Annotated Instructor's Edition

Algebra
A Combined Approach

Fifth Edition

Elayn Martin-Gay

University of New Orleans

PEARSON

Boston   Columbus   Hoboken   Indianapolis   New York   San Francisco
Amsterdam   Cape Town   Dubai   London   Madrid   Milan   Munich   Paris   Montréal   Toronto
Delhi   Mexico City   São Paulo   Sydney   Hong Kong   Seoul   Singapore   Taipei   Tokyo

Editorial Director, Mathematics: *Christine Hoag*
Editor-in-Chief: *Michael Hirsch*
Acquisitions Editor: *Mary Beckwith*
Project Team Lead: *Christina Lepre*
Project Manager: *Lauren Morse*
Assistant Editor: *Matthew Summers*
Editorial Assistant: *Megan Tripp*
Executive Development Editor: *Dawn Nuttall*
Program Team Lead: *Karen Wernholm*
Program Manager: *Patty Bergin*
Cover and Illustration Design: *Tamara Newnam*
Program Design Lead: *Heather Scott*
Executive Content Manager, MathXL: *Rebecca Williams*
Senior Content Developer, TestGen: *John Flanagan*
Director, Course Production: *Ruth Berry*
Media Producer: *Audra Walsh*
Director of Marketing, Mathematics: *Roxanne McCarley*
Senior Marketing Manager: *Rachel Ross*
Marketing Assistant: *Kelly Cross*
Senior Author Support/Technology Specialist: *Joe Vetere*
Senior Procurement Specialist: *Carol Melville*
Interior Design, Production Management, Answer Art, and Composition: *Integra Software
   Services Pvt. Ltd.*
Text Art: *Scientific Illustrators*

For permission to use copyrighted material, grateful acknowledgment is made to the copyright holders on page P1 which is hereby made an extension of this copyright page.

PEARSON, ALWAYS LEARNING, and MYMATHLAB are exclusive trademarks in the U.S. and/or other countries owned by Pearson Education, Inc. or its affiliates.

Unless otherwise indicated herein, any third-party trademarks that may appear in this work are the property of their respective owners and any references to third-party trademarks, logos or other trade dress are for demonstrative or descriptive purposes only. Such references are not intended to imply any sponsorship, endorsement, authorization, or promotion of Pearson's products by the owners of such marks, or any relationship between the owner and Pearson Education, Inc. or its affiliates, authors, licensees or distributors.

Library of Congress Cataloging-in-Publication Data
Martin-Gay, K. Elayn
   Algebra a combined approach/ Elayn Martin-Gay, University of New Orleans.—5th edition.
      pages cm
   ISBN 978-0-321-97753-3 (alk. paper)   ISBN-13: 978-0-321-97807-3
   ISBN 0-321-97753-X (alk. paper)   ISBN-10: 0-321-97807-2 (Annotated Instructor's Edition)
1. Algebra—Textbooks.   I. Title.
QA152.3.M35 2016
512.9–dc23   2014030416

1 2 3 4 5 6 7 8 9 10—CRK—19 18 17 16 15

www.pearsonhighered.com

ISBN-10: 0-321-97807-2 (Annotated Instructor's Edition)
ISBN-13: 978-0-321-97807-3

ISBN-10: 0-321-97753-X (Student Edition)
ISBN-13: 978-0-321-97753-3

Contents

Preface

Algebra: A Combined Approach, **Fifth Edition** is intended for a two-semester course in introductory and intermediate algebra. Specific care was taken to make sure students have the most up-to-date relevant text preparation for their next mathematics course or for nonmathematical courses that require an understanding of algebraic fundamentals. I have tried to achieve this by writing a user-friendly text that is keyed to objectives and contains many worked-out examples. As suggested by AMATYC and the NCTM Standards (plus Addenda), real-life and real-data applications, data interpretation, conceptual understanding, problem solving, writing, cooperative learning, appropriate use of technology, number sense, estimation, critical thinking, and geometric concepts are emphasized and integrated throughout the book.

The many factors that contributed to the success of the previous editions have been retained. In preparing the Fifth Edition, I considered comments and suggestions of colleagues, students, and many users of the prior edition throughout the country.

What's New in the Fifth Edition?

- **The Martin-Gay Program** has been revised and enhanced with a new design in the text and MyMathLab® to actively encourage students to use the text, video program, and Video Organizer as an integrated learning system.

- **The New Video Organizer** is designed to help students take notes and work practice exercises while watching the Interactive Lecture Series videos (available in MyMathLab and on DVD). All content in the Video Organizer is presented in the same order as it is presented in the videos, making it easy for students to create a course notebook and build good study habits.

  — Covers all of the video examples in order.
  — Provides ample space for students to write down key definitions and properties.
  — Includes "Play" and "Pause" button icons to prompt students to follow along with the author for some exercises while they try others on their own.

  The Video Organizer is available in a loose-leaf, notebook-ready format. It is also available for download in MyMathLab.

- **Vocabulary, Readiness & Video Check** questions have been added prior to every section exercise set. These exercises quickly check a student's understanding of new vocabulary words. The **readiness** exercises center on a student's understanding of a concept that is necessary in order to continue to the exercise set. **New Video Check questions for the Martin-Gay Interactive Lecture videos** are now included in every section for each learning objective. **These exercises are all available for assignment in MyMathLab** and are a great way to assess whether students have viewed and understood the key concepts presented in the videos.

- **New Student Success Tips Videos** are 3- to 5-minute video segments designed to be daily reminders to students to continue practicing and maintaining good organizational and study habits. They are organized in three categories and are available in MyMathLab and the Interactive Lecture Series. The categories are:

  1. Success Tips that apply to any course in college in general, such as Time Management.
  2. Success Tips that apply to any mathematics course. One example is based on understanding that mathematics is a course that requires homework to be completed in a timely fashion.

3. Section- or Content-specific Success Tips to help students avoid common mistakes or to better understand concepts that often prove challenging. One example of this type of tip is how to apply the order of operations to simplify an expression.

- **Interactive DVD Lecture Series**, featuring your text author (Elayn Martin-Gay), provides students with active learning at their own pace. The videos offer the following resources and more:

  A complete lecture for each section of the text highlights key examples and exercises from the text. "Pop-ups" reinforce key terms, definitions, and concepts.

  An interface with menu navigation features allows students to quickly find and focus on the examples and exercises they need to review.

  Interactive Concept Check exercises measure students' understanding of key concepts and common trouble spots.

  New Student Success Tips Videos.

- **The Interactive DVD Lecture Series** also includes the following resources for test prep:

  The Chapter Test Prep Videos help students during their most teachable moment—when they are preparing for a test. This innovation provides step-by-step solutions for the exercises found in each Chapter Test. For the Fifth Edition, the chapter test prep videos are also available on YouTube™. The videos are captioned in English and Spanish.

  The Practice Final Exam Videos help students prepare for an end-of-course final. Students can watch full video solutions to each exercise in the Practice Final Exam at the end of this text.

- **The Martin-Gay MyMathLab** course has been updated and revised to provide more exercise coverage, including assignable video check questions and an expanded video program. There are section lecture videos for every section, which students can also access at the specific objective level; Student Success Tips Videos; and an increased number of watch clips at the exercise level to help students while doing homework in MathXL. Suggested homework assignments have been premade for assignment at the instructor's discretion.

- **New MyMathLab Ready to Go Courses** (access code required) provide students with all the same great MyMathLab features that you're used to, but make it easier for instructors to get started. Each course includes preassigned homework and quizzes to make creating your course even simpler. Ask your Pearson representative about the details for this particular course or to see a copy of this course.

Key Pedagogical Features

The following key features have been retained and/or updated for the Fifth Edition of the text:

Problem-Solving Process   This is formally introduced in Chapter 2 with a four-step process that is integrated throughout the text. The four steps are **Understand, Translate, Solve,** and **Interpret.** The repeated use of these steps in a variety of examples shows their wide applicability. Reinforcing the steps can increase students' comfort level and confidence in tackling problems.

Exercise Sets Revised and Updated   The exercise sets have been carefully examined and extensively revised. Special focus was placed on making sure that even- and odd-numbered exercises are paired and that real-life applications are updated.

Examples   Detailed, step-by-step examples were added, deleted, replaced, or updated as needed. Many examples reflect real life. Additional instructional support is provided in the annotated examples.

Practice Exercises   Throughout the text, each worked-out example has a parallel Practice exercise. These invite students to be actively involved in the learning process. Students should try each Practice exercise after finishing the corresponding example. Learning by doing will help students grasp ideas before moving on to other concepts. Answers to the Practice exercises are provided at the bottom of each page.

Helpful Hints   Helpful Hints contain practical advice on applying mathematical concepts. Strategically placed where students are most likely to need immediate reinforcement, Helpful Hints help students avoid common trouble areas and mistakes.

Concept Checks   This feature allows students to gauge their grasp of an idea as it is being presented in the text. Concept Checks stress conceptual understanding at the point-of-use and help suppress misconceived notions before they start. Answers appear at the bottom of the page. Exercises related to Concept Checks are included in the exercise sets.

Mixed Practice Exercises   In the section exercise sets, these exercises require students to determine the problem type and strategy needed to solve it just as they would need to do on a test.

Integrated Reviews   This unique mid-chapter exercise set helps students assimilate new skills and concepts that they have learned separately over several sections. These reviews provide yet another opportunity for students to work with "mixed" exercises as they master the topics.

Vocabulary Check   This feature provides an opportunity for students to become more familiar with the use of mathematical terms as they strengthen their verbal skills. These appear at the end of each chapter before the Chapter Highlights. Vocabulary, Readiness & Video exercises provide practice at the section level.

Chapter Highlights   Found at the end of every chapter, these contain key definitions and concepts with examples to help students understand and retain what they have learned and help them organize their notes and study for tests.

Chapter Review   The end of every chapter contains a comprehensive review of topics introduced in the chapter. The Chapter Review offers exercises keyed to every section in the chapter, as well as Mixed Review exercises that are not keyed to sections.

Chapter Test and Chapter Test Prep Videos   The Chapter Test is structured to include those problems that involve common student errors. The **Chapter Test Prep Videos** give students instant access to a step-by-step video solution of each exercise in the Chapter Test.

Cumulative Review   This review follows every chapter in the text (except Chapters R and 1). Each odd-numbered exercise contained in the Cumulative Review is an earlier worked example in the text that is referenced in the back of the book along with the answer.

Writing Exercises   ✎   These exercises occur in almost every exercise set and require students to provide a written response to explain concepts or justify their thinking.

Applications Real-world and real-data applications have been thoroughly updated, and many new applications are included. These exercises occur in almost every exercise set and show the relevance of mathematics and help students gradually and continuously develop their problem-solving skills.

Review Exercises These exercises occur in each exercise set (except in Chapters R and 1) and are keyed to earlier sections. They review concepts learned earlier in the text that will be needed in the next section or chapter.

Exercise Set Resource Icons Located at the opening of each exercise set, these icons remind students of the resources available for extra practice and support:

See Student Resources descriptions on page xii for details on the individual resources available.

Exercise Icons These icons facilitate the assignment of specialized exercises and let students know what resources can support them.

- DVD Video icon: exercise worked on the Interactive DVD Lecture Series and available in MyMathLab.
- △ Triangle icon: identifies exercises involving geometric concepts.
- Pencil icon: indicates a written response is needed.
- Calculator icon: optional exercises intended to be solved using a scientific or graphing calculator.

Group Activities Found at the end of each chapter, these activities are for individual or group completion, and are usually hands-on or data-based activities that extend the concepts found in the chapter, allowing students to make decisions and interpretations and to think and write about algebra.

Optional: Calculator Exploration Boxes and Calculator Exercises The optional Calculator Explorations provide keystrokes and exercises at appropriate points to give students an opportunity to become familiar with these tools. Section exercises that are best completed by using a calculator are identified by 📱 for ease of assignment.

Student and Instructor Resources

STUDENT RESOURCES

Interactive DVD Lecture Series Videos	Video Organizer	Student Solutions Manual
Provides students with active learning at their pace. The videos offer: • A complete lecture for each text section. The interface allows easy navigation to examples and exercises students need to review. • Interactive Concept Check exercises • Student Success Tips Videos • Practice Final Exam • Chapter Test Prep Videos	Designed to help students take notes and work practice exercises while watching the Interactive Lecture Series Videos. The Video Organizer: • Covers all of the video examples in order • Provides ample space for students to write down key definitions and rules • Includes "Play" and "Pause" button icons to prompt students to follow along with the author for some exercises while they try others on their own Available in loose-leaf, notebook-ready format and in MyMathLab.	Provides completely worked-out solutions to the odd-numbered section exercises; all exercises in the Integrated Reviews, Chapter Reviews, Chapter Tests, and Cumulative Reviews

INSTRUCTOR RESOURCES

Annotated Instructor's Edition	Instructor's Resource Manual with Tests and Mini-Lectures
Contains all the content found in the student edition, plus the following: • Answers to exercises on the same text page • Teaching Tips throughout the text placed at key points • Video Answer Section	• Mini-lectures for each text section • Additional practice worksheets for each section • Several forms of test per chapter—free response and multiple choice • Answers to all items **Instructor's Solutions Manual** **TestGen**® (Available for download from the IRC)
Instructor-to-Instructor Videos—available in the Instructor Resources section of the MyMathLab course.	**Online Resources** **MyMathLab**® (access code required) **MathXL**® (access code required)

Acknowledgments

There are many people who helped me develop this text, and I will attempt to thank some of them here. Courtney Slade and Cindy Trimble were *invaluable* for contributing to the overall accuracy of the text. Dawn Nuttall was *invaluable* for her many suggestions and contributions during the development and writing of this Fifth Edition. Allison Campbell and Lauren Morse provided guidance throughout the production process.

A very special thank you goes to my editor, Mary Beckwith, for being there 24/7/365, as my students say. And my thanks to the staff at Pearson for all their support: Heather Scott, Patty Bergin, Matt Summers, Michelle Renda, Roxanne McCarley, Rachel Ross, Michael Hirsch, Chris Hoag, and Paul Corey.

I would like to thank the following reviewers for their input and suggestions:

Sheila Anderson, *Housatonic Community College*

Lisa Angelo, *Bucks County Community College*

Victoria Baker, *Nicholls State University*

Teri Barnes, *McLennan Community College*

Laurel Berry, *Bryant & Stratton College*

Thomas Blackburn, *Northeastern Illinois University*

Gail Burkett, *Palm Beach State College*

James Butterbach, *Joliet Junior College*

Anita Collins, *Mesa Community College*

Lois Colpo, *Harrisburg Area Community College*

Fay Dang, *Joliet Junior College*

Robert Diaz, *Fullerton College*

Tamie Dickson, *Reading Area Community College*

Laura Dyer, *Southwestern Illinois College*

Sharon Edgemon, *Bakersfield College*

Latonya Ellis, *Gulf Coast Community College*

Hope Essien, *Olive-Harvey College*

Sonia Ford, *Midland College*

Cheryl Gibby, *Cypress College*

Kathryn Gunderson, *Three Rivers Community College*

Elizabeth Hamman, *Cypress College*

Craig Hardesty, *Hillsborough Community College*

Lloyd Harris, *Gulf Coast Community College*

Teresa Hasenauer, *Indian River College*

Julia Hassett, *Oakton Community College*

Todd Hoff, *Wisconsin Indianhead Technical College*

Jeff Koleno, *Lorain County Community College*

Randa Kress, *Idaho State University*

Ted Lai, *Hudson County Community College*

Nicole Lang, *North Hennepin Community College*

Judy Langer, *Westchester Community College*

Lee LaRue, *Paris Junior College*

Jeri Lee, *Des Moines Area Community College*

Sandy Lofstock, *St. Petersburg College*

Stan Mattoon, *Merced College*

Jean McArthur, *Joliet Junior College*

Michael Montano, *Riverside Community College*

Dr. Kris Mudunuri, *Long Beach City College*

Carol Murphy, *San Diego Miramar College*

Lisa J. Music, *Big Sandy Community and Technical College*

Greg Nguyen, *Fullerton College*

Jean Olsen, *Pikes Peak Community College*

Darlene Ornelas, *Fullerton College*

Linda Padilla, *Joliet Junior College*

Scott Perkins, *Lake Sumter State College*

Faith Peters, *Miami Dade*

Marilyn Platt, *Gaston College*

Warren Powell, *Tyler Junior College*

Jeanette Shea, *Central Texas College*

Linda Shoesmith, *Scott Community College*

Mark Shore, *Allegheny College*

Sandy Spears, *Jefferson Community College*

Sue Stokey, *Spartanburg Technical College*

Ping Charlene Tintera, *Texas A & M University*

Katerina Vishnyakova, *Collin County Community College*

Corey Wadlington, *West Kentucky Community and Technical College*

Edward Wagner, *Central Texas College*

Jane Wampler, *Housatonic Community College*

Marjorie Whitmore, *Northwest Arkansas Community College*

Diane Williams, *Northern Kentucky University*

Jenny Wilson, *Tyler Junior College*

Alma Wlazlinski, *McLennan Community College*

Peter Zimmer, *West Chester University*

I would also like to thank the following dedicated group of instructors who participated in our focus groups, Martin-Gay Summits, and our design review for the series. Their feedback and insights have helped to strengthen this edition of the text. These instructors include:

Billie Anderson, *Tyler Junior College*

Cedric Atkins, *Mott Community College*

Lois Beardon, *Schoolcraft College*

Laurel Berry, *Bryant & Stratton College*

John Beyers, *University of Maryland*

Bob Brown, *Community College of Baltimore County–Essex*

Lisa Brown, *Community College of Baltimore County–Essex*

NeKeith Brown, *Richland College*

Gail Burkett, *Palm Beach State College*

Cheryl Cantwell, *Seminole State College*

Jackie Cohen, *Augusta State College*

Julie Dewan, *Mohawk Valley Community College*

Janice Ervin, *Central Piedmont Community College*

Richard Fielding, *Southwestern College*

Cindy Gaddis, *Tyler Junior College*

Nita Graham, *St. Louis Community College*

Pauline Hall, *Iowa State University*

Pat Hussey, *Triton College*

Dorothy Johnson, *Lorain County Community College*

Sonya Johnson, *Central Piedmont Community College*

Irene Jones, *Fullerton College*

Paul Jones, *University of Cincinnati*

Kathy Kopelousous, *Lewis and Clark Community College*

Nancy Lange, *Inver Hills Community College*

Judy Langer, *Westchester Community College*

Lisa Lindloff, *McLennan Community College*

Sandy Lofstock, *St. Petersburg College*

Kathy Lovelle, *Westchester Community College*

Jean McArthur, *Joliet Junior College*

Kevin McCandless, *Evergreen Valley College*

Ena Michael, *State College of Florida*

Daniel Miller, *Niagara County Community College*

Marica Molle, *Metropolitan Community College*

Carol Murphy, *San Diego Miramar College*

Greg Nguyen, *Fullerton College*

Eric Oilila, *Jackson Community College*

Linda Padilla, *Joliet Junior College*

Davidson Pierre, *State College of Florida*

Marilyn Platt, *Gaston College*

Carole Shapero, *Oakton Community College*

Janet Sibol, *Hillsborough Community College*

Anne Smallen, *Mohawk Valley Community College*

Barbara Stoner, *Reading Area Community College*

Jennifer Strehler, *Oakton Community College*

Ellen Stutes, *Louisiana State University Eunice*

Tanomo Taguchi, *Fullerton College*

MaryAnn Tuerk, *Elgin Community College*

Walter Wang, *Baruch College*

Leigh Ann Wheeler, *Greenville Technical Community College*

Valerie Wright, *Central Piedmont Community College*

A special thank you to those students who participated in our design review: Katherine Browne, Mike Bulfin, Nancy Canipe, Ashley Carpenter, Jeff Chojnachi, Roxanne Davis, Mike Dieter, Amy Dombrowski, Kay Herring, Todd Jaycox, Kaleena Levan, Matt Montgomery, Tony Plese, Abigail Polkinghorn, Harley Price, Eli Robinson, Avery Rosen, Robyn Schott, Cynthia Thomas, and Sherry Ward.

Elayn Martin-Gay

This book is dedicated to students everywhere—
and we should all be students. After all, is there anyone among
us who really knows too much? Take that hint and continue
to learn something new every day of your life.

Best of wishes from a fellow student:
Elayn Martin-Gay

About the Author

Elayn Martin-Gay has taught mathematics at the University of New Orleans for more than 25 years. Her numerous teaching awards include the local University Alumni Association's Award for Excellence in Teaching, and Outstanding Developmental Educator at University of New Orleans, presented by the Louisiana Association of Developmental Educators.

Prior to writing textbooks, Elayn Martin-Gay developed an acclaimed series of lecture videos to support developmental mathematics students in their quest for success. These highly successful videos originally served as the foundation material for her texts. Today, the videos are specific to each book in the Martin-Gay series. The author has also created Chapter Test Prep Videos to help students during their most "teachable moment"—as they prepare for a test—along with Instructor-to-Instructor videos that provide teaching tips, hints, and suggestions for each developmental mathematics course, including basic mathematics, prealgebra, beginning algebra, and intermediate algebra.

Elayn is the author of 12 published textbooks as well as multimedia, interactive mathematics, all specializing in developmental mathematics courses. She has also published series in Algebra 1, Algebra 2, and Geometry. She has participated as an author across the broadest range of educational materials: textbooks, videos, tutorial software, and courseware. This provides an opportunity of various combinations for an integrated teaching and learning package offering great consistency for the student.

Applications Index

Prealgebra Review

Check Your Progress

Vocabulary Check

Chapter Highlights

Chapter Review

Chapter Test

This optional review chapter covers basic topics and skills from prealgebra, such as fractions, decimals, and percents. Knowledge of these topics is needed for success in algebra.

What Majors Do College Freshmen Choose?

The following graph is called a circle graph or a pie chart. Each sector (shaped like a piece of pie) shows the fraction of entering college freshmen who choose to major in each discipline shown. In Exercise Set R.2, Exercises 91–94, we show this same circle graph, but in 3-D design. We simplify some of the fractions in it and also study sector size versus fraction value.

College Freshman Majors

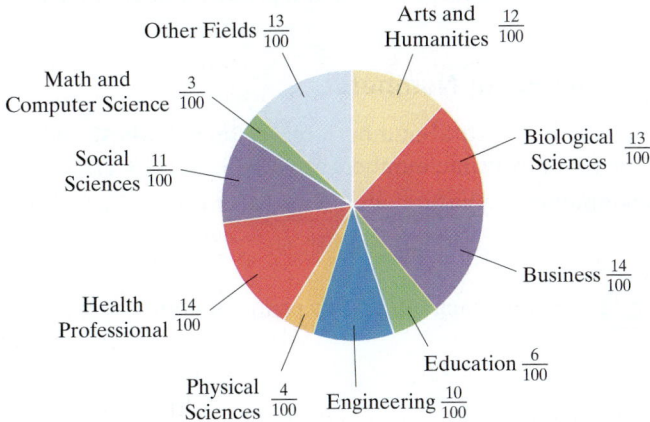

Other Fields $\frac{13}{100}$

Arts and Humanities $\frac{12}{100}$

Math and Computer Science $\frac{3}{100}$

Social Sciences $\frac{11}{100}$

Biological Sciences $\frac{13}{100}$

Business $\frac{14}{100}$

Health Professional $\frac{14}{100}$

Physical Sciences $\frac{4}{100}$

Engineering $\frac{10}{100}$

Education $\frac{6}{100}$

Source: The Higher Education Research Institute

Objectives

A Write the Factors of a Number. ▶

B Write the Prime Factorization of a Number. ▶

C Find the LCM of a List of Numbers. ▶

Objective A Factoring Numbers ▶

In arithmetic we factor numbers, and in algebra we factor expressions containing variables.

> To **factor** means to write as a product.

Throughout this text, you will encounter the word *factor* often. Always remember that factoring means writing as a product.

Since $2 \cdot 3 = 6$, we say that 2 and 3 are **factors** of 6. Also, $2 \cdot 3$ is a **factorization** of 6.

Practice 1

List the factors of 10.

Example 1 List the factors of 6.

Solution: First we write the different factorizations of 6.

$$6 = 1 \cdot 6, \quad 6 = 2 \cdot 3$$

The factors of 6 are 1, 2, 3, and 6.

■ **Work Practice 1**

Practice 2

List the factors of 18.

Example 2 List the factors of 20.

Solution: $20 = 1 \cdot 20, \quad 20 = 2 \cdot 10, \quad 20 = 4 \cdot 5$

The factors of 20 are 1, 2, 4, 5, 10, and 20.

■ **Work Practice 2**

In this section, we will concentrate on **natural numbers** only. The natural numbers (also called counting numbers) are

Natural Numbers: 1, 2, 3, 4, 5, 6, 7, and so on

Every natural number except 1 is either a prime number or a composite number.

> ### Prime and Composite Numbers
>
> A **prime number** is a natural number greater than 1 whose only factors are 1 and itself. The first few prime numbers are 2, 3, 5, 7, 11, 13, 17, . . .
> A **composite number** is a natural number greater than 1 that is not prime.

Practice 3

Identify each number as prime or composite: 5, 16, 23, 42.

Example 3 Identify each number as prime or composite: 3, 28, 19, 35

Solution:

3 is a prime number. Its factors are 1 and 3 only.
28 is a composite number. Its factors are 1, 2, 4, 7, 14, and 28.
19 is a prime number. Its factors are 1 and 19 only.
35 is a composite number. Its factors are 1, 5, 7, and 35.

■ **Work Practice 3**

Answers

1. 1, 2, 5, 10 **2.** 1, 2, 3, 6, 9, 18
3. 5, 23 prime; 16, 42 composite

Objective B Writing Prime Factorizations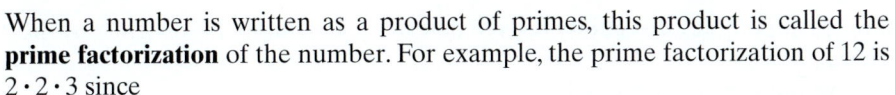

When a number is written as a product of primes, this product is called the **prime factorization** of the number. For example, the prime factorization of 12 is $2 \cdot 2 \cdot 3$ since

$$12 = 2 \cdot 2 \cdot 3$$

and all the factors are prime numbers.

Example 4 Write the prime factorization of 45.

Solution: We can begin by writing 45 as the product of two numbers, say 9 and 5.

$$45 = 9 \cdot 5$$

The number 5 is prime, but 9 is not. So we write 9 as $3 \cdot 3$.

$$45 = 9 \cdot 5$$
$$= 3 \cdot 3 \cdot 5$$

Each factor is now a prime number, so the prime factorization of 45 is $3 \cdot 3 \cdot 5$.

■ **Work Practice 4**

Practice 4

Write the prime factorization of 44.

Helpful Hint

Recall that order is not important when multiplying numbers. For example,

$$3 \cdot 3 \cdot 5 = 3 \cdot 5 \cdot 3 = 5 \cdot 3 \cdot 3 = 45$$

For this reason, any of the products shown can be called *the* prime factorization of 45, and we say that the prime factorization of a number is unique.

Example 5 Write the prime factorization of 80.

Solution: We first write 80 as a product of two numbers. We continue this process until all factors are prime.

$$80 = 8 \cdot 10$$
$$= 4 \cdot 2 \cdot 2 \cdot 5$$
$$= 2 \cdot 2 \cdot 2 \cdot 2 \cdot 5$$

All factors are now prime, so the prime factorization of 80 is

$$2 \cdot 2 \cdot 2 \cdot 2 \cdot 5.$$

■ **Work Practice 5**

Practice 5

Write the prime factorization of 60.

✓**Concept Check** Suppose that you choose $80 = 4 \cdot 20$ as your first step in Example 5 and another student chooses $80 = 5 \cdot 16$. Will you both end up with the same prime factorization as in Example 5? Explain.

Answers

4. $2 \cdot 2 \cdot 11$ **5.** $2 \cdot 2 \cdot 3 \cdot 5$

✓**Concept Check Answer**

yes; answers may vary

Copyright 2016 Pearson Education, Inc.

> ## Helpful Hint
>
> There are a few quick **divisibility tests** to determine if a number is divisible by the primes 2, 3, or 5.
>
> A whole number is divisible by
>
> - **2** if the ones digit is 0, 2, 4, 6, or 8.
>   ↓
>
> 132 is divisible by 2
> - **3** if the sum of the digits is divisible by 3.
>
> 144 is divisible by 3 since $1 + 4 + 4 = 9$ is divisible by 3
> - **5** if the ones digit is 0 or 5.
>   ↓
>
> 1115 is divisible by 5

When finding the prime factorization of larger numbers, you may want to use the procedure shown in Example 6.

Practice 6

Write the prime factorization of 297.

Example 6   Write the prime factorization of 252.

Solution: Since the ones digit of 252 is 2, we know that 252 is divisible by 2.

$$\begin{array}{r} 126 \\ 2\overline{)252} \end{array}$$

126 is divisible by 2 also.

$$\begin{array}{r} 63 \\ 2\overline{)126} \\ 2\overline{)252} \end{array}$$

63 is not divisible by 2 but is divisible by 3. We divide 63 by 3 and continue in this same manner until the quotient is a prime number.

$$\begin{array}{r} 7 \\ 3\overline{)\ 21} \\ 3\overline{)\ 63} \\ 2\overline{)126} \\ 2\overline{)252} \end{array}$$

The prime factorization of 252 is $2 \cdot 2 \cdot 3 \cdot 3 \cdot 7$.

Work Practice 6

Teaching Tip

Consider suggesting that students adopt a procedure that can be used each time they look for a prime factorization. For instance, they might try looking first for factors of 2, then 3, then 5, and so on in order of consecutive primes.

Objective C Finding the Least Common Multiple

A **multiple** of a number is the product of that number and any natural number. For example, the multiples of 3 are

$3 \cdot 1$	$3 \cdot 2$	$3 \cdot 3$	$3 \cdot 4$	$3 \cdot 5$	$3 \cdot 6$	$3 \cdot 7$	
3,	6,	9,	12,	15,	18,	21,	and so on.

The multiples of 2 are

$2 \cdot 1$	$2 \cdot 2$	$2 \cdot 3$	$2 \cdot 4$	$2 \cdot 5$	$2 \cdot 6$	$2 \cdot 7$	
2,	4,	6,	8,	10,	12,	14,	and so on.

Answer

6. $3 \cdot 3 \cdot 3 \cdot 11$

Notice that 2 and 3 have multiples that are common to both.

Multiples of 2:   2, 4, **6**, 8, 10, **12**, 1 4, 16, **18**, and so on

Multiples of 3:   3, **6**, 9, **12**, 15, **18**, 21, and so on

Common multiples of 2 and 3: 6, 12, 18, . . .

The least or smallest common multiple of 2 and 3 is 6. The number 6 is called the **least common multiple** or **LCM** of 2 and 3. It is the smallest number that is a multiple of both 2 and 3.

The **least common multiple (LCM)** of a list of numbers is the smallest number that is a multiple of all the numbers in the list.

Finding the LCM by the method above can sometimes be time-consuming. Let's look at another method that uses prime factorization.

To find the LCM of 4 and 10, for example, we write the prime factorization of each.

$$4 = 2 \cdot 2$$
$$10 = 2 \cdot 5$$

If the LCM is to be a multiple of 4, it must contain the factors $2 \cdot 2$. If the LCM is to be a multiple of 10, it must contain the factors $2 \cdot 5$. Since we decide whether the LCM is a multiple of 4 and 10 separately, the LCM does not need to contain three factors of 2. The LCM only needs to contain a factor the greatest number of times that the factor appears in any **one** prime factorization.

The LCM is a
multiple of 4.

$$LCM = 2 \cdot 2 \cdot 5 = 20$$

The number 2 is a factor twice since that is the greatest number of times that 2 is a factor in either of the prime factorizations.

The LCM is a
multiple of 10.

To Find the LCM of a List of Numbers

Step 1: Write the prime factorization of each number.

Step 2: Write the product containing each different prime factor (from Step 1) the greatest number of times that it appears in any one factorization. This product is the LCM.

Example 7   Find the LCM of 18 and 24.

Solution: First we write the prime factorization of each number.

$$18 = 2 \cdot 3 \cdot 3$$
$$24 = 2 \cdot 2 \cdot 2 \cdot 3$$

Now we write each factor the greatest number of times that it appears in any **one** prime factorization.

The greatest number of times that 2 appears is **3** times.

The greatest number of times that 3 appears is **2** times.

$$LCM = 2 \cdot 2 \cdot 2 \cdot 3 \cdot 3 = 72$$

2 is a factor   3 is a factor
3 times.   2 times.

■ **Work Practice 7**

Practice 7

Find the LCM of 14 and 35.

Answer
7. 70

Practice 8

Find the LCM of 5 and 9.

Example 8 Find the LCM of 11 and 10.

Solution: 11 is a prime number, so we simply rewrite it. Then we write the prime factorization of 10.

$$11 = 11$$
$$10 = 2 \cdot 5$$
$$LCM = 2 \cdot 5 \cdot 11 = 110$$

■ **Work Practice 8**

Teaching Tip

Challenge your students to find all the possible solutions to the following: 60 is the least common multiple of 12 and _____.

Practice 9

Find the LCM of 4, 15, and 10.

Example 9 Find the LCM of 5, 6, and 12.

Solution:

$$5 = 5$$
$$6 = 2 \cdot 3$$
$$12 = 2 \cdot 2 \cdot 3$$
$$LCM = 2 \cdot 2 \cdot 3 \cdot 5 = 60.$$

■ **Work Practice 9**

Answers
8. 45 **9.** 60

Vocabulary, Readiness & Video Check

Use the choices below to fill in each blank.

least common multiple   composite   multiple
prime factorization   prime   factor

1. The number 40 equals $2 \cdot 2 \cdot 2 \cdot 5$. Since each factor is prime, we call $2 \cdot 2 \cdot 2 \cdot 5$ the ____prime factorization____ of 40.

2. A natural number, other than 1, that is not prime is called a(n) ____composite____ number.

3. A natural number that has exactly two different factors, 1 and itself, is called a(n) ____prime____ number.

4. The ____least common multiple____ of a list of numbers is the smallest number that is a multiple of all the numbers in the list.

5. To ____factor____ means to write as a product.

6. A(n) ____multiple____ of a number is the product of that number and any natural number.

Martin-Gay Interactive Videos   Watch the section lecture video and answer the following questions.

Objective A 7. From the lecture before ⊞ Example 2, are all natural numbers either prime or composite? ▶

Objective B 8. Complete this statement based on ⊞ Example 4: We may write factors in different _____, but every natural number has only _____ prime factorization. ▶

Objective C 9. From the lecture before ⊞ Example 7, the least common multiple, LCM, of a list of numbers is the _____ number that is a multiple of each number in the list. ▶

See Video R.1 🍎

See video answer section.

R.1 Exercise Set MyMathLab®

Objective A *List the factors of each number. See Examples 1 and 2.*

1. 9
1, 3, 9

2. 8
1, 2, 4, 8

3. 24
1, 2, 3, 4, 6, 8, 12, 24

4. 36
1, 2, 3, 4, 6, 9, 12, 18, 36

5. 42
1, 2, 3, 6, 7, 14, 21, 42

6. 63
1, 3, 7, 9, 21, 63

7. 80
1, 2, 4, 5, 8, 10, 16, 20, 40, 80

8. 50
1, 2, 5, 10, 25, 50

9. 19
1, 19

10. 31
1, 31

Identify each number as prime or composite. See Example 3.

11. 13 prime

12. 21 composite

13. 39 composite

14. 53 prime

15. 41 prime

16. 51 composite

17. 201 composite

18. 307 prime

19. 2065 composite

20. 1798 composite

Objective B *Write each prime factorization. See Examples 4 through 6.*

21. 18 $2 \cdot 3 \cdot 3$

22. 28 $2 \cdot 2 \cdot 7$

23. 20 $2 \cdot 2 \cdot 5$

24. 30 $2 \cdot 3 \cdot 5$

25. 56 $2 \cdot 2 \cdot 2 \cdot 7$

26. 48 $2 \cdot 2 \cdot 2 \cdot 2 \cdot 3$

27. 81 $3 \cdot 3 \cdot 3 \cdot 3$

28. 64 $2 \cdot 2 \cdot 2 \cdot 2 \cdot 2 \cdot 2$

29. 300 $2 \cdot 2 \cdot 3 \cdot 5 \cdot 5$

30. 500 $2 \cdot 2 \cdot 5 \cdot 5 \cdot 5$

31. 588 $2 \cdot 2 \cdot 3 \cdot 7 \cdot 7$

32. 315 $3 \cdot 3 \cdot 5 \cdot 7$

Multiple choice. Select the best choice to complete each statement.

33. The factors of 48 are
a. $2 \cdot 2 \cdot 2 \cdot 6$
b. $2 \cdot 2 \cdot 2 \cdot 3$
c. $2 \cdot 2 \cdot 2 \cdot 2 \cdot 3$
d. 1, 2, 3, 4, 6, 8, 12, 16, 24, 48   d

34. The prime factorization of 63 is
a. 1, 3, 7, 9, 63
b. 1, 3, 7, 9, 21, 63
c. $3 \cdot 3 \cdot 7$
d. 1, 3, 21, 63   c

Objective C *Find the LCM of each list of numbers. See Examples 7 through 9.*

35. 3, 4   12

36. 4, 5   20

37. 6, 14   42

38. 9, 15   45

39. 20, 30   60

40. 30, 40   120

41. 5, 7   35

42. 2, 11   22

43. 9, 12   36

44. 4, 18   36

45. 16, 20   80

46. 18, 30   90

47. 40, 90   360

48. 50, 70   350

49. 24, 36   72

50. 21, 28   84

51. 2, 8, 15   120

52. 3, 9, 20   180

53. 2, 3, 7   42

54. 3, 5, 7   105

55. 8, 24, 48   48

56. 9, 36, 72   72

57. 8, 18, 30   360

58. 4, 14, 35   140

Concept Extensions

59. Solve. See the Concept Check in the section.

   a. Write the prime factorization of 40 using 2 and 20 as the first pair of factors. $2 \cdot 2 \cdot 2 \cdot 5$

   b. Write the prime factorization of 40 using 4 and 10 as the first pair of factors. $2 \cdot 2 \cdot 2 \cdot 5$

   c. Explain any similarities or differences found in parts a and b.   answers may vary

60. The LCM of 6 and 7 is 42. In general, describe when the LCM of two numbers is equal to their product.   answers may vary

61. Craig Campanella and Edie Hall both have night jobs. Craig has every fifth night off and Edie has every seventh night off. How often will they have the same night off?   every 35 days

62. Elizabeth Kaster and Lori Sypher are both publishing company representatives in Louisiana. Elizabeth spends a day in New Orleans every 35 days, and Lori spends a day in New Orleans every 20 days. How often are they in New Orleans on the same day?   every 140 days

Find the LCM of each pair of numbers.

63. 315, 504   2520

64. 1000, 1125   9000

R.2   Fractions

Objectives

A Discover Fraction Properties Having to Do with 0 and 1. ▶

B Write Equivalent Fractions. ▶

C Write Fractions in Simplest Form. ▶

D Multiply and Divide Fractions. ▶

E Add and Subtract Fractions. ▶

F Perform Operations on Mixed Numbers. ▶

A quotient of two numbers such as $\frac{2}{9}$ is called a **fraction.** The parts of a fraction are:

Fraction bar → $\dfrac{2}{9}$ ← Numerator
     ← Denominator

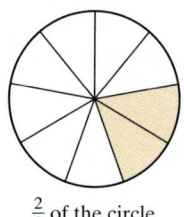

$\frac{2}{9}$ of the circle
is shaded.

A fraction may be used to refer to part of a whole. For example, $\frac{2}{9}$ of the circle is shaded. The denominator 9 tells us how many equal parts the whole circle is divided into, and the numerator 2 tells us how many equal parts are shaded.

In this section, we will use numerators that are **whole numbers** and denominators that are nonzero whole numbers. The whole numbers consist of 0 and the natural numbers.

Whole Numbers: 0, 1, 2, 3, 4, 5, and so on

Objective A   Discovering Fraction Properties with 0 and 1 ▶

Before we continue further, don't forget that the fraction bar indicates division. For example,

$$\frac{8}{4} = 8 \div 4 = 2 \quad \text{since} \quad 2 \cdot 4 = 8$$

Thus, we may simplify some fractions by recalling that the fraction bar means division.

$$\frac{6}{6} = 6 \div 6 = 1 \quad \text{and} \quad \frac{3}{1} = 3 \div 1 = 3$$

Examples Simplify by dividing the numerator by the denominator.

1. $\frac{3}{3} = 1$ Since $3 \div 3 = 1$.

2. $\frac{4}{2} = 2$ Since $4 \div 2 = 2$.

3. $\frac{7}{7} = 1$ Since $7 \div 7 = 1$.

4. $\frac{8}{1} = 8$ Since $8 \div 1 = 8$.

5. $\frac{0}{6} = 0$ Since $0 \cdot 6 = 0$.

6. $\frac{6}{0}$ is undefined because there is no number that when multiplied by 0 gives 6.

▶ Work Practice 1–6

Practice 1–6

Simplify by dividing the numerator by the denominator.

1. $\frac{4}{4}$ **2.** $\frac{9}{3}$ **3.** $\frac{10}{10}$

4. $\frac{5}{1}$ **5.** $\frac{0}{11}$ **6.** $\frac{11}{0}$

From Examples 1 through 6, we can say the following:

> Let a be any number other than 0.
>
> $$\frac{a}{a} = 1, \qquad \frac{0}{a} = 0,$$
>
> $$\frac{a}{1} = a, \qquad \frac{a}{0} \text{ is undefined}$$

Objective B   Writing Equivalent Fractions ▶

More than one fraction can be used to name the same part of a whole. Such fractions are called **equivalent fractions.**

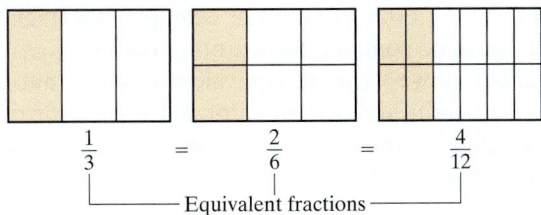

$$\frac{1}{3} \quad = \quad \frac{2}{6} \quad = \quad \frac{4}{12}$$

Equivalent fractions

> ### Equivalent Fractions
>
> Fractions that represent the same portion of a whole are called **equivalent fractions.**

Answers

1. 1   **2.** 3   **3.** 1   **4.** 5   **5.** 0

6. undefined

Copyright 2016 Pearson Education, Inc.

Teaching Tip

When describing equivalent fractions, consider discussing the following problem. Suppose we have two cakes of equal size. One is cut into 6 equal pieces and the other is cut into 12 equal pieces. How many pieces of the 12-piece cake would equal one piece of the 6-piece cake? We can describe this situation using fractions:

$$\frac{1}{6} = \frac{1}{6} \cdot \frac{2}{2} = \frac{1 \cdot 2}{6 \cdot 2} = \frac{2}{12}.$$

For example, let's write $\frac{1}{3}$ as an equivalent fraction with a denominator of 12.

To do so, notice the denominator of 3, multiplied by 4, gives a denominator of 12. Thus let's multiply by 1 in the form of $\frac{4}{4}$.

$$\frac{1}{3} = \frac{1}{3} \cdot 1 = \frac{1}{3} \cdot \frac{4}{4} = \frac{1 \cdot 4}{3 \cdot 4} = \frac{4}{12}$$

$$\frac{4}{4} = 1$$

So $\frac{1}{3} = \frac{4}{12}$.

To Write an Equivalent Fraction

$$\frac{a}{b} = \frac{a}{b} \cdot \frac{c}{c} = \frac{a \cdot c}{b \cdot c}$$

Since $\frac{a}{b} = \frac{a}{b} \cdot 1$

where a, b, and c are nonzero numbers.

Practice 7

Write $\frac{1}{4}$ as an equivalent fraction with a denominator of 20.

Example 7 Write $\frac{2}{5}$ as an equivalent fraction with a denominator of 15.

Solution: In the denominator, since $5 \cdot 3 = 15$, we multiply the fraction $\frac{2}{5}$ by 1 in the form of $\frac{3}{3}$.

$$\frac{2}{5} = \frac{2}{5} \cdot \frac{3}{3} = \frac{2 \cdot 3}{5 \cdot 3} = \frac{6}{15}$$

Then $\frac{2}{5}$ is equivalent to $\frac{6}{15}$. They both represent the same part of a whole.

■ **Work Practice 7**

Objective C Simplifying Fractions ▶

A special equivalent fraction is one that is simplified or in lowest terms. A fraction is said to be **simplified** or in **lowest terms** when the numerator and the denominator have no factors in common other than 1. For example, the fraction $\frac{5}{11}$ is in lowest terms since 5 and 11 have no common factors other than 1.

To simplify a fraction, we write an equivalent fraction, but one with no common factors in the numerator and denominator. Since we are writing an equivalent fraction, we use the same method as before, except we are "removing" factors of 1 instead of "inserting" factors of 1.

To Write a Simplified, Equivalent Fraction

$$\frac{a \cdot c}{b \cdot c} = \frac{a}{b} \cdot \frac{c}{c} = \frac{a}{b}$$

Since $\frac{a}{b} \cdot 1 = \frac{a}{b}$

Answer

7. $\frac{5}{20}$

Thus, we may simplify some fractions by recalling that the fraction bar means division.

$$\frac{6}{6} = 6 \div 6 = 1 \quad \text{and} \quad \frac{3}{1} = 3 \div 1 = 3$$

Examples Simplify by dividing the numerator by the denominator.

1. $\frac{3}{3} = 1$ Since $3 \div 3 = 1$.

2. $\frac{4}{2} = 2$ Since $4 \div 2 = 2$.

3. $\frac{7}{7} = 1$ Since $7 \div 7 = 1$.

4. $\frac{8}{1} = 8$ Since $8 \div 1 = 8$.

5. $\frac{0}{6} = 0$ Since $0 \cdot 6 = 0$.

6. $\frac{6}{0}$ is undefined because there is no number that when multiplied by 0 gives 6.

■ **Work Practice 1–6**

From Examples 1 through 6, we can say the following:

Let a be any number other than 0.

$$\frac{a}{a} = 1, \qquad \frac{0}{a} = 0,$$

$$\frac{a}{1} = a, \qquad \frac{a}{0} \text{ is undefined}$$

Objective B Writing Equivalent Fractions

More than one fraction can be used to name the same part of a whole. Such fractions are called **equivalent fractions.**

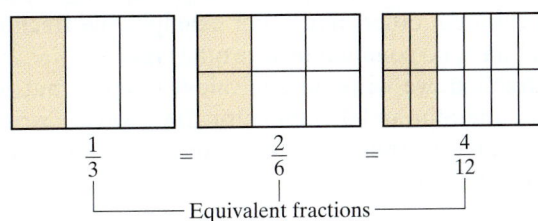

$$\frac{1}{3} \quad = \quad \frac{2}{6} \quad = \quad \frac{4}{12}$$

└──── Equivalent fractions ────┘

Equivalent Fractions

Fractions that represent the same portion of a whole are called **equivalent fractions.**

Practice 1–6

Simplify by dividing the numerator by the denominator.

1. $\frac{4}{4}$ **2.** $\frac{9}{3}$ **3.** $\frac{10}{10}$

4. $\frac{5}{1}$ **5.** $\frac{0}{11}$ **6.** $\frac{11}{0}$

Answers
1. 1 **2.** 3 **3.** 1 **4.** 5 **5.** 0
6. undefined

Copyright 2016 Pearson Education, Inc.

Teaching Tip

When describing equivalent fractions, consider discussing the following problem. Suppose we have two cakes of equal size. One is cut into 6 equal pieces and the other is cut into 12 equal pieces. How many pieces of the 12-piece cake would equal one piece of the 6-piece cake? We can describe this situation using fractions:

$$\frac{1}{6} = \frac{1}{6} \cdot \frac{2}{2} = \frac{1 \cdot 2}{6 \cdot 2} = \frac{2}{12}.$$

For example, let's write $\frac{1}{3}$ as an equivalent fraction with a denominator of 12. To do so, notice the denominator of 3, multiplied by 4, gives a denominator of 12. Thus let's multiply by 1 in the form of $\frac{4}{4}$.

$$\frac{1}{3} = \frac{1}{3} \cdot 1 = \frac{1}{3} \cdot \frac{4}{4} = \frac{1 \cdot 4}{3 \cdot 4} = \frac{4}{12}$$

$$\frac{4}{4} = 1$$

So $\frac{1}{3} = \frac{4}{12}$.

To Write an Equivalent Fraction

$$\frac{a}{b} = \frac{a}{b} \cdot \frac{c}{c} = \frac{a \cdot c}{b \cdot c}$$

Since $\frac{a}{b} = \frac{a}{b} \cdot 1$

where a, b, and c are nonzero numbers.

Practice 7

Write $\frac{1}{4}$ as an equivalent fraction with a denominator of 20.

Example 7 Write $\frac{2}{5}$ as an equivalent fraction with a denominator of 15.

Solution: In the denominator, since $5 \cdot 3 = 15$, we multiply the fraction $\frac{2}{5}$ by 1 in the form of $\frac{3}{3}$.

$$\frac{2}{5} = \frac{2}{5} \cdot \frac{3}{3} = \frac{2 \cdot 3}{5 \cdot 3} = \frac{6}{15}$$

Then $\frac{2}{5}$ is equivalent to $\frac{6}{15}$. They both represent the same part of a whole.

▪ **Work Practice 7**

Objective C Simplifying Fractions ▶

A special equivalent fraction is one that is simplified or in lowest terms. A fraction is said to be **simplified** or in **lowest terms** when the numerator and the denominator have no factors in common other than 1. For example, the fraction $\frac{5}{11}$ is in lowest terms since 5 and 11 have no common factors other than 1.

To simplify a fraction, we write an equivalent fraction, but one with no common factors in the numerator and denominator. Since we are writing an equivalent fraction, we use the same method as before, except we are "removing" factors of 1 instead of "inserting" factors of 1.

To Write a Simplified, Equivalent Fraction

$$\frac{a \cdot c}{b \cdot c} = \frac{a}{b} \cdot \frac{c}{c} = \frac{a}{b}$$

Since $\frac{a}{b} \cdot 1 = \frac{a}{b}$

Answer

7. $\frac{5}{20}$

Example 8 Simplify: $\dfrac{42}{49}$

Solution: To help us see common factors in the numerator and denominator, or factors of 1, we write the numerator and the denominator as products of primes.

$$\frac{42}{49} = \frac{2 \cdot 3 \cdot 7}{7 \cdot 7} = \frac{2 \cdot 3}{7} \cdot \frac{7}{7} = \frac{2 \cdot 3}{7} = \frac{6}{7}$$

■ **Work Practice 8**

✓**Concept Check** Explain the error in the following steps.

a. $\dfrac{15}{55} = \dfrac{\cancel{15}}{\cancel{55}} = \dfrac{1}{5}$ **b.** $\dfrac{6}{7} = \dfrac{5 + 1}{5 + 2} = \dfrac{1}{2}$

Examples Simplify each fraction.

9. $\dfrac{11}{27} = \dfrac{11}{3 \cdot 3 \cdot 3}$ There are no common factors in the numerator and denominator other than 1, so $\dfrac{11}{27}$ is already simplified.

10. $\dfrac{88}{20} = \dfrac{2 \cdot 2 \cdot 2 \cdot 11}{2 \cdot 2 \cdot 5} = \dfrac{2}{2} \cdot \dfrac{2}{2} \cdot \dfrac{2 \cdot 11}{5} = \dfrac{22}{5}$

■ **Work Practice 9–10**

Below are two important notes about simplifying fractions.

Note 1: When simplifying, we can use a shortcut notation if desired. From Example 8,

$$\frac{42}{49} = \frac{2 \cdot 3 \cdot \cancel{7}^{1}}{7 \cdot \cancel{7}_{1}} = \frac{2 \cdot 3}{7} = \frac{6}{7}$$

Note 2: Also, feel free to save time if you immediately notice common factors. In Example 10, notice that the numerator and denominator of $\dfrac{88}{20}$ have a common factor of 4.

$$\frac{88}{20} = \frac{\cancel{4}^{1} \cdot 22}{\cancel{4}_{1} \cdot 5} = \frac{22}{5}$$

A **proper fraction** is a fraction whose numerator is less than its denominator. The fraction $\dfrac{22}{5}$ from Example 10 is called an improper fraction. An **improper fraction** is a fraction whose numerator is greater than or equal to its denominator.

The improper fraction $\dfrac{22}{5}$ may be written as the mixed number $4\dfrac{2}{5}$. Notice that a **mixed number** has a whole number part and a fraction part. We review operations on mixed numbers in objective **F** in this section. First, let's review operations on fractions.

Objective D Multiplying and Dividing Fractions ▶

To multiply two fractions, we multiply numerator times numerator to obtain the numerator of the product. Then we multiply denominator times denominator to obtain the denominator of the product.

Practice 8

Simplify: $\dfrac{20}{35}$

Practice 9–10

Simplify each fraction.

9. $\dfrac{7}{20}$ **10.** $\dfrac{12}{40}$

Teaching Tip

Before Example 11, help students visualize multiplying fractions of the form $\dfrac{1}{a} \cdot \dfrac{1}{b}$. For instance, $\dfrac{1}{3} \cdot \dfrac{1}{4}$ can be visualized by letting a square represent 1 whole. Represent $\dfrac{1}{4}$ by dividing the square into 4 equal vertical strips. Represent $\dfrac{1}{3}$ of $\dfrac{1}{4}$ by dividing one of the vertical strips into 3 equal horizontal strips. Now ask your students how much of the whole is $\dfrac{1}{3}$ of $\dfrac{1}{4}$?

Answers

8. $\dfrac{4}{7}$ **9.** $\dfrac{7}{20}$ **10.** $\dfrac{3}{10}$

✓**Concept Check Answers**

a. $\dfrac{15}{55} = \dfrac{3 \cdot 5}{11 \cdot 5} = \dfrac{3}{11}$

b. $\dfrac{6}{7}$ can't be simplified

Multiplying Fractions

$$\frac{a}{b} \cdot \frac{c}{d} = \frac{a \cdot c}{b \cdot d} \quad \text{if } b \neq 0 \text{ and } d \neq 0$$

Practice 11

Multiply: $\frac{3}{4} \cdot \frac{8}{9}$. Simplify the product if possible.

Example 11 Multiply: $\frac{2}{15} \cdot \frac{5}{13}$. Simplify the product if possible.

Solution:

$$\frac{2}{15} \cdot \frac{5}{13} = \frac{2 \cdot 5}{15 \cdot 13} \quad \begin{array}{l}\text{Multiply numerators.}\\ \text{Multiply denominators.}\end{array}$$

To simplify the product, we divide the numerator and the denominator by any common factors.

$$\frac{2}{15} \cdot \frac{5}{13} = \frac{2 \cdot \overset{1}{\cancel{5}}}{3 \cdot \underset{1}{\cancel{5}} \cdot 13} = \frac{2}{39}$$

■ **Work Practice 11**

Before we divide fractions, we first define **reciprocals.** Two numbers are reciprocals of each other if their product is 1.

The reciprocal of $\frac{2}{3}$ is $\frac{3}{2}$ because $\frac{2}{3} \cdot \frac{3}{2} = \frac{6}{6} = 1$.

The reciprocal of 5 is $\frac{1}{5}$ because $5 \cdot \frac{1}{5} = \frac{5}{1} \cdot \frac{1}{5} = \frac{5}{5} = 1$.

To divide fractions, we multiply the first fraction by the reciprocal of the second fraction. For example,

$$\frac{1}{2} \div \frac{5}{7} = \frac{1}{2} \cdot \frac{7}{5} = \frac{1 \cdot 7}{2 \cdot 5} = \frac{7}{10}$$

To divide, multiply by the reciprocal.

Dividing Fractions

$$\frac{a}{b} \div \frac{c}{d} = \frac{a}{b} \cdot \frac{d}{c}, \quad \text{if } b \neq 0, d \neq 0, \text{ and } c \neq 0$$

Practice 12–14

Divide and simplify.

12. $\frac{2}{9} \div \frac{3}{4}$

13. $\frac{8}{11} \div 24$

14. $\frac{5}{4} \div \frac{15}{8}$

Examples Divide and simplify.

12. $\frac{4}{5} \div \frac{5}{16} = \frac{4}{5} \cdot \frac{16}{5} = \frac{4 \cdot 16}{5 \cdot 5} = \frac{64}{25}$ The numerator and denominator have no common factors.

13. $\frac{7}{10} \div 14 = \frac{7}{10} \div \frac{14}{1} = \frac{7}{10} \cdot \frac{1}{14} = \frac{\overset{1}{\cancel{7}} \cdot 1}{2 \cdot 5 \cdot 2 \cdot \underset{1}{\cancel{7}}} = \frac{1}{20}$

14. $\frac{3}{8} \div \frac{3}{10} = \frac{3}{8} \cdot \frac{10}{3} = \frac{\overset{1}{\cancel{3}} \cdot \overset{1}{\cancel{2}} \cdot 5}{\underset{1}{\cancel{2}} \cdot 2 \cdot 2 \cdot \underset{1}{\cancel{3}}} = \frac{5}{4}$

■ **Work Practice 12–14**

Answers

11. $\frac{2}{3}$ **12.** $\frac{8}{27}$ **13.** $\frac{1}{33}$ **14.** $\frac{2}{3}$

Objective E Adding and Subtracting Fractions

To add or subtract fractions with the same denominator, we combine numerators and place the sum or difference over the common denominator.

Adding and Subtracting Fractions with the Same Denominator

$$\frac{a}{b} + \frac{c}{b} = \frac{a+c}{b}, \qquad \text{if } b \neq 0$$

$$\frac{a}{b} - \frac{c}{b} = \frac{a-c}{b}, \qquad \text{if } b \neq 0$$

Examples   Add or subtract as indicated. Then simplify if possible.

15. $\dfrac{2}{7} + \dfrac{4}{7} = \dfrac{2+4}{7} = \dfrac{6}{7}$ ← Add numerators.
                             ← Keep the common denominator.

16. $\dfrac{3}{10} + \dfrac{2}{10} = \dfrac{3+2}{10} = \dfrac{5}{10} = \dfrac{\overset{1}{\cancel{5}}}{2 \cdot \underset{1}{\cancel{5}}} = \dfrac{1}{2}$

17. $\dfrac{5}{3} - \dfrac{1}{3} = \dfrac{5-1}{3} = \dfrac{4}{3}$ ← Subtract numerators.
                             ← Keep the common denominator.

18. $\dfrac{9}{7} - \dfrac{2}{7} = \dfrac{9-2}{7} = \dfrac{7}{7} = 1$

■ **Work Practice 15–18**

To add or subtract with different denominators, we first write the fractions as **equivalent fractions** with the same denominator. We use the smallest or **least common denominator,** or **LCD.** The LCD is the same as the least common multiple of the denominators (see Section R.1).

Example 19   Add: $\dfrac{2}{5} + \dfrac{1}{4}$

Solution: We first must find the least common denominator before the fractions can be added. The least common multiple of the denominators 5 and 4 is 20. This is the LCD we will use.

We write both fractions as equivalent fractions with denominators of 20. Since

$$\frac{2}{5} = \frac{2}{5} \cdot 1 = \frac{2}{5} \cdot \frac{4}{4} = \frac{2 \cdot 4}{5 \cdot 4} = \frac{8}{20} \quad \text{and} \quad \frac{1}{4} = \frac{1}{4} \cdot 1 = \frac{1}{4} \cdot \frac{5}{5} = \frac{1 \cdot 5}{4 \cdot 5} = \frac{5}{20}$$

then

$$\frac{2}{5} + \frac{1}{4} = \frac{8}{20} + \frac{5}{20} = \frac{13}{20}$$

■ **Work Practice 19**

Practice 20

Subtract and simplify: $\dfrac{8}{15} - \dfrac{1}{3}$

Teaching Tip

Consider ending this lesson with the following activities:

- Will the sum of two fractions that are less than 1 ever be greater than 1? If so, give an example.
- Will the product of two fractions that are less than 1 ever be greater than 1? If so, give an example.
- Will the quotient of two fractions that are less than 1 ever be greater than 1? If so, give an example.

Example 20 Subtract and simplify: $\dfrac{19}{6} - \dfrac{23}{12}$

Solution: The LCD is 12. We write both fractions as equivalent fractions with denominators of 12.

$$\frac{19}{6} - \frac{23}{12} = \frac{19}{6} \cdot \frac{2}{2} - \frac{23}{12}$$

$$= \frac{19 \cdot 2}{6 \cdot 2} - \frac{23}{12}$$

$$= \frac{38}{12} - \frac{23}{12}$$

$$= \frac{15}{12} = \frac{\overset{1}{\cancel{3}} \cdot 5}{2 \cdot 2 \cdot \underset{1}{\cancel{3}}} = \frac{5}{4}$$

■ **Work Practice 20**

Objective F Performing Operations on Mixed Numbers ▶

To perform operations on mixed numbers, first write each mixed number as an improper fraction. To recall how this is done, let's write $3\dfrac{1}{5}$ as an improper fraction.

$$3\frac{1}{5} = 3 + \frac{1}{5} = \frac{15}{5} + \frac{1}{5} = \frac{16}{5}$$

Because of the steps above, notice we can use a shortcut process for writing a mixed number as an improper fraction.

$$3\frac{1}{5} = \frac{5 \cdot 3 + 1}{5} = \frac{16}{5}$$

Practice 21

Multiply: $5\dfrac{1}{6} \cdot 4\dfrac{2}{5}$

Example 21 Divide: $2\dfrac{1}{8} \div 1\dfrac{2}{3}$

Solution: First write each mixed number as an improper fraction.

$$2\frac{1}{8} = \frac{8 \cdot 2 + 1}{8} = \frac{17}{8}; \qquad 1\frac{2}{3} = \frac{3 \cdot 1 + 2}{3} = \frac{5}{3}$$

Now divide as usual.

$$2\frac{1}{8} \div 1\frac{2}{3} = \frac{17}{8} \div \frac{5}{3} = \frac{17}{8} \cdot \frac{3}{5} = \frac{51}{40}$$

The fraction $\dfrac{51}{40}$ is improper. To write it as an equivalent mixed number, remember that the fraction bar means division, and divide.

$$\begin{array}{r} 1\frac{11}{40} \\ 40\overline{)51} \\ -40 \\ \hline 11 \end{array}$$

Thus, the quotient is $\dfrac{51}{40}$ or $1\dfrac{11}{40}$.

■ **Work Practice 21**

Answers

20. $\dfrac{1}{5}$ **21.** $\dfrac{341}{15}$ or $22\dfrac{11}{15}$

As a general rule, if the original exercise contains mixed numbers, write the result as a mixed number, if possible.

Example 22 Add: $2\frac{1}{8} + 1\frac{2}{3}$

Solution: $2\frac{1}{8} + 1\frac{2}{3} = \frac{17}{8} + \frac{5}{3} = \frac{17 \cdot 3}{8 \cdot 3} + \frac{5 \cdot 8}{3 \cdot 8} = \frac{51}{24} + \frac{40}{24} = \frac{91}{24}$ or $3\frac{19}{24}$

■ **Work Practice 22**

When adding or subtracting larger mixed numbers, you might want to use the following method.

Example 23 Subtract: $50\frac{1}{6} - 38\frac{1}{3}$

Solution:

$$50\frac{1}{6} = 50\frac{1}{6} = 49\frac{7}{6}$$
$$-38\frac{1}{3} = -38\frac{2}{6} = -38\frac{2}{6}$$
$$\overline{\hspace{2cm}11\frac{5}{6}}$$

$50\frac{1}{6} = 49 + 1 + \frac{1}{6} = 49\frac{7}{6}$

■ **Work Practice 23**

Practice 22

Add: $7\frac{3}{8} + 6\frac{3}{4}$

Practice 23

Subtract: $76\frac{1}{12} - 35\frac{1}{4}$

Answers

22. $14\frac{1}{8}$ **23.** $40\frac{5}{6}$

Vocabulary, Readiness & Video Check

Use the choices below to fill in each blank.

improper fraction proper reciprocals mixed number least common denominator (LCD) $\dfrac{a \cdot d}{b \cdot c}$ $\dfrac{a \cdot c}{b \cdot d}$

equivalent denominator 24 simplest form numerator $\dfrac{a - c}{b}$ $\dfrac{a + c}{b}$

1. The number $\dfrac{17}{31}$ is called a(n) _____fraction_____. The number 31 is called its _____denominator_____ and 17 is called its _____numerator_____.

2. The fraction $\dfrac{8}{3}$ is called a(n) _____improper_____ fraction, the fraction $\dfrac{3}{8}$ is called a(n) _____proper_____ fraction, and $10\frac{3}{8}$ is called a(n) _____mixed number_____.

3. In $\dfrac{11}{48}$, since 11 and 48 have no common factors other than 1, $\dfrac{11}{48}$ is in _____simplest form_____.

4. Fractions that represent the same portion of a whole are called _____equivalent_____ fractions.

5. To multiply two fractions, we write $\dfrac{a}{b} \cdot \dfrac{c}{d} = $ _____$\dfrac{a \cdot c}{b \cdot d}$_____.

6. Two numbers are _____reciprocals_____ of each other if their product is 1.

7. To divide two fractions, we write $\dfrac{a}{b} \div \dfrac{c}{d} = $ _____$\dfrac{a \cdot d}{b \cdot c}$_____.

8. $\dfrac{a}{b} + \dfrac{c}{b} = $ _____$\dfrac{a + c}{b}$_____ and $\dfrac{a}{b} - \dfrac{c}{b} = $ _____$\dfrac{a - c}{b}$_____.

9. The smallest positive number divisible by all the denominators of a list of fractions is called the _____least common denominator (LCD)_____.

10. The LCD for $\dfrac{1}{6}$ and $\dfrac{5}{8}$ is _____24_____.

Martin-Gay Interactive Videos   Watch the section lecture video and answer the following questions.

See Video R.2

Objective A 11. From the lecture before Example 1, what can we conclude about any fraction where the numerator and denominator are the same nonzero number?

Objective B 12. Complete this statement based on ⊞ Example 4. Equivalent fractions represent the same _____.

Objective C 13. Describe the first step taken to simplify the fraction in ⊞ Example 5.

Objective D 14. What is the reciprocal of the second fraction in ⊞ Example 7?

Objective E 15. In ⊞ Example 8, what is the main difference between adding or subtracting fractions and multiplying or dividing fractions?

Objective F 16. In ⊞ Example 10, why isn't our original sum in proper form?

See video answer section.

R.2 Exercise Set MyMathLab®

Objective A *Simplify by dividing the numerator by the denominator. See Examples 1 through 6.*

1. $\dfrac{14}{14}$ 1

2. $\dfrac{19}{19}$ 1

3. $\dfrac{20}{2}$ 10

4. $\dfrac{30}{5}$ 6

5. $\dfrac{13}{1}$ 13

6. $\dfrac{21}{1}$ 21

7. $\dfrac{0}{9}$ 0

8. $\dfrac{0}{15}$ 0

9. $\dfrac{9}{0}$ undefined

10. $\dfrac{15}{0}$ undefined

Objective B *Write each fraction as an equivalent fraction with the given denominator. See Example 7.*

11. $\dfrac{7}{10}$ with a denominator of 30 $\dfrac{21}{30}$

12. $\dfrac{2}{3}$ with a denominator of 9 $\dfrac{6}{9}$

13. $\dfrac{2}{9}$ with a denominator of 18 $\dfrac{4}{18}$

14. $\dfrac{8}{7}$ with a denominator of 56 $\dfrac{64}{56}$

15. $\dfrac{4}{5}$ with a denominator of 20 $\dfrac{16}{20}$

16. $\dfrac{4}{5}$ with a denominator of 25 $\dfrac{20}{25}$

Objective C *Simplify each fraction. See Examples 8 through 10.*

17. $\dfrac{2}{4}$ $\dfrac{1}{2}$

18. $\dfrac{3}{6}$ $\dfrac{1}{2}$

19. $\dfrac{10}{15}$ $\dfrac{2}{3}$

20. $\dfrac{15}{20}$ $\dfrac{3}{4}$

21. $\dfrac{3}{7}$ $\dfrac{3}{7}$

22. $\dfrac{5}{9}$ $\dfrac{5}{9}$

23. $\dfrac{18}{30}$ $\dfrac{3}{5}$

24. $\dfrac{42}{45}$ $\dfrac{14}{15}$

25. $\dfrac{16}{20}$ $\dfrac{4}{5}$
26. $\dfrac{8}{40}$ $\dfrac{1}{5}$
27. $\dfrac{66}{48}$ $\dfrac{11}{8}$
28. $\dfrac{64}{24}$ $\dfrac{8}{3}$

29. $\dfrac{120}{244}$ $\dfrac{30}{61}$
30. $\dfrac{360}{700}$ $\dfrac{18}{35}$
31. $\dfrac{192}{264}$ $\dfrac{8}{11}$
32. $\dfrac{455}{525}$ $\dfrac{13}{15}$

Objectives D F Mixed Practice *Multiply or divide as indicated. See Examples 11 through 14 and 21.*

33. $\dfrac{1}{2} \cdot \dfrac{3}{4}$ $\dfrac{3}{8}$
34. $\dfrac{7}{11} \cdot \dfrac{3}{5}$ $\dfrac{21}{55}$
▶ 35. $\dfrac{2}{3} \cdot \dfrac{3}{4}$ $\dfrac{1}{2}$
36. $\dfrac{7}{8} \cdot \dfrac{3}{21}$ $\dfrac{1}{8}$

37. $\dfrac{1}{2} \div \dfrac{7}{12}$ $\dfrac{6}{7}$
38. $\dfrac{7}{12} \div \dfrac{1}{2}$ $\dfrac{7}{6}$
▶ 39. $\dfrac{3}{4} \div \dfrac{1}{20}$ 15
40. $\dfrac{3}{5} \div \dfrac{9}{10}$ $\dfrac{2}{3}$

41. $5\dfrac{1}{9} \cdot 3\dfrac{2}{3}$ $18\dfrac{20}{27}$
42. $2\dfrac{3}{4} \cdot 1\dfrac{7}{8}$ $5\dfrac{5}{32}$
43. $8\dfrac{3}{5} \div 2\dfrac{9}{10}$ $2\dfrac{28}{29}$
44. $1\dfrac{7}{8} \div 3\dfrac{8}{9}$ $\dfrac{27}{56}$

Objectives E F Mixed Practice *Add or subtract as indicated. See Examples 15 through 20, 22, and 23.*

45. $\dfrac{4}{5} + \dfrac{1}{5}$ 1
46. $\dfrac{6}{7} + \dfrac{1}{7}$ 1
47. $\dfrac{4}{15} - \dfrac{1}{12}$ $\dfrac{11}{60}$
48. $\dfrac{11}{12} - \dfrac{1}{16}$ $\dfrac{41}{48}$

49. $\dfrac{2}{3} + \dfrac{3}{7}$ $\dfrac{23}{21}$
50. $\dfrac{3}{4} + \dfrac{1}{6}$ $\dfrac{11}{12}$
▶ 51. $\dfrac{10}{3} - \dfrac{5}{21}$ $\dfrac{65}{21}$
52. $\dfrac{11}{7} - \dfrac{3}{35}$ $\dfrac{52}{35}$

53. $8\dfrac{1}{8} - 6\dfrac{3}{8}$ $1\dfrac{3}{4}$
54. $5\dfrac{2}{5} - 3\dfrac{4}{5}$ $1\dfrac{3}{5}$
▶ 55. $1\dfrac{1}{2} + 3\dfrac{2}{3}$ $5\dfrac{1}{6}$
56. $7\dfrac{3}{20} + 2\dfrac{13}{15}$ $10\dfrac{1}{60}$

Objectives D E F Mixed Practice *Perform the indicated operations. See Examples 11 through 23.*

57. $\dfrac{23}{105} + \dfrac{4}{105}$ $\dfrac{9}{35}$
58. $\dfrac{13}{132} + \dfrac{35}{132}$ $\dfrac{4}{11}$
▶ 59. $\dfrac{17}{21} - \dfrac{10}{21}$ $\dfrac{1}{3}$
60. $\dfrac{18}{35} - \dfrac{11}{35}$ $\dfrac{1}{5}$

61. $\dfrac{7}{10} \cdot \dfrac{5}{21}$ $\dfrac{1}{6}$
62. $\dfrac{3}{35} \cdot \dfrac{10}{63}$ $\dfrac{2}{147}$
63. $\dfrac{9}{20} \div 12$ $\dfrac{3}{80}$
64. $\dfrac{25}{36} \div 10$ $\dfrac{5}{72}$

65. $\dfrac{5}{22} - \dfrac{5}{33}$ $\dfrac{5}{66}$
66. $\dfrac{7}{15} - \dfrac{7}{25}$ $\dfrac{14}{75}$
67. $17\dfrac{2}{5} + 30\dfrac{2}{3}$ $48\dfrac{1}{15}$
68. $26\dfrac{11}{20} + 40\dfrac{7}{10}$ $67\dfrac{1}{4}$

69. $7\dfrac{2}{5} \div \dfrac{1}{5}$ 37
70. $9\dfrac{5}{6} \div \dfrac{1}{6}$ 59
71. $4\dfrac{2}{11} \cdot 2\dfrac{1}{2}$ $10\dfrac{5}{11}$
72. $6\dfrac{6}{7} \cdot 3\dfrac{1}{2}$ 24

73. $\dfrac{12}{5} - 1$ $\dfrac{7}{5}$
74. $2 - \dfrac{3}{8}$ $\dfrac{13}{8}$
75. $8\dfrac{11}{12} - 1\dfrac{5}{6}$ $7\dfrac{1}{12}$
76. $4\dfrac{7}{8} - 2\dfrac{3}{16}$ $2\dfrac{11}{16}$

Concept Extensions

Perform the indicated operations.

77. $\dfrac{2}{3} - \dfrac{5}{9} + \dfrac{5}{6}$ $\dfrac{17}{18}$
78. $\dfrac{8}{11} - \dfrac{1}{4} + \dfrac{1}{2}$ $\dfrac{43}{44}$

For Exercises 79–82, determine whether the work is correct or incorrect. If incorrect, find the error and correct. See the Concept Check in this section.

79. $\dfrac{12}{24} = \dfrac{2 + 4 + 6}{2 + 4 + 6 + 12} = \dfrac{1}{12}$

incorrect; $\dfrac{12}{24} = \dfrac{2 \cdot 2 \cdot 3}{2 \cdot 2 \cdot 2 \cdot 3} = \dfrac{1}{2}$

80. $\dfrac{30}{60} = \dfrac{2 \cdot 3 \cdot 5}{2 \cdot 2 \cdot 3 \cdot 5} = \dfrac{1}{2}$

correct

81. $\dfrac{2}{7} + \dfrac{9}{7} = \dfrac{11}{14}$

incorrect; $\dfrac{2}{7} + \dfrac{9}{7} = \dfrac{11}{7}$

82. $\dfrac{16}{28} = \dfrac{2 \cdot 5 + 6 \cdot 1}{2 \cdot 5 + 6 \cdot 3} = \dfrac{1}{3}$

incorrect; $\dfrac{16}{28} = \dfrac{2 \cdot 2 \cdot 2 \cdot 2}{2 \cdot 2 \cdot 7} = \dfrac{4}{7}$

83. In your own words, describe how to divide fractions.   answers may vary

84. In your own words, describe how to add or subtract fractions.   answers may vary

Each circle below represents a whole, or 1. Determine the unknown part of the circle.

85. 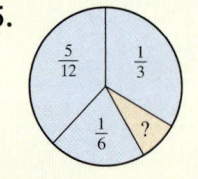 $\dfrac{1}{12}$

86. 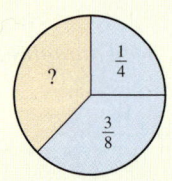 $\dfrac{3}{8}$

87. 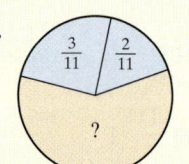 $\dfrac{6}{11}$

88. 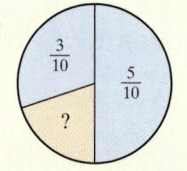 $\dfrac{1}{5}$

89. If Tucson's average rainfall is $11\dfrac{1}{4}$ inches and Yuma's is $3\dfrac{3}{5}$ inches, how much more rain, on average, does Tucson get than Yuma? $7\dfrac{13}{20}$ in.

90. A pair of crutches needs adjustment. One crutch is 43 inches and the other is $41\dfrac{5}{8}$ inches. Find how much the short crutch should be lengthened to make both crutches the same length. $1\dfrac{3}{8}$ in.

The following graph is called a circle graph or pie chart. Each sector (shaped like a piece of pie) shows the fraction of entering college freshmen who choose to major in each discipline shown. The whole circle represents the entire class of college freshmen. Use this graph to answer Exercises 91 through 94. Write each fraction answer in simplest form.

College Freshmen Majors

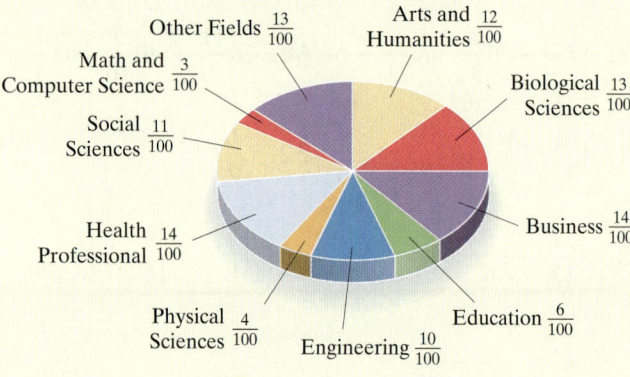

Other Fields $\dfrac{13}{100}$

Arts and Humanities $\dfrac{12}{100}$

Math and Computer Science $\dfrac{3}{100}$

Biological Sciences $\dfrac{13}{100}$

Social Sciences $\dfrac{11}{100}$

Health Professional $\dfrac{14}{100}$

Business $\dfrac{14}{100}$

Physical Sciences $\dfrac{4}{100}$

Engineering $\dfrac{10}{100}$

Education $\dfrac{6}{100}$

Source: The Higher Education Research Institute

91. What fraction of entering college freshmen plan to major in education? $\dfrac{3}{50}$

92. What fraction of entering college freshmen plan to major in engineering? $\dfrac{1}{10}$

93. Why is the Business sector the same size as the Health Professional sector?   answers may vary

94. Why is the Physical Sciences sector smaller than the Business sector?   answers may vary

Use this circle graph to answer Exercises 95 through 98. Write each fraction answer in simplest form.

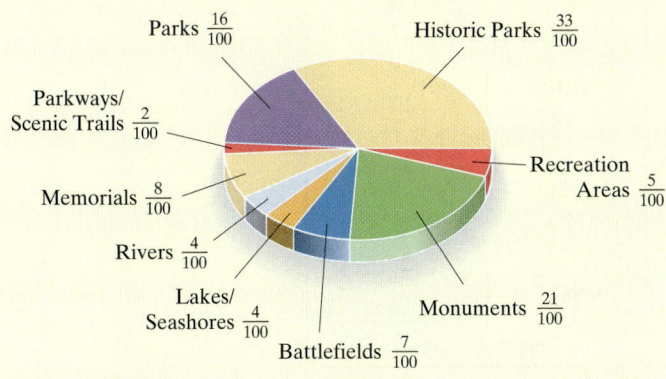

Areas Maintained by the National Park Service

Parks $\frac{16}{100}$ Historic Parks $\frac{33}{100}$

Parkways/ Scenic Trails $\frac{2}{100}$

Memorials $\frac{8}{100}$

Rivers $\frac{4}{100}$

Lakes/ Seashores $\frac{4}{100}$

Battlefields $\frac{7}{100}$

Monuments $\frac{21}{100}$

Recreation Areas $\frac{5}{100}$

Source: National Park Service

95. What fraction of National Park Service areas are National Memorials? $\frac{2}{25}$

96. What fraction of National Park Service areas are National Parks? $\frac{4}{25}$

97. Why is the National Battlefields sector smaller than the National Monuments sector?   answers may vary

98. Why is the National Lakes/National Seashores sector the same size as the National Rivers sector?   answers may vary

The area of a plane figure is a measure of the amount of surface of the figure. Find the area of each figure. (The area of a rectangle is the product of its length and width. The area of a triangle is $\frac{1}{2}$ the product of its base and height. Recall that area is measured in square units.)

△**99.**

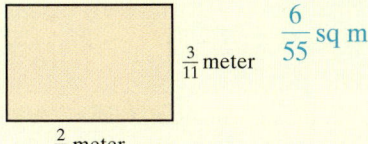

$\frac{3}{11}$ meter

$\frac{2}{5}$ meter

$\frac{6}{55}$ sq m

△**100.**

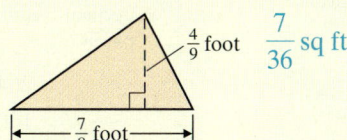

$\frac{4}{9}$ foot $\frac{7}{36}$ sq ft

$\frac{7}{8}$ foot

R.3   Decimals and Percents

Objective A Writing Decimals as Fractions

Like fractional notation, **decimal notation** is used to denote a part of a whole. Below is a **place value chart** that shows the value of each place.

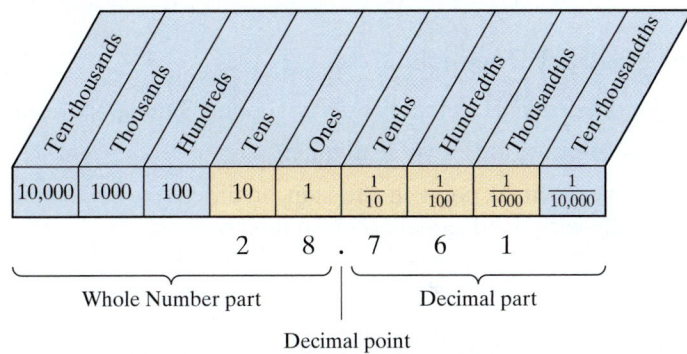

Ten-thousands	Thousands	Hundreds	Tens	Ones	Tenths	Hundredths	Thousandths	Ten-thousandths
10,000	1000	100	10	1	$\frac{1}{10}$	$\frac{1}{100}$	$\frac{1}{1000}$	$\frac{1}{10,000}$

2   8 . 7   6   1

Whole Number part   Decimal part

Decimal point

✓**Concept Check** Fill in the blank: In the number 52.634, the 3 is in the _____ place.

a. Tens   **b.** Ones   **c.** Tenths

d. Hundredths   **e.** Thousandths

Objectives

A Write Decimals as Fractions.

B Add, Subtract, Multiply, and Divide Decimals.

C Round Decimals to a Given Decimal Place.

D Write Fractions as Decimals.

E Write Percents as Decimals and Decimals as Percents.

Teaching Tip

Point out to students the symmetry about the ones column on both sides of the place value chart.

✓**Concept Check Answer**

d

The next chart shows decimals written as fractions.

Decimal Form	Fractional Form
0.1	$\dfrac{1}{10}$
0.07	$\dfrac{7}{100}$
2.31	$\dfrac{231}{100}$
0.9862	$\dfrac{9862}{10,000}$

0.1 ——— tenths ——— $\dfrac{1}{10}$

0.07 ——— hundredths ——— $\dfrac{7}{100}$

2.31 ——— hundredths ——— $\dfrac{231}{100}$

0.9862 ——— ten-thousandths ——— $\dfrac{9862}{10,000}$

To write a decimal as a fraction, use place values.

Practice 1–3

Write each decimal as a fraction. Do not simplify.
1. 0.27 **2.** 5.1 **3.** 7.685

Examples Write each decimal as a fraction. Do not simplify.

1. $0.37 = \dfrac{37}{100}$

2 decimal   2 zeros
places

2. $1.3 = \dfrac{13}{10}$

1 decimal   1 zero
place

3. $2.649 = \dfrac{2649}{1000}$

3 decimal   3 zeros
places

■ **Work Practice 1–3**

Objective B Adding, Subtracting, Multiplying, and Dividing Decimals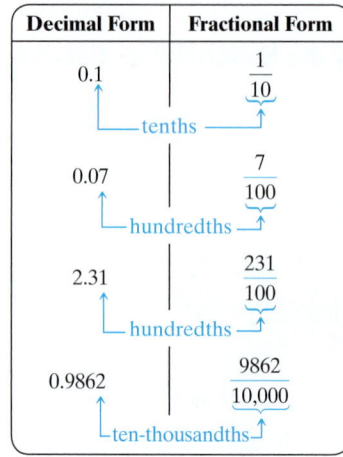

To **add** or **subtract** decimals, follow the steps below.

Teaching Tip

Before discussing adding and subtracting decimals, consider using the following activity: Ask students to explain in writing, as if to someone who knows nothing about our money system, how to add 53 cents and 17 dollars and 12 cents. Ask them to share their explanation. Then ask them to use decimals to represent the problem.

To Add or Subtract Decimals

Step 1: Write the decimals so that the decimal points line up vertically.

Step 2: Add or subtract as for whole numbers.

Step 3: Place the decimal point in the sum or difference so that it lines up vertically with the decimal points in the problem.

Notice that these steps simply ensure that we add or subtract digits with the same place value.

Answers

1. $\dfrac{27}{100}$ **2.** $\dfrac{51}{10}$ **3.** $\dfrac{7685}{1000}$

Example 4 Add.

a. $5.87 + 23.279 + 0.003$

b. $7 + 0.23 + 0.6$

Solution:

a.
```
    5.87
   23.279
 +  0.003
  ───────
   29.152
```

b.
```
   7.
   0.23
  +0.6
  ─────
   7.83
```

■ **Work Practice 4**

Practice 4

Add.

a. $7.19 + 19.782 + 1.006$

b. $12 + 0.79 + 0.03$

Example 5 Subtract.

a. $32.15 - 11.237$

b. $70 - 0.48$

Solution:

a.
```
     1 11 4 10
  3 2. 1̸ 5̸ 0̸
 − 1 1. 2 3 7
 ────────────
   2 0. 9 1 3
```

b.
```
    6 9  9 10
  7̸ 0̸. 0̸ 0̸
 −  0. 4 8
 ──────────
  6 9. 5 2
```

■ **Work Practice 5**

Practice 5

Subtract.

a. $84.23 - 26.982$

b. $90 - 0.19$

Now let's study the following product of decimals. Notice the pattern in the decimal points.

$$0.03 \times 0.6 = \frac{3}{100} \times \frac{6}{10} = \frac{18}{1000} \quad \text{or} \quad 0.018$$

2 decimal places   1 decimal place   3 decimal places

In general, to **multiply** decimals, follow the steps below.

To Multiply Decimals

Step 1: Multiply the decimals as though they are whole numbers.

Step 2: The decimal point in the product is placed so that the number of decimal places in the product is equal to the **sum** of the number of decimal places in the factors.

Example 6 Multiply.

a. 0.072×3.5

b. 0.17×0.02

Solution:

a.
```
    0.072    3 decimal places
  × 3.5      1 decimal place
  ──────
    360
    216
  ──────
  0.2520    4 decimal places
```

b.
```
    0.17     2 decimal places
  × 0.02     2 decimal places
  ──────
  0.0034     4 decimal places
```

■ **Work Practice 6**

Practice 6

Multiply.

a. 0.31×4.6

b. 1.26×0.03

Answers

4. a. 27.978 **b.** 12.82 **5. a.** 57.248
b. 89.81 **6. a.** 1.426 **b.** 0.0378

To divide a decimal by a whole number using long division, we place the decimal point in the quotient directly above the decimal point in the dividend. For example,

$$
\begin{array}{r}
2.47 \\
3\overline{)7.41} \\
-6 \quad\quad \\
\overline{1\,4} \\
-1\,2 \\
\overline{21} \\
-21 \\
\overline{0}
\end{array}
$$

To check, see that $2.47 \times 3 = 7.41$

Helpful Hint   Don't forget the names of the numbers in a division problem.

$$\text{divisor}\overline{)\text{dividend}}^{\text{quotient}}$$

In general, to **divide** decimals, use the steps below.

To Divide Decimals

Step 1: Move the decimal point in the divisor to the right until the divisor is a whole number.

Step 2: Move the decimal point in the dividend to the right the **same number of places** as the decimal point was moved in Step 1.

Step 3: Divide. The decimal point in the quotient is directly over the moved decimal point in the dividend.

Practice 7

Divide.

a. $21.75 \div 0.5$

b. $15.6 \div 0.006$

Example 7   Divide.

a. $9.46 \div 0.04$

b. $31.5 \div 0.007$

Solution:

a.
$$
\begin{array}{r}
236.5 \\
004.\overline{)946.0} \\
-8\quad\quad\quad \\
\overline{14} \\
-12 \\
\overline{26} \\
-24 \\
\overline{2\,0} \\
-2\,0 \\
\overline{0}
\end{array}
$$
A zero is inserted to continue dividing.

b.
$$
\begin{array}{r}
4500. \\
0007.\overline{)31500.} \\
-28\quad\quad \\
\overline{35} \\
-35 \\
\overline{0}
\end{array}
$$
Zeros are inserted in order to move the decimal point three places to the right.

Work Practice 7

Objective C Rounding Decimals

We **round** the decimal part of a decimal number in nearly the same way as we round whole numbers. The only difference is that we drop digits to the right of the rounding place, instead of replacing these digits by 0s. For example,

24.954   rounded to the nearest hundredth is   24.95

↑
hundredths place

Answers

7. a. 43.5   **b.** 2600

To Round Decimals to a Place Value to the Right of the Decimal Point

Step 1: Locate the digit to the right of the given place value.

Step 2: • If this digit is 5 or greater, add 1 to the digit in the given place value and drop all digits to its right.

   • If this digit is less than 5, drop all digits to the right of the given place.

Example 8 Round 7.8265 to the nearest hundredth.

Solution:

hundredths place

7.8265

Step 1. Locate the digit to the right of the hundredths place.
Step 2. This digit is 5 or greater, so we add 1 to the hundredths place digit and drop all digits to its right.

Thus, 7.8265 rounded to the nearest hundredth is 7.83.

■ **Work Practice 8**

Practice 8

Round 12.9187 to the nearest hundredth.

Example 9 Round 19.329 to the nearest tenth.

Solution:

tenths place

19.329

Step 1. Locate the digit to the right of the tenths place.
Step 2. This digit is less than 5, so we drop this digit and all digits to its right.

Thus, 19.329 rounded to the nearest tenth is 19.3.

■ **Work Practice 9**

Practice 9

Round 245.348 to the nearest tenth.

Objective D Writing Fractions as Decimals ▶

To write fractions as decimals, interpret the fraction bar as division and find the quotient.

To Write a Fraction as a Decimal

Divide the numerator by the denominator.

Example 10 Write $\frac{1}{4}$ as a decimal.

Solution:

$$
\begin{array}{r}
0.25 \\
4\overline{)1.00} \\
-8 \\
\hline
20 \\
-20 \\
\hline
0
\end{array}
\qquad \frac{1}{4} = 0.25
$$

■ **Work Practice 10**

Practice 10

Write $\frac{2}{5}$ as a decimal.

Practice 11

Write $\dfrac{5}{6}$ as a decimal.

Example 11 Write $\dfrac{2}{3}$ as a decimal.

Solution:

$$\begin{array}{r} 0.666 \\ 3\overline{)2.000} \\ -18 \\ \hline 20 \\ -18 \\ \hline 20 \\ -18 \\ \hline 2 \end{array}$$

This division pattern will continue so that $\dfrac{2}{3} = 0.6666\ldots$

A bar can be placed over the digit 6 to indicate that it repeats. We call this a **repeating decimal.**

$$\frac{2}{3} = 0.666\ldots = 0.\overline{6}$$

■ **Work Practice 11**

We can also write a decimal approximation for $\dfrac{2}{3}$. For example, $\dfrac{2}{3}$ rounded to the nearest hundredth is 0.67. This can be written as $\dfrac{2}{3} \approx 0.67$. The $\approx$ sign means "is approximately equal to."

✓**Concept Check** The notation $0.5\overline{2}$ is the same as

a. $\dfrac{52}{100}$ **b.** $\dfrac{52\ldots}{100}$ **c.** $0.52222222\ldots$

Practice 12

Write $\dfrac{1}{9}$ as a decimal. Round to the nearest thousandth.

Example 12 Write $\dfrac{22}{7}$ as a decimal. Round to the nearest hundredth.

Solution:

$$\begin{array}{r} 3.142 \approx 3.14 \\ 7\overline{)22.000} \\ -21 \\ \hline 1\,0 \\ -7 \\ \hline 30 \\ -28 \\ \hline 20 \\ -14 \\ \hline 6 \end{array}$$

If rounding to the nearest hundredth, carry the division process out to one more decimal place, the thousandths place.

The fraction $\dfrac{22}{7}$ in decimal form is approximately 3.14. (The fraction $\dfrac{22}{7}$ is an approximation for π.)

■ **Work Practice 12**

Objective E Writing Percents as Decimals and Decimals as Percents ▶

The word **percent** comes from the Latin phrase *per centum*, which means **"per 100."** The % symbol is used to denote percent. Thus, 53% means 53 per 100, or

$$53\% = \frac{53}{100}$$

Answers

11. $0.8\overline{3}$ **12.** 0.111

✓**Concept Check Answer**

c

When solving problems containing percents, it is often necessary to write a percent as a decimal. To see how this is done, study the chart below.

Percent	Fraction	Decimal
7%	$\dfrac{7}{100}$	0.07
63%	$\dfrac{63}{100}$	0.63
109%	$\dfrac{109}{100}$	1.09

To convert directly from a percent to a decimal, notice that

$7\% = 0.07$

To Write a Percent as a Decimal

Drop the percent symbol, %, and move the decimal point two places to the left.

Example 13   Write each percent as a decimal.

a. 25%   **b.** 2.6%   **c.** 195%

Solution: We drop the % and move the decimal point two places to the left. Recall that the decimal point of a whole number is to the right of the ones place digit.

a. $25\% = 25.\% = 0.25$

b. $2.6\% = 02.6\% = 0.026$

c. $195\% = 195.\% = 1.95$

■ **Work Practice 13**

To write a decimal as a percent, we simply reverse the preceding steps. That is, we move the decimal point two places to the right and attach the percent symbol, %.

To Write a Decimal as a Percent

Move the decimal point two places to the right and attach the percent symbol, %.

Example 14   Write each decimal as a percent.

a. 0.85   **b.** 1.25   **c.** 0.012   **d.** 0.6

Solution: We move the decimal point two places to the right and attach the percent symbol, %.

a. $0.85 = 0.85 = 85\%$

b. $1.25 = 1.25 = 125\%$

c. $0.012 = 0.012 = 1.2\%$

d. $0.6 = 0.60 = 60\%$

■ **Work Practice 14**

Practice 13

Write each percent as a decimal.

a. 20%

b. 1.4%

c. 465%

Practice 14

Write each decimal as a percent.

a. 0.42

b. 0.003

c. 2.36

d. 0.7

Answers

13. a. 0.20   **b.** 0.014   **c.** 4.65

14. a. 42%   **b.** 0.3%   **c.** 236%

d. 70%

Vocabulary, Readiness & Video Check

Fill in each blank with one of the choices listed below. Some choices may be used more than once and some not at all.

vertically	decimal	right	100%	percent
left	0.01	sum	denominator	numerator

1. Like fractional notation, _____decimal_____ notation is used to denote a part of a whole.
2. To write fractions as decimals, divide the _____numerator_____ by the _____denominator_____.
3. To add or subtract decimals, write the decimals so that the decimal points line up _____vertically_____.
4. When multiplying decimals, the decimal point in the product is placed so that the number of decimal places in the product is equal to the _____sum_____ of the number of decimal places in the factors.
5. _____Percent_____ means "per hundred."
6. _____100%_____ = 1.
7. The % symbol is read as _____percent_____.
8. To write a percent as a *decimal*, drop the % symbol and move the decimal point two places to the _____left_____.
9. To write a decimal as a *percent*, move the decimal point two places to the _____right_____ and attach the % symbol.

Martin-Gay Interactive Videos   Watch the section lecture video and answer the following questions.

See Video R.3 🍒

See video answer section.

Objective A 10. From ▥ Example 1, how does reading a decimal number correctly help us write it as an equivalent fraction? ▶

Objective B 11. From ▥ Examples 3 and 4, complete each statement with "do" or "do not." When adding or subtracting decimal numbers, we _____ line up decimal points. When multiplying decimal numbers, we _____ need to line up decimal points. ▶

Objective C 12. From the lecture before ▥ Example 7, explain the difference between rounding whole numbers and rounding decimal numbers to a place to the right of the decimal point. ▶

Objective D 13. Complete this statement based on the lecture before ▥ Example 8. To write a fraction as a decimal, we divide the _____ by the _____. ▶

Objective E 14. In ▥ Example 13, 100% equals what natural number? ▶

R.3 Exercise Set MyMathLab®

Objective A *Write each decimal as a fraction. Do not simplify. See Examples 1 through 3.*

▶1. 0.6 $\dfrac{6}{10}$

2. 0.9 $\dfrac{9}{10}$

▶3. 1.86 $\dfrac{186}{100}$

4. 7.23 $\dfrac{723}{100}$

5. 0.114 $\dfrac{114}{1000}$

6. 0.239 $\dfrac{239}{1000}$

7. 123.1 $\dfrac{1231}{10}$

8. 892.7 $\dfrac{8927}{10}$

Objective B *Add or subtract as indicated. See Examples 4 and 5.*

9. 5.7 + 1.13 6.83

10. 2.31 + 6.4 8.71

11. 24.6 + 2.39 + 0.0678 27.0578

12. 32.4 + 1.58 + 0.0934 34.0734

13. 8.8 − 2.3 6.5

14. 7.6 − 2.1 5.5

15. 18 − 2.78 15.22

16. 28 − 3.31 24.69

Multiply or divide as indicated. See Examples 6 and 7.

17.
$$\begin{array}{r} 0.2 \\ \times\ 0.6 \\ \hline 0.12 \end{array}$$

18.
$$\begin{array}{r} 0.7 \\ \times\ 0.9 \\ \hline 0.63 \end{array}$$

19.
$$\begin{array}{r} 0.063 \\ \times\ \ 4.2 \\ \hline 0.2646 \end{array}$$

20.
$$\begin{array}{r} 0.079 \\ \times\ \ 3.6 \\ \hline 0.2844 \end{array}$$

21. $5\overline{)8.4}$ 1.68

22. $2\overline{)11.7}$ 5.85

23. $0.82\overline{)4.756}$ 5.8

24. $0.92\overline{)3.312}$ 3.6

Mixed Practice *Perform the indicated operation. See Examples 4 through 7.*

25.
$$\begin{array}{r} 45.02 \\ 3.006 \\ +\ 8.405 \\ \hline 56.431 \end{array}$$

26.
$$\begin{array}{r} 65.0028 \\ 5.0903 \\ +\ 6.9 \\ \hline 76.9931 \end{array}$$

27.
$$\begin{array}{r} 6.75 \\ \times\ \ 10 \\ \hline 67.5 \end{array}$$

28.
$$\begin{array}{r} 8.91 \\ \times\ 100 \\ \hline 891 \end{array}$$

29. $0.6\overline{)42}$ 70

30. $0.9\overline{)36}$ 40

31.
$$\begin{array}{r} 654.9 \\ -\ 56.67 \\ \hline 598.23 \end{array}$$

32.
$$\begin{array}{r} 863.2 \\ -\ 39.45 \\ \hline 823.75 \end{array}$$

33.
$$\begin{array}{r} 5.62 \\ \times\ 7.7 \\ \hline 43.274 \end{array}$$

34.
$$\begin{array}{r} 8.03 \\ \times\ 5.5 \\ \hline 44.165 \end{array}$$

35. $0.063\overline{)52.92}$ 840

36. $0.054\overline{)51.84}$ 960

37.
$$\begin{array}{r} 16.003 \\ \times\ 5.31 \\ \hline 84.97593 \end{array}$$

38.
$$\begin{array}{r} 31.006 \\ \times\ 3.71 \\ \hline 115.03226 \end{array}$$

Objective C *Round each decimal to the given place value. See Examples 8 and 9.*

39. 0.57, nearest tenth 0.6

40. 0.75, nearest tenth 0.8

41. 0.234, nearest hundredth 0.23

42. 0.452, nearest hundredth 0.45

43. 0.5945, nearest thousandth 0.595

44. 63.4529, nearest thousandth 63.453

45. 98,207.23, nearest tenth 98,207.2

46. 68,936.543, nearest tenth 68,936.5

47. 12.347, nearest hundredth 12.35

48. 42.9878, nearest thousandth 42.988

Objective D *Write each fraction as a decimal. If the decimal is a repeating decimal, write using the bar notation and then round to the nearest hundredth. See Examples 10 through 12.*

49. $\dfrac{3}{4}$ 0.75

50. $\dfrac{9}{25}$ 0.36

51. $\dfrac{1}{3}$
$0.\overline{3} \approx 0.33$

52. $\dfrac{7}{9}$
$0.\overline{7} \approx 0.78$

53. $\dfrac{7}{16}$ 0.4375

54. $\dfrac{5}{8}$ 0.625

55. $\dfrac{6}{11}$
$0.\overline{54} \approx 0.55$

56. $\dfrac{1}{6}$ $0.1\overline{6} \approx 0.17$

57. $\dfrac{29}{6}$
$4.8\overline{3} \approx 4.83$

58. $\dfrac{34}{9}$ $3.\overline{7} \approx 3.78$

Objective E *Write each percent as a decimal. See Example 13.*

▶ **59.** 28%   0.28   **60.** 36%   0.36   ▶ **61.** 3.1%   0.031   **62.** 2.2%   0.022

63. 135%   1.35   **64.** 417%   4.17   **65.** 200%   2   **66.** 700%   7

67. 96.55%   0.9655   **68.** 81.49%   0.8149   **69.** 0.1%   0.001   **70.** 0.6%   0.006

71. In the United States recently, 15.8% of households had no landlines, just cell phones. (*Source:* CTIA— The Wireless Association)   0.158

72. Japan exports 73.2% of all motorcycles manufactured there. (*Source:* Japan Automobile Manufacturers Association)   0.732

Write each decimal as a percent. See Example 14.

▶ **73.** 0.68   68%   **74.** 0.32   32%   **75.** 0.876   87.6%   **76.** 0.521   52.1%

▶ **77.** 1   100%   **78.** 3   300%   **79.** 0.5   50%   **80.** 0.1   10%

81. 1.92   192%   **82.** 2.15   215%   **83.** 0.004   0.4%   **84.** 0.005   0.5%

85. In a recent year, 0.781 of all electricity produced in France was nuclear generated.   78.1%

86. The United States' share of the total world motor vehicle production is 0.142. (*Source:* World Almanac)   14.2%

Concept Extensions

In Exercises 87 through 90, write the percent from the circle graph as a decimal and a fraction.

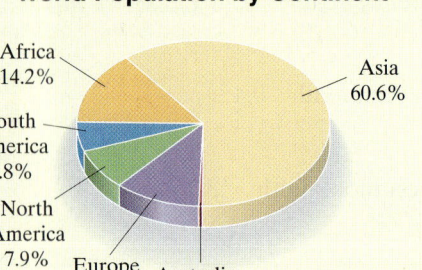

World Population by Continent

Africa 14.2%
Asia 60.6%
South America 5.8%
North America 7.9%
Europe 11.0%
Australia 0.5%

87. Australia: 0.5% $0.005; \dfrac{1}{200}$

88. Europe: 11% $0.11; \dfrac{11}{100}$

89. Africa: 14.2% $0.142; \dfrac{71}{500}$

90. Asia: 60.6% $0.606; \dfrac{303}{500}$

Solve. See the Concept Checks in this section.

91. In the number 3.659, identify the place value of the
  a. 6   tenths   **b.** 9   thousandths   **c.** 3   ones

92. The notation $0.\overline{67}$ is the same as
  a. 0.6777 . . .   **b.** 0.67666 . . .   **c.** 0.6767 . . .   c

93. In your own words, describe how to multiply decimal numbers.   answers may vary

94. In your own words, describe how to add or subtract decimal numbers.   answers may vary

The chart shows the average number of pounds of various dairy products consumed by each U.S. citizen. Use this chart for Exercises 95 and 96. (Source: Dairy Information Center)

Dairy Product	Pounds
Fluid Milk	213.4
Cheese	30.8
Butter	4.4

95. How much more fluid milk products than cheese products does the average U.S. citizen consume?   182.6 lb

96. What is the total amount of these milk products consumed by the average U.S. citizen annually?   248.6 lb

97. Given the percent 52.8647%, round as indicated.
  a. Round to the nearest tenth percent.   52.9%
  b. Round to the nearest hundredth percent.
     52.86%

98. Given the percent 0.5269%, round as indicated.
  a. Round to the nearest tenth percent.   0.5%
  b. Round to the nearest hundredth percent.
     0.53%

99. Which of the following are correct?
  a. 6.5% = 0.65
  b. 7.8% = 0.078
  c. 120% = 0.12
  d. 0.35% = 0.0035   b, d

100. Which of the following are correct?
  a. 0.231 = 23.1%
  b. 5.12 = 0.0512%
  c. 3.2 = 320%
  d. 0.0175 = 0.175%   a, c

Recall that 1 = 100%. This means that 1 whole is 100%. Use this for Exercises 101 and 102. (Source: Some Body, by Dr. Pete Rowen)

101. The four blood types are A, B, O, and AB. (Each blood type can also be further classified as Rh-positive or Rh-negative depending upon whether your blood contains protein or not.) Given the percent blood types for people in the United States below, calculate the percent of the U.S. population with AB blood type.   4%

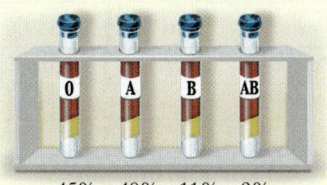

45%   40%   11%   ?%

102. The top four components of bone are below. Find the missing percent.
  1. Minerals—45%
  2. Living tissue—30%
  3. Water—20%
  4. Other—?   5%

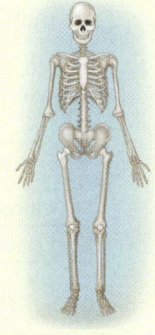

The bar graph shows the predicted fastest-growing occupations. Use this graph for Exercises 103 through 106.

Fastest-Growing Occupations 2006–2016

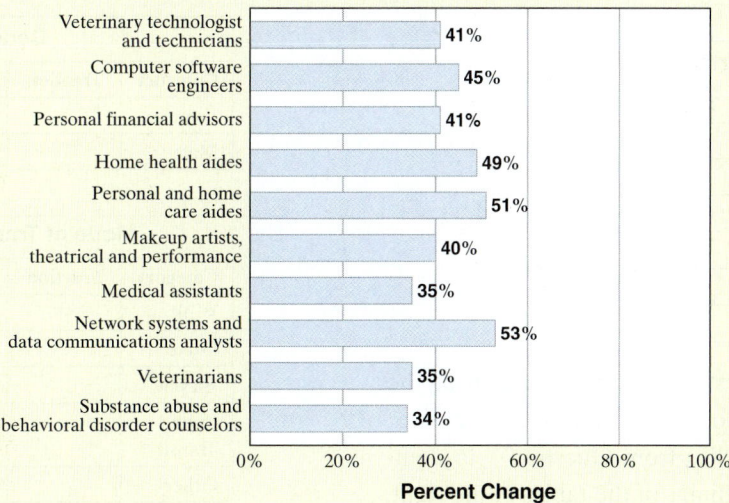

Source: Bureau of Labor Statistics

103. What occupation is predicted to be the fastest growing?   network systems and data communications analysts

104. What occupation is predicted to be the second fastest growing?   personal and home care aides

105. Write the percent change for veterinarians as a decimal.   0.35

106. Write the percent change for makeup artists as a decimal.   0.40

107. In your own words, explain how to write a percent as a decimal.   answers may vary

108. In your own words, explain how to write a decimal as a percent.   answers may vary

Chapter R Group Activity

Additional group activities are available in the *Instructor's Resource Manual with Tests.*

Interpreting Survey Results

This activity may be completed by working in groups or individually.

Conduct the following survey with 12 students in one of your classes and record the results.

a. What is your age?

   Under 20 20s 30s 40s 50s 60 and older

b. What is your gender?

   Female Male

c. How did you arrive on campus today?

   Walked Drove Bicycled

   Took public transportation Other

1. For each survey question, tally the results for each category. answers may vary

Age

Category	Tally
Under 20	
20s	
30s	
40s	
50s	
60+	
Total	

Gender

Category	Tally
Female	
Male	
Total	

Mode of Transportation

Category	Tally
Walk	
Drive	
Bicycle	
Public Transit	
Other	
Total	

2. For each survey question, find the fraction of the total number of responses that fall in each answer category. Use the tallies from Question 1 to complete the Fraction columns of the tables at the right.

answers may vary

3. For each survey question, convert the fraction of the total number of responses that fall in each answer category to a decimal number. Use the fractions from Question 2 to complete the Decimal columns of the tables below.

answers may vary

4. For each survey question, find the percent of the total number of responses that falls in each answer category. Complete the Percent columns of the tables below.

answers may vary

5. Study the tables. What may you conclude from them? What do they tell you about your survey respondents? Write a paragraph summarizing your findings.

answers may vary

Age

Category	Fraction	Decimal	Percent
Under 20			
20s			
30s			
40s			
50s			
60+			

Gender

Category	Fraction	Decimal	Percent
Female			
Male			

Mode of Transportation

Category	Fraction	Decimal	Percent
Walk			
Drive			
Bicycle			
Public Transit			
Other			

Chapter R Vocabulary Check

Fill in each blank with one of the words or phrases listed below.

mixed number	factor	improper fraction	percent
multiple	composite number	proper fraction	simplified
prime number	equivalent		

1. To _____factor_____ means to write as a product.

2. A(n) _____multiple_____ of a number is the product of that number and any natural number.

3. A(n) __composite number__ is a natural number greater than 1 that is not prime.

4. The word _____percent_____ means per 100.

5. Fractions that represent the same portion of a whole are called _____equivalent_____ fractions.

6. A(n) __improper fraction__ is a fraction whose numerator is greater than or equal to its denominator.

7. A(n) ____prime number____ is a natural number greater than 1 whose only factors are 1 and itself.

8. A fraction is _____simplified_____ when the numerator and the denominator have no factors in common other than 1.

9. A(n) ____proper fraction____ is one whose numerator is less than its denominator.

10. A(n) ____mixed number____ contains a whole number part and a fraction part.

> **Helpful Hint**
>
> ▶ Are you preparing for your test? Don't forget to take the Chapter R Test on page R-36. Then check your answers at the back of the text and use the Chapter Test Prep Videos to see the fully worked-out solutions to any of the exercises you want to review.

R Chapter Highlights

Definitions and Concepts	Examples
Section R.1 Factors and the Least Common Multiple	
To **factor** means to write as a product. Since $2 \cdot 6 = 12$, 2 and 6 are **factors** of 12.	The factors of 12 are $1, 2, 3, 4, 6, 12$
When a number is written as a product of primes, this product is called the **prime factorization** of a number.	Write the prime factorization of 60. $60 = 6 \cdot 10$ $\quad\ \downarrow\ \downarrow\quad \searrow$ $= 2 \cdot 3 \cdot 2 \cdot 5$ The prime factorization of 60 is $2 \cdot 2 \cdot 3 \cdot 5$.
The **least common multiple (LCM)** of a list of numbers is the smallest number that is a multiple of all the numbers in the list.	
To Find the LCM of a List of Numbers	Find the LCM of 12 and 40.
Step 1: Write the prime factorization of each number.	$12 = 2 \cdot 2 \cdot 3$
Step 2: Write the product containing each different prime factor (from Step 1) the greatest number of times that it appears in any one factorization. This product is the LCM.	$40 = 2 \cdot 2 \cdot 2 \cdot 5$ $\mathrm{LCM} = 2 \cdot 2 \cdot 2 \cdot 3 \cdot 5 = 120$

Definitions and Concepts	Examples
Section R.2 Fractions	

Definitions and Concepts	Examples
Fractions that represent the same portion of a whole are called **equivalent fractions.**	$\dfrac{1}{5} = \dfrac{1 \cdot 4}{5 \cdot 4} = \dfrac{4}{20}$ $\dfrac{1}{5}$ and $\dfrac{4}{20}$ are equivalent fractions.
To write an equivalent fraction, $\dfrac{a}{b} = \dfrac{a}{b} \cdot \dfrac{c}{c} = \dfrac{a \cdot c}{b \cdot c}$	Write $\dfrac{8}{21}$ as an equivalent fraction with a denominator of 63. $\dfrac{8}{21} = \dfrac{8}{21} \cdot \dfrac{3}{3} = \dfrac{8 \cdot 3}{21 \cdot 3} = \dfrac{24}{63}$
A fraction is **simplified** when the numerator and the denominator have no factors in common other than 1.	$\dfrac{13}{17}$ is simplified.
To simplify a fraction, $\dfrac{a \cdot c}{b \cdot c} = \dfrac{a}{b} \cdot \dfrac{c}{c} = \dfrac{a}{b}$	Simplify. $\dfrac{6}{14} = \dfrac{2 \cdot 3}{2 \cdot 7} = \dfrac{2}{2} \cdot \dfrac{3}{7} = \dfrac{3}{7}$
Two fractions are **reciprocals** if their product is 1. The reciprocal of $\dfrac{a}{b}$ is $\dfrac{b}{a}$, as long as a and b are not 0.	The reciprocal of $\dfrac{6}{25}$ is $\dfrac{25}{6}$.
To multiply fractions, multiply numerator times numerator to find the numerator of the product and denominator times denominator to find the denominator of the product.	$\dfrac{2}{5} \cdot \dfrac{3}{7} = \dfrac{6}{35}$
To divide fractions, multiply the first fraction by the reciprocal of the second fraction.	$\dfrac{5}{9} \div \dfrac{2}{7} = \dfrac{5}{9} \cdot \dfrac{7}{2} = \dfrac{35}{18}$
To add fractions with the same denominator, add the numerators and place the sum over the common denominator.	$\dfrac{5}{11} + \dfrac{3}{11} = \dfrac{8}{11}$
To subtract fractions with the same denominator, subtract the numerators and place the difference over the common denominator.	$\dfrac{13}{15} - \dfrac{3}{15} = \dfrac{10}{15} = \dfrac{2}{3}$
To add or subtract fractions with different denominators, first write each fraction as an equivalent fraction with the LCD as denominator.	$\dfrac{2}{9} + \dfrac{3}{6} = \dfrac{2 \cdot 2}{9 \cdot 2} + \dfrac{3 \cdot 3}{6 \cdot 3} = \dfrac{4 + 9}{18} = \dfrac{13}{18}$

Definitions and Concepts	Examples
Section R.3 Decimals and Percents	

Definitions and Concepts	Examples
To write decimals as fractions, use place values.	$0.11 = \dfrac{11}{100}$
To Add or Subtract Decimals **Step 1:** Write the decimals so that the decimal points line up vertically. **Step 2:** Add or subtract as for whole numbers. **Step 3:** Place the decimal point in the sum or difference so that it lines up vertically with the decimal points in the problem.	Subtract: $2.8 - 1.04$ Add: $25 + 0.02$ $\begin{array}{r} {\scriptstyle 7\ 10} \\ 2.8\llap{/}0 \\ -1.0\ 4 \\ \hline 1.7\ 6 \end{array}$ $\begin{array}{r} 25. \\ +\ 0.02 \\ \hline 25.02 \end{array}$

(continued)

Definitions and Concepts	Examples
Section R.3 Decimals and Percents (*continued*)	

To Multiply Decimals

Step 1: Multiply the decimals as though they are whole numbers.

Step 2: The decimal point in the product is placed so that the number of decimal places in the product is equal to the **sum** of the number of decimal places in the factors.

Multiply: 1.48×5.9

$$
\begin{array}{r}
1.4\,8 \quad \leftarrow 2 \text{ decimal places} \\
\times \quad 5.9 \quad \leftarrow 1 \text{ decimal place} \\
\hline
1\,3\,3\,2 \\
7\,4\,0 \\
\hline
8.7\,3\,2 \quad \leftarrow 3 \text{ decimal places}
\end{array}
$$

To Divide Decimals

Step 1: Move the decimal point in the divisor to the right until the divisor is a whole number.

Step 2: Move the decimal point in the dividend to the right the **same number of places** as the decimal point was moved in Step 1.

Step 3: Divide. The decimal point in the quotient is directly over the moved decimal point in the dividend.

Divide: $1.118 \div 2.6$

$$
\begin{array}{r}
0.43 \\
26\overline{\smash{)}11.18} \\
-10\,4 \\
\hline
78 \\
-78 \\
\hline
0
\end{array}
$$

To write fractions as decimals, divide the numerator by the denominator.

Write $\dfrac{3}{8}$ as a decimal.

$$
\begin{array}{r}
0.375 \\
8\overline{\smash{)}3.000} \\
-2\,4 \\
\hline
60 \\
-56 \\
\hline
40 \\
-40 \\
\hline
0
\end{array}
$$

To write a percent as a decimal, drop the percent symbol, %, and move the decimal point two places to the left.

$25\% = 25.\% = 0.25$

To write a decimal as a percent, move the decimal point two places to the right and attach the percent symbol, %.

$0.7 = 0.70 = 70\%$

Chapter R   Review

(R.1) *Write the prime factorization of each number.*

1. 42 $2 \cdot 3 \cdot 7$

2. 800 $2 \cdot 2 \cdot 2 \cdot 2 \cdot 2 \cdot 5 \cdot 5$

Find the least common multiple (LCM) of each list of numbers.

3. 12, 30   60

4. 7, 42   42

5. 4, 6, 10   60

6. 2, 5, 7   70

(R.2) *Write each fraction as an equivalent fraction with the given denominator.*

7. $\dfrac{5}{8}$ with a denominator of 24 $\dfrac{15}{24}$

8. $\dfrac{2}{3}$ with a denominator of 60 $\dfrac{40}{60}$

Simplify each fraction.

9. $\dfrac{8}{20}$ $\dfrac{2}{5}$

10. $\dfrac{15}{100}$ $\dfrac{3}{20}$

11. $\dfrac{12}{6}$ 2

12. $\dfrac{8}{8}$ 1

Perform each indicated operation and simplify.

13. $\dfrac{1}{7} \cdot \dfrac{8}{11}$ $\dfrac{8}{77}$

14. $\dfrac{5}{12} + \dfrac{2}{15}$ $\dfrac{11}{20}$

15. $\dfrac{3}{10} \div 6$ $\dfrac{1}{20}$

16. $\dfrac{7}{9} - \dfrac{1}{6}$ $\dfrac{11}{18}$

17. $3\dfrac{3}{8} \cdot 4\dfrac{1}{4}$ $14\dfrac{11}{32}$

18. $2\dfrac{1}{3} - 1\dfrac{5}{6}$ $\dfrac{1}{2}$

19. $16\dfrac{9}{10} + 3\dfrac{2}{3}$ $20\dfrac{17}{30}$

20. $6\dfrac{2}{7} \div 2\dfrac{1}{5}$ $2\dfrac{6}{7}$

The area of a plane figure is a measure of the amount of surface of the figure. Find the area of each figure below. (The area of a rectangle is the product of its length and width. The area of a triangle is $\dfrac{1}{2}$ the product of its base and height.)

△ **21.**

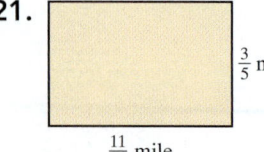

$\dfrac{3}{5}$ mile

$\dfrac{11}{12}$ mile

$\dfrac{11}{20}$ sq mi

△ **22.**

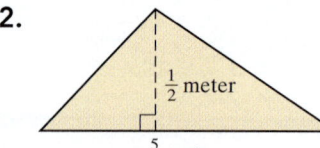

$\dfrac{1}{2}$ meter

$\dfrac{5}{4}$ meters

$\dfrac{5}{16}$ sq m

R-34

(R.3) *Write each decimal as a fraction. Do not simplify.*

23. 1.81 $\dfrac{181}{100}$

24. 0.035 $\dfrac{35}{1000}$

Perform each indicated operation.

25.
$$\begin{array}{r} 76.358 \\ +18.76 \\ \hline 95.118 \end{array}$$

26. $35 + 0.02 + 1.765$
36.785

27. $18 - 4.62$ 13.38

28.
$$\begin{array}{r} 804.062 \\ -112.489 \\ \hline 691.573 \end{array}$$

29.
$$\begin{array}{r} 7.6 \\ \times\ 12 \\ \hline 91.2 \end{array}$$

30.
$$\begin{array}{r} 14.63 \\ \times\ \ 3.2 \\ \hline 46.816 \end{array}$$

31. $27\overline{)772.2}$ 28.6

32. $0.06\overline{)13.8}$ 230

Round each decimal to the given place value.

33. 0.7652, nearest hundredth   0.77

34. 25.6293, nearest tenth   25.6

Write each fraction as a decimal. If the decimal is a repeating decimal, write it using the bar notation and then round to the nearest thousandth.

35. $\dfrac{1}{2}$ 0.5

36. $\dfrac{3}{8}$ 0.375

37. $\dfrac{4}{11}$ $0.\overline{36} \approx 0.364$

38. $\dfrac{5}{6}$ $0.8\overline{3} \approx 0.833$

Write each percent as a decimal.

39. 29%   0.29

40. 1.4%   0.014

Write each decimal as a percent.

41. 0.39   39%

42. 1.2   120%

43. In 2003, the home ownership rate in the United States was 68.3%. Write this percent as a decimal.   0.683

44. Choose the true statement.
 a. $2.3\% = 0.23$
 b. $5 = 500\%$
 c. $40\% = 4$ b

Answers

Note to Instructor: The Chapter R Test file in the TestGen program provides algorithms specifically matched to this test for easy replication for practice or assessment purposes.

1. $2 \cdot 2 \cdot 2 \cdot 3 \cdot 3$

2. 180

3. $\dfrac{25}{60}$

4. $\dfrac{3}{4}$

5. $\dfrac{12}{25}$

6. $\dfrac{13}{10}$

7. $\dfrac{53}{40}$

8. $\dfrac{18}{49}$

9. $\dfrac{1}{20}$

10. $\dfrac{29}{36}$

11. $4\dfrac{8}{9}$

12. $2\dfrac{5}{22}$

13. 55

14. $13\dfrac{13}{20}$

15. 45.11

16. 65.88

17. 12.688

18. 320

19. 23.73

20. 0.875

1. Write the prime factorization of 72.

2. Find the LCM of 5, 18, 20.

3. Write $\dfrac{5}{12}$ as an equivalent fraction with a denominator of 60.

Simplify each fraction.

4. $\dfrac{15}{20}$

5. $\dfrac{48}{100}$

6. Write 1.3 as a fraction.

Perform each indicated operation and simplify.

7. $\dfrac{5}{8} + \dfrac{7}{10}$

8. $\dfrac{2}{3} \cdot \dfrac{27}{49}$

9. $\dfrac{9}{10} \div 18$

10. $\dfrac{8}{9} - \dfrac{1}{12}$

11. $1\dfrac{2}{9} + 3\dfrac{2}{3}$

12. $5\dfrac{6}{11} - 3\dfrac{7}{22}$

13. $6\dfrac{7}{8} \div \dfrac{1}{8}$

14. $2\dfrac{1}{10} \cdot 6\dfrac{1}{2}$

Perform each indicated operation.

15. $43 + 0.21 + 1.9$

16. $123.6 - 57.72$

17. $\begin{array}{r} 7.93 \\ \times\ 1.6 \\ \hline \end{array}$

18. $0.25\overline{)80}$

19. Round 23.7272 to the nearest hundredth.

20. Write $\dfrac{7}{8}$ as a decimal.

(R.3) *Write each decimal as a fraction. Do not simplify.*

23. 1.81 $\dfrac{181}{100}$

24. 0.035 $\dfrac{35}{1000}$

Perform each indicated operation.

25.
```
  76.358
 +18.76
 ───────
  95.118
```

26. 35 + 0.02 + 1.765
36.785

27. 18 − 4.62   13.38

28.
```
  804.062
 −112.489
 ────────
  691.573
```

29.
```
    7.6
 ×  12
 ──────
   91.2
```

30.
```
   14.63
 ×   3.2
 ───────
  46.816
```

31. 27)‾772.2   28.6

32. 0.06)‾13.8   230

Round each decimal to the given place value.

33. 0.7652, nearest hundredth   0.77

34. 25.6293, nearest tenth   25.6

Write each fraction as a decimal. If the decimal is a repeating decimal, write it using the bar notation and then round to the nearest thousandth.

35. $\dfrac{1}{2}$ 0.5

36. $\dfrac{3}{8}$ 0.375

37. $\dfrac{4}{11}$ $0.\overline{36} \approx 0.364$

38. $\dfrac{5}{6}$ $0.8\overline{3} \approx 0.833$

Write each percent as a decimal.

39. 29%   0.29

40. 1.4%   0.014

Write each decimal as a percent.

41. 0.39   39%

42. 1.2   120%

43. In 2003, the home ownership rate in the United States was 68.3%. Write this percent as a decimal.   0.683

44. Choose the true statement.
   a. 2.3% = 0.23
   b. 5 = 500%
   c. 40% = 4   b

Answers

Note to Instructor: The Chapter R Test file in the TestGen program provides algorithms specifically matched to this test for easy replication for practice or assessment purposes.

1. $2 \cdot 2 \cdot 2 \cdot 3 \cdot 3$

2. 180

3. $\dfrac{25}{60}$

4. $\dfrac{3}{4}$

5. $\dfrac{12}{25}$

6. $\dfrac{13}{10}$

7. $\dfrac{53}{40}$

8. $\dfrac{18}{49}$

9. $\dfrac{1}{20}$

10. $\dfrac{29}{36}$

11. $4\dfrac{8}{9}$

12. $2\dfrac{5}{22}$

13. 55

14. $13\dfrac{13}{20}$

15. 45.11

16. 65.88

17. 12.688

18. 320

19. 23.73

20. 0.875

1. Write the prime factorization of 72.

2. Find the LCM of 5, 18, 20.

3. Write $\dfrac{5}{12}$ as an equivalent fraction with a denominator of 60.

Simplify each fraction.

4. $\dfrac{15}{20}$

5. $\dfrac{48}{100}$

6. Write 1.3 as a fraction.

Perform each indicated operation and simplify.

7. $\dfrac{5}{8} + \dfrac{7}{10}$

8. $\dfrac{2}{3} \cdot \dfrac{27}{49}$

9. $\dfrac{9}{10} \div 18$

10. $\dfrac{8}{9} - \dfrac{1}{12}$

11. $1\dfrac{2}{9} + 3\dfrac{2}{3}$

12. $5\dfrac{6}{11} - 3\dfrac{7}{22}$

13. $6\dfrac{7}{8} \div \dfrac{1}{8}$

14. $2\dfrac{1}{10} \cdot 6\dfrac{1}{2}$

Perform each indicated operation.

15. $43 + 0.21 + 1.9$

16. $123.6 - 57.72$

17. $\begin{array}{r} 7.93 \\ \times\ 1.6 \end{array}$

18. $0.25\overline{)80}$

19. Round 23.7272 to the nearest hundredth.

20. Write $\dfrac{7}{8}$ as a decimal.

21. Write $\dfrac{1}{6}$ as a repeating decimal. Then approximate the result to the nearest thousandth.

22. Write 63.2% as a decimal.

23. Write 0.09 as a percent.

24. Write $\dfrac{3}{4}$ as a percent. (*Hint:* Write $\dfrac{3}{4}$ as a decimal, and then write the decimal as a percent.)

Most of the water on Earth is in the form of oceans. Only a small part is fresh water. The graph below is called a circle graph or pie chart. This particular circle graph shows the distribution of fresh water on Earth. Use this graph to answer Exercises 25 through 28. (Source: Philip's World Atlas)

Fresh Water Distribution

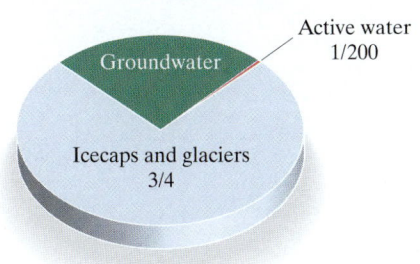

25. What fractional part of fresh water is icecaps and glaciers?

26. What fractional part of fresh water is active water?

27. What fractional part of fresh water is groundwater?

28. What fractional part of fresh water is groundwater or icecaps and glaciers?

Find the area of each figure. (The area of a rectangle is the product of its length and width. The area of a triangle is $\dfrac{1}{2}$ the product of its base and height.)

△ **29.**

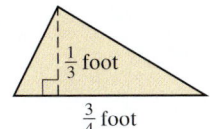

△ **30.**

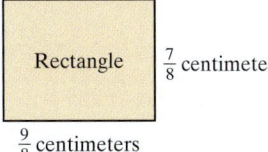

21. $0.1\overline{6} \approx 0.167$

22. 0.632

23. 9%

24. 75%

25. $\dfrac{3}{4}$

26. $\dfrac{1}{200}$

27. $\dfrac{49}{200}$

28. $\dfrac{199}{200}$

29. $\dfrac{1}{8}$ sq ft

30. $\dfrac{63}{64}$ sq cm

Real Numbers and Introduction to Algebra

A Selection of Resources for Success in This Mathematics Course

Textbook

Instructor

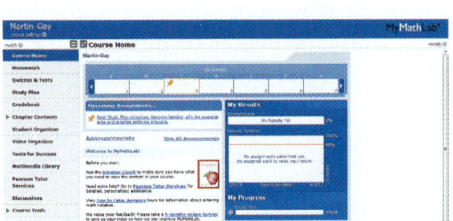

MyMathLab and MathXL

Video Organizer

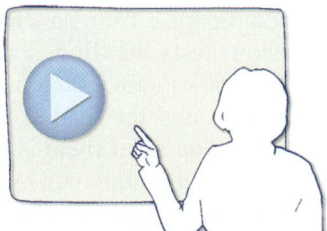

Interactive Lecture Series

For more information about the resources illustrated above, read Section 1.1.

In this chapter, we begin with a review of the basic symbols—the language—of mathematics. We then introduce algebra by using a variable in place of a number. From there, we translate phrases to algebraic expressions and sentences to equations. This is the beginning of problem solving, which we formally study in Chapter 2.

Objectives

A Get Ready for This Course.

B Understand Some General Tips for Success.

C Know How to Use This Text.

D Know How to Use Text Resources.

E Get Help as Soon as You Need It.

F Learn How to Prepare for and Take an Exam.

G Develop Good Time Management.

Before reading Section 1.1, you might want to ask yourself a few questions.

1. When you took your last math course, were you organized? Were your notes and materials from that course easy to find, or were they disorganized and hard to find—if you saved them at all?

2. Were you satisfied—really satisfied—with your performance in that course? In other words, do you feel that your outcome represented your best effort?

If the answer is "no" to these questions, then it is time to make a change. Changing to or resuming good study skill habits is not a process you can start and stop as you please. It is something that you must remember and practice each and every day. To begin, continue reading this section.

Objective A Getting Ready for This Course

Now that you have decided to take this course, remember that a *positive attitude* will make all the difference in the world. Your belief that you can succeed is just as important as your commitment to this course. Make sure you are ready for this course by having the time and positive attitude that it takes to succeed.

Make sure that you are familiar with the way that this course is being taught. Is it a traditional course, in which you have a printed textbook and meet with an instructor? Is it taught totally online, and your textbook is electronic and you e-mail your instructor? Or is your course structured somewhere in between these two methods? (Not all of the tips that follow will apply to all forms of instruction.)

Also make sure that you have scheduled your math course for a time that will give you the best chance for success. For example, if you are also working, you may want to check with your employer to make sure that your work hours will not conflict with your course schedule.

On the day of your first class period, double-check your schedule and allow yourself extra time to arrive on time in case of traffic problems or difficulty locating your classroom. Make sure that you are aware of and bring all necessary class materials.

Objective B General Tips for Success

Below are some general tips that will increase your chance for success in a mathematics class. Many of these tips will also help you in other courses you may be taking.

Most important! Organize your class materials. In the next couple pages, many ideas will be presented to help you organize your class materials—notes, any handouts, completed homework, previous tests, etc. In general, you MUST have these materials organized. All of them will be valuable references throughout your course and when studying for upcoming tests and the final exam. One way to make sure you can locate these materials when you need them is to use a three-ring binder. This binder should be used solely for your mathematics class and should be brought to each and every class or lab. This way, any material can be immediately inserted in a section of this binder and will be there when you need it.

Form study groups and/or exchange names and e-mail addresses. Depending on how your course is taught, you may want to keep in contact with your fellow students. Some ways of doing this are to form a study group—whether in person or through the Internet. Also, you may want to ask if anyone is interested in exchanging e-mail addresses or any other form of contact.

Helpful Hint

MyMathLab® and MathXL®
When assignments are turned in online, keep a hard copy of your complete written work. You will need to refer to your written work to be able to ask questions and to study for tests later.

Choose to attend all class periods. If possible, sit near the front of the classroom. This way, you will see and hear the presentation better. It may also be easier for you to participate in classroom activities.

Do your homework. You've probably heard the phrase "practice makes perfect" in relation to music and sports. It also applies to mathematics. You will find that the more time you spend solving mathematics exercises, the easier the process becomes. Be sure to schedule enough time to complete your assignments before the due date assigned by your instructor.

Check your work. Review the steps you took while working a problem. Learn to check your answers in the original exercises. You may also compare your answers with the "Answers to Selected Exercises" section in the back of the book. If you have made a mistake, try to figure out what went wrong. Then correct your mistake. If you can't find what went wrong, **don't** erase your work or throw it away. Show your work to your instructor, a tutor in a math lab, or a classmate. It is easier for someone to find where you had trouble if he or she looks at your original work.

Learn from your mistakes and be patient with yourself. Everyone, even your instructor, makes mistakes. (That definitely includes me—Elayn Martin-Gay.) Use your errors to learn and to become a better math student. The key is finding and understanding your errors.

Was your mistake a careless one, or did you make it because you can't read your own math writing? If so, try to work more slowly or write more neatly and make a conscious effort to carefully check your work.

Did you make a mistake because you don't understand a concept? Take the time to review the concept or ask questions to better understand it.

Did you skip too many steps? Skipping steps or trying to do too many steps mentally may lead to preventable mistakes.

Know how to get help if you need it. It's all right to ask for help. In fact, it's a good idea to ask for help whenever there is something that you don't understand. Make sure you know when your instructor has office hours and how to find his or her office. Find out whether math tutoring services are available on your campus. Check on the hours, location, and requirements of the tutoring service.

Don't be afraid to ask questions. You are not the only person in class with questions. Other students are normally grateful that someone has spoken up.

Turn in assignments on time. This way, you can be sure that you will not lose points for being late. Show every step of a problem and be neat and organized. Also be sure that you understand which problems are assigned for homework. If allowed, you can always double-check the assignment with another student in your class.

Objective C Knowing and Using Your Text

Flip through the pages of this text or view the e-text pages on a computer screen. Start noticing examples, exercise sets, end-of-chapter material, and so on. Every text is organized in some manner. Learn the way this text is organized by reading about and then finding an example in your text of each type of resource listed below. Finding and using these resources throughout your course will increase your chance of success.

- *Practice Exercises.* Each example in every section has a parallel Practice exercise. As you read a section, try each Practice exercise after you've finished the corresponding example. Answers are at the bottom of the page. This "learn-by-doing" approach will help you grasp ideas before you move on to other concepts.

- *Symbols at the Beginning of an Exercise Set.* If you need help with a particular section, the symbols listed at the beginning of each exercise set will remind you of the resources available.

- *Objectives.* The main section of exercises in each exercise set is referenced by an objective, such as **A** or **B**, and also an example(s). There is also often a section of exercises entitled "Mixed Practice," which is referenced by two or more objectives or sections. These are mixed exercises written to prepare you for your next exam. Use all of this referencing if you have trouble completing an assignment from the exercise set.

Helpful Hint

MyMathLab® and MathXL®
If you are doing your homework online, you can work and re-work those exercises that you struggle with until you master them. Try working through all the assigned exercises twice before the due date.

Helpful Hint

MyMathLab® and MathXL®
If you are completing your homework online, it's important to work each exercise on paper before submitting the answer. That way, you can check your work and follow your steps to find and correct any mistakes.

Helpful Hint

MyMathLab® and MathXL®
Be aware of assignments and due dates set by your instructor. Don't wait until the last minute to submit work online.

- *Icons (Symbols)*. Make sure that you understand the meaning of the icons that are beside many exercises. ▶ tells you that the corresponding exercise may be viewed on the video Lecture Series that corresponds to that section. ✎ tells you that this exercise is a writing exercise in which you should answer in complete sentences. △ tells you that the exercise involves geometry.

- *Integrated Reviews*. Found in the middle of each chapter, these reviews offer you a chance to practice—in one place—the many concepts that you have learned separately over several sections.

- *End-of-Chapter Opportunities*. There are many opportunities at the end of each chapter to help you understand the concepts of the chapter.

  Vocabulary Checks contain key vocabulary terms introduced in the chapter.

  Chapter Highlights contain chapter summaries and examples.

  Chapter Reviews contain review problems. The first part is organized section by section and the second part contains a set of mixed exercises.

  Chapter Tests are sample tests to help you prepare for an exam. The Chapter Test Prep Videos found in the Interactive Lecture Series, MyMathLab, and YouTube provide the video solution to each question on each Chapter Test.

  Cumulative Reviews start at Chapter 2 and are reviews consisting of material from the beginning of the book to the end of that particular chapter.

- *Student Resources in Your Textbook*. You will find a **Student Resources** section at the back of this textbook. It contains the following to help you study and prepare for tests:

  Study Skill Builders contain study skills advice. To increase your chance for success in the course, read these study tips and answer the questions.

  Bigger Picture—Study Guide Outline provides you with a study guide outline of the course, with examples.

  Practice Final provides you with a Practice Final Exam to help you prepare for a final.

- *Resources to Check Your Work*. The **Answers to Selected Exercises** section provides answers to all odd-numbered section exercises and to all integrated review, chapter test, and cumulative review exercises.

Objective D Knowing and Using Video and Notebook Organizer Resources ▶

Video Resources

Below is a list of video resources that are all made by me—the author of your text, Elayn Martin-Gay. By making these videos, I can be sure that the methods presented are consistent with those in the text.

Helpful Hint

MyMathLab®

In MyMathLab, you have access to the following video resources:

- Lecture Videos for each section
- Chapter Test Prep Videos

Use these videos provided by the author to prepare for class, review, and study for tests.

- *Interactive DVD Lecture Series*. Exercises marked with a ▶ are fully worked out by the author on the DVDs and within MyMathLab. The Lecture Series provides approximately 20 minutes of instruction per section and is organized by objective.

- *Chapter Test Prep Videos*. These videos provide solutions to all of the Chapter Test exercises worked out by the author. They can be found in MyMathLab, the Interactive Lecture Series, and You Tube. This supplement is very helpful before a test or exam.

- *Student Success Tips*. These video segments are about 3 minutes long and are daily reminders to help you continue practicing and maintaining good organizational and study habits.

- *Final Exam Videos*. These video segments provide solutions to each question. These videos can be found within MyMathLab and the Interactive Lecture Series.

Notebook Organizer Resources

The resources below are in three-ring notebook ready form. They are to be inserted in a three-ring binder and completed. Both resources are numbered according to the sections in your text to which they refer.

- *Video Organizer.* This organizer is closely tied to the Interactive Lecture (Video) Series. Each section should be completed while watching the lecture video on the same section. Once completed, you will have a set of notes to accompany the Lecture (Video) Series section by section.
- *Student Organizer.* This organizer helps you study effectively through note-taking hints, practice, and homework while referencing examples in the text and examples in the Lecture Series.

Objective E Getting Help

If you have trouble completing assignments or understanding the mathematics, get help as soon as you need it! This tip is presented as an objective on its own because it is so important. In mathematics, usually the material presented in one section builds on your understanding of the previous section. This means that if you don't understand the concepts covered during a class period, there is a good chance that you will not understand the concepts covered during the next class period. If this happens to you, get help as soon as you can.

Where can you get help? Many suggestions have been made in this section on where to get help, and now it is up to you to get it. Try your instructor, a tutoring center, or a math lab, or you may want to form a study group with fellow classmates. If you do decide to see your instructor or go to a tutoring center, make sure that you have a neat notebook and are ready with your questions.

> **Helpful Hint**
>
> **MyMathLab® and MathXL®**
>
> - Use the **Help Me Solve This** button to get step-by-step help for the exercise you are working. You will need to work an additional exercise of the same type before you can get credit for having worked it correctly.
> - Use the **Video** button to view a video clip of the author working a similar exercise.

Objective F Preparing for and Taking an Exam

Make sure that you allow yourself plenty of time to prepare for a test. If you think that you are a little "math anxious," it may be that you are not preparing for a test in a way that will ensure success. The way that you prepare for a test in mathematics is important. To prepare for a test:

1. Review your previous homework assignments.
2. Review any notes from class and section-level quizzes you have taken. (If this is a final exam, also review chapter tests you have taken.)
3. Review concepts and definitions by reading the Chapter Highlights at the end of each chapter.
4. Practice working out exercises by completing the Chapter Review found at the end of each chapter. (If this is a final exam, go through a Cumulative Review. There is one found at the end of each chapter except Chapter 1. Choose the review found at the end of the latest chapter that you have covered in your course.) *Don't stop here!*
5. It is important that you place yourself in conditions similar to test conditions to find out how you will perform. In other words, as soon as you feel that you know the material, get a few blank sheets of paper and take a sample test. There is a Chapter Test available at the end of each chapter, or you can work selected problems from the Chapter Review. Your instructor may also provide you with a review sheet. During this sample test, do not use your notes or your textbook. Then check your sample test. If your sample test is the Chapter Test in the text, don't forget that the video solutions are in MyMathLab, the Interactive Lecture Series, and YouTube. If you are not satisfied with the results, study the areas that you are weak in and try again.

> **Helpful Hint**
>
> **MyMathLab® and MathXL®**
> Review your written work for previous assignments. Then, go back and re-work previous assignments. Open a previous assignment, and click **Similar Exercise** to generate new exercises. Re-work the exercises until you fully understand them and can work them without help features.

6. On the day of the test, allow yourself plenty of time to arrive at where you will be taking your exam.

When taking your test:

1. Read the directions on the test carefully.
2. Read each problem carefully as you take the test. Make sure that you answer the question asked.
3. Watch your time and pace yourself so that you can attempt each problem on your test.
4. If you have time, check your work and answers.
5. Do not turn your test in early. If you have extra time, spend it double-checking your work.

Objective G Managing Your Time

As a college student, you know the demands that classes, homework, work, and family place on your time. Some days you probably wonder how you'll ever get everything done. One key to managing your time is developing a schedule. Here are some hints for making a schedule:

1. Make a list of all of your weekly commitments for the term. Include classes, work, regular meetings, extracurricular activities, etc. You may also find it helpful to list such things as laundry, regular workouts, grocery shopping, etc.
2. Next, estimate the time needed for each item on the list. Also make a note of how often you will need to do each item. Don't forget to include time estimates for the reading, studying, and homework you do outside of your classes. You may want to ask your instructor for help estimating the time needed.
3. In the exercise set that follows, you are asked to block out a typical week on the schedule grid given. Start with items with fixed time slots like classes and work.
4. Next, include the items on your list with flexible time slots. Think carefully about how best to schedule items such as study time.
5. Don't fill up every time slot on the schedule. Remember that you need to allow time for eating, sleeping, and relaxing! You should also allow a little extra time in case some items take longer than planned.
6. If you find that your weekly schedule is too full for you to handle, you may need to make some changes in your workload, classload, or other areas of your life. You may want to talk to your advisor, manager or supervisor at work, or someone in your college's academic counseling center for help with such decisions.

1.1 Exercise Set MyMathLab®

1. What is your instructor's name?

2. What are your instructor's office location and office hours?

3. What is the best way to contact your instructor?

4. Do you have the name and contact information of at least one other student in class?

5. Will your instructor allow you to use a calculator in this class?

6. Why is it important that you write step-by-step solutions to homework exercises and keep a hard copy of all work submitted?

7. Is there a tutoring service available on campus? If so, what are its hours? What services are available?

8. Have you attempted this course before? If so, write down ways that you might improve your chances of success during this attempt.

9. List some steps that you can take if you begin having trouble understanding the material or completing an assignment. If you are completing your homework in MyMathLab® and MathXL®, list the resources you can use for help.

10. How many hours of studying does your instructor advise for each hour of instruction?

11. What does the ＼ icon in this text mean?

12. What does the △ icon in this text mean?

13. What does the ● icon in this text mean?

14. Search the minor columns in your text. What are Practice exercises?

15. When might be the best time to work a Practice exercise?

16. Where are the answers to Practice exercises?

17. What answers are contained in this text and where are they?

18. What are Study Skill Tips of the Day and where are they?

19. What and where are Integrated Reviews?

20. How many times is it suggested that you work through the homework exercises in MathXL® before the submission deadline?

21. How far in advance of the assigned due date is it suggested that homework be submitted online? Why?

22. Chapter Highlights are found at the end of each chapter. Find the Chapter 1 Highlights and explain how you might use it and how it might be helpful.

23. Chapter Reviews are found at the end of each chapter. Find the Chapter 1 Review and explain how you might use it and how it might be helpful.

24. Chapter Tests are found at the end of each chapter. Find the Chapter 1 Test and explain how you might use it and how it might be helpful when preparing for an exam on Chapter 1. Include how the Chapter Test Prep Videos may help. If you are working in MyMathLab® and MathXL®, how can you use previous homework assignments to study?

25. What is the Video Organizer? Explain the contents and how it might be used.

26. What is the Student Organizer? Explain the contents and how it might be used.

27. Read or reread Objective **G** and fill out the schedule grid on the next page.

	Monday	Tuesday	Wednesday	Thursday	Friday	Saturday	Sunday
4:00 a.m.							
5:00 a.m.							
6:00 a.m.							
7:00 a.m.							
8:00 a.m.							
9:00 a.m.							
10:00 a.m.							
11:00 a.m.							
12:00 p.m.							
1:00 p.m.							
2:00 p.m.							
3:00 p.m.							
4:00 p.m.							
5:00 p.m.							
6:00 p.m.							
7:00 p.m.							
8:00 p.m.							
9:00 p.m.							
10:00 p.m.							
11:00 p.m.							
Midnight							
1:00 a.m.							
2:00 a.m.							
3:00 a.m.							

1.2 Symbols and Sets of Numbers

Objectives

A Define the Meaning of the Symbols =, ≠, <, >, ≤, and ≥.

B Translate Sentences into Mathematical Statements.

C Identify Integers, Rational Numbers, Irrational Numbers, and Real Numbers.

D Find the Absolute Value of a Real Number.

We begin with a review of the set of natural numbers and the set of whole numbers and how we use symbols to compare these numbers. A **set** is a collection of objects, each of which is called a **member** or **element** of the set. A pair of brace symbols { } encloses the list of elements and is translated as "the set of" or "the set containing."

Natural Numbers

$$\{1, 2, 3, 4, 5, 6, \ldots\}$$

Whole Numbers

$$\{0, 1, 2, 3, 4, 5, 6, \ldots\}$$

Helpful Hint

The three dots (an ellipsis) at the end of the list of elements of a set means that the list continues in the same manner indefinitely.

Objective A Equality and Inequality Symbols

Picturing natural numbers and whole numbers on a number line helps us to see the order of the numbers. Symbols can be used to describe in writing the order of two quantities. We will use equality symbols and inequality symbols to compare quantities.

Below is a review of these symbols. The letters *a* and *b* are used to represent quantities. Letters such as *a* and *b* that are used to represent numbers or quantities are called **variables.**

Equality and Inequality Symbols

		Meaning
Equality symbol:	$a = b$	*a* is equal to *b*.
Inequality symbols:	$a \neq b$	*a* is not equal to *b*.
	$a < b$	*a* is less than *b*.
	$a > b$	*a* is greater than *b*.
	$a \leq b$	*a* is less than or equal to *b*.
	$a \geq b$	*a* is greater than or equal to *b*.

These symbols may be used to form **mathematical statements** such as

$$2 = 2 \quad \text{and} \quad 2 \neq 6$$

Recall that on a number line, we see that a number **to the right of** another number is **larger.** Similarly, a number **to the left of** another number is **smaller.** For example, 3 is to the left of 5 on the number line, which means that 3 is less than 5, or $3 < 5$. Similarly, 2 is to the right of 0 on the number line, which means that 2 is greater than 0, or $2 > 0$. Since 0 is to the left of 2, we can also say that 0 is less than 2, or $0 < 2$.

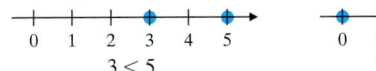
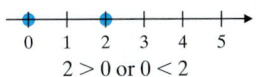

Helpful Hint

Recall that $2 > 0$ has exactly the same meaning as $0 < 2$. Switching the order of the numbers and reversing the direction of the inequality symbol does not change the meaning of the statement.

$6 > 4$ has the same meaning as $4 < 6$.

Also notice that when the statement is true, the inequality arrow points to the smaller number.

Our discussion above can be generalized in the order property below.

Order Property for Real Numbers

For any two real numbers a and b, a is less than b if a is to the left of b on a number line.

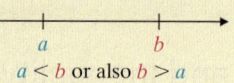

$a < b$ or also $b > a$

Practice 1–6

Determine whether each statement is true or false.

1. $8 < 6$ **2.** $100 > 10$
3. $21 \leq 21$ **4.** $21 \geq 21$
5. $0 \geq 5$ **6.** $25 \geq 22$

Helpful Hint

If either $3 < 3$ or $3 = 3$ is true, then $3 \leq 3$ is true.

Examples   Determine whether each statement is true or false.

1. $2 < 3$ True.   Since 2 is to the left of 3 on a number line
2. $72 < 27$ False.   72 is to the right of 27 on a number line, so $72 > 27$.
3. $8 \geq 8$ True.   Since $8 = 8$ is true
4. $8 \leq 8$ True.   Since $8 = 8$ is true
5. $23 \leq 0$ False.   Since neither $23 < 0$ nor $23 = 0$ is true
6. $0 \leq 23$ True.   Since $0 < 23$ is true

◼ Work Practice 1–6

Objective B Translating Sentences into Mathematical Statements

Now, let's use the symbols discussed above to translate sentences into mathematical statements.

Practice 7

Translate each sentence into a mathematical statement.

a. Fourteen is greater than or equal to fourteen.

b. Zero is less than five.

c. Nine is not equal to ten.

Example 7   Translate each sentence into a mathematical statement.

a. Nine is less than or equal to eleven.
b. Eight is greater than one.
c. Three is not equal to four.

Solution:

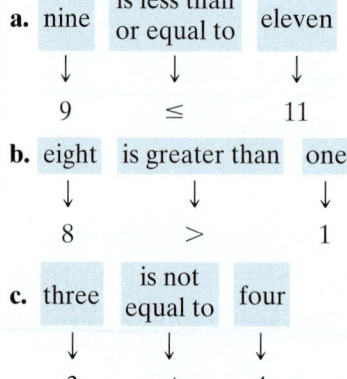

a. nine | is less than or equal to | eleven
         9 | ≤ | 11

b. eight | is greater than | one
         8 | > | 1

c. three | is not equal to | four
         3 | ≠ | 4

◼ Work Practice 7

Objective C Identifying Common Sets of Numbers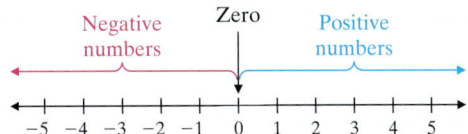

Whole numbers are not sufficient to describe many situations in the real world. For example, quantities smaller than zero must sometimes be represented, such as temperatures less than 0 degrees.

Recall that we can place numbers less than zero on a number line as follows: Numbers less than 0 are to the left of 0 and are labeled −1, −2, −3, and so on. The numbers we have labeled on the number line below are called the set of **integers.**

Negative numbers   Zero   Positive numbers

$$-5 \quad -4 \quad -3 \quad -2 \quad -1 \quad 0 \quad 1 \quad 2 \quad 3 \quad 4 \quad 5$$

Integers to the left of 0 are called **negative integers;** integers to the right of 0 are called **positive integers.** The integer 0 is neither positive nor negative.

Integers

$$\{\ldots, -3, -2, -1, 0, 1, 2, 3, \ldots\}$$

Helpful Hint

A − sign, such as the one in −2, tells us that the number is to the left of 0 on a number line.

   −2 is read "negative two."

A + sign or no sign tells us that the number lies to the right of 0 on a number line. For example, 3 and +3 both mean positive three.

Example 8   Use an integer to express the number in the following. "The lowest temperature ever recorded at South Pole Station, Antarctica, occurred during the month of June. The record-low temperature was 117 degrees below zero." (*Source:* The National Oceanic and Atmospheric Administration)

Solution: The integer −117 represents 117 degrees below zero.

■ **Work Practice 8**

Practice 8

Use an integer to express the number in the following. The elevation of New Orleans, Louisiana, is an average of 8 feet below sea level. (*Source: The World Almanac*)

Answer

8. −8

A problem with integers in real-life settings arises when quantities are smaller than some integer but greater than the next smallest integer. On a number line, these quantities may be visualized by points between integers. Some of these quantities between integers can be represented as a quotient of integers. For example,

The point on the number line halfway between 0 and 1 can be represented by $\frac{1}{2}$, a quotient of integers.

The point on the number line halfway between 0 and -1 can be represented by $-\frac{1}{2}$. Other quotients of integers and their graphs are shown below.

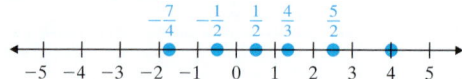

These numbers, each of which can be represented as a quotient of integers, are examples of **rational numbers.** It's not possible to list the set of rational numbers using the notation that we have been using. For this reason, we will use a different notation.

Rational Numbers

$$\left\{ \frac{a}{b} \,\middle|\, a \text{ and } b \text{ are integers and } b \neq 0 \right\}$$

We read this set as "the set of numbers $\frac{a}{b}$ such that a and b are integers and b **is not equal to 0.**"

Helpful Hint

We commonly refer to rational numbers as fractions.

Notice that every integer is also a rational number since each integer can be written as a quotient of integers. For example, the integer 5 is also a rational number since $5 = \frac{5}{1}$. For the rational number $\frac{5}{1}$, recall that the top number, 5, is called the numerator and the bottom number, 1, is called the denominator.

Let's practice **graphing** numbers on a number line.

Example 9 Graph the numbers on a number line.

$$-\frac{4}{3}, \quad \frac{1}{4}, \quad \frac{3}{2}, \quad -2\frac{1}{8}, \quad 3.5$$

Solution: To help graph the improper fractions in the list, we first write them as mixed numbers.

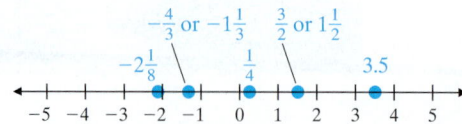

■ **Work Practice 9**

Every rational number has a point on the number line that corresponds to it. But not every point on the number line corresponds to a rational number. Those points that do not correspond to rational numbers correspond instead to **irrational numbers.**

Practice 9

Graph the numbers on the number line.

$$-2\frac{1}{2}, \quad -\frac{2}{3}, \quad \frac{1}{5}, \quad \frac{5}{4}, \quad 2.25$$

<-+--+--+--+--+--+--+--+--+--+->
-5 -4 -3 -2 -1 0 1 2 3 4 5

Answer

9.

-2½   -⅔ ⅕ 5/4 2.25
<-+--+--+--+--+--+--+--+--+--+->
-5 -4 -3 -2 -1 0 1 2 3 4 5

Irrational Numbers

{Nonrational numbers that correspond to points on a number line}

An irrational number that you have probably seen is π. Also, $\sqrt{2}$, the length of the diagonal of the square shown below, is an irrational number.

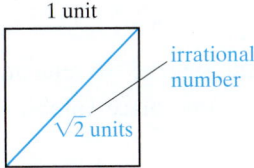

1 unit

irrational number

$\sqrt{2}$ units

Both rational and irrational numbers can be written as decimal numbers. The decimal equivalent of a rational number will either terminate or repeat in a pattern. For example, upon dividing we find that

$$\frac{3}{4} = 0.75 \qquad \text{(Decimal number terminates or ends.)}$$

$$\frac{2}{3} = 0.66666\ldots \qquad \text{(Decimal number repeats in a pattern.)}$$

The decimal representation of an irrational number will neither terminate nor repeat. (For further review of decimals, see Section R.3.)

The set of numbers, each of which corresponds to a point on a number line, is called the set of **real numbers.** One and only one point on a number line corresponds to each real number.

Real Numbers

{All numbers that correspond to points on a number line}

Several different sets of numbers have been discussed in this section. The following diagram shows the relationships among these sets of real numbers. Notice that, together, the rational numbers and the irrational numbers make up the real numbers.

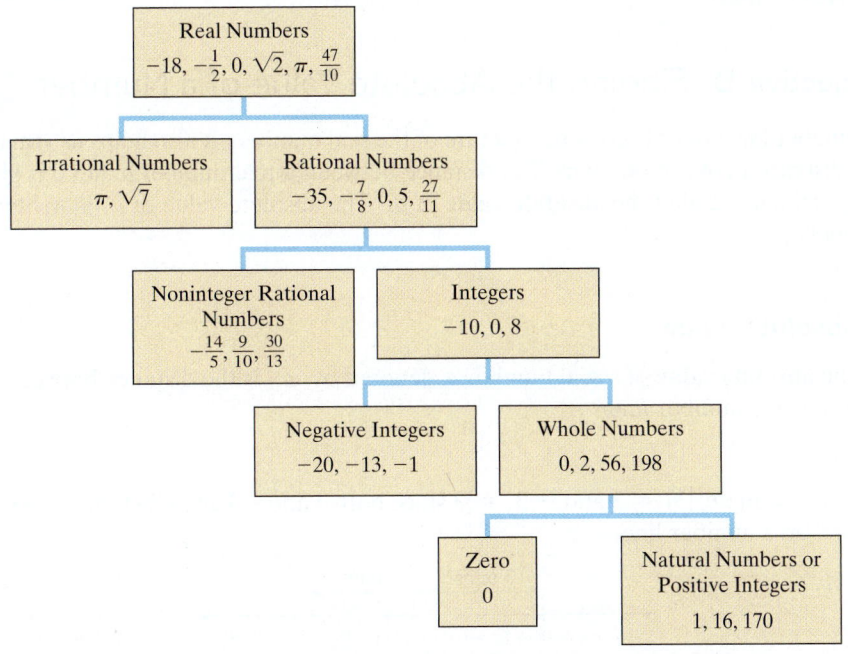

Now that other sets of numbers have been reviewed, let's continue our practice of comparing numbers.

Practice 10

Insert $<$, $>$, or $=$ between the pairs of numbers to form true statements.

a. -11 -9 **b.** 4.511 4.151

c. $\dfrac{7}{8}$ $\dfrac{2}{3}$

Example 10 Insert $<$, $>$, or $=$ between the pairs of numbers to form true statements.

a. -5 -6 **b.** 3.195 3.2 **c.** $\dfrac{1}{4}$ $\dfrac{1}{3}$

Solution:

a. $-5 > -6$ since -5 lies to the right of -6 on a number line.

b. By comparing digits in the same place values, we find that $3.195 < 3.2$, since $0.1 < 0.2$.

c. By dividing, we find that $\dfrac{1}{4} = 0.25$ and $\dfrac{1}{3} = 0.33\ldots$. Since $0.25 < 0.33\ldots$, $\dfrac{1}{4} < \dfrac{1}{3}$.

■ **Work Practice 10**

Practice 11

Given the set

$\left\{-100, -\dfrac{2}{5}, 0, \pi, 6, 913\right\}$, list

the numbers in this set that belong to the set of:

a. Natural numbers

b. Whole numbers

c. Integers

d. Rational numbers

e. Irrational numbers

f. Real numbers

Example 11 Given the set $\left\{-2, 0, \dfrac{1}{4}, 112, -3, 11, \sqrt{2}\right\}$, list the numbers in this set that belong to the set of:

a. Natural numbers **b.** Whole numbers **c.** Integers

d. Rational numbers **e.** Irrational numbers **f.** Real numbers

Solution:

a. The natural numbers are 11 and 112.

b. The whole numbers are $0, 11,$ and 112.

c. The integers are $-3, -2, 0, 11,$ and 112.

d. Recall that integers are rational numbers also. The rational numbers are $-3, -2, 0, \dfrac{1}{4}, 11,$ and 112.

e. The only irrational number is $\sqrt{2}$.

f. All numbers in the given set are real numbers.

■ **Work Practice 11**

Objective D Finding the Absolute Value of a Number

A number line not only gives us a picture of the real numbers, it also helps us visualize the distance between numbers. The distance between a real number a and 0 is given a special name called the **absolute value** of a. "The absolute value of a" is written in symbols as $|a|$.

Helpful Hint Since $|a|$ is a distance, $|a|$ is always either positive or 0. It is never negative. That is, **for any real number a, $|a| \geq 0$.**

Absolute Value

The **absolute value** of a real number a, denoted by $|a|$, is the distance between a and 0 on a number line.

For example, $|3| = 3$ and $|-3| = 3$ since both 3 and -3 are a distance of 3 units from 0 on a number line.

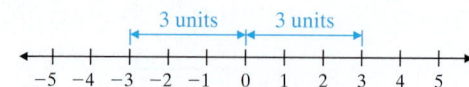

Answers

10. a. $<$ **b.** $>$ **c.** $>$ **11. a.** $6, 913$

b. $0, 6, 913$ **c.** $-100, 0, 6, 913$

d. $-100, -\dfrac{2}{5}, 0, 6, 913$ **e.** π

f. all numbers in the given set

Example 12 Find the absolute value of each number.

a. $|4|$ b. $|-5|$ c. $|0|$

d. $\left|-\dfrac{2}{9}\right|$ e. $|4.93|$

Solution:

a. $|4| = 4$ since 4 is 4 units from 0 on a number line.
b. $|-5| = 5$ since -5 is 5 units from 0 on a number line.
c. $|0| = 0$ since 0 is 0 units from 0 on a number line.
d. $\left|-\dfrac{2}{9}\right| = \dfrac{2}{9}$
e. $|4.93| = 4.93$

🔲 **Work Practice 12**

Practice 12

Find the absolute value of each number.

a. $|7|$ b. $|-8|$ c. $\left|\dfrac{2}{3}\right|$
d. $|0|$ e. $|-3.06|$

Example 13 Insert $<$, $>$, or $=$ in the appropriate space to make each statement true.

a. $|0|$ 2   b. $|-5|$ 5   c. $|-3|$ $|-2|$

d. $|-9|$ $|-9.7|$ e. $\left|-7\dfrac{1}{6}\right|$ $|7|$

Solution:

a. $|0| < 2$ since $|0| = 0$ and $0 < 2$.
b. $|-5| = 5$.
c. $|-3| > |-2|$ since $3 > 2$.
d. $|-9| < |-9.7|$ since $9 < 9.7$.
e. $\left|-7\dfrac{1}{6}\right| > |7|$ since $7\dfrac{1}{6} > 7$.

🔲 **Work Practice 13**

Practice 13

Insert $<$, $>$, or $=$ in the appropriate space to make each statement true.

a. $|-4|$ 4
b. -3 $|0|$
c. $|-2.7|$ $|-2|$
d. $|-6|$ $|-16|$
e. $|10|$ $\left|-10\dfrac{1}{3}\right|$

Answers

12. a. 7   b. 8   c. $\dfrac{2}{3}$ d. 0   e. 3.06

13. a. $=$ b. $<$ c. $>$ d. $<$ e. $<$

Vocabulary, Readiness & Video Check

Use the choices below to fill in each blank. Not all choices will be used.

real	natural	absolute value	$\dfrac{1}{2}$	$\dfrac{1}{4}$	$|a|$	whole
rational	inequality	integers	0	1	$|-1|$	

1. The __whole__ numbers are $\{0, 1, 2, 3, 4, \ldots\}$.
2. The __natural__ numbers are $\{1, 2, 3, 4, 5, \ldots\}$.
3. The symbols $\neq$, $\leq$, and $>$ are called __inequality__ symbols.
4. The __integers__ are $\{\ldots, -3, -2, -1, 0, 1, 2, 3, \ldots\}$.
5. The __real__ numbers are $\{$ all numbers that correspond to points on a number line $\}$.
6. The __rational__ numbers are $\left\{\dfrac{a}{b} \,\middle|\, a \text{ and } b \text{ are integers}, b \neq 0\right\}$.
7. The integer __0__ is neither positive nor negative.

8. The point on a number line halfway between 0 and $\frac{1}{2}$ can be represented by ____$\frac{1}{4}$____.

9. The distance between a real number a and 0 is called the __absolute value__ of a.

10. The absolute value of a is written in symbols as ____$|a|$____.

Martin-Gay Interactive Videos   *Watch the section lecture video and answer the following questions.*

See Video 1.2

Objective A **11.** In Example 2, why is the symbol $<$ inserted between the two numbers? ▶

Objective B **12.** Write the sentence given in ⊞ Example 4 and translate it to a mathematical statement, using symbols. ▶

Objective C **13.** Which sets of numbers does the number in ⊞ Example 6 belong to? Why is this number not an irrational number? ▶

Objective D **14.** Complete this statement based on the lecture given before ⊞ Example 8. The _____ of a real number a, denoted by $|a|$, is the distance between a and 0 on a number line. ▶

See video answer section.

1.2 **Exercise Set** MyMathLab® ▶

Objectives A C Mixed Practice *Insert $<$, $>$, or $=$ in the space between the paired numbers to make each statement true. See Examples 1 through 6 and 10.*

1. $4 < 10$ **2.** $8 > 5$ ▶**3.** $7 > 3$ **4.** $9 < 15$

5. $6.26 = 6.26$ **6.** $1.13 = 1.13$ ▶**7.** $0 < 7$ **8.** $20 > 0$

9. The freezing point of water is 32° Fahrenheit. The boiling point of water is 212° Fahrenheit. Write an inequality statement using $<$ or $>$ comparing the numbers 32 and 212. $32 < 212$

10. The freezing point of water is 0° Celsius. The boiling point of water is 100° Celsius. Write an inequality statement using $<$ or $>$ comparing the numbers 0 and 100. $0 < 100$

△**11.** An angle measuring 30° and an angle measuring 45° are shown. Write an inequality statement using $\leq$ or $\geq$ comparing the numbers 30 and 45. $30 \leq 45$

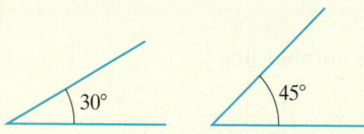

△**12.** The sum of the measures of the angles of a parallelogram is 360°. The sum of the measures of the angles of a triangle is 180°. Write an inequality statement using $\leq$ or $\geq$ comparing the numbers 360 and 180.
$360 \geq 180$

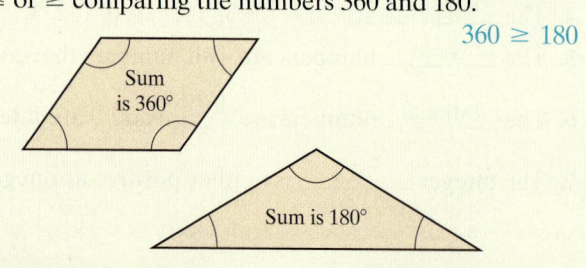

Determine whether each statement is true or false. See Examples 1 through 6 and 10.

13. $11 \leq 11$ true   **14.** $8 \geq 9$ false   **15.** $-11 > -10$ false   **16.** $-16 > -17$ true

17. $5.092 < 5.902$ true   **18.** $1.02 > 1.021$ false   **19.** $\frac{9}{10} \leq \frac{8}{9}$ false   **20.** $\frac{4}{5} \leq \frac{9}{11}$ true

Rewrite each inequality so that the inequality symbol points in the opposite direction and the resulting statement has the same meaning as the given one. See Examples 1 through 6 and 10.

21. $25 \geq 20$ $20 \leq 25$ **22.** $-13 \leq 13$ $13 \geq -13$ **23.** $0 < 6$ $6 > 0$

24. $5 > 3$ $3 < 5$ **25.** $-10 > -12$ $-12 < -10$ **26.** $-4 < -2$ $-2 > -4$

Objectives B C Mixed Practice—Translating *Write each sentence as a mathematical statement. See Example 7.*

27. Seven is less than eleven. $7 < 11$

28. Twenty is greater than two. $20 > 2$

29. Five is greater than or equal to four. $5 \geq 4$

30. Negative ten is less than or equal to thirty-seven. $-10 \leq 37$

31. Fifteen is not equal to negative two. $15 \neq -2$

32. Negative seven is not equal to seven. $-7 \neq 7$

Use integers to represent the value(s) in each statement. See Example 8.

33. The highest elevation in California is Mt. Whitney, with an altitude of 14,494 feet. The lowest elevation in California is Death Valley, with an altitude of 282 feet below sea level. (*Source:* U.S. Geological Survey) 14,494; -282

34. Driskill Mountain, in Louisiana, has an altitude of 535 feet. New Orleans, Louisiana, lies 8 feet below sea level. (*Source:* U.S. Geological Survey) 535; -8

35. The number of graduate students at the University of Texas at Austin was 27,724 fewer than the number of undergraduate students. (*Source:* University of Texas at Austin, 2012) $-27,724$

36. The number of students admitted to the class of 2011 at UCLA is 56,715 fewer students than the number that had applied. (*Source:* UCLA, 2012) $-56,715$

37. A community college student deposited $475 in her savings account. She later withdrew $195. 475; -195

38. A deep-sea diver ascended 17 feet and later descended 15 feet. 17; -15

Graph each set of numbers on the number line. See Example 9.

39. $-4, 0, 2, -2$

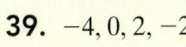

40. $-3, 0, 1, -5$

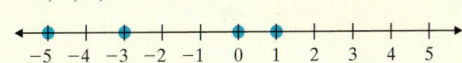

41. $-2, 4, \frac{1}{3}, -\frac{1}{4}$

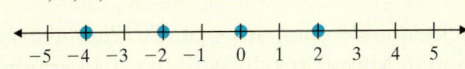

42. $-5, 3, -\frac{1}{3}, \frac{7}{8}$

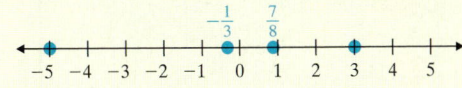

43. $-4.5, \frac{7}{4}, 3.25, -\frac{3}{2}$

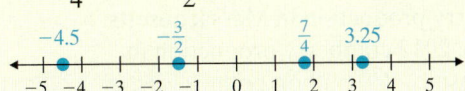

44. $4.5, -\frac{9}{4}, 1.75, -\frac{7}{2}$

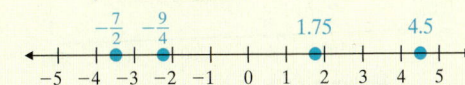

Tell which set or sets each number belongs to: natural numbers, whole numbers, integers, rational numbers, irrational numbers, or real numbers. See Example 11.

▶ **45.** 0   whole, integers, rational, real

46. $\frac{1}{4}$ rational, real

47. −7   integers, rational, real

48. $-\frac{1}{7}$ rational, real

49. 265   natural, whole, integers, rational, real

50. 7941   natural, whole, integers, rational, real

▶ **51.** $\frac{2}{3}$ rational, real

52. $\sqrt{3}$ irrational, real

Determine whether each statement is true or false.

53. Every rational number is also an integer.   false

54. Every natural number is positive.   true

55. 0 is a real number.   true

56. $\frac{1}{2}$ is an integer.   false

57. Every negative number is also a rational number.   false

58. Every rational number is also a real number.   true

59. Every real number is also a rational number.   false

60. Every whole number is an integer.   true

Objective D   *Find each absolute value. See Example 12.*

61. $|8.9|$ 8.9

62. $|11.2|$ 11.2

63. $|-20|$ 20

64. $|-17|$ 17

65. $\left|\frac{9}{2}\right|$ $\frac{9}{2}$

66. $\left|\frac{10}{7}\right|$ $\frac{10}{7}$

67. $\left|-\frac{12}{13}\right|$ $\frac{12}{13}$

68. $\left|-\frac{1}{15}\right|$ $\frac{1}{15}$

Insert <, >, or = in the appropriate space to make each statement true. See Examples 12 and 13.

▶ **69.** $|-5|$ −4   >

70. $|-12|$ $|0|$ >

71. $\left|-\frac{5}{8}\right|$ $\left|\frac{5}{8}\right|$ =

72. $\left|\frac{2}{5}\right|$ $\left|-\frac{2}{5}\right|$ =

73. $|-2|$ $|-2.7|$ <

74. $|-5.01|$ $|-5|$ >

▶ **75.** $|0|$ $|-8|$ <

76. $|-12|$ $\frac{-24}{2}$ >

Concept Extensions

The bar graph shows cranberry production from the top five cranberry-producing states. (Source: National Agricultural Statistics Service)

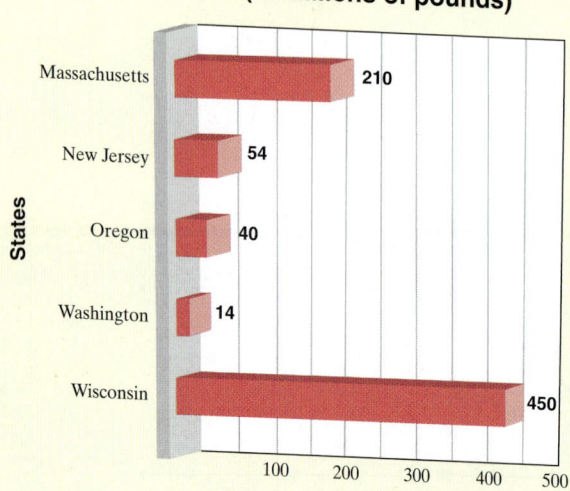

Top Cranberry-Producing States (in millions of pounds)

States

Massachusetts   210
New Jersey   54
Oregon   40
Washington   14
Wisconsin   450

100   200   300   400   500

Millions of lb of Cranberries 2012

Source: National Agricultural Statistics Service

77. Write an inequality comparing the 2012 cranberry production in Oregon with the 2012 cranberry production in Washington.   40 million > 14 million, or 40,000,000 > 14,000,000

78. Write an inequality comparing the 2012 cranberry production in Massachusetts with the 2012 cranberry production in Wisconsin.   210 million < 450 million, or 210,000,000 < 450,000,000

79. Determine the difference between the 2012 cranberry production in Washington and the 2012 cranberry production in New Jersey.   40 million pounds less, or −40 million

80. Determine the difference between the 2012 cranberry production in Massachusetts and the 2012 cranberry production in Wisconsin.   240 million pounds less, or −240 million

The apparent magnitude of a star is the measure of its brightness as seen by someone on Earth. The smaller the apparent magnitude, the brighter the star. Below, the apparent magnitudes of some stars are listed. Use this table to answer Exercises 81 through 86.

Star	Apparent Magnitude	Star	Apparent Magnitude
Arcturus	−0.04	Spica	0.98
Sirius	−1.46	Rigel	0.12
Vega	0.03	Regulus	1.35
Antares	0.96	Canopus	−0.72
Sun (Sol)	−26.7	Hadar	0.61

(*Source: Norton's 2000.0: Star Atlas and Reference Handbooks,* 18th ed., Longman Group, UK, 1989)

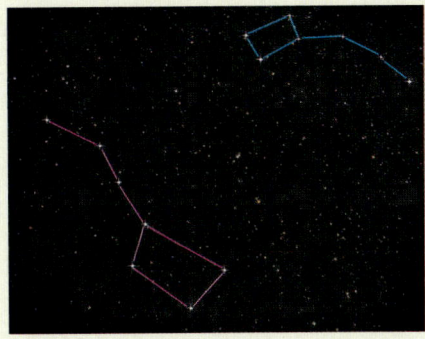

81. The apparent magnitude of the sun is −26.7. The apparent magnitude of the star Arcturus is −0.04. Write an inequality statement comparing the numbers −0.04 and −26.7. $-0.04 > -26.7$

82. The apparent magnitude of Antares is 0.96. The apparent magnitude of Spica is 0.98. Write an inequality statement comparing the numbers 0.96 and 0.98. $0.96 < 0.98$

83. Which is brighter, the sun or Arcturus?   sun

84. Which is dimmer, Antares or Spica?   Spica

85. Which star listed is the brightest?   sun

86. Which star listed is the dimmest?   Regulus

87. In your own words, explain how to find the absolute value of a number.   answers may vary

88. Give an example of a real-life situation that can be described with integers but not with whole numbers.   answers may vary

---

1.3 Exponents, Order of Operations, and Variable Expressions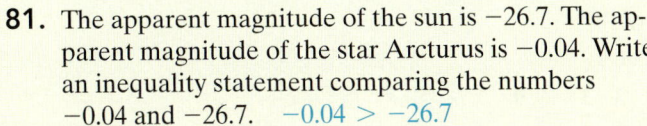

Objective A Exponents and the Order of Operations

Frequently in algebra, products occur that contain repeated multiplication of the same factor. For example, the volume of a cube whose sides each measure 2 centimeters is $(2 \cdot 2 \cdot 2)$ cubic centimeters. We may use **exponential notation** to write such products in a more compact form. For example,

$2 \cdot 2 \cdot 2$ may be written as 2^3.

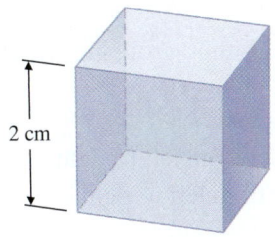

Volume is $(2 \cdot 2 \cdot 2)$
cubic centimeters.

Objectives

A Define and Use Exponents and the Order of Operations.

B Evaluate Algebraic Expressions, Given Replacement Values for Variables.

C Determine Whether a Number Is a Solution of a Given Equation.

D Translate Phrases into Expressions and Sentences into Equations.

The 2 in 2^3 is called the **base;** it is the repeated factor. The 3 in 2^3 is called the **exponent** and is the number of times the base is used as a factor. The expression 2^3 is called an **exponential expression.**

$$\overset{\text{exponent}}{2^3} = 2 \cdot 2 \cdot 2 = 8$$

base ⟶ 2 is a factor 3 times.

Practice 1

Evaluate each expression.

a. 4^2

b. 2^2

c. 3^4

d. 9^1

e. $\left(\dfrac{2}{5}\right)^3$

f. $(0.8)^2$

Helpful Hint

$2^3 \neq 2 \cdot 3$

since 2^3 indicates **repeated multiplication of the same factor.**

$2^3 = 2 \cdot 2 \cdot 2 = 8,$

whereas $2 \cdot 3 = 6$

Example 1   Evaluate (find the value of) each expression.

a. 3^2 [read as "3 squared" or as "3 to the second power"]

b. 5^3 [read as "5 cubed" or as "5 to the third power"]

c. 2^4 [read as "2 to the fourth power"]

d. 7^1

e. $\left(\dfrac{3}{7}\right)^2$

f. $(0.6)^2$

Solution:

a. $3^2 = 3 \cdot 3 = 9$

b. $5^3 = 5 \cdot 5 \cdot 5 = 125$

c. $2^4 = 2 \cdot 2 \cdot 2 \cdot 2 = 16$

d. $7^1 = 7$

e. $\left(\dfrac{3}{7}\right)^2 = \left(\dfrac{3}{7}\right)\left(\dfrac{3}{7}\right) = \dfrac{3 \cdot 3}{7 \cdot 7} = \dfrac{9}{49}$

f. $(0.6)^2 = (0.6)(0.6) = 0.36$

■ **Work Practice 1**

Using symbols for mathematical operations is a great convenience. The more operation symbols presented in an expression, the more careful we must be when performing the indicated operation. For example, in the expression $2 + 3 \cdot 7$, do we add first or multiply first? To eliminate confusion, **grouping symbols** are used. Examples of grouping symbols are parentheses ( ), brackets [ ], braces { }, absolute value bars | |, and the fraction bar. If we wish $2 + 3 \cdot 7$ to be simplified by adding first, we enclose $2 + 3$ in parentheses.

$$(2 + 3) \cdot 7 = 5 \cdot 7 = 35$$

If we wish to multiply first, $3 \cdot 7$ may be enclosed in parentheses.

$$2 + (3 \cdot 7) = 2 + 21 = 23$$

To eliminate confusion when no grouping symbols are present, we use the following agreed-upon order of operations.

Order of Operations

1. Perform all operations within grouping symbols first, starting with the innermost set.
2. Evaluate exponential expressions.
3. Multiply or divide in order from left to right.
4. Add or subtract in order from left to right.

Using this order of operations, we now simplify $2 + 3 \cdot 7$. There are no grouping symbols and no exponents, so we multiply and then add.

$$2 + 3 \cdot 7 = 2 + 21 \quad \text{Multiply.}$$
$$= 23 \quad\quad\quad \text{Add.}$$

Answers

1. a. 16   **b.** 4   **c.** 81   **d.** 9   **e.** $\dfrac{8}{125}$

f. 0.64

Examples   Simplify each expression.

2. $6 \div 3 + 5^2 = 6 \div 3 + 25$ Evaluate 5^2.

$\qquad\qquad\quad = 2 + 25$ Divide.

$\qquad\qquad\quad = 27$ Add.

3. $\underline{20 \div 5} \cdot 4 = 4 \cdot 4$

$\qquad\qquad = 16$

Helpful Hint   Remember to multiply or divide in order from left to right.

4. $\dfrac{3}{2} \cdot \dfrac{1}{2} - \dfrac{1}{2} = \dfrac{3}{4} - \dfrac{1}{2}$ Multiply.

$\qquad\qquad\quad = \dfrac{3}{4} - \dfrac{2}{4}$ The least common denominator is 4.

$\qquad\qquad\quad = \dfrac{1}{4}$ Subtract.

5. $1 + 2[5(2 \cdot 3 + 1) - 10] = 1 + 2[5(7) - 10]$ Simplify the expression in the innermost set of parentheses. $2 \cdot 3 + 1 = 6 + 1 = 7$.

$\qquad\qquad\qquad\qquad\qquad = 1 + 2[35 - 10]$ Multiply 5 and 7.

$\qquad\qquad\qquad\qquad\qquad = 1 + 2[25]$ Subtract inside the brackets.

$\qquad\qquad\qquad\qquad\qquad = 1 + 50$ Multiply 2 and 25.

$\qquad\qquad\qquad\qquad\qquad = 51$ Add.

■ **Work Practice 2–5**

In the next example, the fraction bar serves as a grouping symbol and separates the numerator and denominator. Simplify each separately.

Example 6   Simplify: $\dfrac{3 + |4 - 3| + 2^2}{6 - 3}$

Solution:

$\dfrac{3 + |4 - 3| + 2^2}{6 - 3} = \dfrac{3 + |1| + 2^2}{6 - 3}$ Simplify the expression inside the absolute value bars.

$\qquad\qquad\qquad\quad = \dfrac{3 + 1 + 2^2}{3}$ Find the absolute value and simplify the denominator.

$\qquad\qquad\qquad\quad = \dfrac{3 + 1 + 4}{3}$ Evaluate the exponential expression.

$\qquad\qquad\qquad\quad = \dfrac{8}{3}$ Simplify the numerator.

■ **Work Practice 6**

Helpful Hint

Be careful when evaluating an exponential expression.

$3 \cdot 4^2 = 3 \cdot 16 = 48$ $(3 \cdot 4)^2 = (12)^2 = 144$

$\qquad\uparrow$ $\qquad\qquad\uparrow$

Base is 4.   Base is $3 \cdot 4$.

Practice 2–5

Simplify each expression.

2. $3 \cdot 2 + 4^2$

3. $28 \div 7 \cdot 2$

4. $\dfrac{9}{5} \cdot \dfrac{1}{3} - \dfrac{1}{3}$

5. $5 + 3[2(3 \cdot 4 + 1) - 20]$

Practice 6

Simplify: $\dfrac{1 + |7 - 4| + 3^2}{8 - 5}$

Answers

2. 22   **3.** 8   **4.** $\dfrac{4}{15}$ **5.** 23   **6.** $\dfrac{13}{3}$

Objective B Evaluating Algebraic Expressions

Recall that letters used to represent quantities are called **variables.** An **algebraic expression** is a collection of numbers, variables, operation symbols, and grouping symbols. For example,

$$2x, \quad -3, \quad 2x - 10, \quad 5(p^2 + 1), \quad xy, \quad \text{and} \quad \frac{3y^2 - 6y + 1}{5}$$

are algebraic expressions.

Expressions	Meaning
$2x$	$2 \cdot x$
$5(p^2 + 1)$	$5 \cdot (p^2 + 1)$
$3y^2$	$3 \cdot y^2$
xy	$x \cdot y$

If we give a specific value to a variable, we can **evaluate an algebraic expression.** To evaluate an algebraic expression means to find its numerical value once we know the values of the variables.

Algebraic expressions are often used in problem solving. For example, the expression

$$16t^2$$

gives the distance in feet (neglecting air resistance) that an object will fall in t seconds.

Practice 7

Evaluate each expression when $x = 1$ and $y = 4$.

a. $3y^2$

b. $2y - x$

c. $\dfrac{11x}{3y}$

d. $\dfrac{x}{y} + \dfrac{6}{y}$

e. $y^2 - x^2$

Example 7 Evaluate each expression when $x = 3$ and $y = 2$.

a. $5x^2$ **b.** $2x - y$ **c.** $\dfrac{3x}{2y}$ **d.** $\dfrac{x}{y} + \dfrac{y}{2}$ **e.** $x^2 - y^2$

Solution:

a. Replace x with 3. Then simplify.

$$5x^2 = 5 \cdot (3)^2 = 5 \cdot 9 = 45$$

b. Replace x with 3 and y with 2. Then simplify.

$$\begin{aligned}2x - y &= 2(3) - 2 &&\text{Let } x = 3 \text{ and } y = 2. \\ &= 6 - 2 &&\text{Multiply.} \\ &= 4 &&\text{Subtract.}\end{aligned}$$

c. Replace x with 3 and y with 2. Then simplify.

$$\frac{3x}{2y} = \frac{3 \cdot 3}{2 \cdot 2} = \frac{9}{4} \quad \text{Let } x = 3 \text{ and } y = 2.$$

d. Replace x with 3 and y with 2. Then simplify.

$$\frac{x}{y} + \frac{y}{2} = \frac{3}{2} + \frac{2}{2} = \frac{5}{2}$$

e. Replace x with 3 and y with 2. Then simplify.

$$x^2 - y^2 = 3^2 - 2^2 = 9 - 4 = 5$$

■ **Work Practice 7**

Answers

7. a. 48 **b.** 7 **c.** $\dfrac{11}{12}$ **d.** $\dfrac{7}{4}$ **e.** 15

Objective C Solutions of Equations

Many times a problem-solving situation is modeled by an equation. An **equation** is a mathematical statement that two expressions have equal value. The equal symbol "=" is used to equate the two expressions. For example,

$$3 + 2 = 5, 7x = 35, \frac{2(x - 1)}{3} = 0, \text{ and } I = PRT \text{ are all equations.}$$

Helpful Hint

An equation contains the equal symbol "=". An algebraic expression does not.

✓**Concept Check** Which of the following are equations? Which are expressions?

a. $5x = 8$ **b.** $5x - 8$ **c.** $12y + 3x$ **d.** $12y = 3x$

When an equation contains a variable, deciding which value(s) of the variable make the equation a true statement is called **solving** the equation for the variable. A **solution** of an equation is a value for the variable that makes the equation a true statement. For example, 3 is a solution of the equation $x + 4 = 7$, because if x is replaced with 3 the statement is true.

$$x + 4 = 7$$
$$\downarrow$$
$$3 + 4 \stackrel{?}{=} 7 \quad \text{Replace } x \text{ with 3.}$$
$$7 = 7 \quad \text{True}$$

Similarly, 1 is not a solution of the equation $x + 4 = 7$, because $1 + 4 = 7$ is **not** a true statement.

Example 8 Decide whether 2 is a solution of $3x + 10 = 8x$.

Solution: Replace x with 2 and see if a true statement results.

$$3x + 10 = 8x \quad \text{Original equation}$$
$$3(2) + 10 \stackrel{?}{=} 8(2) \quad \text{Replace } x \text{ with 2.}$$
$$6 + 10 \stackrel{?}{=} 16 \quad \text{Simplify each side.}$$
$$16 = 16 \quad \text{True}$$

Since we arrived at a true statement after replacing x with 2 and simplifying both sides of the equation, 2 is a solution of the equation.

■ **Work Practice 8**

Practice 8

Decide whether 3 is a solution of $5x - 10 = x + 2$.

Objective D Translating Words to Symbols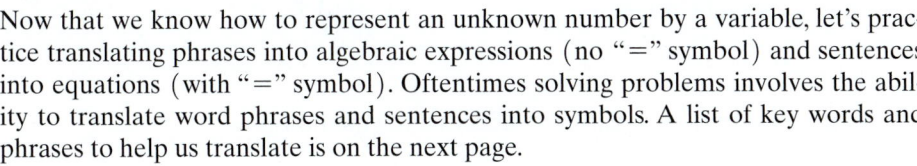

Now that we know how to represent an unknown number by a variable, let's practice translating phrases into algebraic expressions (no "=" symbol) and sentences into equations (with "=" symbol). Oftentimes solving problems involves the ability to translate word phrases and sentences into symbols. A list of key words and phrases to help us translate is on the next page.

Answer

8. It is a solution.

✓**Concept Check Answer**

equations: **a, d**; expressions: **b, c**

Helpful Hint

Order matters when subtracting and also dividing, so be especially careful with these translations.

Addition (+)	Subtraction (−)	Multiplication (·)	Division (÷)	Equality (=)
Sum	Difference of	Product	Quotient	Equals
Plus	Minus	Times	Divide	Gives
Added to	Subtracted from	Multiply	Into	Is/was/should be
More than	Less than	Twice	Ratio	Yields
Increased by	Decreased by	Of	Divided by	Amounts to
Total	Less			Represents
				Is the same as

Practice 9

Write an algebraic expression that represents each phrase. Let the variable x represent the unknown number.

a. The product of 5 and a number

b. A number added to 7

c. A number divided by 11.2

d. A number subtracted from 8

e. Twice a number, plus 1

Example 9 Write an algebraic expression that represents each phrase. Let the variable x represent the unknown number.

a. The sum of a number and 3

b. The product of 3 and a number

c. The quotient of 7.3 and a number

d. 10 decreased by a number

e. 5 times a number, increased by 7

Solution:

a. $x + 3$ since "sum" means to add

b. $3 \cdot x$ and $3x$ are both ways to denote the product of 3 and x

c. $7.3 \div x$ or $\dfrac{7.3}{x}$

d. $10 - x$ because "decreased by" means to subtract

e. $\underbrace{5x}_{\substack{5 \text{ times} \\ \text{a number}}} + 7$

■ **Work Practice 9**

Helpful Hint

Make sure you understand the difference when translating phrases containing "decreased by," "subtracted from," and "less than."

Phrase	Translation
A number decreased by 10	$x - 10$
A number subtracted from 10	$10 - x$
10 less than a number	$x - 10$
A number less 10	$x - 10$

Notice the order.

Now let's practice translating sentences into equations.

Answers

9. a. $5 \cdot x$ or $5x$ **b.** $7 + x$

c. $x \div 11.2$ or $\dfrac{x}{11.2}$ **d.** $8 - x$

e. $2x + 1$

Example 10   Write each sentence as an equation. Let x represent the unknown number.

a. The quotient of 15 and a number is 4.

b. Three subtracted from 12 is a number.

c. 17 added to four times a number is 21.

Solution:

a. In words:

the quotient of 15 and a number	is	4
↓	↓	↓

Translate: $\dfrac{15}{x}$ = 4

b. In words:

three subtracted **from** 12	is	a number
↓	↓	↓

Translate: $12 - 3$ = x

Care must be taken when the operation is subtraction. The expression $3 - 12$ would be incorrect. Notice that $3 - 12 \neq 12 - 3$.

c. In words:

17	added to	four times a number	is	21
↓	↓	↓	↓	↓

Translate: 17   + $4x$ =   21

■ **Work Practice 10**

Practice 10

Write each sentence as an equation. Let x represent the unknown number.

a. The ratio of a number and 6 is 24.

b. The difference of 10 and a number is 18.

c. One less than twice a number is 99.

Answers

10. a. $\dfrac{x}{6} = 24$, **b.** $10 - x = 18$,

c. $2x - 1 = 99$

🖩 Calculator Explorations Exponents

To evaluate exponential expressions on a calculator, find the key marked $\boxed{y^x}$ or $\boxed{\wedge}$. To evaluate, for example, 6^5, press the following keys: $\boxed{6}\boxed{y^x}\boxed{5}\boxed{=}$ or $\boxed{6}\boxed{\wedge}\boxed{5}\boxed{=}$.

⇕ or

$\boxed{\text{ENTER}}$

The display should read $\boxed{\quad 7776 \quad}$

Order of Operations

Some calculators follow the order of operations, and others do not. To see whether or not your calculator has the order of operations built in, use your calculator to find $2 + 3 \cdot 4$. To do this, press the following sequence of keys:

$\boxed{2}\boxed{+}\boxed{3}\boxed{\times}\boxed{4}\boxed{=}$

⇕ or

$\boxed{\text{ENTER}}$

The correct answer is 14 because the order of operations is to multiply before we add. If the calculator displays $\boxed{\quad 14 \quad}$, then it has the order of operations built in.

Even if the order of operations is built in, parentheses must sometimes be inserted. For example, to simplify $\dfrac{5}{12 - 7}$, press the keys

$\boxed{5}\boxed{\div}\boxed{(}\boxed{1}\boxed{2}\boxed{-}\boxed{7}\boxed{)}\boxed{=}$.

⇕ or

$\boxed{\text{ENTER}}$

The display should read $\boxed{\quad 1 \quad}$.

Use a calculator to evaluate each expression.

1. 5^3 125   **2.** 7^4 2401

3. 9^5 59,049   **4.** 8^6 262,144

5. $2(20 - 5)$ 30   **6.** $3(14 - 7) + 21$ 42

7. $24(862 - 455) + 89$ 9857

8. $99 + (401 + 962)$ 1462

9. $\dfrac{4623 + 129}{36 - 34}$ 2376

10. $\dfrac{956 - 452}{89 - 86}$ 168

Vocabulary, Readiness & Video Check

Use the choices below to fill in each blank. Some choices may be used more than once.

addition	multiplication	exponent	expression	solution	evaluating the expression
subtraction	division	base	equation	variable(s)	

1. In 2^5, the 2 is called the ___base___ and the 5 is called the ___exponent___.

2. True or false: 2^5 means $2 \cdot 5$. ___false___.

3. To simplify $8 + 2 \cdot 6$, which operation should be performed first? ___multiplication___

4. To simplify $(8 + 2) \cdot 6$, which operation should be performed first? ___addition___

5. To simplify $9(3 - 2) \div 3 + 6$, which operation should be performed first? ___subtraction___

6. To simplify $8 \div 2 \cdot 6$, which operation should be performed first? ___division___

7. A combination of operations on letters (variables) and numbers is a(n) ___expression___.

8. A letter that represents a number is a(n) ___variable___.

9. $3x - 2y$ is called a(n) ___expression___ and the letters x and y are ___variables___.

10. Replacing a variable in an expression by a number and then finding the value of the expression is called ___evaluating the expression___.

11. A statement of the form "expression = expression" is called a(n) ___equation___.

12. A value for the variable that makes the equation a true statement is called a(n) ___solution___.

Martin-Gay Interactive Videos

See Video 1.3

Watch the section lecture video and answer the following questions.

Objective A 13. In ▣ Example 3 and the lecture before, what is the main point made about the order of operations? ▶

Objective B 14. What happens with the replacement value for z in ▣ Example 6 and why? ▶

Objective C 15. Is the value 0 a solution of the equation given in ▣ Example 9? How is this determined? ▶

Objective D 16. Earlier in this video the point was made that equations have =, while expressions do not. In the lecture before ▣ Example 10, translating from English to math is discussed and another difference between expressions and equations is explained. What is it? ▶

See video answer section.

1.3 Exercise Set MyMathLab®

Objective A *Evaluate. See Example 1.*

1. 3^5 243
2. 5^4 625
3. 3^3 27
4. 4^4 256
5. 1^5 1
6. 1^8 1

7. 5^1 5
8. 8^1 8
9. 7^2 49
10. 9^2 81
11. $\left(\dfrac{2}{3}\right)^4$ $\dfrac{16}{81}$
12. $\left(\dfrac{6}{11}\right)^2$ $\dfrac{36}{121}$

13. $\left(\dfrac{1}{5}\right)^3$ $\dfrac{1}{125}$
14. $\left(\dfrac{1}{2}\right)^5$ $\dfrac{1}{32}$
15. $(1.2)^2$ 1.44
16. $(1.5)^2$ 2.25
17. $(0.7)^3$ 0.343
18. $(0.4)^3$ 0.064

△ **19.** The area of a square whose sides each measure 5 meters is $(5 \cdot 5)$ square meters. Write this area using exponential notation. 5^2 sq m

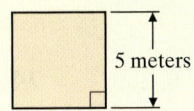

5 meters

△ **20.** The area of a circle whose radius is 9 meters is $(9 \cdot 9 \cdot \pi)$ square meters. Write this area using exponential notation. $9^2 \pi$ sq m

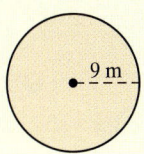

9 m

Simplify each expression. See Examples 2 through 6.

▶ **21.** $5 + 6 \cdot 2$ 17

22. $8 + 5 \cdot 3$ 23

23. $4 \cdot 8 - 6 \cdot 2$ 20

24. $12 \cdot 5 - 3 \cdot 6$ 42

25. $18 \div 3 \cdot 2$ 12

26. $48 \div 6 \cdot 2$ 16

27. $2 + (5 - 2) + 4^2$ 21

28. $6 - 2 \cdot 2 + 2^5$ 34

29. $5 \cdot 3^2$ 45

30. $2 \cdot 5^2$ 50

31. $\frac{1}{4} \cdot \frac{2}{3} - \frac{1}{6}$ 0

32. $\frac{3}{4} \cdot \frac{1}{2} + \frac{2}{3}$ $\frac{25}{24}$

33. $\frac{6 - 4}{9 - 2}$ $\frac{2}{7}$

34. $\frac{8 - 5}{24 - 20}$ $\frac{3}{4}$

▶ **35.** $2[5 + 2(8 - 3)]$ 30

36. $3[4 + 3(6 - 4)]$ 30

37. $\frac{19 - 3 \cdot 5}{6 - 4}$ 2

38. $\frac{14 - 2 \cdot 3}{12 - 8}$ 2

▶ **39.** $\frac{|6 - 2| + 3}{8 + 2 \cdot 5}$ $\frac{7}{18}$

40. $\frac{15 - |3 - 1|}{12 - 3 \cdot 2}$ $\frac{13}{6}$

41. $\frac{3 + 3(5 + 3)}{3^2 + 1}$ $\frac{27}{10}$

42. $\frac{3 + 6(8 - 5)}{4^2 + 2}$ $\frac{7}{6}$

43. $\frac{6 + |8 - 2| + 3^2}{18 - 3}$ $\frac{7}{5}$

44. $\frac{16 + |13 - 5| + 4^2}{17 - 5}$ $\frac{10}{3}$

45. $2 + 3[10(4 \cdot 5 - 16) - 30]$ 32

46. $3 + 4[8(5 \cdot 5 - 20) - 41]$ -1

47. $\left(\frac{2}{3}\right)^3 + \frac{1}{9} + \frac{1}{3} \cdot \frac{4}{3}$ $\frac{23}{27}$

48. $\left(\frac{3}{8}\right)^2 + \frac{1}{4} + \frac{1}{8} \cdot \frac{3}{2}$ $\frac{37}{64}$

Objective B *Evaluate each expression when $x = 1$, $y = 3$, and $z = 5$. See Example 7.*

49. $3y$ 9

50. $4x$ 4

51. $\frac{z}{5x}$ 1

52. $\frac{y}{2z}$ $\frac{3}{10}$

53. $3x - 2$ 1

54. $6y - 8$ 10

▶ **55.** $|2x + 3y|$ 11

56. $|5z - 2y|$ 19

57. $xy + z$ 8

58. $yz - x$ 14

59. $5y^2$ 45

60. $2z^2$ 50

Evaluate each expression when $x = 12$, $y = 8$, and $z = 4$. See Example 7.

61. $\frac{x}{z} + 3y$ 27

62. $\frac{y}{z} + 8x$ 98

63. $x^2 - 3y + x$ 132

64. $y^2 - 3x + y$ 36

▶ **65.** $\frac{x^2 + z}{y^2 + 2z}$ $\frac{37}{18}$

66. $\frac{y^2 + x}{x^2 + 3y}$ $\frac{19}{42}$

Objective **C** *Decide whether the given number is a solution of the given equation. See Example 8.*

67. $3x - 6 = 9; 5$
solution

68. $2x + 7 = 3x; 6$
not a solution

69. $2x + 6 = 5x - 1; 0$
not a solution

70. $4x + 2 = x + 8; 2$
solution

71. $2x - 5 = 5; 8$
not a solution

▶ **72.** $3x - 10 = 8; 6$
solution

73. $x + 6 = x + 6; 2$
solution

74. $x + 6 = x + 6; 10$
solution

▶ **75.** $x = 5x + 15; 0$
not a solution

76. $4 = 1 - x; 1$
not a solution

77. $\frac{1}{3}x = 9; 27$
solution

78. $\frac{2}{7}x = \frac{3}{14}; 6$
not a solution

Objective **D** *Write each phrase as an algebraic expression. Let x represent the unknown number. See Example 9.*

79. Fifteen more than a number $x + 15$

80. A number increased by 9 $x + 9$

81. Five subtracted from a number $x - 5$

82. Five decreased by a number $5 - x$

83. The ratio of a number and 4 $\frac{x}{4}$

84. The quotient of a number and 9 $\frac{x}{9}$

▶ **85.** Three times a number, increased by 22 $3x + 22$

86. Twice a number, decreased by 72 $2x - 72$

Write each sentence as an equation or inequality. Use x to represent any unknown number. See Example 10.

▶ **87.** One increased by two equals the quotient of nine and three. $1 + 2 = 9 \div 3$

88. Four subtracted from eight is equal to two squared. $8 - 4 = 2^2$

▶ **89.** Three is not equal to four divided by two.
$3 \neq 4 \div 2$

90. The difference of sixteen and four is greater than ten. $16 - 4 > 10$

91. The sum of 5 and a number is 20. $5 + x = 20$

92. Seven subtracted from a number is 0. $x - 7 = 0$

93. The product of 7.6 and a number is 17. $7.6x = 17$

94. 9.1 times a number equals 4 $9.1x = 4$

95. Thirteen minus three times a number is 13.
$13 - 3x = 13$

96. Eight added to twice a number is 42. $2x + 8 = 42$

Concept Extensions

97. Are parentheses necessary in the expression $2 + (3 \cdot 5)$? Explain your answer.
no; answers may vary

98. Are parentheses necessary in the expression $(2 + 3) \cdot 5$? Explain your answer.
yes; answers may vary

For Exercises 99 and 100, match each expression in the first column with its value in the second column.

99. **a.** $(6 + 2) \cdot (5 + 3)$ 64   19

   b. $(6 + 2) \cdot 5 + 3$ 43   22

   c. $6 + 2 \cdot 5 + 3$ 19   64

   d. $6 + 2 \cdot (5 + 3)$ 22   43

100. **a.** $(1 + 4) \cdot 6 - 3$ 27   15

   b. $1 + 4 \cdot (6 - 3)$ 13   13

   c. $1 + 4 \cdot 6 - 3$ 22   27

   d. $(1 + 4) \cdot (6 - 3)$ 15   22

△ *Recall that perimeter measures the distance around a plane figure and area measures the amount of surface of a plane figure. The expression $2l + 2w$ gives the perimeter of the rectangle below (measured in units), and the expression lw gives its area (measured in square units). Complete the chart below for the given lengths and widths. Be sure to include units.*

	Length: l	Width: w	Perimeter of Rectangle: $2l + 2w$	Area of Rectangle: lw
101.	4 in.	3 in.	14 in.	12 sq in.
102.	6 in.	1 in.	14 in.	6 sq in.
103.	5.3 in.	1.7 in.	14 in.	9.01 sq in.
104.	4.6 in.	2.4 in.	14 in.	11.04 sq in.

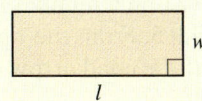

105. Study the perimeters and areas found in the chart to the left. Do you notice any trends?   Rectangles with the same perimeter can have different areas.

106. In your own words, explain the difference between an expression and an equation.   answers may vary

107. Insert one set of parentheses so that the following expression simplifies to 32. $(20 - 4) \cdot 4 \div 2$

$$20 - 4 \cdot 4 \div 2$$

108. Insert parentheses so that the following expression simplifies to 28. $2 \cdot (5 + 3^2)$

$$2 \cdot 5 + 3^2$$

Determine whether each is an expression or an equation. See the Concept Check in this section.

109. **a.** $5x + 6$ expression

   b. $2a = 7$ equation

   c. $3a + 2 = 9$ equation

   d. $4x + 3y - 8z$ expression

   e. $5^2 - 2(6 - 2)$ expression

110. **a.** $3x^2 - 26$ expression

   b. $3x^2 - 26 = 1$ equation

   c. $2x - 5 = 7x - 5$ equation

   d. $9y + x - 8$ expression

   e. $3^2 - 4(5 - 3)$ expression

111. Why is 4^3 usually read as "four cubed"? (*Hint:* What is the volume of the **cube** below?)

△   answers may vary

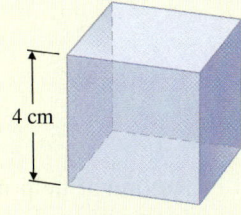

4 cm

112. Why is 8^2 usually read as "eight squared"? (*Hint:* What is the area of the **square** below?)

△   answers may vary

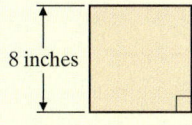

8 inches

113. Write any expression, using 3 or more numbers, that simplifies to -11.

answers may vary; for example, $-2(5) - 1$

114. Write any expression, using 4 or more numbers, that simplifies to 7.

answers may vary; for example, $2(10 - 7) + 1$

1.4 Adding Real Numbers

Objectives

A Add Real Numbers.

B Find the Opposite of a Number.

C Evaluate Algebraic Expressions Using Real Numbers.

D Solve Applications That Involve Addition of Real Numbers.

Real numbers can be added, subtracted, multiplied, divided, and raised to powers, just as whole numbers can.

Objective A Adding Real Numbers

Adding real numbers can be visualized by using a number line. A positive number can be represented on the number line by an arrow of appropriate length pointing to the right, and a negative number by an arrow of appropriate length pointing to the left.

Both arrows represent 2 or +2.

They both point to the right, and they are both 2 units long.

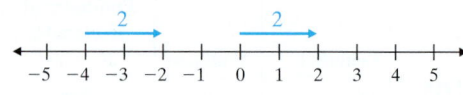

Both arrows represent −3.

They both point to the left, and they are both 3 units long.

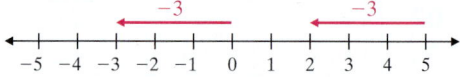

To add signed numbers such as $5 + (-2)$ on a number line, we start at 0 on the number line and draw an arrow representing 5. From the tip of this arrow, we draw another arrow representing −2. The tip of the second arrow ends at their sum, 3.

$$5 + (-2) = 3$$

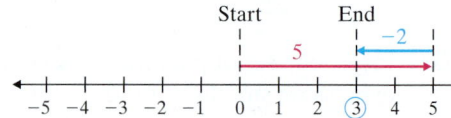

To add $-1 + (-4)$ on the number line, we start at 0 and draw an arrow representing −1. From the tip of this arrow, we draw another arrow representing −4. The tip of the second arrow ends at their sum, −5.

$$-1 + (-4) = -5$$

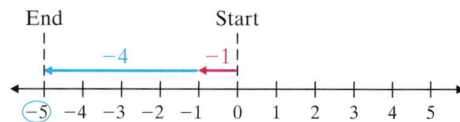

Teaching Tip

We can visualize the addition of integers with + and − signs. For example, $3 + (-5)$ can be represented by

$$\underbrace{+++}_{} \ \underbrace{-----}_{} \Rightarrow \oplus \oplus \oplus \ \underset{}{=} \ \underset{}{=}$$

$$3 + (-5) = -2$$

Using the atom analogy, a positive and a negative will neutralize to a zero. In this problem, the three positives are neutralized by 3 of the negatives and 2 negatives remain so the answer is −2.

Practice 1

Add using a number line:
$-2 + (-4)$

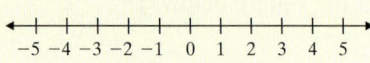

Example 1 Add: $-1 + (-2)$

Solution:

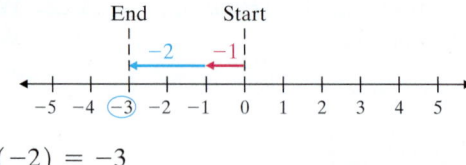

$$-1 + (-2) = -3$$

■ Work Practice 1

Thinking of integers as money earned or lost might help make addition more meaningful. Earnings can be thought of as positive numbers. If $1 is earned and later another $3 is earned, the total amount earned is $4. In other words, $1 + 3 = 4$.

On the other hand, losses can be thought of as negative numbers. If $1 is lost and later another $3 is lost, a total of $4 is lost. In other words, $(-1) + (-3) = -4$.

In Example 1, we added numbers with the same sign. Adding numbers whose signs are not the same can be pictured on a number line also.

Answer

1. −6

30

Example 2   Add: $-4 + 6$

Solution:

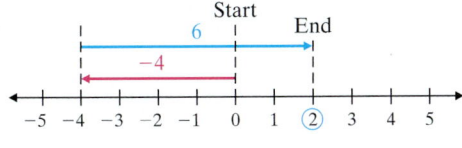

$$-4 + 6 = 2$$

■ **Work Practice 2**

Let's use temperature as an example. If the thermometer registers 4 degrees below 0 degrees and then rises 6 degrees, the new temperature is 2 degrees above 0 degrees. Thus, it is reasonable that $-4 + 6 = 2$. (See the diagram in the margin.)

Example 3   Add: $4 + (-6)$

Solution:

$$4 + (-6) = -2$$

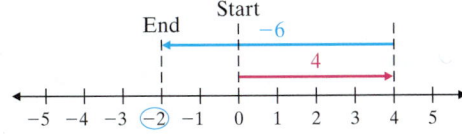

■ **Work Practice 3**

Using a number line each time we add two numbers can be time consuming. Instead, we can notice patterns in the previous examples and write rules for adding real numbers.

Adding Real Numbers

To add two real numbers

1. with the *same sign*, add their absolute values. Use their common sign as the sign of the answer.
2. with *different signs*, subtract their absolute values. Give the answer the same sign as the number with the larger absolute value.

Example 4   Add without using a number line: $(-7) + (-6)$

Solution: Here, we are adding two numbers with the same sign.

$$(-7) + (-6) = -13$$
↑ ↖ sum of absolute values ($|-7| = 7, |-6| = 6, 7 + 6 = 13$)
same sign

■ **Work Practice 4**

Example 5   Add without using a number line: $(-10) + 4$

Solution: Here, we are adding two numbers with different signs.

$$(-10) + 4 = -6$$
↑ ↖ difference of absolute values ($|-10| = 10, |4| = 4, 10 - 4 = 6$)
sign of number with larger absolute value, -10

■ **Work Practice 5**

Practice 2

Add using a number line:
$-5 + 8$

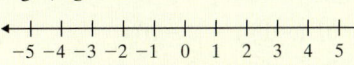

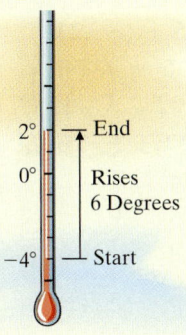

Practice 3

Add using a number line:
$5 + (-4)$

Practice 4

Add without using a number line: $(-8) + (-5)$

Practice 5

Add without using a number line: $(-14) + 6$

Answers
2. 3 **3.** 1 **4.** -13 **5.** -8

Practice 6–11

Add without using a number line.

6. $(-17) + (-10)$

7. $(-4) + 12$

8. $1.5 + (-3.2)$

9. $-\dfrac{5}{12} + \left(-\dfrac{1}{12}\right)$

10. $12.1 + (-3.6)$

11. $-\dfrac{4}{5} + \dfrac{2}{3}$

Examples Add without using a number line.

6. $(-8) + (-11) = -19$

7. $(-2) + 10 = 8$

8. $0.2 + (-0.5) = -0.3$

9. $-\dfrac{7}{10} + \left(-\dfrac{1}{10}\right) = -\dfrac{8}{10} = -\dfrac{\cancel{2} \cdot 4}{\cancel{2} \cdot 5} = -\dfrac{4}{5}$

10. $11.4 + (-4.7) = 6.7$

11. $-\dfrac{3}{8} + \dfrac{2}{5} = -\dfrac{15}{40} + \dfrac{16}{40} = \dfrac{1}{40}$

🔶 Work Practice 6–11

In Example 12a, we add three numbers. Remember that by the associative and commutative properties for addition, we may add numbers in any order that we wish. For Example 12a, let's add the numbers from left to right.

Practice 12

Find each sum.

a. $16 + (-9) + (-9)$

b. $[3 + (-13)] + [-4 + (-7)]$

Example 12 Find each sum.

a. $3 + (-7) + (-8)$

b. $[7 + (-10)] + [-2 + (-4)]$

Solution:

a. Perform the additions from left to right.

$$3 + (-7) + (-8) = -4 + (-8) \quad \text{Adding numbers with different signs}$$
$$= -12 \quad \text{Adding numbers with like signs}$$

b. Simplify inside the brackets first.

$$[7 + (-10)] + [-2 + (-4)] = [-3] + [-6]$$
$$= -9 \quad \text{Add.}$$

🔶 Work Practice 12

Helpful Hint Don't forget that brackets are grouping symbols. We simplify within them first.

Objective B Finding Opposites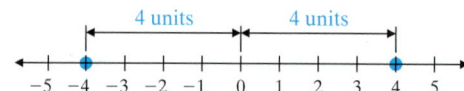

To help us subtract real numbers in the next section, we first review what we mean by opposites. The graphs of 4 and -4 are shown on the number line below.

Notice that the graphs of 4 and -4 lie on opposite sides of 0, and each is 4 units away from 0. Such numbers are known as **opposites** or **additive inverses** of each other.

Answers

6. -27 **7.** 8 **8.** -1.7 **9.** $-\dfrac{1}{2}$

10. 8.5 **11.** $-\dfrac{2}{15}$ **12. a.** -2 **b.** -21

Opposite or Additive Inverse

Two numbers that are the same distance from 0 but lie on opposite sides of 0 are called **opposites** or **additive inverses** of each other.

Examples   Find the opposite of each number.

13. 10   The opposite of 10 is -10.

14. -3 The opposite of -3 is 3.

15. $\dfrac{1}{2}$ The opposite of $\dfrac{1}{2}$ is $-\dfrac{1}{2}$.

16. -4.5 The opposite of -4.5 is 4.5.

■ **Work Practice 13–16**

We use the symbol "$-$" to represent the phrase "the opposite of" or "the additive inverse of." In general, if a is a number, we write the opposite or additive inverse of a as $-a$. We know that the opposite of -3 is 3. Notice that this translates as

the opposite of -3 is   3

$-$ (-3) $=$ 3

This is true in general.

> If a is a number, then $-(-a) = a$.

Example 17   Simplify each expression.

a. $-(-10)$ **b.** $-\left(-\dfrac{1}{2}\right)$ **c.** $-(-2x)$

d. $-|-6|$ **e.** $-|4.1|$

Solution:

a. $-(-10) = 10$

b. $-\left(-\dfrac{1}{2}\right) = \dfrac{1}{2}$

c. $-(-2x) = 2x$

d. $-|-6| = -6$ Since $|-6| = 6$.

e. $-|4.1| = -4.1$ Since $|4.1| = 4.1$

■ **Work Practice 17**

Let's discover another characteristic about opposites. Notice that the sum of a number and its opposite is always 0.

$10 + (-10) = 0$ $-3 + 3 = 0$

opposites   opposites

$\dfrac{1}{2} + \left(-\dfrac{1}{2}\right) = 0$

opposites

In general, we can write the following:

> The sum of a number a and its opposite $-a$ is 0.
>
> $a + (-a) = 0$ Also, $-a + a = 0$.

Notice that this means that the opposite of 0 is then 0 since $0 + 0 = 0$.

Practice 13–16

Find the opposite of each number.

13. -35 **14.** 12

15. $-\dfrac{3}{11}$ **16.** 1.9

Practice 17

Simplify each expression.

a. $-(-22)$

b. $-\left(-\dfrac{2}{7}\right)$

c. $-(-x)$

d. $-|-14|$

e. $-|2.3|$

Practice 18–19

Add.

18. $30 + (-30)$

19. $-81 + 81$

Examples Add.

18. $-56 + 56 = 0$

19. $17 + (-17) = 0$

■ **Work Practice 18–19**

✓**Concept Check** What is wrong with the following calculation?

$$5 + (-22) = 17$$

Objective C Evaluating Algebraic Expressions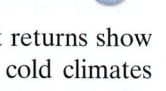

We can continue our work with algebraic expressions by evaluating expressions given real-number replacement values.

Example 20 Evaluate $2x + y$ for $x = 3$ and $y = -5$.

Solution: Replace x with 3 and y with -5 in $2x + y$.

$$2x + y = 2 \cdot 3 + (-5)$$
$$= 6 + (-5)$$
$$= 1$$

■ **Work Practice 20**

Practice 20

Evaluate $x + 3y$ for $x = -6$ and $y = 2$.

Example 21 Evaluate $x + y$ for $x = -2$ and $y = -10$.

Solution: $x + y = (-2) + (-10)$ Replace x with -2 and y with -10.
$$= -12$$

■ **Work Practice 21**

Practice 21

Evaluate $x + y$ for $x = -13$ and $y = -9$.

Objective D Solving Applications That Involve Addition ▶

Positive and negative numbers are used in everyday life. Stock market returns show gains and losses as positive and negative numbers. Temperatures in cold climates often dip into the negative range, commonly referred to as "below zero" temperatures. Bank statements report deposits and withdrawals as positive and negative numbers.

Example 22 Calculating Temperature

In Philadelphia, Pennsylvania, the record extreme high temperature is 104°F. Decrease this temperature by 111 degrees, and the result is the record extreme low temperature. Find this temperature. (*Source:* National Climatic Data Center)

Solution:

In words:	extreme low temperature	=	extreme high temperature	+	decrease of 111°
	↓		↓		↓
Translate:	extreme low temperature	=	104	+	(-111)
		$= -7$			

The record extreme low temperature in Philadelphia, Pennsylvania, is $-7°$F.

■ **Work Practice 22**

Practice 22

If the temperature was $-7°$ Fahrenheit at 6 a.m., and it rose 4 degrees by 7 a.m. and then rose another 7 degrees in the hour from 7 a.m. to 8 a.m., what was the temperature at 8 a.m.?

Answers

18. 0   **19.** 0   **20.** 0   **21.** −22   **22.** 4°F

✓**Concept Check Answer**

$5 + (-22) = -17$

Vocabulary, Readiness & Video Check

Use the choices below to fill in each blank. Not all choices will be used.

$-a$ a 0 commutative   associative

1. If n is a number, then $-n + n = $ _____0_____.

2. Since $x + n = n + x$, we say that addition is __commutative__.

3. If a is a number, then $-(-a) = $ _____a_____.

4. Since $n + (x + a) = (n + x) + a$, we say that addition is ___associative___.

Martin-Gay Interactive Videos   *Watch the section lecture video and answer the following questions.*

See Video 1.4

Objective A **5.** Complete this statement based on the lecture given before
Example 1. To add two numbers with the same sign, add
their _____ and use their common sign as the
sign of the sum.

6. What is the sign of the sum in Example 6 and why?

Objective B **7.** Example 11 illustrates the idea that if a is a real number,
the opposite of $-a$ is a. Example 12 looks similar to
Example 11, but it's actually quite different. Explain the
difference.

Objective C **8.** Explain the difference between the algebraic expression
for Example 13 and the algebraic expression for
Example 14.

Objective D **9.** What is the real-life application of negative numbers used in
Example 15? The answer to Example 15 is -231. What
does this number mean in the context of the problem?

See video answer section.

1.4 **Exercise Set** MyMathLab®

Objectives A B **Mixed Practice** *Add. See Examples 1 through 12, 18, and 19.*

1. $6 + (-3)$ 3

2. $9 + (-12)$ -3

3. $-6 + (-8)$ -14

4. $-6 + (-14)$ -20

5. $8 + (-7)$ 1

6. $16 + (-4)$ 12

7. $-14 + 2$ -12

8. $-10 + 5$ -5

9. $-2 + (-3)$ -5

10. $-7 + (-4)$ -11

11. $-9 + (-3)$ -12

12. $-11 + (-5)$ -16

13. $-7 + 3$ -4

14. $-5 + 9$ 4

15. $10 + (-3)$ 7

16. $8 + (-6)$ 2

17. $5 + (-7)$ -2

18. $3 + (-6)$ -3

19. $-16 + 16$ 0

20. $23 + (-23)$ 0

21. $27 + (-46)$ -19

22. $53 + (-37)$ 16

23. $-18 + 49$ 31

24. $-26 + 14$ -12

25. $-33 + (-14)$ -47 **26.** $-18 + (-26)$ -44 ▶**27.** $6.3 + (-8.4)$ -2.1 **28.** $9.2 + (-11.4)$ -2.2

29. $117 + (-79)$ 38 **30.** $144 + (-88)$ 56 **31.** $-9.6 + (-3.5)$ -13.1 **32.** $-6.7 + (-7.6)$ -14.3

33. $-\dfrac{3}{8} + \dfrac{5}{8}$ $\dfrac{1}{4}$ **34.** $-\dfrac{5}{12} + \dfrac{7}{12}$ $\dfrac{1}{6}$ **35.** $-\dfrac{7}{16} + \dfrac{1}{4}$ $-\dfrac{3}{16}$ **36.** $-\dfrac{5}{9} + \dfrac{1}{3}$ $-\dfrac{2}{9}$

37. $-\dfrac{7}{10} + \left(-\dfrac{3}{5}\right)$ $-\dfrac{13}{10}$ **38.** $-\dfrac{5}{6} + \left(-\dfrac{2}{3}\right)$ $-\dfrac{3}{2}$ **39.** $|-8| + (-16)$ -8 ▶**40.** $|-6| + (-61)$ -55

41. $-15 + 9 + (-2)$ -8 **42.** $-9 + 15 + (-5)$ 1 ▶**43.** $-21 + (-16) + (-22)$ **44.** $-18 + (-6) + (-40)$
                                                                    -59 -64

45. $-23 + 16 + (-2)$ **46.** $-14 + (-3) + 11$ **47.** $|5 + (-10)|$ **48.** $|7 + (-17)|$
        -9 -6 5 10

▶**49.** $6 + (-4) + 9$ **50.** $8 + (-2) + 7$ **51.** $[-17 + (-4)] + [-12 + 15]$
        11 13 -18

52. $[-2 + (-7)] + [-11 + 22]$ 2 **53.** $|9 + (-12)| + |-16|$ 19 **54.** $|43 + (-73)| + |-20|$ 50

55. $-13 + [5 + (-3) + 4]$ -7 **56.** $-30 + [1 + (-6) + 8]$ -27

57. Find the sum of -38 and 12. -26 **58.** Find the sum of -44 and 16. -28

Objective B   *Find each additive inverse or opposite. See Examples 13 through 17.*

▶**59.** 6 -6 **60.** 4 -4 ▶**61.** -2 2 **62.** -8 8

63. 0 0 **64.** $-\dfrac{1}{4}$ $\dfrac{1}{4}$ **65.** $|-6|$ -6 **66.** $|-11|$ -11

Simplify each of the following. See Example 17.

▶**67.** $-|-2|$ **68.** $-|-5|$ ▶**69.** $-(-7)$ **70.** $-(-14)$ **71.** $-(-7.9)$
        -2 -5 7 14 7.9

72. $-(-8.4)$ **73.** $-(-5z)$ **74.** $-(-7m)$ **75.** $\left|-\dfrac{2}{3}\right|$ $\dfrac{2}{3}$ **76.** $-\left|-\dfrac{2}{3}\right|$ $-\dfrac{2}{3}$
        8.4 $5z$ $7m$

Objective C   *Evaluate $x + y$ for the given replacement values. See Examples 20 and 21.*

▶**77.** $x = -20$ and $y = -50$ -70 **78.** $x = -1$ and $y = -29$ -30

Evaluate 3x + y for the given replacement values. See Examples 20 and 21.

▶ **79.** $x = 2$ and $y = -3$ 3

80. $x = 7$ and $y = -11$ 10

Objective D Translating *Translate each phrase; then simplify. See Example 22.*

81. Find the sum of -6 and 25.   19

82. Find the sum of -30 and 15. -15

83. Find the sum of -31, -9, and 30. -10

84. Find the sum of -49, -2, and 40. -11

Solve. See Example 22.

▶ **85.** Suppose a deep-sea diver dives from the surface to 215 feet below the surface. He then dives down 16 more feet. Use positive and negative numbers to represent this situation. Then find the diver's present depth.
$0 + (-215) + (-16) = -231;$
231 ft below the surface

86. Suppose a diver dives from the surface to 248 meters below the surface and then swims up 8 meters, down 16 meters, down another 28 meters, and then up 32 meters. Use positive and negative numbers to represent this situation. Then find the diver's depth after these movements.
$0 + (-248) + 8 + (-16) + (-28) + 32 = -252;$
252 ft below the surface

87. The lowest temperature ever recorded in Massachusetts was $-35°$F. The highest recorded temperature in Massachusetts was $142°$ higher than the record low temperature. Find Massachusetts' highest recorded temperature. (*Source:* National Climatic Data Center) $107°$F

88. On January 2, 1943, the temperature was $-4°$ at 7:30 a.m. in Spearfish, South Dakota. Incredibly, it got $49°$ warmer in the next 2 minutes. To what temperature did it rise by 7:32? $45°$

89. The lowest elevation on Earth is -411 meters (that is, 411 meters below sea level) at the Dead Sea. If you are standing 316 meters above the Dead Sea, what is your elevation? (*Source:* National Geographic Society) -95 m

90. The lowest elevation in Australia is -52 feet at Lake Eyre. If you are standing at a point 439 feet above Lake Eyre, what is your elevation? (*Source:* National Geographic Society)   387 ft

91. During the 2014 PGA Masters Tournament, the winner, Bubba Watson, had scores of −3, +2, −4, and −3 over four rounds of golf. What was his total score for the tournament? (*Source:* Professional Golfer's Association)   −8

92. Lizette Salas won the 2014 LPGA Kingsmill Championship Tournament with scores of −4, −3, −6, and 0 over four rounds of golf. What was her total score for the tournament? (*Source:* LPGA of America)   −13

93. A negative net income results when a company spends more money than it brings in. Johnson Outdoors Inc. had the following quarterly net incomes during its 2013 fiscal year. (*Source:* Market Watch, Inc.)

Quarter of Fiscal 2013	Net Income (in millions)
First	8.9
Second	13.7
Third	−3.5
Fourth	−2.2

What was the total net income for fiscal year 2013?
$16.9 million

94. Barnes & Noble Inc. had the following quarterly net incomes during 2013. (*Source:* Market Watch, Inc.)

Quarter of 2013	Net Income (in millions)
ended January 31	−6.8
ended April 30	−110.2
ended July 31	−87.8
ended October 31	12.6

What was the total net income shown in the table?
−$192.2 million

Concept Extensions

The following bar graph shows each month's average daily low temperature in degrees Fahrenheit for Barrow, Alaska. Use this graph to answer Exercises 95 through 100.

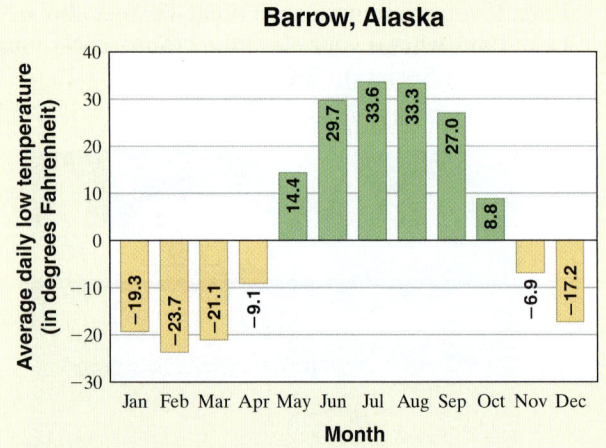

Barrow, Alaska

Average daily low temperature (in degrees Fahrenheit)

Jan −19.3, Feb −23.7, Mar −21.1, Apr −9.1, May 14.4, Jun 29.7, Jul 33.6, Aug 33.3, Sep 27.0, Oct 8.8, Nov −6.9, Dec −17.2

Month

Source: National Climatic Data Center

95. For what month is the graphed temperature the highest?   July

96. For what month is the graphed temperature the lowest?   February

97. For what month is the graphed temperature positive *and* closest to 0°?   October

98. For what month is the graphed temperature negative *and* closest to 0°?   November

99. Find the average of the temperatures shown for the months of April, May, and October. (To find the average of three temperatures, find their sum and divide by 3.)
4.7°F

100. Find the average of the temperatures shown for the months of January, September, and October. (To find the average of three temperatures, find their sum and divide by 3.)   5.5°F

101. Name 2 numbers whose sum is -17.
answers may vary

102. Name 2 numbers whose sum is -30.
answers may vary

Each calculation below is incorrect. Find the error and correct it. See the Concept Check in this section.

103. $7 + (-10) \stackrel{?}{=} 17$ -3

104. $-4 + 14 \stackrel{?}{=} -18$ 10

105. $-10 + (-12) \stackrel{?}{=} -120$ -22

106. $-15 + (-17) \stackrel{?}{=} 32$ -32

For Exercises 107 through 110, determine whether each statement is true or false.

107. The sum of two negative numbers is always a negative number.   true

108. The sum of two positive numbers is always a positive number.   true

109. The sum of a positive number and a negative number is always a negative number.   false

110. The sum of zero and a negative number is always a negative number.   true

111. In your own words, explain how to add two negative numbers.   answers may vary

112. In your own words, explain how to add a positive number and a negative number.   answers may vary

---

1.5 Subtracting Real Numbers

Objective A Subtracting Real Numbers

Now that addition of real numbers has been discussed, we can explore subtraction. We know that $9 - 7 = 2$. Notice that $9 + (-7) = 2$, also. This means that

$$9 - 7 = 9 + (-7)$$

Notice that the *difference* of 9 and 7 is the same as the *sum* of 9 and the opposite of 7. This is how we can subtract real numbers.

> **Subtracting Real Numbers**
>
> If a and b are real numbers, then $a - b = a + (-b)$.

In other words, to find the difference of two numbers, we add the opposite of the number being subtracted.

Objectives

A Subtract Real Numbers.

B Evaluate Algebraic Expressions Using Real Numbers.

C Determine Whether a Number Is a Solution of a Given Equation.

D Solve Applications That Involve Subtraction of Real Numbers.

E Find Complementary and Supplementary Angles.

Practice 1

Subtract.

a. $-20 - 6$

b. $3 - (-5)$

c. $7 - 17$

d. $-4 - (-9)$

Example 1   Subtract.

a. $-13 - 4$ **b.** $5 - (-6)$ **c.** $3 - 6$ **d.** $-1 - (-7)$

Solution:

a. $-13 - 4 = -13 + (-4)$ Add -13 to the opposite of 4, which is -4.

$\qquad = -17$

b. $5 - (-6) = 5 + (6)$ Add 5 to the opposite of -6, which is 6.

$\qquad = 11$

c. $3 - 6 = 3 + (-6)$ Add 3 to the opposite of 6, which is -6.

$\qquad = -3$

d. $-1 - (-7) = -1 + (7) = 6$

■ **Work Practice 1**

Helpful Hint

Study the patterns indicated.

No change ↓   Change to addition. ↓ Change to opposite.

$5 - 11 = \ 5 + (-11) = -6$

$-3 - 4 = -3 + \ (-4) = -7$

$7 - (-1) = \ 7 + \ \ (1) = \ 8$

Practice 2–4

Subtract.

2. $9.6 - (-5.7)$

3. $-\dfrac{4}{9} - \dfrac{2}{9}$

4. $-\dfrac{1}{4} - \left(-\dfrac{2}{5}\right)$

Examples   Subtract.

2. $5.3 - (-4.6) = 5.3 + (4.6) = 9.9$

3. $-\dfrac{3}{10} - \dfrac{5}{10} = -\dfrac{3}{10} + \left(-\dfrac{5}{10}\right) = -\dfrac{8}{10} = -\dfrac{4}{5}$

4. $-\dfrac{2}{3} - \left(-\dfrac{4}{5}\right) = -\dfrac{2}{3} + \left(\dfrac{4}{5}\right) = -\dfrac{10}{15} + \dfrac{12}{15} = \dfrac{2}{15}$

■ **Work Practice 2–4**

Practice 5

Write each phrase as an expression and simplify.

a. Subtract 7 from -11.

b. Decrease 35 by -25.

Example 5   Write each phrase as an expression and simplify.

a. Subtract 8 from -4.   **b.** Decrease 10 by -20.

Solution: Be careful when interpreting these. The order of numbers in subtraction is important.

a. 8 is to be subtracted **from** -4.

$$-4 - 8 = -4 + (-8) = -12$$

b. To decrease 10 by -20, we find 10 **minus** -20.

$$10 - (-20) = 10 + 20 = 30$$

■ **Work Practice 5**

Answers

1. a. -26 **b.** 8   **c.** -10 **d.** 5

2. 15.3   **3.** $-\dfrac{2}{3}$ **4.** $\dfrac{3}{20}$ **5. a.** -18 **b.** 60

If an expression contains additions and subtractions, just write the subtractions as equivalent additions. Then simplify from left to right.

Example 6 Simplify each expression.

a. $-14 - 8 + 10 - (-6)$ **b.** $1.6 - (-10.3) + (-5.6)$

Solution:

a. $-14 - 8 + 10 - (-6) = -14 + (-8) + 10 + 6 = -6$

b. $1.6 - (-10.3) + (-5.6) = 1.6 + 10.3 + (-5.6) = 6.3$

■ **Work Practice 6**

When an expression contains parentheses and brackets, remember the order of operations. Start with the innermost set of parentheses or brackets and work your way outward.

Example 7 Simplify each expression.

a. $-3 + [(-2 - 5) - 2]$ **b.** $2^3 - 10 + [-6 - (-5)]$

Solution:

a. Start with the innermost set of parentheses. Rewrite $-2 - 5$ as an addition.

$$-3 + [(-2 - 5) - 2] = -3 + [(-2 + (-5)) - 2]$$
$$= -3 + [(-7) - 2] \quad \text{Add: } -2 + (-5).$$
$$= -3 + [-7 + (-2)] \quad \text{Write } -7 - 2 \text{ as an addition.}$$
$$= -3 + [-9] \quad \text{Add.}$$
$$= -12 \quad \text{Add.}$$

b. Start simplifying the expression inside the brackets by writing $-6 - (-5)$ as an addition.

$$2^3 - 10 + [-6 - (-5)] = 2^3 - 10 + [-6 + 5]$$
$$= 2^3 - 10 + [-1] \quad \text{Add.}$$
$$= 8 - 10 + (-1) \quad \text{Evaluate } 2^3.$$
$$= 8 + (-10) + (-1) \quad \text{Write } 8 - 10 \text{ as an addition.}$$
$$= -2 + (-1) \quad \text{Add.}$$
$$= -3 \quad \text{Add.}$$

■ **Work Practice 7**

Objective B Evaluating Algebraic Expressions

It is important to be able to evaluate expressions for given replacement values. This helps, for example, when checking solutions of equations.

Example 8 Find the value of each expression when $x = 2$ and $y = -5$.

a. $\dfrac{x - y}{12 + x}$ **b.** $x^2 - y$

Solution:

a. Replace x with 2 and y with -5. Be sure to put parentheses around -5 to separate signs. Then simplify the resulting expression.

$$\frac{x - y}{12 + x} = \frac{2 - (-5)}{12 + 2} = \frac{2 + 5}{14} = \frac{7}{14} = \frac{1}{2}$$

b. Replace x with 2 and y with -5 and simplify.

$$x^2 - y = 2^2 - (-5) = 4 - (-5) = 4 + 5 = 9$$

■ **Work Practice 8**

Practice 6

Simplify each expression.
a. $-20 - 5 + 12 - (-3)$
b. $5.2 - (-4.4) + (-8.8)$

Practice 7

Simplify each expression.
a. $-9 + [(-4 - 1) - 10]$
b. $5^2 - 20 + [-11 - (-3)]$

Teaching Tip

Ask students what makes Example 7 easy to follow. Be sure they observe that each step is shown, that the equal signs are lined up, and that brief comments accompany each step. Encourage them to develop good organizational habits that support their learning of mathematics.

Practice 8

Find the value of each expression when $x = 1$ and $y = -4$.

a. $\dfrac{x - y}{14 + x}$ **b.** $x^2 - y$

Answers
6. a. -10 **b.** 0.8 **7. a.** -24
b. -3 **8. a.** $\dfrac{1}{3}$ **b.** 5

Helpful Hint

For additional help when replacing variables with replacement values, first place parentheses about any variables.

For Example 8b on the previous page, we have

$$x^2 - y = \underbrace{(x)^2 - (y)}_{\substack{\text{Place parentheses} \\ \text{about variables}}} = \underbrace{(2)^2 - (-5)}_{\substack{\text{Replace variables} \\ \text{with values}}} = 4 - (-5) = 4 + 5 = 9$$

Objective C Solutions of Equations

Recall from Section 1.3 that a solution of an equation is a value for the variable that makes the equation true.

Practice 9

Determine whether -2 is a solution of $-1 + x = 1$.

Example 9 Determine whether -4 is a solution of $x - 5 = -9$.

Solution: Replace x with -4 and see if a true statement results.

$$x - 5 = -9 \quad \text{Original equation}$$
$$-4 - 5 \stackrel{?}{=} -9 \quad \text{Replace } x \text{ with } -4.$$
$$-4 + (-5) \stackrel{?}{=} -9$$
$$-9 = -9 \quad \text{True}$$

Thus -4 is a solution of $x - 5 = -9$.

■ **Work Practice 9**

Objective D Solving Applications That Involve Subtraction

Another use of real numbers is in recording altitudes above and below sea level, as shown in the next example.

Practice 10

The highest point in Asia is the top of Mount Everest, at a height of 29,028 feet above sea level. The lowest point is the Dead Sea, which is 1312 feet below sea level. How much higher is Mount Everest than the Dead Sea? (*Source: National Geographic Society*)

Example 10 Finding a Change in Elevation

The highest point in the United States is the top of Mount McKinley, at a height of 20,320 feet above sea level. The lowest point is Death Valley, California, which is 282 feet below sea level. How much higher is Mount McKinley than Death Valley? (*Source:* U.S. Geological Survey)

Solution: To find "how much higher," we subtract. Don't forget that since Death Valley is 282 feet *below* sea level, we represent its height by -282. Draw a diagram to help visualize the problem.

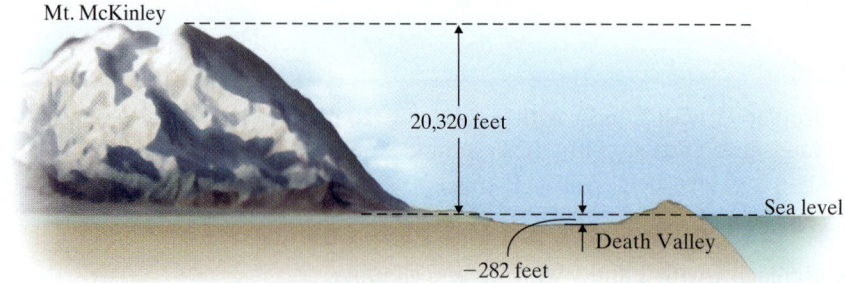

Mt. McKinley

20,320 feet

Sea level

Death Valley

-282 feet

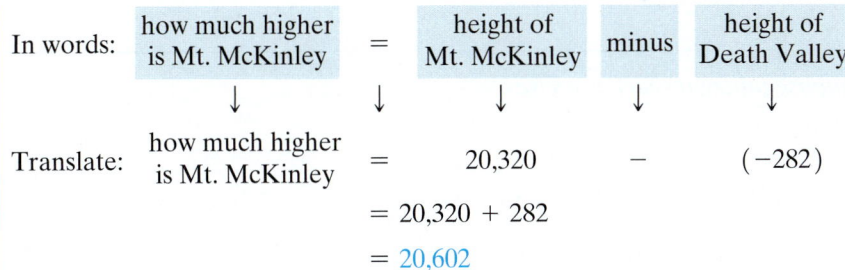

In words:

how much higher is Mt. McKinley	=	height of Mt. McKinley	minus	height of Death Valley
↓	↓	↓	↓	↓

Translate: how much higher is Mt. McKinley $=$ 20,320 $-$ (-282)

$$= 20{,}320 + 282$$

$$= 20{,}602$$

Thus, Mount McKinley is 20,602 feet higher than Death Valley.

■ **Work Practice 10**

Objective E Finding Complementary and Supplementary Angles ▶

A knowledge of geometric concepts is needed by many professionals, such as doctors, carpenters, electronic technicians, gardeners, machinists, and pilots, just to name a few. With this in mind, we review the geometric concepts of **complementary** and **supplementary angles.**

Complementary and Supplementary Angles

Two angles are **complementary** if the sum of their measures is 90°.

Two angles are **supplementary** if the sum of their measures is 180°.

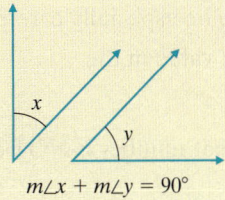

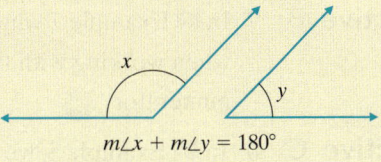

$m\angle x + m\angle y = 90°$

$m\angle x + m\angle y = 180°$

Example 11 Find the measure of each unknown complementary or supplementary angle.

a.

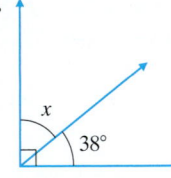

b.

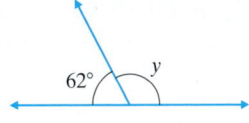

Solution:

a. These angles are complementary, so their sum is 90°. This means that the measure of angle x, $m\angle x$, is $90° - 38°$.

$$m\angle x = 90° - 38° = 52°$$

b. These angles are supplementary, so their sum is 180°. This means that $m\angle y$ is $180° - 62°$.

$$m\angle y = 180° - 62° = 118°$$

■ **Work Practice 11**

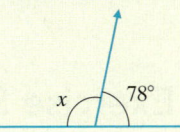
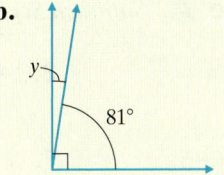

Vocabulary, Readiness & Video Check

Multiple choice: Select the correct lettered response following each exercise.

1. It is true that $a - b =$ ___$a + (-b)$___.

    a. $b - a$ **b.** $a + (-b)$ **c.** $a + b$ b

2. The opposite of n is ___$-n$___.

    a. $-n$ **b.** $-(-n)$ **c.** n a

3. To evaluate $x - y$ for $x = -10$ and $y = -14$, we replace x with -10 and y with -14 and evaluate ___$-10 - (-14)$___.

    a. $10 - 14$ **b.** $-10 - 14$ **c.** $-14 - 10$ **d.** $-10 - (-14)$ d

4. The expression $-5 - 10$ equals ___$-5 + (-10)$___.

    a. $5 - 10$ **b.** $5 + 10$ **c.** $-5 + (-10)$ **d.** $10 - 5$ c

---

Martin-Gay Interactive Videos

See Video 1.5

See video answer section.

Watch the section lecture video and answer the following questions.

Objective A   **5.** Complete this statement based on the lecture given before ⊞ Example 1. To subtract two real numbers, change the operation to _____ and take the _____ of the second number. ▶

        6. When simplifying ⊞ Example 5, what is the result of the first step and why is the expression rewritten in this way? ▶

Objective B   **7.** In ⊞ Example 7, why are you told to be especially careful when working with the replacement value in the numerator? ▶

Objective C   **8.** In ⊞ Example 8, we learned that what number is NOT a solution of what equation? ▶

Objective D   **9.** For ⊞ Example 9, why is the overall vertical change represented as a negative number? ▶

Objective E   **10.** The definition of supplementary angles is given just before ⊞ Example 10. Explain how this definition is used to solve ⊞ Example 10. ▶

1.5   **Exercise Set** MyMathLab®

Objective A *Subtract. See Examples 1 through 4.*

1. $-6 - 4$
    -10

2. $-12 - 8$
    -20

3. $4 - 9$
    -5

4. $8 - 11$
    -3

▶**5.** $16 - (-3)$
    19

6. $12 - (-5)$
    17

7. $7 - (-4)$
    11

8. $3 - (-6)$
    9

9. $-26 - (-18)$
    -8

10. $-60 - (-48)$
    -12

▶**11.** $-6 - 5$
    -11

12. $-8 - 4$
    -12

13. $16 - (-21)$
    37

14. $15 - (-33)$
    48

15. $-6 - (-11)$
    5

16. $-4 - (-16)$
    12

17. $-44 - 27$
    -71

18. $-36 - 51$
    -87

19. $-21 - (-21)$
    0

20. $-17 - (-17)$
    0

21. $-\dfrac{3}{11} - \left(-\dfrac{5}{11}\right)$ **22.** $-\dfrac{4}{7} - \left(-\dfrac{1}{7}\right)$ **23.** $9.7 - 16.1$ **24.** $8.3 - 11.2$ **25.** $-2.6 - (-6.7)$

$\dfrac{2}{11}$ $\dfrac{3}{7}$ -6.4 -2.9 4.1

26. $-6.1 - (-5.3)$ **27.** $\dfrac{1}{2} - \dfrac{2}{3}$ $-\dfrac{1}{6}$ **28.** $\dfrac{3}{4} - \dfrac{7}{8}$ $-\dfrac{1}{8}$ **29.** $-\dfrac{1}{6} - \dfrac{3}{4}$ $-\dfrac{11}{12}$ **30.** $-\dfrac{1}{10} - \dfrac{7}{8}$ $-\dfrac{39}{40}$

-0.8

31. $8.3 - (-0.62)$ **32.** $4.3 - (-0.87)$ **33.** $0 - 8.92$ **34.** $0 - (-4.21)$

8.92 5.17 -8.92 4.21

Translating *Translate each phrase to an expression and simplify. See Example 5.*

35. Subtract -5 from 8.   13

36. Subtract -2 from 3.   5

37. Find the difference between -6 and -1. -5

38. Find the difference between -17 and -1. -16

39. Subtract 8 from 7. -1

40. Subtract 9 from -4. -13

41. Decrease -8 by 15. -23

42. Decrease 11 by -14.   25

Mixed Practice (Sections 1.3, 1.4, 1.5) *Simplify each expression. (Remember the order of operations.) See Examples 6 and 7.*

43. $-10 - (-8) + (-4) - 20$ -26

44. $-16 - (-3) + (-11) - 14$ -38

45. $5 - 9 + (-4) - 8 - 8$ -24

46. $7 - 12 + (-5) - 2 + (-2)$ -14

47. $-6 - (2 - 11)$ 3

48. $-9 - (3 - 8)$ -4

49. $3^3 - 8 \cdot 9$ -45

50. $2^3 - 6 \cdot 3$ -10

51. $2 - 3(8 - 6)$ -4

52. $4 - 6(7 - 3)$ -20

53. $(3 - 6) + 4^2$ 13

54. $(2 - 3) + 5^2$ 24

55. $-2 + [(8 - 11) - (-2 - 9)]$ 6

56. $-5 + [(4 - 15) - (-6) - 8]$ -18

57. $|-3| + 2^2 + [-4 - (-6)]$ 9

58. $|-2| + 6^2 + (-3 - 8)$ 27

Objective B *Evaluate each expression when x = −5, y = 4, and t = 10. See Example 8.*

59. $x - y$ −9

60. $y - x$ 9

▶**61.** $\dfrac{9 - x}{y + 6}$ $\dfrac{7}{5}$

62. $\dfrac{15 - x}{y + 2}$ $\dfrac{10}{3}$

63. $|x| + 2t - 8y$ −7

64. $|y| + 3x - 2t$ −31

65. $y^2 - x$ 21

66. $t^2 - x$ 105

67. $\dfrac{|x - (-10)|}{2t}$ $\dfrac{1}{4}$

68. $\dfrac{|5y - x|}{6t}$ $\dfrac{5}{12}$

Objective C *Decide whether the given number is a solution of the given equation. See Example 9.*

▶**69.** $x - 9 = 5$;   −4
not a solution

70. $x - 10 = -7$;   3
solution

71. $-x + 6 = -x - 1$;   −2
not a solution

72. $-x - 6 = -x - 1$;   −10
not a solution

73. $-x - 13 = -15$;   2
solution

74. $4 = 1 - x$;   5
not a solution

Objectives D E Mixed Practice *Solve. See Examples 10 and 11.*

75. The coldest temperature ever recorded on Earth was −129°F in Antarctica. The warmest temperature ever recorded was 134°F in Death Valley, California. How many degrees warmer is 134°F than −129°F? (*Source: The World Almanac*, 2013)   263°F

76. The coldest temperature ever recorded in the United States was −80°F in Alaska. The warmest temperature ever recorded was 134°F in California. How many degrees warmer is 134°F than −80°F? (*Source: The World Almanac*, 2013)   214°F

77. Mauna Kea in Hawaii has an elevation of 13,796 feet above sea level. The Mid-America Trench in the Pacific Ocean has an elevation of 21,857 feet below sea level. Find the difference in elevation between those two points. (*Source:* National Geographic Society and Defense Mapping Agency)   35,653 ft

78. A woman received a statement of her charge account at Old Navy. She spent $93 on purchases last month. She returned an $18 top because she didn't like the color. She also returned a $26 nightshirt because it was damaged. What does she actually owe on her account? $49

▶**79.** Find *x* if the angles below are complementary angles.

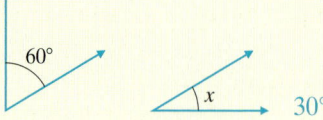

△**80.** Find *y* if the angles below are supplementary angles.

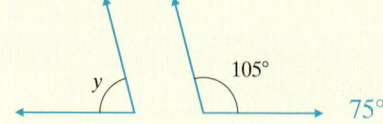

▶**81.** A commercial jetliner hits an air pocket and drops 250 feet. After climbing 120 feet, it drops another 178 feet. What is its overall vertical change?   −308 ft

82. In some card games, it is possible to have a negative score. Lavonne Schultz currently has a score of 15 points. She then loses 24 points. What is her new score?   −9

83. The highest point in Africa is Mt. Kilimanjaro, Tanzania, at an elevation of 19,340 feet. The lowest point is Lake Assal, Djibouti, at 512 feet below sea level. How much higher is Mt. Kilimanjaro than Lake Assal? (*Source:* National Geographic Society)   19,852 ft

84. The airport in Bishop, California, is at an elevation of 4101 feet above sea level. The nearby Furnace Creek Airport in Death Valley, California, is at an elevation of 226 feet below sea level. How much higher in elevation is the Bishop Airport than the Furnace Creek Airport? (*Source:* National Climatic Data Center)   4327 ft

Find each unknown complementary or supplementary angle.

 85.

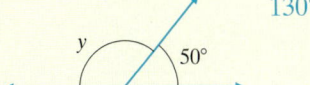   130°

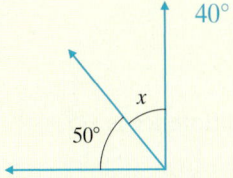 **86.**   40°

Mixed Practice—Translating (*Sections 1.4, 1.5*) *Translate each phrase to an algebraic expression. Use "x" to represent "a number."*

87. The sum of −5 and a number.   −5 + x

88. The difference of −3 and a number.   −3 − x

89. Subtract a number from −20.   −20 − x

90. Add a number and −36.   x + (−36)

Concept Extensions

Recall the bar graph from Section 1.4. It shows each month's average daily low temperature in degrees Fahrenheit for Barrow, Alaska. Use this graph to answer Exercises 91 through 94 on the next page.

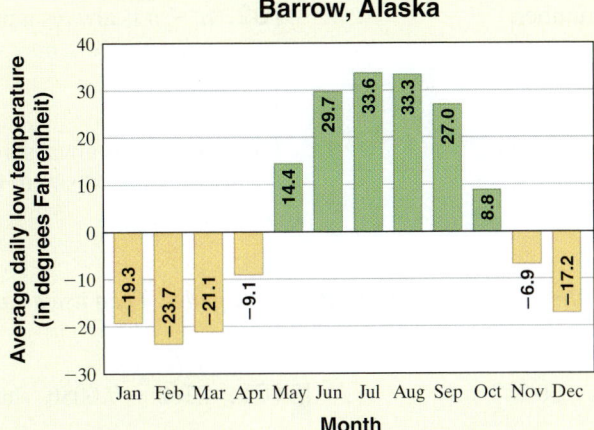

Barrow, Alaska

Source: National Climatic Data Center

91. Record the monthly increases and decreases in the low temperature from the previous month.

Month	Monthly Increase or Decrease (from the previous month)
February	−4.4°
March	2.6°
April	12°
May	23.5°
June	15.3°

92. Record the monthly increases and decreases in the low temperature from the previous month.

Month	Monthly Increase or Decrease (from the previous month)
July	3.9°
August	−0.3°
September	−6.3°
October	−18.2°
November	−15.7°
December	−10.3°

93. Use the tables in Exercises 91 and 92 to determine which month had the greatest increase in temperature?   May

94. Use the tables in Exercises 91 and 92 to determine which month had the greatest decrease in temperature?   October

Solve.

95. Find two numbers whose difference is −5.
answers may vary

96. Find two numbers whose difference is −9.
answers may vary

*Each calculation below is **incorrect**. Find the error and correct it.*

97. $9 - (-7) \overset{?}{=} 2$ 16

98. $-4 - 8 \overset{?}{=} 4$ −12

99. $10 - 30 \overset{?}{=} 20$ −20

100. $-3 - (-10) \overset{?}{=} -13$ 7

If p is a positive number and n is a negative number, determine whether each statement is true or false. Explain your answer.

101. $p - n$ is always a positive number.
true; answers may vary

102. $n - p$ is always a negative number.
true; answers may vary

103. $|n| - |p|$ is always a positive number.
false; answers may vary

104. $|n - p|$ is always a positive number.
true; answers may vary

Without calculating, determine whether each answer is positive or negative. Then use a calculator to find the exact difference.

105. $56{,}875 - 87{,}262$ negative, −30,387

106. $4.362 - 7.0086$ negative, −2.6466

Integrated Review

Operations on Real Numbers

Answer the following with positive, negative, or 0.

1. The opposite of a positive number is a _____ number.

2. The sum of two negative numbers is a _____ number.

3. The absolute value of a negative number is a _____ number.

4. The absolute value of zero is _____.

5. The sum of two positive numbers is a _____ number.

6. The sum of a number and its opposite is _____.

7. The absolute value of a positive number is a _____ number.

8. The opposite of a negative number is a _____ number.

Fill in the chart:

	Number	**Opposite**	**Absolute Value**
9.	$\frac{1}{7}$	$-\frac{1}{7}$	$\frac{1}{7}$
10.	$-\frac{12}{5}$	$\frac{12}{5}$	$\frac{12}{5}$
11.	3	-3	3
12.	$-\frac{9}{11}$	$\frac{9}{11}$	$\frac{9}{11}$

Perform each indicated operation and simplify. Don't forget to use order of operations if needed.

13. $-19 + (-23)$ **14.** $7 - (-3)$ **15.** $-15 + 17$ **16.** $-8 - 10$

17. $18 + (-25)$ **18.** $-2 + (-37)$ **19.** $-14 - (-12)$ **20.** $5 - 14$

21. $4.5 - 7.9$ **22.** $-8.6 - 1.2$ **23.** $-\dfrac{3}{4} - \dfrac{1}{7}$ **24.** $\dfrac{2}{3} - \dfrac{7}{8}$

1. negative

2. negative

3. positive

4. 0

5. positive

6. 0

7. positive

8. positive

9. See chart

10. See chart

11. See chart

12. See chart

13. -42

14. 10

15. 2

16. -18

17. -7

18. -39

19. -2

20. -9

21. -3.4

22. -9.8

23. $-\dfrac{25}{28}$

24. $-\dfrac{5}{24}$

25. $-9 - (-7) + 4 - 6$ **26.** $11 - 20 + (-3) - 12$ **27.** $24 - 6(14 - 11)$

28. $30 - 5(10 - 8)$ **29.** $(7 - 17) + 4^2$ **30.** $9^2 + (10 - 30)$

31. $|-9| + 3^2 + (-4 - 20)$ **32.** $|-4 - 5| + 5^2 + (-50)$

33. $-7 + [(1 - 2) + (-2 - 9)]$ **34.** $-6 + [(-3 + 7) + (4 - 15)]$

35. Subtract 5 from 1.   **36.** Subtract -2 from -3.

37. Subtract $-\dfrac{2}{5}$ from $\dfrac{1}{4}$.   **38.** Subtract $\dfrac{1}{10}$ from $-\dfrac{5}{8}$.

39. $2(19 - 17)^3 - 3(-7 + 9)^2$ **40.** $3(10 - 9)^2 + 6(20 - 19)^3$

Evaluate each expression when $x = -2, y = -1,$ and $z = 9$.

41. $x - y$ **42.** $x + y$

43. $y + z$ **44.** $z - y$

45. $\dfrac{|5z - x|}{y - x}$ **46.** $\dfrac{|-x - y + z|}{2z}$

---

25. -4

26. -24

27. 6

28. 20

29. 6

30. 61

31. -6

32. -16

33. -19

34. -13

35. -4

36. -1

37. $\dfrac{13}{20}$

38. $-\dfrac{29}{40}$

39. 4

40. 9

41. -1

42. -3

43. 8

44. 10

45. 47

46. $\dfrac{2}{3}$

1.6 Multiplying and Dividing Real Numbers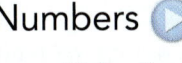

Objective A Multiplying Real Numbers

Multiplication of real numbers is similar to multiplication of whole numbers. We just need to determine when the answer is positive, when it is negative, and when it is zero. To discover sign patterns for multiplication, recall that multiplication is repeated addition. For example, $3(2)$ means that 2 is added to itself three times, or

$$3(2) = 2 + 2 + 2 = 6$$

Also,

$$3(-2) = (-2) + (-2) + (-2) = -6$$

Since $3(-2) = -6$, this suggests that the product of a positive number and a negative number is a negative number.

What about the product of two negative numbers? To find out, consider the following pattern.

Factor decreases by 1 each time.
$$-3 \cdot 2 = -6$$
$$-3 \cdot 1 = -3 \quad \text{Product increases by 3 each time.}$$
$$-3 \cdot 0 = 0$$
$$-3 \cdot -1 = 3$$
$$-3 \cdot -2 = 6$$

This suggests that the product of two negative numbers is a positive number. Our results are given below.

> ### Multiplying Real Numbers
> 1. The product of two numbers with the *same* sign is a positive number.
> 2. The product of two numbers with *different* signs is a negative number.

Examples Multiply.

1. $-7(6) = -42$ Different signs, so the product is negative.
2. $2(-10) = -20$
3. $-2(-14) = 28$ Same sign, so the product is positive.
4. $-\dfrac{2}{3} \cdot \dfrac{4}{7} = -\dfrac{2 \cdot 4}{3 \cdot 7} = -\dfrac{8}{21}$
5. $5(-1.7) = -8.5$
6. $-18(-3) = 54$

▶ **Work Practice 1–6**

We already know that the product of 0 and any whole number is 0. This is true of all real numbers.

> ### Products Involving Zero
> If b is a real number, then $b \cdot 0 = 0$. Also $0 \cdot b = 0$.

Objectives

A Multiply Real Numbers.

B Find the Reciprocal of a Real Number.

C Divide Real Numbers.

D Evaluate Expressions Using Real Numbers.

E Determine Whether a Number Is a Solution of a Given Equation.

F Solve Applications That Involve Multiplication or Division of Real Numbers.

Teaching Tip:
Classroom Group Activity

Consider beginning this lesson with the following activity: Students find and graph the location of a train at a certain time given its velocity (speed and direction). Positive velocities indicate speeds (mph) in the eastward direction, and negative velocities indicate speeds (mph) in the westward direction. Positive times are in the future, and negative times are in the past. All trains are currently at the train depot. Interpret each situation, and find how far the train was/will be from the depot. Graph its position on the number line.

Train	Time (hours)	Velocity (mph)	Miles from Depot (+→ east; −→ west)
A	2	70	
B	5	−45	
C	−4	60	
D	−3	−55	

West   Depot   East

−300 −200 −100   0   100   200   300

Practice 1–6

Multiply.

1. $-8(3)$ 2. $5(-30)$ 3. $-4(-12)$
4. $-\dfrac{5}{6} \cdot \dfrac{1}{4}$ 5. $6(-2.3)$ 6. $-15(-2)$

Answers
1. −24 2. −150 3. 48 4. $-\dfrac{5}{24}$
5. −13.8 6. 30

51

Practice 7

Multiply.
a. $5(0)(-3)$
b. $(-1)(-6)(-7)$
c. $(-2)(4)(-8)(-1)$

Example 7 Multiply.

a. $7(0)(-6)$ **b.** $(-2)(-3)(-4)$ **c.** $(-1)(-5)(-9)(-2)$

Solution:

a. By the order of operations, we multiply from left to right. Notice that because one of the factors is 0, the product is 0.

$$7(0)(-6) = 0(-6) = 0$$

b. Multiply two factors at a time, from left to right.

$$(-2)(-3)(-4) = (6)(-4) \quad \text{Multiply } (-2)(-3).$$
$$= -24$$

c. Multiply from left to right.

$$(-1)(-5)(-9)(-2) = (5)(-9)(-2) \quad \text{Multiply } (-1)(-5).$$
$$= -45(-2) \quad \text{Multiply } 5(-9).$$
$$= 90$$

🔲 **Work Practice 7**

✔**Concept Check** What is the sign of the product of five negative numbers? Explain.

Teaching Tip

Help students notice a pattern between the number of negative numbers in a product and the sign of the product. Try these examples: $7(-1)$; $2(-3)(-5)$; $(-1)(-4)(-2)$; $5(-2)(-1)(-2)$; and so on.

Helpful Hint

Have you noticed a pattern when multiplying signed numbers?

If we let $(-)$ represent a negative number and $(+)$ represent a positive number, then

$$(-)(-) = (+)$$
$$(-)(-)(-) = (-)$$
$$(-)(-)(-)(-) = (+)$$
$$(-)(-)(-)(-)(-) = (-)$$

The product of an even number of negative numbers is a positive result.

The product of an odd number of negative numbers is a negative result.

Teaching Tip

Spend a *lot* of class time reviewing the difference between $(-5)^2$ and -5^2, for example.

Now that we know how to multiply positive and negative numbers, let's see how we find the values of $(-5)^2$ and -5^2, for example. Although these two expressions look similar, the difference between the two is the parentheses. In $(-5)^2$, the parentheses tell us that the base, or repeated factor, is -5. In -5^2, only 5 is the base. Thus,

$$(-5)^2 = (-5)(-5) = 25 \quad \text{The base is } -5.$$
$$-5^2 = -(5 \cdot 5) = -25 \quad \text{The base is } 5.$$

Practice 8

Evaluate.
▶ **a.** $(-2)^4$ ▶ **b.** -2^4
c. $(-1)^5$ **d.** -1^5
e. $\left(-\dfrac{7}{9}\right)^2$

Example 8 Evaluate.

a. $(-2)^3$ **b.** -2^3 **c.** $(-3)^2$ **d.** -3^2 **e.** $\left(-\dfrac{2}{3}\right)^2$

Solution:

a. $(-2)^3 = (-2)(-2)(-2) = -8$ The base is -2.

b. $-2^3 = -(2 \cdot 2 \cdot 2) = -8$ The base is 2.

c. $(-3)^2 = (-3)(-3) = 9$ The base is -3.

d. $-3^2 = -(3 \cdot 3) = -9$ The base is 3.

e. $\left(-\dfrac{2}{3}\right)^2 = \left(-\dfrac{2}{3}\right)\left(-\dfrac{2}{3}\right) = \dfrac{4}{9}$ The base is $-\dfrac{2}{3}$.

🔲 **Work Practice 8**

Answers

7. **a.** 0 **b.** -42 **c.** -64 **8. a.** 16
b. -16 **c.** -1 **d.** -1 **e.** $\dfrac{49}{81}$

✔**Concept Check Answer**
negative

Helpful Hint

Be careful when identifying the base of an exponential expression.

$$(-3)^2 \qquad\qquad\qquad -3^2$$

Base is -3 $\qquad\qquad$ Base is 3

$$(-3)^2 = (-3)(-3) = 9 \qquad\qquad -3^2 = -(3 \cdot 3) = -9$$

Objective B Finding Reciprocals ▶

Addition and subtraction are related. Every difference of two numbers $a - b$ can be written as the sum $a + (-b)$. Multiplication and division are related also. For example, the quotient $6 \div 3$ can be written as the product $6 \cdot \dfrac{1}{3}$. Recall that the pair of numbers 3 and $\dfrac{1}{3}$ has a special relationship. Their product is 1 and they are called **reciprocals** or **multiplicative inverses** of each other.

Reciprocal or Multiplicative Inverse

Two numbers whose product is 1 are called **reciprocals** or **multiplicative inverses** of each other.

Example 9 Find the reciprocal of each number.

a. 22 $\qquad$ Reciprocal is $\dfrac{1}{22}$ since $22 \cdot \dfrac{1}{22} = 1$.

b. $\dfrac{3}{16}$ $\qquad$ Reciprocal is $\dfrac{16}{3}$ since $\dfrac{3}{16} \cdot \dfrac{16}{3} = 1$.

c. -10 $\qquad$ Reciprocal is $-\dfrac{1}{10}$ since $-10 \cdot -\dfrac{1}{10} = 1$.

d. $-\dfrac{9}{13}$ $\qquad$ Reciprocal is $-\dfrac{13}{9}$ since $-\dfrac{9}{13} \cdot -\dfrac{13}{9} = 1$.

e. 1.7 $\qquad$ Reciprocal is $\dfrac{1}{1.7}$ since $1.7 \cdot \dfrac{1}{1.7} = 1$.

■ **Work Practice 9**

Helpful Hint

The fraction $\dfrac{1}{1.7}$ is not simplified since the denominator is a decimal number. For the purpose of finding a reciprocal, we will leave the fraction as is.

Does the number 0 have a reciprocal? If it does, it is a number n such that $0 \cdot n = 1$. Notice that this can never be true since $0 \cdot n = 0$. This means that 0 has no reciprocal.

Quotients Involving Zero

The number 0 does not have a reciprocal.

Practice 9

Find the reciprocal of each number.

a. 13 $\qquad$ **b.** $\dfrac{7}{15}$ $\qquad$ **c.** -5

d. $-\dfrac{8}{11}$ $\qquad$ **e.** 7.9

Answers

9. **a.** $\dfrac{1}{13}$ **b.** $\dfrac{15}{7}$ **c.** $-\dfrac{1}{5}$

d. $-\dfrac{11}{8}$ **e.** $\dfrac{1}{7.9}$

Objective C Dividing Real Numbers ▶

We may now write a quotient as an equivalent product.

> ### Quotient of Two Real Numbers
>
> If a and b are real numbers and b is not 0, then
>
> $$a \div b = \frac{a}{b} = a \cdot \frac{1}{b}$$

In other words, the quotient of two real numbers is the product of the first number and the multiplicative inverse or reciprocal of the second number.

Practice 10

Use the definition of the quotient of two numbers to find each quotient.

a. $-12 \div 4$ **b.** $\dfrac{-20}{-10}$

c. $\dfrac{36}{-4}$

Example 10   Use the definition of the quotient of two numbers to find each quotient. $\left(a \div b = a \cdot \dfrac{1}{b} \right)$

a. $-18 \div 3$ **b.** $\dfrac{-14}{-2}$ **c.** $\dfrac{20}{-4}$

Solution:

a. $-18 \div 3 = -18 \cdot \dfrac{1}{3} = -6$

b. $\dfrac{-14}{-2} = -14 \cdot -\dfrac{1}{2} = 7$

c. $\dfrac{20}{-4} = 20 \cdot -\dfrac{1}{4} = -5$

▪ Work Practice 10

Since the quotient $a \div b$ can be written as the product $a \cdot \dfrac{1}{b}$, it follows that sign patterns for dividing two real numbers are the same as sign patterns for multiplying two real numbers.

> ### Dividing Real Numbers
>
> 1. The quotient of two numbers with the *same* sign is a positive number.
> 2. The quotient of two numbers with *different* signs is a negative number.

Practice 11

Divide.

a. $\dfrac{-25}{5}$ **b.** $\dfrac{-48}{-6}$

c. $\dfrac{50}{-2}$ **d.** $\dfrac{-72}{0.2}$

Example 11   Divide.

a. $\dfrac{-30}{-10} = 3$ Same sign, so the quotient is positive.

b. $\dfrac{-100}{5} = -20$

c. $\dfrac{20}{-2} = -10$ Different signs, so the quotient is negative.

d. $\dfrac{42}{-0.6} = -70$ $0.6\overline{)42.0}$ $70.$

▪ Work Practice 11

Answers
10. a. -3 **b.** 2 **c.** -9 **11. a.** -5
b. 8 **c.** -25 **d.** -360

✓**Concept Check** What is wrong with the following calculation?

$$\cancel{\frac{-36}{-9}} = -4$$

In the examples on the previous page, we divided mentally or by long division. When we divide by a fraction, it is usually easier to multiply by its reciprocal.

Examples Divide.

12. $\dfrac{2}{3} \div \left(-\dfrac{5}{4}\right) = \dfrac{2}{3} \cdot \left(-\dfrac{4}{5}\right) = -\dfrac{8}{15}$

13. $-\dfrac{1}{6} \div \left(-\dfrac{2}{3}\right) = -\dfrac{1}{6} \cdot \left(-\dfrac{3}{2}\right) = \dfrac{3}{12} = \dfrac{\overset{1}{\cancel{3}}}{\underset{1}{\cancel{3}} \cdot 4} = \dfrac{1}{4}$

■ **Work Practice 12–13**

Practice 12–13
Divide.
12. $-\dfrac{5}{9} \div \dfrac{2}{3}$ **13.** $-\dfrac{2}{7} \div \left(-\dfrac{1}{5}\right)$

Our definition of the quotient of two real numbers does not allow for division by 0 because 0 does not have a reciprocal. How then do we interpret $\dfrac{3}{0}$? We say that an expression such as this one is **undefined**. Can we divide 0 by a number other than 0? Yes; for example,

$$\frac{0}{3} = 0 \cdot \frac{1}{3} = 0$$

Division Involving Zero

If a is a nonzero number, then $\dfrac{0}{a} = 0$ and $\dfrac{a}{0}$ is undefined.

Example 14 Divide, if possible.

a. $\dfrac{1}{0}$ is undefined.

b. $\dfrac{0}{-3} = 0$

■ **Work Practice 14**

Practice 14
Divide if possible.
a. $\dfrac{-7}{0}$ **b.** $\dfrac{0}{-2}$

Notice that $\dfrac{12}{-2} = -6$, $-\dfrac{12}{2} = -6$, and $\dfrac{-12}{2} = -6$. This means that

$$\frac{12}{-2} = -\frac{12}{2} = \frac{-12}{2}$$

In other words, a single negative sign in a fraction can be written in the denominator, in the numerator, or in front of the fraction without changing the value of the fraction.

If a and b are real numbers, and $b \neq 0$, then $\dfrac{a}{-b} = \dfrac{-a}{b} = -\dfrac{a}{b}$.

Objective D Evaluating Expressions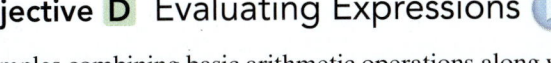

Examples combining basic arithmetic operations along with the principles of the order of operations help us to review these concepts of multiplying and dividing real numbers.

Answers

12. $-\dfrac{5}{6}$ **13.** $\dfrac{10}{7}$ **14. a.** undefined **b.** 0

✓**Concept Check Answer**

$\dfrac{-36}{-9} = 4$

Practice 15

Use order of operations to evaluate each expression.

a. $\dfrac{0(-5)}{3}$

b. $-3(-9) - 4(-4)$

c. $(-3)^2 + 2[(5 - 15) - |-4 - 1|]$

d. $\dfrac{-7(-4) + 2}{-10 - (-5)}$

e. $\dfrac{5(-2)^3 + 52}{-4 + 1}$

Example 15   Use order of operations to evaluate each expression.

a. $\dfrac{0(-8)}{2}$ **b.** $-4(-11) - 5(-2)$

c. $(-2)^2 + 3[(-3 - 2) - |4 - 6|]$ **d.** $\dfrac{(-12)(-3) + 4}{-7 - (-2)}$

e. $\dfrac{2(-3)^2 - 20}{|-5| + 4}$

Solution:

a. $\dfrac{0(-8)}{2} = \dfrac{0}{2} = 0$

b. $(-4)(-11) - 5(-2) = 44 - (-10)$ Find the products.

$\qquad\qquad\qquad\qquad\qquad\quad = 44 + 10$ Add 44 to the opposite of -10.

$\qquad\qquad\qquad\qquad\qquad\quad = 54$ Add.

c. $(-2)^2 + 3[(-3 - 2) - |4 - 6|] = (-2)^2 + 3[(-5) - |-2|]$ Simplify within innermost sets of grouping symbols.

$\qquad\qquad\qquad\qquad\qquad\qquad\quad = (-2)^2 + 3[-5 - 2]$ Write $|-2|$ as 2.

$\qquad\qquad\qquad\qquad\qquad\qquad\quad = (-2)^2 + 3(-7)$ Combine.

$\qquad\qquad\qquad\qquad\qquad\qquad\quad = 4 + (-21)$ Evaluate $(-2)^2$ and multiply $3(-7)$.

$\qquad\qquad\qquad\qquad\qquad\qquad\quad = -17$ Add.

For parts d and e, first simplify the numerator and denominator separately; then divide.

d. $\dfrac{(-12)(-3) + 4}{-7 - (-2)} = \dfrac{36 + 4}{-7 + 2}$

$\qquad\qquad\qquad\quad = \dfrac{40}{-5}$

$\qquad\qquad\qquad\quad = -8$ Divide.

e. $\dfrac{2(-3)^2 - 20}{|-5| + 4} = \dfrac{2 \cdot 9 - 20}{5 + 4} = \dfrac{18 - 20}{9} = \dfrac{-2}{9} = -\dfrac{2}{9}$

■ **Work Practice 15**

Using what we have learned about multiplying and dividing real numbers, we continue to practice evaluating algebraic expressions.

Practice 16

Evaluate each expression when $x = -1$ and $y = -5$.

a. $\dfrac{3y}{45x}$

b. $x^2 - y^3$

c. $\dfrac{x + y}{3x}$

Example 16   Evaluate each expression when $x = -2$ and $y = -4$.

a. $\dfrac{3x}{2y}$ **b.** $x^3 - y^2$ **c.** $\dfrac{x - y}{-x}$

Solution: Replace x with -2 and y with -4 and simplify.

a. $\dfrac{3x}{2y} = \dfrac{3(-2)}{2(-4)} = \dfrac{-6}{-8} = \dfrac{6}{8} = \dfrac{\cancel{2} \cdot 3}{\cancel{2} \cdot 4} = \dfrac{3}{4}$

Answers

15. a. 0   **b.** 43   **c.** -21 **d.** -6

e. -4 **16. a.** $\dfrac{1}{3}$ **b.** 126   **c.** 2

b. $x^3 - y^2 = (-2)^3 - (-4)^2$ Substitute the given values for the variables.

$\qquad = -8 - (16)$ Evaluate $(-2)^3$ and $(-4)^2$.

$\qquad = -8 + (-16)$ Write as a sum.

$\qquad = -24$ Add.

c. $\dfrac{x - y}{-x} = \dfrac{-2 - (-4)}{-(-2)} = \dfrac{-2 + 4}{2} = \dfrac{2}{2} = 1$

🟧 **Work Practice 16**

Helpful Hint

Remember: For additional help when replacing variables with replacement values, first place parentheses about any variables.

Evaluate $3x - y^2$ when $x = 5$ and $y = -4$.

$3x - y^2 = 3(x) - (y)^2$ Place parentheses about variables only.

$\qquad = 3(5) - (-4)^2$ Replace variables with values.

$\qquad = 15 - 16$ Simplify.

$\qquad = -1$

Objective E Solutions of Equations ▶

We use our skills in multiplying and dividing real numbers to check possible solutions of an equation.

Example 17 Determine whether -10 is a solution of $\dfrac{-20}{x} + 15 = 2x$.

Solution:

$\dfrac{-20}{x} + 15 = 2x$ Original equation

$\dfrac{-20}{-10} + 15 \overset{?}{=} 2(-10)$ Replace x with -10.

$2 + 15 \overset{?}{=} -20$ Divide and multiply.

$17 = -20$ False

Since we have a false statement, -10 is *not* a solution of the equation.

🟧 **Work Practice 17**

Practice 17

Determine whether -8 is a solution of $\dfrac{x}{4} - 3 = x + 3$.

Objective F Solving Applications That Involve Multiplying or Dividing Numbers ▶

Many real-life problems involve multiplication and division of numbers.

Answer

17. -8 is a solution

Practice 18

A card player had a score of −13 for each of four games. Find the total score.

Answer

18. −52

Example 18   Calculating a Total Golf Score

A professional golfer finished seven strokes under par (−7) for each of three days of a tournament. What was her total score for the tournament?

Solution: Although the key word is "total," since this is repeated addition of the same number, we multiply.

In words:

golfer's total score	=	number of days	·	score each day

Translate:

$$\text{golfer's total} = 3 \cdot (-7)$$
$$= -21$$

Thus, the golfer's total score was −21, or 21 strokes under par.

■ **Work Practice 18**

Calculator Explorations

Entering Negative Numbers on a Scientific Calculator

To enter a negative number on a scientific calculator, find a key marked $\boxed{+/-}$. (On some calculators, this key is marked $\boxed{\text{CHS}}$ for "change sign.") To enter −8, for example, press the keys $\boxed{8}\,\boxed{+/-}$. The display will read $\boxed{-8}$.

Entering Negative Numbers on a Graphing Calculator

To enter a negative number on a graphing calculator, find a key marked $\boxed{(-)}$. Do not confuse this key with the key $\boxed{-}$, which is used for subtraction. To enter −8, for example, press the keys $\boxed{(-)}\,\boxed{8}$. The display will read $\boxed{-8}$.

Operations with Real Numbers

To evaluate −2(7 − 9) − 20 on a calculator, press the keys

$$\boxed{2}\,\boxed{+/-}\,\boxed{\times}\,\boxed{(}\,\boxed{7}\,\boxed{-}\,\boxed{9}\,\boxed{)}\,\boxed{-}\,\boxed{2}\,\boxed{0}\,\boxed{=},$$

or

$$\boxed{(-)}\,\boxed{2}\,\boxed{(}\,\boxed{7}\,\boxed{-}\,\boxed{9}\,\boxed{)}\,\boxed{-}\,\boxed{2}\,\boxed{0}\,\boxed{\text{ENTER}}.$$

The display will read $\boxed{-16}$ or $\boxed{\begin{array}{l}-2(7-9)-20 \\ \hspace{3em}-16\end{array}}$

Use a calculator to simplify each expression.

1. −38(26 − 27)   38
2. −59(−8) + 1726   2198
3. 134 + 25(68 − 91)   −441
4. 45(32) − 8(218)   −304
5. $\dfrac{-50(294)}{175 - 205}$ 490
6. $\dfrac{-444 - 444.8}{-181 - (-181)}$ undefined
7. $9^5 - 4550$ 54,499
8. $5^8 - 6259$ 384,366
9. $(-125)^2$ (Be careful.)   15,625
10. -125^2 (Be careful.)   −15,625

Vocabulary, Readiness & Video Check

Use the choices below to fill in each blank. Each choice may be used more than once.

negative   0

positive   undefined

1. The product of a negative number and a positive number is a(n) <u>negative</u> number.
2. The product of two negative numbers is a(n) <u>positive</u> number.
3. The quotient of two negative numbers is a(n) <u>positive</u> number.
4. The quotient of a negative number and a positive number is a(n) <u>negative</u> number.
5. The product of a negative number and zero is <u>0</u>.
6. The reciprocal of a negative number is a <u>negative</u> number.
7. The quotient of 0 and a negative number is <u>0</u>.
8. The quotient of a negative number and 0 is <u>undefined</u>.

Martin-Gay Interactive Videos   Watch the section lecture video and answer the following questions.

See Video 1.6

Objective A   9. Explain the significance of the use of parentheses when comparing ⊞ Examples 6 and 7. ▶

Objective B   10. In ⊞ Example 9, why is the reciprocal equal to $\frac{3}{2}$ and not $-\frac{3}{2}$? ▶

Objective C   11. Before ⊞ Example 11, the sign rules for division of real numbers are discussed. Are the sign rules for division the same as for multiplication? Why or why not? ▶

Objective D   12. In ⊞ Example 17, the importance of placing the replacement values in parentheses when evaluating is emphasized. Why? ▶

Objective E   13. In ⊞ Example 18, is 5 a solution of $-3x - 5 = -20$? Why or why not? ▶

Objective F   14. In ⊞ Example 19, explain why each loss of 4 yards is represented by -4 and not 4. ▶

See video answer section.

1.6   Exercise Set MyMathLab®

Objective A   *Multiply. See Examples 1 through 7.*

▶ **1.** $-6(4)$ -24
2. $-8(5)$ -40
▶ **3.** $2(-1)$ -2
4. $7(-4)$ -28

▶ **5.** $-5(-10)$ 50
6. $-6(-11)$ 66
7. $-3 \cdot 15$ -45
8. $-2 \cdot 37$ -74

9. $-\frac{1}{2}\left(-\frac{3}{5}\right)$ $\frac{3}{10}$
10. $-\frac{1}{8}\left(-\frac{1}{3}\right)$ $\frac{1}{24}$
11. $5(-1.4)$ -7
12. $6(-2.5)$ -15

13. $(-1)(-3)(-5)$ -15 **14.** $(-2)(-3)(-4)$ -24 **15.** $(2)(-1)(-3)(0)$ 0 **16.** $(3)(-5)(-2)(0)$ 0

Evaluate. See Example 8.

17. $(-4)^2$ 16

18. $(-3)^3$ -27

19. -4^2 -16

20. -6^2 -36

21. $\left(-\dfrac{3}{4}\right)^2$ $\dfrac{9}{16}$

22. $\left(-\dfrac{2}{7}\right)^2$ $\dfrac{4}{49}$

23. -0.7^2 -0.49

24. -0.8^2 -0.64

Objective B *Find each reciprocal. See Example 9.*

25. $\dfrac{2}{3}$ $\dfrac{3}{2}$

26. $\dfrac{1}{7}$ 7

27. -14 $-\dfrac{1}{14}$

28. -8 $-\dfrac{1}{8}$

29. $-\dfrac{3}{11}$ $-\dfrac{11}{3}$

30. $-\dfrac{6}{13}$ $-\dfrac{13}{6}$

31. 0.2 $\dfrac{1}{0.2}$

32. 1.5 $\dfrac{1}{1.5}$

Objective C *Divide. See Examples 10 through 14.*

33. $\dfrac{18}{-2}$ -9

34. $\dfrac{36}{-9}$ -4

35. $-48 \div 12$ -4

36. $-60 \div 5$ -12

37. $\dfrac{0}{-4}$ 0

38. $\dfrac{0}{-9}$ 0

39. $\dfrac{5}{0}$ undefined

40. $\dfrac{8}{0}$ undefined

41. $\dfrac{6}{7} \div \left(-\dfrac{1}{3}\right)$ $-\dfrac{18}{7}$

42. $\dfrac{4}{5} \div \left(-\dfrac{1}{2}\right)$ $-\dfrac{8}{5}$

43. $-3.2 \div -0.02$ 160

44. $-4.9 \div -0.07$ 70

Objectives A C Mixed Practice *Perform the indicated operation. See Examples 1–14.*

45. $(-8)(-8)$ 64

46. $(-7)(-7)$ 49

47. $\dfrac{2}{3}\left(-\dfrac{4}{9}\right)$ $-\dfrac{8}{27}$

48. $\dfrac{2}{7}\left(-\dfrac{2}{11}\right)$ $-\dfrac{4}{77}$

49. $\dfrac{-12}{-4}$ 3

50. $\dfrac{-45}{-9}$ 5

51. $\dfrac{30}{-2}$ -15

52. $\dfrac{14}{-2}$ -7

53. $(-5)^3$ -125

54. $(-2)^5$ -32

55. $(-0.2)^3$ -0.008

56. $(-0.3)^3$ -0.027

57. $-\dfrac{3}{4}\left(-\dfrac{8}{9}\right)$ $\dfrac{2}{3}$

58. $-\dfrac{5}{6}\left(-\dfrac{3}{10}\right)$ $\dfrac{1}{4}$

59. $-\dfrac{5}{9} \div \left(-\dfrac{3}{4}\right)$ $\dfrac{20}{27}$

60. $-\dfrac{1}{10} \div \left(-\dfrac{8}{11}\right)$ $\dfrac{11}{80}$

61. $-2.1(-0.4)$ 0.84

62. $-1.3(-0.6)$ 0.78

63. $\dfrac{-48}{1.2}$ -40

64. $\dfrac{-86}{2.5}$ -34.4

65. $(-3)^4$ 81

66. -3^4 -81

67. -1^7 -1

68. $(-1)^7$ -1

69. Multiply -11 by 11. -121

70. Multiply -12 by 12. -144

71. Find the quotient of $-\dfrac{4}{9}$ and $\dfrac{4}{9}$. -1

72. Find the quotient of $-\dfrac{5}{12}$ and $\dfrac{5}{12}$. -1

Mixed Practice (*Sections 1.4, 1.5, 1.6*) *Perform the indicated operation.*

73. $-9 - 10$ -19 **74.** $-8 - 11$ -19 **75.** $-9(-10)$ 90 **76.** $-8(-11)$ 88

77. $7(-12)$ -84 **78.** $6(-15)$ -90 **79.** $7 + (-12)$ -5 **80.** $6 + (-15)$ -9

Objective D   *Evaluate each expression. See Example 15.*

81. $\dfrac{-9(-3)}{-6}$ $-\dfrac{9}{2}$

82. $\dfrac{-6(-3)}{-4}$ $-\dfrac{9}{2}$

83. $-3(2 - 8)$ 18

84. $-4(3 - 9)$ 24

85. $-7(-2) - 3(-1)$ 17

86. $-8(-3) - 4(-1)$ 28

87. $2^2 - 3[(2 - 8) - (-6 - 8)]$ -20

88. $3^2 - 2[(3 - 5) - (2 - 9)]$ -1

89. $\dfrac{-6^2 + 4}{-2}$ 16

90. $\dfrac{3^2 + 4}{5}$ $\dfrac{13}{5}$

91. $\dfrac{-3 - 5^2}{2(-7)}$ 2

92. $\dfrac{-2 - 4^2}{3(-6)}$ 1

93. $\dfrac{22 + (3)(-2)^2}{-5 - 2}$ $-\dfrac{34}{7}$

94. $\dfrac{-20 + (-4)^2(3)}{1 - 5}$ -7

95. $\dfrac{(-4)^2 - 16}{4 - 12}$ 0

96. $\dfrac{(-2)^2 - 4}{4 - 9}$ 0

▶ 97. $\dfrac{6 - 2(-3)}{4 - 3(-2)}$ $\dfrac{6}{5}$

98. $\dfrac{8 - 3(-2)}{2 - 5(-4)}$ $\dfrac{7}{11}$

99. $\dfrac{|5 - 9| + |10 - 15|}{|2(-3)|}$ $\dfrac{3}{2}$

100. $\dfrac{|-3 + 6| + |-2 + 7|}{|-2 \cdot 2|}$ 2

101. $\dfrac{-7(-1) + (-3)4}{(-2)(5) + (-6)(-8)}$ $\dfrac{5}{38}$

102. $\dfrac{8(-7) + (-2)(-6)}{(-9)(3) + (-10)(-11)}$ $-\dfrac{44}{83}$

Evaluate each expression when $x = -5$ and $y = -3$. See Example 16.

103. $\dfrac{2x - 5}{y - 2}$ 3

104. $\dfrac{2y - 12}{x - 4}$ 2

105. $\dfrac{6 - y}{x - 4}$ -1

106. $\dfrac{10 - y}{x - 8}$ -1

107. $\dfrac{4 - 2x}{y + 3}$ undefined

108. $\dfrac{2y + 3}{-5 - x}$ undefined

▶ 109. $\dfrac{x^2 + y}{3y}$ $-\dfrac{22}{9}$

110. $\dfrac{y^2 - x}{2x}$ $-\dfrac{7}{5}$

Objective E   *Decide whether the given number is a solution of the given equation. See Example 17.*

▶ 111. $-3x - 5 = -20$; 5 solution

112. $17 - 4x = x + 27$; -2 solution

113. $\dfrac{x}{5} + 2 = -1$; 15 not a solution

114. $\dfrac{x}{6} - 3 = 5$; 48 solution

115. $\dfrac{x - 3}{7} = -2$; -11 solution

116. $\dfrac{x + 4}{5} = -6$; -30 not a solution

Objective **F** Translating *Translate each phrase to an expression. Use x to represent "a number." See Example 18.*

117. The product of -71 and a number
$-71 \cdot x$ or $-71x$

118. The quotient of -8 and a number
$\dfrac{-8}{x}$ or $-8 \div x$

119. Subtract a number from -16. $-16 - x$

120. The sum of a number and -12 $x + (-12)$

121. -29 increased by a number $-29 + x$

122. The difference of a number and -10 $x - (-10)$

123. Divide a number by -33.
$\dfrac{x}{-33}$ or $x \div (-33)$

124. Multiply a number by -17.
$x \cdot (-17)$ or $-17x$

Solve. See Example 18.

▶125. A football team lost four yards on each of three consecutive plays. Represent the total loss as a product of signed numbers and find the total loss.
$3 \cdot (-4) = -12$; a loss of 12 yd

126. A stockbroker lost $400 on each of seven consecutive days in the stock market. Represent his total loss as a product of signed numbers and find his total loss. $7 \cdot (-400) = -2800$; a loss of $2800

127. A deep-sea diver must move up or down in the water in short steps in order to keep from getting a physical condition called the "bends." Suppose a diver moves down from the surface in five steps of 20 feet each. Represent his total movement as a product of signed numbers and find the depth.
$5 \cdot (-20) = -100$; a depth of 100 feet

128. A weather forecaster predicts that the temperature will drop five degrees each hour for the next six hours. Represent this drop as a product of signed numbers and find the total drop in temperature.
$6 \cdot (-5) = -30$; a drop of 30 degrees

Concept Extensions

State whether each statement is true or false.

129. The product of three negative integers is negative.
true

130. The product of three positive integers is positive.
true

131. The product of four negative integers is negative.   false

132. The product of four positive integers is positive.   true

Study the bar graph below showing the average surface temperatures of planets. Use Exercises 133 and 134 to complete the planet temperatures on the graph. (Pluto is now classified as a dwarf planet.)

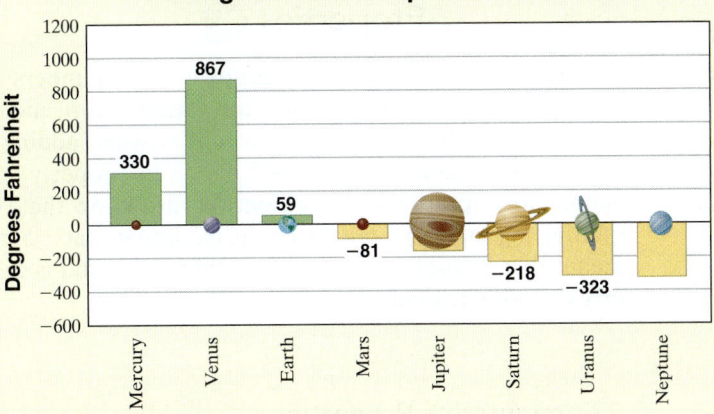

Average Surface Temperature of Planets*

*(For some planets, the temperature given is the temperature where the atmospheric pressure equals 1 Earth atmosphere; Source: The World Almanac)

133. The surface temperature of Jupiter is twice the temperature of Mars. Find this temperature. $-162°F$

134. The surface temperature of Neptune is equal to the temperature of Mercury divided by -1. Find this temperature. $-330°F$

135. For the first quarter of 2013, Wal-Mart Stores, Inc. posted a loss of \$33 million in membership and other income. If this trend was consistent for each month of the quarter, how much would you expect this loss to have been for each month? (*Source:* Wal-Mart Stores, Inc.) $-\$11$ million per month

136. For the first quarter of 2013, Chrysler Group, LLC, maker of Jeep vehicles, posted a loss of about 30,000 Jeep Liberty shipments because they had stopped producing the vehicle in 2012. If this trend was consistent for each month of the quarter, how much would you expect this loss to have been for each month? (*Source:* Chrysler Group, LLC)   loss of 10,000 shipments/month or $-10,000$ shipments per month

137. Explain why the product of an even number of negative numbers is a positive number.
answers may vary

138. If a and b are any real numbers, is the statement $a \cdot b = b \cdot a$ always true? Why or why not?
answers may vary

139. Find two real numbers that are their own reciprocal. Explain why there are only two.
$1, -1$; answers may vary

140. Explain why 0 has no reciprocal.
answers may vary

Mixed Practice (1.4, 1.5, 1.6) *Write each as an algebraic expression. Then simplify the expression.*

141. 7 subtracted from the quotient of 0 and 5
$\dfrac{0}{5} - 7 = -7$

142. Twice the sum of -3 and -4
$2[-3 + (-4)] = -14$

143. -1 added to the product of -8 and -5
$-8(-5) + (-1) = 39$

144. The difference of -9 and the product of -4 and -6
$-9 - (-4)(-6) = -33$

Objectives

A Use the Commutative and Associative Properties.

B Use the Distributive Property.

C Use the Identity and Inverse Properties.

Teaching Tip:
Classroom Activity

Consider the following activity: Together make a list of situations in which the commutative property holds. Then make another list of situations in which the commutative property does not hold. Which list was easier to generate?

Objective A Using the Commutative and Associative Properties

In this section we review properties of real numbers with which we are already familiar. Throughout this section, the variables a, b, and c represent real numbers.

We know that order does not matter when adding numbers. For example, we know that $7 + 5$ is the same as $5 + 7$. This property is given a special name—the **commutative property of addition.** We also know that order does not matter when multiplying numbers. For example, we know that $-5(6) = 6(-5)$. This property means that multiplication is commutative also and is called the **commutative property of multiplication.**

Commutative Properties

Addition: $\qquad\qquad\qquad\qquad a + b = b + a$

Multiplication: $\qquad\qquad\qquad a \cdot b = b \cdot a$

These properties state that the *order* in which any two real numbers are added or multiplied does not change their sum or product. For example, if we let $a = 3$ and $b = 5$, then the commutative properties guarantee that

$$3 + 5 = 5 + 3 \quad \text{and} \quad 3 \cdot 5 = 5 \cdot 3$$

Helpful Hint

Is subtraction also commutative? Try an example. Is $3 - 2 = 2 - 3$? **No!** The left side of this statement equals 1; the right side equals -1. There is no commutative property of subtraction. Similarly, there is no commutative property of division. For example, $10 \div 2$ does not equal $2 \div 10$.

Practice 1

Use a commutative property to complete each statement.

a. $7 \cdot y =$ _____

b. $4 + x =$ _____

Example 1 Use a commutative property to complete each statement.

a. $x + 5 =$ _____ 　　　　　**b.** $3 \cdot x =$ _____

Solution:

a. $x + 5 = 5 + x$ 　　By the commutative property of addition

b. $3 \cdot x = x \cdot 3$ 　　　By the commutative property of multiplication

■ **Work Practice 1**

✓**Concept Check** Which of the following pairs of actions are commutative?

a. "raking the leaves" and "bagging the leaves"

b. "putting on your left glove" and "putting on your right glove"

c. "putting on your coat" and "putting on your shirt"

d. "reading a novel" and "reading a newspaper"

Answers

1. **a.** $y \cdot 7$ **b.** $x + 4$

✓**Concept Check Answer**

b, d

Let's now discuss grouping numbers. When we add three numbers, the way in which they are grouped or associated does not change their sum. For example, we know that $2 + (3 + 4) = 2 + 7 = 9$. This result is the same if we group the numbers differently. In other words, $(2 + 3) + 4 = 5 + 4 = 9$, also. Thus, $2 + (3 + 4) = (2 + 3) + 4$. This property is called the **associative property of addition.**

In the same way, changing the grouping of numbers when multiplying does not change their product. For example, $2 \cdot (3 \cdot 4) = (2 \cdot 3) \cdot 4$ (check it). This is the **associative property of multiplication.**

Associative Properties

Addition: $(a + b) + c = a + (b + c)$

Multiplication: $(a \cdot b) \cdot c = a \cdot (b \cdot c)$

These properties state that the way in which three numbers are *grouped* does not change their sum or their product.

Example 2 Use an associative property to complete each statement.

a. $5 + (4 + 6) = $ _____

b. $(-1 \cdot 2) \cdot 5 = $ _____

c. $(m + n) + 9 = $ _____

d. $(xy) \cdot 12 = $ _____

Solution:

a. $5 + (4 + 6) = (5 + 4) + 6$ By the associative property of addition

b. $(-1 \cdot 2) \cdot 5 = -1 \cdot (2 \cdot 5)$ By the associative property of multiplication

c. $(m + n) + 9 = m + (n + 9)$ By the associative property of addition

d. $(xy) \cdot 12 = x \cdot (y \cdot 12)$ Recall that xy means $x \cdot y$.

■ **Work Practice 2**

Helpful Hint

Remember the difference between the commutative properties and the associative properties. The commutative properties have to do with the *order* of numbers and the associative properties have to do with the *grouping* of numbers.

Examples Determine whether each statement is true by an associative property or a commutative property.

3. $(7 + 10) + 4 = (10 + 7) + 4$ Since the order of two numbers was changed and their grouping was not, this is true by the commutative property of addition.

4. $2 \cdot (3 \cdot 1) = (2 \cdot 3) \cdot 1$ Since the grouping of the numbers was changed and their order was not, this is true by the associative property of multiplication.

■ **Work Practice 3–4**

Let's now illustrate how these properties can help us simplify expressions.

Practice 2

Use an associative property to complete each statement.

a. $5 \cdot (-3 \cdot 6) = $ _____

b. $(-2 + 7) + 3 = $ _____

c. $(q + r) + 17 = $ _____

d. $(ab) \cdot 21 = $ _____

Practice 3–4

Determine whether each statement is true by an associative property or a commutative property.

3. $5 \cdot (4 \cdot 7) = 5 \cdot (7 \cdot 4)$

4. $-2 + (4 + 9)$
$= (-2 + 4) + 9$

Answers

2. a. $(5 \cdot -3) \cdot 6$ **b.** $-2 + (7 + 3)$
c. $q + (r + 17)$ **d.** $a \cdot (b \cdot 21)$
3. commutative **4.** associative

Practice 5–6

Simplify each expression.

5. $(-3 + x) + 17$

6. $4(5x)$

Teaching Tip

You may want to show students that the distributive property can be used to quickly multiply numbers in their head. The product $12 \cdot 104$ can be rewritten as

$$12(100 + 4) = 12 \cdot 100 + 12 \cdot 4$$
$$= 1200 + 48$$
$$= 1248$$

The product $7 \cdot 98$ can be rewritten as

$$7(100 - 2) = 7 \cdot 100 - 7 \cdot 2$$
$$= 700 - 14$$
$$= 686$$

Teaching Tip

It may be helpful to illustrate the distributive property for students.

Let the tile $\boxed{x}$ represent x and the tile $\boxed{1}$ represent 1.

$3(x + 2) = 3x + 3 \cdot 2$ can be illustrated as

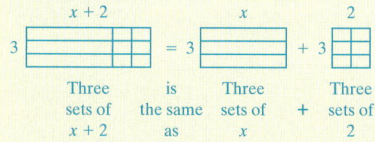

| Three sets of $x + 2$ | is the same as | Three sets of x | + | Three sets of 2 |

Practice 7–12

Use the distributive property to write each expression without parentheses. Then simplify the result.

7. $5(x + y)$

8. $-3(2 + 7x)$

9. $4(x + 6y - 2z)$

10. $-1(3 - a)$

11. $-(8 + a - b)$

12. $\dfrac{1}{2}(2x + 4) + 9$

Answers

5. $14 + x$ **6.** $20x$ **7.** $5x + 5y$
8. $-6 - 21x$ **9.** $4x + 24y - 8z$
10. $-3 + a$ **11.** $-8 - a + b$
12. $x + 11$

Examples Simplify each expression.

5. $10 + (x + 12) = 10 + (12 + x)$ By the commutative property of addition
$$= (10 + 12) + x$$ By the associative property of addition
$$= 22 + x$$ Add.

6. $-3(7x) = (-3 \cdot 7)x$ By the associative property of multiplication
$$= -21x$$ Multiply.

■ **Work Practice 5–6**

Objective B Using the Distributive Property

The **distributive property of multiplication over addition** is used repeatedly throughout algebra. It is useful because it allows us to write a product as a sum or a sum as a product.

We know that $7(2 + 4) = 7(6) = 42$. Compare that with

$$7(2) + 7(4) = 14 + 28 = 42$$

Since both original expressions equal 42, they must equal each other, or

$$7(2 + 4) = 7(2) + 7(4)$$

This is an example of the distributive property. The product on the left side of the equal sign is equal to the sum on the right side. We can think of the 7 as being distributed to each number inside the parentheses.

Distributive Property of Multiplication Over Addition

$$a(b + c) = ab + ac$$

Since multiplication is commutative, this property can also be written as

$$(b + c)a = ba + ca$$

The distributive property can also be extended to more than two numbers inside the parentheses. For example,

$$3(x + y + z) = 3(x) + 3(y) + 3(z)$$
$$= 3x + 3y + 3z$$

Since we define subtraction in terms of addition, the distributive property is also true for subtraction. For example,

$$2(x - y) = 2(x) - 2(y)$$
$$= 2x - 2y$$

Examples Use the distributive property to write each expression without parentheses. Then simplify the result.

7. $2(x + y) = 2(x) + 2(y)$
$$= 2x + 2y$$

8. $-5(-3 + 2z) = -5(-3) + (-5)(2z)$
$$= 15 - 10z$$

9. $5(x + 3y - z) = 5(x) + 5(3y) - 5(z)$
$$= 5x + 15y - 5z$$

10. $-1(2 - y) = (-1)(2) - (-1)(y)$
$$= -2 + y$$

11. $-(3 + x - w) = -1(3 + x - w)$
$$= (-1)(3) + (-1)(x) - (-1)(w)$$
$$= -3 - x + w$$

> **Helpful Hint** Notice in Example 11 that $-(3 + x - w)$ can be rewritten as $-1(3 + x - w)$.

12. $\dfrac{1}{2}(6x + 14) + 10 = \dfrac{1}{2}(6x) + \dfrac{1}{2}(14) + 10$ Apply the distributive property.
$$= 3x + 7 + 10 \qquad\qquad \text{Multiply.}$$
$$= 3x + 17 \qquad\qquad \text{Add.}$$

■ **Work Practice 7–12**

The distributive property can also be used to write a sum as a product.

Examples Use the distributive property to write each sum as a product.

13. $8 \cdot 2 + 8 \cdot x = 8(2 + x)$

14. $7s + 7t = 7(s + t)$

■ **Work Practice 13–14**

Practice 13–14

Use the distributive property to write each sum as a product.

13. $9 \cdot 3 + 9 \cdot y$

14. $4x + 4y$

Objective C Using the Identity and Inverse Properties ▶

Next, we look at the **identity properties.**

The number 0 is called the identity for addition because when 0 is added to any real number, the result is the same real number. In other words, the *identity* of the real number is not changed.

The number 1 is called the identity for multiplication because when a real number is multiplied by 1, the result is the same real number. In other words, the *identity* of the real number is not changed.

Identities for Addition and Multiplication

0 is the identity element for addition.

$$a + 0 = a \quad \text{and} \quad 0 + a = a$$

1 is the identity element for multiplication.

$$a \cdot 1 = a \quad \text{and} \quad 1 \cdot a = a$$

Notice that 0 is the *only* number that can be added to any real number with the result that the sum is the same real number. Also, 1 is the *only* number that can be multiplied by any real number with the result that the product is the same real number.

Additive inverses or **opposites** were introduced in Section 1.4. Two numbers are called additive inverses or opposites if their sum is 0. The additive inverse or opposite of 6 is -6 because $6 + (-6) = 0$. The additive inverse or opposite of -5 is 5 because $-5 + 5 = 0$.

Reciprocals or **multiplicative inverses** were introduced in Section 1.6. Two nonzero numbers are called reciprocals or multiplicative inverses if their product is 1. The reciprocal or multiplicative inverse of $\dfrac{2}{3}$ is $\dfrac{3}{2}$ because $\dfrac{2}{3} \cdot \dfrac{3}{2} = 1$. Likewise, the reciprocal of -5 is $-\dfrac{1}{5}$ because $-5\left(-\dfrac{1}{5}\right) = 1$.

Answers

13. $9(3 + y)$ **14.** $4(x + y)$

Additive or Multiplicative Inverses

The numbers a and $-a$ are additive inverses or opposites of each other because their sum is 0; that is,

$$a + (-a) = 0$$

The numbers b and $\dfrac{1}{b}$ (for $b \neq 0$) are reciprocals or multiplicative inverses of each other because their product is 1; that is,

$$b \cdot \dfrac{1}{b} = 1$$

Practice 15–21

Name the property illustrated by each true statement.

15. $7(a + b) = 7 \cdot a + 7 \cdot b$

16. $12 + y = y + 12$

17. $-4 \cdot (6 \cdot x) = (-4 \cdot 6) \cdot x$

18. $6 + (z + 2) = 6 + (2 + z)$

19. $3\left(\dfrac{1}{3}\right) = 1$

20. $(x + 0) + 23 = x + 23$

21. $(7 \cdot y) \cdot 10 = y \cdot (7 \cdot 10)$

Answers

15. distributive property
16. commutative property of addition
17. associative property of multiplication
18. commutative property of addition
19. multiplicative inverse property
20. identity element for addition
21. commutative and associative properties of multiplication

✔ **Concept Check Answers**

a. $\dfrac{3}{10}$ **b.** $-\dfrac{10}{3}$

✔ **Concept Check** Which of the following is

a. the opposite of $-\dfrac{3}{10}$, and

b. the reciprocal of $-\dfrac{3}{10}$?

$$1, \ -\dfrac{10}{3}, \ \dfrac{3}{10}, \ 0, \ \dfrac{10}{3}, \ -\dfrac{3}{10}$$

Examples   Name the property illustrated by each true statement.

15. $3(x + y) = 3 \cdot x + 3 \cdot y$ Distributive property

16. $(x + 7) + 9 = x + (7 + 9)$ Associative property of addition (grouping changed)

17. $(b + 0) + 3 = b + 3$ Identity element for addition

18. $2 \cdot (z \cdot 5) = 2 \cdot (5 \cdot z)$ Commutative property of multiplication (order changed)

19. $-2 \cdot \left(-\dfrac{1}{2}\right) = 1$ Multiplicative inverse property

20. $-2 + 2 = 0$ Additive inverse property

21. $-6 \cdot (y \cdot 2) = (-6 \cdot 2) \cdot y$ Commutative and associative properties of multiplication (order and grouping changed)

■ **Work Practice 15–21**

Vocabulary, Readiness & Video Check

Use the choices below to fill in each blank.

distributive property   associative property of multiplication   commutative property of addition
opposites or additive inverses   associative property of addition
reciprocals or multiplicative inverses   commutative property of multiplication

1. $x + 5 = 5 + x$ is a true statement by the ___commutative property of addition___.

2. $x \cdot 5 = 5 \cdot x$ is a true statement by the ___commutative property of multiplication___.

3. $3(y + 6) = 3 \cdot y + 3 \cdot 6$ is true by the ___distributive property___.

4. $2 \cdot (x \cdot y) = (2 \cdot x) \cdot y$ is a true statement by the ___associative property of multiplication___.

5. $x + (7 + y) = (x + 7) + y$ is a true statement by the ___associative property of addition___.

6. The numbers $-\dfrac{2}{3}$ and $-\dfrac{3}{2}$ are called ___reciprocals or multiplicative inverses___.

7. The numbers $-\dfrac{2}{3}$ and $\dfrac{2}{3}$ are called ___opposites or additive inverses___.

Martin-Gay Interactive Videos   Watch the section lecture video and answer the following questions.

See Video 1.7

Objective A 8. The commutative properties are discussed in ▣ Examples 1 and 2 and the associative properties are discussed in ▣ Examples 3–7. What's the one word used again and again to describe the commutative property? The associative property? ▶

Objective B 9. In ▣ Example 10, what point is made about the term 2? ▶

Objective C 10. Complete these statements based on the lecture given before ▣ Example 12. ▶

- The identity element for addition is _____ because if we add _____ to any real number, the result is that real number.
- The identity element for multiplication is _____ because any real number times _____ gives a result of that original real number.

See video answer section.

1.7 Exercise Set MyMathLab®

Objective A *Use a commutative property to complete each statement. See Examples 1 and 3.*

▶ **1.** $x + 16 = $ ___$16 + x$___

2. $8 + y = $ ___$y + 8$___

3. $-4 \cdot y = $ ___$y \cdot (-4)$___

4. $-2 \cdot x = $ ___$x \cdot (-2)$___

▶ **5.** $xy = $ ___yx___

6. $ab = $ ___ba___

7. $2x + 13 = $ ___$13 + 2x$___

8. $19 + 3y = $ ___$3y + 19$___

Use an associative property to complete each statement. See Examples 2 and 4.

▶ **9.** $(xy) \cdot z = $ ___$x \cdot (yz)$___

10. $3 \cdot (x \cdot y) = $ ___$(3x) \cdot y$___

11. $2 + (a + b) = $ ___$(2 + a) + b$___

12. $(y + 4) + z = $ ___$y + (4 + z)$___

13. $4 \cdot (ab) = $ ___$(4a) \cdot b$___

14. $(-3y) \cdot z = $ ___$-3 \cdot (yz)$___

▶ **15.** $(a + b) + c = $ ___$a + (b + c)$___

16. $6 + (r + s) = $ ___$(6 + r) + s$___

Use the commutative and associative properties to simplify each expression. See Examples 5 and 6.

▶ **17.** $8 + (9 + b)$ $17 + b$

18. $(r + 3) + 11$ $r + 14$

▶ **19.** $4(6y)$ $24y$

20. $2(42x)$ $84x$

21. $\dfrac{1}{5}(5y)$ y

22. $\dfrac{1}{8}(8z)$ z

23. $(13 + a) + 13$ $26 + a$

24. $7 + (x + 4)$ $11 + x$

25. $-9(8x)$ $-72x$

26. $-3(12y)$ $-36y$

27. $\dfrac{3}{4}\left(\dfrac{4}{3}s\right)$ s

28. $\dfrac{2}{7}\left(\dfrac{7}{2}r\right)$ r

29. $-\dfrac{1}{2}(5x)$ $-\dfrac{5}{2}x$

30. $-\dfrac{1}{3}(7x)$ $-\dfrac{7}{3}x$

Objective B

Use the distributive property to write each expression without parentheses. Then simplify the result, if possible. See Examples 7 through 12.

31. $4(x + y)$ $4x + 4y$

32. $7(a + b)$ $7a + 7b$

33. $9(x - 6)$ $9x - 54$

34. $11(y - 4)$ $11y - 44$

35. $2(3x + 5)$ $6x + 10$

36. $5(7 + 8y)$ $35 + 40y$

37. $7(4x - 3)$ $28x - 21$

38. $3(8x - 1)$ $24x - 3$

39. $3(6 + x)$ $18 + 3x$

40. $2(x + 5)$ $2x + 10$

41. $-2(y - z)$
$-2y + 2z$

42. $-3(z - y)$
$-3z + 3y$

43. $-\dfrac{1}{3}(3y + 5)$ $-y - \dfrac{5}{3}$

44. $-\dfrac{1}{2}(2r + 11)$ $-r - \dfrac{11}{2}$

45. $5(x + 4m + 2)$
$5x + 20m + 10$

46. $8(3y + z - 6)$
$24y + 8z - 48$

47. $-4(1 - 2m + n) + 4$
$8m - 4n$

48. $-4(4 + 2p + 5r) + 16$
$-8p - 20r$

49. $-(5x + 2)$ $-5x - 2$

50. $-(9r + 5)$ $-9r - 5$

51. $-(r - 3 - 7p)$
$-r + 3 + 7p$

52. $-(q - 2 + 6r)$
$-q + 2 - 6r$

53. $\dfrac{1}{2}(6x + 7) + \dfrac{1}{2}$
$3x + 4$

54. $\dfrac{1}{4}(4x - 2) - \dfrac{7}{2}$
$x - 4$

55. $-\dfrac{1}{3}(3x - 9y)$
$-x + 3y$

56. $-\dfrac{1}{5}(10a - 25b)$
$-2a + 5b$

57. $3(2r + 5) - 7$
$6r + 8$

58. $10(4s + 6) - 40$
$40s + 20$

59. $-9(4x + 8) + 2$
$-36x - 70$

60. $-11(5x + 3) + 10$
$-55x - 23$

61. $-0.4(4x + 5) - 0.5$
$-1.6x - 2.5$

62. $-0.6(2x + 1) - 0.1$
$-1.2x - 0.7$

Use the distributive property to write each sum as a product. See Examples 13 and 14.

63. $4 \cdot 1 + 4 \cdot y$
$4(1 + y)$

64. $14 \cdot z + 14 \cdot 5$
$14(z + 5)$

65. $11x + 11y$
$11(x + y)$

66. $9a + 9b$
$9(a + b)$

67. $(-1) \cdot 5 + (-1) \cdot x$
$-1(5 + x)$

68. $(-3)a + (-3)y$
$-3(a + y)$

69. $30a + 30b$
$30(a + b)$

70. $25x + 25y$
$25(x + y)$

Objectives A C Mixed Practice

Name the property illustrated by each true statement. See Examples 15 through 21.

71. $3 \cdot 5 = 5 \cdot 3$
commutative property of multiplication

72. $4(3 + 8) = 4 \cdot 3 + 4 \cdot 8$
distributive property

73. $2 + (x + 5) = (2 + x) + 5$
associative property of addition

74. $9 \cdot (x \cdot 7) = (9 \cdot x) \cdot 7$
associative property of multiplication

75. $(x + 9) + 3 = (9 + x) + 3$
commutative property of addition

76. $1 \cdot 9 = 9$
identity element for multiplication

77. $(4 \cdot y) \cdot 9 = 4 \cdot (y \cdot 9)$
associative property of multiplication

78. $-4 \cdot (8 \cdot 3) = (8 \cdot 3) \cdot (-4)$
commutative property of multiplication

79. $0 + 6 = 6$
identity element for addition

80. $(a + 9) + 6 = a + (9 + 6)$
associative property of addition

81. $-4(y + 7) = -4 \cdot y + (-4) \cdot 7$
distributive property

82. $(11 + r) + 8 = (r + 11) + 8$
commutative property of addition

▶ 83. $6 \cdot \dfrac{1}{6} = 1$ multiplicative inverse property

84. $r + 0 = r$ identity element for addition

85. $-6 \cdot 1 = -6$ identity element for multiplication

86. $-\dfrac{3}{4}\left(-\dfrac{4}{3}\right) = 1$ multiplicative inverse property

Concept Extensions

Fill in the table with the opposite (additive inverse), the reciprocal (multiplicative inverse), or the expression. Assume that the value of each expression is not 0.

	87.	88.	89.	90.	91.	92.
Expression	8	$-\dfrac{2}{3}$	x	$4y$	$2x$	$-7x$
Opposite	-8	$\dfrac{2}{3}$	$-x$	$-4y$	$-2x$	$7x$
Reciprocal	$\dfrac{1}{8}$	$-\dfrac{3}{2}$	$\dfrac{1}{x}$	$\dfrac{1}{4y}$	$\dfrac{1}{2x}$	$-\dfrac{1}{7x}$

Decide whether each statement is true or false. See the second Concept Check in this section.

93. The opposite of $-\dfrac{a}{2}$ is $-\dfrac{2}{a}$.   false

94. The reciprocal of $-\dfrac{a}{2}$ is $\dfrac{a}{2}$.   false

Determine which pairs of actions are commutative. See the first Concept Check in this section.

95. "taking a test" and "studying for the test"   no

96. "putting on your shoes" and "putting on your socks"   no

97. "putting on your left shoe" and "putting on your right shoe"   yes

98. "reading the sports section" and "reading the comics section"   yes

99. "mowing the lawn" and "trimming the hedges"   yes

100. "baking a cake" and "eating the cake"   no

101. "feeding the dog" and "feeding the cat"   yes

102. "dialing a number" and "turning on the cell phone"   no

Name the property illustrated by each step.

103. a. $\triangle + (\square + \bigcirc) = (\square + \bigcirc) + \triangle$
   commutative property of addition
   b. $\qquad\qquad = (\bigcirc + \square) + \triangle$
   commutative property of addition
   c. $\qquad\qquad = \bigcirc + (\square + \triangle)$
   associative property of addition

104. a. $(x + y) + z = x + (y + z)$
   associative property of addition
   b. $\qquad\qquad = (y + z) + x$
   commutative property of addition
   c. $\qquad\qquad = (z + y) + x$
   commutative property of addition

105. Explain why 0 is called the identity element for addition.   answers may vary

106. Explain why 1 is called the identity element for multiplication.   answers may vary

107. Write an example that shows that division is not commutative.   answers may vary

108. Write an example that shows that subtraction is not commutative.   answers may vary

1.8 Simplifying Expressions

As we explore in this section, we will see that an expression such as $3x + 2x$ is not written as simply as possible. This is because—even without replacing x by a value—we can perform the indicated addition.

Objectives

A Identify Terms, Like Terms, and Unlike Terms. ▶

B Combine Like Terms. ▶

C Simplify Expressions Containing Parentheses. ▶

D Write Word Phrases as Algebraic Expressions. ▶

Objective **A** Identifying Terms, Like Terms, and Unlike Terms ▶

Before we practice simplifying expressions, we must learn some new language. A **term** is a number or the product of a number and variables raised to powers.

Terms

$$-y, \quad 2x^3, \quad -5, \quad 3xz^2, \quad \frac{2}{y}, \quad 0.8z$$

The **numerical coefficient** of a term is the numerical factor. The numerical coefficient of $3x$ is 3. Recall that $3x$ means $3 \cdot x$.

Term	Numerical Coefficient
$3x$	3
$\dfrac{y^3}{5}$	$\dfrac{1}{5}$ since $\dfrac{y^3}{5}$ means $\dfrac{1}{5} \cdot y^3$
$-0.7ab^3c^5$	-0.7
z	1
$-y$	-1
-5	-5

Helpful Hint

The term z means $1z$ and thus has a numerical coefficient of 1.
The term $-y$ means $-1y$ and thus has a numerical coefficient of -1.

Practice 1

Identify the numerical coefficient of each term.

a. $-4x$ **b.** $15y^3$ **c.** x

d. $-y$ **e.** $\dfrac{z}{4}$

▶ **Example 1** Identify the numerical coefficient of each term.

a. $-3y$ **b.** $22z^4$ **c.** y **d.** $-x$ **e.** $\dfrac{x}{7}$

Solution:

a. The numerical coefficient of $-3y$ is -3.

b. The numerical coefficient of $22z^4$ is 22.

c. The numerical coefficient of y is 1, since y is $1y$.

d. The numerical coefficient of $-x$ is -1, since $-x$ is $-1x$.

e. The numerical coefficient of $\dfrac{x}{7}$ is $\dfrac{1}{7}$, since $\dfrac{x}{7}$ is $\dfrac{1}{7} \cdot x$.

■ **Work Practice 1**

Answers

1. a. -4 **b.** 15 **c.** 1 **d.** -1 **e.** $\dfrac{1}{4}$

Terms with the same variables raised to exactly the same powers are called **like terms**. Terms that aren't like terms are called **unlike terms**.

Like Terms	Unlike Terms	Reason Why
$3x, 2x$	$5x, 5x^2$	Why? Same variable x, but different powers of x and x^2
$-6x^2y, 2x^2y, 4x^2y$	$7y, 3z, 8x^2$	Why? Different variables
$2ab^2c^3, ac^3b^2$	$6abc^3, 6ab^2$	Why? Different variables and different powers

Helpful Hint

In like terms, each variable and its exponent must match exactly, but these factors don't need to be in the same order.

$2x^2y$ and $3yx^2$ are like terms.

Example 2   Determine whether the terms are like or unlike.

a. $2x, 3x^2$ **b.** $4x^2y, x^2y, -2x^2y$ **c.** $-2yz, -3zy$

d. $-x^4, x^4$ **e.** $-8a^5, 8a^5$

Solution:

a. Unlike terms, since the exponents on x are not the same.
b. Like terms, since each variable and its exponent match.
c. Like terms, since $zy = yz$ by the commutative property of multiplication.
d. Like terms. The variable and its exponent match.
e. Like terms. The variable and its exponent match.

▶ **Work Practice 2**

Objective B Combining Like Terms

An algebraic expression containing the sum or difference of like terms can be simplified by applying the distributive property. For example, by the distributive property, we rewrite the sum of the like terms $6x + 2x$ as

$$6x + 2x = (6 + 2)x = 8x$$

Also,

$$-y^2 + 5y^2 = (-1 + 5)y^2 = 4y^2$$

Simplifying the sum or difference of like terms is called **combining like terms**.

Example 3   Simplify each expression by combining like terms.

a. $7x - 3x$ **b.** $10y^2 + y^2$

c. $8x^2 + 2x - 3x$ **d.** $9n^2 - 5n^2 + n^2$

Solution:

a. $7x - 3x = (7 - 3)x = 4x$
b. $10y^2 + y^2 = (10 + 1)y^2 = 11y^2$
c. $8x^2 + 2x - 3x = 8x^2 + (2 - 3)x = 8x^2 - 1x$ or $8x^2 - x$
d. $9n^2 - 5n^2 + n^2 = (9 - 5 + 1)n^2 = 5n^2$

▶ **Work Practice 3**

Teaching Tip

It may be helpful to point out that the reason students must be able to identify like terms is to simplify expressions. This process is very similar to counting coins. People usually first sort the coins by type and then count how many of each type they have. Counting coins compares to collecting and combining like terms.

Practice 2

Determine whether the terms are like or unlike.
a. $7x^2, -6x^3$
b. $3x^2y^2, -x^2y^2, 4x^2y^2$
c. $-5ab, 3ba$
d. $2x^3, 4y^3$
e. $-7m^4, 7m^4$

Teaching Tip

It may be helpful to give students a visual representation of some terms. This problem illustrates why like terms are combined and unlike terms are not combined.

Use the tiles $\boxed{1}$, $\boxed{x}$, $\boxed{x^2}$.

$2x^2 + 3x + x^2 + 1 + x + 2 = 3x^2 + 4x + 3$

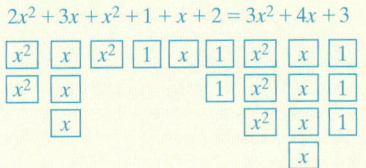

Practice 3

Simplify each expression by combining like terms.
a. $9y - 4y$
b. $11x^2 + x^2$
c. $5y - 3x + 4x$
d. $14m^2 - m^2 + 3m^2$

Answers
2. **a.** unlike **b.** like **c.** like
d. unlike **e.** like 3. **a.** $5y$ **b.** $12x^2$
c. $5y + x$ **d.** $16m^2$

The preceding examples suggest the following.

> ### Combining Like Terms
>
> To **combine like terms**, combine the numerical coefficients and multiply the result by the common variable factors.

Practice 4–7

Simplify each expression by combining like terms.

4. $7y + 2y + 6 + 10$

5. $-2x + 4 + x - 11$

6. $3z - 3z^2$

7. $8.9y + 4.2y - 3$

Examples   Simplify each expression by combining like terms.

4. $2x + 3x + 5 + 2 = (2 + 3)x + (5 + 2)$
$$= 5x + 7$$

5. $-5a - 3 + a + 2 = -5a + 1a + (-3 + 2)$
$$= (-5 + 1)a + (-3 + 2)$$
$$= -4a - 1$$

6. $4y - 3y^2$

These two terms cannot be combined because they are unlike terms.

7. $2.3x + 5x - 6 = (2.3 + 5)x - 6$
$$= 7.3x - 6$$

▨ **Work Practice 4–7**

Objective C Simplifying Expressions Containing Parentheses ▶

In simplifying expressions we make frequent use of the distributive property to remove parentheses.

It may be helpful to study the examples below.

$$+(3a + 2) = +1(3a + 2) = +1(3a) + (+1)(2) = 3a + 2$$
means

$$-(3a + 2) = -1(3a + 2) = -1(3a) + (-1)(2) = -3a - 2$$
means

Practice 8–10

Find each product by using the distributive property to remove parentheses.

8. $3(11y + 6)$

9. $-4(x + 0.2y - 3)$

10. $-(3x + 2y + z - 1)$

Examples   Find each product by using the distributive property to remove parentheses.

8. $5(3x + 2) = 5(3x) + 5(2)$ Apply the distributive property.
$$= 15x + 10$$ Multiply.

9. $-2(y + 0.3z - 1) = -2(y) + (-2)(0.3z) - (-2)(1)$ Apply the distributive property.
$$= -2y - 0.6z + 2$$ Multiply.

10. $-(9x + y - 2z + 6) = -1(9x + y - 2z + 6)$ Distribute -1 over each term.
$$= -1(9x) + (-1)(y) - (-1)(2z) + (-1)(6)$$
$$= -9x - y + 2z - 6$$

▨ **Work Practice 8–10**

Answers

4. $9y + 16$ **5.** $-x - 7$
6. $3z - 3z^2$ **7.** $13.1y - 3$
8. $33y + 18$ **9.** $-4x - 0.8y + 12$
10. $-3x - 2y - z + 1$

Terms with the same variables raised to exactly the same powers are called **like terms**. Terms that aren't like terms are called **unlike terms**.

Like Terms	Unlike Terms	Reason Why
$3x$, $2x$	$5x$, $5x^2$	Why? Same variable x, but different powers of x and x^2
$-6x^2y$, $2x^2y$, $4x^2y$	$7y$, $3z$, $8x^2$	Why? Different variables
$2ab^2c^3$, ac^3b^2	$6abc^3$, $6ab^2$	Why? Different variables and different powers

Helpful Hint

In like terms, each variable and its exponent must match exactly, but these factors don't need to be in the same order.

$2x^2y$ and $3yx^2$ are like terms.

Example 2 Determine whether the terms are like or unlike.

a. $2x$, $3x^2$ **b.** $4x^2y$, x^2y, $-2x^2y$ **c.** $-2yz$, $-3zy$

d. $-x^4$, x^4 **e.** $-8a^5$, $8a^5$

Solution:

a. Unlike terms, since the exponents on x are not the same.
b. Like terms, since each variable and its exponent match.
c. Like terms, since $zy = yz$ by the commutative property of multiplication.
d. Like terms. The variable and its exponent match.
e. Like terms. The variable and its exponent match.

■ **Work Practice 2**

Objective B Combining Like Terms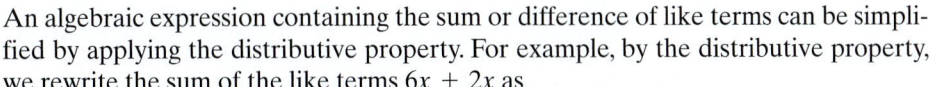

An algebraic expression containing the sum or difference of like terms can be simplified by applying the distributive property. For example, by the distributive property, we rewrite the sum of the like terms $6x + 2x$ as

$$6x + 2x = (6 + 2)x = 8x$$

Also,

$$-y^2 + 5y^2 = (-1 + 5)y^2 = 4y^2$$

Simplifying the sum or difference of like terms is called **combining like terms**.

Example 3 Simplify each expression by combining like terms.

a. $7x - 3x$ **b.** $10y^2 + y^2$

c. $8x^2 + 2x - 3x$ **d.** $9n^2 - 5n^2 + n^2$

Solution:

a. $7x - 3x = (7 - 3)x = 4x$
b. $10y^2 + y^2 = (10 + 1)y^2 = 11y^2$
c. $8x^2 + 2x - 3x = 8x^2 + (2 - 3)x = 8x^2 - 1x$ or $8x^2 - x$
d. $9n^2 - 5n^2 + n^2 = (9 - 5 + 1)n^2 = 5n^2$

■ **Work Practice 3**

Practice 2

Determine whether the terms are like or unlike.

a. $7x^2$, $-6x^3$
b. $3x^2y^2$, $-x^2y^2$, $4x^2y^2$
c. $-5ab$, $3ba$
d. $2x^3$, $4y^3$
e. $-7m^4$, $7m^4$

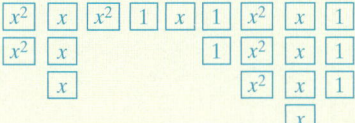

Practice 3

Simplify each expression by combining like terms.

a. $9y - 4y$
b. $11x^2 + x^2$
c. $5y - 3x + 4x$
d. $14m^2 - m^2 + 3m^2$

Answers

2. a. unlike **b.** like **c.** like
d. unlike **e.** like **3. a.** $5y$ **b.** $12x^2$
c. $5y + x$ **d.** $16m^2$

The preceding examples suggest the following.

> ### Combining Like Terms
>
> To **combine like terms,** combine the numerical coefficients and multiply the result by the common variable factors.

Practice 4–7

Simplify each expression by combining like terms.

4. $7y + 2y + 6 + 10$

5. $-2x + 4 + x - 11$

6. $3z - 3z^2$

7. $8.9y + 4.2y - 3$

Examples Simplify each expression by combining like terms.

4. $2x + 3x + 5 + 2 = (2 + 3)x + (5 + 2)$

$\qquad\qquad\qquad\quad = 5x + 7$

5. $-5a - 3 + a + 2 = -5a + 1a + (-3 + 2)$

$\qquad\qquad\qquad\qquad = (-5 + 1)a + (-3 + 2)$

$\qquad\qquad\qquad\qquad = -4a - 1$

6. $4y - 3y^2$

These two terms cannot be combined because they are unlike terms.

7. $2.3x + 5x - 6 = (2.3 + 5)x - 6$

$\qquad\qquad\qquad = 7.3x - 6$

■ **Work Practice 4–7**

Objective C Simplifying Expressions Containing Parentheses ▶

In simplifying expressions we make frequent use of the distributive property to remove parentheses.

It may be helpful to study the examples below.

$$+(3a + 2) = +1(3a + 2) = +1(3a) + (+1)(2) = 3a + 2$$
$$\text{means}$$

$$-(3a + 2) = -1(3a + 2) = -1(3a) + (-1)(2) = -3a - 2$$
$$\text{means}$$

Practice 8–10

Find each product by using the distributive property to remove parentheses.

8. $3(11y + 6)$

9. $-4(x + 0.2y - 3)$

10. $-(3x + 2y + z - 1)$

Examples Find each product by using the distributive property to remove parentheses.

8. $5(3x + 2) = 5(3x) + 5(2)$ Apply the distributive property.

$\qquad\qquad = 15x + 10$ Multiply.

9. $-2(y + 0.3z - 1) = -2(y) + (-2)(0.3z) - (-2)(1)$ Apply the distributive property.

$\qquad\qquad\qquad = -2y - 0.6z + 2$ Multiply.

10. $-(9x + y - 2z + 6) = -1(9x + y - 2z + 6)$ Distribute -1 over each term.

$\qquad\qquad\qquad\quad = -1(9x) + (-1)(y) - (-1)(2z) + (-1)(6)$

$\qquad\qquad\qquad\quad = -9x - y + 2z - 6$

Answers

4. $9y + 16$ **5.** $-x - 7$

6. $3z - 3z^2$ **7.** $13.1y - 3$

8. $33y + 18$ **9.** $-4x - 0.8y + 12$

10. $-3x - 2y - z + 1$

■ **Work Practice 8–10**

If a "−" sign precedes parentheses, the sign of each term inside the parentheses is changed when the distributive property is applied to remove the parentheses.

Examples:

$$-(2x + 1) = -2x - 1$$
$$-(x - 2y) = -x + 2y$$
$$-(-5x + y - z) = 5x - y + z$$
$$-(-3x - 4y - 1) = 3x + 4y + 1$$

When simplifying an expression containing parentheses, we often use the distributive property first to remove parentheses and then again to combine any like terms.

Examples   Simplify each expression.

11. $3(2x - 5) + 1 = 6x - 15 + 1$ Apply the distributive property.

$= 6x - 14$ Combine like terms.

12. $8 - (7x + 2) + 3x = 8 - 7x - 2 + 3x$ Apply the distributive property.

$= -7x + 3x + 8 - 2$

$= -4x + 6$ Combine like terms.

13. $-2(4x + 7) - (3x - 1) = -8x - 14 - 3x + 1$ Apply the distributive property.

$= -11x - 13$ Combine like terms.

14. $9 + 3(4x - 10) = 9 + 12x - 30$ Apply the distributive property.

$= -21 + 12x$ Combine like terms.

or $12x - 21$

■ Work Practice 11–14

Practice 11–14

Simplify each expression.
11. $4(4x - 6) + 20$
12. $5 - (3x + 9) + 6x$
13. $-3(7x + 1) - (4x - 2)$
14. $8 + 11(2y - 9)$

Don't forget to use the distributive property and multiply before adding or subtracting like terms.

Example 15   Subtract $4x - 2$ from $2x - 3$.

Solution: We first note that "subtract $4x - 2$ **from** $2x - 3$" translates to $(2x - 3) - (4x - 2)$. Notice that parentheses were placed around each given expression. This is to ensure that the entire expression after the subtraction sign is subtracted. Next, we simplify the algebraic expression.

$$(2x - 3) - (4x - 2) = 2x - 3 - 4x + 2$$ Apply the distributive property.

$$= -2x - 1$$ Combine like terms.

■ Work Practice 15

Practice 15

Subtract $9x - 10$ from $4x - 3$.

Practice 16–19

Write each phrase as an algebraic expression and simplify if possible. Let x represent the unknown number.

16. Three times a number, subtracted from 10

17. The sum of a number and 2, divided by 5

18. Three times the sum of a number and 6

19. Seven times the difference of a number and 4.

Teaching Tip

After doing the examples, have students write each of the following phrases as an algebraic expression:

a. Four, added to three times a number

b. Three times a number, added to four

c. Four, subtracted from three times a number

d. Three times a number, subtracted from four

Then ask students the following questions: Are the algebraic expressions of **a.** and **b.** equivalent? (Yes) Why or why not? (Addition is commutative)
Are the algebraic expressions of **c.** and **d.** equivalent? (No) Why or why not? (Subtraction is not commutative.)

Answers

16. $10 - 3x$ **17.** $(x + 2) \div 5$ or $\dfrac{x + 2}{5}$

18. $3x + 18$ **19.** $7x - 28$

Objective D Writing Algebraic Expressions

To prepare for problem solving, we next practice writing word phrases as algebraic expressions.

Examples   Write each phrase as an algebraic expression and simplify if possible. Let x represent the unknown number.

16. Twice a number, plus 6

$$2x \qquad + \; 6$$

This expression cannot be simplified.

17. The difference of a number and 4, divided by 7

$$(x - 4) \qquad\qquad \div \qquad 7 \;\; \text{or} \;\; \frac{x - 4}{7}$$

This expression cannot be simplified.

18. Five plus the sum of a number and 1

$$5 \quad + \qquad\qquad (x + 1)$$

We can simplify this expression.

$$5 + (x + 1) = 5 + x + 1$$
$$= 6 + x$$

19. Four times the sum of a number and 3

$$4 \quad \cdot \qquad\qquad (x + 3)$$

Use the distributive property to simplify the expression.

$$4 \cdot (x + 3) = 4(x + 3)$$
$$= 4 \cdot x + 4 \cdot 3$$
$$= 4x + 12$$

🟧 **Work Practice 16–19**

Vocabulary, Readiness & Video Check

Use the choices below to fill in each blank. Some choices may be used more than once.

numerical coefficient	expression	unlike	distributive
combine like terms	like	term	

1. $14y^2 + 2x - 23$ is called a(n) ___expression___ while $14y^2$, $2x$, and -23 are each called a(n) ___term___.

2. To multiply $3(-7x + 1)$, we use the ___distributive___ property.

3. To simplify an expression like $y + 7y$, we ___combine like terms___.

4. The term z has an understood ___numerical coefficient___ of 1.

5. The terms $-x$ and $5x$ are ___like___ terms and the terms $5x$ and $5y$ are ___unlike___ terms.

6. For the term $-3x^2y$, -3 is called the ___numerical coefficient___.

Martin-Gay Interactive Videos   *Watch the section lecture video and answer the following questions.*

Objective A   **7.** ⊞ Example 7 shows two terms with exactly the same variables. Why are these terms not considered like terms? ▶

Objective B   **8.** ⊞ Example 8 shows us that when combining like terms, we are actually applying what property? ▶

Objective C   **9.** The expression in ⊞ Example 11 shows a minus sign before parentheses. When using the distributive property to multiply and remove parentheses, what number are we actually distributing to each term within the parentheses? ▶

Objective D   **10.** Write the phrase given in ⊞ Example 14, translate it to an algebraic expression, then simplify it. Why are we able to simplify it? ▶

See Video 1.8 🍎

See video answer section.

1.8 **Exercise Set** MyMathLab® ▶

Objective A *Identify the numerical coefficient of each term. See Example 1.*

1. $-7y$
   -7

2. $3x$
   3

3. x
   1

4. $-y$
   -1

5. $17x^2y$
   17

6. $1.2xyz$
   1.2

Indicate whether the terms in each list are like or unlike. See Example 2.

▶**7.** $5y, -y$
   like

▶**8.** $-2x^2y, 6xy$
   unlike

9. $2z, 3z^2$
   unlike

10. $ab^2, -7ab^2$
   like

11. $8wz, \frac{1}{7}zw$
   like

12. $7.4p^3q^2, 6.2p^3q^2r$
   unlike

Objective B *Simplify each expression by combining any like terms. See Examples 3 through 7.*

13. $7y + 8y$
   $15y$

▶**14.** $3x + 2x$
   $5x$

15. $8w - w + 6w$
   $13w$

16. $c - 7c + 2c$
   $-4c$

17. $3b - 5 - 10b - 4$
   $-7b - 9$

18. $6g + 5 - 3g - 7$
   $3g - 2$

19. $m - 4m + 2m - 6$
   $-m - 6$

20. $a + 3a - 2 - 7a$
   $-3a - 2$

21. $5g - 3 - 5 - 5g$
   -8

22. $8p + 4 - 8p - 15$
   -11

23. $6.2x - 4 + x - 1.2$
   $7.2x - 5.2$

24. $7.9y - 0.7 - y + 0.2$
   $6.9y - 0.5$

25. $2k - k - 6$
   $k - 6$

26. $7c - 8 - c$
   $6c - 8$

27. $-9x + 4x + 18 - 10x$
   $-15x + 18$

28. $5y - 14 + 7y - 20y$
   $-8y - 14$

29. $6x - 5x + x - 3 + 2x$
   $4x - 3$

30. $8h + 13h - 6 + 7h - h$
   $27h - 6$

31. $7x^2 + 8x^2 - 10x^2$
   $5x^2$

▶**32.** $8x^3 + x^3 - 11x^3$
   $-2x^3$

33. $3.4m - 4 - 3.4m - 7$
   -11

34. $2.8w - 0.9 - 0.5 - 2.8w$
   -1.4

▶**35.** $6x + 0.5 - 4.3x - 0.4x + 3$
   $1.3x + 3.5$

36. $0.4y - 6.7 + y - 0.3 - 2.6y$
   $-1.2y - 7$

Objective C *Simplify each expression. Use the distributive property to remove any parentheses. See Examples 8 through 10.*

37. $5(y + 4)$
$5y + 20$

38. $7(r + 3)$
$7r + 21$

39. $-2(x + 2)$
$-2x - 4$

40. $-4(y + 6)$
$-4y - 24$

41. $-5(2x - 3y + 6)$
$-10x + 15y - 30$

42. $-2(4x - 3z - 1)$
$-8x + 6z + 2$

43. $-(3x - 2y + 1)$
$-3x + 2y - 1$

44. $-(y + 5z - 7)$
$-y - 5z + 7$

Objectives B C Mixed Practice *Remove parentheses and simplify each expression. See Examples 8 through 14.*

45. $7(d - 3) + 10$
$7d - 11$

46. $9(z + 7) - 15$
$9z + 48$

47. $-4(3y - 4) + 12y$ 16

48. $-3(2x + 5) + 6x$
-15

49. $3(2x - 5) - 5(x - 4)$
$x + 5$

50. $2(6x - 1) - (x - 7)$
$11x + 5$

51. $-2(3x - 4) + 7x - 6$
$x + 2$

52. $8y - 2 - 3(y + 4)$
$5y - 14$

53. $5k - (3k - 10)$
$2k + 10$

54. $-11c - (4 - 2c)$
$-9c - 4$

55. $(3x + 4) - (6x - 1)$
$-3x + 5$

56. $(8 - 5y) - (4 + 3y)$
$-8y + 4$

▶57. $5(x + 2) - (3x - 4)$
$2x + 14$

58. $4(2x - 3) - (x + 1)$
$7x - 13$

59. $\frac{1}{3}(7y - 1) + \frac{1}{6}(4y + 7)$
$3y + \frac{5}{6}$

60. $\frac{1}{5}(9y + 2) + \frac{1}{10}(2y - 1)$
$2y + \frac{3}{10}$

61. $2 + 4(6x - 6)$
$-22 + 24x$

62. $8 + 4(3x - 4)$
$-8 + 12x$

63. $0.5(m + 2) + 0.4m$
$0.9m + 1$

64. $0.2(k + 8) - 0.1k$
$0.1k + 1.6$

65. $10 - 3(2x + 3y)$
$10 - 6x - 9y$

66. $14 - 11(5m + 3n)$
$14 - 55m - 33n$

67. $6(3x - 6) - 2(x + 1) - 17x$
$-x - 38$

68. $7(2x + 5) - 4(x + 2) - 20x$
$-10x + 27$

69. $\frac{1}{2}(12x - 4) - (x + 5)$
$5x - 7$

70. $\frac{1}{3}(9x - 6) - (x - 2)$
$2x$

Perform each indicated operation. Don't forget to simplify if possible. See Example 15.

71. Add $6x + 7$ to $4x - 10$.
$10x - 3$

72. Add $3y - 5$ to $y + 16$.
$4y + 11$

73. Subtract $7x + 1$ from $3x - 8$.
$-4x - 9$

74. Subtract $4x - 7$ from $12 + x$.
$-3x + 19$

▶75. Subtract $5m - 6$ from $m - 9$.
$-4m - 3$

76. Subtract $m - 3$ from $2m - 6$.
$m - 3$

Objective **D** *Write each phrase as an algebraic expression and simplify if possible. Let x represent the unknown number. See Examples 16 through 19.*

▶ **77.** Twice a number, decreased by four $2x - 4$

78. The difference of a number and two, divided by five
$$\frac{x - 2}{5}$$

79. Three-fourths of a number, increased by twelve
$$\frac{3}{4}x + 12$$

80. Eight more than triple a number $3x + 8$

▶ **81.** The sum of 5 times a number and −2, added to 7 times the number $12x - 2$

82. The sum of 3 times a number and 10, **subtracted from** 9 times the number $6x - 10$

83. Eight times the sum of a number and six
$8x + 48$

84. Six times the difference of a number and five
$6x - 30$

85. Double a number minus the sum of the number and ten $x - 10$

86. Half a number minus the product of the number and eight $-7.5x$

Concept Extensions

Given the following information, determine whether each scale is balanced or not.

1 cone balances 1 cube

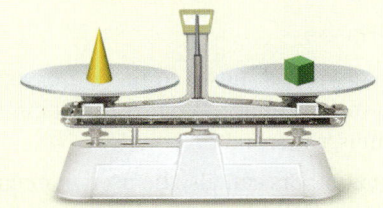

1 cylinder balances 2 cubes

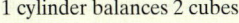

87. 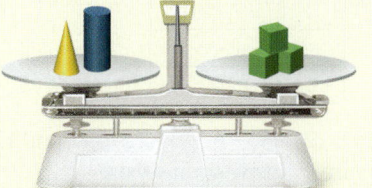   balanced

88. 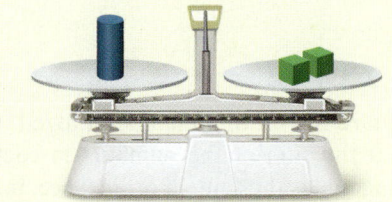   not balanced

89. 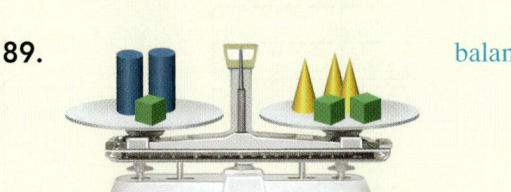   balanced

90. 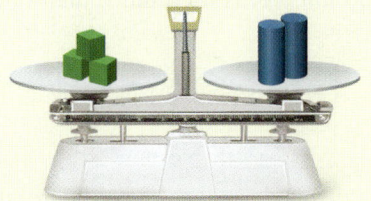   balanced

Write each algebraic expression described.

91. Write an expression with 4 terms that simplifies to $3x - 4$.   answers may vary

92. Write an expression of the form
_____(_____ + _____) whose product is
$6x + 24$.   answers may vary

△**93.** Recall that the perimeter of a figure is the total distance around the figure. Given the following rectangle, express the perimeter as an algebraic expression containing the variable x. $(18x - 2)$ ft

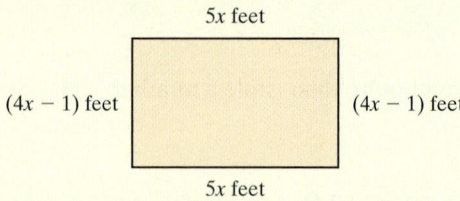

△**94.** Recall that the perimeter of a figure is the total distance around the figure. Given the following triangle, express its perimeter as an algebraic expression containing the variable x. $(5x + 9)$ cm

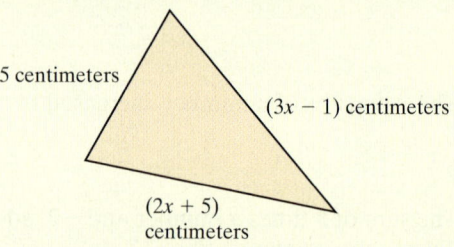

△**95.** To convert from feet to inches, we multiply by 12. For example, the number of inches in 2 feet is $12 \cdot 2$ inches. If one board has a length of $(x + 2)$ *feet* and a second board has a length of $(3x - 1)$ *inches*, express their total length in inches as an algebraic expression. $(15x + 23)$ in.

96. The value of 7 nickels is $5 \cdot 7$ cents. Likewise, the value of x nickels is $5x$ cents. If the money box in a drink machine contains x *nickels*, $3x$ *dimes*, and $(30x - 1)$ *quarters*, express their total value in cents as an algebraic expression. $(785x - 25)$¢

97. In your own words, explain how to combine like terms.   answers may vary

98. Do like terms always contain the same numerical coefficients? Explain your answer.
   no; answers may vary

Chapter 1 Group Activity

Additional group activities are available in the *Instructor's Resource Manual with Tests.*

Magic Squares

Sections 1.3, 1.4, 1.5

A magic square is a set of numbers arranged in a square table so that the sum of the numbers in each column, row, and diagonal is the same. For instance, in the magic square below, the sum of each column, row, and diagonal is 15. Notice that no number is used more than once in the magic square.

2	9	4
7	5	3
6	1	8

The properties of magic squares have been known for a very long time and once were thought to be good luck charms. The ancient Egyptians and Greeks understood their patterns. A magic square even made it into a famous work of art. The engraving titled *Melencolia I*, created by German artist Albrecht Dürer in 1514, features the following four-by-four magic square on the building behind the central figure.

16	3	2	13
5	10	11	8
9	6	7	12
4	15	14	1

Group Exercises

1. Verify that what is shown in the Dürer engraving is, in fact, a magic square. What is the common sum of the columns, rows, and diagonals?   34

2. Negative numbers can also be used in magic squares. Complete the following magic square:

2	−3	−2
−5	−1	3
0	1	−4

3. Use the numbers −12, −9, −6, −3, 0, 3, 6, 9, and 12 to form a magic square.   answers may vary

9	−6	−3
−12	0	12
3	6	−9

Chapter 1 Vocabulary Check

Fill in each blank with one of the words or phrases listed below.

inequality symbols	exponent	term	numerical coefficient
grouping symbols	solution	like terms	unlike terms
equation	absolute value	numerator	denominator
opposites	base	reciprocals	variable

1. The symbols $\neq$, $<$, and $>$ are called _____inequality symbols_____.

2. A mathematical statement that two expressions are equal is called a(n) _____equation_____.

3. The _____absolute value_____ of a number is the distance between that number and 0 on a number line.

4. A symbol used to represent a number is called a(n) _____variable_____.

5. Two numbers that are the same distance from 0 but lie on opposite sides of 0 are called _____opposites_____.

6. The number in a fraction above the fraction bar is called the _____numerator_____.

7. A(n) _____solution_____ of an equation is a value for the variable that makes the equation a true statement.

8. Two numbers whose product is 1 are called _____reciprocals_____.

9. In 2^3, the 2 is called the _____base_____ and the 3 is called the _____exponent_____.

10. The _____numerical coefficient_____ of a term is its numerical factor.

11. The number in a fraction below the fraction bar is called the _____denominator_____.

12. Parentheses and brackets are examples of _____grouping symbols_____.

13. A(n) _____term_____ is a number or the product of a number and variables raised to powers.

14. Terms with the same variables raised to the same powers are called _____like terms_____.

15. If terms are not like terms, then they are _____unlike terms_____.

Teaching Tip

Encourage students to use their notebook along with the Chapter Highlights as a study guide for the test.

> **Helpful Hint**
>
> ● Are you preparing for your test? Don't forget to take the Chapter 1 Test on page 89. Then check your answers at the back of the text and use the Chapter Test Prep Videos to see the fully worked-out solutions to any of the exercises you want to review.

Chapter Highlights

Definitions and Concepts	Examples
Section 1.2 Symbols and Sets of Numbers	

A **set** is a collection of objects, called **elements,** enclosed in braces.	$\{a, c, e\}$ Given the set $\left\{-3.4, \sqrt{3}, 0, \dfrac{2}{3}, 5, -4\right\}$ list the numbers that belong to the set of
Natural numbers: $\{1, 2, 3, 4, \ldots\}$	Natural numbers: 5
Whole numbers: $\{0, 1, 2, 3, 4, \ldots\}$	Whole numbers: 0, 5
Integers: $\{\ldots, -3, -2, -1, 0, 1, 2, 3, \ldots\}$	Integers: $-4, 0, 5$
Rational numbers: {real numbers that can be expressed as quotients of integers}	Rational numbers: $-3.4, 0, \dfrac{2}{3}, 5, -4$
Irrational numbers: {real numbers that cannot be expressed as quotients of integers}	Irrational numbers: $\sqrt{3}$
A line used to picture numbers is called a **number line.**	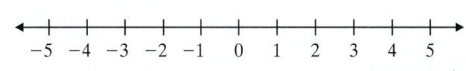

(continued)

Definitions and Concepts	Examples

Section 1.2   Symbols and Sets of Numbers (*continued*)

Real numbers: {all numbers that correspond to points on the number line}

The **absolute value** of a real number a denoted by $|a|$ is the distance between a and 0 on a number line.

Symbols: $=$ is equal to
$\neq$ is not equal to
$>$ is greater than
$<$ is less than
$\leq$ is less than or equal to
$\geq$ is greater than or equal to

Order Property for Real Numbers

For any two real numbers a and b, a is less than b if a is to the left of b on a number line.

Real numbers: $-3.4, \sqrt{3}, 0, \dfrac{2}{3}, 5, -4$

$|5| = 5 \quad |0| = 0 \quad |-2| = 2$

$-7 = -7$
$3 \neq -3$
$4 > 1$
$1 < 4$
$6 \leq 6$
$18 \geq -\dfrac{1}{3}$

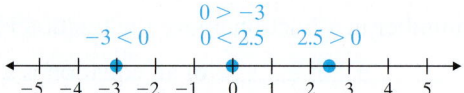
$0 > -3$
$-3 < 0 \qquad 0 < 2.5 \qquad 2.5 > 0$

Section 1.3   Exponents, Order of Operations, and Variable Expressions

The expression a^n is an **exponential expression.** The number a is called the **base;** it is the repeated factor. The number n is called the **exponent;** it is the number of times that the base is a factor.

Order of Operations

1. Perform all operations within grouping symbols first, starting with the innermost set.
2. Evaluate exponential expressions.
3. Multiply or divide in order from left to right.
4. Add or subtract in order from left to right.

A symbol used to represent a number is called a **variable.**

An **algebraic expression** is a collection of numbers, variables, operation symbols, and grouping symbols.

To **evaluate an algebraic expression** containing a variable, substitute a given number for the variable and simplify.

A mathematical statement that two expressions are equal is called an **equation.**

A **solution** of an equation is a value for the variable that makes the equation a true statement.

$4^3 = 4 \cdot 4 \cdot 4 = 64$
$7^2 = 7 \cdot 7 = 49$

$\dfrac{8^2 + 5(7-3)}{3 \cdot 7} = \dfrac{8^2 + 5(4)}{21}$
$= \dfrac{64 + 5(4)}{21}$
$= \dfrac{64 + 20}{21}$
$= \dfrac{84}{21}$
$= 4$

Examples of variables are

q, x, z

Examples of algebraic expressions are

$5x, \quad 2(y-6), \quad \dfrac{q^2 - 3q + 1}{6}$

Evaluate $x^2 - y^2$ when $x = 5$ and $y = 3$.

$x^2 - y^2 = (5)^2 - 3^2$
$= 25 - 9$
$= 16$

Equations:

$3x - 9 = 20$
$A = \pi r^2$

Determine whether 4 is a solution of $5x + 7 = 27$.

$5x + 7 = 27$
$5(4) + 7 \stackrel{?}{=} 27$
$20 + 7 \stackrel{?}{=} 27$
$27 = 27 \quad$ True

4 is a solution.

Definitions and Concepts	**Examples**
Section 1.4 Adding Real Numbers	

To Add Two Numbers with the Same Sign	Add.
1. Add their absolute values.	$10 + 7 = 17$
2. Use their common sign as the sign of the sum.	$-3 + (-8) = -11$
To Add Two Numbers with Different Signs	
1. Subtract their absolute values.	$-25 + 5 = -20$
2. Use the sign of the number whose absolute value is larger as the sign of the sum.	$14 + (-9) = 5$
Two numbers that are the same distance from 0 but lie on opposite sides of 0 are called **opposites** or **additive inverses.** The opposite of a number a is denoted by $-a$.	The opposite of -7 is 7. The opposite of 123 is -123.

Section 1.5 Subtracting Real Numbers	

To subtract two numbers a and b, add the first number a to the opposite of the second number, b. $$a - b = a + (-b)$$	Subtract. $$3 - (-44) = 3 + 44 = 47$$ $$-5 - 22 = -5 + (-22) = -27$$ $$-30 - (-30) = -30 + 30 = 0$$

Section 1.6 Multiplying and Dividing Real Numbers	

Multiplying Real Numbers The product of two numbers with the same sign is a positive number. The product of two numbers with different signs is a negative number.	Multiply. $$7 \cdot 8 = 56 \qquad -7 \cdot (-8) = 56$$ $$-2 \cdot 4 = -8 \qquad 2 \cdot (-4) = -8$$
Products Involving Zero The product of 0 and any number is 0. $$b \cdot 0 = 0 \quad \text{and} \quad 0 \cdot b = 0$$	$$-4 \cdot 0 = 0 \qquad 0 \cdot \left(-\frac{3}{4}\right) = 0$$
Quotient of Two Real Numbers $$\frac{a}{b} = a \cdot \frac{1}{b}$$	Divide. $$\frac{42}{2} = 42 \cdot \frac{1}{2} = 21$$
Dividing Real Numbers The quotient of two numbers with the same sign is a positive number. The quotient of two numbers with different signs is a negative number.	$$\frac{90}{10} = 9 \qquad \frac{-90}{-10} = 9$$ $$\frac{42}{-6} = -7 \qquad \frac{-42}{6} = -7$$
Quotients Involving Zero Let a be a nonzero number. $\dfrac{0}{a} = 0$ and $\dfrac{a}{0}$ is undefined.	$$\frac{0}{18} = 0 \qquad \frac{0}{-47} = 0 \qquad \frac{-85}{0} \text{ is undefined.}$$

Definitions and Concepts	Examples

Section 1.7   Properties of Real Numbers

Commutative Properties

Addition: $a + b = b + a$

Multiplication: $a \cdot b = b \cdot a$

$3 + (-7) = -7 + 3$

$-8 \cdot 5 = 5 \cdot (-8)$

Associative Properties

Addition: $(a + b) + c = a + (b + c)$

Multiplication: $(a \cdot b) \cdot c = a \cdot (b \cdot c)$

$(5 + 10) + 20 = 5 + (10 + 20)$

$(-3 \cdot 2) \cdot 11 = -3 \cdot (2 \cdot 11)$

Two numbers whose product is 1 are called **multiplicative inverses** or **reciprocals.** The reciprocal of a nonzero number a is $\frac{1}{a}$ because $a \cdot \frac{1}{a} = 1$.

The reciprocal of 3 is $\frac{1}{3}$.

The reciprocal of $-\frac{2}{5}$ is $-\frac{5}{2}$.

Distributive Property

$a(b + c) = a \cdot b + a \cdot c$

$5(6 + 10) = 5 \cdot 6 + 5 \cdot 10$

$-2(3 + x) = -2 \cdot 3 + (-2)(x)$

Identities

$a + 0 = a \qquad 0 + a = a$

$a \cdot 1 = a \qquad 1 \cdot a = a$

$5 + 0 = 5 \qquad 0 + (-2) = -2$

$-14 \cdot 1 = -14 \qquad 1 \cdot 27 = 27$

Inverses

Additive or opposite: $a + (-a) = 0$

Multiplicative or reciprocal: $b \cdot \dfrac{1}{b} = 1, \qquad b \neq 0$

$7 + (-7) = 0$

$3 \cdot \dfrac{1}{3} = 1$

Section 1.8   Simplifying Expressions

The **numerical coefficient** of a **term** is its numerical factor.

Term	Numerical Coefficient
$-7y$	-7
x	1
$\dfrac{1}{5}a^2 b$	$\dfrac{1}{5}$

Terms with the same variables raised to exactly the same powers are **like terms.**

Like Terms	Unlike Terms
$12x, -x$	$3y, 3y^2$
$-2xy, 5yx$	$7a^2 b, -2ab^2$

To combine like terms, add the numerical coefficients and multiply the result by the common variable factor.

$9y + 3y = 12y$

$-4z^2 + 5z^2 - 6z^2 = -5z^2$

To remove parentheses, apply the distributive property.

$-4(x + 7) + 10(3x - 1)$

$= -4x - 28 + 30x - 10$

$= 26x - 38$

(1.2) *Insert* <, >, *or* = *in the appropriate space to make each statement true.*

1. $8 < 10$

2. $7 > 2$

3. $-4 > -5$

4. $\dfrac{12}{2} > -8$

5. $|-7| < |-8|$

6. $|-9| > -9$

7. $-|-1| = -1$

8. $|-14| = -(-14)$

9. $1.2 > 1.02$

10. $-\dfrac{3}{2} < -\dfrac{3}{4}$

Translate each statement into symbols.

11. Four is greater than or equal to negative three.
$4 \geq -3$

12. Six is not equal to five. $6 \neq 5$

13. 0.03 is less than 0.3. $0.03 < 0.3$

14. The United States is home to 1729 two-year colleges and 2870 four-year colleges. Write an inequality comparing the numbers 1729 and 2870. (*Source:* National Center for Education Statistics) $1729 < 2870$

Given the sets of numbers below, list the numbers in each set that also belong to the set of:

a. Natural numbers
b. Whole numbers
c. Integers
d. Rational numbers
e. Irrational numbers
f. Real numbers

15. $\left\{ -6, 0, 1, 1\dfrac{1}{2}, 3, \pi, 9.62 \right\}$

    a. $1, 3$ **b.** $0, 1, 3$ **c.** $-6, 0, 1, 3$
    d. $-6, 0, 1, 1\frac{1}{2}, 3, 9.62$ **e.** π **f.** all numbers in set

16. $\left\{ -3, -1.6, 2, 5, \dfrac{11}{2}, 15.1, \sqrt{5}, 2\pi \right\}$

    a. $2, 5$ **b.** $2, 5$ **c.** $-3, 2, 5$ **d.** $-3, -1.6, 2, 5, \frac{11}{2}, 15.1$
    e. $\sqrt{5}, 2\pi$ **f.** all numbers in set

The following chart shows the gains and losses in dollars of Density Oil and Gas stock for a particular week. Use this chart to answer Exercises 17 and 18.

Day	Gain or Loss (in dollars)
Monday	+1
Tuesday	−2
Wednesday	+5
Thursday	+1
Friday	−4

17. Which day showed the greatest loss? Friday

18. Which day showed the greatest gain? Wednesday

(1.3) *Choose the correct answer for each statement.*

19. The expression $6 \cdot 3^2 + 2 \cdot 8$ simplifies to
    a. -52 **b.** 448 **c.** 70 **d.** 64 c

20. The expression $68 - 5 \cdot 2^3$ simplifies to
    a. -232 **b.** 28 **c.** 38 **d.** 504 b

Simplify each expression.

21. $3(1 + 2 \cdot 5) + 4$ 37   **22.** $8 + 3(2 \cdot 6 - 1)$ 41   **23.** $\dfrac{4 + |6 - 2| + 8^2}{4 + 6 \cdot 4}$ $\dfrac{18}{7}$ **24.** $5[3(2 + 5) - 5]$ 80

Translate each word statement to symbols.

25. The difference of twenty and twelve is equal to the product of two and four. $20 - 12 = 2 \cdot 4$

26. The quotient of nine and two is greater than negative five. $\dfrac{9}{2} > -5$

Evaluate each expression when $x = 6$, $y = 2$, and $z = 8$.

27. $2x + 3y$ 18   **28.** $x(y + 2z)$ 108   **29.** $\dfrac{x}{y} + \dfrac{z}{2y}$ 5   **30.** $x^2 - 3y^2$ 24

△ **31.** The expression $180 - a - b$ represents the measure of the unknown angle of the given triangle. Replace a with 37 and b with 80 to find the measure of the unknown angle.   63°

△ **32.** The expression $360 - a - b - c$ represents the measure of the unknown angle of the given quadrilateral. Replace a with 93, b with 80, and c with 82 to find the measure of the unknown angle.   105°

Decide whether the given number is a solution to the given equation.

33. $7x - 3 = 18$;   3   solution

34. $3x^2 + 4 = x - 1$;   1   not a solution

(1.4) *Find the additive inverse or opposite of each number.*

35. -9 9   **36.** $\dfrac{2}{3}$ $-\dfrac{2}{3}$ **37.** $|-2|$ -2 **38.** $-|-7|$ 7

Add.

39. $-15 + 4$ -11 **40.** $-6 + (-11)$ -17 **41.** $\dfrac{1}{16} + \left(-\dfrac{1}{4}\right)$ $-\dfrac{3}{16}$

42. $-8 + |-3|$ -5 **43.** $-4.6 + (-9.3)$ -13.9 **44.** $-2.8 + 6.7$ 3.9

(1.5) *Perform each indicated operation.*

45. $6 - 20$ -14

46. $-3.1 - 8.4$ -11.5

47. $-6 - (-11)$ 5

48. $4 - 15$ -11

49. $-21 - 16 + 3(8 - 2)$ -19

50. $\dfrac{11 - (-9) + 6(8 - 2)}{2 + 3 \cdot 4}$ 4

Evaluate each expression for $x = 3$, $y = -6$, and $z = -9$. Then choose the correct evaluation.

51. $2x^2 - y + z$
   a. 15   **b.** 3   **c.** 27   **d.** -3 a

52. $\dfrac{|y - 4x|}{2x}$
   a. 3   **b.** 1   **c.** -1 **d.** -3 a

53. At the beginning of the week the price of Density Oil and Gas stock from Exercises 17 and 18 is $50 per share. Find the price of a share of stock at the end of the week. $51

54. Find the price of a share of stock by the end of the day on Wednesday. $54

(1.6) *Find each multiplicative inverse or reciprocal.*

55. -6 $-\dfrac{1}{6}$

56. $\dfrac{3}{5}$ $\dfrac{5}{3}$

Simplify each expression.

57. $6(-8)$ -48

58. $(-2)(-14)$ 28

59. $\dfrac{-18}{-6}$ 3

60. $\dfrac{42}{-3}$ -14

61. $-3(-6)(-2)$ -36

62. $(-4)(-3)(0)(-6)$ 0

63. $\dfrac{4(-3) + (-8)}{2 + (-2)}$ undefined

64. $\dfrac{3(-2)^2 - 5}{-14}$ $-\dfrac{1}{2}$

(1.7) *Name the property illustrated in each equation.*

65. $-6 + 5 = 5 + (-6)$
   commutative property of addition

66. $6 \cdot 1 = 6$
   identity element for multiplication

67. $3(8 - 5) = 3 \cdot 8 - 3 \cdot 5$
   distributive property

68. $4 + (-4) = 0$
additive inverse property

69. $2 + (3 + 9) = (2 + 3) + 9$
associative property of addition

70. $2 \cdot 8 = 8 \cdot 2$
commutative property of multiplication

71. $6(8 + 5) = 6 \cdot 8 + 6 \cdot 5$
distributive property

72. $(3 \cdot 8) \cdot 4 = 3 \cdot (8 \cdot 4)$
associative property of multiplication

73. $4 \cdot \dfrac{1}{4} = 1$
multiplicative inverse property

74. $8 + 0 = 8$
identity element for addition

75. $4(8 + 3) = 4(3 + 8)$
commutative property of addition

76. $5(2 + 1) = 5 \cdot 2 + 5 \cdot 1$
distributive property

(1.8) *Simplify each expression.*

77. $5x - x + 2x$ $6x$

78. $0.2z - 4.6z - 7.4z$ $-11.8z$

79. $\dfrac{1}{2}x + 3 + \dfrac{7}{2}x - 5$ $4x - 2$

80. $\dfrac{4}{5}y + 1 + \dfrac{6}{5}y + 2$ $2y + 3$

81. $2(n - 4) + n - 10$ $3n - 18$

82. $3(w + 2) - (12 - w)$ $4w - 6$

83. Subtract $7x - 2$ from $x + 5$.
$-6x + 7$

84. Subtract $1.4y - 3$ from $y - 0.7$.
$-0.4y + 2.3$

Write each phrase as an algebraic expression. Simplify if possible.

85. Three times a number decreased by 7 $3x - 7$

86. Twice the sum of a number and 2.8, added to 3 times the number $5x + 5.6$

Mixed Review

Insert $<, >,$ *or* $=$ *in the space between each pair of numbers.*

87. $-|-11| < |11.4|$

88. $-1\dfrac{1}{2} > -2\dfrac{1}{2}$

Perform the indicated operations.

89. $-7.2 + (-8.1)$ -15.3

90. $14 - 20$ -6

91. $4(-20)$ -80

92. $\dfrac{-20}{4}$ -5

93. $-\dfrac{4}{5}\left(\dfrac{5}{16}\right)$ $-\dfrac{1}{4}$

94. $-0.5(-0.3)$ 0.15

95. $8 \div 2 \cdot 4$ 16

96. $(-2)^4$ 16

97. $\dfrac{-3 - 2(-9)}{-15 - 3(-4)}$ -5

98. $5 + 2[(7 - 5)^2 + (1 - 3)]$ 9

99. $-\dfrac{5}{8} \div \dfrac{3}{4}$ $-\dfrac{5}{6}$

100. $\dfrac{-15 + (-4)^2 + |-9|}{10 - 2 \cdot 5}$
undefined

Remove parentheses and simplify each expression.

101. $7(3x - 3) - 5(x + 4)$
$16x - 41$

102. $8 + 2(9x - 10)$ $18x - 12$

Translate each statement into symbols.

1. The absolute value of negative seven is greater than five.

2. The sum of nine and five is greater than or equal to four.

Simplify each expression.

3. $-13 + 8$

4. $-13 - (-2)$

5. $6 \cdot 3 - 8 \cdot 4$

6. $13(-3)$

7. $(-6)(-2)$

8. $\dfrac{|-16|}{-8}$

9. $\dfrac{-8}{0}$

10. $\dfrac{|-6| + 2}{5 - 6}$

11. $\dfrac{1}{2} - \dfrac{5}{6}$

12. $-1\dfrac{1}{8} + 5\dfrac{3}{4}$

13. $-\dfrac{3}{5} + \dfrac{15}{8}$

14. $3(-4)^2 - 80$

15. $6[5 + 2(3 - 8) - 3]$

16. $\dfrac{-12 + 3 \cdot 8}{4}$

17. $\dfrac{(-2)(0)(-3)}{-6}$

Insert $<$, $>$, or $=$ in the appropriate space to make each statement true.

18. $-3 \quad -7$

19. $4 \quad -8$

20. $|-3| \quad 2$

21. $|-2| \quad -1 - (-3)$

Answers

Note to Instructor: The Chapter 1 Test file in the TestGen program provides algorithms specifically matched to this test for easy replication for practice or assessment purposes.

1. $|-7| > 5$

2. $9 + 5 \geq 4$

3. -5

4. -11

5. -14

6. -39

7. 12

8. -2

9. undefined

10. -8

11. $-\dfrac{1}{3}$

12. $4\dfrac{5}{8}$

13. $\dfrac{51}{40}$

14. -32

15. -48

16. 3

17. 0

18. $>$

19. $>$

20. $>$

21. $=$

22. a. 1, 7

b. 0, 1, 7

c. −5, −1, 0, 1, 7

d. −5, −1, $\frac{1}{4}$, 0, 1, 7, 11.6

e. $\sqrt{7}$, 3π

f. −5, −1, $\frac{1}{4}$, 0, 1, 7, 11.6, $\sqrt{7}$, 3π

23. 40

24. 12

25. 22

26. −1

27. associative property of addition

28. commutative property of multiplication

29. distributive property

30. multiplicative inverse

31. 9

32. −3

33. second down

34. yes

35. 17°

36. $420

37. $y - 10$

38. $5.9x + 1.2$

39. $-2x + 10$

40. $-15y + 1$

---

22. Given $\left\{ -5, -1, \frac{1}{4}, 0, 1, 7, 11.6, \sqrt{7}, 3\pi \right\}$, list the numbers in this set that also belong to the set of:

  a. Natural numbers   **b.** Whole numbers

  c. Integers   **d.** Rational numbers

  e. Irrational numbers   **f.** Real numbers

Evaluate each expression when $x = 6$, $y = -2$, and $z = -3$.

23. $x^2 + y^2$ **24.** $x + yz$ **25.** $2 + 3x - y$ **26.** $\dfrac{y + z - 1}{x}$

Identify the property illustrated by each equation.

27. $8 + (9 + 3) = (8 + 9) + 3$ **28.** $6 \cdot 8 = 8 \cdot 6$

29. $-6(2 + 4) = -6 \cdot 2 + (-6) \cdot 4$ **30.** $\dfrac{1}{6}(6) = 1$

31. Find the opposite of −9.   **32.** Find the reciprocal of $-\dfrac{1}{3}$.

The New Orleans Saints were 22 yards from the goal when the series of gains and losses shown in the chart occurred. Use this chart to answer Exercises 33 and 34.

	Gains and Losses (in yards)
First down	5
Second down	−10
Third down	−2
Fourth down	29

33. During which down did the greatest loss of yardage occur?

34. Was a touchdown scored?

35. The temperature at the Winter Olympics was a frigid 14° below zero in the morning, but by noon it had risen 31°. What was the temperature at noon?

36. A stockbroker decided to sell 280 shares of stock, which decreased in value by $1.50 per share yesterday. How much money did she lose?

Simplify each expression.

37. $2y - 6 - y - 4$ **38.** $2.7x + 6.1 + 3.2x - 4.9$

39. $4(x - 2) - 3(2x - 6)$ **40.** $-5(y + 1) + 2(3 - 5y)$

Equations, Inequalities, and Problem Solving

Top 5 Countries by Number of Internet-Crime Complaints

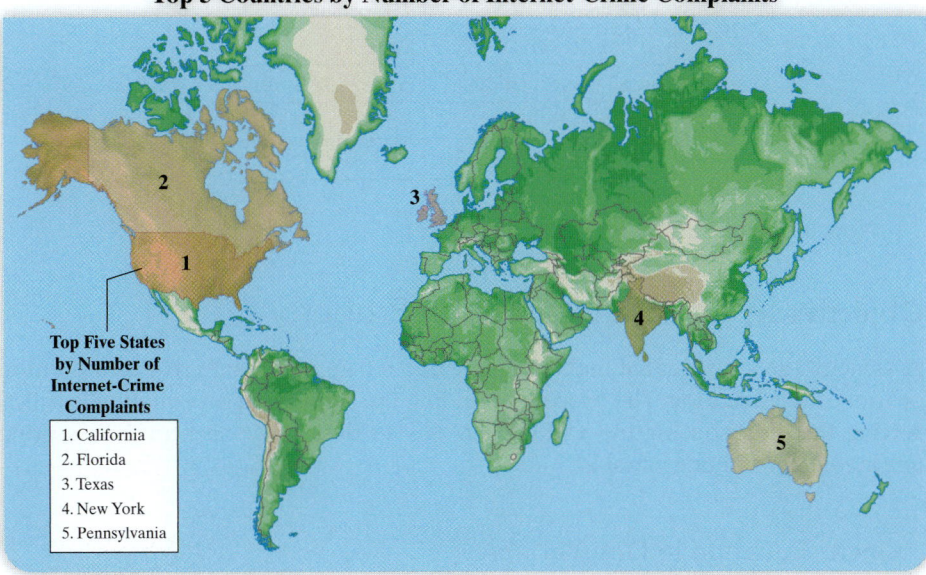

Top Five States by Number of Internet-Crime Complaints
1. California
2. Florida
3. Texas
4. New York
5. Pennsylvania

Internet Crime

The Internet Crime Complaint Center (IC3) is a joint operation between the FBI and the National White-Collar Crime Center. The IC3 receives and refers criminal complaints occurring on the Internet. Of course, nondelivery of merchandise or payment are the highest reported offenses.

In Section 2.6, Exercises 15 and 16, we analyze a bar graph on the yearly number of complaints received by the IC3.

Ages of Persons Filing Complaints

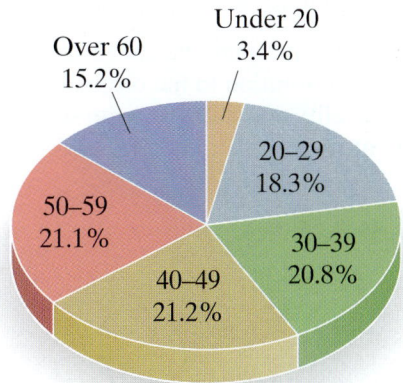

Over 60 15.2%
Under 20 3.4%
20–29 18.3%
30–39 20.8%
40–49 21.2%
50–59 21.1%

In this chapter, we solve equations and inequalities. Once we know how to solve equations and inequalities, we may solve word problems. Of course, problem solving is an integral topic in algebra and its discussion is continued throughout this text.

Objectives

A Use the Addition Property of Equality to Solve Linear Equations.

B Simplify an Equation and Then Use the Addition Property of Equality.

C Write Word Phrases as Algebraic Expressions.

Helpful Hint Simply stated, an equation contains "=" while an expression does not. Also, we *simplify* expressions and *solve* equations.

Let's recall from Section 1.3 the difference between an equation and an expression. A combination of operations on variables and numbers is an expression, and an equation is of the form "expression = expression."

Equations	Expressions
$3x - 1 = -17$	$3x - 1$
area = length · width	$5(20 - 3) + 10$
$8 + 16 = 16 + 8$	y^3
$-9a + 11b = 14b + 3$	$-x^2 + y - 2$

Now, let's concentrate on equations.

Objective **A** Using the Addition Property

A value of the variable that makes an equation a true statement is called a solution or root of the equation. The process of finding the solution of an equation is called **solving** the equation for the variable. In this section, we concentrate on solving *linear equations* in one variable.

Linear Equation in One Variable

A **linear equation in one variable** can be written in the form

$$Ax + B = C$$

where A, B, and C are real numbers and $A \neq 0$.

Evaluating each side of a linear equation for a given value of the variable, as we did in Section 1.3, can tell us whether that value is a solution. But we can't rely on this as our method of solving it—with what value would we start?

Instead, to solve a linear equation in x, we write a series of simpler equations, all *equivalent* to the original equation, so that the final equation has the form

$$x = \text{number} \qquad \text{or} \qquad \text{number} = x$$

Equivalent equations are equations that have the same solution. This means that the "number" above is the solution to the original equation.

The first property of equality that helps us write simpler equivalent equations is the **addition property of equality.**

Addition Property of Equality

Let a, b, and c represent numbers. Then

$$a = b$$
and $a + c = b + c$
are equivalent equations.

Also, $a = b$
and $a - c = b - c$
are equivalent equations.

In other words, **the same number may be added to or subtracted from both sides** of an equation without changing the solution of the equation. (We may subtract the same number from both sides since subtraction is defined in terms of addition.)

Let's visualize how we use the addition property of equality to solve an equation. Picture the equation $x - 2 = 1$ as a balanced scale. The left side of the equation has the same value (weight) as the right side.

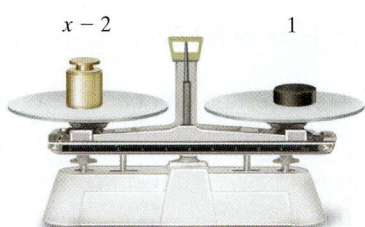

If the same weight is added to each side of a scale, the scale remains balanced. Likewise, if the same number is added to each side of an equation, the left side continues to have the same value as the right side.

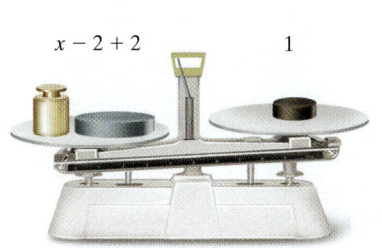

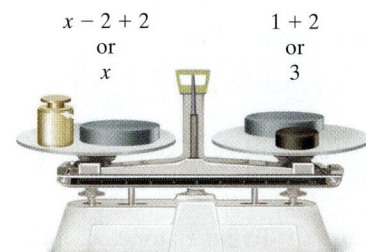

We use the addition property of equality to write equivalent equations until the variable is alone (by itself on one side of the equation) and the equation looks like "x = number" or "number = x."

✓**Concept Check** Use the addition property to fill in the blanks so that the middle equation simplifies to the last equation.

$$x - 5 = 3$$
$$x - 5 + \underline{\hspace{1em}} = 3 + \underline{\hspace{1em}}$$
$$x = 8$$

Example 1 Solve $x - 7 = 10$ for x.

Solution: To solve for x, we first get x alone on one side of the equation. To do this, we add 7 to both sides of the equation.

$$x - 7 = 10$$
$$x - 7 + 7 = 10 + 7 \quad \text{Add 7 to both sides.}$$
$$x = 17 \quad \text{Simplify.}$$

The solution of the equation $x = 17$ is obviously 17.
Since we are writing equivalent equations, the solution of the equation $x - 7 = 10$ is also 17.

Check: To check, replace x with 17 in the original equation.

$$x - 7 = 10 \quad \text{Original equation.}$$
$$17 - 7 \overset{?}{=} 10 \quad \text{Replace } x \text{ with 17.}$$
$$10 = 10 \quad \text{True}$$

Since the statement is true, 17 is the solution,

■ **Work Practice 1**

Practice 1

Solve: $x - 5 = 8$ for x.

Answer
1. $x = 13$

✓**Concept Check Answer**
5

Practice 2

Solve: $y + 1.7 = 0.3$

Example 2 Solve: $y + 0.6 = -1.0$

Solution: To solve for y (get y alone on one side of the equation), we subtract 0.6 from both sides of the equation.

$$y + 0.6 = -1.0$$
$$y + 0.6 - 0.6 = -1.0 - 0.6 \quad \text{Subtract 0.6 from both sides.}$$
$$y = -1.6 \quad \text{Combine like terms.}$$

Check: $y + 0.6 = -1.0$ Original equation.
$$-1.6 + 0.6 \stackrel{?}{=} -1.0 \quad \text{Replace } y \text{ with } -1.6.$$
$$-1.0 = -1.0 \quad \text{True}$$

The solution is -1.6.

■ **Work Practice 2**

Practice 3

Solve: $\dfrac{7}{8} = y - \dfrac{1}{3}$

Example 3 Solve: $\dfrac{1}{2} = x - \dfrac{3}{4}$

Solution: To get x alone, we add $\dfrac{3}{4}$ to both sides.

$$\frac{1}{2} = x - \frac{3}{4}$$
$$\frac{1}{2} + \frac{3}{4} = x - \frac{3}{4} + \frac{3}{4} \quad \text{Add } \frac{3}{4} \text{ to both sides.}$$
$$\frac{1}{2} \cdot \frac{2}{2} + \frac{3}{4} = x \quad \text{The LCD is 4.}$$
$$\frac{2}{4} + \frac{3}{4} = x \quad \text{Add the fractions.}$$
$$\frac{5}{4} = x$$

Check: $\dfrac{1}{2} = x - \dfrac{3}{4}$ Original equation.
$$\frac{1}{2} \stackrel{?}{=} \frac{5}{4} - \frac{3}{4} \quad \text{Replace } x \text{ with } \frac{5}{4}.$$
$$\frac{1}{2} \stackrel{?}{=} \frac{2}{4} \quad \text{Subtract.}$$
$$\frac{1}{2} = \frac{1}{2} \quad \text{True}$$

The solution is $\dfrac{5}{4}$.

■ **Work Practice 3**

> **Helpful Hint**   We may solve an equation so that the variable is alone on *either* side of the equation. For example, $\dfrac{5}{4} = x$ is equivalent to $x = \dfrac{5}{4}$.

Practice 4

Solve: $3x + 10 = 4x$

Example 4 Solve: $5t - 5 = 6t$

Solution: To solve for t, we first want all terms containing t on one side of the equation and numbers on the other side. Notice that if we subtract $5t$ from both sides of the equation, then variable terms will be on one side of the equation and the number -5 will be alone on the other side.

$$5t - 5 = 6t$$
$$5t - 5 - 5t = 6t - 5t \quad \text{Subtract } 5t \text{ from both sides.}$$
$$-5 = t \quad \text{Combine like terms.}$$

Answers

2. $y = -1.4$ **3.** $y = \dfrac{29}{24}$ **4.** $x = 10$

Check:

$$5t - 5 = 6t \quad \text{Original equation.}$$
$$5(-5) - 5 \stackrel{?}{=} 6(-5) \quad \text{Replace } t \text{ with } -5.$$
$$-25 - 5 \stackrel{?}{=} -30$$
$$-30 = -30 \quad \text{True}$$

The solution is -5.

■ **Work Practice 4**

Objective B Simplifying Equations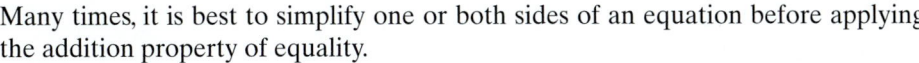

Many times, it is best to simplify one or both sides of an equation before applying the addition property of equality.

Example 5 Solve: $2x + 3x - 5 + 7 = 10x + 3 - 6x - 4$

Solution: First we simplify both sides of the equation.

$$2x + 3x - 5 + 7 = 10x + 3 - 6x - 4$$
$$5x + 2 = 4x - 1 \quad \begin{array}{l}\text{Combine like terms on each} \\ \text{side of the equation.}\end{array}$$

Next, we want all terms with a variable on one side of the equation and all numbers on the other side.

$$5x + 2 - 4x = 4x - 1 - 4x \quad \text{Subtract } 4x \text{ from both sides.}$$
$$x + 2 = -1 \quad \text{Combine like terms.}$$
$$x + 2 - 2 = -1 - 2 \quad \text{Subtract 2 from both sides to get } x \text{ alone.}$$
$$x = -3 \quad \text{Combine like terms.}$$

Check:

$$2x + 3x - 5 + 7 = 10x + 3 - 6x - 4 \quad \text{Original equation.}$$
$$2(-3) + 3(-3) - 5 + 7 \stackrel{?}{=} 10(-3) + 3 - 6(-3) - 4 \quad \text{Replace } x \text{ with } -3.$$
$$-6 - 9 - 5 + 7 \stackrel{?}{=} -30 + 3 + 18 - 4 \quad \text{Multiply.}$$
$$-13 = -13 \quad \text{True}$$

The solution is -3.

■ **Work Practice 5**

If an equation contains parentheses, we use the distributive property to remove them, as before. Then we combine any like terms.

Example 6 Solve: $6(2a - 1) - (11a + 6) = 7$

Solution:

$$6(2a - 1) - 1(11a + 6) = 7$$
$$6(2a) + 6(-1) - 1(11a) - 1(6) = 7 \quad \text{Apply the distributive property.}$$
$$12a - 6 - 11a - 6 = 7 \quad \text{Multiply.}$$
$$a - 12 = 7 \quad \text{Combine like terms.}$$
$$a - 12 + 12 = 7 + 12 \quad \text{Add 12 to both sides.}$$
$$a = 19 \quad \text{Simplify.}$$

Check: Check by replacing a with 19 in the original equation.

■ **Work Practice 6**

Practice 5

Solve:
$$10w + 3 - 4w + 4 = -2w + 3 + 7w$$

Practice 6

Solve:
$$3(2w - 5) - (5w + 1) = -3$$

Answers

5. $w = -4$ **6.** $w = 13$

Practice 7

Solve: $12 - y = 9$

Example 7 Solve: $3 - x = 7$

Solution: First we subtract 3 from both sides.

$$3 - x = 7$$
$$3 - x - 3 = 7 - 3 \quad \text{Subtract 3 from both sides.}$$
$$-x = 4 \quad \text{Simplify.}$$

We have not yet solved for x since x is not alone. However, this equation does say that the opposite of x is 4. If the opposite of x is 4, then x is the opposite of 4, or $x = -4$.

If $\quad -x = 4,$

then $\quad x = -4.$

Check: $\quad 3 - x = 7 \quad$ Original equation.

$$3 - (-4) \overset{?}{=} 7 \quad \text{Replace } x \text{ with } -4.$$
$$3 + 4 \overset{?}{=} 7 \quad \text{Add.}$$
$$7 = 7 \quad \text{True}$$

The solution is -4.

■ **Work Practice 7**

Teaching Tip

After solving Example 7, you may want to point out that there is more than one way to solve this problem.

$$3 - x = 7$$
$$3 - x + x = 7 + x$$
$$3 = 7 + x$$
$$3 - 7 = 7 + x - 7$$
$$-4 = x$$

Objective C Writing Algebraic Expressions

In this section, we continue to practice writing algebraic expressions.

Practice 8

a. The sum of two numbers is 11. If one number is 4, find the other number.

b. The sum of two numbers is 11. If one number is x, write an expression representing the other number.

c. The sum of two numbers is 56. If one number is a, write an expression representing the other number.

Example 8

a. The sum of two numbers is 8. If one number is 3, find the other number.

b. The sum of two numbers is 8. If one number is x, write an expression representing the other number.

Solution:

a. If the sum of two numbers is 8 and one number is 3, we find the other number by subtracting 3 from 8. The other number is $8 - 3$, or 5.

b. If the sum of two numbers is 8 and one number is x, we find the other number by subtracting x from 8. The other number is represented by $8 - x$.

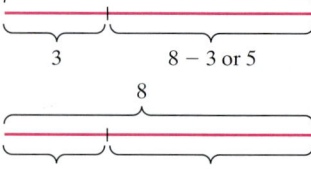

■ **Work Practice 8**

Practice 9

In a recent House of Representatives race in California, Mike Thompson received 100,445 more votes than Zane Starkewolf. If Zane received n votes, how many did Mike receive? (*Source:* Voter News Service)

Example 9 The Verrazano-Narrows Bridge in New York City is the longest suspension bridge in North America. The Golden Gate Bridge in San Francisco is 60 feet shorter than the Verrazano-Narrows Bridge. If the length of the Verrazano-Narrows Bridge is m feet, express the length of the Golden Gate Bridge as an algebraic expression in m. (*Source:* Survey of State Highway Engineers)

Answers

7. $y = 3$ **8. a.** $11 - 4$ or 7 **b.** $11 - x$
c. $56 - a$ **9.** $(n + 100,445)$ votes

Solution: Since the Golden Gate Bridge is 60 feet shorter than the Verrazano-Narrows Bridge, we have that its length is

In words:	Length of Verrazano-Narrows Bridge	minus	60
Translate:	m	$-$	60

The Golden Gate Bridge is $(m - 60)$ feet long.

🟧 **Work Practice 9**

Vocabulary, Readiness & Video Check

Use the choices below to fill in each blank. Some choices may be used more than once or not at all.

equation	multiplication	addition
expression	solution	equivalent

1. A combination of operations on variables and numbers is called a(n) _expression_.
2. A statement of the form "expression = expression" is called a(n) _equation_.
3. A(n) _equation_ contains an equal sign ($=$).
4. A(n) _expression_ does not contain an equal sign ($=$).
5. A(n) _expression_ may be simplified and evaluated while a(n) _equation_ may be solved.
6. A(n) _solution_ of an equation is a number that when substituted for a variable makes the equation a true statement.
7. _Equivalent_ equations have the same solution.
8. By the _addition_ property of equality, the same number may be added to or subtracted from both sides of an equation without changing the solution of the equation.

Solve each equation mentally. See Examples 1 and 2.

9. $x + 4 = 6$ 2
10. $x + 7 = 17$ 10
11. $n + 18 = 30$ 12
12. $z + 22 = 40$ 18
13. $b - 11 = 6$ 17
14. $d - 16 = 5$ 21

Martin-Gay Interactive Videos

See Video 2.1

See video answer section.

Watch the section lecture video and answer the following questions.

Objective A 15. Complete this statement based on the lecture given before 📱 Example 1. The addition property of equality means that if we have an equation, we can add the same real number to _____ of an equation and have an equivalent equation. ▶

Objective B 16. After both sides of 📱 Example 5 are simplified, write down the simplified equation. ▶

Objective C 17. Suppose we were to solve 📱 Example 8 again, this time letting the area of the Sahara Desert be x square miles. Use this to express the area of the Gobi Desert as an algebraic expression in x. ▶

2.1 Exercise Set MyMathLab®

Objective **A** *Solve each equation. Check each solution. See Examples 1 through 4.*

1. $x + 7 = 10$ 3

2. $x + 14 = 25$ 11

3. $x - 2 = -4$ −2

4. $y - 9 = 1$ 10

5. $-11 = 3 + x$ −14

6. $-8 = 8 + z$ −16

7. $r - 8.6 = -8.1$ 0.5

8. $t - 9.2 = -6.8$ 2.4

9. $x - \dfrac{2}{5} = -\dfrac{3}{20}$ $\dfrac{1}{4}$

10. $y - \dfrac{4}{7} = -\dfrac{3}{14}$ $\dfrac{5}{14}$

11. $\dfrac{1}{3} + f = \dfrac{3}{4}$ $\dfrac{5}{12}$

12. $c + \dfrac{1}{6} = \dfrac{3}{8}$ $\dfrac{5}{24}$

Objective **B** *Solve each equation. Don't forget to first simplify each side of the equation, if possible. Check each solution. See Examples 5 through 7.*

13. $7x + 2x = 8x - 3$ −3

14. $3n + 2n = 7 + 4n$ 7

15. $\dfrac{5}{6}x + \dfrac{1}{6}x = -9$ −9

16. $\dfrac{13}{11}y - \dfrac{2}{11}y = -3$ −3

17. $2y + 10 = 5y - 4y$ −10

18. $4x - 4 = 10x - 7x$ 4

19. $-5(n - 2) = 8 - 4n$ 2

20. $-4(z - 3) = 2 - 3z$ 10

21. $\dfrac{3}{7}x + 2 = -\dfrac{4}{7}x - 5$ −7

22. $\dfrac{1}{5}x - 1 = -\dfrac{4}{5}x - 13$ −12

23. $5x - 6 = 6x - 5$ −1

24. $2x + 7 = x - 10$ −17

25. $8y + 2 - 6y = 3 + y - 10$ −9

26. $4p - 11 - p = 2 + 2p - 20$ −7

27. $-3(x - 4) = -4x$ −12

28. $-2(x - 1) = -3x$ −2

29. $\dfrac{3}{8}x - \dfrac{1}{6} = -\dfrac{5}{8}x - \dfrac{2}{3}$ $-\dfrac{1}{2}$

30. $\dfrac{2}{5}x - \dfrac{1}{12} = -\dfrac{3}{5}x - \dfrac{3}{4}$ $-\dfrac{2}{3}$

31. $2(x - 4) = x + 3$ 11

32. $3(y + 7) = 2y - 5$ −26

33. $3(n - 5) - (6 - 2n) = 4n$ 21

34. $5(3 + z) - (8z + 9) = -4z$ −6

35. $-2(x + 6) + 3(2x - 5) = 3(x - 4) + 10$ 25

36. $-5(x + 1) + 4(2x - 3) = 2(x + 2) - 8$ 13

Objectives **A** **B** **Mixed Practice** *Solve. See Examples 1 through 7.*

37. $13x - 3 = 14x$ −3

38. $18x - 9 = 19x$ −9

39. $5b - 0.7 = 6b$ −0.7

40. $9x + 5.5 = 10x$ 5.5

41. $3x - 6 = 2x + 5$ 11

42. $7y + 2 = 6y + 2$ 0

43. $13x - 9 + 2x - 5 = 12x - 1 + 2x$ 13

44. $15x + 20 - 10x - 9 = 25x + 8 - 21x - 7$ −10

45. $7(6 + w) = 6(2 + w)$ −30

46. $6(5 + c) = 5(c - 4)$ −50

47. $n + 4 = 3.6$ −0.4

48. $m + 2 = 7.1$ 5.1

49. $10 - (2x - 4) = 7 - 3x$ −7

50. $15 - (6 - 7k) = 2 + 6k$ −7

51. $\dfrac{1}{3} = x + \dfrac{2}{3}$ $-\dfrac{1}{3}$

52. $\dfrac{1}{11} = y + \dfrac{10}{11}$ $-\dfrac{9}{11}$

53. $-6.5 - 4x - 1.6 - 3x = -6x + 9.8$ −17.9

54. $-1.4 - 7x - 3.6 - 2x = -8x + 4.4$ −9.4

Objective C *Write each algebraic expression described. See Examples 8 and 9.*

55. Two numbers have a sum of 20. If one number is *p*, express the other number in terms of *p*. $20 - p$

56. Two numbers have a sum of 13. If one number is *y*, express the other number in terms of *y*. $13 - y$

57. A 10-foot board is cut into two pieces. If one piece is *x* feet long, express the other length in terms of *x*. $(10 - x)$ ft

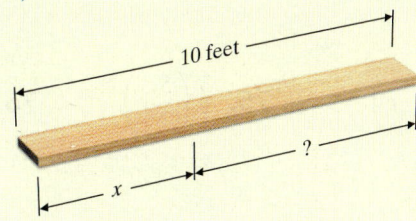

58. A 5-foot piece of string is cut into two pieces. If one piece is *x* feet long, express the other length in terms of *x*. $(5 - x)$ ft

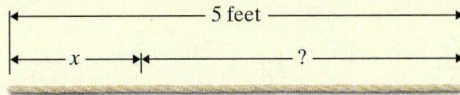

△ **59.** Recall that two angles are *supplementary* if their sum is 180°. If one angle measures *x*°, express the measure of its supplement in terms of *x*. $(180 - x)°$

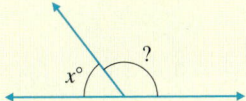

△ **60.** Recall that two angles are *complementary* if their sum is 90°. If one angle measures *x*°, express the measure of its complement in terms of *x*. $(90 - x)°$

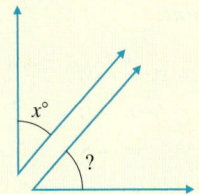

61. In 2013, the number of graduate students at the University of Texas at Austin was approximately 29,000 fewer than the number of undergraduate students. If the number of undergraduate students was *n*, how many graduate students attend UT Austin? (*Source:* University of Texas at Austin) $n - 29,000$

62. The longest interstate highway in the U.S. is I-90, which connects Seattle, Washington, and Boston, Massachusetts. The second longest interstate highway, I-80 (connecting San Francisco, California, and Teaneck, New Jersey), is 121 miles shorter than I-90. If the length of I-80 is *m* miles, express the length of I-90 as an algebraic expression in *m*. (*Source:* U.S. Department of Transportation—Federal Highway Administration) $(m + 121)$ mi

63. The area of the Sahara Desert in Africa is 7 times the area of the Gobi Desert in Asia. If the area of the Gobi Desert is *x* square miles, express the area of the Sahara Desert as an algebraic expression in *x*. $7x$ sq mi

64. The largest meteorite in the world is the Hoba West located in Namibia. Its weight is 3 times the weight of the Armanty meteorite located in Outer Mongolia. If the weight of the Armanty meteorite is *y* kilograms, express the weight of the Hoba West meteorite as an algebraic expression in *y*. $3y$ kg

Review

Find each multiplicative inverse or reciprocal. See Section 1.6.

65. $\dfrac{5}{8}$ $\dfrac{8}{5}$

66. $\dfrac{7}{6}$ $\dfrac{6}{7}$

67. 2 $\dfrac{1}{2}$

68. 5 $\dfrac{1}{5}$

69. $-\dfrac{1}{9}$ -9

70. $-\dfrac{3}{5}$ $-\dfrac{5}{3}$

Perform each indicated operation and simplify. See Sections 1.6 and 1.7.

71. $\dfrac{3x}{3}$ x

72. $\dfrac{-2y}{-2}$ y

73. $-5\left(-\dfrac{1}{5}y\right)$ y

74. $7\left(\dfrac{1}{7}r\right)$ r

75. $\dfrac{3}{5}\left(\dfrac{5}{3}x\right)$ x

76. $\dfrac{9}{2}\left(\dfrac{2}{9}x\right)$ x

Concept Extensions

77. Write two terms whose sum is $-3x$.
answers may vary

78. Write four terms whose sum is $2y - 6$.
answers may vary

Use the addition property to fill in the blank so that the middle equation simplifies to the last equation. See the Concept Check in this section.

79. $x - 4 = -9$
$x - 4 + (\quad) = -9 + (\quad)$
$x = -5$ 4

80. $a + 9 = 15$
$a + 9 + (\quad) = 15 + (\quad)$
$a = 6$ -9

Fill in the blanks with numbers of your choice so that each equation has the given solution. Note: Each blank will be replaced with a different number.

81. _____ $+ x =$ _____; Solution: -3
answers may vary

82. $x -$ _____ $=$ _____; Solution: -10
answers may vary

Solve.

△ **83.** The sum of the angles of a triangle is $180°$. If one angle of a triangle measures $x°$ and a second angle measures $(2x + 7)°$, express the measure of the third angle in terms of x. Simplify the expression. $(173 - 3x)°$

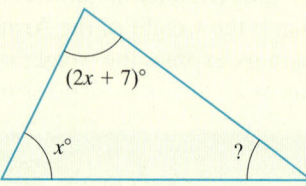

△ **84.** A quadrilateral is a four-sided figure (like the one shown in the figure) whose angle sum is $360°$. If one angle measures $x°$, a second angle measures $3x°$, and a third angle measures $5x°$, express the measure of the fourth angle in terms of x. Simplify the expression. $(360 - 9x)°$

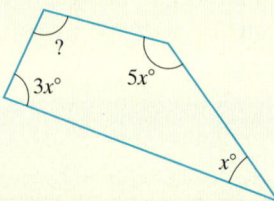

✎ **85.** In your own words, explain what is meant by the solution of an equation.   answers may vary

✎ **86.** In your own words, explain how to check a solution of an equation.   answers may vary

Use a calculator to determine the solution of each equation.

▦ **87.** $36.766 + x = -108.712$ -145.478

▦ **88.** $-85.325 = x - 97.985$ 12.66

2.2 The Multiplication Property of Equality

Objective A Using the Multiplication Property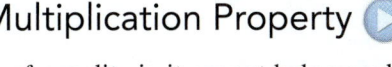

As useful as the addition property of equality is, it cannot help us solve every type of linear equation in one variable. For example, adding or subtracting a value on both sides of the equation does not help solve

$$\frac{5}{2}x = 15$$

because the variable x is being multiplied by a number (other than 1). Instead, we apply another important property of equality, the **multiplication property of equality.**

Objectives

A Use the Multiplication Property of Equality to Solve Linear Equations.

B Use Both the Addition and Multiplication Properties of Equality to Solve Linear Equations.

C Write Word Phrases as Algebraic Expressions.

Multiplication Property of Equality

Let a, b, and c represent numbers and let $c \neq 0$. Then

$a = b$	Also, $a = b$
and $a \cdot c = b \cdot c$	and $\dfrac{a}{c} = \dfrac{b}{c}$
are equivalent equations.	are equivalent equations.

In other words, **both sides** of an equation **may be multiplied or divided by the same nonzero number** without changing the solution of the equation. (We may divide both sides by the same nonzero number since division is defined in terms of multiplication.)

Picturing again our balanced scale, if we multiply or divide the weight on each side by the same nonzero number, the scale (or equation) remains balanced.

$2x$ 6 $\frac{2x}{2}$ or x $\frac{6}{2}$ or 3

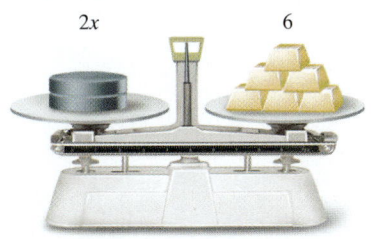 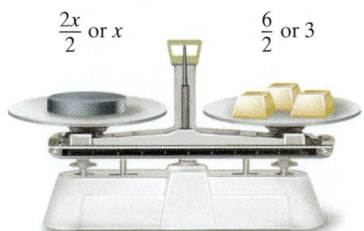

Example 1   Solve: $\dfrac{5}{2}x = 15$

Solution: To get x alone, we multiply both sides of the equation by the reciprocal (or multiplicative inverse) of $\dfrac{5}{2}$, which is $\dfrac{2}{5}$.

$$\frac{5}{2}x = 15$$

$$\frac{2}{5} \cdot \left(\frac{5}{2}x\right) = \frac{2}{5} \cdot 15 \quad \text{Multiply both sides by } \frac{2}{5}.$$

$$\left(\frac{2}{5} \cdot \frac{5}{2}\right)x = \frac{2}{5} \cdot 15 \quad \text{Apply the associative property.}$$

$$1x = 6 \quad \text{Simplify.}$$

or

$$x = 6$$

(*Continued on next page*)

Practice 1

Solve: $\dfrac{3}{7}x = 9$

Answer

1. $x = 21$

Check: Replace x with 6 in the original equation.

$$\frac{5}{2}x = 15 \quad \text{Original equation.}$$

$$\frac{5}{2}(6) \stackrel{?}{=} 15 \quad \text{Replace } x \text{ with 6.}$$

$$15 = 15 \quad \text{True}$$

The solution is 6.

■ **Work Practice 1**

In the equation $\frac{5}{2}x = 15$, $\frac{5}{2}$ is the coefficient of x. When the coefficient of x is a *fraction*, we will get x alone by multiplying by the reciprocal. When the coefficient of x is an integer or a decimal, it is usually more convenient to divide both sides by the coefficient. (Dividing by a number is, of course, the same as multiplying by the reciprocal of the number.)

Practice 2

Solve: $7x = 42$

Example 2   Solve: $5x = 30$

Solution: To get x alone, we divide both sides of the equation by 5, the coefficient of x.

$$5x = 30$$

$$\frac{5x}{5} = \frac{30}{5} \quad \text{Divide both sides by 5.}$$

$$1 \cdot x = 6 \quad \text{Simplify.}$$

$$x = 6$$

Check: $5x = 30$ Original equation.

$$5 \cdot 6 \stackrel{?}{=} 30 \quad \text{Replace } x \text{ with 6.}$$

$$30 = 30 \quad \text{True.}$$

The solution is 6.

■ **Work Practice 2**

Practice 3

Solve: $-4x = 52$

Example 3   Solve: $-3x = 33$

Solution: Recall that $-3x$ means $-3 \cdot x$. To get x alone, we divide both sides by the coefficient of x, that is, -3.

$$-3x = 33$$

$$\frac{-3x}{-3} = \frac{33}{-3} \quad \text{Divide both sides by } -3.$$

$$1x = -11 \quad \text{Simplify.}$$

$$x = -11$$

Check: $-3x = 33$ Original equation.

$$-3(-11) \stackrel{?}{=} 33 \quad \text{Replace } x \text{ with } -11.$$

$$33 = 33 \quad \text{True}$$

The solution is -11.

■ **Work Practice 3**

Teaching Tip

Before beginning Example 4, ask students: If one-seventh of a restaurant bill is $20.00, how much is the entire bill? ($140.00) Discuss how students get their answer.

Recall that 7 and $\frac{1}{7}$ are reciprocals.

Answers

2. $x = 6$ **3.** $x = -13$

Example 4   Solve: $\dfrac{y}{7} = 20$

Solution: Recall that $\dfrac{y}{7} = \dfrac{1}{7}y$. To get y alone, we multiply both sides of the equation by 7, the reciprocal of $\dfrac{1}{7}$.

$$\dfrac{y}{7} = 20$$

$$\dfrac{1}{7}y = 20$$

$$7 \cdot \dfrac{1}{7}y = 7 \cdot 20 \quad \text{Multiply both sides by 7.}$$

$$1y = 140 \quad \text{Simplify.}$$

$$y = 140$$

Check: $\dfrac{y}{7} = 20$ Original equation.

$$\dfrac{140}{7} \stackrel{?}{=} 20 \quad \text{Replace } y \text{ with 140.}$$

$$20 = 20 \quad \text{True}$$

The solution is 140.

■ **Work Practice 4**

Practice 4

Solve: $\dfrac{y}{5} = 13$

Example 5   Solve: $3.1x = 4.96$

Solution: $3.1x = 4.96$

$$\dfrac{3.1x}{3.1} = \dfrac{4.96}{3.1} \quad \text{Divide both sides by 3.1.}$$

$$1x = 1.6 \quad \text{Simplify.}$$

$$x = 1.6$$

Check: Check by replacing x with 1.6 in the original equation. The solution is 1.6.

■ **Work Practice 5**

Practice 5

Solve: $2.6x = 13.52$

Example 6   Solve: $-\dfrac{2}{3}x = -\dfrac{5}{2}$

Solution: To get x alone, we multiply both sides of the equation by $-\dfrac{3}{2}$, the reciprocal of the coefficient of x.

$$-\dfrac{2}{3}x = -\dfrac{5}{2}$$

$$-\dfrac{3}{2} \cdot -\dfrac{2}{3}x = -\dfrac{3}{2} \cdot -\dfrac{5}{2} \quad \text{Multiply both sides by } -\dfrac{3}{2}, \text{ the reciprocal of } -\dfrac{2}{3}.$$

$$x = \dfrac{15}{4} \quad \text{Simplify.}$$

Check: Check by replacing x with $\dfrac{15}{4}$ in the original equation. The solution is $\dfrac{15}{4}$.

■ **Work Practice 6**

Practice 6

Solve: $-\dfrac{5}{6}y = -\dfrac{3}{5}$

Answers

4. $y = 65$ **5.** $x = 5.2$ **6.** $y = \dfrac{18}{25}$

Objective B Using Both the Addition and Multiplication Properties ▶

We are now ready to combine the skills learned in the last section with the skills learned in this section to solve equations by applying more than one property.

Practice 7

Solve: $-x + 7 = -12$

Example 7   Solve: $-z - 4 = 6$

Solution: First, let's get $-z$, the term containing the variable, alone. To do so, we add 4 to both sides of the equation.

$$-z - 4 + 4 = 6 + 4 \quad \text{Add 4 to both sides.}$$
$$-z = 10 \quad \text{Simplify.}$$

Next, recall that $-z$ means $-1 \cdot z$. Thus to get z alone, we either multiply or divide both sides of the equation by -1. In this example, we divide.

$$-z = 10$$
$$\frac{-z}{-1} = \frac{10}{-1} \quad \text{Divide both sides by the coefficient } -1.$$
$$1z = -10 \quad \text{Simplify.}$$
$$z = -10$$

Check: $-z - 4 = 6$ Original equation.
$$-(-10) - 4 \overset{?}{=} 6 \quad \text{Replace } z \text{ with } -10.$$
$$10 - 4 \overset{?}{=} 6$$
$$6 = 6 \quad \text{True}$$

The solution is -10.

◼ **Work Practice 7**

Don't forget to first simplify one or both sides of an equation, if possible.

Practice 8

Solve:
$-7x + 2x + 3 - 20 = -2$

Example 8   Solve: $a + a - 10 + 7 = -13$

Solution: First, we simplify the left side of the equation by combining like terms.

$$a + a - 10 + 7 = -13$$
$$2a - 3 = -13 \quad \text{Combine like terms.}$$
$$2a - 3 + 3 = -13 + 3 \quad \text{Add 3 to both sides.}$$
$$2a = -10 \quad \text{Simplify.}$$
$$\frac{2a}{2} = \frac{-10}{2} \quad \text{Divide both sides by 2.}$$
$$a = -5 \quad \text{Simplify.}$$

Check: To check, replace a with -5 in the original equation. The solution is -5.

◼ **Work Practice 8**

Practice 9

Solve: $10x - 4 = 7x + 14$

Example 9   Solve: $7x - 3 = 5x + 9$

Solution: To get x alone, let's first use the addition property to get variable terms on one side of the equation and numbers on the other side. One way to get variable terms on one side is to subtract $5x$ from both sides.

$$7x - 3 = 5x + 9$$
$$7x - 3 - 5x = 5x + 9 - 5x \quad \text{Subtract } 5x \text{ from both sides.}$$
$$2x - 3 = 9 \quad \text{Simplify.}$$

Answers

7. $x = 19$ **8.** $x = -3$ **9.** $x = 6$

Now, to get numbers on the other side, let's add 3 to both sides.

$$2x - 3 + 3 = 9 + 3 \quad \text{Add 3 to both sides.}$$
$$2x = 12 \quad \text{Simplify.}$$

Use the multiplication property to get x alone.

$$\frac{2x}{2} = \frac{12}{2} \quad \text{Divide both sides by 2.}$$
$$x = 6 \quad \text{Simplify.}$$

Check: To check, replace x with 6 in the original equation to see that a true statement results. The solution is 6.

◼ **Work Practice 9**

If an equation has parentheses, don't forget to use the distributive property to remove them. Then combine any like terms.

Example 10 Solve: $5(2x + 3) = -1 + 7$

Solution:

$$5(2x + 3) = -1 + 7$$
$$5(2x) + 5(3) = -1 + 7 \quad \text{Apply the distributive property.}$$
$$10x + 15 = 6 \quad \text{Multiply and write } -1 + 7 \text{ as 6.}$$
$$10x + 15 - 15 = 6 - 15 \quad \text{Subtract 15 from both sides.}$$
$$10x = -9 \quad \text{Simplify.}$$
$$\frac{10x}{10} = -\frac{9}{10} \quad \text{Divide both sides by 10.}$$
$$x = -\frac{9}{10} \quad \text{Simplify.}$$

Check: To check, replace x with $-\frac{9}{10}$ in the original equation to see that a true statement results. The solution is $-\frac{9}{10}$.

◼ **Work Practice 10**

Objective C Writing Algebraic Expressions

We continue to sharpen our problem-solving skills by writing algebraic expressions.

Example 11 Writing an Expression for Consecutive Integers

If x is the first of three consecutive integers, express the sum of the three integers in terms of x. Simplify if possible.

Solution: An example of three consecutive integers is 7, 8, and 9.

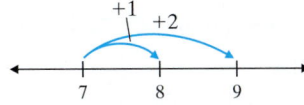

(*Continued on next page*)

Practice 10

Solve: $4(3x - 2) = -1 + 4$

Practice 11

a. If x is the first of two consecutive integers, express the sum of the two integers in terms of x. Simplify if possible.

b. If x is the first of two consecutive odd integers (see next page), express the sum of the two integers in terms of x. Simplify if possible.

Answers

10. $x = \dfrac{11}{12}$ **11. a.** $2x + 1$ **b.** $2x + 2$

The second consecutive integer is always 1 more than the first, and the third consecutive integer is 2 more than the first. If x is the first of three consecutive integers, the three consecutive integers are $x, x + 1,$ and $x + 2$.

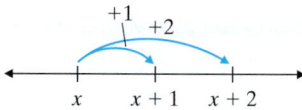

Their sum is shown below.

In words:	first integer	+	second integer	+	third integer
Translate:	x	+	$(x + 1)$	+	$(x + 2)$

This simplifies to $3x + 3$.

■ **Work Practice 11**

Study these examples of consecutive even and consecutive odd integers.

Consecutive even integers:

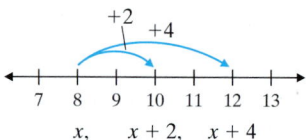

Consecutive odd integers:

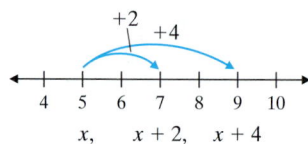

Helpful Hint

If x is an odd integer, then $x + 2$ is the next odd integer. This 2 simply means that odd integers are always 2 units from each other.

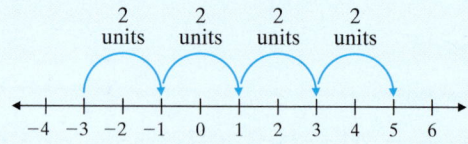

Vocabulary, Readiness & Video Check

Use the choices below to fill in each blank. Some choices may be used more than once. Many of these exercises contain an important review of Section 2.1 also.

equation	multiplication	addition
expression	solution	equivalent

1. By the ___multiplication___ property of equality, both sides of an equation may be multiplied or divided by the same nonzero number without changing the solution of the equation.

2. By the ___addition___ property of equality, the same number may be added to or subtracted from both sides of an equation without changing the solution of the equation.

3. A(n) _____equation_____ may be solved while a(n) _____expression_____ may be simplified and evaluated.

4. A(n) _____equation_____ contains an equal sign (=) while a(n) _____expression_____ does not.

5. _____Equivalent_____ equations have the same solution.

6. A(n) _____solution_____ of an equation is a number that when substituted for the variable makes the equation a true statement.

Solve each equation mentally. See Examples 2 and 3.

7. $3a = 27$ **8.** $9c = 54$ **9.** $5b = 10$ **10.** $7t = 14$ **11.** $6x = -30$ **12.** $8r = -64$
     9   6   2   2   -5   -8

Martin-Gay Interactive Videos   *Watch the section lecture video and answer the following questions.*

Objective A **13.** Complete this statement based on the lecture given before ⊞ Example 1. We can multiply both sides of an equation by the _____ nonzero number and have an equivalent equation.

Objective B **14.** Both the addition and multiplication properties of equality are used to solve ⊞ Examples 4–6. In each of these exercises, what property is applied first? What property is applied last? What conclusion, if any, can you make? ▶

Objective C **15.** Let x be the first of four consecutive integers, as in ⊞ Example 8. Now express the sum of the second integer and the fourth integer as an algebraic expression containing x. ▶

See Video 2.2

See video answer section.

2.2 **Exercise Set** MyMathLab® ▶

Objective A *Solve each equation. Check each solution. See Examples 1 through 6.*

▶**1.** $-5x = -20$ 4   **2.** $-7x = -49$ 7   **3.** $3x = 0$ 0   **4.** $2x = 0$ 0

5. $-x = -12$ 12   **6.** $-y = 8$ -8   ▶**7.** $\frac{2}{3}x = -8$ -12   **8.** $\frac{3}{4}n = -15$ -20

9. $\frac{1}{6}d = \frac{1}{2}$ 3   **10.** $\frac{1}{8}v = \frac{1}{4}$ 2   **11.** $\frac{a}{2} = 1$ 2   ▶**12.** $\frac{d}{15} = 2$ 30

13. $\frac{k}{-7} = 0$ 0   **14.** $\frac{f}{-5} = 0$ 0   **15.** $1.7x = 10.71$ 6.3   **16.** $8.5y = 19.55$ 2.3

Objective B *Solve each equation. Check each solution. See Examples 7 and 8.*

17. $2x - 4 = 16$ 10   **18.** $3x - 1 = 26$ 9   **19.** $-x + 2 = 22$ -20   **20.** $-x + 4 = -24$ 28

▶**21.** $6a + 3 = 3$ 0   **22.** $8t + 5 = 5$ 0   **23.** $\frac{x}{3} - 2 = -5$ -9   **24.** $\frac{b}{4} - 1 = -7$ -24

25. $6z - 8 - z + 3 = 0$
1

26. $4a + 1 + a - 11 = 0$
2

27. $1 = 0.4x - 0.6x - 5$
-30

28. $19 = 0.4x - 0.9x - 6$
-50

29. $\frac{2}{3}y - 11 = -9$ 3

30. $\frac{3}{5}x - 14 = -8$ 10

31. $\frac{3}{4}t - \frac{1}{2} = \frac{1}{3}$ $\frac{10}{9}$

32. $\frac{2}{7}z - \frac{1}{5} = \frac{1}{2}$ $\frac{49}{20}$

Solve each equation. See Examples 9 and 10.

▶ **33.** $8x + 20 = 6x + 18$
-1

34. $11x + 13 = 9x + 9$
-2

35. $3(2x + 5) = -18 + 9$
-4

36. $2(4x + 1) = -12 + 6$
-1

37. $2x - 5 = 20x + 4$
$-\frac{1}{2}$

38. $6x - 4 = -2x - 10$
$-\frac{3}{4}$

39. $2 + 14 = -4(3x - 4)$
0

40. $8 + 4 = -6(5x - 2)$
0

41. $-6y - 3 = -5y - 7$ 4

42. $-17z - 4 = -16z - 20$ 16

43. $\frac{1}{2}(2x - 1) = -\frac{1}{7} - \frac{3}{7}$ $-\frac{1}{14}$

44. $\frac{1}{3}(3x - 1) = -\frac{1}{10} - \frac{2}{10}$ $\frac{1}{30}$

▶ **45.** $-10z - 0.5 = -20z + 1.6$ 0.21

46. $-14y - 1.8 = -24y + 3.9$ 0.57

47. $-4x + 20 = 4x - 20$ 5

48. $-3x + 15 = 3x - 15$ 5

Objectives Ⓐ Ⓑ **Mixed Practice** *See Examples 1 through 10.*

49. $42 = 7x$ 6

50. $81 = 3x$ 27

51. $4.4 = -0.8x$ -5.5

52. $6.3 = -0.6x$ -10.5

53. $6x + 10 = -20$ -5

54. $10y + 15 = -5$ -2

55. $5 - 0.3k = 5$ 0

56. $2 - 0.4p = 2$ 0

57. $13x - 5 = 11x - 11$ -3

58. $20x - 20 = 16x - 40$ -5

▶ **59.** $9(3x + 1) = 4x - 5x$ $-\frac{9}{28}$

60. $7(2x + 1) = 18x - 19x$ $-\frac{7}{15}$

61. $-\frac{3}{7}p = -2$ $\frac{14}{3}$

62. $-\frac{4}{5}r = -5$ $\frac{25}{4}$

63. $-\frac{4}{3}x = 12$ -9

64. $-\frac{10}{3}x = 30$ -9

65. $-2x - \frac{1}{2} = \frac{7}{2}$ -2

66. $-3n - \frac{1}{3} = \frac{8}{3}$ -1

67. $10 = 2x - 1$ $\frac{11}{2}$

68. $12 = 3j - 4$ $\frac{16}{3}$

69. $10 - 3x - 6 - 9x = 7$ $-\frac{1}{4}$

70. $12x + 30 + 8x - 6 = 10$ $-\frac{7}{10}$

71. $z - 5z = 7z - 9 - z$ $\frac{9}{10}$

72. $t - 6t = -13 + t - 3t$ $\frac{13}{3}$

73. $-x - \frac{4}{5} = x + \frac{1}{2} + \frac{2}{5}$ $-\frac{17}{20}$

74. $x + \frac{3}{7} = -x + \frac{1}{3} + \frac{4}{7}$ $\frac{5}{21}$

75. $-15 + 37 = -2(x + 5)$ -16

76. $-19 + 74 = -5(x + 3)$ -14

Objective C *Write each algebraic expression described. Simplify if possible. See Example 11.*

77. If x represents the first of two consecutive odd integers, express the sum of the two integers in terms of x. $2x + 2$

78. If x is the first of three consecutive even integers, write their sum as an algebraic expression in x. $3x + 6$

79. If x is the first of four consecutive integers, express the sum of the first integer and the third integer as an algebraic expression containing the variable x. $2x + 2$

80. If x is the first of two consecutive integers, express the sum of 20 and the second consecutive integer as an algebraic expression containing the variable x. $x + 21$

81. Classrooms on one side of the science building are all numbered with consecutive even integers. If the first room on this side of the building is numbered x, write an expression in x for the sum of five classroom numbers in a row. Then simplify this expression. $5x + 20$

82. Two sides of a quadrilateral have the same length, x, while the other two sides have the same length, both being the next consecutive odd integer. Write the sum of these lengths. Then simplify this expression. $4x + 4$

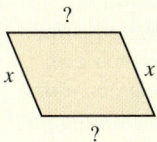

Review

Simplify each expression. See Section 1.8.

83. $5x + 2(x - 6)$
$7x - 12$

84. $-7y + 2y - 3(y + 1)$
$-8y - 3$

85. $6(2z + 4) + 20$
$12z + 44$

86. $-(3a - 3) + 2a - 6$
$-a - 3$

87. $-(x - 1) + x$
1

88. $8(z - 6) + 7z - 1$
$15z - 49$

Concept Extensions

For Exercises 89 and 90, fill in the blank with a number of your choice so that each equation has the given solution.

89. $6x = $ _____; solution: -8 -48

90. _____ $x = 10$; solution: $\dfrac{1}{2}$ 20

91. The equation $3x + 6 = 2x + 10 + x - 4$ is true for all real numbers. Substitute a few real numbers for x to see that this is so and then try solving the equation. Describe what happens.   answers may vary

92. The equation $6x + 2 - 2x = 4x + 1$ has no solution. Try solving this equation for x and describe what happens.   answers may vary

93. From the results of Exercises 91 and 92, when do you think an equation has all real numbers as its solutions?
answers may vary

94. From the results of Exercises 91 and 92, when do you think an equation has no solution?
answers may vary

Solve.

95. $0.07x - 5.06 = -4.92$ 2

96. $0.06y + 2.63 = 2.5562$ -1.23

Objectives

A Apply the General Strategy for Solving a Linear Equation.

B Solve Equations Containing Fractions or Decimals. ▶

C Recognize Identities and Equations with No Solution. ▶

Teaching Tip

Before students are shown the steps for solving a linear equation, let them come up with their own steps for solving a linear equation with your guidance. Then compare their steps with these. They will remember a set of steps better if they are involved in writing them.

Practice 1

Solve:
$5(3x - 1) + 2 = 12x + 6$

Objective A Solving Linear Equations ▶

Let's begin by restating the formal definition of a linear equation in one variable.

A **linear equation in one variable** can be written in the form

$$Ax + B = C$$

where A, B, and C are real numbers and $A \neq 0$.

We now combine our knowledge from the previous sections into a general strategy for solving linear equations.

To Solve Linear Equations in One Variable

Step 1: If an equation contains fractions, multiply both sides by the LCD to clear the equation of fractions.

Step 2: Use the distributive property to remove parentheses if they are present.

Step 3: Simplify each side of the equation by combining like terms.

Step 4: Get all variable terms on one side and all numbers on the other side by using the addition property of equality.

Step 5: Get the variable alone by using the multiplication property of equality.

Step 6: Check the solution by substituting it into the original equation.

We will use these steps as we solve the equations in Examples 1–5.

Example 1 Solve: $4(2x - 3) + 7 = 3x + 5$

Solution: There are no fractions, so we begin with Step 2.

$$4(2x - 3) + 7 = 3x + 5$$

Step 2: $8x - 12 + 7 = 3x + 5$ Use the distributive property.

Step 3: $8x - 5 = 3x + 5$ Combine like terms.

Step 4: Get all variable terms on one side of the equation and all numbers on the other side. One way to do this is by subtracting $3x$ from both sides and then adding 5 to both sides.

$$8x - 5 - 3x = 3x + 5 - 3x \quad \text{Subtract } 3x \text{ from both sides.}$$
$$5x - 5 = 5 \quad \text{Simplify.}$$
$$5x - 5 + 5 = 5 + 5 \quad \text{Add 5 to both sides.}$$
$$5x = 10 \quad \text{Simplify.}$$

Step 5: Use the multiplication property of equality to get x alone.

$$\frac{5x}{5} = \frac{10}{5} \quad \text{Divide both sides by 5.}$$
$$x = 2 \quad \text{Simplify.}$$

Step 6: Check.

$$4(2x - 3) + 7 = 3x + 5 \quad \text{Original equation}$$
$$4[2(2) - 3] + 7 \stackrel{?}{=} 3(2) + 5 \quad \text{Replace } x \text{ with 2.}$$
$$4(4 - 3) + 7 \stackrel{?}{=} 6 + 5$$
$$4(1) + 7 \stackrel{?}{=} 11$$
$$4 + 7 \stackrel{?}{=} 11$$
$$11 = 11 \quad \text{True}$$

The solution is 2.

■ **Work Practice 1**

Answer
1. $x = 3$

Example 2 Solve: $8(2 - t) = -5t$

Solution: First, we apply the distributive property.

$$\overset{\frown}{8(2 - t)} = -5t$$

Step 2: $16 - 8t = -5t$ Use the distributive property.

Step 4: $16 - 8t + 8t = -5t + 8t$ Add $8t$ to both sides.

$$16 = 3t$$ Combine like terms.

Step 5: $\dfrac{16}{3} = \dfrac{3t}{3}$ Divide both sides by 3.

$$\dfrac{16}{3} = t$$ Simplify.

Step 6: Check.

$$8(2 - t) = -5t$$ Original equation

$$8\left(2 - \dfrac{16}{3}\right) \overset{?}{=} -5\left(\dfrac{16}{3}\right)$$ Replace t with $\dfrac{16}{3}$.

$$8\left(\dfrac{6}{3} - \dfrac{16}{3}\right) \overset{?}{=} -\dfrac{80}{3}$$ The LCD is 3.

$$8\left(-\dfrac{10}{3}\right) \overset{?}{=} -\dfrac{80}{3}$$ Subtract fractions.

$$-\dfrac{80}{3} = -\dfrac{80}{3}$$ True

The solution is $\dfrac{16}{3}$.

■ **Work Practice 2**

Objective B Solving Equations Containing Fractions or Decimals ▶

If an equation contains fractions, we can clear the equation of fractions by multiplying both sides by the LCD of all denominators. By doing this, we avoid working with time-consuming fractions.

Example 3 Solve: $\dfrac{x}{2} - 1 = \dfrac{2}{3}x - 3$

Solution: We begin by clearing fractions. To do this, we multiply both sides of the equation by the LCD, which is 6.

$$\dfrac{x}{2} - 1 = \dfrac{2}{3}x - 3$$

Step 1: $6\left(\dfrac{x}{2} - 1\right) = 6\left(\dfrac{2}{3}x - 3\right)$ Multiply both sides by the LCD, 6.

Step 2: $6\left(\dfrac{x}{2}\right) - 6(1) = 6\left(\dfrac{2}{3}x\right) - 6(3)$ Use the distributive property.

$$3x - 6 = 4x - 18$$ Simplify.

There are no longer grouping symbols and no like terms on either side of the equation, so we continue with Step 4.

(Continued on next page)

Practice 2

Solve: $9(5 - x) = -3x$

Helpful Hint When checking solutions, use the original equation.

Practice 3

Solve: $\dfrac{5}{2}x - 1 = \dfrac{3}{2}x - 4$

Helpful Hint Don't forget to multiply *each* term by the LCD.

Answers

2. $x = \dfrac{15}{2}$ **3.** $x = -3$

$$3x - 6 = 4x - 18$$

Step 4: $3x - 6 - 3x = 4x - 18 - 3x$ Subtract $3x$ from both sides.

$$-6 = x - 18$$ Simplify.

$$-6 + 18 = x - 18 + 18$$ Add 18 to both sides.

$$12 = x$$ Simplify.

Step 5: The variable is now alone, so there is no need to apply the multiplication property of equality.

Step 6: Check.

$$\frac{x}{2} - 1 = \frac{2}{3}x - 3$$ Original equation

$$\frac{12}{2} - 1 \overset{?}{=} \frac{2}{3} \cdot 12 - 3$$ Replace x with 12.

$$6 - 1 \overset{?}{=} 8 - 3$$ Simplify.

$$5 = 5$$ True

The solution is 12.

■ **Work Practice 3**

Practice 4

Solve: $\dfrac{3(x - 2)}{5} = 3x + 6$

Example 4   Solve: $\dfrac{2(a + 3)}{3} = 6a + 2$

Solution: We clear the equation of fractions first.

$$\frac{2(a + 3)}{3} = 6a + 2$$

Step 1: $3 \cdot \dfrac{2(a + 3)}{3} = 3(6a + 2)$ Clear the fraction by multiplying both sides by the LCD, 3.

$$2(a + 3) = 3(6a + 2)$$ Simplify.

Step 2: Next, we use the distributive property to remove parentheses.

$$2a + 6 = 18a + 6$$ Use the distributive property.

Step 4: $2a + 6 - 18a = 18a + 6 - 18a$ Subtract $18a$ from both sides.

$$-16a + 6 = 6$$ Simplify.

$$-16a + 6 - 6 = 6 - 6$$ Subtract 6 from both sides.

$$-16a = 0$$

Step 5: $\dfrac{-16a}{-16} = \dfrac{0}{-16}$ Divide both sides by -16.

$$a = 0$$ Simplify.

Step 6: To check, replace a with 0 in the original equation. The solution is 0.

■ **Work Practice 4**

Helpful Hint

Remember: When solving an equation, it makes no difference on which side of the equation variable terms lie. Just make sure that constant terms lie on the other side.

When solving a problem about money, you may need to solve an equation containing decimals. If you choose, you may multiply to clear the equation of decimals.

Answer

4. $x = -3$

Example 5   Solve: $0.25x + 0.10(x - 3) = 1.1$

Solution: First we clear this equation of decimals by multiplying both sides of the equation by 100. Recall that multiplying a decimal number by 100 has the effect of moving the decimal point 2 places to the right.

$$0.25x + 0.10(x - 3) = 1.1$$

Step 1: $0.25x + 0.10(x - 3) = 1.10$ Multiply both sides by 100.

$$25x + 10(x - 3) = 110$$

Step 2: $25x + 10x - 30 = 110$ Apply the distributive property.

Step 3: $35x - 30 = 110$ Combine like terms.

Step 4: $35x - 30 + 30 = 110 + 30$ Add 30 to both sides.

$$35x = 140$$ Combine like terms.

Step 5: $\dfrac{35x}{35} = \dfrac{140}{35}$ Divide both sides by 35.

$$x = 4$$

Step 6: To check, replace x with 4 in the original equation. The solution is 4.

■ **Work Practice 5**

Objective **C** Recognizing Identities and Equations with No Solution ▶

So far, each equation that we have solved has had a single solution. However, not every equation in one variable has a single solution. Some equations have no solution, while others have an infinite number of solutions. For example,

$$x + 5 = x + 7$$

has **no solution** since no matter which real number we replace x with, the equation is false.

    real number + 5 = same real number + 7   FALSE

On the other hand,

$$x + 6 = x + 6$$

has infinitely many solutions since x can be replaced by any real number and the equation will always be true.

    real number + 6 = same real number + 6   TRUE

The equation $x + 6 = x + 6$ is called an **identity.** The next two examples illustrate special equations like these.

Example 6   Solve: $-2(x - 5) + 10 = -3(x + 2) + x$

Solution:

$$-2(x - 5) + 10 = -3(x + 2) + x$$
$$-2x + 10 + 10 = -3x - 6 + x$$ Apply the distributive property on both sides.
$$-2x + 20 = -2x - 6$$ Combine like terms.
$$-2x + 20 + 2x = -2x - 6 + 2x$$ Add $2x$ to both sides.
$$20 = -6$$ Combine like terms.

The final equation contains no variable terms, and the result is the false statement $20 = -6$. This means that there is no value for x that makes $20 = -6$ a true equation. Thus, we conclude that there is **no solution** to this equation.

■ **Work Practice 6**

Practice 5

Solve:
$0.06x - 0.10(x - 2) = -0.16$

Helpful Hint
If you have trouble with this step, try removing parentheses first.

$$0.25x + 0.10(x - 3) = 1.1$$
$$0.25x + 0.10x - 0.3 = 1.1$$
$$0.25x + 0.10x - 0.30 = 1.10$$
$$25x + 10x - 30 = 110$$

Then continue.

Teaching Tip

Help your students understand that each term must be multiplied by 100. In the case of $0.10(x - 3)$, this is accomplished by multiplying 0.10 by 100. Tell your students why.

Practice 6

Solve:
$5(2 - x) + 8x = 3(x - 6)$

Answers

5. $x = 9$ **6.** no solution

Practice 7

Solve: $-6(2x + 1) - 14$
       $= -10(x + 2) - 2x$

Example 7   Solve: $3(x - 4) = 3x - 12$

Solution: $3(x - 4) = 3x - 12$
                 $3x - 12 = 3x - 12$ Apply the distributive property.

The left side of the equation is now identical to the right side. Every real number may be substituted for x and a true statement will result. We arrive at the same conclusion if we continue.

$$3x - 12 = 3x - 12$$
$$3x - 12 - 3x = 3x - 12 - 3x \quad \text{Subtract } 3x \text{ from both sides.}$$
$$-12 = -12 \qquad\qquad \text{Combine like terms.}$$

Again, the final equation contains no variables, but this time the result is the true statement $-12 = -12$. This means that one side of the equation is identical to the other side. Thus, $3(x - 4) = 3x - 12$ is an **identity** and **every real number** is a solution.

■ **Work Practice 7**

Answer

7. Every real number is a solution.

✓ **Concept Check Answers**

a. Every real number is a solution.
b. The solution is 0.
c. There is no solution.

✓**Concept Check** Suppose you have simplified several equations and obtained the following results. What can you conclude about the solutions to the original equation?

a. $7 = 7$ **b.** $x = 0$ **c.** $7 = -4$

🖩 Calculator Explorations Checking Equations

We can use a calculator to check possible solutions of equations. To do this, replace the variable by the possible solution and evaluate each side of the equation separately.

Equation: $3x - 4 = 2(x + 6)$ Solution: $x = 16$
$$3x - 4 = 2(x + 6) \qquad \text{Original equation}$$
$$3(16) - 4 \stackrel{?}{=} 2(16 + 6) \qquad \text{Replace } x \text{ with 16.}$$

Now evaluate each side with your calculator.

Evaluate left side: ③ ☓ ⑯ ⊟ ④ ⊜
                        or
                    ⎣ENTER⎦

Display: ⎣ 44 ⎦

Evaluate right side: ② ⎛ ⑯ ⊞ ⑥ ⎞ ⊜
                         or
                     ⎣ENTER⎦

Display: ⎣ 44 ⎦

Since the left side equals the right side, the equation checks.

Use a calculator to check the possible solutions to each equation.

1. $2x = 48 + 6x$; $x = -12$ solution
2. $-3x - 7 = 3x - 1$; $x = -1$ solution
3. $5x - 2.6 = 2(x + 0.8)$; $x = 4.4$ not a solution
4. $-1.6x - 3.9 = -6.9x - 25.6$; $x = 5$ not a solution
5. $\dfrac{564x}{4} = 200x - 11(649)$; $x = 121$ solution
6. $20(x - 39) = 5x - 432$; $x = 23.2$ solution

Vocabulary, Readiness & Video Check

Throughout algebra, it is important to be able to identify equations and expressions.

Remember,
- an equation contains an equal sign and
- an expression does not.

Among other things,
- we solve equations and
- we simplify or perform operations on expressions.

Identify each as an equation or an expression.

1. $x = -7$ ___equation___

2. $x - 7$ ___expression___

3. $4y - 6 + 9y + 1$ ___expression___

4. $4y - 6 = 9y + 1$ ___equation___

5. $\dfrac{1}{x} - \dfrac{x-1}{8}$ ___expression___

6. $\dfrac{1}{x} - \dfrac{x-1}{8} = 6$ ___equation___

7. $0.1x + 9 = 0.2x$ ___equation___

8. $0.1x^2 + 9y - 0.2x^2$ ___expression___

Martin-Gay Interactive Videos

See Video 2.3

Watch the section lecture video and answer the following questions.

Objective A **9.** The general strategy for solving linear equations in one variable is discussed after ▢ Example 1. How many properties are mentioned in this strategy and what are they? ▶

Objective B **10.** In the first step for solving ▢ Example 2, both sides of the equation are being multiplied by the LCD. Why is the distributive property mentioned? ▶

11. In ▢ Example 3, why is the number of decimal places in each term of the equation important? ▶

Objective C **12.** Complete each statement based on ▢ Examples 4 and 5.

When solving an equation and all variable terms subtract out:
a. If you have a true statement, then the equation has _____ solution(s).
b. If you have a false statement, then the equation has _____ solution(s). ▶

See video answer section.

2.3 Exercise Set MyMathLab® ▶

Objective A *Solve each equation. See Examples 1 and 2.*

1. $-4y + 10 = -2(3y + 1)$ -6

2. $-3x + 1 = -2(4x + 2)$ -1

3. $15x - 8 = 10 + 9x$ 3

4. $15x - 5 = 7 + 12x$ 4

5. $-2(3x - 4) = 2x$ 1

6. $-(5x - 10) = 5x$ 1

▶ **7.** $5(2x - 1) - 2(3x) = 1$ $\dfrac{3}{2}$

8. $3(2 - 5x) + 4(6x) = 12$ $\dfrac{2}{3}$

9. $-6(x - 3) - 26 = -8$ 0

10. $-4(n - 4) - 23 = -7$ 0

11. $8 - 2(a + 1) = 9 + a$ -1

12. $5 - 6(2 + b) = b - 14$ 1

13. $4x + 3 = -3 + 2x + 14$ 4

14. $6y - 8 = -6 + 3y + 13$ 5

15. $-2y - 10 = 5y + 18$ -4

16. $-7n + 5 = 8n - 10$ 1

Objective **B** *Solve each equation. See Examples 3 through 5.*

17. $\dfrac{2}{3}x + \dfrac{4}{3} = -\dfrac{2}{3}$ -3

18. $\dfrac{4}{5}x - \dfrac{8}{5} = -\dfrac{16}{5}$ -2

19. $\dfrac{3}{4}x - \dfrac{1}{2} = 1$ 2

20. $\dfrac{2}{9}x - \dfrac{1}{3} = 1$ 6

▶**21.** $0.50x + 0.15(70) = 35.5$ 50

22. $0.40x + 0.06(30) = 9.8$ 20

23. $\dfrac{2(x + 1)}{4} = 3x - 2$ 1

24. $\dfrac{3(y + 3)}{5} = 2y + 6$ -3

25. $x + \dfrac{7}{6} = 2x - \dfrac{7}{6}$ $\dfrac{7}{3}$

26. $\dfrac{5}{2}x - 1 = x + \dfrac{1}{4}$ $\dfrac{5}{6}$

27. $0.12(y - 6) + 0.06y = 0.08y - 0.70$ 0.2

28. $0.60(z - 300) + 0.05z = 0.70z - 205$ 500

Objective **C** *Solve each equation. See Examples 6 and 7.*

29. $4(3x + 2) = 12x + 8$ all real numbers

30. $14x + 7 = 7(2x + 1)$ all real numbers

31. $\dfrac{x}{4} + 1 = \dfrac{x}{4}$ no solution

32. $\dfrac{x}{3} - 2 = \dfrac{x}{3}$ no solution

33. $3x - 7 = 3(x + 1)$ no solution

34. $2(x - 5) = 2x + 10$ no solution

35. $-2(6x - 5) + 4 = -12x + 14$ all real numbers

36. $-5(4y - 3) + 2 = -20y + 17$ all real numbers

Objectives **A** **B** **C** **Mixed Practice** *Solve. See Examples 1 through 7.*

37. $\dfrac{6(3 - z)}{5} = -z$ 18

38. $\dfrac{4(5 - w)}{3} = -w$ 20

39. $-3(2t - 5) + 2t = 5t - 4$ $\dfrac{19}{9}$

40. $-(4a - 7) - 5a = 10 + a$ $-\dfrac{3}{10}$

41. $5y + 2(y - 6) = 4(y + 1) - 2$ $\dfrac{14}{3}$

42. $9x + 3(x - 4) = 10(x - 5) + 7$ $-\dfrac{31}{2}$

43. $\dfrac{3(x - 5)}{2} = \dfrac{2(x + 5)}{3}$ 13

44. $\dfrac{5(x - 1)}{4} = \dfrac{3(x + 1)}{2}$ -11

45. $0.7x - 2.3 = 0.5$ 4

46. $0.9x - 4.1 = 0.4$ 5

▶**47.** $5x - 5 = 2(x + 1) + 3x - 7$ all real numbers

48. $3(2x - 1) + 5 = 6x + 2$ all real numbers

49. $4(2n + 1) = 3(6n + 3) + 1$ $-\dfrac{3}{5}$

50. $4(4y + 2) = 2(1 + 6y) + 8$ $\dfrac{1}{2}$

51. $x + \dfrac{5}{4} = \dfrac{3}{4}x$ -5

52. $\dfrac{7}{8}x + \dfrac{1}{4} = \dfrac{3}{4}x$ -2

▶**53.** $\dfrac{x}{2} - 1 = \dfrac{x}{5} + 2$ 10

54. $\dfrac{x}{5} - 7 = \dfrac{x}{3} - 5$ -15

55. $2(x + 3) - 5 = 5x - 3(1 + x)$ no solution

56. $4(2 + x) + 1 = 7x - 3(x - 2)$ no solution

57. $0.06 - 0.01(x + 1) = -0.02(2 - x)$ 3

58. $-0.01(5x + 4) = 0.04 - 0.01(x + 4)$ −1

59. $\frac{9}{2} + \frac{5}{2}y = 2y - 4$ −17

60. $3 - \frac{1}{2}x = 5x - 8$ 2

61. $\frac{3}{4}x - 1 + \frac{1}{2}x = \frac{5}{12}x + \frac{1}{6}$ $\frac{7}{5}$

62. $\frac{5}{9}x + 2 - \frac{1}{6}x = \frac{11}{18}x + \frac{1}{3}$ $\frac{15}{2}$

63. $3x + \frac{5}{16} = \frac{3}{4} - \frac{1}{8}x - \frac{1}{2}$ $-\frac{1}{50}$

64. $2x - \frac{1}{10} = \frac{2}{5} - \frac{1}{4}x - \frac{17}{20}$ $-\frac{7}{45}$

Review

Translating *Write each algebraic expression described. See Section 1.8. Recall that the perimeter of a figure is the total distance around the figure.*

△ **65.** A plot of land is in the shape of a triangle. If one side is x meters, a second side is $(2x - 3)$ meters, and a third side is $(3x - 5)$ meters, express the perimeter of the lot as a simplified expression in x.
$(6x - 8)$ m

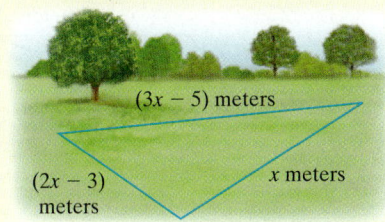

(3x − 5) meters

(2x − 3) meters

x meters

66. A portion of a board has length x feet. The other part has length $(7x - 9)$ feet. Express the total length of the board as a simplified expression in x.
$(8x - 9)$ ft

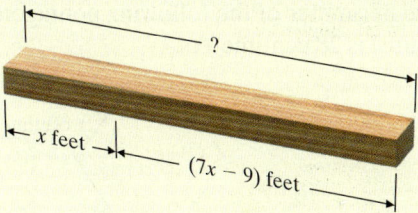

?

x feet

(7x − 9) feet

Translating *Write each phrase as an algebraic expression. Use x for the unknown number. See Section 1.8.*

67. A number subtracted from −8 $-8 - x$

68. Three times a number $3x$

69. The sum of −3 and twice a number $-3 + 2x$

70. The difference of 8 and twice a number $8 - 2x$

71. The product of 9 and the sum of a number and 20 $9(x + 20)$

72. The quotient of −12 and the difference of a number and 3 $\frac{-12}{x - 3}$

Concept Extensions

See the Concept Check in this section.

73. a. Solve: $x + 3 = x + 3$ all real numbers
   b. If you simplify an equation (such as the one in part a) and get a true statement such as $3 = 3$ or $0 = 0$, what can you conclude about the solution(s) of the original equation?   answers may vary
   c. On your own, construct an equation for which every real number is a solution.
   answers may vary

74. a. Solve: $x + 3 = x + 5$ no solution
   b. If you simplify an equation (such as the one in part a) and get a false statement such as $3 = 5$ or $10 = 17$, what can you conclude about the solution(s) of the original equation?   answers may vary
   c. On your own, construct an equation that has no solution.   answers may vary

Match each equation in the first column with its solution in the second column. Items in the second column may be used more than once.

75. $5x + 1 = 5x + 1$ a

76. $3x + 1 = 3x + 2$ b

77. $2x - 6x - 10 = -4x + 3 - 10$ b

78. $x - 11x - 3 = -10x - 1 - 2$ a

79. $9x - 20 = 8x - 20$ c

80. $-x + 15 = x + 15$ c

a. all real numbers

b. no solution

c. 0

81. Explain the difference between simplifying an expression and solving an equation.
answers may vary

82. On your own, write an expression and then an equation. Label each.
answers may vary

For Exercises 83 and 84, **a.** *Write an equation for perimeter. (Recall that the perimeter of a geometric figure is the sum of the lengths of its sides.)* **b.** *Solve the equation in part (a).* **c.** *Find the length of each side.*

△**83.** The perimeter of the following pentagon (five-sided figure) is 28 centimeters.

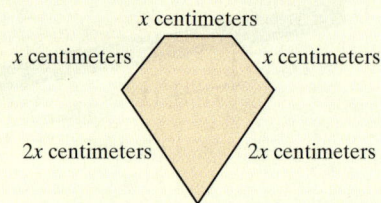

x centimeters

x centimeters   *x* centimeters

2*x* centimeters   2*x* centimeters

a. $x + x + x + 2x + 2x = 28$
b. $x = 4$ **c.** x cm = 4 cm; $2x$ cm = 8 cm

△**84.** The perimeter of the following triangle is 35 meters.

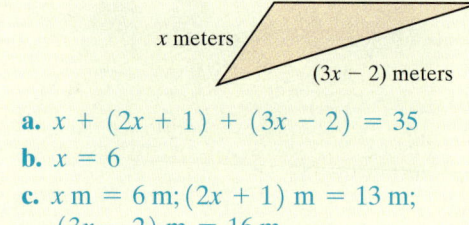

(2*x* + 1) meters

x meters

(3*x* − 2) meters

a. $x + (2x + 1) + (3x - 2) = 35$
b. $x = 6$
c. x m = 6 m; $(2x + 1)$ m = 13 m; $(3x - 2)$ m = 16 m

Fill in the blanks with numbers of your choice so that each equation has the given solution. Note: Each blank will be replaced by a different number.

85. $x +$ ____ $= 2x -$ ____ ; solution: 9   answers may vary

86. $-5x -$ ____ $=$ ____ ; solution: 2   answers may vary

Solve.

87. $1000(7x - 10) = 50(412 + 100x)$ 15.3

88. $1000(x + 40) = 100(16 + 7x)$ −128

89. $0.035x + 5.112 = 0.010x + 5.107$ −0.2

90. $0.127x - 2.685 = 0.027x - 2.38$ 3.05

Solving Linear Equations

Solve. Feel free to use the steps given in Section 2.3.

1. $x - 10 = -4$

2. $y + 14 = -3$

3. $9y = 108$

4. $-3x = 78$

5. $-6x + 7 = 25$

6. $5y - 42 = -47$

7. $\dfrac{2}{3}x = 9$

8. $\dfrac{4}{5}z = 10$

9. $\dfrac{r}{-4} = -2$

10. $\dfrac{y}{-8} = 8$

11. $6 - 2x + 8 = 10$

12. $-5 - 6y + 6 = 19$

13. $2x - 7 = 6x - 27$

14. $3 + 8y = 3y - 2$

15. $9(3x - 1) = -4 + 49$

16. $12(2x + 1) = -6 + 66$

17. $-3a + 6 + 5a = 7a - 8a$

18. $4b - 8 - b = 10b - 3b$

19. $-\dfrac{2}{3}x = \dfrac{5}{9}$

20. $-\dfrac{3}{8}y = -\dfrac{1}{16}$

21. $10 = -6n + 16$

22. $-5 = -2m + 7$

1. 6

2. -17

3. 12

4. -26

5. -3

6. -1

7. $\dfrac{27}{2}$

8. $\dfrac{25}{2}$

9. 8

10. -64

11. 2

12. -3

13. 5

14. -1

15. 2

16. 2

17. -2

18. -2

19. $-\dfrac{5}{6}$

20. $\dfrac{1}{6}$

21. 1

22. 6

23. 4

24. 1

25. $\dfrac{9}{5}$

26. $-\dfrac{6}{5}$

27. all real numbers

28. all real numbers

29. 0

30. -1.6

31. $\dfrac{4}{19}$

32. $-\dfrac{5}{19}$

33. $\dfrac{7}{2}$

34. $-\dfrac{1}{4}$

35. no solution

36. no solution

37. $\dfrac{7}{6}$

38. $\dfrac{1}{15}$

23. $3(5c - 1) - 2 = 13c + 3$

24. $4(3t + 4) - 20 = 3 + 5t$

25. $\dfrac{2(z + 3)}{3} = 5 - z$

26. $\dfrac{3(w + 2)}{4} = 2w + 3$

27. $-2(2x - 5) = -3x + 7 - x + 3$

28. $-4(5x - 2) = -12x + 4 - 8x + 4$

29. $0.02(6t - 3) = 0.04(t - 2) + 0.02$

30. $0.03(m + 7) = 0.02(5 - m) + 0.03$

31. $-3y = \dfrac{4(y - 1)}{5}$

32. $-4x = \dfrac{5(1 - x)}{6}$

33. $\dfrac{5}{3}x - \dfrac{7}{3} = x$

34. $\dfrac{7}{5}n + \dfrac{3}{5} = -n$

35. $\dfrac{1}{10}(3x - 7) = \dfrac{3}{10}x + 5$

36. $\dfrac{1}{7}(2x - 5) = \dfrac{2}{7}x + 1$

37. $5 + 2(3x - 6) = -4(6x - 7)$

38. $3 + 5(2x - 4) = -7(5x + 2)$

2.4 An Introduction to Problem Solving

First, let's review a list of key words and phrases from Section 1.3 to help us translate.

Objectives

A Solve Problems Involving Direct Translations.

B Solve Problems Involving Relationships Among Unknown Quantities.

C Solve Problems Involving Consecutive Integers.

> **Helpful Hint**
>
> Order matters when subtracting and also dividing, so be especially careful with these translations.

Addition (+)	Subtraction (−)	Multiplication (·)	Division (÷)	Equality (=)
Sum	Difference of	Product	Quotient	Equals
Plus	Minus	Times	Divide	Gives
Added to	Subtracted from	Multiply	Into	Is/was/should be
More than	Less than	Twice	Ratio	Yields
Increased by	Decreased by	Of	Divided by	Amounts to
Total	Less			Represents
				Is the same as

We are now ready to put all our translating skills to practical use. To begin, we present a general strategy for problem solving.

General Strategy for Problem Solving

1. UNDERSTAND the problem. During this step, become comfortable with the problem. Some ways of doing this are:

   Read and reread the problem.

   Choose a variable to represent the unknown.

   Construct a drawing.

   Propose a solution and check. Pay careful attention to how you check your proposed solution. This will help when writing an equation to model the problem.

2. TRANSLATE the problem into an equation.

3. SOLVE the equation.

4. INTERPRET the results: *Check* the proposed solution in the stated problem and *state* your conclusion.

Objective A Solving Direct Translation Problems

Much of problem solving involves a direct translation from a sentence to an equation.

Example 1   Finding an Unknown Number

Twice a number, added to seven, is the same as three subtracted from the number. Find the number.

Solution: Translate the sentence into an equation and solve.

In words:	twice a number	added to	seven	is the same as	three subtracted from the number
	↓	↓	↓	↓	↓
Translate:	$2x$	$+$	7	$=$	$x - 3$

(Continued on next page)

Practice 1

Three times a number, minus 6, is the same as two times the number, plus 3. Find the number.

Answer

1. The number is 9.

121

To solve, begin by subtracting x from both sides to isolate the variable term.

$$2x + 7 = x - 3$$
$$2x + 7 - x = x - 3 - x \quad \text{Subtract } x \text{ from both sides.}$$
$$x + 7 = -3 \quad \text{Combine like terms.}$$
$$x + 7 - 7 = -3 - 7 \quad \text{Subtract 7 from both sides.}$$
$$x = -10 \quad \text{Combine like terms.}$$

Check the solution in the problem as it was originally stated. To do so, replace "number" in the sentence with -10. Twice "-10" added to 7 is the same as 3 subtracted from "-10."

$$2(-10) + 7 = -10 - 3$$
$$-13 = -13$$

The unknown number is -10.

■ Work Practice 1

Helpful Hint

When checking solutions, go back to the original stated problem rather than to your equation in case errors have been made in translating to an equation.

Practice 2

Three times the difference of a number and 5 is the same as twice the number decreased by 3. Find the number.

Example 2   Finding an Unknown Number

Twice the sum of a number and 4 is the same as four times the number decreased by 12. Find the number.

Solution:

1. **UNDERSTAND.** Read and reread the problem. If we let $x =$ the unknown number, then

   "the sum of a number and 4" translates to "$x + 4$" and
   "four times the number" translates to "$4x$"

2. **TRANSLATE.**

twice	sum of a number and 4	is the same as	four times the number	decreased by	12
↓	↓	↓	↓	↓	↓
2	$(x + 4)$	$=$	$4x$	$-$	12

3. **SOLVE.**

$$2(x + 4) = 4x - 12$$
$$2x + 8 = 4x - 12 \quad \text{Apply the distributive property.}$$
$$2x + 8 - 4x = 4x - 12 - 4x \quad \text{Subtract } 4x \text{ from both sides.}$$
$$-2x + 8 = -12$$
$$-2x + 8 - 8 = -12 - 8 \quad \text{Subtract 8 from both sides.}$$
$$-2x = -20$$
$$\frac{-2x}{-2} = \frac{-20}{-2} \quad \text{Divide both sides by } -2.$$
$$x = 10$$

4. **INTERPRET.**

Check: Check this solution in the problem as it was originally stated. To do so, replace "number" with 10. Twice the sum of "10" and 4 is 28, which is the same as 4 times "10" decreased by 12.

State: The number is 10.

Copyright 2016 Pearson Education, Inc.

Answer

2. The number is 12.

■ Work Practice 2

Objective B Solving Problems Involving Relationships Among Unknown Quantities

Example 3 Finding the Length of a Board

A 10-foot board is to be cut into two pieces so that the length of the longer piece is 4 times the length of the shorter. Find the length of each piece.

Solution:

1. **UNDERSTAND** the problem. To do so, read and reread the problem. You may also want to propose a solution. For example, if 3 feet represents the length of the shorter piece, then $4(3) = 12$ feet is the length of the longer piece, since it is 4 times the length of the shorter piece. This guess gives a total board length of 3 feet + 12 feet = 15 feet, which is too long. However, the purpose of proposing a solution is not to guess correctly, but to help better understand the problem and how to model it.

In general, if we let

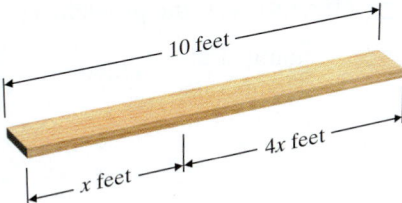

x = length of shorter piece, then
$4x$ = length of longer piece

2. **TRANSLATE** the problem. First, we write the equation in words.

length of shorter piece	added to	length of longer piece	equals	total length of board
↓	↓	↓	↓	↓
x	$+$	$4x$	$=$	10

3. **SOLVE.**

$$x + 4x = 10$$
$$5x = 10 \quad \text{Combine like terms.}$$
$$\frac{5x}{5} = \frac{10}{5} \quad \text{Divide both sides by 5.}$$
$$x = 2$$

4. **INTERPRET.**

Check: Check the solution in the stated problem. If the length of the shorter piece of board is 2 feet, the length of the longer piece is $4 \cdot (2 \text{ feet}) = 8$ feet and the sum of the lengths of the two pieces is 2 feet + 8 feet = 10 feet.

State: The shorter piece of board is 2 feet and the longer piece of board is 8 feet.

◼ **Work Practice 3**

Helpful Hint

Make sure that units are included in your answer, if appropriate.

Example 4 Finding the Number of Republican and Democratic Senators

As of May 2014, the total number of Democrats and Republicans in the U.S. House of Representatives was 432. There were 34 more Republican representatives than Democratic. Find the number of representatives from each party. (*Source:* Office of the Clerk, U.S. Capitol)

(Continued on next page)

Practice 3

An 18-foot wire is to be cut so that the length of the longer piece is 5 times the length of the shorter piece. Find the length of each piece.

Teaching Tip

You may want to show students how to use a chart to help them set up the correct equation for applications. For Example 3. have students propose a few lengths for the short piece and complete the chart for each proposal. After students see a pattern, have them use x for the length of the short piece and complete the chart.

Length of Short Piece	Length of Long Piece	Sum of Lengths
3	$4 \cdot 3$	$3 + 4 \cdot 3 = 3 + 12 = 15$
x	$4x$	$x + 4x = 10$

Practice 4

Through the year 2020, the state of California will have 17 more electoral votes for president than the state of Texas. If the total electoral votes for these two states is 93, find the number of electoral votes for each state. (*Source*: U.S. Census Bureau)

Answers

3. shorter piece: 3 feet; longer piece: 15 feet **4.** Texas: 38 electoral votes; California: 55 electoral votes

Solution:

1. UNDERSTAND the problem. Read and reread the problem. Let's suppose that there are 100 Democratic representatives. Since there are 34 more Republican than Democrats, there must be $100 + 34 = 134$ Republicans. The total number of Republicans and Democrats is then $134 + 100 = 234$. This is incorrect since the total should be 432, but we now have a better understanding of the problem.

In general, if we let

$$x = \text{number of Democrats, then}$$
$$x + 34 = \text{number of Republicans}$$

2. TRANSLATE the problem. First, we write the equation in words.

number of Democrats	added to	number of Republicans	equals	432
↓	↓	↓	↓	↓
x	$+$	$(x + 34)$	$=$	432

3. SOLVE.

$$x + (x + 34) = 432$$
$$2x + 34 = 432 \qquad \text{Combine like terms.}$$
$$2x + 34 - 34 = 432 - 34 \qquad \text{Subtract 34 from both sides.}$$
$$2x = 398$$
$$\frac{2x}{2} = \frac{398}{2} \qquad \text{Divide both sides by 2.}$$
$$x = 199$$

4. INTERPRET.

Check: If there were 199 Democratic representatives, then there were $199 + 34 = 233$ Republican representatives. The total number of representatives is then $199 + 233 = 432$. The results check.

State: There were 199 Democratic and 233 Republican representatives in the U.S. House of Representatives.

■ **Work Practice 4**

Practice 5

A car rental agency charges $28 a day and $0.15 a mile. If you rent a car for a day and your bill (before taxes) is $52, how many miles did you drive?

Example 5 Calculating Hours on the Job

A computer science major at a local university has a part-time job working on computers for his clients. He charges $20 to come to your home or office and then $25 per hour. During one month he visited 10 homes or offices and his total income was $575. How many hours did he spend working on computers?

Solution:

1. UNDERSTAND. Read and reread the problem. Let's propose that the student spent 20 hours working on computers. Pay careful attention as to how his income is calculated. For 20 hours and 10 visits, his income is $20(\$25) + 10(\$20) = \$700$, which is more than $575. We now have a better understanding of the problem and know that the time working on computers is less than 20 hours.

Let's let

$$x = \text{hours working on computers. Then}$$
$$25x = \text{amount of money made while working on computers}$$

Answer

5. 160 miles

2. TRANSLATE.

money made while working on computers	plus	money made for visits	is equal to	575
↓	↓	↓	↓	↓
$25x$	$+$	$10(20)$	$=$	575

3. SOLVE.

$$25x + 200 = 575$$
$$25x + 200 - 200 = 575 - 200 \quad \text{Subtract 200 from both sides.}$$
$$25x = 375 \quad \text{Simplify.}$$
$$\frac{25x}{25} = \frac{375}{25} \quad \text{Divide both sides by 25.}$$
$$x = 15 \quad \text{Simplify.}$$

4. INTERPRET.

Check: If the student works 15 hours and makes 10 visits, his income is $15(\$25) + 10(\$20) = \$575$.

State: The student spent 15 hours working on computers.

■ **Work Practice 5**

 **Example 6** Finding Angle Measures

If the two walls of the Vietnam Veterans Memorial in Washington, D.C., were connected, an isosceles triangle would be formed. The measure of the third angle is 97.5° more than the measure of either of the two equal angles. Find the measure of the third angle. (*Source:* National Park Service)

Solution:

1. UNDERSTAND. Read and reread the problem. We then draw a diagram (recall that an isosceles triangle has two angles with the same measure) and let

$$x = \text{degree measure of one angle}$$
$$x = \text{degree measure of the second equal angle}$$
$$x + 97.5 = \text{degree measure of the third angle}$$

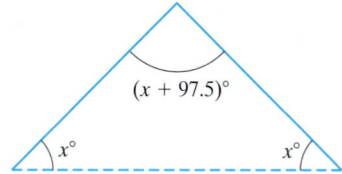

2. TRANSLATE. Recall that the sum of the measures of the angles of a triangle equals 180.

measure of first angle	+	measure of second angle	+	measure of third angle	equal	180
↓		↓		↓	↓	↓
x	$+$	x	$+$	$(x + 97.5)$	$=$	180

(*Continued on next page*)

Practice 6

The measure of the second angle of a triangle is twice the measure of the smallest angle. The measure of the third angle of the triangle is three times the measure of the smallest angle. Find the measures of the angles.

Answer
6. smallest: 30°; second: 60°; third: 90°

3. SOLVE.

$$x + x + (x + 97.5) = 180$$
$$3x + 97.5 = 180 \qquad \text{Combine like terms.}$$
$$3x + 97.5 - 97.5 = 180 - 97.5 \qquad \text{Subtract 97.5 from both sides.}$$
$$3x = 82.5$$
$$\frac{3x}{3} = \frac{82.5}{3} \qquad \text{Divide both sides by 3.}$$
$$x = 27.5$$

4. INTERPRET.

Check: If $x = 27.5$, then the measure of the third angle is $x + 97.5 = 125$. The sum of the angles is then $27.5 + 27.5 + 125 = 180$, the correct sum.

State: The third angle measures $125°$.*

■ **Work Practice 6**

Objective C Solving Consecutive Integer Problems

The next example has to do with consecutive integers. Recall what we have learned thus far about these integers.

	Example	General Representation	
Consecutive Integers	11, 12, 13 +1 +1	Let x be an integer.	x, $x + 1$, $x + 2$ +1 +1
Consecutive Even Integers	38, 40, 42 +2 +2	Let x be an even integer.	x, $x + 2$, $x + 4$ +2 +2
Consecutive Odd Integers	57, 59, 61 +2 +2	Let x be an odd integer.	x, $x + 2$, $x + 4$ +2 +2

The next example has to do with consecutive integers.

Practice 7

The sum of three consecutive even integers is 144. Find the integers.

Helpful Hint
Remember, the 2 here means that odd integers are 2 units apart, for example, the odd integers 13 and $13 + 2 = 15$.

Example 7 Some states have a single area code for the entire state. Two such states have area codes that are consecutive odd integers. If the sum of these integers is 1208, find the two area codes. (*Source: World Almanac*)

Solution:

1. UNDERSTAND. Read and reread the problem. If we let

$$x = \text{the first odd integer, then}$$
$$x + 2 = \text{the next odd integer}$$

2. TRANSLATE.

first odd integer	added to	next odd integer	is	1208
↓	↓	↓		
x	$+$	$(x + 2)$	$=$	1208

Copyright 2016 Pearson Education, Inc.

Answer
7. 46, 48, 50

*The two walls actually meet at an angle of 125 degrees 12 minutes. The measurement of 97.5° given in the problem is an approximation.

3. SOLVE.

$$x + x + 2 = 1208$$
$$2x + 2 = 1208$$
$$2x + 2 - 2 = 1208 - 2$$
$$2x = 1206$$
$$\frac{2x}{2} = \frac{1206}{2}$$
$$x = 603$$

4. INTERPRET.

Check: If $x = 603$, then the next odd integer $x + 2 = 603 + 2 = 605$. Notice their sum, $603 + 605 = 1208$, as needed.

State: The area codes are 603 and 605.

Note: New Hampshire's area code is 603 and South Dakota's area code is 605.

🟧 **Work Practice 7**

Vocabulary, Readiness & Video Check

Fill in the table.

1.	A number: x	→ Double the number: $2x$	→ Double the number, decreased by 31: $2x - 31$
2.	A number: x	→ Three times the number: $3x$	→ Three times the number, increased by 17: $3x + 17$
3.	A number: x	→ The sum of the number and 5: $x + 5$	→ Twice the sum of the number and 5: $2(x + 5)$
4.	A number: x	→ The difference of the number and 11: $x - 11$	→ Seven times the difference of a number and 11: $7(x - 11)$
5.	A number: y	→ The difference of 20 and the number: $20 - y$	→ The difference of 20 and the number, divided by 3: $\frac{20 - y}{3}$ or $(20 - y) \div 3$
6.	A number: y	→ The sum of -10 and the number: $-10 + y$	→ The sum of -10 and the number, divided by 9: $\frac{-10 + y}{9}$ or $(-10 + y) \div 9$

Martin-Gay Interactive Videos

See Video 2.4 🍎

See video answer section.

Watch the section lecture video and answer the following questions.

Objective A **7.** At the end of 🎞 Example 1, where are you told is the best place to check the solution of an application problem? ▶

Objective B **8.** The solution of the equation for 🎞 Example 3 is $x = 43$. Why is this not the solution to the application? ▶

Objective C **9.** What are two things that should be checked to make sure the solution of 🎞 Example 4 is correct? ▶

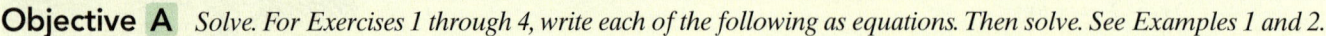

2.4 Exercise Set MyMathLab®

Objective A *Solve. For Exercises 1 through 4, write each of the following as equations. Then solve. See Examples 1 and 2.*

1. The sum of twice a number and 7 is equal to the sum of the number and 6. Find the number.
$2x + 7 = x + 6; -1$

2. The difference of three times a number and 1 is the same as twice the number. Find the number. $3x - 1 = 2x; 1$

3. Three times a number, minus 6, is equal to two times the number, plus 8. Find the number.
$3x - 6 = 2x + 8; 14$

4. The sum of 4 times a number and −2 is equal to the sum of 5 times the number and −2. Find the number. $4x - 2 = 5x - 2; 0$

▶ **5.** Twice the difference of a number and 8 is equal to three times the sum of the number and 3. Find the number. -25

6. Five times the sum of a number and −1 is the same as 6 times the number. Find the number. -5

7. The product of twice a number and three is the same as the difference of five times the number and $\frac{3}{4}$. Find the number. $-\frac{3}{4}$

8. If the difference of a number and four is doubled, the result is $\frac{1}{4}$ less than the number. Find the number. $\frac{31}{4}$

Objective B *Solve. For Exercises 9 and 10, the solutions have been started for you. See Examples 3 and 4.*

9. A 25-inch piece of steel is cut into three pieces so that the second piece is twice as long as the first piece, and the third piece is one inch more than five times the length of the first piece. Find the lengths of the pieces.   3 in.; 6 in.; 16 in.

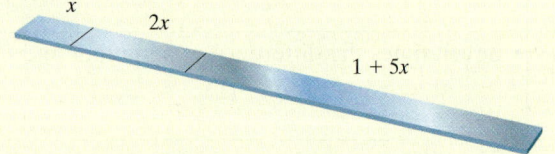

10. A 46-foot piece of rope is cut into three pieces so that the second piece is three times as long as the first piece, and the third piece is two feet more than seven times the length of the first piece. Find the lengths of the pieces.   4 ft; 12 ft; 30 ft

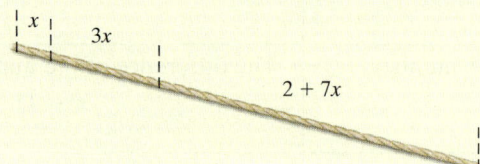

Start the solution:

1. UNDERSTAND the problem. Reread it as many times as needed.
2. TRANSLATE into an equation. (Fill in the blanks below.)

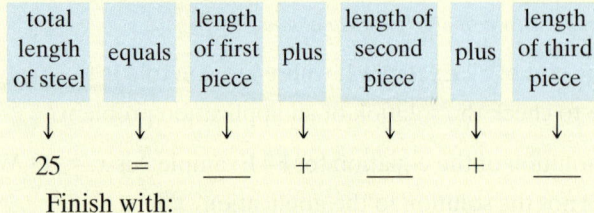

total length of steel	equals	length of first piece	plus	length of second piece	plus	length of third piece
↓	↓	↓	↓	↓	↓	↓
25	=	___	+	___	+	___

Finish with:
3. SOLVE and 4. INTERPRET

Start the solution:

1. UNDERSTAND the problem. Reread it as many times as needed.
2. TRANSLATE into an equation. (Fill in the blanks below.)

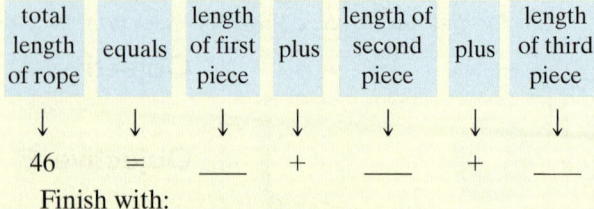

total length of rope	equals	length of first piece	plus	length of second piece	plus	length of third piece
↓	↓	↓	↓	↓	↓	↓
46	=	___	+	___	+	___

Finish with:
3. SOLVE and 4. INTERPRET

11. A 40-inch board is to be cut into three pieces so that the second piece is twice as long as the first piece and the third piece is 5 times as long as the first piece. If x represents the length of the first piece, find the lengths of all three pieces.   1st piece: 5 in.; 2nd piece: 10 in.; 3rd piece: 25 in.

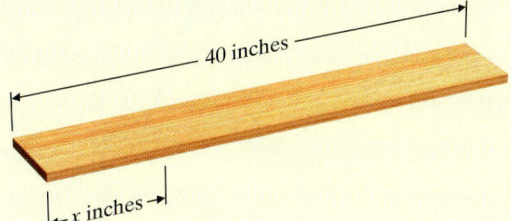

40 inches

x inches

12. A 21-foot beam is to be divided so that the longer piece is 1 foot more than 3 times the length of the shorter piece. If x represents the length of the shorter piece, find the lengths of both pieces.   shorter: 5 ft; longer: 16 ft

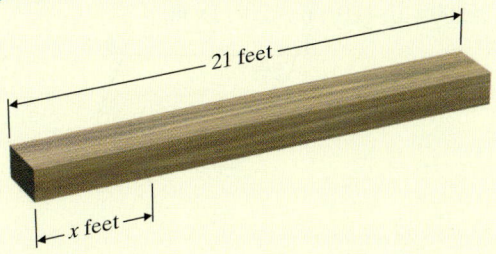

21 feet

x feet

13. In 2012, New York produced 226 million pounds more apples than Pennsylvania. Together, the two states produced 1214 million pounds of apples. Find the amount of apples grown in New York and Pennsylvania in 2012. (*Source:* National Agriculture Statistics Service)   Pennsylvania: 494 million lb; New York: 720 million lb

14. In the 2012 Summer Olympics in London, the U.S. team won 8 more gold medals than the Chinese team. If the total number of gold medals won by both teams was 84, find the number of gold medals won by each team. (*Source:* International Olympic Committee)   U.S.: 46 gold medals; China: 38 gold medals

Solve. See Example 5.

15. A car rental agency advertised renting a Buick Century for $24.95 per day and $0.29 per mile. If you rent this car for 2 days, how many whole miles can you drive on a $100 budget?   172 mi

16. A plumber gave an estimate for the renovation of a kitchen. Her hourly pay is $27 per hour and the plumbing parts will cost $80. If her total estimate is $404, how many hours does she expect this job to take?   12 hr

17. In one U.S. city, the taxi cost is $3 plus $0.80 per mile. If you are traveling from the airport, there is an additional charge of $4.50 for tolls. How far can you travel from the airport by taxi for $27.50?   25 mi

18. A professional carpet cleaning service charges $30 plus $25.50 per hour to come to your home. If your total bill from this company is $119.25 before taxes, for how many hours were you charged?   3.5 hr

Solve. See Example 6.

△ **19.** The flag of Equatorial Guinea contains an isosceles triangle. (Recall that an isosceles triangle contains two angles with the same measure.) If the measure of the third angle of the triangle is 30° more than twice the measure of either of the other two angles, find the measure of each angle of the triangle. (*Hint:* Recall that the sum of the measures of the angles of a triangle is 180°.)   1st angle: 37.5°; 2nd angle: 37.5°; 3rd angle: 105°

△ **20.** The flag of Brazil contains a parallelogram. One angle of the parallelogram is 15° less than twice the measure of the angle next to it. Find the measure of each angle of the parallelogram. (*Hint:* Recall that opposite angles of a parallelogram have the same measure and that the sum of the measures of the angles is 360°.)   1st angle: 65°; 2nd angle: 115°

21. The sum of the measures of the angles of a parallelogram is 360°. In the parallelogram below, angles A and D have the same measure as well as angles C and B. If the measure of angle C is twice the measure of angle A, find the measure of each angle. *A*: 60°; *B*: 120°; *C*: 120°; *D*: 60°

22. Recall that the sum of the measures of the angles of a triangle is 180°. In the triangle below, angle C has the same measure as angle B, and angle A measures 42° less than angle B. Find the measure of each angle. *A*: 32°; *B*: 74°; *C*: 74°

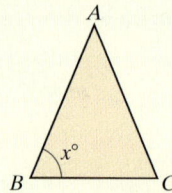

Objective **C** *Solve. See Example 7. Fill in the table. Most of the first row has been completed for you.*

23. Three consecutive integers:

24. Three consecutive integers:

25. Three consecutive even integers:

26. Three consecutive odd integers:

27. Four consecutive integers:

28. Four consecutive integers:

29. Three consecutive odd integers:

30. Three consecutive even integers:

First Integer →	Next Integers	→	Indicated Sum
Integer: x	$x+1$ \ $x+2$		Sum of the three consecutive integers, simplified: $3x+3$
Integer: x	$x+1$ \ $x+2$		Sum of the second and third consecutive integers, simplified: $2x+3$
Even integer: x	$x+2$ \ $x+4$		Sum of the first and third even consecutive integers, simplified: $2x+4$
Odd integer: x	$x+2$ \ $x+4$		Sum of the three consecutive odd integers, simplified: $3x+6$
Integer: x	$x+1$ \ $x+2$ \ $x+3$		Sum of the four consecutive integers, simplified: $4x+6$
Integer: x	$x+1$ \ $x+2$ \ $x+3$		Sum of the first and fourth consecutive integers, simplified: $2x+3$
Odd integer: x	$x+2$ \ $x+4$		Sum of the second and third consecutive odd integers, simplified: $2x+6$
Even integer: x	$x+2$ \ $x+4$		Sum of the three consecutive even integers, simplified: $3x+6$

Solve. See Example 7

31. The left and right page numbers of an open book are two consecutive integers whose sum is 469. Find these page numbers.   234, 235

32. The room numbers of two adjacent classrooms are two consecutive even numbers. If their sum is 654, find the classroom numbers.   326, 328

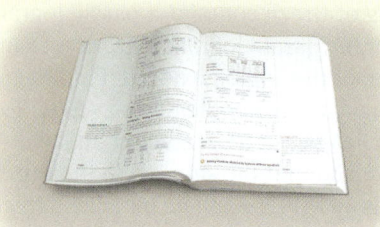

33. To make an international telephone call, you need the code for the country you are calling. The codes for Belgium, France, and Spain are three consecutive integers whose sum is 99. Find the code for each country. (*Source: The World Almanac and Book of Facts*)   Belgium: 32; France: 33; Spain: 34

34. The code to unlock a student's combination lock happens to be three consecutive odd integers whose sum is 51. Find the integers.   15, 17, 19

Objectives A B C **Mixed Practice** *Solve. See Examples 1 through 7.*

35. A 17-foot piece of string is cut into two pieces so that the longer piece is 2 feet longer than twice the length of the shorter piece. Find the lengths of both pieces.   5 ft, 12 ft

36. A 25-foot wire is to be cut so that the longer piece is one foot longer than 5 times the length of the shorter piece. Find the length of each piece.   4 ft, 21 ft

37. Currently, the two fastest trains are the Japanese Maglev and the French TGV. The sum of their fastest speeds is 718.2 miles per hour. If the speed of the Maglev is 3.8 mph faster than the speed of the TGV, find the speeds of each.   Maglev: 361 mph; TGV: 357.2 mph

38. The Pentagon is the world's largest office building in terms of floor space. It has three times the amount of floor space as the Empire State Building. If the total floor space for these two buildings is approximately 8700 thousand square feet, find the floor space of each building.   Empire State Bldg.: 2175 thousand sq ft; Pentagon: 6525 thousand sq ft

39. Two angles are supplementary if their sum is 180°. The larger angle below measures eight degrees more than three times the measure of the smaller angle. If x represents the measure of the smaller angle and these two angles are supplementary, find the measure of each angle.   43°, 137°

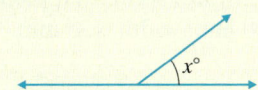

40. Two angles are complementary if their sum is 90°. Given the measures of the complementary angles shown, find the measure of each angle.   31°, 59°

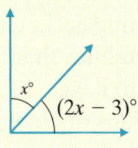

41. The measures of the angles of a triangle are 3 consecutive even integers. Find the measure of each angle.   58°, 60°, 62°

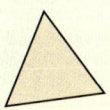

42. A quadrilateral is a polygon with 4 sides. The sum of the measures of the 4 angles in a quadrilateral is 360°. If the measures of the angles of a quadrilateral are consecutive odd integers, find the measures.   87°, 89°, 91°, 93°

43. The sum of $\frac{1}{5}$ and twice a number is equal to $\frac{4}{5}$ subtracted from three times the number. Find the number.   1

44. The sum of $\frac{2}{3}$ and four times a number is equal to $\frac{5}{6}$ subtracted from five times the number. Find the number. $\frac{3}{2}$

45. Hertz Car Rental charges a daily rate of $39 plus $0.20 per mile for a certain car. Suppose that you rent that car for a day and your bill (before taxes) is $95. How many miles did you drive?   280 mi

46. A woman's $15,000 estate is to be divided so that her husband receives twice as much as her son. Find the amount of money that her husband receives and the amount of money that her son receives.   son: $5000; husband: $10,000

47. During the 2013 Rose Bowl, Stanford University beat University of Wisconsin by 6 points. If their combined scores totaled 34, find the individual team scores. (*Source:* Tournament of Roses Association)   Stanford: 20; Wisconsin: 14

48. In June 2013, there were 10 more Republican governors than Democratic governors in the United States. How many Democrats and how many Republicans held governors' offices at that point in time?   Democrats: 20; Republicans: 30

49. The number of counties in California and the number of counties in Montana are consecutive even integers whose sum is 114. If California has more counties than Montana, how many counties does each state have? (*Source: The World Almanac and Book of Facts*)
Montana: 56 counties; California: 58 counties

50. A student is building a bookcase with stepped shelves for her dorm room. She buys a 48-inch board and wants to cut the board into three pieces with lengths equal to three consecutive even integers. Find the three board lengths.   14 in., 16 in., 18 in.

51. Scientists are continually updating information about the planets in our solar system, including the number of satellites orbiting each. Uranus is now believed to have 13 more satellites than Neptune. Also, Saturn is now believed to have 6 more than four times the number of satellites of Neptune. If the total number of satellites for these planets is 103, find the number of satellites for each planet.   Neptune: 14 satellites; Uranus: 27 satellites; Saturn: 62 satellites

52. Apple's iPad Mini tablet computer was launched in 2012. The height of each iPad Mini is 70 millimeters less than twice the width. If the sum of the height and width of an iPad Mini is 335 millimeters, find each dimension. (*Source:* Apple Inc.)
height: 200 mm; width: 135 mm

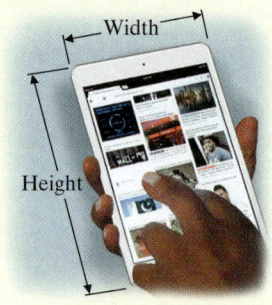

53. If the sum of a number and five is tripled, the result is one less than twice the number. Find the number.   −16

54. Twice the sum of a number and six equals three times the sum of the number and four. Find the number.   0

55. The area of the Sahara Desert is 7 times the area of the Gobi Desert. If the sum of their areas is 4,000,000 square miles, find the area of each desert.   Sahara: 3,500,000 sq mi; Gobi: 500,000 sq mi

56. The largest meteorite in the world is the Hoba West, located in Namibia. Its weight is 3 times the weight of the Armanty meteorite, located in Outer Mongolia. If the sum of their weights is 88 tons, find the weight of each.   Armanty: 22 tons; Hoba West: 66 tons

57. In the 2012 Summer Olympics in London, New Zealand won more gold medals than Cuba, which won more gold medals than Jamaica. If the numbers of gold medals won by these three countries are three consecutive integers whose sum is 15, find the number of gold medals won by each. (*Source:* International Olympic Committee)
Jamaica: 4; Cuba: 5; New Zealand: 6

58. To make an international telephone call, you need the code for the country you are calling. The codes for Mali Republic, Côte d'Ivoire, and Niger are three consecutive odd integers whose sum is 675. Find the code for each country.
Mali Republic: 223; Côte d'Ivoire: 225; Niger: 227

59. In the fall of 2012, there were 1580 more female students enrolled at Rutgers University than male students. If the total student enrollment at Rutgers was 58,788 that fall, find the numbers of female students and male students who were enrolled. (*Source:* Rutgers, The State University of New Jersey)   females: 30,184; males: 28,604

60. The middle-sized car category saw the greatest percentage increase in sales of any vehicle category in the United States in 2012 over the previous year. Approximately 1.1 million fewer pickups were sold than middle-sized cars in 2012. If the total number of pickups and middle-sized cars sold was 4.9 million, find the number of vehicles sold in each category. (*Source:* Alliance of Automobile Manufacturers)   middle-sized cars: 3.0 million; pickups: 1.9 million

61. A geodesic dome, based on the design by Buckminster Fuller, is composed of two different types of triangular panels. One of these is an isosceles triangle. In one geodesic dome, the measure of the third angle is 76.5° more than the measure of either of the two equal angles. Find the measure of the three angles. (*Source:* Buckminster Fuller Institute)   34.5°, 34.5°, 111°

62. The measures of the angles of a particular triangle are such that the second and third angles are each four times the measure of the smallest angle. Find the measures of the angles of this triangle.   20°, 80°, 80°

The graph below shows the states with the highest tourism budgets for the 2012–2013 fiscal year. Use this graph for Exercises 63 through 68.

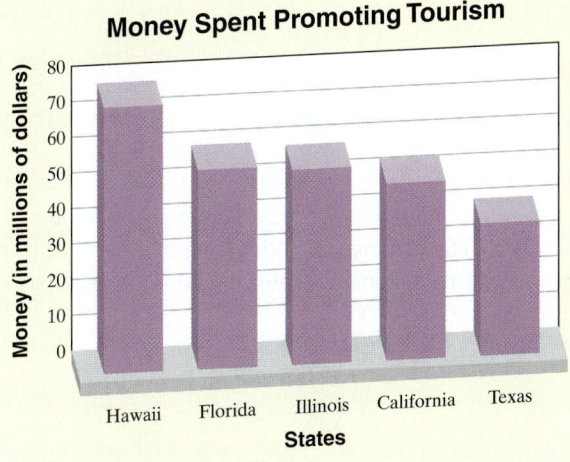

Money Spent Promoting Tourism

Source: U.S. Travel Association

63. Which state spent the most money on tourism?   Hawaii

64. Which state(s) spent between $30 million and $40 million on tourism?   Texas

65. The states of Florida and California spent a total of $106 million on tourism. The state of Florida spent $6 million more than the state of California. Find the amount that each state spent on tourism.   Florida: $56 million; California: $50 million

66. The states of Illinois and Texas spent a total of $92 million on tourism. The state of Illinois spent $19 million less than twice the amount of money that the state of Texas spent. Find the amount that each state spent on tourism.   Illinois: $55 million; Texas: $37 million

Compare the heights of the bars in the graph on the previous page with your results for the exercises below. Are your answers reasonable?

67. Exercise 65   answers may vary

68. Exercise 66   answers may vary

Review

Evaluate each expression for the given values. See Section 1.3.

69. $2W + 2L$; $W = 7$ and $L = 10$ 34

70. $\frac{1}{2}Bh$; $B = 14$ and $h = 22$ 154

71. πr^2; $r = 15$ 225π

72. $r \cdot t$; $r = 15$ and $t = 2$ 30

Concept Extensions

△ **73.** A golden rectangle is a rectangle whose length is approximately 1.6 times its width. The early Greeks thought that a rectangle with these dimensions was the most pleasing to the eye and examples of the golden rectangle are found in many early works of art. For example, the Parthenon in Athens contains many examples of golden rectangles.

Mike Hallahan would like to plant a rectangular garden in the shape of a golden rectangle. If he has 78 feet of fencing available, find the dimensions of the garden.   15 ft by 24 ft

△ **74.** Dr. Dorothy Smith gave the students in her geometry class at the University of New Orleans the following question. Is it possible to construct a triangle such that the second angle of the triangle has a measure that is twice the measure of the first angle and the measure of the third angle is 5 times the measure of the first? If so, find the measure of each angle. (*Hint:* Recall that the sum of the measures of the angles of a triangle is 180°.)   yes; 22.5°, 45°, 112.5°

75. Only male crickets chirp. They chirp at different rates depending on their species and the temperature of their environment. Suppose a certain species is currently chirping at a rate of 90 chirps per minute. At this rate, how many chirps occur in one hour? In one 24-hour day? In one year?   5400 chirps per hour; 129,600 chirps per day; 47,304,000 chirps per year

76. The human eye blinks once every 5 seconds on average. How many times does the average eye blink in one hour? In one 16-hour day while awake? In one year while awake?   720 blinks per hour; 11,520 blinks per day; 4,204,800 blinks per year

77. In your own words, explain why a solution of a word problem should be checked using the original wording of the problem and not the equation written from the wording.   answers may vary

78. Give an example of how you recently solved a problem using mathematics.   answers may vary

Recall from Exercise 73 that a golden rectangle is a rectangle whose length is approximately 1.6 times its width.

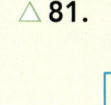 **79.** It is thought that for about 75% of adults, a rectangle in the shape of the golden rectangle is the most pleasing to the eye. Draw three rectangles, one in the shape of the golden rectangle, and poll your class. Do the results agree with the percentage given above?   answers may vary

80. Examples of golden rectangles can be found today in architecture and manufacturing packaging. Find an example of a golden rectangle in your home. A few suggestions: the front face of a book, the floor of a room, the front of a box of food.   answers may vary

For Exercises 81 and 82, measure the dimensions of each rectangle and decide which one best approximates the shape of a golden rectangle.

81.   c

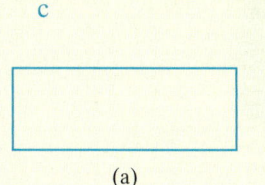
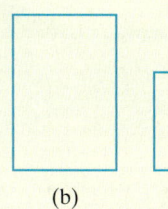
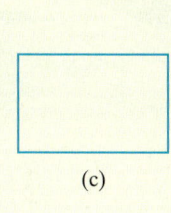

    (a)   (b)   (c)

82.   b

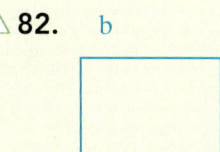
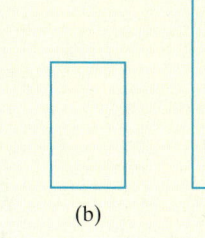

    (a)   (b)   (c)

2.5 Formulas and Problem Solving

Objective A Using Formulas to Solve Problems

A **formula** describes a known relationship among quantities. Many formulas are given as equations. For example, the formula

$$d = r \cdot t$$

stands for the relationship

distance = rate · time

Let's look at one way that we can use this formula.

If we know we traveled a distance of 100 miles at a rate of 40 miles per hour, we can replace the variables d and r in the formula $d = rt$ and find our travel time, t.

$d = rt$ Formula

$100 = 40t$ Replace d with 100 and r with 40.

To solve for t, we divide both sides of the equation by 40.

$\dfrac{100}{40} = \dfrac{40t}{40}$ Divide both sides by 40.

$\dfrac{5}{2} = t$ Simplify.

The travel time was $\dfrac{5}{2}$ hours, or $2\dfrac{1}{2}$ hours, or 2.5 hours.

In this section, we solve problems that can be modeled by known formulas. We use the same problem-solving strategy that was introduced in the previous section.

Objectives

A Use Formulas to Solve Problems.

B Solve a Formula or Equation for One of Its Variables.

Teaching Tip

Remind students to pay attention to the units when working with formulas and to express the answer with the appropriate units.

Practice 1

A family is planning their vacation to visit relatives. They will drive from Cincinnati, Ohio, to Rapid City, South Dakota, a distance of 1180 miles. They plan to average a rate of 50 miles per hour. How much time will they spend driving?

Example 1   Finding Time Given Rate and Distance

A glacier is a giant mass of rocks and ice that flows downhill like a river. Portage Glacier in Alaska is about 6 miles, or 31,680 *feet,* long and moves 400 *feet* per year. Icebergs are created when the front end of the glacier flows into Portage Lake. How long does it take for ice at the head (beginning) of the glacier to reach the lake?

Solution:

1. UNDERSTAND. Read and reread the problem. The appropriate formula needed to solve this problem is the distance formula, $d = rt$. To become familiar with this formula, let's find the distance that ice traveling at a rate of 400 feet per year travels in 100 years. To do so, we let time t be 100 years and rate r be the given 400 feet per year, and substitute these values into the formula $d = rt$. We then have that distance $d = 400(100) = 40,000$ feet. Since we are interested in finding how long it takes ice to travel 31,680 feet, we now know that it is less than 100 years.

   Since we are using the formula $d = rt$, we let

   $t = $ the time in years for ice to reach the lake

   $r = $ rate or speed of ice

   $d = $ distance from beginning of glacier to lake

2. TRANSLATE. To translate to an equation, we use the formula $d = rt$ and let distance $d = 31,680$ feet and rate $r = 400$ feet per year.

$$d = r \cdot t$$
$$31,680 = 400 \cdot t \quad \text{Let } d = 31,680 \text{ and } r = 400.$$

3. SOLVE. Solve the equation for t. To solve for t, we divide both sides by 400.

$$\frac{31,680}{400} = \frac{400 \cdot t}{400} \quad \text{Divide both sides by 400.}$$
$$79.2 = t \quad \text{Simplify.}$$

4. INTERPRET.

Check: To check, substitute 79.2 for t and 400 for r in the distance formula and check to see that the distance is 31,680 feet.

State: It takes 79.2 years for the ice at the head of Portage Glacier to reach the lake.

■ **Work Practice 1**

Helpful Hint   Don't forget to include units, if appropriate.

Answer

1. 23.6 hours

△ **Example 2**   Calculating the Length of a Garden

Charles Pecot can afford enough fencing to enclose a rectangular garden with a perimeter of 140 feet. If the width of his garden is to be 30 feet, find the length.

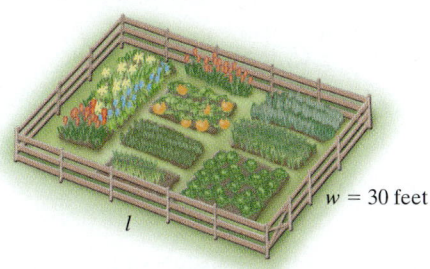

$w = 30$ feet

l

Solution:

1. **UNDERSTAND.** Read and reread the problem. The formula needed to solve this problem is the formula for the perimeter of a rectangle, $P = 2l + 2w$. Before continuing, let's become familar with this formula.

   $l = $ the length of the rectangular garden

   $w = $ the width of the rectangular garden

   $P = $ perimeter of the garden

2. **TRANSLATE.** To translate to an equation, we use the formula $P = 2l + 2w$ and let perimeter $P = 140$ feet and width $w = 30$ feet.

$$P = 2l + 2w \qquad \text{Let } P = 140 \text{ and } w = 30.$$
$$140 = 2l + 2(30)$$

3. **SOLVE.**

$$140 = 2l + 2(30)$$
$$140 = 2l + 60 \qquad \text{Multiply } 2(30).$$
$$140 - 60 = 2l + 60 - 60 \qquad \text{Subtract 60 from both sides.}$$
$$80 = 2l \qquad \text{Combine like terms.}$$
$$40 = l \qquad \text{Divide both sides by 2.}$$

4. **INTERPRET.**

Check: Substitute 40 for l and 30 for w in the perimeter formula and check to see that the perimeter is 140 feet.

State: The length of the rectangular garden is 40 feet.

■ **Work Practice 2**

Example 3   Finding an Equivalent Temperature

The average maximum temperature for the month of January in Algiers, Algeria, is 59° Fahrenheit. Find the equivalent temperature in degrees Celsius.

Solution:

1. **UNDERSTAND.** Read and reread the problem. A formula that can be used to solve this problem is the formula for converting degrees Celsius to degrees Fahrenheit, $F = \dfrac{9}{5}C + 32$. Before continuing, become familiar with this formula. Using this formula, we let

   $C = $ temperature in degrees Celsius, and

   $F = $ temperature in degrees Fahrenheit.

(Continued on next page)

△ **Practice 2**

A wood deck is being built behind a house. The width of the deck must be 18 feet because of the shape of the house. If there is 450 square feet of decking material, find the length of the deck.

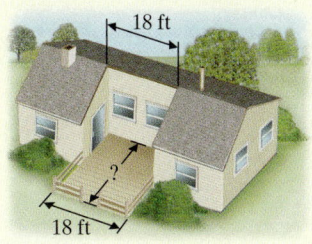

18 ft

?

18 ft

Teaching Tip

You may want to take a moment to remind students that perimeter is measured in units, area in square units, and volume in cubic units.

Practice 3

Convert the temperature 5°C to Fahrenheit.

Answers

2. 25 feet   **3.** 41°F

2. **TRANSLATE.** To translate to an equation, we use the formula $F = \frac{9}{5}C + 32$ and let degrees Fahrenheit $F = 59$.

Formula: $F = \frac{9}{5}C + 32$

Substitute: $59 = \frac{9}{5}C + 32$ Let $F = 59$.

3. **SOLVE.**

$$59 = \frac{9}{5}C + 32$$

$$59 - 32 = \frac{9}{5}C + 32 - 32 \quad \text{Subtract 32 from both sides.}$$

$$27 = \frac{9}{5}C \quad \text{Combine like terms.}$$

$$\frac{5}{9} \cdot 27 = \frac{5}{9} \cdot \frac{5}{9}C \quad \text{Multiply both sides by } \frac{5}{9}.$$

$$15 = C \quad \text{Simplify.}$$

4. **INTERPRET.**

Check: To check, replace C with 15 and F with 59 in the formula and see that a true statement results.

State: Thus, 59° Fahrenheit is equivalent to 15° Celsius.

🟧 **Work Practice 3**

In the next example, we again use the formula for perimeter of a rectangle as in Example 2. In Example 2, we knew the width of the rectangle. In this example, both the length and width are unknown.

Practice 4

The length of a rectangle is one meter more than 4 times its width. Find the dimensions if the perimeter is 52 meters.

⚠️ **Example 4**   Finding Road Sign Dimensions

The length of a rectangular road sign is 2 feet less than three times its width. Find the dimensions if the perimeter is 28 feet.

Solution:

1. **UNDERSTAND.** Read and reread the problem. Recall that the formula for the perimeter of a rectangle is $P = 2l + 2w$. Draw a rectangle and guess the solution. If the width of the rectangular sign is 5 feet, its length is 2 feet less than 3 times the width, or $3(5 \text{ feet}) - 2 \text{ feet} = 13 \text{ feet}$. The perimeter P of the rectangle is then $2(13 \text{ feet}) + 2(5 \text{ feet}) = 36 \text{ feet}$, too much. We now know that the width is less than 5 feet.

Proposed rectangle:

13 feet

5 feet

Answer

4. length: 21 m; width: 5 m

Let

w = the width of the rectangular sign; then

$3w - 2$ = the length of the sign.

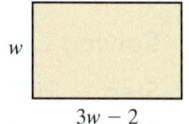

w

$3w - 2$

Draw a rectangle and label it with the assigned variables.

2. TRANSLATE.

Formula: $P = 2l + 2w$

Substitute: $28 = 2(3w - 2) + 2w$

3. SOLVE.

$$28 = 2(3w - 2) + 2w$$

$$28 = 6w - 4 + 2w \qquad \text{Apply the distributive property.}$$

$$28 = 8w - 4$$

$$28 + 4 = 8w - 4 + 4 \qquad \text{Add 4 to both sides.}$$

$$32 = 8w$$

$$\frac{32}{8} = \frac{8w}{8} \qquad \text{Divide both sides by 8.}$$

$$4 = w$$

4. INTERPRET.

Check: If the width of the sign is 4 feet, the length of the sign is $3(4 \text{ feet}) - 2 \text{ feet} = 10 \text{ feet}$. This gives the rectangular sign a perimeter of $P = 2(4 \text{ feet}) + 2(10 \text{ feet}) = 28 \text{ feet}$, the correct perimeter.

State: The width of the sign is 4 feet and the length of the sign is 10 feet.

■ **Work Practice 4**

Objective B Solving a Formula for a Variable ▶

We say that the formula

$$d = rt$$

is solved for d because d is alone on one side of the equation and the other side contains no d's. Suppose that we have a large number of problems to solve where we are given distance d and rate r and asked to find time t. In this case, it may be easier to first solve the formula $d = rt$ for t. To solve for t, we divide both sides of the equation by r.

$$d = rt$$

$$\frac{d}{r} = \frac{rt}{r} \qquad \text{Divide both sides by } r.$$

$$\frac{d}{r} = t \qquad \text{Simplify.}$$

To solve a formula or an equation for a specified variable, we use the same steps as for solving a linear equation except that we treat the specified variable as the only variable in the equation. These steps are listed next.

Solving Equations for a Specified Variable

Step 1: Multiply on both sides to clear the equation of fractions if they appear.

Step 2: Use the distributive property to remove parentheses if they appear.

Step 3: Simplify each side of the equation by combining like terms.

Step 4: Get all terms containing the specified variable on one side and all other terms on the other side by using the addition property of equality.

Step 5: Get the specified variable alone by using the multiplication property of equality.

Practice 5

Solve $C = 2\pi r$ for r.
(This formula is used to find the circumference, C, of a circle given its radius, r.)

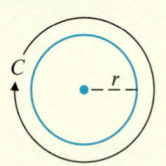

Example 5   Solve $V = lwh$ for l.

Solution: This formula is used to find the volume of a box. To solve for l, divide both sides by wh.

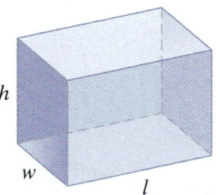

$$V = lwh$$

$$\frac{V}{wh} = \frac{lwh}{wh} \qquad \text{Divide both sides by } wh.$$

$$\frac{V}{wh} = l \qquad \text{Simplify.}$$

Since we have l alone on one side of the equation, we have solved for l in terms of V, w, and h. Remember that it does not matter on which side of the equation we get the variable alone.

■ **Work Practice 5**

Practice 6

Solve $P = 2l + 2w$ for l.

Example 6   Solve $y = mx + b$ for x.

Solution: First we get mx alone by subtracting b from both sides.

$$y = mx + b$$

$$y - b = mx + b - b \qquad \text{Subtract } b \text{ from both sides.}$$

$$y - b = mx \qquad \text{Combine like terms.}$$

Next we solve for x by dividing both sides by m.

$$\frac{y - b}{m} = \frac{mx}{m}$$

$$\frac{y - b}{m} = x \qquad \text{Simplify.}$$

■ **Work Practice 6**

Answers

5. $r = \dfrac{C}{2\pi}$ **6.** $l = \dfrac{P - 2w}{2}$

✓**Concept Check Answer**

a. ⬤ + ▢   **b.** $\dfrac{⬤ \; + \; ▢}{▲}$

✓**Concept Check** Solve:

a. ⬤ = ▢ − ▢ for ▢   **b.** ⬤ = ▢ · ▲ − ▢ for ▢

Example 7 Solve $P = 2l + 2w$ for w.

Solution: This formula relates the perimeter of a rectangle to its length and width. Find the term containing the variable w. To get this term, $2w$, alone, subtract $2l$ from both sides.

$$P = 2l + 2w$$
$$P - 2l = 2l + 2w - 2l \quad \text{Subtract } 2l \text{ from both sides.}$$
$$P - 2l = 2w \quad \text{Combine like terms.}$$
$$\frac{P - 2l}{2} = \frac{2w}{2} \quad \text{Divide both sides by 2.}$$
$$\frac{P - 2l}{2} = w \quad \text{Simplify.}$$

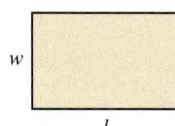

■ **Work Practice 7**

Practice 7

Solve $P = 2a + b - c$ for a.

Helpful Hint

The 2s may *not* be divided out here. Although 2 is a factor of the denominator, 2 is *not* a factor of the numerator since it is not a factor of both terms in the numerator.

The next example has an equation containing a fraction. We will first clear the equation of fractions and then solve for the specified variable.

Example 8 Solve $F = \frac{9}{5}C + 32$ for C.

Solution:
$$F = \frac{9}{5}C + 32$$
$$5(F) = 5\left(\frac{9}{5}C + 32\right) \quad \text{Clear the fraction by multiplying both sides by the LCD.}$$
$$5F = 9C + 160 \quad \text{Distribute the 5.}$$
$$5F - 160 = 9C + 160 - 160 \quad \text{To get the term containing the variable } C \text{ alone, subtract 160 from both sides.}$$
$$5F - 160 = 9C \quad \text{Combine like terms.}$$
$$\frac{5F - 160}{9} = \frac{9C}{9} \quad \text{Divide both sides by 9.}$$
$$\frac{5F - 160}{9} = C \quad \text{Simplify.}$$

■ **Work Practice 8**

Practice 8

Solve $A = \frac{a + b}{2}$ for b.

Answers

7. $a = \dfrac{P - b + c}{2}$ 8. $b = 2A - a$

Vocabulary, Readiness & Video Check

Martin-Gay Interactive Videos

See Video 2.5

See video answer section.

Watch the section lecture video and answer the following questions.

Objective A 1. Complete this statement based on the lecture given before ⊞ Example 1. A formula is an equation that describes known _____ among quantities. ▶

2. In ⊞ Example 2, how are the units for the solution determined? ▶

Objective B 3. During ⊞ Example 4, the equation $5x = 30$ is shown to demonstrate what process? ▶

2.5 Exercise Set MyMathLab®

Objective A *Substitute the given values into each given formula and solve for the unknown variable. See Examples 1 through 4.*

△ **1.** $A = bh$; $A = 45, b = 15$ (Area of a parallelogram) $h = 3$

2. $d = rt$; $d = 195, t = 3$ (Distance formula) $r = 65$

△ **3.** $S = 4lw + 2wh$; $S = 102, l = 7, w = 3$ (Surface area of a special rectangular box) $h = 3$

△ **4.** $V = lwh$; $l = 14, w = 8, h = 3$ (Volume of a rectangular box) $V = 336$

△ **5.** $A = \frac{1}{2}h(B + b)$; $A = 180, B = 11, b = 7$ (Area of a trapezoid) $h = 20$

△ **6.** $A = \frac{1}{2}h(B + b)$; $A = 60, B = 7, b = 3$ (Area of a trapezoid) $h = 12$

△ **7.** $P = a + b + c$; $P = 30, a = 8, b = 10$ (Perimeter of a triangle) $c = 12$

△ **8.** $V = \frac{1}{3}Ah$; $V = 45, h = 5$ (Volume of a pyramid) $A = 27$

△ **9.** $C = 2\pi r$; $C = 15.7$ (Circumference of a circle) (Use the approximation 3.14 for π.) $r = 2.5$

△ **10.** $A = \pi r^2$; $r = 4$ (Area of a circle) (Use the approximation 3.14 for π.) $A = 50.24$

Objective B *Solve each formula for the specified variable. See Examples 5 through 8.*

11. $f = 5gh$ for h

$h = \dfrac{f}{5g}$

△ **12.** $x = 4\pi y$ for y

$y = \dfrac{x}{4\pi}$

△ **13.** $V = lwh$ for w

$w = \dfrac{V}{lh}$

14. $T = mnr$ for n

$n = \dfrac{T}{mr}$

15. $3x + y = 7$ for y

$y = 7 - 3x$

16. $-x + y = 13$ for y

$y = x + 13$

17. $A = P + PRT$ for R

$R = \dfrac{A - P}{PT}$

18. $A = P + PRT$ for T

$T = \dfrac{A - P}{PR}$

△ **19.** $V = \frac{1}{3}Ah$ for A

$A = \dfrac{3V}{h}$

20. $D = \frac{1}{4}fk$ for k

$k = \dfrac{4D}{f}$

△ **21.** $P = a + b + c$ for a

$a = P - b - c$

22. $PR = x + y + z + w$ for z

$z = PR - x - y - w$

△ **23.** $S = 2\pi rh + 2\pi r^2$ for h

$h = \dfrac{S - 2\pi r^2}{2\pi r}$

△ **24.** $S = 4lw + 2wh$ for h

$h = \dfrac{S - 4lw}{2w}$

Objective A *Solve. For Exercises 25 and 26, the solutions have been started for you. See Examples 1 through 4.*

△ **25.** The iconic NASDAQ sign in New York's Times Square has a width of 84 feet and an area of 10,080 square feet. Find the height (or length) of the sign. (*Source:* livedesignonline.com) 120 ft

Start the solution:

1. UNDERSTAND the problem. Reread it as many times as needed.
2. TRANSLATE into an equation. (Fill in the blanks below.)

Area	=	length	times	width
↓	↓	↓	↓	↓
___	=	x	·	___

Finish with:
3. SOLVE and 4. INTERPRET

△ **26.** The world's largest sign for Coca-Cola is located in Arica, Chile. The rectangular sign has a length of 400 feet and an area of 52,400 square feet. Find the width of the sign. (*Source:* Fabulous Facts about Coca-Cola, Atlanta, GA) 131 ft

Start the solution:

1. UNDERSTAND the problem. Reread it as many times as needed.
2. TRANSLATE into an equation. (Fill in the blanks below.)

Area	=	length	times	width
↓	↓	↓	↓	↓
___	=	___	·	x

Finish with:
3. SOLVE and 4. INTERPRET

△ **27.** A frame shop charges according to both the amount of framing needed to surround the picture and the amount of glass needed to cover the picture.

a. Find the area and perimeter of the picture below.   area: 480 sq in.; perimeter: 120 in.
b. Identify whether the frame has to do with perimeter or area and the same with the glass.
frame: perimeter; glass: area

24 in.
20 in.   20 in.
12 in.
56 in.

△ **28.** A decorator is painting and placing a border completely around the parallelogram-shaped wall.

a. Find the area and perimeter of the wall below. ($A = bh$)   area: 65.1 sq ft; perimeter: 42 ft;
b. Identify whether the border has to do with perimeter or area and the same with paint.
border: perimeter; paint: area

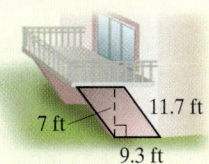

7 ft   11.7 ft
9.3 ft

△ **29.** For the purpose of purchasing new baseboard and carpet,

a. Find the area and perimeter of the room below (neglecting doors).   area: 103.5 sq ft; perimeter: 41 ft;
b. Identify whether baseboard has to do with area or perimeter and the same with carpet.
baseboard: perimeter; carpet: area

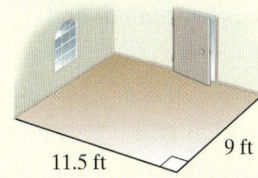

11.5 ft   9 ft

△ **30.** For the purpose of purchasing lumber for a new fence and seed to plant grass,

a. Find the area and perimeter of the yard below.   area: 486 sq ft; perimeter: 108 ft;

$$\left(A = \frac{1}{2}bh \right)$$

b. Identify whether a fence has to do with area or perimeter and the same with grass seed.
fence: perimeter; grass: area

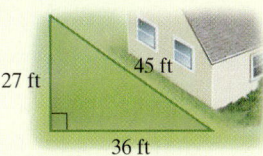

27 ft   45 ft
36 ft

▶ **31.** Convert Nome, Alaska's 14°F high temperature to Celsius.   −10°C

32. Convert Paris, France's low temperature of −5°C to Fahrenheit.   23°F

33. The X-30 is a "space plane" that skims the edge of space at 4000 miles per hour. Neglecting altitude, if the circumference of Earth is approximately 25,000 miles, how long will it take for the X-30 to travel around Earth?   6.25 hr

34. In the United States, a notable hang glider flight was a 303-mile, $8\frac{1}{2}$-hour flight from New Mexico to Kansas. What was the average rate during this flight? $35\frac{11}{17}$ mph

35. An architect designs a rectangular flower garden such that the width is exactly two-thirds of the length. If 260 feet of antique picket fencing are to be used to enclose the garden, find the dimensions of the garden.   length: 78 ft; width: 52 ft

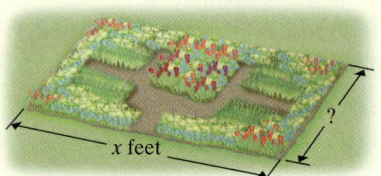

x feet

36. If the length of a rectangular parking lot is 10 meters less than twice its width, and the perimeter is 400 meters, find the length of the parking lot.   130 m

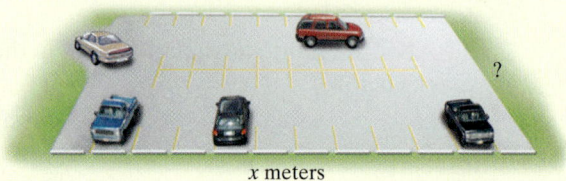

x meters

37. A flower bed is in the shape of a triangle with one side twice the length of the shortest side, and the third side is 30 feet more than the length of the shortest side. Find the dimensions if the perimeter is 102 feet.   18 ft, 36 ft, 48 ft

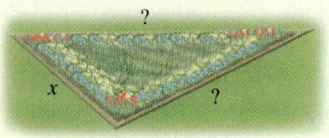

x

38. The perimeter of a yield sign in the shape of an isosceles triangle is 22 feet. If the shortest side is 2 feet less than the other two sides, find the length of the shortest side. (*Hint:* An isosceles triangle has two sides the same length.)   6 ft

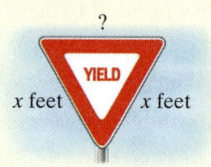

x feet   *x* feet

39. The Cat is a high-speed catamaran auto ferry that operates between Bar Harbor, Maine, and Yarmouth, Nova Scotia. The Cat can make the trip in about $2\frac{1}{2}$ hours at a speed of 55 mph. About how far apart are Bar Harbor and Yarmouth? (*Source:* Bay Ferries)   137.5 mi

40. A family is planning their vacation to Disney World. They will drive from a small town outside New Orleans, Louisiana, to Orlando, Florida, a distance of 700 miles. They plan to average a rate of 55 mph. How long will this trip take? $12\frac{8}{11}$ hr

Dolbear's Law states the relationship between the rate at which Snowy Tree Crickets chirp and the air temperature of their environment. The formula is

$$T = 50 + \frac{N - 40}{4}, where$$ T = temperature in degrees Fahrenheit and
N = number of chirps per minute

41. If $N = 86$, find the temperature in degrees Fahrenheit, T.   61.5°F

42. If $N = 94$, find the temperature in degrees Fahrenheit, T.   63.5°F

43. If $T = 55°F$, find the number of chirps per minute.   60 chirps per minute

44. If $T = 65°F$, find the number of chirps per minute.   100 chirps per minute

Use the results of Exercises 41–44 to complete each sentence with "increases" or "decreases."

45. As the number of cricket chirps per minute increases, the air temperature of their environment ___increases___.

46. As the air temperature of their environment decreases, the number of cricket chirps per minute ___decreases___.

Solve. See Examples 1 through 4.

△ **47.** Piranha fish require 1.5 cubic feet of water per fish to maintain a healthy environment. Find the maximum number of piranhas you could put in a tank measuring 8 feet by 3 feet by 6 feet.   96 piranhas

6 feet

3 feet   8 feet

△ **48.** Find the maximum number of goldfish you can put in a cylindrical tank whose diameter is 8 meters and whose height is 3 meters, if each goldfish needs 2 cubic meters of water. $(V = \pi r^2 h)$ 75 goldfish

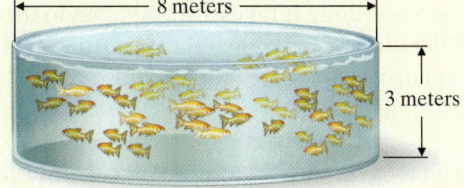

8 meters

3 meters

△ **49.** A lawn is in the shape of a trapezoid with a height of 60 feet and bases of 70 feet and 130 feet. How many bags of fertilizer must be purchased to cover the lawn if each bag covers 4000 square feet?
$\left(A = \frac{1}{2}h(B + b) \right)$ 2 bags

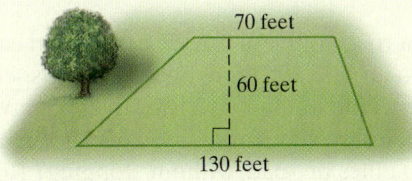

70 feet

60 feet

130 feet

△ **50.** If the area of a right-triangularly shaped sail is 20 square feet and its base is 5 feet, find the height of the sail. $\left(A = \frac{1}{2}bh \right)$ 8 ft

?

5 feet

△ **51.** Maria's Pizza sells one 16-inch cheese pizza or two 10-inch cheese pizzas for $9.99. Determine which size gives more pizza. $(A = \pi r^2)$ one 16-in. pizza

16 inches   10 inches   10 inches

△ **52.** Find how much rope is needed to wrap around Earth at the equator, if the radius of Earth is 4000 miles. (*Hint:* Use 3.14 for π and the formula for circumference.)   25,120 mi

53. A Japanese "bullet" train set a new world record for train speed at 552 kilometers per hour during a manned test run on the Yamanashi Maglev Test Line in April 1999. The Yamanashi Maglev Test Line is 42.8 kilometers long. How many *minutes* would a test run on the Yamanashi Line last at this record-setting speed? Round to the nearest hundredth of a minute. (*Source:* Japan Railways Central Co.)   4.65 min

54. In 1983, the Hawaiian volcano Kilauea began erupting in a series of episodes still occurring at the time of this writing. At times, the lava flows advanced at speeds of up to 0.5 kilometer per hour. In 1983 and 1984 lava flows destroyed 16 homes in the Royal Gardens subdivision, about 6 km away from the eruption site. Roughly how long did it take the lava to reach Royal Gardens? (*Source:* U.S. Geological Survey Hawaiian Volcano Observatory)   12 hr

△ **55.** The perimeter of an equilateral triangle is 7 inches more than the perimeter of a square, and the side of the triangle is 5 inches longer than the side of the square. Find the side of the triangle. (*Hint:* An equilateral triangle has three sides the same length.)   13 in.

△ **56.** A square animal pen and a pen shaped like an equilateral triangle have equal perimeters. Find the length of the sides of each pen if the sides of the triangular pen are fifteen less than twice a side of the square pen. (*Hint:* An equilateral triangle has three sides the same length.)   square's side length: 22.5 units; triangle's side length: 30 units

57. Find how long it takes Tran Nguyen to drive 135 miles on I-10 if he merges onto I-10 at 10 a.m. and drives nonstop with his cruise control set on 60 mph.   2.25 hr

58. Beaumont, Texas, is about 150 miles from Toledo Bend. If Leo Miller leaves Beaumont at 4 a.m. and averages 45 mph, when should he arrive at Toledo Bend?   7:20 a.m.

△ **59.** The longest runway at Los Angeles International Airport has the shape of a rectangle and an area of 1,813,500 square feet. This runway is 150 feet wide. How long is the runway? (*Source:* Los Angeles World Airports)   12,090 ft

60. The return stroke of a bolt of lightning can travel at a speed of 87,000 miles per second (almost half the speed of light). At this speed, how many times can an object travel around the world in one second? (See Exercise 52.) Round to the nearest tenth. (*Source: The Handy Science Answer Book*)   3.5

61. The highest temperature ever recorded in Europe was 122°F in Seville, Spain, in August of 1881. Convert this record high temperature to Celsius. (*Source:* National Climatic Data Center)   50°C

62. The lowest temperature ever recorded in Oceania was −10°C at the Haleakala Summit in Maui, Hawaii, in January 1961. Convert this record low temperature to Fahrenheit. (*Source:* National Climatic Data Center)   14°F

△ **63.** The CART FedEx Championship Series is an open-wheeled race car competition based in the United States. A CART car has a maximum length of 199 inches, a maximum width of 78.5 inches, and a maximum height of 33 inches. When the CART series travels to another country for a grand prix, teams must ship their cars. Find the volume of the smallest shipping crate needed to ship a CART car of maximum dimensions. (*Source:* Championship Auto Racing Teams, Inc.)   515,509.5 cu in.

64. On a road course, a CART car's speed can average up to around 105 mph. Based on this speed, how long would it take a CART driver to travel from Los Angeles to New York City, a distance of about 2810 miles by road, without stopping? Round to the nearest tenth of an hour.   26.8 hr

CART Racing Car

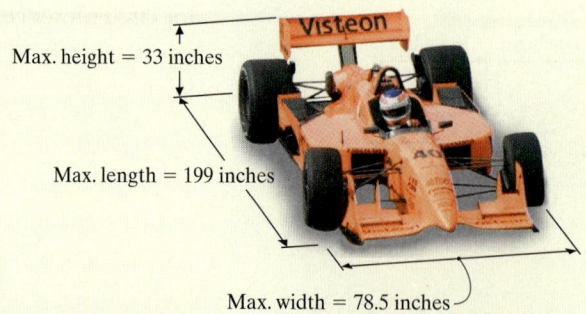

Max. height = 33 inches
Max. length = 199 inches
Max. width = 78.5 inches

△ **65.** The Hoberman Sphere is a toy ball that expands and contracts. When it is completely closed, it has a diameter of 9.5 inches. Find the volume of the Hoberman Sphere when it is completely closed. Use 3.14 for π. Round to the nearest whole cubic inch. (*Hint:* volume of a sphere $= \frac{4}{3}\pi r^3$. *Source:* Hoberman Designs, Inc.) 449 cu in.

△ **66.** When the Hoberman Sphere (see Exercise 65) is completely expanded, its diameter is 30 inches. Find the volume of the Hoberman Sphere when it is completely expanded. Use 3.14 for π. (*Source:* Hoberman Designs, Inc.) 14,130 cu in.

67. The average temperature on the planet Mercury is 167°C. Convert this temperature to degrees Fahrenheit. Round to the nearest degree. (*Source:* National Space Science Data Center) 333°F

68. The average temperature on the planet Jupiter is −227°F. Convert this temperature to degrees Celsius. Round to the nearest degree. (*Source:* National Space Science Data Center) −144°C

Review

Write each percent as a decimal. See Section R.3.

69. 32% 0.32

70. 8% 0.08

71. 200% 2.00 or 2

72. 0.5% 0.005

Write each decimal as a percent. See Section R.3.

73. 0.17 17%

74. 0.03 3%

75. 7.2 720%

76. 5 500%

Concept Extensions

Solve.

77. $N = R + \dfrac{V}{G}$ for V (Urban forestry: tree plantings per year) $V = G(N - R)$

78. $B = \dfrac{F}{P - V}$ for V (Business: break-even point) $V = P - \dfrac{F}{B}$

△ **79.** The formula $V = lwh$ is used to find the volume of a box. If the length of a box is doubled, the width is doubled, and the height is doubled, how does this affect the volume? Explain your answer. multiplies the volume by 8; answers may vary

△ **80.** The formula $A = bh$ is used to find the area of a parallelogram. If the base of a parallelogram is doubled and its height is doubled, how does this affect the area? Explain your answer. multiplies the area by 4; answers may vary

81. Use the Dolbear's Law formula for Exercises 41–46 and calculate when the number of cricket chirps per minute is the same as the temperature in degrees Fahrenheit. (*Hint:* Replace T with N and solve for N or replace N with T and solve for T. $53\frac{1}{3}$

82. Find the temperature at which the Celsius measurement and the Fahrenheit measurement are the same number. −40°

Solve. See the Concept Check in this section.

83. ▲ − ● · ▮ = ▮ for ●   ◯ = $\dfrac{\triangle - \square}{\square}$

84. ⬟ · ▮ + ◣ = ● for ▮   ▮ = $\dfrac{\bigcirc - \triangle}{\pentagon}$

85. Flying fish do not *actually* fly, but glide. They have been known to travel a distance of 1300 feet at a rate of 20 miles per hour. How many seconds would it take to travel this distance? (*Hint:* First convert miles per hour to feet per second. Recall that 1 mile = 5280 feet.) Round to the nearest tenth of a second.   44.3 sec

86. A glacier is a giant mass of rocks and ice that flows downhill like a river. Exit Glacier, near Seward, Alaska, moves at a rate of 20 inches a day. Find the distance in feet the glacier moves in a year. (Assume 365 days a year.) Round to two decimal places.   608.33 ft

Substitute the given values into each given formula and solve for the unknown variable. If necessary, round to one decimal place.

87. $I = PRT$; $I = 1{,}056{,}000, R = 0.055, T = 6$
(Simple interest formula) $P = 3{,}200{,}000$

88. $I = PRT$; $I = 3750, P = 25{,}000, R = 0.05$
(Simple interest formula) $T = 3$

89. $V = \dfrac{4}{3}\pi r^3$; $r = 3$ (Volume of a sphere)
(Use a calculator approximation for π.) $V = 113.1$

90. $V = \dfrac{1}{3}\pi r^2 h$; $V = 565.2, r = 6$ (Volume of a cone)
(Use a calculator approximation for π.) $h = 15.0$

2.6 Percent and Mixture Problem Solving

Objectives

A Solve Percent Equations.

B Solve Discount and Mark-Up Problems. ▶

C Solve Percent of Increase and Percent of Decrease Problems. ▶

D Solve Mixture Problems. ▶

This section is devoted to solving problems in the categories listed. The same problem-solving steps used in previous sections are also followed in this section. They are listed below for review.

General Strategy for Problem Solving

1. **UNDERSTAND** the problem. During this step, become comfortable with the problem. Some ways of doing this are as follows:

   Read and reread the problem.

   Choose a variable to represent the unknown.

   Construct a drawing, whenever possible.

   Propose a solution and check. Pay careful attention to how you check your proposed solution. This will help writing an equation to model the problem.

2. **TRANSLATE** the problem into an equation.

3. **SOLVE** the equation.

4. **INTERPRET** the results: *Check* the proposed solution in the stated problem and *state* your conclusion.

Objective A Solving Percent Equations

Many of today's statistics are given in terms of percent: a basketball player's free throw percent, current interest rates, stock market trends, and nutrition labeling, just to name a few. In this section, we first explore percent, percent equations, and applications involving percents. See Section R.3 if a further review of percents is needed.

Example 1   The number 63 is what percent of 72?

Solution:

1. UNDERSTAND. Read and reread the problem. Next, let's suppose that the percent is 80%. To check, we find 80% of 72.

$$80\% \text{ of } 72 = 0.80(72) = 57.6$$

This is close, but not 63. At this point, though, we have a better understanding of the problem; we know the correct answer is close to and greater than 80%, and we know how to check our proposed solution later.

Let x = the unknown percent.

2. TRANSLATE. Recall that "is" means "equals" and "of" signifies multiplying. Let's translate the sentence directly.

the number 63	is	what percent	of	72
↓	↓	↓	↓	↓
63	=	x	·	72

3. SOLVE.

$$63 = 72x$$
$$0.875 = x \qquad \text{\color{blue}Divide both sides by 72.}$$
$$87.5\% = x \qquad \text{\color{blue}Write as a percent.}$$

4. INTERPRET.

Check: Verify that 87.5% of 72 is 63.

State: The number 63 is 87.5% of 72.

■ **Work Practice 1**

Example 2   The number 120 is 15% of what number?

Solution:

1. UNDERSTAND. Read and reread the problem.

Let x = the unknown number.

2. TRANSLATE.

the number 120	is	15%	of	what number
↓	↓	↓	↓	↓
120	=	15%	·	x

3. SOLVE.

$$120 = 0.15x \qquad \text{\color{blue}Write 15\% as 0.15.}$$
$$800 = x \qquad \text{\color{blue}Divide both sides by 0.15.}$$

4. INTERPRET.

Check: Check the proposed solution by finding 15% of 800 and verifying that the result is 120.

State: Thus, 120 is 15% of 800.

■ **Work Practice 2**

Practice 1

The number 22 is what percent of 40?

Teaching Tips

It may be helpful to point out to students that they can also solve these percent problems using

$$part = percent \cdot total$$

For example, in a class of 125 students, 56% of the students, or 70 students, are female. Here, 125 students is the *total*, 56% is the *percent*, and 70 students is the *part*.

Practice 2

The number 150 is 40% of what number?

Answers

1. 55%   **2.** 375

Practice 3

Use the circle graph to answer each question.

a. What percent of trips made by American travelers are solely for pleasure?

b. What percent of trips made by American travelers are for the purpose of pleasure or combined business/pleasure?

c. On an airplane flight of 250 Americans, how many of these people might we expect to be traveling solely for pleasure?

Example 3 The circle graph below shows the purpose of trips made by American travelers. Use this graph to answer the questions below.

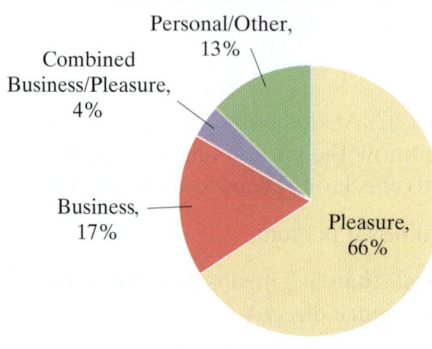

Purpose of Trip

Source: Travel Industry Association of America

a. What percent of trips made by American travelers are solely for the purpose of business?

b. What percent of trips made by American travelers are for the purpose of business or combined business/pleasure?

c. On an airplane flight of 253 Americans, how many of these people might we expect to be traveling solely for business?

Solution:

a. From the circle graph, we see that 17% of trips made by American travelers are solely for the purpose of business.

b. From the circle graph, we know that 17% of trips are solely for business and 4% of trips are for combined business/pleasure. The sum 17% + 4% or 21% of trips made by American travelers are for the purpose of business or combined business/pleasure.

c. Since 17% of trips made by American travelers are for business, we find 17% of 253. Remember that "of" translates to "multiplication."

$$17\% \text{ of } 253 = 0.17(253) \quad \text{Replace "of" with the operation of multiplication.}$$
$$= 43.01$$

We might then expect that about 43 American travelers on the flight are traveling solely for business.

Teaching Tip

You may want to use the following group activity. Give students a true percent equation, like 30% of 120 is 36, or 30% · 120 = 36. Have small groups of students use any two of the three parts of this equation and make up a word problem where the two parts are known and the third part is unknown. Ask the groups to share their word problems. You'll be amazed at their creativity.

■ **Work Practice 3**

Objective B Solving Discount and Mark-Up Problems

The next example has to do with discounting the price of a cell phone.

Practice 4

A surfboard, originally purchased for $400, was sold on eBay at a discount of 40% of the original price. What is the discount and the new price?

Example 4 Cell Phones Unlimited recently reduced the price of a $140 phone by 20%. What is the discount and the new price?

Solution:

1. UNDERSTAND. Read and reread the problem. Make sure you understand the meaning of the word "discount." Discount is the amount of money by which an item has been decreased. To find the discount, we simply find 20% of $140. In other words, we have the formulas,

> discount = percent · original price   Then
>
> new price = original price − discount

Answers

3. a. 66% **b.** 70% **c.** 165 people
4. discount: $160; new price: $240

2, 3. TRANSLATE and SOLVE.

$$\text{discount} = \text{percent} \cdot \text{original price}$$
$$= 20\% \cdot \$140$$
$$= 0.20 \cdot \$140$$
$$= \$28$$

Thus, the discount in price is $28.

$$\text{new price} = \text{original price} - \text{discount}$$
$$= \$140 - \$28$$
$$= \$112$$

4. INTERPRET.

Check: Check your calculations in the formulas, and also see if our results are reasonable. They are.

State: The discount in price is $28 and the new price is $112.

■ Work Practice 4

A concept similar to discount is mark-up. What is the difference between the two? A discount is subtracted from the original price while a mark-up is added to the original price. For mark-ups,

$$\text{mark-up} = \text{percent} \cdot \text{original price}$$

$$\text{new price} = \text{original price} + \text{mark-up}$$

Mark-up exercises can be found in Exercise Set 2.6.

Objective C Solving Percent of Increase and Percent of Decrease Problems

Percent of increase or percent of decrease is a common way to describe how some measurement has increased or decreased. For example, crime increased by 8%, teachers received a 5.5% increase in salary, or a company decreased its employees by 10%. The next example is a review of percent of increase.

Example 5 Calculating the Percent of Increase of Attending College

The average tuition and fees cost of attending a four-year public college rose from $4650 during the 1998–1999 academic year to $8890 during the 2013–2014 year. Find the percent of increase. (*Source:* The College Board)

Solution:

1. UNDERSTAND. Read and reread the problem. Notice that the new tuition, $8890, is almost double the old tuition of $4650. Because of that, we know that the percent of increase is close to 100%. To see this, let's guess that the percent of increase is 100%. To check, we find 100% of $4650 to find the *increase* in cost. Then we add this increase to $4650 to find the *new cost*. In other words, $100\%(\$4650) = 1.00(\$4650) = \$4650$, the *increase* in cost. The *new cost* would be old cost + increase = $4650 + $4650 = $9300, close to the actual new cost of $8890. We now know that the increase is close to, but less than, 100% and we know how to check our proposed solution.

Let x = the percent of increase.

(Continued on next page)

Practice 5

If a number increases from 120 to 200, find the percent of increase. Round to the nearest tenth of a percent.

Answer
5. 66.7%

2. TRANSLATE. First, find the **increase,** and then the **percent of increase.** The increase in cost is found by:

In words:	increase	=	new cost	−	old cost
Translate:	increase	=	$8890	−	$4650
		=	$4240		

Next, find the percent of increase. The percent of increase or percent of decrease is always a percent of the original number or, in this case, the old cost.

In words:	increase	is	what percent	of	old cost
Translate:	$4240	=	x	·	$4650

3. SOLVE.

$$4240 = 4650x$$
$$0.912 \approx x \qquad \text{Divide both sides by 4650 and round to 3 decimal places.}$$
$$91.2\% \approx x \qquad \text{Write as a percent.}$$

4. INTERPRET.

Check: Check the proposed solution

State: The percent of increase in cost is approximately 91.2%.

■ **Work Practice 5**

Percent of decrease is found using a similar method. First find the decrease, then determine what percent of the original or first amount is that decrease.

Read the next example carefully. For Example 5, we were asked to find percent of increase. In Example 6, we are given the percent of increase and asked to find the number before the increase.

Practice 6

Find the original price of a suit if the sale price is $46 after a 20% discount.

Answer

6. $57.50

> **Example 6** Growth in digital 3-D theater screens is fastest in the Asia/Pacific entertainment market. Find the number of digital 3-D screens in Asia/Pacific in 2011 if after a 106% increase, the number in 2013 was 17,726. Round to the nearest whole. (*Source:* MPAA)

Solution:

1. UNDERSTAND. Read and reread the problem. Let's guess a solution and see how we would check our guess. If the number of digital 3-D screens in 2011 was 8000, we would see if 8000 plus the increase is 17,726; that is,

$$8000 + 106\%(8000) = 8000 + 1.06(8000) = 8000 + 8480 = 16,480$$

Since 16,480 is too small, we know that our guess of 8000 is too small. We also have a better understanding of the problem. Let

$$x = \text{number of digital 3-D screens in 2011}$$

2. TRANSLATE. To translate an equation, we remember that

In words:	number of digital 3-D screens in 2011	plus	increase	equals	number of digital 3-D screens in 2013
Translate:	x	+	$1.06x$	=	17,726

3. SOLVE.

$$2.06x = 17,726$$
$$x = \frac{17,726}{2.06}$$
$$x \approx 8605$$

4. INTERPRET.

Check: Recall that x represents the number of digital 3-D screens in 2011. If this number is approximately 8605, let's see if 8605 plus the increase is close to 17,726. (We use the word "close" since 8605 is rounded.)

$$8605 + 106\%(8605) = 8605 + 1.06(8605) = 8605 + 9121.3 = 17,726.3$$

which is close to 17,726.

State: There were approximately 8605 digital 3-D screens in the Asia/Pacific region in 2011.

■ Work Practice 6

Objective D Solving Mixture Problems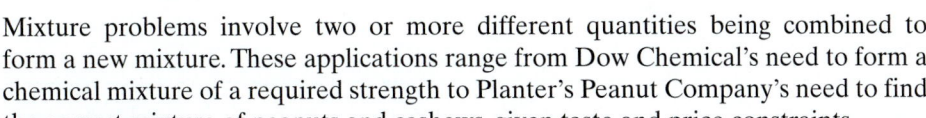

Mixture problems involve two or more different quantities being combined to form a new mixture. These applications range from Dow Chemical's need to form a chemical mixture of a required strength to Planter's Peanut Company's need to find the correct mixture of peanuts and cashews, given taste and price constraints.

Example 7   Calculating Percent for a Lab Experiment

A chemist working on his doctoral degree at Massachusetts Institute of Technology needs 12 liters of a 50% acid solution for a lab experiment. The stockroom has only 40% and 70% solutions. How much of each solution should be mixed together to form 12 liters of a 50% solution?

Solution:

1. UNDERSTAND. First, read and reread the problem a few times. Next, guess a solution. Suppose that we need 7 liters of the 40% solution. Then we need $12 - 7 = 5$ liters of the 70% solution. To see if this is indeed the solution, find the amount of pure acid in 7 liters of the 40% solution, in 5 liters of the 70% solution, and in 12 liters of a 50% solution, the required amount and strength.

number of liters	×	acid strength	=	amount of pure acid
7 liters	×	40%	=	$7(0.40)$ or 2.8 liters
5 liters	×	70%	=	$5(0.70)$ or 3.5 liters
12 liters	×	50%	=	$12(0.50)$ or 6 liters

Since 2.8 liters + 3.5 liters = 6.3 liters and not 6, our guess is incorrect, but we have gained some valuable insight into how to model and check this problem.
   Let

$$x = \text{number of liters of 40\% solution; then}$$

$$12 - x = \text{number of liters of 70\% solution.}$$

2. TRANSLATE. To help us translate to an equation, the following table summarizes the information given. Recall that the amount of acid in each solution is found by multiplying the acid strength of each solution by the number of liters.

	No. of Liters	·	Acid Strength	=	Amount of Acid
40% Solution	x		40%		$0.40x$
70% Solution	$12 - x$		70%		$0.70(12 - x)$
50% Solution Needed	12		50%		$0.50(12)$

The amount of acid in the final solution is the sum of the amounts of acid in the two beginning solutions.

In words: acid in 40% solution   +   acid in 70% solution   =   acid in 50% mixture

Translate: $0.40x$ + $0.70(12 - x)$ = $0.50(12)$

(Continued on next page)

(Continued on next page)

Practice 7

How much 20% dye solution and 50% dye solution should be mixed to obtain 6 liters of a 40% solution?

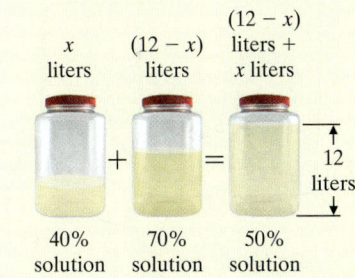

x liters $(12 - x)$ liters $(12 - x)$ liters + x liters

40% solution   70% solution   50% solution   12 liters

Answer

7. 2 liters of the 20% solution; 4 liters of the 50% solution

3. SOLVE.

$$0.40x + 0.70(12 - x) = 0.50(12)$$
$$0.4x + 8.4 - 0.7x = 6 \qquad \text{Apply the distributive property.}$$
$$-0.3x + 8.4 = 6 \qquad \text{Combine like terms.}$$
$$-0.3x = -2.4 \qquad \text{Subtract 8.4 from both sides.}$$
$$x = 8 \qquad \text{Divide both sides by } -0.3.$$

4. INTERPRET.

Check: To check, recall how we checked our guess.

State: If 8 liters of the 40% solution are mixed with $12 - 8$ or 4 liters of the 70% solution, the result is 12 liters of a 50% solution.

■ **Work Practice 7**

Vocabulary, Readiness & Video Check

Tell whether the percent labels in the circle graphs are correct.

1. 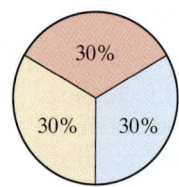 no

2. 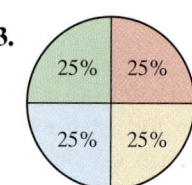 no

3.   25% 25% 25% 25%   yes

4. 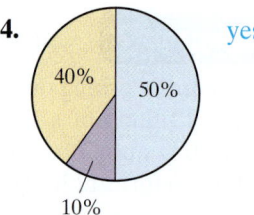 yes

Martin-Gay Interactive Videos

See Video 2.6 🍎

Watch the section lecture video and answer the following questions.

Objective A 5. Answer these questions based on how ⊟ Example 2 was translated to an equation. ▶

    a. What does "is" translate to?

    b. What does "of" translate to?

    c. How do you write a percent as an equivalent decimal?

Objective B 6. At the end of ⊟ Example 3 you are told that the process for finding discount is *almost* the same as finding mark-up. ▶

    a. How is discount similar?

    b. How does discount differ?

Objective C 7. According to ⊟ Example 4, what amount must you find before you can find a percent of increase in price? How do you find this amount? ▶

Objective D 8. The following problem is worded like ⊟ Example 6 in the video, but using different quantities.

How much of an alloy that is 10% copper should be mixed with 400 ounces of an alloy that is 30% copper in order to get an alloy that is 20% copper? Fill in the table and set up an equation that could be used to solve for the unknowns (do not actually solve). Use ⊟ Example 6 in the video as a model for your work. ▶

See video answer section.

Alloy	Ounces	Copper Strength	Amount of Copper
10%	x	0.10	$0.10x$
30%	400	0.30	$0.30(400)$
20%	$x + 400$	0.20	$0.20(x + 400)$

2.6 Exercise Set MyMathLab®

Objective **A** *Find each number described. For Exercises 1 and 2, the solutions have been started for you. See Examples 1 and 2.*

1. What number is 16% of 70?   11.2

Start the solution:

1. UNDERSTAND the problem. Reread it as many times as needed.
2. TRANSLATE into an equation. (Fill in the blanks below.)

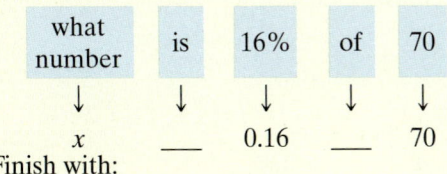

what number	is	16%	of	70
↓	↓	↓	↓	↓
x	___	0.16	___	70

Finish with:

3. SOLVE and
4. INTERPRET

2. What number is 88% of 1000?   880

Start the solution:

1. UNDERSTAND the problem. Reread it as many times as needed.
2. TRANSLATE into an equation. (Fill in the blanks below.)

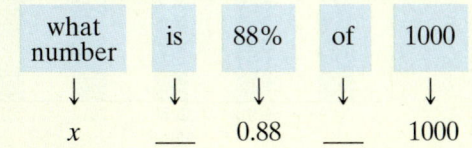

what number	is	88%	of	1000
↓	↓	↓	↓	↓
x	___	0.88	___	1000

Finish with:

3. SOLVE and
4. INTERPRET

3. The number 28.6 is what percent of 52?   55%

4. The number 87.2 is what percent of 436?   20%

▶5. The number 45 is 25% of what number?   180

6. The number 126 is 35% of what number?   360

The circle graph below shows the types of accommodations that overnight visitors to national parks used in 2012. Use this graph for Exercises 7 through 10. See Example 3.

Overnight Stays at National Parks, 2012

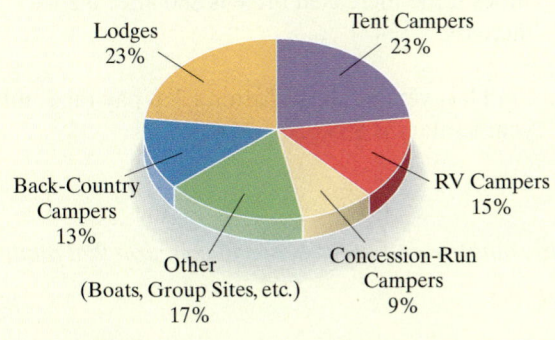

Lodges 23%
Tent Campers 23%
Back-Country Campers 13%
Other (Boats, Group Sites, etc.) 17%
Concession-Run Campers 9%
RV Campers 15%

Source: National Park Service

7. What percent of overnight stays were made in RVs?   15%

8. What percent of overnight stays involved tent camping?   23%

9. In 2012, Yellowstone National Park reported approximately 1,350,000 overnight stays. How many of these stays might you expect were made in lodges?   310,500

10. In 2012, Yosemite National Park reported approximately 1,732,000 overnight stays. How many of these stays might you expect involved back-country camping?   225,160

Objective **B** *Solve. If needed, round answers to the nearest cent. See Example 4.*

11. A used automobile dealership recently reduced the price of a used sports car by 8%. If the price of the car before discount was $18,500, find the discount and the new price.
discount: $1480; new price: $17,020

12. A music store is advertising a 25%-off sale on all new releases. Find the discount and the sale price of a newly released CD that regularly sells for $12.50.
discount: $3.13; new price: $9.37

▶13. A birthday celebration meal is $40.50 including tax. Find the total cost if a 15% tip is added to the cost. $46.58

14. A retirement dinner for two is $65.40 including tax. Find the total cost if a 20% tip is added to the cost. $78.48

Objective C *Solve. Round percents to the nearest tenth. See Example 5.*

Use the graph below for Exercises 15 and 16.

Yearly Complaints Received by the Internet Crime Complaint Center

Source: Data from Internet Crime Complaint Center (www.ic3.gov)

15. The number of Internet-crime complaints decreased from 2012 to 2013. Find the percent of decrease.   9.3%

16. The number of Internet-crime complaints decreased from 2011 to 2012. Find the percent of decrease.   7.6%

17. By increasing each dimension by 1 unit, the area of a rectangle increased from 28 square feet (on the left) to 40 square feet (on the right). Find the percent of increase in area.   42.9%

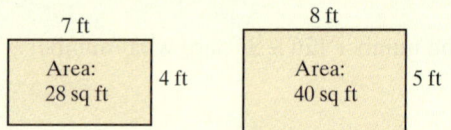

7 ft

Area:
28 sq ft   4 ft

8 ft

Area:
40 sq ft   5 ft

18. By increasing the length of the side by one unit, the area of a square increased from 81 square meters to 100 square meters. Find the percent of increase in area.   23.5%

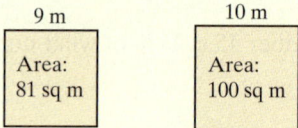

9 m

Area:
81 sq m

10 m

Area:
100 sq m

Solve. See Example 6.

19. Find the original price of a pair of shoes if the sale price is $78 after a 25% discount. $104

20. Find the original price of a popular pair of shoes if the increased price is $80 after a 25% increase. $64

▶**21.** Find last year's salary if after a 4% pay raise, this year's salary is $44,200. $42,500

22. Find last year's salary if after a 3% pay raise, this year's salary is $55,620. $54,000

Objective D *Solve. For each exercise, a table is given for you to complete and use to write an equation that models the situation. See Example 7.*

23. How much pure acid should be mixed with 2 gallons of a 40% acid solution in order to get a 70% acid solution?   2 gal

	Number of Gallons	·	Acid Strength	=	Amount of Acid
Pure Acid			100%		
40% Acid Solution					
70% Acid Solution Needed					

24. How many cubic centimeters (cc) of a 25% antibiotic solution should be added to 10 cubic centimeters of a 60% antibiotic solution in order to get a 30% antibiotic solution?   60 cc

	Number of Cubic cm	·	Antibiotic Strength	=	Amount of Antibiotic
25% Antibiotic Solution					
60% Antibiotic Solution					
30% Antibiotic Solution Needed					

25. Community Coffee Company wants a new flavor of Cajun coffee. How many pounds of coffee worth $7 a pound should be added to 14 pounds of coffee worth $4 a pound to get a mixture worth $5 a pound?   7 lb

	Number of Pounds	·	Cost per Pound	=	Value
$7 per lb Coffee					
$4 per lb Coffee					
$5 per lb Coffee Wanted					

26. Planter's Peanut Company wants to mix 20 pounds of peanuts worth $3 a pound with cashews worth $5 a pound in order to make an experimental mix worth $3.50 a pound. How many pounds of cashews should be added to the peanuts? $6\frac{2}{3}$ lb

	Number of Pounds	·	Cost per Pound	=	Value
$3 per lb Peanuts					
$5 per lb Cashews					
$3.50 per lb Mixture Wanted					

Objectives [A] [B] [C] [D] **Mixed Practice** *Solve. If needed, round money amounts to two decimal places and all other amounts to one decimal place. See Examples 1 through 7.*

▶ 27. Find 23% of 20.   4.6

28. Find 140% of 86.   120.4

29. The number 40 is 80% of what number?   50

30. The number 56.25 is 45% of what number?   125

31. The number 144 is what percent of 480?   30%

32. The number 42 is what percent of 35?   120%

The graph shows the communities in the United States that have the highest percentages of citizens that shop by catalog. Use the graph to answer Exercises 33 through 36.

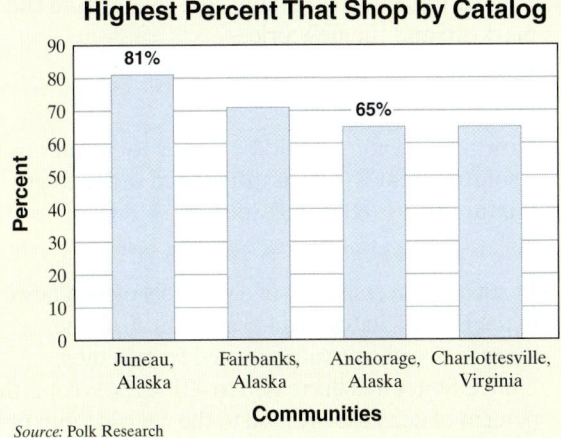

Highest Percent That Shop by Catalog

33. Estimate the percent of the population in Fairbanks, Alaska, who shop by catalog.   71%

34. Estimate the percent of the population in Charlottesville, Virginia, who shop by catalog.   65%

35. According to the U.S. Census Bureau, in 2014, Anchorage had a population of about 291,800. How many catalog shoppers might we predict lived in Anchorage?   189,670

36. According to the U.S. Census Bureau, in 2014, Juneau had a population of 31,275. How many catalog shoppers might we predict lived in Juneau? Round to the nearest whole number.   25,333

For Exercises 37 and 38, fill in the percent column in each table. Each table contains a worked-out example.

37.

Top Cranberry-Producing States in 2012 (in millions of pounds)

	Millions of Pounds	Percent of Total (rounded to nearest percent)
Wisconsin	450	59%
Oregon	40	5%
Massachusetts	210	27%
Washington	14	2%
New Jersey	54	Example: $\frac{54}{768} \approx 7\%$
Total	768	

Source: National Agricultural Statistics Service

38.

The Gap, Inc. Brands North American Stores in 2014

Store Brand	Number of Stores	Percent of Total (rounded to nearest percent)
Gap	968	36%
Athleta	65	2%
Banana Republic	596	Example: $\frac{596}{2670} \approx 22\%$
Intermix	37	1%
Old Navy	1004	38%
Total	2670	

39. Iceberg lettuce is grown and shipped to stores for about 40 cents a head, and consumers purchase it for about 70 cents a head. Find the percent of increase.   75%

40. U.S. macadamia nut production in 2011 was 24,440 tons, and in 2012 the production dropped to 22,000 tons. Find the percent of decrease. (*Source: Agricultural Marketing Resource Center*)   10%

41. A student at the University of New Orleans makes money by buying and selling used cars. Charles bought a used car and later sold it for a 20% profit. If he sold it for $4680, how much did Charles pay for the car? $3900

42. From 2010 to 2020, the number of people employed as physician assistants in the United States is expected to increase by 30%. The number of people employed as physician assistants in 2010 was 83,600. Find the predicted number of physician assistants in 2020. (*Source: Bureau of Labor Statistics*)   108,680 physician assistants

43. By doubling each dimension, the area of a parallelogram increased from 36 square centimeters to 144 square centimeters. Find the percent of increase in area.   300%

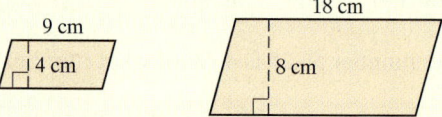

44. By doubling each dimension, the area of a triangle increased from 6 square miles to 24 square miles. Find the percent of increase in area.   300%

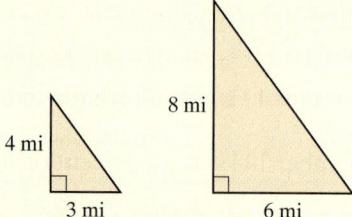

45. A gasoline station recently increased the price of one grade of gasoline by 5%. If this gasoline originally cost $2.20 per gallon, find the mark-up and the new price.   mark-up: $0.11; new price: $2.31

46. The price of a biology book recently increased by 10%. If this book originally cost $89.90, find the mark-up and the new price.   mark-up: $8.99; new price: $98.89

47. How much of an alloy that is 20% copper should be mixed with 200 ounces of an alloy that is 50% copper in order to get an alloy that is 30% copper?   400 oz

48. How much water should be added to 30 gallons of a solution that is 70% antifreeze in order to get a mixture that is 60% antifreeze?   5 gal

49. In 2008, the number of milk cow operations in the United States was 67,000. By 2012, this number had decreased to 58,000. What was the percent of decrease? Round to the nearest tenth of a percent. (*Source: National Agricultural Statistics Service*)   13.4%

50. In 2006, the average size of a privately owned farm in the United States was 443 acres. In 2012, the average size of a privately owned farm in the United States had decreased to 421 acres. What is this percent of decrease? Round to the nearest tenth of a percent. (*Source: National Agricultural Statistics Service*)   5.0%

25. Community Coffee Company wants a new flavor of Cajun coffee. How many pounds of coffee worth $7 a pound should be added to 14 pounds of coffee worth $4 a pound to get a mixture worth $5 a pound?   7 lb

	Number of Pounds	·	Cost per Pound	=	Value
$7 per lb Coffee					
$4 per lb Coffee					
$5 per lb Coffee Wanted					

26. Planter's Peanut Company wants to mix 20 pounds of peanuts worth $3 a pound with cashews worth $5 a pound in order to make an experimental mix worth $3.50 a pound. How many pounds of cashews should be added to the peanuts? $6\frac{2}{3}$ lb

	Number of Pounds	·	Cost per Pound	=	Value
$3 per lb Peanuts					
$5 per lb Cashews					
$3.50 per lb Mixture Wanted					

Objectives Ⓐ Ⓑ Ⓒ Ⓓ **Mixed Practice** *Solve. If needed, round money amounts to two decimal places and all other amounts to one decimal place. See Examples 1 through 7.*

27. Find 23% of 20.   4.6

28. Find 140% of 86.   120.4

29. The number 40 is 80% of what number?   50

30. The number 56.25 is 45% of what number?   125

31. The number 144 is what percent of 480?   30%

32. The number 42 is what percent of 35?   120%

The graph shows the communities in the United States that have the highest percentages of citizens that shop by catalog. Use the graph to answer Exercises 33 through 36.

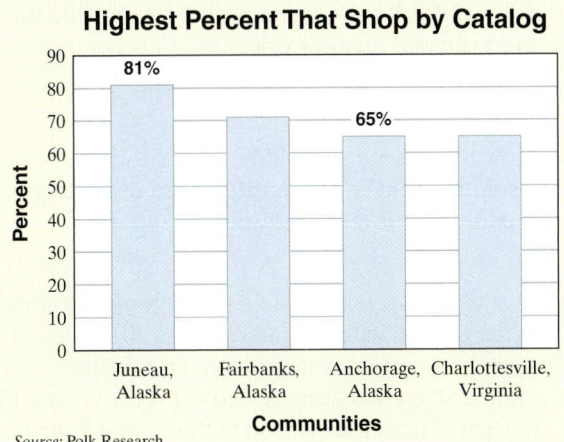

Highest Percent That Shop by Catalog

Source: Polk Research

33. Estimate the percent of the population in Fairbanks, Alaska, who shop by catalog.   71%

34. Estimate the percent of the population in Charlottesville, Virginia, who shop by catalog.   65%

35. According to the U.S. Census Bureau, in 2014, Anchorage had a population of about 291,800. How many catalog shoppers might we predict lived in Anchorage?   189,670

36. According to the U.S. Census Bureau, in 2014, Juneau had a population of 31,275. How many catalog shoppers might we predict lived in Juneau? Round to the nearest whole number.   25,333

For Exercises 37 and 38, fill in the percent column in each table. Each table contains a worked-out example.

37.

Top Cranberry-Producing States in 2012 (in millions of pounds)

	Millions of Pounds	Percent of Total (rounded to nearest percent)
Wisconsin	450	59%
Oregon	40	5%
Massachusetts	210	27%
Washington	14	2%
New Jersey	54	Example: $\frac{54}{768} \approx 7\%$
Total	768	

Source: National Agricultural Statistics Service

38.

The Gap, Inc. Brands North American Stores in 2014

Store Brand	Number of Stores	Percent of Total (rounded to nearest percent)
Gap	968	36%
Athleta	65	2%
Banana Republic	596	Example: $\frac{596}{2670} \approx 22\%$
Intermix	37	1%
Old Navy	1004	38%
Total	2670	

39. Iceberg lettuce is grown and shipped to stores for about 40 cents a head, and consumers purchase it for about 70 cents a head. Find the percent of increase.   75%

40. U.S. macadamia nut production in 2011 was 24,440 tons, and in 2012 the production dropped to 22,000 tons. Find the percent of decrease. (*Source:* Agricultural Marketing Resource Center)   10%

41. A student at the University of New Orleans makes money by buying and selling used cars. Charles bought a used car and later sold it for a 20% profit. If he sold it for $4680, how much did Charles pay for the car? $3900

42. From 2010 to 2020, the number of people employed as physician assistants in the United States is expected to increase by 30%. The number of people employed as physician assistants in 2010 was 83,600. Find the predicted number of physician assistants in 2020. (*Source:* Bureau of Labor Statistics)   108,680 physician assistants

43. By doubling each dimension, the area of a parallelogram increased from 36 square centimeters to 144 square centimeters. Find the percent of increase in area.   300%

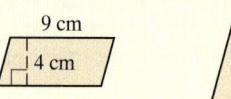 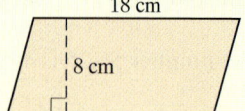

9 cm   4 cm   18 cm   8 cm

44. By doubling each dimension, the area of a triangle increased from 6 square miles to 24 square miles. Find the percent of increase in area.   300%

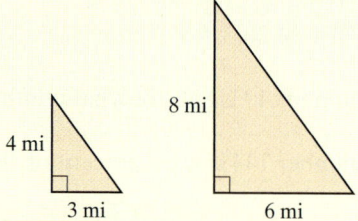

4 mi   3 mi   8 mi   6 mi

45. A gasoline station recently increased the price of one grade of gasoline by 5%. If this gasoline originally cost $2.20 per gallon, find the mark-up and the new price.   mark-up: $0.11; new price: $2.31

46. The price of a biology book recently increased by 10%. If this book originally cost $89.90, find the mark-up and the new price.   mark-up: $8.99; new price: $98.89

47. How much of an alloy that is 20% copper should be mixed with 200 ounces of an alloy that is 50% copper in order to get an alloy that is 30% copper?   400 oz

48. How much water should be added to 30 gallons of a solution that is 70% antifreeze in order to get a mixture that is 60% antifreeze?   5 gal

49. In 2008, the number of milk cow operations in the United States was 67,000. By 2012, this number had decreased to 58,000. What was the percent of decrease? Round to the nearest tenth of a percent. (*Source:* National Agricultural Statistics Service)   13.4%

50. In 2006, the average size of a privately owned farm in the United States was 443 acres. In 2012, the average size of a privately owned farm in the United States had decreased to 421 acres. What is this percent of decrease? Round to the nearest tenth of a percent. (*Source:* National Agricultural Statistics Service)   5.0%

51. A company recently downsized its number of employees by 35%. If there are still 78 employees, how many employees were there prior to the layoffs? 120 employees

52. The average number of children born to each U.S. woman has decreased by 44% since 1920. If this average is now 1.9, find the average in 1920. Round to the nearest tenth.
3.4 children per woman in 1920

53. Nordstrom advertised a 25%-off sale. If a London Fog coat originally sold for $256, find the decrease in price and the sale price.
decrease: $64; sale price: $192

54. A gasoline station decreased the price of a $0.95 cola by 15%. Find the decrease in price and the new price. decrease: $0.14; new price: $0.81

55. Scoville units are used to measure the hotness of a pepper. Measuring 577 thousand Scoville units, the "Red Savina" habañero pepper was known as the hottest chili pepper. That has recently changed with the discovery of Naga Jolokia pepper from India. It measures 48% hotter than the habañero. Find the measure of the Naga Jolokia pepper. Round to the nearest thousand units.
854 thousand Scoville units

56. The number of cell phone tower sites in the United States was 195,613 in 2006. By 2012, the number of cell sites had increased by 54.3%. Find the number of cell towers in 2012. Round to the nearest whole number. (*Source:* CTIA—The Wireless Association) 301,831 cell tower sites

57. In 2013, a survey found that about 55% of all adults in the United States owned a smartphone. There were roughly 240 million U.S. adults at that time. How many U.S. adults owned a smartphone in 2013? (*Source:* Pew Research Center, U.S. Census Bureau) 132 million adults

58. In 2012, there were approximately 225 million moviegoers in the United States and Canada. Of these, about 52% were female. Find the approximate number of females who attended the movies in that year. (*Source:* Motion Picture Association of America) 117 million

59. A new self-tanning lotion for everyday use is to be sold. First, an experimental lotion mixture is made by mixing 800 ounces of everyday moisturizing lotion worth $0.30 an ounce with self-tanning lotion worth $3 per ounce. If the experimental lotion is to cost $1.20 per ounce, how many ounces of the self-tanning lotion should be in the mixture? 400 oz

60. The owner of a local chocolate shop wants to develop a new trail mix. How many pounds of chocolate-covered peanuts worth $5 a pound should be mixed with 10 pounds of granola bites worth $2 a pound to get a mixture worth $3 per pound? 5 lb

Review

Place $<$, $>$, or $=$ in the appropriate space to make each a true statement. See Sections 1.2, 1.3, and 1.6.

61. -5 -7 $>$

62. $\dfrac{12}{3}$ 2^2 $=$

63. $|-5|$ $-(-5)$ $=$

64. -3^3 $(-3)^3$ $=$

65. $(-3)^2$ -3^2 $>$

66. $|-2|$ $-|-2|$ $>$

Concept Extensions

67. Is it possible to mix a 10% acid solution and a 40% acid solution to obtain a 60% acid solution? Why or why not?   no; answers may vary

68. Must the percents in a circle graph have a sum of 100%? Why or why not?   yes; answers may vary

Standardized nutrition labels like the one below have been displayed on food items since 1994. The percent column on the right shows the percent of daily values (based on a 2000-calorie diet) shown at the bottom of the label. For example, a serving of this food contains 4 grams of total fat, where the recommended daily fat based on a 2000-calorie diet is less than 65 grams of fat. This means that $\frac{4}{65}$ or approximately 6% (as shown) of your daily recommended fat is taken in by eating a serving of this food. Use this nutrition label to answer Exercises 69 through 71.

Nutrition Facts

Serving Size 18 Crackers (31g)
Servings Per Container About 9

Amount Per Serving

Calories 130   Calories from Fat 35

	% Daily Value*
Total Fat 4g	6%
Saturated Fat 0.5g	3%
Polyunsaturated Fat 0g	
Monounsaturated Fat 1.5g	
Cholesterol 0mg	0%
Sodium 230mg	*x*
Total Carbohydrate 23g	*y*
Dietary Fiber 2g	8%
Sugars 3g	
Protein 2g	

Vitamin A 0%   •   Vitamin C 0%
Calcium 2%   •   Iron 6%

* Percent Daily Values are based on a 2,000 calorie diet. Your daily values may be higher or lower depending on your calorie needs.

		Calories	2,000	2,500
Total Fat	Less than		65g	80g
Sat. Fat	Less than		20g	25g
Cholesterol	Less than		300mg	300mg
Sodium	Less than		2400mg	2400mg
Total Carbohydrate			300g	375g
Dietary Fiber			25g	30g

69. Based on a 2000-calorie diet, what percent of daily value of sodium is contained in a serving of this food? In other words, find *x* in the label. (Round to the nearest tenth of a percent.)   9.6%

70. Based on a 2000-calorie diet, what percent of daily value of total carbohydrate is contained in a serving of this food? In other words, find *y* in the label. (Round to the nearest tenth of a percent.)   7.7%

71. Notice on the nutrition label that one serving of this food contains 130 calories and 35 of these calories are from fat. Find the percent of calories from fat. (Round to the nearest tenth of a percent.) It is recommended that no more than 30% of calorie intake come from fat. Does this food satisfy this recommendation?   26.9%; yes

Use the nutrition label below to answer Exercises 72 through 74.

NUTRITIONAL INFORMATION PER SERVING

Serving Size: 9.8 oz.   Servings Per Container: 1

Calories 280   Polyunsaturated Fat 1g
Protein . . . . .12g   Saturated Fat 3g
Carbohydrate . . . . .45g   Cholesterol 20mg
Fat . . . . .6g   Sodium 520mg
Percent of Calories from Fat....?   Potassium 220mg

72. If fat contains approximately 9 calories per gram, find the percent of calories from fat in one serving of this food. (Round to the nearest tenth of a percent.)   19.3%

73. If protein contains approximately 4 calories per gram, find the percent of calories from protein from one serving of this food. (Round to the nearest tenth of a percent.)   17.1%

74. Find a food that contains more than 30% of its calories per serving from fat. Analyze the nutrition label and verify that the percents shown are correct.   answers may vary

2.7 Linear Inequalities and Problem Solving

In Chapter 1, we reviewed these inequality symbols and their meanings:

< means "is less than"   ≤ means "is less than or equal to"

> means "is greater than"   ≥ means "is greater than or equal to"

An **inequality** is a statement that contains one of the symbols above.

Equations	Inequalities
$x = 3$	$x \le 3$
$5n - 6 = 14$	$5n - 6 > 14$
$12 = 7 - 3y$	$12 \le 7 - 3y$
$\dfrac{x}{4} - 6 = 1$	$\dfrac{x}{4} - 6 > 1$

Objectives

A Graph Inequalities on a Number Line.

B Use the Addition Property of Inequality to Solve Inequalities.

C Use the Multiplication Property of Inequality to Solve Inequalities.

D Use Both Properties to Solve Inequalities.

E Solve Problems Modeled by Inequalities.

Objective A Graphing Inequalities on a Number Line

Recall that the single solution of the equation $x = 3$ is 3. The solutions of the inequality $x \le 3$ include 3 and *all real numbers less than 3* (for example, -10, $\dfrac{1}{2}$, 2, and 2.9).

Because we can't list all numbers less than 3, we show instead a picture of the solutions by graphing them on a number line.

To graph the solutions of $x \le 3$, we shade the numbers to the left of 3 since they are less than 3. Then we place a closed circle on the point representing 3. The closed circle indicates that 3 *is* a solution: 3 *is* less than or equal to 3.

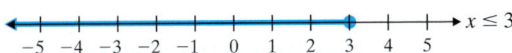

To graph the solutions of $x < 3$, we shade the numbers to the left of 3. Then we place an open circle on the point representing 3. The open circle indicates that 3 *is not* a solution: 3 *is not* less than 3.

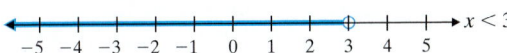

Teaching Tip

You might want to mention to students that the open/closed circle notation is not the only notation in use. Frequently in college algebra, a variation of this notation is used. The notation uses a parenthesis instead of an open circle and a bracket instead of a closed circle.

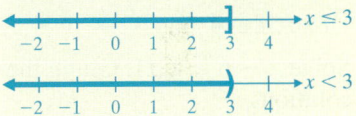

Example 1 Graph: $x \ge -1$

Solution: To graph the solutions of $x \ge -1$, we place a closed circle at -1 since the inequality symbol is $\ge$ and -1 is greater than or equal to -1. Then we shade to the right of -1.

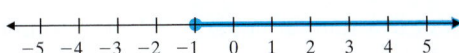

■ Work Practice 1

Practice 1

Graph: $x \ge -2$

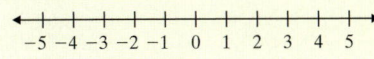

Example 2 Graph: $-1 > x$

Solution: Recall from Section 1.2 that $-1 > x$ means the same as $x < -1$. The graph of the solutions of $x < -1$ is shown below.

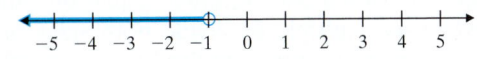

■ Work Practice 2

Practice 2

Graph: $5 > x$

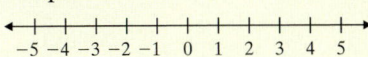

Answers

1.
2.

161

Practice 3

Graph: $-3 \leq x < 1$

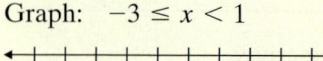

Teaching Tip

Show students how they can easily check the shading on their graph. Substituting any point from the shaded region into the inequality should give a true statement. Substituting any point that has not been shaded should give a false statement.

Example 3   Graph: $-4 < x \leq 2$

Solution: We read as $-4 < x \leq 2$ as "-4 is less than x and x is less than or equal to 2," or as "x is greater than -4 and x is less than or equal to 2." To graph the solutions of this inequality, we place an open circle at -4 (-4 is not part of the graph), a closed circle at 2 (2 is part of the graph), and we shade all numbers between -4 and 2. Why? All numbers between -4 and 2 are greater than -4 *and* also less than 2.

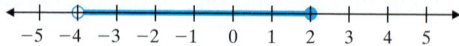

■ **Work Practice 3**

Objective B   Using the Addition Property

When solutions of a linear inequality are not immediately obvious, they are found through a process similar to the one used to solve a linear equation. Our goal is to get the variable alone on one side of the inequality. We use properties of inequality similar to properties of equality.

> ### Addition Property of Inequality
>
> If a, b, and c are real numbers, then
>
> $$a < b \quad \text{and} \quad a + c < b + c$$
>
> are equivalent inequalities.

This property also holds true for subtracting values, since subtraction is defined in terms of addition. In other words, adding or subtracting the same quantity from both sides of an inequality does not change the solutions of the inequality.

Practice 4

Solve $x - 6 \geq -11$. Graph the solutions.

Example 4   Solve $x + 4 \leq -6$. Graph the solutions.

Solution: To solve for x, subtract 4 from both sides of the inequality.

$$
\begin{aligned}
x + 4 &\leq -6 && \text{Original inequality} \\
x + 4 - 4 &\leq -6 - 4 && \text{Subtract 4 from both sides.} \\
x &\leq -10 && \text{Simplify.}
\end{aligned}
$$

The graph of the solutions is shown below.

■ **Work Practice 4**

Helpful Hint

Notice that any number less than or equal to -10 is a solution to $x \leq -10$. For example, solutions include

$$-10, \quad -200, \quad -11\frac{1}{2}, \quad -\sqrt{130}, \quad \text{and} \quad -50.3$$

Objective C   Using the Multiplication Property

An important difference between solving linear equations and solving linear inequalities is shown when we multiply or divide both sides of an inequality by a nonzero real number. For example, start with the true statement $6 < 8$ and multiply both sides by 2. As we see on the next page, the resulting inequality is also true.

Answers

3.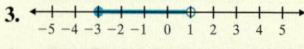

4. $x \geq -5$

$6 < 8$ True

$2(6) < 2(8)$ Multiply both sides by 2.

$12 < 16$ True

But if we start with the same true statement $6 < 8$ and multiply both sides by -2, the resulting inequality is not a true statement.

$6 < 8$ True

$-2(6) < -2(8)$ Multiply both sides by -2.

$-12 < -16$ False

Notice, however, that if we reverse the direction of the inequality symbol, the resulting inequality is true.

$-12 < -16$ False

$-12 > -16$ True

This demonstrates the multiplication property of inequality.

Multiplication Property of Inequality

1. If a, b, and c are real numbers, and c is **positive,** then

$a < b$ and $ac < bc$

are equivalent inequalities.

2. If a, b, and c are real numbers, and c is **negative,** then

$a < b$ and $ac > bc$

are equivalent inequalities.

Because division is defined in terms of multiplication, this property also holds true when dividing both sides of an inequality by a nonzero number: If we multiply or divide both sides of an inequality by a negative number, **the direction of the inequality sign must be reversed for the inequalities to remain equivalent.**

✓**Concept Check** Fill in the box with $<$, $>$, $\leq$, or $\geq$.

a. Since $-8 < -4$, then $3(-8) \,\square\, 3(-4)$.

b. Since $5 \geq -2$, then $\dfrac{5}{-7} \,\square\, \dfrac{-2}{-7}$.

c. If $a < b$, then $2a \,\square\, 2b$.

d. If $a \geq b$, then $\dfrac{a}{-3} \,\square\, \dfrac{b}{-3}$.

Example 5   Solve $-2x \leq -4$. Graph the solutions.

Solution: Remember to reverse the direction of the inequality symbol when dividing by a negative number.

$-2x \leq -4$

$\dfrac{-2x}{-2} \geq \dfrac{-4}{-2}$ Divide both sides by -2 and reverse the inequality sign.

$x \geq 2$ Simplify.

The graph of the solutions is shown.

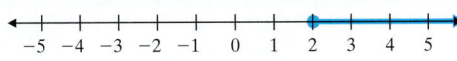

 Work Practice 5

Practice 5

Solve $-3x \leq 12$. Graph the solutions.

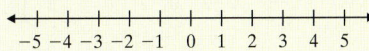

Answer

5. $x \geq -4$

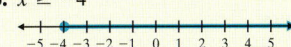

✓**Concept Check Answer**

a. $<$ **b.** $\leq$ **c.** $<$ **d.** $\leq$

Practice 6

Solve $5x > -20$. Graph the solutions.

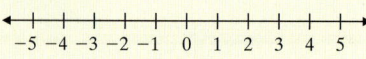

Example 6 Solve $2x < -4$. Graph the solutions.

Solution: $2x < -4$

$$\frac{2x}{2} < \frac{-4}{2}$$ Divide both sides by 2. Do not reverse the inequality sign.

$$x < -2$$ Simplify.

The graph of the solutions is shown.

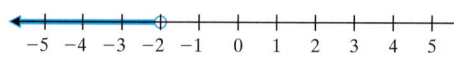

🟧 Work Practice 6

Since we cannot list all solutions to an inequality such as $x < -2$, we will use the set notation $\{x \mid x < -2\}$. Recall from Section 1.2 that this is read "the set of all x such that x is less than -2." We will use this notation when solving inequalities.

Objective D Using Both Properties of Inequality

The following steps may be helpful when solving inequalities in one variable. Notice that these steps are similar to the ones given in Section 2.3 for solving equations.

> **To Solve Linear Inequalities in One Variable**
>
> **Step 1:** If an inequality contains fractions, multiply both sides by the LCD to clear the inequality of fractions.
>
> **Step 2:** Use the distributive property to remove parentheses if they appear.
>
> **Step 3:** Simplify each side of the inequality by combining like terms.
>
> **Step 4:** Get all variable terms on one side and all numbers on the other side by using the addition property of inequality.
>
> **Step 5:** Get the variable alone by using the multiplication property of inequality.

Helpful Hint

Don't forget that if both sides of an inequality are multiplied or divided by a negative number, the direction of the inequality sign must be reversed.

Practice 7

Solve $-3x + 11 \le -13$. Graph the solution set.

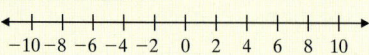

Example 7 Solve $-4x + 7 \ge -9$. Graph the solution set.

Solution: $-4x + 7 \ge -9$

$$-4x + 7 - 7 \ge -9 - 7$$ Subtract 7 from both sides.

$$-4x \ge -16$$ Simplify.

$$\frac{-4x}{-4} \le \frac{-16}{-4}$$ Divide both sides by -4 and reverse the direction of the inequality sign.

$$x \le 4$$ Simplify.

The graph of the solution set $\{x \mid x \le 4\}$ is shown.

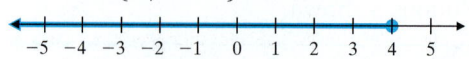

🟧 Work Practice 7

Answers

6. $x > -4$

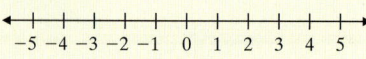

7. $\{x \mid x \ge 8\}$

Example 8   Solve $-5x + 7 < 2(x - 3)$. Graph the solution set.

Solution: $-5x + 7 < 2(x - 3)$

$\qquad -5x + 7 < 2x - 6$ Apply the distributive property.

$\qquad -5x + 7 - 2x < 2x - 6 - 2x$ Subtract $2x$ from both sides.

$\qquad -7x + 7 < -6$ Combine like terms.

$\qquad -7x + 7 - 7 < -6 - 7$ Subtract 7 from both sides.

$\qquad -7x < -13$ Combine like terms.

$\qquad \dfrac{-7x}{-7} > \dfrac{-13}{-7}$ Divide both sides by -7 and reverse the direction of the inequality sign.

$\qquad x > \dfrac{13}{7}$ Simplify.

The graph of the solution set $\left\{x \mid x > \dfrac{13}{7}\right\}$ is shown.

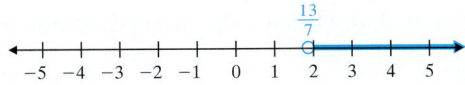

■ **Work Practice 8**

Practice 8

Solve $2x - 3 > 4(x - 1)$.
Graph the solution set.

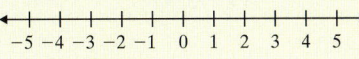

Example 9   Solve: $2(x - 3) - 5 \le 3(x + 2) - 18$

Solution: $2(x - 3) - 5 \le 3(x + 2) - 18$

$\qquad 2x - 6 - 5 \le 3x + 6 - 18$ Apply the distributive property.

$\qquad 2x - 11 \le 3x - 12$ Combine like terms.

$\qquad -x - 11 \le -12$ Subtract $3x$ from both sides.

$\qquad -x \le -1$ Add 11 to both sides.

$\qquad \dfrac{-x}{-1} \ge \dfrac{-1}{-1}$ Divide both sides by -1 and reverse the direction of the inequality sign.

$\qquad x \ge 1$ Simplify.

The solution set is $\{x \mid x \ge 1\}$.

■ **Work Practice 9**

Practice 9

Solve:
$3(x + 5) - 1 \ge 5(x - 1) + 7$

Objective E Solving Problems Modeled by Inequalities ▶

Problems containing words such as "at least," "at most," "between," "no more than," and "no less than" usually indicate that an inequality should be solved instead of an equation. In solving applications involving linear inequalities, we use the same procedure we used to solve applications involving linear equations.

Some Inequality Translations			
≥	≤	<	>
at least	at most	is less than	is greater than
no less than	no more than		

Example 10   12 subtracted from 3 times a number is less than 21. Find all numbers that make this statement true.

Solution:

1. UNDERSTAND. Read and reread the problem. This is a direct translation problem, and let's let

$\qquad x =$ the unknown number   *(Continued on next page)*

Practice 10

Twice a number, subtracted from 35, is greater than 15. Find all numbers that make this true.

Answers

8. $\left\{x \mid x < \dfrac{1}{2}\right\}$

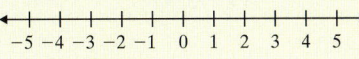

9. $\{x \mid x \le 6\}$

10. all numbers less than 10

2. TRANSLATE.

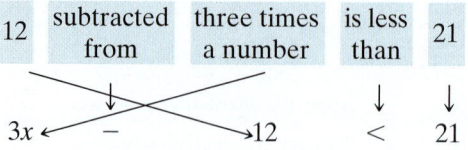

| 12 | subtracted from | three times a number | is less than | 21 |

$$3x \quad - \quad 12 \quad < \quad 21$$

3. SOLVE. $3x - 12 < 21$

$3x < 33$ Add 12 to both sides.

$\dfrac{3x}{3} < \dfrac{33}{3}$ Divide both sides by 3 and do not reverse the direction of the inequality sign.

$x < 11$ Simplify.

4. INTERPRET.

Check: Check the translation; then let's choose a number less than 11 to see if it checks. For example, let's check 10. 12 subtracted from 3 times 10 is 12 subtracted from 30, or 18. Since 18 is less than 21, the number 10 checks.

State: All numbers less than 11 make the original statement true.

■ **Work Practice 10**

Practice 11

Alex earns $600 per month plus 4% of all his sales. Find the minimum sales that will allow Alex to earn at least $3000 per month.

Example 11   Budgeting for a Wedding

Marie Chase and Jonathan Edwards are having their wedding reception at the Gallery reception hall. They may spend at most $1000 for the reception. If the reception hall charges a $100 cleanup fee plus $14 per person, find the greatest number of people that they can invite and still stay within their budget.

Solution:

1. UNDERSTAND. Read and reread the problem. Suppose that 50 people attend the reception. The cost is then $100 + \$14(50) = \$100 + \$700 = \$800$.

   Let x = the number of people who attend the reception.

2. TRANSLATE.

cleanup fee	+	cost per person	times	number of people	must be less than or equal to	$1000
100	+	14	·	x	$\leq$	1000

3. SOLVE.

$100 + 14x \leq 1000$

$14x \leq 900$ Subtract 100 from both sides.

$x \leq 64\dfrac{2}{7}$ Divide both sides by 14.

4. INTERPRET.

Check: Since x represents the number of people, we round down to the nearest whole, or 64. Notice that if 64 people attend, the cost is $\$100 + \$14(64) = \$996$. If 65 people attend, the cost is $\$100 + \$14(65) = \$1010$, which is more than the given $1000.

State: Marie Chase and Jonathan Edwards can invite at most 64 people to the reception.

■ **Work Practice 11**

Answer

11. $60,000

Vocabulary, Readiness & Video Check

Identify each as an equation, expression, or inequality.

1. $6x - 7(x + 9)$ ___expression___

2. $6x = 7(x + 9)$ ___equation___

3. $6x < 7(x + 9)$ ___inequality___

4. $5y - 2 \geq -38$ ___inequality___

5. $\dfrac{9}{7} = \dfrac{x + 2}{14}$ ___equation___

6. $\dfrac{9}{7} - \dfrac{x + 2}{14}$ ___expression___

Decide which number listed is not a solution to each given inequality.

7. $x \geq -3$; $\ -3, 0, -5, \pi$ ___-5___

8. $x < 6$; $\ -6, |-6|, 0, -3.2$ ___$|-6|$___

9. $x < 4.01$; $\ 4, -4.01, 4.1, -4.1$ ___4.1___

10. $x \geq -3$; $\ -4, -3, -2, -(-2)$ ___-4___

Martin-Gay Interactive Videos   *Watch the section lecture video and answer the following questions.*

Objective A   **11.** From ⊞ Example 1, when graphing an inequality, what inequality symbol(s) does an open circle indicate? What inequality symbol(s) does a closed circle indicate? ▶

Objective B   **12.** From the lecture before ⊞ Example 2, which property is the addition property of inequality very similar to? ▶

Objective C   **13.** What is the answer to ⊞ Example 3, written in solution set notation? ▶

Objective D   **14.** When solving ⊞ Example 4, why is special attention given to the coefficient of x in the last step? ▶

Objective E   **15.** What is the phrase in ⊞ Example 5 that tells us to translate to an inequality? What does this phrase translate to? ▶

See Video 2.7

See video answer section.

2.7   Exercise Set MyMathLab®

Objective A   *Graph each inequality on the number line. See Examples 1 and 2.*

▶ 1. $x \leq -1$

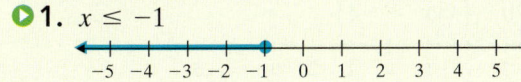

2. $y < 0$

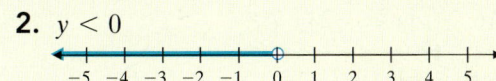

3. $x > \dfrac{1}{2}$

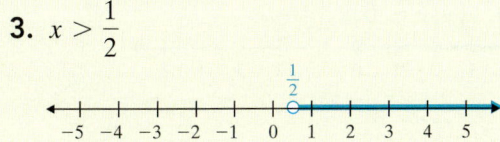

4. $z \geq -\dfrac{2}{3}$

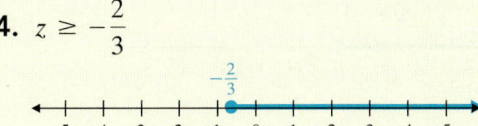

5. $y < 4$

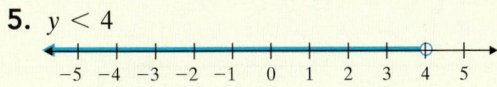

6. $x > 3$

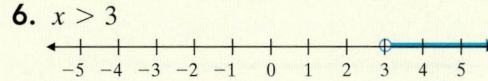

7. $-2 \leq m$

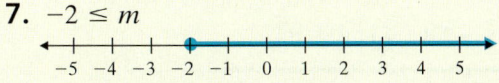

8. $-5 \geq x$

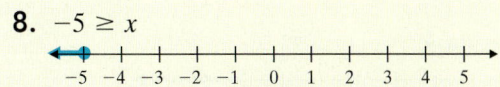

Graph each inequality on the number line. See Example 3.

9. $-1 < x < 3$

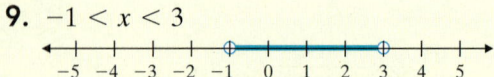

10. $-2 \le x \le 3$

11. $0 \le y < 2$

12. $-4 < x \le 0$

Objective B *Solve each inequality. Graph the solution set. Write each answer using solution set notation. See Example 4.*

13. $x - 2 \ge -7$ $\{x \mid x \ge -5\}$

14. $x + 4 \le 1$ $\{x \mid x \le -3\}$

15. $-9 + y < 0$ $\{y \mid y < 9\}$

16. $-3 + m > 5$ $\{m \mid m > 8\}$

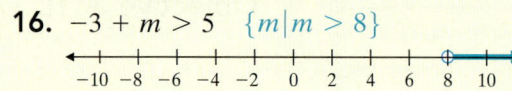

17. $3x - 5 > 2x - 8$ $\{x \mid x > -3\}$

18. $3 - 7x \ge 10 - 8x$ $\{x \mid x \ge 7\}$

19. $4x - 1 \le 5x - 2x$ $\{x \mid x \le 1\}$

20. $7x + 3 < 9x - 3x$ $\{x \mid x < -3\}$

Objective C *Solve each inequality. Graph the solution set. See Examples 5 and 6.*

21. $2x < -6$ $\{x \mid x < -3\}$

22. $3x > -9$ $\{x \mid x > -3\}$

23. $-8x \le 16$ $\{x \mid x \ge -2\}$

24. $-5x < 20$ $\{x \mid x > -4\}$

25. $-x > 0$ $\{x \mid x < 0\}$

26. $-y \ge 0$ $\{y \mid y \le 0\}$

27. $\dfrac{3}{4}y \ge -2$ $\left\{y \mid y \ge -\dfrac{8}{3}\right\}$

28. $\dfrac{5}{6}x \le -8$ $\left\{x \mid x \le -\dfrac{48}{5}\right\}$

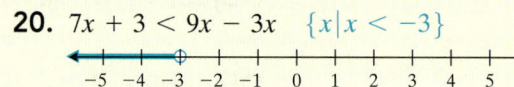

29. $-0.6y < -1.8$ $\{y \mid y > 3\}$

30. $-0.3x > -2.4$ $\{x \mid x < 8\}$

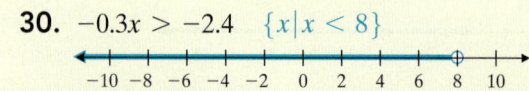

Objectives B C D **Mixed Practice** *Solve each inequality. Write each answer using solution set notation. See Examples 4 through 9.*

31. $-8 < x + 7$
$\{x \mid x > -15\}$

32. $-11 > x + 4$
$\{x \mid x < -15\}$

33. $7(x + 1) - 6x \geq -4$
$\{x \mid x \geq -11\}$

34. $10(x + 2) - 9x \leq -1$
$\{x \mid x \leq -21\}$

35. $4x > 1$ $\left\{x \mid x > \dfrac{1}{4}\right\}$

36. $6x < 5$ $\left\{x \mid x < \dfrac{5}{6}\right\}$

37. $-\dfrac{2}{3}y \leq 8$
$\{y \mid y \geq -12\}$

38. $-\dfrac{3}{4}y \geq 9$
$\{y \mid y \leq -12\}$

39. $4(2z + 1) < 4$
$\{z \mid z < 0\}$

40. $6(2 - z) \geq 12$
$\{z \mid z \leq 0\}$

41. $3x - 7 < 6x + 2$
$\{x \mid x > -3\}$

42. $2x - 1 \geq 4x - 5$
$\{x \mid x \leq 2\}$

43. $5x - 7x \leq x + 2$
$\left\{x \mid x \geq -\dfrac{2}{3}\right\}$

44. $4 - x < 8x + 2x$
$\left\{x \mid x > \dfrac{4}{11}\right\}$

45. $-6x + 2 \geq 2(5 - x)$
$\{x \mid x \leq -2\}$

46. $-7x + 4 > 3(4 - x)$
$\{x \mid x < -2\}$

47. $3(x - 5) < 2(2x - 1)$
$\{x \mid x > -13\}$

48. $5(x - 2) \leq 3(2x - 1)$
$\{x \mid x \geq -7\}$

49. $4(3x - 1) \leq 5(2x - 4)$ $\{x \mid x \leq -8\}$

50. $3(5x - 4) \leq 4(3x - 2)$ $\left\{x \mid x \leq \dfrac{4}{3}\right\}$

▶51. $3(x + 2) - 6 > -2(x - 3) + 14$ $\{x \mid x > 4\}$

52. $7(x - 2) + x \leq -4(5 - x) - 12$ $\left\{x \mid x \leq -\dfrac{9}{2}\right\}$

53. $-5(1 - x) + x \leq -(6 - 2x) + 6$ $\left\{x \mid x \leq \dfrac{5}{4}\right\}$

54. $-2(x - 4) - 3x < -(4x + 1) + 2x$ $\{x \mid x > 3\}$

55. $\dfrac{1}{4}(x + 4) < \dfrac{1}{5}(2x + 3)$ $\left\{x \mid x > \dfrac{8}{3}\right\}$

56. $\dfrac{1}{2}(x - 5) < \dfrac{1}{3}(2x - 1)$ $\{x \mid x > -13\}$

57. $-5x + 4 \leq -4(x - 1)$ $\{x \mid x \geq 0\}$

58. $-6x + 2 < -3(x + 4)$ $\left\{x \mid x > \dfrac{14}{3}\right\}$

Objective E *Solve the following. For Exercises 61 and 62, the solutions have been started for you. See Examples 10 and 11.*

▶59. Six more than twice a number is greater than negative fourteen. Find all numbers that make this statement true.   all numbers greater than -10

60. One more than five times a number is less than or equal to ten. Find all such numbers.

all numbers less than or equal to $\dfrac{9}{5}$

△ **61.** The perimeter of a rectangle is to be no greater than 100 centimeters and the width must be 15 centimeters. Find the maximum length of the rectangle.   35 cm

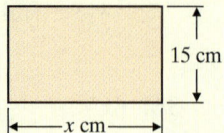

15 cm

x cm

Start the solution:

1. UNDERSTAND the problem. Reread it as many times as needed.
2. TRANSLATE into an equation. (Fill in the blanks below.)

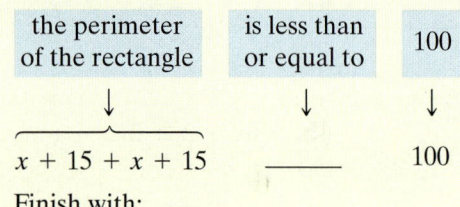

the perimeter of the rectangle	is less than or equal to	100
↓	↓	↓
$x + 15 + x + 15$	_____	100

Finish with:

3. SOLVE and **4.** INTERPRET

△ **62.** One side of a triangle is three times as long as another side, and the third side is 12 inches long. If the perimeter can be no longer than 32 inches, find the maximum lengths of the other two sides.

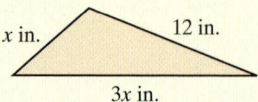

x in.   12 in.

3x in.

5 in.; 15 in. (Since the sum of the lengths of two sides of a triangle must be greater than the length of the third side, some values of x will not form triangles.)

Start the solution:

1. UNDERSTAND the problem. Reread it as many times as needed.
2. TRANSLATE into an equation. (Fill in the blanks below.)

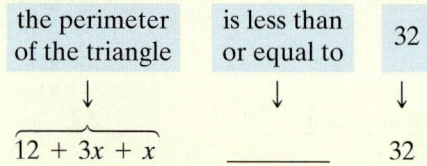

the perimeter of the triangle	is less than or equal to	32
↓	↓	↓
$12 + 3x + x$	_____	32

Finish with:

3. SOLVE and **4.** INTERPRET

63. Ben Holladay bowled 146 and 201 in his first two games. What must he bowl in his third game to have an average of at least 180? (*Hint:* The average of a list of numbers is their sum divided by the number of numbers in the list.)   at least 193

64. On an NBA team the two forwards measure 6′ 8″ and 6′ 6″ tall and the two guards measure 6′ 0″ and 5′ 9″ tall. How tall should the center be if they wish to have a starting team average height of at least 6′ 5″?   at least 7′ 2″

65. Dennis and Nancy Wood are celebrating their 30th wedding anniversary by having a reception at Tiffany Oaks reception hall. They have budgeted $3000 for their reception. If the reception hall charges a $50 cleanup fee plus $34 per person, find the greatest number of people that they may invite and still stay within their budget.   86 people

66. A surprise retirement party is being planned for Pratap Puri. A total of $860 has been collected for the event, which is to be held at a local reception hall. This reception hall charges a cleanup fee of $40 and $15 per person for drinks and light snacks. Find the greatest number of people that may be invited and still stay within the $860 budget.   54 people

67. A 150-pound person uses 5.8 calories per minute when walking at a speed of 4 mph. How long must a person walk at this speed to use at least 200 calories? Round up to the nearest minute. (*Source:* Home & Garden Bulletin No. 72)   at least 35 min

68. A 170-pound person uses 5.3 calories per minute when bicycling at a speed of 5.5 mph. How long must a person ride a bike at this speed in order to use at least 200 calories? Round up to the nearest minute. (*Source:* Same as Exercise 67)   at least 38 min

Review

Evaluate each expression. See Section 1.3.

69. 3^4
81

70. 4^3
64

71. 1^8
1

72. 0^7
0

73. $\left(\dfrac{7}{8}\right)^2$ $\dfrac{49}{64}$

74. $\left(\dfrac{2}{3}\right)^3$ $\dfrac{8}{27}$

The graph shows the number of U.S. Starbucks locations from 2006 to 2013. The height of the graph for each year shown corresponds to the number of Starbucks locations in the United States. Use this graph to answer Exercises 75 through 80. (We study graphs such as this further in Section 6.1.)

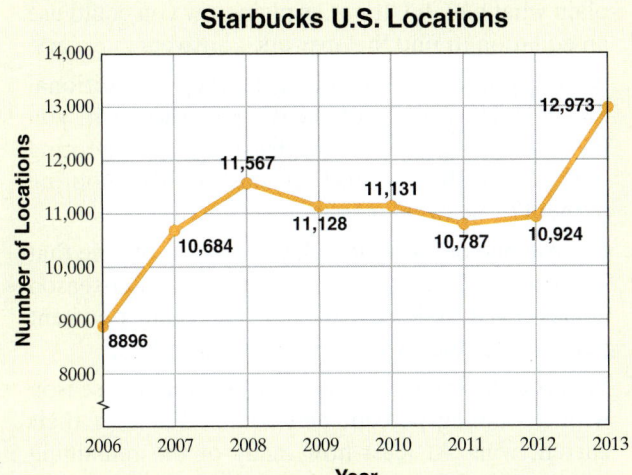

Starbucks U.S. Locations

75. What year shown has the fewest number of Starbucks locations?   2006

76. What year shown has the greatest number of Starbucks locations?   2013

77. Between which two years did the greatest increase in the number of Starbucks locations occur?
2012 and 2013

78. Between what two years did the number of Starbucks locations remain about the same?   2009 and 2010

79. During which year shown did the number of Starbucks locations first rise above 11,500?   2008

80. During which year shown did the number of Starbucks locations rise above 12,000?   2013

Concept Extensions

Fill in the box with $<$, $>$, $\leq$, or $\geq$. See the Concept Check in this section.

81. Since $3 < 5$, then $3(-4) \,\square\, 5(-4)$. $>$

82. If $m \leq n$, then $2m \,\square\, 2n$. $\leq$

83. If $m \leq n$, then $-2m \,\square\, -2n$. $\geq$

84. If $-x < y$, then $x \,\square\, -y$. $>$

85. When solving an inequality, when must you reverse the direction of the inequality symbol?
when multiplying or dividing by a negative number

86. If both sides of the inequality $-3x < -30$ are divided by 3, do you reverse the direction of the inequality symbol? Why or why not?   no; answers may vary

Solve.

87. Eric Daly has scores of 75, 83, and 85 on his history tests. Use an inequality to find the scores he can make on his final exam to receive a B in the class. The final exam counts as **two** tests, and a B is received if the final course average is greater than or equal to 80.   final exam score ≥ 78.5

88. Maria Lipco has scores of 85, 95, and 92 on her algebra tests. Use an inequality to find the scores she can make on her final exam to receive an A in the course. The final exam counts as **three** tests, and an A is received if the final course average is greater than or equal to 90. Round to one decimal place.
final exam score ≥ 89.3

Chapter 2 Group Activity

Additional group activities are available in the *Instructor's Resource Manual with Tests.*

Investigating Averages

Sections 2.1–2.6

Materials:

- small rubber ball or crumpled paper ball
- bucket or waste can

This activity may be completed by working in groups or individually.

1. Try shooting the ball into the bucket or waste can 5 times. Record your results below.

   Shots Made   Shots Missed

2. Find your shooting percent for the 5 shots (that is, the percent of the shots you actually made out of the number you tried).   answers may vary

3. Suppose you are going to try an additional 5 shots. How many of the next 5 shots will you have to make to have a 50% shooting percent for all 10 shots? An 80% shooting percent?   answers may vary

4. Did you solve an equation in Question 3? If so, explain what you did. If not, explain how you could use an equation to find the answers.   answers may vary

5. Now suppose you are going to try an additional 22 shots. How many of the next 22 shots will you have to make to have at least a 50% shooting percent for all 27 shots? At least a 70% shooting percent?   answers may vary

6. Choose one of the sports played at your college that is currently in season. How many regular-season games are scheduled? What is the team's current percent of games won?   answers may vary

7. Suppose the team has a goal of finishing the season with a winning percent better than 110% of their current wins. At least how many of the remaining games must they win to achieve their goal?   answers may vary

Chapter 2 Vocabulary Check

Fill in each blank with one of the words or phrases listed below.

no solution   all real numbers   linear equation in one variable

equivalent equations   formula   reversed

linear inequality in one variable   the same

1. A(n) <u>linear equation in one variable</u> can be written in the form $Ax + B = C$.

2. Equations that have the same solution are called <u>equivalent equations</u>.

3. An equation that describes a known relationship among quantities is called a(n) <u>formula</u>.

4. A(n) <u>linear inequality in one variable</u> can be written in the form $ax + b < c$, (or $>, \le, \ge$).

5. The solution(s) to the equation $x + 5 = x + 5$ is/are <u>all real numbers</u>.

6. The solution(s) to the equation $x + 5 = x + 4$ is/are <u>no solution</u>.

7. If both sides of an inequality are multiplied or divided by the same positive number, the direction of the inequality symbol is <u>the same</u>.

8. If both sides of an inequality are multiplied or divided by the same negative number, the direction of the inequality symbol is <u>reversed</u>.

Helpful Hint

● Are you preparing for your test? Don't forget to take the Chapter 2 Test on page 181. Then check your answers at the back of the text and use the Chapter Test Prep Videos to see the fully worked-out solutions to any of the exercises you want to review.

2 Chapter Highlights

Definitions and Concepts	Examples
Section 2.1 The Addition Property of Equality	

A **linear equation in one variable** can be written in the form $Ax + B = C$ where A, B, and C are real numbers and $A \neq 0$. **Equivalent equations** are equations that have the same solution.	$-3x + 7 = 2$ $3(x - 1) = -8(x + 5) + 4$ $x - 7 = 10$ and $x = 17$ are equivalent equations.
Addition Property of Equality Adding the same number to or subtracting the same number from both sides of an equation does not change its solution.	$$y + 9 = 3$$ $$y + 9 - 9 = 3 - 9$$ $$y = -6$$

Section 2.2 The Multiplication Property of Equality	

Multiplication Property of Equality Multiplying both sides or dividing both sides of an equation by the same nonzero number does not change its solution.	$$\frac{2}{3}a = 18$$ $$\frac{3}{2}\left(\frac{2}{3}a\right) = \frac{3}{2}(18)$$ $$a = 27$$

Section 2.3 Further Solving Linear Equations	

To Solve Linear Equations	*Solve:* $\dfrac{5(-2x + 9)}{6} + 3 = \dfrac{1}{2}$
1. Clear the equation of fractions.	**1.** $6 \cdot \dfrac{5(-2x + 9)}{6} + 6 \cdot 3 = 6 \cdot \dfrac{1}{2}$
2. Remove any grouping symbols such as parentheses.	**2.** $5(-2x + 9) + 18 = 3$ Apply the distributive property. $-10x + 45 + 18 = 3$
3. Simplify each side by combining like terms.	**3.** $-10x + 63 = 3$ Combine like terms.
4. Get all variable terms on one side and all numbers on the other side by using the addition property of equality.	**4.** $-10x + 63 - 63 = 3 - 63$ Subtract 63. $-10x = -60$
5. Get the variable alone by using the multiplication property of equality.	**5.** $\dfrac{-10x}{-10} = \dfrac{-60}{-10}$ Divide by -10. $x = 6$
6. Check the solution by substituting it into the original equation.	

Definitions and Concepts	Examples

Section 2.4   An Introduction to Problem Solving

Problem-Solving Steps

1. UNDERSTAND the problem.

The height of the Hudson volcano in Chile is twice the height of the Kiska volcano in the Aleutian Islands. If the sum of their heights is 12,870 feet, find the height of each.

1. Read and reread the problem. Guess a solution and check your guess.
Let x be the height of the Kiska volcano. Then $2x$ is the height of the Hudson volcano.

 x $2x$

2. TRANSLATE the problem.

2.

height of Kiska	added to	height of Hudson	is	12,870
↓	↓	↓	↓	↓
x	$+$	$2x$	$=$	$12{,}870$

3. SOLVE the equation.

3. $x + 2x = 12{,}870$
$3x = 12{,}870$
$x = 4290$

4. INTERPRET the results.

4. Check: If x is 4290, then $2x$ is $2(4290)$ or 8580. Their sum is $4290 + 8580$ or 12,870, the required amount.

State: The Kiska volcano is 4290 feet tall, and the Hudson volcano is 8580 feet tall.

Section 2.5   Formulas and Problem Solving

An equation that describes a known relationship among quantities is called a **formula.**

To solve a formula for a specified variable, use the same steps as for solving a linear equation. Treat the specified variable as the only variable of the equation.

$A = lw$ (area of a rectangle)
$I = PRT$ (simple interest)

Solve: $P = 2l + 2w$ for l.
$P = 2l + 2w$
$P - 2w = 2l + 2w - 2w$ Subtract $2w$.
$P - 2w = 2l$
$\dfrac{P - 2w}{2} = \dfrac{2l}{2}$ Divide by 2.
$\dfrac{P - 2w}{2} = l$

Section 2.6   Percent and Mixture Problem Solving

Use the same problem-solving steps to solve a problem containing percents.

1. UNDERSTAND.

2. TRANSLATE.

32% of what number is 36.8?

1. Read and reread. Propose a solution and check. Let x = the unknown number.

2.

32%	of	what number	is	36.8
↓	↓	↓	↓	↓
32%	$\cdot$	x	$=$	36.8

Definitions and Concepts	Examples
Section 2.6 Percent and Mixture Problem Solving (*continued*)	

3. SOLVE.

3. *Solve:* $32\% \cdot x = 36.8$

$$0.32x = 36.8$$

$$\frac{0.32x}{0.32} = \frac{36.8}{0.32} \quad \text{Divide by 0.32.}$$

$$x = 115 \quad \text{Simplify.}$$

4. INTERPRET.

4. *Check, then state:* 32% of 115 is 36.8.

How many liters of a 20% acid solution must be mixed with a 50% acid solution in order to obtain 12 liters of a 30% solution?

1. UNDERSTAND.

1. Read and reread. Guess a solution and check.
Let x = number of liters of 20% solution.
Then $12 - x$ = number of liters of 50% solution.

2. TRANSLATE.

2.

	No. of Liters · Acid Strength = Amount of Acid		
20% Solution	x	20%	$0.20x$
50% Solution	$12 - x$	50%	$0.50(12 - x)$
30% Solution Needed	12	30%	$0.30(12)$

In words:

acid in 20% solution	+	acid in 50% solution	=	acid in 30% solution

Translate: $0.20x \quad + 0.50(12 - x) = \quad 0.30(12)$

3. SOLVE.

3. *Solve:* $0.20x + 0.50(12 - x) = 0.30(12)$

$$0.20x + 6 - 0.50x = 3.6 \quad \text{Apply the distributive property.}$$

$$-0.30x + 6 = 3.6 \quad \text{Combine like terms.}$$

$$-0.30x = -2.4 \quad \text{Subtract 6.}$$

$$x = 8 \quad \text{Divide by } -0.30.$$

4. INTERPRET.

4. *Check, then state:*
If 8 liters of a 20% acid solution are mixed with $12 - 8$ or 4 liters of a 50% acid solution, the result is 12 liters of a 30% solution.

Definitions and Concepts	Examples

Section 2.7 Linear Inequalities and Problem Solving

Properties of inequalities are similar to properties of equations. However, if you multiply or divide both sides of an inequality by the same *negative* number, you must reverse the direction of the inequality symbol.

$$-2x \le 4$$

$$\frac{-2x}{-2} \ge \frac{4}{-2} \qquad \text{Divide by } -2; \text{ reverse the inequality symbol.}$$

$$x \ge -2$$

To Solve Linear Inequalities

1. Clear the inequality of fractions.

2. Remove grouping symbols.

3. Simplify each side by combining like terms.

4. Write all variable terms on one side and all numbers on the other side using the addition property of inequality.

5. Get the variable alone by using the multiplication property of inequality.

Solve: $3(x + 2) \le -2 + 8$

1. $3(x + 2) \le -2 + 8$ No fractions to clear.

2. $3x + 6 \le -2 + 8$ Apply the distributive property.

3. $3x + 6 \le 6$ Combine like terms.

4. $3x + 6 - 6 \le 6 - 6$ Subtract 6.

$$3x \le 0$$

5. $\dfrac{3x}{3} \le \dfrac{0}{3}$ Divide by 3.

$$x \le 0$$

The solution set is $\{x \,|\, x \le 0\}$.

Chapter 2   Review

(2.1) *Solve each equation.*

1. $8x + 4 = 9x$ 4

2. $5y - 3 = 6y$ -3

3. $\dfrac{2}{7}x + \dfrac{5}{7}x = 6$ 6

4. $3x - 5 = 4x + 1$ -6

5. $2x - 6 = x - 6$ 0

6. $4(x + 3) = 3(1 + x)$ -9

7. $6(3 + n) = 5(n - 1)$ -23

8. $5(2 + x) - 3(3x + 2) = -5(x - 6) + 2$ 28

Choose the correct algebraic expression.

9. The sum of two numbers is 10. If one number is x, express the other number in terms of x.   b
   a. $x - 10$
   b. $10 - x$
   c. $10 + x$
   d. $10x$

10. Mandy is 5 inches taller than Melissa. If x inches represents the height of Mandy, express Melissa's height in terms of x.   a
   a. $x - 5$
   b. $5 - x$
   c. $5 + x$
   d. $5x$

Definitions and Concepts	**Examples**
Section 2.6 Percent and Mixture Problem Solving (*continued*)	

3. SOLVE.

3. *Solve:* $32\% \cdot x = 36.8$

$$0.32x = 36.8$$

$$\frac{0.32x}{0.32} = \frac{36.8}{0.32} \quad \text{Divide by 0.32.}$$

$$x = 115 \quad \text{Simplify.}$$

4. INTERPRET.

4. *Check, then state:* 32% of 115 is 36.8.

How many liters of a 20% acid solution must be mixed with a 50% acid solution in order to obtain 12 liters of a 30% solution?

1. UNDERSTAND.

1. Read and reread. Guess a solution and check.
Let x = number of liters of 20% solution.
Then $12 - x$ = number of liters of 50% solution.

2. TRANSLATE.

2.

	No. of Liters · Acid Strength = Amount of Acid		
20% Solution	x	20%	$0.20x$
50% Solution	$12 - x$	50%	$0.50(12 - x)$
30% Solution Needed	12	30%	$0.30(12)$

In words: acid in 20% solution   +   acid in 50% solution   =   acid in 30% solution

Translate: $0.20x \quad + 0.50(12 - x) = \quad 0.30(12)$

3. SOLVE.

3. *Solve:* $0.20x + 0.50(12 - x) = 0.30(12)$

$$0.20x + 6 - 0.50x = 3.6 \quad \text{Apply the distributive property.}$$

$$-0.30x + 6 = 3.6 \quad \text{Combine like terms.}$$

$$-0.30x = -2.4 \quad \text{Subtract 6.}$$

$$x = 8 \quad \text{Divide by } -0.30.$$

4. INTERPRET.

4. *Check, then state:*

If 8 liters of a 20% acid solution are mixed with $12 - 8$ or 4 liters of a 50% acid solution, the result is 12 liters of a 30% solution.

Definitions and Concepts	Examples

Section 2.7   Linear Inequalities and Problem Solving

Properties of inequalities are similar to properties of equations. However, if you multiply or divide both sides of an inequality by the same *negative* number, you must reverse the direction of the inequality symbol.

$$-2x \le 4$$

$$\frac{-2x}{-2} \ge \frac{4}{-2} \quad \text{Divide by } -2; \text{ reverse the inequality symbol.}$$

$$x \ge -2$$

To Solve Linear Inequalities

Solve: $3(x + 2) \le -2 + 8$

1. Clear the inequality of fractions.

1. $3(x + 2) \le -2 + 8$ No fractions to clear.

2. Remove grouping symbols.

2. $3x + 6 \le -2 + 8$ Apply the distributive property.

3. Simplify each side by combining like terms.

3. $3x + 6 \le 6$ Combine like terms.

4. Write all variable terms on one side and all numbers on the other side using the addition property of inequality.

4. $3x + 6 - 6 \le 6 - 6$ Subtract 6.

$$3x \le 0$$

5. Get the variable alone by using the multiplication property of inequality.

5. $\dfrac{3x}{3} \le \dfrac{0}{3}$ Divide by 3.

$$x \le 0$$

The solution set is $\{x \,|\, x \le 0\}$.

Chapter 2   Review

(2.1) *Solve each equation.*

1. $8x + 4 = 9x$ 4

2. $5y - 3 = 6y$ -3

3. $\dfrac{2}{7}x + \dfrac{5}{7}x = 6$ 6

4. $3x - 5 = 4x + 1$ -6

5. $2x - 6 = x - 6$ 0

6. $4(x + 3) = 3(1 + x)$ -9

7. $6(3 + n) = 5(n - 1)$ -23

8. $5(2 + x) - 3(3x + 2) = -5(x - 6) + 2$ 28

Choose the correct algebraic expression.

9. The sum of two numbers is 10. If one number is x, express the other number in terms of x.   b
   a. $x - 10$
   b. $10 - x$
   c. $10 + x$
   d. $10x$

10. Mandy is 5 inches taller than Melissa. If x inches represents the height of Mandy, express Melissa's height in terms of x.   a
   a. $x - 5$
   b. $5 - x$
   c. $5 + x$
   d. $5x$

△ **11.** If one angle measures $x°$, express the measure of its complement in terms of x. b
  a. $(180 - x)°$
  b. $(90 - x)°$
  c. $(x - 180)°$
  d. $(x - 90)°$

△ **12.** If one angle measures $(x + 5)°$, express the measure of its supplement in terms of x. c
  a. $(185 + x)°$
  b. $(95 + x)°$
  c. $(175 - x)°$
  d. $(x - 170)°$

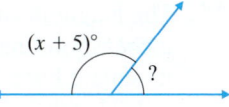

(2.2) *Solve each equation.*

13. $\dfrac{3}{4}x = -9$ -12

14. $\dfrac{x}{6} = \dfrac{2}{3}$ 4

15. $-5x = 0$ 0

16. $-y = 7$ -7

17. $0.2x = 0.15$ 0.75

18. $\dfrac{-x}{3} = 1$ -3

19. $-3x + 1 = 19$ -6

20. $5x + 25 = 20$ -1

21. $7(x - 1) + 9 = 5x$ -1

22. $7x - 6 = 5x - 3$ $\dfrac{3}{2}$

23. $-5x + \dfrac{3}{7} = \dfrac{10}{7}$ $-\dfrac{1}{5}$

24. $5x + x = 9 + 4x - 1 + 6$ 7

25. Write the sum of three consecutive integers as an expression in x. Let x be the first integer. $3x + 3$

26. Write the sum of the first and fourth of four consecutive even integers. Let x be the first even integer. $2x + 6$

(2.3) *Solve each equation.*

27. $\dfrac{5}{3}x + 4 = \dfrac{2}{3}x$ -4

28. $\dfrac{7}{8}x + 1 = \dfrac{5}{8}x$ -4

29. $-(5x + 1) = -7x + 3$ 2

30. $-4(2x + 1) = -5x + 5$ -3

31. $-6(2x - 5) = -3(9 + 4x)$ no solution

32. $3(8y - 1) = 6(5 + 4y)$ no solution

33. $\dfrac{3(2 - z)}{5} = z$ $\dfrac{3}{4}$

34. $\dfrac{4(n + 2)}{5} = -n$ $-\dfrac{8}{9}$

35. $0.5(2n - 3) - 0.1 = 0.4(6 + 2n)$ 20

36. $-9 - 5a = 3(6a - 1)$ $-\dfrac{6}{23}$

37. $\dfrac{5(c + 1)}{6} = 2c - 3$ $\dfrac{23}{7}$

38. $\dfrac{2(8 - a)}{3} = 4 - 4a$ $-\dfrac{2}{5}$

▦ **39.** $200(70x - 3560) = -179(150x - 19{,}300)$ 102

40. $1.72y - 0.04y = 0.42$ 0.25

(2.4) *Solve each of the following.*

41. The height of the Washington Monument is 50.5 inches more than 10 times the length of a side of its square base. If the sum of these two dimensions is 7327 inches, find the height of the Washington Monument. (*Source:* National Park Service)
6665.5 in.

42. A 12-foot board is to be divided into two pieces so that one piece is twice as long as the other. If *x* represents the length of the shorter piece, find the length of each piece.   short piece: 4 ft; long piece: 8 ft

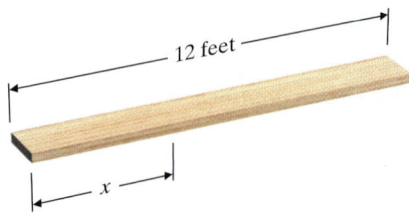

43. The national park system in the United States includes a variety of park unit types. In 2013, there were a total of 40 national battlefields and national memorials. The number of national memorials is four less than three times the number of national battlefields. How many of each park unit was there? (*Source:* National Park System)
national battlefields: 11; national memorials: 29

44. Find three consecutive integers whose sum is −114.
−39, −38, −37

45. The quotient of a number and 3 is the same as the difference of the number and two. Find the number.
3

46. Double the sum of a number and 6 is the opposite of the number. Find the number.   −4

(2.5) *Substitute the given values into the given formulas and solve for the unknown variable.*

47. $P = 2l + 2w$; $P = 46, l = 14$ $w = 9$

48. $V = lwh$; $V = 192, l = 8, w = 6$ $h = 4$

Solve each equation for the indicated variable or constant.

49. $y = mx + b$ for m $m = \dfrac{y - b}{x}$

50. $r = vst - 5$ for s $s = \dfrac{r + 5}{vt}$

51. $2y - 5x = 7$ for x $x = \dfrac{2y - 7}{5}$

52. $3x - 6y = -2$ for y $y = \dfrac{2 + 3x}{6}$

△**53.** $C = \pi D$ for π $\pi = \dfrac{C}{D}$

△**54.** $C = 2\pi r$ for π $\pi = \dfrac{C}{2r}$

△**55.** A swimming pool holds 900 cubic meters of water. If its length is 20 meters and its height is 3 meters, find its width.   15 m

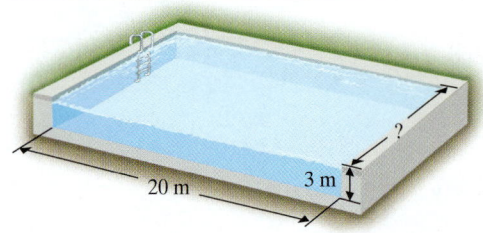

56. The perimeter of a rectangular billboard is 60 feet and the billboard has a length 6 feet longer than its width. Find the dimensions of the billboard.
18 ft by 12 ft

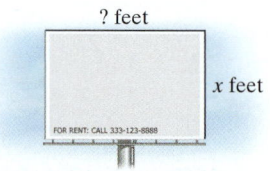

57. A charity 10K race is given annually to benefit a local hospice organization. How long will it take to run/walk a 10K race (10 kilometers or 10,000 meters) if your average pace is 125 **meters** per minute? Give your time in hours and minutes.   1 hr 20 min

58. On September 14, 2013, the highest temperature recorded in the United States was 113°F, which occurred in Death Valley, California. Convert this temperature to degrees Celsius. (*Source:* National Weather Service)   45°C

(2.6) *Find each of the following.*

59. The number 9 is what percent of 45?   20%

60. The number 59.5 is what percent of 85?   70%

61. The number 137.5 is 125% of what number?   110

62. The number 768 is 60% of what number?   1280

63. The price of a small diamond ring was recently increased by 11%. If the ring originally cost $1900, find the mark-up and the new price of the ring.   mark-up: $209; new price: $2109

64. The U.S. motion picture and television industry is made up of over 108,000 businesses. About 85% of these are small businesses with fewer than 10 employees. How many motion picture and television industry businesses have fewer than 10 employees? (*Source:* Motion Picture Association of America)   91,800 businesses

65. Thirty gallons of a 20% acid solution are needed for an experiment. Only 40% and 10% acid solutions are available. How much of each should be mixed to form the needed solution?
40% solution: 10 gal; 10% solution: 20 gal

66. In 2003, the average price of a cinema ticket was $6.03. By 2012, this price had increased to $7.96. What was the percent of increase? (*Source:* National Association of Theatre Owners)   32.0% increase

The graph below shows the percent(s) of cell phone users who have engaged in various behaviors while driving and talking on their cell phones. Use this graph to answer Exercises 67 through 70.

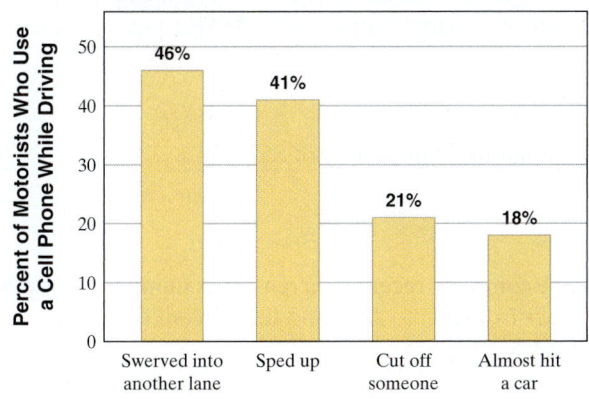

Effects of Cell Phone Use on Driving

Source: Progressive Insurance

67. What percent of motorists who use a cell phone while driving have almost hit another car?   18%

68. What is the most common effect of cell phone use on driving?   swerving into another lane

69. If a cell phone service has an estimated 4600 customers who use their cell phones while driving, how many of these customers would you expect to have cut someone off while driving and talking on their cell phones?   966 customers

70. Do the percents in the graph to the left have a sum of 100%? Why or why not?   no; answers may vary

(2.7) *Graph on a number line.*

71. $x \le -2$

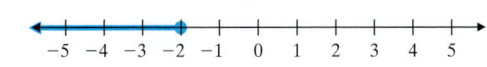

72. $0 < x \le 5$

Solve each inequality.

73. $x - 5 \le -4$
$\{x \mid x \le 1\}$

74. $x + 7 > 2$
$\{x \mid x > -5\}$

75. $-2x \ge -20$
$\{x \mid x \le 10\}$

76. $-3x > 12$
$\{x \mid x < -4\}$

77. $5x - 7 > 8x + 5$ $\{x \mid x < -4\}$ **78.** $x + 4 \geq 6x - 16$ $\{x \mid x \leq 4\}$ **79.** $\dfrac{2}{3}y > 6$ $\{y \mid y > 9\}$

80. $-0.5y \leq 7.5$ $\{y \mid y \geq -15\}$ **81.** $-2(x - 5) > 2(3x - 2)$ **82.** $4(2x - 5) \leq 5x - 1$

$\left\{x \mid x < \dfrac{7}{4}\right\}$ $\left\{x \mid x \leq \dfrac{19}{3}\right\}$

83. Carol Abolafia earns \$175 per week plus a 5% commission on all her sales. Find the minimum amount of sales she must make to ensure that she earns at least \$300 per week.   \$2500

84. Joseph Barrow shot rounds of 76, 82, and 79 golfing. What must he shoot on his next round so that his average will be below 80?   score must be less than 83

Mixed Review

Solve each equation.

85. $6x + 2x - 1 = 5x + 11$
4

86. $2(3y - 4) = 6 + 7y$
-14

87. $4(3 - a) - (6a + 9) = -12a$
$-\dfrac{3}{2}$

88. $\dfrac{x}{3} - 2 = 5$ 21

89. $2(y + 5) = 2y + 10$
all real numbers

90. $7x - 3x + 2 = 2(2x - 1)$
no solution

Solve.

91. The sum of six and twice a number is equal to seven less than the number. Find the number. -13

92. A 23-inch piece of string is to be cut into two pieces so that the length of the longer piece is three more than four times the shorter piece. If x represents the length of the shorter piece, find the lengths of both pieces.   shorter piece: 4 in.; longer piece: 19 in.

93. Solve $V = \dfrac{1}{3}Ah$ for h. $h = \dfrac{3V}{A}$

94. What number is 26% of 85?   22.1

95. The number 72 is 45% of what number?   160

96. A company recently increased its number of employees from 235 to 282. Find the percent of increase.
20%

Solve each inequality. Graph the solution set.

97. $4x - 7 > 3x + 2$ $\{x \mid x > 9\}$

98. $-5x < 20$ $\{x \mid x > -4\}$

99. $-3(1 + 2x) + x \geq -(3 - x)$ $\{x \mid x \leq 0\}$

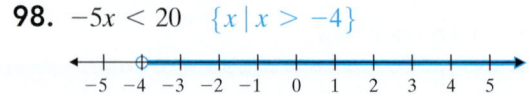

Solve each equation.

1. $-\dfrac{4}{5}x = 4$

2. $4(n - 5) = -(4 - 2n)$

3. $5y - 7 + y = -(y + 3y)$

4. $4z + 1 - z = 1 + z$

5. $\dfrac{2(x + 6)}{3} = x - 5$

6. $\dfrac{4(y - 1)}{5} = 2y + 3$

7. $\dfrac{1}{2} - x + \dfrac{3}{2} = x - 4$

8. $\dfrac{1}{3}(y + 3) = 4y$

9. $-0.3(x - 4) + x = 0.5(3 - x)$

10. $-4(a + 1) - 3a = -7(2a - 3)$

11. $-2(x - 3) = x + 5 - 3x$

Solve each application.

12. A number increased by two-thirds of the number is 35. Find the number.

△ **13.** A gallon of water seal covers 200 square feet. How many gallons are needed to paint two coats of water seal on a deck that measures 20 feet by 35 feet?

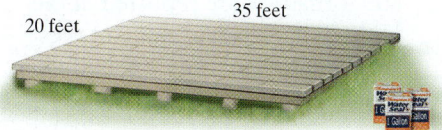

20 feet   35 feet

14. Find the value of x if $y = -14, m = -2,$ and $b = -2$ in the formula $y = mx + b$.

Solve each equation for the indicated variable.

15. $V = \pi r^2 h$ for h

16. $3x - 4y = 10$ for y

1. -5

2. 8

3. $\dfrac{7}{10}$

4. 0

5. 27

6. $-\dfrac{19}{6}$

7. 3

8. $\dfrac{3}{11}$

9. 0.25

10. $\dfrac{25}{7}$

11. no solution

12. 21

13. 7 gal

14. $x = 6$

15. $h = \dfrac{V}{\pi r^2}$

16. $y = \dfrac{3x - 10}{4}$

17. $\{x \mid x \le -2\}$

18. $\{x \mid x < 4\}$

19. $\{x \mid x \le -8\}$

20. $\{x \mid x \ge 11\}$

21. $\left\{x \mid x > \dfrac{2}{5}\right\}$

22. 552

23. 40%

24. 401, 802

25. California: 1107; Ohio: 720

Solve each inequality. Graph the solution set.

17. $3x - 5 \ge 7x + 3$

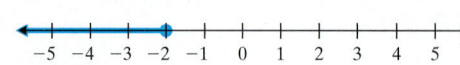

18. $x + 6 > 4x - 6$

Solve each inequality.

19. $-0.3x \ge 2.4$

20. $-5(x - 1) + 6 \le -3(x + 4) + 1$

21. $\dfrac{2(5x + 1)}{3} > 2$

The following graph shows the breakdown of tornadoes occurring in the United States by strength. The corresponding Fujita Tornado Scale categories are shown in parentheses. Use this graph to answer Exercise 22.

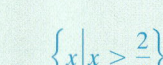

Violent tornadoes (F4–F5) 2%

Strong tornadoes (F2–F3) 29%

Weak tornadoes (F0–F1) 69%

Source: National Climatic Data Center

22. According to the National Climatic Data Center, in an average year, about 800 tornadoes are reported in the United States. How many of these would you expect to be classified as "weak" tornadoes?

23. The number 72 is what percent of 180?

24. Some states have a single area code for the entire state. Two such states have area codes where one is double the other. If the sum of these integers is 1203, find the two area codes.

25. California has more public libraries than any other state. It has 387 more public libraries than Ohio. If the total number of public libraries for these states is 1827, find the number of public libraries in California and the number in Ohio. (*Source:* Institute of Museum and Library Services)

Determine whether each statement is true or false.

1. $8 \geq 8$

2. $-4 < -6$

3. $8 \leq 8$

4. $3 > -3$

5. $23 \leq 0$

6. $-8 \geq -8$

7. $0 \leq 23$

8. $-8 \leq -8$

9. Insert $<$, $>$, or $=$ in the appropriate space to make each statement true.
   a. $|0|$ ___ 2
   b. $|-5|$ ___ 5
   c. $|-3|$ ___ $|-2|$
   d. $|-9|$ ___ $|-9.7|$
   e. $\left|-7\frac{1}{6}\right|$ ___ $|7|$

10. Find the absolute value of each number.
   a. $|5|$
   b. $|-8|$
   c. $\left|-\frac{2}{3}\right|$

Simplify.

11. $\dfrac{3 + |4 - 3| + 2^2}{6 - 3}$

12. $1 + 2(9 - 7)^3 + 4^2$

Add without using number lines.

13. $(-8) + (-11)$

14. $-2 + (-8)$

15. $(-2) + 10$

16. $-10 + 20$

17. $0.2 + (-0.5)$

18. $1.2 + (-1.2)$

Answers

1. true (Sec. 1.2, Ex. 3)

2. false (Sec. 1.2)

3. true (Sec. 1.2, Ex. 4)

4. true (Sec. 1.2)

5. false (Sec. 1.2, Ex. 5)

6. true (Sec. 1.2)

7. true (Sec. 1.2, Ex. 6)

8. true (Sec. 1.2)

9. a. $<$

   b. $=$

   c. $>$

   d. $<$

   e. $>$ (Sec. 1.2, Ex. 13)

10. a. 5

   b. 8

   c. $\frac{2}{3}$ (Sec. 1.2)

11. $\frac{8}{3}$ (Sec. 1.3, Ex. 6)

12. 33 (Sec. 1.3)

13. -19 (Sec. 1.4, Ex. 6)

14. -10 (Sec. 1.4)

15. 8 (Sec. 1.4, Ex. 7)

16. 10 (Sec. 1.4)

17. -0.3 (Sec. 1.4, Ex. 8)

18. 0 (Sec. 1.4)

19. a. -12

   b. -3 (Sec. 1.5, Ex. 7)

20. a. 5

   b. $\dfrac{2}{3}$

   c. a

   d. -3 (Sec. 1.5)

21. a. 0

   b. -24

   c. 90 (Sec. 1.6, Ex. 7)

22. a. -11.1

   b. $-\dfrac{1}{5}$

   c. $\dfrac{3}{4}$ (Sec. 1.5)

23. a. -6

   b. 7

   c. -5 (Sec. 1.6, Ex. 10)

24. a. -0.36

   b. $\dfrac{6}{17}$ (Sec. 1.6)

25. $15 - 10z$ (Sec. 1.7, Ex. 8)

26. $2y - 6x + 8$ (Sec. 1.7)

27. $3x + 17$ (Sec. 1.7, Ex. 12)

28. $2x + 8$ (Sec. 1.7)

29. a. unlike

   b. like

   c. like

   d. like

   e. like (Sec. 1.8, Ex. 2)

---

19. Simplify each expression.
   a. $-3 + [(-2 - 5) - 2]$
   b. $2^3 - 10 + [-6 - (-5)]$

20. Simplify each expression.
   a. $-(-5)$ **b.** $-\left(-\dfrac{2}{3}\right)$
   c. $-(-a)$ **d.** $-|-3|$

21. Multiply.
   a. $7(0)(-6)$
   b. $(-2)(-3)(-4)$
   c. $(-1)(-5)(-9)(-2)$

22. Subtract.
   a. $-2.7 - 8.4$
   b. $-\dfrac{4}{5} - \left(-\dfrac{3}{5}\right)$
   c. $\dfrac{1}{4} - \left(-\dfrac{1}{2}\right)$

23. Use the definition of the quotient of two numbers to find each quotient.
   a. $-18 \div 3$
   b. $\dfrac{-14}{-2}$
   c. $\dfrac{20}{-4}$

24. Find each product.
   a. $(4.5)(-0.08)$
   b. $-\dfrac{3}{4} \cdot -\dfrac{8}{17}$

Use the distributive property to write each expression without parentheses. Then simplify the result.

25. $-5(-3 + 2z)$

26. $2(y - 3x + 4)$

27. $\dfrac{1}{2}(6x + 14) + 10$

28. $-(x + 4) + 3(x + 4)$

29. Determine whether the terms are like or unlike.
   a. $2x, 3x^2$
   b. $4x^2y, x^2y, -2x^2y$
   c. $-2yz, -3zy$
   d. $-x^4, x^4$
   e. $-8a^5, 8a^5$

30. Find each quotient.

   a. $\dfrac{-32}{8}$ **b.** $\dfrac{-108}{-12}$

   c. $-\dfrac{5}{7} \div \left(-\dfrac{9}{2}\right)$

31. Subtract $4x - 2$ from $2x - 3$.

32. Subtract $10x + 3$ from $-5x + 1$.

33. Solve: $x - 7 = 10$

Solve.

34. $\dfrac{5}{6} + x = \dfrac{2}{3}$

35. $-z - 4 = 6$

36. $-3x + 1 - (-4x - 6) = 10$

37. $\dfrac{2(a + 3)}{3} = 6a + 2$

38. $\dfrac{x}{4} = 18$

39. As of May 2014, the total number of Democrats and Republicans in the U.S. House of Representatives was 432. There were 34 more Republican representatives than Democratic. Find the number of representatives from each party. (*Source:* Office of the Clerk, U.S. Capitol)

40. $6x + 5 = 4(x + 4) - 1$

41. A glacier is a giant mass of rocks and ice that flows downhill like a river. Portage Glacier in Alaska is about 6 miles, or 31,680 feet, long and moves 400 feet per year. Icebergs are created when the front end of the glacier flows into Portage Lake. How long does it take for ice at the head (beginning) of the glacier to reach the lake?

42. A number increased by 4 is the same as 3 times the number decreased by 8. Find the number.

43. The number 63 is what percent of 72?

44. Solve: $C = 2\pi r$ for r.

45. Solve: $5(2x + 3) = -1 + 7$

46. Solve: $x - 3 > 2$

47. Graph $-1 > x$.

              $\xleftarrow{\hspace{1cm}}$ $\underset{\substack{-5\ \ -4\ \ -3\ \ -2\ \ -1\ \ \ \ 0\ \ \ \ 1\ \ \ \ 2\ \ \ \ 3\ \ \ \ 4\ \ \ \ 5}}{\overset{\text{\Large◇}}{}}$ $\xrightarrow{\hspace{1cm}}$

48. Solve: $3x - 4 \leq 2x - 14$

49. Solve: $2(x - 3) - 5 \leq 3(x + 2) - 18$

50. Solve: $-3x \geq 9$

---

30. a. -4

    b. 9

    c. $\dfrac{10}{63}$ (Sec. 1.6)

    (Sec. 1.8, Ex. 15)

31. $-2x - 1$

32. $-15x - 2$ (Sec. 1.8)

33. 17 (Sec. 2.1, Ex. 1)

34. $-\dfrac{1}{6}$ (Sec. 2.1)

35. -10 (Sec. 2.2, Ex. 7)

36. 3 (Sec. 2.3)

37. 0 (Sec. 2.3, Ex. 4)

38. 72 (Sec. 2.2)

39. Republicans: 233 Democrats: 199 (Sec. 2.4, Ex. 4)

40. 5 (Sec. 2.3)

41. 79.2 yr (Sec. 2.5, Ex. 1)

42. 6 (Sec. 2.4)

43. 87.5% (Sec. 2.6, Ex. 1)

44. $\dfrac{C}{2\pi} = r$ (Sec. 2.5)

45. $-\dfrac{9}{10}$ (Sec. 2.2, Ex. 10)

46. $\{x \mid x > 5\}$ (Sec. 2.7)

    (Sec. 2.7, Ex. 2)

47. see graph

48. $\{x \mid x \leq -10\}$ (Sec. 2.7)

49. $\{x \mid x \geq 1\}$ (Sec. 2.7, Ex. 9)

50. $\{x \mid x \leq -3\}$ (Sec. 2.7)

3

Graphing Equations and Inequalities

Check Your Progress

In Chapter 2, we learned to solve and graph the solutions of linear equations and inequalities in one variable on number lines. Now we define and present techniques for solving and graphing linear equations and inequalities in two variables on grids.

What Is *Tourism Towards 2030?*

Tourism 2020 Vision was the World Tourism Organization's long-term forecast and study of the growth and economic impact of world tourism through the year 2020. As we approach the year 2020, the World Tourism Organization's new long-term study and forecast is entitled *Tourism Towards 2030*. The broken line graph below shows some actual trends and forecasts from 2000 to 2030. In Chapter 3 Review, Exercises 85–90, we read a bar graph showing the top tourist destinations by country.

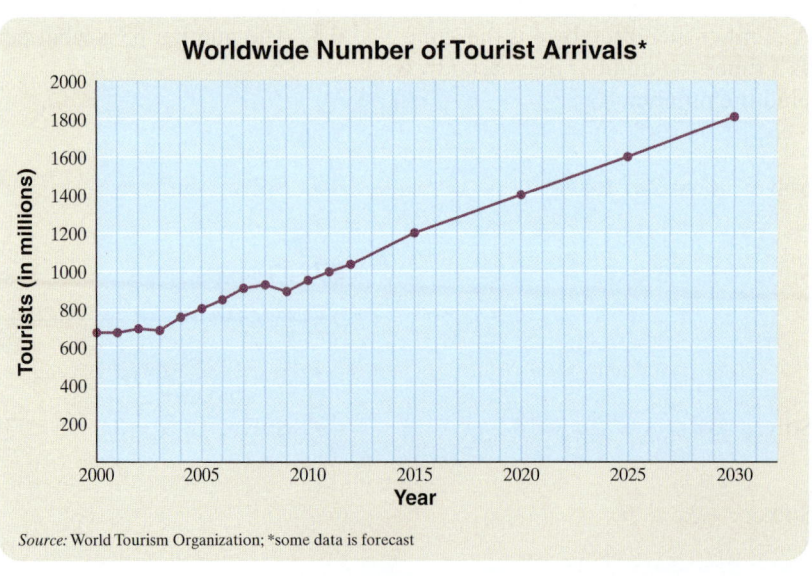

Source: World Tourism Organization; *some data is forecast

Reading Graphs and the Rectangular Coordinate System

In today's world, where the exchange of information must be fast and entertaining, graphs are becoming increasingly popular. They provide a quick visual way of making comparisons, drawing conclusions, and approximating quantities.

Objective A Reading Bar and Line Graphs

One way to visually present data is with a **bar graph.** Bar graphs consist of a series of bars that are arranged vertically or horizontally. Although we have studied bar graphs in previous sections, we now practice reading the height or length of the bars contained in a bar graph. An advantage to using bar graphs is that a scale is usually included for greater accuracy. Care must be taken when reading bar graphs as well as other types of graphs—they may be misleading, as shown later in this section.

Objectives

A Read Bar and Line Graphs.

B Plot Ordered Pairs of Numbers on the Rectangular Coordinate System.

C Graph Paired Data to Create a Scatter Diagram.

D Find the Missing Coordinate of an Ordered Pair Solution, Given One Coordinate of the Pair.

> **Example 1** Finding the Number of Endangered Species

The following bar graph shows the number of endangered species in the United States in 2013. Use this graph to answer the questions.

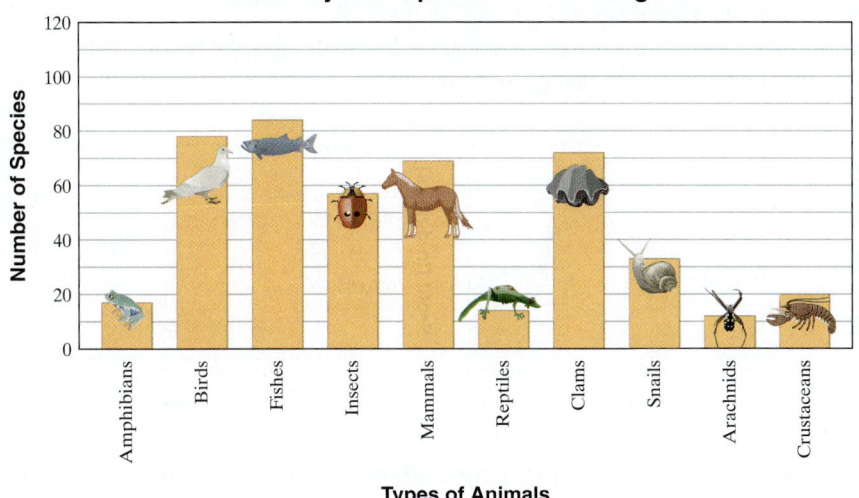

How Many U.S. Species Are Endangered?

Source: U.S. Fish and Wildlife Service

a. Approximate the number of endangered species that are clams.

b. Which category has the most endangered species?

Solution:

a. To approximate the number of endangered species that are clams, we go to the top of the bar that represents clams. From the top of this bar, we move horizontally to the left until the scale is reached. We read the height of the bar on the scale as approximately 72. There are approximately 72 clam species that are endangered, as shown on the next page.

(Continued on next page)

Practice 1

Use the bar graph in Example 1 to answer the following questions:

a. Approximate the number of endangered species that are birds.

b. Which category shows the fewest endangered species?

How Many U.S. Species Are Endangered?

Source: U.S. Fish and Wildlife Service

b. The most endangered species is represented by the tallest (longest) bar. The tallest bar corresponds to fishes.

■ **Work Practice 1**

As mentioned previously, graphs can be misleading. Both graphs below show the same information, but with different scales. Special care should be taken when forming conclusions from the appearance of a graph.

Notice the ⌇ symbol on each vertical scale on the graphs below. This symbol alerts us that numbers are missing from that scale.

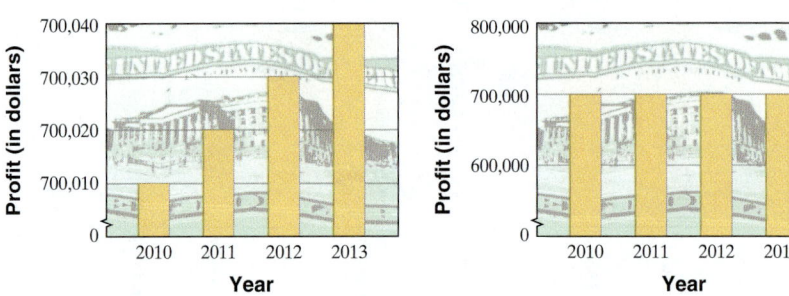

Are the profits shown in the graphs above greatly increasing, or are they remaining about the same?

Next, let's practice reading line graphs.

A **line graph** consists of a series of points connected by a line. The next graph is an example of a line graph. It is also sometimes called a **broken line graph.**

> **Example 2**   The line graph on the next page shows the relationship between time since smoking a cigarette and pulse rate. Time is recorded along the horizontal axis in minutes, with 0 minutes being the moment a smoker lights a cigarette. Pulse is recorded along the vertical axis in heartbeats per minute.

Practice 2

Use the graph from Example 2 to answer the following.

a. What is the pulse rate 40 minutes after lighting a cigarette?

b. What is the pulse rate when the cigarette is being lit?

c. When is the pulse rate the highest?

Answers

2. a. 70 beats per minute   **b.** 60 beats per minute   **c.** 5 minutes after lighting

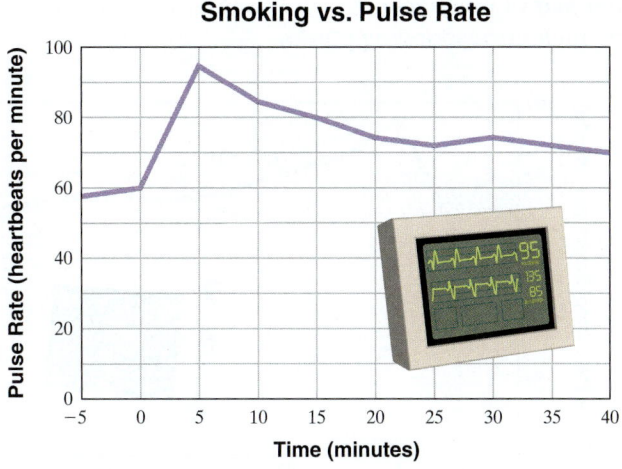

a. What is the pulse rate 15 minutes after a cigarette is lit?

b. When is the pulse rate the lowest?

c. When does the pulse rate show the greatest change?

Solution:

a. We locate the number 15 along the time axis and move vertically upward until the line is reached. From this point on the line, we move horizontally to the left until the pulse rate axis is reached. Reading the number of beats per minute, we find that the pulse rate is 80 beats per minute 15 minutes after a cigarette is lit.

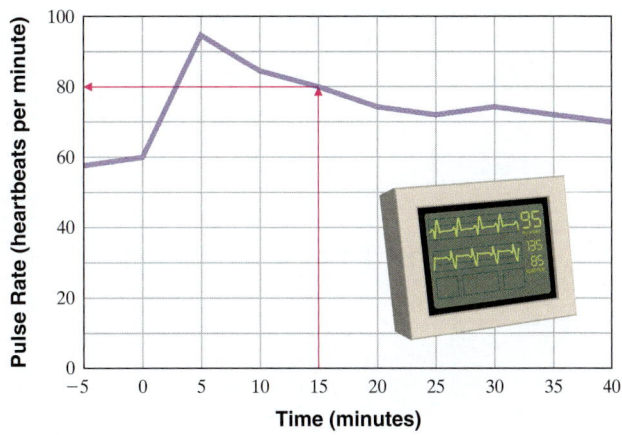

b. We find the lowest point of the line graph, which represents the lowest pulse rate. From this point, we move vertically downward to the time axis. We find that the pulse rate is the lowest at −5 minutes, which means 5 minutes *before* lighting a cigarette.

c. The pulse rate shows the greatest change during the 5 minutes between 0 and 5. Notice that the line graph is *steepest* between 0 and 5 minutes.

■ **Work Practice 2**

Notice in the graph above that there are two numbers associated with each point of the graph. For example, we discussed earlier that 15 minutes after "lighting up," the pulse rate is 80 beats per minute. If we agree to write the time first and the pulse rate second, we can say there is a point on the graph corresponding

Teaching Tip

Have students compare and contrast bar graphs and line graphs.

to the **ordered pair** of numbers (15, 80). A few more ordered pairs are shown below alongside their corresponding points.

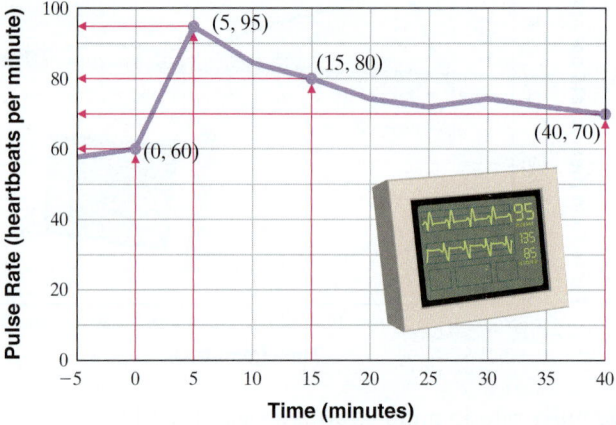

Objective B Plotting Ordered Pairs of Numbers

In general, we use the idea of ordered pairs to describe the location of a point in a plane (such as a piece of paper). We start with a horizontal and a vertical axis. Each axis is a number line, and for the sake of consistency we construct our axes to intersect at the 0 coordinate of both. This point of intersection is called the **origin.** Notice that these two number lines or axes divide the plane into four regions called **quadrants.** The quadrants are usually numbered with Roman numerals as shown. The axes are not considered to be in any quadrant.

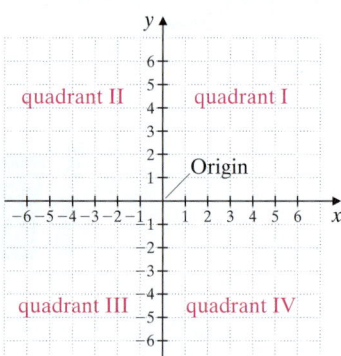

It is helpful to label axes, so we label the horizontal axis the **x-axis** and the vertical axis the **y-axis.** We call the system described above the **rectangular coordinate system,** or the **coordinate plane.** Just as with other graphs shown, we can then describe the locations of points by ordered pairs of numbers. We list the horizontal, **x-axis** measurement first and the vertical, **y-axis** measurement second.

To plot or graph the point corresponding to the ordered pair (a, b) we start at the origin. We then move a units left or right (right if a is positive, left if a is negative). From there, we move b units up or down (up if b is positive, down if b is negative). For example, to plot the point corresponding to the ordered pair (3, 2), we start at the origin, move 3 units right, and from there move 2 units up. (See the figure on the next page.) The x-value, 3, is also called the **x-coordinate,** and the y-value, 2, is also

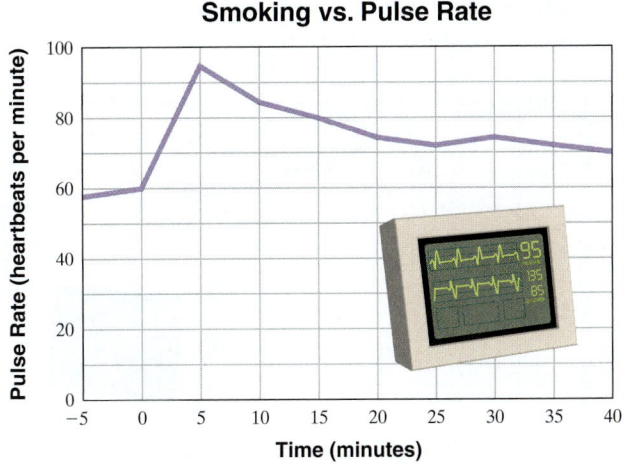

Smoking vs. Pulse Rate

a. What is the pulse rate 15 minutes after a cigarette is lit?

b. When is the pulse rate the lowest?

c. When does the pulse rate show the greatest change?

Teaching Tip

Have students compare and contrast bar graphs and line graphs.

Solution:

a. We locate the number 15 along the time axis and move vertically upward until the line is reached. From this point on the line, we move horizontally to the left until the pulse rate axis is reached. Reading the number of beats per minute, we find that the pulse rate is 80 beats per minute 15 minutes after a cigarette is lit.

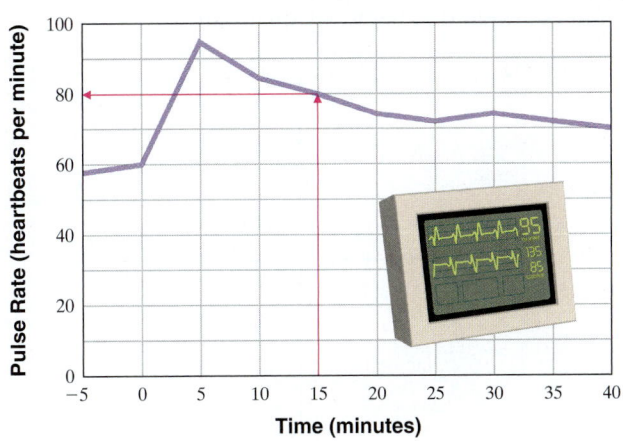

b. We find the lowest point of the line graph, which represents the lowest pulse rate. From this point, we move vertically downward to the time axis. We find that the pulse rate is the lowest at −5 minutes, which means 5 minutes *before* lighting a cigarette.

c. The pulse rate shows the greatest change during the 5 minutes between 0 and 5. Notice that the line graph is *steepest* between 0 and 5 minutes.

■ **Work Practice 2**

Notice in the graph above that there are two numbers associated with each point of the graph. For example, we discussed earlier that 15 minutes after "lighting up," the pulse rate is 80 beats per minute. If we agree to write the time first and the pulse rate second, we can say there is a point on the graph corresponding to the **ordered pair** of numbers (15, 80). A few more ordered pairs are shown below alongside their corresponding points.

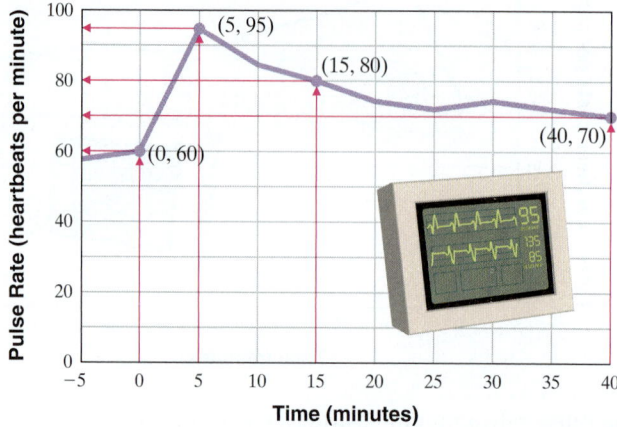

Objective B Plotting Ordered Pairs of Numbers

In general, we use the idea of ordered pairs to describe the location of a point in a plane (such as a piece of paper). We start with a horizontal and a vertical axis. Each axis is a number line, and for the sake of consistency we construct our axes to intersect at the 0 coordinate of both. This point of intersection is called the **origin.** Notice that these two number lines or axes divide the plane into four regions called **quadrants.** The quadrants are usually numbered with Roman numerals as shown. The axes are not considered to be in any quadrant.

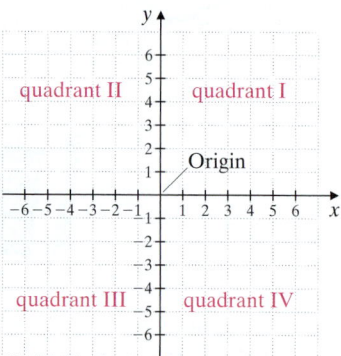

It is helpful to label axes, so we label the horizontal axis the **x-axis** and the vertical axis the **y-axis.** We call the system described above the **rectangular coordinate system,** or the **coordinate plane.** Just as with other graphs shown, we can then describe the locations of points by ordered pairs of numbers. We list the horizontal, **x-axis** measurement first and the vertical, **y-axis** measurement second.

To plot or graph the point corresponding to the ordered pair (*a*, *b*) we start at the origin. We then move *a* units left or right (right if *a* is positive, left if *a* is negative). From there, we move *b* units up or down (up if *b* is positive, down if *b* is negative). For example, to plot the point corresponding to the ordered pair (3, 2), we start at the origin, move 3 units right, and from there move 2 units up. (See the figure on the next page.) The *x*-value, 3, is also called the **x-coordinate,** and the *y*-value, 2, is also called the **y-coordinate.** From now on, we will call the point with coordinates $(3, 2)$ simply the point $(3, 2)$. The point $(-2, 5)$ is also graphed below.

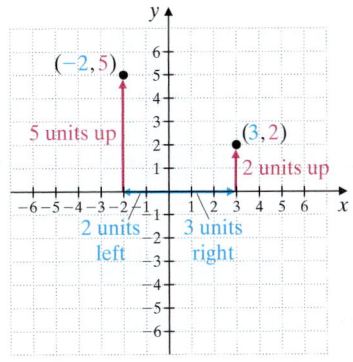

Practice 3

On a single coordinate system, plot each ordered pair. State in which quadrant, or on which axis, each point lies.

a. $(4, 2)$ b. $(-1, -3)$
c. $(2, -2)$ d. $(-5, 1)$
e. $(0, 3)$ f. $(3, 0)$
g. $(0, -4)$ h. $\left(-2\dfrac{1}{2}, 0\right)$
i. $\left(1, -3\dfrac{3}{4}\right)$

Helpful Hint

Don't forget that **each ordered pair corresponds to exactly one point in the plane and that each point in the plane corresponds to exactly one ordered pair.**

✓**Concept Check** Is the graph of the point $(-5, 1)$ in the same location as the graph of the point $(1, -5)$? Explain.

Example 3 On a single coordinate system, plot each ordered pair. State in which quadrant, or on which axis, each point lies.

a. $(5, 3)$ b. $(-2, -4)$ c. $(1, -2)$ d. $(-5, 3)$ e. $(0, 0)$

f. $(0, 2)$ g. $(-5, 0)$ h. $\left(0, -5\dfrac{1}{2}\right)$ i. $\left(4\dfrac{2}{3}, -3\right)$

Solution:

a. Point $(5, 3)$ lies in quadrant I.
b. Point $(-2, -4)$ lies in quadrant III.
c. Point $(1, -2)$ lies in quadrant IV.
d. Point $(-5, 3)$ lies in quadrant II.

e–h. Points $(0, 0)$, $(0, 2)$, and $\left(0, -5\dfrac{1}{2}\right)$ lie on the y-axis. Points $(0, 0)$ and $(-5, 0)$ lie on the x-axis.

i. Point $\left(4\dfrac{2}{3}, -3\right)$ lies in quadrant IV.

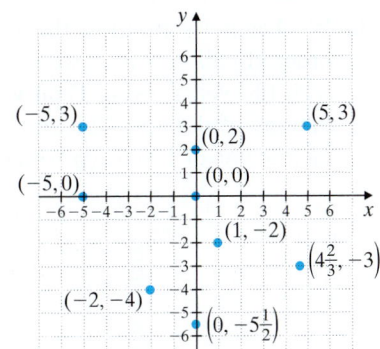

■ **Work Practice 3**

Answers

3.

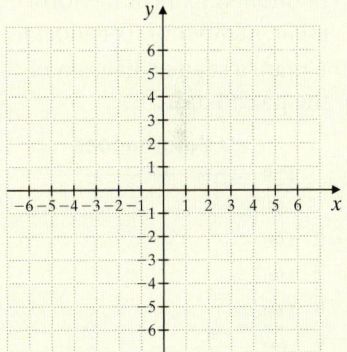

a. Point $(4, 2)$ lies in quadrant I.
b. Point $(-1, -3)$ lies in quadrant III.
c. Point $(2, -2)$ lies in quadrant IV.
d. Point $(-5, 1)$ lies in quadrant II.

e–h. Points $(3, 0)$ and $\left(-2\dfrac{1}{2}, 0\right)$ lie on the x-axis. Points $(0, 3)$ and $(0, -4)$ lie on the y-axis.

i. Point $\left(1, -3\dfrac{3}{4}\right)$ lies in quadrant IV.

✓**Concept Check Answer**

The graph of point $(-5, 1)$ lies in quadrant II, and the graph of point $(1, -5)$ lies in quadrant IV. They are *not* in the same location.

Helpful Hint

In Example 3, notice that the point $(0, 0)$ lies on both the x-axis and the y-axis. It is the only point in the entire rectangular coordinate system that has this feature. Why? It is the only point of intersection of the x-axis and the y-axis.

Practice 4

The table gives the number of tornadoes that occurred in the United States for the years shown. (*Source:* Storm Prediction Center, National Weather Service)

Year	Tornadoes
2008	1692
2009	1156
2010	1282
2011	1693
2012	939
2013	908

a. Write this paired data as a set of ordered pairs of the form (year, number of tornadoes).

b. Create a scatter diagram of the paired data.

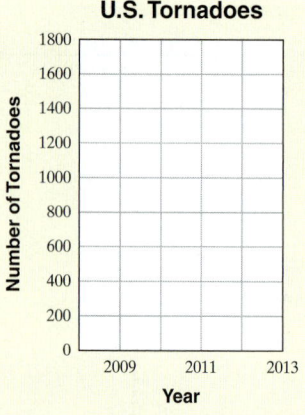
U.S. Tornadoes

c. What trend in the paired data, if any, does the scatter diagram show?

Answers

4. a. (2008, 1692), (2009, 1156), (2010, 1282), (2011, 1693), (2012, 939), (2013, 908)

b.

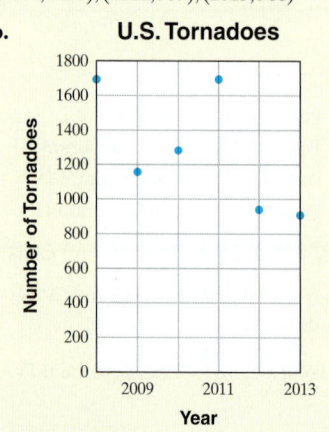
U.S. Tornadoes

c. The number of tornadoes varies greatly from year to year.

✓ **Concept Check Answers**

a. $(-3, 5)$ b. $(0, -6)$

✓**Concept Check** For each description of a point in the rectangular coordinate system, write an ordered pair that represents it.

a. Point *A* is located three units to the left of the *y*-axis and five units above the *x*-axis.

b. Point *B* is located six units below the origin.

From Example 3, notice that the *y*-coordinate of any point on the *x*-axis is 0. For example, the point $(-5, 0)$ lies on the *x*-axis. Also, the *x*-coordinate of any point on the *y*-axis is 0. For example, the point $(0, 2)$ lies on the *y*-axis.

Objective C Creating Scatter Diagrams

Data that can be represented as ordered pairs are called **paired data.** Many types of data collected from the real world are paired data. For instance, the annual measurements of a child's height can be written as ordered pairs of the form (year, height in inches) and are paired data. The graph of paired data as points in the rectangular coordinate system is called a **scatter diagram.** Scatter diagrams can be used to look for patterns and trends in paired data.

Example 4 The table gives the annual net sales for PetSmart for the years shown. (*Source:* PetSmart)

Year	PetSmart Net Sales (in billions of dollars)
2007	4.7
2008	5.1
2009	5.3
2010	5.7
2011	6.1
2012	6.8
2013	6.9

a. Write this paired data as a set of ordered pairs of the form (year, net sales in billions of dollars).

b. Create a scatter diagram of the paired data.

c. What trend in the paired data does the scatter diagram show?

Solution:

a. The ordered pairs are (2007, 4.7), (2008, 5.1), (2009, 5.3), (2010, 5.7), (2011, 6.1), (2012, 6.8), and (2013, 6.9).

b. We begin by plotting the ordered pairs. Because the *x*-coordinate in each ordered pair is a year, we label the *x*-axis "Year" and mark the horizontal axis with the years given. Then we label the *y*-axis, or vertical axis, "Net Sales (in billions of dollars)." In this case, it is convenient to mark the vertical axis in increments of 1, starting with 0.

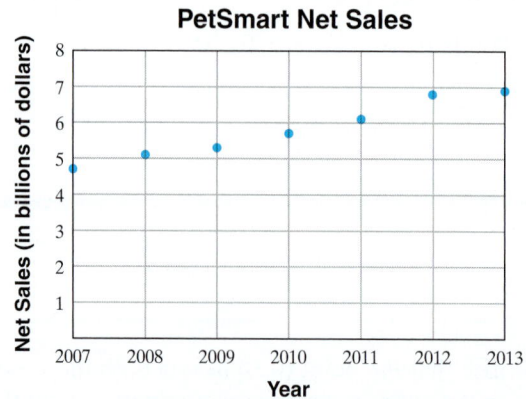
PetSmart Net Sales

c. The scatter diagram shows that PetSmart net sales steadily increased over the years 2007–2013.

■ **Work Practice 4**

Objective D Completing Ordered Pair Solutions ▶

Let's see how we can use ordered pairs to record solutions of equations containing two variables. An equation in one variable such as $x + 1 = 5$ has one solution, 4: The number 4 is the value of the variable x that makes the equation true.

An equation in two variables, such as $2x + y = 8$, has solutions consisting of two values, one for x and one for y. For example, $x = 3$ and $y = 2$ is a solution of $2x + y = 8$ because, if x is replaced with 3 and y with 2, we get a true statement.

$$2x + y = 8$$
$$2(3) + 2 \stackrel{?}{=} 8 \quad \text{Replace } x \text{ with 3 and } y \text{ with 2.}$$
$$8 = 8 \quad \text{True}$$

The solution $x = 3$ and $y = 2$ can be written as $(3, 2)$, an ordered pair of numbers.

> In general, an ordered pair is a **solution** of an equation in two variables if replacing the variables by the values of the ordered pair results in a *true statement*.

For example, another ordered pair solution of $2x + y = 8$ is $(5, -2)$. Replacing x with 5 and y with -2 results in a true statement.

$$2x + y = 8$$
$$2(5) + (-2) \stackrel{?}{=} 8 \quad \text{Replace } x \text{ with 5 and } y \text{ with } -2.$$
$$10 - 2 \stackrel{?}{=} 8$$
$$8 = 8 \quad \text{True}$$

Example 5 Complete each ordered pair so that it is a solution of the equation $3x + y = 12$.

a. $(0, \ )$ **b.** $(\ , 6)$ **c.** $(-1, \ )$

Solution:

a. In the ordered pair $(0, \ )$, the x-value is 0. We let $x = 0$ in the equation and solve for y.

$$3x + y = 12$$
$$3(0) + y = 12 \quad \text{Replace } x \text{ with 0.}$$
$$0 + y = 12$$
$$y = 12$$

The completed ordered pair is $(0, 12)$.

b. In the ordered pair $(\ , 6)$, the y-value is 6. We let $y = 6$ in the equation and solve for x.

$$3x + y = 12$$
$$3x + 6 = 12 \quad \text{Replace } y \text{ with 6.}$$
$$3x = 6 \quad \text{Subtract 6 from both sides.}$$
$$x = 2 \quad \text{Divide both sides by 3.}$$

The ordered pair is $(2, 6)$.

c. In the ordered pair $(-1, \ )$, the x-value is -1. We let $x = -1$ in the equation and solve for y.

$$3x + y = 12$$
$$3(-1) + y = 12 \quad \text{Replace } x \text{ with } -1.$$
$$-3 + y = 12$$
$$y = 15 \quad \text{Add 3 to both sides.}$$

The ordered pair is $(-1, 15)$.

🟧 **Work Practice 5**

Practice 5

Complete each ordered pair so that it is a solution of the equation $x + 2y = 8$.

a. $(0, \ )$

b. $(\ , 3)$

c. $(-4, \ )$

Solutions of equations in two variables can also be recorded in a **table of paired values,** as shown in the next example.

Practice 6

Complete the table for the equation $y = -2x$.

	x	y
a.	−3	
b.		0
c.		10

Example 6

Complete the table for the equation $y = 3x$.

	x	y
a.	−1	
b.		0
c.		−9

Solution:

a. We replace x with -1 in the equation and solve for y.

$$y = 3x$$
$$y = 3(-1) \quad \text{Let } x = -1.$$
$$y = -3$$

The ordered pair is $(-1, -3)$.

b. We replace y with 0 in the equation and solve for x.

$$y = 3x$$
$$0 = 3x \quad \text{Let } y = 0.$$
$$0 = x \quad \text{Divide both sides by 3.}$$

The ordered pair is $(0, 0)$.

c. We replace y with -9 in the equation and solve for x.

$$y = 3x$$
$$-9 = 3x \quad \text{Let } y = -9.$$
$$-3 = x \quad \text{Divide both sides by 3.}$$

The ordered pair is $(-3, -9)$. The completed table is shown to the right.

x	y
−1	−3
0	0
−3	−9

Work Practice 6

Practice 7

Complete the table for the equation $y = \frac{1}{3}x - 1$.

	x	y
a.	−3	
b.	0	
c.		0

Example 7

Complete the table for the equation

$$y = \frac{1}{2}x - 5.$$

	x	y
a.	−2	
b.	0	
c.		0

Solution:

a. Let $x = -2$.

$$y = \frac{1}{2}x - 5$$
$$y = \frac{1}{2}(-2) - 5$$
$$y = -1 - 5$$
$$y = -6$$

b. Let $x = 0$.

$$y = \frac{1}{2}x - 5$$
$$y = \frac{1}{2}(0) - 5$$
$$y = 0 - 5$$
$$y = -5$$

c. Let $y = 0$.

$$y = \frac{1}{2}x - 5$$
$$0 = \frac{1}{2}x - 5 \quad \text{Now, solve for } x.$$
$$5 = \frac{1}{2}x \quad \text{Add 5.}$$
$$10 = x \quad \text{Multiply by 2.}$$

Ordered pairs: $(-2, -6)$ $(0, -5)$ $(10, 0)$

The completed table is

x	−2	0	10
y	−6	−5	0

Work Practice 7

Answers

6.

	x	y
a.	−3	6
b.	0	0
c.	−5	10

7.

	x	y
a.	−3	−2
b.	0	−1
c.	3	0

By now, you have noticed that equations in two variables often have more than one solution. We discuss this more in the next section.

A table showing ordered pair solutions may be written vertically or horizontally, as shown in the next example.

Example 8 A small business purchased a computer for $2000. The business predicts that the computer will be used for 5 years and the value in dollars y of the computer in x years is $y = -300x + 2000$. Complete the table.

x	0	1	2	3	4	5
y						

Solution:

To find the value of y when x is 0, we replace x with 0 in the equation. We use this same procedure to find y when x is 1 and when x is 2.

When $x = 0$,
$y = -300x + 2000$
$y = -300 \cdot 0 + 2000$
$y = 0 + 2000$
$y = 2000$

When $x = 1$,
$y = -300x + 2000$
$y = -300 \cdot 1 + 2000$
$y = -300 + 2000$
$y = 1700$

When $x = 2$,
$y = -300x + 2000$
$y = -300 \cdot 2 + 2000$
$y = -600 + 2000$
$y = 1400$

We have the ordered pairs (0, 2000), (1, 1700), and (2, 1400). This means that in 0 years the value of the computer is $2000, in 1 year the value of the computer is $1700, and in 2 years the value is $1400. To complete the table of values, we continue the procedure for $x = 3$, $x = 4$, and $x = 5$.

When $x = 3$,
$y = -300x + 2000$
$y = -300 \cdot 3 + 2000$
$y = -900 + 2000$
$y = 1100$

When $x = 4$,
$y = -300x + 2000$
$y = -300 \cdot 4 + 2000$
$y = -1200 + 2000$
$y = 800$

When $x = 5$,
$y = -300x + 2000$
$y = -300 \cdot 5 + 2000$
$y = -1500 + 2000$
$y = 500$

The completed table is shown below.

x	0	1	2	3	4	5
y	2000	1700	1400	1100	800	500

Work Practice 8

The ordered pair solutions recorded in the completed table for Example 8 are another set of paired data. They are graphed next. Notice that this scatter diagram gives a visual picture of the decrease in value of the computer.

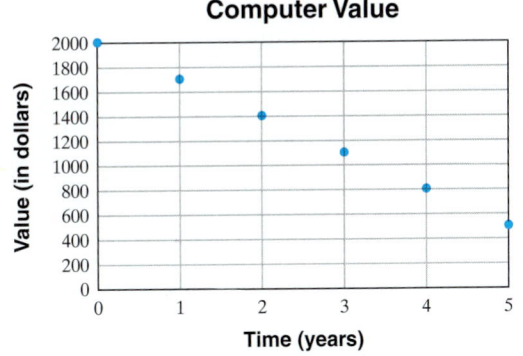

Computer Value

Practice 8

A company purchased a fax machine for $400. The business manager of the company predicts that the fax machine will be used for 7 years and the value in dollars y of the machine in x years is $y = -50x + 400$. Complete the table.

x	1	2	3	4	5	6	7
y							

Answer

8.

x	1	2	3	4	5	6	7
y	350	300	250	200	150	100	50

Vocabulary, Readiness & Video Check

Use the choices below to fill in each blank. The exercises below all have to do with the rectangular coordinate system.

origin	*x*-coordinate	*x*-axis	scatter diagram	four
quadrants	*y*-coordinate	*y*-axis	solution	one

1. The horizontal axis is called the _____*x*-axis_____.

2. The vertical axis is called the _____*y*-axis_____.

3. The intersection of the horizontal axis and the vertical axis is a point called the _____origin_____.

4. The axes divide the plane into regions, called _____quadrants_____. There are _____four_____ of these regions.

5. In the ordered pair of numbers $(-2, 5)$, the number -2 is called the _____*x*-coordinate_____ and the number 5 is called the _____*y*-coordinate_____.

6. Each ordered pair of numbers corresponds to _____one_____ point in the plane.

7. An ordered pair is a(n) _____solution_____ of an equation in two variables if replacing the variables by the coordinates of the ordered pair results in a true statement.

8. The graph of paired data as points in a rectangular coordinate system is called a(n) _____scatter diagram_____.

Martin-Gay Interactive Videos

See Video 3.1

See video answer section.

Watch the section lecture video and answer the following questions.

Objective A **9.** From the line graph in  Examples 5–6, what year had the greatest number of goals per game average and what was this average? ▶

Objective B **10.** Several points are plotted in ⊞ Examples 7–12. Where do we always start when plotting a point? How does the first coordinate tell us to move? How does the second coordinate tell us to move? ▶

Objective C **11.** From ⊞ Example 13, what kind of data can be graphed in a scatter diagram? ▶

Objective D **12.** In ⊞ Example 14, when finding the missing value in an ordered pair solution of a linear equation in two variables, how can we check our solution? ▶

3.1 **Exercise Set** MyMathLab®

Objective A *The National Weather Service has exacting definitions for hurricanes; they are tropical storms with winds in excess of 74 mph. The following bar graph shows the number of hurricanes, by month, that have made landfall on the mainland United States between 1851 and 2012. Use this graph to answer Exercises 1 through 6. See Example 1. (Source: National Weather Service: National Hurricane Center)*

▶ 1. In which month did the most hurricanes make land-fall in the United States?   September

2. In which month did the fewest hurricanes make landfall in the United States?   June

▶ 3. Approximate the number of hurricanes that made landfall in the United States during the month of August.   77

4. Approximate the number of hurricanes that made landfall in the United States in September.   107

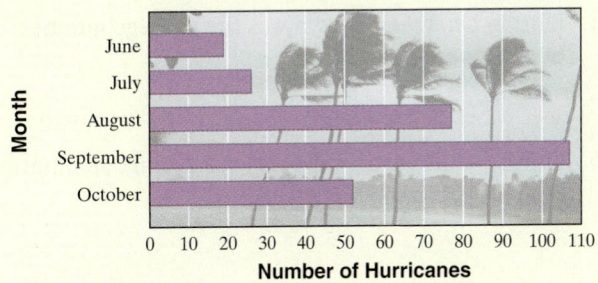

Hurricanes Making Landfall in the United States, by Month, 1851–2012

▶ 5. In 2008 alone, two hurricanes made landfall during the month of August. What fraction of all the 77 hurricanes that made landfall during August is this? $\frac{2}{77}$

6. In 2007, only one hurricane made landfall in the United States during the entire season, in the month of September. If there have been 107 hurricanes to make landfall in the month of September since 1851, what fraction of these arrived in 2007? $\frac{1}{107}$

The following horizontal bar graph shows the approximate 2014 population of the world's largest cities (including their suburbs). Use this graph to answer Exercises 7 through 12. See Example 1. (Source: CityPopulation)

7. Name the city with the largest population, and estimate its population.   Tokyo, Japan; about 39.4 million or 39,400,000

8. Name the cities whose population is between 24 million and 26 million.   Seoul, South Korea; Delhi, India

9. Name the city in Mexico with the largest population, and estimate its population.   Mexico City; 22.2 million or 22,200,000

10. Name the city on the bar graph with a population of about 27 million.   Jakarta, Indonesia

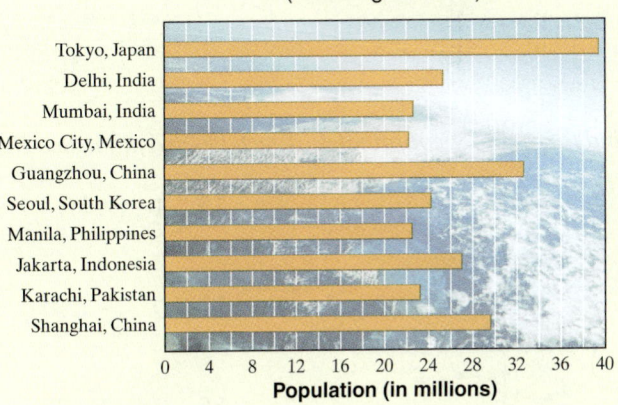

World's Largest Cities
(including Suburbs)

11. How much larger (in terms of population) is Delhi, India, than Seoul, South Korea?   approximately 1 million

12. How much larger (in terms of population) is Shanghai, China, than Mexico City, Mexico?   about 7.5 million

Beach Soccer World Cup is now held every two years. The following line graph shows the World Cup goals per game average for beach soccer during the years shown. Use this graph to answer Exercises 13 through 16. See Example 2.

▶ **13.** Find the average number of goals per game in 2013.
7.6 goals/game

14. Find the average number of goals per game in 2009.
8.7 goals/game

▶ **15.** During what year shown was the average number of goals per game the highest?   2003

16. During what year shown was the average number of goals per game the lowest?   2001

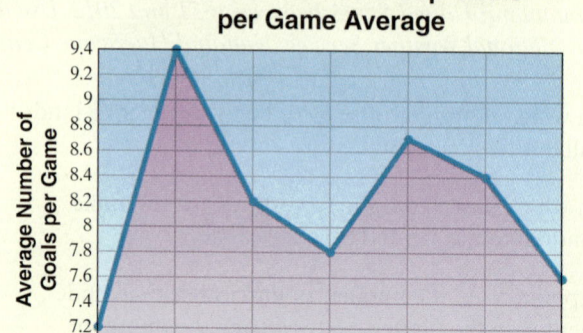

Beach Soccer World Cup Goals per Game Average

Source: Wikipedia

Objective B *Plot each ordered pair. State in which quadrant or on which axis each point lies. See Example 3.*

▶ **17. a.** $(1, 5)$ **b.** $(-5, -2)$ **c.** $(-3, 0)$ **d.** $(0, -1)$ **18. a.** $(2, 4)$ **b.** $(0, 2)$ **c.** $(-2, 1)$ **d.** $(-3, -3)$
e. $(2, -4)$ **f.** $\left(-1, 4\frac{1}{2}\right)$ **g.** $(3.7, 2.2)$ **h.** $\left(\frac{1}{2}, -3\right)$ **e.** $\left(3\frac{3}{4}, 0\right)$ **f.** $(5, -4)$ **g.** $(-3.4, 4.8)$ **h.** $\left(\frac{1}{3}, -5\right)$

$(1, 5)$ and $(3.7, 2.2)$ are in quadrant I, $\left(-1, 4\frac{1}{2}\right)$ is in quadrant II, $(-5, -2)$ is in quadrant III, $(2, -4)$ and $\left(\frac{1}{2}, -3\right)$ are in quadrant IV, $(-3, 0)$ lies on the x-axis, $(0, -1)$ lies on the y-axis

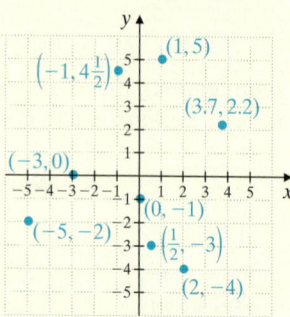

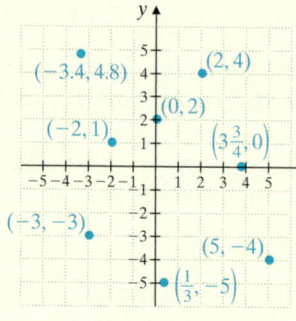

$(2, 4)$ is in quadrant I, $(-2, 1)$ and $(-3.4, 4.8)$ are in quadrant II, $(-3, -3)$ is in quadrant III, $(5, -4)$ and $\left(\frac{1}{3}, -5\right)$ are in quadrant IV, $\left(3\frac{3}{4}, 0\right)$ lies on the x-axis, $(0, 2)$ lies on the y-axis

---

Find the x- and y-coordinates of each labeled point. See Example 3.

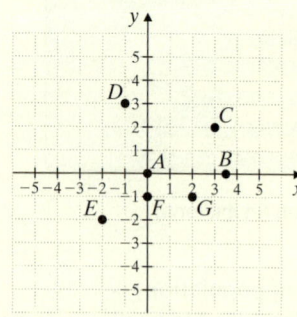

19. A $(0, 0)$ **20.** B $\left(3\frac{1}{2}, 0\right)$ **21.** C $(3, 2)$

22. D $(-1, 3)$ **23.** E $(-2, -2)$ **24.** F $(0, -1)$

25. G $(2, -1)$

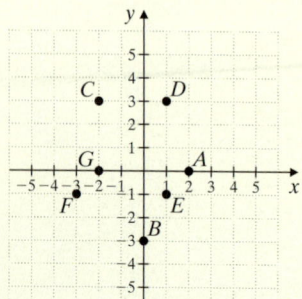

26. A $(2, 0)$ **27.** B $(0, -3)$ **28.** C $(-2, 3)$

29. D $(1, 3)$ **30.** E $(1, -1)$ **31.** F $(-3, -1)$

32. G $(-2, 0)$

Objective C   *Solve. See Example 4.*

33. The table shows the domestic box office (in billions of dollars) for the U.S. movie industry during the years shown. (*Source:* Motion Picture Association of America)

Year	Box Office (in billions of dollars)
2007	9.6
2008	9.6
2009	10.6
2010	10.6
2011	10.2
2012	10.8

a. Write this paired data as a set of ordered pairs of the form (year, box office).   (2007, 9.6), (2008, 9.6), (2009, 10.6), (2010, 10.6), (2011, 10.2), (2012, 10.8)

b. In your own words, write the meaning of the ordered pair (2010, 10.6).   In the year 2010, the domestic box office was $10.6 billion.

c. Create a scatter diagram of the paired data. Be sure to label the axes appropriately.

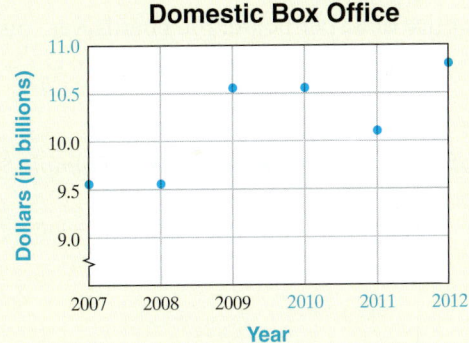

Domestic Box Office

d. What trend in the paired data does the scatter diagram show?   answers may vary

34. The table shows the amount of money (in billions of dollars) that Americans spent on their pets for the years shown. (*Source:* American Pet Products Association, Inc.)

Year	Pet-Related Expenditures (in billions of dollars)
2007	41
2008	43
2009	46
2010	48
2011	51
2012	53

a. Write this paired data as a set of ordered pairs of the form (year, pet-related expenditures).

b. In your own words, write the meaning of the ordered pair (2012, 53).

c. Create a scatter diagram of the paired data. Be sure to label the axes appropriately.

Pet-Related Expenditures

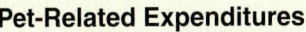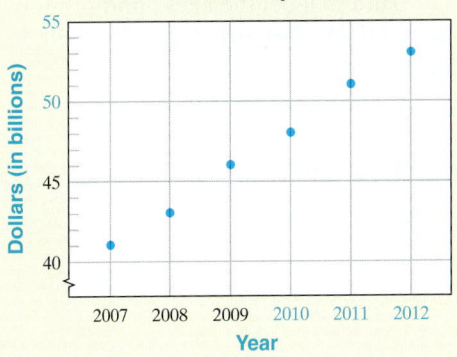

d. What trend in the paired data does the scatter diagram show?

34. a. (2007, 41), (2008, 43), (2009, 46), (2010, 48), (2011, 51), (2012, 53)   **b.** In the year 2012, $53 billion was spent on pet-related expenditures.   **d.** Pet-related expenditures increased every year.

35. Minh, a psychology student, kept a record of how much time she spent studying for each of her 20-point psychology quizzes and her score on each quiz.

Hours Spent Studying	0.50	0.75	1.00	1.25	1.50	1.50	1.75	2.00
Quiz Score	10	12	15	16	18	19	19	20

a. Write the data as ordered pairs of the form (hours spent studying, quiz score).

b. In your own words, write the meaning of the ordered pair (1.25, 16).

c. Create a scatter diagram of the paired data. Be sure to label the axes appropriately.

d. What might Minh conclude from the scatter diagram?

35. a. (0.50, 10), (0.75, 12), (1.00, 15), (1.25, 16), (1.50, 18), (1.50, 19), (1.75, 19), (2.00, 20)
b. When Minh studied 1.25 hours, her quiz score was 16.   **d.** answers may vary

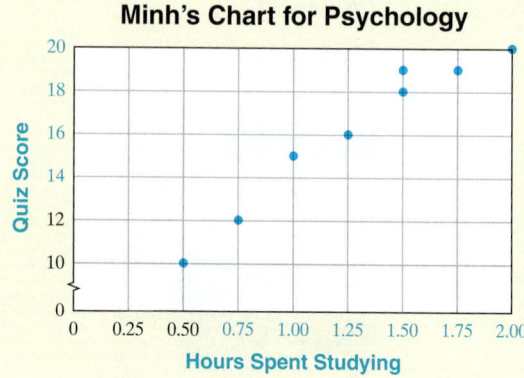

36. A local lumberyard uses quantity pricing. The table shows the price per board for different amounts of lumber purchased.

Price per Board (in dollars)	Number of Boards Purchased
8.00	1
7.50	10
6.50	25
5.00	50
2.00	100

a. Write the data as ordered pairs of the form (price per board, number of boards purchased).

b. In your own words, write the meaning of the ordered pair (2.00, 100).

c. Create a scatter diagram of the paired data. Be sure to label the axes appropriately.

d. What trend in the paired data does the scatter diagram show?

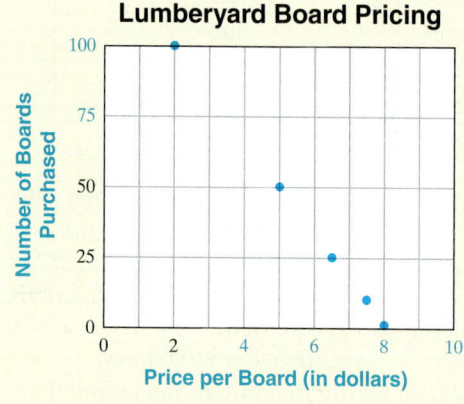

36. a. (8.00, 1), (7.50, 10), (6.50, 25), (5.00, 50), (2.00, 100)   **b.** For the price of $2 per board, 100 boards were purchased.
d. answers may vary

Objective D *Complete each ordered pair so that it is a solution of the given linear equation. See Example 5.*

37. $x - 4y = 4$; (, −2), (4,) $(-4, -2), (4, 0)$

38. $x - 5y = -1$; (, −2), (4,) $(-11, -2), (4, 1)$

39. $y = \frac{1}{4}x - 3$; (−8,), (, 1) $(-8, -5), (16, 1)$

40. $y = \frac{1}{5}x - 2$; (−10,), (, 1) $(-10, -4), (15, 1)$

Complete the table of ordered pairs for each linear equation. See Examples 6 and 7.

41. $y = -7x$

x	y
0	0
−1	7
$-\dfrac{2}{7}$	2

42. $y = -9x$

x	y
0	0
−3	27
$-\dfrac{2}{9}$	2

43. $x = -y + 2$

x	y
0	2
2	0
−3	5

44. $x = -y + 4$

x	y
4	0
0	4
7	−3

45. $y = \dfrac{1}{2}x$

x	y
0	0
−6	−3
2	1

46. $y = \dfrac{1}{3}x$

x	y
0	0
−6	−2
3	1

47. $x + 3y = 6$

x	y
0	2
6	0
3	1

48. $2x + y = 4$

x	y
0	4
2	0
1	2

49. $y = 2x - 12$

x	y
0	−12
5	−2
3	−6

50. $y = 5x + 10$

x	y
−2	0
−1	5
0	10

51. $2x + 7y = 5$

x	y
0	$\dfrac{5}{7}$
$\dfrac{5}{2}$	0
−1	1

52. $x - 6y = 3$

x	y
0	$-\dfrac{1}{2}$
1	$-\dfrac{1}{3}$
−3	−1

Objectives B C D **Mixed Practice** *Complete the table of ordered pairs for each equation. Then plot the ordered pair solutions. See Examples 3 through 7.*

▶ **53.** $x = -5y$

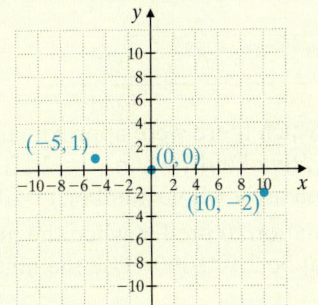

x	y
0	0
−5	1
10	−2

54. $y = -3x$

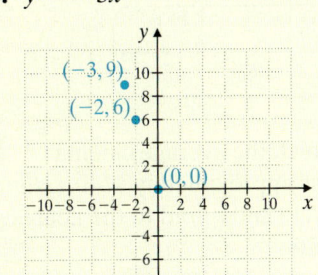

x	y
0	0
−2	6
−3	9

55. $y = \dfrac{1}{3}x + 2$

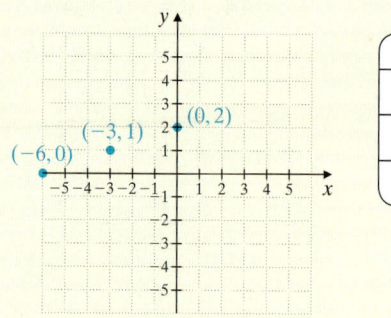

x	y
0	2
−3	1
−6	0

56. $y = \dfrac{1}{2}x + 3$

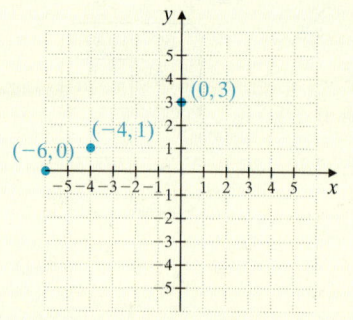

x	y
0	3
−4	1
−6	0

Solve. See Example 8.

57. The cost in dollars y of producing x computer desks is given by $y = 80x + 5000$.

a. Complete the table.

x	100	200	300
y	13,000	21,000	29,000

b. Find the number of computer desks that can be produced for $8600. (*Hint:* Find x when $y = 8600$.)   45 desks

58. The hourly wage y of an employee at a certain production company is given by $y = 0.25x + 9$, where x is the number of units produced by the employee in an hour.

a. Complete the table.

x	0	1	5	10
y	9	9.25	10.25	11.50

b. Find the number of units that the employee must produce each hour to earn an hourly wage of $12.25. (*Hint:* Find x when $y = 12.25$.)   13 units

59. The average annual cinema admission price y (in dollars) from 2003 through 2012 is given by $y = 0.24x + 5.96$. In this equation, x represents the number of years after 2003. (*Source:* Motion Picture Association of America)

a. Complete the table.

x	2	5	8
y	6.44	7.16	7.88

b. Find the year in which the average cinema admission price was approximately $7.40. (*Hint:* Find x when $y = 7.40$ and round to the nearest whole number.)   2009

c. Use the given equation to predict when the cinema admission price might be $9.00. (Use the hint for part **b.**)   2016

60. The number of farms y in the United States from 2009 through 2012 is given by $y = -10,000x + 2,201,000$. In the equation, x represents the number of years after 2009. (*Source:* Based on data from the National Agricultural Statistics Service)

a. Complete the table.

x	0	2	4
y	2,201,000	2,181,000	2,161,000

b. Find the year in which there were approximately 2,170,000 farms. (*Hint:* Find x when $y = 2,170,000$ and round to the nearest whole number.)   2012

c. Use the given equation to predict when the number of farms might be 2,100,000. (Use the hint for part **b.**)   2019

Review

Solve each equation for y. See Section 2.5.

61. $x + y = 5$ $y = 5 - x$

62. $x - y = 3$ $y = x - 3$

63. $2x + 4y = 5$ $y = \dfrac{5 - 2x}{4}$

64. $5x + 2y = 7$ $y = \dfrac{7 - 5x}{2}$

65. $10x = -5y$ $y = -2x$

66. $4y = -8x$ $y = -2x$

Concept Extensions

Answer each exercise with true or false.

67. Point $(-1, 5)$ lies in quadrant IV.   false

68. Point $(3, 0)$ lies on the y-axis.   false

69. For the point $\left(-\dfrac{1}{2}, 1.5\right)$, the first value, $-\dfrac{1}{2}$, is the x-coordinate and the second value, 1.5, is the y-coordinate.   true

70. The ordered pair $\left(2, \dfrac{2}{3}\right)$ is a solution of $2x - 3y = 6$.   false

For Exercises 71 through 75, fill in each blank with "0," "positive," or "negative." For Exercises 76 and 77, fill in each blank with "x" or "y."

	Point	Location
71.	(negative, negative)	quadrant III
72.	(positive, positive)	quadrant I
73.	(positive, negative)	quadrant IV
74.	(negative, positive)	quadrant II
75.	(0 , 0)	origin
76.	(number, 0)	x -axis
77.	(0, number)	y -axis

78. Give an example of an ordered pair whose location is in (or on)

   a. quadrant I   **b.** quadrant II   **c.** quadrant III

   d. quadrant IV   **e.** x-axis   **f.** y-axis   answers may vary

Solve. See the Concept Checks in this section.

79. Is the graph of $(3, 0)$ in the same location as the graph of $(0, 3)$? Explain why or why not.
no; answers may vary

80. Give the coordinates of a point such that if the coordinates are reversed, their location is the same.
answers may vary

81. In general, what points can have coordinates reversed and still have the same location?
answers may vary

82. In your own words, describe how to plot or graph an ordered pair of numbers.
answers may vary

83. Discuss any similarities in the graphs of the ordered pair solutions for Exercises **53–56**.
answers may vary

84. Discuss any differences in the graphs of the ordered pair solutions for Exercises **53–56**.
answers may vary

Write an ordered pair for each point described.

85. Point *C* is four units to the right of the *y*-axis and seven units below the *x*-axis. $(4, -7)$

86. Point *D* is three units to the left of the origin. $(-3, 0)$

Solve.

△**87.** Find the perimeter of the rectangle whose vertices are the points with coordinates $(-1, 5), (3, 5), (3, -4),$ and $(-1, -4)$.   26 units

△**88.** Find the area of the rectangle whose vertices are the points with coordinates $(5, 2), (5, -6), (0, -6),$ and $(0, 2)$.   40 sq units

The scatter diagram below shows the annual number of people enrolled as Gold Star Members at Costco Wholesale. The horizontal axis represents the number of years after 2008.

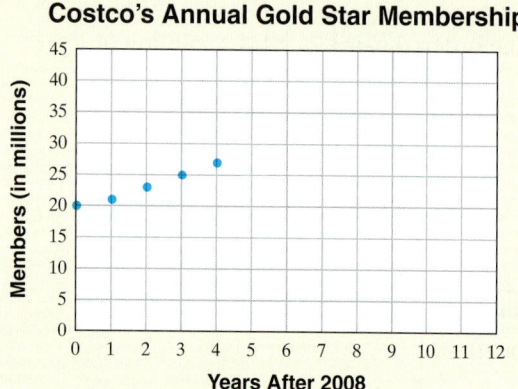

Costco's Annual Gold Star Membership

Source: Costco Wholesale Corporation

89. Estimate the annual Gold Star Membership for years 1, 2, 3, and 4.   21 million, 23 million, 25 million, 27 million

90. Use a straightedge or ruler and this scatter diagram to predict Costco's Gold Star Membership in the year 2018.   approximately 38 million, but answers may vary

The following double line graph shows temperature highs and lows for a week. Use this graph to answer Exercises 91 through 96.

91. What was the high temperature reading on Thursday?   83°F

92. What was the low temperature reading on Thursday?   74°F

93. What day was the temperature the lowest? What was this low temperature?   Sunday; 68°F

94. What day of the week was the temperature the highest? What was this high temperature?   Tuesday; 86°F

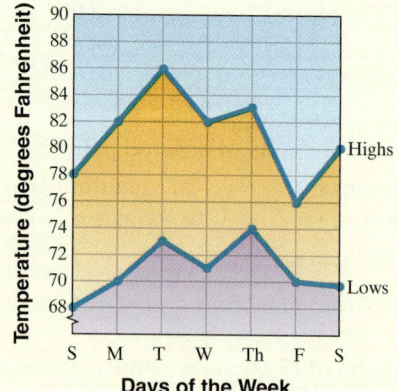

Days of the Week

95. On what day of the week was the difference between the high temperature and the low temperature the greatest? What was this difference in temperature?   Tuesday; 13°F

96. On what day of the week was the difference between the high temperature and the low temperature the least? What was this difference in temperature?   Friday; 6°F

3.2 Graphing Linear Equations

In the previous section, we found that equations in two variables may have more than one solution. For example, both $(2, 2)$ and $(0, 4)$ are solutions of the equation $x + y = 4$. In fact, this equation has an infinite number of solutions. Other solutions include $(-2, 6)$, $(4, 0)$, and $(6, -2)$. Notice the pattern that appears in the graph of these solutions.

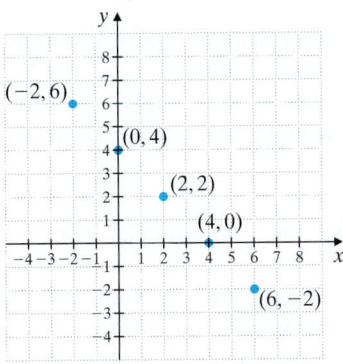

These solutions all appear to lie on the same line, as seen in the second graph. It can be shown that every ordered pair solution of the equation corresponds to a point on this line and that every point on this line corresponds to an ordered pair solution. Thus, we say that this line is the **graph of the equation** $x + y = 4$. Notice that we can show only a part of a line on a graph. The arrowheads on each end of the line below remind us that the line actually extends indefinitely in both directions.

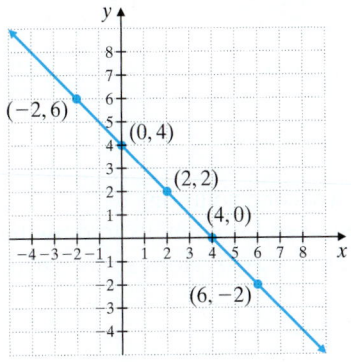

The equation $x + y = 4$ is called a *linear equation in two variables* and *the graph of every linear equation in two variables is a straight line.*

Linear Equation in Two Variables

A **linear equation in two variables** is an equation that can be written in the form

$$Ax + By = C$$

where A, B, and C are real numbers and A and B are not both 0. This form is called **standard form. The graph of a linear equation in two variables is a straight line.**

Objective

A Graph a Linear Equation by Finding and Plotting Ordered Pair Solutions.

Teaching Tip

Begin this section by graphing 4 linear equations on 4 sheets of graph paper using sticky dots. Draw axes on each sheet of paper and label with one of the following equations: $x + y = 2$, $2x - y = 6$, $x = 5$, and $y = -4$. Divide the class into 9 groups. Assign each group an integer from -4 to 4 to use as their x-value. Have each group find the y-value for each equation that corresponds to their x-value and plot each point using a sticky dot. When the graphs are completed, ask what similarities and differences they notice in the graphs.

Teaching Tip

For contrast, you may want to present some equations in two variables that are not linear. For instance,

$$x^2 = 6y + 4$$

$$y + 9 = \sqrt{x}$$

$$3x - \frac{1}{y} = 5$$

A linear equation in two variables may be written in many forms. Standard form, $Ax + By = C$, is just one of these many forms.

Following are examples of linear equations in two variables.

$$2x + y = 8 \qquad -2x = 7y \qquad y = \frac{1}{3}x + 2 \qquad y = 7$$

(Standard form)

Objective A Graphing Linear Equations ▶

From geometry, we know that a straight line is determined by just two points. Thus, to graph a linear equation in two variables, we need to find just two of its infinitely many solutions. Once we do so, we plot the solution points and draw the line connecting the points. Usually, we find a third solution as well, as a check.

Practice 1

Graph the linear equation $x + 3y = 6$.

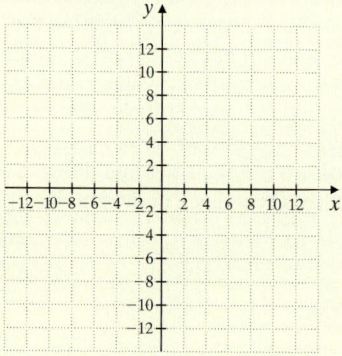

> **Helpful Hint**   All three points should fall on the same straight line. If not, check your ordered pair solutions for a mistake.

Answer

1.

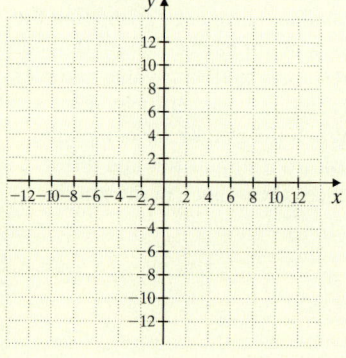

Example 1   Graph the linear equation $2x + y = 5$.

Solution: To graph this equation, we find three ordered pair solutions of $2x + y = 5$. To do this, we choose a value for one variable, x or y, and solve for the other variable. For example, if we let $x = 1$, then $2x + y = 5$ becomes

$$2x + y = 5$$
$$2(1) + y = 5 \quad \text{Replace } x \text{ with 1.}$$
$$2 + y = 5 \quad \text{Multiply.}$$
$$y = 3 \quad \text{Subtract 2 from both sides.}$$

Since $y = 3$ when $x = 1$, the ordered pair $(1, 3)$ is a solution of $2x + y = 5$. Next, we let $x = 0$.

$$2x + y = 5$$
$$2(0) + y = 5 \quad \text{Replace } x \text{ with 0.}$$
$$0 + y = 5$$
$$y = 5$$

The ordered pair $(0, 5)$ is a second solution.

The two solutions found so far allow us to draw the straight line that is the graph of all solutions of $2x + y = 5$. However, we will find a third ordered pair as a check. Let $y = -1$.

$$2x + y = 5$$
$$2x + (-1) = 5 \quad \text{Replace } y \text{ with } -1.$$
$$2x - 1 = 5$$
$$2x = 6 \quad \text{Add 1 to both sides.}$$
$$x = 3 \quad \text{Divide both sides by 2.}$$

The third solution is $(3, -1)$. These three ordered pair solutions are listed in the table and plotted on the coordinate plane. The graph of $2x + y = 5$ is the line through the three points.

x	y
1	3
0	5
3	-1

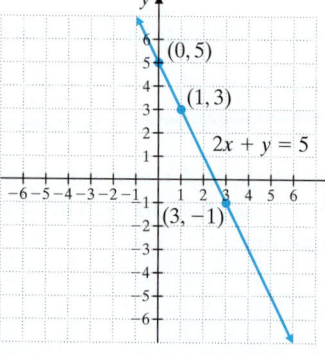

🟧 **Work Practice 1**

Example 2 Graph the linear equation $-5x + 3y = 15$.

Solution: We find three ordered pair solutions of $-5x + 3y = 15$.

Let $x = 0$.	**Let $y = 0$.**	**Let $x = -2$.**
$-5x + 3y = 15$	$-5x + 3y = 15$	$-5x + 3y = 15$
$-5 \cdot 0 + 3y = 15$	$-5x + 3 \cdot 0 = 15$	$-5 \cdot -2 + 3y = 15$
$0 + 3y = 15$	$-5x + 0 = 15$	$10 + 3y = 15$
$3y = 15$	$-5x = 15$	$3y = 5$
$y = 5$	$x = -3$	$y = \dfrac{5}{3}$ or $1\dfrac{2}{3}$

The ordered pairs are $(0, 5)$, $(-3, 0)$, and $\left(-2, 1\dfrac{2}{3}\right)$. The graph of $-5x + 3y = 15$ is the line through the three points.

x	y
0	5
-3	0
-2	$1\dfrac{2}{3}$

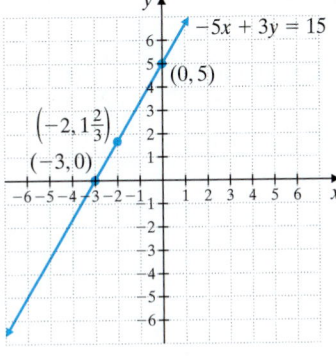

Teaching Tip

Point out the arrowheads on the graph in Example 2. See if the students understand their meaning. Remind them that when they are graphing an equation, they are to graph all the solutions to the equation.

■ **Work Practice 2**

Example 3 Graph the linear equation $y = 3x$.

Solution: We find three ordered pair solutions. Since this equation is solved for y, we'll choose three x-values.

If $x = 2$, $y = 3 \cdot 2 = 6$.
If $x = 0$, $y = 3 \cdot 0 = 0$.
If $x = -1$, $y = 3 \cdot -1 = -3$.

Next, we plot the ordered pair solutions and draw a line through the plotted points. The line is the graph of $y = 3x$.

Think about the following for a moment: A line is made up of an infinite number of points. Every point on the line defined by $y = 3x$ represents an ordered pair solution of the equation, and every ordered pair solution is a point on this line.

x	y
2	6
0	0
-1	-3

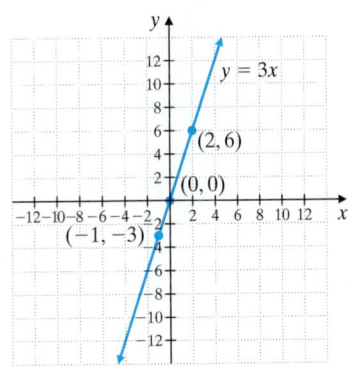

■ **Work Practice 3**

Practice 2

Graph the linear equation $-2x + 4y = 8$.

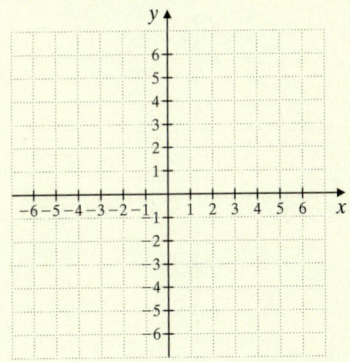

Practice 3

Graph the linear equation $y = 2x$.

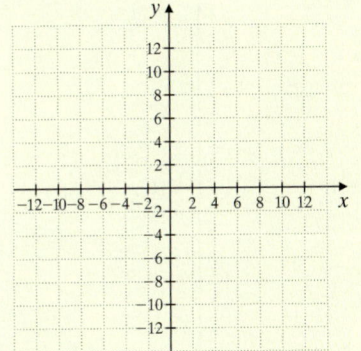

Answers

2.

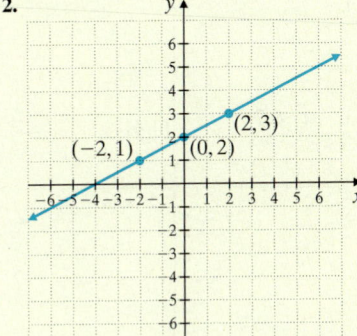

3.

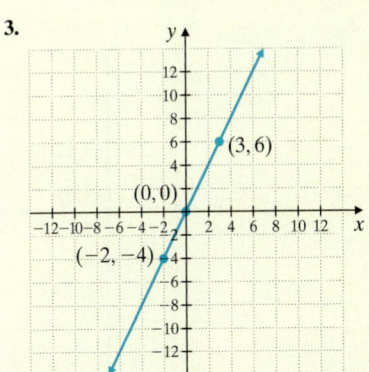

Practice 4

Graph the linear equation
$y = -\dfrac{1}{2}x + 4$.

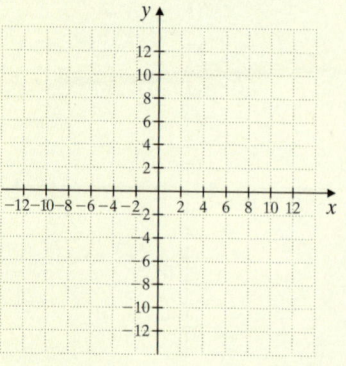

Practice 5

Graph the linear equation
$x = 3$.

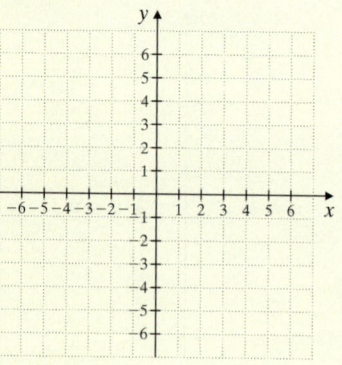

Helpful Hint

When graphing a linear equation in two variables, if it is
- solved for y, it may be easier to find ordered pair solutions by choosing x-values.
- solved for x, it may be easier to find ordered pair solutions by choosing y-values.

Example 4 Graph the linear equation $y = -\dfrac{1}{3}x + 2$.

Solution: We find three ordered pair solutions, plot the solutions, and draw a line through the plotted solutions. To avoid fractions, we'll choose x-values that are multiples of 3 to substitute into the equation.

If $x = 6$, then $y = -\dfrac{1}{3} \cdot 6 + 2 = -2 + 2 = 0$.

If $x = 0$, then $y = -\dfrac{1}{3} \cdot 0 + 2 = 0 + 2 = 2$.

If $x = -3$, then $y = -\dfrac{1}{3} \cdot -3 + 2 = 1 + 2 = 3$.

x	y
6	0
0	2
−3	3

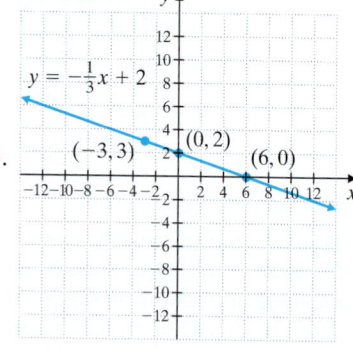

Work Practice 4

Let's take a moment and compare the graphs in Examples 3 and 4. The graph of $y = 3x$ tilts or slants upward (as we follow the line from left to right) and the graph of $y = -\dfrac{1}{3}x + 2$ tilts or slants downward (as we follow the line from left to right). We will learn more about the tilt, or slope, of a line in Section 3.4.

Example 5 Graph the linear equation $y = -2$.

Solution: The equation $y = -2$ can be written in standard form as $0x + y = -2$. No matter what value we replace x with, y is always -2.

x	y
0	−2
3	−2
−2	−2

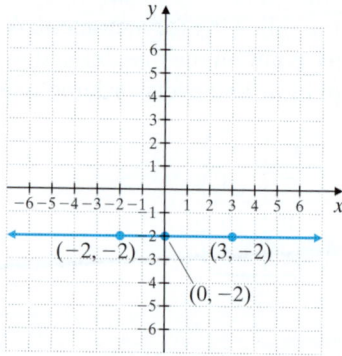

Notice that the graph of $y = -2$ is a horizontal line.

Work Practice 5

Linear equations are often used to model real data, as seen in the next example.

Answers

4.

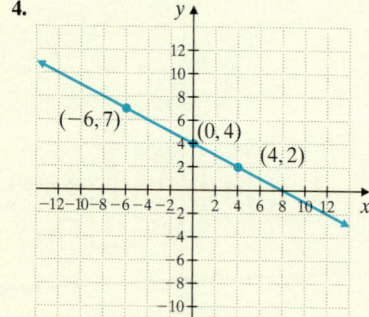

5.

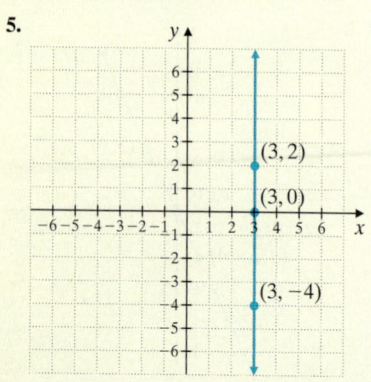

Example 2 Graph the linear equation $-5x + 3y = 15$.

Solution: We find three ordered pair solutions of $-5x + 3y = 15$.

Let $x = 0$.	**Let $y = 0$.**	**Let $x = -2$.**
$-5x + 3y = 15$	$-5x + 3y = 15$	$-5x + 3y = 15$
$-5 \cdot 0 + 3y = 15$	$-5x + 3 \cdot 0 = 15$	$-5 \cdot -2 + 3y = 15$
$0 + 3y = 15$	$-5x + 0 = 15$	$10 + 3y = 15$
$3y = 15$	$-5x = 15$	$3y = 5$
$y = 5$	$x = -3$	$y = \dfrac{5}{3}$ or $1\dfrac{2}{3}$

The ordered pairs are $(0, 5)$, $(-3, 0)$, and $\left(-2, 1\dfrac{2}{3}\right)$. The graph of $-5x + 3y = 15$ is the line through the three points.

x	y
0	5
-3	0
-2	$1\dfrac{2}{3}$

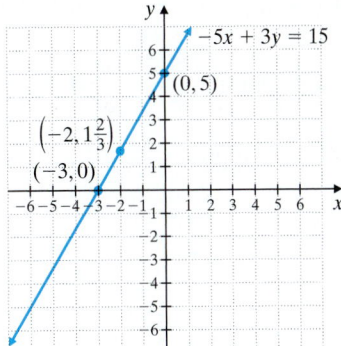

Teaching Tip

Point out the arrowheads on the graph in Example 2. See if the students understand their meaning. Remind them that when they are graphing an equation, they are to graph all the solutions to the equation.

■ **Work Practice 2**

Example 3 Graph the linear equation $y = 3x$.

Solution: We find three ordered pair solutions. Since this equation is solved for y, we'll choose three x-values.

If $x = 2$, $y = 3 \cdot 2 = 6$.
If $x = 0$, $y = 3 \cdot 0 = 0$.
If $x = -1$, $y = 3 \cdot -1 = -3$.

Next, we plot the ordered pair solutions and draw a line through the plotted points. The line is the graph of $y = 3x$.

Think about the following for a moment: A line is made up of an infinite number of points. Every point on the line defined by $y = 3x$ represents an ordered pair solution of the equation, and every ordered pair solution is a point on this line.

x	y
2	6
0	0
-1	-3

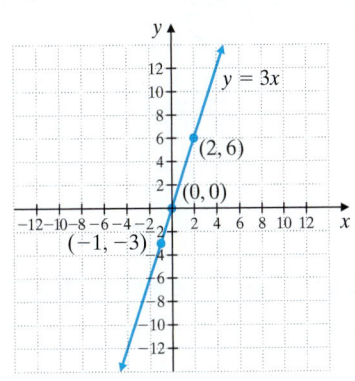

■ **Work Practice 3**

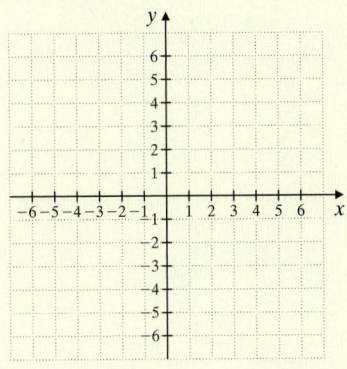
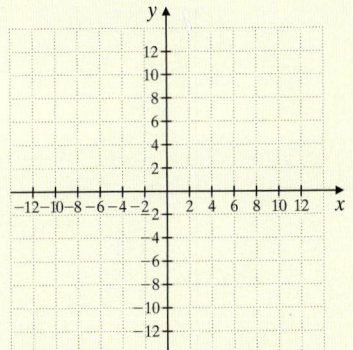
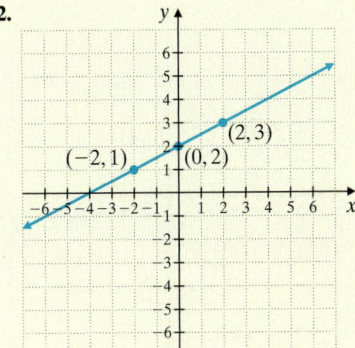
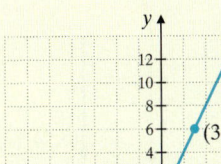
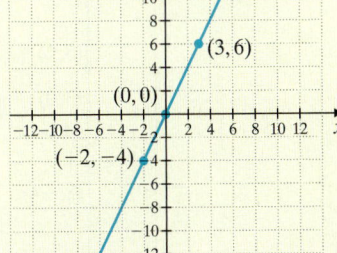

Practice 4

Graph the linear equation $y = -\frac{1}{2}x + 4$.

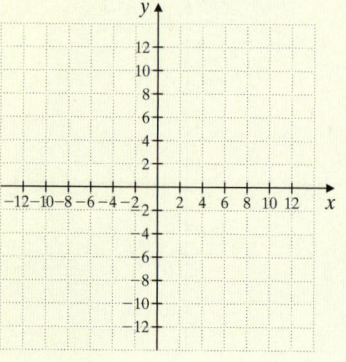

Practice 5

Graph the linear equation $x = 3$.

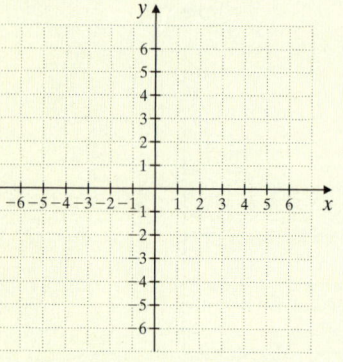

Helpful Hint

When graphing a linear equation in two variables, if it is
- solved for y, it may be easier to find ordered pair solutions by choosing x-values.
- solved for x, it may be easier to find ordered pair solutions by choosing y-values.

Example 4 Graph the linear equation $y = -\frac{1}{3}x + 2$.

Solution: We find three ordered pair solutions, plot the solutions, and draw a line through the plotted solutions. To avoid fractions, we'll choose x-values that are multiples of 3 to substitute into the equation.

If $x = 6$, then $y = -\frac{1}{3} \cdot 6 + 2 = -2 + 2 = 0$.

If $x = 0$, then $y = -\frac{1}{3} \cdot 0 + 2 = 0 + 2 = 2$.

If $x = -3$, then $y = -\frac{1}{3} \cdot -3 + 2 = 1 + 2 = 3$.

x	y
6	0
0	2
−3	3

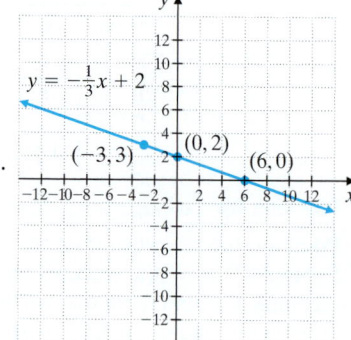

■ Work Practice 4

Let's take a moment and compare the graphs in Examples 3 and 4. The graph of $y = 3x$ tilts or slants upward (as we follow the line from left to right) and the graph of $y = -\frac{1}{3}x + 2$ tilts or slants downward (as we follow the line from left to right). We will learn more about the tilt, or slope, of a line in Section 3.4.

Example 5 Graph the linear equation $y = -2$.

Solution: The equation $y = -2$ can be written in standard form as $0x + y = -2$. No matter what value we replace x with, y is always -2.

x	y
0	−2
3	−2
−2	−2

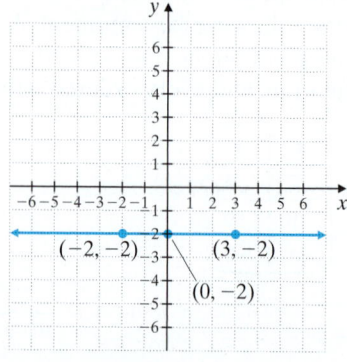

Notice that the graph of $y = -2$ is a horizontal line.

■ Work Practice 5

Linear equations are often used to model real data, as seen in the next example.

Answers

4.

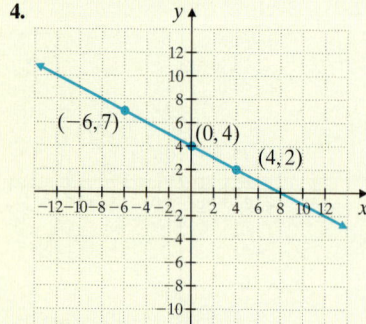

5.

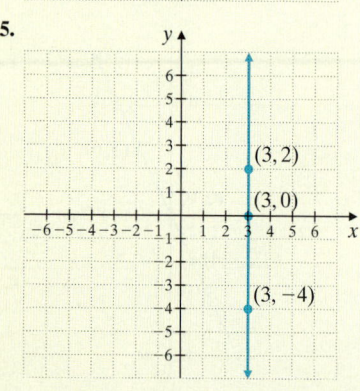

Example 6   Estimating the Number of Registered Nurses

One of the occupations expected to have the most growth in the next few years is registered nurse. The number of people y (in thousands) employed as registered nurses in the United States can be estimated by the linear equation $y = 71.2x + 2734.4$, where x is the number of years after the year 2010. (*Source:* Based on data from the Bureau of Labor Statistics)

a. Graph the equation.

b. Use the graph to predict the number of registered nurses in the year 2021.

Solution:

a. To graph $y = 71.2x + 2734.4$, choose x-values and substitute into the equation.

If $x = 0$, then $y = 71.2(0) + 2734.4 = 2734.4$.

If $x = 2$, then $y = 71.2(2) + 2734.4 = 2876.8$.

If $x = 5$, then $y = 71.2(5) + 2734.4 = 3090.4$.

x	y
0	2734.4
2	2876.8
5	3090.4

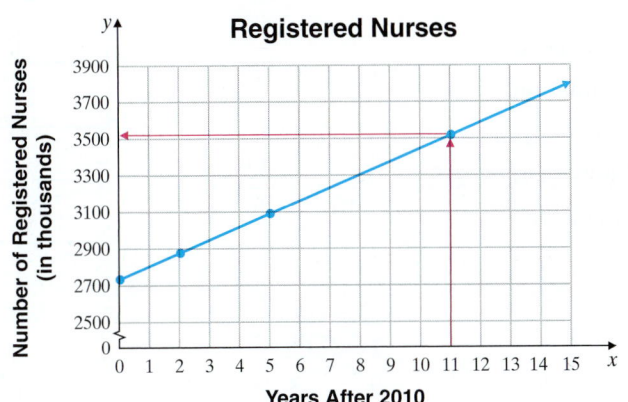

Registered Nurses

Years After 2010

b. To use the graph to *predict* the number of registered nurses in the year 2021, we need to find the y-coordinate that corresponds to $x = 11$. (11 years after 2010 is the year 2021.) To do so, find 11 on the x-axis. Move vertically upward to the graphed line and then horizontally to the left. We approximate the number on the y-axis to be 3500. Thus, in the year 2021, we predict that there will be 3500 thousand registered nurses. (The actual value, using 11 for x, is 3517.6.)

■ **Work Practice 6**

Helpful Hint

Make sure you understand that models are mathematical approximations of the data for the known years. (For example, see the model in Example 6.) Any number of unknown factors can affect future years, so be cautious when using models to make predictions.

Practice 6

Use the graph in Example 6 to predict the number of registered nurses in 2022.

Helpful Hint   From Example 5, we learned that equations such as $y = -2$ are linear equations since $y = -2$ can be written as $0x + y = -2$.

Answer

6. 3600 thousand

Calculator Explorations Graphing

In this section, we begin an optional study of graphing calculators and graphing software packages for computers. These graphers use the same point plotting technique that was introduced in this section. The advantage of this graphing technology is, of course, that graphing calculators and computers can find and plot ordered pair solutions much faster than we can. Note, however, that the features described in these boxes may not be available on all graphing calculators.

The rectangular screen where a portion of the rectangular coordinate system is displayed is called a **window.** We call it a **standard window** for graphing when both the *x*- and *y*-axes show coordinates between −10 and 10. This information is often displayed in the window menu on a graphing calculator as follows.

Xmin = −10
Xmax = 10
Xscl = 1   The scale on the *x*-axis is one unit per tick mark.
Ymin = −10
Ymax = 10
Yscl = 1   The scale on the *y*-axis is one unit per tick mark.

To use a graphing calculator to graph the equation $y = 2x + 3$, press the $\boxed{Y =}$ key and enter the keystrokes $\boxed{2}$ $\boxed{x}$ $\boxed{+}$ $\boxed{3}$. The top row should now read $Y_1 = 2x + 3$. Next press the $\boxed{\text{GRAPH}}$ key, and the display should look like this:

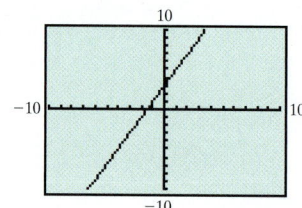

Graph the following linear equations. (Unless otherwise stated, use a standard window when graphing.)

1. $y = -3x + 7$

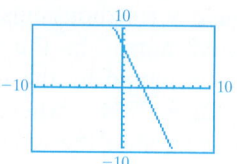

2. $y = -x + 5$

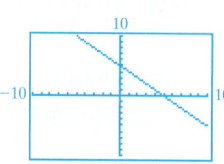

3. $y = 2.5x - 7.9$

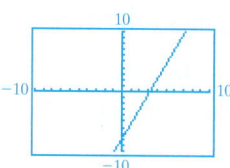

4. $y = -1.3x + 5.2$

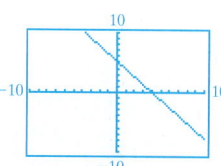

5. $y = -\dfrac{3}{10}x + \dfrac{32}{5}$

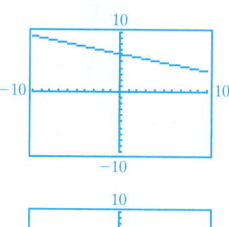

6. $y = \dfrac{2}{9}x - \dfrac{22}{3}$

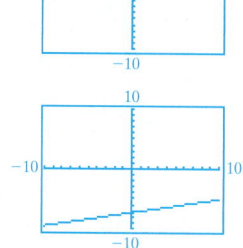

Teaching Tip
Point out that some graphing calculators need to have coefficients that are fractions entered with parentheses. For instance, $y = \dfrac{1}{4}x - 2$ may need to be entered as $y = (1/4)x - 2$.

Vocabulary, Readiness & Video Check

Martin-Gay Interactive Videos

See Video 3.2 🍎

See video answer section.

Watch the section lecture video and answer the following questions.

Objective **A** **1.** In the lecture before 🅗 Example 1, it's mentioned that we need only two points to determine a line. Why, then, are three ordered pair solutions found in 🅗 Examples 1–3? ▶

2. What does a graphed line represent, as discussed at the end of 🅗 Examples 1 and 3? ▶

3.2 Exercise Set MyMathLab®

Objective A *For each equation, find three ordered pair solutions by completing the table. Then use the ordered pairs to graph the equation. See Examples 1 through 5.*

1. $x - y = 6$

x	y
6	0
4	-2
5	-1

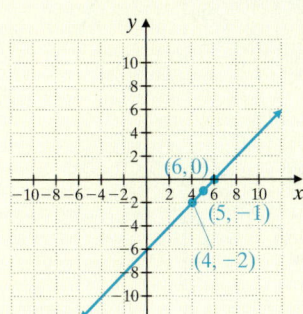

2. $x - y = 4$

x	y
0	-4
6	2
-1	-5

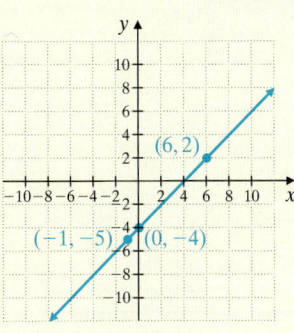

3. $y = -4x$

x	y
1	-4
0	0
-1	4

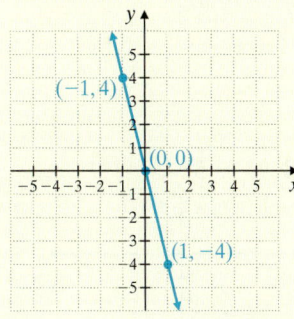

4. $y = -5x$

x	y
1	-5
0	0
-1	5

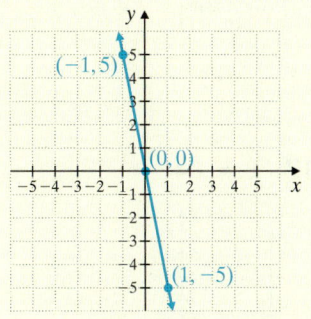

5. $y = \dfrac{1}{3}x$

x	y
0	0
6	2
-3	-1

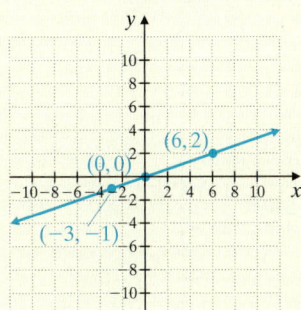

6. $y = \dfrac{1}{2}x$

x	y
0	0
-4	-2
2	1

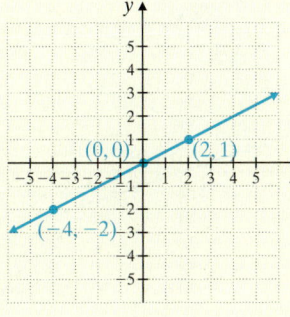

7. $y = -4x + 3$

x	y
0	3
1	-1
2	-5

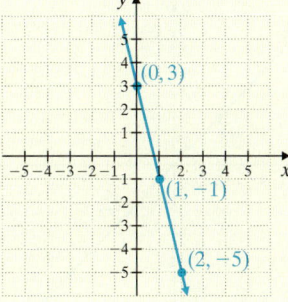

8. $y = -5x + 2$

x	y
0	2
1	-3
2	-8

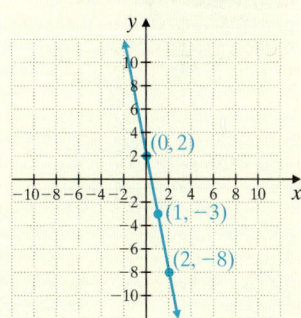

Graph each linear equation. See Examples 1 through 5.

9. $x + y = 1$

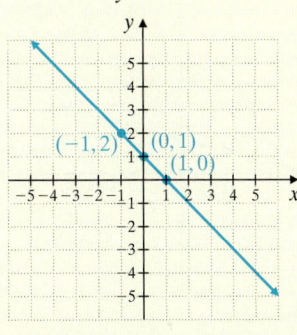

10. $x + y = 7$

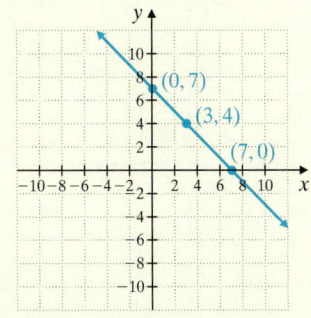

11. $x - y = -2$

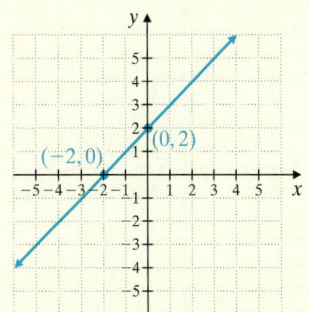

12. $-x + y = 6$

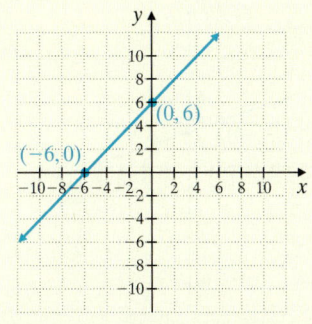

13. $x - 2y = 6$

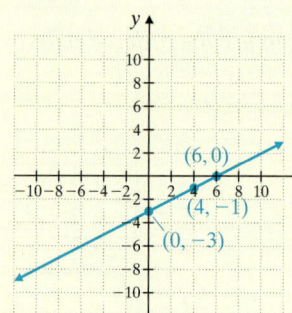

14. $-x + 5y = 5$

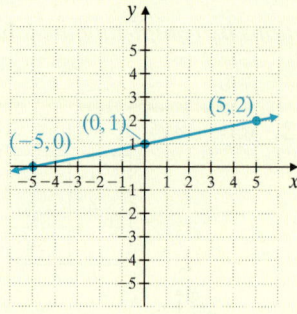

15. $y = 6x + 3$

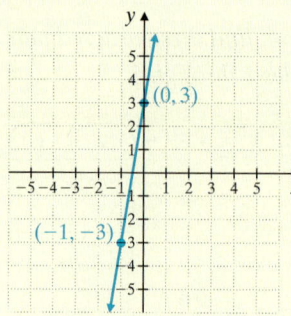

16. $y = -2x + 7$

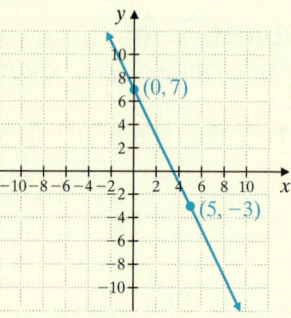

17. $x = -4$

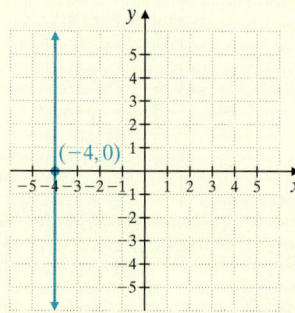

18. $y = 5$

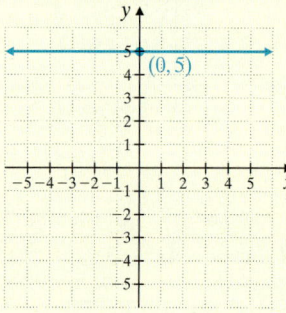

19. $y = 3$

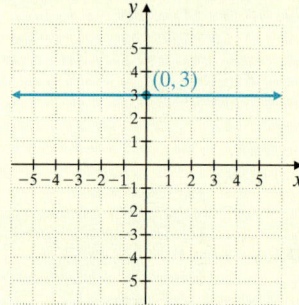

20. $x = -1$

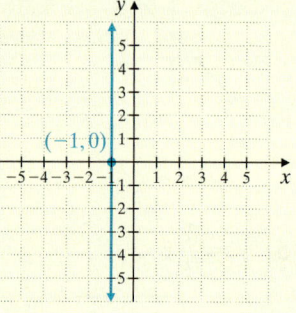

21. $y = x$

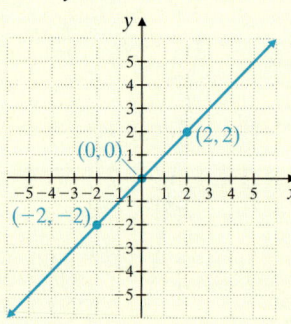

22. $y = -x$

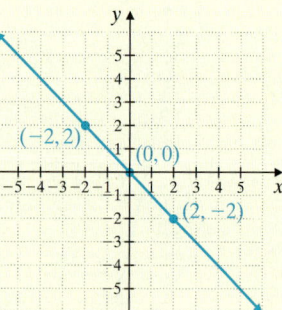

23. $x = -3y$

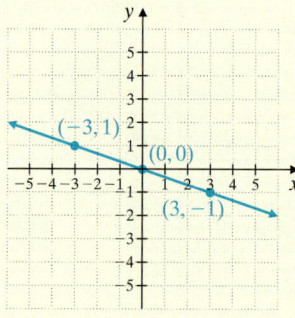

24. $x = 4y$

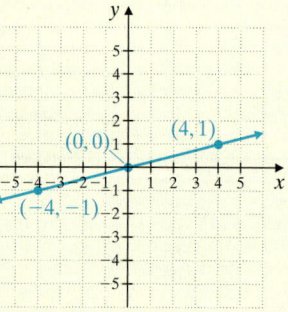

25. $x + 3y = 9$

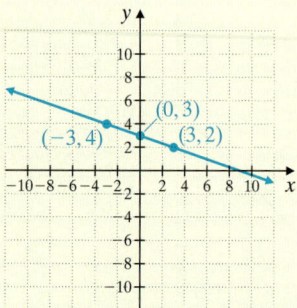

26. $2x + y = 2$

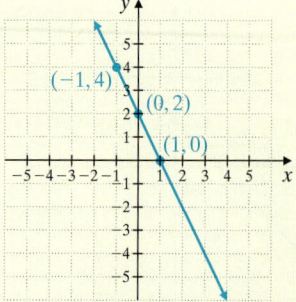

27. $y = \frac{1}{2}x + 2$

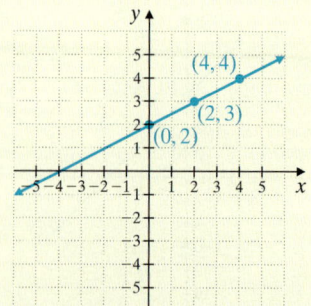

28. $y = \frac{1}{4}x + 3$

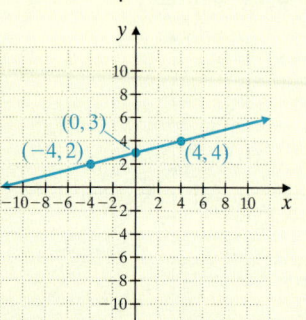

29. $3x - 2y = 12$

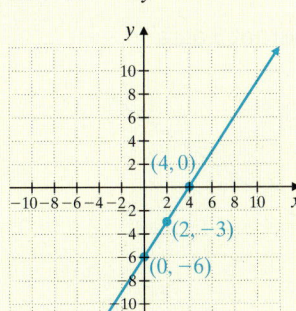

30. $2x - 7y = 14$

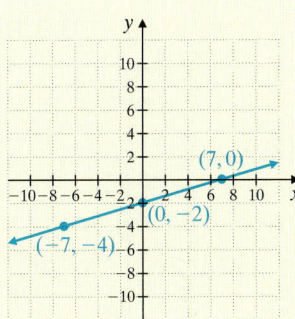

31. $y = -3.5x + 4$

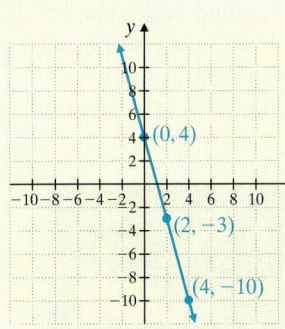

32. $y = -1.5x - 3$

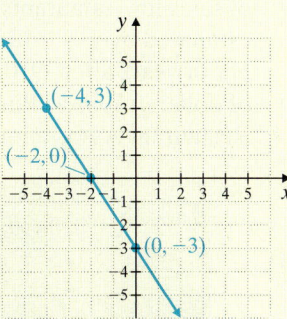

Solve. See Example 6.

33. The number of students y (in thousands) taking the SAT college entrance exam each year from 2008 through 2012 can be approximated by the linear equation $y = 28x + 1552$, where x is the number of years after 2008. (*Source:* Based on data from The College Board)

a. Graph the linear equation.

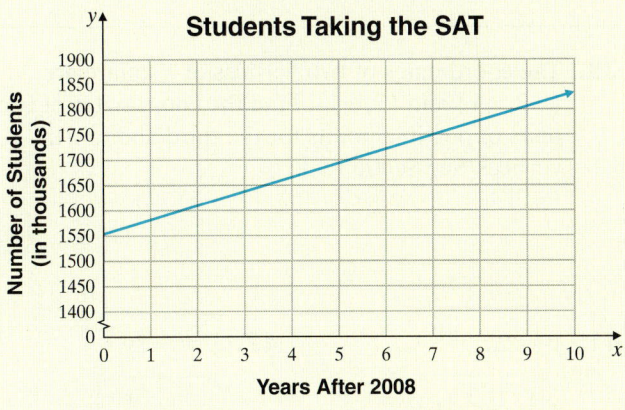

b. Does the point $(7, 1748)$ lie on the line? If so, what does this ordered pair mean?
yes; answers may vary

34. College is getting more expensive every year. The average cost for tuition and fees at a public two-year college y from 1991 through 2012 can be approximated by the linear equation $y = 88x + 973$, where x is the number of years after 1991. (*Source:* The College Board: Trends in College Pricing 2012)

a. Graph the linear equation.

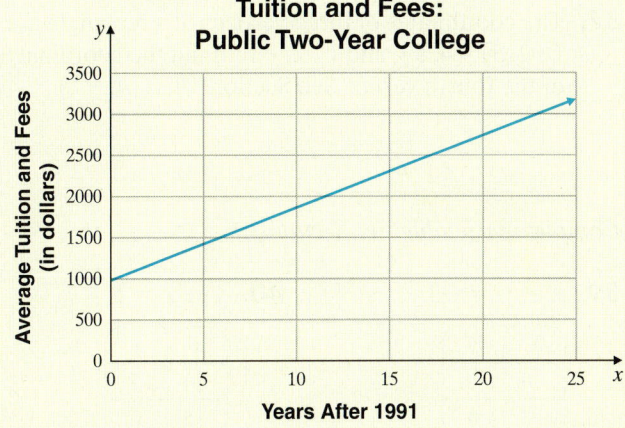

b. Does the point $(21, 2821)$ lie on the line? If so, what does this ordered pair mean?
yes; answers may vary

35. The total annual revenue y (in billions of euros) for IKEA from 2001 through 2012 can be approximated by the equation $y = 1.6x + 9.1$, where x is the number of years after 2001. (*Source:* Based on data from IKEA Group)

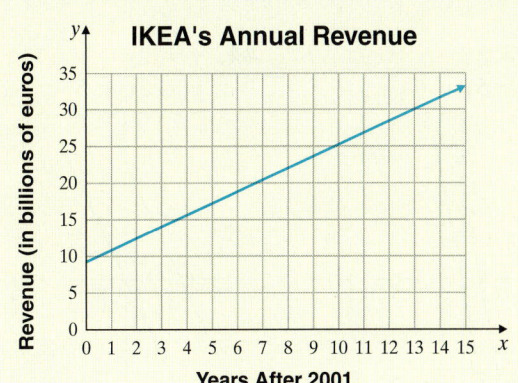

a. Graph the linear equation.
b. Complete the ordered pair $(11,\ )$. $(11, 26.7)$
c. Write a sentence explaining the meaning of the ordered pair found in part **b.**
In 2012, IKEA's total annual revenue was 26.7 billion euros.

36. For the period 1970 through 2013, the annual level of sales for restaurants in the United States can be estimated by $y = 14.3x - 5.3$, where x is the number of years after 1970 and y is the sales in billions of dollars. (*Source:* Based on data from the National Restaurant Association)

a. Graph the linear equation.

b. Complete the ordered pair (43,).   (43, 609.6)

c. Write a sentence explaining the meaning of the ordered pair found in part **b**.
In 2013, U.S. restaurant sales were $609.6 billion.

Review

△ **37.** The coordinates of three vertices of a rectangle are $(-2, 5)$, $(4, 5)$, and $(-2, -1)$. Find the coordinates of the fourth vertex. See Section 3.1. $(4, -1)$

△ **38.** The coordinates of two vertices of a square are $(-3, -1)$ and $(2, -1)$. Find the coordinates of two pairs of possible points for the third and fourth vertices. See Section 3.1.
$(-3, 4), (2, 4); (-3, -6), (2, -6)$

Complete each table. See Section 3.1.

39. $x - y = -3$

x	y
0	3
−3	0

40. $y - x = 5$

x	y
0	5
−5	0

41. $y = 2x$

x	y
0	0
0	0

42. $x = -3y$

x	y
0	0
0	0

Concept Extensions

Graph each pair of linear equations on the same set of axes. Discuss how the graphs are similar and how they are different.

43. $y = 5x$
$y = 5x + 4$

44. $y = 2x$
$y = 2x + 5$

45. $y = -2x$
$y = -2x - 3$

46. $y = x$
$y = x - 7$

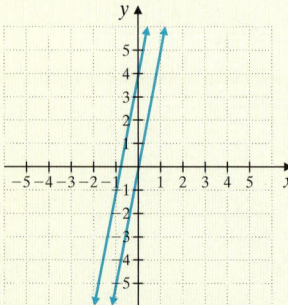

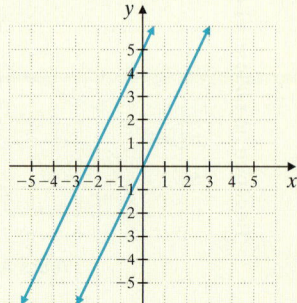

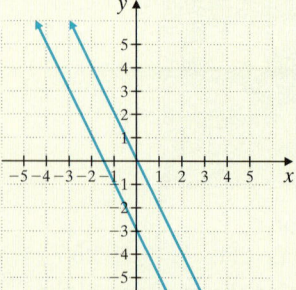

47. Graph the nonlinear equation $y = x^2$ by completing the table shown. Plot the ordered pairs and connect them with a smooth curve.

x	y
0	0
1	1
−1	1
2	4
−2	4

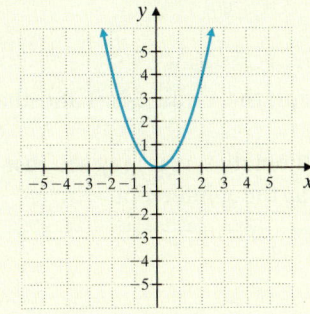

48. Graph the nonlinear equation $y = |x|$ by completing the table shown. Plot the ordered pairs and connect them. This curve is "V" shaped.

x	y
0	0
1	1
−1	1
2	2
−2	2

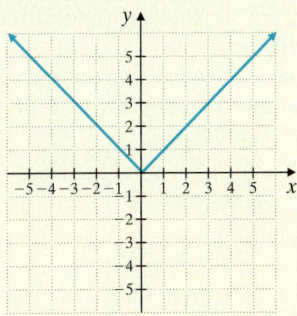

△**49.** The perimeter of the trapezoid below is 22 centimeters. Write a linear equation in two variables for the perimeter. Find y if x is 3 centimeters.
$x + y = 12; 9$ cm

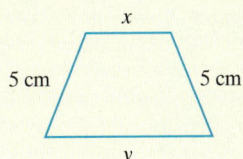

△**50.** The perimeter of the rectangle below is 50 miles. Write a linear equation in two variables for the perimeter. Use this equation to find x when y is 20.
$x + y = 25; x = 5$

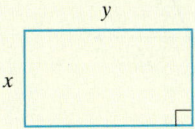

51. If (a, b) is an ordered pair solution of $x + y = 5$, is (b, a) also a solution? Explain why or why not.
yes; answers may vary

52. If (a, b) is an ordered pair solution of $x - y = 5$, is (b, a) also a solution? Explain why or why not.
no; answers may vary

3.3 Intercepts

Objective A Identifying Intercepts

The graph of $y = 4x - 8$ is shown below. Notice that this graph crosses the y-axis at the point $(0, -8)$. This point is called the **y-intercept.** Likewise the graph crosses the x-axis at $(2, 0)$. This point is called the **x-intercept.**

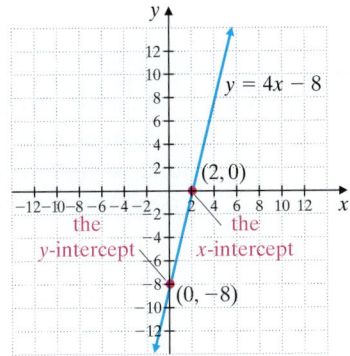

The intercepts are $(2, 0)$ and $(0, -8)$.

Objectives

A Identify Intercepts of a Graph.

B Graph a Linear Equation by Finding and Plotting Intercept Points.

C Identify and Graph Vertical and Horizontal Lines.

Teaching Tip

Start this lesson by showing students a graph of the line $y = 2x + 8$. Remind them that any two points on the line could be used to describe this line. Ask them which two points they would use. Hopefully students will choose the intercepts. Ask what is special about these points.

Teaching Tip

Some questions you may want to ask your students:

- Can the same point ever be the *x*-intercept and the *y*-intercept of a graph?

- What is the maximum number of *x*-intercepts any line (except $y = 0$) will have?

- Will a line always have at least one *y*-intercept?

Practice 1–3

Identify the *x*- and *y*-intercepts.

1.

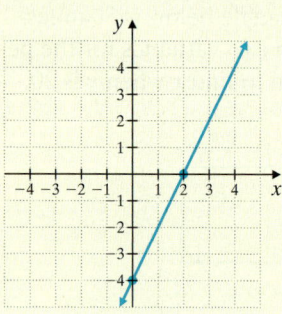

2.

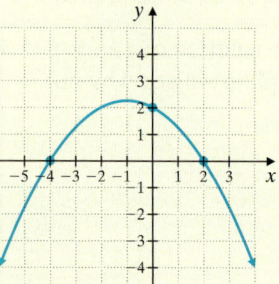

3.

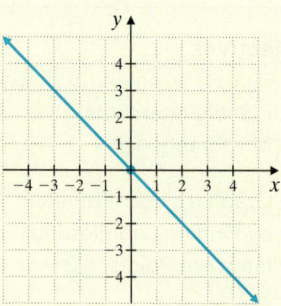

Answers

1. *x*-intercept: $(2, 0)$; *y*-intercept: $(0, -4)$ **2.** *x*-intercepts: $(-4, 0)$, $(2, 0)$; *y*-intercept: $(0, 2)$
3. *x*-intercept and *y*-intercept: $(0, 0)$

Helpful Hint

If a graph crosses the *x*-axis at $(2, 0)$ and the *y*-axis at $(0, -8)$, then

$$(2, 0) \qquad (0, -8)$$

$\uparrow$ *x*-intercept $\uparrow$ *y*-intercept

Notice that for the *x*-intercept, the *y*-value is 0 and that for the *y*-intercept, the *x*-value is 0.

Note: Sometimes in mathematics, you may see just the number -8 stated as the *y*-intercept and 2 stated as the *x*-intercept.

Examples   Identify the *x*- and *y*-intercepts.

1.

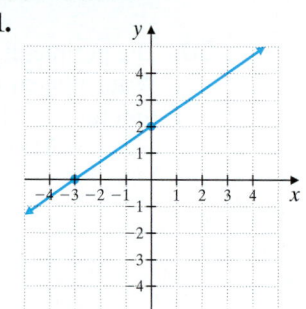

Solution:

x-intercept: $(-3, 0)$
y-intercept: $(0, 2)$

2.

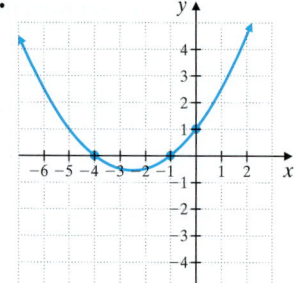

Solution:

x-intercepts: $(-4, 0)$, $(-1, 0)$
y-intercept: $(0, 1)$

3.

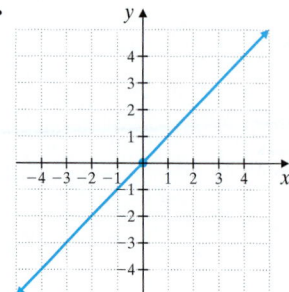

Helpful Hint

Notice that any time $(0, 0)$ is a point of a graph, then it is an *x*-intercept and a *y*-intercept. Why? It is the *only* point that lies on both axes.

Solution:

x-intercept: $(0, 0)$

y-intercept: $(0, 0)$

Here, the x- and y-intercepts happen to be the same point.

🟧 **Work Practice 1–3**

Objective B Finding and Plotting Intercepts ▶

Given an equation of a line, we can usually find intercepts easily since one coordinate is 0.

To find the x-intercept of a line from its equation, let $y = 0$, since a point on the x-axis has a y-coordinate of 0. To find the y-intercept of a line from its equation, let $x = 0$, since a point on the y-axis has an x-coordinate of 0.

> ### Finding x- and y-Intercepts
>
> To find the x-intercept, let $y = 0$ and solve for x.
> To find the y-intercept, let $x = 0$ and solve for y.

Example 4 Graph $x - 3y = 6$ by finding and plotting its intercepts.

Solution: We let $y = 0$ to find the x-intercept and $x = 0$ to find the y-intercept.

$$\text{Let } y = 0. \qquad\qquad \text{Let } x = 0.$$
$$x - 3y = 6 \qquad\qquad x - 3y = 6$$
$$x - 3(0) = 6 \qquad\qquad 0 - 3y = 6$$
$$x - 0 = 6 \qquad\qquad -3y = 6$$
$$x = 6 \qquad\qquad y = -2$$

The x-intercept is $(6, 0)$, and the y-intercept is $(0, -2)$. We find a third ordered pair solution to check our work. If we let $y = -1$, then $x = 3$. We plot the points $(6, 0)$, $(0, -2)$, and $(3, -1)$. The graph of $x - 3y = 6$ is the line drawn through these points, as shown.

x	y
6	0
0	−2
3	−1

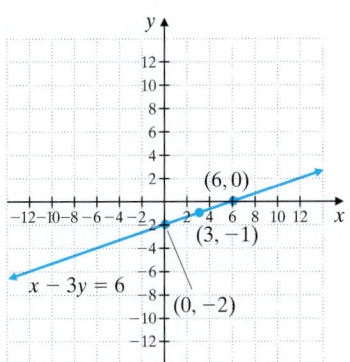

🟧 **Work Practice 4**

Example 5 Graph $x = -2y$ by finding and plotting its intercepts.

Solution: We let $y = 0$ to find the x-intercept and $x = 0$ to find the y-intercept.

$$\text{Let } y = 0. \qquad\qquad \text{Let } x = 0.$$
$$x = -2y \qquad\qquad x = -2y$$
$$x = -2(0) \qquad\qquad 0 = -2y$$
$$x = 0 \qquad\qquad 0 = y$$

(Continued on next page)

Practice 4

Graph $2x - y = 4$ by finding and plotting its intercepts.

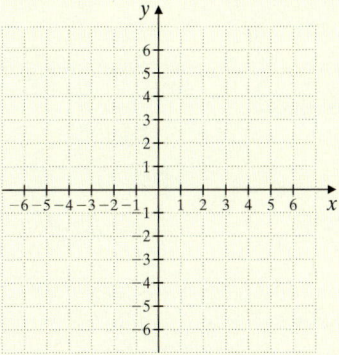

Practice 5

Graph $y = 3x$ by finding and plotting its intercepts.

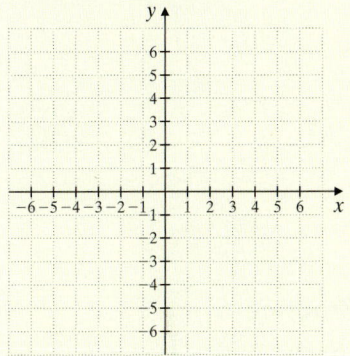

Answers

4.

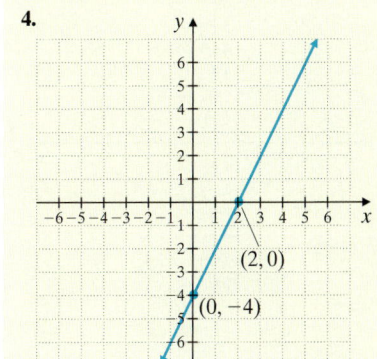

5.

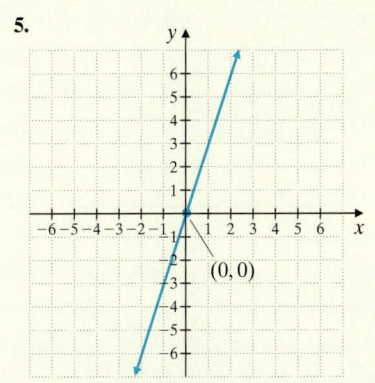

Both the x-intercept and y-intercept are $(0, 0)$. In other words, when $x = 0$, then $y = 0$, which gives the ordered pair $(0, 0)$. Also, when $y = 0$, then $x = 0$, which gives the same ordered pair, $(0, 0)$. This happens when the graph passes through the origin. Since two points are needed to determine a line, we must find at least one more ordered pair that satisfies $x = -2y$. Since the equation is solved for x, we choose y-values so that there is no need to solve to find the corresponding x-value. We let $y = -1$ to find a second ordered pair solution and let $y = 1$ as a check point.

Let $y = -1$.

$x = -2(-1)$

$x = 2$ Multiply.

Let $y = 1$.

$x = -2(1)$

$x = -2$ Multiply.

The ordered pairs are $(0, 0)$, $(2, -1)$, and $(-2, 1)$. We plot these points to graph $x = -2y$.

x	y
0	0
2	-1
-2	1

■ Work Practice 5

Objective C Graphing Vertical and Horizontal Lines

From Section 3.2, recall that the equation $x = 2$, for example, is a linear equation in two variables because it can be written in the form $x + 0y = 2$. The graph of this equation is a vertical line, as reviewed in the next example.

Example 6   Graph: $x = 2$

Solution: The equation $x = 2$ can be written as $x + 0y = 2$. For any y-value chosen, notice that x is 2. No other value for x satisfies $x + 0y = 2$. Any ordered pair whose x-coordinate is 2 is a solution of $x + 0y = 2$. We will use the ordered pair solutions $(2, 3)$, $(2, 0)$, and $(2, -3)$ to graph $x = 2$.

x	y
2	3
2	0
2	-3

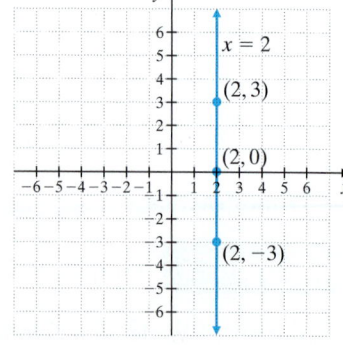

The graph is a vertical line with x-intercept 2. Note that this graph has no y-intercept because x is never 0.

■ Work Practice 6

Practice 6

Graph: $x = -3$

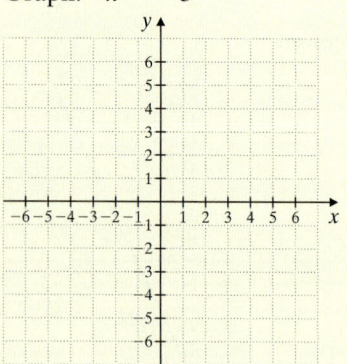

Answer

6.

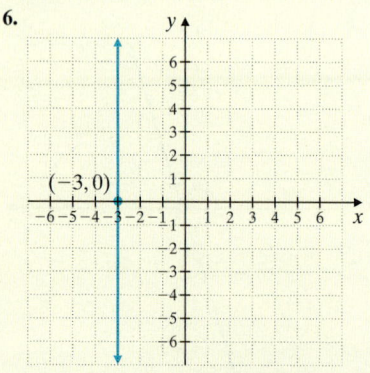

In general, we have the following.

Vertical Lines

The graph of $x = c$, where c is a real number, is a **vertical line** with x-intercept $(c, 0)$.

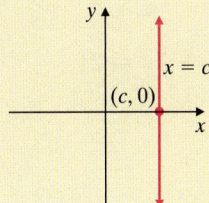

Example 7 Graph: $y = -3$

Solution: The equation $y = -3$ can be written as $0x + y = -3$. For any x-value chosen, y is -3. If we choose 4, 1, and -2 as x-values, the ordered pair solutions are $(4, -3)$, $(1, -3)$, and $(-2, -3)$. We use these ordered pairs to graph $y = -3$. The graph is a horizontal line with y-intercept -3 and no x-intercept.

x	y
4	−3
1	−3
−2	−3

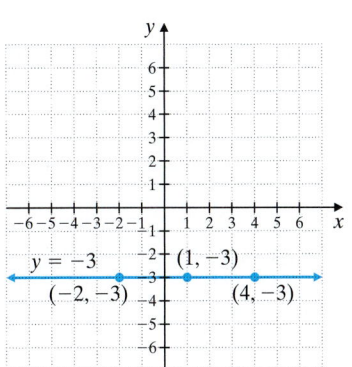

■ **Work Practice 7**

In general, we have the following.

Horizontal Lines

The graph of $y = c$, where c is a real number, is a **horizontal line** with y-intercept $(0, c)$.

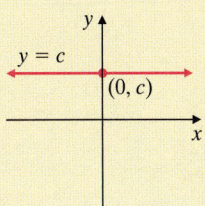

Practice 7

Graph: $y = 4$

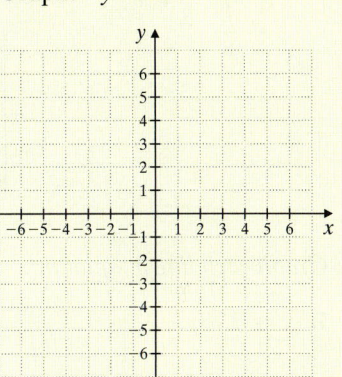

Answer

7.

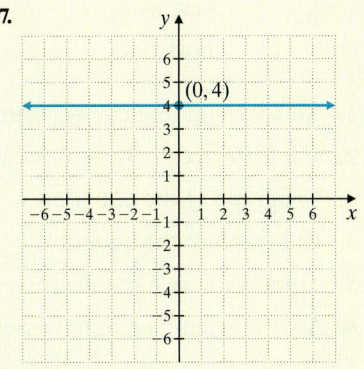

Calculator Explorations Graphing

You may have noticed that to use the $\boxed{Y=}$ key on a graphing calculator to graph an equation, the equation must be solved for y. For example, to graph $2x + 3y = 7$, we solve the equation for y.

$$2x + 3y = 7$$
$$3y = -2x + 7 \qquad \text{Subtract } 2x \text{ from both sides.}$$
$$\frac{3y}{3} = -\frac{2x}{3} + \frac{7}{3} \qquad \text{Divide both sides by 3.}$$
$$y = -\frac{2}{3}x + \frac{7}{3} \qquad \text{Simplify.}$$

To graph $2x + 3y = 7$ or $y = -\frac{2}{3}x + \frac{7}{3}$, press the $\boxed{Y=}$ key and enter

$$Y_1 = -\frac{2}{3}x + \frac{7}{3}$$

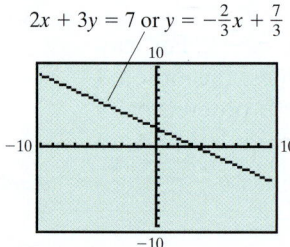

Graph each linear equation.

1. $x = 3.78y$

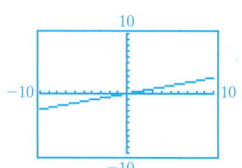

2. $-2.61y = x$

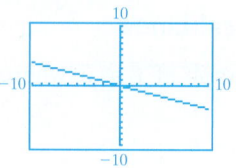

3. $3x + 7y = 21$

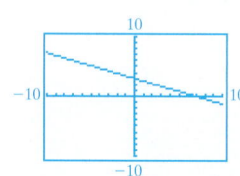

4. $-4x + 6y = 12$

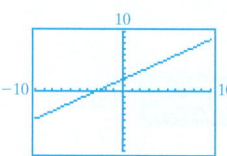

5. $-2.2x + 6.8y = 15.5$

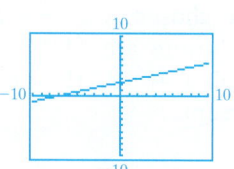

6. $5.9x - 0.8y = -10.4$

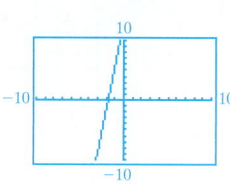

Vocabulary, Readiness & Video Check

Use the choices below to fill in each blank. Some choices may be used more than once. Exercises 1 and 2 come from Section 3.2.

x	vertical	x-intercept	linear
y	horizontal	y-intercept	standard

1. An equation that can be written in the form $Ax + By = C$ is called a(n) ___linear___ equation in two variables.

2. The form $Ax + By = C$ is called ___standard___ form.

3. The graph of the equation $y = -1$ is a(n) ___horizontal___ line.

4. The graph of the equation $x = 5$ is a(n) ___vertical___ line.

5. A point where a graph crosses the y-axis is called a(n) ___y-intercept___.

6. A point where a graph crosses the x-axis is called a(n) ___x-intercept___.

7. Given an equation of a line, to find the x-intercept (if there is one), let ___y___ $= 0$ and solve for ___x___.

8. Given an equation of a line, to find the y-intercept (if there is one), let ___x___ $= 0$ and solve for ___y___.

Answer the following true or false.

9. All lines have an *x*-intercept *and* a *y*-intercept. ___false___

10. The graph of $y = 4x$ contains the point $(0, 0)$. ___true___

11. The graph of $x + y = 5$ has an *x*-intercept of $(5, 0)$ and a *y*-intercept of $(0, 5)$. ___true___

12. The graph of $y = 5x$ contains the point $(5, 1)$. ___false___

Martin-Gay Interactive Videos   Watch the section lecture video and answer the following questions.

See Video 3.3

See video answer section.

Objective A 13. At the end of ⊟ Example 2, patterns are discussed. What reason is given for why *x*-intercepts have *y*-values of 0? For why *y*-intercepts have *x*-values of 0? ▶

Objective B 14. In ⊟ Example 3, the goal is to use the *x*- and *y*-intercepts to graph a line. Yet once the two intercepts are found, a third point is also found before the line is graphed. Why do you think this practice of finding a third point is continued? ▶

Objective C 15. From ⊟ Examples 5 and 6, what is the coefficient of *x* when the equation of a horizontal line is written as $Ax + By = C$? What is the coefficient of *y* when the equation of a vertical line is written as $Ax + By = C$? ▶

3.3 Exercise Set MyMathLab®

Objective A *Identify the intercepts. See Examples 1 through 3.*

▶ **1.**

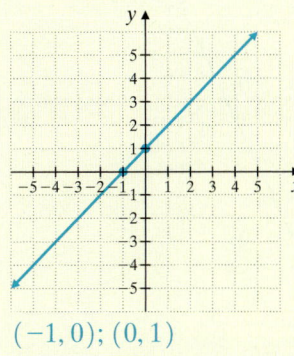

$(-1, 0); (0, 1)$

2.

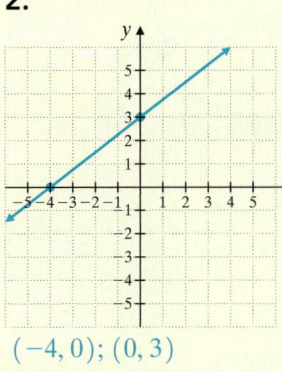

$(-4, 0); (0, 3)$

▶ **3.**

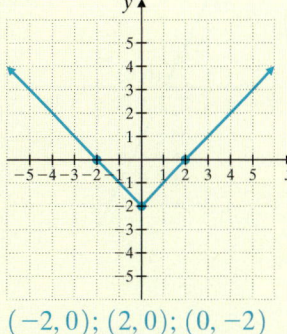

$(-2, 0); (2, 0); (0, -2)$

4.

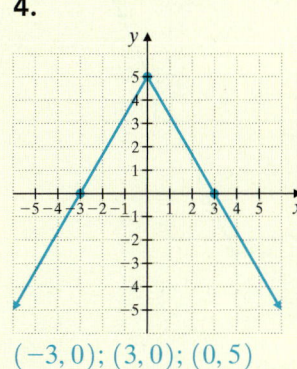

$(-3, 0); (3, 0); (0, 5)$

5.

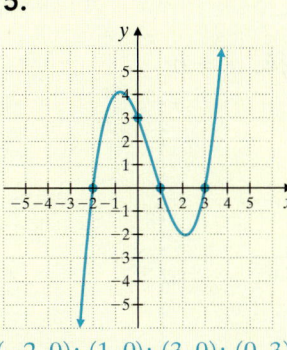

$(-2, 0); (1, 0); (3, 0); (0, 3)$

6.

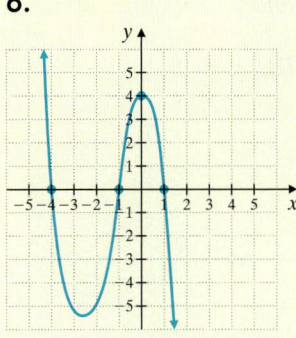

$(-4, 0); (-1, 0); (1, 0); (0, 4)$

7.

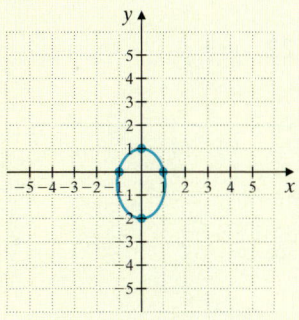

$(-1, 0); (1, 0); (0, 1); (0, -2)$

8.

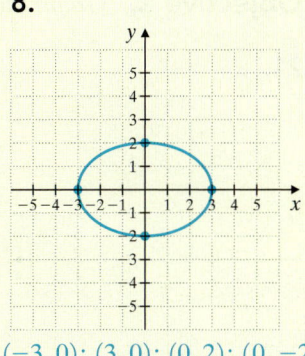

$(-3, 0); (3, 0); (0, 2); (0, -2)$

Objective B *Graph each linear equation by finding and plotting its intercepts. See Examples 4 and 5.*

9. $x - y = 3$

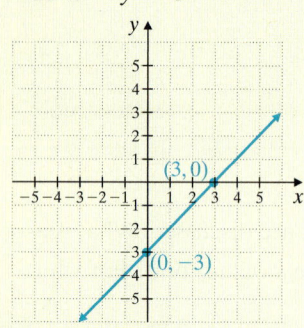

10. $x - y = -4$

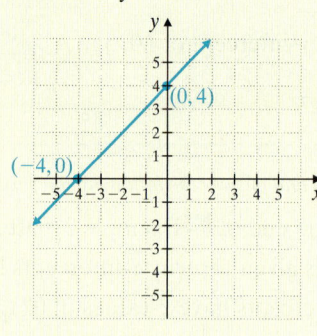

11. $x = 5y$

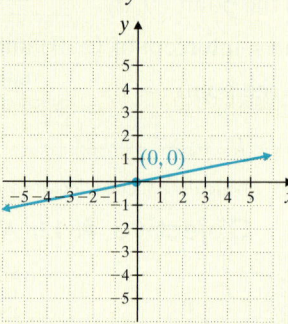

12. $x = 2y$

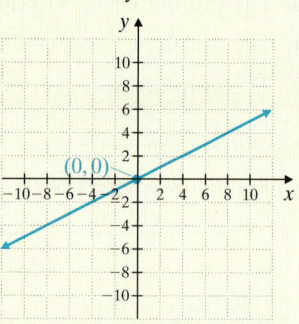

13. $-x + 2y = 6$

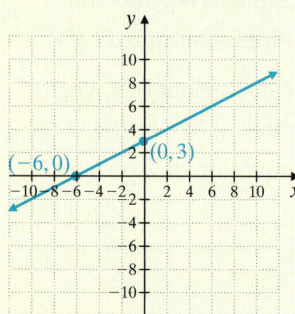

14. $x - 2y = -8$

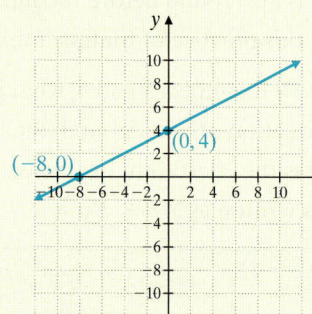

15. $2x - 4y = 8$

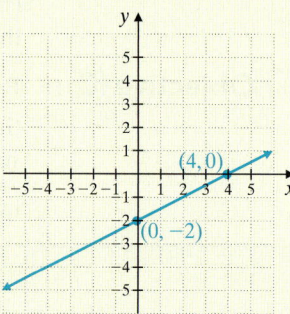

16. $2x + 3y = 6$

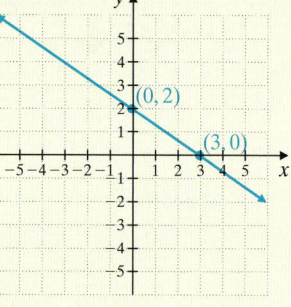

17. $y = 2x$

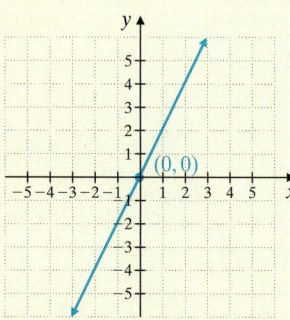

18. $y = -2x$

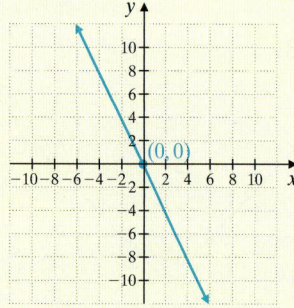

19. $y = 3x + 6$

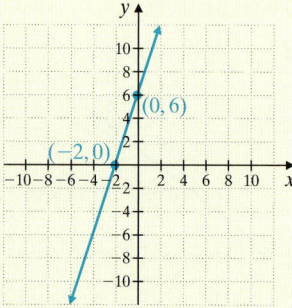

20. $y = 2x + 10$

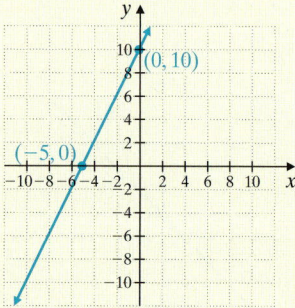

Objective C *Graph each linear equation. See Examples 6 and 7.*

21. $x = -1$

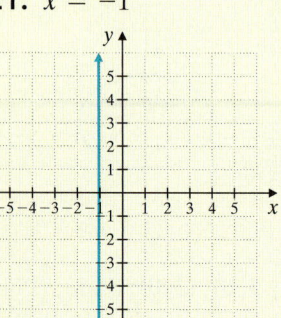

22. $y = 5$

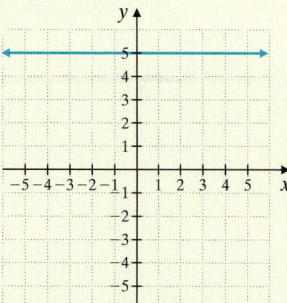

23. $y = 0$

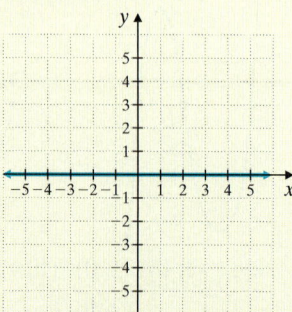

24. $x = 0$

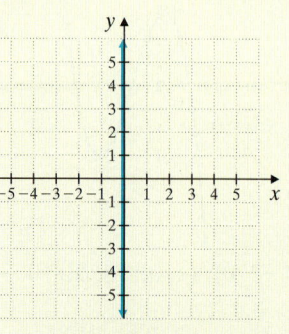

25. $y + 7 = 0$

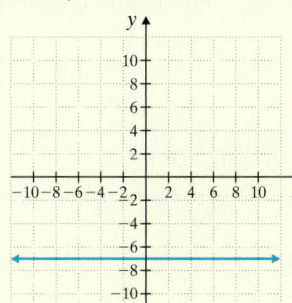

26. $x - 2 = 0$

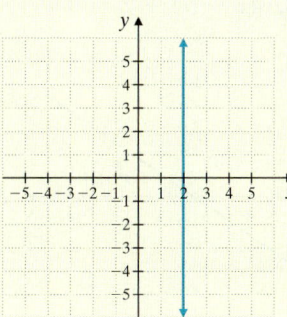

27. $x + 3 = 0$

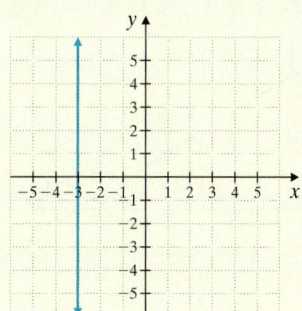

28. $y - 6 = 0$

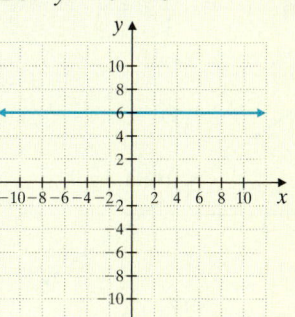

Objectives B C Mixed Practice *Graph each linear equation. See Examples 4 through 7.*

29. $x = y$

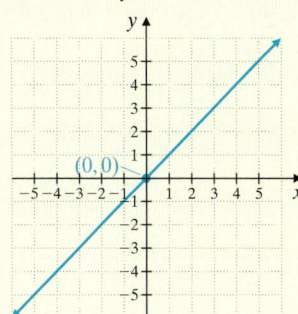

30. $x = -y$

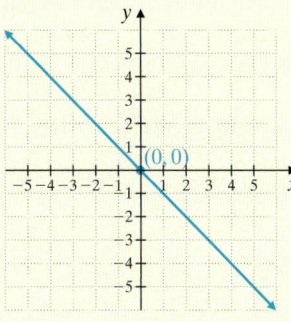

31. $x + 8y = 8$

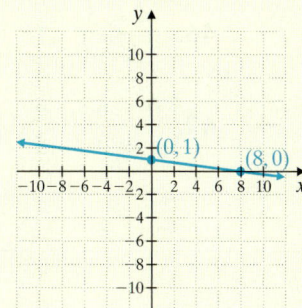

32. $x + 3y = 9$

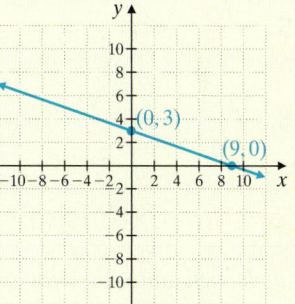

33. $5 = 6x - y$

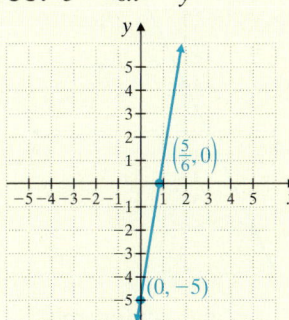

34. $4 = x - 3y$

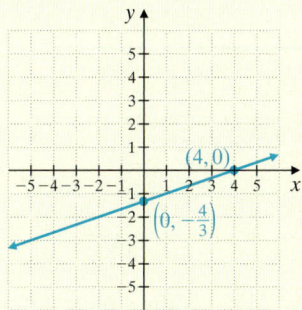

35. $-x + 10y = 11$

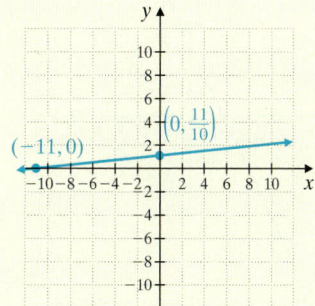

36. $-x + 9y = 10$

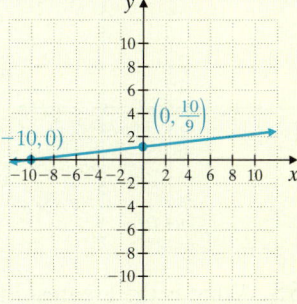

37. $x = -4\dfrac{1}{2}$

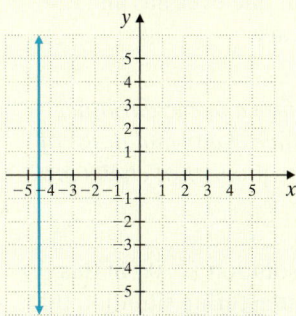

38. $x = -1\dfrac{3}{4}$

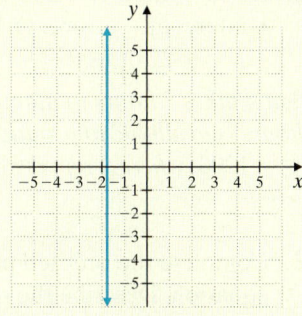

39. $y = 3\dfrac{1}{4}$

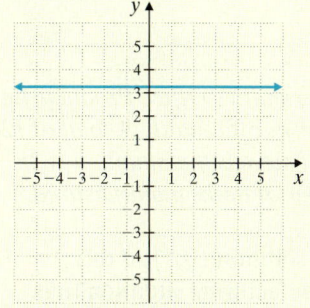

40. $y = 2\dfrac{1}{2}$

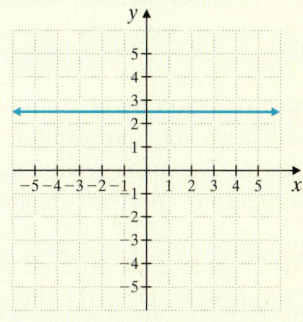

41. $y = -\dfrac{2}{3}x + 1$

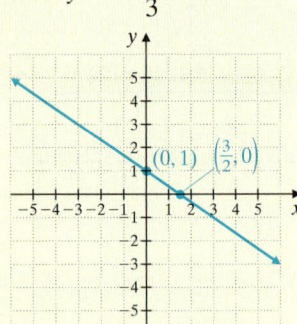

42. $y = -\dfrac{3}{5}x + 3$

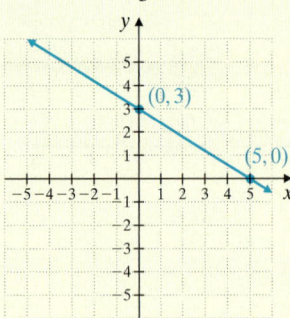

43. $4x - 6y + 2 = 0$

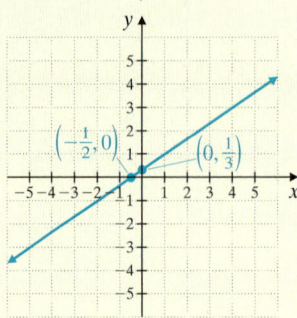

44. $9x - 6y + 3 = 0$

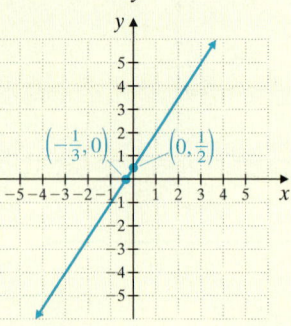

Review

Simplify. See Sections 1.5 and 1.6.

45. $\dfrac{-6 - 3}{2 - 8}$ $\dfrac{3}{2}$

46. $\dfrac{4 - 5}{-1 - 0}$ 1

47. $\dfrac{-8 - (-2)}{-3 - (-2)}$ 6

48. $\dfrac{12 - 3}{10 - 9}$ 9

49. $\dfrac{0 - 6}{5 - 0}$ $-\dfrac{6}{5}$

50. $\dfrac{2 - 2}{3 - 5}$ 0

Concept Extensions

Match each equation with its graph.

51. $y = 3$ c
a.

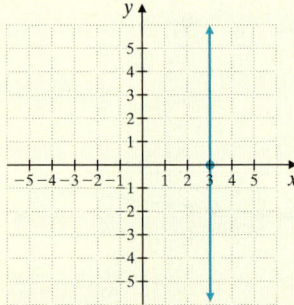

52. $y = 2x + 2$ d
b.

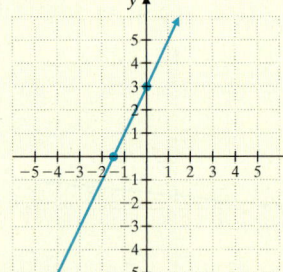

53. $x = 3$ a
c.

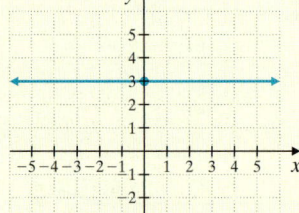

54. $y = 2x + 3$ b
d.

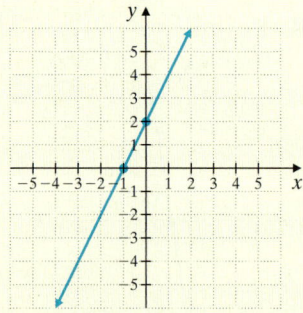

55. What is the greatest number of x- and y-intercepts that a line can have?   infinite

56. What is the smallest number of x- and y-intercepts that a line can have?   1

57. What is the smallest number of x- and y-intercepts that a circle can have?   0

58. What is the greatest number of x- and y-intercepts that a circle can have?   4

59. Discuss whether a vertical line ever has a y-intercept. answers may vary

60. Discuss whether a horizontal line ever has an x-intercept.   answers may vary

The production supervisor at Alexandra's Office Products finds that it takes 3 hours to manufacture a particular office chair and 6 hours to manufacture an office desk. A total of 1200 hours is available to produce office chairs and desks of this style. The linear equation that models this situation is $3x + 6y = 1200$, where x represents the number of chairs produced and y represents the number of desks manufactured.

61. Complete the ordered pair solution $(0, \ )$ of this equation. Describe the manufacturing situation that corresponds to this solution. $(0, 200)$; no chairs and 200 desks are manufactured.

62. Complete the ordered pair solution $(\ , 0)$ of this equation. Describe the manufacturing situation that corresponds to this solution. $(400, 0)$; 400 chairs and no desks are manufactured.

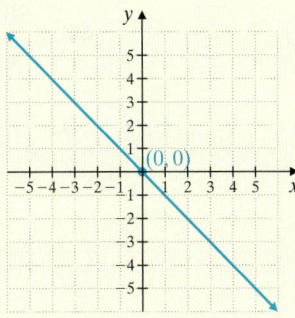

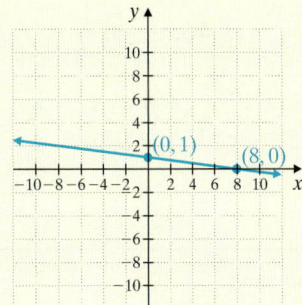

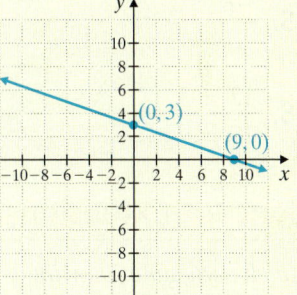

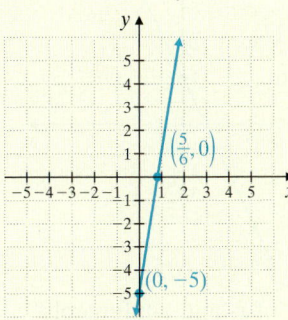

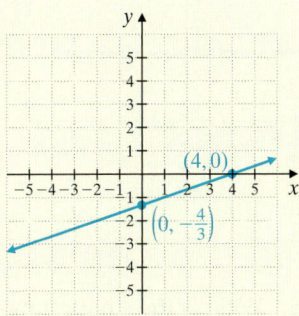

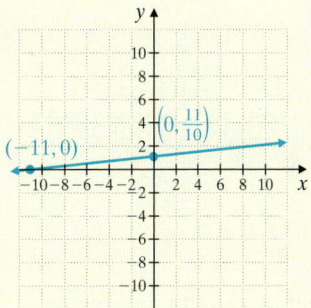

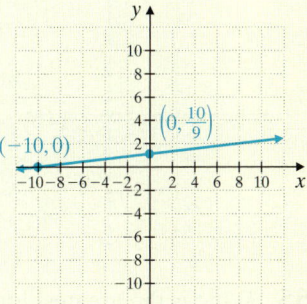

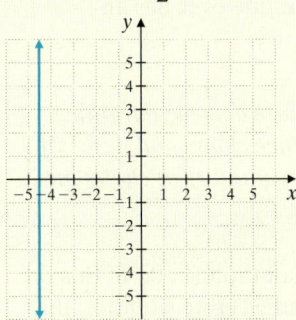

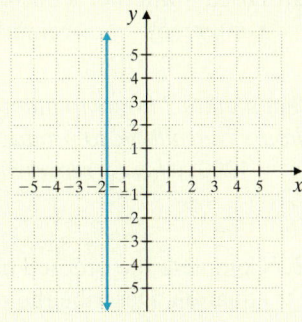

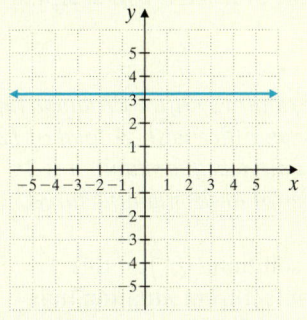

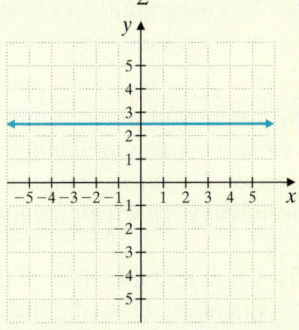

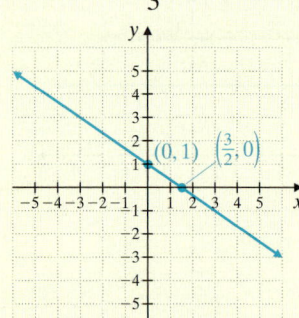

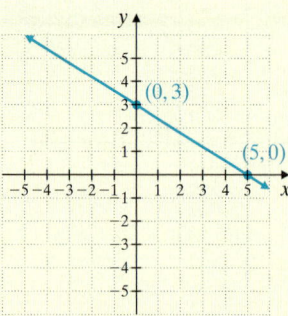

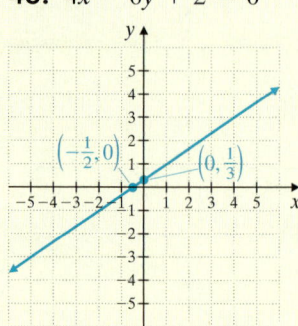

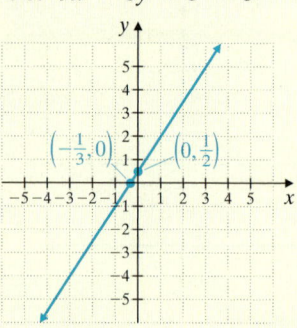

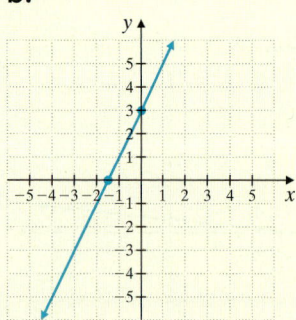

63. If 50 desks are manufactured, find the greatest number of chairs that can be made.   300 chairs

64. If 50 chairs are manufactured, find the greatest number of desks that can be made.   175 desks

Two lines in the same plane that do not intersect are called **parallel lines.**

65. Use your own graph paper to draw a line parallel to the line $y = -1$ that intersects the y-axis at -4. What is the equation of this line? $y = -4$

66. Use your own graph paper to draw a line parallel to the line $x = 5$ that intersects the x-axis at 1. What is the equation of this line? $x = 1$

Solve.

67. It has been said that newspapers are disappearing, replaced by various electronic media. The average circulation of newspapers in the United States y, in millions, from 2006 to 2011 can be modeled by the equation $y = -1.7x + 52$, where x represents the number of years after 2006. (*Source:* Newspaper Association of America)

   a. Find the x-intercept of this equation (round to the nearest tenth). $(30.6, 0)$

   b. What does this x-intercept mean?
   30.6 years after 2006, there may be no newspaper circulation.

68. The number y of Barnes & Noble retail stores in operation for the years 2008–2012 can be modeled by the equation $y = -8.6x + 730$, where x represents the number of years after 2008. (*Source:* Based on data from Barnes & Noble, Inc.)

   a. Find the y-intercept of this equation. $(0, 730)$

   b. What does this y-intercept mean?   In 2008, the number of stores was 730.

3.4   Slope and Rate of Change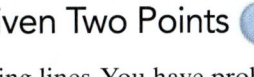

Objective A   Finding the Slope of a Line Given Two Points

Thus far, much of this chapter has been devoted to graphing lines. You have probably noticed by now that a key feature of a line is its slant or steepness. In mathematics, the slant or steepness of a line is formally known as its **slope.** We measure the slope of a line by the ratio of vertical change (rise) to the corresponding horizontal change (run) as we move along the line.

On the line below, for example, suppose that we begin at the point $(1, 2)$ and move to the point $(4, 6)$. The vertical change is the change in y-coordinates: $6 - 2$ or 4 units. The corresponding horizontal change is the change in x-coordinates: $4 - 1 = 3$ units. The ratio of these changes is

$$\text{slope} = \frac{\text{change in } y \text{ (vertical change or rise)}}{\text{change in } x \text{ (horizontal change or run)}} = \frac{4}{3}$$

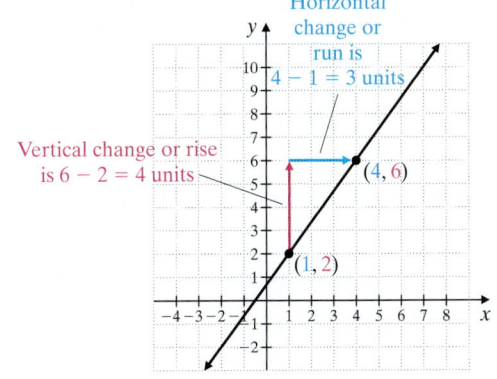

Objectives

A Find the Slope of a Line Given Two Points of the Line.

B Find the Slope of a Line Given Its Equation.

C Find the Slopes of Horizontal and Vertical Lines.

D Compare the Slopes of Parallel and Perpendicular Lines.

E Slope as a Rate of Change.

Teaching Tip

Begin this lesson by having students compare and contrast the graphs of three lines with the same y-intercept and different slopes. Ask how we could describe their differences. Possible lines to use:

$$y = 2x + 3$$
$$y = 3$$
$$y = -4x + 3$$

The slope of this line, then, is $\frac{4}{3}$. This means that for every 4 units of change in y-coordinates, there is a corresponding change of 3 units in x-coordinates.

Helpful Hint

It makes no difference what two points of a line are chosen to find its slope. The slope of a line is the same everywhere on the line.

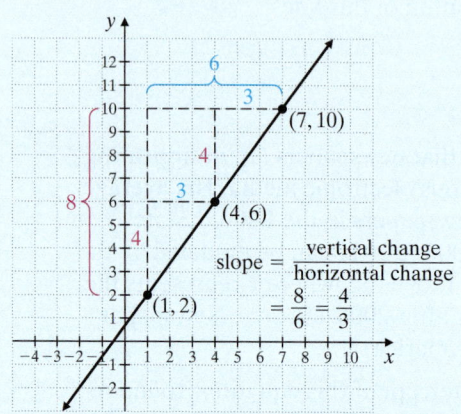

To find the slope of a line, then, choose two points of the line. Label the two x-coordinates of the two points x_1 and x_2 (read "x sub one" and "x sub two"), and label the corresponding y-coordinates y_1 and y_2.

The vertical change or **rise** between these points is the difference in the y-coordinates: $y_2 - y_1$. The horizontal change or **run** between the points is the difference of the x-coordinates: $x_2 - x_1$. The slope of the line is the ratio of $y_2 - y_1$ to $x_2 - x_1$, and we traditionally use the letter m to denote slope: $m = \dfrac{y_2 - y_1}{x_2 - x_1}$.

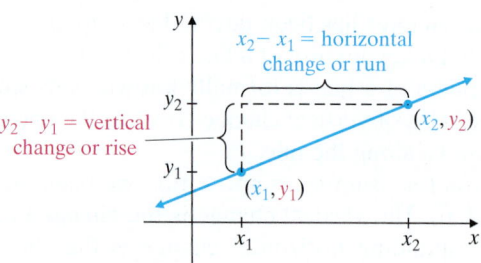

Slope of a Line

The slope m of the line containing the points (x_1, y_1) and (x_2, y_2) is given by

$$ m = \frac{\text{rise}}{\text{run}} = \frac{\text{change in } y}{\text{change in } x} = \frac{y_2 - y_1}{x_2 - x_1}, \qquad \text{as long as } x_2 \neq x_1 $$

Example 1   Find the slope of the line through $(-1, 5)$ and $(2, -3)$. Graph the line.

Solution: Let (x_1, y_1) be $(-1, 5)$ and (x_2, y_2) be $(2, -3)$. Then, by the definition of slope, we have the following.

Copyright 2016 Pearson Education, Inc.

Teaching Tip

For Example 1, have students place their finger on $(-1, 5)$ and make a vertical and horizontal move to $(2, -3)$. Then have them write their move as a ratio of the vertical move to the horizontal move using the following convention: Upward moves are positive, downward moves are negative, rightward moves are positive, leftward moves are negative. Now have them put their finger on $(2, -3)$ and move to $(-1, 5)$ using a vertical and horizontal move. Have them write this move as a ratio. How did the ratios change? Are the ratios equal?

Practice 1

Find the slope of the line through $(-2, 3)$ and $(4, -1)$. Graph the line.

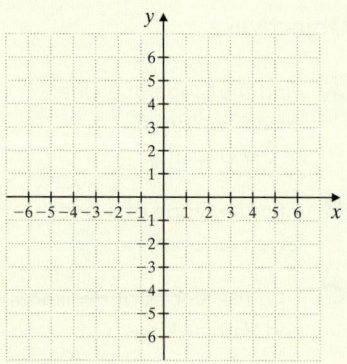

Answer

1. $m = -\dfrac{2}{3}$

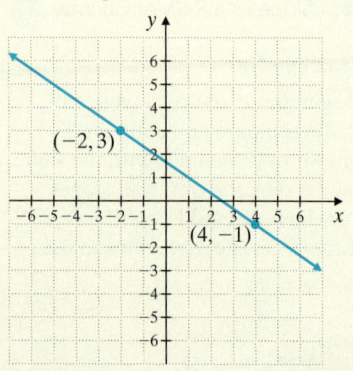

$$m = \frac{y_2 - y_1}{x_2 - x_1}$$

$$= \frac{-3 - 5}{2 - (-1)}$$

$$= \frac{-8}{3} = -\frac{8}{3}$$

The slope of the line is $-\frac{8}{3}$.

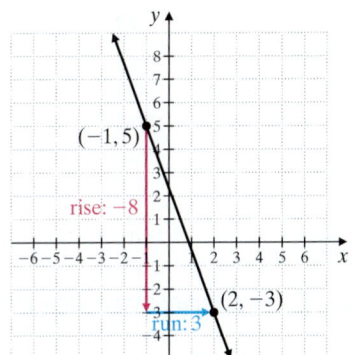

■ **Work Practice 1**

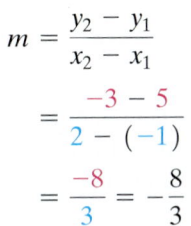**Helpful Hint**

When finding slope, it makes no difference which point is identified as (x_1, y_1) and which is identified as (x_2, y_2). Just remember that whatever y-value is first in the numerator, its corresponding x-value is first in the denominator. Another way to calculate the slope in Example 1 is

$$m = \frac{y_2 - y_1}{x_2 - x_1} = \frac{5 - (-3)}{-1 - 2} = \frac{8}{-3} \text{ or } -\frac{8}{3} \leftarrow \text{Same slope as found in Example 1}$$

✔**Concept Check** The points $(-2, -5)$, $(0, -2)$, $(4, 4)$, and $(10, 13)$ all lie on the same line. Work with a partner and verify that the slope is the same no matter which points are used to find slope.

Example 2 Find the slope of the line through $(-1, -2)$ and $(2, 4)$. Graph the line.

Solution: Let (x_1, y_1) be $(2, 4)$ and (x_2, y_2) be $(-1, -2)$.

$$m = \frac{y_2 - y_1}{x_2 - x_1}$$

$$= \frac{-2 - 4}{-1 - 2} \quad \text{\textcolor{red}{y-value}}$$
$$\text{\textcolor{red}{corresponding x-value}}$$

$$= \frac{-6}{-3} = 2$$

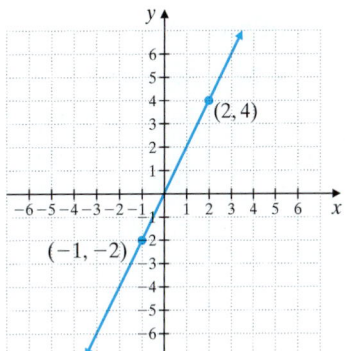

The slope is 2.

■ **Work Practice 2**

✔**Concept Check** What is wrong with the following slope calculation for the points $(3, 5)$ and $(-2, 6)$?

$$m = \frac{5 - 6}{-2 - 3} = \frac{-1}{-5} = \frac{1}{5}$$

Notice that the slope of the line in Example 1 is negative and that the slope of the line in Example 2 is positive. Let your eye follow the line with negative slope from left to right and notice that the line "goes down." If you follow the line with positive slope from left to right, you will notice that the line "goes up." This is true in general.

Practice 2

Find the slope of the line through $(-2, 1)$ and $(3, 5)$. Graph the line.

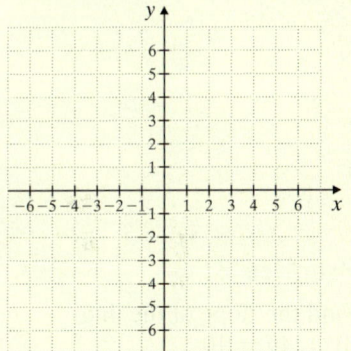

Answer

2. $m = \frac{4}{5}$

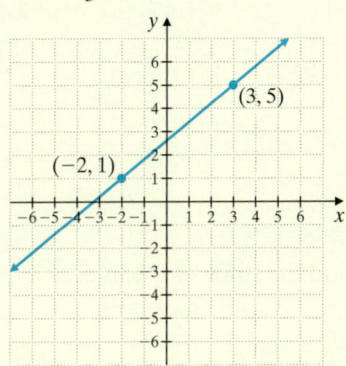

✔**First Concept Check Answer**

$m = \frac{3}{2}$

✔**Second Concept Check Answer**

$$m = \frac{5 - 6}{3 - (-2)} = \frac{-1}{5} = -\frac{1}{5}$$

Negative slope   Positive slope

> **Helpful Hint**
> To decide whether a line "goes up" or "goes down," always follow the line from left to right.

Objective B Finding the Slope of a Line Given Its Equation

As we have seen, the slope of a line is defined by two points on the line. Thus, if we know the equation of a line, we can find its slope by finding two of its points. For example, let's find the slope of the line

$$y = 3x - 2$$

To find two points, we can choose two values for x and substitute to find corresponding y-values. If $x = 0$, for example, $y = 3 \cdot 0 - 2$ or $y = -2$. If $x = 1$, $y = 3 \cdot 1 - 2$ or $y = 1$. This gives the ordered pairs $(0, -2)$ and $(1, 1)$. Using the definition for slope, we have

$$m = \frac{1 - (-2)}{1 - 0} = \frac{3}{1} = 3 \quad \text{The slope is 3.}$$

Notice that the slope, 3, is the same as the coefficient of x in the equation $y = 3x - 2$. This is true in general.

> If a linear equation is solved for y, the coefficient of x is the line's slope. In other words, the slope of the line given by $y = mx + b$ is m, the coefficient of x.
>
> $$y = mx + b$$
> $\uparrow$———— slope

Copyright © 2016 Pearson Education, Inc.

Teaching Tip

Consider having students create tables of values for several pairs of equations such as $y = 5x$ and $y = 5x - 3$; $y = -3x$ and $y = -3x + 2$; and $y = \frac{1}{2}x$ and $y = \frac{1}{2}x + 4$. Then have students use the ordered pairs in each table to find the slope of each equation. Ask them what they notice.

Practice 3

Find the slope of the line $5x + 4y = 10$.

Practice 4

Find the slope of the line $-y = -2x + 7$.

Answers

3. $m = -\frac{5}{4}$ 4. $m = 2$

Example 3   Find the slope of the line $-2x + 3y = 11$.

Solution: When we solve for y, the coefficient of x is the slope.

$$-2x + 3y = 11$$
$$3y = 2x + 11 \quad \text{Add } 2x \text{ to both sides.}$$
$$y = \frac{2}{3}x + \frac{11}{3} \quad \text{Divide both sides by 3.}$$

The slope is $\frac{2}{3}$.

■ Work Practice 3

Example 4   Find the slope of the line $-y = 5x - 2$.

Solution: Remember, the equation must be solved for y (not $-y$) in order for the coefficient of x to be the slope.

To solve for y, let's divide both sides of the equation by -1.

$$-y = 5x - 2$$
$$\frac{-y}{-1} = \frac{5x}{-1} - \frac{2}{-1} \quad \text{Divide both sides by } -1.$$
$$y = -5x + 2 \quad \text{Simplify.}$$

The slope is -5.

■ Work Practice 4

Objective C Finding Slopes of Horizontal and Vertical Lines ▶

Example 5 Find the slope of the line $y = -1$.

Solution: Recall that $y = -1$ is a horizontal line with y-intercept -1. To find the slope, we find two ordered pair solutions of $y = -1$, knowing that solutions of $y = -1$ must have a y-value of -1. We will use $(2, -1)$ and $(-3, -1)$. We let (x_1, y_1) be $(2, -1)$ and (x_2, y_2) be $(-3, -1)$.

$$m = \frac{y_2 - y_1}{x_2 - x_1} = \frac{-1 - (-1)}{-3 - 2} = \frac{0}{-5} = 0$$

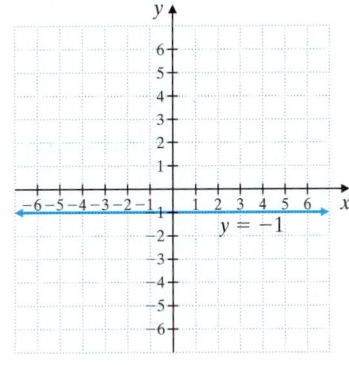

The slope of the line $y = -1$ is 0. Since the y-values will have a difference of 0 for every horizontal line, we can say that all **horizontal lines have a slope of 0.**

🟧 **Work Practice 5**

Example 6 Find the slope of the line $x = 5$.

Solution: Recall that the graph of $x = 5$ is a vertical line with x-intercept 5. To find the slope, we find two ordered pair solutions of $x = 5$. Ordered pair solutions of $x = 5$ must have an x-value of 5. We will use $(5, 0)$ and $(5, 4)$. We let $(x_1, y_1) = (5, 0)$ and $(x_2, y_2) = (5, 4)$.

$$m = \frac{y_2 - y_1}{x_2 - x_1} = \frac{4 - 0}{5 - 5} = \frac{4}{0}$$

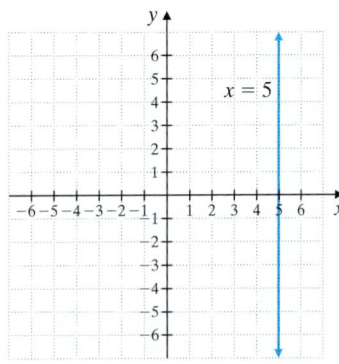

Since $\frac{4}{0}$ is undefined, we say that the slope of the vertical line $x = 5$ is undefined. Since the x-values will have a difference of 0 for every vertical line, we can say that all **vertical lines have undefined slope.**

🟧 **Work Practice 6**

Practice 5

Find the slope of $y = 3$.

Practice 6

Find the slope of the line $x = -2$.

Teaching Tip

Remind students that $\frac{4}{0}$ is undefined because division by zero is undefined.

Helpful Hint Slope of 0 and undefined slope are not the same. Vertical lines have undefined slope, while horizontal lines have a slope of 0.

Answers

5. $m = 0$ **6.** undefined slope

Here is a general review of slope.

Summary of Slope

Slope m of the line through (x_1, y_1) and (x_2, y_2) is given by the equation

$$m = \frac{y_2 - y_1}{x_2 - x_1}.$$

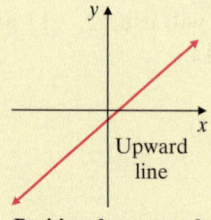

Upward line

Positive slope: $m > 0$

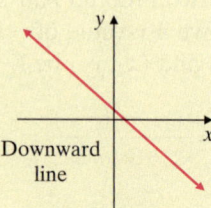

Downward line

Negative slope: $m < 0$

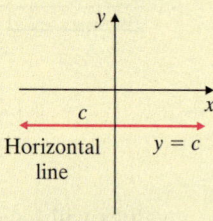

Horizontal line $y = c$

Zero slope: $m = 0$

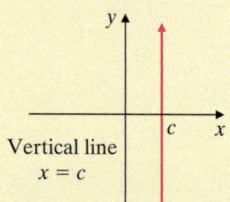

Vertical line $x = c$

No slope or undefined slope

Objective D Slopes of Parallel and Perpendicular Lines

Two lines in the same plane are **parallel** if they do not intersect. Slopes of lines can help us determine whether lines are parallel. Since parallel lines have the same steepness, it follows that they have the same slope.

For example, the graphs of

$$y = -2x + 4$$

and

$$y = -2x - 3$$

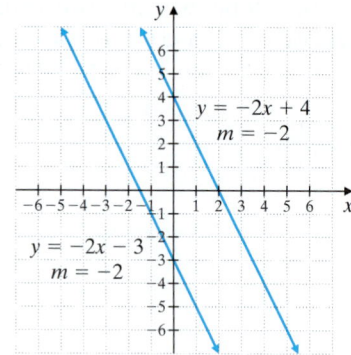

are shown. These lines have the same slope, -2. They also have different y-intercepts, so the lines are parallel. (If the y-intercepts were the same also, the lines would be the same.)

Parallel Lines

Nonvertical parallel lines have the same slope and different y-intercepts.

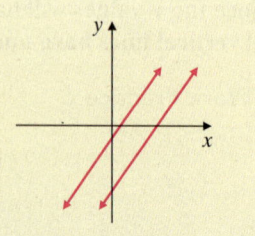

Two lines are **perpendicular** if they lie in the same plane and meet at a 90° (right) angle. How do the slopes of perpendicular lines compare? The product of the slopes of two perpendicular lines is -1.

For example, the graphs of

$$y = 4x + 1$$

and

$$y = -\frac{1}{4}x - 3$$

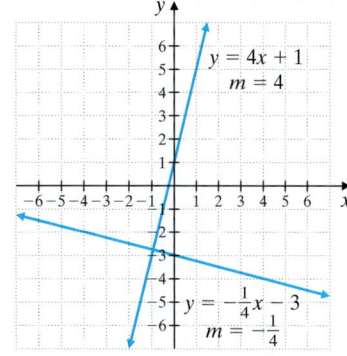

are shown. The slopes of the lines are 4 and $-\frac{1}{4}$. Their product is $4\left(-\frac{1}{4}\right) = -1$, so the lines are perpendicular.

Perpendicular Lines

If the product of the slopes of two lines is -1, then the lines are perpendicular.

(Two nonvertical lines are perpendicular if the slope of one is the negative reciprocal of the slope of the other.)

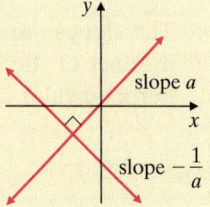

Helpful Hint

Here are examples of numbers that are negative (opposite) reciprocals.

Number	Negative Reciprocal	Their product is -1.
$\frac{2}{3}$	$-\frac{3}{2}$	$\frac{2}{3} \cdot -\frac{3}{2} = -\frac{6}{6} = -1$
-5 or $-\frac{5}{1}$	$\frac{1}{5}$	$-5 \cdot \frac{1}{5} = -\frac{5}{5} = -1$

Here are a few important points about vertical and horizontal lines.

- Two distinct vertical lines are parallel.
- Two distinct horizontal lines are parallel.
- A horizontal line and a vertical line are always perpendicular.

△ **Example 7** Determine whether each pair of lines is parallel, perpendicular, or neither.

a. $y = -\frac{1}{5}x + 1$
$2x + 10y = 3$

b. $x + y = 3$
$-x + y = 4$

c. $3x + y = 5$
$2x + 3y = 6$

(Continued on next page)

Solution:

a. The slope of the line $y = -\dfrac{1}{5}x + 1$ is $-\dfrac{1}{5}$. We find the slope of the second line by solving its equation for y.

$$2x + 10y = 3$$

$$10y = -2x + 3 \qquad \text{Subtract } 2x \text{ from both sides.}$$

$$y = \frac{-2}{10}x + \frac{3}{10} \qquad \text{Divide both sides by 10.}$$

$$y = -\frac{1}{5}x + \frac{3}{10} \qquad \text{Simplify.}$$

The slope of this line is $-\dfrac{1}{5}$ also. Since the lines have the same slope and different y-intercepts, they are parallel, as shown below on the left.

b. To find each slope, we solve each equation for y.

$$x + y = 3 \qquad\qquad\qquad -x + y = 4$$
$$y = -x + 3 \qquad\qquad\qquad y = x + 4$$

The slope is -1.   The slope is 1.

The slopes are not the same, so the lines are not parallel. Next we check the product of the slopes: $(-1)(1) = -1$. Since the product is -1, the lines are perpendicular, as shown below on the right.

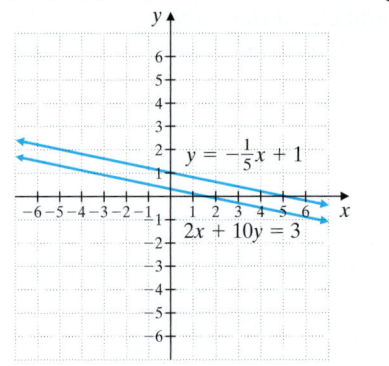
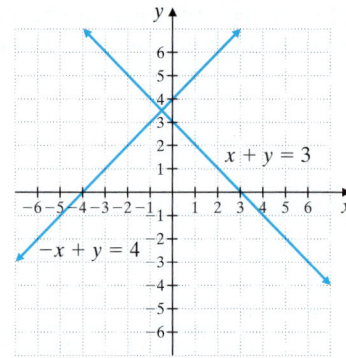

c. We solve each equation for y to find each slope. The slopes are -3 and $-\dfrac{2}{3}$. The slopes are not the same and their product is not -1. Thus, the lines are neither parallel nor perpendicular.

■ **Work Practice 7**

✓**Concept Check** Consider the line $-6x + 2y = 1$.

    a. Write the equations of two lines parallel to this line.

    b. Write the equations of two lines perpendicular to this line.

Objective E Slope as a Rate of Change ▶

Slope can also be interpreted as a rate of change. In other words, slope tells us how fast y is changing with respect to x. To see this, let's look at a few of the many real-world applications of slope. For example, the pitch of a roof, used by builders and architects, is its slope. The pitch of the roof on the left is $\dfrac{7}{10}\left(\dfrac{\text{rise}}{\text{run}}\right)$. This means that the roof rises vertically 7 feet for every horizontal 10 feet. The rate of change for the roof is 7 vertical feet (y) per 10 horizontal feet (x).

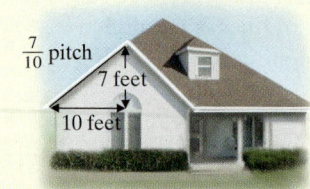

$\dfrac{7}{10}$ pitch

7 feet

10 feet

✓ **Concept Check Answers**

Answers may vary; for example,

a. $y = 3x - 3$, $y = 3x - 1$

b. $y = -\dfrac{1}{3}x$, $y = -\dfrac{1}{3}x + 1$

The grade of a road is its slope written as a percent. A 7% grade, as shown below, means that the road rises (or falls) 7 feet for every horizontal 100 feet. $\left(\text{Recall that } 7\% = \dfrac{7}{100}.\right)$ Here, the slope of $\dfrac{7}{100}$ gives us the rate of change. The road rises (in our diagram) 7 vertical feet (y) for every 100 horizontal feet (x).

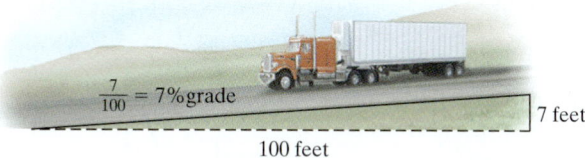

$\dfrac{7}{100} = 7\%$ grade   7 feet

100 feet

Example 8 Finding the Grade of a Road

At one part of the road to the summit of Pike's Peak, the road rises 15 feet for a horizontal distance of 250 feet. Find the grade of the road.

Solution: Recall that the grade of a road is its slope written as a percent.

$$\text{grade} = \frac{\text{rise}}{\text{run}} = \frac{15}{250} = 0.06 = 6\%$$

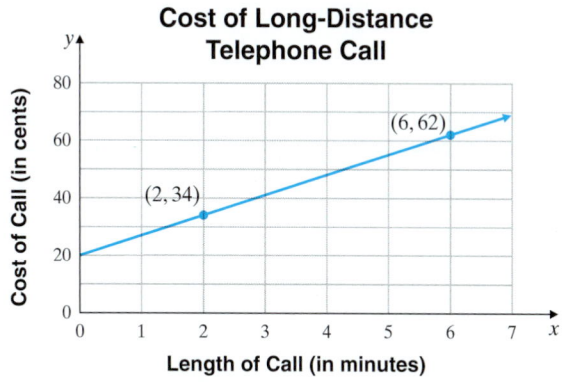

15 feet

250 feet

The grade is 6%.

■ **Work Practice 8**

Example 9 Finding the Slope of a Line

The following graph shows the cost y (in cents) of a nationwide long-distance telephone call from Texas with a certain telephone-calling plan, where x is the length of the call in minutes. Find the slope of the line and attach the proper units for the rate of change. Then write a sentence explaining the meaning of slope in this application.

Solution: Use $(2, 34)$ and $(6, 62)$ to calculate slope.

Cost of Long-Distance Telephone Call

Cost of Call (in cents) vs *Length of Call (in minutes)*

$(2, 34)$ $(6, 62)$

$$m = \frac{62 - 34}{6 - 2} = \frac{28}{4} = \frac{7 \text{ cents}}{1 \text{ minute}}$$

This means that the rate of change of the phone call is 7 cents per 1 minute, or the cost of the phone call is 7 cents per minute.

■ **Work Practice 9**

Practice 8

Find the grade of the road shown.

3 feet

20 feet

Practice 9

Find the slope of the line and write the slope as a rate of change. This graph represents annual restaurant-industry employment y (in millions of workers) for year x. Write a sentence explaining the meaning of slope in this application.

U.S. Restaurant-Industry Employment

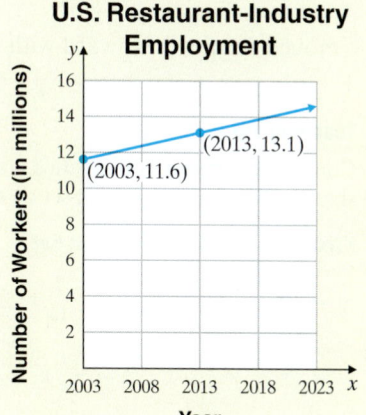

Number of Workers (in millions) vs *Year*

$(2013, 13.1)$

$(2003, 11.6)$

Source: National Restaurant Association

Answers

8. 15%   **9.** $m = 0.15$; Each year the number of workers employed in the U.S. restaurant industry increases by 0.15 million, or 150,000, workers per year.

Calculator Explorations Graphing

It is possible to use a graphing calculator to sketch the graph of more than one equation on the same set of axes. This feature can be used to see parallel lines with the same slope. For example, graph the equations $y = \frac{2}{5}x$, $y = \frac{2}{5}x + 7$, and $y = \frac{2}{5}x - 4$ on the same set of axes. To do so, press the $\boxed{Y =}$ key and enter the equations on the first three lines.

$$Y_1 = \frac{2}{5}x$$

$$Y_2 = \frac{2}{5}x + 7$$

$$Y_3 = \frac{2}{5}x - 4$$

The displayed equations should look like this:

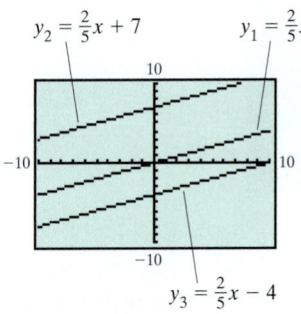

These lines are parallel, as expected, since they all have a slope of $\frac{2}{5}$. The graph of $y = \frac{2}{5}x + 7$ is the graph of $y = \frac{2}{5}x$ moved 7 units upward with a y-intercept of 7. Also, the graph of $y = \frac{2}{5}x - 4$ is the graph of $y = \frac{2}{5}x$ moved 4 units downward with a y-intercept of -4.

Graph the parallel lines on the same set of axes. Describe the similarities and differences in their graphs.

1. $y = 3.8x$, $y = 3.8x - 3$, $y = 3.8x + 9$

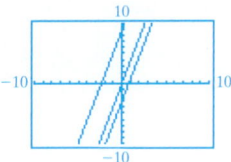

2. $y = -4.9x$, $y = -4.9x + 1$, $y = -4.9x + 8$

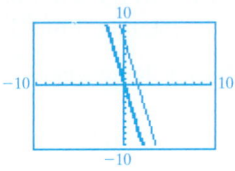

3. $y = \frac{1}{4}x$, $y = \frac{1}{4}x + 5$, $y = \frac{1}{4}x - 8$

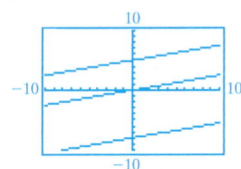

4. $y = -\frac{3}{4}x$, $y = -\frac{3}{4}x - 5$, $y = -\frac{3}{4}x + 6$

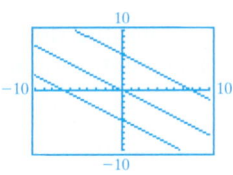

Teaching Tip

Consider exploring slopes with a graphing calculator. Have students graph the following equations on a graphing calculator and sketch all the results of each set on the same axes. Have students compare the first and second sets, then the third and fourth sets.

First Set	Second Set	Third Set	Fourth Set
$y = 10x$	$y = -\frac{1}{10}x$	$y = \frac{1}{2}x$	$y = -2x$
$y = 5x$	$y = -\frac{1}{5}x$	$y = \frac{1}{5}x$	$y = -5x$
$y = 2x$	$y = -\frac{1}{2}x$	$y = \frac{1}{10}x$	$y = -10x$
$y = x$	$y = -1x$		

Then ask how a line of slope 0 would be positioned.

Vocabulary, Readiness & Video Check

Use the choices below to fill in each blank. Not all choices will be used.

m	x	0	positive	undefined
b	y	slope	negative	

1. The measure of the steepness or tilt of a line is called ____slope____.

2. If an equation is written in the form $y = mx + b$, the value of the letter ____m____ is the value of the slope of the graph.

3. The slope of a horizontal line is ____0____.

4. The slope of a vertical line is ____undefined____.

5. If the graph of a line moves upward from left to right, the line has ____positive____ slope.

6. If the graph of a line moves downward from left to right, the line has ____negative____ slope.

7. Given two points of a line, slope $= \dfrac{\text{change in } \underline{\quad y \quad}}{\text{change in } \underline{\quad x \quad}}$.

State whether the slope of the line is positive, negative, 0, or undefined.

8.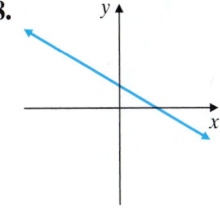

negative

9.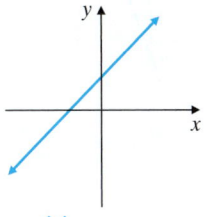

positive

10.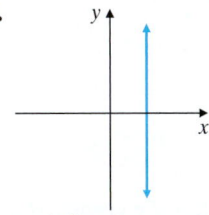

undefined

11.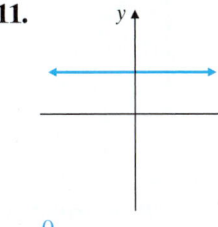

0

Decide whether a line with the given slope slants upward or downward or is horizontal or vertical.

12. $m = \dfrac{7}{6}$ ____upward____   **13.** $m = -3$ ____downward____   **14.** $m = 0$ ____horizontal____   **15.** m is undefined. ____vertical____

Martin-Gay Interactive Videos

See Video 3.4

See video answer section.

Watch the section lecture video and answer the following questions.

Objective A **16.** What important point is made during ▣ Example 1 having to do with the order of the points in the slope formula? ▶

Objective B **17.** From ▣ Example 5, how do we write an equation in "slope-intercept form"? Once the equation is in slope-intercept form, how do we identify the slope? ▶

Objective C **18.** In the lecture after ▣ Example 8, different slopes are summarized. What is the difference between zero slope and undefined slope? What does "no slope" mean? ▶

Objective D **19.** From ▣ Example 10, what form of the equation is best to determine if two lines are parallel or perpendicular? Why? ▶

Objective E **20.** Writing the slope as a rate of change in ▣ Example 11 gave real-life meaning to the slope. What step in the general strategy for problem solving does this correspond to? ▶

3.4 Exercise Set MyMathLab®

Objective A *Find the slope of the line that passes through the given points. See Examples 1 and 2.*

1. $(-1, 5)$ and $(6, -2)$
$m = -1$

2. $(-1, 16)$ and $(3, 4)$
$m = -3$

3. $(1, 4)$ and $(5, 3)$
$m = -\dfrac{1}{4}$

4. $(3, 1)$ and $(2, 6)$
$m = -5$

5. $(5, 1)$ and $(-2, 1)$
$m = 0$

6. $(-8, 3)$ and $(-2, 3)$
$m = 0$

7. $(-4, 3)$ and $(-4, 5)$
undefined slope

8. $(-2, -3)$ and $(-2, 5)$
undefined slope

Use the points shown on each graph to find the slope of each line. See Examples 1 and 2.

9.

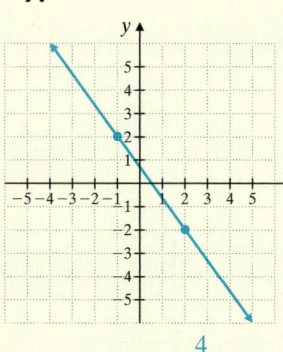

$m = -\dfrac{4}{3}$

10.

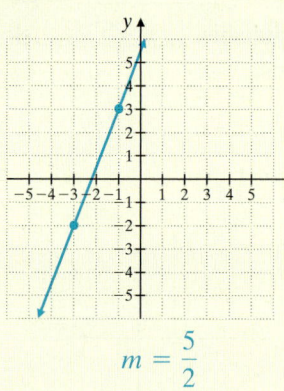

$m = \dfrac{5}{2}$

11.

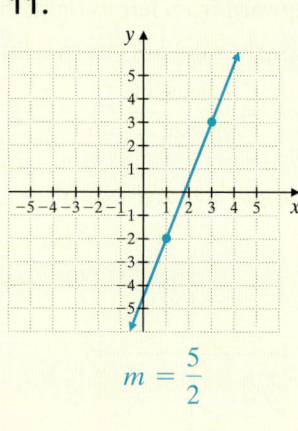

$m = \dfrac{5}{2}$

12.

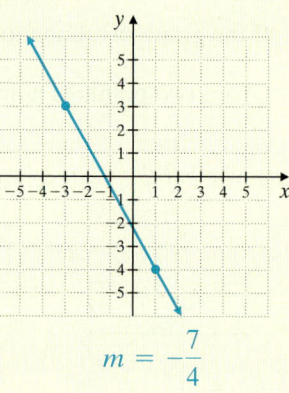

$m = -\dfrac{7}{4}$

For each graph, determine which line has the greater slope.

13.

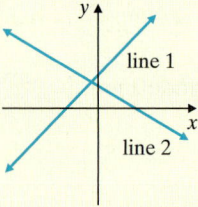

line 1

14.

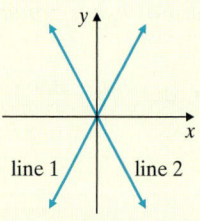

line 1

15.

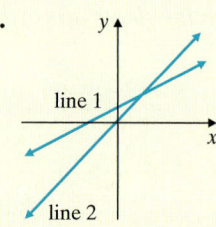

line 2

16.

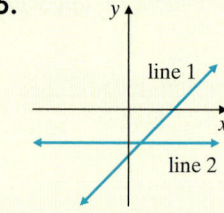

line 1

Objectives B C Mixed Practice *Find the slope of each line. See Examples 3 through 6.*

17. $y = 5x - 2$ $m = 5$

18. $y = -2x + 6$ $m = -2$

19. $y = -0.3x + 2.5$ $m = -0.3$

20. $y = -7.6x - 0.1$ $m = -7.6$

21. $2x + y = 7$ $m = -2$

22. $-5x + y = 10$ $m = 5$

23.

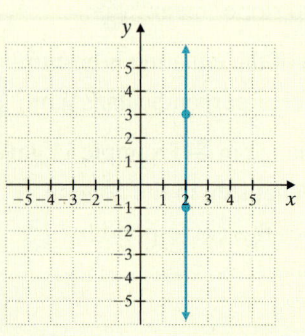

undefined slope

24.

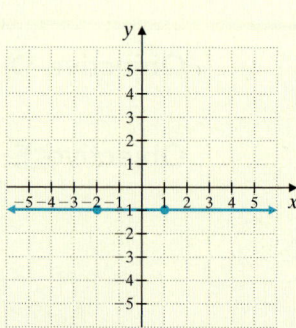

$m = 0$

▶ **25.** $2x - 3y = 10$ **26.** $3x - 5y = 1$ ▶ **27.** $x = 1$ **28.** $y = -2$

    $m = \dfrac{2}{3}$ $m = \dfrac{3}{5}$ undefined slope $m = 0$

29. $x = 2y$ **30.** $x = -4y$ ▶ **31.** $y = -3$ **32.** $x = 5$

    $m = \dfrac{1}{2}$ $m = -\dfrac{1}{4}$ $m = 0$ undefined slope

33. $-3x - 4y = 6$ **34.** $-4x - 7y = 9$ **35.** $20x - 5y = 1.2$ **36.** $24x - 3y = 5.7$

    $m = -\dfrac{3}{4}$ $m = -\dfrac{4}{7}$ $m = 4$ $m = 8$

△ **Objective D** *Determine whether each pair of lines is parallel, perpendicular, or neither. See Example 7.*

▶ **37.** $y = \dfrac{2}{9}x + 3$ **38.** $y = \dfrac{1}{5}x + 20$ **39.** $x - 3y = -6$ **40.** $y = 4x - 2$

    $y = -\dfrac{2}{9}x$ neither $y = -\dfrac{1}{5}x$ neither $y = 3x - 9$ $4x + y = 5$

                                          neither   neither

41. $6x = 5y + 1$ **42.** $-x + 2y = -2$ **43.** $6 + 4x = 3y$ ▶ **44.** $10 + 3x = 5y$

    $-12x + 10y = 1$ $2x = 4y + 3$ $3x + 4y = 8$ $5x + 3y = 1$

    parallel   parallel   perpendicular   perpendicular

△ *Find the slope of a line that is (a) parallel and (b) perpendicular to the line through each pair of points. See Example 7.*

45. $(-3, -3)$ and $(0, 0)$ **46.** $(6, -2)$ and $(1, 4)$ **47.** $(-8, -4)$ and $(3, 5)$ **48.** $(6, -1)$ and $(-4, -10)$

    a. 1 **b.** -1 **a.** $-\dfrac{6}{5}$ **b.** $\dfrac{5}{6}$ **a.** $\dfrac{9}{11}$ **b.** $-\dfrac{11}{9}$ **a.** $\dfrac{9}{10}$ **b.** $-\dfrac{10}{9}$

Objective E *The pitch of a roof is its slope. Find the pitch of each roof shown. See Example 8.*

49. 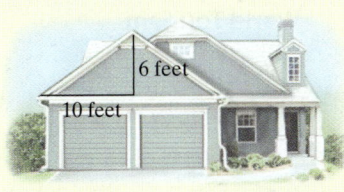 $\dfrac{3}{5}$ **50.**  $\dfrac{1}{2}$

The grade of a road is its slope written as a percent. Find the grade of each road shown. See Example 8.

51.    12.5%   **52.** 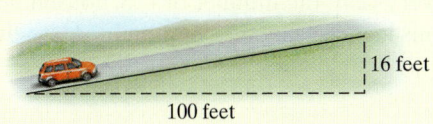   16%

53. One of Japan's superconducting "bullet" trains is researched and tested at the Yamanashi Maglev Test Line near Otsuki City. The steepest section of the track has a rise of 2580 meters for a horizontal distance of 6450 meters. What is the grade (slope written as a percent) of this section of track? (*Source:* Japan Railways Central Co.)   40%

2580 meters

6450 meters

54. Professional plumbers suggest that a sewer pipe rise 0.25 inch for every horizontal foot. Find the recommended slope for a sewer pipe and write the slope as a grade, or percent. Round to the nearest percent.   2%

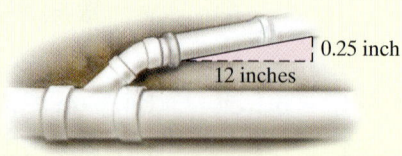

0.25 inch

12 inches

55. There has been controversy over the past few years about the world's steepest street. *The Guinness Book of Records* listed Baldwin Street, in Dunedin, New Zealand, as the world's steepest street, but Canton Avenue in the Pittsburgh neighborhood of Beechview may actually be steeper. Calculate each grade to the nearest percent.

Canton Avenue	For every 30 meters of horizontal distance, the vertical change is 11 meters.	37%
Baldwin Street	For every 2.86 meters of horizontal distance, the vertical change is 1 meter.	35%

56. According to federal regulations, a wheelchair ramp should rise no more than 1 foot for a horizontal distance of 12 feet. Write the slope as a grade. Round to the nearest tenth of a percent.   8.3%

Find the slope of each line and write a sentence using the slope as a rate of change. Don't forget to attach the proper units. See Example 9.

57. This graph approximates the number of U.S. households that have televisions y (in millions) for year x.

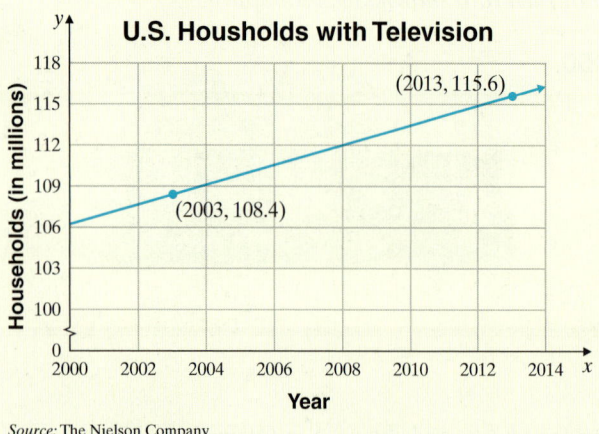

U.S. Housholds with Television

Households (in millions)

(2013, 115.6)

(2003, 108.4)

Year

Source: The Nielson Company

$m = 0.72$; Every year, the number of U.S. households with televisions increases by 0.72 million households.

58. The graph approximates the amount of money y (in billions of dollars) spent worldwide on tourism for year x.

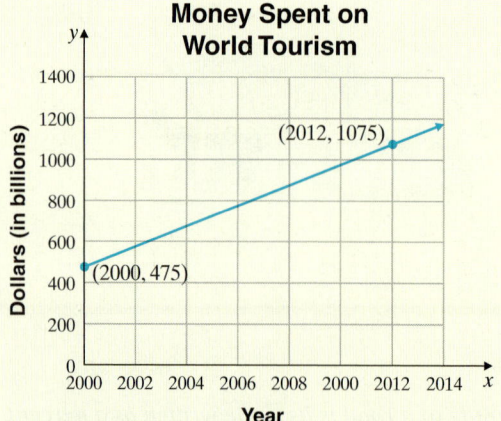

Money Spent on World Tourism

Dollars (in billions)

(2012, 1075)

(2000, 475)

Year

Source: World Tourism Organization

$m = 50$; Every year, an additional $50 billion is spent on tourism worldwide.

59. Americans are keeping their cars longer. The graph below shows the median age y (in years) of automobiles in the United States for the years shown.
m = 0.15; Every year, the median age of automobiles in the United States increases by 0.15 year.

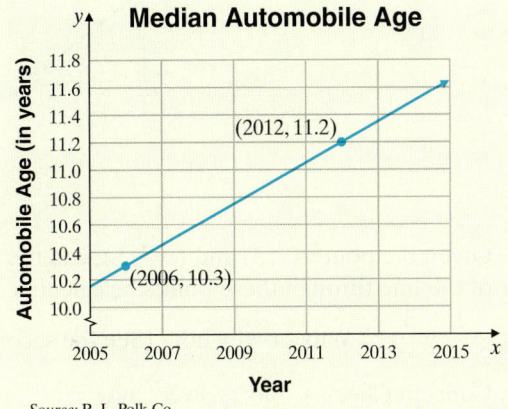

Median Automobile Age

Source: R. L. Polk Co.

○ 60. The graph below shows the total cost y (in dollars) of owning and operating a large sedan in the United States in 2013, where x is the annual number of miles driven. *m = 0.295; The total cost of owning and operating a large sedan increases $0.295 for every additional mile driven.*

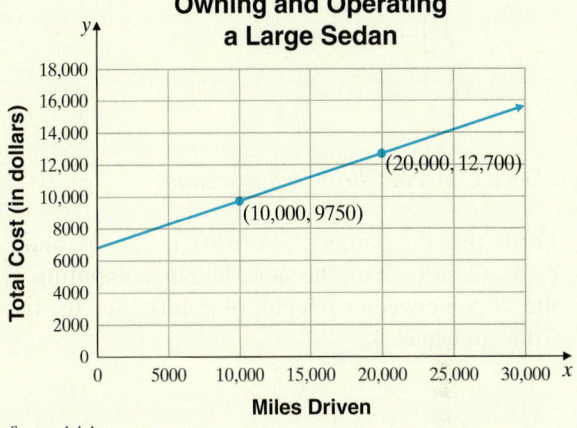

Owning and Operating a Large Sedan

Source: AAA

Review

Solve each equation for y. See Section 2.5.

61. $y - (-6) = 2(x - 4)$ $y = 2x - 14$

62. $y - 7 = -9(x - 6)$ $y = -9x + 61$

63. $y - 1 = -6(x - (-2))$ $y = -6x - 11$

64. $y - (-3) = 4(x - (-5))$ $y = 4x + 17$

Concept Extensions

Match each line with its slope.

a. $m = 0$

b. undefined slope

c. $m = 3$

d. $m = 1$

e. $m = -\dfrac{1}{2}$

f. $m = -\dfrac{3}{4}$

65. d

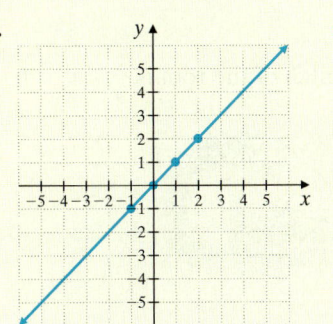

66. a

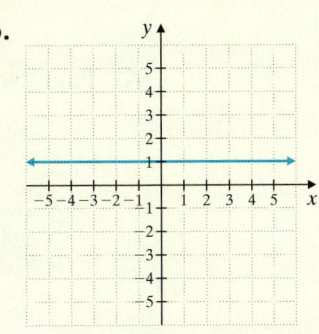

67. b

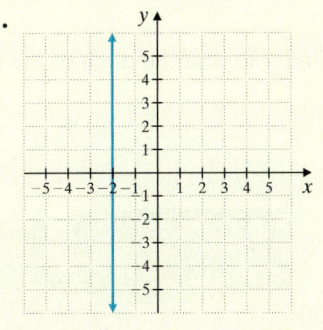

68. 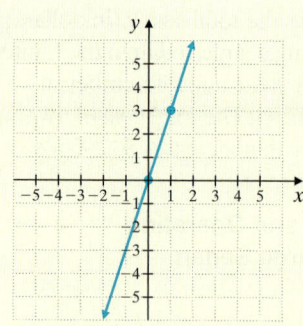   c   **69.** 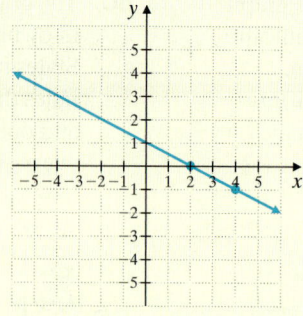   e   **70.** 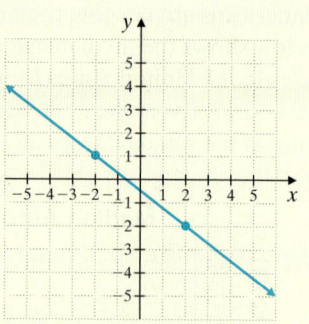   f

Solve. See a Concept Check in this section.

71. Verify that the points $(2, 1), (0, 0), (-2, -1)$, and $(-4, -2)$ are all on the same line by computing the slope between each pair of points. (See the first Concept Check.) $m = \dfrac{1}{2}$

72. Given the points $(2, 3)$ and $(-5, 1)$, can the slope of the line through these points be calculated by $\dfrac{1 - 3}{2 - (-5)}$? Why or why not? (See the second Concept Check.)   no; answers may vary

73. Write the equations of three lines parallel to $10x - 5y = -7$. (See the third Concept Check.)   answers may vary

74. Write the equations of two lines perpendicular to $10x - 5y = -7$. (See the third Concept Check.)   answers may vary

The following line graph shows the average fuel economy (in miles per gallon) of passenger automobiles produced during each of the model years shown. Use this graph to answer Exercises 75 through 80.

75. What was the average fuel economy (in miles per gallon) for automobiles produced during 2008?   31.5 mi per gal

76. Find the decrease in average fuel economy for automobiles between the years 2010 and 2011.   0.8 mi per gal

77. During which of the model year(s) shown was average fuel economy the lowest?   2003 and 2004
What was the average fuel economy for that year/those years?   29.5 mi per gal

78. During which of the model years shown was average fuel economy the highest?   2013
What was the average fuel economy for that year?   36 mi per gal

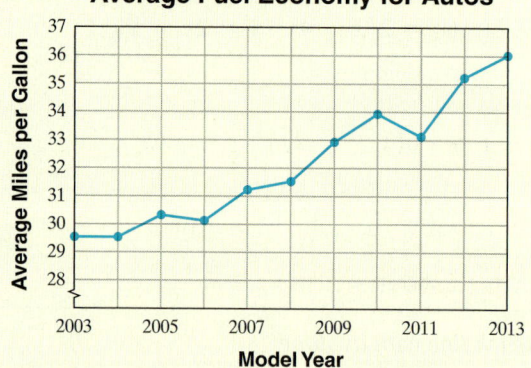

Source: U.S. Department of Transportation

79. Of the following line segments, which has the greatest slope: from 2007 to 2008, from 2009 to 2010, or from 2011 to 2012?   from 2011 to 2012

80. What line segment has a slope of 0?   2003 to 2004

81. Find x so that the pitch of the roof is $\dfrac{2}{5}$. $x = 20$

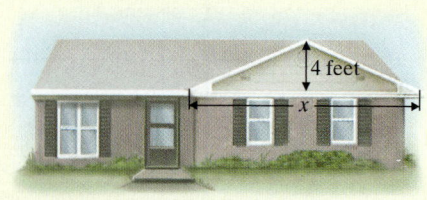

82. Find x so that the pitch of the roof is $\dfrac{1}{3}$. $x = 6$

83. There were approximately 2209 heart transplants performed in the United States in 2007. In 2012, the number of heart transplants in the United States rose to 2378. (*Source:* Organ Procurement and Transplantation Network)

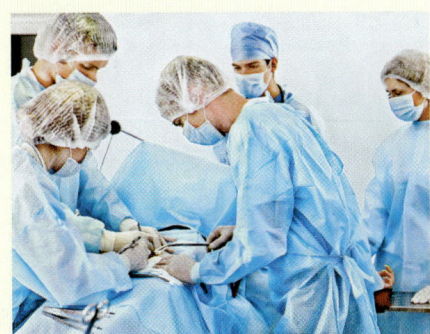

a. Write two ordered pairs of the form (year, number of heart transplants).   (2007, 2209), (2012, 2378)

b. Find the slope of the line between the two points.   33.8

c. Write a sentence explaining the meaning of the slope as a rate of change.   For the years 2007 through 2012, the number of heart transplants increased at a rate of 33.8 per year.

85. Show that the quadrilateral with vertices $(1, 3)$, $(2, 1)$, $(-4, 0)$, and $(-3, -2)$ is a parallelogram. Opposite sides are parallel since their slopes are equal, so the figure is a parallelogram.

84. The average price of an acre of U.S. cropland was $2670 in 2009. In 2013, the price of an acre rose to $4000. (*Source:* National Agricultural Statistics Service)

a. Write two ordered pairs of the form (year, price of acre). (2009, 2670), (2013, 4000)

b. Find the slope of the line through the two points.   332.5

c. Write a sentence explaining the meaning of the slope as a rate of change.   For the years 2009 through 2013, the price per acre of U.S. cropland increased at a rate of $332.50 per acre every year.

86. Show that a triangle with vertices at the points $(1, 1)$, $(-4, 4)$, and $(-3, 0)$ is a right triangle.

The slope through $(-3, 0)$ and $(1, 1)$ is $\frac{1}{4}$. The slope through $(-3, 0)$ and $(-4, 4)$ is -4. The product of the slopes is -1, so the sides are perpendicular.

Find the slope of the line through the given points.

87. $(-3.8, 1.2)$ and $(-2.2, 4.5)$ 2.0625

88. $(2.1, 6.7)$ and $(-8.3, 9.3)$ -0.25

89. $(14.3, -10.1)$ and $(9.8, -2.9)$ -1.6

90. $(2.3, 0.2)$ and $(7.9, 5.1)$ 0.875

91. The graph of $y = \frac{1}{2}x$ has a slope of $\frac{1}{2}$. The graph of $y = 3x$ has a slope of 3. The graph of $y = 5x$ has a slope of 5. Graph all three equations on a single coordinate system. As the slope becomes larger, how does the steepness of the line change?   The line becomes steeper.

92. The graph of $y = -\frac{1}{3}x + 2$ has a slope of $-\frac{1}{3}$. The graph of $y = -2x + 2$ has a slope of -2. The graph of $y = -4x + 2$ has a slope of -4. Graph all three equations on a single coordinate system. As the absolute value of the slope becomes larger, how does the steepness of the line change?   The line becomes steeper.

Integrated Review Sections 3.1–3.4

Summary on Linear Equations

Answers

1. $m = 2$

2. $m = 0$

3. $m = -\dfrac{2}{3}$

4. slope is undefined

5. see graph

6. see graph

7. see graph

8. see graph

Find the slope of each line.

1.

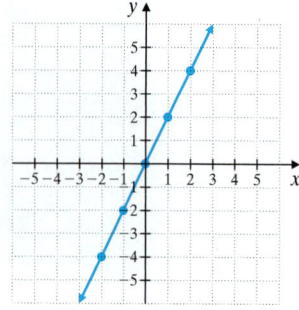

2.

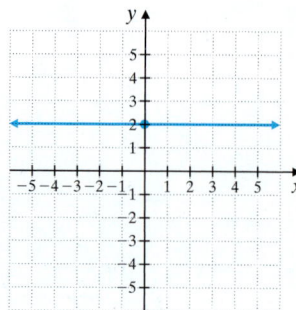

3.

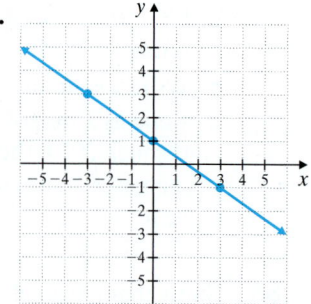

4.

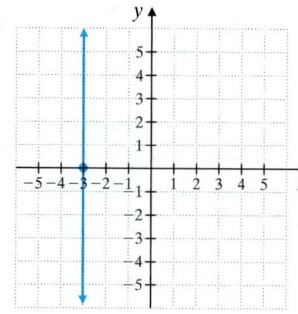

Graph each linear equation. For Exercises 11 and 12, label the intercepts.

5. $y = -2x$

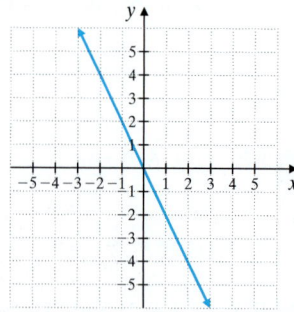

6. $x + y = 3$

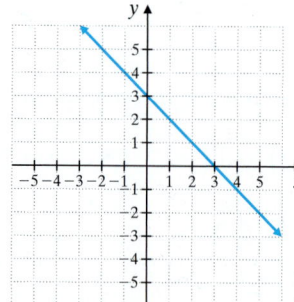

7. $x = -1$

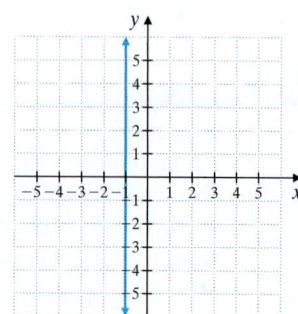

8. $y = 4$

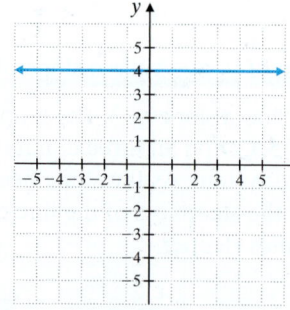

242

Copyright 2016 Pearson Education, Inc.

9. $x - 2y = 6$

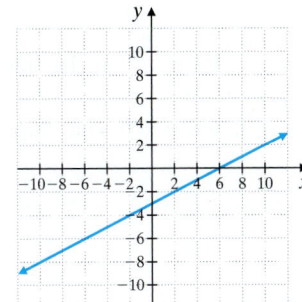

10. $y = 3x + 2$

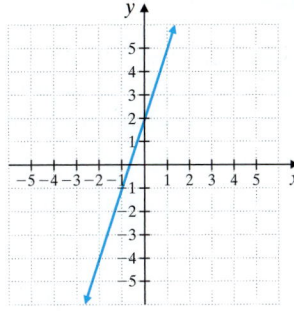

11. $y = -\dfrac{3}{4}x + 3$

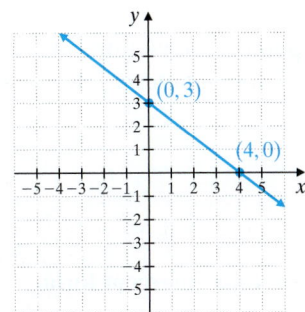

12. $5x - 2y = 8$

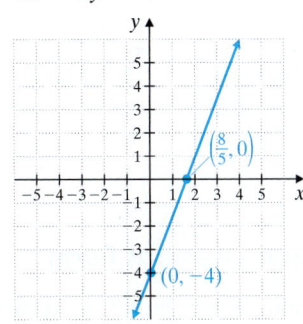

Find the slope of each line by writing the equation in slope-intercept form.

13. $y = 3x - 1$ **14.** $y = -6x + 2$ **15.** $7x + 2y = 11$ **16.** $2x - y = 0$

Find the slope of each line.

17. $x = 2$ **18.** $y = -4$

Determine whether each pair of lines is parallel, perpendicular, or neither.

19. $6x - y = 7$
$2x + 3y = 4$

20. $3x - 6y = 4$
$y = -2x$

21. Yogurt is an ever more popular food item. In 2008, U.S. production of yogurt stood at approximately 3600 million pounds. In 2012, this number rose to 4416 million pounds of yogurt.

   a. Write two ordered pairs of the form (year, millions of pounds of yogurt produced).

   b. Find the slope of the line between these two points.

   c. Write a sentence explaining the meaning of the slope as a rate of change.

---

9. see graph

10. see graph

11. see graph

12. see graph

13. $m = 3$

14. $m = -6$

15. $m = -\dfrac{7}{2}$

16. $m = 2$

17. undefined slope

18. $m = 0$

19. neither

20. perpendicular

21. a. (2008, 3600); (2012, 4416)

   b. 204

   c. For the years 2008 through 2012, the amount of yogurt produced increased at a rate of 204 million pounds per year.

Objectives

A Use the Slope-Intercept Form to Graph a Linear Equation.

B Use the Slope-Intercept Form to Write an Equation of a Line.

C Use the Point-Slope Form to Find an Equation of a Line Given Its Slope and a Point of the Line.

D Use the Point-Slope Form to Find an Equation of a Line Given Two Points of the Line.

E Use the Point-Slope Form to Solve Problems.

Teaching Tip

If you are using graphing calculators, have students make a conjecture about *b* after exploring the following equations on a graphing calculator.

$y = 3x + 0$

$y = 3x + 5$

$y = 3x + 2$

$y = 3x - 5$

Practice 1

Use the slope-intercept form to graph the equation

$$y = \frac{2}{3}x - 4.$$

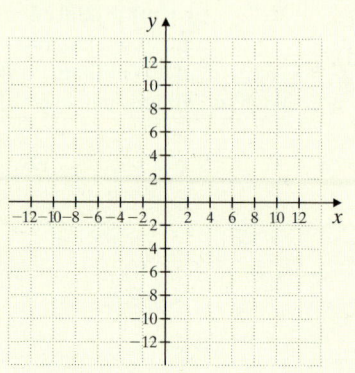

Answer

1. See page 245.

244

We know that when a linear equation is solved for y, the coefficient of x is the slope of the line. For example, the slope of the line whose equation is $y = 3x + 1$ is 3. In the equation $y = 3x + 1$, what does 1 represent? To find out, let $x = 0$ and watch what happens.

$$y = 3x + 1$$
$$y = 3 \cdot 0 + 1 \quad \text{Let } x = 0.$$
$$y = 1$$

We now have the ordered pair $(0, 1)$, which means that 1 is the y-intercept.

This is true in general. To see this, let $x = 0$ and solve for y in $y = mx + b$.

$$y = m \cdot 0 + b \quad \text{Let } x = 0.$$
$$y = b$$

We obtain the ordered pair $(0, b)$, which means that point is the y-intercept.

The form $y = mx + b$ is appropriately called the *slope-intercept form* of a linear equation.

slope y-intercept is $(0, b)$.

Slope-Intercept Form

When a linear equation in two variables is written in **slope-intercept form,**

$$y = mx + b$$

slope $(0, b)$, y-intercept

then m is the slope of the line and $(0, b)$ is the y-intercept of the line.

Objective **A** Using the Slope-Intercept Form to Graph an Equation

We also can use the slope-intercept form of the equation of a line to graph a linear equation.

Example 1 Use the slope-intercept form to graph the equation

$$y = \frac{3}{5}x - 2.$$

Solution: Since the equation $y = \frac{3}{5}x - 2$ is written in slope-intercept form $y = mx + b$, the slope of its graph is $\frac{3}{5}$ and the y-intercept is $(0, -2)$. To graph this equation, we begin by plotting the point $(0, -2)$. From this point, we can find another point of the graph by using the slope $\frac{3}{5}$ and recalling that slope is $\frac{\text{rise}}{\text{run}}$. We start at the y-intercept and move 3 units up since the numerator of the slope is 3; then we move 5 units to the right since the denominator of the slope is 5. We stop at the point $(5, 1)$. The line through $(0, -2)$ and $(5, 1)$ is the graph of $y = \frac{3}{5}x - 2$.

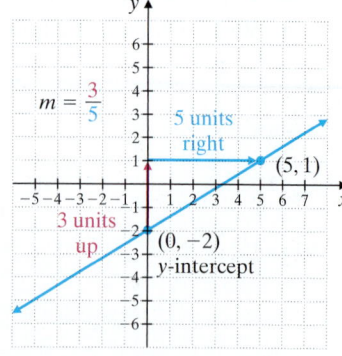

■ **Work Practice 1**

Example 2 Use the slope-intercept form to graph the equation $4x + y = 1$.

Solution: First we write the given equation in slope-intercept form.

$$4x + y = 1$$
$$y = -4x + 1$$

The graph of this equation will have slope -4 and y-intercept $(0, 1)$. To graph this line, we first plot the point $(0, 1)$. To find another point of the graph, we use the slope -4, which can be written as $\dfrac{-4}{1}\left(\dfrac{4}{-1} \text{ could also be used}\right)$. We start at the point $(0, 1)$ and move 4 units down (since the numerator of the slope is -4) and then 1 unit to the right (since the denominator of the slope is 1).

We arrive at the point $(1, -3)$. The line through $(0, 1)$ and $(1, -3)$ is the graph of $4x + y = 1$.

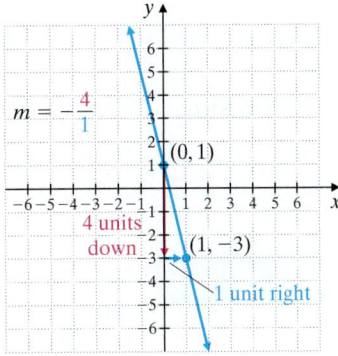

■ **Work Practice 2**

Helpful Hint

In Example 2, if we interpret the slope of -4 as $\dfrac{4}{-1}$, we arrive at $(-1, 5)$ for a second point. Notice that this point is also on the line.

Objective B Using the Slope-Intercept Form to Write an Equation ▶

The slope-intercept form can also be used to write the equation of a line when we know its slope and y-intercept.

Example 3 Find an equation of the line with y-intercept $(0, -3)$ and slope of $\dfrac{1}{4}$.

Solution: We are given the slope and the y-intercept. We let $m = \dfrac{1}{4}$ and $b = -3$ and write the equation in slope-intercept form, $y = mx + b$.

$$y = mx + b$$
$$y = \frac{1}{4}x + (-3) \quad \text{Let } m = \frac{1}{4} \text{ and } b = -3.$$
$$y = \frac{1}{4}x - 3 \quad \text{Simplify.}$$

■ **Work Practice 3**

Objective C Writing an Equation Given Its Slope and a Point ▶

Thus far, we have seen that we can write an equation of a line if we know its slope and y-intercept. We can also write an equation of a line if we know its slope and any

---

Practice 2

Use the slope-intercept form to graph $3x + y = 2$.

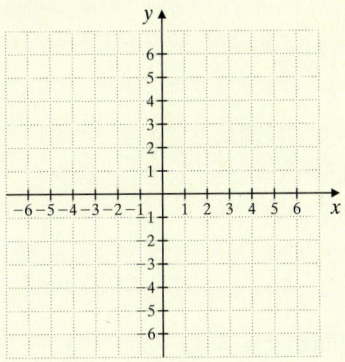

Practice 3

Find an equation of the line with y-intercept $(0, -2)$ and slope of $\dfrac{3}{5}$.

Answers

1.

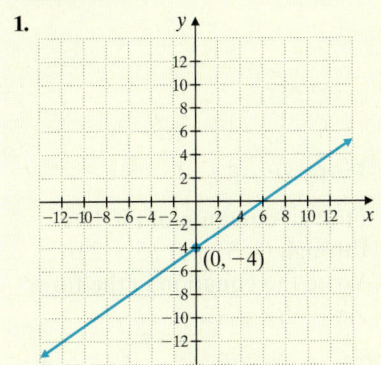

2.

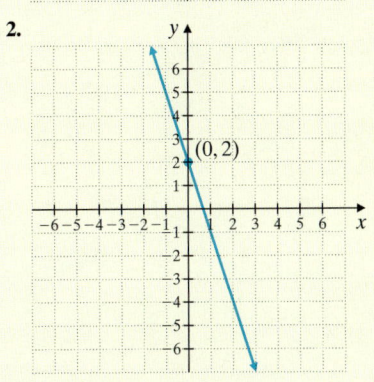

3. $y = \dfrac{3}{5}x - 2$

point on the line. To see how we do this, let m represent slope and (x_1, y_1) represent a point on the line. Then if (x, y) is any other point on the line, we have that

$$\frac{y - y_1}{x - x_1} = m$$

$$y - y_1 = m(x - x_1) \quad \text{Multiply both sides by } (x - x_1).$$

slope

This is the *point-slope form* of the equation of a line.

> ### Point-Slope Form of the Equation of a Line
>
> The **point-slope form** of the equation of a line is $y - y_1 = m(x - x_1)$, where m is the slope of the line and (x_1, y_1) is a point on the line.

Practice 4

Find an equation of the line with slope -3 that passes through $(2, -4)$. Write the equation in slope-intercept form, $y = mx + b$.

Example 4 Find an equation of the line with slope -2 that passes through $(-1, 5)$. Write the equation in slope-intercept form, $y = mx + b$, and in standard form, $Ax + By = C$.

Solution: Since the slope and a point on the line are given, we use point-slope form, $y - y_1 = m(x - x_1)$, to write the equation. Let $m = -2$ and $(-1, 5) = (x_1, y_1)$.

$$y - y_1 = m(x - x_1)$$
$$y - 5 = -2[x - (-1)] \quad \text{Let } m = -2 \text{ and } (x_1, y_1) = (-1, 5).$$
$$y - 5 = -2(x + 1) \quad \text{Simplify.}$$
$$y - 5 = -2x - 2 \quad \text{Use the distributive property.}$$

To write the equation in slope-intercept form, $y = mx + b$, we simply solve the equation for y. To do this, we add 5 to both sides.

$$y - 5 = -2x - 2$$
$$y = -2x + 3 \quad \text{Slope-intercept form}$$
$$2x + y = 3 \quad \text{Add } 2x \text{ to both sides and we have standard form.}$$

■ **Work Practice 4**

Objective D Writing an Equation Given Two Points ▶

We can also find the equation of a line when we are given any two points of the line.

Practice 5

Find an equation of the line through $(1, 3)$ and $(5, -2)$. Write the equation in the form $Ax + By = C$.

Example 5 Find an equation of the line through $(2, 5)$ and $(-3, 4)$. Write the equation in the form $Ax + By = C$.

Solution: First, use the two given points to find the slope of the line.

$$m = \frac{4 - 5}{-3 - 2} = \frac{-1}{-5} = \frac{1}{5}$$

Next we use the slope $\frac{1}{5}$ and either one of the given points to write the equation in point-slope form. We use $(2, 5)$. Let $x_1 = 2$, $y_1 = 5$, and $m = \frac{1}{5}$.

$$y - y_1 = m(x - x_1) \quad \text{Use point-slope form.}$$
$$y - 5 = \frac{1}{5}(x - 2) \quad \text{Let } x_1 = 2, y_1 = 5, \text{ and } m = \frac{1}{5}.$$
$$5(y - 5) = 5 \cdot \frac{1}{5}(x - 2) \quad \text{Multiply both sides by 5 to clear fractions.}$$
$$5y - 25 = x - 2 \quad \text{Use the distributive property and simplify.}$$
$$-x + 5y - 25 = -2 \quad \text{Subtract } x \text{ from both sides.}$$
$$-x + 5y = 23 \quad \text{Add 25 to both sides.}$$

■ **Work Practice 5**

Answers

4. $y = -3x + 2$ **5.** $5x + 4y = 17$

Helpful Hint

When you multiply both sides of the equation from Example 5, $-x + 5y = 23$, by -1, it becomes $x - 5y = -23$.

Both $-x + 5y = 23$ and $x - 5y = -23$ are in the form $Ax + By = C$, and both are equations of the same line.

Objective E Using the Point-Slope Form to Solve Problems

Problems occurring in many fields can be modeled by linear equations in two variables. The next example is from the field of marketing and shows how consumer demand of a product depends on the price of the product.

Example 6 The Whammo Company has learned that by pricing a newly released Frisbee at $6, sales will reach 2000 Frisbees per day. Raising the price to $8 will cause the sales to fall to 1500 Frisbees per day.

a. Assume that the relationship between sales price and number of Frisbees sold is linear, and write an equation describing this relationship. Write the equation in slope-intercept form. Use ordered pairs of the form (sales price, number sold).

b. Predict the daily sales of Frisbees if the price is $7.50.

Solution:

a. We use the given information and write two ordered pairs. Our ordered pairs are $(6, 2000)$ and $(8, 1500)$. To use the point-slope form to write an equation, we find the slope of the line that contains these points.

$$m = \frac{2000 - 1500}{6 - 8} = \frac{500}{-2} = -250$$

Next we use the slope and either one of the points to write the equation in point-slope form. We use $(6, 2000)$.

$y - y_1 = m(x - x_1)$	Use point-slope form.
$y - 2000 = -250(x - 6)$	Let $x_1 = 6$, $y_1 = 2000$, and $m = -250$.
$y - 2000 = -250x + 1500$	Use the distributive property.
$y = -250x + 3500$	Write in slope-intercept form.

b. To predict the sales if the price is $7.50, we find y when $x = 7.50$.

$y = -250x + 3500$	
$y = -250(7.50) + 3500$	Let $x = 7.50$.
$y = -1875 + 3500$	
$y = 1625$	

If the price is $7.50, sales will reach 1625 Frisbees per day.

■ **Work Practice 6**

Practice 6

The Pool Entertainment Company learned that by pricing a new pool toy at $10, local sales will reach 200 a week. Lowering the price to $9 will cause sales to rise to 250 a week.

a. Assume that the relationship between sales price and number of toys sold is linear, and write an equation describing this relationship. Write the equation in slope-intercept form. Use ordered pairs of the form (sales price, number sold).

b. Predict the weekly sales of the toy if the price is $7.50.

Answers

6. a. $y = -50x + 700$ **b.** 325

Teaching Tip

Have students write their strategies for various scenarios. How
would they find the equation if
they knew . . .

• the slope and y-intercept?
• two points?
• a point and the slope?
• a point and the y-intercept?
• the x-intercept and the
  y-intercept?

How much information must be
given in order to determine a line?

We also could have solved Example 6 by using ordered pairs of the form
(number sold, sales price).

Here is a summary of our discussion on linear equations thus far.

Forms of Linear Equations

$Ax + By = C$	**Standard form** of a linear equation. A and B are not both 0.
$y = mx + b$	**Slope-intercept form** of a linear equation. The slope is m, and the y-intercept is $(0, b)$.
$y - y_1 = m(x - x_1)$	**Point-slope form** of a linear equation. The slope is m, and (x_1, y_1) is a point on the line.
$y = c$	**Horizontal line** The slope is 0 and the y-intercept is $(0, c)$.
$x = c$	**Vertical line** The slope is undefined, and the x-intercept is $(c, 0)$.

Parallel and Perpendicular Lines

Nonvertical parallel lines have the same slope.
The product of the slopes of two nonvertical perpendicular lines is -1.

 Calculator Explorations   Graphing

A graphing calculator is a very useful tool for discovering
patterns. To discover the change in the graph of a linear
equation caused by a change in slope, try the following:
Use a standard window and graph a linear equation in
the form $y = mx + b$. Recall that the graph of such an
equation will have slope m and y-intercept $(0, b)$.

First graph $y = x + 3$. To do so, press the $\boxed{Y =}$ key
and enter $Y_1 = x + 3$. Notice that this graph has slope
1 and that the y-intercept is 3. Next, on the same set of
axes, graph $y = 2x + 3$ and $y = 3x + 3$ by pressing $\boxed{Y =}$
and entering $Y_2 = 2x + 3$ and $Y_3 = 3x + 3$.

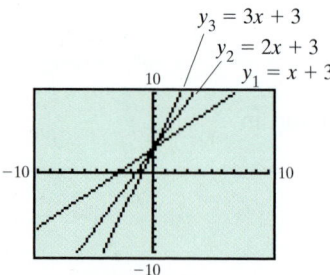

Notice the difference in the graph of each equation
as the slope changes from 1 to 2 to 3. How would the
graph of $y = 5x + 3$ appear? To see the change in the
graph caused by a change to negative slope, try graphing $y = -x + 3$, $y = -2x + 3$, and $y = -3x + 3$ on the
same set of axes.

*Use a graphing calculator to graph the following equa-
tions. For each exercise, graph the first equation and use*
*its graph to predict the appearance of the other equa-
tions. Then graph the other equations on the same set of
axes and check your prediction.*

1. $y = x; y = 6x, y = -6x$

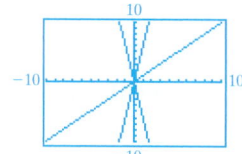

2. $y = -x; y = -5x, y = -10x$

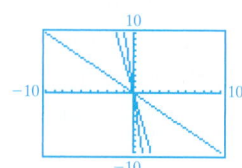

3. $y = \frac{1}{2}x + 2; y = \frac{3}{4}x + 2, y = x + 2$

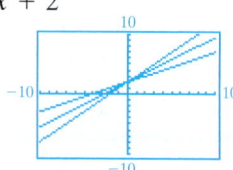

4. $y = x + 1; y = \frac{5}{4}x + 1, y = \frac{5}{2}x + 1$

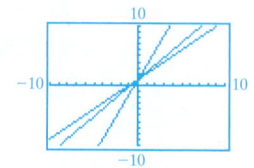

Vocabulary, Readiness & Video Check

Use the choices below to fill in each blank. Some choices may be used more than once and some not at all.

b	(y_1, x_1)	point-slope	vertical	standard
m	(x_1, y_1)	slope-intercept	horizontal	

1. The form $y = mx + b$ is called ___slope-intercept___ form. When a linear equation in two variables is written in this form, ____m____ is the slope of its graph and (0, ____b____) is its y-intercept.

2. The form $y - y_1 = m(x - x_1)$ is called ____point-slope____ form. When a linear equation in two variables is written in this form, ____m____ is the slope of its graph and ____(x_1, y_1)____ is a point on the graph.

For Exercises 3 through 6, identify the form that the linear equation in two variables is written in. For Exercises 7 and 8, identify the appearance of the graph of the equation.

3. $y - 7 = 4(x + 3)$; ____point-slope____ form

4. $5x - 9y = 11$; ____standard____ form

5. $y = \dfrac{3}{4}x - \dfrac{1}{3}$; ____slope-intercept____ form

6. $y + 2 = \dfrac{-1}{3}(x - 2)$ ____point-slope____ form

7. $y = \dfrac{1}{2}$; ____horizontal____ line

8. $x = -17$; ____vertical____ line

Martin-Gay Interactive Videos

See Video 3.5 🍎

Watch the section lecture video and answer the following questions.

Objective A 9. We can use the slope-intercept form to graph a line. Complete these statements based on ▤ Example 1:
Start by graphing the _____.
From this point, find another point by applying the slope — if necessary, rewrite the slope as a(n) _____. ▶

Objective B 10. In ▤ Example 3, what is the y-intercept? ▶

Objective C 11. In ▤ Example 4, we use the point-slope form to find the equation of a line given the slope and a point. How do we then write this equation in standard form? ▶

Objective D 12. The lecture before ▤ Example 5 discusses how to find the equation of a line given two points. Is there any circumstance when we might want to use the slope-intercept form to find the equation of a line given two points? If so, when? ▶

Objective E 13. In ▤ Example 8, we are told to use ordered pairs of the form (time, speed). Explain why it is important to keep track of how we define our ordered pairs and/or our variables. ▶

See video answer section.

3.5 Exercise Set MyMathLab®

Objective A *Use the slope-intercept form to graph each equation. See Examples 1 and 2.*

1. $y = 2x + 1$

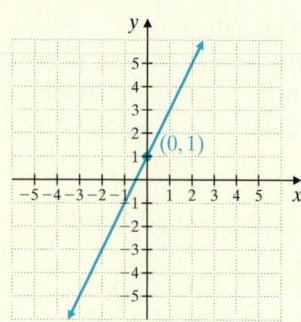

2. $y = -4x - 1$

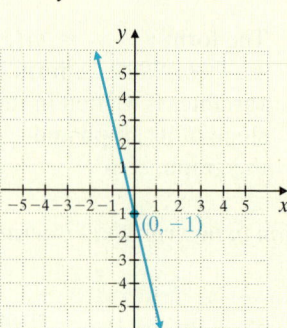

3. $y = \frac{2}{3}x + 5$

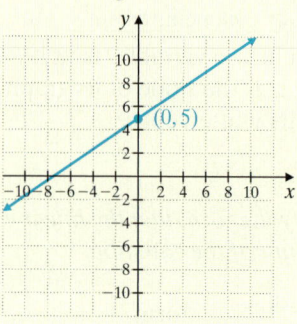

4. $y = \frac{1}{4}x - 3$

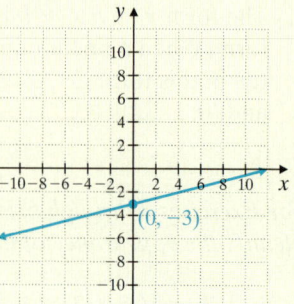

5. $y = -5x$

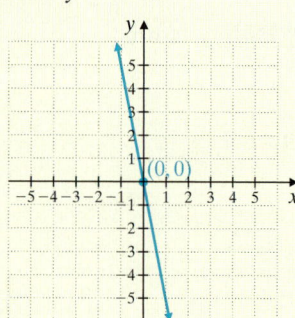

6. $y = -6x$

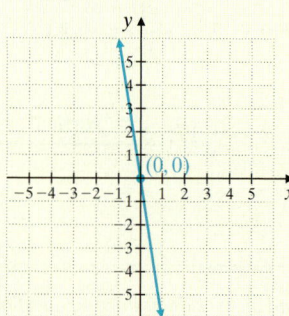

7. $4x + y = 6$

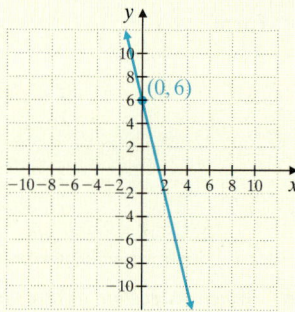

8. $-3x + y = 2$

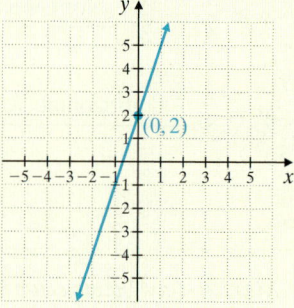

9. $4x - 7y = -14$

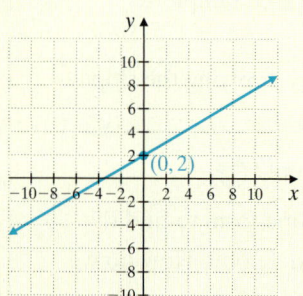

10. $3x - 4y = 4$

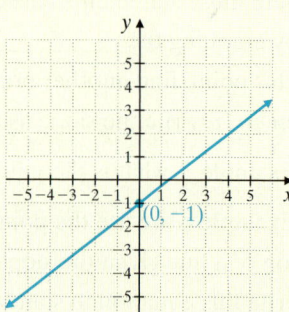

11. $x = \frac{5}{4}y$

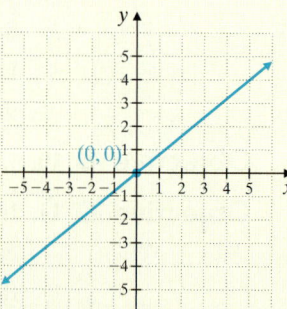

12. $x = \frac{3}{2}y$

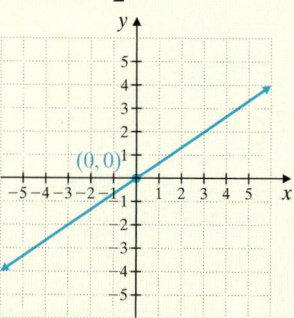

Objective B *Write an equation of the line with each given slope, m, and y-intercept, (0, b). See Example 3.*

13. $m = 5, b = 3$ $y = 5x + 3$

14. $m = -3, b = -3$ $y = -3x - 3$

15. $m = -4, b = -\frac{1}{6}$ $y = -4x - \frac{1}{6}$

16. $m = 2, b = \frac{3}{4}$ $y = 2x + \frac{3}{4}$

17. $m = \frac{2}{3}, b = 0$ $y = \frac{2}{3}x$

18. $m = -\frac{4}{5}, b = 0$ $y = -\frac{4}{5}x$

19. $m = 0, b = -8$ $y = -8$

20. $m = 0, b = -2$ $y = -2$

21. $m = -\frac{1}{5}, b = \frac{1}{9}$ $y = -\frac{1}{5}x + \frac{1}{9}$

22. $m = \frac{1}{2}, b = -\frac{1}{3}$ $y = \frac{1}{2}x - \frac{1}{3}$

Objective C *Find an equation of each line with the given slope that passes through the given point. Write the equation in the form Ax + By = C. See Example 4.*

23. $m = 6;$ $(2, 2)$
$-6x + y = -10$

24. $m = 4;$ $(1, 3)$
$-4x + y = -1$

25. $m = -8;$ $(-1, -5)$
$8x + y = -13$

26. $m = -2;$ $(-11, -12)$
$2x + y = -34$

27. $m = \dfrac{3}{2}$; $(5, -6)$ $3x - 2y = 27$

28. $m = \dfrac{2}{3}$; $(-8, 9)$ $2x - 3y = -43$

29. $m = -\dfrac{1}{2}$; $(-3, 0)$ $x + 2y = -3$

30. $m = -\dfrac{1}{5}$; $(4, 0)$ $x + 5y = 4$

Objective D *Find an equation of the line passing through each pair of points. Write the equation in the form $Ax + By = C$. See Example 5.*

31. $(3, 2)$ and $(5, 6)$ $2x - y = 4$

32. $(6, 2)$ and $(8, 8)$ $3x - y = 16$

33. $(-1, 3)$ and $(-2, -5)$ $8x - y = -11$

34. $(-4, 0)$ and $(6, -1)$ $x + 10y = -4$

▶ **35.** $(2, 3)$ and $(-1, -1)$ $4x - 3y = -1$

36. $(7, 10)$ and $(-1, -1)$ $11x - 8y = -3$

37. $(0, 0)$ and $\left(-\dfrac{1}{8}, \dfrac{1}{13}\right)$ $8x + 13y = 0$

38. $(0, 0)$ and $\left(-\dfrac{1}{2}, \dfrac{1}{3}\right)$ $2x + 3y = 0$

Objectives A C D Mixed Practice *See Examples 1, 4, and 5. Find an equation of each line described. Write each equation in slope-intercept form when possible.*

39. With slope $-\dfrac{1}{2}$, through $\left(0, \dfrac{5}{3}\right)$ $y = -\dfrac{1}{2}x + \dfrac{5}{3}$

40. With slope $\dfrac{5}{7}$, through $(0, -3)$ $y = \dfrac{5}{7}x - 3$

41. Through $(10, 7)$ and $(7, 10)$ $y = -x + 17$

42. Through $(5, -6)$ and $(-6, 5)$ $y = -x - 1$

▶ **43.** With undefined slope, through $\left(-\dfrac{3}{4}, 1\right)$ $x = -\dfrac{3}{4}$

44. With slope 0, through $(6.7, 12.1)$ $y = 12.1$

45. Slope 1, through $(-7, 9)$ $y = x + 16$

46. Slope 5, through $(6, -8)$ $y = 5x - 38$

47. Slope -5, y-intercept $(0, 7)$ $y = -5x + 7$

48. Slope -2, y-intercept $(0, -4)$ $y = -2x - 4$

▶ **49.** Through $(1, 2)$, parallel to $y = 5$ $y = 2$

50. Through $(1, -5)$, parallel to the y-axis $x = 1$

51. Through $(2, 3)$ and $(0, 0)$ $y = \dfrac{3}{2}x$

52. Through $(4, 7)$ and $(0, 0)$ $y = \dfrac{7}{4}x$

53. Through $(-2, -3)$, perpendicular to the y-axis $y = -3$

54. Through $(0, 12)$, perpendicular to the x-axis $x = 0$

55. Slope $-\dfrac{4}{7}$, through $(-1, -2)$ $y = -\dfrac{4}{7}x - \dfrac{18}{7}$

56. Slope $-\dfrac{3}{5}$, through $(4, 4)$ $y = -\dfrac{3}{5}x + \dfrac{32}{5}$

Objective E *Solve. Assume each exercise describes a linear relationship. Write the equations in slope-intercept form. See Example 6.*

57. In 2007, a total of 370 million magazines were sold in the United States. By 2012, this number was 312 million. (*Source:* MPA—The Association of Magazine Media)

   a. Write two ordered pairs of the form (years after 2007, millions of magazines sold) for this situation. $(0, 370), (5, 312)$

   b. Assume the relationship between years after 2007 and millions of magazines sold is linear over this period. Use the ordered pairs from part **a** to write an equation for the line relating year after 2007 to millions of magazines sold. $y = -11.6x + 370$

   c. Use the linear equation in part **b** to estimate the millions of magazines sold in 2010.   335.2 million magazines

58. In 2008, crude oil field production in the United States was 1830 thousand barrels. In 2012, U.S. crude oil field production increased to 2374 thousand barrels. (*Source:* Energy Information Administration)

   a. Write two ordered pairs of the form (years after 2008, crude oil production). $(0, 1830), (4, 2374)$

   b. Assume the relationship between years after 2008 and crude oil production is linear over this period. Use the ordered pairs from part **a** to write an equation of the line relating years after 2008 to crude oil production. $y = 136x + 1830$

   c. Use the linear equation from part **b** to estimate crude oil production in the United States in 2015, if this trend were to continue.   2782 million barrels

59. A rock is dropped from the top of a 400-foot cliff. After 1 second, the rock is traveling 32 feet per second. After 3 seconds, the rock is traveling 96 feet per second.

400 feet

a. Assume that the relationship between time and speed is linear and write an equation describing this relationship. Use ordered pairs of the form (time, speed). $s = 32t$

b. Use this equation to determine the speed of the rock 4 seconds after it is dropped.   128 ft/sec

61. In 2008, there were approximately 314,000 hybrid vehicles sold in the United States. In 2012, there were approximately 434,000 such vehicles sold. (*Source:* HybridCars.com)

a. Write an equation describing the relationship between time and the number of hybrid vehicles sold. Use ordered pairs of the form (years past 2008, number of vehicles sold). $y = 30,000x + 314,000$

b. Use this equation to estimate the number of hybrid sales in 2014.   494,000 vehicles

63. In 2007, there were 5545 indoor cinema sites in the United States. In 2012, there were approximately 5320 indoor cinema sites. (*Source:* National Association of Theater Owners)

a. Write an equation describing this relationship. Use ordered pairs of the form (years past 2007, number of indoor cinema sites). $y = -45x + 5545$

b. Use this equation to predict the number of indoor cinema sites in 2015.   5185 indoor cinema sites

60. A Hawaiian fruit company is studying the sales of a pineapple sauce to see if this product is to be continued. At the end of its first year, profits on this product amounted to $30,000. At the end of the fourth year, profits were $66,000.

a. Assume that the relationship between years on the market and profit is linear and write an equation describing this relationship. Use ordered pairs of the form (years on the market, profit). $p = 12,000t + 18,000$

b. Use this equation to predict the profit at the end of 7 years. $102,000

62. In 2008, there were approximately 945 thousand restaurants in the United States. In 2012, there were 980 thousand restaurants. (*Source:* National Restaurant Association)

a. Write an equation describing the relationship between time and the number of restaurants. Use ordered pairs of the form (years past 2008, number of restaurants in thousands). $y = 8.75x + 945$

b. Use this equation to predict the number of eating establishments in 2016.   1015 thousand eating establishments

64. In 2010, the U.S. population per square mile of land area was approximately 87.4. In 2000, this person-per-square-mile population was 79.7. (*Source:* U.S. Census Bureau)

a. Write an equation describing the relationship between year and person per square mile. Use ordered pairs of the form (years past 2000, person per square mile). $y = 0.77x + 79.7$

b. Use this equation to predict the person-per-square-mile population in 2016.   92.02 persons per sq mi

65. The Pool Fun Company has learned that, by pricing a newly released Fun Noodle at $3, sales will reach 10,000 Fun Noodles per day during the summer. Raising the price to $5 will cause the sales to fall to 8000 Fun Noodles per day.

   a. Assume that the relationship between sales price and number of Fun Noodles sold is linear and write an equation describing this relationship. Use ordered pairs of the form (sales price, number sold). $S = -1000p + 13,000$

   b. Predict the daily sales of Fun Noodles if the price is $3.50.   9500 Fun Noodles

66. The value of a building bought in 1995 may be depreciated (or decreased) as time passes for income tax purposes. Seven years after the building was bought, this value was $225,000 and 12 years after it was bought, this value was $195,000.

   a. If the relationship between number of years past 1995 and the depreciated value of the building is linear, write an equation describing this relationship. Use ordered pairs of the form (years past 1995, value of building). $V = -6000x + 267,000$

   b. Use this equation to estimate the depreciated value of the building in 2013. $159,000

Review

Find the value of $x^2 - 3x + 1$ for each given value of x. See Section 1.3.

67. 2   −1   **68.** 5   11   **69.** −1   5   **70.** −3   19

Concept Extensions

Match each linear equation with its graph.

71. $y = 2x + 1$ b   **72.** $y = -x + 1$ c   **73.** $y = -3x - 2$ d   **74.** $y = \frac{5}{3}x - 2$ a

a. 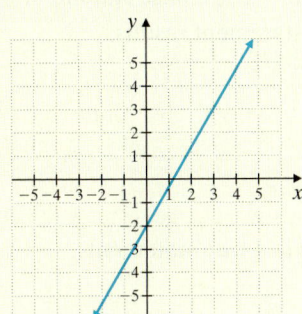   **b.** 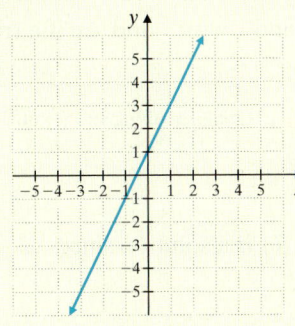   **c.** 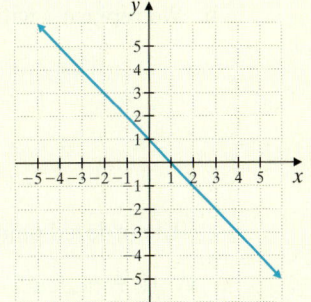   **d.**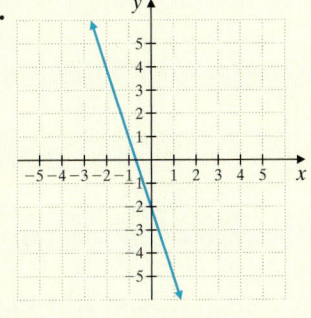

75. Write an equation in standard form of the line that contains the point $(-2, 4)$ and has the same slope as the line $y = 2x + 5$. $2x - y = -8$

76. Write an equation in standard form of the line that contains the point $(3, 0)$ and has the same slope as the line $y = -3x - 1$. $3x + y = 9$

△**77.** Write an equation in standard form of the line that contains the point $(-1, 2)$ and is

   a. parallel to the line $y = 3x - 1$. $3x - y = -5$

   b. perpendicular to the line $y = 3x - 1$.
     $x + 3y = 5$

△**78.** Write an equation in standard form of the line that contains the point $(4, 0)$ and is

   a. parallel to the line $y = -2x + 3$. $2x + y = 8$

   b. perpendicular to the line $y = -2x + 3$.
     $x - 2y = 4$

Graphing Linear Inequalities in Two Variables

Objectives

A Determine Whether an Ordered Pair Is a Solution of a Linear Inequality in Two Variables.

B Graph a Linear Inequality in Two Variables.

Recall that a linear equation in two variables is an equation that can be written in the form $Ax + By = C$, where A, B, and C are real numbers and A and B are not both 0. A **linear inequality in two variables** is an inequality that can be written in one of the forms

$$Ax + By < C \qquad Ax + By \leq C$$
$$Ax + By > C \qquad Ax + By \geq C$$

where A, B, and C are real numbers and A and B are not both 0.

Objective A Determining Solutions of Linear Inequalities in Two Variables

Just as for linear equations in x and y, an ordered pair is a **solution** of an inequality in x and y if replacing the variables with the coordinates of the ordered pair results in a true statement.

Practice 1

Determine whether each ordered pair is a solution of $x - 4y > 8$.

a. $(-3, 2)$ **b.** $(9, 0)$

Example 1 Determine whether each ordered pair is a solution of the inequality $2x - y < 6$.

a. $(5, -1)$ **b.** $(2, 7)$

Solution:

a. We replace x with 5 and y with -1 and see if a true statement results.

$$2x - y < 6$$
$$2(5) - (-1) < 6 \qquad \text{Replace } x \text{ with 5 and } y \text{ with } -1.$$
$$10 + 1 < 6$$
$$11 < 6 \qquad \text{False}$$

The ordered pair $(5, -1)$ is not a solution since $11 < 6$ is a false statement.

b. We replace x with 2 and y with 7 and see if a true statement results.

$$2x - y < 6$$
$$2(2) - (7) < 6 \qquad \text{Replace } x \text{ with 2 and } y \text{ with 7.}$$
$$4 - 7 < 6$$
$$-3 < 6 \qquad \text{True}$$

The ordered pair $(2, 7)$ is a solution since $-3 < 6$ is a true statement.

■ **Work Practice 1**

Objective B Graphing Linear Inequalities in Two Variables

The linear equation $x - y = 1$ is graphed next. Recall that all points on the line correspond to ordered pairs that satisfy the equation $x - y = 1$.

Notice that the line defined by $x - y = 1$ divides the rectangular coordinate system plane into 2 sides. All points on one side of the line satisfy the inequality $x - y < 1$, and all points on the other side satisfy the inequality $x - y > 1$. The graph on the next page shows a few examples of this.

Answers

1. a. no **b.** yes

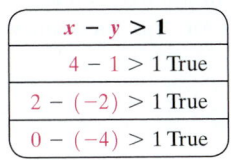

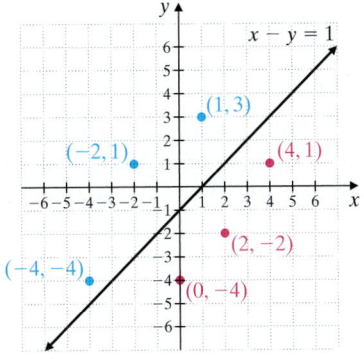

The graph of $x - y < 1$ is the region shaded blue and the graph of $x - y > 1$ is the region shaded red below.

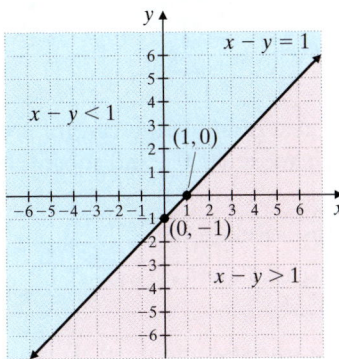

The region to the left of the line and the region to the right of the line are called **half-planes.** Every line divides the plane (similar to a sheet of paper extending indefinitely in all directions) into two half-planes; the line is called the **boundary.**

Recall that the inequality $x - y \le 1$ means

$$x - y = 1 \quad \text{or} \quad x - y < 1$$

Thus, the graph of $x - y \le 1$ is the half-plane $x - y < 1$ along with the boundary line $x - y = 1$.

To Graph a Linear Inequality in Two Variables

Step 1: Graph the boundary line found by replacing the inequality sign with an equal sign. If the inequality sign is $>$ or $<$, graph a dashed boundary line (indicating that the points on the line are not solutions of the inequality). If the inequality sign is $\ge$ or $\le$, graph a solid boundary line (indicating that the points on the line are solutions of the inequality).

Step 2: Choose a point *not* on the boundary line as a test point. Substitute the coordinates of this test point into the *original* inequality.

Step 3: If a true statement is obtained in Step 2, shade the half-plane that contains the test point. If a false statement is obtained, shade the half-plane that does not contain the test point.

Practice 2
Graph: $x - y > 3$

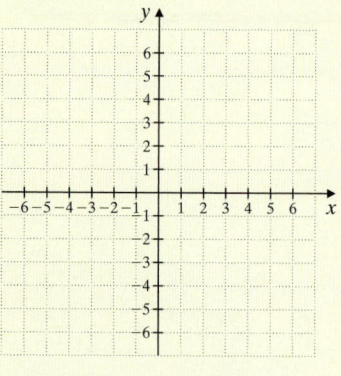

Teaching Tip

Have students use the graph to write the inequality in Example 2 with y isolated on one side of the inequality. Notice that if you look at any vertical slice, the y-values shaded are less than (or below) the line graphed, so the inequality is $y < -x + 7$.

Answer

2.

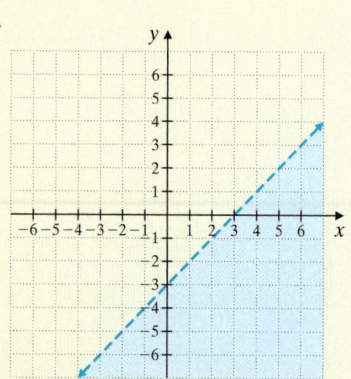

✓**Concept Check Answers**

a. no **b.** yes **c.** yes

Example 2 Graph: $x + y < 7$

Solution:

Step 1: First we graph the boundary line by graphing the equation $x + y = 7$. We graph this boundary as a *dashed line* because the inequality sign is $<$, and thus the points on the line are not solutions of the inequality $x + y < 7$.

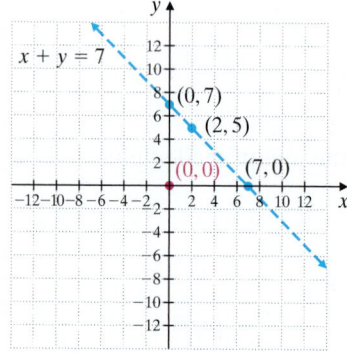

Step 2: Next we choose a test point, being careful *not* to choose a point on the boundary line. We choose $(0, 0)$, and substitute the coordinates of $(0, 0)$ into $x + y < 7$.

$x + y < 7$ Original inequality

$0 + 0 < 7$ Replace x with 0 and y with 0.

$0 < 7$ True

Step 3: Since the result is a true statement, $(0, 0)$ is a solution of $x + y < 7$, and every point in the same half-plane as $(0, 0)$ is also a solution. To indicate this, we shade the entire half-plane containing $(0, 0)$, as shown.

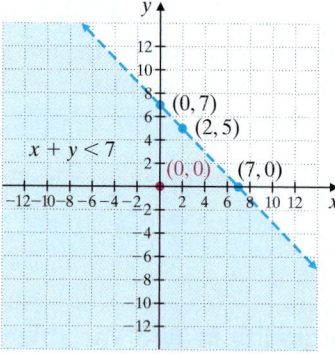

■ **Work Practice 2**

✓**Concept Check** Determine whether $(0, 0)$ is included in the graph of

a. $y \geq 2x + 3$

b. $x < 7$

c. $2x - 3y < 6$

Example 3   Graph: $2x - y \geq 3$

Solution:

Step 1: We graph the boundary line by graphing $2x - y = 3$. We draw this line as a solid line because the inequality sign is $\geq$, and thus the points on the line are solutions of $2x - y \geq 3$.

Step 2: Once again, $(0, 0)$ is a convenient test point since it is not on the boundary line.

We substitute 0 for x and 0 for y into the original inequality.

$$2x - y \geq 3$$
$$2(0) - 0 \geq 3 \quad \text{Let } x = 0 \text{ and } y = 0.$$
$$0 \geq 3 \quad \text{False}$$

Step 3: Since the statement is false, no point in the half-plane containing $(0, 0)$ is a solution. Therefore, we shade the half-plane that does not contain $(0, 0)$. Every point in the shaded half-plane and every point on the boundary line is a solution of $2x - y \geq 3$.

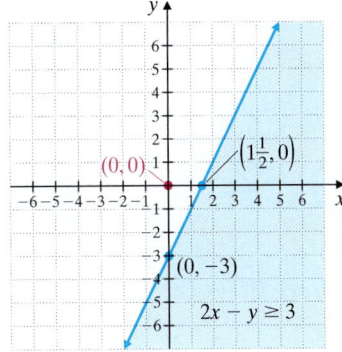

Work Practice 3

> **Helpful Hint**
>
> When graphing an inequality, make sure the test point is substituted into the **original inequality.** For Example 3, we substituted the test point $(0, 0)$ into the **original inequality** $2x - y \geq 3$, *not* $2x - y = 3$.

Example 4   Graph: $x > 2y$

Solution:

Step 1: We find the boundary line by graphing $x = 2y$. The boundary line is a dashed line since the inequality symbol is $>$.

Step 2: We cannot use $(0, 0)$ as a test point because it is a point on the boundary line. We choose instead $(0, 2)$.

$$x > 2y$$
$$0 > 2(2) \quad \text{Let } x = 0 \text{ and } y = 2.$$
$$0 > 4 \quad \text{False}$$

Step 3: Since the statement is false, we shade the half-plane that does not contain the test point $(0, 2)$, as shown.

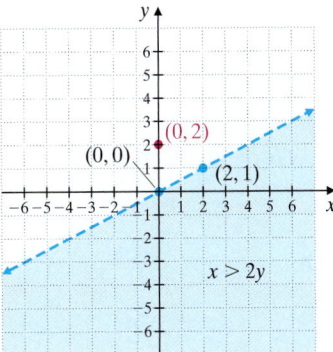

Teaching Tip

For Examples 4 and 5, have your students use the graph to write the inequality with y isolated on one side of the inequality.

Work Practice 4

Practice 3

Graph: $x - 4y \leq 4$

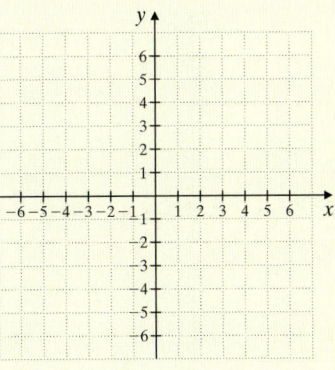

Practice 4

Graph: $y < 3x$

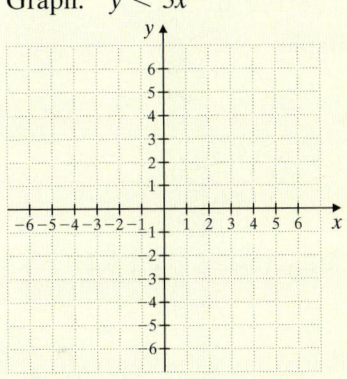

Answers

3.

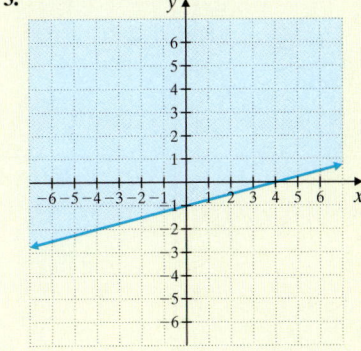

4.

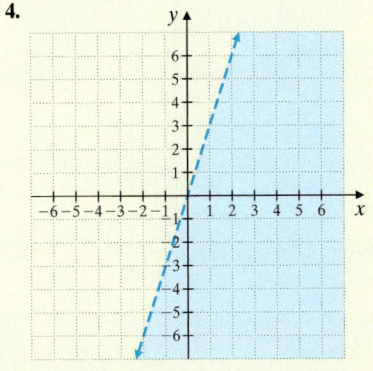

Practice 5

Graph: $3x + 2y \geq 12$

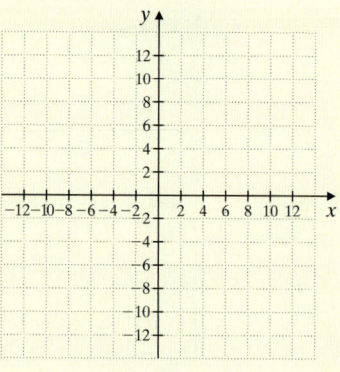

Practice 6

Graph: $x < 2$

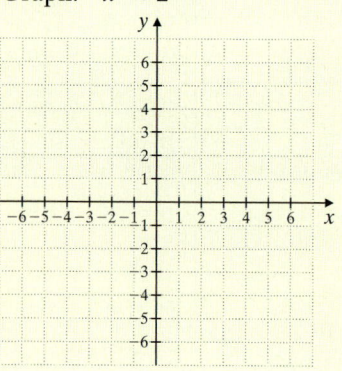

Answers

5.

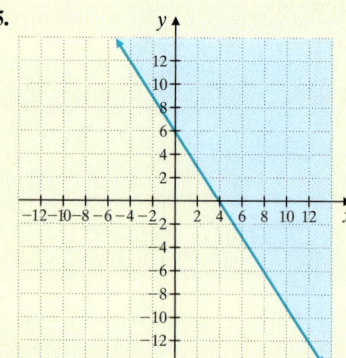

6.

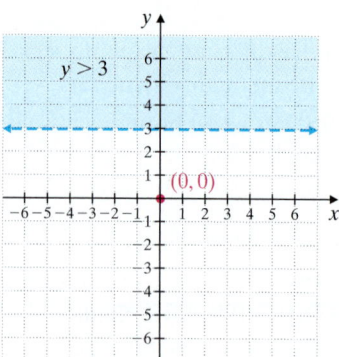

Example 5   Graph: $5x + 4y \leq 20$

Solution: We graph the solid boundary line $5x + 4y = 20$ and choose $(0, 0)$ as the test point.

$$5x + 4y \leq 20$$
$$5(0) + 4(0) \leq 20 \quad \text{Let } x = 0 \text{ and } y = 0.$$
$$0 \leq 20 \quad \text{True}$$

We shade the half-plane that contains $(0, 0)$, as shown.

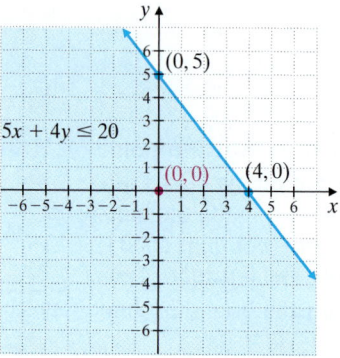

■ **Work Practice 5**

Example 6   Graph: $y > 3$

Solution: We graph the dashed boundary line $y = 3$ and choose $(0, 0)$ as the test point. (Recall that the graph of $y = 3$ is a horizontal line with y-intercept 3.)

$$y > 3$$
$$0 > 3 \quad \text{Let } y = 0.$$
$$0 > 3 \quad \text{False}$$

We shade the half-plane that does not contain $(0, 0)$, as shown.

■ **Work Practice 6**

Example 7 Graph: $y \leq \dfrac{2}{3}x - 4$

Solution: Graph the solid boundary line $y = \dfrac{2}{3}x - 4$. This equation is in slope-intercept form, with slope $\dfrac{2}{3}$ and y-intercept -4.

We use this information to graph the line. Then we choose $(0, 0)$ as our test point.

$$y \leq \dfrac{2}{3}x - 4$$

$$0 \leq \dfrac{2}{3} \cdot 0 - 4$$

$$0 \leq -4 \quad \text{False}$$

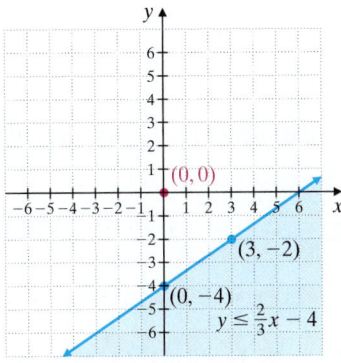

We shade the half-plane that does not contain $(0, 0)$, as shown.

■ **Work Practice 7**

Practice 7

Graph: $y \geq \dfrac{1}{4}x + 3$

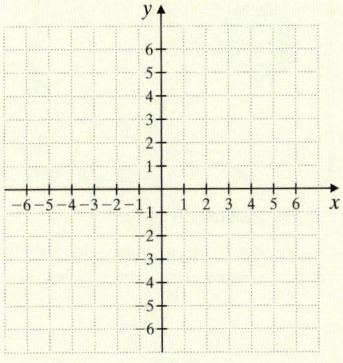

Answer

7.

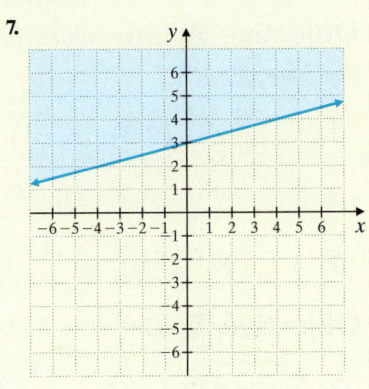

Vocabulary, Readiness & Video Check

Use the choices below to fill in each blank. Some choices may be used more than once, and some not at all.

true	$x < 2$	$y < 2$	half-planes
false	$x \leq 2$	$y \leq 2$	linear inequality in two variables

1. The statement $5x - 6y < 7$ is an example of a ___linear inequality in two variables___.

2. A boundary line divides a plane into two regions called ___half-planes___.

3. True or false? The graph of $5x - 6y < 7$ includes its corresponding boundary line. ___false___

4. True or false? When graphing a linear inequality, to determine which side of the boundary line to shade, choose a point *not* on the boundary line. ___true___

5. True or false? The boundary line for the inequality $5x - 6y < 7$ is the graph of $5x - 6y = 7$.
___true___

6. The graph of ___$y < 2$___ is

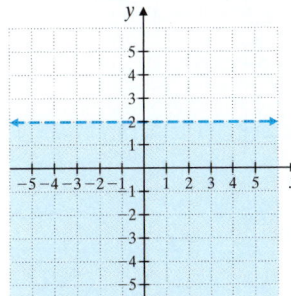

Martin-Gay Interactive Videos   Watch the section lecture video and answer the following questions.

Objective A   **7.** From Example 1, how do we determine whether an ordered pair in *x* and *y* is a solution of an inequality in *x* and *y*? ▶

Objective B   **8.** From Example 3, how do we find the equation of the boundary line? How do we determine if the points on the boundary line are solutions of the inequality? ▶

See Video 3.6 🍎

See video answer section.

3.6   **Exercise Set**   MyMathLab® ▶

Objective A   *Determine whether the ordered pairs given are solutions of the linear inequality in two variables. See Example 1.*

1. $x - y > 3$; $(0, 3)$, $(2, -1)$ no; no

2. $y - x < -2$; $(2, 1)$, $(5, -1)$ no; yes

3. $3x - 5y \leq -4$; $(2, 3)$, $(-1, -1)$ yes; no

4. $2x + y \geq 10$; $(0, 11)$, $(5, 0)$ yes; yes

▶**5.** $x < -y$; $(0, 2)$, $(-5, 1)$ no; yes

6. $y > 3x$; $(0, 0)$, $(1, 4)$ no; yes

Objective B   *Graph each inequality. See Examples 2 through 7.*

7. $x + y \leq 1$

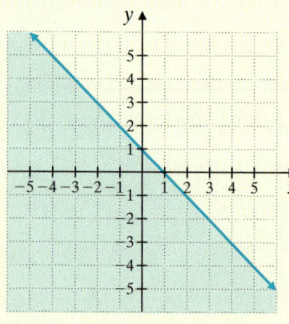

8. $x + y \geq -2$

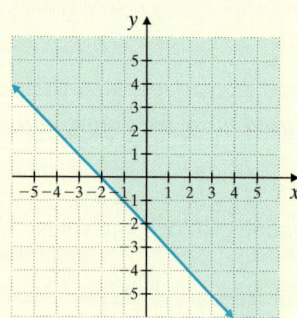

9. $2x - y > -4$

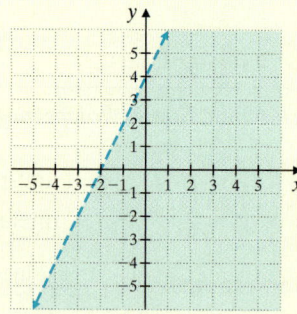

10. $x - 3y < 3$

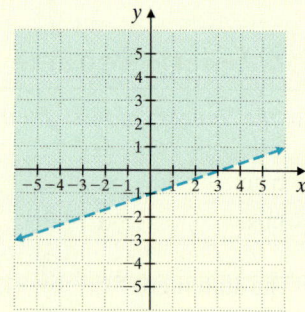

▶**11.** $y \geq 2x$

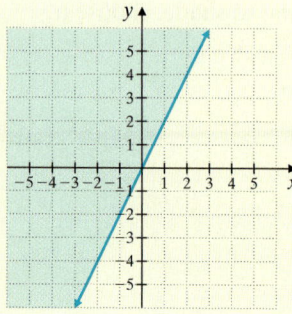

12. $y \leq 3x$

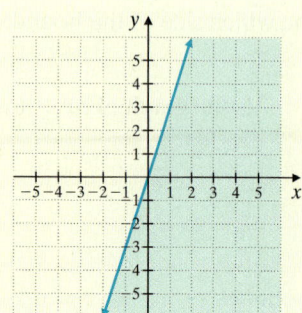

13. $x < -3y$

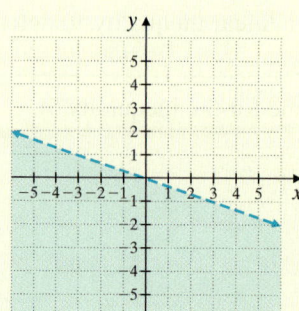

14. $x > -2y$

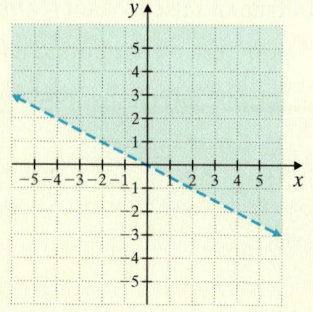

15. $y \geq x + 5$

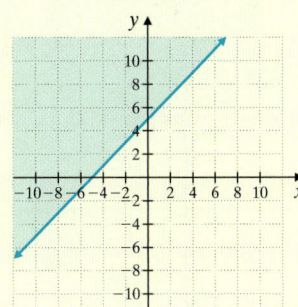

16. $y \leq x + 1$

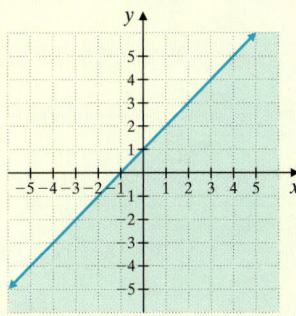

17. $y < 4$

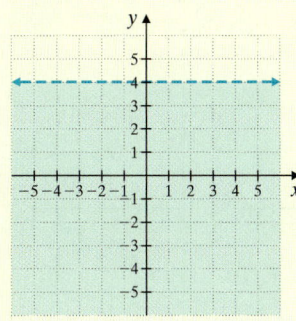

18. $y > 2$

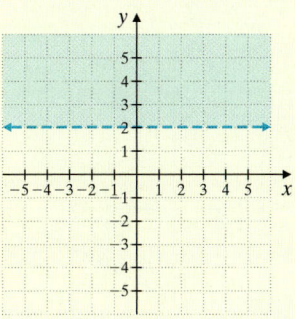

19. $x \geq -3$

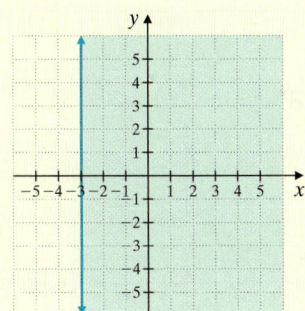

20. $x \leq -1$

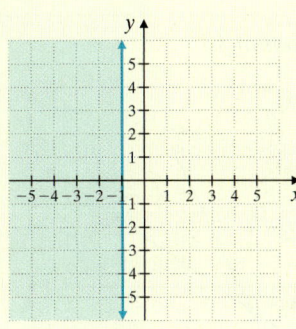

21. $5x + 2y \leq 10$

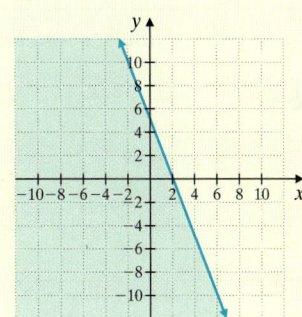

22. $4x + 3y \geq 12$

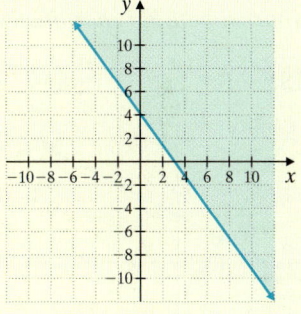

23. $x > y$

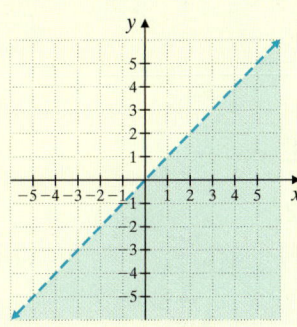

24. $x \leq -y$

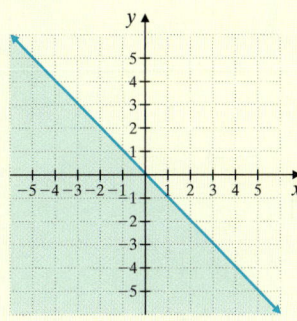

25. $x - y \leq 6$

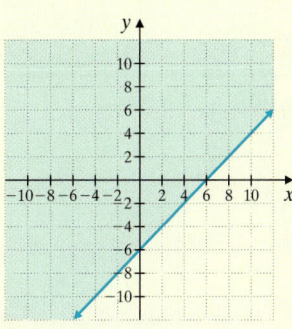

26. $x - y > 10$

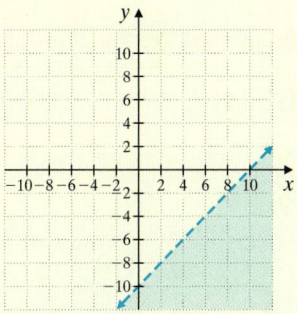

27. $x \geq 0$

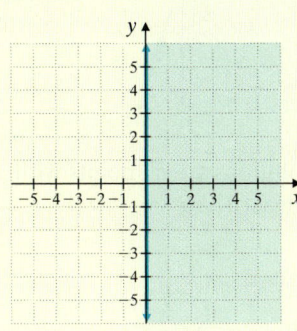

28. $y \leq 0$

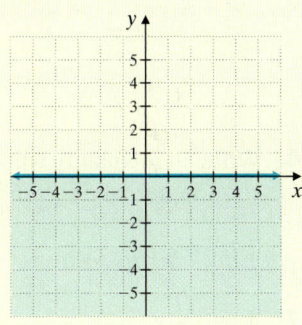

29. $2x + 7y > 5$

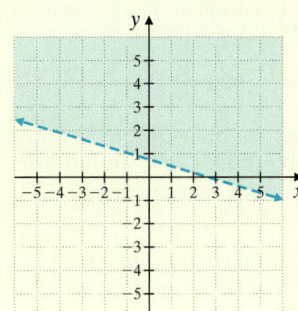

30. $3x + 5y \leq -2$

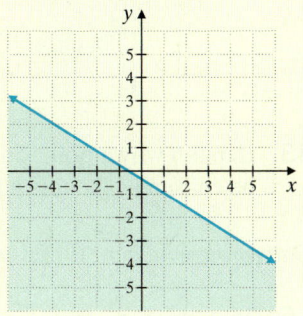

31. $y \geq \dfrac{1}{2}x - 4$

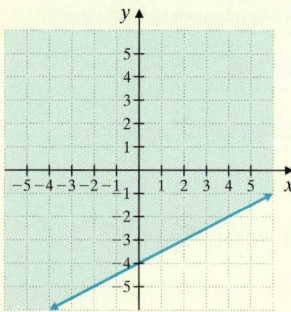

32. $y < \dfrac{2}{5}x - 3$

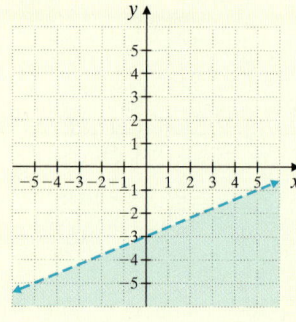

33. $y < -\dfrac{3}{4}x + 2$

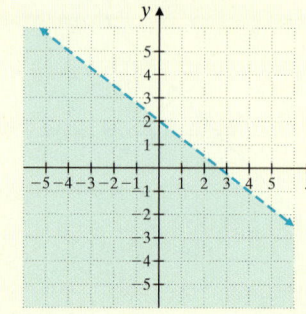

34. $y > -\dfrac{1}{3}x + 4$

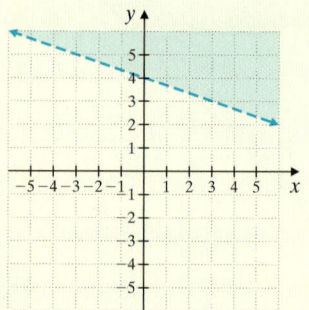

Review

Approximate the coordinates of each point of intersection. See Section 3.1.

35. $(-2, 1)$

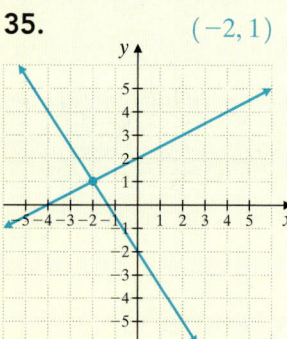

36. $(3, 0)$

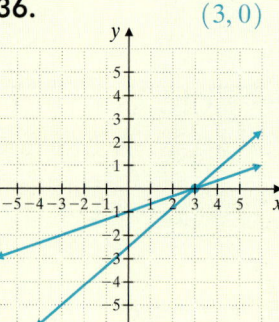

37. $(-3, -1)$

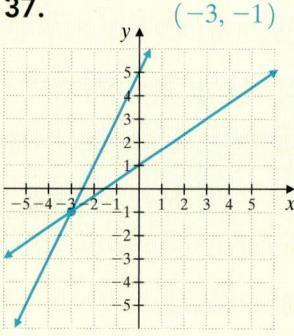

38. $(-3, -3)$

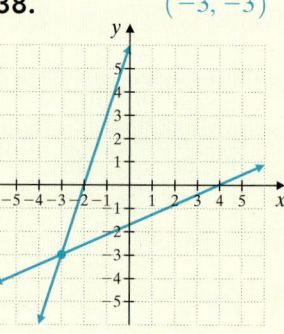

Concept Extensions

Match each inequality with its graph.

a. $x > 2$ **b.** $y < 2$ **c.** $y \leq 2x$ **d.** $y \leq -3x$

39. a

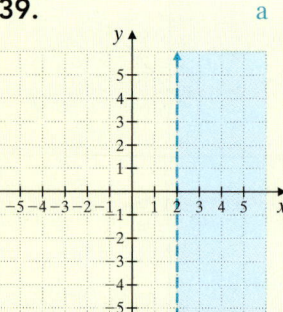

40. c

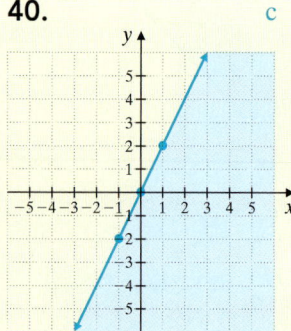

41. b

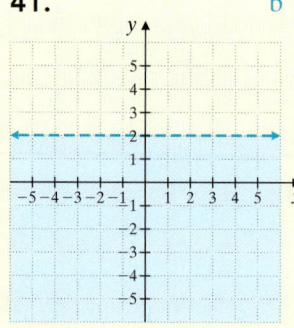

42. d

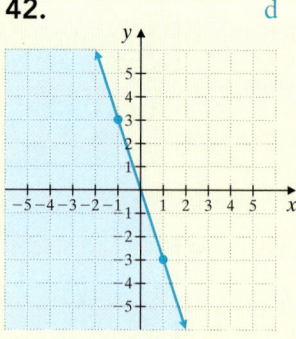

43. Explain why a point on the boundary line should not be chosen as the test point.   answers may vary

44. Write an inequality whose solutions are all points with coordinates whose sum is at least 13. $x + y \geq 13$

Determine whether $(1, 1)$ is included in each graph. See the Concept Check in this section.

45. $3x + 4y < 8$ yes

46. $y > 5x$ no

47. $y \geq -\dfrac{1}{2}x$ yes

48. $x > 3$ no

Solve.

49. It's the end of the budgeting period for Dennis Fernandes and he has $500 left in his budget for car rental expenses. He plans to spend this budget on a sales trip throughout southern Texas. He will rent a car that costs $30 per day and $0.15 per mile and he can spend no more than $500.

a. Write an inequality describing this situation. Let x = number of days and let y = number of miles. $30x + 0.15y \le 500$

b. Graph this inequality below.

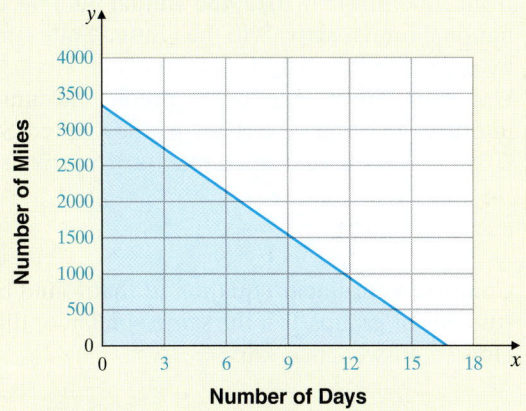

Number of Days

c. Why is the grid showing quadrant I only?
answers may vary

50. Scott Sambracci and Sara Thygeson are planning their wedding. They have calculated that they want the cost of their wedding ceremony x plus the cost of their reception y to be no more than $5000.

a. Write an inequality describing this relationship. $x + y \le 5000$

b. Graph this inequality below.

Wedding Ceremony

c. Why is the grid showing quadrant I only?
answers may vary

Chapter 3 Group Activity

Additional group activities are available in the *Instructor's Resource Manual with Tests.*

Finding a Linear Model

This activity may be completed by working in groups or individually.

The following table shows the actual number of international tourist arrivals to the United States for the years 2009 through 2012.

Year	International Tourist Arrivals to the United States (in millions)
2009	55
2010	60
2011	63
2012	67

Source: World Tourism Organization

1. Make a scatter diagram of the paired data in the table.

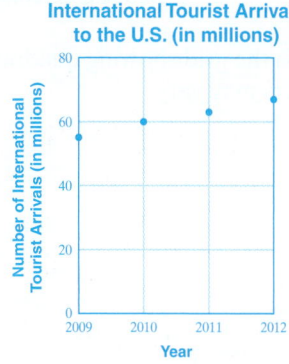

International Tourist Arrivals to the U.S. (in millions)

2. Use what you have learned in this chapter to write an equation of the line representing the paired data in the table. Explain how you found the equation and what each variable represents. $y = 4x + 55$; used a ruler and the first and last points to approximate the line ($y = 3.9x + 55.4$ when linear regression is used on a caculator); x represents the number of years since 2009, y represents international tourist arrivals to the United States in millions

3. What is the slope of your line? What does the slope mean in this context? Slope is 4; this means that from 2009 through 2012, the annual number of international tourist arrivals to the United States increased by approximately 4 million per year

4. Use your linear equation to predict the number of international tourist arrivals to the United States in 2018. 91 million international tourist arrivals

5. Compare your linear equation to that found by other students or groups. Is it the same, similar, or different? How? answers may vary

6. Compare your prediction from question **4** to that of other students or groups. Describe what you find. answers may vary

7. Suppose that the number of international tourist arrivals to the United States for 2013 was estimated to be 65 million. If this data point is added to the chart, how does it affect your results? changes the equation; answers may vary

Chapter 3 Vocabulary Check

Fill in each blank with one of the words listed below.

y-axis	*x*-axis	solution	linear	standard	point-slope
x-intercept	*y*-intercept	*y*	*x*	slope	slope-intercept

1. An ordered pair is a(n) ____solution____ of an equation in two variables if replacing the variables by the coordinates of the ordered pair results in a true statement.

2. The vertical number line in the rectangular coordinate system is called the ____*y*-axis____.

3. A(n) ____linear____ equation can be written in the form $Ax + By = C$.

4. A(n) ____*x*-intercept____ is a point of the graph where the graph crosses the *x*-axis.

5. The form $Ax + By = C$ is called ____standard____ form.

6. A(n) _____y-intercept_____ is a point of the graph where the graph crosses the *y*-axis.

7. The equation $y = 7x - 5$ is written in _____slope-intercept_____ form.

8. To find an *x*-intercept of a graph, let _____*y*_____ = 0.

9. The horizontal number line in the rectangular coordinate system is called the _____*x*-axis_____.

10. To find a *y*-intercept of a graph, let _____*x*_____ = 0.

11. The _____slope_____ of a line measures the steepness or tilt of the line.

Helpful Hint ▶ Are you preparing for your test? Don't forget to take the Chapter 3 Test on page 275. Then check your answers at the back of the text and use the Chapter Test Prep Videos to see the fully worked-out solutions to any of the exercises you want to review.

3 Chapter Highlights

Definitions and Concepts	Examples

Section 3.1 Reading Graphs and the Rectangular Coordinate System

The **rectangular coordinate system** consists of a plane and a vertical and a horizontal number line intersecting at their 0 coordinates. The vertical number line is called the **y-axis** and the horizontal number line is called the **x-axis.** The point of intersection of the axes is called the **origin.**

To **plot** or **graph** an ordered pair means to find its corresponding point on a rectangular coordinate system.

To plot or graph an ordered pair such as $(3, -2)$, start at the origin. Move 3 units to the right and from there, 2 units down.

To plot or graph $(-3, 4)$, start at the origin. Move 3 units to the left and from there, 4 units up.

An ordered pair is a **solution** of an equation in two variables if replacing the variables with the coordinates of the ordered pair results in a true statement.

If one coordinate of an ordered pair solution of an equation is known, the other value can be determined by substitution.

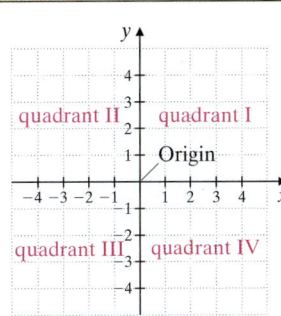 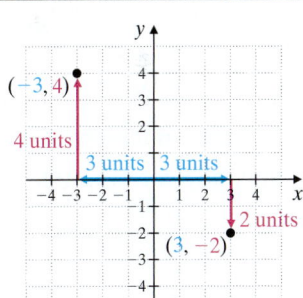

Complete the ordered pair $(0, \ )$ for the equation $x - 6y = 12$.

$$x - 6y = 12$$
$$0 - 6y = 12 \quad \text{Let } x = 0.$$
$$\frac{-6y}{-6} = \frac{12}{-6} \quad \text{Divide by } -6.$$
$$y = -2$$

The ordered pair solution is $(0, -2)$.

Definitions and Concepts	Examples

Section 3.2   Graphing Linear Equations

A **linear equation in two variables** is an equation that can be written in the form $Ax + By = C$, where A and B are not both 0. The form $Ax + By = C$ is called **standard form**.

To graph a linear equation in two variables, find three ordered pair solutions. Plot the solution points and draw the line connecting the points.

$$3x + 2y = -6 \qquad x = -5$$
$$y = 3 \qquad y = -x + 10$$

$x + y = 10$ is in standard form.

Graph: $x - 2y = 5$

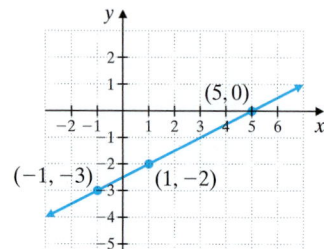

x	y
5	0
1	-2
-1	-3

Section 3.3   Intercepts

An **intercept** of a graph is a point where the graph intersects an axis. If a graph intersects the x-axis at a, then $(a, 0)$ is an **x-intercept**. If a graph intersects the y-axis at b, then $(0, b)$ is a **y-intercept**.

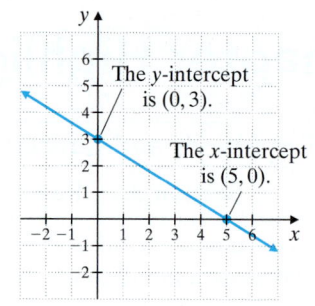

The y-intercept is $(0, 3)$.

The x-intercept is $(5, 0)$.

To find the x-intercept(s), let $y = 0$ and solve for x.
To find the y-intercept(s), let $x = 0$ and solve for y.

Find the intercepts for $2x - 5y = -10$.

If $y = 0$, then
$$2x - 5 \cdot 0 = -10$$
$$2x = -10$$
$$\frac{2x}{2} = \frac{-10}{2}$$
$$x = -5$$

If $x = 0$, then
$$2 \cdot 0 - 5y = -10$$
$$-5y = -10$$
$$\frac{-5y}{-5} = \frac{-10}{-5}$$
$$y = 2$$

The x-intercept is $(-5, 0)$. The y-intercept is $(0, 2)$.

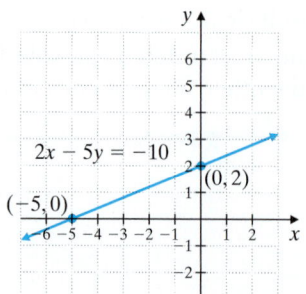

$2x - 5y = -10$

Definitions and Concepts	Examples

Section 3.3   Intercepts (*continued*)

The graph of $x = c$ is a vertical line with x-intercept $(c, 0)$.

The graph of $y = c$ is a horizontal line with y-intercept $(0, c)$.

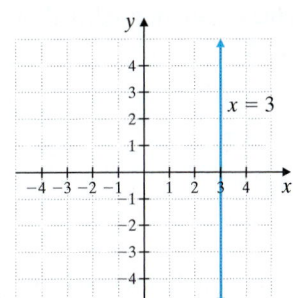
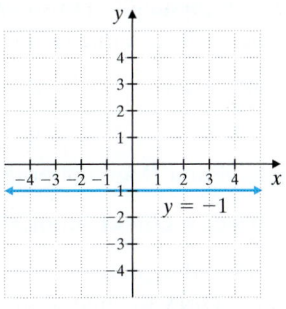

Section 3.4   Slope and Rate of Change

The **slope m** of the line through points (x_1, y_1) and (x_2, y_2) is given by

$$m = \frac{y_2 - y_1}{x_2 - x_1} \quad \text{as long as } x_2 \neq x_1$$

A horizontal line has slope 0.
The slope of a vertical line is undefined.
Nonvertical parallel lines have the same slope.
Two nonvertical lines are perpendicular if the slope of one is the negative reciprocal of the slope of the other.

The slope of the line through points $(-1, 6)$ and $(-5, 8)$ is

$$m = \frac{y_2 - y_1}{x_2 - x_1} = \frac{8 - 6}{-5 - (-1)} = \frac{2}{-4} = -\frac{1}{2}$$

The slope of the line $y = -5$ is 0.
The line $x = 3$ has undefined slope.

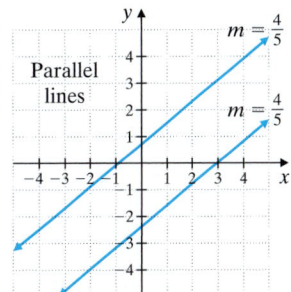
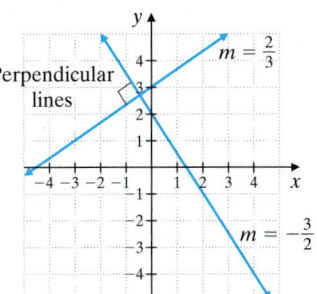

Section 3.5   Equations of Lines

Slope-Intercept Form

$$y = mx + b$$

m is the slope of the line.
$(0, b)$ is the y-intercept.

Find the slope and the y-intercept of the line $2x + 3y = 6$.
Solve for y:

$$2x + 3y = 6$$
$$3y = -2x + 6 \qquad \text{Subtract } 2x.$$
$$y = -\frac{2}{3}x + 2 \qquad \text{Divide by 3.}$$

The slope of the line is $-\dfrac{2}{3}$ and the y-intercept is $(0, 2)$.

Point-Slope Form

$$y - y_1 = m(x - x_1)$$

m is the slope.
(x_1, y_1) is a point of the line.

Find an equation of the line with slope $\dfrac{3}{4}$ that contains the point $(-1, 5)$.

$$y - 5 = \frac{3}{4}[x - (-1)]$$
$$4(y - 5) = 3(x + 1) \qquad \text{Multiply by 4.}$$
$$4y - 20 = 3x + 3 \qquad \text{Distribute.}$$
$$-3x + 4y = 23 \qquad \text{Subtract } 3x \text{ and add 20.}$$

Definitions and Concepts	Examples

Section 3.6   Graphing Linear Inequalities in Two Variables

A **linear inequality in two variables** is an inequality that can be written in one of these forms:

$$Ax + By < C \qquad Ax + By \leq C$$
$$Ax + By > C \qquad Ax + By \geq C$$

where A and B are not both 0.

$$2x - 5y < 6 \qquad\qquad x \geq -5$$
$$y > -8x \qquad\qquad y \leq 2$$

To Graph a Linear Inequality

1. Graph the boundary line by graphing the related equation. Draw the line solid if the inequality symbol is $\leq$ or $\geq$. Draw the line dashed if the inequality symbol is $<$ or $>$.

2. Choose a test point not on the line. Substitute its coordinates into the original inequality.

3. If the resulting inequality is true, shade the half-plane that contains the test point. If the inequality is not true, shade the half-plane that does not contain the test point.

Graph: $2x - y \leq 4$

1. Graph $2x - y = 4$. Draw a solid line because the inequality symbol is $\leq$.

2. Check the test point $(0, 0)$ in the original inequality, $2x - y \leq 4$.

   $2 \cdot 0 - 0 \leq 4$ Let $x = 0$ and $y = 0$.

   $0 \leq 4$ True

3. The inequality is true, so shade the half-plane containing $(0, 0)$, as shown.

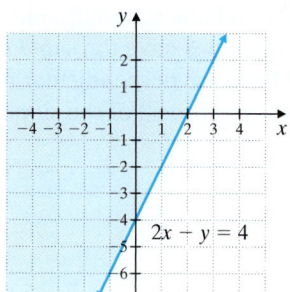

Chapter 3   Review

(3.1) *Plot each point on the same rectangular coordinate system.*

1. $(-7, 0)$

2. $\left(0, 4\dfrac{4}{5}\right)$

3. $(-2, -5)$

4. $(1, -3)$

5. $(0.7, 0.7)$

6. $(-6, 4)$

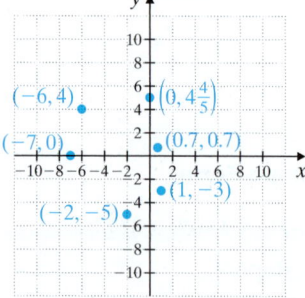

Complete each ordered pair so that it is a solution of the given equation.

7. $-2 + y = 6x$; $(7, \quad)$ $(7, 44)$

8. $y = 3x + 5$; $(\quad, -8)$ $\left(-\dfrac{13}{3}, -8\right)$

Complete the table of values for each given equation.

9. $9 = -3x + 4y$

x	y
−3	0
1	3
9	9

10. $y = 5$

x	y
7	5
−7	5
0	5

11. $x = 2y$

x	y
0	0
10	5
−10	−5

12. The cost in dollars of producing x compact disc holders is given by $y = 5x + 2000$.

    a. Complete the table.

x	1	100	1000
y	2005	2500	7000

b. Find the number of compact disc holders that can be produced for $6430.

886 compact disc holders

(3.2) *Graph each linear equation.*

13. $x - y = 1$

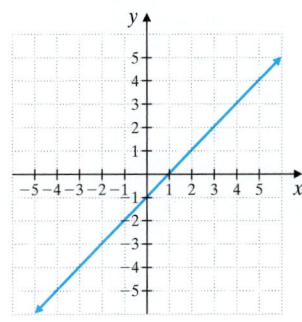

14. $x + y = 6$

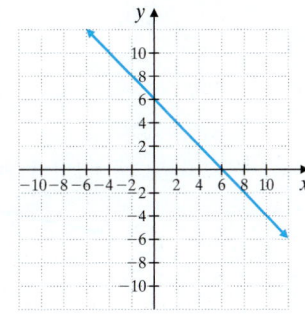

15. $x - 3y = 12$

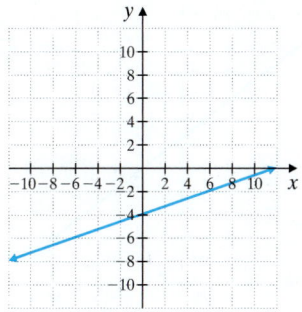

16. $5x - y = -8$

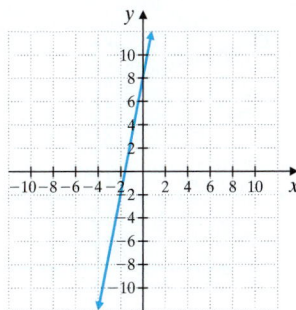

17. $x = 3y$

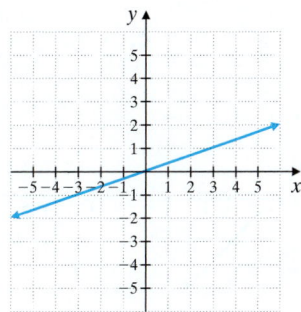

18. $y = -2x$

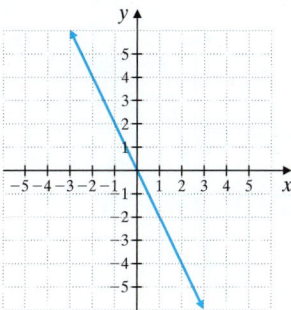

(3.3) *Identify the intercepts in each graph.*

19.

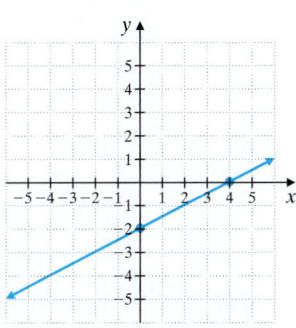

$(4, 0); (0, -2)$

20.

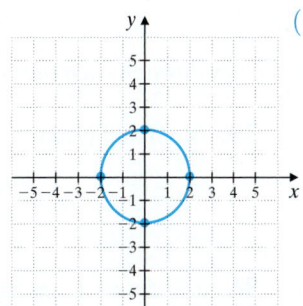

$(-2, 0); (2, 0); (0, 2); (0, -2)$

Graph each linear equation.

21. $y = -3$

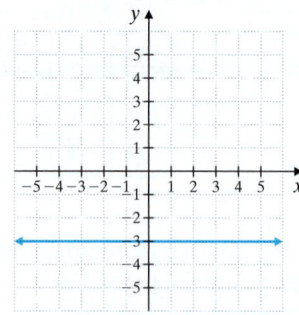

22. $x = 5$

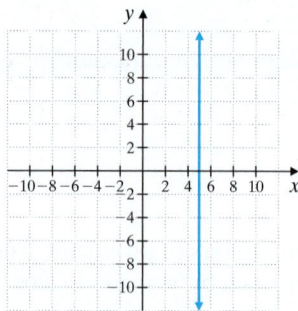

Find the intercepts of each equation.

23. $x - 3y = 12$ $(12, 0), (0, -4)$

24. $-4x + y = 8$ $(-2, 0), (0, 8)$

(3.4) *Find the slope of each line.*

25.

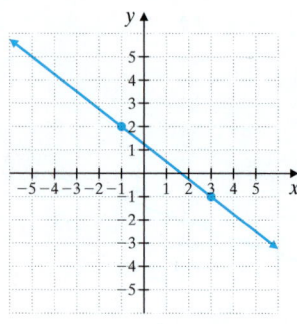

$m = -\dfrac{3}{4}$

26.

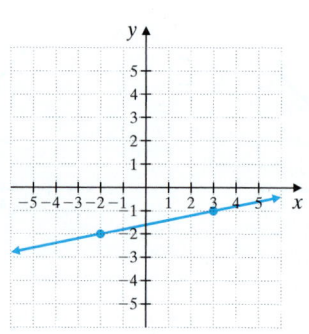

$m = \dfrac{1}{5}$

Match each line with its slope.

a.

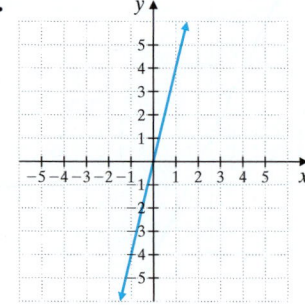

b.

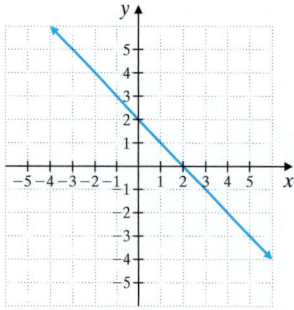

c.

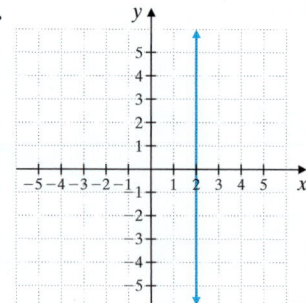

d.

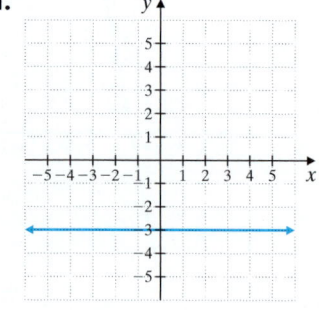

27. $m = 0$ d

28. $m = -1$ b

29. undefined slope   c

30. $m = 4$ a

Find the slope of the line that passes through each pair of points.

31. $(2, 5)$ and $(6, 8)$

$m = \dfrac{3}{4}$

32. $(4, 7)$ and $(1, 2)$

$m = \dfrac{5}{3}$

33. $(1, 3)$ and $(-2, -9)$

$m = 4$

34. $(-4, 1)$ and $(3, -6)$

$m = -1$

Find the slope of each line.

35. $y = 3x + 7$ $m = 3$

36. $x - 2y = 4$ $m = \dfrac{1}{2}$

37. $y = -2$ $m = 0$

38. $x = 0$ undefined slope

Determine whether each pair of lines is parallel, perpendicular, or neither.

39. $x - y = -6$
   $x + y = 3$
   perpendicular

40. $3x + y = 7$
   $-3x - y = 10$
   parallel

41. $y = 4x + \dfrac{1}{2}$
   $4x + 2y = 1$
   neither

42. $y = 6x - \dfrac{1}{3}$
   $x + 6y = 6$
   perpendicular

Find the slope of each line and write the slope as a rate of change. Don't forget to attach the proper units.

43. The graph below approximates the total number of U.S. magazines in print for each year x.

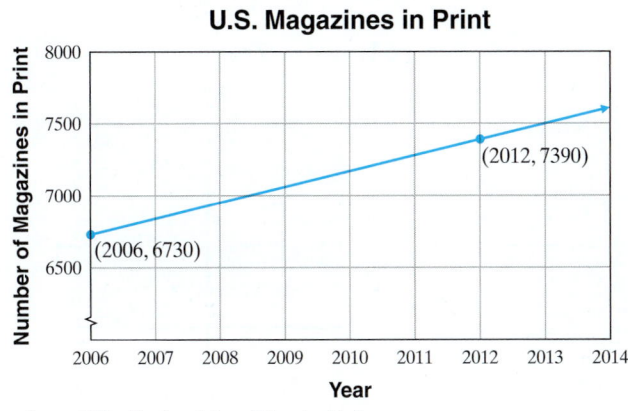

U.S. Magazines in Print

Source: MPA—The Association of Magazine Media

$m = 110$; The total number of U.S. magazines in print increases by 110 magazines per year.

44. The graph below approximates the number of lung transplants y in the United States for year x. Round to the nearest whole.

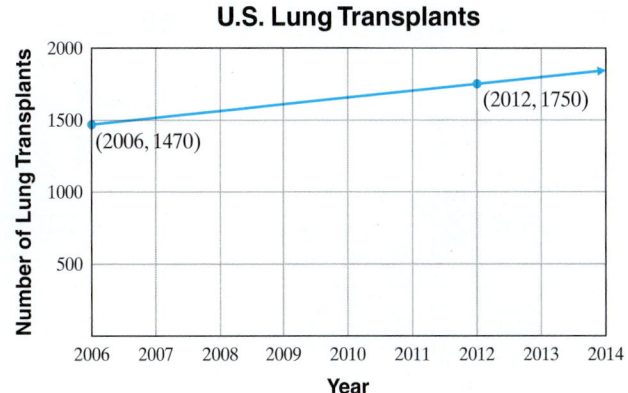

U.S. Lung Transplants

Source: Organ Procurement and Transplantation Network

$m = 47$; The number of U.S. lung transplants increases by 47 transplants per year.

(3.5) *Determine the slope and the y-intercept of the graph of each equation.*

45. $x - 6y = -1$ $m = \dfrac{1}{6}; \left(0, \dfrac{1}{6}\right)$

46. $3x + y = 7$ $m = -3; (0, 7)$

Write an equation of each line.

47. slope -5; y-intercept $\left(0, \dfrac{1}{2}\right)$ $y = -5x + \dfrac{1}{2}$

48. slope $\dfrac{2}{3}$; y-intercept $(0, 6)$ $y = \dfrac{2}{3}x + 6$

Match each equation with its graph.

49. $y = 2x + 1$ d

50. $y = -4x$ c

51. $y = 2x$ a

52. $y = 2x - 1$ b

a.
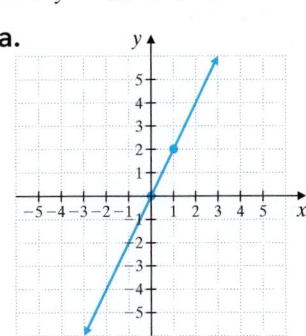

b.
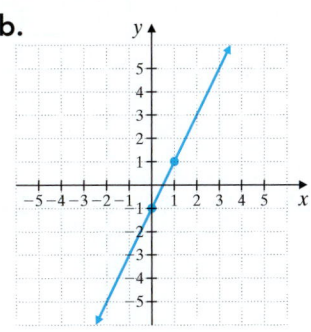

c.
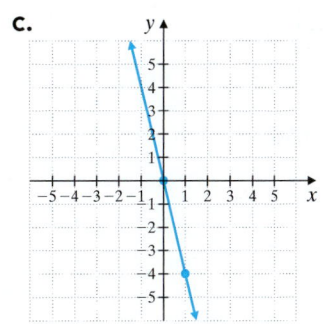

d.
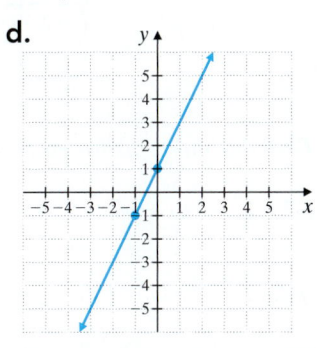

Write an equation of the line with the given slope that passes through the given point. Write the equation in the form
$Ax + By = C$.

53. $m = 4$; $(2, 0)$
$-4x + y = -8$

54. $m = -3$; $(0, -5)$
$3x + y = -5$

55. $m = \dfrac{3}{5}$; $(1, 4)$
$-3x + 5y = 17$

56. $m = -\dfrac{1}{3}$; $(-3, 3)$
$x + 3y = 6$

Write an equation of the line passing through each pair of points. Write the equation in the form $y = mx + b$.

57. $(1, 7)$ and $(2, -7)$ $y = -14x + 21$

58. $(-2, 5)$ and $(-4, 6)$ $y = -\dfrac{1}{2}x + 4$

(3.6) *Graph each inequality.*

59. $x + 6y < 6$

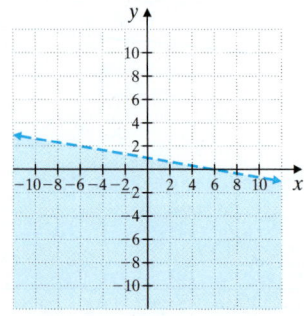

60. $x + y > -2$

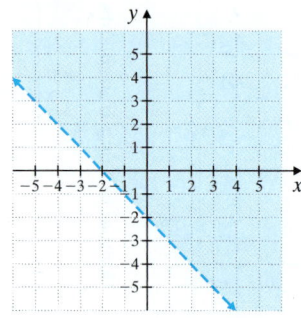

61. $y \geq -7$

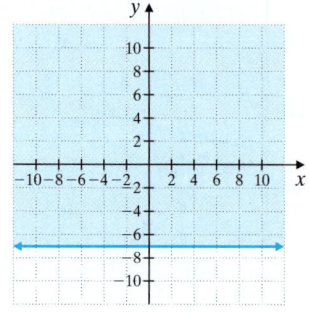

62. $y \leq -4$

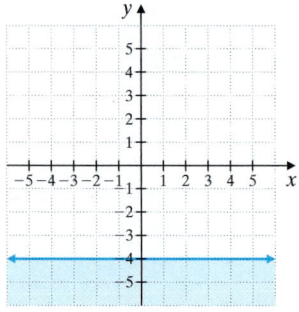

63. $-x \leq y$

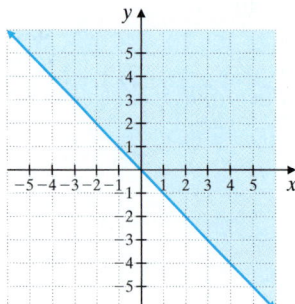

64. $x \geq -y$

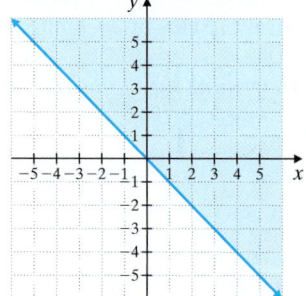

Mixed Review

Complete the table of values for each given equation.

65. $2x - 5y = 9$

x	y
7	1
2	-1
-3	-3

66. $x = -3y$

x	y
0	0
-3	1
6	-2

Find the intercepts for each equation.

67. $2x - 3y = 6$ $(3, 0); (0, -2)$

68. $-5x + y = 10$ $(-2, 0); (0, 10)$

Graph each linear equation.

69. $x - 5y = 10$

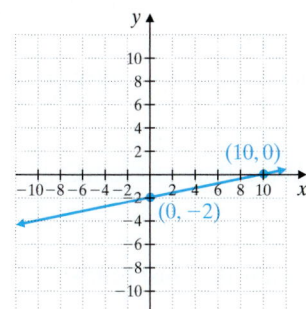

70. $x + y = 4$

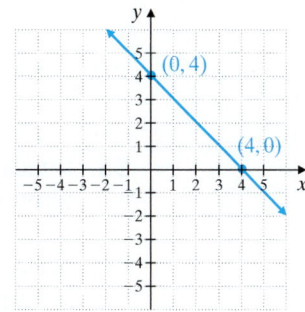

71. $y = -4x$

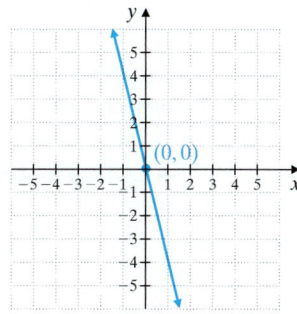

72. $2x + 3y = -6$

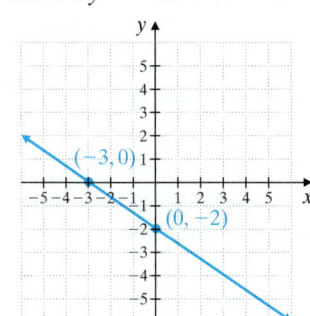

73. $x = 3$

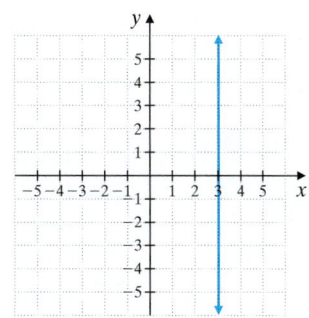

74. $y = -2$

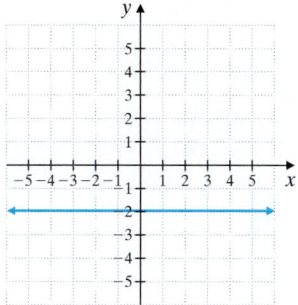

Find the slope of the line that passes through each pair of points.

75. $(3, -5)$ and $(-4, 2)$ $m = -1$

76. $(1, 3)$ and $(-6, -8)$ $m = \dfrac{11}{7}$

Find the slope of each line.

77.

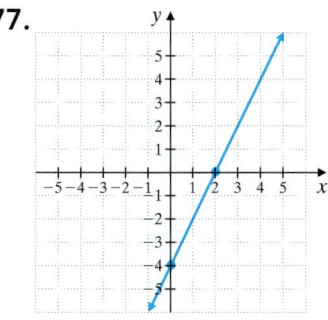

$m = 2$

78.

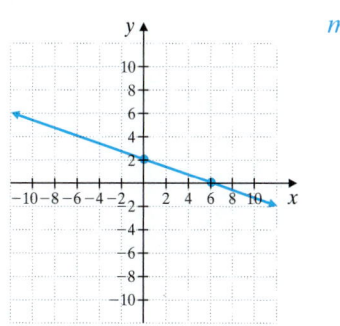

$m = -\dfrac{1}{3}$

Determine the slope and y-intercept of the graph of each equation.

79. $-2x + 3y = -15$ $m = \dfrac{2}{3}; (0, -5)$

80. $6x + y - 2 = 0$ $m = -6; (0, 2)$

Write an equation of the line with the given slope that passes through the given point. Write the equation in the form $Ax + By = C$.

81. $m = -5; (3, -7)$ $5x + y = 8$

82. $m = 3; (0, 6)$ $3x - y = -6$

Write an equation of the line passing through each pair of points. Write the equation in the form Ax + By = C.

83. $(-3, 9)$ and $(-2, 5)$ $4x + y = -3$

84. $(3, 1)$ and $(5, -9)$ $5x + y = 16$

The following bar graph shows the top 10 tourist destinations and the number of tourists that visit each country per year. Use this graph to answer Exercises 85 through 90.

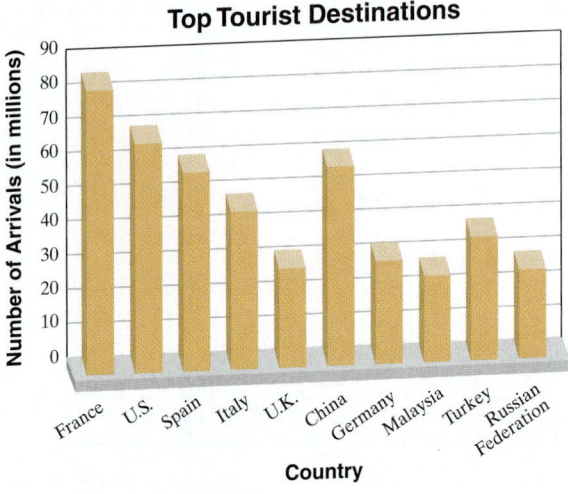

Top Tourist Destinations

Source: World Tourism Organization

85. Which country shown has the most tourist arrivals?   France

86. Which country shown has the least tourist arrivals?   Malaysia

87. Which countries shown have more than 50 million tourists per year?   France, U.S., Spain, China

88. Which countries shown have fewer than 30 million tourists per year?   U.K., Russian Federation, Malaysia

89. Estimate the number of tourists per year whose destination is Germany.   30 million

90. Estimate the number of tourists per year whose destination is Malaysia.   25 million

Complete each ordered pair so that it is a solution of the given equation.

1. $12y - 7x = 5$; $(1, \quad)$

2. $y = 17$; $(-4, \quad)$

Find the slope of each line.

3.
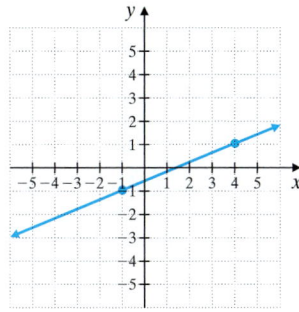

4.
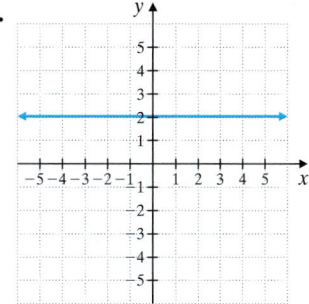

5. Passes through $(6, -5)$ and $(-1, 2)$

6. Passes through $(0, -8)$ and $(-1, -1)$

7. $-3x + y = 5$

8. $x = 6$

Graph.

9. $2x + y = 8$
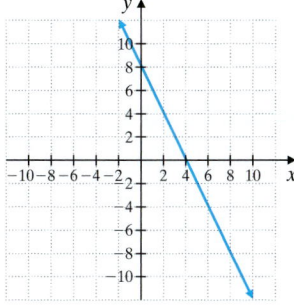

10. $-x + 4y = 5$
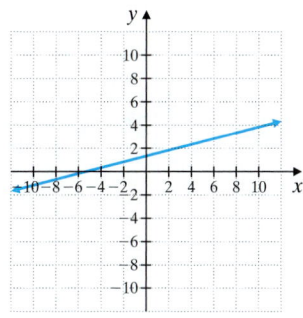

11. $x - y \geq -2$
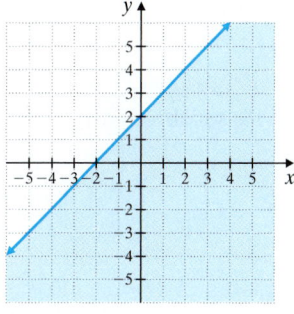

12. $y \geq -4x$
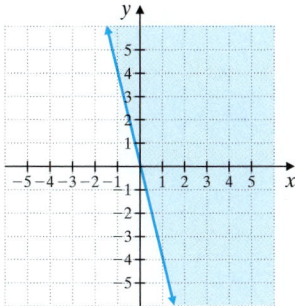

13. $5x - 7y = 10$
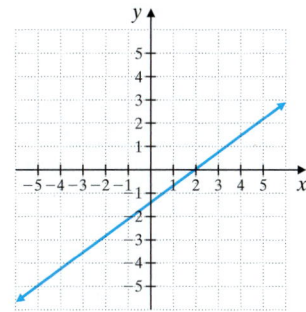

14. $2x - 3y > -6$
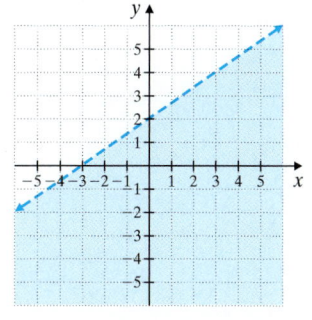

Answers

Note to Instructor: The Chapter 3 Test file in the TestGen program provides algorithms specifically matched to this test for easy replication for practice or assessment purposes.

1. $(1, 1)$

2. $(-4, 17)$

3. $m = \dfrac{2}{5}$

4. $m = 0$

5. $m = -1$

6. $m = -7$

7. $m = 3$

8. undefined slope

9. see graph

10. see graph

11. see graph

12. see graph

13. see graph

14. see graph

15. see graph

15. $6x + y > -1$

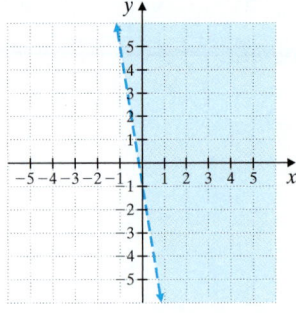

16. see graph

16. $y = -1$

17. neither

17. Determine whether the graphs of $y = 2x - 6$ and $-4x = 2y$ are parallel lines, perpendicular lines, or neither.

18. $x + 4y = 10$

Find the equation of each line. Write the equation in the form $Ax + By = C$.

18. Slope $-\dfrac{1}{4}$, passes through $(2, 2)$

19. $7x + 6y = 0$

19. Passes through the origin and $(6, -7)$

20. $8x + y = 11$

20. Passes through $(2, -5)$ and $(1, 3)$

21. Slope $\dfrac{1}{8}$; y-intercept $(0, 12)$

21. $x - 8y = -96$

△**22.** The perimeter of the parallelogram below is 42 meters. Write a linear equation in two variables for the perimeter. Use this equation to find x when y is 8 meters.

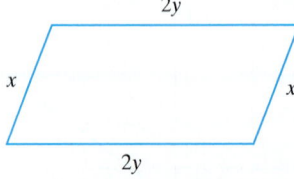

22. $x + 2y = 21; x = 5$ m

23. The table gives the number of cable phone customers (in millions) for the years shown. (*Source:* National Cable and Telecommunications Association)

Year	Cable Phone Customers (in millions)
2008	19.6
2009	22.2
2010	23.9
2011	25.3
2012	26.7

a. Write this data as a set of ordered pairs of the form (year, number of cable phone customers in millions).

b. Create a scatter diagram of the data. Be sure to label the axes properly.

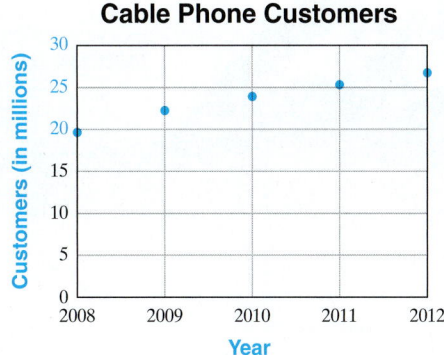

24. This graph approximates the movie ticket sales *y* (in millions) for the year *x*. Find the slope of the line and write the slope as a rate of change. Don't forget to attach the proper units.

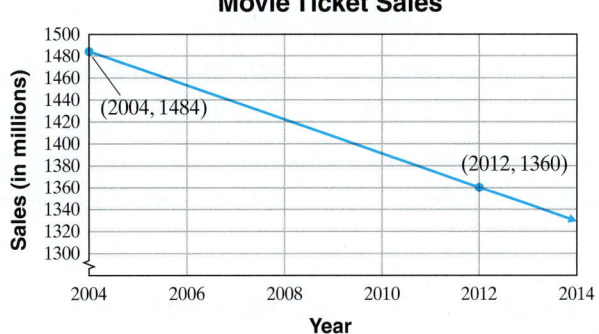

Source: National Association of Theater Owners; Annual U.S./Canada Admissions

(2008, 19.6);
(2009, 22.2);
(2010, 23.9);
(2011, 25.3);
23. a. (2012, 26.7)

b. see graph

$m = -15.5$; For every 1 year, 15.5 million fewer movie tickets
24. are sold.

Answers

Simplify each expression.

1. 27 (Sec. 1.3, Ex. 2)

1. $6 \div 3 + 5^2$

2. $\dfrac{10}{3} + \dfrac{5}{21}$

2. $\dfrac{25}{7}$ (Sec. R.2)

3. 51 (Sec. 1.3, Ex. 5)

3. $1 + 2[5(2 \cdot 3 + 1) - 10]$

4. $16 - 3 \cdot 3 + 2^4$

4. 23 (Sec. 1.3)

5. 20,602 feet (Sec. 1.5, Ex. 10)

5. The highest point in the United States is the top of Mount McKinley, at a height of 20,320 feet above sea level. The lowest point is Death Valley, California, which is 282 feet below sea level. How much higher is Mount McKinley than Death Valley? (*Source:* U.S. Geological Society)

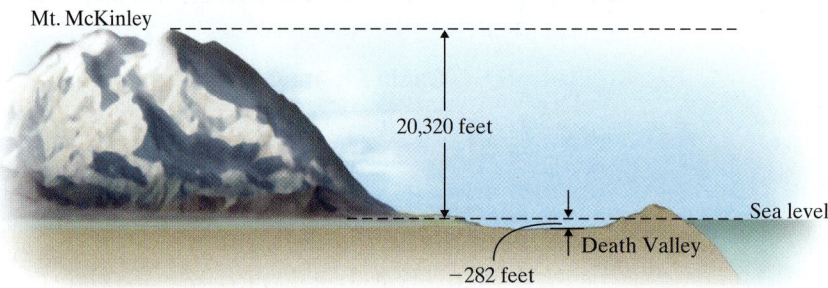

Mt. McKinley

20,320 feet

Sea level

Death Valley

−282 feet

6. $0.8x - 36$ (Sec. 1.8)

7. $2x + 6$ (Sec. 1.8, Ex. 16)

6. Simplify: $1.7x - 11 - 0.9x - 25$

8. $-15\left(x + \dfrac{2}{3}\right)$ $= -15x - 10$ (Sec. 1.8)

9. $(x - 4) \div 7$ or $\dfrac{x - 4}{7}$ (Sec. 1.8, Ex. 17)

Write each phrase as an algebraic expression and simplify if possible. Let x represent the unknown number.

10. $\dfrac{-9}{2x}$ (Sec. 1.8)

7. Twice a number, plus 6.

8. The product of -15 and the sum of a number and $\dfrac{2}{3}$.

11. $5 + (x + 1) = 6 + x$ (Sec. 1.8, Ex. 18)

12. $-86 - x$ (Sec. 1.8)

9. The difference of a number and 4, divided by 7.

10. The quotient of -9 and twice a number.

13. 6 (Sec. 2.2, Ex. 1)

14. -24 (Sec. 2.2)

11. Five plus the sum of a number and 1.

12. A number subtracted from -86.

15. $\{x \mid x < -2\}$ (Sec. 2.7, Ex. 6)

16. $\left\{x \mid x \le \dfrac{8}{3}\right\}$ (Sec. 2.7)

13. Solve for x: $\dfrac{5}{2}x = 15$

14. Solve for x: $\dfrac{x}{4} - 1 = -7$

15. Solve $2x < -4$. Graph the solutions.

16. Solve: $5(x + 4) \ge 4(2x + 3)$

17. $x = \dfrac{y - b}{m}$ (Sec. 2.5, Ex. 6)

18. $y = \dfrac{6 - x}{2}$ (Sec. 2.5)

17. Solve $y = mx + b$ for x.

18. Solve $x + 2y = 6$ for y.

Solve.

19. $-5x + 7 < 2(x - 3)$

20. $-8y + 4 \geq 3(y - 6)$

Complete each table for the given equation.

21. $y = \dfrac{1}{2}x - 5$

	x	y
a.	-2	-6
b.	0	-5
c.	10	0

22. $y = -\dfrac{3}{4}x + 5$

	x	y
a.	4	2
b.	0	5
c.	$\dfrac{20}{3}$	0

23. Complete each ordered pair so that it is a solution of the equation $3x + y = 12$.
   a. $(0, \ \ )$
   b. $(\ \ , 6)$
   c. $(-1, \ \ )$

24. Complete the table for $y = -5x$.

x	y
0	0
-1	5
-2	10

25. Graph the linear equation $2x + y = 5$.

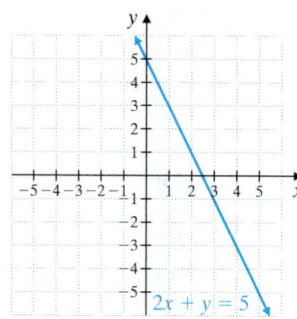

26. Find the slope of the line through $(0, 5)$ and $(-5, 4)$.

27. Find the slope of the line $-2x + 3y = 11$.

28. Find the slope of the line $x = -10$.

29. Find an equation of the line with slope -2 that passes through $(-1, 5)$. Write the equation in slope-intercept form, $y = mx + b$, and in standard form, $Ax + By = C$.

30. Find the slope and y-intercept of the line whose equation is $2x - 5y = 10$.

31. Determine whether each ordered pair is a solution of the inequality $2x - y < 6$.
   a. $(5, -1)$
   b. $(2, 7)$

32. Write an equation of the line through $(2, 3)$ and $(0, 0)$. Write the equation in standard form, $Ax + By = C$.

19. $\left\{ x \middle| x > \dfrac{13}{7} \right\}$ (Sec. 2.7, Ex. 8)

20. $\{y | y \leq 2\}$ (Sec. 2.7)

21. see table (Sec. 3.1, Ex. 7)

22. see table (Sec. 3.1)

23. a. $(0, 12)$

   b. $(2, 6)$

   c. $(-1, 15)$ (Sec. 3.1, Ex. 5)

24. see table (Sec. 3.1)

25. see graph (Sec. 3.2, Ex. 1)

26. $\dfrac{1}{5}$ (Sec. 3.4)

27. $\dfrac{2}{3}$ (Sec. 3.4, Ex. 3)

28. undefined slope (Sec. 3.4)

29. $y = -2x + 3$; $2x + y = 3$ (Sec. 3.5, Ex. 4)

30. $m = \dfrac{2}{5}$; y-intercept: $(0, -2)$ (Sec. 3.5)

31. a. not a solution

   b. is a solution (Sec. 3.6, Ex. 1)

32. $3x - 2y = 0$ (Sec. 3.5)

4

Systems of Equations

In Chapter 3, we graphed equations containing two variables. As we have seen, equations like these are often needed to represent relationships between two different quantities. There are also many opportunities to compare and contrast two such equations, called a *system of equations*. This chapter presents *linear systems* and ways we solve these systems and apply them to real-life situations.

Delivery Methods for Watching Movies Outside the Theater

Movies are not watched just in movie theaters anymore. Consumers can purchase movies for home viewing in such varied formats as DVD, Blu-ray, and digital downloads to television, computer, or other electronic device. Technology is changing so fast that there may well be new ways of obtaining videos that are in the development stream now. The graph below represents consumer spending statistics for two different home entertainment formats, DVD and Blu-ray. In Section 4.2, Exercise 57, we will explore the relationship between consumer spending on these two home entertainment formats.

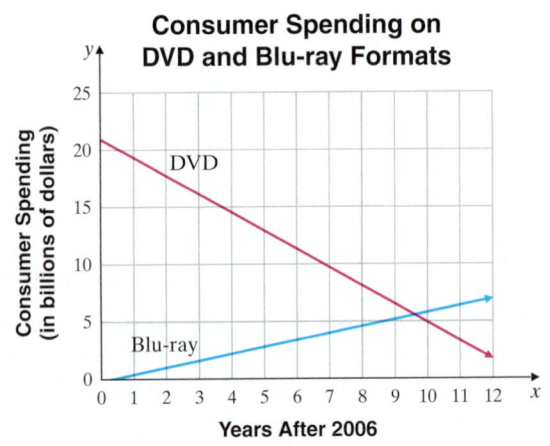

Source: Based on data from DEG: The Digital Entertainment Group

Solve.

19. $-5x + 7 < 2(x - 3)$

20. $-8y + 4 \geq 3(y - 6)$

Complete each table for the given equation.

21. $y = \dfrac{1}{2}x - 5$

	x	y
a.	-2	-6
b.	0	-5
c.	10	0

22. $y = -\dfrac{3}{4}x + 5$

	x	y
a.	4	2
b.	0	5
c.	$\dfrac{20}{3}$	0

23. Complete each ordered pair so that it is a solution of the equation $3x + y = 12$.

a. $(0, \quad)$
b. $(\quad, 6)$
c. $(-1, \quad)$

24. Complete the table for $y = -5x$.

x	y
0	0
-1	5
-2	10

25. Graph the linear equation $2x + y = 5$.

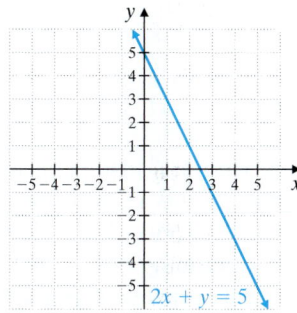

26. Find the slope of the line through $(0, 5)$ and $(-5, 4)$.

27. Find the slope of the line $-2x + 3y = 11$.

28. Find the slope of the line $x = -10$.

29. Find an equation of the line with slope -2 that passes through $(-1, 5)$. Write the equation in slope-intercept form, $y = mx + b$, and in standard form, $Ax + By = C$.

30. Find the slope and y-intercept of the line whose equation is $2x - 5y = 10$.

31. Determine whether each ordered pair is a solution of the inequality $2x - y < 6$.

a. $(5, -1)$
b. $(2, 7)$

32. Write an equation of the line through $(2, 3)$ and $(0, 0)$. Write the equation in standard form, $Ax + By = C$.

4

Systems of Equations

In Chapter 3, we graphed equations containing two variables. As we have seen, equations like these are often needed to represent relationships between two different quantities. There are also many opportunities to compare and contrast two such equations, called a *system of equations*. This chapter presents *linear systems* and ways we solve these systems and apply them to real-life situations.

Delivery Methods for Watching Movies Outside the Theater

Movies are not watched just in movie theaters anymore. Consumers can purchase movies for home viewing in such varied formats as DVD, Blu-ray, and digital downloads to television, computer, or other electronic device. Technology is changing so fast that there may well be new ways of obtaining videos that are in the development stream now. The graph below represents consumer spending statistics for two different home entertainment formats, DVD and Blu-ray. In Section 4.2, Exercise 57, we will explore the relationship between consumer spending on these two home entertainment formats.

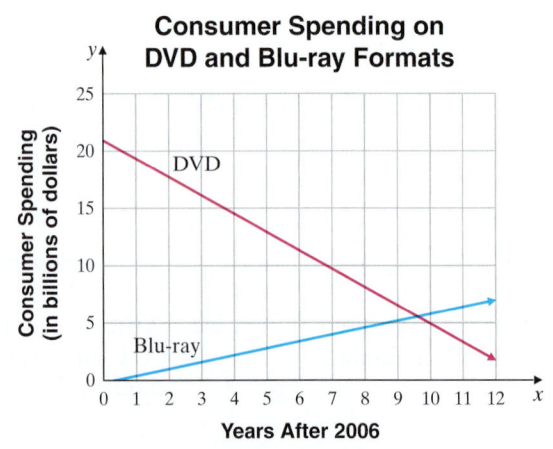

Source: Based on data from DEG: The Digital Entertainment Group

4.1 Solving Systems of Linear Equations by Graphing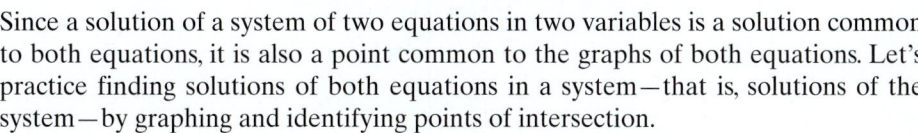

A **system of linear equations** consists of two or more linear equations. In this section, we focus on solving systems of linear equations containing two equations in two variables. Examples of such linear systems are

$$\begin{cases} 3x - 3y = 0 \\ x = 2y \end{cases} \quad \begin{cases} x - y = 0 \\ 2x + y = 10 \end{cases} \quad \begin{cases} y = 7x - 1 \\ y = 4 \end{cases}$$

Objective A Deciding Whether an Ordered Pair Is a Solution

A **solution** of a system of two equations in two variables is an ordered pair of numbers that is a solution of both equations in the system.

Example 1 Determine whether $(12, 6)$ is a solution of the system

$$\begin{cases} 2x - 3y = 6 \\ x = 2y \end{cases}$$

Solution: To determine whether $(12, 6)$ is a solution of the system, we replace x with 12 and y with 6 in both equations.

$2x - 3y = 6$ First equation	$x = 2y$ Second equation
$2(12) - 3(6) \stackrel{?}{=} 6$ Let $x = 12$ and $y = 6$.	$12 \stackrel{?}{=} 2(6)$ Let $x = 12$ and $y = 6$.
$24 - 18 \stackrel{?}{=} 6$ Simplify.	$12 = 12$ True
$6 = 6$ True	

Since $(12, 6)$ is a solution of both equations, it is a solution of the system.

■ Work Practice 1

Example 2 Determine whether $(-1, 2)$ is a solution of the system

$$\begin{cases} x + 2y = 3 \\ 4x - y = 6 \end{cases}$$

Solution: We replace x with -1 and y with 2 in both equations.

$x + 2y = 3$ First equation	$4x - y = 6$ Second equation
$-1 + 2(2) \stackrel{?}{=} 3$ Let $x = -1$ and $y = 2$.	$4(-1) - 2 \stackrel{?}{=} 6$ Let $x = -1$ and $y = 2$.
$-1 + 4 \stackrel{?}{=} 3$ Simplify.	$-4 - 2 \stackrel{?}{=} 6$ Simplify.
$3 = 3$ True	$-6 = 6$ False

$(-1, 2)$ is not a solution of the second equation, $4x - y = 6$, so it is not a solution of the system.

■ Work Practice 2

Objective B Solving Systems of Equations by Graphing

Since a solution of a system of two equations in two variables is a solution common to both equations, it is also a point common to the graphs of both equations. Let's practice finding solutions of both equations in a system—that is, solutions of the system—by graphing and identifying points of intersection.

Objectives

A Decide Whether an Ordered Pair Is a Solution of a System of Linear Equations.

B Solve a System of Linear Equations by Graphing.

C Identify Special Systems of Linear Equations.

D Without Graphing, Determine the Number of Solutions of a System.

Practice 1

Determine whether $(3, 9)$ is a solution of the system

$$\begin{cases} 5x - 2y = -3 \\ y = 3x \end{cases}$$

Teaching Tip

See page 285.

Practice 2

Determine whether $(3, -2)$ is a solution of the system

$$\begin{cases} 2x - y = 8 \\ x + 3y = 4 \end{cases}$$

Teaching Tip

If the desks in your room are arranged in columns and rows, begin with a demonstration. Tell students you will use the room's floor as a coordinate plane with the back right corner as the origin and the rows and columns as unit measures of y and x, respectively. Have students representing $y = x$ stand up (row value equals their column value). While they stand, have students representing $y = 3$ stand up (students in row 3). Is anyone a member of both equations? What happens?

Answers

1. $(3, 9)$ is a solution of the system.

2. $(3, -2)$ is not a solution of the system.

281

Practice 3

Solve the system of equations by graphing.

$$\begin{cases} -3x + y = -10 \\ x - y = 6 \end{cases}$$

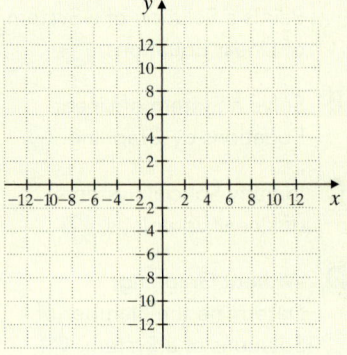

Practice 4

Solve the system of equations by graphing.

$$\begin{cases} x + 3y = -1 \\ y = 1 \end{cases}$$

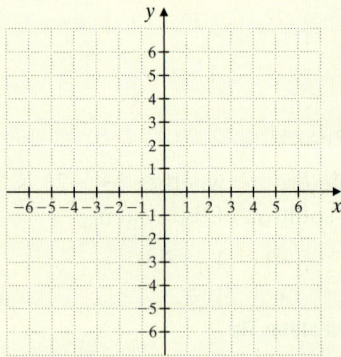

Answers

3. $(2, -4)$;

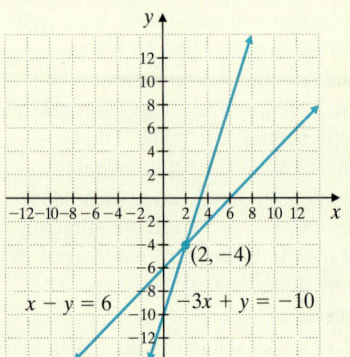

4. See page 284.

Example 3   Solve the system of equations by graphing.

$$\begin{cases} -x + 3y = 10 \\ x + y = 2 \end{cases}$$

Solution: On a single set of axes, graph each linear equation.

$-x + 3y = 10$

x	y
0	$\dfrac{10}{3}$
-4	2
2	4

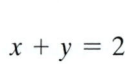

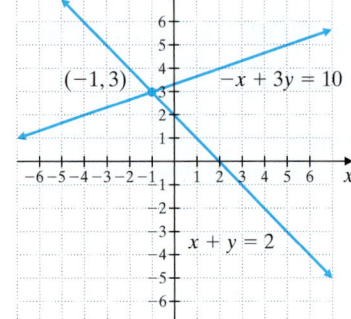

Helpful Hint   The point of intersection gives the solution of the system.

$x + y = 2$

x	y
0	2
2	0
1	1

The two lines appear to intersect at the point $(-1, 3)$. To check, we replace x with -1 and y with 3 in both equations.

$-x + 3y = 10$ First equation $x + y = 2$ Second equation

$-(-1) + 3(3) \stackrel{?}{=} 10$ Let $x = -1$ and $y = 3$. $-1 + 3 \stackrel{?}{=} 2$ Let $x = -1$ and $y = 3$.

$1 + 9 \stackrel{?}{=} 10$ Simplify. $2 = 2$ True

$10 = 10$ True

$(-1, 3)$ checks, so it is the solution of the system.

■ **Work Practice 3**

Helpful Hint

Neatly drawn graphs can help when "guessing" the solution of a system of linear equations by graphing.

Example 4   Solve the system of equations by graphing.

$$\begin{cases} 2x + 3y = -2 \\ x = 2 \end{cases}$$

Solution: We graph each linear equation on a single set of axes.

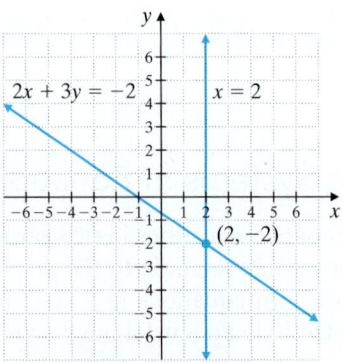

The two lines appear to intersect at the point $(2, -2)$. To determine whether $(2, -2)$ is the solution, we replace x with 2 and y with -2 in both equations.

$2x + 3y = -2$ First equation $x = 2$ Second equation

$2(2) + 3(-2) \stackrel{?}{=} -2$ Let $x = 2$ and $y = -2$. $2 \stackrel{?}{=} 2$ Let $x = 2$.

$4 + (-6) \stackrel{?}{=} -2$ Simplify. $2 = 2$ True

$-2 = -2$ True

Since a true statement results in both equations, $(2, -2)$ is the solution of the system.

■ Work Practice 4

Objective C Identifying Special Systems of Linear Equations ▶

Not all systems of linear equations have a single solution. Some systems have no solution and some have an infinite number of solutions.

Example 5 Solve the system of equations by graphing.

$$\begin{cases} 2x + y = 7 \\ 2y = -4x \end{cases}$$

Solution: We graph the two equations in the system. The equations in slope-intercept form are $y = -2x + 7$ and $y = -2x$. Notice from the equations that the lines have the same slope, -2, and different y-intercepts. This means that the lines are parallel.

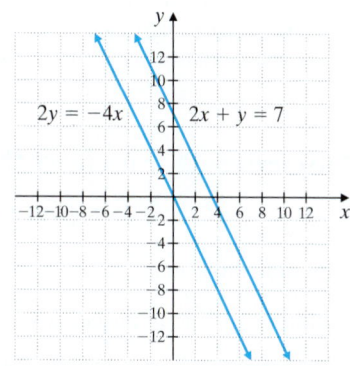

Since the lines are parallel, they do not intersect. This means that the system has *no solution*.

■ Work Practice 5

Example 6 Solve the system of equations by graphing.

$$\begin{cases} x - y = 3 \\ -x + y = -3 \end{cases}$$

Solution: We graph each equation. The graphs of the equations are the same line. To see this, notice that if both sides of the first equation in the system are multiplied by -1, the result is the second equation.

$x - y = 3$ First equation

$-1(x - y) = -1(3)$ Multiply both sides by -1.

$-x + y = -3$ Simplify. This is the second equation.

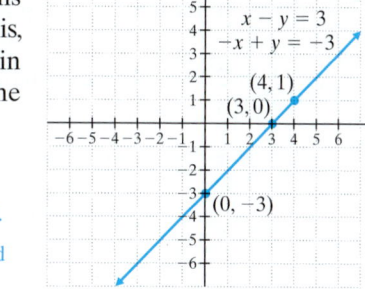

Any ordered pair that is a solution of one equation is a solution of the other equation and is then a solution of the system. This means that the system has an infinite number of solutions.

■ Work Practice 6

Teaching Tip

Before discussing Example 5, do a student demonstration using the floor of the classroom as a coordinate plane with the back right corner as the origin. Have students representing $x = 2$ stand up and students representing $x = 4$ stand up. Then ask, "What is the solution of this system of equations?" Follow up by asking, "Is it possible for these two equations to intersect if we had a bigger room with more desks? Why not?"

Practice 5

Solve the system of equations by graphing.

$$\begin{cases} 3x - y = 6 \\ 6x = 2y \end{cases}$$

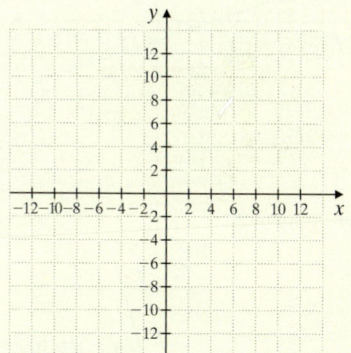

Teaching Tip   See page 285.

Practice 6

Solve the system of equations by graphing.

$$\begin{cases} x + y = -4 \\ -2x - 2y = 8 \end{cases}$$

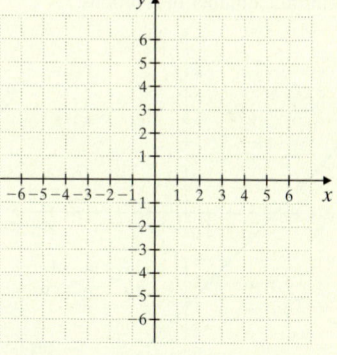

Answers

5. See page 284.   6. See page 284.

Examples 5 and 6 are special cases of systems of linear equations. A system that has no solution is said to be an **inconsistent system.** If the graphs of the two equations of a system are identical, we call the equations **dependent equations.** Thus, the system in Example 5 is an inconsistent system and the equations in the system in Example 6 are dependent equations.

As we have seen, three different situations can occur when graphing the two lines associated with the equations in a linear system. These situations are shown in the figures.

One point of
intersection: one solution

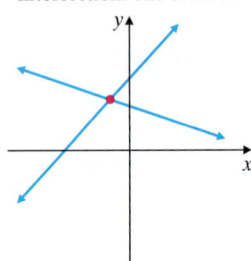

Consistent system
(at least one solution)
Independent equations
(graphs of equations differ)

Parallel lines: no solution

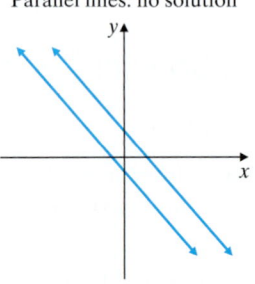

Inconsistent system
(no solution)
Independent equations
(graphs of equations differ)

Same line: infinite number
of solutions

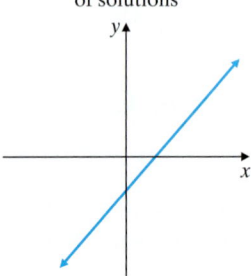

Consistent system
(at least one solution)
Dependent equations
(graphs of
equations identical)

Objective D Finding the Number of Solutions of a System Without Graphing ▶

You may have suspected by now that graphing alone is not an accurate way to solve a system of linear equations. For example, a solution of $\left(\frac{1}{2}, \frac{2}{9}\right)$ is unlikely to be read correctly from a graph. The next two sections present two accurate methods of solving these systems. In the meantime, we can decide how many solutions a system has by writing each equation in slope-intercept form.

Example 7   Without graphing, determine the number of solutions of the system.

$$\begin{cases} \dfrac{1}{2}x - y = 2 \\ x = 2y + 5 \end{cases}$$

Solution: First write each equation in slope-intercept form.

$\dfrac{1}{2}x - y = 2$ First equation

$\dfrac{1}{2}x = y + 2$ Add y to both sides.

$\dfrac{1}{2}x - 2 = y$ Subtract 2 from both sides.

$x = 2y + 5$ Second equation

$x - 5 = 2y$ Subtract 5 from both sides.

$\dfrac{x}{2} - \dfrac{5}{2} = \dfrac{2y}{2}$ Divide both sides by 2.

$\dfrac{1}{2}x - \dfrac{5}{2} = y$ Simplify.

The slope of each line is $\dfrac{1}{2}$, but they have different y-intercepts. This tells us that the lines representing these equations are parallel. Since the lines are parallel, the system has no solution and is inconsistent.

■ **Work Practice 7**

Copyright 2016 Pearson Education, Inc.

Practice 7

Without graphing, determine the number of solutions of the system.

$$\begin{cases} 5x + 4y = 6 \\ x - y = 3 \end{cases}$$

Answers

4. $(-4, 1)$;

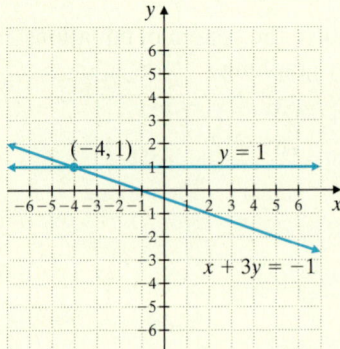

5. no solution;

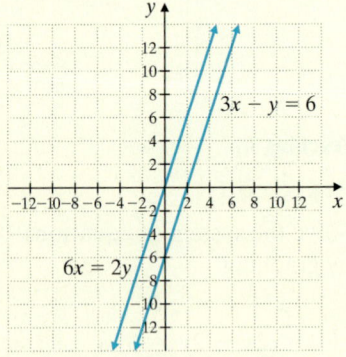

6. infinite number of solutions;

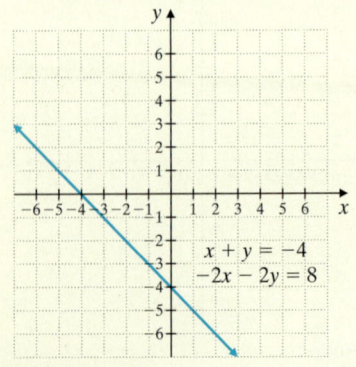

7. one solution

Example 8 Without graphing, determine the number of solutions of the system.

$$\begin{cases} 3x - y = 4 \\ x + 2y = 8 \end{cases}$$

Solution: Once again, the slope-intercept form helps determine how many solutions this system has.

$3x - y = 4$	First equation	$x + 2y = 8$	Second equation
$3x = y + 4$	Add y to both sides.	$x = -2y + 8$	Subtract $2y$ from both sides.
$3x - 4 = y$	Subtract 4 from both sides.	$x - 8 = -2y$	Subtract 8 from both sides.
		$\dfrac{x}{-2} - \dfrac{8}{-2} = \dfrac{-2y}{-2}$	Divide both sides by -2.
		$-\dfrac{1}{2}x + 4 = y$	Simplify.

The slope of the second line is $-\dfrac{1}{2}$, whereas the slope of the first line is 3. Since the slopes are not equal, the two lines are neither parallel nor identical and must intersect. Therefore, this system has one solution and is consistent.

■ **Work Practice 8**

Teaching Tip

Before discussing Example 1, discuss the various scenarios using the graph to the right.
Name a point not on line l or line m. (*S*)
Name a point on line l but not on line m. (*T*)
Name a point on line m but not on line l. (*Q*)
Name a point on both line l and line m. (*R*)
Then state that a point is a solution of a system of equations only when it is on the graphs of all the equations in the system.
Finally, ask, "How can we determine if a point will be on the graph without graphing the equation?" (Substitute the coordinates of the point into the equation and see if we get a true statement.)

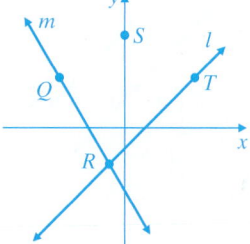

Practice 8

Without graphing, determine the number of solutions of the system.

$$\begin{cases} -\dfrac{2}{3}x + y = 6 \\ 3y = 2x + 5 \end{cases}$$

Teaching Tip

Before discussing Example 6, do a student demonstration using the floor of the classroom as a coordinate plane. Have students representing $3y = 12$ (three times their row value equals twelve) stand up. After asking these students to sit down, have students representing $\dfrac{y}{2} = 2$ (half their row value equals two) stand up. Then ask students to sit down unless they were also a solution of the first equation. (All students remain standing.) Why do we have so many solutions to this system? (The two equations are equivalent.) What is the simplest form of both equations? ($y = 4$)

Answer

8. no solution

📟 **Calculator Explorations Graphing**

A graphing calculator may be used to approximate solutions of systems of equations. For example, to approximate the solution of the system

$$\begin{cases} y = -3.14x - 1.35 \\ y = 4.88x + 5.25, \end{cases}$$

first graph each equation on the same set of axes. Then use the Intersect feature of your calculator to approximate the point of intersection.

  The approximate point of intersection is $(-0.82, 1.23)$.

Solve each system of equations. Approximate the solutions to two decimal places.

1. $\begin{cases} y = -2.68x + 1.21 \\ y = 5.22x - 1.68 \end{cases}$
$(0.37, 0.23)$

2. $\begin{cases} y = 4.25x + 3.89 \\ y = -1.88x + 3.21 \end{cases}$
$(-0.11, 3.42)$

3. $\begin{cases} 4.3x - 2.9y = 5.6 \\ 8.1x + 7.6y = -14.1 \end{cases}$
$(0.03, -1.89)$

4. $\begin{cases} -3.6x - 8.6y = 10 \\ -4.5x + 9.6y = -7.7 \end{cases}$
$(-0.41, -0.99)$

Teaching Tip

Have students approximate the points of intersection using the TRACE feature first and then using the Intersect option.

Vocabulary, Readiness & Video Check

Fill in each blank with one of the words or phrases listed below.

system of linear equations	solution	consistent
dependent	inconsistent	independent

1. In a system of linear equations in two variables, if the graphs of the equations are the same, the equations are _____*dependent*_____ equations.

2. Two or more linear equations are called a(n) ___*system of linear equations*___.

3. A system of equations that has at least one solution is called a(n) _____*consistent*_____ system.

4. A(n) _____*solution*_____ of a system of two equations in two variables is an ordered pair of numbers that is a solution of both equations in the system.

5. A system of equations that has no solution is called a(n) _____*inconsistent*_____ system.

6. In a system of linear equations in two variables, if the graphs of the equations are different, the equations are _____*independent*_____ equations.

Each rectangular coordinate system shows the graph of the equations in a system of equations. Use each graph to determine the number of solutions for each associated system. If the system has only one solution, give its coordinates.

7.
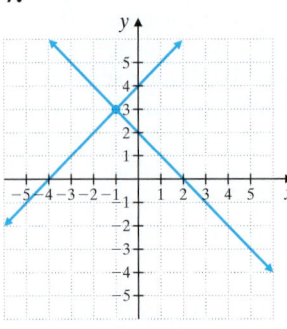
1 solution, $(-1, 3)$

8.
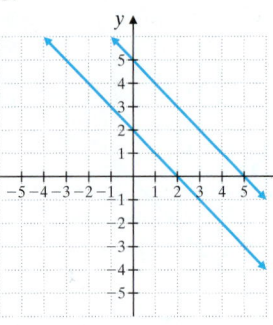
no solution

9.
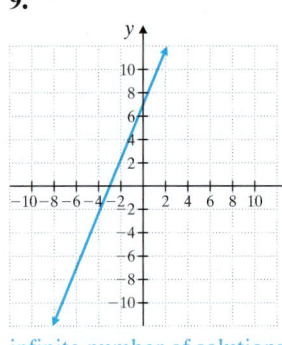
infinite number of solutions

10.
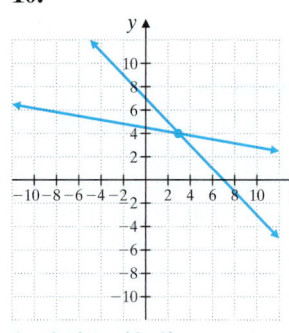
1 solution, $(3, 4)$

Martin-Gay Interactive Videos   *Watch the section lecture video and answer the following questions.*

Objective A 11. In ⊞ Example 1, the first ordered pair is a solution of the first equation of the system. Why is this not enough to determine whether this ordered pair is a solution of the system? ▶

Objectives B C 12. From ⊞ Examples 2 and 3, why is finding the solution of a system of equations from a graph considered "guessing" and this proposed solution checked algebraically? ▶

Objective D 13. From ⊞ Examples 5–7, explain how the slope-intercept form tells us how many solutions a system of equations has. ▶

See Video 4.1 🍎

See video answer section.

Example 8   Without graphing, determine the number of solutions of the system.

$$\begin{cases} 3x - y = 4 \\ x + 2y = 8 \end{cases}$$

Solution: Once again, the slope-intercept form helps determine how many solutions this system has.

$3x - y = 4$ First equation	$x + 2y = 8$ Second equation
$3x = y + 4$ Add y to both sides.	$x = -2y + 8$ Subtract $2y$ from both sides.
$3x - 4 = y$ Subtract 4 from both sides.	$x - 8 = -2y$ Subtract 8 from both sides.
	$\dfrac{x}{-2} - \dfrac{8}{-2} = \dfrac{-2y}{-2}$ Divide both sides by -2.
	$-\dfrac{1}{2}x + 4 = y$ Simplify.

The slope of the second line is $-\dfrac{1}{2}$, whereas the slope of the first line is 3. Since the slopes are not equal, the two lines are neither parallel nor identical and must intersect. Therefore, this system has one solution and is consistent.

■ **Work Practice 8**

Teaching Tip

Before discussing Example 1, discuss the various scenarios using the graph to the right.
Name a point not on line *l* or line *m*. (*S*)
Name a point on line *l* but not on line *m*. (*T*)
Name a point on line *m* but not on line *l*. (*Q*)
Name a point on both line *l* and line *m*. (*R*)
Then state that a point is a solution of a system of equations only when it is on the graphs of all the equations in the system.
Finally, ask, "How can we determine if a point will be on the graph without graphing the equation?" (Substitute the coordinates of the point into the equation and see if we get a true statement.)

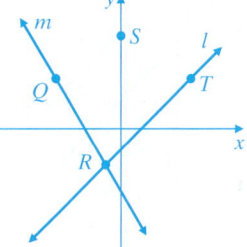

Practice 8

Without graphing, determine the number of solutions of the system.

$$\begin{cases} -\dfrac{2}{3}x + y = 6 \\ 3y = 2x + 5 \end{cases}$$

Teaching Tip

Before discussing Example 6, do a student demonstration using the floor of the classroom as a coordinate plane. Have students representing $3y = 12$ (three times their row value equals twelve) stand up. After asking these students to sit down, have students representing $\dfrac{y}{2} = 2$ (half their row value equals two) stand up. Then ask students to sit down unless they were also a solution of the first equation. (All students remain standing.) Why do we have so many solutions to this system? (The two equations are equivalent.) What is the simplest form of both equations? ($y = 4$)

Answer

8. no solution

📟 **Calculator Explorations Graphing**

A graphing calculator may be used to approximate solutions of systems of equations. For example, to approximate the solution of the system

$$\begin{cases} y = -3.14x - 1.35 \\ y = 4.88x + 5.25, \end{cases}$$

first graph each equation on the same set of axes. Then use the Intersect feature of your calculator to approximate the point of intersection.

The approximate point of intersection is $(-0.82, 1.23)$.

Solve each system of equations. Approximate the solutions to two decimal places.

1. $\begin{cases} y = -2.68x + 1.21 \\ y = 5.22x - 1.68 \end{cases}$
$(0.37, 0.23)$

2. $\begin{cases} y = 4.25x + 3.89 \\ y = -1.88x + 3.21 \end{cases}$
$(-0.11, 3.42)$

3. $\begin{cases} 4.3x - 2.9y = 5.6 \\ 8.1x + 7.6y = -14.1 \end{cases}$
$(0.03, -1.89)$

4. $\begin{cases} -3.6x - 8.6y = 10 \\ -4.5x + 9.6y = -7.7 \end{cases}$
$(-0.41, -0.99)$

Teaching Tip

Have students approximate the points of intersection using the TRACE feature first and then using the Intersect option.

Vocabulary, Readiness & Video Check

Fill in each blank with one of the words or phrases listed below.

system of linear equations	solution	consistent
dependent	inconsistent	independent

1. In a system of linear equations in two variables, if the graphs of the equations are the same, the equations are _____dependent_____ equations.

2. Two or more linear equations are called a(n) __system of linear equations__.

3. A system of equations that has at least one solution is called a(n) _____consistent_____ system.

4. A(n) _____solution_____ of a system of two equations in two variables is an ordered pair of numbers that is a solution of both equations in the system.

5. A system of equations that has no solution is called a(n) _____inconsistent_____ system.

6. In a system of linear equations in two variables, if the graphs of the equations are different, the equations are _____independent_____ equations.

Each rectangular coordinate system shows the graph of the equations in a system of equations. Use each graph to determine the number of solutions for each associated system. If the system has only one solution, give its coordinates.

7.

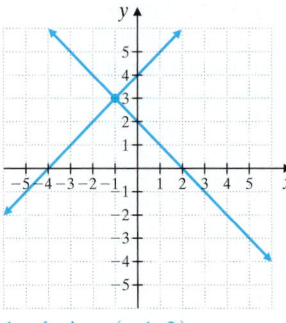

1 solution, $(-1, 3)$

8.

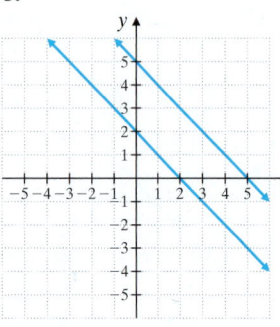

no solution

9.

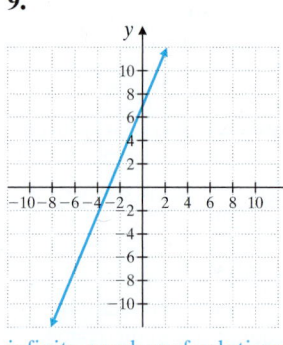

infinite number of solutions

10.

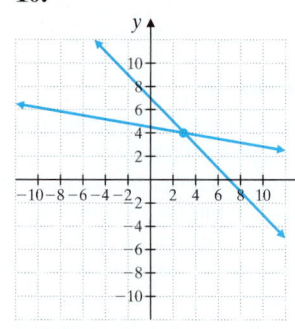

1 solution, $(3, 4)$

Martin-Gay Interactive Videos   *Watch the section lecture video and answer the following questions.*

See Video 4.1

See video answer section.

Objective **A** 11. In  Example 1, the first ordered pair is a solution of the first equation of the system. Why is this not enough to determine whether this ordered pair is a solution of the system? ▶

Objectives **B**
C 12. From  Examples 2 and 3, why is finding the solution of a system of equations from a graph considered "guessing" and this proposed solution checked algebraically? ▶

Objective **D** 13. From 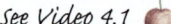 Examples 5–7, explain how the slope-intercept form tells us how many solutions a system of equations has. ▶

4.1 Exercise Set MyMathLab®

Objective A *Determine whether each ordered pair is a solution of the system of linear equations. See Examples 1 and 2.*

1. $\begin{cases} x + y = 8 \\ 3x + 2y = 21 \end{cases}$
a. $(2, 4)$ no
b. $(5, 3)$ yes

2. $\begin{cases} 2x + y = 5 \\ x + 3y = 5 \end{cases}$
a. $(5, 0)$ no
b. $(2, 1)$ yes

⊙ 3. $\begin{cases} 3x - y = 5 \\ x + 2y = 11 \end{cases}$
a. $(3, 4)$ yes
b. $(0, -5)$ no

4. $\begin{cases} 2x - 3y = 8 \\ x - 2y = 6 \end{cases}$
a. $(-2, -4)$ yes
b. $(7, 2)$ no

5. $\begin{cases} 2y = 4x + 6 \\ 2x - y = -3 \end{cases}$
a. $(-3, -3)$ yes
b. $(0, 3)$ yes

6. $\begin{cases} x + 5y = -4 \\ -2x = 10y + 8 \end{cases}$
a. $(-4, 0)$ yes
b. $(6, -2)$ yes

7. $\begin{cases} -2 = x - 7y \\ 6x - y = 13 \end{cases}$
a. $(-2, 0)$ no
b. $\left(\dfrac{1}{2}, \dfrac{5}{14}\right)$ no

8. $\begin{cases} 4x = 1 - y \\ x - 3y = -8 \end{cases}$
a. $(0, 1)$ no
b. $\left(\dfrac{1}{6}, \dfrac{1}{3}\right)$ no

Objectives B C *Solve each system of linear equations by graphing. See Examples 3 through 6.*

9. $\begin{cases} x + y = 4 \\ x - y = 2 \end{cases}$

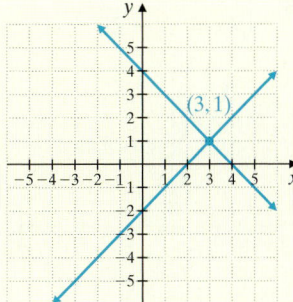

10. $\begin{cases} x + y = 3 \\ x - y = 5 \end{cases}$

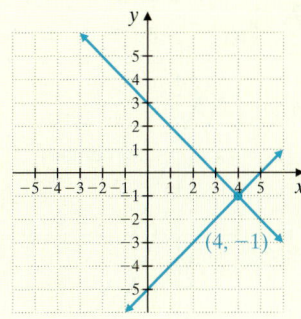

11. $\begin{cases} x + y = 6 \\ -x + y = -6 \end{cases}$

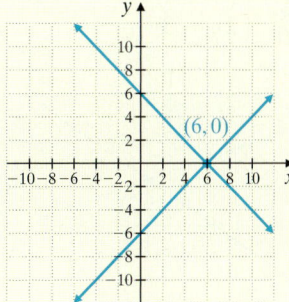

12. $\begin{cases} x + y = 1 \\ -x + y = -3 \end{cases}$

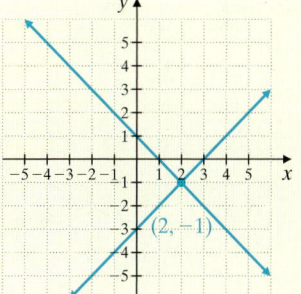

13. $\begin{cases} y = 2x \\ 3x - y = -2 \end{cases}$

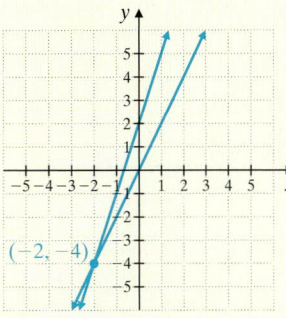

14. $\begin{cases} y = -3x \\ 2x - y = -5 \end{cases}$

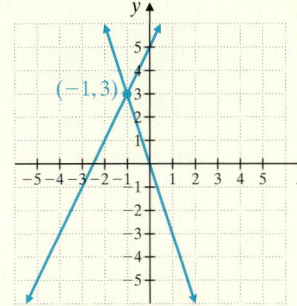

15. $\begin{cases} y = x + 1 \\ y = 2x - 1 \end{cases}$

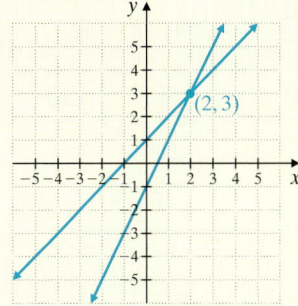

16. $\begin{cases} y = 3x - 4 \\ y = x + 2 \end{cases}$

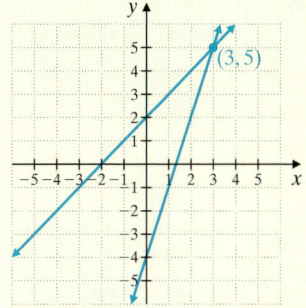

17. $\begin{cases} 2x + y = 0 \\ 3x + y = 1 \end{cases}$

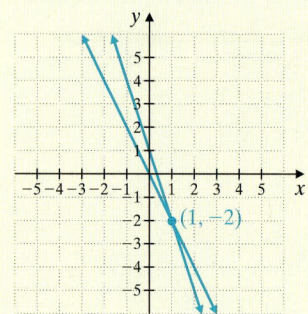

(1, −2)

18. $\begin{cases} 2x + y = 1 \\ 3x + y = 0 \end{cases}$

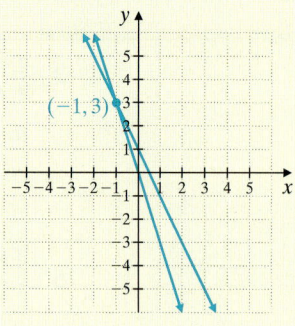

(−1, 3)

19. $\begin{cases} y = -x - 1 \\ y = 2x + 5 \end{cases}$

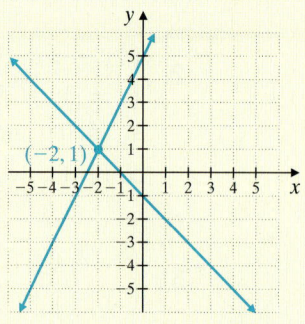

(−2, 1)

20. $\begin{cases} y = x - 1 \\ y = -3x - 5 \end{cases}$

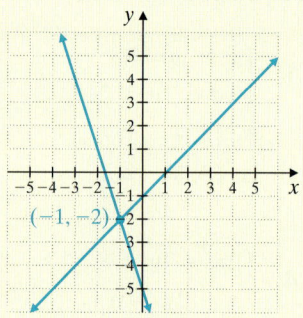

(−1, −2)

21. $\begin{cases} x + y = 5 \\ x + y = 6 \end{cases}$

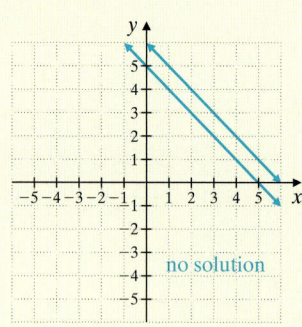

no solution

22. $\begin{cases} x - y = 4 \\ x - y = 1 \end{cases}$

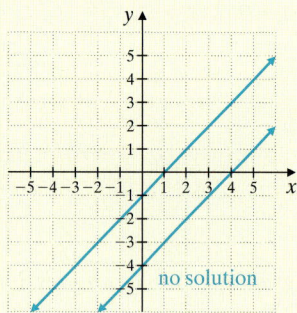

no solution

23. $\begin{cases} 2x - y = 6 \\ y = 2 \end{cases}$

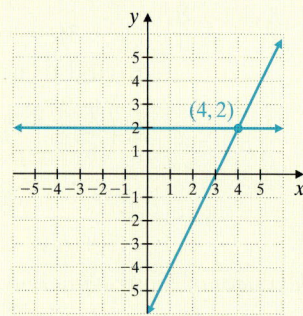

(4, 2)

24. $\begin{cases} x + y = 5 \\ x = 4 \end{cases}$

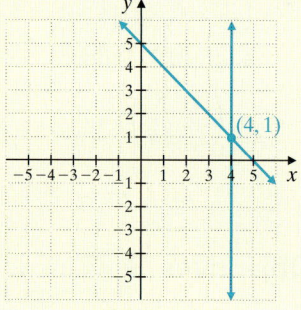

(4, 1)

25. $\begin{cases} x - 2y = 2 \\ 3x + 2y = -2 \end{cases}$

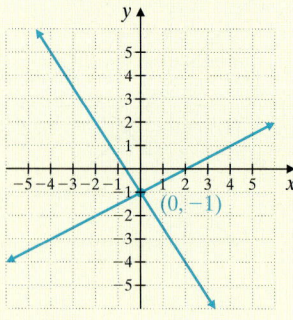

(0, −1)

26. $\begin{cases} x + 3y = 7 \\ 2x - 3y = -4 \end{cases}$

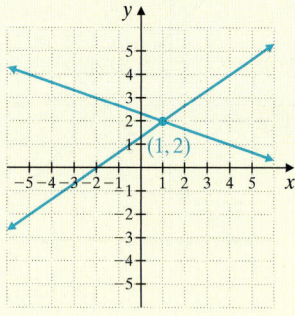

(1, 2)

27. $\begin{cases} 2x + y = 4 \\ 6x = -3y + 6 \end{cases}$

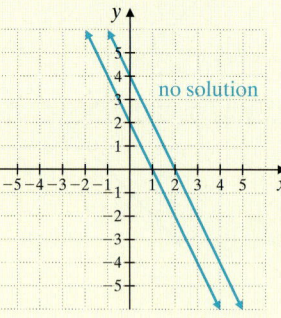

no solution

28. $\begin{cases} y + 2x = 3 \\ 4x = 2 - 2y \end{cases}$

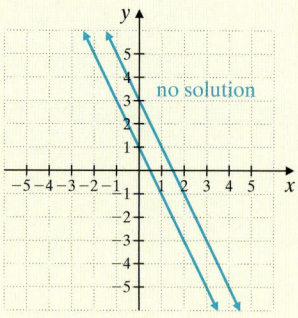

no solution

29. $\begin{cases} y - 3x = -2 \\ 6x - 2y = 4 \end{cases}$

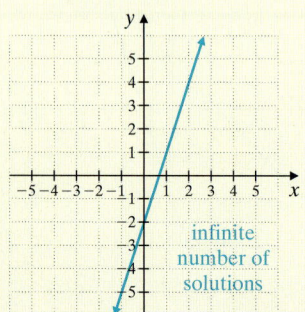

infinite number of solutions

30. $\begin{cases} x - 2y = -6 \\ -2x + 4y = 12 \end{cases}$

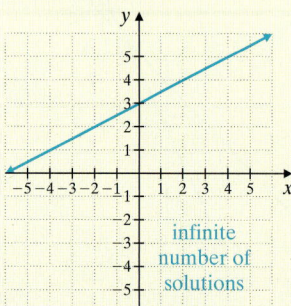

infinite number of solutions

31. $\begin{cases} x = 3 \\ y = -1 \end{cases}$

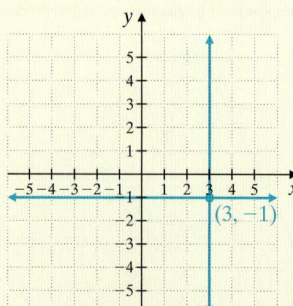

(3, −1)

32. $\begin{cases} x = -5 \\ y = 3 \end{cases}$

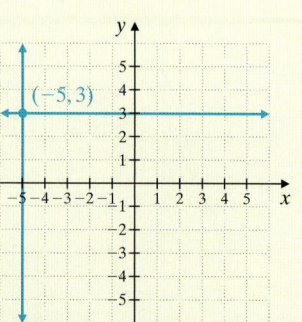

(−5, 3)

33. $\begin{cases} y = x - 2 \\ y = 2x + 3 \end{cases}$

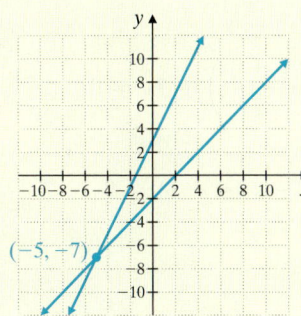

34. $\begin{cases} y = x + 5 \\ y = -2x - 4 \end{cases}$

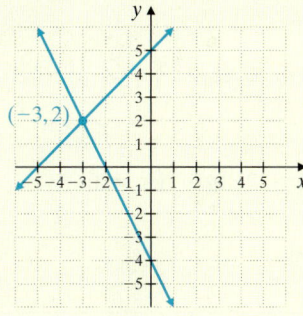

35. $\begin{cases} 2x - 3y = -2 \\ -3x + 5y = 5 \end{cases}$

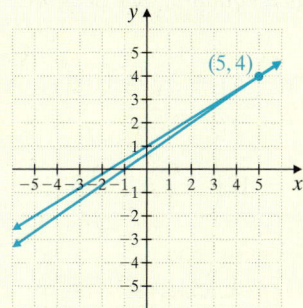

36. $\begin{cases} 4x - y = 7 \\ 2x - 3y = -9 \end{cases}$

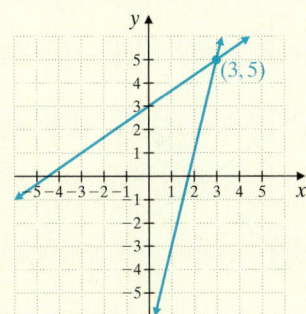

37. $\begin{cases} 6x - y = 4 \\ \dfrac{1}{2}y = -2 + 3x \end{cases}$

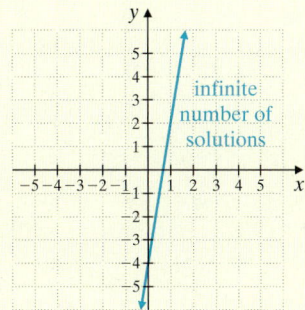

38. $\begin{cases} 3x - y = 6 \\ \dfrac{1}{3}y = -2 + x \end{cases}$

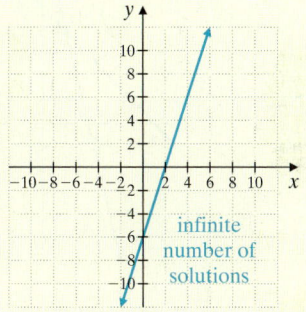

Objective D *Without graphing, decide:*

a. *Are the graphs of the equations identical lines, parallel lines, or lines intersecting at a single point?*
b. *How many solutions does the system have? See Examples 7 and 8.*
For Exercises 39–52, the first answer given is the answer for part **a**, and the second answer given is the answer for part **b**.

39. $\begin{cases} 4x + y = 24 \\ x + 2y = 2 \end{cases}$
intersecting; one solution

40. $\begin{cases} 3x + y = 1 \\ 3x + 2y = 6 \end{cases}$
intersecting; one solution

41. $\begin{cases} 2x + y = 0 \\ 2y = 6 - 4x \end{cases}$
parallel; no solution

42. $\begin{cases} 3x + y = 0 \\ 2y = -6x \end{cases}$
identical lines; infinite number
of solutions

43. $\begin{cases} 6x - y = 4 \\ \dfrac{1}{2}y = -2 + 3x \end{cases}$
identical lines; infinite number
of solutions

44. $\begin{cases} 3x - y = 2 \\ \dfrac{1}{3}y = -2 + 3x \end{cases}$
intersecting; one solution

45. $\begin{cases} x = 5 \\ y = -2 \end{cases}$
intersecting; one solution

46. $\begin{cases} y = 3 \\ x = -4 \end{cases}$
intersecting; one solution

47. $\begin{cases} 3y - 2x = 3 \\ x + 2y = 9 \end{cases}$
intersecting; one solution

48. $\begin{cases} 2y = x + 2 \\ y + 2x = 3 \end{cases}$
intersecting; one solution

49. $\begin{cases} 6y + 4x = 6 \\ 3y - 3 = -2x \end{cases}$
identical lines; infinite number
of solutions

50. $\begin{cases} 8y + 6x = 4 \\ 4y - 2 = 3x \end{cases}$
intersecting; one solution

51. $\begin{cases} x + y = 4 \\ x + y = 3 \end{cases}$
parallel; no solution

52. $\begin{cases} 2x + y = 0 \\ y = -2x + 1 \end{cases}$
parallel; no solution

Review

Solve each equation. See Section 2.3.

53. $5(x - 3) + 3x = 1$ 2

54. $-2x + 3(x + 6) = 17$ -1

55. $4\left(\dfrac{y + 1}{2}\right) + 3y = 0$ $-\dfrac{2}{5}$

56. $-y + 12\left(\dfrac{y - 1}{4}\right) = 3$ 3

57. $8a - 2(3a - 1) = 6$ 2

58. $3z - (4z - 2) = 9$ -7

Concept Extensions

59. Draw a graph of two linear equations whose associated system has the solution $(-1, 4)$.
answers may vary; possible answer:

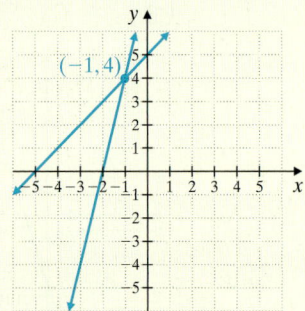

60. Draw a graph of two linear equations whose associated system has the solution $(3, -2)$.
answers may vary; possible answer:

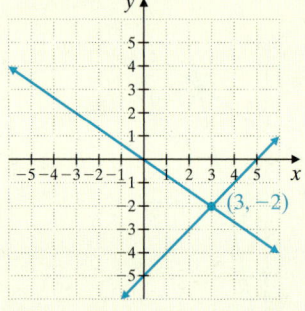

61. Draw a graph of two linear equations whose associated system has no solution.
answers may vary; possible answer:

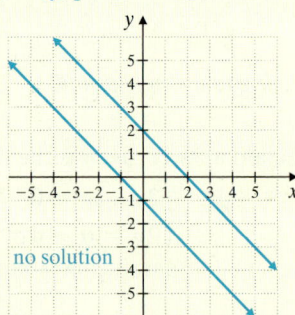

62. Draw a graph of two linear equations whose associated system has an infinite number of solutions.
answers may vary; possible answer:

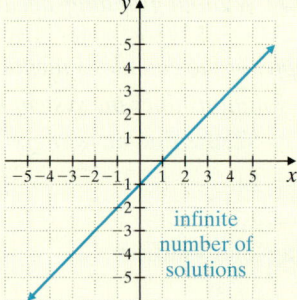

The double line graph below shows the number of digital and analog movie screens in U.S. cinemas for the years shown. Use this graph to answer Exercises 63 and 64. (Source: Motion Picture Association of America, Inc.)

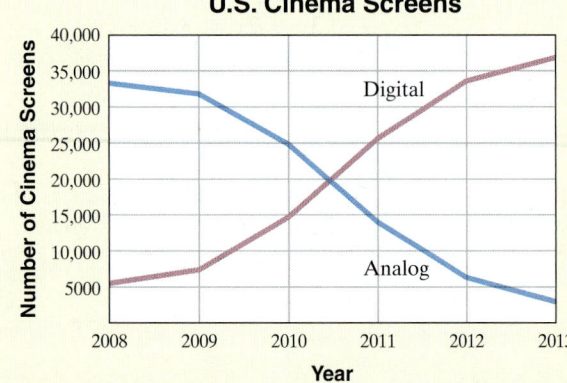

U.S. Cinema Screens

Source: Motion Picture Association of America, Inc.

63. Between what pairs of years did the number of digital cinema screens equal the number of analog cinema screens?   2010–2011

64. For what year(s) was the number of digital cinema screens less than the number of analog cinema screens?   2008, 2009, 2010

The double line graph below shows the average attendance per game for the Cleveland Indians and the Pittsburgh Pirates baseball teams for the years shown. Use this for Exercises 65 and 66. (Source: Baseball Almanac)

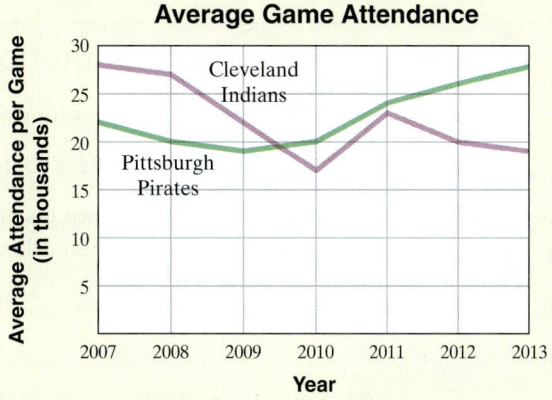

Average Game Attendance

Source: Baseball Almanac

65. In what year(s) was the average attendance per game for the Pittsburgh Pirates greater than the average attendance per game for the Cleveland Indians?   2010, 2011, 2012, 2013

66. In what year was the average attendance per game for the Cleveland Indians closest to the average attendance per game for the Pittsburgh Pirates?   2011

67. Construct a system of two linear equations that has $(2, 5)$ as a solution.   answers may vary

68. Construct a system of two linear equations that has $(0, 1)$ as a solution.   answers may vary

69. The ordered pair $(-2, 3)$ is a solution of the three linear equations below:

$$x + y = 1$$
$$2x - y = -7$$
$$x + 3y = 7$$

If each equation has a distinct graph, describe the graph of all three equations on the same axes.
answers may vary

70. Explain how to use a graph to determine the number of solutions of a system.   answers may vary

71. Below are tables of values for two linear equations.
  a. Find a solution of the corresponding system. $(4, 9)$
  b. Graph several ordered pairs from each table and sketch the two lines.

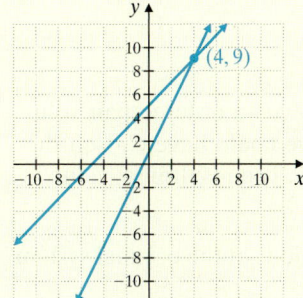

x	y
1	3
2	5
3	7
4	9
5	11

x	y
1	6
2	7
3	8
4	9
5	10

72. Below are tables of values for two linear equations.
  a. Find a solution of the corresponding system. $(-1, 1)$
  b. Graph several ordered pairs from each table and sketch the two lines.

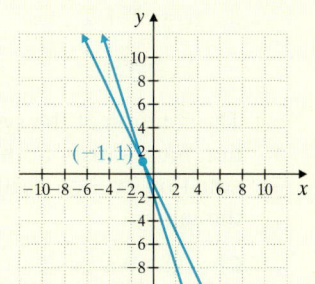

x	y
-3	5
-1	1
0	-1
1	-3
2	-5

x	y
-3	7
-1	1
0	-2
1	-5
2	-8

c. Does your graph confirm the solution from part **a**?
yes

c. Does your graph confirm the solution from part **a**?
yes

4.2 Solving Systems of Linear Equations by Substitution

Objective

A Use the Substitution Method to Solve a System of Linear Equations.

Objective A Using the Substitution Method

You may have suspected by now that graphing alone is not an accurate way to solve a system of linear equations. For example, a solution of $\left(\frac{1}{2}, \frac{2}{9}\right)$ is unlikely to be read correctly from a graph. In this section, we discuss a second, more accurate method for solving systems of equations. This method is called the **substitution method** and is introduced in the next example.

Practice 1

Use the substitution method to solve the system:

$$\begin{cases} 2x + 3y = 13 \\ x = y + 4 \end{cases}$$

Example 1 Solve the system:

$$\begin{cases} 2x + y = 10 & \text{First equation} \\ x = y + 2 & \text{Second equation} \end{cases}$$

Solution: The second equation in this system is $x = y + 2$. This tells us that x and $y + 2$ have the same value. This means that we may substitute $y + 2$ for x in the first equation.

$$2x + y = 10 \quad \text{First equation}$$

$$2(y + 2) + y = 10 \quad \text{Substitute } y + 2 \text{ for } x \text{ since } x = y + 2.$$

Notice that this equation now has one variable, y. Let's now solve this equation for y.

Helpful Hint  Don't forget the distributive property.

$$2(y + 2) + y = 10$$
$$2y + 4 + y = 10 \quad \text{Apply the distributive property.}$$
$$3y + 4 = 10 \quad \text{Combine like terms.}$$
$$3y = 6 \quad \text{Subtract 4 from both sides.}$$
$$y = 2 \quad \text{Divide both sides by 3.}$$

Teaching Tip

Consider beginning this lesson with the following activity: Graph the system of equations

$$\begin{cases} y = x \\ y = 4x - 7 \end{cases}$$

on graph chart paper or an overhead projector. Ask a student to put a sticky dot on the solution(s) to the system. Then ask students to estimate the solution as a point. Ask why it is difficult to give an exact answer. Lead students to notice this is because the intersection point does not occur on intersecting grid lines. Point out that in this section, students will learn to find points of intersection without graphing.

Now we know that the y-value of the ordered pair solution of the system is 2. To find the corresponding x-value, we replace y with 2 in the second equation, $x = y + 2$, and solve for x.

$$x = y + 2 \quad \text{Second equation}$$
$$x = 2 + 2 \quad \text{Let } y = 2.$$
$$x = 4$$

The solution of the system is the ordered pair $(4, 2)$. Since an ordered pair solution must satisfy both linear equations in the system, we could have chosen the equation $2x + y = 10$ to find the corresponding x-value. The resulting x-value is the same.

Check: We check to see that $(4, 2)$ satisfies both equations of the original system.

First Equation	Second Equation
$2x + y = 10$	$x = y + 2$
$2(4) + 2 \stackrel{?}{=} 10$	$4 \stackrel{?}{=} 2 + 2$ Let $x = 4$ and $y = 2.$
$10 = 10$ True	$4 = 4$ True

Answer

1. $(5, 1)$

292

The solution of the system is $(4, 2)$.

A graph of the two equations shows the two lines intersecting at the point $(4, 2)$.

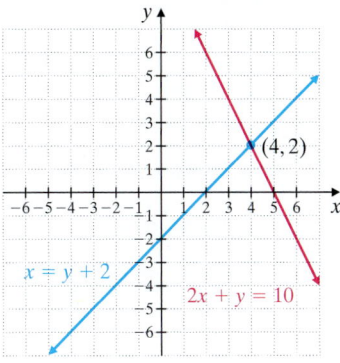

■ **Work Practice 1**

Example 2 Solve the system:

$$\begin{cases} 5x - y = -2 \\ y = 3x \end{cases}$$

Solution: The second equation is solved for y in terms of x. We substitute $3x$ for y in the first equation.

$5x - y = -2$ First equation

$5x - (3x) = -2$ Substitute $3x$ for y.

Now we solve for x.

$5x - 3x = -2$

$2x = -2$ Combine like terms.

$x = -1$ Divide both sides by 2.

The x-value of the ordered pair solution is -1. To find the corresponding y-value, we replace x with -1 in the second equation, $y = 3x$.

$y = 3x$ Second equation

$y = 3(-1)$ Let $x = -1$.

$y = -3$

Check to see that the solution of the system is $(-1, -3)$.

■ **Work Practice 2**

To solve a system of equations by substitution, we first need an equation solved for one of its variables, as in Examples 1 and 2. If neither equation in a system is solved for x or y, this will be our first step.

Example 3 Solve the system:

$$\begin{cases} x + 2y = 7 \\ 2x + 2y = 13 \end{cases}$$

Solution: Notice that neither equation is solved for x or y. Thus, we choose one of the equations and solve for x or y. We will solve the first equation for x so that we will not introduce tedious fractions when solving. To solve the first equation for x, we subtract $2y$ from both sides.

$x + 2y = 7$ First equation

$x = 7 - 2y$ Subtract $2y$ from both sides.

(Continued on next page)

Practice 2

Use the substitution method to solve the system:

$$\begin{cases} 4x - y = 2 \\ y = 5x \end{cases}$$

Teaching Tip

After discussing Example 2, ask students how the substitution helps us solve the first equation for the variable x. Help students see that with substitution, the equation in two variables is written as an equation in one variable.

Practice 3

Solve the system:

$$\begin{cases} 3x + y = 5 \\ 3x - 2y = -7 \end{cases}$$

Answers

2. $(-2, -10)$ **3.** $\left(\frac{1}{3}, 4\right)$

Since $x = 7 - 2y$, we now substitute $7 - 2y$ for x in the second equation and solve for y.

$$2x + 2y = 13 \quad \text{Second equation}$$
$$2(7 - 2y) + 2y = 13 \quad \text{Let } x = 7 - 2y.$$
$$14 - 4y + 2y = 13 \quad \text{Apply the distributive property.}$$
$$14 - 2y = 13 \quad \text{Simplify.}$$
$$-2y = -1 \quad \text{Subtract 14 from both sides.}$$
$$y = \frac{1}{2} \quad \text{Divide both sides by } -2.$$

To find x, we let $y = \frac{1}{2}$ in the equation $x = 7 - 2y$.

$$x = 7 - 2y$$
$$x = 7 - 2\left(\frac{1}{2}\right) \quad \text{Let } y = \frac{1}{2}.$$
$$x = 7 - 1$$
$$x = 6$$

Check the solution in both equations of the original system. The solution is $\left(6, \frac{1}{2}\right)$.

■ **Work Practice 3**

The following steps summarize how to solve a system of equations by the substitution method.

> ### To Solve a System of Two Linear Equations by the Substitution Method
>
> **Step 1:** Solve one of the equations for one of its variables.
>
> **Step 2:** Substitute the expression for the variable found in Step 1 into the other equation.
>
> **Step 3:** Solve the equation from Step 2 to find the value of one variable.
>
> **Step 4:** Substitute the value found in Step 3 into any equation containing both variables to find the value of the other variable.
>
> **Step 5:** Check the proposed solution in the original system.

✔**Concept Check** As you solve the system

$$\begin{cases} 2x + y = -5 \\ x - y = 5 \end{cases}$$

you find that $y = -5$. Is this the solution of the system?

Example 4 Solve the system:

$$\begin{cases} 7x - 3y = -14 \\ -3x + y = 6 \end{cases}$$

Solution: Since the coefficient of y is 1 in the second equation, we will solve the second equation for y. This way, we avoid introducing tedious fractions.

$$-3x + y = 6 \quad \text{Second equation}$$
$$y = 3x + 6$$

Practice 4

Solve the system:
$$\begin{cases} 5x - 2y = 6 \\ -3x + y = -3 \end{cases}$$

Answer

4. $(0, -3)$

✔**Concept Check Answer**

no, the solution will be an ordered pair

Next, we substitute $3x + 6$ for y in the first equation.

$$7x - 3y = -14 \quad \text{First equation}$$
$$7x - 3(3x + 6) = -14 \quad \text{Let } y = 3x + 6.$$
$$7x - 9x - 18 = -14 \quad \text{Use the distributive property.}$$
$$-2x - 18 = -14 \quad \text{Simplify.}$$
$$-2x = 4 \quad \text{Add 18 to both sides.}$$
$$x = -2 \quad \text{Divide both sides by } -2.$$

To find the corresponding y-value, we substitute -2 for x in the equation $y = 3x + 6$. Then $y = 3(-2) + 6$ or $y = 0$. The solution of the system is $(-2, 0)$. Check this solution in both equations of the system.

🟧 **Work Practice 4**

✔**Concept Check** To avoid fractions, which of the equations below would you use to solve for x?

a. $3x - 4y = 15$ **b.** $14 - 3y = 8x$ **c.** $7y + x = 12$

Helpful Hint

When solving a system of equations by the substitution method, begin by solving an equation for one of its variables. If possible, solve for a variable that has a coefficient of 1 or −1 to avoid working with time-consuming fractions.

Example 5 Solve the system: $\begin{cases} \dfrac{1}{2}x - y = 3 \\ x = 6 + 2y \end{cases}$

Solution: The second equation is already solved for x in terms of y. Thus, we substitute $6 + 2y$ for x in the first equation and solve for y.

$$\frac{1}{2}x - y = 3 \quad \text{First equation}$$
$$\frac{1}{2}(6 + 2y) - y = 3 \quad \text{Let } x = 6 + 2y.$$
$$3 + y - y = 3 \quad \text{Apply the distributive property.}$$
$$3 = 3 \quad \text{Simplify.}$$

Arriving at a true statement such as $3 = 3$ indicates that the two linear equations in the original system are equivalent. This means that their graphs are identical, as shown in the figure. There is an infinite number of solutions to the system, and any solution of one equation is also a solution of the other.

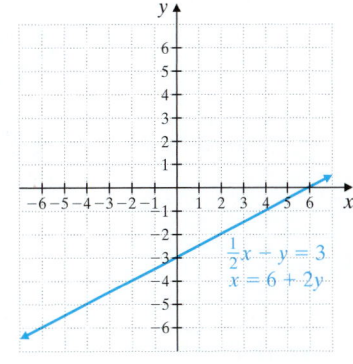

$\frac{1}{2}x - y = 3$
$x = 6 + 2y$

🟧 **Work Practice 5**

Teaching Tip

Ask students to recall the types of solutions to special systems of equations discussed in the previous section. Then ask, "What types of linear systems have an infinite number of solutions?" Have students name an equation that is equivalent to $y = 3x + 5$. Have students use substitution to solve the resulting system. Because a true statement results, any solution of one equation is a solution of the other.

Practice 5

Solve the system:

$$\begin{cases} -x + 3y = 6 \\ y = \dfrac{1}{3}x + 2 \end{cases}$$

Answer

5. infinite number of solutions

✔**Concept Check Answer**

c

Helpful Hint

Know that an infinite number of solutions does *not* mean that any ordered pair is a solution of both equations of the system.

An infinite number of solutions for Example 5 means that any of the infinite number of ordered pairs that is a solution of one equation in the system is also a solution of the other and is thus a solution of the system.

For Example 5,

$(2, 0)$ is *not* a solution of the system, but
$(6, 0)$ is a solution of the system.

Practice 6

Solve the system:

$$\begin{cases} 2x - 3y = 6 \\ -4x + 6y = 12 \end{cases}$$

Teaching Tip

Ask students, "What types of linear systems have no solutions?" Have them name an equation whose graph is a line parallel to $y = 3x + 5$. Now use substitution to solve the resulting system. Because a false statement results, no solution of one equation will satisfy the other.

Example 6   Solve the system:

$$\begin{cases} 6x + 12y = 5 \\ -4x - 8y = 0 \end{cases}$$

Solution: We choose the second equation and solve for y. (*Note:* Although you might not see this beforehand, if you solve the second equation for x, the result is $x = -2y$ and no fractions are introduced. Either way will lead to the correct solution.)

$$-4x - 8y = 0 \qquad \text{Second equation}$$
$$-8y = 4x \qquad \text{Add } 4x \text{ to both sides.}$$
$$\frac{-8y}{-8} = \frac{4x}{-8} \qquad \text{Divide both sides by } -8.$$
$$y = -\frac{1}{2}x \qquad \text{Simplify.}$$

Now we replace y with $-\frac{1}{2}x$ in the first equation.

$$6x + 12y = 5 \qquad \text{First equation}$$
$$6x + 12\left(-\frac{1}{2}x\right) = 5 \qquad \text{Let } y = -\frac{1}{2}x.$$
$$6x + (-6x) = 5 \qquad \text{Simplify.}$$
$$0 = 5 \qquad \text{Combine like terms.}$$

The false statement $0 = 5$ indicates that this system has no solution. The graph of the linear equations in the system is a pair of parallel lines, as shown in the figure.

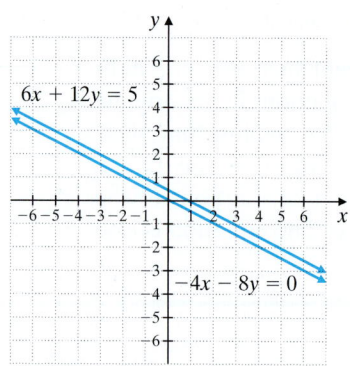

Answer

6. no solution

■ **Work Practice 6**

✓**Concept Check**   Describe how the graphs of the equations in a system appear if the system has

   a. no solution   **b.** one solution   **c.** an infinite number of solutions

Vocabulary, Readiness & Video Check

Give the solution of each system. If the system has no solution or an infinite number of solutions, say so. If the system has one solution, find it.

1. $\begin{cases} y = 4x \\ -3x + y = 1 \end{cases}$

When solving, you obtain $x = 1$. $(1, 4)$

2. $\begin{cases} 4x - y = 17 \\ -8x + 2y = 0 \end{cases}$

When solving, you obtain $0 = 34$.   no solution

3. $\begin{cases} 4x - y = 17 \\ -8x + 2y = -34 \end{cases}$

When solving, you obtain $0 = 0$.   infinite number of solutions

4. $\begin{cases} 5x + 2y = 25 \\ x = y + 5 \end{cases}$

When solving, you obtain $y = 0$. $(5, 0)$

5. $\begin{cases} x + y = 0 \\ 7x - 7y = 0 \end{cases}$

When solving, you obtain $x = 0$. $(0, 0)$

6. $\begin{cases} y = -2x + 5 \\ 4x + 2y = 10 \end{cases}$

When solving, you obtain $0 = 0$.   infinite number of solutions

Martin-Gay Interactive Videos   Watch the section lecture video and answer the following question.

Objective A 7. The systems in ⊟ Examples 2–4 all need one of their equations solved for a variable as a first step. What important part of the substitution method is emphasized in each example?

See Video 4.2 🍎

See video answer section.

4.2 Exercise Set MyMathLab®

Objective A *Solve each system of equations by the substitution method. See Examples 1 and 2.*

1. $\begin{cases} x + y = 3 \\ x = 2y \end{cases}$ $(2, 1)$

2. $\begin{cases} x + y = 20 \\ x = 3y \end{cases}$ $(15, 5)$

▶ **3.** $\begin{cases} x + y = 6 \\ y = -3x \end{cases}$ $(-3, 9)$

4. $\begin{cases} x + y = 6 \\ y = -4x \end{cases}$ $(-2, 8)$

5. $\begin{cases} y = 3x + 1 \\ 4y - 8x = 12 \end{cases}$ $(2, 7)$

6. $\begin{cases} y = 2x + 3 \\ 5y - 7x = 18 \end{cases}$ $(1, 5)$

7. $\begin{cases} y = 2x + 9 \\ y = 7x + 10 \end{cases}$ $\left(-\dfrac{1}{5}, \dfrac{43}{5} \right)$

8. $\begin{cases} y = 5x - 3 \\ y = 8x + 4 \end{cases}$ $\left(-\dfrac{7}{3}, -\dfrac{44}{3} \right)$

Solve each system of equations by the substitution method. See Examples 1 through 6.

9. $\begin{cases} 3x - 4y = 10 \\ y = x - 3 \end{cases}$ $(2, -1)$

10. $\begin{cases} 4x - 3y = 10 \\ y = x - 5 \end{cases}$ $(-5, -10)$

11. $\begin{cases} x + 2y = 6 \\ 2x + 3y = 8 \end{cases}$ $(-2, 4)$

12. $\begin{cases} x + 3y = -5 \\ 2x + 2y = 6 \end{cases}$ $(7, -4)$

13. $\begin{cases} 3x + 2y = 16 \\ x = 3y - 2 \end{cases}$ $(4, 2)$

14. $\begin{cases} 2x + 3y = 18 \\ x = 2y - 5 \end{cases}$ $(3, 4)$

15. $\begin{cases} 2x - 5y = 1 \\ 3x + y = -7 \end{cases}$ $(-2, -1)$

16. $\begin{cases} 3y - x = 6 \\ 4x + 12y = 0 \end{cases}$ $(-3, 1)$

17. $\begin{cases} 4x + 2y = 5 \\ -2x = y + 4 \end{cases}$ no solution

18. $\begin{cases} 2y = x + 2 \\ 6x - 12y = 0 \end{cases}$ no solution

19. $\begin{cases} 4x + y = 11 \\ 2x + 5y = 1 \end{cases}$ $(3, -1)$

20. $\begin{cases} 3x + y = -14 \\ 4x + 3y = -22 \end{cases}$ $(-4, -2)$

21. $\begin{cases} x + 2y + 5 = -4 + 5y - x \\ 2x + x = y + 4 \end{cases}$ $(3, 5)$

(*Hint:* First simplify each equation.)

22. $\begin{cases} 5x + 4y - 2 = -6 + 7y - 3x \\ 3x + 4x = y + 3 \end{cases}$

(*Hint:* See Exercise **21**.) $(1, 4)$

23. $\begin{cases} 6x - 3y = 5 \\ x + 2y = 0 \end{cases}$ $\left(\dfrac{2}{3}, -\dfrac{1}{3}\right)$

24. $\begin{cases} 10x - 5y = -21 \\ x + 3y = 0 \end{cases}$ $\left(-\dfrac{9}{5}, \dfrac{3}{5}\right)$

▶ 25. $\begin{cases} 3x - y = 1 \\ 2x - 3y = 10 \end{cases}$ $(-1, -4)$

26. $\begin{cases} 2x - y = -7 \\ 4x - 3y = -11 \end{cases}$ $(-5, -3)$

27. $\begin{cases} -x + 2y = 10 \\ -2x + 3y = 18 \end{cases}$ $(-6, 2)$

28. $\begin{cases} -x + 3y = 18 \\ -3x + 2y = 19 \end{cases}$ $(-3, 5)$

29. $\begin{cases} 5x + 10y = 20 \\ 2x + 6y = 10 \end{cases}$ $(2, 1)$

30. $\begin{cases} 6x + 3y = 12 \\ 9x + 6y = 15 \end{cases}$ $(3, -2)$

▶ 31. $\begin{cases} 3x + 6y = 9 \\ 4x + 8y = 16 \end{cases}$ no solution

32. $\begin{cases} 2x + 4y = 6 \\ 5x + 10y = 16 \end{cases}$ no solution

▶ 33. $\begin{cases} \dfrac{1}{3}x - y = 2 \\ x - 3y = 6 \end{cases}$ infinite number of solutions

34. $\begin{cases} \dfrac{1}{4}x - 2y = 1 \\ x - 8y = 4 \end{cases}$ infinite number of solutions

35. $\begin{cases} x = \dfrac{3}{4}y - 1 \\ 8x - 5y = -6 \end{cases}$ $\left(\dfrac{1}{2}, 2\right)$

36. $\begin{cases} x = \dfrac{5}{6}y - 2 \\ 12x - 5y = -9 \end{cases}$ $\left(\dfrac{1}{2}, 3\right)$

Review

Write equivalent equations by multiplying both sides of each given equation by the given nonzero number. See Section 2.2.

37. $3x + 2y = 6$ by -2
$-6x - 4y = -12$

38. $-x + y = 10$ by 5
$-5x + 5y = 50$

39. $-4x + y = 3$ by 3
$-12x + 3y = 9$

40. $5a - 7b = -4$ by -4
$-20a + 28b = 16$

Simplify the expressions. See Section 1.8.

41. $(3n + 6m) + (2n - 6m)$ $5n$

42. $(-2x + 5y) + (2x + 11y)$ $16y$

43. $(-5a - 7b) + (5a - 8b)$ $-15b$

44. $(9q + p) + (-9q - p)$ 0

Concept Extensions

Solve each system by the substitution method. First simplify each equation by combining like terms.

45. $\begin{cases} -5y + 6y = 3x + 2(x - 5) - 3x + 5 \\ 4(x + y) - x + y = -12 \end{cases}$ $(1, -3)$

46. $\begin{cases} 5x + 2y - 4x - 2y = 2(2y + 6) - 7 \\ 3(2x - y) - 4x = 1 + 9 \end{cases}$ $(5, 0)$

47. Explain how to identify a system with no solution when using the substitution method.
answers may vary

48. Occasionally, when using the substitution method, we obtain the equation $0 = 0$. Explain how this result indicates that the graphs of the equations in the system are identical.   answers may vary

Solve. See a Concept Check in this section.

49. As you solve the system $\begin{cases} 3x - y = -6 \\ -3x + 2y = 7 \end{cases}$, you find that $y = 1$. Is this the solution of the system?
no

50. As you solve the system $\begin{cases} x = 5y \\ y = 2x \end{cases}$, you find that $x = 0$ and $y = 0$. What is the solution of this system?
$(0, 0)$

51. To avoid fractions, which of the equations below would you use if solving for y? Explain why.

  a. $\dfrac{1}{2}x - 4y = \dfrac{3}{4}$ **b.** $8x - 5y = 13$ **c.** $7x - y = 19$

  c; answers may vary

52. Give the number of solutions for a system if the graphs of the equations in the system are

  a. lines intersecting in one point   1
  b. parallel lines   0
  c. same line   infinite number

Use a graphing calculator to solve each system.

53. $\begin{cases} y = 5.1x + 14.56 \\ y = -2x - 3.9 \end{cases}$
$(-2.6, 1.3)$

54. $\begin{cases} y = 3.1x - 16.35 \\ y = -9.7x + 28.45 \end{cases}$
$(3.5, -5.5)$

55. $\begin{cases} 3x + 2y = 14.04 \\ 5x + y = 18.5 \end{cases}$
$(3.28, 2.1)$

56. $\begin{cases} x + y = -15.2 \\ -2x + 5y = -19.3 \end{cases}$
$(-8.1, -7.1)$

Solve.

57. U.S. consumer spending y (in billions of dollars) on DVD-format home entertainment from 2006 to 2010 is given by $y = -1.6x + 20.9$, where x is the number of years after 2006. U.S. consumer spending y (in billions of dollars) on Blu-ray-format home entertainment from 2006 to 2010 is given by $y = 0.6x - 0.2$, where x is the number of years after 2006. (*Source:* Based on data from DEG: The Digital Entertainment Group)

  a. Use the substitution method to solve this system of equations.

$$\begin{cases} y = -1.6x + 20.9 \\ y = 0.6x - 0.2 \end{cases}$$

  Round x to the nearest tenth and y to the nearest whole. $(9.6, 6)$

  b. Explain the meaning of your answer to part **a.**

  c. Sketch a graph of the system of equations. Write a sentence describing the trends in the popularity of these two types of home entertainment formats.

Consumer Spending on DVD and Blu-ray Formats

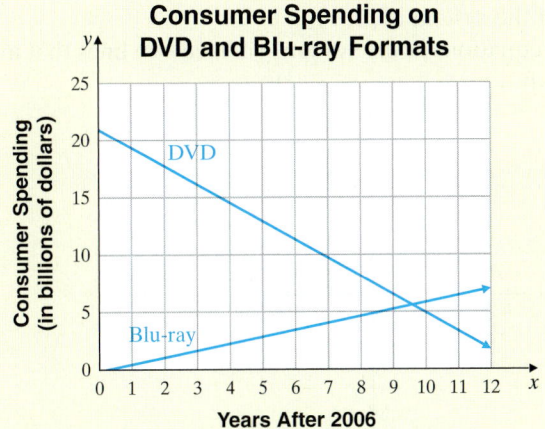

Years After 2006

58. Promoting tourism in a state can significantly boost jobs and tax revenue for that state. For that reason, most states operate a tourism office to attract travelers and their spending. For fiscal years 2008 through 2012, Maryland's state tourism office budget y (in millions of dollars) is given by the equation $y = 0.61x + 7.62$, where x is the number of years after fiscal year 2008. For the same period, Georgia's state tourism office budget y (in millions of dollars) is given by the equation $y = -0.74x + 9.32$, where x is the number of years after fiscal year 2008. (*Source:* Based on data from the U.S. Travel Association)

  a. Use the substitution method to solve this system of equations.

$$\begin{cases} y = 0.61x + 7.62 \\ y = -0.74x + 9.32 \end{cases}$$

  Round x and y to the nearest tenth. $(1.3, 8.4)$

  b. Explain the meaning of your answer to part **a.**

  c. Sketch a graph of the system of equations. Write a sentence describing the trends in Maryland's and Georgia's state tourism office budgets between fiscal years 2008 and 2012.   answers may vary

State Tourism Office Budgets

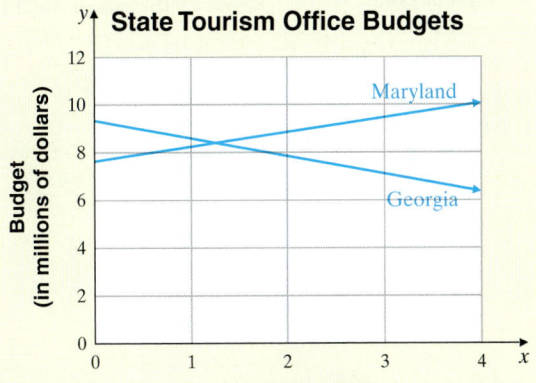

Years After 2008

57. b. In about 9.6 years after 2006, U.S. consumer spending on DVD- and Blu-ray-format home entertainment will be approximately $6 billion for each.
c. answers may vary

58. b. In about 1.3 years after fiscal year 2008, Maryland's and Georgia's state tourism office budgets were the same, approximately $8.4 million.

4.3 Solving Systems of Linear Equations by Addition

Objective

A Use the Addition Method to Solve a System of Linear Equations.

Objective A Using the Addition Method

We have seen that substitution is an accurate method for solving a system of linear equations. Another accurate method is the **addition** or **elimination method.** The addition method is based on the addition property of equality: Adding equal quantities to both sides of an equation does not change the solution of the equation. In symbols,

$$\text{if } A = B \text{ and } C = D, \text{ then } A + C = B + D$$

To see how we use this to solve a system of equations, study Example 1.

Practice 1

Use the addition method to solve the system:
$$\begin{cases} x + y = 13 \\ x - y = 5 \end{cases}$$

Example 1 Solve the system: $\begin{cases} x + y = 7 \\ x - y = 5 \end{cases}$

Solution: Since the left side of each equation is equal to its right side, we are adding equal quantities when we add the left sides of the equations together and add the right sides of the equations together. This adding eliminates the variable y and gives us an equation in one variable, x. We can then solve for x.

$$\begin{array}{ll} x + y = 7 & \text{First equation} \\ \underline{x - y = 5} & \text{Second equation} \\ \quad 2x = 12 & \text{Add the equations to eliminate } y. \\ \quad\ \ x = 6 & \text{Divide both sides by 2.} \end{array}$$

> **Helpful Hint**
> Notice in Example 1 that our goal when solving a system of equations by the addition method is to eliminate a variable when adding the equations.

The x-value of the solution is 6. To find the corresponding y-value, we let $x = 6$ in either equation of the system. We will use the first equation.

$$\begin{array}{ll} x + y = 7 & \text{First equation} \\ 6 + y = 7 & \text{Let } x = 6. \\ \quad\ \ y = 1 & \text{Solve for } y. \end{array}$$

The solution is $(6, 1)$.

Check: Check the solution in both equations of the original system.

First Equation	**Second Equation**
$x + y = 7$	$x - y = 5$
$6 + 1 \overset{?}{=} 7$ Let $x = 6$ and $y = 1$.	$6 - 1 \overset{?}{=} 5$ Let $x = 6$ and $y = 1$.
$7 = 7$ True	$5 = 5$ True

Thus, the solution of the system is $(6, 1)$.

If we graph the two equations in the system, we have two lines that intersect at the point $(6, 1)$, as shown.

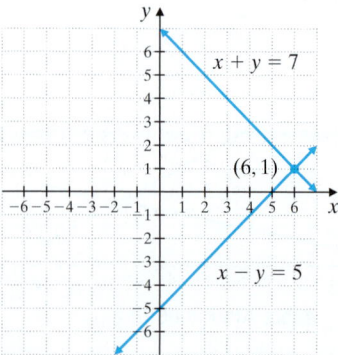

Teaching Tip

Warm up with the following exercise: Find the value for m that makes each statement true:

$$m(3x) - (9x) = 0$$
$$20x + m(5x) = 0$$
$$-12x + m(2x) = 0$$
$$m(-6x) - 30x = 0$$

■ **Work Practice 1**

Answer

1. $(9, 4)$

Example 2 Solve the system: $\begin{cases} -2x + y = 2 \\ -x + 3y = -4 \end{cases}$

Solution: If we simply add these two equations, the result is still an equation in two variables. However, from Example 1, remember that our goal is to eliminate one of the variables so that we have an equation in the other variable. To do this, notice what happens if we multiply *both sides* of the first equation by -3. We are allowed to do this by the multiplication property of equality. Then the system

$$\begin{cases} -3(-2x + y) = -3(2) \\ -x + 3y = -4 \end{cases} \quad \text{simplifies to} \quad \begin{cases} 6x - 3y = -6 \\ -x + 3y = -4 \end{cases}$$

When we add the resulting equations, the y-variable is eliminated.

$$\begin{array}{rl} 6x - 3y &= -6 \\ -x + 3y &= -4 \\ \hline 5x &= -10 \quad \text{Add.} \\ x &= -2 \quad \text{Divide both sides by 5.} \end{array}$$

To find the corresponding y-value, we let $x = -2$ in either of the original equations. We use the first equation of the original system.

$$\begin{array}{rl} -2x + y &= 2 \quad \text{First equation} \\ -2(-2) + y &= 2 \quad \text{Let } x = -2. \\ 4 + y &= 2 \\ y &= -2 \end{array}$$

Check the ordered pair $(-2, -2)$ in both equations of the *original* system. The solution is $(-2, -2)$.

■ **Work Practice 2**

Helpful Hint

When finding the second value of an ordered pair solution, any equation equivalent to one of the original equations in the system may be used.

In Example 2, the decision to multiply the first equation by -3 was no accident. **To eliminate a variable** when adding two equations, **the coefficient of the variable in one equation must be the opposite of its coefficient in the other equation.**

Helpful Hint

Be sure to multiply *both sides* of an equation by the chosen number when solving by the addition method. A common mistake is to multiply only the side containing the variables.

Example 3 Solve the system: $\begin{cases} 2x - y = 7 \\ 8x - 4y = 1 \end{cases}$

Solution: When we multiply both sides of the first equation by -4, the resulting coefficient of x is -8. This is the opposite of 8, the coefficient of x in the second equation. Then the system

$$\begin{cases} -4(2x - y) = -4(7) \\ 8x - 4y = 1 \end{cases} \quad \text{simplifies to}$$

Helpful Hint Don't forget to multiply both sides by -4.

$$\begin{cases} -8x + 4y = -28 \\ \underline{8x - 4y = 1} \end{cases}$$
$$0 = -27 \quad \text{Add the equations.}$$

(Continued on next page)

(Continued on next page)

Practice 2

Solve the system:
$$\begin{cases} 2x - y = -6 \\ -x + 4y = 17 \end{cases}$$

Teaching Tip

Before discussing Example 3, have students predict what will happen when using the addition method to solve a system of two parallel lines. Have students check their prediction with the following system:
$$\begin{cases} 3x - 4y = 12 \\ 3x - 4y = 16 \end{cases}$$

Practice 3

Solve the system:
$$\begin{cases} x - 3y = -2 \\ -3x + 9y = 5 \end{cases}$$

Answers

2. $(-1, 4)$ **3.** no solution

Copyright 2016 Pearson Education, Inc.

Teaching Tip

You may want students to warm up with the following exercises: Find values for m and n that make each statement true.

$m(3x) + n(7x) = 0$

$m(-4x) + n(-5x) = 0$

$m(2x) - n(9x) = 0$

$m(-11x) - n(3x) = 0$

Practice 4

Solve the system:
$$\begin{cases} 2x + 5y = 1 \\ -4x - 10y = -2 \end{cases}$$

Teaching Tip

Before discussing Example 4, have students predict what will happen when they use the addition method to solve a system of two equivalent equations. Have students check their prediction with the following system:
$$\begin{cases} 3x - 4y = 12 \\ -3x + 4y = -12 \end{cases}$$

Practice 5

Solve the system:
$$\begin{cases} 4x + 5y = 14 \\ 3x - 2y = -1 \end{cases}$$

When we add the equations, both variables are eliminated and we have $0 = -27$, a false statement. This means that the system has no solution. The equations, if graphed, would represent parallel lines.

◼ **Work Practice 3**

Example 4 Solve the system: $\begin{cases} 3x - 2y = 2 \\ -9x + 6y = -6 \end{cases}$

Solution: First we multiply both sides of the first equation by 3 and then we add the resulting equations.

$$\begin{cases} 3(3x - 2y) = 3(2) \\ -9x + 6y = -6 \end{cases} \quad \text{simplifies to} \quad \begin{cases} 9x - 6y = 6 \\ \underline{-9x + 6y = -6} \quad \text{Add the equations.} \\ 0 = 0 \end{cases}$$

Both variables are eliminated and we have $0 = 0$, a true statement. This means that the system has an infinite number of solutions. The equations, if graphed, would be the same line.

◼ **Work Practice 4**

✓**Concept Check** Suppose you are solving the system
$$\begin{cases} 3x + 8y = -5 \\ 2x - 4y = 3 \end{cases}$$

You decide to use the addition method by multiplying both sides of the second equation by 2. In which of the following was the multiplication performed correctly? Explain.

a. $4x - 8y = 3$ **b.** $4x - 8y = 6$

In the next example, we multiply both equations by numbers so that coefficients of a variable are opposites.

Example 5 Solve the system: $\begin{cases} 3x + 4y = 13 \\ 5x - 9y = 6 \end{cases}$

Solution: We can eliminate the variable y by multiplying the first equation by 9 and the second equation by 4. Then we add the resulting equations.

$$\begin{cases} 9(3x + 4y) = 9(13) \\ 4(5x - 9y) = 4(6) \end{cases} \quad \text{simplifies to} \quad \begin{cases} 27x + 36y = 117 \\ \underline{20x - 36y = \ \ 24} \\ \begin{aligned} 47x \qquad\quad &= 141 \quad \text{Add the equations.} \\ x &= 3 \quad \text{Solve for } x. \end{aligned} \end{cases}$$

To find the corresponding y-value, we let $x = 3$ in one of the original equations of the system. Doing so in either of these equations will give $y = 1$. Check to see that $(3, 1)$ satisfies each equation in the original system. The solution is $(3, 1)$.

◼ **Work Practice 5**

If we had decided to eliminate x instead of y in Example 5, the first equation could have been multiplied by 5 and the second by -3. Try solving the original system this way to check that the solution is $(3, 1)$.

The following steps summarize how to solve a system of linear equations by the addition method.

Answers

4. infinite number of solutions

5. $(1, 2)$

✓**Concept Check Answer**

b; answers may vary

Example 2 Solve the system: $\begin{cases} -2x + y = 2 \\ -x + 3y = -4 \end{cases}$

Solution: If we simply add these two equations, the result is still an equation in two variables. However, from Example 1, remember that our goal is to eliminate one of the variables so that we have an equation in the other variable. To do this, notice what happens if we multiply *both sides* of the first equation by -3. We are allowed to do this by the multiplication property of equality. Then the system

$$\begin{cases} -3(-2x + y) = -3(2) \\ -x + 3y = -4 \end{cases} \quad \text{simplifies to} \quad \begin{cases} 6x - 3y = -6 \\ -x + 3y = -4 \end{cases}$$

When we add the resulting equations, the y-variable is eliminated.

$$\begin{array}{rl} 6x - 3y = -6 & \\ \underline{-x + 3y = -4} & \\ 5x \phantom{{}- 3y} = -10 & \text{Add.} \\ x = -2 & \text{Divide both sides by 5.} \end{array}$$

To find the corresponding y-value, we let $x = -2$ in either of the original equations. We use the first equation of the original system.

$$\begin{array}{rl} -2x + y = 2 & \text{First equation} \\ -2(-2) + y = 2 & \text{Let } x = -2. \\ 4 + y = 2 & \\ y = -2 & \end{array}$$

Check the ordered pair $(-2, -2)$ in both equations of the *original* system. The solution is $(-2, -2)$.

■ **Work Practice 2**

> **Helpful Hint**
>
> When finding the second value of an ordered pair solution, any equation equivalent to one of the original equations in the system may be used.

In Example 2, the decision to multiply the first equation by -3 was no accident. **To eliminate a variable** when adding two equations, **the coefficient of the variable in one equation must be the opposite of its coefficient in the other equation.**

> **Helpful Hint**
>
> Be sure to multiply *both sides* of an equation by the chosen number when solving by the addition method. A common mistake is to multiply only the side containing the variables.

Example 3 Solve the system: $\begin{cases} 2x - y = 7 \\ 8x - 4y = 1 \end{cases}$

Solution: When we multiply both sides of the first equation by -4, the resulting coefficient of x is -8. This is the opposite of 8, the coefficient of x in the second equation. Then the system

$$\begin{cases} -4(2x - y) = -4(7) \\ 8x - 4y = 1 \end{cases} \quad \text{simplifies to}$$

> **Helpful Hint** Don't forget to multiply both sides by -4.

$$\begin{cases} -8x + 4y = -28 \\ \underline{8x - 4y = 1} \end{cases}$$
$$0 = -27 \quad \text{Add the equations.}$$

(Continued on next page)

Practice 2

Solve the system:
$$\begin{cases} 2x - y = -6 \\ -x + 4y = 17 \end{cases}$$

Practice 3

Solve the system:
$$\begin{cases} x - 3y = -2 \\ -3x + 9y = 5 \end{cases}$$

Answers

2. $(-1, 4)$ **3.** no solution

Practice 4

Solve the system:

$$\begin{cases} 2x + 5y = 1 \\ -4x - 10y = -2 \end{cases}$$

Practice 5

Solve the system:

$$\begin{cases} 4x + 5y = 14 \\ 3x - 2y = -1 \end{cases}$$

When we add the equations, both variables are eliminated and we have $0 = -27$, a false statement. This means that the system has no solution. The equations, if graphed, would represent parallel lines.

■ **Work Practice 3**

Example 4 Solve the system: $\begin{cases} 3x - 2y = 2 \\ -9x + 6y = -6 \end{cases}$

Solution: First we multiply both sides of the first equation by 3 and then we add the resulting equations.

$$\begin{cases} 3(3x - 2y) = 3(2) \\ -9x + 6y = -6 \end{cases} \quad \text{simplifies to} \quad \begin{cases} 9x - 6y = 6 \\ \underline{-9x + 6y = -6} \quad \text{Add the equations.} \\ 0 = 0 \end{cases}$$

Both variables are eliminated and we have $0 = 0$, a true statement. This means that the system has an infinite number of solutions. The equations, if graphed, would be the same line.

■ **Work Practice 4**

✓**Concept Check** Suppose you are solving the system

$$\begin{cases} 3x + 8y = -5 \\ 2x - 4y = 3 \end{cases}$$

You decide to use the addition method by multiplying both sides of the second equation by 2. In which of the following was the multiplication performed correctly? Explain.

a. $4x - 8y = 3$ **b.** $4x - 8y = 6$

In the next example, we multiply both equations by numbers so that coefficients of a variable are opposites.

Example 5 Solve the system: $\begin{cases} 3x + 4y = 13 \\ 5x - 9y = 6 \end{cases}$

Solution: We can eliminate the variable y by multiplying the first equation by 9 and the second equation by 4. Then we add the resulting equations.

$$\begin{cases} 9(3x + 4y) = 9(13) \\ 4(5x - 9y) = 4(6) \end{cases} \quad \text{simplifies to} \quad \begin{cases} 27x + 36y = 117 \\ \underline{20x - 36y = 24} \\ 47x = 141 \quad \text{Add the equations.} \\ x = 3 \quad \text{Solve for } x. \end{cases}$$

To find the corresponding y-value, we let $x = 3$ in one of the original equations of the system. Doing so in either of these equations will give $y = 1$. Check to see that $(3, 1)$ satisfies each equation in the original system. The solution is $(3, 1)$.

■ **Work Practice 5**

If we had decided to eliminate x instead of y in Example 5, the first equation could have been multiplied by 5 and the second by -3. Try solving the original system this way to check that the solution is $(3, 1)$.

The following steps summarize how to solve a system of linear equations by the addition method.

To Solve a System of Two Linear Equations by the Addition Method

Step 1: Rewrite each equation in standard form, $Ax + By = C$.

Step 2: If necessary, multiply one or both equations by a nonzero number so that the coefficients of the chosen variable in the system are opposites.

Step 3: Add the equations.

Step 4: Find the value of one variable by solving the resulting equation from Step 3.

Step 5: Find the value of the second variable by substituting the value found in Step 4 into either of the original equations.

Step 6: Check the proposed solution in the original system.

✔**Concept Check** Suppose you are solving the system

$$\begin{cases} -4x + 7y = 6 \\ x + 2y = 5 \end{cases}$$

by the addition method.

a. What step(s) should you take if you wish to eliminate x when adding the equations?

b. What step(s) should you take if you wish to eliminate y when adding the equations?

Example 6 Solve the system: $\begin{cases} -x - \dfrac{y}{2} = \dfrac{5}{2} \\ \dfrac{x}{6} - \dfrac{y}{2} = 0 \end{cases}$

Solution: We begin by clearing each equation of fractions. To do so, we multiply both sides of the first equation by the LCD, 2, and both sides of the second equation by the LCD, 6. Then the system

$$\begin{cases} 2\left(-x - \dfrac{y}{2}\right) = 2\left(\dfrac{5}{2}\right) \\ 6\left(\dfrac{x}{6} - \dfrac{y}{2}\right) = 6(0) \end{cases} \quad \text{simplifies to} \quad \begin{cases} -2x - y = 5 \\ x - 3y = 0 \end{cases}$$

We can now eliminate the variable x by multiplying the second equation by 2.

$$\begin{cases} -2x - y = 5 \\ 2(x - 3y) = 2(0) \end{cases} \quad \text{simplifies to} \quad \begin{array}{l} -2x - y = 5 \\ \underline{2x - 6y = 0} \\ {-7y = 5} \quad \text{Add the equations.} \\ y = -\dfrac{5}{7} \quad \text{Solve for } y. \end{array}$$

To find x, we could replace y with $-\dfrac{5}{7}$ in one of the equations with two variables.

Instead, let's go back to the simplified system and multiply by appropriate factors to eliminate the variable y and solve for x. To do this, we multiply the first equation by -3. Then the system

$$\begin{cases} -3(-2x - y) = -3(5) \\ x - 3y = 0 \end{cases} \quad \text{simplifies to} \quad \begin{array}{l} 6x + 3y = -15 \\ \underline{x - 3y = 0} \\ 7x = -15 \quad \text{Add the equations.} \\ x = -\dfrac{15}{7} \quad \text{Solve for } x. \end{array}$$

Check the ordered pair $\left(-\dfrac{15}{7}, -\dfrac{5}{7}\right)$ in both equations of the original system. The solution is $\left(-\dfrac{15}{7}, -\dfrac{5}{7}\right)$.

Work Practice 6

Teaching Tip

Before attempting Example 6, remind students to use what they know to make this system as easy as possible to solve. Then ask them how Example 6 could be written as a simpler system.

Practice 6

Solve the system:

$$\begin{cases} -\dfrac{x}{3} + y = \dfrac{4}{3} \\ \dfrac{x}{2} - \dfrac{5}{2}y = -\dfrac{1}{2} \end{cases}$$

Teaching Tip

Conclude this section by having students describe the three methods they have learned for solving systems of linear equations and list the pros and cons of each.

Answer

6. $\left(-\dfrac{17}{2}, -\dfrac{3}{2}\right)$

✔**Concept Check Answers**

a. multiply the second equation by 4

b. possible answer: multiply the first equation by -2 and the second equation by 7

Vocabulary, Readiness & Video Check

Given the system $\begin{cases} 3x - 2y = -9 \\ x + 5y = 14 \end{cases}$ *read each row (Step 1, Step 2, and Result). Then answer whether the result is true or false.*

	Step 1	Step 2	Result	True or False?
1.	Multiply 2nd equation through by -3.	Add the resulting equation to the 1st equation.	The y's are eliminated.	false
2.	Multiply 2nd equation through by -3.	Add the resulting equation to the 1st equation.	The x's are eliminated.	true
3.	Multiply 1st equation by 5 and 2nd equation by 2.	Add the two new equations.	The y's are eliminated.	true
4.	Multiply 1st equation by 5 and 2nd equation by -2.	Add the two new equations.	The y's are eliminated.	false

Martin-Gay Interactive Videos   Watch the section lecture video and answer the following question.

Objective A 5. For the addition/elimination method, sometimes we need to multiply an equation through by a nonzero number so that the coefficients of a variable are opposites, as is shown in Example 2. What property allows us to do this? What important reminder is made at this step?

See Video 4.3

See video answer section.

4.3 Exercise Set MyMathLab®

Objective A *Solve each system of equations by the addition method. See Example 1.*

1. $\begin{cases} 3x + y = 5 \\ 6x - y = 4 \end{cases}$
$(1, 2)$

2. $\begin{cases} 4x + y = 13 \\ 2x - y = 5 \end{cases}$
$(3, 1)$

3. $\begin{cases} x - 2y = 8 \\ -x + 5y = -17 \end{cases}$
$(2, -3)$

4. $\begin{cases} x - 2y = -11 \\ -x + 5y = 23 \end{cases}$
$(-3, 4)$

Solve each system of equations by the addition method. If a system contains fractions or decimals, you may want to first clear each equation of the fractions or decimals. See Examples 2 through 6.

5. $\begin{cases} 3x + y = -11 \\ 6x - 2y = -2 \end{cases}$ $(-2, -5)$

6. $\begin{cases} 4x + y = -13 \\ 6x - 3y = -15 \end{cases}$ $(-3, -1)$

7. $\begin{cases} 3x + 2y = 11 \\ 5x - 2y = 29 \end{cases}$ $(5, -2)$

8. $\begin{cases} 4x + 2y = 2 \\ 3x - 2y = 12 \end{cases}$ $(2, -3)$

9. $\begin{cases} x + 5y = 18 \\ 3x + 2y = -11 \end{cases}$ $(-7, 5)$

10. $\begin{cases} x + 4y = 14 \\ 5x + 3y = 2 \end{cases}$ $(-2, 4)$

11. $\begin{cases} x + y = 6 \\ x - y = 6 \end{cases}$ $(6, 0)$

12. $\begin{cases} x - y = 1 \\ -x + 2y = 0 \end{cases}$ $(2, 1)$

13. $\begin{cases} 2x + 3y = 0 \\ 4x + 6y = 3 \end{cases}$ no solution

14. $\begin{cases} 3x + y = 4 \\ 9x + 3y = 6 \end{cases}$ no solution

15. $\begin{cases} -x + 5y = -1 \\ 3x - 15y = 3 \end{cases}$ infinite number of solutions

16. $\begin{cases} 2x + y = 6 \\ 4x + 2y = 12 \end{cases}$ infinite number of solutions

17. $\begin{cases} 3x - 2y = 7 \\ 5x + 4y = 8 \end{cases}$ $\left(2, -\dfrac{1}{2}\right)$

18. $\begin{cases} 6x - 5y = 25 \\ 4x + 15y = 13 \end{cases}$ $\left(4, -\dfrac{1}{5}\right)$

19. $\begin{cases} 8x = -11y - 16 \\ 2x + 3y = -4 \end{cases}$ $(-2, 0)$

20. $\begin{cases} 10x + 3y = -12 \\ 5x = -4y - 16 \end{cases}$ $(0, -4)$ **21.** $\begin{cases} 4x - 3y = 7 \\ 7x + 5y = 2 \end{cases}$ $(1, -1)$ **22.** $\begin{cases} -2x + 3y = 10 \\ 3x + 4y = 2 \end{cases}$ $(-2, 2)$

23. $\begin{cases} 4x - 6y = 8 \\ 6x - 9y = 12 \end{cases}$ infinite number of solutions   **24.** $\begin{cases} 9x - 3y = 12 \\ 12x - 4y = 18 \end{cases}$ no solution   **25.** $\begin{cases} 2x - 5y = 4 \\ 3x - 2y = 4 \end{cases}$ $\left(\frac{12}{11}, -\frac{4}{11} \right)$

26. $\begin{cases} 6x - 5y = 7 \\ 4x - 6y = 7 \end{cases}$ $\left(\frac{7}{16}, -\frac{7}{8} \right)$ **27.** $\begin{cases} \dfrac{x}{3} + \dfrac{y}{6} = 1 \\ \dfrac{x}{2} - \dfrac{y}{4} = 0 \end{cases}$ $\left(\frac{3}{2}, 3 \right)$ **28.** $\begin{cases} \dfrac{x}{2} + \dfrac{y}{8} = 3 \\ x - \dfrac{y}{4} = 0 \end{cases}$ $(3, 12)$

29. $\begin{cases} \dfrac{10}{3}x + 4y = -4 \\ 5x + 6y = -6 \end{cases}$ infinite number of solutions   **30.** $\begin{cases} \dfrac{3}{2}x + 4y = 1 \\ 9x + 24y = 5 \end{cases}$ no solution   **31.** $\begin{cases} x - \dfrac{y}{3} = -1 \\ -\dfrac{x}{2} + \dfrac{y}{8} = \dfrac{1}{4} \end{cases}$ $(1, 6)$

32. $\begin{cases} 2x - \dfrac{3y}{4} = -3 \\ x + \dfrac{y}{9} = \dfrac{13}{3} \end{cases}$ $(3, 12)$ **33.** $\begin{array}{l} -4(x + 2) = 3y \\ 2x - 2y = 3 \end{array}$ $\left(-\dfrac{1}{2}, -2 \right)$ **34.** $\begin{array}{l} -9(x + 3) = 8y \\ 3x - 3y = 8 \end{array}$ $\left(-\dfrac{1}{3}, -3 \right)$

▶ **35.** $\begin{cases} \dfrac{x}{3} - y = 2 \\ -\dfrac{x}{2} + \dfrac{3y}{2} = -3 \end{cases}$ infinite number of solutions   **36.** $\begin{cases} \dfrac{x}{2} + \dfrac{y}{4} = 1 \\ -\dfrac{x}{4} - \dfrac{y}{8} = 1 \end{cases}$ no solution   **37.** $\begin{cases} \dfrac{3}{5}x - y = -\dfrac{4}{5} \\ 3x + \dfrac{y}{2} = -\dfrac{9}{5} \end{cases}$ $\left(-\dfrac{2}{3}, \dfrac{2}{5} \right)$

38. $\begin{cases} 3x + \dfrac{7}{2}y = \dfrac{3}{4} \\ -\dfrac{x}{2} + \dfrac{5}{3}y = -\dfrac{5}{4} \end{cases}$ $\left(\dfrac{5}{6}, -\dfrac{1}{2} \right)$ ▶ **39.** $\begin{cases} 3.5x + 2.5y = 17 \\ -1.5x - 7.5y = -33 \end{cases}$ $(2, 4)$ **40.** $\begin{cases} -2.5x - 6.5y = 47 \\ 0.5x - 4.5y = 37 \end{cases}$ $(2, -8)$

41. $\begin{cases} 0.02x + 0.04y = 0.09 \\ -0.1x + 0.3y = 0.8 \end{cases}$ $(-0.5, 2.5)$ **42.** $\begin{cases} 0.04x - 0.05y = 0.105 \\ 0.2x - 0.6y = 1.05 \end{cases}$ $(0.75, -1.5)$

Review

Translating *Rewrite each sentence using mathematical symbols. Do not solve the equations. See Section 2.4.*

43. Twice a number, added to 6, is 3 less than the number. $2x + 6 = x - 3$

44. The sum of three consecutive integers is 66. $x + (x + 1) + (x + 2) = 66$

45. Three times a number, subtracted from 20, is 2. $20 - 3x = 2$

46. Twice the sum of 8 and a number is the difference of the number and 20. $2(8 + x) = x - 20$

47. The product of 4 and the sum of a number and 6 is twice the number. $4(x + 6) = 2x$

48. If the quotient of twice a number and 7 is subtracted from the reciprocal of the number, the result is 2. $\dfrac{1}{x} - \dfrac{2x}{7} = 2$

Concept Extensions

Solve. See a Concept Check in this section.

49. To solve this system by the addition method and eliminate the variable y,

$$\begin{cases} 4x + 2y = -7 \\ 3x - y = -12 \end{cases}$$

by what value would you multiply the second equation? What do you get when you complete the multiplication? $2; 6x - 2y = -24$

Given the system of linear equations $\begin{cases} 3x - y = -8 \\ 5x + 3y = 2 \end{cases}$:

50. Use the addition method and
   a. Solve the system by eliminating x.
   b. Solve the system by eliminating y. $\left(-\dfrac{11}{7}, \dfrac{23}{7} \right)$

Solve.

51. Suppose you are solving the system

$$\begin{cases} 3x + 8y = -5 \\ 2x - 4y = 3 \end{cases}$$

You decide to use the addition method by multiplying both sides of the second equation by 2. In which of the following was the multiplication performed correctly? Explain.
   a. $4x - 8y = 3$
   b. $4x - 8y = 6$ b; answers may vary

52. Suppose you are solving the system

$$\begin{cases} -2x - y = 0 \\ -2x + 3y = 6 \end{cases}$$

You decide to use the addition method by multiplying both sides of the first equation by 3, then adding the resulting equation to the second equation. Which of the following is the correct sum? Explain.
   a. $-8x = 6$ a; answers may vary
   b. $-8x = 9$

53. When solving a system of equations by the addition method, how do we know when the system has no solution?   answers may vary

54. Explain why the addition method might be preferred over the substitution method for solving the system $\begin{cases} 2x - 3y = 5 \\ 5x + 2y = 6. \end{cases}$ answers may vary

55. Use the system of linear equations below to answer the questions.

$$\begin{cases} x + y = 5 \\ 3x + 3y = b \end{cases}$$

   a. Find the value of b so that the system has an infinite number of solutions. $b = 15$
   b. Find a value of b so that there are no solutions to the system.   any real number except 15

56. Use the system of linear equations below to answer the questions.

$$\begin{cases} x + y = 4 \\ 2x + by = 8 \end{cases}$$

   a. Find the value of b so that the system has an infinite number of solutions. $b = 2$
   b. Find a value of b so that the system has a single solution.   any real number except 2

Solve each system by the addition method.

57. $\begin{cases} 2x + 3y = 14 \\ 3x - 4y = -69.1 \end{cases}$ $(-8.9, 10.6)$

58. $\begin{cases} 5x - 2y = -19.8 \\ -3x + 5y = -3.7 \end{cases}$ $(-5.6, -4.1)$

59. As the economy and job marketplace change, demand for certain types of workers changes. The number of jobs for postal service mail carriers that is predicted for 2010 through 2020 can be approximated by $38x + 10y = 3167$. The number of jobs for market research analysts that is predicted for the same period can be approximated by $117x - 10y = -2827$. For both equations, x is the number of years since 2010, and y is the number of jobs in the thousands. (*Source*: Based on data from the U.S. Bureau of Labor Statistics)

a. Use the addition method to solve this system of equations.

$$\begin{cases} 38x + 10y = 3167 \\ 117x - 10y = -2827 \end{cases}$$

(Eliminate y first and solve for x. Round this result to the nearest whole number.) $(2, 309)$ or $(2, 306)$

b. Interpret your solution from part **a**.   In about 2012 $(2010 + 2)$, the number of mail carrier jobs was approximately equal to the number of market research analyst jobs.
c. Using the year in your answer to part **b**, estimate the number of mail carrier jobs and market research analyst jobs in that year.   309 thousand or 306 thousand

60. In recent years, the number of newspapers printed as morning editions has been increasing and the number of newspapers printed as evening editions has been decreasing. The number y of daily morning newspapers in existence from 1995 through 2011 is approximated by the equation $153x - 10y = -6720$, where x is the number of years since 1995. The number y of daily evening newspapers in existence from 1995 through 2011 is approximated by $125x + 5y = 4350$, where x is the number of years since 1995. (*Source:* Based on data from Newspaper Association of America)

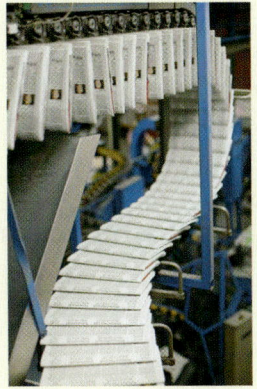

a. Use the addition method to solve this system of equations.

$$\begin{cases} 153x - 10y = -6720 \\ 125x + 5y = 4350 \end{cases}$$

(Round to the nearest whole number. Because of rounding, the y-value of your ordered pair solution may vary.) $(5, 749)$ or $(5, 745)$

b. Interpret your solution from part **a**.   In 2000 $(1995 + 5)$, the number of morning papers equaled the number of evening papers.
c. Use the year in your answer to part **b** to find how many of each type of newspaper were in existence that year.   749 or 745

Summary on Solving Systems of Equations

Answers

Solve each system by either the addition method or the substitution method.

1. $(2, 5)$

2. $(4, 2)$

3. $(5, -2)$

4. $(6, -14)$

5. $(-3, 2)$

6. $(-4, 3)$

7. $(0, 3)$

8. $(-2, 4)$

9. $(5, 7)$

10. $(-3, -23)$

11. $\left(\dfrac{1}{3}, 1\right)$

12. $\left(-\dfrac{1}{4}, 2\right)$

13. no solution

14. infinite number of solutions

15. $(0.5, 3.5)$

16. $(-0.75, 1.25)$

17. infinite number of solutions

18. no solution

19. $(7, -3)$

20. $(-1, -3)$

21. answers may vary

22. answers may vary

1. $\begin{cases} 2x - 3y = -11 \\ y = 4x - 3 \end{cases}$

2. $\begin{cases} 4x - 5y = 6 \\ y = 3x - 10 \end{cases}$

3. $\begin{cases} x + y = 3 \\ x - y = 7 \end{cases}$

4. $\begin{cases} x - y = 20 \\ x + y = -8 \end{cases}$

5. $\begin{cases} x + 2y = 1 \\ 3x + 4y = -1 \end{cases}$

6. $\begin{cases} x + 3y = 5 \\ 5x + 6y = -2 \end{cases}$

7. $\begin{cases} y = x + 3 \\ 3x = 2y - 6 \end{cases}$

8. $\begin{cases} y = -2x \\ 2x - 3y = -16 \end{cases}$

9. $\begin{cases} y = 2x - 3 \\ y = 5x - 18 \end{cases}$

10. $\begin{cases} y = 6x - 5 \\ y = 4x - 11 \end{cases}$

11. $\begin{cases} x + \dfrac{1}{6}y = \dfrac{1}{2} \\ 3x + 2y = 3 \end{cases}$

12. $\begin{cases} x + \dfrac{1}{3}y = \dfrac{5}{12} \\ 8x + 3y = 4 \end{cases}$

13. $\begin{cases} x - 5y = 1 \\ -2x + 10y = 3 \end{cases}$

14. $\begin{cases} -x + 2y = 3 \\ 3x - 6y = -9 \end{cases}$

15. $\begin{cases} 0.2x - 0.3y = -0.95 \\ 0.4x + 0.1y = 0.55 \end{cases}$

16. $\begin{cases} 0.08x - 0.04y = -0.11 \\ 0.02x - 0.06y = -0.09 \end{cases}$

17. $\begin{cases} x = 3y - 7 \\ 2x - 6y = -14 \end{cases}$

18. $\begin{cases} y = \dfrac{x}{2} - 3 \\ 2x - 4y = 0 \end{cases}$

19. $\begin{cases} 2x + 5y = -1 \\ 3x - 4y = 33 \end{cases}$

20. $\begin{cases} 7x - 3y = 2 \\ 6x + 5y = -21 \end{cases}$

21. Which method, substitution or addition, would you prefer to use to solve the system below? Explain your reasoning.

$$\begin{cases} 3x + 2y = -2 \\ y = -2x \end{cases}$$

22. Which method, substitution or addition, would you prefer to use to solve the system below? Explain your reasoning.

$$\begin{cases} 3x - 2y = -3 \\ 6x + 2y = 12 \end{cases}$$

4.4 Systems of Linear Equations and Problem Solving ▶

Objective A Using a System of Equations for Problem Solving ▶

Objective

A Use a System of Equations to Solve Problems.

Many of the word problems solved earlier with one-variable equations can also be solved with two equations in two variables. We use the same problem-solving steps that we have used throughout this text. The only difference is that two variables are assigned to represent the two unknown quantities and that the problem is translated into two equations.

Problem-Solving Steps

1. UNDERSTAND the problem. During this step, become comfortable with the problem. Some ways of doing this are to

Read and reread the problem.

Choose two variables to represent the two unknowns.

Construct a drawing.

Propose a solution and check. Pay careful attention to how you check your proposed solution. This will help when writing equations to model the problem.

2. TRANSLATE the problem into two equations.

3. SOLVE the system of equations.

4. INTERPRET the results: *Check* the proposed solution in the stated problem and *state* your conclusion.

Example 1 Finding Unknown Numbers

Find two numbers whose sum is 37 and whose difference is 21.

Solution:

1. **UNDERSTAND.** Read and reread the problem. Suppose that one number is 20. If their sum is 37, the other number is 17 because $20 + 17 = 37$. Is their difference 21? No; $20 - 17 = 3$. Our proposed solution is incorrect, but we now have a better understanding of the problem.

Since we are looking for two numbers, we let

x = first number and

y = second number

2. **TRANSLATE.** Since we have assigned two variables to this problem, we translate our problem into two equations.

In words: two numbers whose sum is 37
             ↓   ↓   ↓
Translate: $x + y$ =   37

In words: two numbers whose difference is 21
             ↓   ↓   ↓
Translate: $x - y$ =   21

(Continued on next page)

Practice 1

Find two numbers whose sum is 50 and whose difference is 22.

Teaching Tip

Consider beginning Example 1 by having students list some numbers whose sum is 37. Ask them, "How many such pairs exist?" Then have them list some numbers whose difference is 21. Again ask them, "How many such pairs exist?" Point out that finding numbers that satisfy both conditions could take a while using the list approach. Solving the problem with a system of equations can thus be more efficient.

Answer

1. 36 and 14

309

3. SOLVE. Now we solve the system.

$$\begin{cases} x + y = 37 \\ x - y = 21 \end{cases}$$

Notice that the coefficients of the variable y are opposites. Let's then solve by the addition method and begin by adding the equations.

$$\begin{array}{rl} x + y = 37 & \\ \underline{x - y = 21} & \text{Add the equations.} \\ 2x \quad\;\; = 58 & \\ x = 29 & \text{Divide both sides by 2.} \end{array}$$

Now we let $x = 29$ in the first equation to find y.

$$\begin{array}{rl} x + y = 37 & \text{First equation} \\ 29 + y = 37 & \\ y = 8 & \text{Subtract 29 from both sides.} \end{array}$$

4. INTERPRET. The solution of the system is $(29, 8)$.

Check: Notice that the sum of 29 and 8 is $29 + 8 = 37$, the required sum. Their difference is $29 - 8 = 21$, the required difference.

State: The numbers are 29 and 8.

■ **Work Practice 1**

Teaching Tip

After Example 1, point out to students that the system involved 2 equations with 2 unknowns. One equation was the sum, one equation was the difference, and the two unknowns represented the two numbers that were to be found. After reading Example 2, have students identify the unknowns (price of an adult's ticket and price of a child's ticket) and the two equations that can be written using these unknowns.

Practice 2

Admission prices at a local weekend fair were $5 for children and $7 for adults. The total money collected was $3379, and 587 people attended the fair. How many children and how many adults attended the fair?

Example 2   Solving a Problem About Prices

The Cirque du Soleil show Varekai is performing locally. Matinee admission for 4 adults and 2 children is $374, while admission for 2 adults and 3 children is $285.

a. What is the price of an adult's ticket?

b. What is the price of a child's ticket?

c. Suppose that a special rate of $1000 is offered for groups of 20 persons. Should a group of 4 adults and 16 children use the group rate? Why or why not?

Solution:

1. UNDERSTAND. Read and reread the problem and guess a solution. Let's suppose that the price of an adult's ticket is $50 and the price of a child's ticket is $40. To check our proposed solution, let's see if admission for 4 adults and 2 children is $374. Admission for 4 adults is 4($50) or $200 and admission for 2 children is 2($40) or $80. This gives a total admission of $200 + $80 = $280, not the required $374. Again, though, we have accomplished the purpose of this process: We have a better understanding of the problem. To continue, we let

$$A = \text{the price of an adult's ticket and}$$
$$C = \text{the price of a child's ticket}$$

Answer

2. 365 children and 222 adults

4.4 Systems of Linear Equations and Problem Solving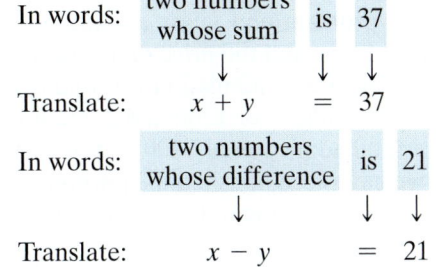

Objective A Using a System of Equations for Problem Solving

Objective

A Use a System of Equations to Solve Problems.

Many of the word problems solved earlier with one-variable equations can also be solved with two equations in two variables. We use the same problem-solving steps that we have used throughout this text. The only difference is that two variables are assigned to represent the two unknown quantities and that the problem is translated into two equations.

Problem-Solving Steps

1. **UNDERSTAND** the problem. During this step, become comfortable with the problem. Some ways of doing this are to

   Read and reread the problem.

   Choose two variables to represent the two unknowns.

   Construct a drawing.

   Propose a solution and check. Pay careful attention to how you check your proposed solution. This will help when writing equations to model the problem.

2. **TRANSLATE** the problem into two equations.

3. **SOLVE** the system of equations.

4. **INTERPRET** the results: *Check* the proposed solution in the stated problem and *state* your conclusion.

Example 1 Finding Unknown Numbers

Find two numbers whose sum is 37 and whose difference is 21.

Solution:

1. **UNDERSTAND.** Read and reread the problem. Suppose that one number is 20. If their sum is 37, the other number is 17 because $20 + 17 = 37$. Is their difference 21? No; $20 - 17 = 3$. Our proposed solution is incorrect, but we now have a better understanding of the problem.

   Since we are looking for two numbers, we let

   x = first number and

   y = second number

2. **TRANSLATE.** Since we have assigned two variables to this problem, we translate our problem into two equations.

   In words: two numbers whose sum   is   37

   $\downarrow$ $\downarrow$ $\downarrow$

   Translate: $x + y$ =   37

   In words: two numbers whose difference   is   21

   $\downarrow$ $\downarrow$ $\downarrow$

   Translate: $x - y$ =   21

(*Continued on next page*)

Practice 1

Find two numbers whose sum is 50 and whose difference is 22.

Teaching Tip

Consider beginning Example 1 by having students list some numbers whose sum is 37. Ask them, "How many such pairs exist?" Then have them list some numbers whose difference is 21. Again ask them, "How many such pairs exist?" Point out that finding numbers that satisfy both conditions could take a while using the list approach. Solving the problem with a system of equations can thus be more efficient.

Answer

1. 36 and 14

3. SOLVE. Now we solve the system.

$$\begin{cases} x + y = 37 \\ x - y = 21 \end{cases}$$

Notice that the coefficients of the variable y are opposites. Let's then solve by the addition method and begin by adding the equations.

$$\begin{array}{ll} x + y = 37 \\ \underline{x - y = 21} & \text{Add the equations.} \\ 2x \phantom{{}- y} = 58 \\ x = 29 & \text{Divide both sides by 2.} \end{array}$$

Now we let $x = 29$ in the first equation to find y.

$$\begin{array}{ll} x + y = 37 & \text{First equation} \\ 29 + y = 37 \\ \phantom{29 +{}} y = 8 & \text{Subtract 29 from both sides.} \end{array}$$

4. INTERPRET. The solution of the system is $(29, 8)$.

Check: Notice that the sum of 29 and 8 is $29 + 8 = 37$, the required sum. Their difference is $29 - 8 = 21$, the required difference.

State: The numbers are 29 and 8.

■ **Work Practice 1**

Practice 2

Admission prices at a local weekend fair were $5 for children and $7 for adults. The total money collected was $3379, and 587 people attended the fair. How many children and how many adults attended the fair?

Example 2   Solving a Problem About Prices

The Cirque du Soleil show Varekai is performing locally. Matinee admission for 4 adults and 2 children is $374, while admission for 2 adults and 3 children is $285.

a. What is the price of an adult's ticket?

b. What is the price of a child's ticket?

c. Suppose that a special rate of $1000 is offered for groups of 20 persons. Should a group of 4 adults and 16 children use the group rate? Why or why not?

Solution:

1. UNDERSTAND. Read and reread the problem and guess a solution. Let's suppose that the price of an adult's ticket is $50 and the price of a child's ticket is $40. To check our proposed solution, let's see if admission for 4 adults and 2 children is $374. Admission for 4 adults is 4($50) or $200 and admission for 2 children is 2($40) or $80. This gives a total admission of $200 + $80 = $280, not the required $374. Again, though, we have accomplished the purpose of this process: We have a better understanding of the problem. To continue, we let

$A = $ the price of an adult's ticket and

$C = $ the price of a child's ticket

Answer

2. 365 children and 222 adults

2. TRANSLATE. We translate the problem into two equations using both variables.

In words:
| admission for 4 adults | and | admission for 2 children | is | $374 |

Translate: $4A + 2C = 374$

In words:
| admission for 2 adults | and | admission for 3 children | is | $285 |

Translate: $2A + 3C = 285$

3. SOLVE. We solve the system.

$$\begin{cases} 4A + 2C = 374 \\ 2A + 3C = 285 \end{cases}$$

Since both equations are written in standard form, we solve by the addition method. First we multiply the second equation by -2 so that when we add the equations, we eliminate the variable A. Then the system

$$\begin{cases} 4A + 2C = 374 \\ -2(2A + 3C) = -2(285) \end{cases}$$

simplifies to

$$\begin{cases} 4A + 2C = 374 \\ -4A - 6C = -570 \end{cases}$$

Add the equations.

$$\begin{aligned} -4C &= -196 \\ C &= 49 \text{ or } \$49, \text{ the child's ticket price} \end{aligned}$$

To find A, we replace C with 49 in the first equation.

$4A + 2C = 374$ First equation

$4A + 2(49) = 374$ Let $C = 49$.

$4A + 98 = 374$

$4A = 276$

$A = 69 \text{ or } \$69, \text{ the adult's ticket price}$

4. INTERPRET.

Check: Notice that 4 adults and 2 children will pay $4(\$69) + 2(\$49) = \$276 + \$98 = \$374$, the required amount. Also, the price for 2 adults and 3 children is $2(\$69) + 3(\$49) = \$138 + \$147 = \$285$, the required amount.

State: Answer the three original questions.

a. Since $A = 69$, the price of an adult's ticket is $69.

b. Since $C = 49$, the price of a child's ticket is $49.

c. The regular admission price for 4 adults and 16 children is

$$4(\$69) + 16(\$49) = \$276 + \$784$$
$$= \$1060$$

This is $60 more than the special group rate of $1000, so they should request the group rate.

Work Practice 2

Practice 3

Two cars are 440 miles apart and traveling toward each other. They meet in 3 hours. If one car's speed is 10 miles per hour faster than the other car's speed, find the speed of each car.

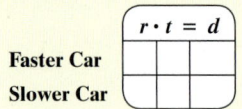

$r \cdot t = d$		
Faster Car		
Slower Car		

Example 3   Finding Rates

As part of an exercise program, two students, Louisa and Alfredo, start walking each morning. They live 15 miles away from each other. They decide to meet one day by walking toward one another. After 2 hours they meet. If Louisa walks one mile per hour faster than Alfredo, find both walking speeds.

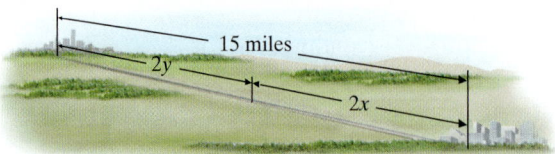

Solution:

1. UNDERSTAND. Read and reread the problem. Let's propose a solution and use the formula $d = r \cdot t$ to check. Suppose that Louisa's rate is 4 miles per hour. Since Louisa's rate is 1 mile per hour faster, Alfredo's rate is 3 miles per hour. To check, see if they can walk a total of 15 miles in 2 hours. Louisa's distance is rate $\cdot$ time $= 4(2) = 8$ miles and Alfredo's distance is rate $\cdot$ time $= 3(2) = 6$ miles. Their total distance is 8 miles + 6 miles = 14 miles, not the required 15 miles. Now that we have a better understanding of the problem, let's model it with a system of equations.

First, we let

x = Alfredo's rate in miles per hour and

y = Louisa's rate in miles per hour

Now we use the facts stated in the problem and the formula $d = rt$ to fill in the following chart.

	r	$\cdot\ t$	$=\ d$
Alfredo	x	2	$2x$
Louisa	y	2	$2y$

2. TRANSLATE. We translate the problem into two equations using both variables.

In words:	Alfredo's distance	+	Louisa's distance	=	15 miles
	↓	↓	↓	↓	↓
Translate:	$2x$	+	$2y$	=	15

In words:	Louisa's rate	is	1 mile per hour faster than Alfredo's
	↓	↓	↓
Translate:	y	=	$x + 1$

3. SOLVE. The system of equations we are solving is

$$\begin{cases} 2x + 2y = 15 \\ y = x + 1 \end{cases}$$

Let's use substitution to solve the system since the second equation is solved for y.

$$2x + 2y = 15 \qquad \text{First equation}$$

$$2x + 2(x + 1) = 15 \qquad \text{Replace } y \text{ with } x + 1.$$

$$2x + 2x + 2 = 15$$

$$4x = 13$$

$$x = \frac{13}{4} = 3\frac{1}{4} \text{ or } 3.25$$

$$y = x + 1 = 3\frac{1}{4} + 1 = 4\frac{1}{4} \text{ or } 4.25$$

4. INTERPRET. Alfredo's proposed rate is $3\frac{1}{4}$ miles per hour, and Louisa's proposed rate is $4\frac{1}{4}$ miles per hour.

Check: Use the formula $d = rt$ and find that in 2 hours, Alfredo's distance is $(3.25)(2)$ miles or 6.5 miles. In 2 hours, Louisa's distance is $(4.25)(2)$ miles or 8.5 miles. The total distance walked is 6.5 miles + 8.5 miles or 15 miles, the given distance.

State: Alfredo walks at a rate of 3.25 miles per hour, and Louisa walks at a rate of 4.25 miles per hour.

■ **Work Practice 3**

Example 4 Finding Amounts of Solutions

A chemistry teaching assistant needs 10 liters of a 20% saline solution (salt water) for his 2 p.m. laboratory class. Unfortunately, the only mixtures on hand are a 5% saline solution and a 25% saline solution. How much of each solution should he mix to produce the 20% solution?

Solution:

1. UNDERSTAND. Read and reread the problem. Suppose that we need 4 liters of the 5% solution. Then we need $10 - 4 = 6$ liters of the 25% solution. To see if this gives us 10 liters of a 20% saline solution, let's find the amount of pure salt in each solution.

	concentration rate	×	amount of solution	=	amount of pure salt
	↓		↓		↓
5% solution:	0.05	×	4 liters	=	0.2 liter
25% solution:	0.25	×	6 liters	=	1.5 liters
20% solution:	0.20	×	10 liters	=	2 liters

Since 0.2 liter + 1.5 liters = 1.7 liters, not 2 liters, our proposed solution is incorrect. But we have gained some insight into how to model and check this problem.

We let

x = number of liters of 5% solution and
y = number of liters of 25% solution

5% saline solution   25% saline solution   20% saline solution

(*Continued on next page*)

Teaching Tip

Before beginning Example 4, you may want to have students make a table showing the amounts of salt and water in 1, 5, 10, 25, 50, and 100 liters of a 5% saline solution.

Practice 4

A pharmacist needs 50 liters of a 60% alcohol solution. She currently has available a 20% solution and a 70% solution. How many liters of each must she use to make the needed 50 liters of 60% alcohol solution?

Answer
4. 10 liters of the 20% alcohol solution and 40 liters of the 70% alcohol solution

Now we use a table to organize the given data.

	Concentration Rate	Liters of Solution	Liters of Pure Salt
First Solution	5%	x	$0.05x$
Second Solution	25%	y	$0.25y$
Mixture Needed	20%	10	$(0.20)(10)$

2. TRANSLATE. We translate into two equations using both variables.

In words: | liters of 5% solution | $+$ | liters of 25% solution | $=$ | 10 liters |

$$\downarrow \qquad\qquad \downarrow \qquad\qquad \downarrow$$

Translate: $x \quad + \quad y \quad = \quad 10$

In words: | salt in 5% solution | $+$ | salt in 25% solution | $=$ | salt in mixture |

$$\downarrow \qquad\qquad \downarrow \qquad\qquad \downarrow$$

Translate: $0.05x \quad + \quad 0.25y \quad = \quad (0.20)(10)$

3. SOLVE. Here we solve the system

$$\begin{cases} x + y = 10 \\ 0.05x + 0.25y = 2 \end{cases}$$

To solve by the addition method, we first multiply the first equation by -25 and the second equation by 100. Then the system

$$\begin{cases} -25(x + y) = -25(10) \\ 100(0.05x + 0.25y) = 100(2) \end{cases} \quad \text{simplifies to} \quad \begin{cases} -25x - 25y = -250 \\ 5x + 25y = 200 \end{cases}$$

$${-20x} = -50 \qquad \text{Add.}$$
$$x = 2.5$$

To find y, we let $x = 2.5$ in the first equation of the original system.

$$x + y = 10$$
$$2.5 + y = 10 \qquad \text{Let } x = 2.5.$$
$$y = 7.5$$

4. INTERPRET. Thus, we propose that he needs to mix 2.5 liters of 5% saline solution with 7.5 liters of 25% saline solution.

Check: Notice that $2.5 + 7.5 = 10$, the required number of liters. Also, the sum of the liters of salt in the two solutions equals the liters of salt in the required mixture:

$$0.05(2.5) + 0.25(7.5) = 0.20(10)$$
$$0.125 + 1.875 = 2$$

State: He needs 2.5 liters of the 5% saline solution and 7.5 liters of the 25% saline solution.

■ **Work Practice 4**

✓**Concept Check** Suppose you mix an amount of a 30% acid solution with an amount of a 50% acid solution. Which of the following acid strengths would be possible for the resulting acid mixture?

a. 22%   **b.** 44%   **c.** 63%

✓**Concept Check Answer**

b

Vocabulary, Readiness & Video Check

Martin-Gay Interactive Videos   *Watch the section lecture video and answer the following question.*

Objective A **1.** In the lecture before Example 1, the problem-solving steps for solving applications involving systems are discussed. How do these steps differ from the general problem-solving strategy steps? ▶

See Video 4.4 🍎

See video answer section.

4.4 Exercise Set MyMathLab® ▶

Without actually solving each problem, choose the correct solution by deciding which choice satisfies the given conditions.

△ **1.** The length of a rectangle is 3 feet longer than the width. The perimeter is 30 feet. Find the dimensions of the rectangle.   c
   a. length = 8 feet; width = 5 feet
   b. length = 8 feet; width = 7 feet
   c. length = 9 feet; width = 6 feet

△ **2.** An isosceles triangle, a triangle with two sides of equal length, has a perimeter of 20 inches. Each of the equal sides is one inch longer than the third side. Find the lengths of the three sides.   b
   a. 6 inches, 6 inches, and 7 inches
   b. 7 inches, 7 inches, and 6 inches
   c. 6 inches, 7 inches, and 8 inches

3. Two computer disks and three notebooks cost $17. However, five computer disks and four notebooks cost $32. Find the price of each.   b
   a. notebook = $4; computer disk = $3
   b. notebook = $3; computer disk = $4
   c. notebook = $5; computer disk = $2

4. Two music CDs and four DVDs cost a total of $40. However, three music CDs and five DVDs cost $55. Find the price of each.   c
   a. CD = $12; DVD = $4
   b. CD = $15; DVD = $2
   c. CD = $10; DVD = $5

5. Kesha has a total of 100 coins, all of which are either dimes or quarters. The total value of the coins is $13.00. Find the number of each type of coin.   a
   a. 80 dimes; 20 quarters
   b. 20 dimes; 44 quarters
   c. 60 dimes; 40 quarters

6. Samuel has 28 gallons of saline solution available in two large containers at his pharmacy. One container holds three times as much as the other container. Find the capacity of each container.   c
   a. 15 gallons; 5 gallons
   b. 20 gallons; 8 gallons
   c. 21 gallons; 7 gallons

Objective A *Write a system of equations describing each situation. Do not solve the system. See Example 1.*

7. Two numbers add up to 15 and have a difference of 7.
$$\begin{cases} x + y = 15 \\ x - y = 7 \end{cases}$$

8. The total of two numbers is 16. The first number plus 2 more than 3 times the second equals 18.
$$\begin{cases} x + y = 16 \\ x + 3y + 2 = 18 \end{cases}$$

9. Keiko has a total of $6500, which she has invested in two accounts. The larger account is $800 greater than the smaller account. $\begin{cases} x + y = 6500 \\ x = y + 800 \end{cases}$

10. Dominique has four times as much money in his savings account as in his checking account. The total amount is $2300. $\begin{cases} x + y = 2300 \\ y = 4x \end{cases}$

Solve. See Examples 1 through 4.

11. Two numbers total 83 and have a difference of 17. Find the two numbers.   33 and 50

12. The sum of two numbers is 76 and their difference is 52. Find the two numbers.   64 and 12

13. A first number plus twice a second number is 8. Twice the first number plus the second totals 25. Find the numbers.   14 and −3

14. One number is 4 more than twice a second number. Their total is 25. Find the numbers.   18 and 7

15. Miguel Cabrera of the Detroit Tigers led Major League Baseball in runs batted in for the 2012 regular season. Josh Hamilton of the Texas Rangers, who came in second to Cabrera, had 11 fewer runs batted in for the 2012 regular season. Together, these two players brought home 267 runs during the 2012 regular season. How many runs batted in each did Cabrera and Hamilton account for? (*Source: Baseball Almanac*)   Cabrera: 139; Hamilton: 128

16. The highest scorer during the WNBA 2013 regular season was Angel McCoughtry of the Atlanta Dream. Over the season, McCoughtry scored 60 more points than the second-highest scorer, Diana Taurasi, of the Phoenix Mercury. Together, McCoughtry and Taurasi scored 1362 points during the 2013 regular season. How many points did each player score over the course of the season? (*Source:* Women's National Basketball Association)   McCoughtry: 711 points; Taurasi: 651 points

17. Ann Marie Jones has been pricing Amtrak train fares for a group trip to New York. Three adults and four children must pay $159. Two adults and three children must pay $112. Find the price of an adult's ticket, and find the price of a child's ticket.   child's ticket: $18; adult's ticket: $29

18. Last month, Jerry Papa purchased five DVDs and two CDs at Wall-to-Wall Sound for $65. This month he bought three DVDs and four CDs for $81. Find the price of each DVD, and find the price of each CD.   DVD: $7; CD: $15

19. Johnston and Betsy Waring have a jar containing 80 coins, all of which are either quarters or nickels. The total value of the coins is $14.60. How many of each type of coin do they have?   quarters: 53; nickels: 27

20. Sarah and Keith Robinson purchased 40 stamps, a mixture of 44¢ and 28¢ stamps. Find the number of each type of stamp if they spent $16.80.   44¢ stamps: 35; 28¢ stamps: 5

21. Norman and Suzanne Scarpulla own 30 shares of McDonald's Corp. stock and 68 shares of Ford Motor Co. stock. As the New York Stock Exchange opened on a day in 2013, their stock portfolio consisting of these two stocks was worth $4107. The McDonald's stock was $80.55 more per share than the Ford stock. What was the price of each stock on that day? (*Source:* SIX Financial Information)   McDonald's: $97.80; Ford: $17.25

22. Saralee Rose has investments in Google and Facebook stock. As the NASDAQ exchange opened on a day in 2013, Google stock was at $886.50 per share, and Facebook stock was at $50 per share. Saralee's portfolio made up of these two stocks was worth $32,964 at that time. If Saralee owns 15 more shares of Google stock than she owns of Facebook stock, how many shares of each type of stock does she own? (*Source:* SIX Financial Information)   Google: 36 shares; Facebook: 21 shares

23. Twice last month, Judy Carter rented a car from Enterprise in Fresno, California, and traveled around the Southwest on business. Enterprise rents this car for a daily fee, plus an additional charge per mile driven. Judy recalls that her first trip lasted 4 days, she drove 450 miles, and the rental cost her $240.50. On her second business trip she drove the same level of car 200 miles in 3 days and paid $146.00 for the rental. Find the daily fee and the mileage charge.   daily fee: $32; mileage charge: $0.25 per mi

24. Joan Gundersen rented the same car model twice from Hertz, which rents this car model for a daily fee plus an additional charge per mile driven. Joan recalls that the car rented for 5 days and driven for 300 miles cost her $178, while the same model car rented for 4 days and driven for 500 miles cost $197. Find the daily fee and find the mileage charge.   daily fee: $23; mileage charge: $0.21 per mi

25. Pratap Puri rowed 18 miles down the Delaware River in 2 hours, but the return trip took him $4\frac{1}{2}$ hours. Find the rate Pratap can row in still water, and find the rate of the current.

Let x = rate Pratap can row in still water and
y = rate of the current
still water: 6.5 mph; current: 2.5 mph

d =	r	$\cdot$	t
Downstream	18	$x + y$	2
Upstream	18	$x - y$	$4\frac{1}{2}$

26. The Jonathan Schultz family took a canoe 10 miles down the Allegheny River in $1\frac{1}{4}$ hours. After lunch it took them 4 hours to return. Find the rate of the current.

Let x = rate the family can row in still water and
y = rate of the current
2.75 mph

d =	r	$\cdot$	t
Downstream	10	$x + y$	$1\frac{1}{4}$
Upstream	10	$x - y$	4

27. Dave and Sandy Hartranft are frequent flyers with Delta Airlines. They often fly from Philadelphia to Chicago, a distance of 780 miles. On one particular trip they fly into the wind, and the flight takes 2 hours. The return trip, with the wind behind them, takes only $1\frac{1}{2}$ hours. Find the speed of the wind and find the speed of the plane in still air.   still air: 455 mph; wind: 65 mph

28. With a strong wind behind it, a United Airlines jet flies 2400 miles from Los Angeles to Orlando in $4\frac{3}{4}$ hours. The return trip takes 6 hours, as the plane flies into the wind. Find the speed of the plane in still air, and find the wind speed to the nearest tenth of a mile per hour.   still air: 452.6 mph; wind: 52.6 mph

29. Kevin Briley began a 186-mile bicycle trip to build up stamina for a triathlon competition. Unfortunately, his bicycle chain broke, so he finished the trip walking. The whole trip took 6 hours. If Kevin walks at a rate of 4 miles per hour and rides at 40 miles per hour, find the amount of time he spent on the bicycle. $4\frac{1}{2}$ hr

30. In Canada, eastbound and westbound trains travel along the same track, with sidings to pull onto to avoid accidents. Two trains are now 150 miles apart, with the westbound train traveling twice as fast as the eastbound train. A warning must be issued to pull one train onto a siding, or else the trains will crash in $1\frac{1}{4}$ hours. Find the speed of the eastbound train and the speed of the westbound train.
westbound: 80 mph; eastbound: 40 mph

31. Dorren Schmidt is a chemist with Gemco Pharmaceutical. She needs to prepare 12 ounces of a 9% hydrochloric acid solution. Find the amount of a 4% solution and the amount of a 12% solution she should mix to get this solution.   12% solution: $7\frac{1}{2}$ oz; 4% solution: $4\frac{1}{2}$ oz

Concentration Rate	Liters of Solution	Liters of Pure Acid
0.04	x	0.04x
0.12	y	?
0.09	12	?

32. Elise Everly is preparing 15 liters of a 25% saline solution. Elise has two other saline solutions with strengths of 40% and 10%. Find the amount of 40% solution and the amount of 10% solution she should mix to get 15 liters of a 25% solution.   40% solution: $7\frac{1}{2}$ liters; 10% solution: $7\frac{1}{2}$ liters

Concentration Rate	Liters of Solution	Liters of Pure Salt
0.40	x	0.40x
0.10	y	?
0.25	15	?

33. Wayne Osby blends coffee for a local coffee café. He needs to prepare 200 pounds of blended coffee beans selling for $3.95 per pound. He intends to do this by blending together a high-quality bean costing $4.95 per pound and a cheaper bean costing $2.65 per pound. To the nearest pound, find how much of the high-quality coffee beans and how much of the cheaper coffee beans he should blend. $4.95 beans: 113 lb; $2.65 beans: 87 lb

34. Macadamia nuts cost an astounding $16.50 per pound, but research by an independent firm says that mixed nuts sell better if macadamias are included. The standard mix costs $9.25 per pound. Find how many pounds of macadamias and how many pounds of the standard mix should be combined to produce 40 pounds that will cost $10 per pound. Find the amounts to the nearest tenth of a pound.   macadamia: 4.1 lb; standard mix: 35.9 lb

35. Recall that two angles are complementary if the sum of their measures is 90°. Find the measures of two complementary angles if one angle is twice the other.   60°, 30°

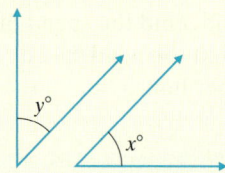

36. Recall that two angles are supplementary if the sum of their measures is 180°. Find the measures of two supplementary angles if one angle is 20° more than four times the other.   32°, 148°

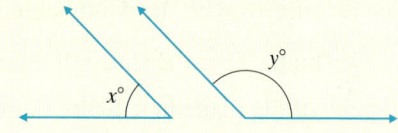

37. Find the measures of two complementary angles if one angle is 10° more than three times the other.   20°, 70°

38. Find the measures of two supplementary angles if one angle is 18° more than twice the other.   54°, 126°

39. Kathi and Robert Hawn had a pottery stand at the annual Skippack Craft Fair. They sold some of their pottery at the original price of $9.50 each, but later decreased the price of each by $2. If they sold all 90 pieces and took in $721, find how many they sold at the original price and how many they sold at the reduced price.   number sold at $9.50: 23; number sold at $7.50: 67

40. A charity fundraiser consisted of a spaghetti supper where a total of 387 people were fed. They charged $6.80 for adults and half price for children. If they took in $2444.60, find how many adults and how many children attended the supper.   adults: 332; children: 55

41. The Santa Fe National Historic Trail is approximately 1200 miles between Old Franklin, Missouri, and Santa Fe, New Mexico. Suppose that a group of hikers start from each town and walk the trail toward each other. They meet after a total hiking time of 240 hours. If one group travels $\frac{1}{2}$ mile per hour slower than the other group, find the rate of each group. (*Source:* National Park Service) $2\frac{1}{4}$ mph and $2\frac{3}{4}$ mph

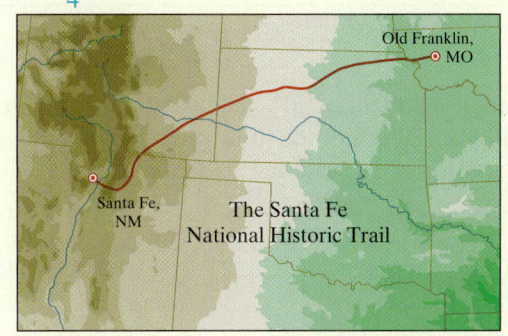

The Santa Fe National Historic Trail

42. California 1 South is a historic highway that stretches 123 miles along the coast from Monterey to Morro Bay. Suppose that two cars start driving this highway, one from each town. They meet after 3 hours. Find the rate of each car if one car travels 1 mile per hour faster than the other car. (*Source: National Geographic*)   20 mph, 21 mph

43. A 30% solution of fertilizer is to be mixed with a 60% solution of fertilizer in order to get 150 gallons of a 50% solution. How many gallons of the 30% solution and 60% solution should be mixed?
30%: 50 gal; 60%: 100 gal

44. A 10% acid solution is to be mixed with a 50% acid solution in order to get 120 ounces of a 20% acid solution. How many ounces of the 10% solution and 50% solution should be mixed?
10%: 90 oz; 50%: 30 oz

45. Traffic signs are regulated by the *Manual on Uniform Traffic Control Devices* (MUTCD). According to this manual, if the sign below is placed on a freeway, its perimeter must be 144 inches. Also, its length must be 12 inches longer than its width. Find the dimensions of this sign.   length: 42 in.; width: 30 in.

46. According to the MUTCD (see Exercise **45**), this sign must have a perimeter of 60 inches. Also, its length must be 6 inches longer than its width. Find the dimensions of this sign.   length: 18 in.; width: 12 in.

Review

Evaluate. See Sections 1.3 and 1.6.

47. 4^2
16

48. 3^2
9

49. $(-5)^2$
25

50. $(-11)^2$
121

51. $10^2 - 5^3$
-25

52. $7^2 - 3^4$
-32

Concept Extensions

Solve. See the Concept Check in this section.

53. Suppose you mix an amount of candy costing $0.49 a pound with candy costing $0.65 a pound. Which of the following costs per pound could result?
  a. $0.58 **b.** $0.72 **c.** $0.29   a

54. Suppose you mix a 50% acid solution with pure acid (100%). Which of the following acid strengths are possible for the resulting acid mixture?
  a. 25% **b.** 150% **c.** 62% **d.** 90%   c, d

Solve.

△ **55.** Dale and Sharon Mahnke have decided to fence off a garden plot behind their house, using their house as the "fence" along one side of the garden. The length (which runs parallel to the house) is 3 feet less than twice the width. Find the dimensions if 33 feet of fencing is used along the three sides requiring it.   width: 9 ft; length: 15 ft

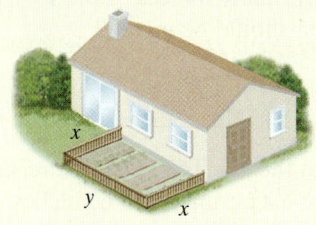

△ **56.** Judy McElroy plans to erect 152 feet of fencing to make a rectangular horse pasture. A river bank serves as one side length of the rectangle. If each width is 4 feet longer than half the length, find the dimensions.   width: 40 ft; length: 72 ft

Chapter 4 Group Activity

Additional group activities are available in the *Instructor's Resource Manual with Tests.*

Break-Even Point

Sections 4.1, 4.2, 4.3, 4.4

When a business sells a new product, it generally does not start making a profit right away. There are usually many expenses associated with creating a new product. These expenses might include an advertising blitz to introduce the product to the public. These start-up expenses might also include the cost of market research and product development or any brand-new equipment needed to manufacture the product. Start-up costs like these are generally called *fixed costs* because they don't depend on the number of items manufactured. Expenses that do depend on the number of items manufactured, such as the cost of materials and shipping, are called *variable costs*. The total cost of manufacturing the new product is given by the cost equation Total cost = Fixed costs + Variable costs.

For instance, suppose a greeting card company is launching a new line of greeting cards. The company spent $7000 doing product research and development for the new line and spent $15,000 advertising the new line. The company does not need to buy any new equipment to manufacture the cards, but the paper and ink needed to make each card will cost $0.20 per card. The total cost y in dollars for manufacturing x cards is $y = 22,000 + 0.20x$.

Once a business sets a price for a new product, the company can find the product's expected *revenue*. Revenue is the amount of money the company takes in from the sales of its product. The revenue from selling a product is given by the revenue equation Revenue = Price per item × Number of items sold.

For instance, suppose that the card company plans to sell its new cards for $1.50 each. The revenue y, in dollars, that the company can expect to receive from the sales of x cards is $y = 1.50x$.

If the total cost and revenue equations are graphed on the same coordinate system, the graphs should intersect. The point of intersection is where total cost equals revenue and is called the *break-even point*. The break-even point gives the number of items x that must be manufactured and sold for the company to recover its expenses. If fewer than this number of items are produced and sold, the company loses money. If more than this number of items are produced and sold, the company makes a profit. In the case of the greeting card company, approximately 16,923 cards must be manufactured and sold for the company to break even on this new card line. The total cost and revenue of producing and selling 16,923 cards is the same. It is approximately $25,385.

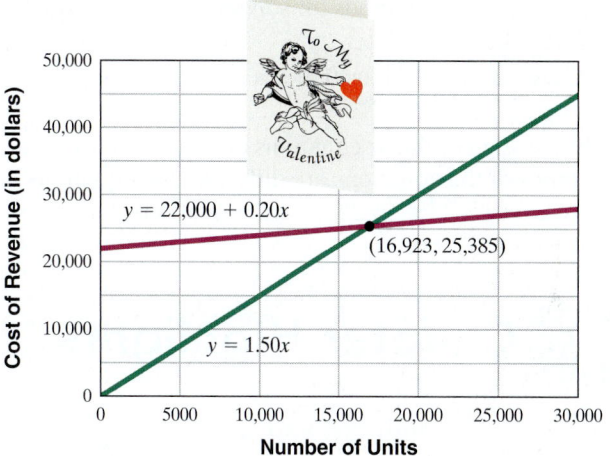

Group Activity

Suppose your group is starting a small business near your campus.

a. Choose a business and decide what campus-related product or service you will provide.

b. Research the fixed costs of starting up such a business.

c. Research the variable costs of producing such a product or providing such a service.

d. Decide how much you will charge per unit of your product or service.

e. Find a system of equations for the total cost and revenue of your product or service.

f. How many units of your product or service must be sold before your business will break even?

Chapter 4 Vocabulary Check

Fill in each blank with one of the words or phrases listed below.

system of linear equations   solution   consistent   independent

dependent   inconsistent   substitution   addition

1. In a system of linear equations in two variables, if the graphs of the equations are the same, the equations are _____dependent_____ equations.

2. Two or more linear equations are called a(n) ___system of linear equations___.

3. A system of equations that has at least one solution is called a(n) _____consistent_____ system.

4. A(n) _____solution_____ of a system of two equations in two variables is an ordered pair of numbers that is a solution of both equations in the system.

5. Two algebraic methods for solving systems of equations are _____addition_____ and _____substitution_____.

6. A system of equations that has no solution is called a(n) _____inconsistent_____ system.

7. In a system of linear equations in two variables, if the graphs of the equations are different, the equations are _____independent_____ equations.

> **Helpful Hint**
>
> ▶ Are you preparing for your test? Don't forget to take the Chapter 4 Test on page 328. Then check your answers at the back of the text and use the Chapter Test Prep Videos to see the fully worked-out solutions to any of the exercises you want to review.

 **Chapter Highlights**

Definitions and Concepts	Examples
Section 4.1 Solving Systems of Linear Equations by Graphing	

Definitions and Concepts	Examples
A **system of linear equations** consists of two or more linear equations. A **solution** of a system of two equations in two variables is an ordered pair of numbers that is a solution of both equations in the system.	$\begin{cases} 2x + y = 6 \\ x = -3y \end{cases}$ $\begin{cases} -3x + 5y = 10 \\ x - 4y = -2 \end{cases}$ Determine whether $(-1, 3)$ is a solution of the system. $\begin{cases} 2x - y = -5 \\ x = 3y - 10 \end{cases}$ Replace x with -1 and y with 3 in both equations. $2x - y = -5$ $2(-1) - 3 \overset{?}{=} -5$ $-5 = -5$ True $x = 3y - 10$ $-1 \overset{?}{=} 3(3) - 10$ $-1 = -1$ True $(-1, 3)$ is a solution of the system.
Graphically, a solution of a system is a point common to the graphs of both equations.	Solve by graphing: $\begin{cases} 3x - 2y = -3 \\ x + y = 4 \end{cases}$ 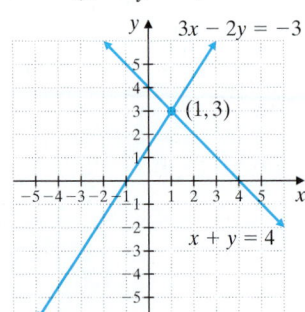

Definitions and Concepts	Examples

Section 4.1 Solving Systems of Linear Equations by Graphing (*continued*)

Three different situations can occur when graphing the two lines associated with the equations in a linear system.

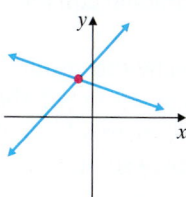

One point of inter-
section; one solution

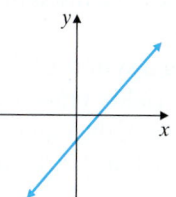

Same line; infinite
number of solutions

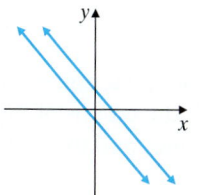

Parallel lines; no solution

Section 4.2 Solving Systems of Linear Equations by Substitution

To Solve a System of Linear Equations by the Substitution Method

Step 1: Solve one equation for a variable.

Step 2: Substitute the expression for the variable into the other equation.

Step 3: Solve the equation from Step 2 to find the value of one variable.

Step 4: Substitute the value from Step 3 in either original equation to find the value of the other variable.

Step 5: Check the solution in both original equations.

Solve by substitution.

$$\begin{cases} 3x + 2y = 1 \\ x = y - 3 \end{cases}$$

Substitute $y - 3$ for x in the first equation.

$$3x + 2y = 1$$
$$3(y - 3) + 2y = 1$$
$$3y - 9 + 2y = 1$$
$$5y = 10$$
$$y = 2 \quad \text{Divide by 5.}$$

To find x, substitute 2 for y in $x = y - 3$ so that $x = 2 - 3$ or -1. The solution $(-1, 2)$ checks.

Section 4.3 Solving Systems of Linear Equations by Addition

To Solve a System of Linear Equations by the Addition Method

Step 1: Rewrite each equation in standard form, $Ax + By = C$.

Step 2: Multiply one or both equations by a nonzero number so that the coefficients of a variable are opposites.

Step 3: Add the equations.

Step 4: Find the value of one variable by solving the resulting equation.

Step 5: Substitute the value from Step 4 into either original equation to find the value of the other variable.

Solve by addition.

$$\begin{cases} x - 2y = 8 \\ 3x + y = -4 \end{cases}$$

Multiply both sides of the first equation by -3.

$$\begin{cases} -3x + 6y = -24 \\ \underline{3x + y = -4} \end{cases}$$
$$7y = -28 \quad \text{Add.}$$
$$y = -4 \quad \text{Divide by 7.}$$

To find x, let $y = -4$ in an original equation.

$$x - 2(-4) = 8 \quad \text{First equation}$$
$$x + 8 = 8$$
$$x = 0$$

(continued)

Definitions and Concepts	Examples

Section 4.3   Solving Systems of Linear Equations by Addition (*continued*)

Step 6: Check the solution in both original equations.

If solving a system of linear equations by substitution or addition yields a true statement such as $-2 = -2$, then the graphs of the equations in the system are identical and the system has an infinite number of solutions.

The solution $(0, -4)$ checks.

Solve: $\begin{cases} 2x - 6y = -2 \\ x = 3y - 1 \end{cases}$

Substitute $3y - 1$ for x in the first equation.

$$2(3y - 1) - 6y = -2$$
$$6y - 2 - 6y = -2$$
$$-2 = -2 \quad \text{True}$$

The system has an infinite number of solutions.

If solving a system of linear equations yields a false statement such as $0 = 3$, the graphs of the equations in the system are parallel lines and the system has no solution.

Solve: $\begin{cases} 5x - 2y = 6 \\ -5x + 2y = -3 \end{cases}$

$$0 = 3 \quad \text{False}$$

The system has no solution.

Section 4.4   Systems of Linear Equations and Problem Solving

Problem-Solving Steps

1. UNDERSTAND. Read and reread the problem.

Two angles are supplementary if the sum of their measures is 180°. The larger of two supplementary angles is three times the smaller, decreased by twelve. Find the measure of each angle. Let

$\quad x = $ measure of smaller angle and

$\quad y = $ measure of larger angle

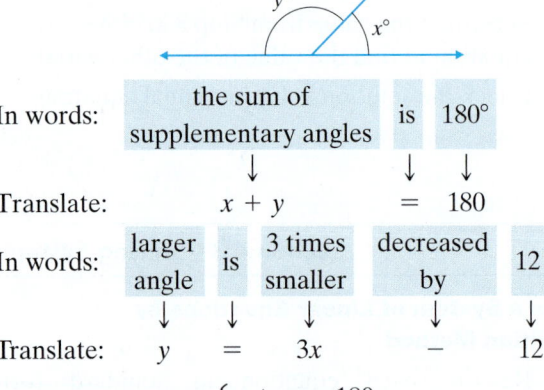

2. TRANSLATE.

In words: the sum of supplementary angles   is   180°

Translate: $x + y$ $=$ 180

In words: larger angle   is   3 times smaller   decreased by   12

Translate: y $=$ $3x$ $-$ 12

3. SOLVE.

Solve the system. $\begin{cases} x + y = 180 \\ y = 3x - 12 \end{cases}$

Use the substitution method and replace y with $3x - 12$ in the first equation.

$$x + y = 180$$
$$x + (3x - 12) = 180$$
$$4x = 192$$
$$x = 48$$

4. INTERPRET.

Since $y = 3x - 12$, then $y = 3 \cdot 48 - 12$ or 132.

The solution checks. The smaller angle measures 48° and the larger angle measures 132°.

(4.1) *Determine whether each ordered pair is a solution of the system of linear equations.*

1. $\begin{cases} 2x - 3y = 12 \\ 3x + 4y = 1 \end{cases}$

   a. $(12, 4)$ no
   b. $(3, -2)$ yes

2. $\begin{cases} 2x + 3y = 1 \\ 3y - x = 4 \end{cases}$

   a. $(2, 2)$ no
   b. $(-1, 1)$ yes

3. $\begin{cases} 5x - 6y = 18 \\ 2y - x = -4 \end{cases}$

   a. $(-6, -8)$ no
   b. $\left(3, \dfrac{5}{2}\right)$ no

4. $\begin{cases} 4x + y = 0 \\ -8x - 5y = 9 \end{cases}$

   a. $\left(\dfrac{3}{4}, -3\right)$ yes
   b. $(-2, 8)$ no

Solve each system of equations by graphing.

5. $\begin{cases} x + y = 5 \\ x - y = 1 \end{cases}$

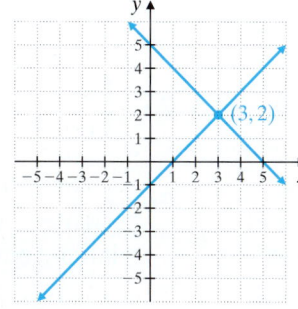

6. $\begin{cases} x + y = 3 \\ x - y = -1 \end{cases}$

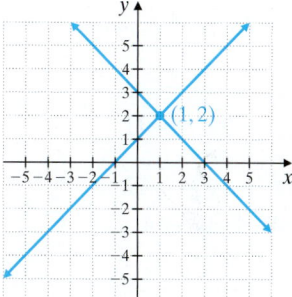

7. $\begin{cases} x = 5 \\ y = -1 \end{cases}$

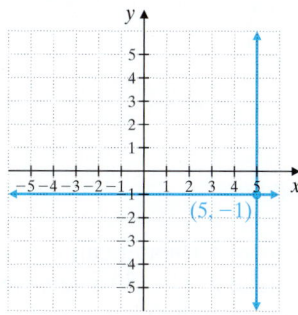

8. $\begin{cases} x = -3 \\ y = 2 \end{cases}$

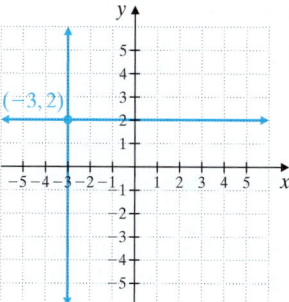

9. $\begin{cases} 2x + y = 5 \\ x = -3y \end{cases}$

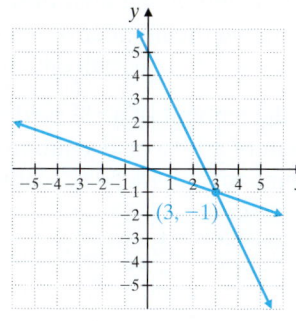

10. $\begin{cases} 3x + y = -2 \\ y = -5x \end{cases}$

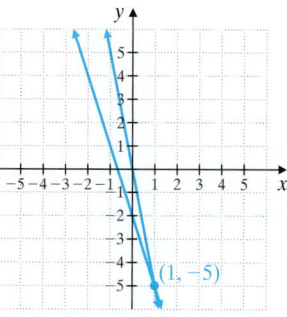

11. $\begin{cases} y = 3x \\ -6x + 2y = 6 \end{cases}$

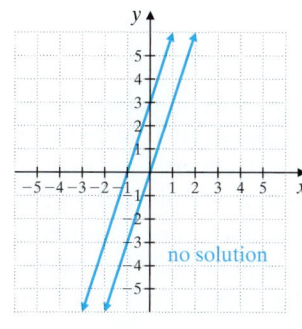

12. $\begin{cases} x - 2y = 2 \\ -2x + 4y = -4 \end{cases}$

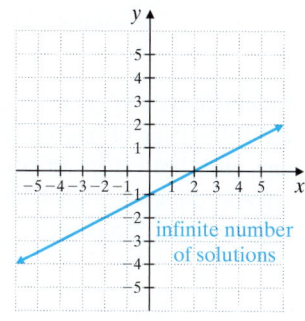

(4.2) *Solve each system of equations by the substitution method.*

13. $\begin{cases} y = 2x + 6 \\ 3x - 2y = -11 \end{cases}$ $(-1, 4)$

14. $\begin{cases} y = 3x - 7 \\ 2x - 3y = 7 \end{cases}$ $(2, -1)$

15. $\begin{cases} x + 3y = -3 \\ 2x + y = 4 \end{cases}$ $(3, -2)$

16. $\begin{cases} 3x + y = 11 \\ x + 2y = 12 \end{cases}$ $(2, 5)$

17. $\begin{cases} 4y = 2x + 6 \\ x - 2y = -3 \end{cases}$ infinite number of solutions

18. $\begin{cases} 9x = 6y + 3 \\ 6x - 4y = 2 \end{cases}$ infinite number of solutions

19. $\begin{cases} x + y = 6 \\ y = -x - 4 \end{cases}$ no solution

20. $\begin{cases} -3x + y = 6 \\ y = 3x + 2 \end{cases}$ no solution

(4.3) *Solve each system of equations by the addition method.*

21. $\begin{cases} 2x + 3y = -6 \\ x - 3y = -12 \end{cases}$ $(-6, 2)$

22. $\begin{cases} 4x + y = 15 \\ -4x + 3y = -19 \end{cases}$ $(4, -1)$

23. $\begin{cases} 2x - 3y = -15 \\ x + 4y = 31 \end{cases}$ $(3, 7)$

24. $\begin{cases} x - 5y = -22 \\ 4x + 3y = 4 \end{cases}$ $(-2, 4)$

25. $\begin{cases} 2x - 6y = -1 \\ -x + 3y = \dfrac{1}{2} \end{cases}$ infinite number of solutions

26. $\begin{cases} 0.6x - 0.3y = -1.5 \\ 0.04x - 0.02y = -0.1 \end{cases}$ infinite number of solutions

27. $\begin{cases} \dfrac{3}{4}x + \dfrac{2}{3}y = 2 \\ x + \dfrac{y}{3} = 6 \end{cases}$ $(8, -6)$

28. $\begin{cases} 10x + 2y = 0 \\ 3x + 5y = 33 \end{cases}$ $\left(-\dfrac{3}{2}, \dfrac{15}{2}\right)$

(4.4) *Solve each problem by writing and solving a system of linear equations.*

29. The sum of two numbers is 16. Three times the larger number decreased by the smaller number is 72. Find the two numbers. -6 and 22

30. The Forrest Theater can seat a total of 360 people. They take in $15,150 when every seat is sold. If orchestra section tickets cost $45 and balcony tickets cost $35, find the number of seats in the orchestra section and the number of seats in the balcony. orchestra: 255 seats; balcony: 105 seats

31. A riverboat can go 340 miles upriver in 19 hours, but the return trip takes only 14 hours. Find the current of the river and find the speed of the riverboat in still water to the nearest tenth of a mile. current of river: 3.2 mph; speed in still water: 21.1 mph

$d =$	r	$\cdot$	t
Upriver	340	$x - y$	19
Downriver	340	$x + y$	14

32. Find the amount of a 6% acid solution and the amount of a 14% acid solution Pat Mayfield should combine to prepare 50 cc (cubic centimeters) of a 12% solution.

6% solution: $12\dfrac{1}{2}$ cc; 14% solution: $37\dfrac{1}{2}$ cc

33. A deli charges $3.80 for a breakfast of three eggs and four strips of bacon. The charge is $2.75 for two eggs and three strips of bacon. Find the cost of each egg and the cost of each strip of bacon. egg: $0.40; strip of bacon: $0.65

34. An exercise enthusiast alternates between jogging and walking. He traveled 15 miles during the past 3 hours. He jogs at a rate of 7.5 miles per hour and walks at a rate of 4 miles per hour. Find how much time, to the nearest hundredth of an hour, he actually spent jogging and how much time he spent walking. jogging: 0.86 hr; walking: 2.14 hr

Mixed Review

Solve each system of equations by graphing.

35. $\begin{cases} x - 2y = 1 \\ 2x + 3y = -12 \end{cases}$

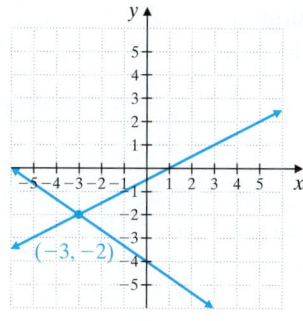

$(-3, -2)$

36. $\begin{cases} 3x - y = -4 \\ 6x - 2y = -8 \end{cases}$

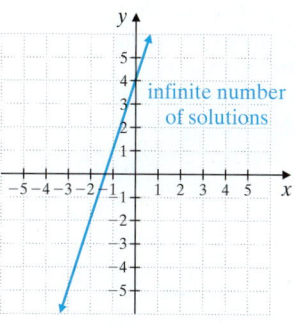

infinite number of solutions

Solve each system of equations.

37. $\begin{cases} x + 4y = 11 \\ 5x - 9y = -3 \end{cases}$ $(3, 2)$

38. $\begin{cases} x + 9y = 16 \\ 3x - 8y = 13 \end{cases}$ $(7, 1)$

39. $\begin{cases} y = -2x \\ 4x + 7y = -15 \end{cases}$ $(1\frac{1}{2}, -3)$

40. $\begin{cases} 3y = 2x + 15 \\ -2x + 3y = 21 \end{cases}$ no solution

41. $\begin{cases} 3x - y = 4 \\ 4y = 12x - 16 \end{cases}$ infinite number of solutions

42. $\begin{cases} x + y = 19 \\ x - y = -3 \end{cases}$ $(8, 11)$

43. $\begin{cases} x - 3y = -11 \\ 4x + 5y = -10 \end{cases}$ $(-5, 2)$

44. $\begin{cases} -x - 15y = 44 \\ 2x + 3y = 20 \end{cases}$ $(16, -4)$

45. $\begin{cases} 2x + y = 3 \\ 6x + 3y = 9 \end{cases}$ infinite number of solutions

46. $\begin{cases} -3x + y = 5 \\ -3x + y = -2 \end{cases}$ no solution

Solve each problem by writing and solving a system of linear equations.

47. The sum of two numbers is 12. Three times the smaller number increased by the larger number is 20. Find the numbers.   4 and 8

48. The difference of two numbers is -18. Twice the smaller decreased by the larger is -23. Find the two numbers. -5 and 13

49. Emma Hodges has a jar containing 65 coins, all of which are either nickels or dimes. The total value of the coins is $5.30. How many of each type does she have?   24 nickels and 41 dimes

50. Sarah and Owen Hebert purchased 26 stamps, a mixture of 13¢ and 22¢ stamps. Find the number of each type of stamp if they spent $4.19.   13¢ stamps: 17; 22¢ stamps: 9

Chapter 4 Test

Step-by-step test solutions are found on the Chapter Test Prep Videos. Where available: **MyMathLab®** or **You Tube**

Answers

Note to Instructor: The Chapter 4 Test file in the TestGen program provides algorithms specifically matched to this test for easy replication for practice or assessment purposes.

1. false

2. false

3. true

4. false

5. no

6. yes

7. see graph

8. see graph

9. $(-4, 1)$

10. $\left(\frac{1}{2}, -2\right)$

11. $(20, 8)$

12. no solution

13. $(4, -5)$

14. $(7, 2)$

Answer each question true or false.

1. A system of two linear equations in two variables can have exactly two solutions.

2. Although $(1, 4)$ is not a solution of $x + 2y = 6$, it can still be a solution of the system $\begin{cases} x + 2y = 6 \\ x + y = 5 \end{cases}$.

3. If the two equations in a system of linear equations are added and the result is $3 = 0$, the system has no solution.

4. If the two equations in a system of linear equations are added and the result is $3x = 0$, the system has no solution.

Is the ordered pair a solution of the given linear system?

5. $\begin{cases} 2x - 3y = 5 \\ 6x + y = 1 \end{cases}$; $(1, -1)$

6. $\begin{cases} 4x - 3y = 24 \\ 4x + 5y = -8 \end{cases}$; $(3, -4)$

Solve each system by graphing.

7. $\begin{cases} x - y = 2 \\ 3x - y = -2 \end{cases}$

8. $\begin{cases} y = -3x \\ 3x + y = 6 \end{cases}$

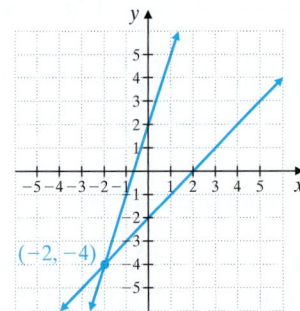

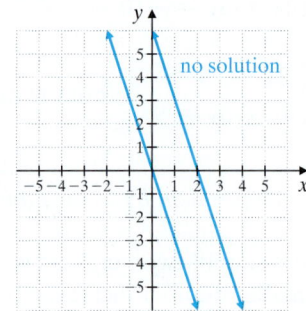

Solve each system by the substitution method.

9. $\begin{cases} 3x - 2y = -14 \\ y = x + 5 \end{cases}$

10. $\begin{cases} \dfrac{1}{2}x + 2y = -\dfrac{15}{4} \\ 4x = -y \end{cases}$

Solve each system by the addition method.

11. $\begin{cases} x + y = 28 \\ x - y = 12 \end{cases}$

12. $\begin{cases} 4x - 6y = 7 \\ -2x + 3y = 0 \end{cases}$

Solve each system using the substitution method or the addition method.

13. $\begin{cases} 3x + y = 7 \\ 4x + 3y = 1 \end{cases}$

14. $\begin{cases} 3(2x + y) = 4x + 20 \\ x - 2y = 3 \end{cases}$

15. $\begin{cases} \dfrac{x-3}{2} = \dfrac{2-y}{4} \\ \dfrac{7-2x}{3} = \dfrac{y}{2} \end{cases}$

16. $\begin{cases} 8x - 4y = 12 \\ y = 2x - 3 \end{cases}$

17. $\begin{cases} 0.01x - 0.06y = -0.23 \\ 0.2x + 0.4y = 0.2 \end{cases}$

18. $\begin{cases} x - \dfrac{2}{3}y = 3 \\ -2x + 3y = 10 \end{cases}$

Solve each problem by writing and using a system of linear equations.

19. Two numbers have a sum of 124 and a difference of 32. Find the numbers.

20. Find the amount of a 12% saline solution a lab assistant should add to 80 cc (cubic centimeters) of a 22% saline solution in order to have a 16% solution.

21. Although the number of farms in the United States is still decreasing, small farms are making a comeback. In 2012, Texas and Missouri were the states with the most number of farms. Texas had 139 thousand more farms than Missouri, and the total number of farms for these two states was 351 thousand. Find the number of farms for each state. (*Source:* National Agricultural Statistics Service)

22. Two hikers start at opposite ends of the St. Tammany Trails and walk toward each other. The trail is 36 miles long and they meet in 4 hours. If one hiker is twice as fast as the other, find both hiking speeds.

The graph below shows the number of music album sales that fell within the alternative or R&B music genres for the years shown. Use this graph to answer Exercises 23 and 24.

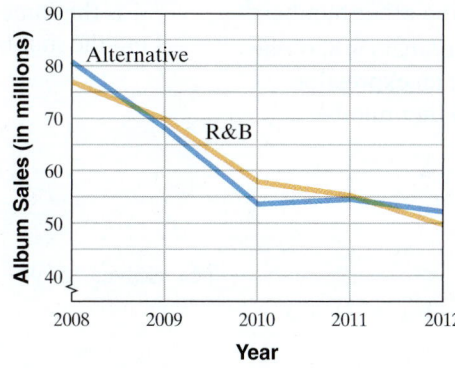

Source: Nielsen Company

23. Between what years were total album sales of R&B music equal to total album sales of alternative music?

24. List each year in which there were more album sales of alternative music than R&B music.

---

15. $(5, -2)$

16. infinite number of solutions

17. $(-5, 3)$

18. $\left(\dfrac{47}{5}, \dfrac{48}{5}\right)$

19. $78, 46$

20. 120 cc

21. Texas: 245 thousand; Missouri: 106 thousand

22. 3 mph; 6 mph

23. 2008–2009; 2011–2012

24. 2008, 2012

Chapters 1–4 Cumulative Review

Answers

1. a. −6

   b. 6.3 (Sec. 1.5, Ex. 6)

2. a. 25

   b. 32 (Sec. 1.3)

3. $\frac{1}{22}$ (Sec. 1.6, Ex. 9a)

4. −22 (Sec. 1.4)

5. $\frac{16}{3}$ (Sec. 1.6, Ex. 9b)

6. $-\frac{3}{16}$ (Sec. 1.4)

7. $-\frac{1}{10}$ (Sec. 1.6, Ex. 9c)

8. 10 (Sec. 1.4)

9. $-\frac{13}{9}$ (Sec. 1.6, Ex. 9d)

10. $\frac{9}{13}$ (Sec. 1.4)

11. $\frac{1}{1.7}$ (Sec. 1.6, Ex. 9e)

12. −1.7 (Sec. 1.4)

13. a. 5

    b. $8 - x$ (Sec. 2.1, Ex. 8)

14. −5 (Sec. 2.1)

15. no solution (Sec. 2.3, Ex. 6)

16. no solution (Sec. 2.3)

17. 12 (Sec. 2.3, Ex. 3)

18. 40 (Sec. 2.3)

330

---

1. Simplify each expression.
   a. $-14 - 8 + 10 - (-6)$
   b. $1.6 - (-10.3) + (-5.6)$

2. Evaluate:
   a. 5^2
   b. 2^5

Find the reciprocal or opposite of each number.

3. reciprocal of 22

4. opposite of 22

5. reciprocal of $\frac{3}{16}$

6. opposite of $\frac{3}{16}$

7. reciprocal of -10

8. opposite of -10

9. reciprocal of $-\frac{9}{13}$

10. opposite of $-\frac{9}{13}$

11. reciprocal of 1.7

12. opposite of 1.7

13. a. The sum of two numbers is 8. If one number is 3, find the other number.
    b. The sum of two numbers is 8. If one number is x, write an expression representing the other number.

14. Five times the sum of a number and -1 is the same as 6 times the number. Find the number.

15. Solve:
    $-2(x - 5) + 10 = -3(x + 2) + x$

16. Solve: $5(y - 5) = 5y + 10$

17. Solve: $\frac{x}{2} - 1 = \frac{2}{3}x - 3$

18. Solve: $7(x - 2) - 6(x + 1) = 20$

19. Solve:
$$2(x - 3) - 5 \leq 3(x + 2) - 18$$

20. Solve $P = a + b + c$ for b.

21. Find the slope of the line through $(-1, 5)$ and $(2, -3)$.

22. Find the slope of a line parallel to the line passing through $(-1, 3)$ and $(2, -8)$.

23. Find the slope of the line $y = -1$.

24. Find the slope of the line $x = 2$.

25. Find an equation of the line through $(2, 5)$ and $(-3, 4)$. Write the equation in the form $Ax + By = C$.

26. Write an equation of the line with slope -5 through $(-2, 3)$.

27. Determine whether $(12, 6)$ is a solution of the system $\begin{cases} 2x - 3y = 6 \\ x = 2y \end{cases}$

28. Determine whether each ordered pair is a solution of the given system.
$$\begin{cases} 2x - y = 6 \\ 3x + 2y = -5 \end{cases}$$
a. $(1, -4)$ **b.** $(0, 6)$ **c.** $(3, 0)$

Solve each system.

29. $\begin{cases} x + 2y = 7 \\ 2x + 2y = 13 \end{cases}$

30. $\begin{cases} 3x - 4y = 10 \\ y = 2x \end{cases}$

31. $\begin{cases} -x - \dfrac{y}{2} = \dfrac{5}{2} \\ \dfrac{x}{6} - \dfrac{y}{2} = 0 \end{cases}$

32. $\begin{cases} x = 5y - 3 \\ x = 8y + 4 \end{cases}$

33. Find two numbers whose sum is 37 and whose difference is 21.

34. Find two numbers whose sum is 67 and whose difference is 29.

$\{x \mid x \geq 1\}$
19. (Sec. 2.7, Ex. 9)

$b = P - a - c$
20. (Sec. 2.5)

$m = -\dfrac{8}{3}$
21. (Sec. 3.4, Ex. 1)

$-\dfrac{11}{3}$ (Sec. 3.4)
22.

$m = 0$
23. (Sec. 3.4, Ex. 5)

24. undefined (Sec. 3.4)

$-x + 5y = 23$
25. (Sec. 3.5, Ex. 5)

$y = -5x - 7$
26. (Sec. 3.5)

It is a solution.
27. (Sec. 4.1, Ex. 1)

28. a. yes

b. no

c. no (Sec. 4.1)

$\left(6, \dfrac{1}{2}\right)$
29. (Sec. 4.2, Ex. 3)

30. $(-2, -4)$ (Sec. 4.2)

$\left(-\dfrac{15}{7}, -\dfrac{5}{7}\right)$
31. (Sec. 4.3, Ex. 6)

$\left(-\dfrac{44}{3}, -\dfrac{7}{3}\right)$
32. (Sec. 4.2)

29 and 8
33. (Sec. 4.4, Ex. 1)

34. 48 and 19 (Sec. 4.4)

5

Exponents and Polynomials

Recall from Chapter 1 that an exponent is a shorthand notation for repeated factors. This chapter explores additional concepts about exponents and exponential expressions. An especially useful type of exponential expression is a polynomial. Polynomials model many real-world phenomena. Our goal in this chapter is to become proficient with operations on polynomials.

Growth of Smartphones

Most wireless phones sold in the United States are smartphone models. As smartphone ownership continues to rise, the wireless network industry will need to expand its capacity for wireless connections. The wireless device category also includes tablet computers, mobile hotspots, standard wireless phones, and so-called feature phones. Below is a graph showing the growth of smartphones as a percent of all types of active wireless devices receiving wireless subscriber service.

In Exercises 25 and 26 of Section 5.3, we will explore some information about the growth of wireless technology. (*Source:* CTIA-the Wireless Association)

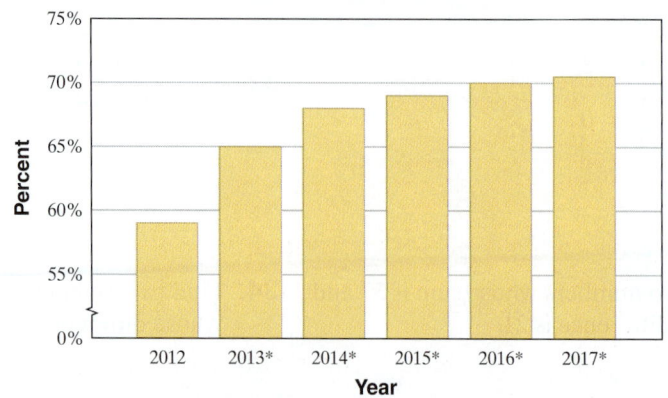

Smartphone Share of All Active Wireless Devices

Source: International Data Corporation

Note: ⌇ means percents missing on vertical bar; * is for forecasted years

5.1 Exponents

Objective A Evaluating Exponential Expressions

In this section, we continue our work with integer exponents. Recall from Section 1.3 that repeated multiplication of the same factor can be written using exponents. For example,

$$2 \cdot 2 \cdot 2 \cdot 2 \cdot 2 = 2^5$$

The exponent 5 tells us how many times 2 is a factor. The expression 2^5 is called an **exponential expression.** It is also called the fifth **power** of 2, or we can say that 2 is **raised** to the fifth power.

$$5^6 = \underbrace{5 \cdot 5 \cdot 5 \cdot 5 \cdot 5 \cdot 5}_{\text{6 factors; each factor is 5}} \quad \text{and} \quad (-3)^4 = \underbrace{(-3) \cdot (-3) \cdot (-3) \cdot (-3)}_{\text{4 factors; each factor is } -3}$$

The base of an exponential expression is the repeated factor. The exponent is the number of times that the base is used as a factor.

$$\overset{\text{exponent or power}}{\underset{\text{base}}{a^n} = \underbrace{a \cdot a \cdot a \cdots a}_{n \text{ factors; each factor is } a}}$$

Examples Evaluate each expression.

1. $2^3 = 2 \cdot 2 \cdot 2 = 8$
2. $3^1 = 3$. To raise 3 to the first power means to use 3 as a factor only once. When no exponent is shown, the exponent is assumed to be 1.
3. $(-4)^2 = (-4)(-4) = 16$
4. $-4^2 = -(4 \cdot 4) = -16$
5. $\left(\dfrac{1}{2}\right)^4 = \dfrac{1}{2} \cdot \dfrac{1}{2} \cdot \dfrac{1}{2} \cdot \dfrac{1}{2} = \dfrac{1}{16}$
6. $4 \cdot 3^2 = 4 \cdot 9 = 36$

Work Practice 1–6

Notice how similar -4^2 is to $(-4)^2$ in the examples above. The difference between the two is the parentheses. In $(-4)^2$, the parentheses tell us that the base, or the repeated factor, is -4. In -4^2, only 4 is the base.

Helpful Hint

Be careful when identifying the base of an exponential expression. Pay close attention to the use of parentheses.

$(-3)^2$	-3^2	$2 \cdot 3^2$
The base is -3.	The base is 3.	The base is 3.
$(-3)^2 = (-3)(-3) = 9$	$-3^2 = -(3 \cdot 3) = -9$	$2 \cdot 3^2 = 2 \cdot 3 \cdot 3 = 18$

An exponent has the same meaning whether the base is a number or a variable. If x is a real number and n is a positive integer, then x^n is the product of n factors, each of which is x.

$$x^n = \underbrace{x \cdot x \cdot x \cdot x \cdot x \cdots x}_{n \text{ factors; each factor is } x}$$

Objectives

A Evaluate Exponential Expressions.

B Use the Product Rule for Exponents.

C Use the Power Rule for Exponents.

D Use the Power Rules for Products and Quotients.

E Use the Quotient Rule for Exponents, and Define a Number Raised to the 0 Power.

F Decide Which Rule(s) to Use to Simplify an Expression.

Practice 1–6

Evaluate each expression.
1. 3^4 2. 7^1 3. $(-2)^3$
4. -2^3 5. $\left(\dfrac{2}{3}\right)^2$ 6. $5 \cdot 6^2$

Teaching Tip

After the examples, ask students to write each of the following expressions as an exponential expression.

	Answers
1. 27	3^3
2. 7	7^1
3. 36	6^2 or $(-6)^2$
4. -36	-6^2
5. $\dfrac{1}{81}$	$\left(\dfrac{1}{9}\right)^2$
6. 75	$3 \cdot 5^2$

Practice 7

Evaluate each expression for the given value of x.

a. $3x^2$ when x is 4

b. $\dfrac{x^4}{-8}$ when x is -2

Teaching Tip

Before doing Example 7, have students fill out the following table.

x	2	-2
x^2	$(2)^2 = 4$	$(-2)^2 = 4$
x^3	$(2)^3 = 8$	$(-2)^3 = -8$
x^4	$(2)^4 = 16$	$(-2)^4 = 16$
x^5	$(2)^5 = 32$	$(-2)^5 = -32$

Example 7   Evaluate each expression for the given value of x.

a. $2x^3$ when x is 5   **b.** $\dfrac{9}{x^2}$ when x is -3

Solution:

a. When x is 5, $2x^3 = 2 \cdot 5^3$

$$= 2 \cdot (5 \cdot 5 \cdot 5)$$
$$= 2 \cdot 125$$
$$= 250$$

b. When x is -3, $\dfrac{9}{x^2} = \dfrac{9}{(-3)^2}$

$$= \dfrac{9}{(-3)(-3)}$$
$$= \dfrac{9}{9} = 1$$

■ **Work Practice 7**

Objective B Using the Product Rule ▶

Exponential expressions can be multiplied, divided, added, subtracted, and themselves raised to powers. Let's see if we can discover a shortcut method for multiplying exponential expressions with the same base. By our definition of an exponent,

$$5^4 \cdot 5^3 = \underbrace{(5 \cdot 5 \cdot 5 \cdot 5)}_{4 \text{ factors of 5}} \cdot \underbrace{(5 \cdot 5 \cdot 5)}_{3 \text{ factors of 5}}$$
$$= \underbrace{5 \cdot 5 \cdot 5 \cdot 5 \cdot 5 \cdot 5 \cdot 5}_{7 \text{ factors of 5}}$$
$$= 5^7$$

Also,

▶ $x^2 \cdot x^3 = (x \cdot x) \cdot (x \cdot x \cdot x)$
$$= x \cdot x \cdot x \cdot x \cdot x$$
$$= x^5$$

In both cases, notice that the result is exactly the same if the exponents are added.

$$5^4 \cdot 5^3 = 5^{4+3} = 5^7 \quad \text{and} \quad x^2 \cdot x^3 = x^{2+3} = x^5$$

This suggests the following rule.

Product Rule for Exponents

If m and n are positive integers and a is a real number, then

$$a^m \cdot a^n = a^{m+n} \leftarrow \text{Add exponents.}$$
$$\uparrow\!\!_____ \text{ Keep common base.}$$

For example,

$$3^5 \cdot 3^7 = 3^{5+7} = 3^{12} \leftarrow \text{Add exponents.}$$
$$\uparrow\!\!_____ \text{ Keep common base.}$$

Answers

7. a. 48 **b.** -2

Helpful Hint

Don't forget that

$3^5 \cdot 3^7 \neq 9^{12}$ ← Add exponents.

└── Common base *not* kept.

$3^5 \cdot 3^7 = \underbrace{3 \cdot 3 \cdot 3 \cdot 3 \cdot 3}_{\text{5 factors of 3}} \cdot \underbrace{3 \cdot 3 \cdot 3 \cdot 3 \cdot 3 \cdot 3 \cdot 3}_{\text{7 factors of 3}}$

$= 3^{12}$ 12 factors of 3, *not* 9

In other words, to multiply two exponential expressions with the **same base,** we keep the base and add the exponents. We call this **simplifying** the exponential expression.

Examples   Use the product rule to simplify each expression.

8. $4^2 \cdot 4^5 = 4^{2+5} = 4^7$ ← Add exponents.

└── Keep common base.

9. $x^2 \cdot x^5 = x^{2+5} = x^7$

10. $y^3 \cdot y = y^3 \cdot y^1$

$= y^{3+1}$

$= y^4$

Helpful Hint   Don't forget that if no exponent is written, it is assumed to be 1.

11. $y^3 \cdot y^2 \cdot y^7 = y^{3+2+7} = y^{12}$

12. $(-5)^7 \cdot (-5)^8 = (-5)^{7+8} = (-5)^{15}$

■ **Work Practice 8–12**

✓**Concept Check** Where possible, use the product rule to simplify the expression.

a. $z^2 \cdot z^{14}$ **b.** $x^2 \cdot z^{14}$ **c.** $9^8 \cdot 9^3$ **d.** $9^8 \cdot 2^7$

Example 13   Use the product rule to simplify $(2x^2)(-3x^5)$.

Solution: Recall that $2x^2$ means $2 \cdot x^2$ and $-3x^5$ means $-3 \cdot x^5$.

$(2x^2)(-3x^5) = (2 \cdot x^2) \cdot (-3 \cdot x^5)$

$= (2 \cdot -3) \cdot (x^2 \cdot x^5)$ Group factors with common bases (using commutative and associative properties).

$= -6x^7$ Simplify.

■ **Work Practice 13**

Examples   Simplify.

14. $(x^2y)(x^3y^2) = (x^2 \cdot x^3) \cdot (y^1 \cdot y^2)$ Group like bases and write y as y^1.

$= x^5 \cdot y^3$ or x^5y^3 Multiply.

15. $(-a^7b^4)(3ab^9) = (-1 \cdot 3) \cdot (a^7 \cdot a^1) \cdot (b^4 \cdot b^9)$

$= -3a^8 b^{13}$

■ **Work Practice 14–15**

Practice 8–12

Use the product rule to simplify each expression.
8. $7^3 \cdot 7^2$ **9.** $x^4 \cdot x^9$
10. $r^5 \cdot r$ **11.** $s^6 \cdot s^2 \cdot s^3$
12. $(-3)^9 \cdot (-3)$

Practice 13

Use the product rule to simplify $(6x^3)(-2x^9)$.

Practice 14–15

Simplify.
14. $(m^5n^{10})(mn^8)$
15. $(-x^9y)(4x^2y^{11})$

Answers

8. 7^5 **9.** x^{13} **10.** r^6 **11.** s^{11}
12. $(-3)^{10}$ **13.** $-12x^{12}$ **14.** $m^6 n^{18}$
15. $-4x^{11}y^{12}$

✓**Concept Check Answers**

a. z^{16} **b.** cannot be simplified

c. 9^{11} **d.** cannot be simplified

Helpful Hint

These examples will remind you of the difference between adding and multiplying terms.

Addition

$$5x^3 + 3x^3 = (5 + 3)x^3 = 8x^3 \qquad \text{By the distributive property}$$
$$7x + 4x^2 = 7x + 4x^2 \qquad \text{Cannot be combined}$$

Multiplication

$$\left(5x^3\right)\left(3x^3\right) = 5 \cdot 3 \cdot x^3 \cdot x^3 = 15x^{3+3} = 15x^6 \quad \text{By the product rule}$$
$$\left(7x\right)\left(4x^2\right) = 7 \cdot 4 \cdot x \cdot x^2 = 28x^{1+2} = 28x^3 \quad \text{By the product rule}$$

Teaching Tip

Consider beginning this objective by having students discover the power rule using a numerical expression.

$$8^4 = \left(2^3\right)^4$$
$$= 2^3 \cdot 2^3 \cdot 2^3 \cdot 2^3$$
$$= 2^{3+3+3+3}$$
$$= 2^{12}$$

Optional: Now have students use a calculator to evaluate both 8^4 and 2^{12}.

Objective C Using the Power Rule

Exponential expressions can themselves be raised to powers. Let's try to discover a rule that simplifies an expression like $\left(x^2\right)^3$. By the definition of a^n,

$$\left(x^2\right)^3 = \left(x^2\right)\left(x^2\right)\left(x^2\right) \quad \left(x^2\right)^3 \text{ means 3 factors of } \left(x^2\right).$$

which can be simplified by the product rule for exponents.

$$\left(x^2\right)^3 = \left(x^2\right)\left(x^2\right)\left(x^2\right) = x^{2+2+2} = x^6$$

Notice that the result is exactly the same if we multiply the exponents.

$$\left(x^2\right)^3 = x^{2 \cdot 3} = x^6$$

The following rule states this result.

Power Rule for Exponents

If m and n are positive integers and a is a real number, then

$$(a^m)^n = a^{mn} \leftarrow \text{Multiply exponents.}$$
$$\uparrow \text{—— Keep the base.}$$

For example,

$$\left(7^2\right)^5 = 7^{2 \cdot 5} = 7^{10} \leftarrow \text{Multiply exponents.}$$
$$\uparrow \text{—— Keep the base.}$$

$$\left[(-5)^3\right]^7 = (-5)^{3 \cdot 7} = (-5)^{21} \leftarrow \text{Multiply exponents.}$$
$$\uparrow \text{—— Keep the base.}$$

In other words, to raise an exponential expression to a power, we keep the base and multiply the exponents.

Examples Use the power rule to simplify each expression.

16. $\left(5^3\right)^6 = 5^{3 \cdot 6} = 5^{18}$ **17.** $\left(y^8\right)^2 = y^{8 \cdot 2} = y^{16}$

◼ **Work Practice 16–17**

Practice 16–17

Use the power rule to simplify each expression.

16. $\left(9^4\right)^{10}$ **17.** $\left(z^6\right)^3$

Helpful Hint

Take a moment to make sure that you understand when to apply the product rule and when to apply the power rule.

Product Rule → Add Exponents	Power Rule → Multiply Exponents
$x^5 \cdot x^7 = x^{5+7} = x^{12}$	$\left(x^5\right)^7 = x^{5 \cdot 7} = x^{35}$
$y^6 \cdot y^2 = y^{6+2} = y^8$	$(y^6)^2 = y^{6 \cdot 2} = y^{12}$

Answers

16. 9^{40} **17.** z^{18}

Objective D Using the Power Rules for Products and Quotients ▶

When the base of an exponential expression is a product, the definition of a^n still applies. For example, simplify $(xy)^3$ as follows.

$$(xy)^3 = (xy)(xy)(xy) \qquad (xy)^3 \text{ means 3 factors of } (xy).$$
$$= x \cdot x \cdot x \cdot y \cdot y \cdot y \quad \text{Group factors with common bases.}$$
$$= x^3 y^3 \qquad \text{Simplify.}$$

Notice that to simplify the expression $(xy)^3$, we raise each factor within the parentheses to a power of 3.

$$(xy)^3 = x^3 y^3$$

In general, we have the following rule.

Power of a Product Rule

If n is a positive integer and a and b are real numbers, then

$$(ab)^n = a^n b^n$$

For example,

$$(3x)^5 = 3^5 x^5$$

In other words, to raise a product to a power, we raise each factor to the power.

Teaching Tip

It may be helpful to emphasize to students that this rule does not apply when a power is taken of a sum. In other words, $(a + b)^{11} \neq a^{11} + b^{11}$. For example,

$$(3 + 4)^2 \neq 3^2 + 4^2$$
$$7^2 \neq 9 + 16$$
$$49 \neq 25$$

Examples Simplify each expression.

18. $(st)^4 = s^4 \cdot t^4 = s^4 t^4$ \qquad Use the power of a product rule.

19. $(2a)^3 = 2^3 \cdot a^3 = 8a^3$ \qquad Use the power of a product rule.

20. $\left(-5x^2 y^3 z\right)^2 = (-5)^2 \cdot \left(x^2\right)^2 \cdot \left(y^3\right)^2 \cdot \left(z^1\right)^2$ \qquad Use the power of a product rule.
$$= 25x^4 y^6 z^2$$

21. $\left(-xy^3\right)^5 = \left(-1xy^3\right)^5 = (-1)^5 \cdot x^5 \cdot \left(y^3\right)^5$ \qquad Use the power of a product rule.
$$= -1x^5 y^{15} \quad \text{or} \quad -x^5 y^{15}$$

■ **Work Practice 18–21**

Practice 18–21

Simplify each expression.

18. $(xy)^7$ \qquad **19.** $(3y)^4$

20. $\left(-2p^4 q^2 r\right)^3$ \qquad **21.** $\left(-a^4 b\right)^7$

Let's see what happens when we raise a quotient to a power. For example, we simplify $\left(\dfrac{x}{y}\right)^3$ as follows.

$$\left(\frac{x}{y}\right)^3 = \left(\frac{x}{y}\right)\left(\frac{x}{y}\right)\left(\frac{x}{y}\right) \quad \left(\frac{x}{y}\right)^3 \text{ means 3 factors of } \left(\frac{x}{y}\right).$$

$$= \frac{x \cdot x \cdot x}{y \cdot y \cdot y} \qquad \text{Multiply fractions.}$$

$$= \frac{x^3}{y^3} \qquad \text{Simplify.}$$

Notice that to simplify the expression $\left(\dfrac{x}{y}\right)^3$, we raise both the numerator and the denominator to a power of 3.

$$\left(\frac{x}{y}\right)^3 = \frac{x^3}{y^3}$$

Answers

18. $x^7 y^7$ **19.** $81y^4$

20. $-8p^{12} q^6 r^3$ **21.** $-a^{28} b^7$

In general, we have the following rule.

Power of a Quotient Rule

If n is a positive integer and a and c are real numbers, then

$$\left(\frac{a}{c}\right)^n = \frac{a^n}{c^n}, \quad c \neq 0$$

For example,

$$\left(\frac{y}{7}\right)^3 = \frac{y^3}{7^3}$$

In other words, to raise a quotient to a power, we raise both the numerator and the denominator to the power.

Practice 22–23

Simplify each expression.

22. $\left(\dfrac{r}{s}\right)^6$ **23.** $\left(\dfrac{5x^6}{9y^3}\right)^2$

Examples   Simplify each expression.

22. $\left(\dfrac{m}{n}\right)^7 = \dfrac{m^7}{n^7}, \quad n \neq 0$ Use the power of a quotient rule.

23. $\left(\dfrac{2x^4}{3y^5}\right)^4 = \dfrac{2^4 \cdot \left(x^4\right)^4}{3^4 \cdot \left(y^5\right)^4}$ Use the power of a quotient rule.

$$= \frac{16x^{16}}{81y^{20}}, \quad y \neq 0 \quad \text{Use the power rule for exponents.}$$

■ Work Practice 22–23

Objective E Using the Quotient Rule and Defining the Zero Exponent ▶

Another pattern for simplifying exponential expressions involves quotients.

$$\frac{x^5}{x^3} = \frac{x \cdot x \cdot x \cdot x \cdot x}{x \cdot x \cdot x}$$

$$= \frac{x \cdot x \cdot x \cdot x \cdot x}{x \cdot x \cdot x}$$

$$= 1 \cdot 1 \cdot 1 \cdot x \cdot x$$

$$= x \cdot x$$

$$= x^2$$

Teaching Tip

Emphasize to students that the quotient rule does not apply when the numerator and denominator are not in factored form. For example,

$$\frac{2^3 + 3^4}{2^2 + 3^1} \neq 2^{3-2} + 3^{4-1}$$

$$\frac{8 + 81}{4 + 3} \neq 2^1 + 3^3$$

$$\frac{89}{7} \neq 2 + 27$$

$$12\frac{5}{7} \neq 29$$

Notice that the result is exactly the same if we subtract exponents of the common bases.

$$\frac{x^5}{x^3} = x^{5-3} = x^2$$

The following rule states this result in a general way.

Quotient Rule for Exponents

If m and n are positive integers and a is a real number, then

$$\frac{a^m}{a^n} = a^{m-n}, \quad a \neq 0$$

For example,

$$\frac{x^6}{x^2} = x^{6-2} = x^4, \quad x \neq 0$$

Answers

22. $\dfrac{r^6}{s^6}, \; s \neq 0$ **23.** $\dfrac{25x^{12}}{81y^6}, \; y \neq 0$

In other words, to divide one exponential expression by another with a common base, we keep the base and subtract the exponents.

Examples Simplify each quotient.

▶ **24.** $\dfrac{x^5}{x^2} = x^{5-2} = x^3$ Use the quotient rule.

25. $\dfrac{4^7}{4^3} = 4^{7-3} = 4^4 = 256$ Use the quotient rule.

26. $\dfrac{(-3)^5}{(-3)^2} = (-3)^3 = -27$ Use the quotient rule.

27. $\dfrac{2x^5y^2}{xy} = 2 \cdot \dfrac{x^5}{x^1} \cdot \dfrac{y^2}{y^1}$

$\phantom{27.\ \dfrac{2x^5y^2}{xy}} = 2 \cdot \left(x^{5-1}\right) \cdot \left(y^{2-1}\right)$ Use the quotient rule.

$\phantom{27.\ \dfrac{2x^5y^2}{xy}} = 2x^4y^1 \quad \text{or} \quad 2x^4y$

🟧 **Work Practice 24–27**

Practice 24–27

Simplify each quotient.

24. $\dfrac{y^7}{y^3}$ **25.** $\dfrac{5^9}{5^6}$

26. $\dfrac{(-2)^{14}}{(-2)^{10}}$ **27.** $\dfrac{7a^4b^{11}}{ab}$

Let's now give meaning to an expression such as x^0. To do so, we will simplify $\dfrac{x^3}{x^3}$ in two ways and compare the results.

▶ $\dfrac{x^3}{x^3} = x^{3-3} = x^0$ Apply the quotient rule.

$\dfrac{x^3}{x^3} = \dfrac{x \cdot x \cdot x}{x \cdot x \cdot x} = 1$ Apply the fundamental principle for fractions.

Since $\dfrac{x^3}{x^3} = x^0$ and $\dfrac{x^3}{x^3} = 1$, we define that $x^0 = 1$ as long as x is not 0.

Zero Exponent

$a^0 = 1$, as long as a is not 0.

For example, $5^0 = 1$.

In other words, a base raised to the 0 power is 1, as long as the base is not 0.

Examples Simplify each expression.

28. $3^0 = 1$
29. $\left(5x^3y^2\right)^0 = 1$
30. $(-4)^0 = 1$
31. $-4^0 = -1 \cdot 4^0 = -1 \cdot 1 = -1$
32. $5x^0 = 5 \cdot x^0 = 5 \cdot 1 = 5$

🟧 **Work Practice 28–32**

Practice 28–32

Simplify each expression.

28. 8^0 **29.** $(2r^2s)^0$
30. $(-7)^0$ **31.** -7^0
32. $7y^0$

Answers

24. y^4 **25.** 125   **26.** 16   **27.** $7a^3b^{10}$
28. 1   **29.** 1   **30.** 1   **31.** -1 **32.** 7

✓**Concept Check** Suppose you are simplifying each expression. Tell whether you would *add* the exponents, *subtract* the exponents, *multiply* the exponents, *divide* the exponents, or *none of these*.

a. $\left(x^{63}\right)^{21}$ **b.** $\dfrac{y^{15}}{y^3}$ **c.** $z^{16} + z^8$ **d.** $w^{45} \cdot w^9$

Objective F Deciding Which Rule to Use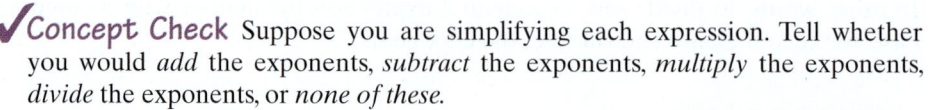

Let's practice deciding which rule to use to simplify an expression. We will continue this discussion with more examples in the next section.

Practice 33

Simplify each expression.

a. $\dfrac{x^7}{x^4}$ **b.** $\left(3y^4\right)^4$ **c.** $\left(\dfrac{x}{4}\right)^3$

Answers

33. **a.** x^3 **b.** $81y^{16}$ **c.** $\dfrac{x^3}{64}$

✓**Concept Check Answers**
a. multiply **b.** subtract
c. none of these **d.** add

Example 33 Simplify each expression.

a. $x^7 \cdot x^4$ **b.** $\left(\dfrac{t}{2}\right)^4$ **c.** $\left(9y^5\right)^2$

Solution:

a. Here, we have a product, so we use the product rule to simplify.
$$x^7 \cdot x^4 = x^{7+4} = x^{11}$$

b. This is a quotient raised to a power, so we use the power of a quotient rule.
$$\left(\dfrac{t}{2}\right)^4 = \dfrac{t^4}{2^4} = \dfrac{t^4}{16}$$

c. This is a product raised to a power, so we use the power of a product rule.
$$\left(9y^5\right)^2 = 9^2\left(y^5\right)^2 = 81y^{10}$$

■ **Work Practice 33**

Vocabulary, Readiness & Video Check

Use the choices below to fill in each blank. Some choices may be used more than once.

0	base	add
1	exponent	multiply

1. Repeated multiplication of the same factor can be written using a(n) ___exponent___ .
2. In 5^2, the 2 is called the ___exponent___ and the 5 is called the ___base___.
3. To simplify $x^2 \cdot x^7$, keep the base and ___add___ the exponents.
4. To simplify $\left(x^3\right)^6$, keep the base and ___multiply___ the exponents.
5. The understood exponent on the term y is ___1___.
6. If $x^{\square} = 1$, the exponent is ___0___.

State the bases and the exponents for each of the following expressions.

7. 3^2 ___base: 3; exponent: 2___

8. $(-3)^6$ ___base: −3; exponent: 6___

9. -4^2 ___base: 4; exponent: 2___

10. $5 \cdot 3^4$ base: 5; exponent: 1; base: 3; exponent: 4

11. $5x^2$ base: 5; exponent: 1; base: x; exponent: 2

12. $(5x)^2$ ___base: 5x; exponent: 2___

Martin-Gay Interactive Videos   Watch the section lecture video and answer the following questions.

See Video 5.1

Objective A 13. ⊞ Examples 3 and 4 illustrate how to find the base of an exponential expression both with and without parentheses. Explain how identifying the base of ⊞ Example 7 is similar to identifying the base of ⊞ Example 4. ▶

Objective B 14. Why were the commutative and associative properties applied in ⊞ Example 12? ▶

Objective C 15. What point is made at the end of ⊞ Example 15? ▶

Objective D 16. Although it's not especially emphasized in ⊞ Example 20, what is helpful to remind ourself about the -2 in the problem? ▶

Objective E 17. In ⊞ Example 24, which exponent rule is used to show that any nonzero base raised to the power of zero is 1? ▶

Objective F 18. When simplifying an exponential expression that's a fraction, will we always use the quotient rule for exponents? Refer to ⊞ Example 30 to support your answer. ▶

See video answer section.

5.1 Exercise Set MyMathLab®

Objective A *Evaluate each expression. See Examples 1 through 6.*

1. 7^2 49   **2.** -3^2 -9 **3.** $(-5)^1$ -5 **4.** $(-3)^2$ 9   **5.** -2^4 -16 **6.** -4^3 -64

7. $(-2)^4$ 16   **8.** $(-4)^3$ -64 **9.** $\left(\frac{1}{3}\right)^3$ $\frac{1}{27}$ **10.** $\left(-\frac{1}{9}\right)^2$ $\frac{1}{81}$ **11.** $7 \cdot 2^4$ 112   **12.** $9 \cdot 2^2$ 36

Evaluate each expression with the given replacement values. See Example 7.

13. x^2 when $x = -2$ 4

14. x^3 when $x = -2$ -8

15. $5x^3$ when $x = 3$ 135

16. $4x^2$ when $x = 5$ 100

17. $2xy^2$ when $x = 3$ and $y = -5$ 150

18. $-4x^2y^3$ when $x = 2$ and $y = -1$ 16

▶ 19. $\frac{2z^4}{5}$ when $z = -2$ $\frac{32}{5}$

20. $\frac{10}{3y^3}$ when $y = -3$ $-\frac{10}{81}$

Objective B *Use the product rule to simplify each expression. Write the results using exponents. See Examples 8 through 15.*

21. $x^2 \cdot x^5$
x^7

22. $y^2 \cdot y$
y^3

23. $(-3)^3 \cdot (-3)^9$
$(-3)^{12}$

24. $(-5)^7 \cdot (-5)^6$
$(-5)^{13}$

▶ 25. $(5y^4)(3y)$
$15y^5$

26. $(-2z^3)(-2z^2)$
$4z^5$

▶ 27. $(x^9y)(x^{10}y^5)$
$x^{19}y^6$

28. $(a^2b)(a^{13}b^{17})$
$a^{15}b^{18}$

29. $(-8mn^6)(9m^2n^2)$
$-72m^3n^8$

30. $(-7a^3b^3)(7a^{19}b)$
$-49a^{22}b^4$

31. $(4z^{10})(-6z^7)(z^3)$
$-24z^{20}$

32. $(12x^5)(-x^6)(x^4)$
$-12x^{15}$

33. The rectangle below has width $4x^2$ feet and length $5x^3$ feet. Find its area as an expression in x. $20x^5$ sq ft

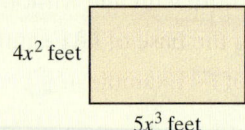

4x² feet

5x³ feet

△ 34. The parallelogram below has base length $9y^7$ meters and height $2y^{10}$ meters. Find its area as an expression in y. $18y^{17}$ sq m

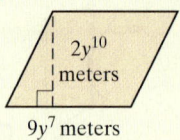

$2y^{10}$ meters

$9y^7$ meters

Objectives C D Mixed Practice *Use the power rule and the power of a product or quotient rule to simplify each expression. See Examples 16 through 23.*

▶ 35. $\left(x^9\right)^4$ x^{36}

36. $\left(y^7\right)^5$ y^{35}

▶ 37. $(pq)^8$ p^8q^8

38. $(ab)^6$ a^6b^6

39. $\left(2a^5\right)^3$ $8a^{15}$

40. $\left(4x^6\right)^2$ $16x^{12}$

▶ 41. $\left(x^2y^3\right)^5$ $x^{10}y^{15}$

42. $\left(a^4b\right)^7$ $a^{28}b^7$

43. $\left(-7a^2b^5c\right)^2$ $49a^4b^{10}c^2$

44. $\left(-3x^7yz^2\right)^3$ $-27x^{21}y^3z^6$

▶ 45. $\left(\dfrac{r}{s}\right)^9$ $\dfrac{r^9}{s^9}$

46. $\left(\dfrac{q}{t}\right)^{11}$ $\dfrac{q^{11}}{t^{11}}$

47. $\left(\dfrac{mp}{n}\right)^9$ $\dfrac{m^9p^9}{n^9}$

48. $\left(\dfrac{xy}{7}\right)^2$ $\dfrac{x^2y^2}{49}$

▶ 49. $\left(\dfrac{-2xz}{y^5}\right)^2$ $\dfrac{4x^2z^2}{y^{10}}$

50. $\left(\dfrac{xy^4}{-3z^3}\right)^3$ $-\dfrac{x^3y^{12}}{27z^9}$

△ 51. The square shown has sides of length $8z^5$ decimeters. Find its area. $64z^{10}$ sq dm

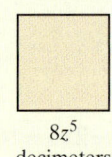

$8z^5$ decimeters

△ 52. Given the circle below with radius $5y$ centimeters, find its area. Do not approximate π. $25y^2\pi$ sq cm

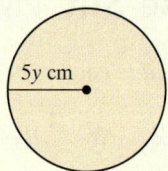

5y cm

53. The vault below is in the shape of a cube. If each side is $3y^4$ feet, find its volume. $27y^{12}$ cu ft

$3y^4$ feet $3y^4$ feet

$3y^4$ feet

54. The silo shown is in the shape of a cylinder. If its radius is $4x$ meters and its height is $5x^3$ meters, find its volume. Do not approximate π. $80x^5\pi$ cu m

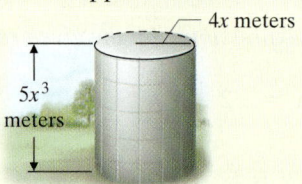

4x meters

$5x^3$ meters

Objective E *Use the quotient rule and simplify each expression. See Examples 24 through 27.*

55. $\dfrac{x^3}{x}$ x^2

56. $\dfrac{y^{10}}{y^9}$ y

▶ 57. $\dfrac{(-4)^6}{(-4)^3}$ -64

58. $\dfrac{(-6)^{13}}{(-6)^{11}}$ 36

59. $\dfrac{p^7q^{20}}{pq^{15}}$ p^6q^5

60. $\dfrac{x^8y^6}{xy^5}$ x^7y

61. $\dfrac{7x^2y^6}{14x^2y^3}$ $\dfrac{y^3}{2}$

▶ 62. $\dfrac{9a^4b^7}{27ab^2}$ $\dfrac{a^3b^5}{3}$

Simplify each expression. See Examples 28 through 32.

63. 7^0 1

64. 23^0 1

▶ 65. $(2x)^0$ 1

66. $(4y)^0$ 1

▶ 67. $-7x^0$ -7

68. $-2x^0$ -2

▶ 69. $5^0 + y^0$ 2

70. $-3^0 + 4^0$ 0

Objectives A B C D E F **Mixed Practice** *Simplify each expression. See Examples 1 through 6 and 8 through 33.*

71. -9^2 -81

72. $(-9)^2$ 81

73. $\left(\dfrac{1}{4}\right)^3$ $\dfrac{1}{64}$

74. $\left(\dfrac{2}{3}\right)^3$ $\dfrac{8}{27}$

75. b^4b^2
b^6

76. y^4y
y^5

77. $a^2a^3a^4$
a^9

78. $x^2x^{15}x^9$
x^{26}

▶ **79.** $(2x^3)(-8x^4)$
$-16x^7$

80. $(3y^4)(-5y)$
$-15y^5$

81. $(a^7b^{12})(a^4b^8)$
$a^{11}b^{20}$

82. $(y^2z^2)(y^{15}z^{13})$
$y^{17}z^{15}$

83. $(-2mn^6)(-13m^8n)$
$26m^9n^7$

84. $(-3s^5t)(-7st^{10})$
$21s^6t^{11}$

85. $(z^4)^{10}$
z^{40}

86. $(t^5)^{11}$
t^{55}

87. $(4ab)^3$
$64a^3b^3$

88. $(2ab)^4$
$16a^4b^4$

89. $(-6xyz^3)^2$
$36x^2y^2z^6$

90. $(-3xy^2a^3)^3$
$-27x^3y^6a^9$

91. $\dfrac{z^{12}}{z^4}$ z^8

92. $\dfrac{b^6}{b^3}$ b^3

▶ **93.** $\dfrac{3x^5}{x}$ $3x^4$

94. $\dfrac{5x^9}{x}$ $5x^8$

95. $(6b)^0$ 1

96. $(5ab)^0$ 1

97. $(9xy)^2$ $81x^2y^2$

98. $(2ab)^5$ $32a^5b^5$

99. $2^3 + 2^5$ 40

100. $7^2 - 7^0$ 48

▶ **101.** $\left(\dfrac{3y^5}{6x^4}\right)^3$ $\dfrac{y^{15}}{8x^{12}}$

102. $\left(\dfrac{2ab}{6yz}\right)^4$ $\dfrac{a^4b^4}{81y^4z^4}$

103. $\dfrac{2x^3y^2z}{xyz}$ $2x^2y$

104. $\dfrac{5x^{12}y^{13}}{x^5y^7}$ $5x^7y^6$

Review

Subtract. See Section 1.5.

105. $5 - 7$ -2

106. $9 - 12$ -3

107. $3 - (-2)$ 5

108. $5 - (-10)$ 15

109. $-11 - (-4)$ -7

110. $-15 - (-21)$ 6

Concept Extensions

Solve. See the Concept Checks in this section. For Exercises 111 through 114, match the expression with the operation needed to simplify each. A letter may be used more than once and a letter may not be used at all.

111. $(x^{14})^{23}$ **c**

112. $x^{14} \cdot x^{23}$ **a**

113. $x^{14} + x^{23}$ **e**

114. $\dfrac{x^{35}}{x^{17}}$ **b**

a. Add the exponents.
b. Subtract the exponents.
c. Multiply the exponents.
d. Divide the exponents.
e. None of these

Fill in the boxes so that each statement is true. (More than one answer is possible for each exercise.)

115. $x^\square \cdot x^\square = x^{12}$ answers may vary

116. $\left(x^\square\right)^\square = x^{20}$ answers may vary

117. $\dfrac{y^\square}{y^\square} = y^7$ answers may vary

118. $\left(y^\square\right)^\square \cdot \left(y^\square\right)^\square = y^{30}$ answers may vary

△ **119.** The formula $V = x^3$ can be used to find the volume V of a cube with side length x. Find the volume of a cube with side length 7 meters. (Volume is the capacity of a solid such as a cube and is measured in cubic units.)   343 cu m

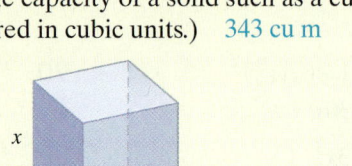

△ **120.** The formula $S = 6x^2$ can be used to find the surface area S of a cube with side length x. Find the surface area of a cube with side length 5 meters. (Surface area is the area of the surface of the cube and is measured in square units.)   150 sq m

△ **121.** To find the amount of water that a swimming pool in the shape of a cube can hold, do we use the formula for volume of the cube or surface area of the cube? (See Exercises **119** and **120**.)   volume

△ **122.** To find the amount of material needed to cover an ottoman in the shape of a cube, do we use the formula for volume of the cube or surface area of the cube? (See Exercises **119** and **120**.)   surface area

123. Explain why $(-5)^4 = 625$, while $-5^4 = -625$. answers may vary

124. Explain why $5 \cdot 4^2 = 80$, while $(5 \cdot 4)^2 = 400$. answers may vary

125. In your own words, explain why $5^0 = 1$. answers may vary

126. In your own words, explain when $(-3)^n$ is positive and when it is negative.   answers may vary

Simplify each expression. Assume that variables represent positive integers.

127. $x^{5a}x^{4a}$ x^{9a}

128. $b^{9a}b^{4a}$ b^{13a}

129. $\left(a^b\right)^5$ a^{5b}

130. $\left(2a^{4b}\right)^4$ $16a^{16b}$

131. $\dfrac{x^{9a}}{x^{4a}}$ x^{5a}

132. $\dfrac{y^{15b}}{y^{6b}}$ y^{9b}

Negative Exponents and Scientific Notation

Objective A Simplifying Expressions Containing Negative Exponents

Our work with exponential expressions so far has been limited to exponents that are positive integers or 0. Here we will also give meaning to an expression like x^{-3}.

Suppose that we wish to simplify the expression $\dfrac{x^2}{x^5}$. If we use the quotient rule for exponents, we subtract exponents:

$$\frac{x^2}{x^5} = x^{2-5} = x^{-3}, \quad x \neq 0$$

But what does x^{-3} mean? Let's simplify $\dfrac{x^2}{x^5}$ using the definition of a^n.

$$\frac{x^2}{x^5} = \frac{x \cdot x}{x \cdot x \cdot x \cdot x \cdot x}$$

$$= \frac{x \cdot x}{x \cdot x \cdot x \cdot x \cdot x} \qquad \text{Divide numerator and denominator by common factors by applying the fundamental principle for fractions.}$$

$$= \frac{1}{x^3}$$

If the quotient rule is to hold true for negative exponents, then x^{-3} must equal $\dfrac{1}{x^3}$. From this example, we state the definition for negative exponents.

Negative Exponents

If a is a real number other than 0 and n is an integer, then

$$a^{-n} = \frac{1}{a^n}$$

For example,

$$x^{-3} = \frac{1}{x^3}$$

In other words, another way to write a^{-n} is to take its reciprocal and change the sign of its exponent.

Examples Simplify by writing each expression with positive exponents only.

1. $3^{-2} = \dfrac{1}{3^2} = \dfrac{1}{9}$ Use the definition of negative exponents.

2. $2x^{-3} = 2^1 \cdot \dfrac{1}{x^3} = \dfrac{2^1}{x^3}$ or $\dfrac{2}{x^3}$ Use the definition of negative exponents.

3. $2^{-1} + 4^{-1} = \dfrac{1}{2} + \dfrac{1}{4} = \dfrac{2}{4} + \dfrac{1}{4} = \dfrac{3}{4}$

4. $(-2)^{-4} = \dfrac{1}{(-2)^4} = \dfrac{1}{(-2)(-2)(-2)(-2)} = \dfrac{1}{16}$

> **Helpful Hint** Don't forget that since there are no parentheses, only x is the base for the exponent -3.

■ Work Practice 1–4

Objectives

A Simplify Expressions Containing Negative Exponents. ▶

B Use the Rules and Definitions for Exponents to Simplify Exponential Expressions. ▶

C Write Numbers in Scientific Notation. ▶

D Convert Numbers in Scientific Notation to Standard Form. ▶

E Perform Operations on Numbers Written in Scientific Notation. ▶

Teaching Tip

Verbalize the meaning of a negative exponent. For instance, 3^{-2} is two factors of the reciprocal of 3 and $2x^{-3}$ is 2 times three factors of the reciprocal of x.

Practice 1–4

Simplify by writing each expression with positive exponents only.

1. 5^{-3} **2.** $7x^{-4}$

3. $5^{-1} + 3^{-1}$ **4.** $(-3)^{-4}$

Answers

1. $\dfrac{1}{125}$ **2.** $\dfrac{7}{x^4}$ **3.** $\dfrac{8}{15}$ **4.** $\dfrac{1}{81}$

> ● **Helpful Hint**
>
> A negative exponent *does not affect* the sign of its base.
> Remember: Another way to write a^{-n} is to take its reciprocal and change the sign of its exponent: $a^{-n} = \dfrac{1}{a^n}$. For example,
>
> $$x^{-2} = \frac{1}{x^2}, \qquad 2^{-3} = \frac{1}{2^3} \text{ or } \frac{1}{8}$$
>
> ● $\dfrac{1}{y^{-4}} = \dfrac{1}{\frac{1}{y^4}} = y^4, \qquad \dfrac{1}{5^{-2}} = 5^2 \text{ or } 25$

From the preceding Helpful Hint, we know that $x^{-2} = \dfrac{1}{x^2}$ and $\dfrac{1}{y^{-4}} = y^4$. We can use this to include another statement in our definition of negative exponents.

Negative Exponents

If a is a real number other than 0 and n is an integer, then

$$a^{-n} = \frac{1}{a^n} \quad \text{and} \quad \frac{1}{a^{-n}} = a^n$$

Practice 5–8

Simplify each expression. Write each result using positive exponents only.

5. $\left(\dfrac{6}{7}\right)^{-2}$ **6.** $\dfrac{x}{x^{-4}}$

7. $\dfrac{y^{-9}}{z^{-5}}$ **8.** $\dfrac{y^{-4}}{y^6}$

Examples   Simplify each expression. Write each result using positive exponents only.

5. $\left(\dfrac{2}{x}\right)^{-3} = \dfrac{2^{-3}}{x^{-3}} = \dfrac{2^{-3}}{1} \cdot \dfrac{1}{x^{-3}} = \dfrac{1}{2^3} \cdot \dfrac{x^3}{1} = \dfrac{x^3}{2^3} = \dfrac{x^3}{8}$ Use the negative exponents rule.

6. $\dfrac{y}{y^{-2}} = \dfrac{y^1}{y^{-2}} = y^{1-(-2)} = y^3$ Use the quotient rule.

7. $\dfrac{p^{-4}}{q^{-9}} = p^{-4} \cdot \dfrac{1}{q^{-9}} = \dfrac{1}{p^4} \cdot q^9 = \dfrac{q^9}{p^4}$ Use the negative exponents rule.

8. $\dfrac{x^{-5}}{x^7} = x^{-5-7} = x^{-12} = \dfrac{1}{x^{12}}$

■ Work Practice 5–8

Objective B Simplifying Exponential Expressions

All the previously stated rules for exponents apply for negative exponents also. Here is a summary of the rules and definitions for exponents.

Teaching Tip

Consider asking students if they see another approach to Example 5. For example,

$$\left(\dfrac{2}{x}\right)^{-3} = \left(\dfrac{x}{2}\right)^3$$

$$= \dfrac{x^3}{2^3}$$

$$= \dfrac{x^3}{8}$$

Summary of Exponent Rules

If m and n are integers and a, b, and c are real numbers, then

Product rule for exponents:	$a^m \cdot a^n = a^{m+n}$
Power rule for exponents:	$(a^m)^n = a^{m \cdot n}$
Power of a product:	$(ab)^n = a^n b^n$
Power of a quotient:	$\left(\dfrac{a}{c}\right)^n = \dfrac{a^n}{c^n}, \quad c \neq 0$
Quotient rule for exponents:	$\dfrac{a^m}{a^n} = a^{m-n}, \quad a \neq 0$
Zero exponent:	$a^0 = 1, \quad a \neq 0$
Negative exponent:	$a^{-n} = \dfrac{1}{a^n}, \quad a \neq 0$

Answers

5. $\dfrac{49}{36}$ **6.** x^5 **7.** $\dfrac{z^5}{y^9}$ **8.** $\dfrac{1}{y^{10}}$

Examples Simplify each expression. Write each result using positive exponents only.

9. $\dfrac{\left(x^3\right)^4 x}{x^7} = \dfrac{x^{12} \cdot x}{x^7} = \dfrac{x^{12+1}}{x^7} = \dfrac{x^{13}}{x^7} = x^{13-7} = x^6$ Use the power rule.

10. $\left(\dfrac{3a^2}{b}\right)^{-3} = \dfrac{3^{-3}\left(a^2\right)^{-3}}{b^{-3}}$ Raise each factor in the numerator and the denominator to the -3 power.

$= \dfrac{3^{-3}a^{-6}}{b^{-3}}$ Use the power rule.

$= \dfrac{b^3}{3^3 a^6}$ Use the negative exponent rule.

$= \dfrac{b^3}{27a^6}$ Write 3^3 as 27.

11. $\left(y^{-3}z^6\right)^{-6} = \left(y^{-3}\right)^{-6}\left(z^6\right)^{-6}$ Raise each factor to the -6 power.

$= y^{18}z^{-36} = \dfrac{y^{18}}{z^{36}}$

12. $\dfrac{(2x)^5}{x^3} = \dfrac{2^5 \cdot x^5}{x^3} = 2^5 \cdot x^{5-3} = 32x^2$ Raise each factor in the numerator to the fifth power.

13. $\dfrac{x^{-7}}{\left(x^4\right)^3} = \dfrac{x^{-7}}{x^{12}} = x^{-7-12} = x^{-19} = \dfrac{1}{x^{19}}$

14. $\left(5y^3\right)^{-2} = 5^{-2}\left(y^3\right)^{-2} = 5^{-2}y^{-6} = \dfrac{1}{5^2y^6} = \dfrac{1}{25y^6}$

15. $-\dfrac{22a^7b^{-5}}{11a^{-2}b^3} = -\dfrac{22}{11} \cdot a^{7-(-2)}b^{-5-3} = -2a^9b^{-8} = -\dfrac{2a^9}{b^8}$

16. $\dfrac{(2xy)^{-3}}{\left(x^2y^3\right)^2} = \dfrac{2^{-3}x^{-3}y^{-3}}{\left(x^2\right)^2\left(y^3\right)^2} = \dfrac{2^{-3}x^{-3}y^{-3}}{x^4y^6} = 2^{-3}x^{-3-4}y^{-3-6}$

$= 2^{-3}x^{-7}y^{-9} = \dfrac{1}{2^3x^7y^9}$ or $\dfrac{1}{8x^7y^9}$

■ **Work Practice 9–16**

Objective C Writing Numbers in Scientific Notation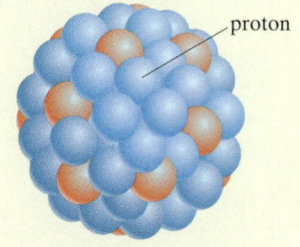

Both very large and very small numbers frequently occur in many fields of science. For example, the distance between the sun and the dwarf planet Pluto is approximately 5,906,000,000 kilometers, and the mass of a proton is approximately 0.00000000000000000000000165 gram. It can be tedious to write these numbers in this standard decimal notation, so **scientific notation** is used as a convenient shorthand for expressing very large and very small numbers.

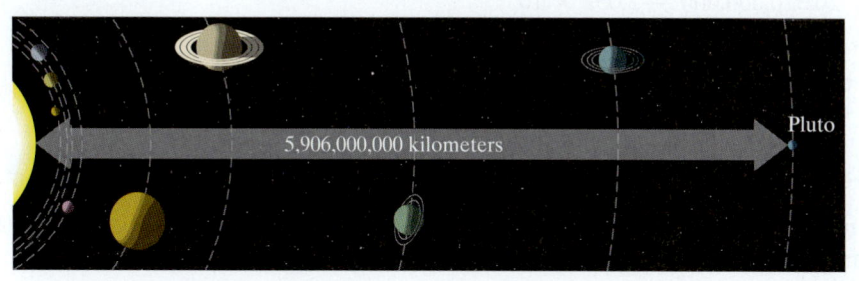

5,906,000,000 kilometers

Pluto

Teaching Tip

Point out that scientific notation is a factored form of a number. For positive numbers, one of the factors is a power of 10 and the other is a number greater than or equal to 1 but less than 10. It is a useful form for multiplying and dividing very large or very small numbers, as will be seen in Example 19.

proton

Mass of proton is approximately
0.000 000 000 000 000 000 000 001 65 gram

Answers

9. x^{12} **10.** $\dfrac{y^2}{81x^6}$ **11.** $\dfrac{a^{20}}{b^{35}}$ **12.** $\dfrac{16}{x^4}$

13. $\dfrac{1}{y^{30}}$ **14.** $\dfrac{1}{64a^6}$ **15.** $-\dfrac{4x^2}{y^4}$

16. $\dfrac{1}{72x^{17}y^5}$

Scientific Notation

A positive number is written in scientific notation if it is written as the product of a number a, where $1 \leq a < 10$, and an integer power r of 10: $a \times 10^r$.

The following numbers are written in scientific notation. The $\times$ sign for multiplication is used as part of the notation.

$$2.03 \times 10^2 \quad 7.362 \times 10^7 \quad 5.906 \times 10^9 \quad \text{(Distance between the sun and Pluto)}$$
$$1 \times 10^{-3} \quad 8.1 \times 10^{-5} \quad 1.65 \times 10^{-24} \quad \text{(Mass of a proton)}$$

The following steps are useful when writing numbers in scientific notation.

To Write a Number in Scientific Notation

Step 1: Move the decimal point in the original number so that the new number has a value greater than or equal to 1 and less than 10.

Step 2: Count the number of decimal places the decimal point is moved in Step 1. If the original number is 10 or greater, the count is positive. If the original number is less than 1, the count is negative.

Step 3: Multiply the new number in Step 1 by 10 raised to an exponent equal to the count found in Step 2.

Practice 17

Write each number in scientific notation.

a. 420,000   **b.** 0.00017

c. 9,060,000,000   **d.** 0.000007

Example 17   Write each number in scientific notation.

a. 367,000,000

b. 0.000003

c. 20,520,000,000

d. 0.00085

Solution:

a. **Step 1:** Move the decimal point until the number is greater than or equal to 1 and less than 10.

$$367,000,000.$$
8 places

Step 2: The decimal point is moved 8 places and the original number is 10 or greater, so the count is positive 8.

Step 3: $367,000,000 = 3.67 \times 10^8$

b. **Step 1:** Move the decimal point until the number is greater than or equal to 1 and less than 10.

$$0.000003$$
6 places

Step 2: The decimal point is moved 6 places and the original number is less than 1, so the count is -6.

Step 3: $0.000003 = 3.0 \times 10^{-6}$

c. $20,520,000,000 = 2.052 \times 10^{10}$

d. $0.00085 = 8.5 \times 10^{-4}$

■ **Work Practice 17**

Objective D Converting Numbers to Standard Form

A number written in scientific notation can be rewritten in standard form. For example, to write 8.63×10^3 in standard form, recall that $10^3 = 1000$.

$$8.63 \times 10^3 = 8.63(1000) = 8630$$

Notice that the exponent on the 10 is positive 3, and we moved the decimal point 3 places to the right.

To write 7.29×10^{-3} in standard form, recall that $10^{-3} = \dfrac{1}{10^3} = \dfrac{1}{1000}$.

$$7.29 \times 10^{-3} = 7.29\left(\dfrac{1}{1000}\right) = \dfrac{7.29}{1000} = 0.00729$$

The exponent on the 10 is negative 3, and we moved the decimal to the left 3 places.

In general, **to write a scientific notation number in standard form,** move the decimal point the same number of places as the exponent on 10. If the exponent is positive, move the decimal point to the right; if the exponent is negative, move the decimal point to the left.

Example 18   Write each number in standard form, without exponents.

a. 1.02×10^5 **b.** 7.358×10^{-3}

c. 8.4×10^7 **d.** 3.007×10^{-5}

Solution:

a. Move the decimal point 5 places to the right.
$1.02 \times 10^5 = 102{,}000.$

b. Move the decimal point 3 places to the left.
$7.358 \times 10^{-3} = 0.007358$

c. $8.4 \times 10^7 = 84{,}000{,}000.$ 7 places to the right

d. $3.007 \times 10^{-5} = 0.00003007$ 5 places to the left

■ **Work Practice 18**

✓**Concept Check** Which number in each pair is larger?
   a. 7.8×10^3 or 2.1×10^5
   b. 9.2×10^{-2} or 2.7×10^4
   c. 5.6×10^{-4} or 6.3×10^{-5}

Objective E Performing Operations with Scientific Notation ▶

Performing operations on numbers written in scientific notation makes use of the rules and definitions for exponents.

Example 19   Perform each indicated operation. Write each result in standard decimal form.

a. $\left(8 \times 10^{-6}\right)\left(7 \times 10^3\right)$

b. $\dfrac{12 \times 10^2}{6 \times 10^{-3}}$

Solution:

a. $\left(8 \times 10^{-6}\right)\left(7 \times 10^3\right) = 8 \cdot 7 \cdot 10^{-6} \cdot 10^3$
$= 56 \times 10^{-3}$
$= 0.056$

b. $\dfrac{12 \times 10^2}{6 \times 10^{-3}} = \dfrac{12}{6} \times 10^{2-(-3)} = 2 \times 10^5 = 200{,}000$

■ **Work Practice 19**

Practice 18

Write the numbers in standard form, without exponents.
a. 3.062×10^{-4}
b. 5.21×10^4
c. 9.6×10^{-5}
d. 6.002×10^6

Practice 19

Perform each indicated operation. Write each result in standard decimal form.
a. $\left(9 \times 10^7\right)\left(4 \times 10^{-9}\right)$
b. $\dfrac{8 \times 10^4}{2 \times 10^{-3}}$

Answers
18. a. 0.0003062   **b.** 52,100
c. 0.000096   **d.** 6,002,000
19. a. 0.36   **b.** 40,000,000

✓**Concept Check Answers**
a. 2.1×10^5 **b.** 2.7×10^4
c. 5.6×10^{-4}

Calculator Explorations Scientific Notation

To enter a number written in scientific notation on a scientific calculator, locate the scientific notation key, which may be marked $\boxed{EE}$ or $\boxed{EXP}$. To enter 3.1×10^7, press $\boxed{3.1}\,\boxed{EE}\,\boxed{7}$. The display should read $\boxed{3.1 \quad 07}$.

Enter each number written in scientific notation on your calculator.

1. 5.31×10^3 5.31 EE 3
2. -4.8×10^{14} −4.8 EE 14
3. 6.6×10^{-9} 6.6 EE −9
4. -9.9811×10^{-2} −9.9811 EE −2

Multiply each of the following on your calculator. Notice the form of the result.

5. $3,000,000 \times 5,000,000$ 1.5×10^{13}
6. $230,000 \times 1,000$ 2.3×10^8

Multiply each of the following on your calculator. Write the product in scientific notation.

7. $(3.26 \times 10^6)(2.5 \times 10^{13})$ 8.15×10^{19}
8. $(8.76 \times 10^{-4})(1.237 \times 10^9)$ 1.083612×10^6

Vocabulary, Readiness & Video Check

Fill in each blank with the correct choice.

1. The expression x^{-3} equals $\underline{\quad \frac{1}{x^3} \quad}$.

   a. $-x^3$ b. $\frac{1}{x^3}$ c. $\frac{-1}{x^3}$ d. $\frac{1}{x^{-3}}$

2. The expression 5^{-4} equals $\underline{\quad \frac{1}{625} \quad}$.

   a. -20 b. -625 c. $\frac{1}{20}$ d. $\frac{1}{625}$

3. The number 3.021×10^{-3} is written in $\underline{\text{scientific notation}}$.

   a. standard form   b. expanded form

   c. scientific notation

4. The number 0.0261 is written in $\underline{\text{standard form}}$.

   a. standard form   b. expanded form

   c. scientific notation

Write each expression using positive exponents only.

5. $5x^{-2}$ $\frac{5}{x^2}$

6. $3x^{-3}$ $\frac{3}{x^3}$

7. $\frac{1}{y^{-6}}$ y^6

8. $\frac{1}{x^{-3}}$ x^3

9. $\frac{4}{y^{-3}}$ $4y^3$

10. $\frac{16}{y^{-7}}$ $16y^7$

Martin-Gay Interactive Videos

See Video 5.2

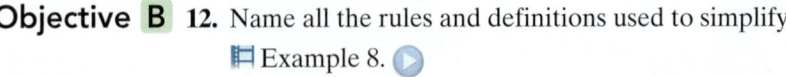

Watch the section lecture video and answer the following questions.

Objective A 11. What important reminder is given at the end of ⊞Example 1? ▶

Objective B 12. Name all the rules and definitions used to simplify ⊞Example 8. ▶

Objective C 13. From ⊞Examples 9 and 10, explain how the movement of the decimal point in Step 1 suggests the sign of the exponent on the number 10. ▶

Objective D 14. From ⊞Example 11, what part of a number written in scientific notation is key in telling us how to write the number in standard form? ▶

Objective E 15. In ⊞Example 13, what exponent rules were needed to evaluate the expression? ▶

See video answer section.

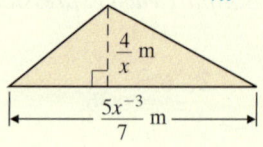

5.2 Exercise Set MyMathLab®

Objective A *Simplify each expression. Write each result using positive exponents only. See Examples 1 through 8.*

1. 4^{-3} $\dfrac{1}{64}$ 2. 6^{-2} $\dfrac{1}{36}$ 3. $7x^{-3}$ $\dfrac{7}{x^3}$ 4. $(7x)^{-3}$ $\dfrac{1}{343x^3}$ 5. $\left(-\dfrac{1}{4}\right)^{-3}$ -64 6. $\left(-\dfrac{1}{8}\right)^{-2}$ 64

7. $3^{-1}+2^{-1}$ $\dfrac{5}{6}$ 8. $4^{-1}+4^{-2}$ $\dfrac{5}{16}$ 9. $\dfrac{1}{p^{-3}}$ p^3 10. $\dfrac{1}{q^{-5}}$ q^5 11. $\dfrac{p^{-5}}{q^{-4}}$ $\dfrac{q^4}{p^5}$ 12. $\dfrac{r^{-5}}{s^{-2}}$ $\dfrac{s^2}{r^5}$

13. $\dfrac{x^{-2}}{x}$ $\dfrac{1}{x^3}$ 14. $\dfrac{y}{y^{-3}}$ y^4 15. $\dfrac{z^{-4}}{z^{-7}}$ z^3 16. $\dfrac{x^{-4}}{x^{-1}}$ $\dfrac{1}{x^3}$ 17. $3^{-2}+3^{-1}$ $\dfrac{4}{9}$ 18. $4^{-2}-4^{-3}$ $\dfrac{3}{64}$

19. $(-3)^{-2}$ $\dfrac{1}{9}$ 20. $(-2)^{-6}$ $\dfrac{1}{64}$ 21. $\dfrac{-1}{p^{-4}}$ $-p^4$ 22. $\dfrac{-1}{y^{-6}}$ $-y^6$ 23. -2^0-3^0 -2 24. $5^0+(-5)^0$ 2

Objective B *Simplify each expression. Write each result using positive exponents only. See Examples 9 through 16.*

25. $\dfrac{x^2x^5}{x^3}$ x^4 26. $\dfrac{y^4y^5}{y^6}$ y^3 27. $\dfrac{p^2p}{p^{-1}}$ p^4 28. $\dfrac{y^3y}{y^{-2}}$ y^6 29. $\dfrac{(m^5)^4m}{m^{10}}$ m^{11} 30. $\dfrac{(x^2)^8x}{x^9}$ x^8

31. $\dfrac{r}{r^{-3}r^{-2}}$ r^6 32. $\dfrac{p}{p^{-3}q^{-5}}$ p^4q^5 33. $(x^5y^3)^{-3}$ $\dfrac{1}{x^{15}y^9}$ 34. $(z^5x^5)^{-3}$ $\dfrac{1}{z^{15}x^{15}}$ 35. $\dfrac{(x^2)^3}{x^{10}}$ $\dfrac{1}{x^4}$ 36. $\dfrac{(y^4)^2}{y^{12}}$ $\dfrac{1}{y^4}$

37. $\dfrac{(a^5)^2}{(a^3)^4}$ $\dfrac{1}{a^2}$ 38. $\dfrac{(x^2)^5}{(x^4)^3}$ $\dfrac{1}{x^2}$ 39. $\dfrac{8k^4}{2k}$ $4k^3$ 40. $\dfrac{27r^6}{3r^4}$ $9r^2$ 41. $\dfrac{-6m^4}{-2m^3}$ $3m$ 42. $\dfrac{15a^4}{-15a^5}$ $-\dfrac{1}{a}$

43. $\dfrac{-24a^6b}{6ab^2}$ $-\dfrac{4a^5}{b}$ 44. $\dfrac{-5x^4y^5}{15x^4y^2}$ $-\dfrac{y^3}{3}$ 45. $\dfrac{6x^2y^3}{-7x^2y^5}$ $-\dfrac{6}{7y^2}$ 46. $\dfrac{-8xa^2b}{-5xa^5b}$ $\dfrac{8}{5a^3}$ 47. $(3a^2b^{-4})^3$ $\dfrac{27a^6}{b^{12}}$ 48. $(5x^3y^{-2})^2$ $\dfrac{25x^6}{y^4}$

49. $(a^{-5}b^2)^{-6}$ $\dfrac{a^{30}}{b^{12}}$ 50. $(4^{-1}x^5)^{-2}$ $\dfrac{16}{x^{10}}$ 51. $\left(\dfrac{x^{-2}y^4}{x^3y^7}\right)^{-2}$ $\dfrac{1}{x^{10}y^6}$ 52. $\left(\dfrac{a^5b}{a^7b^{-2}}\right)^{-3}$ $\dfrac{a^6}{b^9}$ 53. $\dfrac{4^2z^{-3}}{4^3z^{-5}}$ $\dfrac{z^2}{4}$ 54. $\dfrac{5^{-1}z^7}{5^{-2}z^9}$ $\dfrac{5}{z^2}$

55. $\dfrac{3^{-1}x^4}{3^3x^{-7}}$ $\dfrac{x^{11}}{81}$ 56. $\dfrac{2^{-3}x^{-4}}{2^2x}$ $\dfrac{1}{32x^5}$ 57. $\dfrac{7ab^{-4}}{7^{-1}a^{-3}b^2}$ $\dfrac{49a^4}{b^6}$ 58. $\dfrac{6^{-5}x^{-1}y^2}{6^{-2}x^{-4}y^4}$ $\dfrac{x^3}{216y^2}$ 59. $\dfrac{-12m^5n^{-7}}{4m^{-2}n^{-3}}$ $-\dfrac{3m^7}{n^4}$ 60. $\dfrac{-15r^{-6}s}{5r^{-4}s^{-3}}$ $-\dfrac{3s^4}{r^2}$

61. $\left(\dfrac{a^{-5}b}{ab^3}\right)^{-4}$ $a^{24}b^8$ 62. $\left(\dfrac{r^{-2}s^{-3}}{r^{-4}s^{-3}}\right)^{-3}$ $\dfrac{1}{r^6}$ 63. $(5^2)(8)(2^0)$ 200 64. $(3^4)(7^0)(2)$ 162 65. $\dfrac{(xy^3)^5}{(xy)^{-4}}$ x^9y^{19} 66. $\dfrac{(rs)^{-3}}{(r^2s^3)^2}$ $\dfrac{1}{r^7s^9}$

67. $\dfrac{(-2xy^{-3})^{-3}}{(xy^{-1})^{-1}}$ $-\dfrac{y^8}{8x^2}$ 68. $\dfrac{(-3x^2y^2)^{-2}}{(xyz)^{-2}}$ $\dfrac{z^2}{9x^2y^2}$ 69. $\dfrac{(a^4b^{-7})^{-5}}{(5a^2b^{-1})^{-2}}$ $\dfrac{25b^{33}}{a^{16}}$ 70. $\dfrac{(a^6b^{-2})^4}{(4a^{-3}b^{-3})^3}$ $\dfrac{a^{33}b}{64}$

71. Find the volume of the cube. $\dfrac{27}{z^3 x^6}$ cu in.

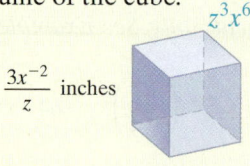

$\dfrac{3x^{-2}}{z}$ inches

72. Find the area of the triangle. $\dfrac{10}{7x^4}$ sq m

$\dfrac{4}{x}$ m

$\dfrac{5x^{-3}}{7}$ m

Objective C *Write each number in scientific notation. See Example 17.*

73. 78,000
7.8×10^4

74. 9,300,000,000
9.3×10^9

75. 0.00000167
1.67×10^{-6}

76. 0.00000017
1.7×10^{-7}

77. 0.00635
6.35×10^{-3}

78. 0.00194
1.94×10^{-3}

79. 1,160,000
1.16×10^6

80. 700,000
7.0×10^5

81. When it is completed in 2022, the Thirty Meter Telescope is expected to be the world's largest optical telescope. Located in an observatory complex at the summit of Mauna Kea in Hawaii, the elevation of the Thirty Meter Telescope will be roughly 4200 meters above sea level. Write 4200 in scientific notation. 4.2×10^3

82. The Thirty Meter Telescope (see Exercise **81**) will have the ability to view objects 13,000,000,000 light-years away. Write 13,000,000,000 in scientific notation. 1.3×10^{10}

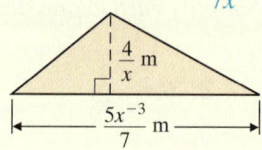

Objective D *Write each number in standard form. See Example 18.*

83. 8.673×10^{-10} 0.0000000008673

84. 9.056×10^{-4} 0.0009056

85. 3.3×10^{-2} 0.033

86. 4.8×10^{-6} 0.0000048

87. 2.032×10^4 20,320

88. 9.07×10^{10} 90,700,000,000

89. Each second, the Sun converts 7.0×10^8 tons of hydrogen into helium and energy in the form of gamma rays. Write this number in standard form. (*Source:* Students for the Exploration and Development of Space) 700,000,000

90. In chemistry, Avogadro's number is the number of atoms in one mole of an element. Avogadro's number is $6.02214199 \times 10^{23}$. Write this number in standard form. (*Source:* National Institute of Standards and Technology)
602,214,199,000,000,000,000,000

Objectives C D Mixed Practice *See Examples 17 and 18. If a number is written in standard form, write it in scientific notation. If a number is written in scientific notation, write it in standard form. (Source: CIA World Factbook) The bar graph below shows estimates of the top six national debts as of December 31, 2012.*

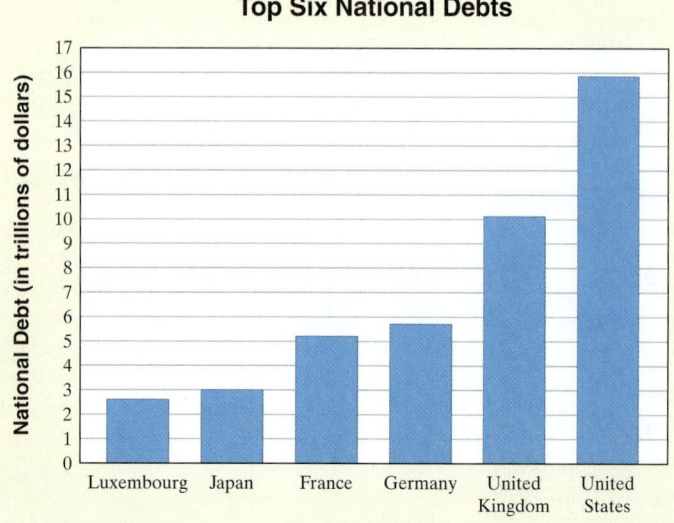

Top Six National Debts

National Debt (in trillions of dollars)

Countries: Luxembourg, Japan, France, Germany, United Kingdom, United States

Source: CIA World Factbook

91. Germany's national debt as of the end of 2012 was $5,700,000,000,000. 5.7×10^{12}

92. Luxembourg's national debt as of the end of 2012 was $2,600,000,000,000. 2.6×10^{12}

93. The United Kingdom's national debt as of the end of 2012 was 1.01×10^{13}. 10,100,000,000,000

94. France's national debt as of the end of 2012 was 5.2×10^{12}. 5,200,000,000,000

95. Use the bar graph to estimate the national debt of Japan and then express it in both standard and scientific notation. 3,000,000,000,000; 3×10^{12}

96. Use the bar graph to estimate the national debt of the United States and then express it in both standard and scientific notation. 15,900,000,000,000; 1.59×10^{13}

Objective **E** *Evaluate each expression using exponential rules. Write each result in standard form. See Example 19.*

97. $(1.2 \times 10^{-3})(3 \times 10^{-2})$ 0.000036

98. $(2.5 \times 10^{6})(2 \times 10^{-6})$ 5

99. $(4 \times 10^{-10})(7 \times 10^{-9})$ 0.0000000000000000028

100. $(5 \times 10^{6})(4 \times 10^{-8})$ 0.2

101. $\dfrac{8 \times 10^{-1}}{16 \times 10^{5}}$ 0.0000005

102. $\dfrac{25 \times 10^{-4}}{5 \times 10^{-9}}$ 500,000

103. $\dfrac{1.4 \times 10^{-2}}{7 \times 10^{-8}}$ 200,000

104. $\dfrac{0.4 \times 10^{5}}{0.2 \times 10^{11}}$ 0.000002

105. Although the actual amount varies by season and time of day, the average volume of water that flows over Niagara Falls (the American and Canadian falls combined) each second is 7.5×10^{5} gallons. How much water flows over Niagara Falls in an hour? Write the result in scientific notation. (*Hint:* 1 hour equals 3600 seconds.) (*Source:* http://niagarafallslive.com) 2.7×10^{9} gal

106. A beam of light travels 9.460×10^{12} kilometers per year. How far does light travel in 10,000 years? Write the result in scientific notation. 9.46×10^{16} km

Review

Simplify each expression by combining any like terms. See Section 1.8.

107. $3x - 5x + 7$ $-2x + 7$

108. $7w + w - 2w$ $6w$

109. $y - 10 + y$ $2y - 10$

110. $-6z + 20 - 3z$ $-9z + 20$

111. $7x + 2 - 8x - 6$ $-x - 4$

112. $10y - 14 - y - 14$ $9y - 28$

Concept Extensions

For Exercises 113–120, write each number in standard form. Then write the number in scientific notation.

113. In September 2013, Google received an estimated 900 million unique monthly visitors. (*Source:* eBizMBA Inc.)   900,000,000; 9×10^{8}

114. In September 2013, YouTube received an estimated 0.45 billion unique monthly visitors. (*Source:* eBizMBA Inc.)   450,000,000; 4.5×10^{8}

115. The surface of the Arctic Ocean encompasses 14.056 million square kilometers of water. (*Source:* CIA World Factbook)   14,056,000; 1.4056×10^{7}

116. The surface of the Pacific Ocean encompasses 155.557 million square kilometers of water. (*Source:* CIA World Factbook)   155,557,000; 1.55557×10^{8}

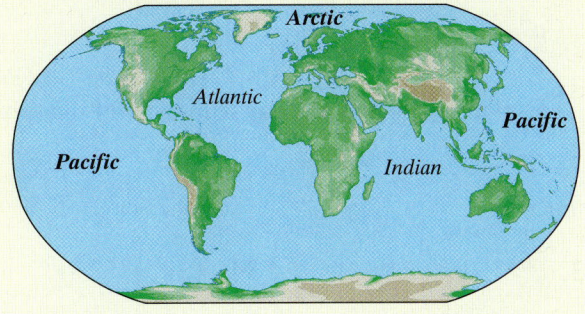

Solve.

117. A nanometer is one-billionth, or 10^{-9}, of a meter. A strand of DNA is about 2.5 nanometers in diameter. Use scientific notation to write the diameter of a DNA strand in terms of meters. (*Source:* United States National Nanotechnology Initiative)
$2.5 \times 10^{-9}\,\text{m}$

118. A micrometer (sometimes referred to as a micron) is one-millionth, or 10^{-6}, of a meter. A single red blood cell is about 7 micrometers in diameter. Use scientific notation to write the diameter of a red blood cell in terms of meters. (*Source:* National Institute of Standards and Technology)
$7 \times 10^{-6}\,\text{m}$

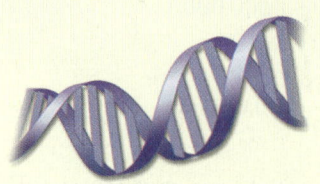

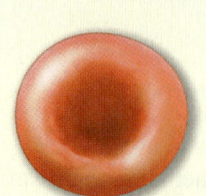

119. The Thirty Meter Telescope, described in Exercises **81–82**, will be capable of observing ultraviolet wavelengths measuring 310 nanometers. Express this wavelength in terms of meters using both standard form and scientific notation. (See Exercise **117** for a definition of nanometer.) (*Source:* TMT Observatory Corporation) $0.00000031\,\text{m}; 3.1 \times 10^{-7}\,\text{m}$

120. The Thirty Meter Telescope, described in Exercises **81–82**, will be capable of observing infrared wavelengths measuring 28 micrometers. Express this wavelength in terms of meters using both standard form and scientific notation. (See Exercise **118** for a definition of micrometer.) (*Source:* TMT Observatory Corporation)
$0.000028\,\text{m}; 2.8 \times 10^{-5}\,\text{m}$

Simplify.

121. $\left(2a^3\right)^3 a^4 + a^5 a^8$ $9a^{13}$

122. $\left(2a^3\right)^3 a^{-3} + a^{11} a^{-5}$ $9a^6$

Fill in the boxes so that each statement is true. (More than one answer is possible for these exercises.)

123. $x^{\square} = \dfrac{1}{x^5}$ -5

124. $7^{\square} = \dfrac{1}{49}$ -2

125. $z^{\square} \cdot z^{\square} = z^{-10}$ answers may vary

126. $\left(x^{\square}\right)^{\square} = x^{-15}$ answers may vary

127. Which is larger? See the Concept Check in this section.
a. 9.7×10^{-2} or 1.3×10^1 1.3×10^1
b. 8.6×10^5 or 4.4×10^7 4.4×10^7
c. 6.1×10^{-2} or 5.6×10^{-4} 6.1×10^{-2}

128. Determine whether each statement is true or false.
a. $5^{-1} < 5^{-2}$ false
b. $\left(\dfrac{1}{5}\right)^{-1} < \left(\dfrac{1}{5}\right)^{-2}$ true
c. $a^{-1} < a^{-2}$ for all nonzero numbers.   false

129. It was stated earlier that for an integer n,

$$x^{-n} = \frac{1}{x^n}, \quad x \neq 0.$$

Explain why x may not equal 0.   answers may vary

130. The quotient rule states that

$$\frac{a^m}{a^n} = a^{m-n}, \quad a \neq 0.$$

Explain why a may not equal 0.   answers may vary

Simplify each expression. Assume that variables represent positive integers.

131. $\left(x^{-3s}\right)^3$ $\dfrac{1}{x^{9s}}$

132. $a^{-4m} \cdot a^{5m}$ a^m

133. $a^{4m+1} \cdot a^4$ a^{4m+5}

134. $\left(3y^{2z}\right)^3$ $27y^{6z}$

Objective A Defining Term and Coefficient

Objectives

A Define Term and Coefficient of a Term.

B Define Polynomial, Monomial, Binomial, Trinomial, and Degree.

C Evaluate a Polynomial for Given Replacement Values.

D Simplify a Polynomial by Combining Like Terms.

E Simplify a Polynomial in Several Variables.

F Write a Polynomial in Descending Powers of the Variable and with No Missing Powers of the Variable.

In this section, we introduce a special algebraic expression called a polynomial. Let's first review some definitions presented in Section 1.8.

Recall that a term is a number or the product of a number and variables raised to powers. The terms of an expression are separated by plus signs. The terms of the expression $4x^2 + 3x$ are $4x^2$ and $3x$. The terms of the expression $9x^4 - 7x - 1$, or $9x^4 + (-7x) + (-1)$, are $9x^4$, $-7x$, and -1.

Expression	Terms
$4x^2 + 3x$	$4x^2, 3x$
$9x^4 - 7x - 1$	$9x^4, -7x, -1$
$7y^3$	$7y^3$

The **numerical coefficient** of a term, or simply the **coefficient,** is the numerical factor of each term. If no numerical factor appears in the term, then the coefficient is understood to be 1. If the term is a number only, it is called a **constant term** or simply a **constant.**

Term	Coefficient
x^5	1
$3x^2$	3
$-4x$	-4
$-x^2 y$	-1
3 (constant)	3

Example 1 Complete the table for the expression $7x^5 - 8x^4 + x^2 - 3x + 5$.

Term	Coefficient
x^2	
	-8
$-3x$	
	7
5	

Solution: The completed table is shown below.

Term	Coefficient
x^2	1
$-8x^4$	-8
$-3x$	-3
$7x^5$	7
5	5

■ **Work Practice 1**

Practice 1

Complete the table for the expression
$-6x^6 + 4x^5 + 7x^3 - 9x^2 - 1$.

Term	Coefficient
$7x^3$	
	-9
$-6x^6$	
	4
-1	

Answer

1. term: $-9x^2$, $4x^5$; coefficient: 7, -6, -1

Objective B Defining Polynomial, Monomial, Binomial, Trinomial, and Degree ▶

Now we are ready to define what we mean by a polynomial.

For example,

$$x^5 - 3x^3 + 2x^2 - 5x + 1$$

is a polynomial in x. Notice that this polynomial is written in **descending powers** of x because the powers of x decrease from left to right. (Recall that the term 1 can be thought of as $1x^0$.)

On the other hand,

$$x^{-5} + 2x - 3$$

is **not** a polynomial because one of its terms contains a variable with an exponent, -5, that is not a whole number.

The following are examples of monomials, binomials, and trinomials. Each of these examples is also a polynomial.

Polynomials			
Monomials	**Binomials**	**Trinomials**	**More than Three Terms**
ax^2	$x + y$	$x^2 + 4xy + y^2$	$5x^3 - 6x^2 + 3x - 6$
$-3z$	$3p + 2$	$x^5 + 7x^2 - x$	$-y^5 + y^4 - 3y^3 - y^2 + y$
4	$4x^2 - 7$	$-q^4 + q^3 - 2q$	$x^6 + x^4 - x^3 + 1$

Each term of a polynomial has a degree. The **degree of a term in one variable** is the exponent on the variable.

Identify the degree of each term of the trinomial $12x^4 - 7x + 3$.

Solution: The term $12x^4$ has degree 4.
The term $-7x$ has degree 1 since $-7x$ is $-7x^1$.
The term 3 has degree 0 since 3 is $3x^0$.

▪ **Work Practice 2**

Each polynomial also has a degree.

Practice 2

Identify the degree of each term of the trinomial $-15x^3 + 2x^2 - 5$.

Answer

2. 3; 2; 0

Example 3 Find the degree of each polynomial and tell whether the polynomial is a monomial, binomial, trinomial, or none of these.

a. $-2t^2 + 3t + 6$ **b.** $15x - 10$ **c.** $7x + 3x^3 + 2x^2 - 1$

Solution:

a. The degree of the trinomial $-2t^2 + 3t + 6$ is 2, the greatest degree of any of its terms.

b. The degree of the binomial $15x - 10$ or $15x^1 - 10$ is 1.

c. The degree of the polynomial $7x + 3x^3 + 2x^2 - 1$ is 3. The polynomial is neither a monomial, binomial, nor trinomial.

■ **Work Practice 3**

Objective C Evaluating Polynomials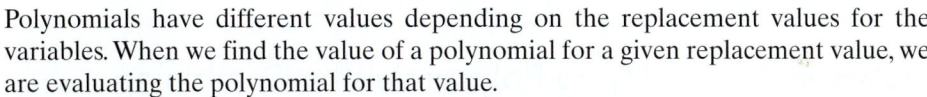

Polynomials have different values depending on the replacement values for the variables. When we find the value of a polynomial for a given replacement value, we are evaluating the polynomial for that value.

Example 4 Evaluate each polynomial when $x = -2$.

a. $-5x + 6$ **b.** $3x^2 - 2x + 1$

Solution:

a. $-5x + 6 = -5(-2) + 6$ Replace x with -2.

$= 10 + 6$

$= 16$

b. $3x^2 - 2x + 1 = 3(-2)^2 - 2(-2) + 1$ Replace x with -2.

$= 3(4) - 2(-2) + 1$

$= 12 + 4 + 1$

$= 17$

■ **Work Practice 4**

Many physical phenomena can be modeled by polynomials.

Example 5 Finding Free-Fall Time

The Swiss Re Building in London is a unique building. Londoners often refer to it as the "pickle building." The building is 592.1 feet tall. An object is dropped from the highest point of this building. Neglecting air resistance, the height in feet of the object above ground at time t seconds is given by the polynomial $-16t^2 + 592.1$. Find the height of the object when $t = 1$ second and when $t = 6$ seconds.

Solution: To find each height, we evaluate the polynomial when $t = 1$ and when $t = 6$.

$-16t^2 + 592.1 = -16(1)^2 + 592.1$ Replace t with 1.

$= -16(1) + 592.1$

$= -16 + 592.1$

$= 576.1$

The height of the object at 1 second is 576.1 feet.

$-16t^2 + 592.1 = -16(6)^2 + 592.1$ Replace t with 6.

$= -16(36) + 592.1$

$= -576 + 592.1 = 16.1$

(Continued on next page)

Practice 3

Find the degree of each polynomial and tell whether the polynomial is a monomial, binomial, trinomial, or none of these.

a. $-6x + 14$

b. $9x - 3x^6 + 5x^4 + 2$

c. $10x^2 - 6x - 6$

Practice 4

Evaluate each polynomial when $x = -1$.

a. $-2x + 10$

b. $6x^2 + 11x - 20$

Practice 5

Find the height of the object in Example 5 when $t = 2$ seconds and $t = 4$ seconds.

Answers

3. a. binomial, 1   **b.** none of these, 6

c. trinomial, 2   **4. a.** 12   **b.** -25

5. 528.1 feet, 336.1 feet

The height of the object at 6 seconds is 16.1 feet.

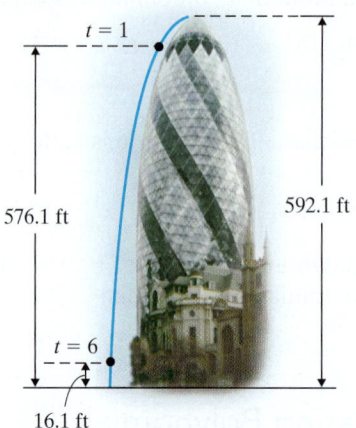

576.1 ft   592.1 ft

$t = 1$

$t = 6$

16.1 ft

🟧 **Work Practice 5**

Objective D Simplifying Polynomials by Combining Like Terms ▶

We can simplify polynomials with like terms by combining the like terms. Recall from Section 1.8 that like terms are terms that contain exactly the same variables raised to exactly the same powers.

Like Terms	Unlike Terms
$5x^2, -7x^2$	$3x, 3y$
$y, 2y$	$-2x^2, -5x$
$\frac{1}{2}a^2 b, -a^2 b$	$6st^2, 4s^2 t$

Only like terms can be combined. We combine like terms by applying the distributive property.

Practice 6–10

Simplify each polynomial by combining any like terms.

6. $-6y + 8y$

7. $14y^2 + 3 - 10y^2 - 9$

8. $7x^3 + x^3$

9. $23x^2 - 6x - x - 15$

10. $\dfrac{2}{7}x^3 - \dfrac{1}{4}x + 2 - \dfrac{1}{2}x^3 + \dfrac{3}{8}x$

🟩 **Examples** Simplify each polynomial by combining any like terms.

6. $-3x + 7x = (-3 + 7)x = 4x$

7. $11x^2 + 5 + 2x^2 - 7 = 11x^2 + 2x^2 + 5 - 7$

$\qquad\qquad\qquad\qquad = 13x^2 - 2$

8. $9x^3 + x^3 = 9x^3 + 1x^3$ Write x^3 as $1x^3$.

$\qquad\qquad = 10x^3$

9. $5x^2 + 6x - 9x - 3 = 5x^2 - 3x - 3$ Combine like terms $6x$ and $-9x$.

10. $\dfrac{2}{5}x^4 + \dfrac{2}{3}x^3 - x^2 + \dfrac{1}{10}x^4 - \dfrac{1}{6}x^3$

$= \left(\dfrac{2}{5} + \dfrac{1}{10}\right)x^4 + \left(\dfrac{2}{3} - \dfrac{1}{6}\right)x^3 - x^2$

$= \left(\dfrac{4}{10} + \dfrac{1}{10}\right)x^4 + \left(\dfrac{4}{6} - \dfrac{1}{6}\right)x^3 - x^2$

$= \dfrac{5}{10}x^4 + \dfrac{3}{6}x^3 - x^2$

$= \dfrac{1}{2}x^4 + \dfrac{1}{2}x^3 - x^2$

🟧 **Work Practice 6–10**

Answers

6. $2y$ **7.** $4y^2 - 6$ **8.** $8x^3$

9. $23x^2 - 7x - 15$

10. $-\dfrac{3}{14}x^3 + \dfrac{1}{8}x + 2$

Example 3 Find the degree of each polynomial and tell whether the polynomial is a monomial, binomial, trinomial, or none of these.

a. $-2t^2 + 3t + 6$ **b.** $15x - 10$ **c.** $7x + 3x^3 + 2x^2 - 1$

Solution:

a. The degree of the trinomial $-2t^2 + 3t + 6$ is 2, the greatest degree of any of its terms.

b. The degree of the binomial $15x - 10$ or $15x^1 - 10$ is 1.

c. The degree of the polynomial $7x + 3x^3 + 2x^2 - 1$ is 3. The polynomial is neither a monomial, binomial, nor trinomial.

■ **Work Practice 3**

Objective C Evaluating Polynomials ▶

Polynomials have different values depending on the replacement values for the variables. When we find the value of a polynomial for a given replacement value, we are evaluating the polynomial for that value.

Example 4 Evaluate each polynomial when $x = -2$.

a. $-5x + 6$ **b.** $3x^2 - 2x + 1$

Solution:

a. $-5x + 6 = -5(-2) + 6$ Replace x with -2.

$\qquad = 10 + 6$

$\qquad = 16$

b. $3x^2 - 2x + 1 = 3(-2)^2 - 2(-2) + 1$ Replace x with -2.

$\qquad = 3(4) - 2(-2) + 1$

$\qquad = 12 + 4 + 1$

$\qquad = 17$

■ **Work Practice 4**

Many physical phenomena can be modeled by polynomials.

Example 5 Finding Free-Fall Time

The Swiss Re Building in London is a unique building. Londoners often refer to it as the "pickle building." The building is 592.1 feet tall. An object is dropped from the highest point of this building. Neglecting air resistance, the height in feet of the object above ground at time t seconds is given by the polynomial $-16t^2 + 592.1$. Find the height of the object when $t = 1$ second and when $t = 6$ seconds.

Solution: To find each height, we evaluate the polynomial when $t = 1$ and when $t = 6$.

$-16t^2 + 592.1 = -16(1)^2 + 592.1$ Replace t with 1.

$\qquad = -16(1) + 592.1$

$\qquad = -16 + 592.1$

$\qquad = 576.1$

The height of the object at 1 second is 576.1 feet.

$-16t^2 + 592.1 = -16(6)^2 + 592.1$ Replace t with 6.

$\qquad = -16(36) + 592.1$

$\qquad = -576 + 592.1 = 16.1$

(Continued on next page)

Practice 3

Find the degree of each polynomial and tell whether the polynomial is a monomial, binomial, trinomial, or none of these.

a. $-6x + 14$

b. $9x - 3x^6 + 5x^4 + 2$

c. $10x^2 - 6x - 6$

Practice 4

Evaluate each polynomial when $x = -1$.

a. $-2x + 10$

b. $6x^2 + 11x - 20$

Practice 5

Find the height of the object in Example 5 when $t = 2$ seconds and $t = 4$ seconds.

Answers

3. **a.** binomial, 1   **b.** none of these, 6

c. trinomial, 2   4. **a.** 12   **b.** -25

5. 528.1 feet, 336.1 feet

The height of the object at 6 seconds is 16.1 feet.

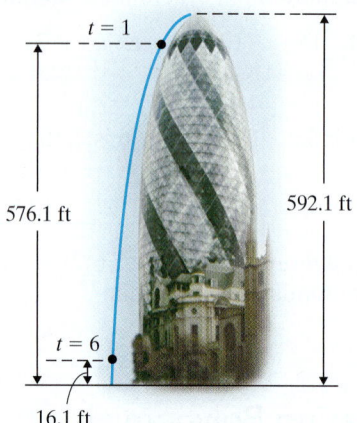

■ **Work Practice 5**

Objective D Simplifying Polynomials by Combining Like Terms

We can simplify polynomials with like terms by combining the like terms. Recall from Section 1.8 that like terms are terms that contain exactly the same variables raised to exactly the same powers.

Like Terms	Unlike Terms
$5x^2, -7x^2$	$3x, 3y$
$y, 2y$	$-2x^2, -5x$
$\frac{1}{2}a^2 b, -a^2 b$	$6st^2, 4s^2 t$

Only like terms can be combined. We combine like terms by applying the distributive property.

Practice 6–10

Simplify each polynomial by combining any like terms.

6. $-6y + 8y$

7. $14y^2 + 3 - 10y^2 - 9$

8. $7x^3 + x^3$

9. $23x^2 - 6x - x - 15$

10. $\frac{2}{7}x^3 - \frac{1}{4}x + 2 - \frac{1}{2}x^3 + \frac{3}{8}x$

Examples Simplify each polynomial by combining any like terms.

6. $-3x + 7x = (-3 + 7)x = 4x$

7. $11x^2 + 5 + 2x^2 - 7 = 11x^2 + 2x^2 + 5 - 7$
$$= 13x^2 - 2$$

8. $9x^3 + x^3 = 9x^3 + 1x^3$ Write x^3 as $1x^3$.
$$= 10x^3$$

9. $5x^2 + 6x - 9x - 3 = 5x^2 - 3x - 3$ Combine like terms $6x$ and $-9x$.

10. $\frac{2}{5}x^4 + \frac{2}{3}x^3 - x^2 + \frac{1}{10}x^4 - \frac{1}{6}x^3$

$$= \left(\frac{2}{5} + \frac{1}{10}\right)x^4 + \left(\frac{2}{3} - \frac{1}{6}\right)x^3 - x^2$$

$$= \left(\frac{4}{10} + \frac{1}{10}\right)x^4 + \left(\frac{4}{6} - \frac{1}{6}\right)x^3 - x^2$$

$$= \frac{5}{10}x^4 + \frac{3}{6}x^3 - x^2$$

$$= \frac{1}{2}x^4 + \frac{1}{2}x^3 - x^2$$

■ **Work Practice 6–10**

Answers

6. $2y$ **7.** $4y^2 - 6$ **8.** $8x^3$

9. $23x^2 - 7x - 15$

10. $-\frac{3}{14}x^3 + \frac{1}{8}x + 2$

Example 11 Write a polynomial that describes the total area of the squares and rectangles shown below. Then simplify the polynomial.

Solution: Recall that the area of a rectangle is length times width.

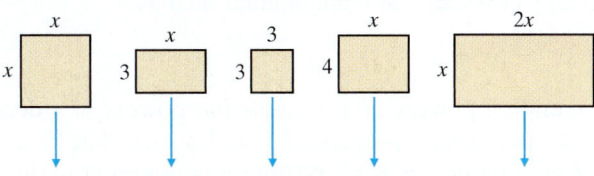

Area: $x \cdot x + 3 \cdot x + 3 \cdot 3 + 4 \cdot x + x \cdot 2x$

$= x^2 + 3x + 9 + 4x + 2x^2$

$= 3x^2 + 7x + 9$ Combine like terms.

■ Work Practice 11

Objective E Simplifying Polynomials Containing Several Variables ▶

A polynomial may contain more than one variable. One example is

$$5x + 3xy^2 - 6x^2y^2 + x^2y - 2y + 1$$

We call this expression a polynomial in several variables.

The **degree of a term** with more than one variable is the sum of the exponents on the variables. The **degree of a polynomial** in several variables is still the greatest degree of any term of the polynomial.

Example 12 Identify the degree of each term and the degree of the polynomial $5x + 3xy^2 - 6x^2y^2 + x^2y - 2y + 1$.

Solution: To organize our work, we use a table.

Terms of Polynomial	Degree of Term	Degree of Polynomial
$5x$	1	
$3xy^2$	1 + 2, or 3	
$-6x^2y^2$	2 + 2, or 4	4 (greatest degree)
x^2y	2 + 1, or 3	
$-2y$	1	
1	0	

■ Work Practice 12

To simplify a polynomial containing several variables, we combine any like terms.

Examples Simplify each polynomial by combining any like terms.

13. $3xy - 5y^2 + 7yx - 9x^2 = (3 + 7)xy - 5y^2 - 9x^2$

$\qquad\qquad\qquad\qquad = 10xy - 5y^2 - 9x^2$

14. $9a^2b - 6a^2 + 5b^2 + a^2b - 11a^2 + 2b^2$

$\qquad = 10a^2b - 17a^2 + 7b^2$

Helpful Hint This term can be written as $7yx$ or $7xy$.

■ Work Practice 13–14

Practice 11

Write a polynomial that describes the total area of the squares and rectangles shown below. Then simplify the polynomial.

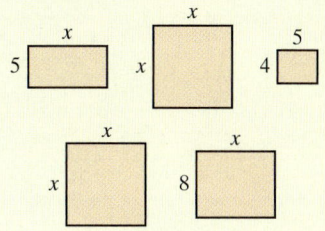

Practice 12

Identify the degrees of the terms and the degree of the polynomial $-2x^3y^2 + 4 - 8xy + 3x^3y + 5xy^2$.

Practice 13–14

Simplify each polynomial by combining any like terms.

13. $11ab - 6a^2 - ba + 8b^2$

14. $7x^2y^2 + 2y^2 - 4y^2x^2 + x^2 - y^2 + 5x^2$

Answers

11. $5x + x^2 + 20 + x^2 + 8x$; $2x^2 + 13x + 20$ **12.** $5, 0, 2, 4, 3; 5$

13. $10ab - 6a^2 + 8b^2$

14. $3x^2y^2 + y^2 + 6x^2$

Objective F Inserting "Missing" Terms

To prepare for dividing polynomials in Section 5.7, let's practice writing a polynomial in descending powers of the variable and with no "missing" powers.

Recall from Objective **B** that a polynomial such as

$$x^5 - 3x^3 + 2x^2 - 5x + 1$$

is written in descending powers of x because the powers of x decrease from left to right. Study the decreasing powers of x and notice that there is a "missing" power of x. This missing power is x^4. Writing a polynomial in decreasing powers of the variable helps you immediately determine important features of the polynomial, such as its degree. It is also sometimes helpful to write a polynomial so that there are no "missing" powers of x. For our polynomial above, if we simply insert a term of $0x^4$, which equals 0, we have an equivalent polynomial with no missing powers of x.

$$x^5 - 3x^3 + 2x^2 - 5x + 1 = x^5 + 0x^4 - 3x^3 + 2x^2 - 5x + 1$$

Practice 15

Write each polynomial in descending powers of the variable with no missing powers.

a. $x^2 + 9$

b. $9m^3 + m^2 - 5$

c. $-3a^3 + a^4$

Example 15 Write each polynomial in descending powers of the variable with no missing powers.

a. $x^2 - 4$ **b.** $3m^3 - m + 1$ **c.** $2x + x^4$

Solution:

a. $x^2 - 4 = x^2 + 0x^1 - 4$ or $x^2 + 0x - 4$ Insert a missing term of $0x^1$ or $0x$.

b. $3m^3 - m + 1 = 3m^3 + 0m^2 - m + 1$ Insert a missing term of $0m^2$.

c. $2x + x^4 = x^4 + 2x$ Write in descending powers of variable.

$\qquad\qquad = x^4 + 0x^3 + 0x^2 + 2x + 0x^0$ Insert missing terms of $0x^3, 0x^2,$ and $0x^0$ (or 0).

■ **Work Practice 15**

Answers

15. a. $x^2 + 0x + 9$

b. $9m^3 + m^2 + 0m - 5$

c. $a^4 - 3a^3 + 0a^2 + 0a + 0a^0$

Helpful Hint

Since there is no constant as a last term, we insert a $0x^0$. This $0x^0$ (or 0) is the final power of x in our polynomial.

Vocabulary, Readiness & Video Check

Use the choices below to fill in each blank. Not all choices will be used.

least   monomial   trinomial   coefficient
greatest   binomial   constant

1. A ___binomial___ is a polynomial with exactly two terms.

2. A ___monomial___ is a polynomial with exactly one term.

3. A ___trinomial___ is a polynomial with exactly three terms.

4. The numerical factor of a term is called the ___coefficient___.

5. A number term is also called a ___constant___.

6. The degree of a polynomial is the ___greatest___ degree of any term of the polynomial.

Martin-Gay Interactive Videos   *Watch the section lecture video and answer the following questions.*

See Video 5.3

Objective A 7. How many terms does the polynomial in ▢ Example 1 have? What are they? ▶

Objective B 8. For ▢ Example 2, why is the degree of each **term** found when the example asks for the degree of the **polynomial** only? ▶

Objective C 9. From ▢ Example 3, what does the value of a polynomial depend on? ▶

Objective D 10. What is another name for combining like terms in a polynomial, as in ▢ Example 5? ▶

Objective E 11. In ▢ Example 6, after combining like terms what is the degree of the binomial? Which term determines this? ▶

Objective F 12. In ▢ Example 7, what power is "missing" from the original polynomial? What term is inserted to replace this missing power? ▶

See video answer section.

5.3 Exercise Set MyMathLab® ▶

Objective A *Complete each table for each polynomial. See Example 1.*

▶ **1.** $x^2 - 3x + 5$

Term	Coefficient
x^2	1
$-3x$	-3
5	5

2. $2x^3 - x + 4$

Term	Coefficient
$2x^3$	2
$-x$	-1
4	4

3. $-5x^4 + 3.2x^2 + x - 5$

Term	Coefficient
$-5x^4$	-5
$3.2x^2$	3.2
x	1
-5	-5

4. $9.7x^7 - 3x^5 + x^3 - \frac{1}{4}x^2$

Term	Coefficient
$9.7x^7$	9.7
$-3x^5$	-3
x^3	1
$-\frac{1}{4}x^2$	$-\frac{1}{4}$

Objective B *Find the degree of each polynomial and determine whether it is a monomial, binomial, trinomial, or none of these. See Examples 2 and 3.*

5. $x + 2$
   1; binomial

6. $-6y + 4$
   1; binomial

▶ **7.** $9m^3 - 5m^2 + 4m - 8$
   3; none of these

8. $a + 5a^2 + 3a^3 - 4a^4$
   4; none of these

9. $12x^4 - x^6 - 12x^2$
   6; trinomial

10. $7r^2 + 2r - 3r^5$
   5; trinomial

11. $3z - 5z^4$
   4; binomial

12. $5y^6 + 2$
   6; binomial

Objective C *Evaluate each polynomial when* **a.** $x = 0$ *and* **b.** $x = -1$. *See Examples 4 and 5.*

13. $5x - 6$
   (a) -6; **(b)** -11

14. $2x - 10$
   (a) -10; **(b)** -12

▶ **15.** $x^2 - 5x - 2$
   (a) -2; **(b)** 4

16. $x^2 + 3x - 4$
   (a) -4; **(b)** -6

17. $-x^3 + 4x^2 - 15$
   (a) -15; **(b)** -10

18. $-2x^3 + 3x^2 - 6$
   (a) -6; **(b)** -1

A rocket is fired upward from the ground with an initial velocity of 200 feet per second. Neglecting air resistance, the height of the rocket at any time t can be described in feet by the polynomial $-16t^2 + 200t$. *Find the height of the rocket at the time given in Exercises 19 through 22. See Example 5.*

	Time, t (in seconds)	Height $-16t^2 + 200t$
19.	1	184 ft
20.	5	600 ft
21.	7.6	595.84 ft
22.	10.3	362.56 ft

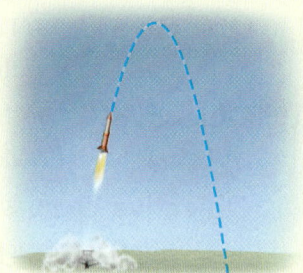

Pictured Rocks National Lakeshore

Scotts Bluff National Monument

23. The polynomial $-15x^2 + 77x + 499$ models the yearly number of visitors (in thousands) x years after 2010 at Pictured Rocks National Lakeshore. Use this polynomial to estimate the number of visitors to the park in 2012 ($x = 2$).   593 thousand

24. The polynomial $-0.5x^2 - 4.5x + 134$ models the yearly number of visitors (in thousands) x years after 2010 at Scotts Bluff National Monument. Use this polynomial to estimate the number of visitors to the monument in 2011 ($x = 1$).   129 thousand

25. The number of cell sites (in thousands) in the United States x years after 2008 is given by the polynomial $3.7x^2 + 0.7x + 241.6$ for 2008 through 2012. Use this model to predict the number of cell sites in the United States in 2015 ($x = 7$). (*Source:* Based on data from CTIA—The Wireless Association)   427.8 thousand

26. The number of active wireless devices (in millions) in the United States x years after 2004 is given by the polynomial $-x^2 + 26x + 184$ for 2004 through 2012. Use this model to predict the number of active wireless devices in the United States in 2016 ($x = 12$). (*Source:* Based on data from CTIA—The Wireless Association)   352 million

Objective D *Simplify each expression by combining like terms. See Examples 6 through 10.*

27. $9x - 20x$
$-11x$

28. $14y - 30y$
$-16y$

29. $14x^3 + 9x^3$
$23x^3$

30. $18x^3 + 4x^3$
$22x^3$

31. $7x^2 + 3 + 9x^2 - 10$
$16x^2 - 7$

32. $8x^2 + 4 + 11x^2 - 20$
$19x^2 - 16$

33. $15x^2 - 3x^2 - 13$
$12x^2 - 13$

34. $12k^3 - 9k^3 + 11$
$3k^3 + 11$

35. $8s - 5s + 4s$
$7s$

36. $5y + 7y - 6y$
$6y$

▶ **37.** $0.1y^2 - 1.2y^2 + 6.7 - 1.9$
$-1.1y^2 + 4.8$

38. $7.6y + 3.2y^2 - 8y - 2.5y^2$
$0.7y^2 - 0.4y$

39. $\frac{2}{3}x^4 + 12x^3 + \frac{1}{6}x^4 - 19x^3 - 19$ $\frac{5}{6}x^4 - 7x^3 - 19$

40. $\frac{2}{5}x^4 - 23x^2 + \frac{1}{15}x^4 + 5x^2 - 5$ $\frac{7}{15}x^4 - 18x^2 - 5$

41. $\frac{3}{20}x^3 + \frac{1}{10} - \frac{3}{10}x - \frac{1}{5} - \frac{7}{20}x + 6x^2$
$\frac{3}{20}x^3 + 6x^2 - \frac{13}{20}x - \frac{1}{10}$

42. $\frac{5}{16}x^3 - \frac{1}{8} + \frac{3}{8}x + \frac{1}{4} - \frac{9}{16}x - 14x^2$
$\frac{5}{16}x^3 - 14x^2 - \frac{3}{16}x + \frac{1}{8}$

Write a polynomial that describes the total area of each set of rectangles and squares shown in Exercises 43 and 44. Then simplify the polynomial. See Example 11.

△ **43.**

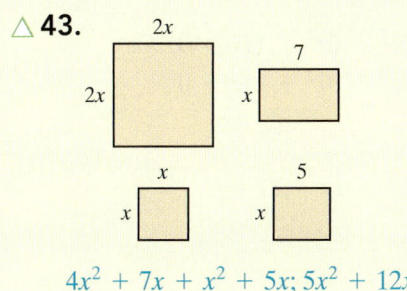

$4x^2 + 7x + x^2 + 5x;\ 5x^2 + 12x$

△ **44.**

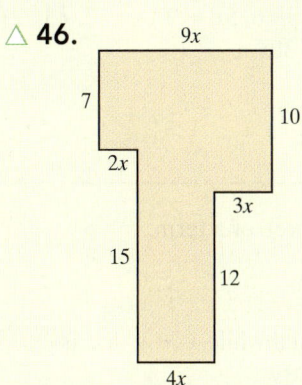

$7x + 4x^2 + 4x^2 + 18 + 16;\ 8x^2 + 7x + 34$

Recall that the perimeter of a figure such as the ones shown in Exercises 45 and 46 is the sum of the lengths of its sides. Write each perimeter as a polynomial. Then simplify the polynomial.

△ **45.**

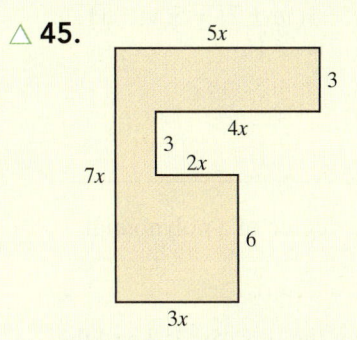

$5x + 3 + 4x + 3 + 2x + 6 + 3x + 7x;\ 21x + 12$

△ **46.**

$9x + 10 + 3x + 12 + 4x + 15 + 2x + 7;\ 18x + 44$

Objective E *Identify the degree of each term and the degree of the polynomial. See Example 12.*

47. $9ab - 6a + 5b - 3$ $2, 1, 1, 0; 2$

48. $y^4 - 6y^3x + 2x^2y^2 - 5y^2 + 3$ $4, 4, 4, 2, 0; 4$

49. $x^3y - 6 + 2x^2y^2 + 5y^3$ $4, 0, 4, 3; 4$

50. $2a^2b + 10a^4b - 9ab + 6$ $3, 5, 2, 0; 5$

Simplify each polynomial by combining any like terms. See Examples 13 and 14.

▶ **51.** $3ab - 4a + 6ab - 7a$
$9ab - 11a$

52. $-9xy + 7y - xy - 6y$
$-10xy + y$

53. $4x^2 - 6xy + 3y^2 - xy$ $4x^2 - 7xy + 3y^2$

54. $3a^2 - 9ab + 4b^2 - 7ab$
$3a^2 - 16ab + 4b^2$

55. $5x^2y + 6xy^2 - 5yx^2 + 4 - 9y^2x$
$-3xy^2 + 4$

56. $17a^2b - 16ab^2 + 3a^3 + 4ba^3 - b^2a$
$17a^2b - 17ab^2 + 3a^3 + 4ba^3$

57. $14y^3 - 9 + 3a^2b^2 - 10 - 19b^2a^2$
$14y^3 - 19 - 16a^2b^2$

58. $18x^4 + 2x^3y^3 - 1 - 2y^3x^3 - 17x^4$
$x^4 - 1$

Objective F *Write each polynomial in descending powers of the variable and with no missing powers. See Example 15.*

59. $7x^2 + 3$
$7x^2 + 0x + 3$

60. $5x^2 - 2$
$5x^2 + 0x - 2$

61. $x^3 - 64$
$x^3 + 0x^2 + 0x - 64$

62. $x^3 - 8$
$x^3 + 0x^2 + 0x - 8$

63. $5y^3 + 2y - 10$
$5y^3 + 0y^2 + 2y - 10$

64. $6m^3 - 3m + 4$
$6m^3 + 0m^2 - 3m + 4$

65. $8y + 2y^4$ $2y^4 + 0y^3 + 0y^2 + 8y + 0y^0$ or
$2y^4 + 0y^3 + 0y^2 + 8y + 0$

66. $11z + 4z^4$ $4z^4 + 0z^3 + 0z^2 + 11z + 0z^0$ or
$4z^4 + 0z^3 + 0z^2 + 11z + 0$

67. $6x^5 + x^3 - 3x + 15$
$6x^5 + 0x^4 + x^3 + 0x^2 - 3x + 15$

68. $9y^5 - y^2 + 2y - 11$
$9y^5 + 0y^4 + 0y^3 - y^2 + 2y - 11$

Review

Simplify each expression. See Section 1.8.

69. $4 + 5(2x + 3)$
$10x + 19$

70. $9 - 6(5x + 1)$
$-30x + 3$

71. $2(x - 5) + 3(5 - x)$
$-x + 5$

72. $-3(w + 7) + 5(w + 1)$
$2w - 16$

Concept Extensions

73. Describe how to find the degree of a term.
answers may vary

74. Describe how to find the degree of a polynomial.
answers may vary

75. Explain why xyz is a monomial while $x + y + z$ is a trinomial.   answers may vary

76. Explain why the degree of the term $5y^3$ is 3 and the degree of the polynomial $2y + y + 2y$ is 1.
answers may vary

Simplify, if possible.

77. $x^4 \cdot x^9$ x^{13}

78. $x^4 + x^9$ $x^4 + x^9$

79. $a \cdot b^3 \cdot a^2 \cdot b^7$ a^3b^{10}

80. $a + b^3 + a^2 + b^7$
$a + b^3 + a^2 + b^7$

81. $(y^5)^4 + (y^2)^{10}$
$2y^{20}$

82. $x^5y^2 + y^2x^5$
$2x^5y^2$ or $2y^2x^5$

Fill in the boxes so that the terms in each expression can be combined. Then simplify. Each exercise has more than one solution.

83. $7x^\square + 2x^\square$ answers may vary

84. $(3y^2)^\square + (4y^3)^\square$ answers may vary

85. Explain why the height of the rocket in Exercises **19** through **22** increases and then decreases as time passes.   answers may vary

86. Approximate (to the nearest tenth of a second) how long before the rocket in Exercises **19** through **22** hits the ground.   12.5 sec

Simplify each polynomial by combining like terms.

87. $1.85x^2 - 3.76x + 9.25x^2 + 10.76 - 4.21x$
$11.1x^2 - 7.97x + 10.76$

88. $7.75x + 9.16x^2 - 1.27 - 14.58x^2 - 18.34$
$-5.42x^2 + 7.75x - 19.61$

5.4 Adding and Subtracting Polynomials

Objective A Adding Polynomials

To add polynomials, we use commutative and associative properties and then combine like terms. To see if you are ready to add polynomials, try the Concept Check.

✓**Concept Check** When combining like terms in the expression $5x - 8x^2 - 8x$, which of the following is the proper result?
 a. $-11x^2$ **b.** $-3x - 8x^2$ **c.** $-11x$ **d.** $-11x^4$

> **To Add Polynomials**
>
> To add polynomials, combine all like terms.

Objectives

A   Add Polynomials.

B   Subtract Polynomials.

C   Add or Subtract Polynomials in One Variable.

D   Add or Subtract Polynomials in Several Variables.

Examples   Add.

1. $\left(4x^3 - 6x^2 + 2x + 7\right) + \left(5x^2 - 2x\right)$
   $= 4x^3 - 6x^2 + 2x + 7 + 5x^2 - 2x$ Remove parentheses.
   $= 4x^3 + \left(-6x^2 + 5x^2\right) + \left(2x - 2x\right) + 7$ Group like terms.
   $= 4x^3 - x^2 + 7$ Simplify.

2. $\left(-2x^2 + 5x - 1\right) + \left(-2x^2 + x + 3\right)$
   $= -2x^2 + 5x - 1 - 2x^2 + x + 3$ Remove parentheses.
   $= \left(-2x^2 - 2x^2\right) + \left(5x + 1x\right) + \left(-1 + 3\right)$ Group like terms.
   $= -4x^2 + 6x + 2$ Simplify.

■ **Work Practice 1–2**

Just as we can add numbers vertically, polynomials can be added vertically if we line up like terms underneath one another.

Example 3   Add $\left(7y^3 - 2y^2 + 7\right)$ and $\left(6y^2 + 1\right)$ using a vertical format.

Solution: Vertically line up like terms and add.

$$\begin{array}{r} 7y^3 - 2y^2 + 7 \\ 6y^2 + 1 \\ \hline 7y^3 + 4y^2 + 8 \end{array}$$

■ **Work Practice 3**

Objective B Subtracting Polynomials

To subtract one polynomial from another, recall the definition of subtraction. To subtract a number, we add its opposite: $a - b = a + (-b)$. To subtract a polynomial, we also add its opposite. Just as $-b$ is the opposite of b, $-(x^2 + 5)$ is the opposite of $(x^2 + 5)$.

> **To Subtract Polynomials**
>
> To subtract two polynomials, change the signs of the terms of the polynomial being subtracted and then add.

Practice 1–2
Add.
1. $\left(3x^5 - 7x^3 + 2x - 1\right)$
    $+ \left(3x^3 - 2x\right)$
2. $\left(5x^2 - 2x + 1\right)$
    $+ \left(-6x^2 + x - 1\right)$

Practice 3
Add $\left(9y^2 - 6y + 5\right)$ and $(4y + 3)$ using a vertical format.

Practice 4

Subtract:

$(9x + 5) - (4x - 3)$

Example 4   Subtract: $(5x - 3) - (2x - 11)$

Solution: From the definition of subtraction, we have

$$(5x - 3) - (2x - 11) = (5x - 3) + [-(2x - 11)] \quad \text{Add the opposite.}$$
$$= (5x - 3) + (-2x + 11) \quad \text{Apply the distributive property.}$$
$$= 5x - 3 - 2x + 11 \quad \text{Remove parentheses.}$$
$$= 3x + 8 \quad \text{Combine like terms.}$$

■ **Work Practice 4**

Practice 5

Subtract:

$(4x^3 - 10x^2 + 1)$
$-(-4x^3 + x^2 - 11)$

Example 5   Subtract: $(2x^3 + 8x^2 - 6x) - (2x^3 - x^2 + 1)$

Solution: First, we change the sign of each term of the second polynomial; then we add.

$$(2x^3 + 8x^2 - 6x) - (2x^3 - x^2 + 1)$$
$$= (2x^3 + 8x^2 - 6x) + (-2x^3 + x^2 - 1)$$
$$= 2x^3 + 8x^2 - 6x - 2x^3 + x^2 - 1$$
$$= 2x^3 - 2x^3 + 8x^2 + x^2 - 6x - 1$$
$$= 9x^2 - 6x - 1 \quad \text{Combine like terms.}$$

■ **Work Practice 5**

Just as polynomials can be added vertically, so can they be subtracted vertically.

Practice 6

Subtract $(6y^2 - 3y + 2)$ from $(2y^2 - 2y + 7)$ using a vertical format.

Example 6   Subtract $(5y^2 + 2y - 6)$ from $(-3y^2 - 2y + 11)$ using a vertical format.

Solution: Arrange the polynomials in a vertical format, lining up like terms.

$$\begin{array}{r} -3y^2 - 2y + 11 \\ -(5y^2 + 2y - 6) \end{array} \qquad \begin{array}{r} -3y^2 - 2y + 11 \\ -5y^2 - 2y + 6 \\ \hline -8y^2 - 4y + 17 \end{array}$$

■ **Work Practice 6**

> **Helpful Hint**
>
> Don't forget to change the sign of each term in the polynomial being subtracted.

Objective C Adding and Subtracting Polynomials in One Variable ▶

Let's practice adding and subtracting polynomials in one variable.

Teaching Tip
Consider having students also try the following problem:
Subtract the sum of $(3x + 6)$ and $(8x - 5)$ from $(5x + 2)$.
$(5x + 2) - [(3x + 6) + (8x - 5)]$
$\quad = 5x + 2 - (11x + 1)$
$\quad = -6x + 1$

Practice 7

Subtract $(3x + 1)$ from the sum of $(4x - 3)$ and $(12x - 5)$.

Example 7   Subtract $(5z - 7)$ from the sum of $(8z + 11)$ and $(9z - 2)$.

Solution: Notice that $(5z - 7)$ is to be subtracted **from** a sum. The translation is

$$[(8z + 11) + (9z - 2)] - (5z - 7)$$
$$= 8z + 11 + 9z - 2 - 5z + 7 \quad \text{Remove grouping symbols.}$$
$$= 8z + 9z - 5z + 11 - 2 + 7 \quad \text{Group like terms.}$$
$$= 12z + 16 \quad \text{Combine like terms.}$$

■ **Work Practice 7**

Answers

4. $5x + 8$ **5.** $8x^3 - 11x^2 + 12$
6. $-4y^2 + y + 5$ **7.** $13x - 9$

Objective D Adding and Subtracting Polynomials in Several Variables

Now that we know how to add or subtract polynomials in one variable, we can also add and subtract polynomials in several variables.

Examples   Add or subtract as indicated.

8. $\left(3x^2 - 6xy + 5y^2\right) + \left(-2x^2 + 8xy - y^2\right)$
$= 3x^2 - 6xy + 5y^2 - 2x^2 + 8xy - y^2$
$= x^2 + 2xy + 4y^2$ Combine like terms.

9. $\left(9a^2b^2 + 6ab - 3ab^2\right) - \left(5b^2a + 2ab - 3 - 9b^2\right)$
$= 9a^2b^2 + 6ab - 3ab^2 - 5b^2a - 2ab + 3 + 9b^2$
$= 9a^2b^2 + 4ab - 8ab^2 + 9b^2 + 3$ Combine like terms.

■ **Work Practice 8–9**

✓**Concept Check** If possible, simplify each expression by performing the indicated operation.
   a. $2y + y$
   b. $2y \cdot y$
   c. $-2y - y$
   d. $(-2y)(-y)$
   e. $2x + y$

Practice 8–9

Add or subtract as indicated.
8. $\left(2a^2 - ab + 6b^2\right)$
$\quad + \left(-3a^2 + ab - 7b^2\right)$
9. $\left(5x^2y^2 + 3 - 9x^2y + y^2\right)$
$\quad -\left(-x^2y^2 + 7 - 8xy^2 + 2y^2\right)$

Answers
8. $-a^2 - b^2$
9. $6x^2y^2 - 4 - 9x^2y + 8xy^2 - y^2$

✓**Concept Check Answers**
a. $3y$ **b.** $2y^2$ **c.** $-3y$ **d.** $2y^2$
e. cannot be simplified

Vocabulary, Readiness & Video Check

Simplify by combining like terms if possible.

1. $-9y - 5y$ $-14y$

2. $6m^5 + 7m^5$ $13m^5$

3. $x + 6x$ $7x$

4. $7z - z$ $6z$

5. $5m^2 + 2m$ $5m^2 + 2m$

6. $8p^3 + 3p^2$ $8p^3 + 3p^2$

Martin-Gay Interactive Videos

See Video 5.4

Watch the section lecture video and answer the following questions.

Objective A **7.** In ▦ Example 1, like terms are combined when adding the polynomials. What are the two sets of like terms? ▶

Objective B **8.** In ▦ Example 2, why can't parentheses just be removed as they were in ▦ Example 1? ▶

Objective C **9.** In ▦ Example 3, why are we told to be careful when translating to an expression? ▶

Objective D **10.** In ▦ Example 5, why aren't any signs changed when parentheses are removed? ▶

See video answer section.

5.4 Exercise Set MyMathLab®

Objective A *Add. See Examples 1 and 2.*

1. $(3x + 7) + (9x + 5)$ $12x + 12$

2. $(-y - 2) + (3y + 5)$ $2y + 3$

3. $(-7x + 5) + (-3x^2 + 7x + 5)$ $-3x^2 + 10$

4. $(3x - 8) + (4x^2 - 3x + 3)$ $4x^2 - 5$

5. $(-5x^2 + 3) + (2x^2 + 1)$ $-3x^2 + 4$

6. $(3x^2 + 7) + (3x^2 + 9)$ $6x^2 + 16$

▶ 7. $(-3y^2 - 4y) + (2y^2 + y - 1)$ $-y^2 - 3y - 1$

8. $(7x^2 + 2x - 9) + (-3x^2 + 5)$ $4x^2 + 2x - 4$

9. $(1.2x^3 - 3.4x + 7.9) + (6.7x^3 + 4.4x^2 - 10.9)$
$7.9x^3 + 4.4x^2 - 3.4x - 3$

10. $(9.6y^3 + 2.7y^2 - 8.6) + (1.1y^3 - 8.8y + 11.6)$
$10.7y^3 + 2.7y^2 - 8.8y + 3$

11. $\left(\frac{3}{4}m^2 - \frac{2}{5}m + \frac{1}{8}\right) + \left(-\frac{1}{4}m^2 - \frac{3}{10}m + \frac{11}{16}\right)$
$\frac{1}{2}m^2 - \frac{7}{10}m + \frac{13}{16}$

12. $\left(-\frac{4}{7}n^2 + \frac{5}{6}m - \frac{1}{20}\right) + \left(\frac{3}{7}n^2 - \frac{5}{12}m - \frac{3}{10}\right)$
$-\frac{1}{7}n^2 + \frac{5}{12}m - \frac{7}{20}$

Add using a vertical format. See Example 3.

13.
$3t^2 + 4$
$5t^2 - 8$
$\overline{8t^2 - 4}$

14.
$7x^3 + 3$
$2x^3 - 7$
$\overline{9x^3 - 4}$

15.
$10a^3 - 8a^2 + 4a + 9$
$15a^3 + 9a^2 - 7a + 7$
$\overline{15a^3 + a^2 - 3a + 16}$

16.
$2x^3 - 3x^2 + x - 4$
$5x^3 + 2x^2 - 3x + 2$
$\overline{7x^3 - x^2 - 2x - 2}$

Objective B *Subtract. See Examples 4 and 5.*

17. $(2x + 5) - (3x - 9)$
$-x + 14$

18. $(4 + 5a) - (-a - 5)$
$6a + 9$

19. $(5x^2 + 4) - (-2x^2 + 4)$
$7x^2$

20. $(-7y^2 + 5) - (-8y^2 + 12)$
$y^2 - 7$

21. $3x - (5x - 9)$
$-2x + 9$

22. $4 - (-y - 4)$
$y + 8$

23. $(2x^2 + 3x - 9) - (-4x + 7)$
$2x^2 + 7x - 16$

24. $(-7x^2 + 4x + 7) - (-8x + 2)$
$-7x^2 + 12x + 5$

▶ 25. $(5x + 8) - (-2x^2 - 6x + 8)$
$2x^2 + 11x$

26. $(-6y^2 + 3y - 4) - (9y^2 - 3y)$
$-15y^2 + 6y - 4$

27. $(0.7x^2 + 0.2x - 0.8) - (0.9x^2 + 1.4)$
$-0.2x^2 + 0.2x - 2.2$

28. $(-0.3y^2 + 0.6y - 0.3) - (0.5y^2 + 0.3)$
$-0.8y^2 + 0.6y - 0.6$

29. $\left(\frac{1}{4}z^2 - \frac{1}{5}z\right) - \left(-\frac{3}{20}z^2 + \frac{1}{10}z - \frac{7}{20}\right)$
$\frac{2}{5}z^2 - \frac{3}{10}z + \frac{7}{20}$

30. $\left(\frac{1}{3}x^2 - \frac{2}{7}x\right) - \left(\frac{4}{21}x^2 + \frac{1}{21}x - \frac{2}{3}\right)$
$\frac{1}{7}x^2 - \frac{1}{3}x + \frac{2}{3}$

Objective D Adding and Subtracting Polynomials in Several Variables

Now that we know how to add or subtract polynomials in one variable, we can also add and subtract polynomials in several variables.

Examples Add or subtract as indicated.

8. $(3x^2 - 6xy + 5y^2) + (-2x^2 + 8xy - y^2)$
$= 3x^2 - 6xy + 5y^2 - 2x^2 + 8xy - y^2$
$= x^2 + 2xy + 4y^2$ Combine like terms.

9. $(9a^2b^2 + 6ab - 3ab^2) - (5b^2a + 2ab - 3 - 9b^2)$
$= 9a^2b^2 + 6ab - 3ab^2 - 5b^2a - 2ab + 3 + 9b^2$
$= 9a^2b^2 + 4ab - 8ab^2 + 9b^2 + 3$ Combine like terms.

■ **Work Practice 8–9**

✓**Concept Check** If possible, simplify each expression by performing the indicated operation.

a. $2y + y$
b. $2y \cdot y$
c. $-2y - y$
d. $(-2y)(-y)$
e. $2x + y$

Practice 8–9

Add or subtract as indicated.
8. $(2a^2 - ab + 6b^2)$
$+ (-3a^2 + ab - 7b^2)$
9. $(5x^2y^2 + 3 - 9x^2y + y^2)$
$- (-x^2y^2 + 7 - 8xy^2 + 2y^2)$

Answers
8. $-a^2 - b^2$
9. $6x^2y^2 - 4 - 9x^2y + 8xy^2 - y^2$

✓**Concept Check Answers**
a. $3y$ **b.** $2y^2$ **c.** $-3y$ **d.** $2y^2$
e. cannot be simplified

Vocabulary, Readiness & Video Check

Simplify by combining like terms if possible.

1. $-9y - 5y$ $-14y$

2. $6m^5 + 7m^5$ $13m^5$

3. $x + 6x$ $7x$

4. $7z - z$ $6z$

5. $5m^2 + 2m$ $5m^2 + 2m$

6. $8p^3 + 3p^2$ $8p^3 + 3p^2$

Martin-Gay Interactive Videos

See Video 5.4

See video answer section.

Watch the section lecture video and answer the following questions.

Objective A **7.** In ▤ Example 1, like terms are combined when adding the polynomials. What are the two sets of like terms? ▶

Objective B **8.** In ▤ Example 2, why can't parentheses just be removed as they were in ▤ Example 1? ▶

Objective C **9.** In ▤ Example 3, why are we told to be careful when translating to an expression? ▶

Objective D **10.** In ▤ Example 5, why aren't any signs changed when parentheses are removed? ▶

5.4 Exercise Set MyMathLab®

Objective A *Add. See Examples 1 and 2.*

1. $(3x + 7) + (9x + 5)$ $12x + 12$

2. $(-y - 2) + (3y + 5)$ $2y + 3$

3. $(-7x + 5) + (-3x^2 + 7x + 5)$ $-3x^2 + 10$

4. $(3x - 8) + (4x^2 - 3x + 3)$ $4x^2 - 5$

5. $(-5x^2 + 3) + (2x^2 + 1)$ $-3x^2 + 4$

6. $(3x^2 + 7) + (3x^2 + 9)$ $6x^2 + 16$

▶ 7. $(-3y^2 - 4y) + (2y^2 + y - 1)$ $-y^2 - 3y - 1$

8. $(7x^2 + 2x - 9) + (-3x^2 + 5)$ $4x^2 + 2x - 4$

9. $(1.2x^3 - 3.4x + 7.9) + (6.7x^3 + 4.4x^2 - 10.9)$
$7.9x^3 + 4.4x^2 - 3.4x - 3$

10. $(9.6y^3 + 2.7y^2 - 8.6) + (1.1y^3 - 8.8y + 11.6)$
$10.7y^3 + 2.7y^2 - 8.8y + 3$

11. $\left(\frac{3}{4}m^2 - \frac{2}{5}m + \frac{1}{8}\right) + \left(-\frac{1}{4}m^2 - \frac{3}{10}m + \frac{11}{16}\right)$
$\frac{1}{2}m^2 - \frac{7}{10}m + \frac{13}{16}$

12. $\left(-\frac{4}{7}n^2 + \frac{5}{6}m - \frac{1}{20}\right) + \left(\frac{3}{7}n^2 - \frac{5}{12}m - \frac{3}{10}\right)$
$-\frac{1}{7}n^2 + \frac{5}{12}m - \frac{7}{20}$

Add using a vertical format. See Example 3.

13.
$$\begin{array}{r} 3t^2 + 4 \\ 5t^2 - 8 \\ \hline 8t^2 - 4 \end{array}$$

14.
$$\begin{array}{r} 7x^3 + 3 \\ 2x^3 - 7 \\ \hline 9x^3 - 4 \end{array}$$

15.
$$\begin{array}{r} 10a^3 - 8a^2 + 4a + 9 \\ 15a^3 + 9a^2 - 7a + 7 \\ \hline 15a^3 + a^2 - 3a + 16 \end{array}$$

16.
$$\begin{array}{r} 2x^3 - 3x^2 + x - 4 \\ 5x^3 + 2x^2 - 3x + 2 \\ \hline 7x^3 - x^2 - 2x - 2 \end{array}$$

Objective B *Subtract. See Examples 4 and 5.*

17. $(2x + 5) - (3x - 9)$
$-x + 14$

18. $(4 + 5a) - (-a - 5)$
$6a + 9$

19. $(5x^2 + 4) - (-2x^2 + 4)$
$7x^2$

20. $(-7y^2 + 5) - (-8y^2 + 12)$
$y^2 - 7$

21. $3x - (5x - 9)$
$-2x + 9$

22. $4 - (-y - 4)$
$y + 8$

23. $(2x^2 + 3x - 9) - (-4x + 7)$
$2x^2 + 7x - 16$

24. $(-7x^2 + 4x + 7) - (-8x + 2)$
$-7x^2 + 12x + 5$

▶ 25. $(5x + 8) - (-2x^2 - 6x + 8)$
$2x^2 + 11x$

26. $(-6y^2 + 3y - 4) - (9y^2 - 3y)$
$-15y^2 + 6y - 4$

27. $(0.7x^2 + 0.2x - 0.8) - (0.9x^2 + 1.4)$
$-0.2x^2 + 0.2x - 2.2$

28. $(-0.3y^2 + 0.6y - 0.3) - (0.5y^2 + 0.3)$
$-0.8y^2 + 0.6y - 0.6$

29. $\left(\frac{1}{4}z^2 - \frac{1}{5}z\right) - \left(-\frac{3}{20}z^2 + \frac{1}{10}z - \frac{7}{20}\right)$
$\frac{2}{5}z^2 - \frac{3}{10}z + \frac{7}{20}$

30. $\left(\frac{1}{3}x^2 - \frac{2}{7}x\right) - \left(\frac{4}{21}x^2 + \frac{1}{21}x - \frac{2}{3}\right)$
$\frac{1}{7}x^2 - \frac{1}{3}x + \frac{2}{3}$

Subtract using a vertical format. See Example 6.

31.
$$\begin{array}{r} 4z^2 - 8z + 3 \\ -(6z^2 + 8z - 3) \\ \hline -2z^2 - 16z + 6 \end{array}$$

32.
$$\begin{array}{r} 7a^2 - 9a + 6 \\ -(11a^2 - 4a + 2) \\ \hline -4a^2 - 5a + 4 \end{array}$$

33.
$$\begin{array}{r} 5u^5 - 4u^2 + 3u - 7 \\ -(3u^5 + 6u^2 - 8u + 2) \\ \hline 2u^5 - 10u^2 + 11u - 9 \end{array}$$

34.
$$\begin{array}{r} 5x^3 - 4x^2 + 6x - 2 \\ -(3x^3 - 2x^2 - x - 4) \\ \hline 2x^3 - 2x^2 + 7x + 2 \end{array}$$

Objectives A B C Mixed Practice *Add or subtract as indicated. See Examples 1 through 7.*

35. $(3x + 5) + (2x - 14)$
$5x - 9$

36. $(2y + 20) + (5y - 30)$
$7y - 10$

37. $(9x - 1) - (5x + 2)$
$4x - 3$

38. $(7y + 7) - (y - 6)$
$6y + 13$

39. $(14y + 12) + (-3y - 5)$
$11y + 7$

40. $(26y + 17) + (-20y - 10)$
$6y + 7$

41. $(x^2 + 2x + 1) - (3x^2 - 6x + 2)$
$-2x^2 + 8x - 1$

42. $(5y^2 - 3y - 1) - (2y^2 + y + 1)$
$3y^2 - 4y - 2$

43. $(3x^2 + 5x - 8) + (5x^2 + 9x + 12) - (8x^2 - 14)$
$14x + 18$

44. $(2x^2 + 7x - 9) + (x^2 - x + 10) - (3x^2 - 30)$
$6x + 31$

45. $(-a^2 + 1) - (a^2 - 3) + (5a^2 - 6a + 7)$
$3a^2 - 6a + 11$

46. $(-m^2 + 3) - (m^2 - 13) + (6m^2 - m + 1)$
$4m^2 - m + 17$

Translating *Perform each indicated operation. See Examples 3, 6, and 7.*

47. Subtract $4x$ from $(7x - 3)$.
$3x - 3$

48. Subtract y from $(y^2 - 4y + 1)$.
$y^2 - 5y + 1$

49. Add $(4x^2 - 6x + 1)$ and $(3x^2 + 2x + 1)$.
$7x^2 - 4x + 2$

50. Add $(-3x^2 - 5x + 2)$ and $(x^2 - 6x + 9)$.
$-2x^2 - 11x + 11$

▸ **51.** Subtract $(5x + 7)$ from $(7x^2 + 3x + 9)$.
$7x^2 - 2x + 2$

52. Subtract $(5y^2 + 8y + 2)$ from $(7y^2 + 9y - 8)$.
$2y^2 + y - 10$

53. Subtract $(4y^2 - 6y - 3)$ from the sum of $(8y^2 + 7)$ and $(6y + 9)$.
$4y^2 + 12y + 19$

54. Subtract $(4x^2 - 2x + 2)$ from the sum of $(x^2 + 7x + 1)$ and $(7x + 5)$. $-3x^2 + 16x + 4$

55. Subtract $(3x^2 - 4)$ from the sum of $(x^2 - 9x + 2)$ and $(2x^2 - 6x + 1)$. $-15x + 7$

56. Subtract $(y^2 - 9)$ from the sum of $(3y^2 + y + 4)$ and $(2y^2 - 6y - 10)$. $4y^2 - 5y + 3$

Objective D *Add or subtract as indicated. See Examples 8 and 9.*

57. $(9a + 6b - 5) + (-11a - 7b + 6)$
$-2a - b + 1$

58. $(3x - 2 + 6y) + (7x - 2 - y)$
$10x - 4 + 5y$

59. $(4x^2 + y^2 + 3) - (x^2 + y^2 - 2)$ $3x^2 + 5$

60. $(7a^2 - 3b^2 + 10) - (-2a^2 + b^2 - 12)$
$9a^2 - 4b^2 + 22$

61. $\left(x^2 + 2xy - y^2\right) + \left(5x^2 - 4xy + 20y^2\right)$
$6x^2 - 2xy + 19y^2$

62. $\left(a^2 - ab + 4b^2\right) + \left(6a^2 + 8ab - b^2\right)$
$7a^2 + 7ab + 3b^2$

63. $\left(11r^2s + 16rs - 3 - 2r^2s^2\right) - \left(3sr^2 + 5 - 9r^2s^2\right)$
$8r^2s + 16rs - 8 + 7r^2s^2$

64. $\left(3x^2y - 6xy + x^2y^2 - 5\right) - \left(11x^2y^2 - 1 + 5yx^2\right)$
$-2x^2y - 6xy - 10x^2y^2 - 4$

Objectives A B C Mixed Practice *For Exercises 65 through 68, find the perimeter of each figure.*

65.

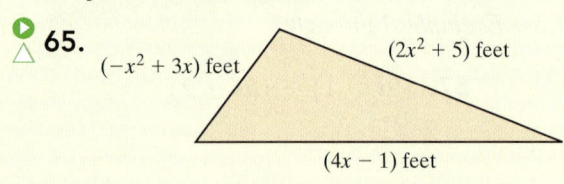

$(-x^2 + 3x)$ feet, $(2x^2 + 5)$ feet, $(4x - 1)$ feet

$\left(x^2 + 7x + 4\right)$ ft

66.

$(-x + 4)$ centimeters, $5x$ centimeters, x^2 centimeters, $(x^2 - 6x - 2)$ centimeters

$\left(2x^2 - 2x + 2\right)$ cm

67.

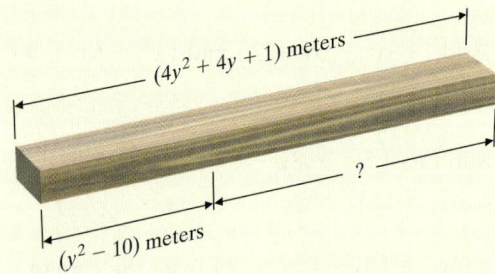

$2x - 3$, $\frac{4}{5}x$, $3x + 5$, $\frac{7}{10}x - 1$, $2x - 2$, $x + 4$

$\left(\frac{19}{2}x + 3\right)$ units

68.

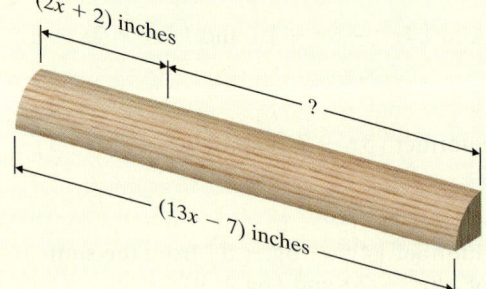

$3y - 4$, $2y - 3$, $3y + 1$, $\frac{3}{2}y$, $\frac{7}{4}y$, $\frac{3}{2}y$

$\left(\frac{51}{4}y - 6\right)$ units

69. A wooden beam is $(4y^2 + 4y + 1)$ meters long. If a piece $(y^2 - 10)$ meters is cut off, express the length of the remaining piece of beam as a polynomial in y. $\left(3y^2 + 4y + 11\right)$ m

70. A piece of quarter-round molding is $(13x - 7)$ inches long. If a piece $(2x + 2)$ inches long is removed, express the length of the remaining piece of molding as a polynomial in x. $(11x - 9)$ in.

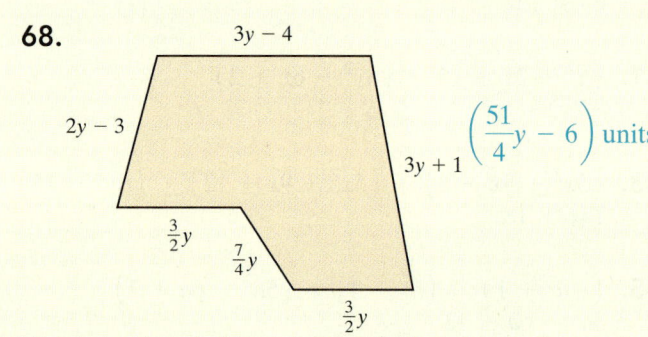

$(4y^2 + 4y + 1)$ meters, $(y^2 - 10)$ meters, ?

$(2x + 2)$ inches, ?, $(13x - 7)$ inches

Perform each indicated operation.

71. $\left[\left(1.2x^2 - 3x + 9.1\right) - \left(7.8x^2 - 3.1 + 8\right)\right] + (1.2x - 6)$ $-6.6x^2 - 1.8x - 1.8$

72. $\left[\left(7.9y^4 - 6.8y^3 + 3.3y\right) + \left(6.1y^3 - 5\right)\right] - \left(4.2y^4 + 1.1y - 1\right)$ $3.7y^4 - 0.7y^3 + 2.2y - 4$

Review

Multiply. See Section 5.1.

73. $3x(2x)$
$6x^2$

74. $-7x(x)$
$-7x^2$

75. $\left(12x^3\right)\left(-x^5\right)$
$-12x^8$

76. $6r^3\left(7r^{10}\right)$
$42r^{13}$

77. $10x^2\left(20xy^2\right)$
$200x^3y^2$

78. $-z^2y(11zy)$
$-11y^2z^3$

Concept Extensions

Fill in the squares so that each is a true statement.

79. $3x^\square + 4x^2 = 7x^\square$ 2; 2

80. $9y^7 + 3y^\square = 12y^7$ 7

81. $2x^\square + 3x^\square - 5x^\square + 4x^\square = 6x^4 - 2x^3$ 4; 3; 3; 4

82. $3y^\square + 7y^\square - 2y^\square - y^\square = 10y^5 - 3y^2$ 5; 5; 2; 2

Match each expression on the left with its simplification on the right. Not all letters on the right must be used and a letter may be used more than once.

83. $10y - 6y^2 - y$ b

84. $5x + 5x$ c

85. $(5x - 3) + (5x - 3)$ e

86. $(15x - 3) - (5x - 3)$ c

a. $3y$
b. $9y - 6y^2$
c. $10x$
d. $25x^2$
e. $10x - 6$
f. none of these

Simplify each expression by performing the indicated operation. Explain how you arrived at each answer. See the second Concept Check in this section.

87. a. $z + 3z$ $4z$
  b. $z \cdot 3z$ $3z^2$
  c. $-z - 3z$ $-4z$
  d. $(-z)(-3z)$ $3z^2$; answers may vary

88. a. $2y + y$ $3y$
  b. $2y \cdot y$ $2y^2$
  c. $-2y - y$ $-3y$
  d. $(-2y)(-y)$ $2y^2$; answers may vary

89. a. $m \cdot m \cdot m$ m^3
  b. $m + m + m$ $3m$
  c. $(-m)(-m)(-m)$ $-m^3$
  d. $-m - m - m$ $-3m$; answers may vary

90. a. $x + x$ $2x$
  b. $x \cdot x$ x^2
  c. $-x - x$ $-2x$
  d. $(-x)(-x)$ x^2; answers may vary

91. The polynomial $377x^2 - 720x + 1003$ represents the electricity generated (in thousand megawatthours) by solar sources in the United States during 2008–2012. The polynomial $538x^2 + 19,421x + 54,762$ represents the electricity generated (in thousand megawatthours) by wind power in the United States during 2008–2012. In both polynomials, x represents the number of years after 2008. Find a polynomial for the total electricity generated by both solar and wind power during 2008–2012. (*Source:* Based on information from the Energy Information Administration) $915x^2 + 18,701x + 55,765$

92. The polynomial $-0.4x^2 + 4.8x + 48.5$ represents the number of Americans (in millions) under age 65 covered by government health insurance during 2007–2012. The polynomial $0.8x^2 - 5.8x + 183.3$ represents the number of Americans (in millions) under age 65 covered by private health insurance during 2007–2012. In both polynomials, x represents the number of years since 2007. Find a polynomial for the total number of Americans (in millions) under age 65 with some form of health insurance during this period. (*Source:* Based on data from U.S. Census Bureau) $0.4x^2 - x + 231.8$

Objectives

A Multiply Monomials.

B Multiply a Monomial by a Polynomial.

C Multiply Two Polynomials.

D Multiply Polynomials Vertically.

Objective A Multiplying Monomials

Recall from Section 5.1 that to multiply two monomials such as $\left(-5x^3\right)$ and $\left(-2x^4\right)$, we use the associative and commutative properties and regroup. Remember also that to multiply exponential expressions with a common base, we use the product rule for exponents and add exponents.

$$\left(-5x^3\right)\left(-2x^4\right) = (-5)(-2)\left(x^3 \cdot x^4\right) \quad \text{Use the commutative and associative properties.}$$
$$= 10x^7 \qquad\qquad \text{Multiply.}$$

Practice 1–3

Multiply.
1. $10x \cdot 9x$
2. $8x^3\left(-11x^7\right)$
3. $\left(-5x^4\right)(-x)$

Examples Multiply.

1. $6x \cdot 4x = (6 \cdot 4)(x \cdot x)$ Use the commutative and associative properties.
$$= 24x^2 \qquad\qquad \text{Multiply.}$$

2. $-7x^2 \cdot 2x^5 = (-7 \cdot 2)\left(x^2 \cdot x^5\right)$
$$= -14x^7$$

3. $\left(-12x^5\right)(-x) = \left(-12x^5\right)(-1x)$
$$= (-12)(-1)\left(x^5 \cdot x\right)$$
$$= 12x^6$$

▪ **Work Practice 1–3**

Teaching Tip

Example 4 can be illustrated with an area diagram. Note that the result of the multiplication is written inside the rectangles.

$$x^2 + 2x + 5$$

	$7x^3$	$14x^2$	$35x$
$7x$			

✓**Concept Check** Simplify.
   a. $3x \cdot 2x$ **b.** $3x + 2x$

Objective B Multiplying Monomials by Polynomials

To multiply a monomial such as $7x$ by a trinomial such as $x^2 + 2x + 5$, we use the distributive property.

Practice 4–6

Multiply.
4. $4x\left(x^2 + 4x + 3\right)$
5. $8x\left(7x^4 + 1\right)$
6. $-2x^3\left(3x^2 - x + 2\right)$

Examples Multiply.

4. $7x\left(x^2 + 2x + 5\right) = 7x\left(x^2\right) + 7x(2x) + 7x(5)$ Apply the distributive property.
$$= 7x^3 + 14x^2 + 35x \qquad\qquad \text{Multiply.}$$

5. $5x\left(2x^3 + 6\right) = 5x\left(2x^3\right) + 5x(6)$ Apply the distributive property.
$$= 10x^4 + 30x \qquad\qquad \text{Multiply.}$$

6. $-3x^2\left(5x^2 + 6x - 1\right)$
$$= \left(-3x^2\right)\left(5x^2\right) + \left(-3x^2\right)(6x) + \left(-3x^2\right)(-1) \quad \text{Apply the distributive property.}$$
$$= -15x^4 - 18x^3 + 3x^2 \qquad\qquad\qquad \text{Multiply.}$$

▪ **Work Practice 4–6**

Answers
1. $90x^2$ **2.** $-88x^{10}$ **3.** $5x^5$
4. $4x^3 + 16x^2 + 12x$ **5.** $56x^5 + 8x$
6. $-6x^5 + 2x^4 - 4x^3$

✓**Concept Check Answers**
a. $6x^2$ **b.** $5x$

Objective C Multiplying Two Polynomials

We also use the distributive property to multiply two binomials.

Example 7   Multiply.

a. $(m + 4)(m + 6)$ **b.** $(3x + 2)(2x - 5)$

Solution:

a. $(m + 4)(m + 6) = m(m + 6) + 4(m + 6)$ Use the distributive property.

$= m \cdot m + m \cdot 6 + 4 \cdot m + 4 \cdot 6$ Use the distributive property.

$= m^2 + 6m + 4m + 24$ Multiply.

$= m^2 + 10m + 24$ Combine like terms.

b. $(3x + 2)(2x - 5) = 3x(2x - 5) + 2(2x - 5)$ Use the distributive property.

$= 3x(2x) + 3x(-5) + 2(2x) + 2(-5)$

$= 6x^2 - 15x + 4x - 10$ Multiply.

$= 6x^2 - 11x - 10$ Combine like terms.

🟧 **Work Practice 7**

This idea can be expanded so that we can multiply any two polynomials.

> **To Multiply Two Polynomials**
>
> Multiply each term of the first polynomial by each term of the second polynomial, and then combine like terms.

Examples   Multiply.

8. $(2x - y)^2$

$= (2x - y)(2x - y)$ Using the meaning of an exponent, we have 2 factors of $(2x - y)$.

$= 2x(2x) + 2x(-y) + (-y)(2x) + (-y)(-y)$

$= 4x^2 - 2xy - 2xy + y^2$ Multiply.

$= 4x^2 - 4xy + y^2$ Combine like terms.

9. $(t + 2)(3t^2 - 4t + 2)$

$= t(3t^2) + t(-4t) + t(2) + 2(3t^2) + 2(-4t) + 2(2)$

$= 3t^3 - 4t^2 + 2t + 6t^2 - 8t + 4$

$= 3t^3 + 2t^2 - 6t + 4$ Combine like terms.

🟧 **Work Practice 8–9**

✓**Concept Check** Square where indicated. Simplify if possible.

    a. $(4a)^2 + (3b)^2$ **b.** $(4a + 3b)^2$

Objective D Multiplying Polynomials Vertically ▶

Another convenient method for multiplying polynomials is to multiply vertically, similar to the way we multiply real numbers. This method is shown in the next examples.

Practice 7

Multiply:

a. $(x + 5)(x + 10)$

b. $(4x + 5)(3x - 4)$

Teaching Tip

Example 7(b) can be illustrated with an area diagram.

	$3x$	$+ \; 2$
$2x$	$6x^2$	$4x$
-5	$-15x$	-10

Practice 8–9

Multiply.

8. $(3x - 2y)^2$

9. $(x + 3)(2x^2 - 5x + 4)$

Teaching Tip

Before doing Example 10, you may wish to review vertical multiplication using the following example:

$$\begin{array}{r} 134 \\ \times 25 \\ \hline 670 \\ 2680 \\ \hline 3350 \end{array}$$

After doing Example 10, have students compare and contrast the numerical example with multiplying polynomials vertically.

Answers

7. a. $x^2 + 15x + 50$

b. $12x^2 - x - 20$

8. $9x^2 - 12xy + 4y^2$

9. $2x^3 + x^2 - 11x + 12$

✓**Concept Check Answers**

a. $16a^2 + 9b^2$ **b.** $16a^2 + 24ab + 9b^2$

Practice 10

Multiply vertically:

$(3y^2 + 1)(y^2 - 4y + 5)$

Example 10   Multiply vertically: $(2y^2 + 5)(y^2 - 3y + 4)$

Solution:

$$
\begin{array}{r}
y^2 - 3y + 4 \\
2y^2 + 5 \\
\hline
5y^2 - 15y + 20 \\
2y^4 - 6y^3 + 8y^2 \\
\hline
2y^4 - 6y^3 + 13y^2 - 15y + 20
\end{array}
$$

Multiply $y^2 - 3y + 4$ by 5.
Multiply $y^2 - 3y + 4$ by $2y^2$.
Combine like terms.

🟧 **Work Practice 10**

Practice 11

Find the product of $(4x^2 - x - 1)$ and $(3x^2 + 6x - 2)$ using a vertical format.

Example 11   Find the product of $(2x^2 - 3x + 4)$ and $(x^2 + 5x - 2)$ using a vertical format.

Solution: First, we arrange the polynomials in a vertical format. Then we multiply each term of the second polynomial by each term of the first polynomial.

$$
\begin{array}{r}
2x^2 - 3x + 4 \\
x^2 + 5x - 2 \\
\hline
-4x^2 + 6x - 8 \\
10x^3 - 15x^2 + 20x \\
2x^4 - 3x^3 + 4x^2 \\
\hline
2x^4 + 7x^3 - 15x^2 + 26x - 8
\end{array}
$$

Multiply $2x^2 - 3x + 4$ by -2.
Multiply $2x^2 - 3x + 4$ by $5x$.
Multiply $2x^2 - 3x + 4$ by x^2.
Combine like terms.

🟧 **Work Practice 11**

Answers

10. $3y^4 - 12y^3 + 16y^2 - 4y + 5$
11. $12x^4 + 21x^3 - 17x^2 - 4x + 2$

Vocabulary, Readiness & Video Check

Fill in each blank with the correct choice.

1. The expression $5x(3x + 2)$ equals $5x \cdot 3x + 5x \cdot 2$ by the _____distributive_____ property.
   a. commutative   **b.** associative   **c.** distributive

2. The expression $(x + 4)(7x - 1)$ equals $x(7x - 1) + 4(7x - 1)$ by the _____distributive_____ property.
   a. commutative   **b.** associative   **c.** distributive

3. The expression $(5y - 1)^2$ equals __$(5y - 1)(5y - 1)$__ .
   a. $2(5y - 1)$ **b.** $(5y - 1)(5y + 1)$ **c.** $(5y - 1)(5y - 1)$

4. The expression $9x \cdot 3x$ equals ____$27x^2$____ .
   a. $27x$ **b.** $27x^2$ **c.** $12x$ **d.** $12x^2$

Perform the indicated operation, if possible.

5. $x^3 \cdot x^5$
   x^8

6. $x^2 \cdot x^6$
   x^8

7. $x^3 + x^5$
   cannot simplify

8. $x^2 + x^6$
   cannot simplify

9. $x^7 \cdot x^7$
   x^{14}

10. $x^{11} \cdot x^{11}$
    x^{22}

11. $x^7 + x^7$
    $2x^7$

12. $x^{11} + x^{11}$
    $2x^{11}$

Martin-Gay Interactive Videos   Watch the section lecture video and answer the following questions.

See Video 5.5 🍎

See video answer section.

Objective A **13.** For 📺 Example 1, we use the product property to multiply two monomials. Is it possible to add the same two monomials? Why or why not? ▶

Objective B **14.** What property and what exponent rule are used in 📺 Examples 3 and 4? ▶

Objective C **15.** In 📺 Example 5, how many times is the distributive property actually applied? Explain. ▶

Objective D **16.** Would you say the vertical format shown in 📺 Example 8 also makes use of the distributive property? Explain. ▶

5.5 Exercise Set MyMathLab®

Objective A *Multiply. See Examples 1 through 3.*

1. $8x^2 \cdot 3x$
$24x^3$

2. $6x \cdot 3x^2$
$18x^3$

3. $(-x^3)(-x)$
x^4

4. $(-x^6)(-x)$
x^7

5. $-4n^3 \cdot 7n^7$
$-28n^{10}$

6. $9t^6(-3t^5)$
$-27t^{11}$

7. $(-3.1x^3)(4x^9)$
$-12.4x^{12}$

8. $(-5.2x^4)(3x^4)$
$-15.6x^8$

9. $\left(-\dfrac{1}{3}y^2\right)\left(\dfrac{2}{5}y\right)$
$-\dfrac{2}{15}y^3$

10. $\left(-\dfrac{3}{4}y^7\right)\left(\dfrac{1}{7}y^4\right)$
$-\dfrac{3}{28}y^{11}$

11. $(2x)(-3x^2)(4x^5)$
$-24x^8$

12. $(x)(5x^4)(-6x^7)$
$-30x^{12}$

Objective B *Multiply. See Examples 4 through 6.*

13. $3x(2x+5)$
$6x^2+15x$

14. $2x(6x+3)$
$12x^2+6x$

15. $7x(x^2+2x-1)$
$7x^3+14x^2-7x$

16. $5y(y^2+y-10)$
$5y^3+5y^2-50y$

17. $-2a(a+4)$
$-2a^2-8a$

18. $-3a(2a+7)$
$-6a^2-21a$

19. $3x(2x^2-3x+4)$
$6x^3-9x^2+12x$

20. $4x(5x^2-6x-10)$
$20x^3-24x^2-40x$

21. $3a^2(4a^3+15)$
$12a^5+45a^2$

22. $9x^3(5x^2+12)$
$45x^5+108x^3$

23. $-2a^2(3a^2-2a+3)$
$-6a^4+4a^3-6a^2$

24. $-4b^2(3b^3-12b^2-6)$
$-12b^5+48b^4+24b^2$

25. $3x^2y(2x^3-x^2y^2+8y^3)$ $6x^5y-3x^4y^3+24x^2y^4$

26. $4xy^2(7x^3+3x^2y^2-9y^3)$ $28x^4y^2+12x^3y^4-36xy^5$

27. $-y(4x^3-7x^2y+xy^2+3y^3)$
$-4x^3y+7x^2y^2-xy^3-3y^4$

28. $-x(6y^3-5xy^2+x^2y-5x^3)$
$-6xy^3+5x^2y^2-x^3y+5x^4$

29. $\dfrac{1}{2}x^2(8x^2-6x+1)$ $4x^4-3x^3+\dfrac{1}{2}x^2$

30. $\dfrac{1}{3}y^2(9y^2-6y+1)$ $3y^4-2y^3+\dfrac{1}{3}y^2$

Objective C *Multiply. See Examples 7 through 9.*

31. $(x + 4)(x + 3)$
$x^2 + 7x + 12$

32. $(x + 2)(x + 9)$
$x^2 + 11x + 18$

33. $(a + 7)(a - 2)$
$a^2 + 5a - 14$

34. $(y - 10)(y + 11)$
$y^2 + y - 110$

35. $\left(x + \dfrac{2}{3}\right)\left(x - \dfrac{1}{3}\right)$
$x^2 + \dfrac{1}{3}x - \dfrac{2}{9}$

36. $\left(x + \dfrac{3}{5}\right)\left(x - \dfrac{2}{5}\right)$
$x^2 + \dfrac{1}{5}x - \dfrac{6}{25}$

37. $(3x^2 + 1)(4x^2 + 7)$
$12x^4 + 25x^2 + 7$

38. $(5x^2 + 2)(6x^2 + 2)$
$30x^4 + 22x^2 + 4$

39. $(4x - 3)(3x - 5)$
$12x^2 - 29x + 15$

40. $(8x - 3)(2x - 4)$
$16x^2 - 38x + 12$

41. $(1 - 3a)(1 - 4a)$
$1 - 7a + 12a^2$

42. $(3 - 2a)(2 - a)$
$6 - 7a + 2a^2$

43. $(2y - 4)^2$
$4y^2 - 16y + 16$

44. $(6x - 7)^2$
$36x^2 - 84x + 49$

45. $(x - 2)(x^2 - 3x + 7)$
$x^3 - 5x^2 + 13x - 14$

46. $(x + 3)(x^2 + 5x - 8)$
$x^3 + 8x^2 + 7x - 24$

47. $(x + 5)(x^3 - 3x + 4)$
$x^4 + 5x^3 - 3x^2 - 11x + 20$

48. $(a + 2)(a^3 - 3a^2 + 7)$
$a^4 - a^3 - 6a^2 + 7a + 14$

49. $(2a - 3)(5a^2 - 6a + 4)$
$10a^3 - 27a^2 + 26a - 12$

50. $(3 + b)(2 - 5b - 3b^2)$
$-3b^3 - 14b^2 - 13b + 6$

51. $(7xy - y)^2$
$49x^2y^2 - 14xy^2 + y^2$

52. $(x^2 - 4)^2$
$x^4 - 8x^2 + 16$

Objective D *Multiply vertically. See Examples 10 and 11.*

53. $(2x - 11)(6x + 1)$
$12x^2 - 64x - 11$

54. $(4x - 7)(5x + 1)$
$20x^2 - 31x - 7$

55. $(x + 3)(2x^2 + 4x - 1)$
$2x^3 + 10x^2 + 11x - 3$

56. $(4x - 5)(8x^2 + 2x - 4)$
$32x^3 - 32x^2 - 26x + 20$

57. $(x^2 + 5x - 7)(2x^2 - 7x - 9)$
$2x^4 + 3x^3 - 58x^2 + 4x + 63$

58. $(3x^2 - x + 2)(x^2 + 2x + 1)$
$3x^4 + 5x^3 + 3x^2 + 3x + 2$

Objectives A B C D **Mixed Practice** *Multiply. See Examples 1 through 11.*

59. $-1.2y(-7y^6)$
$8.4y^7$

60. $-4.2x(-2x^5)$
$8.4x^6$

61. $-3x(x^2 + 2x - 8)$
$-3x^3 - 6x^2 + 24x$

62. $-5x(x^2 - 3x + 10)$
$-5x^3 + 15x^2 - 50x$

63. $(x + 19)(2x + 1)$
$2x^2 + 39x + 19$

64. $(3y + 4)(y + 11)$
$3y^2 + 37y + 44$

65. $\left(x + \dfrac{1}{7}\right)\left(x - \dfrac{3}{7}\right)$
$x^2 - \dfrac{2}{7}x - \dfrac{3}{49}$

66. $\left(m + \dfrac{2}{9}\right)\left(m - \dfrac{1}{9}\right)$
$m^2 + \dfrac{1}{9}m - \dfrac{2}{81}$

67. $(3y + 5)^2$
$9y^2 + 30y + 25$

68. $(7y + 2)^2$
$49y^2 + 28y + 4$

69. $(a + 4)(a^2 - 6a + 6)$
$a^3 - 2a^2 - 18a + 24$

70. $(t + 3)(t^2 - 5t + 5)$
$t^3 - 2t^2 - 10t + 15$

Express as the product of polynomials. Then multiply.

71. Find the area of the rectangle.
$(4x^2 - 25)$ sq yd

(2x + 5) yards
(2x − 5) yards

72. Find the area of the square field.
$(x^2 + 8x + 16)$ sq ft

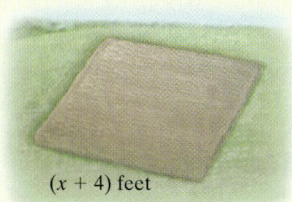

(x + 4) feet

73. Find the area of the triangle.
$(6x^2 - 4x)$ sq in.

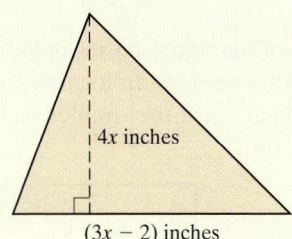

4x inches

(3x − 2) inches

△ **74.** Find the volume of the cube-shaped glass block.
$(y^3 - 3y^2 + 3y - 1)$ cu m

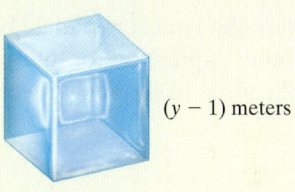

(y − 1) meters

Review

In this section, we review operations on monomials. Study the box below, then proceed. See Sections 1.8, 5.1, and 5.2.

Operations on Monomials	
Multiply	Review the product rule for exponents.
Divide	Review the quotient rule for exponents.
Add or Subtract	Remember, we may only combine like terms.

Perform the operations on the monomials, if possible. The first two rows have been completed for you.

	Monomials	Add	Subtract	Multiply	Divide
	$6x, 3x$	$6x + 3x = 9x$	$6x - 3x = 3x$	$6x \cdot 3x = 18x^2$	$\dfrac{6x}{3x} = 2$
	$-12x^2, 2x$	$-12x^2 + 2x$; can't be simplified	$-12x^2 - 2x$; can't be simplified	$-12x^2 \cdot 2x = -24x^3$	$\dfrac{-12x^2}{2x} = -6x$
75.	$5a, 15a$	$5a + 15a = 20a$	$5a - 15a = -10a$	$5a \cdot 15a = 75a^2$	$\dfrac{5a}{15a} = \dfrac{1}{3}$
76.	$4y^3, 4y^7$	$4y^3 + 4y^7$; can't be simplified	$4y^3 - 4y^7$; can't be simplified	$4y^3 \cdot 4y^7 = 16y^{10}$	$\dfrac{4y^3}{4y^7} = \dfrac{1}{y^4}$
77.	$-3y^5, 9y^4$	$-3y^5 + 9y^4$; can't be simplified	$-3y^5 - 9y^4$; can't be simplified	$-3y^5 \cdot 9y^4 = -27y^9$	$\dfrac{-3y^5}{9y^4} = -\dfrac{y}{3}$
78.	$-14x^2, 2x^2$	$-14x^2 + 2x^2 = -12x^2$	$-14x^2 - 2x^2 = -16x^2$	$-14x^2 \cdot 2x^2 = -28x^4$	$\dfrac{-14x^2}{2x^2} = -7$

Concept Extensions

79. Perform each indicated operation. Explain the difference between the two expressions.
a. $(3x + 5) + (3x + 7)$ $6x + 12$
b. $(3x + 5)(3x + 7)$
$9x^2 + 36x + 35$; answers may vary

80. Perform each indicated operation. Explain the difference between the two expressions.
a. $(8x - 3) - (5x - 2)$ $3x - 1$
b. $(8x - 3)(5x - 2)$
$40x^2 - 31x + 6$; answers may vary

Mixed Practice *Perform the indicated operations. See Sections 5.4 and 5.5.*

81. $(3x - 1) + (10x - 6)$
$13x - 7$

82. $(2x - 1) + (10x - 7)$
$12x - 8$

83. $(3x - 1)(10x - 6)$
$30x^2 - 28x + 6$

84. $(2x - 1)(10x - 7)$
$20x^2 - 24x + 7$

85. $(3x - 1) - (10x - 6)$
$-7x + 5$

86. $(2x - 1) - (10x - 7)$
$-8x + 6$

△87. The area of the largest rectangle below is $x(x + 3)$. Find another expression for this area by finding the sum of the areas of the smaller rectangles.
$x^2 + 3x$

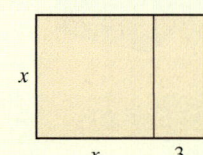

△88. The area of the figure below is $(x + 2)(x + 3)$. Find another expression for this area by finding the sum of the areas of the smaller rectangles.
$x^2 + 5x + 6$

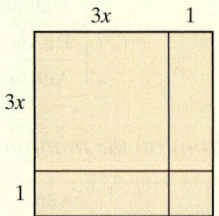

89. Write an expression for the area of the largest rectangle below in two different ways.
$x + 2x^2$; $x(1 + 2x)$

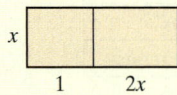

90. Write an expression for the area of the figure below in two different ways.
$(3x + 1)(3x + 1)$; $9x^2 + 6x + 1$

Simplify. See the Concept Checks in this section.

91. $5a + 6a$
$11a$

92. $5a \cdot 6a$
$30a^2$

Square where indicated. Simplify if possible.

93. $(5x)^2 + (2y)^2$
$25x^2 + 4y^2$

94. $(5x + 2y)^2$
$25x^2 + 20xy + 4y^2$

95. Multiply each of the following polynomials.
a. $(a + b)(a - b)$ $a^2 - b^2$
b. $(2x + 3y)(2x - 3y)$ $4x^2 - 9y^2$
c. $(4x + 7)(4x - 7)$ $16x^2 - 49$
d. Can you make a general statement about all products of the form $(x + y)(x - y)$?
answers may vary

96. Evaluate each of the following.
a. $(2 + 3)^2$; $2^2 + 3^2$ $25; 13$
b. $(8 + 10)^2$; $8^2 + 10^2$ $324; 164$
c. Does $(a + b)^2 = a^2 + b^2$ no matter what the values of a and b are? Why or why not?
no; answers may vary

5.6 Special Products

Objective A Using the FOIL Method

In this section, we multiply binomials using special products. First, we introduce a special order for multiplying binomials called the FOIL order or method. This order, or pattern, is a result of the distributive property. We demonstrate by multiplying $(3x + 1)$ by $(2x + 5)$.

Objectives

A Multiply Two Binomials Using the FOIL Method.

B Square a Binomial.

C Multiply the Sum and Difference of Two Terms.

D Use Special Products to Multiply Binomials.

The FOIL Method

F stands for the product of the **First** terms.

$$(3x + 1)(2x + 5)$$
$$(3x)(2x) = 6x^2 \qquad \text{F}$$

O stands for the product of the **Outer** terms.

$$(3x + 1)(2x + 5)$$
$$(3x)(5) = 15x \qquad \text{O}$$

I stands for the product of the **Inner** terms.

$$(3x + 1)(2x + 5)$$
$$(1)(2x) = 2x \qquad \text{I}$$

L stands for the product of the **Last** terms.

$$(3x + 1)(2x + 5)$$
$$(1)(5) = 5 \qquad \text{L}$$

$$
\begin{array}{cccc}
\text{F} & \text{O} & \text{I} & \text{L}
\end{array}
$$
$$(3x + 1)(2x + 5) = 6x^2 + 15x + 2x + 5$$
$$= 6x^2 + 17x + 5 \qquad \text{Combine like terms.}$$

Let's practice multiplying binomials using the FOIL method.

Teaching Tip

Point out that the special products in this section are shortcuts for multiplying binomials. They can all be worked out by the method in the previous section in which each term in the first binomial is multiplied by every term in the second binomial.

Teaching Tip

After doing Example 1, it may be helpful to point out to students that just as 6 and 4 are factors of 24 because $6 \cdot 4 = 24$, $x - 3$ and $x + 4$ are factors of $x^2 + x - 12$ because $(x - 3)(x + 4) = x^2 + x - 12$.

Example 1 Multiply: $(x - 3)(x + 4)$

Solution:

$$(x - 3)(x + 4) = (x)(x) + (x)(4) + (-3)(x) + (-3)(4)$$
$$= x^2 + 4x - 3x - 12$$
$$= x^2 + x - 12 \qquad \text{Combine like terms.}$$

■ **Work Practice 1**

Example 2 Multiply: $(5x - 7)(x - 2)$

Solution:

$$(5x - 7)(x - 2) = 5x(x) + 5x(-2) + (-7)(x) + (-7)(-2)$$
$$= 5x^2 - 10x - 7x + 14$$
$$= 5x^2 - 17x + 14 \qquad \text{Combine like terms.}$$

■ **Work Practice 2**

Practice 1

Multiply: $(x + 7)(x - 5)$

Helpful Hint Remember that the FOIL order for multiplying can be used only for the product of 2 binomials.

Practice 2

Multiply: $(6x - 1)(x - 4)$

Answers
1. $x^2 + 2x - 35$ **2.** $6x^2 - 25x + 4$

Practice 3

Multiply: $(2y^2 + 3)(y - 4)$

Teaching Tip

Consider having students discover patterns for squaring binomials themselves. Before doing Example 4, you may want to have students multiply $(4x + 3)^2$. Then have them multiply $(4x - 3)^2$. Ask if they notice any relationship between the problem and its solution.

Practice 4

Multiply: $(2x + 9)^2$

Example 3   Multiply: $(y^2 + 6)(2y - 1)$

Solution:

$$\overset{\text{F}\qquad\text{O}\qquad\text{I}\qquad\text{L}}{(y^2 + 6)(2y - 1) = 2y^3 - 1y^2 + 12y - 6}$$

Notice in this example that there are no like terms that can be combined, so the product is $2y^3 - y^2 + 12y - 6$.

■ **Work Practice 3**

Objective B Squaring Binomials ▶

An expression such as $(3y + 1)^2$ is called the square of a binomial. Since $(3y + 1)^2 = (3y + 1)(3y + 1)$, we can use the FOIL method to find this product.

Example 4   Multiply: $(3y + 1)^2$

Solution: $(3y + 1)^2 = (3y + 1)(3y + 1)$

$$\overset{\text{F}\qquad\quad\text{O}\qquad\quad\text{I}\qquad\quad\text{L}}{= (3y)(3y) + (3y)(1) + 1(3y) + 1(1)}$$
$$= 9y^2 + 3y + 3y + 1$$
$$= 9y^2 + 6y + 1$$

■ **Work Practice 4**

Notice the pattern that appears in Example 4.

$(3y + 1)^2 = 9y^2 + 6y + 1$

→ $9y^2$ is the first term of the binomial squared: $(3y)^2 = 9y^2$.

→ $6y$ is 2 times the product of both terms of the binomial: $(2)(3y)(1) = 6y$.

→ 1 is the second term of the binomial squared: $(1)^2 = 1$.

This pattern leads to the formulas below, which can be used when squaring a binomial. We call these **special products.**

Squaring a Binomial

A binomial squared is equal to the square of the first term plus or minus twice the product of both terms plus the square of the second term.

$$(a + b)^2 = a^2 + 2ab + b^2$$
$$(a - b)^2 = a^2 - 2ab + b^2$$

Answers

3. $2y^3 - 8y^2 + 3y - 12$

4. $4x^2 + 36x + 81$

This product can be visualized geometrically.

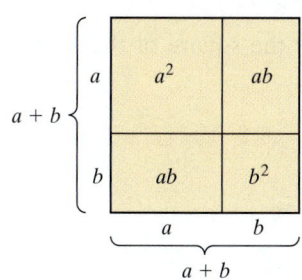

The area of the large square is side · side.
Area $= (a + b)(a + b) = (a + b)^2$
The area of the large square is also the sum of the areas of the smaller rectangles.
Area $= a^2 + ab + ab + b^2 = a^2 + 2ab + b^2$
Thus, $(a + b)^2 = a^2 + 2ab + b^2$.

Examples Use a special product to square each binomial.

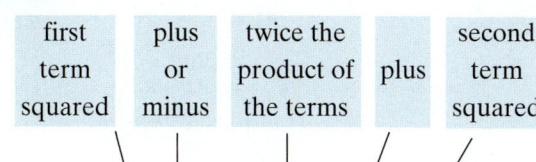

| first term squared | plus or minus | twice the product of the terms | plus | second term squared |

5. $(t + 2)^2 = t^2 + 2(t)(2) + 2^2 = t^2 + 4t + 4$
6. $(p - q)^2 = p^2 - 2(p)(q) + q^2 = p^2 - 2pq + q^2$
7. $(2x + 5)^2 = (2x)^2 + 2(2x)(5) + 5^2 = 4x^2 + 20x + 25$
8. $(x^2 - 7y)^2 = (x^2)^2 - 2(x^2)(7y) + (7y)^2 = x^4 - 14x^2y + 49y^2$

🟧 **Work Practice 5–8**

Helpful Hint

Notice that

$(a + b)^2 \neq a^2 + b^2$ The middle term, $2ab$, is missing.
$(a + b)^2 = (a + b)(a + b) = a^2 + 2ab + b^2$

Likewise,

$(a - b)^2 \neq a^2 - b^2$
$(a - b)^2 = (a - b)(a - b) = a^2 - 2ab + b^2$

Objective C Multiplying the Sum and Difference of Two Terms ▶

Another special product is the product of the sum and difference of the same two terms, such as $(x + y)(x - y)$. Finding this product by the FOIL method, we see a pattern emerge.

$$(x + y)(x - y) = x^2 - xy + xy - y^2$$
$$= x^2 - y^2$$

Notice that the two middle terms subtract out. This is because the **O**uter product is the opposite of the **I**nner product. Only the **difference of squares** remains.

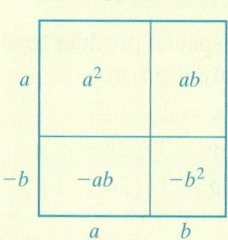

Multiplying the Sum and Difference of Two Terms

The product of the sum and difference of two terms is the square of the first term minus the square of the second term.

$$(a + b)(a - b) = a^2 - b^2$$

Practice 9–13

Use a special product to multiply.

9. $(x + 9)(x - 9)$

10. $(5 + 4y)(5 - 4y)$

11. $\left(x - \dfrac{1}{3}\right)\left(x + \dfrac{1}{3}\right)$

12. $(3a - b)(3a + b)$

13. $\left(2x^2 - 6y\right)\left(2x^2 + 6y\right)$

Examples   Use a special product to multiply.

$$\underset{\downarrow}{\text{first term squared}} \quad \underset{\downarrow}{\text{minus}} \quad \underset{\downarrow}{\text{second term squared}}$$

9. $(x + 4)(x - 4) = x^2 \quad - \quad 4^2 = x^2 - 16$

10. $(6t + 7)(6t - 7) = (6t)^2 \quad - \quad 7^2 = 36t^2 - 49$

11. $\left(x - \dfrac{1}{4}\right)\left(x + \dfrac{1}{4}\right) = x^2 \quad - \quad \left(\dfrac{1}{4}\right)^2 = x^2 - \dfrac{1}{16}$

12. $(2p - q)(2p + q) = (2p)^2 - q^2 = 4p^2 - q^2$

13. $\left(3x^2 - 5y\right)\left(3x^2 + 5y\right) = \left(3x^2\right)^2 - (5y)^2 = 9x^4 - 25y^2$

■ **Work Practice 9–13**

Teaching Tip

You may want to check that students know the differences among the following phrases by having them write an example of each.

• The sum of squares
• The sum squared
• The difference squared
• The difference of squares
• The product of the sum and difference of two terms

✓**Concept Check** Match each expression on the left to the equivalent expression or expressions in the list on the right.

$(a + b)^2$

$(a + b)(a - b)$

    a. $(a + b)(a + b)$

    b. $a^2 - b^2$

    c. $a^2 + b^2$

    d. $a^2 - 2ab + b^2$

    e. $a^2 + 2ab + b^2$

Practice 14–17

Use a special product to multiply, if possible.

14. $(7x - 1)^2$

15. $(5y + 3)(2y - 5)$

16. $(2a - 1)(2a + 1)$

17. $\left(5y - \dfrac{1}{9}\right)^2$

Objective D Using Special Products

Let's now practice using our special products on a variety of multiplication problems. This practice will help us recognize when to apply what special product formula.

Answers

9. $x^2 - 81$ 10. $25 - 16y^2$ 11. $x^2 - \dfrac{1}{9}$

12. $9a^2 - b^2$ 13. $4x^4 - 36y^2$

14. $49x^2 - 14x + 1$

15. $10y^2 - 19y - 15$

16. $4a^2 - 1$

17. $25y^2 - \dfrac{10}{9}y + \dfrac{1}{81}$

✓**Concept Check Answers**

a and **e, b**

Examples   Use a special product to multiply, if possible.

14. $(4x - 9)(4x + 9)$ This is the sum and difference of the same two terms.

$$= (4x)^2 - 9^2 = 16x^2 - 81$$

15. $(3y + 2)^2$ This is a binomial squared.

$$= (3y)^2 + 2(3y)(2) + 2^2$$
$$= 9y^2 + 12y + 4$$

This product can be visualized geometrically.

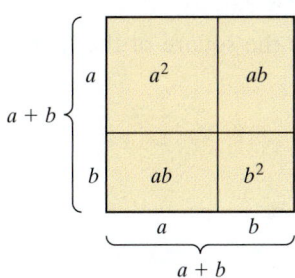

The area of the large square is side $\cdot$ side.
Area $= (a + b)(a + b) = (a + b)^2$
The area of the large square is also the sum of the areas of the smaller rectangles.
Area $= a^2 + ab + ab + b^2 = a^2 + 2ab + b^2$
Thus, $(a + b)^2 = a^2 + 2ab + b^2$.

Examples   Use a special product to square each binomial.

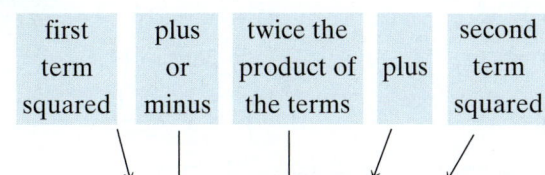

first term squared	plus or minus	twice the product of the terms	plus	second term squared

5. $(t + 2)^2 = \quad t^2 \; + \quad 2(t)(2) \; + \quad 2^2 = t^2 + 4t + 4$

6. $(p - q)^2 = \quad p^2 \; - \quad 2(p)(q) \; + \quad q^2 = p^2 - 2pq + q^2$

7. $(2x + 5)^2 = (2x)^2 \, + \quad 2(2x)(5) \; + \quad 5^2 = 4x^2 + 20x + 25$

8. $(x^2 - 7y)^2 = (x^2)^2 \, - \quad 2(x^2)(7y) \; + \quad (7y)^2 = x^4 - 14x^2y + 49y^2$

■ **Work Practice 5–8**

Helpful Hint

Notice that

$(a + b)^2 \neq a^2 + b^2$ The middle term, $2ab$, is missing.
$(a + b)^2 = (a + b)(a + b) = a^2 + 2ab + b^2$

Likewise,

$(a - b)^2 \neq a^2 - b^2$
$(a - b)^2 = (a - b)(a - b) = a^2 - 2ab + b^2$

Objective C   Multiplying the Sum and Difference of Two Terms ▶

Another special product is the product of the sum and difference of the same two terms, such as $(x + y)(x - y)$. Finding this product by the FOIL method, we see a pattern emerge.

$$(x + y)(x - y) = x^2 - xy + xy - y^2$$
$$= x^2 - y^2$$

Notice that the two middle terms subtract out. This is because the **O**uter product is the opposite of the **I**nner product. Only the **difference of squares** remains.

Practice 5–8

Use a special product to square each binomial.

5. $(y + 3)^2$
6. $(r - s)^2$
7. $(6x + 5)^2$
8. $(x^2 - 3y)^2$

Teaching Tip

It may be helpful for students to draw a square diagram of $(a + b)(a - b)$.

$(a + b)(a - b)$
$= a^2 + ab - ab - b^2$
$= a^2 - b^2$

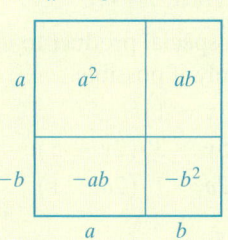

Teaching Tip

Point out to students that the ability to recognize the patterns they discovered in this section will help them factor binomials in the next chapter.

Answers

5. $y^2 + 6y + 9$
6. $r^2 - 2rs + s^2$
7. $36x^2 + 60x + 25$
8. $x^4 - 6x^2y + 9y^2$

Multiplying the Sum and Difference of Two Terms

The product of the sum and difference of two terms is the square of the first term minus the square of the second term.

$$(a + b)(a - b) = a^2 - b^2$$

Practice 9–13

Use a special product to multiply.

9. $(x + 9)(x - 9)$

10. $(5 + 4y)(5 - 4y)$

11. $\left(x - \dfrac{1}{3}\right)\left(x + \dfrac{1}{3}\right)$

12. $(3a - b)(3a + b)$

13. $\left(2x^2 - 6y\right)\left(2x^2 + 6y\right)$

Examples Use a special product to multiply.

	first term squared	minus	second term squared
	↓	↓	↓

9. $(x + 4)(x - 4) = x^2 \quad - \quad 4^2 = x^2 - 16$

10. $(6t + 7)(6t - 7) = (6t)^2 \quad - \quad 7^2 = 36t^2 - 49$

11. $\left(x - \dfrac{1}{4}\right)\left(x + \dfrac{1}{4}\right) = x^2 \quad - \quad \left(\dfrac{1}{4}\right)^2 = x^2 - \dfrac{1}{16}$

12. $(2p - q)(2p + q) = (2p)^2 - q^2 = 4p^2 - q^2$

13. $\left(3x^2 - 5y\right)\left(3x^2 + 5y\right) = \left(3x^2\right)^2 - (5y)^2 = 9x^4 - 25y^2$

■ **Work Practice 9–13**

Teaching Tip

You may want to check that students know the differences among the following phrases by having them write an example of each.

- The sum of squares
- The sum squared
- The difference squared
- The difference of squares
- The product of the sum and difference of two terms

✓**Concept Check** Match each expression on the left to the equivalent expression or expressions in the list on the right.

$(a + b)^2$

$(a + b)(a - b)$

a. $(a + b)(a + b)$

b. $a^2 - b^2$

c. $a^2 + b^2$

d. $a^2 - 2ab + b^2$

e. $a^2 + 2ab + b^2$

Practice 14–17

Use a special product to multiply, if possible.

14. $(7x - 1)^2$

15. $(5y + 3)(2y - 5)$

16. $(2a - 1)(2a + 1)$

17. $\left(5y - \dfrac{1}{9}\right)^2$

Objective D Using Special Products

Let's now practice using our special products on a variety of multiplication problems. This practice will help us recognize when to apply what special product formula.

Answers

9. $x^2 - 81$ **10.** $25 - 16y^2$ **11.** $x^2 - \dfrac{1}{9}$

12. $9a^2 - b^2$ **13.** $4x^4 - 36y^2$

14. $49x^2 - 14x + 1$

15. $10y^2 - 19y - 15$

16. $4a^2 - 1$

17. $25y^2 - \dfrac{10}{9}y + \dfrac{1}{81}$

✓ **Concept Check Answers**

a and e, b

Examples Use a special product to multiply, if possible.

14. $(4x - 9)(4x + 9)$

$= (4x)^2 - 9^2 = 16x^2 - 81$

This is the sum and difference of the same two terms.

15. $(3y + 2)^2$

$= (3y)^2 + 2(3y)(2) + 2^2$

$= 9y^2 + 12y + 4$

This is a binomial squared.

16. $(6a + 1)(a - 7)$ No special product applies.

      F   **O**   **I**   **L**   Use the FOIL method.

$= 6a \cdot a + 6a(-7) + 1 \cdot a + 1(-7)$

$= 6a^2 - 42a + a - 7$

$= 6a^2 - 41a - 7$

17. $\left(4x - \dfrac{1}{11}\right)^2$ This is a binomial squared.

$= (4x)^2 - 2(4x)\left(\dfrac{1}{11}\right) + \left(\dfrac{1}{11}\right)^2$

$= 16x^2 - \dfrac{8}{11}x + \dfrac{1}{121}$

■ **Work Practice 14–17**

Helpful Hint

- When multiplying two binomials, you may always use the FOIL order or method.
- When multiplying any two polynomials, you may always use the distributive property to find the product.

Vocabulary, Readiness & Video Check

Answer each exercise true or false.

1. $(x + 4)^2 = x^2 + 16$ ___false___

2. For $(x + 6)(2x - 1)$, the product of the first terms is $2x^2$. ___true___

3. $(x + 4)(x - 4) = x^2 + 16$ ___false___

4. The product $(x - 1)(x^3 + 3x - 1)$ is a polynomial of degree 5. ___false___

Martin-Gay Interactive Videos   *Watch the section lecture video and answer the following questions.*

See Video 5.6

See video answer section.

Objective A **5.** From  Examples 1–3, for what type of multiplication problem is the FOIL order of multiplication used?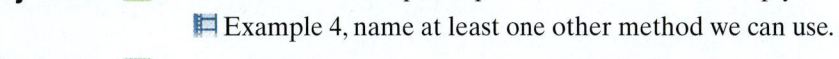

Objective B **6.** In addition to the special product rule used to multiply  Example 4, name at least one other method we can use.

Objective C **7.** From ▦ Example 5, why does multiplying the sum and difference of the same two terms always give us a binomial answer? ▶

Objective D **8.** At the end of ▦ Example 8, what three special products for multiplying binomials are summarized? ▶

5.6 Exercise Set MyMathLab®

Objective A *Multiply using the FOIL method. See Examples 1 through 3.*

1. $(x + 3)(x + 4)$
$x^2 + 7x + 12$

2. $(x + 5)(x + 1)$
$x^2 + 6x + 5$

3. $(x - 5)(x + 10)$
$x^2 + 5x - 50$

4. $(y - 12)(y + 4)$
$y^2 - 8y - 48$

5. $(5x - 6)(x + 2)$
$5x^2 + 4x - 12$

6. $(3y - 5)(2y + 7)$
$6y^2 + 11y - 35$

7. $(y - 6)(4y - 1)$
$4y^2 - 25y + 6$

8. $(2x - 9)(x - 11)$
$2x^2 - 31x + 99$

9. $(2x + 5)(3x - 1)$
$6x^2 + 13x - 5$

10. $(6x + 2)(x - 2)$
$6x^2 - 10x - 4$

11. $(y^2 + 7)(6y + 4)$
$6y^3 + 4y^2 + 42y + 28$

12. $(y^2 + 3)(5y + 6)$
$5y^3 + 6y^2 + 15y + 18$

13. $\left(x - \dfrac{1}{3}\right)\left(x + \dfrac{2}{3}\right)$
$x^2 + \dfrac{1}{3}x - \dfrac{2}{9}$

14. $\left(x - \dfrac{2}{5}\right)\left(x + \dfrac{1}{5}\right)$
$x^2 - \dfrac{1}{5}x - \dfrac{2}{25}$

15. $(0.4 - 3a)(0.2 - 5a)$
$0.08 - 2.6a + 15a^2$

16. $(0.3 - 2a)(0.6 - 5a)$
$0.18 - 2.7a + 10a^2$

17. $(x + 5y)(2x - y)$
$2x^2 + 9xy - 5y^2$

18. $(x + 4y)(3x - y)$
$3x^2 + 11xy - 4y^2$

Objective B *Multiply. See Examples 4 through 8.*

19. $(x + 2)^2$
$x^2 + 4x + 4$

20. $(x + 7)^2$
$x^2 + 14x + 49$

21. $(2a - 3)^2$
$4a^2 - 12a + 9$

22. $(7x - 3)^2$
$49x^2 - 42x + 9$

23. $(3a - 5)^2$
$9a^2 - 30a + 25$

24. $(5a - 2)^2$
$25a^2 - 20a + 4$

25. $\left(x^2 + 0.5\right)^2$
$x^4 + x^2 + 0.25$

26. $(x^2 + 0.3)^2$
$x^4 + 0.6x^2 + 0.09$

27. $\left(y - \dfrac{2}{7}\right)^2$
$y^2 - \dfrac{4}{7}y + \dfrac{4}{49}$

28. $\left(y - \dfrac{3}{4}\right)^2$
$y^2 - \dfrac{3}{2}y + \dfrac{9}{16}$

29. $(2x - 1)^2$
$4x^2 - 4x + 1$

30. $(5b - 4)^2$
$25b^2 - 40b + 16$

31. $(5x + 9)^2$
$25x^2 + 90x + 81$

32. $(6s + 2)^2$
$36s^2 + 24s + 4$

33. $(3x - 7y)^2$
$9x^2 - 42xy + 49y^2$

34. $(4s - 2y)^2$
$16s^2 - 16sy + 4y^2$

35. $(4m + 5n)^2$
$16m^2 + 40mn + 25n^2$

36. $(3n + 5m)^2$
$9n^2 + 30mn + 25m^2$

37. $\left(5x^4 - 3\right)^2$
$25x^8 - 30x^4 + 9$

38. $\left(7x^3 - 6\right)^2$
$49x^6 - 84x^3 + 36$

Objective C *Multiply. See Examples 9 through 13.*

▶ 39. $(a - 7)(a + 7)$
$a^2 - 49$

40. $(b + 3)(b - 3)$
$b^2 - 9$

41. $(x + 6)(x - 6)$
$x^2 - 36$

42. $(x - 8)(x + 8)$
$x^2 - 64$

43. $(3x - 1)(3x + 1)$
$9x^2 - 1$

44. $(7x - 5)(7x + 5)$
$49x^2 - 25$

45. $(x^2 + 5)(x^2 - 5)$
$x^4 - 25$

46. $(a^2 + 6)(a^2 - 6)$
$a^4 - 36$

47. $(2y^2 - 1)(2y^2 + 1)$
$4y^4 - 1$

48. $(3x^2 + 1)(3x^2 - 1)$
$9x^4 - 1$

49. $(4 - 7x)(4 + 7x)$
$16 - 49x^2$

50. $(8 - 7x)(8 + 7x)$
$64 - 49x^2$

51. $\left(3x - \dfrac{1}{2}\right)\left(3x + \dfrac{1}{2}\right)$
$9x^2 - \dfrac{1}{4}$

52. $\left(10x + \dfrac{2}{7}\right)\left(10x - \dfrac{2}{7}\right)$
$100x^2 - \dfrac{4}{49}$

▶ 53. $(9x + y)(9x - y)$
$81x^2 - y^2$

54. $(2x - y)(2x + y)$
$4x^2 - y^2$

55. $(2m + 5n)(2m - 5n)$
$4m^2 - 25n^2$

56. $(5m + 4n)(5m - 4n)$
$25m^2 - 16n^2$

Objective D **Mixed Practice** *Multiply. See Examples 14 through 17.*

57. $(a + 5)(a + 4)$
$a^2 + 9a + 20$

58. $(a + 5)(a + 7)$
$a^2 + 12a + 35$

59. $(a - 7)^2$
$a^2 - 14a + 49$

60. $(b - 2)^2$
$b^2 - 4b + 4$

61. $(4a + 1)(3a - 1)$
$12a^2 - a - 1$

62. $(6a + 7)(6a + 5)$
$36a^2 + 72a + 35$

63. $(x + 2)(x - 2)$
$x^2 - 4$

64. $(x - 10)(x + 10)$
$x^2 - 100$

65. $(3a + 1)^2$
$9a^2 + 6a + 1$

66. $(4a + 2)^2$
$16a^2 + 16a + 4$

67. $(x + y)(4x - y)$
$4x^2 + 3xy - y^2$

68. $(3x + 2)(4x - 2)$
$12x^2 + 2x - 4$

▶ 69. $\left(\dfrac{1}{3}a^2 - 7\right)\left(\dfrac{1}{3}a^2 + 7\right)$
$\dfrac{1}{9}a^4 - 49$

70. $\left(\dfrac{a}{2} + 4y\right)\left(\dfrac{a}{2} - 4y\right)$
$\dfrac{a^2}{4} - 16y^2$

▶ 71. $(3b + 7)(2b - 5)$
$6b^2 - b - 35$

72. $(3y - 13)(y - 3)$
$3y^2 - 22y + 39$

73. $(x^2 + 10)(x^2 - 10)$
$x^4 - 100$

74. $(x^2 + 8)(x^2 - 8)$
$x^4 - 64$

▶ 75. $(4x + 5)(4x - 5)$
$16x^2 - 25$

76. $(3x + 5)(3x - 5)$
$9x^2 - 25$

77. $(5x - 6y)^2$
$25x^2 - 60xy + 36y^2$

78. $(4x - 9y)^2$
$16x^2 - 72xy + 81y^2$

79. $(2r - 3s)(2r + 3s)$
$4r^2 - 9s^2$

80. $(6r - 2x)(6r + 2x)$
$36r^2 - 4x^2$

Express each as a product of polynomials in x. Then multiply and simplify.

81. Find the area of the square rug if its side is $(2x + 1)$ feet. $(4x^2 + 4x + 1)$ sq ft

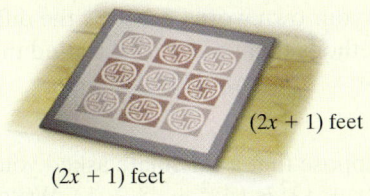

$(2x + 1)$ feet

$(2x + 1)$ feet

82. Find the area of the rectangular canvas if its length is $(3x - 2)$ inches and its width is $(x - 4)$ inches. $(3x^2 - 14x + 8)$ sq in.

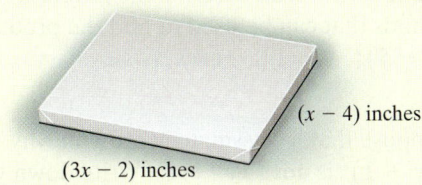

$(x - 4)$ inches

$(3x - 2)$ inches

Review

Simplify each expression. See Sections 5.1 and 5.2.

83. $\dfrac{50b^{10}}{70b^5}$

$\dfrac{5b^5}{7}$

84. $\dfrac{60y^6}{80y^2}$

$\dfrac{3y^4}{4}$

85. $\dfrac{8a^{17}b^5}{-4a^7b^{10}}$

$-\dfrac{2a^{10}}{b^5}$

86. $\dfrac{-6a^8y}{3a^4y}$

$-2a^4$

87. $\dfrac{2x^4y^{12}}{3x^4y^4}$

$\dfrac{2y^8}{3}$

88. $\dfrac{-48ab^6}{32ab^3}$

$-\dfrac{3b^3}{2}$

Concept Extensions

Match each expression on the left to the equivalent expression on the right. See the Concept Check in this section. (Not all choices will be used.)

89. $(a - b)^2$ c

90. $(a - b)(a + b)$ a

91. $(a + b)^2$ d

92. $(a + b)^2(a - b)^2$ e

a. $a^2 - b^2$

b. $a^2 + b^2$

c. $a^2 - 2ab + b^2$

d. $a^2 + 2ab + b^2$

e. none of these

Fill in the squares so that a true statement forms.

93. $(x^{\square} + 7)(x^{\square} + 3) = x^4 + 10x^2 + 21$ 2

94. $(5x^{\square} - 2)^2 = 25x^6 - 20x^3 + 4$ 3

Find the area of the shaded figure. To do so, subtract the area of the smaller square(s) from the area of the larger geometric figure.

△ **95.**

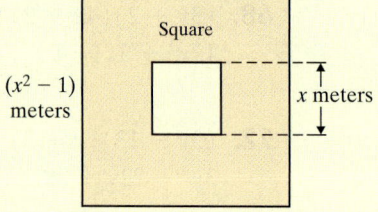

$(x^4 - 3x^2 + 1)$ sq m

△ **96.**

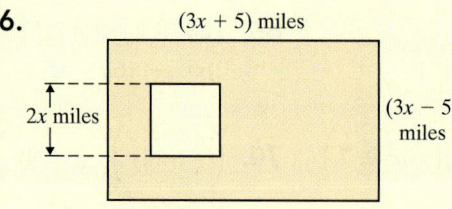

$(5x^2 - 25)$ sq mi

△ **97.**

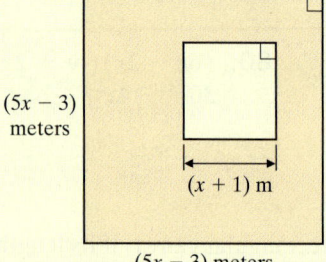

$(24x^2 - 32x + 8)$ sq m

△ **98.**

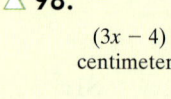

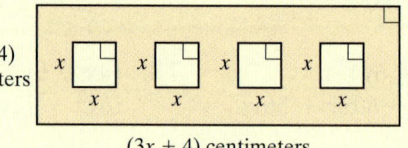

$(5x^2 - 16)$ sq cm

99. In your own words, describe the different methods that can be used to find the product: $(2x - 5)(3x + 1)$.   answers may vary

100. In your own words, describe the different methods that can be used to find the product: $(5x + 1)^2$.   answers may vary

101. Suppose that a classmate asked you why $(2x + 1)^2$ is **not** $(4x^2 + 1)$. Write down your response to this classmate.   answers may vary

102. Suppose that a classmate asked you why $(2x + 1)^2$ **is** $(4x^2 + 4x + 1)$. Write down your response to this classmate.   answers may vary

Exponents and Operations on Polynomials

Perform the operations and simplify.

Answers

1. $(5x^2)(7x^3)$

2. $(4y^2)(-8y^7)$

3. -4^2

4. $(-4)^2$

5. $(x - 5)(2x + 1)$

6. $(3x - 2)(x + 5)$

7. $(x - 5) + (2x + 1)$

8. $(3x - 2) + (x + 5)$

9. $\dfrac{7x^9 y^{12}}{x^3 y^{10}}$

10. $\dfrac{20a^2 b^8}{14a^2 b^2}$

11. $(12m^7 n^6)^2$

12. $(4y^9 z^{10})^3$

13. $(4y - 3)(4y + 3)$

14. $(7x - 1)(7x + 1)$

15. $(x^{-7} y^5)^9$

16. 8^{-2}

17. $(3^{-1} x^9)^3$

18. $\dfrac{(r^7 s^{-5})^6}{(2r^{-4} s^{-4})^4}$

19. $(7x^2 - 2x + 3) - (5x^2 + 9)$

20. $(10x^2 + 7x - 9) - (4x^2 - 6x + 2)$

1. $35x^5$

2. $-32y^9$

3. -16

4. 16

5. $2x^2 - 9x - 5$

6. $3x^2 + 13x - 10$

7. $3x - 4$

8. $4x + 3$

9. $7x^6 y^2$

10. $\dfrac{10b^6}{7}$

11. $144m^{14} n^{12}$

12. $64y^{27} z^{30}$

13. $16y^2 - 9$

14. $49x^2 - 1$

15. $\dfrac{y^{45}}{x^{63}}$

16. $\dfrac{1}{64}$

17. $\dfrac{x^{27}}{27}$

18. $\dfrac{r^{58}}{16s^{14}}$

19. $2x^2 - 2x - 6$

20. $6x^2 + 13x - 11$

21. $2.5y^2 - 6y - 0.2$

21. $0.7y^2 - 1.2 + 1.8y^2 - 6y + 1$ **22.** $7.8x^2 - 6.8x - 3.3 + 0.6x^2 - 0.9$

22. $8.4x^2 - 6.8x - 4.2$

23. $2y^2 - 6y - 1$

23. Subtract $(y^2 + 2)$ from $(3y^2 - 6y + 1)$.   **24.** $(z^2 + 5) - (3z^2 - 1) + \left(8z^2 + 2z - \dfrac{1}{2}\right)$

24. $6z^2 + 2z + \dfrac{11}{2}$

25. $x^2 + 8x + 16$

25. $(x + 4)^2$ **26.** $(y - 9)^2$

26. $y^2 - 18y + 81$

27. $2x + 8$

27. $(x + 4) + (x + 4)$ **28.** $(y - 9) + (y - 9)$

28. $2y - 18$

29. $7x^2 - 10xy + 4y^2$

29. $7x^2 - 6xy + 4(y^2 - xy)$ **30.** $5a^2 - 3ab + 6(b^2 - a^2)$

30. $-a^2 - 3ab + 6b^2$

31. $x^3 + 2x^2 - 16x + 3$

31. $(x - 3)(x^2 + 5x - 1)$ **32.** $(x + 1)(x^2 - 3x - 2)$

32. $x^3 - 2x^2 - 5x - 2$

33. $6x^2 - x - 70$

33. $(2x - 7)(3x + 10)$ **34.** $(5x - 1)(4x + 5)$

34. $20x^2 + 21x - 5$

35. $2x^3 - 19x^2 + 44x - 7$

35. $(2x - 7)(x^2 - 6x + 1)$ **36.** $(5x - 1)(x^2 + 2x - 3)$

36. $5x^3 + 9x^2 - 17x + 3$

37. $4x^2 - \dfrac{25}{81}$

37. $\left(2x + \dfrac{5}{9}\right)\left(2x - \dfrac{5}{9}\right)$ **38.** $\left(12y + \dfrac{3}{7}\right)\left(12y - \dfrac{3}{7}\right)$

38. $144y^2 - \dfrac{9}{49}$

5.7 Dividing Polynomials

Objective A Dividing by a Monomial

To divide a polynomial by a monomial, recall addition of fractions. Fractions that have a common denominator are added by adding the numerators:

$$\frac{a}{c} + \frac{b}{c} = \frac{a+b}{c}$$

If we read this equation from right to left and let a, b, and c be monomials, $c \neq 0$, we have the following.

Objectives

A Divide a Polynomial by a Monomial.

B Use Long Division to Divide a Polynomial by a Polynomial Other than a Monomial.

> **To Divide a Polynomial by a Monomial**
>
> Divide each term of the polynomial by the monomial.
>
> $$\frac{a+b}{c} = \frac{a}{c} + \frac{b}{c}, \quad c \neq 0$$

Throughout this section, we assume that denominators are not 0.

Example 1 Divide: $(6m^2 + 2m) \div 2m$

Solution: We begin by writing the quotient in fraction form. Then we divide each term of the polynomial $6m^2 + 2m$ by the monomial $2m$ and use the quotient rule for exponents to simplify.

$$\frac{6m^2 + 2m}{2m} = \frac{6m^2}{2m} + \frac{2m}{2m}$$

$$= 3m + 1 \qquad \text{Simplify.}$$

Check: To check, we multiply.

$$2m(3m + 1) = 2m(3m) + 2m(1) = 6m^2 + 2m$$

The quotient $3m + 1$ checks.

■ **Work Practice 1**

✓**Concept Check** In which of the following is $\dfrac{x+5}{5}$ simplified correctly?

a. $\dfrac{x}{5} + 1$ **b.** x **c.** $x + 1$

Example 2 Divide: $\dfrac{9x^5 - 12x^2 + 3x}{3x^2}$

Solution: $\dfrac{9x^5 - 12x^2 + 3x}{3x^2} = \dfrac{9x^5}{3x^2} - \dfrac{12x^2}{3x^2} + \dfrac{3x}{3x^2}$ Divide each term by $3x^2$.

$$= 3x^3 - 4 + \frac{1}{x} \qquad \text{Simplify.}$$

Notice that the quotient is not a polynomial because of the term $\dfrac{1}{x}$. This expression is called a rational expression—we will study rational expressions in Chapter 7. Although the quotient of two polynomials is not always a polynomial, we may still check by multiplying.

(Continued on next page)

Practice 1

Divide: $(25x^3 + 5x^2) \div 5x^2$

Teaching Tip

It may be helpful for students to verify that the answer to Example 1 is true for $m = 1, 2, 3$. Then ask if it would be true for all values. Have them verify that it is not true for $m = 0$.

Practice 2

Divide: $\dfrac{24x^7 + 12x^2 - 4x}{4x^2}$

Answers

1. $5x + 1$ **2.** $6x^5 + 3 - \dfrac{1}{x}$

✓**Concept Check Answer**
a

Check: $3x^2\left(3x^3 - 4 + \dfrac{1}{x}\right) = 3x^2(3x^3) - 3x^2(4) + 3x^2\left(\dfrac{1}{x}\right)$

$$= 9x^5 - 12x^2 + 3x$$

■ **Work Practice 2**

Practice 3

Divide: $\dfrac{12x^3y^3 - 18xy + 6y}{3xy}$

Example 3   Divide: $\dfrac{8x^2y^2 - 16xy + 2x}{4xy}$

Solution: $\dfrac{8x^2y^2 - 16xy + 2x}{4xy} = \dfrac{8x^2y^2}{4xy} - \dfrac{16xy}{4xy} + \dfrac{2x}{4xy}$ Divide each term by $4xy$.

$$= 2xy - 4 + \dfrac{1}{2y} \qquad \text{Simplify.}$$

Check: $4xy\left(2xy - 4 + \dfrac{1}{2y}\right) = 4xy(2xy) - 4xy(4) + 4xy\left(\dfrac{1}{2y}\right)$

$$= 8x^2y^2 - 16xy + 2x$$

■ **Work Practice 3**

Objective B   Dividing by a Polynomial Other than a Monomial ▶

To divide a polynomial by a polynomial other than a monomial, we use a process known as long division. Polynomial long division is similar to number long division, so we review long division by dividing 13 into 3660.

$$
\begin{array}{r}
281 \\
13\overline{)3660} \\
-26\downarrow \\
\hline
106 \\
-104\downarrow \\
\hline
20 \\
-13 \\
\hline
7
\end{array}
$$

> **Helpful Hint**   Recall that 3660 is called the dividend.

$2 \cdot 13 = 26$

Subtract and bring down the next digit in the dividend.

$8 \cdot 13 = 104$

Subtract and bring down the next digit in the dividend.

$1 \cdot 13 = 13$

Subtract. There are no more digits to bring down, so the remainder is 7.

The quotient is 281 R 7, which can be written as $281\dfrac{7}{13}$. $\leftarrow$ remainder $\leftarrow$ divisor

Recall that division can be checked by multiplication. To check this division problem, we see that

$$13 \cdot 281 + 7 = 3660, \text{ the dividend.}$$

Now we demonstrate long division of polynomials.

Teaching Tip

In Example 4, you may want to point out that a term of the quotient is placed over the like term in the dividend.

Practice 4

Divide $x^2 + 12x + 35$ by $x + 5$ using long division.

Answers

3. $4x^2y^2 - 6 + \dfrac{2}{x}$ 4. $x + 7$

Example 4   Divide $x^2 + 7x + 12$ by $x + 3$ using long division.

Solution:

> To subtract, change the signs of these terms and add.

How many times does x divide x^2?

$$\dfrac{x^2}{x} = x.$$

$$
\begin{array}{r}
x \\
x + 3\overline{)x^2 + 7x + 12} \\
x^2 + 3x \downarrow \\
\hline
4x + 12
\end{array}
$$

Multiply: $x(x + 3)$

Subtract and bring down the next term.

Now we repeat this process.

$$x + 3 \overline{)x^2 + 7x + 12}$$

$$x + 4$$ How many times does x divide $4x$? $\dfrac{4x}{x} = 4$.

$$\dfrac{x^2 + 3x}{}$$

To subtract, change the signs of these terms and add.

$$4x + 12$$ Multiply: $4(x + 3)$

$$\underrightarrow{} \;\; 4x + 12$$ Subtract. The remainder is 0.

$$0$$

The quotient is $x + 4$.

Check: We check by multiplying.

$$\begin{array}{ccccccc} \text{divisor} & \cdot & \text{quotient} & + & \text{remainder} & = & \text{dividend} \\ \downarrow & & \downarrow & & \downarrow & & \downarrow \\ (x + 3) & \cdot & (x + 4) & + & 0 & = & x^2 + 7x + 12 \end{array}$$

or

The quotient checks.

🟫 **Work Practice 4**

Example 5   Divide $6x^2 + 10x - 5$ by $3x - 1$ using long division.

Solution:

$$3x - 1 \overline{)6x^2 + 10x - 5}$$

$$2x + 4$$ $\dfrac{6x^2}{3x} = 2x$, so $2x$ is a term of the quotient.

$$\underline{ 6x^2 + 2x}$$ Multiply: $2x(3x - 1)$

$$12x - 5$$ Subtract and bring down the next term.

$$\underline{ 12x + 4}$$ $\dfrac{12x}{3x} = 4$. Multiply: $4(3x - 1)$

$$-1$$ Subtract. The remainder is -1.

Thus, $\left(6x^2 + 10x - 5\right)$ divided by $\left(3x - 1\right)$ is $\left(2x + 4\right)$ with a remainder of -1. This can be written as follows.

$$\dfrac{6x^2 + 10x - 5}{3x - 1} = 2x + 4 + \dfrac{-1}{3x - 1} \quad \leftarrow \text{remainder} \\ \leftarrow \text{divisor}$$

$$\text{or } 2x + 4 - \dfrac{1}{3x - 1}$$

Check: To check, we multiply $(3x - 1)(2x + 4)$. Then we add the remainder, -1, to this product.

$$(3x - 1)(2x + 4) + (-1) = \left(6x^2 + 12x - 2x - 4\right) - 1 \\ = 6x^2 + 10x - 5$$

The quotient checks.

🟧 **Work Practice 5**

Notice that the division process is continued until the degree of the remainder polynomial is less than the degree of the divisor polynomial.

Recall that in Section 5.3 we practiced writing polynomials in descending order of powers and with no missing terms. For example, $2 - 4x^2$ written in this form is $-4x^2 + 0x + 2$. Writing the dividend and divisor in this form is helpful when dividing polynomials.

Practice 5

Divide: $8x^2 + 2x - 7$ by $2x - 1$

Answer

5. $4x + 3 + \dfrac{-4}{2x - 1}$ or $4x + 3 - \dfrac{4}{2x - 1}$

Practice 6

Divide: $(15 - 2x^2) \div (x - 3)$

Example 6 Divide: $(2 - 4x^2) \div (x + 1)$

Solution: We use the rewritten form of $2 - 4x^2$ from the previous page.

$$
\begin{array}{r}
-4x + 4 \\
x + 1 \overline{) -4x^2 + 0x + 2} \\
-4x^2 - 4x \\
\hline
4x + 2 \\
4x + 4 \\
\hline
-2
\end{array}
$$

$\dfrac{-4x^2}{x} = -4x$, so $-4x$ is a term of the quotient.

Multiply: $-4x(x + 1)$

Subtract and bring down the next term.

$\dfrac{4x}{x} = 4$. Multiply: $4(x + 1)$

Remainder

Thus, $\dfrac{-4x^2 + 0x + 2}{x + 1}$ or $\dfrac{2 - 4x^2}{x + 1} = -4x + 4 + \dfrac{-2}{x + 1}$ or $-4x + 4 - \dfrac{2}{x + 1}$.

Check: To check, see that $(x + 1)(-4x + 4) + (-2) = 2 - 4x^2$.

■ **Work Practice 6**

Practice 7

Divide: $\dfrac{5 - x + 9x^3}{3x + 2}$

Example 7 Divide: $\dfrac{4x^2 + 7 + 8x^3}{2x + 3}$

Solution: Before we begin the division process, we rewrite $4x^2 + 7 + 8x^3$ as $8x^3 + 4x^2 + 0x + 7$. Notice that we have written the polynomial in descending order and have represented the missing x-term by $0x$.

$$
\begin{array}{r}
4x^2 - 4x + 6 \\
2x + 3 \overline{) 8x^3 + 14x^2 + 0x + 7} \\
-8x^3 + 12x^2 \\
\hline
-8x^2 + 0x \\
8x^2 + 12x \\
\hline
12x + 7 \\
-12x + 18 \\
\hline
-11
\end{array}
$$

Remainder

Thus, $\dfrac{4x^2 + 7 + 8x^3}{2x + 3} = 4x^2 - 4x + 6 + \dfrac{-11}{2x + 3}$ or $4x^2 - 4x + 6 - \dfrac{11}{2x + 3}$.

■ **Work Practice 7**

Practice 8

Divide: $x^3 - 1$ by $x - 1$

Example 8 Divide $x^3 - 8$ by $x - 2$.

Solution: Notice that the polynomial $x^3 - 8$ is missing an x^2-term and an x-term. We'll represent these terms by inserting $0x^2$ and $0x$.

$$
\begin{array}{r}
x^2 + 2x + 4 \\
x - 2 \overline{) x^3 + 0x^2 + 0x - 8} \\
-x^3 - 2x^2 \\
\hline
2x^2 + 0x \\
-2x^2 - 4x \\
\hline
4x - 8 \\
-4x - 8 \\
\hline
0
\end{array}
$$

Thus, $\dfrac{x^3 - 8}{x - 2} = x^2 + 2x + 4$.

Check: To check, see that $(x^2 + 2x + 4)(x - 2) = x^3 - 8$.

■ **Work Practice 8**

Answers

6. $-2x - 6 + \dfrac{-3}{x - 3}$

or $-2x - 6 - \dfrac{3}{x - 3}$

7. $3x^2 - 2x + 1 + \dfrac{3}{3x + 2}$

8. $x^2 + x + 1$

Vocabulary, Readiness & Video Check

Use the choices below to fill in each blank. Choices may be used more than once.

dividend   divisor   quotient

1. In $6\overline{)18}$ with quotient 3, the 18 is the ___dividend___, the 3 is the ___quotient___, and the 6 is the ___divisor___.

2. In $x+1\overline{)x^2+3x+2}$ with quotient $x+2$, the $x+1$ is the ___divisor___, the x^2+3x+2 is the ___dividend___, and the $x+2$ is the ___quotient___.

Simplify each expression mentally.

3. $\dfrac{a^6}{a^4}$ a^2 4. $\dfrac{p^8}{p^3}$ p^5 5. $\dfrac{y^2}{y}$ y 6. $\dfrac{a^3}{a}$ a^2

Martin-Gay Interactive Videos

See Video 5.7

See video answer section.

Watch the section lecture video and answer the following questions.

Objective A 7. The lecture before ⊞ Example 1 begins with adding two fractions with the same denominator. From there, the lecture continues to a method for dividing a polynomial by a monomial. What role does the monomial play in the fraction example? ▶

Objective B 8. In ⊞ Example 5, we're told that although we don't have to fill in missing powers in the divisor and the dividend, it really is a good idea to do so. Why? ▶

5.7 Exercise Set MyMathLab®

Objective A *Perform each division. See Examples 1 through 3.*

▶ 1. $\dfrac{12x^4+3x^2}{x}$
$12x^3+3x$

2. $\dfrac{15x^2-9x^5}{x}$
$15x-9x^4$

3. $\dfrac{20x^3-30x^2+5x+5}{5}$
$4x^3-6x^2+x+1$

4. $\dfrac{8x^3-4x^2+6x+2}{2}$
$4x^3-2x^2+3x+1$

5. $\dfrac{15p^3+18p^2}{3p}$
$5p^2+6p$

6. $\dfrac{6x^5+3x^4}{3x^4}$
$2x+1$

7. $\dfrac{-9x^4+18x^5}{6x^5}$
$-\dfrac{3}{2x}+3$

8. $\dfrac{14m^2-27m^3}{7m}$
$2m-\dfrac{27m^2}{7}$

▶ 9. $\dfrac{-9x^5+3x^4-12}{3x^3}$
$-3x^2+x-\dfrac{4}{x^3}$

10. $\dfrac{6a^2-4a+12}{-2a^2}$
$-3+\dfrac{2}{a}-\dfrac{6}{a^2}$

11. $\dfrac{4x^4-6x^3+7}{-4x^4}$
$-1+\dfrac{3}{2x}-\dfrac{7}{4x^4}$

12. $\dfrac{-12a^3+36a-15}{3a}$
$-4a^2+12-\dfrac{5}{a}$

Objective B *Find each quotient using long division. See Examples 4 and 5.*

▶ **13.** $\dfrac{x^2 + 4x + 3}{x + 3}$

$x + 1$

14. $\dfrac{x^2 + 7x + 10}{x + 5}$

$x + 2$

15. $\dfrac{2x^2 + 13x + 15}{x + 5}$

$2x + 3$

16. $\dfrac{3x^2 + 8x + 4}{x + 2}$

$3x + 2$

17. $\dfrac{2x^2 - 7x + 3}{x - 4}$

$2x + 1 + \dfrac{7}{x - 4}$

18. $\dfrac{3x^2 - x - 4}{x - 1}$

$3x + 2 - \dfrac{2}{x - 1}$

19. $\dfrac{9a^3 - 3a^2 - 3a + 4}{3a + 2}$

$3a^2 - 3a + 1 + \dfrac{2}{3a + 2}$

20. $\dfrac{4x^3 + 12x^2 + x - 14}{2x + 3}$

$2x^2 + 3x - 4 - \dfrac{2}{2x + 3}$

21. $\dfrac{8x^2 + 10x + 1}{2x + 1}$

$4x + 3 - \dfrac{2}{2x + 1}$

22. $\dfrac{3x^2 + 17x + 7}{3x + 2}$

$x + 5 - \dfrac{3}{3x + 2}$

23. $\dfrac{2x^3 + 2x^2 - 17x + 8}{x - 2}$

$2x^2 + 6x - 5 - \dfrac{2}{x - 2}$

24. $\dfrac{4x^3 + 11x^2 - 8x - 10}{x + 3}$

$4x^2 - x - 5 + \dfrac{5}{x + 3}$

Find each quotient using long division. Don't forget to write the polynomials in descending order and fill in any missing terms. See Examples 6 through 8.

25. $\dfrac{x^2 - 36}{x - 6}$

$x + 6$

26. $\dfrac{a^2 - 49}{a - 7}$

$a + 7$

▶ **27.** $\dfrac{x^3 - 27}{x - 3}$

$x^2 + 3x + 9$

28. $\dfrac{x^3 + 64}{x + 4}$

$x^2 - 4x + 16$

29. $\dfrac{1 - 3x^2}{x + 2}$

$-3x + 6 - \dfrac{11}{x + 2}$

30. $\dfrac{7 - 5x^2}{x + 3}$

$-5x + 15 - \dfrac{38}{x + 3}$

31. $\dfrac{-4b + 4b^2 - 5}{2b - 1}$

$2b - 1 - \dfrac{6}{2b - 1}$

32. $\dfrac{-3y + 2y^2 - 15}{2y + 5}$

$y - 4 + \dfrac{5}{2y + 5}$

Objectives A B Mixed Practice *Divide. If the divisor contains 2 or more terms, use long division. See Examples 1 through 8.*

33. $\dfrac{a^2b^2 - ab^3}{ab}$

$ab - b^2$

34. $\dfrac{m^3n^2 - mn^4}{mn}$

$m^2n - n^3$

35. $\dfrac{8x^2 + 6x - 27}{2x - 3}$

$4x + 9$

36. $\dfrac{18w^2 + 18w - 8}{3w + 4}$

$6w - 2$

37. $\dfrac{2x^2y + 8x^2y^2 - xy^2}{2xy}$

$x + 4xy - \dfrac{y}{2}$

38. $\dfrac{11x^3y^3 - 33xy + x^2y^2}{11xy}$

$x^2y^2 - 3 + \dfrac{xy}{11}$

▶ **39.** $\dfrac{2b^3 + 9b^2 + 6b - 4}{b + 4}$

$2b^2 + b + 2 - \dfrac{12}{b + 4}$

40. $\dfrac{2x^3 + 3x^2 - 3x + 4}{x + 2}$

$2x^2 - x - 1 + \dfrac{6}{x + 2}$

41. $\dfrac{y^3 + 3y^2 + 4}{y - 2}$

$y^2 + 5y + 10 + \dfrac{24}{y - 2}$

42. $\dfrac{3x^3 + 11x + 12}{x + 4}$

$3x^2 - 12x + 59 - \dfrac{224}{x + 4}$

43. $\dfrac{5 - 6x^2}{x - 2}$

$-6x - 12 - \dfrac{19}{x - 2}$

44. $\dfrac{3 - 7x^2}{x - 3}$

$-7x - 21 - \dfrac{60}{x - 3}$

45. $\dfrac{x^5 + x^2}{x^2 + x}$ $x^3 - x^2 + x$

46. $\dfrac{x^6 - x^3}{x^3 - x^2}$ $x^3 + x^2 + x$

Review

Fill in each blank. See Section 5.1.

47. $12 = 4 \cdot \underline{3}$

48. $12 = 2 \cdot \underline{6}$

49. $20 = -5 \cdot \underline{-4}$

50. $20 = -4 \cdot \underline{-5}$

51. $9x^2 = 3x \cdot \underline{3x}$

52. $9x^2 = 9x \cdot \underline{x}$

53. $36x^2 = 4x \cdot \underline{9x}$

54. $36x^2 = 2x \cdot \underline{18x}$

Concept Extensions

Solve.

55. The perimeter of a square is $\left(12x^3 + 4x - 16\right)$ feet. Find the length of its side. $\left(3x^3 + x - 4\right)$ ft

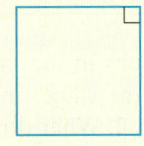

Perimeter is
$(12x^3 + 4x - 16)$ feet

△ **56.** The volume of the swimming pool shown is $\left(36x^5 - 12x^3 + 6x^2\right)$ cubic feet. If its height is $2x$ feet and its width is $3x$ feet, find its length. $\left(6x^3 - 2x + 1\right)$ ft

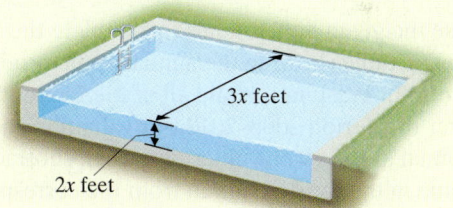

3x feet

2x feet

57. The area of the parallelogram shown is $\left(10x^2 + 31x + 15\right)$ square meters. If its base is $(5x + 3)$ meters, find its height. $(2x + 5)$ m

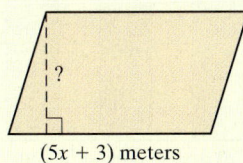

?

$(5x + 3)$ meters

58. The area of the top of the Ping-Pong table shown is $\left(49x^2 + 70x - 200\right)$ square inches. If its length is $(7x + 20)$ inches, find its width. $(7x - 10)$ in.

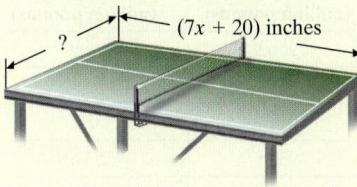

? — $(7x + 20)$ inches

59. Explain how to check a polynomial long division result when the remainder is 0. answers may vary

60. Explain how to check a polynomial long division result when the remainder is not 0.
answers may vary

61. In which of the following is $\dfrac{a + 7}{7}$ simplified correctly? See the Concept Check in this section.

 a. $a + 1$

 b. a

 c. $\dfrac{a}{7} + 1$ c

62. In which of the following is $\dfrac{5x + 15}{5}$ simplified correctly? See the Concept Check in this section.

 a. $x + 15$

 b. $x + 3$

 c. $x + 1$ b

Chapter 5 Group Activity

Modeling with Polynomials

Materials:

• calculator

This activity may be completed by working in groups or individually.

Washington state is the leading producer of apples in the United States. The polynomial model $184x^2 - 545x + 5649$ gives Washington's annual apple production (in million pounds) for the period 2008–2012. The polynomial model $-23x^2 - 50x + 9662$ gives the total U.S. annual apple production (in million pounds) for the same period. In both models, x is the number of years after 2008. (*Source:* Based on data from the National Agricultural Statistics Service)

1. Use the given polynomials to complete the following table showing the annual apple production (both for Washington and all of the United States) over the period 2008–2012 by evaluating each polynomial at the given values of x. Then subtract each value in the fourth column from the corresponding value in the third column. Record the result in the last column, labeled "Difference." What do you think these values represent?

Year	x	Total U.S. Annual Apple Production (million pounds)	Washington's Annual Apple Production (million pounds)	Difference
2008	0	9662	5649	4013
2009	1	9589	5288	4301
2010	2	9470	5295	4175
2011	3	9305	5670	3635
2012	4	9094	6413	2681

2. Use the polynomial models to find a new polynomial model representing the annual apple production of *all other* U.S. states, excluding Washington. Then evaluate your new polynomial model to complete the accompanying table. $-207x^2 + 495x + 4013$

Year	x	Other States' Annual Apple Production (million pounds)
2008	0	4013
2009	1	4301
2010	2	4175
2011	3	3635
2012	4	2681

3. Compare the values in the last column of the table in Question **1** to the values in the last column of the table in Question **2**. What do you notice? What can you conclude?   They are the same; answers may vary.

4. Make a bar graph of the data in the table in Question **2**. Describe what you see.

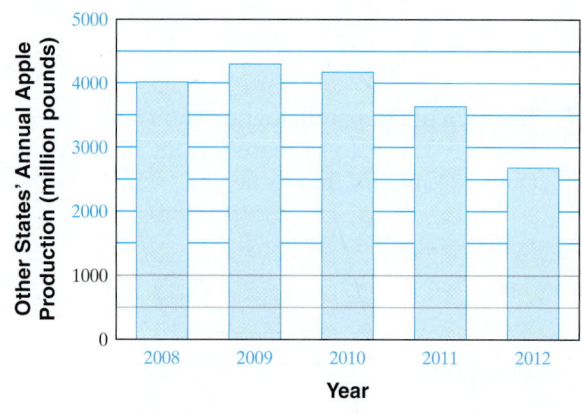

Chapter 5 Vocabulary Check

Fill in each blank with one of the words or phrases listed below.

term	coefficient	monomial	binomial	trinomial
polynomials	degree of a term	degree of a polynomial	distributive	FOIL

1. A ___**term**___ is a number or the product of a number and variables raised to powers.

2. The ___**FOIL**___ method may be used when multiplying two binomials.

3. A polynomial with exactly 3 terms is called a ___**trinomial**___.

4. The ___**degree of a polynomial**___ is the greatest degree of any term of the polynomial.

5. A polynomial with exactly 2 terms is called a ___**binomial**___.

6. The ___**coefficient**___ of a term is its numerical factor.

7. The ___degree of a term___ is the sum of the exponents on the variables in the term.

8. A polynomial with exactly 1 term is called a ___monomial___.

9. Monomials, binomials, and trinomials are all examples of ___polynomials___.

10. The ___distributive___ property is used to multiply $2x(x - 4)$.

Helpful Hint

▶ Are you preparing for your test? Don't forget to take the Chapter 5 Test on page 405. Then check your answers at the back of the text and use the Chapter Test Prep Videos to see the fully worked-out solutions to any of the exercises you want to review.

5 Chapter Highlights

Definitions and Concepts	Examples

Section 5.1   Exponents

a^n means the product of n factors, each of which is a.	$3^2 = 3 \cdot 3 = 9$ $(-5)^3 = (-5)(-5)(-5) = -125$ $\left(\dfrac{1}{2}\right)^4 = \dfrac{1}{2} \cdot \dfrac{1}{2} \cdot \dfrac{1}{2} \cdot \dfrac{1}{2} = \dfrac{1}{16}$
Let m and n be integers and no denominators be 0. **Product Rule:** $a^m \cdot a^n = a^{m+n}$ **Power Rule:** $(a^m)^n = a^{mn}$ **Power of a Product Rule:** $(ab)^n = a^n b^n$ **Power of a Quotient Rule:** $\left(\dfrac{a}{b}\right)^n = \dfrac{a^n}{b^n}$ **Quotient Rule:** $\dfrac{a^m}{a^n} = a^{m-n}$ **Zero Exponent:** $a^0 = 1, a \neq 0$	$x^2 \cdot x^7 = x^{2+7} = x^9$ $\left(5^3\right)^8 = 5^{3 \cdot 8} = 5^{24}$ $(7y)^4 = 7^4 y^4$ $\left(\dfrac{x}{8}\right)^3 = \dfrac{x^3}{8^3}$ $\dfrac{x^9}{x^4} = x^{9-4} = x^5$ $5^0 = 1; x^0 = 1, x \neq 0$

Section 5.2   Negative Exponents and Scientific Notation

If $a \neq 0$ and n is an integer, $a^{-n} = \dfrac{1}{a^n}$	$3^{-2} = \dfrac{1}{3^2} = \dfrac{1}{9}; 5x^{-2} = \dfrac{5}{x^2}$ Simplify: $\left(\dfrac{x^{-2}y}{x^5}\right)^{-2} = \dfrac{x^4 y^{-2}}{x^{-10}}$ $\qquad = x^{4-(-10)}y^{-2}$ $\qquad = \dfrac{x^{14}}{y^2}$
A positive number is written in scientific notation if it is written as the product of a number a, where $1 \leq a < 10$, and an integer power r of 10. $a \times 10^r$	$1200 = 1.2 \times 10^3$ $0.000000568 = 5.68 \times 10^{-7}$

Definitions and Concepts	Examples

Section 5.3   Introduction to Polynomials

A **term** is a number or the product of a number and variables raised to powers.

$$-5x, \; 7a^2b, \; \frac{1}{4}y^4, \; 0.2$$

The **numerical coefficient,** or **coefficient,** of a term is its numerical factor.

Term	Coefficient
$7x^2$	7
y	1
$-a^2b$	-1

A **polynomial** is a finite sum of terms of the form ax^n where a is a real number and n is a whole number.

$$5x^3 - 6x^2 + 3x - 6 \quad \text{(Polynomial)}$$

A **monomial** is a polynomial with exactly 1 term.

$$\frac{5}{6}y^3 \quad \text{(Monomial)}$$

A **binomial** is a polynomial with exactly 2 terms.

$$-0.2a^2b - 5b^2 \quad \text{(Binomial)}$$

A **trinomial** is a polynomial with exactly 3 terms.

$$3x^2 - 2x + 1 \quad \text{(Trinomial)}$$

The **degree of a polynomial** is the greatest degree of any term of the polynomial.

Polynomial	Degree
$5x^2 - 3x + 2$	2
$7y + 8y^2z^3 - 12$	$2 + 3 = 5$

Section 5.4   Adding and Subtracting Polynomials

To add polynomials, combine like terms.

Add.

$$(7x^2 - 3x + 2) + (-5x - 6)$$
$$= 7x^2 - 3x + 2 - 5x - 6$$
$$= 7x^2 - 8x - 4$$

To subtract two polynomials, change the signs of the terms of the second polynomial, and then add.

Subtract.

$$(17y^2 - 2y + 1) - (-3y^3 + 5y - 6)$$
$$= (17y^2 - 2y + 1) + (3y^3 - 5y + 6)$$
$$= 17y^2 - 2y + 1 + 3y^3 - 5y + 6$$
$$= 3y^3 + 17y^2 - 7y + 7$$

Section 5.5   Multiplying Polynomials

To multiply two polynomials, multiply each term of one polynomial by each term of the other polynomial, and then combine like terms.

Multiply.

$$(2x + 1)(5x^2 - 6x + 2)$$

$$= 2x(5x^2 - 6x + 2) + 1(5x^2 - 6x + 2)$$
$$= 10x^3 - 12x^2 + 4x + 5x^2 - 6x + 2$$
$$= 10x^3 - 7x^2 - 2x + 2$$

Definitions and Concepts	**Examples**

Section 5.6   Special Products

The **FOIL method** may be used when multiplying two binomials.	Multiply: $(5x - 3)(2x + 3)$

$$\begin{array}{c}\text{First} \qquad \text{Last} \\ (5x - 3)(2x + 3) \\ \text{Inner} \\ \text{Outer}\end{array}$$

$$\begin{array}{cccc} \text{F} & \text{O} & \text{I} & \text{L} \end{array}$$
$$= (5x)(2x) + (5x)(3) + (-3)(2x) + (-3)(3)$$
$$= 10x^2 + 15x - 6x - 9$$
$$= 10x^2 + 9x - 9$$

Squaring a Binomial

$$(a + b)^2 = a^2 + 2ab + b^2$$

$$(a - b)^2 = a^2 - 2ab + b^2$$

Square each binomial.

$$(x + 5)^2 = x^2 + 2(x)(5) + 5^2$$
$$= x^2 + 10x + 25$$
$$(3x - 2y)^2 = (3x)^2 - 2(3x)(2y) + (2y)^2$$
$$= 9x^2 - 12xy + 4y^2$$

Multiplying the Sum and Difference of Two Terms

$$(a + b)(a - b) = a^2 - b^2$$

Multiply.

$$(6y + 5)(6y - 5) = (6y)^2 - 5^2$$
$$= 36y^2 - 25$$

Section 5.7   Dividing Polynomials

To divide a polynomial by a monomial,	Divide.

$$\frac{a + b}{c} = \frac{a}{c} + \frac{b}{c}, c \neq 0$$

$$\frac{15x^5 - 10x^3 + 5x^2 - 2x}{5x^2}$$

$$= \frac{15x^5}{5x^2} - \frac{10x^3}{5x^2} + \frac{5x^2}{5x^2} - \frac{2x}{5x^2}$$

$$= 3x^3 - 2x + 1 - \frac{2}{5x}$$

To divide a polynomial by a polynomial other than a monomial, use long division.

$$\begin{array}{r} 5x - 1 + \dfrac{-4}{2x + 3} \\ 2x + 3\overline{)10x^2 + 13x - 7} \\ \underline{10x^2 + 15x} \\ -2x - 7 \\ \underline{-2x - 3} \\ -4 \end{array}$$

or $5x - 1 - \dfrac{4}{2x + 3}$

(5.1) *State the base and the exponent for each expression.*

1. 3^2
base: 3; exponent: 2

2. $(-5)^4$
base: -5; exponent: 4

3. -5^4
base: 5; exponent: 4

4. x^6
base: x; exponent: 6

Evaluate each expression.

5. 8^3
512

6. $(-6)^2$
36

7. -6^2
-36

8. $-4^3 - 4^0$
-65

9. $(3b)^0$
1

10. $\dfrac{8b}{8b}$
1

Simplify each expression.

11. $y^2 \cdot y^7$
y^9

12. $x^9 \cdot x^5$
x^{14}

13. $(2x^5)(-3x^6)$
$-6x^{11}$

14. $(-5y^3)(4y^4)$
$-20y^7$

15. $(x^4)^2$
x^8

16. $(y^3)^5$
y^{15}

17. $(3y^6)^4$
$81y^{24}$

18. $(2x^3)^3$
$8x^9$

19. $\dfrac{x^9}{x^4}$ x^5

20. $\dfrac{z^{12}}{z^5}$ z^7

21. $\dfrac{a^5 b^4}{ab}$ $a^4 b^3$

22. $\dfrac{x^4 y^6}{xy}$ $x^3 y^5$

23. $\dfrac{3x^4 y^{10}}{12xy^6}$ $\dfrac{x^3 y^4}{4}$

24. $\dfrac{2x^7 y^8}{8xy^2}$ $\dfrac{x^6 y^6}{4}$

25. $5a^7(2a^4)^3$
$40a^{19}$

26. $(2x)^2(9x)$
$36x^3$

27. $(-5a)^0 + 7^0 + 8^0$
3

28. $8x^0 + 9^0$ 9

Simplify the given expression and choose the correct result.

29. $\left(\dfrac{3x^4}{4y}\right)^3$

a. $\dfrac{27x^{64}}{64y^3}$

b. $\dfrac{27x^{12}}{64y^3}$

c. $\dfrac{9x^{12}}{12y^3}$

d. $\dfrac{3x^{12}}{4y^3}$ **b**

30. $\left(\dfrac{5a^6}{b^3}\right)^2$

a. $\dfrac{10a^{12}}{b^6}$

b. $\dfrac{25a^{36}}{b^9}$

c. $\dfrac{25a^{12}}{b^6}$

d. $25a^{12}b^6$ **c**

(5.2) *Simplify each expression.*

31. 7^{-2} $\dfrac{1}{49}$

32. -7^{-2} $-\dfrac{1}{49}$

33. $2x^{-4}$ $\dfrac{2}{x^4}$

34. $(2x)^{-4}$ $\dfrac{1}{16x^4}$

35. $\left(\dfrac{1}{5}\right)^{-3}$ 125

36. $\left(\dfrac{-2}{3}\right)^{-2}$ $\dfrac{9}{4}$

37. $2^0 + 2^{-4}$ $\dfrac{17}{16}$

38. $6^{-1} - 7^{-1}$ $\dfrac{1}{42}$

400

Simplify each expression. Write each answer using positive exponents only.

39. $\dfrac{x^5}{x^{-3}}$ x^8

40. $\dfrac{z^4}{z^{-4}}$ z^8

41. $\dfrac{r^{-3}}{r^{-4}}$ r

42. $\dfrac{y^{-2}}{y^{-5}}$ y^3

43. $\left(\dfrac{bc^{-2}}{bc^{-3}}\right)^4$ c^4

44. $\left(\dfrac{x^{-3}y^{-4}}{x^{-2}y^{-5}}\right)^{-3}$ $\dfrac{x^3}{y^3}$

45. $\dfrac{x^{-4}y^{-6}}{x^2y^7}$ $\dfrac{1}{x^6y^{13}}$

46. $\dfrac{a^5b^{-5}}{a^{-5}b^5}$ $\dfrac{a^{10}}{b^{10}}$

Write each number in scientific notation.

47. 0.00027
2.7×10^{-4}

48. 0.8868
8.868×10^{-1}

49. 80,800,000
8.08×10^7

50. 868,000
8.68×10^5

51. In November 2012, approximately 130,300,000 people cast ballots in the U.S. presidential election. Write this number in scientific notation. (*Source:* Nonprofit VOTE) 1.303×10^8

52. The approximate diameter of the Milky Way galaxy is 150,000 light-years. Write this number in scientific notation. (*Source:* NASA IMAGE/POETRY Education and Public Outreach Program)
1.5×10^5

150,000 light-years

Write each number in standard form.

53. 8.67×10^5
867,000

54. 3.86×10^{-3}
0.00386

55. 8.6×10^{-4}
0.00086

56. 8.936×10^5
893,600

57. The volume of the planet Jupiter is 1.43128×10^{15} cubic kilometers. Write this number in standard form. (*Source:* National Space Science Data Center) 1,431,280,000,000,000

58. An angstrom is a unit of measure, equal to 1×10^{-10} meter, used for measuring wavelengths or the diameters of atoms. Write this number in standard form. (*Source:* National Institute of Standards and Technology) 0.0000000001

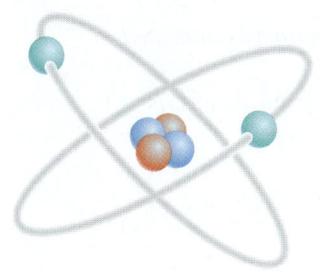

Simplify. Express each result in standard form.

59. $(8 \times 10^4)(2 \times 10^{-7})$ 0.016

60. $\dfrac{8 \times 10^4}{2 \times 10^{-7}}$ 400,000,000,000

(5.3) *Find the degree of each polynomial.*

61. $y^5 + 7x - 8x^4$ 5

62. $9y^2 + 30y + 25$ 2

63. $-14x^2y - 28x^2y^3 - 42x^2y^2$ 5

64. $6x^2y^2z^2 + 5x^2y^3 - 12xyz$ 6

65. The Glass Bridge Skywalk is suspended 4000 feet over the Colorado River at the very edge of the Grand Canyon. Neglecting air resistance, the height of an object dropped from the Skywalk at time t seconds is given by the polynomial $-16t^2 + 4000$. Find the height of the object at the given times below.

t	**0 seconds**	**1 second**	**3 seconds**	**5 seconds**
$-16t^2 + 4000$	4000 ft	3984 ft	3856 ft	3600 ft

△ **66.** The surface area of a box with a square base and a height of 5 units is given by the polynomial $2x^2 + 20x$. Fill in the table below by evaluating $2x^2 + 20x$ for the given values of x.

x	**1**	**3**	**5.1**	**10**
$2x^2 + 20x$	22	78	154.02	400

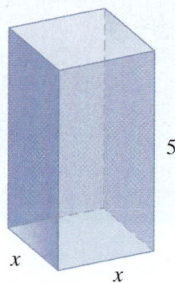

Combine like terms in each expression.

67. $7a^2 - 4a^2 - a^2$ $2a^2$

68. $9y + y - 14y$ $-4y$

69. $6a^2 + 4a + 9a^2$ $15a^2 + 4a$

70. $21x^2 + 3x + x^2 + 6$
   $22x^2 + 3x + 6$

71. $4a^2b - 3b^2 - 8q^2 - 10a^2b + 7q^2$
   $-6a^2b - 3b^2 - q^2$

72. $2s^{14} + 3s^{13} + 12s^{12} - s^{10}$
   cannot be combined

(5.4) *Add or subtract as indicated.*

73. $(3x^2 + 2x + 6) + (5x^2 + x)$
   $8x^2 + 3x + 6$

74. $(2x^5 + 3x^4 + 4x^3 + 5x^2) + (4x^2 + 7x + 6)$
   $2x^5 + 3x^4 + 4x^3 + 9x^2 + 7x + 6$

75. $(-5y^2 + 3) - (2y^2 + 4)$
   $-7y^2 - 1$

76. $(2m^7 + 3x^4 + 7m^6) - (8m^7 + 4m^2 + 6x^4)$
   $-6m^7 - 3x^4 + 7m^6 - 4m^2$

77. $(3x^2 - 7xy + 7y^2) - (4x^2 - xy + 9y^2)$
   $-x^2 - 6xy - 2y^2$

78. $(8x^6 - 5xy - 10y^2) - (7x^6 - 9xy - 12y^2)$
   $x^6 + 4xy + 2y^2$

Translating *Perform the indicated operations.*

79. Add $\left(-9x^2 + 6x + 2\right)$ and $\left(4x^2 - x - 1\right).$ $-5x^2 + 5x + 1$

80. Subtract $\left(4x^2 + 8x - 7\right)$ from the sum of $\left(x^2 + 7x + 9\right)$ and $\left(x^2 + 4\right).$ $-2x^2 - x + 20$

(5.5) *Multiply each expression.*

81. $6(x + 5)$
$6x + 30$

82. $9(x - 7)$
$9x - 63$

83. $4(2a + 7)$
$8a + 28$

84. $9(6a - 3)$
$54a - 27$

85. $-7x(x^2 + 5)$
$-7x^3 - 35x$

86. $-8y(4y^2 - 6)$
$-32y^3 + 48y$

87. $-2(x^3 - 9x^2 + x)$
$-2x^3 + 18x^2 - 2x$

88. $-3a\left(a^2b + ab + b^2\right)$
$-3a^3b - 3a^2b - 3ab^2$

89. $\left(3a^3 - 4a + 1\right)(-2a)$
$-6a^4 + 8a^2 - 2a$

90. $\left(6b^3 - 4b + 2\right)(7b)$
$42b^4 - 28b^2 + 14b$

91. $(2x + 2)(x - 7)$
$2x^2 - 12x - 14$

92. $(2x - 5)(3x + 2)$
$6x^2 - 11x - 10$

93. $(4a - 1)(a + 7)$
$4a^2 + 27a - 7$

94. $(6a - 1)(7a + 3)$
$42a^2 + 11a - 3$

95. $(x + 7)\left(x^3 + 4x - 5\right)$
$x^4 + 7x^3 + 4x^2 + 23x - 35$

96. $(x + 2)\left(x^5 + x + 1\right)$
$x^6 + 2x^5 + x^2 + 3x + 2$

97. $\left(x^2 + 2x + 4\right)\left(x^2 + 2x - 4\right)$
$x^4 + 4x^3 + 4x^2 - 16$

98. $\left(x^3 + 4x + 4\right)\left(x^3 + 4x - 4\right)$
$x^6 + 8x^4 + 16x^2 - 16$

99. $(x + 7)^3$
$x^3 + 21x^2 + 147x + 343$

100. $(2x - 5)^3$
$8x^3 - 60x^2 + 150x - 125$

(5.6) *Use special products to multiply each of the following.*

101. $(x + 7)^2$
$x^2 + 14x + 49$

102. $(x - 5)^2$
$x^2 - 10x + 25$

103. $(3x - 7)^2$
$9x^2 - 42x + 49$

104. $(4x + 2)^2$
$16x^2 + 16x + 4$

105. $(5x - 9)^2$
$25x^2 - 90x + 81$

106. $(5x + 1)(5x - 1)$
$25x^2 - 1$

107. $(7x + 4)(7x - 4)$
$49x^2 - 16$

108. $(a + 2b)(a - 2b)$
$a^2 - 4b^2$

109. $(2x - 6)(2x + 6)$
$4x^2 - 36$

110. $\left(4a^2 - 2b\right)\left(4a^2 + 2b\right)$
$16a^4 - 4b^2$

Express each as a product of polynomials in x. Then multiply and simplify.

111. Find the area of the square if its side is $(3x - 1)$ meters. $\left(9x^2 - 6x + 1\right)$ sq m

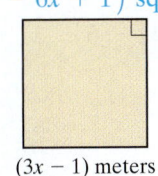

$(3x - 1)$ meters

112. Find the area of the rectangle.
$\left(5x^2 - 3x - 2\right)$ sq mi

$(x - 1)$ miles

$(5x + 2)$ miles

(5.7) *Divide.*

113. $\dfrac{x^2 + 21x + 49}{7x^2}$ $\dfrac{1}{7} + \dfrac{3}{x} + \dfrac{7}{x^2}$

114. $\dfrac{5a^3b - 15ab^2 + 20ab}{-5ab}$ $-a^2 + 3b - 4$

115. $\left(a^2 - a + 4\right) \div (a - 2)$ $a + 1 + \dfrac{6}{a - 2}$

116. $\left(4x^2 + 20x + 7\right) \div (x + 5)$ $4x + \dfrac{7}{x + 5}$

117. $\dfrac{a^3 + a^2 + 2a + 6}{a - 2}$ $\quad a^2 + 3a + 8 + \dfrac{22}{a - 2}$

118. $\dfrac{9b^3 - 18b^2 + 8b - 1}{3b - 2}$ $\quad 3b^2 - 4b - \dfrac{1}{3b - 2}$

119. $\dfrac{4x^4 - 4x^3 + x^2 + 4x - 3}{2x - 1}$

$2x^3 - x^2 + 2 - \dfrac{1}{2x - 1}$

120. $\dfrac{-10x^2 - x^3 - 21x + 18}{x - 6}$

$-x^2 - 16x - 117 - \dfrac{684}{x - 6}$

△ **121.** The area of the rectangle below is $\left(15x^3 - 3x^2 + 60\right)$ square feet. If its length is $3x^2$ feet, find its width.

$\left(5x - 1 + \dfrac{20}{x^2}\right)$ ft

Area is $(15x^3 - 3x^2 + 60)$ sq feet

122. The perimeter of the equilateral triangle below is $\left(21a^3b^6 + 3a - 3\right)$ units. Find the length of a side.
$\left(7a^3b^6 + a - 1\right)$ units

Perimeter is
$(21a^3b^6 + 3a - 3)$ units

Mixed Review

Evaluate.

123. 3^3 $\quad 27$

124. $\left(-\dfrac{1}{2}\right)^3$ $\quad -\dfrac{1}{8}$

Simplify each expression. Write each answer using positive exponents only.

125. $\left(4xy^2\right)\left(x^3y^5\right)$ $4x^4y^7$

126. $\dfrac{18x^9}{27x^3}$ $\dfrac{2x^6}{3}$

127. $\left(\dfrac{3a^4}{b^2}\right)^3$ $\dfrac{27a^{12}}{b^6}$

128. $\left(2x^{-4}y^3\right)^{-4}$ $\dfrac{x^{16}}{16y^{12}}$

129. $\dfrac{a^{-3}b^6}{9^{-1}a^{-5}b^{-2}}$ $9a^2b^8$

Perform the indicated operations and simplify.

130. $\left(-y^2 - 4\right) + \left(3y^2 - 6\right)$ $2y^2 - 10$

131. $(6x + 2) + (5x - 7)$ $11x - 5$

132. $\left(5x^2 + 2x - 6\right) - (-x - 4)$ $5x^2 + 3x - 2$

133. $\left(8y^2 - 3y + 1\right) - \left(3y^2 + 2\right)$ $5y^2 - 3y - 1$

134. $(2x + 5)(3x - 2)$ $6x^2 + 11x - 10$

135. $4x\left(7x^2 + 3\right)$ $28x^3 + 12x$

136. $(7x - 2)(4x - 9)$ $28x^2 - 71x + 18$

137. $(x - 3)\left(x^2 + 4x - 6\right)$ $x^3 + x^2 - 18x + 18$

Use special products to multiply.

138. $(5x + 4)^2$ $25x^2 + 40x + 16$

139. $(6x + 3)(6x - 3)$ $36x^2 - 9$

Divide.

140. $\dfrac{8a^4 - 2a^3 + 4a - 5}{2a^3}$ $4a - 1 + \dfrac{2}{a^2} - \dfrac{5}{2a^3}$

141. $\dfrac{x^2 + 2x + 10}{x + 5}$ $x - 3 + \dfrac{25}{x + 5}$

142. $\dfrac{4x^3 + 8x^2 - 11x + 4}{2x - 3}$ $2x^2 + 7x + 5 + \dfrac{19}{2x - 3}$

Evaluate each expression.

1. 2^5

2. $(-3)^4$

3. -3^4

4. 4^{-3}

Simplify each expression. Write the result using only positive exponents.

5. $(3x^2)(-5x^9)$

6. $\dfrac{y^7}{y^2}$

7. $\dfrac{r^{-8}}{r^{-3}}$

8. $\left(\dfrac{4x^2y^3}{x^3y^{-4}}\right)^2$

9. $\dfrac{6^2x^{-4}y^{-1}}{6^3x^{-3}y^7}$

Express each number in scientific notation.

10. 563,000

11. 0.0000863

Write each number in standard form.

12. 1.5×10^{-3}

13. 6.23×10^4

14. Simplify. Write the answer in standard form.
$(1.2 \times 10^5)(3 \times 10^{-7})$

15. a. Complete the table for the polynomial $4xy^2 + 7xyz + x^3y - 2$.

Term	Numerical Coefficient	Degree of Term
$4xy^2$	4	3
$7xyz$	7	3
x^3y	1	4
-2	-2	0

b. What is the degree of the polynomial?

16. Simplify by combining like terms.
$5x^2 + 4x - 7x^2 + 11 + 8x$

Perform each indicated operation.

17. $(8x^3 + 7x^2 + 4x - 7) + (8x^3 - 7x - 6)$

18. $\begin{array}{r} 5x^3 + x^2 + 5x - 2 \\ -(8x^3 - 4x^2 + 5x - 7) \\ \hline \end{array}$

19. Subtract $(4x + 2)$ from the sum of $(8x^2 + 7x + 5)$ and $(x^3 - 8)$.

Answers

Note to Instructor: The Chapter 5 Test file in the TestGen program provides algorithms specifically matched to this test for easy replication for practice or assessment purposes.

1. 32

2. 81

3. -81

4. $\dfrac{1}{64}$

5. $-15x^{11}$

6. y^5

7. $\dfrac{1}{r^5}$

8. $\dfrac{16y^{14}}{x^2}$

9. $\dfrac{1}{6xy^8}$

10. 5.63×10^5

11. 8.63×10^{-5}

12. 0.0015

13. 62,300

14. 0.036

15. a. see table

b. 4

16. $-2x^2 + 12x + 11$

17. $16x^3 + 7x^2 - 3x - 13$

18. $-3x^3 + 5x^2 + 4x + 5$

19. $x^3 + 8x^2 + 3x - 5$

20. $\dfrac{3x^3 + 22x^2}{+ 41x + 14}$

21. $6x^4 - 9x^3 + 21x^2$

22. $3x^2 + 16x - 35$

23. $9x^2 - \dfrac{1}{25}$

24. $16x^2 - 16x + 4$

25. $64x^2 + 48x + 9$

26. $x^4 - 81b^2$

27. see table

28. $(4x^2 - 9)$ sq in.

29. $\dfrac{x}{2y} + \dfrac{1}{4} - \dfrac{7}{8y}$

30. $x + 2$

31. $\dfrac{9x^2 - 6x + 4}{-\dfrac{16}{3x+2}}$

Multiply in Exercises 20 through 26.

20. $(3x + 7)(x^2 + 5x + 2)$

21. $3x^2(2x^2 - 3x + 7)$

22. $(x + 7)(3x - 5)$

23. $\left(3x - \dfrac{1}{5}\right)\left(3x + \dfrac{1}{5}\right)$

24. $(4x - 2)^2$

25. $(8x + 3)^2$

26. $(x^2 - 9b)(x^2 + 9b)$

27. The height of the Bank of China in Hong Kong is 1001 feet. Neglecting air resistance, the height of an object dropped from this building at time t seconds is given by the polynomial $-16t^2 + 1001$. Find the height of the object at the given times below.

t	0 seconds	1 second	3 seconds	5 seconds
$-16t^2 + 1001$	1001 ft	985 ft	857 ft	601 ft

△ **28.** Find the area of the top of the table. Express the area as a product, then multiply and simplify.

$(2x - 3)$ inches $(2x + 3)$ inches

Divide.

29. $\dfrac{4x^2 + 2xy - 7x}{8xy}$

30. $(x^2 + 7x + 10) \div (x + 5)$

31. $\dfrac{27x^3 - 8}{3x + 2}$

1. Given the set
$$\left\{-2, 0, \frac{1}{4}, 112, -3, 11, \sqrt{2}\right\},$$ list the
numbers in this set that belong to the
set of:
   a. Natural numbers
   b. Whole numbers
   c. Integers
   d. Rational numbers
   e. Irrational numbers
   f. Real numbers

2. Find the absolute value of each number.
   a. $|-7.2|$
   b. $|0|$
   c. $\left|-\frac{1}{2}\right|$

3. Evaluate (find the value of) each
expression.
   a. 3^2
   b. 5^3
   c. 2^4
   d. 7^1
   e. $\left(\dfrac{3}{7}\right)^2$
   f. $(0.6)^2$

4. Multiply. Write products in lowest
terms.
   a. $\dfrac{3}{4} \cdot \dfrac{7}{21}$
   b. $\dfrac{1}{2} \cdot 4\dfrac{5}{6}$

5. Simplify: $\dfrac{3}{2} \cdot \dfrac{1}{2} - \dfrac{1}{2}$

6. Evaluate $\dfrac{2x - 7y}{x^2}$ for $x = 5$ and $y = 1$.

7. Write an algebraic expression that
represents each phrase. Let the variable
x represent the unknown number.
   a. The sum of a number and 3
   b. The product of 3 and a number
   c. The quotient of 7.3 and a number
   d. 10 decreased by a number
   e. 5 times a number, increased by 7

8. Simplify: $8 + 3(2 \cdot 6 - 1)$

Answers

1. a. 11, 112

   b. 0, 11, 112

   c. $-3, -2, 0, 11, 112$

   d. $-3, -2, 0, \dfrac{1}{4}, 11, 112$

   e. $\sqrt{2}$

   f. $-2, 0, \dfrac{1}{4}, 112, -3, 11, \sqrt{2}$ (Sec. 1.2, Ex. 11)

2. a. 7.2

   b. 0

   c. $\dfrac{1}{2}$ (Sec. 1.2)

3. a. 9

   b. 125

   c. 16

   d. 7

   e. $\dfrac{9}{49}$

   f. 0.36 (Sec. 1.3, Ex. 1)

4. a. $\dfrac{1}{4}$

   b. $2\dfrac{5}{12}$ (Sec. R.2)

5. $\dfrac{1}{4}$ (Sec. 1.3, Ex. 4)

6. $\dfrac{3}{25}$ (Sec. 1.3)

7. a. $x + 3$

   b. $3x$

   c. $7.3 \div x$ or $\dfrac{7.3}{x}$

   d. $10 - x$

   e. $5x + 7$ (Sec. 1.3, Ex. 9)

8. 41 (Sec. 1.3)

9. Add: $11.4 + (-4.7)$

10. Is $x = 1$ a solution of $5x^2 + 2 = x - 8$?

11. Find the value of each expression when $x = 2$ and $y = -5$.

   a. $\dfrac{x - y}{12 + x}$

   b. $x^2 - y$

12. Subtract.

   a. $7 - 40$

   b. $-5 - (-10)$

Divide.

13. $\dfrac{-30}{-10}$

14. $\dfrac{-48}{6}$

15. $\dfrac{42}{-0.6}$

16. $\dfrac{-30}{-0.2}$

Find each product by using the distributive property to remove parentheses.

17. $5(3x + 2)$

18. $-3(2x - 3)$

19. $-2(y + 0.3z - 1)$

20. $4(-x^2 + 6x - 1)$

21. $-(9x + y - 2z + 6)$

22. $-(-4xy + 6y - 2)$

23. Solve: $6(2a - 1) - (11a + 6) = 7$

24. Solve: $2x + \dfrac{1}{8} = x - \dfrac{3}{8}$

25. Solve: $\dfrac{y}{7} = 20$

26. Solve: $10 = 5j - 2$

27. Solve: $0.25x + 0.10(x - 3) = 1.1$

28. Solve: $\dfrac{7x + 5}{3} = x + 3$

29. Twice the sum of a number and 4 is the same as four times the number decreased by 12. Find the number.

30. Write the phrase as an algebraic expression and simplify if possible. Double a number, subtracted from the sum of the number and seven.

△ **31.** Charles Pecot can afford enough fencing to enclose a rectangular garden with a perimeter of 140 feet. If the width of his garden is to be 30 feet, find the length.

32. Simplify: $\dfrac{4(-3) + (-8)}{5 + (-5)}$

33. The number 120 is 15% of what number?

34. Graph $x < 5$.

35. Solve: $-4x + 7 \geq -9$. Graph the solution set.

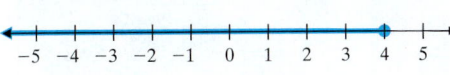

36. Evaluate.

   a. $(-5)^2$ **b.** -5^2 **c.** $2 \cdot 5^2$

37. Simplify each expression.

   a. $x^7 \cdot x^4$

   b. $\left(\dfrac{t}{2}\right)^4$

   c. $(9y^5)^2$

38. Simplify: $\dfrac{(z^2)^3 \cdot z^7}{z^9}$

Simplify each expression. Write each result using positive exponents only.

39. $\left(\dfrac{3a^2}{b}\right)^{-3}$ **40.** $(5x^7)(-3x^9)$ **41.** $(5y^3)^{-2}$ **42.** $(-3)^{-2}$

Perform any indicated operation. Then simplify the expression by combining any like terms.

43. $9x^3 + x^3$ **44.** $(5y^2 - 6) - (y^2 + 2)$ **45.** $5x^2 + 6x - 9x - 3$

46. Multiply: $(10x^2 - 3)(10x^2 + 3)$ **47.** Multiply: $7x(x^2 + 2x + 5)$

48. Multiply: $(10x^2 + 3)^2$ **49.** Divide: $\dfrac{9x^5 - 12x^2 + 3x}{3x^2}$

---

33. 800 (Sec. 2.6, Ex. 2)

34. see graph (Sec. 2.7)

35. $\{x \mid x \leq 4\}$; see graph (Sec. 2.7, Ex. 7)

36. a. 25

   b. -25

   c. 50 (Sec. 5.1)

37. a. x^{11}

   b. $\dfrac{t^4}{16}$

   c. $81y^{10}$ (Sec. 5.1, Ex. 33)

38. z^4 (Sec. 5.1)

39. $\dfrac{b^3}{27a^6}$ (Sec. 5.2, Ex. 10)

40. $-15x^{16}$ (Sec. 5.1)

41. $\dfrac{1}{25y^6}$ (Sec. 5.2, Ex. 14)

42. $\dfrac{1}{9}$ (Sec. 5.2)

43. $10x^3$ (Sec. 5.3, Ex. 8)

44. $4y^2 - 8$ (Sec. 5.4)

45. $5x^2 - 3x - 3$ (Sec. 5.3, Ex. 9)

46. $100x^4 - 9$ (Sec. 5.6)

47. $7x^3 + 14x^2 + 35x$ (Sec. 5.5, Ex. 4)

48. $100x^4 + 60x^2 + 9$ (Sec. 5.6)

49. $3x^3 - 4 + \dfrac{1}{x}$ (Sec. 5.7, Ex. 2)

6

Factoring Polynomials

In Chapter 5, we learned how to multiply polynomials. Now we will deal with an operation that is the reverse process of multiplying—factoring. Factoring is an important algebraic skill because it allows us to write a sum as a product. As we will see in Sections 6.6 and 6.7, factoring can be used to solve equations other than linear equations. In Chapter 7, we will also use factoring to simplify and perform arithmetic operations on rational expressions.

500,000 Units

1,000,000 Units

2,000,000 Units

Downloaded Music Now Calculated in Gold, Platinum, or Multi-Platinum Program

In 2012, digital album sales set a new high. Also, digital track sales were up 5%, and for the first time, two digital songs each had more than 6,000,000 downloads for the calendar year. In 2013, the Recording Industry Association of America (RIAA) announced the integration of downloaded music into the calculation of its Gold, Platinum, or Multi-Platinum Program for artistic achievement. Simply put, units are defined as:

- Each permanent digital download counts as 1 unit for certification purposes.
- 100 on-demand audio and/or video streams will count as 1 unit for certification purposes.

In Section 6.1, Exercises 99 and 100, we continue to study the increase in popularity of digital music.

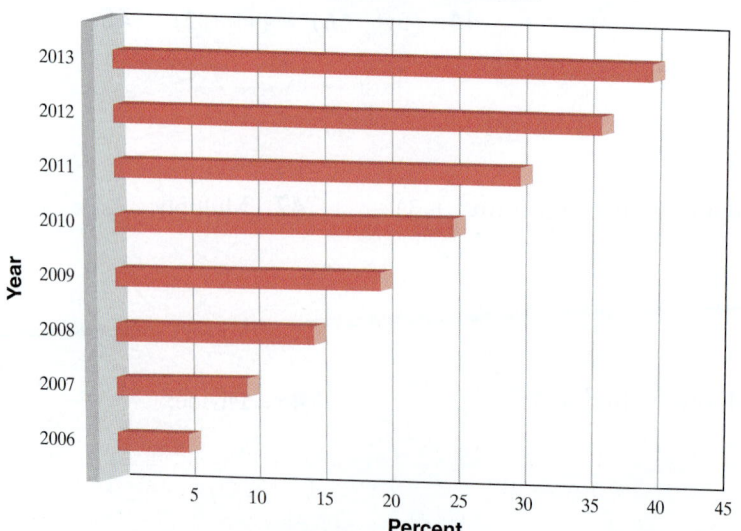

Percent of Digital Album Sales Compared to All Album Purchases

Source: The Nielsen Company & Billboard

6.1 The Greatest Common Factor and Factoring by Grouping

In the product $2 \cdot 3 = 6$, the numbers 2 and 3 are called **factors** of 6 and $2 \cdot 3$ is a **factored form** of 6. This is true of polynomials also. Since $(x + 2)(x + 3) = x^2 + 5x + 6$, then $(x + 2)$ and $(x + 3)$ are factors of $x^2 + 5x + 6$, and $(x + 2)(x + 3)$ is a factored form of the polynomial.

a factored form of 6

$$\underset{\text{factor}}{2} \cdot \underset{\text{factor}}{3} = \underset{\text{product}}{6}$$

a factored form of x^5

$$\underset{\text{factor}}{x^2} \cdot \underset{\text{factor}}{x^3} = \underset{\text{product}}{x^5}$$

a factored form of $x^2 + 5x + 6$

$$\underset{\text{factor}}{(x + 2)} \underset{\text{factor}}{(x + 3)} = \underset{\text{product}}{x^2 + 5x + 6}$$

> The process of writing a polynomial as a product is called **factoring** the polynomial.

Study the examples below and look for a pattern.

Multiplying: $5(x^2 + 3) = 5x^2 + 15$ $2x(x - 7) = 2x^2 - 14x$

Factoring: $5x^2 + 15 = 5(x^2 + 3)$ $2x^2 - 14x = 2x(x - 7)$

Do you see that factoring is the reverse process of multiplying?

$$x^2 + 5x + 6 = (x + 2)(x + 3)$$
factoring
multiplying

✓Concept Check Multiply: $2(x - 4)$
What do you think the result of factoring $2x - 8$ would be? Why?

Objective A Finding the Greatest Common Factor of a List of Numbers

The first step in factoring a polynomial is to see whether the terms of the polynomial have a common factor. If there is one, we can write the polynomial as a product by **factoring out** the common factor. We will usually factor out the *greatest* common factor (GCF).

The GCF of a list of integers is the largest integer that is a factor of all the integers in the list. For example, the GCF of 12 and 20 is 4 because 4 is the largest integer that is a factor of both 12 and 20. With large integers, the GCF may not be easily found by inspection. When this happens, use the following steps.

Teaching Tip
You may want to begin this chapter by pointing out that students have the tools to solve linear equations such as $8x = x - 12$ but have not yet learned the tools to solve equations involving higher-order polynomials such as $x^2 + 8x = x - 12$. The factoring skills students learn in this chapter will give them the tools needed to begin solving these more complicated kinds of equations.

Teaching Tip
This might be a good time to remind students that no matter what number they substitute for x in the equation $x^2 + 5x + 6 = (x + 2)(x + 3)$, it will always be a true statement. Ask students to verify this with several numerical values.

✓Concept Check Answer
$2x - 8$; the result would be $2(x - 4)$ because factoring is the reverse process of multiplying.

> **Finding the GCF of a List of Integers**
>
> **Step 1:** Write each number as a product of prime numbers.
>
> **Step 2:** Identify the common prime factors.
>
> **Step 3:** The product of all common prime factors found in Step 2 is the greatest common factor. If there are no common prime factors, the greatest common factor is 1.

Recall that a prime number is a whole number other than 1 whose only factors are 1 and itself. (See Section R.1 for a review.)

Practice 1

Find the GCF of each list of numbers.

a. 45 and 75

b. 32 and 33

c. 14, 24, and 60

Example 1   Find the GCF of each list of numbers.

a. 28 and 40   **b.** 55 and 21   **c.** 15, 18, and 66

Solution:

a. Write each number as a product of primes.

$$28 = 2 \cdot 2 \cdot 7 = 2^2 \cdot 7$$
$$40 = 2 \cdot 2 \cdot 2 \cdot 5 = 2^3 \cdot 5$$

There are two common factors, each of which is 2, so the GCF is

$$\text{GCF} = 2 \cdot 2 = 4$$

b. $55 = 5 \cdot 11$
$21 = 3 \cdot 7$

There are no common prime factors; thus, the GCF is 1.

c. $15 = 3 \cdot 5$
$18 = 2 \cdot 3 \cdot 3 = 2 \cdot 3^2$
$66 = 2 \cdot 3 \cdot 11$

The only prime factor common to all three numbers is 3, so the GCF is

$$\text{GCF} = 3$$

■ **Work Practice 1**

Objective B Finding the Greatest Common Factor of a List of Terms ▶

The greatest common factor of a list of variables raised to powers is found in a similar way. For example, the GCF of x^2, x^3, and x^5 is x^2 because each term contains a factor of x^2 and no higher power of x is a factor of each term.

$$x^2 = x \cdot x$$
$$x^3 = x \cdot x \cdot x$$
$$x^5 = x \cdot x \cdot x \cdot x \cdot x$$

Practice 2

Find the GCF of each list of terms.

a. $y^4, y^5,$ and y^8

b. x and x^{10}

There are two common factors, each of which is x, so the GCF $= x \cdot x$ or x^2. From this example, we see that **the GCF of a list of common variables raised to powers is the variable raised to the smallest exponent in the list.**

Example 2   Find the GCF of each list of terms.

a. $x^3, x^7,$ and x^5

b. $y, y^4,$ and y^7

Answers

1. a. 15   **b.** 1   **c.** 2

2. a. y^4 **b.** x

Solution:

a. The GCF is x^3, since 3 is the smallest exponent to which x is raised.

b. The GCF is y^1 or y, since 1 is the smallest exponent on y.

■ Work Practice 2

The **greatest common factor (GCF) of a list of terms** is the product of the GCF of the numerical coefficients and the GCF of the variable factors.

$$20x^2y^2 = 2 \cdot 2 \cdot 5 \cdot x \cdot x \cdot y \cdot y$$
$$6xy^3 = 2 \cdot 3 \cdot x \cdot y \cdot y \cdot y$$
$$\text{GCF} = 2 \cdot x \cdot y \cdot y = 2xy^2$$

Helpful Hint

Remember that the GCF of a list of terms contains the smallest exponent on each common variable.

The GCF of x^5y^6, x^2y^7, and x^3y^4 is x^2y^4.
— Smallest exponent on x
— Smallest exponent on y

Example 3 Find the greatest common factor of each list of terms.

a. $6x^2$, $10x^3$, and $-8x$

b. $-18y^2$, $-63y^3$, and $27y^4$

c. a^3b^2, a^5b, and a^6b^2

Solution:

a. $6x^2 = 2 \cdot 3 \cdot x^2$

$10x^3 = 2 \cdot 5 \cdot x^3$ $\quad\longrightarrow$ The GCF of x^2, x^3, and x^1 is x^1 or x.

$-8x = -1 \cdot 2 \cdot 2 \cdot 2 \cdot x^1$

$\text{GCF} = 2 \cdot x^1 \quad \text{or} \quad 2x$

b. $-18y^2 = -1 \cdot 2 \cdot 3 \cdot 3 \cdot y^2$

$-63y^3 = -1 \cdot 3 \cdot 3 \cdot 7 \cdot y^3$ $\quad\longrightarrow$ The GCF of y^2, y^3, and y^4 is y^2.

$27y^4 = 3 \cdot 3 \cdot 3 \cdot y^4$

$\text{GCF} = 3 \cdot 3 \cdot y^2 \quad \text{or} \quad 9y^2$

c. The GCF of a^3, a^5, and a^6 is a^3.

The GCF of b^2, b, and b^2 is b. Thus,

the GCF of a^3b^2, a^5b, and a^6b^2 is a^3b.

■ Work Practice 3

Objective C Factoring Out the Greatest Common Factor ▶

To factor a polynomial such as $8x + 14$, we first see whether the terms have a greatest common factor other than 1. In this case, they do: The GCF of $8x$ and 14 is 2.

We factor out 2 from each term by writing each term as the product of 2 and the term's remaining factors.

$$8x + 14 = 2 \cdot 4x + 2 \cdot 7$$

Using the distributive property, we can write

$$8x + 14 = 2 \cdot 4x + 2 \cdot 7$$
$$= 2(4x + 7)$$

Practice 3

Find the greatest common factor of each list of terms.

a. $6x^2$, $9x^4$, and $-12x^5$

b. $-16y$, $-20y^6$, and $40y^4$

c. a^5b^4, ab^3, and a^3b^2

Answer

3. **a.** $3x^2$ **b.** $4y$ **c.** ab^2

Thus, a factored form of $8x + 14$ is $2(4x + 7)$. We can check by multiplying:

$$2(4x + 7) = 2 \cdot 4x + 2 \cdot 7 = 8x + 14$$

Helpful Hint

A factored form of $8x + 14$ is *not*

$$2 \cdot 4x + 2 \cdot 7$$

Although the *terms* have been factored (written as products), the *polynomial* $8x + 14$ has not been factored. A factored form of $8x + 14$ is the *product* $2(4x + 7)$.

✓**Concept Check** Which of the following is/are factored form(s) of $6t + 18$?

    a. 6 **b.** $6 \cdot t + 6 \cdot 3$ **c.** $6(t + 3)$ **d.** $3(t + 6)$

Practice 4

Factor each polynomial by factoring out the greatest common factor (GCF).

a. $10y + 25$

b. $x^4 - x^9$

Example 4   Factor each polynomial by factoring out the greatest common factor (GCF).

a. $5ab + 10a$ **b.** $y^5 - y^{12}$

Solution:

a. The GCF of terms $5ab$ and $10a$ is $5a$. Thus,

$$5ab + 10a = 5a \cdot b + 5a \cdot 2$$
$$= 5a(b + 2) \qquad \text{Apply the distributive property.}$$

We can check our work by multiplying $5a$ and $(b + 2)$.
$5a(b + 2) = 5a \cdot b + 5a \cdot 2 = 5ab + 10a$, the original polynomial.

b. The GCF of y^5 and y^{12} is y^5. Thus,

$$y^5 - y^{12} = y^5(1) - y^5(y^7)$$
$$= y^5(1 - y^7)$$

Helpful Hint
Don't forget the 1.

■ **Work Practice 4**

Practice 5

Factor: $-10x^3 + 8x^2 - 2x$

Teaching Tip

If students have trouble factoring out a GCF that is a binomial, consider suggesting using substitution to write the expression as one that does not contain a binomial factor. For instance, in Example 9:

Let $z = x + 3$.

$5(x + 3) + y(x + 3) = 5z + yz$
$\qquad\qquad\qquad\qquad = z(5 + y)$

Then substitute $(x + 3)$ back for z to get $(x + 3)(5 + y)$.

Answers

4. a. $5(2y + 5)$ **b.** $x^4(1 - x^5)$

5. $2x(-5x^2 + 4x - 1)$

✓**Concept Check Answer**

c

Example 5   Factor: $-9a^5 + 18a^2 - 3a$

Solution:

$$-9a^5 + 18a^2 - 3a = 3a(-3a^4) + 3a(6a) + 3a(-1)$$
$$= 3a(-3a^4 + 6a - 1)$$

Helpful Hint
Don't forget the -1.

■ **Work Practice 5**

In Example 5, we could have chosen to factor out $-3a$ instead of $3a$. If we factor out $-3a$, we have

$$-9a^5 + 18a^2 - 3a = (-3a)(3a^4) + (-3a)(-6a) + (-3a)(1)$$
$$= -3a(3a^4 - 6a + 1)$$

Helpful Hint

Notice the changes in signs when factoring out $-3a$.

Examples   Factor.

6. $6a^4 - 12a = 6a(a^3 - 2)$

7. $\dfrac{3}{7}x^4 + \dfrac{1}{7}x^3 - \dfrac{5}{7}x^2 = \dfrac{1}{7}x^2(3x^2 + x - 5)$

8. $15p^2q^4 + 20p^3q^5 + 5p^3q^3 = 5p^2q^3(3q + 4pq^2 + p)$

■ **Work Practice 6–8**

Example 9   Factor: $5(x + 3) + y(x + 3)$

Solution: The binomial $(x + 3)$ is present in both terms and is the greatest common factor. We use the distributive property to factor out $(x + 3)$.

$5(x + 3) + y(x + 3) = (x + 3)(5 + y)$

■ **Work Practice 9**

Example 10   Factor: $3m^2n(a + b) - (a + b)$

Solution: The greatest common factor is $(a + b)$.

$3m^2n(a + b) - 1(a + b) = (a + b)(3m^2n - 1)$

■ **Work Practice 10**

Objective D Factoring by Grouping ▶

Once the GCF is factored out, we can often continue to factor the polynomial using a variety of techniques. We discuss here a technique called **factoring by grouping**. This technique can be used to factor some polynomials with four terms.

Example 11   Factor $xy + 2x + 3y + 6$ by grouping.

Solution: Notice that the first two terms of this polynomial have a common factor of x and that the second two terms have a common factor of 3. Because of this, group the first two terms, then the last two terms, and then factor out these common factors.

$$xy + 2x + 3y + 6 = (xy + 2x) + (3y + 6) \quad \text{Group terms.}$$
$$= x(y + 2) + 3(y + 2) \quad \text{Factor out GCF from each grouping.}$$

Next we factor out the common binomial factor, $(y + 2)$.

$$x(y + 2) + 3(y + 2) = (y + 2)(x + 3)$$

Now the result is a factored form because it is a product. We were able to write the polynomial as a product because of the common binomial factor, $(y + 2)$, that appeared. If this does not happen, try rearranging the terms of the original polynomial.

Check: Multiply $(y + 2)$ by $(x + 3)$.

$$(y + 2)(x + 3) = xy + 2x + 3y + 6,$$

the original polynomial.
Thus, a factored form of $xy + 2x + 3y + 6$ is the product $(y + 2)(x + 3)$.

■ **Work Practice 11**

Practice 6–8

Factor.

6. $4x^3 + 12x$

7. $\dfrac{2}{5}a^5 - \dfrac{4}{5}a^3 + \dfrac{1}{5}a^2$

8. $6a^3b + 3a^3b^2 + 9a^2b^4$

Practice 9

Factor: $7(p + 2) + q(p + 2)$

Practice 10

Factor $7xy^3(p + q) - (p + q)$

Practice 11

Factor $ab + 7a + 2b + 14$ by grouping.

Helpful Hint   Notice that this form, $x(y + 2) + 3(y + 2)$, is *not* a factored form of the original polynomial. It is a sum, not a product.

Answers

6. $4x(x^2 + 3)$

7. $\dfrac{1}{5}a^2(2a^3 - 4a + 1)$

8. $3a^2b(2a + ab + 3b^3)$

9. $(p + 2)(7 + q)$

10. $(p + q)(7xy^3 - 1)$

11. $(b + 7)(a + 2)$

You may want to try these steps when factoring by grouping.

> **To Factor a Four-Term Polynomial by Grouping**
>
> **Step 1:** Group the terms in two groups of two terms so that each group has a common factor.
>
> **Step 2:** Factor out the GCF from each group.
>
> **Step 3:** If there is a common binomial factor, factor it out.
>
> **Step 4:** If not, rearrange the terms and try these steps again.

Practice 12–14

Factor by grouping.

12. $28x^3 - 7x^2 + 12x - 3$

13. $2xy + 5y^2 - 4x - 10y$

14. $3x^2 + 4xy + 3x + 4y$

> **Examples**   Factor by grouping.

12. $15x^3 - 10x^2 + 6x - 4$

$= (15x^3 - 10x^2) + (6x - 4)$ Group the terms.

$= 5x^2(3x - 2) + 2(3x - 2)$ Factor each group.

$= (3x - 2)(5x^2 + 2)$ Factor out the common factor, $(3x - 2)$.

13. $3x^2 + 4xy - 3x - 4y$

$= (3x^2 + 4xy) + (-3x - 4y)$

$= x(3x + 4y) - 1(3x + 4y)$ Factor each group. A -1 is factored from the second pair of terms so that there is a common factor, $(3x + 4y)$.

$= (3x + 4y)(x - 1)$ Factor out the common factor, $(3x + 4y)$.

14. $2a^2 + 5ab + 2a + 5b$

$= (2a^2 + 5ab) + (2a + 5b)$ Factor each group. An understood 1 is written before $(2a + 5b)$ to help remember that

$= a(2a + 5b) + 1(2a + 5b)$ $(2a + 5b)$ is $1(2a + 5b)$.

$= (2a + 5b)(a + 1)$ Factor out the common factor, $(2a + 5b)$.

> **Helpful Hint**   Notice that the factor of 1 is written when $(2a + 5b)$ is factored out.

▶ **Work Practice 12–14**

Practice 15–17

Factor by grouping.

15. $4x^3 + x - 20x^2 - 5$

16. $3xy - 4 + x - 12y$

17. $2x - 2 + x^3 - 3x^2$

> **Examples**   Factor by grouping.

15. $3x^3 - 2x - 9x^2 + 6$

$= x(3x^2 - 2) - 3(3x^2 - 2)$ Factor each group. A -3 is factored from the second pair of terms so that there is a common factor, $(3x^2 - 2)$.

$= (3x^2 - 2)(x - 3)$ Factor out the common factor, $(3x^2 - 2)$.

16. $3xy + 2 - 3x - 2y$

Notice that the first two terms have no common factor other than 1. However, if we rearrange these terms, a grouping emerges that does lead to a common factor.

$3xy + 2 - 3x - 2y$

$= (3xy - 3x) + (-2y + 2)$

$= 3x(y - 1) - 2(y - 1)$ Factor -2 from the second group.

$= (y - 1)(3x - 2)$ Factor out the common factor, $(y - 1)$.

17. $5x - 10 + x^3 - x^2 = 5(x - 2) + x^2(x - 1)$

There is no common binomial factor that can now be factored out. No matter how we rearrange the terms, no grouping will lead to a common factor. Thus, this polynomial is not factorable by grouping.

▶ **Work Practice 15–17**

Answers

12. $(4x - 1)(7x^2 + 3)$

13. $(2x + 5y)(y - 2)$

14. $(3x + 4y)(x + 1)$

15. $(4x^2 + 1)(x - 5)$

16. $(3y + 1)(x - 4)$

17. cannot be factored by grouping

> **Helpful Hint**
>
> Throughout this chapter, we will be factoring polynomials. Even when the instructions do not so state, it is always a good idea to check your answers by multiplying.

Vocabulary, Readiness & Video Check

Use the choices below to fill in each blank. Some choices may be used more than once and some may not be used at all.

greatest common factor   factors   factoring   true   false   least   greatest

1. Since $5 \cdot 4 = 20$, the numbers 5 and 4 are called _____ factors _____ of 20.

2. The __ greatest common factor __ of a list of integers is the largest integer that is a factor of all the integers in the list.

3. The greatest common factor of a list of common variables raised to powers is the variable raised to the _____ least _____ exponent in the list.

4. The process of writing a polynomial as a product is called _____ factoring _____.

5. True or false? A factored form of $7x + 21 + xy + 3y$ is $7(x + 3) + y(x + 3)$. _____ false _____

6. True or false? A factored form of $3x^3 + 6x + x^2 + 2$ is $3x(x^2 + 2)$. _____ false _____

Write the prime factorization of the following integers.

7. 14 $2 \cdot 7$ **8.** 15 $3 \cdot 5$

Write the GCF of the following pairs of integers.

9. 18, 3   3   **10.** 7, 35   7   **11.** 20, 15   5   **12.** 6, 15   3

Martin-Gay Interactive Videos

See Video 6.1

Watch the section lecture video and answer the following questions.

Objective **A** **13.** Based on ⊞ Example 1, give a general definition for the greatest common factor (GCF) of a list of numbers. ▶

Objective **B** **14.** When finding the GCF of the terms in ⊞ Example 3, why are the numerical parts of the terms factored out, but not the variable parts? ▶

Objective **C** **15.** From ⊞ Example 5, once we factor out the GCF, how can the number of terms in the other factor help us determine if our factorization is correct? ▶

Objective **D** **16.** In ⊞ Examples 7 and 8, what are we reminded to always do first when factoring a polynomial? Also, a polynomial with how many terms suggests it might be factored by grouping? ▶

See video answer section.

6.1 **Exercise Set** MyMathLab®

Objectives **A** **B** **Mixed Practice** *Find the GCF for each list. See Examples 1 through 3.*

1. 32, 36   4   ▶ **2.** 36, 90   18   **3.** 18, 42, 84   6   **4.** 30, 75, 135   15

5. 24, 14, 21   1   **6.** 15, 25, 27   1   **7.** y^2, y^4, y^7 y^2 ▶ **8.** x^3, x^2, x^5 x^2

9. z^7, z^9, z^{11} z^7 **10.** y^8, y^{10}, y^{12} y^8 **11.** $x^{10}y^2, xy^2, x^3y^3$ xy^2 **12.** p^7q, p^8q^2, p^9q^3 p^7q

13. $14x, 21$ 7   **14.** $20y, 15$ 5   ▶ **15.** $12y^4, 20y^3$ $4y^3$ **16.** $32x^5, 18x^2$ $2x^2$

17. $-10x^2, 15x^3$ $5x^2$

18. $-21x^3, 14x$ $7x$

19. $12x^3, -6x^4, 3x^5$ $3x^3$

20. $15y^2, 5y^7, -20y^3$ $5y^2$

21. $-18x^2y, 9x^3y^3, 36x^3y$
  $9x^2y$

22. $7x^3y^3, -21x^2y^2, 14xy^4$
  $7xy^2$

23. $20a^6b^2c^8, 50a^7b$
  $10a^6b$

24. $40x^7y^2z, 64x^9y$
  $8x^7y$

Objective C *Factor out the GCF from each polynomial. See Examples 4 through 10.*

25. $3a + 6$
  $3(a + 2)$

26. $18a + 12$
  $6(3a + 2)$

▶ **27.** $30x - 15$
  $15(2x - 1)$

28. $42x - 7$
  $7(6x - 1)$

29. $x^3 + 5x^2$
  $x^2(x + 5)$

30. $y^5 + 6y^4$
  $y^4(y + 6)$

31. $6y^4 + 2y^3$
  $2y^3(3y + 1)$

32. $5x^2 + 10x^6$
  $5x^2(1 + 2x^4)$

33. $32xy - 18x^2$
  $2x(16y - 9x)$

34. $10xy - 15x^2$
  $5x(2y - 3x)$

35. $4x - 8y + 4$
  $4(x - 2y + 1)$

36. $7x + 21y - 7$
  $7(x + 3y - 1)$

37. $6x^3 - 9x^2 + 12x$
  $3x(2x^2 - 3x + 4)$

38. $12x^3 + 16x^2 - 8x$
  $4x(3x^2 + 4x - 2)$

39. $a^7b^6 - a^3b^2 + a^2b^5 - a^2b^2$
  $a^2b^2(a^5b^4 - a + b^3 - 1)$

40. $x^9y^6 + x^3y^5 - x^4y^3 + x^3y^3$
  $x^3y^3(x^6y^3 + y^2 - x + 1)$

41. $5x^3y - 15x^2y + 10xy$
  $5xy(x^2 - 3x + 2)$

▶ **42.** $14x^3y + 7x^2y - 7xy$
  $7xy(2x^2 + x - 1)$

43. $8x^5 + 16x^4 - 20x^3 + 12$
  $4(2x^5 + 4x^4 - 5x^3 + 3)$

44. $9y^6 - 27y^4 + 18y^2 + 6$
  $3(3y^6 - 9y^4 + 6y^2 + 2)$

45. $\dfrac{1}{3}x^4 + \dfrac{2}{3}x^3 - \dfrac{4}{3}x^5 + \dfrac{1}{3}x$

46. $\dfrac{2}{5}y^7 - \dfrac{4}{5}y^5 + \dfrac{3}{5}y^2 - \dfrac{2}{5}y$

▶ **47.** $y(x^2 + 2) + 3(x^2 + 2)$
  $(x^2 + 2)(y + 3)$

48. $x(y^2 + 1) - 3(y^2 + 1)$
  $(y^2 + 1)(x - 3)$

49. $z(y + 4) + 3(y + 4)$
  $(y + 4)(z + 3)$

50. $8(x + 2) - y(x + 2)$
  $(x + 2)(8 - y)$

51. $r(z^2 - 6) + (z^2 - 6)$
  $(z^2 - 6)(r + 1)$

52. $q(b^3 - 5) + (b^3 - 5)$
  $(b^3 - 5)(q + 1)$

45. $\dfrac{1}{3}x(x^3 + 2x^2 - 4x^4 + 1)$

46. $\dfrac{1}{5}y(2y^6 - 4y^4 + 3y - 2)$

Factor a negative number or a GCF with a negative coefficient from each polynomial. See Example 5.

53. $-2x - 14$ $-2(x + 7)$

54. $-7y - 21$ $-7(y + 3)$

55. $-2x^5 + x^7$ $-x^5(2 - x^2)$

56. $-5y^3 + y^6$ $-y^3(5 - y^3)$

57. $-6a^4 + 9a^3 - 3a^2$
  $-3a^2(2a^2 - 3a + 1)$

58. $-5m^6 + 10m^5 - 5m^3$
  $-5m^3(m^3 - 2m^2 + 1)$

Objective D *Factor each four-term polynomial by grouping. If this is not possible, write "not factorable by grouping." See Examples 11 through 17.*

59. $x^3 + 2x^2 + 5x + 10$
  $(x + 2)(x^2 + 5)$

60. $x^3 + 4x^2 + 3x + 12$
  $(x + 4)(x^2 + 3)$

61. $5x + 15 + xy + 3y$
  $(x + 3)(5 + y)$

62. $xy + y + 2x + 2$
  $(x + 1)(y + 2)$

63. $6x^3 - 4x^2 + 15x - 10$
  $(3x - 2)(2x^2 + 5)$

64. $16x^3 - 28x^2 + 12x - 21$
  $(4x - 7)(4x^2 + 3)$

65. $5m^3 + 6mn + 5m^2 + 6n$
  $(5m^2 + 6n)(m + 1)$

66. $8w^2 + 7wv + 8w + 7v$
  $(8w + 7v)(w + 1)$

67. $2y - 8 + xy - 4x$
  $(y - 4)(2 + x)$

68. $6x - 42 + xy - 7y$
  $(x - 7)(6 + y)$

69. $2x^3 + x^2 + 8x + 4$
  $(2x + 1)(x^2 + 4)$

70. $2x^3 - x^2 - 10x + 5$
  $(2x - 1)(x^2 - 5)$

71. $3x - 3 + x^3 - 4x^2$
  not factorable by grouping

72. $7x - 21 + x^3 - 2x^2$
  not factorable by grouping

73. $4x^2 - 8xy - 3x + 6y$
  $(x - 2y)(4x - 3)$

▶ **74.** $5xy - 15x - 6y + 18$
  $(y - 3)(5x - 6)$

75. $5q^2 - 4pq - 5q + 4p$
  $(5q - 4p)(q - 1)$

76. $6m^2 - 5mn - 6m + 5n$
  $(6m - 5n)(m - 1)$

Objectives C D Mixed Practice *Factor out the GCF from each polynomial. Then factor by grouping.*

77. $12x^2y - 42x^2 - 4y + 14$ $2(2y - 7)(3x^2 - 1)$

78. $90 + 15y^2 - 18x - 3xy^2$ $3(6 + y^2)(5 - x)$

▶ **79.** $6a^2 + 9ab^2 + 6ab + 9b^3$ $3(2a + 3b^2)(a + b)$

80. $16x^2 + 4xy^2 + 8xy + 2y^3$ $2(4x + y^2)(2x + y)$

Review

Multiply. See Section 5.5.

81. $(x + 2)(x + 5)$
$x^2 + 7x + 10$

82. $(y + 3)(y + 6)$
$y^2 + 9y + 18$

83. $(b + 1)(b - 4)$
$b^2 - 3b - 4$

84. $(x - 5)(x + 10)$
$x^2 + 5x - 50$

Fill in the chart by finding two numbers that have the given product and sum. The first column is filled in for you.

		85.	**86.**	**87.**	**88.**	**89.**	**90.**	**91.**	**92.**
Two Numbers	4, 7	2, 6	4, 5	−1, −8	−2, −8	−2, 5	−3, 3	−8, 3	−9, 4
Their Product	28	12	20	8	16	−10	−9	−24	−36
Their Sum	11	8	9	−9	−10	3	0	−5	−5

Concept Extensions

See the Concept Checks in this section.

93. Which of the following is/are factored form(s) of $-2x + 14$?
   a. $-2(x + 7)$ **b.** $-2 \cdot x + 14$
   c. $-2(x - 14)$ **d.** $-2(x - 7)$ d

94. Which of the following is/are factored form(s) of $8a - 24$?
   a. $8 \cdot a - 24$ **b.** $8(a - 3)$
   c. $4(2a - 12)$ **d.** $8 \cdot a - 2 \cdot 12$ b

Which of the following expressions are factored?

95. $(a + 6)(a + 2)$ factored

96. $(x + 5)(x + y)$ factored

97. $5(2y + z) - b(2y + z)$ not factored

98. $3x(a + 2b) + 2(a + 2b)$ not factored

99. The annual digital music track units sold (in millions) in the United States for the period 2009–2012 can be approximated by the polynomial $15x^2 - 250x + 2200$ where x is the number of years after 2000. (*Source:* The Nielsen Company & Billboard)
   a. Find the approximate U.S. annual digital music track units sold in 2011. To do so, let $x = 11$ and evaluate $15x^2 - 250x + 2200$.   1265 million units
   b. Find the approximate U.S. annual digital music track units sold in 2012.   1360 million units
   c. Suppose the annual digital music track units sold continues to be approximated by the polynomial $15x^2 - 250x + 2200$. Use this polynomial to predict the track music units sold in 2018.   2560 million units
   d. Factor out the GCF from the polynomial $15x^2 - 250x + 2200$. $5(3x^2 - 50x + 440)$

100. The annual percent of digital album sales when compared to all album purchases in the United States for the period 2006–2012 can be approximated by the polynomial $\frac{1}{10}x^2 + \frac{32}{10}x - \frac{180}{10}$ where x is the number of years after 2000. (*Source:* The Nielsen Company & Billboard)
   a. Find the approximate U.S. percent of digital album sales in 2010. To do so, let $x = 10$ and evaluate $\frac{1}{10}x^2 + \frac{32}{10}x - \frac{180}{10}$.   24%
   b. Find the approximate percent of U.S. digital album sales in 2012.   34.8%
   c. Suppose the percent of digital album sales continues to be approximated by the polynomial $\frac{1}{10}x^2 + \frac{32}{10}x - \frac{180}{10}$. Use this polynomial to predict the percent of digital album sales in 2020.   86%
   d. Factor out a factor of $\frac{1}{10}$ from the polynomial $\frac{1}{10}x^2 + \frac{32}{10}x - \frac{180}{10}$. $\frac{1}{10}(x^2 + 32x - 180)$

101. The annual orange production (in thousand tons) in the United States for the period 2010–2013 can be approximated by the polynomial $-322x^2 + 966x + 8372$, where x is the number of years after 2010. (*Source:* Based on data from the National Agricultural Statistics Service)

   a. Find the approximate U.S. orange production in 2011. To do so, let $x = 1$ and evaluate $-322x^2 + 966x + 8372$.   9016 thousand tons

   b. Find the approximate U.S. orange production in 2013.   8372 thousand tons

   c. Factor out the GCF from the polynomial $-322x^2 + 966x + 8372$. $-322(x^2 - 3x - 26)$ or $322(-x^2 + 3x + 26)$

102. The polynomial $4x^2 - 28x + 580$ represents the approximate number of visitors (in thousands) per year to Lowell National Historical Park in Lowell, Massachusetts, during 2008–2012. In this polynomial, x represents the years since 2008. (*Source:* Based on data from the National Park Service)

   a. Find the approximate number of visitors to Lowell National Historical Park in 2010. To do so, let $x = 2$ and evaluate $4x^2 - 28x + 580$.   540 thousand visitors

   b. Find the approximate number of visitors to Lowell National Historical Park in 2012.   532 thousand visitors

   c. Factor out the GCF from the polynomial $4x^2 - 28x + 580$. $4(x^2 - 7x + 145)$

Write an expression for the area of each shaded region. Then write the expression as a factored polynomial.

△**103.** 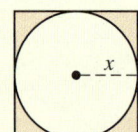 $4x^2 - \pi x^2;\ x^2(4 - \pi)$

△**104.** 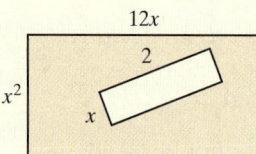 $12x^3 - 2x;\ 2x(6x^2 - 1)$

Write an expression for the length of each rectangle. (Hint: Factor the area binomial and recall that Area = width · length.)

△**105.** 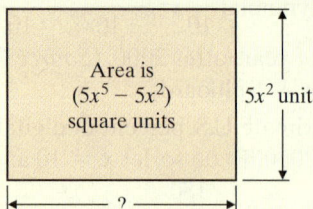 $(x^3 - 1)$ units

Area is $(5x^5 - 5x^2)$ square units $5x^2$ units   ?

△**106.** 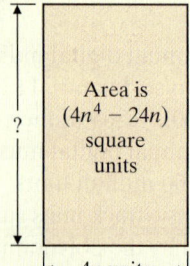 $(n^3 - 6)$ units

?   Area is $(4n^4 - 24n)$ square units   4n units

107. Construct a binomial whose greatest common factor is $5a^3$. (*Hint:* Multiply $5a^3$ by a binomial whose terms contain no common factor other than 1: $5a^3(\Box + \Box)$.)   answers may vary

108. Construct a trinomial whose greatest common factor is $2x^2$. See the hint for Exercise **107**.   answers may vary

109. Explain how you can tell whether a polynomial is written in factored form.   answers may vary

110. Construct a four-term polynomial that can be factored by grouping. Explain how you constructed the polynomial.   answers may vary

6.2 Factoring Trinomials of the Form $x^2 + bx + c$ ▶

Objective A Factoring Trinomials of the Form $x^2 + bx + c$ ▶

In this section, we factor trinomials of the form $x^2 + bx + c$, such as

$$x^2 + 7x + 12, \quad x^2 - 12x + 35, \quad x^2 + 4x - 12, \quad \text{and} \quad r^2 - r - 42$$

Notice that for these trinomials, the coefficient of the squared variable is 1.

Recall that factoring means to write as a product and that factoring and multiplying are reverse processes. Using the FOIL method of multiplying binomials, we have the following.

$$\begin{array}{ccccc} & F & O & I & L \\ (x + 3)(x + 1) = & x^2 & + 1x & + 3x & + 3 \\ & = x^2 + 4x + 3 \end{array}$$

Thus, a factored form of $x^2 + 4x + 3$ is $(x + 3)(x + 1)$.

Notice that the product of the first terms of the binomials is $x \cdot x = x^2$, the first term of the trinomial. Also, the product of the last two terms of the binomials is $3 \cdot 1 = 3$, the third term of the trinomial. The sum of these same terms is $3 + 1 = 4$, the coefficient of the middle, x, term of the trinomial.

The product of these numbers is 3.

$$x^2 + 4x + 3 = (x + 3)(x + 1)$$

The sum of these numbers is 4.

Many trinomials, such as the one above, factor into two binomials. To factor $x^2 + 7x + 10$, let's assume that it factors into two binomials and begin by writing two pairs of parentheses. The first term of the trinomial is x^2, so we use x and x as the first terms of the binomial factors.

$$x^2 + 7x + 10 = (x + \square)(x + \square)$$

To determine the last term of each binomial factor, we look for two integers whose product is 10 and whose sum is 7. The integers are 2 and 5. Thus,

$$x^2 + 7x + 10 = (x + 2)(x + 5)$$

Check: To see if we have factored correctly, we multiply.

$$\begin{array}{ll} (x + 2)(x + 5) = x^2 + 5x + 2x + 10 & \\ = x^2 + 7x + 10 & \text{Combine like terms.} \end{array}$$

Helpful Hint

Since multiplication is commutative, the factored form of $x^2 + 7x + 10$ can be written as either $(x + 2)(x + 5)$ or $(x + 5)(x + 2)$.

To Factor a Trinomial of the Form $x^2 + bx + c$

The product of these numbers is c.

$$x^2 + bx + c = (x + \square)(x + \square)$$

The sum of these numbers is b.

Teaching Tip

Encourage students to look at the problems in this section as brainteasers to solve. Students should look for two numbers such that the constant term of the polynomial is their product and the x-coefficient is their sum.

Teaching Tip

This might be a good time to remind students about positive integers that have no factors other than themselves and 1. Use the following questions.

- Do all positive integers have factors other than themselves and 1? (no)

- What do we call these integers with no other factors? (prime numbers)

- Do you think all polynomials have factors other than themselves and 1? (no)

- What might be a good name to call these polynomials that can't be factored? (prime polynomials)

Practice 1

Factor: $x^2 + 12x + 20$

Teaching Tip

For Examples 1 and 2, you may want to use an area model to help students visualize what they are finding.

Example 1

	x	a
x	x^2	ax
b	bx	ab

$ax + bx = 7x$
$ab = 12$

Example 2

	x	a
x	x^2	ax
b	bx	ab

$ax + bx = -12x$
$ab = 35$

Practice 2

Factor each trinomial.

a. $x^2 - 23x + 22$

b. $x^2 - 27x + 50$

Teaching Tip

Example 3 may be a good opportunity to challenge students. They know that a pair of factors with opposite signs is needed. Since the x-term coefficient is positive, what does this tell us about the factors? Lead them to discover that in this case, the absolute value of the positive factor will be larger than the absolute value of the negative factor. Point out that realizing this fact will cut their factor search in half.

Practice 3

Factor: $x^2 + 5x - 36$

Answers

1. $(x + 10)(x + 2)$

2. a. $(x - 1)(x - 22)$

  b. $(x - 2)(x - 25)$

3. $(x + 9)(x - 4)$

Example 1   Factor: $x^2 + 7x + 12$

Solution: We begin by writing the first terms of the binomial factors.

$$(x + \square)(x + \square)$$

Next we look for two numbers whose product is 12 and whose sum is 7. Since our numbers must have a positive product and a positive sum, we look at pairs of positive factors of 12 only.

Factors of 12	Sum of Factors
1, 12	13
2, 6	8
3, 4	7

Correct sum, so the numbers are 3 and 4.

Thus, $x^2 + 7x + 12 = (x + 3)(x + 4)$

Check: $(x + 3)(x + 4) = x^2 + 4x + 3x + 12 = x^2 + 7x + 12$

■ **Work Practice 1**

Example 2   Factor: $x^2 - 12x + 35$

Solution: Again, we begin by writing the first terms of the binomials.

$$(x + \square)(x + \square)$$

Now we look for two numbers whose product is 35 and whose sum is -12. Since our numbers must have a positive product and a negative sum, we look at pairs of negative factors of 35 only.

Factors of 35	Sum of Factors
$-1, -35$	-36
$-5, -7$	-12

Correct sum, so the numbers are -5 and -7.

$$x^2 - 12x + 35 = (x - 5)(x - 7)$$

Check: To check, multiply $(x - 5)(x - 7)$.

■ **Work Practice 2**

Example 3   Factor: $x^2 + 4x - 12$

Solution: $x^2 + 4x - 12 = (x + \square)(x + \square)$
We look for two numbers whose product is -12 and whose sum is 4. Since our numbers must have a negative product, we look at pairs of factors with opposite signs.

Factors of -12	Sum of Factors
$-1, 12$	11
$1, -12$	-11
$-2, 6$	4
$2, -6$	-4
$-3, 4$	1
$3, -4$	-1

Correct sum, so the numbers are -2 and 6.

$$x^2 + 4x - 12 = (x - 2)(x + 6)$$

■ **Work Practice 3**

Example 4   Factor: $r^2 - r - 42$

Solution: Because the variable in this trinomial is r, the first term of each binomial factor is r.

$$r^2 - r - 42 = (r + \square)(r + \square)$$

Now we look for two numbers whose product is -42 and whose sum is -1, the numerical coefficient of r. The numbers are 6 and -7. Therefore,

$$r^2 - r - 42 = (r + 6)(r - 7)$$

■ **Work Practice 4**

Practice 4

Factor each trinomial.
a. $q^2 - 3q - 40$
b. $y^2 + 2y - 48$

Example 5   Factor: $a^2 + 2a + 10$

Solution: Look for two numbers whose product is 10 and whose sum is 2. Neither 1 and 10 nor 2 and 5 give the required sum, 2. We conclude that $a^2 + 2a + 10$ is not factorable with integers. A polynomial such as $a^2 + 2a + 10$ is called a **prime polynomial.**

■ **Work Practice 5**

Practice 5

Factor: $x^2 + 6x + 15$

Example 6   Factor: $x^2 + 5xy + 6y^2$

Solution: $x^2 + 5xy + 6y^2 = (x + \square)(x + \square)$

Recall that the middle term, $5xy$, is the same as $5yx$. Thus, we can see that $5y$ is the "coefficient" of x. We then look for two terms whose product is $6y^2$ and whose sum is $5y$. The terms are $2y$ and $3y$ because $2y \cdot 3y = 6y^2$ and $2y + 3y = 5y$. Therefore,

$$x^2 + 5xy + 6y^2 = (x + 2y)(x + 3y)$$

■ **Work Practice 6**

Practice 6

Factor each trinomial.
a. $x^2 + 9xy + 14y^2$
b. $a^2 - 13ab + 30b^2$

Example 7   Factor: $x^4 + 5x^2 + 6$

Solution: As usual, we begin by writing the first terms of the binomials. Since the greatest power of x in this polynomial is x^4, we write

$$(x^2 + \square)(x^2 + \square) \quad \text{Since } x^2 \cdot x^2 = x^4$$

Now we look for two factors of 6 whose sum is 5. The numbers are 2 and 3. Thus,

$$x^4 + 5x^2 + 6 = (x^2 + 2)(x^2 + 3)$$

■ **Work Practice 7**

Practice 7

Factor: $x^4 + 8x^2 + 12$

If the terms of a polynomial are not written in descending powers of the variable, you may want to rearrange the terms before factoring.

Example 8   Factor: $40 - 13t + t^2$

Solution: First, we rearrange terms so that the trinomial is written in descending powers of t.

$$40 - 13t + t^2 = t^2 - 13t + 40$$

Next, try to factor.

$$t^2 - 13t + 40 = (t + \square)(t + \square)$$

Now we look for two factors of 40 whose sum is -13. The numbers are -8 and -5. Thus,

$$t^2 - 13t + 40 = (t - 8)(t - 5)$$

■ **Work Practice 8**

Practice 8

Factor: $48 - 14x + x^2$

Answers

4. a. $(q - 8)(q + 5)$
  b. $(y + 8)(y - 6)$
5. prime polynomial
6. a. $(x + 2y)(x + 7y)$
  b. $(a - 3b)(a - 10b)$
7. $(x^2 + 6)(x^2 + 2)$
8. $(x - 6)(x - 8)$

The following sign patterns may be useful when factoring trinomials.

Helpful Hint

A positive constant in a trinomial tells us to look for two numbers with the same sign. The sign of the coefficient of the middle term tells us whether the signs are both positive or both negative.

both positive · same sign
↓   ↓
$$x^2 + 10x + 16 = (x + 2)(x + 8)$$

both negative · same sign
↓   ↓
$$x^2 - 10x + 16 = (x - 2)(x - 8)$$

A negative constant in a trinomial tells us to look for two numbers with opposite signs.

opposite signs
↓
$$x^2 + 6x - 16 = (x + 8)(x - 2)$$

opposite signs
↓
$$x^2 - 6x - 16 = (x - 8)(x + 2)$$

Teaching Tip

Example 9 may be a good time to remind students that when given the instruction to factor, they should factor completely.

Objective B Factoring Out the Greatest Common Factor ▶

Remember that the first step in factoring any polynomial is to factor out the greatest common factor (if there is one other than 1 or -1).

Practice 9

Factor each trinomial.
a. $4x^2 - 24x + 36$
b. $x^3 + 3x^2 - 4x$

Example 9 Factor: $3m^2 - 24m - 60$

Solution: First we factor out the greatest common factor, 3, from each term.

$$3m^2 - 24m - 60 = 3(m^2 - 8m - 20)$$

Now we factor $m^2 - 8m - 20$ by looking for two factors of -20 whose sum is -8. The factors are -10 and 2. Therefore, the complete factored form is

$$3m^2 - 24m - 60 = 3(m + 2)(m - 10)$$

■ **Work Practice 9**

Helpful Hint

Remember to write the common factor, 3, as part of the factored form.

Practice 10

Factor: $5x^5 - 25x^4 - 30x^3$

Example 10 Factor: $2x^4 - 26x^3 + 84x^2$

Solution:

$$2x^4 - 26x^3 + 84x^2 = 2x^2(x^2 - 13x + 42)$$ Factor out common factor, $2x^2$.
$$= 2x^2(x - 6)(x - 7)$$ Factor $x^2 - 13x + 42$.

■ **Work Practice 10**

Answers
9. a. $4(x - 3)(x - 3)$
   b. $x(x + 4)(x - 1)$
10. $5x^3(x + 1)(x - 6)$

Vocabulary, Readiness & Video Check

Fill in each blank with "true" or "false."

1. To factor $x^2 + 7x + 6$, we look for two numbers whose product is 6 and whose sum is 7. __true__

2. We can write the factorization $(y + 2)(y + 4)$ also as $(y + 4)(y + 2)$. __true__

3. The factorization $(4x - 12)(x - 5)$ is completely factored. __false__

4. The factorization $(x + 2y)(x + y)$ may also be written as $(x + 2y)^2$. __false__

Complete each factored form.

5. $x^2 + 9x + 20 = (x + 4)(x + 5)$

6. $x^2 + 12x + 35 = (x + 5)(x + 7)$

7. $x^2 - 7x + 12 = (x - 4)(x - 3)$

8. $x^2 - 13x + 22 = (x - 2)(x - 11)$

9. $x^2 + 4x + 4 = (x + 2)(x + 2)$

10. $x^2 + 10x + 24 = (x + 6)(x + 4)$

 *Martin-Gay Interactive Videos* Watch the section lecture video and answer the following questions.

Objective A **11.** In ▤ Example 2, why are only negative factors of 15 considered?

Objective B **12.** In ▤ Example 5, we know we need a positive and a negative factor of -10. How do we determine which factor is negative?

See Video 6.2 🍎

See video answer section.

6.2 Exercise Set MyMathLab®

Objective A *Factor each trinomial completely. If a polynomial can't be factored, write "prime." See Examples 1 through 8.*

1. $x^2 + 7x + 6$
$(x + 6)(x + 1)$

2. $x^2 + 6x + 8$
$(x + 4)(x + 2)$

3. $y^2 - 10y + 9$
$(y - 9)(y - 1)$

4. $y^2 - 12y + 11$
$(y - 11)(y - 1)$

5. $x^2 - 6x + 9$
$(x - 3)(x - 3)$ or $(x - 3)^2$

6. $x^2 - 10x + 25$
$(x - 5)(x - 5)$ or $(x - 5)^2$

7. $x^2 - 3x - 18$
$(x - 6)(x + 3)$

8. $x^2 - x - 30$
$(x - 6)(x + 5)$

9. $x^2 + 3x - 70$
$(x + 10)(x - 7)$

10. $x^2 + 4x - 32$
$(x + 8)(x - 4)$

11. $x^2 + 5x + 2$
prime

12. $x^2 - 7x + 5$
prime

13. $x^2 + 8xy + 15y^2$
$(x + 5y)(x + 3y)$

14. $x^2 + 6xy + 8y^2$
$(x + 4y)(x + 2y)$

15. $a^4 - 2a^2 - 15$
$(a^2 - 5)(a^2 + 3)$

16. $y^4 - 3y^2 - 70$
$(y^2 - 10)(y^2 + 7)$

17. $13 + 14m + m^2$
$(m + 13)(m + 1)$

18. $17 + 18n + n^2$
$(n + 17)(n + 1)$

19. $10t - 24 + t^2$
$(t - 2)(t + 12)$

20. $6q - 27 + q^2$
$(q - 3)(q + 9)$

21. $a^2 - 10ab + 16b^2$
$(a - 2b)(a - 8b)$

22. $a^2 - 9ab + 18b^2$
$(a - 3b)(a - 6b)$

Objectives A B Mixed Practice *Factor each trinomial completely. Some of these trinomials contain a greatest common factor (other than 1). Don't forget to factor out the GCF first. See Examples 1 through 10.*

23. $2z^2 + 20z + 32$
$2(z + 8)(z + 2)$

24. $3x^2 + 30x + 63$
$3(x + 7)(x + 3)$

25. $2x^3 - 18x^2 + 40x$
$2x(x - 5)(x - 4)$

26. $3x^3 - 12x^2 - 36x$
$3x(x - 6)(x + 2)$

◐ 27. $x^2 - 3xy - 4y^2$
$(x - 4y)(x + y)$

28. $x^2 - 4xy - 77y^2$
$(x - 11y)(x + 7y)$

29. $x^2 + 15x + 36$
$(x + 12)(x + 3)$

30. $x^2 + 19x + 60$
$(x + 4)(x + 15)$

31. $x^4 - x^2 - 2$
$(x^2 - 2)(x^2 + 1)$

32. $x^4 - 5x^2 - 14$
$(x^2 - 7)(x^2 + 2)$

33. $r^2 - 16r + 48$
$(r - 12)(r - 4)$

34. $r^2 - 10r + 21$
$(r - 7)(r - 3)$

35. $x^2 + xy - 2y^2$
$(x + 2y)(x - y)$

36. $x^2 - xy - 6y^2$
$(x - 3y)(x + 2y)$

◐ 37. $3x^2 + 9x - 30$
$3(x + 5)(x - 2)$

38. $4x^2 - 4x - 48$
$4(x - 4)(x + 3)$

39. $3x^4 - 60x^2 + 108$
$3(x^2 - 18)(x^2 - 2)$

40. $2x^4 - 24x^2 + 70$
$2(x^2 - 7)(x^2 - 5)$

41. $x^2 - 18x - 144$
$(x - 24)(x + 6)$

42. $x^2 + x - 42$
$(x + 7)(x - 6)$

43. $r^2 - 3r + 6$
prime

44. $x^2 + 4x - 10$
prime

◐ 45. $x^2 - 8x + 15$
$(x - 5)(x - 3)$

46. $x^2 - 9x + 14$
$(x - 7)(x - 2)$

47. $6x^3 + 54x^2 + 120x$
$6x(x + 4)(x + 5)$

48. $3x^3 + 3x^2 - 126x$
$3x(x + 7)(x - 6)$

49. $4x^2y + 4xy - 12y$
$4y(x^2 + x - 3)$

50. $3x^2y - 9xy + 45y$
$3y(x^2 - 3x + 15)$

51. $x^2 - 4x - 21$
$(x - 7)(x + 3)$

52. $x^2 - 4x - 32$
$(x - 8)(x + 4)$

53. $x^2 + 7xy + 10y^2$
$(x + 5y)(x + 2y)$

54. $x^2 - 3xy - 4y^2$
$(x - 4y)(x + y)$

55. $64 + 24t + 2t^2$
$2(t + 8)(t + 4)$

56. $50 + 20t + 2t^2$
$2(t + 5)(t + 5)$ or $2(t + 5)^2$

57. $x^3 - 2x^2 - 24x$
$x(x - 6)(x + 4)$

58. $x^3 - 3x^2 - 28x$
$x(x - 7)(x + 4)$

59. $2t^5 - 14t^4 + 24t^3$
$2t^3(t - 4)(t - 3)$

60. $3x^6 + 30x^5 + 72x^4$
$3x^4(x + 6)(x + 4)$

◐ 61. $5x^3y - 25x^2y^2 - 120xy^3$
$5xy(x - 8y)(x + 3y)$

62. $7a^3b - 35a^2b^2 + 42ab^3$
$7ab(a - 3b)(a - 2b)$

63. $162 - 45m + 3m^2$
$3(m - 9)(m - 6)$

64. $48 - 20n + 2n^2$
$2(n - 6)(n - 4)$

65. $-x^2 + 12x - 11$
(Factor out −1 first.)
$-1(x - 11)(x - 1)$

66. $-x^2 + 8x - 7$
(Factor out −1 first.)
$-1(x - 7)(x - 1)$

67. $\frac{1}{2}y^2 - \frac{9}{2}y - 11$

(Factor out $\frac{1}{2}$ first.)

$\frac{1}{2}(y - 11)(y + 2)$

68. $\frac{1}{3}y^2 - \frac{5}{3}y - 8$

(Factor out $\frac{1}{3}$ first.)

$\frac{1}{3}(y - 8)(y + 3)$

69. $x^3y^2 + x^2y - 20x$
$x(xy - 4)(xy + 5)$

70. $a^2b^3 + ab^2 - 30b$
$b(ab - 5)(ab + 6)$

Review

Multiply. See Section 5.5.

71. $(2x + 1)(x + 5)$
$2x^2 + 11x + 5$

72. $(3x + 2)(x + 4)$
$3x^2 + 14x + 8$

73. $(5y - 4)(3y - 1)$
$15y^2 - 17y + 4$

74. $(4z - 7)(7z - 1)$
$28z^2 - 53z + 7$

75. $(a + 3b)(9a - 4b)$
$9a^2 + 23ab - 12b^2$

76. $(y - 5x)(6y + 5x)$
$6y^2 - 25xy - 25x^2$

Concept Extensions

77. Write a polynomial that factors as $(x - 3)(x + 8)$.
$x^2 + 5x - 24$

78. To factor $x^2 + 13x + 42$, think of two numbers whose _____ is 42 and whose _____ is 13.
product; sum

Complete each sentence in your own words.

79. If $x^2 + bx + c$ is factorable and c is negative, then the signs of the last-term factors of the binomials are opposite because … answers may vary

80. If $x^2 + bx + c$ is factorable and c is positive, then the signs of the last-term factors of the binomials are the same because … answers may vary

Remember that perimeter means distance around. Write the perimeter of each rectangle as a simplified polynomial. Then factor the polynomial completely.

△**81.**

$4x + 33$

$x^2 + 10x$

$2x^2 + 28x + 66; 2(x + 3)(x + 11)$

△**82.**

$12x^2$

$2x^3 + 16x$ $4x^3 + 24x^2 + 32x; 4x(x + 4)(x + 2)$

83. An object is thrown upward from the top of an 80-foot building with an initial velocity of 64 feet per second. Neglecting air resistance, the height of the object after t seconds is given by $-16t^2 + 64t + 80$. Factor this polynomial.
$-16(t - 5)(t + 1)$

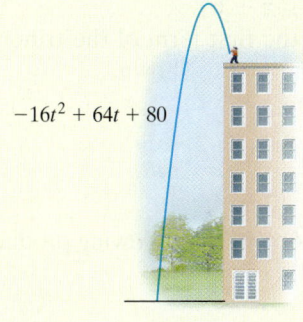

$-16t^2 + 64t + 80$

84. An object is thrown upward from the top of a 112-foot building with an initial velocity of 96 feet per second. Neglecting air resistance, the height of the object after t seconds is given by $-16t^2 + 96t + 112$. Factor this polynomial.
$-16(t - 7)(t + 1)$

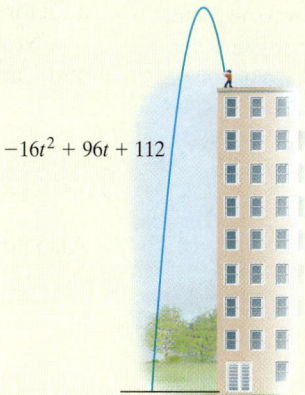

$-16t^2 + 96t + 112$

Factor each trinomial completely.

85. $x^2 + \dfrac{1}{2}x + \dfrac{1}{16}$ $\left(x + \dfrac{1}{4}\right)\left(x + \dfrac{1}{4}\right)$ or $\left(x + \dfrac{1}{4}\right)^2$

86. $x^2 + x + \dfrac{1}{4}$ $\left(x + \dfrac{1}{2}\right)\left(x + \dfrac{1}{2}\right)$ or $\left(x + \dfrac{1}{2}\right)^2$

87. $z^2(x + 1) - 3z(x + 1) - 70(x + 1)$
$(x + 1)(z - 10)(z + 7)$

88. $y^2(x + 1) - 2y(x + 1) - 15(x + 1)$
$(x + 1)(y - 5)(y + 3)$

Find a positive value of c so that each trinomial is factorable.

89. $n^2 - 16n + c$ 15; 28; 39; 48; 55; 60; 63; 64

90. $y^2 - 4y + c$ 3; 4

Find a positive value of b so that each trinomial is factorable.

91. $y^2 + by + 20$ 9; 12; 21

92. $x^2 + bx + 15$ 8; 16

Factor each trinomial. (Hint: Notice that $x^{2n} + 4x^n + 3$ factors as $(x^n + 1)(x^n + 3)$. Remember: $x^n \cdot x^n = x^{n+n}$ or x^{2n}.)

93. $x^{2n} + 8x^n - 20$ $(x^n + 10)(x^n - 2)$

94. $x^{2n} + 5x^n + 6$ $(x^n + 2)(x^n + 3)$

6.3   Factoring Trinomials of the Form $ax^2 + bx + c$

Objectives

A Factor Trinomials of the Form $ax^2 + bx + c$, where $a \neq 1$.

B Factor Out the GCF Before Factoring a Trinomial of the Form $ax^2 + bx + c$.

Teaching Tip

Point out that when the coefficient of the squared term is prime and the constant term is prime, there are only 4 possible ways to factor the trinomial. Have students create the 4 trinomials that can be factored when 7 is the coefficient of the squared term and 3 is the constant:

$(7x + 3)(x + 1) = 7x^2 + 10x + 3$
$(7x + 1)(x + 3) = 7x^2 + 22x + 3$
$(7x - 3)(x - 1) = 7x^2 - 10x + 3$
$(7x - 1)(x - 3) = 7x^2 - 22x + 3$

Objective A Factoring Trinomials of the Form $ax^2 + bx + c$

In this section, we factor trinomials of the form $ax^2 + bx + c$, such as

$$3x^2 + 11x + 6, \quad 8x^2 - 22x + 5, \quad \text{and} \quad 2x^2 + 13x - 7$$

Notice that the coefficient of the squared variable in these trinomials is a number other than 1. We will factor these trinomials using a trial-and-check method based on our work in the last section.

To begin, let's review the relationship between the numerical coefficients of the trinomial and the numerical coefficients of its factored form. For example, since

$$(2x + 1)(x + 6) = 2x^2 + 13x + 6,$$

a factored form of $2x^2 + 13x + 6$ is $(2x + 1)(x + 6)$.

Notice that $2x$ and x are factors of $2x^2$, the first term of the trinomial. Also, 6 and 1 are factors of 6, the last term of the trinomial, as shown:

$$2x^2 + 13x + 6 = (2x + 1)(x + 6)$$

Also notice that $13x$, the middle term, is the sum of the following products:

$$2x^2 + 13x + 6 = (2x + 1)(x + 6)$$

$1x$
$+ 12x$
$13x$ Middle term

Let's use this pattern to factor $5x^2 + 7x + 2$. First, we find factors of $5x^2$. Since all numerical coefficients in this trinomial are positive, we will use factors with positive numerical coefficients only. Thus, the factors of $5x^2$ are $5x$ and x. Let's try these factors as first terms of the binomials. Thus far, we have

$$5x^2 + 7x + 2 = (5x + \square)(x + \square)$$

Next, we need to find positive factors of 2. Positive factors of 2 are 1 and 2. Now we try possible combinations of these factors as second terms of the binomials until we obtain a middle term of $7x$.

$$(5x + 1)(x + 2) = 5x^2 + 11x + 2$$

$$\begin{array}{r} 1x \\ +\, 10x \\ \hline 11x \end{array} \longrightarrow \textbf{Incorrect} \text{ middle term}$$

Let's try switching factors 2 and 1.

$$(5x + 2)(x + 1) = 5x^2 + 7x + 2$$

$$\begin{array}{r} 2x \\ +\, 5x \\ \hline 7x \end{array} \longrightarrow \textbf{Correct} \text{ middle term}$$

Thus, a factored form of $5x^2 + 7x + 2$ is $(5x + 2)(x + 1)$. To check, we multiply $(5x + 2)$ and $(x + 1)$. The product is $5x^2 + 7x + 2$.

Example 1   Factor: $3x^2 + 11x + 6$

Solution: Since all numerical coefficients are positive, we use factors with positive numerical coefficients. We first find factors of $3x^2$.

Factors of $3x^2$: $3x^2 = 3x \cdot x$

If factorable, the trinomial will be of the form

$$3x^2 + 11x + 6 = (3x + \square)(x + \square)$$

Next we factor 6.

Factors of 6: $6 = 1 \cdot 6,$ $6 = 2 \cdot 3$

Now we try combinations of factors of 6 until a middle term of $11x$ is obtained. Let's try 1 and 6 first.

$$(3x + 1)(x + 6) = 3x^2 + 19x + 6$$

$$\begin{array}{r} 1x \\ +\, 18x \\ \hline 19x \end{array} \longrightarrow \textbf{Incorrect} \text{ middle term}$$

Now let's next try 6 and 1.

$$(3x + 6)(x + 1)$$

Before multiplying, notice that the terms of the factor $3x + 6$ have a common factor of 3. The terms of the original trinomial $3x^2 + 11x + 6$ have no common factor other than 1, so the terms of its factors will also contain no common factor other than 1. This means that $(3x + 6)(x + 1)$ is not a factored form.

Next let's try 2 and 3 as last terms.

$$(3x + 2)(x + 3) = 3x^2 + 11x + 6$$

$$\begin{array}{r} 2x \\ +\, 9x \\ \hline 11x \end{array} \longrightarrow \textbf{Correct} \text{ middle term}$$

Thus, a factored form of $3x^2 + 11x + 6$ is $(3x + 2)(x + 3)$.

■ **Work Practice 1**

Practice 1

Factor each trinomial.
a. $5x^2 + 27x + 10$
b. $4x^2 + 12x + 5$

Helpful Hint   This is true in general: If the terms of a trinomial have no common factor (other than 1), then the terms of each of its binomial factors will contain no common factor (other than 1).

Answers
1. a. $(5x + 2)(x + 5)$
    b. $(2x + 5)(2x + 1)$

✔ **Concept Check** Do the terms of $3x^2 + 29x + 18$ have a common factor? Without multiplying, decide which of the following factored forms could not be a factored form of $3x^2 + 29x + 18$.

a. $(3x + 18)(x + 1)$ **b.** $(3x + 2)(x + 9)$

c. $(3x + 6)(x + 3)$ **d.** $(3x + 9)(x + 2)$

Practice 2

Factor each trinomial.

a. $2x^2 - 11x + 12$

b. $6x^2 - 5x + 1$

Example 2   Factor: $8x^2 - 22x + 5$

Solution: Factors of $8x^2$: $8x^2 = 8x \cdot x$, $8x^2 = 4x \cdot 2x$

We'll try $8x$ and x.

$$8x^2 - 22x + 5 = (8x + \square)(x + \square)$$

Since the middle term, $-22x$, has a negative numerical coefficient, we factor 5 into negative factors.

     Factors of 5: $5 = -1 \cdot -5$

Let's try -1 and -5.

$$(8x - 1)(x - 5) = 8x^2 - 41x + 5$$

$-1x$
$+(-40x)$
$-41x$ ⟶ **Incorrect** middle term

Now let's try -5 and -1.

$$(8x - 5)(x - 1) = 8x^2 - 13x + 5$$

$-5x$
$+(-8x)$
$-13x$ ⟶ **Incorrect** middle term

Don't give up yet! We can still try other factors of $8x^2$. Let's try $4x$ and $2x$ with -1 and -5.

$$(4x - 1)(2x - 5) = 8x^2 - 22x + 5$$

$-2x$
$+(-20x)$
$-22x$ ⟶ **Correct** middle term

A factored form of $8x^2 - 22x + 5$ is $(4x - 1)(2x - 5)$.

■ **Work Practice 2**

Teaching Tip

Encourage students to analyze a problem before trying to solve it. Ask them to list some key characteristics about the trinomial in Example 3 before factoring it. (Key characteristics: Squared term is prime, constant term is prime and negative, x-term is positive.)

Practice 3

Factor each trinomial.

a. $3x^2 + 14x - 5$

b. $35x^2 + 4x - 4$

Answers

2. **a.** $(2x - 3)(x - 4)$

   b. $(3x - 1)(2x - 1)$

3. **a.** $(3x - 1)(x + 5)$

   b. $(5x + 2)(7x - 2)$

✔ **Concept Check Answers**

no; a, c, d

Example 3   Factor: $2x^2 + 13x - 7$

Solution: Factors of $2x^2$: $2x^2 = 2x \cdot x$

Factors of -7: $-7 = -1 \cdot 7$, $-7 = 1 \cdot -7$

We try possible combinations of these factors:

$(2x + 1)(x - 7) = 2x^2 - 13x - 7$ **Incorrect** middle term

$(2x - 1)(x + 7) = 2x^2 + 13x - 7$ **Correct** middle term

A factored form of $2x^2 + 13x - 7$ is $(2x - 1)(x + 7)$.

■ **Work Practice 3**

Example 4   Factor: $10x^2 - 13xy - 3y^2$

Solution: Factors of $10x^2$: $10x^2 = 10x \cdot x,$ $10x^2 = 2x \cdot 5x$

Factors of $-3y^2$: $-3y^2 = -3y \cdot y,$ $-3y^2 = 3y \cdot -y$

We try some combinations of these factors:

Correct   Correct
↓   ↓

$$(10x - 3y)(x + y) = 10x^2 + 7xy - 3y^2$$
$$(x + 3y)(10x - y) = 10x^2 + 29xy - 3y^2$$
$$(5x + 3y)(2x - y) = 10x^2 + xy - 3y^2$$
$$(2x - 3y)(5x + y) = 10x^2 - 13xy - 3y^2 \quad \textbf{Correct} \text{ middle term}$$

A factored form of $10x^2 - 13xy - 3y^2$ is $(2x - 3y)(5x + y)$.

🟧 **Work Practice 4**

Example 5   Factor: $3x^4 - 5x^2 - 8$

Solution: Factors of $3x^4$: $3x^4 = 3x^2 \cdot x^2$

Factors of -8: $-8 = -2 \cdot 4, 2 \cdot -4, -1 \cdot 8, 1 \cdot -8$

Try combinations of these factors:

Correct   Correct
↓   ↓

$$(3x^2 - 2)(x^2 + 4) = 3x^4 + 10x^2 - 8$$
$$(3x^2 + 4)(x^2 - 2) = 3x^4 - 2x^2 - 8$$
$$(3x^2 + 8)(x^2 - 1) = 3x^4 + 5x^2 - 8 \quad \textbf{Incorrect sign} \text{ on middle term, so switch}$$
$$\text{signs in binomial factors.}$$
$$(3x^2 - 8)(x^2 + 1) = 3x^4 - 5x^2 - 8 \quad \textbf{Correct} \text{ middle term}$$

🟧 **Work Practice 5**

Helpful Hint

Study the last two lines of Example 5. If a factoring attempt gives you a middle term whose numerical coefficient is the opposite of the desired numerical coefficient, try switching the signs of the last terms in the binomials.

Switched signs $\Big\langle$
$$(3x^2 + 8)(x^2 - 1) = 3x^4 + 5x^2 - 8 \quad \text{Middle term: } +5x$$
$$(3x^2 - 8)(x^2 + 1) = 3x^4 - 5x^2 - 8 \quad \text{Middle term: } -5x$$

Objective B Factoring Out the Greatest Common Factor ▶

Don't forget that the first step in factoring any polynomial is to look for a common factor to factor out.

Example 6   Factor: $24x^4 + 40x^3 + 6x^2$

Solution: Notice that all three terms have a common factor of $2x^2$. Thus, we factor out $2x^2$ first.

$$24x^4 + 40x^3 + 6x^2 = 2x^2(12x^2 + 20x + 3)$$

(Continued on next page)

Practice 4

Factor each trinomial.
a. $14x^2 - 3xy - 2y^2$
b. $12a^2 - 16ab - 3b^2$

Practice 5

Factor: $2x^4 - 5x^2 - 7$

Practice 6

Factor each trinomial.
a. $3x^3 + 17x^2 + 10x$
b. $6xy^2 + 33xy - 18x$

Answers
4. a. $(7x + 2y)(2x - y)$
  b. $(6a + b)(2a - 3b)$
5. $(2x^2 - 7)(x^2 + 1)$
6. a. $x(3x + 2)(x + 5)$
  b. $3x(2y - 1)(y + 6)$

Next we factor $12x^2 + 20x + 3$.

Factors of $12x^2$: $12x^2 = 4x \cdot 3x$, $12x^2 = 12x \cdot x$, $12x^2 = 6x \cdot 2x$

Since all terms in the trinomial have positive numerical coefficients, we factor 3 using positive factors only.

Factors of 3: $3 = 1 \cdot 3$

We try some combinations of the factors.

$$2x^2(4x + 3)(3x + 1) = 2x^2(12x^2 + 13x + 3)$$
$$2x^2(12x + 1)(x + 3) = 2x^2(12x^2 + 37x + 3)$$
$$2x^2(2x + 3)(6x + 1) = 2x^2(12x^2 + 20x + 3) \quad \textbf{Correct} \text{ middle term}$$

A factored form of $24x^4 + 40x^3 + 6x^2$ is $2x^2(2x + 3)(6x + 1)$.

> **Helpful Hint** Don't forget to include the common factor in the factored form.

■ **Work Practice 6**

When the term containing the squared variable has a negative coefficient, you may want to first factor out a common factor of -1.

Practice 7

Factor: $-5x^2 - 19x + 4$

Answer

7. $-1(x + 4)(5x - 1)$

Example 7 Factor: $-6x^2 - 13x + 5$

Solution: We begin by factoring out a common factor of -1.

$$-6x^2 - 13x + 5 = -1(6x^2 + 13x - 5) \quad \text{Factor out } -1.$$
$$= -1(3x - 1)(2x + 5) \quad \text{Factor } 6x^2 + 13x - 5.$$

■ **Work Practice 7**

Vocabulary, Readiness & Video Check

Complete each factorization.

1. $2x^2 + 5x + 3$ factors as $(2x + 3)(\underline{\quad})$. d
    a. $(x + 3)$ **b.** $(2x + 1)$ **c.** $(3x + 4)$ **d.** $(x + 1)$

2. $7x^2 + 9x + 2$ factors as $(7x + 2)(\underline{\quad})$. b
    a. $(3x + 1)$ **b.** $(x + 1)$ **c.** $(x + 2)$ **d.** $(7x + 1)$

3. $3x^2 + 31x + 10$ factors as $\underline{\quad}$. c
    a. $(3x + 2)(x + 5)$ **b.** $(3x + 5)(x + 2)$ **c.** $(3x + 1)(x + 10)$

4. $5x^2 + 61x + 12$ factors as $\underline{\quad}$. a
    a. $(5x + 1)(x + 12)$ **b.** $(5x + 3)(x + 4)$ **c.** $(5x + 2)(x + 6)$

Martin-Gay Interactive Videos   *Watch the section lecture video and answer the following questions.*

Objective A 5. From ⊞ Example 1, explain in general terms how we would go about factoring a trinomial with a first-term coefficient $\neq 1$. ▶

Objective B 6. From ⊞ Examples 3 and 5, how can factoring the GCF from a trinomial help us save time when trying to factor the remaining trinomial? ▶

See Video 6.3 🍎

See video answer section.

6.3 Exercise Set MyMathLab®

Objective **A** *Complete each factored form. See Examples 1 through 5.*

1. $5x^2 + 22x + 8 = (5x + 2)$ $(x + 4)$

2. $2y^2 + 15y + 25 = (2y + 5)$ $(y + 5)$

3. $50x^2 + 15x - 2 = (5x + 2)$ $(10x - 1)$

4. $6y^2 + 11y - 10 = (2y + 5)$ $(3y - 2)$

5. $20x^2 - 7x - 6 = (5x + 2)$ $(4x - 3)$

6. $8y^2 - 2y - 55 = (2y + 5)$ $(4y - 11)$

Factor each trinomial completely. See Examples 1 through 5.

7. $2x^2 + 13x + 15$
$(2x + 3)(x + 5)$

8. $3x^2 + 8x + 4$
$(3x + 2)(x + 2)$

9. $8y^2 - 17y + 9$
$(y - 1)(8y - 9)$

10. $21x^2 - 41x + 10$
$(7x - 2)(3x - 5)$

11. $2x^2 - 9x - 5$
$(2x + 1)(x - 5)$

12. $36r^2 - 5r - 24$
$(9r - 8)(4r + 3)$

13. $20r^2 + 27r - 8$
$(4r - 1)(5r + 8)$

14. $3x^2 + 20x - 63$
$(3x - 7)(x + 9)$

15. $10x^2 + 31x + 3$
$(10x + 1)(x + 3)$

16. $12x^2 + 17x + 5$
$(12x + 5)(x + 1)$

17. $x + 3x^2 - 2$
$(3x - 2)(x + 1)$

18. $y + 8y^2 - 9$
$(y - 1)(8y + 9)$

19. $6x^2 - 13xy + 5y^2$
$(3x - 5y)(2x - y)$

20. $8x^2 - 14xy + 3y^2$
$(4x - y)(2x - 3y)$

21. $15m^2 - 16m - 15$
$(3m - 5)(5m + 3)$

22. $25n^2 - 5n - 6$
$(5n + 2)(5n - 3)$

23. $-9x + 20 + x^2$
$(x - 4)(x - 5)$

24. $-7x + 12 + x^2$
$(x - 3)(x - 4)$

25. $2x^2 - 7x - 99$
$(2x + 11)(x - 9)$

26. $2x^2 + 7x - 72$
$(2x - 9)(x + 8)$

27. $-27t + 7t^2 - 4$
$(7t + 1)(t - 4)$

28. $-3t + 4t^2 - 7$
$(4t - 7)(t + 1)$

29. $3a^2 + 10ab + 3b^2$
$(3a + b)(a + 3b)$

30. $2a^2 + 11ab + 5b^2$
$(2a + b)(a + 5b)$

31. $49p^2 - 7p - 2$
$(7p + 1)(7p - 2)$

32. $3r^2 + 10r - 8$
$(3r - 2)(r + 4)$

33. $18x^2 - 9x - 14$
$(6x - 7)(3x + 2)$

34. $42a^2 - 43a + 6$
$(7a - 6)(6a - 1)$

35. $2m^2 + 17m + 10$
prime

36. $3n^2 + 20n + 5$
prime

37. $24x^2 + 41x + 12$
$(8x + 3)(3x + 4)$

38. $24x^2 - 49x + 15$
$(8x - 3)(3x - 5)$

Objectives **A** **B** **Mixed Practice** *Factor each trinomial completely. See Examples 1 through 7.*

39. $12x^3 + 11x^2 + 2x$
$x(3x + 2)(4x + 1)$

40. $8a^3 + 14a^2 + 3a$
$a(4a + 1)(2a + 3)$

41. $21b^2 - 48b - 45$
$3(7b + 5)(b - 3)$

42. $12x^2 - 14x - 10$
$2(3x - 5)(2x + 1)$

43. $7z + 12z^2 - 12$
$(3z + 4)(4z - 3)$

44. $16t + 15t^2 - 15$
$(5t - 3)(3t + 5)$

45. $6x^2y^2 - 2xy^2 - 60y^2$
$2y^2(3x - 10)(x + 3)$

46. $8x^2y + 34xy - 84y$
$2y(4x - 7)(x + 6)$

47. $4x^2 - 8x - 21$
$(2x - 7)(2x + 3)$

48. $6x^2 - 11x - 10$
$(3x + 2)(2x - 5)$

49. $3x^2 - 42x + 63$
$3(x^2 - 14x + 21)$

50. $5x^2 - 75x + 60$
$5(x^2 - 15x + 12)$

51. $8x^2 + 6xy - 27y^2$
$(4x + 9y)(2x - 3y)$

52. $54a^2 + 39ab - 8b^2$
$(9a + 8b)(6a - b)$

53. $-x^2 + 2x + 24$
$-1(x - 6)(x + 4)$

54. $-x^2 + 4x + 21$
$-1(x + 3)(x - 7)$

55. $4x^3 - 9x^2 - 9x$
$x(4x + 3)(x - 3)$

56. $6x^3 - 31x^2 + 5x$
$x(x - 5)(6x - 1)$

57. $24x^2 - 58x + 9$
$(4x - 9)(6x - 1)$

58. $36x^2 + 55x - 14$
$(4x + 7)(9x - 2)$

59. $40a^2b + 9ab - 9b$
$b(8a - 3)(5a + 3)$

60. $24y^2x + 7yx - 5x$
$x(8y + 5)(3y - 1)$

▶ **61.** $30x^3 + 38x^2 + 12x$
$2x(3x + 2)(5x + 3)$

62. $6x^3 - 28x^2 + 16x$
$2x(3x - 2)(x - 4)$

63. $6y^3 - 8y^2 - 30y$
$2y(3y + 5)(y - 3)$

64. $12x^3 - 34x^2 + 24x$
$2x(3x - 4)(2x - 3)$

65. $10x^4 + 25x^3y - 15x^2y^2$
$5x^2(2x - y)(x + 3y)$

66. $42x^4 - 99x^3y - 15x^2y^2$
$3x^2(2x - 5y)(7x + y)$

▶ **67.** $-14x^2 + 39x - 10$
$-1(2x - 5)(7x - 2)$

68. $-15x^2 + 26x - 8$
$-1(3x - 4)(5x - 2)$

69. $16p^4 - 40p^3 + 25p^2$
$p^2(4p - 5)(4p - 5)$
or $p^2(4p - 5)^2$

70. $9q^4 - 42q^3 + 49q^2$
$q^2(3q - 7)(3q - 7)$
or $q^2(3q - 7)^2$

71. $-2x^2 + 9x + 5$
$-1(2x + 1)(x - 5)$

72. $-3x^2 + 8x + 16$
$-1(3x + 4)(x - 4)$

73. $-4 + 52x - 48x^2$
$-4(12x - 1)(x - 1)$

74. $-5 + 55x - 50x^2$
$-5(10x - 1)(x - 1)$

75. $2t^4 + 3t^2 - 27$
$(2t^2 + 9)(t^2 - 3)$

76. $4r^4 - 17r^2 - 15$
$(4r^2 + 3)(r^2 - 5)$

77. $5x^2y^2 + 20xy + 1$
prime

78. $3a^2b^2 + 12ab + 1$
prime

79. $6a^5 + 37a^3b^2 + 6ab^4$
$a(6a^2 + b^2)(a^2 + 6b^2)$

80. $5m^5 + 26m^3h^2 + 5mh^4$
$m(5m^2 + h^2)(m^2 + 5h^2)$

Review

Multiply. See Section 5.6.

81. $(x - 4)(x + 4)$ $x^2 - 16$

82. $(2x - 9)(2x + 9)$ $4x^2 - 81$

83. $(x + 2)^2$ $x^2 + 4x + 4$

84. $(x + 3)^2$ $x^2 + 6x + 9$

85. $(2x - 1)^2$ $4x^2 - 4x + 1$

86. $(3x - 5)^2$ $9x^2 - 30x + 25$

The following graph shows the percent of text message users in each age group. See Section 1.2.

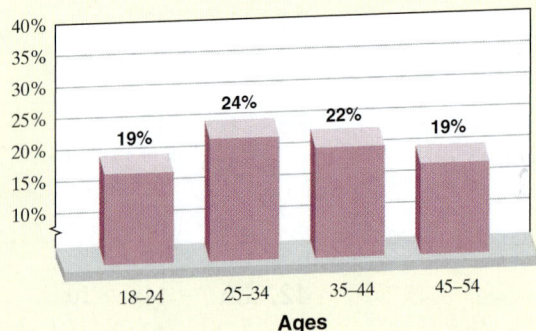

Source: Data from CellSigns, Inc.

87. What range of ages shown has the highest percent of text message users?   25–34

88. What range of ages shown has the lowest percent of text message users?   18–24 and 45–54

89. Describe any trend you see.   answers may vary

90. Why don't the percents shown in the graph add to 100%?   answers may vary

Concept Extensions

See the Concept Check in this section.

91. Do the terms of $4x^2 + 19x + 12$ have a common factor (other than 1)?   no

92. Without multiplying, decide which of the following factored forms is not a factored form of $4x^2 + 19x + 12$.
a. $(2x + 4)(2x + 3)$
b. $(4x + 4)(x + 3)$
c. $(4x + 3)(x + 4)$
d. $(2x + 2)(2x + 6)$ a, b, d

Write the perimeter of each figure as a simplified polynomial. Then factor the polynomial completely.

93.

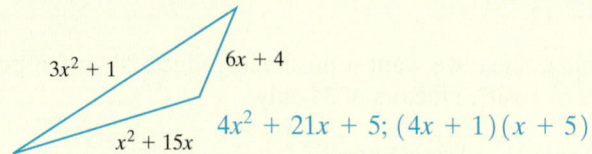

$3x^2 + 1$ $6x + 4$

$x^2 + 15x$ $4x^2 + 21x + 5; (4x + 1)(x + 5)$

94.

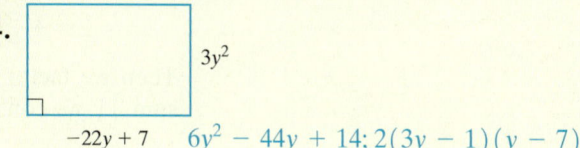

$3y^2$

$-22y + 7$ $6y^2 - 44y + 14; 2(3y - 1)(y - 7)$

Factor each trinomial completely.

95. $4x^2 + 2x + \dfrac{1}{4}$ $\left(2x + \dfrac{1}{2}\right)\left(2x + \dfrac{1}{2}\right)$ or $\left(2x + \dfrac{1}{2}\right)^2$

96. $27x^2 + 2x - \dfrac{1}{9}$ $\left(3x + \dfrac{1}{3}\right)\left(9x - \dfrac{1}{3}\right)$

97. $4x^2(y - 1)^2 + 25x(y - 1)^2 + 25(y - 1)^2$
$(y - 1)^2(4x + 5)(x + 5)$

98. $3x^2(a + 3)^3 - 28x(a + 3)^3 + 25(a + 3)^3$
$(a + 3)^3(3x - 25)(x - 1)$

Find a positive value of b so that each trinomial is factorable.

99. $3x^2 + bx - 5$ $2; 14$

100. $2z^2 + bz - 7$ $5; 13$

Find a positive value of c so that each trinomial is factorable.

101. $5x^2 + 7x + c$ 2

102. $3x^2 - 8x + c$ $4; 5$

103. In your own words, describe the steps you use to factor a trinomial.   answers may vary

104. A student in your class factored $6x^2 + 7x + 1$ as $(3x + 1)(2x + 1)$. Write down how you would explain the student's error.   answers may vary

---

6.4   # Factoring Trinomials of the Form $ax^2 + bx + c$ by Grouping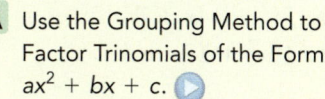

Objective A Using the Grouping Method ▶

There is an alternative method that can be used to factor trinomials of the form $ax^2 + bx + c, a \neq 1$. This method is called the **grouping method** because it uses factoring by grouping as we learned in Section 6.1.

To see how this method works, recall from Section 6.2 that to factor a trinomial such as $x^2 + 11x + 30$, we find two numbers such that

Product is 30.
↓
$x^2 + 11x + 30$
↓
Sum is 11.

To factor a trinomial such as $2x^2 + 11x + 12$ by grouping, we use an extension of the method in Section 6.1. Here we look for two numbers such that

Product is $2 \cdot 12 = 24$.
↓
$2x^2 + 11x + 12$
↓
Sum is 11.

Objective

A Use the Grouping Method to Factor Trinomials of the Form $ax^2 + bx + c$. ▶

Teaching Tip

Consider beginning this lesson with an example reminding students of the grouping method they learned in Section 6.1. For instance, factor $2y^3 + 5y^2 - 4y - 10$ and $3x^2 - x + 6x - 2$.

This time, we use the two numbers to write

$$2x^2 + 11x + 12 \text{ as}$$
$$= 2x^2 + \square x + \square x + 12$$

Then we factor by grouping. Since we want a positive product, 24, and a positive sum, 11, we consider pairs of positive factors of 24 only.

Factors of 24	Sum of Factors	
1, 24	25	
2, 12	14	
3, 8	11	**Correct** sum

The factors are 3 and 8. Now we use these factors to write the middle term, $11x$, as $3x + 8x$ (or $8x + 3x$). We replace $11x$ with $3x + 8x$ in the original trinomial and then we can factor by grouping.

$$
\begin{aligned}
2x^2 + 11x + 12 &= 2x^2 + 3x + 8x + 12 \\
&= (2x^2 + 3x) + (8x + 12) \quad \text{Group the terms.} \\
&= x(2x + 3) + 4(2x + 3) \quad \text{Factor each group.} \\
&= (2x + 3)(x + 4) \quad \text{Factor out } (2x + 3).
\end{aligned}
$$

In general, we have the following procedure.

To Factor Trinomials by Grouping

Step 1: Factor out a greatest common factor, if there is one other than 1.

Step 2: For the resulting trinomial $ax^2 + bx + c$, find two numbers whose product is $a \cdot c$ and whose sum is b.

Step 3: Write the middle term, bx, using the factors found in Step 2.

Step 4: Factor by grouping.

Practice 1

Factor each trinomial by grouping.

a. $3x^2 + 14x + 8$

b. $12x^2 + 19x + 5$

Practice 2

Factor each trinomial by grouping.

a. $30x^2 - 26x + 4$

b. $6x^2y - 7xy - 5y$

Answers

1. a. $(x + 4)(3x + 2)$
   b. $(4x + 5)(3x + 1)$
2. a. $2(5x - 1)(3x - 2)$
   b. $y(2x + 1)(3x - 5)$

Example 1   Factor $8x^2 - 14x + 5$ by grouping.

Solution:

Step 1: The terms of this trinomial contain no greatest common factor other than 1.

Step 2: This trinomial is of the form $ax^2 + bx + c$, with $a = 8, b = -14$, and $c = 5$. Find two numbers whose product is $a \cdot c$ or $8 \cdot 5 = 40$ and whose sum is b or -14.
The numbers are -4 and -10.

Factors of 40	Sum of Factors
$-40, -1$	-41
$-20, -2$	-22
$-10, -4$	-14

Step 3: Write $-14x$ as $-4x - 10x$ so that
$$8x^2 - 14x + 5 = 8x^2 - 4x - 10x + 5$$

Correct sum

Step 4: Factor by grouping.
$$
\begin{aligned}
8x^2 - 4x - 10x + 5 &= 4x(2x - 1) - 5(2x - 1) \\
&= (2x - 1)(4x - 5)
\end{aligned}
$$

■ **Work Practice 1**

Example 2   Factor $6x^2 - 2x - 20$ by grouping.

Solution:

Step 1: First factor out the greatest common factor, 2.
$$6x^2 - 2x - 20 = 2(3x^2 - x - 10)$$

Step 2: Next notice that $a = 3$, $b = -1$, and $c = -10$ in the resulting trinomial. Find two numbers whose product is $a \cdot c$ or $3(-10) = -30$ and whose sum is b, -1. The numbers are -6 and 5.

Step 3: $3x^2 - x - 10 = 3x^2 - 6x + 5x - 10$

Step 4: $3x^2 - 6x + 5x - 10 = 3x(x - 2) + 5(x - 2)$
$$= (x - 2)(3x + 5)$$

A factored form of $6x^2 - 2x - 20 = 2(x - 2)(3x + 5)$.

└─ Don't forget to include the common factor of 2.

🟧 **Work Practice 2**

Teaching Tip

In Example 2, remind students that if they forget to factor out the GCF in Step 1, they may end up with a factored form that is not factored *completely*.

Example 3 Factor $18y^4 + 21y^3 - 60y^2$ by grouping.

Solution:

Step 1: First factor out the greatest common factor, $3y^2$.

$$18y^4 + 21y^3 - 60y^2 = 3y^2(6y^2 + 7y - 20)$$

Step 2: Notice that $a = 6$, $b = 7$, and $c = -20$ in the resulting trinomial. Find two numbers whose product is $a \cdot c$ or $6(-20) = -120$ and whose sum is 7. It may help to factor -120 as a product of primes and -1.

$$-120 = 2 \cdot 2 \cdot 2 \cdot 3 \cdot 5 \cdot (-1)$$

Then choose pairings of factors until you have two pairings whose sum is 7.

$$\underbrace{2 \cdot 2 \cdot 2}_{-8} \cdot \underbrace{3 \cdot 5}_{15} \cdot (-1) \qquad \text{The numbers are } -8 \text{ and } 15.$$

Step 3: $6y^2 + 7y - 20 = 6y^2 - 8y + 15y - 20$

Step 4: $6y^2 - 8y + 15y - 20 = 2y(3y - 4) + 5(3y - 4)$
$$= (3y - 4)(2y + 5)$$

A factored form of $18y^4 + 21y^3 - 60y^2$ is $3y^2(3y - 4)(2y + 5)$.

└─ Don't forget to include the common factor of $3y^2$.

🟧 **Work Practice 3**

Practice 3

Factor $12y^5 + 10y^4 - 42y^3$ by grouping.

Answer

3. $2y^3(3y + 7)(2y - 3)$

Vocabulary, Readiness & Video Check

For each trinomial $ax^2 + bx + c$, choose two numbers whose product is $a \cdot c$ and whose sum is b.

1. $x^2 + 6x + 8$

    a. $4, 2$ **b.** $7, 1$ **c.** $6, 2$ **d.** $6, 8$ a

2. $x^2 + 11x + 24$

    a. $6, 4$ **b.** $12, 2$ **c.** $8, 3$ **d.** $5, 6$ c

3. $2x^2 + 13x + 6$

    a. $2, 6$ **b.** $12, 1$ **c.** $13, 1$ **d.** $3, 4$ b

4. $4x^2 + 8x + 3$

    a. $4, 3$ **b.** $4, 4$ **c.** $12, 1$ **d.** $2, 6$ d

Martin-Gay Interactive Videos   *Watch the section lecture video and answer the following question.*

Objective **A** **5.** In the lecture following 🔲 Example 1, why does writing a term as the sum or difference of two terms suggest we'd then try to factor by grouping?

See Video 6.4 🍎
See video answer section.

6.4 **Exercise Set** MyMathLab®

Objective **A** *Factor each polynomial by grouping. Notice that Step 3 has already been done in these exercises. See Examples 1 through 3.*

1. $x^2 + 3x + 2x + 6$
$(x + 3)(x + 2)$

2. $x^2 + 5x + 3x + 15$
$(x + 5)(x + 3)$

3. $y^2 + 8y - 2y - 16$
$(y + 8)(y - 2)$

4. $z^2 + 10z - 7z - 70$
$(z + 10)(z - 7)$

5. $8x^2 - 5x - 24x + 15$
$(8x - 5)(x - 3)$

6. $4x^2 - 9x - 32x + 72$
$(4x - 9)(x - 8)$

7. $5x^4 - 3x^2 + 25x^2 - 15$
$(5x^2 - 3)(x^2 + 5)$

8. $2y^4 - 10y^2 + 7y^2 - 35$
$(y^2 - 5)(2y^2 + 7)$

Factor each trinomial by grouping. Exercises 9 through 12 are broken into parts to help you get started. See Examples 1 through 3.

9. $6x^2 + 11x + 3$
  a. Find two numbers whose product is $6 \cdot 3 = 18$ and whose sum is 11. $9, 2$
  b. Write $11x$ using the factors from part **a.** $9x + 2x$
  c. Factor by grouping. $(2x + 3)(3x + 1)$

10. $8x^2 + 14x + 3$
  a. Find two numbers whose product is $8 \cdot 3 = 24$ and whose sum is 14. $2, 12$
  b. Write $14x$ using the factors from part **a.** $2x + 12x$
  c. Factor by grouping. $(4x + 1)(2x + 3)$

11. $15x^2 - 23x + 4$
  a. Find two numbers whose product is $15 \cdot 4 = 60$ and whose sum is -23. $-20, -3$
  b. Write $-23x$ using the factors from part **a.**
  c. Factor by grouping. $(3x - 4)(5x - 1)$

12. $6x^2 - 13x + 5$
  a. Find two numbers whose product is $6 \cdot 5 = 30$ and whose sum is -13. $-10, -3$
  b. Write $-13x$ using the factors from part **a.**
  c. Factor by grouping. $(3x - 5)(2x - 1)$

13. $21y^2 + 17y + 2$
$(3y + 2)(7y + 1)$

14. $15x^2 + 11x + 2$
$(3x + 1)(5x + 2)$

15. $7x^2 - 4x - 11$
$(7x - 11)(x + 1)$

16. $8x^2 - x - 9$
$(8x - 9)(x + 1)$

17. $10x^2 - 9x + 2$
$(5x - 2)(2x - 1)$

18. $30x^2 - 23x + 3$
$(5x - 3)(6x - 1)$

19. $2x^2 - 7x + 5$
$(2x - 5)(x - 1)$

20. $2x^2 - 7x + 3$
$(2x - 1)(x - 3)$

21. $12x + 4x^2 + 9$

22. $20x + 25x^2 + 4$

23. $4x^2 - 8x - 21$
$(2x + 3)(2x - 7)$

24. $6x^2 - 11x - 10$
$(3x + 2)(2x - 5)$

25. $10x^2 - 23x + 12$
$(5x - 4)(2x - 3)$

26. $21x^2 - 13x + 2$
$(7x - 2)(3x - 1)$

27. $2x^3 + 13x^2 + 15x$
$x(2x + 3)(x + 5)$

28. $3x^3 + 8x^2 + 4x$
$x(3x + 2)(x + 2)$

29. $16y^2 - 34y + 18$
$2(8y - 9)(y - 1)$

30. $4y^2 - 2y - 12$
$2(2y + 3)(y - 2)$

31. $-13x + 6 + 6x^2$
$(2x - 3)(3x - 2)$

32. $-25x + 12 + 12x^2$
$(3x - 4)(4x - 3)$

11. b. $-20x - 3x$ **12. b.** $-10x - 3x$ **21.** $(2x + 3)(2x + 3)$ or $(2x + 3)^2$ **22.** $(5x + 2)(5x + 2)$ or $(5x + 2)^2$

33. $54a^2 - 9a - 30$
$3(3a + 2)(6a - 5)$

34. $30a^2 + 38a - 20$
$2(3a + 5)(5a - 2)$

35. $20a^3 + 37a^2 + 8a$
$a(4a + 1)(5a + 8)$

36. $10a^3 + 17a^2 + 3a$
$a(2a + 3)(5a + 1)$

▶ 37. $12x^3 - 27x^2 - 27x$
$3x(4x + 3)(x - 3)$

38. $30x^3 - 155x^2 + 25x$
$5x(6x - 1)(x - 5)$

39. $3x^2y + 4xy^2 + y^3$
$y(3x + y)(x + y)$

40. $6r^2t + 7rt^2 + t^3$
$t(6r + t)(r + t)$

41. $20z^2 + 7z + 1$
prime

42. $36z^2 + 6z + 1$
prime

43. $24a^2 - 6ab - 30b^2$
$6(a + b)(4a - 5b)$

44. $30a^2 + 5ab - 25b^2$
$5(a + b)(6a - 5b)$

45. $15p^4 + 31p^3q + 2p^2q^2$
$p^2(15p + q)(p + 2q)$

46. $20s^4 + 61s^3t + 3s^2t^2$
$s^2(20s + t)(s + 3t)$

47. $35 + 12x + x^2$
$(7 + x)(5 + x)$ or $(x + 7)(x + 5)$

48. $33 + 14x + x^2$
$(3 + x)(11 + x)$ or $(x + 3)(x + 11)$

49. $6 - 11x + 5x^2$
$(6 - 5x)(1 - x)$ or $(5x - 6)(x - 1)$

50. $5 - 12x + 7x^2$
$(5 - 7x)(1 - x)$ or $(7x - 5)(x - 1)$

Review

Multiply. See Section 5.6.

51. $(x - 2)(x + 2)$
$x^2 - 4$

52. $(y - 5)(y + 5)$
$y^2 - 25$

53. $(y + 4)(y + 4)$
$y^2 + 8y + 16$

54. $(x + 7)(x + 7)$
$x^2 + 14x + 49$

55. $(9z + 5)(9z - 5)$
$81z^2 - 25$

56. $(8y + 9)(8y - 9)$
$64y^2 - 81$

57. $(4x - 3)^2$
$16x^2 - 24x + 9$

58. $(2z - 1)^2$
$4z^2 - 4z + 1$

Concept Extensions

Write the perimeter of each figure as a simplified polynomial. Then factor the polynomial.

59. $10x^2 + 45x + 45; 5(2x + 3)(x + 3)$

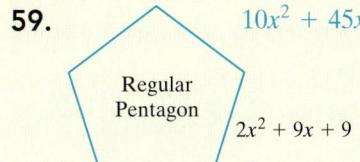

Regular Pentagon

$2x^2 + 9x + 9$

60. $21x^2 + 33xy + 12y^2; 3(7x + 4y)(x + y)$

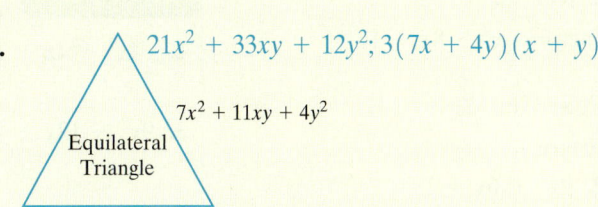

$7x^2 + 11xy + 4y^2$

Equilateral Triangle

Factor each polynomial by grouping.

61. $x^{2n} + 2x^n + 3x^n + 6$
(*Hint:* Don't forget that $x^{2n} = x^n \cdot x^n$.)
$(x^n + 2)(x^n + 3)$

62. $x^{2n} + 6x^n + 10x^n + 60$
$(x^n + 6)(x^n + 10)$

63. $3x^{2n} + 16x^n - 35$
$(3x^n - 5)(x^n + 7)$

64. $12x^{2n} - 40x^n + 25$
$(2x^n - 5)(6x^n - 5)$

✎ 65. In your own words, explain how to factor a trinomial by grouping.
answers may vary

6.5 Factoring by Special Products

Objectives

A Factor Perfect Square Trinomials.

B Factor the Difference of Two Squares.

C Factor the Sum or Difference of Two Cubes.

Teaching Tip

Remind students that learning math can become easier when their experience allows them to recognize certain patterns that occur over and over again. In algebra we extend the concept of perfect square from the specific $(1^2, 2^2, 3^2, 4^2, 5^2, 6^2, 7^2, \ldots)$ to the abstract $((x + 1)^2, (x + 2)^2, (x + 3)^2, \ldots)$. Just as students have come to recognize 1, 4, 9, 16, 25, 36, and 49 as perfect squares, they will come to recognize certain simplified algebraic expressions as perfect squares. Make an area diagram of Example 1 to show that it is a square because the length = width.

	$m + 5$	
m	m^2	$5m$
$+$		
5	$5m$	25

Practice 1

Factor: $x^2 + 8x + 16$

Practice 2–3

Factor.

2. $9x^2 + 6x + 1$

3. $25x^2 - 20x + 4$

Practice 4

Factor: $4x^3 - 32x^2y + 64xy^2$

Answers

1. $(x + 4)^2$ **2.** $(3x + 1)^2$
3. $(5x - 2)^2$ **4.** $4x(x - 4y)^2$

Objective A Factoring Perfect Square Trinomials

A trinomial that is the square of a binomial is called a **perfect square trinomial.** For example,

$$(x + 3)^2 = (x + 3)(x + 3)$$
$$= x^2 + 6x + 9$$

Thus, $x^2 + 6x + 9$ is a perfect square trinomial.

In Chapter 5, we discovered special product formulas for squaring binomials.

$$(a + b)^2 = a^2 + 2ab + b^2 \quad \text{and} \quad (a - b)^2 = a^2 - 2ab + b^2$$

Because multiplication and factoring are reverse processes, we can now use these special products to help us factor perfect square trinomials. If we reverse these equations, we have the following.

> **Factoring Perfect Square Trinomials**
>
> $$a^2 + 2ab + b^2 = (a + b)^2$$
> $$a^2 - 2ab + b^2 = (a - b)^2$$

To use these equations to help us factor, we must first be able to recognize a perfect square trinomial. A trinomial is a perfect square trinomial if it can be written so that its first term is the square of some quantity a, its last term is the square of some quantity b, and its middle term is twice the product of the quantities a and b.

Example 1   Factor: $m^2 + 10m + 25$

Solution: Notice that the first term is a square: $m^2 = (m)^2$, the last term is a square: $25 = 5^2$, and the middle term $10m = 2 \cdot 5 \cdot m$. This is a perfect square trinomial. Thus,

$$m^2 + 10m + 25 = m^2 + 2(m)(5) + 5^2 = (m + 5)^2$$

■ **Work Practice 1**

Examples   Factor each trinomial.

2. $4x^2 + 4x + 1 = (2x)^2 + 2 \cdot 2x \cdot 1 + 1^2$ See whether it is a perfect square trinomial.
$\qquad\qquad\quad = (2x + 1)^2$ Factor.

3. $9x^2 - 12x + 4 = (3x)^2 - 2(3x)(2) + 2^2$ See whether it is a perfect square trinomial.
$\qquad\qquad\quad = (3x - 2)^2$ Factor.

■ **Work Practice 2–3**

Example 4   Factor: $3a^2x - 12abx + 12b^2x$

Solution: The terms of this trinomial have a greatest common factor of $3x$, which we factor out first.

$$3a^2x - 12abx + 12b^2x = 3x(a^2 - 4ab + 4b^2)$$

The polynomial $a^2 - 4ab + 4b^2$ is a perfect square trinomial. Notice that the first term is a square: $a^2 = (a)^2$, the last term is a square: $4b^2 = (2b)^2$, and $4ab = 2(a)(2b)$. The factoring can now be completed as

$$3x(a^2 - 4ab + 4b^2) = 3x(a - 2b)^2$$

■ **Work Practice 4**

Helpful Hint

If you recognize a trinomial as a perfect square trinomial, use the special formulas to factor. However, general methods for factoring trinomials from Sections 6.3 and 6.4 will also result in the correct factored form.

Objective B Factoring the Difference of Two Squares ▶

We now factor special types of binomials, beginning with the **difference of two squares**. The special product pattern presented in Section 5.6 for the product of a sum and a difference of two terms is used again here. However, the emphasis is now on factoring rather than on multiplying.

Difference of Two Squares

$$a^2 - b^2 = (a + b)(a - b)$$

Notice that a binomial is a difference of two squares when it is the difference of the square of some quantity a and the square of some quantity b.

Examples Factor.

5. $x^2 - 9 = x^2 - 3^2$
$$= (x + 3)(x - 3)$$

6. $16y^2 - 9 = (4y)^2 - 3^2$
$$= (4y + 3)(4y - 3)$$

7. $50 - 8y^2 = 2(25 - 4y^2)$ Factor out the common factor of 2.
$$= 2[5^2 - (2y)^2]$$
$$= 2(5 + 2y)(5 - 2y)$$

8. $x^2 - \dfrac{1}{4} = x^2 - \left(\dfrac{1}{2}\right)^2$
$$= \left(x + \dfrac{1}{2}\right)\left(x - \dfrac{1}{2}\right)$$

■ **Work Practice 5–8**

The binomial $x^2 + 9$ is a **sum of two squares** and cannot be factored by using real numbers. *In general, except for factoring out a greatest common factor, the sum of two squares usually cannot be factored by using real numbers.*

Helpful Hint

The sum of two squares whose greatest common factor is 1 usually cannot be factored by using real numbers.

Example 9 Factor: $p^4 - 16$

Solution:

$$p^4 - 16 = (p^2)^2 - 4^2$$
$$= (p^2 + 4)(p^2 - 4)$$

(Continued on next page)

Practice 5–8

Factor:

5. $x^2 - 49$ **6.** $4y^2 - 81$

7. $12 - 3a^2$ **8.** $y^2 - \dfrac{1}{25}$

Practice 9

Factor: $a^4 - 81$

Answers

5. $(x + 7)(x - 7)$

6. $(2y + 9)(2y - 9)$

7. $3(2 + a)(2 - a)$

8. $\left(y + \dfrac{1}{5}\right)\left(y - \dfrac{1}{5}\right)$

9. $(a^2 + 9)(a + 3)(a - 3)$

The binomial factor $p^2 + 4$ cannot be factored by using real numbers, but the binomial factor $p^2 - 4$ is a difference of squares.

$$(p^2 + 4)(p^2 - 4) = (p^2 + 4)(p + 2)(p - 2)$$

■ Work Practice 9

Helpful Hint

1. Don't forget to first see whether there's a greatest common factor (other than 1) that can be factored out.
2. Factor completely. In other words, check to see whether any factors can be factored further (as in Example 9).

Practice 10–12

Factor each binomial.
10. $9x^3 - 25x$
11. $48x^4 - 3$
12. $-9x^2 + 100$

Examples   Factor each binomial.

10. $4x^3 - 49x = x(4x^2 - 49)$ — Factor out the common factor, x.
$\qquad = x[(2x)^2 - 7^2]$
$\qquad = x(2x + 7)(2x - 7)$ — Factor the difference of two squares.

11. $162x^4 - 2 = 2(81x^4 - 1)$ — Factor out the common factor, 2.
$\qquad = 2(9x^2 + 1)(9x^2 - 1)$ — Factor the difference of two squares.
$\qquad = 2(9x^2 + 1)(3x + 1)(3x - 1)$ — Factor the difference of two squares.

12. $-49x^2 + 16 = -1(49x^2 - 16)$ — Factor out -1.
$\qquad = -1(7x + 4)(7x - 4)$ — Factor the difference of two squares.

■ Work Practice 10–12

✔**Concept Check** Is $(x - 4)(y^2 - 9)$ completely factored? Why or why not?

Practice 13

Factor: $(x + 1)^2 - 9$

Example 13 Factor: $(x + 3)^2 - 36$

Solution:

$$(x + 3)^2 - 36 = (x + 3)^2 - 6^2$$
$$\qquad = [(x + 3) + 6][(x + 3) - 6] \quad \text{Factor as the difference of two squares.}$$
$$\qquad = [x + 3 + 6][x + 3 - 6] \quad \text{Remove parentheses.}$$
$$\qquad = (x + 9)(x - 3) \quad \text{Simplify.}$$

■ Work Practice 13

Practice 14

Factor: $a^2 + 2a + 1 - b^2$

Example 14 Factor: $x^2 + 4x + 4 - y^2$

Solution: Factoring by grouping comes to mind since the sum of the first three terms of this polynomial is a perfect square trinomial.

$$x^2 + 4x + 4 - y^2 = (x^2 + 4x + 4) - y^2 \quad \text{Group the first three terms.}$$
$$\qquad = (x + 2)^2 - y^2 \quad \text{Factor the perfect square trinomial.}$$

This is not completely factored yet since we have a *difference*, not a *product*. Since $(x + 2)^2 - y^2$ is a difference of squares, we have

$$(x + 2)^2 - y^2 = [(x + 2) + y][(x + 2) - y]$$
$$\qquad = (x + 2 + y)(x + 2 - y)$$

■ Work Practice 14

Answers

10. $x(3x - 5)(3x + 5)$
11. $3(4x^2 + 1)(2x + 1)(2x - 1)$
12. $-1(3x - 10)(3x + 10)$
13. $(x - 2)(x + 4)$
14. $(a + 1 + b)(a + 1 - b)$

✔**Concept Check Answer**

no; $(y^2 - 9)$ can be factored

Objective C Factoring the Sum or Difference of Two Cubes ▶

Although the sum of two squares usually cannot be factored, the sum of two cubes, as well as the difference of two cubes, can be factored as follows.

Sum and Difference of Two Cubes

$$a^3 + b^3 = (a + b)(a^2 - ab + b^2)$$
$$a^3 - b^3 = (a - b)(a^2 + ab + b^2)$$

To check the first pattern, let's find the product of $(a + b)$ and $(a^2 - ab + b^2)$.

$$(a + b)(a^2 - ab + b^2) = a(a^2 - ab + b^2) + b(a^2 - ab + b^2)$$
$$= a^3 - a^2b + ab^2 + a^2b - ab^2 + b^3$$
$$= a^3 + b^3$$

Example 15 Factor: $x^3 + 8$

Solution: First we write the binomial in the form $a^3 + b^3$. Then we use the formula
$a^3 + b^3 = (a + b)(a^2 - a \cdot b + b^2)$, where a is x and b is 2.

$$x^3 + 8 = x^3 + 2^3 = (x + 2)(x^2 - x \cdot 2 + 2^2)$$

Thus, $x^3 + 8 = (x + 2)(x^2 - 2x + 4)$

▪ **Work Practice 15**

Practice 15

Factor: $x^3 + 27$

Example 16 Factor: $p^3 + 27q^3$

Solution: $p^3 + 27q^3 = p^3 + (3q)^3$
$$= (p + 3q)[p^2 - (p)(3q) + (3q)^2]$$
$$= (p + 3q)(p^2 - 3pq + 9q^2)$$

▪ **Work Practice 16**

Practice 16

Factor: $x^3 + 64y^3$

Example 17 Factor: $y^3 - 64$

Solution: This is a difference of cubes since $y^3 - 64 = y^3 - 4^3$.

From $a^3 - b^3 = (a - b)(a^2 + a \cdot b + b^2)$ we have that

$$y^3 - 4^3 = (y - 4)(y^2 + y \cdot 4 + 4^2)$$
$$= (y - 4)(y^2 + 4y + 16)$$

▪ **Work Practice 17**

Practice 17

Factor: $y^3 - 8$

Answers

15. $(x + 3)(x^2 - 3x + 9)$
16. $(x + 4y)(x^2 - 4xy + 16y^2)$
17. $(y - 2)(y^2 + 2y + 4)$

Helpful Hint

When factoring sums or differences of cubes, be sure to notice the sign patterns.

$$x^3 + y^3 = (x + y)(x^2 - xy + y^2)$$

same sign — opposite signs — always positive

$$x^3 - y^3 = (x - y)(x^2 + xy + y^2)$$

same sign — opposite signs — always positive

Practice 18

Factor: $27a^2 - b^3a^2$

Answer

18. $a^2(3 - b)(9 + 3b + b^2)$

Example 18   Factor: $125q^2 - n^3q^2$

Solution: First we factor out a common factor of q^2.

$$125q^2 - n^3q^2 = q^2(125 - n^3)$$
$$= q^2(5^3 - n^3)$$

Opposite sign   Positive

$$= q^2(5 - n)[5^2 + (5)(n) + (n^2)]$$
$$= q^2(5 - n)(25 + 5n + n^2)$$

Thus, $125q^2 - n^3q^2 = q^2(5 - n)(25 + 5n + n^2)$. The trinomial $25 + 5n + n^2$ cannot be factored further.

Work Practice 18

 Calculator Explorations **Graphing**

A graphing calculator is a convenient tool for evaluating an expression at a given replacement value. For example, let's evaluate $x^2 - 6x$ when $x = 2$. To do so, store the value 2 in the variable x and then enter and evaluate the algebraic expression.

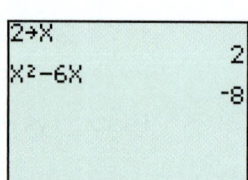

The value of $x^2 - 6x$ when $x = 2$ is -8. You may want to use this method for evaluating expressions as you explore the following.

We can use a graphing calculator to explore factoring patterns numerically. Use your calculator to evaluate

Teaching Tip

Some graphing calculators have a TABLE feature, which allows the user to evaluate an expression for various values. Enter the expressions using the Y= key. Then use the TABLE feature.

$x^2 - 2x + 1, x^2 - 2x - 1$, and $(x - 1)^2$ for each value of x given in the table. What do you observe?

	$x^2 - 2x + 1$	$x^2 - 2x - 1$	$(x - 1)^2$
$x = 5$	16	14	16
$x = -3$	16	14	16
$x = 2.7$	2.89	0.89	2.89
$x = -12.1$	171.61	169.61	171.61
$x = 0$	1	-1	1

Notice in each case that $x^2 - 2x - 1 \neq (x - 1)^2$. Because for each x in the table the value of $x^2 - 2x + 1$ and the value of $(x - 1)^2$ are the same, we might guess that $x^2 - 2x + 1 = (x - 1)^2$. We can verify our guess algebraically with multiplication:

$$(x - 1)(x - 1) = x^2 - x - x + 1 = x^2 - 2x + 1$$

Vocabulary, Readiness & Video Check

Write each number as a square.

1. 1 1^2 **2.** 25 5^2 **3.** 81 9^2 **4.** 64 8^2 **5.** 9 3^2 **6.** 100 10^2

Write each term as a square.

7. $9x^2$ $(3x)^2$ **8.** $16y^2$ $(4y)^2$ **9.** $25a^2$ $(5a)^2$ **10.** $81b^2$ $(9b)^2$ **11.** $36p^4$ $(6p^2)^2$ **12.** $4q^4$ $(2q^2)^2$

Martin-Gay Interactive Videos

See Video 6.5

See video answer section.

Watch the section lecture video and answer the following questions.

Objective A **13.** From ▣ Example 1, what is the first step to see if you have a perfect square trinomial? How do you then finish determining that you do indeed have a perfect square trinomial? ▶

Objective B **14.** In ▣ Example 5, the original binomial is rewritten to write each term as a square. Give two reasons why this is helpful. ▶

Objective C **15.** In ▣ Examples 8 and 9, what tips are given to remember how to factor the sum or difference of two cubes rather than memorizing the formulas? ▶

6.5 Exercise Set MyMathLab®

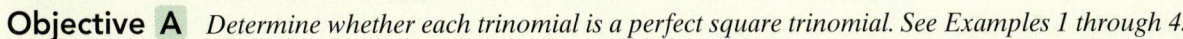

Objective A *Determine whether each trinomial is a perfect square trinomial. See Examples 1 through 4.*

1. $x^2 + 16x + 64$ yes   **2.** $x^2 + 22x + 121$ yes   ▶**3.** $y^2 + 5y + 25$ no   **4.** $y^2 + 4y + 16$ no

5. $4x^2 + 12xy + 8y^2$ no   **6.** $25x^2 + 20xy + 2y^2$ no   **7.** $25a^2 - 40ab + 16b^2$ yes

8. $36a^2 - 12ab + b^2$ yes   **9.** $m^2 - 2m + 1$ yes   **10.** $p^2 - 4p + 4$ yes

Factor each trinomial completely. See Examples 1 through 4.

▶**11.** $x^2 + 22x + 121$
$(x + 11)^2$

12. $x^2 + 18x + 81$
$(x + 9)^2$

13. $x^2 - 16x + 64$
$(x - 8)^2$

14. $x^2 - 12x + 36$
$(x - 6)^2$

15. $16a^2 - 24a + 9$
$(4a - 3)^2$

16. $25x^2 + 20x + 4$
$(5x + 2)^2$

17. $3x^2 - 24x + 48$
$3(x - 4)^2$

18. $2n^2 - 28n + 98$
$2(n - 7)^2$

19. $x^2y^2 - 10xy + 25$
$(xy - 5)^2$

20. $4x^2y^2 - 28xy + 49$
$(2xy - 7)^2$

21. $m^3 + 18m^2 + 81m$
$m(m + 9)^2$

22. $y^3 + 12y^2 + 36y$
$y(y + 6)^2$

23. $1 + 6x^2 + x^4$
prime

24. $1 + 16x^2 + x^4$
prime

▶**25.** $9x^2 - 24xy + 16y^2$
$(3x - 4y)^2$

26. $25x^2 - 60xy + 36y^2$
$(5x - 6y)^2$

27. $x^4 + 4x^2 + 4$
$(x^2 + 2)^2$

28. $m^4 + 10m^2 + 25$
$(m^2 + 5)^2$

Objective **B**   *Factor each completely. See Examples 5 through 14.*

29. $x^2 - 25$
$(x + 5)(x - 5)$

30. $y^2 - 100$
$(y + 10)(y - 10)$

31. $9 - 4z^2$
$(3 + 2z)(3 - 2z)$

32. $16x^2 - y^2$
$(4x + y)(4x - y)$

33. $16r^2 + 1$
prime

34. $49y^2 + 9$
prime

35. $x^3y - 121xy^3$
$xy(x + 11y)(x - 11y)$

36. $25xy^2 - 4x$
$x(5y + 2)(5y - 2)$

37. $(y + 2)^2 - 49$
$(y + 9)(y - 5)$

38. $(x - 1)^2 - z^2$
$(x - 1 + z)(x - 1 - z)$

39. $64x^2 - 100$
$4(4x + 5)(4x - 5)$

40. $4x^2 - 36$
$4(x + 3)(x - 3)$

41. $18x^2y - 2y$
$2y(3x + 1)(3x - 1)$

42. $12xy^2 - 108x$
$12x(y + 3)(y - 3)$

43. $9x^2 - 49$
$(3x + 7)(3x - 7)$

44. $25x^2 - 4$
$(5x + 2)(5x - 2)$

45. $x^4 - 81$
$(x^2 + 9)(x + 3)(x - 3)$

46. $x^4 - 256$
$(x^2 + 16)(x + 4)(x - 4)$

47. $(x + 2y)^2 - 9$
$(x + 2y + 3)(x + 2y - 3)$

48. $(3x + y)^2 - 25$
$(3x + y + 5)(3x + y - 5)$

49. $x^2 + 16x + 64 - x^4$
$(x + 8 + x^2)(x + 8 - x^2)$

50. $x^2 + 20x + 100 - x^4$
$(x + 10 + x^2)(x + 10 - x^2)$

51. $x^2 - 10x + 25 - y^2$
$(x - 5 + y)(x - 5 - y)$

52. $x^2 - 18x + 81 - y^2$
$(x - 9 + y)(x - 9 - y)$

53. $4x^2 + 4x + 1 - z^2$
$(2x + 1 + z)(2x + 1 - z)$

54. $9y^2 + 12y + 4 - x^2$
$(3y + 2 + x)(3y + 2 - x)$

55. $m^4 - 1$
$(m^2 + 1)(m + 1)(m - 1)$

56. $n^4 - 16$
$(n^2 + 4)(n + 2)(n - 2)$

Objective **C**   *Factor. See Examples 15 through 18.*

57. $x^3 + 27$
$(x + 3)(x^2 - 3x + 9)$

58. $y^3 + 1$
$(y + 1)(y^2 - y + 1)$

59. $z^3 - 1$
$(z - 1)(z^2 + z + 1)$

60. $8 - z^3$
$(2 - z)(4 + 2z + z^2)$

61. $m^3 + n^3$
$(m + n)(m^2 - mn + n^2)$

62. $r^3 + 125$
$(r + 5)(r^2 - 5r + 25)$

63. $x^3y^2 - 27y^2$
$y^2(x - 3)(x^2 + 3x + 9)$

64. $64 - p^3$
$(4 - p)(16 + 4p + p^2)$

65. $a^3b + 8b^4$
$b(a + 2b)(a^2 - 2ab + 4b^2)$

66. $8ab^3 + 27a^4$
$a(2b + 3a)(4b^2 - 6ab + 9a^2)$

67. $125y^3 - 8x^3$
$(5y - 2x)(25y^2 + 10xy + 4x^2)$

68. $54y^3 - 128$
$2(3y - 4)(9y^2 + 12y + 16)$

69. $x^6 - y^3$
$(x^2 - y)(x^4 + x^2y + y^2)$

70. $x^3 - y^6$
$(x - y^2)(x^2 + xy^2 + y^4)$

71. $8x^3 + 27y^3$
$(2x + 3y)(4x^2 - 6xy + 9y^2)$

72. $125x^3 + 8y^3$
$(5x + 2y)(25x^2 - 10xy + 4y^2)$

73. $x^3 - 1$
$(x - 1)(x^2 + x + 1)$

74. $x^3 - 8$
$(x - 2)(x^2 + 2x + 4)$

75. $x^3 + 125$
$(x + 5)(x^2 - 5x + 25)$

76. $x^3 + 216$
$(x + 6)(x^2 - 6x + 36)$

77. $3x^6y^2 + 81y^2$
$3y^2(x^2 + 3)(x^4 - 3x^2 + 9)$

78. $x^2y^9 + x^2y^3$
$x^2y^3(y^2 + 1)(y^4 - y^2 + 1)$

Review

Solve each equation. See Section 2.3.

79. $x - 5 = 0$ 5

80. $x + 7 = 0$ -7

81. $3x + 1 = 0$ $-\dfrac{1}{3}$

82. $5x - 15 = 0$ 3

83. $-2x = 0$ 0

84. $3x = 0$ 0

85. $-5x + 25 = 0$ 5

86. $-4x - 16 = 0$ -4

Concept Extensions

Factor each expression completely.

87. $x^2 - \dfrac{2}{3}x + \dfrac{1}{9}$ $\left(x - \dfrac{1}{3}\right)^2$

88. $x^2 - \dfrac{1}{25}$ $\left(x + \dfrac{1}{5}\right)\left(x - \dfrac{1}{5}\right)$

89. $(x + 2)^2 - y^2$
$(x + 2 + y)(x + 2 - y)$

90. $(y - 6)^2 - z^2$
$(y - 6 + z)(y - 6 - z)$

91. $a^2(b - 4) - 16(b - 4)$
$(b - 4)(a + 4)(a - 4)$

92. $m^2(n + 8) - 9(n + 8)$
$(n + 8)(m + 3)(m - 3)$

93. $(x^2 + 6x + 9) - 4y^2$ (*Hint:* Factor the trinomial in parentheses first.) $(x + 3 + 2y)(x + 3 - 2y)$

94. $(x^2 + 2x + 1) - 36y^2$ (*Hint:* Factor the trinomial in parentheses first.) $(x + 1 + 6y)(x + 1 - 6y)$

95. $x^{2n} - 100$
$(x^n + 10)(x^n - 10)$

96. $x^{2n} - 81$
$(x^n + 9)(x^n - 9)$

97. Fill in the blank so that $x^2 +$ ___8___ $x + 16$ is a perfect square trinomial.

98. Fill in the blank so that $9x^2 +$ ___30___ $x + 25$ is a perfect square trinomial.

99. Describe a perfect square trinomial.
answers may vary

100. Write a perfect square trinomial that factors as $(x + 3y)^2$. $x^2 + 6xy + 9y^2$

101. What binomial multiplied by $(x - 6)$ gives the difference of two squares? $(x + 6)$

102. What binomial multiplied by $(5 + y)$ gives the difference of two squares? $(5 - y)$

The area of the largest square in the figure is $(a + b)^2$. Use this figure to answer Exercises 103 and 104.

103. Write the area of the largest square as the sum of the areas of the smaller squares and rectangles. $a^2 + 2ab + b^2$

104. What factoring formula from this section is visually represented by this square?   perfect square trinomial

105. The Toroweap Overlook, on the North Rim of the Grand Canyon, lies 3000 vertical feet above the Colorado River. The view is spectacular, and the sheer drop is dramatic. A film crew creating a documentary about the Grand Canyon has suspended a camera platform 296 feet below the Overlook. A camera filter comes loose and falls to the river below. The height of the filter above the river after t seconds is given by the expression $2704 - 16t^2$.

  a. Find the height of the filter above the river after 3 seconds.　2560 ft

  b. Find the height of the filter above the river after 7 seconds.　1920 ft

  c. To the nearest whole second, estimate when the filter lands in the river.　13 sec

  d. Factor $2704 - 16t^2$.　$16(13 - t)(13 + t)$

106. An object is dropped from the top of Pittsburgh's USX Towers, which is 841 feet tall. (*Source: World Almanac* research) The height of the object after t seconds is given by the expression $841 - 16t^2$.

  a. Find the height of the object after 2 seconds.　777 ft

  b. Find the height of the object after 5 seconds.　441 ft

  c. To the nearest whole second, estimate when the object hits the ground.　7 sec

  d. Factor $841 - 16t^2$.　$(29 + 4t)(29 - 4t)$

841 feet

107. At this writing, the world's tallest building is the Taipei 101 in Taipei, Taiwan, at a height of 1671 feet. (*Source:* Council on Tall Buildings and Urban Habitat) Suppose a worker is suspended 71 feet below the top of the pinnacle atop the building, at a height of 1600 feet above the ground. If the worker accidentally drops a bolt, the height of the bolt after t seconds is given by the expression $1600 - 16t^2$.

  a. Find the height of the bolt after 3 seconds.

  b. Find the height of the bolt after 7 seconds.

  c. To the nearest whole second, estimate when the bolt hits the ground.　10 seconds

  d. Factor $1600 - 16t^2$.　$16(10 + t)(10 - t)$

  107. a. 1456 feet　**b.** 816 feet

108. A performer with the Moscow Circus is planning a stunt involving a free fall from the top of the Moscow State University building, which is 784 feet tall. (*Source:* Council on Tall Buildings and Urban Habitat) Neglecting air resistance, the performer's height above gigantic cushions positioned at ground level after t seconds is given by the expression $784 - 16t^2$.

  a. Find the performer's height after 2 seconds.

  b. Find the performer's height after 5 seconds.

  c. To the nearest whole second, estimate when the performer reaches the cushions positioned at ground level.　7 sec

  d. Factor $784 - 16t^2$.　$16(7 - t)(7 + t)$

  108. a. 720 ft　**b.** 384 ft

Review

Solve each equation. See Section 2.3.

79. $x - 5 = 0$ 5

80. $x + 7 = 0$ -7

81. $3x + 1 = 0$ $-\dfrac{1}{3}$

82. $5x - 15 = 0$ 3

83. $-2x = 0$ 0

84. $3x = 0$ 0

85. $-5x + 25 = 0$ 5

86. $-4x - 16 = 0$ -4

Concept Extensions

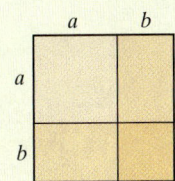

Choosing a Factoring Strategy

The key to proficiency in factoring polynomials is to practice until you are comfortable with each technique. A strategy for factoring polynomials completely is given next.

Factoring a Polynomial

Step 1: Are there any common factors? If so, factor out the greatest common factor.

Step 2: How many terms are in the polynomial?

    a. If there are *two* terms, decide if one of the following formulas may be applied:

      i. Difference of two squares: $a^2 - b^2 = (a + b)(a - b)$

      ii. Difference of two cubes: $a^3 - b^3 = (a - b)(a^2 + ab + b^2)$

      iii. Sum of two cubes: $a^3 + b^3 = (a + b)(a^2 - ab + b^2)$

    b. If there are *three* terms, try one of the following:

      i. Perfect square trinomial: $a^2 + 2ab + b^2 = (a + b)^2$
                                 $a^2 - 2ab + b^2 = (a - b)^2$

      ii. If not a perfect square trinomial, factor by using the methods presented in Sections 6.2 through 6.4.

    c. If there are *four* or more terms, try factoring by grouping.

Step 3: See whether any factors in the factored polynomial can be factored further.

Factor each polynomial completely.

1. $x^2 + x - 12$

2. $x^2 - 10x + 16$

3. $x^2 - x - 6$

4. $x^2 + 2x + 1$

5. $x^2 - 6x + 9$

6. $x^2 + x - 2$

7. $x^2 + x - 6$

8. $x^2 + 7x + 12$

▶ **9.** $x^2 - 7x + 10$

10. $x^2 - x - 30$

▶ **11.** $2x^2 - 98$

12. $3x^2 - 75$

13. $x^2 + 3x + 5x + 15$

14. $3y - 21 + xy - 7x$

15. $x^2 + 6x - 16$

16. $x^2 - 3x - 28$

17. $4x^3 + 20x^2 - 56x$

18. $6x^3 - 6x^2 - 120x$

19. $12x^2 + 34x + 24$

20. $8a^2 + 6ab - 5b^2$

21. $4a^2 - b^2$

22. $x^2 - 25y^2$

23. $28 - 13x - 6x^2$

24. $20 - 3x - 2x^2$

25. $x^2 - 2x + 4$

26. $a^2 + a - 3$

27. $6y^2 + y - 15$

28. $4x^2 - x - 5$

29. $18x^3 - 63x^2 + 9x$

30. $12a^3 - 24a^2 + 4a$

▶ **31.** $16a^2 - 56a + 49$

32. $25p^2 - 70p + 49$

33. $14 + 5x - x^2$

34. $3 - 2x - x^2$

35. $3x^4y + 6x^3y - 72x^2y$

36. $2x^3y + 8x^2y^2 - 10xy^3$

▶ **37.** $12x^3y + 243xy$

38. $6x^3y^2 + 8xy^2$

39. $2xy - 72x^3y$

40. $2x^3 - 18x$

▶ **41.** $x^3 + 6x^2 - 4x - 24$

42. $x^3 - 2x^2 - 36x + 72$

Answers

1. $(x - 3)(x + 4)$

2. $(x - 8)(x - 2)$

3. $(x + 2)(x - 3)$

4. $(x + 1)^2$

5. $(x - 3)^2$

6. $(x + 2)(x - 1)$

7. $(x + 3)(x - 2)$

8. $(x + 3)(x + 4)$

9. $(x - 5)(x - 2)$

10. $(x - 6)(x + 5)$

11. $2(x + 7)(x - 7)$

12. $3(x + 5)(x - 5)$

13. $(x + 3)(x + 5)$

14. $(y - 7)(3 + x)$

15. $(x + 8)(x - 2)$

16. $(x - 7)(x + 4)$

17. $4x(x + 7)(x - 2)$

18. $6x(x - 5)(x + 4)$

19. $2(3x + 4)(2x + 3)$

20. $(2a - b)(4a + 5b)$

21. $(2a + b)(2a - b)$

22. $(x + 5y)(x - 5y)$

23. $(4 - 3x)(7 + 2x)$

24. $(5 - 2x)(4 + x)$

25. prime

26. prime

27. $(3y + 5)(2y - 3)$

28. $(4x - 5)(x + 1)$

29. $9x(2x^2 - 7x + 1)$

30. $4a(3a^2 - 6a + 1)$

31. $(4a - 7)^2$

32. $(5p - 7)^2$

33. $(7 - x)(2 + x)$

34. $(3 + x)(1 - x)$

35. $3x^2y(x + 6)(x - 4)$

36. $2xy(x + 5y)(x - y)$

37. $3xy(4x^2 + 81)$

38. $2xy^2(3x^2 + 4)$

39. $2xy(1 + 6x)(1 - 6x)$

40. $2x(x + 3)(x - 3)$

41. $(x + 6)(x + 2)(x - 2)$

42. $(x - 2)(x + 6)(x - 6)$

449

43. $2a^2(3a + 5)$

44. $2n(2n - 3)$

45. $(3x - 1)(x^2 + 4)$

46. $(x - 2)(x^2 + 3)$

47. $6(x + 2y)(x + y)$

48. $2(x + 4y)(6x - y)$

49. $(x + y)(5 + x)$

50. $(x - y)(7 + y)$

51. $(7t - 1)(2t - 1)$

52. prime

53. $(3x + 5)(x - 1)$

54. $(7x - 2)(x + 3)$

55. $(1 - 10a)(1 + 2a)$

56. $(1 + 5a)(1 - 12a)$

57. $\begin{array}{l}(x + 3)(x - 3) \\ (x + 1)(x - 1)\end{array}$

58. $\begin{array}{l}(x + 3)(x - 3) \\ (x + 2)(x - 2)\end{array}$

59. $(x - 15)(x - 8)$

60. $(y + 16)(y + 6)$

61. prime

62. $(4a - 7b)^2$

63. $(5p - 7q)^2$

64. $(7x + 3y)(x + 3y)$

65. $-1(x - 5)(x + 6)$

66. $-1(x - 2)(x - 4)$

67. $(3r - 1)(s + 4)$

68. $(x - 2)(x^2 + 1)$

69. $(x - 2y)(4x - 3)$

70. $(2x - y)(2x + 7z)$

71. $(x + 12y)(x - 3y)$

72. $\begin{array}{l}(3x - 2y)(x + 4y) \\ (x^2 + 2)(x + 4)\end{array}$

73. $(x - 4)$

74. $\begin{array}{l}(x^2 + 3)(x + 5) \\ (x - 5)\end{array}$

75. $\begin{array}{l}x(x - 1) \\ (x^2 + x + 1)\end{array}$

76. $\begin{array}{l}x^3(x + 1) \\ (x^2 - x + 1)\end{array}$

77. $\begin{array}{l}(2x + 5y) \\ (4x^2 - 10xy + 25y^2)\end{array}$

78. $\begin{array}{l}(3x - 4y) \\ (9x^2 + 12xy + 16y^2)\end{array}$

79. answers may vary

80. yes; $9(x^2 + 9y^2)$

---

43. $6a^3 + 10a^2$　　**44.** $4n^2 - 6n$　　**45.** $3x^3 - x^2 + 12x - 4$

46. $x^3 - 2x^2 + 3x - 6$　　**47.** $6x^2 + 18xy + 12y^2$　　**48.** $12x^2 + 46xy - 8y^2$

49. $5(x + y) + x(x + y)$　　**50.** $7(x - y) + y(x - y)$　　**51.** $14t^2 - 9t + 1$

52. $3t^2 - 5t + 1$　　**53.** $3x^2 + 2x - 5$　　**54.** $7x^2 + 19x - 6$

55. $1 - 8a - 20a^2$　　**56.** $1 - 7a - 60a^2$　　**57.** $x^4 - 10x^2 + 9$

58. $x^4 - 13x^2 + 36$　　**59.** $x^2 - 23x + 120$　　**60.** $y^2 + 22y + 96$

61. $x^2 - 14x - 48$　　**62.** $16a^2 - 56ab + 49b^2$　　**63.** $25p^2 - 70pq + 49q^2$

64. $7x^2 + 24xy + 9y^2$　　**65.** $-x^2 - x + 30$　　**66.** $-x^2 + 6x - 8$

67. $3rs - s + 12r - 4$　　**68.** $x^3 - 2x^2 + x - 2$　　**69.** $4x^2 - 8xy - 3x + 6y$

70. $4x^2 - 2xy - 7yz + 14xz$　　**71.** $x^2 + 9xy - 36y^2$　　**72.** $3x^2 + 10xy - 8y^2$

73. $x^4 - 14x^2 - 32$　　**74.** $x^4 - 22x^2 - 75$　　▶ **75.** $x^4 - x$

76. $x^6 + x^3$　　▶ **77.** $8x^3 + 125y^3$　　**78.** $27x^3 - 64y^3$

79. Explain why it makes good sense to factor out the GCF first, before using other methods of factoring.

80. The sum of two squares usually does not factor. Is the sum of two squares $9x^2 + 81y^2$ factorable?

6.6 Solving Quadratic Equations by Factoring

In this section, we introduce a new type of equation—the **quadratic equation.**

> ### Quadratic Equation
>
> A quadratic equation is one that can be written in the form
>
> $$ax^2 + bx + c = 0$$
>
> where a, b, and c are real numbers and $a \neq 0$.

Objectives

A Solve Quadratic Equations by Factoring.

B Solve Equations with Degree Greater than Two by Factoring.

Some examples of quadratic equations are shown below.

$$x^2 - 9x - 22 = 0 \qquad 4x^2 - 28 = -49 \qquad x(2x - 7) = 4$$

The form $ax^2 + bx + c = 0$ is called the **standard form** of a quadratic equation. The quadratic equation $x^2 - 9x - 22 = 0$ is the only equation above that is in standard form.

Quadratic equations model many real-life situations. For example, let's suppose we want to know how long before a person diving from a 144-foot cliff reaches the ocean. The answer to this question is found by solving the quadratic equation $-16t^2 + 144 = 0$. (See Example 1 in Section 6.7.)

144 feet

Objective **A** Solving Quadratic Equations by Factoring

Some quadratic equations can be solved by making use of factoring and the **zero-factor property.**

> ### Zero-Factor Property
>
> If a and b are real numbers and if $ab = 0$, then $a = 0$ or $b = 0$.

In other words, if the product of two numbers is 0, then at least one of the numbers must be 0.

Teaching Tip

You may want to begin this lesson with the following activity: Ask students to suppose they are told that the product of two factors is 12. Do they know for certain what either factor is? Why or why not? Now tell students the product of two factors is 0. Do they know for certain what either factor is? Why or why not?

Example 1 Solve: $(x - 3)(x + 1) = 0$

Solution: If this equation is to be a true statement, then either the factor $x - 3$ must be 0 or the factor $x + 1$ must be 0. In other words, either

$$x - 3 = 0 \qquad \text{or} \qquad x + 1 = 0$$

If we solve these two linear equations, we have

$$x = 3 \qquad \text{or} \qquad x = -1$$

(Continued on next page)

Practice 1

Solve: $(x - 7)(x + 2) = 0$

Answer
1. 7 and −2

451

Thus, 3 and -1 are both solutions of the equation $(x - 3)(x + 1) = 0$. To check, we replace x with 3 in the original equation. Then we replace x with -1 in the original equation.

Check:

$(x - 3)(x + 1) = 0$ 　　　　　　　　　　　$(x - 3)(x + 1) = 0$

$(3 - 3)(3 + 1) \stackrel{?}{=} 0$ Replace x with 3.　$(-1 - 3)(-1 + 1) \stackrel{?}{=} 0$ Replace x with -1.

　　　　$0(4) = 0$ True　　　　　　　　$(-4)(0) = 0$ True

The solutions are 3 and -1.

■ **Work Practice 1**

Helpful Hint

The zero-factor property says that *if a product is 0, then a factor is 0*.

If $a \cdot b = 0$, then $a = 0$ or $b = 0$.

If $x(x + 5) = 0$, then $x = 0$ or $x + 5 = 0$.

If $(x + 7)(2x - 3) = 0$, then $x + 7 = 0$ or $2x - 3 = 0$.

Use this property only when the product is 0. For example, if $a \cdot b = 8$, we do not know the value of a or b. The values may be $a = 2, b = 4$ or $a = 8, b = 1$, or any other two numbers whose product is 8.

Practice 2

Solve: $(x - 10)(3x + 1) = 0$

Example 2 Solve: $(x - 5)(2x + 7) = 0$

Solution: The product is 0. By the zero-factor property, this is true only when a factor is 0. To solve, we set each factor equal to 0 and solve the resulting linear equations.

$$(x - 5)(2x + 7) = 0$$

$$x - 5 = 0 \quad \text{or} \quad 2x + 7 = 0$$

$$x = 5 \qquad\qquad 2x = -7$$

$$x = -\frac{7}{2}$$

Check: Let $x = 5$.

$$(x - 5)(2x + 7) = 0$$

$$(5 - 5)(2 \cdot 5 + 7) \stackrel{?}{=} 0 \quad \text{Replace } x \text{ with 5.}$$

$$0 \cdot 17 \stackrel{?}{=} 0$$

$$0 = 0 \quad \text{True}$$

Let $x = -\frac{7}{2}$.

$$(x - 5)(2x + 7) = 0$$

$$\left(-\frac{7}{2} - 5\right)\left(2\left(-\frac{7}{2}\right) + 7\right) \stackrel{?}{=} 0 \quad \text{Replace } x \text{ with } -\frac{7}{2}.$$

$$\left(-\frac{17}{2}\right)(-7 + 7) \stackrel{?}{=} 0$$

$$\left(-\frac{17}{2}\right) \cdot 0 \stackrel{?}{=} 0$$

$$0 = 0 \quad \text{True}$$

The solutions are 5 and $-\frac{7}{2}$.

■ **Work Practice 2**

Answer

2. 10 and $-\dfrac{1}{3}$

Example 3 Solve: $x(5x - 2) = 0$

Solution:

$$x(5x - 2) = 0$$

$x = 0$ or $5x - 2 = 0$ Use the zero-factor property.

$$5x = 2$$

$$x = \frac{2}{5}$$

Check these solutions in the original equation. The solutions are 0 and $\frac{2}{5}$.

■ **Work Practice 3**

Example 4 Solve: $x^2 - 9x - 22 = 0$

Solution: One side of the equation is 0. However, to use the zero-factor property, one side of the equation must be 0 *and* the other side must be written as a product (must be factored). Thus, we must first factor this polynomial.

$$x^2 - 9x - 22 = 0$$

$(x - 11)(x + 2) = 0$ Factor.

Now we can apply the zero-factor property.

$x - 11 = 0$ or $x + 2 = 0$

$x = 11$ $x = -2$

Check: Let $x = 11$.   Let $x = -2$.

$$x^2 - 9x - 22 = 0 \qquad\qquad x^2 - 9x - 22 = 0$$

$$11^2 - 9 \cdot 11 - 22 \stackrel{?}{=} 0 \qquad (-2)^2 - 9(-2) - 22 \stackrel{?}{=} 0$$

$$121 - 99 - 22 \stackrel{?}{=} 0 \qquad\qquad 4 + 18 - 22 \stackrel{?}{=} 0$$

$$22 - 22 \stackrel{?}{=} 0 \qquad\qquad\qquad 22 - 22 \stackrel{?}{=} 0$$

$$0 = 0 \quad \text{True} \qquad\qquad\qquad 0 = 0 \quad \text{True}$$

The solutions are 11 and -2.

■ **Work Practice 4**

Example 5 Solve: $4x^2 - 28x = -49$

Solution: First we rewrite the equation in standard form so that one side is 0. Then we factor the polynomial.

$$4x^2 - 28x = -49$$

$4x^2 - 28x + 49 = 0$ Write in standard form by adding 49 to both sides.

$(2x - 7)(2x - 7) = 0$ Factor.

Next we use the zero-factor property and set each factor equal to 0. Since the factors are the same, the related equations will give the same solution.

$2x - 7 = 0$ or $2x - 7 = 0$ Set each factor equal to 0.

$2x = 7$ $2x = 7$ Solve.

$$x = \frac{7}{2} \qquad\qquad x = \frac{7}{2}$$

Check this solution in the original equation. The solution is $\frac{7}{2}$.

■ **Work Practice 5**

Practice 3

Solve each equation.

a. $y(y + 3) = 0$

b. $x(4x - 3) = 0$

Practice 4

Solve: $x^2 - 3x - 18 = 0$

Practice 5

Solve: $9x^2 - 24x = -16$

Teaching Tip

Consider asking your students why we want to rewrite the equation so that one side is equal to zero. What does that accomplish?

Answers

3. a. 0 and -3 **b.** 0 and $\frac{3}{4}$

4. 6 and -3 **5.** $\frac{4}{3}$

The following steps may be used to solve a quadratic equation by factoring.

To Solve Quadratic Equations by Factoring

Step 1: Write the equation in standard form so that one side of the equation is 0.

Step 2: Factor the quadratic equation completely.

Step 3: Set each factor containing a variable equal to 0.

Step 4: Solve the resulting equations.

Step 5: Check each solution in the original equation.

Since it is not always possible to factor a quadratic polynomial, not all quadratic equations can be solved by factoring. Other methods of solving quadratic equations are presented in Chapter 11.

Practice 6

Solve each equation.

a. $x(x - 4) = 5$

b. $x(3x + 7) = 6$

Example 6 Solve: $x(2x - 7) = 4$

Solution: First we write the equation in standard form; then we factor.

$$x(2x - 7) = 4$$
$$2x^2 - 7x = 4 \qquad \text{Multiply.}$$
$$2x^2 - 7x - 4 = 0 \qquad \text{Write in standard form.}$$
$$(2x + 1)(x - 4) = 0 \qquad \text{Factor.}$$
$$2x + 1 = 0 \quad \text{or} \quad x - 4 = 0 \qquad \text{Set each factor equal to zero.}$$
$$2x = -1 \qquad\qquad x = 4 \qquad \text{Solve.}$$
$$x = -\frac{1}{2}$$

Check the solutions in the original equation. The solutions are $-\frac{1}{2}$ and 4.

■ **Work Practice 6**

Helpful Hint

To solve the equation $x(2x - 7) = 4$, do **not** set each factor equal to 4. Remember that to apply the zero-factor property, one side of the equation must be 0 and the other side of the equation must be in factored form.

✓**Concept Check** Explain the error and solve the equation correctly.

$$(x - 3)(x + 1) = 5$$
$$x - 3 = 0 \quad \text{or} \quad x + 1 = 0$$
$$x = 3 \quad \text{or} \qquad x = -1$$

Answers

6. **a.** 5 and −1   **b.** $\frac{2}{3}$ and −3

✓**Concept Check Answer**

To use the zero-factor property, one side of the equation must be 0, not 5. Correctly, $(x - 3)(x + 1) = 5$, $x^2 - 2x - 3 = 5$, $x^2 - 2x - 8 = 0$, $(x - 4)(x + 2) = 0$, $x - 4 = 0$ or $x + 2 = 0, x = 4$ or $x = -2$.

Objective B Solving Equations with Degree Greater than Two by Factoring ▶

Some equations with degree greater than 2 can be solved by factoring and then using the zero-factor property.

Example 7   Solve: $3x^3 - 12x = 0$

Solution: To factor the left side of the equation, we begin by factoring out the greatest common factor, $3x$.

$$3x^3 - 12x = 0$$
$$3x(x^2 - 4) = 0 \quad \text{Factor out the GCF, } 3x.$$
$$3x(x + 2)(x - 2) = 0 \quad \text{Factor } x^2 - 4, \text{ a difference of two squares.}$$
$$3x = 0 \quad \text{or} \quad x + 2 = 0 \quad \text{or} \quad x - 2 = 0 \quad \text{Set each factor equal to 0.}$$
$$x = 0 \qquad\qquad x = -2 \qquad\qquad x = 2 \quad \text{Solve.}$$

Thus, the equation $3x^3 - 12x = 0$ has three solutions: 0, -2, and 2.

Check: Replace x with each solution in the original equation.
Let $x = 0$.

$$3(0)^3 - 12(0) \overset{?}{=} 0$$
$$0 = 0 \quad \text{True}$$

Let $x = -2$.

$$3(-2)^3 - 12(-2) \overset{?}{=} 0$$
$$3(-8) + 24 \overset{?}{=} 0$$
$$0 = 0 \quad \text{True}$$

Let $x = 2$.

$$3(2)^3 - 12(2) \overset{?}{=} 0$$
$$3(8) - 24 \overset{?}{=} 0$$
$$0 = 0 \quad \text{True}$$

The solutions are 0, -2, and 2.

■ **Work Practice 7**

Example 8   Solve: $(5x - 1)(2x^2 + 15x + 18) = 0$

Solution:

$$(5x - 1)(2x^2 + 15x + 18) = 0$$
$$(5x - 1)(2x + 3)(x + 6) = 0 \qquad\qquad \text{Factor the trinomial.}$$
$$5x - 1 = 0 \quad \text{or} \quad 2x + 3 = 0 \quad \text{or} \quad x + 6 = 0 \quad \text{Set each factor equal to 0.}$$
$$5x = 1 \qquad\qquad 2x = -3 \qquad\qquad x = -6 \quad \text{Solve.}$$
$$x = \frac{1}{5} \qquad\qquad x = -\frac{3}{2}$$

Check each solution in the original equation. The solutions are $\frac{1}{5}$, $-\frac{3}{2}$, and -6.

■ **Work Practice 8**

Practice 7

Solve: $2x^3 - 18x = 0$

Teaching Tip

Consider challenging students to notice some patterns. Ask them to make a conjecture about the number of solutions that a quadratic equation has, that a cubic equation has, and so on. Also ask students if they know a solution of a quadratic equation, do they know a factor of the equation?

Practice 8

Solve:
$(x + 3)(3x^2 - 20x - 7) = 0$

Answers

7. $0, 3,$ and -3 **8.** $-3, -\frac{1}{3},$ and 7

Vocabulary, Readiness & Video Check

Use the choices below to fill in each blank. Not all choices will be used.

$-3, 5$	$a = 0 \text{ or } b = 0$	0	linear
$3, -5$	quadratic	1	

1. An equation that can be written in the form $ax^2 + bx + c = 0$ (with $a \neq 0$) is called a _____quadratic_____ equation.

2. If the product of two numbers is 0, then at least one of the numbers must be _____0_____.

3. The solutions of $(x - 3)(x + 5) = 0$ are _____3, −5_____.

4. If $a \cdot b = 0$, then _____$a = 0$ or $b = 0$_____.

Martin-Gay Interactive Videos   Watch the section lecture video and answer the following questions.

See Video 6.6

See video answer section.

Objective A **5.** As shown in ⊞Examples 1–3, what two things have to be true in order to use the zero-factor theorem?

Objective B **6.** ⊞Example 4 implies that the zero-factor theorem can be used with any number of factors on one side of the equation as long as the other side of the equation is zero. Why do you think this is true? ▶

6.6 Exercise Set MyMathLab® ▶

Objective A *Solve each equation. See Examples 1 through 3.*

1. $(x - 2)(x + 1) = 0$
2, −1

2. $(x + 3)(x + 2) = 0$
−3, −2

3. $(x - 6)(x - 7) = 0$
6, 7

4. $(x + 4)(x - 10) = 0$
−4, 10

5. $(x + 9)(x + 17) = 0$
−9, −17

6. $(x - 11)(x - 1) = 0$
11, 1

7. $x(x + 6) = 0$
0, −6

8. $x(x - 7) = 0$
0, 7

9. $3x(x - 8) = 0$
0, 8

10. $2x(x + 12) = 0$ 0, −12

▶ **11.** $(2x + 3)(4x - 5) = 0$ $-\dfrac{3}{2}, \dfrac{5}{4}$

12. $(3x - 2)(5x + 1) = 0$ $\dfrac{2}{3}, -\dfrac{1}{5}$

13. $(2x - 7)(7x + 2) = 0$ $\dfrac{7}{2}, -\dfrac{2}{7}$

14. $(9x + 1)(4x - 3) = 0$ $-\dfrac{1}{9}, \dfrac{3}{4}$

15. $\left(x - \dfrac{1}{2}\right)\left(x + \dfrac{1}{3}\right) = 0$ $\dfrac{1}{2}, -\dfrac{1}{3}$

16. $\left(x + \dfrac{2}{9}\right)\left(x - \dfrac{1}{4}\right) = 0$ $-\dfrac{2}{9}, \dfrac{1}{4}$

17. $(x + 0.2)(x + 1.5) = 0$
−0.2, −1.5

18. $(x + 1.7)(x + 2.3) = 0$
−1.7, −2.3

Solve. See Examples 4 through 6.

19. $x^2 - 13x + 36 = 0$
9, 4

20. $x^2 + 2x - 63 = 0$
−9, 7

▶ **21.** $x^2 + 2x - 8 = 0$
−4, 2

22. $x^2 - 5x + 6 = 0$
3, 2

23. $x^2 - 7x = 0$
0, 7

24. $x^2 - 3x = 0$
0, 3

25. $x^2 + 20x = 0$
0, −20

26. $x^2 + 15x = 0$
0, −15

27. $x^2 = 16$
4, −4

28. $x^2 = 9$
3, −3

29. $x^2 - 4x = 32$
8, −4

30. $x^2 - 5x = 24$
8, −3

31. $(x + 4)(x - 9) = 4x$
−3, 12

32. $(x + 3)(x + 8) = x$
−4, −6

▶ **33.** $x(3x - 1) = 14$
$\dfrac{7}{3}, -2$

34. $x(4x - 11) = 3$
$-\dfrac{1}{4}, 3$

35. $3x^2 + 19x - 72 = 0$ $\quad \frac{8}{3}, -9$

36. $36x^2 + x - 21 = 0$ $\quad \frac{3}{4}, -\frac{7}{9}$

Objectives **A** **B** **and Section 2.3** **Mixed Practice** *Solve each equation. See Examples 1 through 8. (A few exercises are linear equations.)*

37. $4x^3 - x = 0$ $\quad 0, \frac{1}{2}, -\frac{1}{2}$

38. $4y^3 - 36y = 0$ $\quad 0, 3, -3$

39. $4(x - 7) = 6$ $\quad \frac{17}{2}$

40. $5(3 - 4x) = 9$
$\frac{3}{10}$

41. $(4x - 3)(16x^2 - 24x + 9) = 0$
$\frac{3}{4}$

42. $(2x + 5)(4x^2 + 20x + 25) = 0$
$-\frac{5}{2}$

43. $4y^2 - 1 = 0$
$\frac{1}{2}, -\frac{1}{2}$

44. $4y^2 - 81 = 0$
$\frac{9}{2}, -\frac{9}{2}$

▶ 45. $(2x + 3)(2x^2 - 5x - 3) = 0$
$-\frac{3}{2}, -\frac{1}{2}, 3$

46. $(2x - 9)(x^2 + 5x - 36) = 0$
$\frac{9}{2}, -9, 4$

47. $x^2 - 15 = -2x$
$-5, 3$

48. $x^2 - 26 = -11x$
$-13, 2$

49. $30x^2 - 11x = 30$ $\quad -\frac{5}{6}, \frac{6}{5}$

50. $9x^2 + 7x = 2$ $\quad -1, \frac{2}{9}$

51. $5x^2 - 6x - 8 = 0$ $\quad 2, -\frac{4}{5}$

52. $12x^2 + 7x - 12 = 0$ $\quad \frac{3}{4}, -\frac{4}{3}$

53. $6y^2 - 22y - 40 = 0$ $\quad -\frac{4}{3}, 5$

54. $3x^2 - 6x - 9 = 0$ $\quad 3, -1$

55. $(y - 2)(y + 3) = 6$
$-4, 3$

56. $(y - 5)(y - 2) = 28$
$9, -2$

57. $x^3 - 12x^2 + 32x = 0$
$0, 8, 4$

58. $x^3 - 14x^2 + 49x = 0$
$0, 7$

59. $x^2 + 14x + 49 = 0$
-7

60. $x^2 + 22x + 121 = 0$
-11

61. $12y = 8y^2$ $\quad 0, \frac{3}{2}$

62. $9y = 6y^2$ $\quad 0, \frac{3}{2}$

63. $7x^3 - 7x = 0$ $\quad 0, 1, -1$

64. $3x^3 - 27x = 0$
$0, 3, -3$

65. $3x^2 + 8x - 11 = 13 - 6x$
$-6, \frac{4}{3}$

66. $2x^2 + 12x - 1 = 4 + 3x$
$-5, \frac{1}{2}$

67. $3x^2 - 20x = -4x^2 - 7x - 6$
$\frac{6}{7}, 1$

68. $4x^2 - 20x = -5x^2 - 6x - 5$
$\frac{5}{9}, 1$

Review

Perform each indicated operation. Write all results in lowest terms. See Section R.2.

69. $\frac{3}{5} + \frac{4}{9}$ $\quad \frac{47}{45}$

70. $\frac{2}{3} + \frac{3}{7}$ $\quad \frac{23}{21}$

71. $\frac{7}{10} - \frac{5}{12}$ $\quad \frac{17}{60}$

72. $\frac{5}{9} - \frac{5}{12}$ $\quad \frac{5}{36}$

73. $\frac{4}{5} \cdot \frac{7}{8}$ $\quad \frac{7}{10}$

74. $\frac{3}{7} \cdot \frac{12}{17}$ $\quad \frac{36}{119}$

Concept Extensions

For Exercises 75 and 76, see the Concept Check in this section.

75. Explain the error and solve correctly:

$x(x - 2) = 8$

$x = 8$ or $x - 2 = 8$ didn't write equation in
            standard form; should
$x = 10$ be $x = 4$ or $x = -2$

76. Explain the error and solve correctly:

$(x - 4)(x + 2) = 0$ didn't solve the linear
            equations correctly;
$x = -4$ or $x = 2$ should be $x = 4$ or
            $x = -2$

77. Write a quadratic equation that has two solutions, 6 and −1. Leave the polynomial in the equation in factored form.
answers may vary, for example, $(x - 6)(x + 1) = 0$

78. Write a quadratic equation that has two solutions, 0 and −2. Leave the polynomial in the equation in factored form.
answers may vary, for example, $x(x + 2) = 0$

79. Write a quadratic equation in standard form that has two solutions, 5 and 7.   answers may vary, for example, $x^2 - 12x + 35 = 0$

80. Write an equation that has three solutions, 0, 1, and 2.   answers may vary, for example, $x^3 - 3x^2 + 2x = 0$

81. A compass is accidentally thrown upward and out of a hot-air balloon at a height of 300 feet. The height, y, of the compass at time x is given by the equation $y = -16x^2 + 20x + 300$.

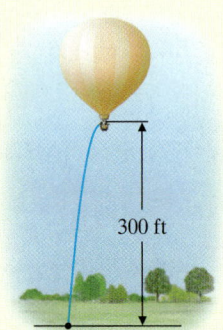

300 ft

a. Find the height of the compass at the given times by filling in the table below.

Time, x (in seconds)	0	1	2	3	4	5	6
Height, y (in feet)	300	304	276	216	124	0	−156

b. Use the table to determine when the compass strikes the ground.   5 sec
c. Use the table to approximate the maximum height of the compass.   304 ft

82. A rocket is fired upward from the ground with an initial velocity of 100 feet per second. The height, y, of the rocket at any time x is given by the equation $y = -16x^2 + 100x$.

y

a. Find the height of the rocket at the given times by filling in the table below.

Time, x (in seconds)	0	1	2	3	4	5	6	7
Height, y (in feet)	0	84	136	156	144	100	24	−84

b. Use the table to determine between what two whole numbered seconds the rocket strikes the ground.   between 6 and 7 sec
c. Use the table to approximate the maximum height of the rocket.   156 ft

Solve each equation.

83. $(x - 3)(3x + 4) = (x + 2)(x - 6)$ $0, \dfrac{1}{2}$

84. $(2x - 3)(x + 6) = (x - 9)(x + 2)$ $0, -16$

85. $(2x - 3)(x + 8) = (x - 6)(x + 4)$ $0, -15$

86. $(x + 6)(x - 6) = (2x - 9)(x + 4)$ $0, 1$

6.7 Quadratic Equations and Problem Solving

Objective A Solving Problems Modeled by Quadratic Equations

Objective

A Solve Problems That Can Be Modeled by Quadratic Equations.

Some problems may be modeled by quadratic equations. To solve these problems, we use the same problem-solving steps that were introduced in Section 2.4. When solving these problems, keep in mind that a solution of an equation that models a problem may not be a solution of the problem. For example, a person's age or the length of a rectangle is always a positive number. Thus, we discard solutions that do not make sense as solutions of the problem.

Example 1 Finding Free-Fall Time

Since the 1940s, one of the top tourist attractions in Acapulco, Mexico, is watching the cliff divers off La Quebrada. The divers' platform is about 144 feet above the sea. These divers must time their descent just right, since they land in the crashing Pacific, in an inlet that is at most $9\frac{1}{2}$ feet deep. Neglecting air resistance, the height h in feet of a cliff diver above the ocean after t seconds is given by the quadratic equation $h = -16t^2 + 144$.

Find out how long it takes the diver to reach the ocean.

Practice 1

Cliff divers also frequent the falls at Waimea Falls Park in Oahu, Hawaii. Here, a diver can jump from a ledge 64 feet up the waterfall into a rocky pool below. Neglecting air resistance, the height of a diver above the pool after t seconds is $h = -16t^2 + 64$. Find how long it takes the diver to reach the pool.

Solution:

1. UNDERSTAND. Read and reread the problem. Then draw a picture of the problem.

   The equation $h = -16t^2 + 144$ models the height of the falling diver at time t. Familiarize yourself with this equation by finding the height of the diver at time $t = 1$ second and $t = 2$ seconds.

   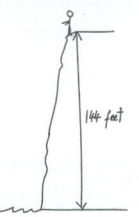

   When $t = 1$ second, the height of the diver is $h = -16(1)^2 + 144 = 128$ feet.

   When $t = 2$ seconds, the height of the diver is $h = -16(2)^2 + 144 = 80$ feet.

2. TRANSLATE. To find out how long it takes the diver to reach the ocean, we want to know the value of t for which $h = 0$.

3. SOLVE. Solve the equation.

$$0 = -16t^2 + 144$$
$$0 = -16(t^2 - 9) \qquad \text{Factor out } -16.$$
$$0 = -16(t - 3)(t + 3) \qquad \text{Factor completely.}$$
$$t - 3 = 0 \quad \text{or} \quad t + 3 = 0 \qquad \text{Set each factor containing a variable equal to 0.}$$
$$t = 3 \quad \text{or} \qquad t = -3 \qquad \text{Solve.}$$

4. INTERPRET. Since the time t cannot be negative, the proposed solution is 3 seconds.

   Check: Verify that the height of the diver when t is 3 seconds is 0.

   When $t = 3$ seconds, $h = -16(3)^2 + 144 = -144 + 144 = 0$.

■ **Work Practice 1**

Answer
1. 2 sec

Practice 2

The square of a number minus twice the number is 63. Find the number.

Example 2 Finding a Number

The square of a number plus three times the number is 70. Find the number.

Solution:

1. **UNDERSTAND. Read and reread the problem.** Suppose that the number is 5. The square of 5 is 5^2 or 25. Three times 5 is 15. Then $25 + 15 = 40$, not 70, so the number must be greater than 5. Remember, the purpose of proposing a number, such as 5, is to better understand the problem. Now that we do, we will let $x =$ the number.

2. **TRANSLATE.**

the square of a number	plus	three times the number	is	70
↓	↓	↓	↓	↓
x^2	$+$	$3x$	$=$	70

3. **SOLVE.**

$$x^2 + 3x = 70$$
$$x^2 + 3x - 70 = 0 \qquad \text{Subtract 70 from both sides.}$$
$$(x + 10)(x - 7) = 0 \qquad \text{Factor.}$$
$$x + 10 = 0 \quad \text{or} \quad x - 7 = 0 \qquad \text{Set each factor equal to 0.}$$
$$x = -10 \qquad\qquad x = 7 \quad \text{Solve.}$$

4. **INTERPRET.**

   Check: The square of -10 is $(-10)^2$, or 100. Three times -10 is $3(-10)$ or -30. Then $100 + (-30) = 70$, the correct sum, so -10 checks.
   The square of 7 is 7^2 or 49. Three times 7 is $3(7)$, or 21. Then $49 + 21 = 70$, the correct sum, so 7 checks.

   State: There are two numbers. They are -10 and 7.

■ **Work Practice 2**

Practice 3

The length of a rectangular garden is 5 feet more than its width. The area of the garden is 176 square feet. Find the length and the width of the garden.

⚠ Example 3 Finding the Dimensions of a Sail

The height of a triangular sail is 2 meters less than twice the length of the base. If the sail has an area of 30 square meters, find the length of its base and the height.

Solution:

1. **UNDERSTAND. Read and reread the problem.** Since we are finding the length of the base and the height, we let

   $x =$ the length of the base

   Since the height is 2 meters less than twice the length of the base,

   $2x - 2 =$ the height

   An illustration is shown in the margin.

2. **TRANSLATE.** We are given that the area of the triangle is 30 square meters, so we use the formula for area of a triangle.

area of triangle	=	$\frac{1}{2}$	·	base	·	height
↓		↓		↓		↓
30	$=$	$\frac{1}{2}$	·	x	·	$(2x - 2)$

Answers

2. 9 and -7

3. length: 16 ft; width: 11 ft

3. SOLVE. Now we solve the quadratic equation.

$$30 = \frac{1}{2}x(2x - 2)$$

$30 = x^2 - x$	Multiply.
$0 = x^2 - x - 30$	Write in standard form.
$0 = (x - 6)(x + 5)$	Factor.
$x - 6 = 0$ or $x + 5 = 0$	Set each factor equal to 0.
$x = 6$ $\qquad x = -5$	

4. INTERPRET. Since x represents the length of the base, we discard the solution -5. The base of a triangle cannot be negative. The base is then 6 meters and the height is $2(6) - 2 = 10$ meters.

Check: To check this problem, we recall that

$$\text{area} = \frac{1}{2}\,\text{base}\cdot\text{height or}$$

$$30 \stackrel{?}{=} \frac{1}{2}(6)(10)$$

$30 = 30$	True

State: The base of the triangular sail is 6 meters and the height is 10 meters.

■ **Work Practice 3**

Teaching Tip
Remind students that although the equation that models an application may have more than one solution, the solution to the application must make sense in the context of the problem. Tell them to always check to see that a solution makes sense for the question being asked.

The next example has to do with consecutive integers. Study the following diagrams for a review of consecutive integers.

Examples

If x is the first integer, then consecutive integers are
$x, x + 1, x + 2, \ldots$.

If x is the first even integer, then consecutive even integers are
$x, x + 2, x + 4, \ldots$.

If x is the first odd integer, then consecutive odd integers are
$x, x + 2, x + 4, \ldots$.

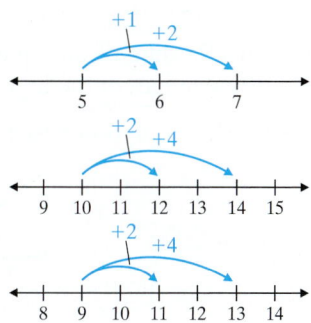

Example 4 Finding Consecutive Even Integers

Find two consecutive even integers whose product is 34 more than their sum.

Solution:

1. UNDERSTAND. Read and reread the problem. Let's just choose two consecutive even integers to help us better understand the problem. Let's choose 10 and 12. Their product is $10(12) = 120$ and their sum is $10 + 12 = 22$. The product is $120 - 22$, or 98 greater than the sum. Thus, our guess is incorrect, but we have a better understanding of this example.

Let's let x and $x + 2$ be the consecutive even integers.

2. TRANSLATE.

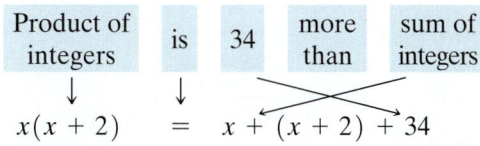

(Continued on next page)

Practice 4

Find two consecutive odd integers whose product is 23 more than their sum.

Teaching Tip
Example 4 may be a good opportunity to point out to students that they are not finished with an application when they have found the values for x. They must also answer the question that was asked.

Answer
4. 5 and 7 or -5 and -3

3. SOLVE. Now we solve the equation.

$$x(x + 2) = x + (x + 2) + 34$$
$$x^2 + 2x = x + x + 2 + 34 \qquad \text{Multiply.}$$
$$x^2 + 2x = 2x + 36 \qquad \text{Combine like terms.}$$
$$x^2 - 36 = 0 \qquad \text{Write in standard form.}$$
$$(x + 6)(x - 6) = 0 \qquad \text{Factor.}$$
$$x + 6 = 0 \quad \text{or} \quad x - 6 = 0 \qquad \text{Set each factor equal to 0.}$$
$$x = -6 \qquad\qquad x = 6 \qquad \text{Solve.}$$

4. INTERPRET. If $x = -6$, then $x + 2 = -6 + 2$, or -4.
If $x = 6$, then $x + 2 = 6 + 2$, or 8.

Check: $-6, -4$ $\qquad\qquad\qquad\qquad\qquad$ $6, 8$

$$-6(-4) \stackrel{?}{=} -6 + (-4) + 34 \qquad\qquad 6(8) \stackrel{?}{=} 6 + 8 + 34$$
$$24 \stackrel{?}{=} -10 + 34 \qquad\qquad\qquad 48 \stackrel{?}{=} 14 + 34$$
$$24 = 24 \qquad \text{True} \qquad\qquad 48 = 48 \qquad \text{True}$$

State: The two consecutive even integers are -6 and -4 or 6 and 8.

■ **Work Practice 4**

The next example makes use of the **Pythagorean theorem.** Before we review this theorem, recall that a **right triangle** is a triangle that contains a 90° or right angle. The **hypotenuse** of a right triangle is the side opposite the right angle and is the longest side of the triangle. The **legs** of a right triangle are the other sides of the triangle.

Pythagorean Theorem

In a right triangle, the sum of the squares of the lengths of the two legs is equal to the square of the length of the hypotenuse.

$$(\text{leg})^2 + (\text{leg})^2 = (\text{hypotenuse})^2 \quad \text{or} \quad a^2 + b^2 = c^2$$

Hypotenuse c

Leg b

Leg a

> **Helpful Hint**
> If you use this formula, don't forget that c represents the length of the hypotenuse.

Practice 5

The length of one leg of a right triangle is 7 meters less than the length of the other leg. The length of the hypotenuse is 13 meters. Find the lengths of the legs.

△ **Example 5**   Finding the Dimensions of a Triangle

Find the lengths of the sides of a right triangle if the lengths can be expressed as three consecutive even integers.

Solution:

1. UNDERSTAND. Read and reread the problem. Let's suppose that the length of one leg of the right triangle is 4 units. Then the other leg is the next even integer, or 6 units, and the hypotenuse of the triangle is the next even integer, or 8 units. Remember that the hypotenuse is the longest side. Let's see if a triangle with sides of these lengths forms a right triangle. To do this, we check to see whether the Pythagorean theorem holds true.

$$4^2 + 6^2 \stackrel{?}{=} 8^2$$
$$16 + 36 \stackrel{?}{=} 64$$
$$52 = 64 \qquad \text{False}$$

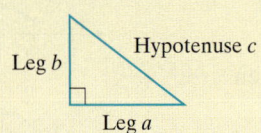

Our proposed numbers do not check, but we now have a better understanding of the problem.

Answer

5. 5 meters, 12 meters

We let x, $x + 2$, and $x + 4$ be three consecutive even integers. Since these integers represent lengths of the sides of a right triangle, we have the following.

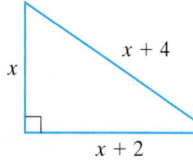

$x = $ one leg

$x + 2 = $ other leg

$x + 4 = $ hypotenuse (longest side)

2. **TRANSLATE.** By the Pythagorean theorem, we have that

$$(\text{leg})^2 + (\text{leg})^2 = (\text{hypotenuse})^2$$
$$(x)^2 + (x + 2)^2 = (x + 4)^2$$

3. **SOLVE.** Now we solve the equation.

$$x^2 + (x + 2)^2 = (x + 4)^2$$

$x^2 + x^2 + 4x + 4 = x^2 + 8x + 16$	Multiply.
$2x^2 + 4x + 4 = x^2 + 8x + 16$	Combine like terms.
$x^2 - 4x - 12 = 0$	Write in standard form.
$(x - 6)(x + 2) = 0$	Factor.
$x - 6 = 0 \quad \text{or} \quad x + 2 = 0$	Set each factor equal to 0.
$x = 6 \qquad\qquad x = -2$	

4. **INTERPRET.** We discard $x = -2$ since length cannot be negative. If $x = 6$, then $x + 2 = 8$ and $x + 4 = 10$.

Check: Verify that

$$(\text{leg})^2 + (\text{leg})^2 = (\text{hypotenuse})^2$$
$$6^2 + 8^2 \overset{?}{=} 10^2$$
$$36 + 64 \overset{?}{=} 100$$
$$100 = 100 \qquad \text{True}$$

State: The sides of the right triangle have lengths 6 units, 8 units, and 10 units.

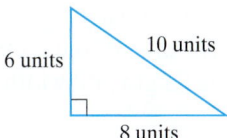

■ **Work Practice 5**

Vocabulary, Readiness & Video Check

Martin-Gay Interactive Videos Watch the section lecture video and answer the following question.

Objective A 1. In each of ▣ Examples 1–3, why aren't both solutions of the translated equation accepted as solutions of the application? ▶

See Video 6.7 🍎

See video answer section.

6.7 Exercise Set MyMathLab® ▶

Objective **A** *See Examples 1 through 5 for all exercises.*

Translating *For Exercises 1 through 6, represent each given condition using a single variable, x.*

△**1.** The length and width of a rectangle whose length is 4 centimeters more than its width
width: x; length: $x + 4$

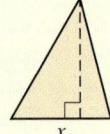

△**2.** The length and width of a rectangle whose length is twice its width   width: x; length: $2x$

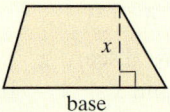

3. Two consecutive odd integers
x and $x + 2$ if x is an odd integer

4. Two consecutive even integers
x and $x + 2$ if x is an even integer

△**5.** The base and height of a triangle whose height is one more than four times its base
base: x; height: $4x + 1$

△**6.** The base and height of a trapezoid whose base is three less than five times its height
height: x; base: $5x - 3$

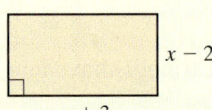

Use the information given to find the dimensions of each figure.

△**7.**

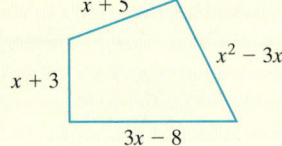

The *area* of the square is 121 square units. Find the length of its sides.   11 units

△**8.**

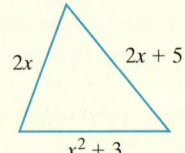

The *area* of the rectangle is 84 square inches. Find its length and width.   length: 12 in.; width: 7 in.

△**9.**

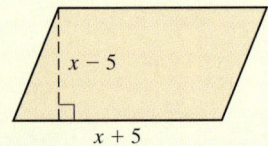

The *perimeter* of the quadrilateral is 120 centimeters. Find the lengths of its sides.
15 cm, 13 cm, 22 cm, 70 cm

▶ **10.**

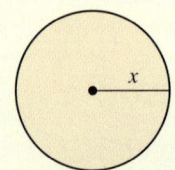

The *perimeter* of the triangle is 85 feet. Find the lengths of its sides.   14 ft, 19 ft, 52 ft

△**11.**

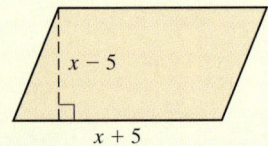

The *area* of the parallelogram is 96 square miles. Find its base and height.   base: 16 mi; height: 6 mi

△**12.**

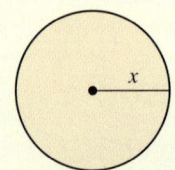

The *area* of the circle is 25π square kilometers. Find its radius.   5 km

Solve.

▶ **13.** An object is thrown upward from the top of an 80-foot building with an initial velocity of 64 feet per second. The height h of the object after t seconds is given by the quadratic equation $h = -16t^2 + 64t + 80$. When will the object hit the ground?　5 sec

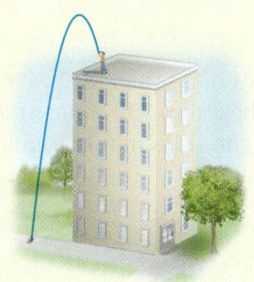

14. A hang glider accidentally drops her compass from the top of a 400-foot cliff. The height h of the compass after t seconds is given by the quadratic equation $h = -16t^2 + 400$. When will the compass hit the ground?　5 sec

15. The width of a rectangle is 7 centimeters less than twice its length. Its area is 30 square centimeters. Find the dimensions of the rectangle.
width: 5 cm; length: 6 cm

16. The length of a rectangle is 9 inches more than its width. Its area is 112 square inches. Find the dimensions of the rectangle.
length: 16 in.; width: 7 in.

△ *The equation* $D = \dfrac{1}{2}n(n-3)$ *gives the number of diagonals D for a polygon with n sides. For example, a polygon with 6 sides has* $D = \dfrac{1}{2}\cdot 6(6-3)$ *or D = 9 diagonals. (See if you can count all 9 diagonals. Some are shown in the figure.)*

Use this equation, $D = \dfrac{1}{2}n(n-3)$, *for Exercises 17 through 20.*

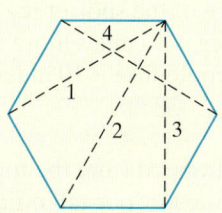

17. Find the number of diagonals for a polygon that has 12 sides.　54 diagonals

18. Find the number of diagonals for a polygon that has 15 sides.　90 diagonals

19. Find the number of sides n for a polygon that has 35 diagonals.　10 sides

20. Find the number of sides n for a polygon that has 14 diagonals.　7 sides

21. The sum of a number and its square is 132. Find the number.　-12 or 11

▶ **22.** The sum of a number and its square is 182. Find the number.　-14 or 13

23. The product of two consecutive room numbers is 210. Find the room numbers.　14, 15

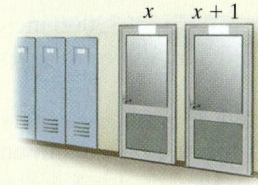

▶ **24.** The product of two consecutive page numbers is 420. Find the page numbers.　20, 21

25. A ladder is leaning against a building so that the distance from the ground to the top of the ladder is one foot less than the length of the ladder. Find the length of the ladder if the distance from the bottom of the ladder to the building is 5 feet.
13 feet

26. Use the given figure to find the length of the guy wire.
50 feet

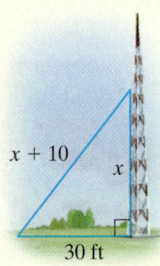

△ **27.** If the sides of a square are increased by 3 inches, the area becomes 64 square inches. Find the length of the sides of the original square.   5 in.

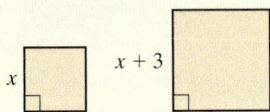

△ **28.** If the sides of a square are increased by 5 meters, the area becomes 100 square meters. Find the length of the sides of the original square.   5 m

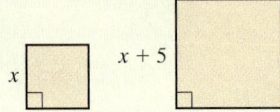

△ **29.** One leg of a right triangle is 4 millimeters longer than the shorter leg and the hypotenuse is 8 millimeters longer than the shorter leg. Find the lengths of the sides of the triangle.   12 mm, 16 mm, 20 mm

△ **30.** One leg of a right triangle is 9 centimeters longer than the other leg and the hypotenuse is 45 centimeters. Find the lengths of the legs of the triangle.   27 cm and 36 cm

△ **31.** The length of the base of a triangle is twice its height. If the area of the triangle is 100 square kilometers, find the height.
10 km

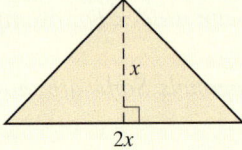

△ **32.** The height of a triangle is 2 millimeters less than the base. If the area is 60 square millimeters, find the base.
12 mm

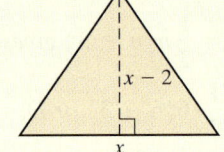

△ **33.** Find the length of the shorter leg of a right triangle if the longer leg is 12 feet more than the shorter leg and the hypotenuse is 12 feet less than twice the shorter leg.   36 ft

△ **34.** Find the length of the shorter leg of a right triangle if the longer leg is 10 miles more than the shorter leg and the hypotenuse is 10 miles less than twice the shorter leg.   30 mi

35. An object is dropped from 39 feet below the tip of the pinnacle atop one of the 1483-foot-tall Petronas Twin Towers in Kuala Lumpur, Malaysia. (*Source:* Council on Tall Buildings and Urban Habitat) The height h of the object after t seconds is given by the equation $h = -16t^2 + 1444$. Find how many seconds pass before the object reaches the ground.
9.5 sec

36. An object is dropped from the top of 311 South Wacker Drive, a 961-foot-tall office building in Chicago. (*Source:* Council on Tall Buildings and Urban Habitat) The height h of the object after t seconds is given by the equation $h = -16t^2 + 961$. Find how many seconds pass before the object reaches the ground.
7.75 sec

37. At the end of 2 years, P dollars invested at an interest rate r compounded annually increases to an amount, A dollars, given by

$$A = P(1 + r)^2$$

Find the interest rate if $100 increased to $144 in 2 years. Write your answer as a percent.   20%

38. At the end of 2 years, P dollars invested at an interest rate r compounded annually increases to an amount, A dollars, given by

$$A = P(1 + r)^2$$

Find the interest rate if $2000 increased to $2420 in 2 years. Write your answer as a percent.   10%

△ **39.** Find the dimensions of a rectangle whose width is 7 miles less than its length and whose area is 120 square miles.   length: 15 mi; width: 8 mi

△ **40.** Find the dimensions of a rectangle whose width is 2 inches less than half its length and whose area is 160 square inches.   width: 8 in.; length: 20 in.

41. If the cost, C, for manufacturing x units of a certain product is given by $C = x^2 - 15x + 50$, find the number of units manufactured at a cost of $9500.   105 units

42. If a switchboard handles n telephones, the number C of telephone connections it can make simultaneously is given by the equation $C = \dfrac{n(n-1)}{2}$. Find how many telephones are handled by a switchboard making 120 telephone connections simultaneously.   16 telephones

Review

The following double line graph shows a comparison of the number of annual visitors (in millions) to Acadia National Park and Cuyahoga Valley National Park for the years shown. Use this graph to answer Exercises 43 through 50. See Section 2.4.

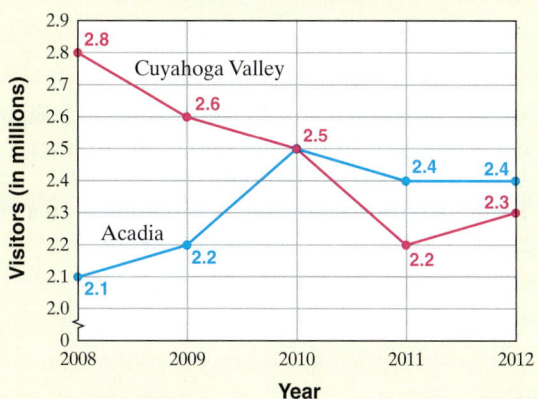

Annual Visitors to Acadia and Cuyahoga Valley National Parks

Source: National Park Service

43. Find the number of visitors to Acadia National Park in 2009.   2.2 million or 2,200,000

44. Find the number of visitors to Cuyahoga Valley National Park in 2009.
2.6 million or 2,600,000

45. Find the number of visitors to Acadia National Park in 2012.   2.4 million or 2,400,000

46. Find the number of visitors to Cuyahoga Valley National Park in 2012.
2.3 million or 2,300,000

47. Determine the year that the colored lines in this graph intersect.   2010

48. For what year(s) on the graph is the number of visitors to Cuyahoga Valley Park greater than the number of visitors to Acadia Park?   2008, 2009

49. In your own words, explain the meaning of the point of intersection in the graph.   answers may vary

50. Describe the trends shown in this graph and speculate as to why these trends have occurred.   answers may vary

Write each fraction in simplest form. See Section R.2.

51. $\dfrac{20}{35}$ $\dfrac{4}{7}$

52. $\dfrac{24}{32}$ $\dfrac{3}{4}$

53. $\dfrac{27}{18}$ $\dfrac{3}{2}$

54. $\dfrac{15}{27}$ $\dfrac{5}{9}$

55. $\dfrac{14}{42}$ $\dfrac{1}{3}$

56. $\dfrac{45}{50}$ $\dfrac{9}{10}$

Concept Extensions

△ **57.** The side of a square equals the width of a rectangle. The length of the rectangle is 6 meters longer than its width. The sum of the areas of the square and the rectangle is 176 square meters. Find the side of the square.   8 m

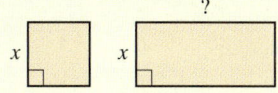

△ **58.** Two boats travel at right angles to each other after leaving the same dock at the same time. One hour later the boats are 17 miles apart. If one boat travels 7 miles per hour faster than the other boat, find the rate of each boat.
slower boat: 8 mph; faster boat: 15 mph

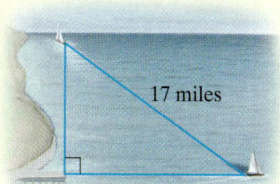

59. The sum of two numbers is 25, and the sum of their squares is 325. Find the numbers.   10 and 15

60. The sum of two numbers is 20, and the sum of their squares is 218. Find the numbers.   13 and 7

△ **61.** A rectangular pool is surrounded by a walk 4 meters wide. The pool is 6 meters longer than its width. If the total area of the pool and walk is 576 square meters more than the area of the pool, find the dimensions of the pool.
width of pool: 29 m; length of pool: 35 m

△ **62.** A rectangular garden is surrounded by a walk of uniform width. The area of the garden is 180 square yards. If the dimensions of the garden plus the walk are 16 yards by 24 yards, find the width of the walk.
3 yd

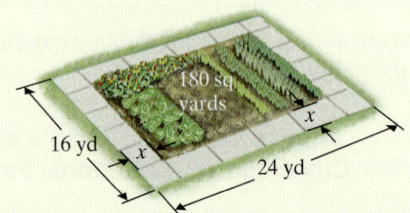

✎ **63.** Write down two numbers whose sum is 10. Square each number and find the sum of the squares. Use this work to write a word problem like Exercise **59.** Then give the word problem to a classmate to solve.   answers may vary

✎ **64.** Write down two numbers whose sum is 12. Square each number and find the sum of the squares. Use this work to write a word problem like Exercise **60.** Then give the word problem to a classmate to solve.   answers may vary

Chapter 6 Group Activity

Additional group activities are available in the Instructor's Resource Manual with Tests.

Factoring polynomials can be visualized using areas of rectangles. To see this, let's first find the areas of the following squares and rectangles. (Recall that Area = Length · Width.)

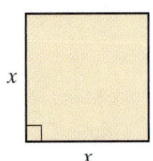 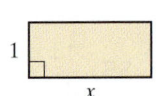

To use these areas to visualize factoring the polynomial $x^2 + 3x + 2$, for example, use the shapes below to form a rectangle. The factored form is found by reading the length and the width of the rectangle, as shown below.

Thus, $x^2 + 3x + 2 = (x + 2)(x + 1)$.

Try using this method to visualize the factored form of each polynomial below.

Work in a group and use tiles to find a factored form for the polynomials below. (Tiles can be handmade from index cards.)

1. $x^2 + 6x + 5$ $(x + 5)(x + 1)$

2. $x^2 + 5x + 6$ $(x + 2)(x + 3)$

3. $x^2 + 5x + 4$ $(x + 1)(x + 4)$

4. $x^2 + 4x + 3$ $(x + 1)(x + 3)$

5. $x^2 + 6x + 9$ $(x + 3)(x + 3)$ or $(x + 3)^2$

6. $x^2 + 4x + 4$ $(x + 2)(x + 2)$ or $(x + 2)^2$

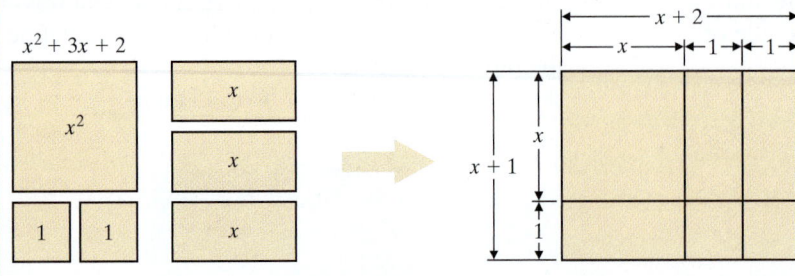

Chapter 6 Vocabulary Check

Fill in each blank with one of the words or phrases listed below. Some words or phrases may be used more than once.

factoring	leg	hypotenuse
greatest common factor	perfect square trinomial	quadratic equation

1. An equation that can be written in the form $ax^2 + bx + c = 0$ (with a not 0) is called a ___quadratic equation___.

2. ___Factoring___ is the process of writing an expression as a product.

3. The ___greatest common factor___ of a list of terms is the product of all common factors.

4. A trinomial that is the square of some binomial is called a ___perfect square trinomial___.

5. In a right triangle, the side opposite the right angle is called the ___hypotenuse___.

6. In a right triangle, each side adjacent to the right angle is called a ___leg___.

7. The Pythagorean theorem states that $(\text{leg})^2 + (\text{leg})^2 = ($___hypotenuse___$)^2$.

> **Helpful Hint**
> ▶ Are you preparing for your test? Don't forget to take the Chapter 6 Test on page 476. Then check your answers at the back of the text and use the Chapter Test Prep Videos to see the fully worked-out solutions to any of the exercises you want to review.

6 Chapter Highlights

Definitions and Concepts	Examples
Section 6.1 The Greatest Common Factor and Factoring by Grouping	

Factoring is the process of writing an expression as a product.	Factor: $6 = 2 \cdot 3$ Factor: $x^2 + 5x + 6 = (x + 2)(x + 3)$
The GCF of a list of variable terms contains the smallest exponent on each common variable.	The GCF of z^5, z^3, and z^{10} is z^3.
The GCF of a list of terms is the product of all common factors.	Find the GCF of $8x^2y$, $10x^3y^2$, and $50x^2y^3$. $\quad 8x^2y = 2 \cdot 2 \cdot 2 \cdot x^2 \cdot y$ $\quad 10x^3y^2 = 2 \cdot 5 \cdot x^3 \cdot y^2$ $\quad 50x^2y^3 = 2 \cdot 5 \cdot 5 \cdot x^2 \cdot y^3$ $\quad\quad \text{GCF} = 2 \cdot x^2 \cdot y \quad \text{or} \quad 2x^2y$
To Factor by Grouping **Step 1:** Group the terms in two groups so that each group has a common factor. **Step 2:** Factor out the GCF from each group. **Step 3:** If there is a common binomial factor, factor it out. **Step 4:** If not, rearrange the terms and try these steps again.	Factor: $10ax + 15a - 6xy - 9y$ **Step 1:** $(10ax + 15a) + (-6xy - 9y)$ **Step 2:** $5a(2x + 3) - 3y(2x + 3)$ **Step 3:** $(2x + 3)(5a - 3y)$

Definitions and Concepts	Examples

Section 6.2   Factoring Trinomials of the Form $x^2 + bx + c$

The product of these numbers is c.

$$x^2 + bx + c = (x + \square)(x + \square)$$

The sum of these numbers is b.

Factor: $x^2 + 7x + 12$

$3 + 4 = 7 \quad 3 \cdot 4 = 12$

$x^2 + 7x + 12 = (x + 3)(x + 4)$

Section 6.3   Factoring Trinomials of the Form $ax^2 + bx + c$

To factor $ax^2 + bx + c$, try various combinations of factors of ax^2 and c until a middle term of bx is obtained when checking.

Factor: $3x^2 + 14x - 5$

Factors of $3x^2$: $3x, x$

Factors of -5: $-1, 5$ and $1, -5$

$(3x - 1)(x + 5)$

$-1x$

$+15x$

$14x$ Correct middle term

Section 6.4   Factoring Trinomials of the Form $ax^2 + bx + c$ by Grouping

To Factor $ax^2 + bx + c$ by Grouping

Step 1:   Find two numbers whose product is $a \cdot c$ and whose sum is b.

Step 2:   Rewrite bx, using the factors found in Step 1.

Step 3:   Factor by grouping.

Factor: $3x^2 + 14x - 5$

Step 1:   Find two numbers whose product is $3 \cdot (-5)$ or -15 and whose sum is 14. They are 15 and -1.

Step 2: $3x^2 + 14x - 5$
$= 3x^2 + 15x - 1x - 5$

Step 3: $= 3x(x + 5) - 1(x + 5)$
$= (x + 5)(3x - 1)$

Section 6.5   Factoring by Special Products

A **perfect square trinomial** is a trinomial that is the square of some binomial.

Perfect Square Trinomial = Square of Binomial

$$x^2 + 4x + 4 = (x + 2)^2$$
$$25x^2 - 10x + 1 = (5x - 1)^2$$

Factoring Perfect Square Trinomials

$$a^2 + 2ab + b^2 = (a + b)^2$$
$$a^2 - 2ab + b^2 = (a - b)^2$$

Factor.

$$x^2 + 6x + 9 = x^2 + 2 \cdot x \cdot 3 + 3^2 = (x + 3)^2$$
$$4x^2 - 12x + 9 = (2x)^2 - 2 \cdot 2x \cdot 3 + 3^2$$
$$= (2x - 3)^2$$

Difference of Two Squares

$$a^2 - b^2 = (a + b)(a - b)$$

Factor.

$$x^2 - 9 = x^2 - 3^2 = (x + 3)(x - 3)$$

Sum and Difference of Two Cubes

$$a^3 + b^3 = (a + b)(a^2 - ab + b^2)$$
$$a^3 - b^3 = (a - b)(a^2 + ab + b^2)$$

Factor.

$$8y^3 + 1 = (2y + 1)(4y^2 - 2y + 1)$$
$$27p^3 - 64q^3 = (3p - 4q)(9p^2 + 12pq + 16q^2)$$

Definitions and Concepts	Examples

Section 6.6   Solving Quadratic Equations by Factoring

A **quadratic equation** is an equation that can be written in the form $ax^2 + bx + c = 0$ with a not 0.

The form $ax^2 + bx + c = 0$ is called the **standard form** of a quadratic equation.

Quadratic Equation	Standard Form
$x^2 = 16$	$x^2 - 16 = 0$
$y = -2y^2 + 5$	$2y^2 + y - 5 = 0$

Zero-Factor Property

If a and b are real numbers and if $ab = 0$, then $a = 0$ or $b = 0$.

If $(x + 3)(x - 1) = 0$, then $x + 3 = 0$ or $x - 1 = 0$.

To Solve Quadratic Equations by Factoring

Step 1: Write the equation in standard form so that one side of the equation is 0.

Step 2: Factor completely.

Step 3: Set each factor containing a variable equal to 0.

Step 4: Solve the resulting equations.

Step 5: Check solutions in the original equation.

Solve: $3x^2 = 13x - 4$

Step 1: $3x^2 - 13x + 4 = 0$

Step 2: $(3x - 1)(x - 4) = 0$

Step 3: $3x - 1 = 0$ or $x - 4 = 0$

Step 4: $3x = 1$ $x = 4$

$$x = \frac{1}{3}$$

Step 5: Check both $\frac{1}{3}$ and 4 in the original equation.

Section 6.7   Quadratic Equations and Problem Solving

Problem-Solving Steps

A garden is in the shape of a rectangle whose length is two feet more than its width. If the area of the garden is 35 square feet, find its dimensions.

1. UNDERSTAND the problem.

1. Read and reread the problem. Guess a solution and check your guess. Draw a diagram.
Let x be the width of the rectangular garden. Then $x + 2$ is the length.

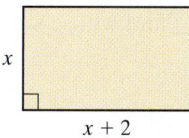

x

$x + 2$

2. TRANSLATE.

2. length · width = area
↓ ↓ ↓
$(x + 2)$ · x = 35

3. SOLVE.

3.

$$(x + 2)x = 35$$
$$x^2 + 2x - 35 = 0$$
$$(x - 5)(x + 7) = 0$$
$$x - 5 = 0 \quad \text{or} \quad x + 7 = 0$$
$$x = 5 \qquad\qquad x = -7$$

4. INTERPRET.

4. Discard the solution $x = -7$ since x represents width.

Check: If x is 5 feet, then $x + 2 = 5 + 2 = 7$ feet. The area of a rectangle whose width is 5 feet and whose length is 7 feet is (5 feet)(7 feet) or 35 square feet.

State: The garden is 5 feet by 7 feet.

(6.1) *Complete each factoring.*

1. $6x^2 - 15x = 3x(2x - 5)$

2. $4x^5 + 2x - 10x^4 = 2x(2x^4 + 1 - 5x^3)$

Factor out the GCF from each polynomial.

3. $5m + 30$ $5(m + 6)$

4. $20x^3 + 12x^2 + 24x$ $4x(5x^2 + 3x + 6)$

5. $3x(2x + 3) - 5(2x + 3)$ $(2x + 3)(3x - 5)$

6. $5x(x + 1) - (x + 1)$ $(x + 1)(5x - 1)$

Factor each polynomial by grouping.

7. $3x^2 - 3x + 2x - 2$ $(x - 1)(3x + 2)$

8. $3a^2 + 9ab + 3b^2 + ab$ $(a + 3b)(3a + b)$

9. $10a^2 + 5ab + 7b^2 + 14ab$ $(2a + b)(5a + 7b)$

10. $6x^2 + 10x - 3x - 5$ $(3x + 5)(2x - 1)$

(6.2) *Factor each trinomial.*

11. $x^2 + 6x + 8$
$(x + 4)(x + 2)$

12. $x^2 - 11x + 24$
$(x - 8)(x - 3)$

13. $x^2 + x + 2$ prime

14. $x^2 - 5x - 6$
$(x - 6)(x + 1)$

15. $x^2 + 2x - 8$
$(x + 4)(x - 2)$

16. $x^2 + 4xy - 12y^2$
$(x + 6y)(x - 2y)$

17. $x^2 + 8xy + 15y^2$
$(x + 5y)(x + 3y)$

18. $72 - 18x - 2x^2$
$2(3 - x)(12 + x)$

19. $32 + 12x - 4x^2$
$4(8 + 3x - x^2)$

20. $5y^3 - 50y^2 + 120y$
$5y(y - 6)(y - 4)$

21. To factor $x^2 + 2x - 48$, think of two numbers whose product is _____ and whose sum is _____. $-48, 2$

22. What is the first step in factoring $3x^2 + 15x + 30$?   factor out the GCF, 3

(6.3) or (6.4) *Factor each trinomial.*

23. $2x^2 + 13x + 6$
$(2x + 1)(x + 6)$

24. $4x^2 + 4x - 3$
$(2x + 3)(2x - 1)$

25. $6x^2 + 5xy - 4y^2$
$(3x + 4y)(2x - y)$

26. $x^2 - x + 2$ prime

27. $2x^2 - 23x - 39$
$(2x + 3)(x - 13)$

28. $18x^2 - 9xy - 20y^2$
$(6x + 5y)(3x - 4y)$

29. $10y^3 + 25y^2 - 60y$
$5y(2y - 3)(y + 4)$

30. $60y^3 - 39y^2 + 6y$
$3y(4y - 1)(5y - 2)$

Write the perimeter of each figure as a simplified polynomial. Then factor each polynomial completely.

△ **31.**

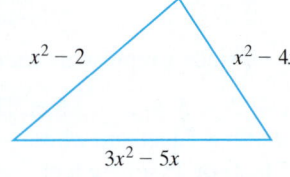

$x^2 - 2$ $x^2 - 4x$
$3x^2 - 5x$

$5x^2 - 9x - 2$; $(5x + 1)(x - 2)$

△ **32.**

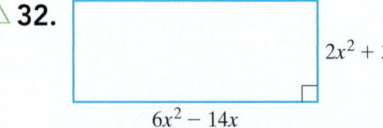

$2x^2 + 3$
$6x^2 - 14x$

$16x^2 - 28x + 6$; $2(4x - 1)(2x - 3)$

(6.5) *Factor each polynomial completely.*

33. $x^2 - 81$
$(x + 9)(x - 9)$

34. $x^2 + 12x + 36$
$(x + 6)^2$

35. $4x^2 - 9$
$(2x + 3)(2x - 3)$

36. $9t^2 - 25s^2$
$(3t + 5s)(3t - 5s)$

37. $16x^2 + y^2$
prime

38. $n^2 - 18n + 81$
$(n - 9)^2$

39. $3r^2 + 36r + 108$
$3(r + 6)^2$

40. $9y^2 - 42y + 49$
$(3y - 7)^2$

41. $5m^8 - 5m^6$
$5m^6(m + 1)(m - 1)$

42. $4x^2 - 28xy + 49y^2$
$(2x - 7y)^2$

43. $3x^2y + 6xy^2 + 3y^3$
$3y(x + y)^2$

44. $16x^4 - 1$
$(4x^2 + 1)(2x + 1)(2x - 1)$

45. $(y + 2)^2 - 25$
$(y + 7)(y - 3)$

46. $(x - 3)^2 - 16$
$(x + 1)(x - 7)$

47. $8 - 27y^3$
$(2 - 3y)(4 + 6y + 9y^2)$

48. $1 - 64y^3$
$(1 - 4y)(1 + 4y + 16y^2)$

49. $6x^4y + 48xy$
$6xy(x + 2)(x^2 - 2x + 4)$

50. $2x^5 + 16x^2y^3$
$2x^2(x + 2y)(x^2 - 2xy + 4y^2)$

51. $x^2 - 2x + 1 - y^2$
$(x - 1 + y)(x - 1 - y)$

△ **52.** The volume of the cylindrical shell is $\pi R^2 h - \pi r^2 h$ cubic units. Write this volume as a factored expression. $\pi h(R + r)(R - r)$ cu units

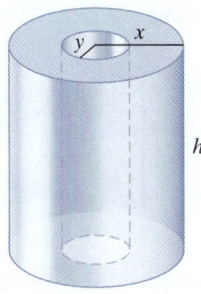

(6.6) *Solve each equation.*

53. $(x + 6)(x - 2) = 0$
$-6, 2$

54. $(x - 7)(x + 11) = 0$
$-11, 7$

55. $3x(x + 1)(7x - 2) = 0$
$0, -1, \dfrac{2}{7}$

56. $4(5x + 1)(x + 3) = 0$
$-\dfrac{1}{5}, -3$

57. $x^2 + 8x + 7 = 0$
$-7, -1$

58. $x^2 - 2x - 24 = 0$
$-4, 6$

59. $x^2 + 10x = -25$
-5

60. $x(x - 10) = -16$
$2, 8$

61. $(3x - 1)(9x^2 + 3x + 1) = 0$
$\dfrac{1}{3}$

62. $56x^2 - 5x - 6 = 0$
$-\dfrac{2}{7}, \dfrac{3}{8}$

63. $m^2 = 6m$ $0, 6$

64. $r^2 = 25$ $5, -5$

65. Write a quadratic equation that has the two solutions 4 and 5.
$x^2 - 9x + 20 = 0$

66. Write a quadratic equation that has two solutions, both -1.
$x^2 + 2x + 1 = 0$

(6.7) *Use the given information to choose the correct dimensions.*

△ **67.** The perimeter of a rectangle is 24 inches. The length is twice the width. Find the dimensions of the rectangle.   **c**
  a. 5 inches by 7 inches   **b.** 5 inches by 10 inches
  c. 4 inches by 8 inches   **d.** 2 inches by 10 inches

△ **68.** The area of a rectangle is 80 meters. The length is one more than three times the width. Find the dimensions of the rectangle.   **d**
  a. 8 meters by 10 meters   **b.** 4 meters by 13 meters
  c. 4 meters by 20 meters   **d.** 5 meters by 16 meters

Use the given information to find the dimensions of each figure.

△ **69.** The *area* of the square is 81 square units. Find the length of a side. 9 units

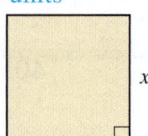

x

△ **70.** The *perimeter* of the quadrilateral is 47 units. Find the lengths of the sides. 8 units, 13 units, 16 units, 10 units

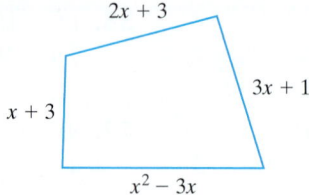

$2x + 3$

$3x + 1$

$x + 3$

$x^2 - 3x$

Solve.

△ **71.** A flag for a local organization is in the shape of a rectangle whose length is 15 inches less than twice its width. If the area of the flag is 500 square inches, find its dimensions. width: 20 in.; length: 25 in.

x

△ **72.** The base of a triangular sail is four times its height. If the area of the triangle is 162 square yards, find the base. 36 yd

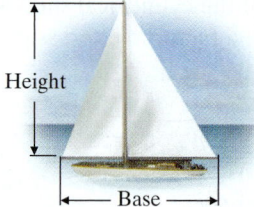

Height

Base

73. Find two consecutive positive integers whose product is 380. 19 and 20

74. Find two consecutive even positive integers whose product is 440. 20 and 22

75. A rocket is fired from the ground with an initial velocity of 440 feet per second. Its height h after t seconds is given by the equation $h = -16t^2 + 440t$.

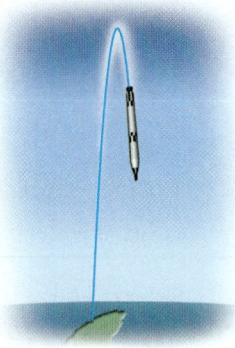

  a. Find how many seconds pass before the rocket reaches a height of 2800 feet. Explain why two answers are obtained. 17.5 sec and 10 sec; answers may vary

  b. Find how many seconds pass before the rocket reaches the ground again. 27.5 sec

△ **76.** An architect's squaring instrument is in the shape of a right triangle. Find the length of the longer leg of the right triangle if the hypotenuse is 8 centimeters longer than the longer leg and the shorter leg is 8 centimeters shorter than the longer leg. 32 cm

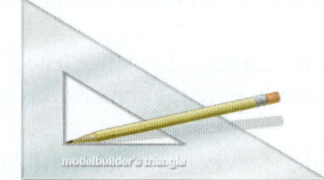

Mixed Review

Factor completely.

77. $6x + 24$
$6(x + 4)$

78. $7x - 63$
$7(x - 9)$

79. $11x(4x - 3) - 6(4x - 3)$
$(4x - 3)(11x - 6)$

80. $2x(x - 5) - (x - 5)$
$(x - 5)(2x - 1)$

81. $3x^3 - 4x^2 + 6x - 8$
$(3x - 4)(x^2 + 2)$

82. $xy + 2x - y - 2$
$(y + 2)(x - 1)$

83. $2x^2 + 2x - 24$
$2(x + 4)(x - 3)$

84. $3x^3 - 30x^2 + 27x$
$3x(x - 9)(x - 1)$

85. $4x^2 - 81$
$(2x + 9)(2x - 9)$

86. $2x^2 - 18$
$2(x + 3)(x - 3)$

87. $16x^2 - 24x + 9$
$(4x - 3)^2$

88. $5x^2 + 20x + 20$
$5(x + 2)^2$

Solve.

89. $2x^2 - x - 28 = 0$
$-\dfrac{7}{2}, 4$

90. $x^2 - 2x = 15$
$-3, 5$

91. $2x(x + 7)(x + 4) = 0$
$0, -7, -4$

92. $x(x - 5) = -6$ $3, 2$

93. $x^2 = 16x$ $0, 16$

94. The perimeter of the following triangle is 48 inches. Find the lengths of its sides.   19 in.; 8 in.; 21 in.

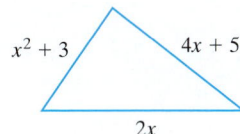

95. The width of a rectangle is 4 inches less than its length. Its area is 12 square inches. Find the dimensions of the rectangle.   length: 6 in.; width: 2 in.

Chapter 6

Test

Step-by-step test solutions are found on the Chapter Test Prep Videos. Where available: MyMathLab® or YouTube™

Answers

Note to Instructor: The Chapter 6 Test file in the TestGen program provides algorithms specifically matched to this test for easy replication for practice or assessment purposes.

1. $3x(3x - 1)$

2. $(x + 7)(x + 4)$

3. $(7 - m)(7 + m)$

4. $(y + 11)^2$

5. $\begin{array}{l}(x^2 + 4)(x + 2)\\(x - 2)\end{array}$

6. $(a + 3)(4 - y)$

7. prime

8. $(y - 12)(y + 4)$

9. $(a + b)(3a - 7)$

10. $(3x - 2)(x - 1)$

11. $5(6 - x)(6 + x)$

12. $3x(3x + 1)(x + 4)$

13. $(6t + 5)(t - 1)$

14. $\begin{array}{l}(x - 7)\\(y - 2)(y + 2)\end{array}$

15. $\begin{array}{l}x(1 - x)(1 + x)\\(1 + x^2)\end{array}$

16. $(x + 12y)(x + 2y)$

17. $\begin{array}{l}(x + 4)\\(x^2 - 4x + 16)\end{array}$

18. $\begin{array}{l}3x(3y - z)\\(9y^2 + 3yz + z^2)\end{array}$

19. $3, -9$

20. $-7, 2$

Factor each polynomial completely. If a polynomial cannot be factored, write "prime."

1. $9x^2 - 3x$

2. $x^2 + 11x + 28$

3. $49 - m^2$

4. $y^2 + 22y + 121$

5. $x^4 - 16$

6. $4(a + 3) - y(a + 3)$

7. $x^2 + 4$

8. $y^2 - 8y - 48$

9. $3a^2 + 3ab - 7a - 7b$

10. $3x^2 - 5x + 2$

11. $180 - 5x^2$

12. $9x^3 + 39x^2 + 12x$

13. $6t^2 - t - 5$

14. $xy^2 - 7y^2 - 4x + 28$

15. $x - x^5$

16. $x^2 + 14xy + 24y^2$

17. $x^3 + 64$

18. $81xy^3 - 3xz^3$

Solve each equation.

19. $(x - 3)(x + 9) = 0$

20. $x^2 + 5x = 14$

21. $x(x + 6) = 7$

22. $3x(2x - 3)(3x + 4) = 0$

23. $t^2 - 2t - 15 = 0$

24. $3x^2 = -12x$

25. $5t^3 - 45t = 0$

26. $(x - 1)(3x^2 - x - 2) = 0$

Solve.

27. A deck for a home is in the shape of a triangle. The length of the base of the triangle is 9 feet longer than its height. If the area of the triangle is 68 square feet, find the length of the base.

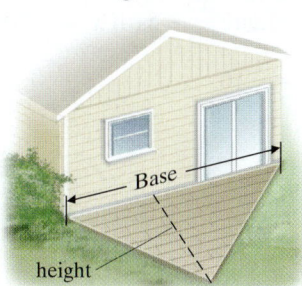

28. An object is dropped from the top of the Woolworth Building on Broadway in New York City. The height h of the object after t seconds is given by the equation

$$h = -16t^2 + 784$$

Find how many seconds pass before the object reaches the ground.

△ **29.** The *area* of the rectangle is 54 square units. Find the dimensions of the rectangle.

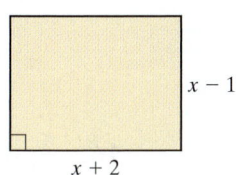

$x - 1$

$x + 2$

△ **30.** Find the lengths of the sides of a right triangle if the hypotenuse is 10 centimeters longer than the shorter leg and 5 centimeters longer than the longer leg.

31. A window washer is suspended 38 feet below the roof of the 1127-foot-tall John Hancock Center in Chicago. (*Source:* Council on Tall Buildings and Urban Habitat) If the window washer drops an object from this height, the object's height h after t seconds is given by the equation $h = -16t^2 + 1089$. Find how many seconds pass before the object reaches the ground.

21. $-7, 1$

22. $0, \dfrac{3}{2}, -\dfrac{4}{3}$

23. $-3, 5$

24. $0, -4$

25. $0, 3, -3$

26. $-\dfrac{2}{3}, 1$

27. 17 ft

28. 7 sec

29. width: 6 units; length: 9 units

30. hypotenuse: 25 cm; legs: 15 cm, 20 cm

31. 8.25 sec

Answers

1. a. $9 \le 11$

b. $8 > 1$

$3 \ne 4$

c. (Sec. 1.2, Ex. 7)

2. a. $>$

b. $<$ (Sec. 1.2)

3. solution (Sec. 1.3, Ex. 8)

4. 102 (Sec. 1.6)

5. -12 (Sec 1.5, Ex. 5a)

6. -102 (Sec. 1.6)

7. a. $\frac{3}{4}$

b. -24

c. 1 (Sec. 1.6, Ex. 16)

8. -98 (Sec. 1.6)

9. $5x + 7$ (Sec. 1.8, Ex. 4)

10. $19 - 6x$ (Sec. 1.8)

11. $-4a - 1$ (Sec. 1.8, Ex. 5)

12. $-13x - 21$ (Sec. 1.8)

13. $7.3x - 6$ (Sec. 1.8, Ex. 7)

14. 2 (Sec. 2.3)

15. -11 (Sec. 2.2, Ex. 3)

16. 28 (Sec. 2.2)

17. every real number (Sec. 2.3, Ex. 7)

18. 33 (Sec. 2.2)

19. $l = \frac{V}{wh}$ (Sec. 2.5, Ex. 5)

20. $y = \frac{-3x - 7}{2}$ or $y = -\frac{3}{2}x - \frac{7}{2}$ (Sec. 2.5)

21. 5^{18} (Sec. 5.1, Ex. 16)

22. 30 (Sec. 5.1)

23. y^{16} (Sec. 5.1, Ex. 17)

24. y^{10} (Sec. 5.1)

1. Translate each sentence into a mathematical statement.
   a. Nine is less than or equal to eleven.
   b. Eight is greater than one.
   c. Three is not equal to four.

2. Insert $<$ or $>$ in the space to make each statement true.
   a. $|-5| \quad |-3|$
   b. $|0| \quad |-2|$

3. Decide whether 2 is a solution of $3x + 10 = 8x$.

4. Evaluate $\frac{x}{y} + 5x$ if $x = 20$ and $y = 10$.

5. Subtract 8 from -4.

6. Evaluate $\frac{x}{y} + 5x$ if $x = -20$ and $y = 10$.

7. Evaluate each expression when $x = -2$ and $y = -4$.
   a. $\frac{3x}{2y}$ **b.** $x^3 - y^2$ **c.** $\frac{x - y}{-x}$

8. Evaluate $\frac{x}{y} + 5x$ if $x = -20$ and $y = -10$.

Simplify each expression by combining like terms.

9. $2x + 3x + 5 + 2$

10. $5 - 2(3x - 7)$

11. $-5a - 3 + a + 2$

12. $5(x - 6) + 9(-2x + 1)$

13. $2.3x + 5x - 6$

Solve each equation.

14. $0.8y + 0.2(y - 1) = 1.8$

15. $-3x = 33$

16. $\frac{x}{-7} = -4$

17. $3(x - 4) = 3x - 12$

18. $-\frac{2}{3}x = -22$

19. Solve $V = lwh$ for l.

20. Solve $3x + 2y = -7$ for y.

Simplify each expression.

21. $(5^3)^6$

22. $5^2 + 5^1$

23. $(y^8)^2$

24. $y^8 \cdot y^2$

Simplify each expression. Write each result using positive exponents only.

25. $\dfrac{(x^3)^4 x}{x^7}$

26. 3^{-2}

27. $(y^{-3}z^6)^{-6}$

28. $\dfrac{x^{-3}}{x^{-7}}$

29. $\dfrac{x^{-7}}{(x^4)^3}$

30. $\dfrac{(5a^7)^2}{a^5}$

Simplify each polynomial by combining any like terms.

31. $-3x + 7x$

32. $\dfrac{2}{3}x + 23 + \dfrac{1}{6}x - 100$

33. $11x^2 + 5 + 2x^2 - 7$

34. $0.2x - 1.1 + 2.3 - 0.7x$

35. Multiply: $(2x - y)^2$

36. Multiply: $(3x - 7y)^2$

Use a special product to square each binomial.

37. $(t + 2)^2$

38. $(x - 13)^2$

39. $(x^2 - 7y)^2$

40. $(7x + y)^2$

41. Divide: $\dfrac{8x^2y^2 - 16xy + 2x}{4xy}$

Factor each polynomial.

42. $z^3 + 7z + z^2 + 7$

43. $5(x + 3) + y(x + 3)$

44. $2x^3 + 2x^2 - 84x$

45. $x^4 + 5x^2 + 6$

46. $-4x^2 - 23x + 6$

47. $6x^2 - 2x - 20$

48. $9xy^2 - 16x$

49. The platform for the cliff divers in Acapulco, Mexico, is about 144 feet above the sea. Neglecting air resistance, the height h in feet of a cliff diver above the ocean after t seconds is given by the quadratic equation $h = -16t^2 + 144$. Find how long it takes the diver to reach the ocean.

50. Solve $x^2 - 13x = -36$.

---

25. x^6 (Sec. 5.2, Ex. 9)

26. $\dfrac{1}{9}$ (Sec. 5.2)

27. $\dfrac{y^{18}}{z^{36}}$ (Sec. 5.2, Ex. 11)

28. x^4 (Sec. 5.2)

29. $\dfrac{1}{x^{19}}$ (Sec. 5.2, Ex. 13)

30. $25a^9$ (Sec. 5.2)

31. $4x$ (Sec. 5.3, Ex. 6)

32. $\dfrac{5}{6}x - 77$ (Sec. 5.3)

33. $13x^2 - 2$ (Sec. 5.3, Ex. 7)

34. $-0.5x + 1.2$ (Sec. 5.3)

35. $4x^2 - 4xy + y^2$ (Sec. 5.5, Ex. 8)

36. $9x^2 - 42xy + 49y^2$ (Sec. 5.5)

37. $t^2 + 4t + 4$ (Sec. 5.6, Ex. 5)

38. $x^2 - 26x + 169$ (Sec. 5.6)

39. $x^4 - 14x^2y + 49y^2$ (Sec. 5.6, Ex. 8)

40. $49x^2 + 14xy + y^2$ (Sec. 5.6)

41. $2xy - 4 + \dfrac{1}{2y}$ (Sec. 5.7, Ex. 3)

42. $(z^2 + 7)(z + 1)$ (Sec. 6.1)

43. $(x + 3)(5 + y)$ (Sec. 6.1, Ex. 9)

44. $2x(x + 7)(x - 6)$ (Sec. 6.2)

45. $(x^2 + 2)(x^2 + 3)$ (Sec. 6.2, Ex. 7)

46. $(-4x + 1)(x + 6)$ or $-1(4x - 1)(x + 6)$ (Sec. 6.3)

47. $2(x - 2)(3x + 5)$ (Sec. 6.4, Ex. 2)

48. $x(3y + 4)(3y - 4)$ (Sec. 6.5)

49. 3 sec (Sec. 6.7, Ex. 1)

50. 9, 4 (Sec. 6.6)

Check Your Progress

Vocabulary Check

Chapter Highlights

Chapter Review

Chapter Test

Cumulative Review

In this chapter, we expand our knowledge of algebraic expressions to include algebraic fractions, called *rational expressions*. We explore the operations of addition, subtraction, multiplication, and division using principles similar to the principles for numerical fractions.

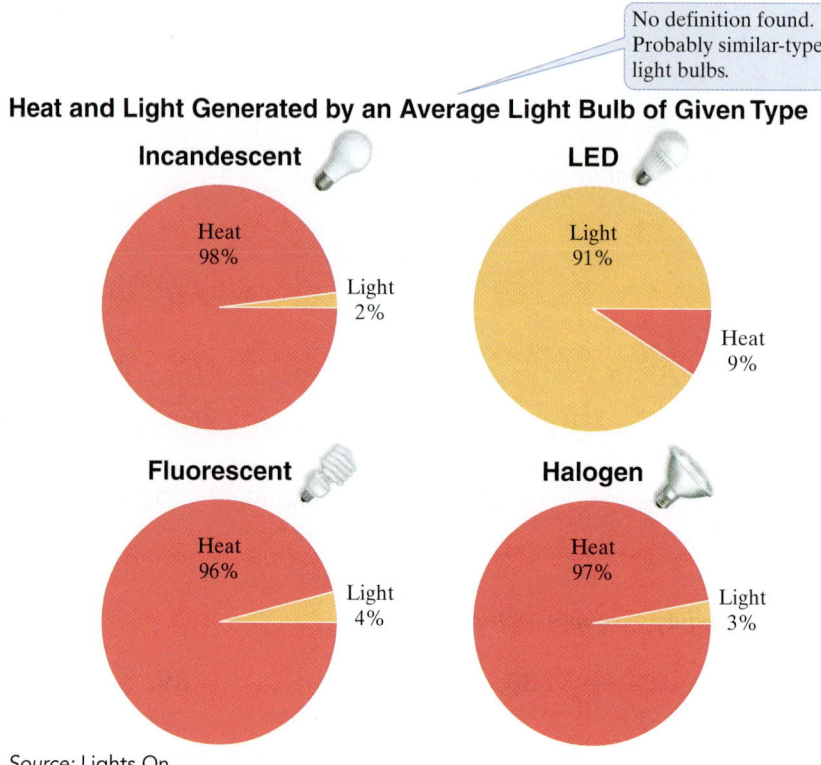

No definition found. Probably similar-type light bulbs.

Heat and Light Generated by an Average Light Bulb of Given Type

Source: Lights On

Is an Average Always an Average?

In mathematics, the average or mean of a set of data values is defined by the sum of the data values divided by the number of data values. For example, the average of two numbers, a and b, is $\frac{a+b}{2}$. This is an excellent example of a rational expression and sometimes a complex rational expression as we see in Section 7.7. Unfortunately, the word *average* is often used and defined in ways that we might not expect, as noted above and below. Sometimes the average is defined and sometimes not. In Section 7.7, Exercises 47–50, we find mathematical averages.

Defined as how long it takes for a percentage in a test batch to fail.
Source: The LIGHTBULB Company

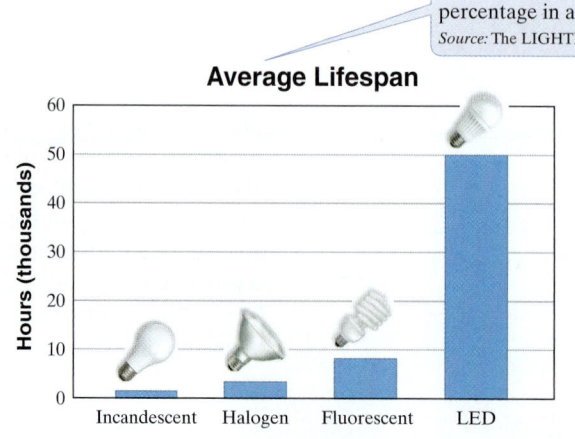

Source: Lights On

Objective A Evaluating Rational Expressions ▶

A rational number is a number that can be written as a quotient of integers. A *rational expression* is also a quotient; it is a quotient of polynomials. Examples are

$$\frac{2}{3}, \quad \frac{3y^3}{8}, \quad \frac{-4p}{p^3 + 2p + 1}, \quad \text{and} \quad \frac{5x^2 - 3x + 2}{3x + 7}$$

Rational Expression

A **rational expression** is an expression that can be written in the form

$$\frac{P}{Q}$$

where P and Q are polynomials and $Q \neq 0$.

Rational expressions have different numerical values depending on what values replace the variables.

Example 1 Find the numerical value of $\dfrac{x + 4}{2x - 3}$ for each replacement value.

a. $x = 5$ **b.** $x = -2$

Solution:

a. We replace each x in the expression with 5 and then simplify.

$$\frac{x + 4}{2x - 3} = \frac{5 + 4}{2(5) - 3} = \frac{9}{10 - 3} = \frac{9}{7}$$

b. We replace each x in the expression with -2 and then simplify.

$$\frac{x + 4}{2x - 3} = \frac{-2 + 4}{2(-2) - 3} = \frac{2}{-7} \quad \text{or} \quad -\frac{2}{7}$$

▪ Work Practice 1

In the example above, we wrote $\dfrac{2}{-7}$ as $-\dfrac{2}{7}$. For a negative fraction such as $\dfrac{2}{-7}$, recall from Section 1.6 that

$$\frac{2}{-7} = \frac{-2}{7} = -\frac{2}{7}$$

In general, for any fraction,

$$\frac{-a}{b} = \frac{a}{-b} = -\frac{a}{b}, \qquad b \neq 0$$

This is also true for rational expressions. For example,

$$\frac{-(x + 2)}{\underbrace{x}} = \frac{x + 2}{-x} = -\frac{x + 2}{x}$$

↑
Notice the parentheses.

Objectives

A Find the Value of a Rational Expression Given a Replacement Number. ▶

B Identify Values for Which a Rational Expression Is Undefined. ▶

C Simplify, or Write Rational Expressions in Lowest Terms. ▶

D Write Equivalent Rational Expressions of the Forms $-\dfrac{a}{b} = \dfrac{-a}{b} = \dfrac{a}{-b}$. ▶

Practice 1

Find the value of $\dfrac{x - 3}{5x + 1}$ for each replacement value.

a. $x = 4$

b. $x = -3$

Answers

1. **a.** $\dfrac{1}{21}$ **b.** $\dfrac{3}{7}$

481

Objective B Identifying When a Rational Expression Is Undefined

In the definition of rational expression (first "box" in this section), notice that we wrote $Q \neq 0$ for the denominator Q. The denominator of a rational expression must not equal 0 since division by 0 is not defined. (See the Helpful Hint to the left.) This means we must be careful when replacing the variable in a rational expression by a number. For example, suppose we replace x with 5 in the rational expression $\frac{3 + x}{x - 5}$. The expression becomes

$$\frac{3 + x}{x - 5} = \frac{3 + 5}{5 - 5} = \frac{8}{0}$$

But division by 0 is undefined. Therefore, in this expression we can allow x to be any real number *except* 5. **A rational expression is undefined for values that make the denominator 0.** Thus,

> To find values for which a rational expression is undefined, find values for which the denominator is 0.

Practice 2

Are there any values for x for which each rational expression is undefined?

a. $\dfrac{x}{x + 8}$

b. $\dfrac{x - 3}{x^2 + 5x + 4}$

c. $\dfrac{x^2 - 3x + 2}{5}$

Example 2 Are there any values for x for which each expression is undefined?

a. $\dfrac{x}{x - 3}$ b. $\dfrac{x^2 + 2}{x^2 - 3x + 2}$ c. $\dfrac{x^3 - 6x^2 - 10x}{3}$

Solution: To find values for which a rational expression is undefined, we find values that make the denominator 0.

a. The denominator of $\dfrac{x}{x - 3}$ is 0 when $x - 3 = 0$ or when $x = 3$. Thus, when $x = 3$, the expression $\dfrac{x}{x - 3}$ is undefined.

b. We set the denominator equal to 0.

$$x^2 - 3x + 2 = 0$$
$$(x - 2)(x - 1) = 0 \qquad \text{Factor.}$$
$$x - 2 = 0 \quad \text{or} \quad x - 1 = 0 \qquad \text{Set each factor equal to 0.}$$
$$x = 2 \qquad\qquad x = 1 \qquad \text{Solve.}$$

Thus, when $x = 2$ or $x = 1$, the denominator $x^2 - 3x + 2$ is 0. So the rational expression $\dfrac{x^2 + 2}{x^2 - 3x + 2}$ is undefined when $x = 2$ or when $x = 1$.

c. The denominator of $\dfrac{x^3 - 6x^2 - 10x}{3}$ is never 0, so there are no values of x for which this expression is undefined.

■ Work Practice 2

Note: Unless otherwise stated, we will now assume that variables in rational expressions are replaced only by values for which the expressions are defined.

Answers

2. **a.** $x = -8$ **b.** $x = -4, x = -1$

c. no

Objective C Simplifying Rational Expressions

A fraction is said to be written in lowest terms or simplest form when the numerator and denominator have no common factors other than 1 (or -1). For example, the fraction $\dfrac{7}{10}$ is written in lowest terms since the numerator and denominator have no common factors other than 1 (or -1).

The process of writing a rational expression in lowest terms or simplest form is called **simplifying** the rational expression.

Simplifying a rational expression is similar to simplifying a fraction. Recall from Section R.2 that to simplify a fraction, we essentially "remove factors of 1." Our ability to do this comes from these facts:

- Any nonzero number over itself simplifies to 1 $\left(\dfrac{5}{5} = 1, \dfrac{-7.26}{-7.26} = 1, \text{and } \dfrac{c}{c} = 1 \right.$ as long as c is not $0 \left.\right)$, and

- The product of any number and 1 is that number $\left(19 \cdot 1 = 19, -8.9 \cdot 1 = -8.9, \right.$ $\dfrac{a}{b} \cdot 1 = \dfrac{a}{b} \left.\right)$.

In other words, we have the following: ⎯⎯⎯

$$\frac{a \cdot c}{b \cdot c} = \frac{a}{b} \cdot \frac{c}{c} = \frac{a}{b}$$

Since $\dfrac{a}{b} \cdot 1 = \dfrac{a}{b}$

Simplify: $\dfrac{15}{20}$

$$\frac{15}{20} = \frac{3 \cdot 5}{2 \cdot 2 \cdot 5} \qquad \text{Factor the numerator and the denominator.}$$

$$= \frac{3 \cdot \boxed{5}}{2 \cdot 2 \cdot \boxed{5}} \qquad \text{Look for common factors.}$$

$$= \frac{3}{2 \cdot 2} \cdot \frac{5}{5} \qquad \text{Common factors in the numerator and denominator form factors of 1.}$$

$$= \frac{3}{2 \cdot 2} \cdot 1 \qquad \text{Write } \frac{5}{5} \text{ as 1.}$$

$$= \frac{3}{2 \cdot 2} = \frac{3}{4} \qquad \text{Multiply to remove a factor of 1.}$$

Before we use the same technique to simplify a rational expression, remember that as long as the denominator is not 0, $\dfrac{a^3 b}{a^3 b} = 1$, $\dfrac{x + 3}{x + 3} = 1$, and $\dfrac{7x^2 + 5x - 100}{7x^2 + 5x - 100} = 1$.

Simplify: $\dfrac{x^2 - 9}{x^2 + x - 6}$

$$\frac{x^2 - 9}{x^2 + x - 6} = \frac{(x - 3)(x + 3)}{(x - 2)(x + 3)} \qquad \text{Factor the numerator and the denominator.}$$

$$= \frac{(x - 3)\boxed{(x + 3)}}{(x - 2)\boxed{(x + 3)}} \qquad \text{Look for common factors.}$$

$$= \frac{x - 3}{x - 2} \cdot \frac{x + 3}{x + 3}$$

$$= \frac{x - 3}{x - 2} \cdot 1 \qquad \text{Write } \frac{x + 3}{x + 3} \text{ as 1.}$$

$$= \frac{x - 3}{x - 2} \qquad \text{Multiply to remove a factor of 1.}$$

Just as for numerical fractions, we can use a shortcut notation. Remember that as long as exact factors in both the numerator and denominator are divided out, we are "removing a factor of 1." We will use the following notation to show this:

$$\frac{x^2 - 9}{x^2 + x - 6} = \frac{(x - 3)(x + 3)}{(x - 2)(x + 3)}$$ A factor of 1 is identified by the shading.

$$= \frac{x - 3}{x - 2}$$ Remove a factor of 1.

Thus, the rational expression $\dfrac{x^2 - 9}{x^2 + x - 6}$ has the same value as the rational expression $\dfrac{x - 3}{x - 2}$ for all values of x except 2 and -3. (Remember that when x is 2, the denominator of both rational expressions is 0 and that when x is -3, the original rational expression has a denominator of 0.)

As we simplify rational expressions, we will assume that the simplified rational expression is equal to the original rational expression for all real numbers except those for which either denominator is 0. The following steps may be used to simplify rational expressions.

To Simplify a Rational Expression

Step 1: Completely factor the numerator and denominator.

Step 2: Divide out factors common to the numerator and denominator. (This is the same as "removing a factor of 1.")

Teaching Tip

It may be useful for students to use the TABLE feature on a graphing calculator to evaluate $\dfrac{x^2 - 9}{x^2 + x - 6}$ and $\dfrac{x - 3}{x - 2}$ for various values of x to verify that the expressions are the same for all values except -3 and 2. Then ask students which expression is easier to evaluate. Finally, point out that one reason to simplify rational expressions before evaluating them is to simplify your work.

Practice 3

Simplify: $\dfrac{x^4 + x^3}{5x + 5}$

Practice 4

Simplify: $\dfrac{x^2 + 11x + 18}{x^2 + x - 2}$

Practice 5

Simplify: $\dfrac{x^2 + 10x + 25}{x^2 + 5x}$

Answers

3. $\dfrac{x^3}{5}$ **4.** $\dfrac{x + 9}{x - 1}$ **5.** $\dfrac{x + 5}{x}$

Example 3 Simplify: $\dfrac{5x - 5}{x^3 - x^2}$

Solution: To begin, we factor the numerator and denominator if possible. Then we look for common factors.

$$\frac{5x - 5}{x^3 - x^2} = \frac{5(x - 1)}{x^2(x - 1)} = \frac{5}{x^2}$$

■ **Work Practice 3**

Example 4 Simplify: $\dfrac{x^2 + 8x + 7}{x^2 - 4x - 5}$

Solution: We factor the numerator and denominator and then look for common factors.

$$\frac{x^2 + 8x + 7}{x^2 - 4x - 5} = \frac{(x + 7)(x + 1)}{(x - 5)(x + 1)} = \frac{x + 7}{x - 5}$$

■ **Work Practice 4**

Example 5 Simplify: $\dfrac{x^2 + 4x + 4}{x^2 + 2x}$

Solution: We factor the numerator and denominator and then look for common factors.

$$\frac{x^2 + 4x + 4}{x^2 + 2x} = \frac{(x + 2)(x + 2)}{x(x + 2)} = \frac{x + 2}{x}$$

■ **Work Practice 5**

Teaching Tip

Give students a concrete example that illustrates why $\dfrac{2x}{x}$ can be simplified to 2 and $\dfrac{x+2}{x}$ cannot be simplified to 2. For example, evaluate each expression for $x = 1$ and then $x = 5$. Have students notice that the first expression is equal to 2 for both values of x, but the second expression is not.

Helpful Hint

When simplifying a rational expression, we look for **common *factors*, not common *terms*.**

$$\frac{x \cdot (x + 2)}{x \cdot x} = \frac{x + 2}{x}$$

Common factors. These can be divided out.

$$\frac{x + 2}{x}$$

Common terms. There is no factor of 1 that can be generated.

✓**Concept Check** Recall that we can remove only *factors* of 1. Which of the following are *not* true? Explain why.

a. $\dfrac{3 - 1}{3 + 5}$ simplifies to $-\dfrac{1}{5}$.

b. $\dfrac{2x + 10}{2}$ simplifies to $x + 5$.

c. $\dfrac{37}{72}$ simplifies to $\dfrac{3}{2}$.

d. $\dfrac{2x + 3}{2}$ simplifies to $x + 3$.

Example 6 Simplify: $\dfrac{x + 9}{x^2 - 81}$

Solution: We factor and then apply the fundamental principle. Remember that this principle allows us to divide the numerator and denominator by all common factors.

$$\frac{x + 9}{x^2 - 81} = \frac{x + 9}{(x + 9)(x - 9)} = \frac{1}{x - 9}$$

■ **Work Practice 6**

Practice 6

Simplify: $\dfrac{x + 5}{x^2 - 25}$

Example 7 Simplify each rational expression.

a. $\dfrac{x + y}{y + x}$

b. $\dfrac{x - y}{y - x}$

Solution:

a. The expression $\dfrac{x + y}{y + x}$ can be simplified by using the commutative property of addition to rewrite the denominator $y + x$ as $x + y$.

$$\frac{x + y}{y + x} = \frac{x + y}{x + y} = 1$$

b. The expression $\dfrac{x - y}{y - x}$ can be simplified by recognizing that $y - x$ and $x - y$ are opposites. In other words, $y - x = -1(x - y)$. We proceed as follows:

$$\frac{x - y}{y - x} = \frac{1 \cdot (x - y)}{(-1)(x - y)} = \frac{1}{-1} = -1$$

■ **Work Practice 7**

Practice 7

Simplify each rational expression.

a. $\dfrac{x + 4}{4 + x}$

b. $\dfrac{x - 4}{4 - x}$

Answers

6. $\dfrac{1}{x - 5}$ **7. a.** 1 **b.** -1

✓**Concept Check Answer**

a, c, d

Practice 8

Simplify: $\dfrac{2x^2 - 5x - 12}{16 - x^2}$

Example 8 Simplify: $\dfrac{4 - x^2}{3x^2 - 5x - 2}$

Solution:

$$\frac{4 - x^2}{3x^2 - 5x - 2} = \frac{(2 - x)(2 + x)}{(x - 2)(3x + 1)} \qquad \text{Factor.}$$

$$= \frac{(-1)(x - 2)(2 + x)}{(x - 2)(3x + 1)} \qquad \text{Write } 2 - x \text{ as } -1(x - 2).$$

$$= \frac{(-1)(2 + x)}{3x + 1} \quad \text{or} \quad \frac{-2 - x}{3x + 1} \qquad \text{Simplify.}$$

■ **Work Practice 8**

Objective D Writing Equivalent Forms of Rational Expressions ▶

From Example 7(a), we have $y + x = x + y$. $y + x$ and $x + y$ are equivalent.

From Example 7(b), we have $y - x = -1(x - y)$. $y - x$ and $x - y$ are opposites.

Thus, $\dfrac{x + y}{y + x} = \dfrac{x + y}{x + y} = 1$ and $\dfrac{x - y}{y - x} = \dfrac{x - y}{-1(x - y)} = \dfrac{1}{-1} = -1.$

When performing operations on rational expressions, equivalent forms of answers often result. For this reason, it is very important to be able to recognize equivalent answers.

Practice 9

List 4 equivalent forms of

$-\dfrac{3x + 7}{x - 6}.$

Example 9 List some equivalent forms of $-\dfrac{5x - 1}{x + 9}.$

Solution: To do so, recall that $-\dfrac{a}{b} = \dfrac{-a}{b} = \dfrac{a}{-b}$. Thus,

$$-\frac{5x - 1}{x + 9} = \frac{-(5x - 1)}{x + 9} = \frac{-5x + 1}{x + 9} \quad \text{or} \quad \frac{1 - 5x}{x + 9}$$

Also,

$$-\frac{5x - 1}{x + 9} = \frac{5x - 1}{-(x + 9)} = \frac{5x - 1}{-x - 9} \quad \text{or} \quad \frac{5x - 1}{-9 - x}$$

Thus, $-\dfrac{5x - 1}{x + 9} = \dfrac{-(5x - 1)}{x + 9} = \dfrac{-5x + 1}{x + 9} = \dfrac{5x - 1}{-(x + 9)} = \dfrac{5x - 1}{-x - 9}$

■ **Work Practice 9**

> ### Helpful Hint
> Remember, a negative sign in front of a fraction or rational expression may be moved to the numerator or the denominator, but *not* both.

Keep in mind that many rational expressions may look different but in fact are equivalent.

Answers

8. $-\dfrac{2x + 3}{x + 4}$ or $\dfrac{-2x - 3}{x + 4}$

9. $\dfrac{-(3x + 7)}{x - 6}$; $\dfrac{-3x - 7}{x - 6}$; $\dfrac{3x + 7}{-(x - 6)}$;

$\dfrac{3x + 7}{-x + 6}$

Vocabulary, Readiness & Video Check

Use the choices below to fill in each blank. Not all choices will be used.

-1	0	simplifying	$\dfrac{-a}{-b}$	$\dfrac{-a}{b}$	$\dfrac{a}{-b}$
1	2	rational expression			

1. A <u>rational expression</u> is an expression that can be written in the form $\dfrac{P}{Q}$, where P and Q are polynomials and $Q \neq 0$.

2. The expression $\dfrac{x+3}{3+x}$ simplifies to ____<u>1</u>____.

3. The expression $\dfrac{x-3}{3-x}$ simplifies to ____<u>-1</u>____.

4. A rational expression is undefined for values that make the denominator ____<u>0</u>____.

5. The expression $\dfrac{7x}{x-2}$ is undefined for $x =$ ____<u>2</u>____.

6. The process of writing a rational expression in lowest terms is called ____<u>simplifying</u>____.

7. For a rational expression, $-\dfrac{a}{b} =$ ____<u>$\dfrac{-a}{b}$</u>____ $=$ ____<u>$\dfrac{a}{-b}$</u>____.

Decide which rational expression(s) can be simplified. (Do not actually simplify.)

8. $\dfrac{x}{x+7}$ no

9. $\dfrac{3+x}{x+3}$ yes

10. $\dfrac{5-x}{x-5}$ yes

11. $\dfrac{x+2}{x+8}$ no

Mrtin-Gay Interactive Videos

See Video 7.1 🔴

See video answer section.

Watch the section lecture video and answer the following questions.

Objective A 12. From the lecture before 🔲 Example 1, what do the different values of a rational expression depend on? How are these different values found? ▶

Objective B 13. Why can't the denominators of rational expressions be zero? How can we find the numbers for which a rational expression is undefined? ▶

Objective C 14. In 🔲 Example 7, why isn't a factor of x divided out of the expression at the end? ▶

Objective D 15. From 🔲 Example 9, if we move a negative sign from in front of a rational expression to either the numerator or the denominator, when would we need to use parentheses and why? ▶

7.1 Exercise Set MyMathLab®

Objective A *Find the value of the following expressions when $x = 2$, $y = -2$, and $z = -5$. See Example 1.*

1. $\dfrac{x+5}{x+2}$ $\dfrac{7}{4}$

2. $\dfrac{x+8}{x+1}$ $\dfrac{10}{3}$

3. $\dfrac{y^3}{y^2-1}$ $-\dfrac{8}{3}$

▶ 4. $\dfrac{z}{z^2-5}$ $-\dfrac{1}{4}$

5. $\dfrac{x^2+8x+2}{x^2-x-6}$ $-\dfrac{11}{2}$

6. $\dfrac{x+5}{x^2+4x-8}$ $\dfrac{7}{4}$

7. The average cost per DVD, in dollars, for a company to produce x DVDs on exercising is given by the formula $A = \dfrac{3x + 400}{x}$, where A is the average cost per DVD and x is the number of DVDs produced.

   a. Find the cost for producing 1 DVD. $403

   b. Find the average cost for producing 100 DVDs. $7

   c. Does the cost per DVD decrease or increase when more DVDs are produced? Explain your answer.   decrease; answers may vary

8. For a certain model of fax machine, the manufacturing cost C per machine is given by the equation

$$C = \frac{250x + 10{,}000}{x}$$

where x is the number of fax machines manufactured and cost C is in dollars per machine.

   a. Find the cost per fax machine when manufacturing 100 fax machines. $350

   b. Find the cost per fax machine when manufacturing 1000 fax machines. $260

   c. Does the cost per machine decrease or increase when more machines are manufactured? Explain why this is so.   decrease; answers may vary

Objective B   *Find any numbers for which each rational expression is undefined. See Example 2.*

9. $\dfrac{7}{2x}$ $x = 0$

10. $\dfrac{3}{5x}$ $x = 0$

11. $\dfrac{x + 3}{x + 2}$ $x = -2$

12. $\dfrac{5x + 1}{x - 9}$ $x = 9$

13. $\dfrac{x - 4}{2x - 5}$ $x = \dfrac{5}{2}$

14. $\dfrac{x + 1}{5x - 2}$ $x = \dfrac{2}{5}$

15. $\dfrac{9x^3 + 4}{15x^2 + 30x}$ $x = 0, x = -2$

16. $\dfrac{19x^3 + 2}{x^2 - x}$ $x = 0, x = 1$

17. $\dfrac{x^2 - 5x - 2}{4}$ none

18. $\dfrac{9y^5 + y^3}{9}$ none

19. $\dfrac{3x^2 + 9}{x^2 - 5x - 6}$ $x = 6, x = -1$

20. $\dfrac{11x^2 + 1}{x^2 - 5x - 14}$ $x = 7, x = -2$

21. $\dfrac{x}{3x^2 + 13x + 14}$ $x = -2, x = -\dfrac{7}{3}$

22. $\dfrac{x}{2x^2 + 15x + 27}$ $x = -3, x = -\dfrac{9}{2}$

Objective C   *Simplify each expression. See Examples 3 through 8.*

23. $\dfrac{x + 7}{7 + x}$ 1

24. $\dfrac{y + 9}{9 + y}$ 1

25. $\dfrac{x - 7}{7 - x}$ -1

26. $\dfrac{y - 9}{9 - y}$ -1

27. $\dfrac{2}{8x + 16}$ $\dfrac{1}{4(x + 2)}$

28. $\dfrac{3}{9x + 6}$ $\dfrac{1}{3x + 2}$

29. $\dfrac{x - 2}{x^2 - 4}$ $\dfrac{1}{x + 2}$

30. $\dfrac{x + 5}{x^2 - 25}$ $\dfrac{1}{x - 5}$

31. $\dfrac{2x - 10}{3x - 30}$ can't simplify

32. $\dfrac{3x - 9}{4x - 16}$ can't simplify

33. $\dfrac{-5a - 5b}{a + b}$ -5

34. $\dfrac{-4x - 4y}{x + y}$ -4

35. $\dfrac{7x + 35}{x^2 + 5x}$ $\dfrac{7}{x}$

36. $\dfrac{9x + 99}{x^2 + 11x}$ $\dfrac{9}{x}$

37. $\dfrac{x + 5}{x^2 - 4x - 45}$ $\dfrac{1}{x - 9}$

38. $\dfrac{x-3}{x^2-6x+9}$ $\quad\dfrac{1}{x-3}$

39. $\dfrac{5x^2+11x+2}{x+2}$ $\quad 5x+1$

40. $\dfrac{12x^2+4x-1}{2x+1}$ $\quad 6x-1$

▶ 41. $\dfrac{x^3+7x^2}{x^2+5x-14}$ $\quad\dfrac{x^2}{x-2}$

42. $\dfrac{x^4-10x^3}{x^2-17x+70}$ $\quad\dfrac{x^3}{x-7}$

43. $\dfrac{14x^2-21x}{2x-3}$ $\quad 7x$

44. $\dfrac{4x^2+24x}{x+6}$ $\quad 4x$

45. $\dfrac{x^2+7x+10}{x^2-3x-10}$ $\quad\dfrac{x+5}{x-5}$

46. $\dfrac{2x^2+7x-4}{x^2+3x-4}$ $\quad\dfrac{2x-1}{x-1}$

47. $\dfrac{3x^2+7x+2}{3x^2+13x+4}$ $\quad\dfrac{x+2}{x+4}$

48. $\dfrac{4x^2-4x+1}{2x^2+9x-5}$ $\quad\dfrac{2x-1}{x+5}$

49. $\dfrac{2x^2-8}{4x-8}$ $\quad\dfrac{x+2}{2}$

50. $\dfrac{5x^2-500}{35x+350}$ $\quad\dfrac{x-10}{7}$

▶ 51. $\dfrac{4-x^2}{x-2}$ $\quad -(x+2)$

52. $\dfrac{49-y^2}{y-7}$ $\quad -(y+7)$

53. $\dfrac{x^2-1}{x^2-2x+1}$ $\quad\dfrac{x+1}{x-1}$

54. $\dfrac{x^2-16}{x^2-8x+16}$ $\quad\dfrac{x+4}{x-4}$

Simplify each expression. Each exercise contains a four-term polynomial that should be factored by grouping. See Examples 3 through 8.

55. $\dfrac{x^2+xy+2x+2y}{x+2}$ $\quad x+y$

56. $\dfrac{ab+ac+b^2+bc}{b+c}$ $\quad a+b$

57. $\dfrac{5x+15-xy-3y}{2x+6}$ $\quad\dfrac{5-y}{2}$

58. $\dfrac{xy-6x+2y-12}{y^2-6y}$ $\quad\dfrac{x+2}{y}$

59. $\dfrac{2xy+5x-2y-5}{3xy+4x-3y-4}$ $\quad\dfrac{2y+5}{3y+4}$

60. $\dfrac{2xy+2x-3y-3}{2xy+4x-3y-6}$ $\quad\dfrac{y+1}{y+2}$

Objective D *Study Example 9. Then list four equivalent forms for each rational expression.*

61. $-\dfrac{x-10}{x+8}$ $\quad\dfrac{-(x-10)}{x+8};\dfrac{-x+10}{x+8};\dfrac{x-10}{-(x+8)};\dfrac{x-10}{-x-8}$

▶ 62. $-\dfrac{x+11}{x-4}$ $\quad\dfrac{-(x+11)}{x-4};\dfrac{-x-11}{x-4};\dfrac{x+11}{-(x-4)};\dfrac{x+11}{-x+4}$

63. $-\dfrac{5y-3}{y-12}$ $\quad\dfrac{-(5y-3)}{y-12};\dfrac{-5y+3}{y-12};\dfrac{5y-3}{-(y-12)};\dfrac{5y-3}{-y+12}$

64. $-\dfrac{8y-1}{y-15}$ $\quad\dfrac{-(8y-1)}{y-15};\dfrac{-8y+1}{y-15};\dfrac{8y-1}{-(y-15)};\dfrac{8y-1}{-y+15}$

Objectives C D Mixed Practice *Simplify each expression. Then determine whether the given answer is correct. See Examples 3 through 9.*

65. $\dfrac{9-x^2}{x-3}$; Answer: $-3-x$? $\quad$ correct

66. $\dfrac{100-x^2}{x-10}$; Answer: $-10-x$? $\quad$ correct

67. $\dfrac{7-34x-5x^2}{25x^2-1}$; Answer: $\dfrac{x+7}{-5x-1}$? $\quad$ correct

68. $\dfrac{2-15x-8x^2}{64x^2-1}$; Answer: $\dfrac{x+2}{-8x-1}$? $\quad$ correct

Review

Perform each indicated operation. See Section R.2.

69. $\dfrac{1}{3}\cdot\dfrac{9}{11}$ $\quad\dfrac{3}{11}$

70. $\dfrac{5}{27}\cdot\dfrac{2}{5}$ $\quad\dfrac{2}{27}$

71. $\dfrac{1}{3}\div\dfrac{1}{4}$ $\quad\dfrac{4}{3}$

72. $\dfrac{7}{8}\div\dfrac{1}{2}$ $\quad\dfrac{7}{4}$

73. $\dfrac{13}{20}\div\dfrac{2}{9}$ $\quad\dfrac{117}{40}$

74. $\dfrac{8}{15}\div\dfrac{5}{8}$ $\quad\dfrac{64}{75}$

Concept Extensions

Which of the following are incorrect and why? See the Concept Check in this section.

75. $\dfrac{5a - 15}{5}$ simplifies to $a - 3$?   correct

76. $\dfrac{7m - 9}{7}$ simplifies to $m - 9$?   incorrect; $\dfrac{7m - 9}{7} = \dfrac{7m}{7} - \dfrac{9}{7} = m - \dfrac{9}{7}$

77. $\dfrac{1 + 2}{1 + 3}$ simplifies to $\dfrac{2}{3}$?   incorrect; $\dfrac{1 + 2}{1 + 3} = \dfrac{3}{4}$

78. $\dfrac{46}{54}$ simplifies to $\dfrac{6}{5}$?   incorrect; $\dfrac{46}{54} = \dfrac{2 \cdot 23}{2 \cdot 27} = \dfrac{23}{27}$

79. Explain how to write a fraction in lowest terms.
answers may vary

80. Explain how to write a rational expression in lowest terms.   answers may vary

81. Explain why the denominator of a fraction or a rational expression must not equal 0.
answers may vary

82. Does $\dfrac{(x - 3)(x + 3)}{x - 3}$ have the same value as $x + 3$ for all real numbers? Explain why or why not.
no; answers may vary

83. The dose of medicine prescribed for a child depends on the child's age A in years and the adult dose D for the medication. Young's Rule is a formula used by pediatricians that gives a child's dose C as

$$C = \dfrac{DA}{A + 12}$$

Suppose that an 8-year-old child needs medication, and the normal adult dose is 1000 mg. What size dose should the child receive?   400 mg

84. Calculating body-mass index is a way to gauge whether a person should lose weight. Doctors recommend that body-mass index values fall between 19 and 25. The formula for body-mass index B is

$$B = \dfrac{705w}{h^2}$$

where w is weight in pounds and h is height in inches. Should a 148-pound person who is 5 feet 6 inches tall lose weight?   no

85. Anthropologists and forensic scientists use a measure called the cephalic index to help classify skulls. The cephalic index of a skull with width W and length L from front to back is given by the formula

$$C = \dfrac{100W}{L}$$

A long skull has an index value less than 75, a medium skull has an index value between 75 and 85, and a broad skull has an index value over 85. Find the cephalic index of a skull that is 5 inches wide and 6.4 inches long. Classify the skull.
$C = 78.125$; medium

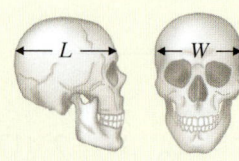

86. A company's gross profit margin P can be computed with the formula $P = \dfrac{R - C}{R}$, where

R = the company's revenue and C = cost of goods sold. For the fiscal year 2012, Ford Motor Company had revenues of \$134.25 billion and cost of goods sold \$115.1 billion. (*Source:* Ford Motor Company) What was Ford's gross profit margin in 2012? Express the answer as a percent, rounded to the nearest tenth of a percent.   14.3%

87. A baseball player's slugging percentage S can be calculated with the following formula:

$$S = \frac{h + d + 2t + 3r}{b},$$ where h = number of hits,

d = number of doubles, t = number of triples, r = number of home runs, and b = number of at bats. In 2012, Giancarlo Stanton of the Miami Marlins led Major League Baseball in slugging percentage. During the 2012 season, Stanton had 449 at bats, 130 hits, 30 doubles, 1 triple, and 37 home runs. (*Source:* Major League Baseball) Calculate Stanton's 2012 slugging percentage. Round to the nearest tenth of a percent. 60.8%

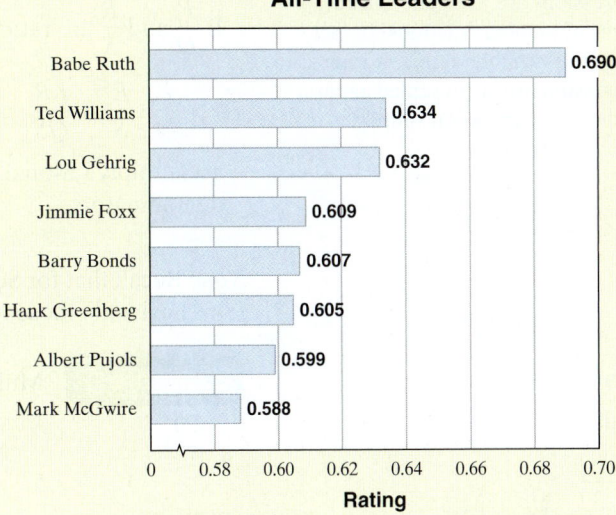

Baseball Slugging Percentage— All-Time Leaders

Player	Rating
Babe Ruth	0.690
Ted Williams	0.634
Lou Gehrig	0.632
Jimmie Foxx	0.609
Barry Bonds	0.607
Hank Greenberg	0.605
Albert Pujols	0.599
Mark McGwire	0.588

Source: Baseball Almanac

88. To calculate a quarterback's rating in NCAA football, you may use the formula

$$\frac{100C + 330T - 200I + 8.4Y}{A},$$ where C = the

number of completed passes, A = the number of attempted passes, T = the number of touchdown passes, Y = the number of yards in the completed passes, and I = the number of interceptions. Jameis Winston of Florida State University was selected as the 2013 winner of the Heisman Memorial Trophy as the Most Outstanding Football Player. Winston, a freshman quarterback with the Seminoles, ended the season with 384 attempted passes, 257 completed passes, 4057 yards, 40 touchdowns, and only 10 interceptions. Calculate Winston's quarterback rating for the 2013 season. (*Source:* NCAA) Round the answer to the nearest tenth. 184.8

7.2 Multiplying and Dividing Rational Expressions

Objective A Multiplying Rational Expressions

Just as simplifying rational expressions is similar to simplifying number fractions, multiplying and dividing rational expressions are similar to multiplying and dividing number fractions.

Objectives

A Multiply Rational Expressions.

B Divide Rational Expressions.

C Multiply and Divide Rational Expressions.

D Convert Between Units of Measure.

Fractions	**Rational Expressions**
Multiply: $\dfrac{3}{5} \cdot \dfrac{10}{11}$	Multiply: $\dfrac{x - 3}{x + 5} \cdot \dfrac{2x + 10}{x^2 - 9}$

Multiply numerators and then multiply denominators.

$\dfrac{3}{5} \cdot \dfrac{10}{11} = \dfrac{3 \cdot 10}{5 \cdot 11}$	$\dfrac{x - 3}{x + 5} \cdot \dfrac{2x + 10}{x^2 - 9} = \dfrac{(x - 3) \cdot (2x + 10)}{(x + 5) \cdot (x^2 - 9)}$

Simplify by factoring numerators and denominators.

$= \dfrac{3 \cdot 2 \cdot 5}{5 \cdot 11}$	$= \dfrac{(x - 3) \cdot 2 \, (x + 5)}{(x + 5) \, (x + 3) \, (x - 3)}$

Apply the fundamental principle.

$= \dfrac{3 \cdot 2}{11}$ or $\dfrac{6}{11}$	$= \dfrac{2}{x + 3}$

Teaching Tip

Point out to students that rational expressions are just generalized forms of number fractions. In fact, when the variable(s) in a rational expression are given replacement values, the expression simplifies to a number fraction.

Multiplying Rational Expressions

If $\dfrac{P}{Q}$ and $\dfrac{R}{S}$ are rational expressions, then

$$\frac{P}{Q} \cdot \frac{R}{S} = \frac{PR}{QS}$$

To multiply rational expressions, multiply the numerators and then multiply the denominators.

Note: Recall that for Sections 7.1 through 7.4, we assume variables in rational expressions have only those replacement values for which the expressions are defined.

Practice 1

Multiply.

a. $\dfrac{16y}{3} \cdot \dfrac{1}{x^2}$

b. $\dfrac{-5a^3}{3b^3} \cdot \dfrac{2b^2}{15a}$

Example 1   Multiply.

a. $\dfrac{25x}{2} \cdot \dfrac{1}{y^3}$

b. $\dfrac{-7x^2}{5y} \cdot \dfrac{3y^5}{14x^2}$

Solution: To multiply rational expressions, we first multiply the numerators and then multiply the denominators of both expressions. Then we write the product in lowest terms.

a. $\dfrac{25x}{2} \cdot \dfrac{1}{y^3} = \dfrac{25x \cdot 1}{2 \cdot y^3} = \dfrac{25x}{2y^3}$

The expression $\dfrac{25x}{2y^3}$ is in lowest terms.

b. $\dfrac{-7x^2}{5y} \cdot \dfrac{3y^5}{14x^2} = \dfrac{-7x^2 \cdot 3y^5}{5y \cdot 14x^2}$ Multiply.

The expression $\dfrac{-7x^2 \cdot 3y^5}{5y \cdot 14x^2}$ is not in lowest terms, so we factor the numerator and the denominator and apply the fundamental principle to "remove factors of 1."

$= \dfrac{-1 \cdot 7 \cdot 3 \cdot x^2 \cdot y \cdot y^4}{5 \cdot 2 \cdot 7 \cdot x^2 \cdot y}$ Common factors in the numerator and denominator form factors of 1.

$= -\dfrac{3y^4}{10}$ Divide out common factors. (This is the same as "removing a factor of 1.")

■ **Work Practice 1**

When multiplying rational expressions, it is usually best to factor each numerator and denominator first. This will help us when we apply the fundamental principle to write the product in lowest terms.

Practice 2

Multiply: $\dfrac{3x + 6}{14} \cdot \dfrac{7x^2}{x^3 + 2x^2}$

Example 2   Multiply: $\dfrac{x^2 + x}{3x} \cdot \dfrac{6}{5x + 5}$

Solution:

$\dfrac{x^2 + x}{3x} \cdot \dfrac{6}{5x + 5} = \dfrac{x(x + 1)}{3x} \cdot \dfrac{2 \cdot 3}{5(x + 1)}$ Factor numerators and denominators.

$= \dfrac{x(x + 1) \cdot 2 \cdot 3}{3x \cdot 5 (x + 1)}$ Multiply.

$= \dfrac{2}{5}$ Divide out common factors.

Answers

1. a. $\dfrac{16y}{3x^2}$ b. $-\dfrac{2a^2}{9b}$ 2. $\dfrac{3}{2}$

■ **Work Practice 2**

The following steps may be used to multiply rational expressions.

> **To Multiply Rational Expressions**
>
> **Step 1:** Completely factor numerators and denominators.
>
> **Step 2:** Multiply numerators and multiply denominators.
>
> **Step 3:** Simplify or write the product in lowest terms by dividing out common factors.

✔**Concept Check** Which of the following is a true statement?

a. $\dfrac{1}{3} \cdot \dfrac{1}{2} = \dfrac{1}{5}$ **b.** $\dfrac{2}{x} \cdot \dfrac{5}{x} = \dfrac{10}{x}$ **c.** $\dfrac{3}{x} \cdot \dfrac{1}{2} = \dfrac{3}{2x}$ **d.** $\dfrac{x}{7} \cdot \dfrac{x+5}{4} = \dfrac{2x+5}{28}$

Example 3 Multiply: $\dfrac{3x+3}{5x^2-5x} \cdot \dfrac{2x^2+x-3}{4x^2-9}$

Solution:

$$\dfrac{3x+3}{5x^2-5x} \cdot \dfrac{2x^2+x-3}{4x^2-9} = \dfrac{3(x+1)}{5x(x-1)} \cdot \dfrac{(2x+3)(x-1)}{(2x-3)(2x+3)} \quad \text{Factor.}$$

$$= \dfrac{3(x+1)\,(2x+3)(x-1)}{5x\,(x-1)\,(2x-3)\,(2x+3)} \quad \text{Multiply.}$$

$$= \dfrac{3(x+1)}{5x(2x-3)} \quad \text{Simplify.}$$

■ **Work Practice 3**

Practice 3

Multiply:

$$\dfrac{4x+8}{7x^2-14x} \cdot \dfrac{3x^2-5x-2}{9x^2-1}$$

Objective B Dividing Rational Expressions ▶

We can divide by a rational expression in the same way we divide by a number fraction. Recall that to divide by a fraction, we multiply by its reciprocal.

For example, to divide $\dfrac{3}{2}$ by $\dfrac{7}{8}$, we multiply $\dfrac{3}{2}$ by $\dfrac{8}{7}$.

$$\dfrac{3}{2} \div \dfrac{7}{8} = \dfrac{3}{2} \cdot \dfrac{8}{7} = \dfrac{3 \cdot 4 \cdot 2}{2 \cdot 7} = \dfrac{12}{7}$$

Helpful Hint

Don't forget how to find reciprocals. The reciprocal of $\dfrac{a}{b}$ is $\dfrac{b}{a}$, $a \neq 0, b \neq 0$.

> **Dividing Rational Expressions**
>
> If $\dfrac{P}{Q}$ and $\dfrac{R}{S}$ are rational expressions and $\dfrac{R}{S}$ is not 0, then
>
> $$\dfrac{P}{Q} \div \dfrac{R}{S} = \dfrac{P}{Q} \cdot \dfrac{S}{R} = \dfrac{PS}{QR}$$

To divide two rational expressions, multiply the first rational expression by the reciprocal of the second rational expression.

Answer

3. $\dfrac{4(x+2)}{7x(3x-1)}$

✔**Concept Check Answer**

c

Practice 4

Divide: $\dfrac{7x^2}{6} \div \dfrac{x}{2y}$

Example 4

Divide: $\dfrac{3x^3}{40} \div \dfrac{4x^3}{y^2}$

Solution:

$$\dfrac{3x^3}{40} \div \dfrac{4x^3}{y^2} = \dfrac{3x^3}{40} \cdot \dfrac{y^2}{4x^3} \qquad \text{Multiply by the reciprocal of } \dfrac{4x^3}{y^2}.$$

$$= \dfrac{3\ x^3 \cdot y^2}{160\ x^3}$$

$$= \dfrac{3y^2}{160} \qquad \text{Simplify.}$$

■ Work Practice 4

Practice 5

Divide: $\dfrac{(x-4)^2}{6} \div \dfrac{3x-12}{2}$

Example 5

Divide: $\dfrac{(x+2)^2}{10} \div \dfrac{2x+4}{5}$

Solution:

$$\dfrac{(x+2)^2}{10} \div \dfrac{2x+4}{5} = \dfrac{(x+2)^2}{10} \cdot \dfrac{5}{2x+4} \qquad \text{Multiply by the reciprocal of } \dfrac{2x+4}{5}.$$

$$= \dfrac{(x+2)\,(x+2)\cdot 5}{5\cdot 2\cdot 2\cdot (x+2)} \qquad \text{Factor and multiply.}$$

$$= \dfrac{x+2}{4} \qquad \text{Simplify.}$$

■ Work Practice 5

> **Helpful Hint**
>
> Remember, **to Divide by a Rational Expression,** multiply by its reciprocal.

Practice 6

Divide: $\dfrac{10x+4}{x^2-4} \div \dfrac{5x^3+2x^2}{x+2}$

Example 6

Divide: $\dfrac{6x+2}{x^2-1} \div \dfrac{3x^2+x}{x-1}$

Solution:

$$\dfrac{6x+2}{x^2-1} \div \dfrac{3x^2+x}{x-1} = \dfrac{6x+2}{x^2-1} \cdot \dfrac{x-1}{3x^2+x} \qquad \text{Multiply by the reciprocal.}$$

$$= \dfrac{2\,(3x+1)(x-1)}{(x+1)\,(x-1)\cdot x\,(3x+1)} \qquad \text{Factor and multiply.}$$

$$= \dfrac{2}{x(x+1)} \qquad \text{Simplify.}$$

■ Work Practice 6

Teaching Tip

Consider telling students that it is OK to leave their final answer in factored form when working with rational expressions.

Practice 7

Divide:

$\dfrac{3x^2-10x+8}{7x-14} \div \dfrac{9x-12}{21}$

Example 7

Divide: $\dfrac{2x^2-11x+5}{5x-25} \div \dfrac{4x-2}{10}$

Solution:

$$\dfrac{2x^2-11x+5}{5x-25} \div \dfrac{4x-2}{10} = \dfrac{2x^2-11x+5}{5x-25} \cdot \dfrac{10}{4x-2} \qquad \text{Multiply by the reciprocal.}$$

$$= \dfrac{(2x-1)(x-5)\cdot 2\cdot 5}{5(x-5)\cdot 2(2x-1)} \qquad \text{Factor and multiply.}$$

$$= \dfrac{1}{1} \quad \text{or} \quad 1 \qquad \text{Simplify.}$$

■ Work Practice 7

Answers

4. $\dfrac{7xy}{3}$ 5. $\dfrac{x-4}{9}$ 6. $\dfrac{2}{x^2(x-2)}$ 7. 1

Objective C Multiplying and Dividing Rational Expressions ▶

Let's make sure that we understand the difference between multiplying and dividing rational expressions.

Rational Expressions	
Multiplication	Multiply the numerators and multiply the denominators.
Division	Multiply by the reciprocal of the divisor.

Example 8 Multiply or divide as indicated.

a. $\dfrac{x-4}{5} \cdot \dfrac{x}{x-4}$

b. $\dfrac{x-4}{5} \div \dfrac{x}{x-4}$

c. $\dfrac{x^2-4}{2x+6} \cdot \dfrac{x^2+4x+3}{2-x}$

Solution:

a. $\dfrac{x-4}{5} \cdot \dfrac{x}{x-4} = \dfrac{(x-4) \cdot x}{5 \cdot (x-4)} = \dfrac{x}{5}$

b. $\dfrac{x-4}{5} \div \dfrac{x}{x-4} = \dfrac{x-4}{5} \cdot \dfrac{x-4}{x} = \dfrac{(x-4)^2}{5x}$

c. $\dfrac{x^2-4}{2x+6} \cdot \dfrac{x^2+4x+3}{2-x} = \dfrac{(x-2)(x+2) \cdot (x+1)(x+3)}{2(x+3) \cdot (2-x)}$ Factor and multiply.

Recall from Section 7.1 that $x-2$ and $2-x$ are opposites. This means that $\dfrac{x-2}{2-x} = -1$. Thus,

$$\dfrac{(x-2)\,(x+2) \cdot (x+1)\,(x+3)}{2\,(x+3) \cdot (2-x)} = \dfrac{-1(x+2)(x+1)}{2}$$

$$= -\dfrac{(x+2)(x+1)}{2}$$

◼ **Work Practice 8**

Objective D Converting Between Units of Measure ▶

How many square inches are in 1 square foot?
How many cubic feet are in a cubic yard?

If you have trouble answering these questions, this section will be helpful to you.

Now that we know how to multiply fractions and rational expressions, we can use this knowledge to help us convert between units of measure. To do so, we will use **unit fractions**. A unit fraction is a fraction that equals 1. For example, since 12 in. = 1 ft, we have the unit fractions

$$\dfrac{12 \text{ in.}}{1 \text{ ft}} = 1 \qquad \text{and} \qquad \dfrac{1 \text{ ft}}{12 \text{ in.}} = 1$$

Practice 8

Multiply or divide as indicated.

a. $\dfrac{x+3}{x} \cdot \dfrac{7}{x+3}$

b. $\dfrac{x+3}{x} \div \dfrac{7}{x+3}$

c. $\dfrac{3-x}{x^2+6x+5} \cdot \dfrac{2x+10}{x^2-7x+12}$

Answers

8. a. $\dfrac{7}{x}$ **b.** $\dfrac{(x+3)^2}{7x}$

c. $-\dfrac{2}{(x+1)(x-4)}$

Practice 9

288 square inches = _____ square feet

Example 9   18 square feet = _____ square yards

Solution: Let's multiply 18 square feet by a unit fraction that has square feet in the denominator and square yards in the numerator. From the diagram, you can see that

$$1 \text{ square yard} = 9 \text{ square feet}$$

Thus,

$$18 \text{ sq ft} = \frac{18 \text{ sq ft}}{1} \cdot 1 = \frac{\overset{2}{\cancel{18 \text{ sq ft}}}}{1} \cdot \frac{1 \text{ sq yd}}{\underset{1}{\cancel{9 \text{ sq ft}}}}$$

$$= \frac{2 \cdot 1}{1 \cdot 1} \text{ sq yd} = 2 \text{ sq yd}$$

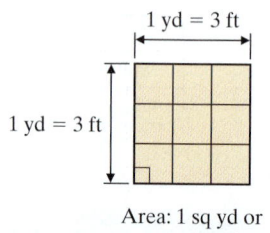

1 yd = 3 ft

1 yd = 3 ft

Area: 1 sq yd or 9 sq ft

Thus, 18 sq ft = 2 sq yd.

Draw a diagram of 18 sq ft to help you see that this is reasonable.

■ **Work Practice 9**

Practice 10

3.5 square feet = _____ square inches

Example 10   5.2 square yards = _____ square feet

Solution:

$$5.2 \text{ sq yd} = \frac{5.2 \text{ sq yd}}{1} \cdot 1 = \frac{5.2 \text{ sq yd}}{1} \cdot \frac{9 \text{ sq ft}}{1 \text{ sq yd}} \quad \leftarrow \text{ Units converting to}$$
$$\leftarrow \text{ Units given}$$

$$= \frac{5.2 \cdot 9}{1 \cdot 1} \text{ sq ft}$$

$$= 46.8 \text{ sq ft}$$

Thus, 5.2 sq yd = 46.8 sq ft.

Draw a diagram to see that this is reasonable.

■ **Work Practice 10**

Practice 11

The largest casino in the world is the Venetian, in Macau, on the southern tip of China. The gaming area for this casino is approximately 61,000 *square yards*. Find the size of the gaming area in *square feet*. (*Source: USA Today*)

Example 11   Converting from Cubic Feet to Cubic Yards

The largest building in the world by volume is The Boeing Company's Everett, Washington, factory complex, where Boeing's wide-body jetliners, the 747, 767, and 777, are built. The volume of this factory complex is 472,370,319 cubic feet. Find the volume of this Boeing facility in cubic yards. (*Source:* The Boeing Company)

Answers

9. 2 sq ft   **10.** 504 sq in.
11. 549,000 sq ft

Solution: There are 27 cubic feet in 1 cubic yard. (See the diagram.)

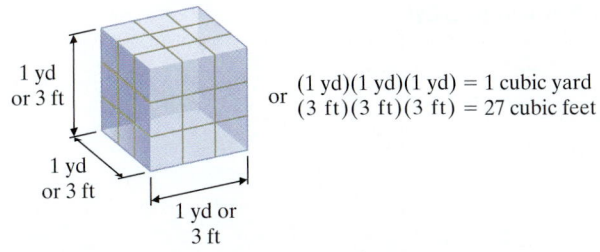

$$472{,}370{,}319 \text{ cu ft} = 472{,}370{,}319 \text{ cu ft} \cdot \frac{1 \text{ cu yd}}{27 \text{ cu ft}}$$

$$= \frac{472{,}370{,}319}{27} \text{ cu yd}$$

$$= 17{,}495{,}197 \text{ cu yd}$$

■ **Work Practice 11**

Helpful Hint

When converting between units of measurement, if possible, write the unit fraction so that **the numerator contains the units you are converting to** and **the denominator contains the original units.**

Unit fraction

$$48 \text{ in.} = \frac{48 \text{ in.}}{1} \cdot \frac{1 \text{ ft}}{12 \text{ in.}} \quad \leftarrow \text{Units converting to} \atop \leftarrow \text{Original units}$$

$$= \frac{48}{12} \text{ ft} = 4 \text{ ft}$$

Example 12

At the Summer Olympics, Jamaican athlete Usain Bolt won the gold medal in the men's 100-meter track event. He ran the distance at an average speed of 34.1 feet per second. Convert this speed to miles per hour. (*Source:* International Olympic Committee)

Solution: Recall that 1 mile = 5280 feet and 1 hour = 3600 seconds $(60 \cdot 60)$.

Unit fractions

$$34.1 \text{ feet/second} = \frac{34.1 \text{ feet}}{1 \text{ second}} \cdot \frac{3600 \text{ seconds}}{1 \text{ hour}} \cdot \frac{1 \text{ mile}}{5280 \text{ feet}}$$

$$= \frac{34.1 \cdot 3600}{5280} \text{ miles/hour}$$

$$= 23.25 \text{ miles/hour}$$

■ **Work Practice 12**

Practice 12

The cheetah is the fastest land animal, being clocked at about 102.7 feet per second. Convert this to miles per hour. Round to the nearest tenth. (*Source: World Almanac and Book of Facts*)

Answer

12. 70.0 miles per hour

Vocabulary, Readiness & Video Check

Use the choices below to fill in each blank. Not all choices will be used.

opposites $\dfrac{a \cdot d}{b \cdot c}$ $\dfrac{a \cdot c}{b \cdot d}$ $\dfrac{x}{42}$ $\dfrac{x^2}{42}$ $\dfrac{2x}{42}$ $\dfrac{6}{7}$ $\dfrac{7}{6}$

reciprocals

1. The expressions $\dfrac{x}{2y}$ and $\dfrac{2y}{x}$ are called ___reciprocals___.

2. $\dfrac{a}{b} \cdot \dfrac{c}{d} = $ ___$\dfrac{a \cdot c}{b \cdot d}$___

3. $\dfrac{a}{b} \div \dfrac{c}{d} = $ ___$\dfrac{a \cdot d}{b \cdot c}$___

4. $\dfrac{x}{7} \cdot \dfrac{x}{6} = $ ___$\dfrac{x^2}{42}$___

5. $\dfrac{x}{7} \div \dfrac{x}{6} = $ ___$\dfrac{6}{7}$___

Martin-Gay Interactive Videos

See Video 7.2

See video answer section.

Watch the section lecture video and answer the following questions.

Objective A **6.** Would you say a person needs to be quite comfortable with factoring polynomials in order to be successful with multiplying rational expressions? Explain, referencing ⊞ Example 2 in your answer. ▶

Objective B **7.** Based on the lecture before ⊞ Example 3, complete the following statements: Dividing rational expressions is exactly like dividing _____. Therefore, to divide by a rational expression, multiply by its _____. ▶

Objective C **8.** In ⊞ Examples 4 and 5, determining the operation is the first step in deciding how to perform the operation. Why is this so? ▶

Objective D **9.** In ⊞ Example 6, why is the unit fraction $\dfrac{27 \text{ cu ft}}{1 \text{ cu yd}}$ used? ▶

7.2 Exercise Set MyMathLab® ▶

Objective A *Find each product and simplify if possible. See Examples 1 through 3.*

1. $\dfrac{3x}{y^2} \cdot \dfrac{7y}{4x}$ $\dfrac{21}{4y}$

2. $\dfrac{9x^2}{y} \cdot \dfrac{4y}{3x^3}$ $\dfrac{12}{x}$

▶ **3.** $\dfrac{8x}{2} \cdot \dfrac{x^5}{4x^2}$ x^4

4. $\dfrac{6x^2}{10x^3} \cdot \dfrac{5x}{12}$ $\dfrac{1}{4}$

5. $-\dfrac{5a^2b}{30a^2b^2} \cdot b^3$ $-\dfrac{b^2}{6}$

6. $-\dfrac{9x^3y^2}{18xy^5} \cdot y^3$ $-\dfrac{x^2}{2}$

7. $\dfrac{x}{2x-14} \cdot \dfrac{x^2-7x}{5}$ $\dfrac{x^2}{10}$

8. $\dfrac{4x-24}{20x} \cdot \dfrac{5}{x-6}$ $\dfrac{1}{x}$

9. $\dfrac{6x+6}{5} \cdot \dfrac{10}{36x+36}$ $\dfrac{1}{3}$

10. $\dfrac{x^2+x}{8} \cdot \dfrac{16}{x+1}$ $2x$

11. $\dfrac{(m+n)^2}{m-n} \cdot \dfrac{m}{m^2+mn}$ $\dfrac{m+n}{m-n}$

12. $\dfrac{(m-n)^2}{m+n} \cdot \dfrac{m}{m^2-mn}$ $\dfrac{m-n}{m+n}$

13. $\dfrac{x^2 - 25}{x^2 - 3x - 10} \cdot \dfrac{x + 2}{x}$ $\dfrac{x + 5}{x}$

14. $\dfrac{a^2 - 4a + 4}{a^2 - 4} \cdot \dfrac{a + 3}{a - 2}$ $\dfrac{a + 3}{a + 2}$

15. $\dfrac{x^2 + 6x + 8}{x^2 + x - 20} \cdot \dfrac{x^2 + 2x - 15}{x^2 + 8x + 16}$ $\dfrac{(x + 2)(x - 3)}{(x - 4)(x + 4)}$

16. $\dfrac{x^2 + 9x + 20}{x^2 - 15x + 44} \cdot \dfrac{x^2 - 11x + 28}{x^2 + 12x + 35}$ $\dfrac{(x + 4)(x - 7)}{(x - 11)(x + 7)}$

Objective B *Find each quotient and simplify. See Examples 4 through 7.*

17. $\dfrac{5x^7}{2x^5} \div \dfrac{15x}{4x^3}$ $\dfrac{2x^4}{3}$

18. $\dfrac{9y^4}{6y} \div \dfrac{y^2}{3}$ $\dfrac{9y}{2}$

19. $\dfrac{8x^2}{y^3} \div \dfrac{4x^2y^3}{6}$ $\dfrac{12}{y^6}$

20. $\dfrac{7a^2b}{3ab^2} \div \dfrac{21a^2b^2}{14ab}$ $\dfrac{14}{9b^2}$

21. $\dfrac{(x - 6)(x + 4)}{4x} \div \dfrac{2x - 12}{8x^2}$ $x(x + 4)$

22. $\dfrac{(x + 3)^2}{5} \div \dfrac{5x + 15}{25}$ $x + 3$

23. $\dfrac{3x^2}{x^2 - 1} \div \dfrac{x^5}{(x + 1)^2}$ $\dfrac{3(x + 1)}{x^3(x - 1)}$

24. $\dfrac{9x^5}{a^2 - b^2} \div \dfrac{27x^2}{3b - 3a}$ $-\dfrac{x^3}{a + b}$

25. $\dfrac{m^2 - n^2}{m + n} \div \dfrac{m}{m^2 + nm}$ $m^2 - n^2$

26. $\dfrac{(m - n)^2}{m + n} \div \dfrac{m^2 - mn}{m}$ $\dfrac{m - n}{m + n}$

▶ 27. $\dfrac{x + 2}{7 - x} \div \dfrac{x^2 - 5x + 6}{x^2 - 9x + 14}$ $-\dfrac{x + 2}{x - 3}$

28. $\dfrac{x - 3}{2 - x} \div \dfrac{x^2 + 3x - 18}{x^2 + 2x - 8}$ $-\dfrac{x + 4}{x + 6}$

29. $\dfrac{x^2 + 7x + 10}{x - 1} \div \dfrac{x^2 + 2x - 15}{x - 1}$ $\dfrac{x + 2}{x - 3}$

30. $\dfrac{x + 1}{2x^2 + 5x + 3} \div \dfrac{20x + 100}{2x + 3}$ $\dfrac{1}{20(x + 5)}$

Objective C Mixed Practice *Multiply or divide as indicated. See Example 8.*

▶ 31. $\dfrac{5x - 10}{12} \div \dfrac{4x - 8}{8}$ $\dfrac{5}{6}$

32. $\dfrac{6x + 6}{5} \div \dfrac{9x + 9}{10}$ $\dfrac{4}{3}$

33. $\dfrac{x^2 + 5x}{8} \cdot \dfrac{9}{3x + 15}$ $\dfrac{3x}{8}$

34. $\dfrac{3x^2 + 12x}{6} \cdot \dfrac{9}{2x + 8}$ $\dfrac{9x}{4}$

35. $\dfrac{7}{6p^2 + q} \div \dfrac{14}{18p^2 + 3q}$ $\dfrac{3}{2}$

36. $\dfrac{3x + 6}{20} \div \dfrac{4x + 8}{8}$ $\dfrac{3}{10}$

37. $\dfrac{3x + 4y}{x^2 + 4xy + 4y^2} \cdot \dfrac{x + 2y}{2}$ $\dfrac{3x + 4y}{2(x + 2y)}$

38. $\dfrac{x^2 - y^2}{3x^2 + 3xy} \cdot \dfrac{3x^2 + 6x}{3x^2 - 2xy - y^2}$ $\dfrac{x + 2}{3x + y}$

39. $\dfrac{(x + 2)^2}{x - 2} \div \dfrac{x^2 - 4}{2x - 4}$ $\dfrac{2(x + 2)}{x - 2}$

40. $\dfrac{x + 3}{x^2 - 9} \div \dfrac{5x + 15}{(x - 3)^2}$ $\dfrac{x - 3}{5(x + 3)}$

41. $\dfrac{x^2 - 4}{24x} \div \dfrac{2 - x}{6xy}$ $-\dfrac{y(x+2)}{4}$

42. $\dfrac{3y}{3 - x} \div \dfrac{12xy}{x^2 - 9}$ $-\dfrac{x+3}{4x}$

43. $\dfrac{a^2 + 7a + 12}{a^2 + 5a + 6} \cdot \dfrac{a^2 + 8a + 15}{a^2 + 5a + 4}$ $\dfrac{(a+5)(a+3)}{(a+2)(a+1)}$

44. $\dfrac{b^2 + 2b - 3}{b^2 + b - 2} \cdot \dfrac{b^2 - 4}{b^2 + 6b + 8}$ $\dfrac{(b+3)(b-2)}{(b+2)(b+4)}$

▶ 45. $\dfrac{5x - 20}{3x^2 + x} \cdot \dfrac{3x^2 + 13x + 4}{x^2 - 16}$ $\dfrac{5}{x}$

46. $\dfrac{9x + 18}{4x^2 - 3x} \cdot \dfrac{4x^2 - 11x + 6}{x^2 - 4}$ $\dfrac{9}{x}$

47. $\dfrac{8n^2 - 18}{2n^2 - 5n + 3} \div \dfrac{6n^2 + 7n - 3}{n^2 - 9n + 8}$ $\dfrac{2(n-8)}{3n-1}$

48. $\dfrac{36n^2 - 64}{3n^2 - 10n + 8} \div \dfrac{3n^2 - 5n - 12}{n^2 - 9n + 14}$ $\dfrac{4(n-7)}{n-3}$

Objective D   *Convert as indicated. See Examples 9 through 12.*

49. 10 square feet = ___1440___ square inches.

50. 1008 square inches = ___7___ square feet.

51. 45 square feet = ___5___ square yards.

52. 2 square yards = ___2592___ square inches.

▶ 53. 3 cubic yards = ___81___ cubic feet.

54. 2 cubic yards = ___93,312___ cubic inches.

55. 50 miles per hour = ___73___ feet per second (round to the nearest whole).

56. 10 feet per second = ___6.8___ miles per hour (round to the nearest tenth).

57. 6.3 square yards = ___56.7___ square feet.

58. 3.6 square yards = ___32.4___ square feet.

59. In January 2010, the Burj Khalifa Tower officially became the tallest building in the world. This tower has a curtain wall (the exterior skin of the building) that is approximately 133,500 square yards. Convert this to square feet. (*Source:* Burj Khalifa)   1,201,500 sq ft

60. The Pentagon, headquarters for the Department of Defense, contains 3,705,793 square feet of office and storage space. Convert this to square yards. Round to the nearest square yard. (*Source:* U.S. Department of Defense)   411,755 sq yd

61. On January 7, 2011, Australian driver Barton Mawer set a new solar-powered-car land speed record of 80.9 feet per second in the Sunswift IV, built by a student team at the University of New South Wales. Convert this speed to miles per hour. Round to the nearest tenth. (*Source:* University of New South Wales)   55.2 miles/hour

62. Peregrine falcons are among the fastest birds in the world. When engaged in a high-speed dive for prey, a peregrine falcon can reach speeds over 200 miles per hour. Find this speed in feet per second. Round to the nearest tenth. (*Source:* Ohio Department of Natural Resources)   293.3 feet per second

Review

Perform each indicated operation. See Section R.2.

63. $\dfrac{1}{5} + \dfrac{4}{5}$ 1

64. $\dfrac{3}{15} + \dfrac{6}{15}$ $\dfrac{3}{5}$

65. $\dfrac{9}{9} - \dfrac{19}{9}$ $-\dfrac{10}{9}$

66. $\dfrac{4}{3} - \dfrac{8}{3}$ $-\dfrac{4}{3}$

67. $\dfrac{6}{5} + \left(\dfrac{1}{5} - \dfrac{8}{5}\right)$ $-\dfrac{1}{5}$

68. $-\dfrac{3}{2} + \left(\dfrac{1}{2} - \dfrac{3}{2}\right)$ $-\dfrac{5}{2}$

Concept Extensions

Identify each statement as true or false. If false, correct the multiplication. See the Concept Check in this section.

69. $\dfrac{4}{a} \cdot \dfrac{1}{b} = \dfrac{4}{ab}$ true

70. $\dfrac{2}{3} \cdot \dfrac{2}{4} = \dfrac{2}{7}$ false; $\dfrac{1}{3}$

71. $\dfrac{x}{5} \cdot \dfrac{x+3}{4} = \dfrac{2x+3}{20}$ false; $\dfrac{x^2+3x}{20}$

72. $\dfrac{7}{a} \cdot \dfrac{3}{a} = \dfrac{21}{a}$ false; $\dfrac{21}{a^2}$

73. Find the area of the rectangle.

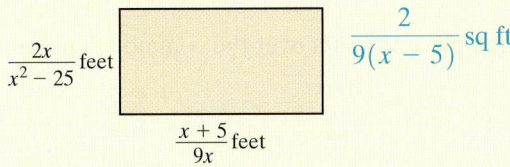

$\dfrac{2x}{x^2-25}$ feet

$\dfrac{x+5}{9x}$ feet

$\dfrac{2}{9(x-5)}$ sq ft

74. Find the area of the square.

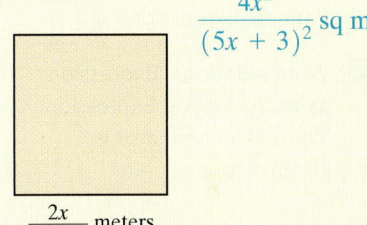

$\dfrac{4x^2}{(5x+3)^2}$ sq m

$\dfrac{2x}{5x+3}$ meters

Multiply or divide as indicated.

75. $\left(\dfrac{x^2-y^2}{x^2+y^2} \div \dfrac{x^2-y^2}{3x}\right) \cdot \dfrac{x^2+y^2}{6}$ $\dfrac{x}{2}$

76. $\left(\dfrac{x^2-9}{x^2-1} \cdot \dfrac{x^2+2x+1}{2x^2+9x+9}\right) \div \dfrac{2x+3}{1-x}$ $-\dfrac{(x-3)(x+1)}{(2x+3)^2}$

77. $\left(\dfrac{2a+b}{b^2} \cdot \dfrac{3a^2-2ab}{ab+2b^2}\right) \div \dfrac{a^2-3ab+2b^2}{5ab-10b^2}$ $\dfrac{5a(2a+b)(3a-2b)}{b^2(a-b)(a+2b)}$

78. $\left(\dfrac{x^2y^2-xy}{4x-4y} \div \dfrac{3y-3x}{8x-8y}\right) \cdot \dfrac{y-x}{8}$ $\dfrac{xy(xy-1)}{12}$

79. In your own words, explain how you multiply rational expressions.   answers may vary

80. Explain how dividing rational expressions is similar to dividing rational numbers.   answers may vary

81. On a day in August 2014, 1 euro was equivalent to 1.3387 American dollars. If you wanted to exchange $2000 U.S. for euros on that day for a European vacation, how many would you have received? Round to the nearest hundredth. (*Source:* Barclays Bank)   1493.99 euros

82. An environmental technician finds that warm water from an industrial process is being discharged into a nearby pond at a rate of 30 gallons per minute. Plant regulations state that the flow rate should be no more than 0.1 cubic foot per second. Is the flow rate of 30 gallons per minute in violation of the plant regulations? (*Hint:* 1 cubic foot is equivalent to 7.48 gallons.)   no

Adding and Subtracting Rational Expressions with the Same Denominator and Least Common Denominator

Objectives

A Add and Subtract Rational Expressions with Common Denominators.

B Find the Least Common Denominator of a List of Rational Expressions.

C Write a Rational Expression as an Equivalent Expression Whose Denominator Is Given.

Teaching Tip

Before beginning this lesson, consider asking students to explain why the denominators must be the same when adding and subtracting fractions. Be sure they understand that the denominator gives the unit to which the numerator is referring. If all fractions are referring to the same unit, the expressions can be added (subtracted) by adding (subtracting) the numerators and writing the sum (difference) over the common denominator. You may want to illustrate this with a diagram.

Practice 1

Add: $\dfrac{8x}{3y} + \dfrac{x}{3y}$

Teaching Tip

Before doing Example 1, ask students to simplify a particular form of this expression. For instance, have them use $m = 4$ and $n = 5$. After they have completed Example 1, have them compare their answers with the simplified form of the sum.

Practice 2

Subtract: $\dfrac{3x}{3x - 7} - \dfrac{7}{3x - 7}$

Answers

1. $\dfrac{3x}{y}$ 2. 1

502

Objective A Adding and Subtracting Rational Expressions with the Same Denominator

Like multiplication and division, addition and subtraction of rational expressions are similar to addition and subtraction of rational numbers. In this section, we add and subtract rational expressions with a common denominator.

Add: $\dfrac{6}{5} + \dfrac{2}{5}$ | Add: $\dfrac{9}{x + 2} + \dfrac{3}{x + 2}$

Add the numerators and place the sum over the common denominator.

$$\dfrac{6}{5} + \dfrac{2}{5} = \dfrac{6 + 2}{5}$$ | $$\dfrac{9}{x + 2} + \dfrac{3}{x + 2} = \dfrac{9 + 3}{x + 2}$$

$$= \dfrac{8}{5} \quad \text{Simplify.}$$ | $$= \dfrac{12}{x + 2} \quad \text{Simplify.}$$

Adding and Subtracting Rational Expressions with Common Denominators

If $\dfrac{P}{R}$ and $\dfrac{Q}{R}$ are rational expressions, then

$$\dfrac{P}{R} + \dfrac{Q}{R} = \dfrac{P + Q}{R} \quad \text{and} \quad \dfrac{P}{R} - \dfrac{Q}{R} = \dfrac{P - Q}{R}$$

To add or subtract rational expressions, add or subtract numerators and place the sum or difference over the common denominator.

Example 1 Add: $\dfrac{5m}{2n} + \dfrac{m}{2n}$

Solution:

$$\dfrac{5m}{2n} + \dfrac{m}{2n} = \dfrac{5m + m}{2n} \qquad \text{Add the numerators.}$$

$$= \dfrac{6m}{2n} \qquad \text{Simplify the numerator by combining like terms.}$$

$$= \dfrac{3m}{n} \qquad \text{Simplify by applying the fundamental principle.}$$

■ **Work Practice 1**

Example 2 Subtract: $\dfrac{2y}{2y - 7} - \dfrac{7}{2y - 7}$

Solution:

$$\dfrac{2y}{2y - 7} - \dfrac{7}{2y - 7} = \dfrac{2y - 7}{2y - 7} \qquad \text{Subtract the numerators.}$$

$$= \dfrac{1}{1} \text{ or } 1 \qquad \text{Simplify.}$$

■ **Work Practice 2**

Example 3 Subtract: $\dfrac{3x^2 + 2x}{x - 1} - \dfrac{10x - 5}{x - 1}$

Solution:

$$\dfrac{3x^2 + 2x}{x - 1} - \dfrac{10x - 5}{x - 1} = \dfrac{3x^2 + 2x - (10x - 5)}{x - 1}$$ Subtract the numerators.
Notice the parentheses.

$$= \dfrac{3x^2 + 2x - 10x + 5}{x - 1}$$ Use the distributive property.

$$= \dfrac{3x^2 - 8x + 5}{x - 1}$$ Combine like terms.

$$= \dfrac{(x - 1)\,(3x - 5)}{x - 1}$$ Factor.

$$= 3x - 5$$ Simplify.

🟧 **Work Practice 3**

Practice 3

Subtract: $\dfrac{2x^2 + 5x}{x + 2} - \dfrac{4x + 6}{x + 2}$

Helpful Hint

Notice how the numerator $10x - 5$ was subtracted in Example 3.

This $-$ sign applies to the   So parentheses are inserted
entire numerator $10x - 5$.   here to indicate this.

$$\dfrac{3x^2 + 2x}{x - 1} - \dfrac{10x - 5}{x - 1} = \dfrac{3x^2 + 2x - (10x - 5)}{x - 1}$$

Objective B Finding the Least Common Denominator ▶

Recall from Section R.2 that to add and subtract fractions with different denominators, we first find the least common denominator (LCD). Then we write all fractions as equivalent fractions with the LCD.

For example, suppose we want to add $\dfrac{3}{8}$ and $\dfrac{1}{6}$. To find the LCD of the denominators, factor 8 and 6. Remember, the LCD is the same as the least common multiple, LCM. It is the smallest number that is a multiple of 6 and also 8.

$$8 = 2 \cdot 2 \cdot 2$$
$$6 = 2 \cdot 3$$

The LCM is a multiple of 6.

$$\text{LCM} = 2 \cdot 2 \cdot 2 \cdot 3 = 24$$

The LCM is a multiple of 8.

In the next section, we will find the sum $\dfrac{3}{8} + \dfrac{1}{6}$, but for now, let's concentrate on the LCD.

To add or subtract rational expressions with different denominators, we also first find the LCD and then write all rational expressions as equivalent expressions with the LCD. The **least common denominator (LCD) of a list of rational expressions** is a polynomial of least degree whose factors include all the factors of the denominators in the list.

To Find the Least Common Denominator (LCD)

Step 1: Factor each denominator completely.

Step 2: The least common denominator (LCD) is the product of all unique factors found in Step 1, each raised to a power equal to the greatest number of times that the factor appears in any one factored denominator.

Answer

3. $2x - 3$

Practice 4

Find the LCD for each pair.

a. $\dfrac{2}{9}, \dfrac{7}{15}$

b. $\dfrac{5}{6x^3}, \dfrac{11}{8x^5}$

Example 4 Find the LCD for each pair.

a. $\dfrac{1}{8}, \dfrac{3}{22}$

b. $\dfrac{7}{5x}, \dfrac{6}{15x^2}$

Solution:

a. We start by finding the prime factorization of each denominator.

$$8 = 2^3 \quad \text{and}$$
$$22 = 2 \cdot 11$$

Next we write the product of all the unique factors, each raised to a power equal to the greatest number of times that the factor appears.

The greatest number of times that the factor 2 appears is 3.
The greatest number of times that the factor 11 appears is 1.

$$\text{LCD} = 2^3 \cdot 11^1 = 8 \cdot 11 = 88$$

b. We factor each denominator.

$$5x = 5 \cdot x \quad \text{and}$$
$$15x^2 = 3 \cdot 5 \cdot x^2$$

The greatest number of times that the factor 5 appears is 1.
The greatest number of times that the factor 3 appears is 1.
The greatest number of times that the factor x appears is 2.

$$\text{LCD} = 3^1 \cdot 5^1 \cdot x^2 = 15x^2$$

■ **Work Practice 4**

Practice 5

Find the LCD of $\dfrac{3a}{a+5}$ and $\dfrac{7a}{a-5}$.

Example 5 Find the LCD of $\dfrac{7x}{x+2}$ and $\dfrac{5x^2}{x-2}$.

Solution: The denominators $x + 2$ and $x - 2$ are completely factored already. The factor $x + 2$ appears once and the factor $x - 2$ appears once.

$$\text{LCD} = (x+2)(x-2)$$

■ **Work Practice 5**

Practice 6

Find the LCD of $\dfrac{7x^2}{(x-4)^2}$ and $\dfrac{5x}{3x-12}$.

Example 6 Find the LCD of $\dfrac{6m^2}{3m+15}$ and $\dfrac{2}{(m+5)^2}$.

Solution: We factor each denominator.

$$3m + 15 = 3(m+5)$$
$$(m+5)^2 = (m+5)^2 \quad \text{This denominator is already factored.}$$

The greatest number of times that the factor 3 appears is 1.
The greatest number of times that the factor $m + 5$ appears *in any one denominator* is 2.

$$\text{LCD} = 3(m+5)^2$$

■ **Work Practice 6**

Answers

4. a. 45 b. $24x^5$

5. $(a+5)(a-5)$ 6. $3(x-4)^2$

✓ **Concept Check Answer**

b

✓**Concept Check** Choose the correct LCD of $\dfrac{x}{(x+1)^2}$ and $\dfrac{5}{x+1}$.

a. $x + 1$ b. $(x+1)^2$ c. $(x+1)^3$ d. $5x(x+1)^2$

Example 7 Find the LCD of $\dfrac{t-10}{2t^2+t-6}$ and $\dfrac{t+5}{t^2+3t+2}$.

Solution:

$$2t^2+t-6 = (2t-3)(t+2)$$
$$t^2+3t+2 = (t+1)(t+2)$$
$$\text{LCD} = (2t-3)(t+2)(t+1)$$

■ **Work Practice 7**

Example 8 Find the LCD of $\dfrac{2}{x-2}$ and $\dfrac{10}{2-x}$.

Solution: The denominators $x-2$ and $2-x$ are opposites. That is, $2-x = -1(x-2)$. We can use either $x-2$ or $2-x$ as the LCD.

$$\text{LCD} = x-2 \quad \text{or} \quad \text{LCD} = 2-x$$

■ **Work Practice 8**

Objective C Writing Equivalent Rational Expressions

Next we practice writing a rational expression as an equivalent rational expression with a given denominator. To do this, we multiply by a form of 1. Recall that multiplying an expression by 1 produces an equivalent expression. In other words,

$$\frac{P}{Q} = \frac{P}{Q} \cdot 1 = \frac{P}{Q} \cdot \frac{R}{R} = \frac{PR}{QR}$$

Example 9 Write each rational expression as an equivalent rational expression with the given denominator.

a. $\dfrac{4b}{9a} = \dfrac{}{27a^2b}$ **b.** $\dfrac{7x}{2x+5} = \dfrac{}{6x+15}$

Solution:

a. We can ask ourselves: "What do we multiply $9a$ by to get $27a^2b$?" The answer is $3ab$, since $9a(3ab) = 27a^2b$. So we multiply by 1 in the form of $\dfrac{3ab}{3ab}$.

$$\frac{4b}{9a} = \frac{4b}{9a} \cdot 1 = \frac{4b}{9a} \cdot \frac{3ab}{3ab}$$
$$= \frac{4b(3ab)}{9a(3ab)} = \frac{12ab^2}{27a^2b}$$

b. First, factor the denominator on the right.

$$\frac{7x}{2x+5} = \frac{}{3(2x+5)}$$

To obtain the denominator on the right from the denominator on the left, we multiply by 1 in the form of $\dfrac{3}{3}$.

$$\frac{7x}{2x+5} = \frac{7x}{2x+5} \cdot \frac{3}{3} = \frac{7x \cdot 3}{(2x+5) \cdot 3} = \frac{21x}{3(2x+5)}$$

■ **Work Practice 9**

Practice 7

Find the LCD of $\dfrac{y+5}{y^2+2y-3}$ and $\dfrac{y+4}{y^2-3y+2}$.

Practice 8

Find the LCD of $\dfrac{6}{x-4}$ and $\dfrac{9}{4-x}$.

Teaching Tip

Before discussing Example 8, have your students find the LCD of

$\dfrac{2}{15}$ and $\dfrac{10}{-15}$

$\dfrac{2}{17-2}$ and $\dfrac{10}{2-17}$

$\dfrac{2}{x-2}$ and $\dfrac{10}{2-x}$ (Example 8)

Practice 9

Write the rational expression as an equivalent rational expression with the given denominator.

$$\frac{2x}{5y} = \frac{}{20x^2y^2}$$

Answers

7. $(y+3)(y-1)(y-2)$

8. $x-4$ or $4-x$

9. $\dfrac{8x^3y}{20x^2y^2}$

Practice 10

Write the rational expression as an equivalent rational expression with the given denominator.

$$\frac{3}{x^2 - 25} = \frac{}{(x+5)(x-5)(x-3)}$$

Answer

10. $\dfrac{3x - 9}{(x+5)(x-5)(x-3)}$

Example 10 Write the rational expression as an equivalent rational expression with the given denominator.

$$\frac{5}{x^2 - 4} = \frac{}{(x-2)(x+2)(x-4)}$$

Solution: First we factor the denominator $x^2 - 4$ as $(x-2)(x+2)$. If we multiply the original denominator $(x-2)(x+2)$ by $x-4$, the result is the new denominator $(x-2)(x+2)(x-4)$. Thus, we multiply by 1 in the form of $\dfrac{x-4}{x-4}$.

$$\frac{5}{x^2 - 4} = \frac{5}{(x-2)(x+2)} = \frac{5}{(x-2)(x+2)} \cdot \frac{x-4}{x-4}$$

$$= \frac{5(x-4)}{(x-2)(x+2)(x-4)}$$

$$= \frac{5x - 20}{(x-2)(x+2)(x-4)}$$

■ **Work Practice 10**

Vocabulary, Readiness & Video Check

Use the choices below to fill in each blank. Not all choices will be used.

$$\frac{9}{22} \qquad \frac{5}{22} \qquad \frac{9}{11} \qquad \frac{5}{11} \qquad \frac{ac}{b} \qquad \frac{a-c}{b} \qquad \frac{a+c}{b} \qquad \frac{5-6+x}{x} \qquad \frac{5-(6+x)}{x}$$

1. $\dfrac{7}{11} + \dfrac{2}{11} = \underline{\dfrac{9}{11}}$

2. $\dfrac{7}{11} - \dfrac{2}{11} = \underline{\dfrac{5}{11}}$

3. $\dfrac{a}{b} + \dfrac{c}{b} = \underline{\dfrac{a+c}{b}}$

4. $\dfrac{a}{b} - \dfrac{c}{b} = \underline{\dfrac{a-c}{b}}$

5. $\dfrac{5}{x} - \dfrac{6+x}{x} = \underline{\dfrac{5-(6+x)}{x}}$

Martin-Gay Interactive Videos

See Video 7.3

See video answer section.

Watch the section lecture video and answer the following questions.

Objective A 6. In ▱ Example 3, why is it important to place parentheses around the second numerator when writing as one expression? ▶

Objective B 7. In ▱ Examples 4 and 5, we factor the denominators completely. How does this help determine the LCD? ▶

Objective C 8. Based on ▱ Example 6, complete the following statements: To write an equivalent rational expression, we multiply the _____ of a rational expression by the same expression as the denominator. This means we're multiplying the original rational expression by a factor of _____ and therefore not changing the _____ of the original expression. ▶

7.3 Exercise Set MyMathLab®

Objective A *Add or subtract as indicated. Simplify the result if possible. See Examples 1 through 3.*

1. $\dfrac{a}{13} + \dfrac{9}{13}$ $\dfrac{a+9}{13}$

2. $\dfrac{x+1}{7} + \dfrac{6}{7}$ $\dfrac{x+7}{7}$

3. $\dfrac{4m}{3n} + \dfrac{5m}{3n}$ $\dfrac{3m}{n}$

4. $\dfrac{3p}{2q} + \dfrac{11p}{2q}$ $\dfrac{7p}{q}$

5. $\dfrac{4m}{m-6} - \dfrac{24}{m-6}$ 4

6. $\dfrac{8y}{y-2} - \dfrac{16}{y-2}$ 8

7. $\dfrac{9}{3+y} + \dfrac{y+1}{3+y}$ $\dfrac{y+10}{3+y}$

8. $\dfrac{9}{y+9} + \dfrac{y-5}{y+9}$ $\dfrac{y+4}{y+9}$

9. $\dfrac{5x^2+4x}{x-1} - \dfrac{6x+3}{x-1}$ $5x+3$

10. $\dfrac{x^2+9x}{x+7} - \dfrac{4x+14}{x+7}$ $x-2$

11. $\dfrac{4a}{a^2+2a-15} - \dfrac{12}{a^2+2a-15}$ $\dfrac{4}{a+5}$

12. $\dfrac{3y}{y^2+3y-10} - \dfrac{6}{y^2+3y-10}$ $\dfrac{3}{y+5}$

13. $\dfrac{2x+3}{x^2-x-30} - \dfrac{x-2}{x^2-x-30}$ $\dfrac{1}{x-6}$

14. $\dfrac{3x-1}{x^2+5x-6} - \dfrac{2x-7}{x^2+5x-6}$ $\dfrac{1}{x-1}$

15. $\dfrac{2x+1}{x-3} + \dfrac{3x+6}{x-3}$ $\dfrac{5x+7}{x-3}$

16. $\dfrac{4p-3}{2p+7} + \dfrac{3p+8}{2p+7}$ $\dfrac{7p+5}{2p+7}$

17. $\dfrac{2x^2}{x-5} - \dfrac{25+x^2}{x-5}$ $x+5$

18. $\dfrac{6x^2}{2x-5} - \dfrac{25+2x^2}{2x-5}$ $2x+5$

19. $\dfrac{5x+4}{x-1} - \dfrac{2x+7}{x-1}$ 3

20. $\dfrac{7x+1}{x-4} - \dfrac{2x+21}{x-4}$ 5

Objective B *Find the LCD for each list of rational expressions. See Examples 4 through 8.*

21. $\dfrac{19}{2x}, \dfrac{5}{4x^3}$ $4x^3$

22. $\dfrac{17x}{4y^5}, \dfrac{2}{8y}$ $8y^5$

23. $\dfrac{9}{8x}, \dfrac{3}{2x+4}$ $8x(x+2)$

24. $\dfrac{1}{6y}, \dfrac{3x}{4y+12}$ $12y(y+3)$

25. $\dfrac{2}{x+3}, \dfrac{5}{x-2}$ $(x+3)(x-2)$

26. $\dfrac{-6}{x-1}, \dfrac{4}{x+5}$ $(x-1)(x+5)$

27. $\dfrac{x}{x+6}, \dfrac{10}{3x+18}$ $3(x+6)$

28. $\dfrac{12}{x+5}, \dfrac{x}{4x+20}$ $4(x+5)$

29. $\dfrac{8x^2}{(x-6)^2}, \dfrac{13x}{5x-30}$ $5(x-6)^2$

30. $\dfrac{9x^2}{7x-14}, \dfrac{6x}{(x-2)^2}$ $7(x-2)^2$

31. $\dfrac{1}{3x+3}, \dfrac{8}{2x^2+4x+2}$ $6(x+1)^2$

32. $\dfrac{19x+5}{4x-12}, \dfrac{3}{2x^2-12x+18}$ $4(x-3)^2$

33. $\dfrac{5}{x-8}, \dfrac{3}{8-x}$ $x-8$ or $8-x$

34. $\dfrac{2x+5}{3x-7}, \dfrac{5}{7-3x}$ $3x-7$ or $7-3x$

35. $\dfrac{5x+1}{x^2+3x-4}, \dfrac{3x}{x^2+2x-3}$ $(x-1)(x+4)(x+3)$

36. $\dfrac{4}{x^2+4x+3}, \dfrac{4x-2}{x^2+10x+21}$ $(x+3)(x+1)(x+7)$

37. $\dfrac{2x}{3x^2 + 4x + 1}, \dfrac{7}{2x^2 - x - 1}$
$(3x + 1)(x + 1)(x - 1)(2x + 1)$

38. $\dfrac{3x}{4x^2 + 5x + 1}, \dfrac{5}{3x^2 - 2x - 1}$
$(4x + 1)(x + 1)(x - 1)(3x + 1)$

39. $\dfrac{1}{x^2 - 16}, \dfrac{x + 6}{2x^3 - 8x^2}$ $2x^2(x + 4)(x - 4)$

40. $\dfrac{5}{x^2 - 25}, \dfrac{x + 9}{3x^3 - 15x^2}$ $3x^2(x + 5)(x - 5)$

Objective C *Rewrite each rational expression as an equivalent rational expression with the given denominator.*
See Examples 9 and 10.

41. $\dfrac{3}{2x} = \dfrac{6x}{4x^2}$

42. $\dfrac{3}{9y^5} = \dfrac{24y^4}{72y^9}$

▶ **43.** $\dfrac{6}{3a} = \dfrac{24b^2}{12ab^2}$

44. $\dfrac{5}{4y^2x} = \dfrac{40yx}{32y^3x^2}$

45. $\dfrac{9}{2x + 6} = \dfrac{9y}{2y(x + 3)}$

46. $\dfrac{4x + 1}{3x + 6} = \dfrac{4xy + y}{3y(x + 2)}$

▶ **47.** $\dfrac{9a + 2}{5a + 10} = \dfrac{9ab + 2b}{5b(a + 2)}$

48. $\dfrac{5 + y}{2x^2 + 10} = \dfrac{10 + 2y}{4(x^2 + 5)}$

49. $\dfrac{x}{x^3 + 6x^2 + 8x} = \dfrac{x^2 + x}{x(x + 4)(x + 2)(x + 1)}$

50. $\dfrac{5x}{x^3 + 2x^2 - 3x} = \dfrac{5x^2 - 25x}{x(x - 1)(x - 5)(x + 3)}$

51. $\dfrac{9y - 1}{15x^2 - 30} = \dfrac{18y - 2}{30x^2 - 60}$

52. $\dfrac{6m - 5}{3x^2 - 9} = \dfrac{24m - 20}{12x^2 - 36}$

Mixed Practice (*Sections 7.2, 7.3*) *Perform the indicated operations.*

53. $\dfrac{5x}{7} + \dfrac{9x}{7}$
$2x$

54. $\dfrac{5x}{7} \cdot \dfrac{9x}{7}$
$\dfrac{45x^2}{49}$

55. $\dfrac{x + 3}{4} \div \dfrac{2x - 1}{4}$
$\dfrac{x + 3}{2x - 1}$

56. $\dfrac{x + 3}{4} - \dfrac{2x - 1}{4}$
$\dfrac{-x + 4}{4}$

57. $\dfrac{x^2}{x - 6} - \dfrac{5x + 6}{x - 6}$
$x + 1$

58. $\dfrac{-2x}{x^3 - 8x} + \dfrac{3x}{x^3 - 8x}$
$\dfrac{1}{x^2 - 8}$

59. $\dfrac{x^2 + 5x}{x^2 - 25} \cdot \dfrac{3x - 15}{x^2} \cdot \dfrac{3}{x}$

60. $\dfrac{-2x}{x^3 - 8x} \div \dfrac{3x}{x^3 - 8x}$
$-\dfrac{2}{3}$

61. $\dfrac{x^3 + 7x^2}{3x^3 - x^2} \div \dfrac{5x^2 + 36x + 7}{9x^2 - 1}$ $\dfrac{3x + 1}{5x + 1}$

62. $\dfrac{12x - 6}{x^2 + 3x} \cdot \dfrac{4x^2 + 13x + 3}{4x^2 - 1}$ $\dfrac{6(4x + 1)}{x(2x + 1)}$

Review

Perform each indicated operation. See Section R.2.

63. $\dfrac{2}{3} + \dfrac{5}{7}$ $\dfrac{29}{21}$

64. $\dfrac{9}{10} - \dfrac{3}{5}$ $\dfrac{3}{10}$

65. $\dfrac{2}{6} - \dfrac{3}{4}$ $-\dfrac{5}{12}$

66. $\dfrac{11}{15} + \dfrac{5}{9}$ $\dfrac{58}{45}$

67. $\dfrac{1}{12} + \dfrac{3}{20}$ $\dfrac{7}{30}$

68. $\dfrac{7}{30} + \dfrac{3}{18}$ $\dfrac{2}{5}$

Concept Extensions

For Exercises 69 and 70, see the Concept Check in this section.

69. Choose the correct LCD of $\dfrac{11a^3}{4a - 20}$ and $\dfrac{15a^3}{(a - 5)^2}$.

    a. $4a(a - 5)(a + 5)$ **b.** $a - 5$
    c. $(a - 5)^2$ **d.** $4(a - 5)^2$
    e. $(4a - 20)(a - 5)^2$ d

70. Choose the correct LCD of $\dfrac{5}{14x^2}$ and $\dfrac{y}{6x^3}$.

    a. $84x^5$ **b.** $84x^3$
    c. $42x^3$ **d.** $42x^5$ c

For Exercises 71 and 72, an algebra student approaches you with each incorrect solution. Find the error and correct the work shown below.

71. $\dfrac{2x-6}{x-5} - \dfrac{x+4}{x-5}$

$= \dfrac{2x-6-x+4}{x-5}$

$= \dfrac{x-2}{x-5}$ answers may vary

72. $\dfrac{x}{x+3} + \dfrac{2}{x+3}$

$= \dfrac{x+2}{x+3}$

$= \dfrac{2}{3}$ answers may vary

△ **73.** A square has a side of length $\dfrac{5}{x-2}$ meters. Express its perimeter as a rational expression. $\dfrac{20}{x-2}$ m

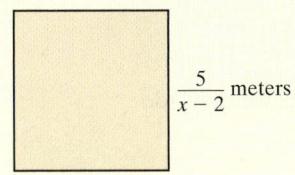

$\dfrac{5}{x-2}$ meters

△ **74.** A trapezoid has sides of the indicated lengths. Find its perimeter. $\dfrac{2x+15}{x+3}$ in.

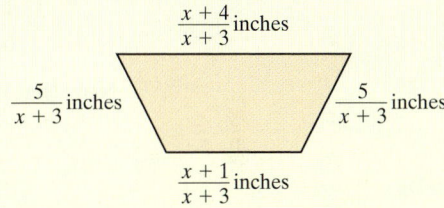

$\dfrac{x+4}{x+3}$ inches

$\dfrac{5}{x+3}$ inches $\dfrac{5}{x+3}$ inches

$\dfrac{x+1}{x+3}$ inches

75. Write two rational expressions with the same denominator whose sum is $\dfrac{5}{3x-1}$.
answers may vary

76. Write two rational expressions with the same denominator whose difference is $\dfrac{x-7}{x^2+1}$.
answers may vary

77. The planet Mercury revolves around the Sun in 88 Earth days. It takes Jupiter 4332 Earth days to make one revolution around the Sun. (*Source:* National Space Science Data Center) If the two planets are aligned as shown in the figure, how long will it take for them to align again? 95,304 Earth days

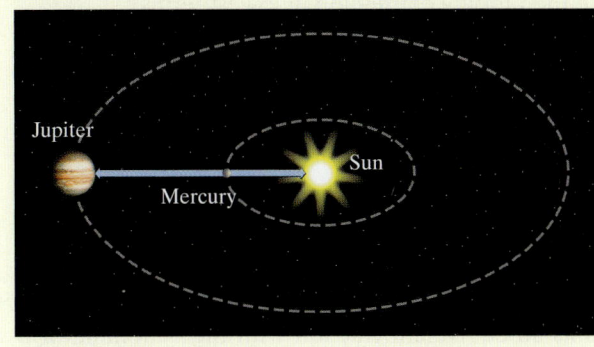

78. You are throwing a barbecue and you want to make sure that you purchase the same number of hot dogs as hot dog buns. Hot dogs come 8 to a package and hot dog buns come 12 to a package. What is the least number of each type of package you should buy? 3 packages hot dogs and 2 packages buns

79. Write some instructions to help a friend who is having difficulty finding the LCD of two rational expressions. answers may vary

80. In your own words, describe how to add or subtract two rational expressions with the same denominator. answers may vary

81. Explain why the LCD of the rational expressions $\dfrac{7}{x+1}$ and $\dfrac{9x}{(x+1)^2}$ is $(x+1)^2$ and not $(x+1)^3$. answers may vary

82. Explain the similarities between subtracting $\dfrac{3}{8}$ from $\dfrac{7}{8}$ and subtracting $\dfrac{6}{x+3}$ from $\dfrac{9}{x+3}$. answers may vary

Objective

A Add and Subtract Rational Expressions with Different Denominators.

Objective A Adding and Subtracting Rational Expressions with Different Denominators

Let's add $\frac{3}{8}$ and $\frac{1}{6}$. In the previous section, we found the LCD of 8 and 6 to be 24. Now let's write equivalent fractions with denominator 24 by multiplying by different forms of 1.

$$\frac{3}{8} = \frac{3}{8} \cdot 1 = \frac{3}{8} \cdot \frac{3}{3} = \frac{3 \cdot 3}{8 \cdot 3} = \frac{9}{24}$$

$$\frac{1}{6} = \frac{1}{6} \cdot 1 = \frac{1}{6} \cdot \frac{4}{4} = \frac{1 \cdot 4}{6 \cdot 4} = \frac{4}{24}$$

Now that the denominators are the same, we may add.

$$\frac{3}{8} + \frac{1}{6} = \frac{9}{24} + \frac{4}{24} = \frac{9 + 4}{24} = \frac{13}{24}$$

We add or subtract rational expressions the same way. You may want to use the steps below.

> **To Add or Subtract Rational Expressions with Different Denominators**
>
> **Step 1:** Find the LCD of the rational expressions.
>
> **Step 2:** Rewrite each rational expression as an equivalent expression whose denominator is the LCD found in Step 1.
>
> **Step 3:** Add or subtract numerators and write the sum or difference over the common denominator.
>
> **Step 4:** Simplify or write the rational expression in lowest terms.

Practice 1

Perform each indicated operation.

a. $\frac{y}{5} - \frac{3y}{15}$ **b.** $\frac{5}{8x} + \frac{11}{10x^2}$

Example 1 Perform each indicated operation.

a. $\frac{a}{4} - \frac{2a}{8}$

b. $\frac{3}{10x^2} + \frac{7}{25x}$

Solution:

a. First, we must find the LCD. Since $4 = 2^2$ and $8 = 2^3$, the LCD $= 2^3 = 8$. Next we write each fraction as an equivalent fraction with the denominator 8, and then we subtract.

$$\frac{a}{4} = \frac{a}{4} \cdot 1 = \frac{a}{4} \cdot \frac{2}{2} = \frac{a \cdot 2}{4 \cdot 2} = \frac{2a}{8}$$

$$\frac{a}{4} - \frac{2a}{8} = \frac{2a}{8} - \frac{2a}{8} = \frac{2a - 2a}{8} = \frac{0}{8} = 0$$

Notice that we wrote $\frac{a}{4}$ as the equivalent expression $\frac{2a}{8}$. Multiplying by a form of 1 means we multiply the numerator and the denominator by the same number. Since this is so, we will start using the shorthand notation on the next page.

$$\frac{a}{4} = \frac{a(2)}{4(2)} = \frac{2a}{8}$$

Multiplying the numerator and denominator by 2 is the same as multiplying by $\frac{2}{2}$ or 1.

b. Since $10x^2 = 2 \cdot 5 \cdot x \cdot x$ and $25x = 5 \cdot 5 \cdot x$, the LCD $= 2 \cdot 5^2 \cdot x^2 = 50x^2$. We write each fraction as an equivalent fraction with a denominator of $50x^2$.

$$\frac{3}{10x^2} + \frac{7}{25x} = \frac{3(5)}{10x^2(5)} + \frac{7(2x)}{25x(2x)}$$

$$= \frac{15}{50x^2} + \frac{14x}{50x^2}$$

$$= \frac{15 + 14x}{50x^2} \qquad \text{Add numerators. Write the sum over the common denominator.}$$

■ **Work Practice 1**

Example 2 Subtract: $\dfrac{6x}{x^2 - 4} - \dfrac{3}{x + 2}$

Solution: Since $x^2 - 4 = (x + 2)(x - 2)$, the LCD $= (x + 2)(x - 2)$. We write equivalent expressions with the LCD as denominators.

$$\frac{6x}{x^2 - 4} - \frac{3}{x + 2} = \frac{6x}{(x + 2)(x - 2)} - \frac{3(x - 2)}{(x + 2)(x - 2)}$$

$$= \frac{6x - 3(x - 2)}{(x + 2)(x - 2)} \qquad \text{Subtract numerators. Write the difference over the common denominator.}$$

$$= \frac{6x - 3x + 6}{(x + 2)(x - 2)} \qquad \text{Apply the distributive property in the numerator.}$$

$$= \frac{3x + 6}{(x + 2)(x - 2)} \qquad \text{Combine like terms in the numerator.}$$

Next we factor the numerator to see if this rational expression can be simplified.

$$\frac{3x + 6}{(x + 2)(x - 2)} = \frac{3\,(x + 2)}{(x + 2)\,(x - 2)} \qquad \text{Factor.}$$

$$= \frac{3}{x - 2} \qquad \text{Apply the fundamental principle to simplify.}$$

■ **Work Practice 2**

Example 3 Add: $\dfrac{2}{3t} + \dfrac{5}{t + 1}$

Solution: The LCD is $3t(t + 1)$. We write each rational expression as an equivalent rational expression with a denominator of $3t(t + 1)$.

$$\frac{2}{3t} + \frac{5}{t + 1} = \frac{2(t + 1)}{3t(t + 1)} + \frac{5(3t)}{(t + 1)(3t)}$$

$$= \frac{2(t + 1) + 5(3t)}{3t(t + 1)} \qquad \text{Add numerators. Write the sum over the common denominator.}$$

$$= \frac{2t + 2 + 15t}{3t(t + 1)} \qquad \text{Apply the distributive property in the numerator.}$$

$$= \frac{17t + 2}{3t(t + 1)} \qquad \text{Combine like terms in the numerator.}$$

■ **Work Practice 3**

Practice 2

Subtract: $\dfrac{10x}{x^2 - 9} - \dfrac{5}{x + 3}$

Practice 3

Add: $\dfrac{5}{7x} + \dfrac{2}{x + 1}$

Answers

2. $\dfrac{5}{x - 3}$ **3.** $\dfrac{19x + 5}{7x(x + 1)}$

Practice 4

Subtract: $\dfrac{10}{x-6} - \dfrac{15}{6-x}$

Teaching Tip

In Example 4, to verify for students that $x-3$ and $3-x$ are opposites, replace x with several values and have students notice the results. To verify that $3-x$ and $-(x-3)$ are equal, have students replace x with several values and notice the results.

Example 4 Subtract: $\dfrac{7}{x-3} - \dfrac{9}{3-x}$

Solution: To find a common denominator, we notice that $x-3$ and $3-x$ are opposites. That is, $3-x = -(x-3)$. We write the denominator $3-x$ as $-(x-3)$ and simplify.

$$\dfrac{7}{x-3} - \dfrac{9}{3-x} = \dfrac{7}{x-3} - \dfrac{9}{-(x-3)}$$

$$= \dfrac{7}{x-3} - \dfrac{-9}{x-3} \qquad \text{Apply } \dfrac{a}{-b} = \dfrac{-a}{b}.$$

$$= \dfrac{7-(-9)}{x-3} \qquad \text{Subtract numerators. Write the difference over the common denominator.}$$

$$= \dfrac{16}{x-3}$$

■ **Work Practice 4**

Practice 5

Add: $2 + \dfrac{x}{x+5}$

Example 5 Add: $1 + \dfrac{m}{m+1}$

Solution: Recall that 1 is the same as $\dfrac{1}{1}$. The LCD of $\dfrac{1}{1}$ and $\dfrac{m}{m+1}$ is $m+1$.

$$1 + \dfrac{m}{m+1} = \dfrac{1}{1} + \dfrac{m}{m+1} \qquad \text{Write 1 as } \dfrac{1}{1}.$$

$$= \dfrac{1(m+1)}{1(m+1)} + \dfrac{m}{m+1} \qquad \text{Multiply both the numerator and the denominator of } \dfrac{1}{1} \text{ by } m+1.$$

$$= \dfrac{m+1+m}{m+1} \qquad \text{Add numerators. Write the sum over the common denominator.}$$

$$= \dfrac{2m+1}{m+1} \qquad \text{Combine like terms in the numerator.}$$

■ **Work Practice 5**

Practice 6

Subtract: $\dfrac{4}{3x^2+2x} - \dfrac{3x}{12x+8}$

Example 6 Subtract: $\dfrac{3}{2x^2+x} - \dfrac{2x}{6x+3}$

Solution: First, we factor the denominators.

$$\dfrac{3}{2x^2+x} - \dfrac{2x}{6x+3} = \dfrac{3}{x(2x+1)} - \dfrac{2x}{3(2x+1)}$$

The LCD is $3x(2x+1)$. We write equivalent expressions with denominator $3x(2x+1)$.

$$\dfrac{3}{x(2x+1)} - \dfrac{2x}{3(2x+1)} = \dfrac{3(3)}{x(2x+1)(3)} - \dfrac{2x(x)}{3(2x+1)(x)}$$

$$= \dfrac{9-2x^2}{3x(2x+1)} \qquad \text{Subtract numerators. Write the difference over the common denominator.}$$

■ **Work Practice 6**

Answers

4. $\dfrac{25}{x-6}$ **5.** $\dfrac{3x+10}{x+5}$ **6.** $\dfrac{16-3x^2}{4x(3x+2)}$

Example 7 Add: $\dfrac{2x}{x^2 + 2x + 1} + \dfrac{x}{x^2 - 1}$

Solution: First we factor the denominators.

$\dfrac{2x}{x^2 + 2x + 1} + \dfrac{x}{x^2 - 1}$

$= \dfrac{2x}{(x + 1)(x + 1)} + \dfrac{x}{(x + 1)(x - 1)}$ Rewrite each expression with LCD $(x + 1)(x + 1)(x - 1)$.

$= \dfrac{2x(x - 1)}{(x + 1)(x + 1)(x - 1)} + \dfrac{x(x + 1)}{(x + 1)(x - 1)(x + 1)}$

$= \dfrac{2x(x - 1) + x(x + 1)}{(x + 1)^2(x - 1)}$ Add numerators. Write the sum over the common denominator.

$= \dfrac{2x^2 - 2x + x^2 + x}{(x + 1)^2(x - 1)}$ Apply the distributive property in the numerator.

$= \dfrac{3x^2 - x}{(x + 1)^2(x - 1)}$ or $\dfrac{x(3x - 1)}{(x + 1)^2(x - 1)}$

The numerator was factored as a last step to see if the rational expression could be simplified further. Since there are no factors common to the numerator and the denominator, we can't simplify further.

■ **Work Practice 7**

Practice 7

Add: $\dfrac{6x}{x^2 + 4x + 4} + \dfrac{x}{x^2 - 4}$

Answer

7. $\dfrac{x(7x - 10)}{(x + 2)^2(x - 2)}$

Vocabulary, Readiness & Video Check

Multiple choice. Choose the correct response.

1. $\dfrac{3}{7x} + \dfrac{5}{7} =$

a. $\dfrac{3}{7x} + \dfrac{5}{7x} = \dfrac{8}{7x}$

b. $\dfrac{3}{7x} + \dfrac{5}{7} \cdot \dfrac{x}{x} = \dfrac{3 + 5x}{7x}$

c. $\dfrac{3}{7x} + \dfrac{5}{7} \cdot \dfrac{x}{x} = \dfrac{8x}{7x}$ or $\dfrac{8}{7}$ b

2. $\dfrac{1}{x} + \dfrac{2}{x^2} =$

a. $\dfrac{1}{x} \cdot \dfrac{x}{x} + \dfrac{2}{x^2} = \dfrac{x + 2}{x^2}$

b. $\dfrac{3}{x^3}$

c. $\dfrac{1}{x} \cdot \dfrac{x}{x} + \dfrac{2}{x^2} = \dfrac{3x}{x^2}$ or $\dfrac{3}{x}$ a

Martin-Gay Interactive Videos Watch the section lecture video and answer the following question.

Objective A 3. What special case is shown in ⊞ Example 2, and what's the purpose of presenting it? ▶

See Video 7.4 🍎

See video answer section.

7.4 Exercise Set MyMathLab®

Objective A *Perform each indicated operation. Simplify if possible. See Examples 1 through 7.*

1. $\dfrac{4}{2x} + \dfrac{9}{3x}$ $\dfrac{5}{x}$

2. $\dfrac{15}{7a} + \dfrac{8}{6a}$ $\dfrac{73}{21a}$

3. $\dfrac{15a}{b} + \dfrac{6b}{5}$ $\dfrac{75a + 6b^2}{5b}$

4. $\dfrac{4c}{d} - \dfrac{8d}{5}$ $\dfrac{20c - 8d^2}{5d}$

5. $\dfrac{3}{x} + \dfrac{5}{2x^2}$ $\dfrac{6x + 5}{2x^2}$

6. $\dfrac{14}{3x^2} + \dfrac{6}{x}$ $\dfrac{14 + 18x}{3x^2}$

7. $\dfrac{6}{x + 1} + \dfrac{10}{2x + 2}$ $\dfrac{11}{x + 1}$

8. $\dfrac{8}{x + 4} - \dfrac{3}{3x + 12}$ $\dfrac{7}{x + 4}$

9. $\dfrac{3}{x + 2} - \dfrac{2x}{x^2 - 4}$ $\dfrac{x - 6}{(x - 2)(x + 2)}$

10. $\dfrac{5}{x - 4} + \dfrac{4x}{x^2 - 16}$ $\dfrac{9x + 20}{(x - 4)(x + 4)}$

11. $\dfrac{3}{4x} + \dfrac{8}{x - 2}$ $\dfrac{35x - 6}{4x(x - 2)}$

12. $\dfrac{5}{y^2} - \dfrac{y}{2y + 1}$ $\dfrac{5 + 10y - y^3}{y^2(2y + 1)}$

13. $\dfrac{6}{x - 3} + \dfrac{8}{3 - x} - \dfrac{2}{x - 3}$

14. $\dfrac{15}{y - 4} + \dfrac{20}{4 - y} - \dfrac{5}{y - 4}$

15. $\dfrac{9}{x - 3} + \dfrac{9}{3 - x}$ 0

16. $\dfrac{5}{a - 7} + \dfrac{5}{7 - a}$ 0

17. $\dfrac{-8}{x^2 - 1} - \dfrac{7}{1 - x^2} - \dfrac{1}{x^2 - 1}$

18. $\dfrac{-9}{25x^2 - 1} + \dfrac{7}{1 - 25x^2} - \dfrac{16}{25x^2 - 1}$

19. $\dfrac{5}{x} + 2$ $\dfrac{5 + 2x}{x}$

20. $\dfrac{7}{x^2} - 5x$ $\dfrac{7 - 5x^3}{x^2}$

21. $\dfrac{5}{x - 2} + 6$ $\dfrac{6x - 7}{x - 2}$

22. $\dfrac{6y}{y + 5} + 1$ $\dfrac{7y + 5}{y + 5}$

23. $\dfrac{y + 2}{y + 3} - 2$ $-\dfrac{y + 4}{y + 3}$

24. $\dfrac{7}{2x - 3} - 3$ $\dfrac{2(8 - 3x)}{2x - 3}$

25. $\dfrac{-x + 2}{x} - \dfrac{x - 6}{4x}$ $\dfrac{-5x + 14}{4x}$ or $-\dfrac{5x - 14}{4x}$

26. $\dfrac{-y + 1}{y} - \dfrac{2y - 5}{3y}$ $\dfrac{-5y + 8}{3y}$ or $-\dfrac{5y - 8}{3y}$

27. $\dfrac{5x}{x + 2} - \dfrac{3x - 4}{x + 2}$ 2

28. $\dfrac{7x}{x - 3} - \dfrac{4x + 9}{x - 3}$ 3

29. $\dfrac{3x^4}{7} - \dfrac{4x^2}{21}$ $\dfrac{9x^4 - 4x^2}{21}$

30. $\dfrac{5x}{6} + \dfrac{11x^2}{2}$ $\dfrac{5x + 33x^2}{6}$

31. $\dfrac{1}{x + 3} - \dfrac{1}{(x + 3)^2}$ $\dfrac{x + 2}{(x + 3)^2}$

32. $\dfrac{5x}{(x - 2)^2} - \dfrac{3}{x - 2}$ $\dfrac{2(x + 3)}{(x - 2)^2}$

33. $\dfrac{4}{5b} + \dfrac{1}{b - 1}$ $\dfrac{9b - 4}{5b(b - 1)}$

34. $\dfrac{1}{y + 5} + \dfrac{2}{3y}$ $\dfrac{5(y + 2)}{3y(y + 5)}$

35. $\dfrac{2}{m} + 1$ $\dfrac{2 + m}{m}$

36. $\dfrac{6}{x} - 1$ $\dfrac{6 - x}{x}$

37. $\dfrac{2x}{x - 7} - \dfrac{x}{x - 2}$ $\dfrac{x(x + 3)}{(x - 7)(x - 2)}$

38. $\dfrac{9x}{x - 10} - \dfrac{x}{x - 3}$ $\dfrac{x(8x - 17)}{(x - 10)(x - 3)}$

39. $\dfrac{6}{1 - 2x} - \dfrac{4}{2x - 1}$ $\dfrac{10}{1 - 2x}$

40. $\dfrac{10}{3n - 4} - \dfrac{5}{4 - 3n}$ $\dfrac{15}{3n - 4}$

41. $\dfrac{7}{(x+1)(x-1)} + \dfrac{8}{(x+1)^2}$ $\dfrac{15x-1}{(x+1)^2(x-1)}$

42. $\dfrac{5}{(x+1)(x+5)} - \dfrac{2}{(x+5)^2}$ $\dfrac{3x+23}{(x+5)^2(x+1)}$

43. $\dfrac{x}{x^2-1} - \dfrac{2}{x^2-2x+1}$ $\dfrac{x^2-3x-2}{(x-1)^2(x+1)}$

44. $\dfrac{x}{x^2-4} - \dfrac{5}{x^2-4x+4}$ $\dfrac{x^2-7x-10}{(x+2)(x-2)^2}$

45. $\dfrac{3a}{2a+6} - \dfrac{a-1}{a+3}$ $\dfrac{a+2}{2(a+3)}$

46. $\dfrac{1}{2x+2y} - \dfrac{y}{x+y}$ $\dfrac{1-2y}{2x+2y}$

47. $\dfrac{y-1}{2y+3} + \dfrac{3}{(2y+3)^2}$ $\dfrac{y(2y+1)}{(2y+3)^2}$

48. $\dfrac{x-6}{5x+1} + \dfrac{6}{(5x+1)^2}$ $\dfrac{x(5x-29)}{(5x+1)^2}$

49. $\dfrac{5}{2-x} + \dfrac{x}{2x-4}$ $\dfrac{x-10}{2(x-2)}$

50. $\dfrac{-1}{a-2} + \dfrac{4}{4-2a}$ $\dfrac{3}{2-a}$

51. $\dfrac{15}{x^2+6x+9} + \dfrac{2}{x+3}$ $\dfrac{2x+21}{(x+3)^2}$

52. $\dfrac{2}{x^2+4x+4} + \dfrac{1}{x+2}$ $\dfrac{x+4}{(x+2)^2}$

53. $\dfrac{13}{x^2-5x+6} - \dfrac{5}{x-3}$ $\dfrac{-5x+23}{(x-2)(x-3)}$

54. $\dfrac{-7}{y^2-3y+2} - \dfrac{2}{y-1}$ $\dfrac{-3-2y}{(y-2)(y-1)}$

55. $\dfrac{70}{m^2-100} + \dfrac{7}{2(m+10)}$ $\dfrac{7}{2(m-10)}$

56. $\dfrac{27}{y^2-81} + \dfrac{3}{2(y+9)}$ $\dfrac{3}{2(y-9)}$

57. $\dfrac{x+8}{x^2-5x-6} + \dfrac{x+1}{x^2-4x-5}$ $\dfrac{2(x^2-x-23)}{(x+1)(x-6)(x-5)}$

58. $\dfrac{x+4}{x^2+12x+20} + \dfrac{x+1}{x^2+8x-20}$ $\dfrac{2x^2+5x-6}{(x+10)(x+2)(x-2)}$

59. $\dfrac{5}{4n^2-12n+8} - \dfrac{3}{3n^2-6n}$ $\dfrac{n+4}{4n(n-1)(n-2)}$

60. $\dfrac{6}{5y^2-25y+30} - \dfrac{2}{4y^2-8y}$ $\dfrac{7y+15}{10y(y-2)(y-3)}$

Mixed Practice (Sections 7.2, 7.3, 7.4) *Perform the indicated operations. Addition, subtraction, multiplication, and division of rational expressions are included here.*

61. $\dfrac{15x}{x+8} \cdot \dfrac{2x+16}{3x}$ 10

62. $\dfrac{9z+5}{15} \cdot \dfrac{5z}{81z^2-25}$ $\dfrac{z}{3(9z-5)}$

63. $\dfrac{8x+7}{3x+5} - \dfrac{2x-3}{3x+5}$ 2

64. $\dfrac{2z^2}{4z-1} - \dfrac{z-2z^2}{4z-1}$ z

65. $\dfrac{5a+10}{18} \div \dfrac{a^2-4}{10a}$ $\dfrac{25a}{9(a-2)}$

66. $\dfrac{9}{x^2-1} \div \dfrac{12}{3x+3}$ $\dfrac{9}{4(x-1)}$

67. $\dfrac{5}{x^2-3x+2} + \dfrac{1}{x-2}$ $\dfrac{x+4}{(x-2)(x-1)}$

68. $\dfrac{4}{2x^2+5x-3} + \dfrac{2}{x+3}$ $\dfrac{4x+2}{(2x-1)(x+3)}$

Review

Solve each linear or quadratic equation. See Sections 2.3 and 6.6.

69. $3x + 5 = 7$ $\dfrac{2}{3}$

70. $5x - 1 = 8$ $\dfrac{9}{5}$

71. $2x^2 - x - 1 = 0$ $-\dfrac{1}{2}, 1$

72. $4x^2 - 9 = 0$ $\dfrac{3}{2}, -\dfrac{3}{2}$

73. $4(x + 6) + 3 = -3$ $-\dfrac{15}{2}$

74. $2(3x + 1) + 15 = -7$ -4

Concept Extensions

Perform each indicated operation.

75. $\dfrac{3}{x} - \dfrac{2x}{x^2 - 1} + \dfrac{5}{x + 1}$ $\dfrac{6x^2 - 5x - 3}{x(x + 1)(x - 1)}$

76. $\dfrac{5}{x - 2} + \dfrac{7x}{x^2 - 4} - \dfrac{11}{x}$ $\dfrac{x^2 + 10x + 44}{x(x - 2)(x + 2)}$

77. $\dfrac{5}{x^2 - 4} + \dfrac{2}{x^2 - 4x + 4} - \dfrac{3}{x^2 - x - 6}$ $\dfrac{4x^2 - 15x + 6}{(x - 2)^2(x + 2)(x - 3)}$

78. $\dfrac{8}{x^2 + 6x + 5} - \dfrac{3x}{x^2 + 4x - 5} + \dfrac{2}{x^2 - 1}$ $\dfrac{-3x^2 + 7x + 2}{(x + 5)(x + 1)(x - 1)}$

79. $\dfrac{9}{x^2 + 9x + 14} - \dfrac{3x}{x^2 + 10x + 21} + \dfrac{x + 4}{x^2 + 5x + 6}$ $\dfrac{-2x^2 + 14x + 55}{(x + 2)(x + 7)(x + 3)}$

80. $\dfrac{x + 10}{x^2 - 3x - 4} - \dfrac{8}{x^2 + 6x + 5} - \dfrac{9}{x^2 + x - 20}$ $\dfrac{x^2 - 2x + 73}{(x - 4)(x + 1)(x + 5)}$

81. A board of length $\dfrac{3}{x + 4}$ inches was cut into two pieces. If one piece is $\dfrac{1}{x - 4}$ inches, express the length of the other piece as a rational expression. $\dfrac{2(x - 8)}{(x - 4)(x + 4)}$ in.

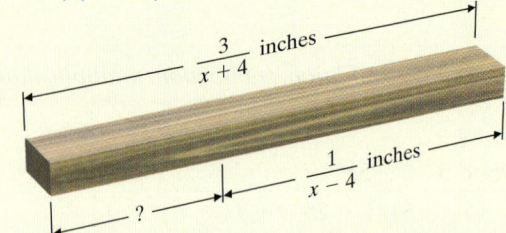

- $\dfrac{3}{x + 4}$ inches
- $\dfrac{1}{x - 4}$ inches
- ?

△ **82.** The length of a rectangle is $\dfrac{3}{y - 5}$ feet, while its width is $\dfrac{2}{y}$ feet. Find its perimeter and then find its area. $\dfrac{10(y - 2)}{y(y - 5)}$ ft; $\dfrac{6}{y^2 - 5y}$ sq ft

- $\dfrac{3}{y - 5}$ feet
- $\dfrac{2}{y}$ feet

83. In ice hockey, penalty killing percentage is a statistic calculated as $1 - \dfrac{G}{P}$, where $G =$ opponent's power play goals and $P =$ opponent's power play opportunities. Simplify this expression. $\dfrac{P - G}{P}$

84. The dose of medicine prescribed for a child depends on the child's age A in years and the adult dose D for the medication. Two expressions that give a child's dose are Young's Rule, $\dfrac{DA}{A + 12}$, and Cowling's Rule, $\dfrac{D(A + 1)}{24}$. Find an expression for the difference in the doses given by these expressions. $\dfrac{11DA - DA^2 - 12D}{24(A + 12)}$

85. Explain when the LCD of the rational expressions in a sum is the product of the denominators.
answers may vary

86. Explain when the LCD is the same as one of the denominators of a rational expression to be added or subtracted. *answers may vary*

△ **87.** Two angles are said to be complementary if the sum of their measures is 90°. If one angle measures $\frac{40}{x}$ degrees, find the measure of its complement. $\left(\frac{90x - 40}{x}\right)^{\circ}$

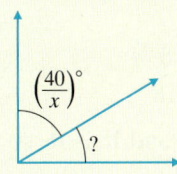

△ **88.** Two angles are said to be supplementary if the sum of their measures is 180°. If one angle measures $\frac{x + 2}{x}$ degrees, find the measure of its supplement. $\left(\frac{179x - 2}{x}\right)^{\circ}$

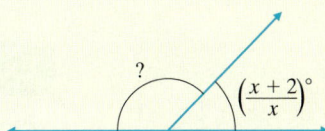

89. In your own words, explain how to add two rational expressions with different denominators. *answers may vary*

90. In your own words, explain how to subtract two rational expressions with different denominators. *answers may vary*

7.5 Solving Equations Containing Rational Expressions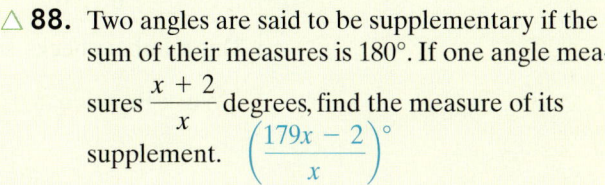

Objective A Solving Equations Containing Rational Expressions

Objectives

A Solve Equations Containing Rational Expressions.

B Solve Equations Containing Rational Expressions for a Specified Variable.

In Chapter 2, we solved equations containing fractions. In this section, we continue the work we began in that chapter by solving equations containing rational expressions. For example,

$$\frac{x}{2} + \frac{8}{3} = \frac{1}{6} \quad \text{and} \quad \frac{4x}{x^2 + x - 30} + \frac{2}{x - 5} = \frac{1}{x + 6}$$

are equations containing rational expressions. To solve equations such as these, we use the multiplication property of equality to clear the equation of fractions by multiplying both sides of the equation by the LCD.

Teaching Tip

You may want to begin this lesson with a review of solving linear equations such as $3x + 16 = 1$.

Example 1 Solve: $\frac{x}{2} + \frac{8}{3} = \frac{1}{6}$

Solution: The LCD of denominators 2, 3, and 6 is 6, so we multiply both sides of the equation by 6.

$$6\left(\frac{x}{2} + \frac{8}{3}\right) = 6\left(\frac{1}{6}\right)$$

$$6\left(\frac{x}{2}\right) + 6\left(\frac{8}{3}\right) = 6\left(\frac{1}{6}\right) \quad \text{Apply the distributive property.}$$

$$3 \cdot x + 16 = 1 \quad \text{Multiply and simplify.}$$

$$3x = -15 \quad \text{Subtract 16 from both sides.}$$

$$x = -5 \quad \text{Divide both sides by 3.}$$

(Continued on next page)

Practice 1

Solve: $\frac{x}{4} + \frac{4}{5} = \frac{1}{20}$

Helpful Hint Make sure that *each* term is multiplied by the LCD.

Answer
1. $x = -3$

Check: To check, we replace x with -5 in the original equation.

$$\frac{-5}{2} + \frac{8}{3} \stackrel{?}{=} \frac{1}{6} \quad \text{Replace } x \text{ with } -5.$$

$$\frac{1}{6} = \frac{1}{6} \quad \text{True}$$

This number checks, so the solution is -5.

■ **Work Practice 1**

Practice 2

Solve: $\dfrac{x + 2}{3} - \dfrac{x - 1}{5} = \dfrac{1}{15}$

Helpful Hint

Multiply *each* term by 18.

Example 2 Solve: $\dfrac{t - 4}{2} - \dfrac{t - 3}{9} = \dfrac{5}{18}$

Solution: The LCD of denominators 2, 9, and 18 is 18, so we multiply both sides of the equation by 18.

$$18\left(\frac{t - 4}{2} - \frac{t - 3}{9}\right) = 18\left(\frac{5}{18}\right)$$

$$18\left(\frac{t - 4}{2}\right) - 18\left(\frac{t - 3}{9}\right) = 18\left(\frac{5}{18}\right) \quad \text{Apply the distributive property.}$$

$$9(t - 4) - 2(t - 3) = 5 \quad \text{Simplify.}$$

$$9t - 36 - 2t + 6 = 5 \quad \text{Use the distributive property.}$$

$$7t - 30 = 5 \quad \text{Combine like terms.}$$

$$7t = 35$$

$$t = 5 \quad \text{Solve for } t.$$

Check: $\dfrac{t - 4}{2} - \dfrac{t - 3}{9} = \dfrac{5}{18}$

$$\frac{5 - 4}{2} - \frac{5 - 3}{9} \stackrel{?}{=} \frac{5}{18} \quad \text{Replace } t \text{ with 5.}$$

$$\frac{1}{2} - \frac{2}{9} \stackrel{?}{=} \frac{5}{18} \quad \text{Simplify.}$$

$$\frac{5}{18} = \frac{5}{18} \quad \text{True}$$

The solution is 5.

■ **Work Practice 2**

Recall from Section 7.1 that a rational expression is defined for all real numbers except those that make the denominator of the expression 0. This means that if an equation contains *rational expressions with variables in the denominator,* we must be certain that the proposed solution does not make the denominator 0. If replacing the variable with the proposed solution makes the denominator 0, the rational expression is undefined and this proposed solution must be rejected.

Answer

2. $x = -6$

Example 3 Solve: $3 - \dfrac{6}{x} = x + 8$

Helpful Hint Notice that Example 3 contains our first equation with a variable in the denominator.

Solution: In this equation, 0 cannot be a solution because if x is 0, the rational expression $\dfrac{6}{x}$ is undefined. The LCD is x, so we multiply both sides of the equation by x.

$$x\left(3 - \dfrac{6}{x}\right) = x(x + 8)$$

$$x(3) - x\left(\dfrac{6}{x}\right) = x \cdot x + x \cdot 8 \quad \text{Apply the distributive property.}$$

$$3x - 6 = x^2 + 8x \quad \text{Simplify.}$$

Helpful Hint Multiply *each* term by x.

Now we write the quadratic equation in standard form and solve for x.

$$0 = x^2 + 5x + 6$$
$$0 = (x + 3)(x + 2) \quad \text{Factor.}$$
$$x + 3 = 0 \quad \text{or} \quad x + 2 = 0 \quad \text{Set each factor equal to 0 and solve.}$$
$$x = -3 \qquad\qquad x = -2$$

Notice that neither -3 nor -2 makes the denominator in the original equation equal to 0.

Check: To check these solutions, we replace x in the original equation by -3, and then by -2.

If $x = -3$:

$$3 - \dfrac{6}{x} = x + 8$$
$$3 - \dfrac{6}{-3} \stackrel{?}{=} -3 + 8$$
$$3 - (-2) \stackrel{?}{=} 5$$
$$5 = 5 \quad \text{True}$$

If $x = -2$:

$$3 - \dfrac{6}{x} = x + 8$$
$$3 - \dfrac{6}{-2} \stackrel{?}{=} -2 + 8$$
$$3 - (-3) \stackrel{?}{=} 6$$
$$6 = 6 \quad \text{True}$$

Both -3 and -2 are solutions.

■ **Work Practice 3**

The following steps may be used to solve an equation containing rational expressions.

> **To Solve an Equation Containing Rational Expressions**
>
> **Step 1:** Multiply both sides of the equation by the LCD of all rational expressions in the equation.
>
> **Step 2:** Remove any grouping symbols and solve the resulting equation.
>
> **Step 3:** Check the solution in the original equation.

Practice 3

Solve: $2 + \dfrac{6}{x} = x + 7$

Answer

3. $x = -6, x = 1$

Practice 4

Solve:

$$\frac{2}{x+3} + \frac{3}{x-3} = \frac{-2}{x^2-9}$$

Example 4 Solve: $\dfrac{4x}{x^2+x-30} + \dfrac{2}{x-5} = \dfrac{1}{x+6}$

Solution: The denominator $x^2 + x - 30$ factors as $(x+6)(x-5)$. The LCD is then $(x+6)(x-5)$, so we multiply both sides of the equation by this LCD.

$$(x+6)(x-5)\left(\frac{4x}{x^2+x-30} + \frac{2}{x-5}\right) = (x+6)(x-5)\left(\frac{1}{x+6}\right) \quad \begin{array}{l}\text{Multiply}\\\text{by the}\\\text{LCD.}\end{array}$$

$$(x+6)(x-5)\cdot\frac{4x}{x^2+x-30} + (x+6)(x-5)\cdot\frac{2}{x-5} \quad \begin{array}{l}\text{Apply the distributive}\\\text{property.}\end{array}$$

$$= (x+6)(x-5)\cdot\frac{1}{x+6}$$

$$4x + 2(x+6) = x - 5 \quad \text{Simplify.}$$

$$4x + 2x + 12 = x - 5 \quad \text{Apply the distributive property.}$$

$$6x + 12 = x - 5 \quad \text{Combine like terms.}$$

$$5x = -17$$

$$x = -\frac{17}{5} \quad \text{Divide both sides by 5.}$$

Check: Check by replacing x with $-\dfrac{17}{5}$ in the original equation. The solution is $-\dfrac{17}{5}$.

◼ **Work Practice 4**

Practice 5

Solve: $\dfrac{5x}{x-1} = \dfrac{5}{x-1} + 3$

Example 5 Solve: $\dfrac{2x}{x-4} = \dfrac{8}{x-4} + 1$

Solution: Multiply both sides by the LCD, $x - 4$.

$$(x-4)\left(\frac{2x}{x-4}\right) = (x-4)\left(\frac{8}{x-4} + 1\right) \quad \text{Multiply by the LCD.}$$

$$(x-4)\cdot\frac{2x}{x-4} = (x-4)\cdot\frac{8}{x-4} + (x-4)\cdot 1 \quad \text{Use the distributive property.}$$

$$2x = 8 + (x-4) \quad \text{Simplify.}$$

$$2x = 4 + x$$

$$x = 4$$

Notice that 4 makes a denominator 0 in the original equation. Therefore, 4 is *not* a solution and this equation has *no solution*.

◼ **Work Practice 5**

✓**Concept Check** When can we clear fractions by multiplying through by the LCD?

a. When adding or subtracting rational expressions

b. When solving an equation containing rational expressions

c. Both of these

d. Neither of these

Answers

4. $x = -1$ **5.** no solution

✓**Concept Check Answer**

b

As we can see from Example 5, it is important to check the proposed solution(s) in the original equation.

Example 6 Solve: $x + \dfrac{14}{x-2} = \dfrac{7x}{x-2} + 1$

Practice 6

Solve:

$$x - \dfrac{6}{x+3} = \dfrac{2x}{x+3} + 2$$

Solution: Notice the denominators in this equation. We can see that 2 can't be a solution. The LCD is $x - 2$, so we multiply both sides of the equation by $x - 2$.

$$(x-2)\left(x + \dfrac{14}{x-2}\right) = (x-2)\left(\dfrac{7x}{x-2} + 1\right)$$

$$(x-2)(x) + (x-2)\left(\dfrac{14}{x-2}\right) = (x-2)\left(\dfrac{7x}{x-2}\right) + (x-2)(1)$$

$$x^2 - 2x + 14 = 7x + x - 2 \quad \text{Simplify.}$$

$$x^2 - 2x + 14 = 8x - 2 \quad \text{Combine like terms.}$$

$$x^2 - 10x + 16 = 0 \quad \begin{array}{l}\text{Write the quadratic}\\\text{equation in standard form.}\end{array}$$

$$(x-8)(x-2) = 0 \quad \text{Factor.}$$

$$x - 8 = 0 \quad \text{or} \quad x - 2 = 0 \quad \text{Set each factor equal to 0.}$$

$$x = 8 \qquad\qquad x = 2 \quad \text{Solve.}$$

As we have already noted, 2 can't be a solution of the original equation. So we need replace x only with 8 in the original equation. We find that 8 is a solution; the only solution is 8.

Teaching Tip

Ask students why Example 6 simply states "Solve" whereas Example 7 states "Solve ... for x." Since Example 7 has more than one unknown, the unknown we need to find must be specified.

■ **Work Practice 6**

Objective B Solving Equations for a Specified Variable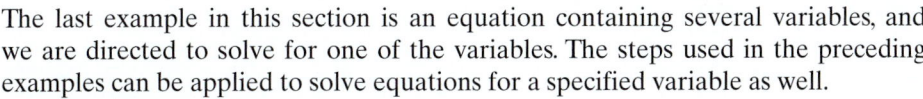

The last example in this section is an equation containing several variables, and we are directed to solve for one of the variables. The steps used in the preceding examples can be applied to solve equations for a specified variable as well.

Example 7 Solve $\dfrac{1}{a} + \dfrac{1}{b} = \dfrac{1}{x}$ for x.

Practice 7

Solve $\dfrac{1}{a} + \dfrac{1}{b} = \dfrac{1}{x}$ for a.

Solution: (This type of equation often models a work problem, as we shall see in the next section.) The LCD is abx, so we multiply both sides by abx.

$$abx\left(\dfrac{1}{a} + \dfrac{1}{b}\right) = abx\left(\dfrac{1}{x}\right)$$

$$abx\left(\dfrac{1}{a}\right) + abx\left(\dfrac{1}{b}\right) = abx \cdot \dfrac{1}{x}$$

$$bx + ax = ab \quad \text{Simplify.}$$

$$x(b + a) = ab \quad \text{Factor out } x \text{ from each term on the left side.}$$

$$\dfrac{x(b+a)}{b+a} = \dfrac{ab}{b+a} \quad \text{Divide both sides by } b + a.$$

$$x = \dfrac{ab}{b+a} \quad \text{Simplify.}$$

This equation is now solved for x.

Teaching Tip

It may be instructional to have students solve Example 7 for b and note the similarities and differences to Example 7. You may want to point out that in both cases, the equation is solved in terms of the other unknowns.

Answers

6. $x = 4$ **7.** $a = \dfrac{bx}{b-x}$

■ **Work Practice 7**

Vocabulary, Readiness & Video Check

Multiple choice. Choose the correct response.

1. Multiply both sides of the equation $\dfrac{3x}{2} + 5 = \dfrac{1}{4}$ by 4. The result is:

   a. $3x + 5 = 1$ **b.** $6x + 5 = 1$ **c.** $6x + 20 = 1$ **d.** $6x + 9 = 1$ c

2. Multiply both sides of the equation $\dfrac{1}{x} - \dfrac{3}{5x} = 2$ by $5x$. The result is:

   a. $1 - 3 = 10x$ **b.** $5 - 3 = 10x$ **c.** $5x - 3 = 10x$ **d.** $5 - 3 = 7x$ b

Choose the correct LCD for the fractions in each equation.

3. Equation: $\dfrac{9}{x} + \dfrac{3}{4} = \dfrac{1}{12}$; LCD: _____

   a. $4x$ **b.** $12x$ **c.** $48x$ **d.** x b

4. Equation: $\dfrac{8}{3x} - \dfrac{1}{x} = \dfrac{7}{9}$; LCD: _____

   a. x **b.** $3x$ **c.** $27x$ **d.** $9x$ d

5. Equation: $\dfrac{9}{x - 1} = \dfrac{7}{(x - 1)^2}$; LCD: _____

   a. $(x - 1)^2$ **b.** $(x - 1)$ **c.** $(x - 1)^3$ **d.** 63 a

6. Equation: $\dfrac{1}{x - 2} - \dfrac{3}{x^2 - 4} = 8$; LCD: _____

   a. $(x - 2)$ **b.** $(x + 2)$ **c.** $(x^2 - 4)$ **d.** $(x - 2)(x^2 - 4)$ c

---

Martin-Gay Interactive Videos   *Watch the section lecture video and answer the following questions.*

See Video 7.5 🍎

See video answer section.

Objective A 7. After multiplying through by the LCD and then simplifying, why is it important to take a moment and determine whether we have a linear or a quadratic equation before we finish solving the problem? ▶

8. From ▥ Examples 2–5, what extra step is needed when checking solutions to an equation containing rational expressions? ▶

Objective B 9. The steps for solving ▥ Example 6 for a specified variable are the same as what other steps? How do we treat this specified variable? ▶

7.5 Exercise Set MyMathLab®

Objective A *Solve each equation and check each solution. See Examples 1 through 3.*

1. $\dfrac{x}{5} + 3 = 9$ 30

2. $\dfrac{x}{5} - 2 = 9$ 55

3. $\dfrac{x}{2} + \dfrac{5x}{4} = \dfrac{x}{12}$ 0

4. $\dfrac{x}{6} + \dfrac{4x}{3} = \dfrac{x}{18}$ 0

5. $2 - \dfrac{8}{x} = 6$ −2

6. $5 + \dfrac{4}{x} = 1$ −1

7. $2 + \dfrac{10}{x} = x + 5$ −5, 2

8. $6 + \dfrac{5}{y} = y - \dfrac{2}{y}$ −1, 7

9. $\dfrac{a}{5} = \dfrac{a-3}{2}$ 5

10. $\dfrac{b}{5} = \dfrac{b+2}{6}$ 10

11. $\dfrac{x-3}{5} + \dfrac{x-2}{2} = \dfrac{1}{2}$ 3

12. $\dfrac{a+5}{4} + \dfrac{a+5}{2} = \dfrac{a}{8}$ −6

Solve each equation and check each proposed solution. See Examples 4 through 6.

13. $\dfrac{3}{2a-5} = -1$ 1

14. $\dfrac{6}{4-3x} = -3$ 2

15. $\dfrac{4y}{y-4} + 5 = \dfrac{5y}{y-4}$ 5

16. $\dfrac{2a}{a+2} - 5 = \dfrac{7a}{a+2}$ −1

17. $2 + \dfrac{3}{a-3} = \dfrac{a}{a-3}$ no solution

18. $\dfrac{2y}{y-2} - \dfrac{4}{y-2} = 4$ no solution

19. $\dfrac{1}{x+3} + \dfrac{6}{x^2-9} = 1$ 4

20. $\dfrac{1}{x+2} + \dfrac{4}{x^2-4} = 1$ 3

21. $\dfrac{2y}{y+4} + \dfrac{4}{y+4} = 3$ −8

22. $\dfrac{5y}{y+1} - \dfrac{3}{y+1} = 4$ 7

23. $\dfrac{2x}{x+2} - 2 = \dfrac{x-8}{x-2}$ 6, −4

24. $\dfrac{4y}{y-3} - 3 = \dfrac{3y-1}{y+3}$ 12, −1

Solve each equation. See Examples 1 through 6.

25. $\dfrac{2}{y} + \dfrac{1}{2} = \dfrac{5}{2y}$ 1

26. $\dfrac{6}{3y} + \dfrac{3}{y} = 1$ 5

27. $\dfrac{a}{a-6} = \dfrac{-2}{a-1}$ 3, −4

28. $\dfrac{5}{x-6} = \dfrac{x}{x-2}$ 10, 1

29. $\dfrac{11}{2x} + \dfrac{2}{3} = \dfrac{7}{2x}$ −3

30. $\dfrac{5}{3} - \dfrac{3}{2x} = \dfrac{3}{2}$ 9

31. $\dfrac{2}{x-2} + 1 = \dfrac{x}{x+2}$ 0

32. $1 + \dfrac{3}{x+1} = \dfrac{x}{x-1}$ 2

33. $\dfrac{x+1}{3} - \dfrac{x-1}{6} = \dfrac{1}{6}$ −2

34. $\dfrac{3x}{5} - \dfrac{x-6}{3} = -\dfrac{2}{5}$ −9

35. $\dfrac{t}{t-4} = \dfrac{t+4}{6}$ 8, −2

36. $\dfrac{15}{x+4} = \dfrac{x-4}{x}$ 16, −1

37. $\dfrac{y}{2y+2} + \dfrac{2y-16}{4y+4} = \dfrac{2y-3}{y+1}$ no solution

38. $\dfrac{1}{x+2} = \dfrac{4}{x^2-4} - \dfrac{1}{x-2}$ no solution

39. $\dfrac{4r - 4}{r^2 + 5r - 14} + \dfrac{2}{r + 7} = \dfrac{1}{r - 2}$ 3

40. $\dfrac{3}{x + 3} = \dfrac{12x + 19}{x^2 + 7x + 12} - \dfrac{5}{x + 4}$ 2

41. $\dfrac{x + 1}{x + 3} = \dfrac{x^2 - 11x}{x^2 + x - 6} - \dfrac{x - 3}{x - 2}$ −11, 1

42. $\dfrac{2t + 3}{t - 1} - \dfrac{2}{t + 3} = \dfrac{5 - 6t}{t^2 + 2t - 3}$ $-\dfrac{1}{2}$, −6

Objective B   *Solve each equation for the indicated variable. See Example 7.*

43. $R = \dfrac{E}{I}$ for I (Electronics: resistance of a circuit)

$I = \dfrac{E}{R}$

44. $T = \dfrac{V}{Q}$ for Q (Water purification: settling time)

$Q = \dfrac{V}{T}$

45. $T = \dfrac{2U}{B + E}$ for B (Merchandising: stock turnover rate)

$B = \dfrac{2U - TE}{T}$

46. $i = \dfrac{A}{t + B}$ for t (Hydrology: rainfall intensity)

$t = \dfrac{A - Bi}{i}$

47. $B = \dfrac{705w}{h^2}$ for w (Health: body-mass index)

$w = \dfrac{Bh^2}{705}$

△ 48. $\dfrac{A}{W} = L$ for W (Geometry: area of a rectangle)

$W = \dfrac{A}{L}$

49. $N = R + \dfrac{V}{G}$ for G (Urban forestry: tree plantings per year) $G = \dfrac{V}{N - R}$

50. $C = \dfrac{D(A + 1)}{24}$ for A (Medicine: Cowling's Rule for child's dose) $A = \dfrac{24C - D}{D}$

△ 51. $\dfrac{C}{\pi r} = 2$ for r (Geometry: circumference of a circle)

$r = \dfrac{C}{2\pi}$

52. $W = \dfrac{CE^2}{2}$ for C (Electronics: energy stored in a capacitor) $C = \dfrac{2W}{E^2}$

53. $\dfrac{1}{y} + \dfrac{1}{3} = \dfrac{1}{x}$ for x $x = \dfrac{3y}{3 + y}$

54. $\dfrac{1}{5} + \dfrac{2}{y} = \dfrac{1}{x}$ for x $x = \dfrac{5y}{y + 10}$

Review

Translating   *Write each phrase as an expression. See Sections R.2 and 1.8.*

55. The reciprocal of x $\dfrac{1}{x}$

56. The reciprocal of $x + 1$ $\dfrac{1}{x + 1}$

57. The reciprocal of x, added to the reciprocal of 2 $\dfrac{1}{x} + \dfrac{1}{2}$

58. The reciprocal of x, subtracted from the reciprocal of 5 $\dfrac{1}{5} - \dfrac{1}{x}$

Answer each question.

59. If a tank is filled in 3 hours, what part of the tank is filled in 1 hour? $\dfrac{1}{3}$

60. If a strip of beach is cleaned in 4 hours, what part of the beach is cleaned in 1 hour? $\dfrac{1}{4}$

Concept Extensions

61. Explain the difference between solving an equation such as $\dfrac{x}{2} + \dfrac{3}{4} = \dfrac{x}{4}$ for x and performing an operation such as adding $\dfrac{x}{2} + \dfrac{3}{4}$.   *answers may vary*

62. When solving an equation such as $\dfrac{y}{4} = \dfrac{y}{2} - \dfrac{1}{4}$, we may multiply all terms by 4. When subtracting two rational expressions such as $\dfrac{y}{2} - \dfrac{1}{4}$, we may not. Explain why.   *answers may vary*

Determine whether each of the following is an equation or an expression. If it is an equation, then solve it for its variable. If it is an expression, perform the indicated operation.

△ **63.** $\dfrac{1}{x} + \dfrac{5}{9}$ $\dfrac{5x + 9}{9x}$

64. $\dfrac{1}{x} + \dfrac{5}{9} = \dfrac{2}{3}$ 9

65. $\dfrac{5}{x - 1} - \dfrac{2}{x} = \dfrac{5}{x(x - 1)}$ *no solution*

66. $\dfrac{5}{x - 1} - \dfrac{2}{x}$ $\dfrac{3x + 2}{x(x - 1)}$

Recall that two angles are supplementary if the sum of their measures is 180°. Find the measures of the supplementary angles.

△ **67.** $100°, 80°$

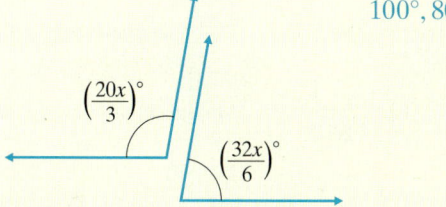

△ **68.** $30°, 150°$

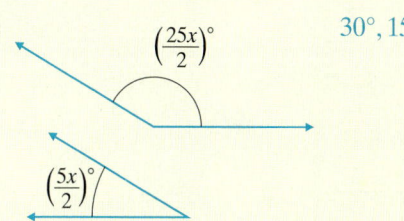

Recall that two angles are complementary if the sum of their measures is 90°. Find the measures of the complementary angles.

△ **69.** $22.5°, 67.5°$

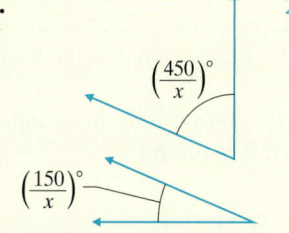

△ **70.** $40°, 50°$

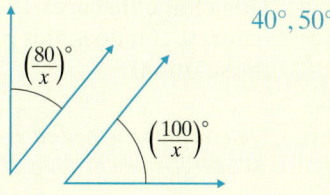

Solve each equation.

71. $\dfrac{4}{a^2 + 4a + 3} + \dfrac{2}{a^2 + a - 6} - \dfrac{3}{a^2 - a - 2} = 0$ 5

72. $\dfrac{-4}{a^2 + 2a - 8} + \dfrac{1}{a^2 + 9a + 20} = \dfrac{-4}{a^2 + 3a - 10}$ 6

Summary on Rational Expressions

1. expression; $\dfrac{3 + 2x}{3x}$

2. expression; $\dfrac{18 + 5a}{6a}$

3. equation; 3

4. equation; 18

5. expression; $\dfrac{x - 1}{x(x + 1)}$

6. expression; $\dfrac{3(x + 1)}{x(x - 3)}$

7. equation; no solution

8. equation; 1

9. expression; 10

10. expression; $\dfrac{z}{3(9z - 5)}$

It is important to know the difference between performing operations with rational expressions and solving an equation containing rational expressions. Study the examples below.

Performing Operations with Rational Expressions

Adding: $\dfrac{1}{x} + \dfrac{1}{x + 5} = \dfrac{1 \cdot (x + 5)}{x(x + 5)} + \dfrac{1 \cdot x}{x(x + 5)} = \dfrac{x + 5 + x}{x(x + 5)} = \dfrac{2x + 5}{x(x + 5)}$

Subtracting: $\dfrac{3}{x} - \dfrac{5}{x^2 y} = \dfrac{3 \cdot xy}{x \cdot xy} - \dfrac{5}{x^2 y} = \dfrac{3xy - 5}{x^2 y}$

Multiplying: $\dfrac{2}{x} \cdot \dfrac{5}{x - 1} = \dfrac{2 \cdot 5}{x(x - 1)} = \dfrac{10}{x(x - 1)}$

Dividing: $\dfrac{4}{2x + 1} \div \dfrac{x - 3}{x} = \dfrac{4}{2x + 1} \cdot \dfrac{x}{x - 3} = \dfrac{4x}{(2x + 1)(x - 3)}$

Solving an Equation Containing Rational Expressions

To solve an equation containing rational expressions, we clear the equation of fractions by multiplying both sides by the LCD.

$$\dfrac{3}{x} - \dfrac{5}{x - 1} = \dfrac{1}{x(x - 1)} \qquad \text{Note that } x \text{ can't be 0 or 1.}$$

$$x(x - 1)\left(\dfrac{3}{x}\right) - x(x - 1)\left(\dfrac{5}{x - 1}\right) = x(x - 1) \cdot \dfrac{1}{x(x - 1)} \qquad \text{Multiply both sides by the LCD.}$$

$$3(x - 1) - 5x = 1 \qquad \text{Simplify.}$$

$$3x - 3 - 5x = 1 \qquad \text{Use the distributive property.}$$

$$-2x - 3 = 1 \qquad \text{Combine like terms.}$$

$$-2x = 4 \qquad \text{Add 3 to both sides.}$$

$$x = -2 \qquad \text{Divide both sides by } -2.$$

Don't forget to check to make sure our proposed solution of -2 does not make any denominators 0. If it does, this proposed solution is *not* a solution of the equation. -2 checks and is the solution.

Determine whether each of the following is an equation or an expression. If it is an equation, solve it for its variable. If it is an expression, perform the indicated operation.

1. $\dfrac{1}{x} + \dfrac{2}{3}$

2. $\dfrac{3}{a} + \dfrac{5}{6}$

3. $\dfrac{1}{x} + \dfrac{2}{3} = \dfrac{3}{x}$

4. $\dfrac{3}{a} + \dfrac{5}{6} = 1$

5. $\dfrac{2}{x + 1} - \dfrac{1}{x}$

6. $\dfrac{4}{x - 3} - \dfrac{1}{x}$

7. $\dfrac{2}{x + 1} - \dfrac{1}{x} = 1$

8. $\dfrac{4}{x - 3} - \dfrac{1}{x} = \dfrac{6}{x(x - 3)}$

9. $\dfrac{15x}{x + 8} \cdot \dfrac{2x + 16}{3x}$

10. $\dfrac{9z + 5}{15} \cdot \dfrac{5z}{81z^2 - 25}$

11. $\dfrac{2x+1}{x-3} + \dfrac{3x+6}{x-3}$

12. $\dfrac{4p-3}{2p+7} + \dfrac{3p+8}{2p+7}$

13. $\dfrac{x+5}{7} = \dfrac{8}{2}$

14. $\dfrac{1}{2} = \dfrac{x+1}{8}$

15. $\dfrac{5a+10}{18} \div \dfrac{a^2-4}{10a}$

16. $\dfrac{9}{x^2-1} \div \dfrac{12}{3x+3}$

17. $\dfrac{x+2}{3x-1} + \dfrac{5}{(3x-1)^2}$

18. $\dfrac{4}{(2x-5)^2} + \dfrac{x+1}{2x-5}$

19. $\dfrac{x-7}{x} - \dfrac{x+2}{5x}$

20. $\dfrac{10x-9}{x} - \dfrac{x-4}{3x}$

21. $\dfrac{3}{x+3} = \dfrac{5}{x^2-9} - \dfrac{2}{x-3}$

22. $\dfrac{9}{x^2-4} + \dfrac{2}{x+2} = \dfrac{-1}{x-2}$

23. Explain the difference between solving an equation such as $\dfrac{x}{3} + \dfrac{1}{6} = \dfrac{x}{6}$ for x and performing an operation such as adding $\dfrac{x}{3} + \dfrac{1}{6}$.

24. When solving an equation such as $\dfrac{y}{6} = \dfrac{y}{3} - \dfrac{1}{6}$, we may multiply all terms by 6. When subtracting two rational expressions such as $\dfrac{y}{3} - \dfrac{1}{6}$, we may not. Explain why.

11. expression; $\dfrac{5x+7}{x-3}$

12. expression; $\dfrac{7p+5}{2p+7}$

13. equation; 23

14. equation; 3

15. expression; $\dfrac{25a}{9(a-2)}$

16. expression; $\dfrac{9}{4(x-1)}$

17. expression; $\dfrac{3x^2+5x+3}{(3x-1)^2}$

18. expression; $\dfrac{2x^2-3x-1}{(2x-5)^2}$

19. expression; $\dfrac{4x-37}{5x}$

20. expression; $\dfrac{29x-23}{3x}$

21. equation; $\dfrac{8}{5}$

22. equation; $-\dfrac{7}{3}$

23. answers may vary

24. answers may vary

7.6 Proportions and Problem Solving with Rational Equations

Objectives

A Solve Proportions.

B Use Proportions to Solve Problems, Including Similar Triangle Problems.

C Solve Problems About Numbers.

D Solve Problems About Work.

E Solve Problems About Distance.

Teaching Tip

Ask students to mentally calculate whether the statements below are true or false by using cross products.

$$\frac{2}{3} = \frac{6}{10}$$

$$\frac{4}{6} = \frac{10}{15}$$

$$\frac{5}{7} = \frac{7}{10}$$

$$\frac{3}{12} = \frac{2}{8}$$

Objective A Solving Proportions

A **ratio** is the quotient of two numbers or two quantities. For example, the ratio of 2 to 5 can be written in fraction form as $\frac{2}{5}$, the quotient of 2 and 5.

A **rate** is a special type of ratio with different kinds of measurement. For example, the ratio "110 miles in 2 hours" written as a fraction in simplest form is $\frac{110 \text{ miles}}{2 \text{ hours}} = \frac{55 \text{ mi}}{1 \text{ hr}}$ or 55 mph.

If two ratios are equal, we say the ratios are **in proportion** to each other. A **proportion** is a mathematical statement that two ratios are equal.

For example, the equation $\frac{1}{2} = \frac{4}{8}$ is a proportion, as is $\frac{x}{5} = \frac{8}{10}$, because both sides of the equations are ratios. When we want to emphasize the equation as a proportion, we

read the proportion $\frac{1}{2} = \frac{4}{8}$ as "one is to two as four is to eight"

In a proportion, cross products are equal. To understand cross products, let's start with the proportion

$$\frac{a}{b} = \frac{c}{d}$$

and multiply both sides by the LCD, bd.

$$bd\left(\frac{a}{b}\right) = bd\left(\frac{c}{d}\right) \qquad \text{Multiply both sides by the LCD, } bd.$$

$$\underset{\text{Cross product}}{\underline{ad}} = \underset{\text{Cross product}}{\underline{bc}} \qquad \text{Simplify.}$$

Notice why ad and bc are called cross products.

$$ad \qquad\qquad bc$$
$$\frac{a}{b} = \frac{c}{d}$$

Cross Products

If $\frac{a}{b} = \frac{c}{d}$, then $ad = bc$.

For example, if

$$\frac{1}{2} = \frac{4}{8}, \quad \text{then} \quad 1 \cdot 8 = 2 \cdot 4 \quad \text{or} \quad 8 = 8$$

Notice that a proportion contains four numbers (or expressions). If any three numbers are known, we can solve and find the fourth number.

Example 1 Solve for x: $\dfrac{45}{x} = \dfrac{5}{7}$

Solution: This is an equation with rational expressions and also a proportion. Below are two ways to solve.

Since this is a rational equation, we can use the methods of the previous section.	Since this is also a proportion, we may set cross products equal.

$$\dfrac{45}{x} = \dfrac{5}{7}$$

$7x \cdot \dfrac{45}{x} = 7x \cdot \dfrac{5}{7}$ Multiply both sides by LCD, $7x$.

$7 \cdot 45 = x \cdot 5$ Divide out common factors.

$315 = 5x$ Multiply.

$\dfrac{315}{5} = \dfrac{5x}{5}$ Divide both sides by 5.

$63 = x$ Simplify.

$$\dfrac{45}{x} = \dfrac{5}{7}$$

$45 \cdot 7 = x \cdot 5$ Set cross products equal.

$315 = 5x$ Multiply.

$\dfrac{315}{5} = \dfrac{5x}{5}$ Divide both sides by 5.

$63 = x$ Simplify.

Check: Both methods give us a solution of 63. To check, substitute 63 for x in the original proportion. The solution is 63.

◼ **Work Practice 1**

In this section, if the rational equation is a proportion, we will use cross products to solve.

Example 2 Solve for x: $\dfrac{x-5}{3} = \dfrac{x+2}{5}$

Solution:

$$\dfrac{x-5}{3} = \dfrac{x+2}{5}$$

$5(x-5) = 3(x+2)$ Set cross products equal.

$5x - 25 = 3x + 6$ Multiply.

$5x = 3x + 31$ Add 25 to both sides.

$2x = 31$ Subtract $3x$ from both sides.

$\dfrac{2x}{2} = \dfrac{31}{2}$ Divide both sides by 2.

$x = \dfrac{31}{2}$

Check: Verify that $\dfrac{31}{2}$ is the solution.

◼ **Work Practice 2**

Objective B Using Proportions to Solve Problems ▶

Proportions can be used to model and solve many real-life problems. When using proportions in this way, it is important to judge whether the solution is reasonable. Doing so helps us to decide if the proportion has been formed correctly. We use the same problem-solving steps that were introduced in Section 2.4.

Practice 1

Solve for x: $\dfrac{3}{8} = \dfrac{63}{x}$

Practice 2

Solve for x: $\dfrac{2x+1}{7} = \dfrac{x-3}{5}$

Answers

1. $x = 168$ **2.** $x = -\dfrac{26}{3}$

Practice 3

To estimate the number of people in Jackson, population 50,000, who have a flu shot, 250 people were polled. Of those polled, 26 had a flu shot. How many people in the city might we expect to have a flu shot?

Example 3 Calculating the Cost of Recordable Compact Discs

Three boxes of CD-Rs (recordable compact discs) cost $37.47. How much should 5 boxes cost?

Solution:

1. UNDERSTAND. Read and reread the problem. We know that the cost of 5 boxes is more than the cost of 3 boxes, or $37.47, and less than the cost of 6 boxes, which is double the cost of 3 boxes, or $2(\$37.47) = \$74.94$. Let's suppose that 5 boxes cost $60.00. To check, we see if 3 boxes is to 5 boxes as the *price* of 3 boxes is to the *price* of 5 boxes. In other words, we see if

$$\frac{3 \text{ boxes}}{5 \text{ boxes}} = \frac{\text{price of 3 boxes}}{\text{price of 5 boxes}}$$

or

$$\frac{3}{5} = \frac{37.47}{60.00}$$

$3(60.00) = 5(37.47)$　　Set cross products equal.

or

$180.00 = 187.35$　　Not a true statement

Thus, $60 is not correct, but we now have a better understanding of the problem.

Let $x =$ price of 5 boxes of CD-Rs.

2. TRANSLATE.

$$\frac{3 \text{ boxes}}{5 \text{ boxes}} = \frac{\text{price of 3 boxes}}{\text{price of 5 boxes}}$$

$$\frac{3}{5} = \frac{37.47}{x}$$

3. SOLVE.

$$\frac{3}{5} = \frac{37.47}{x}$$

$3x = 5(37.47)$　　Set cross products equal.

$3x = 187.35$

$x = 62.45$　　Divide both sides by 3.

4. INTERPRET.

Check: Verify that 3 boxes is to 5 boxes as $37.47 is to $62.45. Also, notice that our solution is a reasonable one as discussed in Step 1.

State: Five boxes of CD-Rs cost $62.45.

■ Work Practice 3

Answer

3. 5200 people

Helpful Hint

The proportion $\dfrac{5 \text{ boxes}}{3 \text{ boxes}} = \dfrac{\text{price of 5 boxes}}{\text{price of 3 boxes}}$ could also have been used to solve Example 3. Notice that the cross products are the same.

Similar triangles have the same shape but not necessarily the same size. In similar triangles, the measures of corresponding angles are equal, and corresponding sides are in proportion.

If triangle ABC and triangle XYZ shown are similar, then we know that the measure of angle A = the measure of angle X, the measure of angle B = the measure of angle Y, and the measure of angle C = the measure of angle Z. We also know that corresponding sides are in proportion: $\dfrac{a}{x} = \dfrac{b}{y} = \dfrac{c}{z}$.

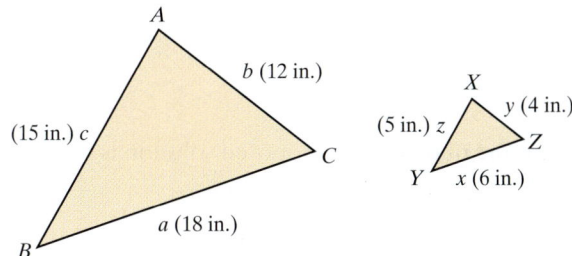

In this section, we will position similar triangles so that they have the same orientation.

To show that corresponding sides are in proportion for the triangles above, we write the ratios of the corresponding sides.

$$\frac{a}{x} = \frac{18}{6} = 3 \qquad \frac{b}{y} = \frac{12}{4} = 3 \qquad \frac{c}{z} = \frac{15}{5} = 3$$

△ **Example 4** Finding the Length of a Side of a Triangle

If the following two triangles are similar, find the missing length x.

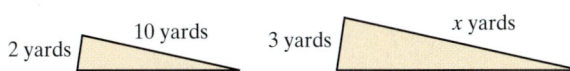

Solution:

1. UNDERSTAND. Read the problem and study the figure.
2. TRANSLATE. Since the triangles are similar, their corresponding sides are in proportion and we have

$$\frac{2}{3} = \frac{10}{x}$$

3. SOLVE. To solve, we multiply both sides by the LCD, $3x$, or cross multiply.

$$2x = 30$$

$$x = 15 \quad \text{Divide both sides by 2.}$$

4. INTERPRET.

Check: To check, replace x with 15 in the original proportion and see that a true statement results.

State: The missing length is 15 yards.

■ **Work Practice 4**

Practice 4

For the similar triangles, find x.

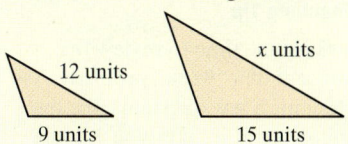

Answer

4. 20 units

Objective C Solving Problems About Numbers ▶

Let's continue to solve problems. The remaining problems are all modeled by rational equations.

Practice 5

The quotient of a number and 2, minus $\frac{1}{3}$, is the quotient of the number and 6. Find the number.

Example 5 Finding an Unknown Number

The quotient of a number and 6, minus $\frac{5}{3}$, is the quotient of the number and 2. Find the number.

Solution:

1. **UNDERSTAND.** Read and reread the problem. Suppose that the unknown number is 2; then we see if the quotient of 2 and 6, or $\frac{2}{6}$, minus $\frac{5}{3}$ is equal to the quotient of 2 and 2, or $\frac{2}{2}$.

$$\frac{2}{6} - \frac{5}{3} = \frac{1}{3} - \frac{5}{3} = -\frac{4}{3}, \text{ not } \frac{2}{2}$$

Don't forget that the purpose of a proposed solution is to better understand the problem.

Let $x =$ the unknown number.

2. **TRANSLATE.**

In words:	the quotient of x and 6	minus	$\frac{5}{3}$	is	the quotient of x and 2
	↓	↓	↓	↓	↓
Translate:	$\frac{x}{6}$	$-$	$\frac{5}{3}$	$=$	$\frac{x}{2}$

3. **SOLVE.** Here, we solve the equation $\frac{x}{6} - \frac{5}{3} = \frac{x}{2}$. We begin by multiplying both sides of the equation by the LCD, 6.

$$6\left(\frac{x}{6} - \frac{5}{3}\right) = 6\left(\frac{x}{2}\right)$$

$$6\left(\frac{x}{6}\right) - 6\left(\frac{5}{3}\right) = 6\left(\frac{x}{2}\right) \quad \text{Apply the distributive property.}$$

$$x - 10 = 3x \quad \text{Simplify.}$$

$$-10 = 2x \quad \text{Subtract } x \text{ from both sides.}$$

$$\frac{-10}{2} = \frac{2x}{2} \quad \text{Divide both sides by 2.}$$

$$-5 = x \quad \text{Simplify.}$$

4. **INTERPRET.**

Check: To check, we verify that "the quotient of -5 and 6 minus $\frac{5}{3}$ is the quotient of -5 and 2," or $-\frac{5}{6} - \frac{5}{3} = -\frac{5}{2}$.

State: The unknown number is -5.

■ **Work Practice 5**

Teaching Tip

Take some time to review the concept of a "work" problem. Ask students a few questions like the following: If it takes you 2 hours to complete a job, what part of the job has been completed in 1 hour? After a few of these, see if you can successfully insert the variable x.

Answer

5. 1

Objective D Solving Problems About Work

The next example is often called a work problem. Work problems usually involve people or machines doing a certain task.

Example 6 Finding Work Rates

Sam Waterton and Frank Schaffer work in a plant that manufactures automobiles. Sam can complete a quality control tour of the plant in 3 hours while his assistant, Frank, needs 7 hours to complete the same job. The regional manager is coming to inspect the plant facilities, so both Sam and Frank are directed to complete a quality control tour together. How long will this take?

Practice 6

Practice 6

Guillaume Beauchesne and Greg Langacker volunteer at a local recycling plant. Guillaume can sort a batch of recyclables in 2 hours alone while his friend Greg needs 3 hours to complete the same job. If they work together, how long will it take them to sort one batch?

Solution:

1. UNDERSTAND. Read and reread the problem. The key idea here is the relationship between the **time** (hours) it takes to complete the job and the **part of the job** completed in 1 unit of time (hour). For example, if the **time** it takes Sam to complete the job is 3 hours, the **part of the job** he can complete in 1 hour is $\frac{1}{3}$. Similarly, Frank can complete $\frac{1}{7}$ of the job in 1 hour.

Let x = the **time** in hours it takes Sam and Frank to complete the job together.

Then $\frac{1}{x}$ = the **part of the job** they complete in 1 hour.

	Hours to Complete Total Job	Part of Job Completed in 1 Hour
Sam	3	$\frac{1}{3}$
Frank	7	$\frac{1}{7}$
Together	x	$\frac{1}{x}$

2. TRANSLATE.

In words:	part of job job Sam completes in 1 hour	added to	part of job Frank completes in 1 hour	is equal to	part of job they complete together in in 1 hour
	↓	↓	↓	↓	↓
Translate:	$\frac{1}{3}$	$+$	$\frac{1}{7}$	$=$	$\frac{1}{x}$

3. SOLVE. Here, we solve the equation $\frac{1}{3} + \frac{1}{7} = \frac{1}{x}$. We begin by multiplying both sides of the equation by the LCD, $21x$.

$$21x\left(\frac{1}{3}\right) + 21x\left(\frac{1}{7}\right) = 21x\left(\frac{1}{x}\right)$$

$$7x + 3x = 21 \qquad \text{Simplify.}$$

$$10x = 21$$

$$x = \frac{21}{10} \quad \text{or} \quad 2\frac{1}{10} \text{ hours}$$

(*Continued on next page*)

Answer

6. $1\frac{1}{5}$ hours

4. INTERPRET.

Check: Our proposed solution is $2\frac{1}{10}$ hours. This proposed solution is reasonable since $2\frac{1}{10}$ hours is more than half of Sam's time and less than half of Frank's time. Check this solution in the originally *stated* problem.

State: Sam and Frank can complete the quality control tour in $2\frac{1}{10}$ hours.

■ **Work Practice 6**

✓**Concept Check**   Solve $E = mc^2$
a. for m **b.** for c^2

Objective E Solving Problems About Distance

Next we look at a problem solved by the distance formula,

$$d = r \cdot t$$

Practice 7

A car travels 600 miles in the same time that a motorcycle travels 450 miles. If the car's speed is 15 miles per hour faster than the motorcycle's, find the speed of the car and the speed of the motorcycle.

Example 7   Finding Speeds of Vehicles

A car travels 180 miles in the same time that a truck travels 120 miles. If the car's speed is 20 miles per hour faster than the truck's, find the car's speed and the truck's speed.

Solution:

1. UNDERSTAND. Read and reread the problem. Suppose that the truck's speed is 45 miles per hour. Then the car's speed is 20 miles per hour faster, or 65 miles per hour.

We are given that the car travels 180 miles in the same time that the truck travels 120 miles. To find the time it takes the car to travel 180 miles, remember that since $d = rt$, we know that $\dfrac{d}{r} = t$.

	Car's Time	**Truck's Time**
	$t = \dfrac{d}{r} = \dfrac{180}{65} = 2\dfrac{50}{65} = 2\dfrac{10}{13}$ hours	$t = \dfrac{d}{r} = \dfrac{120}{45} = 2\dfrac{30}{45} = 2\dfrac{2}{3}$ hours

Since the times are not the same, our proposed solution is not correct. But we have a better understanding of the problem.

Let $x =$ the speed of the truck.

Since the car's speed is 20 miles per hour faster than the truck's, then

$x + 20 =$ the speed of the car

Use the formula $d = r \cdot t$ or **distance** = **rate** · **time**. Prepare a chart to organize the information in the problem.

	Distance	**=**	**Rate**	**·**	**Time**
Truck	120		x		$\dfrac{120}{x}$ ← distance / ← rate
Car	180		$x + 20$		$\dfrac{180}{x + 20}$ ← distance / ← rate

Helpful Hint

If $d = r \cdot t$,
then $t = \dfrac{d}{r}$
or *time* $= \dfrac{distance}{rate}$.

Answer
7. car: 60 mph; motorcycle: 45 mph

✓**Concept Check Answers**

a. $m = \dfrac{E}{c^2}$ **b.** $c^2 = \dfrac{E}{m}$

2. **TRANSLATE.** Since the car and the truck traveled the same amount of time, we have that

In words: car's time   =   truck's time
                  ↓   ↓

Translate: $\dfrac{180}{x + 20}$ = $\dfrac{120}{x}$

3. **SOLVE.** We begin by multiplying both sides of the equation by the LCD, $x(x + 20)$, or cross multiplying.

$$\frac{180}{x + 20} = \frac{120}{x}$$

$$180x = 120(x + 20)$$

$180x = 120x + 2400$ Use the distributive property.

$60x = 2400$ Subtract $120x$ from both sides.

$x = 40$ Divide both sides by 60.

4. **INTERPRET.** The speed of the truck is 40 miles per hour. The speed of the car must then be $x + 20$ or 60 miles per hour.

Check: Find the time it takes the car to travel 180 miles and the time it takes the truck to travel 120 miles.

Car's Time	Truck's Time
$t = \dfrac{d}{r} = \dfrac{180}{60} = 3$ hours	$t = \dfrac{d}{r} = \dfrac{120}{40} = 3$ hours

Since both travel the same amount of time, the proposed solution is correct.

State: The car's speed is 60 miles per hour and the truck's speed is 40 miles per hour.

■ **Work Practice 7**

Vocabulary, Readiness & Video Check

Without solving algebraically, select the best choice for each exercise.

1. One person can complete a job in 7 hours. A second person can complete the same job in 5 hours. How long will it take them to complete the job if they work together?

   a. more than 7 hours
   b. between 5 and 7 hours
   c. less than 5 hours   c

2. One inlet pipe can fill a pond in 30 hours. A second inlet pipe can fill the same pond in 25 hours. How long before the pond is filled if both inlet pipes are on?

   a. less than 25 hours
   b. between 25 and 30 hours
   c. more than 30 hours   a

Fill in a Table *Given the variable in the first column, use the phrase in the second column to translate to an expression, and then continue to the phrase in the third column to translate to another expression.*

3.	A number: x	The reciprocal of the number: $\dfrac{1}{x}$	The reciprocal of the number, decreased by 3: $\dfrac{1}{x} - 3$
4.	A number: y	The reciprocal of the number: $\dfrac{1}{y}$	The reciprocal of the number, increased by 2: $\dfrac{1}{y} + 2$
5.	A number: z	The sum of the number and 5: $z + 5$	The reciprocal of the sum of the number and 5: $\dfrac{1}{z + 5}$
6.	A number: x	The difference of the number and 1: $x - 1$	The reciprocal of the difference of the number and 1: $\dfrac{1}{x - 1}$
7.	A number: y	Twice the number: $2y$	Eleven divided by twice the number: $\dfrac{11}{2y}$
8.	A number: z	Triple the number: $3z$	Negative ten divided by triple the number: $\dfrac{-10}{3z}$

Martin-Gay Interactive Videos   Watch the section lecture video and answer the following questions.

See Video 7.6 🍎

Objective A   **9.** Based on ▣ Examples 1 and 2, can proportions only be solved by using cross products? Explain. ▶

Objective B   **10.** In ▣ Example 3, we are told there are many ways to set up a correct proportion. Why does this fact make it even more important to check that your solution is reasonable? ▶

Objective C   **11.** What words or phrases in ▣ Example 5 told you to translate to an equation containing rational expressions? ▶

Objective D   **12.** From ▣ Example 6, how can you determine a somewhat reasonable answer to a work problem before you even begin to solve it? ▶

Objective E   **13.** The following problem is worded like ▣ Example 7 in the video, but using different quantities.

A car travels 325 miles in the same time that a motorcycle travels 290 miles. If the car's speed is 7 miles per hour more than the motorcycle's, find the speed of the car and the speed of the motorcycle. Fill in the table and set up an equation based on this problem (do not solve). Use ▣ Example 7 in the video as a model for your work. ▶

	d	=	r	·	t
car	325		$x + 7$		$\dfrac{325}{x + 7}$
motorcycle	290		x		$\dfrac{290}{x}$

See video answer section.

7.6   Exercise Set   MyMathLab® ▶

Objective A   *Solve each proportion. See Examples 1 and 2.*

1. $\dfrac{2}{3} = \dfrac{x}{6}$ 4

2. $\dfrac{x}{2} = \dfrac{16}{6}$ $\dfrac{16}{3}$

▶**3.** $\dfrac{x}{10} = \dfrac{5}{9}$ $\dfrac{50}{9}$

4. $\dfrac{9}{4x} = \dfrac{6}{2}$ $\dfrac{3}{4}$

▶**5.** $\dfrac{x + 1}{2x + 3} = \dfrac{2}{3}$ -3

6. $\dfrac{x + 1}{x + 2} = \dfrac{5}{3}$ $-\dfrac{7}{2}$

7. $\dfrac{9}{5} = \dfrac{12}{3x + 2}$ $\dfrac{14}{9}$

8. $\dfrac{6}{11} = \dfrac{27}{3x - 2}$ $\dfrac{103}{6}$

Objective B   *Solve. See Example 3.*

9. The ratio of the weight of an object on Earth to the weight of the same object on Pluto is 100 to 3. If an elephant weighs 4100 pounds on Earth, find the elephant's weight on Pluto.   123 lb

10. If a 170-pound person weighs approximately 65 pounds on Mars, about how much does a 9000-pound satellite weigh? Round your answer to the nearest pound.   3441 lb

11. There are 110 calories per 28.4 grams of Crispy Rice cereal. Find how many calories are in 42.6 grams of this cereal.   165 cal

12. On an architect's blueprint, 1 inch corresponds to 4 feet. Find the length of a wall represented by a line that is $3\frac{7}{8}$ inches long on the blueprint. $15\frac{1}{2}$ ft

Find the unknown length x or y in the following pairs of similar triangles. See Example 4.

13. 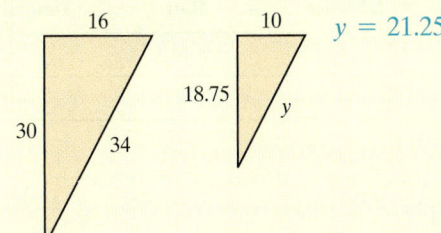 $y = 21.25$

14. 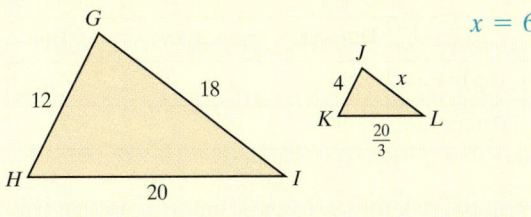 $x = 6$

15.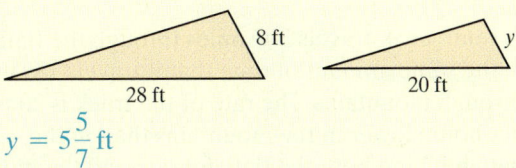

$y = 5\frac{5}{7}$ ft

16. 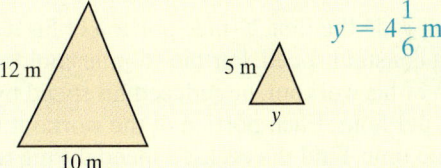 $y = 4\frac{1}{6}$ m

Objective C *Solve the following. See Example 5.*

17. Three times the reciprocal of a number equals 9 times the reciprocal of 6. Find the number.   2

18. Twelve divided by the sum of x and 2 equals the quotient of 4 and the difference of x and 2. Find x.   4

19. If twice a number added to 3 is divided by the number plus 1, the result is three halves. Find the number. -3

20. A number added to the product of 6 and the reciprocal of the number equals -5. Find the number. -3 or -2

Objective D *See Example 6.*

21. Smith Engineering found that an experienced surveyor surveys a roadbed in 4 hours. An apprentice surveyor needs 5 hours to survey the same stretch of road. If the two work together, find how long it takes them to complete the job. $2\frac{2}{9}$ hr

22. An experienced bricklayer constructs a small wall in 3 hours. The apprentice completes the job in 6 hours. Find how long it takes if they work together.   2 hr

23. In 2 minutes, a conveyor belt moves 300 pounds of recyclable aluminum from the delivery truck to a storage area. A smaller belt moves the same quantity of cans the same distance in 6 minutes. If both belts are used, find how long it takes to move the cans to the storage area. $1\frac{1}{2}$ min

24. Find how long it takes the conveyor belts described in Exercise **23** to move 1200 pounds of cans. (*Hint:* Think of 1200 pounds as four 300-pound jobs.)   6 min

Objective E *See Example 7.*

25. A jogger begins her workout by jogging to the park, a distance of 12 miles. She then jogs home at the same speed but along a different route. This return trip is 18 miles and her time is one hour longer. Find her jogging speed. Complete the accompanying chart and use it to find her jogging speed.

	Distance	=	Rate	·	Time
Trip to Park	12				
Return Trip	18				

trip to park rate: r; to park time: $\dfrac{12}{r}$; return trip

rate: r; return time: $\dfrac{18}{r} = \dfrac{12}{r} + 1$; $r = 6$ mph

26. A boat can travel 9 miles upstream in the same amount of time it takes to travel 11 miles downstream. If the current of the river is 3 miles per hour, complete the chart below and use it to find the speed of the boat in still water.

	Distance	=	Rate	·	Time
Upstream	9		$r - 3$		
Downstream	11		$r + 3$		

upstream time: $\dfrac{9}{r-3}$; downstream time: $\dfrac{11}{r+3}$;
$r = 30$ mph

27. A cyclist rode the first 20-mile portion of his workout at a constant speed. For the 16-mile cooldown portion of his workout, he reduced his speed by 2 miles per hour. Each portion of the workout took the same time. Find the cyclist's speed during the first portion and find his speed during the cooldown portion.   1st portion: 10 mph; cooldown: 8 mph

28. A semi-truck travels 300 miles through the flatland in the same amount of time that it travels 180 miles through mountains. The rate of the truck is 20 miles per hour slower in the mountains than in the flatland. Find both the flatland rate and the mountain rate.   flatland: 50 mph; mountains: 30 mph

Objectives A B C D E Mixed Practice *Solve the following. See Examples 1 through 7. (Note: Some exercises can be modeled by equations without rational expressions.)*

29. A human factors expert recommends that there be at least 9 square feet of floor space in a college classroom for every student in the class. Find the minimum floor space that 40 students need.
360 sq ft

30. Due to space problems at a local university, a 20-foot by 12-foot conference room is converted into a classroom. Find the maximum number of students the room can accommodate. (See Exercise **29**.)
26 students

31. One-fourth equals the quotient of a number and 8. Find the number.   2

32. Four times a number added to 5 is divided by 6. The result is $\dfrac{7}{2}$. Find the number.   4

33. Marcus and Tony work for Lombardo's Pipe and Concrete. Mr. Lombardo is preparing an estimate for a customer. He knows that Marcus lays a slab of concrete in 6 hours. Tony lays the same size slab in 4 hours. If both work on the job and the cost of labor is $45.00 per hour, decide what the labor estimate should be. $108.00

34. Mr. Dodson can paint his house by himself in 4 days. His son needs an additional day to complete the job if he works by himself. If they work together, find how long it takes to paint the house. $2\dfrac{2}{9}$ days

35. A pilot can travel 400 miles with the wind in the same amount of time as 336 miles against the wind. Find the speed of the wind if the pilot's speed in still air is 230 miles per hour.   20 mph

36. A fisherman on Pearl River rows 9 miles downstream in the same amount of time he rows 3 miles upstream. If the current is 6 miles per hour, find how long it takes him to cover the 12 miles.   1 hr

37. Find the unknown length y. $y = 37\frac{1}{2}$ ft

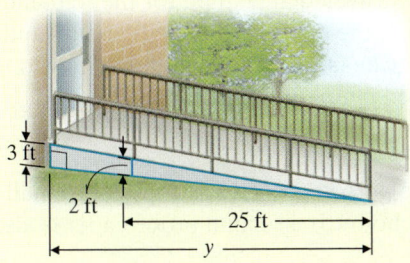

△**38.** Find the unknown length y. $y = 18$ ft

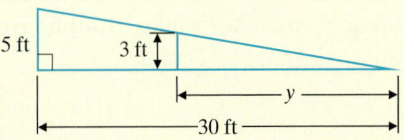

39. Suppose two trains leave Holbrook, Arizona, at the same time, traveling in opposite directions. One train travels 10 mph faster than the other. In 3.5 hours, the trains are 322 miles apart. Find the speed of each train. 41 mph; 51 mph

40. Suppose two cars leave Brinkley, Arkansas, at the same time, traveling in opposite directions. One car travels 8 mph faster than the other car. In 2.5 hours, the cars are 280 miles apart. Find the speed of each car. 52 mph; 60 mph

41. Two divided by the difference of a number and 3 minus 4 divided by the number plus 3, equals 8 times the reciprocal of the difference of the number squared and 9. What is the number? 5

42. If 15 times the reciprocal of a number is added to the ratio of 9 times the number minus 7 and the number plus 2, the result is 9. What is the number? 3

43. A pilot flies 630 miles with a tailwind of 35 miles per hour. Against the wind, he flies only 455 miles in the same amount of time. Find the rate of the plane in still air. 217 mph

44. A marketing manager travels 1080 miles in a corporate jet and then an additional 240 miles by car. If the car ride takes one hour longer than the jet ride takes, and if the rate of the jet is 6 times the rate of the car, find the time the manager travels by jet and find the time the manager travels by car. car: 4 hr; jet: 3 hr

45. To mix weed killer with water correctly, it is necessary to mix 8 teaspoons of weed killer with 2 gallons of water. Find how many gallons of water are needed to mix with the entire box if it contains 36 teaspoons of weed killer. 9 gal

46. The directions for a certain bug spray concentrate are to mix 3 ounces of concentrate with 2 gallons of water. How many ounces of concentrate are needed to mix with 5 gallons of water? 7.5 oz

47. A boater travels 16 miles per hour on the water on a still day. During one particularly windy day, he finds that he travels 48 miles with the wind behind him in the same amount of time that he travels 16 miles into the wind. Find the rate of the wind. 8 mph

Let x be the rate of the wind.

	r	·	t	=	d
with wind	$16 + x$				48
into wind	$16 - x$				16

48. The current on a portion of the Mississippi River is 3 miles per hour. A barge can go 6 miles upstream in the same amount of time it takes to go 10 miles downstream. Find the speed of the boat in still water. 12 mph

Let x be the speed of the boat in still water.

	r	·	t	=	d
upstream	$x - 3$				6
downstream	$x + 3$				10

49. Two hikers are 11 miles apart and walking toward each other. They meet in 2 hours. Find the rate of each hiker if one hiker walks 1.1 mph faster than the other.   2.2 mph; 3.3 mph

50. On a 255-mile trip, Gary Alessandrini traveled at an average speed of 70 mph, got a speeding ticket, and then traveled at 60 mph for the remainder of the trip. If the entire trip took 4.5 hours and the speeding ticket stop took 30 minutes, how long did Gary speed before getting stopped? $1\frac{1}{2}$ hr

51. One custodian cleans a suite of offices in 3 hours. When a second worker is asked to join the regular custodian, the job takes only $1\frac{1}{2}$ hours. How long does it take the second worker to do the same job alone?   3 hr

52. One person proofreads copy for a small newspaper in 4 hours. If a second proofreader is also employed, the job can be done in $2\frac{1}{2}$ hours. How long does it take for the second proofreader to do the same job alone? $6\frac{2}{3}$ hr

△ **53.** An architect is completing the plans for a triangular deck. Use the diagram below to find the missing dimension. $26\frac{2}{3}$ ft

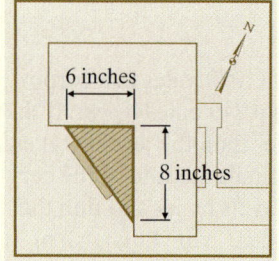
6 inches
8 inches

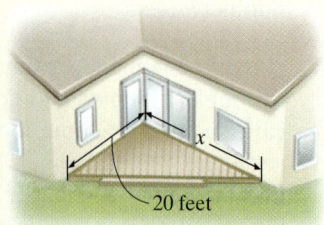
x
20 feet

△ **54.** A student wishes to make a small model of a triangular mainsail in order to study the effects of wind on the sail. The smaller model will be the same shape as a regular-size sailboat's mainsail. Use the following diagram to find the missing dimensions. $x = 4.4$ ft; $y = 5.6$ ft

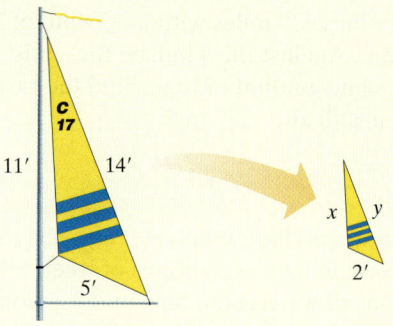
C 17
11'   14'
5'
x   y
2'

55. A manufacturer of cans of salted mixed nuts states that the ratio of peanuts to other nuts is 3 to 2. If 324 peanuts are in a can, find how many other nuts should also be in the can.   216 nuts

56. There are 1280 calories in a 14-ounce portion of Eagle Brand Milk. Find how many calories are in 2 ounces of Eagle Brand Milk. $182\frac{6}{7}$ cal

57. A jet plane traveling at 500 mph overtakes a propeller plane traveling at 200 mph that had a 2-hour head start. How far from the starting point are the planes? $666\frac{2}{3}$ mi

58. How long will it take a bus traveling at 60 miles per hour to overtake a car traveling at 40 miles per hour if the car had a 1.5-hour head start?   3 hr

59. One pipe fills a storage pool in 20 hours. A second pipe fills the same pool in 15 hours. When a third pipe is added and all three are used to fill the pool, it takes only 6 hours. Find how long it takes the third pipe to do the job.   20 hr

60. One pump fills a tank in 9 hours. A second pump fills the same tank in 6 hours. When a third pump is added and all three are used to fill the tank, it takes only 3 hours. Find how long it takes the third pump to fill the tank.   18 hr

▶ **61.** A car travels 280 miles in the same time that a motorcycle travels 240 miles. If the car's speed is 10 miles per hour more than the motorcycle's, find the speed of the car and the speed of the motorcycle.   car: 70 mph; motorcycle: 60 mph

62. A bus traveled on a level road for 3 hours at an average speed 20 miles per hour faster than it traveled on a winding road. The time spent on the winding road was 4 hours. Find the average speed on the level road if the entire trip was 305 miles.   55 mph

63. In 6 hours, an experienced cook prepares enough pies to supply a local restaurant's daily order. Another cook prepares the same number of pies in 7 hours. Together with a third cook, they prepare the pies in 2 hours. Find how long it takes the third cook to prepare the pies alone. $5\frac{1}{4}$ hr

64. Mrs. Smith balances the company books in 8 hours. It takes her assistant 12 hours to do the same job. If they work together, find how long it takes them to balance the books. $4\frac{4}{5}$ hr

65. The quotient of a number and 3, minus 1, equals $\frac{5}{3}$. Find the number.   8

66. The quotient of a number and 5, minus 1, equals $\frac{7}{5}$. Find the number.   12

67. Currently, the Ford Focus is the best-selling car in the world. A driver of this car took a day trip around the California coastline driving at two different speeds. He drove 70 miles at a slower speed and 300 miles at a speed 40 miles per hour faster. If the time spent driving at the faster speed was twice that spent driving at the slower speed, find the two speeds during the trip. (*Source: Top Ten of Everything*)   35 mph; 75 mph

68. The second best-selling car in the world is the Toyota Corolla. Suppose that during a test drive of two Corollas, one car travels 224 miles in the same time that the second car travels 175 miles. If the speed of the first car is 14 miles per hour faster than the speed of the second car, find the speed of both cars. (*Source: R. L. Polk*)   first car: 64 mph; second car: 50 mph

69. A pilot can fly an MD-11 2160 miles with the wind in the same time she can fly 1920 miles against the wind. If the speed of the wind is 30 mph, find the speed of the plane in still air. (*Source: Air Transport Association of America*)   510 mph

70. A pilot can fly a DC-10 1365 miles against the wind in the same time he can fly 1575 miles with the wind. If the speed of the plane in still air is 490 miles per hour, find the speed of the wind. (*Source: Air Transport Association of America*)   35 mph

Given that the following pairs of triangles are similar, find each missing length.

△ **71.** 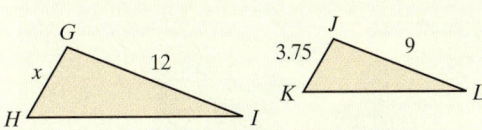 $x = 5$

△ **72.** 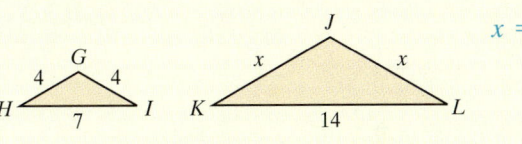 $x = 8$

△ **73.** 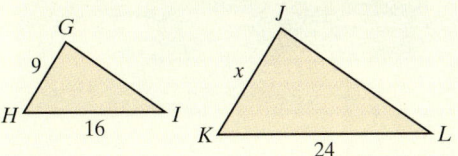 $x = 13.5$

△ **74.** 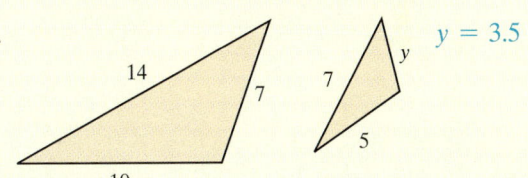 $y = 3.5$

Review

Simplify. Follow the circled steps in the order shown. See Section R.2.

75. $\left.\begin{array}{c}\dfrac{3}{4}+\dfrac{1}{4}\\[2mm]\dfrac{3}{8}+\dfrac{13}{8}\end{array}\right\}$ ←①Add.   ←②Add. $\dfrac{1}{2}$

76. $\left.\begin{array}{c}\dfrac{9}{5}+\dfrac{6}{5}\\[2mm]\dfrac{17}{6}+\dfrac{7}{6}\end{array}\right\}$ ←①Add.   ←②Add. $\dfrac{3}{4}$

77. $\left.\begin{array}{c}\dfrac{2}{5}+\dfrac{1}{5}\\[2mm]\dfrac{7}{10}+\dfrac{7}{10}\end{array}\right\}$ ①Add.   ←③Divide.   ②Add. $\dfrac{3}{7}$

78. $\left.\begin{array}{c}\dfrac{1}{4}+\dfrac{5}{4}\\[2mm]\dfrac{3}{8}+\dfrac{7}{8}\end{array}\right\}$ ①Add.   ←③Divide.   ②Add. $\dfrac{6}{5}$

Concept Extensions

79. One pump fills a tank 3 times as fast as another pump. If the pumps work together, they fill the tank in 21 minutes. How long does it take each pump to fill the tank?   faster pump: 28 min; slower pump: 84 min

80. For which of the following equations can we immediately use cross products to solve for x?   a

   a. $\dfrac{2-x}{5}=\dfrac{1+x}{3}$ **b.** $\dfrac{2}{5}-x=\dfrac{1+x}{3}$

81. Person A can complete a job in 5 hours, and person B can complete the same job in 3 hours. Without solving algebraically, discuss reasonable and unreasonable answers for how long it would take them to complete the job together.   *answers may vary*

82. For what value of x is $\dfrac{x}{x-1}$ in proportion to $\dfrac{x+1}{x}$? Explain your result.
none; answers may vary

Solve. See the Concept Check in this section.

Solve D = RT

83. for R $R = \dfrac{D}{T}$

84. for T $T = \dfrac{D}{R}$

85. A hyena spots a giraffe 0.5 mile away and begins running toward it. The giraffe starts running away from the hyena just as the hyena begins running toward it. A hyena can run at a speed of 40 mph and a giraffe can run at 32 mph. How long will it take the hyena to overtake the giraffe? (*Source: The World Almanac and Book of Facts*)   3.75 min

H   G
|⟵———— 0.5 mile ————⟶|

86. During the 2013 Formula 1 Grand Prix of Spain, Esteban Gutierrez posted the fastest lap of the race while Romain Grosjean's best race lap was the slowest fast lap in the field. The Spanish Grand Prix circuit is 4,655 kilometers long. When traveling at their fastest lap speeds, Grosjean drove 4,404 kilometers in the same time that Gutierrez completed an entire 4,655-kilometer lap. Gutierrez's fastest lap speed was 10.491 km/hr faster than Grosjean's fastest lap speed. Find each driver's fastest lap speed. Round each speed to the nearest tenth. (*Source: Formula One World Championship Limited*)
Gutierrez: 194.6 km/hr; Grosjean: 184.1 km/hr

7.7   Simplifying Complex Fractions

Objectives

A Simplify Complex Fractions Using Method 1.

B Simplify Complex Fractions Using Method 2.

A rational expression whose numerator or denominator or both numerator and denominator contain fractions is called a **complex rational expression** or a **complex fraction.** Some examples are

$$\dfrac{4}{2-\dfrac{1}{2}} \qquad \dfrac{\dfrac{3}{2}}{\dfrac{4}{7}-x} \qquad \dfrac{\dfrac{1}{x+2}}{x+2-\dfrac{1}{x}}$$

← Numerator of complex fraction
← Main fraction bar
← Denominator of complex fraction

Our goal in this section is to write complex fractions in simplest form. A complex fraction is in simplest form when it is in the form $\dfrac{P}{Q}$, where P and Q are polynomials that have no common factors.

Objective A Simplifying Complex Fractions—Method 1

In this section, two methods of simplifying complex fractions are presented. The first method presented uses the fact that the main fraction bar indicates division.

Method 1: To Simplify a Complex Fraction

Step 1: Add or subtract fractions in the numerator or denominator so that the numerator is a single fraction and the denominator is a single fraction.

Step 2: Perform the indicated division by multiplying the numerator of the complex fraction by the reciprocal of the denominator of the complex fraction.

Step 3: Write the rational expression in lowest terms.

Example 1 Simplify the complex fraction $\dfrac{\frac{5}{8}}{\frac{2}{3}}$.

Solution: Since the numerator and denominator of the complex fraction are already single fractions, we proceed to Step 2: Perform the indicated division by multiplying the numerator $\dfrac{5}{8}$ by the reciprocal of the denominator $\dfrac{2}{3}$.

$$\frac{\frac{5}{8}}{\frac{2}{3}} = \frac{5}{8} \div \frac{2}{3} = \frac{5}{8} \cdot \frac{3}{2} = \frac{15}{16}$$

The reciprocal of $\dfrac{2}{3}$ is $\dfrac{3}{2}$.

■ **Work Practice 1**

Practice 1

Simplify the complex fraction $\dfrac{\frac{3}{7}}{\frac{5}{9}}$.

Example 2 Simplify: $\dfrac{\frac{2}{3} + \frac{1}{5}}{\frac{2}{3} - \frac{2}{9}}$

Solution: We simplify the numerator and denominator of the complex fraction separately. First we add $\dfrac{2}{3}$ and $\dfrac{1}{5}$ to obtain a single fraction in the numerator. Then we subtract $\dfrac{2}{9}$ from $\dfrac{2}{3}$ to obtain a single fraction in the denominator.

$$\frac{\frac{2}{3} + \frac{1}{5}}{\frac{2}{3} - \frac{2}{9}} = \frac{\frac{2(5)}{3(5)} + \frac{1(3)}{5(3)}}{\frac{2(3)}{3(3)} - \frac{2}{9}}$$

    The LCD of the numerator's fractions is 15.

    The LCD of the denominator's fractions is 9.

$$= \frac{\frac{10}{15} + \frac{3}{15}}{\frac{6}{9} - \frac{2}{9}}$$

    Simplify.

$$= \frac{\frac{13}{15}}{\frac{4}{9}}$$

    Add the numerator's fractions.

    Subtract the denominator's fractions.

(Continued on next page)

Practice 2

Simplify: $\dfrac{\frac{3}{4} - \frac{2}{3}}{\frac{1}{2} + \frac{3}{8}}$

Answers

1. $\dfrac{27}{35}$ **2.** $\dfrac{2}{21}$

Next we perform the indicated division by multiplying the numerator of the complex fraction by the reciprocal of the denominator of the complex fraction.

$$\frac{\dfrac{13}{15}}{\dfrac{4}{9}} = \frac{13}{15} \cdot \frac{9}{4} \qquad \text{The reciprocal of } \frac{4}{9} \text{ is } \frac{9}{4}.$$

$$= \frac{13 \cdot 3 \cdot 3}{3 \cdot 5 \cdot 4} = \frac{39}{20}$$

■ **Work Practice 2**

Practice 3

Simplify: $\dfrac{\dfrac{2}{5} - \dfrac{1}{x}}{\dfrac{2x}{15} - \dfrac{1}{3}}$

Example 3   Simplify: $\dfrac{\dfrac{1}{z} - \dfrac{1}{2}}{\dfrac{1}{3} - \dfrac{z}{6}}$

Solution: Subtract to get a single fraction in the numerator and a single fraction in the denominator of the complex fraction.

$$\frac{\dfrac{1}{z} - \dfrac{1}{2}}{\dfrac{1}{3} - \dfrac{z}{6}} = \frac{\dfrac{2}{2z} - \dfrac{z}{2z}}{\dfrac{2}{6} - \dfrac{z}{6}} \qquad \begin{array}{l}\text{The LCD of the numerator's fractions is } 2z.\\[1.5em]\text{The LCD of the denominator's fractions is } 6.\end{array}$$

$$= \frac{\dfrac{2 - z}{2z}}{\dfrac{2 - z}{6}}$$

$$= \frac{2 - z}{2z} \cdot \frac{6}{2 - z} \qquad \text{Multiply by the reciprocal of } \frac{2 - z}{6}.$$

$$= \frac{2 \cdot 3 \cdot (2 - z)}{2 \cdot z \cdot (2 - z)} \qquad \text{Factor.}$$

$$= \frac{3}{z} \qquad \text{Write in lowest terms.}$$

■ **Work Practice 3**

Objective B Simplifying Complex Fractions— Method 2 ▶

Next we study a second method for simplifying complex fractions. In this method, we multiply the numerator and the denominator of the complex fraction by the LCD of all fractions in the complex fraction.

Method 2: To Simplify a Complex Fraction

Step 1: Find the LCD of all the fractions in the complex fraction.

Step 2: Multiply both the numerator and the denominator of the complex fraction by the LCD from Step 1.

Step 3: Perform the indicated operations and write the result in lowest terms.

Answer

3. $\dfrac{3}{x}$

We use Method 2 to rework Example 2.

Example 4 Simplify: $\dfrac{\dfrac{2}{3} + \dfrac{1}{5}}{\dfrac{2}{3} - \dfrac{2}{9}}$

Solution: The LCD of $\dfrac{2}{3}, \dfrac{1}{5}, \dfrac{2}{3}$, and $\dfrac{2}{9}$ is 45, so we multiply the numerator and the denominator of the complex fraction by 45. Then we perform the indicated operations and write in lowest terms.

$$\frac{\dfrac{2}{3} + \dfrac{1}{5}}{\dfrac{2}{3} - \dfrac{2}{9}} = \frac{45\left(\dfrac{2}{3} + \dfrac{1}{5}\right)}{45\left(\dfrac{2}{3} - \dfrac{2}{9}\right)}$$

$$= \frac{45\left(\dfrac{2}{3}\right) + 45\left(\dfrac{1}{5}\right)}{45\left(\dfrac{2}{3}\right) - 45\left(\dfrac{2}{9}\right)} \quad \text{Apply the distributive property.}$$

$$= \frac{30 + 9}{30 - 10} = \frac{39}{20} \quad \text{Simplify.}$$

■ **Work Practice 4**

Helpful Hint

The same complex fraction was simplified using two different methods in Examples 2 and 4. Notice that the simplified results are the same.

Example 5 Simplify: $\dfrac{\dfrac{x+1}{y}}{\dfrac{x}{y} + 2}$

Solution: The LCD of $\dfrac{x+1}{y}, \dfrac{x}{y}$ and $\dfrac{2}{1}$ is y, so we multiply the numerator and the denominator of the complex fraction by y.

$$\frac{\dfrac{x+1}{y}}{\dfrac{x}{y} + 2} = \frac{y\left(\dfrac{x+1}{y}\right)}{y\left(\dfrac{x}{y} + 2\right)}$$

$$= \frac{y\left(\dfrac{x+1}{y}\right)}{y\left(\dfrac{x}{y}\right) + y \cdot 2} \quad \text{Apply the distributive property in the denominator.}$$

$$= \frac{x+1}{x+2y} \quad \text{Simplify.}$$

■ **Work Practice 5**

Practice 4

Use Method 2 to simplify the complex fraction in Practice 2:

$$\frac{\dfrac{3}{4} - \dfrac{2}{3}}{\dfrac{1}{2} + \dfrac{3}{8}}$$

Teaching Tip

In Example 4, remind students that multiplying both the numerator and the denominator by 45 does not change the value of the expression because that is equivalent to multiplying the entire expression by 1.

Practice 5

Simplify: $\dfrac{1 + \dfrac{x}{y}}{\dfrac{2x+1}{y}}$

Answers

4. $\dfrac{2}{21}$ 5. $\dfrac{y+x}{2x+1}$

Practice 6

Simplify: $\dfrac{\dfrac{5}{6y}+\dfrac{y}{x}}{\dfrac{y}{3}-x}$

Teaching Tip

Have students verify the simplification of Example 6 by evaluating both expressions for $x = 2, y = 3$. In both cases, students should obtain $\dfrac{17}{48}$ as the result.

Answer

6. $\dfrac{5x+6y^2}{2xy^2-6x^2y}$ or $\dfrac{5x+6y^2}{2xy(y-3x)}$

Example 6 Simplify: $\dfrac{\dfrac{x}{y}+\dfrac{3}{2x}}{\dfrac{x}{2}+y}$

Solution: The LCD of $\dfrac{x}{y}, \dfrac{3}{2x}, \dfrac{x}{2}$, and $\dfrac{y}{1}$ is $2xy$, so we multiply both the numerator and the denominator of the complex fraction by $2xy$.

$$\dfrac{\dfrac{x}{y}+\dfrac{3}{2x}}{\dfrac{x}{2}+y}=\dfrac{2xy\left(\dfrac{x}{y}+\dfrac{3}{2x}\right)}{2xy\left(\dfrac{x}{2}+y\right)}$$

$$=\dfrac{2xy\left(\dfrac{x}{y}\right)+2xy\left(\dfrac{3}{2x}\right)}{2xy\left(\dfrac{x}{2}\right)+2xy(y)} \qquad \text{Apply the distributive property.}$$

$$=\dfrac{2x^2+3y}{x^2y+2xy^2}$$

$$\text{or } \dfrac{2x^2+3y}{xy(x+2y)}$$

■ **Work Practice 6**

Vocabulary, Readiness & Video Check

One method for simplifying a complex fraction is to multiply the fraction's numerator and denominator by the LCD of all fractions in the complex fraction. For each complex fraction, choose the LCD of its fractions.

1. $\dfrac{\dfrac{1}{4}+\dfrac{1}{2}}{\dfrac{1}{3}+\dfrac{1}{2}}$ The LCD for $\dfrac{1}{4}, \dfrac{1}{2}, \dfrac{1}{3}$, and $\dfrac{1}{2}$ is

   a. 4   **b.** 2   **c.** 12   **d.** 6   c

2. $\dfrac{\dfrac{3}{5}+\dfrac{2}{3}}{\dfrac{1}{10}+\dfrac{1}{6}}$ The LCD for $\dfrac{3}{5}, \dfrac{2}{3}, \dfrac{1}{10}$, and $\dfrac{1}{6}$ is

   a. 15   **b.** 30   **c.** 60   **d.** 180   b

3. $\dfrac{\dfrac{5}{2x^2}+\dfrac{3}{16x}}{\dfrac{x}{8}+\dfrac{3}{4x}}$ The LCD for $\dfrac{5}{2x^2}, \dfrac{3}{16x}, \dfrac{x}{8}$, and $\dfrac{3}{4x}$ is

   a. $16x^2$ **b.** $32x^3$ **c.** $16x$ **d.** $16x^3$ a

4. $\dfrac{\dfrac{11}{6}+\dfrac{10}{x^2}}{\dfrac{7}{9}+\dfrac{5}{x}}$ The LCD for $\dfrac{11}{6}, \dfrac{10}{x^2}, \dfrac{7}{9}$, and $\dfrac{5}{x}$ is

   a. 18   **b.** x^2 **c.** $18x^2$ **d.** $54x^3$ c

Martin-Gay Interactive Videos

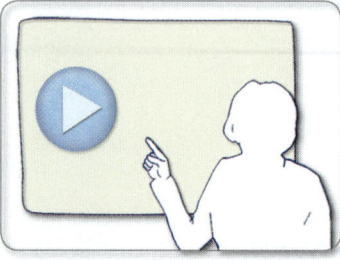

See Video 7.7

See video answer section.

Watch the section lecture video and answer the following questions.

Objective A 5. From ▦ Example 2, before we can rewrite the complex fraction as a division problem, what must we make sure we have? ▶

Objective B 6. How does finding an LCD in Method 2, as in ▦ Examples 4 and 5, differ from finding an LCD in Method 1? Mention the purpose of the LCD in each method. ▶

7.7 **Exercise Set** MyMathLab®

Objectives **A** **B** **Mixed Practice** *Simplify each complex fraction. See Examples 1 through 6.*

1. $\dfrac{\frac{1}{2}}{\frac{3}{4}}$ $\dfrac{2}{3}$

2. $\dfrac{\frac{1}{8}}{-\frac{5}{12}}$ $-\dfrac{3}{10}$

3. $\dfrac{-\frac{4x}{9}}{-\frac{2x}{3}}$ $\dfrac{2}{3}$

4. $\dfrac{-\frac{6y}{11}}{\frac{4y}{9}}$ $-\dfrac{27}{22}$

5. $\dfrac{\frac{1+x}{6}}{\frac{1+x}{3}}$ $\dfrac{1}{2}$

6. $\dfrac{\frac{6x-3}{5x^2}}{\frac{2x-1}{10x}}$ $\dfrac{6}{x}$

7. $\dfrac{\frac{1}{2}+\frac{2}{3}}{\frac{5}{9}-\frac{5}{6}}$ $-\dfrac{21}{5}$

8. $\dfrac{\frac{3}{4}-\frac{1}{2}}{\frac{3}{8}+\frac{1}{6}}$ $\dfrac{6}{13}$

9. $\dfrac{2+\frac{7}{10}}{1+\frac{3}{5}}$ $\dfrac{27}{16}$

10. $\dfrac{4-\frac{11}{12}}{5+\frac{1}{4}}$ $\dfrac{37}{63}$

11. $\dfrac{\frac{1}{3}}{\frac{1}{2}-\frac{1}{4}}$ $\dfrac{4}{3}$

12. $\dfrac{\frac{7}{10}-\frac{3}{5}}{\frac{1}{2}}$ $\dfrac{1}{5}$

13. $\dfrac{-\frac{2}{9}}{-\frac{14}{3}}$ $\dfrac{1}{21}$

14. $\dfrac{\frac{3}{8}}{\frac{4}{15}}$ $\dfrac{45}{32}$

15. $\dfrac{-\frac{5}{12x^2}}{\frac{25}{16x^3}}$ $-\dfrac{4x}{15}$

16. $\dfrac{-\frac{7}{8y}}{\frac{21}{4y}}$ $-\dfrac{1}{6}$

17. $\dfrac{\frac{m}{n}-1}{\frac{m}{n}+1}$ $\dfrac{m-n}{m+n}$

18. $\dfrac{\frac{x}{2}+2}{\frac{x}{2}-2}$ $\dfrac{x+4}{x-4}$

19. $\dfrac{\frac{1}{5}-\frac{1}{x}}{\frac{7}{10}+\frac{1}{x^2}}$ $\dfrac{2x(x-5)}{7x^2+10}$

20. $\dfrac{\frac{1}{y^2}+\frac{2}{3}}{\frac{1}{y}-\frac{5}{6}}$ $\dfrac{2(3+2y^2)}{6y-5y^2}$

21. $\dfrac{1+\frac{1}{y-2}}{y+\frac{1}{y-2}}$ $\dfrac{1}{y-1}$

22. $\dfrac{x-\frac{1}{2x+1}}{1-\frac{x}{2x+1}}$ $2x-1$

23. $\dfrac{\frac{4y-8}{16}}{\frac{6y-12}{4}}$ $\dfrac{1}{6}$

24. $\dfrac{\frac{7y+21}{3}}{\frac{3y+9}{8}}$ $\dfrac{56}{9}$

25. $\dfrac{\frac{x}{y}+1}{\frac{x}{y}-1}$ $\dfrac{x+y}{x-y}$

26. $\dfrac{\frac{3}{5y}+8}{\frac{3}{5y}-8}$ $\dfrac{3+40y}{3-40y}$

27. $\dfrac{1}{2+\frac{1}{3}}$ $\dfrac{3}{7}$

28. $\dfrac{3}{1-\frac{4}{3}}$ -9

29. $\dfrac{\frac{ax+ab}{x^2-b^2}}{\frac{x+b}{x-b}}$ $\dfrac{a}{x+b}$

30. $\dfrac{\frac{m+2}{m-2}}{\frac{2m+4}{m^2-4}}$ $\dfrac{m+2}{2}$

31. $\dfrac{\frac{-3+y}{4}}{\frac{8+y}{28}}$ $\dfrac{7(y-3)}{8+y}$

32. $\dfrac{\frac{-x+2}{18}}{\frac{8}{9}}$ $\dfrac{-x+2}{16}$

33. $\dfrac{3+\frac{12}{x}}{1-\frac{16}{x^2}}$ $\dfrac{3x}{x-4}$

34. $\dfrac{2+\frac{6}{x}}{1-\frac{9}{x^2}}$ $\dfrac{2x}{x-3}$

35. $\dfrac{\frac{8}{x+4}+2}{\frac{12}{x+4}-2}$ $-\dfrac{x+8}{x-2}$

36. $\dfrac{\frac{25}{x+5}+5}{\frac{3}{x+5}-5}$ $-\dfrac{5(x+10)}{5x+22}$

37. $\dfrac{\dfrac{s}{r}+\dfrac{r}{s}}{\dfrac{s}{r}-\dfrac{r}{s}}$ $\dfrac{s^2+r^2}{s^2-r^2}$

38. $\dfrac{\dfrac{2}{x}+\dfrac{x}{2}}{\dfrac{2}{x}-\dfrac{x}{2}}$ $\dfrac{4+x^2}{4-x^2}$

39. $\dfrac{\dfrac{6}{x-5}+\dfrac{x}{x-2}}{\dfrac{3}{x-6}-\dfrac{2}{x-5}}$ $\dfrac{(x-6)(x+4)}{x-2}$

40. $\dfrac{\dfrac{4}{x}+\dfrac{x}{x+1}}{\dfrac{1}{2x}+\dfrac{1}{x+6}}$ $\dfrac{2(x+2)(x+6)}{3(x+1)}$

Review

Use the bar graph below to answer Exercises 41 through 44. See Section R.3. Note: Some of these players are still competing; thus, their total prize money may increase.

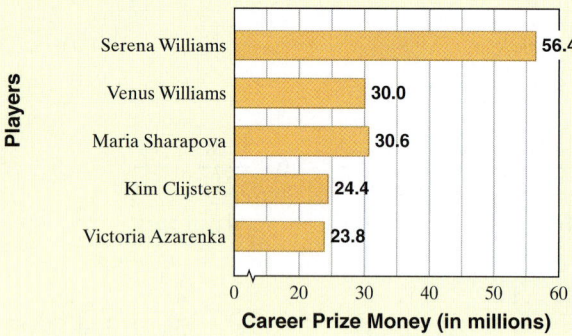

Women's Tennis Career Prize Money Leaders

Players (vertical axis): Serena Williams 56.4, Venus Williams 30.0, Maria Sharapova 30.6, Kim Clijsters 24.4, Victoria Azarenka 23.8

Career Prize Money (in millions) — horizontal axis: 0, 20, 30, 40, 50, 60

Source: WTA Media Information System, as of August 2014

41. Which women's tennis player has earned the most prize money in her career?   Serena Williams

42. How much more prize money has Maria Sharapova earned in her career than Victoria Azarenka? $6.8 million

43. What is the difference in lifetime prize money between Kim Clijsters and Venus Williams? $5.6 million

44. To date in her career, Serena Williams has won 82 doubles and singles tournament titles. Assuming her prize money is earned only for tournament titles, how much prize money has she earned, on average, per tournament title?
about $0.688 million or $688,000 per tournament title

Concept Extensions

45. Explain how to simplify a complex fraction using Method 1.   answers may vary

46. Explain how to simplify a complex fraction using Method 2.   answers may vary

To find the average of two numbers, we find their sum and divide by 2. For example, the average of 65 and 81 is found by simplifying $\dfrac{65+81}{2}$. *This simplifies to* $\dfrac{146}{2}=73$. *Use this for Exercises 47–50.*

47. Find the average of $\dfrac{1}{3}$ and $\dfrac{3}{4}$. $\dfrac{13}{24}$

48. Write the average of $\dfrac{3}{n}$ and $\dfrac{5}{n^2}$ as a simplified rational expression. $\dfrac{3n+5}{2n^2}$

49. A carpenter needs to drill a hole halfway between the two marked points. An intersecting board keeps him from measuring between the marked points, but he does have earlier measurements as shown. How far from the left side of the marked board should he drill?
$4\dfrac{1}{4}$ ft or 4.25 ft

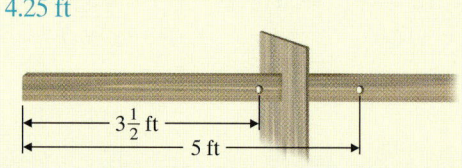

$3\dfrac{1}{2}$ ft

5 ft

50. Use the same diagram as for Exercise **49**. Suppose the measurements are 7.2 inches and 10.3 inches. How far from the left side of the marked board should he drill?   8.75 in. or $8\dfrac{3}{4}$ in.

Solve.

51. In electronics, when two resistors R_1 (read R sub 1) and R_2 (read R sub 2) are connected in parallel, the total resistance is given by the complex fraction

$$\frac{1}{\dfrac{1}{R_1} + \dfrac{1}{R_2}}.$$

Simplify this expression. $\dfrac{R_1 R_2}{R_2 + R_1}$

Resistance R_1 R_2

52. Astronomers occasionally need to know the day of the week a particular date fell on. The complex fraction

$$\frac{J + \dfrac{3}{2}}{7}$$

where J is the *Julian day number,* is used to make this calculation. Simplify this expression. $\dfrac{2J + 3}{14}$

Simplify each of the following. First, write each expression with positive exponents. Then simplify the complex fraction. The first step has been completed for Exercise 53.

53. $\dfrac{x^{-1} + 2^{-1}}{x^{-2} - 4^{-1}} = \dfrac{\dfrac{1}{x} + \dfrac{1}{2}}{\dfrac{1}{x^2} - \dfrac{1}{4}}$

$\dfrac{2x}{2 - x}$

54. $\dfrac{3^{-1} - x^{-1}}{9^{-1} - x^{-2}}$ $\dfrac{3x}{x + 3}$

55. $\dfrac{y^{-2}}{1 - y^{-2}}$ $\dfrac{1}{y^2 - 1}$

56. $\dfrac{4 + x^{-1}}{3 + x^{-1}}$ $\dfrac{4x + 1}{3x + 1}$

57. If the distance formula $d = r \cdot t$ is solved for t, then $t = \dfrac{d}{r}$. Use this formula to find t if distance d is $\dfrac{20x}{3}$ miles and rate r is $\dfrac{5x}{9}$ miles per hour. Write t in simplified form.   12 hr

△ **58.** If the formula for the area of a rectangle, $A = l \cdot w$, is solved for w, then $w = \dfrac{A}{l}$. Use this formula to find w if area A is $\dfrac{4x - 2}{3}$ square meters and length l is $\dfrac{6x - 3}{5}$ meters. Write w in simplified form. $\dfrac{10}{9}$ m

Chapter 7 Group Activity

Additional group activities are available in the *Instructor's Resource Manual with Tests.*

Fast-Growing Careers

According to U.S. Bureau of Labor Statistics projections, the careers listed below will have the largest job growth in the years shown.

Occupation	Employment (number in thousands)		
	2010	**2020**	**Change**
1. Registered nurses	2737.4	3449.3	+711.9
2. Retail salespersons	4261.6	4968.4	+706.8
3. Home health aides	1017.7	1723.9	+706.3
4. Personal care aides	861.0	1468.0	+607.0
5. Office clerks, general	2950.7	3440.2	+489.5
6. Combined food preparation and serving workers, including fast food	2682.1	3080.1	+398.0
7. Customer service representatives	2187.3	2525.6	+338.3
8. Heavy and tractor-trailer truck drivers	1604.8	1934.9	+330.1
9. Laborers and freight, stock, and material movers, hand	2068.2	2387.3	+319.1
10. Postsecondary teachers	1756.0	2061.7	+305.7

What do all of these in-demand occupations have in common? They all require a knowledge of math! For some careers, like nurses, postsecondary teachers, and salespersons, the ways math is used on the job may be obvious. For other occupations, the use of math may not be quite as obvious. However, tasks common to many jobs, such as filling in a time sheet or a medication log, writing up an expense report, planning a budget, figuring a bill, ordering supplies, and even making a work schedule, all require math.

Activity

Suppose that your college placement office is planning to publish an occupational handbook on math in popular occupations. Choose one of the occupations from the given list that interests you. Research the occupation. Then write a brief entry for the occupational handbook that describes how a person in that career would use math in his or her job. Include an example if possible.

Chapter 7 Vocabulary Check

Fill in each blank with one of the words or phrases listed below. Not all choices will be used.

least common denominator   simplifying   reciprocals   numerator $\dfrac{-a}{b}$

rational expression   unit   complex fraction   denominator $\dfrac{-a}{-b}$ $\dfrac{a}{-b}$

1. A ___rational expression___ is an expression that can be written in the form $\dfrac{P}{Q}$, where P and Q are polynomials and Q is not 0.

2. In a ___complex fraction___ the numerator or denominator or both may contain fractions.

3. For a rational expression, $-\dfrac{a}{b} = $ ___$\dfrac{-a}{b}$___ $ = $ ___$\dfrac{a}{-b}$___ .

4. A rational expression is undefined when the ___denominator___ is 0.

5. The process of writing a rational expression in lowest terms is called ___simplifying___ .

6. The expressions $\dfrac{2x}{7}$ and $\dfrac{7}{2x}$ are called ___reciprocals___ .

7. The ___least common denominator___ of a list of rational expressions is a polynomial of least degree whose factors include all factors of the denominators in the list.

8. A ___unit___ fraction is a fraction that equals 1.

Helpful Hint

▶ Are you preparing for your test? Don't forget to take the Chapter 7 Test on page 559. Then check your answers at the back of the text and use the Chapter Test Prep Videos to see the fully worked-out solutions to any of the exercises you want to review.

7 Chapter Highlights

Definitions and Concepts	Examples
Section 7.1 Simplifying Rational Expressions	

A **rational expression** is an expression that can be written in the form $\dfrac{P}{Q}$, where P and Q are polynomials and Q does not equal 0.

$$\frac{7y^3}{4}, \quad \frac{x^2 + 6x + 1}{x - 3}, \quad \frac{-5}{s^3 + 8}$$

To find values for which a rational expression is undefined, find values for which the denominator is 0.

Find any values for which the expression $\dfrac{5y}{y^2 - 4y + 3}$ is undefined.

$$
\begin{aligned}
y^2 - 4y + 3 &= 0 && \text{Set the denominator equal to 0.}\\
(y - 3)(y - 1) &= 0 && \text{Factor.}\\
y - 3 = 0 \ \text{ or } \ y - 1 &= 0 && \text{Set each factor equal to 0.}\\
y = 3 \qquad\quad y &= 1 && \text{Solve.}
\end{aligned}
$$

The expression is undefined when y is 3 and when y is 1.

To Simplify a Rational Expression

Step 1: Factor the numerator and denominator.

Step 2: Divide out factors common to the numerator and denominator. (This is the same as removing a factor of 1.)

Simplify: $\dfrac{4x + 20}{x^2 - 25}$

$$\frac{4x + 20}{x^2 - 25} = \frac{4\,(x + 5)}{(x + 5)\,(x - 5)} = \frac{4}{x - 5}$$

Section 7.2 Multiplying and Dividing Rational Expressions	

To Multiply Rational Expressions

Step 1: Factor numerators and denominators.

Step 2: Multiply numerators and multiply denominators.

Step 3: Write the product in lowest terms.

$$\frac{P}{Q} \cdot \frac{R}{S} = \frac{PR}{QS}$$

Multiply: $\dfrac{4x + 4}{2x - 3} \cdot \dfrac{2x^2 + x - 6}{x^2 - 1}$

$$
\begin{aligned}
&\frac{4x + 4}{2x - 3} \cdot \frac{2x^2 + x - 6}{x^2 - 1}\\[4pt]
&= \frac{4(x + 1)}{2x - 3} \cdot \frac{(2x - 3)(x + 2)}{(x + 1)(x - 1)}\\[4pt]
&= \frac{4\,(x + 1)(2x - 3)\,(x + 2)}{(2x - 3)(x + 1)\,(x - 1)}\\[4pt]
&= \frac{4(x + 2)}{x - 1}
\end{aligned}
$$

(continued)

Definitions and Concepts	Examples

Section 7.2 Multiplying and Dividing Rational Expressions (*continued*)

To divide by a rational expression, multiply by the reciprocal.

$$\frac{P}{Q} \div \frac{R}{S} = \frac{P}{Q} \cdot \frac{S}{R} = \frac{PS}{QR}$$

Divide: $\dfrac{15x + 5}{3x^2 - 14x - 5} \div \dfrac{15}{3x - 12}$

$$\frac{15x + 5}{3x^2 - 14x - 5} \div \frac{15}{3x - 12}$$

$$= \frac{5(3x + 1)}{(3x + 1)(x - 5)} \cdot \frac{3(x - 4)}{3 \cdot 5}$$

$$= \frac{x - 4}{x - 5}$$

Section 7.3 Adding and Subtracting Rational Expressions with the Same Denominator and Least Common Denominator

To add or subtract rational expressions with the same denominator, add or subtract numerators, and place the sum or difference over the common denominator.

$$\frac{P}{R} + \frac{Q}{R} = \frac{P + Q}{R}$$

$$\frac{P}{R} - \frac{Q}{R} = \frac{P - Q}{R}$$

Perform each indicated operation.

$$\frac{5}{x + 1} + \frac{x}{x + 1} = \frac{5 + x}{x + 1}$$

$$\frac{2y + 7}{y^2 - 9} - \frac{y + 4}{y^2 - 9}$$

$$= \frac{2y + 7 - (y + 4)}{y^2 - 9}$$

$$= \frac{2y + 7 - y - 4}{y^2 - 9}$$

$$= \frac{y + 3}{(y + 3)(y - 3)}$$

$$= \frac{1}{y - 3}$$

To Find the Least Common Denominator (LCD)

Step 1: Factor the denominators.

Step 2: The LCD is the product of all unique factors, each raised to a power equal to the greatest number of times that it appears in any one factored denominator.

Find the LCD for

$$\frac{7x}{x^2 + 10x + 25} \text{ and } \frac{11}{3x^2 + 15x}$$

$$x^2 + 10x + 25 = (x + 5)(x + 5)$$

$$3x^2 + 15x = 3x(x + 5)$$

$$\text{LCD} = 3x(x + 5)(x + 5) \text{ or}$$

$$3x(x + 5)^2$$

Definitions and Concepts	Examples

Section 7.4   Adding and Subtracting Rational Expressions with Different Denominators

To Add or Subtract Rational Expressions with Different Denominators

Step 1: Find the LCD.

Step 2: Rewrite each rational expression as an equivalent expression whose denominator is the LCD.

Step 3: Add or subtract numerators and place the sum or difference over the common denominator.

Step 4: Write the result in lowest terms.

Perform the indicated operation.

$$\frac{9x + 3}{x^2 - 9} - \frac{5}{x - 3}$$

$$= \frac{9x + 3}{(x + 3)(x - 3)} - \frac{5}{x - 3}$$

LCD is $(x + 3)(x - 3)$.

$$= \frac{9x + 3}{(x + 3)(x - 3)} - \frac{5(x + 3)}{(x - 3)(x + 3)}$$

$$= \frac{9x + 3 - 5(x + 3)}{(x + 3)(x - 3)}$$

$$= \frac{9x + 3 - 5x - 15}{(x + 3)(x - 3)}$$

$$= \frac{4x - 12}{(x + 3)(x - 3)}$$

$$= \frac{4(x - 3)}{(x + 3)(x - 3)} = \frac{4}{x + 3}$$

Section 7.5   Solving Equations Containing Rational Expressions

To Solve an Equation Containing Rational Expressions

Step 1: Multiply both sides of the equation by the LCD of all rational expressions in the equation.

Step 2: Remove any grouping symbols and solve the resulting equation.

Step 3: Check the solution in the original equation.

Solve: $\dfrac{5x}{x + 2} + 3 = \dfrac{4x - 6}{x + 2}$ The LCD is $x + 2$.

$$(x + 2)\left(\frac{5x}{x + 2} + 3\right) = (x + 2)\left(\frac{4x - 6}{x + 2}\right)$$

$$(x + 2)\left(\frac{5x}{x + 2}\right) + (x + 2)(3) = (x + 2)\left(\frac{4x - 6}{x + 2}\right)$$

$$5x + 3x + 6 = 4x - 6$$

$$4x = -12$$

$$x = -3$$

The solution checks; the solution is -3.

Section 7.6   Proportions and Problem Solving with Rational Equations

A **ratio** is the quotient of two numbers or two quantities. A **proportion** is a mathematical statement that two ratios are equal.

Cross products:

If $\dfrac{a}{b} = \dfrac{c}{d}$, then $ad = bc$.

Proportions

$$\frac{2}{3} = \frac{8}{12} \qquad \frac{x}{7} = \frac{15}{35}$$

Cross Products

$2 \cdot 12$ or 24 $\qquad\qquad$ $3 \cdot 8$ or 24

$$\frac{2}{3} = \frac{8}{12}$$

Solve: $\dfrac{3}{4} = \dfrac{x}{x - 1}$

$$\frac{3}{4} = \frac{x}{x - 1}$$

$$3(x - 1) = 4x \qquad \text{Set cross products equal.}$$

$$3x - 3 = 4x$$

$$-3 = x$$

(continued)

Definitions and Concepts	Examples

Section 7.6   Proportions and Problem Solving with Rational Equations (*continued*)

Problem-Solving Steps

1. UNDERSTAND. Read and reread the problem.

A small plane and a car leave Kansas City, Missouri, and head for Minneapolis, Minnesota, a distance of 450 miles. The speed of the plane is 3 times the speed of the car, and the plane arrives 6 hours ahead of the car. Find the speed of the car.

Let x = the speed of the car.

Then $3x$ = the speed of the plane.

	Distance	= Rate	· Time
Car	450	x	$\dfrac{450}{x}\left(\dfrac{\text{distance}}{\text{rate}}\right)$
Plane	450	$3x$	$\dfrac{450}{3x}\left(\dfrac{\text{distance}}{\text{rate}}\right)$

2. TRANSLATE.

In words: plane's time   +   6 hours   =   car's time

3. SOLVE.

Translate: $\dfrac{450}{3x}$ +   6   = $\dfrac{450}{x}$

$$\frac{450}{3x} + 6 = \frac{450}{x}$$

$$3x\left(\frac{450}{3x}\right) + 3x(6) = 3x\left(\frac{450}{x}\right)$$

$$450 + 18x = 1350$$

$$18x = 900$$

$$x = 50$$

4. INTERPRET.

Check this solution in the originally stated problem. **State** the conclusion: The speed of the car is 50 miles per hour.

Section 7.7   Simplifying Complex Fractions

Method 1: To Simplify a Complex Fraction

Step 1:   Add or subtract fractions in the numerator and the denominator of the complex fraction.

Step 2:   Perform the indicated division.

Step 3:   Write the result in lowest terms.

Simplify:

$$\frac{\dfrac{1}{x} + 2}{\dfrac{1}{x} - \dfrac{1}{y}} = \frac{\dfrac{1}{x} + \dfrac{2x}{x}}{\dfrac{y}{xy} - \dfrac{x}{xy}}$$

$$= \frac{\dfrac{1 + 2x}{x}}{\dfrac{y - x}{xy}}$$

$$= \frac{1 + 2x}{x} \cdot \frac{x\,y}{y - x}$$

$$= \frac{y(1 + 2x)}{y - x}$$

Definitions and Concepts	Examples

Section 7.7 Simplifying Complex Fractions (*continued*)

Method 2: To Simplify a Complex Fraction

Step 1: Find the LCD of all fractions in the complex fraction.

Step 2: Multiply the numerator and the denominator of the complex fraction by the LCD.

Step 3: Perform the indicated operations and write the result in lowest terms.

$$\frac{\dfrac{1}{x} + 2}{\dfrac{1}{x} - \dfrac{1}{y}} = \frac{xy\left(\dfrac{1}{x} + 2\right)}{xy\left(\dfrac{1}{x} - \dfrac{1}{y}\right)}$$

$$= \frac{xy\left(\dfrac{1}{x}\right) + xy(2)}{xy\left(\dfrac{1}{x}\right) - xy\left(\dfrac{1}{y}\right)}$$

$$= \frac{y + 2xy}{y - x} \quad \text{or} \quad \frac{y(1 + 2x)}{y - x}$$

Chapter 7   Review

(7.1) *Find any real number(s) for which each rational expression is undefined.*

1. $\dfrac{x + 5}{x^2 - 4}$ $x = 2, x = -2$

2. $\dfrac{5x + 9}{4x^2 - 4x - 15}$ $x = \dfrac{5}{2}, x = -\dfrac{3}{2}$

Find the value of each rational expression when $x = 5, y = 7,$ and $z = -2$.

3. $\dfrac{2 - z}{z + 5}$ $\dfrac{4}{3}$

4. $\dfrac{x^2 + xy - y^2}{x + y}$ $\dfrac{11}{12}$

Simplify each rational expression.

5. $\dfrac{2x + 6}{x^2 + 3x}$ $\dfrac{2}{x}$

6. $\dfrac{3x - 12}{x^2 - 4x}$ $\dfrac{3}{x}$

7. $\dfrac{x + 2}{x^2 - 3x - 10}$ $\dfrac{1}{x - 5}$

8. $\dfrac{x + 4}{x^2 + 5x + 4}$ $\dfrac{1}{x + 1}$

9. $\dfrac{x^3 - 4x}{x^2 + 3x + 2}$ $\dfrac{x(x - 2)}{x + 1}$

10. $\dfrac{5x^2 - 125}{x^2 + 2x - 15}$ $\dfrac{5(x - 5)}{x - 3}$

11. $\dfrac{x^2 - x - 6}{x^2 - 3x - 10}$ $\dfrac{x - 3}{x - 5}$

12. $\dfrac{x^2 - 2x}{x^2 + 2x - 8}$ $\dfrac{x}{x + 4}$

Simplify each expression. First, factor the four-term polynomials by grouping.

13. $\dfrac{x^2 + xa + xb + ab}{x^2 - xc + bx - bc}$ $\dfrac{x + a}{x - c}$

14. $\dfrac{x^2 + 5x - 2x - 10}{x^2 - 3x - 2x + 6}$ $\dfrac{x + 5}{x - 3}$

(7.2) *Perform each indicated operation and simplify.*

15. $\dfrac{15x^3y^2}{z} \cdot \dfrac{z}{5xy^3}$ $\dfrac{3x^2}{y}$

16. $\dfrac{-y^3}{8} \cdot \dfrac{9x^2}{y^3}$ $-\dfrac{9x^2}{8}$

17. $\dfrac{x^2-9}{x^2-4} \cdot \dfrac{x-2}{x+3}$ $\dfrac{x-3}{x+2}$

18. $\dfrac{2x+5}{x-6} \cdot \dfrac{2x}{-x+6}$ $\dfrac{-2x(2x+5)}{(x-6)^2}$

19. $\dfrac{x^2-5x-24}{x^2-x-12} \div \dfrac{x^2-10x+16}{x^2+x-6}$ $\dfrac{x+3}{x-4}$

20. $\dfrac{4x+4y}{xy^2} \div \dfrac{3x+3y}{x^2y}$ $\dfrac{4x}{3y}$

21. $\dfrac{x^2+x-42}{x-3} \cdot \dfrac{(x-3)^2}{x+7}$ $(x-6)(x-3)$

22. $\dfrac{2a+2b}{3} \cdot \dfrac{a-b}{a^2-b^2}$ $\dfrac{2}{3}$

23. $\dfrac{2x^2-9x+9}{8x-12} \div \dfrac{x^2-3x}{2x}$ $\dfrac{1}{2}$

24. $\dfrac{x^2-y^2}{x^2+xy} \div \dfrac{3x^2-2xy-y^2}{3x^2+6x}$ $\dfrac{3(x+2)}{3x+y}$

(7.3) *Perform each indicated operation and simplify.*

25. $\dfrac{x}{x^2+9x+14} + \dfrac{7}{x^2+9x+14}$ $\dfrac{1}{x+2}$

26. $\dfrac{x}{x^2+2x-15} + \dfrac{5}{x^2+2x-15}$ $\dfrac{1}{x-3}$

27. $\dfrac{4x-5}{3x^2} - \dfrac{2x+5}{3x^2}$ $\dfrac{2(x-5)}{3x^2}$

28. $\dfrac{9x+7}{6x^2} - \dfrac{3x+4}{6x^2}$ $\dfrac{2x+1}{2x^2}$

Find the LCD of each pair of rational expressions.

29. $\dfrac{x+4}{2x}, \dfrac{3}{7x}$ $14x$

30. $\dfrac{x-2}{x^2-5x-24}, \dfrac{3}{x^2+11x+24}$ $(x-8)(x+8)(x+3)$

Rewrite each rational expression as an equivalent expression whose denominator is the given polynomial.

31. $\dfrac{5}{7x} = \dfrac{10x^2y}{14x^3y}$

32. $\dfrac{9}{4y} = \dfrac{36y^2x}{16y^3x}$

33. $\dfrac{x+2}{x^2+11x+18} = \dfrac{x^2-3x-10}{(x+2)(x-5)(x+9)}$

34. $\dfrac{3x-5}{x^2+4x+4} = \dfrac{3x^2+4x-15}{(x+2)^2(x+3)}$

(7.4) *Perform each indicated operation and simplify.*

35. $\dfrac{4}{5x^2} + \dfrac{6}{y}$ $\dfrac{4y+30x^2}{5x^2y}$

36. $\dfrac{2}{x-3} - \dfrac{4}{x-1}$ $\dfrac{-2x+10}{(x-3)(x-1)}$

37. $\dfrac{4}{x+3} - 2$ $\dfrac{-2x-2}{x+3}$

38. $\dfrac{3}{x^2+2x-8} + \dfrac{2}{x^2-3x+2}$ $\dfrac{5(x+1)}{(x+4)(x-2)(x-1)}$

39. $\dfrac{2x-5}{6x+9} - \dfrac{4}{2x^2+3x}$ $\dfrac{x-4}{3x}$

40. $\dfrac{x-1}{x^2-2x+1} - \dfrac{x+1}{x-1}$ $-\dfrac{x}{x-1}$

(7.5) *Solve each equation.*

41. $\dfrac{n}{10} = 9 - \dfrac{n}{5}$ 30

42. $\dfrac{2}{x+1} - \dfrac{1}{x-2} = -\dfrac{1}{2}$ 3, −4

43. $\dfrac{y}{2y+2} + \dfrac{2y-16}{4y+4} = \dfrac{y-3}{y+1}$ no solution

44. $\dfrac{2}{x-3} - \dfrac{4}{x+3} = \dfrac{8}{x^2-9}$ 5

45. $\dfrac{x-3}{x+1} - \dfrac{x-6}{x+5} = 0$ $\dfrac{9}{7}$

46. $x + 5 = \dfrac{6}{x}$ −6, 1

(7.6) *Solve each proportion.*

47. $\dfrac{2}{x-1} = \dfrac{3}{x+3}$ 9

48. $\dfrac{4}{y-3} = \dfrac{2}{y-3}$ no solution

Solve.

49. A machine can process 300 parts in 20 minutes. Find how many parts can be processed in 45 minutes.
675 parts

50. As his consulting fee, Mr. Visconti charges $90.00 per day. Find how much he charges for 3 hours of consulting. Assume an 8-hour workday. $33.75

51. Five times the reciprocal of a number equals the sum of $\dfrac{3}{2}$ the reciprocal of the number and $\dfrac{7}{6}$. What is the number?   3

52. The reciprocal of a number equals the reciprocal of the difference of 4 and the number. Find the number.   2

53. A car travels 90 miles in the same time that a car traveling 10 miles per hour slower travels 60 miles. Find the speed of each car.
fast car speed: 30 mph; slow car speed: 20 mph

54. The current in a bayou near Lafayette, Louisiana, is 4 miles per hour. A paddleboat travels 48 miles upstream in the same amount of time it takes to travel 72 miles downstream. Find the speed of the boat in still water.   20 mph

55. When Mark and Maria manicure Mr. Stergeon's lawn, it takes them 5 hours. If Mark works alone, it takes 7 hours. Find how long it takes Maria alone. $17\dfrac{1}{2}$ hr

56. It takes pipe A 20 days to fill a fish pond. Pipe B takes 15 days. Find how long it takes both pipes together to fill the pond. $8\dfrac{4}{7}$ days

Given that the pairs of triangles are similar, find each missing length x.

△**57.** 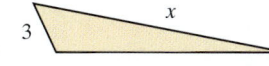   x = 15   △**58.** 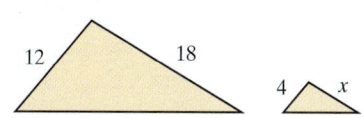   x = 6

(7.7) *Simplify each complex fraction.*

59. $\dfrac{\dfrac{5x}{27}}{-\dfrac{10xy}{21}}$ $-\dfrac{7}{18y}$

60. $\dfrac{\dfrac{3}{5}+\dfrac{2}{7}}{\dfrac{1}{5}+\dfrac{5}{6}}$ $\dfrac{6}{7}$

61. $\dfrac{3-\dfrac{1}{y}}{2-\dfrac{1}{y}}$ $\dfrac{3y-1}{2y-1}$

62. $\dfrac{\dfrac{6}{x+2}+4}{\dfrac{8}{x+2}-4}$ $-\dfrac{7+2x}{2x}$

Mixed Review

Simplify each rational expression.

63. $\dfrac{4x+12}{8x^2+24x}$ $\dfrac{1}{2x}$

64. $\dfrac{x^3-6x^2+9x}{x^2+4x-21}$ $\dfrac{x(x-3)}{x+7}$

Perform the indicated operations and simplify.

65. $\dfrac{x^2+9x+20}{x^2-25}\cdot\dfrac{x^2-9x+20}{x^2+8x+16}$ $\dfrac{x-4}{x+4}$

66. $\dfrac{x^2-x-72}{x^2-x-30}\div\dfrac{x^2+6x-27}{x^2-9x+18}$ $\dfrac{(x-9)(x+8)}{(x+5)(x+9)}$

67. $\dfrac{x}{x^2-36}+\dfrac{6}{x^2-36}$ $\dfrac{1}{x-6}$

68. $\dfrac{5x-1}{4x}-\dfrac{3x-2}{4x}$ $\dfrac{2x+1}{4x}$

69. $\dfrac{4}{3x^2+8x-3}+\dfrac{2}{3x^2-7x+2}$ $\dfrac{2}{(x+3)(x-2)}$

70. $\dfrac{3x}{x^2+9x+14}-\dfrac{6x}{x^2+4x-21}$ $-\dfrac{3x}{(x+2)(x-3)}$

Solve.

71. $\dfrac{4}{a-1}+2=\dfrac{3}{a-1}$ $\dfrac{1}{2}$

72. $\dfrac{x}{x+3}+4=\dfrac{x}{x+3}$ no solution

Solve.

73. The quotient of twice a number and 3, minus one-sixth, is the quotient of the number and 2. Find the number.   1

74. Mr. Crocker can paint his shed by himself in three days. His son will need an additional day to complete the job if he works alone. If they work together, find how long it takes to paint the shed. $1\dfrac{5}{7}$ days

Given that the following pairs of triangles are similar, find each missing length.

75. 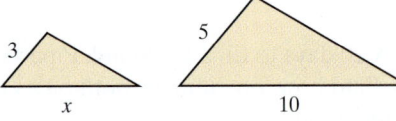 $x=6$

76. 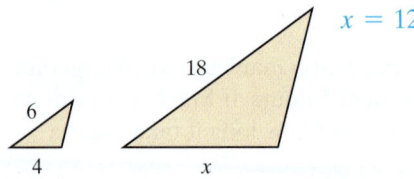 $x=12$

Simplify each complex fraction.

77. $\dfrac{\dfrac{1}{4}}{\dfrac{1}{3}+\dfrac{1}{2}}$ $\dfrac{3}{10}$

78. $\dfrac{4+\dfrac{2}{x}}{6+\dfrac{3}{x}}$ $\dfrac{2}{3}$

1. Find any real numbers for which the following expression is undefined.

$$\frac{x + 5}{x^2 + 4x + 3}$$

2. For a certain computer desk, the average manufacturing cost C per desk (in dollars) is

$$C = \frac{100x + 3000}{x}$$

where x is the number of desks manufactured.
   a. Find the average cost per desk when manufacturing 200 computer desks.
   b. Find the average cost per desk when manufacturing 1000 computer desks.

Simplify each rational expression.

3. $\dfrac{3x - 6}{5x - 10}$

4. $\dfrac{x + 6}{x^2 + 12x + 36}$

5. $\dfrac{7 - x}{x - 7}$

6. $\dfrac{y - x}{x^2 - y^2}$

7. $\dfrac{2m^3 - 2m^2 - 12m}{m^2 - 5m + 6}$

8. $\dfrac{ay + 3a + 2y + 6}{ay + 3a + 5y + 15}$

Perform each indicated operation and simplify if possible.

9. $\dfrac{x^2 - 13x + 42}{x^2 + 10x + 21} \div \dfrac{x^2 - 4}{x^2 + x - 6}$

10. $\dfrac{3}{x - 1} \cdot (5x - 5)$

11. $\dfrac{y^2 - 5y + 6}{2y + 4} \cdot \dfrac{y + 2}{2y - 6}$

12. $\dfrac{5}{2x + 5} - \dfrac{6}{2x + 5}$

13. $\dfrac{5a}{a^2 - a - 6} - \dfrac{2}{a - 3}$

14. $\dfrac{6}{x^2 - 1} + \dfrac{3}{x + 1}$

15. $\dfrac{x^2 - 9}{x^2 - 3x} \div \dfrac{x^2 + 4x + 1}{2x + 10}$

16. $\dfrac{x + 2}{x^2 + 11x + 18} + \dfrac{5}{x^2 - 3x - 10}$

17. $\dfrac{4y}{y^2 + 6y + 5} - \dfrac{3}{y^2 + 5y + 4}$

Answers

Note to Instructor: The Chapter 7 Test file in the TestGen program provides algorithms specifically matched to this test for easy replication for practice or assessment purposes.

1. $x = -1, x = -3$

2. a. $115

    b. $103

3. $\dfrac{3}{5}$

4. $\dfrac{1}{x + 6}$

5. -1

6. $-\dfrac{1}{x + y}$

7. $\dfrac{2m(m + 2)}{m - 2}$

8. $\dfrac{a + 2}{a + 5}$

9. $\dfrac{(x - 6)(x - 7)}{(x + 7)(x + 2)}$

10. 15

11. $\dfrac{y - 2}{4}$

12. $-\dfrac{1}{2x + 5}$

13. $\dfrac{3a - 4}{(a - 3)(a + 2)}$

14. $\dfrac{3}{x - 1}$

15. $\dfrac{2(x + 3)(x + 5)}{x(x^2 + 4x + 1)}$

16. $\dfrac{x^2 + 2x + 35}{(x + 9)(x + 2)(x - 5)}$

17. $\dfrac{4y^2 + 13y - 15}{(y + 5)(y + 1)(y + 4)}$

18. $\dfrac{30}{11}$

19. -6

20. no solution

21. no solution

22. $-2, 5$

23. $\dfrac{xz}{2y}$

24. $b - a$

25. $\dfrac{5y^2 - 1}{y + 2}$

26. 1 or 5

27. 30 mph

28. $6\dfrac{2}{3}$ hr

29. $x = 12$

30. 18 bulbs

Solve each equation.

18. $\dfrac{4}{y} - \dfrac{5}{3} = -\dfrac{1}{5}$

19. $\dfrac{5}{y + 1} = \dfrac{4}{y + 2}$

20. $\dfrac{a}{a - 3} = \dfrac{3}{a - 3} - \dfrac{3}{2}$

21. $\dfrac{10}{x^2 - 25} = \dfrac{3}{x + 5} + \dfrac{1}{x - 5}$

22. $x - \dfrac{14}{x - 1} = 4 - \dfrac{2x}{x - 1}$

Simplify each complex fraction.

23. $\dfrac{\dfrac{5x^2}{yz^2}}{\dfrac{10x}{z^3}}$

24. $\dfrac{\dfrac{b}{a} - \dfrac{a}{b}}{\dfrac{1}{b} + \dfrac{1}{a}}$

25. $\dfrac{5 - \dfrac{1}{y^2}}{\dfrac{1}{y} + \dfrac{2}{y^2}}$

26. One number plus five times its reciprocal is equal to six. Find the number.

27. A pleasure boat traveling down the Red River takes the same time to go 14 miles upstream as it takes to go 16 miles downstream. If the current of the river is 2 miles per hour, find the speed of the boat in still water.

28. An inlet pipe can fill a tank in 12 hours. A second pipe can fill the tank in 15 hours. If both pipes are used, find how long it takes to fill the tank.

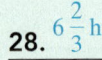 **29.** Given that the two triangles are similar, find x.

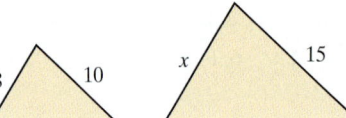

30. In a sample of 85 fluorescent bulbs, 3 were found to be defective. At this rate, how many defective bulbs should be found in 510 bulbs?

1. Write each sentence as an equation. Let x represent the unknown number.
   a. The quotient of 15 and a number is 4.
   b. Three subtracted from 12 is a number.
   c. 17 added to four times a number is 21.

2. Write each sentence as an equation. Let x represent the unknown number.
   a. The difference of 12 and a number is -45.
   b. The product of 12 and a number is -45.
   c. A number less 10 is twice the number.

3. Find each sum.
   a. $3 + (-7) + (-8)$
   b. $[7 + (-10)] + [-2 + (-4)]$

4. Find each difference.
   a. $28 - 6 - 30$
   b. $7 - 2 - 22$

For Exercises 5 through 8, name the property illustrated by each true statement.

5. $3(x + y) = 3 \cdot x + 3 \cdot y$

6. $3 + y = y + 3$

7. $(x + 7) + 9 = x + (7 + 9)$

8. $(x \cdot 7) \cdot 9 = x \cdot (7 \cdot 9)$

9. Solve: $3 - x = 7$

10. Solve: $7x - 6 = 6x - 6$

11. A 10-foot board is to be cut into two pieces so that the length of the longer piece is 4 times the length of the shorter. Find the length of each piece.

12. Find two consecutive even integers whose sum is 382.

13. Graph $y = -3$.

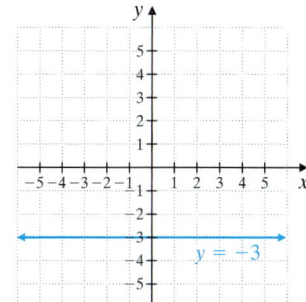

14. Complete the table for the equation $2x + y = 6$.

x	y
0	6
4	-2
3	0

Answers

1. a. $\dfrac{15}{x} = 4$

   b. $12 - 3 = x$

   c. $4x + 17 = 21$ (Sec. 1.3, Ex. 10)

2. a. $12 - x = -45$

   b. $12x = -45$

   c. $x - 10 = 2x$ (Sec. 1.4, 1.5)

3. a. -12

   b. -9 (Sec. 1.4, Ex. 12)

4. a. -8

   b. -17 (Sec. 1.5)

5. distributive property (Sec. 1.7, Ex. 15)

6. commutative property of addition (Sec. 1.7)

7. associative property of addition (Sec. 1.7, Ex. 16)

8. associative property of multiplication (Sec. 1.7)

9. $x = -4$ (Sec. 2.1, Ex. 7)

10. $x = 0$ (Sec. 2.1)

11. shorter piece, 2 ft; longer piece, 8 ft; (Sec. 2.4, Ex. 3)

12. 190, 192 (Sec. 2.4)

13. See graph (Sec. 3.3, Ex. 7)

14. see table (Sec. 3.1)

15. Find an equation of the line with y-intercept $(0, -3)$ and slope of $\frac{1}{4}$.

16. Find an equation of the line perpendicular to $y = 2x + 4$ and passing through $(1, 5)$.

Simplify.

17. $\dfrac{x^5}{x^2}$

18. $\dfrac{y^{14}}{y^{14}}$

19. $\dfrac{4^7}{4^3}$

20. $(x^5y^2)^3$

21. $\dfrac{(-3)^5}{(-3)^2}$

22. $\dfrac{x^{19}y^5}{xy}$

23. $\dfrac{2x^5y^2}{xy}$

24. $(-3a^2b)(5a^3b)$

Simplify by writing each expression with positive exponents only.

25. $2x^{-3}$

26. 7^{-2}

27. $(-2)^{-4}$

28. $5z^{-7}$

Multiply.

29. $5x(2x^3 + 6)$

30. $(x + 9)^2$

31. $-3x^2(5x^2 + 6x - 1)$

32. $(2x + 1)(2x - 1)$

Perform the indicated operations.

33. Divide: $\dfrac{4x^2 + 7 + 8x^3}{2x + 3}$

34. Divide: $(4x^3 - 9x + 2)$ by $(x - 4)$

35. Factor: $x^2 + 7x + 12$

36. Factor: $-2a^2 + 10a + 12$

37. Factor: $x^2 - 9$

38. Factor: $x^2 - 4$

39. Solve: $x^2 - 9x - 22 = 0$

40. Solve: $3x^2 + 5x = 2$

41. Multiply: $\dfrac{x^2 + x}{3x} \cdot \dfrac{6}{5x + 5}$

42. Simplify: $\dfrac{2x^2 - 50}{4x^4 - 20x^3}$

43. Subtract: $\dfrac{3x^2 + 2x}{x - 1} - \dfrac{10x - 5}{x - 1}$

44. Factor: $7x^6 - 7x^5 + 7x^4$

45. Subtract: $\dfrac{6x}{x^2 - 4} - \dfrac{3}{x + 2}$

46. Factor: $4x^2 + 12x + 9$

47. Solve: $\dfrac{t - 4}{2} - \dfrac{t - 3}{9} = \dfrac{5}{18}$

48. Multiply: $\dfrac{6x^2 - 18x}{3x^2 - 2x} \cdot \dfrac{15x - 10}{x^2 - 9}$

49. Sam Waterton and Frank Schaffer work in a plant that manufactures automobiles. Sam can complete a quality control tour of the plant in 3 hours while his assistant, Frank, needs 7 hours to complete the same job. The regional manager is coming to inspect the plant facilities, so both Sam and Frank are directed to complete a quality control tour together. How long will this take?

50. Simplify: $\dfrac{\dfrac{m}{3} + \dfrac{n}{6}}{\dfrac{m + n}{12}}$

37. $(x + 3)(x - 3)$ (Sec. 6.5, Ex. 5)

38. $(x + 2)(x - 2)$ (Sec. 6.5)

39. $11, -2$ (Sec. 6.6, Ex. 4)

40. $-2, \dfrac{1}{3}$ (Sec. 6.6)

41. $\dfrac{2}{5}$ (Sec. 7.2, Ex. 2)

42. $\dfrac{x + 5}{2x^3}$ (Sec. 7.1)

43. $3x - 5$ (Sec. 7.3, Ex. 3)

44. $7x^4(x^2 - x + 1)$ (Sec. 6.1)

45. $\dfrac{3}{x - 2}$ (Sec. 7.4, Ex. 2)

46. $(2x + 3)^2$ (Sec. 6.5)

47. $t = 5$ (Sec. 7.5, Ex. 2)

48. $\dfrac{30}{x + 3}$ (Sec. 7.2)

49. $2\dfrac{1}{10}$ hr (Sec. 7.6, Ex. 6)

50. $\dfrac{4m + 2n}{m + n}$ or $\dfrac{2(2m + n)}{m + n}$ (Sec. 7.7)

8 Graphs and Functions

In this chapter, the discussion we started in Chapter 3 is continued. We examine statements about two variables, linear equations and inequalities in two variables. We focus particularly on graphs of these equations, which leads to the notion of relation and to the notion of function, perhaps the single most important and useful concept in all of mathematics.

London, England, was the first city to host the Olympics three times

Which Has More Participants, Summer or Winter Olympic Games?

The Summer Olympic Games, which were first held in 1896, are an international competition held every four years. The Summer Olympic Games are much larger than the Winter Olympic Games because there are many team sports that require multiple participants. London, England, the host of the 2012 Summer Olympic Games, was the first city to host the Olympics three times. In Section 8.2, Exercise 13, we will explore the number of gold medals won by the United States in the 2012 and previous Summer Olympics.

Gold Medals at 2012 Summer Olympics

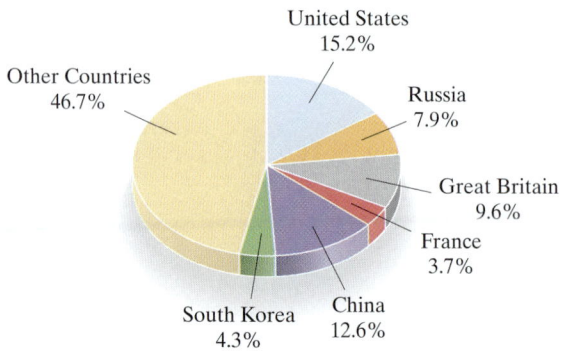

United States 15.2%

Other Countries 46.7%

Russia 7.9%

Great Britain 9.6%

France 3.7%

China 12.6%

South Korea 4.3%

8.1 Review of Equations of Lines and Writing Parallel and Perpendicular Lines

Objective A Using the Point-Slope Form to Write an Equation

Recall that when the slope of a line and a point on the line are known, the equation of the line can also be found. To do this, we use the slope formula to write the slope of a line that passes through points (x, y) and (x_1, y_1). We have

$$m = \frac{y - y_1}{x - x_1}$$

We multiply both sides of this equation by $x - x_1$ to obtain

$$y - y_1 = m(x - x_1)$$

This form is called the *point-slope form* of the equation of a line.

> **Point-Slope Form of the Equation of a Line**
>
> The **point-slope form** of the equation of a line is
>
> $$\overset{\text{slope}}{\underset{\text{point}}{y - y_1 = m(x - x_1)}}$$
>
> where m is the slope of the line and (x_1, y_1) is a point on the line.

Objectives

A Use the Point-Slope Form to Write the Equation of a Line.

B Write Equations of Vertical and Horizontal Lines.

C Write Equations of Parallel and Perpendicular Lines.

D Use the Point-Slope Form in Real-World Applications.

Example 1 Write an equation of the line with slope -3 and containing the point $(1, -5)$. Write the equation in slope-intercept form, $y = mx + b$.

Solution: Because we know the slope and a point on the line, we use the point-slope form with $m = -3$ and $(x_1, y_1) = (1, -5)$.

$$
\begin{array}{ll}
y - y_1 = m(x - x_1) & \text{Point-slope form} \\
y - (-5) = -3(x - 1) & \text{Let } m = -3 \text{ and } (x_1, y_1) = (1, -5). \\
y + 5 = -3x + 3 & \text{Use the distributive property.} \\
y = -3x - 2 & \text{Write in slope-intercept form (solve for } y). \\
\end{array}
$$

In slope-intercept form, the equation is $y = -3x - 2$.

■ **Work Practice 1**

Practice 1

Write an equation of the line with slope -2 and containing the point $(2, -4)$. Write the equation in slope-intercept form, $y = mx + b$.

> **Helpful Hint** Remember, "slope-intercept form" means the equation is "solved for y."

Example 2 Write an equation of the line through points $(4, 0)$ and $(-4, -5)$. Write the equation in standard form, $Ax + By = C$.

Solution: First we find the slope of the line.

$$m = \frac{-5 - 0}{-4 - 4} = \frac{-5}{-8} = \frac{5}{8}$$

Next we make use of the point-slope form. We replace (x_1, y_1) by either $(4, 0)$ or $(-4, -5)$ in the point-slope equation. We will choose the point $(4, 0)$. The line through $(4, 0)$ with slope $\frac{5}{8}$ is as follows.

(*Continued on next page*)

Practice 2

Write an equation of the line through points $(3, 0)$ and $(-2, 4)$. Write the equation in standard form, $Ax + By = C$.

565

$$y - y_1 = m(x - x_1) \quad \text{Point-slope form}$$

$$y - 0 = \frac{5}{8}(x - 4) \quad \text{Let } m = \frac{5}{8} \text{ and } (x_1, y_1) = (4, 0).$$

Let's multiply through by 8 so that the coefficients are integers and are less tedious to work with.

$$8(y - 0) = 8 \cdot \frac{5}{8}(x - 4)$$

$$8y = 5(x - 4) \qquad \text{Simplify.}$$

$$8y = 5x - 20 \qquad \text{Multiply.}$$

$$-5x + 8y = -20 \qquad \text{Write in standard form.}$$

If we multiply both sides of $-5x + 8y = -20$ by -1, we have an equivalent equation in standard form. Both $-5x + 8y = -20$ and $5x - 8y = 20$ are acceptable.

▶ **Work Practice 2**

Objective B Writing Equations of Vertical and Horizontal Lines

A few special types of linear equations are those whose graphs are vertical and horizontal lines.

Practice 3

Write an equation of the horizontal line containing the point $(-1, 6)$.

Example 3 Write an equation of the horizontal line containing the point $(2, 3)$.

Solution: Recall from Section 3.3 that a horizontal line has an equation of the form $y = c$. Since the line contains the point $(2, 3)$, the equation is $y = 3$.

▶ **Work Practice 3**

Practice 4

Write an equation of the line containing the point $(4, 7)$ with undefined slope.

Example 4 Write an equation of the line containing the point $(2, 3)$ with undefined slope.

Solution: Since the line has undefined slope, the line must be vertical. A vertical line has an equation of the form $x = c$, and since the line contains the point $(2, 3)$, the equation is $x = 2$.

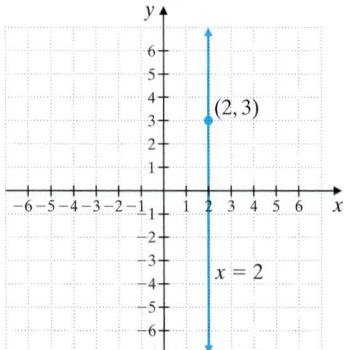

▶ **Work Practice 4**

Answers

3. $y = 6$ **4.** $x = 4$

Objective C Writing Equations of Parallel and Perpendicular Lines ▶

Next, we write equations of parallel and perpendicular lines.

Recall from Section 3.4 that nonvertical parallel lines have the same slope and nonvertical perpendicular lines have slopes whose product is -1.

Example 5 Write an equation of the line containing the point $(4, 4)$ and parallel to the line $2x + y = -6$. Write the equation in slope-intercept form, $y = mx + b$.

Solution: Because the line we want to find is *parallel* to the line $2x + y = -6$, the two lines must have equal slopes. So we first find the slope of $2x + y = -6$ by solving the equation for y to write it in the form $y = mx + b$. Here $y = -2x - 6$, so the slope is -2.

Now we use the point-slope form to write an equation of the line through $(4, 4)$ with slope -2.

$$y - y_1 = m(x - x_1)$$
$$y - 4 = -2(x - 4) \quad \text{Let } m = -2, x_1 = 4, \text{ and } y_1 = 4.$$
$$y - 4 = -2x + 8 \quad \text{Use the distributive property.}$$
$$y = -2x + 12$$

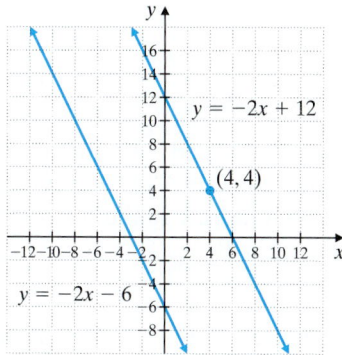

The equation $y = -2x - 6$ and the new equation $y = -2x + 12$ have the same slope but different y-intercepts, so their graphs are parallel. Also, the graph of $y = -2x + 12$ contains the point $(4, 4)$, as desired.

■ **Work Practice 5**

Example 6 Write an equation of the line containing the point $(-2, 1)$ and perpendicular to the line $3x + 5y = 4$. Write the equation in slope-intercept form, $y = mx + b$.

Solution: First we find the slope of $3x + 5y = 4$ by solving the equation for y.

$$5y = -3x + 4$$
$$y = -\frac{3}{5}x + \frac{4}{5}$$

The slope of the given line is $-\frac{3}{5}$. A line perpendicular to this line will have a slope that is the negative reciprocal of $-\frac{3}{5}$, or $\frac{5}{3}$. We use the point-slope form to write an equation of the new line through $(-2, 1)$ with slope $\frac{5}{3}$.

(Continued on next page)

Practice 5

Write an equation of the line containing the point $(-1, 2)$ and parallel to the line $3x + y = 5$. Write the equation in slope-intercept form $y = mx + b$.

Practice 6

Write an equation of the line containing the point $(3, 4)$ and perpendicular to the line $2x + 4y = 5$. Write the equation in slope-intercept form, $y = mx + b$.

Answers

5. $y = -3x - 1$ **6.** $y = 2x - 2$

$$y - y_1 = m(x - x_1)$$

$$y - 1 = \frac{5}{3}[x - (-2)]$$

$$y - 1 = \frac{5}{3}(x + 2) \qquad \text{Simplify.}$$

$$y - 1 = \frac{5}{3}x + \frac{10}{3} \qquad \text{Use the distributive property.}$$

$$y = \frac{5}{3}x + \frac{13}{3} \qquad \text{Add 1 to both sides.}$$

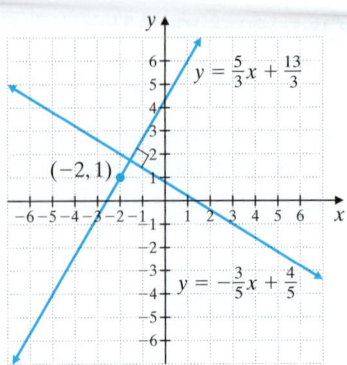

The equation $y = -\frac{3}{5}x + \frac{4}{5}$ and the new equation $y = \frac{5}{3}x + \frac{13}{3}$ have negative reciprocal slopes, so their graphs are perpendicular. Also, the graph of $y = \frac{5}{3}x + \frac{13}{3}$ contains the point $(-2, 1)$, as desired.

■ Work Practice 6

Objective D Using the Point-Slope Form in Applications ▶

The point-slope form of an equation is very useful for solving real-world problems.

Practice 7

Southwest Regional is an established office product maintenance company that has enjoyed constant growth in new maintenance contracts since 2005. In 2007, the company obtained 15 new contracts, and in 2013, the company obtained 33 new contracts. Use these figures to predict the number of new contracts this company can expect in 2022.

Example 7   Predicting Sales

Southern Star Realty is an established real estate company that has enjoyed constant growth in sales since 2005. In 2009, the company sold 250 houses, and in 2013, the company sold 330 houses. Use these figures to predict the number of houses this company will sell in 2018.

Solution:

1. UNDERSTAND. Read and reread the problem. Then let

    $x =$ the number of years after 2005 and

    $y =$ the number of houses sold in the year corresponding to x

The information provided then gives the ordered pairs $(4, 250)$ and $(8, 330)$. To better visualize the sales of Southern Star Realty, we graph the line that passes through the points $(4, 250)$ and $(8, 330)$.

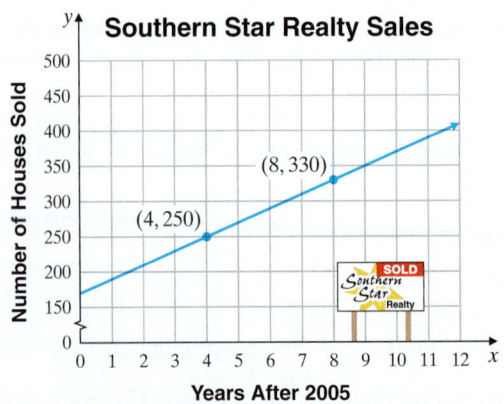

2. TRANSLATE. We write the equation of the line that passes through the points $(4, 250)$ and $(8, 330)$. To do so, we first find the slope of the line.

$$m = \frac{330 - 250}{8 - 4} = \frac{80}{4} = 20$$

Then using the point-slope form to write the equation, we have

$$
\begin{aligned}
y - y_1 &= m(x - x_1) \\
y - 250 &= 20(x - 4) \quad \text{Let } m = 20 \text{ and } (x_1, y_1) = (4, 250). \\
y - 250 &= 20x - 80 \quad \text{Multiply.} \\
y &= 20x + 170 \quad \text{Add 250 to both sides.}
\end{aligned}
$$

3. SOLVE. To predict the number of houses sold in 2018, we use $y = 20x + 170$ and complete the ordered pair $(13, \quad)$, since $2018 - 2005 = 13$.

$$y = 20(13) + 170$$

$$y = 430$$

4. INTERPRET.

Check: Verify that the point $(13, 430)$ is a point on the extended line graphed in Step 1.

State: Southern Star Realty should expect to sell 430 houses in 2018.

■ **Work Practice 7**

📟 Calculator Explorations **Graphing**

Many graphing calculators have a TRACE feature. This feature allows you to trace along a graph and see the corresponding x- and y-coordinates appear on the screen. Use this feature for the following exercises.

Graph each equation and then use the TRACE feature to complete each ordered pair solution. (Many times the tracer will not show the exact x- or y-value asked for. In each case, trace as closely as you can to the given x- or y-coordinate and approximate the other, unknown coordinate to one decimal place.)

1. $y = 2.3x + 6.7$;
   $x = 5.1, y = ?$ 18.4
2. $y = -4.8x + 2.9$;
   $x = -1.8, y = ?$ 11.5
3. $y = -5.9x - 1.6$;
   $x = ?, y = 7.2$ −1.5
4. $y = 0.4x - 8.6$;
   $x = ?, y = -4.4$ 10.5
5. $y = 5.2x - 3.3$;
   $x = 2.3, y = ?$
   $x = ?, y = 36$ 8.7; 7.6
6. $y = -6.2x - 8.3$;
   $x = 3.2, y = ?$
   $x = ?, y = 12$ −28.1; −3.3

Teaching Tip

Also point out the TABLE feature, which is useful for finding the y-value when the x-value is known.

Vocabulary, Readiness & Video Check

Find the slope of and a point on the line described by each equation.

1. $y - 4 = -2(x - 1)$ $m = -2; (1, 4)$
2. $y - 6 = -3(x - 4)$ $m = -3; (4, 6)$
3. $y - 0 = \frac{1}{4}(x - 2)$ $m = \frac{1}{4}; (2, 0)$
4. $y - 1 = -\frac{2}{3}(x - 0)$ $m = -\frac{2}{3}; (0, 1)$
5. $y + 2 = 5(x - 3)$ $m = 5; (3, -2)$
6. $y - 7 = 4(x + 6)$ $m = 4; (-6, 7)$

Martin-Gay Interactive Videos   *Watch the section lecture video and answer the following questions.*

See Video 8.1

See video answer section.

Objective A **7.** Example 2 discusses how to find an equation of a line given two points. Under what circumstances might the slope-intercept form be chosen over the point-slope form to find an equation? ▶

Objective B **8.** Solve ⊞ Examples 3 and 4 again, this time using the point $(-1, 3)$ in each exercise. ▶

Objective C **9.** Solve ⊞ Example 5 again, this time write the equation of the line in slope-intercept form, *parallel* to the given line through the given point. ▶

Objective D **10.** Use the equation found in ⊞ Example 6 to determine the rate of the rock 2 seconds after it was dropped. ▶

8.1 Exercise Set MyMathLab® ▶

Objective A *Write an equation of each line with the given slope and containing the given point. Write the equation in the slope-intercept form $y = mx + b$. See Example 1.*

▶**1.** Slope 3; through $(1, 2)$ $y = 3x - 1$

2. Slope 4; through $(5, 1)$ $y = 4x - 19$

3. Slope -2; through $(1, -3)$ $y = -2x - 1$

4. Slope -4; through $(2, -4)$ $y = -4x + 4$

5. Slope $\frac{1}{2}$; through $(-6, 2)$ $y = \frac{1}{2}x + 5$

6. Slope $\frac{2}{3}$; through $(-9, 4)$ $y = \frac{2}{3}x + 10$

7. Slope $-\frac{9}{10}$; through $(-3, 0)$ $y = -\frac{9}{10}x - \frac{27}{10}$

8. Slope $-\frac{1}{5}$; through $(4, -6)$ $y = -\frac{1}{5}x - \frac{26}{5}$

Write an equation of the line passing through the given points. Write the equation in standard form $Ax + By = C$. See Example 2.

9. $(2, 0)$ and $(4, 6)$ $3x - y = 6$

10. $(3, 0)$ and $(7, 8)$ $2x - y = 6$

11. $(-2, 5)$ and $(-6, 13)$ $2x + y = 1$

12. $(7, -4)$ and $(2, 6)$ $2x + y = 10$

13. $(-2, -4)$ and $(-4, -3)$ $x + 2y = -10$

14. $(-9, -2)$ and $(-3, 10)$ $2x - y = -16$

▶**15.** $(-3, -8)$ and $(-6, -9)$ $x - 3y = 21$

16. $(8, -3)$ and $(4, -8)$ $5x - 4y = 52$

17. $\left(\frac{3}{5}, \frac{4}{10}\right)$ and $\left(-\frac{1}{5}, \frac{7}{10}\right)$ $3x + 8y = 5$

18. $\left(\frac{1}{2}, -\frac{1}{4}\right)$ and $\left(\frac{3}{2}, \frac{3}{4}\right)$ $4x - 4y = 3$

Objective B *Write an equation of each line. See Examples 3 and 4.*

19. Vertical; through $(2, 6)$ $x = 2$

20. Slope 0; through $(-2, -4)$ $y = -4$

▶**21.** Horizontal; through $(-3, 1)$ $y = 1$

22. Vertical; through $(4, 7)$ $x = 4$

▶**23.** Undefined slope; through $(0, 5)$ $x = 0$

24. Horizontal; through $(0, 5)$ $y = 5$

Objective C *Write an equation of each line. Write the equation in the form* $x = a$, $y = b$, *or* $y = mx + b$. *See Examples 5 and 6.*

25. Through $(3, 8)$; parallel to $y = 4x - 2$ $y = 4x - 4$

26. Through $(1, 5)$; parallel to $y = 3x - 4$ $y = 3x + 2$

▶ **27.** Through $(2, -5)$; perpendicular to $3y = x - 6$
$y = -3x + 1$

28. Through $(-4, 8)$; perpendicular to $2x - 3y = 1$
$y = -\dfrac{3}{2}x + 2$

29. Through $(1, 4)$; parallel to $y = 7$ $y = 4$

30. Through $(-2, 6)$; perpendicular to $y = 7$ $x = -2$

31. Through $(-2, -3)$; parallel to $3x + 2y = 5$
$y = -\dfrac{3}{2}x - 6$

32. Through $(-2, -3)$; perpendicular to $3x + 2y = 5$
$y = \dfrac{2}{3}x - \dfrac{5}{3}$

33. Through $(-1, -5)$; perpendicular to $x = 3$
$y = -5$

34. Through $(4, -6)$; parallel to $x = -2$
$x = 4$

35. Through $(-1, 5)$; perpendicular to $x - 4y = 4$
$y = -4x + 1$

36. Through $(2, -3)$; parallel to $x - 5y = 10$
$y = \dfrac{1}{5}x - \dfrac{17}{5}$

Objectives A B C Mixed Practice *Write an equation of each line. Write the equation in standard form unless indicated otherwise. See Examples 1 through 6.*

37. Slope 2; through $(-2, 3)$ $2x - y = -7$

38. Slope 3; through $(-4, 2)$ $3x - y = -14$

39. Through $(1, 6)$ and $(5, 2)$; use slope-intercept form.
$y = -x + 7$

40. Through $(2, 9)$ and $(8, 6)$; use slope-intercept form.
$y = -\dfrac{1}{2}x + 10$

41. With slope $-\dfrac{1}{2}$; y-intercept $\left(0, \dfrac{3}{8}\right)$; use slope-intercept form. $y = -\dfrac{1}{2}x + \dfrac{3}{8}$

42. With slope -4; y-intercept $\left(0, \dfrac{2}{9}\right)$; use slope-intercept form. $y = -4x + \dfrac{2}{9}$

43. Through $(-7, -4)$ and $(0, -6)$ $2x + 7y = -42$

44. Through $(2, -8)$ and $(-4, -3)$ $5x + 6y = -38$

45. Slope $-\dfrac{4}{3}$; through $(-5, 0)$ $4x + 3y = -20$

46. Slope $-\dfrac{3}{5}$; through $(4, -1)$ $3x + 5y = 7$

47. Vertical line; through $(-2, -10)$ $x = -2$

48. Horizontal line; through $(1, 0)$ $y = 0$

49. Through $(6, -2)$; parallel to the line $2x + 4y = 9$
$x + 2y = 2$

50. Through $(8, -3)$; parallel to the line $6x + 2y = 5$
$3x + y = 21$

51. Slope 0; through $(-9, 12)$ $y = 12$

52. Undefined slope; through $(10, -8)$ $x = 10$

53. Through $(6, 1)$; parallel to the line $8x - y = 9$
$8x - y = 47$

54. Through $(3, 5)$; perpendicular to the line $2x - y = 8$
$x + 2y = 13$

55. Through $(5, -6)$; perpendicular to $y = 9$ $x = 5$

56. Through $(-3, -5)$; parallel to $y = 9$ $y = -5$

57. Through $(2, -8)$ and $(-6, -5)$; use slope-intercept form. $y = -\dfrac{3}{8}x - \dfrac{29}{4}$

58. Through $(-4, -2)$ and $(-6, 5)$; use slope-intercept form. $y = -\dfrac{7}{2}x - 16$

Objective D *Solve. See Example 7.*

▶ **59.** A rock is dropped from the top of a 400-foot building. After 1 second, the rock is traveling 32 feet per second. After 3 seconds, the rock is traveling 96 feet per second. Let y be the rate of descent and x be the number of seconds since the rock was dropped.

   a. Write a linear equation that relates time x to rate y. [*Hint:* Use the ordered pairs $(1, 32)$ and $(3, 96)$.] $y = 32x$

   b. Use this equation to determine the rate of travel of the rock 4 seconds after it was dropped.
   128 ft per sec

60. A fruit company recently released a new applesauce. By the end of its first year, profits on this product amounted to $30,000. The anticipated profit for the end of the fourth year is $66,000. The ratio of change in time to change in profit is constant. Let x be years and y be profit.

   a. Write a linear equation that relates profit and time. [*Hint:* Use the ordered pairs $(1, 30,000)$ and $(4, 66,000)$.] $y = 12,000x + 18,000$

   b. Use this equation to predict the company's profit at the end of the seventh year. $102,000

   c. Predict when the profit should reach $126,000.   9 yr

61. The Whammo Company has learned that by pricing a newly released Frisbee at $6, sales will reach 2000 per day. Raising the price to $8 will cause the sales to fall to 1500 per day. Assume that the ratio of change in price to change in daily sales is constant, and let x be the price of the Frisbee and y be number of sales.

   a. Find the linear equation that models the price–sales relationship for this Frisbee. [*Hint:* The line must pass through $(6, 2000)$ and $(8, 1500)$.] $y = -250x + 3500$

   b. Use this equation to predict the daily sales of Frisbees if the price is set at $7.50.   1625 Frisbees

62. The Pool Fun Company has learned that, by pricing a newly released Fun Noodle at $3, sales will reach 10,000 Fun Noodles per day during the summer. Raising the price to $5 will cause the sales to fall to 8000 Fun Noodles per day. Let x be price and y be the number sold.

   a. Assume that the relationship between sales price and number of Fun Noodles sold is linear and write an equation describing this relationship. [*Hint:* The line must pass through $(3, 10,000)$ and $(5, 8000)$.] $y = -1000x + 13,000$

   b. Use this equation to predict the daily sales of Fun Noodles if the price is $3.50.   9500 Fun Noodles

63. In 2012, the average price of a new home sold in the United States was $292,200. In 2004, the average price of a new home in the United States was $297,000. Let y be the price of a new home in the year x, where $x = 0$ represents the year 2004. (*Source:* Based on data from the U.S. Census)

   a. Write a linear equation that models the average price of new home in terms of the year x. [*Hint:* The line must pass through the points $(0, 297,000)$ and $(8, 292,200)$.] $y = -600x + 297,000$

   b. Use this equation to predict the average price of a new home in 2014. $291,000

64. The number of McDonald's restaurants worldwide in 2012 was 34,480. In 2007, there were 31,377 McDonald's restaurants worldwide. Let y be the number of McDonald's restaurants in the year x, where $x = 0$ represents the year 2007. (*Source:* McDonald's Corporation)

   a. Write a linear equation that models the growth in the number of McDonald's restaurants worldwide in terms of the year x. $y = 620.6x + 31,377$

   b. Use this information to predict the number of McDonald's restaurants in 2015. (Round to the nearest whole number.)   36,342

65. The number of theater screens in the United States in 2008 was 38,834. By 2012, the number had grown to 39,918. Let y be the number of theater screens in the United States in year x, where $x = 0$ represents the year 2008. (*Source:* Motion Picture Association of America)

   a. Write a linear equation that models the number of theater screens in the United States in year x. $y = 271x + 38,834$

   b. Use this equation to predict the number of theater screens in 2015.   40,731

66. In 2007, Chrysler Corporation employed 77,778 people worldwide. By 2013, this had shrunk to 70,386. Let y be the number of Chrysler employees worldwide in the year x, where $x = 0$ represents the year 2007. (*Source:* Chrysler Corporation)

   a. Write a linear equation that models the number of Chrysler employees in the year x. $y = -1232x + 77{,}778$

   b. Use this equation to estimate the number of Chrysler employees for the year 2016.   66,690

Review

Complete each ordered pair for the given equation. See Section 3.1.

67. $y = 7x + 3; (4, \quad)$ 31

68. $y = 2x - 6; (2, \quad)$ −2

69. $y = 4.2x; (-2, \quad)$ −8.4

70. $y = -1.3x; (6, \quad)$ −7.8

71. $y = x^2 + 2x + 1; (1, \quad)$ 4

72. $y = x^2 - 6x + 4; (0, \quad)$ 4

Concept Extensions

Find an equation of each line graphed. Write the equation in standard form.

73. $2x + y = 3$

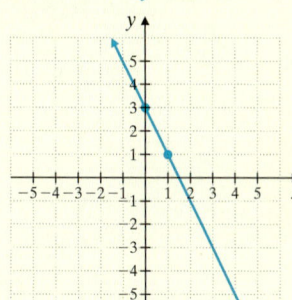

74. $2x - y = 2$

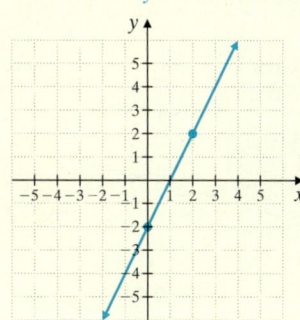

75. $2x - 3y = -7$

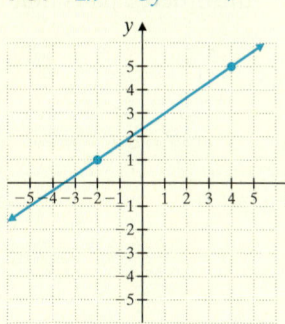

76. $x + 7y = -4$

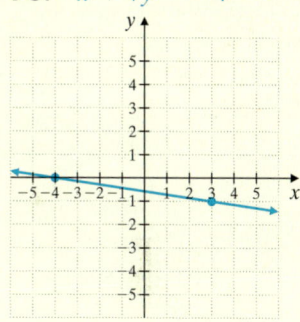

Answer true or false.

77. A vertical line is always perpendicular to a horizontal line.   true

78. A vertical line is always parallel to another vertical line.   true

Solve.

79. Describe how to check to see if the graph of $2x - 4y = 7$ passes through the points $(1.4, -1.05)$ and $(0, -1.75)$. Then follow your directions and check these points.   answers may vary

80. Describe how to use the point-slope form to write an equation of a line if the slope and y-intercept of a line are given.   answers may vary

Use a grapher with a TRACE feature to see the results of each exercise.

81. Exercise 25: Graph the equation and verify that it passes through $(3, 8)$ and is parallel to $y = 4x - 2$.

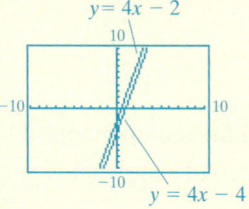

82. Exercise 26: Graph the equation and verify that it passes through $(1, 5)$ and is parallel to $y = 3x - 4$.

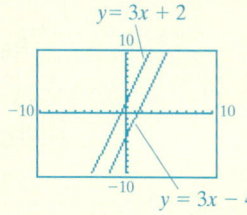

Introduction to Functions

Objectives

A Define Relation, Domain, and Range.

B Identify Functions.

C Use the Vertical Line Test for Functions.

D Use Function Notation.

E Graph a Linear Function.

Objective A Defining Relation, Domain, and Range

Equations in two variables, such as $y = 2x + 1$, describe **relations** between x-values and y-values. For example, if $x = 1$, then this equation describes how to find the y-value related to $x = 1$. In words, the equation $y = 2x + 1$ says that twice the x-value increased by 1 gives the corresponding y-value. The x-value of 1 corresponds to the y-value of $2(1) + 1 = 3$ for this equation, and we have the ordered pair $(1, 3)$. In other words, for the relationship (or relation) between x and y defined by $y = 2x + 1$, the x-value 1 is paired with the y-value 3.

There are other ways of describing relations or correspondences between two numbers or, in general, a set of first components (sometimes called the set of *inputs*) and a set of second components (sometimes called the set of *outputs*). For example,

First Set: Input	Correspondence	Second Set: Output
People in a certain city	Each person's age, to the nearest year	The set of nonnegative integers

A few examples of ordered pairs from this relation might be (Ana, 4); (Bob, 36); (Trey, 21); and so on.

Below are just a few other ways of describing relations between two sets and the ordered pairs that they generate.

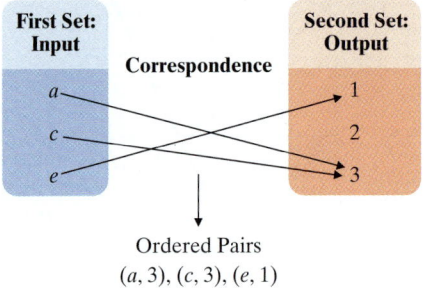

Ordered Pairs
$(a, 3), (c, 3), (e, 1)$

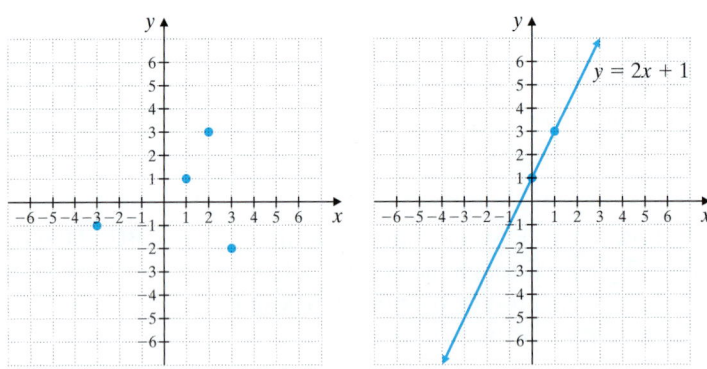

Ordered Pairs

$(-3, -1), (1, 1), (2, 3), (3, -2)$

Some Ordered Pairs

$(0, 1), (1, 3)$, and so on

Relation, Domain, and Range

A **relation** is a set of ordered pairs.

The **domain** of the relation is the set of all first components of the ordered pairs.

The **range** of the relation is the set of all second components of the ordered pairs.

For example, the domain for our middle relation on the previous page is $\{a, c, e\}$ and the range is $\{1, 3\}$. Notice that the range does not include the element 2 of the second set. This is because no element of the first set is assigned to this element. If a relation is defined in terms of x- and y-values, we will agree that the domain corresponds to x-values and that the range corresponds to y-values that are paired with x-values.

Helpful Hint

Remember that the range only includes elements that are paired with domain values. For

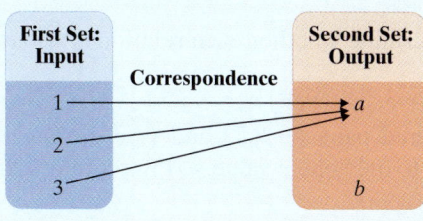

the range is $\{a\}$.

Examples   Determine the domain and range of each relation.

1. $\{(2, 3), (2, 4), (0, -1), (3, -1)\}$

The domain is the set of all first coordinates of the ordered pairs, $\{2, 0, 3\}$. The range is the set of all second coordinates, $\{3, 4, -1\}$.

2.

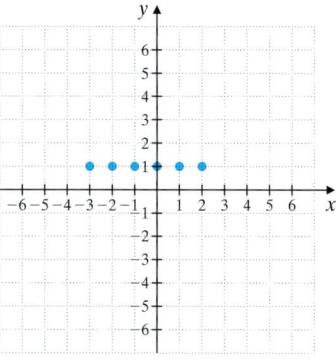

Helpful Hint

Equivalent domain or range elements that occur more than once need to be listed only once.

The relation is $\{(-3, 1), (-2, 1), (-1, 1), (0, 1), (1, 1), (2, 1)\}$.
The domain is $\{-3, -2, -1, 0, 1, 2\}$.
The range is $\{1\}$.

3.

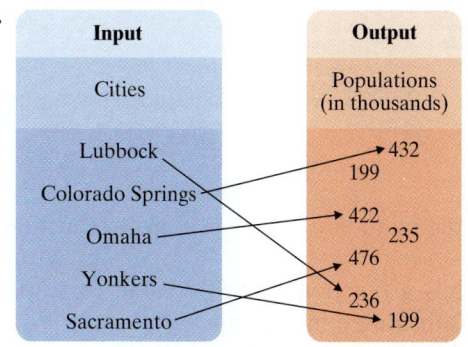

The domain is the set of inputs {Lubbock, Colorado Springs, Omaha, Yonkers, Sacramento}. The range is the numbers in the set of outputs that correspond to elements in the set of inputs {432, 422, 476, 236, 199}.

Work Practice 1–3

Teaching Tip

Before formally defining functions, have students write a response to the following: "Suppose I gave you a list of people and their heights. Would it be easier for you to name the person if I gave you the height or to give the height if I named the person? Explain." Then have several students share their responses with the class.

Practice 1–3

Determine the domain and range of each relation.

1. $\{(1, 6), (2, 8), (0, 3), (0, -2)\}$

2.

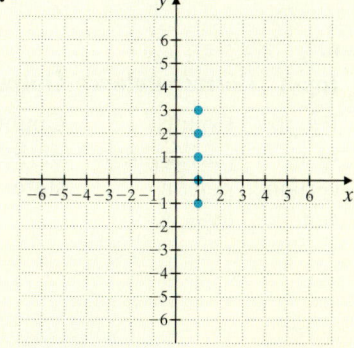

3.

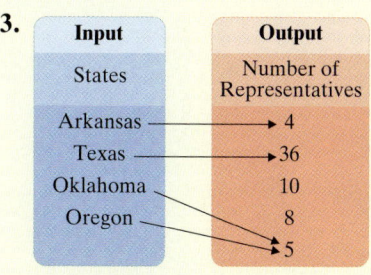

Helpful Hint   A function is a special type of relation, so all functions are relations. But not all relations are functions.

Objective B Identifying Functions ▶

Now we consider a special kind of relation called a *function*.

Function

A **function** is a relation in which each first component in the ordered pairs corresponds to *exactly one* second component.

Practice 4–6

Determine whether each relation is also a function.

4. $\{(-3, 7), (1, 7), (2, 2)\}$

5.

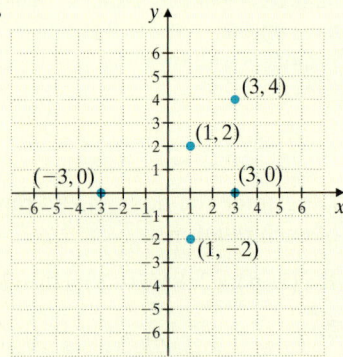

6.

Input	Correspondence	Output
People in a certain state	County/ parish that a person lives in	Counties of that state

Examples   Determine whether each relation is also a function.

4. $\{(-2, 5), (2, 7), (-3, 5), (9, 9)\}$

Although the ordered pairs $(-2, 5)$ and $(-3, 5)$ have the same *y*-value, each *x*-value is assigned to only one *y*-value, so this set of ordered pairs is a function.

5.

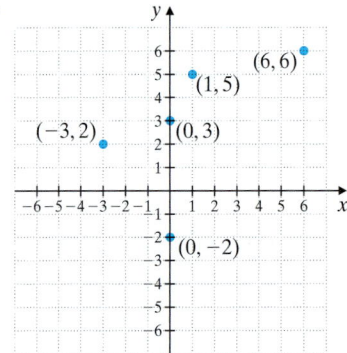

The *x*-value 0 is assigned to two *y*-values, -2 and 3, in this graph, so this relation is not a function.

6.

Input	Correspondence	Output
People in a certain city	Each person's age, to the nearest year	The set of nonnegative integers

This relation is a function because although two different people may have the same age, each person has only one age. This means that each element in the first set is assigned to only one element in the second set.

◼ **Work Practice 4–6**

✓**Concept Check** Explain why a function can contain both the ordered pairs $(1, 3)$ and $(2, 3)$ but not both $(3, 1)$ and $(3, 2)$.

Recall that an equation such as $y = 2x + 1$ is a relation since this equation defines a set of ordered pair solutions.

Practice 7

Determine whether the relation $y = 7x - 3$ is also a function.

Example 7   Determine whether the relation $y = 2x + 1$ is also a function.

Solution: The relation $y = 2x + 1$ is a function if each *x*-value corresponds to just one *y*-value. For each *x*-value substituted in the equation $y = 2x + 1$, the multiplication and addition performed give a single result, so only one *y*-value will be associated with each *x*-value. Thus, $y = 2x + 1$ is a function.

◼ **Work Practice 7**

Answers

4. function **5.** not a function
6. function **7.** function

✓**Concept Check Answer**

In a function, two different ordered pairs can have the same *y*-value, but not the same *x*-value.

Example 8 Determine whether the relation $x = y^2$ is also a function.

Solution: In $x = y^2$, if $y = 3$, then $x = 9$. Also, if $y = -3$, then $x = 9$. In other words, we have the ordered pairs $(9, 3)$ and $(9, -3)$. Since the x-value 9 corresponds to two y-values, 3 and -3, $x = y^2$ is not a function.

■ **Work Practice 8**

Objective C Using the Vertical Line Test ▶

As we have seen, not all relations are functions. Consider the graphs of $y = 2x + 1$ and $x = y^2$ shown next. On the graph of $y = 2x + 1$, notice that each x-value corresponds to only one y-value. Recall from Example 7 that $y = 2x + 1$ is a function.

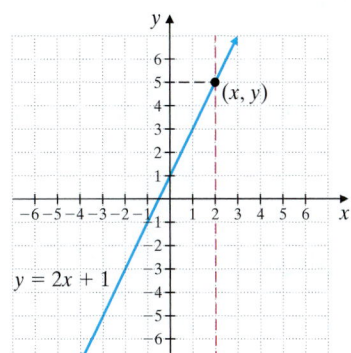
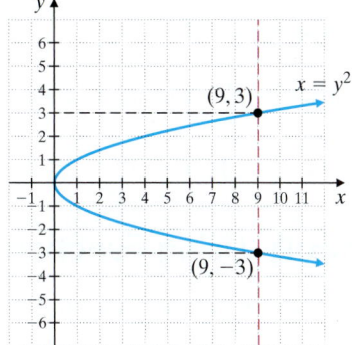

On the graph of $x = y^2$, the x-value 9, for example, corresponds to two y-values, 3 and -3, as shown by the vertical line. Recall from Example 8 that $x = y^2$ is not a function.

Graphs can be used to help determine whether a relation is also a function by the following **vertical line test.**

Vertical Line Test

If no vertical line can be drawn so that it intersects a graph more than once, the graph is the graph of a function. If such a line can be drawn, the graph is not that of a function.

Examples Use the vertical line test to determine which are graphs of functions.

9.
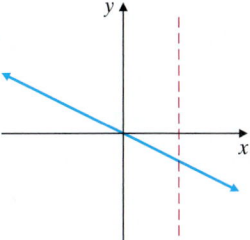

This is the graph of a function since no vertical line will intersect this graph more than once.

10.
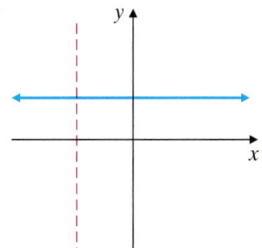

This is the graph of a function.

11.
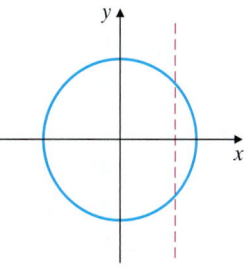

This is not the graph of a function. Note that vertical lines can be drawn that intersect the graph in two points.

(Continued on next page)

Practice 8

Determine whether the relation $x = y^2 + 1$ is also a function.

Practice 9–13

Use the vertical line test to determine which are graphs of functions.

9.
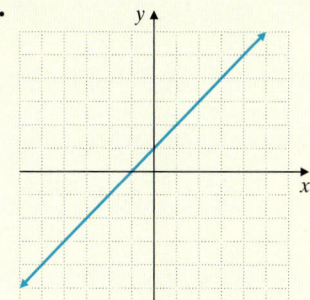

10.
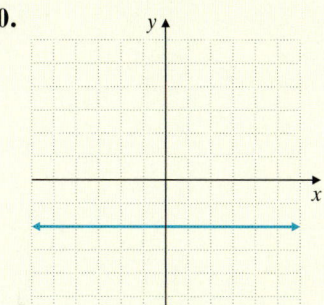

11.
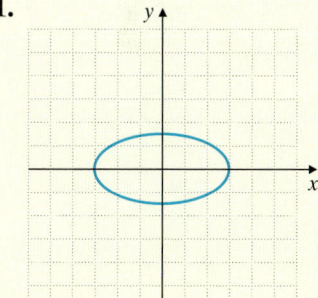

12.
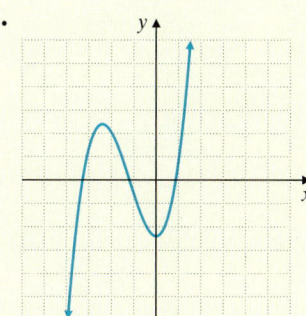

13. See page 578

Answers
8. not a function   **9.** function
10. function   **11.** not a function
12. function   **13.** not a function

13.

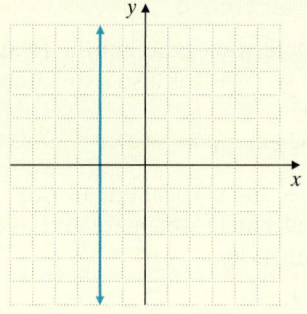

12.

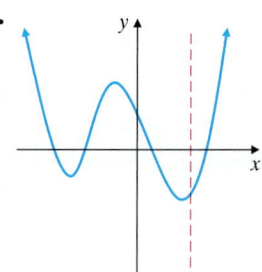

This is the graph of a function.

13.

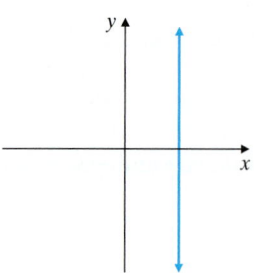

This is not the graph of a function. A vertical line can be drawn that intersects this line at every point.

Work Practice 9–13

✓**Concept Check** Determine which equations represent functions. Explain your answer.

a. $y = 14$ **b.** $x = -5$ **c.** $x + y = 6$

Objective D Using Function Notation

Many times letters such as f, g, and h are used to name functions.

> **Function Notation**
>
> To denote that y is a function of x, we can write
>
> $$y = \underbrace{f(x)}_{\text{Function Notation}} \ (\text{Read "}f\text{ of }x\text{"})$$
>
> This notation means that **y is a function of x** or that y *depends on* x. For this reason, y is called the **dependent variable** and x the **independent variable.**

For example, to use function notation with the function $y = 4x + 3$, we write $f(x) = 4x + 3$. The notation $f(1)$ means to replace x with 1 and find the resulting y- or function value. Since

$$f(x) = 4x + 3$$

then

$$f(1) = 4(1) + 3 = 7$$

This means that when $x = 1$, y or $f(x) = 7$. The corresponding ordered pair is $(1, 7)$. Here, the input is 1 and the output is $f(1)$ or 7. Now let's find $f(2)$, $f(0)$, and $f(-1)$.

$f(x) = 4x + 3$	$f(x) = 4x + 3$	$f(x) = 4x + 3$
$f(2) = 4(2) + 3$	$f(0) = 4(0) + 3$	$f(-1) = 4(-1) + 3$
$= 8 + 3$	$= 0 + 3$	$= -4 + 3$
$= 11$	$= 3$	$= -1$

Ordered Pairs:

$(2, 11)$ $(0, 3)$ $(-1, -1)$

Helpful Hint

Note that $f(x)$ is a special symbol in mathematics used to denote a function. The symbol $f(x)$ is read "f of x." It does *not* mean $f \cdot x$ (f times x).

✓**Concept Check Answers**

a, c

Examples Find each function value.

14. If $g(x) = 3x - 2$, find $g(1)$.

$g(1) = 3(1) - 2 = 1$

15. If $g(x) = 3x - 2$, find $g(0)$.

$g(0) = 3(0) - 2 = -2$

16. If $f(x) = 7x^2 - 3x + 1$, find $f(1)$.

$f(1) = 7(1)^2 - 3(1) + 1 = 5$

17. If $f(x) = 7x^2 - 3x + 1$, find $f(-2)$.

$f(-2) = 7(-2)^2 - 3(-2) + 1 = 35$

Teaching Tip

Typically students need extra time to master function notation. Continue to review the idea that $f(0) = 2$ corresponds to the ordered pair $(0, 2)$, for example. Examples similar to Exercises 59–70 are an excellent way to help students understand this notation.

■ **Work Practice 14–17**

✓**Concept Check** Suppose $y = f(x)$ and we are told that $f(3) = 9$. Which is not true?

a. When $x = 3$, $y = 9$.

b. A possible function is $f(x) = x^2$.

c. A point on the graph of the function is $(3, 9)$.

d. A possible function is $f(x) = 2x + 4$.

If it helps, think of a function, f, as a machine that has been programmed with a certain correspondence or rule. An input value (a member of the domain) is then fed into the machine, the machine does the correspondence or rule, and the result is the output (a member of the range).

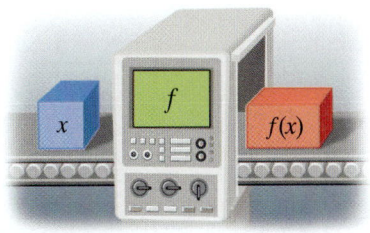 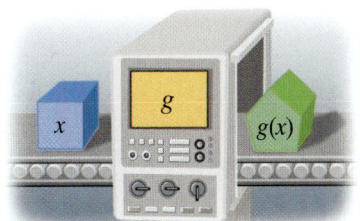

Example 18 Given the graphs of the functions f and g, find each function value by inspecting the graphs.

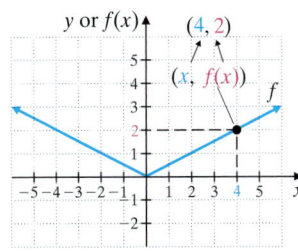 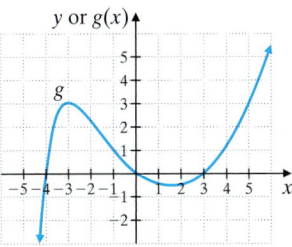

a. $f(4)$ **b.** $f(-2)$ **c.** $g(5)$ **d.** $g(0)$

e. Find all x-values such that $f(x) = 1$.

f. Find all x-values such that $g(x) = 0$.

Solution:

a. To find $f(4)$, find the y-value when $x = 4$. We see from the graph that when $x = 4$, y or $f(x) = 2$. Thus, $f(4) = 2$.

b. $f(-2) = 1$ from the ordered pair $(-2, 1)$.

c. $g(5) = 3$ from the ordered pair $(5, 3)$.

(Continued on next page)

Practice 14–17

Find each function value.

14. If $g(x) = 4x + 5$, find $g(0)$.

15. If $g(x) = 4x + 5$, find $g(-5)$.

16. If $f(x) = 3x^2 - x + 2$, find $f(2)$.

17. If $f(x) = 3x^2 - x + 2$, find $f(-1)$.

Practice 18

Given the graphs of the functions f and g, find each function value by inspecting the graphs.

a. $f(1)$ **b.** $f(0)$

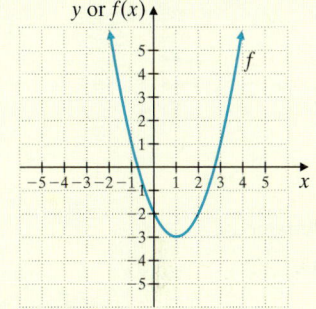

c. $g(-2)$ **d.** $g(0)$

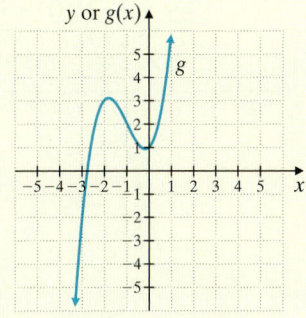

e. Find all x-values such that $f(x) = 1$.

f. Find all x-values such that $g(x) = -2$.

Answers

14. $g(0) = 5$ **15.** $g(-5) = -15$

16. $f(2) = 12$ **17.** $f(-1) = 6$

18. a. -3 **b.** -2 **c.** 3 **d.** 1

e. -1 and 3 **f.** -3

✓**Concept Check Answer**

d

d. $g(0) = 0$ from the ordered pair $(0, 0)$.

e. To find x-values such that $f(x) = 1$, we are looking for any ordered pairs on the graph of f whose $f(x)$ or y-value is 1. They are $(2, 1)$ and $(-2, 1)$. Thus, $f(2) = 1$ and $f(-2) = 1$. The x-values are 2 and -2.

f. Find ordered pairs on the graph of g whose $g(x)$ or y-value is 0. They are $(3, 0)$ $(0, 0)$ and $(-4, 0)$. Thus $g(3) = 0$, $g(0) = 0$, and $g(-4) = 0$. The x-values are $3, 0$ and -4.

■ **Work Practice 18**

Objective E Graphing Linear Functions

Recall that the graph of a linear equation in two variables is a line, and a line that is not vertical will always pass the vertical line test. Thus, *all linear equations are functions except those whose graphs are vertical lines.* We call such functions *linear functions.*

> ### Linear Function
>
> A **linear function** is a function that can be written in the form
>
> $$f(x) = mx + b$$

Practice 19

Graph the function
$f(x) = 3x - 2$.

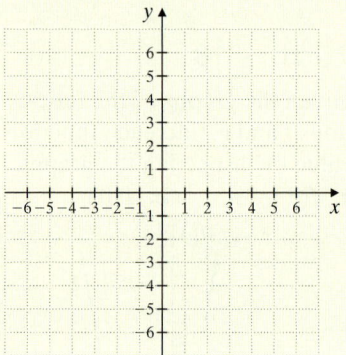

Example 19 Graph the function $f(x) = 4x + 2$.

Solution: Since $y = f(x)$, we can replace $f(x)$ with y and graph as usual. The graph of $y = 4x + 2$ has slope 4 and y-intercept $(0, 2)$. Its graph is shown.

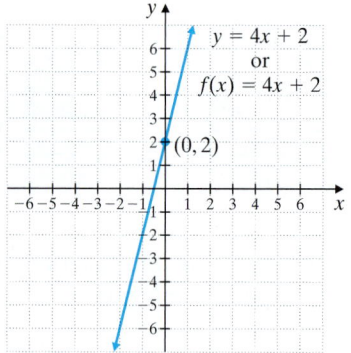

■ **Work Practice 19**

Answer

19.

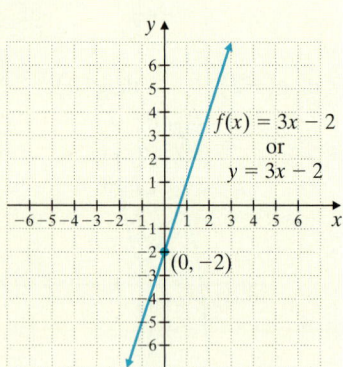

Vocabulary, Readiness & Video Check

Use the choices below to fill in each blank. Not all choices will be used.

domain	vertical	relation	$(1.7, -2)$
range	horizontal	function	$(-2, 1.7)$

1. A ___relation___ is a set of ordered pairs.

2. The ___range___ of a relation is the set of all second components of the ordered pairs.

3. The ___domain___ of a relation is the set of all first components of the ordered pairs.

4. A ___function___ is a relation in which each first component in the ordered pairs corresponds to *exactly* one second component.

5. By the vertical line test, all linear equations are functions except those whose graphs are ___vertical___ lines.

6. If $f(-2) = 1.7$, the corresponding ordered pair is ___$(-2, 1.7)$___ .

Martin-Gay Interactive Videos   *Watch the section lecture video and answer the following questions.*

Objective A **7.** Based on the lecture before ⊞ Example 1, why can an equation in two variables define a relation? ▶

Objective B **8.** Based on the lecture before ⊞ Example 2, can a relation in which a second component corresponds to more than one first component be a function? ▶

Objective C **9.** Based on ⊞ Example 3 and the lecture before, explain why the vertical line test works. ▶

See Video 8.2 🍒

Objective D **10.** From ⊞ Examples 5 and 6, what is the connection between function notation to evaluate a function at certain values and ordered pair solutions of the function? ▶

Objective E **11.** From ⊞ Example 7 and the lecture before, do all linear equations in two variables define functions? Explain. ▶

12. Based on the lecture before ⊞ Example 7, in what form can a linear function be written? ▶

See video answer section.

8.2 **Exercise Set** MyMathLab®

Objectives A B Mixed Practice *Find the domain and the range of each relation. Also determine whether the relation is a function. See Examples 1 through 6.*

▶ 1. $\{(-1, 7), (0, 6), (-2, 2), (5, 6)\}$
domain: $\{-1, 0, -2, 5\}$; range: $\{7, 6, 2\}$; function

2. $\{(4, 9), (-4, 9), (2, 3), (10, -5)\}$
domain: $\{4, -4, 2, 10\}$; range: $\{9, 3, -5\}$; function

3. $\{(-2, 4), (6, 4), (-2, -3), (-7, -8)\}$
domain: $\{-2, 6, -7\}$; range: $\{4, -3, -8\}$; not a function

4. $\{(6, 6), (5, 6), (5, -2), (7, 6)\}$
domain: $\{6, 5, 7\}$; range: $\{6, -2\}$; not a function

5. $\{(1, 1), (1, 2), (1, 3), (1, 4)\}$
domain: $\{1\}$; range: $\{1, 2, 3, 4\}$; not a function

6. $\{(1, 1), (2, 1), (3, 1), (4, 1)\}$
domain: $\{1, 2, 3, 4\}$; range: $\{1\}$; function

7. $\left\{\left(\frac{3}{2}, \frac{1}{2}\right), \left(1\frac{1}{2}, -7\right), \left(0, \frac{4}{5}\right)\right\}$
domain: $\left\{\frac{3}{2}, 0\right\}$; range: $\left\{\frac{1}{2}, -7, \frac{4}{5}\right\}$; not a function

8. $\left\{\left(\frac{1}{2}, \frac{1}{4}\right), \left(0, \frac{7}{8}\right), (0.5, \pi)\right\}$
domain: $\left\{\frac{1}{2}, 0\right\}$; range: $\left\{\frac{1}{4}, \frac{7}{8}, \pi\right\}$; not a function

9. $\{(-3, -3), (0, 0), (3, 3)\}$
domain: $\{-3, 0, 3\}$; range: $\{-3, 0, 3\}$; function

10. $\{(\pi, 0), (0, \pi), (-2, 4), (4, -2)\}$
domain: $\{\pi, 0, -2, 4\}$; range: $\{0, \pi, 4, -2\}$; function

11.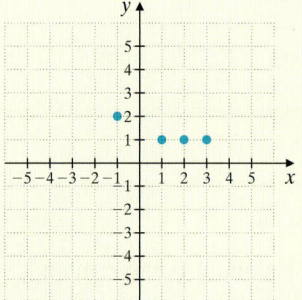

domain: $\{-1, 1, 2, 3\}$; range: $\{2, 1\}$; function

12.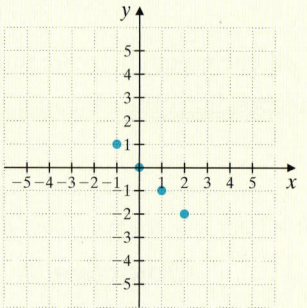

domain: $\{-1, 0, 1, 2\}$; range: $\{1, 0, -1, -2\}$; function

13.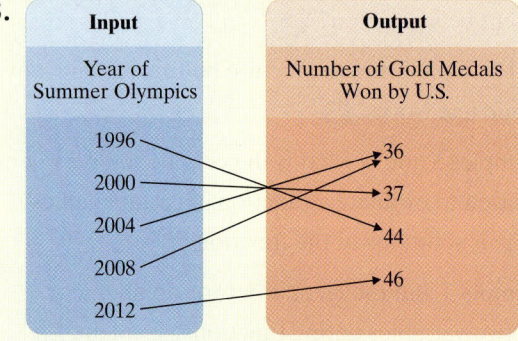

Domain: {1996, 2000, 2004, 2008, 2012}; range: {36, 37, 44, 46}; function

14.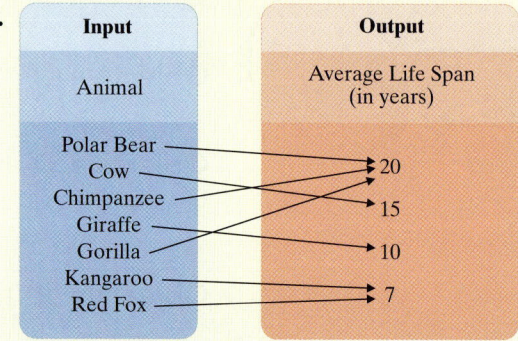

domain: {polar bear, cow, chimpanzee, giraffe, gorilla, kangaroo, red fox}; range: $\{20, 15, 10, 7\}$; function

15.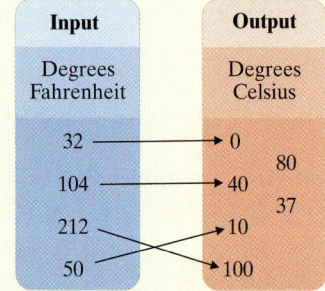

domain: $\{32, 104, 212, 50\}$; range: $\{0, 40, 10, 100\}$; function

16.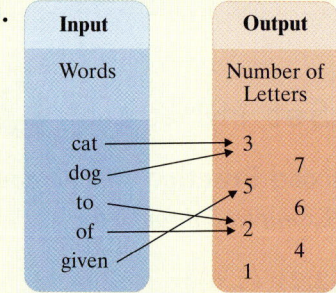

domain: {cat, dog, to, of, given}; range: $\{3, 5, 2\}$; function

17.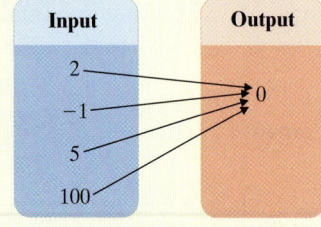

domain: $\{2, -1, 5, 100\}$; range: $\{0\}$; function

18.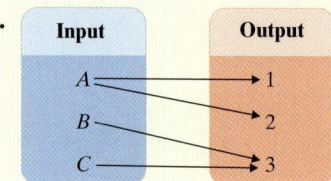

domain: $\{A, B, C\}$; range: $\{1, 2, 3\}$; not a function

Determine whether each relation is a function. See Examples 4 through 6.

	First Set: Input	Correspondence	Second Set: Output
19.	class of algebra students	final grade average	nonnegative numbers
20.	people who live in Cincinnati, Ohio	birth date	days of the year
21.	blue, green, brown	eye color	people who live in Cincinnati, Ohio
22.	whole numbers from 0 to 4	number of children	50 women in a water aerobics class

19. function   **20.** function   **21.** not a function   **22.** not a function

Determine whether each relation is also a function. See Examples 7 and 8.

23. $y = x + 1$ function

24. $y = x - 1$ function

25. $x = 2y^2$ not a function

26. $y = x^2$ function

27. $y - x = 7$ function

28. $2x - 3y = 9$ not a function

Objective C *Use the vertical line test to determine whether each graph is the graph of a function. See Examples 9 through 13.*

29.

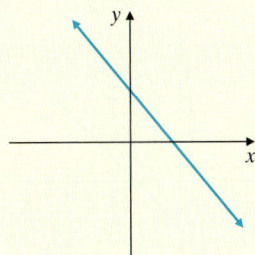

function

30.

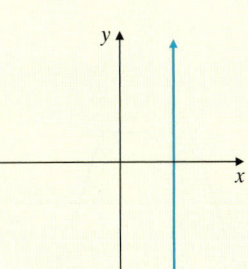

not a function

31.

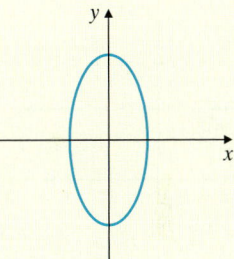

not a function

32.

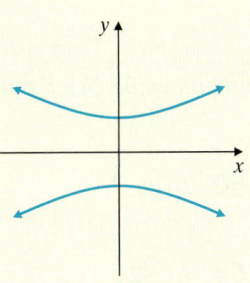

not a function

33.

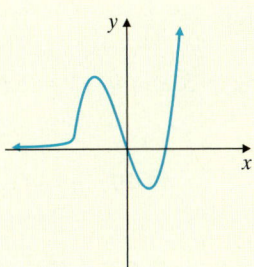

function

34.

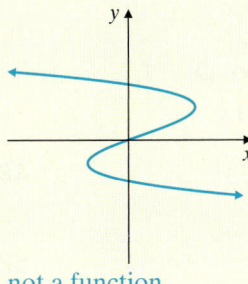

not a function

35.

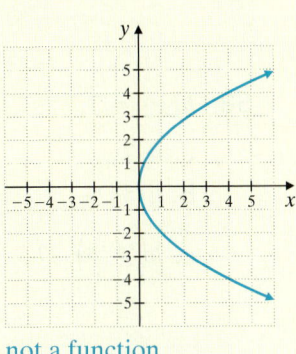

not a function

36.

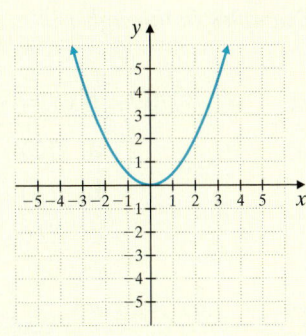

function

37.

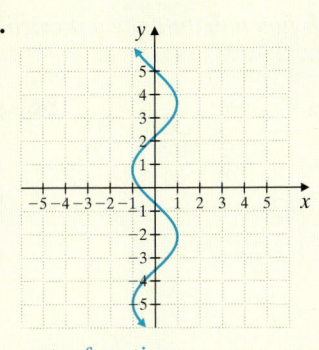

not a function

38.

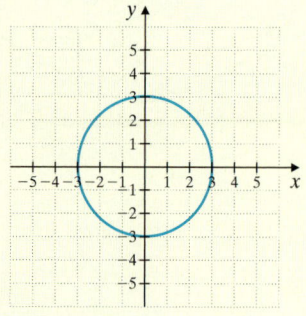

not a function

39.

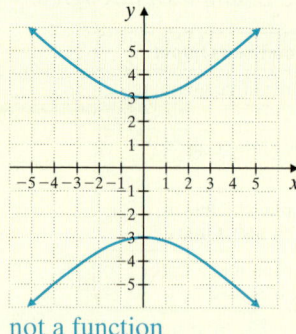

not a function

40.

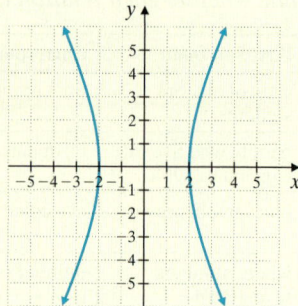

not a function

41.

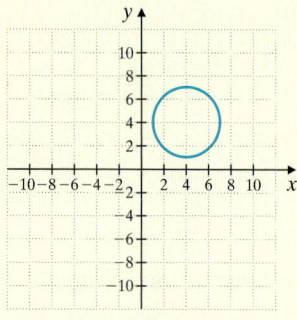

not a function

42.

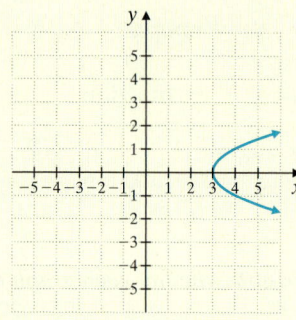

not a function

43.

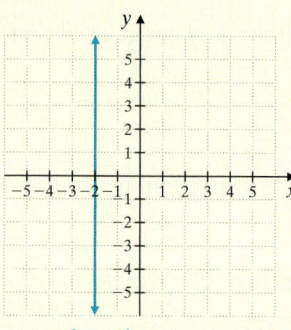

not a function

44.

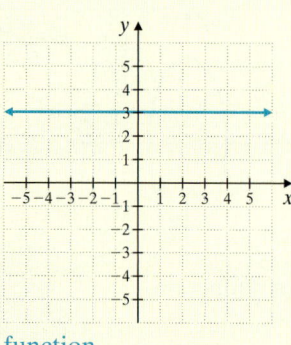

function

45.

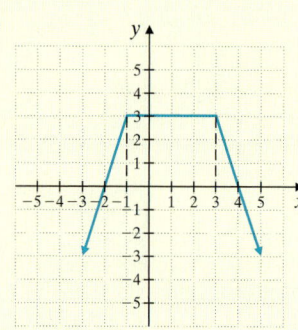

function

46.

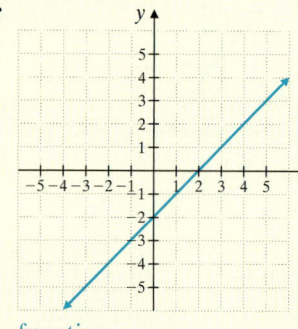

function

Objective D *If $f(x) = 3x + 3$, $g(x) = 4x^2 - 6x + 3$, and $h(x) = 5x^2 - 7$, find each function value. See Examples 14 through 17.*

47. $f(4)$ 15

48. $f(-1)$ 0

▶ **49.** $h(-3)$ 38

50. $h(0)$ −7

51. $g(2)$ 7

52. $g(1)$ 1

▶ **53.** $g(0)$ 3

54. $h(-2)$ 13

For each function, find the indicated values. See Examples 14 through 17.

55. $f(x) = \dfrac{1}{2}x$;

   a. $f(0)$ 0

   b. $f(2)$ 1

   c. $f(-2)$ −1

56. $g(x) = -\dfrac{1}{3}x$;

   a. $g(0)$ 0

   b. $g(-1)$ $\dfrac{1}{3}$

   c. $g(3)$ −1

57. $f(x) = -5$;

   a. $f(2)$ −5

   b. $f(0)$ −5

   c. $f(606)$ −5

58. $h(x) = 7$;

   a. $h(7)$ 7

   b. $h(542)$ 7

   c. $h\left(-\dfrac{3}{4}\right)$ 7

Use the graph of the functions below to answer Exercises 59–70. See Example 18.

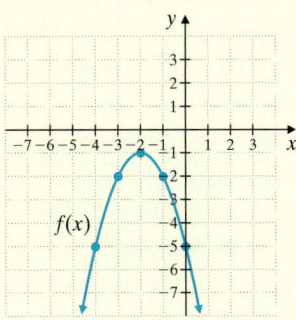
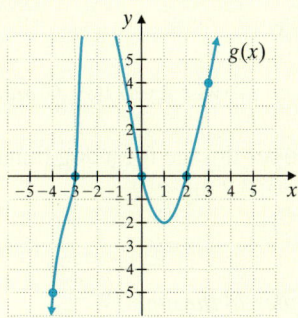

59. If $f(1) = -10$, write the corresponding ordered pair. $(1, -10)$

60. If $f(-5) = -10$, write the corresponding ordered pair. $(-5, -10)$

61. If $g(4) = 56$, write the corresponding ordered pair. $(4, 56)$

62. If $g(-2) = 8$, write the corresponding ordered pair. $(-2, 8)$

63. Find $f(-1)$. -2

64. Find $f(-2)$. -1

65. Find $g(2)$. 0

66. Find $g(-4)$. -5

67. Find all values of x such that $f(x) = -5$. $-4, 0$

68. Find all values of x such that $f(x) = -2$. $-3, -1$

69. Find all positive values of x such that $g(x) = 4$. 3

70. Find all values of x such that $g(x) = 0$. $-3, 0, 2$

The function $A(r) = \pi r^2$ may be used to find the area of a circle if we are given its radius. Use this function to answer Exercises 71 and 72. See Examples 14 through 17.

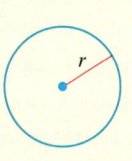

△**71.** Find the area of a circle whose radius is 5 centimeters. (Do not approximate π.) 25π sq cm

△**72.** Find the area of a circular garden whose radius is 8 feet. (Do not approximate π.) 64π sq ft

The function $V(x) = x^3$ may be used to find the volume of a cube if we are given the length x of a side. Use this function to answer Exercises 73 and 74. See Examples 14 through 17.

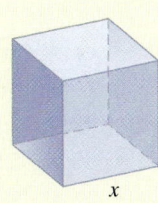

△**73.** Find the volume of a cube whose side is 14 inches. 2744 cu in.

△**74.** Find the volume of a die whose side is 1.7 centimeters. 4.913 cu cm

Forensic scientists use the following functions to find the height of a woman if they are given the height of her femur bone (f) or her tibia bone (t) in centimeters.

$$H(f) = 2.59f + 47.24$$
$$H(t) = 2.72t + 61.28$$

← 46 cm Femur → ← 35 cm Tibia →

Use these functions to answer Exercises 75 and 76. See Examples 14 through 17.

75. Find the height of a woman whose femur measures 46 centimeters. 166.38 cm

76. Find the height of a woman whose tibia measures 35 centimeters. 156.48 cm

The dosage in milligrams D of Ivermectin, a heartworm preventive, for a dog who weighs x pounds is given by

$$D(x) = \frac{136}{25}x$$

Use this function to answer Exercises 77 and 78. See Examples 14 through 17.

77. Find the proper dosage for a dog that weighs 30 pounds.   163.2 mg

78. Find the proper dosage for a dog that weighs 50 pounds.   272 mg

Solve. See Examples 14 through 17.

79. The value of U.S. agricultural exports to China, in billions of dollars, is approximated by the function $E(x) = 1.8x - 1.44$, where x is the number of years since 2000. (*Source:* Based on USDA Economic Research Service data)

   a. Find and interpret $E(5)$.   7.56; Agricultural exports to China in 2005 were approximately $7.56 billion.

   b. Estimate the value of the agricultural exports to China in 2014. $23.76 billion

80. The per capita consumption (in kilograms) of all rice in the United States is given by the function $C(x) = 0.11x + 12.8$, where x is the number of years since 2000. (*Source:* Based on data from USA Rice federation)

   a. Find and interpret $C(6)$.   13.46; In 2006 consumption of rice was 13.46 kilograms of rice per person

   b. Estimate the per capita consumption of rice in the United States in 2015.
   14.45 kilograms per person

Objective E   *Graph each linear function. See Example 19.*

81. $f(x) = 2x + 3$

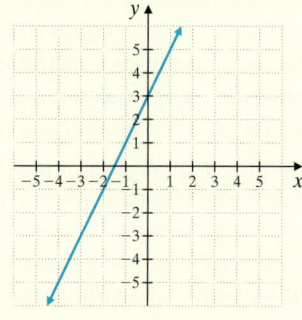

82. $f(x) = 5x - 1$

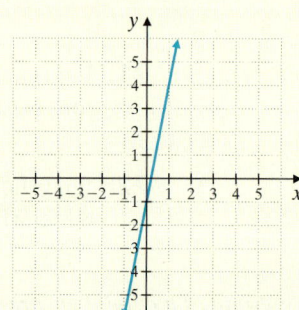

83. $f(x) = -3x$

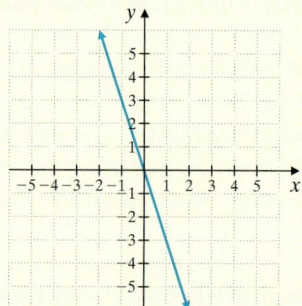

84. $f(x) = -4x$

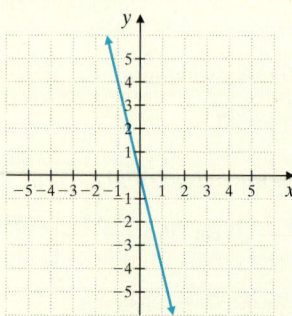

85. $f(x) = -x + 2$

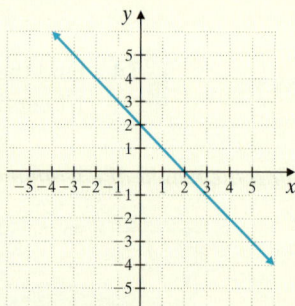

86. $f(x) = -x + 1$

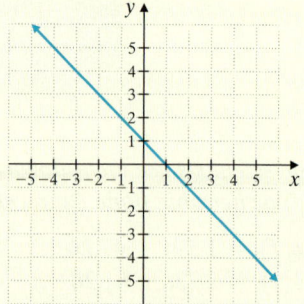

Review

Solve. See Section 2.7.

87. $2x - 7 \leq 21$
$\{x \mid x \leq 14\}$

88. $-3x + 1 > 0$
$\left\{x \mid x < \dfrac{1}{3}\right\}$

89. $5(x - 2) \geq 3(x - 1)$
$\left\{x \mid x \geq \dfrac{7}{2}\right\}$

90. $-2(x + 1) \leq -x + 10$
$\{x \mid x \geq -12\}$

91. $\dfrac{x}{2} + \dfrac{1}{4} < \dfrac{1}{8}$
$\left\{x \mid x < -\dfrac{1}{4}\right\}$

92. $\dfrac{x}{5} - \dfrac{3}{10} \geq \dfrac{x}{2} - 1$
$\left\{x \mid x \leq \dfrac{7}{3}\right\}$

Concept Extensions

Think about the appearance of each graph. Without graphing, determine which equations represent functions. Explain each answer. See the second Concept Check in this section.

93. $x = -1$
no; answers may vary

94. $y = 5$
yes; answers may vary

95. $y = 2x$
yes; answers may vary

96. $x + y = -5$
yes; answers may vary

Suppose that $y = f(x)$ and it is true that $f(7) = 50$. For Exercises 97–100, determine whether each statement is true or false. See the third Concept Check in this section.

97. An ordered-pair solution of the function is $(7, 50)$.
true

98. When x is 50, y is 7.   false

99. A possible function is $f(x) = x^2 + 1$.   true

100. A possible function is $f(x) = 10x - 20$.   true

Solve.

101. What is the greatest number of x-intercepts that a function may have? Explain your answer.
infinite number

102. What is the greatest number of y-intercepts that a function may have? Explain your answer.   1

103. In your own words, explain how to find the domain of a function given its graph.   answers may vary

104. Explain the vertical line test and how it is used.   answers may vary

For each function, find the indicated values.

105. $f(x) = x - 12$
a. $f(12)$ 0
b. $f(a)$ $a - 12$
c. $f(-x)$ $-x - 12$
d. $f(x + h)$ $x + h - 12$

106. $f(x) = 2x + 7$
a. $f(2)$ 11
b. $f(a)$ $2a + 7$
c. $f(-x)$ $-2x + 7$
d. $f(x + h)$ $2x + 2h + 7$

Objectives

A Evaluate Polynomial Functions.

B Find the Domain of a Rational Expression.

C Evaluate Rational Functions.

Objective A Evaluating Polynomial Functions

Recall function notation, first introduced in Section 8.2. At times it is convenient to use function notation to represent polynomials. For example, we may write $P(x)$ to represent the polynomial $3x^4 - 2x^2 - 5$. In symbols, we would write

$$P(x) = 3x^4 - 2x^2 - 5$$

This function is called a **polynomial function** because the expression $3x^4 - 2x^2 - 5$ is a polynomial.

Helpful Hint

Recall that the symbol $P(x)$ **does not mean** P times x. It is a special symbol used to denote a function.

Practice 1–2

If $P(x) = 5x^4 - 3x^2 + 7$, find each function value.

1. $P(2)$

2. $P(-1)$

Examples If $P(x) = 3x^4 - 2x^2 - 5$, find each function value.

1. $P(1) = 3(1)^4 - 2(1)^2 - 5 = -4$ Let $x = 1$ in the function $P(x)$.

2. $P(-2) = 3(-2)^4 - 2(-2)^2 - 5$ Let $x = -2$ in the function $P(x)$.

$\qquad = 3(16) - 2(4) - 5$

$\qquad = 35$

■ Work Practice 1–2

Many real-world phenomena are modeled by polynomial functions. If the polynomial function model is given, we can often find the solution of a problem by evaluating the function at a certain value.

Practice 3

The largest natural bridge is in the canyons at the base of Navajo Mountain, Utah. From the base to the top of the arch, it measures 290 feet. Neglecting air resistance, the height of an object dropped off the bridge is given by the polynomial function $P(t) = -16t^2 + 290$ at time t seconds. Find the height of the object at time $t = 0$ second and $t = 2$ seconds.

Example 3 Finding the Height of an Object

The Millau Viaduct, at 1125 feet, is the highest bridge in the world and overlooks the river Tarn in France. An object is dropped from the top of this bridge. Neglecting air resistance, the height of the object at time t seconds is given by the polynomial function $P(t) = -16t^2 + 1125$. Find the height of the object when $t = 1$ second and when $t = 8$ seconds.

Solution: To find the height of the object at 1 second, we find $P(1)$.

$$P(t) = -16t^2 + 1125$$
$$P(1) = -16(1)^2 + 1125$$
$$\quad = 1109$$

When $t = 1$ second, the height of the object is 1109 feet.

Teaching Tip

After discussing Example 3, ask, "What does the constant term represent in a height problem?" Follow up with "How high is a bridge if the height at time t of an object dropped from the bridge is given by $P(t) = -16t^2 + 934$?"

Answers

1. $P(2) = 75$ **2.** $P(-1) = 9$

3. at 0 sec, height is 290 ft; at 2 sec, height is 226 ft

To find the height of the object at 8 seconds, we find $P(8)$.

$$P(t) = -16t^2 + 1125$$
$$P(8) = -16(8)^2 + 1125$$
$$= -1024 + 1125$$
$$= 101$$

When $t = 8$ seconds, the height of the object is 101 feet. Notice that as time t increases, the height of the object decreases.

■ **Work Practice 3**

Objective B Finding Domains of Rational Functions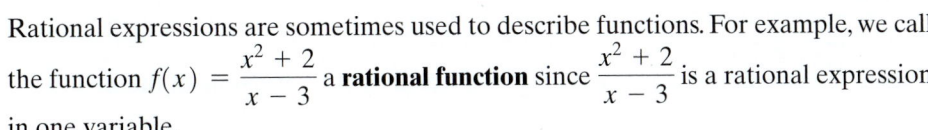

Rational expressions are sometimes used to describe functions. For example, we call the function $f(x) = \dfrac{x^2 + 2}{x - 3}$ a **rational function** since $\dfrac{x^2 + 2}{x - 3}$ is a rational expression in one variable.

The domain of a rational function such as $f(x) = \dfrac{x^2 + 2}{x - 3}$ is the set of all possible replacement values for x. In other words, since the rational expression $\dfrac{x^2 + 2}{x - 3}$ is not defined when $x = 3$, we say that the domain of $f(x) = \dfrac{x^2 + 2}{x - 3}$ is all real numbers except 3.

The domain of f is then

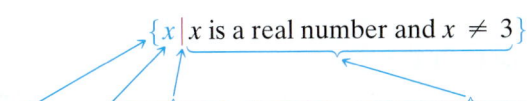

$$\{x \,|\, x \text{ is a real number and } x \neq 3\}$$

"The set of all x such that x is a real number and x is not equal to 3."

In this section, we will use this set builder notation to write domains. Unless told otherwise, we assume that the domain of a function described by an equation is the set of all real numbers for which the equation is defined.

Teaching Tip

Remind students that the domain of a function is a list (usually in set notation) of the values that x can be.

Example 4 Find the domain of each rational function.

a. $f(x) = \dfrac{8x^3 + 7x^2 + 20}{2}$ **b.** $g(x) = \dfrac{5x^2 - 3}{x - 1}$ **c.** $f(x) = \dfrac{7x - 2}{x^2 - 2x - 15}$

Solution: The domain of each function will contain all real numbers except those values that make the denominator 0.

a. No matter what the value of x, the denominator of $f(x) = \dfrac{8x^3 + 7x^2 + 20}{2}$ is never 0, so the domain of f is $\{x \,|\, x \text{ is a real number}\}$.

b. To find the values of x that make the denominator of $g(x)$ equal to 0, we solve the equation "denominator = 0":

$$x - 1 = 0, \quad \text{or} \quad x = 1$$

The domain must exclude 1 since the rational expression is undefined when x is 1. The domain of g is $\{x \,|\, x \text{ is a real number and } x \neq 1\}$.

(Continued on next page)

Practice 4

Find the domain of each rational function.

a. $f(x) = \dfrac{4x^5 - 3x^2 + 2}{-6}$

b. $g(x) = \dfrac{6x^2 + 1}{x + 3}$

c. $h(x) = \dfrac{8x - 3}{x^2 - 5x + 6}$

c. We find the domain by setting the denominator equal to 0.

$$x^2 - 2x - 15 = 0 \quad \text{Set the denominator equal to 0 and solve.}$$
$$(x - 5)(x + 3) = 0$$
$$x - 5 = 0 \quad \text{or} \quad x + 3 = 0$$
$$x = 5 \quad \text{or} \quad x = -3$$

If x is replaced with 5 or with -3, the rational expression is undefined.
The domain of f is $\{x \mid x \text{ is a real number and } x \neq 5, x \neq -3\}$.

■ **Work Practice 4**

✓**Concept Check** For which of these values (if any) is the rational expression $\dfrac{x - 3}{x^2 + 2}$ undefined?

a. 2 　　　 **b.** 3 　　　 **c.** -2 　　　 **d.** 0 　　　 **e.** None of these

See the Calculator Explorations on the next page for further domain exercises.

Objective C Evaluating Rational Functions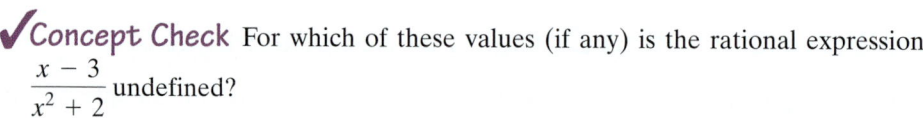

Now let's practice finding function values of rational functions.

Practice 5

A company's cost per book for printing x particular books is given by the rational function

$$C(x) = \frac{0.8x + 5000}{x}.$$

Find the cost per book for printing:
a. 100 books
b. 1000 books

Teaching Tip

For Example 5, have students give the domain of the *cost* function. Then have them give the domain of the rational function $f(x) = \dfrac{2.6x + 10,000}{x}$. Why are the domains different?

Example 5 Finding Unit Cost

For the ICL Production Company, the rational function $C(x) = \dfrac{2.6x + 10,000}{x}$ describes the company's cost per disc of pressing x compact discs. Find the cost per disc for pressing:

a. 100 compact discs

b. 1000 compact discs

Solution:

a. $C(100) = \dfrac{2.6(100) + 10,000}{100} = \dfrac{10,260}{100} = 102.6$

The cost per disc for pressing 100 compact discs is $102.60.

b. $C(1000) = \dfrac{2.6(1000) + 10,000}{1000} = \dfrac{12,600}{1000} = 12.6$

The cost per disc for pressing 1000 compact discs is $12.60.
Notice that as more compact discs are produced, the cost per disc decreases.

■ **Work Practice 5**

Answers

5. **a.** $50.80　**b.** $5.80

✓**Concept Check Answer**

e

Calculator Explorations Graphing

Recall that since the rational expression $\dfrac{7x - 2}{(x - 2)(x + 5)}$ is not defined when $x = 2$ or when $x = -5$, we say that the domain of the rational function $f(x) = \dfrac{7x - 2}{(x - 2)(x + 5)}$ is all real numbers except 2 and -5. This domain can be written as $\{x \mid x \text{ is a real number and } x \neq 2, x \neq -5\}$. This means that the graph of f should not cross the vertical lines $x = 2$ and $x = -5$. The graph of f in *connected* mode is shown below. In connected mode the grapher tries to connect all dots of the graph so that the result is a smooth curve. This is what has happened in the graph. Notice that the graph appears to contain vertical lines at $x = 2$ and at $x = -5$. We know that this cannot happen because the function is not defined at $x = 2$ and at $x = -5$. We also know that this cannot happen because the graph of this function would not pass the vertical line test.

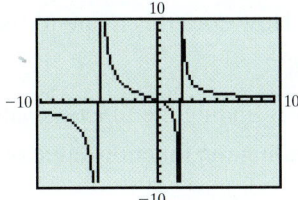

If we graph f in *dot* mode, the graph appears as below. In dot mode the grapher will not connect dots with a smooth curve. Notice that the vertical lines have disappeared, and we have a better picture of the graph. It actually appears more like the hand-drawn graph above on the right. By using a TABLE feature or a CALCULATE VALUE feature or by tracing, we can see that the function is not defined at $x = 2$ and at $x = -5$.

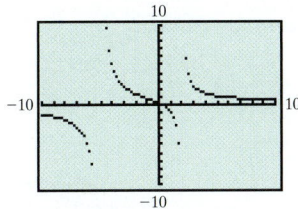

Note: Some calculator manufacturers now offer downloadable operating systems that eliminate the need to use dot mode to graph rational functions.

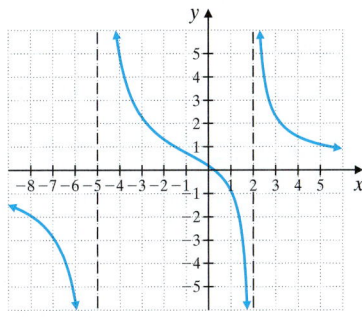

Find the domain of each rational function. Then graph each rational function and use the graph to confirm the domain.

1. $f(x) = \dfrac{5x}{x - 6}$

$\{x \mid x \text{ is a real number and } x \neq 6\}$

2. $f(x) = \dfrac{x}{x + 4}$

$\{x \mid x \text{ is a real number and } x \neq -4\}$

3. $f(x) = \dfrac{x + 1}{x^2 - 4}$

$\{x \mid x \text{ is a real number and } x \neq -2, x \neq 2\}$

4. $g(x) = \dfrac{5x}{x^2 - 9}$

$\{x \mid x \text{ is a real number and } x \neq 3, x \neq -3\}$

5. $h(x) = \dfrac{x^2}{2x^2 + 7x - 4}$

$\left\{x \mid x \text{ is a real number and } x \neq -4, x \neq \dfrac{1}{2}\right\}$

6. $f(x) = \dfrac{3x + 2}{4x^2 - 19x - 5}$

$\left\{x \mid x \text{ is a real number and } x \neq -\dfrac{1}{4}, x \neq 5\right\}$

7. $g(x) = \dfrac{x^2 + x + 1}{5}$

$\{x \mid x \text{ is a real number}\}$

8. $h(x) = \dfrac{x^2 + 25}{2}$

$\{x \mid x \text{ is a real number}\}$

Vocabulary, Readiness & Video Check

Use the choices below to fill in each blank.

polynomial expression   rational expression
polynomial function   rational function

1. $7x^2 - 6x + 2$ is a <u>polynomial expression</u> .

2. $f(x) = \dfrac{2x + 1}{x + 2}$ is a <u>rational function</u> .

3. $\dfrac{2x + 1}{x + 2}$ is a <u>rational expression</u> .

4. $f(x) = 7x - 6x + 2$ is a <u>polynomial function</u> .

Martin-Gay Interactive Videos   *Watch the section lecture video and answer the following questions.*

Objective A 5. From Example 2, what polynomial function value were we finding, and what is that value we found? ▶

Objective B 6. From Example 4, why can't the denominators of rational expressions be zero? How does this relate to the domain of a rational function? ▶

Objective C 7. In Example 5, how is finding the function value of a rational function similar to finding the function value of a polynomial function? ▶

See Video 8.3
See video answer section.

8.3 Exercise Set MyMathLab® ▶

Objective A   *If $P(x) = x^2 + x + 1$ and $Q(x) = 5x^2 - 1$, find each function value. See Examples 1 and 2.*

▶**1.** $P(7)$ **2.** $Q(4)$ ▶**3.** $Q(-10)$ **4.** $P(-4)$ **5.** $P(0)$ **6.** $Q(0)$
   57   79   499   13   1 -1

If $P(x) = x^3 + 2x - 3$ and $Q(x) = 7x + 5$, find each function value. See Examples 1 and 2.

7. $P(2)$ **8.** $Q(6)$ **9.** $Q(-3)$ **10.** $P(-1)$ **11.** $P(-2)$ **12.** $Q(0)$
   9   47 -16 -6 -15 5

Solve. See Example 3.

The surface area of a rectangular box is given by the polynomial
$$2HL + 2LW + 2HW$$
and is measured in square units. In business, surface area is often calculated to help determine cost of materials.

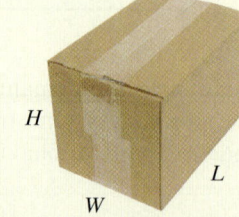

13. A rectangular box is to be constructed to hold a new camcorder. The box is to have dimensions 5 inches by 4 inches by 9 inches. Find the surface area of the box.   202 sq in.

14. Suppose it has been determined that a box of dimensions 4 inches by 4 inches by 8.5 inches can be used to contain the camcorder in Exercise 13. Find the surface area of this box and calculate the square inches of material saved by using this box instead of the box in Exercise 13.   168 sq in.; 34 sq in. saved

▶15. A projectile is fired upward from the ground with an initial velocity of 300 feet per second. Neglecting air resistance, the height of the projectile at any time t can be described by the polynomial function $P(t) = -16t^2 + 300t$. Find the height of the projectile at each given time.

a. $t = 1$ second   284 ft

b. $t = 2$ seconds   536 ft

c. $t = 3$ seconds   756 ft

d. $t = 4$ seconds   944 ft

e. Explain why the height increases and then decreases as time passes.   answers may vary

f. Approximate (to the nearest second) how long before the object hits the ground.   19 sec

16. Two workers at the Hoover Dam bypass bridge were spending their lunch break tossing a football. One of the workers threw the ball upward with an initial velocity of 10 feet per second, but his partner missed the ball and it went over his head and down toward the river. Neglecting air resistance, the height of the football above the Colorado River at any time t can be described by the polynomial function $P(t) = -16t^2 + 10t + 910$. Find the height of the football at each given time.

a. $t = 0$ seconds   910 ft

b. $t = \dfrac{1}{2}$ second   911 ft

c. $t = 3$ seconds   796 ft

d. $t = 5$ seconds   560 ft

e. Explain why the height increases and then decreases as time passes.   answers may vary

f. Approximate (to the nearest second) how long before the football lands in the river.   8 sec

17. The polynomial function $P(x) = 45x - 100{,}000$ models the relationship between the number of computer briefcases x that a company sells and the profit the company makes, $P(x)$. Find $P(4000)$, the profit from selling 4000 computer briefcases. $80,000

18. The total cost (in dollars) for MCD, Inc., Manufacturing Company to produce x blank audio-cassette tapes per week is given by the polynomial function $C(x) = 0.8x + 10{,}000$. Find the total cost of producing 20,000 tapes per week. $26,000

19. The total revenues (in dollars) for an art supply company to sell x boxes of colored pencils per week over the Internet is given by the polynomial function $R(x) = 11x$. Find the total revenue from selling 1500 boxes of colored pencils. $16,500

20. The total revenues (in dollars) for MCD, Inc., Manufacturing Company to sell x blank audiocassette tapes per week is given by the polynomial function $R(x) = 2x$. Find the total revenue from selling 20,000 tapes per week. $40,000

21. Suppose that a movie is being filmed in New York City. An action shot requires an object to be thrown upward with an initial velocity of 80 feet per second off the top of 1 Madison Square Plaza, a height of 576 feet. Neglecting air resistance, the height $h(t)$ in feet of the object after t seconds is given by the function $h(t) = -16t^2 + 80t + 576$. (*Source: The World Almanac*)

a. Find the height of the object at $t = 0$ seconds, $t = 2$ seconds, $t = 4$ seconds, and $t = 6$ seconds.   576 ft; 672 ft; 640 ft; 480 ft

b. Explain why the height of the object increases and then decreases as time passes.   answers may vary

c. Factor the polynomial $-16t^2 + 80t + 576$.
$-16(t + 4)(t - 9)$

22. Suppose that an object is thrown upward with an initial velocity of 64 feet per second off the edge of a 960-foot-cliff. Neglecting air resistance, the height $h(t)$ in feet of the object after t seconds is given by the function.

$$h(t) = -16t^2 + 64t + 960$$

a. Find the height of the object at $t = 0$ seconds, $t = 3$ seconds, $t = 6$ seconds, and $t = 9$ seconds.   960 ft; 1008 ft; 768 ft; 240 ft

b. Explain why the height of the object increases and then decreases as time passes.   answers may vary

c. Factor the polynomial $-16t^2 + 64t + 960$. $-16(t - 10)(t + 6)$

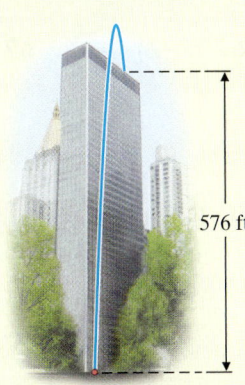

576 ft

Objective B *Find the domain of each rational function. See Example 4.*

23. $f(x) = \dfrac{5x - 7}{4}$
{$x \mid x$ is a real number}

24. $g(x) = \dfrac{4 - 3x}{2}$
{$x \mid x$ is a real number}

25. $s(t) = \dfrac{t^2 + 1}{2t}$
{$t \mid t$ is a real number and $t \neq 0$}

26. $v(t) = -\dfrac{5t + t^2}{3t}$
{$t \mid t$ is a real number and $t \neq 0$}

▶ 27. $f(x) = \dfrac{3x}{7 - x}$
{$x \mid x$ is a real number and $x \neq 7$}

28. $f(x) = \dfrac{-4x}{-2 + x}$
{$x \mid x$ is a real number and $x \neq 2$}

29. $f(x) = \dfrac{x}{3x - 1}$
$\left\{ x \mid x \text{ is a real number and } x \neq \dfrac{1}{3} \right\}$

30. $g(x) = \dfrac{-2}{2x + 5}$
$\left\{ x \mid x \text{ is a real number and } x \neq -\dfrac{5}{2} \right\}$

31. $R(x) = \dfrac{3 + 2x}{x^3 + x^2 - 2x}$
{$x \mid x$ is a real number and $x \neq -2, x \neq 0, x \neq 1$}

32. $h(x) = \dfrac{5 - 3x}{2x^2 - 14x + 20}$
{$x \mid x$ is a real number and $x \neq 5, x \neq 2$}

▶ 33. $C(x) = \dfrac{x + 3}{x^2 - 4}$
{$x \mid x$ is a real number and $x \neq 2, x \neq -2$}

34. $R(x) = \dfrac{5}{x^2 - 7x}$
{$x \mid x$ is a real number and $x \neq 0, x \neq 7$}

Objectives B C **Mixed Practice** *If $f(x) = \dfrac{x + 8}{2x - 1}$ and $g(x) = \dfrac{x - 2}{x - 5}$, find each function value. See Examples 4 and 5.*

35. $f(2)$ $\dfrac{10}{3}$ **36.** $g(10)$ $\dfrac{8}{5}$ **37.** $g(0)$ $\dfrac{2}{5}$ **38.** $f(0)$ -8 **39.** $f(-1)$ $-\dfrac{7}{3}$ **40.** $g(-5)$ $\dfrac{7}{10}$

41. Find the domain of $g(x)$.
{$x \mid x$ is a real number and $x \neq 5$}

42. Find the domain of $f(x)$.
$\left\{ x \mid x \text{ is a real number and } x \neq \dfrac{1}{2} \right\}$

If $f(x) = \dfrac{x^2 + 5}{x}$ and $g(x) = \dfrac{x^2 + 2x}{x + 3}$, find each function value. See Examples 4 and 5.

43. $f(1)$ 6 **44.** $g(3)$ $\dfrac{5}{2}$ **▶ 45.** $g(-6)$ -8 **46.** $f(-2)$ $-\dfrac{9}{2}$ **47.** $g(0)$ 0 **48.** $f(2)$ $\dfrac{9}{2}$

▶ 49. Find the domain of $f(x)$.
{$x \mid x$ is a real number and $x \neq 0$}

50. Find the domain of $g(x)$.
{$x \mid x$ is a real number and $x \neq -3$}

Solve. See Example 5.

51. The total revenue from the sale of a popular book is approximated by the rational function $R(x) = \dfrac{1000x^2}{x^2 + 4}$, where x is the number of years since publication and $R(x)$ is the total revenue in millions of dollars.

a. Find the total revenue at the end of the first year. $200 million

b. Find the total revenue at the end of the second year. $500 million

c. Find the revenue during the second year only. $300 million

52. The function $f(x) = \dfrac{100,000x}{100 - x}$ models the cost in dollars for removing x percent of the pollutants from a bayou in which a nearby company dumped creosol.

a. Find the cost of removing 20% of the pollutants from the bayou. (*Hint:* Find $f(20)$.) $25,000

b. Find the cost of removing 60% of the pollutants and then 80% of the pollutants. $150,000; $400,000

c. Find $f(90)$, then $f(95)$, and then $f(99)$. What happens to the cost as x approaches 100%? $900,000; $1,900,000; $9,900,000; increases without bound

Review

Solve each equation for x. See Section 2.3.

53. $\dfrac{x}{5} = \dfrac{x + 2}{3}$ -5

54. $\dfrac{x}{4} = \dfrac{x + 3}{6}$ 6

55. $\dfrac{x - 3}{2} = \dfrac{x - 5}{6}$ 2

56. $\dfrac{x - 6}{4} = \dfrac{x - 2}{5}$ 22

Concept Extensions

57. An object is thrown upward from the ground with an initial velocity of 64 feet per second. Neglecting air resistance, the height $h(t)$ in feet of the object after t seconds is given by the polynomial function

$$h(t) = -16t^2 + 64t$$

a. Write an equivalent factored expression for the function $h(t)$ by factoring $-16t^2 + 64t$. $h(t) = -16t(t - 4)$

b. Find $h(1)$ by using

$$h(t) = -16t^2 + 64t \quad \text{48 ft}$$

and then by using the factored form of $h(t)$.

c. Explain why the values found in part (b) are the same.   answers may vary

58. An object is dropped from the gondola of a hot-air balloon at a height of 224 feet. Neglecting air resistance, the height $h(t)$ in feet of the object after t seconds is given by the polynomial function

$$h(t) = -16t^2 + 224$$

a. Write an equivalent factored expression for the function $h(t)$ by factoring $-16t^2 + 224$. $h(t) = -16(t^2 - 14)$

b. Find $h(2)$ by using   160 ft

$$h(t) = -16t^2 + 224$$

and then by using the factored form of the function.

c. Explain why the values found in part (b) are the same.   answers may vary

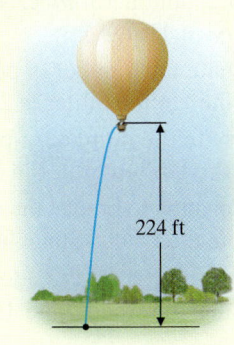

224 ft

If $P(x) = 3x + 3$, $Q(x) = 4x^2 - 6x + 3$, *and* $R(x) = 5x^2 - 7$, *find each function.*

59. $P(x) + Q(x)$ $4x^2 - 3x + 6$

60. $Q(x) - R(x)$ $-x^2 - 6x + 10$

61. If $P(x) = 2x - 3$, find $P(a)$, $P(-x)$, and $P(x + h)$.
$2a - 3; -2x - 3; 2x + 2h - 3$

62. If $P(x) = 5x + 1$, find $P(a)$, $P(-x)$, and $P(x + h)$.
$5a + 1; -5x + 1; 5x + 5h + 1$

If $R(x) = x + 5$, $Q(x) = x^2 - 2$, *and* $P(x) = 5x$, *find each function.*

63. $P(x) \cdot R(x)$ $5x^2 + 25x$

64. $P(x) \cdot Q(x)$ $5x^3 - 10x$

If $f(x) = x^2 - 3x$, *find each function value.*

65. $f(a)$ $a^2 - 3a$

66. $f(a + h)$ $a^2 + 2ah + h^2 - 3a - 3h$

67. Write a polynomial function, $P(x)$, so that $P(0) = 7$.
answers may vary

68. Write a rational function, $R(x)$, so that $R(1) = 2$.
answers may vary

69. The function $f(x) = 0.21x^2 + 5.09x + 45.3$ can be used to approximate the growth of restaurant food-and-drink sales, where x is the number of years since 1970 and $f(x)$ or y is the sales (in billions of dollars.)

  a. Approximate the restaurant food-and-drink sales in 2010. $584.9 billion

  b. Approximate the restaurant food-and-drink sales in 2014. $675.82 billion

  c. Use this function to predict the restaurant food-and-drink sales in 2020. $824.8 billion

  d. From parts (a), (b), and (c), determine whether the restaurant food-and-drink sales are increasing at a steady rate. Explain why or why not.
answers may vary

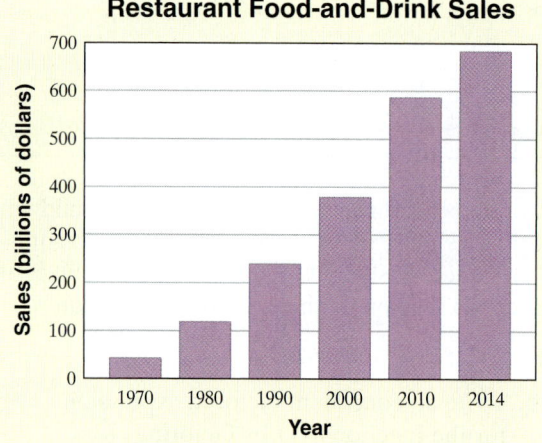

Restaurant Food-and-Drink Sales

Source: National Restaurant Association

70. The function $f(x) = -0.001x^2 + 0.245x + 9.22$ can be used to approximate the total cheese production in the United States from 2005 to 2014, where x is the number of years after 2005 and $f(x)$ or y is pounds of cheese (in billions). Round answers to the nearest hundredth of a billion. (*Source*: National Agricultural Statistics Service, USDA)

  a. Approximate the number of pounds of cheese produced in the United States in 2010. 10.42 billion lb

  b. Approximate the number of pounds of cheese produced in the United States in 2013. 11.12 billion lb

  c. Use this function to predict the pounds of cheese produced in the United States in 2020. 12.67 billion lb

  d. From parts (a), (b), and (c), determine whether the number of pounds of cheese produced in the United States is increasing at a steady rate. Explain why or why not. answers may vary

Total U.S. Cheese Production

1. $y = -x + 7$

2. $x = -2$

3. $y = 0$

5. $y = -5x - 6$

9. $y = 3x - 2$

Calculating body-mass index (BMI) is a way to gauge whether a person should lose weight. Doctors recommend that body-mass index values fall between 19 and 25. The formula for body-mass index B is $B = \dfrac{705w}{h^2}$, where w is weight in pounds and h is height in inches. Use this formula to answer Exercises 71 and 72.

71. A patient is 5 ft 8 in. tall. What should his or her weight be to have a body-mass index of 25? Round to the nearest whole pound.   164 lb

72. A doctor recorded a body-mass index of 47 on a patient's chart. Later, a nurse notices that the doctor recorded the patient's weight as 240 pounds but neglected to record the patient's height. Explain how the nurse can use the information from the chart to find the patient's height. Then find the height.
answers may vary; 60 in. or 5 ft

In physics, when the source of a sound is traveling toward an observer, the relationship between the actual pitch a of the sound and the pitch h that the observer hears due to the Doppler effect is described by the formula $h = \dfrac{a}{1 - \dfrac{s}{770}}$, where s is the speed of the sound source in miles per hour. Use this formula to answer Exercises 73 and 74.

73. An emergency vehicle has a single-tone siren with the pitch of the musical note E. As it approaches an observer standing by the road, the vehicle is traveling 50 mph. Is the pitch that the observer hears due to the Doppler effect lower or higher than the actual pitch? To which musical note is the pitch that the observer hears closest?   higher; F

74. Suppose an emergency van has a single-tone siren with the pitch of the musical note G. If the van is traveling at 80 mph approaching a standing observer, name the pitch the observer hears and the musical note closest to that pitch. Round to the nearest tenth.   437.4; A

Pitch of an Octave of Musical Notes in Hertz (Hz)	
Note	**Pitch**
Middle C	261.63
D	293.66
E	329.63
F	349.23
G	392.00
A	440.00
B	493.88

Note: Greater numbers indicate higher pitches (acoustically).
(*Source:* American Standards Association)

Linear Equations in Two Variables and Functions

Answers

Below is a review of equations of lines.

Teaching Tip

After reviewing the box for Forms of Linear Equations, ask, "How many pieces of information about a line must you know to determine the equation of the line?"

1. $y = -x + 7$

2. $x = -2$

3. $y = 0$

4. $y = -\dfrac{3}{8}x - \dfrac{29}{4}$

5. $y = -5x - 6$

6. $y = -4x + \dfrac{1}{3}$

7. $y = \dfrac{1}{2}x - 1$

8. $y = 3x - \dfrac{3}{2}$

9. $y = 3x - 2$

10. $y = -\dfrac{5}{4}x + 4$

11. $y = \dfrac{1}{4}x - \dfrac{7}{2}$

12. $y = -\dfrac{5}{2}x - \dfrac{5}{2}$

Forms of Linear Equations

$Ax + By = C$	**Standard form** of a linear equation A and B are not both 0.
$y = mx + b$	**Slope-intercept form** of a linear equation The slope is m, and the y-intercept is $(0, b)$.
$y - y_1 = m(x - x_1)$	**Point-slope form** of a linear equation The slope is m, and (x_1, y_1) is a point on the line.
$y = c$	**Horizontal line** The slope is 0, and the y-intercept is $(0, c)$.
$x = c$	**Vertical line** The slope is undefined, and the x-intercept is $(c, 0)$.

Parallel and Perpendicular Lines

Nonvertical parallel lines have the same slope. The product of the slopes of two nonvertical perpendicular lines is -1.

Find an equation of each line. Write the equation in the form $x = a$, $y = b$, or $y = mx + b$.

1. Through $(1, 6)$ and $(5, 2)$

2. Vertical line; through $(-2, -10)$

3. Horizontal line; through $(1, 0)$

4. Through $(2, -8)$ and $(-6, -5)$

5. Through $(-2, 4)$ with slope -5

6. Slope -4; y-intercept $\left(0, \dfrac{1}{3}\right)$

7. Slope $\dfrac{1}{2}$; y-intercept $(0, -1)$

8. Through $\left(\dfrac{1}{2}, 0\right)$ with slope 3

9. Through $(-1, -5)$; parallel to $3x - y = 5$

10. Through $(0, 4)$; perpendicular to $4x - 5y = 10$

11. Through $(2, -3)$; perpendicular to $4x + y = \dfrac{2}{3}$

12. Through $(-1, 0)$; parallel to $5x + 2y = 2$

13. Undefined slope; through $(-1, 3)$ **14.** $m = 0$; through $(-1, 3)$

Find the domain and the range of each relation. Also determine whether the relation is a function.

15. $\{(1, 1), (2, 2), (-3.5, -3.5)\}$

16. $\left\{(-1, 7), \left(\dfrac{3}{4}, 8\right), (-1, 9)\right\}$

17. 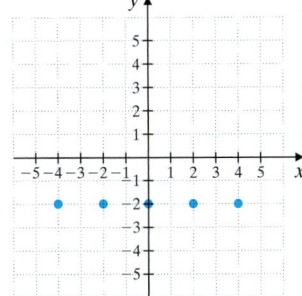   **18.**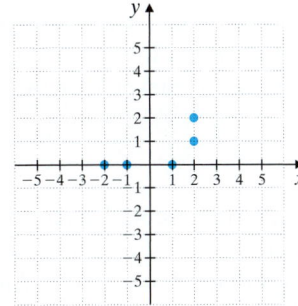

If $f(x) = x^3 - 2x + 7$ and $g(x) = \dfrac{x^2 + 1}{x - 1}$, find each function value.

19. $f(2)$ **20.** $g(2)$

21. $g(0)$ **22.** $f(0)$

23. $f(-1)$ **24.** $g(-5)$

8.4 Interval Notation, Finding Domains and Ranges from Graphs, and Graphing Piecewise-Defined Functions

Objectives

A Use Interval Notation.

B Find the Domain and Range from a Graph.

C Graph Piecewise-Defined Functions.

Objective A Using Interval Notation

Recall that a **solution** of an inequality is a value of the variable that makes the inequality a true statement. The **solution set** of an inequality is the set of all solutions. Notice that the solution set of the inequality $x > 2$, for example, contains all numbers greater than 2. Its graph is an interval on the number line since an infinite number of values satisfies the inequality. If we use open/closed-circle notation, the graph of $\{x \mid x > 2\}$ looks like:

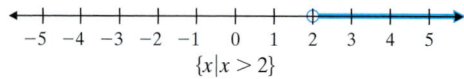

$$\{x \mid x > 2\}$$

In this section, a different graphing notation will be used to help us understand **interval notation.** Instead of an open circle, we use a parenthesis; instead of a closed circle, we use a bracket. With this new notation, the graph of $\{x \mid x > 2\}$ now looks like:

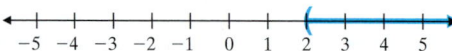

and can be represented in interval notation as $(2, \infty)$. The symbol ∞ is read "infinity" and indicates that the interval includes *all* numbers greater than 2. The left parenthesis indicates that 2 *is not* included in the interval. Using a left bracket, [, would indicate that 2 *is* included in the interval. The following table shows three equivalent ways to describe an interval: in set notation, as a graph, and in interval notation.

Set Notation	Graph	Interval Notation
$\{x \mid x < a\}$		$(-\infty, a)$
$\{x \mid x > a\}$		(a, ∞)
$\{x \mid x \le a\}$		$(-\infty, a]$
$\{x \mid x \ge a\}$		$[a, \infty)$
$\{x \mid a < x < b\}$		(a, b)
$\{x \mid a \le x \le b\}$		$[a, b]$
$\{x \mid a < x \le b\}$		$(a, b]$
$\{x \mid a \le x < b\}$		$[a, b)$
$\{x \mid x \text{ is a real number}\}$		$(-\infty, \infty)$

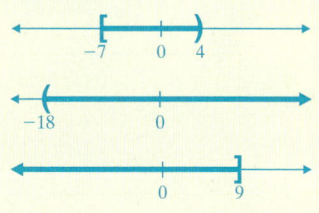

Helpful Hint

Notice that a parenthesis is always used to enclose ∞ and $-\infty$.

Why? The symbols ∞ (or $+\infty$) and $-\infty$ do not represent actual numbers. Although we can certainly approach ∞ and $-\infty$, we will never "reach" them.

Examples   Graph each set on a number line and then write it in interval notation.

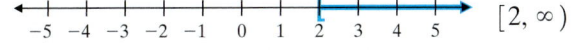

1. $\{x \mid x \geq 2\}$ $[2, \infty)$

2. $\{x \mid x < -1\}$ $(-\infty, -1)$

3. $\{x \mid 0.5 < x \leq 3\}$ $(0.5, 3]$

■ **Work Practice 1–3**

✓**Concept Check**   Explain what is wrong with writing the interval $(5, \infty]$.

Objective B   Finding the Domain and Range from a Graph

Recall from Section 8.2 that the

domain of a relation is the set of all first components of the ordered pairs of the relation and the
range of a relation is the set of all second components of the ordered pairs of the relation.

In this section, we use the graph of a relation to find its domain and range. Let's use interval notation to write these domains and ranges. Remember, we use a parenthesis to indicate that a number is not part of the domain and we use a bracket to indicate that a number is part of the domain. Parentheses are placed about infinity symbols indicating that we approach but never reach infinity.

To find the domain of a function (or relation) from its graph, recall that on the rectangular coordinate system, "domain" is the set of first components of the ordered pairs, so this means the x-values that are graphed. Similarly, "range" is the set of second components of the ordered pairs, so this means the y-values that are graphed.

Examples   Find the domain and range of each relation. Write each in interval notation.

4.
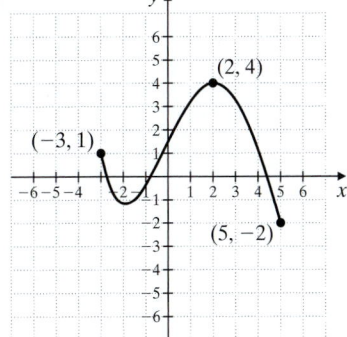

5.
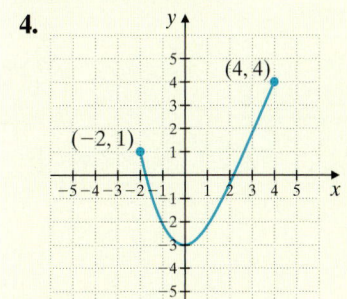

6.
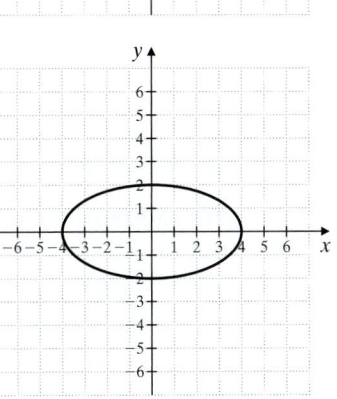

7.
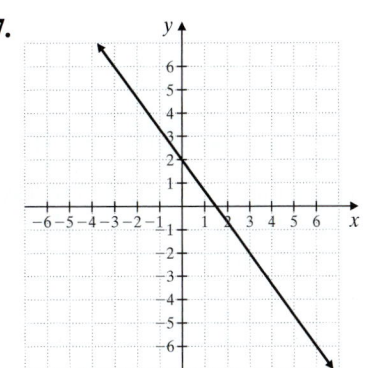

(*Continued on next page*)

(*Continued on next page*)

Practice 1–3

Graph each set on a number line and then write it in interval notation.

1. $\{x \mid x > -3\}$

2. $\{x \mid x \leq 0\}$

3. $\{x \mid -0.5 \leq x < 2\}$

Teaching Tip:
Classroom Activity

Give students a domain and range to use to make four sketches: two of a relation that is not a function and two of a relation that is a function. When they have finished, have them exchange sketches with a partner to determine which ones are sketches of functions.

Practice 4–7

Find the domain and range of each relation. Write each in interval notation.

4.
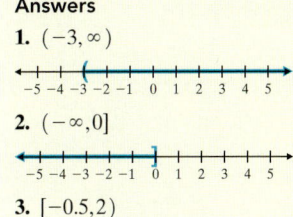

Answers

1. $(-3, \infty)$

2. $(-\infty, 0]$

3. $[-0.5, 2)$
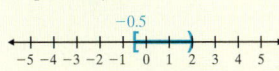

4. domain: $[-2, 4]$; range: $[-3, 4]$

✓**Concept Check Answer**

should be $(5, \infty)$ since a parenthesis is always used to enclose ∞

5.

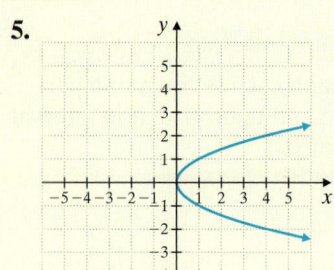

6.

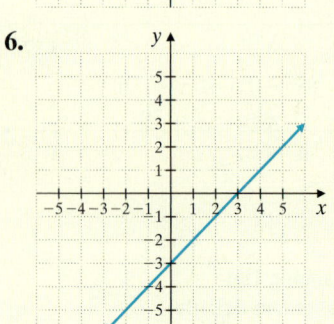

7.

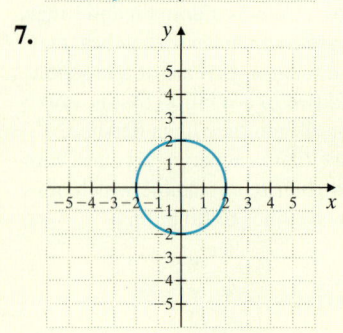

Solution: Notice that the graphs for Examples 4, 5, and 7 are graphs of functions because each passes the vertical line test.

4.

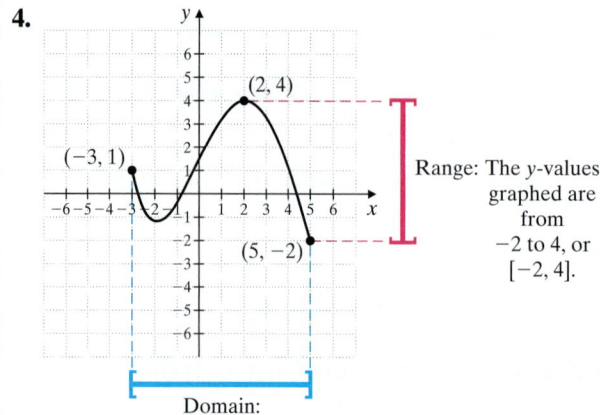

Range: The y-values graphed are from -2 to 4, or $[-2, 4]$.

Domain:
The x-values graphed are from -3 to 5, or $[-3, 5]$.

5.

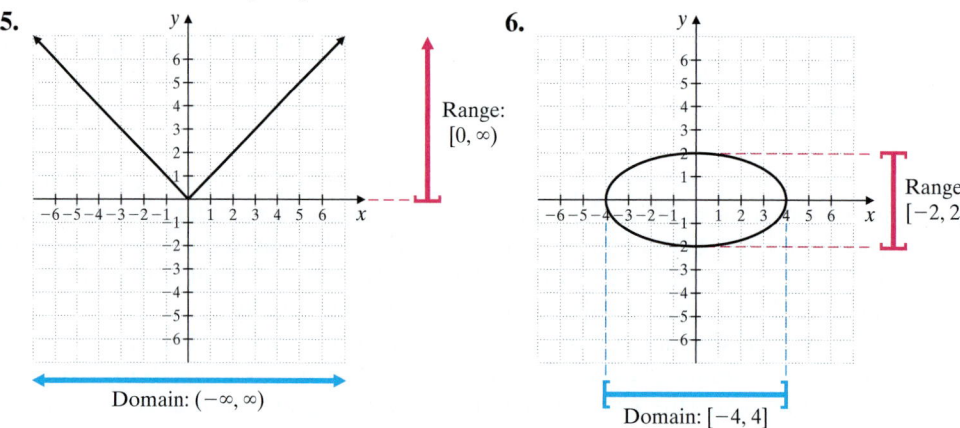

Range: $[0, \infty)$

Domain: $(-\infty, \infty)$

6.

Range: $[-2, 2]$

Domain: $[-4, 4]$

7.

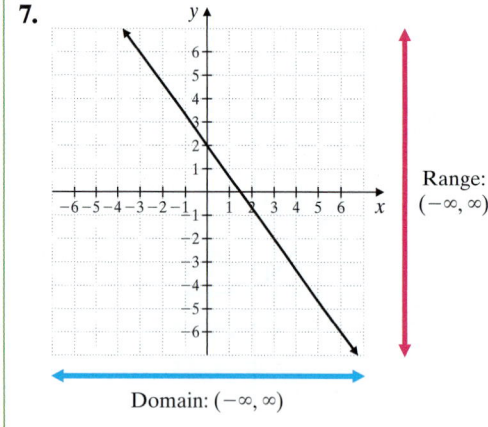

Range: $(-\infty, \infty)$

Domain: $(-\infty, \infty)$

Work Practice 4–7

Objective C Graphing Piecewise-Defined Functions

Sometimes a function is defined by two or more expressions. The expression to use to find the function value depends upon the value of x. Before we actually graph these piecewise-defined functions, let's practice finding function values.

Answers

5. domain: $[0, \infty)$; range: $(-\infty, \infty)$

6. domain: $(-\infty, \infty)$; range: $(-\infty, \infty)$

7. domain: $[-2, 2]$; range: $[-2, 2]$

Example 8   Evaluate $f(2)$, $f(-6)$, and $f(0)$ for the function

$$f(x) = \begin{cases} 2x + 3 & \text{if } x \le 0 \\ -x - 1 & \text{if } x > 0 \end{cases}$$

Then write your results in ordered-pair form.

Solution: Take a moment to study this function. It is a single function defined by two expressions depending on the value of x. From above, if $x \le 0$, use $f(x) = 2x + 3$. If $x > 0$, use $f(x) = -x - 1$. Thus,

$f(2) = -(2) - 1$	$f(-6) = 2(-6) + 3$	$f(0) = 2(0) + 3$
$\quad = -3$ since $2 > 0$	$\quad = -9$ since $-6 \le 0$	$\quad = 3$ since $0 \le 0$
$f(2) = -3$	$f(-6) = -9$	$f(0) = 3$
Ordered pairs: $(2, -3)$	$(-6, -9)$	$(0, 3)$

■ **Work Practice 8**

Now, let's graph a piecewise-defined function.

Example 9   Graph $f(x) = \begin{cases} 2x + 3 & \text{if } x \le 0 \\ -x - 1 & \text{if } x > 0 \end{cases}$

Solution: Let's graph each piece.

If $x \le 0$,   If $x > 0$,

$f(x) = 2x + 3$ $f(x) = -x - 1$

Values ≤ 0
x	$f(x) = 2x + 3$
0	3 Closed circle
−1	1
−2	−1

Values > 0
x	$f(x) = -x - 1$
1	−2
2	−3
3	−4

The graph of the first part of $f(x)$ listed will look like a ray with a closed-circle endpoint at $(0, 3)$. The graph of the second part of $f(x)$ listed will look like a ray with an open-circle endpoint. To find the exact location of the open-circle endpoint, use $f(x) = -x - 1$ and find $f(0)$. Since $f(0) = -0 - 1 = -1$, we graph the values from the second table and place an open circle at $(0, -1)$.

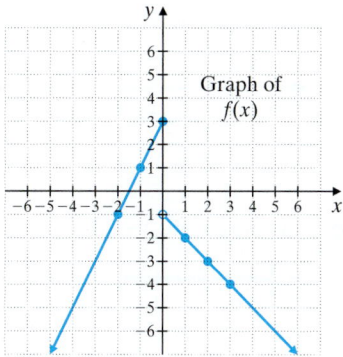

Notice that this graph is the graph of a function because it passes the vertical line test. The domain of this function is $(-\infty, \infty)$ and the range is $(-\infty, 3]$.

■ **Work Practice 9**

In the exercises in the next section, we shall graph piecewise-defined functions whose pieces are not necessarily pieces of lines.

Practice 8

Evaluate $f(-4)$, $f(3)$, and $f(0)$ for the function

$$f(x) = \begin{cases} 3x + 4 & \text{if } x < 0 \\ -x + 2 & \text{if } x \ge 0 \end{cases}$$

Then write your results in ordered-pair solution form.

Practice 9

Graph

$$f(x) = \begin{cases} 3x + 4 & \text{if } x < 0 \\ -x + 2 & \text{if } x \ge 0 \end{cases}$$

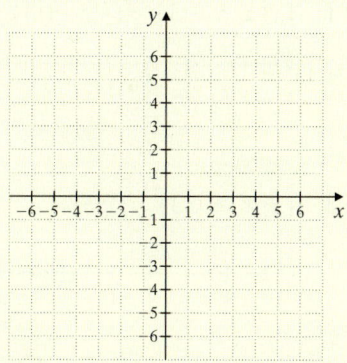

Answers

8. $f(-4) = -8; f(3) = -1; f(0) = 2;$
$(-4, -8); (3, -1); (0, 2)$

9.

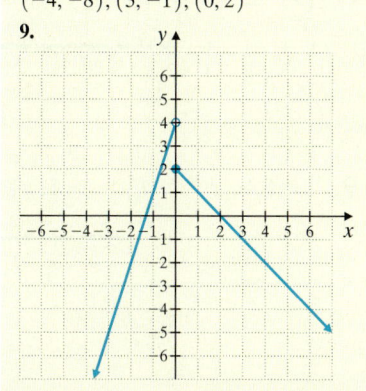

Vocabulary, Readiness & Video Check

Martin-Gay Interactive Videos   *Watch the section lecture video and answer the following questions.*

Objective A 1. In the lecture before ⊞ Example 1, instead of open and closed circles, what symbols are used in interval notation?

Objective B 2. Use ⊞ Examples 3–6 to fill in the blanks. To find the domain of a relation from its graph, we see how far _____ and how far _____ the graph extends. For range, we see how far _____ and how far _____ the graph extends. ▶

Objective C 3. In ⊞ Example 7, only one piece of the function is defined for the value $x = -1$. Why do we find $f(-1)$ for $f(x) = x + 3$? ▶

See Video 8.4 🍎

See video answer section.

8.4 **Exercise Set** MyMathLab® ▶

Objective A *Graph the solution set of each inequality on a number line and then write it in interval notation. See Examples 1 through 3.*

▶**1.** $\{x \mid x < -3\}$ $(-\infty, -3)$

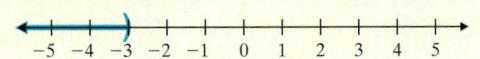

2. $\{x \mid x > 5\}$ $(5, \infty)$

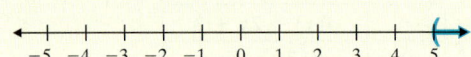

3. $\{x \mid x \geq 0.3\}$ $[0.3, \infty)$

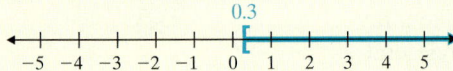

4. $\{x \mid x < -0.2\}$ $(-\infty, -0.2)$

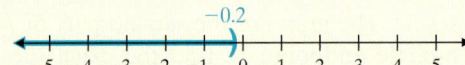

5. $\{x \mid -7 \leq x\}$ $[-7, \infty)$

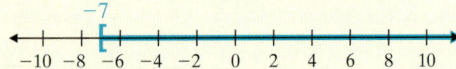

6. $\{x \mid -7 \geq x\}$ $(-\infty, -7]$

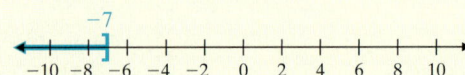

7. $\{x \mid -2 < x < 5\}$ $(-2, 5)$

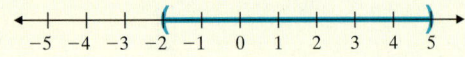

8. $\{x \mid -5 \leq x \leq -1\}$ $[-5, -1]$

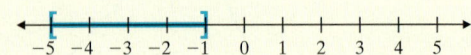

▶**9.** $\{x \mid 5 \geq x > -1\}$ $(-1, 5]$

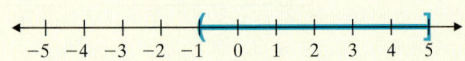

10. $\{x \mid -3 > x \geq -7\}$ $[-7, -3)$

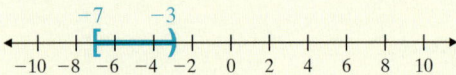

Objective **B** *Find the domain and the range of each relation. Write each in interval notation. See Examples 4 through 7.*

11.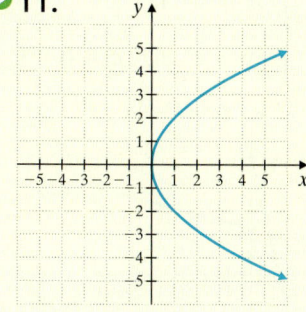

domain: $[0, \infty)$;
range: $(-\infty, \infty)$

12.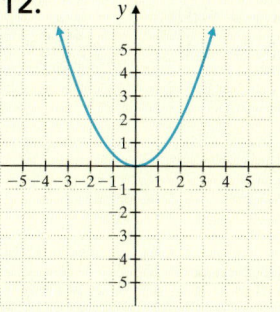

domain: $(-\infty, \infty)$;
range: $[0, \infty)$

13.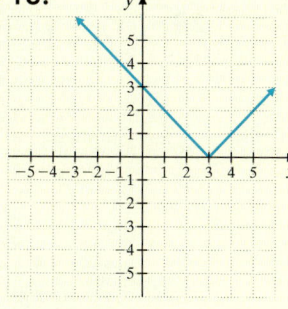

domain: $(-\infty, \infty)$;
range: $[0, \infty)$

14.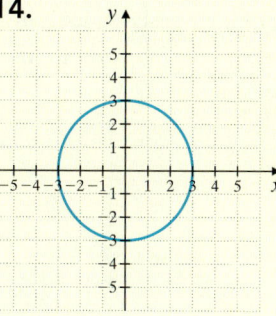

domain: $[-3, 3]$;
range: $[-3, 3]$

15.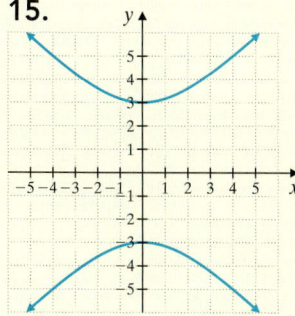

domain: $(-\infty, \infty)$;
range: $(-\infty, -3] \cup [3, \infty)$

16.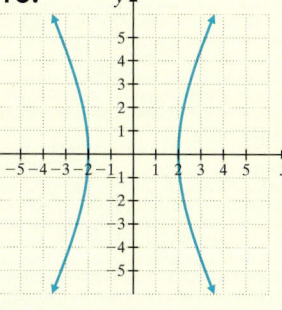

domain: $(-\infty, -2] \cup [2, \infty)$;
range: $(-\infty, \infty)$

17.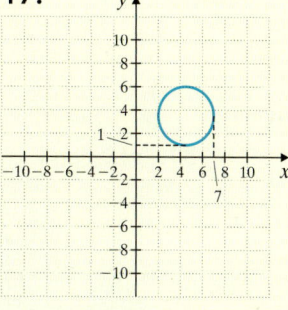

domain: $[2, 7]$;
range: $[1, 6]$

18.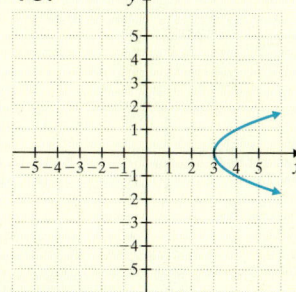

domain: $[3, \infty)$;
range: $(-\infty, \infty)$

19.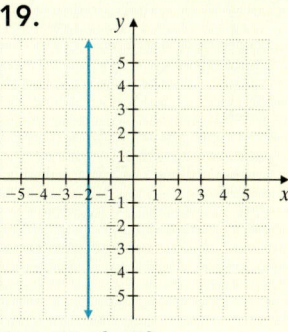

domain: $\{-2\}$;
range: $(-\infty, \infty)$

20.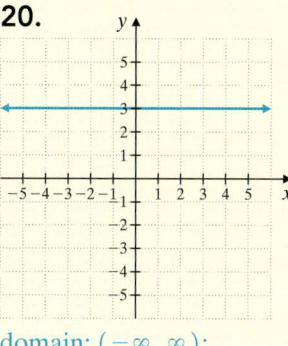

domain: $(-\infty, \infty)$;
range: $\{3\}$

21.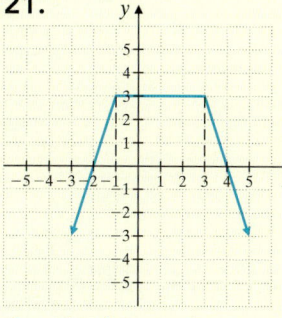

domain: $(-\infty, \infty)$;
range: $(-\infty, 3]$

22.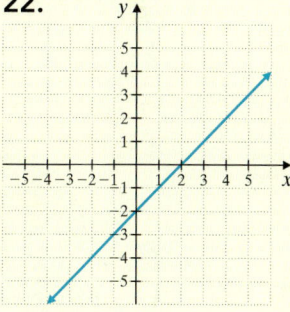

domain: $(-\infty, \infty)$;
range: $(-\infty, \infty)$

23.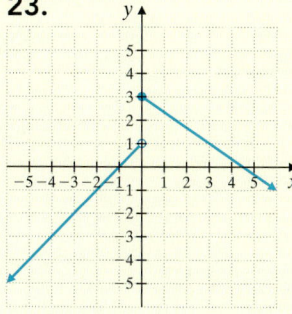

domain: $(-\infty, \infty)$;
range: $(-\infty, 3]$

24.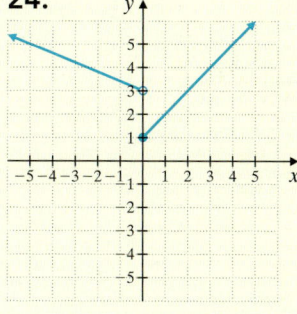

domain: $(-\infty, \infty)$;
range: $[1, \infty)$

25.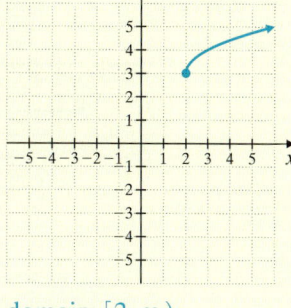

domain: $[2, \infty)$;
range: $[3, \infty)$

26.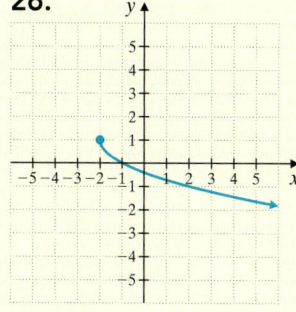

domain: $[-2, \infty)$;
range: $(-\infty, 1]$

Objective C *Graph each piecewise-defined function. See Examples 8 and 9.*

27. $f(x) = \begin{cases} 2x & \text{if } x < 0 \\ x + 1 & \text{if } x \geq 0 \end{cases}$

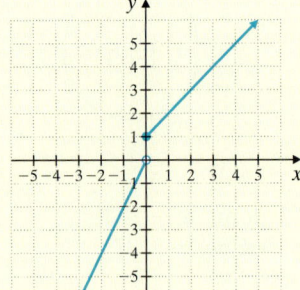

28. $f(x) = \begin{cases} 3x & \text{if } x < 0 \\ x + 2 & \text{if } x \geq 0 \end{cases}$

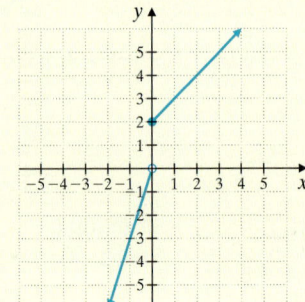

29. $f(x) = \begin{cases} 4x + 5 & \text{if } x \leq 0 \\ \dfrac{1}{4}x + 2 & \text{if } x > 0 \end{cases}$

30. $f(x) = \begin{cases} 5x + 4 & \text{if } x \leq 0 \\ \dfrac{1}{3}x - 1 & \text{if } x > 0 \end{cases}$

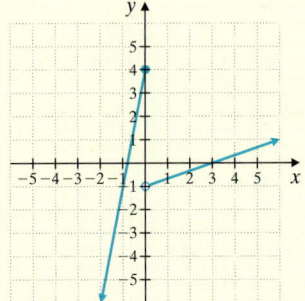

31. $g(x) = \begin{cases} -x & \text{if } x \leq 1 \\ 2x + 1 & \text{if } x > 1 \end{cases}$

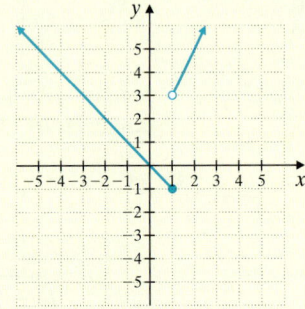

32. $g(x) = \begin{cases} 3x - 1 & \text{if } x \leq 2 \\ -x & \text{if } x > 2 \end{cases}$

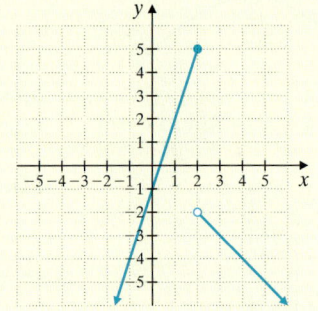

33. $f(x) = \begin{cases} 5 & \text{if } x < -2 \\ 3 & \text{if } x \geq -2 \end{cases}$

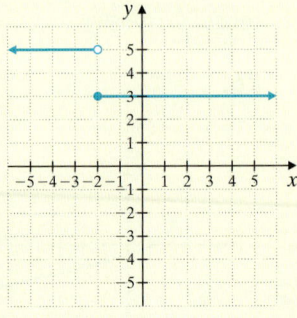

34. $f(x) = \begin{cases} 4 & \text{if } x < -3 \\ -2 & \text{if } x \geq -3 \end{cases}$

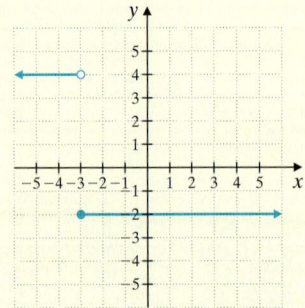

Objectives A B C **Mixed Practice** *Graph each piecewise-defined function. Use the graph to determine the domain and range of the function. See Examples 1 through 9.*

35. $f(x) = \begin{cases} -2x & \text{if } x \le 0 \\ 2x + 1 & \text{if } x > 0 \end{cases}$

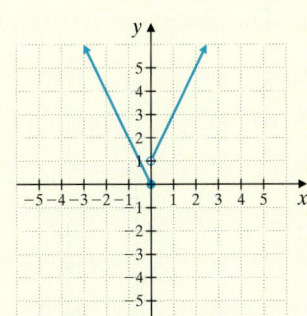

domain: $(-\infty, \infty)$;
range: $[0, \infty)$

36. $g(x) = \begin{cases} -3x & \text{if } x \le 0 \\ 3x + 2 & \text{if } x > 0 \end{cases}$

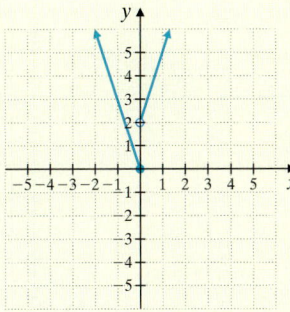

domain: $(-\infty, \infty)$;
range: $[0, \infty)$

37. $h(x) = \begin{cases} 5x - 5 & \text{if } x < 2 \\ -x + 3 & \text{if } x \ge 2 \end{cases}$

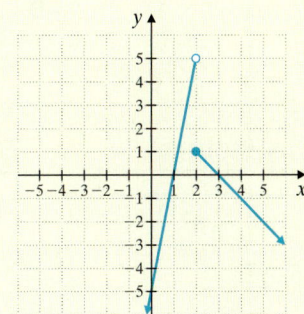

domain: $(-\infty, \infty)$;
range: $(-\infty, 5)$

38. $f(x) = \begin{cases} 4x - 4 & \text{if } x < 2 \\ -x + 1 & \text{if } x \ge 2 \end{cases}$

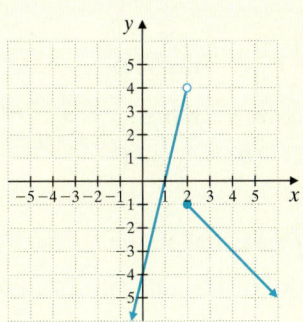

domain: $(-\infty, \infty)$;
range: $(-\infty, 4)$

▶ **39.** $f(x) = \begin{cases} x + 3 & \text{if } x < -1 \\ -2x + 4 & \text{if } x \ge -1 \end{cases}$

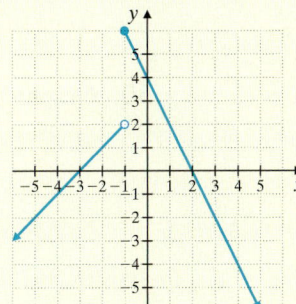

domain: $(-\infty, \infty)$;
range: $(-\infty, 6]$

40. $h(x) = \begin{cases} x + 2 & \text{if } x < 1 \\ 2x - 1 & \text{if } x \ge 1 \end{cases}$

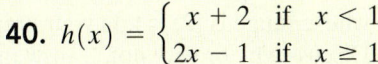
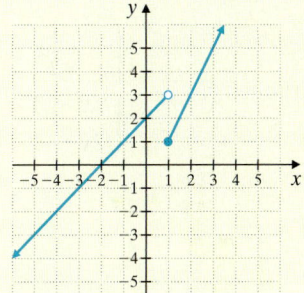

domain: $(-\infty, \infty)$;
range: $(-\infty, \infty)$

41. $g(x) = \begin{cases} -2 & \text{if } x \le 0 \\ -4 & \text{if } x \ge 1 \end{cases}$

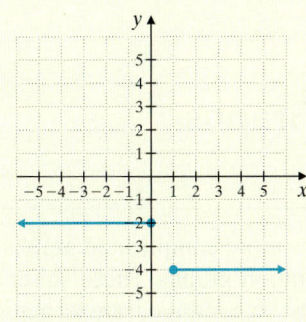

domain: $(-\infty, 0] \cup [1, \infty)$;
range: $\{-4, -2\}$

42. $f(x) = \begin{cases} -1 & \text{if } x \le 0 \\ -3 & \text{if } x \ge 2 \end{cases}$

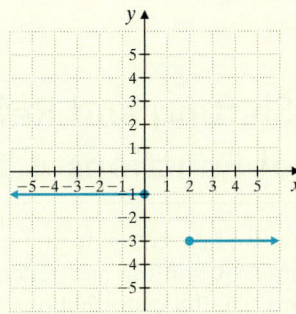

domain: $(-\infty, 0] \cup [2, \infty)$;
range: $\{-3, -1\}$

Review

Match each equation with its graph. See Section 3.3.

43. $y = -1$ A

A

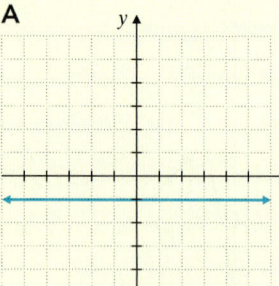

44. $x = -1$ C

B

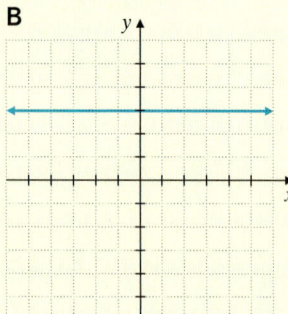

45. $x = 3$ D

C

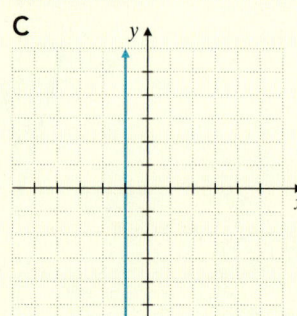

46. $y = 3$ B

D

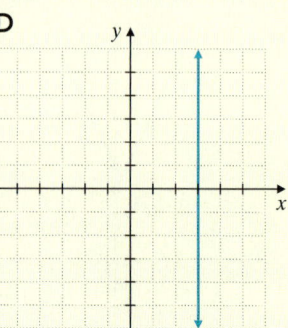

Concept Extensions

47. Draw a graph whose domain is $(-\infty, 5]$ and whose range is $[2, \infty)$. Is your graph a function? Discuss why or why not.   answers may vary

48. In your own words, describe how to graph a piecewise-defined function.   answers may vary

49. Graph: $f(x) = \begin{cases} -\dfrac{1}{2}x & \text{if } x \leq 0 \\ x + 1 & \text{if } 0 < x \leq 2 \\ 2x - 1 & \text{if } x > 2 \end{cases}$

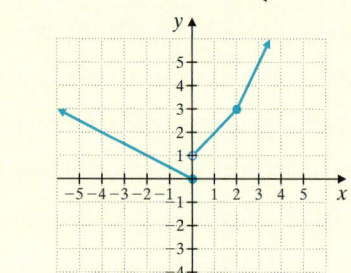

50. Graph: $f(x) = \begin{cases} \dfrac{1}{3}x & \text{if } x < 0 \\ -x + 2 & \text{if } 0 \leq x < 4 \\ 3x - 10 & \text{if } x \geq 4 \end{cases}$

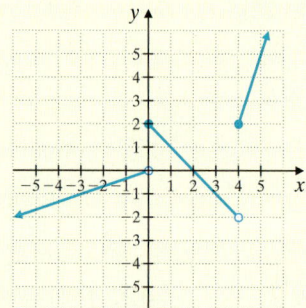

8.5   Shifting and Reflecting Graphs of Functions

Objectives

A Graph Common Equations.

B Vertical and Horizontal Shifts.

C Reflect Graphs.

In this section, we take common graphs and learn how more complicated graphs are actually formed by shifting and reflecting these common graphs. These shifts and reflections are called transformations, and it is possible to combine transformations. A knowledge of these transformations will help you simplify future graphs.

Objective A   Graphing Common Equations

Let's begin with the graphs of four common functions.

First, let's graph the linear function $f(x) = x$, or $y = x$. Ordered-pair solutions of this graph consist of ordered pairs whose x- and y-values are the same.

x	y or $f(x) = x$
-3	-3
0	0
1	1
4	4

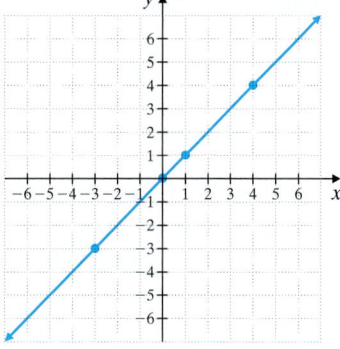

Next, let's graph the nonlinear function $f(x) = x^2$ or $y = x^2$.

This equation is not linear because the x^2 term does not allow us to write it in the form $Ax + By = C$. Its graph is not a line. We begin by finding ordered pair solutions. Because this graph is solved for $f(x)$, or y, we choose x-values and find corresponding $f(x)$, or y-values.

If $x = -3$, then $y = (-3)^2$, or 9.

If $x = -2$, then $y = (-2)^2$, or 4.

If $x = -1$, then $y = (-1)^2$, or 1.

If $x = 0$, then $y = 0^2$, or 0.

If $x = 1$, then $y = 1^2$, or 1.

If $x = 2$, then $y = 2^2$, or 4.

If $x = 3$, then $y = 3^2$, or 9.

x	$f(x)$ or y
-3	9
-2	4
-1	1
0	0
1	1
2	4
3	9

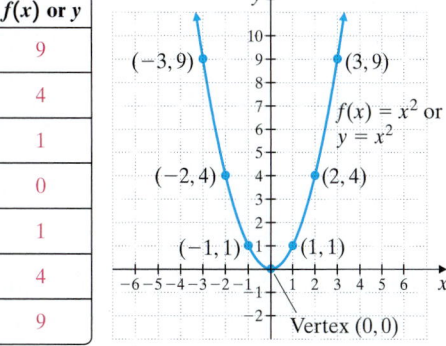

Study the table a moment and look for patterns. Notice that the ordered pair solution $(0, 0)$ contains the smallest y-value because any other x-value squared will give a positive result. This means that the point $(0, 0)$ will be the lowest point on the graph. Also notice that all other y-values correspond to two different x-values. For example, $3^2 = 9$ and also $(-3)^2 = 9$. This means that the graph will be a mirror image of itself across the y-axis. Connect the plotted points with a smooth curve to sketch its graph.

This curve is given a special name, a **parabola.** We will study more about parabolas in later chapters.

Next, let's graph another nonlinear function $f(x) = |x|$ or $y = |x|$.

This is not a linear equation since it cannot be written in the form $Ax + By = C$. Its graph is not a line. Because we do not know the shape of this graph, we find many ordered pair solutions. We will choose x-values and substitute to find corresponding y-values.

If $x = -3$, then $y = |-3|$, or 3.

If $x = -2$, then $y = |-2|$, or 2.

If $x = -1$, then $y = |-1|$, or 1.

If $x = 0$, then $y = |0|$, or 0.

If $x = 1$, then $y = |1|$, or 1.

If $x = 2$, then $y = |2|$, or 2.

If $x = 3$, then $y = |3|$, or 3.

x	$f(x)$ or y
-3	3
-2	2
-1	1
0	0
1	1
2	2
3	3

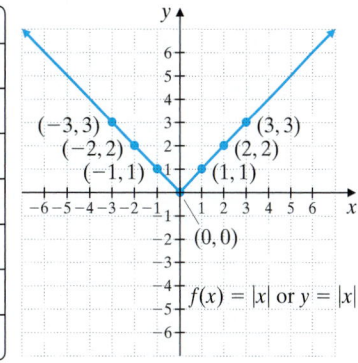

Again, study the table of values for a moment and notice any patterns.

From the plotted ordered pairs, we see that the graph of this absolute value equation is V-shaped.

Finally, a fourth common function, $f(x) = \sqrt{x}$ or $y = \sqrt{x}$. For this graph, you need to recall basic facts about square roots and use your calculator to approximate some square roots to help locate points. Recall also that the square root of a negative number is not a real number, so be careful when finding your domain.

Now let's graph the square root function $f(x) = \sqrt{x}$, or $y = \sqrt{x}$.

To graph, we identify the domain, evaluate the function for several values of x, plot the resulting points, and connect the points with a smooth curve. Since $\sqrt{x}$ is a real number for a nonnegative value of x, the domain of this function is the set of all nonnegative numbers, $\{x \mid x \geq 0\}$, or $[0, \infty)$. We have approximated $\sqrt{3}$ below to help us locate the point corresponding to $(3, \sqrt{3})$.

If $x = 0$, then $y = \sqrt{0}$, or 0.

If $x = 1$, then $y = \sqrt{1}$, or 1.

If $x = 3$, then $y = \sqrt{3}$, or ≈ 1.7.

If $x = 4$, then $y = \sqrt{4}$, or 2.

If $x = 9$, then $y = \sqrt{9}$, or 3.

x	$f(x) = \sqrt{x}$
0	0
1	1
3	$\sqrt{3} \approx 1.7$
4	2
9	3

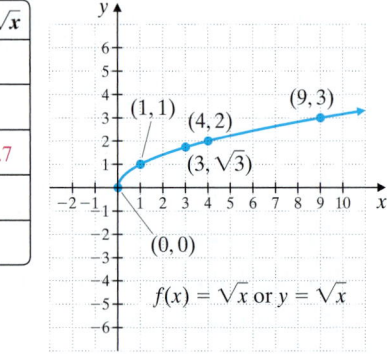

Notice that the graph of this function passes the vertical line test, as expected.

Following is a summary of our four common graphs. Take a moment and study these graphs. Your success in the rest of this section depends on your knowledge of these graphs.

Common Graphs

$f(x) = x$

$f(x) = x^2$

$f(x) = \sqrt{x}$

$f(x) = |x|$

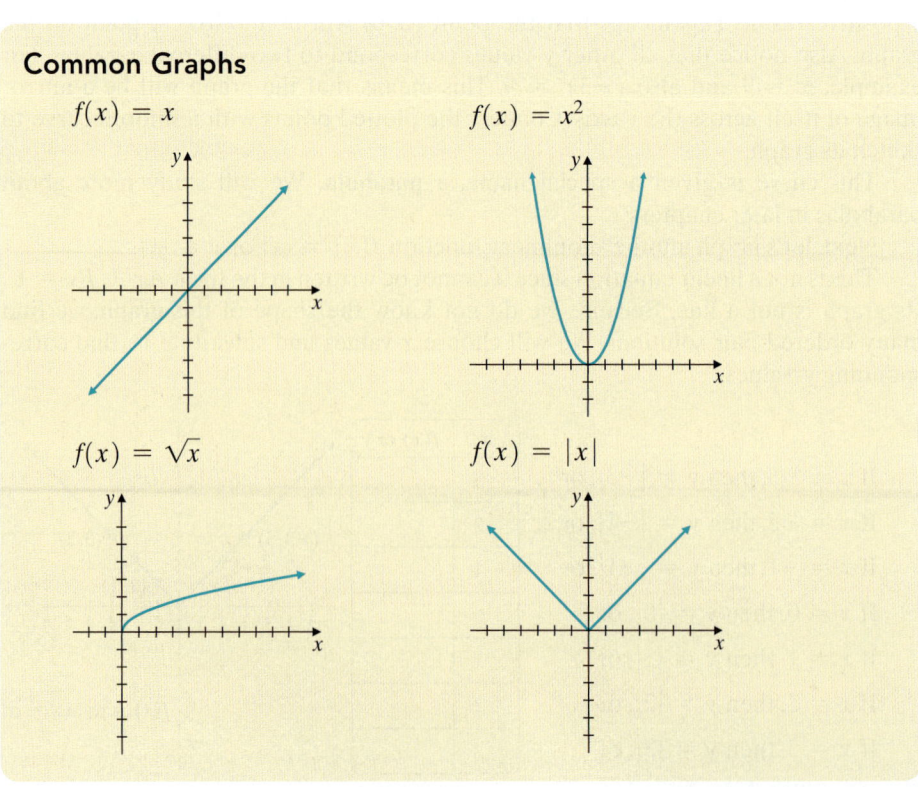

Objective B Vertical and Horizontal Shifting

Our knowledge of the slope-intercept form, $f(x) = mx + b$, will help us understand simple shifting of transformations such as vertical shifts. For example, what is the difference between the graphs of $f(x) = x$ and $g(x) = x + 3$?

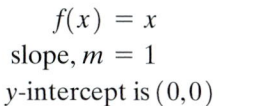

$f(x) = x$
slope, $m = 1$
y-intercept is $(0,0)$

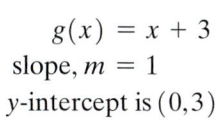

$g(x) = x + 3$
slope, $m = 1$
y-intercept is $(0,3)$

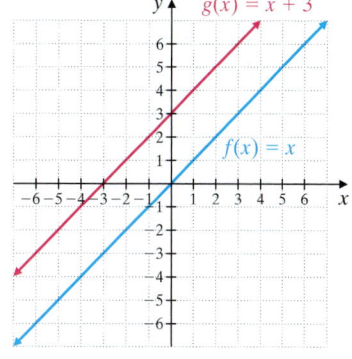

Notice that the graph of $g(x) = x + 3$ is the same as the graph of $f(x) = x$, but moved upward 3 units. This is an example of a **vertical shift** and is true for graphs in general.

Vertical Shifts (Upward and Downward)
Let k Be a Positive Number

Graph of	Same as	Moved
$g(x) = f(x) + k$	$f(x)$	k units upward
$g(x) = f(x) - k$	$f(x)$	k units downward

Examples Without plotting points, sketch the graph of each pair of functions on the same set of axes.

1. $f(x) = x^2$ and $g(x) = x^2 + 2$

2. $f(x) = \sqrt{x}$ and $g(x) = \sqrt{x} - 3$

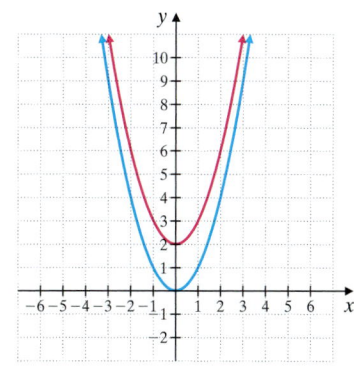

■ Work Practice 1–2

A horizontal shift to the left or right may be slightly more difficult to understand. Let's graph $g(x) = |x - 2|$ and compare it with $f(x) = |x|$.

Practice 1–2

Without plotting points, sketch the graphs of each pair of functions on the same set of axes.

1. $f(x) = x^2$ and
   $g(x) = x^2 - 1$

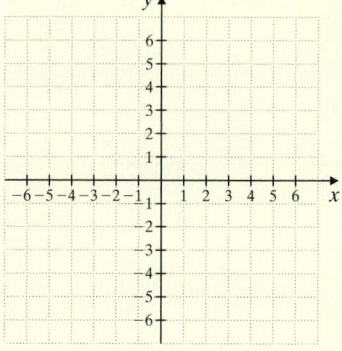

2. $f(x) = \sqrt{x}$ and
   $g(x) = \sqrt{x} + 2$

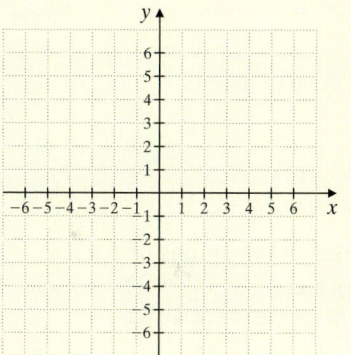

Answers

1.

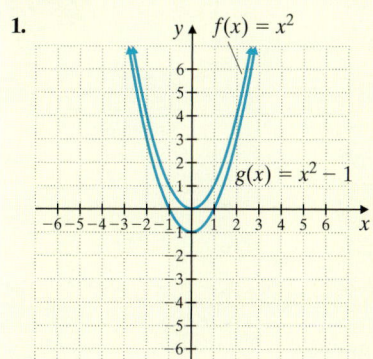

2.

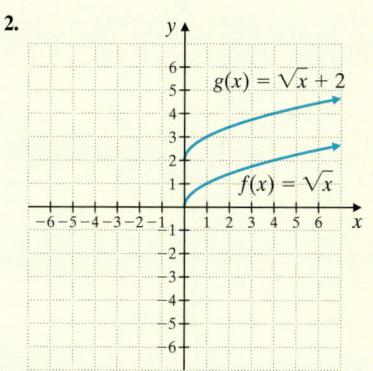

Practice 3

Without plotting points, sketch the graphs of $f(x) = |x|$ and $g(x) = |x + 2|$ on the same set of axes.

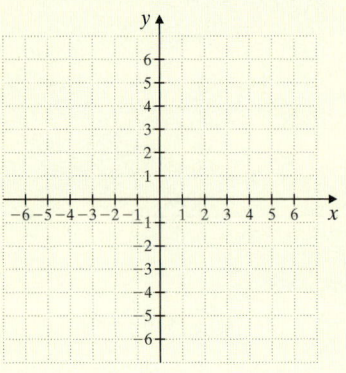

Practice 4

Sketch the graphs of $f(x) = x^2$ and $g(x) = (x + 2)^2 - 1$ on the same set of axes.

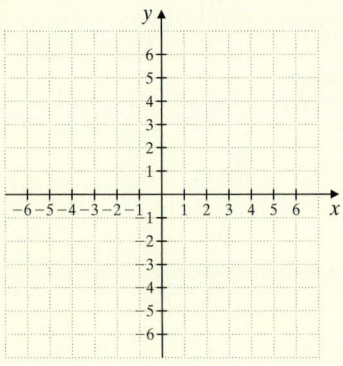

Answers

3.

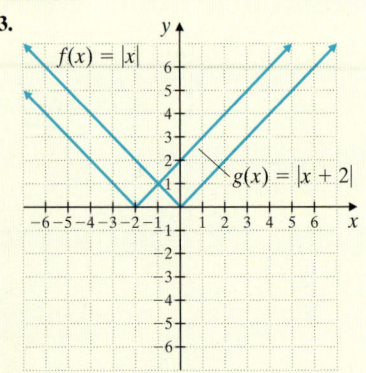

4. See page 613.

Example 3 Without plotting points, sketch the graphs of $f(x) = |x|$ and $g(x) = |x - 2|$ on the same set of axes.

Solution:

x	$f(x) = |x|$	$g(x) = |x - 2|$
-3	3	5
-2	2	4
-1	1	3
0	0	2
1	1	1
2	2	0
3	3	1

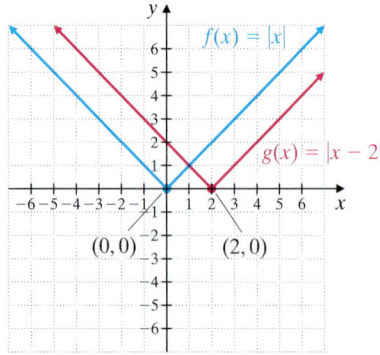

The graph of $g(x) = |x - 2|$ is the same as the graph of $f(x) = |x|$, but moved 2 units to the right. This is an example of a **horizontal shift** and is true for graphs in general. The table is provided to verify the graphs.

▪ **Work Practice 3**

Horizontal Shift (to the Left or Right) Let h Be a Positive Number

Graph of	Same as	Moved
$g(x) = f(x - h)$	$f(x)$	h units to the right
$g(x) = f(x + h)$	$f(x)$	h units to the left

Helpful Hint

Notice that $f(x - h)$ corresponds to a shift to the right and $f(x + h)$ corresponds to a shift to the left.

Vertical and horizontal shifts can be combined.

Example 4 Sketch the graphs of $f(x) = x^2$ and $g(x) = (x - 2)^2 + 1$ on the same set of axes.

Solution: The graph of $g(x)$ is the same as the graph of $f(x)$ shifted 2 units to the right and 1 unit up.

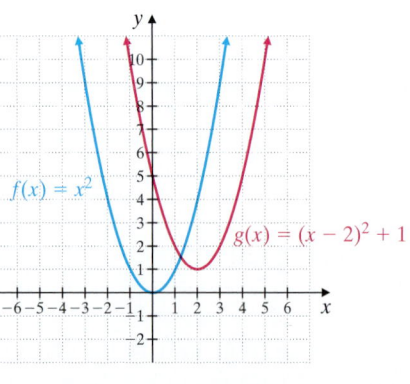

▪ **Work Practice 4**

Objective C Reflecting Graphs

Another type of transformation is called a **reflection**. In this section, we will study reflections (mirror images) about the x-axis only. For example, take a moment and study these two graphs. The graph of $g(x) = -x^2$ can be found, as usual, by plotting points.

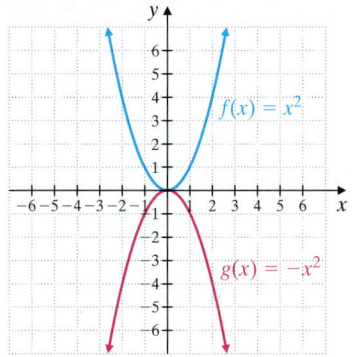

Reflection About the x-axis

The graph of $g(x) = -f(x)$ is the graph of $f(x)$ reflected about the x-axis.

Example 5   Sketch the graph of $h(x) = -|x - 3| + 2$.

Solution: The graph of $h(x) = -|x - 3| + 2$ is the same as the graph of $f(x) = |x|$ reflected about the x-axis, then moved three units to the right and two units upward.

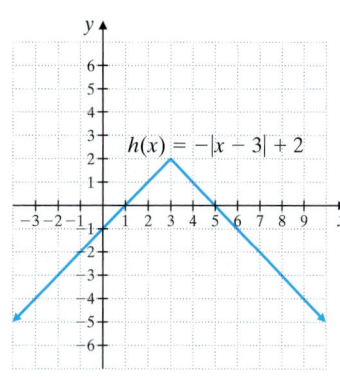

🟧 **Work Practice 5**

There are other transformations, such as stretching, that won't be covered in this section. For a review of this transformation, see the Appendix.

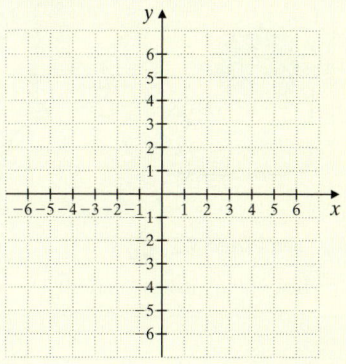
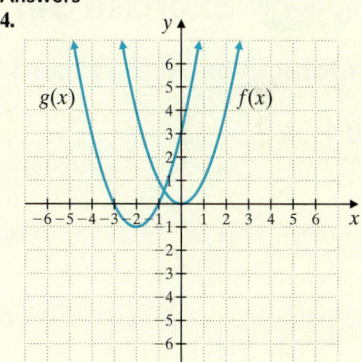
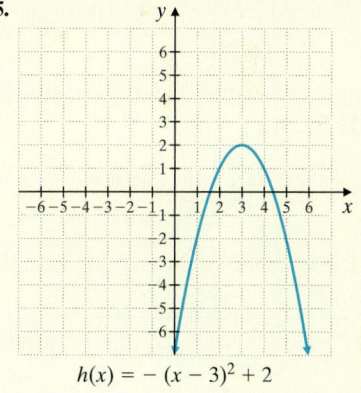
Vocabulary, Readiness & Video Check

Match each equation with its graph.

1. $y = \sqrt{x}$ C   **2.** $y = x^2$ B   **3.** $y = x$ D   **4.** $y = |x|$ A

A

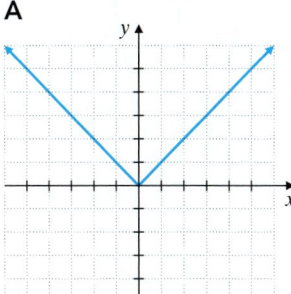

B

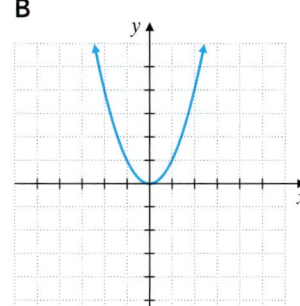

C

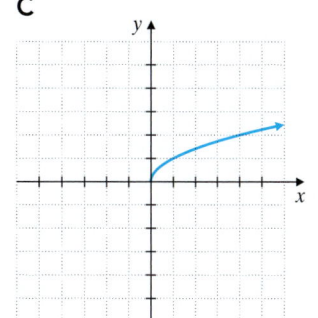

D

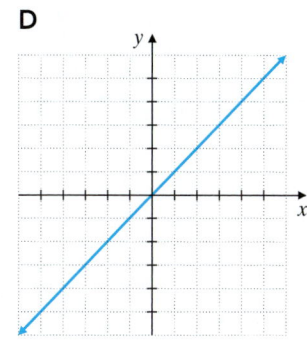

Watch the section lecture video and answer the following questions.

Objective A 5. Based on the lecture before ▣ Example 1, what common function has a graph that is V-shaped? ▶

Objective B 6. For ▣ Examples 1–7, why is it helpful to be familiar with common graphs and their basic shapes? ▶

Objective C 7. Based on the lecture before ▣ Example 8, complete the following statement. The graph of $f(x) = -\sqrt{x} + 6$ has the same shape as the graph of $f(x) = \sqrt{x} + 6$ but is reflected about the _____. ▶

8.5 Exercise Set MyMathLab® ▶

Objectives A B Mixed Practice *Sketch the graph of each function. See Examples 1 through 4.*

▶ 1. $f(x) = |x| + 3$

2. $f(x) = |x| - 2$

▶ 3. $f(x) = \sqrt{x} - 2$

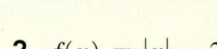

4. $f(x) = \sqrt{x} + 3$

▶ 5. $f(x) = |x - 4|$

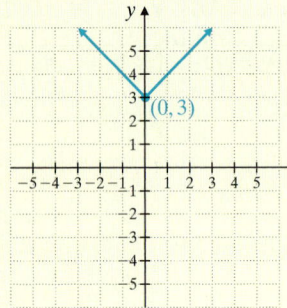

6. $f(x) = |x + 3|$

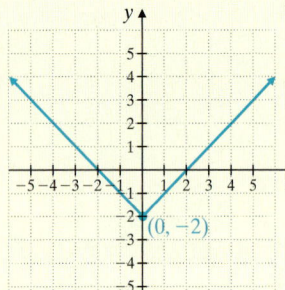

7. $f(x) = \sqrt{x + 2}$

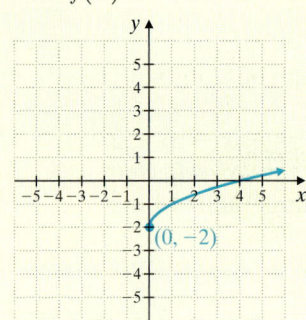

8. $f(x) = \sqrt{x - 2}$

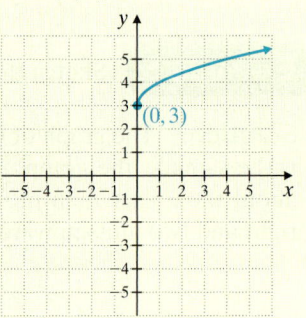

▶ 9. $y = (x - 4)^2$

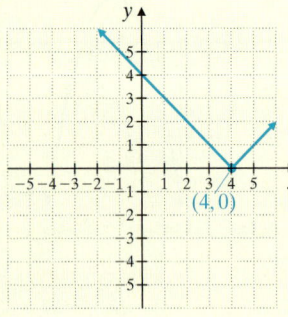

10. $y = (x + 4)^2$

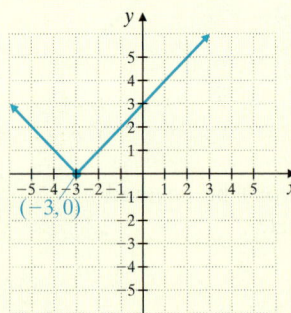

11. $f(x) = x^2 + 4$

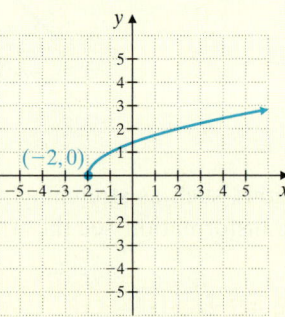

12. $f(x) = x^2 - 4$

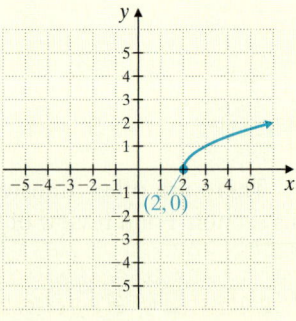

13. $f(x) = \sqrt{x-2} + 3$

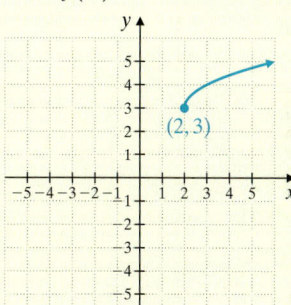

14. $f(x) = \sqrt{x-1} + 3$

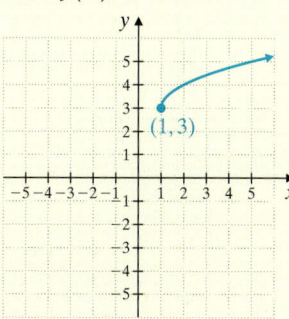

15. $f(x) = |x-1| + 5$

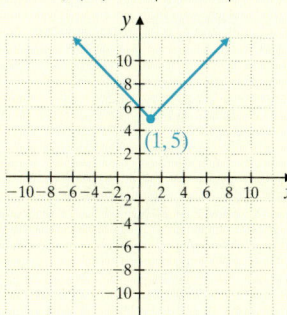

16. $f(x) = |x-3| + 2$

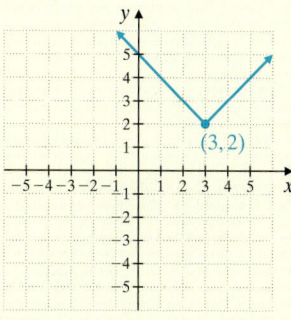

17. $f(x) = \sqrt{x+1} + 1$

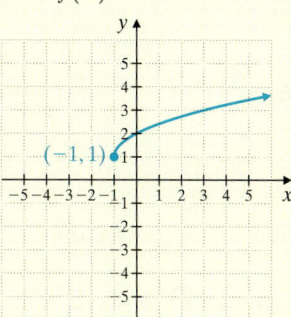

18. $f(x) = \sqrt{x+3} + 2$

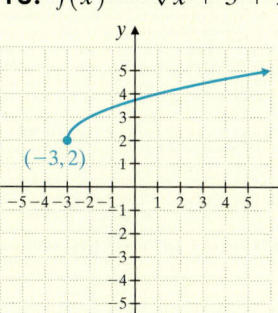

19. $f(x) = |x+3| - 1$

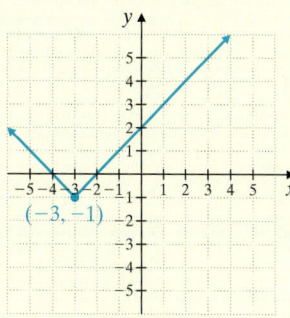

20. $f(x) = |x+1| - 4$

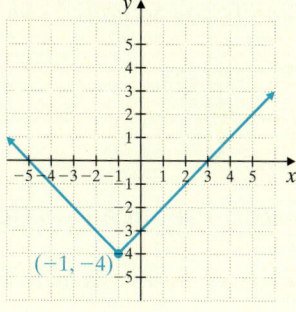

21. $g(x) = (x-1)^2 - 1$

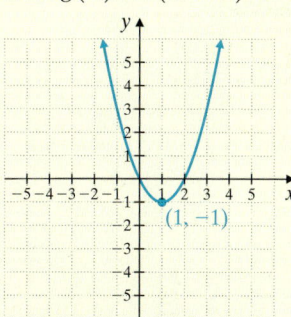

22. $h(x) = (x+2)^2 + 2$

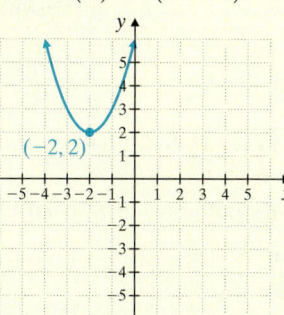

23. $f(x) = (x+3)^2 - 2$

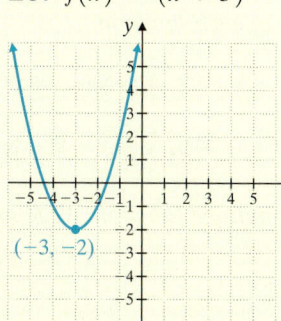

24. $f(x) = (x+2)^2 + 4$

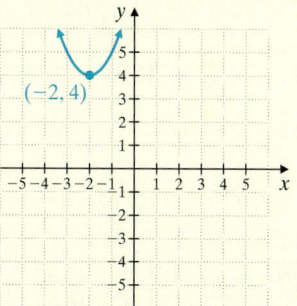

Objectives A B C **Mixed Practice** *Sketch the graph of each function. See Examples 1 through 5.*

25. $f(x) = -(x-1)^2$

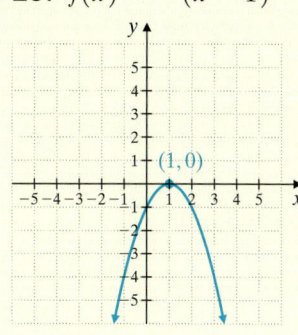

26. $g(x) = -(x+2)^2$

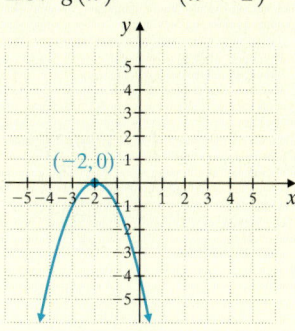

27. $h(x) = -\sqrt{x} + 3$

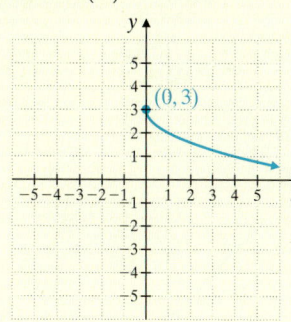

28. $f(x) = -\sqrt{x+3}$

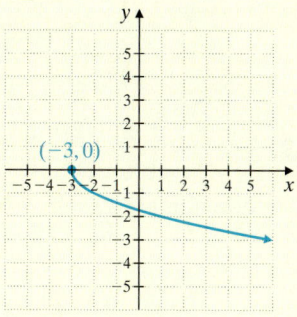

29. $h(x) = -|x + 2| + 3$

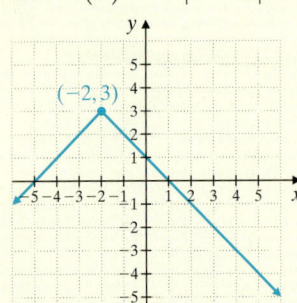

30. $g(x) = -|x + 1| + 1$

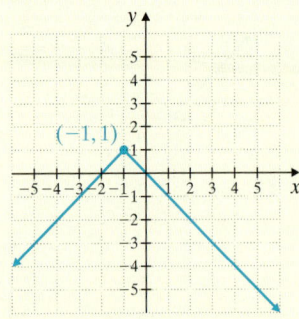

31. $f(x) = (x - 3) + 2$

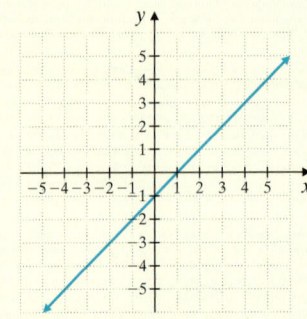

32. $f(x) = (x - 1) + 4$

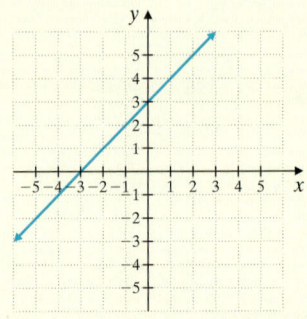

Review

Simplify. See Sections 5.4 and 5.5.

33. $-3x^4 \cdot 5x^4$
$-15x^8$

34. $-3x^4 + 5x^4$
$2x^4$

35. $8(y^7 + y^{11})$
$8y^7 + 8y^{11}$

36. $y^7 \cdot y^{11}$
y^{18}

Concept Extensions

Mixed Practice (Sections 8.2, 8.4, 8.5) *Write the domain and range of the indicated function in this section.*

37. Exercise 13   domain: $[2, \infty)$; range: $[3, \infty)$

38. Exercise 14   domain: $[1, \infty)$; range: $[3, \infty)$

39. Exercise 29   domain: $(-\infty, \infty)$; range: $(-\infty, 3]$

40. Exercise 30   domain: $(-\infty, \infty)$; range: $(-\infty, 1]$

Without graphing, find the domain of each function.

41. $f(x) = 5\sqrt{x - 20} + 1$ $[20, \infty)$

42. $g(x) = -3\sqrt{x + 5}$ $[-5, \infty)$

43. $h(x) = 5|x - 20| + 1$ $(-\infty, \infty)$

44. $f(x) = -3|x + 5.7|$ $(-\infty, \infty)$

45. $g(x) = 9 - \sqrt{x + 103}$ $[-103, \infty)$

46. $h(x) = \sqrt{x - 17} - 3$ $[17, \infty)$

Sketch the graph of each piecewise-defined function. Write the domain and range of each function.

47. $f(x) = \begin{cases} |x| & \text{if } x \le 0 \\ x^2 & \text{if } x > 0 \end{cases}$

domain: $(-\infty, \infty)$;
range: $[0, \infty)$

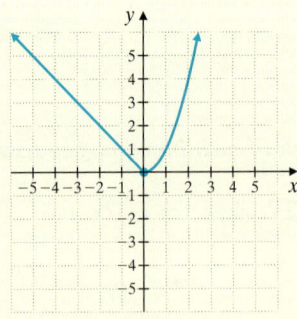

48. $f(x) = \begin{cases} x^2 & \text{if } x < 0 \\ \sqrt{x} & \text{if } x \ge 0 \end{cases}$

domain: $(-\infty, \infty)$;
range: $[0, \infty)$

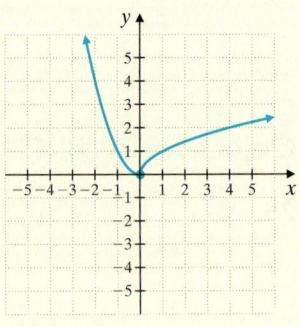

49. $g(x) = \begin{cases} |x - 2| & \text{if } x < 0 \\ -x^2 & \text{if } x \ge 0 \end{cases}$

domain: $(-\infty, \infty)$;
range: $(-\infty, 0] \cup (2, \infty)$

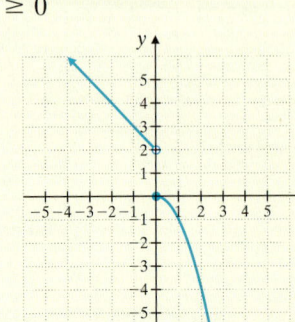

50. $g(x) = \begin{cases} -|x + 1| - 1 & \text{if } x < -2 \\ \sqrt{x + 2} - 4 & \text{if } x \ge -2 \end{cases}$

domain: $(-\infty, \infty)$;
range: $(-\infty, \infty)$

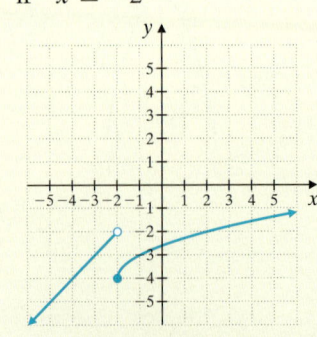

13. $f(x) = \sqrt{x - 2} + 3$

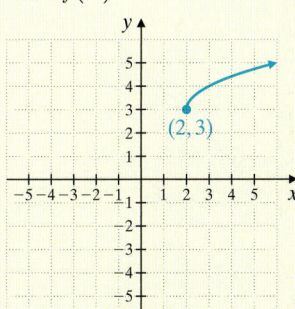

14. $f(x) = \sqrt{x - 1} + 3$

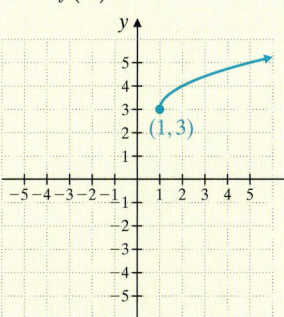

15. $f(x) = |x - 1| + 5$

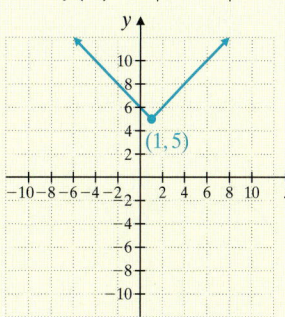

16. $f(x) = |x - 3| + 2$

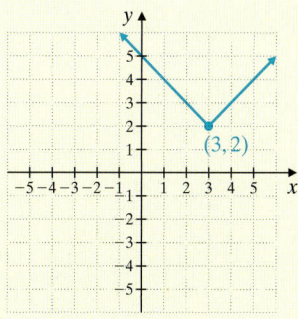

▶ **17.** $f(x) = \sqrt{x + 1} + 1$

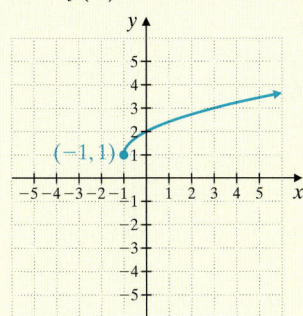

18. $f(x) = \sqrt{x + 3} + 2$

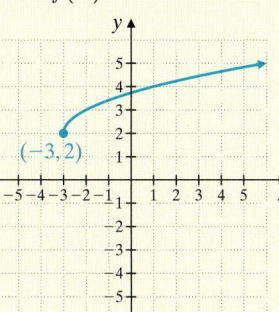

▶ **19.** $f(x) = |x + 3| - 1$

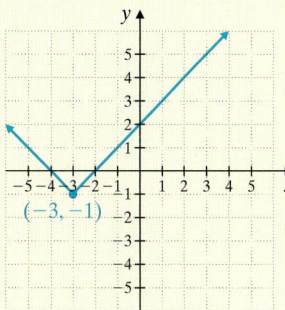

20. $f(x) = |x + 1| - 4$

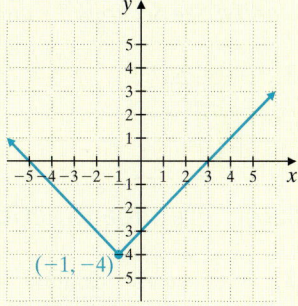

▶ **21.** $g(x) = (x - 1)^2 - 1$

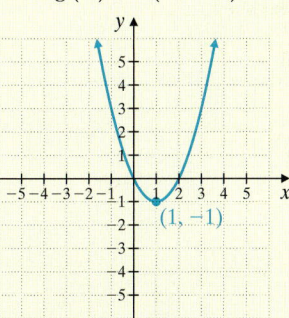

22. $h(x) = (x + 2)^2 + 2$

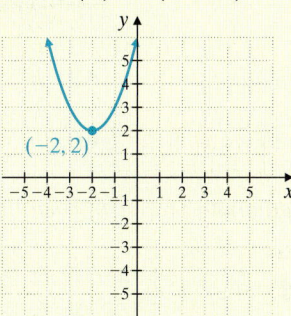

23. $f(x) = (x + 3)^2 - 2$

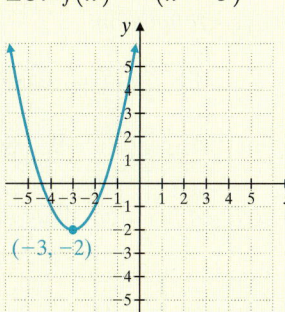

24. $f(x) = (x + 2)^2 + 4$

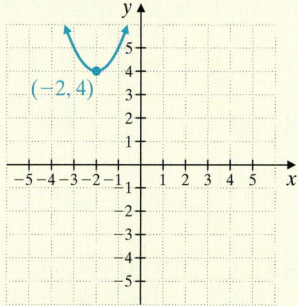

Objectives A B C **Mixed Practice** *Sketch the graph of each function. See Examples 1 through 5.*

▶ **25.** $f(x) = -(x - 1)^2$

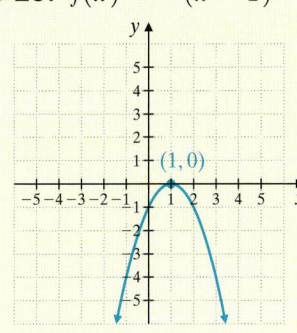

26. $g(x) = -(x + 2)^2$

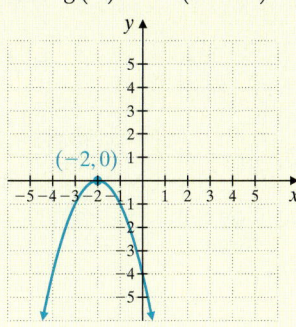

27. $h(x) = -\sqrt{x} + 3$

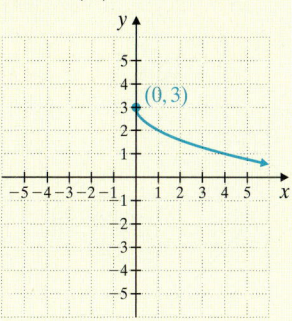

28. $f(x) = -\sqrt{x + 3}$

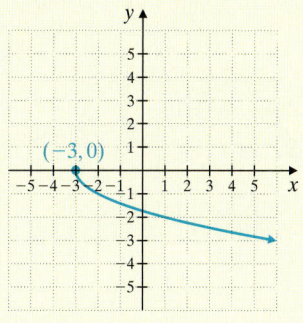

⊙ 29. $h(x) = -|x + 2| + 3$ **30.** $g(x) = -|x + 1| + 1$ **31.** $f(x) = (x - 3) + 2$ **32.** $f(x) = (x - 1) + 4$

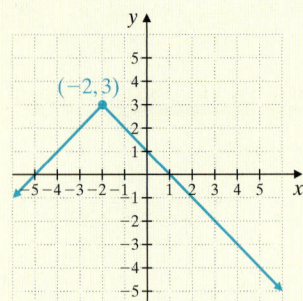   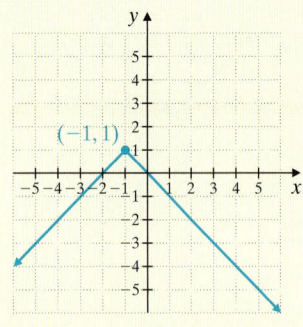   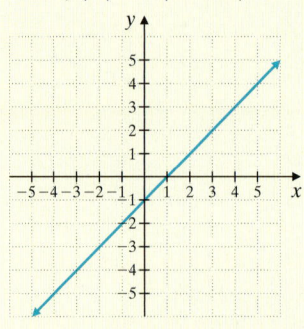   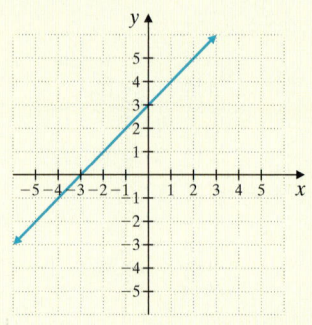

Review

Simplify. See Sections 5.4 and 5.5.

33. $-3x^4 \cdot 5x^4$
$-15x^8$

34. $-3x^4 + 5x^4$
$2x^4$

35. $8(y^7 + y^{11})$
$8y^7 + 8y^{11}$

36. $y^7 \cdot y^{11}$
y^{18}

Concept Extensions

Mixed Practice (Sections 8.2, 8.4, 8.5) *Write the domain and range of the indicated function in this section.*

37. Exercise 13   domain: $[2, \infty)$; range: $[3, \infty)$

38. Exercise 14   domain: $[1, \infty)$; range: $[3, \infty)$

39. Exercise 29   domain: $(-\infty, \infty)$; range: $(-\infty, 3]$

40. Exercise 30   domain: $(-\infty, \infty)$; range: $(-\infty, 1]$

Without graphing, find the domain of each function.

41. $f(x) = 5\sqrt{x - 20} + 1$ $[20, \infty)$ **42.** $g(x) = -3\sqrt{x + 5}$ $[-5, \infty)$ **43.** $h(x) = 5|x - 20| + 1$ $(-\infty, \infty)$

44. $f(x) = -3|x + 5.7|$ $(-\infty, \infty)$ **45.** $g(x) = 9 - \sqrt{x + 103}$
$[-103, \infty)$ **46.** $h(x) = \sqrt{x - 17} - 3$ $[17, \infty)$

Sketch the graph of each piecewise-defined function. Write the domain and range of each function.

47. $f(x) = \begin{cases} |x| & \text{if } x \le 0 \\ x^2 & \text{if } x > 0 \end{cases}$
domain: $(-\infty, \infty)$;
range: $[0, \infty)$

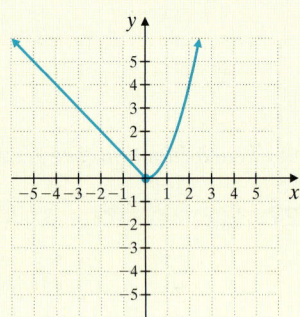

48. $f(x) = \begin{cases} x^2 & \text{if } x < 0 \\ \sqrt{x} & \text{if } x \ge 0 \end{cases}$
domain: $(-\infty, \infty)$;
range: $[0, \infty)$

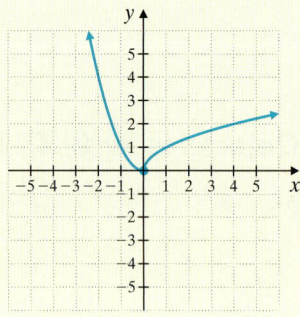

49. $g(x) = \begin{cases} |x - 2| & \text{if } x < 0 \\ -x^2 & \text{if } x \ge 0 \end{cases}$
domain: $(-\infty, \infty)$;
range: $(-\infty, 0] \cup (2, \infty)$

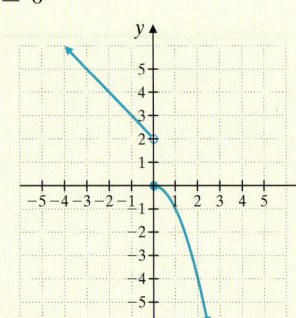

50. $g(x) = \begin{cases} -|x + 1| - 1 & \text{if } x < -2 \\ \sqrt{x + 2} - 4 & \text{if } x \ge -2 \end{cases}$
domain: $(-\infty, \infty)$;
range: $(-\infty, \infty)$

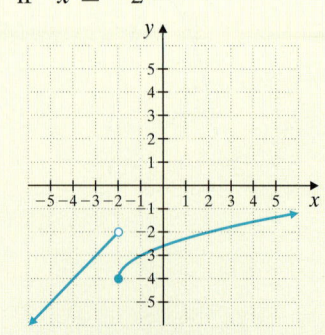

Chapter 8 Group Activity

Additional group activities are available in the *Instructor's Resource Manual with Tests.*

Linear Modeling

As we saw in Sections 8.1, businesses often depend on equations that "closely fit" data. To *model* the data means to find an equation that describes the relationship between the paired data of two variables, such as time in years and profit. A model that accurately summarizes the relationship between two variables can be used to replace a potentially lengthy listing of the raw data. An accurate model might also be used to predict future trends by answering questions such as "If the trend seen in our company's performance in the last several years continues, what level of profit can we reasonably expect in 3 years?"

There are several ways to find a linear equation that models a set of data. If only two ordered pair data points are involved, an exact equation that contains both points can be found using the methods of Section 8.1. When more than two ordered pair data points are involved, it may be impossible to find a linear equation that contains all of the data points. In this case, the graph of the **best fit equation** should have a majority of the plotted ordered pair data points on the graph or close to it. In statistics, a technique called least squares regression is used to determine an equation that best fits a set of data. Various graphing utilities have built-in capabilities for finding an equation (called a regression equation) that best fits a set of ordered pair data points. Regression capabilities are often found with a graphing utility's statistics features.* A best fit equation can also be estimated using an algebraic method, which is outlined in the Group Activity below. In either case, a useful first step when finding a linear equation that models a set of data is creating a scatter diagram of the ordered pair data points to verify that a linear equation is an appropriate model.

Group Activity

The Windows operating system is one of the well-known products of the Microsoft Corporation. This company develops, manufactures, licenses, and supports a wide range of software products for various computing devices. It also provides the MSN network of Internet products and services, and even the Xbox video game system. The table shows Microsoft's total revenues (in billions) for the years 2007–2013. Use the table along with your answers to the questions below to find a linear equation $y = mx + b$ that represents total revenue y (in billions) for Microsoft Corporation, where x represents the number of years after 2007. (Source: Microsoft Corporation)

Year	0	1	2	3	4	5	6
Total Microsoft Revenues (in billions)	51	60	58	62	70	74	78

1. Create a scatter diagram of the paired data given in the table. Does a linear model seem appropriate for the data?   answers may vary

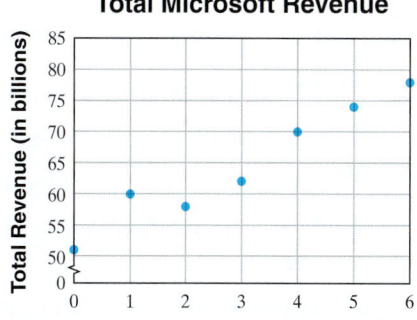

Total Microsoft Revenue

2. Use a straightedge to draw on your graph what appears to be the line that "best fits" the data you plotted.
   answers may vary

3. Estimate the coordinate of the two points that fit on your best fit line. Use these points to find the equation of the line that passes through both points.
   answers may vary

4. Use this equation to find the value of y for $x = 11$. Interpret the meaning of this pair of data.
   answers may vary; this predicts revenue in 2018

5. How could this equation be useful to accountants who work at Microsoft?
   answers may vary

6. Compare your group's linear equation with other groups' equations. Are they the same or different? Explain why.
   answers may vary

7. (Optional) Enter the data from the table into a graphing utility and use the linear regression feature to find a linear equation that models the data. Compare this equation with the one you found in Question 3. How are they alike or different?
   $y = 4.32x + 51.75$; answers may vary

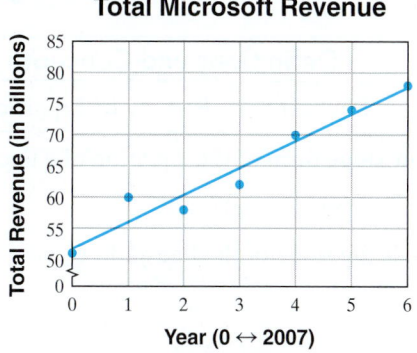

Total Microsoft Revenue

8. (Optional) Using corporation annual reports or articles from magazines or newspapers, search for a set of business-related data that could be modeled with a linear equation. Explain how modeling this data could be useful to a business. Then find the best fit equation for the data.   answers may vary

*To find out more about using a graphing utility to find a regression equation, consult the user's manual for your graphing utility.

Chapter 8 Vocabulary Check

Fill in each blank with one of the words or phrases listed below.

| relation | function | parallel | range | rational | polynomial |
| perpendicular | slope | domain | interval | linear | |

1. A(n) __relation__ is a set of ordered pairs.

2. The __range__ of a relation is the set of all second components of the ordered pairs of the relation.

3. __Parallel__ lines have the same slope and different y-intercepts.

4. A(n) __function__ is a relation in which each first component in the ordered pairs corresponds to exactly one second component.

5. In the equation $y = 4x - 2$, the coefficient of x is the __slope__ of its corresponding graph.

6. Two lines are __perpendicular__ if the product of their slopes is -1.

7. The __domain__ of a relation is the set of all first components of the ordered pairs of the relation.

8. A(n) __linear__ function is a function that can be written in the form $f(x) = mx + b$.

9. The notation $(-\infty, 5]$ is called __interval__ notation.

10. The function $f(x) = \dfrac{x + 3}{x - 1}$ is an example of a(n) __rational__ function.

11. The function $f(x) = x^2 - 3x + 2$ is an example of a(n) __polynomial__ function.

Helpful Hint

▶ Are you preparing for your test? Don't forget to take the Chapter 8 Test on page 626. Then check your answers at the back of the text and use the Chapter Test Prep Videos to see the fully worked-out solutions to any of the exercises you want to review.

8 Chapter Highlights

Definitions and Concepts	Examples
Section 8.1 Review of Equations of Lines and Writing Parallel and Perpendicular Lines	

Definitions and Concepts	Examples
The **point-slope form** of the equation of a line is $$y - y_1 = m(x - x_1)$$ where m is the slope of the line and (x_1, y_1) is a point on the line.	Find an equation of the line with slope 2 containing the point $(1, -4)$. Write the equation in standard form, $Ax + By = C$. $$\begin{aligned} y - y_1 &= m(x - x_1) \\ y - (-4) &= 2(x - 1) \\ y + 4 &= 2x - 2 \\ -2x + y &= -6 \qquad \text{Standard form} \end{aligned}$$

Definitions and Concepts	Examples

Section 8.2   Introduction to Functions

A **relation** is a set of ordered pairs. The **domain** of the relation is the set of all first components of the ordered pairs. The **range** of the relation is the set of all second components of the ordered pairs.

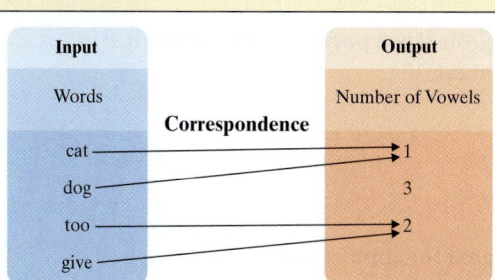

Domain: {cat, dog, too, give}
Range: {1, 2}

A **function** is a relation in which each element of the first set corresponds to exactly one element of the second set.

The previous relation is a function. Each word contains one exact number of vowels.

Find the domain and the range of the relation. Also determine whether the relation is a function.

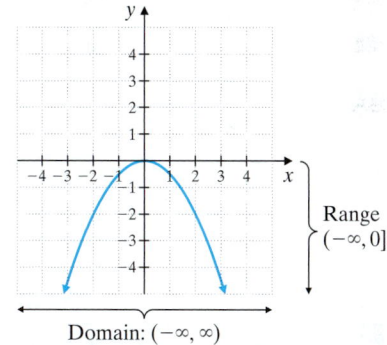

Domain: $(-\infty, \infty)$

Range $(-\infty, 0]$

Vertical Line Test

If no vertical line can be drawn so that it intersects a graph more than once, the graph is the graph of a function. If such a line can be drawn, the graph is not that of a function.

By the vertical line test, this is the graph of a function.

The symbol $f(x)$ means **function of x** and is called **function notation.**

If $f(x) = 2x^2 - 5$, find $f(-3)$.

$$f(-3) = 2(-3)^2 - 5 = 2(9) - 5 = 13$$

A **linear function** is a function that can be written in the form

$$f(x) = mx + b$$

To graph a linear function, use the slope and y-intercept.

Examples of Functions:

$$f(x) = -3, g(x) = 5x, h(x) = -\frac{1}{3}x - 7$$

Graph: $f(x) = -2x$ (or $y = -2x + 0$)

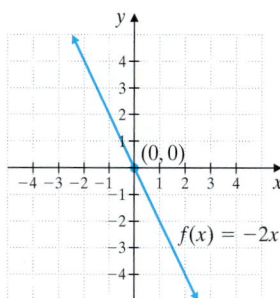

The slope is $\dfrac{2}{-1}$.

The y-intercept is $(0, 0)$.

Definitions and Concepts	Examples

Section 8.3 Polynomial and Rational Functions

A function P is a **polynomial function** if $P(x)$ is a polynomial.

For the polynomial function

$$P(x) = -x^2 + 6x - 12$$

find $P(-2)$.

$$P(-2) = -(-2)^2 + 6(-2) - 12 = -28$$

A **rational function** is a function described by a rational expression.

$$f(x) = \frac{2x - 6}{7}, \quad h(t) = \frac{t^2 - 3t + 5}{t - 1}$$

Section 8.4 Interval Notation, Finding Domains and Ranges from Graphs, and Graphing Piecewise-Defined Functions

To find the domain of a function (or relation) from its graph, recall that on the rectangular coordinate system, "domain" means the x-values that are graphed. Similarly, "range" means the y-values that are graphed.

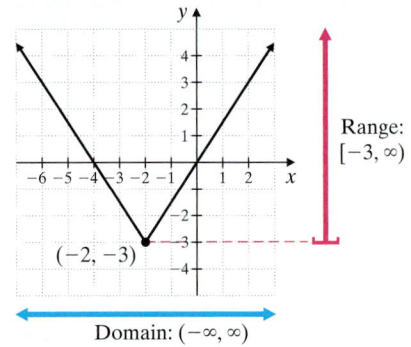

Range: $[-3, \infty)$

$(-2, -3)$

Domain: $(-\infty, \infty)$

Section 8.5 Shifting and Reflecting Graphs of Functions

Vertical shifts (upward and downward)
Let k be a positive number.

Graph of	Same as	Moved
$g(x) = f(x) + k$	$f(x)$	k units upward
$g(x) = f(x) - k$	$f(x)$	k units downward

Horizontal shift (to the left or right)
Let h be a positive number.

Graph of	Same as	Moved
$g(x) = f(x - h)$	$f(x)$	h units to the right
$g(x) = f(x + h)$	$f(x)$	h units to the left

Reflection about the x-axis
The graph of $g(x) = -f(x)$ is the graph of $f(x)$ reflected about the x-axis.

The graph of $h(x) = -|x - 3| + 1$ is the same as the graph of $f(x) = |x|$, reflected about the x-axis, then shifted 3 units right, and 1 unit up.

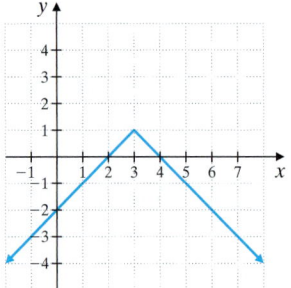

(8.1) *Write an equation of the line satisfying each set of conditions.*

1. Horizontal; through $(3, -1)$ $y = -1$

2. Slope undefined; through $(-4, -3)$ $x = -4$

Write the equation of the line satisfying each set of conditions. Write the equation in the form $y = mx + b$.

3. Through $(-3, 5)$; slope 3 $y = 3x + 14$

4. Through $(-6, -1)$ and $(-4, -2)$ $y = -\frac{1}{2}x - 4$

5. Through $(2, -6)$; parallel to $y = -2x + 3$
$y = -2x - 2$

6. Through $(-6, -1)$; perpendicular to $4x + 3y = 5$
$y = \frac{3}{4}x + \frac{7}{2}$

7. The value of an automobile bought in 2010 continues to decrease as time passes. Two years after the car was bought, it was worth $17,500; four years after it was bought, it was worth $14,300.

  a. Assuming that this relationship between the number of years past 2010 and the value of the car is linear, write an equation describing this relationship. [*Hint*: Use ordered pairs of the form (years past 2010, value of the automobile).] $y = -1600x + 20,700$

  b. Use this equation to estimate the value of the automobile in 2016. $11,100

8. The value of a building bought in 2002 continues to increase as time passes. Seven years after the building was bought, it was worth $210,000; 12 years after it was bought, it was worth $270,000.

  a. Assuming that this relationship between the number of years past 2002 and the value of the building is linear, write an equation describing this relationship. [*Hint*: Use ordered pairs of the form (years past 2002, value of the building).] $y = 12,000x + 126,000$

  b. Use this equation to estimate the value of the building in 2020. $342,000

(8.2) *Find the domain and range of each relation. Then determine whether the relation is also a function. (For Exercises 13–16, write the domain and range using interval notation from Section 8.4.)*

9. $\left\{ \left(-\frac{1}{2}, \frac{3}{4}\right), (6, 0.65), (0, -12), (25, 25) \right\}$

  domain: $\left\{ -\frac{1}{2}, 6, 0, 25 \right\}$;

  range: $\left\{ \frac{3}{4}, 0.65, -12, 25 \right\}$; function

10. $\left\{ \left(\frac{3}{4}, -\frac{1}{2}\right), (0.65, 6), (-12, 0), (25, 25) \right\}$

  domain: $\left\{ \frac{3}{4}, 0.65, -12, 25 \right\}$;

  range: $\left\{ -\frac{1}{2}, 6, 0, 25 \right\}$; function

11.

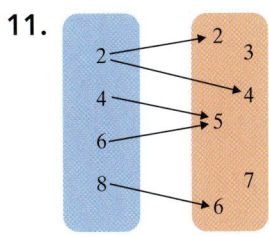

domain: $\{2, 4, 6, 8\}$;
range: $\{2, 4, 5, 6\}$; not a function

12.

Triangle
Square
Rectangle
Parallelogram
3
4

domain: {triangle, square, rectangle, parallelogram}; range: {3, 4}; function

13.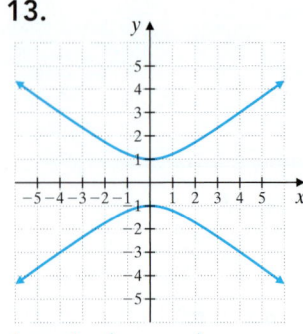

domain: $(-\infty, \infty)$;
range: $(-\infty, -1] \cup [1, \infty)$;
not a function

14.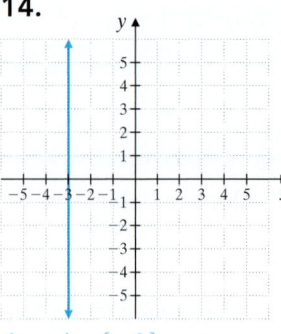

domain: $\{-3\}$; range:
$(-\infty, \infty)$; not a function

15.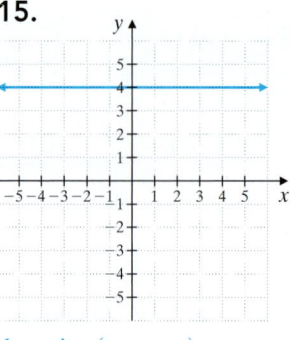

domain: $(-\infty, \infty)$; range:
$\{4\}$; function

16.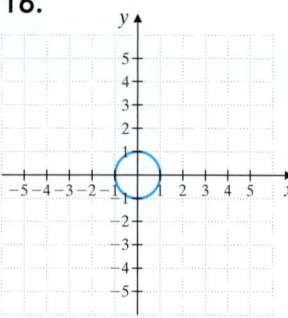

domain: $[-1, 1]$; range:
$[-1, 1]$; not a function

If $f(x) = x - 5, g(x) = -3x,$ and $h(x) = 2x^2 - 6x + 1,$ find each function value.

17. $f(2)$ -3

18. $g(0)$ 0

19. $g(-6)$ 18

20. $h(-1)$ 9

21. $h(1)$ -3

22. $f(5)$ 0

The function $J(x) = 2.54x$ may be used to calculate the weight of an object on Jupiter (J) given its weight on Earth (x).

23. If a person weighs 150 pounds on Earth, find the equivalent weight on Jupiter.   381 lb

24. A 2000-pound probe on Earth weighs how many pounds on Jupiter?   5080 lb

Graph each linear function.

25. $f(x) = x + 2$

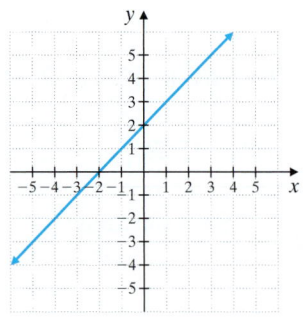

26. $f(x) = -\dfrac{1}{2}x + 3$

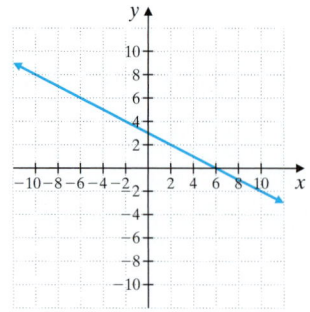

(8.3) *If $P(x) = 9x^2 - 7x + 8,$ find each function value.*

27. $P(6)$ 290

28. $P(-2)$ 58

29. $P(-3)$ 110

30. $P(0)$ 8

If $P(x) = 2x - 1$ and $Q(x) = x^2 + 2x - 5$, find each function.

31. $P(x) + Q(x)$ $x^2 + 4x - 6$

32. $2P(x) - Q(x)$ $-x^2 + 2x + 3$

Find the domain for each rational function.

33. $f(x) = \dfrac{3 - 5x}{7}$ $\{x \mid x \text{ is a real number}\}$

34. $g(x) = \dfrac{2x + 4}{11}$ $\{x \mid x \text{ is a real number}\}$

35. $F(x) = \dfrac{-3x^2}{x - 5}$ $\{x \mid x \text{ is a real number and } x \neq 5\}$

36. $h(x) = \dfrac{4x}{3x - 12}$ $\{x \mid x \text{ is a real number and } x \neq 4\}$

37. $f(x) = \dfrac{x^3 + 2}{x^2 + 8x}$

$\{x \mid x \text{ is a real number and } x \neq 0, x \neq -8\}$

38. $G(x) = \dfrac{20}{3x^2 - 48}$

$\{x \mid x \text{ is a real number and } x \neq -4, x \neq 4\}$

39. The average cost of manufacturing x bookcases is given by the rational function.

$$C(x) = \dfrac{35x + 4200}{x}$$

a. Find the average cost per bookcase of manufacturing 50 bookcases. $119

b. Find the average cost per bookcase of manufacturing 100 bookcases. $77

c. As the number of bookcases increases, does the average cost per bookcase increase or decrease? (See parts (a) and (b)).   decrease

40. If $R(x) = \dfrac{x^2 - 6x}{x - 11}$, find $R(2)$. $\dfrac{8}{9}$

(8.4) *Find the domain and range of each relation. Write each in interval notation.*

41.

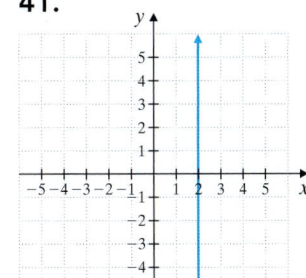

domain: $\{2\}$
range: $(-\infty, \infty)$

42.

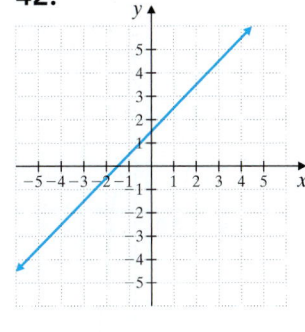

domain: $(-\infty, \infty)$
range: $(-\infty, \infty)$

43.

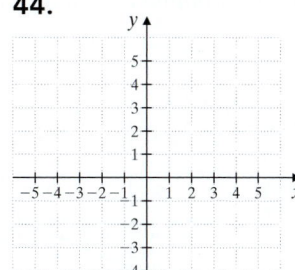

domain: $[-4, 4]$
range: $[-1, 5]$

44.

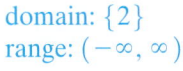

domain: $(-\infty, \infty)$
range: $\{-5\}$

Graph each function.

45. $f(x) = \begin{cases} -3x & \text{if } x < 0 \\ x - 3 & \text{if } x \geq 0 \end{cases}$

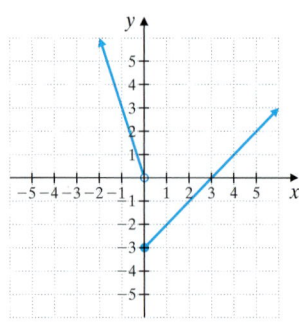

46. $g(x) = \begin{cases} -\dfrac{1}{5}x & \text{if } x \leq -1 \\ -4x + 2 & \text{if } x > -1 \end{cases}$

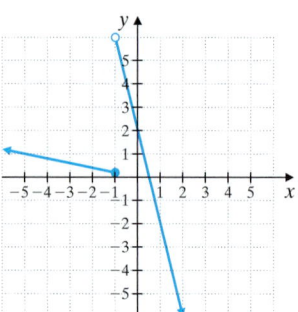

(8.5) *Graph each function.*

47. $y = \sqrt{x} - 4$

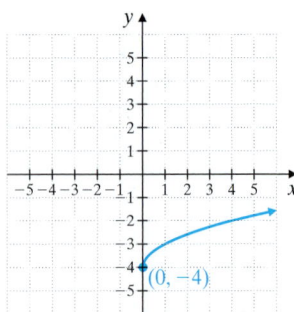

48. $f(x) = \sqrt{x - 4}$

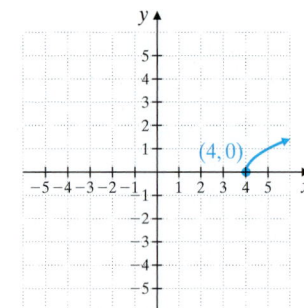

49. $g(x) = |x - 2| - 2$

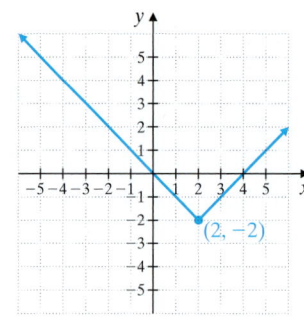

50. $h(x) = -(x + 3)^2 - 1$

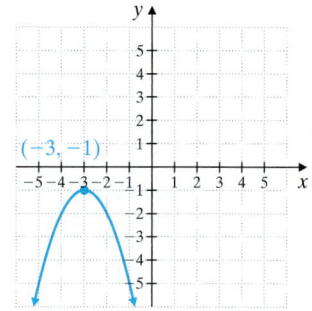

Mixed Review

Write an equation of the line satisfying each set of conditions. If possible, write the equation in the form $y = mx + b$.

51. Vertical; through $(-2, -4)$ $x = -2$

52. Slope 0; through $(2, 5)$ $y = 5$

53. Through $(-4, -2)$; parallel to $y = -\dfrac{3}{2}x + 1$

$y = -\dfrac{3}{2}x - 8$

54. Through $(-4, 5)$; perpendicular to $2x - 3y = 6$

$y = -\dfrac{3}{2}x - 1$

Find the domain and range of each relation.

55.

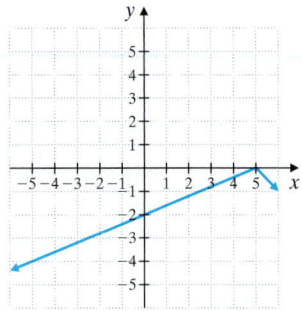

domain: $(-\infty, \infty)$ range: $(-\infty, 0]$

56.

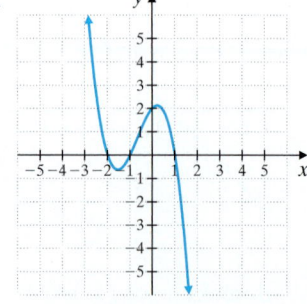

domain: $(-\infty, \infty)$ range: $(-\infty, \infty)$

Graph each piecewise-defined function.

57. $f(x) = \begin{cases} x - 2 & \text{if } x \leq 0 \\ -\dfrac{x}{3} & \text{if } x \geq 3 \end{cases}$

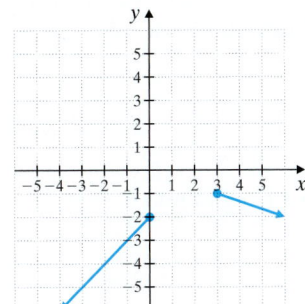

58. $g(x) = \begin{cases} 4x - 3 & \text{if } x \leq 1 \\ 2x & \text{if } x > 1 \end{cases}$

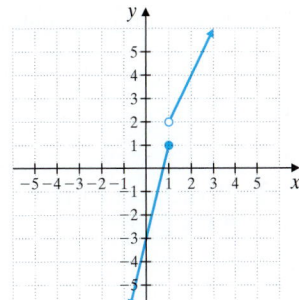

Graph each function.

59. $f(x) = \sqrt{x - 2}$

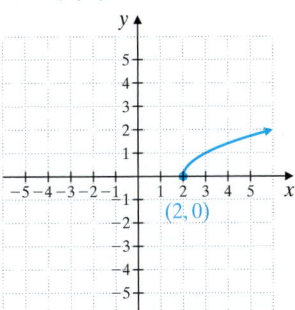

60. $f(x) = |x + 1| - 3$

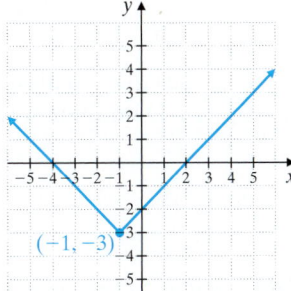

Answers

Note to Instructor: The Chapter 8 Test file in the TestGen program provides algorithms specifically matched to this test for easy replication for practice or assessment purposes.

Find an equation of the line satisfying each set of conditions. Write the equations in the form $x = a$, $y = b$, or $y = mx + b$.

1. Horizontal; through $(2, -8)$

2. Vertical; through $(-4, -3)$

1. $y = -8$

2. $x = -4$

3. Perpendicular to $x = 5$; through $(3, -2)$

4. Through $(4, -2)$ and $(6, -3)$

3. $y = -2$

5. Through $(-1, 2)$; perpendicular to $3x - y = 4$

6. Parallel to $2y + x = 3$; through $(3, -2)$

4. $y = -\dfrac{1}{2}x$

5. $y = -\dfrac{1}{3}x + \dfrac{5}{3}$

6. $y = -\dfrac{1}{2}x - \dfrac{1}{2}$

7. Line L_1 has the equation $2x - 5y = 8$. Line L_2 passes through the points $(1, 4)$ and $(-1, -1)$. Determine whether these lines are parallel, perpendicular, or neither.

7. neither

Find the domain and range of each relation. Also determine whether the relation is a function.

8. domain: $(-\infty, \infty)$, range: $\{5\}$; function

9. domain: $\{-2\}$; range: $(-\infty, \infty)$; not a function

8.
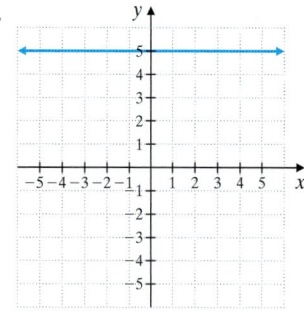

9.
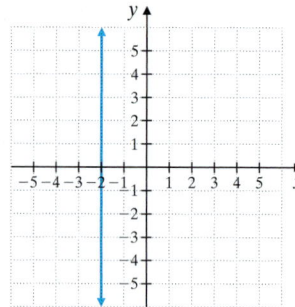

10. domain: $(-\infty, \infty)$; range: $[0, \infty)$; function

11. domain: $(-\infty, \infty)$; range: $(-\infty, \infty)$; function

10.
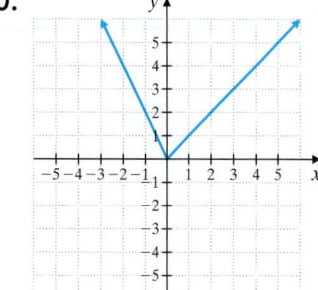

11.
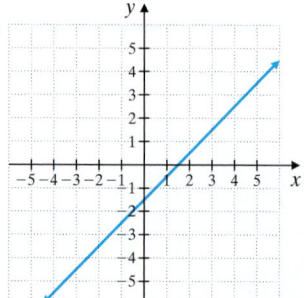

Graph each function. For Exercises 12 and 14, state the domain and the range of the function.

12. $f(x) = \begin{cases} -\dfrac{1}{2}x & \text{if } x \leq 0 \\ 2x - 3 & \text{if } x > 0 \end{cases}$

13. $f(x) = (x - 4)^2$

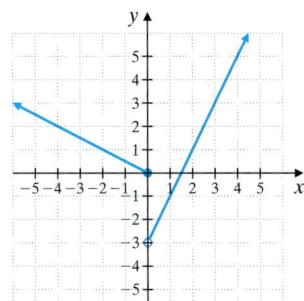

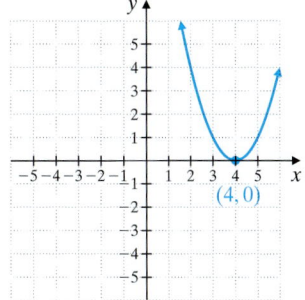

14. $g(x) = -|x + 2| - 1$

15. $h(x) = \sqrt{x} - 1$

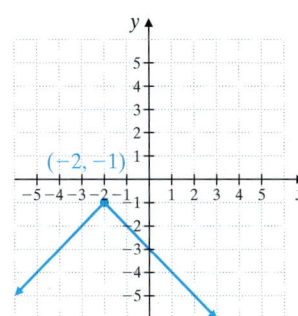

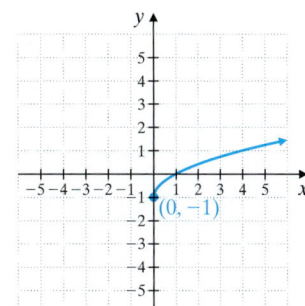

16. A pebble is hurled upward from the top of the 880-foot-tall Canada Trust Tower with an initial velocity of 96 feet per second. Neglecting air resistance, the height of the pebble after t seconds is given by the polynomial function,

$$h(t) = -16t^2 + 96t + 880.$$

Find the height of the pebble at each given time.

a. $t = 1$ second
b. $t = 5.1$ seconds
c. $t = 11$ seconds

Find the domain of each rational function.

▶ **17.** $f(x) = \dfrac{5x^2}{1 - x}$

▶ **18.** $g(x) = \dfrac{9x^2 - 9}{x^2 + 4x + 3}$

19. If $f(x) = \dfrac{5x^2}{1 - x}$, find $f(2)$.

---

domain: $(-\infty, \infty)$;
range: $(-3, \infty)$
12. see graph

13. see graph

domain: $(-\infty, \infty)$;
range: $(-\infty, -1]$
14. see graph

15. see graph

16. a. 960 ft

b. 953.44 ft

c. 0 ft

$\{x \mid x \text{ is a real number}$
17. and $x \neq 1\}$

$\{x \mid x \text{ is a real number}$
18. and $x \neq -3, x \neq -1\}$

19. -20

Answers

1. a. −6

b. 6.3 (Sec. 1.5, Ex. 6)

2. a. 7

b. 0

c. −10 (Sec. 1.6)

3. $\frac{1}{22}$ (Sec. 1.6, Ex. 9a)

4. $-\frac{1}{45}$ (Sec. 1.6)

5. $\frac{16}{3}$ (Sec. 1.6, Ex. 9b)

6. 9 (Sec. 1.6)

7. $-\frac{1}{10}$ (Sec. 1.6, Ex. 9c)

8. no reciprocal (Sec. 1.6)

9. $-\frac{13}{9}$ (Sec. 1.6, Ex. 9d)

10. −1 (Sec. 1.6)

11. $\frac{1}{1.7}$ (Sec. 1.6, Ex. 9e)

12. a. 98

b. 98 (Sec. 1.6)

13. a. 5

b. $8 - x$ (Sec. 2.1, Ex. 8)

14. 22 (Sec. 1.8)

15. no solution (Sec. 2.3, Ex. 6)

16. $\frac{7}{15}$ (Sec. 1.2)

628

---

1. Simplify each expression.
 a. $-14 - 8 + 10 - (-6)$
 b. $1.6 - (-10.3) + (-5.6)$

2. Multiply or divide.
 a. $\frac{-42}{-6}$
 b. $\frac{0}{14}$
 c. $-1(-5)(-2)$

Find the reciprocal of each number, if possible.

3. 22

4. −45

5. $\frac{3}{16}$

6. $\frac{1}{9}$

7. −10

8. 0

9. $-\frac{9}{13}$

10. −1

11. 1.7

12. Evaluate $2x^2$ for
 a. $x = 7$
 b. $x = -7$

13. a. The sum of two numbers is 8. If one number is 3, find the other number.
 b. The sum of two numbers is 8. If one number is x, write an expression representing the other number.

14. Simplify $-2 + 3[5 - (7 - 10)]$.

15. Solve:
 $-2(x - 5) + 10 = -3(x + 2) + x$

16. Find the absolute value: $\left| -\frac{7}{15} \right|$

Graph each inequality on a number line.

17. $x \geq -1$

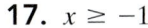

18. $5 < x$

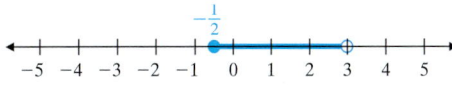

19. $-4 < x \leq 2$

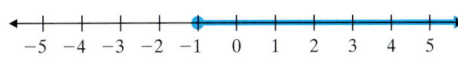

20. $-\dfrac{1}{2} \leq x < 3$

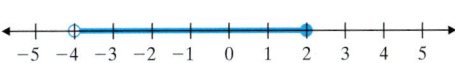

21. Graph the linear equation $y = 3x$.

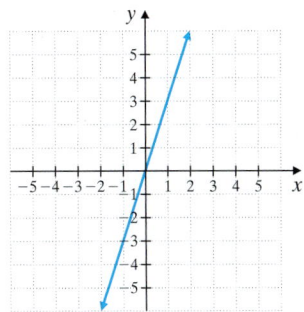

22. Graph the linear equation $-2x + 3y = -6$.

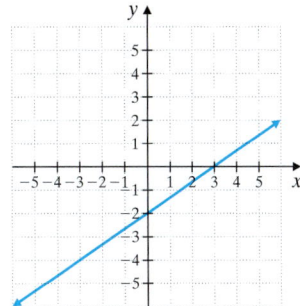

23. Graph the linear equation $y = -2$.

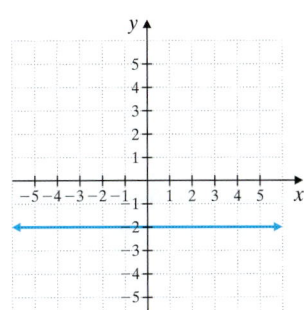

24. Graph the linear equation $x = -4$.

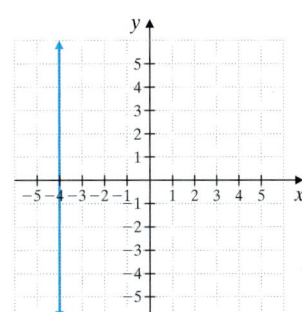

Simplify each expression.

25. $\left(\dfrac{m}{n}\right)^7$

26. $-7x^9 \cdot \dfrac{3}{14} x^4$

27. $\left(\dfrac{2x^4}{3y^5}\right)^4$

28. $(-2ab)(-0.3a^7b^{10})(8.1b^2)$

29. Subtract:
$(2x^3 + 8x^2 - 6x) - (2x^3 - x^2 + 1)$

30. Solve: $5(x - 7) = 4x - 35 + x$.

31. Divide $6x^2 + 10x - 5$ by $3x - 1$ using long division.

32. Perform each indicated operation.
a. $-4 + (-3)$
b. $\dfrac{1}{2} - \left(-\dfrac{1}{3}\right)$
c. $7 - 20$

(Sec. 2.7, Ex. 1)
17. see graph

18. see graph (Sec. 2.7)

(Sec. 2.7, Ex. 3)
19. see graph

20. see graph (Sec. 2.7)

(Sec. 3.2, Ex. 3)
21. see graph

22. see graph (Sec. 3.2)

(Sec. 3.2, Ex. 5)
23. see graph

24. see graph (Sec. 3.2)

25. $\dfrac{m^7}{n^7}, n \neq 0$ (Sec. 5.1, Ex. 22)

26. $-\dfrac{3}{2}x^{13}$ (Sec. 5.1)

27. $\dfrac{16x^{16}}{81y^{20}}, y \neq 0$ (Sec. 5.1, Ex. 23)

28. $4.86a^8b^{13}$ (Sec. 5.1)

29. $9x^2 - 6x - 1$ (Sec. 5.4, Ex. 5)

30. all real numbers (Sec. 2.3)

31. $2x + 4 - \dfrac{1}{3x - 1}$ (Sec. 5.7, Ex. 5)

32. a. -7

b. $\dfrac{5}{6}$

c. -13 (Sec. 1.5)

33. Solve: $x(2x - 7) = 4$

34. Solve: $x(2x - 7) = 0$

35. Find the lengths of the sides of a right triangle if the lengths can be expressed by three consecutive even integers.

36. Find three consecutive odd integers whose sum is 213.

37. Subtract: $\dfrac{2y}{2y - 7} - \dfrac{7}{2y - 7}$

38. Solve. $3x + 10 > \dfrac{5}{2}(x - 1)$

39. Simplify: $\dfrac{\dfrac{x}{y} + \dfrac{3}{2x}}{\dfrac{x}{2} + y}$

40. $\left(\dfrac{5a^3}{b^{-2} c^5} \right)^{-3}$

41. Write an equation of the line containing the point $(-2, 1)$ and perpendicular to the line $3x + 5y = 4$. Write the equation in slope-intercept form, $y = mx + b$.

42. Find an equation of the vertical line through $\left(-2, -\dfrac{3}{4} \right)$.

43. Write an equation of the horizontal line containing the point $(2, 3)$.

44. Find an equation of the line parallel to $x = -4$, through the point $(-1, 7)$.

45. If $f(x) = 7x^2 - 3x + 1$, find $f(1)$.

46. Find the slope of the line defined by $f(x) = -2.3x - 6$.

47. If $g(x) = 3x - 2$, find $g(0)$.

48. Determine whether the graph below is the graph of a function.

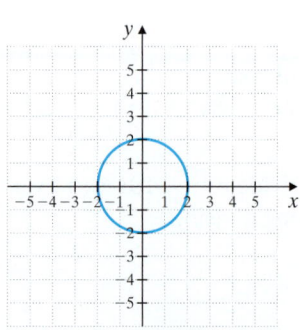

Systems of Equations and Inequalities and Variation

Does Winter Occur Globally During the Same Months?

While North Americans and other citizens of the northern hemisphere, including the English, are enjoying winter and possibly snow, Australians are celebrating the holidays on the beach. Why such a difference in temperature? It is because the seasons are reversed between the northern and southern hemispheres. While winter in the northern hemisphere is cold, it is warm in the southern hemisphere. The winter solstice is our shortest day of the year, while below the equator it is the longest day of the year.

Australia and England are examples of countries on opposite sides of the equator. In Exercise 40 of the Chapter Review, we will find the month (between February and July) when the average high temperature in Australia is the same as the average high temperature in England.

Recall from Chapter 4 that a *system of equations* consists of two or more equations in two or more variables. In this chapter, we use the methods learned in Chapter 4 to solve systems of equations in three variables. Next, we will study new methods for solving systems of equations, then solving systems of inequalities, and then conclude with variation and problem solving.

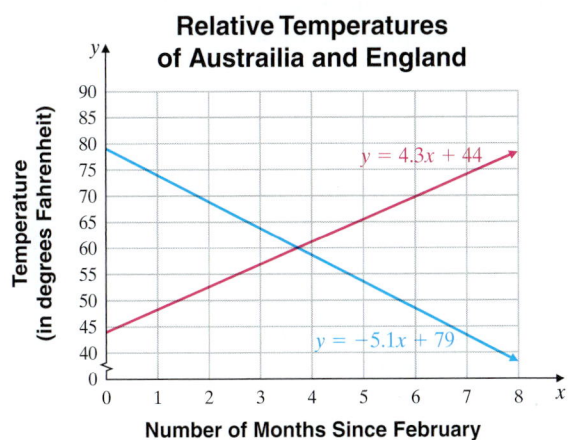

Relative Temperatures of Austrailia and England

$y = 4.3x + 44$

$y = -5.1x + 79$

Temperature (in degrees Fahrenheit)

Number of Months Since February

9.1 Solving Systems of Linear Equations in Three Variables and Problem Solving

Objectives

A Solve a System of Three Linear Equations in Three Variables.

B Solve Problems Modeled by Systems of Three Equations.

Teaching Tip

Use your room to help students visualize solutions, if any, to linear equations in three variables. For instance, what is the intersection, if any, of:

- the floor and two adjacent walls in your room?

- the floor, the ceiling, and the back wall?

In this section, we solve systems of linear equations in three variables. We call the equation $3x - y + z = -15$, for example, a **linear equation in three variables** since there are three variables and each variable is raised only to the power 1. A solution of this equation is an **ordered triple (x, y, z)** that makes the equation a true statement.

For example, the ordered triple $(2, 0, -21)$ is a solution of $3x - y + z = -15$ since replacing x with 2, y with 0, and z with -21 yields the true statement

$$3(2) - 0 + (-21) = -15$$

The graph of this equation is a plane in three-dimensional space, just as the graph of a linear equation in two variables is a line in two-dimensional space.

Although we will not discuss the techniques for graphing equations in three variables, visualizing the possible patterns of intersecting planes gives us insight into the possible patterns of solutions of a system of three three-variable linear equations. There are four possible patterns.

1. Three planes have a single point in common. This point represents the single solution of the system. This system is **consistent.**

2. Three planes intersect at no point common to all three. This system has no solution. A few ways that this can occur are shown. This system is **inconsistent.**

3. Three planes intersect at all the points of a single line. The system has infinitely many solutions. This system is **consistent.**

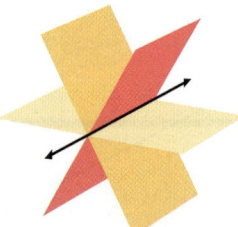

4. Three planes coincide at all points on the plane. The system is **consistent,** and the equations are **dependent.**

Objective A Solving a System of Three Linear Equations in Three Variables ▶

Just as with systems of two equations in two variables, we can use the elimination or substitution method to solve a system of three equations in three variables. To use the elimination method, we eliminate a variable and obtain a system of two equations in two variables. Then we use the methods we learned in Chapter 4 to solve the system of two equations.

Example 1 Solve the system:

$$\begin{cases} 3x - y + z = -15 & \text{Equation (1)} \\ x + 2y - z = 1 & \text{Equation (2)} \\ 2x + 3y - 2z = 0 & \text{Equation (3)} \end{cases}$$

Solution: Let's add equations (1) and (2) to eliminate z.

$$3x - y + z = -15$$
$$\underline{x + 2y - z = \quad 1}$$
$$4x + y \quad = -14 \quad \text{Equation (4)}$$

Next we add two *other* equations and *eliminate z again*. To do so, we multiply both sides of equation (1) by 2 and add the resulting equation to equation (3). Then

$$\begin{cases} 2(3x - y + z) = 2(-15) \\ 2x + 3y - 2z = 0 \end{cases} \quad \text{simplifies to} \quad \begin{cases} 6x - 2y + 2z = -30 \\ \underline{2x + 3y - 2z = \quad 0} \\ 8x + y \quad = -30 \\ \quad \text{Equation (5)} \end{cases}$$

We now have two equations (4 and 5) in the same two variables. This means we can solve equations (4) and (5) for x and y. To solve by elimination, we multiply both sides of equation (4) by -1 and add the resulting equation to equation (5). Then

$$\begin{cases} -1(4x + y) = -1(-14) \\ 8x + y = -30 \end{cases} \quad \text{simplifies to} \quad \begin{cases} -4x - y = \quad 14 \\ \underline{8x + y = -30} \quad \text{Add the} \\ 4x \quad = -16 \quad \text{equations.} \\ \quad x = -4 \quad \text{Solve for } x. \end{cases}$$

We now replace x with -4 in equation (4) or (5).

$$4x + y = -14 \quad \text{Equation (4)}$$
$$4(-4) + y = -14 \quad \text{Let } x = -4.$$
$$y = 2 \quad \text{Solve for } y.$$

Finally, we replace x with -4 and y with 2 in equation (1), (2), or (3).

$$x + 2y - z = 1 \quad \text{Equation (2)}$$
$$-4 + 2(2) - z = 1 \quad \text{Let } x = -4 \text{ and } y = 2.$$
$$-4 + 4 - z = 1$$
$$-z = 1$$
$$z = -1$$

The ordered triple solution is $(-4, 2, -1)$.

(Continued on next page)

Practice 1

Solve the system:

$$\begin{cases} 2x - y + 3z = 13 \\ x + y - z = -2 \\ 3x + 2y + 2z = 13 \end{cases}$$

Helpful Hint Make sure you add two other equations besides equations (1) and (2) and *also* **eliminate the same variable.** You will see why as you follow this example.

Teaching Tip

For Practice 1 ask, "Which variable would you eliminate first?"

Answer

1. $(1, 1, 4)$

Check: To check, we let $x = -4$, $y = 2$, and $z = -1$ in *all three* original equations of the system.

Equation (1)

$$3x - y + z = -15$$
$$3(-4) - 2 + (-1) \overset{?}{=} -15$$
$$-12 - 2 - 1 \overset{?}{=} -15$$
$$-15 = -15 \quad \text{True}$$

Equation (2)

$$x + 2y - z = 1$$
$$-4 + 2(2) - (-1) \overset{?}{=} 1$$
$$-4 + 4 + 1 \overset{?}{=} 1$$
$$1 = 1 \quad \text{True}$$

Equation (3)

$$2x + 3y - 2z = 0$$
$$2(-4) + 3(2) - 2(-1) \overset{?}{=} 0$$
$$-8 + 6 + 2 \overset{?}{=} 0$$
$$0 = 0 \quad \text{True}$$

All three statements are true, so the ordered triple solution is $(-4, 2, -1)$.

🟧 **Work Practice 1**

Practice 2

Solve the system:

$$\begin{cases} 2x + 4y - 2z = 3 \\ -x + y - z = 6 \\ x + 2y - z = 1 \end{cases}$$

Example 2 Solve the system:

$$\begin{cases} 2x - 4y + 8z = 2 & (1) \\ -x - 3y + z = 11 & (2) \\ x - 2y + 4z = 0 & (3) \end{cases}$$

Solution: Add equations (2) and (3) to eliminate x, and the new equation is

$$-5y + 5z = 11 \quad (4)$$

To eliminate x again, we multiply both sides of equation (2) by 2 and add the resulting equation to equation (1). Then

$$\begin{cases} 2x - 4y + 8z = 2 \\ 2(-x - 3y + z) = 2(11) \end{cases} \quad \text{simplifies to} \quad \begin{cases} 2x - 4y + 8z = 2 \\ \underline{-2x - 6y + 2z = 22} \\ \quad -10y + 10z = 24 \quad (5) \end{cases}$$

Next we solve for y and z using equations (4) and (5). To do so, we multiply both sides of equation (4) by -2 and add the resulting equation to equation (5).

$$\begin{cases} -2(-5y + 5z) = -2(11) \\ -10y + 10z = 24 \end{cases} \quad \text{simplifies to} \quad \begin{cases} 10y - 10z = -22 \\ \underline{-10y + 10z = 24} \\ \quad 0 = 2 \quad \text{False} \end{cases}$$

Since the statement is false, this system is inconsistent and has no solution. The solution set is the empty set $\{\ \}$, or $\varnothing$.

🟧 **Work Practice 2**

The elimination method is summarized next.

Answer

2. $\varnothing$

Solving a System of Three Linear Equations by the Elimination Method

Step 1: Write each equation in standard form, $Ax + By + Cz = D$.

Step 2: Choose a pair of equations and use them to eliminate a variable.

Step 3: Choose any other pair of equations and eliminate the *same variable* as in Step 2.

Step 4: Two equations in two variables should be obtained from Step 2 and Step 3. Use methods from Section 4.3 to solve this system for both variables.

Step 5: To solve for the third variable, substitute the values of the variables found in Step 4 into any of the original equations containing the third variable.

Step 6: Check the ordered triple solution in *all three* original equations.

Teaching Tip

The most common student mistake made is not following Step 3. Emphasize this step and its importance.

Helpful Hint Make sure you read closely and follow Step 3.

✓**Concept Check** In the system

$$\begin{cases} x + y + z = 6 & \text{Equation (1)} \\ 2x - y + z = 3 & \text{Equation (2)} \\ x + 2y + 3z = 14 & \text{Equation (3)} \end{cases}$$

equations (1) and (2) are used to eliminate y. Which action could be used to finish solving? Why?

a. Use (1) and (2) to eliminate z.

b. Use (2) and (3) to eliminate y.

c. Use (1) and (3) to eliminate x.

Example 3 Solve the system:

$$\begin{cases} 2x + 4y = 1 & (1) \\ 4x - 4z = -1 & (2) \\ y - 4z = -3 & (3) \end{cases}$$

Solution: Notice that equation (2) has no term containing the variable y. Let's eliminate y using equations (1) and (3). We multiply both sides of equation (3) by -4 and add the resulting equation to equation (1). Then

$$\begin{cases} 2x + 4y = 1 \\ -4(y - 4z) = -4(-3) \end{cases} \quad \text{simplifies to} \quad \begin{cases} 2x + 4y = 1 \\ -4y + 16z = 12 \\ \hline 2x + 16z = 13 \quad (4) \end{cases}$$

Next we solve for z using equations (4) and (2). We multiply both sides of equation (4) by -2 and add the resulting equation to equation (2).

$$\begin{cases} -2(2x + 16z) = -2(13) \\ 4x - 4z = -1 \end{cases} \quad \text{simplifies to} \quad \begin{cases} -4x - 32z = -26 \\ 4x - 4z = -1 \\ \hline -36z = -27 \\ z = \dfrac{3}{4} \end{cases}$$

(*Continued on next page*)

Practice 3

Solve the system:

$$\begin{cases} 3x + 2y = -1 \\ 6x - 2z = 4 \\ y - 3z = 2 \end{cases}$$

Answer

3. $\left(\dfrac{1}{3}, -1, -1\right)$

✓**Concept Check Answer**

b; answers may vary

Now we replace z with $\dfrac{3}{4}$ in equation (3) and solve for y.

$$y - 4\left(\dfrac{3}{4}\right) = -3 \quad \text{Let } z = \dfrac{3}{4} \text{ in equation (3).}$$

$$y - 3 = -3$$

$$y = 0$$

Finally, we replace y with 0 in equation (1) and solve for x.

$$2x + 4(0) = 1 \quad \text{Let } y = 0 \text{ in equation (1).}$$

$$2x = 1$$

$$x = \dfrac{1}{2}$$

The ordered triple solution is $\left(\dfrac{1}{2}, 0, \dfrac{3}{4}\right)$. Check to see that this solution satisfies *all three* equations of the system.

■ **Work Practice 3**

Practice 4

Solve the system:

$$\begin{cases} x - 3y + 4z = 2 \\ -2x + 6y - 8z = -4 \\ \dfrac{1}{2}x - \dfrac{3}{2}y + 2z = 1 \end{cases}$$

Example 4 Solve the system:

$$\begin{cases} x - 5y - 2z = 6 & (1) \\ -2x + 10y + 4z = -12 & (2) \\ \dfrac{1}{2}x - \dfrac{5}{2}y - z = 3 & (3) \end{cases}$$

Solution: We multiply both sides of equation (3) by 2 to eliminate fractions, and we multiply both sides of equation (2) by $-\dfrac{1}{2}$ so that the coefficient of x is 1. The resulting system is then

$$\begin{cases} x - 5y - 2z = 6 & (1) \\ x - 5y - 2z = 6 & \text{Multiply (2) by } -\dfrac{1}{2}. \\ x - 5y - 2z = 6 & \text{Multiply (3) by 2.} \end{cases}$$

All three resulting equations are identical, and therefore equations (1), (2), and (3) are all equivalent. There are infinitely many solutions of this system. The equations are dependent. The solution set can be written as $\{(x, y, z) \mid x - 5y - 2z = 6\}$.

■ **Work Practice 4**

Teaching Tip

After discussing Example 4, ask students to name three points that would be solutions to this system and three points that would not be solutions.

As mentioned earlier, we can also use the substitution method to solve a system of linear equations in three variables.

Practice 5

Solve the system:

$$\begin{cases} 2x + 5y - 3z = 30 & (1) \\ x + y \phantom{{}- 3z} = -3 & (2) \\ 2x \phantom{{}+ 5y} - z = 0 & (3) \end{cases}$$

(*Hint*: Equations (2) and (3) each contain the variable x and have a variable missing.)

Example 5 Solve the system:

$$\begin{cases} x - 4y - 5z = 35 & (1) \\ x - 3y \phantom{{}- 5z} = 0 & (2) \\ -y + z = -55 & (3) \end{cases}$$

Solution: Notice in equations (2) and (3) that a variable is missing. Also notice that both equations contain the variable y. Let's use the substitution method by solving equation (2) for x and equation (3) for z and substituting the results in equation (1).

$$x - 3y = 0 \quad (2)$$

$$x = 3y \quad \text{Solve equation (2) for } x.$$

$$-y + z = -55 \quad (3)$$

$$z = y - 55 \quad \text{Solve equation (3) for } z.$$

Answers

4. $\{(x, y, z) \mid x - 3y + 4z = 2\}$
5. $(-5, 2, -10)$

Now substitute $3y$ for x and $y - 55$ for z in equation (1).

$$x - 4y - 5z = 35 \qquad (1)$$
$$3y - 4y - 5(y - 55) = 35 \qquad \text{Let } x = 3y \text{ and } z = y - 55.$$
$$3y - 4y - 5y + 275 = 35 \qquad \text{Use the distributive property and multiply.}$$
$$-6y + 275 = 35 \qquad \text{Combine like terms.}$$
$$-6y = -240 \qquad \text{Subtract 275 from both sides.}$$
$$y = 40 \qquad \text{Solve.}$$

Helpful Hint Do not forget to distribute.

To find x, recall that $x = 3y$ and substitute 40 for y. Then $x = 3y$ becomes $x = 3 \cdot 40 = 120$. To find z, recall that $z = y - 55$ and also substitute 40 for y. Then $z = y - 55$ becomes $z = 40 - 55 = -15$. The solution is $(120, 40, -15)$.

■ **Work Practice 5**

Objective B Solving Problems Modeled by Systems of Three Equations ▶

To introduce problem solving with systems of three linear equations in three variables, we solve a problem about triangles.

▶ **Example 6** Finding Angle Measures

The measure of the largest angle of a triangle is 80° more than the measure of the smallest angle, and the measure of the remaining angle is 10° more than the measure of the smallest angle. Find the measure of each angle.

Solution:

1. **UNDERSTAND.** Read and reread the problem. Recall that the sum of the measures of the angles of a triangle is 180°. Then guess a solution. If the smallest angle measures 20°, the measure of the largest angle is 80° more, or $20° + 80° = 100°$. The measure of the remaining angle is 10° more than the measure of the smallest angle, or $20° + 10° = 30°$. The sum of these three angles is $20° + 100° + 30° = 150°$, not the required 180°. We now know that the measure of the smallest angle is greater than 20°

To model this problem, we will let

x = degree measure of the smallest angle
y = degree measure of the largest angle
z = degree measure of the remaining angle

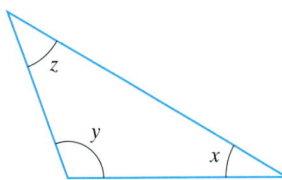

2. **TRANSLATE.** We translate the given information into three equations.

In words: the sum of the measures = 180

Translate: $x + y + z = 180$

In words: the largest angle is 80 more than the smallest angle

Translate: $y = x + 80$

In words: the remaining angle is 10 more than the smallest angle

Translate: $z = x + 10$

(Continued on next page)

Practice 6

The measure of the largest angle of a triangle is 80° more than the measure of the smallest angle, and the measure of the remaining angle is 40° more than the measure of the smallest angle. Find the measure of each angle.

Teaching Tip: Classroom Activity

After reviewing Example 6, have students work in pairs to write their own triangle word problem. Have them begin by deciding the measure of each angle in a triangle. Then have them write a word problem based on a mathematical relationship between one pair of angles and between another pair of angles. Finally, have them give their word problem to another pair to solve.

Answer
6. 20°; 60°; 100°

3. SOLVE. We solve the system

$$\begin{cases} x + y + z = 180 \\ y = x + 80 \\ z = x + 10 \end{cases}$$

Since y and z are both expressed in terms of x, we will solve using the substitution method. We substitute $y = x + 80$ and $z = x + 10$ in the first equation. Then

$$x + y + z = 180 \quad \text{First equation}$$

$$x + (x + 80) + (x + 10) = 180 \quad \text{Let } y = x + 80 \text{ and } z = x + 10.$$

$$3x + 90 = 180$$

$$3x = 90$$

$$x = 30$$

Then $y = x + 80 = 30 + 80 = 110$, and $z = x + 10 = 30 + 10 = 40$. The ordered triple solution is $(30, 110, 40)$.

4. INTERPRET.

Check: Notice that $30° + 40° + 110° = 180°$. Also, the measure of the largest angle, $110°$, is $80°$ more than the measure of the smallest angle, $30°$. The measure of the remaining angle, $40°$, is $10°$ more than the measure of the smallest angle, $30°$.

State: The angles measure $30°, 110°,$ and $40°$.

■ **Work Practice 6**

Vocabulary, Readiness & Video Check

Solve.

1. Choose the equation(s) that has $(-1, 3, 1)$ as a solution.
    a. $x + y + z = 3$ **b.** $-x + y + z = 5$ **c.** $-x + y + 2z = 0$ **d.** $x + 2y - 3z = 2$ a, b, d

2. Choose the equation(s) that has $(2, 1, -4)$ as a solution.
    a. $x + y + z = -1$ **b.** $x - y - z = -3$ **c.** $2x - y + z = -1$ **d.** $-x - 3y - z = -1$ a, c, d

3. Use the result of Exercise 1 to determine whether $(-1, 3, 1)$ is a solution of the system below. Explain your answer.

$$\begin{cases} x + y + z = 3 \\ -x + y + z = 5 \quad \text{yes; answers may vary} \\ x + 2y - 3z = 2 \end{cases}$$

4. Use the result of Exercise 2 to determine whether $(2, 1, -4)$ is a solution of the system below. Explain your answer.

$$\begin{cases} x + y + z = -1 \\ x - y - z = -3 \quad \text{no; answers may vary} \\ 2x - y + z = -1 \end{cases}$$

Martin-Gay Interactive Videos　Watch the section lecture video and answer the following questions.

Objective A 5. From ▤ Example 1 and the lecture before, why does Step 3 stress that the same variable be eliminated from two other equations? ▶

Objective B 6. In ▤ Example 2, why is the ordered triple not the final stated solution to the application? ▶

See Video 9.1

See video answer section.

9.1 **Exercise Set** MyMathLab®

Objective A *Solve each system. See Examples 1 through 5.*

1. $\begin{cases} x - y + z = -4 \\ 3x + 2y - z = 5 \\ -2x + 3y - z = 15 \end{cases}$

$(-1, 5, 2)$

2. $\begin{cases} x + y - z = -1 \\ -4x - y + 2z = -7 \\ 2x - 2y - 5z = 7 \end{cases}$

$(3, -3, 1)$

3. $\begin{cases} x + y = 3 \\ 2y = 10 \\ 3x + 2y - 3z = 1 \end{cases}$

$(-2, 5, 1)$

4. $\begin{cases} 5x = 5 \\ 2x + y = 4 \\ 3x + y - 4z = -15 \end{cases}$

$(1, 2, 5)$

5. $\begin{cases} 2x + 2y + z = 1 \\ -x + y + 2z = 3 \\ x + 2y + 4z = 0 \end{cases}$

$(-2, 3, -1)$

6. $\begin{cases} 2x - 3y + z = 5 \\ x + y + z = 0 \\ 4x + 2y + 4z = 4 \end{cases}$

$(-3, -2, 5)$

7. $\begin{cases} x - 2y + z = -5 \\ -3x + 6y - 3z = 15 \\ 2x - 4y + 2z = -10 \end{cases}$

$\{(x, y, z) \,|\, x - 2y + z = -5\}$

8. $\begin{cases} 3x + y - 2z = 2 \\ -6x - 2y + 4z = -4 \\ 9x + 3y - 6z = 6 \end{cases}$

$\{(x, y, z) \,|\, 3x + y - 2z = 2\}$

9. $\begin{cases} 4x - y + 2z = 5 \\ 2y + z = 4 \\ 4x + y + 3z = 10 \end{cases}$

$\varnothing$

10. $\begin{cases} 5y - 7z = 14 \\ 2x + y + 4z = 10 \\ 2x + 6y - 3z = 30 \end{cases}$

$\varnothing$

11. $\begin{cases} x + 5z = 0 \\ 5x + y = 0 \\ y - 3z = 0 \end{cases}$

$(0, 0, 0)$

12. $\begin{cases} x - 5y = 0 \\ x - z = 0 \\ -x + 5z = 0 \end{cases}$

$(0, 0, 0)$

13. $\begin{cases} 6x - 5z = 17 \\ 5x - y + 3z = -1 \\ 2x + y = -41 \end{cases}$

$(-3, -35, -7)$

14. $\begin{cases} x + 2y = 6 \\ 7x + 3y + z = -33 \\ x - z = 16 \end{cases}$

$(-4, 5, -20)$

15. $\begin{cases} x + y + z = 8 \\ 2x - y - z = 10 \\ x - 2y - 3z = 22 \end{cases}$

$(6, 22, -20)$

16. $\begin{cases} 5x + y + 3z = 1 \\ x - y + 3z = -7 \\ -x + y = 1 \end{cases}$

$(1, 2, -2)$

17. $\begin{cases} x + 2y - z = 5 \\ 6x + y + z = 7 \\ 2x + 4y - 2z = 5 \end{cases}$

$\varnothing$

18. $\begin{cases} 4x - y + 3z = 10 \\ x + y - z = 5 \\ 8x - 2y + 6z = 10 \end{cases}$

$\varnothing$

19. $\begin{cases} 2x - 3y + z = 2 \\ x - 5y + 5z = 3 \\ 3x + y - 3z = 5 \end{cases}$
$(3, 2, 2)$

20. $\begin{cases} 4x + y - z = 8 \\ x - y + 2z = 3 \\ 3x - y + z = 6 \end{cases}$
$(2, 1, 1)$

21. $\begin{cases} -2x - 4y + 6z = -8 \\ x + 2y - 3z = 4 \\ 4x + 8y - 12z = 16 \end{cases}$
$\{(x, y, z) \mid x + 2y - 3z = 4\}$

22. $\begin{cases} -6x + 12y + 3z = -6 \\ 2x - 4y - z = 2 \\ -x + 2y + \dfrac{z}{2} = -1 \end{cases}$
$\{(x, y, z) \mid -6x + 12y + 3z = -6\}$

23. $\begin{cases} 2x + 2y - 3z = 1 \\ y + 2z = -14 \\ 3x - 2y = -1 \end{cases}$
$(-3, -4, -5)$

24. $\begin{cases} 7x + 4y = 10 \\ x - 4y + 2z = 6 \\ y - 2z = -1 \end{cases}$
$(2, -1, 0)$

25. $\begin{cases} x + 2y - z = 5 \\ -3x - 2y - 3z = 11 \\ 4x + 4y + 5z = -18 \end{cases}$
$\left(0, \dfrac{1}{2}, -4\right)$

26. $\begin{cases} 3x - 3y + z = -1 \\ 3x - y - z = 3 \\ -6x + 4y + 3z = -8 \end{cases}$
$\left(\dfrac{1}{3}, 0, -2\right)$

27. $\begin{cases} \dfrac{3}{4}x - \dfrac{1}{3}y + \dfrac{1}{2}z = 9 \\ \dfrac{1}{6}x + \dfrac{1}{3}y - \dfrac{1}{2}z = 2 \\ \dfrac{1}{2}x - y + \dfrac{1}{2}z = 2 \end{cases}$
$(12, 6, 4)$

28. $\begin{cases} \dfrac{1}{3}x - \dfrac{1}{4}y + z = -9 \\ \dfrac{1}{2}x - \dfrac{1}{3}y - \dfrac{1}{4}z = -6 \\ x - \dfrac{1}{2}y - z = -8 \end{cases}$
$(-6, 12, -4)$

Objective B and Section 4.4 Mixed Practice
Solve. See Example 6. For Exercises 29 and 30, the solutions have been started for you. The first few exercises are each modeled by a system of two linear equations in two variables.

29. One number is two more than a second number. Twice the first is 4 less than 3 times the second. Find the numbers.   10 and 8

Start the solution:

1. UNDERSTAND the problem. Since we are looking for two numbers, let
   $x =$ one number
   $y =$ second number

2. TRANSLATE. Since we have assigned two variables, we will translate the facts into two equations. (Fill in the blanks.)
   First equation:

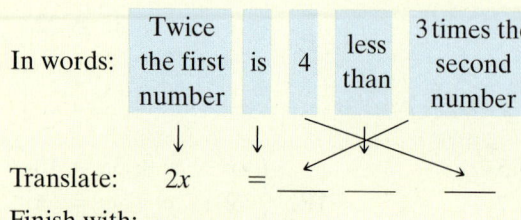

   Translate: x = ___ ___ ___

   Second equation:

   In words: | Twice the first number | is | 4 | less than | 3 times the second number |

   Translate: $2x$ = ___ ___ ___

   Finish with:

3. SOLVE the system and
4. INTERPRET the results.

30. Three times one number minus a second is 8, and the sum of the numbers is 12. Find the numbers.   5 and 7

Start the solution:

1. UNDERSTAND the problem. Since we are looking for two numbers, let
   $x =$ one number
   $y =$ second number

2. TRANSLATE. Since we have assigned two variables, we will translate the facts into two equations. (Fill in the blanks.)
   First equation:

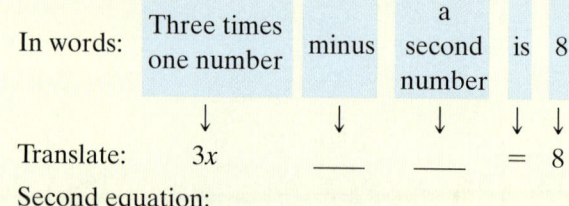

   Translate: $3x$ ___ ___ = 8

   Second equation:

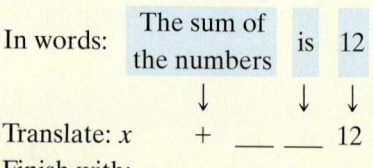

   Translate: x + ___ ___ 12

   Finish with:

3. SOLVE the system and
4. INTERPRET the results.

31. The United States currently has the world's only "large deck" aircraft carriers, which can hold up to 85 fixed-wing aircraft. A Gerald R. Ford class carrier, which is still under production, is longest in length, while a Nimitz class carrier is the second longest. The total length of these two carriers is 2198 feet, while the difference of their lengths is only 14 feet. (*Source:* Naval-technology.com)

 a. Find the length of each class of carrier.
 b. If a football field has a length of 100 yards, determine the length of a Gerald R. Ford class carrier in terms of the number of football fields. (Round to the nearest tenth.)   3.69 football fields
 31. a. Ford class: 1106 ft; Nimitz class: 1092 ft

32. Baseball has often been called "America's game." For the 2013 season, the team with the highest average attendance was the Los Angeles Dodgers, and the team with the lowest average attendance was the Tampa Bay Rays. The average number of fans who attended a home game of the Dodgers was 8926 more than twice the average number of fans who attended a home game of the Rays. If the total average attendance for both teams was 64,861, find the average attendance for each team for the 2013 Major League Baseball season. (*Source:* ESPN)
Tampa Bay: 18,645; Los Angeles: 46,216

33. Rabbits in a lab are to be kept on a strict daily diet that includes 30 grams of protein, 16 grams of fat, and 24 grams of carbohydrates. The scientist has only three food mixes available with the following grams of nutrients per unit.

	Protein	Fat	Carbohydrate
Mix A	4	6	3
Mix B	6	1	2
Mix C	4	1	12

2 units of Mix A:
3 units of Mix B:
1 unit of Mix C

Find how many units of each mix are needed daily to meet each rabbit's dietary need.

34. Gerry Gundersen mixes different solutions with concentrations of 25%, 40%, and 50% to get 200 liters of a 32% solution. If he uses twice as much of the 25% solution as of the 40% solution, find how many liters of each kind he uses.
120 L of 25%; 60 L of 40%; 20 L of 50%

△**35.** The perimeter of a quadrilateral (four-sided polygon) is 29 inches. The longest side is twice as long as the shortest side. The other two sides are equally long and are 2 inches longer than the shortest side. Find the lengths of all four sides.   5 in.; 7 in.; 7 in.; 10 in.

△**36.** The measure of the largest angle of a triangle is 90° more than the measure of the smallest angle, and the measure of the remaining angle is 30° more than the measure of the smallest angle. Find the measure of each angle.   20°, 50°, 110°

37. The sum of three numbers is 40. The first number is five more than the second number. It is also twice the third. Find the numbers.   18, 13, and 9

38. The sum of the digits of a three-digit number is 15. The tens-place digit is twice the hundreds-place digit, and the ones-place digit is 1 less than the hundreds-place digit. Find the three-digit number.   483

39. For 2013, the WNBA top scorer was Maya Moore of the Minnesota Lynx. She scored a total of 628 points during the regular season. The number of two-point field goals that Moore made was 55 fewer than three times the number of three-point field goals that she made. The number of free throws (each worth one point) she made was 71 fewer than the number of two-point field goals she made. Find how many free throws, two-point field goals, and three-point field goals, Maya Moore made during the 2013 regular season. (*Source:* Women's National Basketball Association)   free throws: 90; two-point field goals: 161; three-point field goals: 72

40. During the 2012–2013 regular NBA season, the top-scoring player was Kevin Durant of the Oklahoma City Thunder. Durant scored a total of 2280 points during the regular season. The number of free throws (each worth one point) he made was 123 more than four times the number of three-point field goals he made. The number of two-point field goals that Durant made was 87 less than the number of free throws he made. How many free throws, two-point field goals, and three-point field goals did Kevin Durant make during the 2012–2013 NBA season? (*Source:* National Basketball Association)   free throws: 679; two-point field goals: 592; three-point field goals: 139

△**41.** Find the values of x, y, and z in the following triangle. $x = 60; y = 55; z = 65$

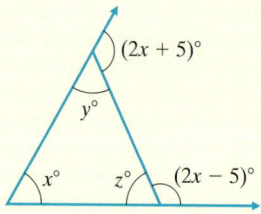

△**42.** The sum of the measures of the angles of a quadrilateral is 360°. Find the values of x, y, and z in the following quadrilateral. $x = 95; y = 123; z = 70$

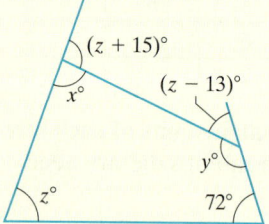

Review

Multiply both sides of equation (1) by 2, and add the resulting equation to equation (2). See Section 4.3.

43. $3x - y + z = 2$ (1)
   $-x + 2y + 3z = 6$ (2) $5x + 5z = 10$

44. $2x + y + 3z = 7$ (1)
   $-4x + y + 2z = 4$ (2) $3y + 8z = 18$

Multiply both sides of equation (1) by −3, and add the resulting equation to equation (2). See Section 4.3.

45. $x + 2y - z = 0$ (1)
   $3x + y - z = 2$ (2) $-5y + 2z = 2$

46. $2x - 3y + 2z = 5$ (1)
   $x - 9y + z = -1$ (2) $-5x - 5z = -16$

Concept Extensions

Solve.

47. Write a single linear equation in three variables that has $(-1, 2, -4)$ as a solution. (There are many possibilities.) Explain the process you used to write your equation.   answers may vary

48. When solving a system of three equations in three unknowns, explain how to determine that a system has no solution.   answers may vary

49. Write a system of linear equations in three variables that has the solution $(-1, 2, -4)$. (There are many possibilities.) Explain the process you used to write your system.   answers may vary

50. Write a system of three linear equations in three variables that has $(2, 1, 5)$ as a solution. (There are many possibilities.) Explain the process you used to write your system.   answers may vary

51. The fraction $\dfrac{1}{24}$ can be written as the following sum:

$$\frac{1}{24} = \frac{x}{8} + \frac{y}{4} + \frac{z}{3}$$

where the numbers $x, y,$ and z are solutions of

$$\begin{cases} x + y + z = 1 \\ 2x - y + z = 0 \\ -x + 2y + 2z = -1 \end{cases}$$

Solve the system and see that the sum of the fractions is $\dfrac{1}{24}$. $(1, 1, -1)$

52. The fraction $\dfrac{1}{18}$ can be written as the following sum:

$$\frac{1}{18} = \frac{x}{2} + \frac{y}{3} + \frac{z}{9}$$

where the numbers $x, y,$ and z are solutions of

$$\begin{cases} x + 3y + z = -3 \\ -x + y + 2z = -14 \\ 3x + 2y - z = 12 \end{cases}$$

Solve the system and see that the sum of the fractions is $\dfrac{1}{18}$. $(1, 1, -7)$

Solving systems involving more than three variables can be accomplished with methods similar to those encountered in this section. For Exercises 53 through 56, apply what you already know to solve each system of equations in four variables.

53. $\begin{cases} x + y \quad\quad - w = 0 \\ \quad\quad y + 2z + w = 3 \\ x \quad\quad - z \quad\quad = 1 \\ 2x - y \quad\quad - w = -1 \end{cases}$ $(1, 1, 0, 2)$

54. $\begin{cases} 5x + 4y \quad\quad\quad = 29 \\ \quad\quad y + z - w = -2 \\ 5x \quad\quad + z \quad\quad = 23 \\ \quad\quad y - z + w = 4 \end{cases}$ $(5, 1, -2, 1)$

55. $\begin{cases} x + y + z + w = 5 \\ 2x + y + z + w = 6 \\ x + y + z \quad\quad = 2 \\ x + y \quad\quad\quad = 0 \end{cases}$ $(1, -1, 2, 3)$

56. $\begin{cases} 2x \quad\quad - z \quad\quad = -1 \\ \quad\quad y + z + w = 9 \\ \quad\quad y \quad\quad - 2w = -6 \\ x + y \quad\quad\quad = 3 \end{cases}$ $(1, 2, 3, 4)$

Solve.

57. Write a system of three linear equations in three variables that are dependent equations.   answers may vary

58. What is the solution to the system in Exercise 57?   infinite number of solutions

59. The 2005 amendments to the bankruptcy laws of the United States made it more difficult for citizens to file for bankruptcy. Since this change in the law, the number of bankruptcy petitions has been decreasing. In 2010, the number of bankruptcy petitions was 622,317 less than twice the number of petitions filed in 2013. This is equivalent to a decrease of 485,382 petitions filed from 2010 to 2013. Find how many personal bankruptcy petitions were filed in each year. (*Source:* Based on data from the Administrative Office of the United States Courts)   2010: 1,593,081; 2013: 1,107,699

60. In 2012, the median weekly earnings for male registered nurses were $103 more than the median weekly earnings for female registered nurses. The median weekly earnings for female nurses were 0.913 times that of their male counterparts. Also in 2012, the median weekly earnings for female network and computer system administrators were $197 less than the median weekly earnings for male network and computer system administrators. The median weekly earnings for male network and computer system administrators were 1.186 times that of their female counterparts. (*Source:* Based on data from the Bureau of Labor Statistics, U.S. Department of Labor)

  a. Find the median weekly earnings for female registered nurses in the United States in 2012. (Round to the nearest dollar.) $1081

  b. Find the median weekly earnings for female network and computer administrators in the United States in 2012. (Round to the nearest dollar.) $1059

  c. Of the four groups of workers described in the problem, which group makes the greatest weekly earnings? Which group makes the least weekly earnings?   greatest: male network and computer administrators   least: female network and computer administrators

61. Find the values of a, b, and c such that the equation $y = ax^2 + bx + c$ has ordered pair solutions $(1, 6), (-1, -2)$, and $(0, -1)$. To do so, substitute each ordered pair solution into the equation. Each time, the result is an equation in three unknowns: a, b, and c. Then solve the resulting system of three linear equations in three unknowns, a, b, and c.
$a = 3, b = 4, c = -1$

62. Find the values of a, b, and c such that the equation $y = ax^2 + bx + c$ has ordered pair solutions $(1, 2), (2, 3)$, and $(-1, 6)$. (*Hint: See Exercise 61.*)
$a = 1, b = -2, c = 3$

63. Data (x, y) for the total number (in thousands) of college-bound students who took the ACT assessment in the year x are approximately $(1, 1179), (9, 1495)$ and $(12, 1666)$ where $x = 1$ represents 2001 and $x = 9$ represents 2009. Find the values a, b, and c such that the equation $y = ax^2 + bx + c$ models these data. According to your model, how many students will take the ACT in 2015? Round all answers to the nearest tenth. (*Source:* ACT, Inc.)
$a = 1.6; b = 23.6; c = 1153.8$
2015: 1867.8 thousand students

64. Monthly normal rainfall data (x, y) for Portland, Oregon, are $(4, 2.47), (7, 0.58), (8, 1.07)$, where x represents time in months (with $x = 1$ representing January) and y represents rainfall in inches. Find the values of a, b, and c such that the equation $y = ax^2 + bx + c$ models this data. According to your model, how much rain should Portland expect during September? (*Source:* National Climatic Data Center)
$a = 0.28, b = -3.71, c = 12.83; 2.12$ in. in Sept.

9.2 Solving Systems of Equations Using Matrices

By now, you may have noticed that the solution of a system of equations depends on the coefficients of the equations in the system and not on the variables. In this section, we introduce how to solve a system of equations using a **matrix.**

Objectives

A Use Matrices to Solve a System of Two Equations.

B Use Matrices to Solve a System of Three Equations.

Objective **A** Using Matrices to Solve a System of Two Equations

A matrix (plural: **matrices**) is a rectangular array of numbers. The following are examples of matrices.

$$\begin{bmatrix} 1 & 0 \\ 0 & 1 \end{bmatrix} \quad \begin{bmatrix} 2 & 1 & 3 & -1 \\ 0 & -1 & 4 & 5 \\ -6 & 2 & 1 & 0 \end{bmatrix} \quad \begin{bmatrix} a & b & c \\ d & e & f \end{bmatrix}$$

The numbers aligned horizontally in a matrix are in the same **row.** The numbers aligned vertically are in the same **column.**

Row 1 → $\begin{bmatrix} 2 & 1 & 0 \\ -1 & 6 & 2 \end{bmatrix}$ ← Row 2

Column 1   Column 2   Column 3

This matrix has 2 rows and 3 columns. It is called a 2×3 (read "two by three") matrix

To see the relationship between systems of equations and matrices, study the example below.

System of Equations
(in Standard Form)

$$\begin{cases} 2x - 3y = 6 & \text{Equation 1} \\ x + y = 0 & \text{Equation 2} \end{cases}$$

Corresponding Matrix

$$\begin{bmatrix} 2 & -3 & \vdots & 6 \\ 1 & 1 & \vdots & 0 \end{bmatrix} \begin{matrix} \text{Row 1} \\ \text{Row 2} \end{matrix}$$

Notice that the rows of the matrix correspond to the equations in the system. The coefficients of the variables are placed to the left of a vertical dashed line. The constants are placed to the right. Each of the numbers in the matrix is called an **element.**

The method of solving systems by matrices is to write this matrix as an equivalent matrix from which we can easily identify the solution. Two matrices are equivalent if they represent systems that have the same solution set. The following **row operations** can be performed on matrices, and the result is an equivalent matrix.

> ### Elementary Row Operations
>
> **1.** Any two rows in a matrix may be interchanged.
> **2.** The elements of any row may be multiplied (or divided) by the same nonzero number.
> **3.** The elements of any row may be multiplied (or divided) by a nonzero number and added to their corresponding elements in any other row.

Helpful Hint

Notice that these *row* operations are the same operations that we can perform on *equations* in a system.

Teaching Tip

If your classroom seating is arranged in rows and columns, let the seats in your classroom represent elements in a matrix and have students determine the dimensions of the matrix they represent.

Helpful Hint   Before writing the corresponding matrix associated with a system of equations, make sure that the equations are written in standard form.

Teaching Tip

Point out that matrices provide a shorthand notation for solving systems by elimination.

To solve a system of two equations in x and y by matrices, write the corresponding matrix associated with the system. Then use elementary row operations to write equivalent matrices until you have a matrix of the form

$$\begin{bmatrix} 1 & a & | & b \\ 0 & 1 & | & c \end{bmatrix},$$

where a, b, and c are constants. Why? If a matrix associated with a system of equations is in this form, we can easily solve for x and y. For example,

Matrix   **System of Equations**

$$\begin{bmatrix} 1 & 2 & | & -3 \\ 0 & 1 & | & 5 \end{bmatrix} \quad \text{corresponds to} \quad \begin{cases} 1x + 2y = -3 \\ 0x + 1y = 5 \end{cases} \quad \text{or} \quad \begin{cases} x + 2y = -3 \\ y = 5 \end{cases}$$

In the second equation, we have $y = 5$. Substituting this in the first equation, we have $x + 2(5) = -3$ or $x = -13$. The solution of the system is the ordered pair $(-13, 5)$.

Practice 1

Use matrices to solve the system:

$$\begin{cases} x + 2y = -4 \\ 2x - 3y = 13 \end{cases}$$

> **Example 1** Use matrices to solve the system:
>
> $$\begin{cases} x + 3y = 5 \\ 2x - y = -4 \end{cases}$$

Solution: The corresponding matrix is $\begin{bmatrix} 1 & 3 & | & 5 \\ 2 & -1 & | & -4 \end{bmatrix}$. We use elementary row

operations to write an equivalent matrix that looks like $\begin{bmatrix} 1 & a & | & b \\ 0 & 1 & | & c \end{bmatrix}$.

For the matrix given, the element in the first row, first column is already 1, as desired. Next we write an equivalent matrix with a 0 below the 1. To do this, we multiply row 1 by -2 and add to row 2. *We will change only row 2.*

$$\begin{bmatrix} 1 & 3 & | & 5 \\ -2(1) + 2 & -2(3) + (-1) & | & -2(5) + (-4) \end{bmatrix} \quad \text{simplifies to}$$

Row 1 Row 2   Row 1 Row 2   Row 1 Row 2
element element element element element element

$$\begin{bmatrix} 1 & 3 & | & 5 \\ 0 & -7 & | & -14 \end{bmatrix}$$

Now we change the -7 to a 1 by use of an elementary row operation. We divide row 2 by -7, then

$$\begin{bmatrix} 1 & 3 & | & 5 \\ \frac{0}{-7} & \frac{-7}{-7} & | & \frac{-14}{-7} \end{bmatrix} \quad \text{simplifies to} \quad \begin{bmatrix} 1 & 3 & | & 5 \\ 0 & 1 & | & 2 \end{bmatrix}$$

This last matrix corresponds to the system

$$\begin{cases} x + 3y = 5 \\ y = 2 \end{cases}$$

Thus we know that y is 2. To find x, we let $y = 2$ in the first equation, $x + 3y = 5$.

$$x + 3y = 5 \qquad \text{First equation}$$
$$x + 3(2) = 5 \qquad \text{Let } y = 2.$$
$$x = -1$$

The ordered pair solution is $(-1, 2)$. Check to see that this ordered pair satisfies both original equations.

◾ **Work Practice 1**

Answer

1. $(2, -3)$

Example 2   Use matrices to solve the system:

$$\begin{cases} 2x - y = 3 \\ 4x - 2y = 5 \end{cases}$$

Solution: The corresponding matrix is $\begin{bmatrix} 2 & -1 & \vdots & 3 \\ 4 & -2 & \vdots & 5 \end{bmatrix}$ To get 1 in the row 1, column 1 position, we divide the elements of row 1 by 2.

$$\begin{bmatrix} \dfrac{2}{2} & \dfrac{-1}{2} & \vdots & \dfrac{3}{2} \\ 4 & -2 & \vdots & 5 \end{bmatrix} \text{ simplifies to } \begin{bmatrix} 1 & -\dfrac{1}{2} & \vdots & \dfrac{3}{2} \\ 4 & -2 & \vdots & 5 \end{bmatrix}$$

To get 0 under the 1, we multiply the elements of row 1 by -4 and add the new elements to the elements of row 2.

$$\begin{bmatrix} 1 & -\dfrac{1}{2} & \vdots & \dfrac{3}{2} \\ -4(1) + 4 & -4\left(-\dfrac{1}{2}\right) - 2 & \vdots & -4\left(\dfrac{3}{2}\right) + 5 \end{bmatrix} \begin{array}{c} \text{simplifies} \\ \text{to} \end{array} \begin{bmatrix} 1 & -\dfrac{1}{2} & \vdots & \dfrac{3}{2} \\ 0 & 0 & \vdots & -1 \end{bmatrix}$$

The corresponding system is $\begin{cases} x - \dfrac{1}{2}y = \dfrac{3}{2} \\ \quad\quad\quad 0 = -1 \end{cases}$. The equation $0 = -1$ is false for

all y or x values; hence the system is inconsistent and has no solution. The solution set is $\varnothing$ or $\{\,\}$.

▪ **Work Practice 2**

✓**Concept Check** Consider the system

$$\begin{cases} 2x - 3y = 8 \\ x + 5y = -3 \end{cases}$$

What is wrong with its corresponding matrix shown below?

$$\begin{bmatrix} 2 & -3 & \vdots & 8 \\ 0 & 5 & \vdots & -3 \end{bmatrix}$$

Objective B Using Matrices to Solve a System of Three Equations ▶

To solve a system of three equations in three variables using matrices, we will write the corresponding matrix in the equivalent form

$$\begin{bmatrix} 1 & a & b & \vdots & d \\ 0 & 1 & c & \vdots & e \\ 0 & 0 & 1 & \vdots & f \end{bmatrix}$$

Example 3   Use matrices to solve the system:

$$\begin{cases} x + 2y + z = 2 \\ -2x - y + 2z = 5 \\ x + 3y - 2z = -8 \end{cases}$$

Solution: The corresponding matrix is $\begin{bmatrix} 1 & 2 & 1 & \vdots & 2 \\ -2 & -1 & 2 & \vdots & 5 \\ 1 & 3 & -2 & \vdots & -8 \end{bmatrix}$.

(Continued on next page)

Practice 2

Use matrices to solve the system:

$$\begin{cases} -3x + y = 0 \\ -6x + 2y = 2 \end{cases}$$

Teaching Tip

Have students use matrices to solve a system of equivalent equations such as

$$\begin{cases} 3x - 4y = 7 \\ -9x + 12y = -21 \end{cases}$$

Practice 3

Use matrices to solve the system:

$$\begin{cases} x + 3y + z = 5 \\ -3x + y - 3z = 5 \\ x + 2y - 2z = 9 \end{cases}$$

Answers

2. $\varnothing$ **3.** $(1, 2, -2)$

✓**Concept Check Answer**

matrix should be $\begin{bmatrix} 2 & -3 & \vdots & 8 \\ 1 & 5 & \vdots & -3 \end{bmatrix}$

Our goal is to write an equivalent matrix with 1s along the diagonal (see the numbers in red in the previous matrix) and 0s below the 1s. The element in row 1, column 1 is already 1. Next we get 0s for each element in the rest of column 1. To do this, first we multiply the elements of row 1 by 2 and add the new elements to row 2. Also, we multiply the elements of row 1 by -1 and add the new elements to the elements of row 3. *We do not change row 1.* Then

$$\begin{bmatrix} 1 & 2 & 1 & \vdots & 2 \\ 2(1)-2 & 2(2)-1 & 2(1)+2 & \vdots & 2(2)+5 \\ -1(1)+1 & -1(2)+3 & -1(1)-2 & \vdots & -1(2)-8 \end{bmatrix} \quad \text{simplifies to}$$

$$\begin{bmatrix} 1 & 2 & 1 & \vdots & 2 \\ 0 & 3 & 4 & \vdots & 9 \\ 0 & 1 & -3 & \vdots & -10 \end{bmatrix}$$

We continue down the diagonal and use elementary row operations to get 1 where the element 3 is now. To do this, we interchange rows 2 and 3.

$$\begin{bmatrix} 1 & 2 & 1 & \vdots & 2 \\ 0 & 3 & 4 & \vdots & 9 \\ 0 & 1 & -3 & \vdots & -10 \end{bmatrix} \quad \text{is equivalent to} \quad \begin{bmatrix} 1 & 2 & 1 & \vdots & 2 \\ 0 & 1 & -3 & \vdots & -10 \\ 0 & 3 & 4 & \vdots & 9 \end{bmatrix}$$

Next we want the new row 3, column 2 element to be 0. We multiply the elements of row 2 by -3 and add the result to the elements of row 3.

$$\begin{bmatrix} 1 & 2 & 1 & \vdots & 2 \\ 0 & 1 & -3 & \vdots & -10 \\ -3(0)+0 & -3(1)+3 & -3(-3)+4 & \vdots & -3(-10)+9 \end{bmatrix} \quad \text{simplifies to}$$

$$\begin{bmatrix} 1 & 2 & 1 & \vdots & 2 \\ 0 & 1 & -3 & \vdots & -10 \\ 0 & 0 & 13 & \vdots & 39 \end{bmatrix}$$

Finally, we divide the elements of row 3 by 13 so that the final diagonal element is 1.

$$\begin{bmatrix} 1 & 2 & 1 & \vdots & 2 \\ 0 & 1 & -3 & \vdots & -10 \\ \dfrac{0}{13} & \dfrac{0}{13} & \dfrac{13}{13} & \vdots & \dfrac{39}{13} \end{bmatrix} \quad \text{simplifies to} \quad \begin{bmatrix} 1 & 2 & 1 & \vdots & 2 \\ 0 & 1 & -3 & \vdots & -10 \\ 0 & 0 & 1 & \vdots & 3 \end{bmatrix}$$

This matrix corresponds to the system

$$\begin{cases} x + 2y + z = 2 \\ \quad\quad y - 3z = -10 \\ \quad\quad\quad\quad z = 3 \end{cases}$$

We identify the z-coordinate of the solution as 3. Next we replace z with 3 in the second equation and solve for y.

$$y - 3z = -10 \quad \text{Second equation}$$
$$y - 3(3) = -10 \quad \text{Let } z = 3.$$
$$y = -1$$

To find x, we let $z = 3$ and $y = -1$ in the first equation.

$$x + 2y + z = 2 \quad \text{First equation}$$
$$x + 2(-1) + 3 = 2 \quad \text{Let } z = 3 \text{ and } y = -1.$$
$$x = 1$$

The ordered triple solution is $(1, -1, 3)$. Check to see that it satisfies all three equations in the original system.

Work Practice 3

Vocabulary, Readiness & Video Check

Use the choices below to fill in each blank.

column   element   row   matrix

1. A(n) _____matrix_____ is a rectangular array of numbers.
2. Each of the numbers in a matrix is called a(n) _____element_____.
3. The numbers aligned horizontally in a matrix are in the same _____row_____.
4. The numbers aligned vertically in a matrix are in the same _____column_____.

Answer true or false for each statement about operations within a matrix forming an equivalent matrix.

5. Any two columns may be interchanged. _____false_____
6. Any two rows may be interchanged. _____true_____
7. The elements in a row may be added to their corresponding elements in another row. _____true_____
8. The elements of a column may be multiplied by any nonzero number. _____false_____

Martin-Gay Interactive Videos

See Video 9.2

See video answer section.

Watch the section lecture video and answer the following questions.

Objective A   9. From the lecture before ▤ Example 1, what elementary row operations can be performed on matrices? Which operation is not performed during Example 1? ▶

Objective B   10. In ▤ Example 2, why do you think the suggestion is made to write neatly when using matrices to solve systems? ▶

9.2   Exercise Set   MyMathLab® ▶

Objective A   *Use matrices to solve each system of linear equations. See Example 1.*

1. $\begin{cases} x + y = 1 \\ x - 2y = 4 \end{cases}$
$(2, -1)$

2. $\begin{cases} 2x - y = 8 \\ x + 3y = 11 \end{cases}$
$(5, 2)$

▶ 3. $\begin{cases} x + 3y = 2 \\ x + 2y = 0 \end{cases}$
$(-4, 2)$

4. $\begin{cases} 4x - y = 5 \\ 3x - 3y = 6 \end{cases}$
$(1, -1)$

Use matrices to solve each system of linear equations. See Example 2.

5. $\begin{cases} x - 2y = 4 \\ 2x - 4y = 4 \end{cases}$
$\varnothing$

6. $\begin{cases} -x + 3y = 6 \\ 3x - 9y = 9 \end{cases}$
$\varnothing$

7. $\begin{cases} 3x - 3y = 9 \\ 2x - 2y = 6 \end{cases}$
$\{(x, y) \mid 3x - 3y = 9\}$

8. $\begin{cases} 9x - 3y = 6 \\ -18x + 6y = -12 \end{cases}$
$\{(x, y) \mid 9x - 3y = 6\}$

Objective B   *Use matrices to solve each system of linear equations. See Example 3.*

9. $\begin{cases} x + y = 3 \\ 2y = 10 \\ 3x + 2y - 4z = 12 \end{cases}$
$(-2, 5, -2)$

10. $\begin{cases} 5x = 5 \\ 2x + y = 4 \\ 3x + y - 5z = -15 \end{cases}$
$(1, 2, 4)$

11. $\begin{cases} 2y - z = -7 \\ x + 4y + z = -4 \\ 5x - y + 2z = 13 \end{cases}$
$(1, -2, 3)$

12. $\begin{cases} 4y + 3z = -2 \\ 5x - 4y = 1 \\ -5x + 4y + z = -3 \end{cases}$
$(1, 1, -2)$

Objectives A B Mixed Practice *Solve each system of linear equations using matrices. See Examples 1 through 3.*

13. $\begin{cases} x - 4 = 0 \\ x + y = 1 \end{cases}$
$(4, -3)$

14. $\begin{cases} 3y = 6 \\ x + y = 7 \end{cases}$
$(5, 2)$

15. $\begin{cases} x + y + z = 2 \\ 2x \quad\;\; - z = 5 \\ 3y + z = 2 \end{cases}$
$(2, 1, -1)$

16. $\begin{cases} x + 2y + z = 5 \\ x - y - z = 3 \\ y + z = 2 \end{cases}$
$(5, -2, 4)$

17. $\begin{cases} 5x - 2y = 27 \\ -3x + 5y = 18 \end{cases}$
$(9, 9)$

18. $\begin{cases} 4x - y = 9 \\ 2x + 3y = -27 \end{cases}$
$(0, -9)$

19. $\begin{cases} 4x - 7y = 7 \\ 12x - 21y = 24 \end{cases}$
$\varnothing$

20. $\begin{cases} 2x - 5y = 12 \\ -4x + 10y = 20 \end{cases}$
$\varnothing$

21. $\begin{cases} 4x - y + 2z = 5 \\ 2y + z = 4 \\ 4x + y + 3z = 10 \end{cases}$
$\varnothing$

22. $\begin{cases} 5y - 7z = 14 \\ 2x + y + 4z = 10 \\ 2x + 6y - 3z = 30 \end{cases}$
$\varnothing$

▶ **23.** $\begin{cases} 4x + y + z = 3 \\ -x + y - 2z = -11 \\ x + 2y + 2z = -1 \end{cases}$
$(1, -4, 3)$

24. $\begin{cases} x + y + z = 9 \\ 3x - y + z = -1 \\ -2x + 2y - 3z = -2 \end{cases}$
$(0, 5, 4)$

Review

Determine whether each graph is the graph of a function. See Section 8.2.

25.
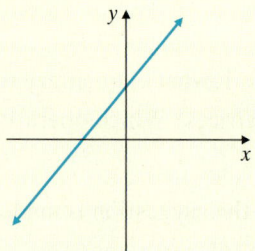
function

26.
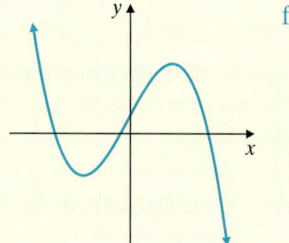
function

27.
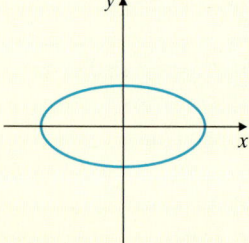
not a function

28.
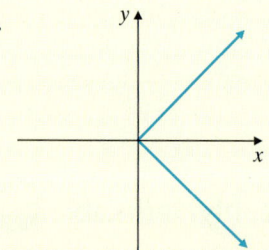
not a function

Concept Extensions

Solve. For Exercises 29 and 30, see the Concept Check in this section.

29. For the system $\begin{cases} x \quad\;\; + z = 7 \\ y + 2z = -6, \\ 3x - y \quad\;\; = 0 \end{cases}$ which is the correct corresponding matrix?

a. $\begin{bmatrix} 1 & 1 & \vdots & 7 \\ 1 & 2 & \vdots & -6 \\ 3 & -1 & \vdots & 0 \end{bmatrix}$

b. $\begin{bmatrix} 1 & 0 & 1 & \vdots & 7 \\ 1 & 2 & 0 & \vdots & -6 \\ 3 & -1 & 0 & \vdots & 0 \end{bmatrix}$

c. $\begin{bmatrix} 1 & 0 & 1 & \vdots & 7 \\ 0 & 1 & 2 & \vdots & -6 \\ 3 & -1 & 0 & \vdots & 0 \end{bmatrix}$ c

30. For the system $\begin{cases} x - 6 = 0 \\ 2x - 3y = 1 \end{cases}$, which is the correct corresponding matrix?

a. $\begin{bmatrix} 1 & -6 & \vdots & 0 \\ 2 & -3 & \vdots & 1 \end{bmatrix}$

b. $\begin{bmatrix} 1 & 0 & \vdots & 6 \\ 2 & -3 & \vdots & 1 \end{bmatrix}$

c. $\begin{bmatrix} 1 & 0 & \vdots & -6 \\ 2 & -3 & \vdots & 1 \end{bmatrix}$ b

31. The amount of electricity generated by geothermal sources (in billions of kilowatts) from 2010 to 2013 can be modeled by the linear equation $y - 0.86x = 15.22$, where x represents the number of years since 2010. Similarly, the amount of electricity y generated by wind power (in billions of kilowatts) during the same time period can be modeled by the linear equation $y = 21.83x + 94.62$. (*Source:* Based on data from the Energy Information Administration, U.S. Department of Energy)

    a. The data used to form these two models were incomplete. It is impossible to tell from the data the year in which the electricity generated by geothermal sources was the same as the electricity generated by wind power. Use matrix methods to estimate the year in which this occurred. 2006

    b. The earliest data for wind power was in 1989, where 2.1 billion kilowatts of electricity was generated. Can this data be determined from the given equation? Why do you think this is? no; answers may vary

    c. According to these models, will the amount of electricity generated by geothermal ever be zero? Why? no; it has a positive slope

    d. Can you think of an explanation why the amount of electricity generated by wind power is increasing so much faster than the amount of electricity generated by geothermal power? answers may vary

32. The most popular amusement park in the world (according to annual attendance) is Walt Disney World's Magic Kingdom, whose yearly attendance in thousands can be approximated by the equation $y = 178x + 15{,}400$, where x is the number of years after 2000. The park that held the distinction of most popular until just recently is Tokyo Disneyland, whose yearly attendance, in thousands, can be approximated by the equation $y = -138x + 16{,}500$. Find the last year when attendance at Tokyo Disneyland was greater than attendance at Walt Disney World's Magic Kingdom. (*Source:* Themed Entertainment Association) 2003

33. For the system $\begin{cases} 2x - 3y = 8 \\ x + 5y = -3 \end{cases}$, explain what is wrong with writing the corresponding matrix as $\begin{bmatrix} 2 & 3 & | & 8 \\ 0 & 5 & | & -3 \end{bmatrix}$.

answers may vary

34. For the system $\begin{cases} 5x + 2y = 0 \\ -y = 2 \end{cases}$, explain what is wrong with writing the corresponding matrix as $\begin{bmatrix} 5 & 2 & | & 0 \\ -1 & 0 & | & 2 \end{bmatrix}$.

answers may vary

Systems of Linear Equations

Answers

The graphs of various systems of equations are shown. Match each graph with the solution of its corresponding system.

1. C

2. D

3. A

4. B

A 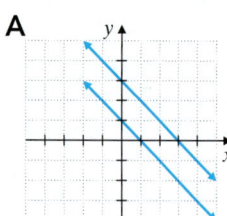   **B** 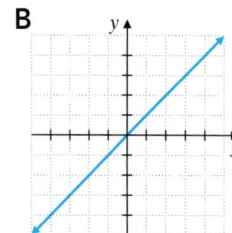   **C** 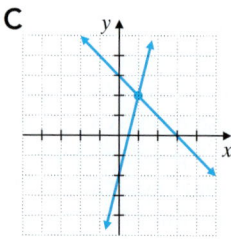   **D**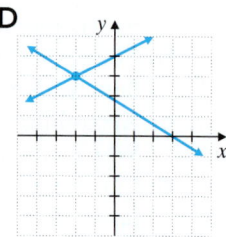

1. Solution: $(1, 2)$ **2.** Solution: $(-2, 3)$

3. No solution   **4.** Infinite number of solutions

5. $(2, -1)$

6. $(5, 2)$

7. $\varnothing$

8. $\{(x, y) \mid 2x - 5y = 3\}$

9. $(-1, 3, 2)$

10. $(1, -3, 0)$

11. $\varnothing$

12. $\{(x, y, z) \mid x - y + 3z = 2\}$

13. $\left(2, 5, \dfrac{1}{2}\right)$

14. $\left(1, 1, \dfrac{1}{3}\right)$

15. 70°; 70°; 100°; 120°

Solve each system by using matrices.

5. $\begin{cases} x + y = 1 \\ x - 2y = 4 \end{cases}$ **6.** $\begin{cases} 2x - y = 8 \\ x + 3y = 11 \end{cases}$

7. $\begin{cases} 4x - 7y = 7 \\ 12x - 21y = 24 \end{cases}$ **8.** $\begin{cases} 2x - 5y = 3 \\ -4x + 10y = -6 \end{cases}$

Solve each system.

9. $\begin{cases} x + y = 2 \\ -3y + z = -7 \\ 2x + y - z = -1 \end{cases}$ **10.** $\begin{cases} y + 2z = -3 \\ x - 2y = 7 \\ 2x - y + z = 5 \end{cases}$ **11.** $\begin{cases} 2x + 4y - 6z = 3 \\ -x + y - z = 6 \\ x + 2y - 3z = 1 \end{cases}$

12. $\begin{cases} x - y + 3z = 2 \\ -2x + 2y - 6z = -4 \\ 3x - 3y + 9z = 6 \end{cases}$ **13.** $\begin{cases} x + y - 4z = 5 \\ x - y + 2z = -2 \\ 3x + 2y + 4z = 18 \end{cases}$ **14.** $\begin{cases} 2x - y + 3z = 2 \\ x + y - 6z = 0 \\ 3x + 4y - 3z = 6 \end{cases}$

15. The sum of the measures of the angles of a quadrilateral is 360°. The two smallest angles of the quadrilateral shown have the same measure. The third angle measures 30° more than the measure of one of the smallest angles, and the fourth angle measures 50° more than the measure of one of the smallest angles. Find the measure of each angle.

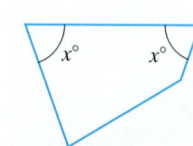

9.3 Systems of Linear Inequalities

Objective A Graphing Systems of Linear Inequalities

In Section 3.6, we graphed linear inequalities in two variables. Just as two linear equations make a system of linear equations, two linear inequalities make a **system of linear inequalities.** Systems of inequalities are very important in a process called linear programming. Many businesses use linear programming to find the most profitable way to use limited resources such as employees, machines, or buildings.

A **solution of a system of linear inequalities** is an ordered pair that satisfies each inequality in the system. The set of all such ordered pairs is the solution set of the system. Graphing this set gives us a picture of the solution set. We can graph a system of inequalities by graphing each inequality in the system and identifying the region of overlap.

> **Graphing the Solutions of a System of Linear Inequalities**
>
> **Step 1:** Graph each inequality in the system on the same set of axes.
>
> **Step 2:** The solutions of the system are the points common to the graphs of all the inequalities in the system.

Example 1 Graph the solutions of the system: $\begin{cases} 3x \geq y \\ x + 2y \leq 8 \end{cases}$

Solution: We begin by graphing each inequality on the *same* set of axes. The graph of the solutions of the system is the region contained in the graphs of both inequalities. In other words, it is their intersection.

First let's graph $3x \geq y$. The boundary line is the graph of $3x = y$. We sketch a solid boundary line since the inequality $3x \geq y$ means $3x > y$ or $3x = y$. The test point $(1, 0)$ satisfies the inequality, so we shade the half-plane that includes $(1, 0)$.

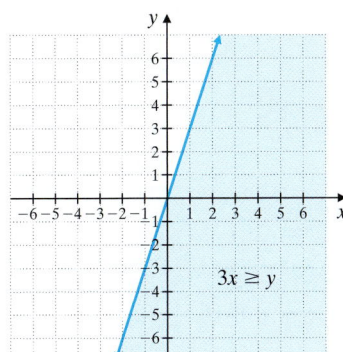

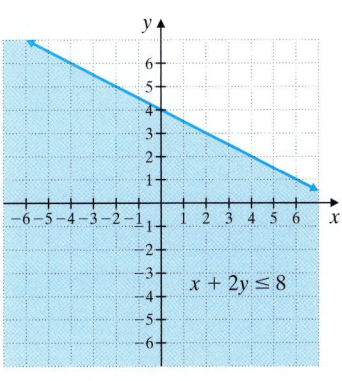

Next we sketch a solid boundary line $x + 2y = 8$ on the same set of axes. The test point $(0, 0)$ satisfies the inequality $x + 2y \leq 8$, so we shade the half-plane that includes $(0, 0)$. (For clarity, the graph of $x + 2y \leq 8$ is shown here on a separate set of axes.) An ordered pair solution of the system must satisfy both inequalities.

(Continued on next page)

(Continued on next page)

Objective

A Graph a System of Linear Inequalities.

Teaching Tip

Suggest that students graph each inequality in a different color.

Teaching Tip

After reviewing Example 1, ask, "When graphing the solution of a system of inequalities, are you graphing the union or intersection of the inequalities?"

Practice 1

Graph the solutions of the system:

$$\begin{cases} 2x \leq y \\ x + 4y \geq 4 \end{cases}$$

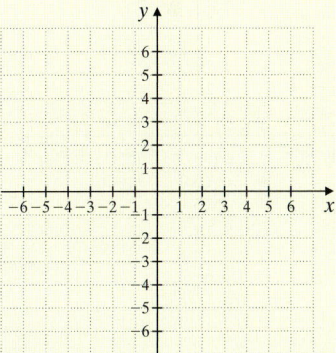

Answer

1.

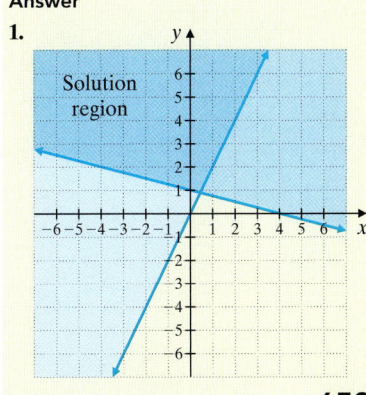

These solutions are points that lie in both shaded regions. The solution of the system is the darkest shaded region. This solution includes parts of both boundary lines.

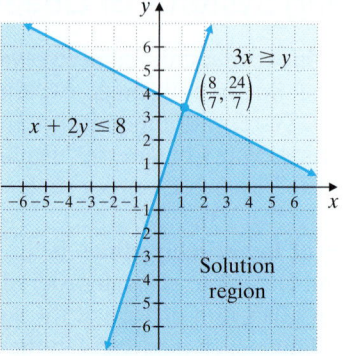

● **Work Practice 1**

In linear programming, it is sometimes necessary to find the coordinates of the **corner point:** the point at which the two boundary lines intersect. To find the corner point for the system of Example 1, we solve the related linear system

$$\begin{cases} 3x = y \\ x + 2y = 8 \end{cases}$$

using either the substitution or the elimination method. The lines intersect at $\left(\dfrac{8}{7}, \dfrac{24}{7}\right)$, the corner point of the graph.

Practice 2

Graph the solutions of the system:

$$\begin{cases} -x + y < 3 \\ y < 1 \\ 2x + y > -2 \end{cases}$$

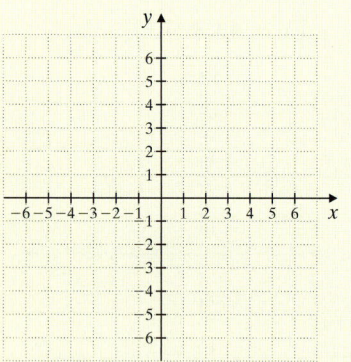

Example 2 Graph the solutions of the system:

$$\begin{cases} x - y < 2 \\ x + 2y > -1 \\ y < 2 \end{cases}$$

Solution: First we graph all three inequalities on the same set of axes. All boundary lines are dashed lines since the inequality symbols are $<$ and $>$. The solution of the system is the region shown by the shading. In this example, the boundary lines are *not* a part of the solution.

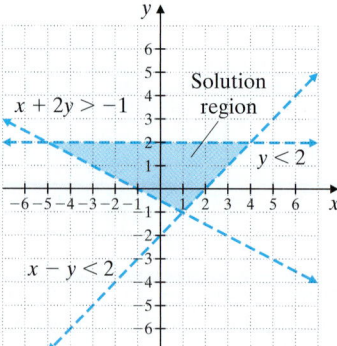

Teaching Tip:
Classroom Activity

End this section by providing students with a shaded region on graph paper and having them work in pairs to determine the system of linear inequalities that is graphed.

Answer

2.

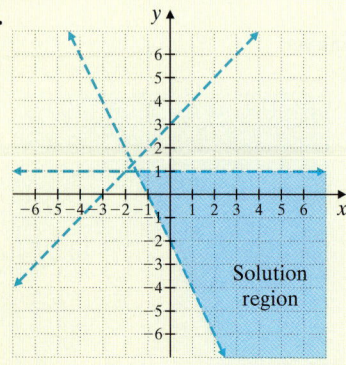

● **Work Practice 2**

✓**Concept Check** Describe the solution of the system of inequalities:

$$\begin{cases} x \le 2 \\ x \ge 2 \end{cases}$$

Example 3 Graph the solutions of the system:
$$\begin{cases} -3x + 4y \le 12 \\ x \le 3 \\ x \ge 0 \\ y \ge 0 \end{cases}$$

Solution: We graph the inequalities on the same set of axes. The intersection of the inequalities is the solution region. It is the only region shaded in this graph and includes the portions of all four boundary lines that border the shaded region.

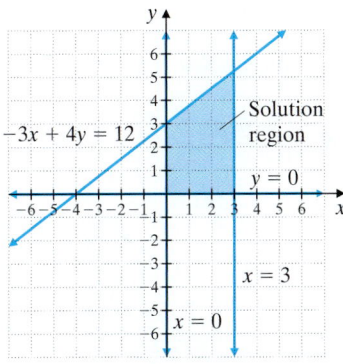

■ **Work Practice 3**

Practice 3
Graph the solutions of the system:

$$\begin{cases} 2x - 3y \le 6 \\ y \ge 0 \\ y \le 4 \\ x \ge 0 \end{cases}$$

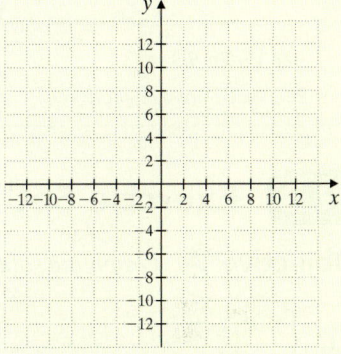

Answer

3.

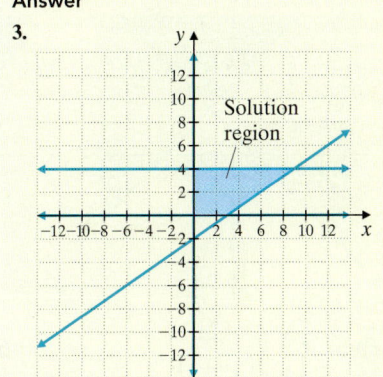

✓**Concept Check Answer**
the line $x = 2$

Vocabulary, Readiness & Video Check

Use the choices below to fill in each blank. Not all choices will be used.

solution	union	system
corner	intersection	

1. Two or more linear inequalities form a(n) ____system____ of linear inequalities.
2. An ordered pair that satisfies each inequality in a system is a(n) ____solution____ of the system.
3. The point where two boundary lines intersect is a(n) ____corner____ point.
4. The solution region of a system of inequalities consists of the ____intersection____ of the solution regions of the inequalities in the system.

Martin-Gay Interactive Videos   *Watch the section lecture video and answer the following question.*

Objective A **5.** In ▶ Example 1, do the solutions of the first inequality of the system limit where we can choose the test point for the second inequality? Why or why not?

See Video 9.3 🍒

9.3 Exercise Set MyMathLab®

Objective A *Graph the solutions of each system of linear inequalities. See Examples 1 through 3.*

1. $\begin{cases} y \geq x + 1 \\ y \geq 3 - x \end{cases}$

2. $\begin{cases} y \geq x - 3 \\ y \geq -1 - x \end{cases}$

3. $\begin{cases} y < 3x - 4 \\ y \leq x + 2 \end{cases}$

4. $\begin{cases} y \leq 2x + 1 \\ y > x + 2 \end{cases}$

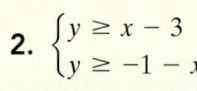
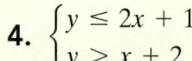

▶ 5. $\begin{cases} y < -2x - 2 \\ y > x + 4 \end{cases}$

6. $\begin{cases} y \leq 2x + 4 \\ y \geq -x - 5 \end{cases}$

7. $\begin{cases} y \geq -x + 2 \\ y \leq 2x + 5 \end{cases}$

8. $\begin{cases} y \geq x - 5 \\ y \leq -3x + 3 \end{cases}$

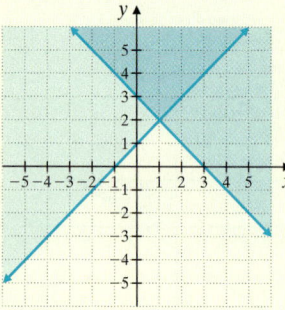
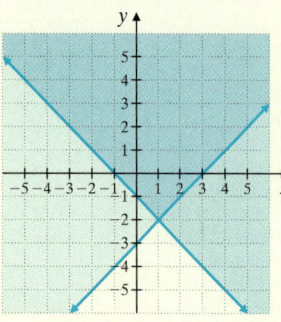
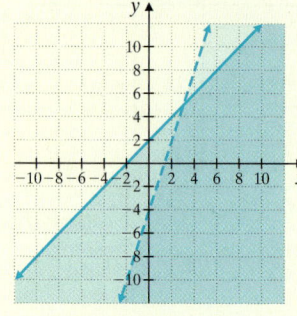
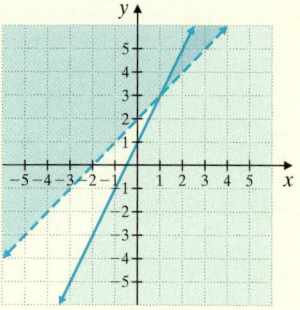

9. $\begin{cases} x \geq 3y \\ x + 3y \leq 6 \end{cases}$

10. $\begin{cases} -2x < y \\ x + 2y < 3 \end{cases}$

11. $\begin{cases} x \leq 2 \\ y \geq -3 \end{cases}$

12. $\begin{cases} x \geq -3 \\ y \geq -2 \end{cases}$

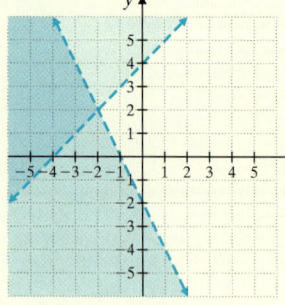
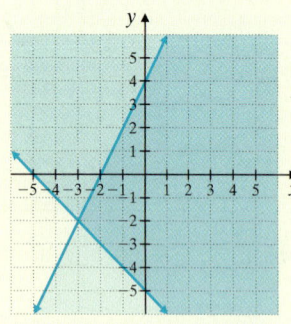
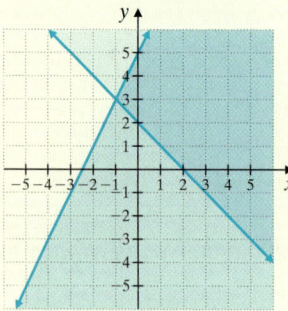
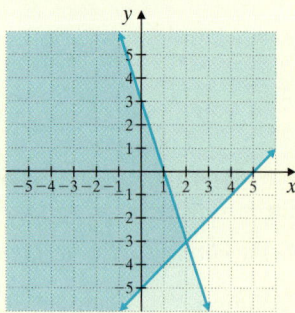

13. $\begin{cases} y \geq 1 \\ x < -3 \end{cases}$

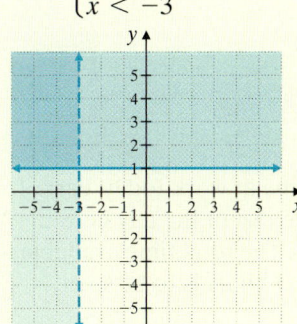

14. $\begin{cases} y > 2 \\ x \geq -1 \end{cases}$

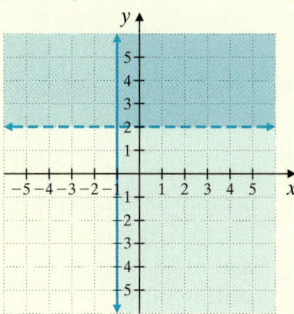

15. $\begin{cases} y + 2x \geq 0 \\ 5x - 3y \leq 12 \\ y \leq 2 \end{cases}$

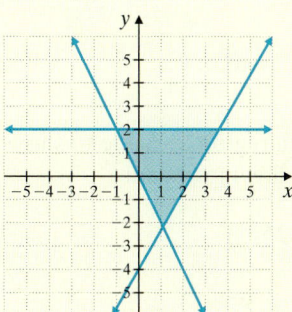

16. $\begin{cases} y + 2x \leq 0 \\ 5x + 3y \geq -2 \\ y \leq 4 \end{cases}$

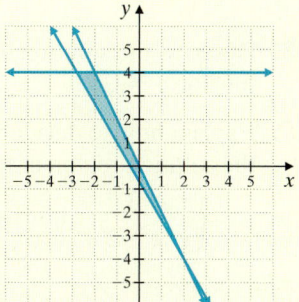

17. $\begin{cases} 3x - 4y \geq -6 \\ 2x + y \leq 7 \\ y \geq -3 \end{cases}$

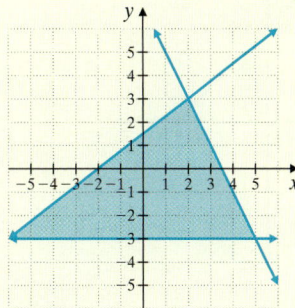

18. $\begin{cases} 4x - y \geq -2 \\ 2x + 3y \leq -8 \\ y \geq -5 \end{cases}$

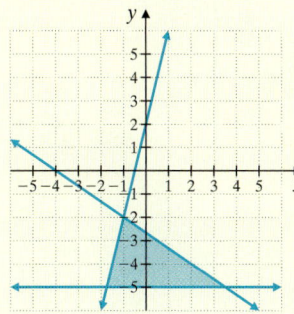

19. $\begin{cases} 2x + y \leq 5 \\ x \leq 3 \\ x \geq 0 \\ y \geq 0 \end{cases}$

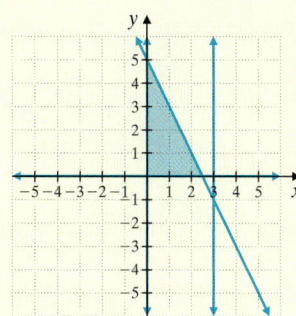

20. $\begin{cases} 3x + y \leq 4 \\ x \leq 4 \\ x \geq 0 \\ y \geq 0 \end{cases}$

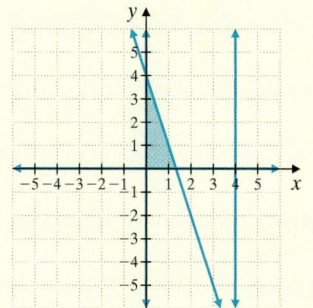

Match each system of inequalities to the corresponding graph.

A

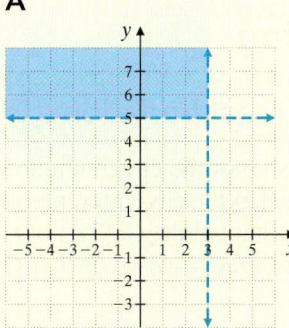

B

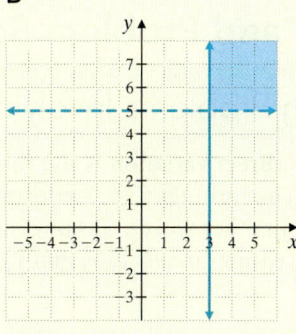

C

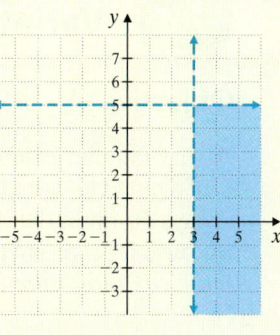

D

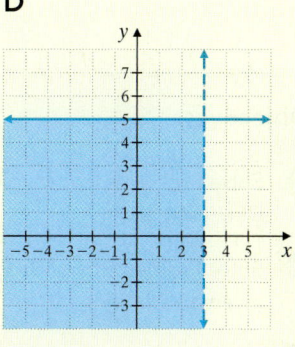

21. $\begin{cases} y < 5 \\ x > 3 \end{cases}$ C

22. $\begin{cases} y > 5 \\ x < 3 \end{cases}$ A

23. $\begin{cases} y \leq 5 \\ x < 3 \end{cases}$ D

24. $\begin{cases} y > 5 \\ x \geq 3 \end{cases}$ B

Review

Evaluate each expression. See Section 1.6.

25. $(-3)^2$ 9

26. $(-5)^3$ -125

27. $\left(\dfrac{2}{3}\right)^2$ $\dfrac{4}{9}$

28. $\left(\dfrac{3}{4}\right)^3$ $\dfrac{27}{64}$

Perform each indicated operation. See Sections 1.3 and 1.6.

29. $(-2)^2 - (-3) + 2(-1)$
5

30. $5^2 - 11 + 3(-5)$
-1

31. $8^2 + (-13) - 4(-2)$
59

32. $(-12)^2 + (-1)(2) - 6$
136

Concept Extensions

Solve. For Exercises 33 and 34, see the Concept Check in this section.

33. Describe the solution of the system: $\begin{cases} y \le 3 \\ y \ge 3 \end{cases}$

the line $y = 3$

34. Describe the solution of the system: $\begin{cases} x \le 5 \\ x \le 3 \end{cases}$

the region described by $x \le 3$

35. Explain how to decide which region to shade to show the solution region of the following system.

$\begin{cases} x \ge 3 \\ y \ge -2 \end{cases}$ answers may vary

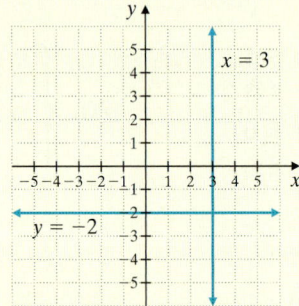

36. Tony Noellert budgets his time at work today. Part of the day he can write bills; the rest of the day he can use to write purchase orders. The total time available is at most 8 hours. Less than 3 hours is to be spent writing bills.

  a. Write a system of inequalities to describe the situation. (Let x = hours available for writing bills and y = hours available for writing purchase orders.)

  b. Graph the solutions of the system.

a. $\begin{cases} x + y \le 8 \\ x < 3 \\ x \ge 0 \\ y \ge 0 \end{cases}$

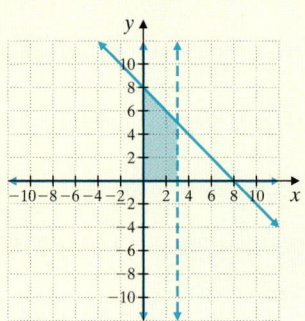

9.4   **Variation and Problem Solving**

Objectives

A Solve Problems Involving Direct Variation.

B Solve Problems Involving Inverse Variation.

C Solve Problems Involving Joint Variation.

D Solve Problems Involving Combined Variation.

Objective A Solving Problems Involving Direct Variation

A very familiar example of **direct variation** is the relationship of the circumference C of a circle to its radius r. The formula $C = 2\pi r$ expresses that the circumference is always 2π times the radius. In other words, C is always a constant multiple (2π) of r. Because it is, we say that *C varies directly as r*, that *C varies directly with r*, or that *C is directly proportional to r*.

$C = 2\pi r$

constant

> ### Direct Variation
>
> **y varies directly as x,** or **y is directly proportional to x,** if there is a nonzero constant k such that
>
> $$y = kx$$
>
> The number k is called the **constant of variation** or the **constant of proportionality.**

In the definition on the previous page, the relationship described between x and y is a linear one. In other words, the graph of $y = kx$ is a line. The slope of the line is k, and the line passes through the origin.

For example, the graph of the direct variation equation $C = 2\pi r$ is shown. The horizontal axis represents the radius r, and the vertical axis is the circumference C. From the graph we can read that when the radius is 6 units, the circumference is approximately 38 units. Also, when the circumference is 45 units, the radius is between 7 and 8 units. Notice that as the radius increases, the circumference increases.

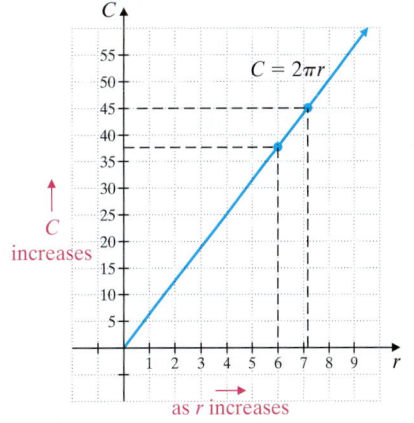

Teaching Tip

After discussing Example 1, have students make a table of ordered pairs that satisfy the direct variation equation. For each ordered pair, ask them to find the ratio, $\frac{y}{x}$, and the product, xy. Ask, "What patterns, if any, are present? Is the ratio defined for all possible ordered pairs?"

Example 1 Suppose that y varies directly as x. If y is 5 when x is 30, find the constant of variation and the direct variation equation.

Solution: Since y varies directly as x, we write $y = kx$. If $y = 5$ when $x = 30$, we have that

$$y = kx$$
$$5 = k(30) \quad \text{Replace } y \text{ with 5 and } x \text{ with 30.}$$
$$\frac{1}{6} = k \quad \text{Solve for } k.$$

The constant of variation is $\frac{1}{6}$.

After finding the constant of variation k, the direct variation equation can be written as $y = \frac{1}{6}x$.

■ **Work Practice 1**

Practice 1

Suppose that y varies directly as x. If y is 24 when x is 8, find the constant of variation and the direct variation equation.

Example 2 Using Direct Variation and Hooke's Law

Hooke's law states that the distance a spring stretches is directly proportional to the weight attached to the spring. If a 40-pound weight attached to the spring stretches the spring 5 inches, find the distance that a 65-pound weight attached to the spring stretches the spring.

Solution:

1. UNDERSTAND. Read and reread the problem. Notice that we are given that the distance a spring stretches is *directly proportional* to the weight attached. We let

   d = the distance stretched.
   w = the weight attached

The constant of variation is represented by k.

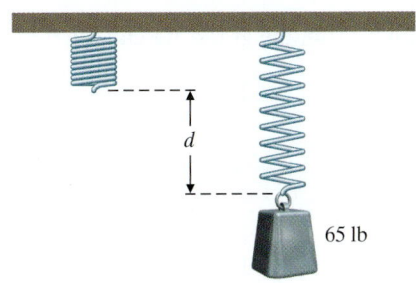

65 lb

(*Continued on next page*)

Practice 2

Use Hooke's law as stated in Example 2. If a 56-pound weight attached to a spring stretches the spring 8 inches, find the distance that an 85-pound weight attached to the spring stretches the spring.

Answers

1. $k = 3; y = 3x$ **2.** $12\frac{1}{7}$ in.

2. **TRANSLATE.** Because d is directly proportional to w, we write

$$d = kw$$

3. **SOLVE.** When a weight of 40 pounds is attached, the spring stretches 5 inches. That is, when $w = 40, d = 5$.

$$5 = k(40) \quad \text{Replace } d \text{ with 5 and } w \text{ with 40.}$$

$$\frac{1}{8} = k \qquad \text{Solve for } k.$$

Now when we replace k with $\frac{1}{8}$ in the equation $d = kw$, we have

$$d = \frac{1}{8}w$$

To find the stretch when a weight of 65 pounds is attached, we replace w with 65 to find d.

$$d = \frac{1}{8}(65)$$

$$= \frac{65}{8} = 8\frac{1}{8} \text{ or } 8.125$$

4. **INTERPRET.**

Check: Check the proposed solution of 8.125 inches in the original problem.

State: The spring stretches 8.125 inches when a 65-pound weight is attached.

■ **Work Practice 2**

Objective B Solving Problems Involving Inverse Variation ▶

When y is proportional to the *reciprocal* of another variable x, we say that y *varies inversely as* x, or that y *is inversely proportional to* x. An example of the **inverse variation** relationship is the relationship between the pressure that a gas exerts and the volume of its container. As the volume of a container decreases, the pressure of the gas it contains increases.

> ### Inverse Variation
>
> y **varies inversely as** x, or y **is inversely proportional to** x, if there is a nonzero constant k such that
>
> $$y = \frac{k}{x}$$
>
> The number k is called the **constant of variation** or the **constant of proportionality.**

Notice that $y = \dfrac{k}{x}$ is an equation containing a rational expression. Its graph for $k > 0$ and $x > 0$ is shown. From the graph, we can see that as x increases, y decreases.

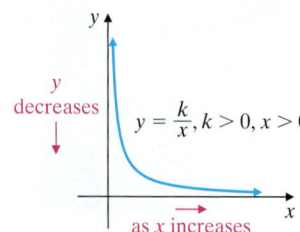

Copyright 2016 Pearson Education, Inc.

Teaching Tip

Before students do Practice 4, discuss the problem in general using a graph. Have students discuss the relationship between time and speed. Then ask, "Is there any speed at which it will take us 0 seconds to get to our destination? What does this tell us about our graph? How long will it take to get to our destination if we travel at 0 mph? What does this tell us about our graph?"

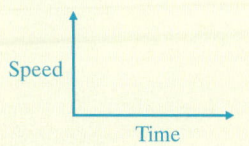

Example 3 Suppose that u varies inversely as w. If u is 3 when w is 5, find the constant of variation and the inverse variation equation.

Solution: Since u varies inversely as w, we have $u = \dfrac{k}{w}$. We let $u = 3$ and $w = 5$, and we solve for k.

$$u = \frac{k}{w}$$

$$3 = \frac{k}{5} \quad \text{Let } u = 3 \text{ and } w = 5.$$

$$15 = k \quad \text{Multiply both sides by 5.}$$

The constant of variation k is 15. This gives the inverse variation equation

$$u = \frac{15}{w}$$

■ **Work Practice 3**

Example 4 Using Inverse Variation and Boyle's Law

Boyle's law says that if the temperature stays the same, the pressure P of a gas is inversely proportional to the volume V. If a cylinder in a steam engine has a pressure of 960 kilopascals when the volume is 1.4 cubic meters, find the pressure when the volume increases to 2.5 cubic meters.

Solution:

1. UNDERSTAND. Read and reread the problem. Notice that we are given that the pressure of a gas is *inversely proportional* to the volume. We will let $P =$ the pressure and $V =$ the volume. The constant of variation is represented by k.

2. TRANSLATE. Because P is inversely proportional to V, we write

$$P = \frac{k}{V}.$$

When $P = 960$ kilopascals, the volume $V = 1.4$ cubic meters. We use this information to find k.

$$960 = \frac{k}{1.4} \quad \text{Let } P = 960 \text{ and } V = 1.4.$$

$$1344 = k \quad \text{Multiply both sides by 1.4.}$$

Thus, the value of k is 1344. Replacing k with 1344 in the variation equation, we have

$$P = \frac{1344}{V}$$

Next we find P when V is 2.5 cubic meters.

(Continued on next page)

Practice 3

Suppose that y varies inversely as x. If y is 6 when x is 3, find the constant of variation and the inverse variation equation.

Teaching Tip

After discussing Example 3, have students make a table of ordered pairs that satisfy the inverse variation equation. For each ordered pair, ask them to find the ratio, $\dfrac{u}{w}$, and the product uw. Ask, "What patterns, if any, are present? Is the ratio defined for all possible ordered pairs?"

Practice 4

The speed r at which one needs to drive in order to travel a constant distance is inversely proportional to the time t. A fixed distance can be driven in 5 hours at a rate of 24 mph. Find the rate needed to drive the same distance in 4 hours.

Answers

3. $k = 18; y = \dfrac{18}{x}$ **4.** 30 mph

Copyright 2016 Pearson Education, Inc.

Teaching Tip:
Classroom Activity

Have students work in pairs to determine whether each of the following sets of ordered pairs represents direct variation, inverse variation, or neither.

a.

x	2	5.5	8	11	19
y	4.5	3.625	3	2.25	0.25

b.

x	0.5	2	8	15	120
y	720	180	45	24	3

c.

x	1	3	7	16	42
y	0.75	2.25	5.25	12	31.5

3. SOLVE.

$$P = \frac{1344}{2.5} \qquad \text{Let } V = 2.5$$

$$= 537.6$$

4. INTERPRET. *Check* the proposed solution in the original problem.

State: When the volume is 2.5 cubic meters, the pressure is 537.6 kilopascals.

■ **Work Practice 4**

Objective C Solving Problems Involving Joint Variation ▶

Sometimes the ratio of a variable to the product of many other variables is constant. For example, the ratio of distance traveled to the product of speed and time traveled is constantly 1:

$$\frac{d}{rt} = 1 \qquad \text{or} \qquad d = rt$$

Such a relationship is called **joint variation.**

> ### Joint Variation
>
> If the ratio of a variable y to the product of two or more variables is constant, then y **varies jointly as,** or **is jointly proportional to,** the other variables. If
>
> $$y = kxz$$
>
> then the number k is the **constant of variation** or the **constant of proportionality.**

✓**Concept Check** Which type of variation is represented by the equation $xy = 8$? Explain.

   a. Direct variation

   b. Inverse variation

   c. Joint variation

Practice 5

The area of a triangle varies jointly as its base and height. Express the area A in terms of base b and height h.

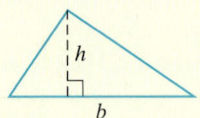

△ **Example 5** The lateral surface area of a cylinder varies jointly as its radius and height. Express surface area S in terms of radius r and height h.

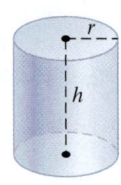

Solution: Because the surface area varies jointly as the radius r and the height h, we equate S to a constant multiple of r and h:

$$S = krh$$

Note: From actual values of $S, r,$ and h, it can be determined that the constant k is 2π, and we then have the formula $S = 2\pi rh$.

■ **Work Practice 5**

Answer

5. $A = kbh$

✓ **Concept Check Answer**

b; answers may vary

Objective D Solving Problems Involving Combined Variation ▶

Some examples of variation involve combinations of direct, inverse, and joint variation. We will call these variations **combined variation.**

Example 6 Suppose that y varies directly as the square of x. If y is 24 when x is 2, find the constant of variation and the variation equation.

Solution: Since y varies directly as the square of x, we have

$$y = kx^2$$

Now let $y = 24$ and $x = 2$ and solve for k.

$$y = kx^2$$
$$24 = k \cdot 2^2$$
$$24 = 4k$$
$$6 = k$$

The constant of variation is 6, so the variation equation is

$$y = 6x^2$$

■ **Work Practice 6**

Example 7 Using Combined Variation

The maximum weight that a circular column can support is directly proportional to the fourth power of its diameter and inversely proportional to the square of its height. A 2-meter-wide column that is 8 meters in height can support 1 ton. Find the weight that a 1-meter-wide column that is 4 meters in height can support.

Solution:

1. UNDERSTAND. Read and reread the problem. Let w = weight, d = diameter, h = height, and k = the constant of variation.

2. TRANSLATE. Since w is directly proportional to d^4 and inversely proportional to h^2, we have

$$w = \frac{kd^4}{h^2}$$

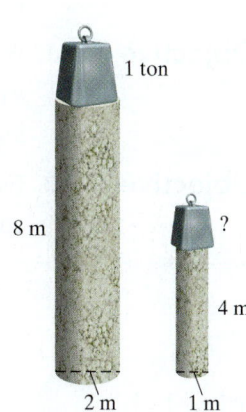

1 ton

8 m

4 m

?

2 m   1 m

3. SOLVE. To find k, we are given that a 2-meter-wide column that is 8 meters in height can support 1 ton. That is, $w = 1$ when $d = 2$ and $h = 8$, or

$$1 = \frac{k \cdot 2^4}{8^2} \qquad \text{Let } w = 1, d = 2, \text{ and } h = 8.$$

$$1 = \frac{k \cdot 16}{64}$$

$$4 = k \qquad \text{Solve for } k.$$

Now we replace k with 4 in the equation $w = \dfrac{kd^4}{h^2}$:

$$w = \frac{4d^4}{h^2}$$

To find weight, w, for a 1-meter-wide column that is 4 meters in height, we let $d = 1$ and $h = 4$.

$$w = \frac{4 \cdot 1^4}{4^2}$$

$$w = \frac{4}{16} = \frac{1}{4}$$

4. INTERPRET. *Check* the proposed solution in the original problem.

State: The 1-meter-wide column that is 4 meters in height can hold $\frac{1}{4}$ ton of weight.

■ **Work Practice 7**

Practice 6

Suppose that y varies inversely as the square of x. If y is 24 when x is 2, find the constant of variation and the variation equation.

Practice 7

The maximum weight that a rectangular beam can support varies jointly as its width and the square of its height and inversely as its length. If a beam $\frac{1}{3}$ foot wide, 1 foot high, and 10 feet long can support 3 tons, find how much weight a similar beam can support if it is 1 foot wide, $\frac{1}{3}$ foot high, and 9 feet long.

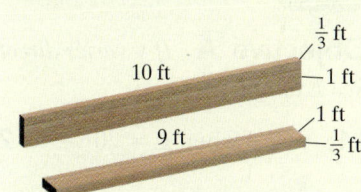

$\frac{1}{3}$ ft

10 ft   1 ft

1 ft

9 ft $\frac{1}{3}$ ft

Answers

6. $k = 96; y = \dfrac{96}{x^2}$ **7.** $1\frac{1}{9}$ tons

Vocabulary, Readiness & Video Check

State whether each equation represents direct, inverse, or joint variation.

1. $y = 5x$ direct

2. $y = \dfrac{700}{x}$ inverse

3. $y = 5xz$ joint

4. $y = \dfrac{1}{2}abc$ joint

5. $y = \dfrac{9.1}{x}$ inverse

6. $y = 2.3x$ direct

7. $y = \dfrac{2}{3}x$ direct

8. $y = 3.1st$ joint

Martin-Gay Interactive Videos   Watch the section lecture video and answer the following questions.

See Video 9.4

See video answer section.

Objective A **9.** Based on the lecture before  Example 1, what kind of equation is a direct variation equation? What does k, the constant of variation, represent in this equation? ▶

Objective B **10.** In  Example 3, why is it not necessary to replace the given values of x and y in the inverse variation equation in order to find k? ▶

Objective C **11.** Based on  Example 5 and the lecture before, what is the variation equation for "y varies jointly as the square of a and the fifth power of b"? ▶

Objective D **12.** From Example 6, what kind of variation does a combined variation application involve? ▶

9.4 Exercise Set MyMathLab® ▶

Objective A *If y varies directly as x, find the constant of variation and the direct variation equation for each situation. See Example 1.*

▶1. $y = 4$ when $x = 20$
$k = \dfrac{1}{5}; y = \dfrac{1}{5}x$

2. $y = 6$ when $x = 30$
$k = \dfrac{1}{5}; y = \dfrac{1}{5}x$

3. $y = 6$ when $x = 4$
$k = \dfrac{3}{2}; y = \dfrac{3}{2}x$

4. $y = 12$ when $x = 8$
$k = \dfrac{3}{2}; y = \dfrac{3}{2}x$

5. $y = 7$ when $x = \dfrac{1}{2}$
$k = 14; y = 14x$

6. $y = 11$ when $x = \dfrac{1}{3}$
$k = 33; y = 33x$

7. $y = 0.2$ when $x = 0.8$
$k = 0.25; y = 0.25x$

8. $y = 0.4$ when $x = 2.5$
$k = 0.16; y = 0.16x$

Objectives A D Mixed Practice *Solve. See Examples 2 and 7.*

▶9. The weight of a synthetic ball varies directly with the cube of its radius. A ball with a radius of 2 inches weighs 1.20 pounds. Find the weight of a ball of the same material with a 3-inch radius.   4.05 lb

10. At sea, the distance to the horizon is directly proportional to the square root of the elevation of the observer. If a person who is 36 feet above the water can see 7.4 miles, find how far a person 64 feet above the water can see. Round to the nearest tenth of a mile.   9.9 mi

11. The amount P of pollution varies directly with the population N of people. Kansas City had a population of 464,000 in 2012 and produced 170,900 tons of pollution. Find how many tons of pollution we should expect St. Louis produced if we know its population was 318,000. Round to the nearest whole ton. (*Source:* Kansas City Government Web site)   117,125 tons

12. Charles's law states that if the pressure P stays the same, the volume V of a gas is directly proportional to its temperature T. If a balloon is filled with 20 cubic meters of a gas at a temperature of 300 K, find the new volume if the temperature rises to 360 K while the pressure stays the same.   24 cu m

Objective **B**   *If y varies inversely as x, find the constant of variation and the inverse variation equation for each situation. See Example 3.*

▶ **13.** $y = 6$ when $x = 5$
$k = 30; y = \dfrac{30}{x}$

14. $y = 20$ when $x = 9$
$k = 180; y = \dfrac{180}{x}$

15. $y = 100$ when $x = 7$
$k = 700; y = \dfrac{700}{x}$

16. $y = 63$ when $x = 3$
$k = 189; y = \dfrac{189}{x}$

17. $y = \dfrac{1}{8}$ when $x = 16$
$k = 2; y = \dfrac{2}{x}$

18. $y = \dfrac{1}{10}$ when $x = 40$
$k = 4; y = \dfrac{4}{x}$

19. $y = 0.2$ when $x = 0.7$
$k = 0.14; y = \dfrac{0.14}{x}$

20. $y = 0.6$ when $x = 0.3$
$k = 0.18; y = \dfrac{0.18}{x}$

Objectives **B** **D** **Mixed Practice** *Solve. See Examples 4 and 7.*

▶ **21.** Pairs of markings a set distance apart are made on highways so that police can detect drivers exceeding the speed limit. Over a fixed distance, the speed R varies inversely with the time T. In one particular pair of markings, R is 45 mph when T is 6 seconds. Find the speed of a car that travels the given distance in 5 seconds.   54 mph

22. The weight of an object on or above the surface of Earth varies inversely as the square of the distance between the object and Earth's center. If a person weighs 160 pounds on Earth's surface, find the individual's weight if he moves 200 miles above Earth. Round to the nearest whole pound. (Assume that Earth's radius is 4000 miles.)   145 lb

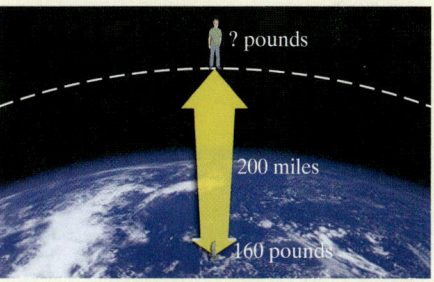

? pounds

200 miles

160 pounds

23. If the voltage V in an electric circuit is held constant, the current I is inversely proportional to the resistance R. If the current is 40 amperes when the resistance is 270 ohms, find the current when the resistance is 150 ohms.   72 amps

24. Because it is more efficient to produce larger numbers of items, the cost of producing a certain computer DVD is inversely proportional to the number produced. If 4000 can be produced at a cost of $1.20 each, find the cost per DVD when 6000 are produced. $0.80

25. The intensity I of light varies inversely as the square of the distance d from the light source. If the distance from the light source is doubled (see the figure), determine what happens to the intensity of light at the new location.   divided by 4

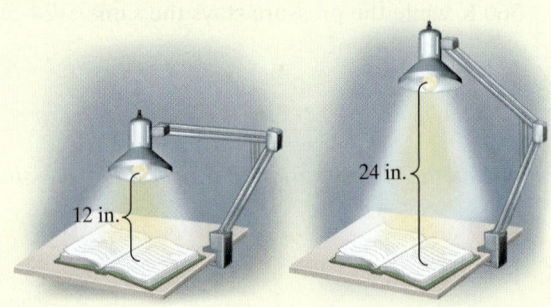

△ **26.** The maximum weight that a circular column can hold is inversely proportional to the square of its height. If an 8-foot column can hold 2 tons, find how much weight a 10-foot column can hold.   1.28 tons

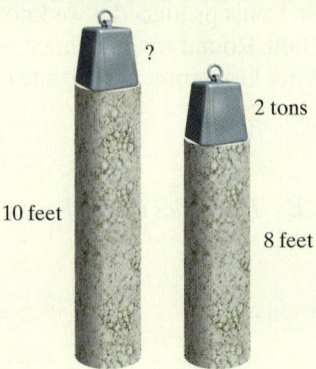

Objectives C D Mixed Practice *Write each statement as an equation. Use k as the constant of variation. See Example 5.*

27. x varies jointly as y and z. $x = kyz$

28. P varies jointly as R and the square of S. $P = kRS^2$

29. r varies jointly as s and the cube of t. $r = kst^3$

30. a varies jointly as b and c. $a = kbc$

For each statement, find the constant of variation and the variation equation. See Examples 5 and 6.

31. y varies directly as the cube of x; $y = 9$ when $x = 3$
$k = \frac{1}{3}; y = \frac{1}{3}x^3$

32. y varies directly as the cube of x; $y = 32$ when $x = 4$
$k = \frac{1}{2}; y = \frac{1}{2}x^3$

33. y varies directly as the square root of x; $y = 0.4$ when $x = 4$ $k = 0.2; y = 0.2\sqrt{x}$

34. y varies directly as the square root of x; $y = 2.1$ when $x = 9$ $k = 0.7; y = 0.7\sqrt{x}$

35. y varies inversely as the square of x; $y = 0.052$ when $x = 5$
$k = 1.3; y = \frac{1.3}{x^2}$

36. y varies inversely as the square of x; $y = 0.011$ when $x = 10$
$k = 1.1; y = \frac{1.1}{x^2}$

▶ **37.** y varies jointly as x and the cube of z; $y = 120$ when $x = 5$ and $z = 2$ $k = 3; y = 3xz^3$

38. y varies jointly as x and the square of z; $y = 360$ when $x = 4$ and $z = 3$ $k = 10; y = 10xz^2$

Solve. See Example 7.

▶△ **39.** The maximum weight that a rectangular beam can support varies jointly as its width and the square of its height and inversely as its length. If a beam $\frac{1}{2}$ foot wide, $\frac{1}{3}$ foot high, and 10 feet long can support 12 tons, find how much a similar beam can support if the beam is $\frac{2}{3}$ foot wide, $\frac{1}{2}$ foot high, and 16 feet long.   22.5 tons

40. The number of cars manufactured on an assembly line at a General Motors plant varies jointly as the number of workers and the time they work. If 200 workers can produce 60 cars in 2 hours, find how many cars 240 workers should be able to make in 3 hours.   108 cars

△ **41.** The volume of a cone varies jointly as its height and the square of its radius. If the volume of a cone is 32π cubic inches when the radius is 4 inches and the height is 6 inches, find the volume of a cone when the radius is 3 inches and the height is 5 inches. 15π cu in.

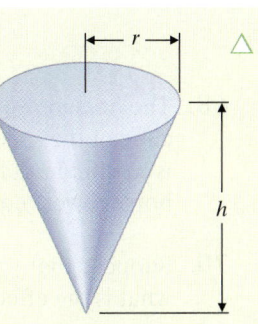

△ **42.** When a wind blows perpendicularly against a flat surface, its force is jointly proportional to the surface area and the speed of the wind. A sail whose surface area is 12 square feet experiences a 20-pound force when the wind speed is 10 miles per hour. Find the force on an 8-square-foot sail if the wind speed is 12 miles per hour.   16 lb

43. The intensity of light (in foot-candles) varies inversely as the square of x, the distance in feet from the light source. The intensity of light 2 feet from the source is 80 foot-candles. How far away is the source if the intensity of light is 5 foot-candles?   8 ft

44. The horsepower that can be safely transmitted to a shaft varies jointly as the shaft's angular speed of rotation (in revolutions per minute) and the cube of its diameter. A 2-inch shaft making 120 revolutions per minute safely transmits 40 horsepower. Find how much horsepower can be safely transmitted by a 3-inch shaft making 80 revolutions per minute.   90 hp

Objectives Ⓐ Ⓑ Ⓒ Ⓓ **Mixed Practice** *Write an equation to describe each variation. Use k for the constant of proportionally. See Examples 1 through 7.*

45. y varies directly as x $y = kx$

46. p varies directly as q $p = kq$

47. a varies inversely as b $a = \dfrac{k}{b}$

48. y varies inversely as x $y = \dfrac{k}{x}$

49. y varies jointly as x and z $y = kxz$

50. y varies jointly as q, r, and t $y = kqrt$

51. y varies inversely as x^3 $y = \dfrac{k}{x^3}$

52. y varies inversely as a^4 $y = \dfrac{k}{a^4}$

53. y varies directly as x and inversely as p^2 $y = \dfrac{kx}{p^2}$

54. y varies directly as a^5 and inversely as b $y = \dfrac{ka^5}{b}$

Review

Find the exact circumference C and area A of each circle. Use the formulas $C = 2\pi r$ and $A = \pi r^2$ where r is the radius of the circle. See Section 2.5.

△ **55.**

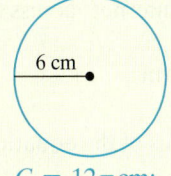

6 cm

$C = 12\pi$ cm;
$A = 36\pi$ sq cm

△ **56.**

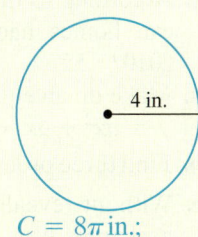

4 in.

$C = 8\pi$ in.;
$A = 16\pi$ sq in.

△ **57.**

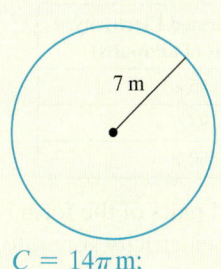

7 m

$C = 14\pi$ m;
$A = 49\pi$ sq m

△ **58.**

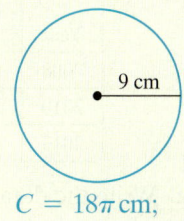

9 cm

$C = 18\pi$ cm;
$A = 81\pi$ sq cm

Find the slope of the line containing each pair of points. See Section 3.4.

59. $(3, 6), (-2, 6)$ 0

60. $(-5, -2), (0, 7)$ $\dfrac{9}{5}$

61. $(4, -1), (5, -2)$ -1

62. $(2, 1), (2, -3)$ undefined

Concept Extensions

Solve. See the Concept Check in this section. Choose the type of variation that each equation represents.
a. *Direct variation*   **b.** *Inverse variation*   **c.** *Joint variation*

63. $y = \dfrac{2}{3}x$ a

64. $y = \dfrac{0.6}{x}$ b

65. $y = 9ab$ c

66. $xy = \dfrac{2}{11}$ b

Solve.

△ **67.** The volume of a cylinder varies jointly as the height and the square of the radius. If the height is halved and the radius is doubled, determine what happens to the volume.   multiplied by 2

68. The horsepower to drive a boat varies directly as the cube of the speed of the boat. If the speed of the boat is to double, determine the corresponding increase in horsepower required.   multiplied by 8

69. Suppose that y varies directly as x^2. If x is doubled, what is the effect on y?   multiplied by 4

70. Suppose that y varies directly as x. If x is doubled, what is the effect on y?   multiplied by 2

Chapter 9 Group Activity

Additional group activities are available in the *Instructor's Resource Manual with Tests.*

Another Mathematical Model

Sometimes mathematical models other than linear models are appropriate for data. Suppose that an equation of the form $y = ax^2 + bx + c$ is an appropriate model for the ordered pairs (x_1, y_1), (x_2, y_2), and (x_3, y_3). Then it is necessary to find the values of a, b, and c such that the given ordered pairs are solutions of the equation $y = ax^2 + bx + c$. To do so, substitute each ordered pair into the equation. Each time, the result is an equation in three unknowns: a, b, and c. Solving the resulting system of three linear equation in three unknowns will give the required values of a, b, and c.

1. The table gives the supply of electricity (in billions of kilowatts) generated in the United States by wind power in each of the years listed. (*Source:* Based on data from the Energy Information Administration, U.S. Department of Energy)

Year	Wind-Generated Electricity (in billions of kilowatts)
2008	55.4
2010	94.6
2013	160.1

a. Write the data as ordered pairs of the form (x, y), where y is the amount of electricity generated (in billions of kilowatts) in the year x ($x = 0$ represents 2008). $(0, 55.4); (2, 94.6); (5, 160.1)$

b. Find the values of a, b, and c rounded to 2 decimal places (where needed) such that the equation $y = ax^2 + bx + c$ models these data.

c. Verify that the model you found in part (b) gives each of the ordered pair solutions from part (a).

d. According to the model, how much electricity was generated by wind power in 2012?   137.44 billion kilowatts

2. The table gives the percent of American households with broadband Internet access for each of the years listed. (*Source:* Pew Internet and American Life Project)

Year	Percent of U.S. Households with Broadband Internet Access
2000	3
2005	33
2013	70

a. Write the data as ordered pairs of the form (x, y), where y is the percent of U.S. households with broadband Internet access in the year x ($x = 0$ represents 2000). $(0, 3); (5, 33); (13, 70)$

b. Find the values of a, b, and c rounded to 2 decimal places such that the equation $y = ax^2 + bx + c$ models these data.

c. According to the model, what percent of American homes had broadband Internet access in 2010?   57.3%

3. a. Make up an equation of the form $y = ax^2 + bx + c$.

b. Find three ordered pair solutions of the equation.

c. Without revealing your equation from part (a), exchange lists of ordered pair solutions with another group.

d. Use the method described above to find the values of a, b, and c such that the equation $y = ax^2 + bx + c$ has the ordered pair solutions you received from the other group.

e. Check with the other group to see if your equation from part (d) is the correct one.

1. b. $a = 0.45$, $b = 18.71$, $c = 55.4$; $y = 0.45x^2 + 18.71x + 55.4$

2. b. $a = -0.11$, $b = 6.53$, $c = 3$; $y = -0.11x^2 + 6.53x + 3$

Chapter 9 Vocabulary Check

Fill in each blank with one of the words or phrases listed below.

matrix	consistent	system of equations	inversely	column	jointly
solution	inconsistent	element	directly	row	

1. Two or more linear equations in two variables form a(n) _system of equations_ .

2. A(n) _solution_ of a system of two equations in two variables is an ordered pair that makes both equations true.

3. A(n) _consistent_ system of equations has at least one solution.

4. A(n) _inconsistent_ system of equations has no solution.

5. A(n) _matrix_ is a rectangular array of numbers.

6. Each of the numbers in a matrix is called a(n) _element_ .

7. The numbers aligned horizontally in a matrix are in the same _row_ .

8. The numbers aligned vertically in a matrix are in the same _column_ .

9. In the equation $y = kx$, y varies _directly_ as x.

10. In the equation $y = \dfrac{k}{x}$, y varies _inversely_ as x.

11. In the equation $y = kxz$, y varies _jointly_ as x and z.

Helpful Hint

▶ Are you preparing for your test? Don't forget to take the Chapter 9 Test on page 675. Then check your answers at the back of the text and use the Chapter Test Prep Videos to see the fully worked-out solutions to any of the exercises you want to review.

9 Chapter Highlights

Definitions and Concepts	Examples
Section 9.1 Solving Systems of Linear Equations in Three Variables and Problem Solving	

A **solution** of an equation in three variables $x, y,$ and z is an **ordered triple** (x, y, z) that makes the equation a true statement.

Verify that $(-2, 1, 3)$ is a solution of $2x + 3y - 2z = -7$.
Replace x with -2, y with 1, and z with 3.

$$2(-2) + 3(1) - 2(3) \stackrel{?}{=} -7$$
$$-4 + 3 - 6 \stackrel{?}{=} -7$$
$$-7 = -7 \quad \text{True}$$

$(-2, 1, 3)$ is a solution.

Solving a System of Three Linear Equations by the Elimination Method

Step 1: Write each equation in standard form, $Ax + By + Cz = D$.

Step 2: Choose a pair of equations and use them to eliminate a variable.

Step 3: Choose any other pair of equations and eliminate the same variable.

Solve:

$$\begin{cases} 2x + y - z = 0 & (1) \\ x - y - 2z = -6 & (2) \\ -3x - 2y + 3z = -22 & (3) \end{cases}$$

1. Each equation is written in standard form.

2. $\begin{aligned} 2x + y - z &= 0 \quad (1) \\ x - y - 2z &= -6 \quad (2) \\ \hline 3x \quad\quad - 3z &= -6 \quad (4) \quad \text{Add.} \end{aligned}$

3. Eliminate y from equations (1) and (3) also.

$\begin{aligned} 4x + 2y - 2z &= 0 \quad\quad \text{Multiply equation (1) by 2.} \\ -3x - 2y + 3z &= -22 \quad (3) \\ \hline x \quad\quad + z &= -22 \quad (5) \quad \text{Add.} \end{aligned}$

(continued)

Definitions and Concepts	Examples

Section 9.1 Solving Systems of Linear Equations in Three Variables and Problem Solving (*continued*)

Step 4: Solve the system of two equations in two variables from Steps 2 and 3.

Step 5: Solve for the third variable by substituting the values of the variables from Step 4 into any of the original equations.

Step 6: Check the solution in all three original equations.

4. Solve.

$$\begin{cases} 3x - 3z = -6 & (4) \\ x + z = -22 & (5) \end{cases}$$

$$\begin{aligned} x - z &= -2 \quad \text{Divide equation (4) by 3.} \\ \underline{x + z} &= \underline{-22} \quad (5) \\ 2x &= -24 \\ x &= -12 \end{aligned}$$

To find z, use equation (5).

$$\begin{aligned} x + z &= -22 \\ -12 + z &= -22 \\ z &= -10 \end{aligned}$$

5. To find y, use equation (1).

$$\begin{aligned} 2x + y - z &= 0 \\ 2(-12) + y - (-10) &= 0 \\ -24 + y + 10 &= 0 \\ y &= 14 \end{aligned}$$

6. The solution $(-12, 14, -10)$ checks.

Section 9.2 Solving Systems of Equations Using Matrices

A **matrix** is a rectangular array of numbers.

$$\begin{bmatrix} -7 & 0 & 3 \\ 1 & 2 & 4 \end{bmatrix} \qquad \begin{bmatrix} a & b & c \\ d & e & f \\ g & h & i \end{bmatrix}$$

The **matrix** corresponding to a system is composed of the coefficients of the variables and the constants of the system.

The matrix corresponding to the system

$$\begin{cases} x - y = 1 \\ 2x + y = 11 \end{cases} \text{ is } \left[\begin{array}{cc|c} 1 & -1 & 1 \\ 2 & 1 & 11 \end{array}\right]$$

The following **row operations** can be performed on matrices, and the result is an equivalent matrix.

Elementary row operations:

1. Interchange any two rows.

2. Multiply (or divide) the elements of one row by the same nonzero number.

3. Multiply (or divide) the elements of one row by the same nonzero number and add them to their corresponding elements in any other row.

Use matrices to solve: $\begin{cases} x - y = 1 \\ 2x + y = 11 \end{cases}$

The corresponding matrix is

$$\left[\begin{array}{cc|c} 1 & -1 & 1 \\ 2 & 1 & 11 \end{array}\right]$$

Use row operations to write an equivalent matrix with 1s along the diagonal and 0s below each 1 in the diagonal. Multiply row 1 by -2 and add to row 2. Change row 2 only.

$$\left[\begin{array}{cc|c} 1 & -1 & 1 \\ -2(1) + 2 & -2(-1) + 1 & -2(1) + 11 \end{array}\right]$$

simplifies to $\left[\begin{array}{cc|c} 1 & -1 & 1 \\ 0 & 3 & 9 \end{array}\right]$

Definitions and Concepts	Examples

Section 9.2   Solving Systems of Equations Using Matrices (*continued*)

Divide row 2 by 3.

$$\begin{bmatrix} 1 & -1 & | & 1 \\ 0 & \dfrac{3}{3} & | & \dfrac{9}{3} \end{bmatrix} \text{ simplifies to } \begin{bmatrix} 1 & -1 & | & 1 \\ 0 & 1 & | & 3 \end{bmatrix}$$

This matrix corresponds to the system

$$\begin{cases} x - y = 1 \\ \quad\ y = 3 \end{cases}$$

Let $y = 3$ in the first equation.

$$x - 3 = 1$$
$$x = 4$$

The ordered pair solution is $(4, 3)$.

Section 9.3   Systems of Linear Inequalities

A **system of linear inequalities** consists of two or more linear inequalities.

To graph a system of inequalities, graph each inequality in the system. The overlapping region is the solution of the system.

$$\begin{cases} x - y \geq 3 \\ \quad\ y \leq -2x \end{cases}$$

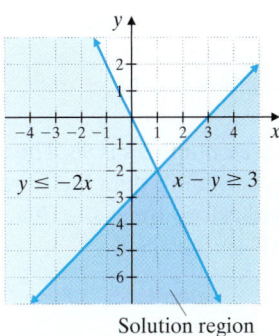

Solution region

Section 9.4   Variation and Problem Solving

y **varies directly** as x, or y is **directly proportional** to x, if there is a nonzero constant k such that

$$y = kx$$

y **varies inversely** as x, or y is **inversely proportional** to x, if there is a nonzero constant k such that

$$y = \frac{k}{x}$$

y **varies jointly** as x and z, or y is **jointly proportional** to x and z, if there is a nonzero constant k such that

$$y = kxz$$

The circumference of a circle C varies directly as its radius r.

$$C = \underset{k}{2\pi} r$$

Pressure P varies inversely with volume V.

$$P = \frac{k}{V}$$

The lateral surface area S of a cylinder varies jointly as its radius r and height h.

$$S = \underset{k}{2\pi} rh$$

(9.1) *Solve each system of equations in three variables.*

1. $\begin{cases} x \qquad + z = 4 \\ 2x - y \qquad = 4 \\ x + y - z = 0 \end{cases}$

$(2, 0, 2)$

2. $\begin{cases} 2x + 5y \qquad = 4 \\ x - 5y + z = -1 \\ 4x \qquad - z = 11 \end{cases}$

$(2, 0, -3)$

3. $\begin{cases} 4y + 2z = 5 \\ 2x + 8y \qquad = 5 \\ 6x + \qquad 4z = 1 \end{cases}$

$\left(-\dfrac{1}{2}, \dfrac{3}{4}, 1 \right)$

4. $\begin{cases} 5x + 7y \qquad = 9 \\ 14y - z = 28 \\ 4x \qquad + 2z = -4 \end{cases}$

$(-1, 2, 0)$

5. $\begin{cases} 3x - 2y + 2z = 5 \\ -x + 6y + z = 4 \\ 3x + 14y + 7z = 20 \end{cases}$

$\varnothing$

6. $\begin{cases} x + 2y + 3z = 11 \\ y + 2z = 3 \\ 2x \qquad + 2z = 10 \end{cases}$

$(5, 3, 0)$

7. $\begin{cases} 7x - 3y + 2z = 0 \\ 4x - 4y - z = 2 \\ 5x + 2y + 3z = 1 \end{cases}$

$(1, 1, -2)$

8. $\begin{cases} x - 3y - 5z = -5 \\ 4x - 2y + 3z = 13 \\ 5x + 3y + 4z = 22 \end{cases}$

$(3, 1, 1)$

Use systems of equations to solve.

9. The sum of three numbers is 98. The sum of the first and second is two more than the third number, and the second is four times the first. Find the numbers.   10, 40, and 48

10. An employee at See's Candy Store needs a special mixture of candy. She has creme-filled chocolates that sell for $3.00 per pound, chocolate-covered nuts that sell for $2.70 per pound, and chocolate-covered raisins that sell for $2.25 per pound. She wants to have twice as many raisins as nuts in the mixture. Find how many pounds of each she should use to make 45 pounds worth $2.80 per pound.   30 lb of creme-filled; 5 lb of chocolate-covered nuts; 10 lb of chocolate-covered raisins

11. Chris Kringler has $2.77 in her coin jar—all in pennies, nickels, and dimes. If she has 53 coins in all and four more nickels than dimes, find how many of each type of coin she has.   17 pennies; 20 nickels; 16 dimes

12. The sum of three numbers is 295. One number is five more than a second and twice the third. Find the numbers.   120, 115, and 60

(9.2) *Use matrices to solve each system.*

13. $\begin{cases} 3x + 10y = 1 \\ x + 2y = -1 \end{cases}$ $(-3, 1)$

14. $\begin{cases} 3x - 6y = 12 \\ 2y = x - 4 \end{cases}$

$\{(x, y) \mid x - 2y = 4\}$

15. $\begin{cases} 3x - 2y = -8 \\ 6x + 5y = 11 \end{cases}$ $\left(-\dfrac{2}{3}, 3 \right)$

16. $\begin{cases} 6x - 6y = -5 \\ 10x - 2y = 1 \end{cases}$ $\left(\dfrac{1}{3}, \dfrac{7}{6} \right)$

17. $\begin{cases} 3x - 6y = 0 \\ 2x + 4y = 5 \end{cases}$ $\left(\dfrac{5}{4}, \dfrac{5}{8} \right)$

18. $\begin{cases} 5x - 3y = 10 \\ -2x + y = -1 \end{cases}$ $(-7, -15)$

19. $\begin{cases} 0.2x - 0.3y = -0.7 \\ 0.5x + 0.3y = 1.4 \end{cases}$
$(1, 3)$

20. $\begin{cases} 3x + 2y = 8 \\ 3x - y = 5 \end{cases}$
$(2, 1)$

21. $\begin{cases} x \quad\quad + z = 4 \\ 2x - y \quad\quad = 0 \\ x + y - z = 0 \end{cases}$
$(1, 2, 3)$

22. $\begin{cases} 2x + 5y \quad\quad = 4 \\ x - 5y + z = -1 \\ 4x \quad\quad - z = 11 \end{cases}$
$(2, 0, -3)$

23. $\begin{cases} 3x - y \quad\quad = 11 \\ x \quad\quad + 2z = 13 \\ y - z = -7 \end{cases}$
$(3, -2, 5)$

24. $\begin{cases} 5x + 7y + 3z = 9 \\ 14y - z = 28 \\ 4x \quad\quad + 2z = -4 \end{cases}$
$(-1, 2, 0)$

25. $\begin{cases} 7x - 3y + 2z = 0 \\ 4x - 4y - z = 2 \\ 5x + 2y + 3z = 1 \end{cases}$
$(1, 1, -2)$

26. $\begin{cases} x + 2y + 3z = 14 \\ y + 2z = 3 \\ 2x \quad\quad - 2z = 10 \end{cases}$
$\varnothing$

(9.3) *Graph the solution of each system of linear inequalities.*

27. $\begin{cases} y \geq 2x - 3 \\ y \leq -2x + 1 \end{cases}$

28. $\begin{cases} y \leq -3x - 3 \\ y \leq 2x + 7 \end{cases}$

29. $\begin{cases} x + 2y > 0 \\ x - y \leq 6 \end{cases}$

30. $\begin{cases} x - 2y \geq 7 \\ x + y \leq -5 \end{cases}$

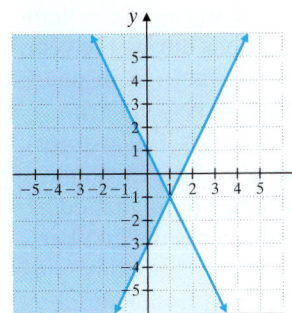 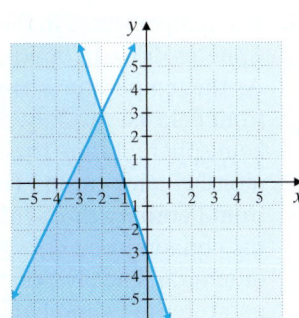 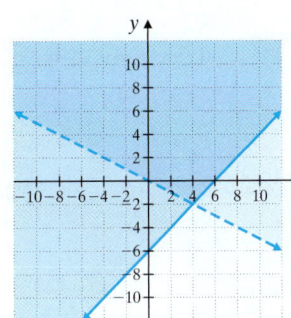 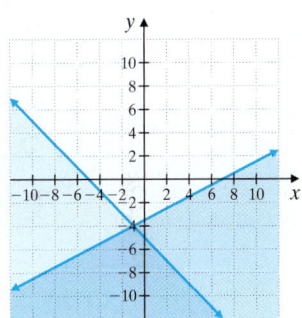

31. $\begin{cases} 3x - 2y \leq 4 \\ 2x + y \geq 5 \\ y \leq 4 \end{cases}$

32. $\begin{cases} 4x - y \leq 0 \\ 3x - 2y \geq -5 \\ y \geq -4 \end{cases}$

33. $\begin{cases} x + 2y \leq 5 \\ x \leq 2 \\ x \geq 0 \\ y \geq 0 \end{cases}$

34. $\begin{cases} x + 3y \leq 7 \\ y \leq 5 \\ x \geq 0 \\ y \geq 0 \end{cases}$

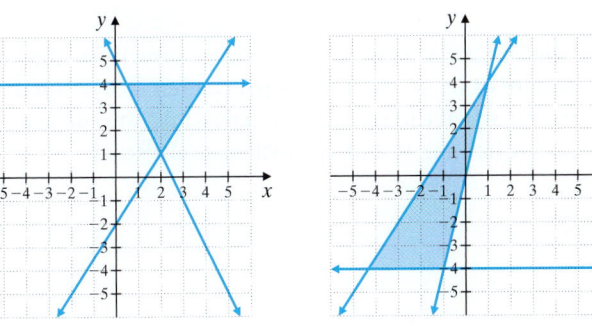 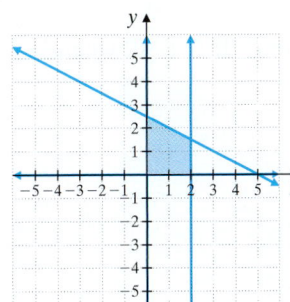 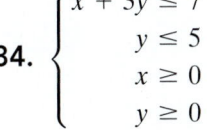 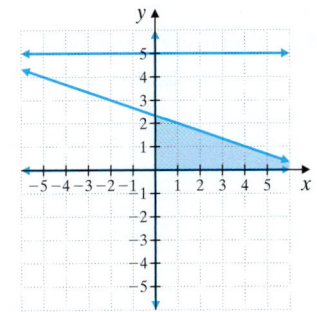

(9.4) *Solve each variation problem.*

35. A is directly proportional to B. If $A = 6$ when $B = 14$, find A when $B = 21$.   9

36. According to Boyle's law, the pressure exerted by a gas is inversely proportional to the volume, as long as the temperature stays the same. If a gas exerts a pressure of 1250 kilopascals when the volume is 2 cubic meters, find the volume when the pressure is 800 kilopascals.   3.125 cu m

Mixed Review

Solve the system.

37. $\begin{cases} x - 3y + 2z = 0 \\ \quad\quad 9y - z = 22 \\ 5x \quad\quad\quad + 3z = 10 \end{cases}$ $(-1, 3, 5)$

38. The perimeter of a triangle is 126 units. The length of one side is twice the length of the shortest side. The length of the third side is 14 more than the length of the shortest side. Find the lengths of the sides of the triangle.   28 units, 42 units, 56 units

39. Graph the solution of the system: $\begin{cases} y \le 3x - \dfrac{1}{2} \\ 3x + 4y \ge 6 \end{cases}$

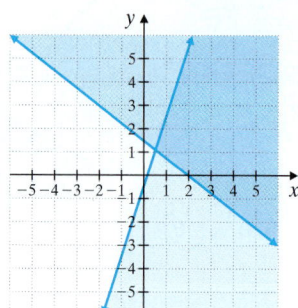

40. The temperatures in the northern and southern hemispheres move in opposite directions during the year. As the temperatures rise in the northern hemisphere from February to July, they drop in the southern hemisphere during these same months. The average monthly temperature (in Fahrenheit) for England from February through July can be approximated by the equation $y = 4.3x + 44$, where x is the number of months since February. The average monthly temperature (in Fahrenheit) in Australia from February through July can be approximated by the equation $y = -5.1x + 79$, where x is the number of months since February. Use these equations to determine the month in which the average monthly temperature of England is the same as the average monthly temperature of Australia.   during May

41. C is inversely proportional to D. If $C = 12$ when $D = 8$, find C when $D = 24$.   4

42. The surface area of a sphere varies directly as the square of its radius. If the surface area is 36π square inches when the radius is 3 inches, find the surface area when the radius is 4 inches. 64π sq in.

Solve each system.

1. $\begin{cases} 2x - 3y = 4 \\ 3y + 2z = 2 \\ x - z = -5 \end{cases}$

2. $\begin{cases} 3x - 2y - z = -1 \\ 2x - 2y = 4 \\ 2x - 2z = -12 \end{cases}$

3. $\begin{cases} x + y + z = 4 \\ 2x + 5y = 1 \\ x - y - 2z = 0 \end{cases}$

4. $\begin{cases} 3x + 2y + 3z = 3 \\ x - z = 9 \\ 4y + z = -4 \end{cases}$

Use matrices to solve each system.

5. $\begin{cases} x - y = -2 \\ 3x - 3y = -6 \end{cases}$

6. $\begin{cases} x + 2y = -1 \\ 2x + 5y = -5 \end{cases}$

7. $\begin{cases} x - y - z = 0 \\ 3x - y - 5z = -2 \\ 2x + 3y = -5 \end{cases}$

8. $\begin{cases} 2x - y + 3z = 4 \\ 3x - 3z = -2 \\ -5x + y = 0 \end{cases}$

9. The measure of the largest angle of a triangle is three less than 5 times the measure of the smallest angle. The measure of the remaining angle is 1 less than twice the measure of the smallest angle. Find the measure of each angle.

Graph the solutions of each system of linear inequalities.

10. $\begin{cases} 2y - x \ge 1 \\ x + y \ge -4 \\ y \le 2 \end{cases}$

11. $\begin{cases} y + 2x \le 4 \\ y \le 2 \\ y \ge 0 \\ x \ge 0 \end{cases}$

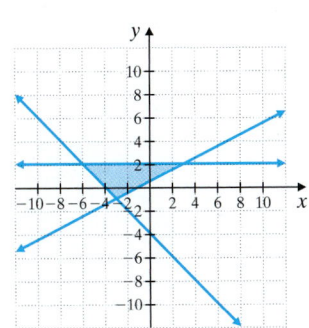

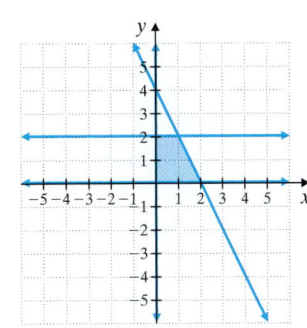

Answers

Note to Instructor: The Chapter 9 Test file in the TestGen program provides algorithms specifically matched to this test for easy replication for practice or assessment purposes.

1. $(-1, -2, 4)$

2. $\varnothing$

3. $(3, -1, 2)$

4. $(5, 0, -4)$

5. $\{(x, y) \mid x - y = -2\}$

6. $(5, -3)$

7. $(-1, -1, 0)$

8. $\varnothing$

9. $23°, 45°, 112°$

10. see graph

11. see graph

675

12. 16

12. Suppose that W is inversely proportional to V. If $W = 20$ when $V = 12$, find W when $V = 15$.

13. Suppose that Q is jointly proportional to R and the square of S. If $Q = 24$ when $R = 3$ and $S = 4$, find Q when $R = 2$ and $S = 3$.

13. 9

14. When an anvil is dropped into a gorge, the speed with which it strikes the ground is directly proportional to the square root of the distance it falls. An anvil that falls 400 feet hits the ground at a speed of 160 feet per second. Find the height of a cliff over the gorge if a dropped anvil hits the ground at a speed of 128 feet per second.

14. 256 ft

1. Solve: $y + 0.6 = -1.0$

2. Solve: $\frac{1}{2}x = -5$

3. Solve: $8(2 - t) = -5t$

4. Solve: $\frac{x}{7} + \frac{x}{5} = \frac{12}{5}$

5. The length of a rectangular road sign is 2 feet less than three times its width. Find the dimensions if the perimeter is 28 feet.

6. Translate each phrase to an algebraic expression. Use the variable x to represent each unknown number.

  a. One third subtracted from a number.

  b. Six less than five times a number.

  c. Three more than eight times a number.

  d. The quotient of seven and the difference of two and a number.

7. Complete the table for the equation $y = \frac{1}{2}x - 5$.

x	y
-2	-6
0	-5
10	0

8. Subtract: $(2x - 7)$ from $(2x^2 + 8x - 3)$

9. Find the slope of the line through $(-1, 5)$ and $(2, -3)$. Graph the line.

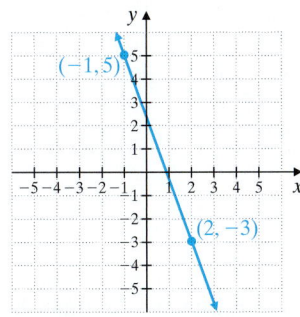

10. Find an equation of the line with slope $\frac{1}{2}$ containing the point $(-1, 3)$. Use function notation to write the equation.

Simplify the following expressions.

11. 3^0 **12.** $\left(\frac{3}{4}\right)^0$ **13.** $(5x^3y^2)^0$ **14.** $7x^0$ **15.** -4^0 **16.** $-8y^0$

17. Multiply: $(3y + 2)^2$

18. Multiply: $(4x - 3z)(4x + 3z)$

Answers

1. $y = -1.6$ (Sec. 2.1, Ex. 2)

2. $x = -10$ (Sec. 2.2)

3. $t = \frac{16}{3}$ (Sec. 2.3, Ex. 2)

4. $x = 7$ (Sec. 2.3)

5. width: 4 ft; length: 10 ft (Sec. 2.5, Ex. 4)

6. a. $x - \frac{1}{3}$ (Sec. 2.1)

   b. $5x - 6$

   c. $8x + 3$

   d. $\frac{7}{2 - x}$ (Sec. 2.1)

7. see table (Sec. 3.1, Ex. 7)

8. $2x^2 + 6x + 4$ (Sec. 5.4)

9. see graph $-\frac{8}{3}$ (Sec. 3.4, Ex. 1)

10. $f(x) = \frac{1}{2}x + \frac{7}{2}$ (Sec. 8.2)

11. 1 (Sec. 5.1, Ex. 28)

12. 1 (Sec. 5.1)

13. 1 (Sec. 5.1, Ex. 29)

14. 7 (Sec. 5.1)

15. -1 (Sec. 5.1, Ex. 31)

16. -8 (Sec. 5.1)

17. $9y^2 + 12y + 4$ (Sec. 5.6, Ex. 15)

18. $16x^2 - 9z^2$ (Sec. 5.6)

19. Divide $x^2 + 7x + 12$ by $x + 3$ using long division.

20. Simplify the following. Write answers with positive exponents.

  a. $2^{-2} + 3^{-1}$

  b. $-6a^0$

  c. $\dfrac{x^{-5}}{x^{-2}}$

21. Factor: $r^2 - r - 42$

22. Factor: $xy + 2x - 5y - 10$

23. Factor: $10x^2 - 13xy - 3y^2$

24. Factor: $6x^2 - x - 35$

25. Factor $8x^2 - 14x + 5$ by grouping.

26. Factor: $4x^2 - 4x + 1 - 9y^2$

27. Factor each binomial.

  a. $4x^3 - 49x$

  b. $162x^4 - 2$

28. Factor: $2x^3 - 16$

29. Solve:

$(5x - 1)(2x^2 + 15x + 18) = 0$

30. Solve: $2x(3x + 1)(x - 3) = 0$

31. Simplify: $\dfrac{x^2 + 8x + 7}{x^2 - 4x - 5}$

32. Simplify: $\dfrac{(3a^{-2}b)^{-2}}{(2ab^{-3})^{-3}}$

33. The quotient of a number and 6, minus $\dfrac{5}{3}$, is the quotient of the number and 2. Find the number.

34. Solve: $\dfrac{3x - 4}{2x} = -\dfrac{8}{x}$

35. Write an equation of the line containing the point $(4, 4)$ and parallel to the line $2x + y = -6$. Write the equation in slope-intercept form, $y = mx + b$.

36. Simplify the following. Assume that a and b are integers and that x and y are not 0.

  a. $3x^{4a}(4x^{-a})^2$

  b. $\dfrac{(y^{4b})^3}{y^{2b-3}}$

If $P(x) = 3x^4 - 2x^2 - 5$, find the following:

37. $P(1)$ **38.** $P(0)$ **39.** $P(-2)$

40. If $f(x) = -x^2 + 3x - 2$, find
    a. $f(0)$
    b. $f(-3)$
    c. $f\left(\dfrac{1}{3}\right)$

41. Hooke's law states that the distance a spring stretches is directly proportional to the weight attached to the spring. If a 40-pound weight attached to a spring stretches the spring 5 inches, find the distance that a 65-pound weight attached to the spring stretches the spring.

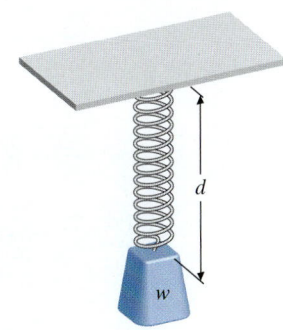

40. a. -2

    b. -20

    c. $-\dfrac{10}{9}$ (Sec. 8.3)

     8.125 in.
41. (Sec. 9.4, Ex. 2)

     Paper, \$3.80; folders,
42. \$5.25 (Sec. 4.4)

42. Kernersville office supply sold three reams of paper and two boxes of manila folders for \$21.90. Also, five reams of paper and one box of manila folders cost \$24.25. Find the price of a ream of paper and a box of manila folders.

43. Use the substitution method to solve the system.
$$\begin{cases} 2x + y = 10 \\ x = y + 2 \end{cases}$$

43. $(4, 2)$ (Sec. 4.2, Ex. 1)

44. Use the substitution method to solve the system.
$$\begin{cases} -2x + 3y = 6 \\ 3x - y = 5 \end{cases}$$

45. Use the addition method to solve the system.
$$\begin{cases} 2x - y = 7 \\ 8x - 4y = 1 \end{cases}$$

44. $(3, 4)$ (Sec. 4.2)

     no solution
45. (Sec. 4.3, Ex. 3)

46. Use the addition method to solve the system.
$$\begin{cases} 5x + y = 5 \\ -7x - 2y = -7 \end{cases}$$

47. Solve the system:
$$\begin{cases} x - 5y - 2z = 6 \\ -2x + 10y + 4z = -12 \\ \dfrac{1}{2}x - \dfrac{5}{2}y - z = 3 \end{cases}$$

46. $(1, 0)$ (Sec. 4.3)

     $\{(x, y, z) \mid$
     $x - 5y - 2z = 6\}$
47. (Sec. 9.1, Ex. 4)

48. Solve the system:
$$\begin{cases} 2x - 2y + 4z = 6 \\ -4x - y + z = -8 \\ 3x - y + z = 6 \end{cases}$$

49. Use matrices to solve the system.
$$\begin{cases} 2x - y = 3 \\ 4x - 2y = 5 \end{cases}$$

48. $(2, 1, 1)$ (Sec. 9.1)

     $\varnothing$ or $\{\ \}$
49. (Sec. 9.2, Ex. 2)

50. Use matrices to solve the system.
$$\begin{cases} x + y + z = 9 \\ 2x - 2y + 3z = 2 \\ -3x + y - z = 1 \end{cases}$$

50. $(0, 5, 4)$ (Sec. 9.2)

10 Rational Exponents, Radicals, and Complex Numbers

Check Your Progress

Vocabulary Check

Chapter Highlights

Chapter Review

Chapter Test

Cumulative Review

In this chapter, radical notation is reviewed, and then rational exponents are introduced. As the name implies, rational exponents are exponents that are rational numbers. We present an interpretation of rational exponents that is consistent with the meaning and rules already established for integer exponents, and we present two forms of notation for roots: radical and exponent. We conclude this chapter with complex numbers, a natural extension of the real number system.

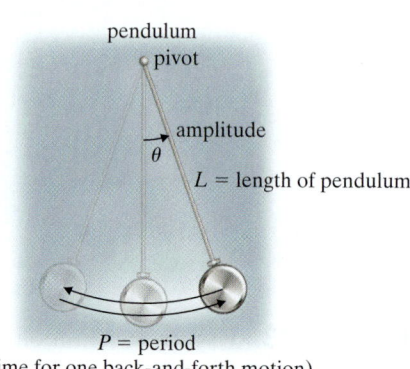

P = period
(time for one back-and-forth motion)

How Do Pendulums Keep Time?

A pendulum is a weight suspended from a pivot. The weight is called a bob, and we call the time for one back-and-forth swing of a pendulum a period. Interestingly enough, the time for one back-and-forth swing of a pendulum depends only on the length of the pendulum. Galileo discovered this around the early 1600s, and it was this regular period (time) that made pendulums useful for timekeeping.

A floor clock, also known as a grandfather clock, is a tall pendulum-driven clock. These early clocks were 6–8 feet tall and were the world's most accurate method of timekeeping until the 1930s. (*Source:* Wikipedia)

For small angles of motion, or amplitudes, the formula $P = 2\pi\sqrt{\dfrac{L}{32}}$ approximates the time, in seconds, for a back-and-forth swing (period). In this formula, P is the period (the time in seconds for one full back-and-forth swing) and L is the length, in feet, of a pendulum. Below is the graph of this formula, and in Sections 10.3 and 10.6, we compute the periods of pendulums of different lengths.

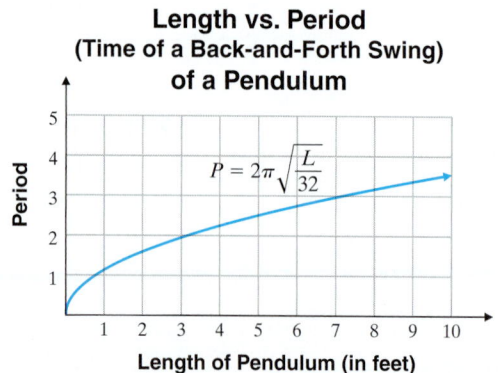

Length vs. Period (Time of a Back-and-Forth Swing) of a Pendulum

$P = 2\pi\sqrt{\dfrac{L}{32}}$

Period

Length of Pendulum (in feet)

Note: This graph has the appearance of a square root function, as it should because of $\sqrt{L}$ in the formula.

10.1 Radical Expressions and Radical Functions

Objective A Finding Square Roots

To find a *square root* of a number a, we find a number that was squared to get a.

Square Root

The number b is a **square root** of a if $b^2 = a$.

Examples Find the real square roots of each number.

1. 25   Since $5^2 = 25$ and $(-5)^2 = 25$, the square roots of 25 are 5 and -5.
2. 49   Since $7^2 = 49$ and $(-7)^2 = 49$, the square roots of 49 are 7 and -7.
3. -4 There is no real number whose square is -4. The number -4 has no real number square root.

Work Practice 1–3

Recall that we denote the *nonnegative*, or *principal*, *square root* with the **radical sign:**

$$\sqrt{25} = 5$$

We denote the *negative square root* with the **negative radical sign:**

$$-\sqrt{25} = -5$$

An expression containing a radical sign is called a **radical expression.** An expression within, or "under," a radical sign is called a **radicand.**

radical expression: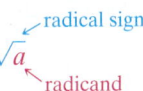

Principal and Negative Square Roots

The **principal square root** of a nonnegative number a is its nonnegative square root. The principal square root is written as $\sqrt{a}$. The **negative square root** of a is written as $-\sqrt{a}$.

Examples Find each square root. Assume that all variables represent non-negative real numbers.

4. $\sqrt{36} = 6$ because $6^2 = 36$.
5. $\sqrt{0} = 0$ because $0^2 = 0$.
6. $\sqrt{\dfrac{4}{49}} = \dfrac{2}{7}$ because $\left(\dfrac{2}{7}\right)^2 = \dfrac{4}{49}$.
7. $\sqrt{0.25} = 0.5$ because $(0.5)^2 = 0.25$.
8. $\sqrt{x^6} = x^3$ because $(x^3)^2 = x^6$.
9. $\sqrt{9x^{12}} = 3x^6$ because $(3x^6)^2 = 9x^{12}$.

(Continued on next page)

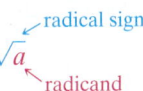

Objectives

A Find Square Roots.

B Approximate Roots Using a Calculator.

C Find Cube Roots.

D Find nth Roots.

E Find $\sqrt[n]{a^n}$ when a Is Any Real Number.

F Find Function Values and Graph Square and Cube Root Functions.

Practice 1–3

Find the real square roots of each number.

1. 36   **2.** 81   **3.** -16

Practice 4–11

Find each square root. Assume that all variables represent nonnegative real numbers.

4. $\sqrt{25}$ **5.** $\sqrt{0}$
6. $\sqrt{\dfrac{9}{25}}$ **7.** $\sqrt{0.36}$
8. $\sqrt{x^{10}}$ **9.** $\sqrt{36x^6}$
10. $-\sqrt{25}$ **11.** $\sqrt{-25}$

Answers

1. $6, -6$ **2.** $9, -9$
3. no real number square root
4. 5   **5.** 0   **6.** $\dfrac{3}{5}$ **7.** 0.6   **8.** x^5
9. $6x^3$ **10.** -5 **11.** not a real number

▶ **10.** $-\sqrt{81} = -9$. The negative in front of the radical indicates the negative square root of 81.

▶ **11.** $\sqrt{-81}$ is not a real number. (There is no real number whose square is -81.)

■ **Work Practice 4–11**

See Example 3 for a discussion of the square root of a negative number. For example, can we simplify $\sqrt{-4}$? That is, can we find a real number whose square is -4? No, there is no real number whose square is -4, and we say that $\sqrt{-4}$ is not a real number. In general:

The square root of a negative number is not a real number.

Helpful Hint

- Remember: $\sqrt{0} = 0$.
- Don't forget that the square root of a negative number is not a real number. For example,

  $\sqrt{-9}$ is not a real number

because there is no real number that when multiplied by itself would give a product of -9. In Section 10.7, we will see what kind of a number $\sqrt{-9}$ is.

Objective B Approximating Roots ▶

Recall that numbers such as 1, 4, 9, and 25 are called **perfect squares,** since $1 = 1^2, 4 = 2^2, 9 = 3^2$, and $25 = 5^2$. Square roots of perfect square radicands simplify to rational numbers. What happens when we try to simplify a root such as $\sqrt{3}$? Since 3 is not a perfect square, $\sqrt{3}$ is not a rational number. It is called an **irrational number,** and we can find a decimal **approximation** of it. To find decimal approximations, we can use a calculator. For example, an approximation for $\sqrt{3}$ is

$$\sqrt{3} \approx 1.732$$
$\uparrow$
approximation symbol

To see if the approximation is reasonable, notice that since

$1 < 3 < 4$,   then
$\sqrt{1} < \sqrt{3} < \sqrt{4}$,   or
$1 < \sqrt{3} < 2$.

We found $\sqrt{3} \approx 1.732$, a number between 1 and 2, so our result is reasonable.

Practice 12

Use a calculator to approximate $\sqrt{30}$. Round the approximation to three decimal places and check to see that your approximation is reasonable.

Example 12   Use a calculator to approximate $\sqrt{20}$. Round the approximation to three decimal places and check to see that your approximation is reasonable.

Solution:

$$\sqrt{20} \approx 4.472$$

Is this reasonable? Since $16 < 20 < 25$, then $\sqrt{16} < \sqrt{20} < \sqrt{25}$, or $4 < \sqrt{20} < 5$. The approximation is between 4 and 5 and is thus reasonable.

■ **Work Practice 12**

Objective C Finding Cube Roots ▶

Finding roots can be extended to other roots such as cube roots. For example, since $2^3 = 8$, we call 2 the *cube root* of 8. In symbols, we write

$$\sqrt[3]{8} = 2$$

Answer

12. 5.477

Cube Root

The **cube root** of a real number a is written as $\sqrt[3]{a}$, and

$$\sqrt[3]{a} = b \quad \text{only if} \quad b^3 = a$$

From this definition, we have

$\sqrt[3]{64} = 4$ since $4^3 = 64$

$\sqrt[3]{-27} = -3$ since $(-3)^3 = -27$

$\sqrt[3]{x^3} = x$ since $x^3 = x^3$

Notice that, unlike with square roots, *it is possible to have a negative radicand when finding a cube root.* This is so because the *cube* of a negative number is a negative number. Therefore, the *cube root* of a negative number is a negative number.

Examples Find each cube root.

13. $\sqrt[3]{1} = 1$ because $1^3 = 1$.

14. $\sqrt[3]{-64} = -4$ because $(-4)^3 = -64$.

15. $\sqrt[3]{\dfrac{8}{125}} = \dfrac{2}{5}$ because $\left(\dfrac{2}{5}\right)^3 = \dfrac{8}{125}$.

16. $\sqrt[3]{x^6} = x^2$ because $(x^2)^3 = x^6$.

17. $\sqrt[3]{-8x^9} = -2x^3$ because $(-2x^3)^3 = -8x^9$.

■ **Work Practice 13–17**

Practice 13–17

Find each cube root.

13. $\sqrt[3]{0}$ **14.** $\sqrt[3]{-8}$

15. $\sqrt[3]{\dfrac{1}{64}}$ **16.** $\sqrt[3]{x^9}$

17. $\sqrt[3]{-64x^6}$

Teaching Tip

For Examples 16 and 17, have students verify that the answer is correct for positive and negative values of x.

Objective D Finding *n*th Roots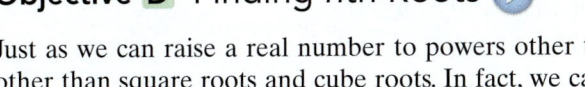

Just as we can raise a real number to powers other than 2 or 3, we can find roots other than square roots and cube roots. In fact, we can find the **nth root** of a number, where n is any natural number. In symbols, the *n*th root of a is written as $\sqrt[n]{a}$, where n is called the **index.** The index 2 is usually omitted for square roots.

Helpful Hint

If the index is even, such as in $\sqrt{}, \sqrt[4]{}, \sqrt[6]{}$, and so on, the radicand must be nonnegative for the root to be a real number. For example,

$\sqrt[4]{16} = 2$, but $\sqrt[4]{-16}$ is not a real number,

$\sqrt[6]{64} = 2$, but $\sqrt[6]{-64}$ is not a real number,

If the index is odd, such as in $\sqrt[3]{}, \sqrt[5]{}$, and so on, the radicand may be any real number. For example,

$\sqrt[3]{64} = 4$ and $\sqrt[3]{-64} = -4$,

$\sqrt[5]{32} = 2$ and $\sqrt[5]{-32} = -2$

✓**Concept Check** Which one is not a real number?

a. $\sqrt[3]{-15}$ **b.** $\sqrt[4]{-15}$ **c.** $\sqrt[5]{-15}$ **d.** $\sqrt{(-15)^2}$

Answers

13. 0 **14.** −2 **15.** $\dfrac{1}{4}$ **16.** x^3 **17.** $-4x^2$

✓**Concept Check Answer**

b

Practice 18–22

Find each root.

18. $\sqrt[4]{16}$ **19.** $\sqrt[5]{-32}$

20. $-\sqrt{36}$ **21.** $\sqrt[4]{-16}$

22. $\sqrt[3]{8x^6}$

Examples   Find each root.

18. $\sqrt[4]{81} = 3$ because $3^4 = 81$ and 3 is positive.

19. $\sqrt[5]{-243} = -3$ because $(-3)^5 = -243$.

20. $-\sqrt{25} = -5$ because -5 is the opposite of $\sqrt{25}$.

21. $\sqrt[4]{-81}$ is not a real number. There is no real number that, when raised to the fourth power, is -81.

22. $\sqrt[3]{64x^3} = 4x$ because $(4x)^3 = 64x^3$.

■ **Work Practice 18–22**

Objective E Finding $\sqrt[n]{a^n}$ when a Is Any Real Number

Recall that the notation $\sqrt{a^2}$ indicates the positive square root of a^2 only. For example,

$$\sqrt{(-5)^2} = \sqrt{25} = 5$$

When variables are present in the radicand and it is *unclear whether the variable represents a positive number or a negative number*, absolute value bars are sometimes needed to ensure that the result is a positive number. For example,

$$\sqrt{x^2} = |x|$$

This ensures that the result is positive. This same situation may occur when the index is any *even* positive integer. When the index is any *odd* positive integer, absolute value bars are not necessary.

> ### Finding $\sqrt[n]{a^n}$
>
> If n is an *even* positive integer, then $\sqrt[n]{a^n} = |a|$.
>
> If n is an *odd* positive integer, then $\sqrt[n]{a^n} = a$.

Practice 23–29

Simplify. Assume that the variables represent any real number.

23. $\sqrt{(-5)^2}$ **24.** $\sqrt{x^6}$

25. $\sqrt[4]{(x+6)^4}$ **26.** $\sqrt[3]{(-3)^3}$

27. $\sqrt[5]{(7x-1)^5}$ **28.** $\sqrt{36x^2}$

29. $\sqrt{x^2 + 6x + 9}$

Teaching Tip

Have students work in pairs to determine the domains of the functions in the examples and practices. Then have students share results with the class. If results vary, have students try to convince others of their solution until agreement is reached.

Answers

18. 2 **19.** -2 **20.** -6

21. not a real number **22.** $2x^2$ **23.** 5

24. $|x^3|$ **25.** $|x+6|$ **26.** -3

27. $7x-1$ **28.** $6|x|$ **29.** $|x+3|$

Examples   Simplify. Assume that the variables represent any real number.

23. $\sqrt{(-3)^2} = |-3| = 3$ When the index is even, the absolute value bars ensure that the result is not negative.

24. $\sqrt{x^2} = |x|$

25. $\sqrt[4]{(x-2)^4} = |x-2|$

26. $\sqrt[3]{(-5)^3} = -5$ Absolute value bars are not needed when the index is odd.

27. $\sqrt[5]{(2x-7)^5} = 2x - 7$

28. $\sqrt{25x^2} = 5|x|$

29. $\sqrt{x^2 + 2x + 1} = \sqrt{(x+1)^2} = |x+1|$

■ **Work Practice 23–29**

Objective F Finding Function Values and Graphing Radical Functions ▶

In general, **radical functions** are functions of the form

$$f(x) = \sqrt[n]{x}.$$

Recall that the domain of a function in x is the set of all possible replacement values of x. This means that if n is even, the domain is the set of all nonnegative numbers, or $\{x \,|\, x \geq 0\}$ or $[0, \infty)$. If n is odd, the domain is the set of all real numbers, or $(-\infty, \infty)$. Keep this in mind as we find function values. In Section 8.5, we graphed square root functions and discussed their domains.

Examples If $f(x) = \sqrt{x - 4}$ and $g(x) = \sqrt[3]{x + 2}$, find each function value.

30. $f(8) = \sqrt{8 - 4} = \sqrt{4} = 2$

31. $f(6) = \sqrt{6 - 4} = \sqrt{2}$

32. $g(-1) = \sqrt[3]{-1 + 2} = \sqrt[3]{1} = 1$

33. $g(1) = \sqrt[3]{1 + 2} = \sqrt[3]{3}$

■ **Work Practice 30–33**

Practice 30–33

If $f(x) = \sqrt{x + 2}$ and $g(x) = \sqrt[3]{x - 1}$, find each function value.

30. $f(7)$ **31.** $g(9)$

32. $f(0)$ **33.** $g(10)$

Notice that for the function $f(x) = \sqrt{x - 4}$, the domain includes all real numbers that make the radicand ≥ 0. To see what numbers these are, solve $x - 4 \geq 0$ and find that $x \geq 4$. The domain is $\{x \mid x \geq 4\}$ or $[4, \infty)$.

The domain of the cube root function $g(x) = \sqrt[3]{x + 2}$ is the set of real numbers or $(-\infty, \infty)$.

Recall from Section 8.5 that the graph of $f(x) = \sqrt{x}$ is

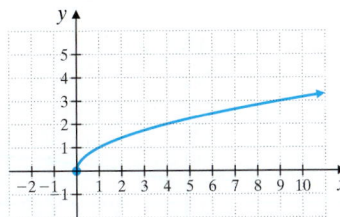

Let's now graph the function $f(x) = \sqrt{x - 2}$.

Example 34 Identify the domain and then graph the function $f(x) = \sqrt{x - 2}$.

Solution: To graph, we identify the domain, evaluate the function for several values of x, plot the resulting points, and connect the points with a smooth curve. The domain of $f(x) = \sqrt{x - 2}$ includes all real numbers that make the radicand ≥ 0. By solving $x - 2 \geq 0$, we find that $x \geq 2$. The domain is $\{x \mid x \geq 2\}$ or $[2, \infty)$.

x	$f(x) = \sqrt{x - 2}$
2	0
3	1
5	$\sqrt{3} \approx 1.7$
6	2
11	3

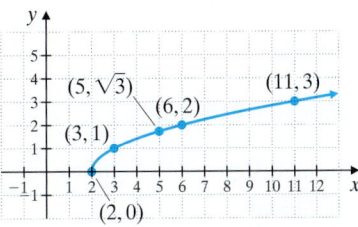

Notice that the graph of this function passes the vertical line test, as expected.

■ **Work Practice 34**

Practice 34

Identify the domain and then graph the function $h(x) = \sqrt{x + 2}$.

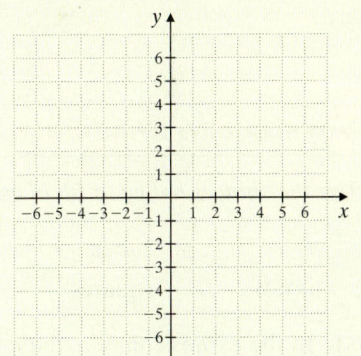

The equation $f(x) = \sqrt[3]{x}$ also describes a function. Here, x may be any real number, so the domain of this function is the set of all real numbers, or $(-\infty, \infty)$. A few function values are given next.

$$f(0) = \sqrt[3]{0} = 0$$
$$f(1) = \sqrt[3]{1} = 1$$
$$f(-1) = \sqrt[3]{-1} = -1$$
$$f(6) = \sqrt[3]{6}$$
$$f(-6) = \sqrt[3]{-6}$$

Here, there is no rational number whose cube is 6. Thus, the radicals do not simplify to rational numbers.

$$f(8) = \sqrt[3]{8} = 2$$
$$f(-8) = \sqrt[3]{-8} = -2$$

Answers

30. 3   **31.** 2   **32.** $\sqrt{2}$ **33.** $\sqrt[3]{9}$

34.

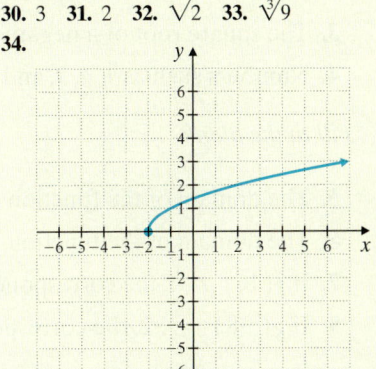

Practice 35

Identify the domain and then graph the function $h(x) = \sqrt[3]{x} + 2$.

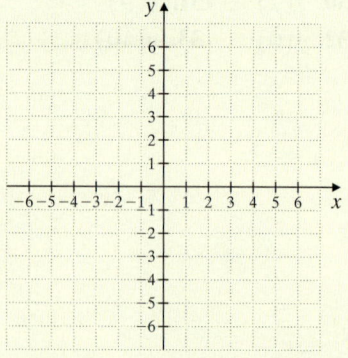

Answer

35.

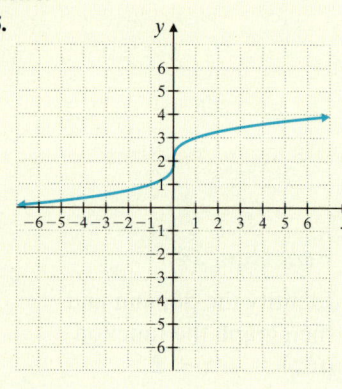

Example 35 Identify the domain and then graph the function $f(x) = \sqrt[3]{x}$.

Solution: To graph, we identify the domain, plot points, and connect the points with a smooth curve. The domain of this function is the set of all real numbers, or $(-\infty, \infty)$. The table comes from the function values obtained earlier. (We have approximated $\sqrt[3]{6}$ and $\sqrt[3]{-6}$ for graphing purposes.)

x	$f(x) = \sqrt[3]{x}$
0	0
1	1
-1	-1
6	$\sqrt[3]{6} \approx 1.8$
-6	$\sqrt[3]{-6} \approx -1.8$
8	2
-8	-2

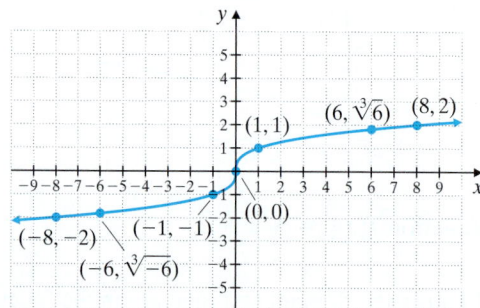

The graph of this function passes the vertical line test, as expected.

🟧 **Work Practice 35**

Vocabulary, Readiness & Video Check

Use the choices below to fill in each blank. Not all choices will be used.

is	cubes	$-\sqrt{a}$	radical sign	index
is not	squares	$\sqrt{-a}$	radicand	

1. In the expression $\sqrt[n]{a}$, the n is called the _____index_____, the $\sqrt{}$ is called the _____radical sign_____, and a is called the _____radicand_____.

2. If $\sqrt{a}$ is the positive square root of a, $a \neq 0$, then _____$-\sqrt{a}$_____ is the negative square root of a.

3. The square root of a negative number _____is not_____ a real number.

4. Numbers such as 1, 4, 9, and 25 are called perfect _____squares_____.

Fill in the blank.

5. The domain of the function $f(x) = \sqrt{x}$ is _____$[0, \infty)$_____.

6. The domain of the function $f(x) = \sqrt[3]{x}$ is _____$(-\infty, \infty)$_____.

7. If $f(16) = 4$, the corresponding ordered pair is _____$(16, 4)$_____.

8. If $g(-8) = -2$, the corresponding ordered pair is _____$(-8, -2)$_____.

Choose the correct letter or letters. No pencil is needed; just think your way through these.

9. Which radical is not a real number?

   a. $\sqrt{3}$ **b.** $-\sqrt{11}$ **c.** $\sqrt[3]{-10}$ **d.** $\sqrt{-10}$ d

10. Which radical(s) simplify to 3?

   a. $\sqrt{9}$ **b.** $\sqrt{-9}$ **c.** $\sqrt[3]{27}$ **d.** $\sqrt[3]{-27}$ a, c

11. Which radical(s) simplify to −3?

   a. $\sqrt{9}$ **b.** $\sqrt{-9}$ **c.** $\sqrt[3]{27}$ **d.** $\sqrt[3]{-27}$ d

12. Which radical does not simplify to a whole number?

   a. $\sqrt{64}$ **b.** $\sqrt[3]{64}$ **c.** $\sqrt{8}$ **d.** $\sqrt[3]{8}$ c

---

Martin-Gay Interactive Videos

See Video 10.1 🍒

Watch the section lecture video and answer the following questions.

Objective A **13.** From ▥ Examples 5 and 6, when simplifying radicals containing variables with exponents, describe a shortcut you can use. ▶

Objective B **14.** From ▥ Example 9, how can you determine a reasonable approximation for a non-perfect square root without using a calculator? ▶

Objective C **15.** From ▥ Example 11, what is an important difference between the square root and the cube root of a negative number? ▶

Objective D **16.** From ▥ Example 12, what conclusion is made about the even root of a negative number? ▶

Objective E **17.** From the lecture before ▥ Example 17, why do you think no absolute value bars are used when *n* is odd? ▶

Objective F **18.** In ▥ Example 19, the domain is found by looking at the graph. How can the domain be found by looking at the function? ▶

See video answer section.

10.1 Exercise Set MyMathLab® ▶

Objective A *Find the real square roots of each number. See Examples 1 through 3.*

1. 4   2, −2

2. 9   3, −3

3. −25   no real number square roots

4. −49   no real number square roots

5. 100   10, −10

6. 64   8, −8

Find each square root. Assume that all variables represent nonnegative real numbers. See Examples 4 through 11.

7. $\sqrt{100}$ 10

8. $\sqrt{400}$ 20

9. $\sqrt{\dfrac{1}{4}}$ $\dfrac{1}{2}$

10. $\sqrt{\dfrac{9}{25}}$ $\dfrac{3}{5}$

11. $\sqrt{0.0001}$ 0.01

12. $\sqrt{0.04}$ 0.2

13. $-\sqrt{36}$ −6

14. $-\sqrt{9}$ −3

15. $\sqrt{x^{10}}$ x^5

16. $\sqrt{x^{16}}$ x^8

17. $\sqrt{16y^6}$ $4y^3$

18. $\sqrt{64y^{20}}$ $8y^{10}$

Objective B *Use a calculator to approximate each square root to three decimal places. Check to see that each approximation is reasonable. See Example 12.*

19. $\sqrt{7}$ 2.646   **20.** $\sqrt{11}$ 3.317 **21.** $\sqrt{38}$ 6.164   **22.** $\sqrt{56}$ 7.483   **23.** $\sqrt{200}$ 14.142 **24.** $\sqrt{300}$ 17.321

Objective C *Find each cube root. See Examples 13 through 17.*

25. $\sqrt[3]{64}$ 4

26. $\sqrt[3]{27}$ 3

▶**27.** $\sqrt[3]{\dfrac{1}{8}}$ $\dfrac{1}{2}$

28. $\sqrt[3]{\dfrac{27}{64}}$ $\dfrac{3}{4}$

29. $\sqrt[3]{-1}$ -1

30. $\sqrt[3]{-125}$ -5

31. $\sqrt[3]{x^{12}}$ x^4

32. $\sqrt[3]{x^{15}}$ x^5

▶**33.** $\sqrt[3]{-27x^9}$ $-3x^3$

34. $\sqrt[3]{-64x^6}$ $-4x^2$

Objective D *Find each root. Assume that all variables represent nonnegative real numbers. See Examples 18 through 22.*

35. $-\sqrt[4]{16}$ -2

36. $\sqrt[5]{-243}$ -3

▶**37.** $\sqrt[4]{-16}$
not a real number

38. $\sqrt{-16}$
not a real number

▶**39.** $\sqrt[5]{-32}$ -2

40. $\sqrt[5]{-1}$ -1

41. $\sqrt[5]{x^{20}}$ x^4

42. $\sqrt[4]{x^{20}}$ x^5

▶**43.** $\sqrt[6]{64x^{12}}$ $2x^2$

44. $\sqrt[5]{-32x^{15}}$
$-2x^3$

45. $\sqrt{81x^4}$ $9x^2$

46. $\sqrt[4]{81x^4}$ $3x$

47. $\sqrt[4]{256x^8}$ $4x^2$

48. $\sqrt{256x^8}$ $16x^4$

Objective E *Simplify. Assume that the variables represent any real number. See Examples 23 through 29.*

▶**49.** $\sqrt{(-8)^2}$ 8

50. $\sqrt{(-7)^2}$ 7

▶**51.** $\sqrt[3]{(-8)^3}$ -8

52. $\sqrt[5]{(-7)^5}$ -7

53. $\sqrt{4x^2}$ $2|x|$

54. $\sqrt[4]{16x^4}$ $2|x|$

55. $\sqrt[3]{x^3}$ x

56. $\sqrt[5]{x^5}$ x

▶**57.** $\sqrt{(x-5)^2}$
$|x-5|$

58. $\sqrt{(y-6)^2}$
$|y-6|$

59. $\sqrt{x^2+4x+4}$
(*Hint:* Factor the polynomial first.) $|x+2|$

60. $\sqrt{x^2-8x+16}$
(*Hint:* Factor the polynomial first.) $|x-4|$

Objectives A B C D Mixed Practice *Simplify each radical. Assume that all variables represent positive real numbers. See Examples 1 through 22.*

61. $-\sqrt{121}$ -11

62. $-\sqrt[3]{125}$ -5

63. $\sqrt[3]{8x^3}$ $2x$

64. $\sqrt{16x^8}$ $4x^4$

65. $\sqrt{y^{12}}$ y^6

66. $\sqrt[3]{y^{12}}$
y^4

67. $\sqrt{25a^2b^{20}}$
$5ab^{10}$

68. $\sqrt{9x^4y^6}$
$3x^2y^3$

69. $\sqrt[3]{-27x^{12}y^9}$
$-3x^4y^3$

70. $\sqrt[3]{-8a^{21}y^6}$
$-2a^7b^2$

71. $\sqrt[4]{a^{16}b^4}$
a^4b

72. $\sqrt[4]{x^8y^{12}}$
x^2y^3

▶**73.** $\sqrt[5]{-32x^{10}y^5}$
$-2x^2y$

74. $\sqrt[5]{-243z^{15}}$
$-3z^3$

75. $\sqrt{\dfrac{25}{49}}$ $\dfrac{5}{7}$

76. $\sqrt{\dfrac{4}{81}}$ $\dfrac{2}{9}$

77. $\sqrt{\dfrac{x^2}{4y^2}}$ $\dfrac{x}{2y}$

78. $\sqrt{\dfrac{y^{10}}{9x^6}}$ $\dfrac{y^5}{3x^3}$

79. $-\sqrt[3]{\dfrac{z^{21}}{27x^3}}$ $-\dfrac{z^7}{3x}$

80. $-\sqrt[3]{\dfrac{64a^3}{b^9}}$ $-\dfrac{4a}{b^3}$

81. $\sqrt[4]{\dfrac{x^4}{16}}$ $\dfrac{x}{2}$

82. $\sqrt[4]{\dfrac{y^4}{81x^4}}$ $\dfrac{y}{3x}$

Choose the correct letter or letters. No pencil is needed; just think your way through these.

9. Which radical is not a real number?

    a. $\sqrt{3}$ **b.** $-\sqrt{11}$ **c.** $\sqrt[3]{-10}$ **d.** $\sqrt{-10}$ d

10. Which radical(s) simplify to 3?

    a. $\sqrt{9}$ **b.** $\sqrt{-9}$ **c.** $\sqrt[3]{27}$ **d.** $\sqrt[3]{-27}$ a, c

11. Which radical(s) simplify to -3?

    a. $\sqrt{9}$ **b.** $\sqrt{-9}$ **c.** $\sqrt[3]{27}$ **d.** $\sqrt[3]{-27}$ d

12. Which radical does not simplify to a whole number?

    a. $\sqrt{64}$ **b.** $\sqrt[3]{64}$ **c.** $\sqrt{8}$ **d.** $\sqrt[3]{8}$ c

---

Martin-Gay Interactive Videos

See Video 10.1 🍒

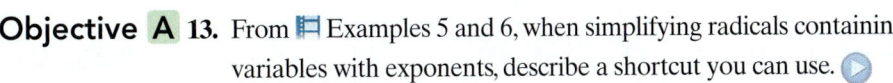
Watch the section lecture video and answer the following questions.

Objective A 13. From ⊞ Examples 5 and 6, when simplifying radicals containing variables with exponents, describe a shortcut you can use. ▶

Objective B 14. From ⊞ Example 9, how can you determine a reasonable approximation for a non-perfect square root without using a calculator? ▶

Objective C 15. From ⊞ Example 11, what is an important difference between the square root and the cube root of a negative number? ▶

Objective D 16. From ⊞ Example 12, what conclusion is made about the even root of a negative number? ▶

Objective E 17. From the lecture before ⊞ Example 17, why do you think no absolute value bars are used when n is odd? ▶

Objective F 18. In ⊞ Example 19, the domain is found by looking at the graph. How can the domain be found by looking at the function? ▶

See video answer section.

10.1   **Exercise Set** MyMathLab® ▶

Objective A   *Find the real square roots of each number. See Examples 1 through 3.*

1. 4   2, -2 **2.** 9   3, -3 **3.** -25
                                                                           no real number square roots

4. -49 **5.** 100   10, -10 **6.** 64   8, -8
     no real number square roots

Find each square root. Assume that all variables represent nonnegative real numbers. See Examples 4 through 11.

7. $\sqrt{100}$ 10   **8.** $\sqrt{400}$ 20   **9.** $\sqrt{\dfrac{1}{4}}$ $\dfrac{1}{2}$ **10.** $\sqrt{\dfrac{9}{25}}$ $\dfrac{3}{5}$

11. $\sqrt{0.0001}$ 0.01   **12.** $\sqrt{0.04}$ 0.2   **13.** $-\sqrt{36}$ -6 **14.** $-\sqrt{9}$ -3

15. $\sqrt{x^{10}}$ x^5 **16.** $\sqrt{x^{16}}$ x^8 **17.** $\sqrt{16y^6}$ $4y^3$ **18.** $\sqrt{64y^{20}}$ $8y^{10}$

Objective B   *Use a calculator to approximate each square root to three decimal places. Check to see that each approximation is reasonable. See Example 12.*

19. $\sqrt{7}$ 2.646   **20.** $\sqrt{11}$ 3.317 ▶ **21.** $\sqrt{38}$ 6.164   **22.** $\sqrt{56}$ 7.483   **23.** $\sqrt{200}$ 14.142   **24.** $\sqrt{300}$ 17.321

Objective C *Find each cube root. See Examples 13 through 17.*

25. $\sqrt[3]{64}$ 4

26. $\sqrt[3]{27}$ 3

27. $\sqrt[3]{\dfrac{1}{8}}$ $\dfrac{1}{2}$

28. $\sqrt[3]{\dfrac{27}{64}}$ $\dfrac{3}{4}$

29. $\sqrt[3]{-1}$ -1

30. $\sqrt[3]{-125}$ -5

31. $\sqrt[3]{x^{12}}$ x^4

32. $\sqrt[3]{x^{15}}$ x^5

33. $\sqrt[3]{-27x^9}$ $-3x^3$

34. $\sqrt[3]{-64x^6}$ $-4x^2$

Objective D *Find each root. Assume that all variables represent nonnegative real numbers. See Examples 18 through 22.*

35. $-\sqrt[4]{16}$ -2

36. $\sqrt[5]{-243}$ -3

37. $\sqrt[4]{-16}$
not a real number

38. $\sqrt{-16}$
not a real number

39. $\sqrt[5]{-32}$ -2

40. $\sqrt[5]{-1}$ -1

41. $\sqrt[5]{x^{20}}$ x^4

42. $\sqrt[4]{x^{20}}$ x^5

43. $\sqrt[6]{64x^{12}}$ $2x^2$

44. $\sqrt[5]{-32x^{15}}$
$-2x^3$

45. $\sqrt{81x^4}$ $9x^2$

46. $\sqrt[4]{81x^4}$ $3x$

47. $\sqrt[4]{256x^8}$ $4x^2$

48. $\sqrt{256x^8}$ $16x^4$

Objective E *Simplify. Assume that the variables represent any real number. See Examples 23 through 29.*

49. $\sqrt{(-8)^2}$ 8

50. $\sqrt{(-7)^2}$ 7

51. $\sqrt[3]{(-8)^3}$ -8

52. $\sqrt[5]{(-7)^5}$ -7

53. $\sqrt{4x^2}$ $2|x|$

54. $\sqrt[4]{16x^4}$ $2|x|$

55. $\sqrt[3]{x^3}$ x

56. $\sqrt[5]{x^5}$ x

57. $\sqrt{(x-5)^2}$
$|x-5|$

58. $\sqrt{(y-6)^2}$
$|y-6|$

59. $\sqrt{x^2+4x+4}$
(*Hint:* Factor the polynomial first.) $|x+2|$

60. $\sqrt{x^2-8x+16}$
(*Hint:* Factor the polynomial first.) $|x-4|$

Objectives A B C D **Mixed Practice** *Simplify each radical. Assume that all variables represent positive real numbers. See Examples 1 through 22.*

61. $-\sqrt{121}$ -11

62. $-\sqrt[3]{125}$ -5

63. $\sqrt[3]{8x^3}$ $2x$

64. $\sqrt{16x^8}$ $4x^4$

65. $\sqrt{y^{12}}$ y^6

66. $\sqrt[3]{y^{12}}$
y^4

67. $\sqrt{25a^2b^{20}}$
$5ab^{10}$

68. $\sqrt{9x^4y^6}$
$3x^2y^3$

69. $\sqrt[3]{-27x^{12}y^9}$
$-3x^4y^3$

70. $\sqrt[3]{-8a^{21}y^6}$
$-2a^7b^2$

71. $\sqrt[4]{a^{16}b^4}$
a^4b

72. $\sqrt[4]{x^8y^{12}}$
x^2y^3

73. $\sqrt[5]{-32x^{10}y^5}$
$-2x^2y$

74. $\sqrt[5]{-243z^{15}}$
$-3z^3$

75. $\sqrt{\dfrac{25}{49}}$ $\dfrac{5}{7}$

76. $\sqrt{\dfrac{4}{81}}$ $\dfrac{2}{9}$

77. $\sqrt{\dfrac{x^2}{4y^2}}$ $\dfrac{x}{2y}$

78. $\sqrt{\dfrac{y^{10}}{9x^6}}$ $\dfrac{y^5}{3x^3}$

79. $-\sqrt[3]{\dfrac{z^{21}}{27x^3}}$ $-\dfrac{z^7}{3x}$

80. $-\sqrt[3]{\dfrac{64a^3}{b^9}}$ $-\dfrac{4a}{b^3}$

81. $\sqrt[4]{\dfrac{x^4}{16}}$ $\dfrac{x}{2}$

82. $\sqrt[4]{\dfrac{y^4}{81x^4}}$ $\dfrac{y}{3x}$

Objective F *If $f(x) = \sqrt{2x + 3}$ and $g(x) = \sqrt[3]{x - 8}$, find each function value. See Examples 30 through 33.*

▶**83.** $f(0)$ $\sqrt{3}$

84. $g(0)$ -2

▶**85.** $g(7)$ -1

86. $f(-1)$ 1

87. $g(-19)$ -3

88. $f(3)$ 3

89. $f(2)$ $\sqrt{7}$

90. $g(1)$ $\sqrt[3]{-7}$

Identify the domain and then graph each function. See Example 34.

▶**91.** $f(x) = \sqrt{x} + 2$

92. $f(x) = \sqrt{x} - 2$

93. $f(x) = \sqrt{x - 3}$

94. $f(x) = \sqrt{x + 1}$

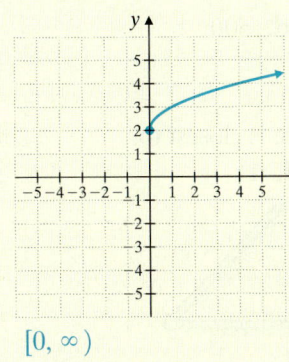

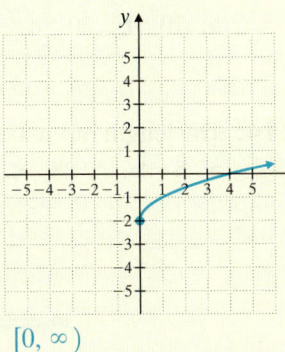

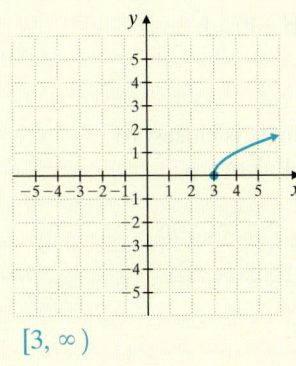

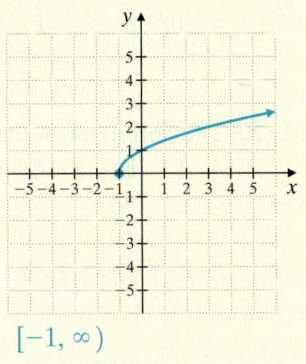

$[0, \infty)$

$[0, \infty)$

$[3, \infty)$

$[-1, \infty)$

Identify the domain and then graph each function. See Example 35.

95. $f(x) = \sqrt[3]{x} + 1$

96. $f(x) = \sqrt[3]{x} - 2$

97. $g(x) = \sqrt[3]{x - 1}$

98. $g(x) = \sqrt[3]{x + 1}$

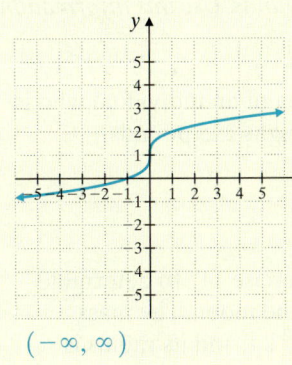

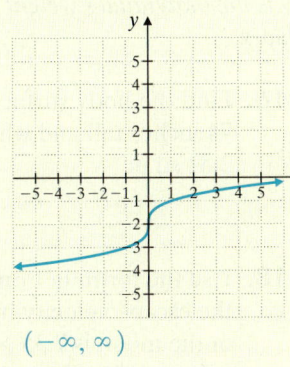

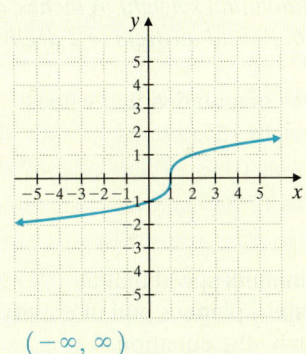

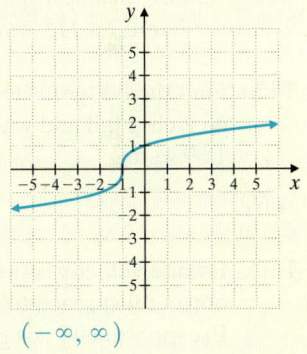

$(-\infty, \infty)$

$(-\infty, \infty)$

$(-\infty, \infty)$

$(-\infty, \infty)$

Review

Simplify each exponential expression. See Sections 5.1 and 5.2.

99. $(-2x^3y^2)^5$ $-32x^{15}y^{10}$

100. $(4y^6z^7)^3$ $64y^{18}z^{21}$

101. $(-3x^2y^3z^5)(20x^5y^7)$
$-60x^7y^{10}z^5$

102. $(-14a^5bc^2)(2abc^4)$
$-28a^6b^2c^6$

103. $\dfrac{7x^{-1}y}{14(x^5y^2)^{-2}}$ $\dfrac{x^9y^5}{2}$

104. $\dfrac{(2a^{-1}b^2)^3}{(8a^2b)^{-2}}$ $512ab^8$

Concept Extensions

Which of the following are not real numbers? See the Concept Check in this section.

105. $\sqrt{-17}$ not a real   **106.** $\sqrt[3]{-17}$
      number   real number

107. $\sqrt[10]{-17}$ **108.** $\sqrt[15]{-17}$
    not a real number   real number

109. Explain why $\sqrt{-64}$ is not a real number.
     answers may vary

110. Explain why $\sqrt[3]{-64}$ is a real number.
     answers may vary

For Exercises 111 through 114, do not use a calculator.

111. $\sqrt{160}$ is closest to
     a. 10   **b.** 13   **c.** 20   **d.** 40   b

112. $\sqrt{1000}$ is closest to
     a. 10   **b.** 30   **c.** 100   **d.** 500   b

△113. The perimeter of the triangle is closest to
     a. 12   **b.** 18   **c.** 66   **d.** 132   b

$\sqrt{30}$ $\sqrt{10}$

$\sqrt{90}$

114. The length of the bent wire is closest to
     a. 5   **b.** $\sqrt{28}$ **c.** 7   **d.** 14   c

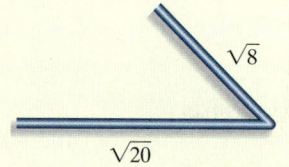

$\sqrt{8}$

$\sqrt{20}$

The Mosteller formula for calculating adult body surface area is $B = \sqrt{\dfrac{hw}{3131}}$, where B is an individual's body surface area in square meters, h is the individual's height in inches, and w is the individual's weight in pounds. Use this information to answer Exercises 115 and 116. Round answers to 2 decimal places.

△115. Find the body surface area of an individual who is 66 inches tall and who weighs 135 pounds.
     1.69 sq m

△116. Find the body surface area of an individual who is 74 inches tall and who weighs 225 pounds.
     2.31 sq m

Solve.

117. Escape velocity is the minimum speed that an object must reach to escape a planet's pull of gravity. Escape velocity v is given by the equation $v = \sqrt{\dfrac{2Gm}{r}}$, where m is the mass of the planet, r is its radius, and G is the universal gravitational constant, which has a value of $G = 6.67 \times 10^{-11}$ m³/kg·s². The mass of Earth is 5.97×10^{24} kg, and its radius is 6.37×10^6 m. Use this information to find the escape velocity for Earth in meters per second. Round to the nearest whole number. (*Source:* National Space Science Data Center)   11,181 m per sec

118. Use the formula from Exercise 117 to determine the escape velocity from the moon. The mass of the moon is 7.35×10^{22} kg, and its radius is 1.74×10^6 m. Round to the nearest whole. (*Source:* National Space Science Data Center)
     2374 m per sec

119. Suppose a classmate tells you that $\sqrt{13} \approx 5.7$. Without a calculator, how can you convince your classmate that he or she must have made an error?
     answers may vary

120. Suppose a classmate tells you that $\sqrt[3]{10} \approx 3.2$. Without a calculator, how can you convince your friend that he or she must have made an error?
     answers may vary

10.2 Rational Exponents

Objective A Understanding $a^{1/n}$

So far in this text, we have not defined expressions with rational exponents such as $3^{1/2}$, $x^{2/3}$, and $-9^{-1/4}$. We will define these expressions so that the rules for exponents shall apply to these rational exponents as well.

Suppose that $x = 5^{1/3}$. Then

$$x^3 = (5^{1/3})^3 = 5^{1/3 \cdot 3} = 5^1 \text{ or } 5$$

$\underset{\text{for exponents}}{\underline{\quad}\text{using rules} \uparrow}$

Since $x^3 = 5$, then x is the number whose cube is 5, or $x = \sqrt[3]{5}$. Notice that we also know that $x = 5^{1/3}$. This means that

$$5^{1/3} = \sqrt[3]{5}$$

> **Definition of $a^{1/n}$**
>
> If n is a positive integer greater than 1 and $\sqrt[n]{a}$ is a real number, then
>
> $$a^{1/n} = \sqrt[n]{a}$$

Notice that the denominator of the rational exponent corresponds to the index of the radical.

Examples   Use radical notation to rewrite each expression. Simplify if possible.

1. $4^{1/2} = \sqrt{4} = 2$
2. $64^{1/3} = \sqrt[3]{64} = 4$
3. $x^{1/4} = \sqrt[4]{x}$
4. $-9^{1/2} = -\sqrt{9} = -3$
5. $(81x^8)^{1/4} = \sqrt[4]{81x^8} = 3x^2$
6. $5y^{1/3} = 5\sqrt[3]{y}$

■ **Work Practice 1–6**

Objective B Understanding $a^{m/n}$

As we expand our use of exponents to include $\frac{m}{n}$, we define their meaning so that rules for exponents still hold true. For example, by properties of exponents,

$$8^{2/3} = (8^{1/3})^2 = (\sqrt[3]{8})^2 \quad \text{or} \quad 8^{2/3} = (8^2)^{1/3} = \sqrt[3]{8^2}$$

> **Definition of $a^{m/n}$**
>
> If m and n are positive integers greater than 1 with $\frac{m}{n}$ in simplest form, then
>
> $$a^{m/n} = \sqrt[n]{a^m} = (\sqrt[n]{a})^m$$
>
> as long as $\sqrt[n]{a}$ is a real number.

Objectives

A Understand the Meaning of $a^{1/n}$.

B Understand the Meaning of $a^{m/n}$.

C Understand the Meaning of $a^{-m/n}$.

D Use Rules for Exponents to Simplify Expressions That Contain Rational Exponents.

E Use Rational Exponents to Simplify Radical Expressions.

Practice 1–6

Use radical notation to rewrite each expression. Simplify if possible.

1. $25^{1/2}$ 2. $27^{1/3}$
3. $x^{1/5}$ 4. $-25^{1/2}$
5. $(-27y^6)^{1/3}$ 6. $7x^{1/5}$

Teaching Tip:
Classroom Activity

Point out the precise definition of $a^{m/n}$. Then ask students to work in pairs to investigate why this definition is only true for $\frac{m}{n}$ in simplest form. If they need a hint, have them try the definition for $a < 0$ and let m and n have a common factor of 2.

Answers
1. 5 2. 3 3. $\sqrt[5]{x}$ 4. -5
5. $-3y^2$ 6. $7\sqrt[5]{x}$

691

Notice that the denominator n of the rational exponent corresponds to the index of the radical. The numerator m of the rational exponent indicates that the base is to be raised to the mth power. This means that

$$8^{2/3} = \sqrt[3]{8^2} = \sqrt[3]{64} = 4 \quad \text{or} \quad 8^{2/3} = (\sqrt[3]{8})^2 = 2^2 = 4$$

Helpful Hint

Most of the time, $(\sqrt[n]{a})^m$ will be easier to calculate than $\sqrt[n]{a^m}$.

Practice 7–11

Use radical notation to rewrite each expression. Simplify if possible.

7. $9^{3/2}$ **8.** $-256^{3/4}$

9. $(-32)^{2/5}$ **10.** $\left(\dfrac{1}{4}\right)^{3/2}$

11. $(2x + 1)^{2/7}$

Examples   Use radical notation to rewrite each expression. Simplify if possible.

7. $4^{3/2} = (\sqrt{4})^3 = 2^3 = 8$

8. $-16^{3/4} = -(\sqrt[4]{16})^3 = -(2)^3 = -8$

9. $(-27)^{2/3} = (\sqrt[3]{-27})^2 = (-3)^2 = 9$

10. $\left(\dfrac{1}{9}\right)^{3/2} = \left(\sqrt{\dfrac{1}{9}}\right)^3 = \left(\dfrac{1}{3}\right)^3 = \dfrac{1}{27}$

11. $(4x - 1)^{3/5} = \sqrt[5]{(4x - 1)^3}$

◼ **Work Practice 7–11**

Helpful Hint

The *denominator* of a rational exponent is the index of the corresponding radical. For example, $x^{1/5} = \sqrt[5]{x}$, and $z^{2/3} = \sqrt[3]{z^2}$ or $z^{2/3} = (\sqrt[3]{z})^2$.

Teaching Tip

Ask students, "What is 3^{-2}? What is a^{-2}? What is a^{-n} when n is an integer? Using this information, what is $a^{-m/n}$ when m and n are integers?"

Objective C Understanding $a^{-m/n}$ ▶

The rational exponents we have given meaning to exclude negative rational numbers. To complete the set of definitions, we define $a^{-m/n}$.

Definition of $a^{-m/n}$

$$a^{-m/n} = \dfrac{1}{a^{m/n}}$$

as long as $a^{m/n}$ is a nonzero real number.

Practice 12–13

Write each expression with a positive exponent. Then simplify.

12. $27^{-2/3}$ **13.** $-256^{-3/4}$

Examples   Write each expression with a positive exponent. Then simplify.

12. $16^{-3/4} = \dfrac{1}{16^{3/4}} = \dfrac{1}{(\sqrt[4]{16})^3} = \dfrac{1}{2^3} = \dfrac{1}{8}$

13. $(-27)^{-2/3} = \dfrac{1}{(-27)^{2/3}} = \dfrac{1}{(\sqrt[3]{-27})^2} = \dfrac{1}{(-3)^2} = \dfrac{1}{9}$

◼ **Work Practice 12–13**

Answers

7. 27 **8.** −64 **9.** 4 **10.** $\dfrac{1}{8}$

11. $\sqrt[7]{(2x + 1)^2}$ **12.** $\dfrac{1}{9}$ **13.** $-\dfrac{1}{64}$

If an expression contains a negative rational exponent, you may want to first write the expression with a positive exponent, then interpret the rational exponent. Notice that the sign of the base is not affected by the sign of its exponent. For example,

$$9^{-3/2} = \frac{1}{9^{3/2}} = \frac{1}{(\sqrt{9})^3} = \frac{1}{27}$$

Also,

$$(-27)^{-1/3} = \frac{1}{(-27)^{1/3}} = -\frac{1}{3}$$

✓ **Concept Check** Which one is correct?

a. $-8^{2/3} = \frac{1}{4}$ **b.** $8^{-2/3} = -\frac{1}{4}$ **c.** $8^{-2/3} = -4$ **d.** $-8^{-2/3} = -\frac{1}{4}$

Objective **D** Using Rules for Exponents

It can be shown that the properties of integer exponents hold for rational exponents. By using these properties and definitions, we can now simplify expressions that contain rational exponents. These rules are repeated here for review.

Summary of Exponent Rules

If m and n are rational numbers, and a, b, and c are numbers for which the expressions below exist, then

Product rule for exponents: $a^m \cdot a^n = a^{m+n}$

Power rule for exponents: $(a^m)^n = a^{m \cdot n}$

Power rules for products and quotients: $(ab)^n = a^n b^n$ and

$$\left(\frac{a}{c}\right)^n = \frac{a^n}{c^n}, \quad c \neq 0$$

Quotient rule for exponents: $\dfrac{a^m}{a^n} = a^{m-n}, \quad a \neq 0$

Zero exponent: $a^0 = 1, \quad a \neq 0$

Negative exponent: $a^{-n} = \dfrac{1}{a^n}, \quad a \neq 0$

Examples   Use the properties of exponents to simplify.

14. $x^{1/2} x^{1/3} = x^{1/2 + 1/3} = x^{3/6 + 2/6} = x^{5/6}$ *Use the product rule.*

15. $\dfrac{7^{1/3}}{7^{4/3}} = 7^{1/3 - 4/3} = 7^{-3/3} = 7^{-1} = \dfrac{1}{7}$ *Use the quotient rule.*

16. $y^{-4/7} \cdot y^{6/7} = y^{-4/7 + 6/7} = y^{2/7}$ *Use the product rule.*

17. $(5^{3/8})^4 = 5^{3/8 \cdot 4} = 5^{12/8} = 5^{3/2}$ *Use the power rule.*

(Continued on next page)

Practice 14–18

Use the properties of exponents to simplify.

14. $x^{1/3} x^{1/4}$ **15.** $\dfrac{9^{2/5}}{9^{12/5}}$

16. $y^{-3/10} \cdot y^{6/10}$ **17.** $(11^{2/9})^3$

18. $\dfrac{(3x^{2/3})^3}{x^2}$

Answers

14. $x^{7/12}$ **15.** $\dfrac{1}{81}$ **16.** $y^{3/10}$

17. $11^{2/3}$ **18.** 27

✓ **Concept Check Answer**

d

18. $\dfrac{(2x^{2/5})^5}{x^2} = \dfrac{2^5(x^{2/5})^5}{x^2}$ Use the power rule.

$= \dfrac{32x^2}{x^2}$ Simplify.

$= 32x^{2-2}$ Use the quotient rule.

$= 32x^0$ Simplify.

$= 32 \cdot 1$ or 32 Substitute 1 for x^0.

■ **Work Practice 14–18**

Objective E Using Rational Exponents to Simplify Radical Expressions ▶

We can simplify some radical expressions by first writing the expression with rational exponents. Then we can use the properties of exponents to simplify, and finally convert the expression back to radical notation.

Practice 19–21

Use rational exponents to simplify. Assume that all variables represent positive real numbers.

19. $\sqrt[10]{y^5}$ **20.** $\sqrt[4]{9}$

21. $\sqrt[9]{a^6 b^3}$

Examples Use rational exponents to simplify. Assume that all variables represent positive real numbers.

19. $\sqrt[8]{x^4} = x^{4/8}$ Write with rational exponents.

$= x^{1/2}$ Simplify the exponent.

$= \sqrt{x}$ Write with radical notation.

20. $\sqrt[6]{25} = 25^{1/6}$ Write with rational exponents.

$= (5^2)^{1/6}$ Write 25 as 5^2.

$= 5^{2/6}$ Use the power rule.

$= 5^{1/3}$ Simplify the exponent.

$= \sqrt[3]{5}$ Write with radical notation.

21. $\sqrt[6]{r^2 s^4} = (r^2 s^4)^{1/6}$ Write with rational exponents.

$= r^{2/6} s^{4/6}$ Use the power rule.

$= r^{1/3} s^{2/3}$ Simplify the exponents.

$= (rs^2)^{1/3}$ Use $a^n b^n = (ab)^n$.

$= \sqrt[3]{rs^2}$ Write with radical notation.

■ **Work Practice 19–21**

Practice 22–24

Use rational exponents to write as a single radical.

22. $\sqrt{y} \cdot \sqrt[3]{y}$ **23.** $\dfrac{\sqrt[3]{x}}{\sqrt[4]{x}}$

24. $\sqrt{5} \cdot \sqrt[3]{2}$

Examples Use rational exponents to write as a single radical expression.

22. $\sqrt{x} \cdot \sqrt[4]{x} = x^{1/2} \cdot x^{1/4} = x^{1/2+1/4}$

$= x^{3/4} = \sqrt[4]{x^3}$

23. $\dfrac{\sqrt{x}}{\sqrt[3]{x}} = \dfrac{x^{1/2}}{x^{1/3}} = x^{1/2-1/3} = x^{3/6-2/6}$

$= x^{1/6} = \sqrt[6]{x}$

24. $\sqrt[3]{3} \cdot \sqrt{2} = 3^{1/3} \cdot 2^{1/2}$ Write with rational exponents.

$= 3^{2/6} \cdot 2^{3/6}$ Write the exponents so that they have the same denominator.

$= (3^2 \cdot 2^3)^{1/6}$ Use $a^n b^n = (ab)^n$.

$= \sqrt[6]{3^2 \cdot 2^3}$ Write with radical notation.

$= \sqrt[6]{72}$ Multiply $3^2 \cdot 2^3$.

■ **Work Practice 22–24**

Answers

19. $\sqrt{y}$ **20.** $\sqrt{3}$ **21.** $\sqrt[3]{a^2 b}$

22. $\sqrt[6]{y^5}$ **23.** $\sqrt[12]{x}$ **24.** $\sqrt[6]{500}$

Vocabulary, Readiness & Video Check

Answer each true or false.

1. $9^{-1/2}$ is a positive number. _____true_____

2. $9^{-1/2}$ is an integer. _____false_____

3. $\dfrac{1}{a^{-m/n}} = a^{m/n}$ (where $a^{m/n}$ is a nonzero real number). _____true_____

Fill in the blank with the correct choice.

4. To simplify $x^{2/3} \cdot x^{1/5}$, _____add_____ the exponents.

    a. add   **b.** subtract   **c.** multiply   **d.** divide   a

5. To simplify $(x^{2/3})^{1/5}$, _____multiply_____ the exponents.

    a. add   **b.** subtract   **c.** multiply   **d.** divide   c

6. To simplify $\dfrac{x^{2/3}}{x^{1/5}}$, _____subtract_____ the exponents.

    a. add   **b.** subtract   **c.** multiply   **d.** divide   b

Choose the correct letter for each exercise. Letters will be used more than once. No pencil is needed; just think about the meaning of each expression.

A = 2, B = −2, C = not a real number

7. $4^{1/2}$ _A_   **8.** $-4^{1/2}$ _B_   **9.** $(-4)^{1/2}$ _C_   **10.** $8^{1/3}$ _A_   **11.** $-8^{1/3}$ _B_   **12.** $(-8)^{1/3}$ _B_

Martin-Gay Interactive Videos

See Video 10.2

Watch the section lecture video and answer the following questions.

Objective A **13.** From looking at ▥ Example 2, what is $-(3x)^{1/5}$ in radical notation? ▶

Objective B **14.** From ▥ Examples 3 and 4, in a fractional exponent, what do the numerator and denominator each represent in radical form? ▶

Objective C **15.** Based on ▥ Example 5, complete the following statements. A negative fractional exponent will move a base from the numerator to the _____ with the fractional exponent becoming _____. ▶

Objective D **16.** Based on ▥ Examples 7–8, complete the following statements. Assume you have an expression with fractional exponents. If applying the product rule of exponents, you _____ the exponents. If applying the quotient rule of exponents, you _____ the exponents. If applying the power rule of exponents, you _____ the exponents. ▶

Objective E **17.** From ▥ Example 10, describe a way to simplify a radical of a variable raised to a power if the index and the power have a common factor. ▶

See video answer section.

10.2 Exercise Set MyMathLab®

Objective A *Use radical notation to rewrite each expression. Simplify if possible. See Examples 1 through 6.*

▶ 1. $49^{1/2}$
7

2. $64^{1/3}$
4

3. $27^{1/3}$
3

4. $8^{1/3}$
2

5. $\left(\dfrac{1}{16}\right)^{1/4}$ $\dfrac{1}{2}$

6. $\left(\dfrac{1}{64}\right)^{1/2}$ $\dfrac{1}{8}$

7. $169^{1/2}$
13

8. $81^{1/4}$
3

▶ 9. $2m^{1/3}$
$2\sqrt[3]{m}$

10. $(2m)^{1/3}$
$\sqrt[3]{2m}$

11. $(9x^4)^{1/2}$
$3x^2$

12. $(16x^8)^{1/2}$
$4x^4$

13. $(-27)^{1/3}$
-3

14. $-64^{1/2}$
-8

15. $-16^{1/4}$
-2

16. $(-32)^{1/5}$
-2

Objective B *Use radical notation to rewrite each expression. Simplify if possible. See Examples 7 through 11.*

▶ 17. $16^{3/4}$
8

18. $4^{5/2}$
32

▶ 19. $(-64)^{2/3}$
16

20. $(-8)^{4/3}$
16

21. $(-16)^{3/4}$
not a real number

22. $(-9)^{3/2}$
not a real number

23. $(2x)^{3/5}$
$\sqrt[5]{(2x)^3}$

24. $2x^{3/5}$
$2\sqrt[5]{x^3}$

25. $(7x+2)^{2/3}$
$\sqrt[3]{(7x+2)^2}$
or $(\sqrt[3]{7x+2})^2$

26. $(x-4)^{3/4}$
$\sqrt[4]{(x-4)^3}$
or $(\sqrt[4]{x-4})^3$

27. $\left(\dfrac{16}{9}\right)^{3/2}$
$\dfrac{64}{27}$

28. $\left(\dfrac{49}{25}\right)^{3/2}$
$\dfrac{343}{125}$

Objective C *Write with positive exponents. Simplify if possible. See Examples 12 and 13.*

▶ 29. $8^{-4/3}$
$\dfrac{1}{16}$

30. $64^{-2/3}$
$\dfrac{1}{16}$

31. $(-64)^{-2/3}$
$\dfrac{1}{16}$

32. $(-8)^{-4/3}$
$\dfrac{1}{16}$

33. $(-4)^{-3/2}$
not a real number

34. $(-16)^{-5/4}$
not a real number

▶ 35. $x^{-1/4}$
$\dfrac{1}{x^{1/4}}$

36. $y^{-1/6}$
$\dfrac{1}{y^{1/6}}$

37. $\dfrac{1}{a^{-2/3}}$
$a^{2/3}$

38. $\dfrac{1}{n^{-8/9}}$
$n^{8/9}$

39. $\dfrac{5}{7x^{-3/4}}$
$\dfrac{5x^{3/4}}{7}$

40. $\dfrac{2}{3y^{-5/7}}$
$\dfrac{2y^{5/7}}{3}$

Objective D *Use the properties of exponents to simplify each expression. Write with positive exponents. See Examples 14 through 18.*

▶ 41. $a^{2/3}a^{5/3}$
$a^{7/3}$

42. $b^{9/5}b^{8/5}$
$b^{17/5}$

43. $x^{-2/5}\cdot x^{7/5}$
x

44. $y^{4/3}\cdot y^{-1/3}$
y

45. $3^{1/4}\cdot 3^{3/8}$
$3^{5/8}$

46. $5^{1/2}\cdot 5^{1/6}$
$5^{2/3}$

47. $\dfrac{y^{1/3}}{y^{1/6}}$
$y^{1/6}$

48. $\dfrac{x^{3/4}}{x^{1/8}}$
$x^{5/8}$

49. $(4u^2)^{3/2}$
$8u^3$

50. $(32^{1/5}x^{2/3})^3$
$8x^2$

51. $\dfrac{b^{1/2}b^{3/4}}{-b^{1/4}}$
$-b$

52. $\dfrac{a^{1/4}a^{-1/2}}{a^{2/3}}$
$\dfrac{1}{a^{11/12}}$

53. $\dfrac{(x^3)^{1/2}}{x^{7/2}}$
$\dfrac{1}{x^2}$

54. $\dfrac{y^{11/3}}{(y^5)^{1/3}}$
y^2

▶ 55. $\dfrac{(3x^{1/4})^3}{x^{1/12}}$
$27x^{2/3}$

56. $\dfrac{(2x^{1/5})^4}{x^{3/10}}$
$16x^{1/2}$

57. $\dfrac{(y^3z)^{1/6}}{y^{-1/2}z^{1/3}}$
$\dfrac{y}{z^{1/6}}$

58. $\dfrac{(m^2n)^{1/4}}{m^{-1/2}n^{5/8}}$
$\dfrac{m}{n^{3/8}}$

59. $\dfrac{(x^3y^2)^{1/4}}{(x^{-5}y^{-1})^{-1/2}}$
$\dfrac{1}{x^{7/4}}$

60. $\dfrac{(a^{-2}b^3)^{1/8}}{(a^{-3}b)^{-1/4}}$
$\dfrac{b^{5/8}}{a}$

Objective E *Use rational exponents to simplify each radical. Assume that all variables represent positive real numbers. See Examples 19 through 21.*

61. $\sqrt[6]{x^3}$
$\sqrt{x}$

62. $\sqrt[9]{a^3}$
$\sqrt[3]{a}$

63. $\sqrt[6]{4}$
$\sqrt[3]{2}$

64. $\sqrt[4]{36}$
$\sqrt{6}$

65. $\sqrt[4]{16x^2}$
$2\sqrt{x}$

66. $\sqrt[8]{4y^2}$
$\sqrt[4]{2y}$

67. $\sqrt[4]{(x+3)^2}$
$\sqrt{x+3}$

68. $\sqrt[8]{(y+1)^4}$
$\sqrt{y+1}$

69. $\sqrt[8]{x^4y^4}$
$\sqrt{xy}$

70. $\sqrt[9]{y^6z^3}$
$\sqrt[3]{y^2z}$

71. $\sqrt[12]{a^8b^4}$
$\sqrt[3]{a^2b}$

72. $\sqrt[10]{a^5b^5}$
$\sqrt{ab}$

Use rational expressions to write as a single radical expression. See Examples 22 through 24.

73. $\sqrt[3]{y} \cdot \sqrt[5]{y^2}$
$\sqrt[15]{y^{11}}$

74. $\sqrt[3]{y^2} \cdot \sqrt[6]{y}$
$\sqrt[6]{y^5}$

75. $\dfrac{\sqrt[3]{b^2}}{\sqrt[4]{b}}$
$\sqrt[12]{b^5}$

76. $\dfrac{\sqrt[4]{a}}{\sqrt[5]{a}}$
$\sqrt[20]{a}$

77. $\sqrt[3]{x} \cdot \sqrt[4]{x} \cdot \sqrt[8]{x^3}$
$\sqrt[24]{x^{23}}$

78. $\sqrt[6]{y} \cdot \sqrt[3]{y} \cdot \sqrt[5]{y^2}$
$\sqrt[10]{y^9}$

79. $\dfrac{\sqrt[3]{a^2}}{\sqrt[6]{a}}$
$\sqrt{a}$

80. $\dfrac{\sqrt[5]{b^2}}{\sqrt[10]{b^3}}$
$\sqrt[10]{b}$

81. $\sqrt{3} \cdot \sqrt[3]{4}$
$\sqrt[6]{432}$

82. $\sqrt[3]{5} \cdot \sqrt{2}$
$\sqrt[6]{200}$

83. $\sqrt[5]{7} \cdot \sqrt[3]{y}$
$\sqrt[15]{343y^5}$

84. $\sqrt[4]{5} \cdot \sqrt[3]{x}$
$\sqrt[12]{125x^4}$

85. $\sqrt{5r} \cdot \sqrt[3]{s}$
$\sqrt[6]{125r^3s^2}$

86. $\sqrt[3]{b} \cdot \sqrt[5]{4a}$
$\sqrt[15]{64a^3b^5}$

Review

Write each integer as a product of two integers such that one of the factors is a perfect square. For example, write 18 as $9 \cdot 2$, because 9 is a perfect square. See Section 10.1.

87. 75 $25 \cdot 3$

88. 20 $4 \cdot 5$

89. 48 $16 \cdot 3$ or $4 \cdot 12$

90. 45 $9 \cdot 5$

Write each integer as a product of two integers such that one of the factors is a perfect cube. For example, write 24 as $8 \cdot 3$, because 8 is a perfect cube. See Section 10.1.

91. 16 $8 \cdot 2$

92. 56 $8 \cdot 7$

93. 54 $27 \cdot 2$

94. 80 $8 \cdot 10$

Concept Extensions

Basal metabolic rate (BMR) is the number of calories per day a person needs to maintain life. A person's basal metabolic rate $B(w)$ in calories per day can be estimated with the function $B(w) = 70w^{3/4}$, where w is the person's weight in kilograms. Use this information to answer Exercises 95 and 96.

95. Estimate the BMR for a person who weighs 60 kilograms. Round to the nearest calorie. (*Note:* 60 kilograms is approximately 132 pounds.)   1509 calories

96. Estimate the BMR for a person who weighs 90 kilograms. Round to the nearest calorie. (*Note:* 90 kilograms is approximately 198 pounds.)   2045 calories

The number of tablet computer users worldwide from 2010 to 2017 can be predicted by $f(x) = 76x^{13/10}$, where $f(x)$ is the number of tablet computer users worldwide in millions, x years after 2010. (Source: Forrester Research: World Tablet Adoption Forecast 2012–2017) Use this information to answer Exercises 97 and 98.

97. Use this model to estimate the number of tablet computer users worldwide in 2015. Round to the nearest tenth of a million.   615.8 million

98. Predict the number of tablet computer users worldwide in 2020. Round to the nearest tenth of a million.   1516.4 million

99. Explain how writing x^{-7} with positive exponents is similar to writing $x^{-1/4}$ with positive exponents.   answers may vary

100. Explain how writing $2x^{-5}$ with positive exponents is similar to writing $2x^{-3/4}$ with positive exponents.   answers may vary

Fill in each box with the correct expression.

101. $\square \cdot a^{2/3} = a^{3/3}$, or a $a^{1/3}$

102. $\square \cdot x^{1/8} = x^{4/8}$, or $x^{1/2}$ $x^{3/8}$

103. $\dfrac{\square}{x^{-2/5}} = x^{3/5}$ $x^{1/5}$

104. $\dfrac{\square}{y^{-3/4}} = y^{4/4}$, or y $y^{1/4}$

Use a calculator to write a four-decimal-place approximation of each number.

105. $8^{1/4}$ 1.6818

106. $18^{3/5}$ 5.6645

Solve.

107. In physics, the speed of a wave traveling over a stretched string with tension t and density u is given by the expression $\dfrac{\sqrt{t}}{\sqrt{u}}$. Write this expression with rational exponents. $\dfrac{t^{1/2}}{u^{1/2}}$

108. In electronics, the angular frequency of oscillations in a certain type of circuit is given by the expression $(LC)^{-1/2}$. Use radical notation to write this expression. $\dfrac{1}{\sqrt{LC}}$

10.3   Simplifying Radical Expressions

Objectives

A Use the Product Rule for Radicals. ▶

B Use the Quotient Rule for Radicals. ▶

C Simplify Radicals. ▶

D Use the Distance and Midpoint Formulas. ▶

Objective **A** Using the Product Rule

It is possible to simplify some radicals that do not evaluate to rational numbers. To do so, we use a product rule and a quotient rule for radicals. To discover the product rule, notice the following pattern:

$$\sqrt{9} \cdot \sqrt{4} = 3 \cdot 2 = 6$$
$$\sqrt{9 \cdot 4} = \sqrt{36} = 6$$

Since both expressions simplify to 6, it is true that

$$\sqrt{9} \cdot \sqrt{4} = \sqrt{9 \cdot 4}$$

This pattern suggests the following product rule for radicals.

Product Rule for Radicals

If $\sqrt[n]{a}$ and $\sqrt[n]{b}$ are real numbers, then

$$\sqrt[n]{a} \cdot \sqrt[n]{b} = \sqrt[n]{ab}$$

Notice that the product rule is the relationship $a^{1/n} \cdot b^{1/n} = (ab)^{1/n}$ stated in radical notation.

Examples Use the product rule to multiply.

1. $\sqrt{3} \cdot \sqrt{5} = \sqrt{3 \cdot 5} = \sqrt{15}$
2. $\sqrt{21} \cdot \sqrt{x} = \sqrt{21x}$
3. $\sqrt[3]{4} \cdot \sqrt[3]{2} = \sqrt[3]{4 \cdot 2} = \sqrt[3]{8} = 2$
4. $\sqrt[4]{5} \cdot \sqrt[4]{2x^3} = \sqrt[4]{5 \cdot 2x^3} = \sqrt[4]{10x^3}$
5. $\sqrt{\dfrac{2}{a}} \cdot \sqrt{\dfrac{b}{3}} = \sqrt{\dfrac{2}{a} \cdot \dfrac{b}{3}} = \sqrt{\dfrac{2b}{3a}}$

🟧 **Work Practice 1–5**

Objective B Using the Quotient Rule

To discover the quotient rule for radicals, notice the following pattern:

$$\sqrt{\dfrac{4}{9}} = \dfrac{2}{3}$$

$$\dfrac{\sqrt{4}}{\sqrt{9}} = \dfrac{2}{3}$$

Since both expressions simplify to $\dfrac{2}{3}$, it is true that

$$\sqrt{\dfrac{4}{9}} = \dfrac{\sqrt{4}}{\sqrt{9}}$$

This pattern suggests the following quotient rule for radicals.

Quotient Rule for Radicals

If $\sqrt[n]{a}$ and $\sqrt[n]{b}$ are real numbers and $\sqrt[n]{b}$ is not zero, then

$$\sqrt[n]{\dfrac{a}{b}} = \dfrac{\sqrt[n]{a}}{\sqrt[n]{b}}$$

Notice that the quotient rule is the relationship $\left(\dfrac{a}{b}\right)^{1/n} = \dfrac{a^{1/n}}{b^{1/n}}$ stated in radical notation. We can use the quotient rule to simplify radical expressions by reading the rule from left to right or to divide radicals by reading the rule from right to left.

For example.

$$\sqrt{\dfrac{x}{16}} = \dfrac{\sqrt{x}}{\sqrt{16}} = \dfrac{\sqrt{x}}{4} \qquad \text{Using } \sqrt[n]{\dfrac{a}{b}} = \dfrac{\sqrt[n]{a}}{\sqrt[n]{b}}$$

$$\dfrac{\sqrt{50}}{\sqrt{2}} = \sqrt{\dfrac{50}{2}} = \sqrt{25} = 5 \qquad \text{Using } \dfrac{\sqrt[n]{a}}{\sqrt[n]{b}} = \sqrt[n]{\dfrac{a}{b}}$$

Note: For the remainder of this chapter, we will assume that variables represent positive real numbers. If this is so, we need not insert absolute value bars when we simplify even roots.

Practice 6–9

Use the quotient rule to simplify. Assume that all variables represent positive real numbers.

6. $\sqrt{\dfrac{9}{25}}$ **7.** $\sqrt{\dfrac{y}{36}}$

8. $\sqrt[3]{\dfrac{27}{64}}$ **9.** $\sqrt[5]{\dfrac{7}{32x^5}}$

Examples   Use the quotient rule to simplify.

6. $\sqrt{\dfrac{25}{49}} = \dfrac{\sqrt{25}}{\sqrt{49}} = \dfrac{5}{7}$

7. $\sqrt{\dfrac{x}{9}} = \dfrac{\sqrt{x}}{\sqrt{9}} = \dfrac{\sqrt{x}}{3}$

8. $\sqrt[3]{\dfrac{8}{27}} = \dfrac{\sqrt[3]{8}}{\sqrt[3]{27}} = \dfrac{2}{3}$

9. $\sqrt[4]{\dfrac{3}{16y^4}} = \dfrac{\sqrt[4]{3}}{\sqrt[4]{16y^4}} = \dfrac{\sqrt[4]{3}}{2y}$

■ **Work Practice 6–9**

Objective C Simplifying Radicals ▶

Both the product and quotient rules can be used to simplify a radical. If the product rule is read from right to left, we have that $\sqrt[n]{ab} = \sqrt[n]{a} \cdot \sqrt[n]{b}$. We use this to simplify the following radicals.

Practice 10

Simplify: $\sqrt{18}$

Example 10   Simplify: $\sqrt{50}$

Solution: We factor 50 such that one factor is the largest perfect square that divides 50. The largest perfect square factor of 50 is 25, so we write 50 as $25 \cdot 2$ and use the product rule for radicals to simplify.

$$\sqrt{50} = \sqrt{25 \cdot 2} = \sqrt{25} \cdot \sqrt{2} = 5\sqrt{2}$$

the largest perfect square factor of 50

Helpful Hint   Don't forget that, for example, $5\sqrt{2}$ means $5 \cdot \sqrt{2}$.

■ **Work Practice 10**

Practice 11–13

Simplify.

11. $\sqrt[3]{40}$ **12.** $\sqrt{14}$

13. $\sqrt[4]{162}$

Examples   Simplify.

11. $\sqrt[3]{24} = \sqrt[3]{8 \cdot 3} = \sqrt[3]{8} \cdot \sqrt[3]{3} = 2\sqrt[3]{3}$

the largest perfect cube factor of 24

12. $\sqrt{26}$ The largest perfect square factor of 26 is 1, so $\sqrt{26}$ cannot be simplified further.

13. $\sqrt[4]{32} = \sqrt[4]{16 \cdot 2} = \sqrt[4]{16} \cdot \sqrt[4]{2} = 2\sqrt[4]{2}$

the largest 4th power factor of 32

■ **Work Practice 11–13**

Teaching Tip

For large numbers, suggest that students use the prime factorization of the radicand. For instance,

$$\sqrt[3]{185,220} = \sqrt[3]{2^2 \cdot 3^3 \cdot 5 \cdot 7^3}$$
$$= 3 \cdot 7 \cdot \sqrt[3]{2^2 \cdot 5}$$
$$= 21\sqrt[3]{20}$$

After simplifying a radical such as a square root, always check the radicand to see that it contains no other perfect square factors. It may, if the largest perfect square factor of the radicand was not originally recognized. For example,

$$\sqrt{200} = \sqrt{4 \cdot 50} = \sqrt{4} \cdot \sqrt{50} = 2\sqrt{50}$$

Notice that the radicand 50 still contains the perfect square factor 25. This is because 4 is not the largest perfect square factor of 200. We continue as follows:

$$2\sqrt{50} = 2\sqrt{25 \cdot 2} = 2 \cdot \sqrt{25} \cdot \sqrt{2} = 2 \cdot 5 \cdot \sqrt{2} = 10\sqrt{2}$$

The radical is now simplified since 2 contains no perfect square factors (other than 1).

Answers

6. $\dfrac{3}{5}$ **7.** $\dfrac{\sqrt{y}}{6}$ **8.** $\dfrac{3}{4}$ **9.** $\dfrac{\sqrt[5]{7}}{2x}$

10. $3\sqrt{2}$ **11.** $2\sqrt[3]{5}$ **12.** $\sqrt{14}$

13. $3\sqrt[4]{2}$

Helpful Hint

To recognize the largest perfect power factors of a radicand, it will help if you are familiar with some perfect powers. A few are listed below.

Perfect Squares	1, 4, 9, 16, 25, 36, 49, 64, 81, 100, 121, 144
	1^2 2^2 3^2 4^2 5^2 6^2 7^2 8^2 9^2 10^2 11^2 12^2

Perfect Cubes	1, 8, 27, 64, 125
	1^3 2^3 3^3 4^3 5^3

Perfect 4th powers	1, 16, 81, 256
	1^4 2^4 3^4 4^4

Helpful Hint

We say that a radical of the form $\sqrt[n]{a}$ is simplified when the radicand a contains no factors that are perfect nth powers (other than 1 or −1).

Examples Simplify. Assume that all variables represent positive real numbers.

14.
$$\sqrt{25x^3} = \sqrt{25 \cdot x^2 \cdot x} \qquad \text{Find the largest perfect square factor.}$$
$$= \sqrt{25 \cdot x^2} \cdot \sqrt{x} \qquad \text{Use the product rule.}$$
$$= 5x\sqrt{x} \qquad \text{Simplify.}$$

15.
$$\sqrt[3]{54x^6y^8} = \sqrt[3]{27 \cdot 2 \cdot x^6 \cdot y^6 \cdot y^2} \qquad \text{Factor the radicand and identify perfect cube factors.}$$
$$= \sqrt[3]{27 \cdot x^6 \cdot y^6 \cdot 2y^2}$$
$$= \sqrt[3]{27 \cdot x^6 \cdot y^6} \cdot \sqrt[3]{2y^2} \qquad \text{Use the product rule.}$$
$$= 3x^2y^2\sqrt[3]{2y^2} \qquad \text{Simplify.}$$

16.
$$\sqrt[4]{81z^{11}} = \sqrt[4]{81 \cdot z^8 \cdot z^3} \qquad \text{Factor the radicand and identify perfect 4th power factors.}$$
$$= \sqrt[4]{81 \cdot z^8} \cdot \sqrt[4]{z^3} \qquad \text{Use the product rule.}$$
$$= 3z^2\sqrt[4]{z^3} \qquad \text{Simplify.}$$

■ **Work Practice 14–16**

Examples Use the quotient rule to divide. Then simplify if possible. Assume that all variables represent positive real numbers.

17.
$$\frac{\sqrt{20}}{\sqrt{5}} = \sqrt{\frac{20}{5}} \qquad \text{Use the quotient rule.}$$
$$= \sqrt{4} \qquad \text{Simplify.}$$
$$= 2 \qquad \text{Simplify.}$$

18.
$$\frac{\sqrt{50x}}{2\sqrt{2}} = \frac{1}{2} \cdot \sqrt{\frac{50x}{2}} \qquad \text{Use the quotient rule.}$$
$$= \frac{1}{2} \cdot \sqrt{25x} \qquad \text{Simplify.}$$
$$= \frac{1}{2} \cdot \sqrt{25} \cdot \sqrt{x} \qquad \text{Factor } 25x.$$
$$= \frac{1}{2} \cdot 5 \cdot \sqrt{x} \qquad \text{Simplify.}$$
$$= \frac{5}{2}\sqrt{x}$$

(Continued on next page)

Practice 14–16

Simplify. Assume that all variables represent positive real numbers.

14. $\sqrt{49a^5}$ **15.** $\sqrt[3]{24x^9y^7}$

16. $\sqrt[4]{16z^9}$

Practice 17–20

Use the quotient rule to divide. Then simplify if possible. Assume that all variables represent positive real numbers.

17. $\dfrac{\sqrt{75}}{\sqrt{3}}$ **18.** $\dfrac{\sqrt{80y}}{3\sqrt{5}}$

19. $\dfrac{5\sqrt[3]{162x^8}}{\sqrt[3]{3x^2}}$ **20.** $\dfrac{3\sqrt[4]{243x^9y^6}}{\sqrt[4]{x^{-3}y}}$

Answers

14. $7a^2\sqrt{a}$ **15.** $2x^3y^2\sqrt[3]{3y}$ **16.** $2z^2\sqrt[4]{z}$

17. 5 **18.** $\dfrac{4}{3}\sqrt{y}$ **19.** $15x^2\sqrt[3]{2}$

20. $9x^3y\sqrt[4]{3y}$

19. $\dfrac{7\sqrt[3]{48y^4}}{\sqrt[3]{2y}} = 7\sqrt[3]{\dfrac{48y^4}{2y}} = 7\sqrt[3]{24y^3} = 7\sqrt[3]{8 \cdot y^3 \cdot 3}$

$= 7\sqrt[3]{8 \cdot y^3} \cdot \sqrt[3]{3} = 7 \cdot 2y\sqrt[3]{3} = 14y\sqrt[3]{3}$

20. $\dfrac{2\sqrt[4]{32a^8b^6}}{\sqrt[4]{a^{-1}b^2}} = 2\sqrt[4]{\dfrac{32a^8b^6}{a^{-1}b^2}} = 2\sqrt[4]{32a^9b^4} = 2\sqrt[4]{16 \cdot a^8 \cdot b^4 \cdot 2 \cdot a}$

$= 2\sqrt[4]{16 \cdot a^8 \cdot b^4} \cdot \sqrt[4]{2 \cdot a} = 2 \cdot 2a^2b \cdot \sqrt[4]{2a} = 4a^2b\sqrt[4]{2a}$

■ **Work Practice 17–20**

✓**Concept Check** Find and correct the error:

$$\dfrac{\sqrt[3]{27}}{\sqrt{9}} \cancel{=} \sqrt[3]{\dfrac{27}{9}} = \sqrt[3]{3}$$

Objective D Using the Distance and Midpoint Formulas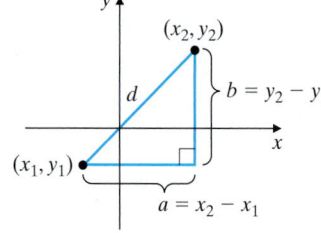

Now that we know how to simplify radicals, we can derive and use the distance formula. The midpoint formula is often confused with the distance formula, so to clarify both, we will also review the midpoint formula.

The Cartesian coordinate system helps us visualize the distance between points. To find the distance between two points, we use the distance formula, which is derived from the Pythagorean theorem.

To find the distance d between two points (x_1, y_1) and (x_2, y_2), draw vertical and horizontal lines so that a right triangle is formed, as shown. Notice that the length of leg a is $x_2 - x_1$ and that the length of leg b is $y_2 - y_1$. Thus, the Pythagorean theorem tells us that

$$d^2 = a^2 + b^2$$

or

$$d^2 = (x_2 - x_1)^2 + (y_2 - y_1)^2$$

or

$$d = \sqrt{(x_2 - x_1)^2 + (y_2 - y_1)^2}$$

This formula gives us the distance between any two points on the real plane.

Distance Formula

The distance d between two points (x_1, y_1) and (x_2, y_2) is given by

$$d = \sqrt{(x_2 - x_1)^2 + (y_2 - y_1)^2}$$

✓**Concept Check Answer**

$$\dfrac{\sqrt[3]{27}}{\sqrt{9}} = \dfrac{3}{3} = 1$$

Example 21   Find the distance between $(2, -5)$ and $(1, -4)$. Give the exact distance and a three-decimal-place approximation.

Solution: To use the distance formula, it makes no difference which point we call (x_1, y_1) and which point we call (x_2, y_2). We will let $(x_1, y_1) = (2, -5)$ and $(x_2, y_2) = (1, -4)$.

$$
\begin{aligned}
d &= \sqrt{(x_2 - x_1)^2 + (y_2 - y_1)^2} \\
&= \sqrt{(1 - 2)^2 + [-4 - (-5)]^2} \\
&= \sqrt{(-1)^2 + (1)^2} \\
&= \sqrt{1 + 1} \\
&= \sqrt{2} \approx 1.414
\end{aligned}
$$

The distance between the two points is exactly $\sqrt{2}$ units, or approximately 1.414 units.

■ **Work Practice 21**

Practice 21

Find the distance between $(-1, 3)$ and $(-2, 6)$. Give the exact distance and a three-decimal-place approximation.

Teaching Tip

For Example 21, have students explain why it does not matter which of the two points is (x_1, y_1).

The **midpoint** of a line segment is the **point** located exactly halfway between the two endpoints of the line segment. On the following graph, the point M is the midpoint of line segment PQ. Thus, the distance between M and P equals the distance between M and Q. *Note:* We usually need no knowledge of roots to calculate the midpoint of a line segment. We review midpoint here only because it is often confused with the distance between two points.

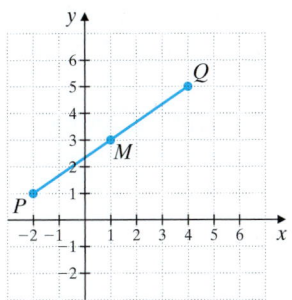

The x-coordinate of M is at half the distance between the x-coordinates of P and Q, and the y-coordinate of M is at half the distance between the y-coordinates of P and Q. That is, the x-coordinate of M is the average of the x-coordinates of P and Q; the y-coordinate of M is the average of the y-coordinates of P and Q.

Midpoint Formula

The midpoint of the line segment whose endpoints are (x_1, y_1) and (x_2, y_2) is the point with coordinates

$$
\left(\frac{x_1 + x_2}{2}, \frac{y_1 + y_2}{2} \right)
$$

Teaching Tip

Have students show that the distance from the midpoint to each endpoint is half of the distance between the endpoints.

Answer

21. $\sqrt{10} \approx 3.162$

Practice 22

Find the midpoint of the line segment that joins points $P(-2, 5)$ and $Q(4, -6)$.

Example 22   Find the midpoint of the line segment that joins points $P(-3, 3)$ and $Q(1, 0)$.

Solution: To use the midpoint formula, it makes no difference which point we call (x_1, y_1) and which point we call (x_2, y_2). We will let $(x_1, y_1) = (-3, 3)$ and $(x_2, y_2) = (1, 0)$.

$$\text{midpoint} = \left(\frac{x_1 + x_2}{2}, \frac{y_1 + y_2}{2} \right)$$
$$= \left(\frac{-3 + 1}{2}, \frac{3 + 0}{2} \right)$$
$$= \left(\frac{-2}{2}, \frac{3}{2} \right)$$
$$= \left(-1, \frac{3}{2} \right)$$

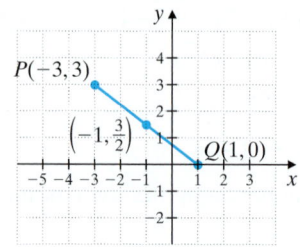

The midpoint of the segment is $\left(-1, \frac{3}{2} \right)$.

■ **Work Practice 22**

Answer

22. $\left(1, -\dfrac{1}{2} \right)$

Helpful Hint

The distance **between two points is a distance**. The **midpoint of a line segment is the point** halfway between the endpoints of the segment.

distance—measured in units

midpoint—it is a point

Vocabulary, Readiness & Video Check

Use the choices below to fill in each blank. Some choices may be used more than once.

distance   midpoint   point

1. The __midpoint__ of a line segment is a __point__ exactly halfway between the two endpoints of the line segment.
2. The __distance__ formula is $d = \sqrt{(x_2 - x_1)^2 + (y_2 - y_1)^2}$.
3. The __midpoint__ formula is $\left(\dfrac{x_1 + x_2}{2}, \dfrac{y_1 + y_2}{2} \right)$.

Answer true or false. Assume all radicals represent nonzero real numbers.

4. $\sqrt[n]{a} \cdot \sqrt[n]{b} = \sqrt[n]{ab}$, __true__

5. $\sqrt[3]{7} \cdot \sqrt[3]{11} = \sqrt[3]{18}$, __false__

6. $\sqrt[3]{7} \cdot \sqrt{11} = \sqrt{77}$, __false__

7. $\sqrt{x^7 y^8} = \sqrt{x^7} \cdot \sqrt{y^8}$, __true__

8. $\dfrac{\sqrt[n]{a}}{\sqrt[n]{b}} = \sqrt[n]{\dfrac{a}{b}}$, __true__

9. $\dfrac{\sqrt[3]{12}}{\sqrt[3]{4}} = \sqrt[3]{8}$, __false__

10. $\dfrac{\sqrt[n]{x^7}}{\sqrt[n]{x}} = \sqrt[n]{x^6}$, __true__

Martin-Gay Interactive Videos

See Video 10.3 🍎

Watch the section lecture video and answer the following questions.

Objective A **11.** From Example 1 and the lecture before, in order to apply the product rule for radicals, what must be true about the indexes of the radicals being multiplied? ▶

Objective B **12.** From Example 2–6, when might you apply the quotient rule (in either direction) in order to simplify a fractional radical expression? ▶

Objective C **13.** From Example 8, we know that an even power of a variable is a perfect square factor of the variable, leaving no factor in the radicand once simplified. Therefore, what must be true about the power of any variable left in the radicand of a simplified square root? Explain. ▶

Objective D **14.** From Example 10, the formula uses the coordinates of two points similar to the slope formula. What caution should you take when replacing values in the formula? ▶

15. Based on Example 11, complete the following statement. The x-value of the midpoint is the _____ of the x-values of the endpoints and the y-value of the midpoint is the _____ of the y-values of the endpoints. ▶

See video answer section.

10.3 **Exercise Set** **MyMathLab®** ▶

Objective A *Use the product rule to multiply. Assume that all variables represent positive real numbers. See Examples 1 through 5.*

1. $\sqrt{7} \cdot \sqrt{2}$ $\sqrt{14}$

2. $\sqrt{11} \cdot \sqrt{10}$ $\sqrt{110}$

3. $\sqrt[4]{8} \cdot \sqrt[4]{2}$ 2

4. $\sqrt[4]{27} \cdot \sqrt[4]{3}$ 3

5. $\sqrt[3]{4} \cdot \sqrt[3]{9}$ $\sqrt[3]{36}$

6. $\sqrt[3]{10} \cdot \sqrt[3]{5}$ $\sqrt[3]{50}$

▶ **7.** $\sqrt{2} \cdot \sqrt{3x}$ $\sqrt{6x}$

8. $\sqrt{3y} \cdot \sqrt{5x}$ $\sqrt{15xy}$

9. $\sqrt{\dfrac{7}{x}} \cdot \sqrt{\dfrac{2}{y}}$ $\sqrt{\dfrac{14}{xy}}$

10. $\sqrt{\dfrac{6}{m}} \cdot \sqrt{\dfrac{n}{5}}$ $\sqrt{\dfrac{6n}{5m}}$

11. $\sqrt[4]{4x^3} \cdot \sqrt[4]{5}$ $\sqrt[4]{20x^3}$

12. $\sqrt[4]{ab^2} \cdot \sqrt[4]{27ab}$ $\sqrt[4]{27a^2b^3}$

Objective B *Use the quotient rule to simplify. Assume that all variables represent positive real numbers. See Examples 6 through 9.*

▶ **13.** $\sqrt{\dfrac{6}{49}}$ $\dfrac{\sqrt{6}}{7}$

14. $\sqrt{\dfrac{10}{81}}$ $\dfrac{\sqrt{10}}{9}$

15. $\sqrt{\dfrac{2}{49}}$ $\dfrac{\sqrt{2}}{7}$

16. $\sqrt{\dfrac{5}{121}}$ $\dfrac{\sqrt{5}}{11}$

▶ **17.** $\sqrt[4]{\dfrac{x^3}{16}}$ $\dfrac{\sqrt[4]{x^3}}{2}$

18. $\sqrt[4]{\dfrac{y}{81x^4}}$ $\dfrac{\sqrt[4]{y}}{3x}$

19. $\sqrt[3]{\dfrac{4}{27}}$ $\dfrac{\sqrt[3]{4}}{3}$

20. $\sqrt[3]{\dfrac{3}{64}}$ $\dfrac{\sqrt[3]{3}}{4}$

21. $\sqrt[4]{\dfrac{8}{x^8}}$ $\dfrac{\sqrt[4]{8}}{x^2}$

22. $\sqrt[4]{\dfrac{a^3}{81}}$ $\dfrac{\sqrt[4]{a^3}}{3}$

23. $\sqrt[3]{\dfrac{2x}{27y^{12}}}$ $\dfrac{\sqrt[3]{2x}}{3y^4}$

24. $\sqrt[3]{\dfrac{3y}{8x^6}}$ $\dfrac{\sqrt[3]{3y}}{2x^2}$

25. $\sqrt{\dfrac{x^2y}{169}}$

$\dfrac{x\sqrt{y}}{13}$

26. $\sqrt{\dfrac{y^2z}{225}}$

$\dfrac{y\sqrt{z}}{15}$

27. $\sqrt{\dfrac{5x^2}{4y^2}}$

$\dfrac{x\sqrt{5}}{2y}$

28. $\sqrt{\dfrac{y^{10}}{9x^6}}$

$\dfrac{y^5}{3x^3}$

29. $-\sqrt[3]{\dfrac{z^7}{125x^3}}$

$-\dfrac{z^2\sqrt[3]{z}}{5x}$

30. $-\sqrt[3]{\dfrac{1000a}{b^9}}$

$-\dfrac{10\sqrt[3]{a}}{b^3}$

Objective C   *Simplify. Assume that all variables represent positive real numbers. See Examples 10 through 16.*

31. $\sqrt{32}$

$4\sqrt{2}$

32. $\sqrt{27}$

$3\sqrt{3}$

33. $\sqrt[3]{192}$

$4\sqrt[3]{3}$

34. $\sqrt[3]{108}$

$3\sqrt[3]{4}$

35. $5\sqrt{75}$

$25\sqrt{3}$

36. $3\sqrt{8}$

$6\sqrt{2}$

37. $\sqrt{24}$

$2\sqrt{6}$

38. $\sqrt{20}$

$2\sqrt{5}$

39. $\sqrt{100x^5}$

$10x^2\sqrt{x}$

40. $\sqrt{64y^9}$

$8y^4\sqrt{y}$

41. $\sqrt[3]{16y^7}$

$2y^2\sqrt[3]{2y}$

42. $\sqrt[3]{128y^{10}}$

$4y^3\sqrt[3]{2y}$

43. $\sqrt[4]{a^8b^7}$

$a^2b\sqrt[4]{b^3}$

44. $\sqrt[5]{32z^{12}}$

$2z^2\sqrt[5]{z^2}$

45. $\sqrt{y^5}$

$y^2\sqrt{y}$

46. $\sqrt[3]{y^5}$

$y\sqrt[3]{y^2}$

47. $\sqrt{25a^2b^3}$

$5ab\sqrt{b}$

48. $\sqrt{9x^5y^7}$

$3x^2y^3\sqrt{xy}$

49. $\sqrt[5]{-32x^{10}y}$

$-2x^2\sqrt[5]{y}$

50. $\sqrt[5]{-243z^9}$

$-3z\sqrt[5]{z^4}$

51. $\sqrt[3]{50x^{14}}$

$x^4\sqrt[3]{50x^2}$

52. $\sqrt[3]{40y^{10}}$

$2y^3\sqrt[3]{5y}$

53. $-\sqrt{32a^8b^7}$

$-4a^4b^3\sqrt{2b}$

54. $-\sqrt{20ab^6}$

$-2b^3\sqrt{5a}$

55. $\sqrt{9x^7y^9}$

$3x^3y^4\sqrt{xy}$

56. $\sqrt{12r^9s^{12}}$

$2r^4s^6\sqrt{3r}$

57. $\sqrt[3]{125r^9s^{12}}$

$5r^3s^4$

58. $\sqrt[3]{8a^6b^9}$

$2a^2b^3$

59. $\sqrt[4]{32x^{12}y^5}$

$2x^3y\sqrt[4]{2y}$

60. $\sqrt[4]{162x^7y^{20}}$

$3xy^5\sqrt[4]{2x^3}$

Use the quotient rule to divide. Then simplify if possible. Assume that all variables represent positive real numbers. See Examples 17 through 20.

61. $\dfrac{\sqrt{14}}{\sqrt{7}}$ $\sqrt{2}$

62. $\dfrac{\sqrt{45}}{\sqrt{9}}$ $\sqrt{5}$

63. $\dfrac{\sqrt[3]{24}}{\sqrt[3]{3}}$ 2

64. $\dfrac{\sqrt[3]{10}}{\sqrt[3]{2}}$ $\sqrt[3]{5}$

65. $\dfrac{5\sqrt[4]{48}}{\sqrt[4]{3}}$ 10

66. $\dfrac{7\sqrt[4]{162}}{\sqrt[4]{2}}$ 21

67. $\dfrac{\sqrt{x^5y^3}}{\sqrt{xy}}$ x^2y

68. $\dfrac{\sqrt{a^7b^6}}{\sqrt{a^3b^2}}$ a^2b^2

69. $\dfrac{8\sqrt[3]{54m^7}}{\sqrt[3]{2m}}$

$24m^2$

70. $\dfrac{\sqrt[3]{128x^3}}{-3\sqrt[3]{2x}}$

$-\dfrac{4\sqrt[3]{x^2}}{3}$

71. $\dfrac{3\sqrt{100x^2}}{2\sqrt{2x^{-1}}}$

$\dfrac{15x\sqrt{2x}}{2}$ or $\dfrac{15x}{2}\sqrt{2x}$

72. $\dfrac{\sqrt{270y^2}}{5\sqrt{3y^{-4}}}$

$\dfrac{3y^3\sqrt{10}}{5}$ or $\dfrac{3y^3}{5}\sqrt{10}$

73. $\dfrac{\sqrt[4]{96a^{10}b^3}}{\sqrt[4]{3a^2b^3}}$

$2a^2\sqrt[4]{2}$

74. $\dfrac{\sqrt[4]{160x^{10}y^5}}{\sqrt[4]{2x^2y^2}}$

$2x^2\sqrt[4]{5y^3}$

75. $\dfrac{\sqrt[5]{64x^{10}y^3}}{\sqrt[5]{2x^3y^{-7}}}$

$2xy^2\sqrt[5]{x^2}$

76. $\dfrac{\sqrt[5]{192x^6y^{12}}}{\sqrt[5]{2x^{-1}y^{-3}}}$

$2xy^3\sqrt[5]{3x^2}$

Copyright 2016 Pearson Education, Inc.

Objective D *Find the distance between each pair of points. Give the exact distance and a three-decimal-place approximation. See Example 21.*

77. $(5, 1)$ and $(8, 5)$
5 units

78. $(2, 3)$ and $(14, 8)$
13 units

▶ **79.** $(-3, 2)$ and $(1, -3)$
$\sqrt{41}$ units ≈ 6.403 units

80. $(3, -2)$ and $(-4, 1)$
$\sqrt{58}$ units ≈ 7.616 units

81. $(-9, 4)$ and $(-8, 1)$
$\sqrt{10}$ units ≈ 3.162 units

82. $(-5, -2)$ and $(-6, -6)$
$\sqrt{17}$ units ≈ 4.123 units

83. $(0, -\sqrt{2})$ and $(\sqrt{3}, 0)$
$\sqrt{5}$ units ≈ 2.236 units

84. $(-\sqrt{5}, 0)$ and $(0, \sqrt{7})$
$2\sqrt{3}$ units ≈ 3.464 units

85. $(1.7, -3.6)$ and $(-8.6, 5.7)$
$\sqrt{192.58}$ units ≈ 13.877 units

86. $(9.6, 2.5)$ and $(-1.9, -3.7)$
$\sqrt{170.69}$ units ≈ 13.065 units

Find the midpoint of each line segment whose endpoints are given. See Example 22.

87. $(6, -8); (2, 4)$
$(4, -2)$

88. $(3, 9); (7, 11)$
$(5, 10)$

▶ **89.** $(-2, -1); (-8, 6)$
$\left(-5, \dfrac{5}{2}\right)$

90. $(-3, -4); (6, -8)$
$\left(\dfrac{3}{2}, -6\right)$

91. $(6, 3); (-1, -3)$
$\left(\dfrac{5}{2}, 0\right)$

92. $(-2, 5); (2, 6)$
$\left(0, \dfrac{11}{2}\right)$

93. $\left(\dfrac{1}{2}, \dfrac{3}{8}\right); \left(-\dfrac{3}{2}, \dfrac{5}{8}\right)$
$\left(-\dfrac{1}{2}, \dfrac{1}{2}\right)$

94. $\left(-\dfrac{2}{5}, \dfrac{7}{15}\right); \left(-\dfrac{2}{5}, -\dfrac{4}{15}\right)$
$\left(-\dfrac{2}{5}, \dfrac{1}{10}\right)$

95. $(\sqrt{2}, 3\sqrt{5}); (\sqrt{2}, -2\sqrt{5})$
$\left(\sqrt{2}, \dfrac{\sqrt{5}}{2}\right)$

96. $(\sqrt{8}, -\sqrt{12}); (3\sqrt{2}, 7\sqrt{3})$
$\left(\dfrac{5\sqrt{2}}{2}, \dfrac{5\sqrt{3}}{2}\right)$

97. $(4.6, -3.5); (7.8, -9.8)$
$(6.2, -6.65)$

98. $(-4.6, 2.1); (-6.7, 1.9)$
$(-5.65, 2)$

Review

Perform each indicated operation. See Sections 1.8 and 5.5.

99. $6x + 8x$
$14x$

100. $(6x)(8x)$
$48x^2$

101. $(2x + 3)(x - 5)$
$2x^2 - 7x - 15$

102. $(2x + 3) + (x - 5)$
$3x - 2$

103. $9y^2 - 8y^2$
y^2

104. $(9y^2)(-8y^2)$
$-72y^4$

105. $(x - 4)^2$
$x^2 - 8x + 16$

106. $(2x + 1)^2$
$4x^2 + 4x + 1$

Concept Extensions

Find and correct the error. See the Concept Check in this section.

107. $\dfrac{\sqrt[3]{64}}{\sqrt{64}} = \cancel{\sqrt[3]{\dfrac{64}{64}} = \sqrt[3]{1} = 1} \qquad \dfrac{\sqrt[3]{64}}{\sqrt{64}} = \dfrac{4}{8} = \dfrac{1}{2}$

108. $\dfrac{\sqrt[4]{16}}{\sqrt{4}} = \cancel{\sqrt[4]{\dfrac{16}{4}} = \sqrt[4]{4}} \qquad \dfrac{\sqrt[4]{16}}{\sqrt{4}} = \dfrac{2}{2} = 1$

Solve.

109. The formula for the radius r of a sphere with surface area A is given by $r = \sqrt{\dfrac{A}{4\pi}}$. Calculate the radius of a standard zorb whose outside surface area is 32.17 sq m. Round to the nearest tenth. (A zorb is a large inflated ball within a ball in which a person, strapped inside, may choose to roll down a hill. *Source:* Zorb, Ltd.)   1.6 m

110. Before Mount Vesuvius, a volcano in Italy, erupted violently in 79 A.D., its height was 4190 feet. Vesuvius was roughly cone shaped, and its base had a radius of approximately 25,200 feet. Use the formula $A = \pi r \sqrt{r^2 + h^2}$ for the lateral surface area A of a cone with radius r and height h to approximate the surface area of this volcano before it erupted. (*Source:* Global Volcanism Network) 2,022,426,050 sq ft

111. The time it takes a pendulum to swing back and forth is dependent on the length of the pendulum. The formula for this is $P = 2\pi \sqrt{\dfrac{L}{32}}$, where $P(L)$ is the time in seconds it takes the pendulum to swing back and forth, L is the length in feet of the pendulum, and 32 represents the force due to gravity.

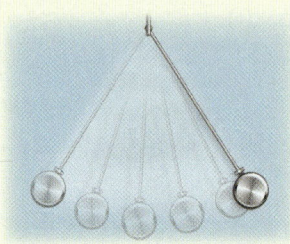

   a. Approximate to one decimal place the time of the pendulum's swing if the length of the pendulum is 4 feet.   2.2 sec

   b. Approximate to one decimal place the time of the pendulum's swing if the length of the pendulum is 1.5 feet.   1.4 sec

   c. Explain how the formula can be used to determine the length of the pendulum if the time of the swing is known.   answers may vary

---

10.4   Adding, Subtracting, and Multiplying Radical Expressions ▶

Objectives

A   Add or Subtract Radical Expressions. ▶

B   Multiply Radical Expressions. ▶

Objective **A**   Adding or Subtracting Radical Expressions ▶

We have learned that the sum or difference of like terms can be simplified. To simplify these sums or differences, we use the distributive property. For example,

$$2x + 3x = (2 + 3)x = 5x$$

The distributive property can also be used to add *like radicals*.

Like Radicals

Radicals with the same index and the same radicand are **like radicals.** The example below shows how to use the distributive property to simplify an expression containing like radicals.

$$2\sqrt{7} + 3\sqrt{7} = (2 + 3)\sqrt{7} = 5\sqrt{7}$$

Like radicals

Helpful Hint

The expression

$$5\sqrt{7} - 3\sqrt{6}$$

does not contain like radicals and cannot be simplified further.

Examples   Add or subtract as indicated.

1. $4\sqrt{11} + 8\sqrt{11} = (4 + 8)\sqrt{11} = 12\sqrt{11}$

2. $5\sqrt[3]{3x} - 7\sqrt[3]{3x} = (5 - 7)\sqrt[3]{3x} = -2\sqrt[3]{3x}$

3. $2\sqrt{7} + 2\sqrt[3]{7}$ This expression cannot be simplified since $2\sqrt{7}$ and $2\sqrt[3]{7}$ do not contain like radicals.

▶ **Work Practice 1–3**

✔**Concept Check** True or false:

$$\sqrt{a} + \sqrt{b} = \sqrt{a + b}$$

Explain your answer.

When adding or subtracting radicals, always check first to see whether any radicals can be simplified.

Examples   Add or subtract as indicated. Assume that all variables represent positive real numbers.

4. $\sqrt{20} + 2\sqrt{45} = \sqrt{4 \cdot 5} + 2\sqrt{9 \cdot 5}$ Factor 20 and 45.

$\qquad\qquad = \sqrt{4} \cdot \sqrt{5} + 2 \cdot \sqrt{9} \cdot \sqrt{5}$ Use the product rule.

$\qquad\qquad = 2 \cdot \sqrt{5} + 2 \cdot 3 \cdot \sqrt{5}$ Simplify $\sqrt{4}$ and $\sqrt{9}$.

$\qquad\qquad = 2\sqrt{5} + 6\sqrt{5}$ Add like radicals.

$\qquad\qquad = 8\sqrt{5}$

5. $\sqrt[3]{54} - 5\sqrt[3]{16} + \sqrt[3]{2}$

$\qquad = \sqrt[3]{27} \cdot \sqrt[3]{2} - 5 \cdot \sqrt[3]{8} \cdot \sqrt[3]{2} + \sqrt[3]{2}$ Factor and use the product rule.

$\qquad = 3 \cdot \sqrt[3]{2} - 5 \cdot 2 \cdot \sqrt[3]{2} + \sqrt[3]{2}$ Simplify $\sqrt[3]{27}$ and $\sqrt[3]{8}$.

$\qquad = 3\sqrt[3]{2} - 10\sqrt[3]{2} + \sqrt[3]{2}$ Write $5 \cdot 2$ as 10.

$\qquad = -6\sqrt[3]{2}$ Combine like radicals.

6. $\sqrt{27x} - 2\sqrt{9x} + \sqrt{72x}$

$\qquad = \sqrt{9} \cdot \sqrt{3x} - 2 \cdot \sqrt{9} \cdot \sqrt{x} + \sqrt{36} \cdot \sqrt{2x}$ Factor and use the product rule.

$\qquad = 3 \cdot \sqrt{3x} - 2 \cdot 3 \cdot \sqrt{x} + 6 \cdot \sqrt{2x}$ Simplify $\sqrt{9}$ and $\sqrt{36}$.

$\qquad = 3\sqrt{3x} - 6\sqrt{x} + 6\sqrt{2x}$ Write $2 \cdot 3$ as 6.

(Continued on next page)

Teaching Tip

Before discussing combining like radicals, have students evaluate $2x + 3x$ for $x = \sqrt{7}$.

Teaching Tip

Show that radical expressions can be rewritten as simple algebraic expressions by letting a variable represent the radical portion. In Example 2, let $y = \sqrt[3]{3x}$.

$$5\sqrt[3]{3x} - 7\sqrt[3]{3x} = 5y - 7y$$
$$= -2y$$
$$= -2\sqrt[3]{3x}$$

Practice 1–3

Add or subtract as indicated.

1. $5\sqrt{15} + 2\sqrt{15}$

2. $9\sqrt[3]{2y} - 15\sqrt[3]{2y}$

3. $6\sqrt{10} - 3\sqrt[3]{10}$

Practice 4–8

Add or subtract as indicated. Assume that all variables represent positive real numbers.

4. $\sqrt{50} + 5\sqrt{18}$

5. $\sqrt[3]{24} - 4\sqrt[3]{192} + \sqrt[3]{3}$

6. $\sqrt{20x} - 6\sqrt{16x} + \sqrt{32x}$

7. $\sqrt[4]{32} + \sqrt{32}$

8. $\sqrt[3]{8y^5} + \sqrt[3]{27y^5}$

Helpful Hint

None of these terms contains like radicals. We can simplify no further.

Answers

1. $7\sqrt{15}$ **2.** $-6\sqrt[3]{2y}$ **3.** $6\sqrt{10} - 3\sqrt[3]{10}$
4. $20\sqrt{2}$ **5.** $-13\sqrt[3]{3}$
6. $2\sqrt{5x} - 24\sqrt{x} + 4\sqrt{2x}$
7. $2\sqrt[4]{2} + 4\sqrt{2}$ **8.** $5y\sqrt[3]{y^2}$

✔**Concept Check Answer**

false; answers may vary

7. $\sqrt[3]{98} + \sqrt{98} = \sqrt[3]{98} + \sqrt{49} \cdot \sqrt{2}$ Factor and use the product rule.

$= \sqrt[3]{98} + 7\sqrt{2}$ No further simplification is possible.

8. $\sqrt[3]{48y^4} + \sqrt[3]{6y^4} = \sqrt[3]{8y^3} \cdot \sqrt[3]{6y} + \sqrt[3]{y^3} \cdot \sqrt[3]{6y}$ Factor and use the product rule.

$= 2y\sqrt[3]{6y} + y\sqrt[3]{6y}$ Simplify $\sqrt[3]{8y^3}$ and $\sqrt[3]{y^3}$.

$= 3y\sqrt[3]{6y}$ Combine like radicals.

■ **Work Practice 4–8**

Practice 9–10

Add or subtract as indicated. Assume that all variables represent positive real numbers.

9. $\dfrac{\sqrt{75}}{9} - \dfrac{\sqrt{3}}{2}$

10. $\sqrt[3]{\dfrac{5x}{27}} + 4\sqrt[3]{5x}$

Examples   Add or subtract as indicated. Assume that all variables represent positive real numbers.

9. $\dfrac{\sqrt{45}}{4} - \dfrac{\sqrt{5}}{3} = \dfrac{3\sqrt{5}}{4} - \dfrac{\sqrt{5}}{3}$ To subtract, notice that the LCD is 12.

$= \dfrac{3\sqrt{5} \cdot 3}{4 \cdot 3} - \dfrac{\sqrt{5} \cdot 4}{3 \cdot 4}$ Write each expression as an equivalent expression with a denominator of 12.

$= \dfrac{9\sqrt{5}}{12} - \dfrac{4\sqrt{5}}{12}$ Multiply factors in the numerators and the denominators.

$= \dfrac{5\sqrt{5}}{12}$ Subtract.

10. $\sqrt[3]{\dfrac{7x}{8}} + 2\sqrt[3]{7x} = \dfrac{\sqrt[3]{7x}}{\sqrt[3]{8}} + 2\sqrt[3]{7x}$ Use the quotient rule for radicals.

$= \dfrac{\sqrt[3]{7x}}{2} + 2\sqrt[3]{7x}$ Simplify.

$= \dfrac{\sqrt[3]{7x}}{2} + \dfrac{2\sqrt[3]{7x} \cdot 2}{2}$ Write each expression as an equivalent expression with a denominator of 2.

$= \dfrac{\sqrt[3]{7x}}{2} + \dfrac{4\sqrt[3]{7x}}{2}$

$= \dfrac{5\sqrt[3]{7x}}{2}$ Add.

■ **Work Practice 9–10**

Objective B Multiplying Radical Expressions

We can multiply radical expressions by using many of the same properties used to multiply polynomial expressions. For instance, to multiply $\sqrt{2}(\sqrt{6} - 3\sqrt{2})$, we use the distributive property and multiply $\sqrt{2}$ by each term inside the parentheses.

$\sqrt{2}(\sqrt{6} - 3\sqrt{2}) = \sqrt{2}(\sqrt{6}) - \sqrt{2}(3\sqrt{2})$ Use the distributive property.

$= \sqrt{2 \cdot 6} - 3\sqrt{2 \cdot 2}$

$= \sqrt{2 \cdot 2 \cdot 3} - 3 \cdot 2$ Use the product rule for radicals.

$= 2\sqrt{3} - 6$

Answers

9. $\dfrac{\sqrt{3}}{18}$ **10.** $\dfrac{13\sqrt[3]{5x}}{3}$

Example 11 Multiply: $\sqrt{3}(5 + \sqrt{30})$

Solution:

$$\sqrt{3}(5 + \sqrt{30}) = \sqrt{3}(5) + \sqrt{3}(\sqrt{30})$$
$$= 5\sqrt{3} + \sqrt{3 \cdot 30}$$
$$= 5\sqrt{3} + \sqrt{3 \cdot 3 \cdot 10}$$
$$= 5\sqrt{3} + 3\sqrt{10}$$

■ **Work Practice 11**

Examples Multiply. Assume that all variables represent positive real numbers.

12. $(\sqrt{5} - \sqrt{6})(\sqrt{7} + 1) = \overset{\text{First}}{\sqrt{5} \cdot \sqrt{7}} + \overset{\text{Outer}}{\sqrt{5} \cdot 1} - \overset{\text{Inner}}{\sqrt{6} \cdot \sqrt{7}} - \overset{\text{Last}}{\sqrt{6} \cdot 1}$

                              Using the FOIL order

                 $= \sqrt{35} + \sqrt{5} - \sqrt{42} - \sqrt{6}$ Simplify.

13. $(\sqrt{2x} + 5)(\sqrt{2x} - 5) = (\sqrt{2x})^2 - 5^2$ Multiply the sum and difference of

                               two terms: $(a + b)(a - b) = a^2 - b^2$

                 $= 2x - 25$

14. $(\sqrt{3} - 1)^2 = (\sqrt{3})^2 - 2 \cdot \sqrt{3} \cdot 1 + 1^2$ Square the binomial:

                             $(a - b)^2 = a^2 - 2ab + b^2$

            $= 3 - 2\sqrt{3} + 1$

            $= 4 - 2\sqrt{3}$

15. $(\underbrace{\sqrt{x-3}}_{a} + \underbrace{5}_{b})^2 = \underbrace{(\sqrt{x-3})^2}_{a^2} + \underbrace{2 \cdot}_{+\,2\,\cdot} \underbrace{\sqrt{x-3}}_{a} \cdot \underbrace{5}_{\cdot\, b} + \underbrace{5^2}_{+\, b^2}$ Square the binomial:

                               $(a + b)^2 = a^2 + 2ab + b^2$

            $= x - 3 + 10\sqrt{x-3} + 25$ Simplify.

            $= x + 22 + 10\sqrt{x-3}$ Combine like terms.

■ **Work Practice 12–15**

Practice 11

Multiply: $\sqrt{2}(6 + \sqrt{10})$

Practice 12–15

Multiply. Assume that all variables represent positive real numbers.

12. $(\sqrt{3} - \sqrt{5})(\sqrt{2} + 7)$

13. $(\sqrt{5y} + 2)(\sqrt{5y} - 2)$

14. $(\sqrt{3} - 7)^2$

15. $(\sqrt{x+1} + 2)^2$

Teaching Tip

While discussing Example 13, point out that the product does not contain radical signs. Then ask, "What is special about the factors?"

Answers

11. $6\sqrt{2} + 2\sqrt{5}$
12. $\sqrt{6} + 7\sqrt{3} - \sqrt{10} - 7\sqrt{5}$
13. $5y - 4$ **14.** $52 - 14\sqrt{3}$
15. $x + 5 + 4\sqrt{x+1}$

Vocabulary, Readiness & Video Check

Complete the table with "Like" or "Unlike."

	Terms	Like or Unlike Radical Terms?
1.	$\sqrt{7}, \sqrt[3]{7}$	Unlike
2.	$\sqrt[3]{x^2 y}, \sqrt[3]{yx^2}$	Like
3.	$\sqrt[3]{abc}, \sqrt[3]{cba}$	Like
4.	$2x\sqrt{5}, 2x\sqrt{10}$	Unlike

Simplify. Assume that all variables represent positive real numbers.

5. $2\sqrt{3} + 4\sqrt{3} = $ _____ $6\sqrt{3}$ _____

6. $5\sqrt{7} + 3\sqrt{7} = $ _____ $8\sqrt{7}$ _____

7. $8\sqrt{x} - \sqrt{x} = $ _____ $7\sqrt{x}$ _____

8. $3\sqrt{y} - \sqrt{y} = $ _____ $2\sqrt{y}$ _____

9. $7\sqrt[3]{x} + \sqrt[3]{x} = $ _____ $8\sqrt[3]{x}$ _____

10. $8\sqrt[3]{z} + \sqrt[3]{z} = $ _____ $9\sqrt[3]{z}$ _____

Add or subtract if possible.

11. $\sqrt{11} + \sqrt[3]{11} = $ _____ $\sqrt{11} + \sqrt[3]{11}$ _____

12. $9\sqrt{13} - \sqrt[4]{13} = $ _____ $9\sqrt{13} - \sqrt[4]{13}$ _____

13. $8\sqrt[3]{2x} + 3\sqrt[3]{2x} - \sqrt[3]{2x} = $ _____ $10\sqrt[3]{2x}$ _____

14. $8\sqrt[3]{2x} + 3\sqrt[3]{2x^2} - \sqrt[3]{2x} = $ _____ $7\sqrt[3]{2x} + 3\sqrt[3]{2x^2}$ _____

Martin-Gay Interactive Videos   *Watch the section lecture video and answer the following questions.*

Objective A **15.** From Examples 1 and 2, why should you always check to see if all terms in your expression are simplified before attempting to add or subtract radicals? ▶

Objective B **16.** In Example 4, what are you told to remember about the square root of a positive number times the square root of the same positive number? ▶

See Video 10.4

See video answer section.

10.4 **Exercise Set** MyMathLab® ▶

Objective A *Add or subtract as indicated. Assume that all variables represent positive real numbers. See Examples 1 through 10.*

1. $\sqrt{8} - \sqrt{32}$
$-2\sqrt{2}$

2. $\sqrt{27} - \sqrt{75}$
$-2\sqrt{3}$

3. $2\sqrt{2x^3} + 4x\sqrt{8x}$
$10x\sqrt{2x}$

4. $3\sqrt{45x^3} + x\sqrt{5x}$
$10x\sqrt{5x}$

5. $2\sqrt{50} - 3\sqrt{125} + \sqrt{98}$
$17\sqrt{2} - 15\sqrt{5}$

6. $4\sqrt{32} - \sqrt{18} + 2\sqrt{128}$
$29\sqrt{2}$

7. $\sqrt[3]{16x} - \sqrt[3]{54x}$
$-\sqrt[3]{2x}$

8. $2\sqrt[3]{3a^4} - 3a\sqrt[3]{81a}$
$-7a\sqrt[3]{3a}$

9. $\sqrt{9b^3} - \sqrt{25b^3} + \sqrt{49b^3}$
$5b\sqrt{b}$

10. $\sqrt{4x^7} + 9x^2\sqrt{x^3} - 5x\sqrt{x^5}$
$6x^3\sqrt{x}$

11. $\dfrac{5\sqrt{2}}{3} + \dfrac{2\sqrt{2}}{5}$
$\dfrac{31\sqrt{2}}{15}$

12. $\dfrac{\sqrt{3}}{2} + \dfrac{4\sqrt{3}}{3}$
$\dfrac{11\sqrt{3}}{6}$

13. $\sqrt[3]{\dfrac{11}{8}} - \dfrac{\sqrt[3]{11}}{6}$
$\dfrac{\sqrt[3]{11}}{3}$

14. $\dfrac{2\sqrt[3]{4}}{7} - \dfrac{\sqrt[3]{4}}{14}$
$\dfrac{3\sqrt[3]{4}}{14}$

15. $\dfrac{\sqrt{20x}}{9} + \sqrt{\dfrac{5x}{9}}$
$\dfrac{5\sqrt{5x}}{9}$

16. $\dfrac{3x\sqrt{7}}{5} + \sqrt{\dfrac{7x^2}{100}}$
$\dfrac{7x\sqrt{7}}{10}$

17. $7\sqrt{9} - 7 + \sqrt{3}$
$14 + \sqrt{3}$

18. $\sqrt{16} - 5\sqrt{10} + 7$
$11 - 5\sqrt{10}$

19. $2 + 3\sqrt{y^2} - 6\sqrt{y^2} + 5$
$7 - 3y$

20. $3\sqrt{7} - \sqrt[3]{x} + 4\sqrt{7} - 3\sqrt[3]{x}$
$7\sqrt{7} - 4\sqrt[3]{x}$

21. $3\sqrt{108} - 2\sqrt{18} - 3\sqrt{48}$
$6\sqrt{3} - 6\sqrt{2}$

22. $-\sqrt{75} + \sqrt{12} - 3\sqrt{3}$
$-6\sqrt{3}$

23. $-5\sqrt[3]{625} + \sqrt[3]{40}$
$-23\sqrt[3]{5}$

24. $-2\sqrt[3]{108} - \sqrt[3]{32}$
$-8\sqrt[3]{4}$

25. $3\sqrt{a^5b^7} - ab\sqrt{25a^3b^5} + a^2\sqrt{16ab^7}$
$2a^2b^3\sqrt{ab}$

26. $\sqrt{4x^7y^5} + 9x^2\sqrt{x^3y^5} - 5xy\sqrt{x^5y^3}$
$6x^3y^2\sqrt{xy}$

27. $5y\sqrt{8y} + 2\sqrt{50y^3}$
$20y\sqrt{2y}$

28. $3\sqrt{8x^2y^3} - 2x\sqrt{32y^3}$
$-2xy\sqrt{2y}$

29. $\sqrt[3]{54xy^3} - 5\sqrt[3]{2xy^3} + y\sqrt[3]{128x}$
$2y\sqrt[3]{2x}$

30. $2\sqrt[3]{24x^3y^4} + 4x\sqrt[3]{81y^4}$
$16xy\sqrt[3]{3y}$

Example 11 Multiply: $\sqrt{3}(5 + \sqrt{30})$

Solution:

$$\sqrt{3}(5 + \sqrt{30}) = \sqrt{3}(5) + \sqrt{3}(\sqrt{30})$$
$$= 5\sqrt{3} + \sqrt{3 \cdot 30}$$
$$= 5\sqrt{3} + \sqrt{3 \cdot 3 \cdot 10}$$
$$= 5\sqrt{3} + 3\sqrt{10}$$

■ **Work Practice 11**

Practice 11

Multiply: $\sqrt{2}(6 + \sqrt{10})$

Examples Multiply. Assume that all variables represent positive real numbers.

12. $(\sqrt{5} - \sqrt{6})(\sqrt{7} + 1) = \overset{\text{First}}{\sqrt{5} \cdot \sqrt{7}} + \overset{\text{Outer}}{\sqrt{5} \cdot 1} - \overset{\text{Inner}}{\sqrt{6} \cdot \sqrt{7}} - \overset{\text{Last}}{\sqrt{6} \cdot 1}$

Using the FOIL order

$$= \sqrt{35} + \sqrt{5} - \sqrt{42} - \sqrt{6} \qquad \text{Simplify.}$$

13. $(\sqrt{2x} + 5)(\sqrt{2x} - 5) = (\sqrt{2x})^2 - 5^2$ Multiply the sum and difference of two terms: $(a + b)(a - b) = a^2 - b^2$

$$= 2x - 25$$

14. $(\sqrt{3} - 1)^2 = (\sqrt{3})^2 - 2 \cdot \sqrt{3} \cdot 1 + 1^2$ Square the binomial: $(a - b)^2 = a^2 - 2ab + b^2$

$$= 3 - 2\sqrt{3} + 1$$
$$= 4 - 2\sqrt{3}$$

15. $(\underbrace{\sqrt{x - 3}}_{a} + \underbrace{5}_{b})^2 = (\underbrace{\sqrt{x - 3}}_{a^2})^2 + \underbrace{2 \cdot}_{+ 2 \cdot} \underbrace{\sqrt{x - 3}}_{a} \cdot \underbrace{5}_{\cdot b} + \underbrace{5^2}_{+ b^2}$ Square the binomial: $(a + b)^2 = a^2 + 2ab + b^2$

$$= x - 3 + 10\sqrt{x - 3} + 25 \qquad \text{Simplify.}$$
$$= x + 22 + 10\sqrt{x - 3} \qquad \text{Combine like terms.}$$

■ **Work Practice 12–15**

Practice 12–15

Multiply. Assume that all variables represent positive real numbers.

12. $(\sqrt{3} - \sqrt{5})(\sqrt{2} + 7)$
13. $(\sqrt{5y} + 2)(\sqrt{5y} - 2)$
14. $(\sqrt{3} - 7)^2$
15. $(\sqrt{x + 1} + 2)^2$

Teaching Tip

While discussing Example 13, point out that the product does not contain radical signs. Then ask, "What is special about the factors?"

Answers
11. $6\sqrt{2} + 2\sqrt{5}$
12. $\sqrt{6} + 7\sqrt{3} - \sqrt{10} - 7\sqrt{5}$
13. $5y - 4$ **14.** $52 - 14\sqrt{3}$
15. $x + 5 + 4\sqrt{x + 1}$

Vocabulary, Readiness & Video Check

Complete the table with "Like" or "Unlike."

Terms	Like or Unlike Radical Terms?
1. $\sqrt{7}, \sqrt[3]{7}$	Unlike
2. $\sqrt[3]{x^2 y}, \sqrt[3]{yx^2}$	Like
3. $\sqrt[3]{abc}, \sqrt[3]{cba}$	Like
4. $2x\sqrt{5}, 2x\sqrt{10}$	Unlike

Simplify. Assume that all variables represent positive real numbers.

5. $2\sqrt{3} + 4\sqrt{3} =$ _____ $6\sqrt{3}$

6. $5\sqrt{7} + 3\sqrt{7} =$ _____ $8\sqrt{7}$

7. $8\sqrt{x} - \sqrt{x} =$ _____ $7\sqrt{x}$

8. $3\sqrt{y} - \sqrt{y} =$ _____ $2\sqrt{y}$

9. $7\sqrt[3]{x} + \sqrt[3]{x} =$ _____ $8\sqrt[3]{x}$

10. $8\sqrt[3]{z} + \sqrt[3]{z} =$ _____ $9\sqrt[3]{z}$

Add or subtract if possible.

11. $\sqrt{11} + \sqrt[3]{11} =$ _____ $\sqrt{11} + \sqrt[3]{11}$

12. $9\sqrt{13} - \sqrt[4]{13} =$ _____ $9\sqrt{13} - \sqrt[4]{13}$

13. $8\sqrt[3]{2x} + 3\sqrt[3]{2x} - \sqrt[3]{2x} =$ _____ $10\sqrt[3]{2x}$

14. $8\sqrt[3]{2x} + 3\sqrt[3]{2x^2} - \sqrt[3]{2x} =$ _____ $7\sqrt[3]{2x} + 3\sqrt[3]{2x^2}$

Martin-Gay Interactive Videos   Watch the section lecture video and answer the following questions.

See Video 10.4

See video answer section.

Objective A **15.** From ⊟ Examples 1 and 2, why should you always check to see if all terms in your expression are simplified before attempting to add or subtract radicals?

Objective B **16.** In ⊟ Example 4, what are you told to remember about the square root of a positive number times the square root of the same positive number? ▶

10.4 Exercise Set MyMathLab®

Objective A *Add or subtract as indicated. Assume that all variables represent positive real numbers. See Examples 1 through 10.*

1. $\sqrt{8} - \sqrt{32}$
$-2\sqrt{2}$

2. $\sqrt{27} - \sqrt{75}$
$-2\sqrt{3}$

3. $2\sqrt{2x^3} + 4x\sqrt{8x}$
$10x\sqrt{2x}$

4. $3\sqrt{45x^3} + x\sqrt{5x}$
$10x\sqrt{5x}$

5. $2\sqrt{50} - 3\sqrt{125} + \sqrt{98}$
$17\sqrt{2} - 15\sqrt{5}$

6. $4\sqrt{32} - \sqrt{18} + 2\sqrt{128}$
$29\sqrt{2}$

7. $\sqrt[3]{16x} - \sqrt[3]{54x}$
$-\sqrt[3]{2x}$

8. $2\sqrt[3]{3a^4} - 3a\sqrt[3]{81a}$
$-7a\sqrt[3]{3a}$

9. $\sqrt{9b^3} - \sqrt{25b^3} + \sqrt{49b^3}$
$5b\sqrt{b}$

10. $\sqrt{4x^7} + 9x^2\sqrt{x^3} - 5x\sqrt{x^5}$
$6x^3\sqrt{x}$

11. $\dfrac{5\sqrt{2}}{3} + \dfrac{2\sqrt{2}}{5}$
$\dfrac{31\sqrt{2}}{15}$

12. $\dfrac{\sqrt{3}}{2} + \dfrac{4\sqrt{3}}{3}$
$\dfrac{11\sqrt{3}}{6}$

13. $\sqrt[3]{\dfrac{11}{8}} - \dfrac{\sqrt[3]{11}}{6}$
$\dfrac{\sqrt[3]{11}}{3}$

14. $\dfrac{2\sqrt[3]{4}}{7} - \dfrac{\sqrt[3]{4}}{14}$
$\dfrac{3\sqrt[3]{4}}{14}$

15. $\dfrac{\sqrt{20x}}{9} + \sqrt{\dfrac{5x}{9}}$
$\dfrac{5\sqrt{5x}}{9}$

16. $\dfrac{3x\sqrt{7}}{5} + \sqrt{\dfrac{7x^2}{100}}$
$\dfrac{7x\sqrt{7}}{10}$

17. $7\sqrt{9} - 7 + \sqrt{3}$
$14 + \sqrt{3}$

18. $\sqrt{16} - 5\sqrt{10} + 7$
$11 - 5\sqrt{10}$

19. $2 + 3\sqrt{y^2} - 6\sqrt{y^2} + 5$
$7 - 3y$

20. $3\sqrt{7} - \sqrt[3]{x} + 4\sqrt{7} - 3\sqrt[3]{x}$
$7\sqrt{7} - 4\sqrt[3]{x}$

21. $3\sqrt{108} - 2\sqrt{18} - 3\sqrt{48}$
$6\sqrt{3} - 6\sqrt{2}$

22. $-\sqrt{75} + \sqrt{12} - 3\sqrt{3}$
$-6\sqrt{3}$

23. $-5\sqrt[3]{625} + \sqrt[3]{40}$
$-23\sqrt[3]{5}$

24. $-2\sqrt[3]{108} - \sqrt[3]{32}$
$-8\sqrt[3]{4}$

25. $3\sqrt{a^5b^7} - ab\sqrt{25a^3b^5} + a^2\sqrt{16ab^7}$
$2a^2b^3\sqrt{ab}$

26. $\sqrt{4x^7y^5} + 9x^2\sqrt{x^3y^5} - 5xy\sqrt{x^5y^3}$
$6x^3y^2\sqrt{xy}$

27. $5y\sqrt{8y} + 2\sqrt{50y^3}$
$20y\sqrt{2y}$

28. $3\sqrt{8x^2y^3} - 2x\sqrt{32y^3}$
$-2xy\sqrt{2y}$

29. $\sqrt[3]{54xy^3} - 5\sqrt[3]{2xy^3} + y\sqrt[3]{128x}$
$2y\sqrt[3]{2x}$

30. $2\sqrt[3]{24x^3y^4} + 4x\sqrt[3]{81y^4}$
$16xy\sqrt[3]{3y}$

31. $6\sqrt[3]{11} + 8\sqrt{11} - 12\sqrt{11}$
$6\sqrt[3]{11} - 4\sqrt{11}$

32. $3\sqrt[3]{5} + 4\sqrt{5} - 2\sqrt{5}$
$3\sqrt[3]{5} + 2\sqrt{5}$

33. $-2\sqrt[4]{x^7} + 3\sqrt[4]{16x^7}$
$4x\sqrt[4]{x^3}$

34. $6\sqrt[3]{24x^3} - 2\sqrt[3]{81x^3} - x\sqrt[3]{3}$
$5x\sqrt[3]{3}$

35. $\dfrac{4\sqrt{3}}{3} - \dfrac{\sqrt{12}}{3}$
$\dfrac{2\sqrt{3}}{3}$

36. $\dfrac{\sqrt{45}}{10} + \dfrac{7\sqrt{5}}{10}$
$\sqrt{5}$

37. $\dfrac{\sqrt[3]{8x^4}}{7} + \dfrac{3x\sqrt[3]{x}}{7}$
$\dfrac{5x\sqrt[3]{x}}{7}$

38. $\dfrac{\sqrt[4]{48}}{5x} - \dfrac{2\sqrt[4]{3}}{10x}$
$\dfrac{\sqrt[4]{3}}{5x}$

39. $\sqrt{\dfrac{28}{x^2}} + \sqrt{\dfrac{7}{4x^2}}$
$\dfrac{5\sqrt{7}}{2x}$

40. $\dfrac{\sqrt{99}}{5x} - \sqrt{\dfrac{44}{x^2}}$
$\dfrac{7\sqrt{11}}{5x}$

41. $\sqrt[3]{\dfrac{16}{27}} - \dfrac{\sqrt[3]{54}}{6}$
$\dfrac{\sqrt[3]{2}}{6}$

42. $\dfrac{\sqrt[3]{3}}{10} + \sqrt[3]{\dfrac{24}{125}}$
$\dfrac{\sqrt[3]{3}}{2}$

43. $-\dfrac{\sqrt[3]{2x^4}}{9} + \sqrt[3]{\dfrac{250x^4}{27}}$
$\dfrac{14x\sqrt[3]{2x}}{9}$

44. $\dfrac{\sqrt[3]{y^5}}{8} + \dfrac{5y\sqrt[3]{y^2}}{4}$
$\dfrac{11y\sqrt[3]{y^2}}{8}$

△ **45.** Find the perimeter of the trapezoid.

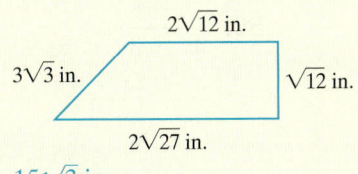

2√12 in.
3√3 in.
√12 in.
2√27 in.
$15\sqrt{3}$ in.

△ **46.** Find the perimeter of the triangle.

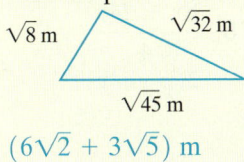

√8 m
√32 m
√45 m
$(6\sqrt{2} + 3\sqrt{5})$ m

Objective B *Multiply. Then simplify if possible. Assume that all variables represent positive real numbers. See Examples 11 through 15.*

▶ **47.** $\sqrt{7}(\sqrt{5} + \sqrt{3})$
$\sqrt{35} + \sqrt{21}$

48. $\sqrt{2}(\sqrt{15} - \sqrt{35})$
$\sqrt{30} - \sqrt{70}$

49. $(\sqrt{5} - \sqrt{2})^2$
$7 - 2\sqrt{10}$

50. $(3x - \sqrt{2})(3x - \sqrt{2})$
$9x^2 - 6x\sqrt{2} + 2$

51. $\sqrt{3x}(\sqrt{3} - \sqrt{x})$
$3\sqrt{x} - x\sqrt{3}$

52. $\sqrt{5y}(\sqrt{y} + \sqrt{5})$
$y\sqrt{5} + 5\sqrt{y}$

53. $(2\sqrt{x} - 5)(3\sqrt{x} + 1)$
$6x - 13\sqrt{x} - 5$

54. $(8\sqrt{y} + z)(4\sqrt{y} - 1)$
$32y - 8\sqrt{y} + 4z\sqrt{y} - z$

55. $(\sqrt[3]{a} - 4)(\sqrt[3]{a} + 5)$
$\sqrt[3]{a^2} + \sqrt[3]{a} - 20$

56. $(\sqrt[3]{a} + 2)(\sqrt[3]{a} + 7)$
$\sqrt[3]{a^2} + 9\sqrt[3]{a} + 14$

57. $6(\sqrt{2} - 2)$
$6\sqrt{2} - 12$

58. $\sqrt{5}(6 - \sqrt{5})$
$6\sqrt{5} - 5$

59. $\sqrt{2}(\sqrt{2} + x\sqrt{6})$
$2 + 2x\sqrt{3}$

60. $\sqrt{3}(\sqrt{3} - 2\sqrt{5x})$
$3 - 2\sqrt{15x}$

▶ **61.** $(2\sqrt{7} + 3\sqrt{5})(\sqrt{7} - 2\sqrt{5})$
$-16 - \sqrt{35}$

62. $(\sqrt{x} - y)(\sqrt{x} + y)$
$x - y^2$

63. $(\sqrt{6} - 4\sqrt{2})(3\sqrt{6} + 1)$
$18 + \sqrt{6} - 24\sqrt{3} - 4\sqrt{2}$

64. $(3\sqrt{x} + 2)(\sqrt{3x} - 2)$
$3x\sqrt{3} - 6\sqrt{x} + 2\sqrt{3x} - 4$

65. $(\sqrt{3} + x)^2$
$3 + 2x\sqrt{3} + x^2$

66. $(\sqrt{y} - 3x)^2$
$y - 6x\sqrt{y} + 9x^2$

67. $(\sqrt{5x} - 3\sqrt{2})(\sqrt{5x} - 3\sqrt{3})$
$5x - 3\sqrt{15x} - 3\sqrt{10x} + 9\sqrt{6}$

68. $(5\sqrt{3x} - \sqrt{y})(4\sqrt{x} + 1)$
$20x\sqrt{3} + 5\sqrt{3x} - 4\sqrt{xy} - \sqrt{y}$

69. $(\sqrt[3]{4} + 2)(\sqrt[3]{2} - 1)$
$-\sqrt[3]{4} + 2\sqrt[3]{2}$

70. $(\sqrt[3]{3} + \sqrt[3]{2})(\sqrt[3]{9} - \sqrt[3]{4})$
$1 - \sqrt[3]{12} + \sqrt[3]{18}$

71. $(\sqrt[3]{x} + 1)(\sqrt[3]{x} - 4\sqrt{x} + 7)$
$\sqrt[3]{x^2} - 4\sqrt[6]{x^5} + 8\sqrt[3]{x} - 4\sqrt{x} + 7$

72. $(\sqrt[3]{3x} + 3)(\sqrt[3]{2x} - 3x - 1)$
$\sqrt[3]{6x^2} - 3x\sqrt[3]{3x} - \sqrt[3]{3x} + 3\sqrt[3]{2x} - 9x - 3$

73. $(\sqrt{x - 1} + 5)^2$
$x + 24 + 10\sqrt{x - 1}$

74. $(\sqrt{3x + 1} + 2)^2$
$3x + 5 + 4\sqrt{3x + 1}$

75. $(\sqrt{2x + 5} - 1)^2$
$2x + 6 - 2\sqrt{2x + 5}$

76. $(\sqrt{x - 6} - 7)^2$
$x + 43 - 14\sqrt{x - 6}$

Review

Factor each numerator and denominator. Then simplify if possible. See Section 7.1.

77. $\dfrac{2x - 14}{2}$
$x - 7$

78. $\dfrac{8x - 24y}{4}$
$2(x - 3y)$

79. $\dfrac{7x - 7y}{x^2 - y^2}$
$\dfrac{7}{x + y}$

80. $\dfrac{x^3 - 8}{4x - 8}$
$\dfrac{x^2 + 2x + 4}{4}$

81. $\dfrac{6a^2b - 9ab}{3ab}$
$2a - 3$

82. $\dfrac{14r - 28r^2s^2}{7rs}$
$\dfrac{2}{s} - 4rs$

83. $\dfrac{-4 + 2\sqrt{3}}{6}$
$\dfrac{-2 + \sqrt{3}}{3}$

84. $\dfrac{-5 + 10\sqrt{7}}{5}$
$-1 + 2\sqrt{7}$

Concept Extensions

△ **85.** Find the perimeter and area of the rectangle.
$22\sqrt{5}$ ft; 150 sq ft

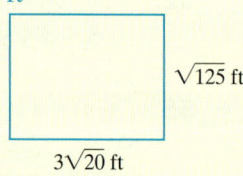

$\sqrt{125}$ ft

$3\sqrt{20}$ ft

△ **86.** Find the perimeter and area of the trapezoid. (*Hint:* The area of a trapezoid is the product of half the height $6\sqrt{3}$ meters and the sum of the bases $2\sqrt{63}$ and $7\sqrt{7}$ meters.) $(12\sqrt{3} + 13\sqrt{7})$ m; $39\sqrt{21}$ sq m

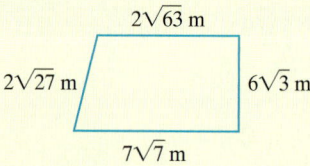

$2\sqrt{63}$ m

$2\sqrt{27}$ m $6\sqrt{3}$ m

$7\sqrt{7}$ m

87. a. Add: $\sqrt{3} + \sqrt{3}$ $2\sqrt{3}$
b. Multiply: $\sqrt{3} \cdot \sqrt{3}$ 3
c. Describe the differences in parts (a) and (b).
answers may vary

88. a. Add: $2\sqrt{5} + \sqrt{5}$ $3\sqrt{5}$
b. Multiply: $2\sqrt{5} \cdot \sqrt{5}$ 10
c. Describe the differences in parts (a) and (b).
answers may vary

89. Multiply: $(\sqrt{2} + \sqrt{3} - 1)^2$
$2\sqrt{6} - 2\sqrt{2} - 2\sqrt{3} + 6$

90. Multiply: $(\sqrt{5} - \sqrt{2} + 1)^2$
$8 - 2\sqrt{10} + 2\sqrt{5} - 2\sqrt{2}$

10.5 Rationalizing Numerators and Denominators of Radical Expressions

Objective A Rationalizing Denominators

Objectives

A Rationalize Denominators.

B Rationalize Denominators Having Two Terms.

C Rationalize Numerators.

Often in mathematics it is helpful to write a radical expression such as $\frac{\sqrt{3}}{\sqrt{2}}$ either without a radical in the denominator or without a radical in the numerator. The process of writing this expression as an equivalent expression but without a radical in the denominator is called **rationalizing the denominator.** To rationalize the denominator of $\frac{\sqrt{3}}{\sqrt{2}}$, we multiply the numerator and the denominator by $\sqrt{2}$. Recall that this is the same as multiplying by $\frac{\sqrt{2}}{\sqrt{2}}$, which simplifies to 1.

$$\frac{\sqrt{3}}{\sqrt{2}} = \frac{\sqrt{3} \cdot \sqrt{2}}{\sqrt{2} \cdot \sqrt{2}} = \frac{\sqrt{6}}{\sqrt{4}} = \frac{\sqrt{6}}{2}$$

Example 1 Rationalize the denominator of $\frac{2}{\sqrt{5}}$.

Solution: To rationalize the denominator, we multiply the numerator and denominator by a factor that makes the radicand in the denominator a perfect square.

$$\frac{2}{\sqrt{5}} = \frac{2 \cdot \sqrt{5}}{\sqrt{5} \cdot \sqrt{5}} = \frac{2\sqrt{5}}{5} \quad \text{The denominator is now rationalized.}$$

■ Work Practice 1

Practice 1

Rationalize the denominator of $\frac{7}{\sqrt{2}}$.

Example 2 Rationalize the denominator of $\frac{2\sqrt{16}}{\sqrt{9x}}$.

Solution: First we simplify the radicals; then we rationalize the denominator.

$$\frac{2\sqrt{16}}{\sqrt{9x}} = \frac{2(4)}{\sqrt{9} \cdot \sqrt{x}} = \frac{8}{3\sqrt{x}}$$

To rationalize the denominator, we multiply the numerator and the denominator by $\sqrt{x}$.

$$\frac{8}{3\sqrt{x}} = \frac{8 \cdot \sqrt{x}}{3\sqrt{x} \cdot \sqrt{x}} = \frac{8\sqrt{x}}{3x}$$

■ Work Practice 2

Practice 2

Rationalize the denominator of $\frac{2\sqrt{9}}{\sqrt{16y}}$.

Example 3 Rationalize the denominator of $\sqrt[3]{\frac{1}{2}}$.

Solution: $\sqrt[3]{\frac{1}{2}} = \frac{\sqrt[3]{1}}{\sqrt[3]{2}} = \frac{1}{\sqrt[3]{2}}$

Now we rationalize the denominator. Since $\sqrt[3]{2}$ is a cube root, we want to multiply by a value that will make the radicand 2 a perfect cube. If we multiply by $\sqrt[3]{2^2}$, we get $\sqrt[3]{2^3} = 2$. Thus,

$$\frac{1 \cdot \sqrt[3]{2^2}}{\sqrt[3]{2} \cdot \sqrt[3]{2^2}} = \frac{\sqrt[3]{4}}{\sqrt[3]{2^3}} = \frac{\sqrt[3]{4}}{2} \quad \text{Multiply numerator and denominator by } \sqrt[3]{2^2} \text{ and then simplify.}$$

■ Work Practice 3

Practice 3

Rationalize the denominator of $\sqrt[3]{\frac{2}{25}}$.

Answers

1. $\frac{7\sqrt{2}}{2}$ 2. $\frac{3\sqrt{y}}{2y}$ 3. $\frac{\sqrt[3]{10}}{5}$

✓ **Concept Check** Determine by which number both the numerator and denominator should be multiplied to rationalize the denominator of the radical expression.

a. $\dfrac{1}{\sqrt[3]{7}}$ **b.** $\dfrac{1}{\sqrt[4]{8}}$

Practice 4

Rationalize the denominator of $\sqrt{\dfrac{5m}{11n}}$. Assume that all variables represent positive real numbers.

Example 4 Rationalize the denominator of $\sqrt{\dfrac{7x}{3y}}$. Assume that all variables represent positive real numbers.

Solution:

$$\sqrt{\dfrac{7x}{3y}} = \dfrac{\sqrt{7x}}{\sqrt{3y}}$$ Use the quotient rule. No radical may be simplified further.

$$= \dfrac{\sqrt{7x} \cdot \sqrt{3y}}{\sqrt{3y} \cdot \sqrt{3y}}$$ Multiply numerator and denominator by $\sqrt{3y}$ so that the radicand in the denominator is a perfect square.

$$= \dfrac{\sqrt{21xy}}{3y}$$ Use the product rule in the numerator and denominator. Remember that $\sqrt{3y} \cdot \sqrt{3y} = 3y$.

■ **Work Practice 4**

Practice 5

Rationalize the denominator of $\dfrac{\sqrt[5]{a^2}}{\sqrt[5]{32b^{12}}}$. Assume that all variables represent positive real numbers.

Example 5 Rationalize the denominator of $\dfrac{\sqrt[4]{x}}{\sqrt[4]{81y^5}}$. Assume that all variables represent positive real numbers.

Solution: First we simplify each radical if possible.

$$\dfrac{\sqrt[4]{x}}{\sqrt[4]{81y^5}} = \dfrac{\sqrt[4]{x}}{\sqrt[4]{81y^4} \cdot \sqrt[4]{y}}$$ Use the product rule in the denominator.

$$= \dfrac{\sqrt[4]{x}}{3y\sqrt[4]{y}}$$ Write $\sqrt[4]{81y^4}$ as $3y$.

$$= \dfrac{\sqrt[4]{x} \cdot \sqrt[4]{y^3}}{3y\sqrt[4]{y} \cdot \sqrt[4]{y^3}}$$ Multiply numerator and denominator by $\sqrt[4]{y^3}$ so that the radicand in the denominator is a perfect 4th power.

$$= \dfrac{\sqrt[4]{xy^3}}{3y\sqrt[4]{y^4}}$$ Use the product rule in the numerator and denominator.

$$= \dfrac{\sqrt[4]{xy^3}}{3y^2}$$ In the denominator, $\sqrt[4]{y^4} = y$ and $3y \cdot y = 3y^2$.

■ **Work Practice 5**

Objective B Rationalizing Denominators Having Two Terms ▶

Remember the product of the sum and difference of two terms?

$$(a + b)(a - b) = a^2 - b^2$$

These two expressions are called **conjugates** of each other.

Answers

4. $\dfrac{\sqrt{55mn}}{11n}$ **5.** $\dfrac{\sqrt[5]{a^2b^3}}{2b^3}$

✓ **Concept Check Answers**

a. $\sqrt[3]{7^2}$ or $\sqrt[3]{49}$ **b.** $\sqrt[4]{2}$

To rationalize a denominator that is a sum or difference of two terms, we use conjugates. To see how and why this works, let's rationalize the denominator of the expression $\dfrac{5}{\sqrt{3} - 2}$. To do so, we multiply both the numerator and the denominator by $\sqrt{3} + 2$, the *conjugate* of the denominator $\sqrt{3} - 2$, and see what happens.

$$\dfrac{5}{\sqrt{3} - 2} = \dfrac{5(\sqrt{3} + 2)}{(\sqrt{3} - 2)(\sqrt{3} + 2)}$$

$$= \dfrac{5(\sqrt{3} + 2)}{(\sqrt{3})^2 - 2^2} \quad \text{\small Multiply the sum and difference of two terms:}$$
$$\text{\small} \quad (a + b)(a - b) = a^2 - b^2.$$

$$= \dfrac{5(\sqrt{3} + 2)}{3 - 4}$$

$$= \dfrac{5(\sqrt{3} + 2)}{-1}$$

$$= -5(\sqrt{3} + 2) \text{ or } -5\sqrt{3} - 10$$

Notice in the denominator that the product of $(\sqrt{3} - 2)$ and its conjugate, $(\sqrt{3} + 2)$, is -1. In general, the product of an expression and its conjugate will contain no radical terms. This is why, when rationalizing a denominator or a numerator containing two terms, we multiply by its conjugate. Examples of conjugates are

$$\sqrt{a} - \sqrt{b} \quad \text{and} \quad \sqrt{a} + \sqrt{b}$$
$$x + \sqrt{y} \quad \text{and} \quad x - \sqrt{y}$$

Example 6 Rationalize the denominator of $\dfrac{2}{3\sqrt{2} + 4}$.

Solution: We multiply the numerator and the denominator by the conjugate of $3\sqrt{2} + 4$.

$$\dfrac{2}{3\sqrt{2} + 4} = \dfrac{2(3\sqrt{2} - 4)}{(3\sqrt{2} + 4)(3\sqrt{2} - 4)}$$

$$= \dfrac{2(3\sqrt{2} - 4)}{(3\sqrt{2})^2 - 4^2} \quad \text{\small Multiply the sum and difference}$$
$$\text{\small of two terms: } (a + b)(a - b) = a^2 - b^2.$$

$$= \dfrac{2(3\sqrt{2} - 4)}{18 - 16} \quad \text{\small Write } (3\sqrt{2})^2 \text{ as } 9 \cdot 2 \text{ or } 18 \text{ and } 4^2 \text{ as } 16.$$

$$= \dfrac{2(3\sqrt{2} - 4)}{2} = 3\sqrt{2} - 4$$

Work Practice 6

As we saw in Example 6, it is often helpful to leave a numerator in factored form to help determine whether the expression can be simplified.

Practice 6

Rationalize the denominator of $\dfrac{3}{2\sqrt{5} + 1}$.

Answer

6. $\dfrac{3(2\sqrt{5} - 1)}{19}$

Practice 7

Rationalize the denominator of $\dfrac{\sqrt{5}+3}{\sqrt{3}-\sqrt{2}}$.

Example 7 Rationalize the denominator of $\dfrac{\sqrt{6}+2}{\sqrt{5}-\sqrt{3}}$.

Solution: We multiply the numerator and the denominator by the conjugate of $\sqrt{5}-\sqrt{3}$.

$$\frac{\sqrt{6}+2}{\sqrt{5}-\sqrt{3}} = \frac{(\sqrt{6}+2)(\sqrt{5}+\sqrt{3})}{(\sqrt{5}-\sqrt{3})(\sqrt{5}+\sqrt{3})}$$

$$= \frac{\sqrt{6}\sqrt{5}+\sqrt{6}\sqrt{3}+2\sqrt{5}+2\sqrt{3}}{(\sqrt{5})^2-(\sqrt{3})^2}$$

$$= \frac{\sqrt{30}+\sqrt{18}+2\sqrt{5}+2\sqrt{3}}{5-3}$$

$$= \frac{\sqrt{30}+3\sqrt{2}+2\sqrt{5}+2\sqrt{3}}{2}$$

■ **Work Practice 7**

Practice 8

Rationalize the denominator of $\dfrac{3}{2-\sqrt{x}}$. Assume that all variables represent positive real numbers.

Example 8 Rationalize the denominator of $\dfrac{2\sqrt{m}}{3\sqrt{x}+\sqrt{m}}$. Assume that all variables represent positive real numbers.

Solution: We multiply by the conjugate of $3\sqrt{x}+\sqrt{m}$ to eliminate the radicals from the denominator.

$$\frac{2\sqrt{m}}{3\sqrt{x}+\sqrt{m}} = \frac{2\sqrt{m}(3\sqrt{x}-\sqrt{m})}{(3\sqrt{x}+\sqrt{m})(3\sqrt{x}-\sqrt{m})} = \frac{6\sqrt{mx}-2m}{(3\sqrt{x})^2-(\sqrt{m})^2}$$

$$= \frac{6\sqrt{mx}-2m}{9x-m}$$

■ **Work Practice 8**

Teaching Tip

After discussing rationalizing the numerator, ask students, "Is it possible to rationalize both the denominator and numerator? Explain."

Objective C Rationalizing Numerators ▶

As mentioned earlier, it is also often helpful to write an expression such as $\dfrac{\sqrt{3}}{\sqrt{2}}$ as an equivalent expression without a radical in the numerator. This process is called **rationalizing the numerator.** To rationalize the numerator of $\dfrac{\sqrt{3}}{\sqrt{2}}$, we multiply the numerator and the denominator by $\sqrt{3}$.

$$\frac{\sqrt{3}}{\sqrt{2}} = \frac{\sqrt{3}\cdot\sqrt{3}}{\sqrt{2}\cdot\sqrt{3}} = \frac{\sqrt{9}}{\sqrt{6}} = \frac{3}{\sqrt{6}}$$

Practice 9

Rationalize the numerator of $\dfrac{\sqrt{18}}{\sqrt{75}}$.

Example 9 Rationalize the numerator of $\dfrac{\sqrt{7}}{\sqrt{45}}$.

Solution: First we simplify $\sqrt{45}$.

$$\frac{\sqrt{7}}{\sqrt{45}} = \frac{\sqrt{7}}{\sqrt{9\cdot5}} = \frac{\sqrt{7}}{3\sqrt{5}}$$

Next we rationalize the numerator by multiplying the numerator and the denominator by $\sqrt{7}$.

$$\frac{\sqrt{7}}{3\sqrt{5}} = \frac{\sqrt{7}\cdot\sqrt{7}}{3\sqrt{5}\cdot\sqrt{7}} = \frac{7}{3\sqrt{5\cdot7}} = \frac{7}{3\sqrt{35}}$$

■ **Work Practice 9**

Answers

7. $\sqrt{15}+\sqrt{10}+3\sqrt{3}+3\sqrt{2}$

8. $\dfrac{6+3\sqrt{x}}{4-x}$ 9. $\dfrac{6}{5\sqrt{6}}$

Example 10 Rationalize the numerator of $\dfrac{\sqrt[3]{2x^2}}{\sqrt[3]{5y}}$.

Solution:

$$\dfrac{\sqrt[3]{2x^2}}{\sqrt[3]{5y}} = \dfrac{\sqrt[3]{2x^2} \cdot \sqrt[3]{2^2x}}{\sqrt[3]{5y} \cdot \sqrt[3]{2^2x}}$$ Multiply the numerator and denominator by $\sqrt[3]{2^2x}$ so that the radicand in the numerator is a perfect cube.

$$= \dfrac{\sqrt[3]{2^3x^3}}{\sqrt[3]{5y \cdot 2^2x}}$$ Use the product rule in the numerator and denominator.

$$= \dfrac{2x}{\sqrt[3]{20xy}}$$ Simplify.

■ **Work Practice 10**

Just as for denominators, to rationalize a numerator that is a sum or difference of two terms, we use conjugates.

Example 11 Rationalize the numerator of $\dfrac{\sqrt{x} + 2}{5}$. Assume that all variables represent positive real numbers.

Solution: We multiply the numerator and the denominator by the conjugate of $\sqrt{x} + 2$, the numerator.

$$\dfrac{\sqrt{x} + 2}{5} = \dfrac{(\sqrt{x} + 2)(\sqrt{x} - 2)}{5(\sqrt{x} - 2)}$$ Multiply by $\sqrt{x} - 2$, the conjugate of $\sqrt{x} + 2$.

$$= \dfrac{(\sqrt{x})^2 - 2^2}{5(\sqrt{x} - 2)}$$ $(a + b)(a - b) = a^2 - b^2$.

$$= \dfrac{x - 4}{5(\sqrt{x} - 2)}$$

■ **Work Practice 11**

Practice 10

Rationalize the numerator of $\dfrac{\sqrt[3]{3a}}{\sqrt[3]{7b}}$.

Practice 11

Rationalize the numerator of $\dfrac{\sqrt{x} + 5}{3}$. Assume that all variables represent positive real numbers.

Answers

10. $\dfrac{3a}{\sqrt[3]{63a^2b}}$ 11. $\dfrac{x - 25}{3(\sqrt{x} - 5)}$

Vocabulary, Readiness & Video Check

Use the choices below to fill in each blank. Not all choices will be used.

rationalizing the numerator   conjugate $\dfrac{\sqrt{3}}{\sqrt{3}}$

rationalizing the denominator $\dfrac{5}{5}$

1. The _____ conjugate _____ of $a + b$ is $a - b$.

2. The process of writing an equivalent expression, but without a radical in the denominator, is called _____ rationalizing the denominator _____ .

3. The process of writing an equivalent expression, but without a radical in the numerator, is called _____ rationalizing the numerator _____ .

4. To rationalize the denominator of $\dfrac{5}{\sqrt{3}}$, we multiply by _____ $\dfrac{\sqrt{3}}{\sqrt{3}}$ _____ .

Find the conjugate of each expression.

5. $\sqrt{2} + x$ $\sqrt{2} - x$ 6. $\sqrt{3} + y$ $\sqrt{3} - y$ 7. $5 - \sqrt{a}$ $5 + \sqrt{a}$ 8. $6 - \sqrt{b}$ $6 + \sqrt{b}$

9. $-7\sqrt{5} + 8\sqrt{x}$ $-7\sqrt{5} - 8\sqrt{x}$ 10. $-9\sqrt{2} - 6\sqrt{y}$ $-9\sqrt{2} + 6\sqrt{y}$

Martin-Gay Interactive Videos   Watch the section lecture video and answer the following questions.

Objective A **11.** From Examples 1–3, what is the goal of rationalizing a denominator? ▶

Objective B **12.** From Examples 4, why will multiplying a denominator by its conjugate always rationalize the denominator? ▶

Objective C **13.** From Example 5, is the process of rationalizing a numerator any different from rationalizing a denominator? ▶

See Video 10.5 🍎

See video answer section.

10.5 Exercise Set MyMathLab® ▶

Objective A *Rationalize each denominator. Assume that all variables represent positive real numbers. See Examples 1 through 5.*

▶**1.** $\dfrac{\sqrt{2}}{\sqrt{7}}$ $\dfrac{\sqrt{14}}{7}$

2. $\dfrac{\sqrt{5}}{\sqrt{2}}$ $\dfrac{\sqrt{10}}{2}$

3. $\sqrt{\dfrac{1}{5}}$ $\dfrac{\sqrt{5}}{5}$

4. $\sqrt{\dfrac{1}{2}}$ $\dfrac{\sqrt{2}}{2}$

▶**5.** $\dfrac{4}{\sqrt[3]{3}}$ $\dfrac{4\sqrt[3]{9}}{3}$

6. $\dfrac{6}{\sqrt[3]{9}}$ $2\sqrt[3]{3}$

▶**7.** $\dfrac{3}{\sqrt{8x}}$ $\dfrac{3\sqrt{2x}}{4x}$

8. $\dfrac{5}{\sqrt{27a}}$ $\dfrac{5\sqrt{3a}}{9a}$

9. $\dfrac{3}{\sqrt[3]{4x^2}}$ $\dfrac{3\sqrt[3]{2x}}{2x}$

10. $\dfrac{5}{\sqrt[3]{3y}}$ $\dfrac{5\sqrt[3]{9y^2}}{3y}$

11. $\dfrac{9}{\sqrt{3a}}$ $\dfrac{3\sqrt{3a}}{a}$

12. $\dfrac{x}{\sqrt{5}}$ $\dfrac{x\sqrt{5}}{5}$

13. $\dfrac{3}{\sqrt[3]{2}}$ $\dfrac{3\sqrt[3]{4}}{2}$

14. $\dfrac{5}{\sqrt[3]{9}}$ $\dfrac{5\sqrt[3]{3}}{3}$

15. $\dfrac{2\sqrt{3}}{\sqrt{7}}$ $\dfrac{2\sqrt{21}}{7}$

16. $\dfrac{-5\sqrt{2}}{\sqrt{11}}$ $\dfrac{-5\sqrt{22}}{11}$

17. $\sqrt{\dfrac{2x}{5y}}$ $\dfrac{\sqrt{10xy}}{5y}$

18. $\sqrt{\dfrac{13a}{2b}}$ $\dfrac{\sqrt{26ab}}{2b}$

19. $\sqrt[3]{\dfrac{3}{5}}$ $\dfrac{\sqrt[3]{75}}{5}$

20. $\sqrt[3]{\dfrac{7}{10}}$ $\dfrac{\sqrt[3]{700}}{10}$

21. $\sqrt{\dfrac{3x}{50}}$ $\dfrac{\sqrt{6x}}{10}$

22. $\sqrt{\dfrac{11y}{45}}$ $\dfrac{\sqrt{55y}}{15}$

23. $\dfrac{1}{\sqrt{12z}}$ $\dfrac{\sqrt{3z}}{6z}$

24. $\dfrac{1}{\sqrt{32x}}$ $\dfrac{\sqrt{2x}}{8x}$

25. $\dfrac{\sqrt[3]{2y^2}}{\sqrt[3]{9x^2}}$ $\dfrac{\sqrt[3]{6xy^2}}{3x}$

26. $\dfrac{\sqrt[3]{3x}}{\sqrt[3]{4y^4}}$ $\dfrac{\sqrt[3]{6xy^2}}{2y^2}$

27. $\sqrt[4]{\dfrac{16}{9x^7}}$ $\dfrac{2\sqrt[4]{9x}}{3x^2}$

28. $\sqrt[5]{\dfrac{32}{m^6 n^{13}}}$ $\dfrac{2\sqrt[5]{m^4 n^2}}{m^2 n^3}$

29. $\dfrac{5a}{\sqrt[5]{8a^9 b^{11}}}$ $\dfrac{5\sqrt[5]{4ab^4}}{2ab^3}$

30. $\dfrac{9y}{\sqrt[4]{4y^9}}$ $\dfrac{9\sqrt[4]{4y^3}}{2y^2}$

Objective B *Rationalize each denominator. Assume that all variables represent positive real numbers. See Examples 6 through 8.*

31. $\dfrac{6}{2-\sqrt{7}}$ $-2(2+\sqrt{7})$

32. $\dfrac{3}{\sqrt{7}-4}$ $-\dfrac{\sqrt{7}+4}{3}$

▶**33.** $\dfrac{-7}{\sqrt{x}-3}$ $\dfrac{7(\sqrt{x}+3)}{9-x}$

34. $\dfrac{-8}{\sqrt{y}+4}$ $\dfrac{32-8\sqrt{y}}{y-16}$

35. $\dfrac{\sqrt{2}-\sqrt{3}}{\sqrt{2}+\sqrt{3}}$ $-5+2\sqrt{6}$

36. $\dfrac{\sqrt{3}+\sqrt{4}}{\sqrt{2}+\sqrt{3}}$ $-\sqrt{6}+3-2\sqrt{2}+2\sqrt{3}$

37. $\dfrac{\sqrt{a}+1}{2\sqrt{a}-\sqrt{b}}$ $\dfrac{2a+\sqrt{ab}+2\sqrt{a}+\sqrt{b}}{4a-b}$

38. $\dfrac{2\sqrt{a}-3}{2\sqrt{a}-\sqrt{b}}$ $\dfrac{4a+2\sqrt{ab}-6\sqrt{a}-3\sqrt{b}}{4a-b}$

39. $\dfrac{8}{1+\sqrt{10}}$ $-\dfrac{8(1-\sqrt{10})}{9}$

40. $\dfrac{-3}{\sqrt{6}-2}$ $\dfrac{-3(\sqrt{6}+2)}{2}$

41. $\dfrac{\sqrt{x}}{\sqrt{x}+\sqrt{y}}$ $\dfrac{x-\sqrt{xy}}{x-y}$

42. $\dfrac{2\sqrt{a}}{2\sqrt{x}-\sqrt{y}}$ $\dfrac{4\sqrt{ax}+2\sqrt{ay}}{4x-y}$

43. $\dfrac{2\sqrt{3}+\sqrt{6}}{4\sqrt{3}-\sqrt{6}}$ $\dfrac{5+3\sqrt{2}}{7}$

44. $\dfrac{4\sqrt{5}+\sqrt{2}}{2\sqrt{5}-\sqrt{2}}$ $\dfrac{7+\sqrt{10}}{3}$

Objective C *Rationalize each numerator. Assume that all variables represent positive real numbers. See Examples 9 and 10.*

45. $\sqrt{\dfrac{5}{3}}$ $\dfrac{5}{\sqrt{15}}$

46. $\sqrt{\dfrac{3}{2}}$ $\dfrac{3}{\sqrt{6}}$

▶ **47.** $\sqrt{\dfrac{18}{5}}$ $\dfrac{6}{\sqrt{10}}$

48. $\sqrt{\dfrac{12}{7}}$ $\dfrac{6}{\sqrt{21}}$

49. $\dfrac{\sqrt{4x}}{7}$ $\dfrac{2x}{7\sqrt{x}}$

50. $\dfrac{\sqrt{3x^5}}{6}$ $\dfrac{x^3}{2\sqrt{3x}}$

51. $\dfrac{\sqrt[3]{5y^2}}{\sqrt[3]{4x}}$ $\dfrac{5y}{\sqrt[3]{100xy}}$

52. $\dfrac{\sqrt[3]{4x}}{\sqrt[3]{z^4}}$ $\dfrac{2x}{z\sqrt[3]{2x^2z}}$

53. $\sqrt{\dfrac{2}{5}}$ $\dfrac{2}{\sqrt{10}}$

54. $\sqrt{\dfrac{3}{7}}$ $\dfrac{3}{\sqrt{21}}$

55. $\dfrac{\sqrt{2x}}{11}$ $\dfrac{2x}{11\sqrt{2x}}$

56. $\dfrac{\sqrt{y}}{7}$ $\dfrac{y}{7\sqrt{y}}$

57. $\sqrt[3]{\dfrac{7}{8}}$ $\dfrac{7}{2\sqrt[3]{49}}$

58. $\sqrt[3]{\dfrac{25}{2}}$ $\dfrac{5}{\sqrt[3]{10}}$

59. $\dfrac{\sqrt[3]{3x^5}}{10}$ $\dfrac{3x^2}{10\sqrt[3]{9x}}$

60. $\sqrt[3]{\dfrac{9y}{7}}$ $\dfrac{3y}{\sqrt[3]{21y^2}}$

61. $\sqrt{\dfrac{18x^4y^6}{3z}}$ $\dfrac{6x^2y^3}{\sqrt{6z}}$

62. $\sqrt{\dfrac{8x^5y}{2z}}$ $\dfrac{2x^3y}{\sqrt{xyz}}$

✎ **63.** When rationalizing the denominator of $\dfrac{\sqrt{5}}{\sqrt{7}}$, explain why both the numerator and the denominator must be multiplied by $\sqrt{7}$.   answers may vary

✎ **64.** When rationalizing the numerator of $\dfrac{\sqrt{5}}{\sqrt{7}}$, explain why both the numerator and the denominator must be multiplied by $\sqrt{5}$.   answers may vary

Rationalize each numerator. Assume that all variables represent positive real numbers. See Example 11.

65. $\dfrac{2-\sqrt{11}}{6}$ $\dfrac{-7}{12+6\sqrt{11}}$

66. $\dfrac{\sqrt{15}+1}{2}$ $\dfrac{7}{\sqrt{15}-1}$

67. $\dfrac{2-\sqrt{7}}{-5}$ $\dfrac{3}{10+5\sqrt{7}}$

68. $\dfrac{\sqrt{5}+2}{\sqrt{2}}$ $\dfrac{1}{\sqrt{10}-2\sqrt{2}}$

69. $\dfrac{\sqrt{x}+3}{\sqrt{x}}$ $\dfrac{x-9}{x-3\sqrt{x}}$

70. $\dfrac{5+\sqrt{2}}{\sqrt{2x}}$ $\dfrac{23}{5\sqrt{2x}-2\sqrt{x}}$

71. $\dfrac{\sqrt{x}+1}{\sqrt{x}-1}$ $\dfrac{x-1}{x-2\sqrt{x}+1}$

72. $\dfrac{\sqrt{x}+\sqrt{y}}{\sqrt{x}-\sqrt{y}}$ $\dfrac{x-y}{x-2\sqrt{xy}+y}$

Review

Solve each equation. See Sections 2.3 and 6.6.

73. $2x-7=3(x-4)$ $\{5\}$

74. $9x-4=7(x-2)$ $\{-5\}$

75. $(x-6)(2x+1)=0$ $\left\{-\dfrac{1}{2},6\right\}$

76. $(y+2)(5y+4)=0$ $\left\{-2,-\dfrac{4}{5}\right\}$

77. $x^2-8x=-12$ $\{2,6\}$

78. $x^3=x$ $\{0,1,-1\}$

Concept Extensions

△ **79.** The formula of the radius r of a sphere with surface area A is

$$r = \sqrt{\frac{A}{4\pi}}$$

Rationalize the denominator of the radical expression in this formula. $r = \dfrac{\sqrt{A\pi}}{2\pi}$

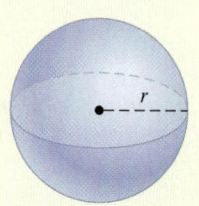

△ **80.** The formula for the radius r of a cone with height 7 centimeters and volume V is

$$r = \sqrt{\frac{3V}{7\pi}}$$

Rationalize the numerator of the radical expression in this formula. $r = \dfrac{3V}{\sqrt{21\pi V}}$

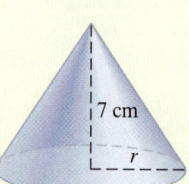

7 cm

81. Explain why rationalizing the denominator does not change the value of the original expression.
answers may vary

82. Explain why rationalizing the numerator does not change the value of the original expression.
answers may vary

83. Given $\dfrac{\sqrt{5y^3}}{\sqrt{12x^3}}$, rationalize the denominator by following parts (a) and (b).

    a. Multiply the numerator and denominator by $\sqrt{12x^3}$. $\dfrac{y\sqrt{15xy}}{6x^2}$

    b. Multiply the numerator and denominator by $\sqrt{3x}$. $\dfrac{y\sqrt{15xy}}{6x^2}$

    c. What can you conclude from parts (a) and (b)?   answers may vary

84. Given $\dfrac{\sqrt[3]{5y}}{\sqrt[3]{4}}$, rationalize the denominator by following parts (a) and (b).

    a. Multiply the numerator and denominator by $\sqrt[3]{16}$. $\dfrac{\sqrt[3]{10y}}{2}$

    b. Multiply the numerator and denominator by $\sqrt[3]{2}$. $\dfrac{\sqrt[3]{10y}}{2}$

    c. What can you conclude from parts (a) and (b)?   answers may vary

Determine the smallest number both the numerator and denominator should be multiplied by to rationalize the denominator of the radical expression. See the Concept Check in this section.

85. $\dfrac{9}{\sqrt[3]{5}}$ $\sqrt[3]{25}$

86. $\dfrac{5}{\sqrt{27}}$ $\sqrt{3}$

Radicals and Rational Exponents

Answers

Throughout this review, assume that all variables represent positive real numbers.
Find each root.

1. $\sqrt{81}$ **2.** $\sqrt[3]{-8}$ **3.** $\sqrt[4]{\dfrac{1}{16}}$ **4.** $\sqrt{x^6}$

5. $\sqrt[3]{y^9}$ **6.** $\sqrt{4y^{10}}$ **7.** $\sqrt[5]{-32y^5}$ **8.** $\sqrt[4]{81b^{12}}$

Use radical notation to rewrite each expression. Simplify if possible.

9. $36^{1/2}$ **10.** $(3y)^{1/4}$ **11.** $64^{-2/3}$ **12.** $(x+1)^{3/5}$

Use the properties of exponents to simplify each expression. Write with positive exponents.

13. $y^{-1/6} \cdot y^{7/6}$ **14.** $\dfrac{(2x^{1/3})^4}{x^{5/6}}$ **15.** $\dfrac{x^{1/4}x^{3/4}}{x^{-1/4}}$ **16.** $4^{1/3} \cdot 4^{2/5}$

Use rational exponents to simplify each radical.

17. $\sqrt[3]{8x^6}$ **18.** $\sqrt[12]{a^9b^6}$

Use rational exponents to write each as a single radical expression.

19. $\sqrt[4]{x} \cdot \sqrt{x}$ **20.** $\sqrt{5} \cdot \sqrt[3]{2}$

Simplify.

21. $\sqrt{40}$ **22.** $\sqrt[4]{16x^7y^{10}}$ **23.** $\sqrt[3]{54x^4}$ **24.** $\sqrt[5]{-64b^{10}}$

1. 9

2. -2

3. $\dfrac{1}{2}$

4. x^3

5. y^3

6. $2y^5$

7. $-2y$

8. $3b^3$

9. 6

10. $\sqrt[4]{3y}$

11. $\dfrac{1}{16}$

12. $\sqrt[5]{(x+1)^3}$

13. y

14. $16x^{1/2}$

15. $x^{5/4}$

16. $4^{11/15}$

17. $2x^2$

18. $\sqrt[4]{a^3b^2}$

19. $\sqrt[4]{x^3}$

20. $\sqrt[6]{500}$

21. $2\sqrt{10}$

22. $2xy^2\sqrt[4]{x^3y^2}$

23. $3x\sqrt[3]{2x}$

24. $-2b^2\sqrt[5]{2}$

723

25. $\sqrt{5x}$

26. $4x$

27. $7y^2\sqrt{y}$

28. $2a^2\sqrt[4]{3}$

29. $2\sqrt{5} - 5\sqrt{3} + 5\sqrt{7}$

30. $y\sqrt[3]{2y}$

31. $\sqrt{15} - \sqrt{6}$

32. $10 + 2\sqrt{21}$

33. $4x^2 - 5$

34. $x + 2 - 2\sqrt{x+1}$

35. $\dfrac{\sqrt{21}}{3}$

36. $\dfrac{5\sqrt[3]{4x}}{2x}$

37. $\dfrac{13 - 3\sqrt{21}}{5}$

38. $\dfrac{7}{\sqrt{21}}$

39. $\dfrac{3y}{\sqrt[3]{33y^2}}$

40. $\dfrac{x-4}{x + 2\sqrt{x}}$

Multiply or divide. Then simplify if possible.

25. $\sqrt{5} \cdot \sqrt{x}$

26. $\sqrt[3]{8x} \cdot \sqrt[3]{8x^2}$

27. $\dfrac{\sqrt{98y^6}}{\sqrt{2y}}$

28. $\dfrac{\sqrt[4]{48a^9b^3}}{\sqrt[4]{ab^3}}$

Perform each indicated operation.

29. $\sqrt{20} - \sqrt{75} + 5\sqrt{7}$

30. $\sqrt[3]{54y^4} - y\sqrt[3]{16y}$

31. $\sqrt{3}(\sqrt{5} - \sqrt{2})$

32. $(\sqrt{7} + \sqrt{3})^2$

33. $(2x - \sqrt{5})(2x + \sqrt{5})$

34. $(\sqrt{x+1} - 1)^2$

Rationalize each denominator.

35. $\sqrt{\dfrac{7}{3}}$

36. $\dfrac{5}{\sqrt[3]{2x^2}}$

37. $\dfrac{\sqrt{3} - \sqrt{7}}{2\sqrt{3} + \sqrt{7}}$

Rationalize each numerator.

38. $\sqrt{\dfrac{7}{3}}$

39. $\sqrt[3]{\dfrac{9y}{11}}$

40. $\dfrac{\sqrt{x} - 2}{\sqrt{x}}$

10.6 Radical Equations and Problem Solving

Objective A Solving Equations That Contain Radical Expressions

Objectives

A Solve Equations That Contain Radical Expressions.

B Use the Pythagorean Theorem to Model Problems.

In this section, we present techniques to solve equations containing radical expressions such as

$$\sqrt{2x - 3} = 9$$

We use the power rule to help us solve these radical equations.

Power Rule

If both sides of an equation are raised to the same power, *all* solutions of the original equation are *among* the solutions of the new equation.

This property *does not* say that raising both sides of an equation to a power yields an equivalent equation. A solution of the new equation *may* or *may not* be a solution of the original equation. Thus, *each solution of the new equation must be checked* to make sure it is a solution of the original equation. Recall that a proposed solution that is not a solution of the original equation is called an extraneous solution.

Example 1 Solve: $\sqrt{2x - 3} = 9$

Solution: We use the power rule to square both sides of the equation to eliminate the radical.

$$\sqrt{2x - 3} = 9$$
$$(\sqrt{2x - 3})^2 = 9^2$$
$$2x - 3 = 81$$
$$2x = 84$$
$$x = 42$$

Now we check the solution in the original equation.

Check:
$$\sqrt{2x - 3} = 9$$
$$\sqrt{2(42) - 3} \stackrel{?}{=} 9 \quad \text{Let } x = 42.$$
$$\sqrt{84 - 3} \stackrel{?}{=} 9$$
$$\sqrt{81} \stackrel{?}{=} 9$$
$$9 = 9 \quad \text{True}$$

The solution checks, so we conclude that the solution set is $\{42\}$.

■ **Work Practice 1**

To solve a radical equation, first isolate a radical on one side of the equation.

Practice 1

Solve: $\sqrt{3x - 2} = 5$

Answer
1. $\{9\}$

The following steps may be used to solve a radical equation.

Solving a Radical Equation

Step 1: Isolate one radical on one side of the equation.

Step 2: Raise each side of the equation to a power equal to the index of the radical and simplify.

Step 3: If the equation still contains a radical term, repeat Steps 1 and 2. If not, solve the equation.

Step 4: Check all proposed solutions in the original equation.

Practice 2

Solve: $\sqrt{9x - 2} - 2x = 0$

Example 2 Solve: $\sqrt{-10x - 1} + 3x = 0$

Solution: First we isolate the radical on one side of the equation. To do this, we subtract $3x$ from both sides.

$$\sqrt{-10x - 1} + 3x = 0$$
$$\sqrt{-10x - 1} + 3x - 3x = 0 - 3x$$
$$\sqrt{-10x - 1} = -3x$$

Next we use the power rule to eliminate the radical.

$$(\sqrt{-10x - 1})^2 = (-3x)^2$$
$$-10x - 1 = 9x^2$$

Since this is a quadratic equation, we can set the equation equal to 0 and try to solve by factoring.

$$9x^2 + 10x + 1 = 0$$
$$(9x + 1)(x + 1) = 0 \qquad \text{Factor.}$$
$$9x + 1 = 0 \quad \text{or} \quad x + 1 = 0 \qquad \text{Set each factor equal to 0.}$$
$$x = -\frac{1}{9} \qquad\qquad x = -1$$

Teaching Tip:

Classroom Activity

After discussing Example 2, have students work in pairs to create an equation involving a square root by starting with desired solutions and then reversing the steps in Example 2. Have them check that their solutions are solutions to the equation they created. Finally, have them give their equation to another pair to solve.

Check: Let $x = -\frac{1}{9}$.

$$\sqrt{-10x - 1} + 3x = 0$$
$$\sqrt{-10\left(-\frac{1}{9}\right) - 1} + 3\left(-\frac{1}{9}\right) \stackrel{?}{=} 0$$
$$\sqrt{\frac{10}{9} - \frac{9}{9}} - \frac{3}{9} \stackrel{?}{=} 0$$
$$\sqrt{\frac{1}{9}} - \frac{1}{3} \stackrel{?}{=} 0$$
$$\frac{1}{3} - \frac{1}{3} = 0 \quad \text{True}$$

Let $x = -1$.

$$\sqrt{-10x - 1} + 3x = 0$$
$$\sqrt{-10(-1) - 1} + 3(-1) \stackrel{?}{=} 0$$
$$\sqrt{10 - 1} - 3 \stackrel{?}{=} 0$$
$$\sqrt{9} - 3 \stackrel{?}{=} 0$$
$$3 - 3 = 0 \quad \text{True}$$

Both solutions check. The solution set is $\left\{-\frac{1}{9}, -1\right\}$.

■ **Work Practice 2**

Practice 3

Solve: $\sqrt[3]{x - 5} + 2 = 1$

Example 3 Solve: $\sqrt[3]{x + 1} + 5 = 3$

Solution: First we isolate the radical by subtracting 5 from both sides of the equation.

$$\sqrt[3]{x + 1} + 5 = 3$$
$$\sqrt[3]{x + 1} = -2$$

Answers

2. $\left\{\frac{1}{4}, 2\right\}$ **3.** $\{4\}$

Next we raise both sides of the equation to the third power to eliminate the radical.

$$(\sqrt[3]{x + 1})^3 = (-2)^3$$
$$x + 1 = -8$$
$$x = -9$$

The solution checks in the original equation, so the solution set is $\{-9\}$.

■ **Work Practice 3**

Example 4   Solve: $\sqrt{4 - x} = x - 2$

Solution:

$$\sqrt{4 - x} = x - 2$$
$$(\sqrt{4 - x})^2 = (x - 2)^2$$
$$4 - x = x^2 - 4x + 4 \quad \text{Write the quadratic equation in standard form.}$$
$$x^2 - 3x = 0$$
$$x(x - 3) = 0 \quad \text{Factor.}$$
$$x = 0 \quad \text{or} \quad x - 3 = 0 \quad \text{Set each factor equal to 0.}$$
$$x = 3$$

Check:

$$\sqrt{4 - x} = x - 2 \qquad\qquad \sqrt{4 - x} = x - 2$$
$$\sqrt{4 - 0} \stackrel{?}{=} 0 - 2 \quad \text{Let } x = 0. \qquad \sqrt{4 - 3} \stackrel{?}{=} 3 - 2 \quad \text{Let } x = 3.$$
$$2 = -2 \quad \text{False} \qquad\qquad 1 = 1 \quad \text{True}$$

The proposed solution 3 checks, but 0 does not. Since 0 is an extraneous solution, the solution set is $\{3\}$.

■ **Work Practice 4**

Helpful Hint

In Example 4, notice that $(x - 2)^2 = x^2 - 4x + 4$. Make sure binomials are squared correctly.

✓**Concept Check** How can you immediately tell that the equation $\sqrt{2y + 3} = -4$ has no real solution?

Example 5   Solve: $\sqrt{2x + 5} + \sqrt{2x} = 3$

Solution: We get one radical alone by subtracting $\sqrt{2x}$ from both sides.

$$\sqrt{2x + 5} + \sqrt{2x} = 3$$
$$\sqrt{2x + 5} = 3 - \sqrt{2x}$$

Now we use the power rule to begin eliminating the radicals. First we square both sides.

$$(\sqrt{2x + 5})^2 = (3 - \sqrt{2x})^2$$
$$2x + 5 = 9 - 6\sqrt{2x} + 2x \quad \text{Multiply:} \quad (3 - \sqrt{2x})(3 - \sqrt{2x})$$

(Continued on next page)

Practice 4
Solve: $\sqrt{9 + x} = x + 3$

Practice 5
Solve: $\sqrt{3x + 1} + \sqrt{3x} = 2$

Answers

4. $\{0\}$ **5.** $\left\{\dfrac{3}{16}\right\}$

✓**Concept Check Answer**
answers may vary

Teaching Tip:
Classroom Activity

After discussing Example 5, have students work in pairs to create an equation involving two square roots by starting with a desired solution and then reversing the steps in Example 5. Have them check that their solution is a solution to the equation they created. Finally, have them give their equation to another pair to solve.

There is still a radical in the equation, so we get the radical alone again. Then we square both sides.

$$2x + 5 = 9 - 6\sqrt{2x} + 2x$$
$$6\sqrt{2x} = 4 \qquad \text{Get the radical alone.}$$
$$(6\sqrt{2x})^2 = 4^2 \qquad \text{Square both sides of the equation to eliminate the radical.}$$
$$36(2x) = 16$$
$$72x = 16 \qquad \text{Multiply.}$$
$$x = \frac{16}{72} \qquad \text{Solve.}$$
$$x = \frac{2}{9} \qquad \text{Simplify.}$$

The proposed solution $\frac{2}{9}$ checks in the original equation. The solution set is $\left\{\frac{2}{9}\right\}$.

■ **Work Practice 5**

Helpful Hint

Make sure expressions are squared correctly. In Example 5, we squared $(3 - \sqrt{2x})$ as

$$(3 - \sqrt{2x})^2 = (3 - \sqrt{2x})(3 - \sqrt{2x})$$
$$= 3 \cdot 3 - 3\sqrt{2x} - 3\sqrt{2x} + \sqrt{2x} \cdot \sqrt{2x}$$
$$= 9 - 6\sqrt{2x} + 2x$$

✓**Concept Check** What is wrong with the following solution?

$$\sqrt{2x + 5} + \sqrt{4 - x} = 8$$
$$(\sqrt{2x + 5} + \sqrt{4 - x})^2 = 8^2$$
$$(2x + 5) + (4 - x) = 64$$
$$x + 9 = 64$$
$$x = 55$$

Objective B Using the Pythagorean Theorem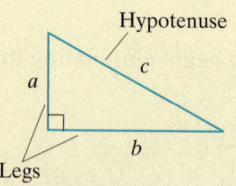

Recall that the Pythagorean theorem states that in a right triangle, the length of the hypotenuse squared equals the sum of the lengths of each of the legs squared.

Pythagorean Theorem

If a and b are the lengths of the legs of a right triangle and c is the length of the hypotenuse, then $a^2 + b^2 = c^2$.

Hypotenuse

c

a

b

Legs

✓**Concept Check Answer**

From the second line of the solution to the third line of the solution, the left side of the equation is squared incorrectly.

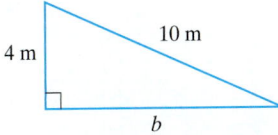 **Example 6**   Find the length of the unknown leg of the right triangle.

Find the length of the unknown leg of the right triangle.

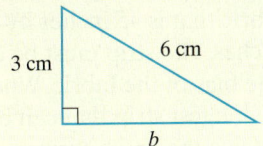

Solution: In the formula $a^2 + b^2 = c^2$, c is the hypotenuse. Here, $c = 10$, the length of the hypotenuse, and $a = 4$. We solve for b. Then $a^2 + b^2 = c^2$ becomes

$$4^2 + b^2 = 10^2$$
$$16 + b^2 = 100$$
$$b^2 = 84 \quad \text{Subtract 16 from both sides.}$$

Recall from Section 10.1 our definition of square root that if $b^2 = a$, then b is a square root of a. Since b is a length and thus is positive, we have that

$$b = \sqrt{84} = \sqrt{4 \cdot 21} = 2\sqrt{21}$$

The unknown leg of the triangle is exactly $2\sqrt{21}$ meters long. Using a calculator, this is approximately 9.2 meters.

■ **Work Practice 6**

Example 7   Calculating Distance Across a Quadrangle

Find the length of a sidewalk that is the diagonal of a square whose side length is 60 feet.

The current route of the Great Court Run at Trinity College runs along the perimeter of the grassy rectangle. If the dimensions of the grassy rectangle are 70 meters by 80 meters, how far would a student travel by walking across the grass from corner to corner?

70 m

80 m

Solution: In the formula $a^2 + b^2 = c^2$, a and b represent the legs of the triangle, and c represents the hypotenuse. Here, $a = 80$ and $b = 70$, and we wish to solve for c, the length of the hypotenuse.

Then $a^2 + b^2 = c^2$ becomes
$$80^2 + 70^2 = c^2$$
$$6400 + 4900 = c^2$$
$$11{,}300 = c^2$$

Recall from Section 10.1 our definition of square root that if $c^2 = a$, then c is a square root of a. Since c is a length, and thus is positive, we have that

$$c = \sqrt{11{,}300} = \sqrt{100 \cdot 113} = 10\sqrt{113}$$

The distance across the grassy rectangle is therefore $10\sqrt{113}$ meters. Using a calculator, this is approximately 106.3 meters.

■ **Work Practice 7**

Practice 8

A furniture upholsterer wishes to cut a strip from a piece of fabric that is 45 inches by 45 inches. The strip must be cut on the bias of the fabric. What is the longest strip that can be cut? Give the exact answer and a two-decimal-place approximation.

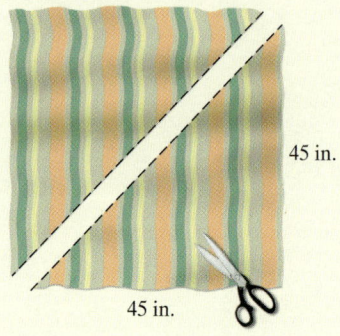

45 in.

45 in.

△ **Example 8**   Calculating Placement of a Wire

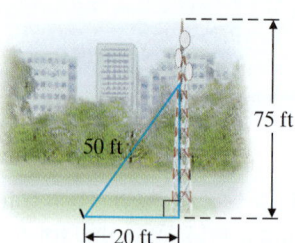

75 ft

50 ft

←20 ft→

A 50-foot supporting wire is to be attached to a 75-foot antenna. Because of surrounding buildings, sidewalks, and roadways, the wire must be anchored exactly 20 feet from the base of the antenna.

a. How high from the base of the antenna must the wire be attached?

b. Local regulations require that a supporting wire be attached at a height no less than $\frac{3}{5}$ of the total height the antenna. From part (a), have local regulations been met?

Solution:

1. UNDERSTAND. Read and reread the problem. From the diagram we notice that a right triangle is formed with hypotenuse 50 feet and one leg 20 feet. We let x = the height from the base of the antenna to the attached wire.

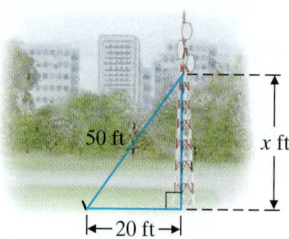

50 ft

x ft

←20 ft→

2. TRANSLATE. We'll use the Pythagorean theorem.

$$a^2 + b^2 = c^2$$
$$20^2 + x^2 = 50^2 \quad a = 20, c = 50$$

3. SOLVE.

$$20^2 + x^2 = 50^2$$
$$400 + x^2 = 2500$$
$$x^2 = 2100 \qquad \text{Subtract 400 from both sides.}$$
$$x = \sqrt{2100}$$
$$= 10\sqrt{21}$$

4. INTERPRET. *Check* the work and *state* the solution.

   a. The wire is attached exactly $10\sqrt{21}$ feet from the base of the pole, or approximately 45.8 feet.

   b. The supporting wire must be attached at a height no less than $\frac{3}{5}$ of the total height of the antenna. This height is $\frac{3}{5}$ (75 feet), or 45 feet.

   Since we know from part (a) that the wire is to be attached at a height of approximately 45.8 feet, local regulations have been met.

■ **Work Practice 8**

Answer

8. $45\sqrt{2}$ in. ≈ 63.64 in.

Calculator Explorations **Graphing**

We can use a graphing calculator to solve radical equations. For example, to use a graphing calculator to approximate the solutions of the equation solved in Example 4, we graph the following:

$$Y_1 = \sqrt{4 - x} \quad \text{and} \quad Y_2 = x - 2$$

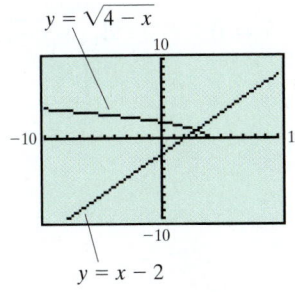

The x-value of the point of intersection is the solution. Use the INTERSECT feature or the ZOOM and TRACE features of your graphing calculator to see that the solution is 3.

Use a graphing calculator to solve each radical equation. Round all solutions to the nearest hundredth.

1. $\sqrt{x + 7} = x$ $\{3.19\}$

2. $\sqrt{3x + 5} = 2x$ $\{1.55\}$

3. $\sqrt{2x + 1} = \sqrt{2x + 2}$ $\varnothing$

4. $\sqrt{10x - 1} = \sqrt{-10x + 10} - 1$ $\{0.34\}$

5. $1.2x = \sqrt{3.1x + 5}$ $\{3.23\}$

6. $\sqrt{1.9x^2 - 2.2} = -0.8x + 3$ $\{-5.44, 1.63\}$

Vocabulary, Readiness & Video Check

Use the choices below to fill in each blank. Not all choices will be used.

hypotenuse	right	$x^2 + 25$	$16 - 8\sqrt{7x} + 7x$
extraneous solution	legs	$x^2 - 10x + 25$	$16 + 7x$

1. A proposed solution that is not a solution of the original equation is called a(n) ___extraneous solution___ .

2. The Pythagorean theorem states that $a^2 + b^2 = c^2$ where a and b are the lengths of the ___legs___ of a(n) ___right___ triangle and c is the length of the ___hypotenuse___ .

3. The square of $x - 5$, or $(x - 5)^2 = $ ___$x^2 - 10x + 25$___ .

4. The square of $4 - \sqrt{7x}$, or $(4 - \sqrt{7x})^2 = $ ___$16 - 8\sqrt{7x} + 7x$___ .

Martin-Gay Interactive Videos

See Video 10.6

See video answer section.

Watch the section lecture video and answer the following questions.

Objective A **5.** From ⊞ Examples 1–4, why must you be careful to check your proposed solution(s) in the original equation? ▶

Objective B **6.** From ⊞ Example 5, when solving problems using the Pythagorean theorem, what two things must you remember? ▶

7. What important reminder is given as the final answer to ⊞ Example 5 is being found? ▶

10.6 Exercise Set MyMathLab®

Objective A *Solve. See Examples 1 and 2.*

1. $\sqrt{2x} = 4$
{8}

2. $\sqrt{3x} = 3$
{3}

▶**3.** $\sqrt{x-3} = 2$
{7}

4. $\sqrt{x+1} = 5$
{24}

5. $\sqrt{2x} = -4$
∅

6. $\sqrt{5x} = -5$
∅

7. $\sqrt{4x-3} - 5 = 0$
{7}

8. $\sqrt{x-3} - 1 = 0$
{4}

9. $\sqrt{2x-3} - 2 = 1$
{6}

10. $\sqrt{3x+3} - 4 = 8$
{47}

Solve. See Example 3.

11. $\sqrt[3]{6x} = -3$ $\left\{-\dfrac{9}{2}\right\}$

12. $\sqrt[3]{4x} = -2$
{−2}

13. $\sqrt[3]{x-2} - 3 = 0$
{29}

14. $\sqrt[3]{2x-6} - 4 = 0$
{35}

Solve. See Examples 4 and 5.

15. $\sqrt{13-x} = x - 1$
{4}

16. $\sqrt{2x-3} = 3 - x$
{2}

▶**17.** $x - \sqrt{4-3x} = -8$
{−4}

18. $2x + \sqrt{x+1} = 8$
{3}

19. $\sqrt{y+5} = 2 + \sqrt{y-2}$ $\left\{\dfrac{41}{16}\right\}$

20. $\sqrt{x+3} + \sqrt{x-5} = 3$ $\left\{\dfrac{181}{36}\right\}$

21. $\sqrt{x-3} + \sqrt{x+2} = 5$
{7}

22. $\sqrt{2x-4} - \sqrt{3x+4} = -2$
{4, 20}

Solve. See Examples 1 through 5.

23. $\sqrt{3x-2} = 5$
{9}

24. $\sqrt{5x-4} = 9$
{17}

25. $-\sqrt{2x} + 4 = -6$
{50}

26. $-\sqrt{3x} + 2 = -10$
{48}

27. $\sqrt{3x+1} + 2 = 0$
∅

28. $\sqrt{3x+1} - 2 = 0$
{1}

29. $\sqrt[4]{4x+1} - 2 = 0$ $\left\{\dfrac{15}{4}\right\}$

30. $\sqrt[4]{2x-9} - 3 = 0$
{45}

31. $\sqrt{3x+4} = 5$
{7}

32. $\sqrt{3x+9} = 12$
{45}

33. $\sqrt[3]{6x-3} - 3 = 0$
{5}

34. $\sqrt[3]{3x} + 4 = 7$
{9}

⊙ 35. $\sqrt[3]{2x - 3} - 2 = -5$
{−12}

36. $\sqrt[3]{x - 4} - 5 = -7$
{−4}

37. $\sqrt{x + 4} = \sqrt{2x - 5}$
{9}

38. $\sqrt{3y + 6} = \sqrt{7y - 6}$
{3}

39. $x - \sqrt{1 - x} = -5$
{−3}

40. $x - \sqrt{x - 2} = 4$
{6}

41. $\sqrt[3]{-6x - 1} = \sqrt[3]{-2x - 5}$
{1}

42. $x + \sqrt{x + 5} = 7$
{4}

⊙ 43. $\sqrt{5x - 1} - \sqrt{x + 2} = 3$
{1}

44. $\sqrt{2x - 1} - 4 = -\sqrt{x - 4}$
{5}

45. $\sqrt{2x - 1} = \sqrt{1 - 2x}$ $\left\{\dfrac{1}{2}\right\}$

46. $\sqrt{7x - 4} = \sqrt{4 - 7x}$ $\left\{\dfrac{4}{7}\right\}$

47. $\sqrt{3x + 4} - 1 = \sqrt{2x + 1}$
{0, 4}

48. $\sqrt{x - 2} + 3 = \sqrt{4x + 1}$
{2, 6}

49. $\sqrt{y + 3} - \sqrt{y - 3} = 1$ $\left\{\dfrac{37}{4}\right\}$

50. $\sqrt{x + 1} - \sqrt{x - 1} = 2$
∅

Objective B *Find the length of the unknown side of each triangle. See Example 6.*

△ **51.**
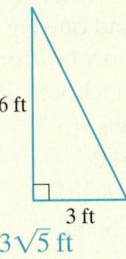
6 ft
3 ft
$3\sqrt{5}$ ft

△ **52.**
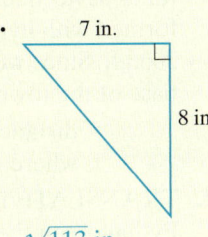
7 in.
8 in.
$\sqrt{113}$ in.

⊙ 53.
△
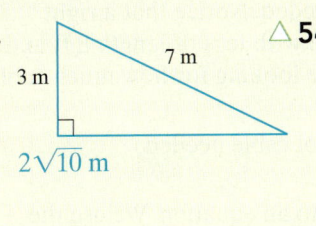
3 m
7 m
$2\sqrt{10}$ m

△ **54.**
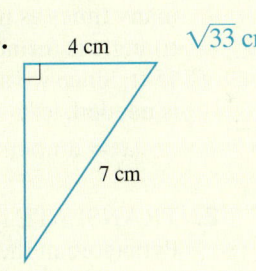
4 cm
$\sqrt{33}$ cm
7 cm

Find the length of the unknown side of each triangle. Give the exact length and a one-decimal-place approximation. See Example 6.

▣ **55.**
△
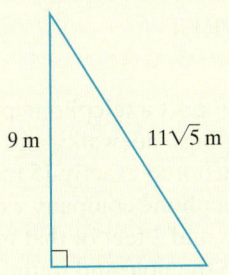
9 m
$11\sqrt{5}$ m
$2\sqrt{131}$ m ≈ 22.9 m

▣ **56.**
△
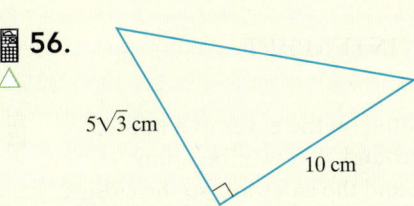
$5\sqrt{3}$ cm
10 cm
$5\sqrt{7}$ cm ≈ 13.2 cm

▣ **57.**
△
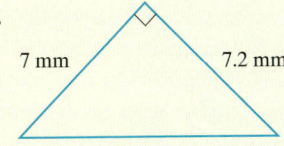
7 mm
7.2 mm
$\sqrt{100.84}$ mm ≈ 10.0 mm

▣ **58.**
△
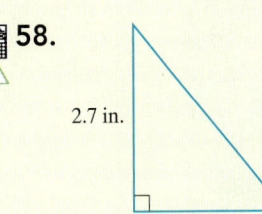
2.7 in.
2.3 in.
$\sqrt{12.58}$ in. ≈ 3.5 in.

Solve. Give exact answers and two-decimal-place approximations where appropriate. For Exercises 59 and 60, the solutions have been started for you. See Examples 7 and 8.

59. A wire is needed to support a vertical pole 15 feet tall. The cable will be anchored to a stake 8 feet from the base of the pole. How much cable is needed?   17 ft

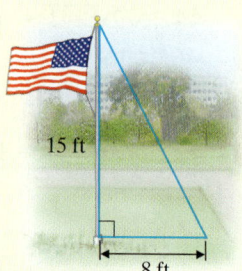

15 ft

8 ft

Start the solution:

1. **UNDERSTAND** the problem. Reread it as many times as needed. Notice that a right triangle is formed with legs of length 8 ft and 15 ft. Since we are looking for how much cable is needed, let

   x = amount of cable needed

2. **TRANSLATE** into an equation. We use the Pythagorean theorem. (Fill in the blanks below.)

$$a^2 \quad + \quad b^2 \quad = \quad c^2$$
$$\downarrow \qquad\qquad \downarrow \qquad\qquad\qquad$$
$$\underline{\quad}^2 \quad + \quad \underline{\quad}^2 \quad = \quad x^2$$

Finish with:
3. **SOLVE** and 4. **INTERPRET**

61. A spotlight is mounted on the eaves of a house, 12 feet above the ground. A flower bed runs between the house and the sidewalk, so the closest the ladder can be placed to the house is 5 feet. How long a ladder is needed so that an electrician can reach the place where the light is mounted?   13 ft

12 ft

5 ft

60. The tallest structure in the United States is a TV tower in Blanchard, North Dakota. Its height is 2063 feet. A 2382-foot length of wire is to be used as a guy wire attached to the top of the tower. Approximate to the nearest foot how far from the base of the tower the guy wire must be anchored. (*Source:* U.S. Geological Survey)   1191 ft

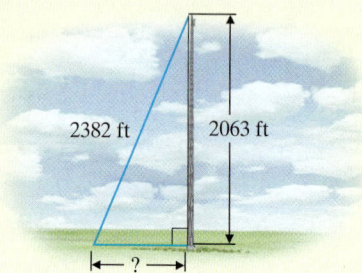

2382 ft   2063 ft

?

Start the solution:

1. **UNDERSTAND** the problem. Reread it as many times as needed. Notice that a right triangle is formed with hypotenuse 2382 ft and one leg 2063 ft. Since we are looking for how far from the base of the tower the guy wire is anchored, let

   x = distance from base of tower to where guy wire is anchored.

2. **TRANSLATE** into an equation. We use the Pythagorean theorem. (Fill in the blanks below.)

$$a^2 \quad + \quad b^2 \quad = \quad c^2$$
$$\downarrow \qquad\qquad \downarrow \qquad\qquad\qquad \downarrow$$
$$\underline{\quad}^2 \quad + \quad x^2 \quad = \quad \underline{\quad}^2$$

Finish with:
3. **SOLVE** and 4. **INTERPRET**

62. A wire is to be attached to support a telephone pole. Because of surrounding buildings, sidewalks, and roadway, the wire must be anchored exactly 15 feet from the base of the pole. Telephone company workers have only 30 feet of cable, and 2 feet of that must be used to attach the cable to the pole and to the stake on the ground. How high from the base of the pole can the wire be attached? $\sqrt{559}$ ft $\approx$ 23.64 ft

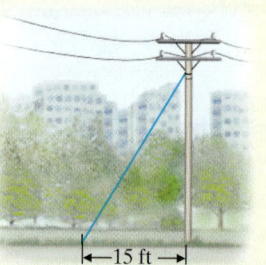

15 ft

△ **63.** The radius of the moon is 1080 miles. Use the formula for the radius r of a sphere given its surface area A.

$$r = \sqrt{\frac{A}{4\pi}}$$

to find the surface area of the moon. Round to the nearest square mile. (*Source:* National Space Science Data Center)   14,657,415 sq mi

64. Police departments find it very useful to be able to approximate driving speeds in skidding accidents. If the road surface is wet concrete, the function $S(x) = \sqrt{10.5x}$ is used, where $S(x)$ is the speed of the car in miles per hour and x is the distance skidded in feet. Find how fast a car was moving if it skidded 280 feet on wet concrete. $14\sqrt{15}$ mph ≈ 54.22 mph

The formula $v = \sqrt{2gh}$ relates the velocity v, in feet per second, of an object after it falls h feet accelerated by gravity g, in feet per second squared. If g is approximately 32 feet per second squared, find how far an object has fallen for each given velocity.

65. Velocity of 80 feet per second   100 ft

66. Velocity of 40 feet per second   25 ft

67. Harvard Yard is known as the heart of Harvard University. It is a great rectangular lawn area in the midst of Cambridge, Massachusetts. This "old yard" has dimensions of 750 feet by 250 feet. How far does a student walk to travel across the diagonal of the yard? $\sqrt{625,000}$ ft ≈ 790.57 ft

68. Two tractors are pulling a tree stump from a field. If two forces A and B pull at right angles (90°) to each other, the resulting force R is given by the formula $R = \sqrt{A^2 + B^2}$. If tractor A is exerting 600 pounds of force and the resulting force is 850 pounds, find how much force tractor B is exerting. $50\sqrt{145}$ lb ≈ 602.08 lb

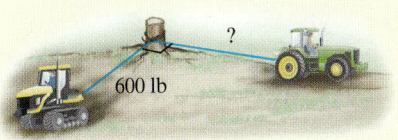

600 lb

In psychology, it has been suggested that the number S of nonsense syllables that a person can repeat consecutively depends on his or her IQ score I according to the equation $S = 2\sqrt{I} - 9$.

69. Use this relationship to estimate the IQ of a person who can repeat 11 nonsense syllables consecutively.   100

70. Use this relationship to estimate the IQ of a person who can repeat 15 nonsense syllables consecutively.   144

*The **period** of a pendulum is the time it takes for the pendulum to make one full back-and-forth swing. The period of a pendulum depends on the length of the pendulum. The formula for the period P, in seconds, is $P = 2\pi\sqrt{\dfrac{L}{32}}$, where L is the length of the pendulum in feet. Use this formula for Exercises 71 through 76.*

71. Find the period of a pendulum whose length is 2 feet. Give the exact answer and a two-decimal-place approximation. $\dfrac{\pi}{2}$ sec ≈ 1.57 sec

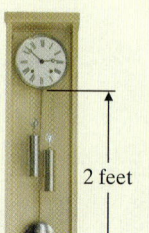

2 feet

72. Klockit sells a 43-inch lyre pendulum. Find the period of this pendulum. Round your answer to 2 decimal places. (*Hint:* First convert inches to feet.)   2.10 sec

73. Find the length of a pendulum whose period is 4 seconds. Round your answer to 2 decimal places.   12.97 ft

74. Find the length of a pendulum whose period is 3 seconds. Round your answer to 3 decimal places.   7.295 ft

75. Study the relationship between period and pendulum length in Exercises 71 through 74 and make a conjecture about this relationship.
answers may vary

76. Galileo experimented with pendulums. He supposedly made conjectures about pendulums of equal length with different bob weights. Try this experiment. Make two pendulums 3 feet long. Attach a heavy weight (lead) to one and a light weight (a cork) to the other. Pull both pendulums back the same angle measure and release. Make a conjecture from your observations.   answers may vary

If the three lengths of the sides of a triangle are known, Heron's formula can be used to find its area. If a, b, and c are the three lengths of the sides, Heron's formula for area is

$$A = \sqrt{s(s-a)(s-b)(s-c)}$$

where s is half the perimeter of the triangle, or $s = \dfrac{1}{2}(a + b + c)$. Use this formula to find the area of each triangle. Give an exact answer and then a two-decimal-place approximation.

△ **77.**

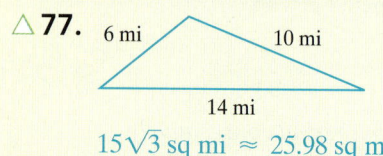

6 mi   10 mi
14 mi

$15\sqrt{3}$ sq mi ≈ 25.98 sq mi

△ **78.**

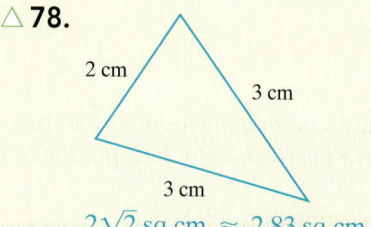

2 cm   3 cm
3 cm

$2\sqrt{2}$ sq cm ≈ 2.83 sq cm

79. Describe when Heron's formula might be useful.
answers may vary

80. In your own words, explain why you think s in Heron's formula is called the *semiperimeter*.
answers may vary

The maximum distance $D(h)$ in kilometers that a person can see from a height h kilometers above the ground is given by the function $D(h) = 111.7\sqrt{h}$. Use this function for Exercises 81 and 82. Round your answers to two decimal places.

81. Find the height that would allow a person to see 80 kilometers.   0.51 km

82. Find the height that would allow a person to see 40 kilometers.   0.13 km

Review

Simplify. See Section 7.7.

83. $\dfrac{\dfrac{x}{6}}{\dfrac{2x}{3} + \dfrac{1}{2}}$ $\dfrac{x}{4x+3}$

84. $\dfrac{\dfrac{1}{y} + \dfrac{4}{5}}{\dfrac{-3}{20}}$ $-\dfrac{20+16y}{3y}$

85. $\dfrac{\dfrac{z}{5} + \dfrac{1}{10}}{\dfrac{z}{20} - \dfrac{z}{5}}$ $-\dfrac{4z+2}{3z}$

86. $\dfrac{\dfrac{1}{y} + \dfrac{1}{x}}{\dfrac{1}{y} - \dfrac{1}{x}}$ $\dfrac{x+y}{x-y}$

Concept Extensions

87. Consider the equations $\sqrt{2x} = 4$ and $\sqrt[3]{2x} = 4$.
 a. Explain the difference in solving these equations.
 b. Explain the similarity in solving these equations.
  a.–b. answers may vary

88. Explain why proposed solutions of radical equations must be checked.   answers may vary

89. Find and correct the error in the following solution. See the second Concept Check in this section.

$$\sqrt{5x-1} + 4 = 7$$
$$(\sqrt{5x-1} + 4)^2 = 7^2$$
$$5x - 1 + 16 = 49$$
$$5x = 34$$
$$x = \frac{34}{5}$$

$$\sqrt{5x-1} + 4 = 7$$
$$\sqrt{5x-1} = 3$$
$$(\sqrt{5x-1})^2 = 3^2$$
$$5x - 1 = 9$$
$$5x = 10$$
$$x = 2$$

90. Solve: $\sqrt{\sqrt{x+3} + \sqrt{x}} = \sqrt{3}$ $\{1\}$

91. The cost $C(x)$ in dollars per day to operate a small delivery service is given by $C(x) = 80\sqrt[3]{x} + 500$, where x is the number of deliveries per day. In July, the manager decides that it is necessary to keep delivery costs below $1620.00. Find the greatest number of deliveries this company can make per day and still keep overhead below $1620.00.   2743 deliveries

Objectives

A Write Square Roots of Negative Numbers in the Form bi.

B Add or Subtract Complex Numbers.

C Multiply Complex Numbers.

D Divide Complex Numbers.

E Raise i to Powers.

Objective A Writing Numbers in the Form bi

Our work with radical expressions has excluded expressions such as $\sqrt{-16}$ because $\sqrt{-16}$ is not a real number; there is no real number whose square is -16. In this section, we discuss a number system that includes roots of negative numbers. This number system is the **complex number system,** and it includes the set of real numbers as a subset. The complex number system allows us to solve equations such as $x^2 + 1 = 0$ that have no real number solutions. The set of complex numbers includes the *imaginary unit.*

> **Imaginary Unit**
>
> The **imaginary unit,** written i, is the number whose square is -1. That is,
>
> $$i^2 = -1 \quad \text{and} \quad i = \sqrt{-1}$$

To write the square root of a negative number in terms of i, we use the property that if a is a positive number, then

$$\sqrt{-a} = \sqrt{-1} \cdot \sqrt{a}$$
$$= i \cdot \sqrt{a}$$

Using i, we can write $\sqrt{-16}$ as

$$\sqrt{-16} = \sqrt{-1 \cdot 16} = \sqrt{-1} \cdot \sqrt{16} = i \cdot 4 \text{ or } 4i$$

Examples Write using i notation.

1. $\sqrt{-36} = \sqrt{-1 \cdot 36} = \sqrt{-1} \cdot \sqrt{36} = i \cdot 6$ or $6i$

2. $\sqrt{-5} = \sqrt{-1(5)} = \sqrt{-1} \cdot \sqrt{5} = i\sqrt{5}$

> **Helpful Hint**
> Since $\sqrt{5}i$ can easily be confused with $\sqrt{5i}$, we write $\sqrt{5}i$ as $i\sqrt{5}$.

3. $-\sqrt{-20} = -\sqrt{-1 \cdot 20} = -\sqrt{-1} \cdot \sqrt{4 \cdot 5} = -i \cdot 2\sqrt{5} = -2i\sqrt{5}$

■ **Work Practice 1–3**

The product rule for radicals does not necessarily hold true for imaginary numbers. *To multiply square roots of negative numbers, first we write each number in terms of the imaginary unit i.* For example, to multiply $\sqrt{-4}$ and $\sqrt{-9}$, we first write each number in the form bi:

$$\sqrt{-4} \cdot \sqrt{-9} = 2i(3i) = 6i^2 = 6(-1) = -6 \quad \text{Correct.}$$

Make sure you notice that the product rule does not work for this example. In other words, $\sqrt{-4} \cdot \sqrt{-9} = \sqrt{(-4)(-9)} = \sqrt{36} = 6$ is incorrect!

Examples Multiply or divide as indicated.

4. $\sqrt{-3} \cdot \sqrt{-5} = i\sqrt{3}(i\sqrt{5}) = i^2\sqrt{15} = -1\sqrt{15} = -\sqrt{15}$

5. $\sqrt{-36} \cdot \sqrt{-1} = 6i(i) = 6i^2 = 6(-1) = -6$

6. $\sqrt{8} \cdot \sqrt{-2} = 2\sqrt{2}(i\sqrt{2}) = 2i(\sqrt{2}\sqrt{2}) = 2i(2) = 4i$

7. $\dfrac{\sqrt{-125}}{\sqrt{5}} = \dfrac{i\sqrt{125}}{\sqrt{5}} = i\sqrt{25} = 5i$

■ **Work Practice 4–7**

Practice 1–3

Write using i notation.

1. $\sqrt{-25}$

2. $\sqrt{-17}$

3. $-\sqrt{-50}$

Teaching Tip

Before discussing how to multiply the square roots of negative numbers, have students work in pairs to multiply a pair of square roots of negative numbers first by using the power rule and then by using the definition of imaginary numbers. Have students report their findings and discuss which method is correct. If necessary, point out that rules can have limits while definitions always hold.

Practice 4–7

Multiply or divide as indicated.

4. $\sqrt{-3} \cdot \sqrt{-7}$

5. $\sqrt{-25} \cdot \sqrt{-1}$

6. $\sqrt{27} \cdot \sqrt{-3}$

7. $\dfrac{\sqrt{-8}}{\sqrt{2}}$

Answers

1. $5i$ **2.** $i\sqrt{17}$ **3.** $-5i\sqrt{2}$

4. $-\sqrt{21}$ **5.** -5 **6.** $9i$ **7.** $2i$

Now that we have practiced working with the imaginary unit, we define *complex numbers*.

Complex Numbers

A **complex number** is a number that can be written in the form $a + bi$, where a and b are real numbers.

Notice that the set of real numbers is a subset of the complex numbers since any real number can be written in the form of a complex number. For example,

$$16 = 16 + 0i$$

In general, a complex number $a + bi$ is a real number if $b = 0$. Also, a complex number is called an **imaginary number** if $a = 0$. For example,

$$3i = 0 + 3i \quad \text{and} \quad i\sqrt{7} = 0 + i\sqrt{7}$$

are imaginary numbers.

The following diagram shows the relationship between complex numbers and their subsets.

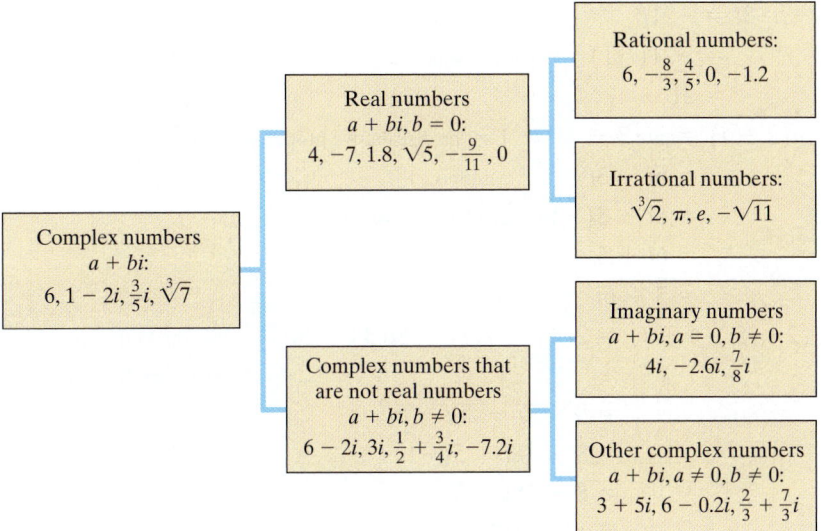

✔**Concept Check** True or false? Every complex number is also a real number.

Teaching Tip

Point out that imaginary numbers are not just an exercise in theoretical mathematics. They have many real-world applications in engineering fields such as signal analysis.

Objective B Adding or Subtracting Complex Numbers ▶

Two complex numbers $a + bi$ and $c + di$ are equal if and only if $a = c$ and $b = d$. Complex numbers can be added or subtracted by adding or subtracting their real parts and then adding or subtracting their imaginary parts.

Teaching Tip

Tell students that adding complex numbers is similar to adding algebraic expressions in that you add like parts.

Sum or Difference of Complex Numbers

If $a + bi$ and $c + di$ are complex numbers, then their sum is

$$(a + bi) + (c + di) = (a + c) + (b + d)i$$

Their difference is

$$(a + bi) - (c + di) = a + bi - c - di = (a - c) + (b - d)i$$

✔**Concept Check Answer**

false

Practice 8–10

Add or subtract as indicated.

8. $(5 + 2i) + (4 - 3i)$

9. $6i - (2 - i)$

10. $(-2 - 4i) - (-3)$

Examples Add or subtract as indicated.

8. $(2 + 3i) + (-3 + 2i) = (2 - 3) + (3 + 2)i = -1 + 5i$

9. $5i - (1 - i) = 5i - 1 + i$
$$= -1 + (5 + 1)i$$
$$= -1 + 6i$$

10. $(-3 - 7i) - (-6) = -3 - 7i + 6$
$$= (-3 + 6) - 7i$$
$$= 3 - 7i$$

■ **Work Practice 8–10**

Objective C Multiplying Complex Numbers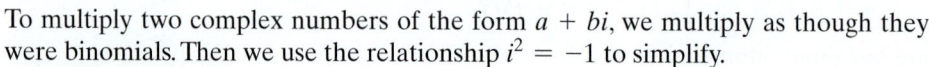

To multiply two complex numbers of the form $a + bi$, we multiply as though they were binomials. Then we use the relationship $i^2 = -1$ to simplify.

Practice 11–15

Multiply.

11. $-5i \cdot 3i$

12. $-2i(6 - 2i)$

13. $(3 - 4i)(6 + i)$

14. $(1 - 2i)^2$

15. $(6 + 5i)(6 - 5i)$

Examples Multiply.

11. $-7i \cdot 3i = -21i^2$
$$= -21(-1) \quad \text{Replace } i^2 \text{ with } -1.$$
$$= 21$$

12. $3i(2 - i) = 3i \cdot 2 - 3i \cdot i \quad \text{Use the distributive property.}$
$$= 6i - 3i^2 \qquad \text{Multiply.}$$
$$= 6i - 3(-1) \quad \text{Replace } i^2 \text{ with } -1.$$
$$= 6i + 3$$
$$= 3 + 6i$$

13. $(2 - 5i)(4 + i) = 2(4) + 2(i) - 5i(4) - 5i(i) \quad \text{Use the FOIL order.}$

 F $\quad$ O $\quad$ I $\quad$ L $\quad$ (First, Outer, Inner, Last)
$$= 8 + 2i - 20i - 5i^2$$
$$= 8 - 18i - 5(-1) \quad i^2 = -1$$
$$= 8 - 18i + 5$$
$$= 13 - 18i$$

14. $(2 - i)^2 = (2 - i)(2 - i)$
$$= 2(2) - 2(i) - 2(i) + i^2$$
$$= 4 - 4i + (-1) \qquad\qquad i^2 = -1$$
$$= 3 - 4i$$

15. $(7 + 3i)(7 - 3i) = 7(7) - 7(3i) + 3i(7) - 3i(3i)$
$$= 49 - 21i + 21i - 9i^2$$
$$= 49 - 9(-1) \qquad\qquad i^2 = -1$$
$$= 49 + 9$$
$$= 58$$

■ **Work Practice 11–15**

Notice that if you add, subtract, or multiply two complex numbers, the result is a complex number.

Objective D Dividing Complex Numbers

From Example 15, notice that the product of $7 + 3i$ and $7 - 3i$ is a real number. These two complex numbers are called *complex conjugates* of one another. In general, we have the following definition.

Teaching Tip

After discussing Example 15, challenge students to find two complex numbers that multiply to give an imaginary number.

Answers

8. $9 - i$ **9.** $-2 + 7i$ **10.** $1 - 4i$
11. 15 **12.** $-4 - 12i$ **13.** $22 - 21i$
14. $-3 - 4i$ **15.** 61

Complex Conjugates

The complex numbers $(a + bi)$ and $(a - bi)$ are called **complex conjugates** of each other, and

$$(a + bi)(a - bi) = a^2 + b^2$$

To see that the product of a complex number $a + bi$ and its conjugate $a - bi$ is the real number $a^2 + b^2$, we multiply:

$$
\begin{aligned}
(a + bi)(a - bi) &= a^2 - abi + abi - b^2 i^2 \\
&= a^2 - b^2(-1) \\
&= a^2 + b^2
\end{aligned}
$$

We use complex conjugates to divide by a complex number.

Example 16 Divide and write in the form $a + bi$: $\dfrac{2 + i}{1 - i}$

Solution: We multiply the numerator and the denominator by the complex conjugate of $1 - i$ to get a real number in the denominator.

$$
\begin{aligned}
\frac{2 + i}{1 - i} &= \frac{(2 + i)(1 + i)}{(1 - i)(1 + i)} \\
&= \frac{2(1) + 2(i) + 1(i) + i^2}{1^2 - i^2} \\
&= \frac{2 + 3i - 1}{1 + 1} \\
&= \frac{1 + 3i}{2} = \frac{1}{2} + \frac{3}{2}i
\end{aligned}
$$

■ **Work Practice 16**

Example 17 Divide and write in the form $a + bi$: $\dfrac{7}{3i}$

Solution: We multiply the numerator and the denominator by the conjugate of $3i$. Note that $3i = 0 + 3i$, so its conjugate is $0 - 3i$ or $-3i$.

$$
\frac{7}{3i} = \frac{7(-3i)}{(3i)(-3i)} = \frac{-21i}{-9i^2} = \frac{-21i}{-9(-1)} = \frac{-21i}{9} = \frac{-7i}{3} = -\frac{7}{3}i
$$

■ **Work Practice 17**

Objective E Finding Powers of i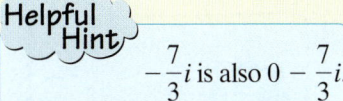

We can use the fact that $i^2 = -1$ to simplify i^3 and i^4.

$$i^3 = i^2 \cdot i = (-1)i = -i$$
$$i^4 = i^2 \cdot i^2 = (-1) \cdot (-1) = 1$$

We continue this process and use the fact that $i^4 = 1$ and $i^2 = -1$ to simplify i^5 and i^6.

$$i^5 = i^4 \cdot i = 1 \cdot i = i$$
$$i^6 = i^4 \cdot i^2 = 1 \cdot (-1) = -1$$

Teaching Tip

Have students compare the following products:

$(x + 5)(x - 5)$
$(x + \sqrt{5})(x - \sqrt{5})$
$(x + 5i)(x - 5i)$

Practice 16

Divide and write in the form

$$a + bi: \frac{3 + i}{2 - 3i}$$

Practice 17

Divide and write in the form

$$a + bi: \frac{6}{5i}$$

Helpful Hint

$-\dfrac{7}{3}i$ is also $0 - \dfrac{7}{3}i$.

When writing in $a + bi$ form, if a is 0, we will write $a + bi$ as bi and if $b = 0$, we will write $a + bi$ as a.

Answers

16. $\dfrac{3}{13} + \dfrac{11}{13}i$ **17.** $-\dfrac{6}{5}i$

If we continue finding powers of i, we generate the following pattern. Notice that the values i, -1, $-i$, and 1 repeat as i is raised to higher and higher powers.

$$i^1 = i \qquad i^5 = i \qquad i^9 = i$$
$$i^2 = -1 \qquad i^6 = -1 \qquad i^{10} = -1$$
$$i^3 = -i \qquad i^7 = -i \qquad i^{11} = -i$$
$$i^4 = 1 \qquad i^8 = 1 \qquad i^{12} = 1$$

This pattern allows us to find other powers of i. To do so, we will use the fact that $i^4 = 1$ and rewrite a power of i in terms of i^4. For example,

$$i^{22} = i^{20} \cdot i^2 = (i^4)^5 \cdot i^2 = 1^5 \cdot (-1) = 1 \cdot (-1) = -1$$

Practice 18–21

Find the powers of i.
18. i^{11} **19.** i^{40}
20. i^{50} **21.** i^{-10}

Answers
18. $-i$ **19.** 1 **20.** -1 **21.** -1

Examples Find each power of i.

18. $i^7 = i^4 \cdot i^3 = 1(-i) = -i$

19. $i^{20} = (i^4)^5 = 1^5 = 1$

20. $i^{46} = i^{44} \cdot i^2 = (i^4)^{11} \cdot i^2 = 1^{11}(-1) = -1$

21. $i^{-12} = \dfrac{1}{i^{12}} = \dfrac{1}{(i^4)^3} = \dfrac{1}{(1)^3} = \dfrac{1}{1} = 1$

▪ **Work Practice 18–21**

Vocabulary, Readiness & Video Check

Use the choices below to fill in each blank. Not all choices will be used.

    -1 $\sqrt{-1}$ real   1 $\sqrt{1}$ complex   imaginary   imaginary unit

1. A(n) ___complex___ number is one that can be written in the form $a + bi$ where a and b are real numbers.

2. In the complex number system, i denotes the ___imaginary unit___.

3. $i^2 =$ ___-1___

4. $i =$ ___$\sqrt{-1}$___

5. A complex number, $a + bi$, is a(n) ___real___ number if $b = 0$.

6. A complex number, $a + bi$, is a(n) ___imaginary___ number if $a = 0$ and $b \neq 0$.

Martin-Gay Interactive Videos

See Video 10.7

See video answer section.

Watch the section lecture video and answer the following questions.

Objective A   **7.** From ▥ Example 4, with what rule must you be especially careful when working with imaginary numbers and why? ▶

Objective B   **8.** In ▥ Examples 5 and 6, what is the process of adding and subtracting complex numbers compared to? What important reminder is given about i? ▶

Objective C   **9.** In ▥ Examples 7 and 8, what part of the definition of the imaginary unit i may be used during the multiplication of complex numbers to help simplify products? ▶

Objective D   **10.** In ▥ Example 9, using complex conjugates to divide complex numbers is compared to what process? ▶

Objective E   **11.** From the lecture before ▥ Example 10, what are the first four powers of i whose values keep repeating? ▶

Complex Conjugates

The complex numbers $(a + bi)$ and $(a - bi)$ are called **complex conjugates** of each other, and

$$(a + bi)(a - bi) = a^2 + b^2$$

To see that the product of a complex number $a + bi$ and its conjugate $a - bi$ is the real number $a^2 + b^2$, we multiply:

$$(a + bi)(a - bi) = a^2 - abi + abi - b^2i^2$$
$$= a^2 - b^2(-1)$$
$$= a^2 + b^2$$

We use complex conjugates to divide by a complex number.

Example 16 Divide and write in the form $a + bi$: $\dfrac{2 + i}{1 - i}$

Solution: We multiply the numerator and the denominator by the complex conjugate of $1 - i$ to get a real number in the denominator.

$$\frac{2 + i}{1 - i} = \frac{(2 + i)(1 + i)}{(1 - i)(1 + i)}$$

$$= \frac{2(1) + 2(i) + 1(i) + i^2}{1^2 - i^2}$$

$$= \frac{2 + 3i - 1}{1 + 1}$$

$$= \frac{1 + 3i}{2} = \frac{1}{2} + \frac{3}{2}i$$

■ **Work Practice 16**

Example 17 Divide and write in the form $a + bi$: $\dfrac{7}{3i}$

Solution: We multiply the numerator and the denominator by the conjugate of $3i$. Note that $3i = 0 + 3i$, so its conjugate is $0 - 3i$ or $-3i$.

$$\frac{7}{3i} = \frac{7(-3i)}{(3i)(-3i)} = \frac{-21i}{-9i^2} = \frac{-21i}{-9(-1)} = \frac{-21i}{9} = \frac{-7i}{3} = -\frac{7}{3}i$$

■ **Work Practice 17**

Objective E Finding Powers of i ▶

We can use the fact that $i^2 = -1$ to simplify i^3 and i^4.

$$i^3 = i^2 \cdot i = (-1)i = -i$$
$$i^4 = i^2 \cdot i^2 = (-1) \cdot (-1) = 1$$

We continue this process and use the fact that $i^4 = 1$ and $i^2 = -1$ to simplify i^5 and i^6.

$$i^5 = i^4 \cdot i = 1 \cdot i = i$$
$$i^6 = i^4 \cdot i^2 = 1 \cdot (-1) = -1$$

Teaching Tip

Have students compare the following products:
$(x + 5)(x - 5)$
$(x + \sqrt{5})(x - \sqrt{5})$
$(x + 5i)(x - 5i)$

Practice 16

Divide and write in the form

$$a + bi: \frac{3 + i}{2 - 3i}$$

Practice 17

Divide and write in the form

$$a + bi: \frac{6}{5i}$$

Helpful Hint

$-\dfrac{7}{3}i$ is also $0 - \dfrac{7}{3}i$.

When writing in $a + bi$ form, if a is 0, we will write $a + bi$ as bi and if $b = 0$, we will write $a + bi$ as a.

Answers

16. $\dfrac{3}{13} + \dfrac{11}{13}i$ **17.** $-\dfrac{6}{5}i$

If we continue finding powers of i, we generate the following pattern. Notice that the values i, -1, $-i$, and 1 repeat as i is raised to higher and higher powers.

$$i^1 = i \qquad\qquad i^5 = i \qquad\qquad i^9 = i$$
$$i^2 = -1 \qquad\quad i^6 = -1 \qquad\quad i^{10} = -1$$
$$i^3 = -i \qquad\quad i^7 = -i \qquad\quad i^{11} = -i$$
$$i^4 = 1 \qquad\qquad i^8 = 1 \qquad\qquad i^{12} = 1$$

This pattern allows us to find other powers of i. To do so, we will use the fact that $i^4 = 1$ and rewrite a power of i in terms of i^4. For example,

$$i^{22} = i^{20} \cdot i^2 = (i^4)^5 \cdot i^2 = 1^5 \cdot (-1) = 1 \cdot (-1) = -1$$

Practice 18–21

Find the powers of i.

18. i^{11} **19.** i^{40}

20. i^{50} **21.** i^{-10}

Answers

18. $-i$ **19.** 1 **20.** -1 **21.** -1

Examples   Find each power of i.

18. $i^7 = i^4 \cdot i^3 = 1(-i) = -i$

19. $i^{20} = (i^4)^5 = 1^5 = 1$

20. $i^{46} = i^{44} \cdot i^2 = (i^4)^{11} \cdot i^2 = 1^{11}(-1) = -1$

21. $i^{-12} = \dfrac{1}{i^{12}} = \dfrac{1}{(i^4)^3} = \dfrac{1}{(1)^3} = \dfrac{1}{1} = 1$

🟧 **Work Practice 18–21**

Vocabulary, Readiness & Video Check

Use the choices below to fill in each blank. Not all choices will be used.

   -1 $\sqrt{-1}$ real 1 $\sqrt{1}$ complex   imaginary   imaginary unit

1. A(n) __complex__ number is one that can be written in the form $a + bi$ where a and b are real numbers.

2. In the complex number system, i denotes the __imaginary unit__.

3. $i^2 = $ ____-1____

4. $i = $ ____$\sqrt{-1}$____

5. A complex number, $a + bi$, is a(n) ____real____ number if $b = 0$.

6. A complex number, $a + bi$, is a(n) ____imaginary____ number if $a = 0$ and $b \neq 0$.

Martin-Gay Interactive Videos

See Video 10.7

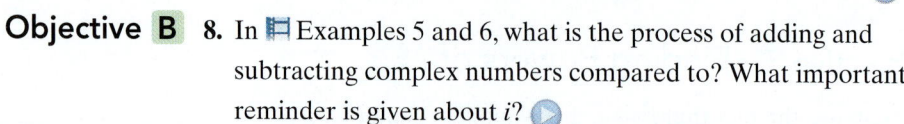

Watch the section lecture video and answer the following questions.

Objective A   **7.** From ▦ Example 4, with what rule must you be especially careful when working with imaginary numbers and why? ▶

Objective B   **8.** In ▦ Examples 5 and 6, what is the process of adding and subtracting complex numbers compared to? What important reminder is given about i? ▶

Objective C   **9.** In ▦ Examples 7 and 8, what part of the definition of the imaginary unit i may be used during the multiplication of complex numbers to help simplify products? ▶

Objective D   **10.** In ▦ Example 9, using complex conjugates to divide complex numbers is compared to what process? ▶

Objective E   **11.** From the lecture before ▦ Example 10, what are the first four powers of i whose values keep repeating? ▶

10.7 Exercise Set MyMathLab®

Objective **A** *Simplify and write using i notation, if possible. See Examples 1 through 3.*

▶ **1.** $\sqrt{-81}$ $9i$ **2.** $\sqrt{-49}$ $7i$ ▶ **3.** $\sqrt{-7}$ $i\sqrt{7}$ **4.** $\sqrt{-3}$ $i\sqrt{3}$

5. $-\sqrt{16}$ -4 **6.** $-\sqrt{4}$ -2 **7.** $\sqrt{-64}$ $8i$ **8.** $\sqrt{-100}$ $10i$

▶ **9.** $\sqrt{-24}$ $2i\sqrt{6}$ **10.** $\sqrt{-32}$ $4i\sqrt{2}$ **11.** $-\sqrt{-36}$ $-6i$ **12.** $-\sqrt{-121}$ $-11i$

13. $8\sqrt{-63}$ $24i\sqrt{7}$ **14.** $4\sqrt{-20}$ $8i\sqrt{5}$ **15.** $-\sqrt{54}$ $-3\sqrt{6}$ **16.** $-\sqrt{63}$ $-3\sqrt{7}$

Multiply or divide as indicated. See Examples 4 through 7.

17. $\sqrt{-2} \cdot \sqrt{-7}$ $-\sqrt{14}$ **18.** $\sqrt{-11} \cdot \sqrt{-3}$ $-\sqrt{33}$ ▶ **19.** $\sqrt{-5} \cdot \sqrt{-10}$ $-5\sqrt{2}$ **20.** $\sqrt{-2} \cdot \sqrt{-6}$ $-2\sqrt{3}$

21. $\sqrt{16} \cdot \sqrt{-1}$ $4i$ **22.** $\sqrt{3} \cdot \sqrt{-27}$ $9i$ **23.** $\dfrac{\sqrt{-9}}{\sqrt{3}}$ $i\sqrt{3}$ **24.** $\dfrac{\sqrt{49}}{\sqrt{-10}}$ $-\dfrac{7i\sqrt{10}}{10}$

25. $\dfrac{\sqrt{-80}}{\sqrt{-10}}$ $2\sqrt{2}$ **26.** $\dfrac{\sqrt{-40}}{\sqrt{-8}}$ $\sqrt{5}$

Objective **B** *Add or subtract as indicated. Write your answers in the form a + bi. See Examples 8 through 10.*

▶ **27.** $(4 - 7i) + (2 + 3i)$ $6 - 4i$ **28.** $(2 - 4i) - (2 - i)$ $-3i$ **29.** $(6 + 5i) - (8 - i)$ $-2 + 6i$

30. $(8 - 3i) + (-8 + 3i)$ 0 ▶ **31.** $6 - (8 + 4i)$ $-2 - 4i$ **32.** $(9 - 4i) - 9$ $-4i$

33. $(6 - 3i) - (4 - 2i)$ $2 - i$ **34.** $(-2 - 4i) - (6 - 8i)$ $-8 + 4i$ **35.** $(5 - 6i) - 4i$ $5 - 10i$

36. $(6 - 2i) + 7i$ $6 + 5i$ **37.** $(2 + 4i) + (6 - 5i)$ $8 - i$ **38.** $(5 - 3i) + (7 - 8i)$ $12 - 11i$

Objective **C** *Multiply. Write your answers in the form a + bi. See Examples 11 through 15.*

39. $6i \cdot 2i$ -12 **40.** $5i \cdot 7i$ -35 **41.** $-9i \cdot 7i$ 63

42. $-6i \cdot 4i$ 24 ▶ **43.** $-10i \cdot -4i$ -40 **44.** $-2i \cdot -11i$ -22

45. $6i(2 - 3i)$ $18 + 12i$ **46.** $5i(4 - 7i)$ $35 + 20i$ **47.** $-3i(-1 + 9i)$ $27 + 3i$

48. $-5i(-2 + i)$ $5 + 10i$

49. $(4 + i)(5 + 2i)$ $18 + 13i$

50. $(3 + i)(2 + 4i)$ $2 + 14i$

51. $(\sqrt{3} + 2i)(\sqrt{3} - 2i)$ 7

52. $(\sqrt{5} - 5i)(\sqrt{5} + 5i)$ 30

53. $(4 - 2i)^2$ $12 - 16i$

54. $(6 - 3i)^2$ $27 - 36i$

▶**55.** $(6 - 2i)(3 + i)$ 20

56. $(2 - 4i)(2 - i)$ $-10i$

57. $(1 - i)(1 + i)$ 2

58. $(6 + 2i)(6 - 2i)$ 40

59. $(9 + 8i)^2$ $17 + 144i$

60. $(4 + 7i)^2$ $-33 + 56i$

61. $(1 - i)^2$ $-2i$

62. $(2 - 2i)^2$ $-8i$

Objective **D**   *Divide. Write your answers in the form a + bi. See Examples 16 and 17.*

63. $\dfrac{4}{i}$ $-4i$

64. $\dfrac{5}{6i}$ $-\dfrac{5}{6}i$

65. $\dfrac{7}{4 + 3i}$ $\dfrac{28}{25} - \dfrac{21}{25}i$

66. $\dfrac{9}{1 - 2i}$ $\dfrac{9}{5} + \dfrac{18}{5}i$

67. $\dfrac{6i}{1 - 2i}$ $-\dfrac{12}{5} + \dfrac{6}{5}i$

68. $\dfrac{3i}{5 + i}$ $\dfrac{3}{26} + \dfrac{15}{26}i$

▶**69.** $\dfrac{3 + 5i}{1 + i}$ $4 + i$

70. $\dfrac{6 + 2i}{4 - 3i}$ $\dfrac{18}{25} + \dfrac{26}{25}i$

71. $\dfrac{4 - 5i}{2i}$ $-\dfrac{5}{2} - 2i$

72. $\dfrac{6 + 8i}{3i}$ $\dfrac{8}{3} - 2i$

73. $\dfrac{16 + 15i}{-3i}$ $-5 + \dfrac{16}{3}i$

74. $\dfrac{2 - 3i}{-7i}$ $\dfrac{3}{7} + \dfrac{2}{7}i$

75. $\dfrac{2}{3 + i}$ $\dfrac{3}{5} - \dfrac{1}{5}i$

76. $\dfrac{5}{3 - 2i}$ $\dfrac{15}{13} + \dfrac{10}{13}i$

77. $\dfrac{2 - 3i}{2 + i}$ $\dfrac{1}{5} - \dfrac{8}{5}i$

78. $\dfrac{6 + 5i}{6 - 5i}$ $\dfrac{11}{61} + \dfrac{60}{61}i$

Objective **E**   *Find each power of i. See Examples 18 through 21.*

▶**79.** i^8 1

80. i^{10} -1

81. i^{21} i

82. i^{15} $-i$

83. i^{11} $-i$

84. i^{40} 1

85. i^{-6} -1

86. i^{-9} $-i$

▶**87.** $(2i)^6$ -64

88. $(5i)^4$ 625

89. $(-3i)^5$ $-243i$

90. $(-2i)^7$ $128i$

Review

Thirty people were recently polled about the average monthly balance in their checking account. The results of this poll are shown in the bar graph. Use this graph to answer Exercises 91 through 96. See Section 3.1.

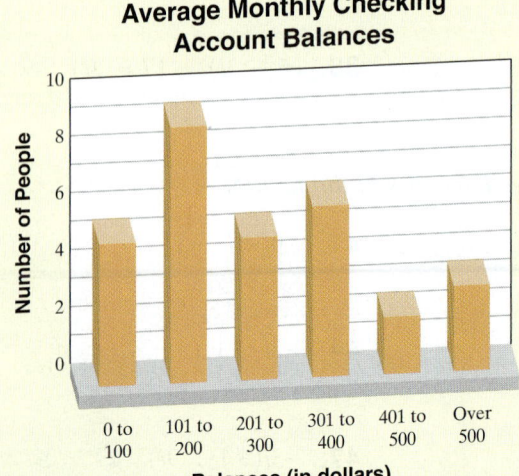

Average Monthly Checking Account Balances

91. How many people polled reported an average checking balance of $201 to $300?   5 people

92. How many people polled reported an average checking balance of $0 to $100?   5 people

93. How many people polled reported an average checking balance of $200 or less?   14 people

94. How many people polled reported an average checking balance of $301 or more?   11 people

95. What percent of people polled reported an average checking balance of $201 to $300?   16.7%

96. What percent of people polled reported an average checking balance of 0 to $100?   16.7%

Concept Extensions

Write each expression in the form a + bi.

97. $i^3 + i^4$
$1 - i$

98. $i^8 - i^7$
$1 + i$

99. $i^6 + i^8$
0

100. $i^4 + i^{12}$
2

101. $2 + \sqrt{-9}$
$2 + 3i$

102. $7 - \sqrt{-25}$
$7 - 5i$

103. $5 - \sqrt{-16}$
$5 - 4i$

104. $\dfrac{6 + \sqrt{-18}}{3}$
$2 + i\sqrt{2}$

105. $\dfrac{4 - \sqrt{-8}}{2}$
$2 - i\sqrt{2}$

106. $\dfrac{5 - \sqrt{-75}}{10}$ $\dfrac{1}{2} - \dfrac{\sqrt{3}}{2}i$

107. Describe how to find the conjugate of a complex number.   answers may vary

108. Explain why the product of a complex number and its complex conjugate is a real number.
answers may vary

Simplify.

109. $(8 - \sqrt{-3}) - (2 + \sqrt{-12})$ $6 - 3i\sqrt{3}$

110. $(8 - \sqrt{-4}) - (2 + \sqrt{-16})$ $6 - 6i$

111. Determine whether $2i$ is a solution of $x^2 + 4 = 0$.
yes

112. Determine whether $-1 + i$ is a solution of $x^2 + 2x = -2$.   yes

---

Chapter 10 Group Activity

Additional group activities are available in the *Instructor's Resource Manual with Tests.*

Heron of Alexandria

Heron (also Hero) was a Greek mathematician and engineer. He lived and worked in Alexandria, Egypt, around 75 A.D. During his prolific work life, Heron developed a rotary steam engine called an aeolipile and a surveying tool called a dioptra as well as a wind organ and a fire engine. As an engineer, he must have had the need to approximate square roots because he described an iterative method for doing so in his work *Metrica*. Heron's method for approximating a square root can be summarized as follows:

Suppose that x is not a perfect square and a^2 is the nearest perfect square to x. For a rough estimate of the value of $\sqrt{x}$, find the value of $y_1 = \dfrac{1}{2}\left(a + \dfrac{x}{a}\right)$. This estimate can be improved by calculating a second estimate using the first estimate y_1 in place of a:

$$y_2 = \frac{1}{2}\left(y_1 + \frac{x}{y_1}\right).$$

Repeating this process several times will give more and more accurate estimates of $\sqrt{x}$.

2 a. 25   **b.** $\dfrac{11}{2}$ **c.** $\dfrac{241}{44}$ **d.** 5.477225575

e. answers may vary   **f.** 3

3 a. $68^2 = 4624$ **b.** $\dfrac{2299}{34}$ **c.** $\dfrac{10,570,633}{156,332}$

d. 67.61656602   **e.** answers may vary   **f.** 3

Critical Thinking

1. a. Which perfect square is closest to 80?   81

b. Use Heron's method for approximating square roots to calculate the first estimate of the square root of 80. $\dfrac{161}{18}$

c. Use the first estimate of the square root of 80 to find a more refined second estimate. $\dfrac{51,841}{5796}$

d. Use a calculator to find the actual value of the square root of 80. List all digits shown on your calculator's display.   8.94427191

e. Compare the actual value from part (d) to the values of the first and second estimates. What do you notice?   answers may vary

f. How many iterations of this process are necessary to get an estimate that differs no more than one digit from the actual value recorded in part (d)?   2

2. Repeat Question 1 for finding an estimate of the square root of 30.

3. Repeat Question 1 for finding an estimate of the square root of 4572.

4. Why would this iterative method have been important to people of Heron's era? Would you say that this method is as important today? Why or why not?   answers may vary

Chapter 10 Vocabulary Check

Fill in each blank with one of the words or phrases listed below.

index   rationalizing   conjugate   principal square root   cube root   midpoint

complex number   like radicals   radicand   imaginary unit   distance

1. The _____conjugate_____ of $\sqrt{3} + 2$ is $\sqrt{3} - 2$.

2. The ___principal square root___ of a nonnegative number a is written as $\sqrt{a}$.

3. The process of writing a radical expression as an equivalent expression but without a radical in the denominator is called _____rationalizing_____ the denominator.

4. The ___imaginary unit___, written i, is the number whose square is -1.

5. The _____cube root_____ of a number is written as $\sqrt[3]{a}$.

6. In the notation $\sqrt[n]{a}$, n is called the _____index_____ and a is called the _____radicand_____.

7. Radicals with the same index and the same radicand are called _____like radicals_____.

8. A(n) ___complex number___ is a number that can be written in the form $a + bi$, where a and b are real numbers.

9. The _____distance_____ formula is $d = \sqrt{(x_2 - x_1)^2 + (y_2 - y_1)^2}$.

10. The _____midpoint_____ formula is $\left(\dfrac{x_1 + x_2}{2}, \dfrac{y_1 + y_2}{2} \right)$.

Helpful Hint

▶ Are you preparing for your test? Don't forget to take the Chapter 10 Test on page 754. Then check your answers at the back of the text and use the Chapter Test Prep Videos to see the fully worked-out solutions to any of the exercises you want to review.

10 Chapter Highlights

Definitions and Concepts	Examples
Section 10.1 Radical Expressions and Radical Functions	

Definitions and Concepts	Examples
The **positive**, or **principal, square root** of a nonnegative number a is written as $\sqrt{a}$. $$\sqrt{a} = b \text{ only if } b^2 = a \text{ and } b \geq 0$$	$\sqrt{36} = 6 \qquad \sqrt{\dfrac{9}{100}} = \dfrac{3}{10}$
The **negative square root** of a is written as $-\sqrt{a}$.	$-\sqrt{36} = -6 \quad -\sqrt{0.04} = -0.2$
The **cube root** of a real number a is written as $\sqrt[3]{a}$. $$\sqrt[3]{a} = b \text{ only if } b^3 = a$$	$\sqrt[3]{27} = 3 \qquad \sqrt[3]{-\dfrac{1}{8}} = -\dfrac{1}{2}$ $\sqrt[3]{y^6} = y^2 \quad \sqrt[3]{64x^9} = 4x^3$
If n is an even positive integer, then $\sqrt[n]{a^n} = \lvert a \rvert$.	$\sqrt{(-3)^2} = \lvert -3 \rvert = 3$
If n is an odd positive integer, then $\sqrt[n]{a^n} = a$.	$\sqrt{(-7)^3} = -7$
A **radical function** in x is a function defined by an expression containing a root of x.	If $f(x) = \sqrt{x} + 2$, $$f(1) = \sqrt{1} + 2 = 1 + 2 = 3$$ $$f(3) = \sqrt{3} + 2 \approx 3.73$$

📱 *Use a calculator and write a three-decimal-place approximation of each number.*

51. $\sqrt{20}$ 4.472

52. $\sqrt[3]{-39}$ -3.391

53. $\sqrt[4]{726}$ 5.191

54. $56^{1/3}$ 3.826

55. $-78^{3/4}$ -26.246

56. $105^{-2/3}$ 0.045

Use rational exponents to write each as a single radical.

57. $\sqrt[3]{2} \cdot \sqrt{7}$ $\sqrt[6]{1372}$

58. $\sqrt[3]{3} \cdot \sqrt[4]{x}$ $\sqrt[12]{81x^3}$

(10.3) *Perform each indicated operation and then simplify if possible. Assume that all variables represent positive real numbers.*

59. $\sqrt{3} \cdot \sqrt{8}$ $2\sqrt{6}$

60. $\sqrt[3]{7y} \cdot \sqrt[3]{x^2z}$ $\sqrt[3]{7x^2yz}$

61. $\dfrac{\sqrt{44x^3}}{\sqrt{11x}}$ $2x$

62. $\dfrac{\sqrt[4]{a^6b^{13}}}{\sqrt[4]{a^2b}}$ ab^3

Simplify.

63. $\sqrt{60}$ $2\sqrt{15}$

64. $-\sqrt{75}$ $-5\sqrt{3}$

65. $\sqrt[3]{162}$ $3\sqrt[3]{6}$

66. $\sqrt[3]{-32}$ $-2\sqrt[3]{4}$

67. $\sqrt{36x^7}$ $6x^3\sqrt{x}$

68. $\sqrt[3]{24a^5b^7}$ $2ab^2\sqrt[3]{3a^2b}$

69. $\sqrt{\dfrac{p^{17}}{121}}$ $\dfrac{p^8\sqrt{p}}{11}$

70. $\sqrt[3]{\dfrac{y^5}{27x^6}}$ $\dfrac{y\sqrt[3]{y^2}}{3x^2}$

71. $\sqrt[4]{\dfrac{xy^6}{81}}$ $\dfrac{y\sqrt[4]{xy^2}}{3}$

72. $\sqrt{\dfrac{2x^3}{49y^4}}$ $\dfrac{x\sqrt{2x}}{7y^2}$

△ *The formula for the radius r of a circle of area A is* $r = \sqrt{\dfrac{A}{\pi}}$. *Use this for Exercises 73 and 74.*

73. Find the exact radius of a circle whose area is 25 square meters. $\dfrac{5}{\sqrt{\pi}}$ m or $\dfrac{5\sqrt{\pi}}{\pi}$ m

📱 **74.** Approximate to two decimal places the radius of a circle whose area is 104 square inches.   5.75 in.

Find the distance between each pair of points. Give the exact value and a three-decimal-place approximation.

75. $(-6, 3)$ and $(8, 4)$ $\sqrt{197}$ units ≈ 14.036 units

76. $(-4, -6)$ and $(-1, 5)$ $\sqrt{130}$ units ≈ 11.402 units

77. $(-1, 5)$ and $(2, -3)$ $\sqrt{73}$ units ≈ 8.544 units

78. $(-\sqrt{2}, 0)$ and $(0, -4\sqrt{6})$ $7\sqrt{2}$ units ≈ 9.899 units

79. $(-\sqrt{5}, -\sqrt{11})$ and $(-\sqrt{5}, -3\sqrt{11})$
$2\sqrt{11}$ units ≈ 6.633 units

📱 **80.** $(7.4, -8.6)$ and $(-1.2, 5.6)$
$\sqrt{275.6}$ units ≈ 16.601 units

Find the midpoint of each line segment whose endpoints are given.

81. $(2, 6); (-12, 4)$ $(-5, 5)$

82. $(-6, -5); (-9, 7)$ $\left(-\dfrac{15}{2}, 1\right)$

83. $(4, -6); (-15, 2)$ $\left(-\dfrac{11}{2}, -2\right)$

84. $\left(0, -\dfrac{3}{8}\right); \left(\dfrac{1}{10}, 0\right)$ $\left(\dfrac{1}{20}, -\dfrac{3}{16}\right)$

85. $\left(\dfrac{3}{4}, -\dfrac{1}{7}\right); \left(-\dfrac{1}{4}, -\dfrac{3}{7}\right)$ $\left(\dfrac{1}{4}, -\dfrac{2}{7}\right)$

86. $(\sqrt{3}, -2\sqrt{6})$ and $(\sqrt{3}, -4\sqrt{6})$
$(\sqrt{3}, -3\sqrt{6})$

(10.4) *Perform each indicated operation. Assume that all variables represent positive real numbers.*

87. $\sqrt{20} + \sqrt{45} - 7\sqrt{5}$ $-2\sqrt{5}$

88. $x\sqrt{75x} - \sqrt{27x^3}$ $2x\sqrt{3x}$

89. $\sqrt[3]{128} + \sqrt[3]{250}$ $9\sqrt[3]{2}$

90. $3\sqrt[4]{32a^5} - a\sqrt[4]{162a}$ $3a\sqrt[4]{2a}$

91. $\dfrac{5}{\sqrt{4}} + \dfrac{\sqrt{3}}{3}$ $\dfrac{15 + 2\sqrt{3}}{6}$

92. $\sqrt{\dfrac{8}{x^2}} - \sqrt{\dfrac{50}{16x^2}}$ $\dfrac{3\sqrt{2}}{4x}$

93. $2\sqrt{50} - 3\sqrt{125} + \sqrt{98}$
$17\sqrt{2} - 15\sqrt{5}$

94. $2a\sqrt[4]{32b^5} - 3b\sqrt[4]{162a^4b} + \sqrt[4]{2a^4b^5}$
$-4ab\sqrt[4]{2b}$

Multiply and then simplify if possible. Assume that all variables represent positive real numbers.

95. $\sqrt{3}(\sqrt{27} - \sqrt{3})$ 6

96. $(\sqrt{x} - 3)^2$ $x - 6\sqrt{x} + 9$

97. $(\sqrt{5} - 5)(2\sqrt{5} + 2)$ $-8\sqrt{5}$

98. $(2\sqrt{x} - 3\sqrt{y})(2\sqrt{x} + 3\sqrt{y})$
$4x - 9y$

99. $(\sqrt{a} + 3)(\sqrt{a} - 3)$ $a - 9$

100. $(\sqrt[3]{a} + 2)^2$
$\sqrt[3]{a^2} + 4\sqrt[3]{a} + 4$

101. $(\sqrt[3]{5x} + 9)(\sqrt[3]{5x} - 9)$
$\sqrt[3]{25x^2} - 81$

102. $(\sqrt[3]{a} + 4)(\sqrt[3]{a^2} - 4\sqrt[3]{a} + 16)$
$a + 64$

(10.5) *Rationalize each denominator. Assume that all variables represent positive real numbers.*

103. $\dfrac{3}{\sqrt{7}}$ $\dfrac{3\sqrt{7}}{7}$

104. $\sqrt{\dfrac{x}{12}}$ $\dfrac{\sqrt{3x}}{6}$

105. $\dfrac{5}{\sqrt[3]{4}}$ $\dfrac{5\sqrt[3]{2}}{2}$

106. $\sqrt{\dfrac{24x^5}{3y}}$ $\dfrac{2x^2\sqrt{2xy}}{y}$

107. $\sqrt[3]{\dfrac{15x^6y^7}{z^2}}$ $\dfrac{x^2y^2\sqrt[3]{15yz}}{z}$

108. $\sqrt[4]{\dfrac{81}{8x^{10}}}$ $\dfrac{3\sqrt[4]{2x^2}}{2x^3}$

109. $\dfrac{3}{\sqrt{y} - 2}$ $\dfrac{3\sqrt{y} + 6}{y - 4}$

110. $\dfrac{\sqrt{2} - \sqrt{3}}{\sqrt{2} + \sqrt{3}}$ $-5 + 2\sqrt{6}$

Rationalize each numerator. Assume that all variables represent positive real numbers.

111. $\dfrac{\sqrt{11}}{3}$ $\dfrac{11}{3\sqrt{11}}$

112. $\sqrt{\dfrac{18}{y}}$ $\dfrac{6}{\sqrt{2y}}$

113. $\dfrac{\sqrt[3]{9}}{7}$ $\dfrac{3}{7\sqrt[3]{3}}$

114. $\sqrt{\dfrac{24x^5}{3y^2}}$ $\dfrac{4x^3}{y\sqrt{2x}}$

115. $\sqrt[3]{\dfrac{xy^2}{10z}}$ $\dfrac{xy}{\sqrt[3]{10x^2yz}}$

116. $\dfrac{\sqrt{x} + 5}{-3}$ $\dfrac{x - 25}{-3\sqrt{x} + 15}$

(10.6) *Solve each equation.*

117. $\sqrt{y - 7} = 5$ $\{32\}$

118. $\sqrt{2x} + 10 = 4$ $\varnothing$

119. $\sqrt[3]{2x - 6} = 4$ $\{35\}$

120. $\sqrt{x + 6} = \sqrt{x + 2}$ $\varnothing$

121. $2x - 5\sqrt{x} = 3$ $\{9\}$

122. $\sqrt{x + 9} = 2 + \sqrt{x - 7}$
$\{16\}$

Find each unknown length.

△**123.**

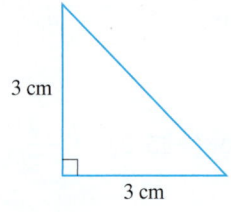

3 cm

3 cm

$3\sqrt{2}$ cm

△**124.**

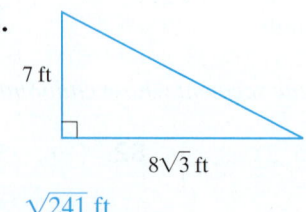

7 ft

$8\sqrt{3}$ ft

$\sqrt{241}$ ft

△ **125.** Craig and Daniel Cantwell want to determine the distance x across a pond on their property. They are able to measure the distances shown on the following diagram. Find how wide the lake is at the crossing point indicated by the triangle to the nearest tenth of a foot.   51.2 ft

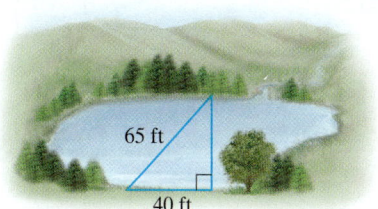

65 ft

40 ft

△ **126.** Andrea Roberts, a pipefitter, needs to connect two underground pipelines that are offset by 3 feet, as pictured in the diagram. Neglecting the joints needed to join the pipes, find the length of the shortest possible connecting pipe rounded to the nearest hundredth of a foot.   4.24 ft

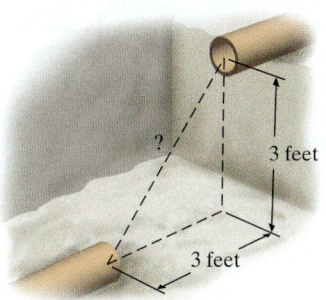

? 3 feet

3 feet

(10.7) *Perform each indicated operation and simplify. Write the results in the form a + bi.*

127. $\sqrt{-8}$ $2i\sqrt{2}$

128. $-\sqrt{-6}$ $-i\sqrt{6}$

129. $\sqrt{-4} + \sqrt{-16}$ $6i$

130. $\sqrt{-2} \cdot \sqrt{-5}$ $-\sqrt{10}$

131. $(12 - 6i) + (3 + 2i)$ $15 - 4i$

132. $(-8 - 7i) - (5 - 4i)$ $-13 - 3i$

133. $(2i)^6$ -64

134. $(3i)^4$ 81

135. $-3i(6 - 4i)$ $-12 - 18i$

136. $(3 + 2i)(1 + i)$ $1 + 5i$

137. $(2 - 3i)^2$ $-5 - 12i$

138. $(\sqrt{6} - 9i)(\sqrt{6} + 9i)$ 87

139. $\dfrac{2 + 3i}{2i}$ $\dfrac{3}{2} - i$

140. $\dfrac{1 + i}{-3i}$ $-\dfrac{1}{3} + \dfrac{1}{3}i$

Mixed Review

Simplify. Use absolute value bars when necessary.

141. $\sqrt[3]{x^3}$ x

142. $\sqrt{(x + 2)^2}$ $|x + 2|$

Simplify. Assume that all variables represent positive real numbers. If necessary, write answers with positive exponents only.

143. $-\sqrt{100}$ -10

144. $\sqrt[3]{-x^{12}y^3}$ $-x^4y$

145. $\sqrt[4]{\dfrac{y^{20}}{16x^{12}}}$ $\dfrac{y^5}{2x^3}$

146. $9^{1/2}$ 3

147. $64^{-1/2}$ $\dfrac{1}{8}$

148. $\left(\dfrac{27}{64}\right)^{-2/3}$ $\dfrac{16}{9}$

149. $\dfrac{(x^{2/3}x^{-3})^3}{x^{-1/2}}$ $\dfrac{1}{x^{13/2}}$

150. $\sqrt{200x^9}$ $10x^4\sqrt{2x}$

151. $\sqrt{\dfrac{3n^3}{121m^{10}}}$ $\dfrac{n\sqrt{3n}}{11m^5}$

152. $3\sqrt{20} - 7x\sqrt[3]{40} + 3\sqrt[3]{5x^3}$ $6\sqrt{5} - 11x\sqrt[3]{5}$

153. $(2\sqrt{x} - 5)^2$ $4x - 20\sqrt{x} + 25$

154. Find the distance between $(-3, 5)$ and $(-8, 9)$. $\sqrt{41}$ units

155. Find the midpoint of the line segment joining $(-3, 8)$ and $(11, 24)$. $(4, 16)$

Rationalize each denominator.

156. $\dfrac{7}{\sqrt{13}}$ $\dfrac{7\sqrt{13}}{13}$

157. $\dfrac{2}{\sqrt{x} + 3}$ $\dfrac{2\sqrt{x} - 6}{x - 9}$

Solve.

158. $\sqrt{x} + 2 = x$ $\{4\}$

159. $\sqrt{2x - 1} + 2 = x$ $\{5\}$

Chapter 10   Test

Step-by-step test solutions are found on the Chapter Test Prep Videos. Where available: MyMathLab® or YouTube™

Answers

Note to Instructor: The Chapter 10 Test file in the TestGen program provides algorithms specifically matched to this test for easy replication for practice or assessment purposes.

1. $6\sqrt{6}$

2. $-x^{16}$

3. $\dfrac{1}{5}$

4. 5

5. $\dfrac{4x^2}{9}$

6. $-a^6b^3$

7. $\dfrac{8a^{1/3}c^{2/3}}{b^{5/12}}$

8. $a^{7/12} - a^{7/3}$

9. $4|xy|$

10. -27

11. $\dfrac{3\sqrt{y}}{y}$

12. $\dfrac{8 - 6\sqrt{x} + x}{8 - 2x}$

13. $\dfrac{2\sqrt[3]{3x^2}}{3x}$

14. $\dfrac{6 - x^2}{8(\sqrt{6} - x)}$

15. $-x\sqrt{5x}$

16. $4\sqrt{3} - \sqrt{6}$

17. $x + 2\sqrt{x} + 1$

18. $-4\sqrt{3} - 4$

19. -20

754

Raise to the power or find the root. Assume that all variables represent positive real numbers. Write with only positive exponents.

1. $\sqrt{216}$

2. $-\sqrt[4]{x^{64}}$

3. $\left(\dfrac{1}{125}\right)^{1/3}$

4. $\left(\dfrac{1}{125}\right)^{-1/3}$

5. $\left(\dfrac{8x^3}{27}\right)^{2/3}$

6. $\sqrt[3]{-a^{18}b^9}$

7. $\left(\dfrac{64c^{4/3}}{a^{-2/3}b^{5/6}}\right)^{1/2}$

8. $a^{-2/3}(a^{5/4} - a^3)$

Find each root. Use absolute value bars when necessary.

9. $\sqrt[4]{(4xy)^4}$

10. $\sqrt[3]{(-27)^3}$

Rationalize each denominator. Assume that all variables represent positive real numbers.

11. $\sqrt{\dfrac{9}{y}}$

12. $\dfrac{4 - \sqrt{x}}{4 + 2\sqrt{x}}$

13. $\sqrt[3]{\dfrac{8}{9x}}$

14. Rationalize the numerator of $\dfrac{\sqrt{6} + x}{8}$ and simplify.

Perform each indicated operation. Assume that all variables represent positive real numbers.

15. $\sqrt{125x^3} - 3\sqrt{20x^3}$

16. $\sqrt{3}(\sqrt{16} - \sqrt{2})$

17. $(\sqrt{x} + 1)^2$

18. $(\sqrt{2} - 4)(\sqrt{3} + 1)$

19. $(\sqrt{5} + 5)(\sqrt{5} - 5)$

Use a calculator to approximate each number to three decimal places.

20. $\sqrt{561}$

21. $386^{-2/3}$

Solve.

22. $x = \sqrt{x-2} + 2$

23. $\sqrt{x^2 - 7} + 3 = 0$

24. $\sqrt{x+5} = \sqrt{2x-1}$

Perform each indicated operation and simplify. Write the results in the form $a + bi$.

25. $\sqrt{-2}$

26. $-\sqrt{-8}$

27. $(12 - 6i) - (12 - 3i)$

28. $(6 - 2i)(6 + 2i)$

29. $(4 + 3i)^2$

30. $\dfrac{1 + 4i}{1 - i}$

△ **31.** Find x.

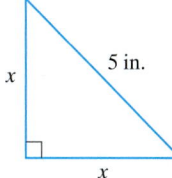

5 in.

x

x

32. If $g(x) = \sqrt{x+2}$, find $g(0)$ and $g(23)$.

33. Find the distance between the points $(-6, 3)$ and $(-8, -7)$.

34. Find the distance between the points $(-2\sqrt{5}, \sqrt{10})$ and $(-\sqrt{5}, 4\sqrt{10})$.

35. Find the midpoint of the line segment whose endpoints are $(-2, -5)$ and $(-6, 12)$.

36. Find the midpoint of the line segment whose endpoints are $\left(-\dfrac{2}{3}, -\dfrac{1}{5}\right)$ and $\left(-\dfrac{1}{3}, \dfrac{4}{5}\right)$.

Solve.

37. The function $V = \sqrt{2.5r}$ can be used to estimate the maximum safe velocity, V, in miles per hour, at which a car can travel if it is driven along a curved road with a *radius of curvature*, r, in feet. To the nearest whole number, find the maximum safe speed if a cloverleaf exit on an expressway has a radius of curvature of 300 feet.

38. Use the formula from Exercise 37 to find the radius of curvature if the safe velocity is 30 miles per hour.

20. 23.685

21. 0.019

22. $\{2, 3\}$

23. $\varnothing$

24. $\{6\}$

25. $i\sqrt{2}$

26. $-2i\sqrt{2}$

27. $-3i$

28. 40

29. $7 + 24i$

30. $-\dfrac{3}{2} + \dfrac{5}{2}i$

31. $x = \dfrac{5\sqrt{2}}{2}$ in.

32. $\sqrt{2}$; 5

33. $2\sqrt{26}$ units

34. $\sqrt{95}$ units

35. $\left(-4, \dfrac{7}{2}\right)$

36. $\left(-\dfrac{1}{2}, \dfrac{3}{10}\right)$

37. 27 mph

38. 360 ft

Multiply.

1. $2(-10)$

2. $-1.7(-3.1)$

3. $-\dfrac{2}{3} \cdot \dfrac{4}{7}$

4. $\dfrac{10}{11} \cdot \left(-\dfrac{5}{8}\right)$

5. Solve: $4(2x - 3) + 7 = 3x + 5$

6. Solve: $\dfrac{a - 1}{2} + a = 2 - \dfrac{2a + 7}{8}$

7. The circle graph below shows the purpose of trips made by American travelers. Use this graph to answer the questions below.

Purpose of Trip

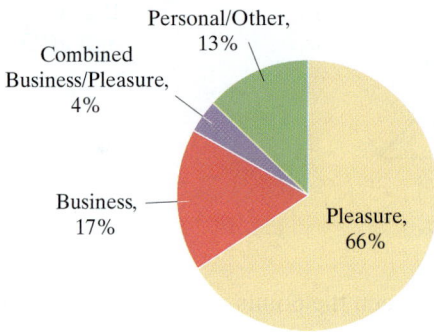

Source: Travel Industry Association of America

   a. What percent of trips made by American travelers are solely for the purpose of business?

   b. What percent of trips made by American travelers are for the purpose of business or combined business/pleasure?

   c. On an airplane flight of 253 Americans, how many of these people might we expect to be traveling solely for business?

8. Simplify each expression.
   a. $2(x - 3) + (5x + 3)$
   b. $4(3x + 2) - 3(5x - 1)$
   c. $7x + 2(x - 7) - 3x$

9. Graph: $y = -3$

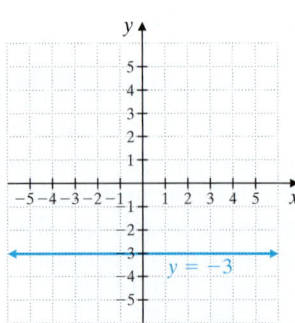

10. Find the slope of $y = -3$.

11. Graph: $5x + 4y \le 20$

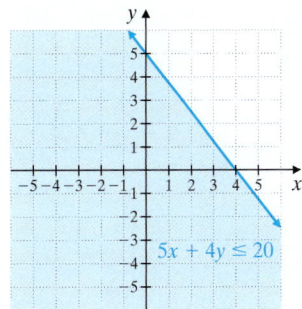

$5x + 4y \le 20$

12. Multiply: $(y - 2)(3y + 4)$

13. Write each number in standard form, without exponents.
  a. 1.02×10^5
  b. 7.358×10^{-3}
  c. 8.4×10^7
  d. 3.007×10^{-5}

14. Use scientific notation to simplify and write the answer in scientific notation.
$$\frac{0.0000035 \times 4000}{0.28}$$

15. Multiply.
$$(3x + 2)(2x - 5)$$

16. Multiply.
$$(3y - 1)(2y^2 + 3y - 1)$$

17. Factor $xy + 2x + 3y + 6$ by grouping.

18. Factor $x^3 - x^2 + 4x - 4$ by grouping.

19. Factor: $3x^2 + 11x + 6$

20. Factor: $2x^3 + 16$

21. Are there any values for x for which each expression is undefined?
  a. $\dfrac{x}{x - 3}$
  b. $\dfrac{x^2 + 2}{x^2 - 3x + 2}$
  c. $\dfrac{x^3 - 6x^2 - 10x}{3}$

22. Factor: $x^4 - 1$

23. Simplify: $\dfrac{x^2 + 4x + 4}{x^2 + 2x}$

24. Simplify: $\dfrac{a^3 - 8}{2 - a}$

---

10. 0 (Sec. 3.4)

11. see graph   (Sec. 3.6, Ex. 5)

12. $3y^2 - 2y - 8$ (Sec. 5.5)

13. a. 102,000

  b. 0.007358

  c. 84,000,000

  d. 0.00003007 (Sec. 5.2, Ex. 18)

14. 5×10^{-2} (Sec. 5.2)

15. $6x^2 - 11x - 10$ (Sec. 5.5, Ex. 7b)

16. $6y^3 + 7y^2 - 6y + 1$ (Sec. 5.5)

17. $(y + 2)(x + 3)$ (Sec. 6.1, Ex. 11)

18. $(x - 1)(x^2 + 4)$ (Sec. 6.1)

19. $(3x + 2)(x + 3)$ (Sec. 6.3, Ex. 1)

20. $2(x + 2)(x^2 - 2x + 4)$ (Sec. 6.5)

21. a. $x = 3$

  b. $x = 2, x = 1$

  c. none (Sec. 7.1, Ex. 2)

22. $(x^2 + 1)(x + 1)(x - 1)$ (Sec. 6.5)

23. $\dfrac{x + 2}{x}$ (Sec. 7.1, Ex. 5)

24. $-a^2 - 2a - 4$ (Sec. 7.1)

Perform each indicated operation.

25. a. $\dfrac{a}{4} - \dfrac{2a}{8}$

    b. $\dfrac{3}{10x^2} + \dfrac{7}{25x}$

26. a. $\dfrac{3}{xy^2} - \dfrac{2}{3x^2y}$

    b. $\dfrac{5x}{x + 3} - \dfrac{2x}{x - 3}$

    c. $\dfrac{x}{x - 2} - \dfrac{5}{2 - x}$

Solve.

27. $\dfrac{4x}{x^2 + x - 30} + \dfrac{2}{x - 5} = \dfrac{1}{x + 6}$

28. $\dfrac{28}{9 - a^2} = \dfrac{2a}{a - 3} + \dfrac{6}{a + 3}$

29. Write an equation of the line with slope -3 and containing the point $(1, -5)$. Write the equation in slope-intercept form, $y = mx + b$.

Write each expression with a positive exponent, and then simplify.

30. $81^{-3/4}$

31. $(-27)^{-2/3}$

32. $16^{-3/4}$

33. $(-125)^{-2/3}$

34. Determine whether the relation $y = 2x + 1$ is also a function.

35. Solve the system: $\begin{cases} 3x + 4y = 13 \\ 5x - 9y = 6 \end{cases}$

36. Use the substitution method to solve the system.

$$\begin{cases} \dfrac{x}{6} - \dfrac{y}{2} = 1 \\ \dfrac{x}{3} - \dfrac{y}{4} = 2 \end{cases}$$

37. Solve the system:

$$\begin{cases} 2x - 4y + 8z = 2 \\ -x - 3y + z = 11 \\ x - 2y + 4z = 0 \end{cases}$$

38. Divide $x^3 - 2x^2 + 3x - 6$ by $x - 2$.

Use matrices to solve.

39. $\begin{cases} 2x - y = 3 \\ 4x - 2y = 5 \end{cases}$

40. $\begin{cases} x - 3y = -21 \\ 3x + 2y = -8 \end{cases}$

Quadratic Equations and Functions

What Are Supertall and Megatall Buildings?

The term *supertall* refers to a building over 300 meters. According to the Council on Tall Buildings and Urban Habitat (CTBUH), this term is no longer adequate to rank the tallest buildings in the world. By 2020, it is predicted that eight buildings will have a height over 600 meters tall. We now use the term *megatall* to describe these buildings over 600 meters. By 2020, it is predicted that the tallest building in the world, the Kingdom Tower, will have a height of 1000 meters (about 3280 feet).

In Section 11.1, Exercises 75 through 78, we will explore the heights of some tall buildings. (*Source:* Council on Tall Buildings and Urban Habitat)

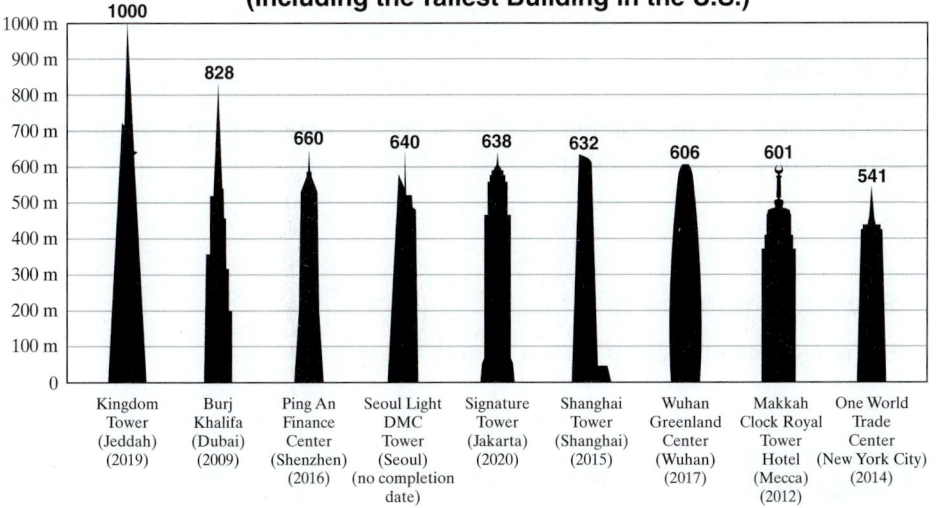

World's Megatall Buildings in 2020 (Projected)
(Including the Tallest Building in the U.S.)

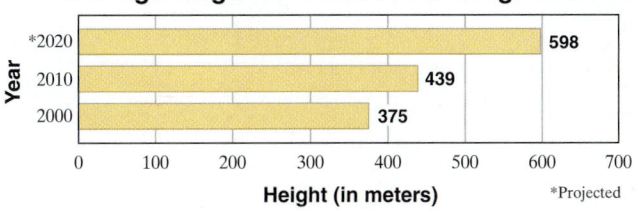

Average Height of Tallest 20 Buildings in World

An important part of algebra is learning to model and solve problems. Often, the model of a problem is a quadratic equation or a function containing a second-degree polynomial. In this chapter, we continue the work from Chapter 6, solving polynomial equations in one variable by factoring. Two other methods of solving quadratic equations are analyzed in this chapter, along with methods of solving nonlinear inequalities in one variable and the graphs of quadratic functions.

Objectives

A Use the Square Root Property to Solve Quadratic Equations.

B Write Perfect Square Trinomials.

C Solve Quadratic Equations by Completing the Square.

D Use Quadratic Equations to Solve Problems.

11.1 Solving Quadratic Equations by Completing the Square

Objective A Using the Square Root Property

In Chapter 6, we solved quadratic equations by factoring. Recall that a **quadratic,** or **second-degree, equation** is an equation that can be written in the form $ax^2 + bx + c = 0$, where a, b, and c are real numbers and a is not 0. To solve a quadratic equation such as $x^2 = 9$ by factoring, we use the zero-factor property. To use the zero-factor property, the equation must first be written in the standard form $ax^2 + bx + c = 0$.

$$x^2 = 9$$
$$x^2 - 9 = 0 \qquad \text{Subtract 9 from both sides to write in standard form.}$$
$$(x + 3)(x - 3) = 0 \qquad \text{Factor.}$$
$$x + 3 = 0 \quad \text{or} \quad x - 3 = 0 \qquad \text{Set each factor equal to 0.}$$
$$x = -3 \qquad\qquad x = 3 \qquad \text{Solve.}$$

The solution set is $\{-3, 3\}$, the positive and negative square roots of 9.

Not all quadratic equations can be solved by factoring, so we need to explore other methods. Notice that the solutions of the equation $x^2 = 9$ are two numbers whose square is 9:

$$3^2 = 9 \qquad \text{and} \qquad (-3)^2 = 9$$

Thus, we can solve the equation $x^2 = 9$ by taking the square root of both sides. Be sure to include both $\sqrt{9}$ and $-\sqrt{9}$ as solutions since both $\sqrt{9}$ and $-\sqrt{9}$ are numbers whose square is 9.

$$x^2 = 9$$
$$x = \pm\sqrt{9} \qquad \text{The notation } \pm\sqrt{9} \text{ (read as "plus or minus } \sqrt{9}\text{") indicates the pair of}$$
$$x = \pm 3 \qquad\qquad \text{numbers } +\sqrt{9} \text{ and } -\sqrt{9}.$$

This illustrates the square root property.

Helpful Hint

The notation ± 3, for example, is read as "plus or minus 3." It is a shorthand notation for the pair of numbers $+3$ and -3.

Square Root Property

If b is a real number and if $a^2 = b$, then $a = \pm\sqrt{b}$.

Teaching Tip

Point out that $\sqrt{9}$ evaluates to positive 3 only, but when solving a problem such as $x^2 = 9$, solutions are the positive and negative square roots of 9, or $\pm\sqrt{9} = \pm 3$.

Example 1 Use the square root property to solve $x^2 = 50$.

Solution:

$$x^2 = 50$$
$$x = \pm\sqrt{50} \qquad \text{Use the square root property.}$$
$$x = \pm 5\sqrt{2} \qquad \text{Simplify the radical.}$$

Practice 1

Use the square root property to solve $x^2 = 45$.

Answer

1. $\{3\sqrt{5}, -3\sqrt{5}\}$

Check: Let $x = 5\sqrt{2}$.   Let $x = -5\sqrt{2}$.
$$x^2 = 50 \qquad\qquad\qquad x^2 = 50$$
$$(5\sqrt{2})^2 \overset{?}{=} 50 \qquad\qquad (-5\sqrt{2})^2 \overset{?}{=} 50$$
$$25 \cdot 2 \overset{?}{=} 50 \qquad\qquad\quad 25 \cdot 2 \overset{?}{=} 50$$
$$50 = 50 \quad \text{True} \qquad\qquad 50 = 50 \quad \text{True}$$

The solution set is $\{5\sqrt{2}, -5\sqrt{2}\}$.

■ **Work Practice 1**

Example 2   Use the square root property to solve $2x^2 = 14$.

Solution: First we get the squared variable alone on one side of the equation.

$$2x^2 = 14$$
$$x^2 = 7 \qquad \text{Divide both sides by 2.}$$
$$x = \pm\sqrt{7} \qquad \text{Use the square root property.}$$

Check: Let $x = \sqrt{7}$.   Let $x = -\sqrt{7}$.
$$2x^2 = 14 \qquad\qquad\qquad 2x^2 = 14$$
$$2(\sqrt{7})^2 \overset{?}{=} 14 \qquad\qquad 2(-\sqrt{7})^2 \overset{?}{=} 14$$
$$2 \cdot 7 \overset{?}{=} 14 \qquad\qquad\quad 2 \cdot 7 \overset{?}{=} 14$$
$$14 = 14 \quad \text{True} \qquad\qquad 14 = 14 \quad \text{True}$$

The solution set is $\{\sqrt{7}, -\sqrt{7}\}$.

■ **Work Practice 2**

Example 3   Use the square root property to solve $(x + 1)^2 = 12$.

Solution: $(x + 1)^2 = 12$
$$x + 1 = \pm\sqrt{12} \qquad \text{Use the square root property.}$$
$$x + 1 = \pm 2\sqrt{3} \qquad \text{Simplify the radical.}$$
$$x = -1 \pm 2\sqrt{3} \qquad \text{Subtract 1 from both sides.}$$

Check: Below is a check for $-1 + 2\sqrt{3}$. The check for $-1 - 2\sqrt{3}$ is almost the same and is left for you to do on your own.

$$(x + 1)^2 = 12$$
$$(-1 + 2\sqrt{3} + 1)^2 \overset{?}{=} 12$$
$$(2\sqrt{3})^2 \overset{?}{=} 12$$
$$4 \cdot 3 \overset{?}{=} 12$$
$$12 = 12 \quad \text{True}$$

The solution set is $\{-1 + 2\sqrt{3}, -1 - 2\sqrt{3}\}$.

■ **Work Practice 3**

Practice 2

Use the square root property to solve $5x^2 = 55$.

Teaching Tip

For Examples 1 and 2, have students use the solution to write the factored form of the original equation.

Practice 3

Use the square root property to solve $(x + 2)^2 = 18$.

Helpful Hint   Don't forget that $-1 \pm 2\sqrt{3}$, for example, means $-1 + 2\sqrt{3}$ and $-1 - 2\sqrt{3}$. In other words, the equation in Example 3 has two solutions.

Answers
2. $\{\sqrt{11}, -\sqrt{11}\}$
3. $\{-2 + 3\sqrt{2}, -2 - 3\sqrt{2}\}$

Practice 4

Use the square root property to solve $(3x - 1)^2 = -4$.

Teaching Tip

Before beginning Objective B, have students work in pairs to determine the relationship between coefficients b and c (in $ax^2 + bx + c = 0$) for equations of the form $(x + n)^2 = 0$, where n is an integer. Have them share their conclusions with the class.

Example 4   Use the square root property to solve $(2x - 5)^2 = -16$.

Solution:
$$(2x - 5)^2 = -16$$
$$2x - 5 = \pm\sqrt{-16} \quad \text{Use the square root property.}$$
$$2x - 5 = \pm 4i \quad \text{Simplify the radical.}$$
$$2x = 5 \pm 4i \quad \text{Add 5 to both sides.}$$
$$x = \frac{5 \pm 4i}{2} \quad \text{Divide both sides by 2.}$$

Check each proposed solution in the original equation to see that the solution set is $\left\{ \dfrac{5 + 4i}{2}, \dfrac{5 - 4i}{2} \right\}$.

■ **Work Practice 4**

✓**Concept Check** How do you know just by looking that $(x - 2)^2 = -81$ has complex solutions?

Objective B Writing Perfect Square Trinomials

Notice from Examples 3 and 4 that, if we write a quadratic equation so that one side is the square of a binomial, we can solve by using the square root property. To write the square of a binomial, we must have a perfect square trinomial. Recall that a perfect square trinomial is a trinomial that can be factored into two identical binomial factors, that is, as a binomial squared.

Perfect Square Trinomials	**Factored Form**
$x^2 + 8x + 16$	$(x + 4)^2$
$x^2 - 6x + 9$	$(x - 3)^2$
$x^2 + 3x + \dfrac{9}{4}$	$\left(x + \dfrac{3}{2}\right)^2$

Notice that for each perfect square trinomial, **the constant term of the trinomial is the square of half the coefficient of the x-term.** For example,

$$x^2 + 8x + 16 \qquad\qquad x^2 - 6x + 9$$

$$\frac{1}{2}(8) = 4 \text{ and } 4^2 = 16 \qquad \frac{1}{2}(-6) = -3 \text{ and } (-3)^2 = 9$$

Practice 5

Add the proper constant to $x^2 + 12x$ so that the result is a perfect square trinomial. Then factor.

Answers

4. $\left\{ \dfrac{1 - 2i}{3}, \dfrac{1 + 2i}{3} \right\}$

5. $x^2 + 12x + 36 = (x + 6)^2$

✓**Concept Check Answer**

answers may vary

Example 5   Add the proper constant to $x^2 + 6x$ so that the result is a perfect square trinomial. Then factor.

Solution: We add the square of half the coefficient of x.

$$x^2 + 6x + 9 \quad = \quad (x + 3)^2 \quad \text{In factored form}$$

$$\frac{1}{2}(6) = 3 \text{ and } 3^2 = 9$$

■ **Work Practice 5**

Example 6 Add the proper constant to $x^2 - 3x$ so that the result is a perfect square trinomial. Then factor.

Solution: We add the square of half the coefficient of x.

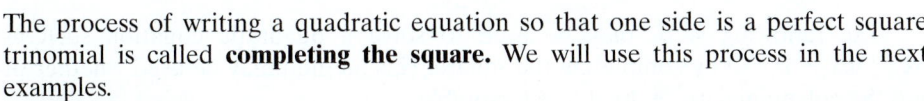

$$x^2 - 3x + \frac{9}{4} = \left(x - \frac{3}{2}\right)^2 \quad \text{In factored form}$$

$$\frac{1}{2}(-3) = -\frac{3}{2} \text{ and } \left(-\frac{3}{2}\right)^2 = \frac{9}{4}$$

■ **Work Practice 6**

Objective C Solving by Completing the Square

The process of writing a quadratic equation so that one side is a perfect square trinomial is called **completing the square.** We will use this process in the next examples.

Example 7 Solve $p^2 + 2p = 4$ by completing the square.

Solution: First we add the square of half the coefficient of p to both sides so that the resulting trinomial will be a perfect square trinomial. The coefficient of p is 2.

$$\frac{1}{2}(2) = 1 \quad \text{and} \quad 1^2 = 1$$

Now we add 1 to both sides of the original equation.

$$p^2 + 2p = 4$$
$$p^2 + 2p + 1 = 4 + 1 \quad \text{Add 1 to both sides.}$$
$$(p + 1)^2 = 5 \quad \text{Factor the trinomial; simplify the right side.}$$

We may now use the square root property and solve for p.

$$p + 1 = \pm\sqrt{5} \quad \text{Use the square root property.}$$
$$p = -1 \pm \sqrt{5} \quad \text{Subtract 1 from both sides.}$$

Don't forget that there are two solutions: $-1 + \sqrt{5}$ and $-1 - \sqrt{5}$. The solution set is $\{-1 + \sqrt{5}, -1 - \sqrt{5}\}$.

■ **Work Practice 7**

Example 8 Solve $m^2 - 7m - 1 = 0$ by completing the square.

Solution: First we add 1 to both sides of the equation so that the left side has no constant term. We can then add the constant term on both sides that will make the left side a perfect square trinomial.

$$m^2 - 7m - 1 = 0$$
$$m^2 - 7m = 1$$

Now we find the constant term that makes the left side a perfect square trinomial by squaring half the coefficient of m. We add this constant to both sides of the equation.

$$\frac{1}{2}(-7) = -\frac{7}{2} \quad \text{and} \quad \left(-\frac{7}{2}\right)^2 = \frac{49}{4}$$

(Continued on next page)

Practice 6

Add the proper constant to $y^2 - 5y$ so that the result is a perfect square trinomial. Then factor.

Practice 7

Solve $x^2 + 8x = 1$ by completing the square.

Practice 8

Solve $y^2 - 5y + 2 = 0$ by completing the square.

Answers

6. $y^2 - 5y + \dfrac{25}{4} = \left(y - \dfrac{5}{2}\right)^2$

7. $\{-4 - \sqrt{17}, -4 + \sqrt{17}\}$

8. $\left\{\dfrac{5 - \sqrt{17}}{2}, \dfrac{5 + \sqrt{17}}{2}\right\}$

$$m^2 - 7m + \frac{49}{4} = 1 + \frac{49}{4}$$ Add $\frac{49}{4}$ to both sides of the equation.

$$\left(m - \frac{7}{2}\right)^2 = \frac{53}{4}$$ Factor the perfect square trinomial and simplify the right side.

$$m - \frac{7}{2} = \pm\sqrt{\frac{53}{4}}$$ Use the square root property.

$$m = \frac{7}{2} \pm \frac{\sqrt{53}}{2}$$ Add $\frac{7}{2}$ to both sides and simplify $\sqrt{\frac{53}{4}}$.

$$m = \frac{7 \pm \sqrt{53}}{2}$$ Simplify.

The solution set is $\left\{\dfrac{7 + \sqrt{53}}{2}, \dfrac{7 - \sqrt{53}}{2}\right\}$.

■ **Work Practice 8**

The following steps may be used to solve a quadratic equation such as $ax^2 + bx + c = 0$ by completing the square. This method may be used whether or not the polynomial $ax^2 + bx + c$ is factorable.

Solving a Quadratic Equation in x by Completing the Square

Step 1: If the coefficient of x^2 is 1, go to Step 2. Otherwise, divide both sides of the equation by the coefficient of x^2.

Step 2: Get all variable terms alone on one side of the equation.

Step 3: Complete the square for the resulting binomial by adding the square of half of the coefficient of x to both sides of the equation.

Step 4: Factor the resulting perfect square trinomial and write it as the square of a binomial.

Step 5: Use the square root property to solve for x.

Practice 9

Solve $2x^2 - 2x + 7 = 0$ by completing the square.

Example 9 Solve $4x^2 - 24x + 41 = 0$ by completing the square.

Solution: First we divide both sides of the equation by 4 so that the coefficient of x^2 is 1.

$$4x^2 - 24x + 41 = 0$$

Step 1: $x^2 - 6x + \dfrac{41}{4} = 0$ Divide both sides of the equation by 4.

Step 2: $x^2 - 6x = -\dfrac{41}{4}$ Subtract $\dfrac{41}{4}$ from both sides.

Since $\dfrac{1}{2}(-6) = -3$ and $(-3)^2 = 9$, we add 9 to both sides of the equation.

Step 3: $x^2 - 6x + 9 = -\dfrac{41}{4} + 9$ Add 9 to both sides.

Step 4: $(x - 3)^2 = -\dfrac{41}{4} + \dfrac{36}{4}$ Factor the perfect square trinomial.

$$(x - 3)^2 = -\dfrac{5}{4}$$

Answer

9. $\left\{\dfrac{1 + i\sqrt{13}}{2}, \dfrac{1 - i\sqrt{13}}{2}\right\}$

Step 5: $x - 3 = \pm\sqrt{-\dfrac{5}{4}}$ Use the square root property.

$\qquad x - 3 = \pm\dfrac{i\sqrt{5}}{2}$ Simplify the radical.

$\qquad\qquad x = 3 \pm \dfrac{i\sqrt{5}}{2}$ Add 3 to both sides.

$\qquad\qquad\quad = \dfrac{6}{2} \pm \dfrac{i\sqrt{5}}{2}$ Find a common denominator.

$\qquad\qquad\quad = \dfrac{6 \pm i\sqrt{5}}{2}$ Simplify.

The solution set is $\left\{\dfrac{6 + i\sqrt{5}}{2}, \dfrac{6 - i\sqrt{5}}{2}\right\}$.

🔲 **Work Practice 9**

Objective D Solving Problems Modeled by Quadratic Equations

Recall the **simple interest** formula $I = Prt$, where I is the interest earned, P is the principal, r is the rate of interest, and t is time. If \$100 is invested at a simple interest rate of 5% annually, at the end of 3 years the total interest I earned is

$\qquad I = P \cdot r \cdot t$

or

$\qquad I = 100 \cdot 0.05 \cdot 3 = \$15$

and the new principal is

$\qquad \$100 + \$15 = \$115$

Most of the time, the interest computed on money borrowed or money deposited is **compound interest.** Unlike simple interest, compound interest is computed on original principal *and* on interest already earned. To see the difference between simple interest and compound interest, suppose that \$100 is invested at a rate of 5% compounded annually. To find the total amount of money at the end of 3 years, we calculate as follows:

$\qquad I = P \cdot r \cdot t$

First year: Interest = $\$100 \cdot 0.05 \cdot 1 = \$5.00$
     New principal = $\$100.00 + \$5.00 = \$105.00$

Second year: Interest = $\$105.00 \cdot 0.05 \cdot 1 = \$5.25$
     New principal = $105.00 + \$5.25 = \$110.25$

Third year: Interest = $\$110.25 \cdot 0.05 \cdot 1 \approx \$5.51$
     New principal = $\$110.25 + \$5.51 = \$115.76$

At the end of the third year, the total compound interest earned is \$15.76, whereas the total simple interest earned is \$15.

It is tedious to calculate compound interest as we did above, so we use a compound interest formula. The formula for calculating the total amount of money when interest is compounded annually is

$\qquad A = P(1 + r)^t$

where P is the original investment, r is the interest rate per compounding period, and t is the number of periods. For example, the amount of money A at the end of 3 years if \$100 is invested at 5% compounded annually is

$\qquad A = \$100(1 + 0.05)^3 \approx 100(1.1576) = \$115.76$

as we previously calculated.

Practice 10

Use the formula from Example 10 to find the interest rate r if $1600 compounded annually grows to $1764 in 2 years.

Finding an Interest Rate

Find the interest rate r if $2000 compounded annually grows to $2420 in 2 years.

Solution:

1. UNDERSTAND the problem. For this example, make sure that you understand the formula for compounding interest annually.
2. TRANSLATE. We substitute the given values into the formula.

$$A = P(1 + r)^t$$
$$2420 = 2000(1 + r)^2 \qquad \text{Let } A = 2420, P = 2000, \text{ and } t = 2.$$

3. SOLVE. We now solve the equation for r.

$$2420 = 2000(1 + r)^2$$

$$\frac{2420}{2000} = (1 + r)^2 \qquad \text{Divide both sides by 2000.}$$

$$\frac{121}{100} = (1 + r)^2 \qquad \text{Simplify the fraction.}$$

$$\pm\sqrt{\frac{121}{100}} = 1 + r \qquad \text{Use the square root property.}$$

$$\pm\frac{11}{10} = 1 + r \qquad \text{Simplify.}$$

$$-1 \pm \frac{11}{10} = r$$

$$-\frac{10}{10} \pm \frac{11}{10} = r$$

$$\frac{1}{10} = r \quad \text{or} \quad -\frac{21}{10} = r$$

4. INTERPRET. The rate cannot be negative, so we reject $-\dfrac{21}{10}$.

Check: $\dfrac{1}{10} = 0.10 = 10\%$ per year. If we invest $2000 at 10% compounded annually, in 2 years the amount in the account would be $2000(1 + 0.10)^2 = 2420$ dollars, the desired amount.

State: The interest rate is 10% compounded annually.

Answer

10. 5%

■ **Work Practice 10**

Vocabulary, Readiness & Video Check

Use the choices below to fill in each blank. Not all choices will be used.

binomial	$\sqrt{b}$	$\pm\sqrt{b}$	b^2	9	25	completing the square
quadratic	$-\sqrt{b}$	$\dfrac{b}{2}$	$\left(\dfrac{b}{2}\right)^2$	3	5	

1. By the square root property, if b is a real number, and $a^2 = b$, then $a = \underline{\quad \pm\sqrt{b} \quad}$.

2. A $\underline{\quad \text{quadratic} \quad}$ equation can be written in the form $ax^2 + bx + c = 0, a \neq 0$.

3. The process of writing a quadratic equation so that one side is a perfect square trinomial is called <u>completing the square</u>.

4. A perfect square trinomial is one that can be factored as a _____binomial_____ squared.

5. To solve $x^2 + 6x = 10$ by completing the square, add _____9_____ to both sides.

6. To solve $x^2 + bx = c$ by completing the square, add _____$\left(\frac{b}{2}\right)^2$_____ to both sides.

Martin-Gay Interactive Videos   *Watch the section lecture video and answer the following questions.*

See Video 11.1 🍎

See video answer section.

Objective A **7.** From Examples 2 and 3, explain a step you can perform so that you may easily apply the square root property to $2x^2 = 16$. Explain why you perform this step. ▶

Objective B **8.** Use Example 5 to explain why $r^2 - r + 1$ isn't a perfect square trinomial. ▶

Objective C **9.** In Example 7, why is the equation first divided through by 3? ▶

Objective D **10.** In Example 8, why is the negative solution not considered? ▶

Teaching Tip

If you prefer to assign exercises with real number solutions only, omit Exercises 15, 16, 19–22, 25, 26, 57–64, 67, and 68.

11.1 **Exercise Set** MyMathLab® ▶

Objective A *Use the square root property to solve each equation. See Examples 1 through 4.*

▶**1.** $x^2 = 16$
$\{-4, 4\}$

2. $x^2 = 49$
$\{-7, 7\}$

3. $x^2 - 7 = 0$
$\{-\sqrt{7}, \sqrt{7}\}$

4. $x^2 - 11 = 0$
$\{-\sqrt{11}, \sqrt{11}\}$

5. $x^2 = 18$
$\{-3\sqrt{2}, 3\sqrt{2}\}$

6. $y^2 = 20$
$\{-2\sqrt{5}, 2\sqrt{5}\}$

7. $3z^2 - 30 = 0$
$\{-\sqrt{10}, \sqrt{10}\}$

8. $2x^2 = 4$
$\{-\sqrt{2}, \sqrt{2}\}$

9. $(x + 5)^2 = 9$
$\{-8, -2\}$

10. $(y - 3)^2 = 4$
$\{1, 5\}$

▶**11.** $(z - 6)^2 = 18$
$\{6 - 3\sqrt{2}, 6 + 3\sqrt{2}\}$

12. $(y + 4)^2 = 27$
$\{-4 - 3\sqrt{3}, -4 + 3\sqrt{3}\}$

13. $(2x - 3)^2 = 8$
$\left\{\dfrac{3 - 2\sqrt{2}}{2}, \dfrac{3 + 2\sqrt{2}}{2}\right\}$

14. $(4x + 9)^2 = 6$
$\left\{\dfrac{-9 - \sqrt{6}}{4}, \dfrac{-9 + \sqrt{6}}{4}\right\}$

15. $x^2 + 9 = 0$
$\{-3i, 3i\}$

16. $x^2 + 4 = 0$
$\{-2i, 2i\}$

▶**17.** $x^2 - 6 = 0$
$\{-\sqrt{6}, \sqrt{6}\}$

18. $y^2 - 10 = 0$
$\{-\sqrt{10}, \sqrt{10}\}$

19. $2z^2 + 16 = 0$
$\{-2i\sqrt{2}, 2i\sqrt{2}\}$

20. $3p^2 + 36 = 0$
$\{-2i\sqrt{3}, 2i\sqrt{3}\}$

21. $(3x - 1)^2 = -16$
$\left\{\dfrac{1 - 4i}{3}, \dfrac{1 + 4i}{3}\right\}$

22. $(4y + 2)^2 = -25$
$\left\{\dfrac{-2 - 5i}{4}, \dfrac{-2 + 5i}{4}\right\}$

23. $(z + 7)^2 = 5$
$\{-7 - \sqrt{5}, -7 + \sqrt{5}\}$

24. $(x + 10)^2 = 11$
$\{-10 - \sqrt{11}, -10 + \sqrt{11}\}$

25. $(x + 3)^2 + 8 = 0$
$\{-3 - 2i\sqrt{2}, -3 + 2i\sqrt{2}\}$

26. $(y - 4)^2 + 18 = 0$
$\{4 - 3i\sqrt{2}, 4 + 3i\sqrt{2}\}$

Objective B

Fill in the blank with the number needed to make the expression a perfect square trinomial. See Examples 5 and 6.

27. $m^2 + 2m + \underline{1}$ **28.** $m^2 - 2m + \underline{1}$ **29.** $y^2 - 14y + \underline{49}$ **30.** $z^2 + z + \underline{\frac{1}{4}}$

Add the proper constant to each binomial so that the resulting trinomial is a perfect square trinomial. Then factor the trinomial. See Examples 5 and 6.

▶ 31. $x^2 + 16x$
$x^2 + 16x + 64 = (x + 8)^2$

32. $y^2 + 2y$
$y^2 + 2y + 1 = (y + 1)^2$

33. $z^2 - 12z$
$z^2 - 12z + 36 = (z - 6)^2$

34. $x^2 - 8x$
$x^2 - 8x + 16 = (x - 4)^2$

35. $p^2 + 9p$
$p^2 + 9p + \dfrac{81}{4} = \left(p + \dfrac{9}{2}\right)^2$

36. $n^2 + 5n$
$n^2 + 5n + \dfrac{25}{4} = \left(n + \dfrac{5}{2}\right)^2$

▶ 37. $r^2 - r$
$r^2 - r + \dfrac{1}{4} = \left(r - \dfrac{1}{2}\right)^2$

38. $p^2 - 7p$
$p^2 - 7p + \dfrac{49}{4} = \left(p - \dfrac{7}{2}\right)^2$

Objective C

Solve each equation by completing the square. See Examples 7 through 9.

39. $x^2 + 8x = -15$
$\{-5, -3\}$

40. $y^2 + 6y = -8$
$\{-4, -2\}$

▶ 41. $x^2 + 6x + 2 = 0$
$\{-3 - \sqrt{7}, -3 + \sqrt{7}\}$

42. $x^2 - 2x - 2 = 0$
$\{1 - \sqrt{3}, 1 + \sqrt{3}\}$

43. $x^2 + x - 1 = 0$
$\left\{\dfrac{-1 - \sqrt{5}}{2}, \dfrac{-1 + \sqrt{5}}{2}\right\}$

44. $x^2 + 3x - 2 = 0$
$\left\{\dfrac{-3 - \sqrt{17}}{2}, \dfrac{-3 + \sqrt{17}}{2}\right\}$

45. $x^2 + 2x - 5 = 0$
$\{-1 - \sqrt{6}, -1 + \sqrt{6}\}$

46. $x^2 - 6x + 3 = 0$
$\{3 - \sqrt{6}, 3 + \sqrt{6}\}$

47. $y^2 + y - 7 = 0$
$\left\{\dfrac{-1 - \sqrt{29}}{2}, \dfrac{-1 + \sqrt{29}}{2}\right\}$

48. $x^2 - 7x - 1 = 0$
$\left\{\dfrac{7 - \sqrt{53}}{2}, \dfrac{7 + \sqrt{53}}{2}\right\}$

49. $x^2 + 8x + 1 = 0$
$\{-4 - \sqrt{15}, -4 + \sqrt{15}\}$

50. $x^2 - 10x + 2 = 0$
$\{5 - \sqrt{23}, 5 + \sqrt{23}\}$

51. $3p^2 - 12p + 2 = 0$
$\left\{\dfrac{6 - \sqrt{30}}{3}, \dfrac{6 + \sqrt{30}}{3}\right\}$

52. $2x^2 + 14x - 1 = 0$
$\left\{\dfrac{-7 - \sqrt{51}}{2}, \dfrac{-7 + \sqrt{51}}{2}\right\}$

53. $2x^2 + 7x = 4$
$\left\{-4, \dfrac{1}{2}\right\}$

54. $3x^2 - 4x = 4$
$\left\{-\dfrac{2}{3}, 2\right\}$

▶ 55. $3y^2 + 6y - 4 = 0$
$\left\{\dfrac{-3 - \sqrt{21}}{3}, \dfrac{-3 + \sqrt{21}}{3}\right\}$

56. $2y^2 + 12y + 3 = 0$
$\left\{\dfrac{-6 - \sqrt{30}}{2}, \dfrac{-6 + \sqrt{30}}{2}\right\}$

57. $y^2 + 2y + 2 = 0$
$\{-1 - i, -1 + i\}$

58. $x^2 + 4x + 6 = 0$
$\{-2 - i\sqrt{2}, -2 + i\sqrt{2}\}$

59. $2a^2 + 8a = -12$
$\{-2 - i\sqrt{2}, -2 + i\sqrt{2}\}$

60. $3x^2 + 12x = -14$
$$\left\{\frac{-6 - i\sqrt{6}}{3}, \frac{-6 + i\sqrt{6}}{3}\right\}$$

61. $2x^2 - x + 6 = 0$
$$\left\{\frac{1 - i\sqrt{47}}{4}, \frac{1 + i\sqrt{47}}{4}\right\}$$

62. $4x^2 - 2x + 5 = 0$
$$\left\{\frac{1 - i\sqrt{19}}{4}, \frac{1 + i\sqrt{19}}{4}\right\}$$

63. $x^2 + 10x + 28 = 0$
$\{-5 - i\sqrt{3}, -5 + i\sqrt{3}\}$

64. $y^2 + 8y + 18 = 0$
$\{-4 - i\sqrt{2}, -4 + i\sqrt{2}\}$

65. $z^2 + 3z - 4 = 0$
$\{-4, 1\}$

66. $y^2 + y - 2 = 0$
$\{-2, 1\}$

67. $2x^2 - 4x + 3 = 0$
$$\left\{\frac{2 - i\sqrt{2}}{2}, \frac{2 + i\sqrt{2}}{2}\right\}$$

68. $9x^2 - 36x = -40$
$$\left\{\frac{6 - 2i}{3}, \frac{6 + 2i}{3}\right\}$$

69. $3x^2 + 3x = 5$
$$\left\{\frac{-3 - \sqrt{69}}{6}, \frac{-3 + \sqrt{69}}{6}\right\}$$

70. $5y^2 - 15y = 1$
$$\left\{\frac{15 - 7\sqrt{5}}{10}, \frac{15 + 7\sqrt{5}}{10}\right\}$$

Objective D *Use the formula $A = P(1 + r)^t$ to solve Exercises 71 through 74. See Example 10.*

▶ **71.** Find the rate r at which \$3000 grows to \$4320 in 2 years.   20%

72. Find the rate r at which \$800 grows to \$882 in 2 years.   5%

73. Find the rate r at which \$810 grows to approximately \$1000 in 2 years.   11%

74. Find the rate r at which \$2000 grows to \$2880 in 2 years.   20%

Neglecting air resistance, the distance $s(t)$ in feet traveled by a freely falling object is given by the function $s(t) = 16t^2$, where t is time in seconds. For Exercises 75 through 78, use this formula to find the time it takes for an object to fall to the ground from the top of each building. Round answers to two decimal places. (Source: Council on Tall Buildings and Urban Habitat)

75. The Petronas Towers are located in Kuala Lumpur, Malaysia. Each tower is 1483 feet tall.   9.63 sec

76. The Burj Khalifa is located in Dubai, United Arab Emirates. It is estimated to be 2717 feet tall.   13.03 sec

77. The Kingdom Tower in Jeddah, Saudi Arabia, has a projected completion date of 2019. It is estimated to be 3281 feet tall.   14.32 sec

78. One World Trade Center in New York City was completed in 2014. It has a height of 1776 feet.   10.54 sec

Use the formula $s(t) = 16t^2$ to solve. See Exercises 75 through 78.

79. The Rogun Dam in Tajikistan (part of the former USSR that borders Afghanistan) is the tallest dam in the world at 1100 feet. How long would it take an object to fall from the top to the base of the dam? (*Source:* U.S. Committee on Large Dams of the International Commission on Large Dams)   8.29 sec

80. The Hoover Dam, located on the Colorado River on the border of Nevada and Arizona near Las Vegas, is 725 feet tall. How long would it take an object to fall from the top to the base of the dam? (*Source:* U.S. Committee on Large Dams of the International Commission on Large Dams)   6.73 sec

Solve.

△ **81.** The area of a square room is 225 square feet. Find the dimensions of the room.   15 ft by 15 ft

△ **82.** The area of a circle is 36π square inches. Find the radius of the circle.   6 in.

△ **83.** An isosceles right triangle has legs of equal length. If the hypotenuse is 20 centimeters long, find the length of each leg. $10\sqrt{2}$ cm

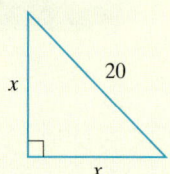

△ **84.** The top of a square coffee table has a diagonal that measures 30 inches. Find the length of each side of the top of the coffee table. $15\sqrt{2}$ in.

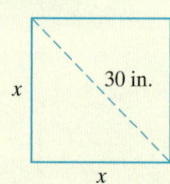

Review

Simplify each expression. See Section 7.1.

85. $\dfrac{6 + 4\sqrt{5}}{2}$

$3 + 2\sqrt{5}$

86. $\dfrac{10 - 20\sqrt{3}}{2}$

$5 - 10\sqrt{3}$

87. $\dfrac{3 - 9\sqrt{2}}{6}$

$\dfrac{1 - 3\sqrt{2}}{2}$

88. $\dfrac{12 - 8\sqrt{7}}{16}$

$\dfrac{3 - 2\sqrt{7}}{4}$

Evaluate $\sqrt{b^2 - 4ac}$ for each set of values. See Section 10.3.

89. $a = 2, b = 4, c = -1$ $2\sqrt{6}$

90. $a = 1, b = 6, c = 2$ $2\sqrt{7}$

91. $a = 3, b = -1, c = -2$ 5

92. $a = 1, b = -3, c = -1$ $\sqrt{13}$

Concept Extensions

For Exercises 93 through 98, without solving, determine whether the solutions of each equation are real numbers or complex, but not real numbers. See the Concept Check in this section.

93. $(x + 1)^2 = -1$ complex, but not real numbers

94. $(y - 5)^2 = -9$ complex, but not real numbers

95. $3z^2 = 10$ real numbers

96. $4x^2 = 17$ real numbers

97. $(2y - 5)^2 + 7 = 3$ complex, but not real numbers

98. $(3m + 2)^2 + 4 = 1$ complex, but not real numbers

99. In your own words, explain how to calculate the number that will complete the square on an expression such as $x^2 - 5x$.   answers may vary

100. In your own words, what is the difference between simple interest and compound interest?
answers may vary

101. If you are depositing money in an account that pays 4%, would you prefer the interest to be simple or compound? Explain your answer.
compound; answers may vary

102. If you are borrowing money at a rate of 10%, would you prefer the interest to be simple or compound? Explain your answer.
simple; answers may vary

Find two possible missing terms so that each is a perfect square trinomial.

103. $x^2 + \square + 16$ $-8x, 8x$

104. $y^2 + \square + 9$ $-6y, 6y$

A common equation used in business is a demand equation. It expresses the relationship between the unit price of some commodity and the quantity demanded. For Exercises 105 and 106, p represents the unit price and x represents the quantity demanded in thousands.

105. A manufacturing company has found that the demand equation for a certain type of scissors is given by the equation $p = -x^2 + 47$. Find the demand for the scissors if the price is $11 per pair.
6 thousand scissors

106. Acme, Inc., sells desk lamps and has found that the demand equation for a certain style of desk lamp is given by the equation $p = -x^2 + 15$. Find the demand for the desk lamp if the price is $7 per lamp.   2.828 thousand units

Objectives

A Solve Quadratic Equations by Using the Quadratic Formula.

B Determine the Number and Type of Solutions of a Quadratic Equation by Using the Discriminant.

C Solve Problems Modeled by Quadratic Equations.

Objective A Solving Equations by Using the Quadratic Formula

Any quadratic equation can be solved by completing the square. Since the same sequence of steps is repeated each time we complete the square, let's complete the square for a general quadratic equation, $ax^2 + bx + c = 0$. By doing so, we will find a pattern for the solutions of a quadratic equation known as the **quadratic formula.**

Recall that to complete the square for an equation such as $ax^2 + bx + c = 0$, $a \neq 0$, we first divide both sides by the coefficient of x^2.

$$ax^2 + bx + c = 0$$

$$x^2 + \frac{b}{a}x + \frac{c}{a} = 0 \qquad \text{Divide both sides by } a \text{, the coefficient of } x^2.$$

$$x^2 + \frac{b}{a}x = -\frac{c}{a} \qquad \text{Subtract the constant } \frac{c}{a} \text{ from both sides.}$$

Next we find the square of half $\frac{b}{a}$, the coefficient of x.

$$\frac{1}{2}\left(\frac{b}{a}\right) = \left(\frac{b}{2a}\right) \qquad \text{and} \qquad \left(\frac{b}{2a}\right)^2 = \frac{b^2}{4a^2}$$

Now we add this result to both sides of the equation.

$$x^2 + \frac{b}{a}x + \frac{b^2}{4a^2} = -\frac{c}{a} + \frac{b^2}{4a^2} \qquad \text{Add } \frac{b^2}{4a^2} \text{ to both sides.}$$

$$x^2 + \frac{b}{a}x + \frac{b^2}{4a^2} = \frac{-c \cdot 4a}{a \cdot 4a} + \frac{b^2}{4a^2} \qquad \text{Find a common denominator on the right side.}$$

$$x^2 + \frac{b}{a}x + \frac{b^2}{4a^2} = \frac{b^2 - 4ac}{4a^2} \qquad \text{Simplify the right side.}$$

$$\left(x + \frac{b}{2a}\right)^2 = \frac{b^2 - 4ac}{4a^2} \qquad \text{Factor the perfect square trinomial on the left side.}$$

$$x + \frac{b}{2a} = \pm\sqrt{\frac{b^2 - 4ac}{4a^2}} \qquad \text{Use the square root property.}$$

$$x + \frac{b}{2a} = \pm\frac{\sqrt{b^2 - 4ac}}{2a} \qquad \text{Simplify the radical.}$$

$$x = -\frac{b}{2a} \pm \frac{\sqrt{b^2 - 4ac}}{2a} \qquad \text{Subtract } \frac{b}{2a} \text{ from both sides.}$$

$$x = \frac{-b \pm \sqrt{b^2 - 4ac}}{2a} \qquad \text{Simplify.}$$

The resulting equation identifies the solutions of the general quadratic equation in standard form and is called the quadratic formula. It can be used to solve any equation written in standard form $ax^2 + bx + c = 0$ as long as a is not 0.

Quadratic Formula

A quadratic equation written in the form $ax^2 + bx + c = 0$, $a \neq 0$, has the solutions

$$x = \frac{-b \pm \sqrt{b^2 - 4ac}}{2a}$$

Example 1 Solve: $3x^2 + 16x + 5 = 0$

Solution: This equation is in standard form with $a = 3$, $b = 16$, and $c = 5$. We substitute these values into the quadratic formula.

$$x = \frac{-b \pm \sqrt{b^2 - 4ac}}{2a} \qquad \text{Quadratic formula}$$

$$= \frac{-16 \pm \sqrt{16^2 - 4(3)(5)}}{2(3)} \qquad \text{Let } a = 3, b = 16, \text{ and } c = 5.$$

$$= \frac{-16 \pm \sqrt{256 - 60}}{6}$$

$$= \frac{-16 \pm \sqrt{196}}{6} = \frac{-16 \pm 14}{6}$$

$$x = \frac{-16 + 14}{6} = -\frac{1}{3} \quad \text{or} \quad x = \frac{-16 - 14}{6} = -\frac{30}{6} = -5$$

The solution set is $\left\{ -\frac{1}{3}, -5 \right\}$.

■ **Work Practice 1**

Helpful Hint

To replace a, b, and c correctly in the quadratic formula, write the quadratic equation in standard form, $ax^2 + bx + c = 0$.

Example 2 Solve: $2x^2 - 4x = 3$

Solution: First we write the equation in standard form by subtracting 3 from both sides.

$$2x^2 - 4x - 3 = 0$$

Now $a = 2$, $b = -4$, and $c = -3$. We substitute these values into the quadratic formula.

$$x = \frac{-b \pm \sqrt{b^2 - 4ac}}{2a}$$

$$= \frac{-(-4) \pm \sqrt{(-4)^2 - 4(2)(-3)}}{2(2)}$$

$$= \frac{4 \pm \sqrt{16 + 24}}{4}$$

$$= \frac{4 \pm \sqrt{40}}{4}$$

$$= \frac{4 \pm 2\sqrt{10}}{4}$$

$$= \frac{2(2 \pm \sqrt{10})}{2 \cdot 2}$$

$$= \frac{2 \pm \sqrt{10}}{2}$$

The solution set is $\left\{ \dfrac{2 + \sqrt{10}}{2}, \dfrac{2 - \sqrt{10}}{2} \right\}$.

■ **Work Practice 2**

Practice 1

Solve: $2x^2 + 9x + 10 = 0$

Practice 2

Solve: $2x^2 - 6x = 1$

Teaching Tip

Have students work in pairs to solve the equation $2x^2 + 7x + 5 = 0$ using factoring, completing the square, and the quadratic formula to illustrate that the solutions are the same no matter which method is used. Have them show each step. Then have them verify that each solution is correct.

Answers

1. $\left\{ -\dfrac{5}{2}, -2 \right\}$

2. $\left\{ \dfrac{3 + \sqrt{11}}{2}, \dfrac{3 - \sqrt{11}}{2} \right\}$

✓**Concept Check** For the quadratic equation $x^2 = 7$, which substitution is correct?

a. $a = 1, b = 0,$ and $c = -7$ **b.** $a = 1, b = 0,$ and $c = 7$

c. $a = 0, b = 0,$ and $c = 7$ **d.** $a = 1, b = 1,$ and $c = -7$

Helpful Hint

To simplify the expression $\dfrac{4 \pm 2\sqrt{10}}{4}$ in Example 2, note that we factored 2 out of both terms of the numerator *before* simplifying.

$$\frac{4 \pm 2\sqrt{10}}{4} = \frac{2\,(2 \pm \sqrt{10})}{2 \cdot 2} = \frac{2 \pm \sqrt{10}}{2}$$

Practice 3

Solve: $\dfrac{1}{6}x^2 - \dfrac{1}{3}x - 1 = 0$

Example 3   Solve: $\dfrac{1}{4}m^2 - m + \dfrac{1}{2} = 0$

Solution: We could use the quadratic formula with $a = \dfrac{1}{4}, b = -1,$ and $c = \dfrac{1}{2}$. Instead, let's find a simpler, equivalent, standard-form equation whose coefficients are not fractions.

First we multiply both sides of the equation by 4 to clear the fractions.

$$4\left(\frac{1}{4}m^2 - m + \frac{1}{2}\right) = 4 \cdot 0$$

$$m^2 - 4m + 2 = 0 \qquad \text{Simplify.}$$

Now we can substitute $a = 1, b = -4,$ and $c = 2$ into the quadratic formula and simplify.

$$m = \frac{-(-4) \pm \sqrt{(-4)^2 - 4(1)(2)}}{2(1)}$$

$$= \frac{4 \pm \sqrt{16 - 8}}{2}$$

$$= \frac{4 \pm \sqrt{8}}{2} = \frac{4 \pm 2\sqrt{2}}{2} = \frac{2\,(2 \pm \sqrt{2})}{2} = 2 \pm \sqrt{2}$$

The solution set is $\{2 + \sqrt{2}, 2 - \sqrt{2}\}$.

■ **Work Practice 3**

Practice 4

Solve: $x = -4x^2 - 4$

Example 4   Solve: $p = -3p^2 - 3$

Solution: The equation in standard form is $3p^2 + p + 3 = 0$. Thus, $a = 3, b = 1,$ and $c = 3$ in the quadratic formula.

$$p = \frac{-1 \pm \sqrt{1^2 - 4(3)(3)}}{2(3)} = \frac{-1 \pm \sqrt{1 - 36}}{6}$$

$$= \frac{-1 \pm \sqrt{-35}}{6} = \frac{-1 \pm i\sqrt{35}}{6}$$

The solution set is $\left\{\dfrac{-1 + i\sqrt{35}}{6}, \dfrac{-1 - i\sqrt{35}}{6}\right\}$.

■ **Work Practice 4**

Answers

3. $\{1 + \sqrt{7}, 1 - \sqrt{7}\}$

4. $\left\{\dfrac{-1 - 3i\sqrt{7}}{8}, \dfrac{-1 + 3i\sqrt{7}}{8}\right\}$

✓**Concept Check Answer**

a

✓**Concept Check** What is the first step in solving $-3x^2 = 5x - 4$ using the quadratic formula?

Objective B Using the Discriminant

In the quadratic formula $x = \dfrac{-b \pm \sqrt{b^2 - 4ac}}{2a}$, the radicand $b^2 - 4ac$ is called the **discriminant** because when we know its value, we can **discriminate** among the possible number and type of solutions of a quadratic equation. Possible values of the discriminant and their meanings are summarized next.

Discriminant

The following table relates the discriminant $b^2 - 4ac$ of a quadratic equation of the form $ax^2 + bx + c = 0$ with the number and type of solutions of the equation.

$b^2 - 4ac$	Number and Type of Solutions
Positive	Two real solutions
Zero	One real solution
Negative	Two complex but not real solutions

Example 5 Use the discriminant to determine the number and type of solutions of $x^2 + 2x + 1 = 0$.

Solution: In $x^2 + 2x + 1 = 0$, $a = 1$, $b = 2$, and $c = 1$. Thus,

$$b^2 - 4ac = 2^2 - 4(1)(1) = 0$$

Since $b^2 - 4ac = 0$, this quadratic equation has one real solution.

■ **Work Practice 5**

Example 6 Use the discriminant to determine the number and type of solutions of $3x^2 + 2 = 0$.

Solution: In this equation, $a = 3$, $b = 0$, and $c = 2$. Then

$$b^2 - 4ac = 0^2 - 4(3)(2) = -24$$

Since $b^2 - 4ac$ is negative, this quadratic equation has two complex but not real solutions.

■ **Work Practice 6**

Example 7 Use the discriminant to determine the number and type of solutions of $2x^2 - 7x - 4 = 0$.

Solution: In this equation, $a = 2$, $b = -7$, and $c = -4$. Then

$$b^2 - 4ac = (-7)^2 - 4(2)(-4) = 81$$

Since $b^2 - 4ac$ is positive, this quadratic equation has two real solutions.

■ **Work Practice 7**

Practice 5

Use the discriminant to determine the number and type of solutions of $x^2 + 4x + 4 = 0$.

Practice 6

Use the discriminant to determine the number and type of solutions of $5x^2 + 7 = 0$.

Practice 7

Use the discriminant to determine the number and type of solutions of $3x^2 - 2x - 2 = 0$.

Answers

5. one real solution
6. two complex but not real solutions
7. two real solutions

✓**Concept Check Answer**
Write the equation in standard form.

Teaching Tip:
Classroom Activity

After discussing Example 7, have students work in pairs to write an equation with two real solutions, an equation with one real solution, and an equation with two complex solutions. Then have them give their equations to another pair to solve.

The discriminant helps us determine the number and type of solutions of a quadratic equation, $ax^2 + bx + c = 0$. To see this, notice that the solutions of this equation are the same as the x-intercepts of its related graph $f(x) = ax^2 + bx + c$. This means that the discriminant of $ax^2 + bx + c = 0$ also tells us the number of x-intercepts for the graph of $f(x) = ax^2 + bx + c$, or, equivalently, $y = ax^2 + bx + c$.

Graph of $f(x) = ax^2 + bx + c$ or $y = ax^2 + bx + c$

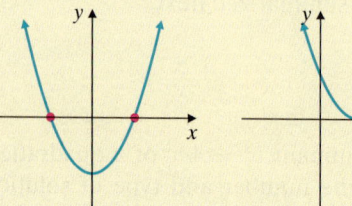

$b^2 - 4ac > 0$,
$f(x)$ has two x-intercepts

$b^2 - 4ac = 0$,
$f(x)$ has one x-intercept

$b^2 - 4ac < 0$,
$f(x)$ has no x-intercepts

Objective C Solving Problems Modeled by Quadratic Equations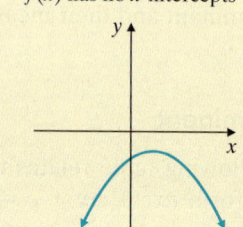

The quadratic formula is useful in solving problems that are modeled by quadratic equations.

Practice 8

Given the diagram below, approximate to the nearest foot how many feet of walking distance a person saves by cutting across the lawn instead of walking on the sidewalk.

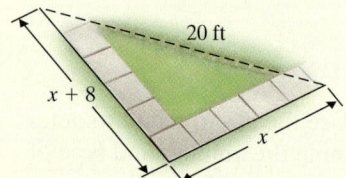

20 ft

$x + 8$

x

Example 8   Calculating Distance Saved

At a local university, students often leave the sidewalk and cut across the lawn to save walking distance. Given the diagram below of a favorite place to cut across the lawn, approximate to the nearest foot how many feet of walking distance a student saves by cutting across the lawn instead of walking on the sidewalk.

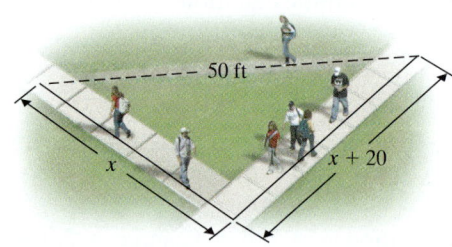

50 ft

x

$x + 20$

Solution:

1. UNDERSTAND. Read and reread the problem. You may want to review the Pythagorean theorem.
2. TRANSLATE. By the Pythagorean theorem, we have

   In words: $(\text{leg})^2 + (\text{leg})^2 = (\text{hypotenuse})^2$

   Translate: $x^2 + (x + 20)^2 = 50^2$

3. SOLVE. Use the quadratic formula to solve.

   $x^2 + x^2 + 40x + 400 = 2500$ Square $(x + 20)$ and 50.

   $2x^2 + 40x - 2100 = 0$ Write the equation in standard form.

We can use the quadratic formula right now with $a = 2, b = 40,$ and $c = -2100$. Instead, just as in Example 3, you may want to find a simpler, equivalent equation by dividing both sides of the equation by 2.

   $x^2 + 20x - 1050 = 0$ Divide by 2.

Answer

8. 8 ft

Here, $a = 1$, $b = 20$, and $c = -1050$. By the quadratic formula,

$$x = \frac{-20 \pm \sqrt{20^2 - 4(1)(-1050)}}{2 \cdot 1}$$

$$= \frac{-20 \pm \sqrt{400 + 4200}}{2} = \frac{-20 \pm \sqrt{4600}}{2}$$

$$= \frac{-20 \pm \sqrt{100 \cdot 46}}{2} = \frac{-20 \pm 10\sqrt{46}}{2}$$

$$= -10 \pm 5\sqrt{46} \qquad \text{Simplify.}$$

Check:

4. INTERPRET. We check our calculations from the quadratic formula. The length of a side of a triangle can't be negative, so we reject $-10 - 5\sqrt{46}$. Since $-10 + 5\sqrt{46} \approx 24$ feet, the walking distance along the sidewalk is

$$x + (x + 20) \approx 24 + (24 + 20) = 68 \text{ feet.}$$

State: A person saves about $68 - 50$ or 18 feet of walking distance by cutting across the lawn.

■ **Work Practice 8**

Example 9 Calculating Landing Time

An object is thrown upward from the top of a 200-foot cliff with a velocity of 12 feet per second. The height above ground h in feet of the object after t seconds is

$$h = -16t^2 + 12t + 200$$

How long after the object is thrown will it strike the ground? Round to the nearest tenth of a second.

200 ft

Solution:

1. UNDERSTAND. Read and reread the problem.

2. TRANSLATE. Since we want to know when the object strikes the ground, we want to know when the height $h = 0$, or

$$0 = -16t^2 + 12t + 200$$

3. SOLVE. First we divide both sides of the equation by -4.

$$0 = 4t^2 - 3t - 50 \qquad \text{Divide both sides by } -4.$$

Here, $a = 4$, $b = -3$, and $c = -50$. By the quadratic formula,

$$t = \frac{-(-3) \pm \sqrt{(-3)^2 - 4(4)(-50)}}{2 \cdot 4}$$

$$= \frac{3 \pm \sqrt{9 + 800}}{8}$$

$$= \frac{3 \pm \sqrt{809}}{8}$$

(Continued on next page)

Practice 9

How long after the object in Example 9 is thrown will it be 100 feet from the ground? Round to the nearest tenth of a second.

Check:

4. INTERPRET. We check our calculations from the quadratic formula. Since the time won't be negative, we reject the proposed solution $\dfrac{3 - \sqrt{809}}{8}$.

State: The time it takes for the object to strike the ground is exactly $\dfrac{3 + \sqrt{809}}{8}$ seconds ≈ 3.9 seconds.

■ **Work Practice 9**

Calculator Explorations Graphing

We can use a grapher to approximate real number solutions of a quadratic equation. For example, to solve $(x + 1)^2 = 12$, the quadratic equation in Example 3 of Section 11.1, we graph the following on the same set of axes. We use Xmin = −10, Xmax = 10, Ymin = −13, and Ymax = 13.

$$Y_1 = (x + 1)^2 \quad \text{and} \quad Y_2 = 12$$

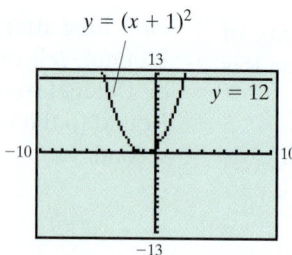

We use the INTERSECT feature or the ZOOM and TRACE features to locate the points of intersection of the graphs. The x-values of these points are the solutions of $(x + 1)^2 = 12$. The solutions, rounded to two decimal points, are 2.46 and −4.46.

Check to see that these numbers are approximations of the exact solutions, $-1 \pm 2\sqrt{3}$.

Use a grapher to solve each quadratic equation. Round all solutions to the nearest hundredth.

1. $x(x - 5) = 8$ $\{-1.27, 6.27\}$
2. $x(x + 2) = 5$ $\{-3.45, 1.45\}$
3. $x^2 + 0.5x = 0.3x + 1$ $\{-1.10, 0.90\}$
4. $x^2 - 2.6x = -2.2x + 3$ $\{-1.54, 1.94\}$
5. Use a grapher to solve $(2x - 5)^2 = -16$, (Example 4, Section 11.1) using the window

$$\begin{aligned} \text{Xmin} &= -20 \\ \text{Xmax} &= 20 \\ \text{Xscl} &= 1 \\ \text{Ymin} &= -20 \\ \text{Ymax} &= 20 \\ \text{Yscl} &= 1 \end{aligned}$$

Explain the results. Compare your results with the solution found in Example 4 of Section 11.1.

6. What are the advantages and disadvantages of using a grapher to solve quadratic equations?
answers may vary

5. $\varnothing$; answers may vary

Vocabulary, Readiness & Video Check

Fill in each blank.

1. The quadratic formula is ___ $x = \dfrac{-b \pm \sqrt{b^2 - 4ac}}{2a}$ ___.

2. For $2x^2 + x + 1 = 0$, if $a = 2$, then $b = $ ___ 1 ___ and $c = $ ___ 1 ___.

3. For $5x^2 - 5x - 7 = 0$, if $a = 5$, then $b = $ ___ −5 ___ and $c = $ ___ −7 ___.

4. For $7x^2 - 4 = 0$, if $a = 7$, then $b = $ ___ 0 ___ and $c = $ ___ −4 ___.

5. For $x^2 + 9 = 0$, if $c = 9$, then $a = $ ___ 1 ___ and $b = $ ___ 0 ___.

6. The correct simplified form of $\dfrac{5 \pm 10\sqrt{2}}{5}$ is ___ $1 \pm 2\sqrt{2}$ ___. c

   a. $1 \pm 10\sqrt{2}$ **b.** $2\sqrt{2}$ **c.** $1 \pm 2\sqrt{2}$ **d.** $\pm 5\sqrt{2}$

Martin-Gay Interactive Videos

See Video 11.2 🍎

Watch the section lecture video and answer the following questions.

Objective A **7.** Based on ▥ Examples 1–3, answer the following. ▸

  a. Must a quadratic equation be written in standard form in order to use the quadratic formula? Why or why not?

  b. Must fractions be cleared from an equation before using the quadratic formula? Why or why not?

Objective B **8.** Based on ▥ Example 4 and the lecture before, complete the following statements. The discriminant is the _____ in the quadratic formula and can be used to find the number and type of solutions of a quadratic equation without _____ the equation. To use the discriminant, the quadratic equation needs to be written in _____ form. ▸

Objective C **9.** In ▥ Example 5, the value of x is found, which is then used to find the dimensions of the triangle. Yet all this work still does solve the problem. Explain. ▸

See video answer section.

Teaching Tip

If you prefer to assign exercises with real number solutions only, omit Exercises 23–40.

11.2 Exercise Set MyMathLab® ▸

Objective A *Use the quadratic formula to solve each equation. These equations have real number solutions only. See Examples 1 through 3.*

1. $m^2 + 5m - 6 = 0$ $\{-6, 1\}$

2. $p^2 + 11p - 12 = 0$ $\{-12, 1\}$

3. $2y = 5y^2 - 3$ $\left\{-\dfrac{3}{5}, 1\right\}$

4. $5x^2 - 3 = 14x$ $\left\{-\dfrac{1}{5}, 3\right\}$

5. $x^2 - 6x + 9 = 0$ $\{3\}$

6. $y^2 + 10y + 25 = 0$ $\{-5\}$

▸ **7.** $x^2 + 7x + 4 = 0$
$\left\{\dfrac{-7 - \sqrt{33}}{2}, \dfrac{-7 + \sqrt{33}}{2}\right\}$

8. $y^2 + 5y + 3 = 0$
$\left\{\dfrac{-5 - \sqrt{13}}{2}, \dfrac{-5 + \sqrt{13}}{2}\right\}$

9. $8m^2 - 2m = 7$
$\left\{\dfrac{1 - \sqrt{57}}{8}, \dfrac{1 + \sqrt{57}}{8}\right\}$

10. $11n^2 - 9n = 1$
$\left\{\dfrac{9 - 5\sqrt{5}}{22}, \dfrac{9 + 5\sqrt{5}}{22}\right\}$

11. $3m^2 - 7m = 3$
$\left\{\dfrac{7 - \sqrt{85}}{6}, \dfrac{7 + \sqrt{85}}{6}\right\}$

12. $x^2 - 13 = 5x$
$\left\{\dfrac{5 - \sqrt{77}}{2}, \dfrac{5 + \sqrt{77}}{2}\right\}$

13. $\dfrac{1}{2}x^2 - x - 1 = 0$
$\{1 - \sqrt{3}, 1 + \sqrt{3}\}$

14. $\dfrac{1}{6}x^2 + x + \dfrac{1}{3} = 0$
$\{-3 - \sqrt{7}, -3 + \sqrt{7}\}$

15. $\dfrac{2}{5}y^2 + \dfrac{1}{5}y = \dfrac{3}{5}$
$\left\{-\dfrac{3}{2}, 1\right\}$

16. $\dfrac{1}{8}x^2 + x = \dfrac{5}{2}$
$\{-10, 2\}$

17. $\dfrac{1}{3}y^2 - y - \dfrac{1}{6} = 0$
$\left\{\dfrac{3 - \sqrt{11}}{2}, \dfrac{3 + \sqrt{11}}{2}\right\}$

18. $\dfrac{1}{2}y^2 = y + \dfrac{1}{2}$
$\{1 - \sqrt{2}, 1 + \sqrt{2}\}$

19. $x^2 + 5x = -2$ $\left\{\dfrac{-5 - \sqrt{17}}{2}, \dfrac{-5 + \sqrt{17}}{2}\right\}$

20. $y^2 - 8 = 4y$ $\{2 - 2\sqrt{3}, 2 + 2\sqrt{3}\}$

21. $(m + 2)(2m - 6) = 5(m - 1) - 12$ $\left\{\dfrac{5}{2}, 1\right\}$

22. $7p(p - 2) + 2(p + 4) = 3$ $\left\{\dfrac{5}{7}, 1\right\}$

Objective A Mixed Practice *Use the quadratic formula to solve each equation. These equations have real solutions and complex, but not real, solutions. See Examples 1 through 4.*

23. $x^2 + 6x + 13 = 0$
$\{-3 - 2i, -3 + 2i\}$

24. $x^2 + 2x + 2 = 0$
$\{-1 - i, -1 + i\}$

⊳ 25. $(x + 5)(x - 1) = 2$
$\{-2 - \sqrt{11}, -2 + \sqrt{11}\}$

26. $x(x + 6) = 2$
$\{-3 - \sqrt{11}, -3 + \sqrt{11}\}$

27. $6 = -4x^2 + 3x$
$\left\{\dfrac{3 - i\sqrt{87}}{8}, \dfrac{3 + i\sqrt{87}}{8}\right\}$

28. $9x^2 + x + 2 = 0$
$\left\{\dfrac{-1 - i\sqrt{71}}{18}, \dfrac{-1 + i\sqrt{71}}{18}\right\}$

29. $\dfrac{x^2}{3} - x = \dfrac{5}{3}$
$\left\{\dfrac{3 - \sqrt{29}}{2}, \dfrac{3 + \sqrt{29}}{2}\right\}$

30. $\dfrac{x^2}{2} - 3 = -\dfrac{9}{2}x$
$\left\{\dfrac{-9 - \sqrt{105}}{2}, \dfrac{-9 + \sqrt{105}}{2}\right\}$

31. $10y^2 + 10y + 3 = 0$
$\left\{\dfrac{-5 - i\sqrt{5}}{10}, \dfrac{-5 + i\sqrt{5}}{10}\right\}$

32. $3y^2 + 6y + 5 = 0$
$\left\{\dfrac{-3 - i\sqrt{6}}{3}, \dfrac{-3 + i\sqrt{6}}{3}\right\}$

33. $x(6x + 2) = 3$
$\left\{\dfrac{-1 - \sqrt{19}}{6}, \dfrac{-1 + \sqrt{19}}{6}\right\}$

34. $x(7x + 1) = 2$
$\left\{\dfrac{-1 - \sqrt{57}}{14}, \dfrac{-1 + \sqrt{57}}{14}\right\}$

⊳ 35. $\dfrac{2}{5}y^2 + \dfrac{1}{5}y + \dfrac{3}{5} = 0$
$\left\{\dfrac{-1 - i\sqrt{23}}{4}, \dfrac{-1 + i\sqrt{23}}{4}\right\}$

36. $\dfrac{1}{8}x^2 + x + \dfrac{5}{2} = 0$
$\{-4 - 2i, -4 + 2i\}$

37. $\dfrac{1}{2}y^2 = y - \dfrac{1}{2}$
$\{1\}$

38. $\dfrac{2}{3}x^2 - \dfrac{20}{3}x = -\dfrac{100}{6}$ $\{5\}$

39. $(n - 2)^2 = 2n$
$\{3 + \sqrt{5}, 3 - \sqrt{5}\}$

40. $\left(p - \dfrac{1}{2}\right)^2 = \dfrac{p}{2}$
$\left\{\dfrac{3 - \sqrt{5}}{4}, \dfrac{3 + \sqrt{5}}{4}\right\}$

Objective B *Use the discriminant to determine the number and types of solutions of each equation. See Examples 5 through 7.*

41. $x^2 - 5 = 0$
two real solutions

42. $x^2 - 7 = 0$
two real solutions

43. $4x^2 + 12x = -9$
one real solution

44. $9x^2 + 1 = 6x$
one real solution

45. $3x = -2x^2 + 7$
two real solutions

46. $3x^2 = 5 - 7x$
two real solutions

⊳ 47. $6 = 4x - 5x^2$
two complex but not
real solutions

48. $5 - 4x + 12x^2 = 0$
two complex but not
real solutions

49. $9x - 2x^2 + 5 = 0$
two real solutions

50. $8x = 3 - 9x^2$
two real solutions

Objective C *Solve. See Examples 8 and 9.*

▶ **51.** Nancy, Thelma, and John Varner live on a corner lot. Often, neighborhood children cut across their lot to save walking distance. Given the diagram below, approximate to the nearest foot how many feet of walking distance children save by cutting across their property instead of walking around the lot. 14 ft

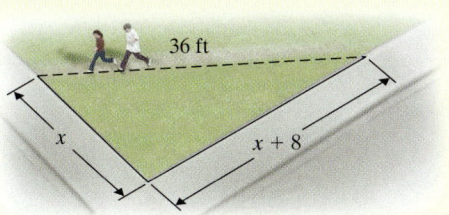

△ **52.** Given the diagram below, approximate to the nearest foot how many feet of walking distance a person saves by cutting across the lawn instead of walking on the sidewalk. 16 ft

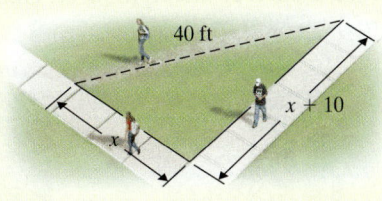

△ **53.** The hypotenuse of an isosceles right triangle is 2 centimeters longer than either of its legs. Find the exact length of each side. (*Hint:* An isosceles right triangle is a right triangle whose legs are the same length.)
$(2 + 2\sqrt{2})$ cm, $(2 + 2\sqrt{2})$ cm, $(4 + 2\sqrt{2})$ cm

△ **54.** The hypotenuse of an isosceles right triangle is one meter longer than either of its legs. Find the length of each side. (See the hint given in Exercise 53.)
$(1 + \sqrt{2})$ m, $(1 + \sqrt{2})$ m, $(2 + \sqrt{2})$ m

△ **55.** Bailey's rectangular dog pen for her Irish setter must have an area of 400 square feet. Also, the length must be 10 feet longer than the width. Find the dimensions of the pen.
width: $(-5 + 5\sqrt{17})$ ft; length: $(5 + 5\sqrt{17})$ ft;

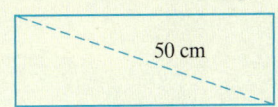

△ **56.** An entry in the Peach Festival Poster Contest must be rectangular and have an area of 1200 square inches. Furthermore, its length must be 20 inches longer than its width. Find the dimensions each entry must have.
width: $(-10 + 10\sqrt{13})$ in.; length: $(10 + 10\sqrt{13})$ in.

△ **57.** A holding pen for cattle must be square and have a diagonal length of 100 meters.

　a. Find the length of a side of the pen. $50\sqrt{2}$ m

　b. Find the area of the pen. 5000 sq m

△ **58.** A rectangle is three times longer than it is wide. It has a diagonal of length 50 centimeters.

　a. Find the dimensions of the rectangle.
　width: $5\sqrt{10}$ cm; length: $15\sqrt{10}$ cm

　b. Find the perimeter of the rectangle. $40\sqrt{10}$ cm

59. The heaviest reported door in the world is the 708.6-ton radiation shield door in the National Institute for Fusion Science at Toki, Japan. If the height of the door is 1.1 feet longer than its width, and its front area (neglecting depth) is 1439.9 square feet, find its width and height. [Interesting note: The door is 6.6 feet thick.] (*Source: Guiness World Records*)   37.4 ft by 38.5 ft

60. Christi and Robbie Wegmann are constructing a rectangular stained glass window whose length is 7.3 inches longer than its width. If the area of the window is 569.9 square inches, find its width and length.   20.5 in. by 27.8 in.

61. The base of a triangle is four more than twice its height. If the area of the triangle is 42 square centimeters, find its base and height.
base: $(2 + 2\sqrt{43})$ cm; height: $(-1 + \sqrt{43})$ cm

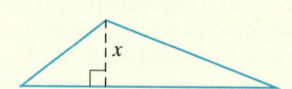

62. If a point B divides a line segment such that the smaller portion is to the larger portion as the larger is to the whole, the whole is the length of the *golden ratio*.

$$x \text{ (whole)}$$

$$\overbrace{\underset{A}{\bullet} \overset{1}{\rule{2cm}{0pt}} \underset{B}{\bullet} \overset{x-1}{\rule{2.5cm}{0pt}} \underset{C}{\bullet}}$$

The golden ratio was thought by the Greeks to be the most pleasing to the eye, and many of their buildings contained numerous examples of the golden ratio. The value of the golden ratio is the positive solution of the following equation.

$$\text{(smaller)} \quad \frac{x-1}{1} = \frac{1}{x} \quad \text{(larger)}$$
$$\text{(larger)} \qquad\qquad\quad \text{(whole)}$$

Find this value. $\dfrac{1 + \sqrt{5}}{2}$

The Wollomombi Falls in Australia have a height of 1100 feet. A pebble is thrown upward from the top of the falls with an initial velocity of 20 feet per second. The height of the pebble h in feet after t seconds is given by the equation $h = -16t^2 + 20t + 1100$. Use this equation for Exercises 63 and 64.

63. How long after the pebble is thrown will it hit the ground? Round to the nearest tenth of a second.   8.9 sec

64. How long after the pebble is thrown will it be 550 feet from the ground? Round to the nearest tenth of a second.   6.5 sec

A ball is thrown downward from the top of a 180-foot building with an initial velocity of 20 feet per second. The height of the ball h in feet after t seconds is given by the equation $h = -16t^2 - 20t + 180$. Use this equation to answer Exercises 65 and 66.

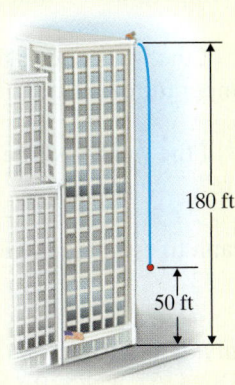

180 ft

50 ft

65. How long after the ball is thrown will it strike the ground? Round the result to the nearest tenth of a second.   2.8 sec

66. How long after the ball is thrown will it be 50 feet from the ground? Round the result to the nearest tenth of a second.   2.3 sec

Review

Solve each equation. See Sections 7.5 and 10.6.

67. $\sqrt{5x - 2} = 3$ $\left\{\dfrac{11}{5}\right\}$

68. $\sqrt{y + 2} + 7 = 12$ $\{23\}$

69. $\dfrac{1}{x} + \dfrac{2}{5} = \dfrac{7}{x}$ $\{15\}$

70. $\dfrac{10}{z} = \dfrac{5}{z} - \dfrac{1}{3}$ $\{-15\}$

Factor. See Sections 6.4 and 6.5.

71. $x^4 + x^2 - 20$ $(x^2 + 5)(x + 2)(x - 2)$

72. $2y^4 + 11y^2 - 6$ $(2y^2 - 1)(y^2 + 6)$

73. $z^4 - 13z^2 + 36$ $(z + 3)(z - 3)(z + 2)(z - 2)$

74. $x^4 - 1$ $(x^2 + 1)(x + 1)(x - 1)$

Concept Extensions

For each quadratic equation in Exercises 75 and 76, choose the correct substitution for a, b, and c in the standard form $ax^2 + bx + c = 0$. See the first Concept Check in this section.

75. $x^2 = -10$
    a. $a = 1, b = 0, c = -10$
    b. $a = 1, b = 0, c = 10$
    c. $a = 0, b = 1, c = -10$
    d. $a = 1, b = 1, c = 10$ b

76. $x^2 + 5 = -x$
    a. $a = 1, b = 5, c = -1$
    b. $a = 1, b = -1, c = 5$
    c. $a = 1, b = 5, c = 1$
    d. $a = 1, b = 1, c = 5$ d

77. Solve Exercise 1 by factoring. Explain the result.   answers may vary

78. Solve Exercise 2 by factoring. Explain the result.   answers may vary

Use the quadratic formula and a calculator to approximate each solution to the nearest tenth.

79. $2x^2 - 6x + 3 = 0$ $\{0.6, 2.4\}$

80. $3.6x^2 + 1.8x - 4.3 = 0$ $\{-1.4, 0.9\}$

The graph shows the daily low temperatures for one week in New Orleans, Louisiana. Use this graph to answer Exercises 81 through 84.

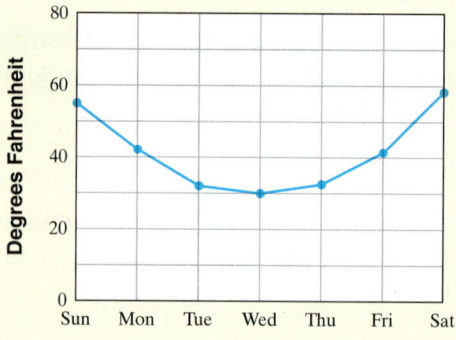

81. Which day of the week shows the greatest decrease in the low temperature?   Sunday to Monday

82. Which day of the week shows the greatest increase in the low temperature?   Friday to Saturday

83. Which day of the week had the lowest temperature?   Wednesday

84. Use the graph to estimate the low temperature on Thursday.   33°F

Notice that the shape of the temperature graph for Exercises 81 through 84 is similar to a parabola (see Section 8.5). In fact, this graph can be approximated by the quadratic function $f(x) = 3x^2 - 18x + 57$, where $f(x)$ is the temperature in degrees Fahrenheit and x is the number of days from Sunday. Use this function to answer Exercises 85 and 86.

85. Use the given quadratic function to approximate the low temperature on Thursday. Does your answer agree with the graph?
$f(4) = 33$; answers may vary

86. Use the given function and the quadratic formula to find when the low temperature was 35°F. [*Hint:* Let $f(x) = 35$ and solve for x.] Round your answer to one decimal place and interpret your result. Does your answer agree with the graph above?
Tuesday and Thursday; answers may vary

Solve.

87. The amount of electricity generated by wind power in the United States can be modeled by the quadratic function $f(x) = 0.45x^2 + 18.71x + 55.42$, where $f(x)$ is the amount of electricity in billions of kilowatts and x is the number of years after 2008. (*Source:* Based on data from the Energy Information Administration, U.S. Department of Energy)
   a. Find the amount of electricity generated by wind power in 2010.   94.64 billion kilowatts
   b. If the trend described by this model continues, find the year after 2008 when the amount of electricity generated by wind power tops 200 billion kilowatts.   2015

88. The function $f(x) = 0.0054x^2 - 1.46x + 95$ can be used to estimate the expected number of years of life remaining for a person of age x years, when $30 \le x \le 100$. (*Source:* Based on data from the U.S. Census Bureau)
   a. Estimate the expected remaining years of life for a person of age 50.   35.5 years
   b. Estimate the expected remaining years of life for a person of 30.   56.06 years
   c. Why do you think that this function is valid only for ages 30 through 100?   answers may vary

89. The number of tablet computer users worldwide (in millions) can be projected by the function $f(x) = 4x^2 + 107x - 15$, where x is the number of years since 2010. (*Source:* Forrester Research: World Tablet Adoption Forecast 2012–2017)
   a. Find the projected number of tablet computer users in 2012.   215 million
   b. According to this model, find the projected number of tablet computer users in 2015.   620 million
   c. If the trend represented by this model continues, in what year will the projected number of tablet computer users reach 1000 million?   2018

90. The relationship between body weight and the Recommended Dietary Allowance (RDA) for vitamin A in children up to age 10 is modeled by the quadratic equation $y = 0.149x^2 - 4.475x + 406.478$, where y is the RDA for vitamin A in micrograms for a child whose weight is x pounds. (*Source:* Based on data from the Food and Nutrition Board, National Academy of Sciences—Institute of Medicine, 1989)
   a. Determine the vitamin A requirements of a child who weighs 35 pounds.   432.378 micrograms
   b. What is the weight of a child whose RDA of vitamin A is 600 micrograms? Round your answer to the nearest pound.   54 lb

91. Use a grapher to solve Exercise 79.   {0.6, 2.4}

92. Use a grapher to solve Exercise 80.   {−1.4, 0.9}

Solving Equations by Using Quadratic Methods

Objective A Solving Equations That Are Quadratic in Form

In this section, we discuss various types of equations that can be solved in part by using the methods for solving quadratic equations.

Once each equation is simplified, you may want to use these steps when deciding what method to use to solve the quadratic equation.

Solving a Quadratic Equation

Step 1: If the equation is in the form $(ax + b)^2 = c$, use the square root property and solve. If not, go to Step 2.

Step 2: Write the equation in standard form by setting it equal to 0: $ax^2 + bx + c = 0$.

Step 3: Try to solve the equation by the factoring method. If not possible, go to Step 4.

Step 4: Solve the equation by the quadratic formula.

The first example is a radical equation that becomes a quadratic equation once we square both sides.

Example 1 Solve: $x - \sqrt{x} - 6 = 0$

Solution: Recall that to solve a radical equation, we first get the radical alone on one side of the equation. Then we square both sides.

$$x - 6 = \sqrt{x} \quad \text{Add } \sqrt{x} \text{ to both sides.}$$
$$x^2 - 12x + 36 = x \quad \text{Square both sides.}$$
$$x^2 - 13x + 36 = 0 \quad \text{Set the equation equal to 0.}$$
$$(x - 9)(x - 4) = 0 \quad \text{Factor.}$$
$$x - 9 = 0 \quad \text{or} \quad x - 4 = 0 \quad \text{Set each factor equal to 0.}$$
$$x = 9 \qquad\qquad x = 4 \quad \text{Solve.}$$

Check: Let $x = 9$.

$$x - \sqrt{x} - 6 = 0$$
$$9 - \sqrt{9} - 6 \stackrel{?}{=} 0$$
$$9 - 3 - 6 \stackrel{?}{=} 0$$
$$0 = 0 \quad \text{True}$$

Let $x = 4$.

$$x - \sqrt{x} - 6 = 0$$
$$4 - \sqrt{4} - 6 \stackrel{?}{=} 0$$
$$4 - 2 - 6 \stackrel{?}{=} 0$$
$$-4 = 0 \quad \text{False}$$

The solution set is $\{9\}$.

Work Practice 1

Practice 1

Solve: $x - \sqrt{x - 1} - 3 = 0$

Answer

1. $\{5\}$

Practice 2

Solve:

$$\frac{2x}{x - 1} - \frac{x + 2}{x} = \frac{5}{x(x - 1)}$$

Example 2   Solve: $\dfrac{3x}{x - 2} - \dfrac{x + 1}{x} = \dfrac{6}{x(x - 2)}$

Solution: In this equation, x cannot be either 2 or 0 because these values cause denominators to equal zero. To solve for x, we first multiply both sides of the equation by $x(x - 2)$ to clear the fractions. By the distributive property, this means that we multiply each term by $x(x - 2)$.

$$x(x - 2)\left(\frac{3x}{x - 2}\right) - x(x - 2)\left(\frac{x + 1}{x}\right) = x(x - 2)\left[\frac{6}{x(x - 2)}\right]$$

$$3x^2 - (x - 2)(x + 1) = 6 \quad \text{Simplify.}$$
$$3x^2 - (x^2 - x - 2) = 6 \quad \text{Multiply.}$$
$$3x^2 - x^2 + x + 2 = 6$$
$$2x^2 + x - 4 = 0 \quad \text{Simplify.}$$

This equation cannot be factored using integers, so we solve by the quadratic formula.

$$x = \frac{-1 \pm \sqrt{1^2 - 4(2)(-4)}}{2 \cdot 2} \quad \text{Let } a = 2, b = 1, \text{ and } c = -4, \text{ in the quadratic formula.}$$

$$= \frac{-1 \pm \sqrt{1 + 32}}{4} \quad \text{Simplify.}$$

$$= \frac{-1 \pm \sqrt{33}}{4}$$

Neither proposed solution will make the denominators 0.

The solution set is $\left\{\dfrac{-1 + \sqrt{33}}{4}, \dfrac{-1 - \sqrt{33}}{4}\right\}$.

■ **Work Practice 2**

Practice 3

Solve: $x^4 + 5x^2 - 36 = 0$

Example 3   Solve: $p^4 - 3p^2 - 4 = 0$

Solution: First we factor the trinomial.

$$p^4 - 3p^2 - 4 = 0$$
$$(p^2 - 4)(p^2 + 1) = 0 \quad \text{Factor.}$$
$$(p - 2)(p + 2)(p^2 + 1) = 0 \quad \text{Factor further.}$$
$$p - 2 = 0 \quad \text{or} \quad p + 2 = 0 \quad \text{or} \quad p^2 + 1 = 0 \quad \text{Set each factor equal}$$
$$p = 2 \qquad\qquad p = -2 \qquad\qquad p^2 = -1 \quad \text{to 0 and solve.}$$
$$p = \pm\sqrt{-1} = \pm i$$

The solution set is $\{2, -2, i, -i\}$.

■ **Work Practice 3**

Helpful Hint

Example 3 can be solved using substitution also. Think of $p^4 - 3p^2 - 4 = 0$ as

$$(p^2)^2 - 3p^2 - 4 = 0 \quad \text{Then let } x = p^2, \text{ and solve and substitute back.}$$

The solution set will be the same.

$$x^2 - 3x - 4 = 0$$

Answers

2. $\left\{\dfrac{1 + \sqrt{13}}{2}, \dfrac{1 - \sqrt{13}}{2}\right\}$

3. $\{2, -2, 3i, -3i\}$

✓Concept Check

a. True or false? The maximum number of solutions that a quadratic equation can have is 2.

b. True or false? The maximum number of solutions that an equation in quadratic form can have is 2.

Teaching Tip

Before discussing Example 4, have students use $y = x + 2$ to rewrite $y^2 + 6y + 7 = 0$ in terms of x.

Example 4 Solve: $(x - 3)^2 - 3(x - 3) - 4 = 0$

Solution: Notice that the quantity $(x - 3)$ is repeated in this equation. Sometimes it is helpful to substitute a variable (in this case other than x) for the repeated quantity. We will let $u = x - 3$. Then

$$(x - 3)^2 - 3(x - 3) - 4 = 0$$

becomes

$$u^2 - 3u - 4 = 0 \quad \text{Let } x - 3 = u.$$
$$(u - 4)(u + 1) = 0 \quad \text{Factor.}$$

To solve, we use the zero-factor property.

$$u - 4 = 0 \quad \text{or} \quad u + 1 = 0 \quad \text{Set each factor equal to 0.}$$
$$u = 4 \qquad\qquad u = -1 \quad \text{Solve.}$$

To find values of x, we substitute back. That is, we substitute $x - 3$ for u.

$$x - 3 = 4 \quad \text{or} \quad x - 3 = -1$$
$$x = 7 \qquad\qquad x = 2$$

Both 2 and 7 check. The solution is $\{2, 7\}$.

■ **Work Practice 4**

Practice 4

Solve:
$(x + 4)^2 - (x + 4) - 6 = 0$

Helpful Hint When using substitution, don't forget to substitute back to the original variable.

Teaching Tip

After discussing Example 4, have students solve the equation by first simplifying it. Then have them compare their results to those in the example.

Example 5 Solve: $x^{2/3} - 5x^{1/3} + 6 = 0$

Solution: The key to solving this equation is recognizing that $x^{2/3} = (x^{1/3})^2$. We replace $x^{1/3}$ with m so that

$$(x^{1/3})^2 - 5x^{1/3} + 6 = 0$$

becomes

$$m^2 - 5m + 6 = 0$$

Now we solve by factoring.

$$m^2 - 5m + 6 = 0$$
$$(m - 3)(m - 2) = 0 \qquad\qquad \text{Factor.}$$
$$m - 3 = 0 \quad \text{or} \quad m - 2 = 0 \quad \text{Set each factor equal to 0.}$$
$$m = 3 \qquad\qquad m = 2$$

Since $m = x^{1/3}$, we have

$$x^{1/3} = 3 \qquad \text{or} \quad x^{1/3} = 2$$
$$x = 3^3 = 27 \quad \text{or} \quad x = 2^3 = 8$$

Both 8 and 27 check. The solution set is $\{8, 27\}$.

■ **Work Practice 5**

Practice 5

Solve: $x^{2/3} - 7x^{1/3} + 10 = 0$

Teaching Tip

Before discussing Example 5, have students use $y = x^{1/3}$ to rewrite $y^2 + 8y + 15 = 0$ in terms of x.

Answers
4. $\{-1, -6\}$ 5. $\{8, 125\}$

✓**Concept Check Answers**
a. true   **b.** false

Objective B Solving Problems That Lead to Quadratic Equations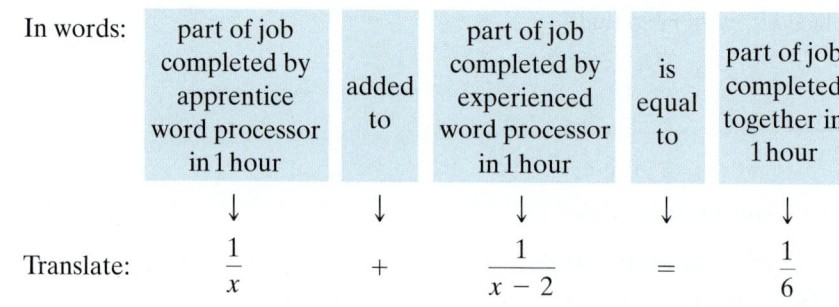

The next example is a work problem. This problem is modeled by a rational equation that simplifies to a quadratic equation.

Practice 6

Together, Karen and Doug Lewis can clean a strip of beach in 5 hours. Alone, Karen can clean the strip of beach 1 hour faster than Doug. Find the time that each person can clean the strip of beach alone. Give an exact answer and a one-decimal-place approximation.

Example 6 Finding Work Time

Together, an experienced word processor and an apprentice word processor can create a document in 6 hours. Alone, the experienced word processor can process the document 2 hours faster than the apprentice word processor can. Find the time in which each person can create the document alone.

Solution:

1. UNDERSTAND. Read and reread the problem. The key idea here is the relationship between the *time* (hours) it takes to complete the job and the *part of the job* completed in one unit of time (hour). For example, because they can complete the job together in 6 hours, the *part of the job* they can complete in

1 hour is $\frac{1}{6}$. We let

$x =$ the *time* in hours it takes the apprentice word processor to complete the job alone, and

$x - 2 =$ the *time* in hours it takes the experienced word processor to complete the job alone

We can summarize in a chart the information discussed.

	Total Hours to Complete Job	Part of Job Completed in 1 Hour
Apprentice Word Processor	x	$\frac{1}{x}$
Experienced Word Processor	$x - 2$	$\frac{1}{x - 2}$
Together	6	$\frac{1}{6}$

2. TRANSLATE.

In words:

part of job completed by apprentice word processor in 1 hour	added to	part of job completed by experienced word processor in 1 hour	is equal to	part of job completed together in 1 hour
↓	↓	↓	↓	↓

Translate: $\dfrac{1}{x} \quad + \quad \dfrac{1}{x-2} \quad = \quad \dfrac{1}{6}$

Answer

6. Doug: $\dfrac{11 + \sqrt{101}}{2} \approx 10.5$ hr;

Karen: $\dfrac{9 + \sqrt{101}}{2} \approx 9.5$ hr

3. SOLVE.

$$\frac{1}{x} + \frac{1}{x-2} = \frac{1}{6}$$

$$6x(x-2)\left(\frac{1}{x} + \frac{1}{x-2}\right) = 6x(x-2)\cdot\frac{1}{6} \qquad \text{Multiply both sides by the LCD } 6x(x-2).$$

$$6x(x-2)\cdot\frac{1}{x} + 6x(x-2)\cdot\frac{1}{x-2} = 6x(x-2)\cdot\frac{1}{6} \qquad \text{Use the distributive property.}$$

$$6(x-2) + 6x = x(x-2)$$

$$6x - 12 + 6x = x^2 - 2x$$

$$0 = x^2 - 14x + 12$$

Now we can substitute $a = 1$, $b = -14$, and $c = 12$ into the quadratic formula and simplify.

$$x = \frac{-(-14) \pm \sqrt{(-14)^2 - 4(1)(12)}}{2\cdot 1} = \frac{14 \pm \sqrt{148}}{2}$$

Using a calculator or a square root table, we see that $\sqrt{148} \approx 12.2$ rounded to one decimal place. Thus,

$$x \approx \frac{14 \pm 12.2}{2}$$

$$x \approx \frac{14 + 12.2}{2} = 13.1 \quad \text{or} \quad x \approx \frac{14 - 12.2}{2} = 0.9$$

4. INTERPRET.

Check: If the apprentice word processor completes the job alone in 0.9 hours, the experienced word processor completes the job alone in $x - 2 = 0.9 - 2 = -1.1$ hours. Since this is not possible, we reject the solution of 0.9. The approximate solution thus is 13.1 hours.

State: The apprentice word processor can complete the job alone in approximately 13.1 hours, and the experienced word processor can complete the job alone in approximately

$$x - 2 = 13.1 - 2 = 11.1 \text{ hours}$$

■ **Work Practice 6**

Example 7 Finding Driving Speeds

Beach and Fargo are about 400 miles apart. A salesperson travels from Fargo to Beach one day at a certain speed. She returns to Fargo the next day and drives 10 mph faster. Her total travel time was $14\frac{2}{3}$ hours. Find her speed to Beach and the return speed to Fargo.

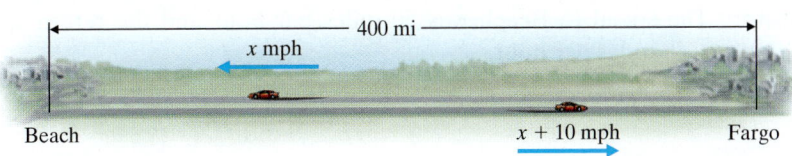

(Continued on next page)

Practice 7

A family drives 500 miles to the beach for a vacation. The return trip was made at a speed that was 10 miles per hour faster. The total traveling time was $18\frac{1}{3}$ hours. Find the speed to the beach and the return speed.

Answer

7. 50 mph to the beach; 60 mph returning

Solution:

1. **UNDERSTAND.** Read and reread the problem. Let

$$x = \text{the speed to Beach, so}$$
$$x + 10 = \text{the return speed to Fargo}$$

Then organize the given information in a table.

> **Helpful Hint**
> Since $d = rt$, then $t = \dfrac{d}{r}$. The time column was completed using $\dfrac{d}{r}$.

	Distance =	Rate ·	Time	
To Beach	400	x	$\dfrac{400}{x}$	← distance ← rate
Return to Fargo	400	$x + 10$	$\dfrac{400}{x + 10}$	← distance ← rate

2. **TRANSLATE.**

In words: time to Beach + return time to Fargo = $14\dfrac{2}{3}$ hours

Translate: $\dfrac{400}{x} + \dfrac{400}{x + 10} = \dfrac{44}{3}$

3. **SOLVE.**

$$\frac{400}{x} + \frac{400}{x + 10} = \frac{44}{3}$$

This next step is optional. Notice that all three numerators in our equation are divisible by 4. To keep the numbers in our equation as simple as possible, we will take a step and divide through by 4.

$$\frac{100}{x} + \frac{100}{x + 10} = \frac{11}{3} \qquad \text{Divide both sides by 4.}$$

$$3x(x + 10)\left(\frac{100}{x} + \frac{100}{x + 10}\right) = 3x(x + 10) \cdot \frac{11}{3} \qquad \begin{array}{l}\text{Multiply both sides by the LCD,} \\ 3x(x + 10).\end{array}$$

$$3x(x + 10)\frac{100}{x} + 3x(x + 10)\frac{100}{x + 10} = 3x(x + 10) \cdot \frac{11}{3} \qquad \begin{array}{l}\text{Use the distribu-} \\ \text{tive property.}\end{array}$$

$$3(x + 10)100 + 3x(100) = x(x + 10)11$$
$$300x + 3000 + 300x = 11x^2 + 110x$$
$$0 = 11x^2 - 490x - 3000 \qquad \begin{array}{l}\text{Set equation} \\ \text{equal to 0.}\end{array}$$
$$0 = (11x + 60)(x - 50) \qquad \text{Factor.}$$

$$11x + 60 = 0 \qquad \text{or} \qquad x - 50 = 0 \qquad \text{Set each factor equal to 0.}$$

$$x = -\frac{60}{11} = -5\frac{5}{11} \qquad\qquad x = 50$$

4. **INTERPRET.**

Check: The speed is not negative, so it's not $-5\dfrac{5}{11}$. The number 50 does check.

State: The speed to Beach was 50 miles per hour and the return speed to Fargo was 60 miles per hour.

■ **Work Practice 7**

Vocabulary, Readiness & Video Check

Martin-Gay Interactive Videos   *Watch the section lecture video and answer the following questions.*

Objective **A** **1.** From ⊞ Examples 1 and 2, what's the main thing to remember when using a substitution in order to solve an equation by quadratic methods?

Objective **B** **2.** In ⊞ Example 4, the translated equation is actually a rational equation. Explain how we end up solving it using quadratic methods. ▶

See Video 11.3 🍒

See video answer section.

Teaching Tip
If you prefer to assign exercises with real number solutions only, omit Exercises 13–18, 49, 53, and 54.

11.3 Exercise Set MyMathLab® ▶

Objective **A** *Solve. See Example 1.*

1. $2x = \sqrt{10 + 3x}$ $\{2\}$

2. $3x = \sqrt{8x + 1}$ $\{1\}$

3. $x - 2\sqrt{x} = 8$ $\{16\}$

4. $x - \sqrt{2x} = 4$ $\{8\}$

5. $\sqrt{9x} = x + 2$ $\{1, 4\}$

6. $\sqrt{16x} = x + 3$ $\{1, 9\}$

Solve. See Example 2.

▶7. $\dfrac{2}{x} + \dfrac{3}{x - 1} = 1$ $\{3 - \sqrt{7}, 3 + \sqrt{7}\}$

8. $\dfrac{6}{x^2} = \dfrac{3}{x + 1}$ $\{1 - \sqrt{3}, 1 + \sqrt{3}\}$

9. $\dfrac{3}{x} + \dfrac{4}{x + 2} = 2$ $\left\{\dfrac{3 - \sqrt{57}}{4}, \dfrac{3 + \sqrt{57}}{4}\right\}$

10. $\dfrac{5}{x - 2} + \dfrac{4}{x + 2} = 1$ $\left\{\dfrac{9 - \sqrt{105}}{2}, \dfrac{9 + \sqrt{105}}{2}\right\}$

11. $\dfrac{7}{x^2 - 5x + 6} = \dfrac{2x}{x - 3} - \dfrac{x}{x - 2}$
$\left\{\dfrac{1 - \sqrt{29}}{2}, \dfrac{1 + \sqrt{29}}{2}\right\}$

12. $\dfrac{11}{2x^2 + x - 15} = \dfrac{5}{2x - 5} - \dfrac{x}{x + 3}$
$\left\{\dfrac{5 - \sqrt{33}}{2}, \dfrac{5 + \sqrt{33}}{2}\right\}$

Solve. See Example 3.

13. $p^4 - 16 = 0$ $\{-2, 2, -2i, 2i\}$

14. $z^4 = 81$ $\{-3, 3, -3i, 3i\}$

15. $z^4 - 5z^2 - 36 = 0$ $\{-3, 3, -2i, 2i\}$

16. $x^4 + 2x^2 - 3 = 0$ $\{-1, 1, -i\sqrt{3}, i\sqrt{3}\}$

17. $4x^4 + 11x^2 = 3$ $\left\{-\dfrac{1}{2}, \dfrac{1}{2}, -i\sqrt{3}, i\sqrt{3}\right\}$

18. $9x^4 + 5x^2 - 4 = 0$ $\left\{-\dfrac{2}{3}, \dfrac{2}{3}, -i, i\right\}$

Solve. See Examples 4 and 5.

▶19. $x^{2/3} - 3x^{1/3} - 10 = 0$ $\{125, -8\}$

20. $x^{2/3} + 2x^{1/3} + 1 = 0$ $\{-1\}$

21. $(5n + 1)^2 + 2(5n + 1) - 3 = 0$ $\left\{-\dfrac{4}{5}, 0\right\}$

22. $(m - 6)^2 + 5(m - 6) + 4 = 0$ $\{2, 5\}$

23. $2x^{2/3} - 5x^{1/3} = 3$ $\left\{-\dfrac{1}{8}, 27\right\}$

24. $3x^{2/3} + 11x^{1/3} = 4$ $\left\{-64, \dfrac{1}{27}\right\}$

25. $1 + \dfrac{2}{3t - 2} = \dfrac{8}{(3t - 2)^2}$ $\left\{-\dfrac{2}{3}, \dfrac{4}{3}\right\}$

26. $2 - \dfrac{7}{x + 6} = \dfrac{15}{(x + 6)^2}$ $\left\{-\dfrac{15}{2}, -1\right\}$

27. $20x^{2/3} - 6x^{1/3} - 2 = 0$ $\left\{-\dfrac{1}{125}, \dfrac{1}{8}\right\}$

28. $4x^{2/3} + 16x^{1/3} = -15$ $\left\{-\dfrac{125}{8}, -\dfrac{27}{8}\right\}$

Objective A Mixed Practice *Solve. See Examples 1 through 5.*

29. $a^4 - 5a^2 + 6 = 0$ $\{-\sqrt{2}, \sqrt{2}, -\sqrt{3}, \sqrt{3}\}$

30. $x^4 - 12x^2 + 11 = 0$ $\{-1, 1, -\sqrt{11}, \sqrt{11}\}$

31. $\dfrac{2x}{x - 2} + \dfrac{x}{x + 3} = \dfrac{-5}{x + 3}$ $\left\{\dfrac{-9 - \sqrt{201}}{6}, \dfrac{-9 + \sqrt{201}}{6}\right\}$

32. $\dfrac{5}{x - 3} + \dfrac{x}{x + 3} = \dfrac{19}{x^2 - 9}$ $\{-1 - \sqrt{5}, -1 + \sqrt{5}\}$

▶ 33. $(p + 2)^2 = 9(p + 2) - 20$ $\{2, 3\}$

34. $2(4m - 3)^2 - 9(4m - 3) = 5$ $\left\{\dfrac{5}{8}, 2\right\}$

35. $2x = \sqrt{11x + 3}$ $\{3\}$

36. $4x = \sqrt{2x + 3}$ $\left\{\dfrac{1}{2}\right\}$

37. $x^{2/3} - 8x^{1/3} + 15 = 0$ $\{27, 125\}$

38. $x^{2/3} - 2x^{1/3} - 8 = 0$ $\{-8, 64\}$

39. $x - \sqrt{19 - 2x} - 2 = 0$ $\{5\}$

40. $x - \sqrt{17 - 4x} - 3 = 0$ $\{4\}$

41. $2x^{2/3} + 3x^{1/3} - 2 = 0$ $\left\{\dfrac{1}{8}, -8\right\}$

42. $6x^{2/3} - 25x^{1/3} - 25 = 0$ $\left\{-\dfrac{125}{216}, 125\right\}$

43. $(t + 3)^2 - 2(t + 3) - 8 = 0$ $\{-5, 1\}$

44. $(2n - 3)^2 - 7(2n - 3) + 12 = 0$ $\left\{\dfrac{7}{2}, 3\right\}$

45. $x - \sqrt{x} = 2$ $\{4\}$

46. $x - \sqrt{3x} = 6$ $\{12\}$

47. $\dfrac{x}{x - 1} + \dfrac{1}{x + 1} = \dfrac{2}{x^2 - 1}$ $\{-3\}$

48. $\dfrac{x}{x - 5} + \dfrac{5}{x + 5} = \dfrac{-1}{x^2 - 25}$ $\{-12, 2\}$

49. $p^4 - p^2 - 20 = 0$ $\{-\sqrt{5}, \sqrt{5}, -2i, 2i\}$

50. $x^4 - 10x^2 + 9 = 0$ $\{-1, 1, -3, 3\}$

51. $1 = \dfrac{4}{x - 7} + \dfrac{5}{(x - 7)^2}$ $\{6, 12\}$

52. $3 + \dfrac{1}{2p + 4} = \dfrac{10}{(2p + 4)^2}$ $\left\{-\dfrac{7}{6}, -3\right\}$

53. $27y^4 + 15y^2 = 2$ $\left\{-\dfrac{1}{3}, \dfrac{1}{3}, -\dfrac{i\sqrt{6}}{3}, \dfrac{i\sqrt{6}}{3}\right\}$

54. $8z^4 + 14z^2 = -5$ $\left\{-\dfrac{i\sqrt{2}}{2}, \dfrac{i\sqrt{2}}{2}, -\dfrac{i\sqrt{5}}{2}, \dfrac{i\sqrt{5}}{2}\right\}$

Objective B *Solve. For Exercises 59 and 60, the solutions have been started for you. See Examples 6 and 7.*

55. A jogger ran 3 miles, decreased her speed by 1 mile per hour, and then ran another 4 miles. If her total time jogging was $1\frac{3}{5}$ hours, find her speed for each part of her run. 5 mph, then 4 mph

56. Mark Keaton's workout consists of jogging for 3 miles and then riding his bike for 5 miles at a speed 4 miles per hour faster than he jogs. If his total workout time is 1 hour, find his jogging speed and his biking speed. jogging: 6 mph; biking: 10 mph

57. A Chinese restaurant in Mandeville, Louisiana, has a large goldfish pond around the restaurant. Suppose that an inlet pipe and a hose together can fill the pond in 8 hours. The inlet pipe alone can complete the job in 1 hour less time than the hose alone. Find the time that the hose can complete the job alone and the time that the inlet pipe can complete the job alone. Round each to the nearest tenth of an hour. inlet pipe: 15.5 hr; hose: 16.5 hr

58. A water tank on a farm in Flatonia, Texas, can be filled with a large inlet pipe and a small inlet pipe in 3 hours. The large inlet pipe alone can fill the tank in 2 hours less time than the small inlet pipe alone. Find the time to the nearest tenth of an hour each pipe can fill the tank alone. small pipe: 7.2 hr; large pipe: 5.2 hr

59. Roma Sherry drove 330 miles from her home town to Tucson. During her return trip, she was able to increase her speed by 11 miles per hour. If her return trip took 1 hour less time, find her original speed and her speed returning home. 55 mph; 66 mph

Start the solution:

1. UNDERSTAND the problem. Reread it as many times as needed. Let

$$x = \text{original speed}$$
$$x + 11 = \text{return-trip speed}$$

Organize the information in a table.

	distance	= rate	· time	
To Tucson	330	x	$\dfrac{330}{x}$	← distance ← rate
Return trip	330	___	$\dfrac{330}{}$	← distance ← rate

2. TRANSLATE into an equation. (Fill in the blanks below.)

Time to Tucson	equals	Return trip time	plus	1 hour
↓	↓	↓	↓	↓
___	=	___	+	1

Finish with:

3. SOLVE and **4.** INTERPRET

60. A salesperson drove to Portland, a distance of 300 miles. During the last 80 miles of his trip, heavy rainfall forced him to decrease his speed by 15 miles per hour. If his total driving time was 6 hours, find his original speed and his speed during the rainfall. 55 mph; 40 mph

Start the solution:

1. UNDERSTAND the problem. Reread it as many times as needed. Let

$$x = \text{original speed}$$
$$x - 15 = \text{rainfall speed}$$

Organize the information in a table.

	distance	= rate	· time	
First part of trip	300 − 80, or 220	x	$\dfrac{220}{x}$	← distance ← rate
Heavy rainfall part of trip	80	$x - 15$	$\dfrac{80}{}$	← distance ← rate

2. TRANSLATE into an equation. (Fill in the blanks below.)

Time during first part of trip	plus	Time during heavy rainfall	equals	6 hr
↓	↓	↓	↓	↓
___	+	___	=	6

Finish with:

3. SOLVE and **4.** INTERPRET

61. Bill Shaughnessy and his son Billy can clean the house together in 4 hours. When the son works alone, it takes him an hour longer to clean than it takes his dad alone. Find how long to the nearest tenth of an hour it takes the son to clean alone.   8.5 hr

62. Together, Noodles and Freckles eat a 50-pound bag of dog food in 30 days. Noodles by herself eats a 50-pound bag in 2 weeks less time than Freckles does by himself. How many days to the nearest whole day would a 50-pound bag of dog food last Freckles?   68 days

63. The product of a number and 4 less than the number is 96. Find the number.   12 or −8

64. A whole number increased by its square is two more than twice itself. Find the number.   2

△ **65.** Suppose that we want to make an open box from a square sheet of cardboard by cutting out squares from each corner as shown and then folding along the dotted lines. If the box is to have a volume of 300 cubic inches, find the original dimensions of the sheet of cardboard.

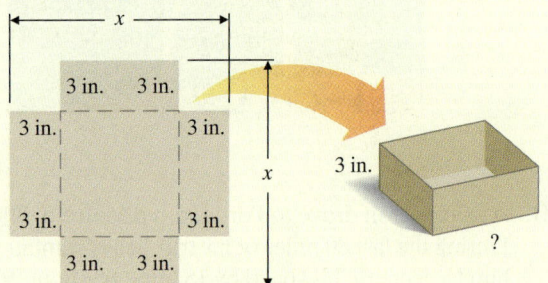

a. The ? in the drawing to the right will be the length (and also the width) of the box as shown in the drawing to the left. Represent this length in terms of x.
$(x - 6)$ in.
b. Use the formula for volume of a box, $V = l \cdot w \cdot h$, to write an equation in x.
$300 = (x - 6) \cdot (x - 6) \cdot 3$
c. Solve the equation for x and give the dimensions of the sheet of cardboard. Check your solution.
16 in. by 16 in.

△ **66.** Suppose that we want to make an open box from a square sheet of cardboard by cutting out squares from each corner as shown and then folding along the dotted lines. If the box is to have a volume of 128 cubic inches, find the original dimensions of the sheet of cardboard. (*Hint:* Use Exercise 65 parts (a), (b), and (c) to help you.)   12 in. by 12 in.

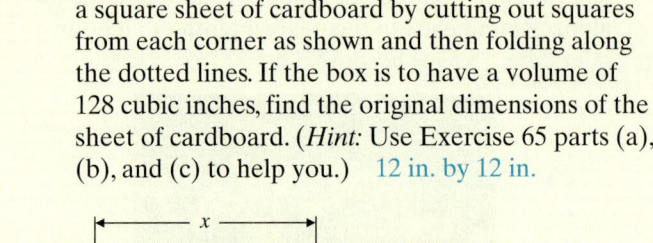
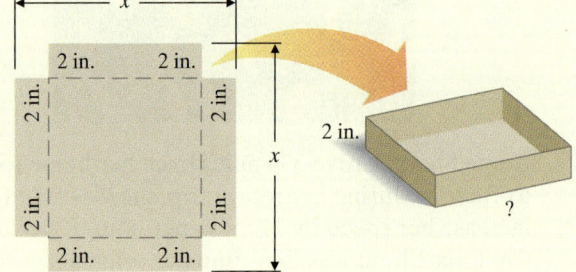

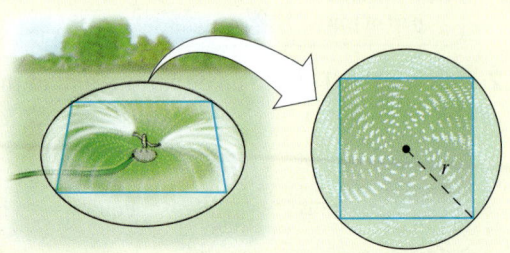

△ **67.** A sprinkler that sprays water in a circular pattern is to be used to water a square garden. If the area of the garden is 920 square feet, find the smallest whole number *radius* that the sprinkler can be adjusted to so that the entire garden is watered.   22 ft

△ **68.** Suppose that a square field has an area of 6270 square feet. See Exercise 67 and find the required sprinkler radius.   56 ft

Objective B *Solve. For Exercises 59 and 60, the solutions have been started for you. See Examples 6 and 7.*

55. A jogger ran 3 miles, decreased her speed by 1 mile per hour, and then ran another 4 miles. If her total time jogging was $1\frac{3}{5}$ hours, find her speed for each part of her run.   5 mph, then 4 mph

56. Mark Keaton's workout consists of jogging for 3 miles and then riding his bike for 5 miles at a speed 4 miles per hour faster than he jogs. If his total workout time is 1 hour, find his jogging speed and his biking speed.   jogging: 6 mph; biking: 10 mph

57. A Chinese restaurant in Mandeville, Louisiana, has a large goldfish pond around the restaurant. Suppose that an inlet pipe and a hose together can fill the pond in 8 hours. The inlet pipe alone can complete the job in 1 hour less time than the hose alone. Find the time that the hose can complete the job alone and the time that the inlet pipe can complete the job alone. Round each to the nearest tenth of an hour.   inlet pipe: 15.5 hr; hose: 16.5 hr

58. A water tank on a farm in Flatonia, Texas, can be filled with a large inlet pipe and a small inlet pipe in 3 hours. The large inlet pipe alone can fill the tank in 2 hours less time than the small inlet pipe alone. Find the time to the nearest tenth of an hour each pipe can fill the tank alone.   small pipe: 7.2 hr; large pipe: 5.2 hr

59. Roma Sherry drove 330 miles from her home town to Tucson. During her return trip, she was able to increase her speed by 11 miles per hour. If her return trip took 1 hour less time, find her original speed and her speed returning home.   55 mph; 66 mph

Start the solution:

1. UNDERSTAND the problem. Reread it as many times as needed. Let

x = original speed

$x + 11$ = return-trip speed

Organize the information in a table.

	distance	= rate	· time	
To Tucson	330	x	$\dfrac{330}{x}$	← distance ← rate
Return trip	330	____	$\dfrac{330}{}$	← distance ← rate

2. TRANSLATE into an equation. (Fill in the blanks below.)

Time to Tucson	equals	Return trip time	plus	1 hour
↓	↓	↓	↓	↓
____	=	____	+	1

Finish with:

3. SOLVE and **4.** INTERPRET

60. A salesperson drove to Portland, a distance of 300 miles. During the last 80 miles of his trip, heavy rainfall forced him to decrease his speed by 15 miles per hour. If his total driving time was 6 hours, find his original speed and his speed during the rainfall.   55 mph; 40 mph

Start the solution:

1. UNDERSTAND the problem. Reread it as many times as needed. Let

x = original speed

$x - 15$ = rainfall speed

Organize the information in a table.

	distance	= rate	· time	
First part of trip	300 − 80, or 220	x	$\dfrac{220}{x}$	← distance ← rate
Heavy rainfall part of trip	80	$x - 15$	$\dfrac{80}{}$	← distance ← rate

2. TRANSLATE into an equation. (Fill in the blanks below.)

Time during first part of trip	plus	Time during heavy rainfall	equals	6 hr
↓	↓	↓	↓	↓
____	+	____	=	6

Finish with:

3. SOLVE and **4.** INTERPRET

61. Bill Shaughnessy and his son Billy can clean the house together in 4 hours. When the son works alone, it takes him an hour longer to clean than it takes his dad alone. Find how long to the nearest tenth of an hour it takes the son to clean alone.   8.5 hr

62. Together, Noodles and Freckles eat a 50-pound bag of dog food in 30 days. Noodles by herself eats a 50-pound bag in 2 weeks less time than Freckles does by himself. How many days to the nearest whole day would a 50-pound bag of dog food last Freckles?   68 days

63. The product of a number and 4 less than the number is 96. Find the number.   12 or −8

64. A whole number increased by its square is two more than twice itself. Find the number.   2

△ **65.** Suppose that we want to make an open box from a square sheet of cardboard by cutting out squares from each corner as shown and then folding along the dotted lines. If the box is to have a volume of 300 cubic inches, find the original dimensions of the sheet of cardboard.

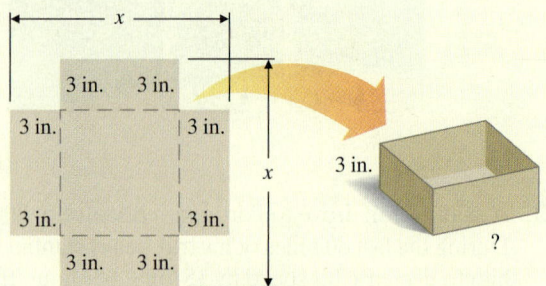

a. The ? in the drawing to the right will be the length (and also the width) of the box as shown in the drawing to the left. Represent this length in terms of x. $(x - 6)$ in.

b. Use the formula for volume of a box, $V = l \cdot w \cdot h$, to write an equation in x.
   $300 = (x - 6) \cdot (x - 6) \cdot 3$

c. Solve the equation for x and give the dimensions of the sheet of cardboard. Check your solution.
   16 in. by 16 in.

△ **66.** Suppose that we want to make an open box from a square sheet of cardboard by cutting out squares from each corner as shown and then folding along the dotted lines. If the box is to have a volume of 128 cubic inches, find the original dimensions of the sheet of cardboard. (*Hint:* Use Exercise 65 parts (a), (b), and (c) to help you.)   12 in. by 12 in.

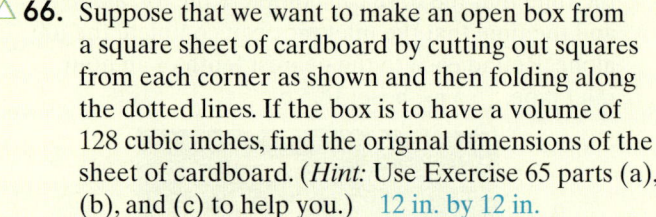

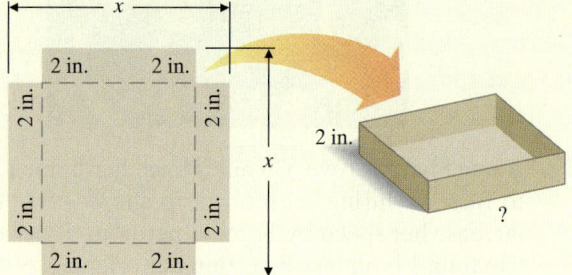

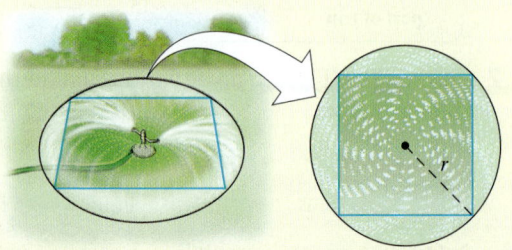

△ **67.** A sprinkler that sprays water in a circular pattern is to be used to water a square garden. If the area of the garden is 920 square feet, find the smallest whole number *radius* that the sprinkler can be adjusted to so that the entire garden is watered.   22 ft

△ **68.** Suppose that a square field has an area of 6270 square feet. See Exercise 67 and find the required sprinkler radius.   56 ft

Objective B *Solve. For Exercises 59 and 60, the solutions have been started for you. See Examples 6 and 7.*

55. A jogger ran 3 miles, decreased her speed by 1 mile per hour, and then ran another 4 miles. If her total time jogging was $1\frac{3}{5}$ hours, find her speed for each part of her run.   5 mph, then 4 mph

56. Mark Keaton's workout consists of jogging for 3 miles and then riding his bike for 5 miles at a speed 4 miles per hour faster than he jogs. If his total workout time is 1 hour, find his jogging speed and his biking speed.   jogging: 6 mph; biking: 10 mph

57. A Chinese restaurant in Mandeville, Louisiana, has a large goldfish pond around the restaurant. Suppose that an inlet pipe and a hose together can fill the pond in 8 hours. The inlet pipe alone can complete the job in 1 hour less time than the hose alone. Find the time that the hose can complete the job alone and the time that the inlet pipe can complete the job alone. Round each to the nearest tenth of an hour.   inlet pipe: 15.5 hr; hose: 16.5 hr

58. A water tank on a farm in Flatonia, Texas, can be filled with a large inlet pipe and a small inlet pipe in 3 hours. The large inlet pipe alone can fill the tank in 2 hours less time than the small inlet pipe alone. Find the time to the nearest tenth of an hour each pipe can fill the tank alone.   small pipe: 7.2 hr; large pipe: 5.2 hr

59. Roma Sherry drove 330 miles from her home town to Tucson. During her return trip, she was able to increase her speed by 11 miles per hour. If her return trip took 1 hour less time, find her original speed and her speed returning home.   55 mph; 66 mph

Start the solution:
1. UNDERSTAND the problem. Reread it as many times as needed. Let

$$x = \text{original speed}$$
$$x + 11 = \text{return-trip speed}$$

Organize the information in a table.

	distance =	rate ·	time	
To Tucson	330	x	$\dfrac{330}{x}$	← distance ← rate
Return trip	330	___	$\dfrac{330}{\_\_}$	← distance ← rate

2. TRANSLATE into an equation. (Fill in the blanks below.)

Time to Tucson	equals	Return trip time	plus	1 hour
↓	↓	↓	↓	↓
___	=	___	+	1

Finish with:
3. SOLVE and 4. INTERPRET

60. A salesperson drove to Portland, a distance of 300 miles. During the last 80 miles of his trip, heavy rainfall forced him to decrease his speed by 15 miles per hour. If his total driving time was 6 hours, find his original speed and his speed during the rainfall.   55 mph; 40 mph

Start the solution:
1. UNDERSTAND the problem. Reread it as many times as needed. Let

$$x = \text{original speed}$$
$$x - 15 = \text{rainfall speed}$$

Organize the information in a table.

	distance =	rate ·	time	
First part of trip	300 − 80, or 220	x	$\dfrac{220}{x}$	← distance ← rate
Heavy rainfall part of trip	80	$x - 15$	$\dfrac{80}{\_\_}$	← distance ← rate

2. TRANSLATE into an equation. (Fill in the blanks below.)

Time during first part of trip	plus	Time during heavy rainfall	equals	6 hr
↓	↓	↓	↓	↓
___	+	___	=	6

Finish with:
3. SOLVE and 4. INTERPRET

61. Bill Shaughnessy and his son Billy can clean the house together in 4 hours. When the son works alone, it takes him an hour longer to clean than it takes his dad alone. Find how long to the nearest tenth of an hour it takes the son to clean alone.   8.5 hr

62. Together, Noodles and Freckles eat a 50-pound bag of dog food in 30 days. Noodles by herself eats a 50-pound bag in 2 weeks less time than Freckles does by himself. How many days to the nearest whole day would a 50-pound bag of dog food last Freckles?   68 days

63. The product of a number and 4 less than the number is 96. Find the number.   12 or −8

64. A whole number increased by its square is two more than twice itself. Find the number.   2

△ **65.** Suppose that we want to make an open box from a square sheet of cardboard by cutting out squares from each corner as shown and then folding along the dotted lines. If the box is to have a volume of 300 cubic inches, find the original dimensions of the sheet of cardboard.

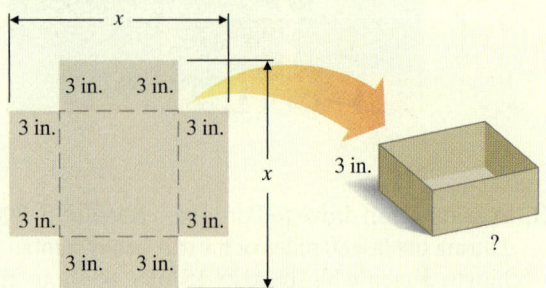

a. The ? in the drawing to the right will be the length (and also the width) of the box as shown in the drawing to the left. Represent this length in terms of x.
$(x − 6)$ in.

b. Use the formula for volume of a box, $V = l \cdot w \cdot h$, to write an equation in x.
$300 = (x − 6) \cdot (x − 6) \cdot 3$

c. Solve the equation for x and give the dimensions of the sheet of cardboard. Check your solution.
16 in. by 16 in.

△ **66.** Suppose that we want to make an open box from a square sheet of cardboard by cutting out squares from each corner as shown and then folding along the dotted lines. If the box is to have a volume of 128 cubic inches, find the original dimensions of the sheet of cardboard. (*Hint:* Use Exercise 65 parts (a), (b), and (c) to help you.)   12 in. by 12 in.

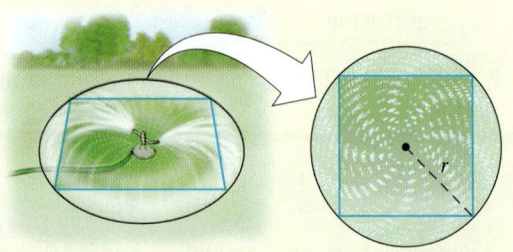

△ **67.** A sprinkler that sprays water in a circular pattern is to be used to water a square garden. If the area of the garden is 920 square feet, find the smallest whole number *radius* that the sprinkler can be adjusted to so that the entire garden is watered.   22 ft

△ **68.** Suppose that a square field has an area of 6270 square feet. See Exercise 67 and find the required sprinkler radius.   56 ft

Review

Solve each inequality. See Section 2.7.

69. $\frac{5x}{3} + 2 \le 7$ $(-\infty, 3]$

70. $\frac{2x}{3} + \frac{1}{6} \ge 2$ $\left[\frac{11}{4}, \infty\right)$

71. $\frac{y-1}{15} > -\frac{2}{5}$ $(-5, \infty)$

72. $\frac{z-2}{12} < \frac{1}{4}$ $(-\infty, 5)$

Concept Extensions

Solve.

73. $y^3 + 9y - y^2 - 9 = 0$ $\{1, -3i, 3i\}$

74. $x^3 + x - 3x^2 - 3 = 0$ $\{3, -i, i\}$

75. $x^{-2} - x^{-1} - 6 = 0$ $\left\{-\frac{1}{2}, \frac{1}{3}\right\}$

76. $y^{-2} - 8y^{-1} + 7 = 0$ $\left\{\frac{1}{7}, 1\right\}$

77. $2x^3 = -54$ $\left\{-3, \frac{3 - 3i\sqrt{3}}{2}, \frac{3 + 3i\sqrt{3}}{2}\right\}$

78. $y^3 - 216 = 0$ $\{6, -3 - 3i\sqrt{3}, -3 + 3i\sqrt{3}\}$

79. Write a polynomial equation that has three solutions: $2, 5$, and -7.   answers may vary

80. Write a polynomial equation that has three solutions: $0, 2i$, and $-2i$.   answers may vary

81. At the FIS World Cup Slalom event in Flachau, Austria, on January 14, 2014, Mikaela Shiffrin of the United States beat Frida Hansdotter of Sweden. During the first run, Shiffrin posted a time that was 1.09 seconds faster than Hansdotter's. On the second run, however, Hansdotter beat Shiffrin's time by 0.26 second. The total of the two runs for Shiffrin was 105.83 seconds, and the first run for Hansdotter was 52.81 seconds.

   a. Find the time for Shiffrin's first run.   51.72 sec

   b. Find the time for Shiffrin's second run.   54.11 sec

   c. Find the final time for Hansdotter's two runs.   106.66 sec

   d. Find the time for Hansdotter's second run.   53.85 sec

82. Use a grapher to solve Exercise 29. Compare the solution with the solution from Exercise 29. Explain any differences.   answers may vary

Summary on Solving Quadratic Equations

Answers

1. $\{-\sqrt{10}, \sqrt{10}\}$

2. $\{-2i\sqrt{2}, 2i\sqrt{2}\}$

3. $\{1 - 2\sqrt{2}, 1 + 2\sqrt{2}\}$

4. $\left\{\dfrac{-5 - 2\sqrt{3}}{2}, \dfrac{-5 + 2\sqrt{3}}{2}\right\}$

5. $\{-1 - \sqrt{13}, -1 + \sqrt{13}\}$

6. $\{1, 11\}$

7. $\left\{\dfrac{-3 - \sqrt{17}}{2}, \dfrac{-3 + \sqrt{17}}{2}\right\}$

8. $\left\{\dfrac{-2 - \sqrt{5}}{4}, \dfrac{-2 + \sqrt{5}}{4}\right\}$

9. $\left\{\dfrac{2 - \sqrt{2}}{2}, \dfrac{2 + \sqrt{2}}{2}\right\}$

10. $\{-3 - \sqrt{5}, -3 + \sqrt{5}\}$

11. $\{-2 + i\sqrt{3}, -2 - i\sqrt{3}\}$

12. $\left\{\dfrac{-3 - i\sqrt{6}}{5}, \dfrac{-3 + i\sqrt{6}}{5}\right\}$

13. $\left\{\dfrac{-3 + i\sqrt{15}}{2}, \dfrac{-3 - i\sqrt{15}}{2}\right\}$

14. $\{3i, -3i\}$

15. $\{0, -17\}$

16. $\left\{\dfrac{1 + \sqrt{13}}{4}, \dfrac{1 - \sqrt{13}}{4}\right\}$

17. $\{2 + 3\sqrt{3}, 2 - 3\sqrt{3}\}$

18. $\{2 + \sqrt{3}, 2 - \sqrt{3}\}$

Teaching Tip

If you prefer to assign exercises with real number solutions only, omit Exercises 2, 11–14, and 24.

Use the square root property to solve each equation.

1. $x^2 - 10 = 0$

2. $3x^2 + 24 = 0$

3. $(x - 1)^2 = 8$

4. $(2x + 5)^2 = 12$

Solve each equation by completing the square.

5. $x^2 + 2x - 12 = 0$

6. $x^2 - 12x + 11 = 0$

7. $4x^2 + 12x = 8$

8. $16y^2 + 16y = 1$

Use the quadratic formula to solve each equation.

9. $2x^2 - 4x + 1 = 0$

10. $\dfrac{1}{2}x^2 + 3x + 2 = 0$

11. $x^2 + 4x = -7$

12. $5x^2 + 6x = -3$

Solve each equation. Use the method of your choice.

13. $x^2 + 3x + 6 = 0$

14. $2x^2 + 18 = 0$

15. $x^2 + 17x = 0$

16. $4x^2 - 2x - 3 = 0$

17. $(x - 2)^2 = 27$

18. $\dfrac{1}{2}x^2 - 2x + \dfrac{1}{2} = 0$

19. $3x^2 + 2x = 8$

20. $2x^2 = -5x - 1$

21. $x(x - 2) = 5$

22. $x^2 - 31 = 0$

23. $4x^2 - 48 = 0$

24. $5x^2 + 55 = 0$

25. $x(x + 5) = 66$

26. $5x^2 + 6x - 2 = 0$

27. $2x^2 + 3x = 1$

28. $x - \sqrt{13 - 3x} - 3 = 0$

29. $\dfrac{5x}{x - 2} - \dfrac{x + 1}{x} = \dfrac{3}{x(x - 2)}$

△ **30.** The diagonal of a square room measures 20 feet. Find the exact length of a side of the room. Then approximate the length to the nearest tenth of a foot.

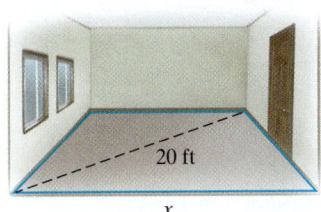

31. Jack and Lucy Hoag together can prepare a crawfish boil for a large party in 4 hours. Lucy alone can complete the job in 2 hours less time than Jack alone. Find the time in which each person can prepare the crawfish boil alone. Round each time to the nearest tenth of an hour.

32. Kraig Blackwelder exercises at Total Body Gym. On the treadmill, he runs 5 miles, then increases his speed by 1 mile per hour and runs an additional 2 miles. If his total time on the treadmill is $1\dfrac{1}{3}$ hours, find his speed during each part of his run.

19. $\left\{-2, \dfrac{4}{3}\right\}$

20. $\left\{\dfrac{-5 + \sqrt{17}}{4}, \dfrac{-5 - \sqrt{17}}{4}\right\}$

21. $\{1 - \sqrt{6}, 1 + \sqrt{6}\}$

22. $\{-\sqrt{31}, \sqrt{31}\}$

23. $\{-2\sqrt{3}, 2\sqrt{3}\}$

24. $\{-i\sqrt{11}, i\sqrt{11}\}$

25. $\{-11, 6\}$

26. $\left\{\dfrac{-3 + \sqrt{19}}{5}, \dfrac{-3 - \sqrt{19}}{5}\right\}$

27. $\left\{\dfrac{-3 + \sqrt{17}}{4}, \dfrac{-3 - \sqrt{17}}{4}\right\}$

28. $\{4\}$

29. $\left\{\dfrac{-1 + \sqrt{17}}{8}, \dfrac{-1 - \sqrt{17}}{8}\right\}$

30. $10\sqrt{2}$ ft ≈ 14.1 ft

31. Jack: 9.1 hr; Lucy: 7.1 hr

32. 5 mph during the first part, then 6 mph

Objectives

A Solve Polynomial Inequalities of Degree 2 or Greater. ▶

B Solve Inequalities That Contain Rational Expressions with Variables in the Denominator. ▶

Teaching Tip

Have students use a graphing calculator to check solutions in this section.

Objective A Solving Polynomial Inequalities ▶

Just as we can solve linear inequalities in one variable, we can also solve quadratic and higher-degree inequalities in one variable. Let's begin with quadratic inequalities. A **quadratic inequality** is an inequality that can be written so that one side is a quadratic expression and the other side is 0. Here are examples of quadratic inequalities in one variable. Each is written in **standard form.**

$$x^2 - 10x + 7 \leq 0 \qquad 3x^2 + 2x - 6 > 0$$
$$2x^2 + 9x - 2 < 0 \qquad x^2 - 3x + 11 \geq 0$$

A solution of a quadratic inequality in one variable is a value of the variable that makes the inequality a true statement.

The value of an expression such as $x^2 - 3x - 10$ will sometimes be positive, sometimes negative, and sometimes 0, depending on the value substituted for x. To solve the inequality $x^2 - 3x - 10 < 0$, we look for all values of x that make the expression $x^2 - 3x - 10$ less than 0, or negative. To understand how we find these values, we'll study the graph of the quadratic function $y = x^2 - 3x - 10$.

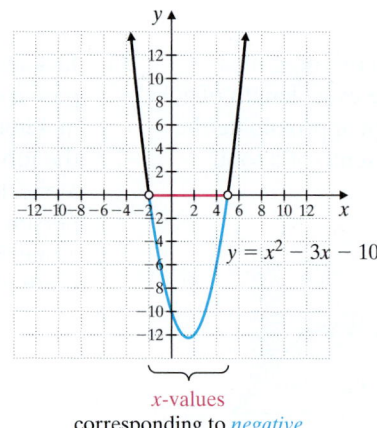

x-values
corresponding to *negative*
y-values

Notice that the x-values for which y or $x^2 - 3x - 10$ is positive are separated from the x-values for which y or $x^2 - 3x - 10$ is negative by the values for which y or $x^2 - 3x - 10$ is 0, the x-intercepts. Thus, the solution set of $x^2 - 3x - 10 < 0$ consists of all real numbers from -2 to 5 or, in interval notation, $(-2, 5)$.

It is not necessary to graph $y = x^2 - 3x - 10$ to solve the related inequality $x^2 - 3x - 10 < 0$. Instead, we can draw a number line representing the x-axis and keep the following in mind: *A region on the number line for which the value of $x^2 - 3x - 10$ is positive is separated from a region on the number line for which the value of $x^2 - 3x - 10$ is negative by a value for which the expression is 0.*

Let's find these values for which the expression is 0 by solving the related equation, $x^2 - 3x - 10 = 0$.

$$x^2 - 3x - 10 = 0$$
$$(x - 5)(x + 2) = 0 \qquad \text{Factor.}$$
$$x - 5 = 0 \quad \text{or} \quad x + 2 = 0 \qquad \text{Set each factor equal to 0.}$$
$$x = 5 \qquad\qquad x = -2 \qquad \text{Solve.}$$

These two numbers -2 and 5 divide the number line into three regions. We will call the regions A, B, and C.

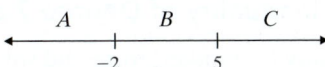

These regions are important because if the value of $x^2 - 3x - 10$ is negative when a number from a region is substituted for x, then $x^2 - 3x - 10$ is negative when any number in that region is substituted for x. Similarly, if the value of $x^2 - 3x - 10$ is positive when a number from a region is substituted for x, then $x^2 - 3x - 10$ is positive when any number in that region is substituted for x.

To see whether the inequality $x^2 - 3x - 10 < 0$ is true or false in each region, we choose a test point from each region and substitute its value for x in the inequality $x^2 - 3x - 10 < 0$. If the resulting inequality is true, the region containing the test point is a solution region.

Region	Test Point Value	$(x - 5)(x + 2) < 0$	Result
A	-3	$(-8)(-1) < 0$	False
B	0	$(-5)(2) < 0$	True
C	6	$(1)(8) < 0$	False

The values in region B satisfy the inequality. The numbers -2 and 5 are not included in the solution set since the inequality symbol is $<$. The solution set is $(-2, 5)$, and its graph is shown.

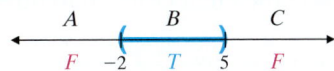

Example 1   Solve: $(x + 3)(x - 3) > 0$

Solution: First we solve the related equation, $(x + 3)(x - 3) = 0$.

$$(x + 3)(x - 3) = 0$$
$$x + 3 = 0 \quad \text{or} \quad x - 3 = 0$$
$$x = -3 \qquad\qquad x = 3$$

The two numbers -3 and 3 separate the number line into three regions, A, B, and C.

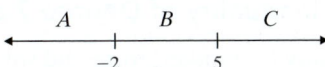

Now we substitute the value of a test point from each region. If the test value satisfies the inequality, every value in the region containing the test value is a solution.

Region	Test Point Value	$(x + 3)(x - 3) > 0$	Result
A	-4	$(-1)(-7) > 0$	True
B	0	$(3)(-3) > 0$	False
C	4	$(7)(1) > 0$	True

The points in regions A and C satisfy the inequality. The numbers -3 and 3 are not included in the solution since the inequality symbol is $>$. The solution set is $(-\infty, -3) \cup (3, \infty)$, and its graph is shown.

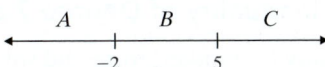

Work Practice 1

Practice 1

Solve: $(x - 2)(x + 4) > 0$

Answer

1. $(-\infty, -4) \cup (2, \infty)$

The steps below may be used to solve a polynomial inequality of degree 2 or greater.

Solving a Polynomial Inequality of Degree 2 or Greater

Step 1: Write the inequality in standard form and solve the related equation.

Step 2: Separate the number line into regions using the solutions from Step 1.

Step 3: For each region, choose a test point and determine whether its value satisfies the *original inequality*.

Step 4: The solution set includes the regions whose test point value is a solution. If the inequality symbol is $\leq$ or $\geq$, the values from Step 1 are solutions; if $<$ or $>$, they are not.

✔**Concept Check** When choosing a test point in Step 4, why would the solutions from Step 1 not make good choices for test points?

Practice 2

Solve: $x^2 - 6x \leq 0$

Example 2 Solve: $x^2 - 4x \leq 0$

Solution: First we solve the related equation, $x^2 - 4x = 0$.

$$x^2 - 4x = 0$$
$$x(x - 4) = 0$$
$$x = 0 \quad \text{or} \quad x = 4$$

The numbers 0 and 4 separate the number line into three regions, A, B, and C.

We check a test value in each region in the original inequality. Values in region B satisfy the inequality. The numbers 0 and 4 are included in the solution since the inequality symbol is $\leq$. The solution set is $[0, 4]$, and its graph is shown.

Teaching Tip

After discussing Example 2, ask, "If an inequality is true in a region, is it always false in the regions on either side of that region?" Have students work in pairs to investigate the answer. Then have several pairs share their conclusions and how they reached them with the class.

■ **Work Practice 2**

Practice 3

Solve:
$(x - 2)(x + 1)(x + 5) \leq 0$

Example 3 Solve: $(x + 2)(x - 1)(x - 5) \leq 0$

Solution: First we solve $(x + 2)(x - 1)(x - 5) = 0$. By inspection, we see that the solutions are -2, 1, and 5. They separate the number line into four regions, A, B, C, and D. Next we check test points from each region.

Region	Test Point Value	$(x + 2)(x - 1)(x - 5) \leq 0$	Result
A	-3	$(-1)(-4)(-8) \leq 0$	True
B	0	$(2)(-1)(-5) \leq 0$	False
C	2	$(4)(1)(-3) \leq 0$	True
D	6	$(8)(5)(1) \leq 0$	False

The solution set is $(-\infty, -2] \cup [1, 5]$, and its graph is shown. We include the numbers -2, 1, and 5 because the inequality symbol is $\leq$.

Answers

2. $[0, 6]$ **3.** $(-\infty, -5] \cup [-1, 2]$

✔**Concept Check Answer**

The solutions found in Step 1 have a value of 0 in the original inequality.

■ **Work Practice 3**

Objective B Solving Rational Inequalities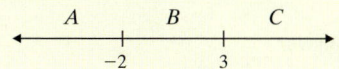

Inequalities containing rational expressions with variables in the denominator are solved by using a similar procedure. Notice as we solve an example that unlike quadratic inequalities, we must also consider values for which the rational inequality is undefined. Why? As usual, these values may not be solution values for the inequality.

Example 4 Solve: $\dfrac{x + 2}{x - 3} \le 0$

Solution: First we find all values that make the denominator equal to 0. To do this, we solve $x - 3 = 0$, or $x = 3$.

Next, we solve the related equation, $\dfrac{x + 2}{x - 3} = 0$.

$$\dfrac{x + 2}{x - 3} = 0 \qquad \text{Multiply both sides by the LCD, } x - 3.$$

$$x + 2 = 0$$

$$x = -2$$

Now we place these numbers on a number line and proceed as before, checking test point values in the original inequality.

Choose -3 from region A	Choose 0 from region B	Choose 4 from region C
$\dfrac{x + 2}{x - 3} \le 0$	$\dfrac{x + 2}{x - 3} \le 0$	$\dfrac{x + 2}{x - 3} \le 0$
$\dfrac{-3 + 2}{-3 - 3} \le 0$	$\dfrac{0 + 2}{0 - 3} \le 0$	$\dfrac{4 + 2}{4 - 3} \le 0$
$\dfrac{-1}{-6} \le 0$	$-\dfrac{2}{3} \le 0$ True	$6 \le 0$ False
$\dfrac{1}{6} \le 0$ False		

The solution set is $[-2, 3)$. This interval includes -2 because -2 satisfies the original inequality. This interval does not include 3 because 3 would make the denominator 0.

```
        A        B        C
←———————[————————)————————→
     F  -2    T   3    F
```

■ **Work Practice 4**

The steps below may be used to solve a rational inequality with variables in the denominator.

Solving a Rational Inequality

Step 1: Solve for values that make any denominator 0.

Step 2: Solve the related equation.

Step 3: Separate the number line into regions using the solutions from Steps 1 and 2.

Step 4: For each region, choose a test point and determine whether its value satisfies the *original inequality*.

Step 5: The solution set includes the regions whose test point value is a solution. Check whether to include values from Step 2. Be sure *not* to include values that make any denominator 0.

Practice 5

Solve: $\dfrac{3}{x-2} < 2$

Teaching Tip

After discussing Example 5, have students verify the solution by graphing $y = 5/(x+1)$ and $y = -2$ on a graphing calculator.

Answer

5. $(-\infty, 2) \cup \left(\dfrac{7}{2}, \infty\right)$

Example 5 Solve: $\dfrac{5}{x+1} < -2$

Solution: First we find values for x that make the denominator equal to 0.

$$x + 1 = 0$$
$$x = -1$$

Next we solve $\dfrac{5}{x+1} = -2$.

$$(x+1) \cdot \dfrac{5}{x+1} = (x+1) \cdot -2 \quad \text{Multiply both sides by the LCD, } x + 1.$$
$$5 = -2x - 2 \quad \text{Simplify.}$$
$$7 = -2x$$
$$-\dfrac{7}{2} = x$$

We use these two solutions to divide a number line into three regions and choose test points. Only a test point value from region B satisfies the *original inequality*. The solution set is $\left(-\dfrac{7}{2}, -1\right)$, and its graph is shown.

Work Practice 5

Vocabulary, Readiness & Video Check

Write the graphed solution set in interval notation.

1.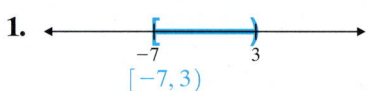
$[-7, 3)$

2.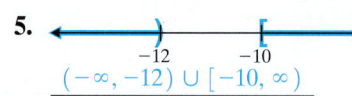
$(-1, 5]$

3.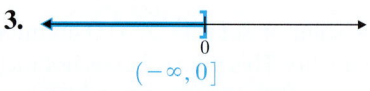
$(-\infty, 0]$

4.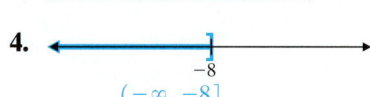
$(-\infty, -8]$

5.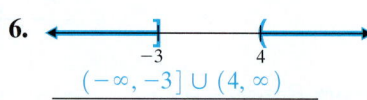
$(-\infty, -12) \cup [-10, \infty)$

6.
$(-\infty, -3] \cup (4, \infty)$

Martin-Gay Interactive Videos

See Video 11.4

See video answer section.

Watch the section lecture video and answer the following questions.

Objective A 7. From ▣ Examples 1–3, how does solving a related equation help you solve a polynomial inequality? Are the solutions to the related equation ever solutions to the inequality? ▶

Objective B 8. In ▣ Example 4, one of the values that separates the number line into regions is 4. The inequality is $\geq$, so why isn't 4 included in the solution set? ▶

11.4 Exercise Set MyMathLab®

Objective A *Solve. Write the solution set in interval notation. See Examples 1 through 3.*

1. $(x + 1)(x + 5) > 0$
$(-\infty, -5) \cup (-1, \infty)$

2. $(x + 1)(x + 5) \leq 0$
$[-5, -1]$

▶**3.** $(x - 3)(x + 4) \leq 0$
$[-4, 3]$

4. $(x + 4)(x - 1) > 0$
$(-\infty, -4) \cup (1, \infty)$

5. $x^2 + 8x + 15 \geq 0$
$(-\infty, -5] \cup [-3, \infty)$

6. $x^2 - 7x + 10 \leq 0$
$[2, 5]$

7. $3x^2 + 16x < -5$
$\left(-5, -\dfrac{1}{3}\right)$

8. $2x^2 - 5x < 7$
$\left(-1, \dfrac{7}{2}\right)$

9. $(x - 6)(x - 4)(x - 2) > 0$
$(2, 4) \cup (6, \infty)$

10. $(x - 6)(x - 4)(x - 2) \leq 0$
$(-\infty, 2] \cup [4, 6]$

11. $x(x - 1)(x + 4) \leq 0$
$(-\infty, -4] \cup [0, 1]$

12. $x(x - 6)(x + 2) > 0$
$(-2, 0) \cup (6, \infty)$

13. $(x^2 - 9)(x^2 - 4) > 0$
$(-\infty, -3) \cup (-2, 2) \cup (3, \infty)$

14. $(x^2 - 16)(x^2 - 1) \leq 0$
$[-4, -1] \cup [1, 4]$

Objective B *Solve. Write the solution set in interval notation. See Examples 4 and 5.*

15. $\dfrac{x + 7}{x - 2} < 0$
$(-7, 2)$

16. $\dfrac{x - 5}{x - 6} > 0$
$(-\infty, 5) \cup (6, \infty)$

17. $\dfrac{5}{x + 1} > 0$
$(-1, \infty)$

18. $\dfrac{3}{y - 5} < 0$
$(-\infty, 5)$

▶**19.** $\dfrac{x + 1}{x - 4} \geq 0$
$(-\infty, -1] \cup (4, \infty)$

20. $\dfrac{x + 1}{x - 4} \leq 0$
$[-1, 4)$

21. $\dfrac{3}{x - 2} < 4$
$(-\infty, 2) \cup \left(\dfrac{11}{4}, \infty\right)$

22. $\dfrac{-2}{y + 3} > 2$
$(-4, -3)$

23. $\dfrac{x^2 + 6}{5x} \geq 1$
$(0, 2] \cup [3, \infty)$

24. $\dfrac{y^2 + 15}{8y} \leq 1$
$(-\infty, 0) \cup [3, 5]$

25. $\dfrac{x + 2}{x - 3} < 1$
$(-\infty, 3)$

26. $\dfrac{x - 1}{x + 4} > 2$
$(-9, -4)$

Objectives A B Mixed Practice *Solve each inequality. Write the solution set in interval notation. See Examples 1 through 5.*

27. $(2x - 3)(4x + 5) \leq 0$
$\left[-\dfrac{5}{4}, \dfrac{3}{2}\right]$

28. $(6x + 7)(7x - 12) > 0$
$\left(-\infty, -\dfrac{7}{6}\right) \cup \left(\dfrac{12}{7}, \infty\right)$

▶**29.** $x^2 > x$
$(-\infty, 0) \cup (1, \infty)$

30. $x^2 < 25$
$(-5, 5)$

31. $\dfrac{x}{x - 10} < 0$
$(0, 10)$

32. $\dfrac{x + 10}{x - 10} > 0$
$(-\infty, -10) \cup (10, \infty)$

33. $(2x - 8)(x + 4)(x - 6) \leq 0$
$(-\infty, -4] \cup [4, 6]$

34. $(3x - 12)(x + 5)(2x - 3) \geq 0$
$\left[-5, \dfrac{3}{2}\right] \cup [4, \infty)$

35. $6x^2 - 5x \geq 6$
$\left(-\infty, -\dfrac{2}{3}\right] \cup \left[\dfrac{3}{2}, \infty\right)$

36. $12x^2 + 11x \leq 15$

$\left[-\dfrac{5}{3}, \dfrac{3}{4} \right]$

37. $\dfrac{x-5}{x+4} \geq 0$

$(-\infty, -4) \cup [5, \infty)$

38. $\dfrac{x-3}{x+2} \leq 0$

$(-2, 3]$

39. $\dfrac{-1}{x-1} > -1$

$(-\infty, 1) \cup (2, \infty)$

40. $\dfrac{4}{y+2} < -2$

$(-4, -2)$

41. $4x^3 + 16x^2 - 9x - 36 > 0$

$\left(-4, -\dfrac{3}{2} \right) \cup \left(\dfrac{3}{2}, \infty \right)$

42. $x^3 + 2x^2 - 4x - 8 < 0$

$(-\infty, -2) \cup (-2, 2)$

▶ 43. $x^4 - 26x^2 + 25 \geq 0$

$(-\infty, -5] \cup [-1, 1] \cup [5, \infty)$

44. $16x^4 - 40x^2 + 9 \leq 0$

$\left[-\dfrac{3}{2}, -\dfrac{1}{2} \right] \cup \left[\dfrac{1}{2}, \dfrac{3}{2} \right]$

45. $\dfrac{x(x+6)}{(x-7)(x+1)} \geq 0$

$(-\infty, -6] \cup (-1, 0] \cup (7, \infty)$

46. $\dfrac{(x-2)(x+2)}{(x+1)(x-4)} \leq 0$

$[-2, -1) \cup [2, 4)$

47. $\dfrac{x}{x+4} \leq 2$

$(-\infty, -8] \cup (-4, \infty)$

48. $\dfrac{4x}{x-3} \geq 5$

$(3, 15]$

49. $(2x-7)(3x+5) > 0$

$\left(-\infty, -\dfrac{5}{3} \right) \cup \left(\dfrac{7}{2}, \infty \right)$

50. $(4x-9)(2x+5) < 0$

$\left(-\dfrac{5}{2}, \dfrac{9}{4} \right)$

51. $\dfrac{z}{z-5} \geq 2z$

$(-\infty, 0] \cup \left(5, \dfrac{11}{2} \right]$

52. $\dfrac{p}{p+4} \leq 3p$

$\left(-4, -\dfrac{11}{3} \right] \cup [0, \infty)$

53. $\dfrac{(x+1)^2}{5x} > 0$

$(0, \infty)$

54. $\dfrac{(2x-3)^2}{x} < 0$

$(-\infty, 0)$

Review

Fill in each table so that each ordered pair is a solution of the given function. See Section 8.2.

55. $f(x) = x^2$

x	y
0	0
1	1
−1	1
2	4
−2	4

56. $f(x) = 2x^2$

x	y
0	0
1	2
−1	2
2	8
−2	8

57. $f(x) = -x^2$

x	y
0	0
1	−1
−1	−1
2	−4
−2	−4

58. $f(x) = -3x^2$

x	y
0	0
1	−3
−1	−3
2	−12
−2	−12

Concept Extensions

59. Explain why $\dfrac{x+2}{x-3} > 0$ and $(x+2)(x-3) > 0$ have the same solution sets.   answers may vary

60. Explain why $\dfrac{x+2}{x-3} \geq 0$ and $(x+2)(x-3) \geq 0$ do not have the same solution sets.   answers may vary

Find all numbers that satisfy each statement.

61. A number minus its reciprocal is less than zero. Find the numbers. $(-\infty, -1) \cup (0, 1)$

62. Twice a number, added to its reciprocal is nonnegative. Find the numbers. $(0, \infty)$

Solve.

63. The total profit $P(x)$ for a company producing x thousand units is given by the function $P(x) = -2x^2 + 26x - 44$. Find the values of x for which the company makes a profit. [*Hint:* The company makes a profit when $P(x) > 0$.] when x is between 2 and 11

64. A projectile is fired straight up from the ground with an initial velocity of 80 feet per second. Its height $s(t)$ in feet at any time t in seconds is given by the function $s(t) = -16t^2 + 80t$. Find the interval of time for which the height of the projectile is greater than 96 feet. between 2 and 3 sec

Solve each inequality, then use a graphing calculator to check.

65. $x^2 - x - 56 > 0$ $(-\infty, -7) \cup (8, \infty)$

66. $x^2 - 4x - 5 < 0$ $(-1, 5)$

Note to Instructor: If you covered Section 8.5, this section is mostly a review. In Section 8.5, the graph of $f(x) = ax^2$ was not covered, but it is in this section.

11.5 Quadratic Functions and Their Graphs

Objective A Graphing $f(x) = x^2 + k$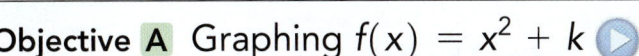

We first graphed the quadratic function $f(x) = x^2$ in Section 8.5. In that section, we discovered that the graph of a quadratic function is a parabola opening upward or downward. Now, as we continue our study, we will discover more details about quadratic functions and their graphs.

First, let's recall the definition of a *quadratic function*.

> **Quadratic Function**
>
> A **quadratic function** is a function that can be written in the form $f(x) = ax^2 + bx + c$, where $a, b,$ and c are real numbers and $a \neq 0$.

Notice that equations of the form $y = ax^2 + bx + c$, where $a \neq 0$, also define quadratic functions since y is a function of x or $y = f(x)$.

Recall that if $a > 0$, the parabola opens upward, and if $a < 0$, the parabola opens downward. Also, the vertex of a parabola is the lowest point if the parabola opens upward and the highest point if the parabola opens downward. The axis of symmetry is the vertical line that passes through the vertex.

Objectives

A Graph Quadratic Functions of the Form $f(x) = x^2 + k$.

B Graph Quadratic Functions of the Form $f(x) = (x - h)^2$.

C Graph Quadratic Functions of the Form $f(x) = (x - h)^2 + k$.

D Graph Quadratic Functions of the Form $f(x) = ax^2$.

E Graph Quadratic Functions of the Form $f(x) = a(x - h)^2 + k$.

Teaching Tip

Have students use a graphing calculator to graph several equations of the form $y = x^2 + k$ or $y = x^2 - k$, where k is any number. Suggest that they sketch the graph first and then use the graphing calculator to verify.

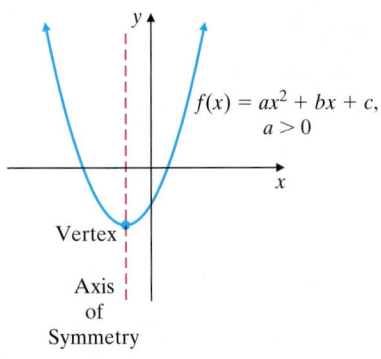

$f(x) = ax^2 + bx + c,$
$a > 0$

Vertex
Axis of Symmetry

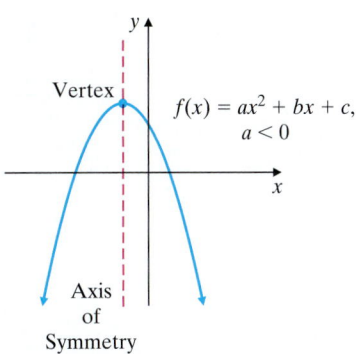

Vertex
$f(x) = ax^2 + bx + c,$
$a < 0$

Axis of Symmetry

Practice 1

Graph $f(x) = x^2$ and $g(x) = x^2 + 4$ on the same set of axes.

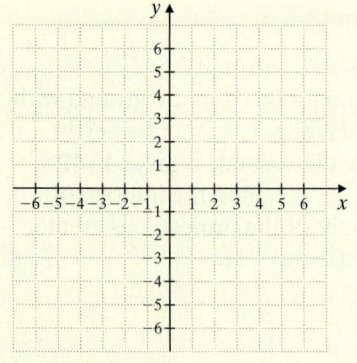

Practice 2–3

Graph each function.

2. $F(x) = x^2 + 1$

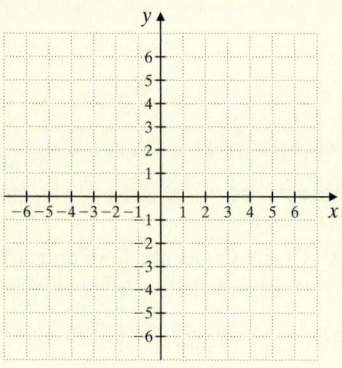

3. $g(x) = x^2 - 2$

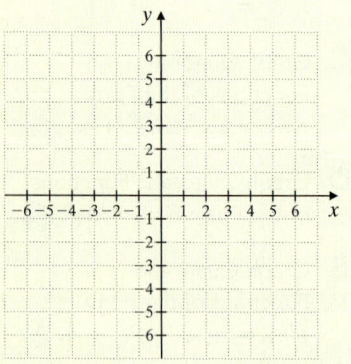

Answers

1.

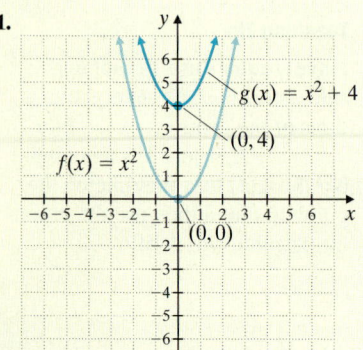

2–3. See answers on page 807.

Example 1 Graph $f(x) = x^2$ and $g(x) = x^2 + 3$ on the same set of axes.

Solution: First we construct a table of values for f and plot the points. Notice that for each x-value, the corresponding value of $g(x)$ must be 3 more than the corresponding value of $f(x)$, since $f(x) = x^2$ and $g(x) = x^2 + 3$. In other words, the graph of $g(x) = x^2 + 3$ is the same as the graph of $f(x) = x^2$ shifted upward 3 units. The axis of symmetry for both graphs is the y-axis.

x	$f(x) = x^2$	$g(x) = x^2 + 3$
-2	4	7
-1	1	4
0	0	3
1	1	4
2	4	7

Each y-value is increased by 3.

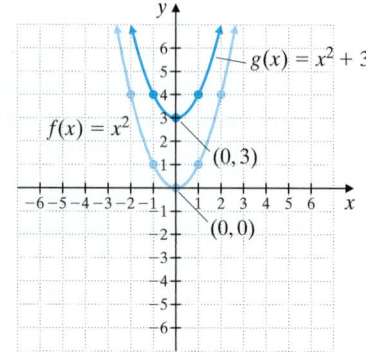

■ **Work Practice 1**

In general, we have the following properties.

> ### Graphing the Parabola Defined by $f(x) = x^2 + k$
>
> If k is positive, the graph of $f(x) = x^2 + k$ is the graph of $y = x^2$ shifted upward k units.
>
> If k is negative, the graph of $f(x) = x^2 + k$ is the graph of $y = x^2$ shifted downward $|k|$ units.
>
> The vertex is $(0, k)$, and the axis of symmetry is the y-axis.

Examples Graph each function.

2. $F(x) = x^2 + 2$

The graph of $F(x) = x^2 + 2$ is obtained by shifting the graph of $y = x^2$ upward 2 units.

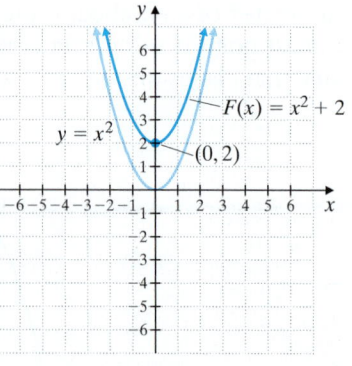

3. $g(x) = x^2 - 3$

The graph of $g(x) = x^2 - 3$ is obtained by shifting the graph of $y = x^2$ downward 3 units.

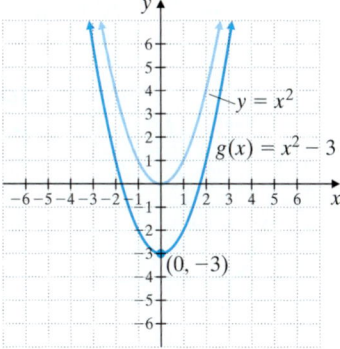

Work Practice 2–3

Objective B Graphing $f(x) = (x - h)^2$ ▶

Now we will graph functions of the form $f(x) = (x - h)^2$.

Example 4 Graph $f(x) = x^2$ and $g(x) = (x - 2)^2$ on the same set of axes.

Solution: By plotting points, we see that for each x-value, the corresponding value of $g(x)$ is the same as the value of $f(x)$ when the x-value is increased by 2. Thus, the graph of $g(x) = (x - 2)^2$ is the graph of $f(x) = x^2$ shifted to the right 2 units. The axis of symmetry for the graph of $g(x) = (x - 2)^2$ is also shifted 2 units to the right and is the line $x = 2$.

x	$f(x) = x^2$	x	$g(x) = (x - 2)^2$
-2	4	0	4
-1	1	1	1
0	0	2	0
1	1	3	1
2	4	4	4

Each x-value increased by 2 corresponds to same y-value.

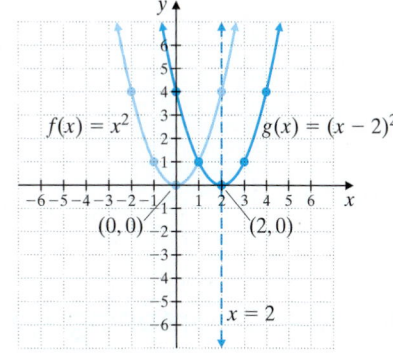

Work Practice 4

Teaching Tip

Before discussing the solution to Example 4, ask, "What value of x makes $x - 2 = 0$?"

In general, we have the following properties.

> ## Graphing the Parabola Defined by $f(x) = (x - h)^2$
>
> If h is positive, the graph of $f(x) = (x - h)^2$ is the graph of $y = x^2$ shifted to the right h units.
>
> If h is negative, the graph of $f(x) = (x - h)^2$ is the graph of $y = x^2$ shifted to the left $|h|$ units.
>
> The vertex is $(h, 0)$, and the axis of symmetry is the vertical line $x = h$.

Practice 4

Graph $f(x) = x^2$ and $g(x) = (x - 1)^2$ on the same set of axes.

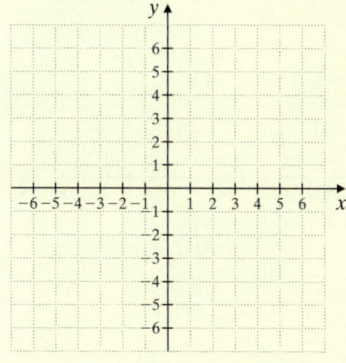

Answers

2.

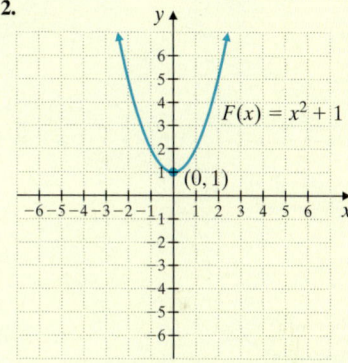

3.

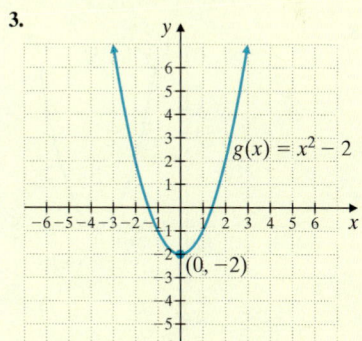

4.

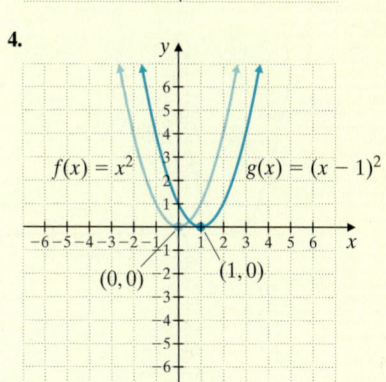

Practice 5–6

Graph each function.

5. $G(x) = (x - 4)^2$

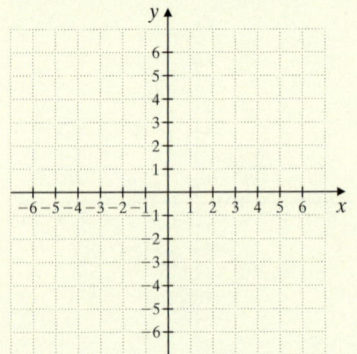

6. $F(x) = (x + 2)^2$

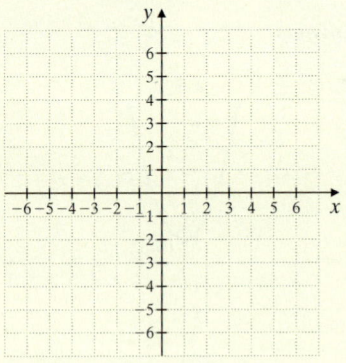

Answers

5.

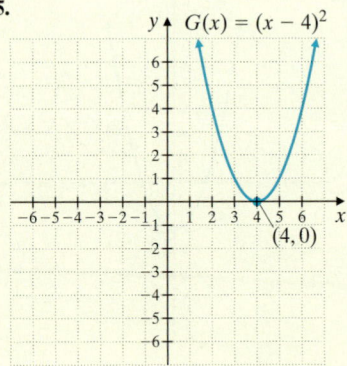

6.

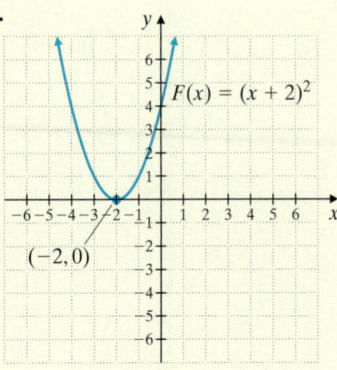

Examples Graph each function.

5. $G(x) = (x - 3)^2$

The graph of $G(x) = (x - 3)^2$ is obtained by shifting the graph of $y = x^2$ to the right 3 units.

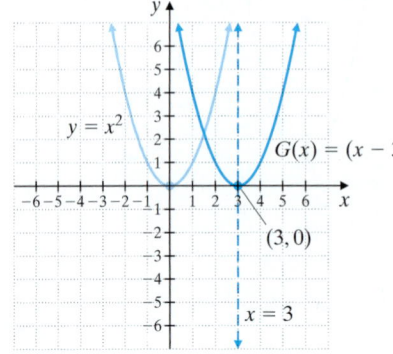

Teaching Tip

Before discussing the solutions to Examples 5 and 6, suggest, "When graphing functions of the form $y = (x - h)^2$, ask yourself what value of x makes the value within parentheses equal to 0. What question do we need to ask ourselves when graphing the functions in Examples 5 and 6?"

6. $F(x) = (x + 1)^2$

The equation $F(x) = (x + 1)^2$ can be written as $F(x) = [x - (-1)]^2$. The graph of $F(x) = [x - (-1)]^2$ is obtained by shifting the graph of $y = x^2$ to the left 1 unit.

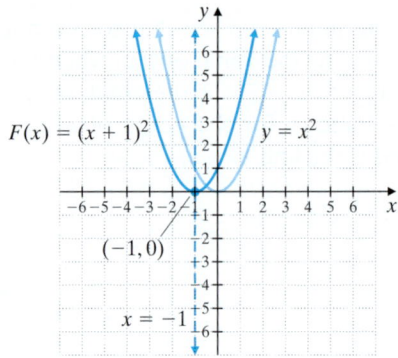

■ **Work Practice 5–6**

Objective C Graphing $f(x) = (x - h)^2 + k$ ▶

As we will see in graphing functions of the form $f(x) = (x - h)^2 + k$, it is possible to combine vertical and horizontal shifts.

Graphing the Parabola Defined by $f(x) = (x - h)^2 + k$

The parabola has the same shape as $y = x^2$.

The vertex is (h, k), and the axis of symmetry is the vertical line $x = h$.

Example 7 Graph: $F(x) = (x - 3)^2 + 1$

Solution: The graph of $F(x) = (x - 3)^2 + 1$ is the graph of $y = x^2$ shifted 3 units to the right and 1 unit up. The vertex is then $(3, 1)$, and the axis of symmetry is $x = 3$. A few ordered pair solutions are plotted to aid in graphing.

x	$F(x) = (x - 3)^2 + 1$
1	5
2	2
4	2
5	5

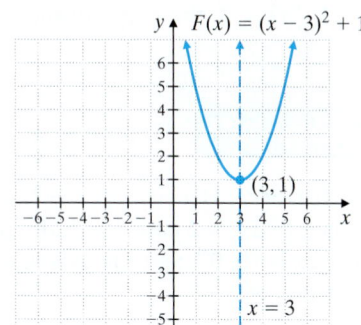

Teaching Tip

Have students use a graphing calculator to graph several equations of the form $y = (x - h)^2$ or $y = (x + h)^2$, where h is any number. Suggest that they sketch the graph first and then use a graphing calculator to verify.

■ **Work Practice 7**

Objective D Graphing $f(x) = ax^2$

Next, we discover the change in the shape of the graph when the coefficient of x^2 is not 1.

Example 8 Graph $f(x) = x^2$, $g(x) = 3x^2$, and $h(x) = \frac{1}{2}x^2$ on the same set of axes.

Solution: Comparing the table of values, we see that for each x-value, the corresponding value of $g(x)$ is triple the corresponding value of $f(x)$. Similarly, the value of $h(x)$ is half the value of $f(x)$.

x	$f(x) = x^2$
-2	4
-1	1
0	0
1	1
2	4

x	$g(x) = 3x^2$
-2	12
-1	3
0	0
1	3
2	12

x	$h(x) = \frac{1}{2}x^2$
-2	2
-1	$\frac{1}{2}$
0	0
1	$\frac{1}{2}$
2	2

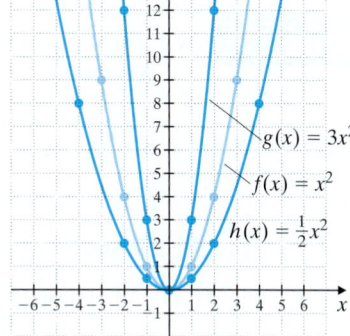

Teaching Tip:
Classroom Activity

Have each student use graph paper and draw a parabola in the shape of $y = x^2$ but in different locations. Have them give their paper to another student to write equations for each parabola.

The result is that the graph of $g(x) = 3x^2$ is narrower than the graph of $f(x) = x^2$ and the graph of $h(x) = \frac{1}{2}x^2$ is wider. The vertex for each graph is $(0, 0)$, and the axis of symmetry is the y-axis.

■ **Work Practice 8**

Practice 7

Graph: $F(x) = (x - 2)^2 + 3$

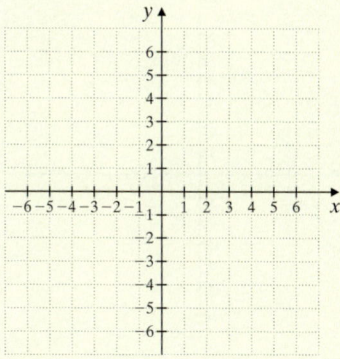

Practice 8

Graph $f(x) = x^2$, $g(x) = 2x^2$, and $h(x) = \frac{1}{3}x^2$ on the same set of axes.

Answers

7.

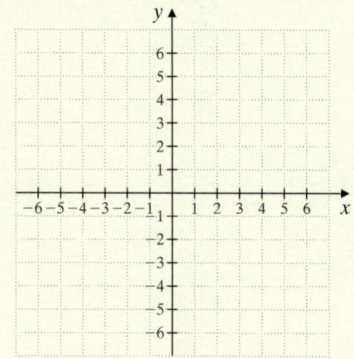

8.

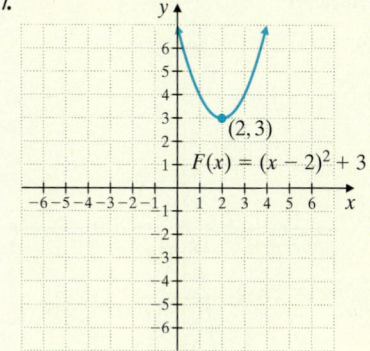
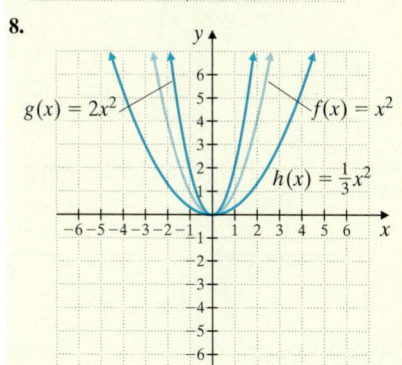

Practice 9

Graph: $f(x) = -3x^2$

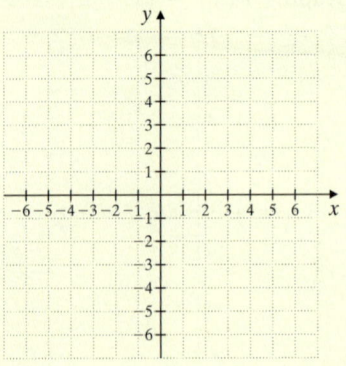

Practice 10

Graph: $f(x) = 2(x + 3)^2 - 4$. Find the vertex and axis of symmetry.

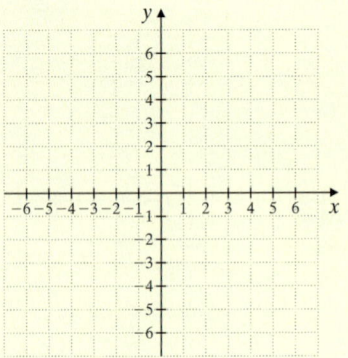

Graphing the Parabola Defined by $f(x) = ax^2$

If a is positive, the parabola opens upward, and if a is negative, the parabola opens downward.

If $|a| > 1$, the graph of the parabola is narrower than the graph of $y = x^2$.

If $|a| < 1$, the graph of the parabola is wider than the graph of $y = x^2$.

Example 9 Graph: $f(x) = -2x^2$

Solution: Because $a = -2$, a negative value, this parabola opens downward. Since $|-2| = 2$ and $2 > 1$, the parabola is narrower than the graph of $y = x^2$. The vertex is $(0, 0)$, and the axis of symmetry is the y-axis. We verify this by plotting a few points.

x	$f(x) = -2x^2$
-2	-8
-1	-2
0	0
1	-2
2	-8

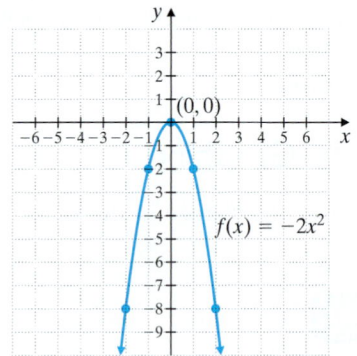

■ **Work Practice 9**

Objective E Graphing $f(x) = a(x - h)^2 + k$ ▶

Now we will see the shape of the graph of a quadratic function of the form $f(x) = a(x - h)^2 + k$.

Example 10 Graph: $g(x) = \frac{1}{2}(x + 2)^2 + 5$. Find the vertex and the axis of symmetry.

Solution: The function $g(x) = \frac{1}{2}(x + 2)^2 + 5$ may be written as $g(x) = \frac{1}{2}[x - (-2)]^2 + 5$. Thus, this graph is the same as the graph of $y = x^2$ shifted 2 units to the left and 5 units upward and widened because a is $\frac{1}{2}$. The vertex is $(-2, 5)$, and the axis of symmetry is $x = -2$. We plot a few points to verify.

x	$g(x) = \frac{1}{2}(x + 2)^2 + 5$
-4	7
-3	$5\frac{1}{2}$
-2	5
-1	$5\frac{1}{2}$
0	7

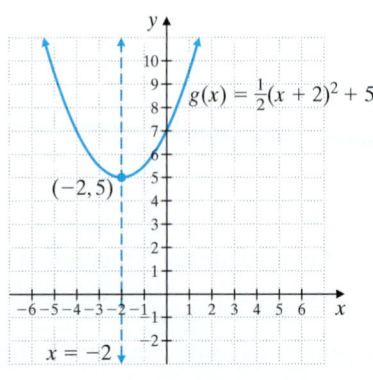

■ **Work Practice 10**

Answers

9.

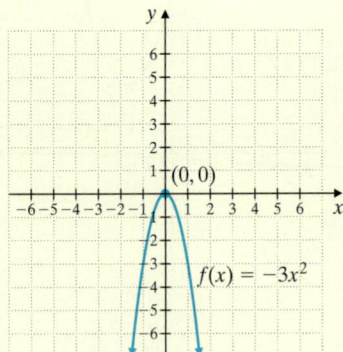

10.

$f(x) = 2(x + 3)^2 - 4$

$(-3, -4)$

$x = -3$

In general, the following holds.

> ## Graphing a Quadratic Function
>
> The graph of a quadratic function written in the form $f(x) = a(x - h)^2 + k$ is a parabola with vertex (h, k).
>
> If $a > 0$, the parabola opens upward.
> If $a < 0$, the parabola opens downward. The axis of symmetry is the line whose equation is $x = h$.
>
>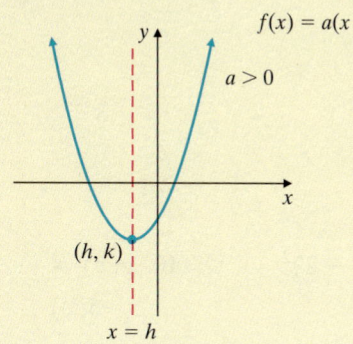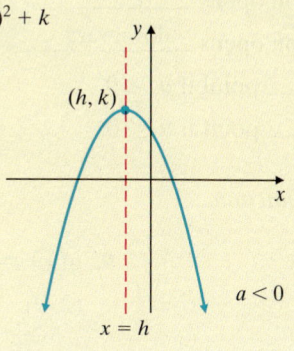

Teaching Tip

Have students write an equation that satisfies given criteria such as "narrow parabola with maximum in third quadrant" or "wide parabola with line of symmetry of $x = -5$." Then have them give their equation to another student to check.

✓**Concept Check** Which description of the graph of $f(x) = -0.35(x + 3)^2 - 4$ is correct?

a. The graph opens downward and has its vertex at $(-3, 4)$.

b. The graph opens upward and has its vertex at $(-3, 4)$.

c. The graph opens downward and has its vertex at $(-3, -4)$.

d. The graph is narrower than the graph of $y = x^2$.

✓**Concept Check Answer**

c

▣ Calculator Explorations   Graphing

Use a graphing calculator to graph the first function of each pair. Then use its graph to predict the graph of the second function. Check your prediction by graphing both on the same set of axes. See this section and Section 8.5.

1. $F(x) = \sqrt{x}$;
$G(x) = \sqrt{x} + 1$

2. $g(x) = x^3$;
$H(x) = x^3 - 2$

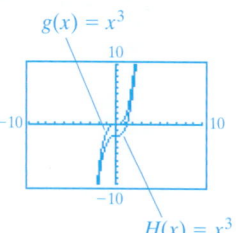

3. $H(x) = |x|$;
$f(x) = |x - 5|$

4. $h(x) = x^3 + 2; g(x) = (x - 3)^3 + 2$

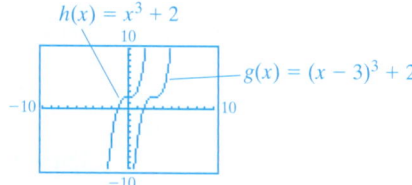

5. $f(x) = |x + 4|; F(x) = |x + 4| + 3$

6. $G(x) = \sqrt{x} - 2; g(x) = \sqrt{x - 4} - 2$

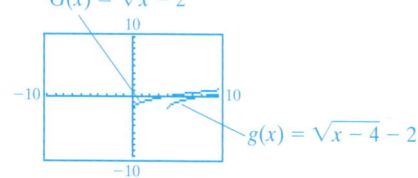

Vocabulary, Readiness & Video Check

Use the choices below to fill in each blank. Some choices will be used more than once.

> upward   highest   parabola   downward   lowest   quadratic

1. A ____quadratic____ function is one that can be written in the form $f(x) = ax^2 + bx + c, a \neq 0$.

2. The graph of a quadratic function is a ____parabola____ opening ____upward____ or ____downward____.

3. If $a > 0$, the graph of the quadratic function opens ____upward____.

4. If $a < 0$, the graph of the quadratic function opens ____downward____.

5. The vertex of a parabola is the ____lowest____ point if $a > 0$.

6. The vertex of a parabola is the ____highest____ point if $a < 0$.

State the vertex of the graph of each quadratic function.

7. $f(x) = x^2$

$(0, 0)$

8. $f(x) = -5x^2$

$(0, 0)$

9. $g(x) = (x - 2)^2$

$(2, 0)$

10. $g(x) = (x + 5)^2$

$(-5, 0)$

11. $f(x) = 2x^2 + 3$

$(0, 3)$

12. $h(x) = x^2 - 1$

$(0, -1)$

13. $g(x) = (x + 1)^2 + 5$

$(-1, 5)$

14. $h(x) = (x - 10)^2 - 7$

$(10, -7)$

Martin-Gay Interactive Videos

See Video 11.5

See video answer section.

Watch the section lecture video and answer the following questions.

Objective A 15. From ▦ Examples 1 and 2 and the lecture before, how do graphs of the form $f(x) = x^2 + k$ differ from $y = x^2$? Consider the location of the vertex $(0, k)$ on these graphs of the form $f(x) = x^2 + k$. By what other name do we call this point on a graph?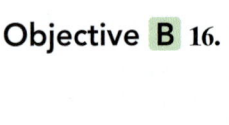

Objective B 16. From ▦ Example 3 and the lecture before, how do graphs of the form $f(x) = (x - h)^2$ differ from $y = x^2$? Consider the location of the vertex $(h, 0)$ on these graphs of the form $f(x) = (x - h)^2$. By what other name do we call this point on a graph? ▶

Objective C 17. From ▦ Example 4 and the lecture before, what general information does the equation $f(x) = (x - h)^2 + k$ tell us about its graph? ▶

Objective D 18. From the lecture before ▦ Example 5, besides the direction a parabola opens, what other graphing information can the value of a tell us? ▶

Objective E 19. In ▦ Examples 6 and 7, what four properties of the graph did we learn from the equations that help us locate and draw the general shape of each parabola? ▶

11.5 Exercise Set MyMathLab®

Objectives A B **Mixed Practice** *Graph each quadratic function. Label the vertex and sketch and label the axis of symmetry. See Examples 1 through 6.*

1. $f(x) = x^2 - 1$

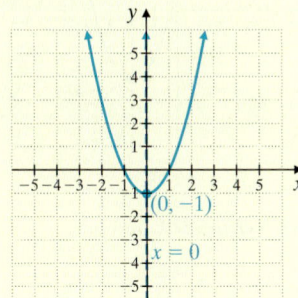

2. $h(x) = x^2 + 3$

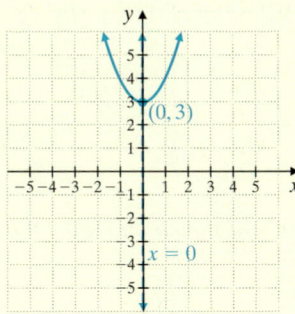

3. $h(x) = x^2 + 5$

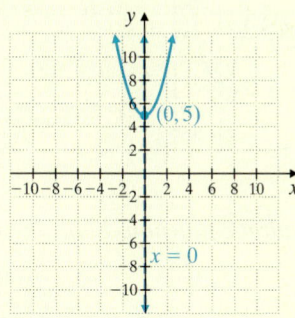

4. $h(x) = x^2 - 4$

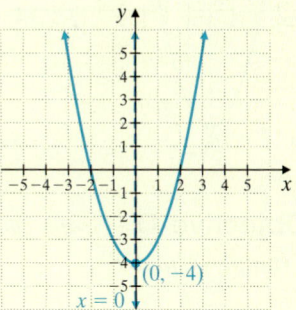

5. $f(x) = (x - 5)^2$

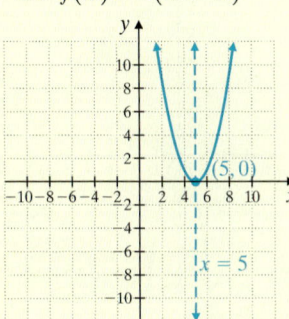

6. $g(x) = (x + 5)^2$

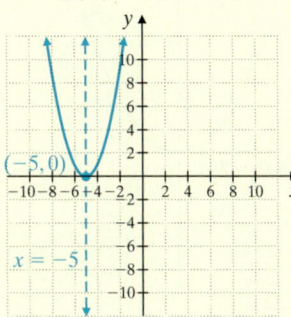

7. $h(x) = (x + 2)^2$

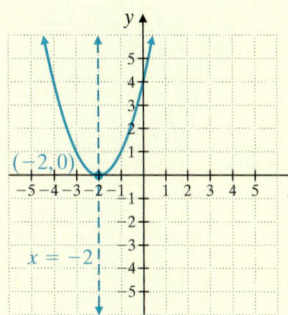

8. $H(x) = (x - 1)^2$

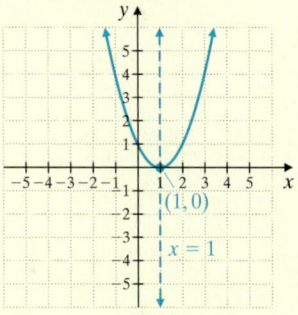

9. $g(x) = x^2 + 7$

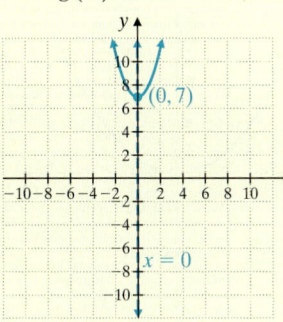

10. $f(x) = x^2 - 2$

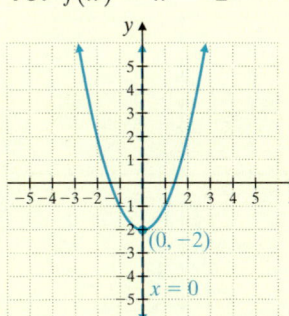

11. $G(x) = (x + 3)^2$

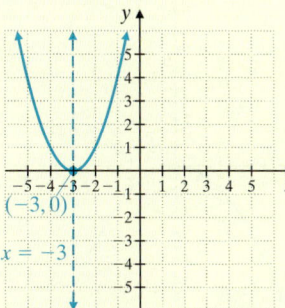

12. $f(x) = (x - 6)^2$

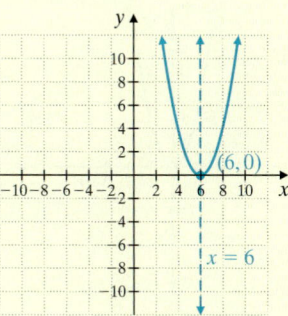

Objective C *Graph each quadratic function. Label the vertex and sketch and label the axis of symmetry. See Example 7.*

13. $f(x) = (x - 2)^2 + 5$

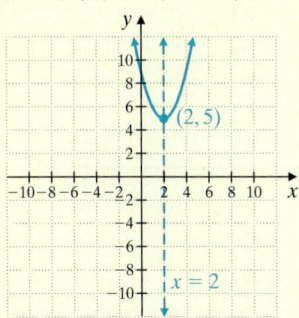

14. $g(x) = (x - 6)^2 + 1$

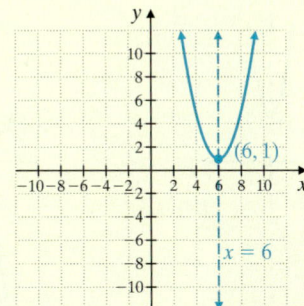

15. $h(x) = (x + 1)^2 + 4$

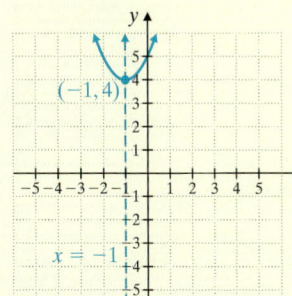

16. $G(x) = (x + 3)^2 + 3$

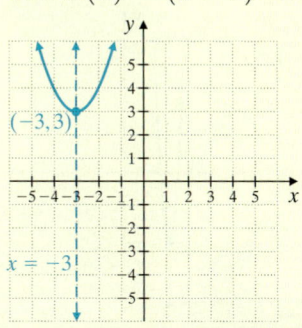

17. $g(x) = (x + 2)^2 - 5$ **18.** $h(x) = (x + 4)^2 - 6$ **19.** $h(x) = (x - 3)^2 + 2$ **20.** $F(x) = (x - 2)^2 - 3$

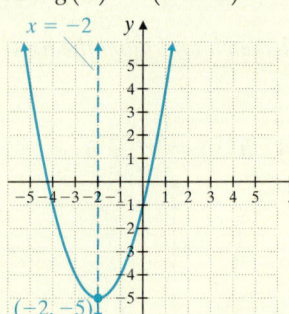   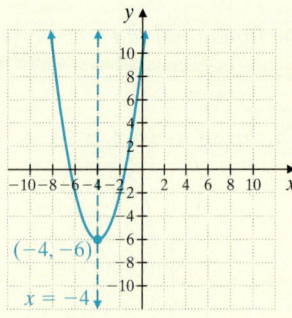   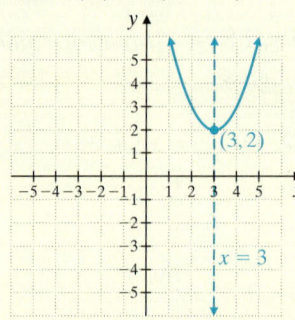   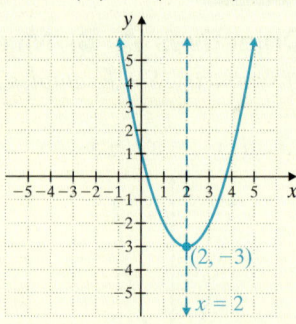

Objective D
Graph each quadratic function. Label the vertex and sketch and label the axis of symmetry. See Examples 8 and 9.

21. $g(x) = -x^2$ **22.** $f(x) = 5x^2$ **23.** $h(x) = \frac{1}{3}x^2$ **24.** $g(x) = -3x^2$

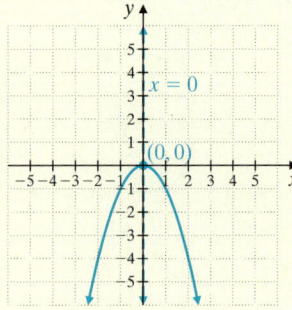   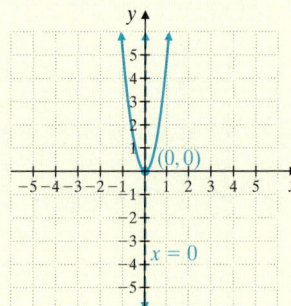   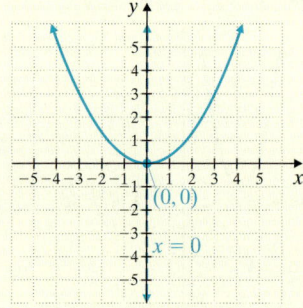   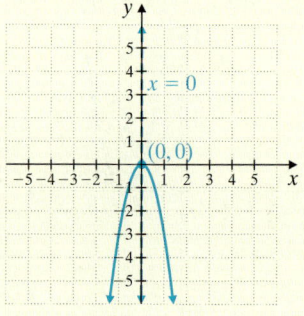

25. $H(x) = 2x^2$ **26.** $f(x) = -\frac{1}{4}x^2$ **27.** $F(x) = -4x^2$ **28.** $G(x) = \frac{1}{5}x^2$

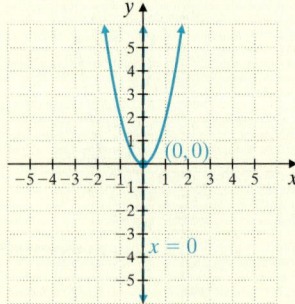   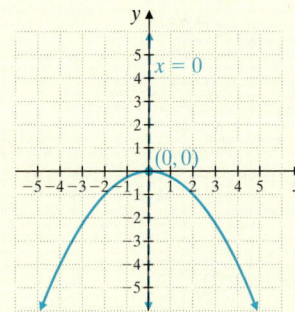   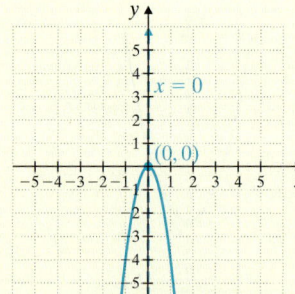   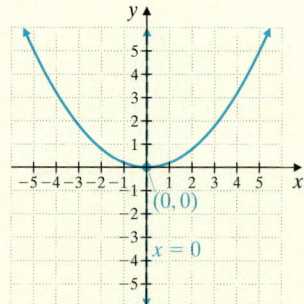

Objective E
Graph each quadratic function. Label the vertex and sketch and label the axis of symmetry. See Example 10.

29. $f(x) = 10(x + 4)^2 - 6$ **30.** $g(x) = 4(x - 4)^2 + 2$ **31.** $h(x) = -3(x + 3)^2 + 1$

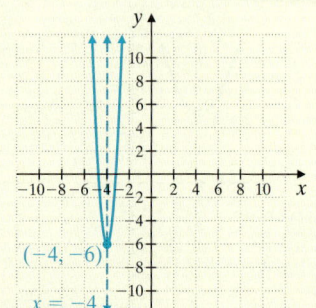   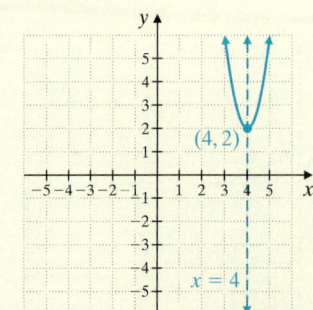   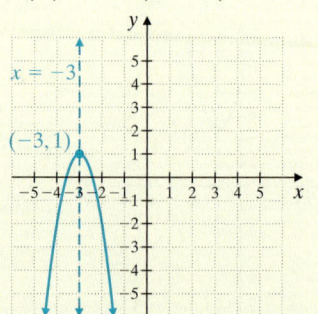

32. $f(x) = -(x - 2)^2 - 6$

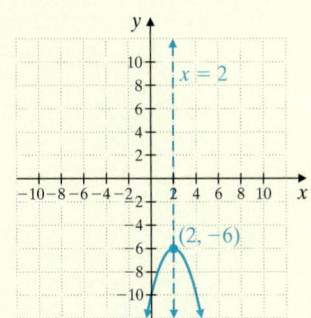

33. $H(x) = \dfrac{1}{2}(x - 6)^2 - 3$

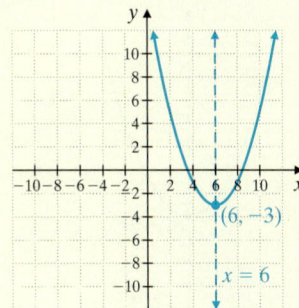

34. $G(x) = \dfrac{1}{5}(x + 4)^2 + 3$

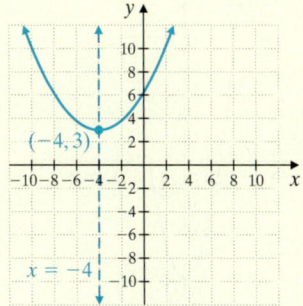

35. $f(x) = -(x - 1)^2$

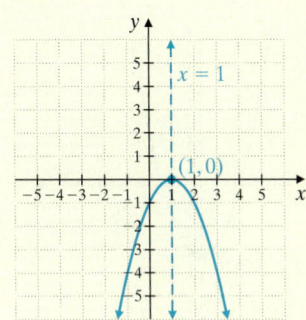

36. $f(x) = 2(x + 3)^2$

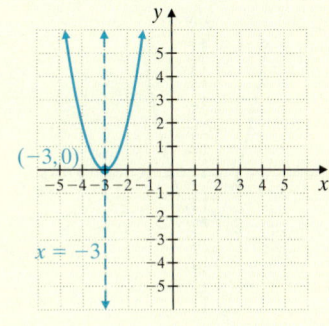

37. $F(x) = \left(x + \dfrac{1}{2}\right)^2 - 2$

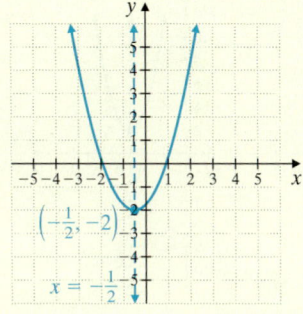

38. $H(x) = \left(x + \dfrac{1}{4}\right)^2 - 3$

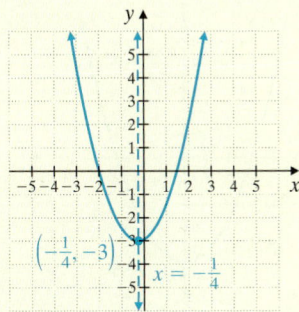

39. $F(x) = -x^2 + 2$

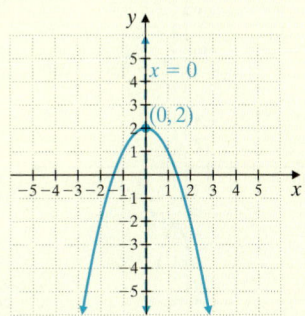

40. $G(x) = 3x^2 + 1$

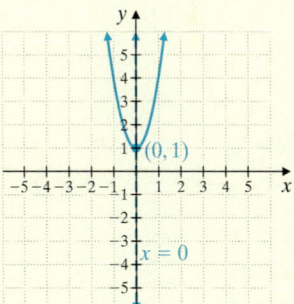

Objectives A B C D E **Mixed Practice** *Sketch the graph of each quadratic function. Label the vertex and sketch and label the axis of symmetry. See Examples 1 through 10.*

41. $f(x) = -(x - 2)^2$

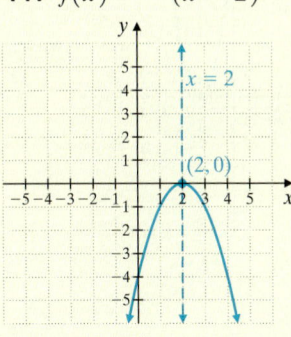

42. $g(x) = -(x + 6)^2$

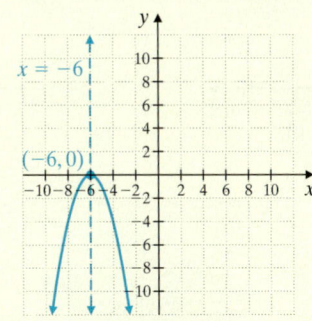

43. $F(x) = -x^2 + 4$

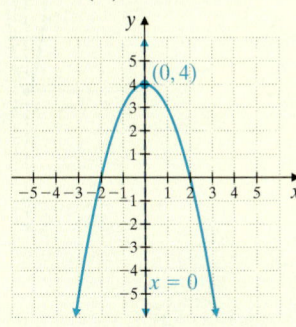

44. $H(x) = -x^2 + 10$

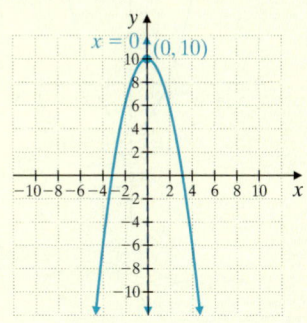

45. $F(x) = 2x^2 - 5$

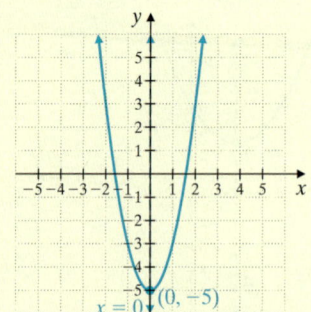

46. $g(x) = \frac{1}{2}x^2 - 2$

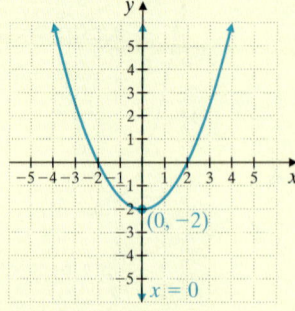

47. $h(x) = (x - 6)^2 + 4$

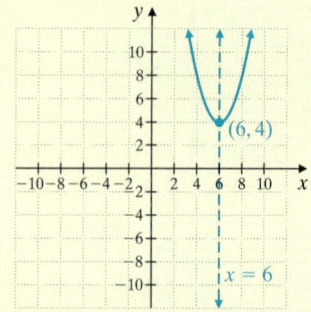

48. $f(x) = (x - 5)^2 + 2$

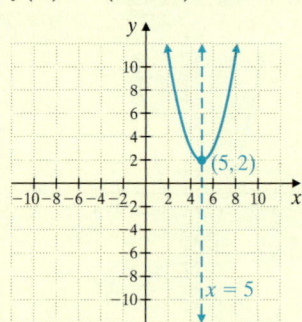

49. $F(x) = \left(x + \frac{1}{2}\right)^2 - 2$

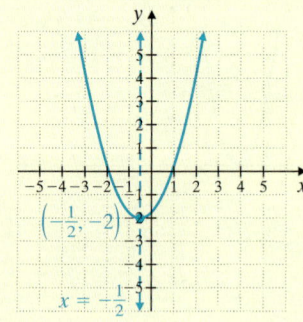

50. $H(x) = \left(x + \frac{1}{2}\right)^2 - 3$

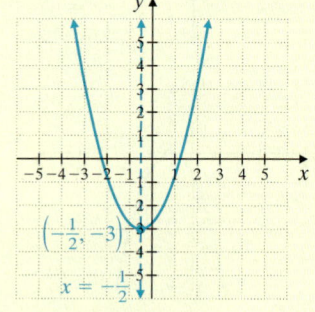

51. $F(x) = \frac{3}{2}(x + 7)^2 + 1$

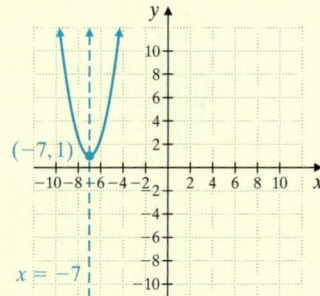

52. $g(x) = -\frac{3}{2}(x - 1)^2 - 5$

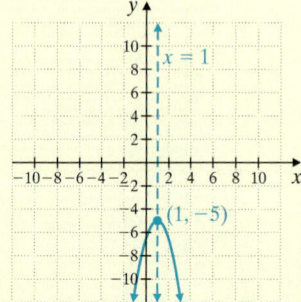

53. $f(x) = \frac{1}{4}x^2 - 9$

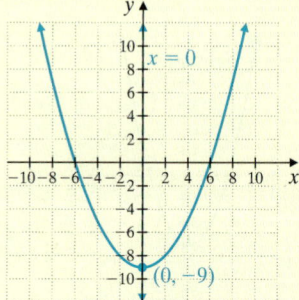

54. $H(x) = \frac{3}{4}x^2 - 2$

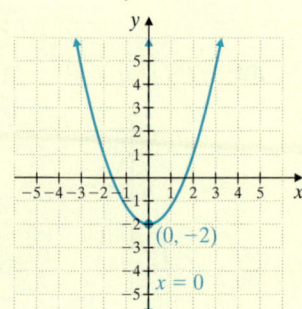

55. $G(x) = 5\left(x + \frac{1}{2}\right)^2$

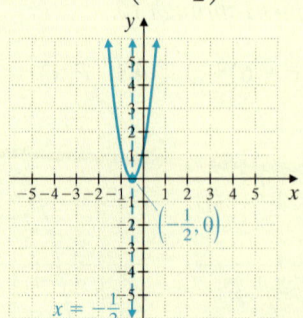

56. $F(x) = 3\left(x - \frac{3}{2}\right)^2$

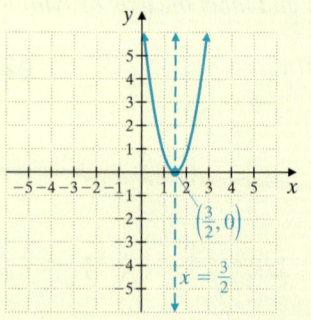

57. $h(x) = -(x - 1)^2 - 1$

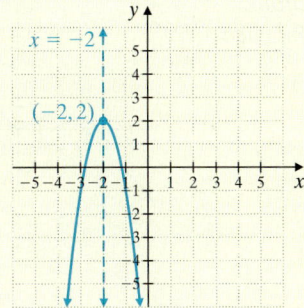

58. $f(x) = -3(x + 2)^2 + 2$

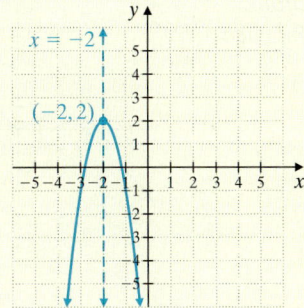

59. $f(x) = 2(x - 1)^2 + 3$

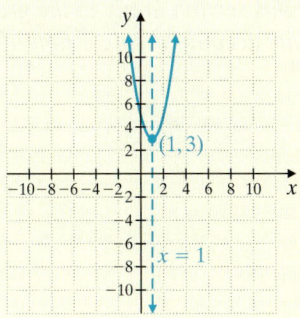

60. $h(x) = 8(x + 1)^2 + 9$

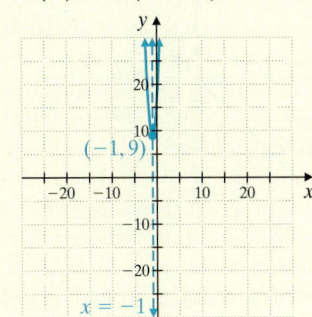

61. $f(x) = -2(x - 4)^2 + 5$

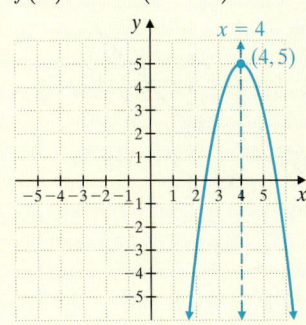

62. $G(x) = -4(x + 9)^2 - 1$

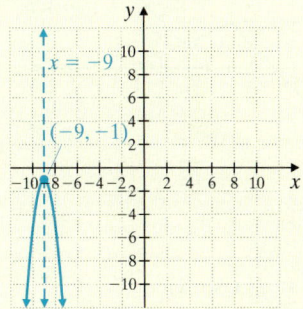

Review

Add the proper constant to each binomial so that the resulting trinomial is a perfect square trinomial. See Section 11.1.

63. $x^2 + 8x$ $x^2 + 8x + 16$

64. $y^2 + 4y$ $y^2 + 4y + 4$

65. $z^2 - 16z$ $z^2 - 16z + 64$

66. $x^2 - 10x$ $x^2 - 10x + 25$

67. $y^2 + y$ $y^2 + y + \dfrac{1}{4}$

68. $z^2 - 3z$ $z^2 - 3z + \dfrac{9}{4}$

Concept Extensions

Write the equation of the parabola that has the same shape as $f(x) = 5x^2$ but with the given vertex. Call each function $g(x)$.

69. $(2, 3)$
    $g(x) = 5(x - 2)^2 + 3$

70. $(1, 6)$
    $g(x) = 5(x - 1)^2 + 6$

71. $(-3, 6)$
    $g(x) = 5(x + 3)^2 + 6$

72. $(4, -1)$
    $g(x) = 5(x - 4)^2 - 1$

Recall from Section 8.5 that the shifting properties covered in this section apply to the graphs of all functions. Given the accompanying graph of $y = f(x)$, graph each function.

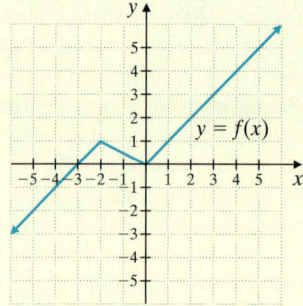

73. $y = f(x) + 1$

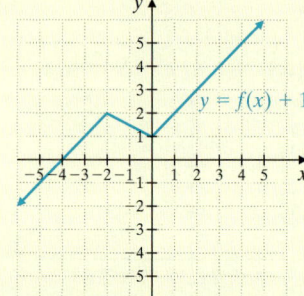

74. $y = f(x) - 2$

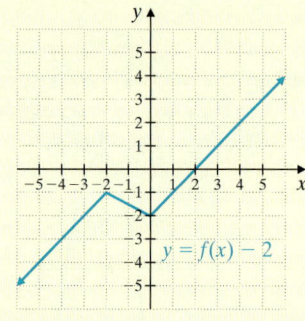

75. $y = f(x - 3)$

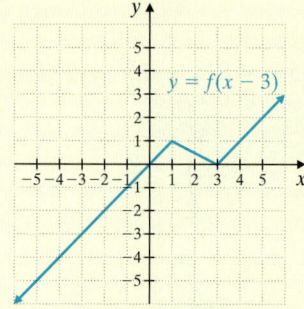

76. $y = f(x + 3)$

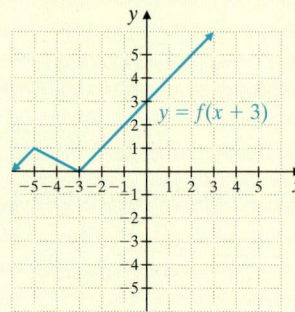

77. $y = f(x + 2) + 2$

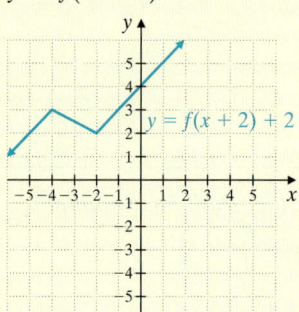

78. $y = f(x - 1) + 1$

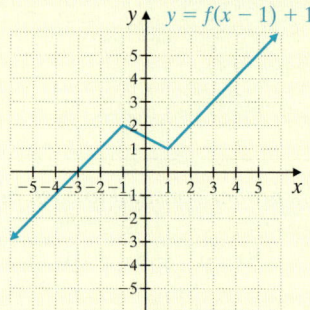

Solve. See the Concept Check in this section.

79. Which description of $f(x) = -213(x - 0.1)^2 + 3.6$ is correct? c

Graph Opens	Vertex
a. upward	$(0.1, 3.6)$
b. upward	$(-213, 3.6)$
c. downward	$(0.1, 3.6)$
d. downward	$(-0.1, 3.6)$

80. Which description of $f(x) = 5\left(x + \dfrac{1}{2}\right)^2 + \dfrac{1}{2}$ is correct? b

Graph Opens	Vertex
a. upward	$\left(\dfrac{1}{2}, \dfrac{1}{2}\right)$
b. upward	$\left(-\dfrac{1}{2}, \dfrac{1}{2}\right)$
c. downward	$\left(\dfrac{1}{2}, -\dfrac{1}{2}\right)$
d. downward	$\left(-\dfrac{1}{2}, -\dfrac{1}{2}\right)$

11.6 Further Graphing of Quadratic Functions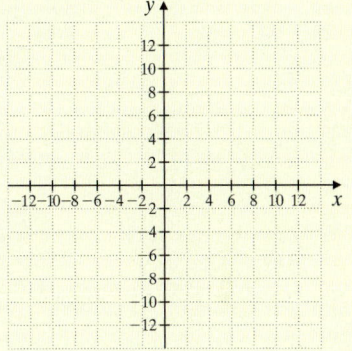

Objective A Writing Quadratic Functions in the Form
$$y = a + (x - h)^2 + k$$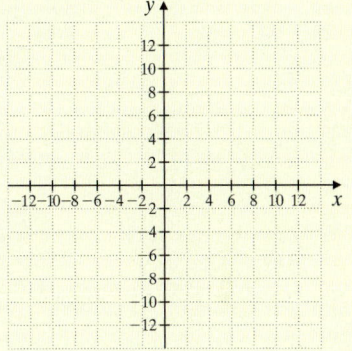

We know that the graph of a quadratic function is a parabola. If a quadratic function is written in the form

$$f(x) = a(x - h)^2 + k$$

we can easily find the vertex (h, k) and graph the parabola. To write a quadratic function in this form, we need to complete the square. (See Section 11.1 for a review of completing the square.)

Objectives

A Write Quadratic Functions in the Form $y = a + (x - h)^2 + k$.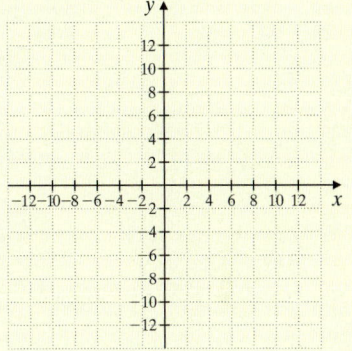

B Derive a Formula for Finding the Vertex of a Parabola.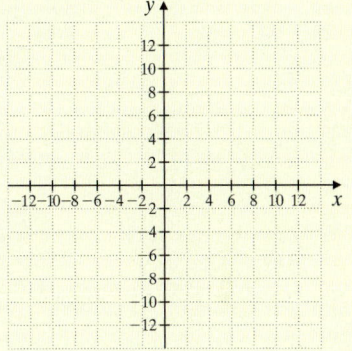

C Find the Minimum or Maximum Value of a Quadratic Function.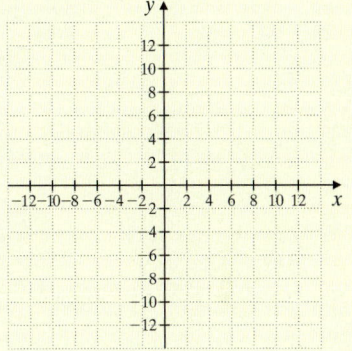

Example 1 Graph: $f(x) = x^2 - 4x - 12$. Find the vertex and any intercepts.

Solution: The graph of this quadratic function is a parabola. To find the vertex of the parabola, we complete the square on the binomial $x^2 - 4x$. To simplify our work, we let $f(x) = y$.

$$y = x^2 - 4x - 12 \qquad \text{Let } f(x) = y.$$
$$y + 12 = x^2 - 4x \qquad \text{Add 12 to both sides to get the } x\text{-variable terms alone.}$$

Now we add the square of half of -4 to both sides.

$$\frac{1}{2}(-4) = -2 \quad \text{and} \quad (-2)^2 = 4$$

$$y + 12 + 4 = x^2 - 4x + 4 \qquad \text{Add 4 to both sides.}$$
$$y + 16 = (x - 2)^2 \qquad \text{Factor the trinomial.}$$
$$y = (x - 2)^2 - 16 \qquad \text{Subtract 16 from both sides.}$$
$$f(x) = (x - 2)^2 - 16 \qquad \text{Replace } y \text{ with } f(x).$$

From this equation, we can see that the vertex of the parabola is $(2, -16)$, a point in quadrant IV, and the axis of symmetry is the line $x = 2$.

Notice that $a = 1$. Since $a > 0$, the parabola opens upward. This parabola opening upward with vertex $(2, -16)$ will have two x-intercepts.

To find the x-intercepts, we let $f(x)$ or $y = 0$.

$$0 = x^2 - 4x - 12$$
$$0 = (x - 6)(x + 2)$$
$$0 = x - 6 \quad \text{or} \quad 0 = x + 2$$
$$6 = x \qquad\qquad -2 = x$$

The two x-intercepts are $(6, 0)$ and $(-2, 0)$. To find the y-intercept, we let $x = 0$.

$$f(0) = 0^2 - 4 \cdot 0 - 12 = -12$$

The y-intercept is $(0, -12)$. The sketch of $f(x) = x^2 - 4x - 12$ is shown.

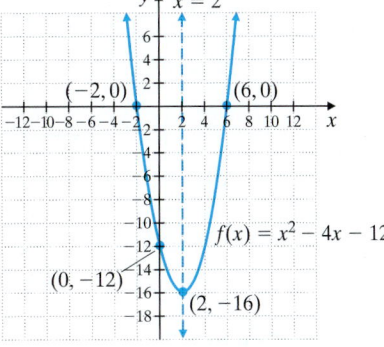

■ **Work Practice 1**

Practice 1

Graph: $f(x) = x^2 - 4x - 5$. Find the vertex and any intercepts.

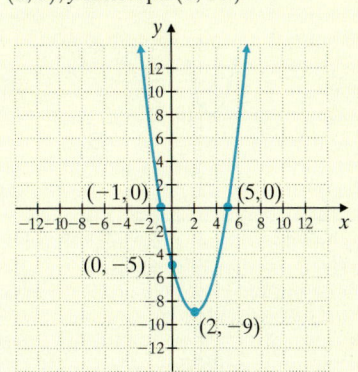

Teaching Tip

For Example 1, have students use the y-intercept and the axis of symmetry to identify another point on the graph.

Answer

1. vertex: $(2, -9)$; x-intercepts: $(-1, 0)$, $(5, 0)$; y-intercept: $(0, -5)$

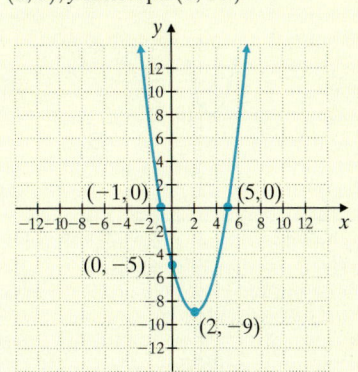

Practice 2

Graph: $f(x) = 2x^2 + 2x + 5$.
Find the vertex and any intercepts.

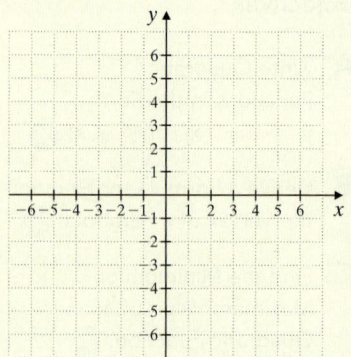

Example 2 Graph: $f(x) = 3x^2 + 3x + 1$. Find the vertex and any intercepts.

Solution: We replace $f(x)$ with y and complete the square on x to write the equation in the form $y = a(x - h)^2 + k$.

$$y = 3x^2 + 3x + 1 \quad \text{Replace } f(x) \text{ with } y.$$
$$y - 1 = 3x^2 + 3x \quad \text{Get the } x\text{-variable terms alone.}$$

Next we factor 3 from the terms $3x^2 + 3x$ so that the coefficient of x^2 is 1.

$$y - 1 = 3(x^2 + x) \quad \text{Factor out 3.}$$

The coefficient of x is 1. Then $\frac{1}{2}(1) = \frac{1}{2}$ and $\left(\frac{1}{2}\right)^2 = \frac{1}{4}$. Since we are adding $\frac{1}{4}$ inside the parentheses, we are really adding $3\left(\frac{1}{4}\right)$, so we *must* add $3\left(\frac{1}{4}\right)$ to the left side.

$$y - 1 + 3\left(\frac{1}{4}\right) = 3\left(x^2 + x + \frac{1}{4}\right)$$

$$y - \frac{1}{4} = 3\left(x + \frac{1}{2}\right)^2 \quad \text{Simplify the left side and factor the right side.}$$

$$y = 3\left(x + \frac{1}{2}\right)^2 + \frac{1}{4} \quad \text{Add } \frac{1}{4} \text{ to both sides.}$$

$$f(x) = 3\left(x + \frac{1}{2}\right)^2 + \frac{1}{4} \quad \text{Replace } y \text{ with } f(x).$$

Then $a = 3$, $h = -\frac{1}{2}$, and $k = \frac{1}{4}$. This means that the parabola opens upward with vertex $\left(-\frac{1}{2}, \frac{1}{4}\right)$ and that the axis of symmetry is the line $x = -\frac{1}{2}$. This parabola has no x-intercepts since the vertex is in the second quadrant and it opens upward.

To find the y-intercept, we let $x = 0$. Then

$$f(0) = 3(0)^2 + 3(0) + 1 = 1$$

We use the vertex, axis of symmetry, and y-intercept to graph the parabola.

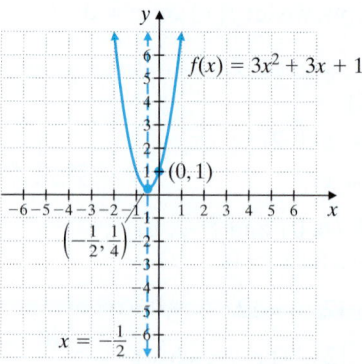

Answer

2. vertex: $\left(-\frac{1}{2}, \frac{9}{2}\right)$; y-intercept: $(0, 5)$

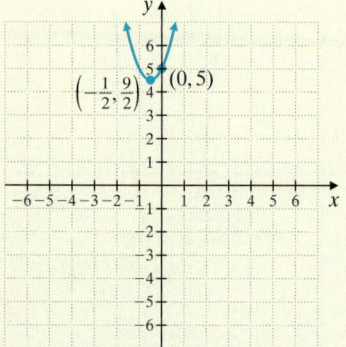

■ **Work Practice 2**

Helpful Hint

Parabola Opens Upward
Vertex in quadrant I or II: no x-intercepts
Vertex in quadrant III or IV: 2 x-intercepts

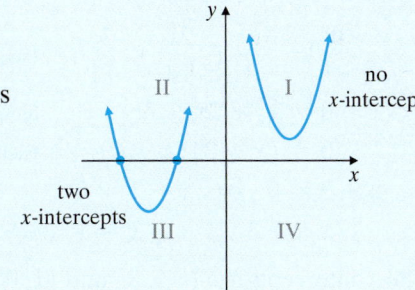

Parabola Opens Downward
Vertex in quadrant I or II: 2 x-intercepts
Vertex in quadrant III or IV: no x-intercepts

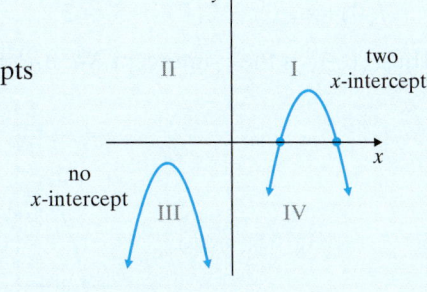

Example 3   Graph: $f(x) = -x^2 - 2x + 3$. Find the vertex and any intercepts.

Solution: We write $f(x)$ in the form $a(x - h)^2 + k$ by completing the square. First we replace $f(x)$ with y.

$$f(x) = -x^2 - 2x + 3$$
$$y = -x^2 - 2x + 3$$
$$y - 3 = -x^2 - 2x \quad \text{Subtract 3 from both sides to get the } x\text{-variable terms alone.}$$
$$y - 3 = -1(x^2 + 2x) \quad \text{Factor } -1 \text{ from the terms } -x^2 - 2x.$$

The coefficient of x is 2. Then $\dfrac{1}{2}(2) = 1$ and $1^2 = 1$. We add 1 to the right side inside the parentheses and add $-1(1)$ to the left side.

$$y - 3 - 1(1) = -1(x^2 + 2x + 1)$$
$$y - 4 = -1(x + 1)^2 \quad \text{Simplify the left side and factor the right side.}$$
$$y = -1(x + 1)^2 + 4 \quad \text{Add 4 to both sides.}$$
$$\underline{f(x) = -1(x + 1)^2 + 4} \quad \text{Replace } y \text{ with } f(x).$$

Since $a = -1$, the parabola opens downward with vertex $(-1, 4)$ and axis of symmetry $x = -1$.
    To find the x-intercepts, we let y or $f(x) = 0$ and solve for x.

$$f(x) = -x^2 - 2x + 3$$
$$0 = -x^2 - 2x + 3 \quad \text{Let } f(x) = 0.$$

(*Continued on next page*)

Practice 3

Graph: $f(x) = -x^2 - 2x + 8$.
Find the vertex and any intercepts.

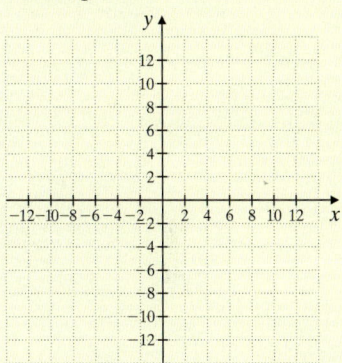

Helpful Hint   This can be written as $f(x) = -1[x - (-1)]^2 + 4$. Notice that the vertex is $(-1, 4)$.

Answer

3. vertex: $(-1, 9)$; x-intercepts: $(-4, 0)$, $(2, 0)$; y-intercept: $(0, 8)$

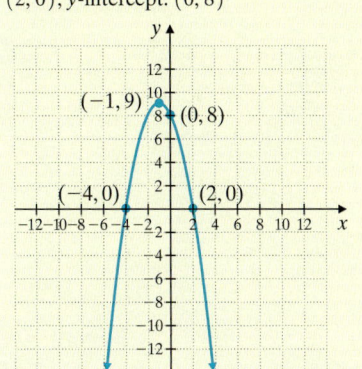

Now we divide both sides by -1 so that the coefficient of x^2 is 1. (If you prefer, you may factor -1 from the trinomial on the right side.)

$$\frac{0}{-1} = \frac{-x^2}{-1} - \frac{2x}{-1} + \frac{3}{-1} \qquad \text{Divide both sides by } -1.$$

$$0 = x^2 + 2x - 3 \qquad \text{Simplify.}$$

$$0 = (x + 3)(x - 1) \qquad \text{Factor.}$$

$$x + 3 = 0 \quad \text{or} \quad x - 1 = 0 \qquad \text{Set each factor equal to 0.}$$

$$x = -3 \qquad\qquad x = 1 \qquad \text{Solve.}$$

The x-intercepts are $(-3, 0)$ and $(1, 0)$.

To find the y-intercept, we let $x = 0$ and solve for y. Then

$$f(0) = -0^2 - 2(0) + 3 = 3$$

Thus, $(0, 3)$ is the y-intercept. We use these points to graph the parabola.

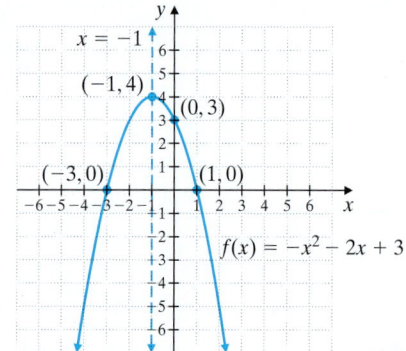

■ Work Practice 3

Objective B Deriving a Formula for Finding the Vertex

As you have seen in previous examples, it can sometimes be tedious to find the vertex of a parabola by completing the square. There is a formula for finding the vertex of a parabola. Now that we have practiced completing the square, we will show that the x-coordinate of the vertex of the graph of $f(x)$ or $y = ax^2 + bx + c$ can be found by the formula $x = \frac{-b}{2a}$. To do so, we complete the square on x and write the equation in the form $y = (x - h)^2 + k$.

First we get the x-variable terms alone by subtracting c from both sides.

$$y = ax^2 + bx + c$$

$$y - c = ax^2 + bx$$

$$y - c = a\left(x^2 + \frac{b}{a}x\right) \qquad \text{Factor } a \text{ from the terms } ax^2 + bx.$$

Now we add the square of half of $\frac{b}{a}$, or $\left(\frac{b}{2a}\right)^2 = \frac{b^2}{4a^2}$, to the right side inside the parentheses. Because of the factor a, what we really added is $a\left(\frac{b^2}{4a^2}\right)$ and this must be added to the left side as well.

$$y - c + a\left(\frac{b^2}{4a^2}\right) = a\left(x^2 + \frac{b}{a}x + \frac{b^2}{4a^2}\right)$$

$$y - c + \frac{b^2}{4a} = a\left(x + \frac{b}{2a}\right)^2$$

$$y = a\left(x + \frac{b}{2a}\right)^2 + c - \frac{b^2}{4a} \qquad \text{Simplify the left side and factor the right side. Add } c \text{ to both sides and subtract } \frac{b^2}{4a} \text{ from both sides.}$$

Compare this form with $f(x)$ or $y = a(x - h)^2 + k$ and see that h is $\dfrac{-b}{2a}$, which means that the x-coordinate of the vertex of the graph of $f(x) = ax^2 + bx + c$ is $\dfrac{-b}{2a}$.

Let's use the vertex formula below to find the vertex of the parabola we graphed in Example 1.

Vertex Formula

The graph of $f(x) = ax^2 + bx + c$, when $a \neq 0$, is a parabola with vertex

$$\left(\frac{-b}{2a}, f\!\left(\frac{-b}{2a} \right) \right)$$

Example 4   Find the vertex of the graph of $f(x) = x^2 - 4x - 12$.

Solution: In the quadratic function $f(x) = x^2 - 4x - 12$, notice that $a = 1$, $b = -4$, and $c = -12$.

$$\frac{-b}{2a} = \frac{-(-4)}{2(1)} = 2$$

The x-value of the vertex is 2. To find the corresponding $f(x)$ or y-value, find $f(2)$. Then

$$f(2) = 2^2 - 4(2) - 12 = 4 - 8 - 12 = -16$$

The vertex is $(2, -16)$. These results agree with our findings in Example 1.

🟨 **Work Practice 4**

Practice 4

Find the vertex of the graph of $f(x) = x^2 - 4x - 5$. Compare your result with the result of Practice 1.

Objective C Finding Minimum and Maximum Values ▶

The quadratic function whose graph is a parabola that opens upward has a minimum value, and the quadratic function whose graph is a parabola that opens downward has a maximum value. The $f(x)$- or y-value of the vertex is the minimum or maximum value of the function.

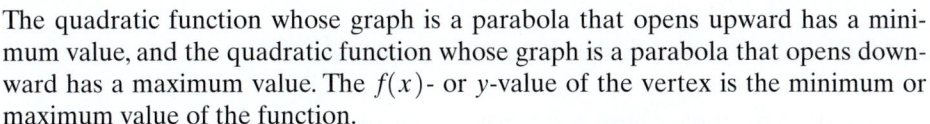

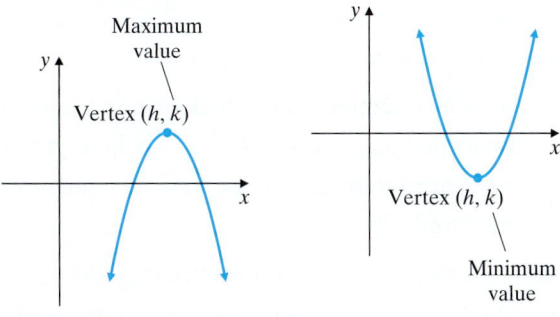

✓**Concept Check** Without making any calculations, tell whether the graph of $f(x) = 7 - x - 0.3x^2$ has a maximum value or a minimum value. Explain your reasoning.

Answer
4. $(2, -9)$

✓**Concept Check Answer**
$f(x)$ has a maximum value since it opens downward.

Practice 5

An object is thrown upward from the top of a 100-foot cliff. Its height in feet above ground after t seconds is given by the function $f(t) = -16t^2 + 10t + 100$. Find the maximum height of the object and the number of seconds it took for the object to reach its maximum height.

Teaching Tip

After discussing Example 5, ask students to determine when the rock will reach a height of 12 feet. Have them find their answer in two ways.

Answer

5. maximum height: $101\dfrac{9}{16}$ ft in $\dfrac{5}{16}$ sec

Example 5 Finding Maximum Height

A rock is thrown upward from the ground. Its height in feet above ground after t seconds is given by the function $f(t) = -16t^2 + 20t$. Find the maximum height of the rock and the number of seconds it took for the rock to reach its maximum height.

Solution:

1. UNDERSTAND. The maximum height of the rock is the largest value of $f(t)$. Since the function $f(t) = -16t^2 + 20t$ is a quadratic function, its graph is a parabola. It opens downward since $-16 < 0$. Thus, the maximum value of $f(t)$ is the $f(t)$- or y-value of the vertex of its graph.

2. TRANSLATE. To find the vertex (h, k), we notice that for $f(t) = -16t^2 + 20t$, $a = -16, b = 20$, and $c = 0$. We will use these values and the vertex formula

$$\left(\frac{-b}{2a}, f\left(\frac{-b}{2a} \right) \right)$$

3. SOLVE. $h = \dfrac{-b}{2a} = \dfrac{-20}{-32} = \dfrac{5}{8}$

$$f\left(\frac{5}{8} \right) = -16\left(\frac{5}{8} \right)^2 + 20\left(\frac{5}{8} \right) = -16\left(\frac{25}{64} \right) + \frac{25}{2} = -\frac{25}{4} + \frac{50}{4} = \frac{25}{4}$$

4. INTERPRET. The graph of $f(t)$ is a parabola opening downward with vertex $\left(\dfrac{5}{8}, \dfrac{25}{4} \right)$. This means that the rock's maximum height is $\dfrac{25}{4}$ feet, or $6\dfrac{1}{4}$ feet, which was reached in $\dfrac{5}{8}$ second.

▪ **Work Practice 5**

Vocabulary, Readiness & Video Check

Fill in each blank.

1. If a quadratic function is in the form $f(x) = a(x - h)^2 + k$, the vertex of its graph is ___(h, k)___.

2. The graph of $f(x) = ax^2 + bx + c, a \neq 0$ is a parabola whose vertex has x-value of ___$\dfrac{-b}{2a}$___.

Martin-Gay Interactive Videos   Watch the section lecture video and answer the following questions.

See Video 11.6

See video answer section.

Objective A 3. From ⊟ Example 1, how does writing a quadratic function in the form $f(x) = a(x - h)^2 + k$ help us graph the function? What procedure can we use to write a quadratic function in this form? ▶

Objective B 4. From ⊟ Example 2, how can locating the vertex and knowing whether parabola opens upward or downward potentially help save unnecessary work? Explain. ▶

Objective C 5. From ⊟ Example 4, when an application involving a quadratic function asks for the maximum or minimum, what part of a parabola should we find? ▶

11.6 **Exercise Set** MyMathLab®

Objective **A**   *Fill in each blank. See Examples 1 through 3.*

	Parabola Opens	Vertex Location	Number of *x*-intercept(s)	Number of *y*-intercept(s)
1.	up	Q I	0	1
2.	up	Q III	2	1
3.	down	Q II	2	1
4.	down	Q IV	0	1
5.	up	*x*-axis	1	1
6.	down	*x*-axis	1	1
7.	down	Q III	0	
8.	down	Q I	2	
9.	up	Q IV	2	
10.	up	Q II	0	

Objectives **A** **B**   **Mixed Practice** *Find the vertex of the graph of each quadratic function by completing the square or using the vertex formula. See Examples 1 through 4.*

11. $f(x) = x^2 + 8x + 7$
$(-4, -9)$

12. $f(x) = x^2 + 6x + 5$
$(-3, -4)$

13. $f(x) = -x^2 + 10x + 5$
$(5, 30)$

14. $f(x) = -x^2 - 8x + 2$
$(-4, 18)$

15. $f(x) = 5x^2 - 10x + 3$
$(1, -2)$

16. $f(x) = -3x^2 + 6x + 4$
$(1, 7)$

17. $f(x) = -x^2 + x + 1$
$\left(\dfrac{1}{2}, \dfrac{5}{4}\right)$

18. $f(x) = x^2 - 9x + 8$
$\left(\dfrac{9}{2}, -\dfrac{49}{4}\right)$

Match each function with its graph. See Examples 1 through 4.

19. $f(x) = x^2 - 4x + 3$
D

20. $f(x) = x^2 + 2x - 3$
A

21. $f(x) = x^2 - 2x - 3$
B

22. $f(x) = x^2 + 4x + 3$
C

A.

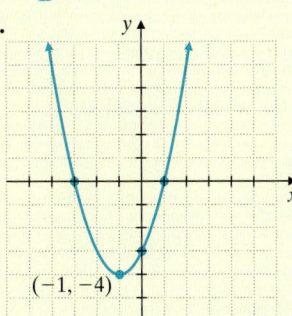
$(-1, -4)$

B.

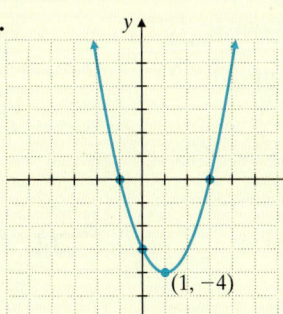
$(1, -4)$

C.

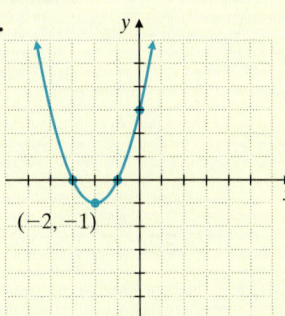
$(-2, -1)$

D.

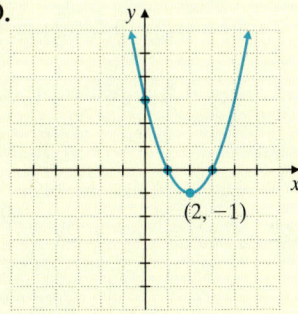
$(2, -1)$

Find the vertex of the graph of each quadratic function. Determine whether the graph opens upward or downward, find any intercepts, and graph the function. See Examples 1 through 4.

23. $f(x) = x^2 + 4x - 5$

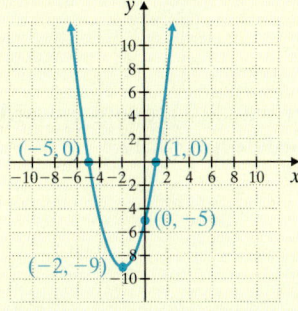
$(-5, 0)$ $(1, 0)$
$(0, -5)$
$(-2, -9)$

vertex: $(-2, -9)$; opens upward;
x-intercepts: $(-5, 0), (1, 0)$;
y-intercept: $(0, -5)$

24. $f(x) = x^2 + 2x - 3$

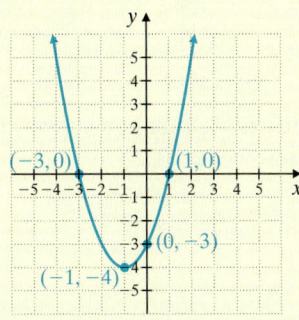
$(-3, 0)$ $(1, 0)$
$(0, -3)$
$(-1, -4)$

vertex: $(-1, -4)$; opens upward;
x-intercepts: $(-3, 0), (1, 0)$;
y-intercept: $(0, -3)$

▶ **25.** $f(x) = -x^2 + 2x - 1$

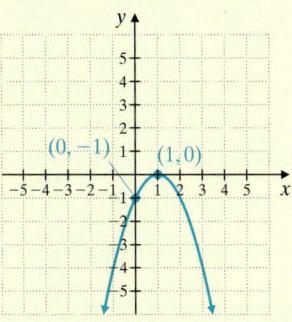
$(0, -1)$ $(1, 0)$

vertex: $(1, 0)$; opens downward;
x-intercept: $(1, 0)$;
y-intercept: $(0, -1)$

26. $f(x) = -x^2 + 4x - 4$

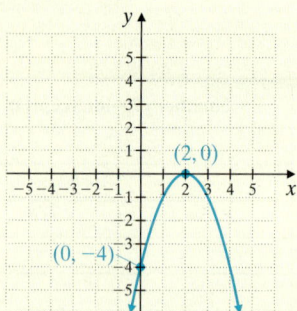

vertex: $(2, 0)$; opens downward;
x-intercept: $(2, 0)$;
y-intercept: $(0, -4)$

27. $f(x) = x^2 - 4$

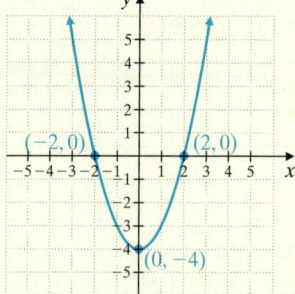

vertex: $(0, -4)$; opens upward;
x-intercepts: $(-2, 0), (2, 0)$;
y-intercept: $(0, -4)$

28. $f(x) = x^2 - 1$

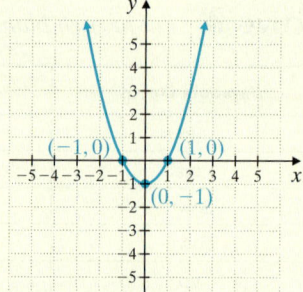

vertex: $(0, -1)$; opens upward;
x-intercepts: $(-1, 0), (1, 0)$;
y-intercept: $(0, -1)$

29. $f(x) = 4x^2 + 4x - 3$

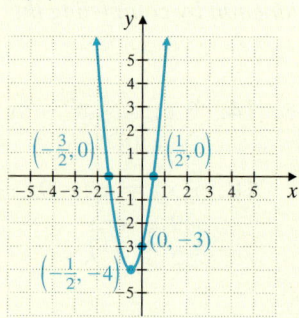

vertex: $\left(-\dfrac{1}{2}, -4\right)$; opens

upward; x-intercepts: $\left(-\dfrac{3}{2}, 0\right)$,

$\left(\dfrac{1}{2}, 0\right)$; y-intercept: $(0, -3)$

30. $f(x) = 2x^2 - x - 3$

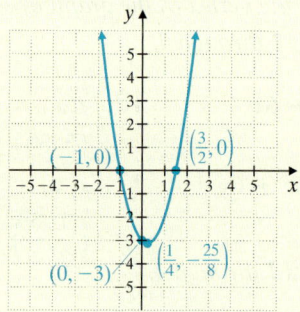

vertex: $\left(\dfrac{1}{4}, -\dfrac{25}{8}\right)$; opens

upward; x-intercepts: $(-1, 0)$,

$\left(\dfrac{3}{2}, 0\right)$; y-intercept: $(0, -3)$

31. $f(x) = \dfrac{1}{2}x^2 + 4x + \dfrac{15}{2}$

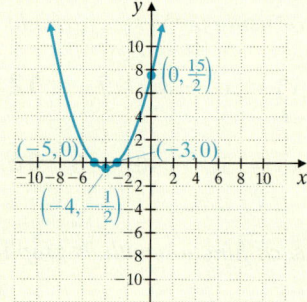

vertex: $\left(-4, -\dfrac{1}{2}\right)$; opens

upward; x-intercepts: $(-5, 0)$,

$(-3, 0)$; y-intercept: $\left(0, \dfrac{15}{2}\right)$

32. $f(x) = \dfrac{1}{5}x^2 + 2x + \dfrac{9}{5}$

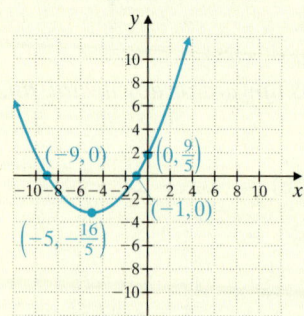

vertex: $\left(-5, -\dfrac{16}{5}\right)$; opens

upward; x-intercepts: $(-9, 0)$,

$(-1, 0)$; y-intercept: $\left(0, \dfrac{9}{5}\right)$

33. $f(x) = x^2 - 4x + 5$

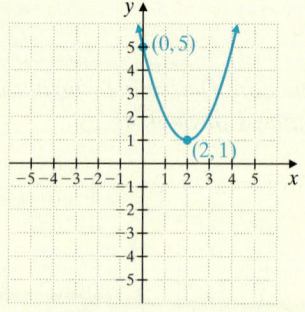

vertex: $(2, 1)$; opens upward;
y-intercept: $(0, 5)$

34. $f(x) = x^2 - 6x + 11$

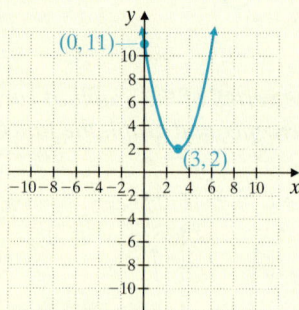

vertex: $(3, 2)$; opens upward;
y-intercept: $(0, 11)$

35. $f(x) = 2x^2 + 4x + 5$

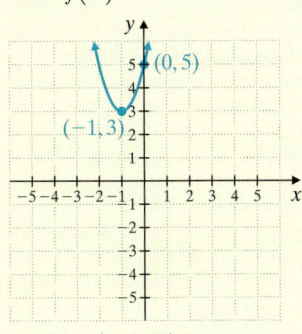

vertex: $(-1, 3)$; opens upward; y-intercept: $(0, 5)$

36. $f(x) = 3x^2 + 12x + 16$

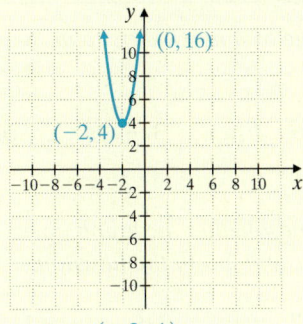

vertex: $(-2, 4)$; opens upward; y-intercept: $(0, 16)$

37. $f(x) = -2x^2 + 12x$

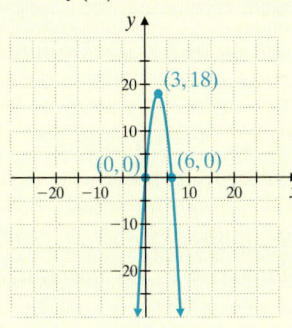

vertex: $(3, 18)$; opens downward; x-intercepts: $(0, 0)$, $(6, 0)$; y-intercept: $(0, 0)$

38. $f(x) = -4x^2 + 8x$

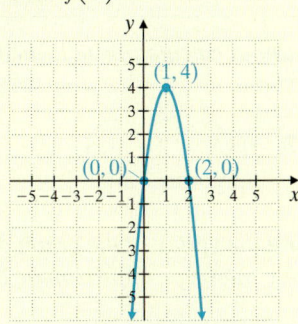

vertex: $(1, 4)$; opens downward; x-intercepts: $(0, 0)$, $(2, 0)$; y-intercept: $(0, 0)$

Objective C *Solve. See Example 5.*

▶ **39.** If a projectile is fired straight upward from the ground with an initial speed of 96 feet per second, then its height h in feet after t seconds is given by the function $h(t) = -16t^2 + 96t$. Find the maximum height of the projectile.   144 ft

40. If Rheam Gaspar throws a ball upward with an initial speed of 32 feet per second, then its height h in feet after t seconds is given by the function $h(t) = -16t^2 + 32t$. Find the maximum height of the ball.   16 ft

41. The cost C in dollars of manufacturing x bicycles at Holladay's Production Plant is given by the function $C(x) = 2x^2 - 800x + 92{,}000$.

   a. Find the number of bicycles that must be manufactured to minimize the cost.   200 bicycles

   b. Find the minimum cost. $12,000

42. The Utah Ski Club sells calendars to raise money. The profit P, in cents, from selling x calendars is given by the function $P(x) = 360x - x^2$.

   a. Find how many calendars must be sold to maximize profit.   180 calendars

   b. Find the maximum profit. $32,400

43. Find two numbers whose sum is 60 and whose product is as large as possible. [*Hint:* Let x and $60 - x$ be the two positive numbers. Their product can be described by the function $f(x) = x(60 - x)$.]   30, 30

44. Find two numbers whose sum is 11 and whose product is as large as possible. (Use the hint for Exercise 43.)   5.5, 5.5

45. Find two numbers whose difference is 10 and whose product is as small as possible. (Use the hint for Exercise 43.)   5, −5

46. Find two numbers whose difference is 8 and whose product is as small as possible.   4, −4

△ **47.** The length and width of a rectangle must have a sum of 40. Find the dimensions of the rectangle that will have the maximum area. (Use the hint for Exercise 43.)   length: 20 units; width: 20 units

△ **48.** The length and width of a rectangle must have a sum of 50. Find the dimensions of the rectangle that will have maximum area.   length: 25; width: 25

Review

Find the vertex of the graph of each function. See Section 11.5.

49. $f(x) = x^2 + 2$
$(0, 2)$

50. $f(x) = (x - 3)^2$
$(3, 0)$

51. $g(x) = (x + 2)^2$
$(-2, 0)$

52. $h(x) = x^2 - 3$
$(0, -3)$

53. $f(x) = (x + 5)^2 + 2$
$(-5, 2)$

54. $f(x) = 2(x - 3)^2 + 2$
$(3, 2)$

55. $f(x) = 3(x - 4)^2 + 1$
$(4, 1)$

56. $f(x) = (x + 1)^2 + 4$
$(-1, 4)$

Concept Extensions

Without calculating, tell whether each graph has a minimum value or a maximum value. See the Concept Check in the section.

57. $f(x) = 2x^2 - 5$
minimum value

58. $g(x) = -7x^2 + x + 1$
maximum value

59. $F(x) = 3 - \frac{1}{2}x^2$
maximum value

60. $G(x) = 3 - \frac{1}{2}x + 0.8x^2$
minimum value

Find the vertex of the graph of each quadratic function. Determine whether the graph opens upward or downward, find the y-intercept, approximate the x-intercepts to one decimal place, and graph the function.

61. $f(x) = x^2 + 10x + 15$

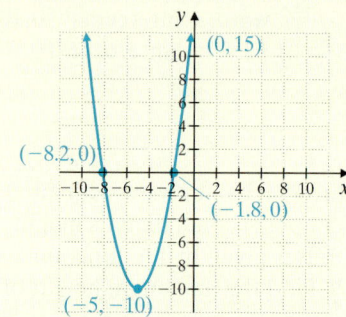

vertex: $(-5, -10)$; opens upward; y-intercept:
$(0, 15)$; x-intercepts: $(-1.8, 0)$, $(-8.2, 0)$

62. $f(x) = 2x^2 + 4x - 1$

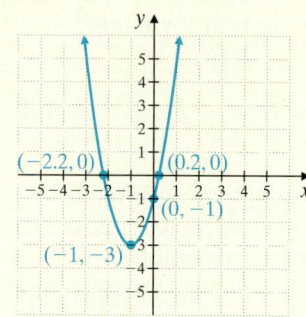

vertex: $(-1, -3)$; opens upward;
y-intercept: $(0, -1)$; x-intercepts: $(-2.2, 0)$, $(0.2, 0)$

Use a graphing calculator to verify the graph of each exercise.

63. Exercise 31.

64. Exercise 32.

Find the maximum or minimum value of each function. Approximate to two decimal places.

65. $f(x) = 2.3x^2 - 6.1x + 3.2$ -0.84

66. $f(x) = 7.6x^2 + 9.8x - 2.1$ -5.26

67. The percent of U.S. households with broadband Internet access can be modeled by the function $f(x) = -0.11x^2 + 6.53x + 3$, where $f(x)$ is the projected percent of American households with broadband Internet access and x is the number of years since 2000. (*Source:* Based on data from Pew Internet and American Life Project)

a. Will this function have a maximum or a minimum? How can you tell?   maximum, answers may vary

b. According to this model, in what year will the percent of American homes with broadband Internet access be at its maximum or minimum? Round to the nearest whole year.   2030

c. What is the predicted maximum/minimum percent of homes with broadband Internet access? (Use your rounded answer from part b.)   99.9%

68. Methane is a gas produced by landfills, natural gas systems, and coal mining that contributes to the greenhouse effect and global warming. Projected methane emissions in the United States can be modeled by the quadratic function

$$f(x) = -0.072x^2 + 1.93x + 173.9$$

where $f(x)$ is the amount of methane produced in millions of metric tons and x is the number of years after 2000. (*Source:* Based on data from the U.S. Environmental Protection Agency, 2000–2020)

a. According to this model, what will U.S. emissions of methane be in 2018?   185.312 million metric tons

b. Will this function have a maximum or a minimum? How can you tell?   maximum; answers may vary

c. In what year will methane emissions in the United States be at their maximum or minimum? Round to the nearest whole year.   2013

d. What is the level of methane emissions for that year? (Use your rounded answer from part c.)   186.822 million metric tons

Chapter 11 Group Activity

Additional group activities are available in the Instructor's Resource Manual with Tests.

Recognizing Linear and Quadratic Models

This activity may be completed by working in groups or individually.

We have seen in this and previous chapters that data can be modeled by both linear models and quadratic models. However, when we are given a set of data to model, how can we tell if a linear or quadratic model is appropriate? The best answer requires looking at a scatter diagram of the data. If the plotted data points fall roughly on a line, a linear model is usually the better choice. If the plotted data points seem to fall on a definite curve or if a maximum or minimum point is apparent, a quadratic model is usually the better choice.

One of the sets of data shown in the tables is best modeled by a linear function and one is best modeled by a quadratic function. In each case, the variable x represents the number of years after 2008.

North American Total Hybrid Car Sales					
Year	2009	2010	2011	2012	2013
x	1	2	3	4	5
Number of Hybrids Sold, y	290,271	274,210	268,752	434,498	495,685

(*Source: The Christian Science Monitor*)

Total Number of Target Stores					
Year	2009	2010	2011	2012	2013
x	1	2	3	4	5
Number of Stores, y	1740	1750	1763	1778	1790

(*Source: Target, Inc.*)

1. Make a scatter diagram for each set of data. Which type of model should be used for each set of data?
   Hybrid sales: quadratic; Target stores: linear

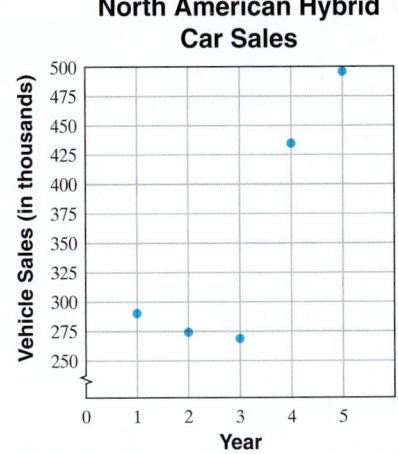

North American Hybrid Car Sales

Number of Target Stores

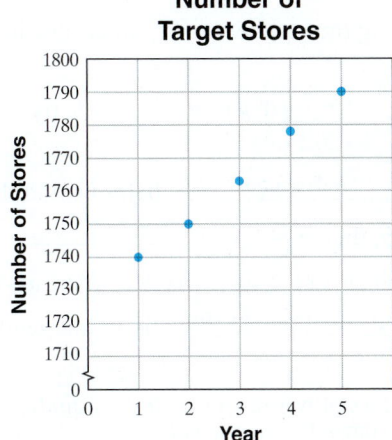

2. For the set of data that you have determined to be linear, find a linear function that fits the data points. Explain the method that you used. $y = 13x + 1726$; answers may vary

3. For the set of data that you have determined to be quadratic, identify the point on your scatter diagram that appears to be the vertex of the parabola. Use the coordinates of this vertex in the quadratic model $f(x) = a(x - h)^2 + k$.   vertex: $(3, 268{,}752)$ $f(x) = a(x - 3)^2 + 268{,}752$

4. Solve for the remaining unknown constant in the quadratic model by substituting the coordinates for another data point into the function. Write the final form of the quadratic model for this data set.
   $f(x) = 56{,}733(x - 3)^2 + 268{,}752$; answers may vary

5. Use your models to estimate the number of hybrid cars sold and the number of Target stores in 2014.
   Hybrid cars: 779,349; answers may vary
   Target stores: 1804; answers may vary

6. (Optional) For each set of data, enter the data from the table into a graphing calculator and use either the linear regression feature or the quadratic regression feature to find an appropriate function that models the data.* Compare these functions with the ones you found in Exercise 2. How are they alike or different?   answers may vary

*To find out more about using your graphing calculator to find a regression equation, consult your user's manual.

Chapter 11 Vocabulary Check

Fill in each blank with one of the words or phrases listed below.

quadratic formula   quadratic   discriminant

completing the square   quadratic inequality

(h, k) $(0, k)$ $(h, 0)$ $\dfrac{-b}{2a}$ $\pm\sqrt{b}$

1. The _____discriminant_____ helps us find the number and type of solutions of a quadratic equation.

2. If $a^2 = b$, then $a =$ _____$\pm\sqrt{b}$_____.

3. The graph of $f(x) = ax^2 + bx + c$, where a is not 0, is a parabola whose vertex has an x-value of _____$\dfrac{-b}{2a}$_____.

4. A(n) _____quadratic inequality_____ is an inequality that can be written so that one side is a quadratic expression and the other side is 0.

5. The process of writing a quadratic equation so that one side is a perfect square trinomial is called _____completing the square_____.

6. The graph of $f(x) = x^2 + k$ has vertex _____$(0, k)$_____.

7. The graph of $f(x) = (x - h)^2$ has vertex _____$(h, 0)$_____.

8. The graph of $f(x) = (x - h)^2 + k$ has vertex _____(h, k)_____.

9. The formula $x = \dfrac{-b \pm \sqrt{b^2 - 4ac}}{2a}$ is called the _____quadratic formula_____.

10. A(n) _____quadratic_____ equation is one that can be written in the form $ax^2 + bx + c = 0$, where a, b, and c are real numbers and a is not 0.

Helpful Hint

▶ Are you preparing for your test? Don't forget to take the Chapter 11 Test on page 837. Then check your answers at the back of the text and use the Chapter Test Prep Videos to see the fully worked-out solutions to any of the exercises you want to review.

11 Chapter Highlights

Definitions and Concepts	Examples
Section 11.1 Solving Quadratic Equations by Completing the Square	

Square Root Property

If b is a real number and if $a^2 = b$, then $a = \pm\sqrt{b}$.

Solving a Quadratic Equation in x by Completing the Square

Step 1: If the coefficient of x^2 is not 1, divide both sides of the equation by the coefficient of x^2.

Step 2: Get the variable terms alone.

Step 3: Complete the square by adding the square of half of the coefficient of x to both sides.

Step 4: Write the resulting trinomial as the square of a binomial.

Step 5: Use the square root property.

Solve: $(x + 3)^2 = 14$
$$x + 3 = \pm\sqrt{14}$$
$$x = -3 \pm \sqrt{14}$$

Solve: $3x^2 - 12x - 18 = 0$

1. $x^2 - 4x - 6 = 0$

2. $x^2 - 4x = 6$

3. $\dfrac{1}{2}(-4) = -2$ and $(-2)^2 = 4$

    $x^2 - 4x + 4 = 6 + 4$

4. $(x - 2)^2 = 10$

5. $x - 2 = \pm\sqrt{10}$
    $$x = 2 \pm \sqrt{10}$$

Definitions and Concepts	Examples

Section 11.2 Solving Quadratic Equations by Using the Quadratic Formula

Quadratic Formula

A quadratic equation written in the form $ax^2 + bx + c = 0$ has solutions

$$x = \frac{-b \pm \sqrt{b^2 - 4ac}}{2a}$$

Solve: $x^2 - x - 3 = 0$

$$a = 1, b = -1, c = -3$$

$$x = \frac{-(-1) \pm \sqrt{(-1)^2 - 4(1)(-3)}}{2 \cdot 1}$$

$$x = \frac{1 \pm \sqrt{13}}{2}$$

Section 11.3 Solving Equations by Using Quadratic Methods

Substitution is often helpful in solving an equation that contains a repeated variable expression.

Solve: $(2x + 1)^2 - 5(2x + 1) + 6 = 0$

Let $m = 2x + 1$, then the equation is

$$m^2 - 5m + 6 = 0 \qquad \text{Let } m = 2x + 1$$

$$(m - 3)(m - 2) = 0$$

$$m = 3 \quad \text{or} \quad m = 2$$

$$2x + 1 = 3 \quad 2x + 1 = 2 \quad \text{Substitute back.}$$

$$x = 1 \qquad x = \frac{1}{2}$$

Section 11.4 Nonlinear Inequalities in One Variable

Solving a Polynomial Inequality

Step 1: Write the inequality in standard form and solve the related equation.

Step 2: Use solutions from Step 1 to separate the number line into regions.

Step 3: Use a test point to determine whether values in each region satisfy the original inequality.

Step 4: Write the solution set as the union of regions whose test point values are solutions.

Solve: $x^2 \geq 6x$

1. $x^2 - 6x \geq 0$

$$x^2 - 6x = 0$$
$$x(x - 6) = 0$$
$$x = 0 \quad \text{or} \quad x = 6$$

2.

Region	Test Point Value	$x^2 \geq 6x$	Result
A	-2	$(-2)^2 \geq 6(-2)$	True
B	1	$1^2 \geq 6(1)$	False
C	7	$7^2 \geq 6(7)$	True

3.

4.

The solution set is $(-\infty, 0] \cup [6, \infty)$.

Solving a Rational Inequality

Step 1: Solve for values that make all denominators 0.

Step 2: Solve the related equation.

Solve: $\dfrac{6}{x - 1} < -2$

1. $x - 1 = 0$ Set the denominator equal to 0.

$$x = 1$$

2. $\dfrac{6}{x - 1} = -2$

$$6 = -2(x - 1) \quad \text{Multiply by } (x - 1).$$
$$6 = -2x + 2$$
$$4 = -2x$$
$$-2 = x$$

(continued)

Definitions and Concepts	Examples

Section 11.4 Nonlinear Inequalities in One Variable (*continued*)

Step 3: Use solutions from Steps 1 and 2 to separate the number line into regions.

3.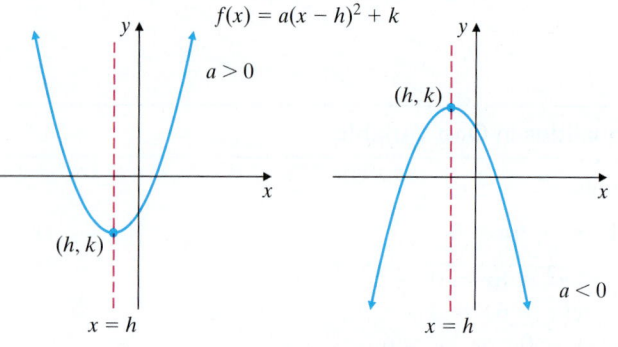

Step 4: Use a test point to determine whether values in each region satisfy the original inequality.

4. Only a test value from region B satisfies the original inequality.

Step 5: Write the solution set as the union of regions whose test point values are solutions.

5.

The solution set is $(-2, 1)$.

Section 11.5 Quadratic Functions and Their Graphs

Graphing a Quadratic function

The graph of a quadratic function written in the form $f(x) = a(x - h)^2 + k$ is a parabola with vertex (h, k).

If $a > 0$, the parabola opens upward.

If $a < 0$, the parabola opens downward.

The axis of symmetry is the line whose equation is $x = h$.

Graph: $g(x) = 3(x - 1)^2 + 4$

The graph is a parabola with vertex $(1, 4)$ and axis of symmetry $x = 1$. Since $a = 3$ is positive, the graph opens upward.

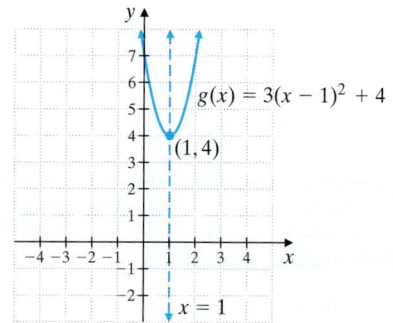

Section 11.6 Further Graphing of Quadratic Functions

The graph of $f(x) = ax^2 + bx + c, a \neq 0$, is a parabola with vertex

$$\left(\frac{-b}{2a}, f\left(\frac{-b}{2a} \right) \right)$$

Graph: $f(x) = x^2 - 2x - 8$. Find the vertex and x- and y-intercepts.

$$\frac{-b}{2a} = \frac{-(-2)}{2 \cdot 1} = 1$$

$$f(1) = 1^2 - 2(1) - 8 = -9$$

The vertex is $(1, -9)$.

$$0 = x^2 - 2x - 8$$

$$0 = (x - 4)(x + 2)$$

$$x = 4 \quad \text{or} \quad x = -2$$

The x-intercepts are $(4, 0)$ and $(-2, 0)$.

$$f(0) = 0^2 - 2 \cdot 0 - 8 = -8$$

The y-intercept is $(0, -8)$.

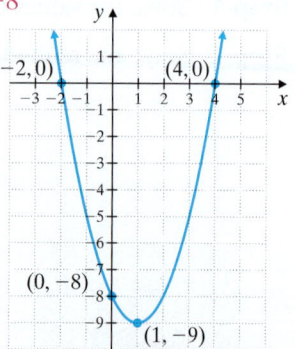

Definitions and Concepts	Examples

Section 11.2 Solving Quadratic Equations by Using the Quadratic Formula

Quadratic Formula

A quadratic equation written in the form $ax^2 + bx + c = 0$ has solutions

$$x = \frac{-b \pm \sqrt{b^2 - 4ac}}{2a}$$

Solve: $x^2 - x - 3 = 0$

$$a = 1, b = -1, c = -3$$

$$x = \frac{-(-1) \pm \sqrt{(-1)^2 - 4(1)(-3)}}{2 \cdot 1}$$

$$x = \frac{1 \pm \sqrt{13}}{2}$$

Section 11.3 Solving Equations by Using Quadratic Methods

Substitution is often helpful in solving an equation that contains a repeated variable expression.

Solve: $(2x + 1)^2 - 5(2x + 1) + 6 = 0$

Let $m = 2x + 1$, then the equation is

$$m^2 - 5m + 6 = 0 \qquad \text{Let } m = 2x + 1$$

$$(m - 3)(m - 2) = 0$$

$$m = 3 \quad \text{or} \quad m = 2$$

$$2x + 1 = 3 \qquad 2x + 1 = 2 \quad \text{Substitute back.}$$

$$x = 1 \qquad x = \frac{1}{2}$$

Section 11.4 Nonlinear Inequalities in One Variable

Solving a Polynomial Inequality

Step 1: Write the inequality in standard form and solve the related equation.

Step 2: Use solutions from Step 1 to separate the number line into regions.

Step 3: Use a test point to determine whether values in each region satisfy the original inequality.

Step 4: Write the solution set as the union of regions whose test point values are solutions.

Solve: $x^2 \geq 6x$

1. $x^2 - 6x \geq 0$

$$x^2 - 6x = 0$$
$$x(x - 6) = 0$$
$$x = 0 \quad \text{or} \quad x = 6$$

2.

3.

Region	Test Point Value	$x^2 \geq 6x$	Result
A	-2	$(-2)^2 \geq 6(-2)$	True
B	1	$1^2 \geq 6(1)$	False
C	7	$7^2 \geq 6(7)$	True

4.

The solution set is $(-\infty, 0] \cup [6, \infty)$.

Solving a Rational Inequality

Step 1: Solve for values that make all denominators 0.

Step 2: Solve the related equation.

Solve: $\dfrac{6}{x - 1} < -2$

1. $x - 1 = 0$ Set the denominator equal to 0.

     $x = 1$

2. $\dfrac{6}{x - 1} = -2$

$$6 = -2(x - 1) \quad \text{Multiply by } (x - 1).$$
$$6 = -2x + 2$$
$$4 = -2x$$
$$-2 = x$$

(continued)

Definitions and Concepts	Examples

Section 11.4   Nonlinear Inequalities in One Variable (*continued*)

Step 3: Use solutions from Steps 1 and 2 to separate the number line into regions.

Step 4: Use a test point to determine whether values in each region satisfy the original inequality.

Step 5: Write the solution set as the union of regions whose test point values are solutions.

3.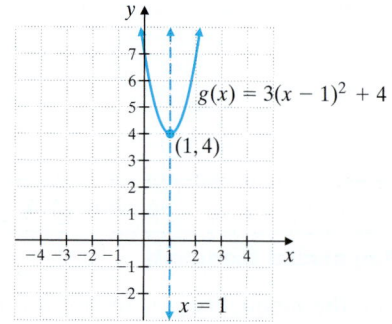

4. Only a test value from region B satisfies the original inequality.

5.

The solution set is $(-2, 1)$.

Section 11.5   Quadratic Functions and Their Graphs

Graphing a Quadratic function

The graph of a quadratic function written in the form $f(x) = a(x - h)^2 + k$ is a parabola with vertex (h, k).

If $a > 0$, the parabola opens upward.

If $a < 0$, the parabola opens downward.

The axis of symmetry is the line whose equation is $x = h$.

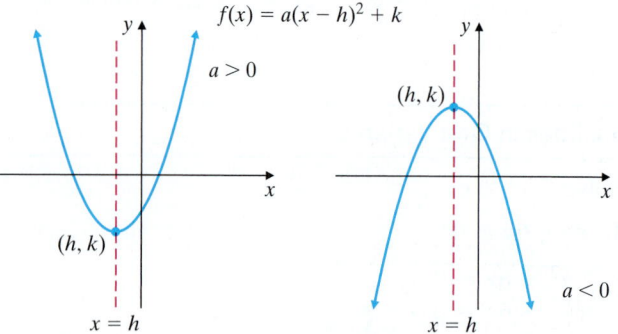

Graph: $g(x) = 3(x - 1)^2 + 4$

The graph is a parabola with vertex $(1, 4)$ and axis of symmetry $x = 1$. Since $a = 3$ is positive, the graph opens upward.

Section 11.6   Further Graphing of Quadratic Functions

The graph of $f(x) = ax^2 + bx + c, a \neq 0$, is a parabola with vertex

$$\left(\frac{-b}{2a}, f\left(\frac{-b}{2a} \right) \right)$$

Graph: $f(x) = x^2 - 2x - 8$. Find the vertex and x- and y-intercepts.

$$\frac{-b}{2a} = \frac{-(-2)}{2 \cdot 1} = 1$$

$$f(1) = 1^2 - 2(1) - 8 = -9$$

The vertex is $(1, -9)$.

$$0 = x^2 - 2x - 8$$
$$0 = (x - 4)(x + 2)$$
$$x = 4 \quad \text{or} \quad x = -2$$

The x-intercepts are $(4, 0)$ and $(-2, 0)$.

$$f(0) = 0^2 - 2 \cdot 0 - 8 = -8$$

The y-intercept is $(0, -8)$.

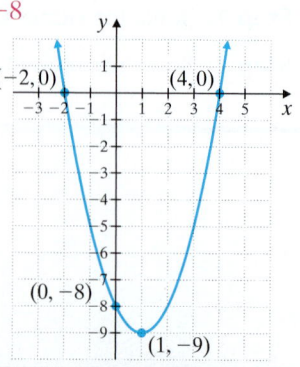

(11.1) *Solve by factoring.*

1. $x^2 - 15x + 14 = 0$ $\{14, 1\}$

2. $7a^2 = 29a + 30$ $\left\{-\dfrac{6}{7}, 5\right\}$

Use the square root property to solve each equation.

3. $4m^2 = 196$ $\{-7, 7\}$

4. $(5x - 2)^2 = 2$ $\left\{\dfrac{2 - \sqrt{2}}{5}, \dfrac{2 + \sqrt{2}}{5}\right\}$

Solve by completing the square.

5. $z^2 + 3z + 1 = 0$ $\left\{\dfrac{-3 - \sqrt{5}}{2}, \dfrac{-3 + \sqrt{5}}{2}\right\}$

6. $(2x + 1)^2 = x$ $\left\{\dfrac{-3 - i\sqrt{7}}{8}, \dfrac{-3 + i\sqrt{7}}{8}\right\}$

7. If P dollars are invested, the formula $A = P(1 + r)^2$ gives the amount A in an account paying interest rate r compounded annually after 2 years. Find the interest rate r such that \$2500 increases to \$2717 in 2 years. Round the result to the nearest hundredth of a percent.   4.25%

8. Two ships leave a port at the same time and travel at the same speed. One ship is traveling due north and the other due east. In a few hours, the ships are 150 miles apart. How many miles has each ship traveled? Give the exact answer and a one-decimal-place approximation. $75\sqrt{2}$ mi; 106.1 mi

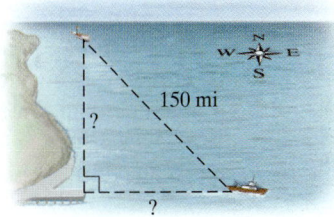

(11.2) *If the discriminant of a quadratic equation has the given value, determine the number and type of solutions of the equation.*

9. -8 two complex but not real solutions

10. 48   two real solutions

11. 100   two real solutions

12. 0   one real solution

Use the quadratic formula to solve each equation.

13. $x^2 - 16x + 64 = 0$ $\{8\}$

14. $x^2 + 5x = 0$ $\{-5, 0\}$

15. $2x^2 + 3x = 5$ $\left\{-\dfrac{5}{2}, 1\right\}$

16. $6x^2 + 7 = 5x$ $\left\{\dfrac{5 - i\sqrt{143}}{12}, \dfrac{5 + i\sqrt{143}}{12}\right\}$

17. $9x^2 + 4 = 2x$ $\left\{\dfrac{1 - i\sqrt{35}}{9}, \dfrac{1 + i\sqrt{35}}{9}\right\}$

18. $(2x - 3)^2 = x$ $\left\{1, \dfrac{9}{4}\right\}$

19. Cadets graduating from military school usually toss their hats high into the air at the end of the ceremony. One cadet threw his hat so that its distance $d(t)$ in feet above the ground t seconds after it was thrown was $d(t) = -16t^2 + 30t + 6$.

   a. Find the distance above the ground of the hat 1 second after it was thrown.   20 ft

   b. Find the time it took that hat to hit the ground. Give the exact time and a one-decimal-place approximation. $\dfrac{15 + \sqrt{321}}{16}$ sec; 2.1 sec

△ **20.** The hypotenuse of an isosceles right triangle is 6 centimeters longer than either of the legs. Find the length of the legs. (*Hint:* Don't forget that an isosceles triangle has two sides of equal length.) $(6 + 6\sqrt{2})$ cm

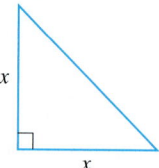

(11.3) *Solve each equation.*

21. $x^3 = 27$ $\left\{3, \dfrac{-3 + 3i\sqrt{3}}{2}, \dfrac{-3 - 3i\sqrt{3}}{2}\right\}$

22. $y^3 = -64$ $\{-4, 2 - 2i\sqrt{3}, 2 + 2i\sqrt{3}\}$

23. $\dfrac{5}{x} + \dfrac{6}{x - 2} = 3$ $\left\{\dfrac{2}{3}, 5\right\}$

24. $x^4 - 21x^2 - 100 = 0$ $\{-5, 5, -2i, 2i\}$

25. $5(x + 3)^2 - 19(x + 3) = 4$ $\left\{-\dfrac{16}{5}, 1\right\}$

26. $x^{2/3} - 6x^{1/3} + 5 = 0$ $\{1, 125\}$

27. $a^6 - a^2 = a^4 - 1$ $\{-1, 1, -i, i\}$

28. $y^{-2} + y^{-1} = 20$ $\left\{-\dfrac{1}{5}, \dfrac{1}{4}\right\}$

29. Two postal workers, Jerome Grant and Tim Bozik, can sort a stack of mail in 5 hours. Working alone, Tim can sort the mail in 1 hour less time than Jerome can. Find the time in which each postal worker can sort the mail alone. Round the result to one decimal place. Jerome: 10.5 hr; Tim: 9.5 hr

30. A negative number decreased by its reciprocal is $-\dfrac{24}{5}$. Find the number. -5

(11.4) *Solve each inequality for x. Write each solution set in interval notation.*

31. $2x^2 - 50 \le 0$ $[-5, 5]$

32. $\dfrac{1}{4}x^2 < \dfrac{1}{16}$ $\left(-\dfrac{1}{2}, \dfrac{1}{2}\right)$

33. $(x^2 - 16)(x^2 - 1) > 0$
$(-\infty, -4) \cup (-1, 1) \cup (4, \infty)$

34. $(x^2 - 4)(y^2 - 25) \le 0$
$[-5, -2] \cup [2, 5]$

35. $\dfrac{x - 5}{x - 6} < 0$ $(5, 6)$

36. $\dfrac{(4x + 3)(x - 5)}{x(x + 6)} > 0$

$(-\infty, -6) \cup \left(-\dfrac{3}{4}, 0\right) \cup (5, \infty)$

37. $(x + 5)(x - 6)(x + 2) \le 0$ $(-\infty, -5] \cup [-2, 6]$

38. $x^3 + 3x^2 - 25x - 75 > 0$ $(-5, -3) \cup (5, \infty)$

39. $\dfrac{x^2 + 4}{3x} \le 1$ $(-\infty, 0)$

40. $\dfrac{(5x + 6)(x - 3)}{x(6x - 5)} < 0$ $\left(-\dfrac{6}{5}, 0\right) \cup \left(\dfrac{5}{6}, 3\right)$

(11.5) *Graph each function. Label the vertex and the axis of symmetry of each graph.*

41. $f(x) = x^2 - 4$

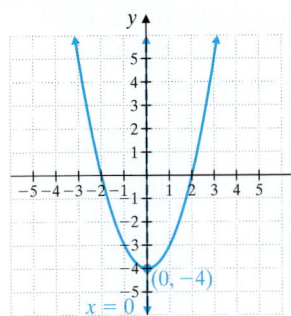

42. $g(x) = x^2 + 7$

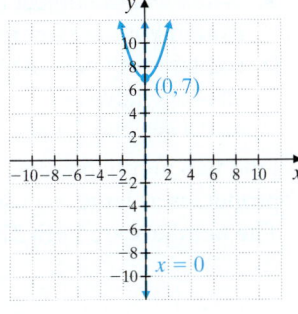

43. $H(x) = 2x^2$

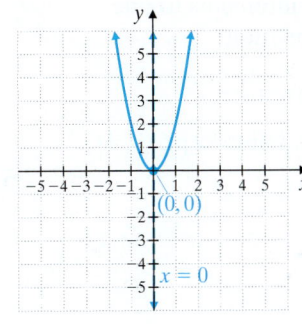

44. $h(x) = -\dfrac{1}{3}x^2$

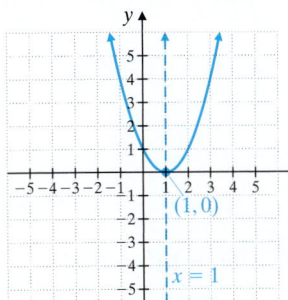

45. $F(x) = (x - 1)^2$

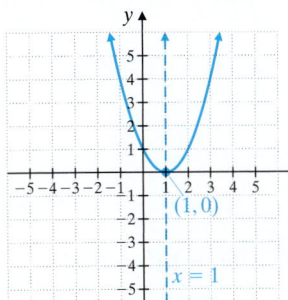

46. $G(x) = (x + 5)^2$

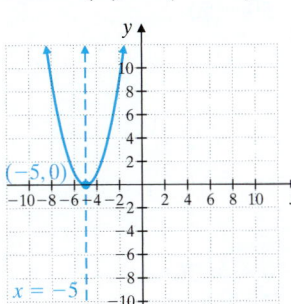

47. $f(x) = (x - 4)^2 - 2$

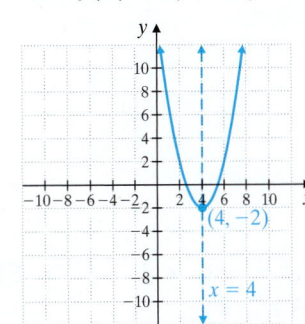

48. $f(x) = -3(x - 1)^2 + 1$

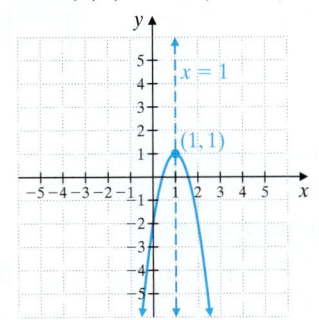

(11.6) *Graph each function. Find the vertex and any intercepts of each graph.*

49. $f(x) = x^2 + 10x + 25$

vertex: $(-5, 0)$; x-intercept: $(-5, 0)$; y-intercept: $(0, 25)$

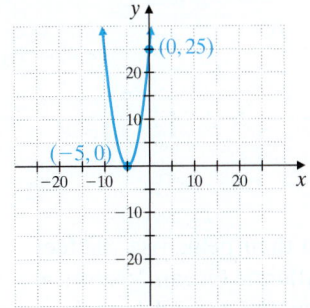

50. $f(x) = -x^2 + 6x - 9$

vertex: $(3, 0)$; x-intercept: $(3, 0)$; y-intercept: $(0, -9)$

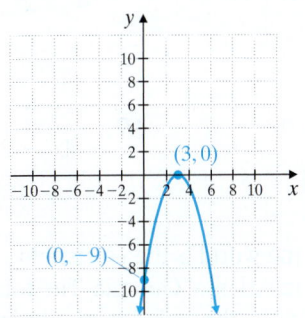

51. $f(x) = 4x^2 - 1$

vertex: $(0, -1)$; x-intercepts: $\left(-\dfrac{1}{2}, 0\right), \left(\dfrac{1}{2}, 0\right)$; y-intercept: $(0, -1)$

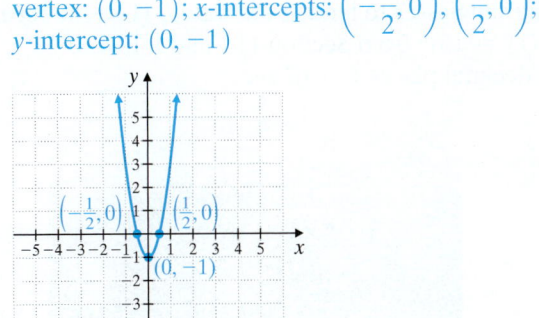

52. $f(x) = -5x^2 + 5$

vertex: $(0, 5)$; x-intercepts: $(-1, 0), (1, 0)$; y-intercept: $(0, 5)$

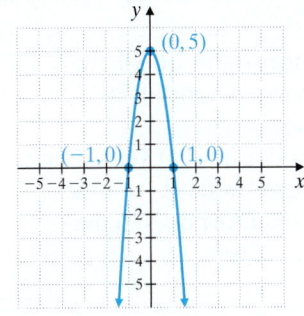

53. Find the vertex of the graph of $f(x) = -3x^2 - 5x + 4$. Determine whether the graph opens upward or downward, find the y-intercept, approximate the x-intercepts to one decimal place, and graph the function.

vertex: $\left(-\dfrac{5}{6}, \dfrac{73}{12}\right)$; opens downward;
x-intercepts: $(-2.3, 0)$, $(0.6, 0)$; y-intercept: $(0, 4)$

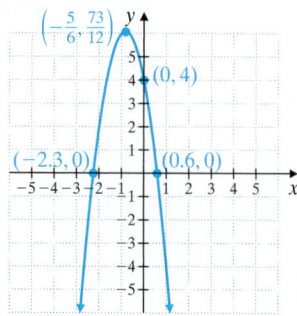

54. The function $h(t) = -16t^2 + 120t + 300$ gives the height in feet of a projectile fired from the top of a building at t seconds.

  a. When will the object reach a height of 350 feet? Round your answer to one decimal place.

  b. Explain why part (**a**) has two answers.
  answers may vary
  a. 0.4 sec and 7.1 sec

55. Find two numbers whose sum is 420 and whose product is as large as possible.
The numbers are both 210.

56. Find two numbers whose difference is 100 and whose product is as small as possible.
$50, -50$

Mixed Review

Solve each equation.

57. $x^2 - x - 30 = 0$
$\{-5, 6\}$

58. $10x^2 = 3x + 4$
$\left\{\dfrac{4}{5}, -\dfrac{1}{2}\right\}$

59. $9y^2 = 36$
$\{-2, 2\}$

60. $(9n + 1)^2 = 9$
$\left\{-\dfrac{4}{9}, \dfrac{2}{9}\right\}$

61. $x^2 + x + 7 = 0$
$\left\{\dfrac{-1 - 3i\sqrt{3}}{2}, \dfrac{-1 + 3i\sqrt{3}}{2}\right\}$

62. $(3x - 4)^2 = 10x$
$\left\{\dfrac{17 - \sqrt{145}}{9}, \dfrac{17 + \sqrt{145}}{9}\right\}$

63. $x^2 + 11 = 0$
$\{-i\sqrt{11}, i\sqrt{11}\}$

64. $x^2 + 7 = 0$
$\{-i\sqrt{7}, i\sqrt{7}\}$

65. $(5x - 2)^2 - x = 0$
$\left\{\dfrac{21 - \sqrt{41}}{50}, \dfrac{21 + \sqrt{41}}{50}\right\}$

66. $\dfrac{7}{8} = \dfrac{8}{x^2}$
$\left\{-\dfrac{8\sqrt{7}}{7}, \dfrac{8\sqrt{7}}{7}\right\}$

67. $x^{2/3} - 6x^{1/3} = -8$ $\{8, 64\}$

68. $(2x - 3)(4x + 5) \ge 0$
$\left(-\infty, -\dfrac{5}{4}\right] \cup \left[\dfrac{3}{2}, \infty\right)$

69. $\dfrac{x(x + 5)}{4x - 3} \ge 0$
$[-5, 0] \cup \left(\dfrac{3}{4}, \infty\right)$

70. $\dfrac{3}{x - 2} > 2$ $\left(2, \dfrac{7}{2}\right)$

71. The busiest airport in the world is the Hartsfield International Airport in Atlanta, Georgia. The total amount of passenger traffic through Atlanta during the period 2008 through 2013 can be modeled by the equation $y = 592x^2 - 1002x + 89,401$, where y is the number of passengers enplaned and deplaned, in thousands, and x is the number of years after 2008. (*Source:* Based on data from Airports Council International)

  a. Estimate the passenger traffic at Atlanta's Hartsfield International Airport in 2015.
  111,395 thousand passengers

  b. According to this model, in what year after 2008 will the passenger traffic at Atlanta's Hartsfield International Airport surpass 150,000 passengers?   2019

72. The Cliffs of Moher, at their peak, rise 702 feet straight up from the Atlantic Ocean. Visitors are often portrayed lying on the ground, looking over the edge, straight down into the cold water. If a visitor's sunglasses were to drop off while he or she was looking over the edge, how long would it take for the sunglasses to land in the water? (Use the formula $s(t) = 16t^2$ from Section 11.1 and round to 2 decimal places.)   6.62 sec

Solve each equation.

1. $5x^2 - 2x = 7$

2. $(x + 1)^2 = 10$

3. $m^2 - m + 8 = 0$

4. $u^2 - 6u + 2 = 0$

5. $7x^2 + 8x + 1 = 0$

6. $y^2 - 3y = 5$

7. $\dfrac{4}{x + 2} + \dfrac{2x}{x - 2} = \dfrac{6}{x^2 - 4}$

8. $x^4 - 8x^2 - 9 = 0$

9. $x^6 + 1 = x^4 + x^2$

10. $(x + 1)^2 - 15(x + 1) + 56 = 0$

Solve by completing the square.

11. $x^2 - 6x = -2$

12. $2a^2 + 5 = 4a$

Solve each inequality. Write each solution set in interval notation.

13. $2x^2 - 7x > 15$

14. $(x^2 - 16)(x^2 - 25) \geq 0$

15. $\dfrac{5}{x + 3} < 1$

16. $\dfrac{7x - 14}{x^2 - 9} \leq 0$

Graph each function. Label the vertex for each graph.

17. $f(x) = 3x^2$

18. $G(x) = -2(x - 1)^2 + 5$

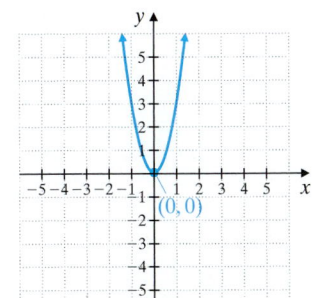

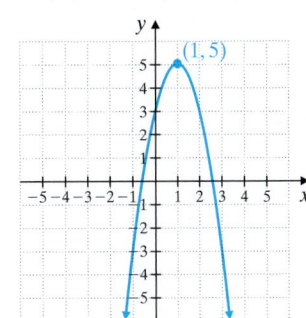

Answers

Note to Instructor: The Chapter 11 Test file in the TestGen program provides algorithms specifically matched to this test for easy replication for practice or assessment purposes.

1. $\left\{\dfrac{7}{5}, -1\right\}$

2. $\{-1 - \sqrt{10}, -1 + \sqrt{10}\}$

3. $\left\{\dfrac{1 + i\sqrt{31}}{2}, \dfrac{1 - i\sqrt{31}}{2}\right\}$

4. $\{3 - \sqrt{7}, 3 + \sqrt{7}\}$

5. $\left\{-\dfrac{1}{7}, -1\right\}$

6. $\left\{\dfrac{3 + \sqrt{29}}{2}, \dfrac{3 - \sqrt{29}}{2}\right\}$

7. $\{-2 - \sqrt{11}, -2 + \sqrt{11}\}$

8. $\{-3, 3, -i, i\}$

9. $\{-1, 1, -i, i\}$

10. $\{6, 7\}$

11. $\{3 - \sqrt{7}, 3 + \sqrt{7}\}$

12. $\left\{\dfrac{2 - i\sqrt{6}}{2}, \dfrac{2 + i\sqrt{6}}{2}\right\}$

13. $\left(-\infty, -\dfrac{3}{2}\right) \cup (5, \infty)$

14. $(-\infty, -5) \cup (-4, 4) \cup (5, \infty)$

15. $(-\infty, -3) \cup (2, \infty)$

16. $(-\infty, -3) \cup [2, 3)$

17. see graph

18. see graph

837

Graph each function. Find and label the vertex, y-intercept, and x-intercepts (if any) for each graph.

19. $h(x) = x^2 - 4x + 4$

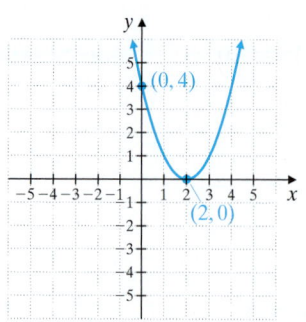

20. $F(x) = 2x^2 - 8x + 9$

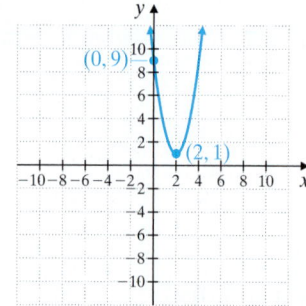

△ **21.** Given the diagram shown, approximate to the nearest foot how many feet of walking distance a person saves by cutting across the lawn instead of walking on the sidewalk.

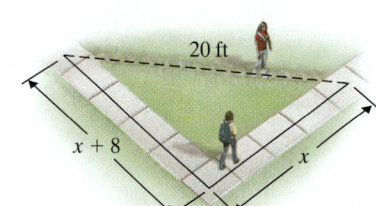

20 ft

$x + 8$

x

22. Dave and Sandy Hartranft can paint a room together in 4 hours. Working alone, Dave can paint the room in 2 hours less time than Sandy can. Find how long it takes Sandy to paint the room alone. Give the exact answer and a two-decimal-place approximation.

23. A stone is thrown upward from a bridge. The stone's height $s(t)$ in feet, above the water t seconds after the stone is thrown is given by the function $s(t) = -16t^2 + 32t + 256$.

　a. Find the maximum height of the stone.

　b. Find the time it takes the stone to hit the water. Round to the nearest hundredth of a second.

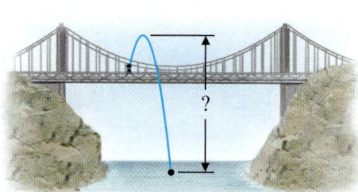

?

1. Write an equation of the line through points $(4, 0)$ and $(-4, -5)$. Write the equation in standard form $Ax + By = C$.

2. Write 2.068×10^{-3} in standard form.

Find the domain and range of each relation.

3.

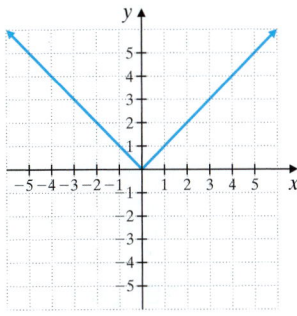

4.

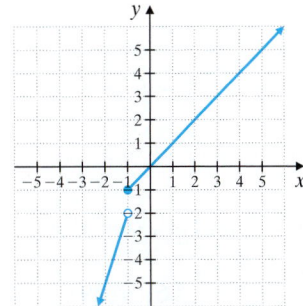

5.

6. Use the substitution method to solve the system.
$$\begin{cases} -6x + y = 5 \\ 4x - 2y = 6 \end{cases}$$

7. Solve the system:
$$\begin{cases} 2x + 4y \quad = 1 \\ 4x \quad - 4z = -1 \\ \quad y - 4z = -3 \end{cases}$$

8. Simplify. Use positive exponents to write each answer.
   a. $(a^{-2}bc^3)^{-3}$
   b. $\left(\dfrac{a^{-4}b^2}{c^3}\right)^{-2}$
   c. $\left(\dfrac{3a^8b^2}{12a^5b^5}\right)^{-2}$

9. Use matrices to solve the system:
$$\begin{cases} x + 2y + z = 2 \\ -2x - y + 2z = 5 \\ x + 3y - 2z = -8 \end{cases}$$

10. Multiply: $(4a - 3)(7a - 2)$

11. Solve the system by the substitution method:
$$\begin{cases} 5x - y = -2 \\ \quad y = 3x \end{cases}$$

12. Factor: $9x^3 + 27x^2 - 15x$

Answers

1. $-5x + 8y = -20$
   (Sec. 8.1, Ex. 2)

2. 0.002068 (Sec. 5.2)

3. domain: $(-\infty, \infty)$;
   range: $[0, \infty)$
   (Sec. 8.4, Ex. 5)

4. domain: $(-\infty, \infty)$;
   range: $(-\infty, -2) \cup [-1, \infty)$
   (Sec. 8.4)

5. domain: $[-4, 4]$;
   range: $[-2, 2]$
   (Sec. 8.4, Ex. 6)

6. $(-2, -7)$ (Sec. 4.2)

7. $\left(\dfrac{1}{2}, 0, \dfrac{3}{4}\right)$
   (Sec. 9.1, Ex. 3)

8. a. $\dfrac{a^6}{b^3c^9}$
   b. $\dfrac{a^8c^6}{b^4}$
   c. $\dfrac{16b^6}{a^6}$ (Sec. 5.2)

9. $(1, -1, 3)$
   (Sec. 9.2, Ex. 3)

10. $28a^2 - 29a + 6$
    (Sec. 5.5)

11. $(-1, -3)$
    (Sec. 4.2, Ex. 2)

12. $3x(3x^2 + 9x - 5)$
    (Sec. 6.1)

13. Graph the solutions of the system:
$$\begin{cases} 3x \geq y \\ x + 2y \leq 8 \end{cases}$$

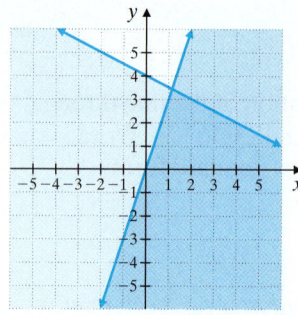

14. Solve: $2(a^2 + 2) - 8 = -2a(a - 2) - 5$

If $f(x) = \sqrt{x - 4}$ and $g(x) = \sqrt[3]{x + 2}$, find each function value.

15. $f(8)$ **16.** $f(29)$ **17.** $g(-1)$ **18.** $g(6)$

19. Use the properties of exponents to simplify: $x^{1/2}x^{1/3}$

20. Simplify: $\dfrac{(2a)^{-1} + b^{-1}}{a^{-1} + (2b)^{-1}}$

21. Use the properties of exponents to simplify: $\dfrac{(2x^{2/5})^5}{x^2}$

22. Divide $x^3 - 3x^2 - 10x + 24$ by $x + 3$.

Use the quotient rule to simplify.

23. $\sqrt{\dfrac{x}{9}}$ **24.** $\sqrt{\dfrac{y^3}{25}}$ **25.** $\sqrt[4]{\dfrac{3}{16y^4}}$ **26.** $\sqrt[3]{\dfrac{5}{27x^3}}$

Multiply. Assume that all variables represent positive real numbers.

27. $(\sqrt{2x} + 5)(\sqrt{2x} - 5)$ **28.** $(\sqrt{3} - 4)(2\sqrt{3} + 2)$

29. $(\sqrt{3} - 1)^2$ **30.** $(2a + b)(3a - 5b)$

31. Rationalize the numerator of $\dfrac{\sqrt{7}}{\sqrt{45}}$.   **32.** Rationalize the denominator of $\dfrac{2\sqrt{16}}{\sqrt{9x}}$.

33. Solve: $\sqrt{4-x} = x - 2$

34. Solve: $\sqrt{x-2} = \sqrt{4x+1} - 3$

35. Add: $(2 + 3i) + (-3 + 2i)$

36. Subtract: $\dfrac{3}{x^2 - 4} - \dfrac{1-x}{x^2-4}$

37. Solve $4x^2 - 24x + 41 = 0$ by completing the square.

38. Use the square root property to solve: $(y-1)^2 = 24$

39. Solve: $\dfrac{1}{4}m^2 - m + \dfrac{1}{2} = 0$

40. Use the quadratic formula to solve: $m^2 = 4m + 8$

41. Solve: $x^{2/3} - 5x^{1/3} + 6 = 0$

42. Solve:

$$\dfrac{x+3}{x^2+5x+6} = \dfrac{3}{2x+4} - \dfrac{1}{x+3}$$

43. Solve: $\dfrac{5}{x+1} < -2$

44. Find the vertex and any intercepts of $f(x) = x^2 + x - 12$.

45. Graph $f(x) = 3x^2 + 3x + 1$. Find the vertex and any intercepts.

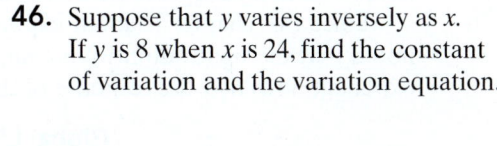

46. Suppose that y varies inversely as x. If y is 8 when x is 24, find the constant of variation and the variation equation.

---

33. $\{3\}$ (Sec. 10.6, Ex. 4)

34. $\{2, 6\}$ (Sec. 10.6)

35. $-1 + 5i$ (Sec. 10.7, Ex. 8)

36. $\dfrac{1}{x-2}$ (Sec. 7.3)

37. $\left\{ \dfrac{6 + i\sqrt{5}}{2}, \dfrac{6 - i\sqrt{5}}{2} \right\}$ (Sec. 11.1, Ex. 9)

38. $\{1 + 2\sqrt{6}, 1 - 2\sqrt{6}\}$ (Sec. 11.1)

39. $\{2 + \sqrt{2}, 2 - \sqrt{2}\}$ (Sec. 11.2, Ex. 3)

40. $\{2 + 2\sqrt{3}, 2 - 2\sqrt{3}\}$ (Sec. 11.2)

41. $\{8, 27\}$ (Sec. 11.3, Ex. 5)

42. $\{-1\}$ (Sec. 7.5)

43. $\left(-\dfrac{7}{2}, -1 \right)$ (Sec. 11.4, Ex. 5)

44. vertex: $\left(-\dfrac{1}{2}, -\dfrac{49}{4} \right)$; y-intercept: $(0, -12)$; x-intercepts: $(3, 0)$, $(-4, 0)$ (Sec. 11.6)

45. (Sec. 11.6, Ex. 2) see graph

46. $k = 192; y = \dfrac{192}{x}$ (Sec. 9.4)

12 Exponential and Logarithmic Functions

Check Your Progress

Vocabulary Check

Chapter Highlights

Chapter Review

Chapter Test

Cumulative Review

In this chapter, we discuss two closely related types of functions: exponential and logarithmic functions. These functions are vital to applications in economics, finance, engineering, the sciences, education, and other fields. Models of tumor growth and learning curves are two examples of the uses of exponential and logarithmic functions.

Are LED Lightbulbs the Lightbulbs of Our Immediate Future?

An LED lightbulb is a light-emitting diode (LED) product that has a predicted future in lighting fixtures. LED lightbulbs have a life span much greater than current rivals in the market. The graph on the bottom of this page shows a comparison of some current lightbulbs available. The other graph on this page shows a predicted exponential growth of LED products. There is also ongoing research in the production of organic LEDs (OLED).

Since this chapter is about exponential and logarithmic functions, it should be noted that the light output of a compact fluorescent lamp (or light) (CFL) decays exponentially. Also, the response of the human eye to light is logarithmic.

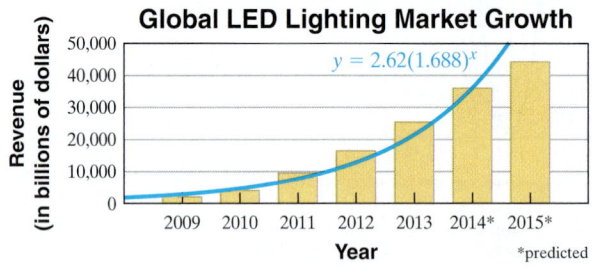

Global LED Lighting Market Growth

$y = 2.62(1.688)^x$

Revenue (in billions of dollars) — Year

2009, 2010, 2011, 2012, 2013, 2014*, 2015* *predicted

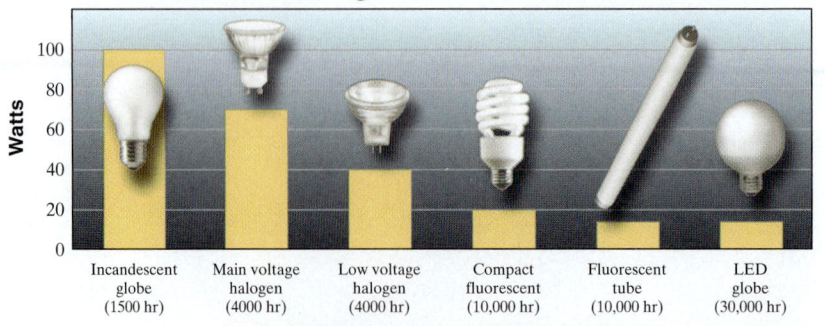

Approximate Watts Used by Different Globes to Deliver the Same Light as a 100 Watt Incandescent Bulb

Watts — Type of Bulb (approximate life span in hours)

| Incandescent globe (1500 hr) | Main voltage halogen (4000 hr) | Low voltage halogen (4000 hr) | Compact fluorescent (10,000 hr) | Fluorescent tube (10,000 hr) | LED globe (30,000 hr) |

12.1 The Algebra of Functions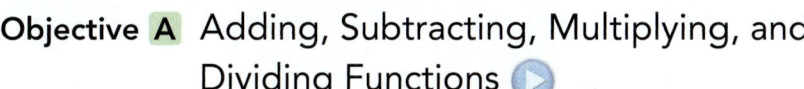

Objective A Adding, Subtracting, Multiplying, and Dividing Functions

As we have seen in earlier chapters, it is possible to add, subtract, multiply, and divide functions. Although we have not stated them as such, the sums, differences, products, and quotients of functions are themselves functions. For example, if $f(x) = 3x$ and $g(x) = x + 1$, their product, $f(x) \cdot g(x) = 3x(x + 1) = 3x^2 + 3x$, is a new function. We can use the notation $(f \cdot g)(x)$ to denote this new function. Using the sum, difference, product, and quotient of functions to generate new functions is called the **algebra of functions.**

Objectives

A Add, Subtract, Multiply, and Divide Functions.

B Construct Composite Functions.

Algebra of Functions

Let f and g be functions. New functions from f and g are defined as follows:

Sum	$(f + g)(x) = f(x) + g(x)$
Difference	$(f - g)(x) = f(x) - g(x)$
Product	$(f \cdot g)(x) = f(x) \cdot g(x)$
Quotient	$\left(\dfrac{f}{g}\right)(x) = \dfrac{f(x)}{g(x)},\ g(x) \neq 0$

Example 1 If $f(x) = x - 1$ and $g(x) = 2x - 3$, find the following.

a. $(f + g)(x)$ **b.** $(f - g)(x)$ **c.** $(f \cdot g)(x)$ **d.** $\left(\dfrac{f}{g}\right)(x)$

Solution: Use the algebra of functions and replace $f(x)$ by $x - 1$ and $g(x)$ by $2x - 3$. Then simplify.

a. $(f + g)(x) = f(x) + g(x)$
$\qquad\qquad\quad = (x - 1) + (2x - 3)$
$\qquad\qquad\quad = 3x - 4$

b. $(f - g)(x) = f(x) - g(x)$
$\qquad\qquad\quad = (x - 1) - (2x - 3)$
$\qquad\qquad\quad = x - 1 - 2x + 3$
$\qquad\qquad\quad = -x + 2$

c. $(f \cdot g)(x) = f(x) \cdot g(x)$
$\qquad\qquad\quad = (x - 1)(2x - 3)$
$\qquad\qquad\quad = 2x^2 - 5x + 3$

d. $\left(\dfrac{f}{g}\right)(x) = \dfrac{f(x)}{g(x)} = \dfrac{x - 1}{2x - 3}$, where $x \neq \dfrac{3}{2}$

■ **Work Practice 1**

Practice 1

If $f(x) = x + 3$ and $g(x) = 3x - 1$, find

a. $(f + g)(x)$

b. $(f - g)(x)$

c. $(f \cdot g)(x)$

d. $\left(\dfrac{f}{g}\right)(x)$

Teaching Tip

For Example 1, have students confirm that the sum (difference, product, quotient) of $f(7)$ and $g(7)$ gives the same result as the functions summed (subtracted, multiplied, divided) and then evaluated at $x = 7$.

There is an interesting but not surprising relationship between the graphs of functions and the graphs of their sum, difference, product, and quotient. For example, the graph of $(f + g)$ can be found by adding the graph of f to the graph of g. We add two graphs by adding corresponding y-values. (*Note:* These graphs on the top of the next page are not from Example 1.)

Answers

1. a. $4x + 2$ **b.** $-2x + 4$
c. $3x^2 + 8x - 3$
d. $\dfrac{x + 3}{3x - 1}$, where $x \neq \dfrac{1}{3}$

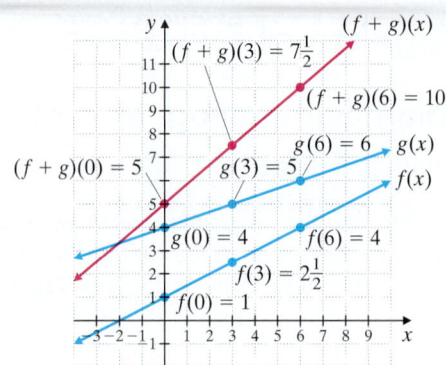

Objective B Constructing Composite Functions

Another way to combine functions is called **function composition.** To understand this new way of combining functions, study the diagrams below. The left diagram (in the margin) shows an illustration by thermometers, and the right diagram (below) is the same illustration but by tables. In both illustrations, we show degrees Celsius $f(x)$ as a function of degrees Fahrenheit x, then Kelvins $g(x)$ as a function of degrees Celsius x. (The Kelvin scale is a temperature scale devised by Lord Kelvin in 1848.) The first function we will call f, and the second function we will call g.

Thermometer Illustration

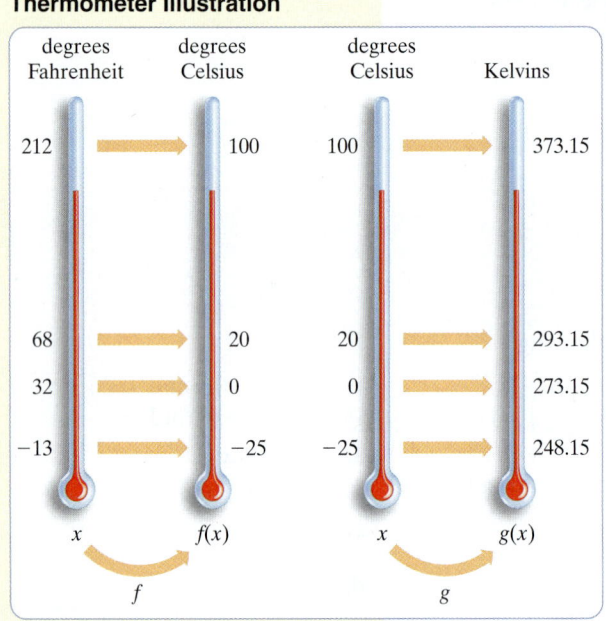

Table Illustration

x = Degrees Fahrenheit (Input)	−13	32	68	212
$f(x)$ = Degrees Celsius (Output)	−25	0	20	100

x = Degrees Celsius (Input)	−25	0	20	100
$g(x)$ = Kelvins (Output)	248.15	273.15	293.15	373.15

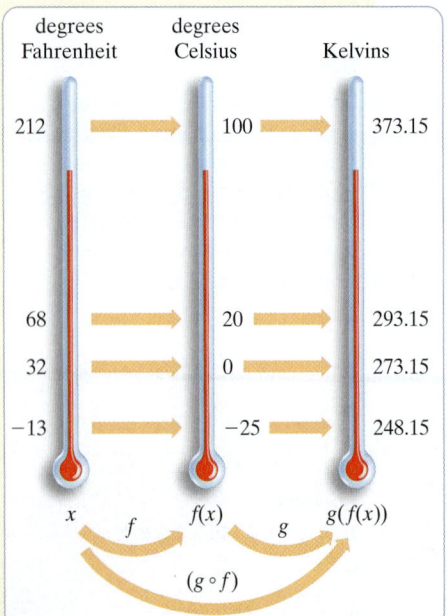

Suppose that we want a function that shows a direct conversion from degrees Fahrenheit to Kelvins. In other words, suppose that a function is needed that shows Kelvins as a function of degrees Fahrenheit. This can easily be done because the output of the first function $f(x)$ is the same as the input of the second function. If we use $g(f(x))$ to represent this, then we get the diagram to the left.

x = Degrees Fahrenheit (Input)	−13	32	68	212
$g(f(x))$ = Kelvins (Output)	248.15	273.15	293.15	373.15

For example $g(f(-13)) = 248.15$, and so on.

Since the output of the first function is used as the input of the second function, we write the new function as $g(f(x))$. The new function is formed from the composition of the other two functions. The mathematical symbol for this composition is $(g \circ f)(x)$. Thus, $(g \circ f)(x) = g(f(x))$.

It is possible to find an equation for the composition of the two functions f and g. In other words, we can find a function that converts degrees Fahrenheit directly to Kelvins. The function $f(x) = \dfrac{5}{9}(x - 32)$ converts degrees Fahrenheit to degrees Celsius, and the function $g(x) = x + 273.15$ converts degrees Celsius to Kelvins. Thus,

$$(g \circ f)(x) = g(f(x)) = g\left(\frac{5}{9}(x - 32)\right) = \frac{5}{9}(x - 32) + 273.15$$

In general, the notation **g(f(x))** means "g composed with f" and can be written as **(g ∘ f) (x).** Also $f(g(x))$, or $(f \circ g)(x)$, means "f composed with g."

Composition of Functions

The composition of functions f and g is

$$(f \circ g)(x) = f(g(x))$$

Helpful Hint

$(f \circ g)(x)$ does not mean the same as $(f \cdot g)(x)$.

$$(f \circ g)(x) = f(g(x)) \text{ while } (f \cdot g)(x) = f(x) \cdot g(x)$$
$$\uparrow \qquad\qquad\qquad\qquad \uparrow$$
Composition of functions   Multiplication of functions

Example 2   If $f(x) = x^2$ and $g(x) = x + 3$, find each composition.

a. $(f \circ g)(2)$ and $(g \circ f)(2)$
b. $(f \circ g)(x)$ and $(g \circ f)(x)$

Solution:

a. $(f \circ g)(2) = f(g(2))$
$\qquad\qquad = f(5)$ Since $g(x) = x + 3$, then $g(2) = 2 + 3 = 5$.
$\qquad\qquad = 5^2 = 25$
$\quad (g \circ f)(2) = g(f(2))$
$\qquad\qquad = g(4)$ Since $f(x) = x^2$, then $f(2) = 2^2 = 4$.
$\qquad\qquad = 4 + 3 = 7$

b. $(f \circ g)(x) = f(g(x))$
$\qquad\qquad = f(x + 3)$ Replace $g(x)$ with $x + 3$.
$\qquad\qquad = (x + 3)^2$ $f(x + 3) = (x + 3)^2$
$\qquad\qquad = x^2 + 6x + 9$ Square $(x + 3)$.
$\quad (g \circ f)(x) = g(f(x))$
$\qquad\qquad = g(x^2)$ Replace $f(x)$ with x^2.
$\qquad\qquad = x^2 + 3$ $g(x^2) = x^2 + 3$

■ **Work Practice 2**

Example 3   If $f(x) = |x|$ and $g(x) = x - 2$, find each composition.

a. $(f \circ g)(x)$ **b.** $(g \circ f)(x)$

Solution:

a. $(f \circ g)(x) = f(g(x)) = f(x - 2) = |x - 2|$
b. $(g \circ f)(x) = g(f(x)) = g(|x|) = |x| - 2$

■ **Work Practice 3**

Practice 2

If $f(x) = x^2$ and $g(x) = 2x + 1$, find each composition.
a. $(f \circ g)(3)$ and $(g \circ f)(3)$
b. $(f \circ g)(x)$ and $(g \circ f)(x)$

Practice 3

If $f(x) = \sqrt{x}$ and $g(x) = x + 1$, find each composition.
a. $(f \circ g)(x)$ **b.** $(g \circ f)(x)$

Answers

2. a. 49; 19   **b.** $4x^2 + 4x + 1$; $2x^2 + 1$
3. a. $\sqrt{x + 1}$ **b.** $\sqrt{x} + 1$

Helpful Hint

In Examples 2 and 3, notice that $(g \circ f)(x) \neq (f \circ g)(x)$. In general, $(g \circ f)(x)$ *may* or *may not* equal $(f \circ g)(x)$.

Practice 4

If $f(x) = 2x$, $g(x) = x + 5$, and $h(x) = |x|$, write each function as a composition of f, g, or h.

a. $F(x) = |x + 5|$

b. $G(x) = 2x + 5$

Teaching Tip

Have students work in pairs to write the function $f(x) = x^2 - 6x + 13$ as a composition of functions in two ways.

Answers

4. a. $(h \circ g)(x)$ **b.** $(g \circ f)(x)$

Example 4 If $f(x) = 5x$, $g(x) = x - 2$, and $h(x) = \sqrt{x}$, write each function as a composition with f, g, or h.

a. $F(x) = \sqrt{x - 2}$

b. $G(x) = 5x - 2$

Solution:

a. Notice the order in which the function F operates on an input value x. First, 2 is subtracted from x, and then the square root of that result is taken. This means that $F(x) = (h \circ g)(x)$. To check, we find $(h \circ g)(x)$.

$$(h \circ g)(x) = h(g(x)) = h(x - 2) = \sqrt{x - 2}$$

b. Notice the order in which the function G operates on an input value x. First, x is multiplied by 5, and then 2 is subtracted from the result. This means that $G(x) = (g \circ f)(x)$. To check, we find $(g \circ f)(x)$.

$$(g \circ f)(x) = g(f(x)) = g(5x) = 5x - 2$$

■ **Work Practice 4**

 Calculator Explorations Graphing

If $f(x) = \dfrac{1}{2}x + 2$ and $g(x) = \dfrac{1}{3}x^2 + 4$, then

$$
\begin{aligned}
(f + g)(x) &= f(x) + g(x) \\
&= \left(\frac{1}{2}x + 2\right) + \left(\frac{1}{3}x^2 + 4\right) \\
&= \frac{1}{3}x^2 + \frac{1}{2}x + 6
\end{aligned}
$$

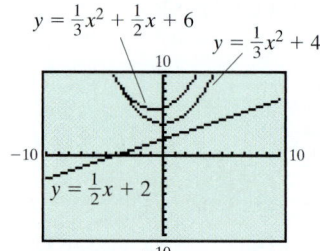

To visualize this addition of functions with a grapher, graph

$$Y_1 = \frac{1}{2}x + 2, \quad Y_2 = \frac{1}{3}x^2 + 4, \quad \text{and} \quad Y_3 = \frac{1}{3}x^2 + \frac{1}{2}x + 6$$

Use a TABLE feature to verify that for a given x value, $Y_1 + Y_2 = Y_3$. For example, verify that when $x = 0$, $Y_1 = 2$, $Y_2 = 4$, and $Y_3 = 2 + 4 = 6$.

Vocabulary, Readiness & Video Check

Match each function (Exercises 1–6) with its definition (Choices A–F).

1. $(f \circ g)(x)$ C

2. $(f \cdot g)(x)$ E

3. $(f - g)(x)$ F

4. $(g \circ f)(x)$ A

5. $\left(\dfrac{f}{g}\right)(x)$ D

6. $(f + g)(x)$ B

A. $g(f(x))$

B. $f(x) + g(x)$

C. $f(g(x))$

D. $\dfrac{f(x)}{g(x)}, g(x) \neq 0$

E. $f(x) \cdot g(x)$

F. $f(x) - g(x)$

Martin-Gay Interactive Videos   *Watch the section lecture video and answer the following questions.*

Objective A 7. From ▦ Example 1 and the lecture before, we know that $(f + g)(x) = f(x) + g(x)$. Use this fact to explain two ways you can find $(f + g)(2)$. ▶

Objective B 8. From ▦ Example 3, given two functions $f(x)$ and $g(x)$, can $f(g(x))$ ever equal $g(f(x))$? ▶

See Video 12.1 🍒

See video answer section.

12.1   Exercise Set MyMathLab® ▶

Objective A *For the functions f and g, find* **a.** $(f + g)(x)$, **b.** $(f - g)(x)$, **c.** $(f \cdot g)(x)$, *and* **d.** $\left(\dfrac{f}{g}\right)(x)$. *See Example 1.*

1. $f(x) = x - 7; g(x) = 2x + 1$
   a. $3x - 6$
   b. $-x - 8$
   c. $2x^2 - 13x - 7$
   d. $\dfrac{x - 7}{2x + 1}$, where $x \neq -\dfrac{1}{2}$

2. $f(x) = x + 4; g(x) = 5x - 2$
   a. $6x + 2$
   b. $-4x + 6$
   c. $5x^2 + 18x - 8$
   d. $\dfrac{x + 4}{5x - 2}$, where $x \neq \dfrac{2}{5}$

▶**3.** $f(x) = x^2 + 1; g(x) = 5x$
   a. $x^2 + 5x + 1$
   b. $x^2 - 5x + 1$
   c. $5x^3 + 5x$
   d. $\dfrac{x^2 + 1}{5x}$, where $x \neq 0$

4. $f(x) = x^2 - 2; g(x) = 3x$
   a. $x^2 + 3x - 2$
   b. $x^2 - 3x - 2$
   c. $3x^3 - 6x$
   d. $\dfrac{x^2 - 2}{3x}$, where $x \neq 0$

5. $f(x) = \sqrt{x}; g(x) = x + 5$
   a. $\sqrt{x} + x + 5$
   b. $\sqrt{x} - x - 5$
   c. $x\sqrt{x} + 5\sqrt{x}$
   d. $\dfrac{\sqrt{x}}{x + 5}$, where $x \neq -5$

6. $f(x) = \sqrt[3]{x}; g(x) = x - 3$
   a. $\sqrt[3]{x} + x - 3$
   b. $\sqrt[3]{x} - x + 3$
   c. $x\sqrt[3]{x} - 3\sqrt[3]{x}$
   d. $\dfrac{\sqrt[3]{x}}{x - 3}$, where $x \neq 3$

7. $f(x) = -3x; g(x) = 5x^2$
   a. $5x^2 - 3x$
   b. $-5x^2 - 3x$
   c. $-15x^3$
   d. $-\dfrac{3}{5x}$, where $x \neq 0$

8. $f(x) = 4x^3; g(x) = -6x$
   a. $4x^3 - 6x$
   b. $4x^3 + 6x$
   c. $-24x^4$
   d. $-\dfrac{2}{3}x^2$, where $x \neq 0$, since before simplification we have $\dfrac{4x^3}{-6x}$

Objective B *If* $f(x) = x^2 - 6x + 2$, $g(x) = -2x$, *and* $h(x) = \sqrt{x}$, *find each composition. See Example 2.*

9. $(f \circ g)(2)$ 42

10. $(h \circ f)(-2)$ $3\sqrt{2}$

▶**11.** $(g \circ f)(-1)$ -18

12. $(f \circ h)(1)$ -3

13. $(g \circ h)(0)$ 0

14. $(h \circ g)(0)$ 0

Find $(f \circ g)(x)$ *and* $(g \circ f)(x)$. *See Examples 2 and 3.*

▶**15.** $f(x) = x^2 + 1; g(x) = 5x$
   $(f \circ g)(x) = 25x^2 + 1; (g \circ f)(x) = 5x^2 + 5$

16. $f(x) = x - 3; g(x) = x^2$
   $(f \circ g)(x) = x^2 - 3; (g \circ f)(x) = x^2 - 6x + 9$

17. $f(x) = 2x - 3; g(x) = x + 7$
   $(f \circ g)(x) = 2x + 11; (g \circ f)(x) = 2x + 4$

18. $f(x) = x + 10; g(x) = 3x + 1$
   $(f \circ g)(x) = 3x + 11; (g \circ f)(x) = 3x + 31$

19. $f(x) = x^3 + x - 2; g(x) = -2x$
$(f \circ g)(x) = -8x^3 - 2x - 2;$
$(g \circ f)(x) = -2x^3 - 2x + 4$

20. $f(x) = -4x; g(x) = x^3 + x^2 - 6$
$(f \circ g)(x) = -4x^3 - 4x^2 + 24;$
$(g \circ f)(x) = -64x^3 + 16x^2 - 6$

21. $f(x) = |x|; g(x) = 10x - 3$
$(f \circ g)(x) = |10x - 3|; (g \circ f)(x) = 10|x| - 3$

22. $f(x) = |x|; g(x) = 14x - 8$
$(f \circ g)(x) = |14x - 8|; (g \circ f)(x) = 14|x| - 8$

23. $f(x) = \sqrt{x}; g(x) = -5x + 2$
$(f \circ g)(x) = \sqrt{-5x + 2}; (g \circ f)(x) = -5\sqrt{x} + 2$

24. $f(x) = 7x - 1; g(x) = \sqrt[3]{x}$
$(f \circ g)(x) = 7\sqrt[3]{x} - 1; (g \circ f)(x) = \sqrt[3]{7x - 1}$

If $f(x) = 3x, g(x) = \sqrt{x}$, and $h(x) = x^2 + 2$, write each function as a composition with f, g, or h. See Example 4.

▶ 25. $H(x) = \sqrt{x^2 + 2}$
$H(x) = (g \circ h)(x)$

26. $G(x) = \sqrt{3x}$
$G(x) = (g \circ f)(x)$

27. $F(x) = 9x^2 + 2$
$F(x) = (h \circ f)(x)$

28. $H(x) = 3x^2 + 6$
$H(x) = (f \circ h)(x)$

29. $G(x) = 3\sqrt{x}$
$G(x) = (f \circ g)(x)$

30. $F(x) = x + 2$
$F(x) = (h \circ g)(x)$

Find f(x) and g(x) so that the given function $h(x) = (f \circ g)(x)$.

31. $h(x) = (x + 2)^2$
answers may vary

32. $h(x) = |x - 1|$
answers may vary

33. $h(x) = \sqrt{x + 5} + 2$
answers may vary

34. $h(x) = (3x + 4)^2 + 3$
answers may vary

35. $h(x) = \dfrac{1}{2x - 3}$
answers may vary

36. $h(x) = \dfrac{1}{x + 10}$
answers may vary

Review

Solve each equation for y. See Section 2.5.

37. $x = y + 2$
$y = x - 2$

38. $x = y - 5$
$y = x + 5$

39. $x = 3y$
$y = \dfrac{x}{3}$

40. $x = -6y$
$y = -\dfrac{x}{6}$

41. $x = -2y - 7$
$y = -\dfrac{x + 7}{2}$

42. $x = 4y + 7$
$y = \dfrac{x - 7}{4}$

Concept Extensions

Solve.

43. Businesspeople are concerned with cost functions, revenue functions, and profit functions. Recall that the profit $P(x)$ obtained from selling x units of a product is equal to the revenue $R(x)$ from selling the x units minus the cost $C(x)$ of manufacturing the x units. Write an equation expressing this relationship among $C(x), R(x)$, and $P(x)$. $P(x) = R(x) - C(x)$

44. Suppose the revenue $R(x)$ for x units of a product can be described by $R(x) = 25x$ and the cost $C(x)$ can be described by $C(x) = 50 + x^2 + 4x$. Find the profit $P(x)$ for x units. $P(x) = 21x - x^2 - 50$

45. If you are given $f(x)$ and $g(x)$, explain in your own words how to find $(f \circ g)(x)$, and then how to find $(g \circ f)(x)$.   answers may vary

46. Given $f(x)$ and $g(x)$, describe in your own words the difference between $(f \circ g)(x)$ and $(f \cdot g)(x)$. answers may vary

12.2 Inverse Functions

In the next sections, we begin a study of two new functions: exponential and logarithmic functions. As we learn more about these functions, we will discover that they share a special relation to each other; they are inverses of each other.

Before we study these functions, we need to learn about inverses. We begin by defining one-to-one functions.

Objective A Determining Whether a Function Is One-to-One

Study the following table.

Degrees Fahrenheit (Input)	−31	−13	32	68	149	212
Degrees Celsius (Output)	−35	−25	0	20	65	100

Recall that since each degrees Fahrenheit (input) corresponds to exactly one degrees Celsius (output), this table of inputs and outputs does describe a function. Also notice that each output corresponds to a different input. This type of function is given a special name—a *one-to-one function*.

Does the set $f = \{(0,1),(2,2),(-3,5),(7,6)\}$ describe a one-to-one function? It is a function since each x-value corresponds to a unique y-value. For this particular function f, each y-value corresponds to a unique x-value. Thus, this function is also a one-to-one function.

> ### One-to-One Function
>
> For a **one-to-one function,** each x-value (input) corresponds to only one y-value (output) and each y-value (output) corresponds to only one x-value (input).

Examples   Determine whether each function described is one-to-one.

1. $f = \{(6,2),(5,4),(-1,0),(7,3)\}$

The function f is one-to-one since each y-value corresponds to only one x-value.

2. $g = \{(3,9),(-4,2),(-3,9),(0,0)\}$

The function g is not one-to-one because the y-value 9 in $(3,9)$ and $(-3,9)$ corresponds to two different x-values.

3. $h = \{(1,1),(2,2),(10,10),(-5,-5)\}$

The function h is one-to-one since each y-value corresponds to only one x-value.

4.

Mineral (Input)	Talc	Gypsum	Diamond	Topaz	Stibnite
Hardness on the Mohs Scale (Output)	1	2	10	8	2

This table does not describe a one-to-one function since the output 2 corresponds to two different inputs, gypsum and stibnite.

(Continued on next page)

Objectives

A Determine Whether a Function Is a One-to-One Function.

B Use the Horizontal Line Test to Decide Whether a Function Is a One-to-One Function.

C Find the Inverse of a Function.

D Find an Equation of the Inverse of a Function.

E Graph Functions and Their Inverses.

F Determine Whether Two Functions Are Inverses of Each Other.

Practice 1–5

Determine whether each function described is one-to-one.

1. $f = \{(7,3),(-1,1),$
$(5,0),(4,-2)\}$

2. $g = \{(-3,2),(6,3),$
$(2,14),(-6,2)\}$

3. $h = \{(0,0),(1,2),$
$(3,4),(5,6)\}$

4.

State (Input)	Colorado	Mississippi	Nevada	New Mexico	Utah
Number of Colleges and Universities (Output)	16	7	4	10	7

Source: American Educational Guidance Center, 2005.

5.

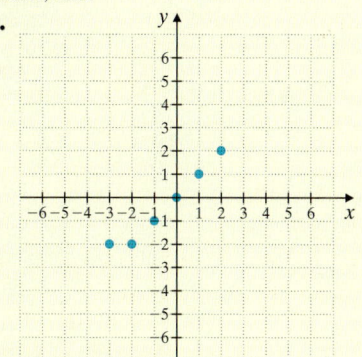

Answers

1. one-to-one   **2.** not one-to-one
3. one-to-one   **4.** not one-to-one
5. not one-to-one

5.

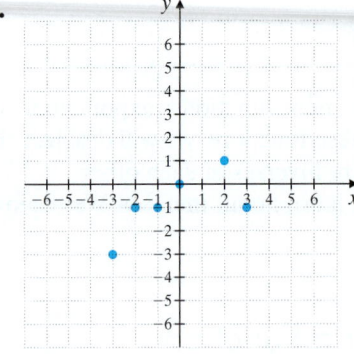

This graph does not describe a one-to-one function since the y-value -1 corresponds to three different x-values, -2, -1 and 3, as shown to the right.

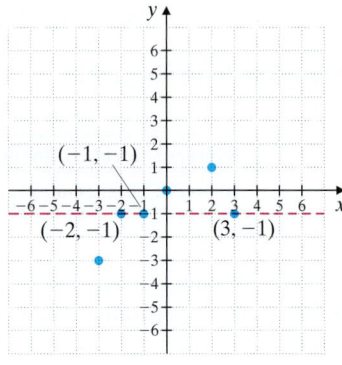

Work Practice 1–5

Objective B Using the Horizontal Line Test

Recall that we recognize the graph of a function when it passes the vertical line test. Since every x-value of the function corresponds to exactly one y-value, each vertical line intersects the function's graph at most once. The graph shown next, for instance, is the graph of a function.

Is this function a *one-to-one* function? The answer is no. To see why not, notice that the y-value of the ordered pair $(-3, 3)$, for example, is the same as the y-value of the ordered pair $(3, 3)$. This function is therefore not one-to-one.

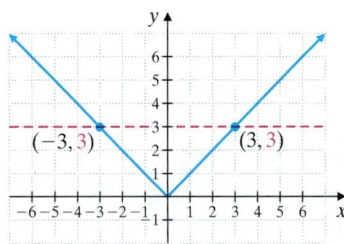

To test whether a graph is the graph of a one-to-one function, we can apply the vertical line test to see whether it is a function, and then apply a similar **horizontal line test** to see whether it is a one-to-one function.

Horizontal Line Test

If every horizontal line intersects the graph of a function at most once, then the function is a one-to-one function.

Example 6	Use the vertical and horizontal line tests to determine whether each graph is the graph of a one-to-one function.

a.

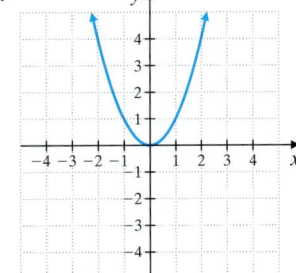

b.

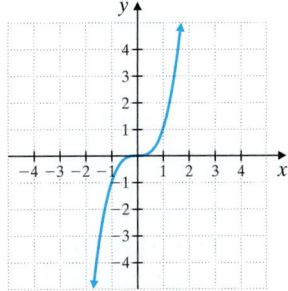

c.

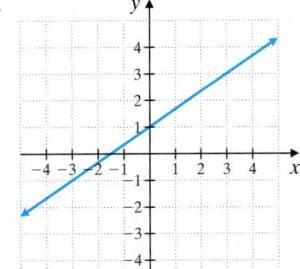

d.

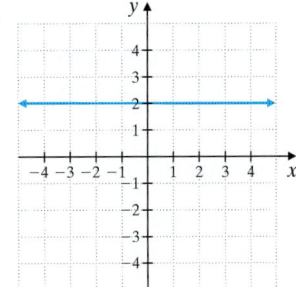

e.

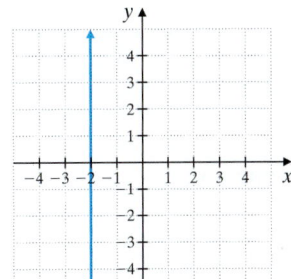

Solution: Graphs **a, b, c,** and **d** all pass the vertical line test, so only these graphs are graphs of functions. But, of these, only **b** and **c** pass the horizontal line test, so only **b** and **c** are graphs of one-to-one functions.

■ **Work Practice 6**

Helpful Hint

All linear equations are one-to-one functions except those whose graphs are horizontal or vertical lines. A vertical line does not pass the vertical line test and hence is not the graph of a function. A horizontal line is the graph of a function but does not pass the horizontal line test and hence is not the graph of a one-to-one function.

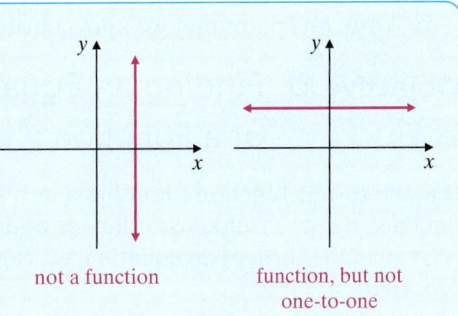

not a function   function, but not one-to-one

Objective **C** Finding the Inverse of a Function

One-to-one functions are special in that their graphs pass the vertical and horizontal line tests. They are special, too, in another sense: We can find the **inverse function** for any one-to-one function by switching the coordinates of the ordered pairs of the

Practice 6

Use the vertical and horizontal line tests to determine whether each graph is the graph of a one-to-one function.

a.

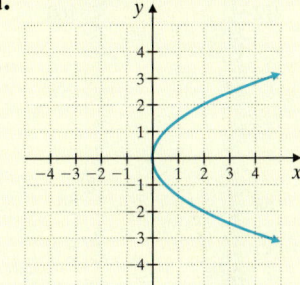

b.

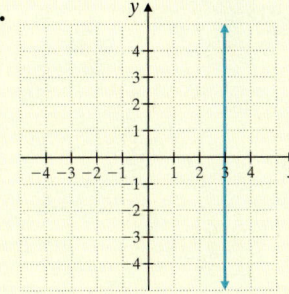

c.

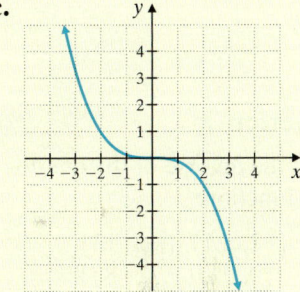

d.

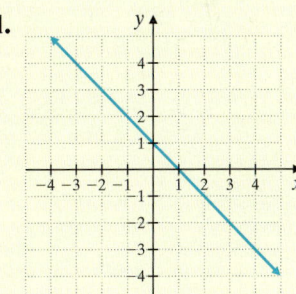

e.

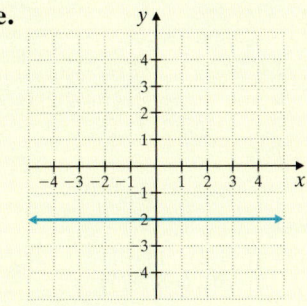

Answers

6. **a.** not a function **b.** not a function
c. one-to-one **d.** one-to-one
e. not one-to-one

function, or the inputs and the outputs. For example, the inverse of the one-to-one function

f

Input: Output:
degrees   degrees
Fahrenheit   Celsius

212 ────────▶ 100

68 ────────▶ 20
32 ────────▶ 0

−13 ────────▶ −25

f⁻¹

Degrees Fahrenheit (Input)	−31	−13	32	68	149	212
Degrees Celsius (Output)	−35	−25	0	20	65	100

is the function

Degrees Celsius (Input)	−35	−25	0	20	65	100
Degrees Fahrenheit (Output)	−31	−13	32	68	149	212

Notice that the ordered pair $(-31, -35)$ of the function, for example, becomes the ordered pair $(-35, -31)$ of its inverse.

Also, the inverse of the one-to-one function $f = \{(2, -3), (5, 10), (9, 1)\}$ is $\{(-3, 2), (10, 5), (1, 9)\}$. For a function f, we use the notation f^{-1}, read "f inverse," to denote its inverse function. Notice that since the coordinates of each ordered pair have been switched, the domain (set of inputs) of f is the range (set of outputs) of f^{-1}, and the range of f is the domain of f^{-1}.

Inverse Function

The inverse of a one-to-one function f is the one-to-one function f^{-1} that consists of the set of all ordered pairs (y, x) where (x, y) belongs to f.

Helpful Hint

If a function is not one-to-one, it does not have an inverse function.

Practice 7

Find the inverse of the one-to-one function:

$f = \{(2, -4), (-1, 13),$
$\quad (0, 0), (-7, -8)\}$

Example 7

Find the inverse of the one-to-one function:

$f = \{(0, 1), (-2, 7), (3, -6), (4, 4)\}$

Solution:

$f^{-1} = \{(1, 0), (7, -2), (-6, 3), (4, 4)\}$

Switch coordinates of each ordered pair.

■ Work Practice 7

✓ Concept Check

Suppose that f is a one-to-one function and that $f(1) = 5$.

a. Write the corresponding ordered pair.

b. Write one point that we know must belong to the inverse function f^{-1}.

Objective D Finding an Equation of the Inverse of a Function ▶

If a one-to-one function f is defined as a set of ordered pairs, we can find f^{-1} by interchanging the x- and y-coordinates of the ordered pairs. If a one-to-one function f is given in the form of an equation, we can find an equation of f^{-1} by using a similar procedure.

Answer

7. $f^{-1} = \{(-4, 2), (13, -1),$
$\quad (0, 0), (-8, -7)\}$

✓ Concept Check Answers

a. $(1, 5)$

b. $(5, 1)$

Finding an Equation of the Inverse of a One-to-One Function f

Step 1: Replace $f(x)$ with y.

Step 2: Interchange x and y.

Step 3: Solve the equation for y.

Step 4: Replace y with the notation $f^{-1}(x)$.

Helpful Hint

The symbol f^{-1} is the single symbol used to denote the inverse of the function f. It is read as "f inverse." This symbol *does not mean* $\dfrac{1}{f}$.

Example 8 Find the equation of the inverse of $f(x) = x + 3$.

Solution: $f(x) = x + 3$

Step 1:	$y = x + 3$	Replace $f(x)$ with y.
Step 2:	$x = y + 3$	Interchange x and y.
Step 3:	$x - 3 = y$	Solve for y.
Step 4:	$f^{-1}(x) = x - 3$	Replace y with $f^{-1}(x)$.

Teaching Tip

Before discussing the solution to Example 8, ask, "If the function adds 3 to the input. What would you expect the inverse of the function to do to the input?"

The inverse of $f(x) = x + 3$ is $f^{-1}(x) = x - 3$. Notice that, for example,

$$f(1) = 1 + 3 = 4 \quad \text{and} \quad f^{-1}(4) = 4 - 3 = 1$$

Ordered pair: $(1, 4)$ Ordered pair: $(4, 1)$

The coordinates are switched, as expected.

■ Work Practice 8

Example 9 Find the equation of the inverse of $f(x) = 3x - 5$. Graph f and f^{-1} on the same set of axes.

Solution: $f(x) = 3x - 5$

Step 1:	$y = 3x - 5$	Replace $f(x)$ with y.
Step 2:	$x = 3y - 5$	Interchange x and y.
Step 3:	$x + 5 = 3y$	Solve for y.

$$\frac{x + 5}{3} = y \quad \text{or} \quad y = \frac{x + 5}{3}$$

| Step 4: | $f^{-1}(x) = \dfrac{x + 5}{3}$ | Replace y with $f^{-1}(x)$. |

Now we graph f and f^{-1} on the same set of axes. Both $f(x) = 3x - 5$ and $f^{-1}(x) = \dfrac{x + 5}{3}$ are linear functions, so each graph is a line.

$f(x) = 3x - 5$	
x	$y = f(x)$
1	-2
0	-5
$\dfrac{5}{3}$	0

$f^{-1}(x) = \dfrac{x + 5}{3}$	
x	$y = f^{-1}(x)$
-2	1
-5	0
0	$\dfrac{5}{3}$

(Continued on next page)

Practice 8

Find the equation of the inverse of $f(x) = x - 6$.

Teaching Tip

Before discussing the solution to Example 9, ask, "What does f do to the input? What would you expect the inverse of f to do to the input?" After discussing the solution, state "Notice that the inverse function does the inverse operations of the function in reverse order."

Practice 9

Find the equation of the inverse of $f(x) = 2x + 3$. Graph f and f^{-1} on the same set of axes.

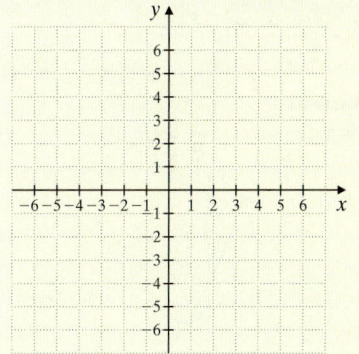

Answers

8. $f^{-1}(x) = x + 6$

9. $f^{-1}(x) = \dfrac{x - 3}{2}$

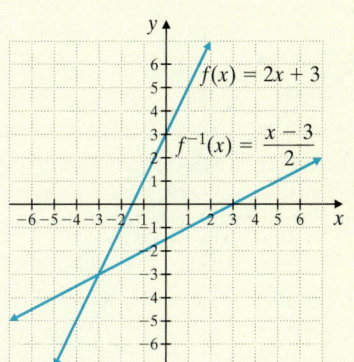

Practice 10

Graph the inverse of each function.

a.

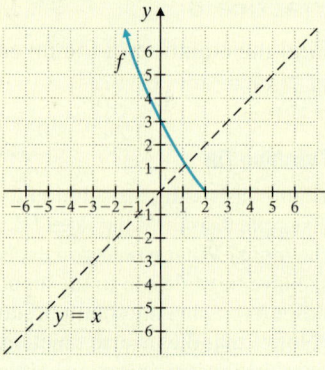

b.

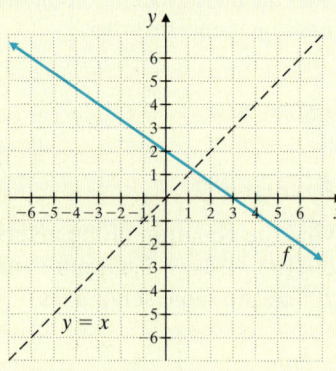

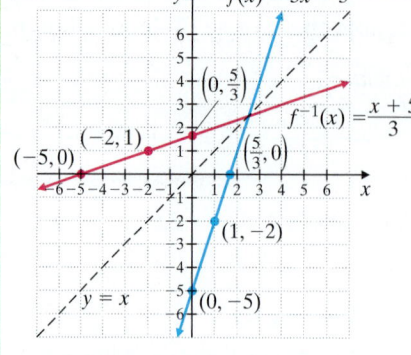

■ **Work Practice 9**

Objective **E** Graphing Inverse Functions

Notice that the graphs of f and f^{-1} in Example 9 are mirror images of each other, and the "mirror" is the dashed line $y = x$. This is true for every function and its inverse. For this reason, we say that the *graphs of f and f^{-1} are symmetric about the line $y = x$.*

To see why this happens, study the graph of a few ordered pairs and their switched coordinates.

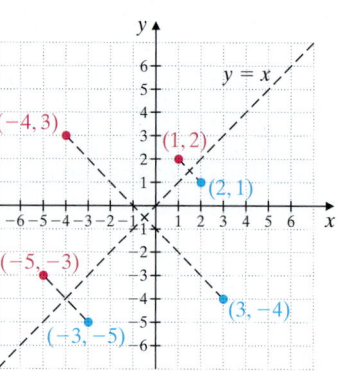

Answers
10. a.

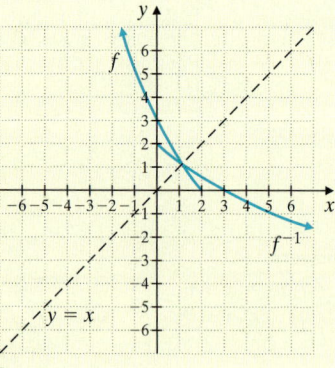

b.

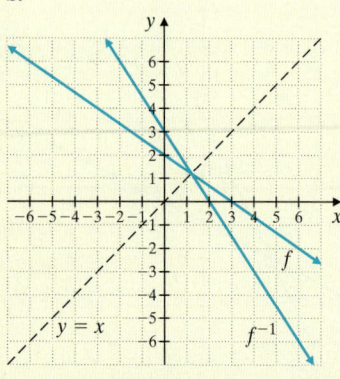

Example 10 Graph the inverse of each function.

Solution: The function is graphed in blue and the inverse is graphed in red.

a.

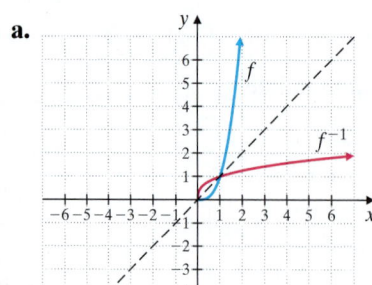

b.

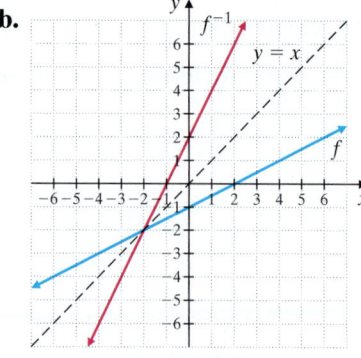

■ **Work Practice 10**

Objective F Determining Whether Functions Are Inverses of Each Other ▶

Notice in the table of values in Example 9 that $f(0) = -5$ and $f^{-1}(-5) = 0$, as expected. Also, for example, $f(1) = -2$ and $f^{-1}(-2) = 1$. In words, we say that for some input x, the function f^{-1} takes the output of x, called $f(x)$, back to x.

$$x \rightarrow f(x) \quad \text{and} \quad f^{-1}(f(x)) \rightarrow x$$
$$\downarrow \quad\quad \downarrow \quad\quad\quad\quad \downarrow \quad\quad \downarrow$$
$$f(0) = -5 \quad \text{and} \quad f^{-1}(-5) = 0$$
$$f(1) = -2 \quad \text{and} \quad f^{-1}(-2) = 1$$

In general,

If f is a one-to-one function, then the inverse of f is the function f^{-1} such that

$$(f^{-1} \circ f)(x) = x \quad \text{and} \quad (f \circ f^{-1})(x) = x$$

Example 11 Show that if $f(x) = 3x + 2$, then $f^{-1}(x) = \dfrac{x - 2}{3}$.

Solution: See that $(f^{-1} \circ f)(x) = x$ and $(f \circ f^{-1})(x) = x$.

$$
\begin{aligned}
(f^{-1} \circ f)(x) &= f^{-1}(f(x)) \\
&= f^{-1}(3x + 2) \qquad \text{Replace } f(x) \text{ with } 3x + 2. \\
&= \frac{3x + 2 - 2}{3} \\
&= \frac{3x}{3} \\
&= x
\end{aligned}
$$

$$
\begin{aligned}
(f \circ f^{-1})(x) &= f(f^{-1}(x)) \\
&= f\left(\frac{x - 2}{3}\right) \qquad \text{Replace } f^{-1}(x) \text{ with } \frac{x-2}{3}. \\
&= 3\left(\frac{x - 2}{3}\right) + 2 \\
&= x - 2 + 2 \\
&= x
\end{aligned}
$$

■ **Work Practice 11**

Practice 11

Show that if $f(x) = 4x - 1$, then $f^{-1}(x) = \dfrac{x + 1}{4}$.

Answer

11.
$$
\begin{aligned}
f(f^{-1}(x)) &= 4\left(\frac{x + 1}{4}\right) - 1 \\
&= x + 1 - 1 = x \\
f^{-1}(f(x)) &= \frac{(4x - 1) + 1}{4} \\
&= \frac{4x}{4} = x
\end{aligned}
$$

▦ Calculator Explorations Graphing

A grapher can be used to visualize functions and their inverses. Recall that the graph of a function f and its inverse f^{-1} are mirror images of each other across the line $y = x$. To see this for the function $f(x) = 3x + 2$, use a square window (see Appendix H) and graph

the given function: $Y_1 = 3x + 2$

its inverse: $\qquad Y_2 = \dfrac{x - 2}{3}$

and the line $\qquad Y_3 = x$

Exercises will follow in Exercise Set 12.2.

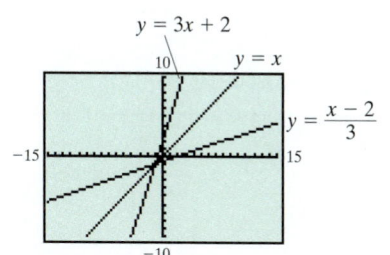

Vocabulary, Readiness & Video Check

Use the choices below to fill in each blank. Some choices will not be used, and some will be used more than once.

vertical	$(3, 7)$	$(11, 2)$	$y = x$	x	true
horizontal	$(7, 3)$	$(2, 11)$	$\dfrac{1}{f}$	the inverse of f	false

1. If $f(2) = 11$, the corresponding ordered pair is _____(2, 11)_____.
2. If $(7, 3)$ is an ordered pair solution of $f(x)$, and $f(x)$ has an inverse, then an ordered pair solution
   of $f^{-1}(x)$ is _____(3, 7)_____.
3. The symbol f^{-1} means _the inverse of f_.
4. True or false: The function notation $f^{-1}(x)$ means $\dfrac{1}{f(x)}$. ____false____
5. To tell whether a graph is the graph of a function, use the ___vertical___ line test.
6. To tell whether the graph of a function is also a one-to-one function, use the ___horizontal___ line test.
7. The graphs of f and f^{-1} are symmetric about the line ___$y = x$___.
8. Two functions are inverses of each other if $(f \circ f^{-1})(x) =$ ____x____ and $(f^{-1} \circ f)(x) =$ ____x____.

Martin-Gay Interactive Videos

See Video 12.2

Watch the section lecture video and answer the following questions.

Objective A   9. From ▥ Example 1 and the definition before, what makes a one-to-one function different from other types of functions? ▶

Objective B   10. From ▥ Examples 2 and 3, if a graph passes the horizontal line test but not the vertical line test, is it a one-to-one function? Explain. ▶

Objective C   11. From ▥ Example 4 and the lecture before, if you find the inverse of a one-to-one function, is this inverse function also a one-to-one function? How do you know? ▶

Objective D   12. From ▥ Examples 5 and 6, explain why the interchanging of x and y when finding an inverse equation makes sense given the definition of an inverse function. ▶

Objective E   13. From ▥ Example 7, if you have the equation or graph of a one-to-one function, how can you graph its inverse without finding the inverse's equation? ▶

Objective F   14. Based on ▥ Example 8 and the lecture before, what's wrong with the following statement? "If f is a one-to-one function, you can prove that f and f^{-1} are inverses of each other by showing that $f(f^{-1}(x)) = f^{-1}(f(x))$." ▶

See video answer section.

Objective F Determining Whether Functions Are Inverses of Each Other

Notice in the table of values in Example 9 that $f(0) = -5$ and $f^{-1}(-5) = 0$, as expected. Also, for example, $f(1) = -2$ and $f^{-1}(-2) = 1$. In words, we say that for some input x, the function f^{-1} takes the output of x, called $f(x)$, back to x.

$$x \rightarrow f(x) \quad \text{and} \quad f^{-1}(f(x)) \rightarrow x$$
$$\downarrow \quad \downarrow \qquad\qquad \downarrow \quad \downarrow$$
$$f(0) = -5 \quad \text{and} \quad f^{-1}(-5) = 0$$
$$f(1) = -2 \quad \text{and} \quad f^{-1}(-2) = 1$$

In general,

If f is a one-to-one function, then the inverse of f is the function f^{-1} such that

$$(f^{-1} \circ f)(x) = x \quad \text{and} \quad (f \circ f^{-1})(x) = x$$

Example 11 Show that if $f(x) = 3x + 2$, then $f^{-1}(x) = \dfrac{x-2}{3}$.

Solution: See that $(f^{-1} \circ f)(x) = x$ and $(f \circ f^{-1})(x) = x$.

$$(f^{-1} \circ f)(x) = f^{-1}(f(x))$$
$$= f^{-1}(3x + 2) \qquad \text{Replace } f(x) \text{ with } 3x + 2.$$
$$= \frac{3x + 2 - 2}{3}$$
$$= \frac{3x}{3}$$
$$= x$$
$$(f \circ f^{-1})(x) = f(f^{-1}(x))$$
$$= f\left(\frac{x-2}{3}\right) \qquad \text{Replace } f^{-1}(x) \text{ with } \frac{x-2}{3}.$$
$$= 3\left(\frac{x-2}{3}\right) + 2$$
$$= x - 2 + 2$$
$$= x$$

■ **Work Practice 11**

Practice 11

Show that if $f(x) = 4x - 1$, then $f^{-1}(x) = \dfrac{x+1}{4}$.

Answer

11. $f(f^{-1}(x)) = 4\left(\dfrac{x+1}{4}\right) - 1$
$$= x + 1 - 1 = x$$
$$f^{-1}(f(x)) = \dfrac{(4x-1)+1}{4}$$
$$= \dfrac{4x}{4} = x$$

📷 Calculator Explorations Graphing

A grapher can be used to visualize functions and their inverses. Recall that the graph of a function f and its inverse f^{-1} are mirror images of each other across the line $y = x$. To see this for the function $f(x) = 3x + 2$, use a square window (see Appendix H) and graph

the given function: $Y_1 = 3x + 2$

its inverse: $Y_2 = \dfrac{x-2}{3}$

and the line $Y_3 = x$

Exercises will follow in Exercise Set 12.2.

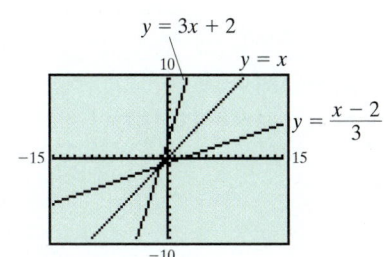

Vocabulary, Readiness & Video Check

Use the choices below to fill in each blank. Some choices will not be used, and some will be used more than once.

vertical	$(3, 7)$	$(11, 2)$	$y = x$	x	true
horizontal	$(7, 3)$	$(2, 11)$	$\dfrac{1}{f}$	the inverse of f	false

1. If $f(2) = 11$, the corresponding ordered pair is ____(2, 11)____.
2. If $(7, 3)$ is an ordered pair solution of $f(x)$, and $f(x)$ has an inverse, then an ordered pair solution of $f^{-1}(x)$ is ____(3, 7)____.
3. The symbol f^{-1} means __the inverse of f__.
4. True or false: The function notation $f^{-1}(x)$ means $\dfrac{1}{f(x)}$. ____false____
5. To tell whether a graph is the graph of a function, use the ____vertical____ line test.
6. To tell whether the graph of a function is also a one-to-one function, use the ____horizontal____ line test.
7. The graphs of f and f^{-1} are symmetric about the line ____$y = x$____.
8. Two functions are inverses of each other if $(f \circ f^{-1})(x) = $ ____x____ and $(f^{-1} \circ f)(x) = $ ____x____.

Martin-Gay Interactive Videos

See Video 12.2

Watch the section lecture video and answer the following questions.

Objective A 9. From ▤ Example 1 and the definition before, what makes a one-to-one function different from other types of functions? ▶

Objective B 10. From ▤ Examples 2 and 3, if a graph passes the horizontal line test but not the vertical line test, is it a one-to-one function? Explain. ▶

Objective C 11. From ▤ Example 4 and the lecture before, if you find the inverse of a one-to-one function, is this inverse function also a one-to-one function? How do you know? ▶

Objective D 12. From ▤ Examples 5 and 6, explain why the interchanging of x and y when finding an inverse equation makes sense given the definition of an inverse function. ▶

Objective E 13. From ▤ Example 7, if you have the equation or graph of a one-to-one function, how can you graph its inverse without finding the inverse's equation? ▶

Objective F 14. Based on ▤ Example 8 and the lecture before, what's wrong with the following statement? "If f is a one-to-one function, you can prove that f and f^{-1} are inverses of each other by showing that $f(f^{-1}(x)) = f^{-1}(f(x))$." ▶

See video answer section.

12.2 Exercise Set MyMathLab®

Objectives **A** **C** **Mixed Practice** *Determine whether each function is a one-to-one function. If it is one-to-one, list the inverse function by switching coordinates, or inputs and outputs. See Examples 1 through 5 and 7.*

1. $g = \{(0, 3), (3, 7), (6, 7), (-2, -2)\}$
not one-to-one

2. $g = \{(8, 6), (9, 6), (3, 4), (-4, 4)\}$
not one-to-one

3. $h = \{(10, 10)\}$
one-to-one; $h^{-1} = \{(10, 10)\}$

4. $r = \{(1, 2), (3, 4), (5, 6), (6, 7)\}$
one-to-one; $r^{-1} = \{(2, 1), (4, 3), (6, 5), (7, 6)\}$

5. $f = \{(11, 12), (4, 3), (3, 4), (6, 6)\}$
one-to-one; $f^{-1} = \{(12, 11), (3, 4), (4, 3), (6, 6)\}$

6. $f = \{(-1, -1), (1, 1), (0, 2), (2, 0)\}$
one-to-one; $f^{-1} = \{(-1, -1), (1, 1), (2, 0), (0, 2)\}$

7.

Month of 2014 (Input)	May	June	July	August	September	October
Unemployment Rate in Percent (Output)	6.3	6.1	6.2	6.1	5.9	5.8

(*Source:* U.S. Bureau of Labor Statistics)

not one-to-one

8.

State (Input)	Maine	South Dakota	New Hampshire	Idaho	Michigan
Number of Two-Year Colleges (Output)	8	6	8	3	30

(*Source:* U.S. University of Texas at Austin)

not one-to-one

9.

State (Input)	California	Alaska	Indiana	Louisiana	New Mexico	Ohio
Rank in Population (Output)	1	47	16	25	36	7

(*Source:* U.S. Bureau of the Census)

one-to-one

Rank in Population (Input)	1	47	16	25	36	7
State (Output)	California	Alaska	Indiana	Louisiana	New Mexico	Ohio

10.

Shape (Input)	Triangle	Pentagon	Quadrilateral	Hexagon	Decagon
Number of Sides (Output)	3	5	4	6	10

one-to-one

Number of Sides (Input)	3	5	4	6	10
Shape (Output)	Triangle	Pentagon	Quadrilateral	Hexagon	Decagon

Given the one-to-one function $f(x) = x^3 + 2$, find the following. (Hint: You do not need to find the equation for f^{-1}.)

11. a. $f(1)$
b. $f^{-1}(3)$
a. 3 **b.** 1

12. a. $f(0)$
b. $f^{-1}(2)$
a. 2 **b.** 0

13. a. $f(-1)$
b. $f^{-1}(1)$
a. 1 **b.** −1

14. a. $f(-2)$
b. $f^{-1}(-6)$
a. −6 **b.** −2

Objective B *Determine whether the graph of each function is the graph of a one-to-one function. See Example 6.*

15.

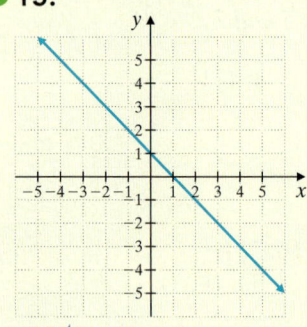

one-to-one

16.

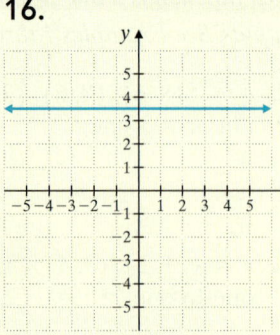

not one-to-one

17.

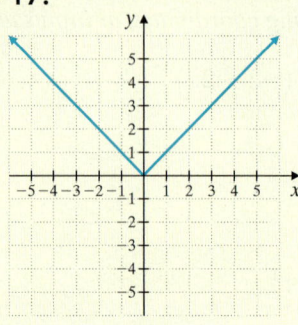

not one-to-one

18.

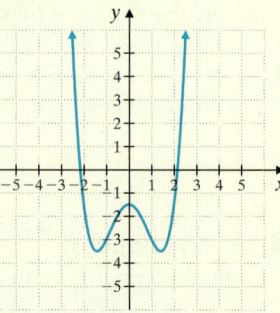

not one-to-one

19.

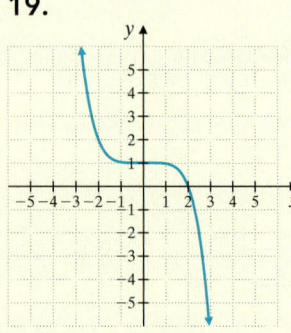

one-to-one

20.

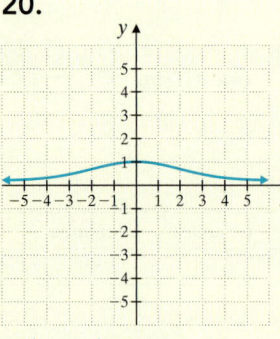

not one-to-one

21.

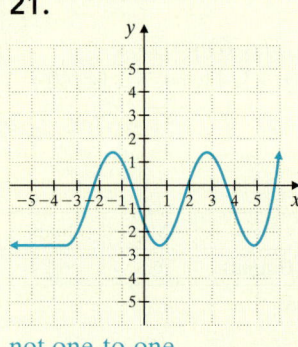

not one-to-one

22.

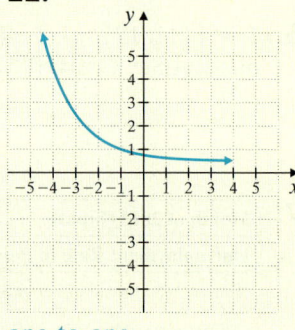

one-to-one

Objectives D E **Mixed Practice** *Each of the following functions is one-to-one. Find the inverse of each function and graph the function and its inverse on the same set of axes. See Examples 8 through 10.*

23. $f(x) = x + 4$

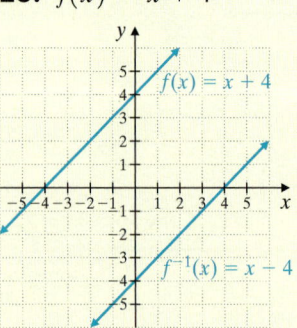

24. $f(x) = x - 5$

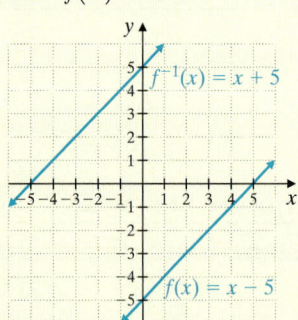

25. $f(x) = 2x - 3$

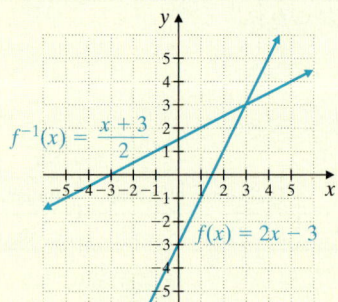

26. $f(x) = 4x + 9$

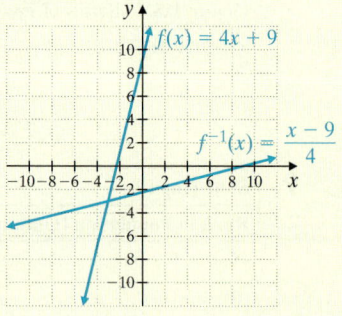

27. $f(x) = \dfrac{1}{2}x - 1$

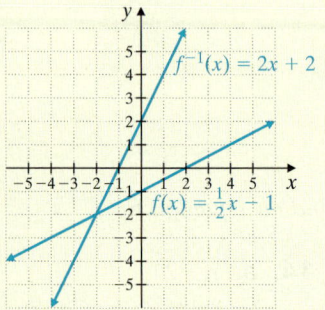

28. $f(x) = -\dfrac{1}{2}x + 2$

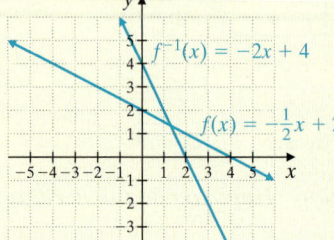

29. $f(x) = x^3$

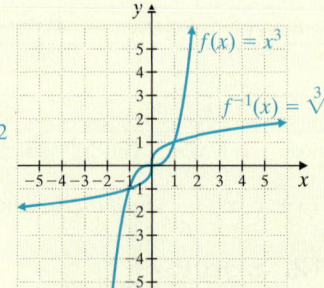

30. $f(x) = x^3 - 1$

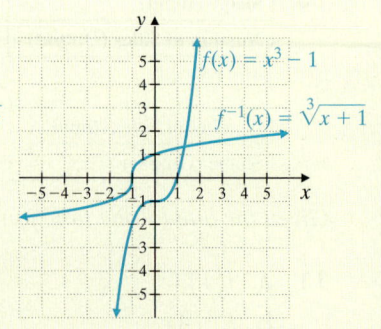

Find the inverse of each one-to-one function. See Examples 8 and 9.

31. $f(x) = 5x + 2$
$f^{-1}(x) = \dfrac{x-2}{5}$

32. $f(x) = 6x - 1$
$f^{-1}(x) = \dfrac{x+1}{6}$

33. $f(x) = \dfrac{x-2}{5}$
$f^{-1}(x) = 5x + 2$

34. $f(x) = \dfrac{4x-3}{2}$
$f^{-1}(x) = \dfrac{2x+3}{4}$

35. $f(x) = \sqrt[3]{x}$
$f^{-1}(x) = x^3$

36. $f(x) = \sqrt[3]{x+1}$
$f^{-1}(x) = x^3 - 1$

 37. $f(x) = \dfrac{5}{3x+1}$
$f^{-1}(x) = \dfrac{\frac{5}{x}-1}{3}$

38. $f(x) = \dfrac{7}{2x+4}$
$f^{-1}(x) = \dfrac{\frac{7}{x}-4}{2}$

39. $f(x) = (x+2)^3$
$f^{-1}(x) = \sqrt[3]{x} - 2$

40. $f(x) = (x-5)^3$
$f^{-1}(x) = \sqrt[3]{x} + 5$

Graph the inverse of each function on the same set of axes. See Example 10.

41.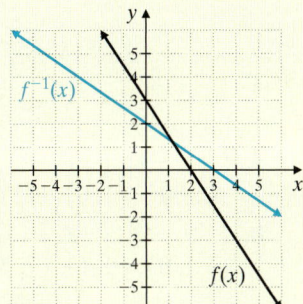

42.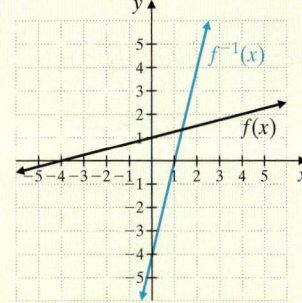

43.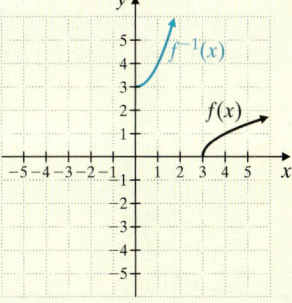

44.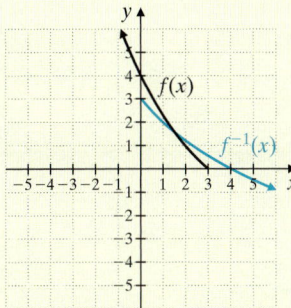

45.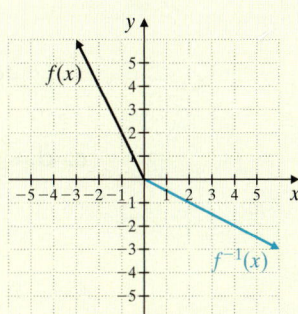

46.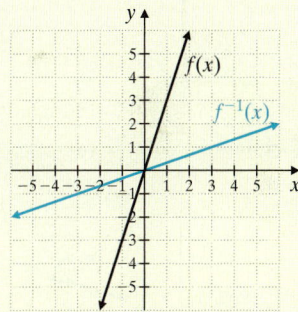

Objective F *Solve. See Example 11.*

 47. If $f(x) = 2x + 1$, show that $f^{-1}(x) = \dfrac{x-1}{2}$.
$(f \circ f^{-1})(x) = x; (f^{-1} \circ f)(x) = x$

48. If $f(x) = 3x - 10$, show that $f^{-1}(x) = \dfrac{x+10}{3}$.
$(f \circ f^{-1})(x) = x; (f^{-1} \circ f)(x) = x$

49. If $f(x) = x^3 + 6$, show that $f^{-1}(x) = \sqrt[3]{x-6}$.
$(f \circ f^{-1})(x) = x; (f^{-1} \circ f)(x) = x$

50. If $f(x) = x^3 - 5$, show that $f^{-1}(x) = \sqrt[3]{x+5}$.
$(f \circ f^{-1})(x) = x; (f^{-1} \circ f)(x) = x$

Review

Evaluate each of the following. See Section 10.2.

51. $25^{1/2}$ 5

52. $49^{1/2}$ 7

53. $16^{3/4}$ 8

54. $27^{2/3}$ 9

55. $9^{-3/2}$ $\dfrac{1}{27}$

56. $81^{-3/4}$ $\dfrac{1}{27}$

If $f(x) = 3^x$, find the following. In Exercises 59 and 60, give the exact answer and a two-decimal-place approximation. See Sections 8.2 and 10.2.

57. $f(2)$ 9

58. $f(0)$ 1

59. $f\left(\dfrac{1}{2}\right)$ $3^{1/2} \approx 1.73$

60. $f\left(\dfrac{2}{3}\right)$ $3^{2/3} \approx 2.08$

Concept Extensions

Solve. See the Concept Check in this section.

61. Suppose that *f* is a one-to-one function and that $f(2) = 9$.
    a. Write the corresponding ordered pair. $(2, 9)$
    b. Name one ordered pair that we know is a solution of the inverse of *f*, or f^{-1}. $(9, 2)$

62. Suppose that *F* is a one-to-one function and that $F\left(\dfrac{1}{2}\right) = -0.7$.
    a. Write the corresponding ordered pair. $\left(\dfrac{1}{2}, -0.7\right)$
    b. Name one ordered pair that we know is a solution of the inverse of *F*, or F^{-1}. $\left(-0.7, \dfrac{1}{2}\right)$

For Exercises 63 and 64.

    a. *Write the ordered pairs for f whose points are highlighted. (Include the points whose coordinates are given.)*
    b. *Write the corresponding ordered pairs for the inverse of f, f^{-1}.*
    c. *Graph the ordered pairs for f^{-1} found in part (b).*
    d. *Graph f^{-1} by drawing a smooth curve through the plotted points.*

63. a.

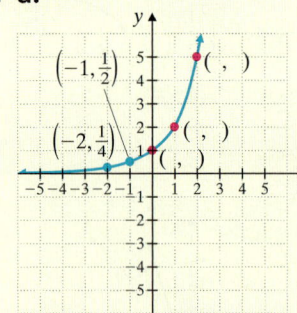

    a. $\left(-2, \dfrac{1}{4}\right), \left(-1, \dfrac{1}{2}\right),$
        $(0, 1), (1, 2), (2, 5)$
    b. $\left(\dfrac{1}{4}, -2\right), \left(\dfrac{1}{2}, -1\right),$
        $(1, 0), (2, 1), (5, 2)$

64. a.

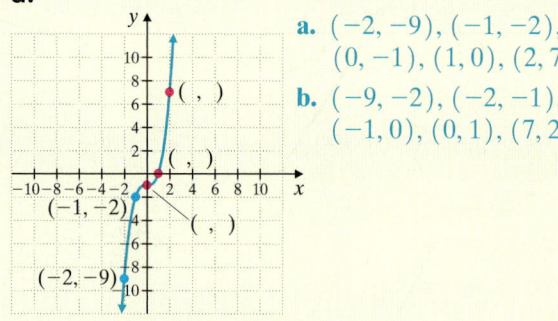

    a. $(-2, -9), (-1, -2),$
        $(0, -1), (1, 0), (2, 7)$
    b. $(-9, -2), (-2, -1),$
        $(-1, 0), (0, 1), (7, 2)$

c. d.

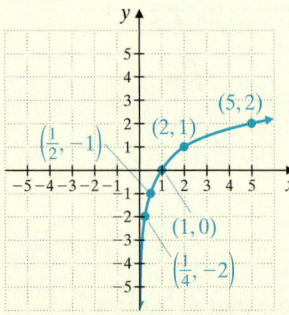

c. d.

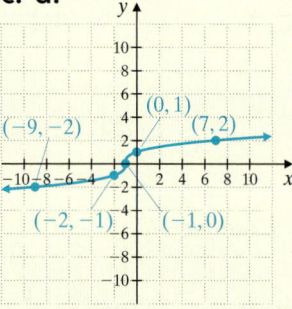

65. If you are given the graph of a function, describe how you can tell from the graph whether the function has an inverse.   answers may vary

66. Describe the appearance of the graphs of a function and its inverse.   answers may vary

Find the inverse of each one-to-one function. Then graph the function and its inverse in a square window.

67. $f(x) = 3x + 1$
$f^{-1}(x) = \dfrac{x - 1}{3}$

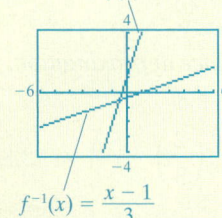

$f(x) = 3x + 1$
$f^{-1}(x) = \dfrac{x - 1}{3}$

68. $f(x) = -2x - 6$
$f^{-1}(x) = -\dfrac{x + 6}{2}$

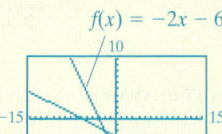

$f(x) = -2x - 6$
$f^{-1}(x) = -\dfrac{x + 6}{2}$

69. $f(x) = \sqrt[3]{x + 3}$
$f^{-1}(x) = x^3 - 3$

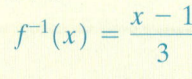

$f(x) = \sqrt[3]{x + 3}$
$f^{-1}(x) = x^3 - 3$

70. $f(x) = x^3 - 3$
$f^{-1}(x) = \sqrt[3]{x + 3}$

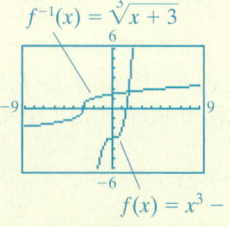

$f^{-1}(x) = \sqrt[3]{x + 3}$
$f(x) = x^3 - 3$

12.3 Exponential Functions

In earlier chapters, we gave meaning to exponential expressions such as 2^x, where x is a rational number. Recall the following examples.

$2^3 = 2 \cdot 2 \cdot 2$ Three factors; each factor is 2

$2^{3/2} = (2^{1/2})^3 = \sqrt{2} \cdot \sqrt{2} \cdot \sqrt{2}$ Three factors; each factor is $\sqrt{2}$

When x is an irrational number (for example, $\sqrt{3}$), what meaning can we give to $2^{\sqrt{3}}$?

It is beyond the scope of this book to give precise meaning to 2^x if x is irrational. We can confirm your intuition and say that $2^{\sqrt{3}}$ is a real number, and since $1 < \sqrt{3} < 2$, then $2^1 < 2^{\sqrt{3}} < 2^2$. We can also use a calculator and approximate $2^{\sqrt{3}}$: $2^{\sqrt{3}} \approx 3.321997$. In fact, as long as the base b is positive, b^x is a real number for all real numbers x. Finally, the rules of exponents apply whether x is rational or irrational, as long as b is positive.

In this section, we are interested in functions of the form $f(x) = b^x$, or $y = b^x$, where $b > 0$. A function of this form is called an *exponential function*.

> ### Exponential Function
>
> A function of the form
>
> $$f(x) = b^x$$
>
> is called an **exponential function**, where $b > 0$, b is not 1, and x is a real number.

Objectives

A Graph Exponential Functions.

B Solve Equations of the Form $b^x = b^y$.

C Solve Problems Modeled by Exponential Equations.

Objective A Graphing Exponential Functions

Now let's practice graphing exponential functions.

Example 1   Graph the exponential functions $f(x) = 2^x$ and $g(x) = 3^x$ on the same set of axes.

Solution: To graph these functions, we find some ordered pair solutions, plot the points, and connect them with a smooth curve. Remember throughout that $y = f(x)$.

$f(x) = 2^x$	x	0	1	2	3	-1	-2
	$f(x)$	1	2	4	8	$\frac{1}{2}$	$\frac{1}{4}$

$g(x) = 3^x$	x	0	1	2	3	-1	-2
	$g(x)$	1	3	9	27	$\frac{1}{3}$	$\frac{1}{9}$

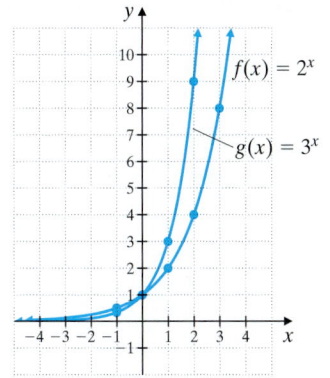

Work Practice 1

Practice 1

Graph the exponential function $f(x) = 6^x$.

Answer

1.

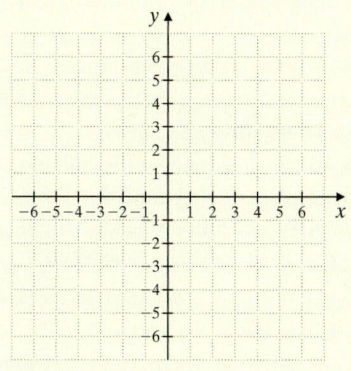

A number of things should be noted about the two graphs of exponential functions in Example 1. First, the graphs show that $f(x) = 2^x$ and $g(x) = 3^x$ are one-to-one functions since each graph passes the vertical and horizontal line tests. The y-intercept of each graph is $(0, 1)$, but neither graph has an x-intercept. From the graph, we can also see that the domain of each function is all real numbers and that the range is $(0, \infty)$. We can also see that as x-values are increasing, y-values are increasing also.

Practice 2

Graph the exponential function

$$f(x) = \left(\frac{1}{6}\right)^x.$$

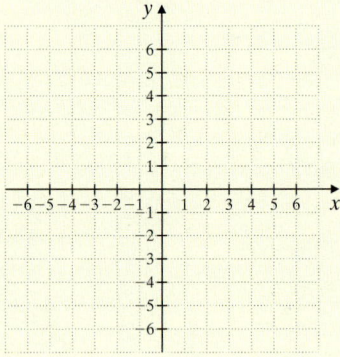

Example 2 Graph the exponential functions $f(x) = \left(\frac{1}{2}\right)^x$ and $g(x) = \left(\frac{1}{3}\right)^x$ on the same set of axes.

Solution: As before, we find some ordered pair solutions, plot the points, and connect them with a smooth curve.

$f(x) = \left(\frac{1}{2}\right)^x$	x	0	1	2	3	−1	−2
	y	1	$\frac{1}{2}$	$\frac{1}{4}$	$\frac{1}{8}$	2	4

$g(x) = \left(\frac{1}{3}\right)^x$	x	0	1	2	3	−1	−2
	y	1	$\frac{1}{3}$	$\frac{1}{9}$	$\frac{1}{27}$	3	9

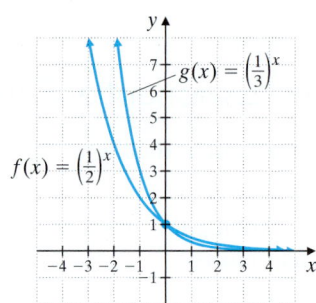

■ **Work Practice 2**

Each function in Example 2 again is a one-to-one function. The y-intercept of both is $(0, 1)$. The domain is the set of all real numbers, and the range is $(0, \infty)$.

Notice the difference between the graphs of Example 1 and the graphs of Example 2. An exponential function is always increasing if the base is greater than 1. When the base is between 0 and 1, the graph is always decreasing. The following figures summarize these characteristics of exponential functions.

Answer

2.

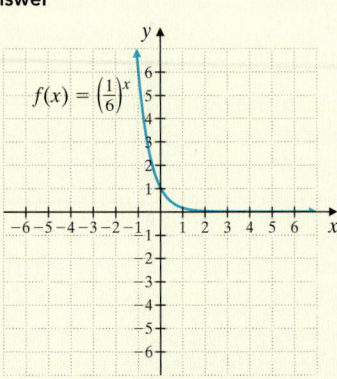

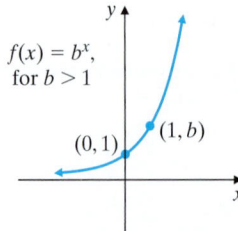

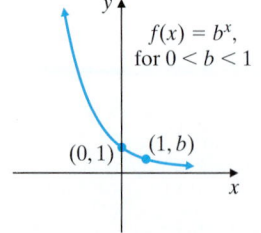

Example 3 Graph the exponential function $f(x) = 3^{x+2}$.

Solution: As before, we find and plot a few ordered pair solutions. Then we connect the points with a smooth curve.

$f(x) = 3^{x+2}$	x	0	-1	-2	-3	-4
	y	9	3	1	$\dfrac{1}{3}$	$\dfrac{1}{9}$

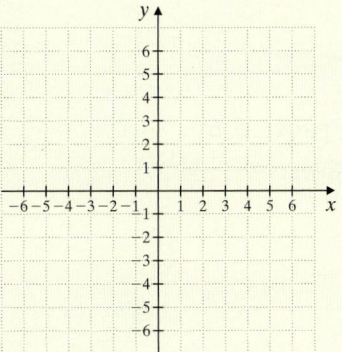

■ **Work Practice 3**

✓**Concept Check** Which functions are exponential functions?

a. $f(x) = x^3$ **b.** $g(x) = \left(\dfrac{2}{3}\right)^x$ **c.** $h(x) = 5^{x-2}$ **d.** $w(x) = (2x)^2$

Objective B Solving Equations of the Form $b^x = b^y$

We have seen that an exponential function $y = b^x$ is a one-to-one function. Another way of stating this fact is a property that we can use to solve exponential equations.

> ### Uniqueness of b^x
>
> Let $b > 0$ and $b \neq 1$. Then $b^x = b^y$ is equivalent to $x = y$.

Thus, one way to solve an exponential equation depends on whether it's possible to write each side of the equation with the same base; that is, $b^x = b^y$. We solve by this method first.

Example 4 Solve: $2^x = 16$

Solution: We write 16 as a power of 2 so that each side of the equation has the same base. Then we use the uniqueness of b^x to solve.

$2^x = 16$
$2^x = 2^4$

Since the bases are the same and are nonnegative, by the uniqueness of b^x we then have that the exponents are equal. Thus,

$x = 4$

To check, we replace x with 4 in the original equation. The solution set is {4}.

■ **Work Practice 4**

Example 5 Solve: $25^x = 125$

Solution: Since both 25 and 125 are powers of 5, we can use the uniqueness of b^x.

$25^x = 125$
$(5^2)^x = 5^3$ Write 25 and 125 as powers of 5.
$5^{2x} = 5^3$
$2x = 3$ Use the uniqueness of b^x.
$x = \dfrac{3}{2}$ Divide both sides by 2.

(Continued on next page)

Practice 3

Graph the exponential function $f(x) = 2^{x-1}$.

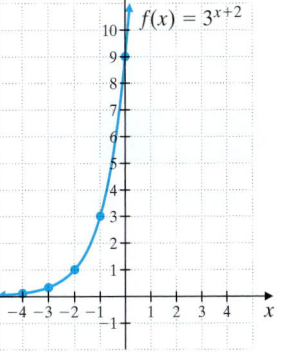

Practice 4

Solve: $5^x = 125$

Practice 5

Solve: $4^x = 8$

Answers

3.

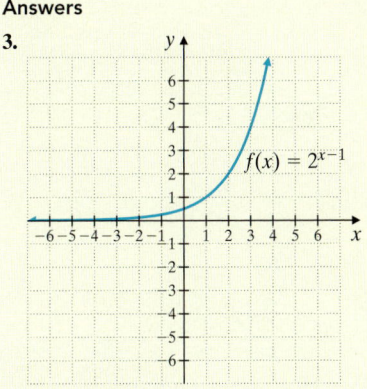

4. {3}   5. $\left\{\dfrac{3}{2}\right\}$

✓**Concept Check Answers**

b and **c**

To check, we replace x with $\dfrac{3}{2}$ in the original equation.

The solution set is $\left\{\dfrac{3}{2}\right\}$.

■ Work Practice 5

Practice 6

Solve: $9^{x-1} = 27^x$

Example 6   Solve: $4^{x+3} = 8^x$

Solution: We write both 4 and 8 as powers of 2, and then use the uniqueness of b^x.

$$4^{x+3} = 8^x$$
$$(2^2)^{x+3} = (2^3)^x$$
$$2^{2x+6} = 2^{3x}$$
$$2x + 6 = 3x \qquad \text{Use the uniqueness of } b^x.$$
$$6 = x \qquad \text{Subtract } 2x \text{ from both sides.}$$

Check to see that the solution set is $\{6\}$.

■ Work Practice 6

There is one major problem with the preceding technique. Often the two sides of an equation, $4 = 3^x$ for example, cannot easily be written as powers of a common base. We explore how to solve such an equation with the help of *logarithms* later.

Objective C Solving Problems Modeled by Exponential Equations

The bar graph below shows the increase in the number of cellular phone users. Notice that the graph of the exponential function $y = 246.98(1.05)^x$ approximates the heights of the bars. This is just one example of how the world abounds with patterns that can be modeled by exponential functions. To make these applications realistic, we use numbers that warrant a calculator.

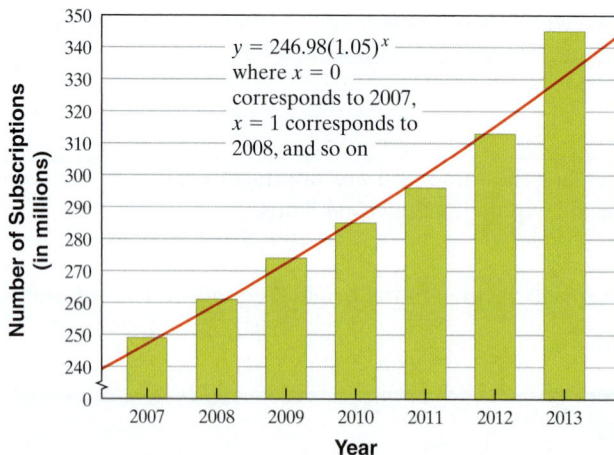

Cellular Phone Subscriptions

$y = 246.98(1.05)^x$ where $x = 0$ corresponds to 2007, $x = 1$ corresponds to 2008, and so on

Source: ICT database of ITU/United Nations

Another application of an exponential function has to do with interest rates on loans. The exponential function defined by $A = P\left(1 + \dfrac{r}{n}\right)^{nt}$ models the pattern relating the dollars A accrued (or owed) after P dollars are invested (or loaned) at an annual rate of interest r compounded n times each year for t years. This function is known as the *compound interest formula.*

Answer

6. $\{-2\}$

Example 7   Using the Compound Interest Formula

Find the amount owed at the end of 5 years if $1600 is loaned at a rate of 9% compounded monthly.

Solution: Use the formula $A = P\left(1 + \dfrac{r}{n}\right)^{nt}$, with the following values:

$P = \$1600$ (the amount of the loan)

$r = 9\% = 0.09$ (the annual rate of interest)

$n = 12$ (the number of times interest is compounded each year)

$t = 5$ (the duration of the loan, in years)

$$A = P\left(1 + \frac{r}{n}\right)^{nt} \qquad \text{Compound interest formula}$$

$$= 1600\left(1 + \frac{0.09}{12}\right)^{12(5)} \qquad \text{Substitute known values.}$$

$$= 1600(1.0075)^{60}$$

To approximate A, use the $\boxed{y^x}$ or $\boxed{\wedge}$ key on your calculator.

$\boxed{2505.0896}$

Thus, the amount A owed is approximately $2505.09.

 Work Practice 7

Teaching Tip

Have students repeat Example 7 assuming the interest is compounded
a. yearly.   **b.** quarterly.   **c.** daily.

Practice 7

a. As a result of the Chernobyl nuclear accident, radioactive debris was carried through the atmosphere. One immediate concern was the impact that the debris had on the milk supply. The percent y of radioactive material in raw milk t days after the accident is estimated by $y = 100(2.7)^{-0.1t}$. Estimate the expected percent of radioactive material in the milk after 30 days.

b. Find the amount owed at the end of 6 years if $23,000 is loaned at a rate of 12% compounded quarterly (4 times a year). Round your answer to the nearest cent.

Answers

7. a. approximately 5.08%

b. $46,754.26

📟 Calculator Explorations   Graphing

We can use a graphing calculator and its TRACE feature to solve Practice 7a graphically.

    To estimate the percent of radioactive material in the milk after 30 days, enter $Y_1 = 100(2.7)^{-0.1x}$. The graph does not appear on a standard viewing window, so we need to determine an appropriate viewing window. Because it doesn't make sense to look at radioactivity *before* the Chernobyl nuclear accident, we use Xmin = 0. We are interested in finding the percent of radioactive material in the milk when $x = 30$, so we choose Xmax = 35 to leave enough space to see the graph at $x = 30$. Because the values of y are percents, it seems appropriate that $0 \le y \le 100$. (We also use Xscl = 1 and Yscl = 10.) Now we graph the function.

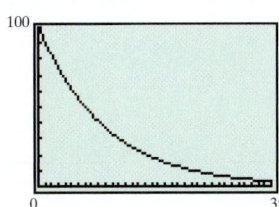

    We can use the TRACE feature to obtain an approximation of the expected percent of radioactive material in the milk when $x = 30$. (A TABLE feature may also be used to approximate the percent.) To obtain a better approximation, let's use the ZOOM feature several times to zoom in near $x = 30$.

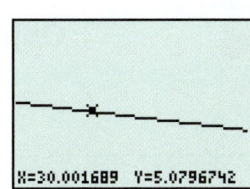

X=30.001689   Y=5.0796742

    The percent of radioactive material in the milk 30 days after the Chernobyl accident was 5.08%, accurate to two decimal places.

Use a graphing calculator to find each percent. Round your solutions to two decimal places.

1. Estimate the percent of radioactive material in the milk 2 days after the Chernobyl nuclear accident.
81.98%

2. Estimate the percent of radioactive material in the milk 10 days after the Chernobyl nuclear accident.
37.04%

3. Estimate the percent of radioactive material in the milk 15 days after the Chernobyl nuclear accident.
22.54%

4. Estimate the percent of radioactive material in the milk 25 days after the Chernobyl nuclear accident.
8.35%

Vocabulary, Readiness & Video Check

Use the choices to fill in each blank.

1. A function such as $f(x) = 2^x$ is a(n) ___exponential___ function.

   a. linear   **b.** quadratic   **c.** exponential   c

2. If $7^x = 7^y$, then ___$x = y$___.

   a. $x = 7^y$ **b.** $x = y$ **c.** $y = 7^x$ **d.** $7 = 7^y$ b

Answer the questions about the graph of $y = 2^x$, shown to the right.

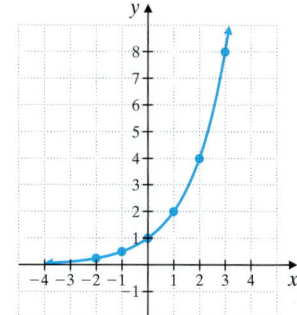

3. Is this a function? ___yes___

4. Is this a one-to-one function? ___yes___

5. Is there an x-intercept? ___no___ If so, name the coordinates. ___none___

6. Is there a y-intercept? ___yes___ If so, name the coordinates. ___$(0, 1)$___

7. The domain of this function, in interval notation, is ___$(-\infty, \infty)$___.

8. The range of this function, in interval notation, is ___$(0, \infty)$___.

Martin-Gay Interactive Videos

See Video 12.3

See video answer section.

Watch the section lecture video and answer the following questions.

Objective A   9. From the lecture before ▣ Example 1, what's the main difference between a polynomial function and an exponential function? ▶

Objective B   10. From ▣ Examples 2 and 3, you can only apply the uniqueness of b^x to solve an exponential equation if you're able to do what? ▶

Objective B   11. For ▣ Example 4, write the equation and find how much uranium will remain after 101 days. Round your answer to the nearest tenth. ▶

12.3 Exercise Set MyMathLab® ▶

Objective A   *Graph each exponential function. See Examples 1 through 3.*

1. $y = 5^x$

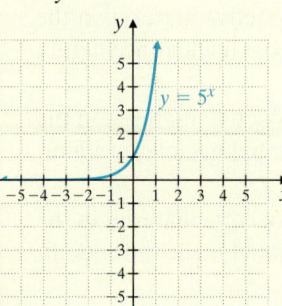

2. $y = 4^x$

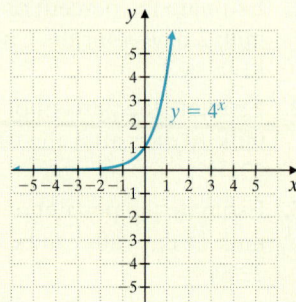

3. $y = 1 + 2^x$

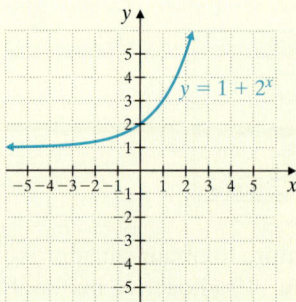

4. $y = 3^x - 1$

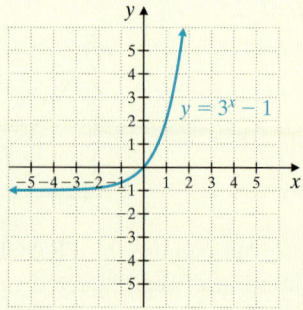

▶ 5. $y = \left(\frac{1}{4}\right)^x$

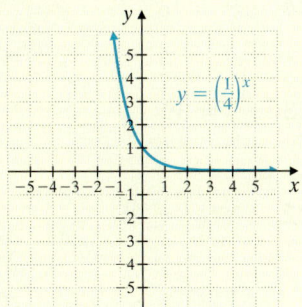

6. $y = \left(\frac{1}{5}\right)^x$

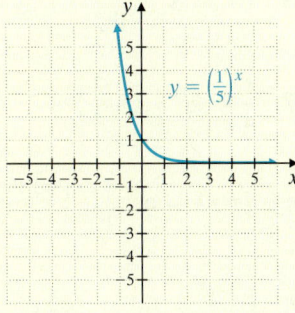

7. $y = \left(\frac{1}{2}\right)^x - 2$

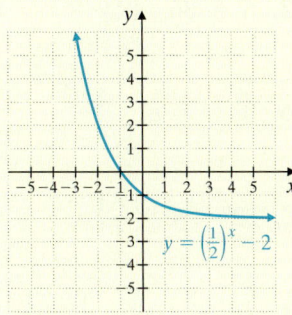

8. $y = \left(\frac{1}{3}\right)^x + 2$

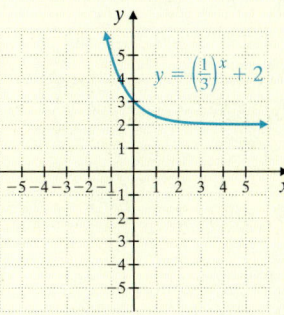

9. $y = -2^x$

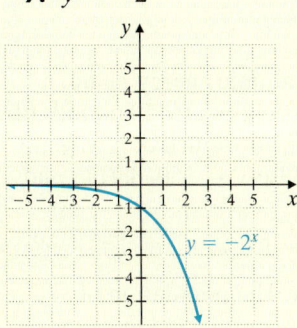

10. $y = -3^x$

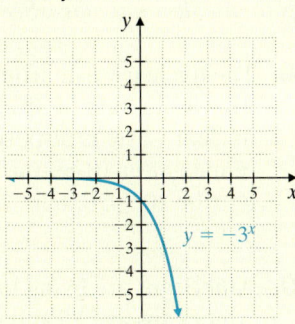

11. $y = 3^x - 2$

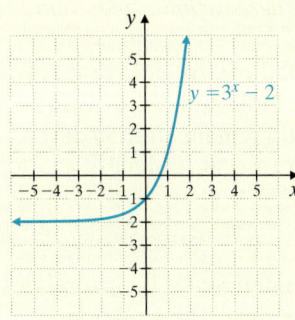

12. $y = 2^x - 3$

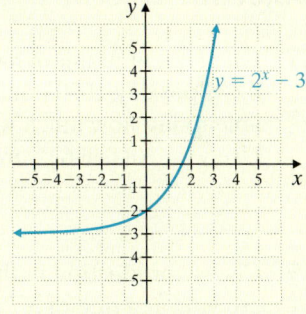

13. $y = -\left(\frac{1}{4}\right)^x$

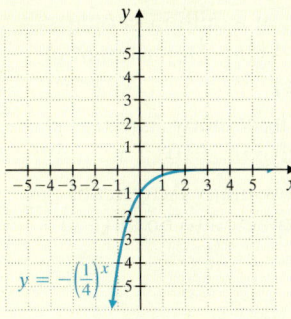

14. $y = -\left(\frac{1}{5}\right)^x$

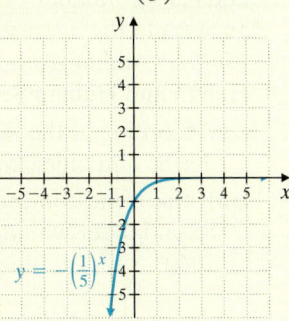

15. $y = \left(\frac{1}{3}\right)^x + 1$

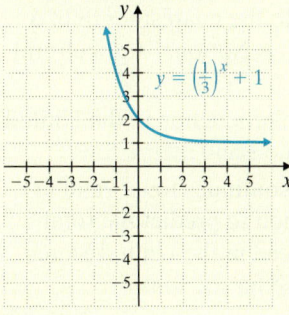

16. $y = \left(\frac{1}{2}\right)^x + 2$

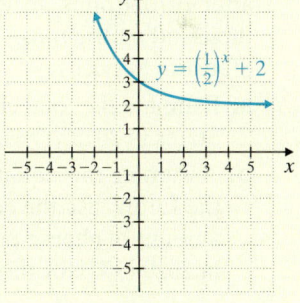

17. $f(x) = 2^{x-2}$

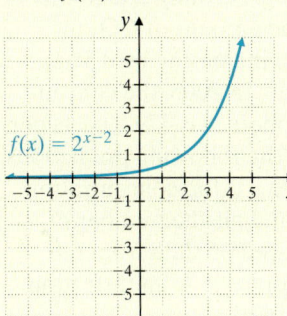

18. $g(x) = 2^{x+1}$

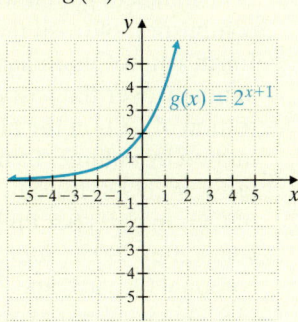

19. $F(x) = 5^{x+1}$

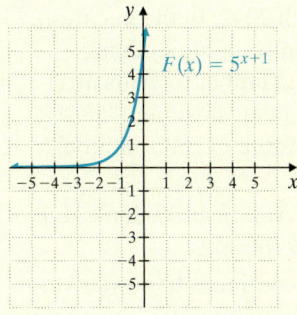

20. $G(x) = 3^{x-2}$

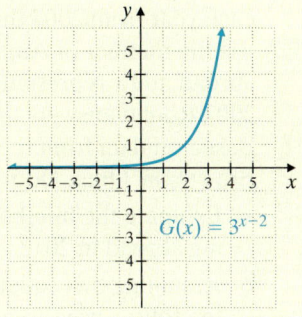

Objective B *Solve. See Examples 4 through 6.*

21. $3^x = 27$ $\{3\}$

22. $6^x = 36$ $\{2\}$

● 23. $16^x = 8$ $\left\{\dfrac{3}{4}\right\}$

24. $64^x = 16$ $\left\{\dfrac{2}{3}\right\}$

25. $32^{2x-3} = 2$ $\left\{\dfrac{8}{5}\right\}$

26. $9^{2x+1} = 81$ $\left\{\dfrac{1}{2}\right\}$

27. $\dfrac{1}{4} = 2^{3x}$ $\left\{-\dfrac{2}{3}\right\}$

28. $\dfrac{1}{27} = 3^{2x}$ $\left\{-\dfrac{3}{2}\right\}$

29. $9^x = 27$ $\left\{\dfrac{3}{2}\right\}$

30. $32^x = 4$ $\left\{\dfrac{2}{5}\right\}$

● 31. $27^{x+1} = 9$ $\left\{-\dfrac{1}{3}\right\}$

32. $125^{x-2} = 25$ $\left\{\dfrac{8}{3}\right\}$

33. $81^{x-1} = 27^{2x}$ $\{-2\}$

34. $4^{3x-7} = 32^{2x}$ $\left\{-\dfrac{7}{2}\right\}$

35. $\left(\dfrac{1}{8}\right)^x = 16^{1-x}$ $\{4\}$

36. $\left(\dfrac{1}{9}\right)^x = 27^{2-x}$ $\{6\}$

Objective C *Solve. Unless otherwise indicated, round results to one decimal place. See Example 7.*

● 37. One type of uranium has a radioactive decay rate of 0.4% per day. If 30 pounds of this uranium is available today, how much will still remain after 50 days? Use $y = 30(0.996)^x$, and let x be 50.   24.6 lb

38. The nuclear waste from an atomic energy plant decays at a rate of 3% each century. If 150 pounds of nuclear waste is disposed of, how much of it will still remain after 10 centuries? Use $y = 150(0.97)^x$, and let x be 10.   110.6 lb

39. Cheese production in the United States is currently growing at a rate of 3.7% per year. The equation $y = 1158.6(1.037)^x$ models the cheese production in the United States from 1970 to 2012. In this equation, y is the amount of cheese produced, in thousands of metric tons, and x represents the number of years after 1970. Round answers to the nearest tenth of a thousand metric tons. (*Source:* National Agricultural Statistics Service)

    a. Estimate the total cheese production in the United States in 2012.   5328.9 thousand metric tons

    b. Assuming this equation continues to be valid in the future, use the equation to predict the total amount of cheese produced in the United States in 2020.   7126.4 thousand metric tons

40. Americans are living longer. The population of Americans age 85 and older is projected to keep growing at a rate of 3.2%. Answer the following questions using $y = 1.09(1.032)^t$, where y is the population, in millions, of people age 85 and older and t is the number of years after 1960. Round answers to the nearest tenth of a million people. (*Source:* Administration on Aging, Department of Health and Human Services)

    a. According to the model, how many Americans, age 85 and older, were alive in 2010?   5.3 million

    b. If this trend continues, how many Americans age 85 and older would be expected to be alive in 2040?   13.5 million

41. The equation $y = 259{,}184(1.02)^x$ models the number of American college students who studied abroad each year from 2007 through 2012. In the equation, y is the number of American students studying abroad and x represents the number of years after 2007. Round answers to the nearest whole. (*Source:* Based on data from the Institute of International Education, Open Doors)

    a. Estimate the number of American students studying abroad in 2012.   286,160 students

    b. Assuming this equation continues to be valid in the future, use this equation to predict the number of American college students studying abroad in 2020.   335,282 students

42. Carbon dioxide (CO_2) is a greenhouse gas that contributes to global warming. Partially due to the combustion of fossil fuels, the amount of CO_2 in Earth's atmosphere has been increasing by 0.5% annually over the past decade. In 2005, the concentration of CO_2 in the atmosphere was 379.8 parts per million by volume. To make the following predictions, use $y = 379.8(1.005)^t$, where y is the concentration of CO_2 in parts per million and t is the number of years after 2005. (*Source:* Based on data from the National Oceanic and Atmospheric Administration, U.S. Department of Commerce)

    a. Predict the concentration of CO_2 in the atmosphere in the year 2015.   399.2 parts per million

    b. Predict the concentration of CO_2 in the atmosphere in the year 2030.   430.2 parts per million

The equation $y = 246.98(1.05)^x$ gives the number of cellular phone subscriptions y (in millions) in the United States for the years 2007 through 2013. In this equation, $x = 0$ corresponds to 2007, $x = 1$ corresponds to 2008, and so on. Use this model to solve Exercises 43 and 44. Round answers to the nearest tenth of a million.

43. Predict the number of cellular subscriptions in the year 2017.   402.3 million

44. Predict the number of cellular subscriptions in 2020.   465.7 million

Solve. Use $A = P\left(1 + \dfrac{r}{n}\right)^{nt}$. Round answers to two decimal places. See Example 7.

45. Find the amount a college student owes at the end of 3 years if $6000 is loaned to her at a rate of 8% compounded monthly. $7621.42

46. Find the amount owed at the end of 5 years if $3000 is loaned at a rate of 10% compounded quarterly. $4915.85

47. Find the total amount a college student has in a savings account if $2000 was invested and earned 6% compounded semiannually for 12 years. $4065.59

48. Find the amount accrued if $500 is invested and earns 7% compounded monthly for 4 years. $661.03

Review

Solve each equation. See Section 2.3.

49. $5x - 2 = 18$ $\{4\}$

50. $3x - 7 = 11$ $\{6\}$

51. $3x - 4 = 3(x + 1)$ $\varnothing$

52. $2 - 6x = 6(1 - x)$ $\varnothing$

Concept Extensions

Is the given function an exponential function? See the Concept Check in this section.

53. $f(x) = 1.5x^2$ no

54. $g(x) = 3^x$ yes

55. $h(x) = \left(\dfrac{1}{2}x\right)^2$ no

56. $F(x) = 0.4^{x+1}$ yes

Match each exponential function with its graph.

57. $f(x) = \left(\dfrac{1}{2}\right)^x$ C

58. $f(x) = 2^x$ B

59. $f(x) = \left(\dfrac{1}{4}\right)^x$ D

60. $f(x) = 3^x$ A

A

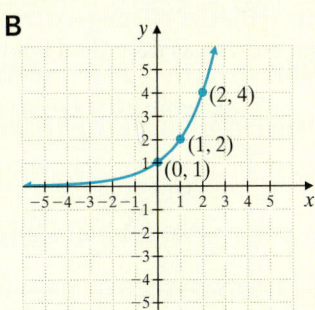

B

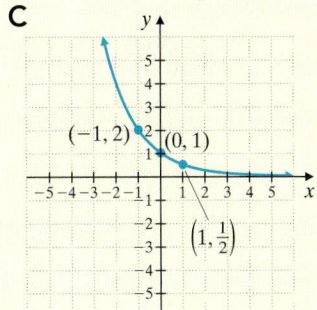

C

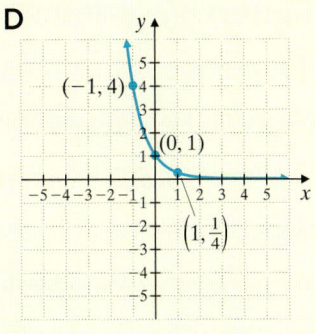

D

61. Explain why the graph of an exponential function $y = b^x$ contains the point $(1, b)$.   answers may vary

62. Explain why an exponential function $y = b^x$ has a y-intercept of $(0, 1)$.   answers may vary

Use a graphing calculator to solve. Estimate your results to two decimal places.

63. Verify the results of Exercise 37.   24.55 lb

64. Verify the results of Exercise 38.   110.61 lb

65. From Exercise 37, estimate the number of pounds of uranium that will remain after 100 days. 20.09 lb

66. From Exercise 37, estimate the number of pounds of uranium that will remain after 120 days. 18.55 lb

12.4 Exponential Growth and Decay Functions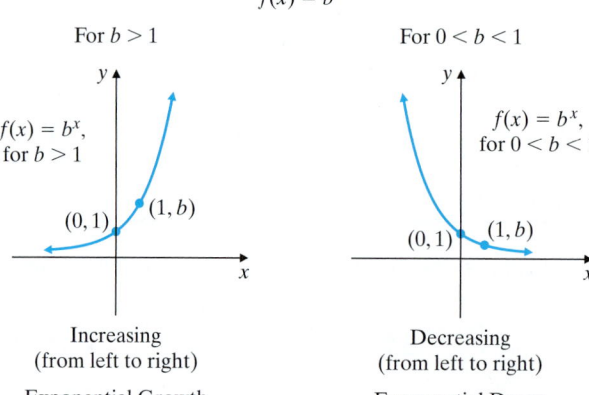

Objectives

A Model Exponential Growth.

B Model Exponential Decay.

Now that we can graph exponential functions, let's learn about exponential growth and exponential decay.

A quantity that grows or decays by the same percent at regular time periods is said to have **exponential growth** or **exponential decay.** There are many real-life examples of exponential growth and decay, such as population, bacteria, viruses, and radioactive substances, just to name a few.

Recall the graphs of exponential functions.

Exponential Functions
$f(x) = b^x$

For $b > 1$

$f(x) = b^x,$
for $b > 1$

$(0, 1)$ $(1, b)$

Increasing
(from left to right)

Exponential Growth

For $0 < b < 1$

$f(x) = b^x,$
for $0 < b < 1$

$(0, 1)$ $(1, b)$

Decreasing
(from left to right)

Exponential Decay

Objective A Modeling Exponential Growth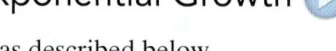

We begin with exponential growth, as described below.

Exponential Growth

initial amount

$$y = C(1 + r)^x$$

number of time intervals

$(1 + r)$ is growth factor; r is growth rate (often a percent)

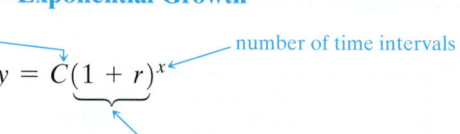

Example 1 In 2005, let's suppose a town named La Combe had a population of 15,500 and was consistently increasing by 10% per year. If this yearly increase continues, predict the city's population in 2025. (Round to the nearest whole.)

Solution: Let's begin to understand by calculating the city's population each year:

Time Interval	$x = 1$	$x = 2$	3	4	5	and so on ...
Year	1996	1997	1998	1999	2000	
Population	17,050	18,755	20,631	22,694	24,963	

$$\boxed{15,500 + 0.10(15,500)} \qquad \boxed{17,050 + 0.10(17,050)}$$

This is an example of exponential growth, so let's use our formula with

$$C = 15,500; r = 0.10; x = 2015 - 1995 = 20$$

Then,

$$y = C(1 + r)^x$$
$$= 15,500(1 + 0.10)^{20}$$
$$= 15,500(1.1)^{20}$$
$$\approx 104,276$$

In 2025, we predict the population of La Combe to be 104,276.

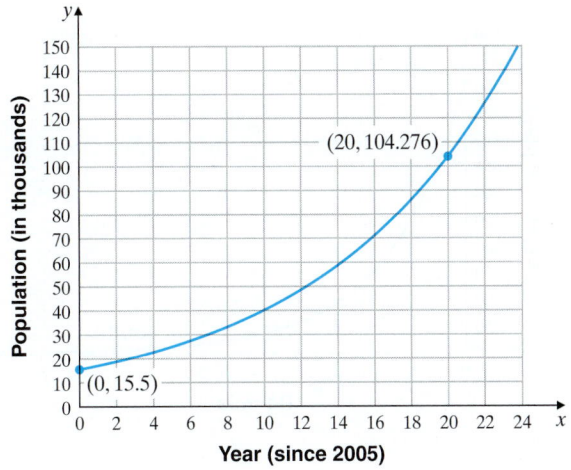

Work Practice 1

Note: The exponential growth formula, $y = C(1 + r)^x$, should remind you of the compound interest formula from the previous section, $A = P(1 + \frac{r}{n})^{nt}$. In fact, if the number of compoundings per year, n, is 1, the interest formula becomes $A = P(1 + r)^t$, which is the exponential growth formula written with different variables.

Objective B Modeling Exponential Decay

Now let's study exponential decay.

Exponential Decay

$$y = C(1 - r)^x$$

initial amount

number of time intervals

$(1 - r)$ is decay factor; r is decay rate (often a percent)

Practice 1

In 2010, the town of La Combe (from Example 1) had a population of 25,000 and started consistently increasing by 12% per year. If this yearly increase continues, predict the city's population in 2025. Round to the nearest whole.

Answer
1. 136,839

Copyright 2016 Pearson Education, Inc.

Practice 2

A tournament with 800 persons is played so that after each round the number of players decreases by 30%. Find the number of players after round 9. Round your answer to the nearest whole.

Example 2 A large golf country club holds a singles tournament each year. At the start of the tournament for a particular year there are 512 players. After each round, half the players are eliminated. How many players remain after 6 rounds?

Solution: This is an example of exponential decay.

Let's begin to understand by calculating the number of players after a few rounds.

Round (same as interval)	1	2	3	4	and so on …
Players (at end of round)	256	128	64	32	

$512 - 0.50(512)$ ↑ $256 - 0.50(256)$ ↑

Here, $C = 512$; $r = \dfrac{1}{2}$ or $50\% = 0.50$; $x = 6$

Thus,

$$y = 512(1 - 0.50)^6$$
$$= 512(0.50)^6$$
$$= 8$$

After 6 rounds, there are 8 players remaining.

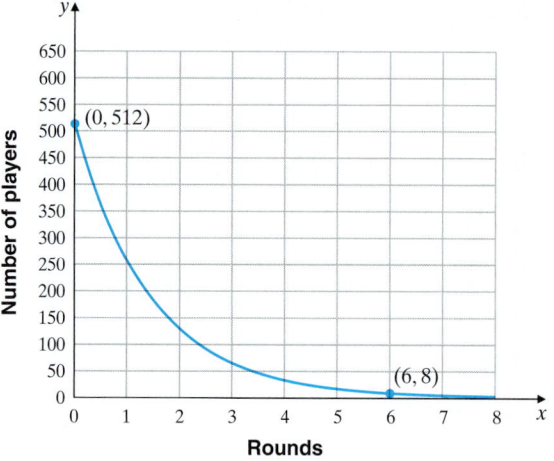

Work Practice 2

The **half-life** of a substance is the amount of time it takes for half of the substance to decay.

Practice 3

Use the information from Example 3 and calculate how much of a 500-gram sample of DDT will remain after 51 years. Round to the nearest tenth.

Example 3 A form of DDT pesticide (banned in 1972) has a half-life of approximately 15 years. If a storage unit had 400 pounds of DDT, find how much DDT is remaining after 72 years. Round to the nearest tenth.

Solution: Here, we need to be careful because each time interval is 15 years, the half-life.

Time Interval	1	2	3	4	5	and so on …
Years Passed	15	$2 \cdot 15 = 30$	45	60	75	
Pounds of DDT	200	100	50	25	12.5	

From the table we see that after 72 years, between 4 and 5 intervals, there should be between 12.5 and 25 pounds of DDT remaining.

Answers

2. 32   **3.** 47.4 g

Let's calculate x, the number of time intervals.

$$x = \frac{72 \text{ (years)}}{15 \text{ (half-life)}} = 4.8$$

Now, using our exponential decay formula and the definition of half-life, for each time interval x, the decay rate r is $\frac{1}{2}$ or 50% or 0.50.

$$y = 400(1 - 0.50)^{4.8} \text{—time intervals for 72 years}$$

$\qquad$ | $\qquad\qquad$ |
original amount $\quad$ decay rate

$$y = 400(0.50)^{4.8}$$
$$y \approx 14.4$$

In 72 years, 14.4 pounds of DDT remain.

■ **Work Practice 3**

Vocabulary, Readiness & Video Check

Martin-Gay Interactive Videos

See Video 12.4

See video answer section.

Watch the section lecture video and answer the following questions.

Objective A **1.** ⊞ Example 1 reviews exponential growth. Explain how you find the growth rate and the correct number of time intervals. ▶

Objective B **2.** Explain how you know that ⊞ Example 2 has to do with exponential decay and not exponential growth. ▶

3. For ⊞ Example 3, which has to do with half-life, explain how to calculate the number of time intervals. Also, what is the decay rate for half-life and why? ▶

12.4 Exercise Set MyMathLab®

Objective A *Practice using the exponential growth formula by completing the table below. Round final amounts to the nearest whole. See Example 1.*

	Original Amount	Growth Rate per Year	Number of Years, x	Final Amount After x Years of Growth
▶ 1.	305	5%	8	451
2.	402	7%	5	564
3.	2000	11%	41	144,302
4.	1000	47%	19	1,510,182
5.	17	29%	28	21,231
6.	29	61%	12	8796

Objective **B** *Practice using the exponential decay formula by completing the table below. Round final amounts to the nearest whole. See Example 2.*

	Original Amount	Decay Rate per Year	Number of Years, x	Final Amount After x Years of Decay
7.	305	5%	8	202
8.	402	7%	5	280
9.	10,000	12%	15	1470
10.	15,000	16%	11	2204
11.	207,000	32%	25	13
12.	325,000	29%	31	8

Objectives **A** **B** **Mixed Practice** *Solve. Unless noted otherwise, round answers to the nearest whole. See Examples 1 and 2.*

13. Suppose a city with population 500,000 has been growing at a rate of 3% per year. If this rate continues, find the population of this city in 12 years.
712,880

14. Suppose a city with population 320,000 has been growing at a rate of 4% per year. If this rate continues, find the population of this city in 20 years.
701,159

15. The number of employees for a certain company has been decreasing each year by 5%. If the company currently has 640 employees and this rate continues, find the number of employees in 10 years.   383

16. The number of students attending summer school at a local community college has been decreasing each year by 7%. If 984 students currently attend summer school and this rate continues, find the number of students attending summer school in 5 years.   685

17. National Park Service personnel are trying to increase the size of the bison population of Theodore Roosevelt National Park. If 260 bison currently live in the park and if the population's rate of growth is 2.5% annually, find how many bison there should be in 10 years.   333 bison

18. The size of the rat population of a wharf area grows at a rate of 8% monthly. If there are 200 rats in January, find how many rats should be expected by next January.   504 rats

19. A rare isotope of a nuclear material is very unstable, decaying at a rate of 15% each second. Find how much isotope remains 10 seconds after 5 grams of the isotope is created.   1 g

20. An accidental spill of 75 grams of radioactive material in a local stream has led to the presence of radioactive debris decaying at a rate of 4% each day. Find how much debris still remains after 14 days.   42 g

Practice using the exponential decay formula with half-lives by completing the table below. The first row has been completed for you. See Example 3.

	Original Amount	Half-Life (in years)	Number of Years	Time Intervals, $x \left(\dfrac{\text{Years}}{\text{Half-Life}} \right)$ Rounded to Tenths If Needed	Final Amount After x Time Intervals (rounded to tenths)	Is Your Final Amount Reasonable?
	60	8	10	$\dfrac{10}{8} = 1.25$	25.2	yes
21. a.	40	7	14	$\dfrac{14}{7} = 2$	10	yes
b.	40	7	11	$\dfrac{11}{7} \approx 1.6$	13.2	yes
22. a.	200	12	36	$\dfrac{36}{12} = 3$	25	yes
b.	200	12	40	$\dfrac{40}{12} \approx 3.3$	20.3	yes
▶ **23.**	21	152	500	$\dfrac{500}{152} \approx 3.3$	2.1	yes
24.	35	119	500	$\dfrac{500}{119} \approx 4.2$	1.9	yes

Solve. Round answers to the nearest tenth.

25. A form of nickel has a half-life of 96 years. How much of a 30-gram sample is left after 250 years?
4.9 g

26. A form of uranium has a half-life of 72 years. How much of a 100-gram sample is left after 500 years?
0.8 g

Review

By inspection, find the value for x that makes each statement true. See Section 12.3.

27. $2^x = 8$ 3

28. $3^x = 9$ 2

29. $5^x = \dfrac{1}{5}$ −1

30. $4^x = 1$ 0

Concept Extensions

31. An item is on sale for 40% off its original price. If it is then marked down an additional 60%, does this mean the item is free? Discuss why or why not.
no; answers may vary

32. Uranium U-232 has a half-life of 72 years. What eventually happens to a 10 gram sample? Does it ever completely decay and disappear? Discuss why or why not.   no; answers may vary

Objectives

A Write Exponential Equations with Logarithmic Notation and Write Logarithmic Equations with Exponential Notation.

B Solve Logarithmic Equations by Using Exponential Notation.

C Identify and Graph Logarithmic Functions.

Objective A Using Logarithmic Notation

Since the exponential function $f(x) = 2^x$ is a one-to-one function, it has an inverse. We can create a table of values for f^{-1} by switching the coordinates in the accompanying table of values for $f(x) = 2^x$.

x	$y = f(x)$
-3	$\frac{1}{8}$
-2	$\frac{1}{4}$
-1	$\frac{1}{2}$
0	1
1	2
2	4
3	8

x	$y = f^{-1}(x)$
$\frac{1}{8}$	-3
$\frac{1}{4}$	-2
$\frac{1}{2}$	-1
1	0
2	1
4	2
8	3

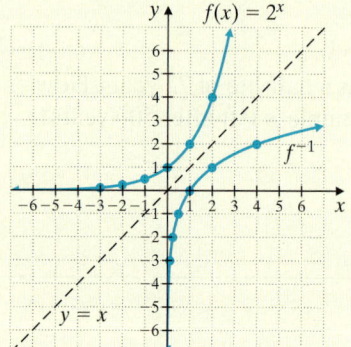

The graphs of f and its inverse are shown in the margin. Notice that the graphs of f and f^{-1} are symmetric about the line $y = x$, as expected.

Now we would like to be able to write an equation for f^{-1}. To do so, we follow the steps for finding the equation of an inverse.

$$f(x) = 2^x$$

Step 1: Replace $f(x)$ by y. $\qquad y = 2^x$

Step 2: Interchange x and y. $\qquad x = 2^y$

Step 3: Solve for y.

At this point, we are stuck. To solve this equation for y, a new notation, **logarithmic notation,** is needed.

The symbol $\log_b x$ means "the power to which b is raised to produce a result of x." In other words,

$$\log_b x = y \quad \text{means} \quad b^y = x$$

We say that $\log_b x$ is "the logarithm of x to the base b" or "the log of x to the base b."

Logarithmic Definition

If $b > 0$, and $b \neq 1$, then

$$y = \log_b x \quad \text{means} \quad x = b^y$$

for every $x > 0$ and every real number y.

Before returning to the function $x = 2^y$ and solving it for y in terms of x, let's practice using the new notation $\log_b x$.

It is important to be able to write exponential equations with logarithmic notation, and vice versa. The following table shows examples of both forms.

876

Logarithmic Equation	Corresponding Exponential Equation
$\log_3 9 = 2$	$3^2 = 9$
$\log_6 1 = 0$	$6^0 = 1$
$\log_2 8 = 3$	$2^3 = 8$
$\log_4 \dfrac{1}{16} = -2$	$4^{-2} = \dfrac{1}{16}$
$\log_8 2 = \dfrac{1}{3}$	$8^{1/3} = 2$

Helpful Hint Notice that a *logarithm* is an *exponent*. In other words, $\log_3 9$ is the *power* that we raise 3 to in order to get 9.

Examples Write as an exponential equation.

1. $\log_5 25 = 2$ means $5^2 = 25$.

2. $\log_6 \dfrac{1}{6} = -1$ means $6^{-1} = \dfrac{1}{6}$.

3. $\log_2 \sqrt{2} = \dfrac{1}{2}$ means $2^{1/2} = \sqrt{2}$.

☛ **Work Practice 1–3**

Examples Write as a logarithmic equation.

4. $9^3 = 729$ means $\log_9 729 = 3$.

5. $6^{-2} = \dfrac{1}{36}$ means $\log_6 \dfrac{1}{36} = -2$.

6. $5^{1/3} = \sqrt[3]{5}$ means $\log_5 \sqrt[3]{5} = \dfrac{1}{3}$.

☛ **Work Practice 4–6**

Example 7 Find the value of each logarithmic expression.

a. $\log_4 16$ **b.** $\log_{10} \dfrac{1}{10}$ **c.** $\log_9 3$

Solution:

a. $\log_4 16 = 2$ because $4^2 = 16$.

b. $\log_{10} \dfrac{1}{10} = -1$ because $10^{-1} = \dfrac{1}{10}$.

c. $\log_9 3 = \dfrac{1}{2}$ because $9^{1/2} = \sqrt{9} = 3$.

☛ **Work Practice 7**

Objective B Solving Logarithmic Equations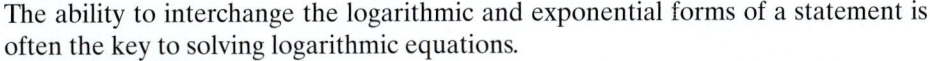

The ability to interchange the logarithmic and exponential forms of a statement is often the key to solving logarithmic equations.

Example 8 Solve: $\log_5 x = 3$

Solution: $\log_5 x = 3$

$\qquad 5^3 = x$ Write as an exponential equation.

$\qquad 125 = x$

The solution set is $\{125\}$.

☛ **Work Practice 8**

Practice 9

Solve: $\log_x 9 = 2$

> **Example 9** Solve: $\log_x 25 = 2$
>
> **Solution:**
> $$\log_x 25 = 2$$
> $$x^2 = 25 \quad \text{Write as an exponential equation.}$$
> $$x = 5$$
>
> Even though $(-5)^2 = 25$, the base b of a logarithm must be positive. The solution set is $\{5\}$.
>
> ■ **Work Practice 9**

Practice 10

Solve: $\log_2 1 = x$

> **Example 10** Solve: $\log_3 1 = x$
>
> **Solution:**
> $$\log_3 1 = x$$
> $$3^x = 1 \quad \text{Write as an exponential equation.}$$
> $$3^x = 3^0 \quad \text{Write 1 as } 3^0.$$
> $$x = 0 \quad \text{Use the uniqueness of } b^x.$$
>
> The solution set is $\{0\}$.
>
> ■ **Work Practice 10**

In Example 10, we illustrated an important property of logarithms. That is, $\log_b 1$ is always 0. This property as well as two important others are given below.

Teaching Tip

For property 1, ask, "b raised to what power is 1?"

> **Properties of Logarithms**
>
> If b is a real number, $b > 0$ and $b \neq 1$, then
>
> 1. $\log_b 1 = 0$
> 2. $\log_b b^x = x$
> 3. $b^{\log_b x} = x$

To see that $\log_b b^x = x$, we change the logarithmic form to exponential form. Then, $\log_b b^x = x$ means $b^x = b^x$. In exponential form, the statement is true, so in logarithmic form, the statement is also true.

Practice 11

Simplify.

a. $\log_6 6^3$ b. $\log_{11} 11^{-4}$

c. $7^{\log_7 13}$ d. $3^{\log_3 10}$

> **Example 11** Simplify.
>
> a. $\log_3 3^2$ b. $\log_7 7^{-1}$
>
> c. $5^{\log_5 3}$ d. $2^{\log_2 6}$
>
> **Solution:**
>
> a. From property 2, $\log_3 3^2 = 2$.
> b. From property 2, $\log_7 7^{-1} = -1$.
> c. From property 3, $5^{\log_5 3} = 3$.
> d. From property 3, $2^{\log_2 6} = 6$.
>
> ■ **Work Practice 11**

Teaching Tip

For property 2 ask, "b raised to what power is b^x?" Then give the example," What power of 2 is 2^5? In other words, what power of 2 is 32?"

Teaching Tip

For property 3, have students convert the equation to logarithmic form.

Objective C Graphing Logarithmic Functions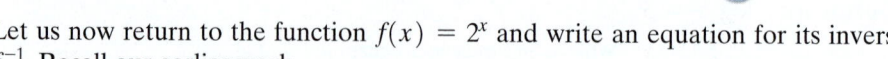

Let us now return to the function $f(x) = 2^x$ and write an equation for its inverse, f^{-1}. Recall our earlier work.

$$f(x) = 2^x$$

Step 1: Replace $f(x)$ by y. $\qquad\qquad y = 2^x$

Step 2: Interchange x and y. $\qquad\qquad x = 2^y$

Answers

9. $\{3\}$ 10. $\{0\}$ 11. **a.** 3   **b.** -4
c. 13   **d.** 10

Having gained proficiency with the notation $\log_b x$, we can now complete the steps for writing the inverse equation.

Step 3: Solve for y. $y = \log_2 x$

Step 4: Replace y with $f^{-1}(x)$. $f^{-1}(x) = \log_2 x$

Thus, $f^{-1}(x) = \log_2 x$ defines a function that is the inverse function of the function $f(x) = 2^x$. The function $f^{-1}(x)$ or $y = \log_2 x$ is called a *logarithmic function*.

> ### Logarithmic Function
>
> If x is a positive real number, b is a constant positive real number, and b is not 1, then a **logarithmic function** is a function that can be defined by
>
> $$f(x) = \log_b x$$
>
> The domain of f is the set of positive real numbers, and the range of f is the set of real numbers.

✔**Concept Check** Let $f(x) = \log_3 x$ and $g(x) = 3^x$. These two functions are inverses of each other. Since $(2, 9)$ is an ordered pair solution of $g(x)$, what ordered pair do we know to be a solution of $f(x)$? Explain why.

We can explore logarithmic functions by graphing them.

Example 12 Graph the logarithmic function $y = \log_2 x$.

Solution: First we write the equation with exponential notation as $2^y = x$. Then we find some ordered pair solutions that satisfy this equation. Finally, we plot the points and connect them with a smooth curve. The domain of this function is $(0, \infty)$, and the range is all real numbers.

Since $x = 2^y$ is solved for x, we choose y-values and compute corresponding x-values.

If $y = 0, x = 2^0 = 1$.

If $y = 1, x = 2^1 = 2$.

If $y = 2, x = 2^2 = 4$

If $y = -1, x = 2^{-1} = \dfrac{1}{2}$.

$x = 2^y$	y
1	0
2	1
4	2
$\dfrac{1}{2}$	-1

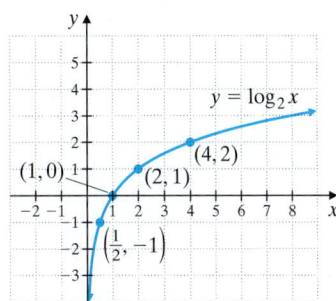

■ **Work Practice 12**

Example 13 Graph the logarithmic function $f(x) = \log_{1/3} x$.

Solution: We can replace $f(x)$ with y, and write the result with exponential notation.

$f(x) = \log_{1/3} x$

$y = \log_{1/3} x$ Replace $f(x)$ with y.

$\left(\dfrac{1}{3}\right)^y = x$ Write in exponential form.

(Continued on next page)

Practice 12

Graph the logarithmic function $y = \log_4 x$.

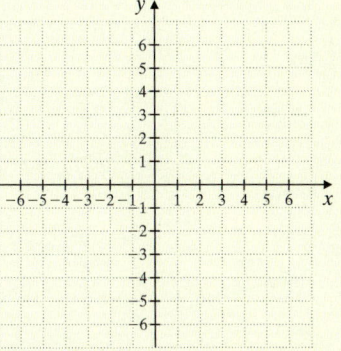

Practice 13

Graph the logarithmic function $f(x) = \log_{1/2} x$.

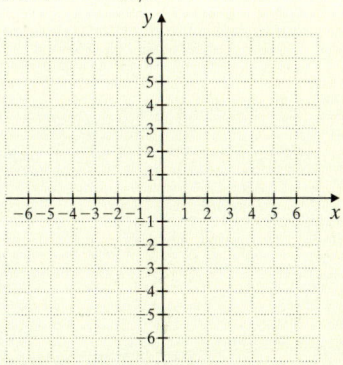

Answers

12.

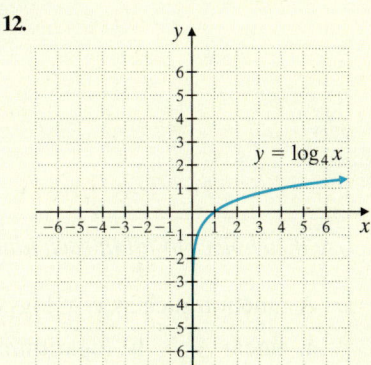

13.

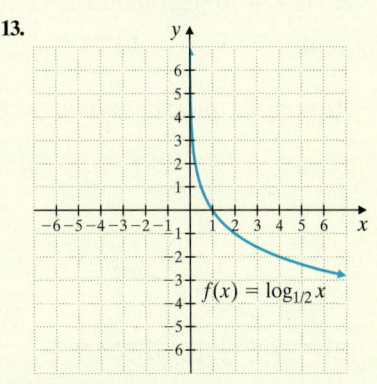

✔**Concept Check Answer**
$(9, 2)$; answers may vary

Having gained proficiency with the notation $\log_b x$, we can now complete the steps for writing the inverse equation.

Step 3: Solve for y. $\qquad\qquad y = \log_2 x$

Step 4: Replace y with $f^{-1}(x)$. $\quad f^{-1}(x) = \log_2 x$

Thus, $f^{-1}(x) = \log_2 x$ defines a function that is the inverse function of the function $f(x) = 2^x$. The function $f^{-1}(x)$ or $y = \log_2 x$ is called a *logarithmic function*.

Logarithmic Function

If x is a positive real number, b is a constant positive real number, and b is not 1, then a **logarithmic function** is a function that can be defined by

$$f(x) = \log_b x$$

The domain of f is the set of positive real numbers, and the range of f is the set of real numbers.

✓**Concept Check** Let $f(x) = \log_3 x$ and $g(x) = 3^x$. These two functions are inverses of each other. Since $(2, 9)$ is an ordered pair solution of $g(x)$, what ordered pair do we know to be a solution of $f(x)$? Explain why.

We can explore logarithmic functions by graphing them.

Example 12 Graph the logarithmic function $y = \log_2 x$.

Solution: First we write the equation with exponential notation as $2^y = x$. Then we find some ordered pair solutions that satisfy this equation. Finally, we plot the points and connect them with a smooth curve. The domain of this function is $(0, \infty)$, and the range is all real numbers.

Since $x = 2^y$ is solved for x, we choose y-values and compute corresponding x-values.

If $y = 0, x = 2^0 = 1$.

If $y = 1, x = 2^1 = 2$.

If $y = 2, x = 2^2 = 4$

If $y = -1, x = 2^{-1} = \dfrac{1}{2}$.

$x = 2^y$	y
1	0
2	1
4	2
$\dfrac{1}{2}$	-1

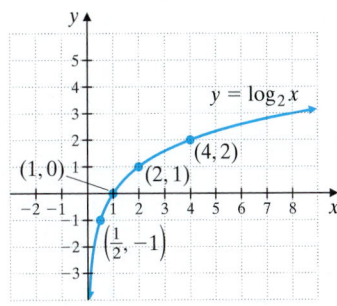

■ **Work Practice 12**

Example 13 Graph the logarithmic function $f(x) = \log_{1/3} x$.

Solution: We can replace $f(x)$ with y, and write the result with exponential notation.

$$f(x) = \log_{1/3} x$$

$$y = \log_{1/3} x \quad \text{Replace } f(x) \text{ with } y.$$

$$\left(\dfrac{1}{3}\right)^y = x \quad \text{Write in exponential form.}$$

(Continued on next page)

(Continued on next page)

Practice 12

Graph the logarithmic function $y = \log_4 x$.

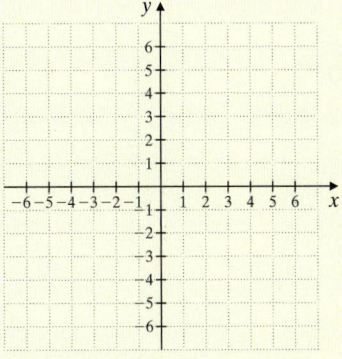

Practice 13

Graph the logarithmic function $f(x) = \log_{1/2} x$.

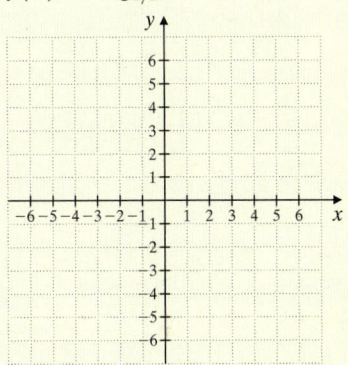

Answers

12.

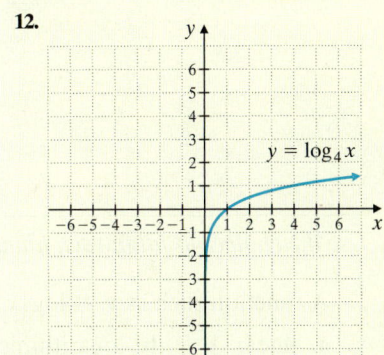

13.

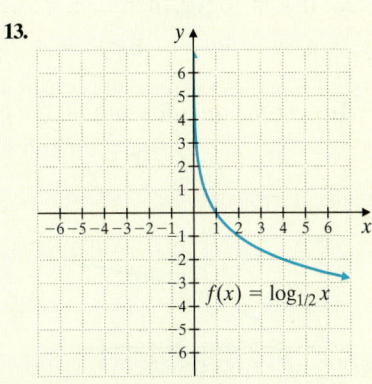

✓**Concept Check Answer**
$(9, 2)$; answers may vary

Now we can find ordered pair solutions that satisfy $\left(\frac{1}{3}\right)^y = x$, plot these points, and connect them with a smooth curve.

If $y = 0, x = \left(\frac{1}{3}\right)^0 = 1$.

If $y = 1, x = \left(\frac{1}{3}\right)^1 = \frac{1}{3}$.

If $y = -1, x = \left(\frac{1}{3}\right)^{-1} = 3$.

If $y = -2, x = \left(\frac{1}{3}\right)^{-2} = 9$.

$x = \left(\frac{1}{3}\right)^y$	y
1	0
$\frac{1}{3}$	1
3	-1
9	-2

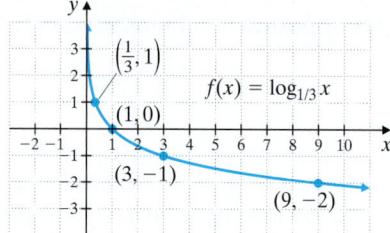

The domain of this function is $(0, \infty)$, and the range is the set of all real numbers.

Work Practice 13

The following figures summarize characteristics of logarithmic functions.

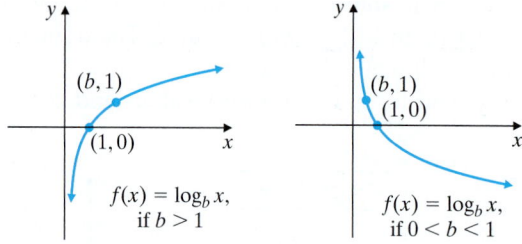

Vocabulary, Readiness & Video Check

Use the choices to fill in each blank.

1. A function such as $y = \log_2 x$ is a(n) ___logarithmic___ function. b
   a. linear   **b.** logarithmic   **c.** quadratic   **d.** exponential

2. If $y = \log_2 x$, then ___$2^y = x$___. c
   a. $x = y$ **b.** $2^x = y$ **c.** $2^y = x$ **d.** $2y = x$

Answer the questions about the graph of $y = \log_2 x$, shown to the left.

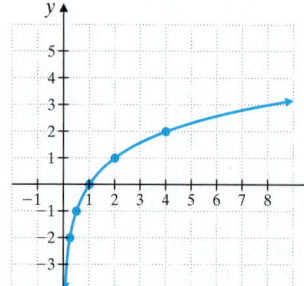

3. Is this a one-to-one function? ___yes___

4. Is there an x-intercept? ___yes___ If so, name the coordinates. ___(1, 0)___

5. Is there a y-intercept? ___no___ If so, name the coordinates. ___none___

6. The domain of this function, in interval notation, is ___$(0, \infty)$___.

7. The range of this function, in interval notation, is ___$(-\infty, \infty)$___.

Martin-Gay Interactive Videos   Watch the section lecture video and answer the following questions.

See Video 12.5

Objective A   **8.** Notice from the definition of a logarithm and from ▦ Examples 1–4 that a logarithmic statement equals the power in the exponent statement, such as $b^y = x$. What conclusion can you make about logarithms and exponents? ▶

Objective B   **9.** From ▦ Examples 8 and 9, how do you solve these types of logarithmic equations? ▶

Objective C   **10.** In ▦ Example 12, why is it easier to choose values for y when finding ordered pairs for the graph? ▶

See video answer section.

12.5 Exercise Set MyMathLab®

Objective A   *Write each as an exponential equation. See Examples 1 through 3.*

▶1. $\log_6 36 = 2$ $6^2 = 36$

2. $\log_2 32 = 5$ $2^5 = 32$

3. $\log_3 \frac{1}{27} = -3$ $3^{-3} = \frac{1}{27}$

4. $\log_5 \frac{1}{25} = -2$ $5^{-2} = \frac{1}{25}$

5. $\log_{10} 1000 = 3$ $10^3 = 1000$

6. $\log_{10} 10 = 1$ $10^1 = 10$

7. $\log_e x = 4$ $e^4 = x$

8. $\log_e y = 7$ $e^7 = y$

9. $\log_e \frac{1}{e^2} = -2$ $e^{-2} = \frac{1}{e^2}$

10. $\log_e \frac{1}{e} = -1$ $e^{-1} = \frac{1}{e}$

▶11. $\log_7 \sqrt{7} = \frac{1}{2}$ $7^{1/2} = \sqrt{7}$

12. $\log_{11} \sqrt[4]{11} = \frac{1}{4}$ $11^{1/4} = \sqrt[4]{11}$

13. $\log_{0.7} 0.343 = 3$ $0.7^3 = 0.343$

14. $\log_{1.2} 1.44 = 2$ $1.2^2 = 1.44$

15. $\log_3 \frac{1}{81} = -4$ $3^{-4} = \frac{1}{81}$

16. $\log_{1/4} 16 = -2$ $\left(\frac{1}{4}\right)^{-2} = 16$

Write each as a logarithmic equation. See Examples 4 through 6.

▶17. $2^4 = 16$ $\log_2 16 = 4$

18. $5^3 = 125$ $\log_5 125 = 3$

19. $10^2 = 100$ $\log_{10} 100 = 2$

20. $10^4 = 10,000$ $\log_{10} 10,000 = 4$

21. $e^3 = x$ $\log_e x = 3$

22. $e^5 = y$ $\log_e y = 5$

▶23. $10^{-1} = \frac{1}{10}$ $\log_{10} \frac{1}{10} = -1$

24. $10^{-2} = \frac{1}{100}$ $\log_{10} \frac{1}{100} = -2$

25. $4^{-2} = \frac{1}{16}$ $\log_4 \frac{1}{16} = -2$

26. $3^{-4} = \frac{1}{81}$ $\log_3 \frac{1}{81} = -4$

27. $5^{1/2} = \sqrt{5}$ $\log_5 \sqrt{5} = \frac{1}{2}$

28. $4^{1/3} = \sqrt[3]{4}$ $\log_4 \sqrt[3]{4} = \frac{1}{3}$

Find the value of each logarithmic expression. See Example 7.

▶29. $\log_2 8$ 3

30. $\log_3 9$ 2

31. $\log_2 \frac{1}{4}$ -2

32. $\log_2 \frac{1}{32}$ -5

33. $\log_{25} 5$ $\frac{1}{2}$

34. $\log_8 \frac{1}{2}$ $-\frac{1}{3}$

▶35. $\log_{1/2} 2$ -1

36. $\log_{2/3} \frac{4}{9}$ 2

▶37. $\log_6 1$ 0

38. $\log_9 9$ 1

39. $\log_{10} 100$ 2

40. $\log_{10} \frac{1}{10}$ -1

41. $\log_3 81$ 4

42. $\log_2 16$ 4

43. $\log_4 \dfrac{1}{64}$ -3

44. $\log_3 \dfrac{1}{9}$ -2

Objective B   *Solve. See Examples 8 through 10.*

45. $\log_3 9 = x$ $\{2\}$

46. $\log_2 8 = x$ $\{3\}$

47. $\log_3 x = 4$ $\{81\}$

48. $\log_2 x = 3$ $\{8\}$

49. $\log_x 49 = 2$ $\{7\}$

50. $\log_x 8 = 3$ $\{2\}$

51. $\log_2 \dfrac{1}{8} = x$ $\{-3\}$

52. $\log_3 \dfrac{1}{81} = x$ $\{-4\}$

53. $\log_3 \dfrac{1}{27} = x$ $\{-3\}$

54. $\log_5 \dfrac{1}{125} = x$ $\{-3\}$

▶**55.** $\log_8 x = \dfrac{1}{3}$ $\{2\}$

56. $\log_9 x = \dfrac{1}{2}$ $\{3\}$

57. $\log_4 16 = x$ $\{2\}$

58. $\log_2 16 = x$ $\{4\}$

59. $\log_{3/4} x = 3$ $\left\{\dfrac{27}{64}\right\}$

60. $\log_{2/3} x = 2$ $\left\{\dfrac{4}{9}\right\}$

▶**61.** $\log_x 100 = 2$ $\{10\}$

62. $\log_x 27 = 3$ $\{3\}$

63. $\log_2 2^4 = x$ $\{4\}$

64. $\log_6 6^{-2} = x$ $\{-2\}$

65. $3^{\log_3 5} = x$ $\{5\}$

66. $5^{\log_5 7} = x$ $\{7\}$

67. $\log_x \dfrac{1}{7} = \dfrac{1}{2}$ $\left\{\dfrac{1}{49}\right\}$

68. $\log_x 2 = -\dfrac{1}{3}$ $\left\{\dfrac{1}{8}\right\}$

Simplify. See Example 11.

69. $\log_5 5^3$ 3

70. $\log_6 6^2$ 2

▶**71.** $2^{\log_2 3}$ 3

72. $7^{\log_7 4}$ 4

▶**73.** $\log_9 9$ 1

74. $\log_2 2$ 1

75. $\log_8 (8)^{-1}$ -1

76. $\log_{11} (11)^{-1}$ -1

Objective C   *Graph each logarithmic function. See Examples 12 and 13.*

▶**77.** $y = \log_3 x$

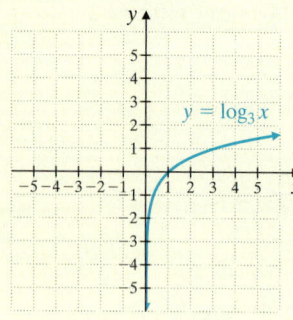

78. $y = \log_2 x$

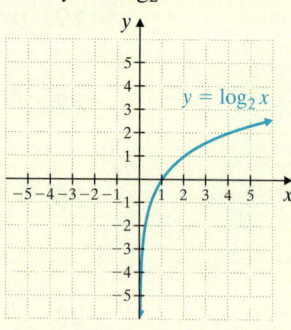

79. $f(x) = \log_{1/4} x$

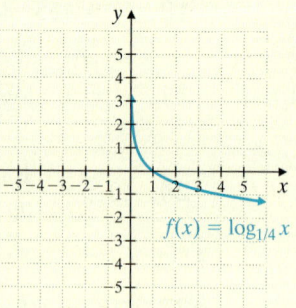

80. $f(x) = \log_{1/2} x$

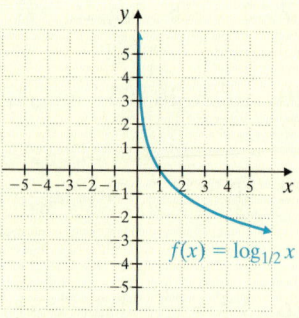

81. $f(x) = \log_5 x$

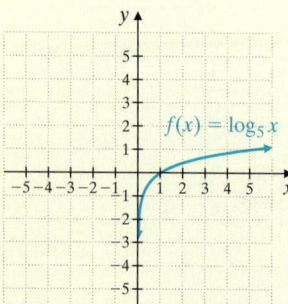

82. $f(x) = \log_6 x$

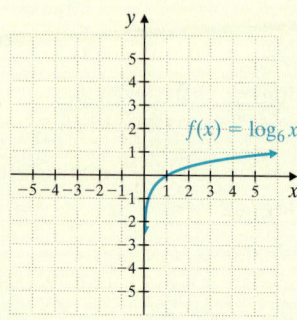

83. $f(x) = \log_{1/6} x$

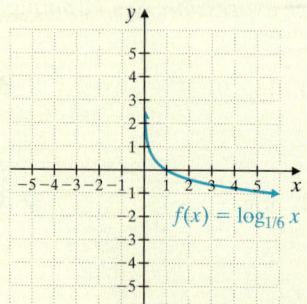

84. $f(x) = \log_{1/5} x$

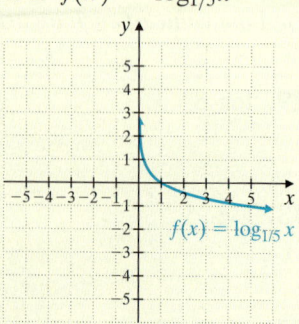

Review

Simplify each rational expression. See Section 7.1.

85. $\dfrac{x+3}{3+x}$ 1

86. $\dfrac{x-5}{5-x}$ −1

87. $\dfrac{x^2-8x+16}{2x-8}$ $\dfrac{x-4}{2}$

88. $\dfrac{x^2-3x-10}{2+x}$ $x-5$

Concept Extensions

Solve. See the Concept Check in this section.

89. Let $f(x) = \log_5 x$. Then $g(x) = 5^x$ is the inverse of $f(x)$. The ordered pair $(2, 25)$ is a solution of the function $g(x)$.
   a. Write this solution using function notation. $g(2) = 25$
   b. Write an ordered pair that we know to be a solution of $f(x)$. $(25, 2)$
   c. Use the answer to part (b) and write the solution using function notation. $f(25) = 2$

90. Let $f(x) = \log_{0.3} x$. Then $g(x) = 0.3^x$ is the inverse of $f(x)$. The ordered pair $(3, 0.027)$ is a solution of the function $g(x)$.
   a. Write this solution using function notation. $g(3) = 0.027$
   b. Write an ordered pair that we know to be a solution of $f(x)$. $(0.027, 3)$
   c. Use the answer to part (b) and write the solution using function notation. $f(0.027) = 3$

91. Explain why negative numbers are not included as logarithmic bases.   answers may vary

92. Explain why 1 is not included as a logarithmic base.   answers may vary

For Exercises 93 through 96, graph each function and its inverse on the same set of axes.

93. $y = 4^x$; $y = \log_4 x$

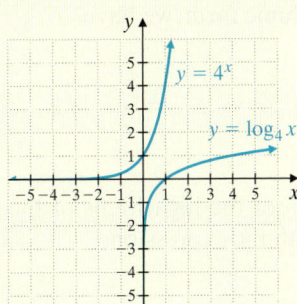

94. $y = 3^x$; $y = \log_3 x$

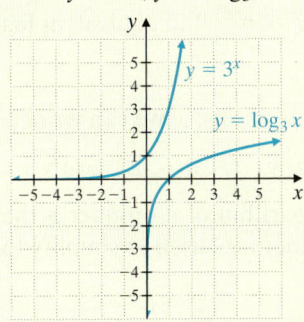

95. $y = \left(\dfrac{1}{3}\right)^x$; $y = \log_{1/3} x$

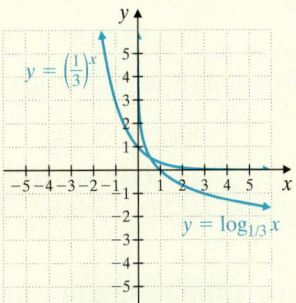

96. $y = \left(\dfrac{1}{2}\right)^x$; $y = \log_{1/2} x$

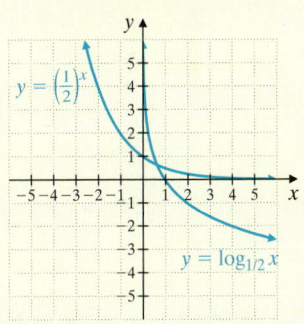

97. Explain why the graph of the function $y = \log_b x$ contains the point $(1, 0)$ no matter what b is.   answers may vary

98. $\log_3 10$ is between which two integers? Explain your answer.   2 and 3

Solve. Round Exercises 99 and 100 to four decimal places.

99. The formula $\log_{10}(1-k) = \dfrac{-0.3}{H}$ models the relationship between the half-life H of a radioactive material and its rate of decay k. Find the rate of decay of the iodine isotope I-131 if its half-life is 8 days.   0.0827

100. The formula $\text{pH} = -\log_{10}(\text{H}^+)$ gives the pH for a liquid, where H^+ stands for the concentration of hydronium ions in moles per liter. Find the concentration of hydronium ions of lemonade, whose pH is 2.3.   0.0050

12.6 Properties of Logarithms

In the previous section, we explored some basic properties of logarithms. We now introduce and study additional properties. Because a logarithm is an exponent, logarithmic properties are just restatements of exponential properties.

Objectives

A Use the Product Property of Logarithms.

B Use the Quotient Property of Logarithms.

C Use the Power Property of Logarithms.

D Use the Properties of Logarithms Together.

Objective A Using the Product Property

The first of these properties is called the **product property of logarithms** because it deals with the logarithm of a product.

> **Product Property of Logarithms**
>
> If x, y, and b are positive real numbers and $b \neq 1$, then
>
> $$\log_b xy = \log_b x + \log_b y$$

To prove this, we let $\log_b x = M$ and $\log_b y = N$. Now we write each logarithm with exponential notation.

$$\log_b x = M \quad \text{is equivalent to} \quad b^M = x$$
$$\log_b y = N \quad \text{is equivalent to} \quad b^N = y$$

When we multiply the left sides and the right sides of the exponential equations, we have that

$$xy = (b^M)(b^N) = b^{M+N}$$

If we write the equation $xy = b^{M+N}$ in equivalent logarithmic form, we have

$$\log_b xy = M + N$$

But since $M = \log_b x$ and $N = \log_b y$, we can write

$$\log_b xy = \log_b x + \log_b y \quad \text{Let } M = \log_b x \text{ and } N = \log_b y.$$

In other words, the logarithm of a product is the sum of the logarithms of the factors. This property is sometimes used to simplify logarithmic expressions.

Practice 1

Write as a single logarithm:
$\log_2 7 + \log_2 5$

Example 1 Write as a single logarithm: $\log_{11} 10 + \log_{11} 3$

Solution: $\log_{11} 10 + \log_{11} 3 = \log_{11}(10 \cdot 3)$ Use the product property.
$$= \log_{11} 30$$

◼ Work Practice 1

Practice 2

Write as a single logarithm:
$\log_3 x + \log_3(x - 9)$

Example 2 Write as a single logarithm: $\log_2(x + 2) + \log_2 x$

Solution: $\log_2(x + 2) + \log_2 x = \log_2[(x + 2) \cdot x] = \log_2(x^2 + 2x)$

◼ Work Practice 2

Objective B Using the Quotient Property

The second property is the **quotient property of logarithms.**

Answers

1. $\log_2 35$ **2.** $\log_3(x^2 - 9x)$

Quotient Property of Logarithms

If x, y, and b are positive real numbers and $b \neq 1$, then

$$\log_b \frac{x}{y} = \log_b x - \log_b y$$

The proof of the quotient property of logarithms is similar to the proof of the product property. Notice that the quotient property says that the logarithm of a quotient is the difference of the logarithms of the dividend and divisor.

✓**Concept Check** Which of the following is the correct way to rewrite $\log_5 \frac{7}{2}$?

a. $\log_5 7 - \log_5 2$ **b.** $\log_5(7 - 2)$ **c.** $\dfrac{\log_5 7}{\log_5 2}$ **d.** $\log_5 14$

Example 3 Write as a single logarithm: $\log_{10} 27 - \log_{10} 3$

Solution: $\log_{10} 27 - \log_{10} 3 = \log_{10} \dfrac{27}{3}$ Use the quotient property.

$$= \log_{10} 9$$

■ **Work Practice 3**

Example 4 Write as a single logarithm: $\log_3(x^2 + 5) - \log_3(x^2 + 1)$

Solution: $\log_3(x^2 + 5) - \log_3(x^2 + 1) = \log_3 \dfrac{x^2 + 5}{x^2 + 1}$ Use the quotient property.

■ **Work Practice 4**

Objective C Using the Power Property

The third and final property we introduce is the **power property of logarithms.**

Power Property of Logarithms

If x and b are positive real numbers, $b \neq 1$, and r is a real number, then

$$\log_b x^r = r \log_b x$$

Examples Use the power property to rewrite each expression.

5. $\log_5 x^3 = 3 \log_5 x$

6. $\log_4 \sqrt{2} = \log_4 2^{1/2} = \dfrac{1}{2}\log_4 2$

■ **Work Practice 5–6**

Objective D Using More Than One Property

Many times we must use more than one property of logarithms to simplify logarithmic expressions.

Practice 3

Write as a single logarithm:
$\log_7 40 - \log_7 8$

Practice 4

Write as a single logarithm:
$\log_3(x^3 + 4) - \log_3(x^2 + 2)$

Teaching Tip

Show how the power property of logarithms is related to the product property of logarithms. For instance,
$\log_2 5^3 = \log_2(5 \cdot 5 \cdot 5) =$
$\log_2 5 + \log_2 5 + \log_2 5 = 3 \log_2 5.$

Practice 5–6

Use the power property to rewrite each expression.
5. $\log_3 x^5$ **6.** $\log_7 \sqrt[3]{4}$

Answers

3. $\log_7 5$ **4.** $\log_3 \dfrac{x^3 + 4}{x^2 + 2}$ **5.** $5 \log_3 x$

6. $\dfrac{1}{3}\log_7 4$

✓**Concept Check Answer**

a

Practice 7–8

Write as a single logarithm.

7. $3\log_4 2 + 2\log_4 5$

8. $5\log_2(2x - 1) - \log_2 x$

Examples   Write as a single logarithm.

7. $2\log_5 3 + 3\log_5 2 = \log_5 3^2 + \log_5 2^3$ Use the power property.

$\qquad\qquad\qquad\quad = \log_5 9 + \log_5 8$

$\qquad\qquad\qquad\quad = \log_5(9 \cdot 8)$ Use the product property.

$\qquad\qquad\qquad\quad = \log_5 72$

8. $3\log_9 x - \log_9(x + 1) = \log_9 x^3 - \log_9(x + 1)$ Use the power property.

$\qquad\qquad\qquad\qquad\quad = \log_9 \dfrac{x^3}{x + 1}$ Use the quotient property.

■ **Work Practice 7–8**

Practice 9–10

Write each expression as sums or differences of logarithms.

9. $\log_7 \dfrac{6 \cdot 2}{5}$

10. $\log_3 \dfrac{x^4}{y^3}$

Examples   Write each expression as sums or differences of logarithms.

9. $\log_3 \dfrac{5 \cdot 7}{4} = \log_3(5 \cdot 7) - \log_3 4$ Use the quotient property.

$\qquad\qquad\quad = \log_3 5 + \log_3 7 - \log_3 4$ Use the product property.

10. $\log_2 \dfrac{x^5}{y^2} = \log_2(x^5) - \log_2(y^2)$ Use the quotient property.

$\qquad\qquad\quad = 5\log_2 x - 2\log_2 y$ Use the power property.

■ **Work Practice 9–10**

> ### Helpful Hint
>
> Notice that we are not able to simplify further a logarithmic expression such as $\log_5(2x - 1)$. None of the basic properties gives a way to write the logarithm of a difference (or sum) in some equivalent form.

✔**Concept Check**   What is wrong with the following?

$$\log_{10}(x^2 + 5) \ \cancel{=}\ \log_{10} x^2 + \log_{10} 5$$
$$\cancel{=}\ 2\log_{10} x + \log_{10} 5$$

Use a numerical example to demonstrate that the result is incorrect.

Practice 11–13

If $\log_b 4 = 0.86$ and $\log_b 7 = 1.21$, use the properties of logarithms to evaluate each expression.

11. $\log_b 28$ **12.** $\log_b 49$

13. $\log_b \sqrt[3]{4}$

Examples   If $\log_b 2 = 0.43$ and $\log_b 3 = 0.68$, use the properties of logarithms to evaluate each expression.

11. $\log_b 6 = \log_b(2 \cdot 3)$ Write 6 as $2 \cdot 3$.

$\qquad\quad = \log_b 2 + \log_b 3$ Use the product property.

$\qquad\quad = 0.43 + 0.68$ Substitute given values.

$\qquad\quad = 1.11$ Simplify.

12. $\log_b 9 = \log_b 3^2$ Write 9 as 3^2.

$\qquad\quad = 2\log_b 3$ Use the power property.

$\qquad\quad = 2(0.68)$ Substitute the given value.

$\qquad\quad = 1.36$ Simplify.

13. $\log_b \sqrt{2} = \log_b 2^{1/2}$ Write $\sqrt{2}$ as $2^{1/2}$.

$\qquad\quad = \dfrac{1}{2}\log_b 2$ Use the power property.

$\qquad\quad = \dfrac{1}{2}(0.43)$ Substitute the given value.

$\qquad\quad = 0.215$ Simplify.

■ **Work Practice 11–13**

Answers

7. $\log_4 200$ **8.** $\log_2 \dfrac{(2x - 1)^5}{x}$

9. $\log_7 6 + \log_7 2 - \log_7 5$

10. $4\log_3 x - 3\log_3 y$ **11.** 2.07

12. 2.42 **13.** $0.28\overline{6}$

✔**Concept Check Answer**

The properties do not give any way to simplify the logarithm of a sum; answers may vary.

Quotient Property of Logarithms

If x, y, and b are positive real numbers and $b \neq 1$, then

$$\log_b \frac{x}{y} = \log_b x - \log_b y$$

The proof of the quotient property of logarithms is similar to the proof of the product property. Notice that the quotient property says that the logarithm of a quotient is the difference of the logarithms of the dividend and divisor.

✓**Concept Check** Which of the following is the correct way to rewrite $\log_5 \frac{7}{2}$?

a. $\log_5 7 - \log_5 2$ **b.** $\log_5(7 - 2)$ **c.** $\dfrac{\log_5 7}{\log_5 2}$ **d.** $\log_5 14$

Example 3   Write as a single logarithm: $\log_{10} 27 - \log_{10} 3$

Solution: $\log_{10} 27 - \log_{10} 3 = \log_{10} \dfrac{27}{3}$ Use the quotient property.

$$= \log_{10} 9$$

■ **Work Practice 3**

Practice 3

Write as a single logarithm:
$\log_7 40 - \log_7 8$

Example 4   Write as a single logarithm: $\log_3(x^2 + 5) - \log_3(x^2 + 1)$

Solution: $\log_3(x^2 + 5) - \log_3(x^2 + 1) = \log_3 \dfrac{x^2 + 5}{x^2 + 1}$ Use the quotient property.

■ **Work Practice 4**

Practice 4

Write as a single logarithm:
$\log_3(x^3 + 4) - \log_3(x^2 + 2)$

Objective C   Using the Power Property ▶

The third and final property we introduce is the **power property of logarithms.**

Power Property of Logarithms

If x and b are positive real numbers, $b \neq 1$, and r is a real number, then

$$\log_b x^r = r \log_b x$$

Teaching Tip

Show how the power property of logarithms is related to the product property of logarithms. For instance,

$\log_2 5^3 = \log_2(5 \cdot 5 \cdot 5) =$

$\log_2 5 + \log_2 5 + \log_2 5 = 3 \log_2 5.$

Examples   Use the power property to rewrite each expression.

5. $\log_5 x^3 = 3 \log_5 x$

6. $\log_4 \sqrt{2} = \log_4 2^{1/2} = \dfrac{1}{2} \log_4 2$

■ **Work Practice 5–6**

Practice 5–6

Use the power property to rewrite each expression.
5. $\log_3 x^5$ **6.** $\log_7 \sqrt[3]{4}$

Answers

3. $\log_7 5$ **4.** $\log_3 \dfrac{x^3 + 4}{x^2 + 2}$ **5.** $5 \log_3 x$

6. $\dfrac{1}{3} \log_7 4$

✓**Concept Check Answer**

a

Objective D   Using More Than One Property ▶

Many times we must use more than one property of logarithms to simplify logarithmic expressions.

Practice 7–8

Write as a single logarithm.

7. $3\log_4 2 + 2\log_4 5$

8. $5\log_2(2x - 1) - \log_2 x$

Examples   Write as a single logarithm.

7. $2\log_5 3 + 3\log_5 2 = \log_5 3^2 + \log_5 2^3$ Use the power property.

$\quad\quad\quad\quad\quad\quad\quad\quad = \log_5 9 + \log_5 8$

$\quad\quad\quad\quad\quad\quad\quad\quad = \log_5(9 \cdot 8)$ Use the product property.

$\quad\quad\quad\quad\quad\quad\quad\quad = \log_5 72$

8. $3\log_9 x - \log_9(x + 1) = \log_9 x^3 - \log_9(x + 1)$ Use the power property.

$\quad\quad\quad\quad\quad\quad\quad\quad\quad\quad = \log_9 \dfrac{x^3}{x + 1}$ Use the quotient property.

■ Work Practice 7–8

Practice 9–10

Write each expression as sums or differences of logarithms.

9. $\log_7 \dfrac{6 \cdot 2}{5}$

10. $\log_3 \dfrac{x^4}{y^3}$

Examples   Write each expression as sums or differences of logarithms.

9. $\log_3 \dfrac{5 \cdot 7}{4} = \log_3(5 \cdot 7) - \log_3 4$ Use the quotient property.

$\quad\quad\quad\quad\quad = \log_3 5 + \log_3 7 - \log_3 4$ Use the product property.

10. $\log_2 \dfrac{x^5}{y^2} = \log_2(x^5) - \log_2(y^2)$ Use the quotient property.

$\quad\quad\quad\quad\quad = 5\log_2 x - 2\log_2 y$ Use the power property.

■ Work Practice 9–10

Helpful Hint

Notice that we are not able to simplify further a logarithmic expression such as $\log_5(2x - 1)$. None of the basic properties gives a way to write the logarithm of a difference (or sum) in some equivalent form.

✓**Concept Check** What is wrong with the following?

$\log_{10}(x^2 + 5) = \log_{10} x^2 + \log_{10} 5$

$\quad\quad\quad\quad\quad\quad = 2\log_{10} x + \log_{10} 5$

Use a numerical example to demonstrate that the result is incorrect.

Practice 11–13

If $\log_b 4 = 0.86$ and $\log_b 7 = 1.21$, use the properties of logarithms to evaluate each expression.

11. $\log_b 28$ **12.** $\log_b 49$

13. $\log_b \sqrt[3]{4}$

Examples   If $\log_b 2 = 0.43$ and $\log_b 3 = 0.68$, use the properties of logarithms to evaluate each expression.

11. $\log_b 6 = \log_b(2 \cdot 3)$ Write 6 as $2 \cdot 3$.

$\quad\quad\quad = \log_b 2 + \log_b 3$ Use the product property.

$\quad\quad\quad = 0.43 + 0.68$ Substitute given values.

$\quad\quad\quad = 1.11$ Simplify.

12. $\log_b 9 = \log_b 3^2$ Write 9 as 3^2.

$\quad\quad\quad = 2\log_b 3$ Use the power property.

$\quad\quad\quad = 2(0.68)$ Substitute the given value.

$\quad\quad\quad = 1.36$ Simplify.

13. $\log_b \sqrt{2} = \log_b 2^{1/2}$ Write $\sqrt{2}$ as $2^{1/2}$.

$\quad\quad\quad = \dfrac{1}{2}\log_b 2$ Use the power property.

$\quad\quad\quad = \dfrac{1}{2}(0.43)$ Substitute the given value.

$\quad\quad\quad = 0.215$ Simplify.

■ Work Practice 11–13

Answers

7. $\log_4 200$ **8.** $\log_2 \dfrac{(2x - 1)^5}{x}$

9. $\log_7 6 + \log_7 2 - \log_7 5$

10. $4\log_3 x - 3\log_3 y$ **11.** 2.07

12. 2.42   **13.** $0.28\overline{6}$

✓**Concept Check Answer**

The properties do not give any way to simplify the logarithm of a sum; answers may vary.

Vocabulary, Readiness & Video Check

Select the correct choice.

1. $\log_b 12 + \log_b 3 = \log_b \underline{\quad 36 \quad}$ a

    a. 36   **b.** 15   **c.** 4   **d.** 9

2. $\log_b 12 - \log_b 3 = \log_b \underline{\quad 4 \quad}$ c

    a. 36   **b.** 15   **c.** 4   **d.** 9

3. $7 \log_b 2 = \underline{\quad \log_b 2^7 \quad}$ b

    a. $\log_b 14$ **b.** $\log_b 2^7$ **c.** $\log_b 7^2$ **d.** $(\log_b 2)^7$

4. $\log_b 1 = \underline{\quad 0 \quad}$ c

    a. b **b.** 1   **c.** 0   **d.** no answer

5. $b^{\log_b x} = \underline{\quad x \quad}$ a

    a. x **b.** b **c.** 1   **d.** 0

6. $\log_5 5^2 = \underline{\quad 2 \quad}$ b

    a. 25   **b.** 2   **c.** 5^{5^2} **d.** 32

Martin-Gay Interactive Videos   Watch the section lecture video and answer the following questions.

See Video 12.6 🍎

See video answer section.

Objective A   **7.** Can the product property of logarithms be used again on the bottom line of ⊞ Example 2 to write $\log_{10}(10x^2 + 20)$ as a sum of logarithms, $\log_{10} 10x^2 + \log_{10} 20$? Explain. ▶

Objective B   **8.** From ⊞ Example 3 and the lecture before, what must be true about bases before you can apply the quotient property of logarithms? ▶

Objective C   **9.** Based on ⊞ Example 5, explain why $\log_2 \dfrac{1}{x} = -\log_2 x$. ▶

Objective D   **10.** From the lecture before ⊞ Example 6, where do the logarithmic properties come from? ▶

12.6 Exercise Set MyMathLab® ▶

Objective A *Write each sum as a single logarithm. Assume that variables represent positive numbers. See Examples 1 and 2.*

▶ **1.** $\log_5 2 + \log_5 7$
    $\log_5 14$

2. $\log_3 8 + \log_3 4$
    $\log_3 32$

3. $\log_4 9 + \log_4 x$
    $\log_4 9x$

4. $\log_2 x + \log_2 y$
    $\log_2 xy$

5. $\log_6 x + \log_6 (x + 1)$
    $\log_6 (x^2 + x)$

6. $\log_5 y^3 + \log_5 (y - 7)$
    $\log_5 (y^4 - 7y^3)$

▶ **7.** $\log_{10} 5 + \log_{10} 2 + \log_{10}(x^2 + 2)$
    $\log_{10} (10x^2 + 20)$

8. $\log_6 3 + \log_6(x + 4) + \log_6 5$
    $\log_6 (15x + 60)$

Objective B *Write each difference as a single logarithm. Assume that variables represent positive numbers. See Examples 3 and 4.*

9. $\log_5 12 - \log_5 4$ $\log_5 3$

10. $\log_7 20 - \log_7 4$ $\log_7 5$

▶**11.** $\log_3 8 - \log_3 2$ $\log_3 4$

12. $\log_5 12 - \log_5 3$ $\log_5 4$

13. $\log_2 x - \log_2 y$ $\log_2 \dfrac{x}{y}$

14. $\log_3 12 - \log_3 z$ $\log_3 \dfrac{12}{z}$

15. $\log_2 (x^2 + 6) - \log_2 (x^2 + 1)$ $\log_2 \dfrac{x^2 + 6}{x^2 + 1}$

16. $\log_7 (x + 9) - \log_7 (x^2 + 10)$ $\log_7 \dfrac{x + 9}{x^2 + 10}$

Objective C *Use the power property to rewrite each expression. See Example 5 and 6.*

▶**17.** $\log_3 x^2$ $2\log_3 x$

18. $\log_2 x^5$ $5\log_2 x$

19. $\log_4 5^{-1}$ $-1\log_4 5 = -\log_4 5$

20. $\log_6 7^{-2}$ $-2\log_6 7$

▶**21.** $\log_5 \sqrt{y}$ $\dfrac{1}{2}\log_5 y$

22. $\log_5 \sqrt[3]{x}$ $\dfrac{1}{3}\log_5 x$

Objective D *Write each as a single logarithm. Assume that variables represent positive numbers. See Examples 7 and 8.*

23. $\log_2 5 + \log_2 x^3$
$\log_2 5x^3$

24. $\log_5 2 + \log_5 y^2$
$\log_5 2y^2$

25. $3 \log_4 2 + \log_4 6$
$\log_4 48$

26. $2 \log_3 5 + \log_3 2$
$\log_3 50$

▶**27.** $3 \log_5 x + 6 \log_5 z$
$\log_5 x^3 z^6$

28. $2 \log_7 y + 6 \log_7 z$
$\log_7 y^2 z^6$

▶**29.** $\log_4 2 + \log_4 10 - \log_4 5$
$\log_4 4$, or 1

30. $\log_6 21 + \log_6 2 - \log_6 7$
$\log_6 6$, or 1

31. $\log_7 6 + \log_7 3 - \log_7 4$
$\log_7 \dfrac{9}{2}$

32. $\log_8 5 + \log_8 15 - \log_8 20$
$\log_8 \dfrac{15}{4}$

33. $\log_{10} x - \log_{10} (x + 1) + \log_{10} (x^2 - 2)$
$\log_{10} \dfrac{x^3 - 2x}{x + 1}$

34. $\log_9 (4x) - \log_9 (x - 3) + \log_9 (x^3 + 1)$
$\log_9 \dfrac{4x^4 + 4x}{x - 3}$

35. $3 \log_2 x + \dfrac{1}{2} \log_2 x - 2 \log_2 (x + 1)$
$\log_2 \dfrac{x^{7/2}}{(x + 1)^2}$

36. $2 \log_5 x + \dfrac{1}{3} \log_5 x - 3 \log_5 (x + 5)$
$\log_5 \dfrac{x^{7/3}}{(x + 5)^3}$

37. $2 \log_8 x - \dfrac{2}{3} \log_8 x + 4 \log_8 x$
$\log_8 x^{16/3}$

38. $5 \log_6 x - \dfrac{3}{4} \log_6 x + 3 \log_6 x$
$\log_6 x^{29/4}$

Write each expression as a sum or difference of logarithms. Assume that variables represent positive numbers. See Examples 9 and 10.

39. $\log_3 \dfrac{4y}{5}$ $\log_3 4 + \log_3 y - \log_3 5$

40. $\log_7 \dfrac{5x}{4}$ $\log_7 5 + \log_7 x - \log_7 4$

41. $\log_4 \dfrac{2}{9z}$ $\log_4 2 - \log_4 9 - \log_4 z$

42. $\log_9 \dfrac{7}{8y}$ $\log_9 7 - \log_9 8 - \log_9 y$

▶ **43.** $\log_2 \dfrac{x^3}{y}$ $3 \log_2 x - \log_2 y$

44. $\log_5 \dfrac{x}{y^4}$ $\log_5 x - 4 \log_5 y$

45. $\log_b \sqrt{7x}$ $\dfrac{1}{2} \log_b 7 + \dfrac{1}{2} \log_b x$

46. $\log_b \sqrt{\dfrac{3}{y}}$ $\dfrac{1}{2} \log_b 3 - \dfrac{1}{2} \log_b y$

47. $\log_6 x^4 y^5$ $4 \log_6 x + 5 \log_6 y$

48. $\log_2 y^3 z$
   $3 \log_2 y + \log_2 z$

49. $\log_5 x^3(x + 1)$
   $3 \log_5 x + \log_5 (x + 1)$

50. $\log_3 x^2(x - 9)$
   $2 \log_3 x + \log_3 (x - 9)$

51. $\log_6 \dfrac{x^2}{x + 3}$ $2 \log_6 x - \log_6 (x + 3)$

52. $\log_3 \dfrac{(x + 5)^2}{x}$ $2 \log_3 (x + 5) - \log_3 x$

If $\log_b 3 = 0.5$ and $\log_b 5 = 0.7$, evaluate each expression. See Examples 11 through 13.

53. $\log_b \dfrac{5}{3}$ 0.2

54. $\log_b 25$ 1.4

55. $\log_b 15$ 1.2

56. $\log_b \dfrac{3}{5}$ -0.2

57. $\log_b \sqrt{5}$ 0.35

58. $\log_b \sqrt[4]{3}$ 0.125

If $\log_b 2 = 0.43$ and $\log_b 3 = 0.68$, evaluate each expression. See Examples 11 through 13.

59. $\log_b 8$
   1.29

60. $\log_b 81$
   2.72

61. $\log_b \dfrac{3}{9}$
   -0.68

62. $\log_b \dfrac{4}{32}$
   -1.29

63. $\log_b \sqrt{\dfrac{2}{3}}$
   -0.125

64. $\log_b \sqrt{\dfrac{3}{2}}$
   0.125

Review

Graph each function on the same set of axes. See Sections 12.3 and 12.5.

65. $y = 10^x$

66. $y = \log_{10} x$

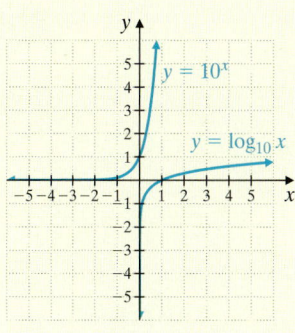

Evaluate each expression. See Section 12.5.

67. $\log_{10} 100$ 2

68. $\log_{10} \dfrac{1}{10}$ -1

69. $\log_7 7^2$ 2

70. $\log_7 \sqrt{7}$ $\dfrac{1}{2}$

Concept Extensions

Solve. See the Concept Checks in this section.

71. Which of the following is the correct way to rewrite $\log_3 \dfrac{14}{11}$?

   a. $\dfrac{\log_3 14}{\log_3 11}$

   b. $\log_3 14 - \log_3 11$

   c. $\log_3 (14 - 11)$

   d. $\log_3 154$ b

72. Which of the following is the correct way to rewrite $\log_9 \dfrac{21}{3}$?

   a. $\log_9 7$

   b. $\log_9 (21 - 3)$

   c. $\dfrac{\log_9 21}{\log_9 3}$

   d. $\log_9 21 - \log_9 3$ a and d

Determine whether each statement is true or false.

73. $\log_2 x^3 = 3 \log_2 x$ true

74. $\log_3 (x + y) = \log_3 x + \log_3 y$ false

75. $\dfrac{\log_7 10}{\log_7 5} = \log_7 2$ false

76. $\log_7 \dfrac{14}{8} = \log_7 14 - \log_7 8$ true

77. $\dfrac{\log_7 x}{\log_7 y} = \log_7 x - \log_7 y$ false

78. $(\log_3 6) \cdot (\log_3 4) = \log_3 24$ false

79. It is true that $\log_b 8 = \log_b (8 \cdot 1) = \log_b 8 + \log_b 1$. Explain how $\log_b 8$ can equal $\log_b 8 + \log_b 1$.
because $\log_b 1 = 0$

80. It is true that $\log_b 7 = \log_b \dfrac{7}{1} = \log_b 7 - \log_b 1$. Explain how $\log_b 7$ can equal $\log_b 7 - \log_b 1$.
because $\log_b 1 = 0$

Integrated Review

Functions and Properties of Logarithms

Answers

If $f(x) = x - 6$ and $g(x) = x^2 + 1$, find each value.

1. $(f + g)(x)$ **2.** $(f - g)(x)$ **3.** $(f \cdot g)(x)$ **4.** $\left(\dfrac{f}{g}\right)(x)$

1. $x^2 + x - 5$

2. $-x^2 + x - 7$

If $f(x) = \sqrt{x}$ and $g(x) = 3x - 1$, find each value.

5. $(f \circ g)(x)$ **6.** $(g \circ f)(x)$

3. $x^3 - 6x^2 + x - 6$

4. $\dfrac{x - 6}{x^2 + 1}$

Determine whether each is a one-to-one function. If it is, find its inverse.

7. $f = \{(-2, 6), (4, 8), (2, -6), (3, 3)\}$ **8.** $g = \{(4, 2), (-1, 3), (5, 3), (7, 1)\}$

5. $\sqrt{3x - 1}$

6. $3\sqrt{x} - 1$

Determine whether the graph of each function is the graph of a one-to-one function.

9.

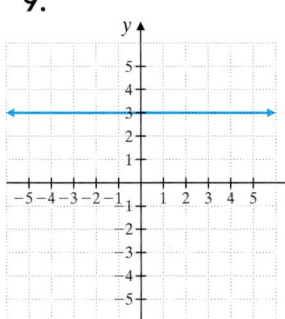

10.

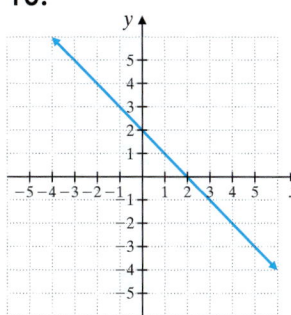

11.

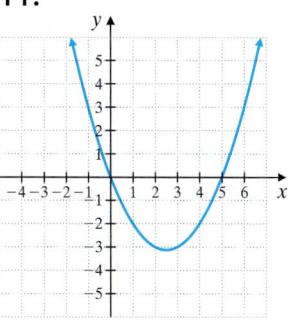

7. one-to-one; $\{(6, -2), (8, 4), (-6, 2), (3, 3)\}$

8. not one-to-one

9. not one-to-one

10. one-to-one

Each function listed is one-to-one. Find the inverse of each function.

12. $f(x) = 3x$ **13.** $f(x) = x + 4$

11. not one-to-one

14. $f(x) = 5x - 1$ **15.** $f(x) = 3x + 2$

12. $f^{-1}(x) = \dfrac{x}{3}$

13. $f^{-1}(x) = x - 4$

Graph each function.

16. $y = \left(\dfrac{1}{2}\right)^x$ **17.** $y = 2^x + 1$

14. $f^{-1}(x) = \dfrac{x + 1}{5}$

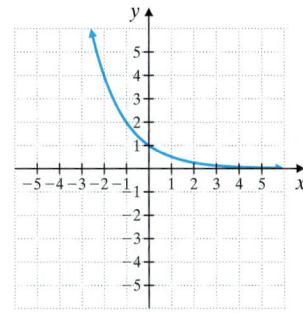

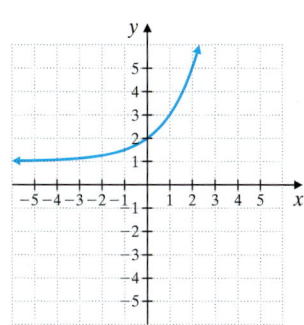

15. $f^{-1}(x) = \dfrac{x - 2}{3}$

16. see graph

17. see graph

18. see graph

19. see graph

20. $\{3\}$

21. $\{7\}$

22. $\{-8\}$

23. $\{3\}$

24. $\{2\}$

25. $\left\{\dfrac{1}{2}\right\}$

26. $\{32\}$

27. $\{4\}$

28. $\{5\}$

29. $\left\{\dfrac{1}{9}\right\}$

30. $\log_2 14x$

31. $\log_2 (5^x \cdot 8)$

32. $\log_5 \dfrac{x^3}{y^5}$

33. $\log_5 x^9 y^3$

34. $\log_2 \dfrac{x^2 - 3x}{x^2 + 4}$

35. $\log_3 \dfrac{y^4 + 11y}{y + 2}$

36. $\log_7 9 + 2\log_7 x - \log_7 y$

37. $\log_6 5 + \log_6 y - 2\log_6 z$

38. 544,000 mosquitoes

---

18. $y = \log_3 x$

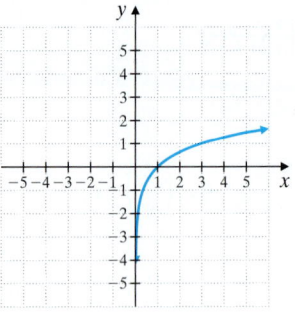

19. $y = \log_{1/3} x$

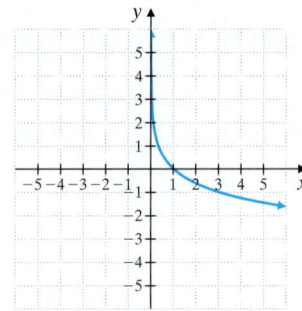

Solve.

20. $2^x = 8$

21. $9 = 3^{x-5}$

22. $4^{x-1} = 8^{x+2}$

23. $25^x = 125^{x-1}$

24. $\log_4 16 = x$

25. $\log_{49} 7 = x$

26. $\log_2 x = 5$

27. $\log_x 64 = 3$

28. $\log_x \dfrac{1}{125} = -3$

29. $\log_3 x = -2$

Write each as a single logarithm.

30. $\log_2 x + \log_2 14$

31. $x\log_2 5 + \log_2 8$

32. $3\log_5 x - 5\log_5 y$

33. $9\log_5 x + 3\log_5 y$

34. $\log_2 x + \log_2(x - 3) - \log_2(x^2 + 4)$

35. $\log_3 y - \log_3(y + 2) + \log_3 (y^3 + 11)$

Write each expression as a sum or difference of logarithms.

36. $\log_7 \dfrac{9x^2}{y}$

37. $\log_6 \dfrac{5y}{z^2}$

38. An unusually wet spring has caused the size of the Cape Cod mosquito population to increase by 8% each day. If an estimated 200,000 mosquitoes are on Cape Cod on May 12, find how many mosquitoes will inhabit the Cape on May 25. Round to the nearest thousand.

In this section, we look closely at two particular logarithmic bases. These two logarithmic bases are used so frequently that logarithms to their bases are given special names. **Common logarithms** are logarithms to base 10. **Natural logarithms** are logarithms to base e, which we define in this section. The work in this section is based on the use of a calculator that has both the "common log" $\boxed{\text{LOG}}$ and the "natural log" $\boxed{\text{LN}}$ keys.

Objective A Approximating Common Logarithms

Logarithms to base 10—**common logarithms**—are used frequently because our number system is a base 10 decimal system. The notation $\log x$ means the same as $\log_{10} x$.

> **Common Logarithm**
>
> $\log x$ means $\log_{10} x$

Example 1 Use a calculator to approximate $\log 7$ to four decimal places.

Solution: Press the following sequence of keys:

$\boxed{7}\,\boxed{\text{LOG}}$ or $\boxed{\text{LOG}}\,\boxed{7}\,\boxed{\text{ENTER}}$

To four decimal places,

$\log 7 \approx 0.8451$

■ **Work Practice 1**

Objective B Evaluating Common Logarithms of Powers of 10

To evaluate the common log of a power of 10, a calculator is not needed. According to the property of logarithms,

$\log_b b^x = x$

It follows that if b is replaced with 10, we have

$\log 10^x = x$

> **Helpful Hint**
> Remember that the base of this logarithm is understood to be 10.

Examples Find the exact value of each logarithm.

2. $\log 10 = \log 10^1 = 1$

3. $\log \dfrac{1}{10} = \log 10^{-1} = -1$

4. $\log 100{,}000 = \log 10^5 = 5$

5. $\log \sqrt[4]{10} = \log 10^{1/4} = \dfrac{1}{4}$

■ **Work Practice 2–5**

Practice 1

Use a calculator to approximate $\log 21$ to four decimal places.

Teaching Tip

Point out that just as the radical symbol represents square root unless another index is given, log represents $\log_{10}$ unless another base is given.

Teaching Tip

Have students estimate answers before using a calculator.

Practice 2–5

Find the exact value of each logarithm.

2. $\log 1000$

3. $\log \dfrac{1}{100}$

4. $\log 10{,}000$

5. $\log \sqrt[3]{10}$

Answers

1. 1.3222 **2.** 3 **3.** −2 **4.** 4 **5.** $\dfrac{1}{3}$

As we will soon see, equations containing common logs are useful models of many natural phenomena.

Practice 6

Solve: $\log x = 2.9$. Give the exact solution and then approximate the solution to four decimal places.

Example 6   Solve: $\log x = 1.2$. Give the exact solution and then approximate the solution to four decimal places.

Solution: Remember that the base of a common log is understood to be 10.

$$\log x = 1.2$$
$$10^{1.2} = x \qquad \text{Write with exponential notation.}$$

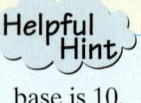
Helpful Hint   The understood base is 10.

The exact solution is $10^{1.2}$ or the solution set is $\{10^{1.2}\}$. To four decimal places, $x \approx 15.8489$.

■ **Work Practice 6**

Objective C Approximating Natural Logarithms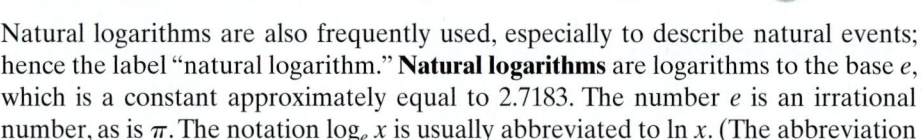

Natural logarithms are also frequently used, especially to describe natural events; hence the label "natural logarithm." **Natural logarithms** are logarithms to the base e, which is a constant approximately equal to 2.7183. The number e is an irrational number, as is π. The notation $\log_e x$ is usually abbreviated to $\ln x$. (The abbreviation ln is read "el en.")

> **Natural Logarithm**
>
> $\ln x$ means $\log_e x$

Practice 7

Use a calculator to approximate ln 11 to four decimal places.

Example 7   Use a calculator to approximate ln 8 to four decimal places.

Solution: Press the following sequence of keys:

| 8 | ln | or | ln | 8 | ENTER |

To four decimal places,
$$\ln 8 \approx 2.0794$$

■ **Work Practice 7**

Objective D Evaluating Natural Logarithms of Powers of e

As a result of the property $\log_b b^x = x$, we know that $\log_e e^x = x$, or $\ln e^x = x$.

Practice 8–9

Find the exact value of each natural logarithm.

8. $\ln e^9$

9. $\ln \sqrt[3]{e}$

Examples   Find the exact value of each natural logarithm.

8. $\ln e^3 = 3$

9. $\ln \sqrt[7]{e} = \ln e^{1/7} = \dfrac{1}{7}$

■ **Work Practice 8–9**

Answers

6. $\{10^{2.9}\}$; $\{794.3282\}$ **7.** 2.3979

8. 9   **9.** $\dfrac{1}{3}$

Example 10   Solve: ln $3x = 5$. Give the exact solution and then approximate the solution to four decimal places.

Solution: Remember that the base of a natural logarithm is understood to be e.

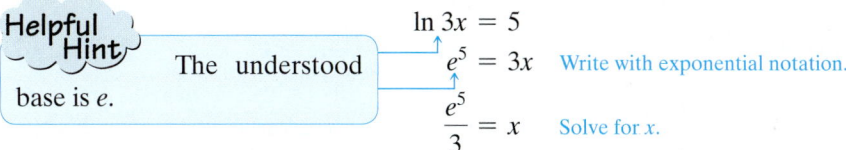

> **Helpful Hint**
> The understood base is e.

$$\ln 3x = 5$$
$$e^5 = 3x \quad \text{Write with exponential notation.}$$
$$\frac{e^5}{3} = x \quad \text{Solve for } x.$$

The exact solution is $\dfrac{e^5}{3}$ or the solution set is $\left\{\dfrac{e^5}{3}\right\}$. To four decimal places, $x \approx 49.4711$.

■ **Work Practice 10**

Practice 10

Solve: ln $7x = 10$. Give the exact solution and then approximate the solution to four decimal places.

Recall from Section 12.3 the formula $A = P\left(1 + \dfrac{r}{n}\right)^{nt}$ for compound interest, where n represents the number of compoundings per year. When interest is compounded continuously, we use the formula $A = Pe^{rt}$, where r is the annual interest rate and interest is compounded continuously for t years.

Example 11   Finding the Amount Owed on a Loan

Find the amount owed at the end of 5 years if $1600 is loaned at a rate of 9% compounded continuously.

Solution: We use the formula $A = Pe^{rt}$ and the following values of the variables.

$P = \$1600$ (the amount of the loan)
$r = 9\% = 0.09$ (the rate of interest)
$t = 5$ (the 5-year duration of the loan)
$A = Pe^{rt}$
$\quad = 1600e^{0.09(5)}$ Substitute known values.
$\quad = 1600e^{0.45}$

Now we can use a calculator to approximate the solution.

$A \approx 2509.30$

The total amount of money owed is approximately $2509.30.

■ **Work Practice 11**

Practice 11

Find the amount owed at the end of 3 years if $1200 is loaned at a rate of 8% compounded continuously.

Teaching Tip

Have students repeat Example 11 assuming that interest is compounded

a. annually.
b. quarterly.
c. monthly.
d. daily.

Objective E Using the Change of Base Formula ▶

Calculators are handy tools for approximating natural and common logarithms. Unfortunately, most calculators cannot be used to approximate logarithms to bases other than e or 10—at least not directly. In such cases, we use the **change of base formula.**

> **Change of Base**
>
> If a, b, and c are positive real numbers and neither b nor c is 1, then
>
> $$\log_b a = \frac{\log_c a}{\log_c b}$$

Answers

10. $\left\{\dfrac{e^{10}}{7}\right\}$; $\{3146.6380\}$

11. $1525.50

Practice 12

Approximate $\log_7 5$ to four decimal places.

Example 12 Approximate $\log_5 3$ to four decimal places.

Solution: We use the change of base property to write $\log_5 3$ as a quotient of logarithms to base 10.

$$\log_5 3 = \frac{\log 3}{\log 5} \quad \text{Use the change of base property.}$$

$$\approx \frac{0.4771213}{0.69897} \quad \text{Approximate the logarithms by calculator.}$$

$$\approx 0.6826063 \quad \text{Simplify by calculator.}$$

To four decimal places, $\log_5 3 \approx 0.6826$.

Answer

12. 0.8271

✔ **Concept Check Answer**

$$f(x) = \frac{\log x}{\log 5}$$

■ **Work Practice 12**

✔**Concept Check** If a graphing calculator cannot directly evaluate logarithms to base 5, describe how you could use the graphing calculator to graph the function $f(x) = \log_5 x$.

Vocabulary, Readiness & Video Check

Use the choices to fill in each blank.

1. The base of $\log 7$ is _____10_____. c

   a. e **b.** 7   **c.** 10   **d.** no answer

2. The base of $\ln 7$ is _____e_____. a

   a. e **b.** 7   **c.** 10   **d.** no answer

3. $\log_{10} 10^7 = $ _____7_____. b

   a. e **b.** 7   **c.** 10   **d.** no answer

4. $\log_7 1 = $ _____0_____. d

   a. e **b.** 7   **c.** 10   **d.** 0

5. $\log_e e^5 = $ _____5_____. b

   a. e **b.** 5   **c.** 0   **d.** 1

6. Study exercise 5 to the left. Then answer:

   $\ln e^5 = $ _____5_____. b

   a. e **b.** 5   **c.** 0   **d.** 1

7. $\log_2 7 = $ _____$\dfrac{\log 7}{\log 2} = \dfrac{\ln 7}{\ln 2}$_____ (There may be more than one answer.)   a, b

   a. $\dfrac{\log 7}{\log 2}$ **b.** $\dfrac{\ln 7}{\ln 2}$ **c.** $\dfrac{\log 2}{\log 7}$ **d.** $\log \dfrac{7}{2}$

Martin-Gay Interactive Videos   *Watch the section lecture video and answer the following questions.*

Objective A **8.** From ▣ Example 1 and the lecture before, what is the understood base of a common logarithm? ▶

Objective B **9.** From ▣ Example 2, why can you find exact values of common logarithms of powers of 10? ▶

Objective C **10.** From ▣ Example 4 and the lecture before, what is the understood base of a natural logarithm? ▶

Objective D **11.** In ▣ Examples 5 and 6, consider how the expression is rewritten and the resulting answer. What logarithm property is actually used here? ▶

Objective E **12.** From ▣ Example 8, what two equivalent fractions will give you the exact value of $\log_6 4$? ▶

See Video 12.7 🍎

See video answer section.

Example 10 Solve: $\ln 3x = 5$. Give the exact solution and then approximate the solution to four decimal places.

Solution: Remember that the base of a natural logarithm is understood to be e.

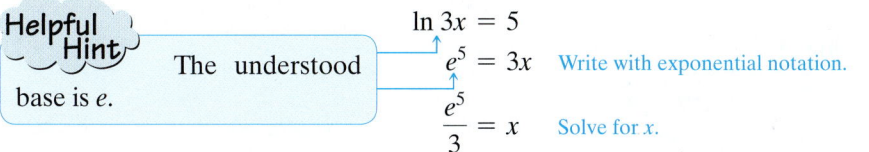

$$\ln 3x = 5$$
$$e^5 = 3x \quad \text{Write with exponential notation.}$$
$$\frac{e^5}{3} = x \quad \text{Solve for } x.$$

Helpful Hint
The understood base is e.

The exact solution is $\dfrac{e^5}{3}$ or the solution set is $\left\{\dfrac{e^5}{3}\right\}$. To four decimal places, $x \approx 49.4711$.

■ **Work Practice 10**

Practice 10
Solve: $\ln 7x = 10$. Give the exact solution and then approximate the solution to four decimal places.

Recall from Section 12.3 the formula $A = P\left(1 + \dfrac{r}{n}\right)^{nt}$ for compound interest, where n represents the number of compoundings per year. When interest is compounded continuously, we use the formula $A = Pe^{rt}$, where r is the annual interest rate and interest is compounded continuously for t years.

Example 11 Finding the Amount Owed on a Loan

Find the amount owed at the end of 5 years if $1600 is loaned at a rate of 9% compounded continuously.

Solution: We use the formula $A = Pe^{rt}$ and the following values of the variables.

$$P = \$1600 \quad \text{(the amount of the loan)}$$
$$r = 9\% = 0.09 \quad \text{(the rate of interest)}$$
$$t = 5 \quad \text{(the 5-year duration of the loan)}$$
$$A = Pe^{rt}$$
$$= 1600e^{0.09(5)} \quad \text{Substitute known values.}$$
$$= 1600e^{0.45}$$

Now we can use a calculator to approximate the solution.

$$A \approx 2509.30$$

The total amount of money owed is approximately $2509.30.

■ **Work Practice 11**

Practice 11
Find the amount owed at the end of 3 years if $1200 is loaned at a rate of 8% compounded continuously.

Teaching Tip
Have students repeat Example 11 assuming that interest is compounded

a. annually.

b. quarterly.

c. monthly.

d. daily.

Objective E Using the Change of Base Formula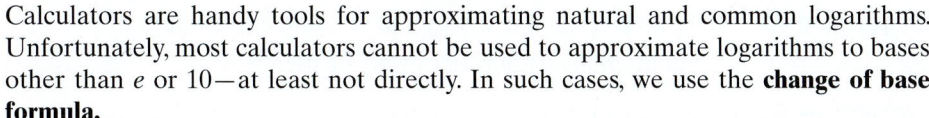

Calculators are handy tools for approximating natural and common logarithms. Unfortunately, most calculators cannot be used to approximate logarithms to bases other than e or 10—at least not directly. In such cases, we use the **change of base formula.**

Change of Base

If a, b, and c are positive real numbers and neither b nor c is 1, then

$$\log_b a = \frac{\log_c a}{\log_c b}$$

Answers

10. $\left\{\dfrac{e^{10}}{7}\right\}$; $\{3146.6380\}$

11. $1525.50

Practice 12

Approximate $\log_7 5$ to four decimal places.

Answer

12. 0.8271

✓ **Concept Check Answer**

$$f(x) = \frac{\log x}{\log 5}$$

Example 12 Approximate $\log_5 3$ to four decimal places.

Solution: We use the change of base property to write $\log_5 3$ as a quotient of logarithms to base 10.

$$\log_5 3 = \frac{\log 3}{\log 5} \qquad \text{Use the change of base property.}$$

$$\approx \frac{0.4771213}{0.69897} \qquad \text{Approximate the logarithms by calculator.}$$

$$\approx 0.6826063 \qquad \text{Simplify by calculator.}$$

To four decimal places, $\log_5 3 \approx 0.6826$.

■ **Work Practice 12**

✓ **Concept Check** If a graphing calculator cannot directly evaluate logarithms to base 5, describe how you could use the graphing calculator to graph the function $f(x) = \log_5 x$.

Vocabulary, Readiness & Video Check

Use the choices to fill in each blank.

1. The base of log 7 is ____10____. c
    a. e **b.** 7   **c.** 10   **d.** no answer

2. The base of ln 7 is ____e____. a
    a. e **b.** 7   **c.** 10   **d.** no answer

3. $\log_{10} 10^7 = $ ____7____. b
    a. e **b.** 7   **c.** 10   **d.** no answer

4. $\log_7 1 = $ ____0____. d
    a. e **b.** 7   **c.** 10   **d.** 0

5. $\log_e e^5 = $ ____5____. b
    a. e **b.** 5   **c.** 0   **d.** 1

6. Study exercise 5 to the left. Then answer:
    $\ln e^5 = $ ____5____. b
    a. e **b.** 5   **c.** 0   **d.** 1

7. $\log_2 7 = $ ____$\dfrac{\log 7}{\log 2} = \dfrac{\ln 7}{\ln 2}$____ (There may be more than one answer.) a, b

    a. $\dfrac{\log 7}{\log 2}$ **b.** $\dfrac{\ln 7}{\ln 2}$ **c.** $\dfrac{\log 2}{\log 7}$ **d.** $\log \dfrac{7}{2}$

Martin-Gay Interactive Videos

See Video 12.7

See video answer section.

Watch the section lecture video and answer the following questions.

Objective A 8. From Example 1 and the lecture before, what is the understood base of a common logarithm?

Objective B 9. From Example 2, why can you find exact values of common logarithms of powers of 10?

Objective C 10. From Example 4 and the lecture before, what is the understood base of a natural logarithm?

Objective D 11. In Examples 5 and 6, consider how the expression is rewritten and the resulting answer. What logarithm property is actually used here?

Objective E 12. From Example 8, what two equivalent fractions will give you the exact value of $\log_6 4$?

12.7 Exercise Set MyMathLab®

Objectives A C Mixed Practice *Use a calculator to approximate each logarithm to four decimal places. See Examples 1 and 7.*

▶ **1.** log 8   0.9031

2. log 6   0.7782

3. log 2.31   0.3636

4. log 4.86   0.6866

5. ln 2   0.6931

6. ln 3   1.0986

7. ln 0.0716   −2.6367

8. ln 0.0032   −5.7446

9. log 12.6   1.1004

10. log 25.9   1.4133

▶ **11.** ln 5   1.6094

12. ln 7   1.9459

13. log 41.5   1.6180

14. ln 41.5   3.7257

Objectives B D Mixed Practice *Find the exact value of each logarithm. See Examples 2 through 5, 8, and 9.*

▶ **15.** log 100   2

16. log 10,000   4

17. $\log \frac{1}{1000}$ −3

18. $\log \frac{1}{10}$ −1

▶ **19.** ln e^2 2

20. ln e^4 4

21. ln $\sqrt[4]{e}$ $\frac{1}{4}$

22. ln $\sqrt[5]{e}$ $\frac{1}{5}$

23. log 10^3 3

24. ln e^5 5

25. ln $e^{3.1}$ 3.1

26. log 10^7 7

27. log 0.0001   −4

28. log 0.001   −3

▶ **29.** ln $\sqrt{e}$ $\frac{1}{2}$

30. log $\sqrt{10}$ $\frac{1}{2}$

Solve each equation. Give the exact solution and a four-decimal-place approximation. See Examples 6 and 10.

31. log x = 1.3 $\{10^{1.3}\}$; $\{19.9526\}$

32. log x = 2.1 $\{10^{2.1}\}$; $\{125.8925\}$

33. ln x = 1.4 $\{e^{1.4}\}$; $\{4.0552\}$

34. ln x = 2.1 $\{e^{2.1}\}$; $\{8.1662\}$

35. log x = 2.3 $\{10^{2.3}\}$; $\{199.5262\}$

36. log x = 3.1 $\{10^{3.1}\}$; $\{1258.9254\}$

37. ln x = −2.3 $\{e^{-2.3}\}$; $\{0.1003\}$

38. ln x = −3.7 $\{e^{-3.7}\}$; $\{0.0247\}$

▶ **39.** log 2x = 1.1 $\left\{\frac{10^{1.1}}{2}\right\}$; $\{6.2946\}$

40. log 3x = 1.3 $\left\{\frac{10^{1.3}}{3}\right\}$; $\{6.6509\}$

41. ln 4x = 0.18 $\left\{\frac{e^{0.18}}{4}\right\}$; $\{0.2993\}$

42. ln 3x = 0.76 $\left\{\frac{e^{0.76}}{3}\right\}$; $\{0.7128\}$

43. ln(3x − 4) = 2.3 $\left\{\frac{4 + e^{2.3}}{3}\right\}$; $\{4.6581\}$

44. ln(2x + 5) = 3.4 $\left\{\frac{e^{3.4} - 5}{2}\right\}$; $\{12.4821\}$

45. log(2x + 1) = −0.5 $\left\{\frac{10^{-0.5} - 1}{2}\right\}$; $\{-0.3419\}$

46. log(3x − 2) = −0.8 $\left\{\frac{10^{-0.8} + 2}{3}\right\}$; $\{0.7195\}$

Use the formula $A = Pe^{rt}$ to solve. See Example 11.

▶ **47.** How much money does Dana Jones have after 12 years if she invests $1400 at 8% interest compounded continuously? $3656.38

48. Determine the size of an account in which $3500 earns 6% interest compounded continuously for 1 year. $3716.43

49. How much money does Barbara Mack owe at the end of 4 years if 6% interest is compounded continuously on her $2000 debt? $2542.50

50. Find the amount of money for which a $2500 certificate of deposit is redeemable if it has been earning 10% interest compounded continuously for 3 years. $3374.65

Objective E *Approximate each logarithm to four decimal places. See Example 12.*

51. $\log_2 3$ 1.5850   **52.** $\log_3 2$ 0.6309   **53.** $\log_8 6$ 0.8617   **54.** $\log_6 8$ 1.1606   ▶**55.** $\log_4 9$ 1.5850

56. $\log_9 4$ 0.6309   **57.** $\log_3 \dfrac{1}{6}$ −1.6309   **58.** $\log_6 \dfrac{2}{3}$ −0.2263   **59.** $\log_{1/2} 5$ −2.3219   **60.** $\log_{1/3} 2$ −0.6309

Review

Solve for x. See Sections 2.3, 2.5, and 6.6.

61. $6x - 3(2 - 5x) = 6$ $\left\{\dfrac{4}{7}\right\}$

62. $2x + 3 = 5 - 2(3x - 1)$ $\left\{\dfrac{1}{2}\right\}$

63. $2x + 3y = 6x$ $x = \dfrac{3y}{4}$

64. $4x - 8y = 10x$ $x = -\dfrac{4y}{3}$

65. $x^2 + 7x = -6$ $\{-6, -1\}$

66. $x^2 + 4x = 12$ $\{-6, 2\}$

Concept Extensions

67. Use a calculator to try to approximate log 0. Describe what happens and explain why.   answers may vary

68. Use a calculator to try to approximate ln 0. Describe what happens and explain why.   answers may vary

Graph each function by finding ordered pair solutions, plotting the solutions, and then drawing a smooth curve through the plotted points.

69. $f(x) = e^x$

70. $f(x) = e^{2x}$

71. $f(x) = \ln x$

72. $f(x) = \log x$

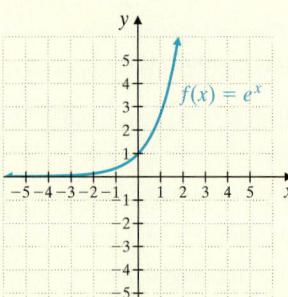
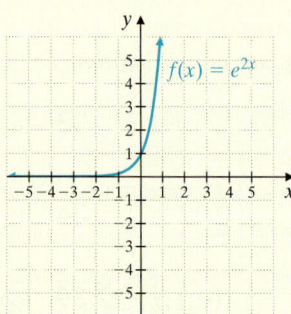
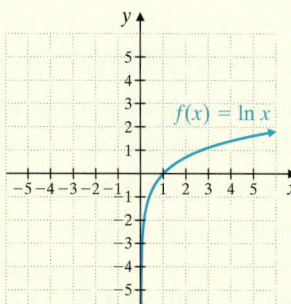
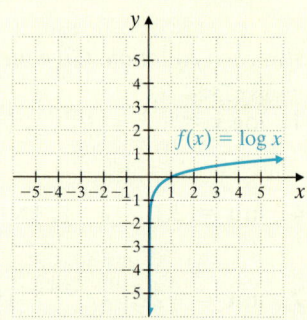

Solve.

73. Without using a calculator, explain which of log 50 or ln 50 must be larger.   answers may vary

74. Without using a calculator, explain which of log 50^{-1} or ln 50^{-1} must be larger.   answers may vary

The Richter scale measures the intensity, or magnitude, of an earthquake. The formula for the magnitude R of an earthquake is $R = \log\left(\dfrac{a}{T}\right) + B$, where a is the amplitude in micrometers of the vertical motion of the ground at the recording station, T is the number of seconds between successive seismic waves, and B is an adjustment factor that takes into account the weakening of the seismic wave as the distance increases from the epicenter of the earthquake.

Use the Richter scale formula to find the magnitude R of the earthquake that fits the description given. Round answers to one decimal place.

75. Amplitude *a* is 200 micrometers, time *T* between waves is 1.6 seconds, and *B* is 2.1.   4.2

76. Amplitude *a* is 150 micrometers, time *T* between waves is 3.6 seconds, and *B* is 1.9.   3.5

77. Amplitude *a* is 400 micrometers, time *T* between waves is 2.6 seconds, and *B* is 3.1.   5.3

78. Amplitude *a* is 450 micrometers, time *T* between waves is 4.2 seconds, and *B* is 2.7.   4.7

12.8 Exponential and Logarithmic Equations and Problem Solving

Objective A Solving Exponential Equations

Objectives

A Solve Exponential Equations.

B Solve Logarithmic Equations.

C Solve Problems That Can Be Modeled by Exponential and Logarithmic Equations.

In Section 12.3, we solved exponential equations such as $2^x = 16$ by writing both sides in terms of the same base. Here, we write 16 as a power of 2 and using the uniqueness of b^x.

$$2^x = 16$$
$$2^x = 2^4 \quad \text{Write 16 as } 2^4.$$
$$x = 4 \quad \text{Use the uniqueness of } b^x.$$

How do we solve an exponential equation when the bases cannot easily be written the same? For example, how do we solve an equation such as $3^x = 7$? We use the fact that $f(x) = \log_b x$ is a one-to-one function. Another way of stating this fact is as a property of equality.

> **Logarithm Property of Equality**
>
> Let a, b, and c be real numbers such that $\log_b a$ and $\log_b c$ are real numbers and b is not 1. Then
>
> $$\log_b a = \log_b c \quad \text{is equivalent to} \quad a = c$$

Example 1 Solve: $3^x = 7$. Give the exact answer and a four-decimal-place approximation.

Solution: We use the logarithm property of equality and take the logarithm of both sides. For this example, we use the common logarithm.

$$3^x = 7$$
$$\log 3^x = \log 7 \quad \text{Take the common log of both sides.}$$
$$x \log 3 = \log 7 \quad \text{Use the power property of logarithms.}$$
$$x = \frac{\log 7}{\log 3} \quad \text{Divide both sides by } \log 3.$$

The exact solution is $\dfrac{\log 7}{\log 3}$. When we approximate to four decimal places, we have

$$\frac{\log 7}{\log 3} \approx \frac{0.845098}{0.477121} \approx 1.7712$$

The solution set is $\left\{ \dfrac{\log 7}{\log 3} \right\}$, or approximately $\{1.7712\}$.

■ **Work Practice 1**

Practice 1

Solve: $2^x = 5$. Give the exact answer and a four-decimal-place approximation.

Teaching Tip

Point out that logarithms are used to solve equations in which the variable is in the exponent.

Teaching Tip

Point out that Example 1 can be solved by taking the logarithm in any base. Then have the students solve for x using the natural log.

Objective B Solving Logarithmic Equations

By applying the appropriate properties of logarithms, we can solve a broad variety of logarithmic equations.

Answer

1. $\left\{ \dfrac{\log 5}{\log 2} \right\}$; $\{2.3219\}$

Practice 2

Solve: $\log_3(x + 5) = 2$

Example 2   Solve: $\log_4(x - 2) = 2$

Solution: Notice that $x - 2$ must be positive, so x must be greater than 2. With this in mind, we first write the equation with exponential notation.

$$\log_4(x - 2) = 2$$
$$4^2 = x - 2$$
$$16 = x - 2$$
$$18 = x \qquad \text{Add 2 to both sides.}$$

To check, we replace x with 18 in the original equation.

$$\log_4(x - 2) = 2$$
$$\log_4(18 - 2) \stackrel{?}{=} 2 \qquad \text{Let } x = 18.$$
$$\log_4 16 \stackrel{?}{=} 2$$
$$4^2 = 16 \qquad \text{True}$$

The solution set is $\{18\}$.

■ **Work Practice 2**

Practice 3

Solve: $\log_6 x + \log_6(x + 1) = 1$

Example 3   Solve: $\log_2 x + \log_2(x - 1) = 1$

Solution: Notice that $x - 1$ must be positive, so x must be greater than 1. We use the product property on the left side of the equation.

$$\log_2 x + \log_2(x - 1) = 1$$
$$\log_2[x(x - 1)] = 1 \qquad \text{Use the product property.}$$
$$\log_2(x^2 - x) = 1$$

Next we write the equation with exponential notation and solve for x.

$$2^1 = x^2 - x$$
$$0 = x^2 - x - 2 \qquad\qquad \text{Subtract 2 from both sides.}$$
$$0 = (x - 2)(x + 1) \qquad\qquad \text{Factor.}$$
$$0 = x - 2 \quad \text{or} \quad 0 = x + 1 \quad \text{Set each factor equal to 0.}$$
$$2 = x \qquad\qquad -1 = x$$

Recall that -1 cannot be a solution because x must be greater than 1. If we forgot this, we would still reject -1 after checking. To see this, we replace x with -1 in the original equation.

$$\log_2 x + \log_2(x - 1) = 1$$
$$\log_2(-1) + \log_2(-1 - 1) \stackrel{?}{=} 1 \quad \text{Let } x = -1.$$

Because the logarithm of a negative number is undefined, -1 is rejected. Check to see that the solution set is $\{2\}$.

■ **Work Practice 3**

Answers

2. $\{4\}$ **3.** $\{2\}$

Example 4 Solve: $\log(x + 2) - \log x = 2$

Solution: We use the quotient property of logarithms on the left side of the equation.

$$\log(x + 2) - \log x = 2$$

$$\log\frac{x + 2}{x} = 2 \qquad \text{Use the quotient property.}$$

$$10^2 = \frac{x + 2}{x} \qquad \text{Write using exponential notation.}$$

$$100 = \frac{x + 2}{x}$$

$$100x = x + 2 \qquad \text{Multiply both sides by } x.$$

$$99x = 2 \qquad \text{Subtract } x \text{ from both sides.}$$

$$x = \frac{2}{99} \qquad \text{Divide both sides by 99.}$$

Check to see that the solution set is $\left\{\frac{2}{99}\right\}$.

■ **Work Practice 4**

Objective C Solving Problems Modeled by Exponential and Logarithmic Equations

Logarithmic and exponential functions are used in a variety of scientific, technical, and business settings. A few examples follow.

Example 5 Estimating Population Size

The population size y of a community of lemmings varies according to the relationship $y = y_0 e^{0.15t}$. In this formula, t is time in months, and y_0 is the initial population at time 0. Estimate the population after 6 months if there were originally 5000 lemmings.

Solution: We substitute 5000 for y_0 and 6 for t.

$$y = y_0 e^{0.15t}$$

$$= 5000 e^{0.15(6)} \qquad \text{Let } t = 6 \text{ and } y_0 = 5000.$$

$$= 5000 e^{0.9} \qquad \text{Multiply.}$$

Using a calculator, we find that $y \approx 12{,}298.016$. In 6 months the population will be approximately 12,300 lemmings.

■ **Work Practice 5**

Example 6 Doubling an Investment

How long does it take an investment of \$2000 to double if it is invested at 5% interest compounded quarterly? The necessary formula is $A = P\left(1 + \frac{r}{n}\right)^{nt}$, where

A is the accrued amount, P is the principal invested, r is the annual rate of interest, n is the number of compounding periods per year, and t is the number of years.

(Continued on next page)

Practice 4

Solve:
$\log(x + 1) - \log x = 1$

Practice 5

Use the equation in Example 5 to estimate the lemming population in 8 months.

Practice 6

How long does it take an investment of \$1000 to double if it is invested at 6% interest compounded quarterly?

Answers

4. $\left\{\frac{1}{9}\right\}$ **5.** approximately 16,600

lemmings **6.** $11\frac{3}{4}$ yr

Teaching Tip

After discussing the solution to Example 6, have students find how long it takes an investment of $8000 to double at the given rate. Then ask, "How long will it take any investment to double at the given rate?"

Solution: We are given that $P = \$2000$ and $r = 5\% = 0.05$. Compounding quarterly means 4 times a year, so $n = 4$. The investment is to double, so A must be $4000. We substitute these values and solve for t.

$$A = P\left(1 + \frac{r}{n}\right)^{nt}$$

$$4000 = 2000\left(1 + \frac{0.05}{4}\right)^{4t}$$ Substitute known values.

$$4000 = 2000(1.0125)^{4t}$$ Simplify $1 + \frac{0.05}{4}$.

$$2 = (1.0125)^{4t}$$ Divide both sides by 2000.

$$\log 2 = \log 1.0125^{4t}$$ Take the logarithm of both sides.

$$\log 2 = 4t(\log 1.0125)$$ Use the power property.

$$\frac{\log 2}{4 \log 1.0125} = t$$ Divide both sides by 4 log 1.0125.

$$13.949408 \approx t$$ Approximate by calculator.

It takes approximately 14 years for the money to double in value.

■ **Work Practice 6**

 Calculator Explorations **Graphing**

Use a grapher to find how long it takes an investment of $1500 to triple if it is invested at 8% interest compounded monthly. First, let $P = \$1500$, $r = 0.08$, and $n = 12$ (for monthly compounding) in the formula

$$A = P\left(1 + \frac{r}{n}\right)^{nt}$$

Notice that when the investment has tripled, the accrued amount A is $4500. Thus,

$$4500 = 1500\left(1 + \frac{0.08}{12}\right)^{12t}$$

Determine an appropriate viewing window and enter and graph the equations

$$Y_1 = 1500\left(1 + \frac{0.08}{12}\right)^{12x}$$

and

$$Y_2 = 4500$$

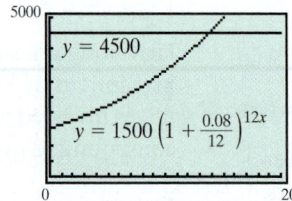

The point of intersection of the two curves is the solution. The x-coordinate tells how long it takes for the investment to triple.

Use a TRACE feature or an INTERSECT feature to approximate the coordinates of the point of intersection of the two curves. It takes approximately 13.78 years, or 13 years and 10 months, for the investment to triple in value to $4500.

Use this graphical solution method to solve each problem. Round each answer to the nearest hundredth. Because of rounding and the timing of interest earned, the actual answers may be slightly different from the answers found using this method.

1. Find how long it takes an investment of $5000 to grow to $6000 if it is invested at 5% interest compounded quarterly.
    3.67 yr, or 3 yr and 8 mo

2. Find how long it takes an investment of $1000 to double if it is invested at 4.5% interest compounded daily. (Use 365 days in a year.)
    15.40 yr, or 15 yr and 5 mo

3. Find how long it takes an investment of $10,000 to quadruple if it is invested at 6% interest compounded monthly.
    23.16 yr, or 23 yr and 2 mo

4. Find how long it takes $500 to grow to $800 if it is invested at 4% interest compounded semiannually.
    11.87 yr, or 11 yr and 11 mo

Vocabulary, Readiness & Video Check

Martin-Gay Interactive Videos   *Watch the section lecture video and answer the following questions.*

Objective A 1. From the lecture before ⊞ Example 1, explain why
ln (4x − 2) = ln 3 is equivalent to 4x − 2 = 3.

Objective B 2. Why is the possible solution of −8 rejected in
⊞ Example 3? ▶

Objective C 3. For ⊞ Example 4, write the equation and find the number of
years it takes $1000 to double at 7% interest compounded
monthly. Explain the similarity to the answer to Example 4.
Round your answer to the nearest tenth. ▶

See Video 12.8 🍒

See video answer section.

12.8 Exercise Set MyMathLab® ▶

Objective A *Solve each equation. Give the exact solution and a four-decimal-place approximation. See Example 1.*

1. $3^x = 6$
$\left\{\dfrac{\log 6}{\log 3}\right\}$; {1.6309}

2. $4^x = 7$
$\left\{\dfrac{\log 7}{\log 4}\right\}$; {1.4037}

3. $9^x = 5$
$\left\{\dfrac{\log 5}{\log 9}\right\}$; {0.7325}

4. $3^x = 11$
$\left\{\dfrac{\log 11}{\log 3}\right\}$; {2.1827}

5. $3^{2x} = 3.8$
$\left\{\dfrac{\log 3.8}{2 \log 3}\right\}$; {0.6076}

6. $5^{3x} = 5.6$
$\left\{\dfrac{\log 5.6}{3 \log 5}\right\}$; {0.3568}

7. $e^{6x} = 5$
$\left\{\dfrac{\ln 5}{6}\right\}$; {0.2682}

8. $e^{2x} = 8$
$\left\{\dfrac{\ln 8}{2}\right\}$; {1.0397}

▶**9.** $2^{x-3} = 5$
$\left\{3 + \dfrac{\log 5}{\log 2}\right\}$; {5.3219}

10. $8^{x-2} = 12$
$\left\{\dfrac{\log 12}{\log 8} + 2\right\}$; {3.1950}

11. $4^{x+7} = 3$
$\left\{\dfrac{\log 3}{\log 4} - 7\right\}$; {−6.2075}

12. $6^{x+3} = 2$
$\left\{\dfrac{\log 2}{\log 6} - 3\right\}$; {−2.6131}

13. $7^{3x-4} = 11$
$\left\{\dfrac{1}{3}\left(4 + \dfrac{\log 11}{\log 7}\right)\right\}$; {1.7441}

14. $5^{2x-6} = 12$
$\left\{\dfrac{1}{2}\left(6 + \dfrac{\log 12}{\log 5}\right)\right\}$; {3.7720}

Objective B *Solve each equation. See Examples 2 through 4.*

▶**15.** $\log_2(x + 5) = 4$
{11}

16. $\log_2(x - 5) = 3$
{13}

17. $\log_4 2 + \log_4 x = 0$
$\left\{\dfrac{1}{2}\right\}$

18. $\log_3 5 + \log_3 x = 1$
$\left\{\dfrac{3}{5}\right\}$

19. $\log_2 6 - \log_2 x = 3$
$\left\{\dfrac{3}{4}\right\}$

20. $\log_4 10 - \log_4 x = 2$
$\left\{\dfrac{5}{8}\right\}$

21. $\log_2(x^2 + x) = 1$
{−2, 1}

22. $\log_6(x^2 - x) = 1$
{−2, 3}

▶**23.** $\log_4 x + \log_4(x + 6) = 2$
{2}

24. $\log_3 x + \log_3(x + 6) = 3$
{3}

25. $\log_5(x + 3) - \log_5 x = 2$
$\left\{\dfrac{1}{8}\right\}$

26. $\log_6(x + 2) - \log_6 x = 2$
$\left\{\dfrac{2}{35}\right\}$

27. $\log_4(x^2 - 3x) = 1$
{4, −1}

28. $\log_8(x^2 - 2x) = 1$
{−2, 4}

29. $\log_2 x + \log_2(3x + 1) = 1$
$\left\{\dfrac{2}{3}\right\}$

30. $\log_3 x + \log_3(x - 8) = 2$ {9}

Objective C *Solve. See Example 5.*

31. The size of the wolf population at Isle Royale National Park increases according to the formula $y = y_0 e^{0.043t}$. In this formula, t is time in years and y_0 is the initial population at time 0. If the size of the current population is 83 wolves, find how many there should be in 5 years. Round to the nearest whole number.   103 wolves

32. The number of victims of a flu epidemic is increasing according to the formula $y = y_0 e^{0.075t}$. In this formula, t is time in weeks and y_0 is the infected population at time 0. If 20,000 people are currently infected, how many might be infected in 3 weeks? Round to the nearest whole number.   25,046

33. The population of Brazil is increasing according to the formula $y = y_0 e^{0.009t}$. In this formula, t is time in years and y_0 is the initial population at time 0. If the size of the population in 2009 was 191 million, use the formula to predict the population of Brazil in year 2020. Round to the nearest tenth of a million. (*Source:* World Bank)   210.9 million

34. The population of the Faeroe Islands is decreasing according to the formula $y = y_0 e^{-0.001t}$. In this formula, t is the time in years and y_0 is the initial population at time 0. If the size of the population in 2009 was 48,778, use the formula to predict the population of the Faeroe Islands in the year 2025. Round to the nearest whole number. (*Source:* World Bank and ArcticStat)   48,004

35. The population of the Cook Islands is decreasing according to the formula $y = y_0 e^{-0.0277t}$. In this formula, t is time in years and y_0 is the initial population at time 0. If the size of the population in 2009 was 11,870, use the formula to predict the population of Cook Islands in the year 2025. Round to the nearest whole number. (*Source:* The World Almanac)   7620

36. The population of Saint Barthelemy is decreasing according to the formula $y = y_0 e^{-0.0034t}$. In this formula, t is time in years and y_0 is the initial population at time 0. If the size of the population in 2009 was 7448, use the formula to predict the population of Saint Barthelemy in the year 2025. Round to the nearest whole number. (*Source:* The World Almanac)   7054

Use the formula $A = P\left(1 + \dfrac{r}{n}\right)^{nt}$ *to solve these compound interest problems. Round to the nearest tenth. See Example 6.*

▶ **37.** How long does it take for $600 to double if it is invested at 7% interest compounded monthly?   9.9 yr

38. How long does it take for $600 to double if it is invested at 12% interest compounded monthly?   5.8 yr

39. How long does it take for a $1200 investment to earn $200 interest if it is invested at 9% interest compounded quarterly?   1.7 yr

40. How long does it take for a $1500 investment to earn $200 interest if it is invested at 10% interest compounded semiannually?   1.3 yr

41. How long does it take for $1000 to double if it is invested at 8% interest compounded semiannually?   8.8 yr

42. How long does it take for $1000 to double if it is invested at 8% interest compounded monthly?   8.7 yr

The formula $w = 0.00185h^{2.67}$ *is used to estimate the normal weight w in pounds of a boy h inches tall. Use this formula to solve Exercises 43 and 44. Round to the nearest tenth.*

43. Find the expected height of a boy who weighs 85 pounds.   55.7 in.

44. Find the expected height of a boy who weighs 140 pounds.   67.2 in.

The formula $P = 14.7e^{-0.21x}$ gives the average atmospheric pressure P, in pounds per square inch, at an altitude x, in miles above sea level. Use this formula to solve Exercises 45 through 48. Round to the nearest tenth.

45. Find the average atmospheric pressure of Denver, which is 1 mile above sea level.   11.9 lb per sq in.

46. Find the average atmospheric pressure of Pikes Peak, which is 2.7 miles above sea level.   8.3 lb per sq in.

47. Find the elevation of a Delta jet if the atmospheric pressure outside the jet is 7.5 pounds per square inch.   3.2 mi

48. Find the elevation of a remote Himalayan peak if the atmospheric pressure atop the peak is 6.5 pounds per square inch.   3.9 mi

Psychologists call the graph of the formula $t = \dfrac{1}{c} \ln\left(\dfrac{A}{A - N}\right)$ the learning curve since the formula relates time t passed, in weeks, to a measure N of learning achieved, to a measure A of maximum learning possible, and to a measure c of an individual's learning style. Use this formula to answer Exercises 49 through 52. Round to the nearest whole number.

49. Norman Weidner is learning to type. If he wants to type at a rate of 50 words per minute ($N = 50$) and his expected maximum rate is 75 words per minute ($A = 75$), how many weeks should it take him to achieve his goal? Assume that c is 0.09.   12 weeks

50. An experiment of teaching chimpanzees sign language shows that a typical chimp can master a maximum of 65 signs. How many weeks should it take a chimpanzee to master 30 signs if c is 0.03?   21 weeks

51. Janine Jenkins is working on her dictation skills. She wants to take dictation at a rate of 150 words per minute and believes that the maximum rate she can hope for is 210 words per minute. How many weeks should it take her to achieve the 150-word level if c is 0.07?   18 weeks

52. A psychologist is measuring human capability to memorize nonsense syllables. How many weeks should it take a subject to learn 15 nonsense syllables if the maximum possible to learn is 24 syllables and c is 0.17?   6 weeks

Review

If $x = -2, y = 0,$ and $z = 3,$ find the value of each expression. See Section 1.6.

53. $\dfrac{x^2 - y + 2z}{3x}$ $-\dfrac{5}{3}$

54. $\dfrac{x^3 - 2y + z}{2z}$ $-\dfrac{5}{6}$

55. $\dfrac{3z - 4x + y}{x + 2z}$ $\dfrac{17}{4}$

56. $\dfrac{4y - 3x + z}{2x + y}$ $-\dfrac{9}{4}$

Concept Extensions

The formula $y = y_0 e^{kt}$ gives the population size y of a population that experiences a relative growth rate k (k is positive if growth is increasing and k is negative if growth is decreasing). In this formula, t is time in years and y_0 is the initial population at time 0. Use this formula to solve Exercises 57 and 58. Round answers to the nearest year.

57. In 2009, the population of Michigan was approximately 9,970,000 and decreasing according to the formula $y = y_0 e^{-0.003t}$. Assume that the population continues to decrease according to the given formula and predict how many years after which the population of Michigan will be 9,500,000. (*Hint:* Let $y_0 = 9,970,000; y = 9,500,000,$ and solve for t.)   16 yr

58. In 2009, the population of Illinois was approximately 12,910,000 and increasing according to the formula $y = y_0 e^{0.005t}$. Assume that the population continues to increase according to the given formula and predict how many years after which the population of Illinois will be 13,500,000. (See the Hint for Exercise 57.) (*Source:* U.S. Census Bureau and Federal Reserve Bank of Chicago)   9 yr

Solve.

59. When solving a logarithmic equation, explain why you must check possible solutions in the original equation.   answers may vary

60. Solve $5^x = 9$ by taking the common logarithm of both sides of the equation. Next, solve this equation by taking the natural logarithm of both sides. Compare your solutions. Are they the same? Why or why not?   answers may vary

Use a graphing calculator to solve. Round your answers to two decimal places.

61. $e^{0.3x} = 8$ 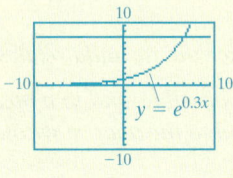 $\{6.93\}$

62. $10^{0.5x} = 7$ 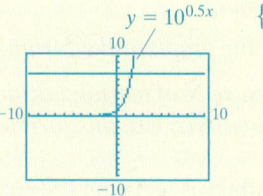 $\{1.69\}$

Chapter 12 Group Activity

Additional group activities are available in the *Instructor's Resource Manual with Tests.*

Sound Intensity

The decibel (dB) measures sound intensity, or the relative loudness or strength of a sound. One decibel is the smallest difference in sound levels that is detectable by humans. The decibel is a logarithmic unit. This means that for approximately every 3-decibel increase in sound intensity, the relative loudness of the sound is doubled. For example, a 35 dB sound is twice as loud as a 32 dB sound.

In the modern world, noise pollution has increasingly become a concern. Sustained exposure to high sound intensities can lead to hearing loss. Regular exposure to 90 dB sounds can eventually lead to loss of hearing. Sounds of 130 dB and more can cause permanent loss of hearing instantaneously.

The relative loudness of a sound D in decibels is given by the equation

$$D = 10 \log_{10} \frac{I}{10^{-16}}$$

where I is the intensity of a sound given in watts per square centimeter. Some sound intensities of common noises are listed in the table in order of increasing sound intensity.

Group Activity

1. Work together to create a table of the relative loudness (in decibels) of the sounds listed in the table.

2. Research the loudness of other common noises. Add these sounds and their decibel levels to your table. Be sure to list the sounds in order of increasing sound intensity.

Some Sound Intensities of Common Noises	
Noise	**Intensity (watts/cm²)**
Whispering	10^{-15}
Rustling leaves	$10^{-14.2}$
Normal conversation	10^{-13}
Background noise in a quiet residence	$10^{-12.2}$
Normal office noise or a quiet stream	10^{-11}
Air conditioning	10^{-10}
Freight train at 50 feet	$10^{-8.5}$
Vacuum cleaner	10^{-8}
Nearby thunder	10^{-7}
Air hammer	$10^{-6.5}$
Jet plane at takeoff	10^{-6}
Threshold of pain	10^{-4}

Chapter 12 Vocabulary Check

Fill in each blank with one of the words or phrases listed below. Some words or phrases may be used more than once.

inverse	common	composition	symmetric	exponential
vertical	logarithmic	natural	half-life	horizontal

1. For a one-to-one function, we can find its ___inverse___ function by switching the coordinates of the ordered pairs of the function.

2. The ___composition___ of functions f and g is $(f \circ g)(x) = f(g(x))$.

3. A function of the form $f(x) = b^x$ is called a(n) ___exponential___ function if $b > 0$, b is not 1, and x is a real number.

4. The graphs of f and f^{-1} are ___symmetric___ about the line $y = x$.

5. ___Natural___ logarithms are logarithms to base e.

6. ___Common___ logarithms are logarithms to base 10.

7. To see whether a graph is the graph of a one-to-one function, apply the ___vertical___ line test to see whether it is a function, and then apply the ___horizontal___ line test to see whether it is a one-to-one function.

8. A(n) ___logarithmic___ function is a function that can be defined by $f(x) = \log_b x$ where x is a positive real number, b is a constant positive real number, and b is not 1.

9. ___Half-life___ is the amount of time it takes for half of the amount of a substance to decay.

10. A quantity that grows or decays by the same percent at regular time periods is said to have ___exponential___ growth or decay.

Helpful Hint

▶ Are you preparing for your test? Don't forget to take the Chapter 12 Test on page 917. Then check your answers at the back of the text and use the Chapter Test Prep Videos to see the fully worked-out solutions to any of the exercises you want to review.

12 Chapter Highlights

Definitions and Concepts	Examples
Section 12.1 The Algebra of Functions	

Algebra of Functions	If $f(x) = 7x$ and $g(x) = x^2 + 1$,
Let f and g be functions.	$(f + g)(x) = f(x) + g(x) = 7x + x^2 + 1$
Sum $\quad (f + g)(x) = f(x) + g(x)$	$(f - g)(x) = f(x) - g(x) = 7x - (x^2 + 1)$
Difference $\quad (f - g)(x) = f(x) - g(x)$	$\qquad\qquad\qquad\qquad\quad = 7x - x^2 - 1$
Product $\quad (f \cdot g)(x) = f(x) \cdot g(x)$	$(f \cdot g)(x) = f(x) \cdot g(x) = 7x(x^2 + 1)$
Quotient $\quad \left(\dfrac{f}{g}\right)(x) = \dfrac{f(x)}{g(x)}, g(x) \neq 0$	$\qquad\qquad\qquad\qquad = 7x^3 + 7x^2$
	$\left(\dfrac{f}{g}\right)(x) = \dfrac{f(x)}{g(x)} = \dfrac{7x}{x^2 + 1}$
Composite Functions	If $f(x) = x^2 + 1$ and $g(x) = x - 5$, find $(f \circ g)(x)$.
The notation $(f \circ g)(x)$ means "f composed with g."	$(f \circ g)(x) = f(g(x))$
$\quad (f \circ g)(x) = f(g(x))$	$\qquad\qquad = f(x - 5)$
$\quad (g \circ f)(x) = g(f(x))$	$\qquad\qquad = (x - 5)^2 + 1$
	$\qquad\qquad = x^2 - 10x + 26$

Definitions and Concepts	Examples

Section 12.2   Inverse Functions

One-to-One Function

If f is a function, then f is a **one-to-one function** only if each y-value (output) corresponds to only one x-value (input).

Horizontal Line Test

If every horizontal line intersects the graph of a function at most once, then the function is a one-to-one function.

Determine whether each graph is a one-to-one function.

A 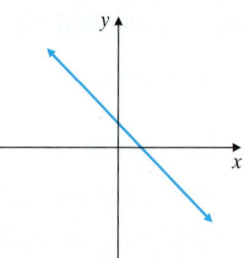   **B**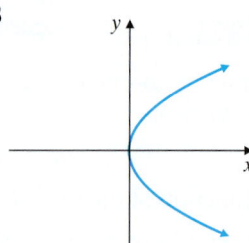

C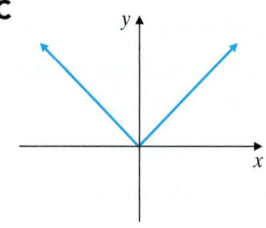

Graphs A and C pass the vertical line test, so only these are graphs of functions. Of graphs A and C, only graph A passes the horizontal line test, so only graph A is the graph of a one-to-one function.

The **inverse** of a one-to-one function f is the one-to-one function f^{-1} that is the set of all ordered pairs (b, a) such that (a, b) belongs to f.

Finding the Inverse of a One-to-One Function f

Step 1: Replace $f(x)$ with y.

Step 2: Interchange x and y.

Step 3: Solve for the equation for y.

Step 4: Replace y with the notation $f^{-1}(x)$.

Find the inverse of $f(x) = 2x + 7$.

$$y = 2x + 7 \quad \text{Replace } f(x) \text{ with } y.$$
$$x = 2y + 7 \quad \text{Interchange } x \text{ and } y.$$
$$2y = x - 7 \quad \text{Solve for } y.$$
$$y = \frac{x - 7}{2}$$
$$f^{-1}(x) = \frac{x - 7}{2} \quad \text{Replace } y \text{ with } f^{-1}(x).$$

The inverse of $f(x) = 2x + 7$ is $f^{-1}(x) = \dfrac{x - 7}{2}$.

Section 12.3   Exponential Functions

Exponential Function

A function of the form $f(x) = b^x$ is an **exponential function,** where $b > 0$, $b \neq 1$, and x is a real number.

Graph the exponential function $y = 4^x$.

x	y
-2	$\dfrac{1}{16}$
-1	$\dfrac{1}{4}$
0	1
1	4
2	16

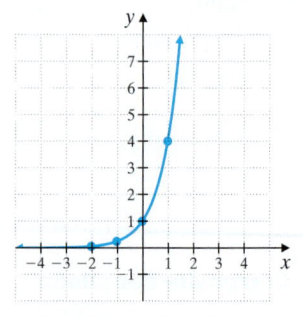

Definitions and Concepts	Examples

Section 12.3　Exponential Functions (*continued*)

Uniqueness of b^x

If $b > 0$ and $b \neq 1$, then $b^x = b^y$ is equivalent to $x = y$.

Solve:　$2^{x+5} = 8$

$\quad 2^{x+5} = 2^3$　Write 8 as 2^3.

$\quad\quad x + 5 = 3$　Use the uniqueness of b^x.

$\quad\quad\quad\quad x = -2$　Subtract 5 from both sides.

Section 12.4　Exponential Growth and Decay Functions

A quantity that grows or decays by the same percent at regular time periods is said to have **exponential growth** or **exponential decay**.

Exponential Growth

initial amount　number of time intervals

$$y = C(1 + r)^x$$

$(1 + r)$ is growth factor; r is growth rate (often a percent)

Exponential Decay

initial amount　number of time intervals

$$y = C(1 - r)^x$$

$(1 - r)$ is decay factor; r is decay rate (often a percent)

A city has a current population of 37,000 that has been increasing at a rate of 3% per year. At this rate, find the city's population in 20 years.

$$y = C(1 + r)^x$$
$$y = 37{,}000(1 + 0.03)^{20}$$
$$y \approx 66{,}826.12$$

In 20 years, the predicted population of the city is 66,826.

A city has a current population of 37,000 that has been decreasing at a rate of 3% per year. At this rate, find the city's population in 20 years.

$$y = C(1 - r)^x$$
$$y = 37{,}000(1 - 0.03)^{20}$$
$$y \approx 20{,}120.39$$

In 20 years, predicted population of the city is 20,120.

Section 12.5　Logarithmic Functions

Logarithmic Definition

If $b > 0$ and $b \neq 1$, then

$$y = \log_b x \quad \text{means} \quad x = b^y$$

for any positive number x and real number y.

Properties of Logarithms

If b is a real number, $b > 0$ and $b \neq 1$, then

$$\log_b 1 = 0, \quad \log_b b^x = x, \quad \text{and} \quad b^{\log_b x} = x$$

Logarithmic function

If $b > 0$ and $b \neq 1$, then a **logarithmic function** is a function that can be defined as

$$f(x) = \log_b x$$

The domain of f is the set of positive real numbers, and the range of f is the set of real numbers.

Logarithmic Form	Corresponding Exponential Statement
$\log_5 25 = 2$	$5^2 = 25$
$\log_9 3 = \dfrac{1}{2}$	$9^{1/2} = 3$

$$\log_5 1 = 0, \quad \log_7 7^2 = 2, \quad \text{and} \quad 3^{\log_3 6} = 6$$

Graph:　$y = \log_3 x$

Write $y = \log_3 x$ as $3^y = x$. Plot the ordered pair solutions listed in the table, and connect them with a smooth curve.

x	y
3	1
1	0
$\dfrac{1}{3}$	-1
$\dfrac{1}{9}$	-2

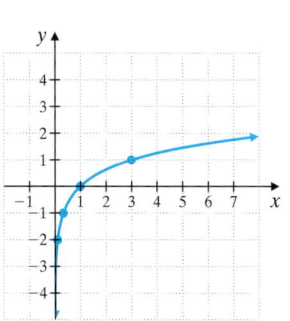

Definitions and Concepts	Examples

Section 12.6   Properties of Logarithms

Let x, y, and b be positive numbers, $b \neq 1$, and r be a real number.

Product Property

$$\log_b xy = \log_b x + \log_b y$$

Quotient Property

$$\log_b \frac{x}{y} = \log_b x - \log_b y$$

Power Property

$$\log_b x^r = r \log_b x$$

Write as a single logarithm:

$$2\log_5 6 + \log_5 x - \log_5(y + 2)$$
$$= \log_5 6^2 + \log_5 x - \log_5(y + 2) \quad \text{Power property}$$
$$= \log_5 36 \cdot x - \log_5(y + 2) \quad \text{Product property}$$
$$= \log_5 \frac{36x}{y + 2} \quad \text{Quotient property}$$

Section 12.7   Common Logarithms, Natural Logarithms, and Change of Base

Common Logarithms

$\log x$ means $\log_{10} x$

Natural Logarithms

$\ln x$ means $\log_e x$

Continuously Compounded Interest Formula

$$A = Pe^{rt}$$

where r is the annual interest rate for P dollars invested for t years.

Change of Base Formula

If a, b, and c are positive real numbers and neither b nor c is 1, then

$$\log_b a = \frac{\log_c a}{\log_c b}$$

$$\log 5 = \log_{10} 5 \approx 0.6990$$

$$\ln 7 = \log_e 7 \approx 1.9459$$

Find the amount in an account at the end of 3 years if $1000 is invested at an interest rate of 4% compounded continuously.

Here, $t = 3$ years, $P = \$1000$, and $r = 0.04$.

$$A = Pe^{rt}$$
$$= \$1000e^{0.04(3)}$$
$$\approx \$1127.50$$

Section 12.8   Exponential and Logarithmic Equations and Problem Solving

Logarithm Property of Equality

Let $\log_b a$ and $\log_b c$ be real numbers and $b \neq 1$. Then

$\log_b a = \log_b c$ is equivalent to $a = c$

Solve: $2^x = 5$

$$\log 2^x = \log 5 \quad \text{Log property of equality}$$
$$x \log 2 = \log 5 \quad \text{Power property}$$
$$x = \frac{\log 5}{\log 2} \quad \text{Divide both sides by log 2.}$$
$$x \approx 2.3219 \quad \text{Use a calculator.}$$

(12.1) *If $f(x) = x - 5$ and $g(x) = 2x + 1$, find the following.*

1. $(f + g)(x)$
$3x - 4$

2. $(f - g)(x)$
$-x - 6$

3. $(f \cdot g)(x)$
$2x^2 - 9x - 5$

4. $\left(\dfrac{g}{f}\right)(x)$

$\dfrac{2x + 1}{x - 5}$ where $x \neq 5$

If $f(x) = x^2 - 2$, $g(x) = x + 1$, and $h(x) = x^3 - x^2$, find each composition.

5. $(f \circ g)(x)$ $x^2 + 2x - 1$

6. $(g \circ f)(x)$ $x^2 - 1$

7. $(h \circ g)(2)$ 18

8. $(f \circ f)(x)$ $x^4 - 4x^2 + 2$

9. $(f \circ g)(-1)$ -2

10. $(h \circ h)(2)$ 48

(12.2) *Determine whether each function is a one-to-one function. If it is one-to-one, list the elements of its inverse.*

11. $h = \{(-9, 14), (6, 8), (-11, 12), (15, 15)\}$
one-to-one;
$h^{-1} = \{(14, -9), (8, 6), (12, -11), (15, 15)\}$

12. $f = \{(-5, 5), (0, 4), (13, 5), (11, -6)\}$
not one-to-one

13.

U.S. Region (Input)	Northeast	Midwest	South	West
Rank in Housing Starts for 2013 (Output)	4	3	1	2

one-to-one

Rank in Housing Starts for 2013 (Input)	4	3	1	2
U.S. Region (Output)	Northeast	Midwest	South	West

14.

Shape (Input)	Square	Triangle	Parallelogram	Rectangle
Number of Sides (Output)	4	3	4	4

not one-to-one

Determine whether each function is a one-to-one function.

15.

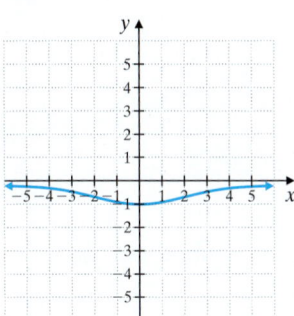

not one-to-one

16.

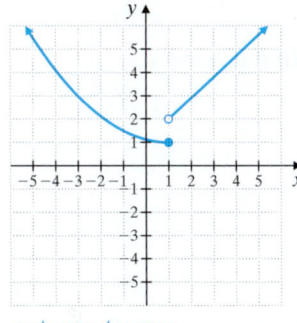

not one-to-one

17.

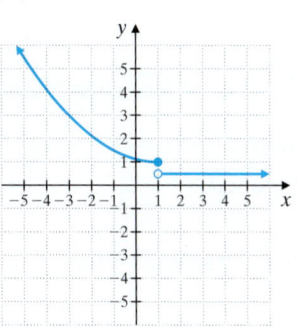

not one-to-one

18.

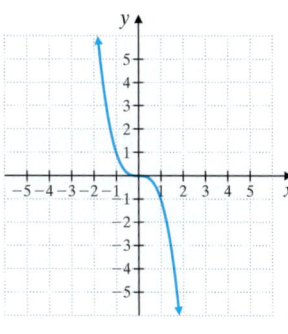

one-to-one

Find an equation defining the inverse function of each one-to-one function.

19. $f(x) = 6x + 11$

$f^{-1}(x) = \dfrac{x - 11}{6}$

20. $f(x) = 12x$

$f^{-1}(x) = \dfrac{x}{12}$

21. $f(x) = 3x - 5$

$f^{-1}(x) = \dfrac{x + 5}{3}$

22. $f(x) = 2x + 1$

$f^{-1}(x) = \dfrac{x - 1}{2}$

Graph each one-to-one function and its inverse on the same set of axes.

23. $f(x) = -2x + 3$

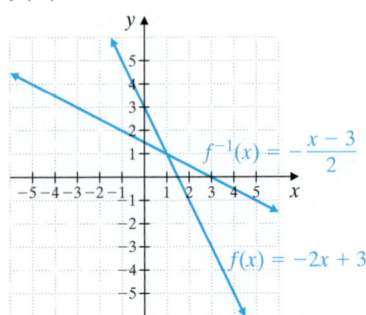

24. $f(x) = 5x - 5$

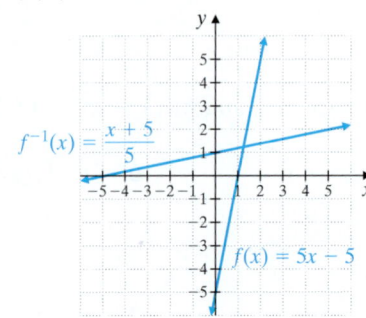

(12.3) *Solve each equation.*

25. $4^x = 64$ $\quad \{3\}$

26. $2^{3x} = \dfrac{1}{16}$ $\quad \left\{-\dfrac{4}{3}\right\}$

27. $9^{x+1} = 243$ $\quad \left\{\dfrac{3}{2}\right\}$

28. $25^{x-1} = 125$ $\quad \left\{\dfrac{5}{2}\right\}$

Graph each exponential function.

29. $y = 3^x$

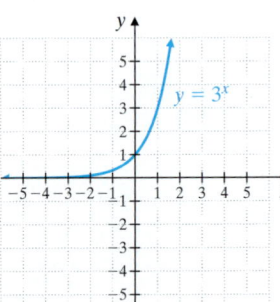

30. $y = \left(\dfrac{1}{3}\right)^x$

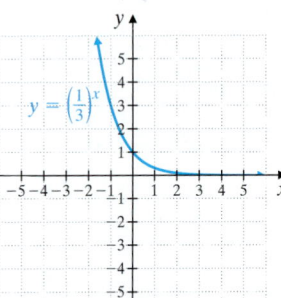

31. $y = 2^{x-4}$

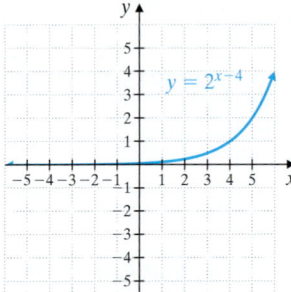

32. $y = 2^x + 4$

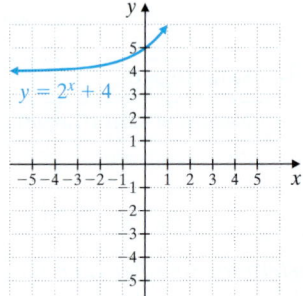

Use the formula $A = P\left(1 + \dfrac{r}{n}\right)^{nt}$ to solve Exercises 33 and 34. In this formula,

A = amount accrued (or owed)

P = principal invested (or loaned)

r = rate of interest

n = number of compounding periods per year

t = time in years

33. A total of $800 is invested in a 7% certificate of deposit for which interest is compounded quarterly. Find the value that this certificate will have at the end of 5 years. $1131.82

34. Find the amount accrued if $1600 is invested at 9% interest compounded semiannually for 7 years. $2963.11

(12.4) *Solve. Round each answer to the nearest whole.*

35. The city of Henderson, Nevada, has been growing at a rate of 4.8% per year since the year 2000. If the population of Henderson was 179,087 in 2000 and this rate continues, predict the city's population in 2020.   457,393

36. The city of Raleigh, North Carolina, has been growing at a rate of 3.9% per year since the year 2000. If the population of Raleigh was 287,370 in 2000 and this rate continues, predict the city's population in 2021.   641,753

37. A summer camp tournament starts with 1024 players. After each round, half the players are eliminated. How many players remain after 7 rounds?   8 players

38. The bear population in a certain national park is decreasing by 11% each year. If this rate continues, and there is currently an estimated bear population of 1280, find the bear population in 6 years.   636 bears

(12.5) *Write each exponential equation with logarithmic notation.*

39. $49 = 7^2$ $\log_7 49 = 2$

40. $2^{-4} = \dfrac{1}{16}$ $\log_2 \dfrac{1}{16} = -4$

Write each logarithmic equation with exponential notation.

41. $\log_{1/2} 16 = -4$ $\left(\dfrac{1}{2}\right)^{-4} = 16$

42. $\log_{0.4} 0.064 = 3$ $0.4^3 = 0.064$

Solve.

43. $\log_4 x = -3$ $\left\{\dfrac{1}{64}\right\}$

44. $\log_3 x = 2$ $\{9\}$

45. $\log_3 1 = x$ $\{0\}$

46. $\log_x 64 = 2$ $\{8\}$

47. $\log_4 4^5 = x$ $\{5\}$

48. $\log_7 7^{-2} = x$ $\{-2\}$

49. $5^{\log_5 4} = x$ $\{4\}$

50. $2^{\log_2 9} = x$ $\{9\}$

51. $\log_3 (2x + 5) = 2$ $\{2\}$

52. $\log_8 (x^2 + 7x) = 1$ $\{-8, 1\}$

Graph each pair of equations on the same set of axes.

53. $y = 2^x;\ y = \log_2 x$

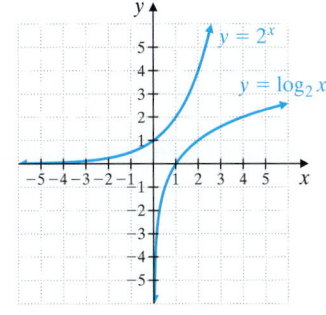

54. $y = \left(\dfrac{1}{2}\right)^x;\ y = \log_{1/2} x$

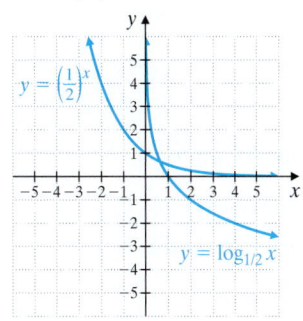

(12.6) *Write each expression as a single logarithm.*

55. $\log_3 8 + \log_3 4$ $\log_3 32$

56. $\log_2 6 + \log_2 3$ $\log_2 18$

57. $\log_7 15 - \log_7 20$ $\log_7 \dfrac{3}{4}$

58. $\log_e 18 - \log_e 12$ $\log_e \dfrac{3}{2}$

59. $\log_{11} 8 + \log_{11} 3 - \log_{11} 6$
$\log_{11} 4$

60. $\log_5 14 + \log_5 3 - \log_5 21$
$\log_5 2$

61. $2\log_5 x - 2\log_5 (x + 1) + \log_5 x$ $\log_5 \dfrac{x^3}{(x + 1)^2}$

62. $4\log_3 x - \log_3 x + \log_3 (x + 2)$ $\log_3 (x^4 + 2x^3)$

Use properties of logarithms to write each expression as a sum or difference of logarithms.

63. $\log_3 \dfrac{x^3}{x + 2}$ $3\log_3 x - \log_3 (x + 2)$

64. $\log_4 \dfrac{x + 5}{x^2}$ $\log_4 (x + 5) - 2\log_4 x$

65. $\log_2 \dfrac{3x^2 y}{z}$ $\log_2 3 + 2\log_2 x + \log_2 y - \log_2 z$

66. $\log_7 \dfrac{yz^3}{x}$ $\log_7 y + 3\log_7 z - \log_7 x$

If $\log_b 2 = 0.36$ and $\log_b 5 = 0.83$, evaluate each expression.

67. $\log_b 50$ 2.02

68. $\log_b \dfrac{4}{5}$ -0.11

(12.7) *Use a calculator to approximate each logarithm to four decimal places.*

69. $\log 3.6$ 0.5563

70. $\log 0.15$ -0.8239

71. $\ln 1.25$ 0.2231

72. $\ln 4.63$ 1.5326

Find the exact value of each logarithm.

73. $\log 1000$ 3

74. $\log \dfrac{1}{10}$ -1

75. $\ln \dfrac{1}{e}$ -1

76. $\ln e^4$ 4

Solve each equation. Give the exact solution and a four-decimal approximation where necessary.

77. $\log 2x = 2$
$\{50\}$

78. $\ln (3x) = 1.6$
$\left\{\dfrac{e^{1.6}}{3}\right\}$; $\{1.6510\}$

79. $\ln (2x - 3) = -1$
$\left\{\dfrac{e^{-1} + 3}{2}\right\}$; $\{1.6839\}$

80. $\ln (3x + 1) = 2$
$\left\{\dfrac{e^2 - 1}{3}\right\}$; $\{2.1297\}$

Approximate each logarithm to four decimal places.

81. $\log_5 1.6$ 0.2920

82. $\log_3 4$ 1.2619

Use the formula $A = Pe^{rt}$ to solve Exercises 83 and 84, in which interest is compounded continuously. In this formula,

A = amount accrued (or owed)
P = principal invested (or loaned)
r = rate of interest
t = time in years

83. Bank of New York offers a 5-year 6% continuously compounded investment option. Find the amount accrued if $1450 is invested. $1957.30

84. Find the amount to which a $940 investment grows if it is invested at 11% interest compounded continuously for 3 years. $1307.51

(12.8) *Solve each exponential equation. Give the exact solution and a four-decimal-place approximation.*

85. $7^x = 20$ $\left\{\dfrac{\log 20}{\log 7}\right\}$; $\{1.5395\}$

86. $3^{2x} = 7$ $\left\{\dfrac{\log 7}{2\log 3}\right\}$; $\{0.8856\}$

87. $3^{2x+1} = 6$ $\left\{\dfrac{1}{2}\left(\dfrac{\log 6}{\log 3} - 1\right)\right\}$; $\{0.3155\}$

88. $8^{4x-2} = 3$ $\left\{\dfrac{1}{4}\left(\dfrac{\log 3}{\log 8} + 2\right)\right\}$; $\{0.6321\}$

Solve each equation.

89. $\log_5 2 + \log_5 x = 2$ $\left\{\dfrac{25}{2}\right\}$

90. $\log 5x - \log(x + 1) = 4$ $\varnothing$

91. $\log_2 x + \log_2 2x - 3 = 1$ $\{2\sqrt{2}\}$

92. $\log_3(x^2 - 8x) = 2$ $\{9, -1\}$

Use the formula $y = y_0 e^{kt}$ to solve Exercises 93 through 96. In this formula,

> y = size of population
> y_0 = initial count of population
> k = rate of growth, expressed as a decimal
> t = time

Round each answer to the nearest tenth.

93. Use this formula to calculate the rate of growth of the population of California condors. In 1987, the population of California condors was only 87 birds. They were all brought in from the wild and an intensive breeding program was instituted. By 2013, there were 416 California condors. *(Source: National Parks Conservation Association)* 6.01%

94. France is experiencing an annual growth rate of 0.5%. In 2012, the population of France was approximately 65,300,000. How long will it take for the population to reach 67,000,000? Round to the nearest tenth. *(Source: Population Reference Bureau)*   5.1 yr

95. In 2013, the population of Australia was approximately 23,400,000. How long will it take Australia to double its population if its growth rate is 1.5% annually? *(Source: Population Reference Bureau)*   46.2 yr

96. Monaco's population is increasing at a rate of 0.6% per year. How long will it take for its 2013 population of 30,429 to double in size? *(Source: Population Reference Bureau)*   115.5 yr

Use the compound interest equation $A = P\left(1 + \dfrac{r}{n}\right)^{nt}$ to solve Exercises 97 and 98. (See the directions for Exercises 33 and 34 for an explanation of this formula.) Round answers to the nearest tenth.

97. How long does it take for a $5000 investment to grow to $10,000 if it is invested at 8% interest compounded quarterly?   8.8 yr

98. An investment of $6000 has grown to $10,000 while the money was invested at 6% interest compounded monthly. How long was it invested?   8.5 yr

Mixed Review

Solve each equation. Give exact answers.

99. $3^x = \dfrac{1}{9}$ $\{-2\}$

100. $5^{2x} = 125$ $\left\{\dfrac{3}{2}\right\}$

101. $8^{3x-2} = 4$ $\left\{\dfrac{8}{9}\right\}$

102. $9^{x-2} = 27$ $\left\{\dfrac{7}{2}\right\}$

103. $\log_4 64 = x$ $\{3\}$

104. $\log_x 81 = 4$ $\{3\}$

105. $\log_4(x^2 - 3x) = 1$ $\{-1, 4\}$

106. $\log_2(3x - 1) = 4$ $\left\{\dfrac{17}{3}\right\}$

107. $\ln x = -1.2$ $\{e^{-1.2}\}$

108. $\log_3 x + \log_3 10 = 2$ $\left\{\dfrac{9}{10}\right\}$

109. $\ln 3x - \ln(x - 3) = 2$
$\left\{\dfrac{3e^2}{e^2 - 3}\right\}$

110. $\log_6 x - \log_6(4x + 7) = 1$
$\varnothing$

If $f(x) = x$ and $g(x) = 2x - 3$, find the following.

1. $(f \cdot g)(x)$

2. $(f - g)(x)$

If $f(x) = x, g(x) = x - 7$, and $h(x) = x^2 - 6x + 5$, find each composition.

3. $(f \circ h)(0)$

4. $(g \circ f)(x)$

5. $(g \circ h)(x)$

Graph the one-to-one function and its inverse on the same set of axes.

6. $f(x) = 7x - 14$

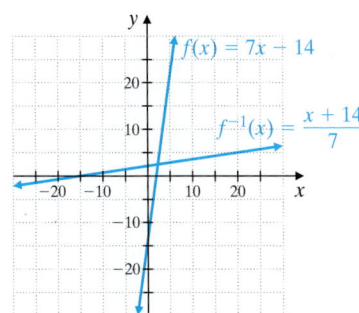

Determine whether each graph is the graph of a one-to-one function.

7.

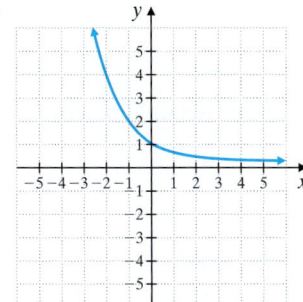

8.

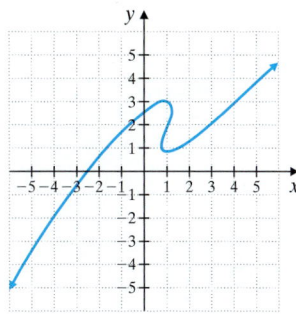

Determine whether each function is one-to-one. If it is one-to-one, find an equation or a set of ordered pairs that defines the inverse function of the given function.

9. $f(x) = 6 - 2x$

10. $f = \{(0,0), (2,3), (-1,5)\}$

11.

Word (Input)	Dog	Cat	House	Desk	Circle
First Letter of Word (Output)	d	c	h	d	c

Answers

Note to Instructor: The Chapter 12 Test file in the TestGen program provides algorithms specifically matched to this test for easy replication for practice or assessment purposes.

1. $2x^2 - 3x$

2. $-x + 3$

3. 5

4. $x - 7$

5. $x^2 - 6x - 2$

6. see graph

7. one-to-one

8. not a function

9. one-to-one; $f^{-1}(x) = \dfrac{-x + 6}{2}$

10. one-to-one; $f^{-1} = \{(0,0),(3,2)\ (5,-1)\}$

11. not one-to-one

12. $\log_3 24$

_____

13. $\log_5 \dfrac{x^4}{x+1}$

_____

14. $\log_6 2 + \log_6 x$
$- 3\log_6 y$

_____

15. -1.53

_____

16. 1.0686

_____

17. $\{-1\}$

_____

18. $\left\{\dfrac{1}{2}\left(\dfrac{\log 4}{\log 3} - 5\right)\right\};$
$\{-1.8691\}$

_____

19. $\left\{\dfrac{1}{9}\right\}$

_____

20. $\left\{\dfrac{1}{2}\right\}$

_____

21. $\{22\}$

_____

22. $\left\{\dfrac{25}{3}\right\}$

_____

23. $\left\{\dfrac{43}{21}\right\}$

_____

24. $\{-1.0979\}$

_____

25. see graph

_____

26. see graph

_____

Use the properties of logarithms to write each expression as a single logarithm.

12. $\log_3 6 + \log_3 4$

13. $\log_5 x + 3\log_5 x - \log_5 (x+1)$

14. Write the expression $\log_6 \dfrac{2x}{y^3}$ as the sum or difference of logarithms.

15. If $\log_b 3 = 0.79$ and $\log_b 5 = 1.16$, find the value of $\log_b \dfrac{3}{25}$.

16. Approximate $\log_7 8$ to four decimal places.

17. Solve $8^{x-1} = \dfrac{1}{64}$ for x. Give the exact solution.

18. Solve $3^{2x+5} = 4$ for x. Give the exact solution and a four-decimal-place approximation.

Solve each logarithmic equation. Give the exact solution.

19. $\log_3 x = -2$

20. $\ln \sqrt{e} = x$

21. $\log_8 (3x - 2) = 2$

22. $\log_5 x + \log_5 3 = 2$

23. $\log_4 (x+1) - \log_4 (x-2) = 3$

24. Solve $\ln(3x + 7) = 1.31$. Round the solution to four decimal places.

25. Graph $f(x) = \left(\dfrac{1}{2}\right)^x + 1$.

26. Graph the functions $y = 3^x$ and $y = \log_3 x$ on the same set of axes.

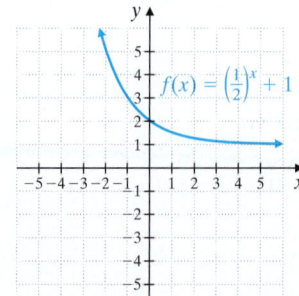

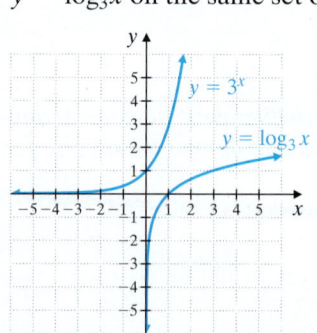

Use the formula $A = P\left(1 + \dfrac{r}{n}\right)^{nt}$ or $A = Pe^{rt}$ to solve Exercises 27 through 29.

27. Find the amount in an account in which $4000 is invested for 3 years at 9% interest compounded monthly.

28. How long will it take $2000 to grow to $3000 if the money is invested at 7% interest compounded semiannually? Round to the nearest whole year.

29. Suppose you have $3000 to invest. Which investment, rounded to the nearest dollar, yields the greater return over 10 years: 6.5% compounded semiannually or 6% compounded monthly? How much more is yielded by the better investment?

Solve. Round answers to the nearest whole.

30. Suppose a city with population of 150,000 has been decreasing at a rate of 2% per year. If this rate continues, predict the population of the city in 20 years.

31. The prairie dog population of the Grand Forks area now stands at 57,000 animals. If the population is growing at a rate of 2.6% annually, how many prairie dogs will there be in that area 5 years from now?

32. In an attempt to save an endangered species of wood duck, naturalists would like to increase the wood duck population from 400 to 1000 ducks. If the annual population growth rate is 6.2%, how long will it take the naturalists to reach their goal? Round to the nearest whole year.

The reliability of a new model of CD player can be described by the exponential function $R(t) = 2.7^{-(1/3)t}$, where the reliability R is the probability (as a decimal) that the CD player is still working t years after it is manufactured. Round answers to the nearest hundredth. Then write your answers as percents.

33. What is the probability that the CD player will still work half a year after it is manufactured?

34. What is the probability that the CD player will still work 2 years after it is manufactured?

Answers

1. $\dfrac{5}{x^2}$ (Sec. 7.1, Ex. 3)

2. $\dfrac{1}{1-x}$ (Sec. 7.3)

3. $\dfrac{1}{x-9}$ (Sec. 7.1, Ex. 6)

4. $\dfrac{-x^2 + 23x + 38}{(x-2)(x+2)^2}$ (Sec. 7.4)

5. $\dfrac{2}{5}$ (Sec. 7.2, Ex. 2)

6. 11 (Sec. 2.3)

7. $\dfrac{2}{x(x+1)}$ (Sec. 7.2, Ex. 6)

8. $2 - x$ (Sec. 7.1)

9. $y = \dfrac{1}{4}x - 3$ (Sec. 3.5, Ex. 1)

10. $(3y - 2)(2x - 5)$ (Sec. 6.1)

11. $x = 2$ (Sec. 8.1, Ex. 4)

12. $(y + 3)(2x - 1)$ (Sec. 6.1)

13. domain: $[-3, 5]$; range: $[-2, 4]$ (Sec. 8.4, Ex. 4)

14. $f(x) = \dfrac{x + 20}{3}$ (Sec. 8.2)

15. $(-1, 3)$ (Sec. 4.1, Ex. 3) see graph

16. $\dfrac{1}{4}x^2 - 9$ (Sec. 5.6)

17. no solution (Sec. 4.2, Ex. 6)

18. $m = \dfrac{3}{2}$ (Sec. 3.4)

1. Simplify: $\dfrac{5x - 5}{x^3 - x^2}$

2. Add: $\dfrac{1}{1 - x^2} + \dfrac{x}{1 - x^2}$

3. Simplify: $\dfrac{x + 9}{x^2 - 81}$

4. Perform the indicated operation and simplify if possible.

$$\dfrac{5}{x - 2} + \dfrac{3}{x^2 + 4x + 4} - \dfrac{6}{x + 2}$$

5. Multiply: $\dfrac{x^2 + x}{3x} \cdot \dfrac{6}{5x + 5}$

6. Solve: $\dfrac{1}{3}(x - 2) = \dfrac{1}{4}(x + 1)$

7. Divide: $\dfrac{6x + 2}{x^2 - 1} \div \dfrac{3x^2 + x}{x - 1}$

8. Simplify: $\dfrac{x^2 - 4x + 4}{2 - x}$

9. Find an equation of the line with y-intercept $(0, -3)$ and slope of $\dfrac{1}{4}$.

10. Factor: $2x(3y - 2) - 5(3y - 2)$

11. Write an equation of the line containing the point $(2, 3)$ with undefined slope.

12. Factor: $2xy + 6x - y - 3$

13. Find the domain and range of the relation.

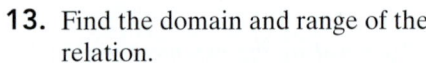
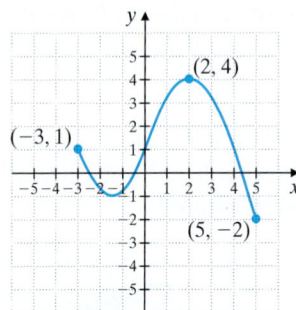

14. Find an equation of the line through $(-2, 6)$ and perpendicular to the graph of $f(x) = -3x + 4$. Write the equation using function notation.

15. Solve the system of equations by graphing.
$$\begin{cases} -x + 3y = 10 \\ x + y = 2 \end{cases}$$

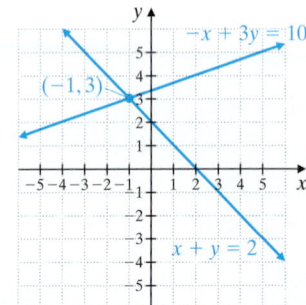

16. Multiply: $\left(\dfrac{1}{2}x + 3\right)\left(\dfrac{1}{2}x - 3\right)$

17. Solve the system:
$$\begin{cases} 6x + 12y = 5 \\ -4x - 8y = 0 \end{cases}$$

18. Find the slope of the line containing $(-2, 6)$ and $(0, 9)$.

19. The Cirque du Soleil show Varekai is performing locally. Matinee admission for 4 adults and 2 children is $374, while admission for 2 adults and 3 children is $285.

    a. What is the price of an adult's ticket?

    b. What is the price of a child's ticket?

    c. Suppose that a special rate of $1000 is offered for groups of 20 persons. Should a group of 4 adults and 16 children use the group rate? Why or why not?

20. Solve the system by the addition method:
$$\begin{cases} 5x + y = -2 \\ 4x - 2y = -10 \end{cases}$$

Simplify. Assume that all variables represent positive real numbers.

21. $\sqrt{0}$

22. $\sqrt[3]{64x^7y^2}$

23. $\sqrt{0.25}$

24. $\sqrt[4]{162a^4b^5}$

Use rational exponents to simplify. Assume that all variables represent positive real numbers.

25. $\sqrt[8]{x^4}$

26. $\sqrt[15]{y^5}$

27. $\sqrt[6]{r^2s^4}$

28. $\sqrt[20]{x^8y^{16}}$

Simplify.

29. $\sqrt[3]{24}$

30. $\sqrt[3]{x^{10}}$

31. $\sqrt[4]{32}$

32. $\sqrt{200x}$

33. Rationalize the denominator of $\dfrac{2}{\sqrt{5}}$.

34. Rationalize the denominator of $\sqrt[3]{\dfrac{27}{m^4n^8}}$.

35. Solve: $\sqrt{-10x - 1} + 3x = 0$

36. Multiply.

    a. $a^{1/4}(a^{3/4} - a^8)$

    b. $(x^{1/2} - 3)(x^{1/2} + 5)$

37. Multiply: $(2 - 5i)(4 + i)$

38. Add or subtract as indicated.

    a. $\dfrac{\sqrt{20}}{3} + \dfrac{\sqrt{5}}{4}$

    b. $\sqrt[3]{\dfrac{24x}{27}} - \dfrac{\sqrt[3]{3x}}{2}$

39. Solve $p^2 + 2p = 4$ by completing the square.

40. Suppose that y varies directly as x. If $y = \dfrac{1}{2}$ when $x = 12$, find the constant of variation and the variation equation.

---

19. a. $69

    b. $49

    c. yes (Sec. 4.4, Ex. 2)

20. $(-1, 3)$ (Sec. 4.3)

21. 0 (Sec. 10.1, Ex. 5)

22. $4x^2\sqrt[3]{xy^2}$ (Sec. 10.3)

23. 0.5 (Sec. 10.1, Ex. 7)

24. $3ab\sqrt[4]{2b}$ (Sec. 10.3)

25. $\sqrt{x}$ (Sec. 10.2, Ex. 19)

26. $\sqrt[3]{y}$ (Sec. 10.2)

27. $\sqrt[3]{rs^2}$ (Sec. 10.2, Ex. 21)

28. $\sqrt[5]{x^2y^4}$ (Sec. 10.2)

29. $2\sqrt[3]{3}$ (Sec. 10.3, Ex. 11)

30. $x^3\sqrt[3]{x}$ (Sec. 10.3)

31. $2\sqrt[4]{2}$ (Sec. 10.3, Ex. 13)

32. $10\sqrt{2x}$ (Sec. 10.3)

33. $\dfrac{2\sqrt{5}}{5}$ (Sec. 10.5, Ex. 1)

34. $\dfrac{3\sqrt[3]{m^2n}}{m^2n^3}$ (Sec. 10.5)

35. $\left\{-\dfrac{1}{9}, -1\right\}$ (Sec. 10.6, Ex. 2)

36. a. $a - a^{33/4}$

    b. $x + 2x^{1/2} - 15$ (Sec. 10.2)

37. $13 - 18i$ (Sec. 10.7, Ex. 13)

38. a. $\dfrac{11\sqrt{5}}{12}$

    b. $\dfrac{\sqrt[3]{3x}}{6}$ (Sec. 10.4)

39. $\{-1 + \sqrt{5}, -1 - \sqrt{5}\}$ (Sec. 11.1, Ex. 7)

40. $k = \dfrac{1}{24}; y = \dfrac{1}{24}x$ (Sec. 9.4)

41. $\left\{\dfrac{2 + \sqrt{10}}{2}, \dfrac{2 - \sqrt{10}}{2}\right\}$ (Sec. 11.2, Ex. 2)

42. $\left\{\dfrac{-2 + \sqrt{5}}{2}, \dfrac{-2 - \sqrt{5}}{2}\right\}$ (Sec. 11.1)

43. $\{9\}$ (Sec. 11.3, Ex. 1)

44. $\left\{\dfrac{3 + \sqrt{5}}{4}, \dfrac{3 - \sqrt{5}}{4}\right\}$ (Sec. 11.2)

45. $(-\infty, -2] \cup [1, 5]$ (Sec. 11.4, Ex. 3)

46. $x^2 + 2x + 4$ (Sec. 5.7)

47. $\{6\}$ (Sec. 12.3, Ex. 6)

48. a. $\dfrac{2a}{a - 1}$

b. $\dfrac{-3(a + 6)}{4(a - 3)}$

c. $\dfrac{y + x}{x^2 y^2}$ (Sec. 7.7)

49. $\{5\}$ (Sec. 12.5, Ex. 9)

50. $\log_3 x^7 y^9$ (Sec. 12.6)

51. $\log_5 72$ (Sec. 12.6, Ex. 7)

52. $\left\{\dfrac{55}{8}\right\}$ (Sec. 12.8)

---

41. Solve: $2x^2 - 4x = 3$

42. Solve $4x^2 + 8x - 1 = 0$ by completing the square.

43. Solve: $x - \sqrt{x} - 6 = 0$

44. Solve by using the quadratic formula.
$$\left(x - \frac{1}{2}\right)^2 = \frac{x}{2}$$

45. Solve: $(x + 2)(x - 1)(x - 5) \le 0$

46. Divide $x^3 - 8$ by $x - 2$.

47. Solve: $4^{x+3} = 8^x$

48. Simplify each complex fraction.

a. $\dfrac{\dfrac{a}{5}}{\dfrac{a - 1}{10}}$

b. $\dfrac{\dfrac{3}{2 + a} + \dfrac{6}{2 - a}}{\dfrac{5}{a + 2} - \dfrac{1}{a - 2}}$

c. $\dfrac{x^{-1} + y^{-1}}{xy}$

49. Solve: $\log_x 25 = 2$

50. Write as a single logarithm:
$7 \log_3 x + 9 \log_3 y$

51. Write as a single logarithm.
$2 \log_5 3 + 3 \log_5 2$

52. Solve: $\log_7 (8x - 6) = 2$

Conic Sections

The circular Baptistery in Pisa, Italy

What Are the Conic Sections, and How Are They Used?

Applications of conic sections—the parabola, the circle, the ellipse, and the hyperbola—are too numerous to mention. For example, a parabolic orbit is often seen in comets. Many telescopes contain a parabolic dish because of the reflective properties of parabolas. The orbits of planets are in the shape of an ellipse. For this chapter opener, we study a building whose base is in the shape of a circle.

Located in the Piazza Dei Miracoli, in Italy, are the Duomo di Pisa, the great cathedral, and the Baptistery building. The circular Baptistery is the largest such building in Italy, and it is even taller than the Leaning Tower of Pisa, which is also in the Piazza.

In Section 13.1, Exercise 100, we will explore the dimensions of this interesting building.

In Chapter 11, we analyzed some of the important connections between a parabola and its equation. Parabolas are interesting in their own right but are more interesting still because they are part of a collection of curves known as conic sections. This chapter is devoted to quadratic equations in two variables and their conic section graphs: the parabola, circle, ellipse, and hyperbola.

Objectives

A Graph Parabolas of the Forms $y = a(x - h)^2 + k$ and $x = a(y - k)^2 + h$. ▶

B Graph Circles of the Form $(x - h)^2 + (y - k)^2 = r^2$. ▶

C Find the Center and the Radius of a Circle, Given Its Equation. ▶

D Write the Equation of a Circle, Given Its Center and Radius. ▶

Conic sections are called such because each conic section is the intersection of a right circular cone and a plane. The circle, parabola, ellipse, and hyperbola are the conic sections.

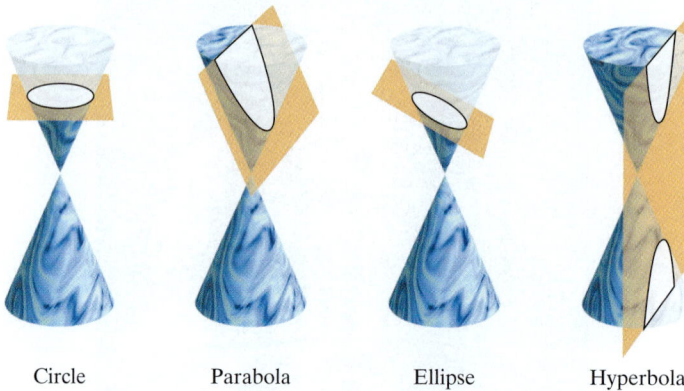

Circle   Parabola   Ellipse   Hyperbola

Objective **A** Graphing Parabolas ▶

Thus far, we have seen that $f(x)$ or $y = a(x - h)^2 + k$ is the equation of a parabola that opens upward if $a > 0$ or downward if $a < 0$. Parabolas can also open left or right or even on a slant. Equations of these parabolas are not functions of x, of course, since a parabola opening any way other than upward or downward fails the vertical line test. In this section, we introduce parabolas that open to the left and to the right. Parabolas opening on a slant will not be developed in this book.

Just as $y = a(x - h)^2 + k$ is the equation of a parabola that opens upward or downward, $x = a(y - k)^2 + h$ is the equation of a parabola that opens to the right or to the left. The parabola opens to the right if $a > 0$ and to the left if $a < 0$. The parabola has vertex (h, k), and its axis of symmetry is the line $y = k$.

Parabolas

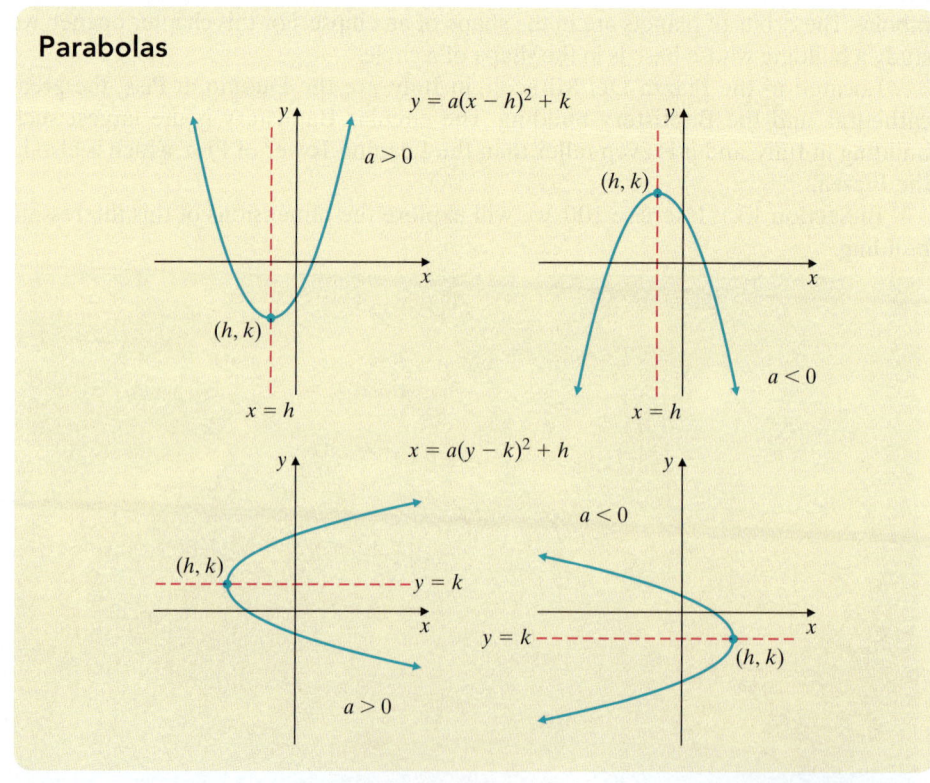

The forms $y = a(x - h)^2 + k$ and $x = a(y - k)^2 + h$ are called **standard forms.**

✓**Concept Check** Does the graph of the parabola given by the equation $x = -3y^2$ open to the left, to the right, upward, or downward?

Example 1 Graph: $x = 2y^2$

Solution: Written in standard form, the equation $x = 2y^2$ is $x = 2(y - 0)^2 + 0$ with $a = 2, h = 0$, and $k = 0$. Its graph is a parabola with vertex $(0, 0)$, and its axis of symmetry is the line $y = 0$. Since $a > 0$, this parabola opens to the right. We use a table to obtain a few more ordered pair solutions to help us graph $x = 2y^2$.

x	y
8	−2
2	−1
0	0
2	1
8	2

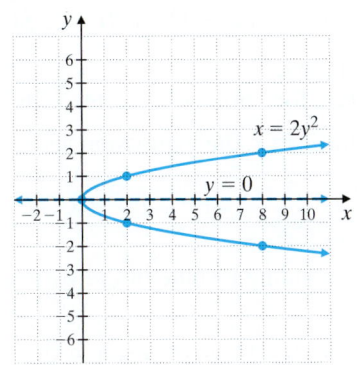

🟧 **Work Practice 1**

Example 2 Graph: $x = -3(y - 1)^2 + 2$

Solution: The equation $x = -3(y - 1)^2 + 2$ is in the form $x = a(y - k)^2 + h$ with $a = -3, k = 1$, and $h = 2$. Since $a < 0$, the parabola opens to the left. The vertex (h, k) is $(2, 1)$, and the axis of symmetry is the horizontal line $y = 1$. When $y = 0$, the x-value is -1, so the x-intercept is $(-1, 0)$.

Again, we use a table to obtain a few ordered pair solutions and then graph the parabola.

x	y
2	1
−1	0
−1	2

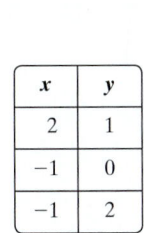
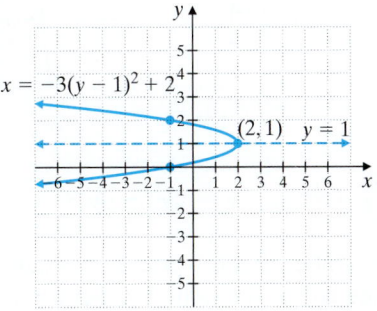

🟧 **Work Practice 2**

Practice 1

Graph: $x = 4y^2$

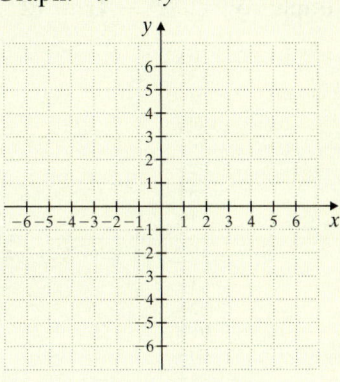

Practice 2

Graph: $x = -2(y - 3)^2 + 1$

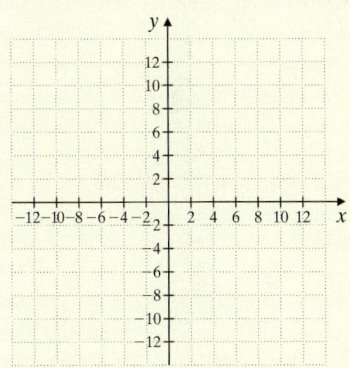

Answers

1.

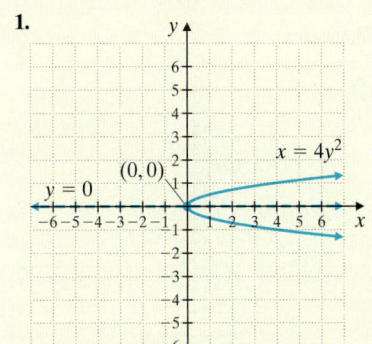

2.

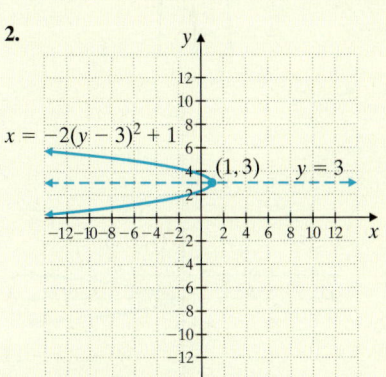

✓**Concept Check Answer**

to the left

Practice 3

Graph: $y = -x^2 - 4x + 12$

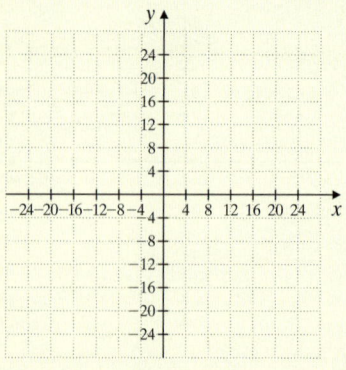

Practice 4

Graph: $x = 3y^2 + 12y + 13$

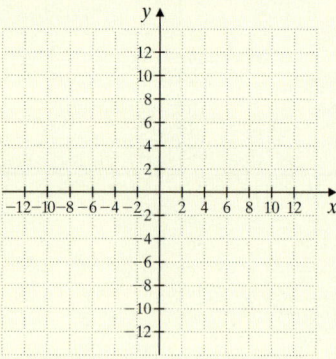

Answers

3.

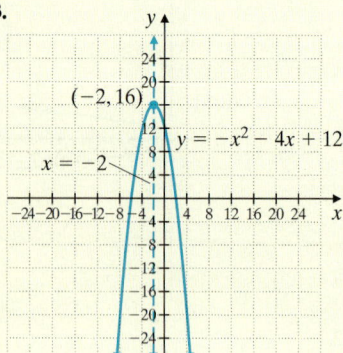

4.

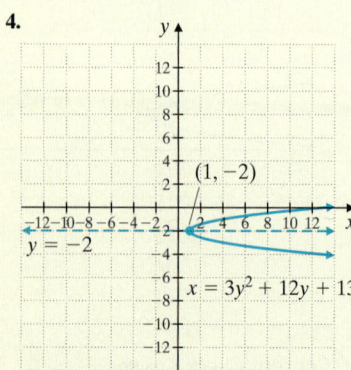

Example 3 Graph: $y = -x^2 - 2x + 15$

Solution: Notice that this equation is not written in standard form, $y = a(x - h)^2 + k$. There are two methods that we can use to find the vertex. The first method is completing the square.

$$y - 15 = -x^2 - 2x \qquad \text{Subtract 15 from both sides.}$$
$$y - 15 = -1(x^2 + 2x) \qquad \text{Factor } -1 \text{ from the terms } -x^2 - 2x.$$

The coefficient of x is 2, so we find the square of half of 2.

$$\frac{1}{2}(2) = 1 \quad \text{and} \quad 1^2 = 1$$

$$y - 15 - 1(1) = -1(x^2 + 2x + 1) \qquad \text{Add } -1(1) \text{ to both sides.}$$
$$y - 16 = -1(x + 1)^2 \qquad \text{Simplify the left side, and factor the right side.}$$
$$y = -(x + 1)^2 + 16 \qquad \text{Add 16 to both sides.}$$

The vertex is $(-1, 16)$.

The second method for finding the vertex is by using the expression $\frac{-b}{2a}$. Since the equation is quadratic in x, the expression gives us the x-value of the vertex.

$$x = \frac{-(-2)}{2(-1)} = \frac{2}{-2} = -1$$

To find the corresponding y-value of the vertex, replace x with -1 in the original equation.

$$y = -(-1)^2 - 2(-1) + 15 = -1 + 2 + 15 = 16$$

Again, we see that the vertex is $(-1, 16)$, and the axis of symmetry is the vertical line $x = -1$. The y-intercept is $(0, 15)$. Now we can use a few more ordered pair solutions to graph the parabola.

x	y
-5	0
-3	12
-2	15
-1	16
0	15
1	12
3	0

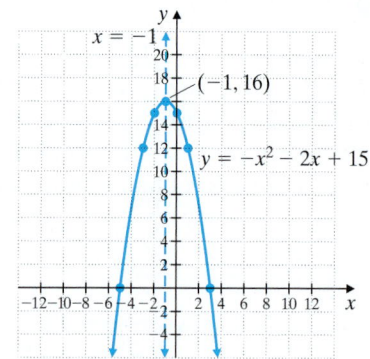

■ **Work Practice 3**

Example 4 Graph: $x = 2y^2 + 4y + 5$

Solution: We notice that this equation is quadratic in y so its graph is a parabola that opens to the left or the right. We can complete the square on y or we can use the expression $\frac{-b}{2a}$ to find the vertex.

Since the equation is quadratic in y, the expression gives us the y-value of the vertex.

$$y = \frac{-4}{2 \cdot 2} = \frac{-4}{4} = -1$$

$$x = 2(-1)^2 + 4(-1) + 5 = 2 \cdot 1 - 4 + 5 = 3$$

The vertex is $(3, -1)$, and the axis of symmetry is the line $y = -1$. The parabola opens to the right since $a > 0$. The x-intercept is $(5, 0)$.

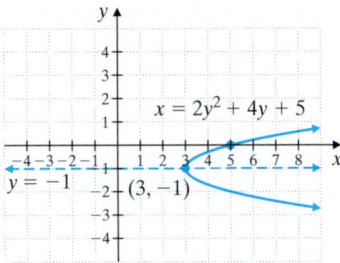

$x = 2y^2 + 4y + 5$

$y = -1$

$(3, -1)$

■ **Work Practice 4**

Objective B Graphing Circles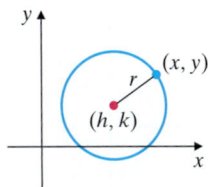

Another conic section is the **circle.** A circle is the set of all points in a plane that are the same distance from a fixed point called the **center.** The distance is called the **radius** of the circle. To find a standard equation for a circle, let (h, k) represent the center of the circle, and let (x, y) represent any point on the circle. The distance between (h, k) and (x, y) is defined to be the radius, r units. We can find this distance r by using the distance formula. (For a review of the distance formula, see Section 10.3.)

$r = \sqrt{(x - h)^2 + (y - k)^2}$ The distance formula.

$r^2 = (x - h)^2 + (y - k)^2$ Square both sides.

Circle

The graph of $(x - h)^2 + (y - k)^2 = r^2$ is a circle with center (h, k) and radius r.

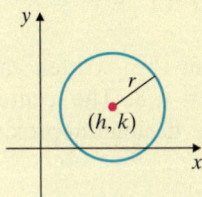

The form $(x - h)^2 + (y - k)^2 = r^2$ is called **standard form.**

If an equation can be written in the standard form

$(x - h)^2 + (y - k)^2 = r^2$

then its graph is a circle, which we can draw by graphing the center (h, k) and using the radius r.

Practice 5

Graph: $x^2 + y^2 = 36$

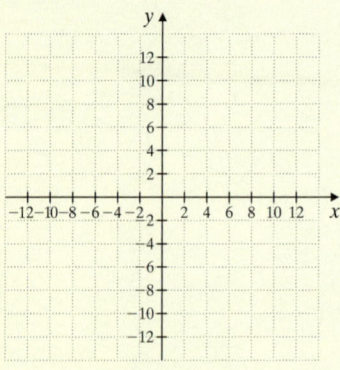

Practice 6

Graph: $x^2 + (y + 2)^2 = 6$

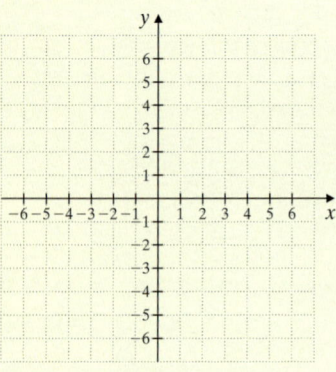

Answers

5.

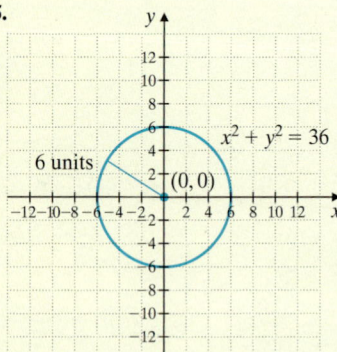

6.

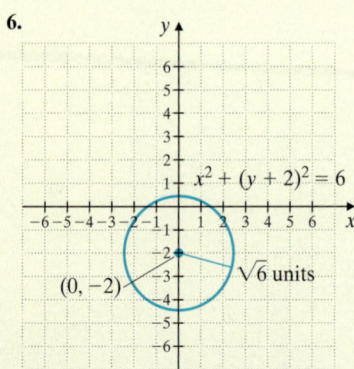

Helpful Hint

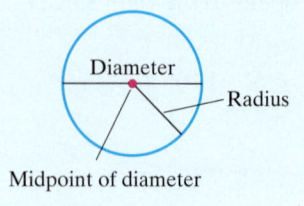

Notice that the radius is the *distance* from the center of the circle to any point of the circle. Also notice that the *midpoint* of a diameter of a circle is the center of the circle.

Example 5   Graph: $x^2 + y^2 = 4$

Solution: The equation can be written in standard form as

$$(x - 0)^2 + (y - 0)^2 = 2^2$$

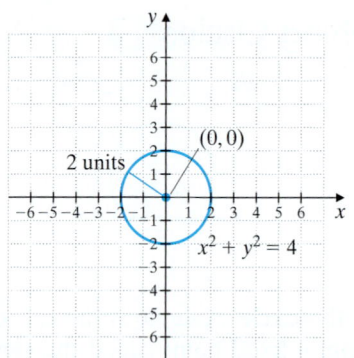

The center of the circle is $(0, 0)$, and the radius is 2. The graph of the circle is shown above.

■ **Work Practice 5**

Helpful Hint

Notice the difference between the equation of a circle and the equation of a parabola. The equation of a circle contains both x^2- and y^2-terms on the same side of the equation with equal coefficients. The equation of a parabola has either an x^2-term or a y^2-term but not both.

Example 6   Graph: $(x + 1)^2 + y^2 = 8$

Solution: The equation can be written as $(x - (-1))^2 + (y - 0)^2 = 8$ with $h = -1$, $k = 0$, and $r = \sqrt{8}$. The center is $(-1, 0)$, and the radius is $\sqrt{8} = 2\sqrt{2} \approx 2.8$. We use the decimal approximation to approximate the radius when graphing.

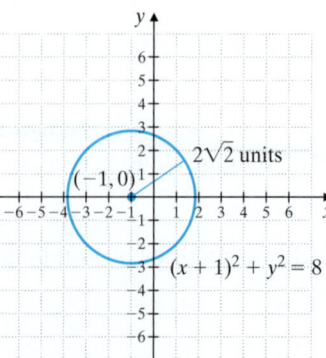

■ **Work Practice 6**

✓**Concept Check** In the graph of the equation $(x - 3)^2 + (y - 2)^2 = 5$, what is the distance between the center of the circle and any point on the circle?

Objective C Finding the Center and the Radius of a Circle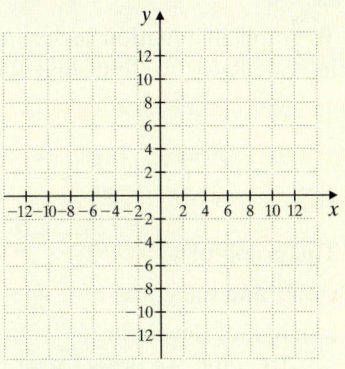

To find the center and the radius of a circle from its equation, we write the equation in standard form. To write the equation of a circle in standard form, we complete the square on both x and y.

Example 7 Graph: $x^2 + y^2 + 4x - 8y = 16$

Solution: Since this equation contains x^2- and y^2-terms on the same side of the equation with equal coefficients, its graph is a circle. To write the equation in standard form, we group the terms involving x and the terms involving y, and then complete the square on each variable.

$$(x^2 + 4x) + (y^2 - 8y) = 16$$

Now, $\frac{1}{2}(4) = 2$ and $2^2 = 4$. Also, $\frac{1}{2}(-8) = -4$ and $(-4)^2 = 16$. We add 4 and then 16 to both sides.

$$(x^2 + 4x + 4) + (y^2 - 8y + 16) = 16 + 4 + 16$$
$$(x + 2)^2 + (y - 4)^2 = 36 \qquad \text{Factor.}$$

This circle has the center $(-2, 4)$ and radius 6, as shown.

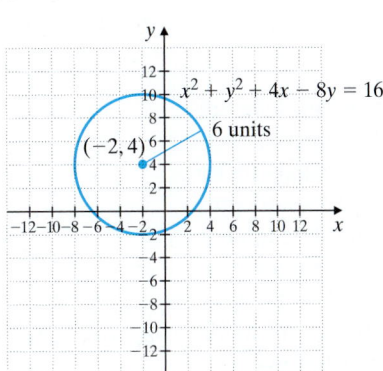

■ **Work Practice 7**

Objective D Writing Equations of Circles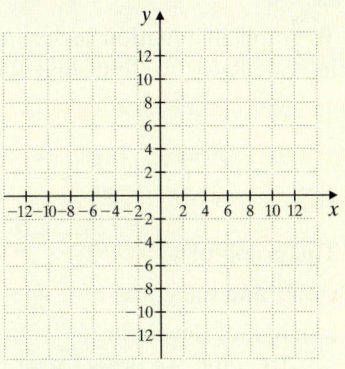

Since a circle is determined entirely by its center and radius, this information is all we need to write an equation of a circle.

Example 8 Write an equation of the circle with center $(-7, 3)$ and radius 10.

Solution: Using the given values $h = -7$, $k = 3$, and $r = 10$, we write the equation

$$(x - h)^2 + (y - k)^2 = r^2$$

or

$$(x - (-7))^2 + (y - 3)^2 = 10^2 \qquad \text{Substitute the given values.}$$

or

$$(x + 7)^2 + (y - 3)^2 = 100$$

■ **Work Practice 8**

Practice 7

Graph: $x^2 + y^2 - 2x + 6y = 6$

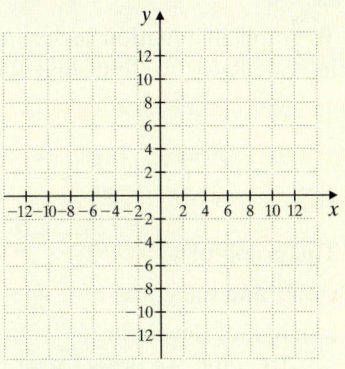

Teaching Tip:
Classroom Activity

Have students arrange desks in groups of four. Assign a quadrant to each member of the group. Have each student use a compass and a rectangular coordinate grid to draw a circle with radius equal to a whole unit and center with x- and y-values equal to whole units in their assigned quadrant. Then have the students pass their drawing clockwise for another member of their group to write the equation of the circle in standard form. Next, have them pass the drawing with the equation clockwise for another member to rewrite the equation in expanded form. Have the group leader write the four expanded equations from the group on one sheet of paper and give the paper to another group to graph. Finally, have the group check that the equations were graphed accurately by the other group.

Practice 8

Write an equation of the circle with the center $(2, -5)$ and radius 7.

Answers

7.

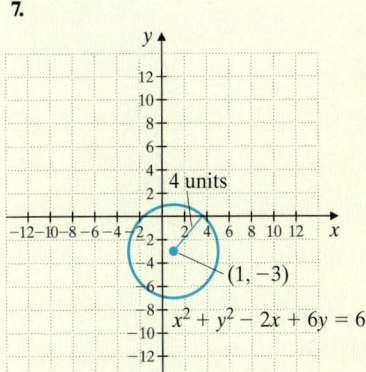

8. $(x - 2)^2 + (y + 5)^2 = 49$

✓**Concept Check Answer**

$\sqrt{5}$ units

Calculator Explorations Graphing

To graph an equation such as $x^2 + y^2 = 25$ with a graphing calculator, we first solve the equation for y.

$$x^2 + y^2 = 25$$
$$y^2 = 25 - x^2$$
$$y = \pm\sqrt{25 - x^2}$$

The graph of $y = \sqrt{25 - x^2}$ will be the top half of the circle, and the graph of $y = -\sqrt{25 - x^2}$ will be the bottom half of the circle.

To graph, we press $\boxed{Y=}$ and enter $Y_1 = \sqrt{25 - x^2}$ and $Y_2 = -\sqrt{25 - x^2}$. We insert parentheses about $25 - x^2$ so that $\sqrt{25 - x^2}$ and not $\sqrt{25} - x^2$ is graphed.

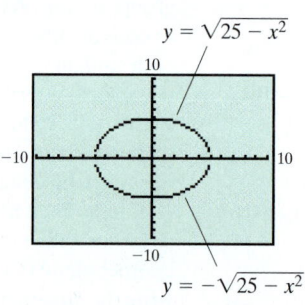

The graph does not appear to be a circle because we are currently using a standard window and the screen is rectangular. This causes the tick marks on the x-axis to be farther apart than the tick marks on the y-axis and thus creates the distorted circle. If we want the graph to appear circular, we define a square window by using a feature of the graphing calculator or redefine the window to show the x-axis from -15 to 15 and the y-axis from -10 to 10. Using a square window, the graph appears as follows:

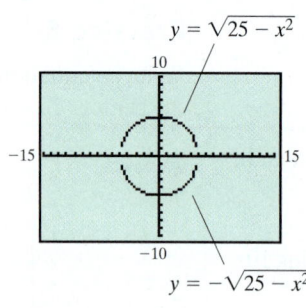

Use a graphing calculator to graph each circle.

1. $x^2 + y^2 = 55$

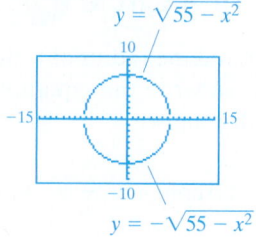

2. $x^2 + y^2 = 20$

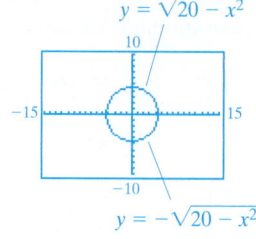

3. $7x^2 + 7y^2 - 89 = 0$

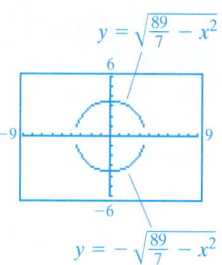

4. $3x^2 + 3y^2 - 35 = 0$

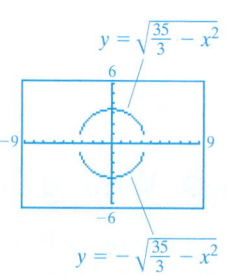

Vocabulary, Readiness & Video Check

Use the choices below to fill in each blank. Some choices may be used more than once.

radius	center	vertex
diameter	circle	conic sections

1. The circle, parabola, ellipse, and hyperbola are called the ___conic sections___.
2. For a parabola that opens upward, the lowest point is the ___vertex___.
3. A ___circle___ is the set of all points in a plane that are the same distance from a fixed point. The fixed point is called the ___center___.
4. The midpoint of a diameter of a circle is the ___center___.
5. The distance from the center of a circle to any point of the circle is called the ___radius___.
6. Twice a circle's radius is its ___diameter___.

Martin-Gay Interactive Videos

See Video 13.1

See video answer section.

Watch the section lecture video and answer the following questions.

Objective A 7. Based on ⊟ Example 1 and the lecture before, would you say that parabolas of the form $x = a(y - k)^2 + h$ are functions? Why or why not? ▶

Objective B 8. Based on the lecture before ⊟ Example 2, what would be the standard form of a circle with its center at the origin? Simplify your answer. ▶

Objective C 9. From ⊟ Example 3, why do we need to complete the square twice when writing this equation of a circle in standard form? ▶

Objective D 10. From ⊟ Example 4, if you know the center and radius of a circle, how can you write that circle's equation? ▶

13.1 Exercise Set MyMathLab®

Objective A *The graph of each equation is a parabola. Determine whether the parabola opens upward, downward, to the left, or to the right. See Examples 1 through 4.*

1. $y = x^2 - 7x + 5$ upward

2. $y = -x^2 + 16$ downward

3. $x = -y^2 - y + 2$ to the left

4. $x = 3y^2 + 2y - 5$ to the right

5. $y = -x^2 + 2x + 1$ downward

6. $x = -y^2 + 2y - 6$ to the left

The graph of each equation is a parabola. Find the vertex of the parabola and then graph it. See Examples 1 through 4.

7. $x = 3y^2$

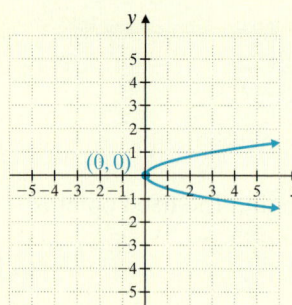

8. $x = 5y^2$

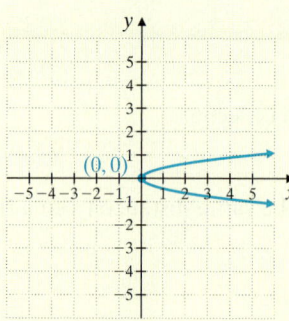

9. $x = -2y^2$

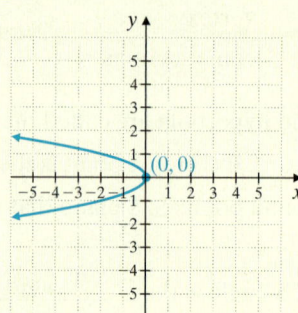

10. $x = -4y^2$

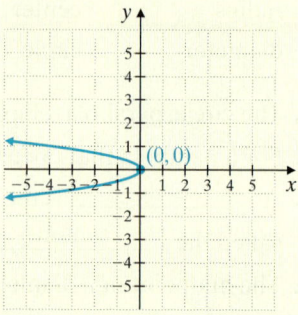

11. $y = -4x^2$

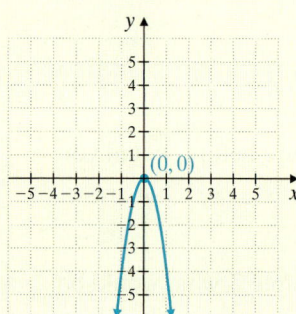

12. $y = -2x^2$

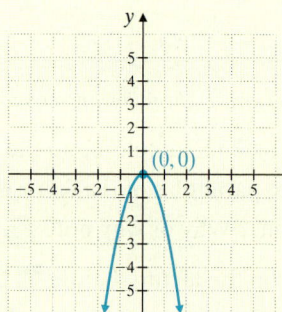

▶ **13.** $x = (y - 2)^2 + 3$

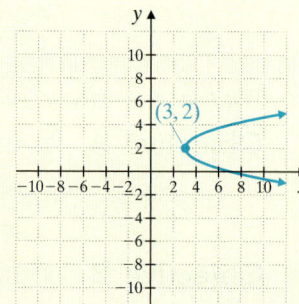

14. $x = (y - 4)^2 - 1$

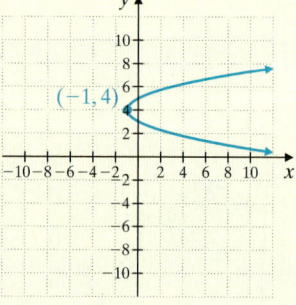

15. $y = -3(x - 1)^2 + 5$

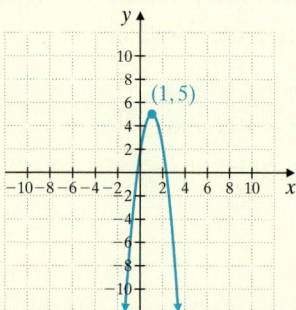

16. $y = -4(x - 2)^2 + 2$

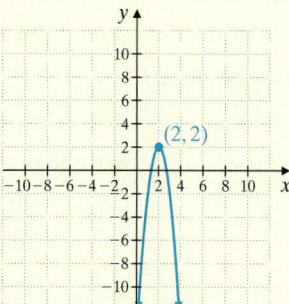

17. $x = y^2 + 6y + 8$

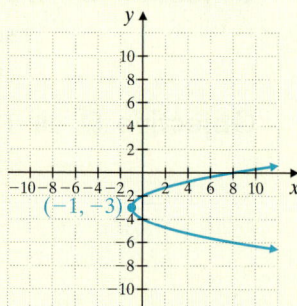

18. $x = y^2 - 6y + 6$

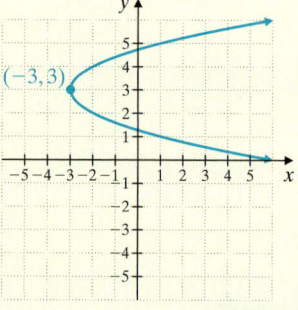

19. $y = x^2 + 10x + 20$

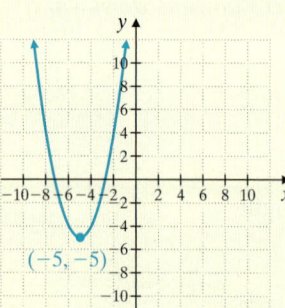

20. $y = x^2 + 4x - 5$

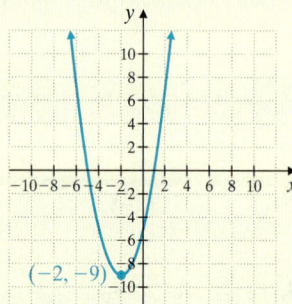

21. $x = -2y^2 + 4y + 6$

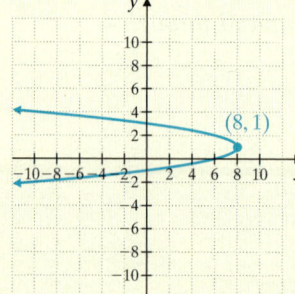

22. $x = 3y^2 + 6y + 7$

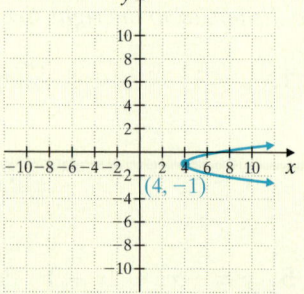

Objectives B C Mixed Practice *The graph of each equation is a circle. Find the center and the radius, and then graph the circle. See Examples 5 through 7.*

23. $x^2 + y^2 = 9$

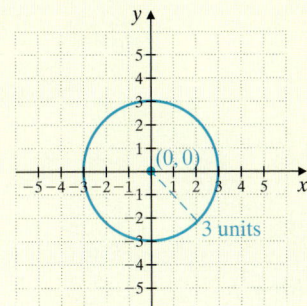

24. $x^2 + y^2 = 25$

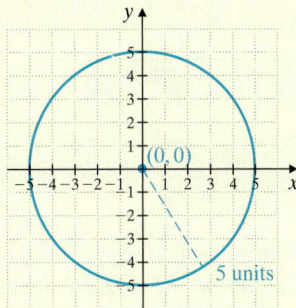

25. $x^2 + (y - 2)^2 = 1$

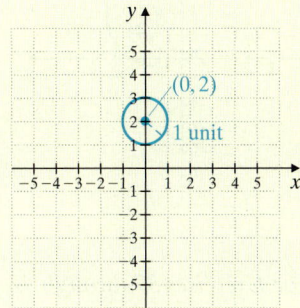

26. $(x - 3)^2 + y^2 = 9$

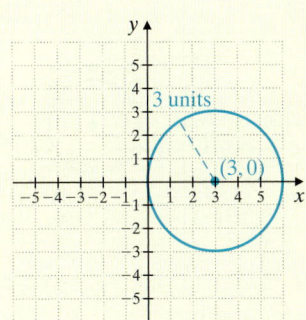

▶ 27. $(x - 5)^2 + (y + 2)^2 = 1$

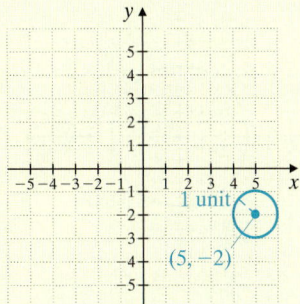

28. $(x + 3)^2 + (y + 3)^2 = 4$

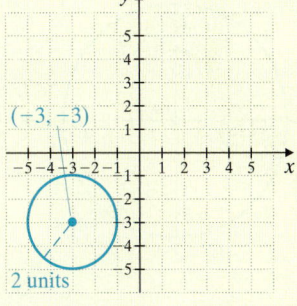

29. $x^2 + y^2 + 6y = 0$

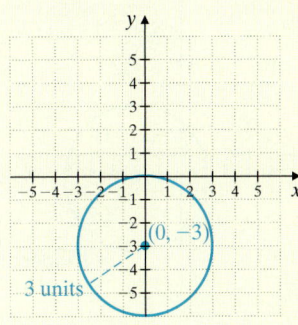

30. $x^2 + 10x + y^2 = 0$

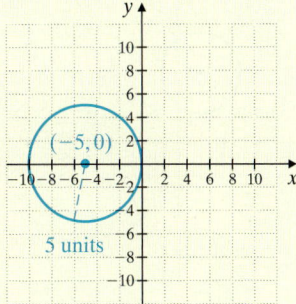

31. $x^2 + y^2 + 2x - 4y = 4$

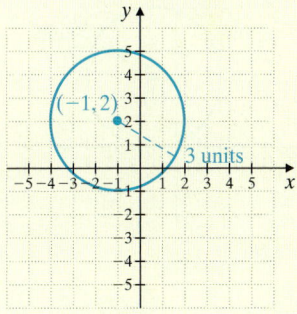

32. $x^2 + y^2 + 6x - 4y = 3$

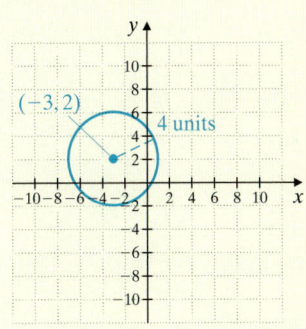

33. $(x + 2)^2 + (y - 3)^2 = 7$

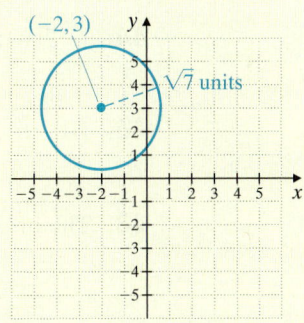

34. $(x + 1)^2 + (y - 2)^2 = 5$

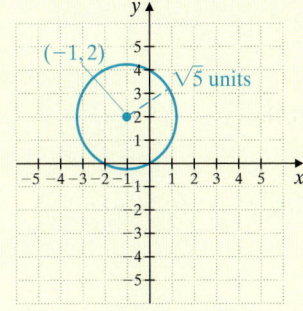

35. $x^2 + y^2 - 4x - 8y - 2 = 0$ **36.** $x^2 + y^2 - 2x - 6y - 5 = 0$

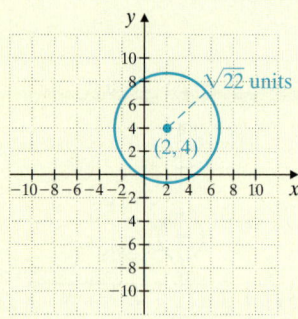   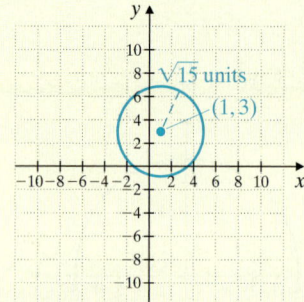

Hint: For Exercises 37 through 42, first divide the equation through by the coefficient of x^2 (or y^2).

37. $3x^2 + 3y^2 = 75$ **38.** $2x^2 + 2y^2 = 18$ **39.** $6(x - 4)^2 + 6(y - 1)^2 = 24$

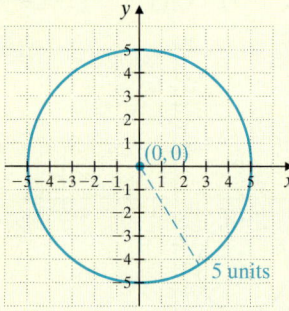   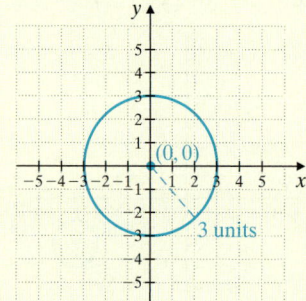   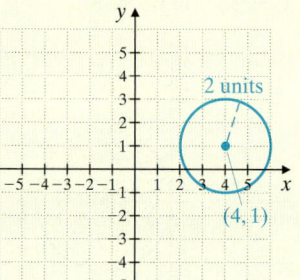

40. $7(x - 1)^2 + 7(y - 3)^2 = 63$ **41.** $4(x + 1)^2 + 4(y - 3)^2 = 12$ **42.** $5(x - 2)^2 + 5(y + 1) = 50$

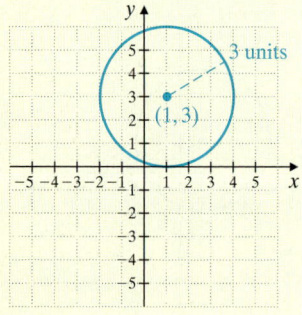   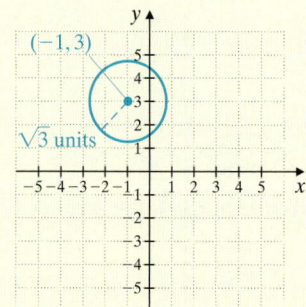   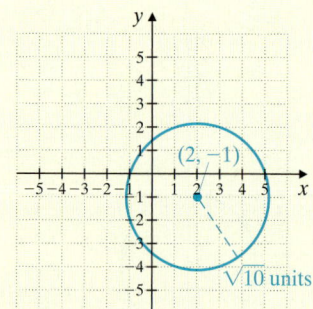

Objective D *Write an equation of the circle with the given center and radius. See Example 8.*

43. $(2, 3); 6$ $(x - 2)^2 + (y - 3)^2 = 36$ **44.** $(-7, 6); 2$ $(x + 7)^2 + (y - 6)^2 = 4$

45. $(0, 0); \sqrt{3}$ $x^2 + y^2 = 3$ **46.** $(0, -6); \sqrt{2}$ $x^2 + (y + 6)^2 = 2$

▶ 47. $(-5, 4); 3\sqrt{5}$ $(x + 5)^2 + (y - 4)^2 = 45$ **48.** The origin; $4\sqrt{7}$ $x^2 + y^2 = 112$

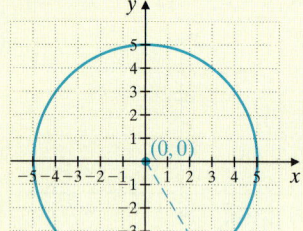

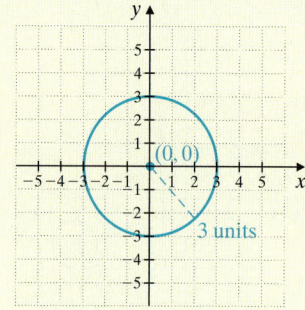

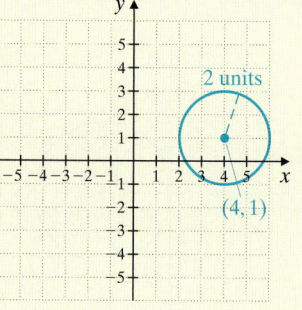

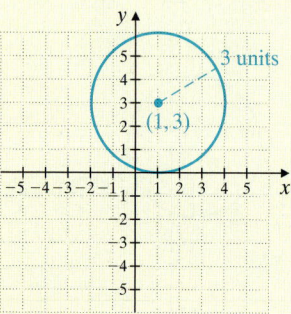

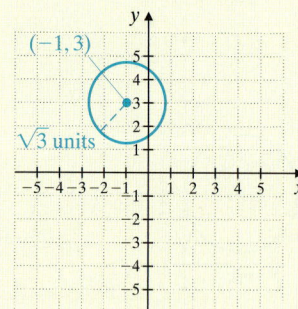

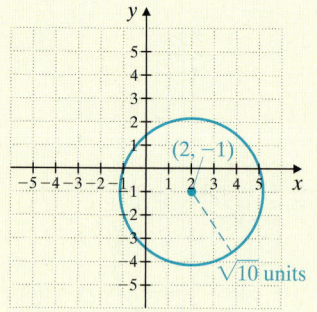

43. $(2, 3); 6$ $(x - 2)^2 + (y - 3)^2 = 36$

44. $(-7, 6); 2$ $(x + 7)^2 + (y - 6)^2 = 4$

45. $(0, 0); \sqrt{3}$ $x^2 + y^2 = 3$

46. $(0, -6); \sqrt{2}$ $x^2 + (y + 6)^2 = 2$

▶**47.** $(-5, 4); 3\sqrt{5}$ $(x + 5)^2 + (y - 4)^2 = 45$

48. The origin; $4\sqrt{7}$ $x^2 + y^2 = 112$

Objectives A B C Mixed Practice *Sketch the graph of each equation. If the graph is a parabola, find its vertex. If the graph is a circle, find its center and radius.*

49. $x = y^2 - 3$

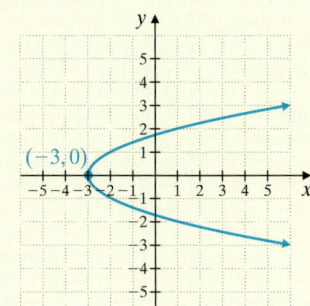

50. $x = y^2 + 2$

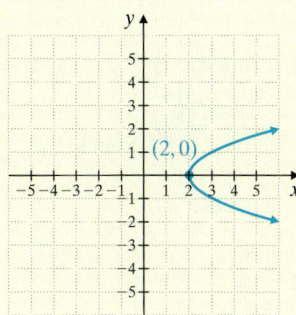

51. $y = (x - 2)^2 - 2$

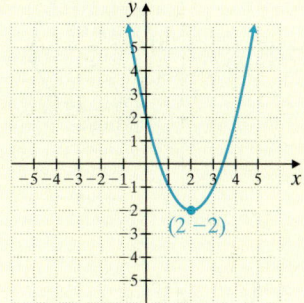

52. $y = (x + 3)^2 + 3$

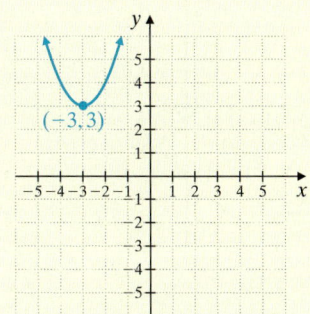

53. $x^2 + y^2 = 1$

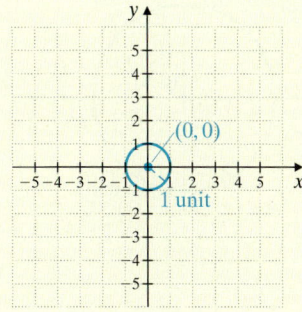

54. $x^2 + y^2 = 49$

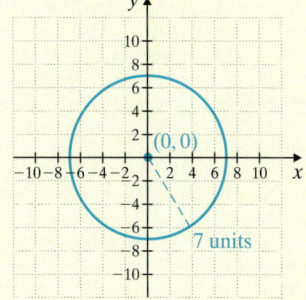

55. $x = (y + 3)^2 - 1$

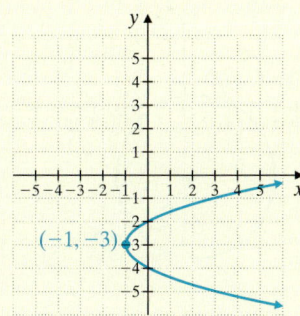

56. $x = (y - 1)^2 + 4$

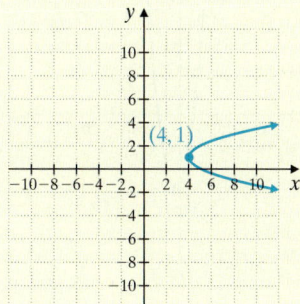

57. $(x - 2)^2 + (y - 2)^2 = 16$

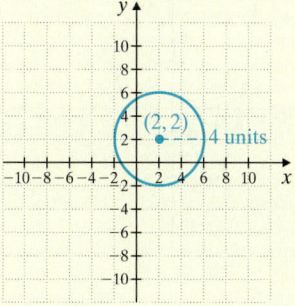

58. $(x + 3)^2 + (y - 1)^2 = 9$

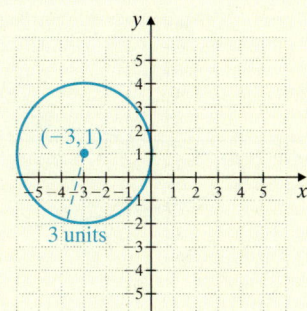

59. $x = -(y - 1)^2$

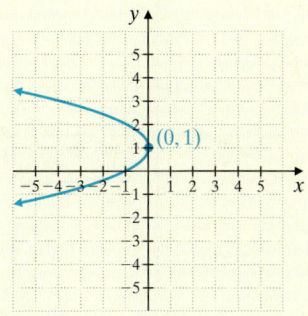

60. $x = -2(y + 5)^2$

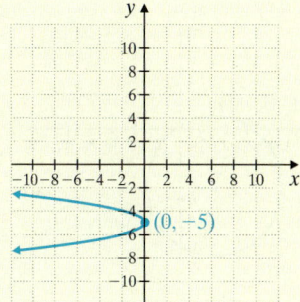

61. $(x - 4)^2 + y^2 = 7$

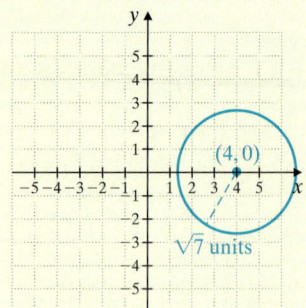

62. $x^2 + (y + 5)^2 = 5$

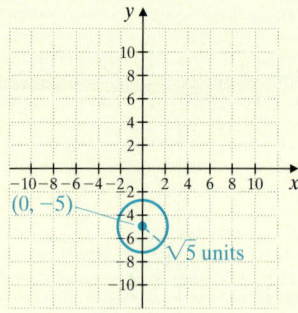

63. $y = 5(x + 5)^2 + 3$

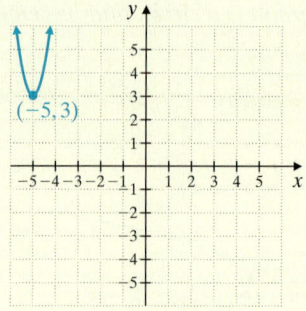

64. $y = 3(x - 4)^2 + 2$

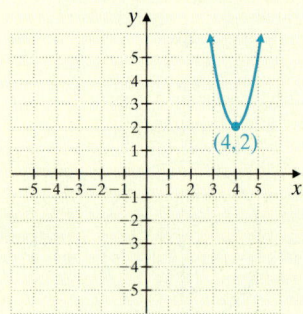

65. $\dfrac{x^2}{8} + \dfrac{y^2}{8} = 2$

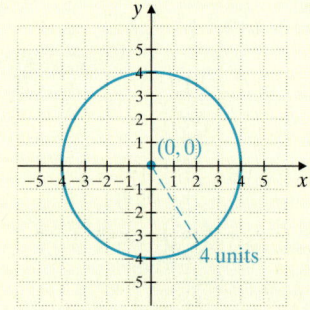

66. $2x^2 + 2y^2 = \dfrac{1}{2}$

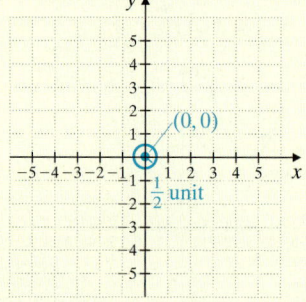

67. $y = x^2 + 7x + 6$

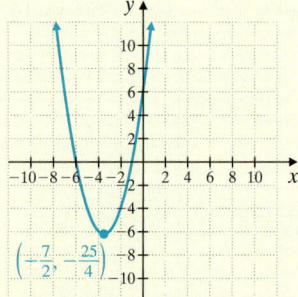

68. $y = x^2 - 2x - 15$

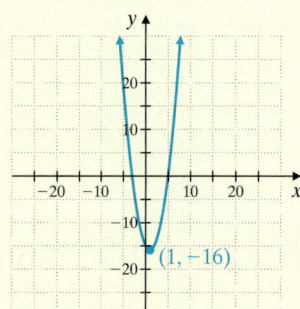

69. $x^2 + y^2 + 2x + 12y - 12 = 0$

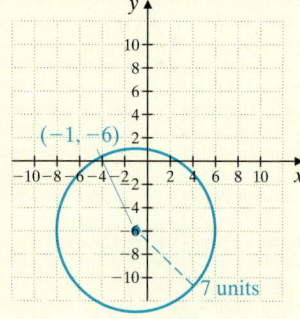

70. $x^2 + y^2 + 6x + 10y - 2 = 0$

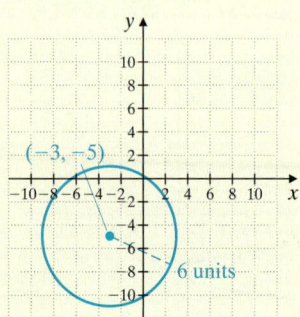

71. $x = y^2 + 8y - 4$

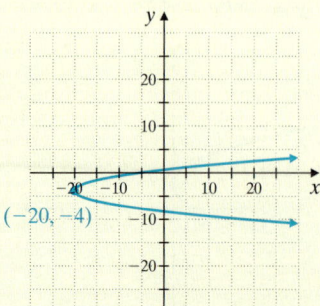

72. $x = y^2 + 6y + 2$

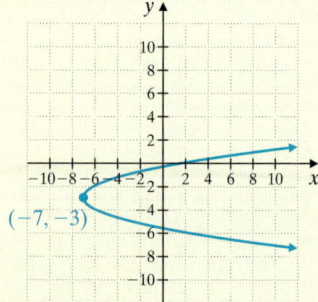

73. $x^2 - 10y + y^2 + 4 = 0$

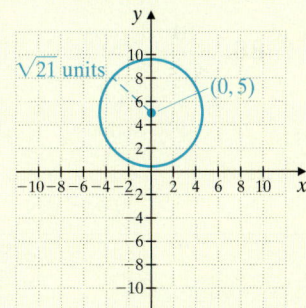

74. $x^2 + y^2 - 8y + 5 = 0$

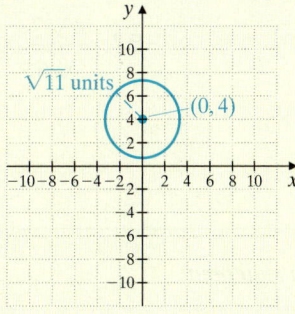

75. $x = -3y^2 + 30y$

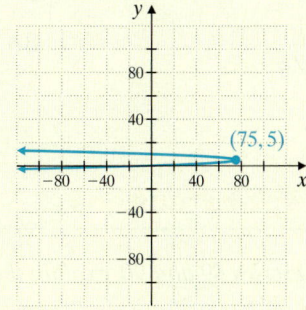

76. $x = -2y^2 - 4y$

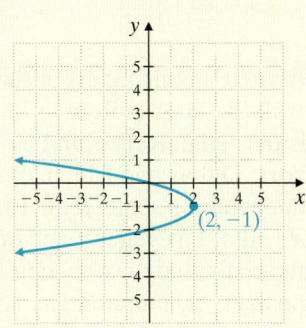

77. $5x^2 + 5y^2 = 25$

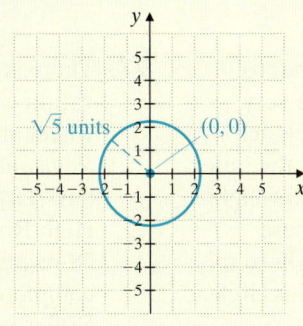

78. $\dfrac{x^2}{3} + \dfrac{y^2}{3} = 2$

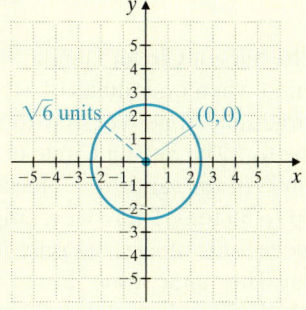

79. $y = 5x^2 - 20x + 16$

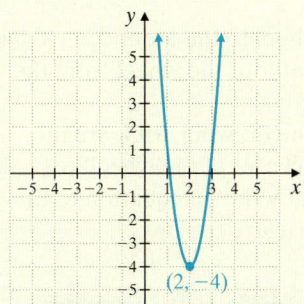

80. $y = 4x^2 - 40x + 105$

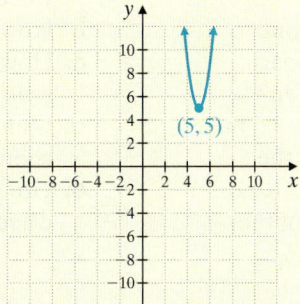

Review

Graph each equation. See Sections 3.2 and 3.3.

81. $y = 2x + 5$

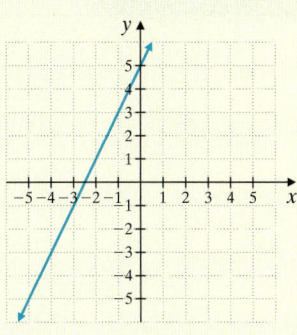

82. $y = -3x + 3$

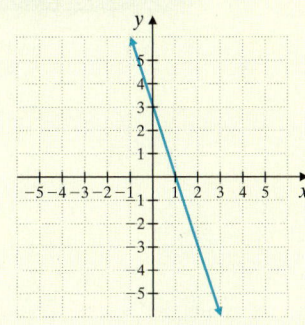

83. $y = 3$

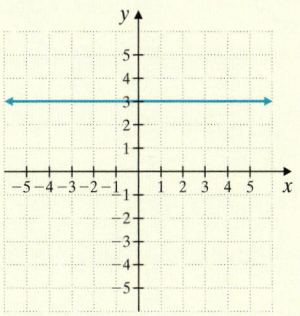

84. $x = -2$

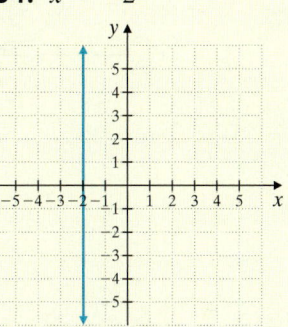

Rationalize each denominator and simplify if possible. See Section 10.5.

85. $\dfrac{1}{\sqrt{3}}$ $\dfrac{\sqrt{3}}{3}$ **86.** $\dfrac{\sqrt{5}}{\sqrt{8}}$ $\dfrac{\sqrt{10}}{4}$ **87.** $\dfrac{4\sqrt{7}}{\sqrt{6}}$ $\dfrac{2\sqrt{42}}{3}$ **88.** $\dfrac{10}{\sqrt{5}}$ $2\sqrt{5}$

Concept Extensions

For Exercises 89 and 90, explain the error in each statement.

89. The graph of $x = 5(y + 5)^2 + 1$ is a parabola with vertex $(-5, 1)$ and opening to the right.
The vertex is $(1, -5)$

90. The graph of $x^2 + (y + 3)^2 = 10$ is a circle with center $(0, -3)$ and radius 5.
The radius is $\sqrt{10}$.

91. The Sarsen Circle: The first image that comes to mind when one thinks of Stonehenge is the very large sandstone blocks with sandstone lintels across the top. The Sarsen Circle of Stonehenge is the outer circle of the sandstone blocks, each of which weighs up to 50 tons. There were originally 30 of these monolithic blocks, but only 17 remain upright to this day. The "altar stone" lies at the center of this circle, which has a diameter of 33 meters.

a. What is the radius of the Sarsen Circle?   16.5 m
b. What is the circumference of the Sarsen Circle? Round your result to 2 decimal places.   103.67 m
c. Since there were originally 30 Sarsen stones located on the circumference, how far apart would the centers of the stones have been? Round to the nearest tenth of a meter.   3.5 m
d. Using the axes in the drawing, what are the coordinates of the center of the circle?   (0, 16.5)
e. Use parts **a** and **d** to write an equation of the Sarsen Circle. $x^2 + (y - 16.5)^2 = (16.5)^2$

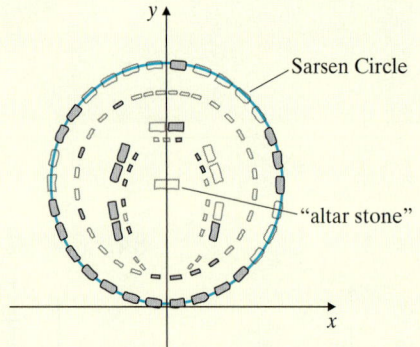

92. There is heavy competition to be the tallest observation wheel in the world. The largest observation wheel at the moment is the "High Roller" in Las Vegas, although New York, Dubai, and China all have larger wheels in the planning stage. From the High Roller, you can see all of Las Vegas. Each of its 28-passenger cabins can accommodate 40 people and completes a full rotation every 30 minutes. The High Roller has a diameter of 520 feet and a height of 550 feet. (*Source:* NBCnews.com)

a. What is the radius of this observation wheel?   260 feet
b. How close is the wheel to the ground?   30 feet
c. How high is the center of the wheel from the ground?   290 feet
d. Using the axes in the drawing, what are the coordinates of the center of the wheel?   (0, 290)
e. Use parts **a** and **d** to write an equation of the wheel. $x^2 + (y - 290)^2 = 260^2$

550 ft   520 ft

93. In 1893, Pittsburgh bridge builder George Ferris designed and built a gigantic revolving steel wheel whose height was 264 feet and diameter was 250 feet. This Ferris wheel opened at the 1893 exposition in Chicago. It had 36 wooden cars, each capable of holding 60 passengers. (*Source: The Handy Science Answer Book*)

a. What was the radius of this Ferris wheel?   125 ft
b. How close is the wheel to the ground?   14 ft
c. How high is the center of the wheel from the ground?   139 ft
d. Using the axes in the drawing, what are the coordinates of the center of the wheel?   (0, 139)
e. Use parts **a** and **d** to write an equation of the wheel. $x^2 + (y - 139)^2 = 125^2$

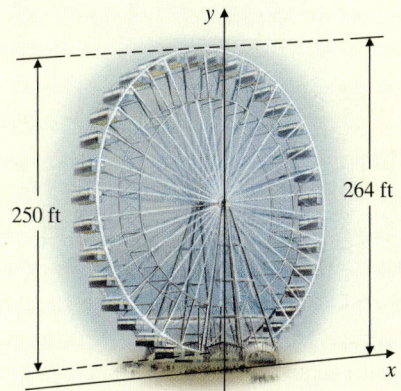

94. Cosmo Clock 21 is a large Ferris wheel currently operating in Yokohama City, Japan. It has a 60-armed wheel, its diameter is 100 meters, and it has a height of 105 meters. (*Source: The Handy Science Answer Book*)

a. What is the radius of this Ferris wheel?   50 m
b. How close is the wheel to the ground?   5 m
c. How high is the center of the wheel from the ground?   55 m
d. Using the axes in the drawing, what are the coordinates of the center of the wheel?   (0, 55)
e. Use parts **a** and **d** to write an equation of the wheel. $x^2 + (y - 55)^2 = 50^2$

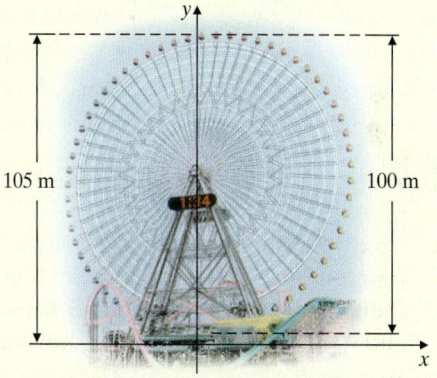

95. If you are given a list of equations of circles and parabolas and none are in standard form, explain how you would determine which is an equation of a circle and which is an equation of a parabola. Explain also how you would distinguish the upward or downward parabolas from the left-opening or right-opening parabolas.   answers may vary

△ **96.** Determine whether the triangle with vertices (2, 6), (0, −2), and (5, 1) is an isosceles triangle.   yes, it is

Solve.

97. Two surveyors need to find the distance across a lake. They place a reference pole at point *A* in the diagram. Point *B* is 3 meters east and 1 meter north of the reference point *A*. Point *C* is 19 meters east and 13 meters north of point *A*. Find the distance across the lake, from *B* to *C*.   20 m

98. A bridge constructed over a bayou has a supporting arch in the shape of a parabola. Find an equation of the parabolic arch if the length of the road over the arch is 100 meters and the maximum height of the arch is 40 meters. (*Hint:* Use the lower horizontal line in the figure as the *x*-axis.)

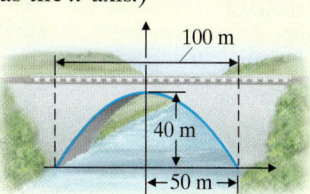

$$y = -\frac{2}{125}x^2 + 40$$

99. Cindy Brown, an architect, is drawing plans on grid paper for a circular pool with a fountain in the middle. The paper is marked off in centimeters, and each centimeter represents 1 foot. On the paper, the diameter of the "pool" is 20 centimeters, and "fountain" is the point $(0, 0)$.

a. Sketch the architect's drawing. Be sure to label the axes.

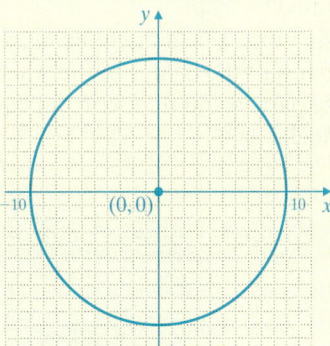

b. Write an equation that describes the circular pool. $x^2 + y^2 = 100$

c. Cindy plans to place a circle of lights around the fountain such that each light is 5 feet from the fountain. Write an equation for the circle of lights and sketch the circle on your drawing. $x^2 + y^2 = 25$

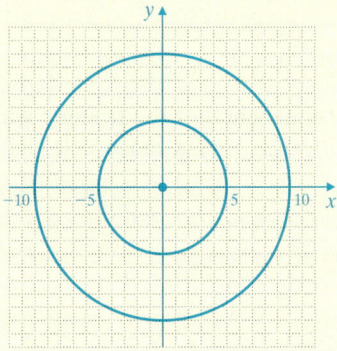

100. The circular Baptistery building in Pisa is the largest such building in Italy, with a height of 180 feet and a circumference of approximately 352 feet. (*Source:* towerofpisa.net)

a. What is the diameter of the base floor (circular imprint) of the Baptistery? Round to two decimal places.
112.05 feet

b. What is the radius of the circular footprint of the Baptistery? Round to two decimal places.
56.03 feet

c. Using the axes in the drawing, what are the coordinates of the center of the circular imprint?
$(0, 56.03)$

d. Write an equation for the circular footprint of the Baptistry. $x^2 + (y - 56.03)^2 = 56.03^2$

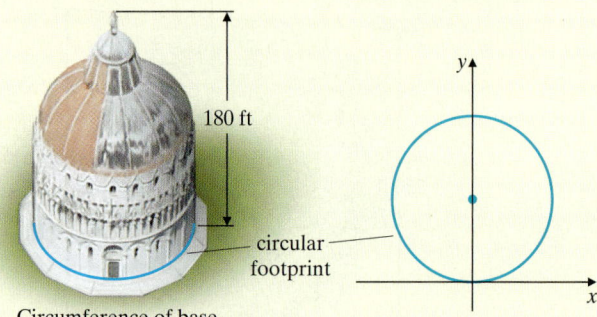

180 ft

circular footprint

Circumference of base
(circular imprint): 352 ft

13.2 **The Ellipse and the Hyperbola**

Objectives

A Define and Graph Ellipses.

B Define and Graph Hyperbolas.

Objective A Graphing Ellipses

An **ellipse** can be thought of as the set of points in a plane such that the sum of the distances of each of those points from two fixed points is constant. Each of the two fixed points is called a **focus.** The plural of focus is **foci.** The point midway between the foci is called the **center.**

An ellipse may be drawn by hand by using two tacks, a piece of string, and a pencil. Secure the two tacks into a piece of cardboard, for example, and tie each end of the string to a tack. Use your pencil to pull the string tight and draw the ellipse. The two tacks are the foci of the drawn ellipse.

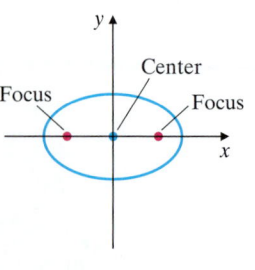

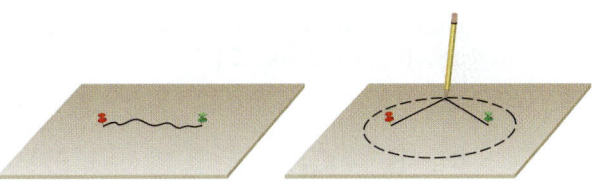

Ellipse with Center (0, 0)

The graph of an equation of the form $\dfrac{x^2}{a^2} + \dfrac{y^2}{b^2} = 1$ is an ellipse with center $(0, 0)$.

The x-intercepts are $(a, 0)$ and $(-a, 0)$, and the y-intercepts are $(0, b)$ and $(0, -b)$.

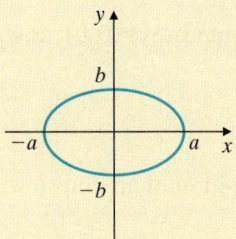

The **standard form** of the equation of an ellipse with center $(0, 0)$ is $\dfrac{x^2}{a^2} + \dfrac{y^2}{b^2} = 1$.

Example 1 Graph: $\dfrac{x^2}{9} + \dfrac{y^2}{16} = 1$

Solution: The equation is of the form $\dfrac{x^2}{a^2} + \dfrac{y^2}{b^2} = 1$ with $a = 3$ and $b = 4$, so its graph is an ellipse with center $(0, 0)$, x-intercepts $(3, 0)$ and $(-3, 0)$, and y-intercepts $(0, 4)$ and $(0, -4)$.

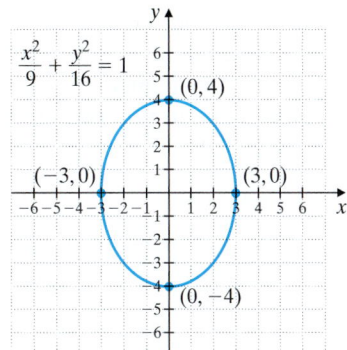

■ **Work Practice 1**

Example 2 Graph: $4x^2 + 16y^2 = 64$

Solution: Although this equation contains a sum of squared terms in x and y on the same side of an equation, this is not the equation of a circle since the coefficients of x^2 and y^2 are not the same. When this happens, the graph is an ellipse. Since the standard form of the equation of an ellipse has 1 on one side, we divide both sides of this equation by 64 to get it in standard form.

$$4x^2 + 16y^2 = 64$$

$$\frac{4x^2}{64} + \frac{16y^2}{64} = \frac{64}{64} \qquad \text{Divide both sides by 64.}$$

$$\frac{x^2}{16} + \frac{y^2}{4} = 1 \qquad \text{Simplify.}$$

(Continued on next page)

Practice 1

Graph: $\dfrac{x^2}{9} + \dfrac{y^2}{4} = 1$

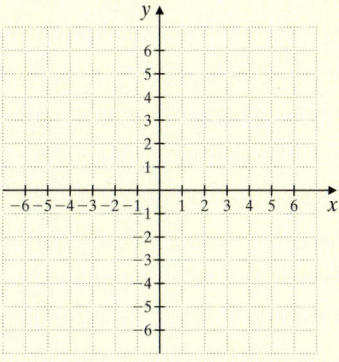

Practice 2

Graph: $4x^2 + 36y^2 = 144$

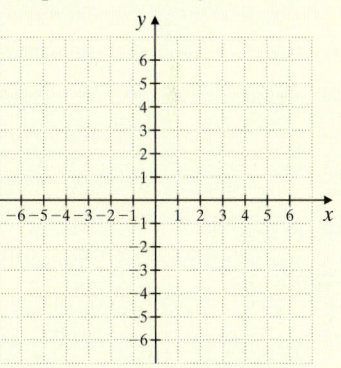

Answers

1.

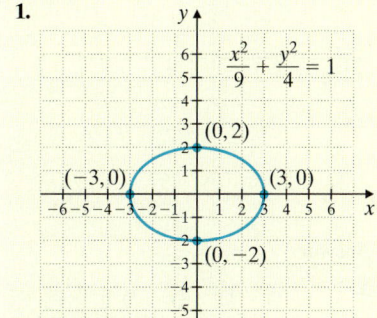

2.

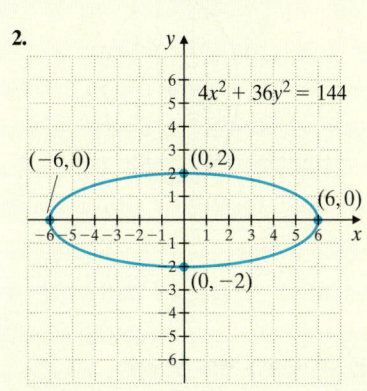

We now recognize the equation of an ellipse with center $(0, 0)$, x-intercepts $(4, 0)$ and $(-4, 0)$, and y-intercepts $(0, 2)$ and $(0, -2)$.

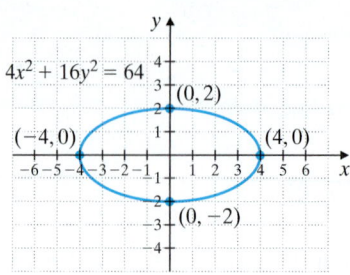

$4x^2 + 16y^2 = 64$

■ **Work Practice 2**

The center of an ellipse is not always $(0, 0)$, as shown in the next example.

> ### Ellipse with Center (h, k)
>
> The standard form of the equation of an ellipse with center (h, k) is
> $$\frac{(x - h)^2}{a^2} + \frac{(y - k)^2}{b^2} = 1$$

Practice 3

Graph:
$$\frac{(x - 1)^2}{9} + \frac{(y - 3)^2}{16} = 1$$

Example 3 Graph: $\dfrac{(x + 3)^2}{25} + \dfrac{(y - 2)^2}{36} = 1$

Solution: This ellipse has center $(-3, 2)$. Notice that $a = 5$ and $b = 6$. To find four points on the graph of the ellipse, we first graph the center, $(-3, 2)$. Since $a = 5$, we count 5 units right and then 5 units left of the point with coordinates $(-3, 2)$. Next, since $b = 6$, we start at $(-3, 2)$ and count 6 units up and then 6 units down to find two more points on the ellipse.

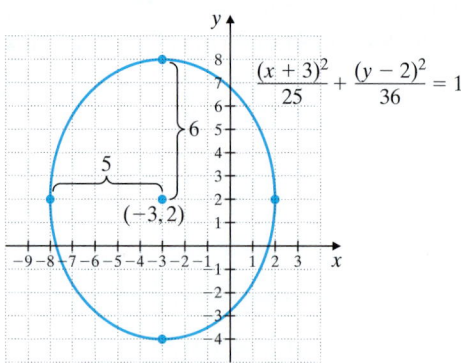

$\dfrac{(x + 3)^2}{25} + \dfrac{(y - 2)^2}{36} = 1$

■ **Work Practice 3**

✓**Concept Check** In the graph of the equation $\dfrac{x^2}{64} + \dfrac{y^2}{36} = 1$, which distance is longer: the distance between the x-intercepts or the distance between the y-intercepts? How much longer? Explain.

Objective B Graphing Hyperbolas ▶

The final conic section is the **hyperbola.** A hyperbola is the set of points in a plane such that for each point in the set, the absolute value of the difference of the

Answer

3.
$$\frac{(x - 1)^2}{9} + \frac{(y - 3)^2}{16} = 1$$

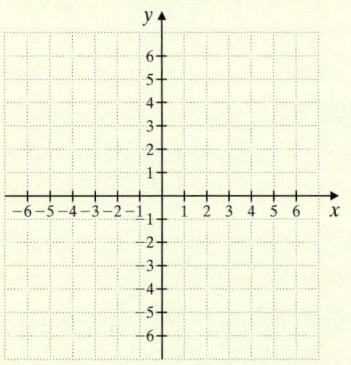

(1, 3)

✓**Concept Check Answer**

x-intercepts, by 4 units

distances from two fixed points is constant. Each of the two fixed points is called a **focus.** The point midway between the foci is called the **center.**

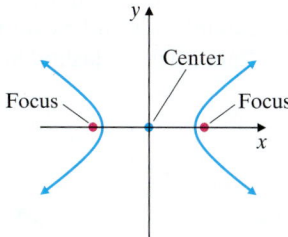

Using the distance formula, we can show that the graph of $\dfrac{x^2}{a^2} - \dfrac{y^2}{b^2} = 1$ is a hyperbola with center $(0, 0)$ and x-intercepts $(a, 0)$ and $(-a, 0)$. Also, the graph of $\dfrac{y^2}{b^2} - \dfrac{x^2}{a^2} = 1$ is a hyperbola with center $(0, 0)$ and y-intercepts $(0, b)$ and $(0, -b)$.

Hyperbola with Center (0, 0)

The graph of an equation of the form $\dfrac{x^2}{a^2} - \dfrac{y^2}{b^2} = 1$ is a hyperbola with center $(0, 0)$ and x-intercepts $(a, 0)$ and $(-a, 0)$.

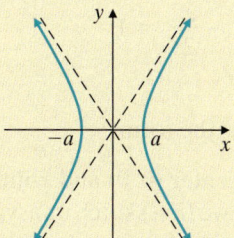

The graph of an equation of the form $\dfrac{y^2}{b^2} - \dfrac{x^2}{a^2} = 1$ is a hyperbola with center $(0, 0)$ and y-intercepts $(0, b)$ and $(0, -b)$.

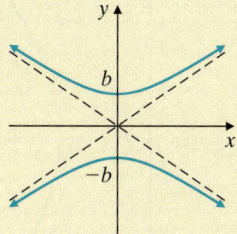

The equations $\dfrac{x^2}{a^2} - \dfrac{y^2}{b^2} = 1$ and $\dfrac{y^2}{b^2} - \dfrac{x^2}{a^2} = 1$ are the **standard forms** for the equation of a hyperbola.

Helpful Hint

Notice the difference between the equation of an ellipse and a hyperbola. The equation of an ellipse contains x^2- and y^2-terms on the same side of the equation with same-sign coefficients. For a hyperbola, the coefficients on the same side of the equation have different signs.

Teaching Tip

After discussing Example 2, ask, "What would we need to do in order to graph this equation using a graphing calculator?" Solve the equation for y with the class, pointing out that the equation turns into two equations when it is solved for y. Then have students graph the equations using graphing calculators.

Teaching Tip

After discussing Example 3, have students expand the equation by clearing the fractions, multiplying the binomials, and collecting like terms. Then have them work in pairs to answer, "If an equation was in this form, what steps would you use to write it in the standard ellipse form?"

Teaching Tip

Have students solve the general forms of the hyperbola with center at $(0, 0)$ for $x = 0$ and again for $y = 0$. Then ask, "Where are these ordered pairs shown on the graph?"

Teaching Tip

Have students find the slope of each asymptote of the general form for the hyperbola with center at $(0, 0)$.

Graphing a hyperbola such as $\dfrac{y^2}{b^2} - \dfrac{x^2}{a^2} = 1$ is made easier by recognizing one of its important characteristics. Examining the figure below, notice how the sides of the branches of the hyperbola extend indefinitely and seem to approach, but not intersect, the dashed lines in the figure. These dashed lines are called the **asymptotes** of the hyperbola.

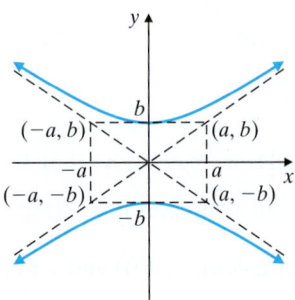

To sketch these lines, or asymptotes, draw a rectangle with vertices (a, b), $(-a, b)$, $(a, -b)$, and $(-a, -b)$. The asymptotes of the hyperbola are the extended diagonals of this rectangle.

Practice 4

Graph: $\dfrac{x^2}{9} - \dfrac{y^2}{4} = 1$

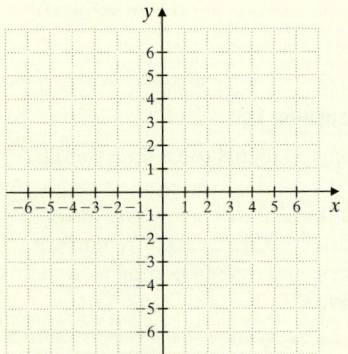

Example 4   Graph: $\dfrac{x^2}{16} - \dfrac{y^2}{25} = 1$

Solution: This equation has the form $\dfrac{x^2}{a^2} - \dfrac{y^2}{b^2} = 1$, with $a = 4$ and $b = 5$. Thus, its graph is a hyperbola with center $(0, 0)$ and x-intercepts of $(4, 0)$ and $(-4, 0)$. To aid in graphing the hyperbola, we first sketch its asymptotes. The extended diagonals of the rectangle with coordinates $(4, 5)$, $(4, -5)$, $(-4, 5)$, and $(-4, -5)$ are the asymptotes of the hyperbola. Then we use the asymptotes to aid in graphing the hyperbola.

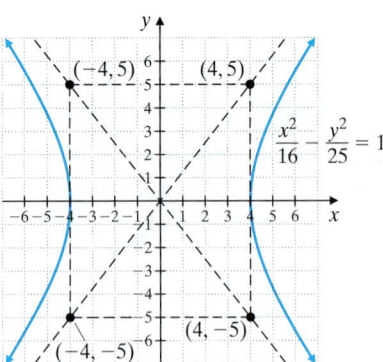

Work Practice 4

Answer

4.

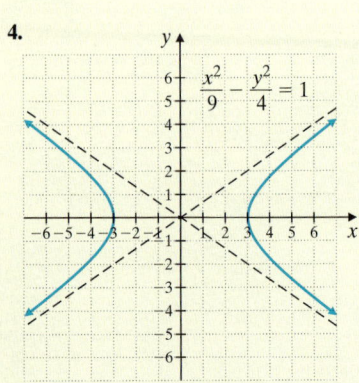

Example 5   Graph: $4y^2 - 9x^2 = 36$

Solution: Since this is a difference of squared terms in x and y on the same side of the equation, its graph is a hyperbola, as opposed to an ellipse or a circle. The standard form of the equation of a hyperbola has a 1 on one side, so we divide both sides of the equation by 36 to get it in standard form.

$$4y^2 - 9x^2 = 36$$

$$\frac{4y^2}{36} - \frac{9x^2}{36} = \frac{36}{36} \qquad \text{Divide both sides by 36.}$$

$$\frac{y^2}{9} - \frac{x^2}{4} = 1 \qquad \text{Simplify.}$$

The equation is of the form $\dfrac{y^2}{b^2} - \dfrac{x^2}{a^2} = 1$ with $a = 2$ and $b = 3$, so the hyperbola is centered at $(0, 0)$ with y-intercepts $(0, 3)$ and $(0, -3)$.

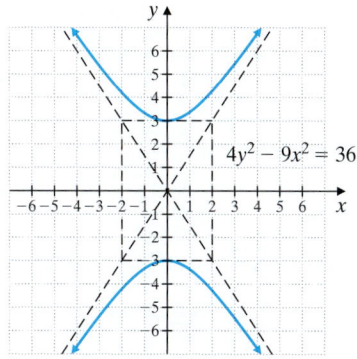

▨ Work Practice 5

Although this is beyond the scope of this text, the standard forms of the equations of hyperbolas with center (h, k) are given below. The Concept Extensions section in this Exercise Set (13.2) contains some hyperbolas of this form.

Hyperbola with Center (h, k)

Standard forms of the equations of hyperbolas with centers (h, k) are:

$$\frac{(x - h)^2}{a^2} - \frac{(y - k)^2}{b^2} = 1 \qquad \frac{(y - k)^2}{b^2} - \frac{(x - h)^2}{a^2} = 1$$

Practice 5

Graph: $9y^2 - 16x^2 = 144$

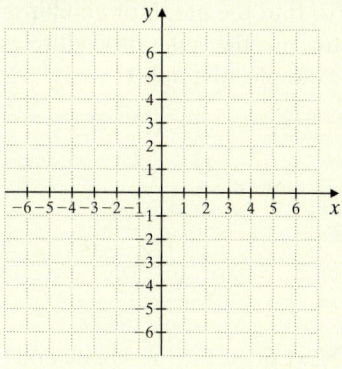

Answer

5.

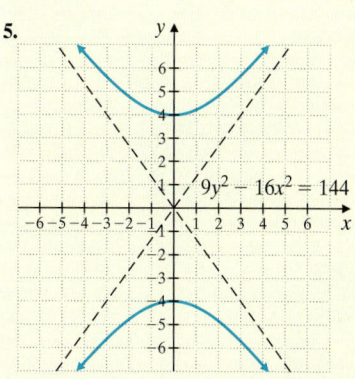

Calculator Explorations Graphing

To find the graph of an ellipse using a graphing calculator, use the same procedure as for graphing a circle. For example, to graph $x^2 + 3y^2 = 22$, first solve for y.

$$3y^2 = 22 - x^2$$

$$y^2 = \frac{22 - x^2}{3}$$

$$y = \pm\sqrt{\frac{22 - x^2}{3}}$$

Next press the $\boxed{Y =}$ key and enter $Y_1 = \sqrt{\frac{22 - x^2}{3}}$ and $Y_2 = -\sqrt{\frac{22 - x^2}{3}}$. (Insert two sets of parentheses in the radicand as in $\sqrt{((22 - x^2)/3)}$ so that the desired graph is obtained.) The graph appears as follows:

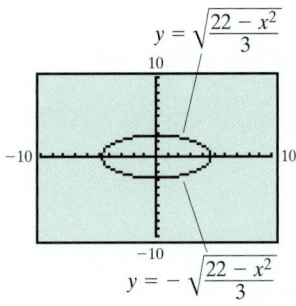

Note: Use a square window for a better view of the shape of each ellipse.

Use a graphing calculator to graph each ellipse.

1. $10x^2 + y^2 = 32$

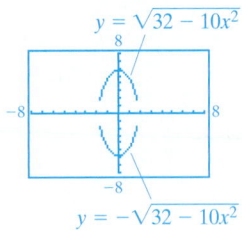

2. $20x^2 + 5y^2 = 100$

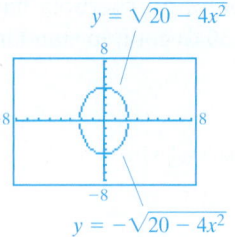

3. $7.3x^2 + 15.5y^2 = 95.2$

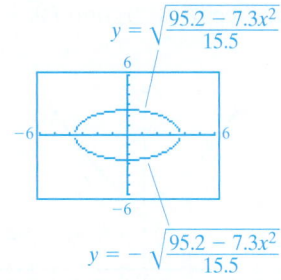

4. $18.8x^2 + 36.1y^2 = 205.8$

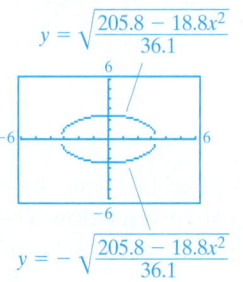

Vocabulary, Readiness & Video Check

Use the choices below to fill in each blank. Some choices will be used more than once and some not at all.

ellipse	$(0, 0)$	x	$(a, 0)$ and $(-a, 0)$	$(0, a)$ and $(0, -a)$	focus
hyperbola	center	y	$(b, 0)$ and $(-b, 0)$	$(0, b)$ and $(0, -b)$	

1. A(n) ____hyperbola____ is the set of points in a plane such that the absolute value of the differences of their distances from two fixed points is constant.

2. A(n) ____ellipse____ is the set of points in a plane such that the sum of their distances from two fixed points is constant.

For Exercises 1 and 2 above,

3. The two fixed points are each called a(n) ____focus____.

4. The point midway between the foci is called the ____center____.

5. The graph of $\dfrac{x^2}{a^2} - \dfrac{y^2}{b^2} = 1$ is a(n) ____hyperbola____ with center ____$(0, 0)$____ and ____x____-intercepts of ____$(a, 0)$ and $(-a, 0)$____.

6. The graph of $\dfrac{x^2}{a^2} + \dfrac{y^2}{b^2} = 1$ is a(n) ____ellipse____ with center ____$(0, 0)$____ and y-intercepts of ____$(0, b)$ and $(0, -b)$____.

Martin-Gay Interactive Videos   Watch the section lecture video and answer the following questions.

Objective A 7. From Example 1, what information do the values of a and b give us about the graph of an ellipse? Answer this same question for Example 2. ▶

Objective B 8. From Example 3, we know the points (a, b), $(a, -b)$, $(-a, b)$, and $(-a, -b)$ are not part of the graph. Explain the role of these points. ▶

See Video 13.2 🍒

See video answer section.

13.2 Exercise Set MyMathLab® ▶

Objective A *Graph each ellipse. See Examples 1 and 2.*

1. $\dfrac{x^2}{4} + \dfrac{y^2}{25} = 1$

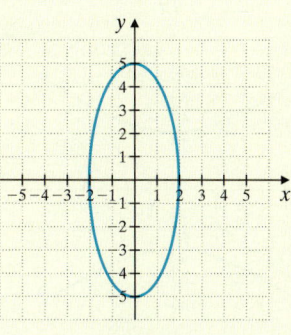

2. $\dfrac{x^2}{16} + \dfrac{y^2}{9} = 1$

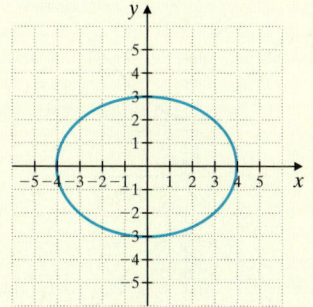

3. $\dfrac{x^2}{9} + y^2 = 1$

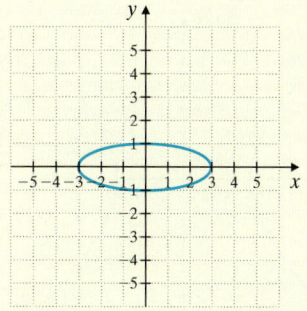

4. $x^2 + \dfrac{y^2}{4} = 1$

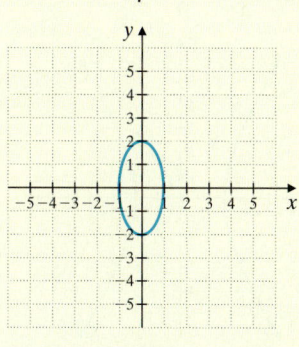

5. $9x^2 + 4y^2 = 36$

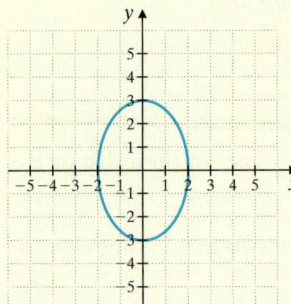

6. $x^2 + 4y^2 = 16$

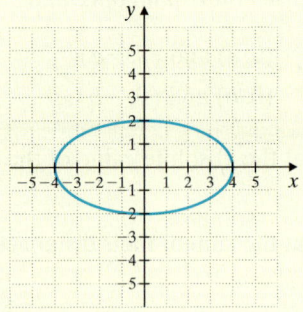

7. $4x^2 + 25y^2 = 100$

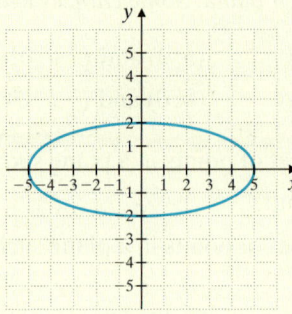

8. $36x^2 + y^2 = 36$

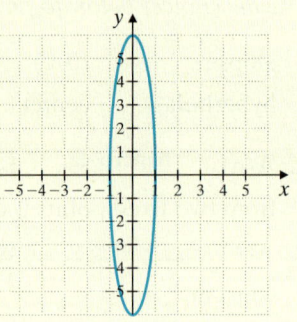

Graph each ellipse. See Example 3.

9. $\dfrac{(x + 1)^2}{36} + \dfrac{(y - 2)^2}{49} = 1$

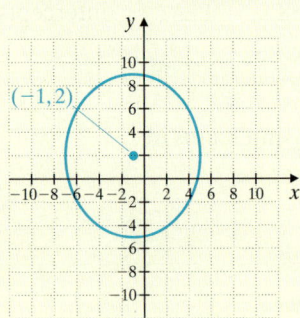

10. $\dfrac{(x - 3)^2}{9} + \dfrac{(y + 3)^2}{16} = 1$

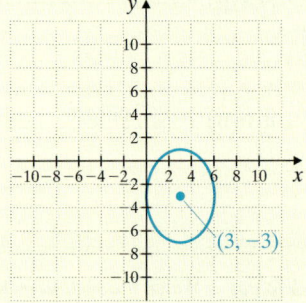

11. $\dfrac{(x - 1)^2}{4} + \dfrac{(y - 1)^2}{25} = 1$

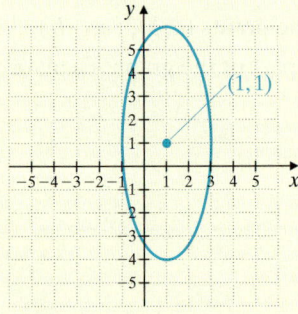

12. $\dfrac{(x + 3)^2}{16} + \dfrac{(y + 2)^2}{4} = 1$

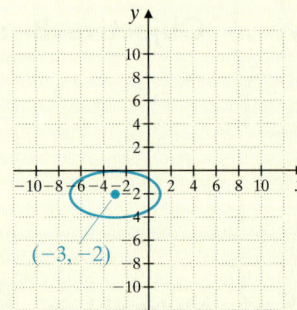

Objective B *Graph each hyperbola. See Examples 4 and 5.*

13. $\dfrac{x^2}{4} - \dfrac{y^2}{9} = 1$

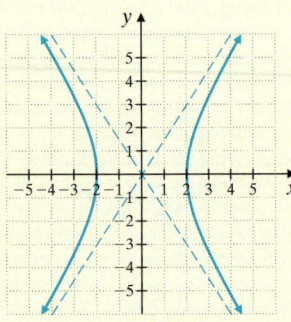

14. $\dfrac{x^2}{36} - \dfrac{y^2}{36} = 1$

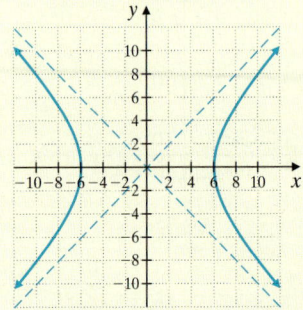

15. $\dfrac{y^2}{25} - \dfrac{x^2}{16} = 1$

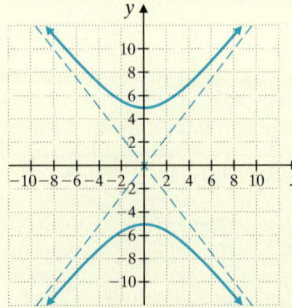

16. $\dfrac{y^2}{25} - \dfrac{x^2}{49} = 1$

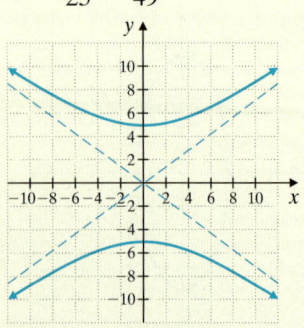

17. $x^2 - 4y^2 = 16$

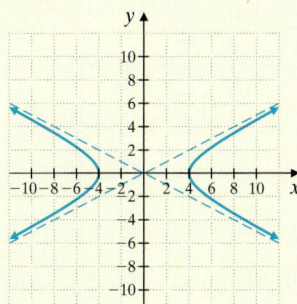

18. $4x^2 - y^2 = 36$

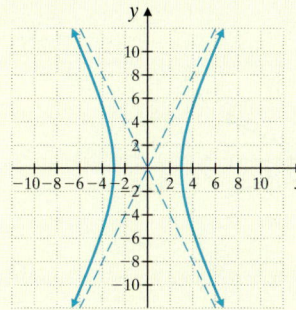

19. $16y^2 - x^2 = 16$

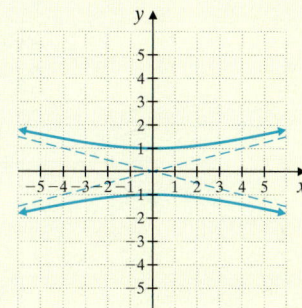

20. $4y^2 - 25x^2 = 100$

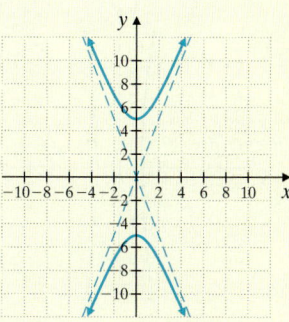

Objectives A B **Mixed Practice** *Identify the graph of each equation as an ellipse or a hyperbola. See Examples 1 through 5.*

21. $\dfrac{x^2}{16} + \dfrac{y^2}{4} = 1$ ellipse

22. $\dfrac{x^2}{16} - \dfrac{y^2}{4} = 1$ hyperbola

23. $x^2 - 5y^2 = 3$ hyperbola

24. $-x^2 + 5y^2 = 3$ hyperbola

25. $-\dfrac{y^2}{25} + \dfrac{x^2}{36} = 1$ hyperbola

26. $\dfrac{y^2}{25} + \dfrac{x^2}{36} = 1$ ellipse

Graph each equation. See Examples 1 through 5.

27. $\dfrac{y^2}{36} = 1 - x^2$

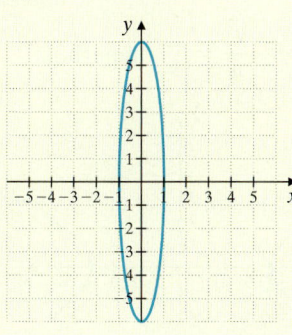

28. $\dfrac{x^2}{36} = 1 - y^2$

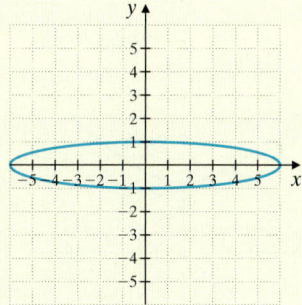

29. $4(x - 1)^2 + 9(y + 2)^2 = 36$

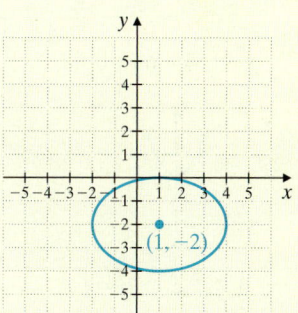

30. $25(x + 3)^2 + 4(y - 3)^2 = 100$

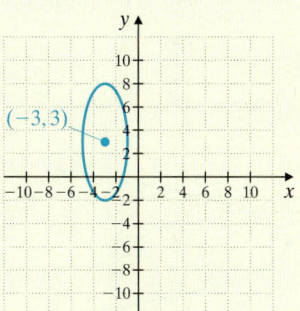

31. $8x^2 + 2y^2 = 32$

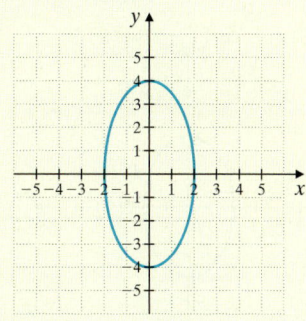

32. $3x^2 + 12y^2 = 48$

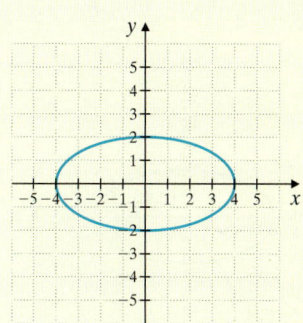

33. $25x^2 - y^2 = 25$

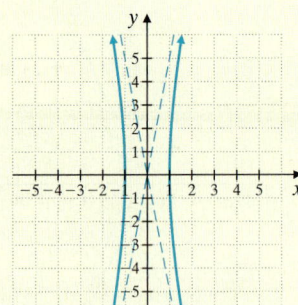

34. $x^2 - 9y^2 = 9$

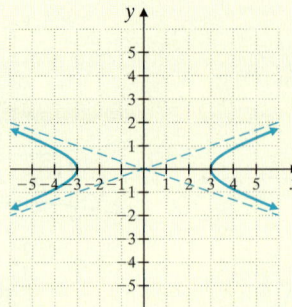

Mixed Practice—Sections 13.1, 13.2 *Identify whether each equation, when graphed, will be a parabola, circle, ellipse, or hyperbola. Sketch the graph of each equation.*

If a parabola, label the vertex.
If a circle, label the center note the radius.
If an ellipse, label the center.
If a hyperbola, label the x- or y-intercepts.

35. $(x - 7)^2 + (y - 2)^2 = 4$ circle

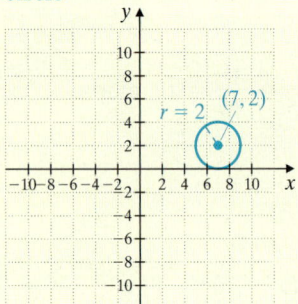

36. $x^2 + y^2 = 16$ circle

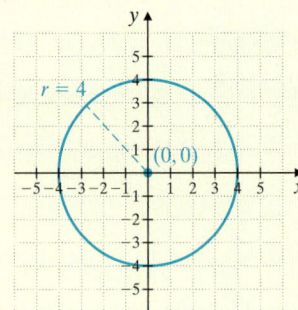

37. $y = x^2 + 12x + 36$ parabola

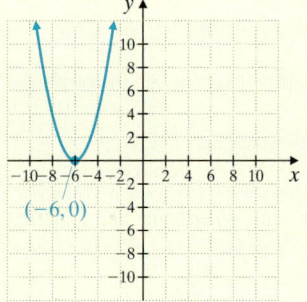

38. $y = x^2 + 4$ parabola

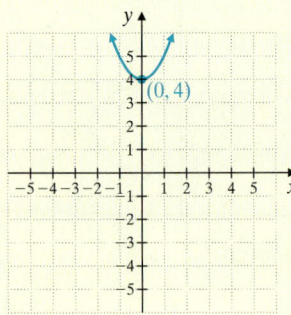

39. $\dfrac{y^2}{9} - \dfrac{x^2}{9} = 1$ hyperbola

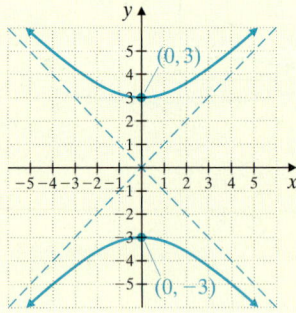

40. $\dfrac{x^2}{16} - \dfrac{y^2}{4} = 1$ hyperbola

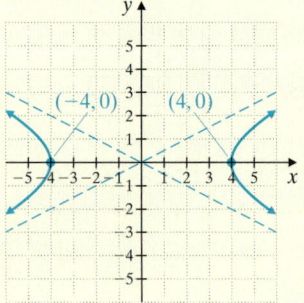

41. $\dfrac{x^2}{16} + \dfrac{y^2}{4} = 1$ ellipse

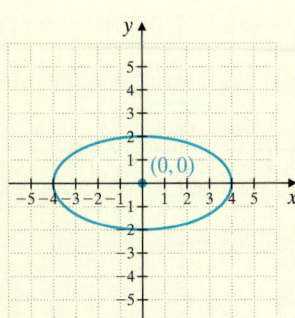

42. $\dfrac{x^2}{4} + \dfrac{y^2}{9} = 1$ ellipse

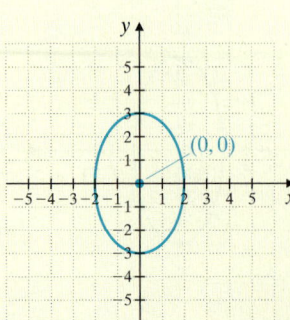

43. $x = y^2 + 4y - 1$ parabola

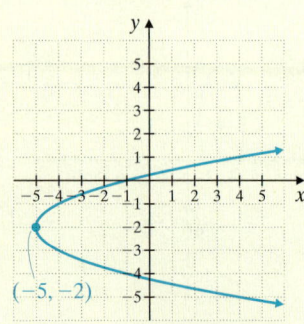

44. $x = -y^2 + 6y$ parabola

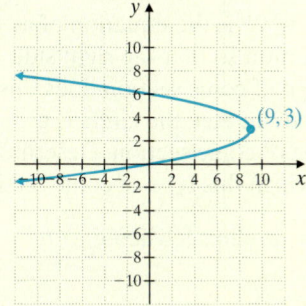

45. $9x^2 - 4y^2 = 36$ hyperbola

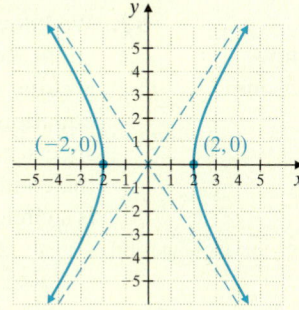

46. $y^2 = x^2 + 16$ hyperbola

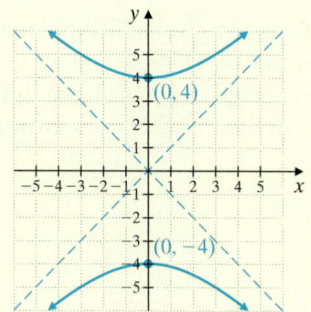

47. $\dfrac{(x-1)^2}{49} + \dfrac{(y+2)^2}{25} = 1$

ellipse

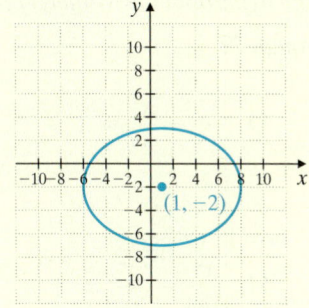

48. $\dfrac{(x+2)^2}{36} + \dfrac{(y-3)^2}{16} = 1$

ellipse

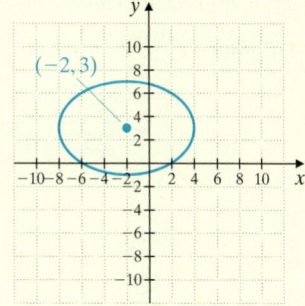

49. $\left(x + \dfrac{1}{2}\right)^2 + \left(y - \dfrac{1}{2}\right)^2 = 1$

circle

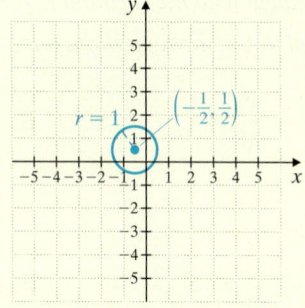

50. $\left(x - \dfrac{3}{2}\right)^2 + \left(y + \dfrac{3}{2}\right)^2 = 9$

circle

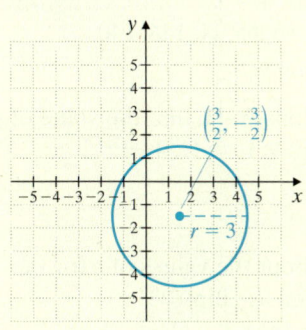

51. $y = -2x^2 + 4x - 3$

parabola

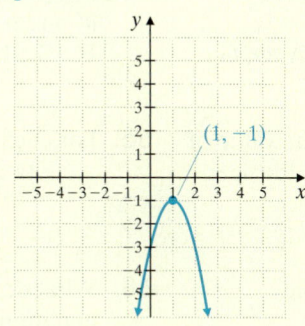

52. $y = -\dfrac{1}{2}x^2 + 2x - 4$

parabola

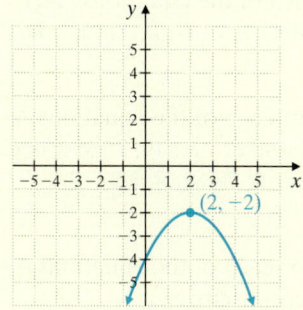

Review

Perform each indicated operation. See Sections 5.1 and 5.4.

53. $(2x^3)(-4x^2)$ $-8x^5$

54. $2x^3 - 4x^3$ $-2x^3$

55. $-5x^2 + x^2$ $-4x^2$

56. $(-5x^2)(x^2)$ $-5x^4$

Concept Extensions

The graph of each equation is an ellipse. Determine which distance is longer, the distance between the x-intercepts or the distance between the y-intercepts. How much longer? See the Concept Check in this section.

57. $\dfrac{x^2}{16} + \dfrac{y^2}{25} = 1$

y-intercepts; 2 units

58. $\dfrac{x^2}{100} + \dfrac{y^2}{49} = 1$

x-intercepts; 6 units

59. $4x^2 + y^2 = 16$

y-intercepts; 4 units

60. $x^2 + 4y^2 = 36$

x-intercepts; 6 units

Solve.

61. If you are given a list of equations of circles, parabolas, ellipses, and hyperbolas, explain how you could distinguish the different conic sections from their equations.   answers may vary

62. We know that $x^2 + y^2 = 25$ is the equation of a circle. Rewrite the equation so that the right side is equal to 1. Which type of conic section does this equation form resemble? In fact, the circle is a special case of this type of conic section. Describe the conditions under which this type of conic section is a circle. $\dfrac{x^2}{25} + \dfrac{y^2}{25} = 1$; ellipse; when $a = b$

The orbits of stars, planets, comets, asteroids, and satellites all have the shape of one of the conic sections. Astronomers use a measure called eccentricity to describe the shape and elongation of an orbital path. For the circle and ellipse, eccentricity e is calculated with the formula $e = \dfrac{c}{d}$, where $c^2 = |a^2 - b^2|$ and d is the larger value of a or b. For a hyperbola, eccentricity e is calculated with the formula $e = \dfrac{c}{d}$, where $c^2 = a^2 + b^2$ and the value of d is equal to a if the hyperbola has x-intercepts or equal to b if the hyperbola has y-intercepts. Use equations A–H to answer Exercises 63 through 72.

A. $\dfrac{x^2}{36} - \dfrac{y^2}{13} = 1$

B. $\dfrac{x^2}{4} + \dfrac{y^2}{4} = 1$

C. $\dfrac{x^2}{25} + \dfrac{y^2}{16} = 1$

D. $\dfrac{y^2}{25} - \dfrac{x^2}{39} = 1$

E. $\dfrac{x^2}{17} + \dfrac{y^2}{81} = 1$

F. $\dfrac{x^2}{36} + \dfrac{y^2}{36} = 1$

G. $\dfrac{x^2}{16} - \dfrac{y^2}{65} = 1$

H. $\dfrac{x^2}{144} + \dfrac{y^2}{140} = 1$

63. Identify the type of conic section represented by each of the equations A–H.
ellipses: C, E, H; circles: B, F; hyperbolas: A, D, G

64. For each of the equations A–H, identify the values of a^2 and b^2.   A: 36, 13; B: 4, 4; C: 25, 16; D: 39, 25; E: 17, 81; F: 36, 36; G: 16, 65; H: 144, 140

65. For each of the equations A–H, calculate the value of c^2 and c.   A: 49, 7; B: 0, 0; C: 9, 3; D: 64, 8; E: 64, 8; F: 0, 0; G: 81, 9; H: 4, 2

66. For each of the equations A–H, find the value of d.
A: 6; B: 2; C: 5; D: 5; E: 9; F: 6; G: 4; H: 12

67. For each of the equations A–H, calculate the eccentricity e.
A: $\dfrac{7}{6}$; B: 0; C: $\dfrac{3}{5}$; D: $\dfrac{8}{5}$; E: $\dfrac{8}{9}$; F: 0; G: $\dfrac{9}{4}$; H: $\dfrac{1}{6}$

68. What do you notice about the values of e for the equations you identified as ellipses?
greater than 0 and less than 1

69. What do you notice about the values of e for the equations you identified as circles?   equal to 0

70. What do you notice about the values of e for the equations you identified as hyperbolas?
greater than 1

71. The eccentricity of a parabola is exactly 1. Use this information and the observations you made in Exercises 68, 69, and 70 to describe a way that could be used to identify the type of conic section based on its eccentricity value.   answers may vary

72. Graph each of the conic sections given in equations A–H. What do you notice about the shape of the ellipses for increasing values of eccentricity? Which is the most elliptical? Which is the least elliptical, that is, the most circular?   answers may vary

A.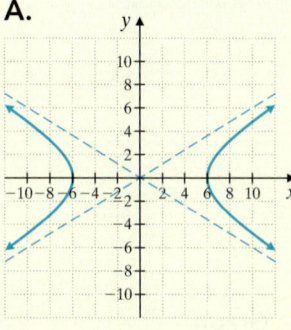

B.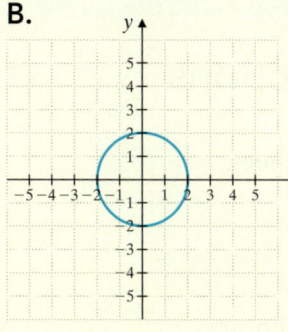

C.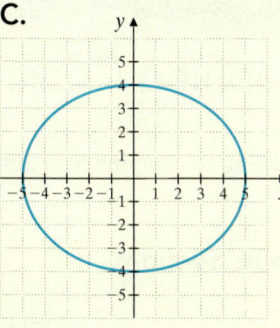

D.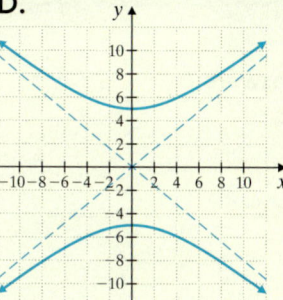

E.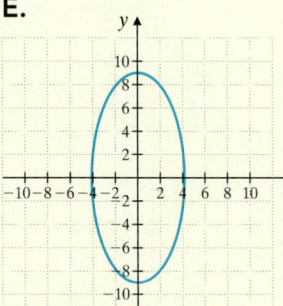

F.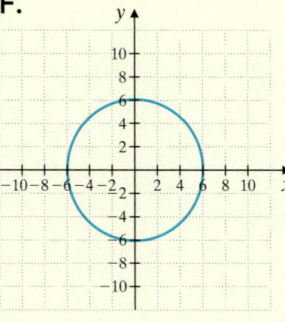

G.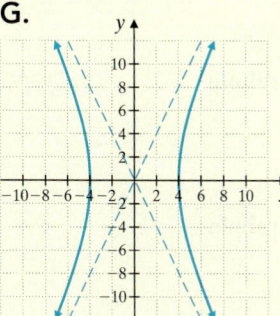

H.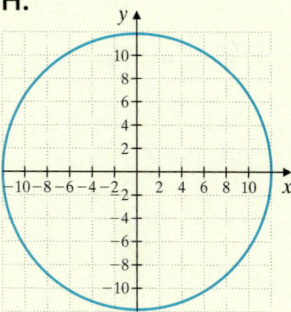

For Exercises 73 through 78, see the example below.

Example

Sketch the graph of $\dfrac{(x-2)^2}{25} - \dfrac{(y-1)^2}{9} = 1$.

Solution

This hyperbola has center $(2, 1)$. Notice that $a = 5$ and $b = 3$.

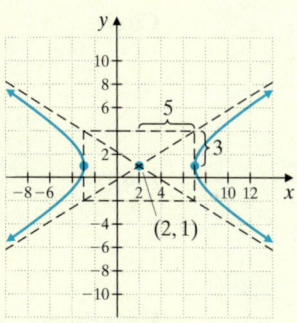

Sketch the graph of each equation.

73. $\dfrac{(x-1)^2}{4} - \dfrac{(y+1)^2}{25} = 1$

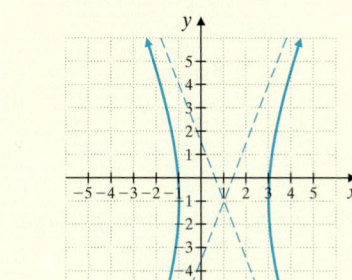

74. $\dfrac{(x+2)^2}{9} - \dfrac{(y-1)^2}{4} = 1$

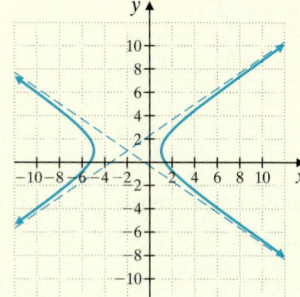

75. $\dfrac{y^2}{16} - \dfrac{(x+3)^2}{9} = 1$

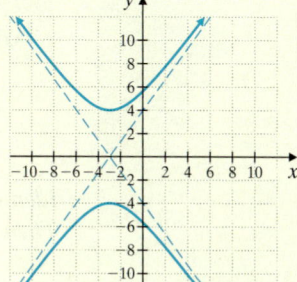

76. $\dfrac{(y+4)^2}{4} - \dfrac{x^2}{25} = 1$

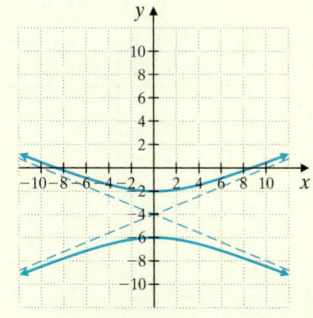

77. $\dfrac{(x+5)^2}{16} - \dfrac{(y+2)^2}{25} = 1$

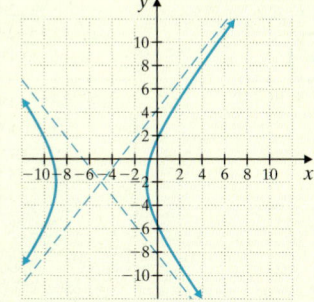

78. $\dfrac{(x-3)^2}{9} - \dfrac{(y-2)^2}{4} = 1$

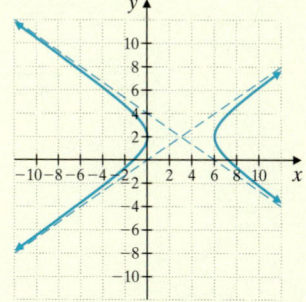

Graphing Conic Sections

Answers

1. circle

2. parabola

3. parabola

Following is a summary of conic sections.

Conic Sections

	Standard Form	**Graph**
Parabola	$y = a(x - h)^2 + k$	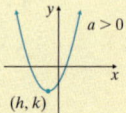
Parabola	$x = a(y - k)^2 + h$	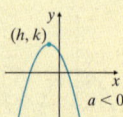
Circle	$(x - h)^2 + (y - k)^2 = r^2$	
Ellipse	$\dfrac{x^2}{a^2} + \dfrac{y^2}{b^2} = 1$	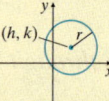
Hyperbola	$\dfrac{x^2}{a^2} - \dfrac{y^2}{b^2} = 1$	
Hyperbola	$\dfrac{y^2}{b^2} - \dfrac{x^2}{a^2} = 1$	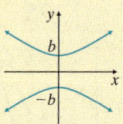

Identify whether each equation, when graphed, will be a parabola, circle, ellipse, or hyperbola. Then graph each equation.

1. $(x - 7)^2 + (y - 2)^2 = 4$ **2.** $y = x^2 + 4$ **3.** $y = x^2 + 12x + 36$

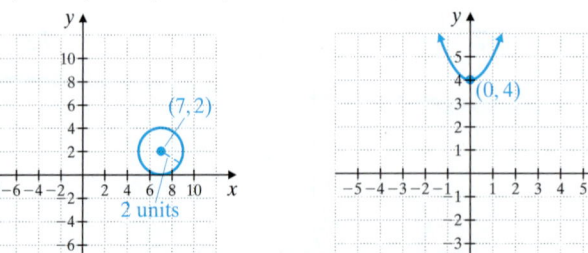

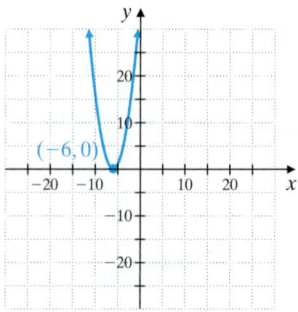

4. $\dfrac{x^2}{4} + \dfrac{y^2}{9} = 1$

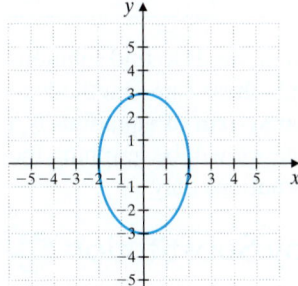

5. $\dfrac{y^2}{9} - \dfrac{x^2}{9} = 1$

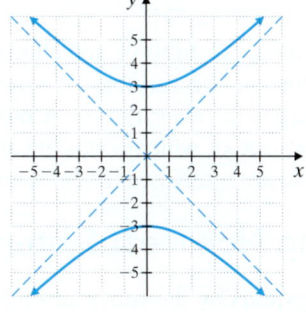

6. $\dfrac{x^2}{16} - \dfrac{y^2}{4} = 1$

7. $2x^2 + 8y^2 = 32$

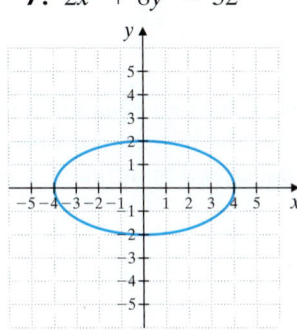

8. $x^2 + y^2 = 16$

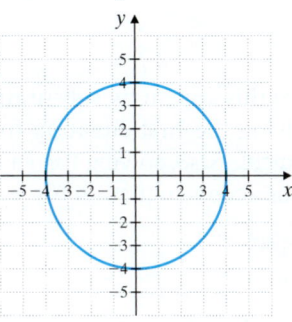

9. $x = y^2 + 4y - 1$

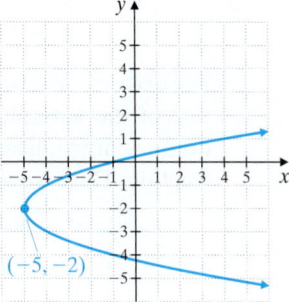

$(-5, -2)$

10. $x = -y^2 + 6y$

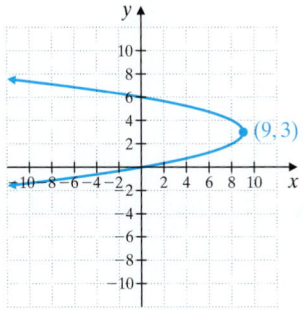

$(9, 3)$

11. $9x^2 - 4y^2 = 36$

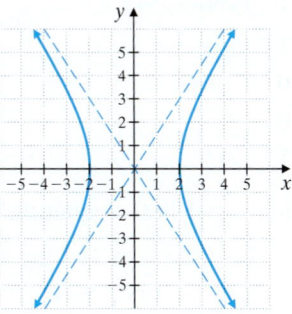

12. $18x^2 + 8y^2 = 72$

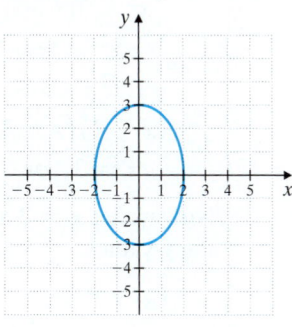

13. $\dfrac{(x-1)^2}{49} + \dfrac{(y+2)^2}{25} = 1$ **14.** $y^2 = x^2 + 16$

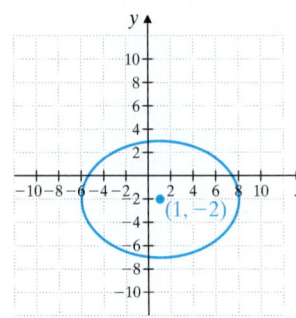

$(1, -2)$

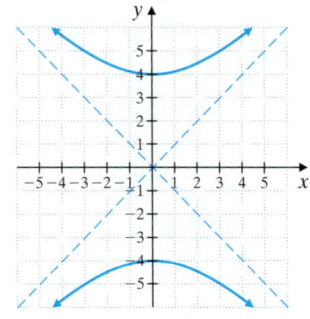

15. $\left(x + \dfrac{1}{2}\right)^2 + \left(y - \dfrac{1}{2}\right)^2 = 1$

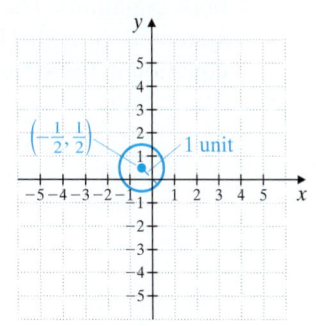

$\left(-\dfrac{1}{2}, \dfrac{1}{2}\right)$ 1 unit

4. ellipse

5. hyperbola

6. hyperbola

7. ellipse

8. circle

9. parabola

10. parabola

11. hyperbola

12. ellipse

13. ellipse

14. hyperbola

15. circle

Solving Nonlinear Systems of Equations

Objectives

A Solve a Nonlinear System by Substitution.

B Solve a Nonlinear System by Elimination.

In Chapter 4, we used graphing, substitution, and elimination methods to find solutions of systems of linear equations in two variables. We now apply these same methods to nonlinear systems of equations in two variables. A **nonlinear system of equations** is a system of equations at least one of which is not linear. Since we will be graphing the equations in each system, we are interested in real number solutions only.

Objective A Solving Nonlinear Systems by Substitution

First we solve nonlinear systems by the substitution method.

Example 1 Solve the system:

$$\begin{cases} x^2 - 3y = 1 \\ x - y = 1 \end{cases}$$

Solution: We can solve this system by substitution if we solve one equation for one of the variables. Solving the first equation for x is not the best choice since doing so introduces a radical. Also, solving for y in the first equation introduces a fraction. Thus, we solve the second equation for y.

$$x - y = 1 \quad \text{Second equation}$$
$$x - 1 = y \quad \text{Solve for } y.$$

Now we replace y with $x - 1$ in the first equation and then solve for x.

$$x^2 - 3y = 1 \quad \text{First equation}$$
$$x^2 - 3(x - 1) = 1 \quad \text{Replace } y \text{ with } x - 1.$$
$$x^2 - 3x + 3 = 1$$
$$x^2 - 3x + 2 = 0$$
$$(x - 2)(x - 1) = 0$$
$$x = 2 \quad \text{or} \quad x = 1$$

Now we let $x = 2$ and then $x = 1$ in the equation $y = x - 1$ to find corresponding y-values.

Let $x = 2$.

$$y = x - 1$$
$$y = 2 - 1 = 1$$

Let $x = 1$.

$$y = x - 1$$
$$y = 1 - 1 = 0$$

When we check $(2, 1)$ and $(1, 0)$ in the equations, we find that both ordered pairs satisfy both equations. Thus, the solution set for the system is $\{(2, 1), (1, 0)\}$. The graph of each equation in the system is shown.

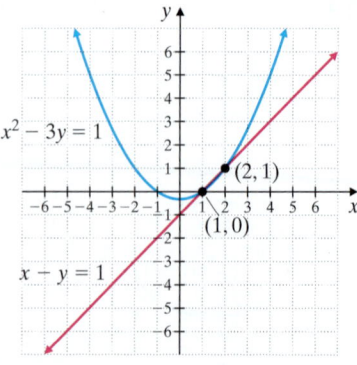

Practice 1

Solve the system:

$$\begin{cases} x^2 - 2y = 5 \\ x + y = -1 \end{cases}$$

Teaching Tip

Before going over the solution for Example 1, have students identify and sketch a generic graph for each equation. Then have them sketch possible intersections for these kinds of graphs.

Answer

1. $\{(-3, 2), (1, -2)\}$

■ **Work Practice 1**

Example 2 Solve the system:

$$\begin{cases} y = \sqrt{x} \\ x^2 + y^2 = 6 \end{cases}$$

Solution: This system is ideal for the substitution method since y is expressed in terms of x in the first equation. Notice that if $y = \sqrt{x}$, then both x and y must be nonnegative if they are real numbers. Let's substitute $\sqrt{x}$ for y in the second equation and solve for x.

$$x^2 + y^2 = 6$$
$$x^2 + (\sqrt{x})^2 = 6 \quad \text{Let } y = \sqrt{x}.$$
$$x^2 + x = 6$$
$$x^2 + x - 6 = 0$$
$$(x + 3)(x - 2) = 0$$
$$x = -3 \quad \text{or} \quad x = 2$$

The solution -3 is discarded because we have noted that x must be nonnegative. To see this, we let $x = -3$ and $x = 2$ in the first equation to find the corresponding y-values.

Let $x = -3$.   Let $x = 2$.
$$y = \sqrt{x} \qquad\qquad y = \sqrt{x}$$
$$y = \sqrt{-3} \quad \text{Not a real number} \qquad y = \sqrt{2}$$

Since we are interested only in real number solutions, the only solution is $(2, \sqrt{2})$. The solution set is $\{(2, \sqrt{2})\}$. Check to see that this solution satisfies both equations. The graph of each equation in this system is shown.

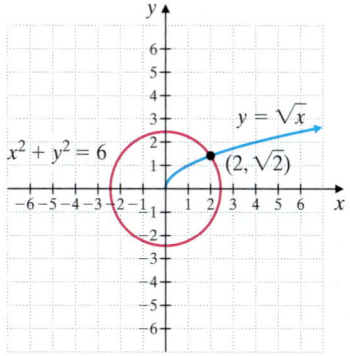

■ Work Practice 2

Example 3 Solve the system:

$$\begin{cases} x^2 + y^2 = 4 \\ x + y = 3 \end{cases}$$

Solution: We use the substitution method and solve the second equation for x

$$x + y = 3 \qquad \text{Second equation}$$
$$x = 3 - y$$

(Continued on next page)

(Continued on next page)

Practice 2

Solve the system:

$$\begin{cases} y = \sqrt{x} \\ x^2 + y^2 = 30 \end{cases}$$

Teaching Tip

Before going over the solution for Example 2, have students identify and sketch a generic graph for each equation. Then have them sketch possible intersections for these kinds of graphs.

Teaching Tip

After discussing the solution to Example 2, have students work in pairs to modify the first equation so that the system would have no solutions. Then have them show algebraically and graphically that their system has no solution.

Teaching Tip

After discussing the solution for Example 3, have students work in pairs to modify the linear equation so that the system has two solutions. Then have them show algebraically and graphically that their system has two solutions.

Practice 3

Solve the system:

$$\begin{cases} x^2 + y^2 = 1 \\ x + y = 4 \end{cases}$$

Answers

2. $\{(5, \sqrt{5})\}$ **3.** no solutions

Now we let $x = 3 - y$ in the first equation.

$$x^2 + y^2 = 4 \quad \text{First equation}$$

$$(3 - y)^2 + y^2 = 4 \quad \text{Let } x = 3 - y.$$

$$9 - 6y + y^2 + y^2 = 4$$

$$2y^2 - 6y + 5 = 0$$

By the quadratic formula, where $a = 2$, $b = -6$, and $c = 5$, we have

$$y = \frac{6 \pm \sqrt{(-6)^2 - 4 \cdot 2 \cdot 5}}{2 \cdot 2} = \frac{6 \pm \sqrt{-4}}{4}$$

Since $\sqrt{-4}$ is not a real number, there is no solution. Graphically, the circle and the line do not intersect, as shown.

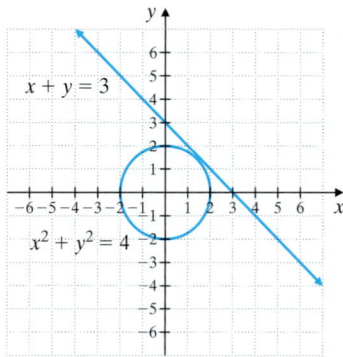

■ **Work Practice 3**

✓**Concept Check** Without solving, how can you tell that the graphs of $x^2 + y^2 = 9$ and $x^2 + y^2 = 16$ do not have any points of intersection?

Objective B Solving Nonlinear Systems by Elimination

Some nonlinear systems may be solved by the elimination method.

Example 4   Solve the system:

$$\begin{cases} x^2 + 2y^2 = 10 \\ x^2 - y^2 = 1 \end{cases}$$

Solution: We will use the elimination, or addition, method to solve this system. To eliminate x^2 when we add the two equations, we multiply both sides of the second equation by -1. Then

$$\begin{array}{l} x^2 + 2y^2 = 10 \\ (-1)(x^2 - y^2) = -1 \cdot 1 \end{array} \quad \begin{array}{l} \text{is} \\ \text{equivalent} \\ \text{to} \end{array} \quad \begin{cases} x^2 + 2y^2 = 10 \\ -x^2 + y^2 = -1 \end{cases}$$

$$3y^2 = 9 \quad \text{Add.}$$

$$y^2 = 3$$

$$y = \pm\sqrt{3} \quad \text{Divide both sides by 3.}$$

Practice 4

Solve the equation:

$$\begin{cases} x^2 + 3y^2 = 21 \\ x^2 - y^2 = 1 \end{cases}$$

Teaching Tip

Ask students, "Can any of the systems in Examples 1–3 be solved using elimination? Explain." Then have them solve Example 1 using elimination.

Answer

4. $\{(\sqrt{6}, \sqrt{5}), (\sqrt{6}, -\sqrt{5}),$
$(-\sqrt{6}, \sqrt{5}), (-\sqrt{6}, -\sqrt{5})\}$

✓ **Concept Check Answer**

$x^2 + y^2 = 9$ is a circle inside the circle $x^2 + y^2 = 16$; therefore, they do not have any points of intersection.

To find the corresponding *x*-values, we let $y = \sqrt{3}$ and $y = -\sqrt{3}$ in either original equation. We choose the second equation.

Let $y = \sqrt{3}$.

$$x^2 - y^2 = 1$$
$$x^2 - (\sqrt{3})^2 = 1$$
$$x^2 - 3 = 1$$
$$x^2 = 4$$
$$x = \pm\sqrt{4} = \pm 2$$

Let $y = -\sqrt{3}$.

$$x^2 - y^2 = 1$$
$$x^2 - (-\sqrt{3})^2 = 1$$
$$x^2 - 3 = 1$$
$$x^2 = 4$$
$$x = \pm\sqrt{4} = \pm 2$$

The solution set is $\{(2, \sqrt{3}), (-2, \sqrt{3}), (2, -\sqrt{3}), (-2, -\sqrt{3})\}$. Check all four ordered pairs in both equations of the system. The graph of each equation in this system is shown.

Teaching Tip

After discussing the solution for Example 4, have students work in pairs to solve it using substitution.

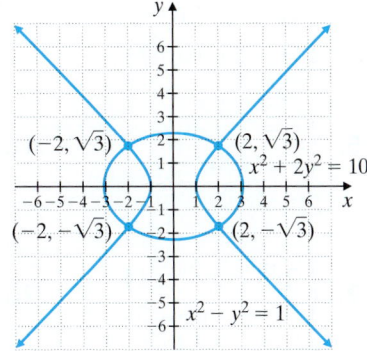

Work Practice 4

Vocabulary, Readiness & Video Check

Martin-Gay Interactive Videos

Watch the section lecture video and answer the following questions.

Objective A **1.** In ▣ Example 1, why do we choose not to solve either equation for *y*? ▶

Objective B **2.** In ▣ Example 2, what important reminder is made as the second equation is multiplied by a number to get opposite coefficients of *x*? ▶

See Video 13.3 🍎

See video answer section.

13.3 Exercise Set MyMathLab®

Objectives **A** **B** **Mixed Practice** *Solve each nonlinear system of equations. See Examples 1 through 4.*

1. $\begin{cases} x^2 + y^2 = 25 \\ 4x + 3y = 0 \end{cases}$
   $\{(3, -4), (-3, 4)\}$

2. $\begin{cases} x^2 + y^2 = 25 \\ 3x + 4y = 0 \end{cases}$
   $\{(-4, 3), (4, -3)\}$

3. $\begin{cases} x^2 + 4y^2 = 10 \\ y = x \end{cases}$
   $\{(\sqrt{2}, \sqrt{2}), (-\sqrt{2}, -\sqrt{2})\}$

4. $\begin{cases} 4x^2 + y^2 = 10 \\ y = x \end{cases}$
   $\{(-\sqrt{2}, -\sqrt{2}), (\sqrt{2}, \sqrt{2})\}$

5. $\begin{cases} y^2 = 4 - x \\ x - 2y = 4 \end{cases}$
   $\{(4, 0), (0, -2)\}$

6. $\begin{cases} x^2 + y^2 = 4 \\ x + y = -2 \end{cases}$
   $\{(-2, 0), (0, -2)\}$

7. $\begin{cases} x^2 + y^2 = 9 \\ 16x^2 - 4y^2 = 64 \end{cases}$
   $\{(-\sqrt{5}, -2), (-\sqrt{5}, 2), (\sqrt{5}, -2), (\sqrt{5}, 2)\}$

8. $\begin{cases} 4x^2 + 3y^2 = 35 \\ 5x^2 + 2y^2 = 42 \end{cases}$
   $\{(2\sqrt{2}, 1), (-2\sqrt{2}, 1), (2\sqrt{2}, -1), (-2\sqrt{2}, -1)\}$

9. $\begin{cases} x^2 + 2y^2 = 2 \\ x - y = 2 \end{cases}$
   $\varnothing$

10. $\begin{cases} x^2 + 2y^2 = 2 \\ x^2 - 2y^2 = 6 \end{cases}$
    $\varnothing$

11. $\begin{cases} y = x^2 - 3 \\ 4x - y = 6 \end{cases}$
    $\{(1, -2), (3, 6)\}$

12. $\begin{cases} y = x + 1 \\ x^2 - y^2 = 1 \end{cases}$
    $\{(-1, 0)\}$

13. $\begin{cases} y = x^2 \\ 3x + y = 10 \end{cases}$
    $\{(2, 4), (-5, 25)\}$

14. $\begin{cases} 6x - y = 5 \\ xy = 1 \end{cases}$
    $\left\{\left(-\dfrac{1}{6}, -6\right), (1, 1)\right\}$

15. $\begin{cases} y = 2x^2 + 1 \\ x + y = -1 \end{cases}$
    $\varnothing$

16. $\begin{cases} x^2 + y^2 = 9 \\ x + y = 5 \end{cases}$
    $\varnothing$

17. $\begin{cases} y = x^2 - 4 \\ y = x^2 - 4x \end{cases}$
    $\{(1, -3)\}$

18. $\begin{cases} x = y^2 - 3 \\ x = y^2 - 3y \end{cases}$
    $\{(-2, 1)\}$

19. $\begin{cases} 2x^2 + 3y^2 = 14 \\ -x^2 + y^2 = 3 \end{cases}$
    $\{(-1, -2), (-1, 2), (1, -2), (1, 2)\}$

20. $\begin{cases} 4x^2 - 2y^2 = 2 \\ -x^2 + y^2 = 2 \end{cases}$
    $\{(\sqrt{3}, \sqrt{5}), (\sqrt{3}, -\sqrt{5}), (-\sqrt{3}, \sqrt{5}), (-\sqrt{3}, -\sqrt{5})\}$

21. $\begin{cases} x^2 + y^2 = 1 \\ x^2 + (y + 3)^2 = 4 \end{cases}$
    $\{(0, -1)\}$

22. $\begin{cases} x^2 + 2y^2 = 4 \\ x^2 - y^2 = 4 \end{cases}$
    $\{(-2, 0), (2, 0)\}$

23. $\begin{cases} y = x^2 + 2 \\ y = -x^2 + 4 \end{cases}$
    $\{(-1, 3), (1, 3)\}$

24. $\begin{cases} x = -y^2 - 3 \\ x = y^2 - 5 \end{cases}$
    $\{(-4, -1), (-4, 1)\}$

25. $\begin{cases} 3x^2 + y^2 = 9 \\ 3x^2 - y^2 = 9 \end{cases}$
    $\{(\sqrt{3}, 0), (-\sqrt{3}, 0)\}$

26. $\begin{cases} x^2 + y^2 = 25 \\ x = y^2 - 5 \end{cases}$
    $\{(-5, 0), (4, -3), (4, 3)\}$

27. $\begin{cases} x^2 + 3y^2 = 6 \\ x^2 - 3y^2 = 10 \end{cases}$
    $\varnothing$

28. $\begin{cases} x^2 + y^2 = 1 \\ y = x^2 - 9 \end{cases}$
    $\varnothing$

29. $\begin{cases} x^2 + y^2 = 36 \\ y = \dfrac{1}{6}x^2 - 6 \end{cases}$
    $\{(-6, 0), (6, 0), (0, -6)\}$

30. $\begin{cases} x^2 + y^2 = 16 \\ y = -\dfrac{1}{4}x^2 + 4 \end{cases}$
    $\{(0, 4), (-4, 0), (4, 0)\}$

31. $\begin{cases} y = \sqrt{x} \\ x^2 + y^2 = 12 \end{cases}$
    $\{(3, \sqrt{3})\}$

32. $\begin{cases} y = \sqrt{x} \\ x^2 + y^2 = 20 \end{cases}$
    $\{(4, 2)\}$

Review

Graph each inequality in two variables. See Section 3.6.

33. $x > -3$

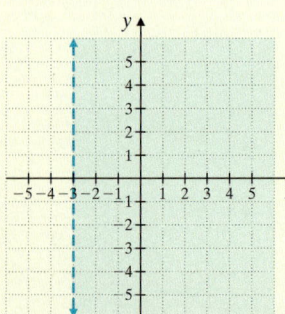

34. $y \le 1$

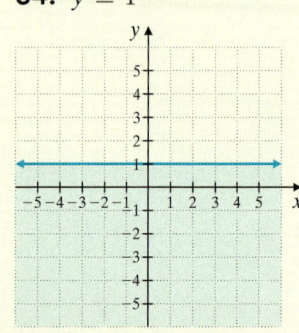

35. $y < 2x - 1$

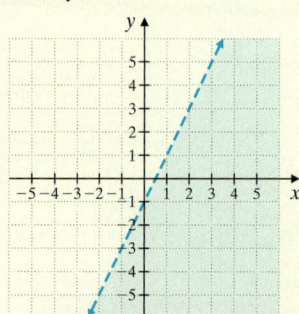

36. $3x - y \le 4$

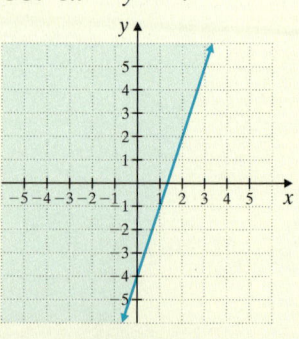

Find the perimeter of each geometric figure. See Section 5.1.

△ **37.**

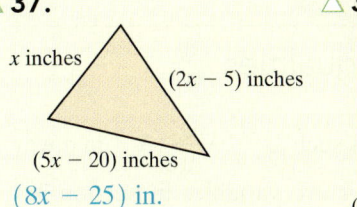

x inches
$(2x - 5)$ inches
$(5x - 20)$ inches
$(8x - 25)$ in.

△ **38.**

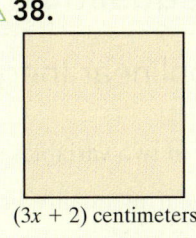

$(3x + 2)$ centimeters
$(12x + 8)$ cm

△ **39.**

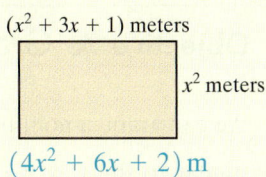

$(x^2 + 3x + 1)$ meters
x^2 meters
$(4x^2 + 6x + 2)$ m

△ **40.**

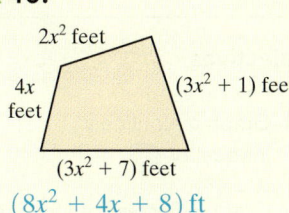

$2x^2$ feet
$4x$ feet
$(3x^2 + 1)$ feet
$(3x^2 + 7)$ feet
$(8x^2 + 4x + 8)$ ft

Concept Extensions

For the exercises below, see the Concept Check in this section.

41. Without solving, how can you tell that the graphs of $x^2 + y^2 = 1$ and $x^2 + y^2 = 4$ do not have any points of intersection?   answers may vary

42. Without solving, how can you tell that the graphs of $y = 2x + 3$ and $y = 2x + 7$ do not have any points of intersection?   answers may vary

43. How many real solutions are possible for a system of equations whose graphs are a circle and a parabola? 0, 1, 2, 3, or 4

44. How many real solutions are possible for a system of equations whose graphs are an ellipse and a line? 0, 1, or 2

Solve.

45. The sum of the squares of two numbers is 130. The difference of the squares of the two numbers is 32. Find the two numbers.
-9 and 7; 9 and -7; 9 and 7; -9 and -7

46. The sum of the squares of two numbers is 20. Their product is 8. Find the two numbers.
2 and 4; -2 and -4

△ **47.** During the development stage of a new rectangular keypad for a security system, it was decided that the area of the rectangle should be 285 square centimeters and the perimeter should be 68 centimeters. Find the dimensions of the keypad.   15 cm by 19 cm

△ **48.** A rectangular holding pen for cattle is to be designed so that its perimeter is 92 feet and its area is 525 feet. Find the dimensions of the holding pen.
21 ft by 25 ft

Recall that in business, a demand function expresses the quantity of a commodity demanded as a function of the commodity's unit price. A supply function expresses the quantity of a commodity supplied as a function of the commodity's unit price. When the quantity produced and supplied is equal to the quantity demanded, then we have what is called **market equilibrium.** *Use this information for Exercises 49 and 50.*

49. The demand function for a certain compact disc is given by the function $p(x) = -0.01x^2 - 0.2x + 9$ and the corresponding supply function is given by $p(x) = 0.01x^2 - 0.1x + 3$, where $p(x)$ is in dollars and x is in thousands of units. Find the equilibrium quantity and the corresponding price by solving the system consisting of the two given equations.
15 thousand compact discs; price: $3.75

50. The demand function for a certain style of picture frame is given by the function $p(x) = -2x^2 + 90$ and the corresponding supply function is given by $p(x) = 9x + 34$, where $p(x)$ is in dollars and x is in thousands of units. Find the equilibrium quantity and the corresponding price by solving the system consisting of the two given equations.
3.5 thousand frames; price: $65.50

Use a grapher to verify the results of each exercise.

51. Exercise 3

52. Exercise 4

53. Exercise 23

54. Exercise 24

13.4 Nonlinear Inequalities and Systems of Inequalities

Objectives

A Graph a Nonlinear Inequality.

B Graph a System of Nonlinear Inequalities.

Objective A Graphing Nonlinear Inequalities

We can graph a nonlinear inequality in two variables such as $\dfrac{x^2}{9} + \dfrac{y^2}{16} \le 1$ in a way similar to the way we graphed a linear inequality in two variables in Section 3.6. First, we graph the related equation $\dfrac{x^2}{9} + \dfrac{y^2}{16} = 1$. The graph of the equation is our boundary. Then, using test points, we determine and shade the region whose points satisfy the inequality.

Practice 1

Graph: $\dfrac{x^2}{25} + \dfrac{y^2}{4} \le 1$

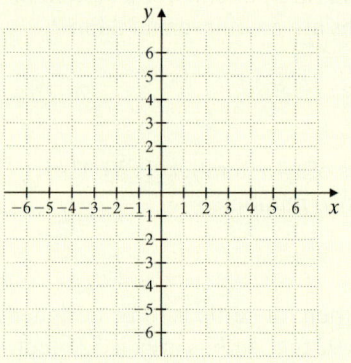

Example 1 Graph: $\dfrac{x^2}{9} + \dfrac{y^2}{16} \le 1$

Solution: First we graph the equation $\dfrac{x^2}{9} + \dfrac{y^2}{16} = 1$. We sketch a solid curve because of the inequality symbol $\le$. It means that the graph of $\dfrac{x^2}{9} + \dfrac{y^2}{16} \le 1$ includes the graph of $\dfrac{x^2}{9} + \dfrac{y^2}{16} = 1$. The graph is an ellipse, and it divides the plane into two regions, the "inside" and the "outside" of the ellipse. Recall from Section 3.6 that to determine which region contains the solutions, we select a test point in either region and determine whether the coordinates of the point satisfy the inequality. We choose $(0, 0)$ as the test point.

$$\frac{x^2}{9} + \frac{y^2}{16} \le 1$$

$$\frac{0^2}{9} + \frac{0^2}{16} \le 1 \quad \text{Let } x = 0 \text{ and } y = 0.$$

$$0 \le 1 \quad \text{True}$$

Since this statement is true, the solution set is the region containing $(0, 0)$. The graph of the solution set includes the points on and inside the ellipse, as shaded in the figure.

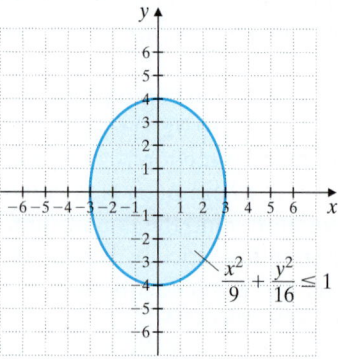

■ **Work Practice 1**

Teaching Tip

Ask students to determine how many regions the graphs of the following types of equations divide a plane into: line, parabola, circle, ellipse, hyperbola. Then have them sketch a picture illustrating the regions.

Answer

1.

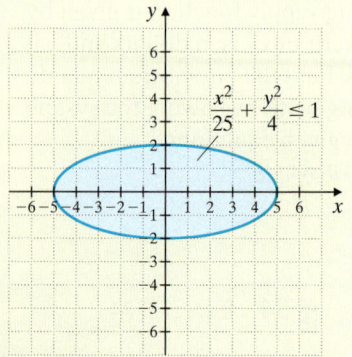

Example 2   Graph: $4y^2 > x^2 + 16$

Solution: The related equation is $4y^2 = x^2 + 16$, or $\dfrac{y^2}{4} - \dfrac{x^2}{16} = 1$, which is a hyperbola. We graph the hyperbola as a dashed curve because of the inequality symbol $>$. It means that the graph of $4y^2 > x^2 + 16$ does *not* include the graph of $4y^2 = x^2 + 16$. The hyperbola divides the plane into three regions. We select a test point in each region—not on a boundary line—to determine whether that region contains solutions of the inequality.

Test Region A with $(0, 4)$	Test Region B with $(0, 0)$	Test Region C with $(0, -4)$
$4y^2 > x^2 + 16$	$4y^2 > x^2 + 16$	$4y^2 > x^2 + 16$
$4(4)^2 > 0^2 + 16$	$4(0)^2 > 0^2 + 16$	$4(-4)^2 > 0^2 + 16$
$64 > 16$ True	$0 > 16$ False	$64 > 16$ True

The graph of the solution set includes the shaded regions *A* and *C* only, not the boundary.

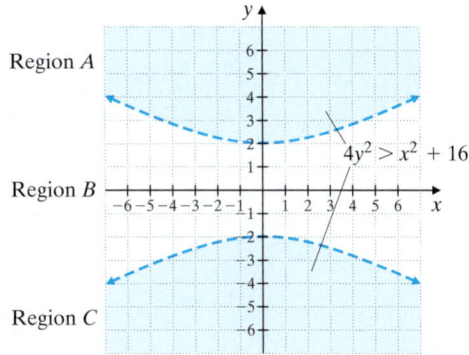

Work Practice 2

Objective B Graphing Systems of Nonlinear Inequalities ▶

In Section 9.3, we graphed systems of linear inequalities. Recall that the graph of a system of inequalities is the intersection of the graphs of the inequalities.

Example 3   Graph the system:

$$\begin{cases} x \le 1 - 2y \\ y \le x^2 \end{cases}$$

Solution: We graph each inequality on the same set of axes. The intersection is the darkest shaded region along with its boundary lines. The coordinates of the points of intersection can be found by solving the related system

$$\begin{cases} x = 1 - 2y \\ y = x^2 \end{cases}$$

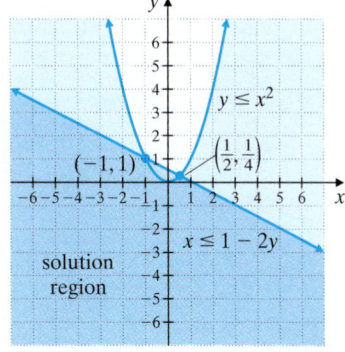

Work Practice 3

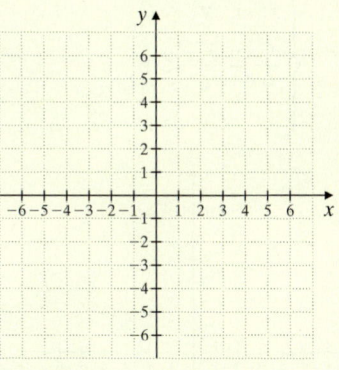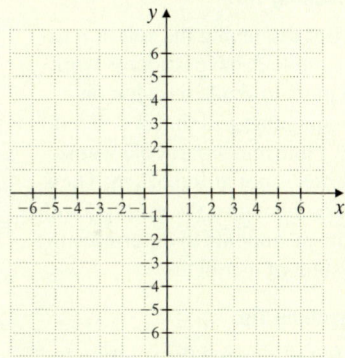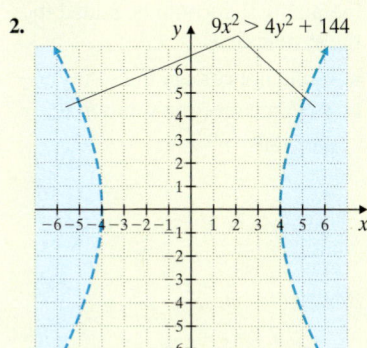
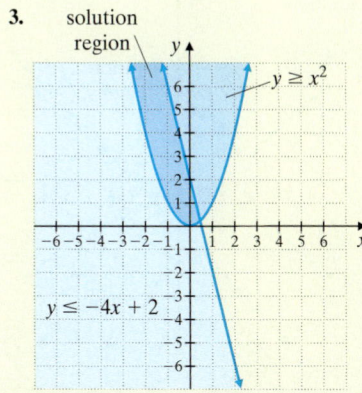

Practice 4

Graph the system:

$$\begin{cases} x^2 + y^2 < 9 \\ \dfrac{x^2}{9} - \dfrac{y^2}{4} < 1 \\ y > x - 2 \end{cases}$$

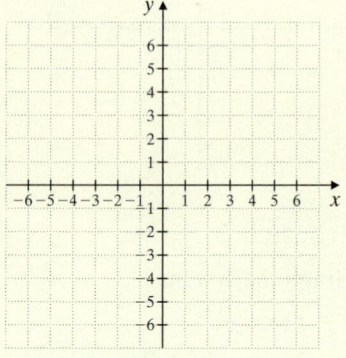

Teaching Tip

After discussing the solution to Example 3, have students work in pairs to shade a region on a rectangular grid that is defined by a line and a parabola whose vertex is not at the origin. Then have them give the grid to another pair of students to determine the system of inequalities.

Teaching Tip

After discussing the solution to Example 4, have students find an inequality that could be added to the system that

a. would not change the solution.

b. would change the solution to the empty set.

Answer

4.

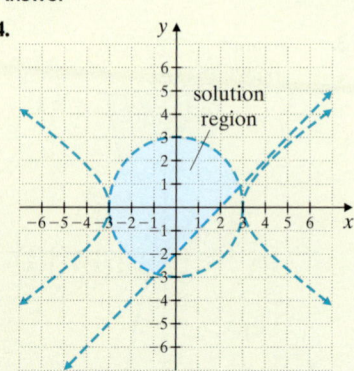

Example 4 Graph the system:

$$\begin{cases} x^2 + y^2 < 25 \\ \dfrac{x^2}{9} - \dfrac{y^2}{25} < 1 \\ y < x + 3 \end{cases}$$

Solution: We graph each inequality. The graph of $x^2 + y^2 < 25$ contains points "inside" the circle that has center $(0,0)$ and radius 5. The graph of $\dfrac{x^2}{9} - \dfrac{y^2}{25} < 1$ is the region between the two branches of the hyperbola with x-intercepts $(-3,0)$ and $(3,0)$ and center $(0,0)$. The graph of $y < x + 3$ is the region "below" the line with slope 1 and y-intercept $(0,3)$. The graph of the solution set of the system is the intersection of all the graphs. This intersection region is shown as the shaded region on the fourth graph. The boundary of this region is not part of the solution.

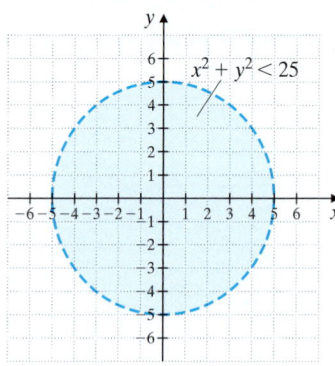

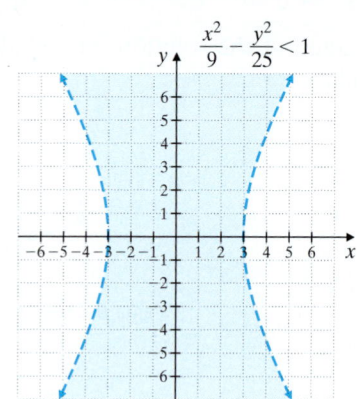

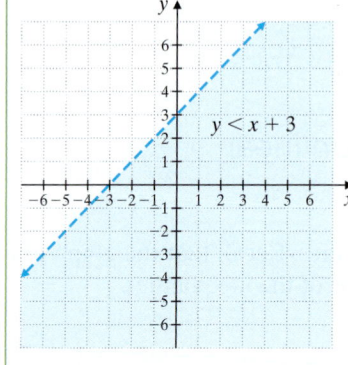

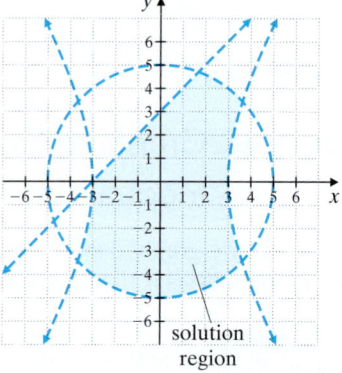

■ **Work Practice 4**

Vocabulary, Readiness & Video Check

Martin-Gay Interactive Videos *Watch the section lecture video and answer the following questions.*

Objective A 1. From Example 1, explain the similarities between graphing linear inequalities and graphing nonlinear inequalities. ▶

Objective B 2. From Example 2, describe one possible illustration of graphs of two circle inequalities in which the system has no solution—that is, the graph of the inequalities in the system do not overlap. ▶

See Video 13.4
See video answer section.

13.4 Exercise Set MyMathLab® ▶

Objective A *Graph each inequality. See Examples 1 and 2.*

1. $y < x^2$

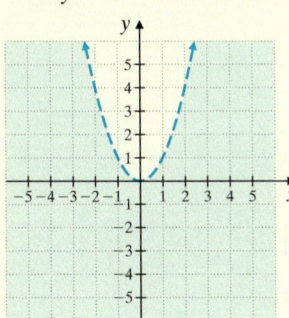

2. $y < -x^2$

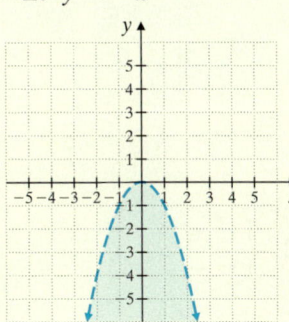

3. $x^2 + y^2 \geq 16$

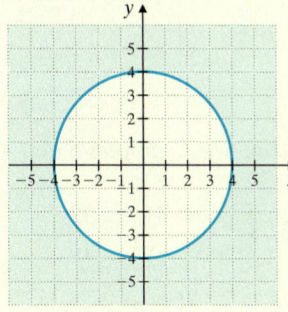

4. $x^2 + y^2 < 36$

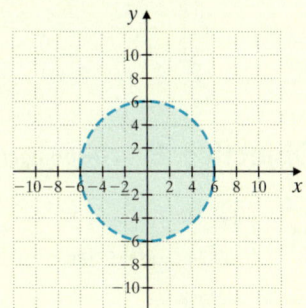

5. $\dfrac{x^2}{4} - y^2 < 1$

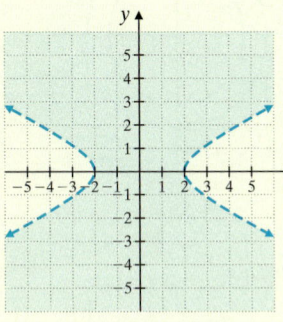

6. $x^2 - \dfrac{y^2}{9} \geq 1$

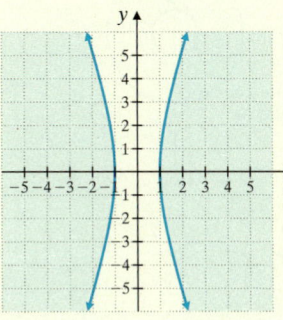

7. $y > (x - 1)^2 - 3$

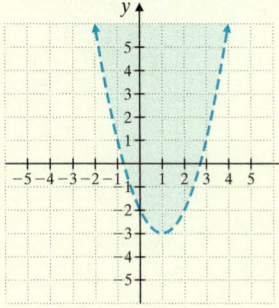

8. $y > (x + 3)^2 + 2$

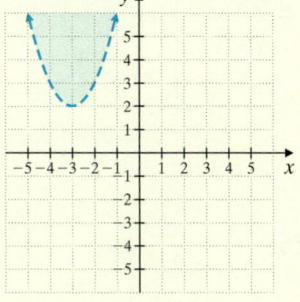

9. $x^2 + y^2 \leq 9$

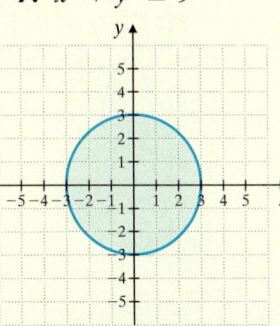

10. $x^2 + y^2 > 4$

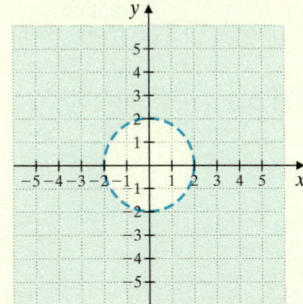

11. $y > -x^2 + 5$

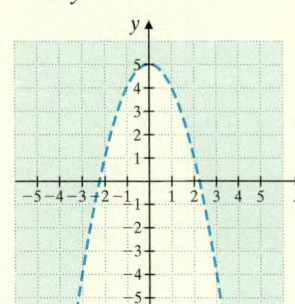

12. $y < -x^2 + 5$

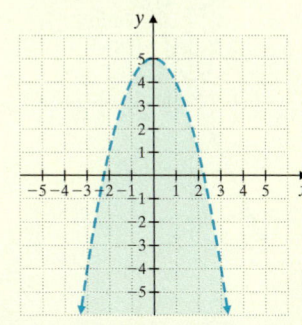

13. $\dfrac{x^2}{4} + \dfrac{y^2}{9} \le 1$

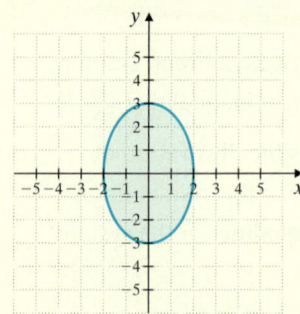

14. $\dfrac{x^2}{25} + \dfrac{y^2}{4} \ge 1$

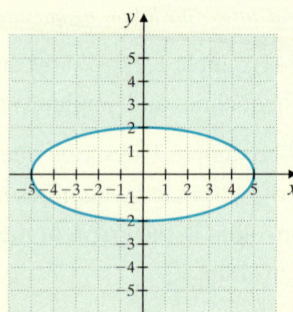

15. $\dfrac{y^2}{4} - x^2 \le 1$

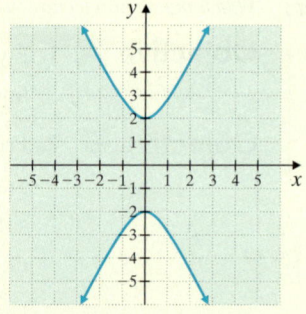

16. $\dfrac{y^2}{16} - \dfrac{x^2}{9} > 1$

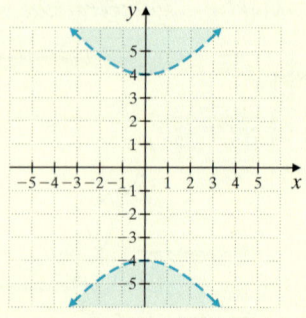

17. $y < (x - 2)^2 + 1$

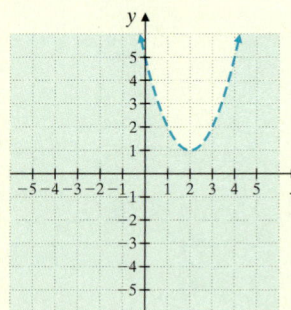

18. $y > (x - 2)^2 + 1$

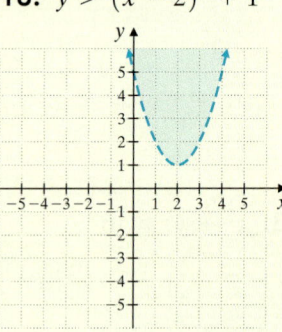

19. $y \le x^2 + x - 2$

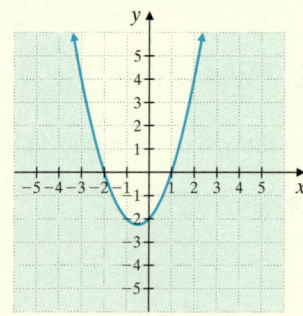

20. $y > x^2 + x - 2$

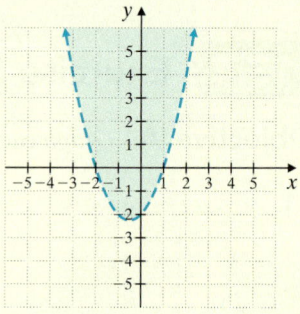

Objective B *Graph each system. See Examples 3 and 4.*

21. $\begin{cases} 4x + 3y \ge 12 \\ x^2 + y^2 < 16 \end{cases}$

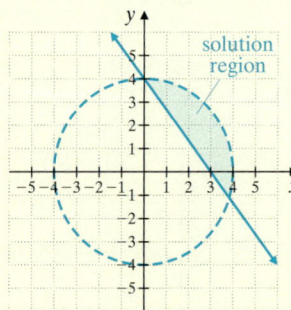

22. $\begin{cases} 3x - 4y \le 12 \\ x^2 + y^2 < 16 \end{cases}$

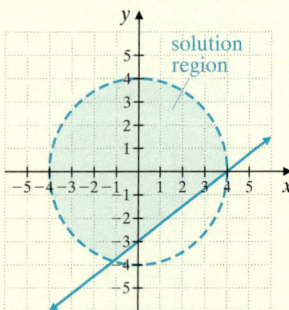

23. $\begin{cases} x^2 + y^2 \le 9 \\ x^2 + y^2 \ge 1 \end{cases}$

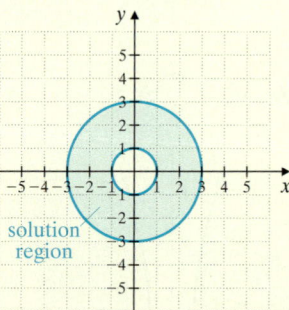

24. $\begin{cases} x^2 + y^2 \ge 9 \\ x^2 + y^2 \ge 16 \end{cases}$

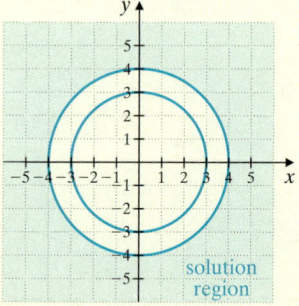

25. $\begin{cases} y > x^2 \\ y \ge 2x + 1 \end{cases}$

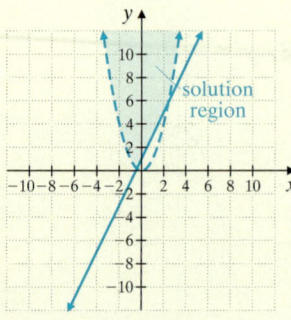

26. $\begin{cases} y \le -x^2 + 3 \\ y \le 2x - 1 \end{cases}$

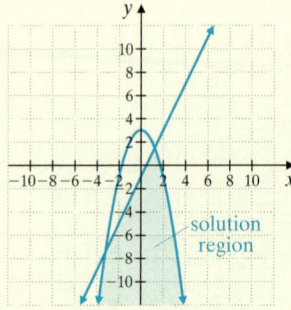

27. $\begin{cases} x^2 + y^2 > 9 \\ y > x^2 \end{cases}$

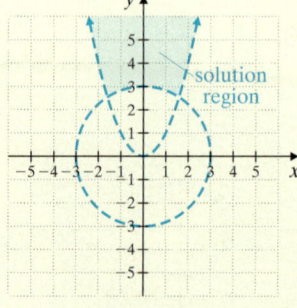

28. $\begin{cases} x^2 + y^2 \le 9 \\ y < x^2 \end{cases}$

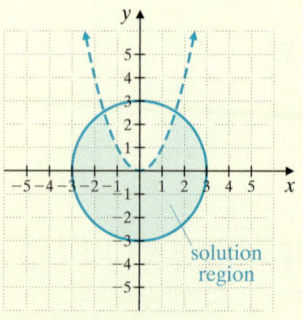

29. $\begin{cases} \dfrac{x^2}{4} + \dfrac{y^2}{9} \geq 1 \\ x^2 + y^2 \geq 4 \end{cases}$

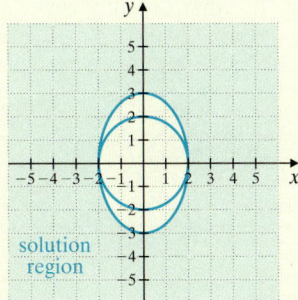

30. $\begin{cases} x^2 + (y-2)^2 \geq 9 \\ \dfrac{x^2}{4} + \dfrac{y^2}{25} < 1 \end{cases}$

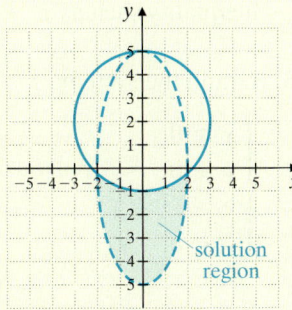

31. $\begin{cases} x^2 - y^2 \geq 1 \\ y \geq 0 \end{cases}$

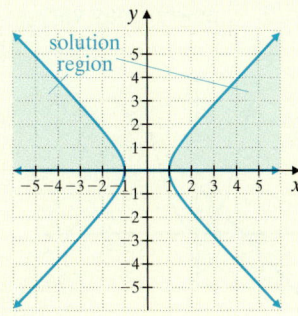

32. $\begin{cases} x^2 - y^2 \geq 1 \\ x \geq 0 \end{cases}$

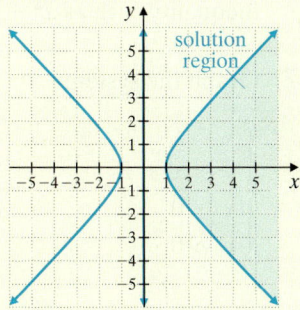

33. $\begin{cases} x + y \geq 1 \\ 2x + 3y < 1 \\ x > -3 \end{cases}$

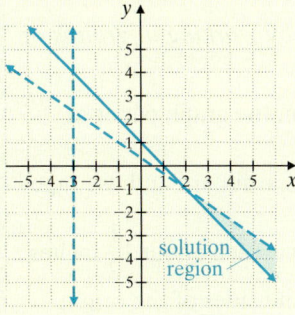

34. $\begin{cases} x - y < -1 \\ 4x - 3y > 0 \\ y > 0 \end{cases}$

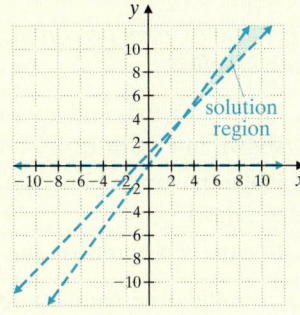

35. $\begin{cases} x^2 - y^2 < 1 \\ \dfrac{x^2}{16} + y^2 \leq 1 \\ x \geq -2 \end{cases}$

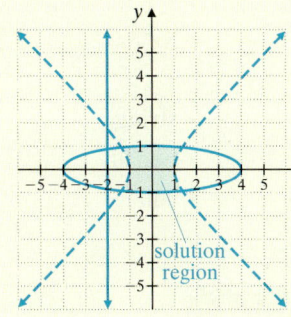

36. $\begin{cases} x^2 - y^2 \geq 1 \\ \dfrac{x^2}{16} + \dfrac{y^2}{4} \leq 1 \\ y \geq 1 \end{cases}$

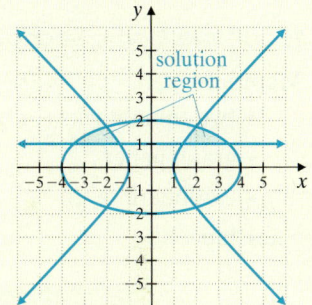

Review

Determine whether each graph is the graph of a function. See Section 8.2

37.

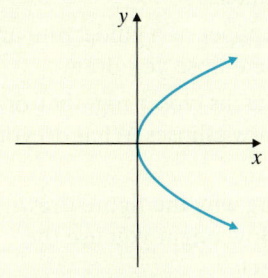

not a function

38.

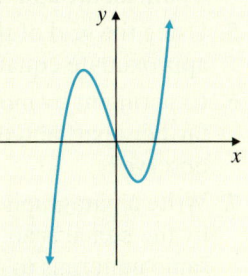

function

39.

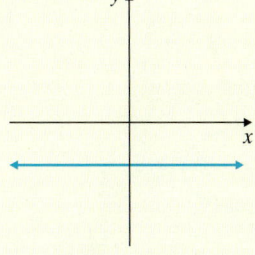

function

40.

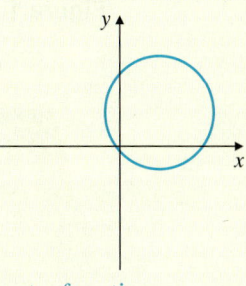

not a function

Find each function value if $f(x) = 3x^2 - 2$. See Section 8.2.

41. $f(-1)$ 1

42. $f(-3)$ 25

43. $f(a)$ $3a^2 - 2$

44. $f(b)$ $3b^2 - 2$

Concept Extensions

45. Discuss how graphing a linear inequality such as $x + y < 9$ is similar to graphing a nonlinear inequality such as $x^2 + y^2 < 9$.   answers may vary

46. Discuss how graphing a linear inequality such as $x + y < 9$ is different from graphing a nonlinear inequality such as $x^2 + y^2 < 9$.   answers may vary

47. Graph the system:

$$\begin{cases} y \le x^2 \\ y \ge x + 2 \\ x \ge 0 \\ y \ge 0 \end{cases}$$

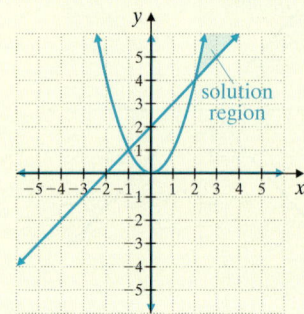

48. Graph the system:

$$\begin{cases} x \ge 0 \\ y \ge 0 \\ y \ge x^2 + 1 \\ y \le 4 - x \end{cases}$$

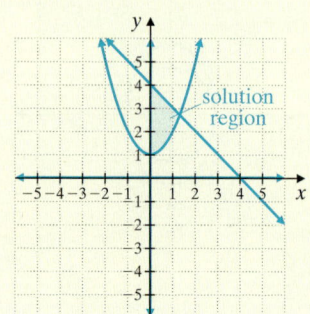

Chapter 13 Group Activity

Modeling Conic Sections

Materials

- two thumbtacks (or nails)
- graph paper
- cardboard
- tape
- string
- pencil
- ruler

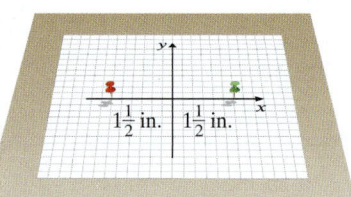

Figure 1

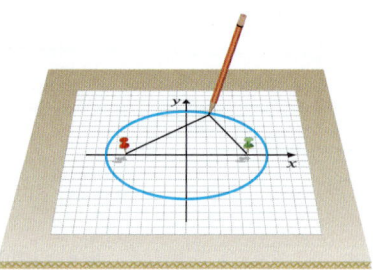

Figure 2

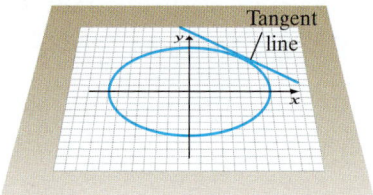

Figure 3

Additional group activities are available in the *Instructor's Resource Manual with Tests*.

This activity may be completed by working in groups or individually.

1. Draw an x-axis and a y-axis on the graph paper as shown in Figure 1.

2. Place the graph paper on the cardboard and use tape to attach.

3. Locate two points on the x-axis each about $1\frac{1}{2}$ inches from the origin and on opposite sides of the origin (see Figure 1). Insert thumbtacks (or nails) at each of these locations.

4. Fasten a 9-inch piece of string to the thumbtacks as shown in Figure 2. Use your pencil to draw and keep the string taut while you carefully move the pencil in a path all around the thumbtacks.

5. Using the grid of the graph paper as a guide, find an approximate equation of the ellipse you drew.

6. Experiment by moving the tacks closer together or farther apart and drawing new ellipses. What do you observe?

7. Write a paragraph explaining why the figure drawn by the pencil is an ellipse. How might you use the same materials to draw a circle?

8. (Optional) Choose one of the ellipses you drew with the string and pencil. Use a ruler to draw any six tangent lines to the ellipse. (A line is tangent to the ellipse if it intersects, or just touches, the ellipse at only one point. See Figure 3.) Extend the tangent lines to yield six points of intersection among the tangents. Use a straightedge to draw lines connecting each pair of opposite points of intersection. What do you observe? Repeat with a different ellipse. Can you make a conjecture about the relationship among the lines that connect opposite points of intersection?

Chapter 13 Vocabulary Check

Fill in each blank with one of the words or phrases listed below.

circle	ellipse	hyperbola
conic sections	vertex	diameter
center	radius	nonlinear system of equations

1. A(n) _____circle_____ is the set of all points in a plane that are the same distance from a fixed point, called the _____center_____ .

2. A(n) __nonlinear system of equations__ is a system of equations at least one of which is not linear.

3. A(n) _____ellipse_____ is the set of points in a plane such that the sum of the distances of those points from two fixed points is a constant.

4. In a circle, the distance from the center to a point of the circle is called its _____radius_____ .

5. A(n) _____hyperbola_____ is the set of points in a plane such that the absolute value of the difference of the distance from two fixed points is constant.

6. The circle, parabola, ellipse, and hyperbola are called the _____conic sections_____ .

7. For a parabola that opens upward, the lowest point is the _____vertex_____ .

8. Twice a circle's radius is its _____diameter_____ .

> **Helpful Hint**
> ▶ Are you preparing for your test? Don't forget to take the Chapter 13 Test on page 977. Then check your answers at the back of the text and use the Chapter Test Prep Videos to see the fully worked-out solutions to any of the exercises you want to review.

13 Chapter Highlights

Definitions and Concepts	Examples
Section 13.1 The Parabola and the Circle	

Parabolas

$$y = a(x - h)^2 + k$$

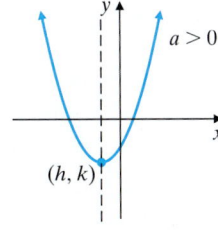

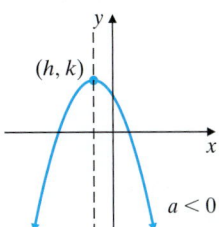

$$x = a(y - k)^2 + h$$

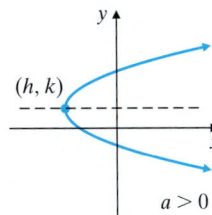

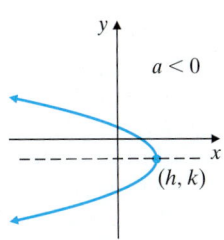

Graph: $x = 3y^2 - 12y + 13$

$$x - 13 = 3(y^2 - 4y)$$
$$x - 13 + 3(4) = 3(y^2 - 4y + 4)$$
$$x = 3(y - 2)^2 + 1$$

Since $a = 3$, this parabola opens to the right with vertex $(1, 2)$. Its axis of symmetry is $y = 2$. The x-intercept is $(13, 0)$.

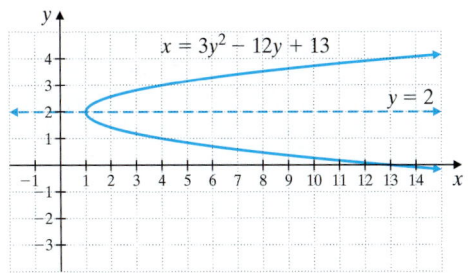

(continued)

Definitions and Concepts	**Examples**

Section 13.1   The Parabola and the Circle (*continued*)

Circle

The graph $(x - h)^2 + (y - k)^2 = r^2$ is a circle with center (h, k) and radius r.

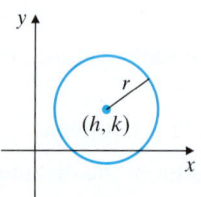

Graph: $x^2 + (y + 3)^2 = 5$

This equation can be written as

$$(x - 0)^2 + (y + 3)^2 = (\sqrt{5})^2$$

with $h = 0, k = -3$, and $r = \sqrt{5}$. The center of this circle is $(0, -3)$, and the radius is $\sqrt{5}$.

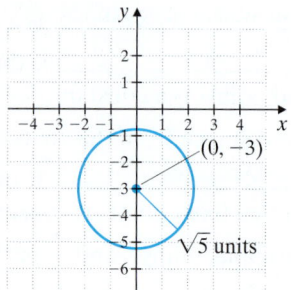

Section 13.2   The Ellipse and the Hyperbola

Ellipse with center (0, 0)

The graph of an equation of the form $\dfrac{x^2}{a^2} + \dfrac{y^2}{b^2} = 1$ is an ellipse with center $(0, 0)$. The x-intercepts are $(a, 0)$ and $(-a, 0)$, and the y-intercepts are $(0, b)$ and $(0, -b)$.

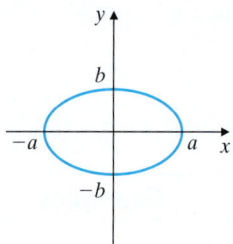

Graph: $4x^2 + 9y^2 = 36$

$$\dfrac{x^2}{9} + \dfrac{y^2}{4} = 1 \quad \text{Divide both sides by 36.}$$

$$\dfrac{x^2}{3^2} + \dfrac{y^2}{2^2} = 1$$

The ellipse has center $(0, 0)$, x-intercepts $(3, 0)$ and $(-3, 0)$, and y-intercepts $(0, 2)$ and $(0, -2)$.

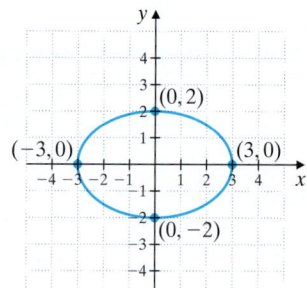

Definitions and Concepts	Examples

Section 13.2　The Ellipse and the Hyperbola (continued)

Hyperbola with center (0, 0)

The graph of an equation of the form $\dfrac{x^2}{a^2} - \dfrac{y^2}{b^2} = 1$ is a hyperbola with center $(0,0)$ and x-intercepts $(a,0)$ and $(-a,0)$.

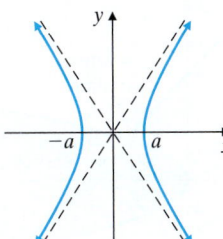

The graph of an equation of the form $\dfrac{y^2}{b^2} - \dfrac{x^2}{a^2} = 1$ is a hyperbola with center $(0,0)$ and y-intercepts $(0,b)$ and $(0,-b)$.

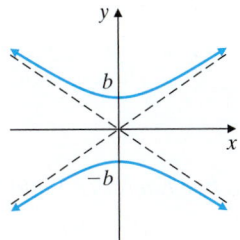

Graph: $\dfrac{x^2}{9} - \dfrac{y^2}{4} = 1$. Here $a = 3$ and $b = 2$.

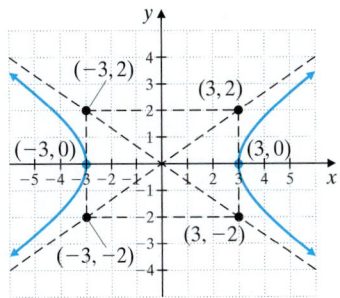

Section 13.3　Solving Nonlinear Systems of Equations

A **nonlinear system of equations** is a system of equations at least one of which is not linear. Both the substitution method and the elimination method may be used to solve a nonlinear system of equations.

Solve the nonlinear system:　$\begin{cases} y = x + 2 \\ 2x^2 + y^2 = 3 \end{cases}$

Substitute $x + 2$ for y in the second equation:

$$2x^2 + y^2 = 3$$
$$2x^2 + (x + 2)^2 = 3$$
$$2x^2 + x^2 + 4x + 4 = 3$$
$$3x^2 + 4x + 1 = 0$$
$$(3x + 1)(x + 1) = 0$$
$$x = -\frac{1}{3}, \quad \text{or} \quad x = -1$$

If $x = -\dfrac{1}{3}, y = x + 2 = -\dfrac{1}{3} + 2 = \dfrac{5}{3}$.

If $x = -1, y = x + 2 = -1 + 2 = 1$.

The solution set is $\left\{ \left(-\dfrac{1}{3}, \dfrac{5}{3} \right), (-1, 1) \right\}$

Definitions and Concepts	Examples

Section 13.4   Nonlinear Inequalities and Systems of Inequalities

The **graph of a system of inequalities** is the intersection of the graphs of the inequalities.

Graph the system: $\begin{cases} x \geq y^2 \\ x + y \leq 4 \end{cases}$

The graph of the system is the darkest shaded region along with its boundary lines.

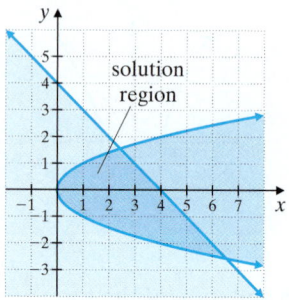

Chapter 13   Review

(13.1) *Write an equation of the circle with the given center and radius or diameter. For Exercises 3 and 4, begin by writing diameters as radii.*

1. Center $(-4, 4)$, radius 3
$(x + 4)^2 + (y - 4)^2 = 9$

2. Center $(-7, -9)$, radius $\sqrt{11}$
$(x + 7)^2 + (y + 9)^2 = 11$

3. Center $(5, 0)$, diameter 10
$(x - 5)^2 + y^2 = 25$

4. Center $(0, 0)$, diameter 7
$x^2 + y^2 = \dfrac{49}{4}$

Graph each equation. If the graph is a circle, find its center and radius. If the graph is a parabola, find its vertex.

5. $x^2 + y^2 = 4$

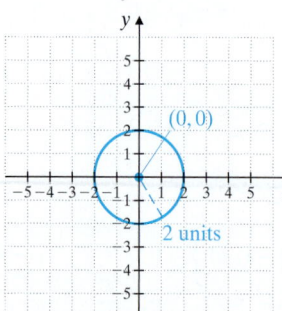

6. $x = 2(y - 5)^2 + 4$

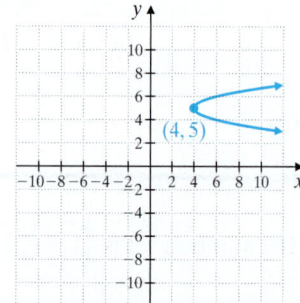

7. $x = -(y + 2)^2 + 3$

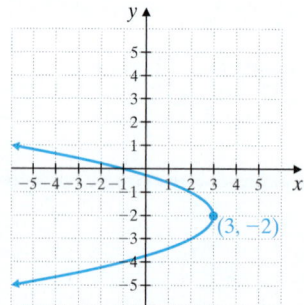

8. $(x - 1)^2 + (y - 2)^2 = 4$

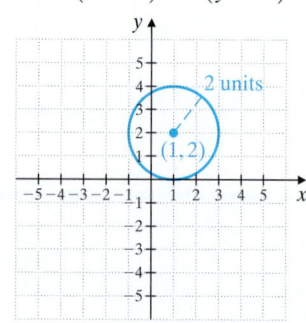

9. $y = -x^2 + 4x + 10$

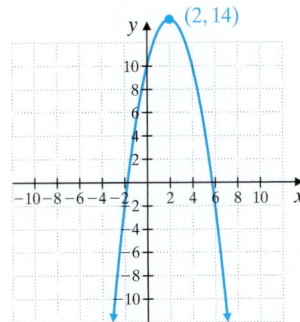

10. $x = -y^2 - 4y + 6$

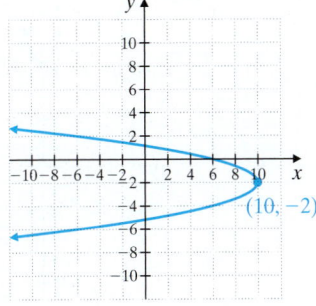

11. $x = \dfrac{1}{2}y^2 + 2y + 1$

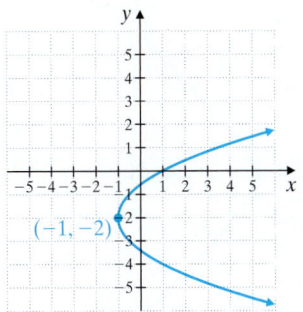

12. $y = -3x^2 + \dfrac{1}{2}x + 4$

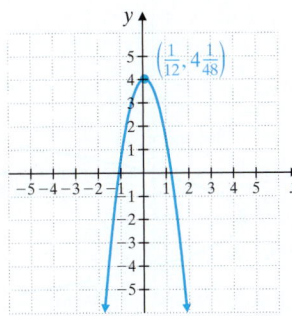

13. $x^2 + y^2 + 2x + y = \dfrac{3}{4}$

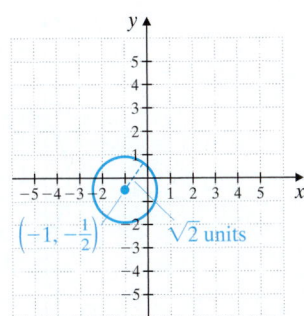

14. $x^2 + y^2 - 3y = \dfrac{7}{4}$

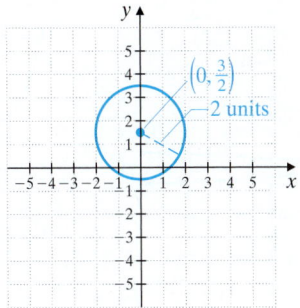

15. $4x^2 + 4y^2 + 16x + 8y = 1$

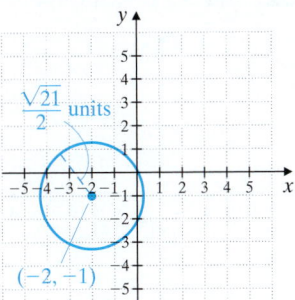

16. $3x^2 + 3y^2 + 18x - 12y = -12$

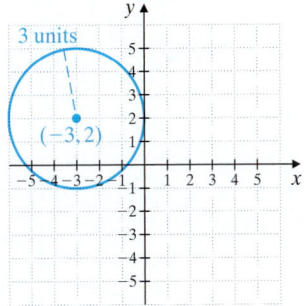

(13.1, 13.2) *Graph each equation.*

17. $x^2 - \dfrac{y^2}{4} = 1$

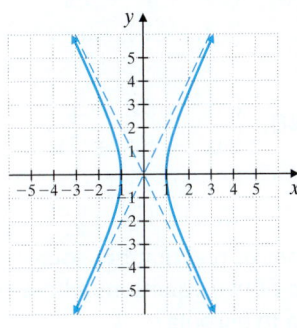

18. $x^2 + \dfrac{y^2}{4} = 1$

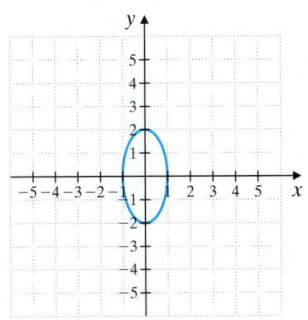

19. $4y^2 + 9x^2 = 36$

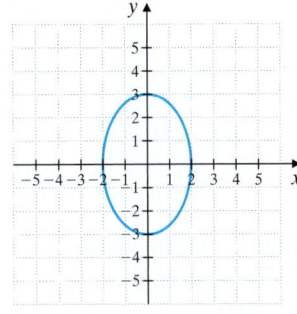

20. $-5x^2 + 25y^2 = 125$

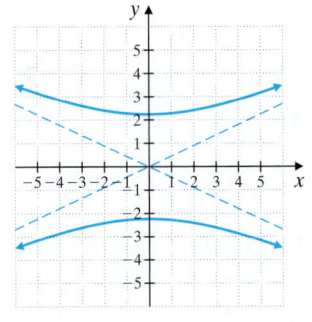

21. $x^2 - y^2 = 1$

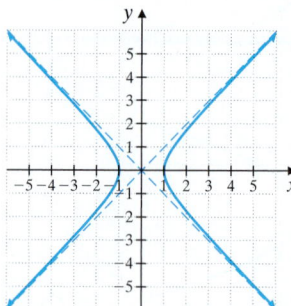

22. $\dfrac{(x+3)^2}{9} + \dfrac{(y-4)^2}{25} = 1$

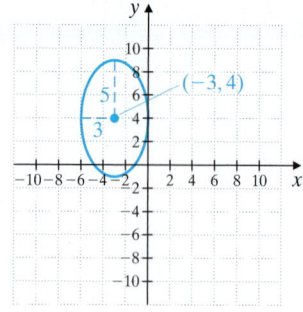

23. $y = x^2 + 9$

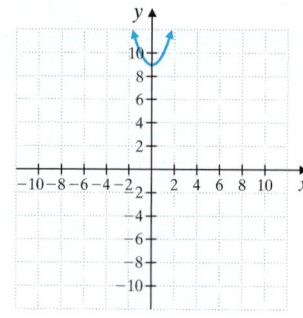

24. $36y^2 - 49x^2 = 1764$

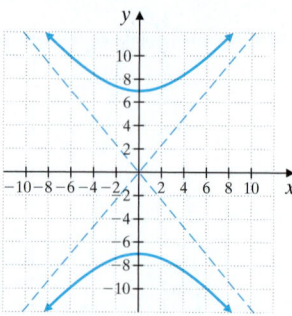

25. $x = 4y^2 - 16$

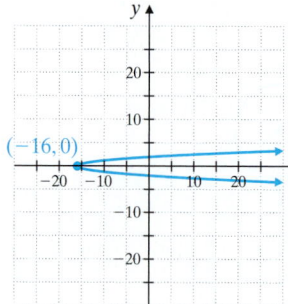

26. $y = x^2 + 4x + 6$

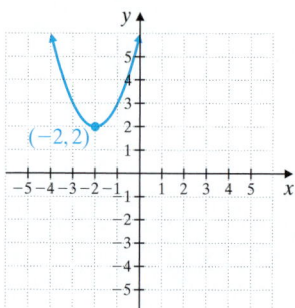

27. $y^2 + 2(x-1)^2 = 8$

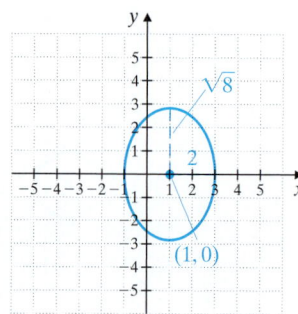

28. $x - 4y = y^2$

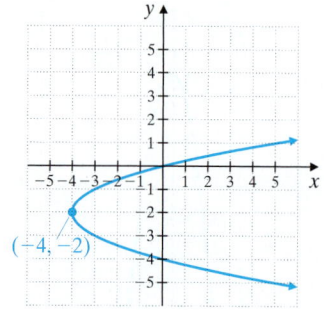

29. $x^2 - 4 = y^2$

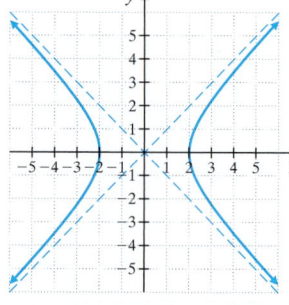

30. $x^2 = 4 - y^2$

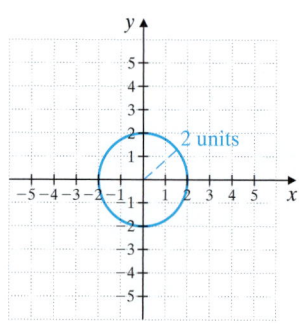

31. $36y^2 = 576 + 16x^2$

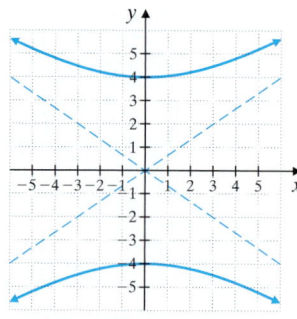

32. $3(x-7)^2 + 3(y+4)^2 = 1$

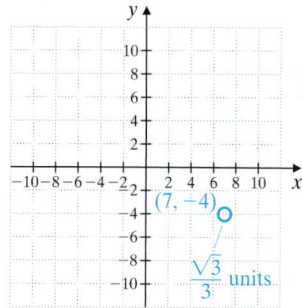

(13.3) *Solve each system of equations.*

33. $\begin{cases} y = 2x - 4 \\ y^2 = 4x \end{cases}$
$\{(1, -2), (4, 4)\}$

34. $\begin{cases} x^2 + y^2 = 4 \\ x - y = 4 \end{cases}$
$\varnothing$

35. $\begin{cases} y = x + 2 \\ y = x^2 \end{cases}$
$\{(-1, 1), (2, 4)\}$

36. $\begin{cases} 4x - y^2 = 0 \\ 2x^2 + y^2 = 16 \end{cases}$
$\{(2, 2\sqrt{2}), (2, -2\sqrt{2})\}$

37. $\begin{cases} x^2 + 4y^2 = 16 \\ x^2 + y^2 = 4 \end{cases}$
$\{(0, 2), (0, -2)\}$

38. $\begin{cases} x^2 + 2y = 9 \\ 5x - 2y = 5 \end{cases}$
$\left\{\left(2, \dfrac{5}{2}\right), (-7, -20)\right\}$

39. $\begin{cases} y = 3x^2 + 5x - 4 \\ y = 3x^2 - x + 2 \end{cases}$
$\{(1, 4)\}$

40. $\begin{cases} x^2 - 3y^2 = 1 \\ 4x^2 + 5y^2 = 21 \end{cases}$
$\{(-2, -1), (-2, 1), (2, -1), (2, 1)\}$

△ **41.** Find the length and the width of a room whose area is 150 square feet and whose perimeter is 50 feet.   length: 15 ft; width: 10 ft

42. What is the greatest number of real number solutions possible for a system of two equations whose graphs are an ellipse and a hyperbola?   4

(13.4) *Graph each inequality or system of inequalities.*

43. $y \le -x^2 + 3$

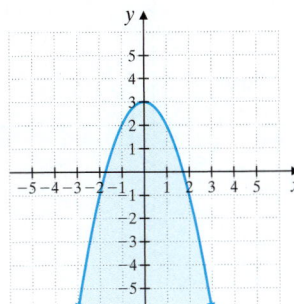

44. $x < y^2 - 1$

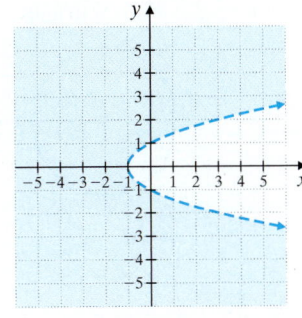

45. $x^2 + y^2 < 9$

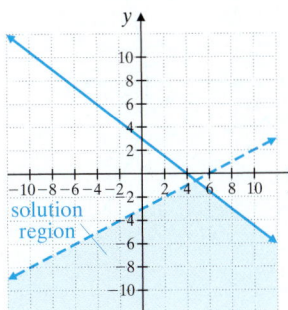

46. $\dfrac{x^2}{4} + \dfrac{y^2}{9} \ge 1$

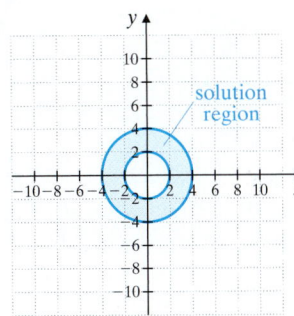

47. $\begin{cases} 3x + 4y \le 12 \\ x - 2y > 6 \end{cases}$

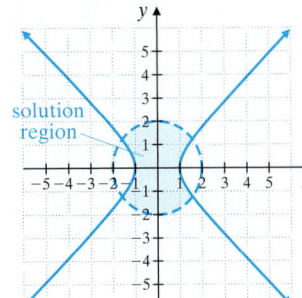

48. $\begin{cases} x^2 + y^2 \le 16 \\ x^2 + y^2 \ge 4 \end{cases}$

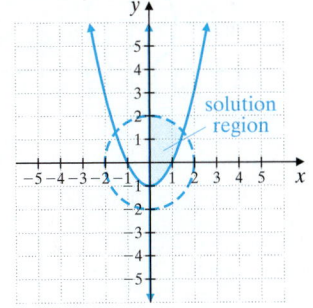

49. $\begin{cases} x^2 + y^2 < 4 \\ x^2 - y^2 \le 1 \end{cases}$

50. $\begin{cases} x^2 + y^2 < 4 \\ y \ge x^2 - 1 \\ x \ge 0 \end{cases}$

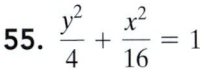

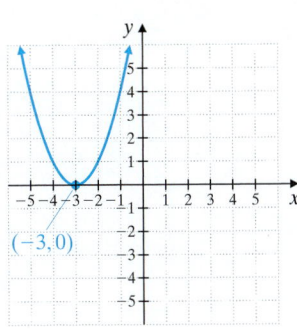

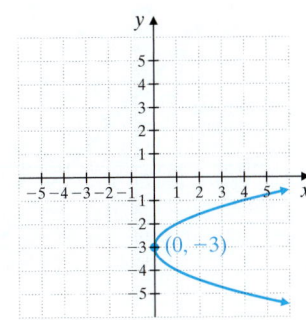

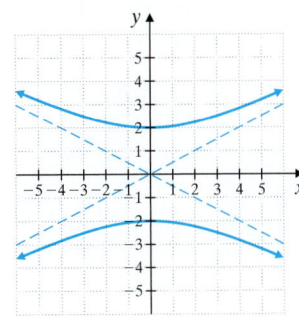

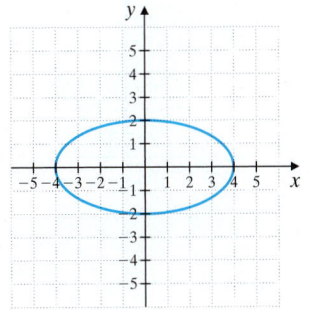

Mixed Review

51. Write an equation of the circle with center $(-7, 8)$ and radius 5. $(x + 7)^2 + (y - 8)^2 = 25$

Graph each equation.

52. $y = x^2 + 6x + 9$

53. $x = y^2 + 6y + 9$

54. $\dfrac{y^2}{4} - \dfrac{x^2}{16} = 1$

55. $\dfrac{y^2}{4} + \dfrac{x^2}{16} = 1$

56. $\dfrac{(x-2)^2}{4} + (y-1)^2 = 1$ **57.** $y^2 = x^2 + 6$ **58.** $y^2 + (x-2)^2 = 10$ **59.** $3x^2 + 6x + 3y^2 = 9$

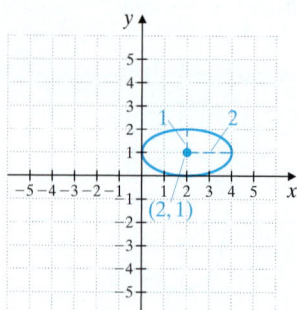   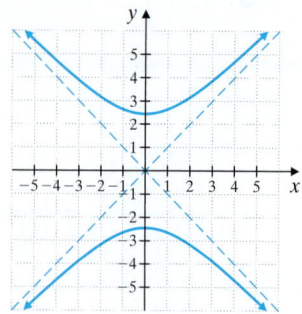   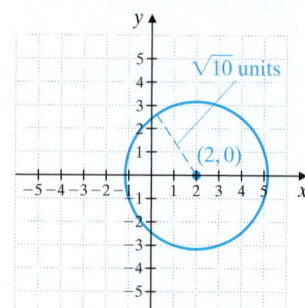   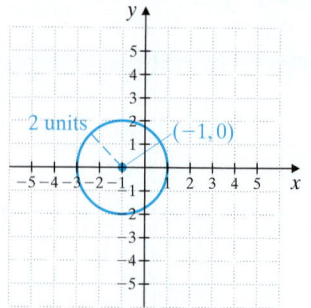

60. $x^2 + y^2 - 8y = 0$ **61.** $6(x-2)^2 + 9(y+5)^2 = 36$ **62.** $\dfrac{x^2}{16} - \dfrac{y^2}{25} = 1$

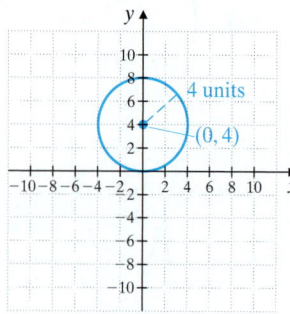   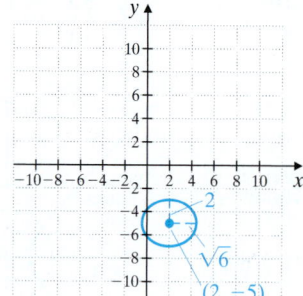   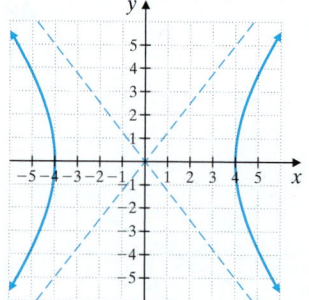

Solve each system of equations.

63. $\begin{cases} y = x^2 - 5x + 1 \\ y = -x + 6 \end{cases}$

$\{(5,1), (-1,7)\}$

64. $\begin{cases} x^2 + y^2 = 10 \\ 9x^2 + y^2 = 18 \end{cases}$

$\{(-1,3), (-1,-3), (1,3), (1,-3)\}$

Graph each inequality or system of inequalities.

65. $x^2 - y^2 < 1$

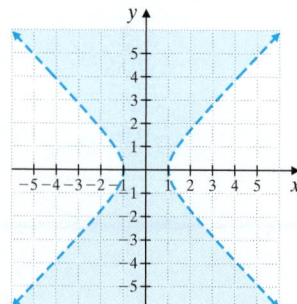

66. $\begin{cases} y > x^2 \\ x + y \geq 3 \end{cases}$

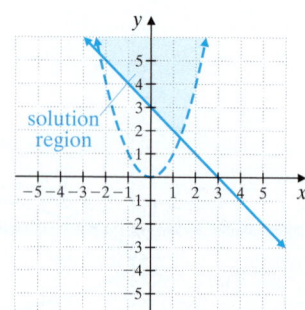

Graph each equation.

1. $x^2 + y^2 = 36$

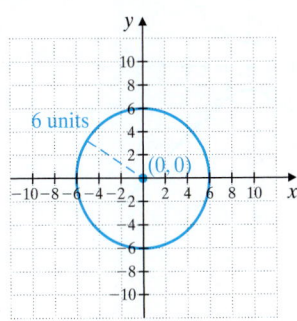

2. $x^2 - y^2 = 36$

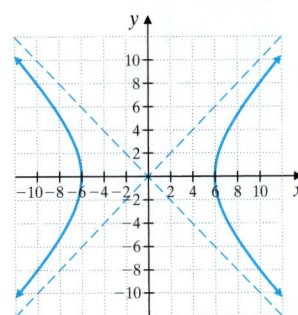

3. $16x^2 + 9y^2 = 144$

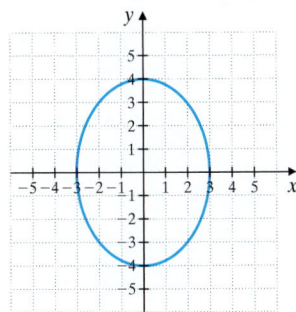

4. $y = x^2 - 8x + 16$

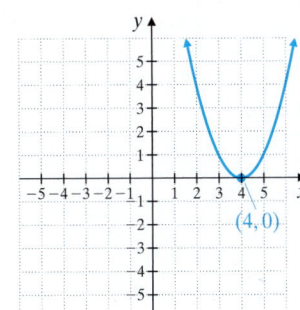

5. $x^2 + y^2 + 6x = 16$

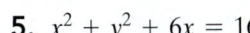
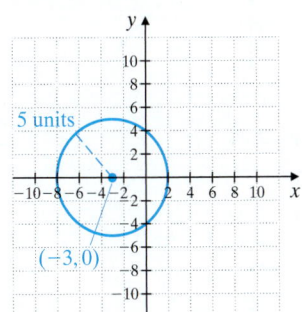

6. $x = y^2 + 8y - 3$

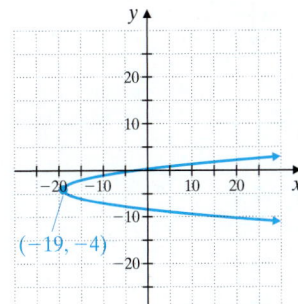

Answers

Note to Instructor: The Chapter 13 Test file in the TestGen program provides algorithms specifically matched to this test for easy replication for practice or assessment purposes.

1. see graph

2. see graph

3. see graph

4. see graph

5. see graph

6. see graph

7. see graph

7. $\dfrac{(x-4)^2}{16} + \dfrac{(y-3)^2}{9} = 1$

8. $y^2 - x^2 = 1$

8. see graph

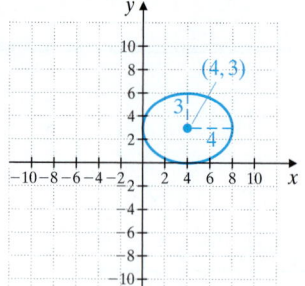

9. $\{(-12, 5), (12, -5)\}$

Solve each system.

$\{(-5,-1),(-5,1),$
10. $(5,-1),(5,1)\}$

9. $\begin{cases} x^2 + y^2 = 169 \\ 5x + 12y = 0 \end{cases}$

10. $\begin{cases} x^2 + y^2 = 26 \\ x^2 - 2y^2 = 23 \end{cases}$

11. $\begin{cases} y = x^2 - 5x + 6 \\ y = 2x \end{cases}$

12. $\begin{cases} x^2 + 4y^2 = 5 \\ y = x \end{cases}$

11. $\{(6, 12), (1, 2)\}$

Graph each system.

12. $\{(1, 1), (-1, -1)\}$

13. $\begin{cases} 2x + 5y \geq 10 \\ y \geq x^2 + 1 \end{cases}$

14. $\begin{cases} \dfrac{x^2}{4} + y^2 \leq 1 \\ x + y > 1 \end{cases}$

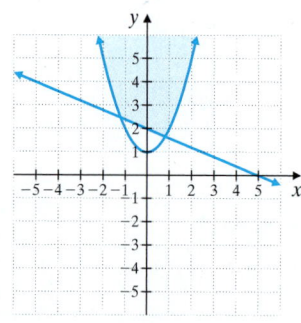

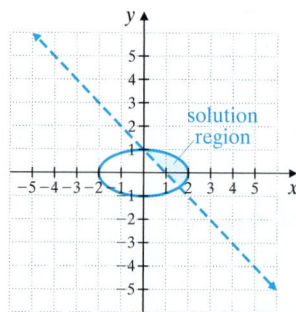

13. see graph

14. see graph

15. $\begin{cases} x^2 + y^2 > 1 \\ \dfrac{x^2}{4} - y^2 \geq 1 \end{cases}$

16. $\begin{cases} x^2 + y^2 \geq 4 \\ x^2 + y^2 < 16 \\ y \geq 0 \end{cases}$

15. see graph

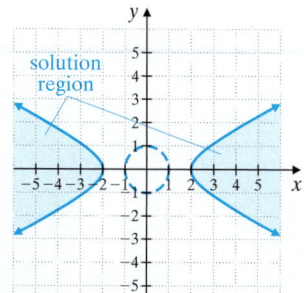

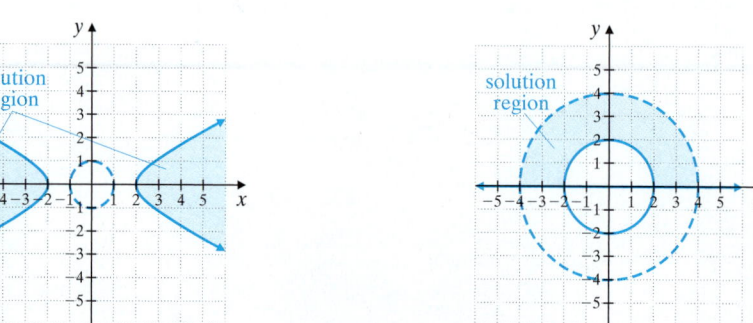

16. see graph

17. Which graph best resembles the graph of $x = a(y - k)^2 + h$ if $a > 0, h < 0,$ and $k > 0$?

17. B

A.

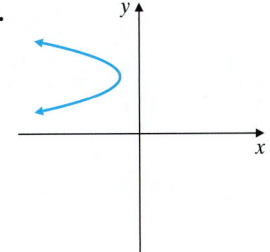

B.

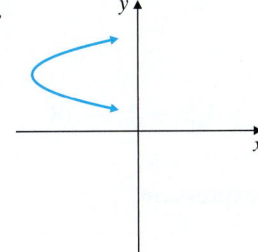

C.

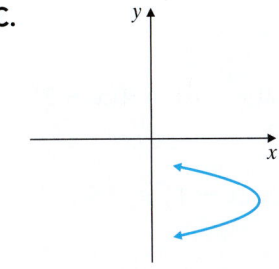

D.

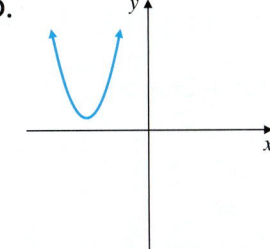

18. A bridge has an arch in the shape of half an ellipse. If the equation of the ellipse, measured in feet, is $100x^2 + 225y^2 = 22{,}500,$ find the height of the arch from the road and the width of the arch.

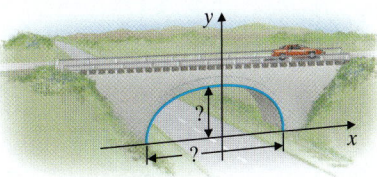

18. height: 10 ft; width: 30 ft

Simplify.

1. Solve:
$2(x - 3) - 5 \leq 3(x + 2) - 18$

2. Solve: $\dfrac{4}{x - 2} - \dfrac{x}{x + 2} = \dfrac{16}{x^2 - 4}$

3. Simplify each expression.
a. $x^7 \cdot x^4$
b. $\left(\dfrac{t}{2}\right)^4$
c. $(9y^5)^2$

4. Simplify: 9^{-2}

5. Multiply: $7x(x^2 + 2x + 5)$

6. Solve: $2(x - 6) = 4(x - 3) - 2x$

7. Divide: $\dfrac{9x^5 - 12x^2 + 3x}{3x^2}$

8. Solve: $5(2x - 1) > -5$

Factor.

9. $5(x + 3) + y(x + 3)$

10. $3y^2 + 14y + 15$

11. $x^4 + 5x^2 + 6$

12. $20a^5 + 54a^4 + 10a^3$

13. Multiply: $\dfrac{-7x^2}{5y} \cdot \dfrac{3y^5}{14x^2}$

14. Simplify: $\dfrac{8x^3 - 1}{2x - 1}$

15. Solve: $\dfrac{t - 4}{2} - \dfrac{t - 3}{9} = \dfrac{5}{18}$

16. Solve: $\dfrac{2}{x + 3} = \dfrac{1}{x^2 - 9} - \dfrac{1}{x - 3}$.

17. Simplify: $\dfrac{\dfrac{1}{z} - \dfrac{1}{2}}{\dfrac{1}{3} - \dfrac{z}{6}}$

18. Simplify: $\dfrac{2}{3a - 15} - \dfrac{a}{25 - a^2}$

19. Graph the linear equation $y = -\dfrac{1}{3}x + 2$.

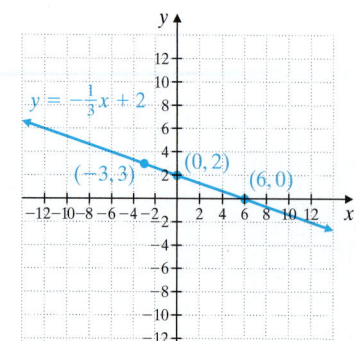

20. Suppose that y varies inversely as x. If $y = 3$ when $x = \dfrac{2}{3}$, find the constant of variation and the variation equation.

21. Write an equation of the line containing the point $(-2, 1)$ and perpendicular to the line $3x + 5y = 4$. Write the equation in slope-intercept form, $y = mx + b$.

22. Find the slope of the line that goes through $(3, 2)$ and $(1, -4)$.

23. Determine the domain and range of the relation:
$$\{(2, 3), (2, 4), (0, -1), (3, -1)\}$$

24. Two planes leave Greensboro, one traveling north and the other south. After 2 hours they are 650 miles apart. If one plane is flying 25 mph faster than the other, what is the speed of each?

25. If $g(x) = 3x - 2$, find $g(1)$.

26. If $f(x) = x^2 - 3x + 2$ and $g(x) = -3x + 5$, find
a. $(f \circ g)(x)$
b. $(f \circ g)(-2)$
c. $(g \circ f)(x)$
d. $(g \circ f)(5)$

27. If $f(x) = 7x^2 - 3x + 1$, find $f(-2)$.

28. Solve the system $\begin{cases} x^2 + y^2 = 36 \\ \quad\; y = x + 6 \end{cases}$

29. Solve the system:
$$\begin{cases} 3x - y + z = -15 \\ x + 2y - z = 1 \\ 2x + 3y - 2z = 0 \end{cases}$$

30. Simplify: $\dfrac{x}{3x^2 - 13x - 10} + \dfrac{9}{9x^2 - 4}$

31. Use matrices to solve the system:
$$\begin{cases} x + 3y = 5 \\ 2x - y = -4 \end{cases}$$

32. Solve the system by the addition method:
$$\begin{cases} -6x + 8y = 0 \\ 9x - 12y = 2 \end{cases}$$

33. Graph the solutions of the system:
$$\begin{cases} x - y < 2 \\ x + 2y > -1 \\ \quad\; y < 2 \end{cases}$$

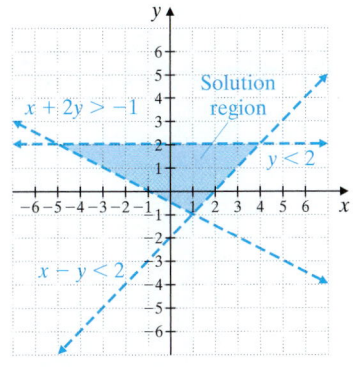

34. Solve $x^3 + 2x^2 - 4x \geq 8$.

Find the roots. Assume that all variables represent positive real numbers.

35. $\sqrt[3]{-64}$

36. $\sqrt{\dfrac{x^7}{49}}$

37. $\sqrt[3]{\dfrac{8}{125}}$

38. $\sqrt{90a^5b^2}$

---

21. $y = \dfrac{5}{3}x + \dfrac{13}{3}$ (Sec. 8.1, Ex. 6)

22. 3 (Sec. 3.4)

23. domain: $\{2, 0, 3\}$; range: $\{3, 4, -1\}$ (Sec. 8.2, Ex. 1)

24. 150 mph, 175 mph (Sec. 4.4)

25. 1 (Sec. 8.2, Ex. 14)

26. a. $9x^2 - 21x + 12$

b. 90

c. $-3x^2 + 9x - 1$

d. -31 (Sec. 12.1)

27. 35 (Sec. 8.2, Ex. 17)

28. $\{(-6, 0), (0, 6)\}$ (Sec. 13.3)

29. $(-4, 2, -1)$ (Sec. 9.1, Ex. 1)

30. $\dfrac{3x^2 + 7x - 45}{(3x + 2)(x - 5)(3x - 2)}$ (Sec. 7.4)

31. $(-1, 2)$ (Sec. 9.2, Ex. 1)

32. $\varnothing$ (Sec. 4.3)

33. see graph (Sec. 9.3, Ex. 2)

34. $\{-2\} \cup [2, \infty)$ (Sec. 11.4)

35. -4 (Sec. 10.1, Ex. 14)

36. $\dfrac{x^3\sqrt{x}}{7}$ (Sec. 10.3)

37. $\dfrac{2}{5}$ (Sec. 10.1, Ex. 15)

38. $3a^2b\sqrt{10a}$ (Sec. 10.3)

Multiply.

39. $\sqrt[3]{4} \cdot \sqrt[3]{2}$

40. $\sqrt{5}(2 + \sqrt{15})$

41. $\sqrt{\dfrac{2}{a}} \cdot \sqrt{\dfrac{b}{3}}$

42. $(2\sqrt{5} - 1)^2$

Add or subtract. Assume that all variables represent positive real numbers

43. $\sqrt[3]{54} - 5\sqrt[3]{16} + \sqrt[3]{2}$

44. $\sqrt{45} + \sqrt{20}$

45. $\sqrt[3]{\dfrac{7x}{8}} + 2\sqrt[3]{7x}$

46. $-3\sqrt[3]{2x} + \sqrt[3]{16x}$

47. Solve: $p = -3p^2 - 3$

48. Solve $\sqrt{2x - 3} = x - 3$.

49. Solve: $\log(x + 2) - \log x = 2$

50. Solve each equation for x.
   a. $64^x = 4$
   b. $125^{x-3} = 25$
   c. $\dfrac{1}{81} = 3^{2x}$

51. Find the midpoint of the line segment that joins points $P(-3, 3)$ and $Q(1, 0)$.

52. Find the length of the segment that joins points $P(-3, 3)$ and $Q(1, 0)$.

53. Graph: $4y^2 - 9x^2 = 36$

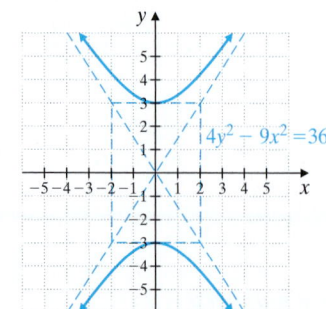

54. Find the inverse of $f(x) = \dfrac{x + 1}{2}$.

55. Solve the system $\begin{cases} y = \sqrt{x} \\ x^2 + y^2 = 6 \end{cases}$

56. Rationalize the denominator. $\dfrac{-2}{\sqrt{3} + 3}$

Transition Review: Exponents, Polynomials, and Factoring Strategies

Appendix

A

Objective A Reviewing Rules for Exponents

The following is a summary of rules for exponents from Sections 5.1 and 5.2.

Summary of Rules for Exponents

If a and b are real numbers and m and n are integers, and no denominator is 0, then

Product rule	$a^m \cdot a^n = a^{m+n}$	$x^2 \cdot x^3 = x^5$
Zero exponent	$a^0 = 1, a \neq 0$	$7^0 = 1, (-10)^0 = 1$
Negative exponent	$a^{-n} = \dfrac{1}{a^n}$	$3^{-2} = \dfrac{1}{3^2} = \dfrac{1}{9}$
Quotient rule	$\dfrac{a^m}{a^n} = a^{m-n}$	$\dfrac{y^{10}}{y^4} = y^{10-4} = y^6$
Power rule	$(a^m)^n = a^{m \cdot n}$	$(7^8)^2 = 7^{16}$
Power of a product	$(ab)^m = a^m \cdot b^m$	$(2y)^3 = 2^3 y^3 = 8y^3$
Power of a quotient	$\left(\dfrac{a}{b}\right)^m = \dfrac{a^m}{b^m}$	$\left(\dfrac{5x^{-3}}{x^2}\right)^{-2} = \dfrac{5^{-2}x^6}{x^{-4}}$
		$= 5^{-2} \cdot x^{6-(-4)}$
		$= \dfrac{x^{10}}{5^2} \text{ or } \dfrac{x^{10}}{25}$

Examples Use the rules for exponents to simplify each expression. Write each answer using positive exponents only.

1. $\left(\dfrac{3p^4}{q^5}\right)^2 = \dfrac{(3p^4)^2}{(q^5)^2} = \dfrac{3^2 \cdot (p^4)^2}{(q^5)^2} = \dfrac{9p^8}{q^{10}}$

2. $(-2x^3p^2)(4xp^{10}) = -2(4)x^3x^1p^2p^{10} = -8x^4p^{12}$

3. $(2x^0y^{-3})^{-2} = 2^{-2}(x^0)^{-2}(y^{-3})^{-2} = 2^{-2}x^0y^6 = \dfrac{1(y^6)}{2^2} = \dfrac{y^6}{4}$ Write x^0 as 1.

4. $\dfrac{5^{-2}x^{-3}y^{11}}{x^2y^{-5}} = 5^{-2}x^{-3-2}y^{11-(-5)} = 5^{-2}x^{-5}y^{16} = \dfrac{y^{16}}{5^2x^5} = \dfrac{y^{16}}{25x^5}$

5. $\left(\dfrac{3a^2}{2x^{-1}}\right)^3\left(\dfrac{x^{-3}}{4a^{-2}}\right)^{-1} = \dfrac{27a^6}{8x^{-3}} \cdot \dfrac{x^3}{4^{-1}a^2} = \dfrac{27 \cdot 4 \cdot a^6x^3x^3}{8 \cdot a^2} = \dfrac{27a^4x^6}{2}$

■ **Work Practice 1–5**

Objective B Reviewing Operations on Polynomials

In Section 5.4, we added and subtracted polynomials. Below is a review of those operations.

Objectives

A Review Rules for Exponents.

B Review Addition, Subtraction, and Multiplication of Polynomials.

C Review Factoring Strategies.

Practice 1–5

Use the rules for exponents to simplify each expression. Write each answer using positive exponents only.

1. $\left(\dfrac{4m^5}{n^3}\right)^3$ **2.** $(-3x^2y^7)(5xy^6)$

3. $(7xy^{-2})^{-2}$ **4.** $\dfrac{6^{-2}x^{-4}y^{10}}{x^2y^{-6}}$

5. $\left(\dfrac{4x^3}{3y^{-1}}\right)^3\left(\dfrac{y^{-2}}{3x^{-1}}\right)^{-1}$

Answers

1. $\dfrac{64m^{15}}{n^9}$ **2.** $-15x^3y^{13}$ **3.** $\dfrac{y^4}{49x^2}$

4. $\dfrac{y^{16}}{36x^6}$ **5.** $\dfrac{64x^8y^5}{9}$

Adding and Subtracting Polynomials

To **add** polynomials, combine all like terms. To **subtract** polynomials, change the signs of the terms of the polynomial being subtracted, and then add.

Practice 6

Add $14x^4 - 6x^3 + x^2 - 6$ and $x^3 - 5x^2 + 1$.

Example 6   Add $11x^3 - 12x^2 + x - 3$ and $x^3 - 10x + 5$.

Solution:

$$(11x^3 - 12x^2 + x - 3) + (x^3 - 10x + 5)$$
$$= 11x^3 + x^3 - 12x^2 + x - 10x - 3 + 5 \quad \text{Group like terms.}$$
$$= 12x^3 - 12x^2 - 9x + 2 \quad \text{Combine like terms.}$$

▪ **Work Practice 6**

Practice 7

Subtract:

$(7x^4 - 8x^2 + x) - (9x^4 + x^2 - 18)$

Example 7   Subtract: $(12z^5 - 12z^3 + z) - (-3z^4 + z^3 + 12z)$

Solution: First we change the sign of each term of the second polynomial, and then we add the result to the first polynomial.

$$(12z^5 - 12z^3 + z) - (-3z^4 + z^3 + 12z)$$
$$= 12z^5 - 12z^3 + z + 3z^4 - z^3 - 12z \quad \text{Change signs and add.}$$
$$= 12z^5 + 3z^4 - 12z^3 - z^3 + z - 12z \quad \text{Group like terms.}$$
$$= 12z^5 + 3z^4 - 13z^3 - 11z \quad \text{Combine like terms.}$$

▪ **Work Practice 7**

Practice 8

Subtract $3b^3 - 4b^2 + 6$ from $7b^3 - b^2$.

Example 8   Subtract $4x^3 - 3x^2 + 2$ from $10x^3 - 7x^2$.

Solution:

$$(10x^3 - 7x^2) - (4x^3 - 3x^2 + 2)$$
$$= 10x^3 - 7x^2 - 4x^3 + 3x^2 - 2 \quad \text{Remove parentheses.}$$
$$= 6x^3 - 4x^2 - 2 \quad \text{Combine like terms.}$$

▪ **Work Practice 8**

To multiply any two polynomials, we can use the following from Section 5.6

Multiplying Any Two Polynomials

To multiply any two polynomials, use the distributive property and multiply each term of one polynomial by each term of the other polynomial. Then combine any like terms.

Practice 9–10

Multiply.

9. $(5x - 1)(2x^2 - x + 4)$

10. $(4x - 3)(x - 6)$

Examples   Multiply:

9. $(2x - 3)(5x^2 - 6x + 7) = 2x(5x^2 - 6x + 7) + (-3)(5x^2 - 6x + 7)$
$$= 10x^3 - 12x^2 + 14x - 15x^2 + 18x - 21$$
$$= 10x^3 - 27x^2 + 32x - 21 \quad \text{Combine like terms.}$$

10. $(2x - 7)(3x - 4) = 2x(3x) + 2x(-4) + (-7)(3x) + (-7)(-4)$
$$= 6x^2 - 8x - 21x + 28 \quad \text{FOIL order}$$
$$= 6x^2 - 29x + 28$$

▪ **Work Practice 9–10**

Answers

6. $14x^4 - 5x^3 - 4x^2 - 5$
7. $-2x^4 - 9x^2 + x + 18$
8. $4b^3 + 3b^2 - 6$
9. $10x^3 - 7x^2 + 21x - 4$
10. $4x^2 - 27x + 18$

Special products from Section 5.6 may also be used to multiply.

Square of a Binomial

$$(a + b)^2 = a^2 + 2ab + b^2$$
$$(a - b)^2 = a^2 - 2ab + b^2$$

Product of Sum and Difference of Two Terms

$$(a + b)(a - b) = a^2 - b^2$$

Examples Multiply.

$$\begin{array}{ccccccc}(a & + & b)^2 & = & a^2 & + & 2\cdot a\cdot b & + & b^2\end{array}$$

11. $(x + 5)^2 = x^2 + 2\cdot x\cdot 5 + 5^2 = x^2 + 10x + 25$

12. $(4m^2 - 3n)^2 = (4m^2)^2 - 2(4m^2)(3n) + (3n)^2 = 16m^4 - 24m^2n + 9n^2$

$$\begin{array}{ccccccc}(a & + & b)(a & - & b) & = & a^2 & - & b^2\end{array}$$

13. $(x + 3)(x - 3) = x^2 - 3^2 = x^2 - 9$

14. $(4y - 1)(4y + 1) = (4y)^2 - 1^2 = 16y^2 - 1$

🟧 **Work Practice 11–14**

Objective C Reviewing Factoring Strategies

The key to proficiency in factoring polynomials is to practice until you are comfortable with each technique. A strategy for factoring polynomials completely is given next. This strategy can also be found in the Chapter 6 Integrated Review.

Factoring a Polynomial

Step 1: Are there any common factors? If so, factor out the greatest common factor.

Step 2: How many terms are in the polynomial?

    a. If there are *two* terms, decide if one of the following formulas may be applied:

        i. Difference of two squares: $a^2 - b^2 = (a - b)(a + b)$

        ii. Difference of two cubes: $a^3 - b^3 = (a - b)(a^2 + ab + b^2)$

        iii. Sum of two cubes: $a^3 + b^3 = (a + b)(a^2 - ab + b^2)$

    b. If there are *three* terms, try one of the following:

        i. Perfect square trinomial: $a^2 + 2ab + b^2 = (a + b)^2$
$$a^2 - 2ab + b^2 = (a - b)^2$$

        ii. If not a perfect square trinomial, factor by using the methods presented in Sections 6.2 through 6.4.

    c. If there are *four* or more terms, try factoring by grouping.

Step 3: See whether any factors in the factored polynomial can be factored further.

Practice 11–14

Multiply.

11. $(x + 3)^2$

12. $(6a^2 - 2b)^2$

13. $(x + 4)(x - 4)$

14. $(3m - 6)(3m + 6)$

Answers
11. $x^2 + 6x + 9$
12. $36a^4 - 24a^2b + 4b^2$
13. $x^2 - 16$ **14.** $9m^2 - 36$

Practice 15

Factor each polynomial completely.

a. $21xy^2 + 6xy$

b. $45x^2 - 20$

c. $4y^2 - 2y - 6$

d. $a^2 + a^2b + 5 + 5b$

e. $8x^3 + y^3$

Example 15   Factor each polynomial completely.

a. $8a^2b - 4ab$

b. $36x^2 - 9$

c. $2x^2 - 5x - 7$

d. $5p^2 + 5 + qp^2 + q$

e. $27a^3 - b^3$

Solution:

a. Step 1: The terms have a common factor of $4ab$, which we factor out.

$$8a^2b - 4ab = 4ab(2a - 1)$$

Step 2: The factor $2a - 1$ has two terms, but it is not the difference of two squares or the sum or difference of two cubes.

Step 3: The factor $2a - 1$ cannot be factored further.

b. Step 1: Factor out a common factor of 9.

$$36x^2 - 9 = 9(4x^2 - 1)$$

Step 2: The factor $4x^2 - 1$ has two terms, and it is the difference of two squares.

$$9(4x^2 - 1) = 9(2x + 1)(2x - 1)$$

Step 3: No factor can be factored further.

c. Step 1: The terms of $2x^2 - 5x - 7$ contain no common factor other than 1 or -1.

Step 2: There are three terms. The trinomial is not a perfect square, so we factor by methods from Section 6.3 or 6.4.

$$2x^2 - 5x - 7 = (2x - 7)(x + 1)$$

Step 3: No factor can be factored further.

d. Step 1: There is no common factor of all terms of $5p^2 + 5 + qp^2 + q$.

Step 2: The polynomial has four terms, so try factoring by grouping.

$$5p^2 + 5 + qp^2 + q = (5p^2 + 5) + (qp^2 + q) \quad \textcolor{blue}{\text{Group the terms.}}$$
$$= 5(p^2 + 1) + q(p^2 + 1)$$
$$= (p^2 + 1)(5 + q)$$

Step 3: No factor can be factored further.

e. Step 1: The terms of $27a^3 - b^3$ contain no common factor.

Step 2: There are two terms and $27a^3 - b^3$ is the difference of cubes.

$$27a^3 - b^3 = (3a)^3 - b^3$$
$$= (3a - b)[(3a)^2 + (3a)(b) + b^2]$$
$$= (3a - b)(9a^2 + 3ab + b^2)$$

Step 3: No factor can be factored further.

■ **Work Practice 15**

Answers

15. a. $3xy(7y + 2)$

b. $5(3x - 2)(3x + 2)$

c. $2(2y - 3)(y + 1)$

d. $(1 + b)(a^2 + 5)$

e. $(2x + y)(4x^2 - 2xy + y^2)$

A Exercise Set MyMathLab®

Objective A *Use rules for exponents to simplify each expression. See Examples 1 through 5.*

1. $(-4x^3p^2)(4y^3x^3)$
$-16x^6y^3p^2$

2. $(-6a^2b^3)(-3ab^3)$
$18a^3b^6$

3. $8x^0 + 1$
9

4. $(5x)^0 + 5x^0$
6

5. $\dfrac{a^{12}b^2}{a^9b}$ a^3b

6. $-\dfrac{26z^{11}}{2z^7}$ $-13z^4$

7. 4^{-2} $\dfrac{1}{16}$

8. 2^{-3} $\dfrac{1}{8}$

9. $4^{-1} + 3^{-2}$ $\dfrac{13}{36}$

10. $1^{-3} - 4^{-2}$ $\dfrac{15}{16}$

11. $\dfrac{y^{-3}}{y^{-7}}$
y^4

12. $\dfrac{z^{-12}}{z^{10}}$
$\dfrac{1}{z^{22}}$

13. $\dfrac{(24x^8)(x)}{20x^{-7}}$
$\dfrac{6x^{16}}{5}$

14. $\dfrac{(30z^2)(z^5)}{55z^{-4}}$
$\dfrac{6z^{11}}{11}$

15. $\left(\dfrac{2x^5}{y^{-3}}\right)^4$
$16x^{20}y^{12}$

16. $\left(\dfrac{3a^{-4}}{b^7}\right)^3$
$\dfrac{27}{a^{12}b^{21}}$

17. $\left(\dfrac{2a^{-2}b^5}{4a^2b^7}\right)^{-2}$
$4a^8b^4$

18. $\left(\dfrac{5x^7y^4}{10x^3y^{-2}}\right)^{-3}$
$\dfrac{8}{x^{12}y^{18}}$

19. $\left(\dfrac{2x^2}{y^4}\right)^3\left(\dfrac{2x^5}{y}\right)^{-2}$
$\dfrac{2}{x^4y^{10}}$

20. $\left(\dfrac{3z^{-2}}{y}\right)^2\left(\dfrac{9y^{-4}}{z^{-3}}\right)^{-1}$
$\dfrac{y^2}{z^7}$

Objective B *Add or subtract as indicated. See Examples 6 through 8.*

21. $(5y^4 - 7y^2 + x^2 - 3) + (-3y^4 + 2y^2 + 4)$
$2y^4 - 5y^2 + x^2 + 1$

22. $(8x^4 - 14x^2 + 6) + (-12x^6 - 21x^4 - 9x^2)$
$-12x^6 - 13x^4 - 23x^2 + 6$

23. $(9y^2 - 7y + 5) - (8y^2 - 7y + 2)$
$y^2 + 3$

24. $(2x^2 + 3x + 12) - (5x - 7)$
$2x^2 - 2x + 19$

25. $(3x^3 - b + 2a - 6) + (-4x^3 + b + 6a - 6)$
$-x^3 + 8a - 12$

26. $(5x^2 - 6) + (2x^2 - 4x + 8)$
$7x^2 - 4x + 2$

27. $(4x^2 - 6x + 2) - (-x^2 + 3x + 5)$
$5x^2 - 9x - 3$

28. $(5x^2 + x + 9) - (2x^2 - 9)$
$3x^2 + x + 18$

29.
$$
\begin{array}{r}
3x^2 - 4x + 8 \\
-\ \ \ \ (5x^2 - 7) \\
\hline
-2x^2 - 4x + 15
\end{array}
$$

30.
$$
\begin{array}{r}
-3x^2 - 4x + 8 \\
-\ \ \ \ (5x + 12) \\
\hline
-3x^2 - 9x - 4
\end{array}
$$

31.
$$
\begin{array}{r}
6y^2 - 6y + 4 \\
-(-y^2 - 6y + 7) \\
\hline
7y^2 - 3
\end{array}
$$

32.
$$
\begin{array}{r}
-4x^3 + 4x^2 - 4x \\
-(2x^3 - 2x^2 + 3x) \\
\hline
-6x^3 + 6x^2 - 7x
\end{array}
$$

33. Subtract $(y^2 + 4yx + 7)$ from $(-19y^2 + 7yx + 7)$.
$-20y^2 + 3yx$

34. Subtract $(x^2y - 4)$ from $(3x^2 - 4x^2y + 5)$.
$3x^2 - 5x^2y + 9$

Multiply. See Examples 9 through 14.

35. $(3x + 1)(3x + 5)$
$9x^2 + 18x + 5$

36. $(4x - 5)(5x + 6)$
$20x^2 - x - 30$

37. $\left(4x + \dfrac{1}{3}\right)\left(4x - \dfrac{1}{2}\right)$
$16x^2 - \dfrac{2}{3}x - \dfrac{1}{6}$

38. $\left(4y - \dfrac{1}{3}\right)\left(3y - \dfrac{1}{8}\right)$
$12y^2 - \dfrac{3}{2}y + \dfrac{1}{24}$

39. $(x + 4)^2$
$x^2 + 8x + 16$

40. $(x - 5)^2$
$x^2 - 10x + 25$

41. $(3x - y)^2$
$9x^2 - 6xy + y^2$

42. $(4x - z)^2$
$16x^2 - 8xz + z^2$

43. $(3b - 6y)(3b + 6y)$
$9b^2 - 36y^2$

44. $(2x - 4y)(2x + 4y)$
$4x^2 - 16y^2$

45. $\left(3x + \dfrac{1}{2}\right)\left(3x - \dfrac{1}{2}\right)$
$9x^2 - \dfrac{1}{4}$

46. $\left(2x - \dfrac{1}{3}\right)\left(2x + \dfrac{1}{3}\right)$
$4x^2 - \dfrac{1}{9}$

Objective C *Factor completely. See Example 15.*

47. $14x^2y - 2xy$
$2xy(7x - 1)$

48. $24ab^2 - 6ab$
$6ab(4b - 1)$

49. $4x^2 - 16$
$4(x + 2)(x - 2)$

50. $9x^2 - 81$
$9(x + 3)(x - 3)$

51. $3x^2 - 8x - 11$
$(3x - 11)(x + 1)$

52. $5x^2 - 2x - 3$
$(5x + 3)(x - 1)$

53. $8x^3 + 125y^3$
$(2x + 5y)$
$(4x^2 - 10xy + 25y^2)$

54. $27x^3 - 64y^3$
$(3x - 4y)$
$(9x^2 + 12xy + 16y^2)$

55. $7x^2 - 63x$
$7x(x - 9)$

56. $15x^2 - 20x$
$5x(3x - 4)$

57. $20x^2 + 23x + 6$
$(4x + 3)(5x + 2)$

58. $20x^2 - 220x + 600$
$20(x - 6)(x - 5)$

59. $ab - 6a + 7b - 42$
$(b - 6)(a + 7)$

60. $2sr + 10s - r - 5$
$(r + 5)(2s - 1)$

61. $x^4 - 1$
$(x^2 + 1)(x - 1)(x + 1)$

62. $y^4 - 16$
$(y^2 + 4)(y + 2)(y - 2)$

63. $2x^3 - 54$
$2(x - 3)(x^2 + 3x + 9)$

64. $250x^4 - 16x$
$2x(5x - 2)$
$(25x^2 + 10x + 4)$

Transition Review: Solving Linear and Quadratic Equations

Appendix

B

Objective A Solving Equations ▶

Recall that an **equation** is a statement that two expressions are equal. When a variable in an equation is replaced by a number and the resulting equation is true, then that number is called a **solution.** The set of solutions of an equation is called its **solution set.**

In this section, we review solving linear and quadratic equations. Study the table below to help you identify these types of equations. Here, a, b, and c are real numbers and a is not 0.

Objective

A Solve Linear and Quadratic Equations. ▶

Linear: Can be written in form $ax + b = c$ (Sections 2.1–2.3)	Quadratic: Can be written in form $ax^2 + bx + c = 0$ (Section 6.6)
$3x = -15$	$-3x^2 + 7 = x^2 - 9$
$2.7 - y = 3y$	$p^2 + 2.6 = p - 4.3$
$\dfrac{4n}{5} - \dfrac{9n}{7} + 1 = 0$	$\dfrac{y^2}{5} - \dfrac{1}{7} + \dfrac{y}{9} = 0$

You may want to use the steps below to help you solve linear and quadratic equations. (In this appendix, only the method of solving quadratic equations by factoring will be reviewed.)

Solving Linear and Quadratic Equations in One Variable

Step 1: Multiply on both sides to clear the equation of fractions if they occur.

Step 2: Use the distributive property to remove parentheses if they occur.

Step 3: Simplify each side of the equation by combining like terms.

Step 4: Decide whether the equation is linear or quadratic.

If linear $(ax + b = c)$,	If quadratic $(ax^2 + bx + c = 0)$,
Step 5: Get all variable terms on one side and all numbers on the other side by using the addition property of equality.	**Step 5:** Write the equation in standard form: $ax^2 + bx + c = 0$.
	Step 6: Factor completely.
Step 6: Get the variable alone by using the multiplication property of equality.	**Step 7:** Set each factor containing a variable equal to 0.
	Step 8: Solve.

Final Step: Check each solution in the original equation.

Example 1 Solve: $2(x - 3) = 5x - 9$

Solution:

Step 1: is not needed since there are no fractions.

$$2(x - 3) = 5x - 9$$

Step 2: $2x - 6 = 5x - 9$ Use the distributive property.

Step 3: is not needed since no simplifying can be done on either side of the equation.

(Continued on next page)

Practice 1

Solve: $4(x - 2) = 6x - 10$

Answer

1. $x = 1$

989

Step 4: $2x - 6 = 5x - 9$ The equation is linear.

Step 5: Next we get variable terms on the same side of the equation by using the addition property of equality.

$$2x - 6 - 5x = 5x - 9 - 5x$$ Subtract $5x$ from both sides.

$$-3x - 6 = -9$$ Simplify.

$$-3x - 6 + 6 = -9 + 6$$ Add 6 to both sides.

$$-3x = -3$$ Simplify.

Step 6:
$$\frac{-3x}{-3} = \frac{-3}{-3}$$ Divide both sides by -3.

$$x = 1$$

Final Step: Check to see that 1 is the solution.

■ **Work Practice 1**

Practice 2

Solve:

$$y + \frac{y-4}{4} = \frac{1}{2} - \frac{y-6}{4}$$

Example 2 Solve:

$$x - \frac{x-2}{6} = \frac{x-7}{3} + \frac{2}{3}$$

Solution:

Step 1: $6\left(x - \dfrac{x-2}{6}\right) = 6\left(\dfrac{x-7}{3} + \dfrac{2}{3}\right)$ Multiply both sides by 6.

$$6x - (x - 2) = 2(x - 7) + 2(2)$$

Step 2: $6x - x + 2 = 2x - 14 + 4$ Remove grouping symbols.

Step 3: $5x + 2 = 2x - 10$ Simplify.

Step 4: $5x + 2 = 2x - 10$ This equation is linear.

Step 5: $5x + 2 - 2 = 2x - 10 - 2$ Subtract 2.

$$5x = 2x - 12$$

$$5x - 2x = 2x - 12 - 2x$$ Subtract $2x$.

$$3x = -12$$

Step 6:
$$\frac{3x}{3} = \frac{-12}{3}$$ Divide by 3.

$$x = -4$$

Final Step: $-4 - \dfrac{-4-2}{6} \overset{?}{=} \dfrac{-4-7}{3} + \dfrac{2}{3}$ Replace x with -4 in the original equation.

$$-4 - \frac{-6}{6} \overset{?}{=} \frac{-11}{3} + \frac{2}{3}$$

$$-4 - (-1) \overset{?}{=} \frac{-9}{3}$$

$$-3 = -3$$ True

The solution is -4.

■ **Work Practice 2**

Answer

2. $y = 2$

Example 3 Solve: $2x^2 = \dfrac{17}{3}x + 1$

Solution:

$$2x^2 = \frac{17}{3}x + 1$$

Step 1: $3(2x^2) = 3\left(\dfrac{17}{3}x + 1\right)$ Clear the equation of fractions.

Step 2: $6x^2 = 17x + 3$ Use the distributive property.

Step 3: is not needed since no simplifying can be done on either side of the equation.

Step 4: $6x^2 = 17x + 3$ The equation is quadratic.

Step 5: $6x^2 - 17x - 3 = 0$ Rewrite the equation in standard form.

Step 6: $(6x + 1)(x - 3) = 0$ Factor.

Step 7: $6x + 1 = 0$ or $x - 3 = 0$ Set each factor equal to 0.

Step 8: $6x = -1$ $x = 3$ Solve each equation.

$$x = -\frac{1}{6}$$

Final Step: Check by substituting into the original equation. The solutions are $-\dfrac{1}{6}$ and 3.

■ **Work Practice 3**

Practice 3

Solve: $2x^2 + \dfrac{5}{2}x = 3$

Answer

3. $-2, \dfrac{3}{4}$

Vocabulary and Readiness Check

Solve each equation.

1. $3x = 18$ 6

2. $2x = 60$ 30

3. $x - 7 = 10$ 17

4. $x - 2 = 15$ 17

5. $\dfrac{x}{2} = 4$ 8

6. $\dfrac{x}{3} = 5$ 15

7. $x + 1 = 11$ 10

8. $x + 4 = 20$ 16

B   **Exercise Set**   MyMathLab®

Objective A *Solve each equation. See Examples 1 through 3.*

1. $x + 2.8 = 1.9$ -0.9

2. $y - 8.6 = -6.3$ 2.3

3. $5x - 4 = 26$ 6

4. $2y - 3 = 11$ 7

5. $-4.1 - 7z = 3.6$ -1.1

6. $10.3 - 6x = -2.3$ 2.1

7. $5y + 12 = 2y - 3$ -5

8. $4x + 14 = 6x + 8$ 3

9. $(x + 3)(3x - 4) = 0$ $-3, \dfrac{4}{3}$

10. $(5x + 1)(x - 2) = 0$ $-\dfrac{1}{5}, 2$

11. $8x - 5x + 3 = x - 7 + 10$ 0

12. $6 + 3x + x = -x + 2 - 26$ -6

13. $3(2x - 5)(4x + 3) = 0$ $-\dfrac{3}{4}, \dfrac{5}{2}$

14. $8(3x - 4)(2x - 7) = 0$ $\dfrac{4}{3}, \dfrac{7}{2}$

15. $x^2 + 11x + 24 = 0$ $-3, -8$

16. $y^2 - 10y + 24 = 0$ 4, 6

▶**17.** $5x + 12 = 2(2x + 7)$ 2

18. $2(x + 3) = x + 5$ -1

19. $12x^2 + 5x - 2 = 0$ $\dfrac{1}{4}, -\dfrac{2}{3}$

20. $3y^2 - y - 14 = 0$ $-2, \dfrac{7}{3}$

21. $z^2 + 9 = 10z$ $1, 9$

22. $n^2 + n = 72$ $-9, 8$

23. $3(x - 6) = 5x$ -9

24. $6x = 4(5 + x)$ 10

25. $\dfrac{x}{2} + \dfrac{2}{3} = \dfrac{3}{4}$ $\dfrac{1}{6}$

26. $\dfrac{x}{2} + \dfrac{x}{3} = \dfrac{5}{2}$ 3

27. $\dfrac{n - 3}{4} + \dfrac{n + 5}{7} = \dfrac{5}{14}$ 1

28. $\dfrac{2 + h}{9} + \dfrac{h - 1}{3} = \dfrac{1}{3}$ 1

▶ **29.** $x(5x + 2) = 3$ $\dfrac{3}{5}, -1$

30. $n(2n - 3) = 2$ $-\dfrac{1}{2}, 2$

31. $x^2 - 6x = x(8 + x)$ 0

32. $n(3 + n) = n^2 + 4n$ 0

33. $\dfrac{z^2}{6} - \dfrac{z}{2} - 3 = 0$ $6, -3$

34. $\dfrac{c^2}{20} - \dfrac{c}{4} + \dfrac{1}{5} = 0$ $1, 4$

35. $2y + 5(y - 4) = 4y - 2(y - 10)$ 8

36. $9c - 3(6 - 5c) = c - 2(3c + 9)$ 0

37. $2(x - 8) + x = 3(x - 6) + 2$ $\{x \mid x \text{ is a real number}\}$

38. $4(x + 5) = 3(x - 4) + x$ $\varnothing$

39. $\dfrac{x^2}{2} + \dfrac{x}{20} = \dfrac{1}{10}$ $\dfrac{2}{5}, -\dfrac{1}{2}$

40. $\dfrac{y^2}{30} = \dfrac{y}{15} + \dfrac{1}{2}$ $-3, 5$

41. $\dfrac{4t^2}{5} = \dfrac{t}{5} + \dfrac{3}{10}$ $\dfrac{3}{4}, -\dfrac{1}{2}$

42. $\dfrac{5x^2}{6} - \dfrac{7x}{2} + \dfrac{2}{3} = 0$ $\dfrac{1}{5}, 4$

▶ **43.** $\dfrac{m - 4}{3} - \dfrac{3m - 1}{5} = 1$ -8

44. $\dfrac{n + 1}{8} - \dfrac{2 - n}{3} = \dfrac{5}{6}$ 3

45. $-(3x - 5) - (2x - 6) + 1 = -5(x - 1) - (3x + 2) + 3$ -2

46. $-4(2x - 3) - (10x + 7) - 2 = -(12x - 5) - (4x + 9) - 1$ 4

47. $3x(x - 5) = 0$ $0, 5$

48. $4x(2x + 3) = 0$ $-\dfrac{3}{2}, 0$

49. $12x^2 + 2x - 2 = 0$ $-\dfrac{1}{2}, \dfrac{1}{3}$

50. $8x^2 + 13x + 5 = 0$ $-1, -\dfrac{5}{8}$

51. $w^2 - 5w = 36$ $-4, 9$

52. $x^2 + 32 = 12x$ $4, 8$

▶ **53.** $2z(z + 6) = 2z^2 + 12z - 8$ $\varnothing$

54. $3c^2 - 8c + 2 = c(3c - 8)$ $\varnothing$

55. $-3(x - 4) + x = 5(3 - x)$ 1

56. $-4(a + 1) - 3a = -7(2a - 3)$ $\dfrac{25}{7}$

41. $x \le -4$ or $x \ge 1$

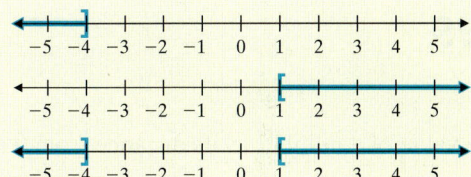

42. $x < 0$ or $x < 1$

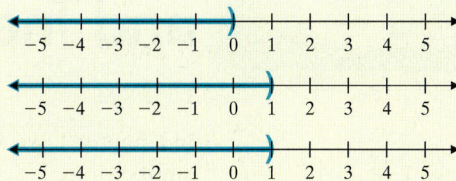

43. $x > 0$ or $x < 3$

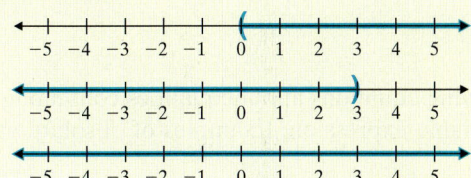

44. $x \ge -3$ or $x \le -4$

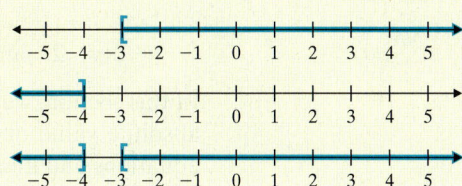

Solve each compound inequality. See Examples 7 and 8.

45. $x < -1$ or $x > 0$
$(-\infty, -1) \cup (0, \infty)$

46. $x \le 1$ or $x \le -3$
$(-\infty, 1]$

47. $-2x \le -4$ or $5x - 20 \ge 5$
$[2, \infty)$

48. $x + 4 < 0$ or $6x > -12$
$(-\infty, -4) \cup (-2, \infty)$

49. $3(x - 1) < 12$ or $x + 7 > 10$
$(-\infty, \infty)$

50. $5(x - 1) \ge -5$ or $5 - x \le 11$
$[-6, \infty)$

▶ 51. $3x + 2 \le 5$ or $7x > 29$
$(-\infty, 1] \cup \left(\dfrac{29}{7}, \infty\right)$

52. $-x < 7$ or $3x + 1 < -20$
$(-\infty, -7) \cup (-7, \infty)$

53. $3x \ge 5$ or $-x - 6 < 1$
$(-7, \infty)$

54. $\dfrac{3}{8}x + 1 \le 0$ or $-2x < -4$
$\left(-\infty, -\dfrac{8}{3}\right] \cup (2, \infty)$

55. $6x - 4 > 2x$ or $4x - 1 < x + 5$
$(-\infty, \infty)$

56. $6x - 2 > 5x + 3$ or $4x - 3 < x$
$(-\infty, 1) \cup (5, \infty)$

Absolute Value Equations and Inequalities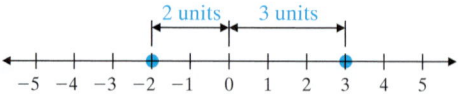

Objectives

A Solve Absolute Value Equations.

B Solve Absolute Value Inequalities.

In Chapter 1, we defined the absolute value of a number as its distance from 0 on a number line.

$$|-2| = 2 \quad \text{and} \quad |3| = 3$$

In this section, we concentrate on solving equations and inequalities containing the absolute value of a variable or a variable expression. Examples of absolute value equations and inequalities are

$$|x| = 3 \qquad -5 \geq |2y + 7| \qquad |z - 6.7| = |3z + 1.2| \qquad |x - 3| > 7$$

Absolute value equations and inequalities are extremely useful in data analysis, especially for calculating acceptable measurement error and errors that result from the way numbers are sometimes represented in computers.

Objective A Solving Absolute Value Equations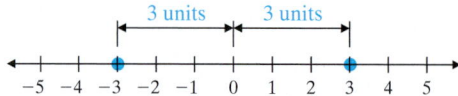

To begin, let's solve a few absolute value equations by inspection.

Practice 1
Solve: $|y| = 5$

Example 1 Solve: $|x| = 3$

Solution: The solution set of this equation will contain all numbers whose distance from 0 is 3 units. Two numbers are 3 units away from 0 on the number line: 3 and -3.

Check: To check, let $x = 3$ and $x = -3$ in the original equation.

$$|x| = 3 \qquad\qquad\qquad |x| = 3$$
$$|3| \overset{?}{=} 3 \quad \text{Let } x = 3. \qquad |-3| \overset{?}{=} 3 \quad \text{Let } x = -3.$$
$$3 = 3 \quad \text{True.} \qquad\qquad 3 = 3 \quad \text{True.}$$

Both solutions check. Thus, the solution set of the equation $|x| = 3$ is $\{3, -3\}$.

■ **Work Practice 1**

Practice 2
Solve: $|p| = -4$

Example 2 Solve: $|x| = -2$

Solution: The absolute value of a number is never negative, so this equation has no solution. The solution set is $\{\ \}$ or $\varnothing$.

■ **Work Practice 2**

Answers

1. $\{-5, 5\}$ **2.** $\varnothing$

41. $x \le -4$ or $x \ge 1$

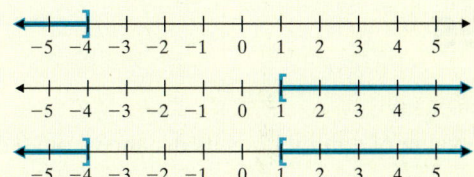

42. $x < 0$ or $x < 1$

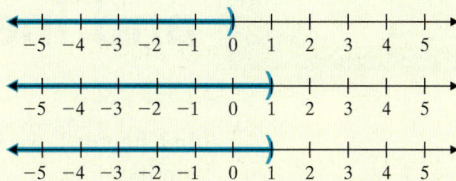

43. $x > 0$ or $x < 3$

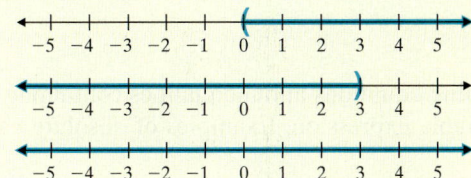

44. $x \ge -3$ or $x \le -4$

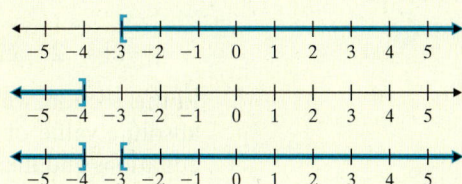

Solve each compound inequality. See Examples 7 and 8.

45. $x < -1$ or $x > 0$
$(-\infty, -1) \cup (0, \infty)$

46. $x \le 1$ or $x \le -3$
$(-\infty, 1]$

47. $-2x \le -4$ or $5x - 20 \ge 5$
$[2, \infty)$

48. $x + 4 < 0$ or $6x > -12$
$(-\infty, -4) \cup (-2, \infty)$

49. $3(x - 1) < 12$ or $x + 7 > 10$
$(-\infty, \infty)$

50. $5(x - 1) \ge -5$ or $5 - x \le 11$
$[-6, \infty)$

▶ **51.** $3x + 2 \le 5$ or $7x > 29$
$(-\infty, 1] \cup \left(\dfrac{29}{7}, \infty\right)$

52. $-x < 7$ or $3x + 1 < -20$
$(-\infty, -7) \cup (-7, \infty)$

53. $3x \ge 5$ or $-x - 6 < 1$
$(-7, \infty)$

54. $\dfrac{3}{8}x + 1 \le 0$ or $-2x < -4$
$\left(-\infty, -\dfrac{8}{3}\right] \cup (2, \infty)$

55. $6x - 4 > 2x$ or $4x - 1 < x + 5$
$(-\infty, \infty)$

56. $6x - 2 > 5x + 3$ or $4x - 3 < x$
$(-\infty, 1) \cup (5, \infty)$

Absolute Value Equations and Inequalities

Objectives

A Solve Absolute Value Equations.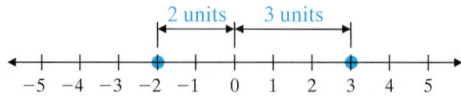

B Solve Absolute Value Inequalities. ▶

In Chapter 1, we defined the absolute value of a number as its distance from 0 on a number line.

$$|-2| = 2 \quad \text{and} \quad |3| = 3$$

In this section, we concentrate on solving equations and inequalities containing the absolute value of a variable or a variable expression. Examples of absolute value equations and inequalities are

$$|x| = 3 \qquad -5 \geq |2y + 7| \qquad |z - 6.7| = |3z + 1.2| \qquad |x - 3| > 7$$

Absolute value equations and inequalities are extremely useful in data analysis, especially for calculating acceptable measurement error and errors that result from the way numbers are sometimes represented in computers.

Objective A Solving Absolute Value Equations ▶

To begin, let's solve a few absolute value equations by inspection.

Practice 1

Solve: $|y| = 5$

Example 1 Solve: $|x| = 3$

Solution: The solution set of this equation will contain all numbers whose distance from 0 is 3 units. Two numbers are 3 units away from 0 on the number line: 3 and -3.

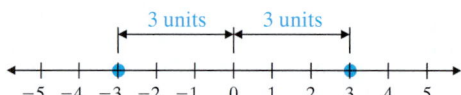

Check: To check, let $x = 3$ and $x = -3$ in the original equation.

$\|x\| = 3$	$\|x\| = 3$
$\|3\| \stackrel{?}{=} 3$ Let $x = 3$.	$\|-3\| \stackrel{?}{=} 3$ Let $x = -3$.
$3 = 3$ True.	$3 = 3$ True.

Both solutions check. Thus, the solution set of the equation $|x| = 3$ is $\{3, -3\}$.

■ **Work Practice 1**

Practice 2

Solve: $|p| = -4$

Example 2 Solve: $|x| = -2$

Solution: The absolute value of a number is never negative, so this equation has no solution. The solution set is $\{ \ \}$ or $\varnothing$.

■ **Work Practice 2**

Answers

1. $\{-5, 5\}$ **2.** $\varnothing$

Example 3 Solve: $|y| = 0$

Solution: We are looking for all numbers whose distance from 0 is zero units. The only number is 0. The solution set is $\{0\}$.

■ **Work Practice 3**

From the above examples, we have the following.

Absolute Value Property

Solve $|X| = a$ as follows.

If a is positive, then solve $X = a$ or $X = -a$.

If a is 0, then $X = 0$.

If a is negative, the equation $|X| = a$ has no solution.

Helpful Hint

For the equation $|X| = a$ in the box above, X can be a single variable or a variable expression.

When we are solving absolute value equations, if $|X|$ is not alone on one side of the equation we first use properties of equality to get $|X|$ alone.

Example 4 Solve: $2|x| + 25 = 37$

Solution: First we get $|x|$ alone.

$$2|x| + 25 = 37$$
$$2|x| = 12 \quad \text{Subtract 25 from both sides.}$$
$$|x| = 6 \quad \text{Divide both sides by 2.}$$
$$x = 6 \quad \text{or} \quad x = -6 \quad \text{Use the absolute value property.}$$

The solution set is $\{-6, 6\}$.

■ **Work Practice 4**

If the expression inside the absolute value bars is more complicated than a single variable x, we can still use the absolute value property.

Example 5 Solve: $|w + 3| = 7$

Solution: If we think of the expression $w + 3$ as X in the absolute value property, we have that

$$|w + 3| = 7$$
$$w + 3 = 7 \quad \text{or} \quad w + 3 = -7 \quad \text{Use the absolute value property.}$$
$$w = 4 \quad \text{or} \qquad w = -10$$

The solution set is $\{4, -10\}$.

■ **Work Practice 5**

Don't forget that to use the absolute value property you must first make sure that the absolute value expression is alone on one side of the equation.

Helpful Hint

If the equation has a single absolute value expression containing variables, get the absolute value expression alone. Then use the absolute value property.

Practice 3

Solve: $|x| = 0$

Teaching Tip

Point out the use of the word *or* in the definition of the absolute value property and then ask why "or" was used rather than "and."

Practice 4

Solve: $3|y| - 4 = 17$

Practice 5

Solve: $|x - 4| = 11$

Answers

3. $\{0\}$ **4.** $\{-7, 7\}$ **5.** $\{15, -7\}$

Practice 6

Solve: $|4x + 2| + 1 = 7$

Example 6 Solve: $|2x - 1| + 5 = 6$

Solution: We want the absolute value expression alone on one side of the equation, so we begin by subtracting 5 from both sides. Then we use the absolute value property.

$$|2x - 1| + 5 = 6$$
$$|2x - 1| = 1 \qquad \text{Subtract 5 from both sides}$$
$$2x - 1 = 1 \quad \text{or} \quad 2x - 1 = -1 \qquad \text{Use the absolute value property.}$$
$$2x = 2 \quad \text{or} \qquad 2x = 0$$
$$x = 1 \quad \text{or} \qquad x = 0 \qquad \text{Solve.}$$

The solution set is $\{0, 1\}$.

■ **Work Practice 6**

Given two absolute value expressions, we might ask, when are the absolute values of two expressions equal? To see the answer, notice that

$$|2| = |2| \quad |-2| = |-2| \quad |-2| = |2| \quad |2| = |-2|$$

same   same   opposites   opposites

Two absolute value expressions are equal when the expressions inside the absolute value bars are equal to or are opposites of each other.

Practice 7

Solve: $|4x - 5| = |3x + 5|$

Example 7 Solve: $|3x + 2| = |5x - 8|$

Solution: This equation is true if the expressions inside the absolute value bars are equal to or are opposites of each other.

$$3x + 2 = 5x - 8 \quad \text{or} \quad 3x + 2 = -(5x - 8)$$

Next we solve each equation.

$$3x + 2 = 5x - 8 \quad \text{or} \quad 3x + 2 = -5x + 8$$
$$-2x + 2 = -8 \qquad \text{or} \quad 8x + 2 = 8$$
$$-2x = -10 \qquad \text{or} \qquad 8x = 6$$
$$x = 5 \qquad \text{or} \qquad x = \frac{3}{4}$$

Check to see that replacing x with 5 or with $\frac{3}{4}$ results in a true statement.

The solution set is $\left\{\frac{3}{4}, 5\right\}$.

■ **Work Practice 7**

Practice 8

Solve: $|x + 2| = |4 - x|$

Example 8 Solve: $|x - 3| = |5 - x|$

Solution:

$$x - 3 = 5 - x \quad \text{or} \qquad x - 3 = -(5 - x)$$
$$2x - 3 = 5 \qquad \text{or} \qquad x - 3 = -5 + x$$
$$2x = 8 \qquad \text{or} \quad x - 3 - x = -5 + x - x$$
$$x = 4 \qquad \text{or} \qquad -3 = -5 \qquad \text{False.}$$

Recall from Section 2.3 that when an equation simplifies to a false statement, the equation has no solution. Thus, the only solution for the original absolute value equation is 4, and the solution set is $\{4\}$.

■ **Work Practice 8**

Answers

6. $\{1, -2\}$ **7.** $\{0, 10\}$ **8.** $\{1\}$

✓**Concept Check** True or false? Absolute value equations always have two solutions. Explain your answer.

Objective B Solving Absolute Value Inequalities ▶

To begin, let's solve a few absolute value inequalities by inspection.

Example 9 Solve $|x| < 2$ using a number line.

Solution: The solution set contains all numbers whose distance from 0 is less than 2 units on the number line.

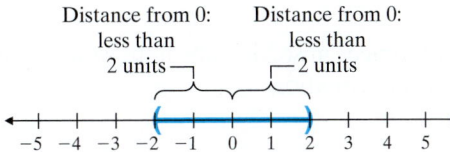

The solution set is $\{x \mid -2 < x < 2\}$, or $(-2, 2)$ in interval notation.

◼ **Work Practice 9**

Example 10 Solve $|x| \geq 3$ using a number line.

Solution: The solution set contains all numbers whose distance from 0 is 3 or more units. Thus, the graph of the solution set contains 3 and all points to the right of 3 on the number line or -3 and all points to the left of -3 on the number line.

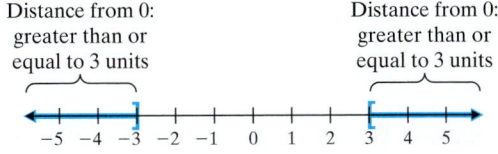

This solution set is $\{x \mid x \leq -3 \text{ or } x \geq 3\}$. In interval notation, the solution set is $(-\infty, -3] \cup [3, \infty)$, since **or** means union.

◼ **Work Practice 10**

The following box summarizes solving absolute value equations and inequalities.

> ### Solving Absolute Value Equations and Inequalities
>
> If a is a positive number,
>
> To solve $|X| = a$, solve $X = a$ or $X = -a$.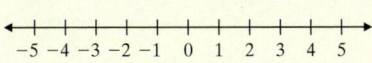
>
> To solve $|X| < a$, solve $-a < X < a$.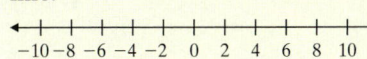
>
> To solve $|X| > a$, solve $X < -a$ or $X > a$.

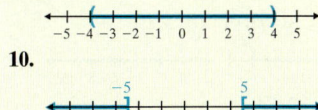

Practice 11

Solve: $|x + 2| > 4$. Graph the solution set.

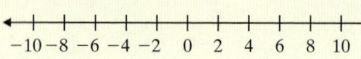

Example 11   Solve: $|x - 3| > 7$

Solution: Since 7 is positive, to solve $|x - 3| > 7$, we solve the compound inequality $x - 3 < -7$ or $x - 3 > 7$.

$$x - 3 < -7 \quad \text{or} \quad x - 3 > 7$$
$$x < -4 \quad \text{or} \quad x > 10 \quad \text{Add 3 to both sides.}$$

The solution set is $\{x \mid x < -4 \text{ or } x > 10\}$ or $(-\infty, -4) \cup (10, \infty)$ in interval notation. Its graph is shown.

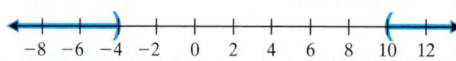

Work Practice 11

Let's review the differences in solving absolute value equations and inequalities by solving an absolute value equation.

Practice 12

Solve: $|x - 3| = 5$. Graph the solution set.

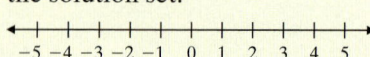

Example 12   Solve: $|x + 1| = 6$

Solution: This is an equation, so we solve

$$x + 1 = 6 \quad \text{or} \quad x + 1 = -6$$
$$x = 5 \quad \text{or} \quad x = -7$$

The solution set is $\{-7, 5\}$. Its graph is shown.

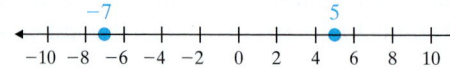

Work Practice 12

Practice 13

Solve: $|x - 2| \le 1$. Graph the solution set.

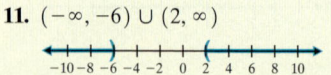

Example 13   Solve: $|x - 6| \le 2$

Solution: To solve $|x - 6| \le 2$, we solve

$$-2 \le x - 6 \le 2$$
$$-2 + 6 \le x - 6 + 6 \le 2 + 6 \quad \text{Add 6 to all three parts.}$$
$$4 \le x \le 8 \quad \text{Simplify.}$$

The solution set is $\{x \mid 4 \le x \le 8\}$, or $[4, 8]$ in interval notation. Its graph is shown.

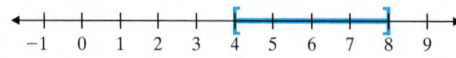

Work Practice 13

Answers

11. $(-\infty, -6) \cup (2, \infty)$

12. $\{-2, 8\}$

13. $[1, 3]$

Helpful Hint

Before using an absolute value inequality property, get an absolute value expression alone on one side of the inequality.

Example 14 Solve: $|5x + 1| + 1 \le 10$

Solution: First we get the absolute value expression alone by subtracting 1 from both sides.

$$|5x + 1| + 1 \le 10$$
$$|5x + 1| \le 10 - 1 \qquad \text{Subtract 1 from both sides.}$$
$$|5x + 1| \le 9 \qquad \text{Simplify.}$$

Since 9 is positive, to solve $|5x + 1| \le 9$, we solve

$$-9 \le 5x + 1 \le 9$$
$$-9 - 1 \le 5x + 1 - 1 \le 9 - 1 \qquad \text{Subtract 1 from all three parts.}$$
$$-10 \le 5x \le 8 \qquad \text{Simplify.}$$
$$-2 \le x \le \frac{8}{5} \qquad \text{Divide all three parts by 5.}$$

The solution set is $\left[-2, \dfrac{8}{5} \right]$.

■ **Work Practice 14**

The next few examples are special cases of absolute value inequalities.

Example 15 Solve: $|x| \le -3$

Solution: The absolute value of a number is never negative. Thus, it will then never be less than or equal to -3. The solution set is $\{\ \}$ or $\varnothing$.

■ **Work Practice 15**

Example 16 Solve: $|x - 1| > -2$

Solution: The absolute value of a number is always nonnegative. Thus, it will always be greater than -2. The solution set contains all real numbers, or $(-\infty, \infty)$.

■ **Work Practice 16**

✓**Concept Check** Without taking any solution steps, how do you know that the absolute value inequality $|3x - 2| > -9$ has a solution? What is its solution?

Practice 14
Solve: $|2x - 5| + 2 \le 9$

Practice 15
Solve: $|x| < -1$

Practice 16
Solve: $|x + 1| \ge -3$

Answers
14. $[-1, 6]$ **15.** $\varnothing$ **16.** $(-\infty, \infty)$

✓**Concept Check Answer**
$(-\infty, \infty)$ since the absolute value is always nonnegative

D Exercise Set MyMathLab®

Objective A *Solve. See Examples 1 through 6.*

▶1. $|x| = 7$
$\{7, -7\}$

2. $|y| = 15$
$\{-15, 15\}$

▶3. $|x| = -4$
$\varnothing$

4. $|x| = -20$
$\varnothing$

5. $|3x| = 12.6$
$\{4.2, -4.2\}$

6. $|6n| = 12.6$
$\{2.1, -2.1\}$

7. $3|x| - 5 = 7$
$\{-4, 4\}$

8. $5|x| - 12 = 8$
$\{-4, 4\}$

9. $-6|x| + 44 = -10$
$\{-9, 9\}$

10. $-4|x| + 18 = -22$
$\{-10, 10\}$

11. $|x - 9| = 14$
$\{-5, 23\}$

12. $|x + 2| = 8$
$\{-10, 6\}$

13. $|2x - 5| = 9$
$\{7, -2\}$

14. $|6 + 2n| = 4$
$\{-5, -1\}$

▶15. $\left|\dfrac{x}{2} - 3\right| = 1$
$\{8, 4\}$

16. $\left|\dfrac{n}{3} + 2\right| = 4$
$\{-18, 6\}$

17. $|z| + 4 = 9$
$\{5, -5\}$

18. $|x| + 1 = 3$
$\{2, -2\}$

19. $|3x| + 5 = 14$
$\{3, -3\}$

20. $|2x| - 6 = 4$
$\{5, -5\}$

21. $\left|\dfrac{4x - 6}{3}\right| = 6$
$\{-3, 6\}$

22. $\left|\dfrac{2x + 1}{5}\right| = 7$
$\{-18, 17\}$

23. $|2x| = 0$
$\{0\}$

24. $|7z| = 0$
$\{0\}$

25. $|4n + 1| + 10 = 4$
$\varnothing$

26. $|3z - 2| + 8 = 1$
$\varnothing$

27. $3|x - 1| + 19 = 23$
$\left\{-\dfrac{1}{3}, \dfrac{7}{3}\right\}$

28. $5|x + 1| - 1 = 3$
$\left\{-\dfrac{9}{5}, -\dfrac{1}{5}\right\}$

Solve. See Examples 7 and 8.

29. $|5x - 7| = |3x + 11|$
$\left\{-\dfrac{1}{2}, 9\right\}$

30. $|9y + 1| = |6y + 4|$
$\left\{-\dfrac{1}{3}, 1\right\}$

31. $|z + 8| = |z - 3|$
$\left\{-\dfrac{5}{2}\right\}$

32. $|2x - 5| = |2x + 5|$
$\{0\}$

▶33. $|2y - 3| = |9 - 4y|$
$\{3, 2\}$

34. $|5z - 1| = |7 - z|$
$\left\{-\dfrac{3}{2}, \dfrac{4}{3}\right\}$

35. $\left|\dfrac{3}{4}x - 2\right| = \left|\dfrac{1}{4}x + 6\right|$
$\{-4, 16\}$

36. $\left|\dfrac{2}{3}x - 5\right| = \left|\dfrac{1}{3}x + 4\right|$
$\{1, 27\}$

37. $|2x - 6| = |10 - 2x|$
$\{4\}$

38. $|4n + 5| = |4n + 3|$
$\{-1\}$

39. $|x + 4| = |7 - x|$
$\left\{\dfrac{3}{2}\right\}$

40. $|8 - y| = |y + 2|$
$\{3\}$

41. $\left|\dfrac{2x + 1}{5}\right| = \left|\dfrac{3x - 7}{3}\right|$

$\left\{\dfrac{32}{21}, \dfrac{38}{9}\right\}$

42. $\left|\dfrac{5x - 1}{2}\right| = \left|\dfrac{4x + 5}{6}\right|$

$\left\{-\dfrac{2}{19}, \dfrac{8}{11}\right\}$

43. $|5x + 1| = |4x - 7|$

$\left\{-8, \dfrac{2}{3}\right\}$

44. $|3 + 6n| = |4n + 11|$

$\left\{-\dfrac{7}{5}, 4\right\}$

Objective **B**   *Solve. Graph the solution set. See Examples 9 through 16.*

45. $|x| \le 4$

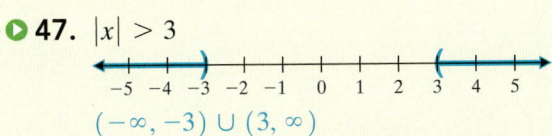

$[-4, 4]$

46. $|x| < 6$

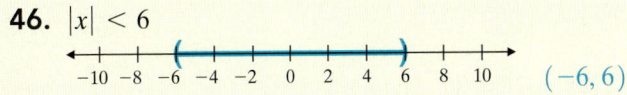

$(-6, 6)$

47. $|x| > 3$

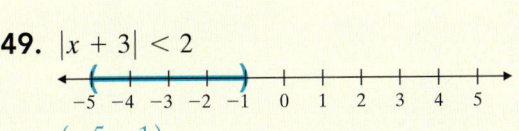

$(-\infty, -3) \cup (3, \infty)$

48. $|y| \ge 4$

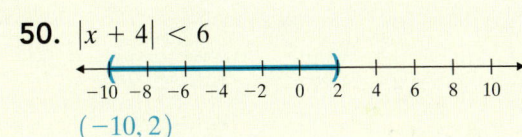

$(-\infty, -4] \cup [4, \infty)$

49. $|x + 3| < 2$

$(-5, -1)$

50. $|x + 4| < 6$

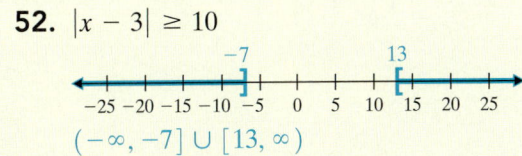

$(-10, 2)$

51. $|y - 6| \ge 7$

$(-\infty, -1] \cup [13, \infty)$

52. $|x - 3| \ge 10$

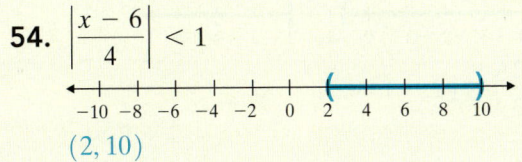

$(-\infty, -7] \cup [13, \infty)$

53. $\left|\dfrac{x + 2}{3}\right| < 1$

$(-5, 1)$

54. $\left|\dfrac{x - 6}{4}\right| < 1$

$(2, 10)$

55. $|x| + 7 \le 12$

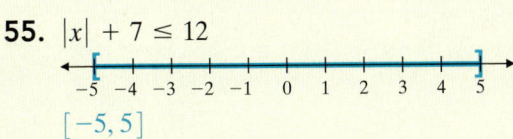

$[-5, 5]$

56. $|x| + 6 \le 7$

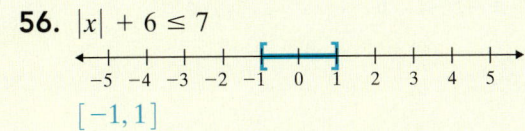

$[-1, 1]$

57. $|x| + 2 > 6$

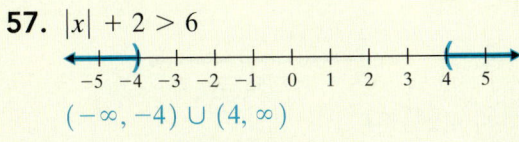

$(-\infty, -4) \cup (4, \infty)$

58. $|x| - 1 > 3$

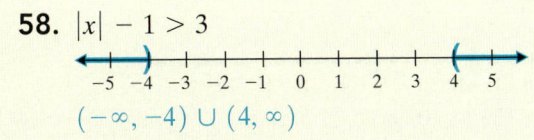

$(-\infty, -4) \cup (4, \infty)$

59. $|2x + 7| \le 13$

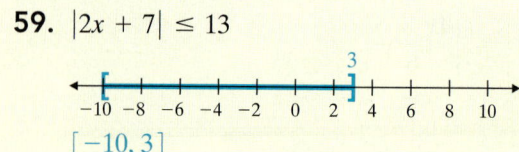

$[-10, 3]$

60. $|5x - 3| \le 18$

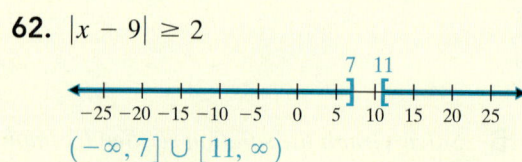

$\left[-3, \dfrac{21}{5}\right]$

61. $|x + 10| \ge 14$

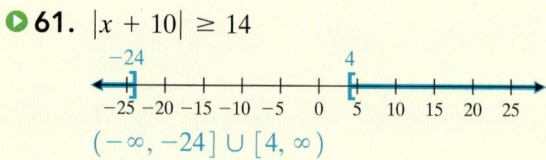

$(-\infty, -24] \cup [4, \infty)$

62. $|x - 9| \ge 2$

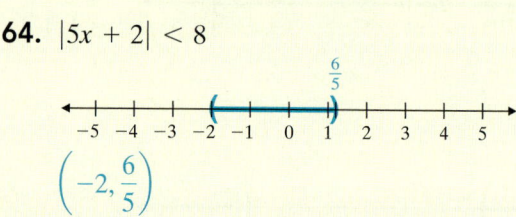

$(-\infty, 7] \cup [11, \infty)$

63. $|2x - 7| \le 11$

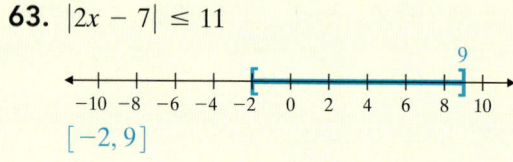

$[-2, 9]$

64. $|5x + 2| < 8$

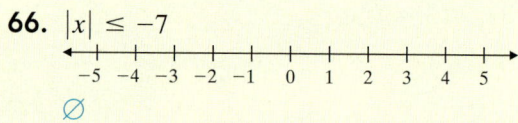

$\left(-2, \dfrac{6}{5}\right)$

65. $|x| > -4$

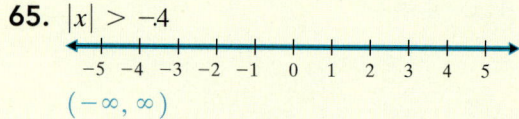

$(-\infty, \infty)$

66. $|x| \le -7$

$\varnothing$

67. $6 + |4x - 1| \le 9$

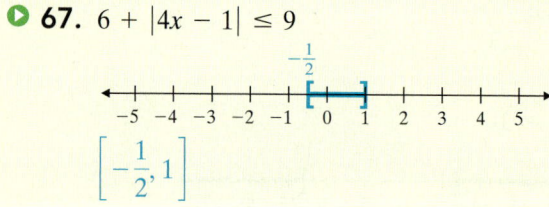

$\left[-\dfrac{1}{2}, 1\right]$

68. $-3 + |5x - 2| \le 4$

$\left[-1, \dfrac{9}{5}\right]$

69. $|6x - 8| + 3 > 7$

$\left(-\infty, \dfrac{2}{3}\right) \cup (2, \infty)$

70. $|10 + 3x| + 1 > 2$

$\left(-\infty, -\dfrac{11}{3}\right) \cup (-3, \infty)$

71. $|5x + 3| < -6$

$\varnothing$

72. $|4 + 9x| \ge -6$

$(-\infty, \infty)$

73. $\left|\dfrac{x + 6}{3}\right| > 2$

$(-\infty, -12) \cup (0, \infty)$

74. $\left|\dfrac{7 + x}{2}\right| \ge 4$

$(-\infty, -15] \cup [1, \infty)$

Mixed Practice *Solve each equation or inequality for x. See Examples 1 through 16.*

75. $|x| = 13$
$\{-13, 13\}$

76. $|x| < 13$
$(-13, 13)$

77. $|x| > 13$
$(-\infty, -13) \cup (13, \infty)$

78. $|3x| = 12$
$\{-4, 4\}$

79. $|x| + 12 = 9$
$\varnothing$

80. $|x| - 4 = -9$
$\varnothing$

81. $2|x| - 9 \le 11$
$[-10, 10]$

82. $4|x| - 2 \ge 6$
$(-\infty, -2] \cup [2, \infty)$

83. $|2x - 3| < 7$
$(-2, 5)$

84. $|2x - 3| > 7$
$(-\infty, -2) \cup (5, \infty)$

85. $|2x - 3| = 7$
$\{5, -2\}$

86. $|5 - 6x| = 29$
$\left\{-4, \dfrac{17}{3}\right\}$

87. $|x - 5| \ge 12$
$(-\infty, -7] \cup [17, \infty)$

88. $|x + 4| \ge 20$
$(-\infty, -24] \cup [16, \infty)$

89. $|9 + 4x| = 0$
$\left\{-\dfrac{9}{4}\right\}$

90. $|9 + 4x| \ge 0$
$(-\infty, \infty)$

91. $|2x + 1| + 4 < 7$
$(-2, 1)$

92. $8 + |5x - 3| \ge 11$
$(-\infty, 0] \cup \left[\dfrac{6}{5}, \infty\right)$

93. $\left|\dfrac{1}{3}x + 1\right| > 5$
$(-\infty, -18) \cup (12, \infty)$

94. $\left|\dfrac{1}{4}x - 2\right| < 1$
$(4, 12)$

95. $|3x - 5| + 4 = 5$
$\left\{2, \dfrac{4}{3}\right\}$

96. $|x - 1| + 7 = 11$
$\{5, -3\}$

97. $|x + 11| = -1$
$\varnothing$

98. $|4x - 4| = -3$
$\varnothing$

99. $\left|\dfrac{2x - 1}{3}\right| = 6$
$\left\{-\dfrac{17}{2}, \dfrac{19}{2}\right\}$

100. $\left|\dfrac{6 - x}{4}\right| = 5$
$\{-14, 26\}$

101. $\left|\dfrac{3x - 5}{6}\right| > 5$
$\left(-\infty, -\dfrac{25}{3}\right) \cup \left(\dfrac{35}{3}, \infty\right)$

102. $\left|\dfrac{4x - 7}{5}\right| < 2$
$\left(-\dfrac{3}{4}, \dfrac{17}{4}\right)$

103. $|6x - 3| = |4x + 5|$
$\left\{4, -\dfrac{1}{5}\right\}$

104. $|3x + 1| = |4x + 10|$
$\left\{-9, -\dfrac{11}{7}\right\}$

Determinants and Cramer's Rule

Objectives

A Define and Evaluate a 2 × 2 Determinant.

B Use Cramer's Rule to Solve a System of Two Linear Equations in Two Variables.

C Define and Evaluate a 3 × 3 Determinant.

D Use Cramer's Rule to Solve a System of Three Linear Equations in Three Variables.

We have solved systems of two linear equations in two variables in four different ways: graphically, by substitution, by elimination, and by matrices. Now we analyze another method, called **Cramer's rule.**

Objective A Evaluating 2 × 2 Determinants

Recall that a matrix is a rectangular array of numbers. If a matrix has the same number of rows and columns, it is called a **square matrix.** Examples of square matrices are

$$\begin{bmatrix} 1 & 6 \\ 5 & 2 \end{bmatrix} \qquad \begin{bmatrix} 2 & 4 & 1 \\ 0 & 5 & 2 \\ 3 & 6 & 9 \end{bmatrix}$$

A **determinant** is a real number associated with a square matrix. The determinant of a square matrix is denoted by placing vertical bars about the array of numbers. Thus,

The determinant of the square matrix $\begin{bmatrix} 1 & 6 \\ 5 & 2 \end{bmatrix}$ is $\begin{vmatrix} 1 & 6 \\ 5 & 2 \end{vmatrix}$.

The determinant of the square matrix $\begin{bmatrix} 2 & 4 & 1 \\ 0 & 5 & 2 \\ 3 & 6 & 9 \end{bmatrix}$ is $\begin{vmatrix} 2 & 4 & 1 \\ 0 & 5 & 2 \\ 3 & 6 & 9 \end{vmatrix}$.

We define the determinant of a 2 × 2 matrix first. (Recall that 2 × 2 is read "two by two." It means that the matrix has 2 rows and 2 columns.)

Determinant of a 2 × 2 Matrix

$$\begin{vmatrix} a & b \\ c & d \end{vmatrix} = ad - bc$$

Practice 1

Evaluate each determinant.

a. $\begin{vmatrix} -3 & 6 \\ 2 & 1 \end{vmatrix}$

b. $\begin{vmatrix} 4 & 5 \\ 0 & -5 \end{vmatrix}$

Example 1 Evaluate each determinant.

a. $\begin{vmatrix} -1 & 2 \\ 3 & -4 \end{vmatrix}$

b. $\begin{vmatrix} 2 & 0 \\ 7 & -5 \end{vmatrix}$

Solution: First we identify the values of a, b, c, and d. Then we perform the evaluation.

a. Here $a = -1, b = 2, c = 3$, and $d = -4$.

$$\begin{vmatrix} -1 & 2 \\ 3 & -4 \end{vmatrix} = ad - bc = (-1)(-4) - (2)(3) = -2$$

b. In this example, $a = 2, b = 0, c = 7$, and $d = -5$.

$$\begin{vmatrix} 2 & 0 \\ 7 & -5 \end{vmatrix} = ad - bc = 2(-5) - (0)(7) = -10$$

■ **Work Practice 1**

Answers

1. a. −15 **b.** −20

Objective B Using Cramer's Rule to Solve a System of Two Linear Equations ▶

To develop Cramer's rule, we solve the system $\begin{cases} ax + by = h \\ cx + dy = k \end{cases}$ using elimination.

First, we eliminate y by multiplying both sides of the first equation by d and both sides of the second equation by $-b$ so that the coefficients of y are opposites. The result is that

$$\begin{cases} d(ax + by) = d \cdot h \\ -b(cx + dy) = -b \cdot k \end{cases} \text{ simplifies to } \begin{cases} adx + bdy = hd \\ -bcx - bdy = -bk \end{cases}$$

We now add the two equations and solve for x.

$$\begin{aligned} adx + bdy &= hd \\ \underline{-bcx - bdy} &= \underline{-bk} \\ adx - bcx &= hd - bk \quad \text{Add the equations.} \\ (ad - bc)x &= hd - bk \\ x &= \frac{hd - bk}{ad - bc} \quad \text{Solve for } x. \end{aligned}$$

When we replace x with $\frac{hd - bk}{ad - bc}$ in the equation $ax + by = h$ and solve for y, we find that $y = \frac{ak - hc}{ad - bc}$.

Notice that the numerator of the value of x is the determinant of

$$\begin{vmatrix} h & b \\ k & d \end{vmatrix} = hd - bk$$

Also, the numerator of the value of y is the determinant of

$$\begin{vmatrix} a & h \\ c & k \end{vmatrix} = ak - hc$$

Finally, the denominators of the values of x and y are the same and are the determinant of

$$\begin{vmatrix} a & b \\ c & d \end{vmatrix} = ad - bc$$

This means that the values of x and y can be written in determinant notation:

$$x = \frac{\begin{vmatrix} h & b \\ k & d \end{vmatrix}}{\begin{vmatrix} a & b \\ c & d \end{vmatrix}} \quad \text{and} \quad y = \frac{\begin{vmatrix} a & h \\ c & k \end{vmatrix}}{\begin{vmatrix} a & b \\ c & d \end{vmatrix}}$$

For convenience, we label the determinants D, D_x, and D_y.

$$\begin{vmatrix} a & b \\ c & d \end{vmatrix} = D \quad \begin{vmatrix} h & b \\ k & d \end{vmatrix} = D_x \quad \begin{vmatrix} a & h \\ c & k \end{vmatrix} = D_y$$

x-column replaced by constants   *y*-column replaced by constants

These determinant formulas for the coordinates of the solution of a system are known as **Cramer's rule.**

> ## Cramer's Rule for Two Linear Equations in Two Variables
>
> The solution of the system $\begin{cases} ax + by = h \\ cx + dy = k \end{cases}$ is given by
>
> $$x = \frac{\begin{vmatrix} h & b \\ k & d \end{vmatrix}}{\begin{vmatrix} a & b \\ c & d \end{vmatrix}} = \frac{D_x}{D} \qquad y = \frac{\begin{vmatrix} a & h \\ c & k \end{vmatrix}}{\begin{vmatrix} a & b \\ c & d \end{vmatrix}} = \frac{D_y}{D}$$
>
> as long as $D = ad - bc$ is not 0.

When $D = 0$, the system is either inconsistent or the equations are dependent. When this happens, we need to use another method to see which is the case.

Practice 2

Use Cramer's rule to solve the system.

$$\begin{cases} x - y = -4 \\ 2x + 3y = 2 \end{cases}$$

Example 2   Use Cramer's rule to solve the system:

$$\begin{cases} 3x + 4y = -7 \\ x - 2y = -9 \end{cases}$$

Solution: First we find D, D_x, and D_y.

$$\begin{array}{ccc} a & b & h \\ \downarrow & \downarrow & \downarrow \end{array}$$
$$\begin{cases} 3x + 4y = -7 \\ x - 2y = -9 \end{cases}$$
$$\begin{array}{ccc} \uparrow & \uparrow & \uparrow \\ c & d & k \end{array}$$

$$D = \begin{vmatrix} a & b \\ c & d \end{vmatrix} = \begin{vmatrix} 3 & 4 \\ 1 & -2 \end{vmatrix} = 3(-2) - 4(1) = -10$$

$$D_x = \begin{vmatrix} h & b \\ k & d \end{vmatrix} = \begin{vmatrix} -7 & 4 \\ -9 & -2 \end{vmatrix} = (-7)(-2) - 4(-9) = 50$$

$$D_y = \begin{vmatrix} a & h \\ c & k \end{vmatrix} = \begin{vmatrix} 3 & -7 \\ 1 & -9 \end{vmatrix} = 3(-9) - (-7)(1) = -20$$

Then $x = \dfrac{D_x}{D} = \dfrac{50}{-10} = -5$ and $y = \dfrac{D_y}{D} = \dfrac{-20}{-10} = 2$.

The ordered pair solution is $(-5, 2)$.
As always, check the solution in both original equations.

■ **Work Practice 2**

Practice 3

Use Cramer's rule to solve the system.

$$\begin{cases} 4x + y = 3 \\ 2x - 3y = -9 \end{cases}$$

Example 3   Use Cramer's rule to solve the system:

$$\begin{cases} 5x + y = 5 \\ -7x - 2y = -7 \end{cases}$$

Solution: First we find D, D_x, and D_y.

$$D = \begin{vmatrix} 5 & 1 \\ -7 & -2 \end{vmatrix} = 5(-2) - (-7)(1) = -3$$

$$D_x = \begin{vmatrix} 5 & 1 \\ -7 & -2 \end{vmatrix} = 5(-2) - (-7)(1) = -3$$

$$D_y = \begin{vmatrix} 5 & 5 \\ -7 & -7 \end{vmatrix} = 5(-7) - 5(-7) = 0$$

Answers

2. $(-2, 2)$ **3.** $(0, 3)$

Then

$$x = \frac{D_x}{D} = \frac{-3}{-3} = 1 \qquad y = \frac{D_y}{D} = \frac{0}{-3} = 0$$

The ordered pair solution is $(1, 0)$.

■ **Work Practice 3**

Objective C Evaluating 3 × 3 Determinants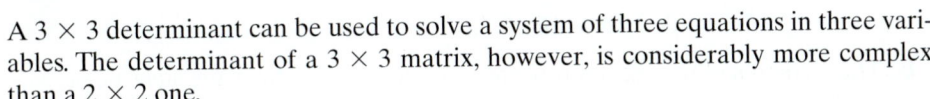

A 3 × 3 determinant can be used to solve a system of three equations in three variables. The determinant of a 3 × 3 matrix, however, is considerably more complex than a 2 × 2 one.

Determinant of a 3 × 3 Matrix

$$\begin{vmatrix} a_1 & b_1 & c_1 \\ a_2 & b_2 & c_2 \\ a_3 & b_3 & c_3 \end{vmatrix} = a_1 \cdot \begin{vmatrix} b_2 & c_2 \\ b_3 & c_3 \end{vmatrix} - a_2 \cdot \begin{vmatrix} b_1 & c_1 \\ b_3 & c_3 \end{vmatrix} + a_3 \cdot \begin{vmatrix} b_1 & c_1 \\ b_2 & c_2 \end{vmatrix}$$

Notice that the determinant of a 3 × 3 matrix is related to the determinants of three 2 × 2 matrices. Each determinant of these 2 × 2 matrices is called a **minor**, and every element of a 3 × 3 matrix has a minor associated with it. For example, the minor of c_2 is the determinant of the 2 × 2 matrix found by deleting the row and column containing c_2.

$$\begin{array}{ccc} a_1 & b_1 & c_1 \\ a_2 & b_2 & c_2 \\ a_3 & b_3 & c_3 \end{array} \qquad \text{The minor of } c_2 \text{ is} \qquad \begin{vmatrix} a_1 & b_1 \\ a_3 & b_3 \end{vmatrix}$$

Also, the minor of element a_1 is the determinant of the 2 × 2 matrix that has no row or column containing a_1.

$$\begin{array}{ccc} a_1 & b_1 & c_1 \\ a_2 & b_2 & c_2 \\ a_3 & b_3 & c_3 \end{array} \qquad \text{The minor of } a_1 \text{ is} \qquad \begin{vmatrix} b_2 & c_2 \\ b_3 & c_3 \end{vmatrix}$$

So the determinant of a 3 × 3 matrix can be written as

$$a_1 \cdot (\text{minor of } a_1) - a_2 \cdot (\text{minor of } a_2) + a_3 \cdot (\text{minor of } a_3)$$

Finding the determinant by using minors of elements in the first column is called **expanding** by the minors of the first column. *The value of a determinant can be found by expanding by the minors of any row or column.* The following **array of signs** is helpful in determining whether to add or subtract the product of an element and its minor.

$$\begin{array}{ccc} + & - & + \\ - & + & - \\ + & - & + \end{array}$$

If an element is in a position marked +, we add. If marked −, we subtract.

✔**Concept Check** Suppose you are interested in finding the determinant of a 4 × 4 matrix. Study the pattern shown in the array of signs for a 3 × 3 matrix. Use the pattern to expand the array of signs for use with a 4 × 4 matrix.

Example 4 Evaluate by expanding by the minors of the given row or column.

$$\begin{vmatrix} 0 & 5 & 1 \\ 1 & 3 & -1 \\ -2 & 2 & 4 \end{vmatrix}$$

a. First column **b.** Second row *(Continued on next page)*

Practice 4

Evaluate by expanding by the minors of the given row or column.

a. First column **b.** Third row

$$\begin{vmatrix} 2 & 0 & 1 \\ -1 & 3 & 2 \\ 5 & 1 & 4 \end{vmatrix}$$

Answers
4. a. 4 **b.** 4

✔**Concept Check Answer**

$$\begin{array}{cccc} + & - & + & - \\ - & + & - & + \\ + & - & + & - \\ - & + & - & + \end{array}$$

Solution:

a. The elements of the first column are $0, 1,$ and -2. The first column of the array of signs is $+, -, +$.

$$\begin{vmatrix} 0 & 5 & 1 \\ 1 & 3 & -1 \\ -2 & 2 & 4 \end{vmatrix} = 0 \cdot \begin{vmatrix} 3 & -1 \\ 2 & 4 \end{vmatrix} - 1 \cdot \begin{vmatrix} 5 & 1 \\ 2 & 4 \end{vmatrix} + (-2) \cdot \begin{vmatrix} 5 & 1 \\ 3 & -1 \end{vmatrix}$$

$$= 0(12 - (-2)) - 1(20 - 2) + (-2)(-5 - 3)$$

$$= 0 - 18 + 16 = -2$$

b. The elements of the second row are $1, 3,$ and -1. This time, the signs begin with $-$ and again alternate.

$$\begin{vmatrix} 0 & 5 & 1 \\ 1 & 3 & -1 \\ -2 & 2 & 4 \end{vmatrix} = -1 \cdot \begin{vmatrix} 5 & 1 \\ 2 & 4 \end{vmatrix} + 3 \cdot \begin{vmatrix} 0 & 1 \\ -2 & 4 \end{vmatrix} - (-1) \cdot \begin{vmatrix} 0 & 5 \\ -2 & 2 \end{vmatrix}$$

$$= -1(20 - 2) + 3(0 - (-2)) - (-1)(0 - (-10))$$

$$= -18 + 6 + 10 = -2$$

Notice that the determinant of the 3×3 matrix is the same regardless of the row or column you select to expand by.

■ **Work Practice 4**

✓**Concept Check** Why would expanding by minors of the second row be a good choice for the determinant $\begin{vmatrix} 3 & 4 & -2 \\ 5 & 0 & 0 \\ 6 & -3 & 7 \end{vmatrix}$?

Objective D Using Cramer's Rule to Solve a System of Three Linear Equations ▶

A system of three equations in three variables may be solved with Cramer's rule also. Using the elimination process to solve a system with unknown constants as coefficients leads to the following.

Cramer's Rule for Three Equations in Three Variables

The solution of the system $\begin{cases} a_1x + b_1y + c_1z = k_1 \\ a_2x + b_2y + c_2z = k_2 \\ a_3x + b_3y + c_3z = k_3 \end{cases}$ is given by

$$x = \frac{D_x}{D} \qquad y = \frac{D_y}{D} \qquad \text{and} \qquad z = \frac{D_z}{D}$$

where

$$D = \begin{vmatrix} a_1 & b_1 & c_1 \\ a_2 & b_2 & c_2 \\ a_3 & b_3 & c_3 \end{vmatrix} \qquad D_x = \begin{vmatrix} k_1 & b_1 & c_1 \\ k_2 & b_2 & c_2 \\ k_3 & b_3 & c_3 \end{vmatrix}$$

$$D_y = \begin{vmatrix} a_1 & k_1 & c_1 \\ a_2 & k_2 & c_2 \\ a_3 & k_3 & c_3 \end{vmatrix} \qquad D_z = \begin{vmatrix} a_1 & b_1 & k_1 \\ a_2 & b_2 & k_2 \\ a_3 & b_3 & k_3 \end{vmatrix}$$

as long as D is not 0.

✓**Concept Check Answer**

Two elements of the second row are 0, which makes calculations easier.

Example 5   Use Cramer's rule to solve the system:

$$\begin{cases} x - 2y + z = 4 \\ 3x + y - 2z = 3 \\ 5x + 5y + 3z = -8 \end{cases}$$

Solution: First we find D, D_x, D_y, and D_z. Beginning with D, we expand by the minors of the first column.

$$D = \begin{vmatrix} 1 & -2 & 1 \\ 3 & 1 & -2 \\ 5 & 5 & 3 \end{vmatrix} = 1 \cdot \begin{vmatrix} 1 & -2 \\ 5 & 3 \end{vmatrix} - 3 \cdot \begin{vmatrix} -2 & 1 \\ 5 & 3 \end{vmatrix} + 5 \cdot \begin{vmatrix} -2 & 1 \\ 1 & -2 \end{vmatrix}$$

$$= 1(3 - (-10)) - 3(-6 - 5) + 5(4 - 1)$$
$$= 13 + 33 + 15 = 61$$

$$D_x = \begin{vmatrix} 4 & -2 & 1 \\ 3 & 1 & -2 \\ -8 & 5 & 3 \end{vmatrix} = 4 \cdot \begin{vmatrix} 1 & -2 \\ 5 & 3 \end{vmatrix} - 3 \cdot \begin{vmatrix} -2 & 1 \\ 5 & 3 \end{vmatrix} + (-8) \cdot \begin{vmatrix} -2 & 1 \\ 1 & -2 \end{vmatrix}$$

$$= 4(3 - (-10)) - 3(-6 - 5) + (-8)(4 - 1)$$
$$= 52 + 33 - 24 = 61$$

$$D_y = \begin{vmatrix} 1 & 4 & 1 \\ 3 & 3 & -2 \\ 5 & -8 & 3 \end{vmatrix} = 1 \cdot \begin{vmatrix} 3 & -2 \\ -8 & 3 \end{vmatrix} - 3 \cdot \begin{vmatrix} 4 & 1 \\ -8 & 3 \end{vmatrix} + 5 \cdot \begin{vmatrix} 4 & 1 \\ 3 & -2 \end{vmatrix}$$

$$= 1(9 - 16) - 3(12 - (-8)) + 5(-8 - 3)$$
$$= -7 - 60 - 55 = -122$$

$$D_z = \begin{vmatrix} 1 & -2 & 4 \\ 3 & 1 & 3 \\ 5 & 5 & -8 \end{vmatrix} = 1 \cdot \begin{vmatrix} 1 & 3 \\ 5 & -8 \end{vmatrix} - 3 \cdot \begin{vmatrix} -2 & 4 \\ 5 & -8 \end{vmatrix} + 5 \cdot \begin{vmatrix} -2 & 4 \\ 1 & 3 \end{vmatrix}$$

$$= 1(-8 - 15) - 3(16 - 20) + 5(-6 - 4)$$
$$= -23 + 12 - 50 = -61$$

From these determinants, we calculate the solution:

$$x = \frac{D_x}{D} = \frac{61}{61} = 1 \quad y = \frac{D_y}{D} = \frac{-122}{61} = -2 \quad z = \frac{D_z}{D} = \frac{-61}{61} = -1$$

The ordered triple solution is $(1, -2, -1)$. Check this solution by verifying that it satisfies each equation of the system.

■ **Work Practice 5**

Practice 5

Use Cramer's rule to solve the system:

$$\begin{cases} x + 2y - z = 3 \\ 2x - 3y + z = -9 \\ -x + y - 2z = 0 \end{cases}$$

Teaching Tip

Mention that larger linear systems can be solved using a similar strategy. While these would be time consuming to solve by hand, computers and powerful calculators can be programmed to solve them quickly and accurately because the solution is found by following a set of instructions involving only the coefficients and constants of the equations.

Answer
5. $(-1, 3, 2)$

Vocabulary and Readiness Check

Evaluate each determinant mentally.

1. $\begin{vmatrix} 7 & 2 \\ 0 & 8 \end{vmatrix}$ 56

2. $\begin{vmatrix} 6 & 0 \\ 1 & 2 \end{vmatrix}$ 12

3. $\begin{vmatrix} -4 & 2 \\ 0 & 8 \end{vmatrix}$ -32

4. $\begin{vmatrix} 5 & 0 \\ 3 & -5 \end{vmatrix}$ -25

5. $\begin{vmatrix} -2 & 0 \\ 3 & -10 \end{vmatrix}$ 20

6. $\begin{vmatrix} -1 & 4 \\ 0 & -18 \end{vmatrix}$ 18

E   Exercise Set   MyMathLab®

Objective A *Evaluate each determinant. See Example 1.*

1. $\begin{vmatrix} 3 & 5 \\ -1 & 7 \end{vmatrix}$ 26

2. $\begin{vmatrix} -5 & 1 \\ 1 & -4 \end{vmatrix}$ 19

3. $\begin{vmatrix} 9 & -2 \\ 4 & -3 \end{vmatrix}$ -19

4. $\begin{vmatrix} 4 & -1 \\ 9 & 8 \end{vmatrix}$ 41

5. $\begin{vmatrix} -2 & 9 \\ 4 & -18 \end{vmatrix}$ 0

6. $\begin{vmatrix} -40 & 8 \\ 70 & -14 \end{vmatrix}$ 0

7. $\begin{vmatrix} \frac{3}{4} & \frac{5}{2} \\ -\frac{1}{6} & \frac{7}{3} \end{vmatrix}$ $\frac{13}{6}$

8. $\begin{vmatrix} \frac{5}{7} & \frac{1}{3} \\ \frac{6}{7} & \frac{2}{3} \end{vmatrix}$ $\frac{4}{21}$

Objective B *Use Cramer's rule, if possible, to solve each system of linear equations. See Examples 2 and 3.*

9. $\begin{cases} 2y - 4 = 0 \\ x + 2y = 5 \end{cases}$
$(1, 2)$

10. $\begin{cases} 4x - y = 5 \\ 3x - 3 = 0 \end{cases}$
$(1, -1)$

11. $\begin{cases} 3x + y = 1 \\ 2y = 2 - 6x \end{cases}$
$\{(x, y) \mid 3x + y = 1\}$

12. $\begin{cases} y = 2x - 5 \\ 8x - 4y = 20 \end{cases}$
$\{(x, y) \mid y = 2x - 5\}$

13. $\begin{cases} 5x - 2y = 27 \\ -3x + 5y = 18 \end{cases}$
$(9, 9)$

14. $\begin{cases} 4x - y = 9 \\ 2x + 3y = -27 \end{cases}$
$(0, -9)$

15. $\begin{cases} 2x - 5y = 4 \\ x + 2y = -7 \end{cases}$
$(-3, -2)$

16. $\begin{cases} 3x - y = 2 \\ -5x + 2y = 0 \end{cases}$
$(4, 10)$

17. $\begin{cases} \dfrac{2}{3}x - \dfrac{3}{4}y = -1 \\ -\dfrac{1}{6}x + \dfrac{3}{4}y = \dfrac{5}{2} \end{cases}$
$(3, 4)$

18. $\begin{cases} \dfrac{1}{2}x - \dfrac{1}{3}y = -3 \\ \dfrac{1}{8}x + \dfrac{1}{6}y = 0 \end{cases}$
$(-4, 3)$

Objective C *Evaluate. See Example 4.*

19. $\begin{vmatrix} 2 & 1 & 0 \\ 0 & 5 & -3 \\ 4 & 0 & 2 \end{vmatrix}$ 8

20. $\begin{vmatrix} -6 & 4 & 2 \\ 1 & 0 & 5 \\ 0 & 3 & 1 \end{vmatrix}$ 92

21. $\begin{vmatrix} 4 & -6 & 0 \\ -2 & 3 & 0 \\ 4 & -6 & 1 \end{vmatrix}$ 0

22. $\begin{vmatrix} 5 & 2 & 1 \\ 3 & -6 & 0 \\ 2 & 8 & 0 \end{vmatrix}$ 36

23. $\begin{vmatrix} 1 & 0 & 4 \\ 1 & -1 & 2 \\ 3 & 2 & 1 \end{vmatrix}$ 15

24. $\begin{vmatrix} 0 & 1 & 2 \\ 3 & -1 & 2 \\ 3 & 2 & -2 \end{vmatrix}$ 30

25. $\begin{vmatrix} 3 & 6 & -3 \\ -1 & -2 & 3 \\ 4 & -1 & 6 \end{vmatrix}$ 54

26. $\begin{vmatrix} 2 & -2 & 1 \\ 4 & 1 & 3 \\ 3 & 1 & 2 \end{vmatrix}$ -3

Objective **D** *Use Cramer's rule, if possible, to solve each system of linear equations. See Example 5.*

27. $\begin{cases} 3x \quad\quad + z = -1 \\ -x - 3y + z = \quad 7 \\ \quad\quad 3y + z = \quad 5 \end{cases}$
$(-2, 1, 5)$

28. $\begin{cases} \quad\quad 4y - 3z = -2 \\ 8x - 4y \quad\quad = \quad 4 \\ -8x + 4y + \quad z = -2 \end{cases}$
$(1, 1, 2)$

29. $\begin{cases} x + \quad y + \quad z = \quad 8 \\ 2x - \quad y - \quad z = 10 \\ x - 2y + 3z = 22 \end{cases}$
$(6, -2, 4)$

30. $\begin{cases} 5x + y + 3z = \quad 1 \\ x - y - 3z = -7 \\ -x + y \quad\quad = \quad 1 \end{cases}$
$(-1, 0, 2)$

31. $\begin{cases} 2x + 2y + \quad z = 1 \\ -x + \quad y + 2z = 3 \\ x + 2y + 4z = 0 \end{cases}$
$(-2, 3, -1)$

32. $\begin{cases} 2x - 3y + \quad z = 5 \\ x + \quad y + \quad z = 0 \\ 4x + 2y + 4z = 4 \end{cases}$
$(-3, -2, 5)$

33. $\begin{cases} x - 2y + \quad z = -5 \\ \quad\quad 3y + 2z = \quad 4 \\ 3x - \quad y \quad\quad = -2 \end{cases}$
$(0, 2, -1)$

34. $\begin{cases} 4x + 5y \quad\quad = 10 \\ \quad\quad 3y + 2z = -6 \\ x + \quad y + \quad z = 3 \end{cases}$
$(5, -2, 0)$

Concept Extensions

Find the value of x that will make each a true statement.

35. $\begin{vmatrix} 1 & x \\ 2 & 7 \end{vmatrix} = -3$ 5

36. $\begin{vmatrix} 6 & 1 \\ -2 & x \end{vmatrix} = 26$ 4

37. If all the elements in a single row of a determinant are zero, what is the value of the determinant? Explain your answer.   0; answers may vary

38. If all the elements in a single column of a determinant are 0, what is the value of the determinant? Explain your answer.   0; answers may vary

Review of Angles, Lines, and Special Triangles

The word **geometry** is formed from the Greek words, **geo,** meaning earth, and **metron,** meaning measure. Geometry literally means to measure the earth.

This appendix contains a review of some basic geometric ideas. It will be assumed that fundamental ideas of geometry such as point, line, ray, and angle are known. In this appendix, the notation $\angle 1$ is read "angle 1" and the notation $m\angle 1$ is read "the measure of angle 1."

We first review types of angles.

Angles

An angle whose measure is greater than 0° but less than 90° is called an **acute angle.**

A **right angle** is an angle whose measure is 90°. A right angle can be indicated by a square drawn at the vertex of the angle, as shown below.

An angle whose measure is greater than 90° but less than 180° is called an **obtuse angle.**

An angle whose measure is 180° is called a **straight angle.**

Two angles are said to be **complementary** if the sum of their measures is 90°. Each angle is called the **complement** of the other.

Two angles are said to be **supplementary** if the sum of their measures is 180°. Each angle is called the **supplement** of the other.

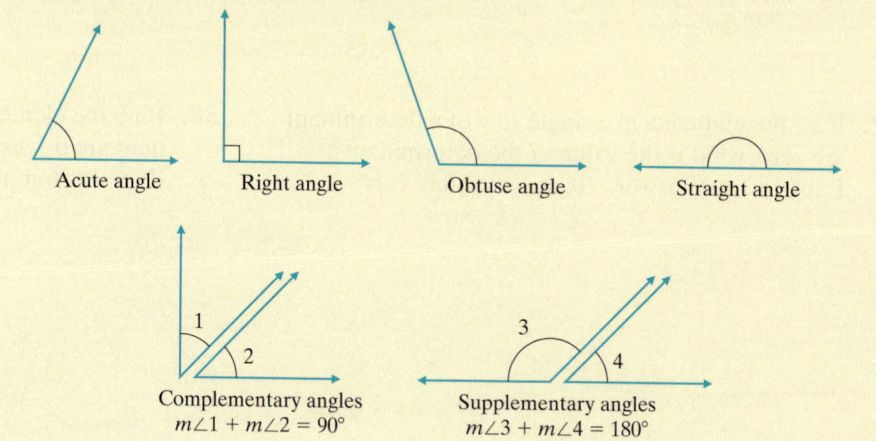

Acute angle   Right angle   Obtuse angle   Straight angle

Complementary angles
$m\angle 1 + m\angle 2 = 90°$

Supplementary angles
$m\angle 3 + m\angle 4 = 180°$

Example 1 If an angle measures 28°, find its complement.

Solution: Two angles are complementary if the sum of their measures is 90°. The complement of a 28° angle is an angle whose measure is $90° - 28° = 62°$. To check, notice that $28° + 62° = 90°$.

Plane is an undefined term that we will describe. A plane can be thought of as a flat surface with infinite length and width, but no thickness. A plane is two dimensional.

The arrows in the following diagram indicate that a plane extends indefinitely and has no boundaries.

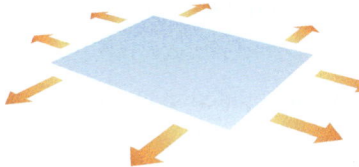

Figures that lie on a plane are called **plane figures.** Lines that lie in the same plane are called **coplanar.**

Lines

Two lines are **parallel** if they lie in the same plane but never meet. **Intersecting lines** meet or cross in one point.

Two lines that form right angles when they intersect are said to be **perpendicular.**

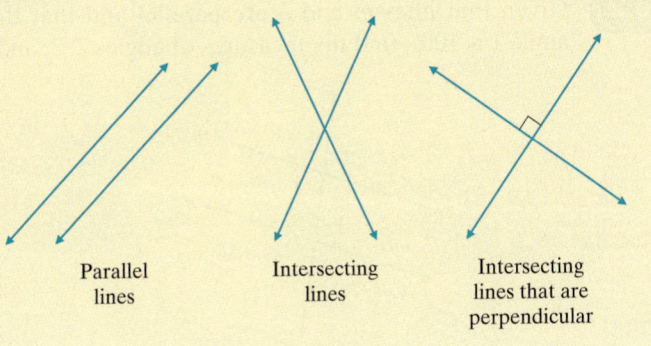

Parallel lines   Intersecting lines   Intersecting lines that are perpendicular

Two intersecting lines form **vertical angles.** Angles 1 and 3 are vertical angles. Also angles 2 and 4 are vertical angles. It can be shown that **vertical angles have equal measures.**

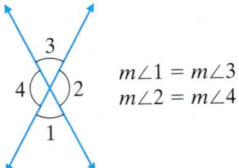

$$m\angle 1 = m\angle 3$$
$$m\angle 2 = m\angle 4$$

Adjacent angles have the same vertex and share a side. Angles 1 and 2 are adjacent angles. Other pairs of adjacent angles are angles 2 and 3, angles 3 and 4, and angles 4 and 1.

A **transversal** is a line that intersects two or more lines in the same plane. Line l is a transversal that intersects lines m and n. The eight angles formed are numbered and certain pairs of these angles are given special names.

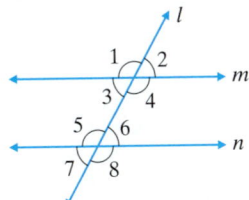

Corresponding angles: $\angle 1$ and $\angle 5$, $\angle 3$ and $\angle 7$, $\angle 2$ and $\angle 6$, and $\angle 4$ and $\angle 8$.

Exterior angles: $\angle 1$, $\angle 2$, $\angle 7$, and $\angle 8$.

Interior angles: $\angle 3$, $\angle 4$, $\angle 5$, and $\angle 6$.

Alternate interior angles: $\angle 3$ and $\angle 6$, $\angle 4$ and $\angle 5$.

These angles and parallel lines are related in the following manner.

> ### Parallel Lines Cut by a Transversal
>
> **1.** If two parallel lines are cut by a transversal, then
>   - **a.** **corresponding angles are equal** and
>   - **b.** **alternate interior angles are equal.**
> **2.** If corresponding angles formed by two lines and a transversal are equal, then the lines are parallel.
> **3.** If alternate interior angles formed by two lines and a transversal are equal, then the lines are parallel.

Example 2   Given that lines m and n are parallel and that the measure of angle 1 is 100°, find the measures of angles 2, 3, and 4.

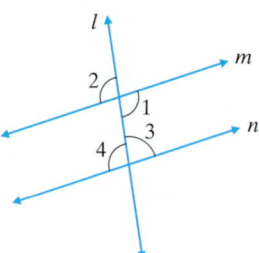

Solution:

$$m\angle 2 = 100° \qquad \text{Since angles 1 and 2 are vertical angles}$$
$$m\angle 4 = 100° \qquad \text{Since angles 1 and 4 are alternate interior angles}$$
$$m\angle 3 = 180° - 100° = 80° \qquad \text{Since angles 4 and 3 are supplementary angles}$$

A **polygon** is the union of three or more coplanar line segments that intersect each other only at each endpoint, with each endpoint shared by exactly two segments.

A **triangle** is a polygon with three sides. The sum of the measures of the three angles of a triangle is 180°. In the following figure, $m\angle 1 + m\angle 2 + m\angle 3 = 180°$.

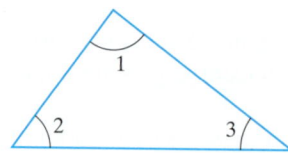

Example 3   Find the measure of the third angle of the triangle shown.

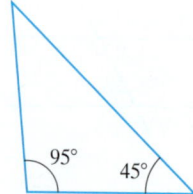

Solution: The sum of the measures of the angles of a triangle is 180°. Since one angle measures 45° and the other angle measures 95°, the third angle measures $180° - 45° - 95° = 40°$.

Two triangles are **congruent** if they have the same size and the same shape. In congruent triangles, the measures of corresponding angles are equal and the lengths of corresponding sides are equal. The following triangles are congruent.

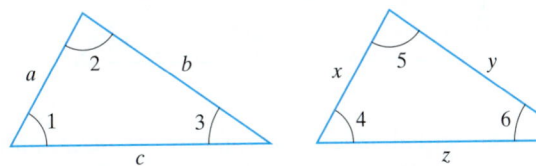

Corresponding angles are equal: $m\angle 1 = m\angle 4$, $m\angle 2 = m\angle 5$, and $m\angle 3 = m\angle 6$. Also, lengths of corresponding sides are equal: $a = x$, $b = y$, and $c = z$.

Any one of the following may be used to determine whether two triangles are congruent.

Congruent Triangles

1. If the measures of two angles of a triangle equal the measures of two angles of another triangle and the lengths of the sides between each pair of angles are equal, the triangles are congruent.

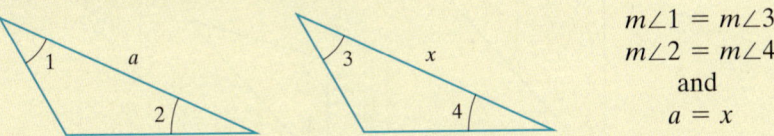

$$m\angle 1 = m\angle 3$$
$$m\angle 2 = m\angle 4$$
$$\text{and}$$
$$a = x$$

2. If the lengths of the three sides of a triangle equal the lengths of corresponding sides of another triangle, the triangles are congruent.

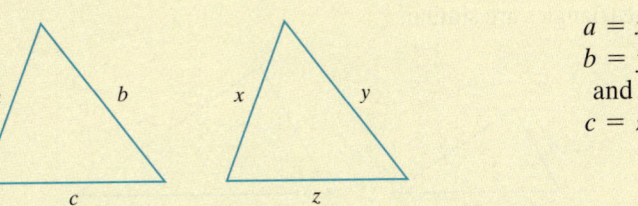

$$a = x$$
$$b = y$$
$$\text{and}$$
$$c = z$$

3. If the lengths of two sides of a triangle equal the lengths of corresponding sides of another triangle, and the measures of the angles between each pair of sides are equal, the triangles are congruent.

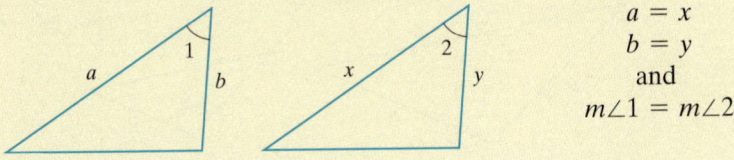

$$a = x$$
$$b = y$$
$$\text{and}$$
$$m\angle 1 = m\angle 2$$

Two triangles are **similar** if they have the same shape but not necessarily the same size. In similar triangles, the measures of corresponding angles are equal and corresponding sides are in proportion. The following triangles are similar. (All similar triangles drawn in this appendix will be oriented the same.)

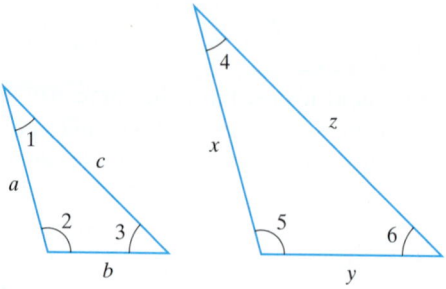

Corresponding angles are equal: $m\angle 1 = m\angle 4$, $m\angle 2 = m\angle 5$, and $m\angle 3 = m\angle 6$. Also, corresponding sides are proportional: $\dfrac{a}{x} = \dfrac{b}{y} = \dfrac{c}{z}$.

Any one of the following may be used to determine whether two triangles are similar.

Similar Triangles

1. If the measures of two angles of a triangle equal the measures of two angles of another triangle, the triangles are similar.

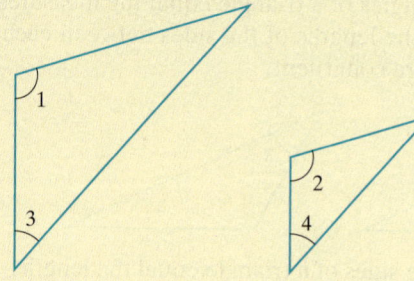

$m\angle 1 = m\angle 2$
and
$m\angle 3 = m\angle 4$

2. If three sides of one triangle are proportional to three sides of another triangle, the triangles are similar.

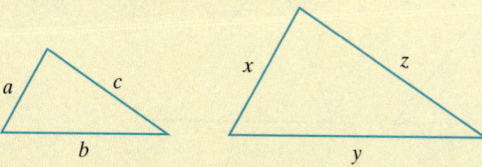

$\dfrac{a}{x} = \dfrac{b}{y} = \dfrac{c}{z}$

3. If two sides of a triangle are proportional to two sides of another triangle and the measures of the included angles are equal, the triangles are similar.

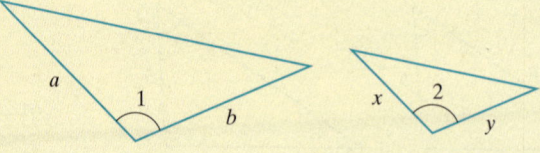

$m\angle 1 = m\angle 2$
and
$\dfrac{a}{x} = \dfrac{b}{y}$

Example 4 Given that the following triangles are similar, find the missing length x.

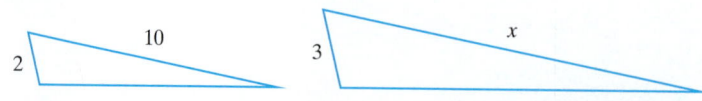

Solution: Since the triangles are similar, corresponding sides are in proportion. Thus, $\frac{2}{3} = \frac{10}{x}$. To solve this equation for x, we cross multiply.

$$\frac{2}{3} = \frac{10}{x}$$

$$2x = 30$$

$$x = 15$$

The missing length is 15 units.

A **right triangle** contains a right angle. The side opposite the right angle is called the **hypotenuse,** and the other two sides are called the **legs.** The **Pythagorean theorem** gives a formula that relates the lengths of the three sides of a right triangle.

The Pythagorean Theorem

If a and b are the lengths of the legs of a right triangle and c is the length of the hypotenuse, then $a^2 + b^2 = c^2$.

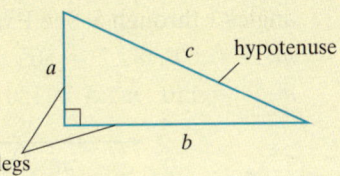

Example 5 Find the length of the hypotenuse of a right triangle whose legs have lengths of 3 centimeters and 4 centimeters.

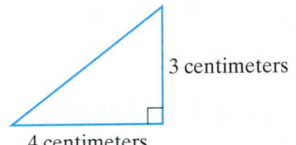

Solution: Because we have a right triangle, we use the Pythagorean theorem. The legs are 3 centimeters and 4 centimeters, so let $a = 3$ and $b = 4$ in the formula.

$$a^2 + b^2 = c^2$$

$$3^2 + 4^2 = c^2$$

$$9 + 16 = c^2$$

$$25 = c^2$$

Since c represents a length, we assume that c is positive. Thus, if c^2 is 25, c must be 5. The hypotenuse has a length of 5 centimeters.

F   **Exercise Set** MyMathLab®

Find the complement of each angle. See Example 1.

1. 19°   71°

2. 65°   25°

3. 70.8°   19.2°

4. $45\frac{2}{3}°$ $44\frac{1}{3}°$

5. $11\frac{1}{4}°$ $78\frac{3}{4}°$

6. 19.6°   70.4°

Find the supplement of each angle.

7. 150°   30°

8. 90°   90°

9. 30.2°   149.8°

10. 81.9°   98.1°

11. $79\frac{1}{2}°$ $100\frac{1}{2}°$

12. $165\frac{8}{9}°$ $14\frac{1}{9}°$

13. If lines *m* and *n* are parallel, find the measures of angles 1 through 7. See Example 2.
$m\angle 1 = m\angle 5 = m\angle 7 = 110°,$
$m\angle 2 = m\angle 3 = m\angle 4 = m\angle 6 = 70°$

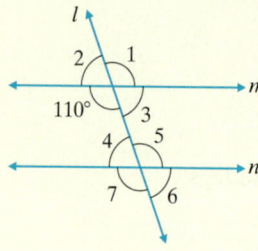

14. If lines *m* and *n* are parallel, find the measures of angles 1 through 5. See Example 2.
$m\angle 1 = 60°, m\angle 2 = 50°, m\angle 3 = 70°,$
$m\angle 4 = 110°, m\angle 5 = 120°$

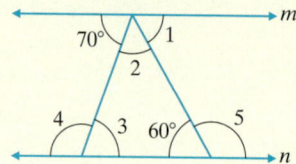

In each of the following, the measures of two angles of a triangle are given. Find the measure of the third angle. See Example 3.

15. 11°, 79°   90°

16. 8°, 102°   70°

17. 25°, 65°   90°

18. 44°, 19°   117°

19. 30°, 60°   90°

20. 67°, 23°   90°

In each of the following, the measure of one angle of a right triangle is given. Find the measures of the other two angles.

21. 45°   45°, 90°

22. 60°   30°, 90°

23. 17°   73°, 90°

24. 30°   60°, 90°

25. $39\frac{3}{4}°$ $50\frac{1}{4}°, 90°$

26. 72.6°   17.4°, 90°

Given that each of the following pairs of triangles is similar, find the missing length x. See Example 4.

27.

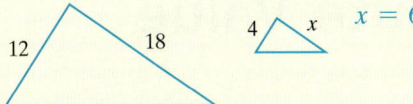

$x = 6$

28.

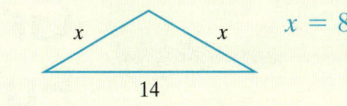

$x = 8$

29.

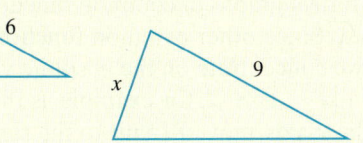

$x = 4.5$

30.

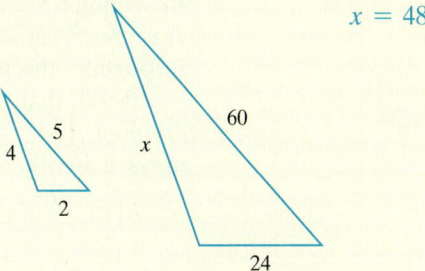

$x = 48$

Use the Pythagorean theorem to find the missing lengths in the right triangles. See Example 5.

31.

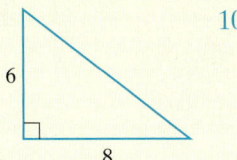

10

32.

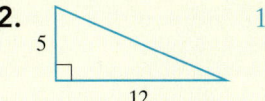

13

33.

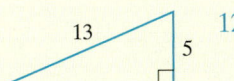

12

34.

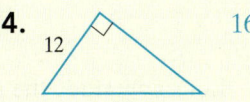

16

Stretching and Compressing Graphs of Absolute Value Functions

In Section 8.5, we learned to shift and reflect graphs of common functions: $f(x) = x$, $f(x) = x^2$, $f(x) = |x|$, and $f(x) = \sqrt{x}$. Since other common functions are studied throughout this text, in this appendix we concentrate on the absolute value function.

Recall that the graph of $h(x) = -|x - 1| + 2$, for example, is the same as the graph of $f(x) = |x|$ reflected about the x-axis, moved 1 unit to the right and 2 units upward. In other words,

$$h(x) = -|x - 1| + 2$$

opens downward $(1, 2)$ location of vertex of V-shape

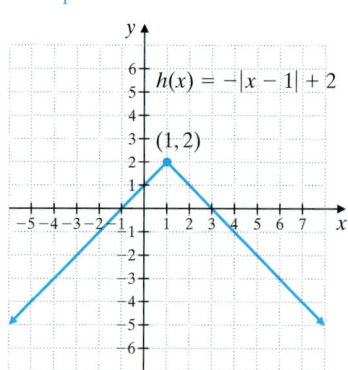

Let's now study the graphs of a few other absolute value functions.

Practice 1

Graph $h(x) = 4|x|$ and $g(x) = \dfrac{1}{5}|x|$ on the same set of axes.

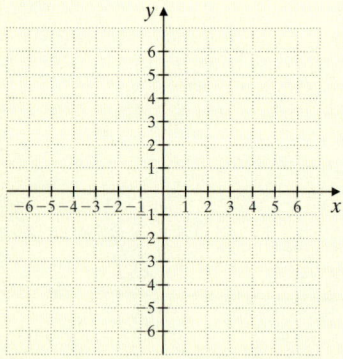

Answer

1.

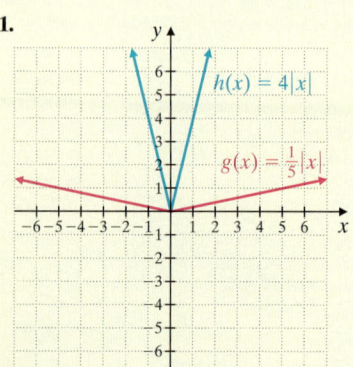

Example 1   Graph $h(x) = 2|x|$, and $g(x) = \dfrac{1}{2}|x|$

Solution: Let's find and plot ordered-pair solutions for the functions.

x	$h(x)$	$g(x)$
-2	4	1
-1	2	$\dfrac{1}{2}$
0	0	0
1	2	$\dfrac{1}{2}$
2	4	1

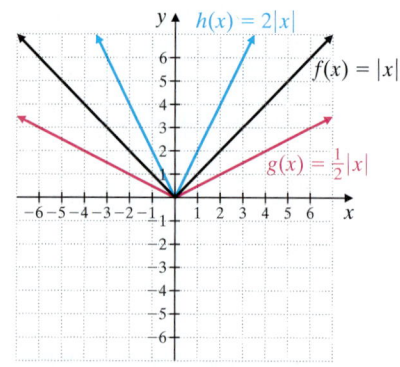

Notice that the graph of $h(x) = 2|x|$ is narrower than the graph of $f(x) = |x|$ and the graph of $g(x) = \dfrac{1}{2}|x|$ is wider than the graph of $f(x) = |x|$.

■ **Work Practice 1**

In general, for the absolute function, we have the following:

The Graph of the Absolute Value Function

The graph of $f(x) = a|x - h| + k$

- Has vertex (h, k) and is V-shaped.
- Opens upward if $a > 0$ and downward if $a < 0$.
- If $|a| < 1$, the graph is wider than the graph of $y = |x|$.
- If $|a| > 1$, the graph is narrower than a graph of $y = |x|$.

Example 2 Graph $f(x) = -\dfrac{1}{3}|x + 2| + 4$

Solution: Let's write this function in the form $f(x) = a|x - h| + k$. For our function, we have $f(x) = -\dfrac{1}{3}|x - (-2)| + 4$. Thus:

- vertex is $(-2, 4)$
- since $a < 0$, V-shape opens down
- since $|a| = \left| -\dfrac{1}{3} \right| = \dfrac{1}{3} < 1$, the graph is wider than $y = |x|$

We will also find and plot ordered-pair solutions.

If $x = -5, f(-5) = -\dfrac{1}{3}|-5 + 2| + 4$, or 3

If $x = 1, f(1) = -\dfrac{1}{3}|1 + 2| + 4$, or 3

If $x = 3, f(3) = -\dfrac{1}{3}|3 + 2| + 4$, or $\dfrac{7}{3}$, or $2\dfrac{1}{3}$

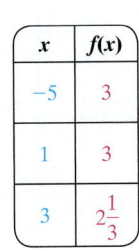

x	$f(x)$
-5	3
1	3
3	$2\dfrac{1}{3}$

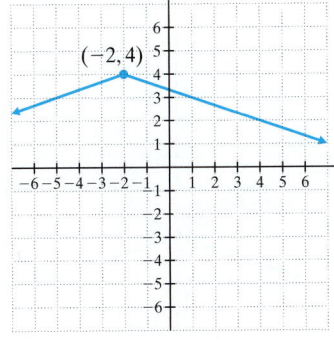

Work Practice 2

Practice 2

Graph $f(x) = -\dfrac{1}{2}|x + 1| + 3$

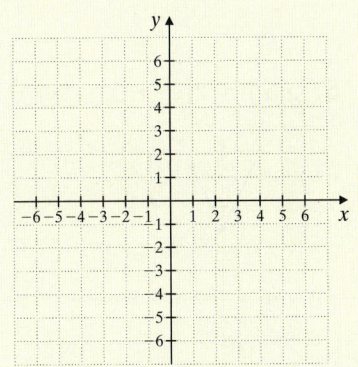

Answer

2.

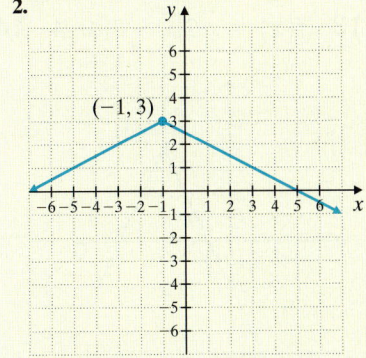

G **Exercise Set** MyMathLab®

Sketch the graph of each function. Label the vertex of the V-shape. See Examples 1 and 2.

1. $f(x) = 3|x|$

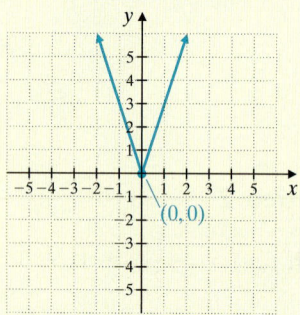

2. $f(x) = 5|x|$

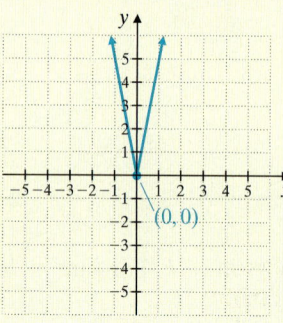

3. $f(x) = \frac{1}{4}|x|$

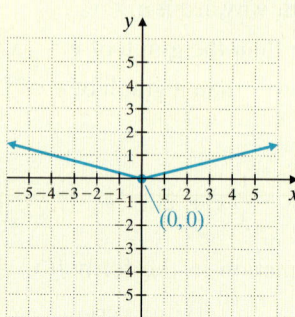

4. $f(x) = \frac{1}{3}|x|$

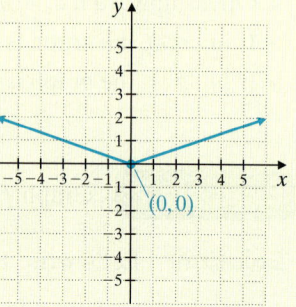

5. $g(x) = 2|x| + 3$

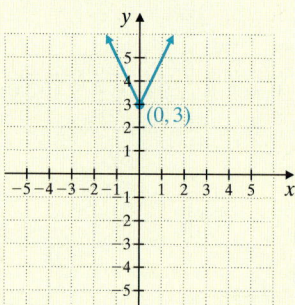

6. $g(x) = 3|x| + 2$

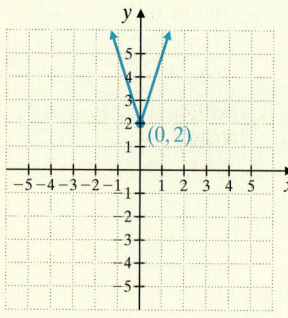

7. $h(x) = -\frac{1}{2}|x|$

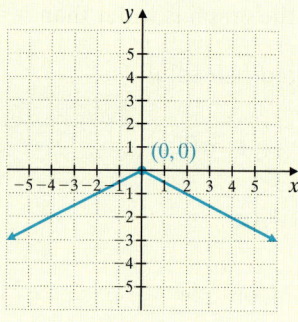

8. $h(x) = -\frac{1}{3}|x|$

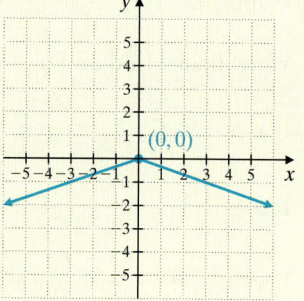

9. $f(x) = 4|x - 1|$

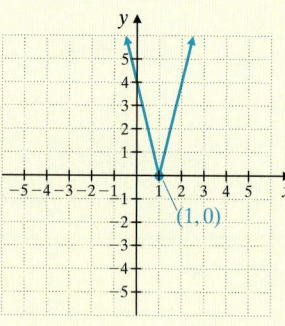

10. $f(x) = 3|x - 2|$

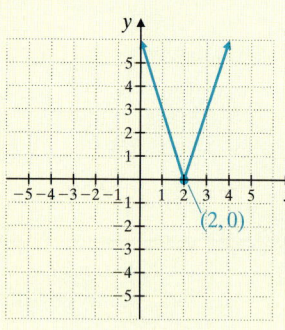

11. $g(x) = -\frac{1}{3}|x| - 2$

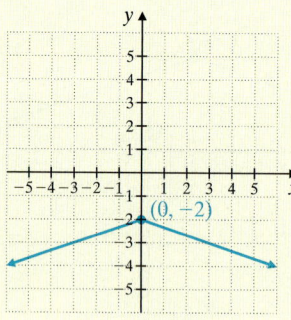

12. $g(x) = -\frac{1}{2}|x| - 3$

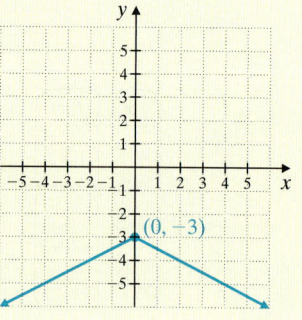

13. $f(x) = -2|x - 3| + 4$

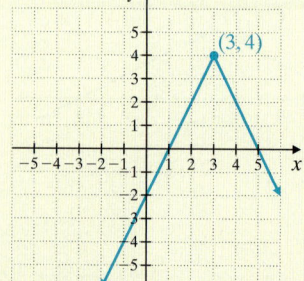

14. $f(x) = -3|x - 1| + 5$

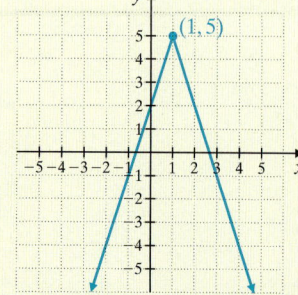

15. $f(x) = \frac{2}{3}|x + 2| - 5$

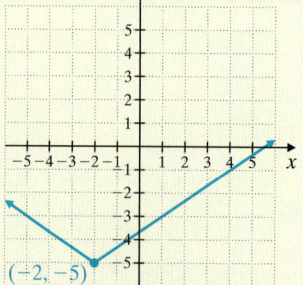

16. $f(x) = \frac{3}{4}|x + 1| - 4$

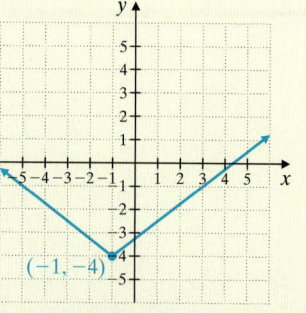

In general, for the absolute function, we have the following:

> ### The Graph of the Absolute Value Function
>
> The graph of $f(x) = a|x - h| + k$
> - Has vertex (h, k) and is V-shaped.
> - Opens upward if $a > 0$ and downward if $a < 0$.
> - If $|a| < 1$, the graph is wider than the graph of $y = |x|$.
> - If $|a| > 1$, the graph is narrower than a graph of $y = |x|$.

Example 2 Graph $f(x) = -\dfrac{1}{3}|x + 2| + 4$

Solution: Let's write this function in the form $f(x) = a|x - h| + k$. For our function, we have $f(x) = -\dfrac{1}{3}|x - (-2)| + 4$. Thus:

- vertex is $(-2, 4)$
- since $a < 0$, V-shape opens down

- since $|a| = \left|-\dfrac{1}{3}\right| = \dfrac{1}{3} < 1$, the graph is wider than $y = |x|$

We will also find and plot ordered-pair solutions.

If $x = -5, f(-5) = -\dfrac{1}{3}|-5 + 2| + 4$, or 3

If $x = 1, f(1) = -\dfrac{1}{3}|1 + 2| + 4$, or 3

If $x = 3, f(3) = -\dfrac{1}{3}|3 + 2| + 4$, or $\dfrac{7}{3}$, or $2\dfrac{1}{3}$

x	$f(x)$
-5	3
1	3
3	$2\dfrac{1}{3}$

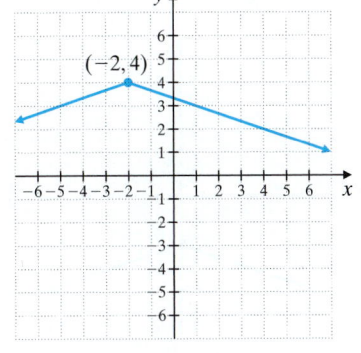

■ **Work Practice 2**

Practice 2

Graph $f(x) = -\dfrac{1}{2}|x + 1| + 3$

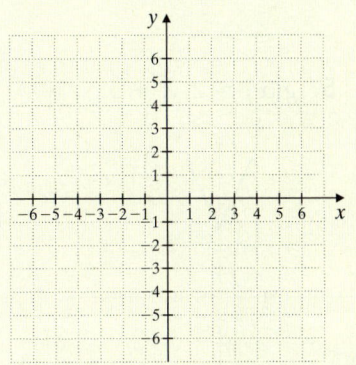

Answer

2.

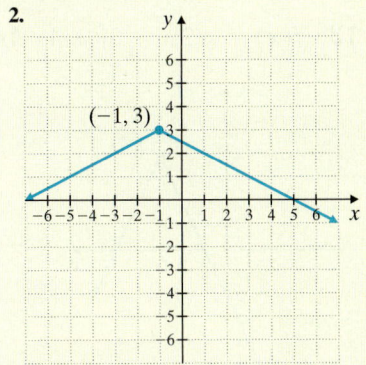

G   Exercise Set MyMathLab®

Sketch the graph of each function. Label the vertex of the V-shape. See Examples 1 and 2.

1. $f(x) = 3|x|$

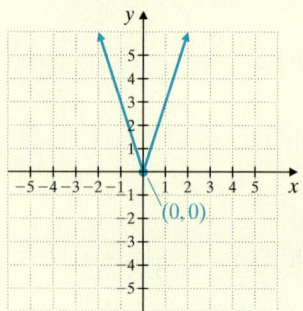

2. $f(x) = 5|x|$

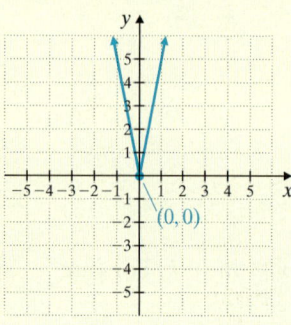

3. $f(x) = \frac{1}{4}|x|$

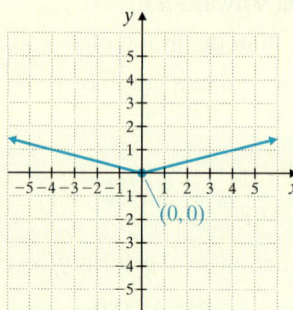

4. $f(x) = \frac{1}{3}|x|$

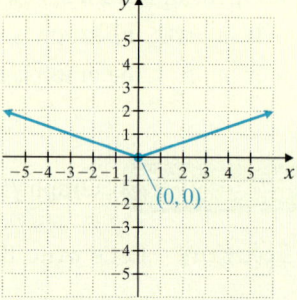

5. $g(x) = 2|x| + 3$

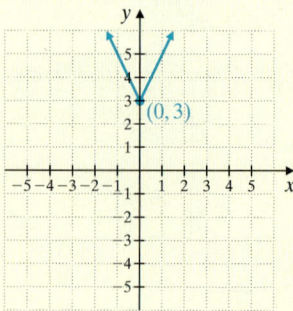

6. $g(x) = 3|x| + 2$

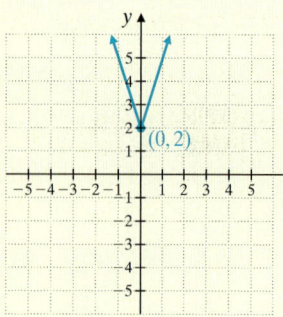

7. $h(x) = -\frac{1}{2}|x|$

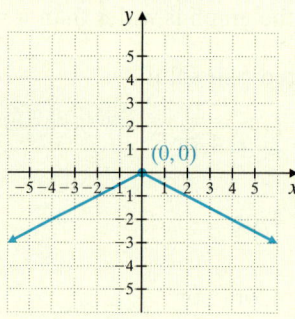

8. $h(x) = -\frac{1}{3}|x|$

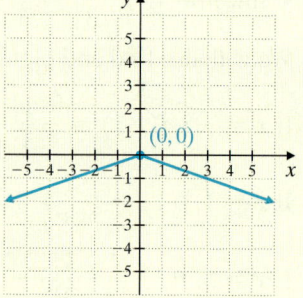

9. $f(x) = 4|x - 1|$

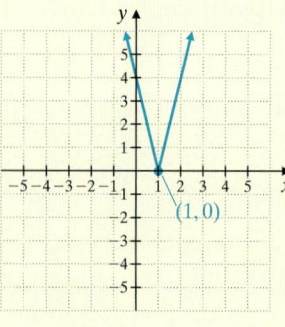

10. $f(x) = 3|x - 2|$

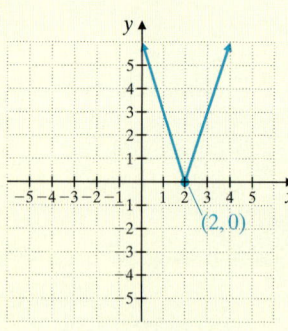

11. $g(x) = -\frac{1}{3}|x| - 2$

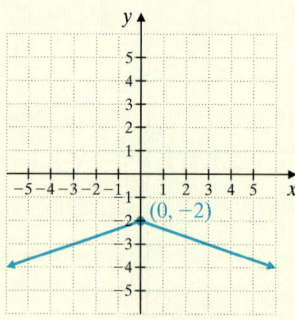

12. $g(x) = -\frac{1}{2}|x| - 3$

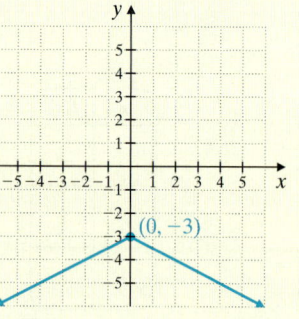

13. $f(x) = -2|x - 3| + 4$

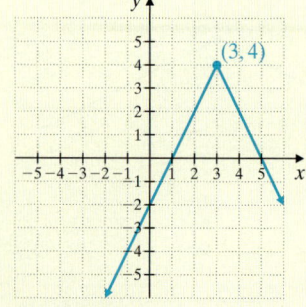

14. $f(x) = -3|x - 1| + 5$

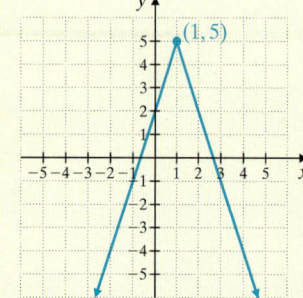

15. $f(x) = \frac{2}{3}|x + 2| - 5$

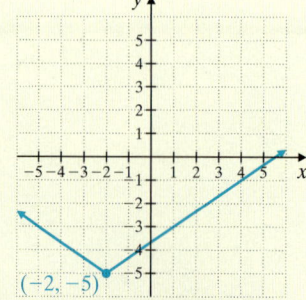

16. $f(x) = \frac{3}{4}|x + 1| - 4$

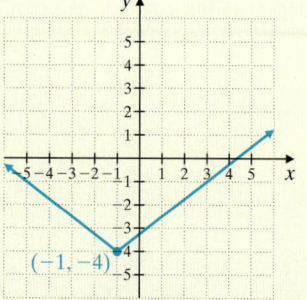

An Introduction to Using a Graphing Utility

Objective A Viewing Window and Interpreting Window Settings

In this appendix, we will use the term **graphing utility** to mean a graphing calculator or a computer software graphing package. All graphing utilities graph equations by plotting points on a screen. While plotting several points can be slow and sometimes tedious for us, a graphing utility can quickly and accurately plot hundreds of points. How does a graphing utility show plotted points? A computer or calculator screen is made up of a grid of small rectangular areas called **pixels.** If a pixel contains a point to be plotted, the pixel is turned "on"; otherwise, the pixel remains "off." The graph of an equation is then a collection of pixels turned "on." The graph of $y = 3x + 1$ from a graphing calculator is shown in Figure H-1. Notice the irregular shape of the line caused by the rectangular pixels.

The portion of the coordinate plane shown on the screen in Figure H-1 is called the **viewing window** or the **viewing rectangle.** Notice the x-axis and the y-axis on the graph. While tick marks are shown on the axes, they are not labeled. This means that from this screen alone, we do not know how many units each tick mark represents. To see what each tick mark represents and the minimum and maximum values on the axes, check the *window setting* of the graphing utility. It defines the viewing window. The window of the graph of $y = 3x + 1$ shown in Figure H-1 has the following setting (Figure H-2):

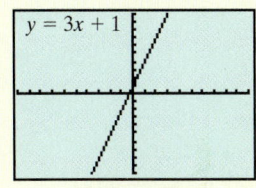

Figure H-1

$$\begin{aligned}
\text{Xmin} &= -10 && \text{The minimum } x\text{-value is } -10. \\
\text{Xmax} &= 10 && \text{The maximum } x\text{-value is } 10. \\
\text{Xscl} &= 1 && \text{The } x\text{-axis scale is 1 unit per tick mark.} \\
\text{Ymin} &= -10 && \text{The minimum } y\text{-value is } -10. \\
\text{Ymax} &= 10 && \text{The maximum } y\text{-value is } 10. \\
\text{Yscl} &= 1 && \text{The } y\text{-axis scale is 1 unit per tick mark.}
\end{aligned}$$

By knowing the scale, we can find the minimum and the maximum values on the axes simply by counting tick marks. For example, if both the Xscl (x-axis scale) and the Yscl are 1 unit per tick mark on the graph in Figure H-3, we can count the tick marks and find that the minimum x-value is -10 and the maximum x-value is 10. Also, the minimum y-value is -10 and the maximum y-value is 10. If the Xscl (x-axis scale) changes to 2 units per tick mark (shown in Figure H-4), by counting tick marks, we see that the minimum x-value is now -20 and the maximum x-value is now 20.

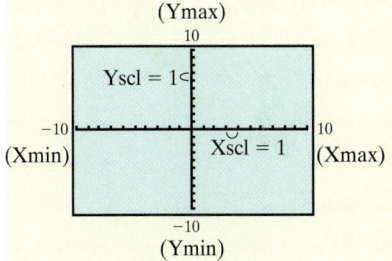

Figure H-2

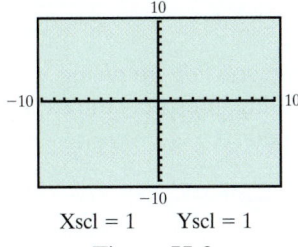

Xscl = 1   Yscl = 1

Figure H-3

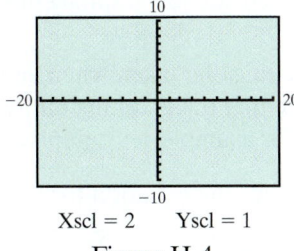

Xscl = 2   Yscl = 1

Figure H-4

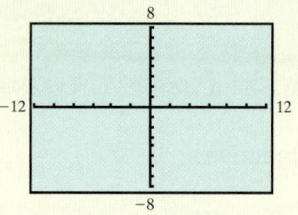

Figure H-5

It is also true that if we know the Xmin and the Xmax values, we can calculate the Xscl by the displayed axes. For example, the Xscl of the graph in Figure H-5 must be 2 units per tick mark for the maximum and minimum x-values to be as shown. Also, the Yscl of that graph must be 1 unit per tick mark for the maximum and minimum y-values to be as shown.

We will call the viewing window in Figure H-3 a *standard* viewing window or rectangle. Although a standard viewing window is sufficient for much of this text, special care must be taken to ensure that all key features of a graph are shown. Figures H-6, H-7, and H-8 show the graph of $y = x^2 + 11x - 1$ on three different viewing windows. Note that certain viewing windows for this equation are misleading.

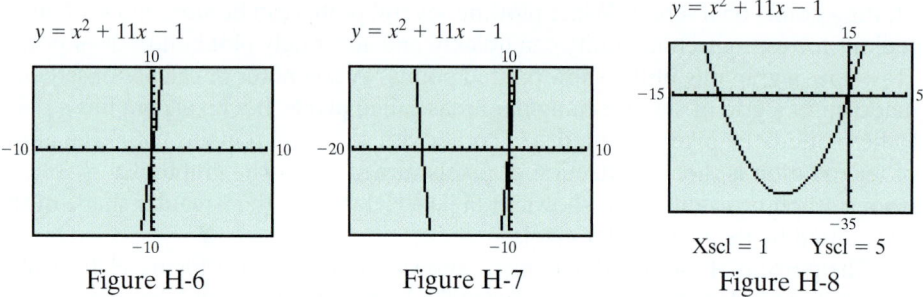

Figure H-6   Figure H-7   Figure H-8

How do we ensure that all distinguishing features of the graph of an equation are shown? It helps to know about the equation that is being graphed. For example, the equation $y = x^2 + 11x - 1$ is not a linear equation, and its graph is not a line. This equation is a quadratic equation, and therefore its graph is a parabola. By knowing this information, we know that the graph shown in Figure H-6, although correct, is misleading. Of the three viewing rectangles shown, the graph in Figure H-8 is best because it shows more of the distinguishing features of the parabola. Properties of equations needed for graphing are studied in this text.

Objective B Graphing Equations and Square Viewing Windows ▶

In general, the following steps may be used to graph an equation on a standard viewing window.

> **To Graph an Equation in x and y with a Graphing Utility on a Standard Viewing Window**
>
> **Step 1:** Solve the equation for y.
> **Step 2:** Use your graphing utility and enter the equation in the form $Y = expression\ involving\ x$
> **Step 3:** Activate the graphing utility.

Special care must be taken when entering the *expression involving x* in Step 2. You must be sure that the graphing utility you are using interprets the expression as you want it to. For example, let's graph $3y = 4x$. To do so,

Step 1: Solve the equation for y.

$$3y = 4x \qquad \frac{3y}{3} = \frac{4x}{3} \qquad y = \frac{4}{3}x$$

Step 2: Using your graphing utility, enter the expression $\frac{4}{3}x$ after the $Y =$ prompt.

In order for your graphing utility to correctly interpret the expression, you may need to enter $(4/3)x$ or $(4 \div 3)x$.

Step 3: Activate the graphing utility. The graph should appear as in Figure H-9.

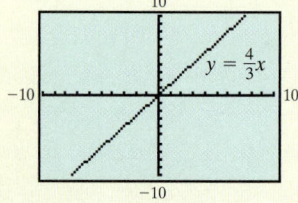

Figure H-9

Distinguishing features of the graph of a line include showing all the intercepts of the line. For example, the window of the graph of the line in Figure H-10 does not show both intercepts of the line, but the window of the graph of the same line in Figure H-11 does show both intercepts.

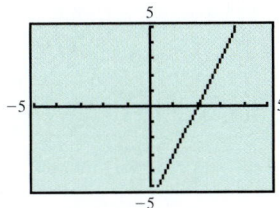

Figure H-10

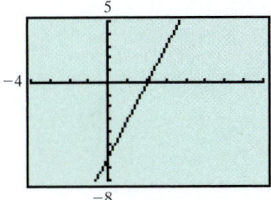

Figure H-11

On a standard viewing window, the tick marks on the *y*-axis are closer than the tick marks on the *x*-axis. This happens because the viewing window is a rectangle, and so 10 equally spaced tick marks on the positive *y*-axis will be closer together than 10 equally spaced tick marks on the positive *x*-axis. This causes the appearance of graphs to be distorted.

For example, notice the different appearances of the same line graphed using different viewing windows. The line in Figure H-12 is distorted because the tick marks along the *x*-axis are farther apart than the tick marks along the *y*-axis. The graph of the same line in Figure H-13 is not distorted because the viewing rectangle has been selected so that there is equal spacing between tick marks on both axes.

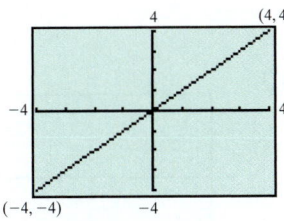

Figure H-12

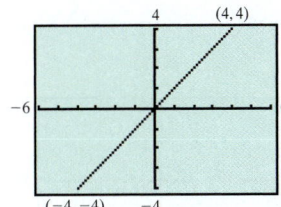

Figure H-13

We say that the line in Figure H-13 is graphed on a *square* setting. Some graphing utilities have a built-in program that, if activated, will automatically provide a square setting. A square setting is especially helpful when we are graphing perpendicular lines, circles, or when a true geometric perspective is desired. Some examples of square screens are shown in Figures H-14 and H-15.

Other features of a graphing utility such as Trace, Zoom, Intersect, and Table are discussed in appropriate Graphing Calculator Explorations in this text.

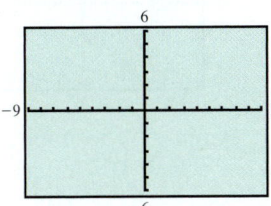

Figure H-14

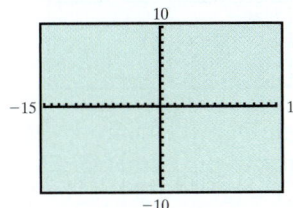

Figure H-15

H Exercise Set MyMathLab®

Objective A *In Exercises 1 through 4, determine whether all ordered pairs listed will lie within a standard viewing rectangle.*

1. $(-9, 0), (5, 8), (1, -8)$ yes

2. $(4, 7), (0, 0), (-8, 9)$ yes

3. $(-11, 0), (2, 2), (7, -5)$ no

4. $(3, 5), (-3, -5), (15, 0)$ no

In Exercises 5 through 10, choose an Xmin, Xmax, Ymin, and Ymax so that all ordered pairs listed will lie within the viewing rectangle.

5. $(-90, 0), (55, 80), (0, -80)$
answers may vary

6. $(4, 70), (20, 20), (-18, 90)$
answers may vary

7. $(-11, 0), (2, 2), (7, -5)$
answers may vary

8. $(3, 5), (-3, -5), (15, 0)$
answers may vary

9. $(200, 200), (50, -50), (70, -50)$
answers may vary

10. $(40, 800), (-30, 500), (15, 0)$
answers may vary

Write the window setting for each viewing window shown. Use the following format:

> Xmin =   Ymin =
> Xmax =   Ymax =
> Xscl =   Yscl =

11.

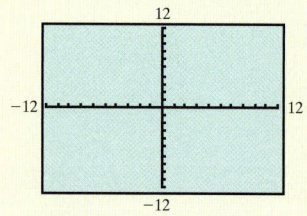

Xmin = -12 Ymin = -12
Xmax = 12 Ymax = 12
Xscl = $\dfrac{6}{5}$ Yscl = $\dfrac{6}{5}$

12.

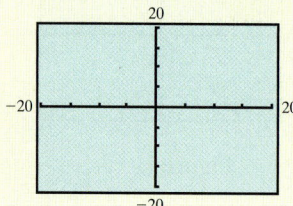

Xmin = -20 Ymin = -20
Xmax = 20 Ymax = 20
Xscl = 5 Yscl = 5

13.

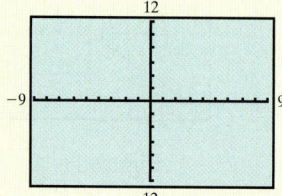

Xmin = -9 Ymin = -12
Xmax = 9 Ymax = 12
Xscl = 1 Yscl = 2

14.

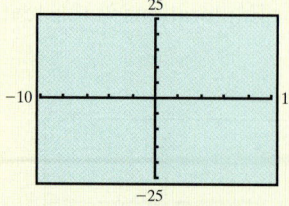

Xmin = -27 Ymin = -6
Xmax = 27 Ymax = 6
Xscl = 3 Yscl = 1

15.

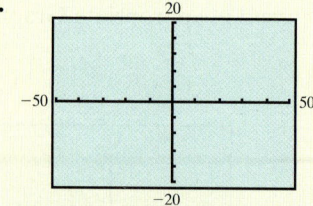

Xmin = -10 Ymin = -25
Xmax = 10 Ymax = 25
Xscl = 2 Yscl = 5

16.

Xmin = -50 Ymin = -20
Xmax = 50 Ymax = 20
Xscl = 10 Yscl = 4

17.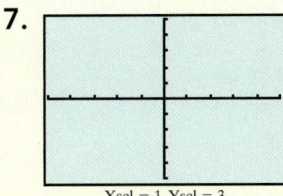
$$Xmin = -5 \quad Ymin = -15$$
$$Xmax = 5 \quad Ymax = 15$$
$$Xscl = 1 \quad Yscl = 3$$

18.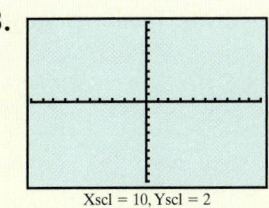
$$Xmin = -100 \quad Ymin = -20$$
$$Xmax = 100 \quad Ymax = 20$$
$$Xscl = 10 \quad Yscl = 2$$

19.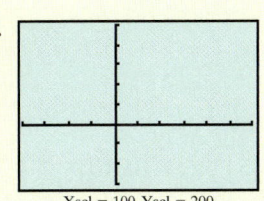
$$Xmin = -20 \quad Ymin = -30$$
$$Xmax = 30 \quad Ymax = 50$$
$$Xscl = 5 \quad Yscl = 10$$

20.
$$Xmin = -400 \quad Ymin = -600$$
$$Xmax = 600 \quad Ymax = 1000$$
$$Xscl = 100 \quad Yscl = 200$$

Objective **B** *Graph each linear equation in two variables, using the two different window settings given. Determine which setting shows all intercepts of the line.*

21. $y = 2x + 12$
Setting A: $[-10, 10]$ by $[-10, 10]$
Setting B: $[-10, 10]$ by $[-10, 15]$ Setting B

22. $y = -3x + 25$
Setting A: $[-5, 5]$ by $[-30, 10]$
Setting B: $[-10, 10]$ by $[-10, 30]$ Setting B

23. $y = -x - 41$
Setting A: $[-50, 10]$ by $[-10, 10]$
Setting B: $[-50, 10]$ by $[-50, 15]$ Setting B

24. $y = 6x - 18$
Setting A: $[-10, 10]$ by $[-20, 10]$
Setting B: $[-10, 10]$ by $[-10, 10]$ Setting A

25. $y = \dfrac{1}{2}x - 15$
Setting A: $[-10, 10]$ by $[-20, 10]$
Setting B: $[-10, 35]$ by $[-20, 15]$ Setting B

26. $y = -\dfrac{2}{3}x - \dfrac{29}{3}$
Setting A: $[-10, 10]$ by $[-10, 10]$
Setting B: $[-15, 5]$ by $[-15, 5]$ Setting B

The graph of each equation is a line. Use a graphing utility and a standard viewing window to graph each equation.

27. $3x = 5y$
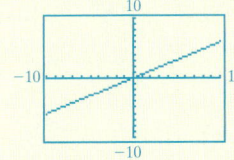

28. $7y = -3x$
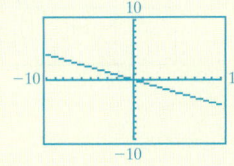

29. $9x - 5y = 30$
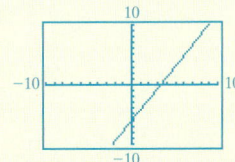

30. $4x + 6y = 20$
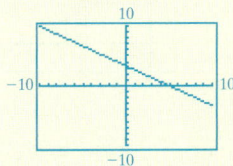

31. $y = -7$
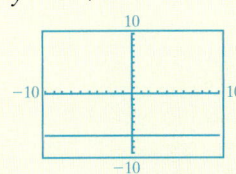

32. $y = 2$
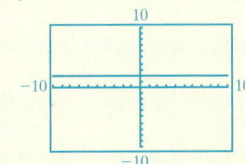

33. $x + 10y = -5$
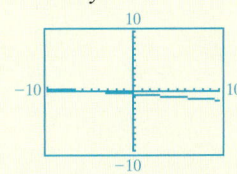

34. $x - 5y = 9$
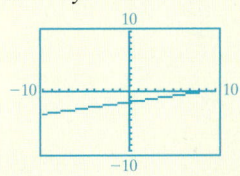

Graph the following equations using the square setting given. Some keystrokes that may be helpful are given.

35. $y = \sqrt{x}$ $[-12, 12]$ by $[-8, 8]$
Suggested keystrokes: $\sqrt{x}$

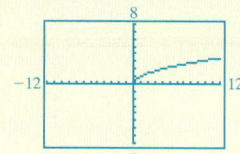

36. $y = \sqrt{2x}$ $[-12, 12]$ by $[-8, 8]$
Suggested keystrokes: $\sqrt{(2x)}$

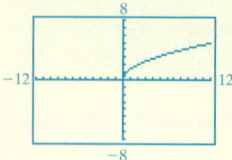

37. $y = x^2 + 2x + 1$ $[-15, 15]$ by $[-10, 10]$
Suggested keystrokes: $x \wedge 2 + 2x + 1$

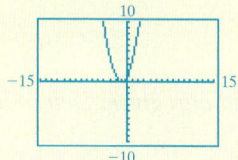

38. $y = x^2 - 5$ $[-15, 15]$ by $[-10, 10]$
Suggested keystrokes: $x \wedge 2 - 5$

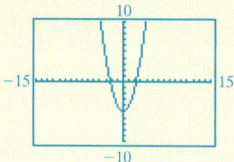

39. $y = |x|$ $[-9, 9]$ by $[-6, 6]$
Suggested keystrokes: ABS (x)

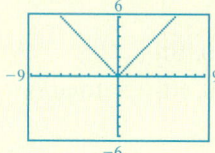

40. $y = |x - 2|$ $[-9, 9]$ by $[-6, 6]$
Suggested keystrokes: ABS $(x - 2)$

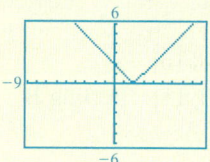

Graph the line on a single set of axes. Use a standard viewing window; then, if necessary, change the viewing window so that all intercepts of the line show.

41. $x + 2y = 30$

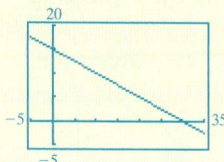

42. $1.5x - 3.7y = 40.3$

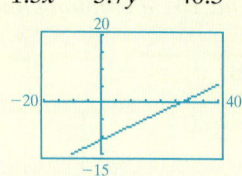

17.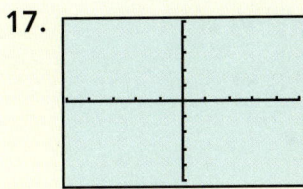

$\text{Xmin} = -5 \quad \text{Ymin} = -15$
$\text{Xmax} = 5 \quad \text{Ymax} = 15$
$\text{Xscl} = 1 \quad \text{Yscl} = 3$

18.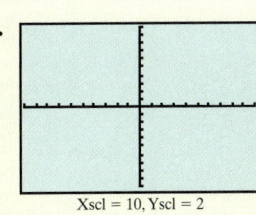

$\text{Xmin} = -100 \quad \text{Ymin} = -20$
$\text{Xmax} = 100 \quad \text{Ymax} = 20$
$\text{Xscl} = 10 \quad \text{Yscl} = 2$

19.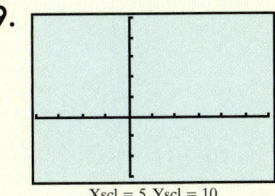

$\text{Xmin} = -20 \quad \text{Ymin} = -30$
$\text{Xmax} = 30 \quad \text{Ymax} = 50$
$\text{Xscl} = 5 \quad \text{Yscl} = 10$

20.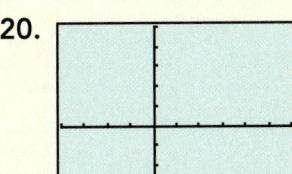

$\text{Xmin} = -400 \quad \text{Ymin} = -600$
$\text{Xmax} = 600 \quad \text{Ymax} = 1000$
$\text{Xscl} = 100 \quad \text{Yscl} = 200$

Objective B *Graph each linear equation in two variables, using the two different window settings given. Determine which setting shows all intercepts of the line.*

21. $y = 2x + 12$
Setting A: $[-10, 10]$ by $[-10, 10]$
Setting B: $[-10, 10]$ by $[-10, 15]$ Setting B

22. $y = -3x + 25$
Setting A: $[-5, 5]$ by $[-30, 10]$
Setting B: $[-10, 10]$ by $[-10, 30]$ Setting B

23. $y = -x - 41$
Setting A: $[-50, 10]$ by $[-10, 10]$
Setting B: $[-50, 10]$ by $[-50, 15]$ Setting B

24. $y = 6x - 18$
Setting A: $[-10, 10]$ by $[-20, 10]$
Setting B: $[-10, 10]$ by $[-10, 10]$ Setting A

25. $y = \dfrac{1}{2}x - 15$
Setting A: $[-10, 10]$ by $[-20, 10]$
Setting B: $[-10, 35]$ by $[-20, 15]$ Setting B

26. $y = -\dfrac{2}{3}x - \dfrac{29}{3}$
Setting A: $[-10, 10]$ by $[-10, 10]$
Setting B: $[-15, 5]$ by $[-15, 5]$ Setting B

The graph of each equation is a line. Use a graphing utility and a standard viewing window to graph each equation.

27. $3x = 5y$

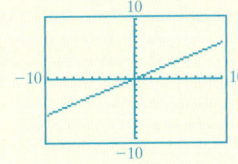

28. $7y = -3x$

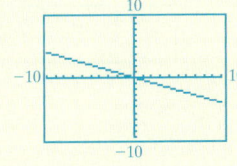

29. $9x - 5y = 30$

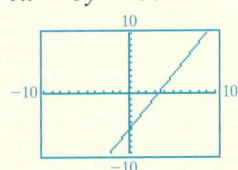

30. $4x + 6y = 20$

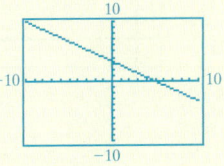

31. $y = -7$

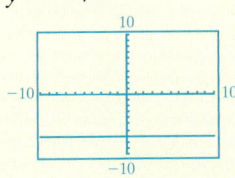

32. $y = 2$

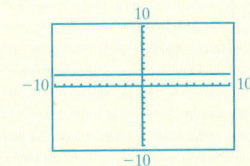

33. $x + 10y = -5$

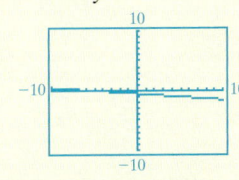

34. $x - 5y = 9$

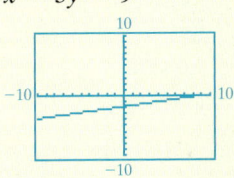

Graph the following equations using the square setting given. Some keystrokes that may be helpful are given.

35. $y = \sqrt{x}$ $[-12, 12]$ by $[-8, 8]$
Suggested keystrokes: $\sqrt{x}$

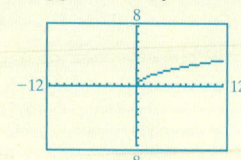

36. $y = \sqrt{2x}$ $[-12, 12]$ by $[-8, 8]$
Suggested keystrokes: $\sqrt{(2x)}$

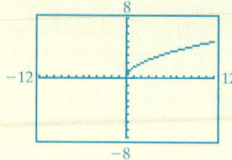

37. $y = x^2 + 2x + 1$ $[-15, 15]$ by $[-10, 10]$
Suggested keystrokes: $x \wedge 2 + 2x + 1$

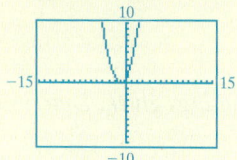

38. $y = x^2 - 5$ $[-15, 15]$ by $[-10, 10]$
Suggested keystrokes: $x \wedge 2 - 5$

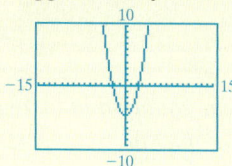

39. $y = |x|$ $[-9, 9]$ by $[-6, 6]$
Suggested keystrokes: ABS (x)

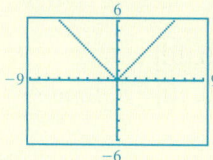

40. $y = |x - 2|$ $[-9, 9]$ by $[-6, 6]$
Suggested keystrokes: ABS $(x - 2)$

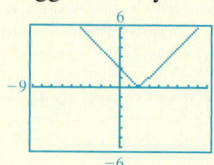

Graph the line on a single set of axes. Use a standard viewing window; then, if necessary, change the viewing window so that all intercepts of the line show.

41. $x + 2y = 30$

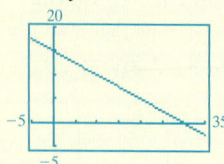

42. $1.5x - 3.7y = 40.3$

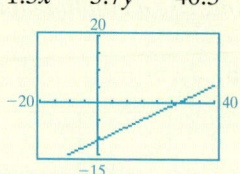

Contents of Student Resources

Student Resources

Study Skills Builders

Attitude and Study Tips

Study Skills Builder 1

Have You Decided to Complete This Course Successfully?

Ask yourself if one of your current goals is to complete this course successfully.

If it is not a goal of yours, ask yourself why. One common reason is fear of failure. Amazingly enough, fear of failure alone can be strong enough to keep many of us from doing our best in any endeavor.

Another common reason is that you simply haven't taken the time to think about or write down your goals for this course. To help accomplish this, answer the questions below.

Exercises

1. Write down your goal(s) for this course.

2. Now list steps you will take to make sure your goal(s) in Exercise 1 are accomplished.

3. Rate your commitment to this course with a number between 1 and 5. Use the diagram below to help.

High Commitment	Average Commitment	Not Committed at All
5 4	3	2 1

4. If you have rated your personal commitment level (from the exercise above) as a 1, 2, or 3, list the reasons why this is so. Then determine whether it is possible to increase your commitment level to a 4 or 5.

Good luck, and don't forget that a positive attitude will make a big difference.

Study Skills Builder 2

Tips for Studying for an Exam

To prepare for an exam, try the following study techniques:

- Start the study process days before your exam.
- Make sure that you are up to date on your assignments.
- If there is a topic that you are unsure of, use one of the many resources that are available to you. For example,

  See your instructor.
  View a lecture video on the topic.
  Visit a learning resource center on campus.
  Read the textbook material and examples on the topic.

- Reread your notes and carefully review the Chapter Highlights at the end of any chapter.
- Work the review exercises at the end of the chapter.
- Find a quiet place to take the Chapter Test found at the end of the chapter. Do not use any resources when taking this sample test. This way, you will have a clear indication of how prepared you are for your exam. Check your answers and use the Chapter Test Prep Videos to make sure that you correct any missed exercises.

Good luck, and keep a positive attitude.

Exercises

Let's see how you did on your last exam.

1. How many days before your last exam did you start studying for that exam?

2. Were you up to date on your assignments at that time, or did you need to catch up on assignments?

3. List the most helpful text supplement (if you used one).

4. List the most helpful campus supplement (if you used one).

5. List your process for preparing for a mathematics test.

6. Was this process helpful? In other words, were you satisfied with your performance on your exam?

7. If not, what changes can you make in your process that will make it more helpful to you?

Study Skills Builder 3

What to Do the Day of an Exam

Your first exam may be soon. On the day of an exam, don't forget to try the following:

- Allow yourself plenty of time to arrive.
- Read the directions on the test carefully.
- Read each problem carefully as you take your test. Make sure that you answer the question asked.
- Watch your time and pace yourself so that you may attempt each problem on your test.
- Check your work and answers.
- ***Do not turn your test in early.*** If you have extra time, spend it double-checking your work.

Good luck!

Exercises

Answer the following questions based on your most recent mathematics exam, whenever that was.

1. How soon before class did you arrive?

2. Did you read the directions on the test carefully?

3. Did you make sure you answered the question asked for each problem on the exam?

4. Were you able to attempt each problem on your exam?

5. If your answer to Exercise 4 is no, list reasons why.

6. Did you have extra time on your exam?

7. If your answer to Exercise 6 is yes, describe how you spent that extra time.

Study Skills Builder 4

Are You Satisfied with Your Performance on a Particular Quiz or Exam?

If not, don't forget to analyze your quiz or exam and look for common errors. Were most of your errors a result of:

- *Carelessness?* Did you turn in your quiz or exam before the allotted time expired? If so, resolve to use any extra time to check your work.

- *Running out of time?* Answer the questions you are sure of first. Then attempt the questions you are unsure of, and delay checking your work until all questions have been answered.

- *Not understanding a concept?* If so, review that concept and correct your work so that you make sure you understand it before the next quiz or the final exam.

- *Test conditions?* When studying for a quiz or exam, make sure you place yourself in conditions similar to test conditions. For example, before your next quiz or exam, take a sample test without the aid of your notes or text.

(For a sample test, see your instructor or use the Chapter Test at the end of each chapter.)

Exercises

1. Have you corrected all your previous quizzes and exams?

2. List any errors you have found common to two or more of your graded papers.

3. Is one of your common errors not understanding a concept? If so, are you making sure you understand all the concepts for the next quiz or exam?

4. Is one of your common errors making careless mistakes? If so, are you now taking all the time allotted to check over your work so that you can minimize the number of careless mistakes?

5. Are you satisfied with your grades thus far on quizzes and tests?

6. If your answer to Exercise 5 is no, are there any more suggestions you can make to your instructor or yourself to help? If so, list them here and share these with your instructor.

Study Skills Builder 5

How Are You Doing?

If you haven't done so yet, take a few moments to think about how you are doing in this course. Are you working toward your goal of successfully completing this course? Is your performance on homework, quizzes, and tests satisfactory? If not, you might want to see your instructor to see if he/she has any suggestions on how you can improve your performance. Reread Section 1.1 for ideas on places to get help with your mathematics course.

Exercises

Answer the following.

1. List any textbook supplements you are using to help you through this course.

2. List any campus resources you are using to help you through this course.

3. Write a short paragraph describing how you are doing in your mathematics course.

4. If improvement is needed, list ways that you can work toward improving your situation as described in Exercise 3.

Study Skills Builder 6

Are You Preparing for Your Final Exam?

To prepare for your final exam, try the following study techniques:

- Review the material that you will be responsible for on your exam. This includes material from your textbook, your notebook, and any handouts from your instructor.

- Review any formulas that you may need to memorize.

- Check to see if your instructor or mathematics department will be conducting a final exam review.

- Check with your instructor to see whether final exams from previous semesters/quarters are available to students for review.

- Use your previously taken exams as a practice final exam. To do so, rewrite the test questions in mixed order on blank sheets of paper. This will help you prepare for exam conditions.

- If you are unsure of a few concepts, see your instructor or visit a learning lab for assistance. Also, view the video segment of any troublesome sections.

- If you need further exercises to work, try the Cumulative Reviews at the end of the chapters.

Once again, good luck! I hope you are enjoying this textbook and your mathematics course.

Organizing Your Work

Study Skills Builder 7

Learning New Terms

Many of the terms used in this text may be new to you. It will be helpful to make a list of new mathematical terms and symbols as you encounter them and to review them frequently. Placing these new terms (including page references) on 3×5 index cards might help you later when you're preparing for a quiz.

Exercises

1. Name one way you might place a word and its definition on a 3×5 card.

2. How do new terms stand out in this text so that they can be found?

Study Skills Builder 8

Are You Organized?

Have you ever had trouble finding a completed assignment? When it's time to study for a test, are your notes neat and organized? Have you ever had trouble reading your own mathematics handwriting? (Be honest—I have.)

When any of these things happens, it's time to get organized. Here are a few suggestions:

- Write your notes and complete your homework assignments in a notebook with pockets (spiral or ring binder).
- Take class notes in this notebook, and then follow the notes with your completed homework assignment.
- When you receive graded papers or handouts, place them in the notebook pocket so that you will not lose them.
- Mark (possibly with an exclamation point) any note(s) that seem extra important to you.
- Mark (possibly with a question mark) any notes or homework that you are having trouble with.
- See your instructor or a math tutor to help you with the concepts or exercises that you are having trouble understanding.
- If you are having trouble reading your own handwriting, *slow down* and write your mathematics work clearly!

Exercises

1. Have you been completing your assignments on time?

2. Have you been correcting any exercises you may be having difficulty with?

3. If you are having trouble with a mathematical concept or correcting any homework exercises, have you visited your instructor, a tutor, or your campus math lab?

4. Are you taking lecture notes in your mathematics course? (By the way, these notes should include worked-out examples solved by your instructor.)

5. Is your mathematics course material (handouts, graded papers, lecture notes) organized?

6. If your answer to Exercise 5 is no, take a moment to review your course material. List at least two ways that you might better organize it.

Study Skills Builder 9

Organizing a Notebook

It's never too late to get organized. If you need ideas about organizing a notebook for your mathematics course, try some of these:

- Use a spiral or ring binder notebook with pockets and use it for mathematics only.
- Start each page by writing the book's section number you are working on at the top.
- When your instructor is lecturing, take notes. *Always* include any examples your instructor works for you.
- Place your worked-out homework exercises in your notebook immediately after the lecture notes from that section. This way, a section's worth of material is together.
- Homework exercises: Attempt and check all assigned homework.
- Place graded quizzes in the pockets of your notebook or a special section of your binder.

Exercises

Check your notebook organization by answering the following questions.

1. Do you have a spiral or ring binder notebook for your mathematics course only?

2. Have you ever had to flip through several sheets of notes and work in your mathematics notebook to determine what section's work you are in?

3. Are you now writing the textbook's section number at the top of each notebook page?

4. Have you ever lost or had trouble finding a graded quiz or test?

5. Are you now placing all your graded work in a dedicated place in your notebook?

6. Are you attempting all of your homework and placing all of your work in your notebook?

7. Are you checking and correcting your homework in your notebook? If not, why not?

8. Are you writing in your notebook the examples your instructor works for you in class?

Study Skills Builder 10

How Are Your Homework Assignments Going?

It is very important in mathematics to keep up with homework. Why? Many concepts build on each other. Often your understanding of a day's concepts depends on an understanding of the previous day's material.

Remember that completing your homework assignment involves a lot more than attempting a few of the problems assigned.

To complete a homework assignment, remember these four things:

- Attempt all of it.
- Check it.
- Correct it.
- If needed, ask questions about it.

Exercises

Take a moment to review your completed homework assignments. Answer the questions below based on this review.

1. Approximate the fraction of your homework you have attempted.

2. Approximate the fraction of your homework you have checked (if possible).

3. If you are able to check your homework, have you corrected it when errors have been found?

4. When working homework, if you do not understand a concept, what do you do?

MyMathLab and MathXL

Study Skills Builder 11

Tips for Turning In Your Homework on Time

It is very important to keep up with your mathematics homework assignments. Why? Many concepts in mathematics build upon each other.

Remember these four tips to help ensure your work is completed on time:

- Know the assignments and due dates set by your instructor.
- Do not wait until the last minute to submit your homework.
- Set a goal to submit your homework 6–8 hours before the scheduled due date in case you have unexpected technology trouble.
- Schedule enough time to complete each assignment.

Following the tips above will also help you avoid potentially losing points for late or missed assignments.

Exercises

Take a moment to consider your work on your homework assignments to date and answer the following questions.

1. What percentage of your assignments have you turned in on time?

2. Why might it be a good idea to submit your homework 6–8 hours before the scheduled deadline?

3. If you have missed submitting any homework by the due date, list some of the reasons why this occurred.

4. What steps do you plan to take in the future to ensure your homework is submitted on time?

Study Skills Builder 12

Tips for Doing Your Homework Online

Practice is one of the main keys to success in any mathematics course. Did you know that MyMathLab/MathXL provides you with **immediate feedback** for each exercise? If you are incorrect, you are given hints to work the exercise correctly. You have **unlimited practice opportunities** and can rework any exercises you have trouble with until you master them, and submit homework assignments unlimited times before the deadline.

Remember these success tips when doing your homework online:

- Attempt all assigned exercises.
- Write down (neatly) your step-by-step work for each exercise before entering your answer.
- Use the immediate feedback provided by the program to help you check and correct your work for each exercise.
- Rework any exercises you have trouble with until you master them.
- Work through your homework assignment as many times as necessary until you are satisfied.

Exercises

Take a moment to think about your homework assignments to date and answer the following questions.

1. Have you attempted all assigned exercises?

2. Of the exercises attempted, have you also written out your work before entering your answer—so that you can check it?

3. Are you familiar with how to enter answers using the MathXL player so that you avoid answer entry type errors?

4. List some ways the immediate feedback and practice supports have helped you with your homework. If you have not used these supports, how do you plan to use them with the success tips above on your next assignment?

Study Skills Builder 13

Organizing Your Work

Have you ever used any readily available paper (such as the back of a flyer, another course assignment, Post-it notes, etc.) to work out homework exercises before entering the answer in MathXL? To save time, have you ever entered answers directly into MathXL without working the exercises on paper? When it's time to study, have you ever been unable to find your completed work or read and follow your own mathematics handwriting?

When any of these things happen, it's time to get organized. Here are some suggestions:

- Write your step-by-step work for each homework exercise, (neatly) on lined, loose-leaf paper and keep this in a 3-ring binder.
- Refer to your step-by-step work when you receive feedback that your answer is incorrect in MathXL. Double-check against the steps and hints provided by the program and correct your work accordingly.
- Keep your written homework with your class notes for that section.

- Identify any exercises you are having trouble with and ask questions about them.
- Keep all graded quizzes and tests in this binder as well to study later.

If you follow the suggestions above, you and your instructor or tutor will be able to follow your steps and correct any mistakes. You will have a written copy of your work to refer to later to ask questions and study for tests.

Exercises

1. Why is it important that you write out your step-by-step work on homework exercises and keep a hard copy of all work submitted online?

2. If you have gotten an incorrect answer, are you able to follow your steps and find your error?

3. If you were asked today to review your previous homework assignments and first test, could you find them? If not, list some ways you might better organize your work.

Study Skills Builder 14

Getting Help with Your Homework Assignments

There are many helpful resources available to you through MathXL to help you work through any homework exercises you may have trouble with. It is important that you know what these resources are and know when and how to use them.

Let's review these features found in the homework exercises:

- **Help Me Solve This**—provides step-by-step help for the exercise you are working. You must work an additional exercise of the same type (without this help) before you can get credit for having worked it correctly.
- **View an Example**—allows you to view a correctly worked exercise similar to the one you are having trouble with. You can go back to your original exercise and work it on your own.
- **E-Book**—allows you to read examples from your text and find similar exercises.

- **Video****—your text author, Elayn Martin-Gay, works an exercise similar to the one you need help with. **Not all exercises have an accompanying video clip.
- **Ask My Instructor**—allows you to e-mail your instructor for help with an exercise.

Exercises

1. How does the "Help Me Solve This" feature work?

2. If the "View an Example" feature is used, is it necessary to work an additional problem before continuing the assignment?

3. When might be a good time to use the "Video" feature? Do all exercises have an accompanying video clip?

4. Which of the features above have you used? List those you found the most helpful to you.

5. If you haven't used the features discussed, list those you plan to try on your next homework assignment.

Study Skills Builder 15

Tips for Preparing for an Exam

Did you know that you can rework your previous homework assignments in MyMathLab and MathXL? This is a great way to prepare for tests. To do this, open a previous homework assignment and click "similar exercise." This will generate new exercises similar to the homework you have submitted. You can then rework the exercises and assignments until you feel confident that you understand them.

To prepare for an exam, follow these tips:

- Review your written work for your previous homework assignments along with your class notes.
- Identify any exercises or topics that you have questions on or have difficulty understanding.
- Rework your previous assignments in MyMathLab and MathXL until you fully understand them and can do them without help.
- Get help for any topics you feel unsure of or for which you have questions.

Exercises

1. Are your current homework assignments up to date and is your written work for them organized in a binder or notebook? If the answer is no, it's time to get organized. For tips on this, see Study Skills Builder 13—Organizing Your Work.

2. How many days in advance of an exam do you usually start studying?

3. List some ways you think that practicing previous homework assignments can help you prepare for your test.

4. List two or three resources you can use to get help for any topics you are unsure of or have questions on.

Good luck!

Study Skills Builder 16

How Well Do You Know the Resources Available to You in MyMathLab?

There are many helpful resources available to you in MyMathLab. Let's take a moment to locate and explore a few of them now. Go into your MyMathLab course, and visit the Multimedia Library, Tools for Success, and E-Book.

Let's see what you found.

Exercises

1. List the resources available to you in the Multimedia Library.

2. List the resources available to you in the Tools for Success folder.

3. Where did you find the English/Spanish Audio Glossary?

4. Can you view videos from the E-Book?

5. Did you find any resources you did not know about? If so, which ones?

6. Which resources have you used most often or found most helpful?

Additional Help Inside and Outside Your Textbook

Study Skills Builder 17

How Well Do You Know Your Textbook?

The questions below will help determine whether you are familiar with your textbook. For additional information, see Section 1.1 in this text.

Exercises

1. What does the ▶ icon mean?

2. What does the ✎ icon mean?

3. What does the △ icon mean?

4. Where can you find a review for each chapter? What answers to this review can be found in the back of your text?

5. Each chapter contains an overview of the chapter along with examples. What is this feature called?

6. Each chapter contains a review of vocabulary. What is this feature called?

7. There are practice exercises that are contained in this text. What are they, and how can they be used?

8. This text contains a student section in the back titled Student Resources. List the contents of this section and how they might be helpful.

9. What exercise answers are available in this text? Where are they located?

Study Skills Builder 18

Are You Familiar with Your Textbook Supplements?

Below is a review of some of the student supplements available for additional study. Check to see if you are using the ones most helpful to you.

- Chapter Test Prep Videos. These videos provide video clip solutions to the Chapter Test exercises in this text. You will find this extremely useful when studying for tests or exams.
- Interactive DVD Lecture Series. These are keyed to each section of the text. The material is presented by me, Elayn Martin-Gay, and I have placed a ▶ by the exercises in the text that I have worked on the video.
- The *Student Solutions Manual*. This contains worked-out solutions to odd-numbered exercises as well as every exercise in the Integrated Reviews, Chapter Reviews, Chapter Tests, Cumulative Reviews, and every Practice exercise.
- Pearson Tutor Center. Mathematics questions may be phoned, faxed, or e-mailed to this center.
- MyMathLab is a text-specific online course. MathXL is an online homework, tutorial, and assessment system.

Take a moment to determine whether these are available to you.

As usual, your instructor is your best source of information.

Exercises

Let's see how you are doing with textbook supplements.

1. Name one way the Lecture Videos can be helpful to you.

2. Name one way the Chapter Test Prep Video can help you prepare for a chapter test.

3. List any textbook supplements that you have found useful.

4. Have you located and visited a learning resource lab located on your campus?

5. List the textbook supplements that are currently housed in your campus's learning resource lab.

Study Skills Builder 19

Are You Getting All the Mathematics Help That You Need?

Remember that, in addition to your instructor, there are many places to get help with your mathematics course. For example:

- This text has an accompanying video lesson by the author for every section. There are also worked-out video solutions by the author to every Chapter Test exercise.
- The back of the book contains answers to odd-numbered exercises.
- A *Student Solutions Manual* is available that contains worked-out solutions to odd-numbered exercises as well as solutions to every exercise in the Integrated Reviews, Chapter Reviews, Chapter Tests, Cumulative Reviews, and every Practice exercise.
- Don't forget to check with your instructor for other local resources available to you, such as a tutor center.

Exercises

1. List items you find helpful in the text and all student supplements to this text.

2. List all the campus help that is available to you for this course.

3. List any help (besides the textbook) from Exercises 1 and 2 above that you are using.

4. List any help (besides the textbook) that you feel you should try.

5. Write a goal for yourself that includes trying everything you listed in Exercise 4 during the next week.

Student Resources

Bigger Picture—
Study Guide Outline

Simplifying Expressions and Solving Equations and Inequalities

I. Simplifying Expressions

  A. Real Numbers

    1. Add: (Sec. 1.4)

$$-1.7 + (-0.21) = -1.91 \qquad \text{Adding like signs.}$$

$$-7 + 3 = -4 \qquad \text{Adding different signs.}$$

Subtract absolute values. Attach the sign of the number with the larger absolute value.

    2. Subtract: Add the first number to the opposite of the second number. (Sec. 1.5)

$$17 - 25 = 17 + (-25) = -8$$

    3. Multiply or divide: Multiply or divide the two numbers as usual. If the signs are the same, the answer is positive. If the signs are different, the answer is negative. (Sec. 1.6)

$$-10 \cdot 3 = -30, \qquad -81 \div (-3) = 27$$

  B. Exponents (Sec. 5.1 and 5.2)

$$x^7 \cdot x^5 = x^{12}; \ (x^7)^5 = x^{35}, \frac{x^7}{x^5} = x^2; \ x^0 = 1; \ 8^{-2} = \frac{1}{8^2} = \frac{1}{64}$$

  C. Polynomials

    1. Add: Combine like terms. (Sec. 5.4)

$$(3y^2 + 6y + 7) + (9y^2 - 11y - 15) = 3y^2 + 6y + 7 + 9y^2 - 11y - 15$$
$$= 12y^2 - 5y - 8$$

    2. Subtract: Change the sign of the terms of the polynomial being subtracted, then add. (Sec. 5.4)

$$(3y^2 + 6y + 7) - (9y^2 - 11y - 15) = 3y^2 + 6y + 7 - 9y^2 + 11y + 15$$
$$= -6y^2 + 17y + 22$$

    3. Multiply: Multiply each term of one polynomial by each term of the other polynomial. (Sec. 5.5)

$$(x + 5)(2x^2 - 3x + 4) = x(2x^2 - 3x + 4) + 5(2x^2 - 3x + 4)$$
$$= 2x^3 - 3x^2 + 4x + 10x^2 - 15x + 20$$
$$= 2x^3 + 7x^2 - 11x + 20$$

    4. Divide: (Sec. 5.7)

      a. To divide by a monomial, divide each term of the polynomial by the monomial.

$$\frac{8x^2 + 2x - 6}{2x} = \frac{8x^2}{2x} + \frac{2x}{2x} - \frac{6}{2x} = 4x + 1 - \frac{3}{x}$$

b. To divide by a polynomial other than a monomial, use long division.

$$
\begin{array}{r}
x - 6 + \dfrac{40}{2x + 5} \\[4pt]
2x + 5 \overline{)\,2x^2 - 7x + 10} \\
\underline{2x^2 + 5x} \\
-12x + 10 \\
\underline{12x - 30} \\
40
\end{array}
$$

D. Factoring Polynomials

See the Chapter 6 Integrated Review for steps.

$$
\begin{aligned}
3x^4 - 78x^2 + 75 &= 3(x^4 - 26x^2 + 25) \quad \text{Factor out GCF—always first step.}\\
&= 3(x^2 - 25)(x^2 - 1) \quad \text{Factor trinomial.}\\
&= 3(x + 5)(x - 5)(x + 1)(x - 1) \quad \text{Factor further—each}\\
&\qquad\qquad\qquad\qquad\qquad\qquad\qquad \text{difference of squares.}
\end{aligned}
$$

E. Rational Expressions

1. Simplify: Factor the numerator and denominator. Then remove factors of 1 by dividing out common factors in the numerator and denominator. (Sec. 7.1)

$$
\frac{x^2 - 9}{7x^2 - 21x} = \frac{(x + 3)(x - 3)}{7x(x - 3)} = \frac{x + 3}{7x}
$$

2. Multiply: Multiply numerators, then multiply denominators. (Sec. 7.2)

$$
\frac{5z}{2z^2 - 9z - 18} \cdot \frac{22z + 33}{10z} = \frac{5 \cdot z}{(2z + 3)(z - 6)} \cdot \frac{11(2z + 3)}{2 \cdot 5 \cdot z} = \frac{11}{2(z - 6)}
$$

3. Divide: First fraction times the reciprocal of the second fraction. (Sec. 7.2)

$$
\frac{14}{x + 5} \div \frac{x + 1}{2} = \frac{14}{x + 5} \cdot \frac{2}{x + 1} = \frac{28}{(x + 5)(x + 1)}
$$

4. Add or subtract: Must have same denominator. If not, find the LCD and write each fraction as an equivalent fraction with the LCD as denominator. (Sec. 7.4)

$$
\begin{aligned}
\frac{9}{10} - \frac{x + 1}{x + 5} &= \frac{9(x + 5)}{10(x + 5)} - \frac{10(x + 1)}{10(x + 5)}\\[4pt]
&= \frac{9x + 45 - 10x - 10}{10(x + 5)} = \frac{-x + 35}{10(x + 5)}
\end{aligned}
$$

F. Radicals

1. Simplify square roots: If possible, factor the radicand so that one factor is a perfect square. Then use the product rule and simplify. (Sec. 10.3)

$$
\sqrt{75} = \sqrt{25 \cdot 3} = \sqrt{25} \cdot \sqrt{3} = 5\sqrt{3}
$$

2. Add or subtract: Only like radicals (same index and radicand) can be added or subtracted. (Sec. 10.4)

$$
8\sqrt{10} - \sqrt{40} + \sqrt{5} = 8\sqrt{10} - 2\sqrt{10} + \sqrt{5} = 6\sqrt{10} + \sqrt{5}
$$

3. Multiply or divide: $\sqrt{a} \cdot \sqrt{b} = \sqrt{ab}$; $\dfrac{\sqrt{a}}{\sqrt{b}} = \sqrt{\dfrac{a}{b}}$. (Sec. 10.4, 10.5)

$$
\sqrt{11} \cdot \sqrt{3} = \sqrt{33}; \quad \frac{\sqrt{140}}{\sqrt{7}} = \sqrt{\frac{140}{7}} = \sqrt{20} = \sqrt{4 \cdot 5} = 2\sqrt{5}
$$

4. Rationalizing the denominator: (Sec. 10.5)

a. If denominator is one term,

$$\frac{5}{\sqrt{11}} = \frac{5 \cdot \sqrt{11}}{\sqrt{11} \cdot \sqrt{11}} = \frac{5\sqrt{11}}{11}$$

b. If denominator is two terms, multiply by 1 in the form of $\dfrac{\text{conjugate of denominator}}{\text{conjugate of denominator}}$.

$$\frac{13}{3 + \sqrt{2}} = \frac{13}{3 + \sqrt{2}} \cdot \frac{3 - \sqrt{2}}{3 - \sqrt{2}} = \frac{13(3 - \sqrt{2})}{9 - 2} = \frac{13(3 - \sqrt{2})}{7}$$

II. Solving Equations and Inequalities

A. Linear Equations: Power on variable is 1 and there are no variables in denominator. (Sec. 2.3)

$7(x - 3) = 4x + 6$	Linear equation. (If fractions, multiply by LCD.)
$7x - 21 = 4x + 6$	Use the distributive property.
$7x = 4x + 27$	Add 21 to both sides.
$3x = 27$	Subtract $4x$ from both sides.
$x = 9$	Divide both sides by 3.

B. Linear Inequalities: Same as linear equation except if you multiply or divide by a negative number, then reverse direction of inequality. (Sec. 2.7)

$-4x + 11 \le -1$	Linear inequality.
$-4x \le -12$	Subtract 11 from both sides.
$\dfrac{-4x}{-4} \ge \dfrac{-12}{-4}$	Divide both sides by -4 and reverse the direction of the inequality symbol.
$x \ge 3$	Simplify.

C. Quadratic and Higher Degree Equations: Solve: first write the equation in standard form (one side is 0).

1. If the polynomial on one side factors, solve by factoring. (Sec. 6.6)

2. If the polynomial does not factor, solve by the quadratic formula. (Sec. 11.2)

By factoring:	**By quadratic formula:**
$x^2 + x = 6$	$x^2 + x = 5$
$x^2 + x - 6 = 0$	$x^2 + x - 5 = 0$
$(x - 2)(x + 3) = 0$	$a = 1, b = 1, c = -5$
$x - 2 = 0 \text{ or } x + 3 = 0$	$x = \dfrac{-1 \pm \sqrt{1^2 - 4(1)(-5)}}{2 \cdot 1}$
$x = 2 \text{ or } \quad x = -3$	$x = \dfrac{-1 \pm \sqrt{21}}{2}$

D. Equations with Rational Expressions: Make sure the proposed solution does not make any denominator 0. (Sec. 7.5)

$$\frac{3}{x} - \frac{1}{x - 1} = \frac{4}{x - 1} \qquad \text{Equation with rational expressions}$$

$$x(x - 1) \cdot \frac{3}{x} - x(x - 1) \cdot \frac{1}{x - 1} = x(x - 1) \cdot \frac{4}{x - 1} \qquad \begin{array}{l}\text{Multiply through by} \\ x(x - 1).\end{array}$$

$3(x - 1) - x \cdot 1 = x \cdot 4$	Simplify.
$3x - 3 - x = 4x$	Use the distributive property.
$-3 = 2x$	Simplify and move variable terms to right side.
$-\dfrac{3}{2} = x$	Divide both sides by 2.

E. Proportions: An equation with two ratios equal. Set cross products equal, then solve. Make sure the proposed solution does not make any denominator 0. (Sec. 7.6)

$$\frac{5}{x} \nwarrow\nearrow \frac{9}{2x - 3}$$

$5(2x - 3) = 9 \cdot x$ Set cross products equal.

$10x - 15 = 9x$ Multiply.

$x = 15$ Write equation with variable terms on one side and constants on the other.

F. Equations with Radicals: To solve, isolate a radical, then square both sides. You may have to repeat this. Check possible solution in the original equation. (Sec. 10.6)

$\sqrt{x + 49} + 7 = x$

$\sqrt{x + 49} = x - 7$ Subtract 7 from both sides.

$x + 49 = x^2 - 14x + 49$ Square both sides.

$0 = x^2 - 15x$ Set terms equal to 0.

$0 = x(x - 15)$ Factor.

$\cancel{x = 0}$ or $x = 15$ Set each factor equal to 0 and solve.

G. Nonlinear Inequalities (Sec. 11.4)

Polynomial Inequality

$x^2 - x < 6$

$x^2 - x - 6 < 0$

$(x - 3)(x + 2) < 0$

$(-2, 3)$

Rational Inequality with variable in denominator

$\frac{x - 5}{x + 1} \geq 0$

$(-\infty, -1) \cup [5, \infty)$

H. Exponential Equations (Sec. 12.8)

1. If we can write with the same base, then set the exponents equal to each other and solve

$9^x = 27^{x+1}$

$(3^2)^x = (3^3)^{x+1}$

$3^{2x} = 3^{3x+3}$

$2x = 3x + 3$

$-3 = x$

2. If we can't write with the same base, then solve using logarithms

$5^x = 7$

$\log 5^x = \log 7$

$x \log 5 = \log 7$

$x = \frac{\log 7}{\log 5}$

I. Logarithmic Equations (Sec. 12.8)

$\log 7 + \log(x + 3) = \log 5$ Write equation so that single logarithm on one side and constant on the other side.

$\log 7(x + 3) = \log 5$

$7(x + 3) = 5$ Use definition of logarithm.

$7x + 21 = 5$ Multiply.

$7x = -16$

$x = -\frac{16}{7}$ Solve.

Practice Final Exam

Answers (left column)

1. -48

2. -81

3. $\dfrac{1}{64}$

4. $-3x^3 + 5x^2 + 4x + 5$

5. $16x^2 - 16x + 4$

6. $3x^3 + 22x^2 + 41x + 14$

7. $(y - 12)(y + 4)$

8. $3x(3x + 1)(x + 4)$

9. $5(6 + x)(6 - x)$

10. $(a + b)(3a - 7)$

11. $\begin{array}{l} 3x(3y - z) \\ (9y^2 + 3yz + z^2) \end{array}$

12. $\dfrac{16y^{14}}{x^2}$

13. $\dfrac{25}{7}$

14. $\{x \mid x \le -2\}$

15. $-7, 1$

16. see graph

17. see graph

18. see graph

Evaluate.

1. $6[5 + 2(3 - 8) - 3]$ **2.** -3^4 **3.** 4^{-3}

Perform the indicated operations and simplify if possible.

4. $(5x^3 + x^2 + 5x - 2) - (8x^3 - 4x^2 + x - 7)$

5. $(4x - 2)^2$ **6.** $(3x + 7)(x^2 + 5x + 2)$

Factor.

7. $y^2 - 8y - 48$ **8.** $9x^3 + 39x^2 + 12x$ **9.** $180 - 5x^2$

10. $3a^2 + 3ab - 7a - 7b$ **11.** $81xy^3 - 3xz^3$

Simplify. Write answers using only positive exponents.

12. $\left(\dfrac{4x^2y^3}{x^3y^{-4}}\right)^2$

Solve each equation or inequality.

13. $-4(a + 1) - 3a = -7(2a - 3)$ **14.** $3x - 5 \ge 7x + 3$

15. $x(x + 6) = 7$

Graph the following.

16. $5x - 7y = 10$ **17.** $y = -1$ **18.** $y \ge -4x$

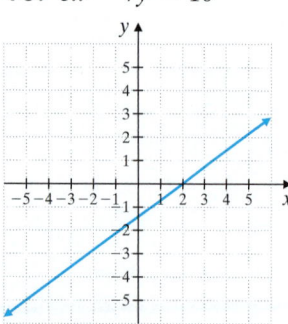   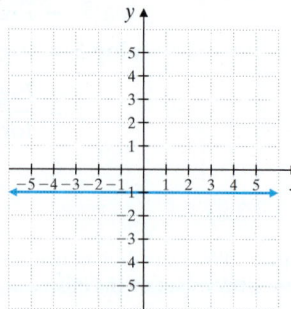   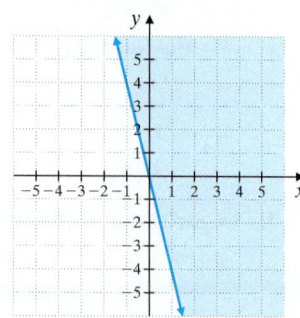

Helpful Hint

- If your course in this text ended with Chapter 6, work Exercises 1–31.
- If your course in this text ended with Chapter 7, work Exercises 1–39.
- If your course in this text started with Chapter 7, start with Exercise 32.
- If your course in this text started with Chapter 8, start with Exercise 40.

1052

Find the slope of each line.

19. through $(6, -5)$ and $(-1, 2)$

20. $-3x + y = 5$

Write an equation of the following lines. Write each equation in standard form, $Ax + By = C$.

21. Passes through $(2, -5)$ and $(1, 3)$

22. slope $\frac{1}{8}$; y-intercept $(0, 12)$

Solve each system of equations.

23. $\begin{cases} \frac{1}{2}x + 2y = -\frac{15}{4} \\ \quad\quad 4x = -y \end{cases}$

24. $\begin{cases} 4x - 6y = 7 \\ -2x + 3y = 0 \end{cases}$

25. Divide by long division: $\dfrac{27x^3 - 8}{3x + 2}$

26. A number increased by two-thirds of the number is 35. Find the number.

27. A gallon of water seal covers 200 square feet. How many gallons are needed to paint two coats of water seal on a deck that measures 20 feet by 35 feet?

20 feet | 35 feet

28. Some states have a single area code for the entire state. Two such states have area codes where one is double the other. If the sum of these integers is 1203, find the two area codes.

29. California has more public libraries than any other state. It has 387 more public libraries than Ohio. If the total number of public libraries for these states is 1827, find the number of public libraries in California and the number in Ohio. (*Source: Institute of Museum and Library Services*)

30. Find the amount of a 12% saline solution a lab assistant should add to 80 cc (cubic centimeters) of a 22% saline solution in order to have a 16% solution.

31. Two hikers start at opposite ends of the St. Tammany Trails and walk toward each other. The trail is 36 miles long and they meet in 4 hours. If one hiker is twice as fast as the other, find both hiking speeds.

Perform the indicated operations and simplify if possible.

32. $\dfrac{5}{2x + 5} - \dfrac{6}{2x + 5}$

33. $\dfrac{x^2 - 9}{x^2 - 3x} \div \dfrac{x^2 + 4x + 1}{2x + 10}$

34. $\dfrac{5a}{a^2 - a - 6} - \dfrac{2}{a - 3}$

35. $\dfrac{5 - \dfrac{1}{y^2}}{\dfrac{1}{y} + \dfrac{2}{y^2}}$

19. $m = -1$

20. $m = 3$

21. $8x + y = 11$

22. $x - 8y = -96$

23. $\left(\frac{1}{2}, -2\right)$

24. no solution

25. $9x^2 - 6x + 4 - \dfrac{16}{3x + 2}$

26. 21

27. 7 gal

28. $401, 802$

29. California: 1107
Ohio: 720

30. 120 cc

31. 3 mph; 6 mph

32. $-\dfrac{1}{2x + 5}$

33. $\dfrac{2(x + 3)(x + 5)}{x(x^2 + 4x + 1)}$

34. $\dfrac{3a - 4}{(a - 3)(a + 2)}$

35. $\dfrac{5y^2 - 1}{y + 2}$

36. $\dfrac{30}{11}$

37. -6

38. no solution

39. 5 or 1

40. $6\sqrt{6}$

41. 5

42. $\dfrac{8a^{1/3}c^{2/3}}{b^{5/12}}$

43. $-x\sqrt{5x}$

44. -20

45. -20

46. domain: $\{-2\}$, range: $(-\infty, \infty)$; not a function

47. domain: $(-\infty, \infty)$; range: $[0, \infty)$; function

48. $\dfrac{3 \pm \sqrt{29}}{2}$

49. 2, 3

50. $\left(-\infty, -\dfrac{3}{2}\right) \cup (5, \infty)$

Solve each equation.

36. $\dfrac{4}{y} - \dfrac{5}{3} = -\dfrac{1}{5}$

37. $\dfrac{5}{y + 1} = \dfrac{4}{y + 2}$

38. $\dfrac{a}{a - 3} = \dfrac{3}{a - 3} - \dfrac{3}{2}$

Solve.

39. One number plus five times its reciprocal is equal to six. Find the number.

Simplify. If needed, write answers with positive exponents only.

40. $\sqrt{216}$

41. $\left(\dfrac{1}{125}\right)^{-1/3}$

42. $\left(\dfrac{64c^{4/3}}{a^{-2/3}b^{5/6}}\right)^{1/2}$

Perform the indicated operations and simplify if possible.

43. $\sqrt{125x^3} - 3\sqrt{20x^3}$

44. $(\sqrt{5} + 5)(\sqrt{5} - 5)$

Answer the question about functions.

45. If $f(x) = \dfrac{5x^2}{1 - x}$, find $f(2)$.

Find the domain and range of each relation. Also determine whether the relation is a function.

46.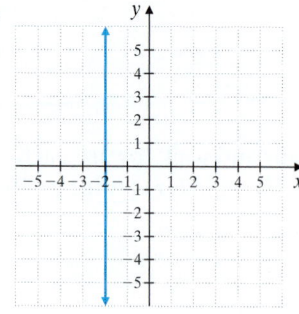

47.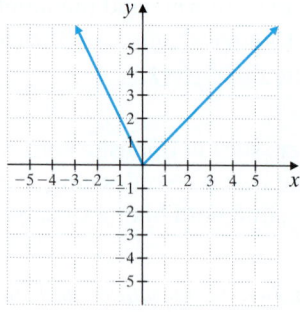

Solve each equation or inequality. Write inequality solutions using interval notation.

48. $y^2 - 3y = 5$

49. $x = \sqrt{x - 2} + 2$

50. $2x^2 - 7x > 15$

Graph each function. For Exercises 52 and 54, state the domain and the range of the function.

51. $h(x) = \sqrt{x} - 1$

52. $g(x) = -|x + 2| - 1$

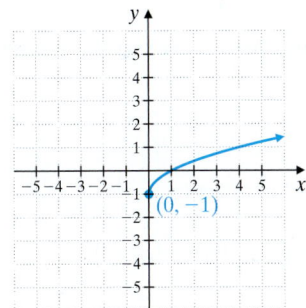

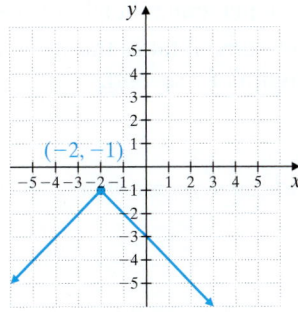

53. Label the vertex and any intercepts for the graph of $h(x) = x^2 - 4x + 4$.

54. $f(x) = \begin{cases} -\dfrac{1}{2}x \text{ if } x \le 0 \\ 2x - 3 \text{ if } x > 0 \end{cases}$

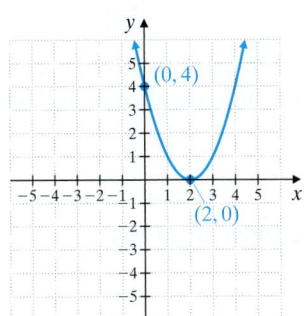

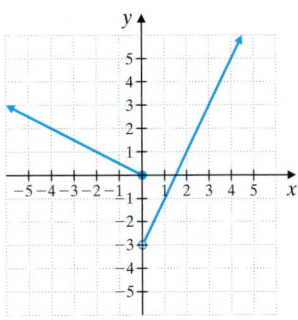

Write equations of the following lines. Write each equation in the form $y = mx + b$.

55. through $(4, -2)$ and $(6, -3)$

56. through $(-1, 2)$ and perpendicular to $3x - y = 4$

Find the distance or midpoint.

57. Find the distance between the points $(-6, 3)$ and $(-8, -7)$.

58. Find the midpoint of the line segment whose endpoints are $(-2, -5)$ and $(-6, 12)$.

Rationalize each denominator. Assume that variables represent positive numbers.

59. $\sqrt{\dfrac{9}{y}}$

60. $\dfrac{4 - \sqrt{x}}{4 + 2\sqrt{x}}$

Solve.

61. Suppose that W is inversely proportional to V. If $W = 20$ when $V = 12$, find W when $V = 15$.

62. Given the diagram shown, approximate to the nearest foot how many feet of walking distance a person saves by cutting across the lawn instead of walking on the sidewalk.

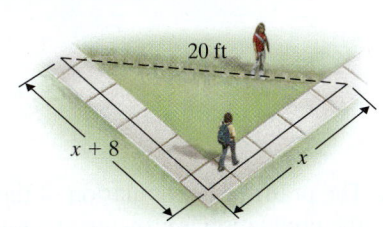

51. see graph

domain:$(-\infty, \infty)$; range:$(-\infty, -1]$

52. see graph

53. see graph

domain:$(-\infty, \infty)$; range:$(-3, \infty)$

54. see graph

55. $y = -\dfrac{1}{2}x$

56. $y = -\dfrac{1}{3}x + \dfrac{5}{3}$

57. $2\sqrt{26}$ units

58. $\left(-4, \dfrac{7}{2}\right)$

59. $\dfrac{3\sqrt{y}}{y}$

60. $\dfrac{8 - 6\sqrt{x} + x}{8 - 2x}$

61. 16

62. 7 ft

63. a. 272 ft

b. 5.12 sec

64. $0 - 2i\sqrt{2}$

65. $0 - 3i$

66. $7 + 24i$

67. $-\dfrac{3}{2} + \dfrac{5}{2}i$

68. $(g \circ h)(x) = x^2 - 6x - 2$

69. $f(x)$ is one-to-one; $f^{-1}(x) = \dfrac{-x + 6}{2}$

70. $\log_5 \dfrac{x^4}{x + 1}$

71. -1

72. $\dfrac{1}{2}\left(\dfrac{\log 4}{\log 3} - 5\right)$; -1.8691

73. 22

74. $\dfrac{43}{21}$

75. $\dfrac{1}{2}$

76. see graph

77. 64,805 prairie dogs

---

63. A stone is thrown upward from a bridge. The stone's height, $s(t)$, in feet above the water t seconds after the stone is thrown is given by the function

$$s(t) = -16t^2 + 32t + 256$$

a. Find the maximum height of the stone.

b. Find the time it takes the stone to hit the water. Round the answer to the nearest hundredth of a second.

Perform the indicated operation and simplify. Write the result in the form $a + bi$.

64. $-\sqrt{-8}$

65. $(12 - 6i) - (12 - 3i)$

66. $(4 + 3i)^2$

67. $\dfrac{1 + 4i}{1 - i}$

68. If $g(x) = x - 7$ and $h(x) = x^2 - 6x + 5$, find $(g \circ h)(x)$.

69. Decide whether $f(x) = 6 - 2x$ is a one-to-one function. If it is, find its inverse.

70. Use properties of logarithms to write the expression as a single logarithm.

$$\log_5 x + 3\log_5 x - \log_5(x + 1)$$

Solve. Give exact solutions.

71. $8^{x-1} = \dfrac{1}{64}$

72. $3^{2x+5} = 4$ Give the exact solution and a 4-decimal place approximation

73. $\log_8(3x - 2) = 2$

74. $\log_4(x + 1) - \log_4(x - 2) = 3$

75. $\ln\sqrt{e} = x$

76. Graph $f(x) = \left(\dfrac{1}{2}\right)^x + 1$.

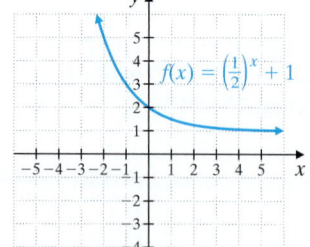

77. The prairie dog population of the Grand Forks area now stands at 57,000 animals. If the population is growing at a rate of 2.6% annually, how many prairie dogs will there be in that area 5 years from now?

Sketch the graph of each equation.

78. $x^2 - y^2 = 36$

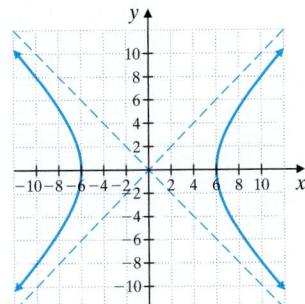

79. $16x^2 + 9y^2 = 144$

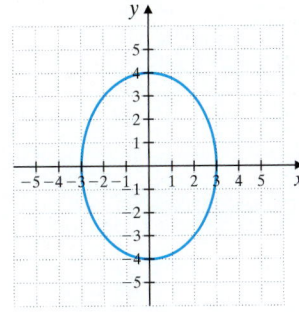

80. $x^2 + y^2 + 6x = 16$

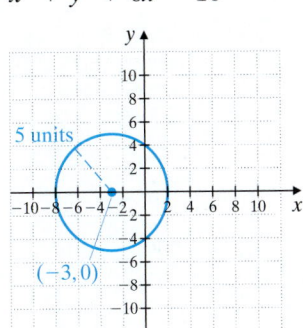

81. Solve the system:

$$\begin{cases} x^2 + y^2 = 26 \\ x^2 - 2y^2 = 23 \end{cases}$$

Answers to Selected Exercises

Chapter R Prealgebra Review

Section R.1

Vocabulary, Readiness & Video Check **1.** prime factorization **3.** prime **5.** factor **7.** No, the natural number 1 is neither prime nor composite. **9.** smallest

Exercise Set R.1 **1.** 1, 3, 9 **3.** 1, 2, 3, 4, 6, 8, 12, 24 **5.** 1, 2, 3, 6, 7, 14, 21, 42 **7.** 1, 2, 4, 5, 8, 10, 16, 20, 40, 80 **9.** 1, 19 **11.** prime
13. composite **15.** prime **17.** composite **19.** composite **21.** $2 \cdot 3 \cdot 3$ **23.** $2 \cdot 2 \cdot 5$ **25.** $2 \cdot 2 \cdot 2 \cdot 7$ **27.** $3 \cdot 3 \cdot 3 \cdot 3$ **29.** $2 \cdot 2 \cdot 3 \cdot 5 \cdot 5$
31. $2 \cdot 2 \cdot 3 \cdot 7 \cdot 7$ **33.** d **35.** 12 **37.** 42 **39.** 60 **41.** 35 **43.** 36 **45.** 80 **47.** 360 **49.** 72 **51.** 120 **53.** 42 **55.** 48 **57.** 360
59. a. $2 \cdot 2 \cdot 2 \cdot 5$ **b.** $2 \cdot 2 \cdot 2 \cdot 5$ **c.** answers may vary **61.** every 35 days **63.** 2520

Section R.2

Vocabulary, Readiness & Video Check **1.** fraction; denominator; numerator **3.** simplest form **5.** $\dfrac{a \cdot c}{b \cdot d}$ **7.** $\dfrac{a \cdot d}{b \cdot c}$ **9.** least common
denominator (LCD) **11.** The fraction is equal to 1. **13.** wrote both the numerator and denominator as products of prime numbers
15. When adding or subtracting fractions, we must have common denominators. When multiplying or dividing fractions, we do not.

Exercise Set R.2 **1.** 1 **3.** 10 **5.** 13 **7.** 0 **9.** undefined **11.** $\dfrac{21}{30}$ **13.** $\dfrac{4}{18}$ **15.** $\dfrac{16}{20}$ **17.** $\dfrac{1}{2}$ **19.** $\dfrac{2}{3}$ **21.** $\dfrac{3}{7}$ **23.** $\dfrac{3}{5}$ **25.** $\dfrac{4}{5}$

27. $\dfrac{11}{8}$ **29.** $\dfrac{30}{61}$ **31.** $\dfrac{8}{11}$ **33.** $\dfrac{3}{8}$ **35.** $\dfrac{1}{2}$ **37.** $\dfrac{6}{7}$ **39.** 15 **41.** $18\dfrac{20}{27}$ **43.** $2\dfrac{28}{29}$ **45.** 1 **47.** $\dfrac{11}{60}$ **49.** $\dfrac{23}{21}$ **51.** $\dfrac{65}{21}$ **53.** $1\dfrac{3}{4}$ **55.** $5\dfrac{1}{6}$

57. $\dfrac{9}{35}$ **59.** $\dfrac{1}{3}$ **61.** $\dfrac{1}{6}$ **63.** $\dfrac{3}{80}$ **65.** $\dfrac{5}{66}$ **67.** $48\dfrac{1}{15}$ **69.** 37 **71.** $10\dfrac{5}{11}$ **73.** $\dfrac{7}{5}$ **75.** $7\dfrac{1}{12}$ **77.** $\dfrac{17}{18}$ **79.** incorrect; $\dfrac{12}{24} = \dfrac{2 \cdot 2 \cdot 3}{2 \cdot 2 \cdot 2 \cdot 3} = \dfrac{1}{2}$

81. incorrect; $\dfrac{2}{7} + \dfrac{9}{7} = \dfrac{11}{7}$ **83.** answers may vary **85.** $\dfrac{1}{12}$ **87.** $\dfrac{6}{11}$ **89.** $7\dfrac{13}{20}$ in. **91.** $\dfrac{3}{50}$ **93.** answers may vary **95.** $\dfrac{2}{25}$

97. answers may vary **99.** $\dfrac{6}{55}$ sq m

Section R.3

Vocabulary, Readiness & Video Check **1.** decimal **3.** vertically **5.** Percent **7.** percent **9.** right **11.** do; do not
13. numerator; denominator

Exercise Set R.3 **1.** $\dfrac{6}{10}$ **3.** $\dfrac{186}{100}$ **5.** $\dfrac{114}{1000}$ **7.** $\dfrac{1231}{10}$ **9.** 6.83 **11.** 27.0578 **13.** 6.5 **15.** 15.22 **17.** 0.12 **19.** 0.2646 **21.** 1.68
23. 5.8 **25.** 56.431 **27.** 67.5 **29.** 70 **31.** 598.23 **33.** 43.274 **35.** 840 **37.** 84.97593 **39.** 0.6 **41.** 0.23 **43.** 0.595 **45.** 98,207.2
47. 12.35 **49.** 0.75 **51.** $0.\overline{3} \approx 0.33$ **53.** 0.4375 **55.** $0.\overline{54} \approx 0.55$ **57.** $4.8\overline{3} \approx 4.83$ **59.** 0.28 **61.** 0.031 **63.** 1.35 **65.** 2
67. 0.9655 **69.** 0.001 **71.** 0.158 **73.** 68% **75.** 87.6% **77.** 100% **79.** 50% **81.** 192% **83.** 0.4% **85.** 78.1%
87. $0.005; \dfrac{1}{200}$ **89.** $0.142; \dfrac{71}{500}$ **91. a.** tenths **b.** thousandths **c.** ones **93.** answers may vary **95.** 182.6 lb **97. a.** 52.9%
b. 52.86% **99.** b, d **101.** 4% **103.** network systems and data communication analysts **105.** 0.35 **107.** answers may vary

Chapter R Vocabulary Check **1.** factor **2.** multiple **3.** composite number **4.** percent **5.** equivalent **6.** improper fraction
7. prime number **8.** simplified **9.** proper fraction **10.** mixed number

Chapter R Review **1.** $2 \cdot 3 \cdot 7$ **2.** $2 \cdot 2 \cdot 2 \cdot 2 \cdot 2 \cdot 5 \cdot 5$ **3.** 60 **4.** 42 **5.** 60 **6.** 70 **7.** $\dfrac{15}{24}$ **8.** $\dfrac{40}{60}$ **9.** $\dfrac{2}{5}$ **10.** $\dfrac{3}{20}$ **11.** 2 **12.** 1
13. $\dfrac{8}{77}$ **14.** $\dfrac{11}{20}$ **15.** $\dfrac{1}{20}$ **16.** $\dfrac{11}{18}$ **17.** $14\dfrac{11}{32}$ **18.** $\dfrac{1}{2}$ **19.** $20\dfrac{17}{30}$ **20.** $2\dfrac{6}{7}$ **21.** $\dfrac{11}{20}$ sq mi **22.** $\dfrac{5}{16}$ sq m **23.** $\dfrac{181}{100}$ **24.** $\dfrac{35}{1000}$
25. 95.118 **26.** 36.785 **27.** 13.38 **28.** 691.573 **29.** 91.2 **30.** 46.816 **31.** 28.6 **32.** 230 **33.** 0.77 **34.** 25.6 **35.** 0.5 **36.** 0.375
37. $0.\overline{36} \approx 0.364$ **38.** $0.8\overline{3} \approx 0.833$ **39.** 0.29 **40.** 0.014 **41.** 39% **42.** 120% **43.** 0.683 **44.** b

Chapter R Test **1.** $2 \cdot 2 \cdot 2 \cdot 3 \cdot 3$ **2.** 180 **3.** $\dfrac{25}{60}$ **4.** $\dfrac{3}{4}$ **5.** $\dfrac{12}{25}$ **6.** $\dfrac{13}{10}$ **7.** $\dfrac{53}{40}$ **8.** $\dfrac{18}{49}$ **9.** $\dfrac{1}{20}$ **10.** $\dfrac{29}{36}$ **11.** $4\dfrac{8}{9}$ **12.** $2\dfrac{5}{22}$ **13.** 55
14. $13\dfrac{13}{20}$ **15.** 45.11 **16.** 65.88 **17.** 12.688 **18.** 320 **19.** 23.73 **20.** 0.875 **21.** $0.1\overline{6} \approx 0.167$ **22.** 0.632 **23.** 9% **24.** 75%
25. $\dfrac{3}{4}$ **26.** $\dfrac{1}{200}$ **27.** $\dfrac{49}{200}$ **28.** $\dfrac{199}{200}$ **29.** $\dfrac{1}{8}$ sq ft **30.** $\dfrac{63}{64}$ sq cm

Chapter 1 Real Numbers and Introduction to Algebra

Section 1.2

Vocabulary, Readiness & Video Check   **1.** whole   **3.** inequality   **5.** real   **7.** 0   **9.** absolute value   **11.** To form a true statement: $0 < 7$.   **13.** 0 belongs to the whole numbers, the integers, the rational numbers, and the real numbers; since 0 is a rational number, it cannot also be an irrational number.

Exercise Set 1.2   **1.** $<$ **3.** $>$ **5.** $=$ **7.** $<$ **9.** $32 < 212$ **11.** $30 \le 45$ **13.** true   **15.** false   **17.** true   **19.** false   **21.** $20 \le 25$ **23.** $6 > 0$ **25.** $-12 < -10$ **27.** $7 < 11$ **29.** $5 \ge 4$ **31.** $15 \ne -2$ **33.** $14{,}494; -282$ **35.** $-27{,}724$ **37.** $475; -195$
39. 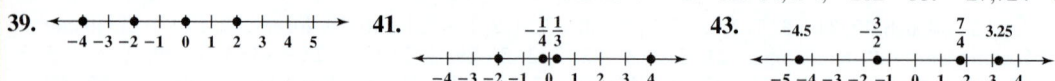   **41.**   **43.**

45. whole, integers, rational, real   **47.** integers, rational, real   **49.** natural, whole, integers, rational, real   **51.** rational, real
53. false   **55.** true   **57.** false   **59.** false   **61.** 8.9   **63.** 20   **65.** $\dfrac{9}{2}$ **67.** $\dfrac{12}{13}$ **69.** $>$ **71.** $=$ **73.** $<$ **75.** $<$
77. 40 million $>$ 14 million, or $40{,}000{,}000 > 14{,}000{,}000$ **79.** 40 million pounds less, or -40 million   **81.** $-0.04 > -26.7$
83. sun   **85.** sun   **87.** answers may vary

Section 1.3

Calculator Explorations   **1.** 125   **3.** 59,049   **5.** 30   **7.** 9857   **9.** 2376

Vocabulary, Readiness & Video Check   **1.** base; exponent   **3.** multiplication   **5.** subtraction   **7.** expression   **9.** expression; variables   **11.** equation   **13.** The order in which we perform operations does matter! We came up with an order of operations to avoid getting more than one answer when evaluating an expression.   **15.** No; the variable was replaced with 0 in the equation to see if a true statement occurred, and it did not.

Exercise Set 1.3   **1.** 243   **3.** 27   **5.** 1   **7.** 5   **9.** 49   **11.** $\dfrac{16}{81}$ **13.** $\dfrac{1}{125}$ **15.** 1.44   **17.** 0.343   **19.** 5^2 sq m   **21.** 17   **23.** 20   **25.** 12
27. 21   **29.** 45   **31.** 0   **33.** $\dfrac{2}{7}$ **35.** 30   **37.** 2   **39.** $\dfrac{7}{18}$ **41.** $\dfrac{27}{10}$ **43.** $\dfrac{7}{5}$ **45.** 32   **47.** $\dfrac{23}{27}$ **49.** 9   **51.** 1   **53.** 1   **55.** 11   **57.** 8
59. 45   **61.** 27   **63.** 132   **65.** $\dfrac{37}{18}$ **67.** solution   **69.** not a solution   **71.** not a solution   **73.** solution   **75.** not a solution
77. solution   **79.** $x + 15$ **81.** $x - 5$ **83.** $\dfrac{x}{4}$ **85.** $3x + 22$ **87.** $1 + 2 = 9 \div 3$ **89.** $3 \ne 4 \div 2$ **91.** $5 + x = 20$ **93.** $7.6x = 17$
95. $13 - 3x = 13$ **97.** no; answers may vary   **99. a.** 64   **b.** 43   **c.** 19   **d.** 22   **101.** 14 in.; 12 sq in.   **103.** 14 in.; 9.01 sq in.
105. Rectangles with the same perimeter can have different areas.   **107.** $(20 - 4) \cdot 4 \div 2$ **109. a.** expression   **b.** equation
c. equation   **d.** expression   **e.** expression   **111.** answers may vary   **113.** answers may vary, for example, $-2(5) - 1$.

Section 1.4

Vocabulary, Readiness & Video Check   **1.** 0   **3.** a **5.** absolute values   **7.** Example 12 is an example of the opposite of the *absolute value* of $-a$, not the opposite of $-a$. The absolute value of $-a$ is positive, so its opposite is negative. Therefore the answers to Examples 11 and 12 have different signs.   **9.** Depths below the surface; the diver's position is 231 feet below the surface.

Exercise Set 1.4   **1.** 3   **3.** -14 **5.** 1   **7.** -12 **9.** -5 **11.** -12 **13.** -4 **15.** 7   **17.** -2 **19.** 0   **21.** -19 **23.** 31   **25.** -47
27. -2.1 **29.** 38   **31.** -13.1 **33.** $\dfrac{1}{4}$ **35.** $-\dfrac{3}{16}$ **37.** $-\dfrac{13}{10}$ **39.** -8 **41.** -8 **43.** -59 **45.** -9 **47.** 5   **49.** 11   **51.** -18 **53.** 19
55. -7 **57.** -26 **59.** -6 **61.** 2   **63.** 0   **65.** -6 **67.** -2 **69.** 7   **71.** 7.9   **73.** $5z$ **75.** $\dfrac{2}{3}$ **77.** -70 **79.** 3   **81.** 19   **83.** -10
85. $0 + (-215) + (-16) = -231$; 231 ft below the surface   **87.** $107°F$ **89.** -95 m   **91.** -8 **93.** \$16.9 million   **95.** July
97. October   **99.** $4.7°F$ **101.** answers may vary   **103.** -3 **105.** -22 **107.** true   **109.** false   **111.** answers may vary

Section 1.5

Vocabulary, Readiness & Video Check   **1.** $a + (-b)$; b   **3.** $-10 - (-14)$; d   **5.** addition, opposite   **7.** There's a minus sign in the numerator and the replacement value is negative (notice parentheses are used around the replacement value), and it's always good to be careful when working with negative signs.   **9.** This means that the overall vertical altitude change of the jet is actually a decrease in altitude from when the Example started.

Exercise Set 1.5   **1.** -10 **3.** -5 **5.** 19   **7.** 11   **9.** -8 **11.** -11 **13.** 37   **15.** 5   **17.** -71 **19.** 0   **21.** $\dfrac{2}{11}$ **23.** -6.4 **25.** 4.1
27. $-\dfrac{1}{6}$ **29.** $-\dfrac{11}{12}$ **31.** 8.92   **33.** -8.92 **35.** 13   **37.** -5 **39.** -1 **41.** -23 **43.** -26 **45.** -24 **47.** 3   **49.** -45 **51.** -4

53. 13 **55.** 6 **57.** 9 **59.** −9 **61.** $\dfrac{7}{5}$ **63.** −7 **65.** 21 **67.** $\dfrac{1}{4}$ **69.** not a solution **71.** not a solution **73.** solution **75.** 263°F

77. 35,653 ft **79.** 30° **81.** −308 ft **83.** 19,852 ft **85.** 130° **87.** −5 + x **89.** −20 − x **91.** −4.4°, 2.6°, 12°, 23.5°, 15.3° **93.** May **95.** answers may vary **97.** 16 **99.** −20 **101.** true; answers may vary **103.** false; answers may vary **105.** negative, −30,387

Integrated Review **1.** negative **2.** negative **3.** positive **4.** 0 **5.** positive **6.** 0 **7.** positive **8.** positive **9.** $-\dfrac{1}{7}; \dfrac{1}{7}$

10. $\dfrac{12}{5}; \dfrac{12}{5}$ **11.** 3; 3 **12.** $-\dfrac{9}{11}; \dfrac{9}{11}$ **13.** −42 **14.** 10 **15.** 2 **16.** −18 **17.** −7 **18.** −39 **19.** −2 **20.** −9 **21.** −3.4 **22.** −9.8

23. $-\dfrac{25}{28}$ **24.** $-\dfrac{5}{24}$ **25.** −4 **26.** −24 **27.** 6 **28.** 20 **29.** 6 **30.** 61 **31.** −6 **32.** −16 **33.** −19 **34.** −13 **35.** −4 **36.** −1

37. $\dfrac{13}{20}$ **38.** $-\dfrac{29}{40}$ **39.** 4 **40.** 9 **41.** −1 **42.** −3 **43.** 8 **44.** 10 **45.** 47 **46.** $\dfrac{2}{3}$

Section 1.6

Calculator Explorations **1.** 38 **3.** −441 **5.** 490 **7.** 54,499 **9.** 15,625

Vocabulary, Readiness & Video Check **1.** negative **3.** positive **5.** 0 **7.** 0 **9.** The parentheses, or lack of them, determine the base of the expression. In Example 6, $(-2)^4$, the base is −2 and all of −2 is raised to the 4th power. In Example 7, -2^4, the base is 2 and only 2 is raised to the 4th power. **11.** Yes; because division of real numbers is defined in terms of multiplication. **13.** Yes; a true statement results when x is replaced with 5.

Exercise Set 1.6 **1.** −24 **3.** −2 **5.** 50 **7.** −45 **9.** $\dfrac{3}{10}$ **11.** −7 **13.** −15 **15.** 0 **17.** 16 **19.** −16 **21.** $\dfrac{9}{16}$ **23.** −0.49

25. $\dfrac{3}{2}$ **27.** $-\dfrac{1}{14}$ **29.** $-\dfrac{11}{3}$ **31.** $\dfrac{1}{0.2}$ **33.** −9 **35.** −4 **37.** 0 **39.** undefined **41.** $-\dfrac{18}{7}$ **43.** 160 **45.** 64 **47.** $-\dfrac{8}{27}$ **49.** 3

51. −15 **53.** −125 **55.** −0.008 **57.** $\dfrac{2}{3}$ **59.** $\dfrac{20}{27}$ **61.** 0.84 **63.** −40 **65.** 81 **67.** −1 **69.** −121 **71.** −1 **73.** −19 **75.** 90

77. −84 **79.** −5 **81.** $-\dfrac{9}{2}$ **83.** 18 **85.** 17 **87.** −20 **89.** 16 **91.** 2 **93.** $-\dfrac{34}{7}$ **95.** 0 **97.** $\dfrac{6}{5}$ **99.** $\dfrac{3}{2}$ **101.** $-\dfrac{5}{38}$ **103.** 3

105. −1 **107.** undefined **109.** $-\dfrac{22}{9}$ **111.** solution **113.** not a solution **115.** solution **117.** $-71 \cdot x$ or $-71x$ **119.** $-16 - x$

121. $-29 + x$ **123.** $\dfrac{x}{-33}$ or $x \div (-33)$ **125.** $3 \cdot (-4) = -12$; a loss of 12 yd **127.** $5(-20) = -100$; a depth of 100 ft

129. true **131.** false **133.** −162°F **135.** −$11 million per month **137.** answers may vary **139.** 1, −1; answers may vary

141. $\dfrac{0}{5} - 7 = -7$ **143.** $-8(-5) + (-1) = 39$

Section 1.7

Vocabulary, Readiness & Video Check **1.** commutative property of addition **3.** distributive property **5.** associative property of addition **7.** opposites or additive inverses **9.** 2 is outside the parentheses, so the point is made that you should only distribute the −9 to the terms within the parentheses and not also to the 2.

Exercise Set 1.7 **1.** $16 + x$ **3.** $y \cdot (-4)$ **5.** yx **7.** $13 + 2x$ **9.** $x \cdot (yz)$ **11.** $(2 + a) + b$ **13.** $(4a) \cdot b$ **15.** $a + (b + c)$

17. $17 + b$ **19.** $24y$ **21.** y **23.** $26 + a$ **25.** $-72x$ **27.** s **29.** $-\dfrac{5}{2}x$ **31.** $4x + 4y$ **33.** $9x - 54$ **35.** $6x + 10$ **37.** $28x - 21$

39. $18 + 3x$ **41.** $-2y + 2z$ **43.** $-y - \dfrac{5}{3}$ **45.** $5x + 20m + 10$ **47.** $8m - 4n$ **49.** $-5x - 2$ **51.** $-r + 3 + 7p$ **53.** $3x + 4$

55. $-x + 3y$ **57.** $6r + 8$ **59.** $-36x - 70$ **61.** $-1.6x - 2.5$ **63.** $4(1 + y)$ **65.** $11(x + y)$ **67.** $-1(5 + x)$ **69.** $30(a + b)$

71. commutative property of multiplication **73.** associative property of addition **75.** commutative property of addition **77.** associative property of multiplication **79.** identity element for addition **81.** distributive property **83.** multiplicative inverse property **85.** identity element for multiplication **87.** $-8; \dfrac{1}{8}$ **89.** $-x; \dfrac{1}{x}$ **91.** $2x; -2x$ **93.** false **95.** no **97.** yes **99.** yes

101. yes **103. a.** commutative property of addition **b.** commutative property of addition **c.** associative property of addition **105.** answers may vary **107.** answers may vary

Section 1.8

Vocabulary, Readiness & Video Check **1.** expression; term **3.** combine like terms **5.** like; unlike **7.** Although these terms have exactly the same variables, the exponents on each are not exactly the same—the exponents on x differ in each term. **9.** −1

Exercise Set 1.8 **1.** -7 **3.** 1 **5.** 17 **7.** like **9.** unlike **11.** like **13.** $15y$ **15.** $13w$ **17.** $-7b - 9$ **19.** $-m - 6$ **21.** -8
23. $7.2x - 5.2$ **25.** $k - 6$ **27.** $-15x + 18$ **29.** $4x - 3$ **31.** $5x^2$ **33.** -11 **35.** $1.3x + 3.5$ **37.** $5y + 20$ **39.** $-2x - 4$
41. $-10x + 15y - 30$ **43.** $-3x + 2y - 1$ **45.** $7d - 11$ **47.** 16 **49.** $x + 5$ **51.** $x + 2$ **53.** $2k + 10$ **55.** $-3x + 5$
57. $2x + 14$ **59.** $3y + \dfrac{5}{6}$ **61.** $-22 + 24x$ **63.** $0.9m + 1$ **65.** $10 - 6x - 9y$ **67.** $-x - 38$ **69.** $5x - 7$ **71.** $10x - 3$
73. $-4x - 9$ **75.** $-4m - 3$ **77.** $2x - 4$ **79.** $\dfrac{3}{4}x + 12$ **81.** $12x - 2$ **83.** $8x + 48$ **85.** $x - 10$ **87.** balanced **89.** balanced
91. answers may vary **93.** $(18x - 2)$ ft **95.** $(15x + 23)$ in. **97.** answers may vary

Chapter 1 Vocabulary Check **1.** inequality symbols **2.** equation **3.** absolute value **4.** variable **5.** opposites **6.** numerator
7. solution **8.** reciprocals **9.** base; exponent **10.** numerical coefficient **11.** denominator **12.** grouping symbols **13.** term
14. like terms **15.** unlike terms

Chapter 1 Review **1.** $<$ **2.** $>$ **3.** $>$ **4.** $>$ **5.** $<$ **6.** $>$ **7.** $=$ **8.** $=$ **9.** $>$ **10.** $<$ **11.** $4 \geq -3$ **12.** $6 \neq 5$ **13.** $0.03 < 0.3$
14. $1729 < 2870$ **15. a.** $1, 3$ **b.** $0, 1, 3$ **c.** $-6, 0, 1, 3$ **d.** $-6, 0, 1, 1\dfrac{1}{2}, 3, 9.62$ **e.** π **f.** all numbers in set **16. a.** $2, 5$ **b.** $2, 5$
c. $-3, 2, 5$ **d.** $-3, -1.6, 2, 5, \dfrac{11}{2}, 15.1$ **e.** $\sqrt{5}, 2\pi$ **f.** all numbers in set **17.** Friday **18.** Wednesday **19.** c **20.** b **21.** 37
22. 41 **23.** $\dfrac{18}{7}$ **24.** 80 **25.** $20 - 12 = 2 \cdot 4$ **26.** $\dfrac{9}{2} > -5$ **27.** 18 **28.** 108 **29.** 5 **30.** 24 **31.** $63°$ **32.** $105°$ **33.** solution
34. not a solution **35.** 9 **36.** $-\dfrac{2}{3}$ **37.** -2 **38.** 7 **39.** -11 **40.** -17 **41.** $-\dfrac{3}{16}$ **42.** -5 **43.** -13.9 **44.** 3.9 **45.** -14
46. -11.5 **47.** 5 **48.** -11 **49.** -19 **50.** 4 **51.** a **52.** a **53.** $\$51$ **54.** $\$54$ **55.** $-\dfrac{1}{6}$ **56.** $\dfrac{5}{3}$ **57.** -48 **58.** 28 **59.** 3
60. -14 **61.** -36 **62.** 0 **63.** undefined **64.** $-\dfrac{1}{2}$ **65.** commutative property of addition **66.** identity element for multiplication
67. distributive property **68.** additive inverse property **69.** associative property of addition **70.** commutative property of
multiplication **71.** distributive property **72.** associative property of multiplication **73.** multiplicative inverse property
74. identity element for addition **75.** commutative property of addition **76.** distributive property **77.** $6x$ **78.** $-11.8z$
79. $4x - 2$ **80.** $2y + 3$ **81.** $3n - 18$ **82.** $4w - 6$ **83.** $-6x + 7$ **84.** $-0.4y + 2.3$ **85.** $3x - 7$ **86.** $5x + 5.6$ **87.** $<$ **88.** $>$
89. -15.3 **90.** -6 **91.** -80 **92.** -5 **93.** $-\dfrac{1}{4}$ **94.** 0.15 **95.** 16 **96.** 16 **97.** -5 **98.** 9 **99.** $-\dfrac{5}{6}$ **100.** undefined **101.** $16x - 41$
102. $18x - 12$

Chapter 1 Test **1.** $|-7| > 5$ **2.** $9 + 5 \geq 4$ **3.** -5 **4.** -11 **5.** -14 **6.** -39 **7.** 12 **8.** -2 **9.** undefined **10.** -8 **11.** $-\dfrac{1}{3}$
12. $4\dfrac{5}{8}$ **13.** $\dfrac{51}{40}$ **14.** -32 **15.** -48 **16.** 3 **17.** 0 **18.** $>$ **19.** $>$ **20.** $>$ **21.** $=$ **22. a.** $1, 7$ **b.** $0, 1, 7$ **c.** $-5, -1, 0, 1, 7$
d. $-5, -1, \dfrac{1}{4}, 0, 1, 7, 11.6$ **e.** $\sqrt{7}, 3\pi$ **f.** $-5, -1, \dfrac{1}{4}, 0, 1, 7, 11.6, \sqrt{7}, 3\pi$ **23.** 40 **24.** 12 **25.** 22 **26.** -1 **27.** associative property
of addition **28.** commutative property of multiplication **29.** distributive property **30.** multiplicative inverse **31.** 9
32. -3 **33.** second down **34.** yes **35.** $17°$ **36.** $\$420$ **37.** $y - 10$ **38.** $5.9x + 1.2$ **39.** $-2x + 10$ **40.** $-15y + 1$

Chapter 2 Equations, Inequalities, and Problem Solving

Section 2.1

Vocabulary, Readiness & Video Check **1.** expression **3.** equation **5.** expression; equation **7.** Equivalent **9.** 2 **11.** 12
13. 17 **15.** both sides **17.** $\dfrac{1}{7}x$

Exercise Set 2.1 **1.** 3 **3.** -2 **5.** -14 **7.** 0.5 **9.** $\dfrac{1}{4}$ **11.** $\dfrac{5}{12}$ **13.** -3 **15.** -9 **17.** -10 **19.** 2 **21.** -7 **23.** -1 **25.** -9 **27.** -12
29. $-\dfrac{1}{2}$ **31.** 11 **33.** 21 **35.** 25 **37.** -3 **39.** -0.7 **41.** 11 **43.** 13 **45.** -30 **47.** -0.4 **49.** -7 **51.** $-\dfrac{1}{3}$ **53.** -17.9
55. $20 - p$ **57.** $(10 - x)$ ft **59.** $(180 - x)°$ **61.** $n - 29{,}000$ **63.** $7x$ sq mi **65.** $\dfrac{8}{5}$ **67.** $\dfrac{1}{2}$ **69.** -9 **71.** x **73.** y **75.** x
77. answers may vary **79.** 4 **81.** answers may vary **83.** $(173 - 3x)°$ **85.** answers may vary **87.** -145.478

Section 2.2

Vocabulary, Readiness & Video Check **1.** multiplication **3.** equation; expression **5.** Equivalent **7.** 9 **9.** 2 **11.** -5
13. same **15.** $(x + 1) + (x + 3) = 2x + 4$

Exercise Set 2.2 **1.** 4 **3.** 0 **5.** 12 **7.** -12 **9.** 3 **11.** 2 **13.** 0 **15.** 6.3 **17.** 10 **19.** -20 **21.** 0 **23.** -9 **25.** 1 **27.** -30
29. 3 **31.** $\dfrac{10}{9}$ **33.** -1 **35.** -4 **37.** $-\dfrac{1}{2}$ **39.** 0 **41.** 4 **43.** $-\dfrac{1}{14}$ **45.** 0.21 **47.** 5 **49.** 6 **51.** -5.5 **53.** -5 **55.** 0 **57.** -3
59. $-\dfrac{9}{28}$ **61.** $\dfrac{14}{3}$ **63.** -9 **65.** -2 **67.** $\dfrac{11}{2}$ **69.** $-\dfrac{1}{4}$ **71.** $\dfrac{9}{10}$ **73.** $-\dfrac{17}{20}$ **75.** -16 **77.** $2x + 2$ **79.** $2x + 2$ **81.** $5x + 20$
83. $7x - 12$ **85.** $12z + 44$ **87.** 1 **89.** -48 **91.** answers may vary **93.** answers may vary **95.** 2

Section 2.3

Calculator Explorations **1.** solution **3.** not a solution **5.** solution

Vocabulary, Readiness & Video Check **1.** equation **3.** expression **5.** expression **7.** equation **9.** 3; distributive property, addition property of equality, multiplication property of equality **11.** The number of decimal places in each number helps you determine what power of 10 you can multiply through by so you are no longer dealing with decimals.

Exercise Set 2.3 **1.** -6 **3.** 3 **5.** 1 **7.** $\dfrac{3}{2}$ **9.** 0 **11.** -1 **13.** 4 **15.** -4 **17.** -3 **19.** 2 **21.** 50 **23.** 1 **25.** $\dfrac{7}{3}$ **27.** 0.2
29. all real numbers **31.** no solution **33.** no solution **35.** all real numbers **37.** 18 **39.** $\dfrac{19}{9}$ **41.** $\dfrac{14}{3}$ **43.** 13 **45.** 4
47. all real numbers **49.** $-\dfrac{3}{5}$ **51.** -5 **53.** 10 **55.** no solution **57.** 3 **59.** -17 **61.** $\dfrac{7}{5}$ **63.** $-\dfrac{1}{50}$ **65.** $(6x - 8)$ m
67. $-8 - x$ **69.** $-3 + 2x$ **71.** $9(x + 20)$ **73. a.** all real numbers **b.** answers may vary **c.** answers may vary **75.** a
77. b **79.** c **81.** answers may vary **83. a.** $x + x + x + 2x + 2x = 28$ **b.** $x = 4$ **c.** x cm $= 4$ cm; $2x$ cm $= 8$ cm
85. answers may vary **87.** 15.3 **89.** -0.2

Integrated Review **1.** 6 **2.** -17 **3.** 12 **4.** -26 **5.** -3 **6.** -1 **7.** $\dfrac{27}{2}$ **8.** $\dfrac{25}{2}$ **9.** 8 **10.** -64 **11.** 2 **12.** -3 **13.** 5
14. -1 **15.** 2 **16.** 2 **17.** -2 **18.** -2 **19.** $-\dfrac{5}{6}$ **20.** $\dfrac{1}{6}$ **21.** 1 **22.** 6 **23.** 4 **24.** 1 **25.** $\dfrac{9}{5}$ **26.** $-\dfrac{6}{5}$ **27.** all real numbers
28. all real numbers **29.** 0 **30.** -1.6 **31.** $\dfrac{4}{19}$ **32.** $-\dfrac{5}{19}$ **33.** $\dfrac{7}{2}$ **34.** $-\dfrac{1}{4}$ **35.** no solution **36.** no solution **37.** $\dfrac{7}{6}$ **38.** $\dfrac{1}{15}$

Section 2.4

Vocabulary, Readiness & Video Check **1.** $2x$; $2x - 31$ **3.** $x + 5$; $2(x + 5)$ **5.** $20 - y$; $\dfrac{20 - y}{3}$ or $(20 - y) \div 3$ **7.** in the statement of the application **9.** That the 3 angle measures are consecutive even integers and that they sum to $180°$.

Exercise Set 2.4 **1.** $2x + 7 = x + 6$; -1 **3.** $3x - 6 = 2x + 8$; 14 **5.** -25 **7.** $-\dfrac{3}{4}$ **9.** 3 in.; 6 in.; 16 in. **11.** 1st piece: 5 in.;
2nd piece: 10 in.; 3rd piece: 25 in. **13.** Pennsylvania: 494 million lb; New York: 720 million lb **15.** 172 mi **17.** 25 mi
19. 1st angle: $37.5°$; 2nd angle: $37.5°$; 3rd angle: $105°$ **21.** A: $60°$; B: $120°$; C: $120°$; D: $60°$ **23.** $3x + 3$ **25.** $x + 2$; $x + 4$; $2x + 4$
27. $x + 1$; $x + 2$; $x + 3$; $4x + 6$ **29.** $x + 2$; $x + 4$; $2x + 6$ **31.** 234, 235 **33.** Belgium: 32; France: 33; Spain: 34 **35.** 5 ft, 12 ft
37. Maglev: 361 mph; TGV: 357.2 mph **39.** $43°$, $137°$ **41.** $58°$, $60°$, $62°$ **43.** 1 **45.** 280 mi **47.** Stanford: 20; Wisconsin: 14
49. Montana: 56 counties; California: 58 counties **51.** Neptune: 14 satellites; Uranus: 27 satellites; Saturn: 62 satellites **53.** -16
55. Sahara: 3,500,000 sq mi; Gobi: 500,000 sq mi **57.** Jamaica: 4; Cuba: 5; New Zealand: 6 **59.** females: 30,184; males: 28,604
61. $34.5°$; $34.5°$; $111°$ **63.** Hawaii **65.** Florida: \$56 million; California: \$50 million **67.** answers may vary **69.** 34 **71.** 225π
73. 15 ft by 24 ft **75.** 5400 chirps per hour; 129,600 chirps per day; 47,304,000 chirps per year **77.** answers may vary
79. answers may vary **81.** c

Section 2.5

Vocabulary, Readiness & Video Check **1.** relationships **3.** To show that the process of solving this equation for x—dividing both sides by 5, the coefficient of x—is the same process used to solve a formula for a specific variable. Treat whatever is multiplied by that specific variable as the coefficient—the coefficient is all the factors except that specific variable.

Exercise Set 2.5 **1.** $h = 3$ **3.** $h = 3$ **5.** $h = 20$ **7.** $c = 12$ **9.** $r = 2.5$ **11.** $h = \dfrac{f}{5g}$ **13.** $w = \dfrac{V}{lh}$ **15.** $y = 7 - 3x$
17. $R = \dfrac{A - P}{PT}$ **19.** $A = \dfrac{3V}{h}$ **21.** $a = P - b - c$ **23.** $h = \dfrac{S - 2\pi r^2}{2\pi r}$ **25.** 120 ft **27. a.** area: 480 sq in.; perimeter: 120 in.
b. frame: perimeter; glass: area **29. a.** area: 103.5 sq ft; perimeter: 41 ft **b.** baseboard: perimeter; carpet: area **31.** $-10°$C
33. 6.25 hr **35.** length: 78 ft; width: 52 ft **37.** 18 ft, 36 ft, 48 ft **39.** 137.5 mi **41.** $61.5°$F **43.** 60 chirps per minute **45.** increases
47. 96 piranhas **49.** 2 bags **51.** one 16-in. pizza **53.** 4.65 min **55.** 13 in. **57.** 2.25 hr **59.** 12,090 ft **61.** $50°$C **63.** 515,509.5 cu in.
65. 449 cu in. **67.** $333°$F **69.** 0.32 **71.** 2.00 or 2 **73.** 17% **75.** 720% **77.** $V = G(N - R)$ **79.** multiplies the volume by 8;
answers may vary **81.** $53\dfrac{1}{3}$ **83.** $\bigcirc = \dfrac{\triangle - \square}{\blacksquare}$ **85.** 44.3 sec **87.** $P = 3,200,000$ **89.** $V = 113.1$

Section 2.6

Vocabulary, Readiness & Video Check   **1.** no   **3.** yes   **5. a.** equals; $=$ **b.** multiplication; $\cdot$ **c.** Drop the percent symbol and move the decimal point two places to the left.   **7.** You must first find the actual amount of increase in price by subtracting the original price from the new price.

Exercise Set 2.6   **1.** 11.2   **3.** 55%   **5.** 180   **7.** 15%   **9.** 310,500   **11.** discount: $1480; new price: $17,020   **13.** $46.58   **15.** 9.3%   **17.** 42.9%   **19.** $104   **21.** $42,500   **23.** 2 gal   **25.** 7 lb   **27.** 4.6   **29.** 50   **31.** 30%   **33.** 71%   **35.** 189,670   **37.** 59%, 5%, 27%, 2%   **39.** 75%   **41.** $3900   **43.** 300%   **45.** mark-up: $0.11; new price: $2.31   **47.** 400 oz   **49.** 13.4%   **51.** 120 employees   **53.** decrease: $64; sale price: $192   **55.** 854 thousand Scoville units   **57.** 132 million adults   **59.** 400 oz   **61.** $>$ **63.** $=$ **65.** $>$ **67.** no; answers may vary   **69.** 9.6%   **71.** 26.9%; yes   **73.** 17.1%

Section 2.7

Vocabulary, Readiness & Video Check   **1.** expression   **3.** inequality   **5.** equation   **7.** -5 **9.** 4.1   **11.** An open circle indicates $>$ or $<$; a closed circle indicates $\geq$ or $\leq$.   **13.** $\{x \mid x \geq -2\}$ **15.** is greater than; $>$

Exercise Set 2.7   **1.**   **3.**   **5.**   **7.**

9.   **11.**   **13.** $\{x \mid x \geq -5\}$

15. $\{y \mid y < 9\}$ **17.** $\{x \mid x > -3\}$ **19.** $\{x \mid x \leq 1\}$

21. $\{x \mid x < -3\}$ **23.** $\{x \mid x \geq -2\}$ **25.** $\{x \mid x < 0\}$

27. $\left\{y \mid y \geq -\dfrac{8}{3}\right\}$ **29.** $\{y \mid y > 3\}$ **31.** $\{x \mid x > -15\}$ **33.** $\{x \mid x \geq -11\}$

35. $\left\{x \mid x > \dfrac{1}{4}\right\}$ **37.** $\{y \mid y \geq -12\}$ **39.** $\{z \mid z < 0\}$ **41.** $\{x \mid x > -3\}$ **43.** $\left\{x \mid x \geq -\dfrac{2}{3}\right\}$ **45.** $\{x \mid x \leq -2\}$

47. $\{x \mid x > -13\}$ **49.** $\{x \mid x \leq -8\}$ **51.** $\{x \mid x > 4\}$ **53.** $\left\{x \mid x \leq \dfrac{5}{4}\right\}$ **55.** $\left\{x \mid x > \dfrac{8}{3}\right\}$ **57.** $\{x \mid x \geq 0\}$ **59.** all numbers greater than -10 **61.** 35 cm   **63.** at least 193   **65.** 86 people   **67.** 35 min   **69.** 81   **71.** 1   **73.** $\dfrac{49}{64}$ **75.** 2006   **77.** 2012 and 2013   **79.** 2008   **81.** $>$ **83.** $\geq$ **85.** when multiplying or dividing by a negative number   **87.** final exam score ≥ 78.5

Chapter 2 Vocabulary Check   **1.** linear equation in one variable   **2.** equivalent equations   **3.** formula   **4.** linear inequality in one variable   **5.** all real numbers   **6.** no solution   **7.** the same   **8.** reversed

Chapter 2 Review   **1.** 4   **2.** -3 **3.** 6   **4.** -6 **5.** 0   **6.** -9 **7.** -23 **8.** 28   **9.** b   **10.** a   **11.** b   **12.** c   **13.** -12 **14.** 4   **15.** 0   **16.** -7 **17.** 0.75   **18.** -3 **19.** -6 **20.** -1 **21.** -1 **22.** $\dfrac{3}{2}$ **23.** $-\dfrac{1}{5}$ **24.** 7   **25.** $3x + 3$ **26.** $2x + 6$ **27.** -4 **28.** -4 **29.** 2   **30.** -3 **31.** no solution   **32.** no solution   **33.** $\dfrac{3}{4}$ **34.** $-\dfrac{8}{9}$ **35.** 20   **36.** $-\dfrac{6}{23}$ **37.** $\dfrac{23}{7}$ **38.** $-\dfrac{2}{5}$ **39.** 102   **40.** 0.25   **41.** 6665.5 in.   **42.** short piece: 4 ft; long piece: 8 ft   **43.** national battlefields: 11; national memorials: 29   **44.** $-39, -38, -37$ **45.** 3   **46.** -4 **47.** $w = 9$ **48.** $h = 4$ **49.** $m = \dfrac{y - b}{x}$ **50.** $s = \dfrac{r + 5}{vt}$ **51.** $x = \dfrac{2y - 7}{5}$ **52.** $y = \dfrac{2 + 3x}{6}$ **53.** $\pi = \dfrac{C}{D}$ **54.** $\pi = \dfrac{C}{2r}$ **55.** 15 m   **56.** 18 ft by 12 ft   **57.** 1 hr 20 min   **58.** 45°C   **59.** 20%   **60.** 70%   **61.** 110   **62.** 1280   **63.** mark-up: $209; new price: $2109   **64.** 91,800 businesses   **65.** 40% solution: 10 gal; 10% solution: 20 gal   **66.** 32.0% increase   **67.** 18%   **68.** swerving into another lane   **69.** 966 customers   **70.** no; answers may vary   **71.**   **72.**   **73.** $\{x \mid x \leq 1\}$ **74.** $\{x \mid x > -5\}$ **75.** $\{x \mid x \leq 10\}$ **76.** $\{x \mid x < -4\}$ **77.** $\{x \mid x < -4\}$ **78.** $\{x \mid x \leq 4\}$ **79.** $\{y \mid y > 9\}$ **80.** $\{y \mid y \geq -15\}$ **81.** $\left\{x \mid x < \dfrac{7}{4}\right\}$ **82.** $\left\{x \mid x \leq \dfrac{19}{3}\right\}$ **83.** $2500   **84.** score must be less than 83   **85.** 4   **86.** -14 **87.** $-\dfrac{3}{2}$ **88.** 21   **89.** all real numbers   **90.** no solution   **91.** -13 **92.** shorter piece: 4 in.; longer piece: 19 in.   **93.** $h = \dfrac{3V}{A}$ **94.** 22.1   **95.** 160   **96.** 20%   **97.** $\{x \mid x > 9\}$ **98.** $\{x \mid x > -4\}$ **99.** $\{x \mid x \leq 0\}$

Chapter 2 Test   **1.** -5 **2.** 8   **3.** $\dfrac{7}{10}$ **4.** 0   **5.** 27   **6.** $-\dfrac{19}{6}$ **7.** 3   **8.** $\dfrac{3}{11}$ **9.** 0.25   **10.** $\dfrac{25}{7}$ **11.** no solution   **12.** 21   **13.** 7 gal   **14.** $x = 6$ **15.** $h = \dfrac{V}{\pi r^2}$ **16.** $y = \dfrac{3x - 10}{4}$ **17.** $\{x \mid x \leq -2\}$ **18.** $\{x \mid x < 4\}$ **19.** $\{x \mid x \leq -8\}$ **20.** $\{x \mid x \geq 11\}$ **21.** $\left\{x \mid x > \dfrac{2}{5}\right\}$ **22.** 552   **23.** 40%   **24.** 401,802   **25.** California: 1107; Ohio: 720

Cumulative Review   1. true; Sec. 1.2, Ex. 3   **2.** false; Sec. 1.2   **3.** true; Sec. 1.2, Ex. 4   **4.** true; Sec. 1.2   **5.** false; Sec. 1.2, Ex. 5
6. true; Sec. 1.2   **7.** true; Sec. 1.2, Ex. 6   **8.** true; Sec. 1.2   **9. a.** $<$ **b.** $=$ **c.** $>$ **d.** $<$ **e.** $>$; Sec. 1.2, Ex. 13   **10. a.** 5   **b.** 8
c. $\frac{2}{3}$; Sec. 1.2   **11.** $\frac{8}{3}$; Sec. 1.3, Ex. 6   **12.** 33; Sec. 1.3   **13.** -19; Sec. 1.4, Ex. 6   **14.** -10; Sec. 1.4   **15.** 8; Sec. 1.4, Ex. 7   **16.** 10; Sec. 1.4
17. -0.3; Sec. 1.4, Ex. 8   **18.** 0; Sec. 1.4   **19. a.** -12 **b.** -3; Sec. 1.5, Ex. 7   **20. a.** 5   **b.** $\frac{2}{3}$ **c.** a **d.** -3; Sec. 1.5   **21. a.** 0   **b.** -24
c. 90; Sec. 1.6, Ex. 7   **22. a.** -11.1 **b.** $-\frac{1}{5}$ **c.** $\frac{3}{4}$; Sec. 1.5   **23. a.** -6 **b.** 7   **c.** -5; Sec. 1.6, Ex. 10   **24. a.** -0.36 **b.** $\frac{6}{17}$; Sec. 1.6
25. $15 - 10z$; Sec. 1.7, Ex. 8   **26.** $2y - 6x + 8$; Sec. 1.7   **27.** $3x + 17$; Sec. 1.7, Ex. 12   **28.** $2x + 8$; Sec. 1.7   **29. a.** unlike   **b.** like
c. like   **d.** like   **e.** like; Sec. 1.8, Ex. 2   **30. a.** -4 **b.** 9   **c.** $\frac{10}{63}$; Sec. 1.6   **31.** $-2x - 1$; Sec. 1.8, Ex. 15   **32.** $-15x - 2$; Sec. 1.8
33. 17; Sec. 2.1, Ex. 1   **34.** $-\frac{1}{6}$; Sec. 2.1   **35.** -10; Sec. 2.2, Ex. 7   **36.** 3; Sec. 2.3   **37.** 0; Sec. 2.3, Ex. 4   **38.** 72; Sec. 2.2
39. Republicans: 233; Democrats: 199; Sec. 2.4, Ex. 4   **40.** 5; Sec. 2.3   **41.** 79.2 yr; Sec. 2.5, Ex. 1   **42.** 6; Sec. 2.4   **43.** 87.5%;
Sec. 2.6, Ex. 1   **44.** $\frac{C}{2\pi} = r$; Sec. 2.5   **45.** $-\frac{9}{10}$; Sec. 2.2, Ex. 10   **46.** $\{x \mid x > 5\}$; Sec. 2.7   **47.** 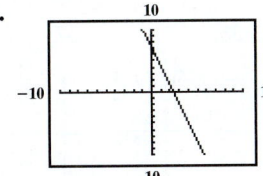; Sec. 2.7,
Ex. 2   **48.** $\{x \mid x \le -10\}$; Sec. 2.7   **49.** $\{x \mid x \ge 1\}$; Sec. 2.7, Ex. 9   **50.** $\{x \mid x \le -3\}$; Sec. 2.7

Chapter 3 Graphing Equations and Inequalities

Section 3.1

Vocabulary, Readiness & Video Check   1. x-axis   **3.** origin   **5.** x-coordinate; y-coordinate   **7.** solution   **9.** 2003; 9.4
11. Paired data, which can be written as ordered pairs and graphed.

Exercise Set 3.1   1. September   **3.** 77   **5.** $\frac{2}{77}$ **7.** Tokyo, Japan; about 39.4 million or 39,400,000   **9.** Mexico City: 22.2 million or
22,200,000   **11.** approximately 1 million   **13.** 7.6 goals/game   **15.** 2003

17. 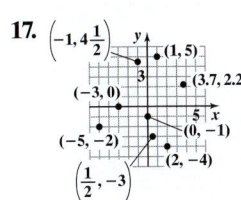 $(1, 5)$ and $(3.7, 2.2)$ are in quadrant I, $\left(-1, 4\frac{1}{2}\right)$ is in quadrant II, $(-5, -2)$ is in quadrant III, $(2, -4)$ and
$\left(\frac{1}{2}, -3\right)$ are in quadrant IV, $(-3, 0)$ lies on the x-axis, $(0, -1)$ lies on the y-axis   **19.** $(0, 0)$ **21.** $(3, 2)$
23. $(-2, -2)$ **25.** $(2, -1)$ **27.** $(0, -3)$ **29.** $(1, 3)$ **31.** $(-3, -1)$ **33. a.** $(2007, 9.6), (2008, 9.6), (2009,$
$10.6), (2010, 10.6), (2011, 10.2), (2012, 10.8)$ **b.** In the year 2010, the domestic box office was $10.6 billion.

c. 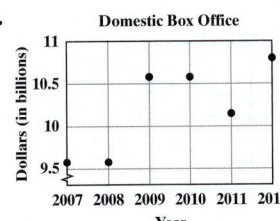   **d.** answers may vary   **35. a.** $(0.50, 10), (0.75, 12), (1.00, 15), (1.25, 16), (1.50, 18), (1.50, 19), (1.75, 19),$
$(2.00, 20)$ **b.** When Minh studied 1.25 hours, her quiz score was 16.
c. 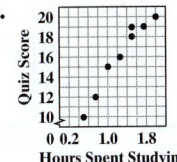   **d.** answers may vary   **37.** $(-4, -2), (4, 0)$ **39.** $(-8, -5), (16, 1)$
41. $0; 7; -\frac{2}{7}$ **43.** $2; 2; 5$ **45.** $0; -3; 2$ **47.** $2; 6; 3$ **49.** $-12; 5; -6$ **51.** $\frac{5}{7}; \frac{5}{2}; -1$

53. $0; -5; -2$ 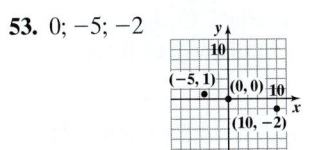   **55.** $2; 1; -6$ 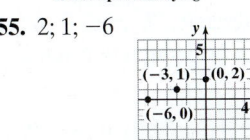   **57. a.** 13,000; 21,000; 29,000   **b.** 45 desks   **59. a.** 6.44; 7.16; 7.88
b. 2009   **c.** 2016   **61.** $y = 5 - x$ **63.** $y = \dfrac{5 - 2x}{4}$ **65.** $y = -2x$
67. false   **69.** true   **71.** negative; negative   **73.** positive; negative
75. $0; 0$ **77.** y **79.** no; answers may vary   **81.** answers may vary
83. answers may vary   **85.** $(4, -7)$ **87.** 26 units   **89.** 21 million; 23 million; 25 million; 27 million   **91.** 83°F   **93.** Sunday; 68°F
95. Tuesday; 13°F

Section 3.2

Calculator Explorations   1. 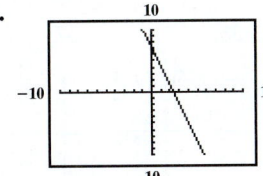   **3.** 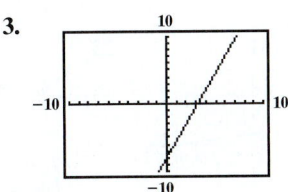   **5.**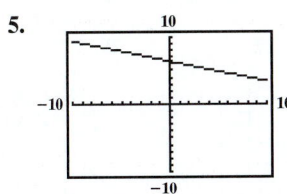

Vocabulary, Readiness & Video Check   1. It is always good practice to use a third point as a check to see that your points lie along a
straight line.

Exercise Set 3.2   **1.** $6; -2; 5$ **3.** $-4; 0; 4$ **5.** $0; 2; -1$ **7.** $3; -1; -5$ **9.**   **11.**

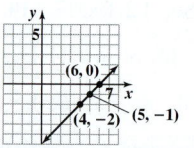 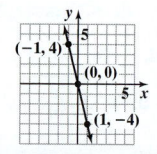 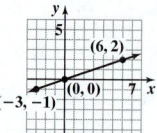 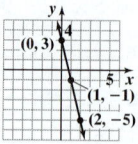 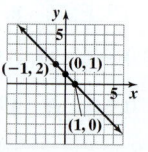 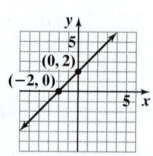

13.   **15.**   **17.**   **19.**   **21.**   **23.**   **25.**

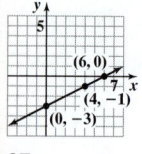 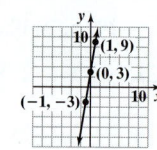 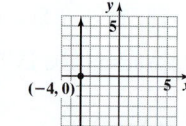 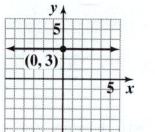 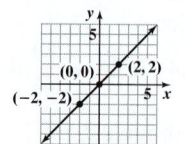 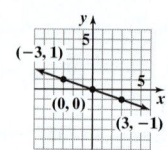 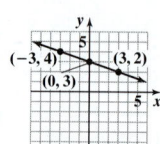

27.   **29.**   **31.**   **33. a.**   **b.** yes; answers may vary

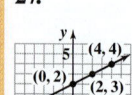 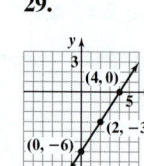 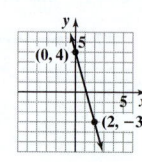 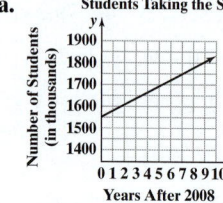

35. a.   **b.** $(11, 26.7)$ **c.** In 2012, IKEA's total annual revenue was 26.7 billion euros.

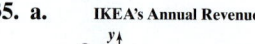

37. $(4, -1)$ **39.** $3; -3$ **41.** $0; 0$ **43.**   **45.**

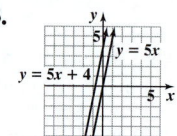 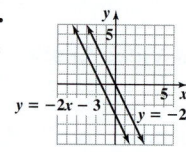

47. $0; 1; 1; 4; 4$ **49.** $x + y = 12; 9$ cm   **51.** yes; answers may vary

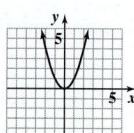

Section 3.3

Calculator Explorations   **1.**   **3.**   **5.**

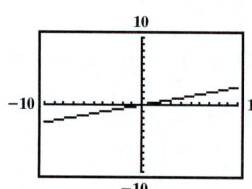 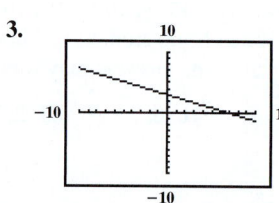 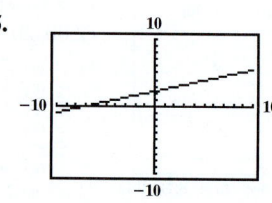

Vocabulary, Readiness & Video Check   **1.** linear   **3.** horizontal   **5.** y-intercept   **7.** $y; x$ **9.** false   **11.** true   **13.** because x-intercepts lie on the x-axis; because y-intercepts lie on the y-axis.   **15.** For a horizontal line, the coefficient of x will be 0; for a vertical line, the coefficient of y will be 0.

Exercise Set 3.3   **1.** $(-1, 0); (0, 1)$ **3.** $(-2, 0); (2, 0); (0, -2)$ **5.** $(-2, 0); (1, 0); (3, 0); (0, 3)$ **7.** $(-1, 0); (1, 0); (0, 1); (0, -2)$

9.   **11.**   **13.**   **15.**   **17.**   **19.**   **21.**   **23.**

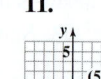

25.   **27.**   **29.**   **31.**   **33.**   **35.**   **37.**   **39.**

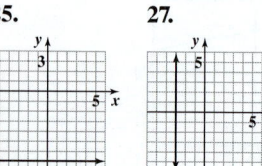 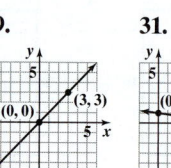 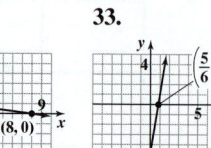 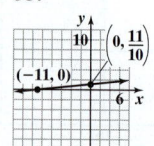 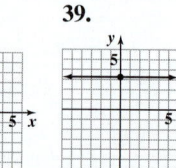

41. 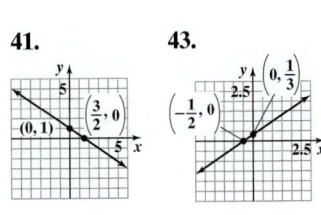 **43.** 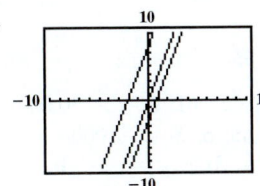 **45.** $\frac{3}{2}$ **47.** 6 **49.** $-\frac{6}{5}$ **51.** c **53.** a **55.** infinite **57.** 0 **59.** answers may vary **61.** $(0, 200)$; no chairs and 200 desks are manufactured. **63.** 300 chairs **65.** $y = -4$ **67. a.** $(30.6, 0)$ **b.** 30.6 years after 2006, there may be no newspaper circulation.

Section 3.4

Calculator Explorations **1.** 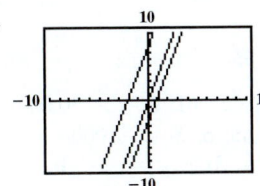 **3.**

Vocabulary, Readiness & Video Check **1.** slope **3.** 0 **5.** positive **7.** $y; x$ **9.** positive **11.** 0 **13.** downward **15.** vertical **17.** Solve the equation for y; the slope is the coefficient of x. **19.** Slope-intercept form; this form makes the slope easy to see, and we need to compare slopes to determine if two lines are parallel or perpendicular.

Exercise Set 3.4 **1.** $m = -1$ **3.** $m = -\frac{1}{4}$ **5.** $m = 0$ **7.** undefined slope **9.** $m = -\frac{4}{3}$ **11.** $m = \frac{5}{2}$ **13.** line 1 **15.** line 2

17. $m = 5$ **19.** $m = -0.3$ **21.** $m = -2$ **23.** undefined slope **25.** $m = \frac{2}{3}$ **27.** undefined slope **29.** $m = \frac{1}{2}$ **31.** $m = 0$

33. $m = -\frac{3}{4}$ **35.** $m = 4$ **37.** neither **39.** neither **41.** parallel **43.** perpendicular **45. a.** 1 **b.** -1 **47. a.** $\frac{9}{11}$ **b.** $-\frac{11}{9}$

49. $\frac{3}{5}$ **51.** 12.5% **53.** 40% **55.** 37%; 35% **57.** $m = 0.72$; Every year, the number of U.S. households with televisions increases by 0.72 million households. **59.** $m = 0.15$; Every year, the median age of automobiles in the United States increases by 0.15 year.

61. $y = 2x - 14$ **63.** $y = -6x - 11$ **65.** d **67.** b **69.** e **71.** $m = \frac{1}{2}$ **73.** answers may vary **75.** 31.5 mi per gal

77. 2003 and 2004; 29.5 mi per gallon **79.** from 2011 to 2012 **81.** $x = 20$ **83. a.** $(2007, 2209), (2012, 2378)$ **b.** 33.8 **c.** For the years 2007 through 2012, the number of heart transplants increased at a rate of 33.8 per year. **85.** Opposite sides are parallel since their slopes are equal, so the figure is a parallelogram. **87.** 2.0625 **89.** -1.6 **91.** The line becomes steeper.

Integrated Review **1.** $m = 2$ **2.** $m = 0$ **3.** $m = -\frac{2}{3}$ **4.** slope is undefined

5. 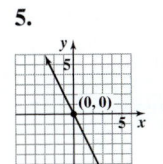 **6.** 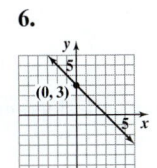 **7.** 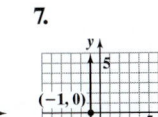 **8.** 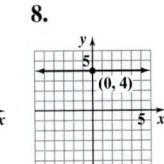 **9.** 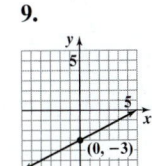 **10.** 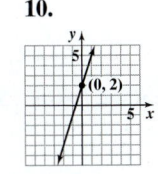 **11.** 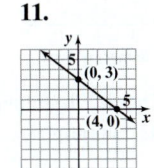 **12.**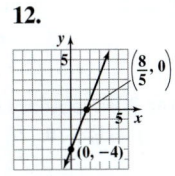

13. $m = 3$ **14.** $m = -6$ **15.** $m = -\frac{7}{2}$ **16.** $m = 2$ **17.** undefined slope **18.** $m = 0$ **19.** neither **20.** perpendicular

21. a. $(2008, 3600); (2012, 4416)$ **b.** 204 **c.** For the years 2008 through 2012, the amount of yogurt produced increased at a rate of 204 million pounds per year.

Section 3.5

Calculator Explorations **1.** 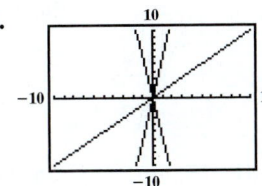 **3.**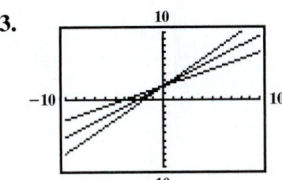

Vocabulary, Readiness & Video Check **1.** slope-intercept; $m; b$ **3.** point-slope **5.** slope-intercept **7.** horizontal **9.** y-intercept, fraction **11.** Write the equation with x- and y-terms on one side of the equal sign and a constant on the other side. **13.** You need to know what your variables stand for in order to solve part (b) of the example, and that depends on how you set up your ordered pairs in part (a).

Exercise Set 3.5 **1.** 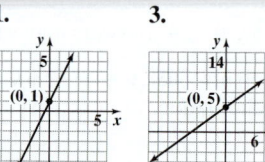 **3.** **5.** **7.** **9.** **11.**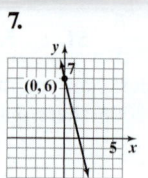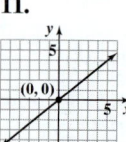

13. $y = 5x + 3$

15. $y = -4x - \dfrac{1}{6}$

17. $y = \dfrac{2}{3}x$

19. $y = -8$

21. $y = -\dfrac{1}{5}x + \dfrac{1}{9}$ **23.** $-6x + y = -10$ **25.** $8x + y = -13$ **27.** $3x - 2y = 27$ **29.** $x + 2y = -3$ **31.** $2x - y = 4$

33. $8x - y = -11$ **35.** $4x - 3y = -1$ **37.** $8x + 13y = 0$ **39.** $y = -\dfrac{1}{2}x + \dfrac{5}{3}$ **41.** $y = -x + 17$ **43.** $x = -\dfrac{3}{4}$

45. $y = x + 16$ **47.** $y = -5x + 7$ **49.** $y = 2$ **51.** $y = \dfrac{3}{2}x$ **53.** $y = -3$ **55.** $y = -\dfrac{4}{7}x - \dfrac{18}{7}$ **57. a.** $(0, 370), (5, 312)$

b. $y = -11.6x + 370$ **c.** 335.2 million magazines **59. a.** $s = 32t$ **b.** 128 ft/sec **61. a.** $y = 30{,}000x + 314{,}000$
b. 494,000 vehicles **63. a.** $y = -45x + 5545$ **b.** 5185 indoor cinema sites **65. a.** $S = -1000p + 13{,}000$
b. 9500 Fun Noodles **67.** -1 **69.** 5 **71.** b **73.** d **75.** $2x - y = -8$ **77. a.** $3x - y = -5$ **b.** $x + 3y = 5$

Section 3.6

Vocabulary, Readiness & Video Check **1.** linear inequality in two variables **3.** false **5.** true **7.** An ordered pair is a solution of an inequality if replacing the variables with the coordinates of the ordered pair results in a true statement.

Exercise Set 3.6 **1.** no; no **3.** yes; no **5.** no; yes **7.** **9.** **11.** **13.** **15.**

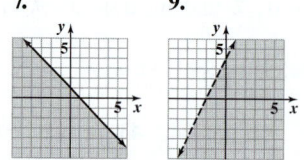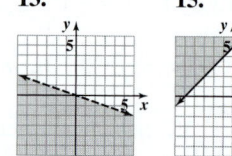

17. **19.** **21.** **23.** **25.** **27.** **29.** **31.**

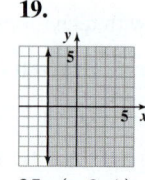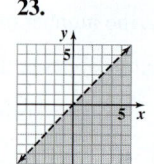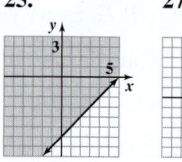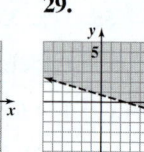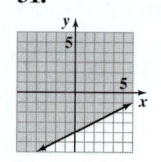

33. **35.** $(-2, 1)$ **37.** $(-3, -1)$ **39.** a **41.** b **43.** answers may vary **45.** yes **47.** yes **49. a.** $30x + 0.15y \leq 500$
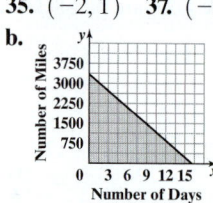
b. **c.** answers may vary

Chapter 3 Vocabulary Check **1.** solution **2.** y-axis **3.** linear **4.** x-intercept **5.** standard **6.** y-intercept **7.** slope-intercept
8. y **9.** x-axis **10.** x **11.** slope

Chapter 3 Review **1.–6.** 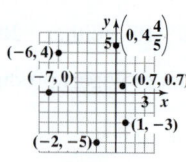 **7.** $(7, 44)$ **8.** $\left(-\dfrac{13}{3}, -8\right)$ **9.** $-3; 1; 9$ **10.** $5; 5; 5$ **11.** $0; 10; -10$

12. a. $2005; 2500; 7000$ **b.** 886 compact disc holders

13. **14.** **15.** **16.** **17.** **18.** **19.** $(4, 0); (0, -2)$

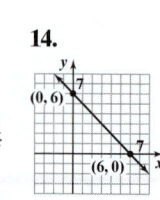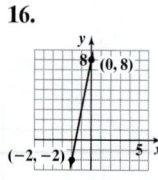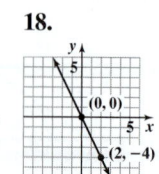

20. $(-2, 0); (2, 0); (0, 2); (0, -2)$

21. **22.** 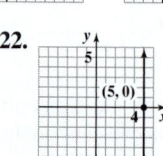 **23.** $(12, 0), (0, -4)$ **24.** $(-2, 0), (0, 8)$ **25.** $m = -\dfrac{3}{4}$ **26.** $m = \dfrac{1}{5}$ **27.** d **28.** b

29. c **30.** a **31.** $m = \dfrac{3}{4}$ **32.** $m = \dfrac{5}{3}$ **33.** $m = 4$ **34.** $m = -1$ **35.** $m = 3$ **36.** $m = \dfrac{1}{2}$

37. $m = 0$ **38.** undefined slope **39.** perpendicular **40.** parallel **41.** neither

42. perpendicular **43.** $m = 110;$ The total number of U.S. magazines in print increases by 110 magazines per year.

44. $m = 47$; The number of U.S. lung transplants increases by 47 transplants per year. **45.** $m = \frac{1}{6}; \left(0, \frac{1}{6}\right)$

46. $m = -3; (0, 7)$ **47.** $y = -5x + \frac{1}{2}$ **48.** $y = \frac{2}{3}x + 6$ **49.** d **50.** c **51.** a **52.** b **53.** $-4x + y = -8$ **54.** $3x + y = -5$

55. $-3x + 5y = 17$ **56.** $x + 3y = 6$ **57.** $y = -14x + 21$ **58.** $y = -\frac{1}{2}x + 4$ **59.** **60.** **61.**

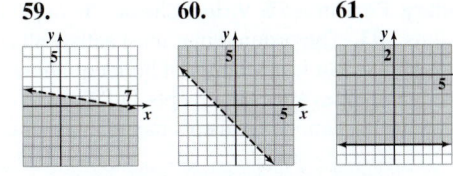

62. **63.** **64.** **65.** $7; -1; -3$ **66.** $0; -3; -2$ **67.** $(3, 0); (0, -2)$ **68.** $(-2, 0); (0, 10)$
69. **70.** **71.** **72.**

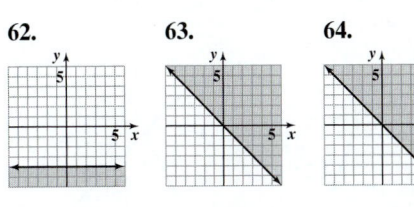
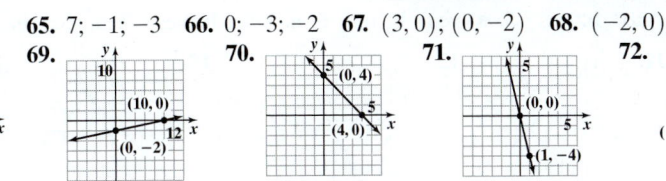

73. **74.** **75.** $m = -1$ **76.** $m = \frac{11}{7}$ **77.** $m = 2$ **78.** $m = -\frac{1}{3}$ **79.** $m = \frac{2}{3}; (0, -5)$

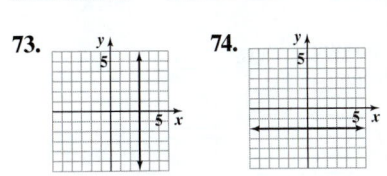
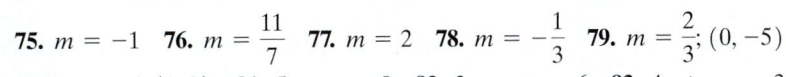

80. $m = -6; (0, 2)$ **81.** $5x + y = 8$ **82.** $3x - y = -6$ **83.** $4x + y = -3$
84. $5x + y = 16$ **85.** France **86.** Malaysia **87.** France, U.S., Spain, China
88. U.K., Russian Federation, Malaysia **89.** 30 million **90.** 25 million

Chapter 3 Test 1. $(1, 1)$ **2.** $(-4, 17)$ **3.** $m = \frac{2}{5}$ **4.** $m = 0$ **5.** $m = -1$ **6.** $m = -7$ **7.** $m = 3$ **8.** undefined slope

9. **10.** **11.** **12.** **13.** **14.** **15.** **16.**

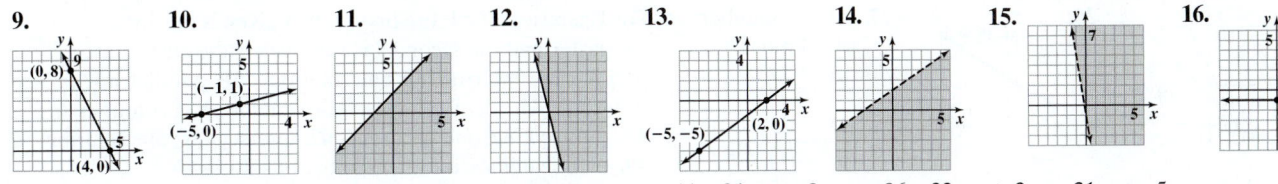
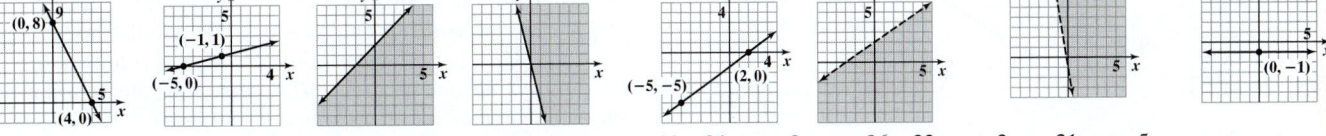

17. neither **18.** $x + 4y = 10$ **19.** $7x + 6y = 0$ **20.** $8x + y = 11$ **21.** $x - 8y = -96$ **22.** $x + 2y = 21; x = 5$ m
23. a. $(2008, 19.6); (2009, 22.2); (2010, 23.9); (2011, 25.3); (2012, 26.7)$ **b.** **24.** $m = -15.5$; For every 1 year,
15.5 million fewer movie tickets
are sold.

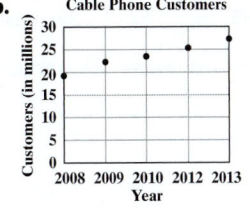

Cumulative Review 1. 27; Sec. 1.3, Ex. 2 **2.** $\frac{25}{7}$; Sec. R.2 **3.** 51; Sec. 1.3, Ex. 5 **4.** 23; Sec. 1.3 **5.** $20{,}602$ feet; Sec. 1.5, Ex. 10

6. $0.8x - 36$; Sec. 1.8 **7.** $2x + 6$; Sec. 1.8, Ex. 16 **8.** $-15\left(x + \frac{2}{3}\right) = -15x - 10$; Sec. 1.8 **9.** $(x - 4) \div 7$ or $\frac{x - 4}{7}$; Sec. 1.8, Ex. 17

10. $\frac{-9}{2x}$; Sec. 1.8 **11.** $5 + (x + 1) = 6 + x$; Sec. 1.8, Ex. 18 **12.** $-86 - x$; Sec. 1.8 **13.** 6; Sec. 2.2, Ex. 1 **14.** -24; Sec. 2.2

15. 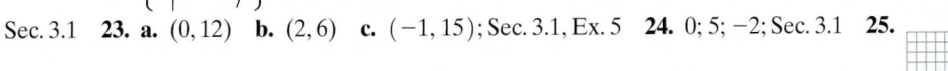 $\{x \mid x < -2\}$; Sec. 2.7, Ex. 6 **16.** $\left\{x \mid x \le \frac{8}{3}\right\}$; Sec. 2.7 **17.** $x = \frac{y - b}{m}$; Sec. 2.5, Ex. 6 **18.** $y = \frac{6 - x}{2}$;

Sec. 2.5 **19.** $\left\{x \mid x > \frac{13}{7}\right\}$; Sec. 2.7, Ex. 8 **20.** $\{y \mid y \le 2\}$; Sec. 2.7 **21. a.** -6 **b.** -5 **c.** 10; Sec. 3.1, Ex. 7 **22. a.** 2 **b.** 5 **c.** $\frac{20}{3}$;

Sec. 3.1 **23. a.** $(0, 12)$ **b.** $(2, 6)$ **c.** $(-1, 15)$; Sec. 3.1, Ex. 5 **24.** $0; 5; -2$; Sec. 3.1 **25.** Sec. 3.2, Ex. 1

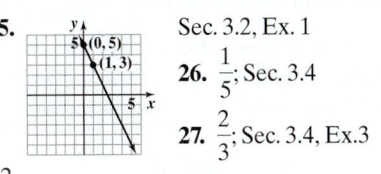

26. $\frac{1}{5}$; Sec. 3.4

27. $\frac{2}{3}$; Sec. 3.4, Ex.3

28. undefined slope; Sec. 3.4 **29.** $y = -2x + 3; 2x + y = 3$; Sec. 3.5, Ex. 4 **30.** $m = \frac{2}{5}$, y-intercept; $(0, -2)$; Sec. 3.5
31. a. not a solution **b.** is a solution; Sec. 3.6, Ex. 1 **32.** $3x - 2y = 0$; Sec. 3.5

Chapter 4 Systems of Equations

Section 4.1

Calculator Explorations **1.** $(0.37, 0.23)$ **3.** $(0.03, -1.89)$

Vocabulary, Readiness & Video Check **1.** dependent **3.** consistent **5.** inconsistent **7.** 1 solution, $(-1, 3)$ **9.** infinite number of solutions **11.** The ordered pair must satisfy all equations of the system in order to be a solution of the system, so we must check that the ordered pair is a solution of both equations. **13.** Writing the equations of a system in slope-intercept form lets us see and compare their slopes and y-intercepts. Different slopes mean one solution; same slopes with different y-intercepts mean no solution; same slopes with same y-intercepts mean an infinite number of solutions.

Exercise Set 4.1 **1. a.** no **b.** yes **3. a.** yes **b.** no **5. a.** yes **b.** yes **7. a.** no **b.** no

9. 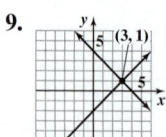 **11.** 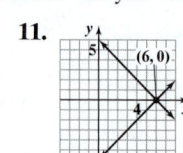 **13.** 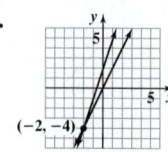 **15.** 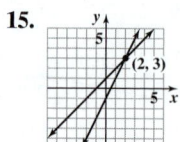 **17.** 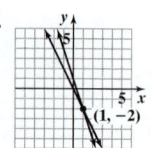 **19.**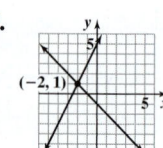

21. no solution 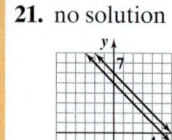 **23.** 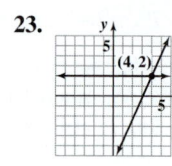 **25.** 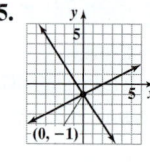 **27.** no solution 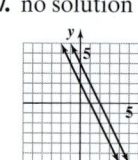 **29.** infinite number of solutions 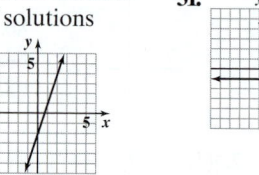 **31.**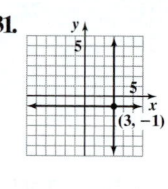

33. 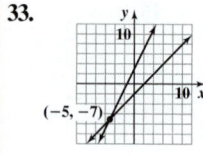 **35.** 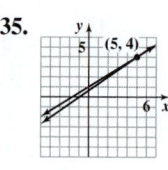 **37.** infinite number of solutions 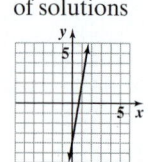 For Exercises 39–51, the first answer given is the answer for part **a**, and the second answer given is the answer for part **b**. **39.** intersecting; one solution **41.** parallel; no solution **43.** identical lines; infinite number of solutions **45.** intersecting; one solution **47.** intersecting; one solution **49.** identical lines; infinite number of solutions

51. parallel; no solution **53.** 2 **55.** $-\dfrac{2}{5}$ **57.** 2 **59.** answers may vary **61.** answers may vary **63.** 2010–2011 **65.** 2010, 2011, 2012, 2013 **67.** answers may vary **69.** answers may vary **71. a.** $(4, 9)$ **b.** 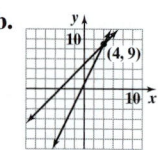 **c.** yes

Section 4.2

Vocabulary, Readiness & Video Check **1.** $(1, 4)$ **3.** infinite number of solutions **5.** $(0, 0)$ **7.** We solved one equation for a variable. Next, be sure to substitute this expression for the variable into the *other* equation.

Exercise Set 4.2 **1.** $(2, 1)$ **3.** $(-3, 9)$ **5.** $(2, 7)$ **7.** $\left(-\dfrac{1}{5}, \dfrac{43}{5}\right)$ **9.** $(2, -1)$ **11.** $(-2, 4)$ **13.** $(4, 2)$ **15.** $(-2, -1)$

17. no solution **19.** $(3, -1)$ **21.** $(3, 5)$ **23.** $\left(\dfrac{2}{3}, -\dfrac{1}{3}\right)$ **25.** $(-1, -4)$ **27.** $(-6, 2)$ **29.** $(2, 1)$ **31.** no solution

33. infinite number of solutions **35.** $\left(\dfrac{1}{2}, 2\right)$ **37.** $-6x - 4y = -12$ **39.** $-12x + 3y = 9$ **41.** $5n$ **43.** $-15b$ **45.** $(1, -3)$

47. answers may vary **49.** no **51.** c; answers may vary **53.** $(-2.6, 1.3)$ **55.** $(3.28, 2.1)$ **57. a.** $(9.6, 6)$ **b.** In about 9.6 years after 2006, U.S. consumer spending on DVD- and Blu-ray-format home entertainment will be approximately $6 billion for each. **c.** answers may vary;

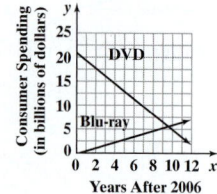

44. $m = 47$; The number of U.S. lung transplants increases by 47 transplants per year. **45.** $m = \dfrac{1}{6}; \left(0, \dfrac{1}{6}\right)$

46. $m = -3; (0, 7)$ **47.** $y = -5x + \dfrac{1}{2}$ **48.** $y = \dfrac{2}{3}x + 6$ **49.** d **50.** c **51.** a **52.** b **53.** $-4x + y = -8$ **54.** $3x + y = -5$

55. $-3x + 5y = 17$ **56.** $x + 3y = 6$ **57.** $y = -14x + 21$ **58.** $y = -\dfrac{1}{2}x + 4$ **59.** **60.** **61.**

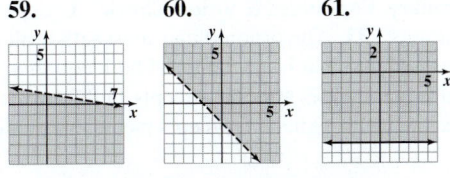

62. **63.** **64.** **65.** $7; -1; -3$ **66.** $0; -3; -2$ **67.** $(3, 0); (0, -2)$ **68.** $(-2, 0); (0, 10)$
69. **70.** **71.** **72.**

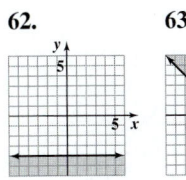 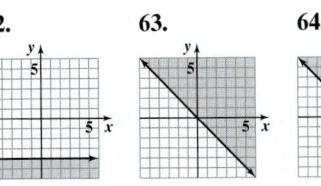 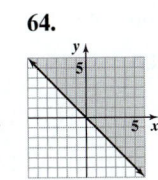 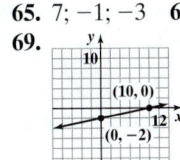 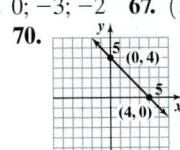 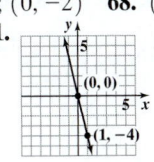 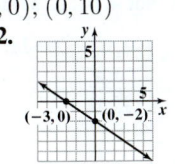

73. **74.** **75.** $m = -1$ **76.** $m = \dfrac{11}{7}$ **77.** $m = 2$ **78.** $m = -\dfrac{1}{3}$ **79.** $m = \dfrac{2}{3}; (0, -5)$

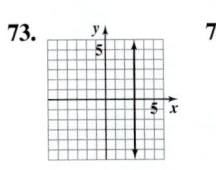 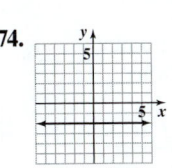

80. $m = -6; (0, 2)$ **81.** $5x + y = 8$ **82.** $3x - y = -6$ **83.** $4x + y = -3$
84. $5x + y = 16$ **85.** France **86.** Malaysia **87.** France, U.S., Spain, China
88. U.K., Russian Federation, Malaysia **89.** 30 million **90.** 25 million

Chapter 3 Test **1.** $(1, 1)$ **2.** $(-4, 17)$ **3.** $m = \dfrac{2}{5}$ **4.** $m = 0$ **5.** $m = -1$ **6.** $m = -7$ **7.** $m = 3$ **8.** undefined slope

9. **10.** **11.** **12.** **13.** **14.** **15.** **16.**

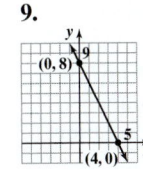 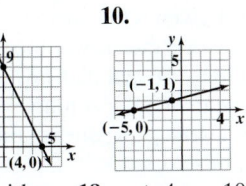 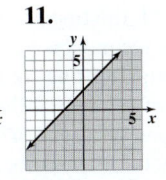 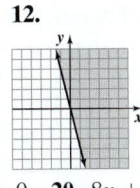 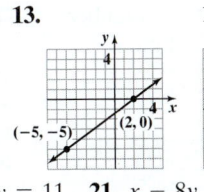 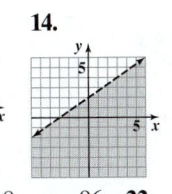 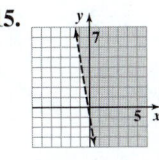 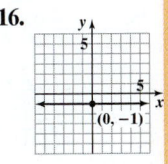

17. neither **18.** $x + 4y = 10$ **19.** $7x + 6y = 0$ **20.** $8x + y = 11$ **21.** $x - 8y = -96$ **22.** $x + 2y = 21; x = 5$ m
23. a. $(2008, 19.6); (2009, 22.2); (2010, 23.9); (2011, 25.3); (2012, 26.7)$ **b.** **24.** $m = -15.5$; For every 1 year,
15.5 million fewer movie tickets
are sold.

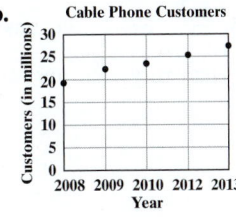

Cumulative Review **1.** 27; Sec. 1.3, Ex. 2 **2.** $\dfrac{25}{7}$; Sec. R.2 **3.** 51; Sec. 1.3, Ex. 5 **4.** 23; Sec. 1.3 **5.** $20{,}602$ feet; Sec. 1.5, Ex. 10

6. $0.8x - 36$; Sec. 1.8 **7.** $2x + 6$; Sec. 1.8, Ex. 16 **8.** $-15\left(x + \dfrac{2}{3}\right) = -15x - 10$; Sec. 1.8 **9.** $(x - 4) \div 7$ or $\dfrac{x - 4}{7}$; Sec. 1.8, Ex. 17

10. $\dfrac{-9}{2x}$; Sec. 1.8 **11.** $5 + (x + 1) = 6 + x$; Sec. 1.8, Ex. 18 **12.** $-86 - x$; Sec. 1.8 **13.** 6; Sec. 2.2, Ex. 1 **14.** -24; Sec. 2.2

15. $\{x \mid x < -2\}$; Sec. 2.7, Ex. 6 **16.** $\left\{x \mid x \le \dfrac{8}{3}\right\}$; Sec. 2.7 **17.** $x = \dfrac{y - b}{m}$; Sec. 2.5, Ex. 6 **18.** $y = \dfrac{6 - x}{2}$;

Sec. 2.5 **19.** $\left\{x \mid x > \dfrac{13}{7}\right\}$; Sec. 2.7, Ex. 8 **20.** $\{y \mid y \le 2\}$; Sec. 2.7 **21. a.** -6 **b.** -5 **c.** 10; Sec. 3.1, Ex. 7 **22. a.** 2 **b.** 5 **c.** $\dfrac{20}{3}$;

Sec. 3.1 **23. a.** $(0, 12)$ **b.** $(2, 6)$ **c.** $(-1, 15)$; Sec. 3.1, Ex. 5 **24.** $0; 5; -2$; Sec. 3.1 **25.** Sec. 3.2, Ex. 1

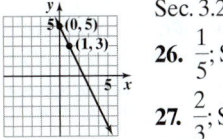

26. $\dfrac{1}{5}$; Sec. 3.4

27. $\dfrac{2}{3}$; Sec. 3.4, Ex. 3

28. undefined slope; Sec. 3.4 **29.** $y = -2x + 3; 2x + y = 3$; Sec. 3.5, Ex. 4 **30.** $m = \dfrac{2}{5}$, y-intercept; $(0, -2)$; Sec. 3.5
31. a. not a solution **b.** is a solution; Sec. 3.6, Ex. 1 **32.** $3x - 2y = 0$; Sec. 3.5

Chapter 4 Systems of Equations

Section 4.1

Calculator Explorations   1. $(0.37, 0.23)$ **3.** $(0.03, -1.89)$

Vocabulary, Readiness & Video Check   1. dependent   **3.** consistent   **5.** inconsistent   **7.** 1 solution, $(-1, 3)$ **9.** infinite number of solutions   **11.** The ordered pair must satisfy all equations of the system in order to be a solution of the system, so we must check that the ordered pair is a solution of both equations.   **13.** Writing the equations of a system in slope-intercept form lets us see and compare their slopes and y-intercepts. Different slopes mean one solution; same slopes with different y-intercepts mean no solution; same slopes with same y-intercepts mean an infinite number of solutions.

Exercise Set 4.1   1. a. no   **b.** yes   **3. a.** yes   **b.** no   **5. a.** yes   **b.** yes   **7. a.** no   **b.** no

9. 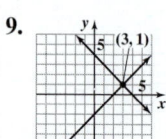   **11.** 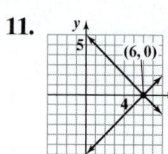   **13.** 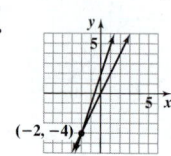   **15.** 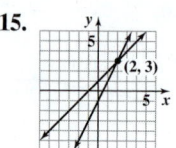   **17.** 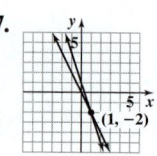   **19.**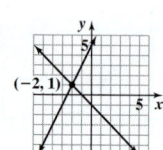

21. no solution   **23.** 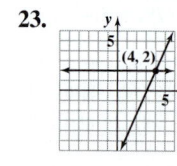   **25.** 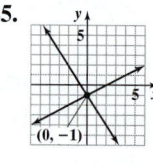   **27.** no solution   **29.** infinite number of solutions   **31.**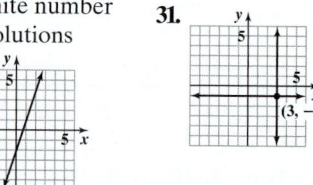

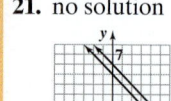

33. 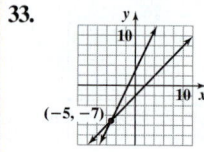   **35.** 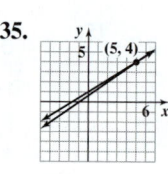   **37.** infinite number of solutions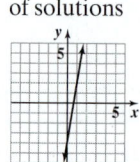

For Exercises 39–51, the first answer given is the answer for part **a**, and the second answer given is the answer for part **b**.   **39.** intersecting; one solution   **41.** parallel; no solution   **43.** identical lines; infinite number of solutions   **45.** intersecting; one solution   **47.** intersecting; one solution   **49.** identical lines; infinite number of solutions

51. parallel; no solution   **53.** 2   **55.** $-\dfrac{2}{5}$ **57.** 2   **59.** answers may vary   **61.** answers may vary   **63.** 2010–2011   **65.** 2010, 2011, 2012, 2013   **67.** answers may vary   **69.** answers may vary   **71. a.** $(4, 9)$ **b.** 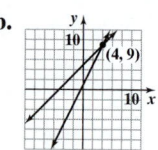   **c.** yes

Section 4.2

Vocabulary, Readiness & Video Check   1. $(1, 4)$ **3.** infinite number of solutions   **5.** $(0, 0)$ **7.** We solved one equation for a variable. Next, be sure to substitute this expression for the variable into the *other* equation.

Exercise Set 4.2   1. $(2, 1)$ **3.** $(-3, 9)$ **5.** $(2, 7)$ **7.** $\left(-\dfrac{1}{5}, \dfrac{43}{5}\right)$ **9.** $(2, -1)$ **11.** $(-2, 4)$ **13.** $(4, 2)$ **15.** $(-2, -1)$

17. no solution   **19.** $(3, -1)$ **21.** $(3, 5)$ **23.** $\left(\dfrac{2}{3}, -\dfrac{1}{3}\right)$ **25.** $(-1, -4)$ **27.** $(-6, 2)$ **29.** $(2, 1)$ **31.** no solution

33. infinite number of solutions   **35.** $\left(\dfrac{1}{2}, 2\right)$ **37.** $-6x - 4y = -12$ **39.** $-12x + 3y = 9$ **41.** $5n$ **43.** $-15b$ **45.** $(1, -3)$

47. answers may vary   **49.** no   **51.** c; answers may vary   **53.** $(-2.6, 1.3)$ **55.** $(3.28, 2.1)$ **57. a.** $(9.6, 6)$ **b.** In about 9.6 years after 2006, U.S. consumer spending on DVD- and Blu-ray-format home entertainment will be approximately $6 billion for each.   **c.** answers may vary;

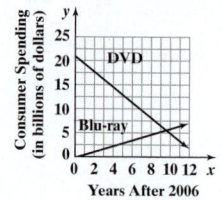

Section 4.3

Vocabulary, Readiness & Video Check **1.** false **3.** true **5.** The multiplication property of equality; be sure to multiply *both* sides of the equation by the nonzero number chosen.

Exercise Set 4.3 **1.** $(1, 2)$ **3.** $(2, -3)$ **5.** $(-2, -5)$ **7.** $(5, -2)$ **9.** $(-7, 5)$ **11.** $(6, 0)$ **13.** no solution **15.** infinite number of solutions **17.** $\left(2, -\frac{1}{2}\right)$ **19.** $(-2, 0)$ **21.** $(1, -1)$ **23.** infinite number of solutions **25.** $\left(\frac{12}{11}, -\frac{4}{11}\right)$ **27.** $\left(\frac{3}{2}, 3\right)$

29. infinite number of solutions **31.** $(1, 6)$ **33.** $\left(-\frac{1}{2}, -2\right)$ **35.** infinite number of solutions **37.** $\left(-\frac{2}{3}, \frac{2}{5}\right)$ **39.** $(2, 4)$

41. $(-0.5, 2.5)$ **43.** $2x + 6 = x - 3$ **45.** $20 - 3x = 2$ **47.** $4(x + 6) = 2x$ **49.** $2; 6x - 2y = -24$ **51.** b; answers may vary **53.** answers may vary **55. a.** $b = 15$ **b.** any real number except 15 **57.** $(-8.9, 10.6)$ **59. a.** $(2, 309)$ or $(2, 306)$ **b.** In about 2012 $(2010 + 2)$, the number of mail carrier jobs was approximately equal to the number of market research analyst jobs. **c.** 309 thousand or 306 thousand

Integrated Review **1.** $(2, 5)$ **2.** $(4, 2)$ **3.** $(5, -2)$ **4.** $(6, -14)$ **5.** $(-3, 2)$ **6.** $(-4, 3)$ **7.** $(0, 3)$ **8.** $(-2, 4)$ **9.** $(5, 7)$ **10.** $(-3, -23)$ **11.** $\left(\frac{1}{3}, 1\right)$ **12.** $\left(-\frac{1}{4}, 2\right)$ **13.** no solution **14.** infinite number of solutions **15.** $(0.5, 3.5)$ **16.** $(-0.75, 1.25)$ **17.** infinite number of solutions **18.** no solution **19.** $(7, -3)$ **20.** $(-1, -3)$ **21.** answers may vary **22.** answers may vary

Section 4.4

Vocabulary, Readiness & Video Check **1.** Up to now we've been working with one variable/unknown and one equation. Because these systems involve two equations with two unknowns, for these applications we need to choose two variables to represent two unknowns and translate the problem into two equations.

Exercise Set 4.4 **1.** c **3.** b **5.** a **7.** $\begin{cases} x + y = 15 \\ x - y = 7 \end{cases}$ **9.** $\begin{cases} x + y = 6500 \\ x = y + 800 \end{cases}$ **11.** 33 and 50 **13.** 14 and -3 **15.** Cabrera: 139; Hamilton: 128 **17.** child's ticket: $18; adult's ticket: $29 **19.** quarters: 53; nickels: 27 **21.** McDonald's: $97.80; Ford: $17.25 **23.** daily fee: $32; mileage charge: $0.25 per mi **25.** distance downstream = distance upstream = 18 mi; time downstream: 2 hr; time upstream: $4\frac{1}{2}$ hr; still water: 6.5 mph; current: 2.5 mph **27.** still air: 455 mph; wind: 65 mph **29.** $4\frac{1}{2}$ hr **31.** 12% solution: $7\frac{1}{2}$ oz; 4% solution: $4\frac{1}{2}$ oz **33.** $4.95 beans: 113 lb; $2.65 beans: 87 lb **35.** $60°, 30°$ **37.** $20°, 70°$ **39.** number sold at $9.50: 23; number sold at $7.50: 67 **41.** $2\frac{1}{4}$ mph and $2\frac{3}{4}$ mph **43.** 30%: 50 gal; 60%: 100 gal **45.** length: 42 in.; width: 30 in. **47.** 16 **49.** 25 **51.** -25 **53.** a **55.** width: 9 ft; length: 15 ft

Chapter 4 Vocabulary Check **1.** dependent **2.** system of linear equations **3.** consistent **4.** solution **5.** addition; substitution **6.** inconsistent **7.** independent

Chapter 4 Review **1. a.** no **b.** yes **2. a.** no **b.** yes **3. a.** no **b.** no **4. a.** yes **b.** no

5. 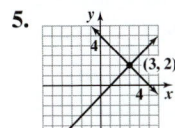 **6.** 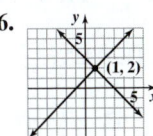 **7.** 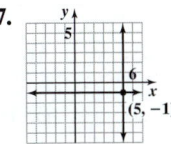 **8.** 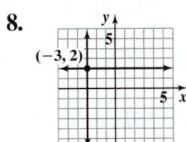 **9.**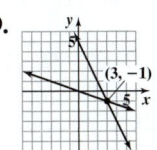

10. 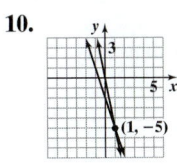 **11.** no solution **12.** infinite number of solutions 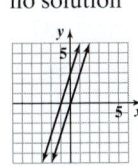 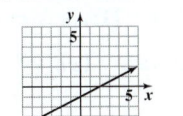 **13.** $(-1, 4)$ **14.** $(2, -1)$ **15.** $(3, -2)$ **16.** $(2, 5)$ **17.** infinite number of solutions **18.** infinite number of solutions **19.** no solution **20.** no solution **21.** $(-6, 2)$ **22.** $(4, -1)$ **23.** $(3, 7)$ **24.** $(-2, 4)$ **25.** infinite number of solutions **26.** infinite number of solutions

27. $(8, -6)$ **28.** $\left(-\frac{3}{2}, \frac{15}{2}\right)$ **29.** -6 and 22 **30.** orchestra: 255 seats; balcony: 105 seats **31.** current of river: 3.2 mph; speed in still water: 21.1 mph **32.** 6% solution: $12\frac{1}{2}$ cc; 14% solution: $37\frac{1}{2}$ cc **33.** egg: $0.40; strip of bacon: $0.65 **34.** jogging: 0.86 hr; walking: 2.14 hr

35.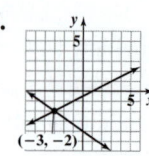

36. infinite number of solutions **37.** $(3, 2)$ **38.** $(7, 1)$ **39.** $\left(1\frac{1}{2}, -3\right)$ **40.** no solution **41.** infinite number of solutions **42.** $(8, 11)$ **43.** $(-5, 2)$ **44.** $(16, -4)$ **45.** infinite number of solutions **46.** no solution **47.** 4 and 8 **48.** -5 and 13 **49.** 24 nickels and 41 dimes **50.** 13¢ stamps: 17; 22¢ stamps: 9

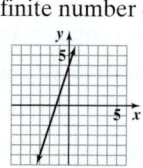

Chapter 4 Test **1.** false **2.** false **3.** true **4.** false **5.** no **6.** yes

7. 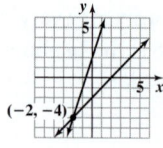 **8.** no solution **9.** $(-4, 1)$ **10.** $\left(\frac{1}{2}, -2\right)$ **11.** $(20, 8)$ **12.** no solution **13.** $(4, -5)$ **14.** $(7, 2)$ **15.** $(5, -2)$ **16.** infinite number of solutions **17.** $(-5, 3)$ **18.** $\left(\frac{47}{5}, \frac{48}{5}\right)$ **19.** 78, 46 **20.** 120 cc **21.** Texas: 245 thousand; Missouri: 106 thousand **22.** 3 mph; 6 mph **23.** 2008–2009, 2011–2012 **24.** 2008, 2012

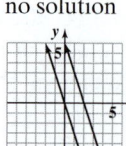

Cumulative Review **1. a.** -6 **b.** 6.3; Sec. 1.5, Ex. 6 **2. a.** 25 **b.** 32; Sec. 1.3 **3.** $\frac{1}{22}$; Sec. 1.6, Ex. 9a **4.** -22; Sec. 1.4 **5.** $\frac{16}{3}$; Sec. 1.6, Ex. 9b **6.** $-\frac{3}{16}$; Sec. 1.4 **7.** $-\frac{1}{10}$; Sec. 1.6, Ex. 9c **8.** 10; Sec. 1.4 **9.** $-\frac{13}{9}$; Sec. 1.6, Ex. 9d **10.** $\frac{9}{13}$; Sec. 1.4 **11.** $\frac{1}{1.7}$; Sec. 1.6, Ex. 9e **12.** -1.7; Sec. 1.4 **13. a.** 5 **b.** $8 - x$; Sec. 2.1, Ex. 8 **14.** -5; Sec. 2.1 **15.** no solution; Sec. 2.3, Ex. 6 **16.** no solution; Sec. 2.3 **17.** 12; Sec. 2.3, Ex. 3 **18.** 40; Sec. 2.3 **19.** $\{x \mid x \geq 1\}$; Sec. 2.7, Ex. 9 **20.** $b = P - a - c$; Sec. 2.5 **21.** $m = -\frac{8}{3}$ Sec. 3.4, Ex. 1 **22.** $-\frac{11}{3}$; Sec. 3.4 **23.** $m = 0$; Sec. 3.4, Ex. 5 **24.** undefined; Sec. 3.4 **25.** $-x + 5y = 23$; Sec. 3.5, Ex. 5 **26.** $y = -5x - 7$; Sec. 3.5

27. It is a solution; Sec. 4.1, Ex. 1 **28. a.** yes **b.** no **c.** no; Sec. 4.1 **29.** $\left(6, \frac{1}{2}\right)$; Sec. 4.2, Ex. 3 **30.** $(-2, -4)$; Sec. 4.2

31. $\left(-\frac{15}{7}, -\frac{5}{7}\right)$; Sec. 4.3, Ex. 6 **32.** $\left(-\frac{44}{3}, -\frac{7}{3}\right)$; Sec. 4.2 **33.** 29 and 8; Sec. 4.4, Ex. 1 **34.** 48 and 19; Sec. 4.4

Chapter 5 Exponents and Polynomials

Section 5.1

Vocabulary, Readiness & Video Check **1.** exponent **3.** add **5.** 1 **7.** base: 3; exponent: 2 **9.** base: 4; exponent: 2 **11.** base: 5; exponent: 1; base: x; exponent: 2 **13.** Example 4 can be written as $-4^2 = -1 \cdot 4^2$, which is similar to Example 7, $4 \cdot 3^2$, and shows why the negative sign should not be considered part of the base when there are no parentheses. **15.** Be careful not to confuse the power rule with the product rule. The power rule involves a power raised to a power (exponents are multiplied), and the product rule involves a product (exponents are added). **17.** the quotient rule

Exercise Set 5.1 **1.** 49 **3.** -5 **5.** -16 **7.** 16 **9.** $\frac{1}{27}$ **11.** 112 **13.** 4 **15.** 135 **17.** 150 **19.** $\frac{32}{5}$ **21.** x^7 **23.** $(-3)^{12}$ **25.** $15y^5$

27. $x^{19}y^6$ **29.** $-72m^3n^8$ **31.** $-24z^{20}$ **33.** $20x^5$ sq ft **35.** x^{36} **37.** p^8q^8 **39.** $8a^{15}$ **41.** $x^{10}y^{15}$ **43.** $49a^4b^{10}c^2$ **45.** $\frac{r^9}{s^9}$ **47.** $\frac{m^9p^9}{n^9}$

49. $\frac{4x^2z^2}{y^{10}}$ **51.** $64z^{10}$ sq dm **53.** $27y^{12}$ cu ft **55.** x^2 **57.** -64 **59.** p^6q^5 **61.** $\frac{y^3}{2}$ **63.** 1 **65.** 1 **67.** -7 **69.** 2 **71.** -81 **73.** $\frac{1}{64}$

75. b^6 **77.** a^9 **79.** $-16x^7$ **81.** $a^{11}b^{20}$ **83.** $26m^9n^7$ **85.** z^{40} **87.** $64a^3b^3$ **89.** $36x^2y^2z^6$ **91.** z^8 **93.** $3x^4$ **95.** 1 **97.** $81x^2y^2$

99. 40 **101.** $\frac{y^{15}}{8x^{12}}$ **103.** $2x^2y$ **105.** -2 **107.** 5 **109.** -7 **111.** c **113.** e **115.** answers may vary **117.** answers may vary **119.** 343 cu m **121.** volume **123.** answers may vary **125.** answers may vary **127.** x^{9a} **129.** a^{5b} **131.** x^{5a}

Section 5.2

Calculator Explorations **1.** 5.31 EE 3 **3.** 6.6 EE -9 **5.** 1.5×10^{13} **7.** 8.15×10^{19}

Vocabulary, Readiness & Video Check **1.** $\frac{1}{x^3}$ **3.** scientific notation **5.** $\frac{5}{x^2}$ **7.** y^6 **9.** $4y^3$ **11.** A negative exponent has nothing to do with the sign of the simplified result. **13.** When you move the decimal point to the left, the sign of the exponent will be positive; when you move the decimal point to the right, the sign of the exponent will be negative. **15.** the quotient rule

Section 4.3

Vocabulary, Readiness & Video Check **1.** false **3.** true **5.** The multiplication property of equality; be sure to multiply *both* sides of the equation by the nonzero number chosen.

Exercise Set 4.3 **1.** $(1, 2)$ **3.** $(2, -3)$ **5.** $(-2, -5)$ **7.** $(5, -2)$ **9.** $(-7, 5)$ **11.** $(6, 0)$ **13.** no solution **15.** infinite number of solutions **17.** $\left(2, -\dfrac{1}{2}\right)$ **19.** $(-2, 0)$ **21.** $(1, -1)$ **23.** infinite number of solutions **25.** $\left(\dfrac{12}{11}, -\dfrac{4}{11}\right)$ **27.** $\left(\dfrac{3}{2}, 3\right)$

29. infinite number of solutions **31.** $(1, 6)$ **33.** $\left(-\dfrac{1}{2}, -2\right)$ **35.** infinite number of solutions **37.** $\left(-\dfrac{2}{3}, \dfrac{2}{5}\right)$ **39.** $(2, 4)$

41. $(-0.5, 2.5)$ **43.** $2x + 6 = x - 3$ **45.** $20 - 3x = 2$ **47.** $4(x + 6) = 2x$ **49.** 2; $6x - 2y = -24$ **51.** b; answers may vary **53.** answers may vary **55. a.** $b = 15$ **b.** any real number except 15 **57.** $(-8.9, 10.6)$ **59. a.** $(2, 309)$ or $(2, 306)$ **b.** In about 2012 $(2010 + 2)$, the number of mail carrier jobs was approximately equal to the number of market research analyst jobs. **c.** 309 thousand or 306 thousand

Integrated Review **1.** $(2, 5)$ **2.** $(4, 2)$ **3.** $(5, -2)$ **4.** $(6, -14)$ **5.** $(-3, 2)$ **6.** $(-4, 3)$ **7.** $(0, 3)$ **8.** $(-2, 4)$ **9.** $(5, 7)$
10. $(-3, -23)$ **11.** $\left(\dfrac{1}{3}, 1\right)$ **12.** $\left(-\dfrac{1}{4}, 2\right)$ **13.** no solution **14.** infinite number of solutions **15.** $(0.5, 3.5)$ **16.** $(-0.75, 1.25)$
17. infinite number of solutions **18.** no solution **19.** $(7, -3)$ **20.** $(-1, -3)$ **21.** answers may vary **22.** answers may vary

Section 4.4

Vocabulary, Readiness & Video Check **1.** Up to now we've been working with one variable/unknown and one equation. Because these systems involve two equations with two unknowns, for these applications we need to choose two variables to represent two unknowns and translate the problem into two equations.

Exercise Set 4.4 **1.** c **3.** b **5.** a **7.** $\begin{cases} x + y = 15 \\ x - y = 7 \end{cases}$ **9.** $\begin{cases} x + y = 6500 \\ x = y + 800 \end{cases}$ **11.** 33 and 50 **13.** 14 and -3 **15.** Cabrera: 139; Hamilton: 128 **17.** child's ticket: \$18; adult's ticket: \$29 **19.** quarters: 53; nickels: 27 **21.** McDonald's: \$97.80; Ford: \$17.25 **23.** daily fee: \$32; mileage charge: \$0.25 per mi **25.** distance downstream = distance upstream = 18 mi; time downstream: 2 hr; time upstream: $4\dfrac{1}{2}$ hr; still water: 6.5 mph; current: 2.5 mph **27.** still air: 455 mph; wind: 65 mph **29.** $4\dfrac{1}{2}$ hr **31.** 12% solution: $7\dfrac{1}{2}$ oz; 4% solution: $4\dfrac{1}{2}$ oz **33.** \$4.95 beans: 113 lb; \$2.65 beans: 87 lb **35.** $60°, 30°$ **37.** $20°, 70°$ **39.** number sold at \$9.50: 23; number sold at \$7.50: 67 **41.** $2\dfrac{1}{4}$ mph and $2\dfrac{3}{4}$ mph **43.** 30%: 50 gal; 60%: 100 gal **45.** length: 42 in.; width: 30 in. **47.** 16 **49.** 25 **51.** -25 **53.** a **55.** width: 9 ft; length: 15 ft

Chapter 4 Vocabulary Check **1.** dependent **2.** system of linear equations **3.** consistent **4.** solution **5.** addition; substitution **6.** inconsistent **7.** independent

Chapter 4 Review **1. a.** no **b.** yes **2. a.** no **b.** yes **3. a.** no **b.** no **4. a.** yes **b.** no

5.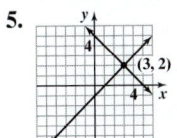
6.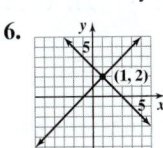
7.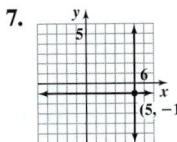
8.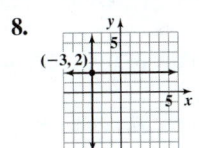
9.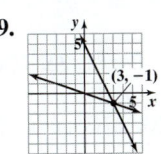

10. 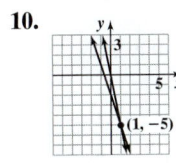 **11.** no solution 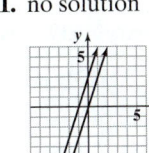 **12.** infinite number of solutions 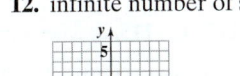 **13.** $(-1, 4)$ **14.** $(2, -1)$ **15.** $(3, -2)$ **16.** $(2, 5)$
17. infinite number of solutions **18.** infinite number of solutions **19.** no solution **20.** no solution **21.** $(-6, 2)$
22. $(4, -1)$ **23.** $(3, 7)$ **24.** $(-2, 4)$ **25.** infinite number of solutions **26.** infinite number of solutions

27. $(8, -6)$ **28.** $\left(-\dfrac{3}{2}, \dfrac{15}{2}\right)$ **29.** -6 and 22 **30.** orchestra: 255 seats; balcony: 105 seats **31.** current of river: 3.2 mph; speed in still water: 21.1 mph **32.** 6% solution: $12\dfrac{1}{2}$ cc; 14% solution: $37\dfrac{1}{2}$ cc **33.** egg: \$0.40; strip of bacon: \$0.65 **34.** jogging: 0.86 hr; walking: 2.14 hr

35. 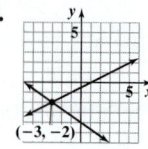   **36.** infinite number of solutions   **37.** $(3, 2)$ **38.** $(7, 1)$ **39.** $\left(1\frac{1}{2}, -3\right)$ **40.** no solution   **41.** infinite

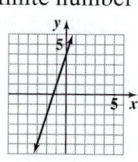 number of solutions   **42.** $(8, 11)$ **43.** $(-5, 2)$ **44.** $(16, -4)$

45. infinite number of solutions   **46.** no solution   **47.** 4 and 8   **48.** -5

and 13   **49.** 24 nickels and 41 dimes   **50.** 13¢ stamps: 17; 22¢ stamps: 9

Chapter 4 Test   **1.** false   **2.** false   **3.** true   **4.** false   **5.** no   **6.** yes

7. 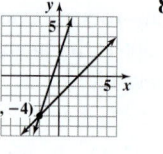   **8.** no solution   **9.** $(-4, 1)$ **10.** $\left(\frac{1}{2}, -2\right)$ **11.** $(20, 8)$ **12.** no solution   **13.** $(4, -5)$ **14.** $(7, 2)$

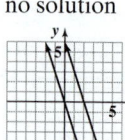   **15.** $(5, -2)$ **16.** infinite number of solutions   **17.** $(-5, 3)$ **18.** $\left(\frac{47}{5}, \frac{48}{5}\right)$ **19.** 78, 46

20. 120 cc   **21.** Texas: 245 thousand; Missouri: 106 thousand   **22.** 3 mph; 6 mph

23. 2008–2009, 2011–2012   **24.** 2008, 2012

Cumulative Review   **1. a.** -6 **b.** 6.3; Sec. 1.5, Ex. 6   **2. a.** 25   **b.** 32; Sec. 1.3   **3.** $\frac{1}{22}$; Sec. 1.6, Ex. 9a   **4.** -22; Sec. 1.4   **5.** $\frac{16}{3}$;

Sec. 1.6, Ex. 9b   **6.** $-\frac{3}{16}$; Sec. 1.4   **7.** $-\frac{1}{10}$; Sec. 1.6, Ex. 9c   **8.** 10; Sec. 1.4   **9.** $-\frac{13}{9}$; Sec. 1.6, Ex. 9d   **10.** $\frac{9}{13}$; Sec. 1.4   **11.** $\frac{1}{1.7}$; Sec. 1.6,

Ex. 9e   **12.** -1.7; Sec. 1.4   **13. a.** 5   **b.** $8 - x$; Sec. 2.1, Ex. 8   **14.** -5; Sec. 2.1   **15.** no solution; Sec. 2.3, Ex. 6   **16.** no solution; Sec. 2.3

17. 12; Sec. 2.3, Ex. 3   **18.** 40; Sec. 2.3   **19.** $\{x \mid x \geq 1\}$; Sec. 2.7, Ex. 9   **20.** $b = P - a - c$; Sec. 2.5   **21.** $m = -\frac{8}{3}$ Sec. 3.4, Ex. 1

22. $-\frac{11}{3}$; Sec. 3.4   **23.** $m = 0$; Sec. 3.4, Ex. 5   **24.** undefined; Sec. 3.4   **25.** $-x + 5y = 23$; Sec. 3.5, Ex. 5   **26.** $y = -5x - 7$; Sec. 3.5

27. It is a solution; Sec. 4.1, Ex. 1   **28. a.** yes   **b.** no   **c.** no; Sec. 4.1   **29.** $\left(6, \frac{1}{2}\right)$; Sec. 4.2, Ex. 3   **30.** $(-2, -4)$; Sec. 4.2

31. $\left(-\frac{15}{7}, -\frac{5}{7}\right)$; Sec. 4.3, Ex. 6   **32.** $\left(-\frac{44}{3}, -\frac{7}{3}\right)$; Sec. 4.2   **33.** 29 and 8; Sec. 4.4, Ex. 1   **34.** 48 and 19; Sec. 4.4

Chapter 5 Exponents and Polynomials

Section 5.1

Vocabulary, Readiness & Video Check   **1.** exponent   **3.** add   **5.** 1   **7.** base: 3; exponent: 2   **9.** base: 4; exponent: 2   **11.** base: 5; exponent: 1; base: x; exponent: 2   **13.** Example 4 can be written as $-4^2 = -1 \cdot 4^2$, which is similar to Example 7, $4 \cdot 3^2$, and shows why the negative sign should not be considered part of the base when there are no parentheses.   **15.** Be careful not to confuse the power rule with the product rule. The power rule involves a power raised to a power (exponents are multiplied), and the product rule involves a product (exponents are added).   **17.** the quotient rule

Exercise Set 5.1   **1.** 49   **3.** -5 **5.** -16 **7.** 16   **9.** $\frac{1}{27}$ **11.** 112   **13.** 4   **15.** 135   **17.** 150   **19.** $\frac{32}{5}$ **21.** x^7 **23.** $(-3)^{12}$ **25.** $15y^5$

27. $x^{19}y^6$ **29.** $-72m^3n^8$ **31.** $-24z^{20}$ **33.** $20x^5$ sq ft   **35.** x^{36} **37.** p^8q^8 **39.** $8a^{15}$ **41.** $x^{10}y^{15}$ **43.** $49a^4b^{10}c^2$ **45.** $\frac{r^9}{s^9}$ **47.** $\frac{m^9p^9}{n^9}$

49. $\frac{4x^2z^2}{y^{10}}$ **51.** $64z^{10}$ sq dm   **53.** $27y^{12}$ cu ft   **55.** x^2 **57.** -64 **59.** p^6q^5 **61.** $\frac{y^3}{2}$ **63.** 1   **65.** 1   **67.** -7 **69.** 2   **71.** -81 **73.** $\frac{1}{64}$

75. b^6 **77.** a^9 **79.** $-16x^7$ **81.** $a^{11}b^{20}$ **83.** $26m^9n^7$ **85.** z^{40} **87.** $64a^3b^3$ **89.** $36x^2y^2z^6$ **91.** z^8 **93.** $3x^4$ **95.** 1   **97.** $81x^2y^2$

99. 40   **101.** $\frac{y^{15}}{8x^{12}}$ **103.** $2x^2y$ **105.** -2 **107.** 5   **109.** -7 **111.** c   **113.** e   **115.** answers may vary   **117.** answers may vary

119. 343 cu m   **121.** volume   **123.** answers may vary   **125.** answers may vary   **127.** x^{9a} **129.** a^{5b} **131.** x^{5a}

Section 5.2

Calculator Explorations   **1.** 5.31 EE 3   **3.** 6.6 EE -9 **5.** 1.5×10^{13} **7.** 8.15×10^{19}

Vocabulary, Readiness & Video Check   **1.** $\frac{1}{x^3}$ **3.** scientific notation   **5.** $\frac{5}{x^2}$ **7.** y^6 **9.** $4y^3$ **11.** A negative exponent has nothing to do with the sign of the simplified result.   **13.** When you move the decimal point to the left, the sign of the exponent will be positive; when you move the decimal point to the right, the sign of the exponent will be negative.   **15.** the quotient rule

Exercise Set 5.2 **1.** $\dfrac{1}{64}$ **3.** $\dfrac{7}{x^3}$ **5.** -64 **7.** $\dfrac{5}{6}$ **9.** p^3 **11.** $\dfrac{q^4}{p^5}$ **13.** $\dfrac{1}{x^3}$ **15.** z^3 **17.** $\dfrac{4}{9}$ **19.** $\dfrac{1}{9}$ **21.** $-p^4$ **23.** -2 **25.** x^4 **27.** p^4

29. m^{11} **31.** r^6 **33.** $\dfrac{1}{x^{15}y^9}$ **35.** $\dfrac{1}{x^4}$ **37.** $\dfrac{1}{a^2}$ **39.** $4k^3$ **41.** $3m$ **43.** $-\dfrac{4a^5}{b}$ **45.** $-\dfrac{6}{7y^2}$ **47.** $\dfrac{27a^6}{b^{12}}$ **49.** $\dfrac{a^{30}}{b^{12}}$ **51.** $\dfrac{1}{x^{10}y^6}$ **53.** $\dfrac{z^2}{4}$

55. $\dfrac{x^{11}}{81}$ **57.** $\dfrac{49a^4}{b^6}$ **59.** $-\dfrac{3m^7}{n^4}$ **61.** $a^{24}b^8$ **63.** 200 **65.** x^9y^{19} **67.** $-\dfrac{y^8}{8x^2}$ **69.** $\dfrac{25b^{33}}{a^{16}}$ **71.** $\dfrac{27}{z^3x^6}$ cu in. **73.** 7.8×10^4

75. 1.67×10^{-6} **77.** 6.35×10^{-3} **79.** 1.16×10^6 **81.** 4.2×10^3 **83.** 0.0000000008673 **85.** 0.033 **87.** 20,320 **89.** 700,000,000

91. 5.7×10^{12} **93.** 10,100,000,000,000 **95.** 3,000,000,000,000; 3×10^{12} **97.** 0.000036 **99.** 0.0000000000000000028

101. 0.0000005 **103.** 200,000 **105.** 2.7×10^9 gal **107.** $-2x + 7$ **109.** $2y - 10$ **111.** $-x - 4$ **113.** 900,000,000; 9×10^8

115. 14,056,000; 1.4056×10^7 **117.** 2.5×10^{-9} m **119.** 0.00000031 m; 3.1×10^{-7} m **121.** $9a^{13}$ **123.** -5 **125.** answers may vary

127. a. 1.3×10^1 **b.** 4.4×10^7 **c.** 6.1×10^{-2} **129.** answers may vary **131.** $\dfrac{1}{x^{9s}}$ **133.** a^{4m+5}

Section 5.3

Vocabulary, Readiness & Video Check **1.** binomial **3.** trinomial **5.** constant **7.** $3; x^2, -3x, 5$ **9.** the replacement value for the variable **11.** $2; 9ab$

Exercise Set 5.3 **1.** $1; -3x; 5$ **3.** $-5; 3.2; 1; -5$ **5.** 1; binomial **7.** 3; none of these **9.** 6; trinomial **11.** 4; binomial **13. a.** -6 **b.** -11 **15. a.** -2 **b.** 4 **17. a.** -15 **b.** -10 **19.** 184 ft **21.** 595.84 ft **23.** 593 thousand **25.** 427.8 thousand **27.** $-11x$

29. $23x^3$ **31.** $16x^2 - 7$ **33.** $12x^2 - 13$ **35.** $7s$ **37.** $-1.1y^2 + 4.8$ **39.** $\dfrac{5}{6}x^4 - 7x^3 - 19$ **41.** $\dfrac{3}{20}x^3 + 6x^2 - \dfrac{13}{20}x - \dfrac{1}{10}$

43. $4x^2 + 7x + x^2 + 5x; 5x^2 + 12x$ **45.** $5x + 3 + 4x + 3 + 2x + 6 + 3x + 7x; 21x + 12$ **47.** 2, 1, 1, 0; 2 **49.** 4, 0, 4, 3; 4
51. $9ab - 11a$ **53.** $4x^2 - 7xy + 3y^2$ **55.** $-3xy^2 + 4$ **57.** $14y^3 - 19 - 16a^2b^2$ **59.** $7x^2 + 0x + 3$ **61.** $x^3 + 0x^2 + 0x - 64$
63. $5y^3 + 0y^2 + 2y - 10$ **65.** $2y^4 + 0y^3 + 0y^2 + 8y + 0y^0$ or $2y^4 + 0y^3 + 0y^2 + 8y + 0$ **67.** $6x^5 + 0x^4 + x^3 + 0x^2 - 3x + 15$
69. $10x + 19$ **71.** $-x + 5$ **73.** answers may vary **75.** answers may vary **77.** x^{13} **79.** a^3b^{10} **81.** $2y^{20}$ **83.** answers may vary
85. answers may vary **87.** $11.1x^2 - 7.97x + 10.76$

Section 5.4

Vocabulary, Readiness & Video Check **1.** $-14y$ **3.** $7x$ **5.** $5m^2 + 2m$ **7.** $-3y^2$ and $2y^2$; $-4y$ and y **9.** We're translating a subtraction problem. Order matters when subtracting, so we need to be careful that the order of the expressions is correct.

Exercise Set 5.4 **1.** $12x + 12$ **3.** $-3x^2 + 10$ **5.** $-3x^2 + 4$ **7.** $-y^2 - 3y - 1$ **9.** $7.9x^3 + 4.4x^2 - 3.4x - 3$

11. $\dfrac{1}{2}m^2 - \dfrac{7}{10}m + \dfrac{13}{16}$ **13.** $8t^2 - 4$ **15.** $15a^3 + a^2 - 3a + 16$ **17.** $-x + 14$ **19.** $7x^2$ **21.** $-2x + 9$ **23.** $2x^2 + 7x - 16$

25. $2x^2 + 11x$ **27.** $-0.2x^2 + 0.2x - 2.2$ **29.** $\dfrac{2}{5}z^2 - \dfrac{3}{10}z + \dfrac{7}{20}$ **31.** $-2z^2 - 16z + 6$ **33.** $2u^5 - 10u^2 + 11u - 9$ **35.** $5x - 9$

37. $4x - 3$ **39.** $11y + 7$ **41.** $-2x^2 + 8x - 1$ **43.** $14x + 18$ **45.** $3a^2 - 6a + 11$ **47.** $3x - 3$ **49.** $7x^2 - 4x + 2$
51. $7x^2 - 2x + 2$ **53.** $4y^2 + 12y + 19$ **55.** $-15x + 7$ **57.** $-2a - b + 1$ **59.** $3x^2 + 5$ **61.** $6x^2 - 2xy + 19y^2$

63. $8r^2s + 16rs - 8 + 7r^2s^2$ **65.** $(x^2 + 7x + 4)$ ft **67.** $\left(\dfrac{19}{2}x + 3\right)$ units **69.** $(3y^2 + 4y + 11)$ m **71.** $-6.6x^2 - 1.8x - 1.8$

73. $6x^2$ **75.** $-12x^8$ **77.** $200x^3y^2$ **79.** $2; 2$ **81.** $4; 3; 3; 4$ **83.** b **85.** e **87. a.** $4z$ **b.** $3z^2$ **c.** $-4z$ **d.** $3z^2$; answers may vary
89. a. m^3 **b.** $3m$ **c.** $-m^3$ **d.** $-3m$; answers may vary **91.** $915x^2 + 18,701x + 55,765$

Section 5.5

Vocabulary, Readiness & Video Check **1.** distributive **3.** $(5y - 1)(5y - 1)$ **5.** x^8 **7.** cannot simplify **9.** x^{14} **11.** $2x^7$
13. No. The monomials are unlike terms. **15.** Three times: First $(a - 2)$ is distributed to a and 7, and then a is distributed to $(a - 2)$ and 7 is distributed to $(a - 2)$.

Exercise Set 5.5 **1.** $24x^3$ **3.** x^4 **5.** $-28n^{10}$ **7.** $-12.4x^{12}$ **9.** $-\dfrac{2}{15}y^3$ **11.** $-24x^8$ **13.** $6x^2 + 15x$ **15.** $7x^3 + 14x^2 - 7x$

17. $-2a^2 - 8a$ **19.** $6x^3 - 9x^2 + 12x$ **21.** $12a^5 + 45a^2$ **23.** $-6a^4 + 4a^3 - 6a^2$ **25.** $6x^5y - 3x^4y^3 + 24x^2y^4$

27. $-4x^3y + 7x^2y^2 - xy^3 - 3y^4$ **29.** $4x^4 - 3x^3 + \dfrac{1}{2}x^2$ **31.** $x^2 + 7x + 12$ **33.** $a^2 + 5a - 14$ **35.** $x^2 + \dfrac{1}{3}x - \dfrac{2}{9}$

37. $12x^4 + 25x^2 + 7$ **39.** $12x^2 - 29x + 15$ **41.** $1 - 7a + 12a^2$ **43.** $4y^2 - 16y + 16$ **45.** $x^3 - 5x^2 + 13x - 14$
47. $x^4 + 5x^3 - 3x^2 - 11x + 20$ **49.** $10a^3 - 27a^2 + 26a - 12$ **51.** $49x^2y^2 - 14xy^2 + y^2$ **53.** $12x^2 - 64x - 11$
55. $2x^3 + 10x^2 + 11x - 3$ **57.** $2x^4 + 3x^3 - 58x^2 + 4x + 63$ **59.** $8.4y^7$ **61.** $-3x^3 - 6x^2 + 24x$ **63.** $2x^2 + 39x + 19$

65. $x^2 - \dfrac{2}{7}x - \dfrac{3}{49}$ **67.** $9y^2 + 30y + 25$ **69.** $a^3 - 2a^2 - 18a + 24$ **71.** $(4x^2 - 25)$ sq yd **73.** $(6x^2 - 4x)$ sq in.

75. $5a + 15a = 20a$; $5a - 15a = -10a$; $5a \cdot 15a = 75a^2$; $\dfrac{5a}{15a} = \dfrac{1}{3}$ **77.** $-3y^5 + 9y^4$, cannot be simplified; $-3y^5 - 9y^4$,

cannot be simplified; $-3y^5 \cdot 9y^4 = -27y^9$; $\dfrac{-3y^5}{9y^4} = -\dfrac{y}{3}$ **79. a.** $6x + 12$ **b.** $9x^2 + 36x + 35$; answers may vary   **81.** $13x - 7$

83. $30x^2 - 28x + 6$ **85.** $-7x + 5$ **87.** $x^2 + 3x$ **89.** $x + 2x^2$; $x(1 + 2x)$ **91.** $11a$ **93.** $25x^2 + 4y^2$ **95. a.** $a^2 - b^2$
b. $4x^2 - 9y^2$ **c.** $16x^2 - 49$ **d.** answers may vary

Section 5.6

Vocabulary, Readiness & Video Check   1. false   **3.** false   **5.** a binomial times a binomial   **7.** Multiplying gives you four terms,
and the two like terms will always subtract out.

Exercise Set 5.6   1. $x^2 + 7x + 12$ **3.** $x^2 + 5x - 50$ **5.** $5x^2 + 4x - 12$ **7.** $4y^2 - 25y + 6$ **9.** $6x^2 + 13x - 5$

11. $6y^3 + 4y^2 + 42y + 28$ **13.** $x^2 + \dfrac{1}{3}x - \dfrac{2}{9}$ **15.** $0.08 - 2.6a + 15a^2$ **17.** $2x^2 + 9xy - 5y^2$ **19.** $x^2 + 4x + 4$

21. $4a^2 - 12a + 9$ **23.** $9a^2 - 30a + 25$ **25.** $x^4 + x^2 + 0.25$ **27.** $y^2 - \dfrac{4}{7}y + \dfrac{4}{49}$ **29.** $4x^2 - 4x + 1$ **31.** $25x^2 + 90x + 81$

33. $9x^2 - 42xy + 49y^2$ **35.** $16m^2 + 40mn + 25n^2$ **37.** $25x^8 - 30x^4 + 9$ **39.** $a^2 - 49$ **41.** $x^2 - 36$ **43.** $9x^2 - 1$ **45.** $x^4 - 25$

47. $4y^4 - 1$ **49.** $16 - 49x^2$ **51.** $9x^2 - \dfrac{1}{4}$ **53.** $81x^2 - y^2$ **55.** $4m^2 - 25n^2$ **57.** $a^2 + 9a + 20$ **59.** $a^2 - 14a + 49$

61. $12a^2 - a - 1$ **63.** $x^2 - 4$ **65.** $9a^2 + 6a + 1$ **67.** $4x^2 + 3xy - y^2$ **69.** $\dfrac{1}{9}a^4 - 49$ **71.** $6b^2 - b - 35$ **73.** $x^4 - 100$

75. $16x^2 - 25$ **77.** $25x^2 - 60xy + 36y^2$ **79.** $4r^2 - 9s^2$ **81.** $(4x^2 + 4x + 1)$ sq ft   **83.** $\dfrac{5b^5}{7}$ **85.** $-\dfrac{2a^{10}}{b^5}$ **87.** $\dfrac{2y^8}{3}$ **89.** c

91. d   **93.** 2   **95.** $(x^4 - 3x^2 + 1)$ sq m   **97.** $(24x^2 - 32x + 8)$ sq m   **99.** answers may vary   **101.** answers may vary

Integrated Review   1. $35x^5$ **2.** $-32y^9$ **3.** -16 **4.** 16 **5.** $2x^2 - 9x - 5$ **6.** $3x^2 + 13x - 10$ **7.** $3x - 4$ **8.** $4x + 3$ **9.** $7x^6y^2$

10. $\dfrac{10b^6}{7}$ **11.** $144m^{14}n^{12}$ **12.** $64y^{27}z^{30}$ **13.** $16y^2 - 9$ **14.** $49x^2 - 1$ **15.** $\dfrac{y^{45}}{x^{63}}$ **16.** $\dfrac{1}{64}$ **17.** $\dfrac{x^{27}}{27}$ **18.** $\dfrac{r^{58}}{16s^{14}}$ **19.** $2x^2 - 2x - 6$

20. $6x^2 + 13x - 11$ **21.** $2.5y^2 - 6y - 0.2$ **22.** $8.4x^2 - 6.8x - 4.2$ **23.** $2y^2 - 6y - 1$ **24.** $6z^2 + 2z + \dfrac{11}{2}$

25. $x^2 + 8x + 16$ **26.** $y^2 - 18y + 81$ **27.** $2x + 8$ **28.** $2y - 18$ **29.** $7x^2 - 10xy + 4y^2$ **30.** $-a^2 - 3ab + 6b^2$
31. $x^3 + 2x^2 - 16x + 3$ **32.** $x^3 - 2x^2 - 5x - 2$ **33.** $6x^2 - x - 70$ **34.** $20x^2 + 21x - 5$ **35.** $2x^3 - 19x^2 + 44x - 7$

36. $5x^3 + 9x^2 - 17x + 3$ **37.** $4x^2 - \dfrac{25}{81}$ **38.** $144y^2 - \dfrac{9}{49}$

Section 5.7

Vocabulary, Readiness & Video Check   1. dividend; quotient; divisor   **3.** a^2 **5.** y **7.** the common denominator

Exercise Set 5.7   1. $12x^3 + 3x$ **3.** $4x^3 - 6x^2 + x + 1$ **5.** $5p^2 + 6p$ **7.** $-\dfrac{3}{2x} + 3$ **9.** $-3x^2 + x - \dfrac{4}{x^3}$ **11.** $-1 + \dfrac{3}{2x} - \dfrac{7}{4x^4}$

13. $x + 1$ **15.** $2x + 3$ **17.** $2x + 1 + \dfrac{7}{x - 4}$ **19.** $3a^2 - 3a + 1 + \dfrac{2}{3a + 2}$ **21.** $4x + 3 - \dfrac{2}{2x + 1}$ **23.** $2x^2 + 6x - 5 - \dfrac{2}{x - 2}$

25. $x + 6$ **27.** $x^2 + 3x + 9$ **29.** $-3x + 6 - \dfrac{11}{x + 2}$ **31.** $2b - 1 - \dfrac{6}{2b - 1}$ **33.** $ab - b^2$ **35.** $4x + 9$ **37.** $x + 4xy - \dfrac{y}{2}$

39. $2b^2 + b + 2 - \dfrac{12}{b + 4}$ **41.** $y^2 + 5y + 10 + \dfrac{24}{y - 2}$ **43.** $-6x - 12 - \dfrac{19}{x - 2}$ **45.** $x^3 - x^2 + x$ **47.** 3 **49.** -4 **51.** $3x$

53. $9x$ **55.** $(3x^3 + x - 4)$ ft   **57.** $(2x + 5)$ m   **59.** answers may vary   **61.** c

Chapter 5 Vocabulary Check   1. term   **2.** FOIL   **3.** trinomial   **4.** degree of a polynomial   **5.** binomial   **6.** coefficient
7. degree of a term   **8.** monomial   **9.** polynomials   **10.** distributive

Chapter 5 Review   1. base: 3; exponent: 2   **2.** base: -5; exponent: 4   **3.** base: 5; exponent: 4   **4.** base: x; exponent: 6   **5.** 512
6. 36   **7.** -36 **8.** -65 **9.** 1   **10.** 1   **11.** y^9 **12.** x^{14} **13.** $-6x^{11}$ **14.** $-20y^7$ **15.** x^8 **16.** y^{15} **17.** $81y^{24}$ **18.** $8x^9$ **19.** x^5

20. z^7 **21.** a^4b^3 **22.** x^3y^5 **23.** $\dfrac{x^3y^4}{4}$ **24.** $\dfrac{x^6y^6}{4}$ **25.** $40a^{19}$ **26.** $36x^3$ **27.** 3   **28.** 9   **29.** b   **30.** c   **31.** $\dfrac{1}{49}$ **32.** $-\dfrac{1}{49}$ **33.** $\dfrac{2}{x^4}$

34. $\dfrac{1}{16x^4}$ **35.** 125   **36.** $\dfrac{9}{4}$ **37.** $\dfrac{17}{16}$ **38.** $\dfrac{1}{42}$ **39.** x^8 **40.** z^8 **41.** r **42.** y^3 **43.** c^4 **44.** $\dfrac{x^3}{y^3}$ **45.** $\dfrac{1}{x^6y^{13}}$ **46.** $\dfrac{a^{10}}{b^{10}}$ **47.** 2.7×10^{-4}

48. 8.868×10^{-1} **49.** 8.08×10^7 **50.** 8.68×10^5 **51.** 1.303×10^8 **52.** 1.5×10^5 **53.** 867,000   **54.** 0.00386
55. 0.00086   **56.** 893,600   **57.** 1,431,280,000,000,000   **58.** 0.0000000001   **59.** 0.016   **60.** 400,000,000,000   **61.** 5   **62.** 2

63. 5 **64.** 6 **65.** 4000 ft; 3984 ft; 3856 ft; 3600 ft **66.** 22; 78; 154.02; 400 **67.** $2a^2$ **68.** $-4y$ **69.** $15a^2 + 4a$ **70.** $22x^2 + 3x + 6$
71. $-6a^2b - 3b^2 - q^2$ **72.** cannot be combined **73.** $8x^2 + 3x + 6$ **74.** $2x^5 + 3x^4 + 4x^3 + 9x^2 + 7x + 6$
75. $-7y^2 - 1$ **76.** $-6m^7 - 3x^4 + 7m^6 - 4m^2$ **77.** $-x^2 - 6xy - 2y^2$ **78.** $x^6 + 4xy + 2y^2$ **79.** $-5x^2 + 5x + 1$
80. $-2x^2 - x + 20$ **81.** $6x + 30$ **82.** $9x - 63$ **83.** $8a + 28$ **84.** $54a - 27$ **85.** $-7x^3 - 35x$ **86.** $-32y^3 + 48y$
87. $-2x^3 + 18x^2 - 2x$ **88.** $-3a^3b - 3a^2b - 3ab^2$ **89.** $-6a^4 + 8a^2 - 2a$ **90.** $42b^4 - 28b^2 + 14b$ **91.** $2x^2 - 12x - 14$
92. $6x^2 - 11x - 10$ **93.** $4a^2 + 27a - 7$ **94.** $42a^2 + 11a - 3$ **95.** $x^4 + 7x^3 + 4x^2 + 23x - 35$
96. $x^6 + 2x^5 + x^2 + 3x + 2$ **97.** $x^4 + 4x^3 + 4x^2 - 16$ **98.** $x^6 + 8x^4 + 16x^2 - 16$ **99.** $x^3 + 21x^2 + 147x + 343$
100. $8x^3 - 60x^2 + 150x - 125$ **101.** $x^2 + 14x + 49$ **102.** $x^2 - 10x + 25$ **103.** $9x^2 - 42x + 49$ **104.** $16x^2 + 16x + 4$
105. $25x^2 - 90x + 81$ **106.** $25x^2 - 1$ **107.** $49x^2 - 16$ **108.** $a^2 - 4b^2$ **109.** $4x^2 - 36$ **110.** $16a^4 - 4b^2$ **111.** $(9x^2 - 6x + 1)$ sq m
112. $(5x^2 - 3x - 2)$ sq mi **113.** $\dfrac{1}{7} + \dfrac{3}{x} + \dfrac{7}{x^2}$ **114.** $-a^2 + 3b - 4$ **115.** $a + 1 + \dfrac{6}{a-2}$ **116.** $4x + \dfrac{7}{x+5}$
117. $a^2 + 3a + 8 + \dfrac{22}{a-2}$ **118.** $3b^2 - 4b - \dfrac{1}{3b-2}$ **119.** $2x^3 - x^2 + 2 - \dfrac{1}{2x-1}$ **120.** $-x^2 - 16x - 117 - \dfrac{684}{x-6}$
121. $\left(5x - 1 + \dfrac{20}{x^2}\right)$ ft **122.** $(7a^3 b^6 + a - 1)$ units **123.** 27 **124.** $-\dfrac{1}{8}$ **125.** $4x^4 y^7$ **126.** $\dfrac{2x^6}{3}$ **127.** $\dfrac{27a^{12}}{b^6}$ **128.** $\dfrac{x^{16}}{16y^{12}}$
129. $9a^2 b^8$ **130.** $2y^2 - 10$ **131.** $11x - 5$ **132.** $5x^2 + 3x - 2$ **133.** $5y^2 - 3y - 1$ **134.** $6x^2 + 11x - 10$ **135.** $28x^3 + 12x$
136. $28x^2 - 71x + 18$ **137.** $x^3 + x^2 - 18x + 18$ **138.** $25x^2 + 40x + 16$ **139.** $36x^2 - 9$ **140.** $4a - 1 + \dfrac{2}{a^2} - \dfrac{5}{2a^3}$
141. $x - 3 + \dfrac{25}{x+5}$ **142.** $2x^2 + 7x + 5 + \dfrac{19}{2x-3}$

Chapter 5 Test **1.** 32 **2.** 81 **3.** -81 **4.** $\dfrac{1}{64}$ **5.** $-15x^{11}$ **6.** y^5 **7.** $\dfrac{1}{r^5}$ **8.** $\dfrac{16y^{14}}{x^2}$ **9.** $\dfrac{1}{6xy^8}$ **10.** 5.63×10^5 **11.** 8.63×10^{-5}
12. 0.0015 **13.** 62,300 **14.** 0.036 **15. a.** 4, 3; 7, 3; 1, 4; -2, 0 **b.** 4 **16.** $-2x^2 + 12x + 11$ **17.** $16x^3 + 7x^2 - 3x - 13$
18. $-3x^3 + 5x^2 + 4x + 5$ **19.** $x^3 + 8x^2 + 3x - 5$ **20.** $3x^3 + 22x^2 + 41x + 14$ **21.** $6x^4 - 9x^3 + 21x^2$ **22.** $3x^2 + 16x - 35$
23. $9x^2 - \dfrac{1}{25}$ **24.** $16x^2 - 16x + 4$ **25.** $64x^2 + 48x + 9$ **26.** $x^4 - 81b^2$ **27.** 1001 ft; 985 ft; 857 ft; 601 ft **28.** $(4x^2 - 9)$ sq in.
29. $\dfrac{x}{2y} + \dfrac{1}{4} - \dfrac{7}{8y}$ **30.** $x + 2$ **31.** $9x^2 - 6x + 4 - \dfrac{16}{3x+2}$

Cumulative Review **1. a.** 11, 112 **b.** 0, 11, 112 **c.** $-3, -2, 0, 11, 112$ **d.** $-3, -2, 0, \dfrac{1}{4}, 11, 112$ **e.** $\sqrt{2}$ **f.** $-2, 0, \dfrac{1}{4}, 112, -3, 11, \sqrt{2}$;
Sec. 1.2, Ex. 11 **2. a.** 7.2 **b.** 0 **c.** $\dfrac{1}{2}$; Sec. 1.2 **3. a.** 9 **b.** 125 **c.** 16 **d.** 7 **e.** $\dfrac{9}{49}$ **f.** 0.36; Sec. 1.3, Ex. 1 **4. a.** $\dfrac{1}{4}$ **b.** $2\dfrac{5}{12}$; Sec. R.2
5. $\dfrac{1}{4}$; Sec. 1.3, Ex. 4 **6.** $\dfrac{3}{25}$; Sec. 1.3 **7. a.** $x + 3$ **b.** $3x$ **c.** $7.3 \div x$ or $\dfrac{7.3}{x}$ **d.** $10 - x$ **e.** $5x + 7$; Sec. 1.3, Ex. 9 **8.** 41; Sec. 1.3
9. 6.7; Sec. 1.4, Ex. 10 **10.** no; Sec. 1.5 **11. a.** $\dfrac{1}{2}$ **b.** 9; Sec. 1.5, Ex. 8 **12. a.** -33 **b.** 5; Sec. 1.5 **13.** 3; Sec. 1.6, Ex. 11a **14.** -8;
Sec. 1.6 **15.** -70; Sec. 1.6, Ex. 11d **16.** 150; Sec. 1.6 **17.** $15x + 10$; Sec. 1.8, Ex. 8 **18.** $-6x + 9$; Sec. 1.8 **19.** $-2y - 0.6z + 2$;
Sec. 1.8, Ex. 9 **20.** $-4x^2 + 24x - 4$; Sec. 1.8 **21.** $-9x - y + 2z - 6$; Sec. 1.8, Ex. 10 **22.** $4xy - 6y + 2$; Sec. 1.8 **23.** $a = 19$;
Sec. 2.1, Ex. 6 **24.** $x = -\dfrac{1}{2}$; Sec. 2.1 **25.** $y = 140$; Sec. 2.2, Ex. 4 **26.** $j = \dfrac{12}{5}$; Sec. 2.2 **27.** $x = 4$; Sec. 2.3, Ex. 5 **28.** $x = 1$; Sec. 2.3
29. 10; Sec. 2.4, Ex. 2 **30.** $(x + 7) - 2x$ or $-x + 7$; Sec. 2.1 **31.** 40 feet; Sec. 2.5, Ex. 2 **32.** undefined; Sec. 1.6 **33.** 800; Sec. 2.6,
Ex. 2 **34.** ◄————○————►; Sec. 2.7 **35.** ◄————○————►; $\{x \mid x \le 4\}$; Sec. 2.7, Ex. 7 **36. a.** 25 **b.** -25 **c.** 50;
          5   4
Sec. 5.1 **37. a.** x^{11} **b.** $\dfrac{t^4}{16}$ **c.** $81y^{10}$; Sec. 5.1, Ex. 33 **38.** z^4; Sec. 5.1 **39.** $\dfrac{b^3}{27a^6}$; Sec. 5.2, Ex. 10 **40.** $-15x^{16}$; Sec. 5.1 **41.** $\dfrac{1}{25y^6}$;
Sec. 5.2, Ex. 14 **42.** $\dfrac{1}{9}$; Sec. 5.2 **43.** $10x^3$; Sec. 5.3, Ex. 8 **44.** $4y^2 - 8$; Sec. 5.4 **45.** $5x^2 - 3x - 3$; Sec. 5.3, Ex. 9 **46.** $100x^4 - 9$;
Sec. 5.6 **47.** $7x^3 + 14x^2 + 35x$; Sec. 5.5, Ex. 4 **48.** $100x^4 + 60x^2 + 9$; Sec. 5.6 **49.** $3x^3 - 4 + \dfrac{1}{x}$; Sec. 5.7, Ex. 2

Chapter 6 Factoring Polynomials

Section 6.1

Vocabulary, Readiness & Video Check **1.** factors **3.** least **5.** false **7.** $2 \cdot 7$ **9.** 3 **11.** 5 **13.** The GCF of a list of numbers is
the largest number that is a factor of all numbers in the list. **15.** When factoring out a GCF, the number of terms in the other factor
should be the same as the number of terms as your original polynomial.

Exercise Set 6.1 **1.** 4 **3.** 6 **5.** 1 **7.** y^2 **9.** z^7 **11.** xy^2 **13.** 7 **15.** $4y^3$ **17.** $5x^2$ **19.** $3x^3$ **21.** $9x^2y$ **23.** $10a^6b$ **25.** $3(a + 2)$
27. $15(2x - 1)$ **29.** $x^2(x + 5)$ **31.** $2y^3(3y + 1)$ **33.** $2x(16y - 9x)$ **35.** $4(x - 2y + 1)$ **37.** $3x(2x^2 - 3x + 4)$

39. $a^2b^2(a^5b^4 - a + b^3 - 1)$ **41.** $5xy(x^2 - 3x + 2)$ **43.** $4(2x^5 + 4x^4 - 5x^3 + 3)$ **45.** $\dfrac{1}{3}x(x^3 + 2x^2 - 4x^4 + 1)$

47. $(x^2 + 2)(y + 3)$ **49.** $(y + 4)(z + 3)$ **51.** $(z^2 - 6)(r + 1)$ **53.** $-2(x + 7)$ **55.** $-x^5(2 - x^2)$ **57.** $-3a^2(2a^2 - 3a + 1)$
59. $(x + 2)(x^2 + 5)$ **61.** $(x + 3)(5 + y)$ **63.** $(3x - 2)(2x^2 + 5)$ **65.** $(5m^2 + 6n)(m + 1)$ **67.** $(y - 4)(2 + x)$
69. $(2x + 1)(x^2 + 4)$ **71.** not factorable by grouping   **73.** $(x - 2y)(4x - 3)$ **75.** $(5q - 4p)(q - 1)$ **77.** $2(2y - 7)(3x^2 - 1)$
79. $3(2a + 3b^2)(a + b)$ **81.** $x^2 + 7x + 10$ **83.** $b^2 - 3b - 4$ **85.** $2, 6$ **87.** $-1, -8$ **89.** $-2, 5$ **91.** $-8, 3$ **93.** d   **95.** factored
97. not factored   **99. a.** 1265 million units   **b.** 1360 million units   **c.** 2560 million units   **d.** $5(3x^2 - 50x + 440)$
101. a. 9016 thousand tons   **b.** 8372 thousand tons   **c.** $-322(x^2 - 3x - 26)$ or $322(-x^2 + 3x + 26)$ **103.** $4x^2 - \pi x^2; x^2(4 - \pi)$
105. $(x^3 - 1)$ units   **107.** answers may vary   **109.** answers may vary

Section 6.2

Vocabulary, Readiness & Video Check   **1.** true   **3.** false   **5.** $+5$ **7.** -3 **9.** $+2$ **11.** 15 is positive, so its factors would have to
be either both positive or both negative. Since the factors need to sum to -8, both factors must be negative.

Exercise Set 6.2   **1.** $(x + 6)(x + 1)$ **3.** $(y - 9)(y - 1)$ **5.** $(x - 3)(x - 3)$ or $(x - 3)^2$ **7.** $(x - 6)(x + 3)$
9. $(x + 10)(x - 7)$ **11.** prime   **13.** $(x + 5y)(x + 3y)$ **15.** $(a^2 - 5)(a^2 + 3)$ **17.** $(m + 13)(m + 1)$ **19.** $(t - 2)(t + 12)$
21. $(a - 2b)(a - 8b)$ **23.** $2(z + 8)(z + 2)$ **25.** $2x(x - 5)(x - 4)$ **27.** $(x - 4y)(x + y)$ **29.** $(x + 12)(x + 3)$
31. $(x^2 - 2)(x^2 + 1)$ **33.** $(r - 12)(r - 4)$ **35.** $(x + 2y)(x - y)$ **37.** $3(x + 5)(x - 2)$ **39.** $3(x^2 - 18)(x^2 - 2)$
41. $(x - 24)(x + 6)$ **43.** prime   **45.** $(x - 5)(x - 3)$ **47.** $6x(x + 4)(x + 5)$ **49.** $4y(x^2 + x - 3)$ **51.** $(x - 7)(x + 3)$
53. $(x + 5y)(x + 2y)$ **55.** $2(t + 8)(t + 4)$ **57.** $x(x - 6)(x + 4)$ **59.** $2t^3(t - 4)(t - 3)$ **61.** $5xy(x - 8y)(x + 3y)$

63. $3(m - 9)(m - 6)$ **65.** $-1(x - 11)(x - 1)$ **67.** $\dfrac{1}{2}(y - 11)(y + 2)$ **69.** $x(xy - 4)(xy + 5)$ **71.** $2x^2 + 11x + 5$

73. $15y^2 - 17y + 4$ **75.** $9a^2 + 23ab - 12b^2$ **77.** $x^2 + 5x - 24$ **79.** answers may vary   **81.** $2x^2 + 28x + 66; 2(x + 3)(x + 11)$

83. $-16(t - 5)(t + 1)$ **85.** $\left(x + \dfrac{1}{4}\right)\left(x + \dfrac{1}{4}\right)$ or $\left(x + \dfrac{1}{4}\right)^2$ **87.** $(x + 1)(z - 10)(z + 7)$ **89.** $15; 28; 39; 48; 55; 60; 63; 64$

91. $9; 12; 21$ **93.** $(x^n + 10)(x^n - 2)$

Section 6.3

Vocabulary, Readiness & Video Check   **1.** d   **3.** c   **5.** Consider the factors of the first and last terms and the signs of the
trinomial. Continue to check by multiplying until we get the middle term of the trinomial.

Exercise Set 6.3   **1.** $x + 4$ **3.** $10x - 1$ **5.** $4x - 3$ **7.** $(2x + 3)(x + 5)$ **9.** $(y - 1)(8y - 9)$ **11.** $(2x + 1)(x - 5)$
13. $(4r - 1)(5r + 8)$ **15.** $(10x + 1)(x + 3)$ **17.** $(3x - 2)(x + 1)$ **19.** $(3x - 5y)(2x - y)$ **21.** $(3m - 5)(5m + 3)$
23. $(x - 4)(x - 5)$ **25.** $(2x + 11)(x - 9)$ **27.** $(7t + 1)(t - 4)$ **29.** $(3a + b)(a + 3b)$ **31.** $(7p + 1)(7p - 2)$
33. $(6x - 7)(3x + 2)$ **35.** prime   **37.** $(8x + 3)(3x + 4)$ **39.** $x(3x + 2)(4x + 1)$ **41.** $3(7b + 5)(b - 3)$
43. $(3z + 4)(4z - 3)$ **45.** $2y^2(3x - 10)(x + 3)$ **47.** $(2x - 7)(2x + 3)$ **49.** $3(x^2 - 14x + 21)$ **51.** $(4x + 9y)(2x - 3y)$
53. $-1(x - 6)(x + 4)$ **55.** $x(4x + 3)(x - 3)$ **57.** $(4x - 9)(6x - 1)$ **59.** $b(8a - 3)(5a + 3)$ **61.** $2x(3x + 2)(5x + 3)$
63. $2y(3y + 5)(y - 3)$ **65.** $5x^2(2x - y)(x + 3y)$ **67.** $-1(2x - 5)(7x - 2)$ **69.** $p^2(4p - 5)(4p - 5)$ or $p^2(4p - 5)^2$
71. $-1(2x + 1)(x - 5)$ **73.** $-4(12x - 1)(x - 1)$ **75.** $(2t^2 + 9)(t^2 - 3)$ **77.** prime   **79.** $a(6a^2 + b^2)(a^2 + 6b^2)$
81. $x^2 - 16$ **83.** $x^2 + 4x + 4$ **85.** $4x^2 - 4x + 1$ **87.** $25-34$ **89.** answers may vary   **91.** no   **93.** $4x^2 + 21x + 5; (4x + 1)(x + 5)$

95. $\left(2x + \dfrac{1}{2}\right)\left(2x + \dfrac{1}{2}\right)$ or $\left(2x + \dfrac{1}{2}\right)^2$ **97.** $(y - 1)^2(4x + 5)(x + 5)$ **99.** $2; 14$ **101.** 2   **103.** answers may vary

Section 6.4

Vocabulary, Readiness & Video Check   **1.** a   **3.** b   **5.** This gives us a four-term polynomial, which may be factored by grouping.

Exercise Set 6.4   **1.** $(x + 3)(x + 2)$ **3.** $(y + 8)(y - 2)$ **5.** $(8x - 5)(x - 3)$ **7.** $(5x^2 - 3)(x^2 + 5)$ **9. a.** $9, 2$ **b.** $9x + 2x$
c. $(2x + 3)(3x + 1)$ **11. a.** $-20, -3$ **b.** $-20x - 3x$ **c.** $(3x - 4)(5x - 1)$ **13.** $(3y + 2)(7y + 1)$ **15.** $(7x - 11)(x + 1)$
17. $(5x - 2)(2x - 1)$ **19.** $(2x - 5)(x - 1)$ **21.** $(2x + 3)(2x + 3)$ or $(2x + 3)^2$ **23.** $(2x + 3)(2x - 7)$ **25.** $(5x - 4)(2x - 3)$
27. $x(2x + 3)(x + 5)$ **29.** $2(8y - 9)(y - 1)$ **31.** $(2x - 3)(3x - 2)$ **33.** $3(3a + 2)(6a - 5)$ **35.** $a(4a + 1)(5a + 8)$
37. $3x(4x + 3)(x - 3)$ **39.** $y(3x + y)(x + y)$ **41.** prime   **43.** $6(a + b)(4a - 5b)$ **45.** $p^2(15p + q)(p + 2q)$
47. $(7 + x)(5 + x)$ or $(x + 7)(x + 5)$ **49.** $(6 - 5x)(1 - x)$ or $(5x - 6)(x - 1)$ **51.** $x^2 - 4$ **53.** $y^2 + 8y + 16$ **55.** $81z^2 - 25$
57. $16x^2 - 24x + 9$ **59.** $10x^2 + 45x + 45; 5(2x + 3)(x + 3)$ **61.** $(x^n + 2)(x^n + 3)$ **63.** $(3x^n - 5)(x^n + 7)$
65. answers may vary

Section 6.5

Calculator Explorations

	$x^2 - 2x + 1$	$x^2 - 2x - 1$	$(x - 1)^2$
$x = 5$	16	14	16
$x = -3$	16	14	16
$x = 2.7$	2.89	0.89	2.89
$x = -12.1$	171.61	169.61	171.61
$x = 0$	1	−1	1

Vocabulary, Readiness & Video Check   1. 1^2 **3.** 9^2 **5.** 3^2 **7.** $(3x)^2$ **9.** $(5a)^2$ **11.** $(6p^2)^2$ **13.** See if the first term is a square, say a^2, and the last term is a square, say b^2. Check to see if the middle term is $2 \cdot a \cdot b$ or $-2 \cdot a \cdot b$. **15.** First rewrite the original binomial so that each term is some quantity cubed. Your answers will then vary, depending on your interpretation.

Exercise Set 6.5   1. yes **3.** no **5.** no **7.** yes **9.** yes **11.** $(x + 11)^2$ **13.** $(x - 8)^2$ **15.** $(4a - 3)^2$ **17.** $3(x - 4)^2$
19. $(xy - 5)^2$ **21.** $m(m + 9)^2$ **23.** prime **25.** $(3x - 4y)^2$ **27.** $(x^2 + 2)^2$ **29.** $(x + 5)(x - 5)$ **31.** $(3 + 2z)(3 - 2z)$
33. prime **35.** $xy(x + 11y)(x - 11y)$ **37.** $(y + 9)(y - 5)$ **39.** $4(4x + 5)(4x - 5)$ **41.** $2y(3x + 1)(3x - 1)$
43. $(3x + 7)(3x - 7)$ **45.** $(x^2 + 9)(x + 3)(x - 3)$ **47.** $(x + 2y + 3)(x + 2y - 3)$ **49.** $(x + 8 + x^2)(x + 8 - x^2)$
51. $(x - 5 + y)(x - 5 - y)$ **53.** $(2x + 1 + z)(2x + 1 - z)$ **55.** $(m^2 + 1)(m + 1)(m - 1)$ **57.** $(x + 3)(x^2 - 3x + 9)$
59. $(z - 1)(z^2 + z + 1)$ **61.** $(m + n)(m^2 - mn + n^2)$ **63.** $y^2(x - 3)(x^2 + 3x + 9)$ **65.** $b(a + 2b)(a^2 - 2ab + 4b^2)$
67. $(5y - 2x)(25y^2 + 10xy + 4x^2)$ **69.** $(x^2 - y)(x^4 + x^2y + y^2)$ **71.** $(2x + 3y)(4x^2 - 6xy + 9y^2)$ **73.** $(x - 1)(x^2 + x + 1)$
75. $(x + 5)(x^2 - 5x + 25)$ **77.** $3y^2(x^2 + 3)(x^4 - 3x^2 + 9)$ **79.** 5 **81.** $-\dfrac{1}{3}$ **83.** 0 **85.** 5 **87.** $\left(x - \dfrac{1}{3}\right)^2$

89. $(x + 2 + y)(x + 2 - y)$ **91.** $(b - 4)(a + 4)(a - 4)$ **93.** $(x + 3 + 2y)(x + 3 - 2y)$ **95.** $(x^n + 10)(x^n - 10)$
97. 8 **99.** answers may vary **101.** $(x + 6)$ **103.** $a^2 + 2ab + b^2$ **105. a.** 2560 ft **b.** 1920 ft **c.** 13 sec **d.** $16(13 - t)(13 + t)$
107. a. 1456 feet **b.** 816 feet **c.** 10 seconds **d.** $16(10 + t)(10 - t)$

Integrated Review   1. $(x - 3)(x + 4)$ **2.** $(x - 8)(x - 2)$ **3.** $(x + 2)(x - 3)$ **4.** $(x + 1)^2$ **5.** $(x - 3)^2$ **6.** $(x + 2)(x - 1)$
7. $(x + 3)(x - 2)$ **8.** $(x + 3)(x + 4)$ **9.** $(x - 5)(x - 2)$ **10.** $(x - 6)(x + 5)$ **11.** $2(x + 7)(x - 7)$ **12.** $3(x + 5)(x - 5)$
13. $(x + 3)(x + 5)$ **14.** $(y - 7)(3 + x)$ **15.** $(x + 8)(x - 2)$ **16.** $(x - 7)(x + 4)$ **17.** $4x(x + 7)(x - 2)$
18. $6x(x - 5)(x + 4)$ **19.** $2(3x + 4)(2x + 3)$ **20.** $(2a - b)(4a + 5b)$ **21.** $(2a + b)(2a - b)$ **22.** $(x + 5y)(x - 5y)$
23. $(4 - 3x)(7 + 2x)$ **24.** $(5 - 2x)(4 + x)$ **25.** prime **26.** prime **27.** $(3y + 5)(2y - 3)$ **28.** $(4x - 5)(x + 1)$
29. $9x(2x^2 - 7x + 1)$ **30.** $4a(3a^2 - 6a + 1)$ **31.** $(4a - 7)^2$ **32.** $(5p - 7)^2$ **33.** $(7 - x)(2 + x)$ **34.** $(3 + x)(1 - x)$
35. $3x^2y(x + 6)(x - 4)$ **36.** $2xy(x + 5y)(x - y)$ **37.** $3xy(4x^2 + 81)$ **38.** $2xy^2(3x^2 + 4)$ **39.** $2xy(1 + 6x)(1 - 6x)$
40. $2x(x + 3)(x - 3)$ **41.** $(x + 6)(x + 2)(x - 2)$ **42.** $(x - 2)(x + 6)(x - 6)$ **43.** $2a^2(3a + 5)$ **44.** $2n(2n - 3)$
45. $(3x - 1)(x^2 + 4)$ **46.** $(x - 2)(x^2 + 3)$ **47.** $6(x + 2y)(x + y)$ **48.** $2(x + 4y)(6x - y)$ **49.** $(x + y)(5 + x)$
50. $(x - y)(7 + y)$ **51.** $(7t - 1)(2t - 1)$ **52.** prime **53.** $(3x + 5)(x - 1)$ **54.** $(7x - 2)(x + 3)$ **55.** $(1 - 10a)(1 + 2a)$
56. $(1 + 5a)(1 - 12a)$ **57.** $(x + 3)(x - 3)(x + 1)(x - 1)$ **58.** $(x + 3)(x - 3)(x + 2)(x - 2)$ **59.** $(x - 15)(x - 8)$
60. $(y + 16)(y + 6)$ **61.** prime **62.** $(4a - 7b)^2$ **63.** $(5p - 7q)^2$ **64.** $(7x + 3y)(x + 3y)$ **65.** $-1(x - 5)(x + 6)$
66. $-1(x - 2)(x - 4)$ **67.** $(3r - 1)(s + 4)$ **68.** $(x - 2)(x^2 + 1)$ **69.** $(x - 2y)(4x - 3)$ **70.** $(2x - y)(2x + 7z)$
71. $(x + 12y)(x - 3y)$ **72.** $(3x - 2y)(x + 4y)$ **73.** $(x^2 + 2)(x + 4)(x - 4)$ **74.** $(x^2 + 3)(x + 5)(x - 5)$
75. $x(x - 1)(x^2 + x + 1)$ **76.** $x^3(x + 1)(x^2 - x + 1)$ **77.** $(2x + 5y)(4x^2 - 10xy + 25y^2)$
78. $(3x - 4y)(9x^2 + 12xy + 16y^2)$ **79.** answers may vary **80.** yes; $9(x^2 + 9y^2)$

Section 6.6

Vocabulary, Readiness & Video Check   1. quadratic **3.** $3, -5$ **5.** One side of the equation must be a factored polynomial and the other side must be zero.

Exercise Set 6.6   1. $2, -1$ **3.** $6, 7$ **5.** $-9, -17$ **7.** $0, -6$ **9.** $0, 8$ **11.** $-\dfrac{3}{2}, \dfrac{5}{4}$ **13.** $\dfrac{7}{2}, -\dfrac{2}{7}$ **15.** $\dfrac{1}{2}, -\dfrac{1}{3}$ **17.** $-0.2, -1.5$ **19.** $9, 4$

21. $-4, 2$ **23.** $0, 7$ **25.** $0, -20$ **27.** $4, -4$ **29.** $8, -4$ **31.** $-3, 12$ **33.** $\dfrac{7}{3}, -2$ **35.** $\dfrac{8}{3}, -9$ **37.** $0, -\dfrac{1}{2}, \dfrac{1}{2}$ **39.** $\dfrac{17}{2}$ **41.** $\dfrac{3}{4}$

43. $-\dfrac{1}{2}, \dfrac{1}{2}$ **45.** $-\dfrac{3}{2}, -\dfrac{1}{2}, 3$ **47.** $-5, 3$ **49.** $-\dfrac{5}{6}, \dfrac{6}{5}$ **51.** $2, -\dfrac{4}{5}$ **53.** $-\dfrac{4}{3}, 5$ **55.** $-4, 3$ **57.** $0, 8, 4$ **59.** -7 **61.** $0, \dfrac{3}{2}$ **63.** $0, 1, -1$

65. $-6, \dfrac{4}{3}$ **67.** $\dfrac{6}{7}, 1$ **69.** $\dfrac{47}{45}$ **71.** $\dfrac{17}{60}$ **73.** $\dfrac{7}{10}$ **75.** didn't write equation in standard form; should be $x = 4$ or $x = -2$

77. answers may vary, for example, $(x - 6)(x + 1) = 0$ **79.** answers may vary, for example, $x^2 - 12x + 35 = 0$ **81. a.** $300; 304;$
$276; 216; 124; 0; -156$ **b.** 5 sec **c.** 304 ft **83.** $0, \dfrac{1}{2}$ **85.** $0, -15$

Section 6.7

Vocabulary, Readiness & Video Check   **1.** In applications, the context of the stated application needs to be considered. Each translated equation resulted in both a positive and a negative solution, and a negative solution is not appropriate for any of the stated application.

Exercise Set 6.7   **1.** width: x; length: $x + 4$ **3.** x and $x + 2$ if x is an odd integer   **5.** base: x; height: $4x + 1$ **7.** 11 units
9. 15 cm, 13 cm, 22 cm, 70 cm   **11.** base: 16 mi; height: 6 mi   **13.** 5 sec   **15.** width: 5 cm; length: 6 cm   **17.** 54 diagonals   **19.** 10 sides
21. -12 or 11   **23.** 14, 15   **25.** 13 feet   **27.** 5 in.   **29.** 12 mm, 16 mm, 20 mm   **31.** 10 km   **33.** 36 ft   **35.** 9.5 sec   **37.** 20%
39. length: 15 mi; width: 8 mi   **41.** 105 units   **43.** 2.2 million or 2,200,000   **45.** 2.4 million or 2,400,000   **47.** 2010   **49.** answers may vary
51. $\dfrac{4}{7}$ **53.** $\dfrac{3}{2}$ **55.** $\dfrac{1}{3}$ **57.** 8 m   **59.** 10 and 15   **61.** width of pool: 29 m; length of pool: 35 m   **63.** answers may vary

Chapter 6 Vocabulary Check   **1.** quadratic equation   **2.** Factoring   **3.** greatest common factor   **4.** perfect square trinomial
5. hypotenuse   **6.** leg   **7.** hypotenuse

Chapter 6 Review   **1.** $2x - 5$ **2.** $2x^4 + 1 - 5x^3$ **3.** $5(m + 6)$ **4.** $4x(5x^2 + 3x + 6)$ **5.** $(2x + 3)(3x - 5)$
6. $(x + 1)(5x - 1)$ **7.** $(x - 1)(3x + 2)$ **8.** $(a + 3b)(3a + b)$ **9.** $(2a + b)(5a + 7b)$ **10.** $(3x + 5)(2x - 1)$
11. $(x + 4)(x + 2)$ **12.** $(x - 8)(x - 3)$ **13.** prime   **14.** $(x - 6)(x + 1)$ **15.** $(x + 4)(x - 2)$ **16.** $(x + 6y)(x - 2y)$
17. $(x + 5y)(x + 3y)$ **18.** $2(3 - x)(12 + x)$ **19.** $4(8 + 3x - x^2)$ **20.** $5y(y - 6)(y - 4)$ **21.** $-48, 2$ **22.** factor out the
GCF, 3   **23.** $(2x + 1)(x + 6)$ **24.** $(2x + 3)(2x - 1)$ **25.** $(3x + 4y)(2x - y)$ **26.** prime   **27.** $(2x + 3)(x - 13)$
28. $(6x + 5y)(3x - 4y)$ **29.** $5y(2y - 3)(y + 4)$ **30.** $3y(4y - 1)(5y - 2)$ **31.** $5x^2 - 9x - 2; (5x + 1)(x - 2)$
32. $16x^2 - 28x + 6; 2(4x - 1)(2x - 3)$ **33.** $(x + 9)(x - 9)$ **34.** $(x + 6)^2$ **35.** $(2x + 3)(2x - 3)$ **36.** $(3t + 5s)(3t - 5s)$
37. prime   **38.** $(n - 9)^2$ **39.** $3(r + 6)^2$ **40.** $(3y - 7)^2$ **41.** $5m^6(m + 1)(m - 1)$ **42.** $(2x - 7y)^2$ **43.** $3y(x + y)^2$
44. $(4x^2 + 1)(2x + 1)(2x - 1)$ **45.** $(y + 7)(y - 3)$ **46.** $(x + 1)(x - 7)$ **47.** $(2 - 3y)(4 + 6y + 9y^2)$
48. $(1 - 4y)(1 + 4y + 16y^2)$ **49.** $6xy(x + 2)(x^2 - 2x + 4)$ **50.** $2x^2(x + 2y)(x^2 - 2xy + 4y^2)$ **51.** $(x - 1 + y)(x - 1 - y)$
52. $\pi h(R + r)(R - r)$ cu units   **53.** $-6, 2$ **54.** $-11, 7$ **55.** $0, -1, \dfrac{2}{7}$ **56.** $-\dfrac{1}{5}, -3$ **57.** $-7, -1$ **58.** $-4, 6$ **59.** -5 **60.** $2, 8$
61. $\dfrac{1}{3}$ **62.** $-\dfrac{2}{7}, \dfrac{3}{8}$ **63.** $0, 6$ **64.** $5, -5$ **65.** $x^2 - 9x + 20 = 0$ **66.** $x^2 + 2x + 1 = 0$ **67.** c   **68.** d   **69.** 9 units   **70.** 8 units,
13 units, 16 units, 10 units   **71.** width: 20 in.; length: 25 in.   **72.** 36 yd   **73.** 19 and 20   **74.** 20 and 22   **75. a.** 17.5 sec and 10 sec;
answers may vary   **b.** 27.5 sec   **76.** 32 cm   **77.** $6(x + 4)$ **78.** $7(x - 9)$ **79.** $(4x - 3)(11x - 6)$ **80.** $(x - 5)(2x - 1)$
81. $(3x - 4)(x^2 + 2)$ **82.** $(y + 2)(x - 1)$ **83.** $2(x + 4)(x - 3)$ **84.** $3x(x - 9)(x - 1)$ **85.** $(2x + 9)(2x - 9)$
86. $2(x + 3)(x - 3)$ **87.** $(4x - 3)^2$ **88.** $5(x + 2)^2$ **89.** $-\dfrac{7}{2}, 4$ **90.** $-3, 5$ **91.** $0, -7, -4$ **92.** $3, 2$ **93.** $0, 16$
94. 19 in.; 8 in.; 21 in.   **95.** length: 6 in.; width: 2 in.

Chapter 6 Test   **1.** $3x(3x - 1)$ **2.** $(x + 7)(x + 4)$ **3.** $(7 + m)(7 - m)$ **4.** $(y + 11)^2$ **5.** $(x^2 + 4)(x + 2)(x - 2)$
6. $(a + 3)(4 - y)$ **7.** prime   **8.** $(y - 12)(y + 4)$ **9.** $(a + b)(3a - 7)$ **10.** $(3x - 2)(x - 1)$ **11.** $5(6 + x)(6 - x)$
12. $3x(3x + 1)(x + 4)$ **13.** $(6t + 5)(t - 1)$ **14.** $(x - 7)(y - 2)(y + 2)$ **15.** $x(1 + x^2)(1 + x)(1 - x)$
16. $(x + 12y)(x + 2y)$ **17.** $(x + 4)(x^2 - 4x + 16)$ **18.** $3x(3y - z)(9y^2 + 3yz + z^2)$ **19.** $3, -9$ **20.** $-7, 2$ **21.** $-7, 1$
22. $0, \dfrac{3}{2}, -\dfrac{4}{3}$ **23.** $-3, 5$ **24.** $0, -4$ **25.** $0, 3, -3$ **26.** $-\dfrac{2}{3}, 1$ **27.** 17 ft   **28.** 7 sec   **29.** width: 6 units; length: 9 units
30. hypotenuse: 25 cm; legs: 15 cm, 20 cm   **31.** 8.25 sec

Cumulative Review   **1. a.** $9 \le 11$ **b.** $8 > 1$ **c.** $3 \ne 4$; Sec. 1.2, Ex. 7   **2. a.** $>$ **b.** $<$; Sec. 1.2   **3.** solution; Sec. 1.3, Ex. 8
4. 102; Sec. 1.6   **5.** -12; Sec. 1.5, Ex. 5a   **6.** -102; Sec. 1.6   **7. a.** $\dfrac{3}{4}$ **b.** -24 **c.** 1; Sec. 1.6, Ex. 16   **8.** -98; Sec. 1.6   **9.** $5x + 7$;
Sec. 1.8, Ex. 4   **10.** $19 - 6x$; Sec. 1.8   **11.** $-4a - 1$; Sec. 1.8, Ex. 5   **12.** $-13x - 21$; Sec. 1.8   **13.** $7.3x - 6$; Sec. 1.8, Ex. 7   **14.** 2; Sec. 2.3
15. -11; Sec. 2.3, Ex. 3   **16.** 28; Sec. 2.2   **17.** every real number; Sec. 2.3, Ex. 7   **18.** 33; Sec. 2.2   **19.** $l = \dfrac{V}{wh}$; Sec. 2.5, Ex. 5
20. $y = \dfrac{-3x - 7}{2}$ or $y = -\dfrac{3}{2}x - \dfrac{7}{2}$; Sec. 2.5   **21.** 5^{18}; Sec. 5.1, Ex. 16   **22.** 30; Sec. 5.1   **23.** y^{16}; Sec. 5.1, Ex. 17   **24.** y^{10}; Sec. 5.1
25. x^6; Sec. 5.2, Ex. 9   **26.** $\dfrac{1}{9}$; Sec. 5.2   **27.** $\dfrac{y^{18}}{z^{36}}$; Sec. 5.2, Ex. 11   **28.** x^4; Sec. 5.2   **29.** $\dfrac{1}{x^{19}}$; Sec. 5.2, Ex. 13   **30.** $25a^9$; Sec. 5.2
31. $4x$; Sec. 5.3, Ex. 6   **32.** $\dfrac{5}{6}x - 77$; Sec. 5.3   **33.** $13x^2 - 2$; Sec. 5.3, Ex. 7   **34.** $-0.5x + 1.2$; Sec. 5.3   **35.** $4x^2 - 4xy + y^2$; Sec. 5.5,
Ex. 8   **36.** $9x^2 - 42xy + 49y^2$; Sec. 5.5   **37.** $t^2 + 4t + 4$; Sec. 5.6, Ex. 5   **38.** $x^2 - 26x + 169$; Sec. 5.6   **39.** $x^4 - 14x^2y + 49y^2$; Sec.
5.6, Ex. 8   **40.** $49x^2 + 14xy + y^2$; Sec. 5.6   **41.** $2xy - 4 + \dfrac{1}{2y}$; Sec. 5.7, Ex. 3   **42.** $(z^2 + 7)(z + 1)$; Sec. 6.1   **43.** $(x + 3)(5 + y)$;
Sec. 6.1, Ex. 9   **44.** $2x(x + 7)(x - 6)$; Sec. 6.2   **45.** $(x^2 + 2)(x^2 + 3)$; Sec. 6.2, Ex. 7   **46.** $(-4x + 1)(x + 6)$ or $-1(4x - 1)(x + 6)$;
Sec. 6.3   **47.** $2(x - 2)(3x + 5)$; Sec. 6.4, Ex. 2   **48.** $x(3y + 4)(3y - 4)$; Sec. 6.5   **49.** 3 sec; Sec. 6.7, Ex. 1   **50.** 9, 4; Sec. 6.6

Chapter 7 Rational Expressions

Section 7.1

Vocabulary, Readiness & Video Check **1.** rational expression **3.** -1 **5.** 2 **7.** $\dfrac{-a}{b}, \dfrac{a}{-b}$ **9.** yes **11.** no **13.** Rational expressions are fractions and are therefore undefined if the denominator is zero; if a denominator contains variables, set it equal to zero and solve. **15.** We would need to write parentheses around the numerator or denominator if it had more than one term because the negative sign needs to apply to the entire numerator or denominator.

Exercise Set 7.1 **1.** $\dfrac{7}{4}$ **3.** $-\dfrac{8}{3}$ **5.** $-\dfrac{11}{2}$ **7. a.** \$403 **b.** \$7 **c.** decrease; answers may vary **9.** $x = 0$ **11.** $x = -2$ **13.** $x = \dfrac{5}{2}$

15. $x = 0, x = -2$ **17.** none **19.** $x = 6, x = -1$ **21.** $x = -2, x = -\dfrac{7}{3}$ **23.** 1 **25.** -1 **27.** $\dfrac{1}{4(x + 2)}$ **29.** $\dfrac{1}{x + 2}$

31. can't simplify **33.** -5 **35.** $\dfrac{7}{x}$ **37.** $\dfrac{1}{x - 9}$ **39.** $5x + 1$ **41.** $\dfrac{x^2}{x - 2}$ **43.** $7x$ **45.** $\dfrac{x + 5}{x - 5}$ **47.** $\dfrac{x + 2}{x + 4}$ **49.** $\dfrac{x + 2}{2}$

51. $-(x + 2)$ **53.** $\dfrac{x + 1}{x - 1}$ **55.** $x + y$ **57.** $\dfrac{5 - y}{2}$ **59.** $\dfrac{2y + 5}{3y + 4}$ **61.** $\dfrac{-(x - 10)}{x + 8}; \dfrac{-x + 10}{x + 8}; \dfrac{x - 10}{-(x + 8)}; \dfrac{x - 10}{-x - 8}$

63. $\dfrac{-(5y - 3)}{y - 12}; \dfrac{-5y + 3}{y - 12}; \dfrac{5y - 3}{-(y - 12)}; \dfrac{5y - 3}{-y + 12}$ **65.** correct **67.** correct **69.** $\dfrac{3}{11}$ **71.** $\dfrac{4}{3}$ **73.** $\dfrac{117}{40}$ **75.** correct

77. incorrect; $\dfrac{1 + 2}{1 + 3} = \dfrac{3}{4}$ **79.** answers may vary **81.** answers may vary **83.** 400 mg **85.** $C = 78.125$; medium **87.** 60.8%

Section 7.2

Vocabulary, Readiness & Video Check **1.** reciprocals **3.** $\dfrac{a \cdot d}{b \cdot c}$ **5.** $\dfrac{6}{7}$ **7.** fractions; reciprocal **9.** We're converting to cubic feet, so we want cubic feet in the numerator. We want cubic yards to divide out so cubic yards is in the denominator.

Exercise Set 7.2 **1.** $\dfrac{21}{4y}$ **3.** x^4 **5.** $-\dfrac{b^2}{6}$ **7.** $\dfrac{x^2}{10}$ **9.** $\dfrac{1}{3}$ **11.** $\dfrac{m + n}{m - n}$ **13.** $\dfrac{x + 5}{x}$ **15.** $\dfrac{(x + 2)(x - 3)}{(x - 4)(x + 4)}$ **17.** $\dfrac{2x^4}{3}$ **19.** $\dfrac{12}{y^6}$

21. $x(x + 4)$ **23.** $\dfrac{3(x + 1)}{x^3(x - 1)}$ **25.** $m^2 - n^2$ **27.** $-\dfrac{x + 2}{x - 3}$ **29.** $\dfrac{x + 2}{x - 3}$ **31.** $\dfrac{5}{6}$ **33.** $\dfrac{3x}{8}$ **35.** $\dfrac{3}{2}$ **37.** $\dfrac{3x + 4y}{2(x + 2y)}$ **39.** $\dfrac{2(x + 2)}{x - 2}$

41. $-\dfrac{y(x + 2)}{4}$ **43.** $\dfrac{(a + 5)(a + 3)}{(a + 2)(a + 1)}$ **45.** $\dfrac{5}{x}$ **47.** $\dfrac{2(n - 8)}{3n - 1}$ **49.** 1440 **51.** 5 **53.** 81 **55.** 73 **57.** 56.7 **59.** 1,201,500 sq ft

61. 55.2 miles/hour **63.** 1 **65.** $-\dfrac{10}{9}$ **67.** $-\dfrac{1}{5}$ **69.** true **71.** false; $\dfrac{x^2 + 3x}{20}$ **73.** $\dfrac{2}{9(x - 5)}$ sq ft **75.** $\dfrac{x}{2}$

77. $\dfrac{5a(2a + b)(3a - 2b)}{b^2(a - b)(a + 2b)}$ **79.** answers may vary **81.** 1493.99 euros

Section 7.3

Vocabulary, Readiness & Video Check **1.** $\dfrac{9}{11}$ **3.** $\dfrac{a + c}{b}$ **5.** $\dfrac{5 - (6 + x)}{x}$ **7.** We completely factor denominators—including coefficients—so we can determine the greatest number of times each unique factor occurs in any one denominator for the LCD.

Exercise Set 7.3 **1.** $\dfrac{a + 9}{13}$ **3.** $\dfrac{3m}{n}$ **5.** 4 **7.** $\dfrac{y + 10}{3 + y}$ **9.** $5x + 3$ **11.** $\dfrac{4}{a + 5}$ **13.** $\dfrac{1}{x - 6}$ **15.** $\dfrac{5x + 7}{x - 3}$ **17.** $x + 5$ **19.** 3

21. $4x^3$ **23.** $8x(x + 2)$ **25.** $(x + 3)(x - 2)$ **27.** $3(x + 6)$ **29.** $5(x - 6)^2$ **31.** $6(x + 1)^2$ **33.** $x - 8$ or $8 - x$

35. $(x - 1)(x + 4)(x + 3)$ **37.** $(3x + 1)(x + 1)(x - 1)(2x + 1)$ **39.** $2x^2(x + 4)(x - 4)$ **41.** $\dfrac{6x}{4x^2}$ **43.** $\dfrac{24b^2}{12ab^2}$

45. $\dfrac{9y}{2y(x + 3)}$ **47.** $\dfrac{9ab + 2b}{5b(a + 2)}$ **49.** $\dfrac{x^2 + x}{x(x + 4)(x + 2)(x + 1)}$ **51.** $\dfrac{18y - 2}{30x^2 - 60}$ **53.** $2x$ **55.** $\dfrac{x + 3}{2x - 1}$ **57.** $x + 1$ **59.** $\dfrac{3}{x}$

61. $\dfrac{3x + 1}{5x + 1}$ **63.** $\dfrac{29}{21}$ **65.** $-\dfrac{5}{12}$ **67.** $\dfrac{7}{30}$ **69.** d **71.** answers may vary **73.** $\dfrac{20}{x - 2}$ m **75.** answers may vary **77.** 95,304 Earth days

79. answers may vary **81.** answers may vary

Section 7.4

Vocabulary, Readiness & Video Check **1.** b **3.** The exercise is adding two rational expressions with denominators that are opposites of each other. Recognizing this special case can save us time and effort. If we recognize that one denominator is -1 times the other denominator, we may save many steps.

Exercise Set 7.4 **1.** $\dfrac{5}{x}$ **3.** $\dfrac{75a + 6b^2}{5b}$ **5.** $\dfrac{6x + 5}{2x^2}$ **7.** $\dfrac{11}{x + 1}$ **9.** $\dfrac{x - 6}{(x - 2)(x + 2)}$ **11.** $\dfrac{35x - 6}{4x(x - 2)}$ **13.** $-\dfrac{2}{x - 3}$ **15.** 0

17. $-\dfrac{1}{x^2 - 1}$ **19.** $\dfrac{5 + 2x}{x}$ **21.** $\dfrac{6x - 7}{x - 2}$ **23.** $-\dfrac{y + 4}{y + 3}$ **25.** $\dfrac{-5x + 14}{4x}$ or $-\dfrac{5x - 14}{4x}$ **27.** 2 **29.** $\dfrac{9x^4 - 4x^2}{21}$ **31.** $\dfrac{x + 2}{(x + 3)^2}$

33. $\dfrac{9b - 4}{5b(b - 1)}$ **35.** $\dfrac{2 + m}{m}$ **37.** $\dfrac{x(x + 3)}{(x - 7)(x - 2)}$ **39.** $\dfrac{10}{1 - 2x}$ **41.** $\dfrac{15x - 1}{(x + 1)^2(x - 1)}$ **43.** $\dfrac{x^2 - 3x - 2}{(x - 1)^2(x + 1)}$ **45.** $\dfrac{a + 2}{2(a + 3)}$

47. $\dfrac{y(2y + 1)}{(2y + 3)^2}$ **49.** $\dfrac{x - 10}{2(x - 2)}$ **51.** $\dfrac{2x + 21}{(x + 3)^2}$ **53.** $\dfrac{-5x + 23}{(x - 2)(x - 3)}$ **55.** $\dfrac{7}{2(m - 10)}$ **57.** $\dfrac{2(x^2 - x - 23)}{(x + 1)(x - 6)(x - 5)}$

59. $\dfrac{n + 4}{4n(n - 1)(n - 2)}$ **61.** 10 **63.** 2 **65.** $\dfrac{25a}{9(a - 2)}$ **67.** $\dfrac{x + 4}{(x - 2)(x - 1)}$ **69.** $\dfrac{2}{3}$ **71.** $-\dfrac{1}{2}, 1$ **73.** $-\dfrac{15}{2}$ **75.** $\dfrac{6x^2 - 5x - 3}{x(x + 1)(x - 1)}$

77. $\dfrac{4x^2 - 15x + 6}{(x - 2)^2(x + 2)(x - 3)}$ **79.** $\dfrac{-2x^2 + 14x + 55}{(x + 2)(x + 7)(x + 3)}$ **81.** $\dfrac{2(x - 8)}{(x + 4)(x - 4)}$ in. **83.** $\dfrac{P - G}{P}$ **85.** answers may vary

87. $\left(\dfrac{90x - 40}{x}\right)^\circ$ **89.** answers may vary

Section 7.5

Vocabulary, Readiness & Video Check **1.** c **3.** b **5.** a **7.** These equations are solved in very different ways, so we need to determine the next correct step to make. For a linear equation, you first "move" variable terms to one side and numbers to the other; for a quadratic equation, we first set the equation equal to 0. **9.** the steps for solving an equation containing rational expressions; as if it's the only variable in the equation

Exercise Set 7.5 **1.** 30 **3.** 0 **5.** -2 **7.** $-5, 2$ **9.** 5 **11.** 3 **13.** 1 **15.** 5 **17.** no solution **19.** 4 **21.** -8 **23.** $6, -4$ **25.** 1

27. $3, -4$ **29.** -3 **31.** 0 **33.** -2 **35.** $8, -2$ **37.** no solution **39.** 3 **41.** $-11, 1$ **43.** $I = \dfrac{E}{R}$ **45.** $B = \dfrac{2U - TE}{T}$

47. $w = \dfrac{Bh^2}{705}$ **49.** $G = \dfrac{V}{N - R}$ **51.** $r = \dfrac{C}{2\pi}$ **53.** $x = \dfrac{3y}{3 + y}$ **55.** $\dfrac{1}{x}$ **57.** $\dfrac{1}{x} + \dfrac{1}{2}$ **59.** $\dfrac{1}{3}$ **61.** answers may vary **63.** $\dfrac{5x + 9}{9x}$

65. no solution **67.** $100°, 80°$ **69.** $22.5°, 67.5°$ **71.** 5

Integrated Review **1.** expression; $\dfrac{3 + 2x}{3x}$ **2.** expression; $\dfrac{18 + 5a}{6a}$ **3.** equation; 3 **4.** equation; 18 **5.** expression; $\dfrac{x - 1}{x(x + 1)}$

6. expression; $\dfrac{3(x + 1)}{x(x - 3)}$ **7.** equation; no solution **8.** equation; 1 **9.** expression; 10 **10.** expression; $\dfrac{z}{3(9z - 5)}$

11. expression; $\dfrac{5x + 7}{x - 3}$ **12.** expression; $\dfrac{7p + 5}{2p + 7}$ **13.** equation; 23 **14.** equation; 3 **15.** expression; $\dfrac{25a}{9(a - 2)}$

16. expression; $\dfrac{9}{4(x - 1)}$ **17.** expression; $\dfrac{3x^2 + 5x + 3}{(3x - 1)^2}$ **18.** expression; $\dfrac{2x^2 - 3x - 1}{(2x - 5)^2}$ **19.** expression; $\dfrac{4x - 37}{5x}$

20. expression; $\dfrac{29x - 23}{3x}$ **21.** equation; $\dfrac{8}{5}$ **22.** equation; $-\dfrac{7}{3}$ **23.** answers may vary **24.** answers may vary

Section 7.6

Vocabulary, Readiness & Video Check **1.** c **3.** $\dfrac{1}{x}; \dfrac{1}{x} - 3$ **5.** $z + 5; \dfrac{1}{z + 5}$ **7.** $2y; \dfrac{11}{2y}$ **9.** No. Proportions are actually equations containing rational expressions, so they can also be solved by using the steps to solve those equations. **11.** divided by, quotient

13. $\dfrac{325}{x + 7} = \dfrac{290}{x}$

Exercise Set 7.6 **1.** 4 **3.** $\dfrac{50}{9}$ **5.** -3 **7.** $\dfrac{14}{9}$ **9.** 123 lb **11.** 165 cal **13.** $y = 21.25$ **15.** $y = 5\dfrac{5}{7}$ ft **17.** 2 **19.** -3 **21.** $2\dfrac{2}{9}$ hr

23. $1\dfrac{1}{2}$ min **25.** trip to park rate: r; to park time: $\dfrac{12}{r}$; return trip rate: r; return time: $\dfrac{18}{r} = \dfrac{12}{r} + 1$; $r = 6$ mph **27.** 1st portion:

10 mph; cooldown: 8 mph **29.** 360 sq ft **31.** 2 **33.** $\$108.00$ **35.** 20 mph **37.** $y = 37\dfrac{1}{2}$ ft **39.** 41 mph; 51 mph **41.** 5 **43.** 217 mph

45. 9 gal **47.** 8 mph **49.** 2.2 mph; 3.3 mph **51.** 3 hr **53.** $26\dfrac{2}{3}$ ft **55.** 216 nuts **57.** $666\dfrac{2}{3}$ mi **59.** 20 hr **61.** car: 70 mph;

motorcycle: 60 mph **63.** $5\dfrac{1}{4}$ hr **65.** 8 **67.** 35 mph; 75 mph **69.** 510 mph **71.** $x = 5$ **73.** $x = 13.5$ **75.** $\dfrac{1}{2}$ **77.** $\dfrac{3}{7}$

79. faster pump: 28 min; slower pump: 84 min **81.** answers may vary **83.** $R = \dfrac{D}{T}$ **85.** 3.75 min

Section 7.7

Vocabulary, Readiness & Video Check   **1.** c   **3.** a   **5.** a single fraction in the numerator and in the denominator

Exercise Set 7.7   **1.** $\dfrac{2}{3}$ **3.** $\dfrac{2}{3}$ **5.** $\dfrac{1}{2}$ **7.** $-\dfrac{21}{5}$ **9.** $\dfrac{27}{16}$ **11.** $\dfrac{4}{3}$ **13.** $\dfrac{1}{21}$ **15.** $-\dfrac{4x}{15}$ **17.** $\dfrac{m-n}{m+n}$ **19.** $\dfrac{2x(x-5)}{7x^2+10}$ **21.** $\dfrac{1}{y-1}$ **23.** $\dfrac{1}{6}$
25. $\dfrac{x+y}{x-y}$ **27.** $\dfrac{3}{7}$ **29.** $\dfrac{a}{x+b}$ **31.** $\dfrac{7(y-3)}{8+y}$ **33.** $\dfrac{3x}{x-4}$ **35.** $-\dfrac{x+8}{x-2}$ **37.** $\dfrac{s^2+r^2}{s^2-r^2}$ **39.** $\dfrac{(x-6)(x+4)}{x-2}$ **41.** Serena Williams
43. \$5.6 million   **45.** answers may vary   **47.** $\dfrac{13}{24}$ **49.** $4\dfrac{1}{4}$ ft or 4.25 ft   **51.** $\dfrac{R_1R_2}{R_2+R_1}$ **53.** $\dfrac{2x}{2-x}$ **55.** $\dfrac{1}{y^2-1}$ **57.** 12 hr

Chapter 7 Vocabulary Check

1. rational expression   **2.** complex fraction   **3.** $\dfrac{-a}{b}; \dfrac{a}{-b}$ **4.** denominator   **5.** simplifying
6. reciprocals   **7.** least common denominator   **8.** unit

Chapter 7 Review

1. $x=2, x=-2$ **2.** $x=\dfrac{5}{2}, x=-\dfrac{3}{2}$ **3.** $\dfrac{4}{3}$ **4.** $\dfrac{11}{12}$ **5.** $\dfrac{2}{x}$ **6.** $\dfrac{3}{x}$ **7.** $\dfrac{1}{x-5}$ **8.** $\dfrac{1}{x+1}$ **9.** $\dfrac{x(x-2)}{x+1}$
10. $\dfrac{5(x-5)}{x-3}$ **11.** $\dfrac{x-3}{x-5}$ **12.** $\dfrac{x}{x+4}$ **13.** $\dfrac{x+a}{x-c}$ **14.** $\dfrac{x+5}{x-3}$ **15.** $\dfrac{3x^2}{y}$ **16.** $-\dfrac{9x^2}{8}$ **17.** $\dfrac{x-3}{x+2}$ **18.** $\dfrac{-2x(2x+5)}{(x-6)^2}$ **19.** $\dfrac{x+3}{x-4}$
20. $\dfrac{4x}{3y}$ **21.** $(x-6)(x-3)$ **22.** $\dfrac{2}{3}$ **23.** $\dfrac{1}{2}$ **24.** $\dfrac{3(x+2)}{3x+y}$ **25.** $\dfrac{1}{x+2}$ **26.** $\dfrac{1}{x-3}$ **27.** $\dfrac{2(x-5)}{3x^2}$ **28.** $\dfrac{2x+1}{2x^2}$ **29.** $14x$
30. $(x-8)(x+8)(x+3)$ **31.** $\dfrac{10x^2y}{14x^3y}$ **32.** $\dfrac{36y^2x}{16y^3x}$ **33.** $\dfrac{x^2-3x-10}{(x+2)(x-5)(x+9)}$ **34.** $\dfrac{3x^2+4x-15}{(x+2)^2(x+3)}$ **35.** $\dfrac{4y+30x^2}{5x^2y}$
36. $\dfrac{-2x+10}{(x-3)(x-1)}$ **37.** $\dfrac{-2x-2}{x+3}$ **38.** $\dfrac{5(x+1)}{(x+4)(x-2)(x-1)}$ **39.** $\dfrac{x-4}{3x}$ **40.** $-\dfrac{x}{x-1}$ **41.** 30   **42.** 3, -4 **43.** no solution
44. 5   **45.** $\dfrac{9}{7}$ **46.** $-6, 1$ **47.** 9   **48.** no solution   **49.** 675 parts   **50.** \$33.75   **51.** 3   **52.** 2   **53.** faster car speed: 30 mph;
slower car speed: 20 mph   **54.** 20 mph   **55.** $17\dfrac{1}{2}$ hr   **56.** $8\dfrac{4}{7}$ days   **57.** $x=15$ **58.** $x=6$ **59.** $-\dfrac{7}{18y}$ **60.** $\dfrac{6}{7}$ **61.** $\dfrac{3y-1}{2y-1}$
62. $-\dfrac{7+2x}{2x}$ **63.** $\dfrac{1}{2x}$ **64.** $\dfrac{x(x-3)}{x+7}$ **65.** $\dfrac{x-4}{x+4}$ **66.** $\dfrac{(x-9)(x+8)}{(x+5)(x+9)}$ **67.** $\dfrac{1}{x-6}$ **68.** $\dfrac{2x+1}{4x}$ **69.** $\dfrac{2}{(x+3)(x-2)}$
70. $-\dfrac{3x}{(x+2)(x-3)}$ **71.** $\dfrac{1}{2}$ **72.** no solution   **73.** 1   **74.** $1\dfrac{5}{7}$ days   **75.** $x=6$ **76.** $x=12$ **77.** $\dfrac{3}{10}$ **78.** $\dfrac{2}{3}$

Chapter 7 Test

1. $x=-1, x=-3$ **2. a.** \$115   **b.** \$103   **3.** $\dfrac{3}{5}$ **4.** $\dfrac{1}{x+6}$ **5.** -1 **6.** $-\dfrac{1}{x+y}$ **7.** $\dfrac{2m(m+2)}{m-2}$ **8.** $\dfrac{a+2}{a+5}$
9. $\dfrac{(x-6)(x-7)}{(x+7)(x+2)}$ **10.** 15   **11.** $\dfrac{y-2}{4}$ **12.** $-\dfrac{1}{2x+5}$ **13.** $\dfrac{3a-4}{(a-3)(a+2)}$ **14.** $\dfrac{3}{x-1}$ **15.** $\dfrac{2(x+3)(x+5)}{x(x^2+4x+1)}$
16. $\dfrac{x^2+2x+35}{(x+9)(x+2)(x-5)}$ **17.** $\dfrac{4y^2+13y-15}{(y+5)(y+1)(y+4)}$ **18.** $\dfrac{30}{11}$ **19.** -6 **20.** no solution   **21.** no solution   **22.** $-2, 5$
23. $\dfrac{xz}{2y}$ **24.** $b-a$ **25.** $\dfrac{5y^2-1}{y+2}$ **26.** 1 or 5   **27.** 30 mph   **28.** $6\dfrac{2}{3}$ hr   **29.** $x=12$ **30.** 18 bulbs

Cumulative Review

1. a. $\dfrac{15}{x}=4$ **b.** $12-3=x$ **c.** $4x+17=21$; Sec. 1.3, Ex. 10   **2. a.** $12-x=-45$ **b.** $12x=-45$
c. $x-10=2x$; Sec. 1.4, 1.5   **3. a.** -12 **b.** -9; Sec. 1.4, Ex. 12   **4. a.** -8 **b.** -17; Sec. 1.5   **5.** distributive property; Sec. 1.7, Ex. 15
6. commutative property of addition; Sec. 1.7   **7.** associative property of addition; Sec. 1.7, Ex. 16   **8.** associative property of
multiplication; Sec. 1.7   **9.** $x=-4$; Sec. 2.1, Ex. 7   **10.** $x=0$; Sec. 2.1   **11.** shorter piece, 2 ft; longer piece, 8 ft; Sec. 2.4, Ex. 3
12. 190, 192; Sec. 2.4   **13.**

; Sec. 3.3, Ex. 7   **14.** 6; 4; 0; Sec 3.1   **15.** $y=\dfrac{1}{4}x-3$; Sec. 3.5, Ex. 1   **16.** $y=-\dfrac{1}{2}x+\dfrac{11}{2}$;
Sec. 3.5   **17.** x^3; Sec. 5.1, Ex. 24   **18.** 1; Sec. 5.1   **19.** 256; Sec. 5.1, Ex. 25   **20.** $x^{15}y^6$;
Sec. 5.1   **21.** -27; Sec. 5.1, Ex. 26   **22.** $x^{18}y^4$; Sec. 5.1   **23.** $2x^4y$; Sec. 5.1, Ex. 27
24. $-15a^5b^2$; Sec. 5.1   **25.** $\dfrac{2}{x^3}$; Sec. 5.2, Ex. 2   **26.** $\dfrac{1}{49}$; Sec. 5.2   **27.** $\dfrac{1}{16}$; Sec. 5.2, Ex. 4
28. $\dfrac{5}{z^7}$; Sec. 5.2   **29.** $10x^4+30x$; Sec. 5.5, Ex. 5   **30.** $x^2+18x+81$; Sec. 5.5   **31.** $-15x^4-18x^3+3x^2$; Sec. 5.5, Ex. 6   **32.** $4x^2-1$;
Sec. 5.6   **33.** $4x^2-4x+6-\dfrac{11}{2x+3}$; Sec. 5.7, Ex. 7   **34.** $4x^2+16x+55+\dfrac{222}{x-4}$; Sec. 5.7   **35.** $(x+3)(x+4)$; Sec. 6.2, Ex. 1
36. $-2(a+1)(a-6)$; Sec. 6.2   **37.** $(x+3)(x-3)$; Sec. 6.5, Ex. 5   **38.** $(x+2)(x-2)$; Sec. 6.5   **39.** 11, -2; Sec. 6.6, Ex. 4

40. $-2, \dfrac{1}{3}$; Sec. 6.6   **41.** $\dfrac{2}{5}$; Sec. 7.2, Ex. 2   **42.** $\dfrac{x+5}{2x^3}$; Sec. 7.1   **43.** $3x-5$; Sec. 7.3, Ex. 3   **44.** $7x^4(x^2-x+1)$; Sec. 6.1

45. $\dfrac{3}{x-2}$; Sec. 7.4, Ex. 2   **46.** $(2x+3)^2$; Sec. 6.5   **47.** $t=5$; Sec. 7.5, Ex. 2   **48.** $\dfrac{30}{x+3}$; Sec. 7.2   **49.** $2\dfrac{1}{10}$ hr; Sec. 7.6, Ex. 6

50. $\dfrac{4m+2n}{m+n}$ or $\dfrac{2(2m+n)}{m+n}$; Sec. 7.7

Chapter 8 Graphs and Functions

Section 8.1

Calculator Explorations   **1.** 18.4   **3.** -1.5 **5.** 8.7; 7.6

Vocabulary, Readiness & Video Check   **1.** $m=-2$; $(1,4)$ **3.** $m=\dfrac{1}{4}$; $(2,0)$ **5.** $m=5$; $(3,-2)$ **7.** if one of the two given points is the y-intercept   **9.** $y=\dfrac{1}{3}x-\dfrac{17}{3}$

Exercise Set 8.1   **1.** $y=3x-1$ **3.** $y=-2x-1$ **5.** $y=\dfrac{1}{2}x+5$ **7.** $y=-\dfrac{9}{10}x-\dfrac{27}{10}$ **9.** $3x-y=6$ **11.** $2x+y=1$

13. $x+2y=-10$ **15.** $x-3y=21$ **17.** $3x+8y=5$ **19.** $x=2$ **21.** $y=1$ **23.** $x=0$ **25.** $y=4x-4$ **27.** $y=-3x+1$

29. $y=4$ **31.** $y=-\dfrac{3}{2}x-6$ **33.** $y=-5$ **35.** $y=-4x+1$ **37.** $2x-y=-7$ **39.** $y=-x+7$ **41.** $y=-\dfrac{1}{2}x+\dfrac{3}{8}$

43. $2x+7y=-42$ **45.** $4x+3y=-20$ **47.** $x=-2$ **49.** $x+2y=2$ **51.** $y=12$ **53.** $8x-y=47$ **55.** $x=5$

57. $y=-\dfrac{3}{8}x-\dfrac{29}{4}$ **59. a.** $y=32x$ **b.** 128 ft per sec   **61. a.** $y=-250x+3500$ **b.** 1625 Frisbees   **63. a.** $y=-600x+297{,}000$

b. \$291,000   **65. a.** $y=271x+38{,}834$ **b.** 40,731   **67.** 31   **69.** -8.4 **71.** 4   **73.** $2x+y=3$ **75.** $2x-3y=-7$ **77.** true
79. answers may vary   **81.**

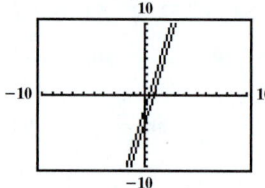

Section 8.2

Vocabulary, Readiness & Video Check   **1.** relation   **3.** domain   **5.** vertical   **7.** An equation is one way of writing down a correspondence that produces ordered pair solutions; a relation between two sets of coordinates, x's and y's.   **9.** If a vertical line intersects a graph two (or more) times, then there's an x-value corresponding to two (or more) different y-values and the graph is thus not the graph of a function.   **11.** No, equations of the form $x=c$, whose graphs are vertical lines, are not functions.

Exercise Set 8.2   **1.** domain; $\{-1,0,-2,5\}$; range: $\{7,6,2\}$; function   **3.** domain; $\{-2,6,-7\}$; range: $\{4,-3,-8\}$; not a function

5. domain: $\{1\}$; range: $\{1,2,3,4\}$; not a function   **7.** domain: $\left\{\dfrac{3}{2},0\right\}$; range: $\left\{\dfrac{1}{2},-7,\dfrac{4}{5}\right\}$; not a function   **9.** domain: $\{-3,0,3\}$;

range: $\{-3,0,3\}$; function   **11.** domain: $\{-1,1,2,3\}$; range: $\{2,1\}$; function   **13.** domain: $\{1996,2000,2004,2008,2012\}$;
range: $\{36,37,44,46\}$; function   **15.** domain: $\{32,104,212,50\}$; range: $\{0,40,10,100\}$; function   **17.** domain: $\{2,-1,5,100\}$;
range: $\{0\}$; function   **19.** function   **21.** not a function   **23.** function   **25.** not a function   **27.** function   **29.** function
31. not a function   **33.** function   **35.** not a function   **37.** not a function   **39.** not a function   **41.** not a function   **43.** not a function
45. function   **47.** 15   **49.** 38   **51.** 7   **53.** 3   **55. a.** 0   **b.** 1   **c.** -1 **57. a.** -5 **b.** -5 **c.** -5 **59.** $(1,-10)$ **61.** $(4,56)$ **63.** -2
65. 0   **67.** $-4,0$ **69.** 3   **71.** 25π sq cm   **73.** 2744 cu in.   **75.** 166.38 cm   **77.** 163.2 mg   **79. a.** 7.56; Agricultural exports to China
in 2005 were approximately \$7.56 billion.   **b.** \$23.76 billion.   **81.**   **83.**   **85.**

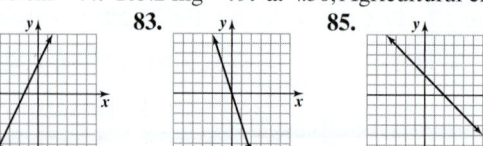

87. $\{x\,|\,x\le 14\}$ **89.** $\left\{x\,\middle|\,x\ge\dfrac{7}{2}\right\}$ **91.** $\left\{x\,\middle|\,x<-\dfrac{1}{4}\right\}$ **93.** no; answers may vary   **95.** yes; answers may vary

97. true   **99.** true   **101.** infinite number   **103.** answers may vary   **105. a.** 0   **b.** $a-12$ **c.** $-x-12$ **d.** $x+h-12$

Section 8.3

Calculator Explorations   **1.** $\{x\,|\,x$ is a real number and $x\ne 6\}$ **3.** $\{x\,|\,x$ is a real number and $x\ne -2, x\ne 2\}$
5. $\left\{x\,\middle|\,x$ is a real number and $x\ne -4, x\ne \dfrac{1}{2}\right\}$ **7.** $\{x\,|\,x$ is a real number$\}$

Vocabulary, Readiness & Video Check   1. polynomial expression   **3.** rational expression   **5.** We are finding $Q(-10)$, and that value is 499. Thus, $Q(-10) = 499$.   **7.** Each time, we replace the variable in the expression with the given value, then simplify.

Exercise Set 8.3   1. 57   **3.** 499   **5.** 1   **7.** 9   **9.** -16 **11.** -15 **13.** 202 sq in.   **15. a.** 284 ft   **b.** 536 ft   **c.** 756 ft   **d.** 944 ft
e. answers may vary   **f.** 19 sec   **17.** \$80,000   **19.** \$16,500   **21. a.** 576 ft; 672 ft; 640 ft; 480 ft   **b.** answers may vary
c. $-16(t + 4)(t - 9)$ **23.** $\{x \mid x \text{ is a real number}\}$ **25.** $\{t \mid t \text{ is a real number and } t \neq 0\}$ **27.** $\{x \mid x \text{ is a real number and } x \neq 7\}$
29. $\left\{x \mid x \text{ is a real number and } x \neq \dfrac{1}{3}\right\}$ **31.** $\{x \mid x \text{ is a real number and } x \neq -2, x \neq 0, x \neq 1\}$ **33.** $\{x \mid x \text{ is a real number and }$
$x \neq 2, x \neq -2\}$ **35.** $\dfrac{10}{3}$ **37.** $\dfrac{2}{5}$ **39.** $-\dfrac{7}{3}$ **41.** $\{x \mid x \text{ is a real number and } x \neq 5\}$ **43.** 6   **45.** -8 **47.** 0   **49.** $\{x \mid x \text{ is a real}$
number and $x \neq 0\}$ **51. a.** \$200 million   **b.** \$500 million   **c.** \$300 million   **53.** -5 **55.** 2   **57. a.** $h(t) = -16t(t - 4)$
b. 48 ft   **c.** answers may vary   **59.** $4x^2 - 3x + 6$ **61.** $2a - 3; -2x - 3; 2x + 2h - 3$ **63.** $5x^2 + 25x$ **65.** $a^2 - 3a$
67. answers may vary   **69. a.** \$584.9 billion   **b.** \$675.82 billion   **c.** \$824.8 billion   **d.** answers may vary   **71.** 164 lb   **73.** higher; F

Integrated Review   1. $y = -x + 7$ **2.** $x = -2$ **3.** $y = 0$ **4.** $y = -\dfrac{3}{8}x - \dfrac{29}{4}$ **5.** $y = -5x - 6$ **6.** $y = -4x + \dfrac{1}{3}$

7. $y = \dfrac{1}{2}x - 1$ **8.** $y = 3x - \dfrac{3}{2}$ **9.** $y = 3x - 2$ **10.** $y = -\dfrac{5}{4}x + 4$ **11.** $y = \dfrac{1}{4}x - \dfrac{7}{2}$ **12.** $y = -\dfrac{5}{2}x - \dfrac{5}{2}$ **13.** $x = -1$

14. $y = 3$ **15.** domain: $\{1, 2, -3.5\}$; range: $\{1, 2, -3.5\}$; function   **16.** domain: $\left\{-1, \dfrac{3}{4}\right\}$; range: $\{7, 8, 9\}$; not a function

17. domain: $\{-4, -2, 0, 2, 4\}$; range: $\{-2\}$; function   **18.** domain: $\{-2, -1, 1, 2\}$; range: $\{0, 1, 2\}$; not a function   **19.** 11   **20.** 5

21. -1 **22.** 7   **23.** 8   **24.** $-\dfrac{13}{3}$

Section 8.4

Vocabulary, Readiness & Video Check   1. Instead of an open circle, use **(** or **)**. Instead of a closed circle, use **[** or **]**.   **3.** Although $f(x) = x + 3$ isn't defined for $x = -1$, we need to clearly indicate the point where this piece of the graph ends. Therefore, we find this point and graph it as an open circle.

Exercise Set 8.4   1. 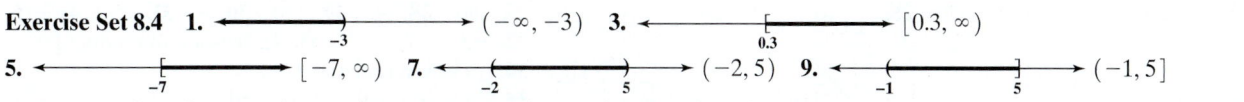 $(-\infty, -3)$ **3.** $[0.3, \infty)$
5. $[-7, \infty)$ **7.** $(-2, 5)$ **9.** $(-1, 5]$
11. domain; $[0, \infty)$; range: $(-\infty, \infty)$ **13.** domain: $(-\infty, \infty)$; range: $[0, \infty)$ **15.** domain: $(-\infty, \infty)$; range: $(-\infty, -3] \cup [3, \infty)$
17. domain: $[2, 7]$; range: $[1, 6]$ **19.** domain: $\{-2\}$; range: $(-\infty, \infty)$ **21.** domain: $(-\infty, \infty)$; range: $(-\infty, 3]$
23. domain: $(-\infty, \infty)$; range: $(-\infty, 3]$ **25.** domain: $[2, \infty)$; range: $[3, \infty)$ **27.** 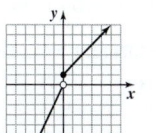   **29.** 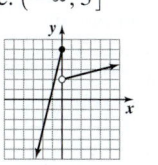   **31.**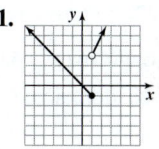

33. 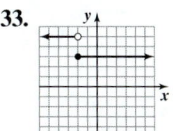   **35.** 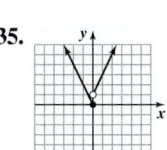 domain: $(-\infty, \infty)$;   **37.** 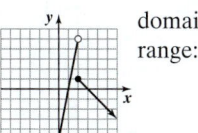 domain: $(-\infty, \infty)$ **39.** 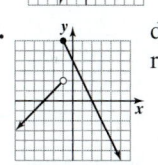 domain: $(-\infty, \infty)$;
range: $[0, \infty)$ range: $(-\infty, 5)$ range: $(-\infty, 6]$

41. 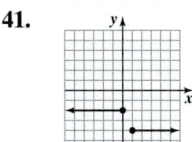 domain: $(-\infty, 0] \cup [1, \infty)$;   **43.** A   **45.** D   **47.** answers may vary   **49.**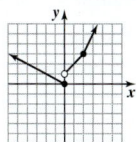
range: $\{-4, -2\}$

Section 8.5

Vocabulary, Readiness & Video Check   1. C   **3.** D   **5.** $f(x) = |x|$ **7.** x-axis

Exercise Set 8.5   1. 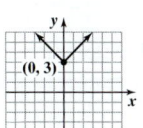   **3.** 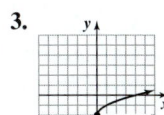   **5.** 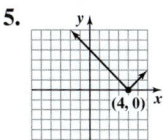   **7.** 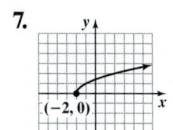   **9.** 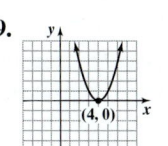   **11.**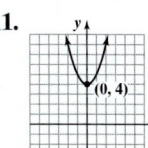

13. 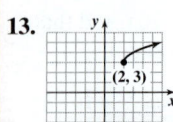 **15.** 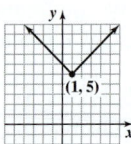 **17.** 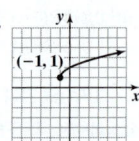 **19.** 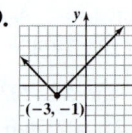 **21.** 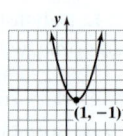 **23.** 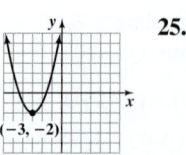 **25.**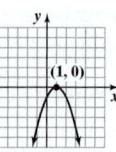

27. 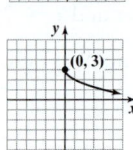 **29.** 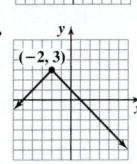 **31.** 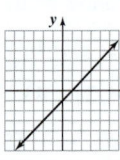 **33.** $-15x^8$ **35.** $8y^7 + 8y^{11}$ **37.** domain: $[2, \infty)$; range: $[3, \infty)$
39. domain: $(-\infty, \infty)$; range: $(-\infty, 3]$ **41.** $[20, \infty)$
43. $(-\infty, \infty)$ **45.** $[-103, \infty)$

47. 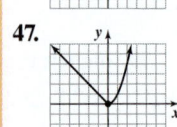 domain: $(-\infty, \infty)$; range: $[0, \infty)$ **49.** 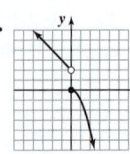 domain: $(-\infty, \infty)$; range: $(-\infty, 0] \cup (2, \infty)$

Chapter 8 Vocabulary Check **1.** relation **2.** range **3.** Parallel **4.** function **5.** slope **6.** perpendicular **7.** domain
8. linear **9.** interval **10.** rational **11.** polynomial

Chapter 8 Review **1.** $y = -1$ **2.** $x = -4$ **3.** $y = 3x + 14$ **4.** $y = -\frac{1}{2}x - 4$ **5.** $y = -2x - 2$ **6.** $y = \frac{3}{4}x + \frac{7}{2}$

7. a. $y = -1600x + 20{,}700$ **b.** \$11,100 **8. a.** $y = 12{,}000x + 126{,}000$ **b.** \$342,000 **9.** domain: $\left\{-\frac{1}{2}, 6, 0, 25\right\}$;

range: $\left\{\frac{3}{4}, 0.65, -12, 25\right\}$; function **10.** domain: $\left\{\frac{3}{4}, 0.65, -12, 25\right\}$; range: $\left\{-\frac{1}{2}, 6, 0, 25\right\}$; function **11.** domain: $\{2, 4, 6, 8\}$;

range: $\{2, 4, 5, 6\}$; not a function **12.** domain: $\{$triangle, square, rectangle, parallelogram$\}$; range: $\{3, 4\}$; function
13. domain: $(-\infty, \infty)$; range: $(-\infty - 1] \cup [1, \infty)$; not a function **14.** domain: $\{-3\}$; range: $(-\infty, \infty)$; not a function
15. domain: $(-\infty, \infty)$; range: $\{4\}$; function **16.** domain: $[-1, 1]$; range: $[-1, 1]$; not a function **17.** -3 **18.** 0 **19.** 18 **20.** 9
21. -3 **22.** 0 **23.** 381 lb **24.** 5080 lb **25.** 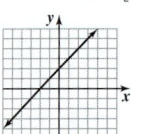 **26.** 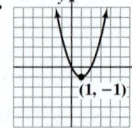 **27.** 290 **28.** 58 **29.** 110 **30.** 8 **31.** $x^2 + 4x - 6$
32. $-x^2 + 2x + 3$ **33.** $\{x \mid x \text{ is a real number}\}$
34. $\{x \mid x \text{ is a real number}\}$
35. $\{x \mid x \text{ is a real number and } x \neq 5\}$
36. $\{x \mid x \text{ is a real number and } x \neq 4\}$

37. $\{x \mid x \text{ is a real number and } x \neq 0, x \neq -8\}$ **38.** $\{x \mid x \text{ is a real number and } x \neq -4, x \neq 4\}$ **39. a.** \$119 **b.** \$77
c. decrease **40.** $\frac{8}{9}$ **41.** domain: $\{2\}$; range: $(-\infty, \infty)$ **42.** domain: $(-\infty, \infty)$; range: $(-\infty, \infty)$ **43.** domain: $[-4, 4]$; range: $[-1, 5]$
44. domain: $(-\infty, \infty)$; range: $\{-5\}$ **45.** 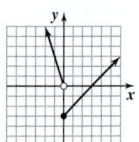 **46.** 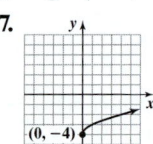 **47.** 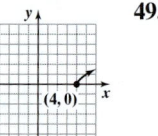 **48.** 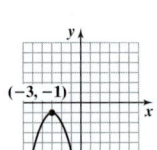 **49.**

50. 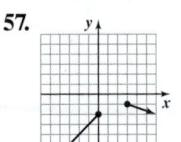 **51.** $x = -2$ **52.** $y = 5$ **53.** $y = -\frac{3}{2}x - 8$ **54.** $y = -\frac{3}{2}x - 1$ **55.** domain: $(-\infty, \infty)$; range: $(\infty, 0]$
56. domain: $(-\infty, \infty)$; range: $(-\infty, \infty)$

57. 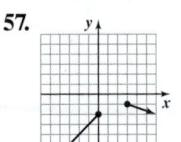 **58.** 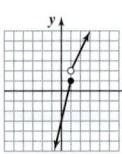 **59.** 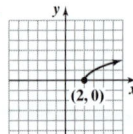 **60.**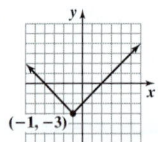

Chapter 8 Test **1.** $y = -8$ **2.** $x = -4$ **3.** $y = -2$ **4.** $y = -\frac{1}{2}x$ **5.** $y = -\frac{1}{3}x + \frac{5}{3}$ **6.** $y = -\frac{1}{2}x - \frac{1}{2}$ **7.** neither

8. domain: $(-\infty, \infty)$; range: $\{5\}$; function **9.** domain: $\{-2\}$; range: $(-\infty, \infty)$; not a function **10.** domain: $(-\infty, \infty)$; range: $[0, \infty)$;
function **11.** domain: $(-\infty, \infty)$; range: $(-\infty, \infty)$; function **12.** 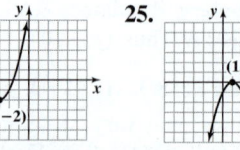 domain: $(-\infty, \infty)$; range: $(-3, \infty)$

13. 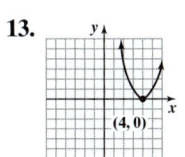 **14.** 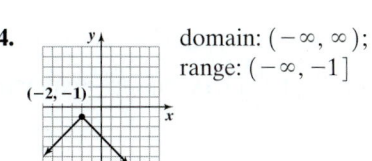 domain: $(-\infty, \infty)$; **15.** 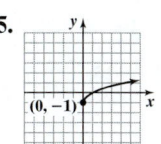 **16. a.** 960 ft **b.** 953.44 ft **c.** 0 ft
range: $(-\infty, -1]$ **17.** $\{x\,|\,x \text{ is a real number and } x \neq 1\}$
18. $\{x\,|\,x \text{ is a real number and } x \neq -3, x \neq -1\}$
19. -20

Cumulative Review 1. a. -6 **b.** 6.3; Sec. 1.5, Ex. 6 **2. a.** 7 **b.** 0 **c.** -10; Sec. 1.6 **3.** $\dfrac{1}{22}$; Sec. 1.6, Ex. 9a **4.** $-\dfrac{1}{45}$; Sec. 1.6

5. $\dfrac{16}{3}$; Sec. 1.6, Ex. 9b **6.** 9; Sec. 1.6 **7.** $-\dfrac{1}{10}$; Sec. 1.6; Ex. 9c **8.** no reciprocal; Sec. 1.6 **9.** $-\dfrac{13}{9}$; Sec. 1.6, Ex. 9d **10.** -1; Sec. 1.6

11. $\dfrac{1}{1.7}$; Sec. 1.6, Ex. 9e **12. a.** 98 **b.** 98; Sec. 1.6 **13. a.** 5 **b.** $8 - x$; Sec. 2.1, Ex. 8 **14.** 22; Sec. 1.8 **15.** no solution; Sec. 2.3, Ex. 6

16. $\dfrac{7}{15}$; Sec. 1.2 **17.** 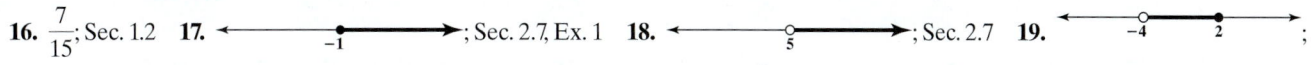; Sec. 2.7, Ex. 1 **18.** ; Sec. 2.7 **19.** ;

Sec. 2.7, Ex. 3 **20.** 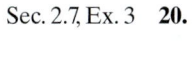; Sec. 2.7 **21.** 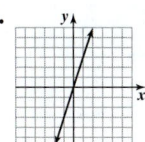; Sec. 3.2, Ex. 3 **22.** 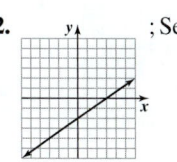; Sec. 3.2

23. 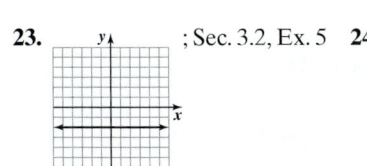; Sec. 3.2, Ex. 5 **24.** 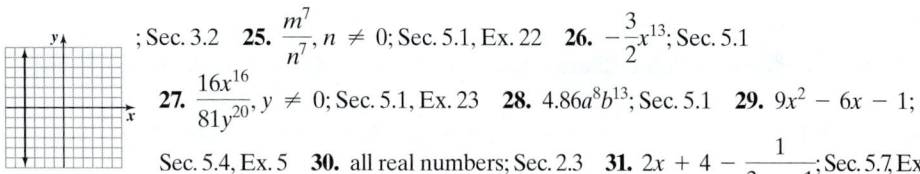; Sec. 3.2 **25.** $\dfrac{m^7}{n^7}, n \neq 0$; Sec. 5.1, Ex. 22 **26.** $-\dfrac{3}{2}x^{13}$; Sec. 5.1

27. $\dfrac{16x^{16}}{81y^{20}}, y \neq 0$; Sec. 5.1, Ex. 23 **28.** $4.86a^8b^{13}$; Sec. 5.1 **29.** $9x^2 - 6x - 1$;

Sec. 5.4, Ex. 5 **30.** all real numbers; Sec. 2.3 **31.** $2x + 4 - \dfrac{1}{3x - 1}$; Sec. 5.7, Ex. 5

32. a. -7 **b.** $\dfrac{5}{6}$ **c.** -13; Sec. 1.5 **33.** $x = -\dfrac{1}{2}; x = 4$; Sec. 6.6, Ex. 6 **34.** $x = 0; x = \dfrac{7}{2}$; Sec. 6.6 **35.** 6, 8, 10; Sec. 6.7, Ex. 5

36. 69, 71, 73; Sec. 2.4 **37.** 1; Sec. 7.3, Ex. 2 **38.** $\{x\,|\,x > -25\}$; Sec. 2.7 **39.** $\dfrac{2x^2 + 3y}{x^2y + 2xy^2}$; Sec. 7.7, Ex. 6 **40.** $\dfrac{c^{15}}{125a^9b^6}$; Sec. 5.2

41. $y = \dfrac{5}{3}x + \dfrac{13}{3}$; Sec. 8.1, Ex. 6 **42.** $x = -2$; Sec. 8.1 **43.** $y = 3$; Sec. 8.1, Ex. 3 **44.** $x = -1$; Sec. 8.1 **45.** 5; Sec. 8.2, Ex. 16

46. $m = -2.3$; Sec. 8.2 **47.** -2; Sec. 8.2, Ex. 15 **48.** no; Sec. 8.2

Chapter 9 Systems of Equations and Inequalities and Variation
Section 9.1

Vocabulary, Readiness & Video Check 1. a, b, d **3.** yes; answers may vary **5.** Once we have one equation in two variables, we need to get another equation in the *same* two variables, giving us a system of two equations in two variables. We solve this new system to find the value of two variables. We then substitute these values into an original equation to find the value of the third.

Exercise Set 9.1 1. $(-1, 5, 2)$ **3.** $(-2, 5, 1)$ **5.** $(-2, 3, -1)$ **7.** $\{(x, y, z)\,|\,x - 2y + z = -5\}$ **9.** $\varnothing$ **11.** $(0, 0, 0)$
13. $(-3, -35, -7)$ **15.** $(6, 22, -20)$ **17.** $\varnothing$ **19.** $(3, 2, 2)$ **21.** $\{(x, y, z)\,|\,x + 2y - 3z = 4\}$ **23.** $(-3, -4, -5)$ **25.** $\left(0, \dfrac{1}{2}, -4\right)$
27. $(12, 6, 4)$ **29.** 10 and 8 **31. a.** Ford class: 1106 ft; Nimitz class: 1092 ft **b.** 3.69 football fields **33.** 2 units of Mix A; 3 units of Mix B; 1 unit of Mix C **35.** 5 in.; 7 in.; 7 in.; 10 in. **37.** 18, 13, and 9 **39.** free throws: 90; 2-pt field goals: 161; 3-pt field goals: 72
41. $x = 60; y = 55; z = 65$ **43.** $5x + 5z = 10$ **45.** $-5y + 2z = 2$ **47.** answers may vary **49.** answers may vary **51.** $(1, 1, -1)$
53. $(1, 1, 0, 2)$ **55.** $(1, -1, 2, 3)$ **57.** answers may vary **59.** 2010: 1,593,081; 2013: 1,107,699 **61.** $a = 3, b = 4, c = -1$
63. $a = 1.6; b = 23.6; c = 1153.8; 1867.8$ thousand students

Section 9.2

Vocabulary, Readiness & Video Check 1. matrix **3.** row **5.** false **7.** true **9.** Two rows may be interchanged, the elements of any row may be multiplied/divided by the same nonzero number, the elements of any row may be multiplied/divided by the same nonzero number and added to their corresponding elements in any other row; rows were not interchanged in Example 1.

Exercise Set 9.2 1. $(2, -1)$ **3.** $(-4, 2)$ **5.** $\varnothing$ **7.** $\{(x, y)\,|\,3x - 3y = 9\}$ **9.** $(-2, 5, -2)$ **11.** $(1, -2, 3)$ **13.** $(4, -3)$
15. $(2, 1, -1)$ **17.** $(9, 9)$ **19.** $\varnothing$ **21.** $\varnothing$ **23.** $(1, -4, 3)$ **25.** function **27.** not a function **29.** c **31. a.** 2006
b. no; answers may vary **c.** no, it has a positive slope **d.** answers may vary **33.** answers may vary

Integrated Review **1.** C **2.** D **3.** A **4.** B **5.** $(2, -1)$ **6.** $(5, 2)$ **7.** $\varnothing$ **8.** $\{(x, y) \mid 2x - 5y = 3\}$ **9.** $(-1, 3, 2)$
10. $(1, -3, 0)$ **11.** $\varnothing$ **12.** $\{(x, y, z) \mid x - y + 3z = 2\}$ **13.** $\left(2, 5, \dfrac{1}{2}\right)$ **14.** $\left(1, 1, \dfrac{1}{3}\right)$ **15.** $70°; 70°; 100°; 120°$

Section 9.3

Vocabulary, Readiness & Video Check **1.** system **3.** corner **5.** No; we can choose any test point except a point on the second inequality's own boundary line.

Exercise Set 9.3 **1.** 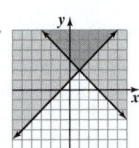 **3.** 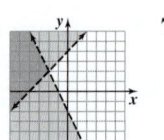 **5.** 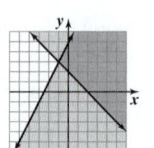 **7.** 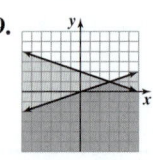 **9.** **11.**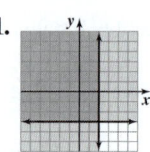

13. 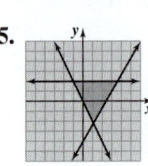 **15.** 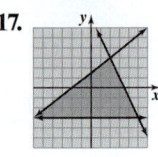 **17.** 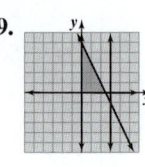 **19.** **21.** C **23.** D **25.** 9 **27.** $\dfrac{4}{9}$ **29.** 5 **31.** 59
33. the line $y = 3$ **35.** answers may vary

Section 9.4

Vocabulary, Readiness & Video Check **1.** direct **3.** joint **5.** inverse **7.** direct **9.** linear; slope **11.** $y = ka^2b^5$

Exercise Set 9.4 **1.** $k = \dfrac{1}{5}; y = \dfrac{1}{5}x$ **3.** $k = \dfrac{3}{2}; y = \dfrac{3}{2}x$ **5.** $k = 14; y = 14x$ **7.** $k = 0.25; y = 0.25x$ **9.** 4.05 lb **11.** 117,125 tons
13. $k = 30; y = \dfrac{30}{x}$ **15.** $k = 700; y = \dfrac{700}{x}$ **17.** $k = 2; y = \dfrac{2}{x}$ **19.** $k = 0.14; y = \dfrac{0.14}{x}$ **21.** 54 mph **23.** 72 amps
25. divided by 4 **27.** $x = kyz$ **29.** $r = kst^3$ **31.** $k = \dfrac{1}{3}; y = \dfrac{1}{3}x^3$ **33.** $k = 0.2; y = 0.2\sqrt{x}$ **35.** $k = 1.3; y = \dfrac{1.3}{x^2}$
37. $k = 3; y = 3xz^3$ **39.** 22.5 tons **41.** 15π cu in. **43.** 8 ft **45.** $y = kx$ **47.** $a = \dfrac{k}{b}$ **49.** $y = kxz$ **51.** $y = \dfrac{k}{x^3}$ **53.** $y = \dfrac{kx}{p^2}$
55. $C = 12\pi$ cm; $A = 36\pi$ sq cm **57.** $C = 14\pi$ m; $A = 49\pi$ sq m **59.** 0 **61.** -1 **63.** a **65.** c **67.** multiplied by 2
69. multiplied by 4

Vocabulary Check **1.** system of equations **2.** solution **3.** consistent **4.** inconsistent **5.** matrix **6.** element
7. row **8.** column **9.** directly **10.** inversely **11.** jointly

Chapter 9 Review **1.** $(2, 0, 2)$ **2.** $(2, 0, -3)$ **3.** $\left(-\dfrac{1}{2}, \dfrac{3}{4}, 1\right)$ **4.** $(-1, 2, 0)$ **5.** $\varnothing$ **6.** $(5, 3, 0)$ **7.** $(1, 1, -2)$ **8.** $(3, 1, 1)$
9. 10, 40, and 48 **10.** 30 lb of creme-filled; 5 lb of chocolate-covered nuts; 10 lb of chocolate-covered raisins **11.** 17 pennies;
20 nickels; 16 dimes **12.** 120, 115, and 60 **13.** $(-3, 1)$ **14.** $\{(x, y) \mid x - 2y = 4\}$ **15.** $\left(-\dfrac{2}{3}, 3\right)$ **16.** $\left(\dfrac{1}{3}, \dfrac{7}{6}\right)$ **17.** $\left(\dfrac{5}{4}, \dfrac{5}{8}\right)$
18. $(-7, -15)$ **19.** $(1, 3)$ **20.** $(2, 1)$ **21.** $(1, 2, 3)$ **22.** $(2, 0, -3)$ **23.** $(3, -2, 5)$ **24.** $(-1, 2, 0)$ **25.** $(1, 1, -2)$ **26.** $\varnothing$
27. 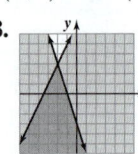 **28.** 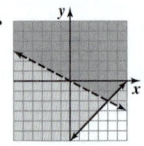 **29.** 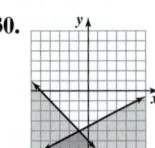 **30.** 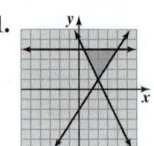 **31.** 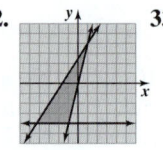 **32.** **33.**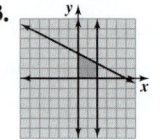

34. **35.** 9 **36.** 3.125 cu m **37.** $(-1, 3, 5)$ **38.** 28 units, 42 units, 56 units **39.**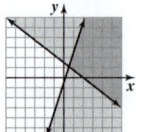

40. during May **41.** 4 **42.** 64π sq in.

Chapter 9 Test **1.** $(-1, -2, 4)$ **2.** $\varnothing$ **3.** $(3, -1, 2)$ **4.** $(5, 0, -4)$ **5.** $\{(x, y) \mid x - y = -2\}$ **6.** $(5, -3)$ **7.** $(-1, -1, 0)$
8. $\varnothing$ **9.** $23°, 45°, 112°$ **10.** 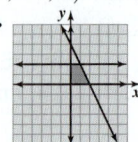 **11.** **12.** 16 **13.** 9 **14.** 256 ft

Cumulative Review   **1.** $y = -1.6$; Sec. 2.1, Ex. 2   **2.** $x = -10$; Sec. 2.2   **3.** $t = \dfrac{16}{3}$; Sec. 2.3, Ex. 2   **4.** $x = 7$; Sec. 2.3

5. width: 4 ft; length: 10 ft; Sec. 2.5, Ex. 4   **6. a.** $x - \dfrac{1}{3}$ **b.** $5x - 6$ **c.** $8x + 3$ **d.** $\dfrac{7}{2 - x}$; Sec. 2.1   **7.** -6; -5; 10; Sec. 3.1, Ex. 7

8. $2x^2 + 6x + 4$; Sec. 5.4   **9.** $m = -\dfrac{8}{3}$

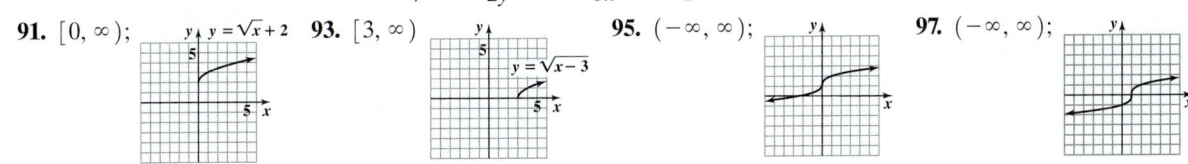

; Sec. 3.4, Ex. 1   **10.** $f(x) = \dfrac{1}{2}x + \dfrac{7}{2}$; Sec. 8.2   **11.** 1; Sec. 5.1, Ex. 28

12. 1; Sec. 5.1   **13.** 1; Sec. 5.1, Ex. 29   **14.** 7; Sec. 5.1   **15.** -1; Sec. 5.1, Ex. 31   **16.** -8; Sec. 5.1   **17.** $9y^2 + 12y + 4$; Sec. 5.6, Ex. 15

18. $16x^2 - 9z^2$; Sec. 5.6   **19.** $x + 4$; Sec. 5.7, Ex. 4   **20. a.** $\dfrac{7}{12}$ **b.** -6 **c.** $\dfrac{1}{x^3}$; Sec. 5.2   **21.** $(r + 6)(r - 7)$; Sec. 6.2, Ex. 4

22. $(y + 2)(x - 5)$; Sec. 6.1   **23.** $(2x - 3y)(5x + y)$; Sec. 6.3, Ex. 4   **24.** $(2x - 5)(3x + 7)$; Sec. 6.3
25. $(2x - 1)(4x - 5)$; Sec. 6.4, Ex. 1   **26.** $(2x - 1 + 3y)(2x - 1 - 3y)$; Sec. 6.5   **27. a.** $x(2x + 7)(2x - 7)$; Sec. 6.5, Ex. 10

b. $2(9x^2 + 1)(3x + 1)(3x - 1)$; Sec. 6.5, Ex. 11   **28.** $2(x - 2)(x^2 + 2x + 4)$; Sec. 6.5   **29.** $x = \dfrac{1}{5}, -\dfrac{3}{2}, -6$; Sec. 6.6, Ex. 8

30. $x = 0, -\dfrac{1}{3}, 3$; Sec. 6.6   **31.** $\dfrac{x + 7}{x - 5}$; Sec. 7.1, Ex. 4   **32.** $\dfrac{8a^7}{9b^{11}}$; Sec. 5.2   **33.** -5; Sec. 7.6, Ex. 5   **34.** -4; Sec. 7.5   **35.** $y = -2x + 12$;

Sec. 8.1, Ex. 5   **36. a.** $48x^{2a}$ **b.** y^{10b+3}; Sec. 5.2   **37.** -4; Sec. 8.3, Ex. 1   **38.** -5; Sec. 8.3   **39.** 35; Sec. 8.3; Ex. 2   **40. a.** -2

b. -20 **c.** $-\dfrac{10}{9}$; Sec. 8.3   **41.** 8.125 in.; Sec. 9.4, Ex. 2   **42.** Paper: \$3.80, folders: \$5.25; Sec. 4.4   **43.** $(4, 2)$; Sec. 4.2, Ex. 1

44. $(3, 4)$; Sec. 4.2   **45.** no solution; Sec. 4.3, Ex. 3   **46.** $(1, 0)$; Sec. 4.3   **47.** $\{(x, y, z) \mid x - 5y - 2z = 6\}$; Sec. 9.1, Ex. 4
48. $(2, 1, 1)$; Sec. 9.1   **49.** $\varnothing$ or $\{\ \}$; Sec. 9.2, Ex. 2   **50.** $(0, 5, 4)$; Sec. 9.2

Chapter 10 Rational Exponents, Radicals, and Complex Numbers
Section 10.1

Vocabulary, Readiness & Video Check   **1.** index; radical sign; radicand   **3.** is not   **5.** $[0, \infty)$ **7.** $(16, 4)$ **9.** d   **11.** d
13. Divide the index into each exponent in the radicand.   **15.** The square root of a negative number is not a real number, but the cube root of a negative number is a real number.   **17.** For odd roots, $\sqrt[n]{a^n} = a$, whether a is positive, 0, or negative, so absolute value bars aren't needed.

Exercise Set 10.1   **1.** $2, -2$ **3.** no real number square roots   **5.** $10, -10$ **7.** 10   **9.** $\dfrac{1}{2}$ **11.** 0.01   **13.** -6 **15.** x^5 **17.** $4y^3$

19. 2.646   **21.** 6.164   **23.** 14.142   **25.** 4   **27.** $\dfrac{1}{2}$ **29.** -1 **31.** x^4 **33.** $-3x^3$ **35.** -2 **37.** not a real number   **39.** -2 **41.** x^4

43. $2x^2$ **45.** $9x^2$ **47.** $4x^2$ **49.** 8   **51.** -8 **53.** $2|x|$ **55.** x **57.** $|x - 5|$ **59.** $|x + 2|$ **61.** -11 **63.** $2x$ **65.** y^6 **67.** $5ab^{10}$

69. $-3x^4y^3$ **71.** a^4b **73.** $-2x^2y$ **75.** $\dfrac{5}{7}$ **77.** $\dfrac{x}{2y}$ **79.** $-\dfrac{z^7}{3x}$ **81.** $\dfrac{x}{2}$ **83.** $\sqrt{3}$ **85.** -1 **87.** -3 **89.** $\sqrt{7}$

91. $[0, \infty)$;   **93.** $[3, \infty)$ **95.** $(-\infty, \infty)$;   **97.** $(-\infty, \infty)$;

99. $-32x^{15}y^{10}$ **101.** $-60x^7y^{10}z^5$ **103.** $\dfrac{x^9y^5}{2}$ **105.** not a real number   **107.** not a real number   **109.** answers may vary   **111.** b

113. b   **115.** 1.69 sq m   **117.** 11,181 m per sec   **119.** answers may vary

Section 10.2

Vocabulary, Readiness & Video Check   **1.** true   **3.** true   **5.** multiply; c   **7.** A   **9.** C   **11.** B   **13.** $-\sqrt[5]{3x}$ **15.** denominator; positive   **17.** Write the radical in equivalent fractional exponent form, use rules for exponents to write a single fractional exponent, simplify the fraction, then write in radical form again.

Exercise Set 10.2   **1.** 7   **3.** 3   **5.** $\dfrac{1}{2}$ **7.** 13   **9.** $2\sqrt[3]{m}$ **11.** $3x^2$ **13.** -3 **15.** -2 **17.** 8   **19.** 16   **21.** not a real number   **23.** $\sqrt[5]{(2x)^3}$

25. $\sqrt[3]{(7x + 2)^2}$ or $(\sqrt[3]{7x + 2})^2$ **27.** $\dfrac{64}{27}$ **29.** $\dfrac{1}{16}$ **31.** $\dfrac{1}{16}$ **33.** not a real number   **35.** $\dfrac{1}{x^{1/4}}$ **37.** $a^{2/3}$ **39.** $\dfrac{5x^{3/4}}{7}$ **41.** $a^{7/3}$

43. x **45.** $3^{5/8}$ **47.** $y^{1/6}$ **49.** $8u^3$ **51.** $-b$ **53.** $\dfrac{1}{x^2}$ **55.** $27x^{2/3}$ **57.** $\dfrac{y}{z^{1/6}}$ **59.** $\dfrac{1}{x^{7/4}}$ **61.** $\sqrt{x}$ **63.** $\sqrt[3]{2}$ **65.** $2\sqrt{x}$ **67.** $\sqrt{x+3}$

69. $\sqrt{xy}$ **71.** $\sqrt[3]{a^2b}$ **73.** $\sqrt[15]{y^{11}}$ **75.** $\sqrt[12]{b^5}$ **77.** $\sqrt[24]{x^{23}}$ **79.** $\sqrt{a}$ **81.** $\sqrt[6]{432}$ **83.** $\sqrt[15]{343y^5}$ **85.** $\sqrt[6]{125r^3s^2}$ **87.** $25\cdot 3$

89. $16\cdot 3$ or $4\cdot 12$ **91.** $8\cdot 2$ **93.** $27\cdot 2$ **95.** 1509 calories **97.** 615.8 million **99.** answers may vary **101.** $a^{1/3}$ **103.** $x^{1/5}$

105. 1.6818 **107.** $\dfrac{t^{1/2}}{u^{1/2}}$

Section 10.3

Vocabulary, Readiness & Video Check **1.** midpoint; point **3.** midpoint **5.** false **7.** true **9.** false **11.** The indexes must be the same. **13.** The power must be 1. Any even power is a perfect square and will leave no factor in the radicand; any higher odd power can have an even power factored from it, leaving one factor remaining in the radicand. **15.** average; average

Exercise Set 10.3 **1.** $\sqrt{14}$ **3.** 2 **5.** $\sqrt[3]{36}$ **7.** $\sqrt{6x}$ **9.** $\sqrt{\dfrac{14}{xy}}$ **11.** $\sqrt[4]{20x^3}$ **13.** $\dfrac{\sqrt{6}}{7}$ **15.** $\dfrac{\sqrt{2}}{7}$ **17.** $\dfrac{\sqrt[4]{x^3}}{2}$ **19.** $\dfrac{\sqrt[3]{4}}{3}$ **21.** $\dfrac{\sqrt[4]{8}}{x^2}$

23. $\dfrac{\sqrt[3]{2x}}{3y^4}$ **25.** $\dfrac{x\sqrt{y}}{13}$ **27.** $\dfrac{x\sqrt{5}}{2y}$ **29.** $-\dfrac{z^2\sqrt[3]{z}}{5x}$ **31.** $4\sqrt{2}$ **33.** $4\sqrt[3]{3}$ **35.** $25\sqrt{3}$ **37.** $2\sqrt{6}$ **39.** $10x^2\sqrt{x}$ **41.** $2y^2\sqrt[3]{2y}$

43. $a^2b\sqrt[4]{b^3}$ **45.** $y^2\sqrt{y}$ **47.** $5ab\sqrt{b}$ **49.** $-2x^2\sqrt[5]{y}$ **51.** $x^4\sqrt[3]{50x^2}$ **53.** $-4a^4b^3\sqrt{2b}$ **55.** $3x^3y^4\sqrt{xy}$ **57.** $5r^3s^4$ **59.** $2x^3y\sqrt[4]{2y}$

61. $\sqrt{2}$ **63.** 2 **65.** 10 **67.** x^2y **69.** $24m^2$ **71.** $\dfrac{15x\sqrt{2x}}{2}$ or $\dfrac{15x}{2}\sqrt{2x}$ **73.** $2a^2\sqrt[4]{2}$ **75.** $2xy^2\sqrt[5]{x^2}$ **77.** 5 units

79. $\sqrt{41}$ units ≈ 6.403 units **81.** $\sqrt{10}$ units ≈ 3.162 units **83.** $\sqrt{5}$ units ≈ 2.236 units **85.** $\sqrt{192.58}$ units ≈ 13.877 units

87. $(4,-2)$ **89.** $\left(-5,\dfrac{5}{2}\right)$ **91.** $\left(\dfrac{5}{2},0\right)$ **93.** $\left(-\dfrac{1}{2},\dfrac{1}{2}\right)$ **95.** $\left(\sqrt{2},\dfrac{\sqrt{5}}{2}\right)$ **97.** $(6.2,-6.65)$ **99.** $14x$ **101.** $2x^2-7x-15$

103. y^2 **105.** $x^2-8x+16$ **107.** $\dfrac{\sqrt[3]{64}}{\sqrt{64}}=\dfrac{4}{8}=\dfrac{1}{2}$ **109.** 1.6 m **111. a.** 2.2 sec **b.** 1.4 sec **c.** answers may vary

Section 10.4

Vocabulary, Readiness & Video Check **1.** Unlike **3.** Like **5.** $6\sqrt{3}$ **7.** $7\sqrt{x}$ **9.** $8\sqrt[3]{x}$ **11.** $\sqrt{11}+\sqrt[3]{11}$ **13.** $10\sqrt[3]{2x}$
15. Sometimes you can't see that there are like radicals until you simplify, so you may incorrectly think you cannot add or subtract if you don't simplify first.

Exercise Set 10.4 **1.** $-2\sqrt{2}$ **3.** $10x\sqrt{2x}$ **5.** $17\sqrt{2}-15\sqrt{5}$ **7.** $-\sqrt[3]{2x}$ **9.** $5b\sqrt{b}$ **11.** $\dfrac{31\sqrt{2}}{15}$ **13.** $\dfrac{\sqrt[3]{11}}{3}$ **15.** $\dfrac{5\sqrt{5x}}{9}$

17. $14+\sqrt{3}$ **19.** $7-3y$ **21.** $6\sqrt{3}-6\sqrt{2}$ **23.** $-23\sqrt[3]{5}$ **25.** $2a^2b^3\sqrt{ab}$ **27.** $20y\sqrt{2y}$ **29.** $2y\sqrt[3]{2x}$ **31.** $6\sqrt[3]{11}-4\sqrt{11}$

33. $4x\sqrt[4]{x^3}$ **35.** $\dfrac{2\sqrt{3}}{3}$ **37.** $\dfrac{5x\sqrt[3]{x}}{7}$ **39.** $\dfrac{5\sqrt{7}}{2x}$ **41.** $\dfrac{\sqrt[3]{2}}{6}$ **43.** $\dfrac{14x\sqrt[3]{2x}}{9}$ **45.** $15\sqrt{3}$ in. **47.** $\sqrt{35}+\sqrt{21}$ **49.** $7-2\sqrt{10}$

51. $3\sqrt{x}-x\sqrt{3}$ **53.** $6x-13\sqrt{x}-5$ **55.** $\sqrt[3]{a^2}+\sqrt[3]{a}-20$ **57.** $6\sqrt{2}-12$ **59.** $2+2x\sqrt{3}$ **61.** $-16-\sqrt{35}$

63. $18+\sqrt{6}-24\sqrt{3}-4\sqrt{2}$ **65.** $3+2x\sqrt{3}+x^2$ **67.** $5x-3\sqrt{15x}-3\sqrt{10x}+9\sqrt{6}$ **69.** $-\sqrt[3]{4}+2\sqrt[3]{2}$

71. $\sqrt[3]{x^2}-4\sqrt[6]{x^5}+8\sqrt[3]{x}-4\sqrt{x}+7$ **73.** $x+24+10\sqrt{x-1}$ **75.** $2x+6-2\sqrt{2x+5}$ **77.** $x-7$ **79.** $\dfrac{7}{x+y}$ **81.** $2a-3$

83. $\dfrac{-2+\sqrt{3}}{3}$ **85.** $22\sqrt{5}$ ft; 150 sq ft **87. a.** $2\sqrt{3}$ **b.** 3 **c.** answers may vary **89.** $2\sqrt{6}-2\sqrt{2}-2\sqrt{3}+6$

Section 10.5

Vocabulary, Readiness & Video Check **1.** conjugate **3.** rationalizing the numerator **5.** $\sqrt{2}-x$ **7.** $5+\sqrt{a}$ **9.** $-7\sqrt{5}-8\sqrt{x}$
11. to write an equivalent expression without a radical in the denominator **13.** No, except for the fact you're working with numerators, the process is the same.

Exercise Set 10.5 **1.** $\dfrac{\sqrt{14}}{7}$ **3.** $\dfrac{\sqrt{5}}{5}$ **5.** $\dfrac{4\sqrt[3]{9}}{3}$ **7.** $\dfrac{3\sqrt{2x}}{4x}$ **9.** $\dfrac{3\sqrt[3]{2x}}{2x}$ **11.** $\dfrac{3\sqrt{3a}}{a}$ **13.** $\dfrac{3\sqrt[3]{4}}{2}$ **15.** $\dfrac{2\sqrt{21}}{7}$ **17.** $\dfrac{\sqrt{10xy}}{5y}$ **19.** $\dfrac{\sqrt[3]{75}}{5}$

21. $\dfrac{\sqrt{6x}}{10}$ **23.** $\dfrac{\sqrt{3z}}{6z}$ **25.** $\dfrac{\sqrt[3]{6xy^2}}{3x}$ **27.** $\dfrac{2\sqrt[4]{9x}}{3x^2}$ **29.** $\dfrac{5\sqrt[5]{4ab^4}}{2ab^3}$ **31.** $-2(2+\sqrt{7})$ **33.** $\dfrac{7(\sqrt{x}+3)}{9-x}$ **35.** $-5+2\sqrt{6}$

37. $\dfrac{2a+\sqrt{ab}+2\sqrt{a}+\sqrt{b}}{4a-b}$ **39.** $-\dfrac{8(1-\sqrt{10})}{9}$ **41.** $\dfrac{x-\sqrt{xy}}{x-y}$ **43.** $\dfrac{5+3\sqrt{2}}{7}$ **45.** $\dfrac{5}{\sqrt{15}}$ **47.** $\dfrac{6}{\sqrt{10}}$ **49.** $\dfrac{2x}{7\sqrt{x}}$

51. $\dfrac{5y}{\sqrt[3]{100xy}}$ **53.** $\dfrac{2}{\sqrt{10}}$ **55.** $\dfrac{2x}{11\sqrt{2x}}$ **57.** $\dfrac{7}{2\sqrt[3]{49}}$ **59.** $\dfrac{3x^2}{10\sqrt[3]{9x}}$ **61.** $\dfrac{6x^2y^3}{\sqrt{6z}}$ **63.** answers may vary **65.** $\dfrac{-7}{12+6\sqrt{11}}$

67. $\dfrac{3}{10 + 5\sqrt{7}}$ **69.** $\dfrac{x - 9}{x - 3\sqrt{x}}$ **71.** $\dfrac{x - 1}{x - 2\sqrt{x} + 1}$ **73.** $\{5\}$ **75.** $\left\{-\dfrac{1}{2}, 6\right\}$ **77.** $\{2, 6\}$ **79.** $r = \dfrac{\sqrt{A\pi}}{2\pi}$ **81.** answers may vary

83. a. $\dfrac{y\sqrt{15xy}}{6x^2}$ **b.** $\dfrac{y\sqrt{15xy}}{6x^2}$ **c.** answers may vary **85.** $\sqrt[3]{25}$

Integrated Review 1. 9 **2.** -2 **3.** $\dfrac{1}{2}$ **4.** x^3 **5.** y^3 **6.** $2y^5$ **7.** $-2y$ **8.** $3b^3$ **9.** 6 **10.** $\sqrt[4]{3y}$ **11.** $\dfrac{1}{16}$ **12.** $\sqrt[5]{(x + 1)^3}$
13. y **14.** $16x^{1/2}$ **15.** $x^{5/4}$ **16.** $4^{11/15}$ **17.** $2x^2$ **18.** $\sqrt[4]{a^3b^2}$ **19.** $\sqrt[4]{x^3}$ **20.** $\sqrt[6]{500}$ **21.** $2\sqrt{10}$ **22.** $2xy^2\sqrt[4]{x^3y^2}$ **23.** $3x\sqrt[3]{2x}$
24. $-2b^2\sqrt[5]{2}$ **25.** $\sqrt{5x}$ **26.** $4x$ **27.** $7y^2\sqrt{y}$ **28.** $2a^2\sqrt[4]{3}$ **29.** $2\sqrt{5} - 5\sqrt{3} + 5\sqrt{7}$ **30.** $y\sqrt[3]{2y}$ **31.** $\sqrt{15} - \sqrt{6}$ **32.** $10 + 2\sqrt{21}$
33. $4x^2 - 5$ **34.** $x + 2 - 2\sqrt{x + 1}$ **35.** $\dfrac{\sqrt{21}}{3}$ **36.** $\dfrac{5\sqrt[3]{4x}}{2x}$ **37.** $\dfrac{13 - 3\sqrt{21}}{5}$ **38.** $\dfrac{7}{\sqrt{21}}$ **39.** $\dfrac{3y}{\sqrt[3]{33y^2}}$ **40.** $\dfrac{x - 4}{x + 2\sqrt{x}}$

Section 10.6

Graphing Calculator Explorations 1. $\{3.19\}$ **3.** $\varnothing$ **5.** $\{3.23\}$

Vocabulary, Readiness & Video Check 1. extraneous solution **3.** $x^2 - 10x + 25$ **5.** Applying the power rule can result in an equation with more solutions than the original equation, so you need to check all proposed solutions in the original equation.
7. Our answer is either a positive square root of a value or a negative square root of a value. We're looking for a length, which must be positive, so our answer must be the positive square root.

Exercise Set 10.6 1. $\{8\}$ **3.** $\{7\}$ **5.** $\varnothing$ **7.** $\{7\}$ **9.** $\{6\}$ **11.** $\left\{-\dfrac{9}{2}\right\}$ **13.** $\{29\}$ **15.** $\{4\}$ **17.** $\{-4\}$ **19.** $\left\{\dfrac{41}{16}\right\}$ **21.** $\{7\}$
23. $\{9\}$ **25.** $\{50\}$ **27.** $\varnothing$ **29.** $\left\{\dfrac{15}{4}\right\}$ **31.** $\{7\}$ **33.** $\{5\}$ **35.** $\{-12\}$ **37.** $\{9\}$ **39.** $\{-3\}$ **41.** $\{1\}$ **43.** $\{1\}$ **45.** $\left\{\dfrac{1}{2}\right\}$
47. $\{0, 4\}$ **49.** $\left\{\dfrac{37}{4}\right\}$ **51.** $3\sqrt{5}$ ft **53.** $2\sqrt{10}$ m **55.** $2\sqrt{131}$ m ≈ 22.9 m **57.** $\sqrt{100.84}$ mm ≈ 10.0 mm **59.** 17 ft **61.** 13 ft
63. 14,657,415 sq mi **65.** 100 ft **67.** $\sqrt{625,000}$ ft ≈ 790.57 ft **69.** 100 **71.** $\dfrac{\pi}{2}$ sec ≈ 1.57 sec **73.** 12.97 ft **75.** answers may vary
77. $15\sqrt{3}$ sq mi ≈ 25.98 sq mi **79.** answers may vary **81.** 0.51 km **83.** $\dfrac{x}{4x + 3}$ **85.** $-\dfrac{4z + 2}{3z}$ **87. a.-b.** answers may vary
89. $\sqrt{5x - 1} + 4 = 7$ **91.** 2743 deliveries
$\qquad \sqrt{5x - 1} = 3$
$\qquad (\sqrt{5x - 1})^2 = 3^2$
$\qquad 5x - 1 = 9$
$\qquad 5x = 10$
$\qquad x = 2$

Section 10.7

Vocabulary, Readiness & Video Check 1. complex **3.** -1 **5.** real **7.** the product rule for radicals; you need to first simplify each radical separately and have nonnegative radicands before applying the product rule. **9.** the fact that $i^2 = -1$
11. $i, i^2 = -1, i^3 = -i, i^4 = 1$

Exercise Set 10.7 1. $9i$ **3.** $i\sqrt{7}$ **5.** -4 **7.** $8i$ **9.** $2i\sqrt{6}$ **11.** $-6i$ **13.** $24i\sqrt{7}$ **15.** $-3\sqrt{6}$ **17.** $-\sqrt{14}$ **19.** $-5\sqrt{2}$ **21.** $4i$
23. $i\sqrt{3}$ **25.** $2\sqrt{2}$ **27.** $6 - 4i$ **29.** $-2 + 6i$ **31.** $-2 - 4i$ **33.** $2 - i$ **35.** $5 - 10i$ **37.** $8 - i$ **39.** -12 **41.** 63 **43.** -40
45. $18 + 12i$ **47.** $27 + 3i$ **49.** $18 + 13i$ **51.** 7 **53.** $12 - 16i$ **55.** 20 **57.** 2 **59.** $17 + 144i$ **61.** $-2i$ **63.** $-4i$ **65.** $\dfrac{28}{25} - \dfrac{21}{25}i$
67. $-\dfrac{12}{5} + \dfrac{6}{5}i$ **69.** $4 + i$ **71.** $-\dfrac{5}{2} - 2i$ **73.** $-5 + \dfrac{16}{3}i$ **75.** $\dfrac{3}{5} - \dfrac{1}{5}i$ **77.** $\dfrac{1}{5} - \dfrac{8}{5}i$ **79.** 1 **81.** i **83.** $-i$ **85.** -1 **87.** -64
89. $-243i$ **91.** 5 people **93.** 14 people **95.** 16.7% **97.** $1 - i$ **99.** 0 **101.** $2 + 3i$ **103.** $5 - 4i$ **105.** $2 - i\sqrt{2}$
107. answers may vary **109.** $6 - 3i\sqrt{3}$ **111.** yes

Chapter 10 Vocabulary Check 1. conjugate **2.** principal square root **3.** rationalizing **4.** imaginary unit **5.** cube root
6. index; radicand **7.** like radicals **8.** complex number **9.** distance **10.** midpoint

Chapter 10 Review 1. 9 **2.** 3 **3.** -2 **4.** not a real number **5.** $-\dfrac{1}{7}$ **6.** x^{32} **7.** -6 **8.** 4 **9.** $-a^2b^3$ **10.** $4a^2b^6$ **11.** $2ab^2$
12. $-2x^3y^4$ **13.** $\dfrac{x^6}{6y}$ **14.** $\dfrac{3y}{z^4}$ **15.** $|x|$ **16.** $|x^2 - 4|$ **17.** -27 **18.** -5 **19.** $-x$ **20.** $-x$ **21.** $2|2y + z|$ **22.** $5|x - y|$
23. y **24.** $|x|$ **25.** 3, 6 **26.** 2, $\sqrt[3]{17}$ **27.** $\dfrac{1}{3}$ **28.** $-\dfrac{1}{3}$ **29.** $-\dfrac{1}{3}$ **30.** $-\dfrac{1}{4}$ **31.** -27 **32.** $\dfrac{1}{4}$ **33.** not a real number **34.** $\dfrac{343}{125}$

35. $\dfrac{9}{4}$ **36.** not a real number **37.** $x^{2/3}$ **38.** $5^{1/5}x^{2/5}y^{3/5}$ **39.** $\sqrt[5]{y^4}$ **40.** $5\sqrt[3]{xy^2z^5}$ **41.** $\dfrac{1}{\sqrt[3]{x+2}}$ **42.** $\dfrac{1}{\sqrt{x+2y}}$ **43.** $a^{13/6}$

44. $\dfrac{1}{b}$ **45.** $\dfrac{1}{a^{9/2}}$ **46.** $\dfrac{y^2}{x}$ **47.** a^4b^6 **48.** $\dfrac{1}{x^{11/12}}$ **49.** $\dfrac{b^{5/6}}{49a^{1/4}c^{5/3}}$ **50.** $a - a^2$ **51.** 4.472 **52.** -3.391 **53.** 5.191 **54.** 3.826

55. -26.246 **56.** 0.045 **57.** $\sqrt[6]{1372}$ **58.** $\sqrt[12]{81x^3}$ **59.** $2\sqrt{6}$ **60.** $\sqrt[3]{7x^2yz}$ **61.** $2x$ **62.** ab^3 **63.** $2\sqrt{15}$ **64.** $-5\sqrt{3}$ **65.** $3\sqrt[3]{6}$

66. $-2\sqrt[3]{4}$ **67.** $6x^3\sqrt{x}$ **68.** $2ab^2\sqrt[3]{3a^2b}$ **69.** $\dfrac{p^8\sqrt{p}}{11}$ **70.** $\dfrac{y\sqrt[3]{y^2}}{3x^2}$ **71.** $\dfrac{y\sqrt[4]{xy^2}}{3}$ **72.** $\dfrac{x\sqrt{2x}}{7y^2}$ **73.** $\dfrac{5}{\sqrt{\pi}}$ m or $\dfrac{5\sqrt{\pi}}{\pi}$ m **74.** 5.75 in.

75. $\sqrt{197}$ units ≈ 14.036 units **76.** $\sqrt{130}$ units ≈ 11.402 units **77.** $\sqrt{73}$ units ≈ 8.544 units **78.** $7\sqrt{2}$ units ≈ 9.899 units

79. $2\sqrt{11}$ units ≈ 6.633 units **80.** $\sqrt{275.6}$ units ≈ 16.601 units **81.** $(-5, 5)$ **82.** $\left(-\dfrac{15}{2}, 1\right)$ **83.** $\left(-\dfrac{11}{2}, -2\right)$ **84.** $\left(\dfrac{1}{20}, -\dfrac{3}{16}\right)$

85. $\left(\dfrac{1}{4}, -\dfrac{2}{7}\right)$ **86.** $(\sqrt{3}, -3\sqrt{6})$ **87.** $-2\sqrt{5}$ **88.** $2x\sqrt{3x}$ **89.** $9\sqrt[3]{2}$ **90.** $3a\sqrt[4]{2a}$ **91.** $\dfrac{15 + 2\sqrt{3}}{6}$ **92.** $\dfrac{3\sqrt{2}}{4x}$

93. $17\sqrt{2} - 15\sqrt{5}$ **94.** $-4ab\sqrt[4]{2b}$ **95.** 6 **96.** $x - 6\sqrt{x} + 9$ **97.** $-8\sqrt{5}$ **98.** $4x - 9y$ **99.** $a - 9$ **100.** $\sqrt[3]{a^2} + 4\sqrt[3]{a} + 4$

101. $\sqrt[3]{25x^2} - 81$ **102.** $a + 64$ **103.** $\dfrac{3\sqrt{7}}{7}$ **104.** $\dfrac{\sqrt{3x}}{6}$ **105.** $\dfrac{5\sqrt[3]{2}}{2}$ **106.** $\dfrac{2x^2\sqrt[3]{2xy}}{y}$ **107.** $\dfrac{x^2y^2\sqrt[3]{15yz}}{z}$ **108.** $\dfrac{3\sqrt[4]{2x^2}}{2x^3}$

109. $\dfrac{3\sqrt{y} + 6}{y - 4}$ **110.** $-5 + 2\sqrt{6}$ **111.** $\dfrac{11}{3\sqrt{11}}$ **112.** $\dfrac{6}{\sqrt{2y}}$ **113.** $\dfrac{3}{7\sqrt[3]{3}}$ **114.** $\dfrac{4x^3}{y\sqrt{2x}}$ **115.** $\dfrac{xy}{\sqrt[3]{10x^2yz}}$ **116.** $\dfrac{x - 25}{-3\sqrt{x} + 15}$

117. $\{32\}$ **118.** $\varnothing$ **119.** $\{35\}$ **120.** $\varnothing$ **121.** $\{9\}$ **122.** $\{16\}$ **123.** $3\sqrt{2}$ cm **124.** $\sqrt{241}$ ft **125.** 51.2 ft **126.** 4.24 ft
127. $2i\sqrt{2}$ **128.** $-i\sqrt{6}$ **129.** $6i$ **130.** $-\sqrt{10}$ **131.** $15 - 4i$ **132.** $-13 - 3i$ **133.** -64 **134.** 81 **135.** $-12 - 18i$

136. $1 + 5i$ **137.** $-5 - 12i$ **138.** 87 **139.** $\dfrac{3}{2} - i$ **140.** $-\dfrac{1}{3} + \dfrac{1}{3}i$ **141.** x **142.** $|x + 2|$ **143.** -10 **144.** $-x^4y$ **145.** $\dfrac{y^5}{2x^3}$

146. 3 **147.** $\dfrac{1}{8}$ **148.** $\dfrac{16}{9}$ **149.** $\dfrac{1}{x^{13/2}}$ **150.** $10x^4\sqrt{2x}$ **151.** $\dfrac{n\sqrt{3n}}{11m^5}$ **152.** $6\sqrt{5} - 11x\sqrt[3]{5}$ **153.** $4x - 20\sqrt{x} + 25$ **154.** $\sqrt{41}$ units

155. $(4, 16)$ **156.** $\dfrac{7\sqrt{13}}{13}$ **157.** $\dfrac{2\sqrt{x} - 6}{x - 9}$ **158.** $\{4\}$ **159.** $\{5\}$

Chapter 10 Test 1. $6\sqrt{6}$ **2.** $-x^{16}$ **3.** $\dfrac{1}{5}$ **4.** 5 **5.** $\dfrac{4x^2}{9}$ **6.** $-a^6b^3$ **7.** $\dfrac{8a^{1/3}c^{2/3}}{b^{5/12}}$ **8.** $a^{7/12} - a^{7/3}$ **9.** $|4xy|$ or $4|xy|$

10. -27 **11.** $\dfrac{3\sqrt{y}}{y}$ **12.** $\dfrac{8 - 6\sqrt{x} + x}{8 - 2x}$ **13.** $\dfrac{2\sqrt[3]{3x^2}}{3x}$ **14.** $\dfrac{6 - x^2}{8(\sqrt{6} - x)}$ **15.** $-x\sqrt{5x}$ **16.** $4\sqrt{3} - \sqrt{6}$ **17.** $x + 2\sqrt{x} + 1$

18. $\sqrt{6} + \sqrt{2} - 4\sqrt{3} - 4$ **19.** -20 **20.** 23.685 **21.** 0.019 **22.** $\{2, 3\}$ **23.** $\varnothing$ **24.** $\{6\}$ **25.** $i\sqrt{2}$ **26.** $-2i\sqrt{2}$ **27.** $-3i$
28. 40 **29.** $7 + 24i$ **30.** $-\dfrac{3}{2} + \dfrac{5}{2}i$ **31.** $x = \dfrac{5\sqrt{2}}{2}$ in. **32.** $\sqrt{2}; 5$ **33.** $2\sqrt{26}$ units **34.** $\sqrt{95}$ units **35.** $\left(-4, \dfrac{7}{2}\right)$ **36.** $\left(-\dfrac{1}{2}, \dfrac{3}{10}\right)$
37. 27 mph **38.** 360 ft

Cumulative Review 1. -20; Sec. 1.6, Ex. 2 **2.** 5.27; Sec. 1.6 **3.** $-\dfrac{8}{21}$; Sec. 1.6, Ex. 4 **4.** $-\dfrac{25}{44}$; Sec. 1.6 **5.** $x = 2$; Sec. 2.3, Ex. 1
6. $a = \dfrac{13}{14}$; Sec. 2.3 **7. a.** 17% **b.** 21% **c.** 43 American travelers; Sec. 2.6, Ex. 3 **8. a.** $7x - 3$ **b.** $-3x + 11$ **c.** $6x - 14$; Sec. 5.5
9. ; Sec. 3.3, Ex. 7 **10.** 0; Sec. 3.4 **11.** ; Sec. 3.6, Ex. 5 **12.** $3y^2 - 2y - 8$; Sec. 5.5 **13. a.** 102,000
b. 0.007358 **c.** 84,000,000 **d.** 0.00003007; Sec. 5.2, Ex. 18
14. 5×10^{-2}; Sec. 5.2 **15.** $6x^2 - 11x - 10$; Sec. 5.5, Ex. 7b
16. $6y^3 + 7y^2 - 6y + 1$; Sec. 5.5 **17.** $(y + 2)(x + 3)$; Sec. 6.1, Ex. 11 **18.** $(x - 1)(x^2 + 4)$; Sec. 6.1
19. $(3x + 2)(x + 3)$; Sec. 6.3, Ex. 1 **20.** $2(x + 2)(x^2 - 2x + 4)$; Sec. 6.5 **21. a.** $x = 3$ **b.** $x = 2, x = 1$ **c.** none; Sec. 7.1, Ex. 2
22. $(x^2 + 1)(x + 1)(x - 1)$; Sec. 6.5 **23.** $\dfrac{x + 2}{x}$; Sec. 7.1, Ex. 5 **24.** $-a^2 - 2a - 4$; Sec. 7.1 **25. a.** 0 **b.** $\dfrac{15 + 14x}{50x^2}$; Sec. 7.4, Ex. 1
26. a. $\dfrac{9x - 2y}{3x^2y^2}$ **b.** $\dfrac{3x(x - 7)}{(x + 3)(x - 3)}$ **c.** $\dfrac{x + 5}{x - 2}$; Sec. 7.4 **27.** $x = -\dfrac{17}{5}$; Sec. 7.5, Ex. 4 **28.** $a = -1, a = -5$; Sec. 7.5
29. $y = -3x - 2$; Sec. 8.1, Ex. 1 **30.** $\dfrac{1}{27}$; Sec. 10.2 **31.** $\dfrac{1}{9}$; Sec. 10.2, Ex. 13 **32.** $\dfrac{1}{8}$; Sec. 10.2, Ex 12 **33.** $\dfrac{1}{25}$; Sec. 10.2
34. function; Sec. 8.2 **35.** $(3, 1)$; Sec. 4.3, Ex. 5 **36.** $(6, 0)$; Sec. 4.2 **37.** $\varnothing$; Sec. 9.1, Ex. 2 **38.** $x^2 + 3$; Sec. 5.7 **39.** $\varnothing$; Sec. 9.2, Ex. 2
40. $(-6, 5)$; Sec. 9.2

Chapter 11 Quadratic Equations and Functions

Section 11.1

Vocabulary, Readiness & Video Check   1. $\pm\sqrt{b}$ **3.** completing the square   **5.** 9   **7.** We need a quantity shown squared by itself on one side of the equation. The only quantity squared is x, so divide both sides by 2 before applying the square root property.
9. The coefficient of y^2 is 3. To use the completing the square method, the coefficient of the squared variable must be 1, so we first divide through by 3.

Exercise Set 11.1   1. $\{-4, 4\}$ **3.** $\{-\sqrt{7}, \sqrt{7}\}$ **5.** $\{-3\sqrt{2}, 3\sqrt{2}\}$ **7.** $\{-\sqrt{10}, \sqrt{10}\}$ **9.** $\{-8, -2\}$ **11.** $\{6 - 3\sqrt{2}, 6 + 3\sqrt{2}\}$
13. $\left\{\dfrac{3 - 2\sqrt{2}}{2}, \dfrac{3 + 2\sqrt{2}}{2}\right\}$ **15.** $\{-3i, 3i\}$ **17.** $\{-\sqrt{6}, \sqrt{6}\}$ **19.** $\{-2i\sqrt{2}, 2i\sqrt{2}\}$ **21.** $\left\{\dfrac{1 - 4i}{3}, \dfrac{1 + 4i}{3}\right\}$ **23.** $\{-7 - \sqrt{5}, -7 + \sqrt{5}\}$
25. $\{-3 - 2i\sqrt{2}, -3 + 2i\sqrt{2}\}$ **27.** 1   **29.** 49   **31.** $x^2 + 16x + 64 = (x + 8)^2$ **33.** $z^2 - 12z + 36 = (z - 6)^2$
35. $p^2 + 9p + \dfrac{81}{4} = \left(p + \dfrac{9}{2}\right)^2$ **37.** $r^2 - r + \dfrac{1}{4} = \left(r - \dfrac{1}{2}\right)^2$ **39.** $\{-5, -3\}$ **41.** $\{-3 - \sqrt{7}, -3 + \sqrt{7}\}$
43. $\left\{\dfrac{-1 - \sqrt{5}}{2}, \dfrac{-1 + \sqrt{5}}{2}\right\}$ **45.** $\{-1 - \sqrt{6}, -1 + \sqrt{6}\}$ **47.** $\left\{\dfrac{-1 - \sqrt{29}}{2}, \dfrac{-1 + \sqrt{29}}{2}\right\}$ **49.** $\{-4 - \sqrt{15}, -4 + \sqrt{15}\}$
51. $\left\{\dfrac{6 - \sqrt{30}}{3}, \dfrac{6 + \sqrt{30}}{3}\right\}$ **53.** $\left\{-4, \dfrac{1}{2}\right\}$ **55.** $\left\{\dfrac{-3 - \sqrt{21}}{3}, \dfrac{-3 + \sqrt{21}}{3}\right\}$ **57.** $\{-1 - i, -1 + i\}$ **59.** $\{-2 - i\sqrt{2}, -2 + i\sqrt{2}\}$
61. $\left\{\dfrac{1 - i\sqrt{47}}{4}, \dfrac{1 + i\sqrt{47}}{4}\right\}$ **63.** $\{-5 - i\sqrt{3}, -5 + i\sqrt{3}\}$ **65.** $\{-4, 1\}$ **67.** $\left\{\dfrac{2 - i\sqrt{2}}{2}, \dfrac{2 + i\sqrt{2}}{2}\right\}$ **69.** $\left\{\dfrac{-3 - \sqrt{69}}{6}, \dfrac{-3 + \sqrt{69}}{6}\right\}$
71. 20%   **73.** 11%   **75.** 9.63 sec   **77.** 14.32 sec   **79.** 8.29 sec   **81.** 15 ft by 15 ft   **83.** $10\sqrt{2}$ cm   **85.** $3 + 2\sqrt{5}$ **87.** $\dfrac{1 - 3\sqrt{2}}{2}$
89. $2\sqrt{6}$ **91.** 5   **93.** complex, but not real numbers   **95.** real numbers   **97.** complex, but not real numbers   **99.** answers may vary
101. compound; answers may vary   **103.** $-8x, 8x$ **105.** 6 thousand scissors

Section 11.2

Calculator Explorations   1. $\{-1.27, 6.27\}$ **3.** $\{-1.10, 0.90\}$ **5.** $\varnothing$; answers may vary

Vocabulary, Readiness & Video Check   1. $x = \dfrac{-b \pm \sqrt{b^2 - 4ac}}{2a}$ **3.** $-5; -7$ **5.** $1; 0$ **7. a.** Yes, in order to make sure we have correct values for $a, b,$ and c.   **b.** No; clearing fractions makes the work less tedious, but it's not a necessary step.   **9.** With applications, we need to make sure we answer the question(s) asked. Here we're asked how much distance is saved, so once the dimensions of the triangle are known, further calculations are needed to answer this question and solve the problem.

Exercise Set 11.2   1. $\{-6, 1\}$ **3.** $\left\{-\dfrac{3}{5}, 1\right\}$ **5.** $\{3\}$ **7.** $\left\{\dfrac{-7 - \sqrt{33}}{2}, \dfrac{-7 + \sqrt{33}}{2}\right\}$ **9.** $\left\{\dfrac{1 - \sqrt{57}}{8}, \dfrac{1 + \sqrt{57}}{8}\right\}$
11. $\left\{\dfrac{7 - \sqrt{85}}{6}, \dfrac{7 + \sqrt{85}}{6}\right\}$ **13.** $\{1 - \sqrt{3}, 1 + \sqrt{3}\}$ **15.** $\left\{-\dfrac{3}{2}, 1\right\}$ **17.** $\left\{\dfrac{3 - \sqrt{11}}{2}, \dfrac{3 + \sqrt{11}}{2}\right\}$ **19.** $\left\{\dfrac{-5 - \sqrt{17}}{2}, \dfrac{-5 + \sqrt{17}}{2}\right\}$
21. $\left\{\dfrac{5}{2}, 1\right\}$ **23.** $\{-3 - 2i, -3 + 2i\}$ **25.** $\{-2 - \sqrt{11}, -2 + \sqrt{11}\}$ **27.** $\left\{\dfrac{3 - i\sqrt{87}}{8}, \dfrac{3 + i\sqrt{87}}{8}\right\}$ **29.** $\left\{\dfrac{3 - \sqrt{29}}{2}, \dfrac{3 + \sqrt{29}}{2}\right\}$
31. $\left\{\dfrac{-5 - i\sqrt{5}}{10}, \dfrac{-5 + i\sqrt{5}}{10}\right\}$ **33.** $\left\{\dfrac{-1 - \sqrt{19}}{6}, \dfrac{-1 + \sqrt{19}}{6}\right\}$ **35.** $\left\{\dfrac{-1 - i\sqrt{23}}{4}, \dfrac{-1 + i\sqrt{23}}{4}\right\}$ **37.** $\{1\}$ **39.** $\{3 + \sqrt{5}, 3 - \sqrt{5}\}$
41. two real solutions   **43.** one real solution   **45.** two real solutions   **47.** two complex but not real solutions   **49.** two real solutions
51. 14 ft   **53.** $(2 + 2\sqrt{2})$ cm, $(2 + 2\sqrt{2})$ cm, $(4 + 2\sqrt{2})$ cm   **55.** width: $(-5 + 5\sqrt{17})$ ft; length: $(5 + 5\sqrt{17})$ ft   **57. a.** $50\sqrt{2}$ m
b. 5000 sq m   **59.** 37.4 ft by 38.5 ft   **61.** base: $(2 + 2\sqrt{43})$ cm; height: $(-1 + \sqrt{43})$ cm   **63.** 8.9 sec   **65.** 2.8 sec   **67.** $\left\{\dfrac{11}{5}\right\}$
69. $\{15\}$ **71.** $(x^2 + 5)(x + 2)(x - 2)$ **73.** $(z + 3)(z - 3)(z + 2)(z - 2)$ **75.** b   **77.** answers may vary   **79.** $\{0.6, 2.4\}$
81. Sunday to Monday   **83.** Wednesday   **85.** $f(4) = 33$; answers may vary   **87. a.** 94.64 billion kilowatts   **b.** 2015   **89. a.** 215
million   **b.** 620 million   **c.** 2018   **91.** $\{0.6, 2.4\}$

Section 11.3

Vocabulary, Readiness & Video Check   1. The values we get for the substituted variable are *not* our final answers. Remember to always substitute back to the original variable and solve for it if necessary.

Exercise Set 11.3   1. $\{2\}$ **3.** $\{16\}$ **5.** $\{1, 4\}$ **7.** $\{3 - \sqrt{7}, 3 + \sqrt{7}\}$ **9.** $\left\{\dfrac{3 - \sqrt{57}}{4}, \dfrac{3 + \sqrt{57}}{4}\right\}$ **11.** $\left\{\dfrac{1 - \sqrt{29}}{2}, \dfrac{1 + \sqrt{29}}{2}\right\}$
13. $\{-2, 2, -2i, 2i\}$ **15.** $\{-3, 3, -2i, 2i\}$ **17.** $\left\{-\dfrac{1}{2}, \dfrac{1}{2}, -i\sqrt{3}, i\sqrt{3}\right\}$ **19.** $\{125, -8\}$ **21.** $\left\{-\dfrac{4}{5}, 0\right\}$ **23.** $\left\{-\dfrac{1}{8}, 27\right\}$ **25.** $\left\{-\dfrac{2}{3}, \dfrac{4}{3}\right\}$

27. $\left\{-\dfrac{1}{125}, \dfrac{1}{8}\right\}$ **29.** $\{-\sqrt{2}, \sqrt{2}, -\sqrt{3}, \sqrt{3}\}$ **31.** $\left\{\dfrac{-9 - \sqrt{201}}{6}, \dfrac{-9 + \sqrt{201}}{6}\right\}$ **33.** $\{2, 3\}$ **35.** $\{3\}$ **37.** $\{27, 125)$ **39.** $\{5\}$

41. $\left\{\dfrac{1}{8}, -8\right\}$ **43.** $\{-5, 1\}$ **45.** $\{4\}$ **47.** $\{-3\}$ **49.** $\{-\sqrt{5}, \sqrt{5}, -2i, 2i\}$ **51.** $\{6, 12\}$ **53.** $\left\{-\dfrac{1}{3}, \dfrac{1}{3}, -\dfrac{i\sqrt{6}}{3}, \dfrac{i\sqrt{6}}{3}\right\}$ **55.** 5 mph,

then 4 mph **57.** inlet pipe: 15.5 hr; hose: 16.5 hr **59.** 55 mph; 66 mph **61.** 8.5 hr **63.** 12 or -8 **65. a.** $(x - 6)$ in.

b. $300 = (x - 6) \cdot (x - 6) \cdot 3$ **c.** 16 in. by 16 in. **67.** 22 ft **69.** $(-\infty, 3]$ **71.** $(-5, \infty)$ **73.** $\{1, -3i, 3i\}$ **75.** $\left\{-\dfrac{1}{2}, \dfrac{1}{3}\right\}$

77. $\left\{-3, \dfrac{3 - 3i\sqrt{3}}{2}, \dfrac{3 + 3i\sqrt{3}}{2}\right\}$ **79.** answers may vary **81. a.** 51.72 sec **b.** 54.11 sec **c.** 106.66 sec **d.** 53.85 sec

Integrated Review 1. $\{-\sqrt{10}, \sqrt{10}\}$ **2.** $\{-2i\sqrt{2}, 2i\sqrt{2}\}$ **3.** $\{1 - 2\sqrt{2}, 1 + 2\sqrt{2}\}$ **4.** $\left\{\dfrac{-5 - 2\sqrt{3}}{2}, \dfrac{-5 + 2\sqrt{3}}{2}\right\}$

5. $\{-1 - \sqrt{13}, -1 + \sqrt{13}\}$ **6.** $\{1, 11\}$ **7.** $\left\{\dfrac{-3 - \sqrt{17}}{2}, \dfrac{-3 + \sqrt{17}}{2}\right\}$ **8.** $\left\{\dfrac{-2 - \sqrt{5}}{4}, \dfrac{-2 + \sqrt{5}}{4}\right\}$ **9.** $\left\{\dfrac{2 - \sqrt{2}}{2}, \dfrac{2 + \sqrt{2}}{2}\right\}$

10. $\{-3 - \sqrt{5}, -3 + \sqrt{5}\}$ **11.** $\{-2 + i\sqrt{3}, -2 - i\sqrt{3}\}$ **12.** $\left\{\dfrac{-3 - i\sqrt{6}}{5}, \dfrac{-3 + i\sqrt{6}}{5}\right\}$ **13.** $\left\{\dfrac{-3 + i\sqrt{15}}{2}, \dfrac{-3 - i\sqrt{15}}{2}\right\}$

14. $\{3i, -3i\}$ **15.** $\{0, -17\}$ **16.** $\left\{\dfrac{1 + \sqrt{13}}{4}, \dfrac{1 - \sqrt{13}}{4}\right\}$ **17.** $\{2 + 3\sqrt{3}, 2 - 3\sqrt{3}\}$ **18.** $\{2 + \sqrt{3}, 2 - \sqrt{3}\}$ **19.** $\left\{-2, \dfrac{4}{3}\right\}$

20. $\left\{\dfrac{-5 + \sqrt{17}}{4}, \dfrac{-5 - \sqrt{17}}{4}\right\}$ **21.** $\{1 - \sqrt{6}, 1 + \sqrt{6}\}$ **22.** $\{-\sqrt{31}, \sqrt{31}\}$ **23.** $\{-2\sqrt{3}, 2\sqrt{3}\}$ **24.** $\{-i\sqrt{11}, i\sqrt{11}\}$

25. $\{-11, 6\}$ **26.** $\left\{\dfrac{-3 + \sqrt{19}}{5}, \dfrac{-3 - \sqrt{19}}{5}\right\}$ **27.** $\left\{\dfrac{-3 + \sqrt{17}}{4}, \dfrac{-3 - \sqrt{17}}{4}\right\}$ **28.** $\{4\}$ **29.** $\left\{\dfrac{-1 + \sqrt{17}}{8}, \dfrac{-1 - \sqrt{17}}{8}\right\}$

30. $10\sqrt{2}$ ft ≈ 14.1 ft **31.** Jack: 9.1 hr; Lucy: 7.1 hr **32.** 5 mph during the first part, then 6 mph

Section 11.4

Vocabulary, Readiness & Video Check 1. $[-7, 3)$ **3.** $(-\infty, 0]$ **5.** $(-\infty, -12) \cup [-10, \infty)$ **7.** We use the solutions to the
related equation to divide the number line into regions that either entirely are or entirely are not solution regions; the solutions to
the related equation are solutions to the inequality only if the inequality symbol is $\leq$ or $\geq$.

Exercise Set 11.4 1. $(-\infty, -5) \cup (-1, \infty)$ **3.** $[-4, 3]$ **5.** $(-\infty, -5] \cup [-3, \infty)$ **7.** $\left(-5, -\dfrac{1}{3}\right)$ **9.** $(2, 4) \cup (6, \infty)$

11. $(-\infty, -4] \cup [0, 1]$ **13.** $(-\infty, -3) \cup (-2, 2) \cup (3, \infty)$ **15.** $(-7, 2)$ **17.** $(-1, \infty)$ **19.** $(-\infty, -1] \cup (4, \infty)$

21. $(-\infty, 2) \cup \left(\dfrac{11}{4}, \infty\right)$ **23.** $(0, 2] \cup [3, \infty)$ **25.** $(-\infty, 3)$ **27.** $\left[-\dfrac{5}{4}, \dfrac{3}{2}\right]$ **29.** $(-\infty, 0) \cup (1, \infty)$ **31.** $(0, 10)$

33. $(-\infty, -4] \cup [4, 6]$ **35.** $\left(-\infty, -\dfrac{2}{3}\right] \cup \left[\dfrac{3}{2}, \infty\right)$ **37.** $(-\infty, -4) \cup [5, \infty)$ **39.** $(-\infty, 1) \cup (2, \infty)$ **41.** $\left(-4, -\dfrac{3}{2}\right) \cup \left(\dfrac{3}{2}, \infty\right)$

43. $(-\infty, -5] \cup [-1, 1] \cup [5, \infty)$ **45.** $(-\infty, -6] \cup (-1, 0] \cup (7, \infty)$ **47.** $(-\infty, -8] \cup (-4, \infty)$ **49.** $\left(-\infty, -\dfrac{5}{3}\right) \cup \left(\dfrac{7}{2}, \infty\right)$

51. $(-\infty, 0] \cup \left(5, \dfrac{11}{2}\right]$ **53.** $(0, \infty)$ **55.** 0; 1; 1; 4; 4 **57.** 0; -1; -1; -4; -4 **59.** answers may vary **61.** $(-\infty, -1) \cup (0, 1)$

63. when x is between 2 and 11 **65.** $(-\infty, -7) \cup (8, \infty)$

Section 11.5

Calculator Explorations 1.
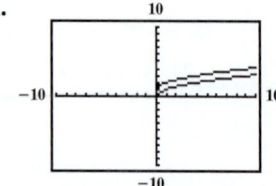
3.
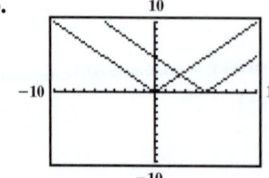
5.
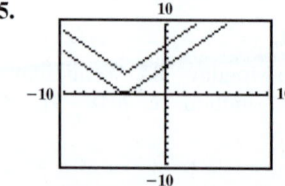

Vocabulary, Readiness & Video Check 1. quadratic **3.** upward **5.** lowest **7.** $(0, 0)$ **9.** $(2, 0)$ **11.** $(0, 3)$ **13.** $(-1, 5)$
15. Graphs of the form $f(x) = x^2 + k$ shift up or down the y-axis k units from $y = x^2$; the y-intercept. **17.** The vertex, (h, k) and
the axis of symmetry, $x = h$; the basic shape of $y = x^2$ does not change. **19.** the coordinates of the vertex, whether the graph opens
upward or downward, whether the graph is narrower or wider than $y = x^2$, and the graph's axis of symmetry

Exercise Set 11.5

1. 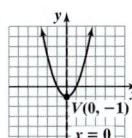 $V(0, -1)$ $x = 0$

3. 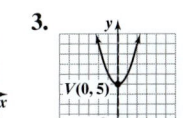 $V(0, 5)$ $x = 0$

5. 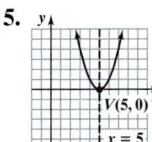 $V(5, 0)$ $x = 5$

7. 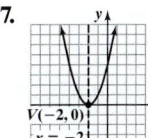 $V(-2, 0)$ $x = -2$

9. 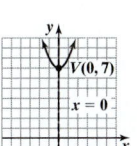 $V(0, 7)$ $x = 0$

11. 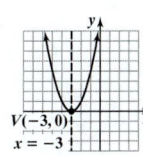 $V(-3, 0)$ $x = -3$

13. 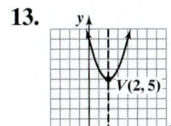 $V(2, 5)$ $x = 2$

15. 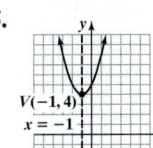 $V(-1, 4)$ $x = -1$

17. 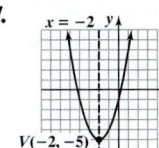 $x = -2$ $V(-2, -5)$

19. 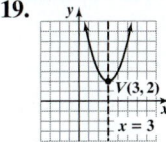 $V(3, 2)$ $x = 3$

21. 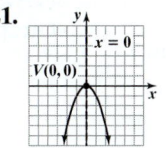 $x = 0$ $V(0, 0)$

23. 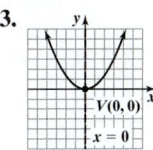 $V(0, 0)$ $x = 0$

25. 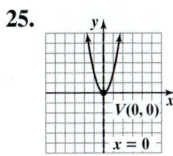 $V(0, 0)$ $x = 0$

27. 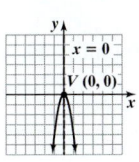 $x = 0$ $V(0, 0)$

29. 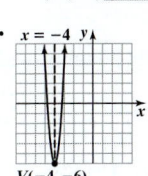 $x = -4$ $V(-4, -6)$

31. 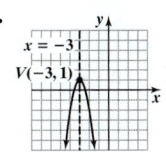 $x = -3$ $V(-3, 1)$

33. 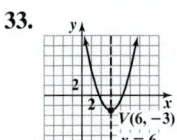 $V(6, -3)$ $x = 6$

35. 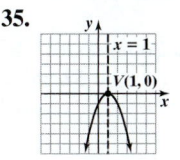 $x = 1$ $V(1, 0)$

37. 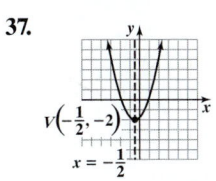 $V\left(-\frac{1}{2}, -2\right)$ $x = -\frac{1}{2}$

39. 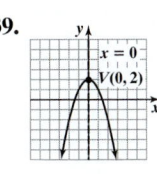 $x = 0$ $V(0, 2)$

41. 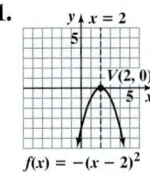 $x = 2$ $V(2, 0)$ $f(x) = -(x - 2)^2$

43. 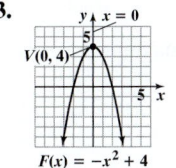 $x = 0$ $V(0, 4)$ $F(x) = -x^2 + 4$

45. 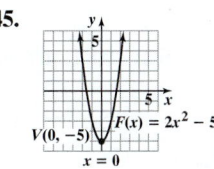 $F(x) = 2x^2 - 5$ $V(0, -5)$ $x = 0$

47. 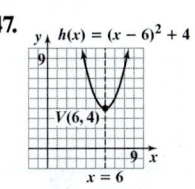 $h(x) = (x - 6)^2 + 4$ $V(6, 4)$ $x = 6$

49. 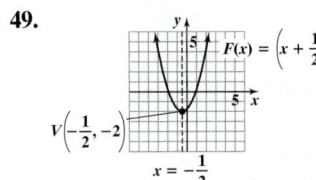 $F(x) = \left(x + \frac{1}{2}\right)^2 - 2$ $V\left(-\frac{1}{2}, -2\right)$ $x = -\frac{1}{2}$

51. 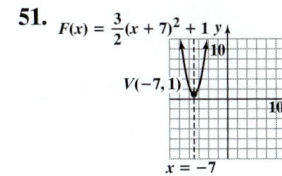 $F(x) = \frac{3}{2}(x + 7)^2 + 1$ $V(-7, 1)$ $x = -7$

53. 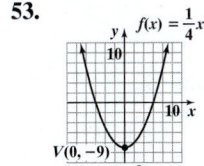 $f(x) = \frac{1}{4}x^2 - 9$ $V(0, -9)$ $x = 0$

55. 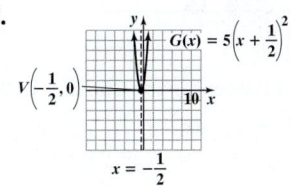 $G(x) = 5\left(x + \frac{1}{2}\right)^2$ $V\left(-\frac{1}{2}, 0\right)$ $x = -\frac{1}{2}$

57. 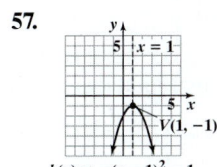 $x = 1$ $V(1, -1)$ $h(x) = -(x - 1)^2 - 1$

59. 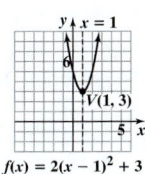 $x = 1$ $V(1, 3)$ $f(x) = 2(x - 1)^2 + 3$

61. 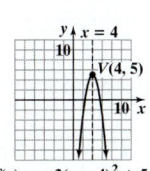 $x = 4$ $V(4, 5)$ $f(x) = -2(x - 4)^2 + 5$

63. $x^2 + 8x + 16$ **65.** $z^2 - 16z + 64$ **67.** $y^2 + y + \frac{1}{4}$

69. $g(x) = 5(x - 2)^2 + 3$ **71.** $g(x) = 5(x + 3)^2 + 6$

73.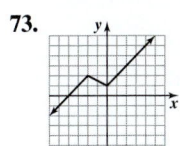

75.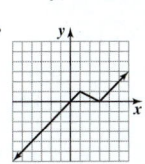

77.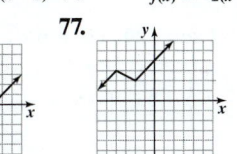

79. c

Section 11.6

Vocabulary, Readiness & Video Check **1.** (h, k) **3.** We can immediately identify the vertex (h, k), whether the parabola opens upward or downward, and know its axis of symmetry; completing the square. **5.** the vertex

Exercise Set 11.6 **1.** $0; 1$ **3.** $2; 1$ **5.** $1; 1$ **7.** down **9.** up **11.** $(-4, -9)$ **13.** $(5, 30)$ **15.** $(1, -2)$ **17.** $\left(\frac{1}{2}, \frac{5}{4}\right)$ **19.** D **21.** B

23. 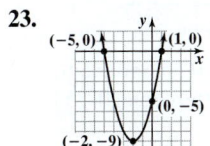 $(-5, 0)$ $(1, 0)$ $(0, -5)$ $(-2, -9)$

vertex: $(-2, -9)$; opens upward;
x-intercepts: $(-5, 0), (1, 0)$;
y-intercept: $(0, -5)$

25. 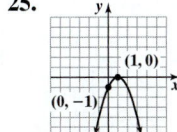 $(1, 0)$ $(0, -1)$

vertex: $(1, 0)$;
opens downward;
x-intercept: $(1, 0)$;
y-intercept: $(0, -1)$

27. 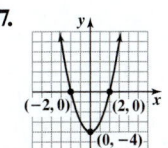 $(-2, 0)$ $(2, 0)$ $(0, -4)$

vertex: $(0, -4)$; opens upward;
x-intercepts: $(-2, 0), (2, 0)$;
y-intercept: $(0, -4)$

29. 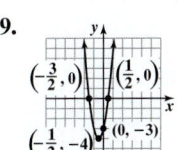 $\left(-\frac{3}{2}, 0\right)$ $\left(\frac{1}{2}, 0\right)$ $\left(-\frac{1}{2}, -4\right)$ $(0, -3)$

vertex: $\left(-\frac{1}{2}, -4\right)$; opens upward;

x-intercepts: $\left(-\frac{3}{2}, 0\right), \left(\frac{1}{2}, 0\right)$;

y-intercept: $(0, -3)$

31.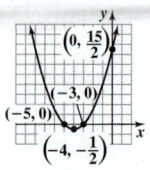

vertex: $\left(-4, -\dfrac{1}{2}\right)$; opens upward;

x-intercepts: $(-5, 0), (-3, 0)$;

y-intercept: $\left(0, \dfrac{15}{2}\right)$

33.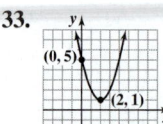

vertex: $(2, 1)$; opens upward;
y-intercept: $(0, 5)$

35.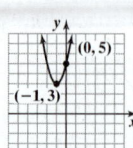

vertex: $(-1, 3)$; opens upward;
y-intercept: $(0, 5)$

37.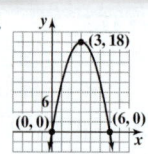

vertex: $(3, 18)$; opens
downward; x-intercepts:
$(0, 0), (6, 0)$; y-intercept:
$(0, 0)$

39. 144 ft   **41. a.** 200 bicycles   **b.** $12,000   **43.** 30, 30   **45.** 5, −5   **47.** length: 20 units; width: 20 units   **49.** $(0, 2)$ **51.** $(-2, 0)$
53. $(-5, 2)$ **55.** $(4, 1)$ **57.** minimum value   **59.** maximum value   **61.**

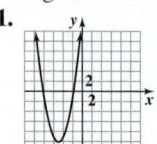

vertex: $(-5, -10)$; opens upward;
y-intercept: $(0, 15)$; x-intercepts:
$(-1.8, 0), (-8.2, 0)$

63.

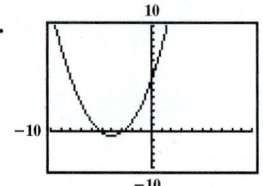

65. −0.84   **67. a.** maximum, answers may vary   **b.** 2030   **c.** 99.9%

Chapter 11 Vocabulary Check   1. discriminant   **2.** $\pm\sqrt{b}$ **3.** $\dfrac{-b}{2a}$ **4.** quadratic inequality   **5.** completing the square
6. $(0, k)$ **7.** $(h, 0)$ **8.** (h, k) **9.** quadratic formula   **10.** quadratic

Chapter 11 Review   1. $\{14, 1\}$ **2.** $\left\{-\dfrac{6}{7}, 5\right\}$ **3.** $\{-7, 7\}$ **4.** $\left\{\dfrac{2 - \sqrt{2}}{5}, \dfrac{2 + \sqrt{2}}{5}\right\}$ **5.** $\left\{\dfrac{-3 - \sqrt{5}}{2}, \dfrac{-3 + \sqrt{5}}{2}\right\}$
6. $\left\{\dfrac{-3 - i\sqrt{7}}{8}, \dfrac{-3 + i\sqrt{7}}{8}\right\}$ **7.** 4.25%   **8.** $75\sqrt{2}$ mi; 106.1 mi   **9.** two complex but not real solutions   **10.** two real solutions
11. two real solutions   **12.** one real solution   **13.** $\{8\}$ **14.** $\{-5, 0\}$ **15.** $\left\{-\dfrac{5}{2}, 1\right\}$ **16.** $\left\{\dfrac{5 - i\sqrt{143}}{12}, \dfrac{5 + i\sqrt{143}}{12}\right\}$
17. $\left\{\dfrac{1 - i\sqrt{35}}{9}, \dfrac{1 + i\sqrt{35}}{9}\right\}$ **18.** $\left\{1, \dfrac{9}{4}\right\}$ **19. a.** 20 ft   **b.** $\dfrac{15 + \sqrt{321}}{16}$ sec; 2.1 sec   **20.** $(6 + 6\sqrt{2})$ cm
21. $\left\{3, \dfrac{-3 + 3i\sqrt{3}}{2}, \dfrac{-3 - 3i\sqrt{3}}{2}\right\}$ **22.** $\{-4, 2 - 2i\sqrt{3}, 2 + 2i\sqrt{3}\}$ **23.** $\left\{\dfrac{2}{3}, 5\right\}$ **24.** $\{-5, 5, -2i, 2i\}$ **25.** $\left\{-\dfrac{16}{5}, 1\right\}$
26. $\{1, 125\}$ **27.** $\{-1, 1, -i, i\}$ **28.** $\left\{-\dfrac{1}{5}, \dfrac{1}{4}\right\}$ **29.** Jerome: 10.5 hr; Tim: 9.5 hr   **30.** −5   **31.** $[-5, 5]$ **32.** $\left(-\dfrac{1}{2}, \dfrac{1}{2}\right)$
33. $(-\infty, -4) \cup (-1, 1) \cup (4, \infty)$ **34.** $[-5, -2] \cup [2, 5]$ **35.** $(5, 6)$ **36.** $(-\infty, -6) \cup \left(-\dfrac{3}{4}, 0\right) \cup (5, \infty)$ **37.** $(-\infty, -5] \cup [-2, 6]$
38. $(-5, -3) \cup (5, \infty)$ **39.** $(-\infty, 0)$ **40.** $\left(-\dfrac{6}{5}, 0\right) \cup \left(\dfrac{5}{6}, 3\right)$ **41.** 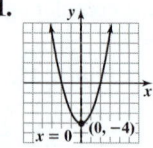   **42.** 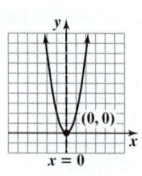   **43.**

44. 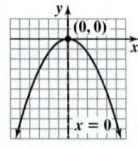   **45.** 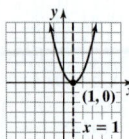   **46.** 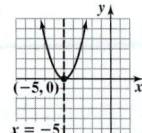   **47.** 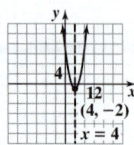   **48.** 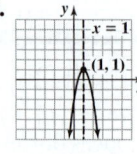   **49.** vertex: $(-5, 0)$; x-intercept:
$(-5, 0)$; y-intercept: $(0, 25)$

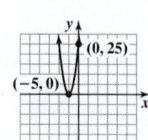

50. vertex: $(3, 0)$; x-intercept: $(3, 0)$; **51.** vertex: $(0, -1)$; x-intercepts: **52.** vertex: $(0, 5)$; x-intercepts: **53.** vertex: $\left(-\dfrac{5}{6}, \dfrac{73}{12}\right)$;
y-intercept: $(0, -9)$

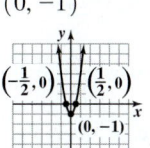

51. $\left(-\dfrac{1}{2}, 0\right), \left(\dfrac{1}{2}, 0\right)$; y-intercept: $(0, -1)$

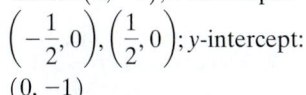

52. $(-1, 0), (1, 0)$; y-intercept: $(0, 5)$

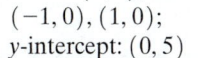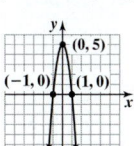

53. opens downward; x-intercepts: $(-2.3, 0)$, $(0.6, 0)$; y-intercept: $(0, 4)$

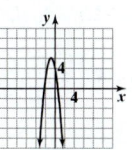

54. a. 0.4 sec and 7.1 sec **b.** answers may vary **55.** The numbers are both 210. **56.** $50, -50$ **57.** $\{-5, 6)$ **58.** $\left\{\dfrac{4}{5}, -\dfrac{1}{2}\right\}$

59. $\{-2, 2\}$ **60.** $\left\{-\dfrac{4}{9}, \dfrac{2}{9}\right\}$ **61.** $\left\{\dfrac{-1 - 3i\sqrt{3}}{2}, \dfrac{-1 + 3i\sqrt{3}}{2}\right\}$ **62.** $\left\{\dfrac{17 - \sqrt{145}}{9}, \dfrac{17 + \sqrt{145}}{9}\right\}$ **63.** $\{-i\sqrt{11}, i\sqrt{11}\}$

64. $\{-i\sqrt{7}, i\sqrt{7}\}$ **65.** $\left\{\dfrac{21 - \sqrt{41}}{50}, \dfrac{21 + \sqrt{41}}{50}\right\}$ **66.** $\left\{-\dfrac{8\sqrt{7}}{7}, \dfrac{8\sqrt{7}}{7}\right\}$ **67.** $\{8, 64\}$ **68.** $\left(-\infty, -\dfrac{5}{4}\right] \cup \left[\dfrac{3}{2}, \infty\right)$

69. $[-5, 0] \cup \left(\dfrac{3}{4}, \infty\right)$ **70.** $\left(2, \dfrac{7}{2}\right)$ **71. a.** 111,395 thousand passengers **b.** 2019 **72.** 6.62 sec

Chapter 11 Test 1. $\left\{\dfrac{7}{5}, -1\right\}$ **2.** $\{-1 - \sqrt{10}, -1 + \sqrt{10}\}$ **3.** $\left\{\dfrac{1 + i\sqrt{31}}{2}, \dfrac{1 - i\sqrt{31}}{2}\right\}$ **4.** $\{3 - \sqrt{7}, 3 + \sqrt{7}\}$

5. $\left\{-\dfrac{1}{7}, -1\right\}$ **6.** $\left\{\dfrac{3 + \sqrt{29}}{2}, \dfrac{3 - \sqrt{29}}{2}\right\}$ **7.** $\{-2 - \sqrt{11}, -2 + \sqrt{11}\}$ **8.** $\{-3, 3, -i, i\}$ **9.** $\{-1, 1, -i, i\}$ **10.** $\{6, 7\}$

11. $\{3 - \sqrt{7}, 3 + \sqrt{7}\}$ **12.** $\left\{\dfrac{2 - i\sqrt{6}}{2}, \dfrac{2 + i\sqrt{6}}{2}\right\}$ **13.** $\left(-\infty, -\dfrac{3}{2}\right) \cup (5, \infty)$ **14.** $(-\infty, -5) \cup (-4, 4) \cup (5, \infty)$

15. $(-\infty, -3) \cup (2, \infty)$ **16.** $(-\infty, -3) \cup [2, 3)$ **17.** **18.** **19.** **20.**

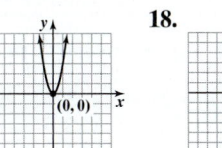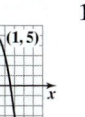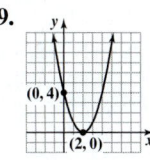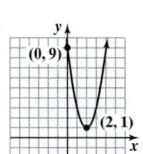

21. 7 ft **22.** $(5 + \sqrt{17})$ hr $\approx$ 9.12 hr **23. a.** 272 ft **b.** 5.12 sec

Cumulative Review 1. $-5x + 8y = -20$; Sec. 8.1, Ex. 2 **2.** 0.002068; Sec. 5.2 **3.** domain: $(-\infty, \infty)$; range: $[0, \infty)$; Sec. 8.4, Ex. 5
4. domain: $(-\infty, \infty)$; range: $(-\infty, -2) \cup [-1, \infty)$; Sec. 8.4 **5.** domain: $[-4, 4]$; range: $[-2, 2]$; Sec. 8.4, Ex. 6 **6.** $(-2, -7)$; Sec. 4.2

7. $\left(\dfrac{1}{2}, 0, \dfrac{3}{4}\right)$; Sec. 9.1, Ex. 3 **8. a.** $\dfrac{a^6}{b^3 c^9}$ **b.** $\dfrac{a^8 c^6}{b^4}$ **c.** $\dfrac{16 b^6}{a^6}$; Sec. 5.2 **9.** $(1, -1, 3)$; Sec. 9.2, Ex. 3 **10.** $28a^2 - 29a + 6$; Sec. 5.5

11. $(-1, -3)$; Sec. 4.2, Ex. 2 **12.** $3x(3x^2 + 9x - 5)$; Sec. 6.1 **13.** 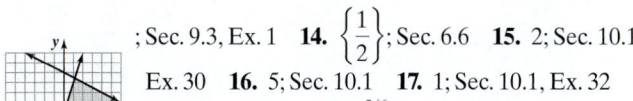; Sec. 9.3, Ex. 1 **14.** $\left\{\dfrac{1}{2}\right\}$; Sec. 6.6 **15.** 2; Sec. 10.1,
Ex. 30 **16.** 5; Sec. 10.1 **17.** 1; Sec. 10.1, Ex. 32 **18.** 2; Sec. 10.1 **19.** $x^{5/6}$; Sec. 10.2, Ex. 14
20. $\dfrac{2a + b}{a + 2b}$; Sec. 7.7 **21.** 32; Sec. 10.2, Ex. 18

22. $x^2 - 6x + 8$; Sec. 5.7 **23.** $\dfrac{\sqrt{x}}{3}$; Sec. 10.3, Ex. 7 **24.** $\dfrac{y\sqrt{y}}{5}$; Sec. 10.3 **25.** $\dfrac{\sqrt[4]{3}}{2y}$; Sec. 10.3, Ex. 9 **26.** $\dfrac{\sqrt[3]{5}}{3x}$; Sec. 10.3 **27.** $2x - 25$;
Sec. 10.4, Ex. 13 **28.** $-2 - 6\sqrt{3}$; Sec. 10.4 **29.** $4 - 2\sqrt{3}$; Sec. 10.4, Ex. 14 **30.** $6a^2 - 7ab - 5b^2$; Sec. 5.5 **31.** $\dfrac{7}{3\sqrt{35}}$; Sec. 10.5, Ex. 9

32. $\dfrac{8\sqrt{x}}{3x}$; Sec. 10.5 **33.** $\{3\}$; Sec. 10.6, Ex. 4 **34.** $\{2, 6\}$; Sec. 10.6 **35.** $-1 + 5i$; Sec. 10.7, Ex. 8 **36.** $\dfrac{1}{x - 2}$; Sec. 7.3

37. $\left\{\dfrac{6 + i\sqrt{5}}{2}, \dfrac{6 - i\sqrt{5}}{2}\right\}$; Sec. 11.1, Ex. 9 **38.** $\{1 + 2\sqrt{6}, 1 - 2\sqrt{6}\}$; Sec. 11.1 **39.** $\{2 + \sqrt{2}, 2 - \sqrt{2}\}$; Sec. 11.2, Ex. 3

40. $\{2 + 2\sqrt{3}, 2 - 2\sqrt{3}\}$; Sec. 11.2 **41.** $\{8, 27\}$; Sec. 11.3, Ex. 5 **42.** $\{-1\}$; Sec. 7.5 **43.** $\left(-\dfrac{7}{2}, -1\right)$; Sec. 11.4, Ex. 5

44. vertex: $\left(-\dfrac{1}{2}, -\dfrac{49}{4}\right)$; y-intercept: $(0, -12)$; x-intercepts: $(3, 0), (-4, 0)$; Sec. 11.6 **45.** 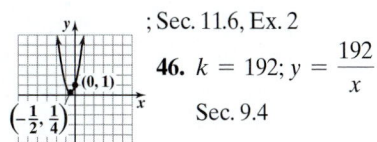; Sec. 11.6, Ex. 2 **46.** $k = 192$; $y = \dfrac{192}{x}$; Sec. 9.4

Chapter 12 Exponential and Logarithmic Functions

Section 12.1

Vocabulary, Readiness & Video Check   **1.** C   **3.** F   **5.** D   **7.** You can find $(f + g)(x)$ and then find $(f + g)(2)$, or you can find $f(2)$ and $g(2)$ and then add those results.

Exercise Set 12.1   **1. a.** $3x - 6$ **b.** $-x - 8$ **c.** $2x^2 - 13x - 7$ **d.** $\dfrac{x - 7}{2x + 1}$, where $x \neq -\dfrac{1}{2}$ **3. a.** $x^2 + 5x + 1$ **b.** $x^2 - 5x + 1$

c. $5x^3 + 5x$ **d.** $\dfrac{x^2 + 1}{5x}$, where $x \neq 0$ **5. a.** $\sqrt{x} + x + 5$ **b.** $\sqrt{x} - x - 5$ **c.** $x\sqrt{x} + 5\sqrt{x}$ **d.** $\dfrac{\sqrt{x}}{x + 5}$, where $x \neq -5$ **7. a.** $5x^2 - 3x$

b. $-5x^2 - 3x$ **c.** $-15x^3$ **d.** $-\dfrac{3}{5x}$, where $x \neq 0$ **9.** 42   **11.** -18 **13.** 0   **15.** $(f \circ g)(x) = 25x^2 + 1$; $(g \circ f)(x) = 5x^2 + 5$

17. $(f \circ g)(x) = 2x + 11$; $(g \circ f)(x) = 2x + 4$ **19.** $(f \circ g)(x) = -8x^3 - 2x - 2$; $(g \circ f)(x) = -2x^3 - 2x + 4$

21. $(f \circ g)(x) = |10x - 3|$; $(g \circ f)(x) = 10|x| - 3$ **23.** $(f \circ g)(x) = \sqrt{-5x + 2}$; $(g \circ f)(x) = -5\sqrt{x} + 2$ **25.** $H(x) = (g \circ h)(x)$

27. $F(x) = (h \circ f)(x)$ **29.** $G(x) = (f \circ g)(x)$ **31.** answers may vary   **33.** answers may vary   **35.** answers may vary   **37.** $y = x - 2$

39. $y = \dfrac{x}{3}$ **41.** $y = -\dfrac{x + 7}{2}$ **43.** $P(x) = R(x) - C(x)$ **45.** answers may vary

Section 12.2

Vocabulary, Readiness & Video Check   **1.** $(2, 11)$ **3.** the inverse of f **5.** vertical   **7.** $y = x$ **9.** Every function must have each x-value correspond to only one y-value. A one-to-one function must also have each y-value correspond to only one x-value.   **11.** Yes; by the definition of an inverse function.   **13.** Once you know some points of the original equation or graph, you can switch the x's and y's of these points to find points that satisfy the inverse and then graph it. You can also check that the two graphs (the original and the inverse) are symmetric about the line $y = x$.

Exercise Set 12.2   **1.** not one-to-one   **3.** one-to-one; $h^{-1} = \{(10, 10)\}$ **5.** one-to-one; $f^{-1} = \{(12, 11), (3, 4), (4, 3), (6, 6)\}$
7. not one-to-one   **9.** one-to-one;

Rank in Population (Input)	1	47	16	25	36	7
State (Output)	California	Alaska	Indiana	Louisiana	New Mexico	Ohio

11. a. 3   **b.** 1   **13. a.** 1   **b.** -1 **15.** one-to-one   **17.** not one-to-one   **19.** one-to-one   **21.** not one-to-one

23. $f^{-1}(x) = x - 4$ **25.** $f^{-1}(x) = \dfrac{x + 3}{2}$ **27.** $f^{-1}(x) = 2x + 2$ **29.** $f^{-1}(x) = \sqrt[3]{x}$ **31.** $f^{-1}(x) = \dfrac{x - 2}{5}$ **33.** $f^{-1}(x) = 5x + 2$

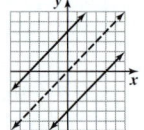   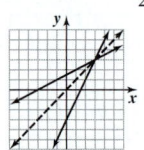   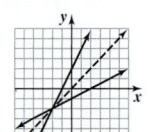   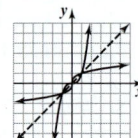

35. $f^{-1}(x) = x^3$ **37.** $f^{-1}(x) = \dfrac{\dfrac{5}{x} - 1}{3}$

39. $f^{-1}(x) = \sqrt[3]{x} - 2$

41. 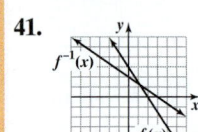   **43.** 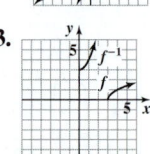   **45.** 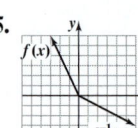   **47.** $(f \circ f^{-1})(x) = x$; $(f^{-1} \circ f)(x) = x$ **49.** $(f \circ f^{-1})(x) = x$; $(f^{-1} \circ f)(x) = x$

51. 5   **53.** 8   **55.** $\dfrac{1}{27}$ **57.** 9   **59.** $3^{1/2} \approx 1.73$ **61. a.** $(2, 9)$ **b.** $(9, 2)$

63. a. $\left(-2, \dfrac{1}{4}\right), \left(-1, \dfrac{1}{2}\right), (0, 1), (1, 2), (2, 5)$

b. $\left(\dfrac{1}{4}, -2\right), \left(\dfrac{1}{2}, -1\right), (1, 0), (2, 1)(5, 2)$ **c.** 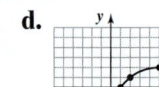   **d.** 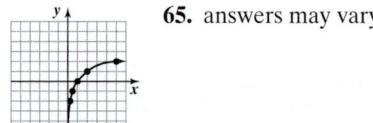   **65.** answers may vary

67. $f^{-1}(x) = \dfrac{x - 1}{3}$; 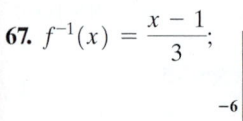   **69.** $f^{-1}(x) = x^3 - 3$;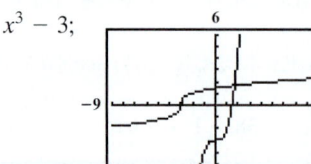

Section 12.3

Calculator Explorations   1. 81.98%   **3.** 22.54%

Vocabulary, Readiness & Video Check   1. exponential; c   **3.** yes   **5.** no; none   **7.** $(-\infty, \infty)$ **9.** In a polynomial function, the base is the variable and the exponent is the constant; in an exponential function, the base is the constant and the exponent is the variable.   **11.** $y = 30(0.996)^{101} \approx 20.0$ lb

Exercise Set 12.3   1. 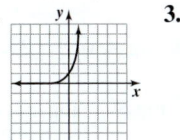   **3.** 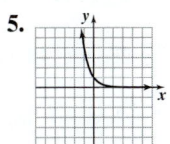   **5.** 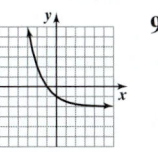   **7.**   **9.**   **11.**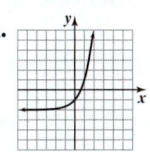

13. 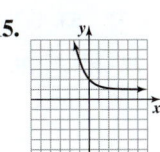   **15.** 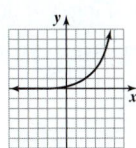   **17.** 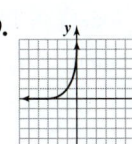   **19.**   **21.** $\{3\}$ **23.** $\left\{\dfrac{3}{4}\right\}$ **25.** $\left\{\dfrac{8}{5}\right\}$ **27.** $\left\{-\dfrac{2}{3}\right\}$ **29.** $\left\{\dfrac{3}{2}\right\}$

31. $\left\{-\dfrac{1}{3}\right\}$ **33.** $\{-2\}$ **35.** $\{4\}$ **37.** 24.6 lb

39. a. 5328.9 thousand metric tons

b. 7126.4 thousand metric tons   **41. a.** 286,160 students   **b.** 335,282 students   **43.** 402.3 million   **45.** $7621.42   **47.** $4065.59
49. $\{4\}$ **51.** $\varnothing$ **53.** no   **55.** no   **57.** C   **59.** D   **61.** answers may vary   **63.** 24.55 lb   **65.** 20.09 lb

Section 12.4

Vocabulary, Readiness & Video Check   1. The growth rate is given as 5% per year. Since this is "per year," the number of time intervals is the "number of years," or 8.   **3.** time intervals = years/half-life; the decay rate is 50% or $\dfrac{1}{2}$ because half-life is the amount of time it takes half of a substance to decay

Exercise Set 12.4   1. 451   **3.** 144,302   **5.** 21,231   **7.** 202   **9.** 1470   **11.** 13   **13.** 712,880   **15.** 383   **17.** 333 bison   **19.** 1 g
21. a. $\dfrac{14}{7} = 2$; 10; yes   **b.** $\dfrac{11}{7} \approx 1.6$; 13.2; yes   **23.** $\dfrac{500}{152} \approx 3.3$; 2.1; yes   **25.** 4.9 g   **27.** 3   **29.** -1 **31.** no; answers may vary

Section 12.5

Vocabulary, Readiness & Video Check   1. logarithmic; b   **3.** yes   **5.** no; none   **7.** $(-\infty, \infty)$ **9.** First write the equation as an equivalent exponential equation. Then solve.

Exercise Set 12.5   1. $6^2 = 36$ **3.** $3^{-3} = \dfrac{1}{27}$ **5.** $10^3 = 1000$ **7.** $e^4 = x$ **9.** $e^{-2} = \dfrac{1}{e^2}$ **11.** $7^{1/2} = \sqrt{7}$ **13.** $0.7^3 = 0.343$

15. $3^{-4} = \dfrac{1}{81}$ **17.** $\log_2 16 = 4$ **19.** $\log_{10} 100 = 2$ **21.** $\log_e x = 3$ **23.** $\log_{10}\dfrac{1}{10} = -1$ **25.** $\log_4 \dfrac{1}{16} = -2$ **27.** $\log_5 \sqrt{5} = \dfrac{1}{2}$ **29.** 3

31. -2 **33.** $\dfrac{1}{2}$ **35.** -1 **37.** 0   **39.** 2   **41.** 4   **43.** -3 **45.** $\{2\}$ **47.** $\{81\}$ **49.** $\{7\}$ **51.** $\{-3\}$ **53.** $\{-3\}$ **55.** $\{2\}$ **57.** $\{2\}$

59. $\left\{\dfrac{27}{64}\right\}$ **61.** $\{10\}$ **63.** $\{4\}$ **65.** $\{5\}$ **67.** $\left\{\dfrac{1}{49}\right\}$ **69.** 3   **71.** 3   **73.** 1   **75.** -1 **77.** 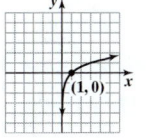   **79.**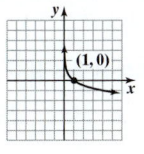

81. 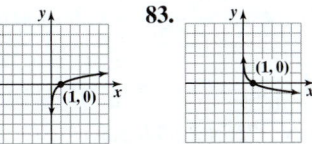   **83.** 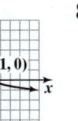   **85.** 1   **87.** $\dfrac{x-4}{2}$ **89. a.** $g(2) = 25$ **b.** $(25, 2)$ **c.** $f(25) = 2$ **91.** answers may vary

93. 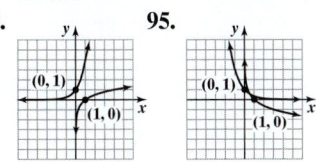   **95.**   **97.** answers may vary   **99.** 0.0827

Section 12.6

Vocabulary, Readiness & Video Check **1.** 36; a **3.** $\log_b 2^7$; b **5.** x; a **7.** No, the product property says the logarithm of a product can be written as a sum of logarithms—the expression in this example is a logarithm of a sum. **9.** Since $\frac{1}{x} = x^{-1}$, this gives us $\log_2 x^{-1}$. Using the power property, we get $-1\log_2 x$ or $-\log_2 x$.

Exercise Set 12.6 **1.** $\log_5 14$ **3.** $\log_4 9x$ **5.** $\log_6 (x^2 + x)$ **7.** $\log_{10}(10x^2 + 20)$ **9.** $\log_5 3$ **11.** $\log_3 4$ **13.** $\log_2 \frac{x}{y}$ **15.** $\log_2 \frac{x^2 + 6}{x^2 + 1}$

17. $2\log_3 x$ **19.** $-1\log_4 5 = -\log_4 5$ **21.** $\frac{1}{2}\log_5 y$ **23.** $\log_2 5x^3$ **25.** $\log_4 48$ **27.** $\log_5 x^3 z^6$ **29.** $\log_4 4$, or 1 **31.** $\log_7 \frac{9}{2}$

33. $\log_{10} \frac{x^3 - 2x}{x + 1}$ **35.** $\log_2 \frac{x^{7/2}}{(x + 1)^2}$ **37.** $\log_8 x^{16/3}$ **39.** $\log_3 4 + \log_3 y - \log_3 5$ **41.** $\log_4 2 - \log_4 9 - \log_4 z$ **43.** $3\log_2 x - \log_2 y$

45. $\frac{1}{2}\log_b 7 + \frac{1}{2}\log_b x$ **47.** $4\log_6 x + 5\log_6 y$ **49.** $3\log_5 x + \log_5(x + 1)$ **51.** $2\log_6 x - \log_6(x + 3)$ **53.** 0.2 **55.** 1.2 **57.** 0.35

59. 1.29 **61.** -0.68 **63.** -0.125 **65.** 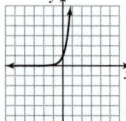 **67.** 2 **69.** 2 **71.** b **73.** true **75.** false **77.** false **79.** because $\log_b 1 = 0$

Integrated Review **1.** $x^2 + x - 5$ **2.** $-x^2 + x - 7$ **3.** $x^3 - 6x^2 + x - 6$ **4.** $\frac{x - 6}{x^2 + 1}$ **5.** $\sqrt{3x - 1}$ **6.** $3\sqrt{x} - 1$

7. one-to-one; $\{(6, -2), (8, 4), (-6, 2), (3, 3)\}$ **8.** not one-to-one **9.** not one-to-one **10.** one-to-one **11.** not one-to-one

12. $f^{-1}(x) = \frac{x}{3}$ **13.** $f^{-1}(x) = x - 4$ **14.** $f^{-1}(x) = \frac{x + 1}{5}$ **15.** $f^{-1}(x) = \frac{x - 2}{3}$ **16.** 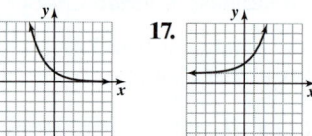 **17.**

18. 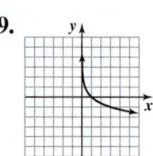 **19.**   **20.** $\{3\}$ **21.** $\{7\}$ **22.** $\{-8\}$ **23.** $\{3\}$ **24.** $\{2\}$ **25.** $\left\{\frac{1}{2}\right\}$ **26.** $\{32\}$ **27.** $\{4\}$ **28.** $\{5\}$

29. $\left\{\frac{1}{9}\right\}$ **30.** $\log_2 14x$ **31.** $\log_2(5^x \cdot 8)$ **32.** $\log_5 \frac{x^3}{y^5}$ **33.** $\log_5 x^9 y^3$ **34.** $\log_2 \frac{x^2 - 3x}{x^2 + 4}$

35. $\log_3 \frac{y^4 + 11y}{y + 2}$ **36.** $\log_7 9 + 2\log_7 x - \log_7 y$ **37.** $\log_6 5 + \log_6 y - 2\log_6 z$

38. 544,000 mosquitoes

Section 12.7

Vocabulary, Readiness & Video Check **1.** 10; c **3.** 7; b **5.** 5; b **7.** $\frac{\log 7}{\log 2} = \frac{\ln 7}{\ln 2}$; a or b **9.** The understood base of a common logarithm is 10. If you're finding the common logarithm of a known power of 10, then the common logarithm is the known power of 10. **11.** $\log_b b^x = x$

Exercise Set 12.7 **1.** 0.9031 **3.** 0.3636 **5.** 0.6931 **7.** -2.6367 **9.** 1.1004 **11.** 1.6094 **13.** 1.6180 **15.** 2 **17.** -3 **19.** 2

21. $\frac{1}{4}$ **23.** 3 **25.** 3.1 **27.** -4 **29.** $\frac{1}{2}$ **31.** $\{10^{1.3}\}$; $\{19.9526\}$ **33.** $\{e^{1.4}\}$; $\{4.0552\}$ **35.** $\{10^{2.3}\}$; $\{199.5262\}$

37. $\{e^{-2.3}\}$; $\{0.1003\}$ **39.** $\left\{\frac{10^{1.1}}{2}\right\}$; $\{6.2946\}$ **41.** $\left\{\frac{e^{0.18}}{4}\right\}$; $\{0.2993\}$ **43.** $\left\{\frac{4 + e^{2.3}}{3}\right\}$; $\{4.6581\}$ **45.** $\left\{\frac{10^{-0.5} - 1}{2}\right\}$; $\{-0.3419\}$

47. \$3656.38 **49.** \$2542.50 **51.** 1.5850 **53.** 0.8617 **55.** 1.5850 **57.** -1.6309 **59.** -2.3219 **61.** $\left\{\frac{4}{7}\right\}$ **63.** $x = \frac{3y}{4}$

65. $\{-6, -1\}$ **67.** answers may vary **69.** 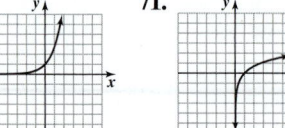 **71.**   **73.** answers may vary **75.** 4.2 **77.** 5.3

Section 12.8

Calculator Explorations **1.** 3.67 yr, or 3 yr and 8 mo **3.** 23.16 yr, or 23 yr and 2 mo

Vocabulary, Readiness & Video Check **1.** $\ln(4x-2) = \ln 3$ is the same as $\log_e(4x-2) = \log_e 3$. Therefore, from the logarithm property of equality, we know that $4x - 2 = 3$. **3.** $2000 = 1000\left(1 + \dfrac{0.07}{12}\right)^{12 \cdot t} \approx 9.9$ yr; as long as the interest rate and compounding are the same, it takes any amount of money the same time to double.

Exercise Set 12.8 **1.** $\left\{\dfrac{\log 6}{\log 3}\right\}$; $\{1.6309\}$ **3.** $\left\{\dfrac{\log 5}{\log 9}\right\}$; $\{0.7325\}$ **5.** $\left\{\dfrac{\log 3.8}{2\log 3}\right\}$; $\{0.6076\}$ **7.** $\left\{\dfrac{\ln 5}{6}\right\}$; $\{0.2682\}$

9. $\left\{3 + \dfrac{\log 5}{\log 2}\right\}$; $\{5.3219\}$ **11.** $\left\{\dfrac{\log 3}{\log 4} - 7\right\}$; $\{-6.2075\}$ **13.** $\left\{\dfrac{1}{3}\left(4 + \dfrac{\log 11}{\log 7}\right)\right\}$; $\{1.7441\}$ **15.** $\{11\}$ **17.** $\left\{\dfrac{1}{2}\right\}$

19. $\left\{\dfrac{3}{4}\right\}$ **21.** $\{-2, 1\}$ **23.** $\{2\}$ **25.** $\left\{\dfrac{1}{8}\right\}$ **27.** $\{4, -1\}$ **29.** $\left\{\dfrac{2}{3}\right\}$ **31.** 103 wolves **33.** 210.9 million **35.** 7620 **37.** 9.9 yr

39. 1.7 yr **41.** 8.8 yr **43.** 55.7 in. **45.** 11.9 lb per sq in. **47.** 3.2 mi **49.** 12 weeks **51.** 18 weeks **53.** $-\dfrac{5}{3}$ **55.** $\dfrac{17}{4}$ **57.** 16 yr
59. answers may vary **61.** 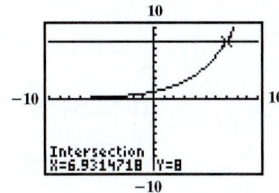 ;$\{6.93\}$

Chapter 12 Vocabulary Check **1.** inverse **2.** composition **3.** exponential **4.** symmetric **5.** Natural **6.** Common
7. vertical; horizontal **8.** logarithmic **9.** Half-life **10.** exponential

Chapter 12 Review **1.** $3x - 4$ **2.** $-x - 6$ **3.** $2x^2 - 9x - 5$ **4.** $\dfrac{2x+1}{x-5}$ where $x \neq 5$ **5.** $x^2 + 2x - 1$ **6.** $x^2 - 1$ **7.** 18
8. $x^4 - 4x^2 + 2$ **9.** -2 **10.** 48 **11.** one-to-one; $h^{-1} = \{(14, -9), (8, 6), (12, -11), (15, 15)\}$ **12.** not one-to-one
13. one-to-one; **14.** not one-to-one;

Rank in Housing Starts for 2013 (Input)	4	3	1	2
U.S. Region (Output)	Northeast	Midwest	South	West

15. not one-to-one **16.** not one-to-one **17.** not one-to-one **18.** one-to-one **19.** $f^{-1}(x) = \dfrac{x-11}{6}$ **20.** $f^{-1}(x) = \dfrac{x}{12}$

21. $f^{-1}(x) = \dfrac{x+5}{3}$ **22.** $f^{-1}(x) = \dfrac{x-1}{2}$ **23.** $f^{-1}(x) = -\dfrac{x-3}{2}$ **24.** $f^{-1}(x) = \dfrac{x+5}{5}$ **25.** $\{3\}$ **26.** $\left\{-\dfrac{4}{3}\right\}$

27. $\left\{\dfrac{3}{2}\right\}$ **28.** $\left\{\dfrac{5}{2}\right\}$

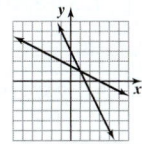 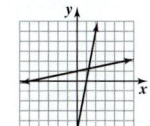

29. 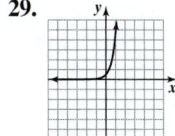 **30.** 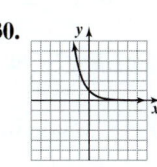 **31.** 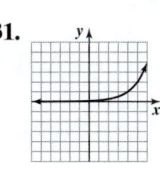 **32.** 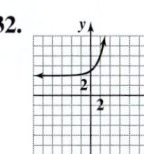 **33.** \$1131.82 **34.** \$2963.11 **35.** 457,393 **36.** 572,165
37. 8 players **38.** 636 bears **39.** $\log_7 49 = 2$
40. $\log_2 \dfrac{1}{16} = -4$ **41.** $\left(\dfrac{1}{2}\right)^{-4} = 16$
42. $0.4^3 = 0.064$ **43.** $\left\{\dfrac{1}{64}\right\}$ **44.** $\{9\}$ **45.** $\{0\}$

46. $\{8\}$ **47.** $\{5\}$ **48.** $\{-2\}$ **49.** $\{4\}$ **50.** $\{9\}$ **51.** $\{2\}$ **52.** $\{-8, 1\}$ **53.** 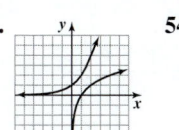 **54.** 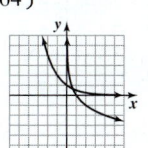 **55.** $\log_3 32$
56. $\log_2 18$

57. $\log_7 \dfrac{3}{4}$ **58.** $\log_e \dfrac{3}{2}$ **59.** $\log_{11} 4$ **60.** $\log_5 2$ **61.** $\log_5 \dfrac{x^3}{(x+1)^2}$ **62.** $\log_3(x^4 + 2x^3)$ **63.** $3\log_3 x - \log_3(x+2)$

64. $\log_4(x+5) - 2\log_4 x$ **65.** $\log_2 3 + 2\log_2 x + \log_2 y - \log_2 z$ **66.** $\log_7 y + 3\log_7 z - \log_7 x$ **67.** 2.02 **68.** -0.11

69. 0.5563 **70.** -0.8239 **71.** 0.2231 **72.** 1.5326 **73.** 3 **74.** -1 **75.** -1 **76.** 4 **77.** $\{50\}$ **78.** $\left\{\dfrac{e^{1.6}}{3}\right\}$; $\{1.6510\}$

79. $\left\{\dfrac{e^{-1}+3}{2}\right\}$; $\{1.6839\}$ **80.** $\left\{\dfrac{e^2-1}{3}\right\}$; $\{2.1297\}$ **81.** 0.2920 **82.** 1.2619 **83.** \$1957.30 **84.** \$1307.51 **85.** $\left\{\dfrac{\log 20}{\log 7}\right\}$; $\{1.5395\}$

86. $\left\{\dfrac{\log 7}{2\log 3}\right\}$; $\{0.8856\}$ **87.** $\left\{\dfrac{1}{2}\left(\dfrac{\log 6}{\log 3} - 1\right)\right\}$; $\{0.3155\}$ **88.** $\left\{\dfrac{1}{4}\left(\dfrac{\log 3}{\log 8} + 2\right)\right\}$; $\{0.6321\}$ **89.** $\left\{\dfrac{25}{2}\right\}$ **90.** $\varnothing$

91. $\{2\sqrt{2}\}$ **92.** $\{9, -1\}$ **93.** 6.01% **94.** 5.1 yr **95.** 46.2 yr **96.** 115.5 yr **97.** 8.8 yr **98.** 8.5 yr **99.** $\{-2\}$ **100.** $\left\{\dfrac{3}{2}\right\}$

101. $\left\{\dfrac{8}{9}\right\}$ **102.** $\left\{\dfrac{7}{2}\right\}$ **103.** $\{3\}$ **104.** $\{3\}$ **105.** $\{-1, 4\}$ **106.** $\left\{\dfrac{17}{3}\right\}$ **107.** $\{e^{-1.2}\}$ **108.** $\left\{\dfrac{9}{10}\right\}$ **109.** $\left\{\dfrac{3e^2}{e^2 - 3}\right\}$ **110.** $\varnothing$

Chapter 12 Test **1.** $2x^2 - 3x$ **2.** $-x + 3$ **3.** 5 **4.** $x - 7$ **5.** $x^2 - 6x - 2$ **6.** $f^{-1}(x) = \dfrac{x + 14}{7}$ **7.** one-to-one
8. not a function
9. one-to-one;

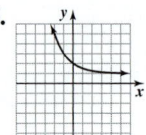

$f^{-1}(x) = \dfrac{-x + 6}{2}$

10. one-to-one; $f^{-1} = \{(0, 0), (3, 2), (5, -1)\}$ **11.** not one-to-one **12.** $\log_3 24$ **13.** $\log_5 \dfrac{x^4}{x + 1}$

14. $\log_6 2 + \log_6 x - 3\log_6 y$ **15.** -1.53 **16.** 1.0686 **17.** $\{-1\}$ **18.** $\left\{\dfrac{1}{2}\left(\dfrac{\log 4}{\log 3} - 5\right)\right\}$; $\{-1.8691\}$ **19.** $\left\{\dfrac{1}{9}\right\}$ **20.** $\left\{\dfrac{1}{2}\right\}$

21. $\{22\}$ **22.** $\left\{\dfrac{25}{3}\right\}$ **23.** $\left\{\dfrac{43}{21}\right\}$ **24.** $\{-1.0979\}$ **25.** **26.** **27.** \$5234.58 **28.** 6 yr **29.** 6.5%; \$230
30. 100,141 **31.** 64,805 prairie dogs
32. 15 yr **33.** 85% **34.** 52%

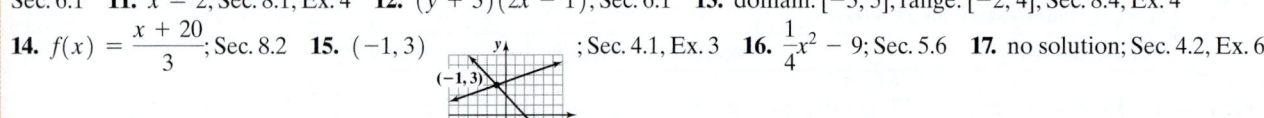

Cumulative Review **1.** $\dfrac{5}{x^2}$; Sec. 7.1, Ex. 3 **2.** $\dfrac{1}{1 - x}$; Sec. 7.3 **3.** $\dfrac{1}{x - 9}$; Sec. 7.1, Ex. 6 **4.** $\dfrac{-x^2 + 23x + 38}{(x - 2)(x + 2)^2}$; Sec. 7.4 **5.** $\dfrac{2}{5}$;

Sec. 7.2, Ex. 2 **6.** 11; Sec. 2.3 **7.** $\dfrac{2}{x(x + 1)}$; Sec. 7.2, Ex. 6 **8.** $2 - x$; Sec. 7.1 **9.** $y = \dfrac{1}{4}x - 3$; Sec. 3.5, Ex. 1 **10.** $(3y - 2)(2x - 5)$;

Sec. 6.1 **11.** $x = 2$; Sec. 8.1, Ex. 4 **12.** $(y + 3)(2x - 1)$; Sec. 6.1 **13.** domain: $[-3, 5]$; range: $[-2, 4]$; Sec. 8.4, Ex. 4

14. $f(x) = \dfrac{x + 20}{3}$; Sec. 8.2 **15.** $(-1, 3)$; Sec. 4.1, Ex. 3 **16.** $\dfrac{1}{4}x^2 - 9$; Sec. 5.6 **17.** no solution; Sec. 4.2, Ex. 6

18. $m = \dfrac{3}{2}$; Sec. 3.4 **19. a.** \$69 **b.** \$49 **c.** yes; Sec. 4.4, Ex. 2 **20.** $(-1, 3)$; Sec. 4.3 **21.** 0; Sec. 10.1, Ex. 5 **22.** $4x^2\sqrt[3]{xy^2}$;

Sec. 10.3 **23.** 0.5; Sec. 10.1, Ex. 7 **24.** $3ab\sqrt[4]{2b}$; Sec. 10.3 **25.** $\sqrt{x}$; Sec. 10.2, Ex. 19 **26.** $\sqrt[5]{y}$; Sec. 10.2 **27.** $\sqrt[3]{rs^2}$; Sec. 10.2, Ex. 21

28. $\sqrt[5]{x^2y^4}$; Sec. 10.2 **29.** $2\sqrt[3]{3}$; Sec. 10.3, Ex. 11 **30.** $x^3\sqrt[3]{x}$; Sec. 10.3 **31.** $2\sqrt[4]{2}$; Sec. 10.3, Ex. 13 **32.** $10\sqrt{2x}$; Sec. 10.3

33. $\dfrac{2\sqrt{5}}{5}$; Sec. 10.5, Ex. 1 **34.** $\dfrac{3\sqrt[3]{m^2n}}{m^2n^3}$; Sec. 10.5 **35.** $\left\{-\dfrac{1}{9}, -1\right\}$; Sec. 10.6, Ex. 2 **36. a.** $a - a^{33/4}$ **b.** $x + 2x^{1/2} - 15$; Sec. 10.2

37. $13 - 18i$; Sec. 10.7, Ex. 13 **38. a.** $\dfrac{11\sqrt{5}}{12}$ **b.** $\dfrac{\sqrt[3]{3x}}{6}$; Sec. 10.4 **39.** $\{-1 + \sqrt{5}, -1 - \sqrt{5}\}$; Sec. 11.1, Ex. 7 **40.** $k = \dfrac{1}{24}$; $y = \dfrac{1}{24}x$;

Sec. 9.4 **41.** $\left\{\dfrac{2 + \sqrt{10}}{2}, \dfrac{2 - \sqrt{10}}{2}\right\}$; Sec. 11.2, Ex. 2 **42.** $\left\{\dfrac{-2 + \sqrt{5}}{2}, \dfrac{-2 - \sqrt{5}}{2}\right\}$; Sec. 11.1 **43.** $\{9\}$; Sec. 11.3, Ex. 1

44. $\left\{\dfrac{3 + \sqrt{5}}{4}, \dfrac{3 - \sqrt{5}}{4}\right\}$; Sec. 11.2 **45.** $(-\infty, -2] \cup [1, 5]$; Sec. 11.4, Ex. 3 **46.** $x^2 + 2x + 4$; Sec. 5.7 **47.** $\{6\}$; Sec. 12.3, Ex. 6

48. a. $\dfrac{2a}{a - 1}$ **b.** $\dfrac{-3(a + 6)}{4(a - 3)}$ **c.** $\dfrac{y + x}{x^2y^2}$; Sec. 7.7 **49.** $\{5\}$; Sec. 12.5, Ex. 9 **50.** $\log_3 x^7y^9$; Sec. 12.6 **51.** $\log_5 72$; Sec. 12.6, Ex. 7

52. $\left\{\dfrac{55}{8}\right\}$; Sec. 12.8

Chapter 13 Conic Sections

Section 13.1

Calculator Explorations **1.**

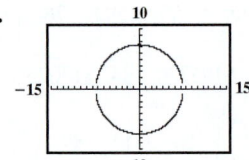

3.

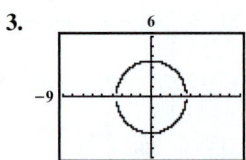

Vocabulary, Readiness & Video Check   1. conic sections   **3.** circle; center   **5.** radius   **7.** No, their graphs don't pass the vertical line test.   **9.** Since the standard form of a circle involves a squared binomial for both x and y, we need to complete the square on both x and y.

Exercise Set 13.1   1. upward   **3.** to the left   **5.** downward   **7.** 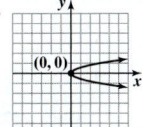   **9.** 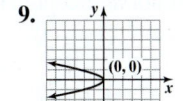   **11.**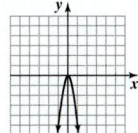

13. 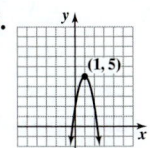   **15.** 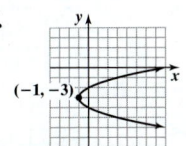   **17.** 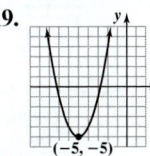   **19.** 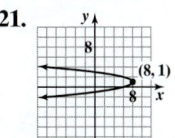   **21.** 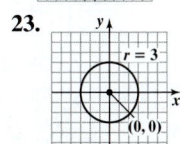   **23.**

25. 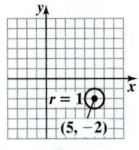   **27.** 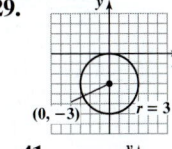   **29.** 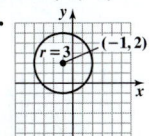   **31.** 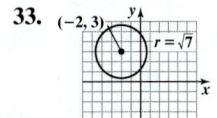   **33.** 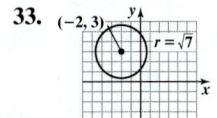   **35.**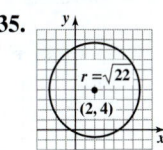

37. 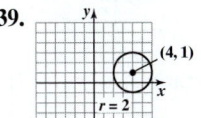   **39.** 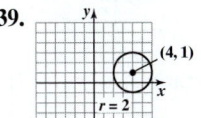   **41.** 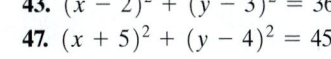   **43.** $(x - 2)^2 + (y - 3)^2 = 36$ **45.** $x^2 + y^2 = 3$
47. $(x + 5)^2 + (y - 4)^2 = 45$

49. 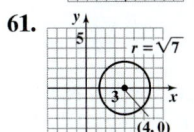   **51.** 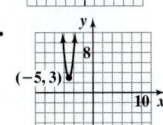   **53.** 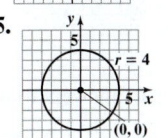   **55.** 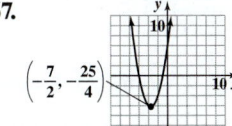   **57.** 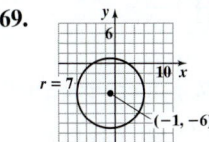   **59.**

61. 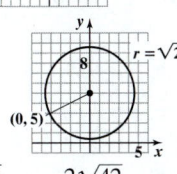   **63.** 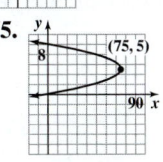   **65.** 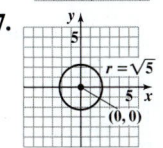   **67.** 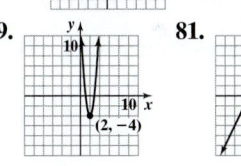   **69.**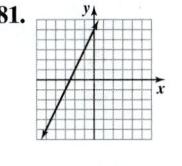

71. 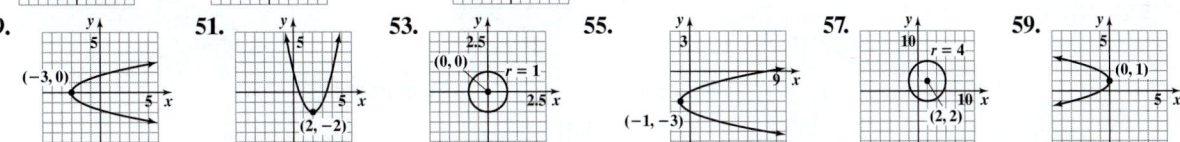   **73.**   **75.**   **77.**   **79.**   **81.**

83. 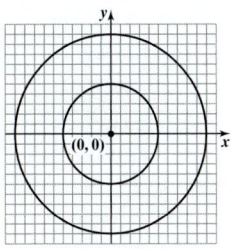   **85.** $\dfrac{\sqrt{3}}{3}$ **87.** $\dfrac{2\sqrt{42}}{3}$ **89.** The vertex is $(1, -5)$.   **91. a.** 16.5 m   **b.** 103.67 m   **c.** 3.5 m   **d.** $(0, 16.5)$
e. $x^2 + (y - 16.5)^2 = (16.5)^2$ **93. a.** 125 ft   **b.** 14 ft   **c.** 139 ft   **d.** $(0, 139)$ **e.** $x^2 + (y - 139)^2 = 125^2$
95. answers may vary   **97.** 20 m

99. a. 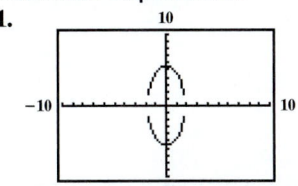   **b.** $x^2 + y^2 = 100$ **c.** $x^2 + y^2 = 25$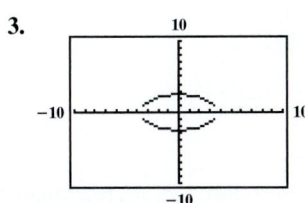

Section 13.2

Calculator Explorations

1. **3.**

A44

Vocabulary, Readiness & Video Check   1. hyperbola   **3.** focus   **5.** hyperbola; $(0,0)$; x; $(a,0)$ and $(-a,0)$ **7.** a and b give us the location of 4 intercepts—$(a,0),(-a,0),(0,b)$, and $(0,-b)$ for $\dfrac{x^2}{a^2}+\dfrac{y^2}{b^2}=1$ with center $(0,0)$. For Example 2, the values of a and b also give us 4 points of the graph, just not intercepts. Here we move a distance of a units horizontally to the left and right of the center and b units above and below the center.

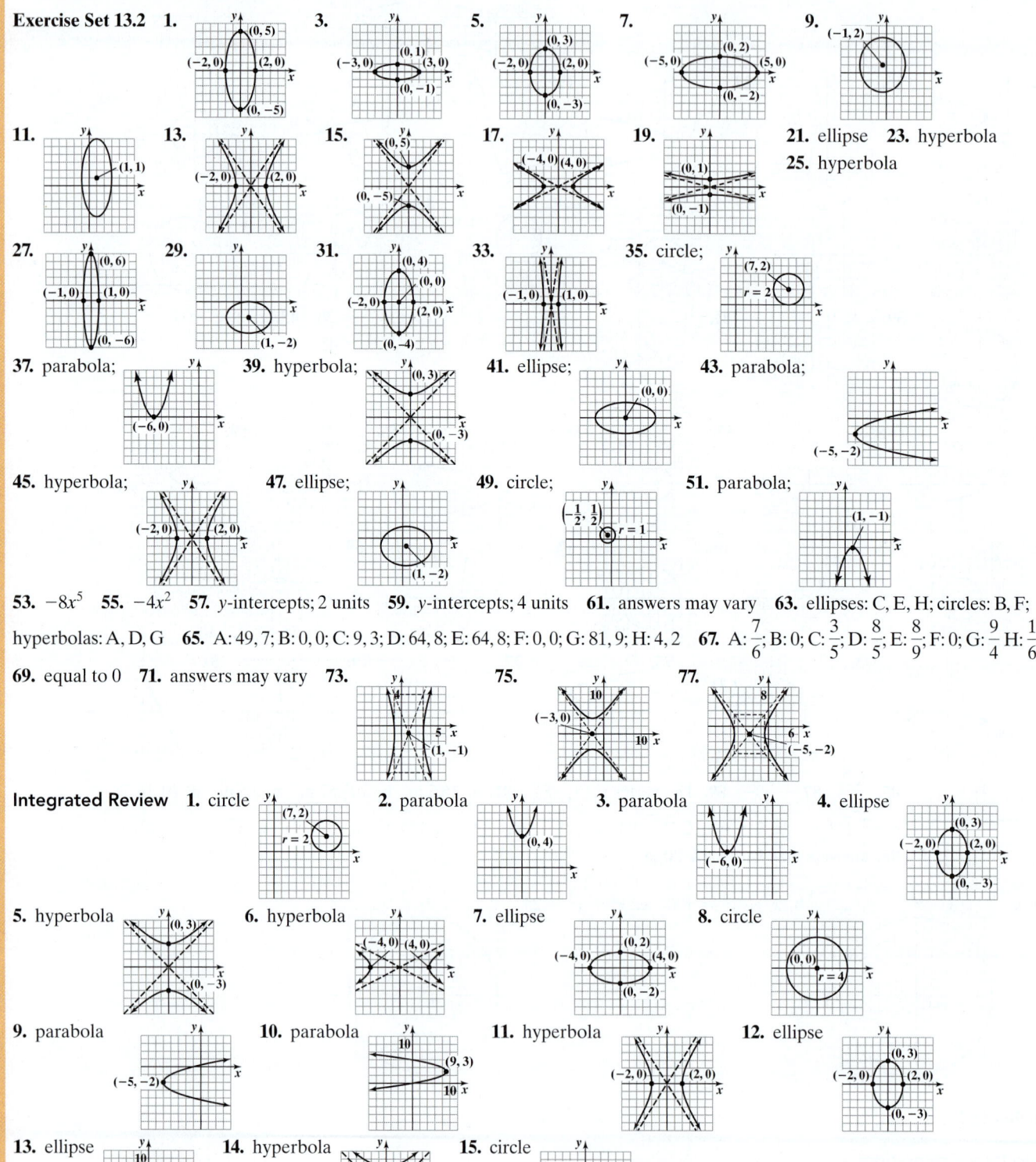

53. $-8x^5$ **55.** $-4x^2$ **57.** y-intercepts; 2 units   **59.** y-intercepts; 4 units   **61.** answers may vary   **63.** ellipses: C, E, H; circles: B, F; hyperbolas: A, D, G   **65.** A: 49, 7; B: 0, 0; C: 9, 3; D: 64, 8; E: 64, 8; F: 0, 0; G: 81, 9; H: 4, 2   **67.** A: $\dfrac{7}{6}$; B: 0; C: $\dfrac{3}{5}$; D: $\dfrac{8}{5}$; E: $\dfrac{8}{9}$; F: 0; G: $\dfrac{9}{4}$ H: $\dfrac{1}{6}$

69. equal to 0   **71.** answers may vary

Section 13.3

Vocabulary, Readiness & Video Check   1. Solving for y would either introduce tedious fractions (2nd equation) or a square root (1st equation) into the calculations.

Exercise Set 13.3 **1.** $\{(3, -4), (-3, 4)\}$ **3.** $\{(\sqrt{2}, \sqrt{2}), (-\sqrt{2}, -\sqrt{2})\}$ **5.** $\{(4, 0), (0, -2)\}$ **7.** $\{(-\sqrt{5}, -2), (-\sqrt{5}, 2), (\sqrt{5}, -2)$ $(\sqrt{5}, 2)\}$ **9.** $\varnothing$ **11.** $\{(1, -2), (3, 6)\}$ **13.** $\{(2, 4), (-5, 25)\}$ **15.** $\varnothing$ **17.** $\{(1, -3)\}$ **19.** $\{(-1, -2), (-1, 2), (1, -2), (1, 2)\}$ **21.** $\{(0, -1)\}$ **23.** $\{(-1, 3), (1, 3)\}$ **25.** $\{(\sqrt{3}, 0), (-\sqrt{3}, 0)\}$ **27.** $\varnothing$ **29.** $\{(-6, 0), (6, 0), (0, -6)\}$ **31.** $\{(3, \sqrt{3})\}$

33. 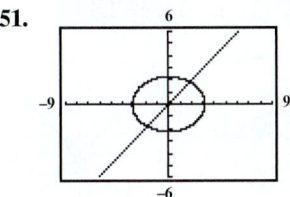 **35.** 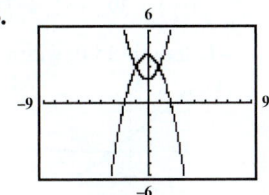 **37.** $(8x - 25)$ in. **39.** $(4x^2 + 6x + 2)$ m **41.** answers may vary **43.** 0, 1, 2, 3, or 4
45. -9 and 7; 9 and -7; 9 and 7; -9 and -7 **47.** 15 cm by 19 cm
49. 15 thousand compact discs; price: $3.75

51. 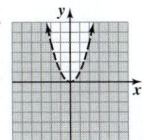 **53.**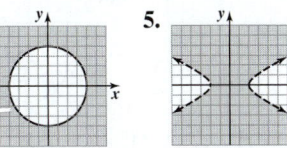

Section 13.4

Vocabulary, Readiness & Video Check **1.** For both, we graph the related equation to find the boundary and sketch it as a solid boundary for $\leq$ or $\geq$ and a dashed boundary for $<$ or $>$; also, we choose a test point not on the boundary and shade the region containing the test point if the test point is a solution of the original inequality or shade the other region if not.

Exercise Set 13.4 **1.** 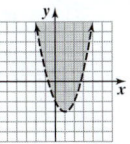 **3.** **5.** **7.** **9.** **11.**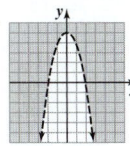

13. 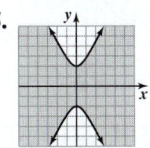 **15.** **17.** **19.** **21.** **23.**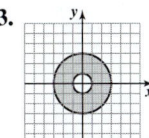

25. 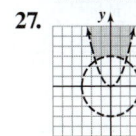 **27.** 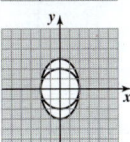 **29.** 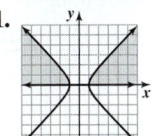 **31.** **33.** 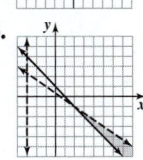 **35.**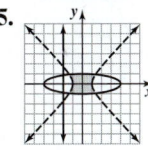

37. not a function **39.** function **41.** 1 **43.** $3a^2 - 2$ **45.** answers may vary **47.**

Chapter 13 Vocabulary Check **1.** circle; center **2.** nonlinear system of equations **3.** ellipse **4.** radius **5.** hyperbola
6. conic sections **7.** vertex **8.** diameter

Chapter 13 Review **1.** $(x + 4)^2 + (y - 4)^2 = 9$ **2.** $(x + 7)^2 + (y + 9)^2 = 11$ **3.** $(x - 5)^2 + y^2 = 25$ **4.** $x^2 + y^2 = \dfrac{49}{4}$

5. 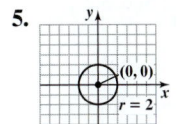 **6.** **7.** **8.** **9.** **10.** **11.**

12. **13.** **14.** **15.** **16.** **17.**

18. 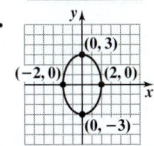 **19.** 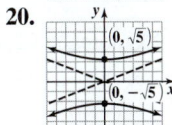 **20.** 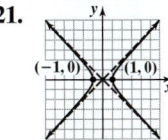 **21.** 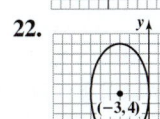 **22.** 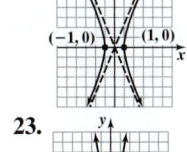 **23.**

24. 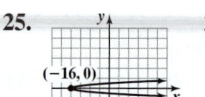 **25.** 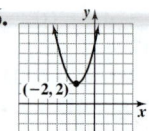 **26.** 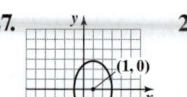 **27.** 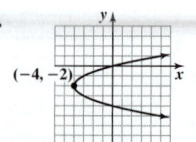 **28.** 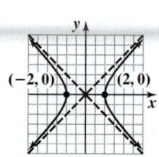 **29.**

30. 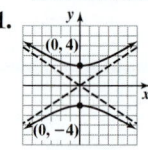 **31.** 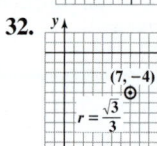 **32.**

33. $\{(1, -2), (4, 4)\}$ **34.** $\varnothing$ **35.** $\{(-1, 1), (2, 4)\}$

36. $\{(2, 2\sqrt{2}), (2, -2\sqrt{2})\}$ **37.** $\{(0, 2), (0, -2)\}$

38. $\left\{\left(2, \frac{5}{2}\right), (-7, -20)\right\}$ **39.** $\{(1, 4)\}$ **40.** $\{(-2, -1), (-2, 1), (2, -1), (2, 1)\}$ **41.** length: 15 ft; width: 10 ft **42.** 4

43. 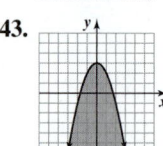 **44.** 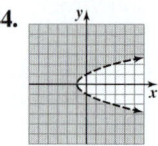 **45.** 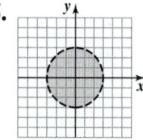 **46.** 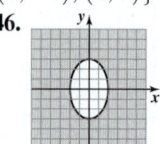 **47.** 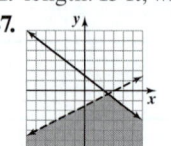 **48.** 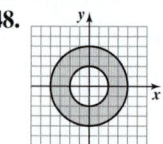 **49.**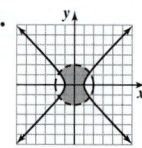

50. 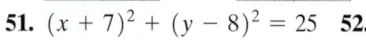 **51.** $(x + 7)^2 + (y - 8)^2 = 25$ **52.** 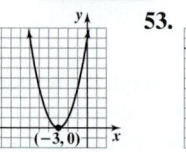 **53.** 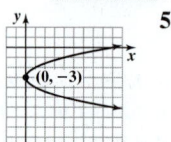 **54.** 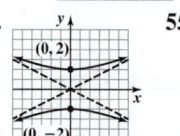 **55.**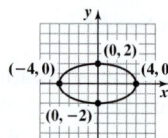

56. 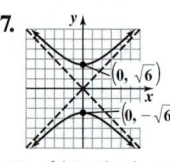 **57.** 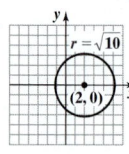 **58.** 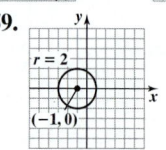 **59.** 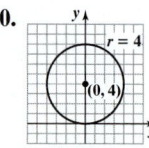 **60.** 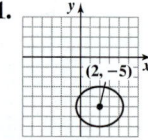 **61.**

62. 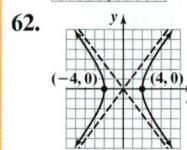 **63.** $\{(5, 1), (-1, 7)\}$ **64.** $\{(-1, 3), (-1, -3), (1, 3), (1, -3)\}$ **65.** 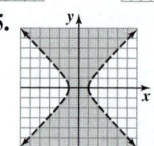 **66.**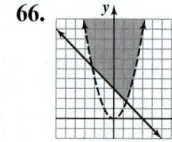

Chapter 13 Test **1.** 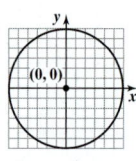 **2.** 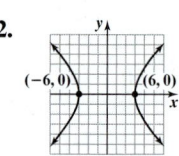 **3.** 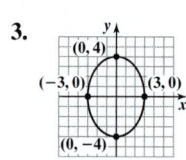 **4.** 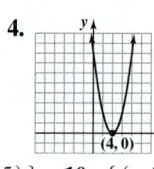 **5.**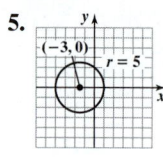

6. 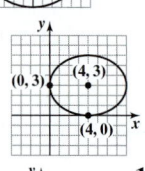 **7.** 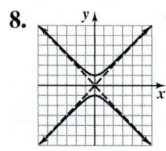 **8.** 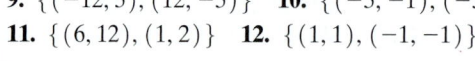 **9.** $\{(-12, 5), (12, -5)\}$ **10.** $\{(-5, -1), (-5, 1), (5, -1), (5, 1)\}$

11. $\{(6, 12), (1, 2)\}$ **12.** $\{(1, 1), (-1, -1)\}$

13. 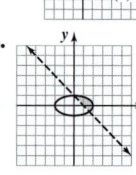 **14.** 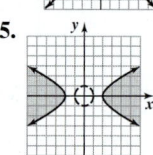 **15.** 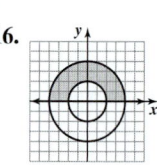 **16.** **17.** B **18.** height: 10 ft; width: 30 ft

Cumulative Review **1.** $\{x \mid x \geq 1\}$; Sec. 2.7, Ex. 9 **2.** 4; Sec. 7.5 **3. a.** x^{11} **b.** $\frac{t^4}{16}$ **c.** $81y^{10}$; Sec. 5.1, Ex. 33 **4.** $\frac{1}{81}$; Sec. 5.2

5. $7x^3 + 14x^2 + 35x$; Sec. 5.5, Ex. 4 **6.** all real numbers; Sec. 2.3 **7.** $3x^3 - 4 + \frac{1}{x}$; Sec. 5.7, Ex. 2 **8.** $(0, \infty)$; Sec. 2.7

9. $(x + 3)(5 + y)$; Sec. 6.1, Ex. 9 **10.** $(3y + 5)(y + 3)$; Sec. 6.3, 6.4 **11.** $(x^2 + 2)(x^2 + 3)$; Sec. 6.2, Ex. 7

12. $2a^3(2a + 5)(5a + 1)$; Sec. 6.3, 6.4 **13.** $-\frac{3y^4}{10}$; Sec. 7.2, Ex. 1b **14.** $4x^2 + 2x + 1$; Sec. 7.1 **15.** 5; Sec. 7.5, Ex. 2 **16.** $\frac{4}{3}$; Sec. 7.5

17. $\frac{3}{z}$; Sec. 7.7, Ex. 3 **18.** $\frac{5(a + 2)}{3(a + 5)(a - 5)}$; Sec. 7.4 **19.** 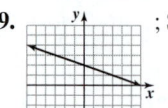 ; Sec. 3.2, Ex. 4 **20.** $k = 2$; $y = \frac{2}{x}$; Sec. 9.4

21. $y = \frac{5}{3}x + \frac{13}{3}$; Sec. 8.1, Ex. 6 **22.** 3; Sec. 3.4

23. domain: $\{2, 0, 3\}$; range: $\{3, 4, -1\}$; Sec. 8.2, Ex. 1

24. 150 mph, 175 mph; Sec. 4.4 **25.** 1; Sec. 8.2, Ex. 14 **26. a.** $9x^2 - 21x + 12$ **b.** 90 **c.** $-3x^2 + 9x - 1$ **d.** -31; Sec. 12.1

27. 35; Sec. 8.2, Ex. 17 **28.** $\{(-6,0),(0,6)\}$; Sec. 13.3 **29.** $(-4,2,-1)$; Sec. 9.1, Ex. 1 **30.** $\dfrac{3x^2 + 7x - 45}{(3x+2)(x-5)(3x-2)}$; Sec. 7.4

31. $(-1,2)$; Sec. 9.2, Ex. 1 **32.** $\varnothing$; Sec. 4.3 **33.** 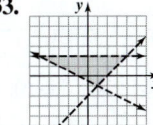 ; Sec. 9.3, Ex. 2 **34.** $\{-2\} \cup [2, \infty)$; Sec 11.4 **35.** -4; Sec. 10.1, Ex. 14

36. $\dfrac{x^3 \sqrt{x}}{7}$; Sec. 10.3 **37.** $\dfrac{2}{5}$; Sec. 10.1, Ex. 15 **38.** $3a^2 b\sqrt{10a}$; Sec. 10.3

39. 2; Sec. 10.3, Ex. 3 **40.** $2\sqrt{5} + 5\sqrt{3}$; Sec. 10.4

41. $\sqrt{\dfrac{2b}{3a}}$; Sec. 10.3, Ex. 5 **42.** $21 - 4\sqrt{5}$; Sec. 10.4 **43.** $-6\sqrt[3]{2}$; Sec. 10.4, Ex. 5 **44.** $5\sqrt{5}$; Sec. 10.4 **45.** $\dfrac{5\sqrt[3]{7x}}{2}$; Sec. 10.4, Ex. 10

46. $-\sqrt[3]{2x}$; Sec. 10.4 **47.** $\left\{\dfrac{-1 + i\sqrt{35}}{6}, \dfrac{-1 - i\sqrt{35}}{6}\right\}$; Sec. 11.2, Ex. 4 **48.** 6; Sec. 10.6 **49.** $\left\{\dfrac{2}{99}\right\}$; Sec. 12.8, Ex. 4 **50. a.** $\left\{\dfrac{1}{3}\right\}$

b. $\left\{\dfrac{11}{3}\right\}$ **c.** $\{-2\}$; Sec. 12.3 **51.** $\left(-1, \dfrac{3}{2}\right)$; Sec. 10.3, Ex. 22 **52.** 5; Sec. 10.3 **53.** 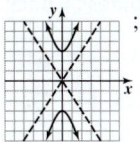 ; Sec. 13.2, Ex. 5

54. $f^{-1}(x) = 2x - 1$; Sec. 12.2 **55.** $\{(2, \sqrt{2})\}$; Sec. 13.3, Ex. 2 **56.** $\dfrac{\sqrt{3} - 3}{3}$; Sec. 10.5

Appendices

Exercise Set Appendix A 1. $-16x^6 y^3 p^2$ **3.** 9 **5.** $a^3 b$ **7.** $\dfrac{1}{16}$ **9.** $\dfrac{13}{36}$ **11.** y^4 **13.** $\dfrac{6x^{16}}{5}$ **15.** $16x^{20} y^{12}$ **17.** $4a^8 b^4$ **19.** $\dfrac{2}{x^4 y^{10}}$

21. $2y^4 - 5y^2 + x^2 + 1$ **23.** $y^2 + 3$ **25.** $-x^3 + 8a - 12$ **27.** $5x^2 - 9x - 3$ **29.** $-2x^2 - 4x + 15$ **31.** $7y^2 - 3$

33. $-20y^2 + 3yx$ **35.** $9x^2 + 18x + 5$ **37.** $16x^2 - \dfrac{2}{3}x - \dfrac{1}{6}$ **39.** $x^2 + 8x + 16$ **41.** $9x^2 - 6xy + y^2$ **43.** $9b^2 - 36y^2$ **45.** $9x^2 - \dfrac{1}{4}$

47. $2xy(7x - 1)$ **49.** $4(x + 2)(x - 2)$ **51.** $(3x - 11)(x + 1)$ **53.** $(2x + 5)(4x^2 - 10xy + 25y^2)$ **55.** $7x(x - 9)$

57. $(4x + 3)(5x + 2)$ **59.** $(b - 6)(a + 7)$ **61.** $(x^2 + 1)(x - 1)(x + 1)$ **63.** $2(x - 3)(x^2 + 3x + 9)$

Appendix B Vocabulary and Readiness Check 1. 6 **3.** 17 **5.** 8 **7.** 10

Exercise Set Appendix B 1. -0.9 **3.** 6 **5.** -1.1 **7.** -5 **9.** $-3, \dfrac{4}{3}$ **11.** 0 **13.** $-\dfrac{3}{4}, \dfrac{5}{2}$ **15.** $-3, -8$ **17.** 2 **19.** $\dfrac{1}{4}, -\dfrac{2}{3}$ **21.** 1, 9

23. -9 **25.** $\dfrac{1}{6}$ **27.** 1 **29.** $\dfrac{3}{5}, -1$ **31.** 0 **33.** 6, -3 **35.** 8 **37.** $\{x \mid x \text{ is a real number}\}$ **39.** $\dfrac{2}{5}, -\dfrac{1}{2}$ **41.** $\dfrac{3}{4}, -\dfrac{1}{2}$ **43.** -8 **45.** -2

47. 0, 5 **49.** $-\dfrac{1}{2}, \dfrac{1}{3}$ **51.** $-4, 9$ **53.** $\varnothing$ **55.** 1

Exercise Set Appendix C 1. $\{2, 4\}$ **3.** $\varnothing$ **5.** $\{3, 5\}$ **7.** ⟵( )⟶ (−3 to 1) **9.** ⟵⟶

11. ⟵( )⟶ (−1) **13.** $(-2, 5)$ **15.** $[6, \infty)$ **17.** $(-\infty, -3]$ **19.** $\varnothing$ **21.** $(11, 17)$ **23.** $[1, 4]$ **25.** $\left[-3, \dfrac{3}{2}\right]$

27. $[-21, -9]$ **29.** $\left[\dfrac{3}{2}, 6\right]$ **31.** $\left(0, \dfrac{14}{3}\right)$ **33.** $\{1, 2, 3, 4, 5, 6, 7, 8\}$ **35.** $\{1, 5, 6\}$ **37.** $\{2, 4, 6, 8\}$ **39.** ⟵⟶ (5)

41. ⟵( )⟶ (−4 to 1) **43.** ⟵⟶ **45.** $(-\infty, -1) \cup (0, \infty)$ **47.** $[2, \infty)$ **49.** $(-\infty, \infty)$

51. $(-\infty, 1] \cup \left(\dfrac{29}{7}, \infty\right)$ **53.** $(-7, \infty)$ **55.** $(-\infty, \infty)$

Exercise Set Appendix D 1. $\{7, -7\}$ **3.** $\varnothing$ **5.** $\{4.2, -4.2\}$ **7.** $\{-4, 4\}$ **9.** $\{-9, 9\}$ **11.** $\{-5, 23\}$ **13.** $\{7, -2\}$

15. $\{8, 4\}$ **17.** $\{5, -5\}$ **19.** $\{3, -3\}$ **21.** $\{-3, 6\}$ **23.** $\{0\}$ **25.** $\varnothing$ **27.** $\left\{-\dfrac{1}{3}, \dfrac{7}{3}\right\}$ **29.** $\left\{-\dfrac{1}{2}, 9\right\}$ **31.** $\left\{-\dfrac{5}{2}\right\}$ **33.** $\{3, 2\}$

35. $\{-4, 16\}$ **37.** $\{4\}$ **39.** $\left\{\dfrac{3}{2}\right\}$ **41.** $\left\{\dfrac{32}{21}, \dfrac{38}{9}\right\}$ **43.** $\left\{-8, \dfrac{2}{3}\right\}$ **45.** ⟵[ ]⟶ (−4 to 4); $[-4, 4]$

47. ⟵( )⟶ (−3 to 3); $(-\infty, -3) \cup (3, \infty)$ **49.** ⟵( )⟶ (−5 to −1); $(-5, -1)$ **51.** ⟵( )⟶ (−1 to 13);

$(-\infty, -1] \cup [13, \infty)$ **53.** ⟵( )⟶ (−5 to 1); $(-5, 1)$ **55.** ⟵[ ]⟶ (−5 to 5); $[-5, 5]$

57. ⟵( )⟶ (−4 to 4); $(-\infty, -4) \cup (4, \infty)$ **59.** ⟵[ ]⟶ (−10 to 3); $[-10, 3]$ **61.** ⟵[ ]⟶ (−24 to 4);

$(-\infty, -24] \cup [4, \infty)$ **63.** 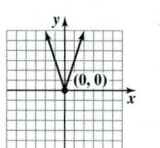 $;[-2, 9]$ **65.** ———————— $;(-\infty, \infty)$ **67.** 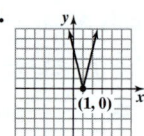 $;$

$\left[-\dfrac{1}{2}, 1\right]$ **69.** ———————— $;\left(-\infty, \dfrac{2}{3}\right) \cup (2, \infty)$ **71.** ———————— $;\varnothing$ **73.** ———————— $;$

$(-\infty, -12) \cup (0, \infty)$ **75.** $\{-13, 13\}$ **77.** $(-\infty, -13) \cup (13, \infty)$ **79.** $\varnothing$ **81.** $[-10, 10]$ **83.** $(-2, 5)$ **85.** $\{5, -2\}$

87. $(-\infty, -7] \cup [17, \infty)$ **89.** $\left\{-\dfrac{9}{4}\right\}$ **91.** $(-2, 1)$ **93.** $(-\infty, -18) \cup (12, \infty)$ **95.** $\left\{2, \dfrac{4}{3}\right\}$ **97.** $\varnothing$ **99.** $\left\{-\dfrac{17}{2}, \dfrac{19}{2}\right\}$

101. $\left(-\infty, -\dfrac{25}{3}\right) \cup \left(\dfrac{35}{3}, \infty\right)$ **103.** $\left\{4, -\dfrac{1}{5}\right\}$

Appendix E Vocabulary and Readiness Check   **1.** 56   **3.** -32 **5.** 20

Exercise Set Appendix E   **1.** 26   **3.** -19 **5.** 0   **7.** $\dfrac{13}{6}$ **9.** $(1, 2)$ **11.** $\{(x, y) \mid 3x + y = 1\}$ **13.** $(9, 9)$ **15.** $(-3, -2)$
17. $(3, 4)$ **19.** 8   **21.** 0   **23.** 15   **25.** 54   **27.** $(-2, 0, 5)$ **29.** $(6, -2, 4)$ **31.** $(-2, 3, -1)$ **33.** $(0, 2, -1)$ **35.** 5
37. 0; answers may vary

Exercise Set Appendix F   **1.** $71°$ **3.** $19.2°$ **5.** $78\frac{3}{4}°$ **7.** $30°$ **9.** $149.8°$ **11.** $100\frac{1}{2}°$ **13.** $m\angle 1 = m\angle 5 = m\angle 7 = 110°$,
$m\angle 2 = m\angle 3 = m\angle 4 = m\angle 6 = 70°$ **15.** $90°$ **17.** $90°$ **19.** $90°$ **21.** $45°, 90°$ **23.** $73°, 90°$ **25.** $50\frac{1}{4}°, 90°$ **27.** $x = 6$
29. $x = 4.5$ **31.** 10   **33.** 12

Exercise Set Appendix G   **1.**   **3.**   **5.**   **7.**   **9.**

11.   **13.**   **15.**

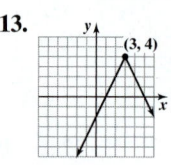
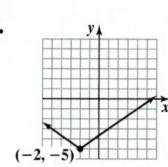

Exercise Set Appendix H   **1.** yes   **3.** no   **5.** answers may vary   **7.** answers may vary   **9.** answers may vary
11. Xmin $= -12$ Ymin $= -12$ **13.** Xmin $= -9$ Ymin $= -12$ **15.** Xmin $= -10$ Ymin $= -25$ **17.** Xmin $= -5$ Ymin $= -15$
   Xmax $= 12$ Ymax $= 12$ Xmax $= 9$ Ymax $= 12$ Xmax $= 10$ Ymax $= 25$ Xmax $= 5$ Ymax $= 15$
   Xscl $= \dfrac{6}{5}$ Yscl $= \dfrac{6}{5}$ Xscl $= 1$ Yscl $= 2$ Xscl $= 2$ Yscl $= 5$ Xscl $= 1$ Yscl $= 3$

19. Xmin $= -20$ Ymin $= -30$ **21.** Setting B   **23.** Setting B   **25.** Setting B
   Xmax $= 30$ Ymax $= 50$
   Xscl $= 5$ Yscl $= 10$

27.   **29.**   **31.**   **33.**

35.   **37.**   **39.**   **41.**

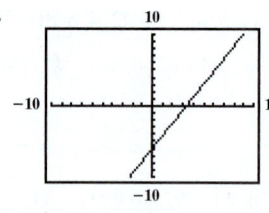
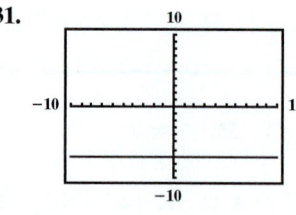
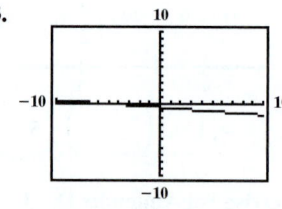
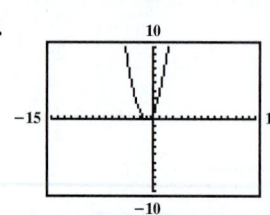
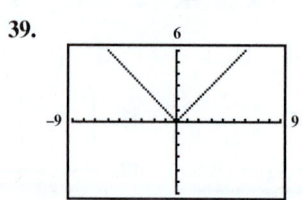
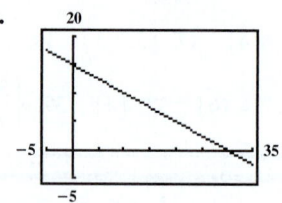

Practice Final Exam   **1.** -48 **2.** -81 **3.** $\dfrac{1}{64}$ **4.** $-3x^3 + 5x^2 + 4x + 5$ **5.** $16x^2 - 16x + 4$ **6.** $3x^3 + 22x^2 + 41x + 14$
7. $(y - 12)(y + 4)$ **8.** $3x(3x + 1)(x + 4)$ **9.** $5(6 + x)(6 - x)$ **10.** $(a + b)(3a - 7)$ **11.** $3x(3y - z)(9y^2 + 3yz + z^2)$

12. $\dfrac{16y^{14}}{x^2}$ **13.** $\dfrac{25}{7}$ **14.** $\{x \mid x \le -2\}$ **15.** $-7, 1$ **16.** 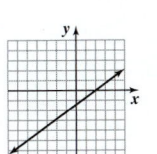   **17.** 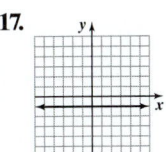   **18.** 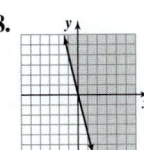   **19.** $m = -1$ **20.** $m = 3$

21. $8x + y = 11$ **22.** $x - 8y = -96$ **23.** $\left(\dfrac{1}{2}, -2\right)$ **24.** no solution   **25.** $9x^2 - 6x + 4 - \dfrac{16}{3x + 2}$ **26.** 21   **27.** 7 gal

28. $401{,}802$ **29.** California: 1107   **30.** 120 cc   **31.** 3 mph; 6 mph   **32.** $-\dfrac{1}{2x + 5}$ **33.** $\dfrac{2(x + 3)(x + 5)}{x(x^2 + 4x + 1)}$ **34.** $\dfrac{3a - 4}{(a - 3)(a + 2)}$
Ohio: 720

35. $\dfrac{5y^2 - 1}{y + 2}$ **36.** $\dfrac{30}{11}$ **37.** -6 **38.** no solution   **39.** 5 or 1   **40.** $6\sqrt{6}$ **41.** 5   **42.** $\dfrac{8a^{1/3}c^{2/3}}{b^{5/12}}$ **43.** $-x\sqrt{5x}$ **44.** -20 **45.** -20

46. domain: $\{-2\}$; range: $(-\infty, \infty)$; not a function   **47.** domain: $(-\infty, \infty)$; range: $[0, \infty)$; function   **48.** $\dfrac{3 \pm \sqrt{29}}{2}$ **49.** $2, 3$

50. $\left(-\infty, -\dfrac{3}{2}\right) \cup (5, \infty)$ **51.** 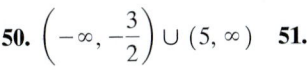   **52.** 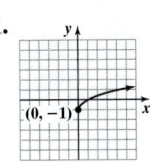   domain: $(-\infty, \infty)$; range: $(-\infty, -1]$ **53.**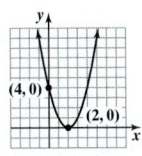

54. 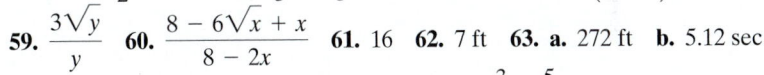   domain: $(-\infty, \infty)$; range: $(-3, \infty)$ **55.** $y = -\dfrac{1}{2}x$ **56.** $y = -\dfrac{1}{3}x + \dfrac{5}{3}$ **57.** $2\sqrt{26}$ units   **58.** $\left(-4, \dfrac{7}{2}\right)$

59. $\dfrac{3\sqrt{y}}{y}$ **60.** $\dfrac{8 - 6\sqrt{x} + x}{8 - 2x}$ **61.** 16   **62.** 7 ft   **63. a.** 272 ft   **b.** 5.12 sec

64. $0 - 2i\sqrt{2}$ **65.** $0 - 3i$ **66.** $7 + 24i$ **67.** $-\dfrac{3}{2} + \dfrac{5}{2}i$

68. $(g \circ h)(x) = x^2 - 6x - 2$ **69.** $f(x)$ is one-to-one; $f^{-1}(x) = \dfrac{-x + 6}{2}$ **70.** $\log_5 \dfrac{x^4}{x + 1}$ **71.** -1 **72.** $\dfrac{1}{2}\left(\dfrac{\log 4}{\log 3} - 5\right)$; -1.8691

73. 22   **74.** $\dfrac{43}{21}$ **75.** $\dfrac{1}{2}$ **76.** 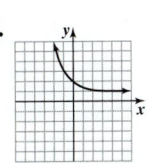   **77.** 64,805 prairie dogs   **78.** 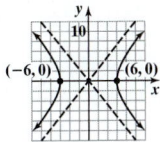   **79.** 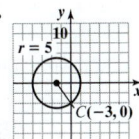   **80.**

81. $(-5, -1), (-5, 1), (5, -1), (5, 1)$

Video Answer Section

Chapter R Prealgebra Review

Section R.1

7. No, the natural number 1 is neither prime nor composite.   **8.** order; one   **9.** smallest

Section R.2

11. The fraction is equal to 1.   **12.** quantity   **13.** wrote both the numerator and denominator as products of prime numbers
14. $\frac{20}{1}$ or 20   **15.** When adding or subtracting fractions, we must have common denominators. When multiplying or dividing fractions, we do not.   **16.** Our original sum, $4\frac{7}{6}$, is not in proper form because the fraction part, $\frac{7}{6}$, is an improper fraction.

Section R.3

10. Reading a decimal correctly gives us the correct place value, which tells us the denominator of our equivalent fraction.
11. do; do not   **12.** when rounding whole numbers, digits to the right of the rounding place are replaced by zeros; when rounding decimal numbers to the right of the decimal point, digits to the right of the rounding place are not replaced by zeros
13. numerator; denominator   **14.** 1

Chapter 1 Real Numbers and Introduction to Algebra

Section 1.2

11. To form a true statement: $0 < 7$.   **12.** Five is greater than or equal to four; $5 \geq 4$ **13.** 0 belongs to the whole numbers, the integers, the rational numbers, and the real numbers; since 0 is a rational number, it cannot also be an irrational number.   **14.** absolute value

Section 1.3

13. The order in which we perform operations does matter! We came up with an order of operations to avoid getting more than one answer when evaluating an expression.   **14.** The replacement value for z is not used because it's not needed—there is no variable z in the given algebraic expression.   **15.** No; the variable was replaced with 0 in the equation to see if a true statement occurred, and it did not.   **16.** We translate phrases to mathematical expressions and sentences to mathematical equations.

Section 1.4

5. absolute values   **6.** Negative; when you add two numbers with different signs, the sign of the sum is the same as the sign of the number with the larger absolute value and -8.4 has a larger absolute value than 6.3.   **7.** Example 12 is an example of the opposite of the *absolute value* of $-a$, not the opposite of $-a$. The absolute value of $-a$ is positive, so its opposite is negative. Therefore the answers to Examples 11 and 12 have different signs.   **8.** The algebraic expressions are $x + y$ and $3x + y$. For $3x + y$, the variable x is multiplied by 3.   **9.** Depths below the surface; the diver's position is 231 feet below the surface.

Section 1.5

5. addition, opposite   **6.** $-10 + (8) + (-4) + (-20)$; it's rewritten to change the subtraction operations to addition and turn the expression into an addition of real numbers problem   **7.** There's a minus sign in the numerator and the replacement value is negative (notice parentheses are used around the replacement value), and it's always good to be careful when working with negative signs.   **8.** -4 is NOT a solution of $x - 9 = 5$ **9.** This means that the overall vertical altitude change of the jet is actually a decrease in altitude from when the Example started.   **10.** In Example 10, we have two supplementary angles and know the measure of one of them. From the definition, we know that the two supplementary angles must sum to $180°$. Therefore we can subtract the known angle measure from $180°$ to get the measure of the other angle.

Section 1.6

9. The parentheses, or lack of them, determine the base of the expression. In Example 6, $(-2)^4$, the base is -2 and all of -2 is raised to the 4th power. In Example 7, -2^4, the base is 2 and only 2 is raised to the 4th power.   **10.** Remember, the product of a number and its reciprocal is 1, *not* -1. $\frac{2}{3} \cdot \frac{3}{2} = 1$, as needed.   **11.** Yes; because division of real numbers is defined in terms of multiplication.

12. The replacement values are negative and one of them will be squared. If a negative number is squared, it must be placed in parentheses so the entire value, including the negative, is squared.   **13.** Yes; a true statement results when x is replaced with 5.
14. The football team lost 4 yards on each play and a loss of yardage is represented by a negative number.

Section 1.7

8. order; grouping　**9.** 2 is outside the parentheses, so the point is made that we should only distribute the -9 to the terms within the parentheses and not also to the 2.　**10.** 0; 0; 1; 1

Section 1.8

7. Although these terms have exactly the same variables, the exponents on each are not exactly the same—the exponents on x differ in each term.　**8.** distributive property　**9.** -1　**10.** The sum of 5 times a number and -2, added to 7 times the number; $5x + (-2) + 7x$; because there are like terms

Chapter 2 Equations, Inequalities, and Problem Solving

Section 2.1

15. both sides　**16.** $15x - 14 = 14x - 1$　**17.** $\frac{1}{7}x$

Section 2.2

13. same　**14.** addition property; multiplication property; answers may vary　**15.** $(x + 1) + (x + 3) = 2x + 4$

Section 2.3

9. 3; distributive property, addition property of equality, multiplication property of equality　**10.** Since both sides have more than one term, you need to apply the distributive property to make sure you multiply every single term in the equation by the LCD. **11.** The number of decimal places in each number helps you determine what power of 10 you can multiply through by so you are no longer dealing with decimals.　**12. a.** all real numbers as a　**b.** no

Section 2.4

7. in the statement of the application　**8.** The original application asks for the measure of two supplementary angles. The solution of $x = 43$ only gives us the measure of one of the angles.　**9.** That the 3 angle measures are consecutive even integers and that they sum to 180°.

Section 2.5

1. relationships　**2.** This is a distance, rate, and time problem. The rate is given in miles per hour (mph) and the time is given in hours, so the distance that we are finding must be in miles.　**3.** To show that the process of solving this equation for x—dividing both sides by 5, the coefficient of x—is the same process used to solve a formula for a specific variable. Treat whatever is multiplied by that specific variable as the coefficient—the coefficient is all the factors except that specific variable.

Section 2.6

5. a. equals; $=$　**b.** multiplication; $\cdot$　**c.** Drop the percent symbol and move the decimal point two places to the left.　**6. a.** You also find a discount amount by multiplying the (discount) percent by the original price.　**b.** For discount, the new price is the original price minus the discount amount, so you *subtract* from the original price rather than *add* as with mark-up.　**7.** You must first find the actual amount of increase in price by subtracting the original price from the new price.　**8.** 10%; x; 0.10; 0.10x; 30%; 400; 0.30; 0.30(400); 20%; $x + 400$; 0.20; 0.20$(x + 400)$; 0.10$x + 0.30(400) = 0.20(x + 400)$

Section 2.7

11. An open circle indicates $>$ or $<$; a closed circle indicates $\geq$ or $\leq$.　**12.** addition property of equality　**13.** $\{x \mid x \geq -2\}$ **14.** The multiplication property of inequality is applied at this step when we divide by the coefficient of x. The coefficient is positive and so the inequality symbol remains the same, but if the coefficient had been negative, the direction of the inequality symbol would have been reversed.　**15.** is greater than; $>$

Chapter 3 Graphing Equations and Inequalities

Section 3.1

9. 2003; 9.4　**10.** origin; left or right; up or down　**11.** Paired data, which can be written as ordered pairs and graphed.　**12.** Replace both values of the ordered pair in the linear equation and see if a true statement results.

Section 3.2

1. It is always good practice to use a third point as a check to see that your points lie along a straight line.　**2.** An infinite number of points make up the line, and every point corresponds to an ordered pair that is a solution of the linear equation in two variables.

Section 3.3

13. because *x*-intercepts lie on the *x*-axis; because *y*-intercepts lie on the *y*-axis **14.** It is always good practice to use a third point as a check to see that your points lie along a straight line. **15.** For a horizontal line, the coefficient of *x* will be 0; for a vertical line, the coefficient of *y* will be 0.

Section 3.4

16. Whatever *y*-value we decide to start with in the numerator, we *must* start with the corresponding *x*-value in the denominator. **17.** Solve the equation for *y*; the slope is the coefficient of *x*. **18.** Zero slope indicates $m = 0$ and a horizontal line; undefined slope indicates *m* is undefined and a vertical line; "no slope" refers to an undefined slope. **19.** Slope-intercept form; this form makes the slope easy to see, and we need to compare slopes to determine if two lines are parallel or perpendicular. **20.** Step 4: INTERPRET the results.

Section 3.5

9. *y*-intercept; fraction **10.** $\left(0, -\dfrac{1}{6}\right)$ **11.** Write the equation with *x*- and *y*-terms on one side of the equal sign and a constant on the other side. **12.** Yes, if one of the points given is the *y*-intercept. We will still need to use the slope formula to find the slope, but then we'll have the slope and *y*-intercept for the slope-intercept form. **13.** We need to know what our variables stand for in order to solve part (b) of the example, and that depends on how we set up our ordered pairs in part (a).

Section 3.6

7. An ordered pair is a solution of an inequality if replacing the variables with the coordinates of the ordered pair results in a true statement. **8.** We find the boundary line equation by replacing the inequality symbol with =. The points on this line are solutions (line is solid) if the inequality is $\geq$ or $\leq$; they are not solutions (line is dashed) if the inequality is $>$ or $<$.

Chapter 4 Systems of Equations

Section 4.1

11. The ordered pair must satisfy all equations of the system in order to be a solution of the system, so we must check that the ordered pair is a solution of both equations. **12.** Graphing is not the most accurate method, especially if your graph is off just slightly, or the point of intersection does not have integer coordinates. **13.** Writing the equations of a system in slope-intercept form lets us see and compare their slopes and *y*-intercepts. Different slopes mean one solution; same slopes with different *y*-intercepts mean no solution; same slopes with same *y*-intercepts mean an infinite number of solutions.

Section 4.2

7. We solved one equation for a variable. Next, be sure to substitute this expression for the variable into the *other* equation.

Section 4.3

5. The multiplication property of equality; be sure to multiply *both* sides of the equation by the nonzero number chosen.

Section 4.4

1. Up to now we've been working with one variable/unknown and one equation. Because these systems involve two equations with two unknowns, for these applications we need to choose two variables to represent two unknowns and translate the problem into two equations.

Chapter 5 Exponents and Polynomials

Section 5.1

13. Example 4 can be written as $-4^2 = -1 \cdot 4^2$, which is similar to Example 7, $4 \cdot 3^2$, and shows why the negative sign should not be considered part of the base when there are no parentheses. **14.** The properties allow us to reorder and regroup factors and put factors with common bases together, making it easier to apply the product rule. **15.** Be careful not to confuse the power rule with the product rule. The power rule involves a power raised to a power (exponents are multiplied), and the product rule involves a product (exponents are added). **16.** Remember to raise the -2 (or any number) to the power along with the variables. **17.** the quotient rule **18.** No, Example 30 is a fraction cubed and the quotient rule for exponents is not needed to simplify.

Section 5.2

11. A negative exponent has nothing to do with the sign of the simplified result. **12.** power of a product rule, power rule for exponents, negative exponent definition, quotient rule for exponents **13.** When you move the decimal point to the left, the sign of the exponent will be positive; when you move the decimal point to the right, the sign of the exponent will be negative. **14.** the exponent on 10 **15.** the quotient rule

Section 5.3

7. $3; x^2, -3x, 5$ **8.** The degree of the polynomial is the greatest degree of any of its terms, so we need to find the degree of each term first.   **9.** the replacement value for the variable   **10.** simplifying it   **11.** $2; 9ab$ **12.** $2; 0y^2$

Section 5.4

7. $-3y^2$ and $2y^2$; $-4y$ and y **8.** Example 2 is a subtraction example, so the signs of the polynomial being subtracted must change when parentheses are removed.   **9.** We're translating a subtraction problem. Order matters when subtracting, so we need to be careful that the order of the expressions is correct.   **10.** Because the operation is addition.

Section 5.5

13. No. The monomials are unlike terms.   **14.** distributive property, product rule   **15.** Three times: First $(a - 2)$ is distributed to a and 7, and then a is distributed to $(a - 2)$ and 7 is distributed to $(a - 2)$.   **16.** Yes. The parentheses have been removed for the vertical format, but every term in the first polynomial is still distributed to every term in the second polynomial.

Section 5.6

5. a binomial times a binomial   **6.** FOIL order for multiplication, distributive property   **7.** Multiplying gives you four terms, and the two like terms will always subtract out.   **8.** Multiplying the sum and difference of the same two terms, squaring a binomial, and the FOIL order for multiplication when multiplying a binomial and a binomial.

Section 5.7

7. the common denominator   **8.** Filling in missing powers helps us keep like terms lined up and our work clear and neat.

Chapter 6 Factoring Polynomials

Section 6.1

13. The GCF of a list of numbers is the largest number that is a factor of all numbers in the list.   **14.** There is no need to factor the variable parts. The GCF of common variable factors is the variable raised to the smallest exponent.   **15.** When factoring out a GCF, the number of terms in the other factor should be the same as the number of terms as your original polynomial.   **16.** Look for a GCF other than 1 or -1; four terms.

Section 6.2

11. 15 is positive, so its factors would have to be either both positive or both negative. Since the factors need to sum to -8, both factors must be negative.   **12.** Since the sum of the factors is 3, the factors are -2 and 5 ($-2 + 5 = 3$). (In other words, the factor with the smaller absolute value is negative so that the sum is positive.) If we accidentally choose factors whose sum is -3, simply "switch" the signs of the factors.

Section 6.3

5. Consider the factors of the first and last terms and the signs of the trinomial. Continue to check possible factors by multiplying until we get the middle term of the trinomial.   **6.** If the GCF has been factored out, then neither binomial can contain a common factor other than 1 or -1. This helps limit our choice of factors for one or both binomials since we cannot choose factors that would give the terms in either binomial a common factor.

Section 6.4

5. This gives us a four-term polynomial, which may be factored by grouping.

Section 6.5

13. See if the first term is a square, say a^2, and the last term is a square, say b^2. Check to see if the middle term is $2 \cdot a \cdot b$ or $-2 \cdot a \cdot b$.   **14.** In order to help you see that the binomial is a difference of squares and also to identify the terms to use in the special factoring formula.   **15.** First rewrite the original binomial so that each term is some quantity cubed. Your answers will then vary, depending on your interpretation.

Section 6.6

5. One side of the equation must be a factored polynomial and the other side must be zero.   **6.** Because no matter how many factors you have in a multiplication problem, it's still true that for a product to be zero, at least one of the factors must be zero.

Section 6.7

1. In applications, the context of the stated application needs to be considered. Each translated equation resulted in both a positive and a negative solution, and a negative solution is not appropriate for any of the stated application.

Chapter 7 Rational Expressions

Section 7.1

12. replacement values of the variables; by evaluating the expression for different replacement values—variables are replaced with these values and the expression is simplified   **13.** Rational expressions are fractions and are therefore undefined if the denominator is zero; if a denominator contains variables, set it equal to zero and solve.   **14.** Although x is a factor in the numerator, it is not a factor in the denominator—factor means to write as a product, and the denominator is a difference, not a product.   **15.** We would need to write parentheses around the numerator or denominator if it had more than one term because the negative sign needs to apply to the entire numerator or denominator.

Section 7.2

6. Yes, multiplying and simplifying rational expressions often require polynomial factoring. Example 2 alone involves factoring out a GCF, factoring a trinomial with $a \neq 1$, and factoring a difference of squares.   **7.** fractions; reciprocal   **8.** Multiplication and division of rational expressions are performed similarly—both involve multiplication—but there are important differences. Note the operation first to see whether you multiply by the reciprocal or not.   **9.** We're converting to cubic feet, so we want cubic feet in the numerator. We want cubic yards to divide out so cubic yards is in the denominator.

Section 7.3

6. In order to carry out the subtraction properly—parentheses make sure each term in the numerator is affected by the subtraction, not just the first term.   **7.** We completely factor denominators—including coefficients—so we can determine the greatest number of times each unique factor occurs in any one denominator for the LCD.   **8.** numerator; one; value

Section 7.4

3. The exercise is adding two rational expressions with denominators that are opposites of each other. Recognizing this special case can save us time and effort. If we recognize that one denominator is -1 times the other denominator, we may save many steps.

Section 7.5

7. These equations are solved in very different ways, so we need to determine the next correct step to make. For a linear equation, we first "move" variable terms to one side and numbers to the other; for a quadratic equation, we first set the equation equal to 0.   **8.** If there are variables in any denominators, we should first check to see if the proposed solutions make these denominators zero in the original equation, giving us an undefined rational expression. If so, that solution is an extraneous solution and is not a solution to the equation.   **9.** the steps for solving an equation containing rational expressions; as if it's the only variable in the equation

Section 7.6

9. No. Proportions are actually equations containing rational expressions, so they can also be solved by using the steps to solve those equations.   **10.** There are also many ways to set up an incorrect proportion, so just checking your solution in your proportion isn't enough. You need to determine if your solution is reasonable from the relationships given in the problem.   **11.** divided by; quotient   **12.** Two machines (or people) take different amounts of time to complete the task, one faster and one slower than the other. When working together, they will complete the task in less time than the faster machine, so your answer must be less than the time of the faster machine.   **13.** $\dfrac{325}{x+7} = \dfrac{290}{x}$

Section 7.7

5. a single fraction in the numerator and in the denominator   **6.** In Method 2, we find the LCD of all fractions in the complex fraction, then multiply both the numerator and the denominator by this LCD so that both will no longer contain fractions; in Method 1, we find the LCD of the fractions only in the numerator and then the LCD of the fractions only in the denominator in order to perform the addition and/or subtraction and get single fractions in the numerator and in the denominator.

Chapter 8 Graphs and Functions

Section 8.1

7. if one of the two points given is the y-intercept   **8.** Example 3: $y = 3$; Example 4: $x = -1$ **9.** $y = \dfrac{1}{3}x - \dfrac{17}{3}$ **10.** 64 ft per sec

Section 8.2

7. An equation is one way of writing down a correspondence that produces ordered pair solutions; a relation between two sets of coordinates, x's and y's.   **8.** Yes. The function definition restricts each first component to correspond to exactly one second component, but it makes no such restriction on each second component.   **9.** If a vertical line intersects a graph two (or more) times, then there's an *x*-value corresponding to two (or more) different *y*-values, and the graph is thus not the graph of a function.   **10.** Using function notation, the replacement value for *x* and the resulting *f*(*x*) or *y*-value corresponds to an ordered pair (x, y) solution to the function.   **11.** No, equations of the form $x = c$, whose graphs are vertical lines, are not functions.   **12.** $f(x) = mx + b$, or slope-intercept form

Section 8.3

5. We are finding $Q(-10)$, and that value is 499. Thus, $Q(-10) = 499$.   **6.** Rational expressions are fractions and are therefore undefined if the denominator is zero; the domain of a rational function consists of all real numbers except those for which the rational expression is undefined.   **7.** Each time, we replace the variable in the expression with the given value, then simplify.

Section 8.4

1. Instead of an open circle, use (or). Instead of a closed circle, use [or].   **2.** left; right; up; down   **3.** Although $f(x) = x + 3$ isn't defined for $x = -1$, we need to clearly indicate the point where this piece of the graph ends. Therefore, we find this point and graph it as an open circle.

Section 8.5

5. $f(x) = |x|$ **6.** Once you know the shapes, use the shifting rules to tell you where to move the vertex or starting point for each graph; then you can easily draw in the appropriate basic shape.   **7.** *x*-axis

Chapter 9 Systems of Equations and Inequalities and Variation

Section 9.1

5. Once we have one equation in two variables, we need to get another equation in the *same* two variables, giving us a system of two equations in two variables. We solve this new system to find the value of two variables. We then substitute these values into an original equation to find the value of the third.   **6.** The ordered triple still needs to be interpreted in the context of the application. Each value actually represents the angle measure of a triangle, in degrees.

Section 9.2

9. Two rows may be interchanged, the elements of any row may be multiplied/divided by the same nonzero number, the elements of any row may be multiplied/divided by the same nonzero number and added to their corresponding elements in any other row; rows were not interchanged in Example 1.   **10.** Consider the possible confusion or errors that might occur if you're careless and it's unclear which row or column numbers belong in, especially when working with larger matrices and/or with a lot of calculations.

Section 9.3

5. No; we can choose any test point except a point on the second inequality's own boundary line.

Section 9.4

9. linear; slope   **10.** When *x* and *y* vary inversely, the product of *x* and *y* is the constant of variation, *k*.   **11.** $y = ka^2b^5$
12. A combined variation problem involves combinations of direct, inverse, and/or joint variation.

Chapter 10 Rational Exponents, Radicals, and Complex Numbers

Section 10.1

13. Divide the index into each exponent in the radicand.   **14.** Find the nearest perfect squares both less than and greater than the radicand. The square root of the radicand falls between the square roots of these two perfect squares.   **15.** The square root of a negative number is not a real number, but the cube root of a negative number is a real number.   **16.** The even root of a negative number is not a real number.   **17.** For odd roots, $\sqrt[n]{a^n} = a$, whether *a* is positive, 0, or negative, so absolute value bars aren't needed.
18. Since the variable *x* is the radicand of a square root and a square root cannot be negative, we know that *x* cannot be negative.

Section 10.2

13. $-\sqrt[5]{3x}$ **14.** The numerator is the power; the denominator is the index.   **15.** denominator; positive   **16.** add; subtract; multiply
17. Write the radical in equivalent fractional exponent form, use rules for exponents to write a single fractional exponent, simplify the fraction, then write in radical form again.

Section 10.3

11. The indexes must be the same.   **12.** if you see that simplifying can be done by separating a fractional radical into separate numerator and denominator radicands or by combining separate numerator and denominator radicands under one radical
13. The power must be 1. Any even power is a perfect square and will leave no factor in the radicand; any higher odd power can

have an even power factored from it, leaving one factor remaining in the radicand.   **14.** Be careful of signs since you're dealing with subtraction.   **15.** average; average

Section 10.4

15. Sometimes you can't see that there are like radicals until you simplify, so you may incorrectly think you cannot add or subtract if you don't simplify first.   **16.** The square root of a positive number times the square root of the same positive number is that positive number.

Section 10.5

11. to write an equivalent expression without a radical in the denominator   **12.** Using the FOIL order to multiply, the Outer product and the Inner product are opposites and they will subtract out.   **13.** No, except for the fact you're working with numerators, the process is the same.

Section 10.6

5. Applying the power rule can result in an equation with more solutions than the original equation, so you need to check all proposed solutions in the original equation.   **6.** The Pythagorean theorem works for a right triangle only, and the side opposite the right angle is the hypotenuse, which is c in the formula $a^2 + b^2 = c^2$.   **7.** Our answer is either a positive square root of a value or a negative square root of a value. We're looking for a length, which must be positive, so our answer must be the positive square root.

Section 10.7

7. the product rule for radicals; you need to first simplify each radical separately and have nonnegative radicands before applying the product rule.   **8.** combining like terms; i is *not* a variable, but a constant, $\sqrt{-1}$ **9.** the fact that $i^2 = -1$ **10.** using conjugates to rationalize denominators with two terms   **11.** $i, i^2 = -1, i^3 = -i, i^4 = 1$

Chapter 11 Quadratic Equations and Functions

Section 11.1

7. We need a quantity shown squared by itself on one side of the equation. The only quantity squared is x, so divide both sides by 2 before applying the square root property.   **8.** $r^2 - r + 1$ is not a perfect square trinomial because it does not factor as a binomial squared. In fact, $r^2 - r + 1$ is prime.   **9.** The coefficient of y^2 is 3. To use the completing the square method, the coefficient of the squared variable must be 1, so we first divide through by 3.   **10.** We're looking for an interest rate, so a negative value does not make sense.

Section 11.2

7. a. Yes, in order to make sure we have correct values for a, b, and c.   **b.** No; clearing fractions makes the work less tedious, but it's not a necessary step.   **8.** radicand; solving; standard   **9.** With applications, we need to make sure we answer the question(s) asked. Here we're asked how much distance is saved, so once the dimensions of the triangle are known, further calculations are needed to answer this question and solve the problem.

Section 11.3

1. The values we get for the substituted variable are *not* our final answers. Remember to always substitute back to the original variable and solve for it if necessary.   **2.** The rational equation simplifies to a quadratic equation once you multiply through by the LCD to rid the equation of fractions.

Section 11.4

7. We use the solutions to the related equation to divide the number line into regions that either entirely are or entirely are not solution regions; the solutions to the related equation are solutions to the inequality only if the inequality symbol is $\leq$ or $\geq$. **8.** The solution set cannot include values that make the denominator zero.

Section 11.5

15. Graphs of the form $f(x) = x^2 + k$ shift up or down the y-axis k units from $y = x^2$; the y-intercept.   **16.** Graphs of the form $f(x) = (x - h)^2$ shift right or left on the x-axis h units from $y = x^2$; the x-intercept.   **17.** The vertex, (h, k) and the axis of symmetry, $x = h$; the basic shape of $y = x^2$ does not change.   **18.** whether the graph is wider or narrower than $y = x^2$ **19.** the coordinates of the vertex, whether the graph opens upward or downward, whether the graph is narrower or wider than $y = x^2$, and the graph's axis of symmetry

Section 11.6

3. We can immediately identify the vertex (h, k), whether the parabola opens upward or downward, and know its axis of symmetry; completing the square.   **4.** This information tells us whether or not the graph has x-intercepts. For example, if the vertex is in quadrant III or IV and the parabola opens downward, then there aren't any x-intercepts and there's no need to go through the steps to locate any. **5.** the vertex

Chapter 12 Exponential and Logarithmic Functions

Section 12.1

7. You can find $(f + g)(x)$ and then find $(f + g)(2)$, or you can find $f(2)$ and $g(2)$ and then add those results.   **8.** Yes, sometimes they can be equal.

Section 12.2

9. Every function must have each x-value correspond to only one y-value. A one-to-one function must also have each y-value correspond to only one x-value.   **10.** No, a graph must pass the vertical line test to even be a function—a graph must pass *both* the vertical and horizontal line tests to be a one-to-one function.   **11.** Yes; by the definition of an inverse function.   **12.** The definition of inverse function tells us that f^{-1} consists of the ordered pairs (y, x) when (x, y) belongs to f. So it makes sense that switching x and y in the equations would result in switching the x and y values in the ordered pairs.   **13.** Once you know some points of the original equation or graph, you can switch the x's and y's of these points to find points that satisfy the inverse and then graph it. You can also check that the two graphs (the original and the inverse) are symmetric about the line $y = x$.   **14.** You must show that $f(f^{-1}(x))$ and $f^{-1}(f(x))$ both equal x in order to prove they are inverses of each other.

Section 12.3

9. In a polynomial function, the base is the variable and the exponent is the constant; in an exponential function, the base is the constant and the exponent is the variable.   **10.** rewrite the equation so the bases are the same   **11.** $y = 30(0.996)^{101} \approx 20.0 \text{ lb}$

Section 12.4

1. The growth rate is given as 5% per year. Since this is "per year," the number of time intervals is the "number of years," or 8. **2.** The number of employees is decreasing and not increasing. It is exponential decay because the decrease is the same percent per year.   **3.** time intervals $=$ years/half-life; the decay rate is 50% or $\frac{1}{2}$ because half-life is the amount of time it takes half of a substance to decay

Section 12.5

8. Logarithms are exponents.   **9.** First write the equation as an equivalent exponential equation. Then solve.   **10.** The exponential equation is solved for x, and y is the exponent in the equation. Since the exponential equation is solved for x, it is easier to choose a y-value and simplify the expression containing y, which is then the x-value.

Section 12.6

7. No, the product property says the logarithm of a product can be written as a sum of logarithms—the expression in this example is a logarithm of a sum.   **8.** The bases must be the same.   **9.** Since $\frac{1}{x} = x^{-1}$, this gives us $\log_2 x^{-1}$. Using the power property, we get $-1 \log_2 x$ or $-\log_2 x$.   **10.** From writing logarithms as equivalent exponents and then using the rules for exponents.

Section 12.7

8. 10   **9.** The understood base of a common logarithm is 10. If you're finding the common logarithm of a known power of 10, then the common logarithm is the known power of 10.   **10.** e **11.** $\log_b b^x = x$ **12.** $\frac{\ln 4}{\ln 6}$ or $\frac{\log 4}{\log 6}$

Section 12.8

1. $\ln (4x - 2) = \ln 3$ is the same as $\log_e(4x - 2) = \log_e 3$. Therefore, from the logarithm property of equality, we know that $4x - 2 = 3$.   **2.** Substituting -8 in the original equation gives us the logarithm of a negative number, which does not exist—we can only take the logarithm of a positive number.   **3.** $2000 = 1000\left(1 + \frac{0.07}{12}\right)^{12 \cdot t} \approx 9.9 \text{ yr}$; as long as the interest rate and compounding are the same, it takes any amount of money the same time to double.

Chapter 13 Conic Sections

Section 13.1

7. No, their graphs don't pass the vertical line test.   **8.** $x^2 + y^2 = r^2$ **9.** Since the standard form of a circle involves a squared binomial for both x and y, we need to complete the square on both x and y.   **10.** The formula for the standard form of a circle identifies the center and radius, so you just need to substitute these values into this formula and simplify.

Section 13.2

7. a and b give us the location of 4 intercepts—$(a, 0), (-a, 0), (0, b)$, and $(0, -b)$ for $\dfrac{x^2}{a^2} + \dfrac{y^2}{b^2} = 1$ with center $(0, 0)$. For Example 2, the values of a and b also give us 4 points of the graph, just not intercepts. Here we move a distance of a units horizontally to the left and right of the center and b units above and below the center.   **8.** We use these points to draw asymptotes (also not part of the graph), which help us draw the correct shape of the hyperbola. The graph of a hyperbola gets closer and closer to the asymptotes without crossing them.

Section 13.3

1. Solving for y would either introduce tedious fractions (2nd equation) or a square root (1st equation) into the calculations.
2. When you multiply the left side of the equation by this number, do not forget to also multiply the right side.

Section 13.4

1. For both, we graph the related equation to find the boundary and sketch it as a solid boundary for $\leq$ or $\geq$ and a dashed boundary for $<$ or $>$; also, we choose a test point not on the boundary and shade the region containing the test point if the test point is a solution of the original inequality or shade the other region if not.   **2.** A circle within a circle (either circle solid or dashed) where the inner circle is shaded inside and the outer circle is shaded outside; also, two nonintersecting circles (either circle solid or dashed), both shaded inside—just to name a few examples.

Index

Index

A

Absolute value
  explanation of, 14, 82, 1000
  method to find, 14–15
Absolute value bars, 20, 1000–1002
Absolute value equations, 1000–1002
Absolute value functions, 1026–1027
Absolute value inequalities, 1003–1005
Acute angles, 1018
Addition
  associative property of, 65, 84
  commutative property of, 64, 84
  of complex numbers, 739–740, 749
  of decimals, R-20–R-21, R-32
  distributive property of multiplication
    over, 66–67
  of fractions, R-13, R-32
  of functions, 843
  identities for, 67
  of polynomials, 365–367, 398, 984
  problem solving with, 34
  of radical expressions, 708–710,
    747–748
  of rational expressions, 502–503, 510–513,
    552, 553
  of real numbers, 30–34, 83
  of signed numbers, 30, 31
  symbol for, 24
  words/phrases for, 24
Addition method
  explanation of, 300
  to solve systems of linear equations,
    300–303, 323–324
Addition property of equality
  explanation of, 92–93, 173
  use of, 93–95, 104–105
Addition property of inequality, 162
Additive inverse. *See also* Opposites
  explanation of, 32, 67, 68, 84
  method to find, 32–34
Adjacent angles, 1019
Algebra, of functions, 843, 907

Algebraic expressions. *See also*
    Expressions
  evaluation of, 22, 34, 41, 55–57, 82, 444
  explanation of, 22, 82
  method to write, 22, 76, 96–97, 105–106
Alternate interior angles, 1020
$a^{m/n}$, 691–692
$a^{-m/n}$, 692–693
$a^{1/n}$, 691
Angles
  acute, 1018
  adjacent, 1019
  alternate interior, 1020
  complementary, 43, 517, 1018
  corresponding, 1020, 1022
  explanation of, 1018
  exterior, 1020
  finding measures of, 125–126, 637–638,
    1018, 1020–1021
  interior, 1020
  obtuse, 1018
  right, 1018
  straight, 1018
  supplementary, 43, 517, 1018
  vertical, 1019
Approximately equal to sign, R-24
Approximation
  decimal, R-24
  of logarithms, 893–896
  of roots, 682, 745
Array of signs, 1013–1014
Associative property
  of addition, 65, 84
  of multiplication, 65, 84
Asymptotes, of hyperbola, 944
Average, 172, 480

B

Bar graphs, 187–188
Base, of exponential expressions, 19,
    333, 335
Best fit equations, 617

Photo Credits